Introduction to Theoretical and Computational Fluid Dynamics

Introduction to Theoretical and Computational Fluid Dynamics

Second Edition

C. Pozrikidis

OXFORD UNIVERSITY PRESS

2011

OXFORD UNIVERSITY PRESS

Oxford University Press, Inc., publishes works that further Oxford University's objective of excellence in research, scholarship, and education.

1006520928
Published by Oxford University Press, Inc.
198 Madison Avenue, New York, New York 10016

www.oup.com

Library of Congress Cataloging-in-Publication Data
Pozrikidis, C.
Introduction to theoretical and computational fluid dynamics / C. Pozrikidis. – 2nd ed.
    p. cm.
ISBN 978-0-19-975207-2 (hardcover : alk. paper) 1. Fluid dynamics. I. Title.
QA911.P65 2011
532'.05–dc22

                    2011002954

Preface

My goal in this book entitled *Introduction to Theoretical and Computational Fluid Dynamics* is to provide a comprehensive and rigorous introduction to the fundamental concepts and basic equations of fluid dynamics, and simultaneously illustrate the application of numerical methods for solving a broad range of fundamental and practical problems involving incompressible Newtonian fluids. The intended audience includes advanced undergraduate students, graduate students, and researchers in most fields of science and engineering, applied mathematics, and scientific computing. Prerequisites are a basic knowledge of classical mechanics, intermediate calculus, elementary numerical methods, and some familiarity with computer programming. The chapters can be read sequentially, randomly, or in parts, according to the reader's experience, interest, and needs.

Scope

This book differs from a typical text on theoretical fluid dynamics in that the discourse is carried into the realm of numerical methods and into the discipline of computational fluid dynamics (CFD). Specific algorithms for computing incompressible flows under diverse conditions are developed, and computer codes encapsulated in the public software library FDLIB are discussed in Appendix C. This book also differs from a typical text on computational fluid dynamics in that a full discussion of the theory with minimal external references is provided, and no experience in computational fluid dynamics or knowledge of its terminology is assumed. Contemporary numerical methods and computational schemes are developed and references for specialized and advanced topics are provided.

Content

The material covered in this text has been selected according to what constitutes essential knowledge of theoretical and computational fluid dynamics. This intent explains the absence of certain specialized and advanced topics, such as turbulent motion and non-Newtonian flow. Although asymptotic and perturbation methods are discussed in several places, emphasis is placed on analytical and numerical computation. The discussion makes extensive usage of the powerful concept of Green's functions and integral representations.

Use as a text

This book is suitable as a text in an advanced undergraduate or introductory graduate course on fluid mechanics, Stokes flow, hydrodynamic stability, computational fluid dynamics, vortex dynamics, or a special topics course, as indicated in the *Note to the Instructor*. Each section is followed by a set of problems that should be solved by hand and another set of problems that should be tackled with the help of a computer. Both categories of problems are suitable for self-study, homework, and project assignment. Some computer problems are coordinated so that a function or subroutine written for one problem can be used as a module in a subsequent problem.

Preface to the Second Edition

The Second Edition considerably extends the contents of the First Edition to include contemporary topics and some new and original material. Clarifications, further explanations, detailed proofs, original derivations, and solved problems have been added in numerous places. Chapter 1 on kinematics, Chapter 8 on hydrodynamic stability, and Chapter 11 on vortex methods have been considerably expanded. Numerous schematic depictions and graphs have been included as visual guides to illustrate the results of theoretical derivations. Expanded appendices containing useful background material have been added for easy reference. These additions underscore the intended purpose of the Second Edition as a teaching, research, and reference resource.

FDLIB

The numerical methods presented in the text are implemented in computer codes contained in the software library FDLIB, as discussed in Appendix C. The directories of FDLIB include a variety of programs written in FORTRAN 77 (compatible with FORTRAN 90), Matlab, and C++. The codes are suitable for self-study, classroom instruction, and fundamental or applied research. Appendix D contains the User Guide of the eighth directory of FDLIB on hydrodynamic stability, complementing Chapter 8.

Acknowledgments

I thankfully acknowledge the support of Todd Porteous and appreciate useful comments by Jeffrey M. Davis and A. I. Hill on a draft of the Second Edition.

C. Pozrikidis

Note to the Instructor

This book is suitable for teaching several general and special-topics courses in theoretical and computational fluid dynamics, applied mathematics, and scientific computing.

Course on fluid mechanics

The first eight chapters combined with selected sections from subsequent chapters can be used in an upper-level undergraduate or entry-level introductory graduate core course on fluid mechanics. The course syllabus includes essential mathematics and numerical methods, flow kinematics, stresses, the equation of motion and flow dynamics, hydrostatics, exact solutions, Stokes flow, irrotational flow, and boundary-layer analysis. The following lecture plan is recommended:

Appendix A	Essential mathematics	Reading assignment
Appendix B	Primer of numerical methods	Reading assignment
Chapter 1	Kinematics	
Chapter 2	Kinematic description	Sections 2.2–2.8 and 2.10–2.13 are optional
Chapter 3	Equation of motion	
Chapter 4	Hydrostatics	
Chapter 5	Exact solutions	
Chapter 6	Stokes flow	Sections 6.8–6.18 are optional
Chapter 7	Irrotational flow	

Some sections can be taught as a guided reading assignment at the instructor's discretion.

Course on Stokes flow

Chapter 6 can be used in its entirety as a text in a course on theoretical and computational Stokes flow. The course syllabus includes governing equations and fundamental properties of Stokes flow, local solutions, particulate microhydrodynamics, singularity methods, boundary-integral formulations, boundary-element methods, unsteady Stokes flow, and unsteady particle motion. The students are assumed to have a basic undergraduate-level knowledge of fluid mechanics. Some topics from previous chapters can be reviewed at the beginning of the course.

Course on hydrodynamic stability

Chapter 9 combined with Appendix D can be used in a course on hydrodynamic stability. The course syllabus includes formulation of the linear stability problem, normal-mode analysis, stability of unidirectional flows, the Rayleigh equation, the Orr–Sommerfeld equation, stability of rotating flows, and stability of inviscid and viscous interfacial flows. The students are assumed to have a basic knowledge of the continuity equation, the Navier–Stokes equation, and the vorticity transport equation. These topics can be reviewed from previous chapters at the beginning of the course.

Course on computational fluid dynamics (CFD)

The following lecture plan is recommended in a course on numerical methods and computational fluid dynamics, following a graduate course on fluid mechanics:

Appendix A	Essential mathematics	Reading assignment
Appendix B	Primer of numerical methods	Reading assignment
Chapter 2	Theory of potential flow	Sections 2.1–2.5
Chapter 6	Boundary-integral methods for Stokes flow	Sections 6.5–6.10
Chapter 10	Boundary-integral methods for potential flow	
Chapter 11	Vortex motion	Selected topics
Chapter 12	Finite-difference methods	
Chapter 13	Finite-difference methods for incompressible flow	

Short course on vortex dynamics

Chapter 11 is suitable as a text in a short course on vortex dynamics. The material can be preceded or supplemented with selected sections from previous chapters to establish the necessary theoretical framework.

Special topics in fluid mechanics

Selected material from Chapters 9–13 can be used in a special-topics course in fluid mechanics, applied mathematics, computational fluids dynamics, and scientific computing. The choice of topics will depend on the students' interests and field of study.

Note to the Reader

For self-study, follow the roadmap outlined in the *Note to the Instructor,* choosing your preferred area of concentration. In the absence of a preferred area, study the text from page one onward, skipping sections that seem specialized, but keeping in mind the material contained in Appendices A and B on essential mathematics and numerical methods. Before embarking on a course of study, familiarize yourself with the entire contents of this book, including the appendices.

Notation

In the text, an italicized variable, such as a, is a scalar, and a bold-faced variable, such as $\mathbf{a}$, is a vector or matrix. Matrices are represented by upper case and bold faced symbols. Matrix–vector and matrix–matrix multiplication is indicated explicitly with a centered dot, such as $\mathbf{A} \cdot \mathbf{B}$. With this convention, a vector, $\mathbf{a}$, can be horizontal or vertical, as the need arises. It is perfectly acceptable to formulate the product $\mathbf{A} \cdot \mathbf{a}$ as well as the product $\mathbf{a} \cdot \mathbf{A}$, where $\mathbf{A}$ is an appropriate square matrix. Index notation and other conventions are defined in Appendix A.

The fluid velocity is denoted by $\mathbf{u}$ or $\mathbf{U}$. The boundary velocity is denoted by $\mathbf{v}$ or $\mathbf{V}$. Exceptions are stated, as required. Dimensionless variables are denoted by a hat (caret). We strongly advocate working with physical dimensional variables and nondimensionalizing at the end, if necessary.

Polymorphism

Occasionally in the analysis, we run out of symbols. A bold faced variable may then be used to represent a vector or a matrix with two or more indices. A mental note should be made that the variable may have different meanings, depending on the current context. This practice is consistent with the concept of polymorphism in object-oriented programming where a symbol or function may represent different entities depending on the data type supplied in the input and requested in the output. The language compiler is trained to pick up the appropriate structure.

Physical entities expressed by vectors and tensors

The velocity of an object, $\mathbf{v}$, is a physical entity characterized by magnitude and direction. In the analysis, the velocity is described by three scalar components referring to Cartesian, polar, or other orthogonal or nonorthogonal coordinates. The Cartesian components, v_x, v_y, and v_z, can be conveniently collected into a Cartesian vector,

$$\mathbf{v} = [v_x, v_y, v_z].$$

Accordingly, $\mathbf{v}$ admits a dual interpretation as a physical entity that is independent of the chosen coordinate system, and as a mathematical vector. In conceptual analysis, we refer to the physical interpretation; in practical analysis and calculations, we invoke the mathematical interpretation. To prevent confusion, the components of a vector in non-Cartesian coordinates should *never* be collected into a vector. Similar restrictions apply to matrices representing physical entities that qualify as Cartesian tensors.

Units

In engineering, all physical variables have physical dimensions such as length, length over time, mass, mass over length cubed. Units must be chosen consistently from a chosen system, such as the cgs (cm, g, s) or the mks (m, kg, s) system. It is not appropriate to take the logarithm or exponential of a length a, $\ln a$ or e^a, because the units of the logarithm or exponential are not defined. Instead, we must always write $\ln(a/b)$ or $e^{a/b}$, where b is another length introduced so that the argument of the logarithm or exponential is a dimensionless variable or number. The logarithm or exponential are then dimensionless variables or numbers.

Computer languages

Basic computer programming skills are necessary for the thorough understanding of contemporary applied sciences and engineering. Applications of computer programming in fluid mechanics range from preparing graphs and producing animations, to computing fluid flows under a broad range of conditions. Recommended general-purpose computer languages include FORTRAN, C, and C++. Free compilers for these languages are available for many operating systems. Helpful tutorials can be found on the Internet and a number of excellent texts are available for self-study.

Units in a computer code

In writing a computer code, always use physical variables and dimensional equations. To scale the results, set the value of one chosen length to unity and, depending on the problem, one viscosity, density, or surface tension equal to unity. For example, setting the length of a pipe, L, to 1.0, renders all lengths dimensionless with respect to L.

Fluid, solid, and continuum mechanics

The union of fluid and solid mechanics comprises the field of continuum mechanics. The basic theory of fluid mechanics derives from the theory of solid mechanics, and *vice versa*, by straightforward substitutions:

fluid		*solid*
velocity		displacement
velocity gradient tensor		deformation gradient tensor
rate of strain or rotation	$\rightarrow$	strain or rotation
surface force		traction
stress		Cauchy stress

However, there are some important differences. The traction in a fluid refers to an infinitesimal surface fixed in space, whereas the traction in a solid refers to a material surface before or after deformation. Consequently, there are several alternative, albeit inter-related, definitions for the stress tensor in a solid. Another important difference is that the velocity gradient tensor in fluid mechanics is the transpose of the deformation gradient tensor in solid mechanics. This difference is easy to identify in Cartesian coordinates, but is lurking, sometimes unnoticed, in curvilinear coordinates. These important subtleties should be born in mind when referring to texts on continuum mechanics.

Contents

Kinematic structure of a flow

1

The study of fluid mechanics is divided broadly into two main themes, kinematics and dynamics. Kinematics analyzes the motion of a fluid with reference to the structure of the velocity field, whereas dynamics examines the forces developing in a fluid as the result of the motion. Fundamental concepts of kinematics and dynamics are combined with the fundamental principle of mass conservation and Newton's second law of motion to derive a system of differential equations governing the structure of a steady flow and the evolution of an unsteady flow.

1.1 Fluid velocity and motion of fluid parcels

Referring to a frame of reference that is fixed in space, we observe the flow of a homogeneous fluid consisting of a single chemical species. We consider, in particular, the motion of a fluid parcel which, at a certain observation time, t, has a spherical shape of radius ϵ centered at a point, $\mathbf{x}$. The velocity of translation of the parcel in a particular direction is defined as the average value of the instantaneous velocity of all molecules residing inside the parcel in that direction.

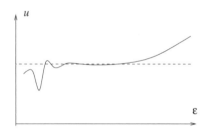

The fluid velocity is defined as the outer limit of the mean parcel velocity.

It is clear that the average parcel velocity depends on the parcel radius, ϵ. Taking the limit as ϵ tends to zero, we find that the average velocity approaches a well-defined asymptotic limit, until ϵ becomes comparable to the molecular size. At that stage, we observe strong oscillations that are manifestations of random molecular motions. We define the velocity of the fluid, $\mathbf{u}$, at position, $\mathbf{x}$, and time, t, as the apparent or outer limit of the velocity of the parcel as ϵ tends to zero, just before the discrete nature of the fluid becomes apparent. This definition is the cornerstone of the *continuum approximation* in fluid mechanics. Under normal conditions, the velocity, $\mathbf{u}(\mathbf{x}, t)$, is an infinitely differentiable function of position, $\mathbf{x}$, and time, t. However, spatial discontinuities may arise under extreme conditions in high-speed flows, or else emerge due to mathematical idealization. If a flow is steady, $\mathbf{u}$ is independent of time, $\partial \mathbf{u}/\partial t = \mathbf{0}$, and the velocity at a certain position in space remains constant in time.

The velocity of a two-dimensional flow in the xy plane is independent of the z coordinate, $\partial \mathbf{u}/\partial z = \mathbf{0}$. The velocity component along the z axis, u_z, may have a constant value that is usually made to vanish by an appropriate choice of the frame of reference.

1.1.1   Subparcels and point particles

To analyze the motion of a fluid parcel, it is helpful to divide the parcel into a collection of subparcels with smaller dimensions and observe that the rate of rotation and deformation of the parcel is determined by the relative motion of the subparcels. For example, if all subparcels move with the same velocity, the parental parcel translates as a rigid body, that is, the rate of rotation and deformation of the parental parcel are both zero. It is conceivable that the velocity of the subparcels may be coordinated so that the undivided parental parcel translates and rotates as a rigid body without suffering deformation, that is, without change in shape.

If we continue to subdivide a parcel into subparcels with decreasingly small size, we will eventually encounter subparcels with infinitesimal dimensions called point particles. Each point particle occupies an infinitesimal volume in space, but an infinite collection of point particles that belongs to a finite (noninfinitesimal) parcel occupies a finite volume in space. By definition, a point particle located at the position $\mathbf{x}$ moves with the local and current fluid velocity, $\mathbf{u}(\mathbf{x}, t)$.

1.1.2   Velocity gradient

To describe the motion of a fluid parcel in quantitative terms, we introduce Cartesian coordinates and consider the spatial variation of the velocity, $\mathbf{u}$, in the neighborhood of a chosen point, $\mathbf{x}_0$, which is located somewhere inside the parcel. Expanding the jth component of $\mathbf{u}(\mathbf{x}, t)$ in a Taylor series with respect to $\mathbf{x}$ about $\mathbf{x}_0$ and retaining only the linear terms, we find that

$$u_j(\mathbf{x}, t) \simeq u_j(\mathbf{x}_0, t) + \widehat{x}_i \, L_{ij}(\mathbf{x}_0, t), \qquad (1.1.1)$$

where $\widehat{\mathbf{x}} = \mathbf{x} - \mathbf{x}_0$ is the vector connecting the point $\mathbf{x}_0$ to the point $\mathbf{x}$, and summation over the repeated index i is implied on the right-hand side. We have introduced the velocity gradient or rate of relative displacement matrix, $\mathbf{L} = \nabla \mathbf{u}$, with components

$$L_{ij} = \frac{\partial u_j}{\partial x_i}. \qquad (1.1.2)$$

Explicitly, the velocity gradient is given by

$$\mathbf{L} \equiv \nabla \mathbf{u} = \begin{bmatrix} \dfrac{\partial u_x}{\partial x} & \dfrac{\partial u_y}{\partial x} & \dfrac{\partial u_z}{\partial x} \\[2mm] \dfrac{\partial u_x}{\partial y} & \dfrac{\partial u_y}{\partial y} & \dfrac{\partial u_z}{\partial y} \\[2mm] \dfrac{\partial u_x}{\partial z} & \dfrac{\partial u_y}{\partial z} & \dfrac{\partial u_z}{\partial z} \end{bmatrix}. \qquad (1.1.3)$$

In physical terms, the velocity gradient expresses the spatial variation of the velocity across a collection of subparcels or point particles composing a parcel. In Section 1.1.8, we will show that $\mathbf{L}$ satisfies a transformation rule that qualifies it as a second-order Cartesian tensor.

In formal mathematics, the velocity gradient is the tensor product of the gradient vector operator, ∇, and the velocity vector field, $\mathbf{u}$. To simplify the notation, we have set

$$\nabla \mathbf{u} \equiv \nabla \otimes \mathbf{u}, \qquad (1.1.4)$$

where $\otimes$ denotes the tensor product of two vectors defined in Section A.4, Appendix A. Similar simplified notation will be used for other tensor products involving the gradient operator; for example, $\nabla\nabla = \nabla \otimes \nabla$ is the two-dimensional operator of the second derivatives.

In vector notation, equation (1.1.1) takes the form

$$\mathbf{u}(\mathbf{x},t) \simeq \mathbf{u}(\mathbf{x}_0,t) + \widehat{\mathbf{x}} \cdot \mathbf{L}(\mathbf{x}_0,t). \qquad (1.1.5)$$

Note the left-to-right vector–matrix multiplication in the second term on the right-hand side.

Divergence of the velocity

The trace of the velocity gradient tensor, $\mathbf{L}$, defined as the sum of the diagonal elements of $\mathbf{L}$, is equal to the divergence of the velocity,

$$\alpha \equiv \text{trace}(\mathbf{L}) = \frac{\partial u_k}{\partial x_k} = \nabla \cdot \mathbf{u}, \qquad (1.1.6)$$

where summation is implied over the repeated index, k. In Section 1.1.6, we will see that the divergence of the velocity expresses the rate of expansion of the fluid.

1.1.3 Dyadic base

Let $\mathbf{e}_m$ be the unit vector along the mth Cartesian axis for $m = 1, 2, 3$. By definition, $\mathbf{e}_i \cdot \mathbf{e}_j = \delta_{ij}$, where δ_{ij} is Kronecker's delta representing the identity matrix, as discussed in Section A.4, Appendix A. The velocity can be expressed in terms of its Cartesian components, u_i, as

$$\mathbf{u} = u_i \, \mathbf{e}_i, \qquad (1.1.7)$$

where summation is implied over the repeated index, i. The velocity gradient can be expressed in the dyadic form

$$\mathbf{L} = L_{ij} \, \mathbf{e}_i \otimes \mathbf{e}_j, \qquad (1.1.8)$$

where summation is implied over the repeated indices, i and j. The kl component of the matrix $\mathbf{e}_i \otimes \mathbf{e}_j$ is

$$[\mathbf{e}_i \otimes \mathbf{e}_j]_{kl} = [\mathbf{e}_i]_k [\mathbf{e}_j]_l. \qquad (1.1.9)$$

Expression (1.1.8) is the algebraic counterpart of the matrix depiction shown in (1.1.3).

Projecting equation (1.1.8) from the right onto the unit vector $\mathbf{e}_q$, where q is a free index, and using the orthogonality property of the unit vectors, we obtain

$$\mathbf{L} \cdot \mathbf{e}_q = L_{ij} \, \mathbf{e}_i \, (\mathbf{e}_j \cdot \mathbf{e}_q) = L_{ij} \, \mathbf{e}_i \, \delta_{jq} = L_{iq} \, \mathbf{e}_i. \qquad (1.1.10)$$

Formulating the inner product of this equation with the unit vector $\mathbf{e}_p$, where p is another free index, and working in a similar fashion, we extract the components of the velocity gradient,

$$L_{pq} = \mathbf{e}_p \cdot \mathbf{L} \cdot \mathbf{e}_q = \mathbf{L} : (\mathbf{e}_p \otimes \mathbf{e}_q), \qquad (1.1.11)$$

where the double dot product of two matrices with matching dimensions is defined in Section A.4, Appendix A. This expression may also be regarded as a consequence of the orthogonality property

$$(\mathbf{e}_i \otimes \mathbf{e}_j) : (\mathbf{e}_p \otimes \mathbf{e}_q) = (\mathbf{e}_i \cdot \mathbf{e}_p)(\mathbf{e}_j \cdot \mathbf{e}_q) = \delta_{ip}\,\delta_{jq}, \tag{1.1.12}$$

where i, j, p, and q are found independent indices.

The identity matrix admits the dyadic decomposition

$$\mathbf{I} = \mathbf{e}_1 \otimes \mathbf{e}_1 + \mathbf{e}_2 \otimes \mathbf{e}_2 + \mathbf{e}_3 \otimes \mathbf{e}_3. \tag{1.1.13}$$

In matrix notation underlying this dyadic expansion, all elements of $\mathbf{I}$ are zero, except for the three diagonal elements that are equal to unity.

Without loss of generality, we may assume that the kth component of the unit vector $\mathbf{e}_i$ is equal to 1 if $k = i$ or 0 if $k \neq i$, that is, $[\mathbf{e}_i]_k = \delta_{ik}$. The kl component of the matrix $\mathbf{e}_i \otimes \mathbf{e}_j$ is $[\mathbf{e}_i \otimes \mathbf{e}_j]_{kl} = [\mathbf{e}_i]_k[\mathbf{e}_j]_l = \delta_{ik}\delta_{jl}$. All elements of the matrix $\mathbf{e}_i \otimes \mathbf{e}_j$ are zero, except for the element located at the intersection of the ith row and jth column that is equal to unity, $[\mathbf{e}_i \otimes \mathbf{e}_j]_{ij} = 1$, where summation is *not* implied over the indices i and j. For example,

$$\mathbf{e}_1 \otimes \mathbf{e}_2 = \begin{bmatrix} 0 & 1 & 0 \\ 0 & 0 & 0 \\ 0 & 0 & 0 \end{bmatrix}. \tag{1.1.14}$$

This expression illustrates that the dyadic base provides us with a natural framework for representing a two-index matrix.

1.1.4 Fundamental decomposition of the velocity gradient

It is useful to decompose the velocity gradient, $\mathbf{L}$, into an antisymmetric component, Ξ, a symmetric component with vanishing trace, $\mathbf{E}$, and an isotropic component with nonzero trace,

$$\mathbf{L} = \Xi + \mathbf{E} + \frac{1}{3}\alpha\,\mathbf{I}, \tag{1.1.15}$$

where $\mathbf{I}$ is the identity matrix. We have introduced the vorticity tensor,

$$\Xi \equiv \frac{1}{2}(\mathbf{L} - \mathbf{L}^T), \qquad \Xi_{ij} = \frac{1}{2}\left(\frac{\partial u_j}{\partial x_i} - \frac{\partial u_i}{\partial x_j}\right), \tag{1.1.16}$$

and the rate-of-deformation tensor, also called the rate-of-strain tensor,

$$\mathbf{E} \equiv \frac{1}{2}(\mathbf{L} + \mathbf{L}^T) - \frac{1}{3}\alpha\,\mathbf{I}, \qquad E_{ij} = \frac{1}{2}\left(\frac{\partial u_j}{\partial x_i} + \frac{\partial u_i}{\partial x_j}\right) - \frac{1}{3}\alpha\,\delta_{ij}, \tag{1.1.17}$$

where the superscript T denotes the matrix transpose. Explicitly, the components of Ξ and $\mathbf{E}$ are given in Table 1.1.1. In Section 1.1.8, we will demonstrate that Ξ and $\mathbf{E}$ obey transformation rules that qualify them as second-order Cartesian tensors.

$$\Xi \equiv \frac{1}{2} \begin{bmatrix} 0 & \dfrac{\partial u_y}{\partial x} - \dfrac{\partial u_x}{\partial y} & \dfrac{\partial u_z}{\partial x} - \dfrac{\partial u_x}{\partial z} \\[3ex] \dfrac{\partial u_x}{\partial y} - \dfrac{\partial u_y}{\partial x} & 0 & \dfrac{\partial u_z}{\partial y} - \dfrac{\partial u_y}{\partial z} \\[3ex] \dfrac{\partial u_x}{\partial z} - \dfrac{\partial u_z}{\partial x} & \dfrac{\partial u_y}{\partial z} - \dfrac{\partial u_z}{\partial y} & 0 \end{bmatrix}$$

$$\mathbf{E} \equiv \begin{bmatrix} \dfrac{\partial u_x}{\partial x} - \dfrac{1}{3}\alpha & \dfrac{1}{2}\Big(\dfrac{\partial u_y}{\partial x} + \dfrac{\partial u_x}{\partial y}\Big) & \dfrac{1}{2}\Big(\dfrac{\partial u_z}{\partial x} + \dfrac{\partial u_x}{\partial z}\Big) \\[3ex] \dfrac{1}{2}\Big(\dfrac{\partial u_x}{\partial y} + \dfrac{\partial u_y}{\partial x}\Big) & \dfrac{\partial u_y}{\partial y} - \dfrac{1}{3}\alpha & \dfrac{1}{2}\Big(\dfrac{\partial u_z}{\partial y} + \dfrac{\partial u_y}{\partial z}\Big) \\[3ex] \dfrac{1}{2}\Big(\dfrac{\partial u_x}{\partial z} + \dfrac{\partial u_z}{\partial x}\Big) & \dfrac{1}{2}\Big(\dfrac{\partial u_y}{\partial z} + \dfrac{\partial u_z}{\partial y}\Big) & \dfrac{\partial u_z}{\partial z} - \dfrac{1}{3}\alpha \end{bmatrix}$$

TABLE 1.1.1 Cartesian components of the vorticity tensor, Ξ, and rate-of-deformation tensor, $\mathbf{E}$, where $\alpha \equiv \nabla \cdot \mathbf{u}$ is the rate of expansion. The trace of the rate-of-deformation tensor, $\mathbf{E}$, is zero by construction.

1.1.5 Vorticity

Because the vorticity tensor is antisymmetric, it has only three independent components that can be accommodated into a vector. To implement this simplification, we introduce the vorticity vector, $\boldsymbol{\omega}$, and set

$$\Xi_{ij} = \frac{1}{2}\,\epsilon_{ijk}\,\omega_k, \tag{1.1.18}$$

where ϵ_{ijk} is the alternating tensor defined in Section A.4, Appendix A, and summation is implied over the repeated index k. Explicitly,

$$\Xi = \frac{1}{2} \begin{bmatrix} 0 & \omega_z & -\omega_y \\ -\omega_z & 0 & \omega_x \\ \omega_y & -\omega_x & 0 \end{bmatrix}. \tag{1.1.19}$$

Conversely, the vorticity vector derives from the vorticity tensor as

$$\omega_k = \epsilon_{kij}\,\Xi_{ij} = \frac{1}{2}\,\epsilon_{kij}\Big(\frac{\partial u_j}{\partial x_i} - \frac{\partial u_i}{\partial x_j}\Big) = \epsilon_{kij}\,\frac{\partial u_j}{\partial x_i}, \tag{1.1.20}$$

where summation is implied over the repeated indices, i and j. The last expression shows that the vorticity is the curl of the velocity,

$$\boldsymbol{\omega} = \nabla \times \mathbf{u}. \tag{1.1.21}$$

Explicitly,

$$\boldsymbol{\omega} = \left(\frac{\partial u_z}{\partial y} - \frac{\partial u_y}{\partial z} \right) \mathbf{e}_x + \left(\frac{\partial u_x}{\partial z} - \frac{\partial u_z}{\partial x} \right) \mathbf{e}_y + \left(\frac{\partial u_y}{\partial x} - \frac{\partial u_x}{\partial y} \right) \mathbf{e}_z, \tag{1.1.22}$$

where $\mathbf{e}_x$, $\mathbf{e}_y$, and $\mathbf{e}_z$ are unit vectors along the x, y, and z axes.

Observing that the vorticity of a two-dimensional flow in the xy plane is oriented along the z axis, we write

$$\boldsymbol{\omega}(x, y, t) = \omega_z(x, y, t) \, \mathbf{e}_z, \tag{1.1.23}$$

where ω_z is the z vorticity component.

1.1.6   Fluid parcel motion

Substituting (1.1.18) into (1.1.15) and the resulting expression into (1.1.1), we derive an expression for the spatial distribution of the velocity in the neighborhood of a point, $\mathbf{x}_0$, in terms of the vorticity vector, $\boldsymbol{\omega}$, the rate-of-deformation tensor, $\mathbf{E}$, and the divergence of the velocity field, α,

$$u_j(\mathbf{x}, t) \simeq u_j(\mathbf{x}_0, t) + \frac{1}{2} \epsilon_{jki} \, \omega_k(\mathbf{x}_0, t) \, \widehat{x}_i + \widehat{x}_i \, E_{ij}(\mathbf{x}_0, t) + \frac{1}{3} \alpha \, \widehat{x}_j, \tag{1.1.24}$$

where $\widehat{\mathbf{x}} = \mathbf{x} - \mathbf{x}_0$. Next, we proceed to analyze the motion of a fluid parcel based on this local representation, recognizing that point particles execute sequential motions under the influence of each term on the right-hand side.

Translation

The first term on the right-hand side of (1.1.24) expresses rigid-body translation. Under the influence of this term, a fluid parcel translates with the fluid velocity evaluated at the designated particle center, $\mathbf{x}_0$.

Vorticity and rotation

The second term on the right-hand side of (1.1.24) can be written as $\boldsymbol{\Omega}(\mathbf{x}_0, t) \times \widehat{\mathbf{x}}$, where

$$\boldsymbol{\Omega} = \frac{1}{2} \boldsymbol{\omega}. \tag{1.1.25}$$

This expression shows that point particles rotate about the point $\mathbf{x}_0$ with angular velocity that is equal to half the centerpoint vorticity. Conversely, the vorticity vector is parallel to the point-particle angular velocity vector, and the magnitude of the vorticity vector is equal to twice that of the point-particle angular velocity vector.

Rate of strain and rate of deformation

We return to (1.1.24) and examine the nature of the motion associated with the third term on the right-hand side. Because the matrix $\mathbf{E}$ is real and symmetric, it has three real eigenvalues, λ_1, λ_2, and λ_3, and three mutually orthogonal eigenvectors. Under the action of this term, three infinitesimal fluid parcels resembling slender needles initially aligned with the eigenvectors elongate or compress in their respective directions while remaining mutually orthogonal (Problem 1.1.2). An ellipsoidal fluid parcel with three axes aligned with the eigenvectors deforms, increasing or decreasing its aspect ratios, while maintaining its initial orientation. These observations suggest that the third term on the right-hand side of (1.1.24) expresses deformation that preserves orientation.

To examine the change of volume of a parcel under the action of this term, we consider a parcel in the shape of a rectangular parallelepiped whose edges are aligned with the eigenvectors of $\mathbf{E}$. Let the initial lengths of the edges be $\mathrm{d}x_1$, $\mathrm{d}x_2$, and $\mathrm{d}x_3$. After a small time interval $\mathrm{d}t$ has elapsed, the lengths of the sides have become $(1 + \lambda_1\,\mathrm{d}t)\,\mathrm{d}x_1$, $(1 + \lambda_2\,\mathrm{d}t)\,\mathrm{d}x_2$, and $(1 + \lambda_3\,\mathrm{d}t)\,\mathrm{d}x_3$. Formulating the product of these lengths, we obtain the new volume,

$$(1 + \lambda_1\mathrm{d}t)(1 + \lambda_2\mathrm{d}t)(1 + \lambda_3\mathrm{d}t)\,\mathrm{d}x_1\,\mathrm{d}x_2\,\mathrm{d}x_3. \qquad (1.1.26)$$

Multiplying the three factors, we find that, to first order in $\mathrm{d}t$, the volume of the parcel has been modified by the factor $1 + (\lambda_1 + \lambda_2 + \lambda_3)\mathrm{d}t$. However, because the sum of the eigenvalues of $\mathbf{E}$ is equal to the trace of $\mathbf{E}$, which is zero by construction, the volume of the parcel is preserved during the deformation.

If the rate of strain is everywhere zero, the flow must necessarily express rigid-body motion, including translation and rotation (Problem 1.1.3).

Expansion

The last term on the right-hand side of (1.1.24) represents isotropic expansion or contraction. Under the action of this term, a small spherical parcel of radius a and volume $\delta V = \frac{4\pi}{3}a^3$ centered at the point $\mathbf{x}_0$ expands or contracts isotropically, undergoing neither translation, nor rotation, nor deformation. To compute the rate of expansion, we note that, after a small time interval $\mathrm{d}t$ has elapsed, the radius of the spherical parcel has become $(1 + \frac{1}{3}\alpha\,\mathrm{d}t)\,a$ and the parcel volume has been changed by the differential amount

$$\mathrm{d}\,\delta V = \frac{4\pi}{3}\left(1 + \frac{1}{3}\alpha\,\mathrm{d}t\right)^3 a^3 - \frac{4\pi}{3}\,a^3. \qquad (1.1.27)$$

Expanding the cubic power of the binomial, linearizing the resulting expression with respect to $\mathrm{d}t$, and rearranging, we find that

$$\frac{1}{\delta V}\frac{\mathrm{d}\,\delta V}{\mathrm{d}t} = \alpha. \qquad (1.1.28)$$

This expression justifies calling the divergence of the velocity, $\alpha \equiv \nabla \cdot \mathbf{u}$, the rate of expansion or rate of dilatation of the fluid.

Summary

We have found that a small fluid parcel translates, rotates, expands, and deforms by rates that are determined by the local velocity, vorticity, rate of expansion, and rate-of-strain tensor, as illustrated in Figure 1.1.1. To first order in $\mathrm{d}t$, the sequence by which these motions occur is inconsequential.

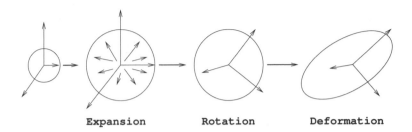

Expansion   **Rotation**   **Deformation**

FIGURE 1.1.1   Illustration of the expansion, rotation, and deformation of a small spherical fluid parcel during an infinitesimal period of time in a three-dimensional flow.

1.1.7 Irrotational and rotational flows

If the vorticity is everywhere zero in a flow, the flow and the velocity field are called irrotational. For example, since the curl of the gradient of any continuous scalar function is identically zero, as shown in identity (A.6.13), Appendix A, the velocity field $\mathbf{u} = \nabla\phi$ is irrotational for any differentiable function, ϕ. The rate of rotation of slender fluid parcels resembling needles averaged all possible orientations is zero in an irrotational flow (e.g., [370]). If the vorticity is nonzero at least in some part of a flow, the flow and the velocity field are called rotational. Sometimes the qualifiers "rotational" and "irrotational" are casually attributed to the fluid, and we speak of an irrotational fluid to describe a fluid that executes irrotational motion. However, although this may be an acceptable simplification, we should keep in mind that, strictly speaking, irrotationality is not a property of the fluid but a kinematic attribute of the flow.

A number of common flows consist of adjacent regions of nearly rotational and nearly irrotational flow. For example, high-speed flow past a streamlined body is irrotational everywhere except inside a thin layer lining the surface of the body and inside a slender wake behind the body, as discussed in Chapter 8. The flow between two parallel streams that merge at different velocities is irrotational everywhere except inside a shear layer along the interface. Other examples of irrotational flows will be presented in later chapters.

Vorticity and circulatory motion

It is important to bear in mind that the occurrence of global circulatory motion does not necessarily mean that individual fluid parcels undergo rotation. For example, the circulatory flow generated by the steady rotation of an infinite cylinder in an ambient fluid of infinite expanse is irrotational, which means that small elongated fluid particles parallel to the principal axes of the rate-of-deformation tensor retain their orientation during any infinitesimal time interval as they circulate around the cylinder.

1.1.8 Cartesian tensors

Consider two Cartesian coordinate systems with common origin, (x_1, x_2, x_3) and (x'_1, x'_2, x'_3), as shown in Figure 1.1.2. The Cartesian unit vectors in the unprimed and primed systems are denoted,

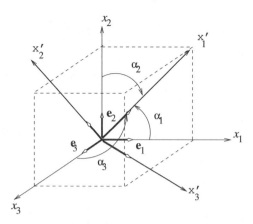

FIGURE 1.1.2   Illustration of two coordinate systems with shared origin. The cosines of the angles, α_1, α_2, and α_3, are the direction cosines of the x_1' axis, subject to the convention that $0 \le \alpha_i \le \pi$. The bold arrows are Cartesian unit vectors.

respectively, by $(\mathbf{e}_1, \mathbf{e}_2, \mathbf{e}_3)$ and $(\mathbf{e}_1', \mathbf{e}_2', \mathbf{e}_3')$. The position of a point particle in physical space can be expressed in the unprimed or primed coordinates and corresponding unit vectors as

$$\mathbf{X} = x_i\,\mathbf{e}_i = x_i'\,\mathbf{e}_i', \tag{1.1.29}$$

where summation is implied over the repeated index, i. The primed coordinates are related to the unprimed coordinates, and *vice versa*, by an orthogonal transformation.

Orthogonal transformations

It is useful to introduce a transformation matrix, $\mathbf{A}$, with elements

$$A_{ij} = \mathbf{e}_i' \cdot \mathbf{e}_j. \tag{1.1.30}$$

Using the geometrical interpretation of the inner vector product, we find that the first row of $\mathbf{A}$ contains the cosines of the angles subtended between each unit vector along the x_1, x_2, and x_3 axes, and the unit vector in the direction of the x_1' axis,

$$A_{11} = \cos\alpha_1, \qquad A_{12} = \cos\alpha_2, \qquad A_{13} = \cos\alpha_3, \tag{1.1.31}$$

called the direction cosines of $\mathbf{e}_1'$, as shown in Figure 1.1.1. The second and third rows of $\mathbf{A}$ contain the corresponding direction cosines for the unit vectors along the x_2' and x_3' axes, $\mathbf{e}_2'$ and $\mathbf{e}_3'$.

   Projecting (1.1.29) on $\mathbf{e}_p$ or $\mathbf{e}_p'$, where p is a free index, and using the orthogonality of the primed and unprimed unit vectors, we find that the $\mathbf{x}' = (x_1', x_2', x_3')$ and $\mathbf{x} = (x_1, x_2, x_3)$ coordinates are related by the linear transformations

$$\mathbf{x}' = \mathbf{A} \cdot \mathbf{x} \qquad \mathbf{x} = \mathbf{A}^T \cdot \mathbf{x}', \tag{1.1.32}$$

where the superscript T indicates the matrix transpose. In index notation, these equations read

$$x'_i = A_{ij}x_j, \qquad x_i = x'_j A_{ji}. \tag{1.1.33}$$

Similar transformation rules can be written for the components of a physical vector, such as the velocity vector, the vorticity vector, and the angular velocity vector.

Relations (1.1.32) demonstrate that the transformation matrix, $\mathbf{A}$, is orthogonal, meaning that its inverse, indicated by the superscript -1, is equal to its transpose, $\mathbf{A}^{-1} = \mathbf{A}^T$. The determinant of $\mathbf{A}$ and the length of any vector represented by any column or row are equal to unity. The projection of any column or row onto a different column or row is zero.

Rotation matrices

In practice, the primed axes can be generated from the unprimed axes by three sequential rotations about the x_1, x_2, and x_3 axes, by respective angles equal to φ_1, φ_2, and φ_3. We may say that the primed system is rotated with respect to the unprimed system, and *vice versa*. To develop the pertinent transformations, we introduce the rotation matrices

$$\mathbf{R}^{(1)} = \begin{bmatrix} 1 & 0 & 0 \\ 0 & \cos\varphi_1 & \sin\varphi_1 \\ 0 & -\sin\varphi_1 & \cos\varphi_1 \end{bmatrix}, \qquad \mathbf{R}^{(2)} = \begin{bmatrix} \cos\varphi_2 & 0 & -\sin\varphi_2 \\ 0 & 1 & 0 \\ \sin\varphi_2 & 0 & \cos\varphi_2 \end{bmatrix},$$

$$\tag{1.1.34}$$

$$\mathbf{R}^{(3)} = \begin{bmatrix} \cos\varphi_3 & \sin\varphi_3 & 0 \\ -\sin\varphi_3 & \cos\varphi_3 & 0 \\ 0 & 0 & 1 \end{bmatrix}.$$

Each rotation matrix is orthogonal, meaning that its transpose is equal to its inverse. The determinant of each rotation matrix and the length of any vector represented by a column or row are equal to unity. The mutual projection of any two different columns or rows is zero. The orthogonal transformation matrix of interest is

$$\mathbf{A} = \mathbf{R}^{(3)} \cdot \mathbf{R}^{(2)} \cdot \mathbf{R}^{(1)}, \tag{1.1.35}$$

where the order of multiplication is consequential.

Dyadic base expansion

Now we consider a two-dimensional (two-index) matrix, $\mathbf{T}$, whose elements are the components of a physical variable, $\boldsymbol{\tau}$, in a Cartesian dyadic base, $\mathbf{e}_i \otimes \mathbf{e}_j$, so that

$$\boldsymbol{\tau} = T_{ij}\,\mathbf{e}_i \otimes \mathbf{e}_j, \tag{1.1.36}$$

where summation is implied over the repeated indices, i and j. The elements of the matrix $\mathbf{T}$ depend on position in space and possibly time. The same physical variable can be expanded in the rotated dyadic base defined with respect to the primed unit vectors as

$$\boldsymbol{\tau} = T'_{ij}\,\mathbf{e}'_i \otimes \mathbf{e}'_j, \tag{1.1.37}$$

where T'_{ij} are the component of $\boldsymbol{\tau}$ in the primed axes.

Projecting (1.1.36) and (1.1.37) from the right onto a unit vector, $\mathbf{e}_m$, noting that $\mathbf{e}_j \cdot \mathbf{e}_m = \delta_{jm}$, where δ_{jm} is the Kronecker delta, and using (1.1.29), we find that

$$\boldsymbol{\tau} \cdot \mathbf{e}_m = T_{im}\,\mathbf{e}_i = T'_{ij}\,A_{jm}\,\mathbf{e}'_i. \tag{1.1.38}$$

Projecting this expression onto a unit vector, $\mathbf{e}_p$, and working in a similar fashion, we obtain

$$\mathbf{e}_p \cdot \boldsymbol{\tau} \cdot \mathbf{e}_m = T_{pm} = T'_{ij}\,A_{jm}\,A_{ip}. \tag{1.1.39}$$

Renaming the indices and working in a similar fashion, we derive the distinguishing properties of a second-order Cartesian tensor,

$$T_{ij} = T'_{kl}A_{ki}A_{lj}, \qquad T'_{ij} = A_{ik}A_{jl}T_{kl}. \tag{1.1.40}$$

In vector notation,

$$\mathbf{T} = \mathbf{A}^T \cdot \mathbf{T}' \cdot \mathbf{A}, \qquad \mathbf{T}' = \mathbf{A} \cdot \mathbf{T} \cdot \mathbf{A}^T. \tag{1.1.41}$$

The transformations (1.1.40) and (1.1.41) are special cases of similarity transformations encountered in matrix calculus (Problem 1.1.5). Because the matrix $\mathbf{A}$ is orthogonal, its transpose can be replaced by its matrix inverse.

Invariants

The characteristic polynomial of a Cartesian tensor, $\mathcal{P}(\lambda) = \det(\mathbf{T} - \lambda \mathbf{I})$, is independent of the coordinate system where the tensor is evaluated. Consequently, the coefficients and hence the roots of the characteristic polynomial defining the eigenvalues of $\mathbf{T}$ are invariant under a change of coordinates. To demonstrate the invariance of the characteristic polynomial, we recall that the transformation matrix $\mathbf{A}$ is orthogonal and write

$$T_{ij} - \lambda \delta_{ij} = T'_{kl}A_{ki}A_{lj} - \lambda A_{ki}A_{kj} = T'_{kl}A_{ki}A_{lj} - \lambda A_{ki}\delta_{kl}A_{lj} = A_{ki}\left(T'_{kl} - \lambda\,\delta_{kl}\right)A_{lj}. \tag{1.1.42}$$

In vector notation, we obtain the statement

$$\mathbf{T} - \lambda\,\mathbf{I} = \mathbf{A}^T \cdot (\mathbf{T}' - \lambda\mathbf{I}) \cdot \mathbf{A}. \tag{1.1.43}$$

Taking the determinant of both sides, expanding the determinant of the product, and noting that $\det(\mathbf{A}^T) = 1/\det(\mathbf{A})$, we find that

$$\det[\mathbf{T} - \lambda\mathbf{I}] = \det(\mathbf{A}^T)\det[\mathbf{T}' - \lambda\mathbf{I}]\det(\mathbf{A}) = \det[\mathbf{T}' - \lambda\,\mathbf{I}], \tag{1.1.44}$$

which completes the proof.

3×3 *tensors*

In the case of 3×3 three tensors of central interest in fluid mechanics, we recast the characteristic polynomial into the form

$$\det(\mathbf{T} - \lambda\mathbf{I}) = -\lambda^3 + I_3\lambda^2 - I_2\lambda + I_1, \tag{1.1.45}$$

involving the three invariants

$$I_1 = \det(\mathbf{T}) = \lambda_1\lambda_2\lambda_3, \qquad I_2 = \lambda_1\lambda_2 + \lambda_2\lambda_3 + \lambda_3\lambda_1 = \frac{1}{2}\,\text{trace}^2(\mathbf{T}) - \text{trace}(\mathbf{T}^2),$$

$$I_3 = \text{trace}(\mathbf{T}) = \lambda_1 + \lambda_2 + \lambda_3, \tag{1.1.46}$$

defined in terms of the roots of the characteristic polynomial, λ_1, λ_2, and λ_3, which are the eigenvalues of $\mathbf{T}$. In Chapter 3, we will see that the invariants (1.1.46) play an important role in developing constitutive equations that provide us with expressions for the stresses developing inside a fluid as the result of the motion.

Kinematic tensors

The Cartesian components of the velocity transform like the components of the position vector, as shown in (1.1.32),

$$u_i' = A_{ij}u_j, \qquad u_i = u_j'A_{ji}. \tag{1.1.47}$$

Using the chain rule of differentiation, we transform the velocity gradient matrix $\mathbf{L}$ introduced in (1.1.2) as

$$L_{ij} = \frac{\partial u_j}{\partial x_i} = \frac{\partial x_k'}{\partial x_i}\frac{\partial u_j}{\partial x_k'} = A_{ki}\frac{\partial u_j}{\partial x_k'} = A_{ki}\,A_{lj}\,\frac{\partial u_l'}{\partial x_k'}, \tag{1.1.48}$$

which is tantamount to

$$L_{ij} = L_{kl}'A_{ki}A_{lj}. \tag{1.1.49}$$

Comparing (1.1.49) with the first equation in (1.1.40), we conclude that $\mathbf{L}$ is a second-order Cartesian tensor. We note that the third invariant I_3 defined in (1.1.46) is equal to the rate of expansion, $\alpha \equiv \nabla \cdot \mathbf{u}$, and confirm that the rate of dilatation of a fluid parcel is independent of the coordinate system where the flow is described.

The symmetric and antisymmetric parts of $\mathbf{L}$ also obey the transformation rules (1.1.40). Thus, the rate-of-strain matrix, $\mathbf{E}$, and vorticity matrix, $\boldsymbol{\Xi}$, are both second-order Cartesian tensors. It is a straightforward exercise to show that the sequence of matrices, $\mathbf{L}\cdot\mathbf{L}$, $\mathbf{L}\cdot\mathbf{L}\cdot\mathbf{L}$, ..., $\mathbf{E}\cdot\mathbf{E}$, $\mathbf{E}\cdot\mathbf{E}\cdot\mathbf{E}$, ..., are all second-order Cartesian tensors.

Problems

1.1.1 *Relative velocity near a point*

(*a*) Confirm the validity of the decomposition (1.1.15).

(*b*) Show that (1.1.20) is consistent with (1.1.18).

1.1.2 *Eigenvalues of a real and symmetric matrix*

Prove that a real and symmetric matrix has real eigenvalues and an orthogonal set of eigenvectors (e.g., [18, 317]).

1.1.3 *Rigid-body motion*

Prove that, if the rate-of-deformation tensor, $\mathbf{E}$, and rate of expansion, α, are zero everywhere in the domain of a flow, the flow must necessarily express rigid-body motion, including translation and rotation.

1.1.4 *Momentum tensor*

Show that the matrix $\rho\, u_i u_j$ is a second-order tensor, called the momentum tensor, where ρ is the fluid density defined in Section 1.3.

1.1.5 *Similarity transformations*

Consider a square matrix, $\mathbf{A}$, select a nonsingular square matrix, $\mathbf{P}$, whose dimensions match those of $\mathbf{A}$, and compute the new matrix $\mathbf{B} = \mathbf{P}^{-1}\!\cdot\!\mathbf{A}\!\cdot\!\mathbf{P}$. This operation is called a similarity transformation, and we say that the matrix $\mathbf{B}$ is similar to $\mathbf{A}$. Show that the eigenvalues of the matrix $\mathbf{B}$ are identical to those of $\mathbf{A}$; thus, similarity transformations preserve the eigenvalues (e.g., [18, 317]).

1.1.6 *Tensor properties*

(*a*) Derive the first relation in (1.1.41).

(*b*) Show that a two-dimensional matrix, $\mathbf{T}$, qualifies as a second-order tensor if $\mathbf{u}(\mathbf{x}) = \mathbf{T}(\mathbf{x}) \cdot \mathbf{x}$ transforms like a vector according to (1.1.32) and (1.1.47).

(*c*) Consider a function $f(x)$ defined through its Taylor series expansion. Show that if $\mathbf{T}$ is a tensor, then $\mathbf{f}(\mathbf{T})$ is also a tensor.

1.2   Curvilinear coordinates

The position vector, velocity field, vorticity field, or any other vector field of interest in fluid mechanics can be described by its components in orthogonal or nonorthogonal curvilinear coordinates, as discussed in Sections A.8–A.17, Appendix A. The velocity gradient tensor, or any other Cartesian tensor, can be represented by its components in a corresponding dyadic base.

1.2.1   Orthogonal curvilinear coordinates

In the case of orthogonal curvilinear coordinates, $(\alpha_1, \alpha_2, \alpha_3)$, with corresponding unit vectors $\mathbf{e}_1, \mathbf{e}_2$, and $\mathbf{e}_3$, the velocity is resolved into corresponding components, u_1, u_2, and u_3, so that

$$\mathbf{u} = u_1\,\mathbf{e}_1 + u_2\,\mathbf{e}_2 + u_3\,\mathbf{e}_3, \tag{1.2.1}$$

as discussed in Section A.8. The gradient operator takes the form

$$\nabla \equiv \mathbf{e}_1\,\frac{1}{h_1}\,\frac{\partial}{\partial\alpha_1} + \mathbf{e}_2\,\frac{1}{h_2}\,\frac{\partial}{\partial\alpha_2} + \mathbf{e}_3\,\frac{1}{h_3}\,\frac{\partial}{\partial\alpha_3}, \tag{1.2.2}$$

where h_1, h_2, and h_3 are metric coefficients. The velocity gradient admits the dyadic representation

$$\mathbf{L} \equiv \nabla\mathbf{u} = L_{ij}\,\mathbf{e}_i \otimes \mathbf{e}_j, \tag{1.2.3}$$

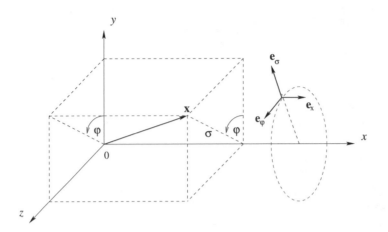

FIGURE 1.2.1   Illustration of cylindrical polar coordinates, (x, σ, φ), defined with respect to companion Cartesian coordinates, (x, y, z).

where summation is implied over the repeated indices, i and j. To prevent misinterpretation, the curvilinear components of the velocity gradient tensor, L_{ij}, should not be collected into a matrix.

The rate of expansion is the trace of the velocity gradient tensor. We note that the trace of the matrix $\mathbf{e}_i \otimes \mathbf{e}_j$ is zero if $i \neq j$ or unity if $i = j$, and obtain the expression

$$\nabla \cdot \mathbf{u} = L_{11} + L_{22} + L_{33}. \tag{1.2.4}$$

The components of the vorticity are given by

$$\omega_1 = L_{23} - L_{32}, \qquad \omega_2 = L_{31} - L_{13}, \qquad \omega_3 = L_{12} - L_{21}. \tag{1.2.5}$$

Other properties are discussed in Section A.8, Appendix A.

Cylindrical, spherical, and plane polar coordinates discussed in Sections A.9–A.11, Appendix A, are used extensively in theoretical analysis and engineering applications. Expressions for the velocity gradient in the corresponding dyadic base are derived in this section.

1.2.2 Cylindrical polar coordinates

Cylindrical polar coordinates, (x, σ, φ), are defined in Figure 1.2.1 (see also Section A.9, Appendix A). Using elementary trigonometry, we derive relations between the polar cylindrical and associated Cartesian coordinates,

$$y = \sigma \cos \varphi, \qquad z = \sigma \sin \varphi. \tag{1.2.6}$$

The inverse relations providing us with the cylindrical polar coordinates in terms of Cartesian coordinates are

$$\sigma = \sqrt{y^2 + z^2}, \qquad \varphi = \arccos \frac{y}{\sigma}. \tag{1.2.7}$$

Velocity gradient

To derive the components of the velocity gradient tensor in cylindrical polar coordinates, we express the gradient operator and velocity field as

$$\nabla = \mathbf{e}_x \frac{\partial}{\partial x} + \mathbf{e}_\sigma \frac{\partial}{\partial \sigma} + \mathbf{e}_\varphi \frac{1}{\sigma} \frac{\partial}{\partial \varphi}, \qquad \mathbf{u} = u_x \mathbf{e}_x + u_\sigma \mathbf{e}_\sigma + u_\varphi \mathbf{e}_\varphi, \qquad (1.2.8)$$

where $\mathbf{e}_x$, $\mathbf{e}_\sigma$, and $\mathbf{e}_\varphi$ are the unit vectors defined in Figure 1.2.1, and u_x, u_σ, and u_φ are the corresponding velocity components. The cylindrical polar unit vectors are related to the Cartesian unit vectors by

$$\mathbf{e}_\sigma = \cos\varphi\, \mathbf{e}_y + \sin\varphi\, \mathbf{e}_z, \qquad \mathbf{e}_\varphi = -\sin\varphi\, \mathbf{e}_y + \cos\varphi\, \mathbf{e}_z. \qquad (1.2.9)$$

The inverse relations are

$$\mathbf{e}_y = \cos\varphi\, \mathbf{e}_\sigma - \sin\varphi\, \mathbf{e}_\varphi, \qquad \mathbf{e}_z = \sin\varphi\, \mathbf{e}_\sigma + \cos\varphi\, \mathbf{e}_\varphi. \qquad (1.2.10)$$

All derivatives $\partial \mathbf{e}_\alpha / \partial \beta$ are zero, except for two derivatives,

$$\frac{\partial \mathbf{e}_\sigma}{\partial \varphi} = \mathbf{e}_\varphi, \qquad \frac{d\mathbf{e}_\varphi}{d\varphi} = -\mathbf{e}_\sigma, \qquad (1.2.11)$$

where Greek variables stand for x, σ, or φ. Using the representations (1.2.8), we find that

$$\mathbf{L} = \left(\mathbf{e}_x \otimes \frac{\partial}{\partial x} + \mathbf{e}_\sigma \otimes \frac{\partial}{\partial \sigma} + \frac{1}{\sigma} \mathbf{e}_\varphi \otimes \frac{\partial}{\partial \varphi} \right) (u_x \mathbf{e}_x + u_\sigma \mathbf{e}_\sigma + u_\varphi \mathbf{e}_\varphi). \qquad (1.2.12)$$

Carrying out the differentiation and using (1.2.11), we obtain

$$\mathbf{L} \equiv \nabla \mathbf{u} = L_{\alpha\beta}\, \mathbf{e}_\alpha \otimes \mathbf{e}_\beta, \qquad (1.2.13)$$

where summation is implied over the repeated indices, α and β. The cylindrical polar components of the velocity gradient tensor, $L_{\alpha\beta}$, are given in Table 1.2.1(a).

To generate the matrix $\mathbf{e}_\sigma \otimes \mathbf{e}_\varphi$, we write $\mathbf{e}_\sigma = \begin{bmatrix} 0, \cos\varphi, \sin\varphi \end{bmatrix}$ and $\mathbf{e}_\varphi = \begin{bmatrix} 0, -\sin\varphi, \cos\varphi \end{bmatrix}$, and formulate their tensor product,

$$\mathbf{e}_\sigma \otimes \mathbf{e}_\varphi = \begin{bmatrix} 0 & 0 & 0 \\ 0 & -\cos\varphi\sin\varphi & \cos^2\varphi \\ 0 & -\sin^2\varphi & \cos\varphi\sin\varphi \end{bmatrix}. \qquad (1.2.14)$$

The rest of the matrices, $\mathbf{e}_\alpha \otimes \mathbf{e}_\beta$, are formulated in a similar fashion.

Rate of expansion and vorticity

The rate of expansion is the trace of the velocity gradient tensor. Making substitutions, we obtain

$$\nabla \cdot \mathbf{u} = L_{xx} + L_{\sigma\sigma} + L_{\varphi\varphi} = \frac{\partial u_x}{\partial x} + \frac{1}{\sigma} \frac{\partial(\sigma u_\sigma)}{\partial \sigma} + \frac{1}{\sigma} \frac{\partial u_\varphi}{\partial \varphi}. \qquad (1.2.15)$$

(*a*)

$$L_{xx} = \frac{\partial u_x}{\partial x} \qquad L_{x\sigma} = \frac{\partial u_\sigma}{\partial x} \qquad L_{x\varphi} = \frac{\partial u_\varphi}{\partial x}$$

$$L_{\sigma x} = \frac{\partial u_x}{\partial \sigma} \qquad L_{\sigma\sigma} = \frac{\partial u_\sigma}{\partial \sigma} \qquad L_{\sigma\varphi} = \frac{\partial u_\varphi}{\partial \sigma}$$

$$L_{\varphi x} = \frac{1}{\sigma}\frac{\partial u_x}{\partial \varphi} \qquad L_{\varphi\sigma} = \frac{1}{\sigma}\frac{\partial u_\sigma}{\partial \varphi} - \frac{u_\varphi}{\sigma} \qquad L_{\varphi\varphi} = \frac{1}{\sigma}\frac{\partial u_\varphi}{\partial \varphi} + \frac{u_\sigma}{\sigma}$$

(*b*)

$$L_{rr} = \frac{\partial u_r}{\partial r} \qquad L_{r\theta} = \frac{\partial u_\theta}{\partial r} \qquad L_{r\varphi} = \frac{\partial u_\varphi}{\partial r}$$

$$L_{\theta r} = \frac{1}{r}\frac{\partial u_r}{\partial \theta} - \frac{u_\theta}{r} \qquad L_{\theta\theta} = \frac{1}{r}\frac{\partial u_\theta}{\partial \theta} + \frac{u_r}{r} \qquad L_{\theta\varphi} = \frac{u_\theta}{r}\frac{\partial u_\varphi}{\partial \theta}$$

$$L_{\varphi r} = \frac{1}{r\sin\theta}\frac{\partial u_r}{\partial \varphi} - \frac{u_\varphi}{r} \qquad L_{\varphi\theta} = \frac{1}{r\sin\theta}\frac{\partial u_\theta}{\partial \varphi} - \frac{\cot\theta}{r}u_\varphi \qquad L_{\varphi\varphi} = \frac{1}{r\sin\theta}\frac{\partial u_\varphi}{\partial \varphi} + \frac{u_r + u_\theta\cot\theta}{r}$$

(*c*)

$$L_{rr} = \frac{\partial u_r}{\partial r} \qquad L_{r\theta} = \frac{\partial u_\theta}{\partial r}$$

$$L_{\theta r} = \frac{1}{r}\frac{\partial u_r}{\partial \theta} - \frac{u_\theta}{r} \qquad L_{\theta\theta} = \frac{1}{r}\frac{\partial u_\theta}{\partial \theta} + \frac{u_r}{r}$$

TABLE 1.2.1   Components of the velocity gradient tensor in (*a*) cylindrical, (*b*) spherical, and (*c*) plane polar coordinates.

The cylindrical polar components of the vorticity are

$$\omega_x = L_{\sigma\varphi} - L_{\varphi\sigma} = \frac{1}{\sigma}\left(\frac{\partial(\sigma u_\varphi)}{\partial \sigma} - \frac{1}{\sigma}\frac{\partial u_\sigma}{\partial \varphi}\right), \qquad \omega_\sigma = L_{\varphi x} - L_{x\varphi} = \frac{1}{\sigma}\frac{\partial u_x}{\partial \varphi} - \frac{\partial u_\varphi}{\partial x},$$

$$\omega_\varphi = L_{x\sigma} - L_{\sigma x} = \frac{\partial u_\sigma}{\partial x} - \frac{\partial u_x}{\partial \sigma}. \tag{1.2.16}$$

If a fluid rotates as a rigid body around the x axis with angular velocity Ω, the velocity components are $u_x = 0$, $u_\sigma = 0$, and $u_\varphi = \Omega\sigma$, and the only surviving vorticity component is $\omega_x = 2\Omega$.

1.2.3   Spherical polar coordinates

Spherical polar coordinates, (r, θ, φ), are defined in Figure 1.2.2 (see also Section A.10, Appendix A). Using elementary trigonometry, we derive the following relations between the Cartesian, cylindrical,

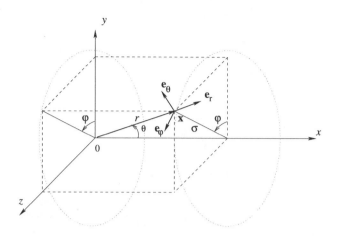

FIGURE 1.2.2   Illustration of spherical polar coordinates, (r, θ, φ), defined with respect to companion Cartesian coordinates, (x, y, z).

and spherical polar coordinates,

$$x = r \cos \theta, \qquad \sigma = r \sin \theta, \tag{1.2.17}$$

and

$$y = \sigma \cos \varphi = r \sin \theta \cos \varphi, \qquad z = \sigma \sin \varphi = r \sin \theta \sin \varphi. \tag{1.2.18}$$

The inverse relations are

$$r = \sqrt{x^2 + y^2 + z^2} = \sqrt{x^2 + \sigma^2}, \qquad \theta = \arccos \frac{x}{r}, \qquad \varphi = \arccos \frac{y}{\sigma}. \tag{1.2.19}$$

Velocity gradient

To derive the components of the velocity gradient in spherical polar coordinates, we express the gradient operator and velocity field as

$$\nabla = \mathbf{e}_x \frac{\partial}{\partial r} + \mathbf{e}_x \frac{1}{r} \frac{\partial}{\partial \theta} + \mathbf{e}_\varphi \frac{1}{r \sin \theta} \frac{\partial}{\partial \varphi}, \qquad \mathbf{u} = u_r \mathbf{e}_r + u_\theta \mathbf{e}_\theta + u_\varphi \mathbf{e}_\varphi, \tag{1.2.20}$$

where $\mathbf{e}_r$, $\mathbf{e}_\theta$, and $\mathbf{e}_\varphi$ are unit vectors defined in Figure 1.2.2, and u_r, u_θ, and u_φ are the corresponding velocity components. The Cartesian, spherical, and cylindrical polar unit vectors are related by

$$\begin{aligned} \mathbf{e}_r &= \cos \theta \, \mathbf{e}_x + \sin \theta \cos \varphi \, \mathbf{e}_y + \sin \theta \sin \varphi \, \mathbf{e}_z = \cos \theta \, \mathbf{e}_x + \sin \theta \, \mathbf{e}_\sigma, \\ \mathbf{e}_\theta &= -\sin \theta \, \mathbf{e}_x + \cos \theta \cos \varphi \, \mathbf{e}_y + \cos \theta \sin \varphi \, \mathbf{e}_z = -\sin \theta \, \mathbf{e}_x + \cos \theta \, \mathbf{e}_\sigma, \\ \mathbf{e}_\varphi &= -\sin \varphi \, \mathbf{e}_y + \cos \varphi \, \mathbf{e}_z. \end{aligned} \tag{1.2.21}$$

All derivatives $\partial \mathbf{e}_\alpha / \partial \beta$ are zero, except for five derivatives,

$$\frac{\mathrm{d}\mathbf{e}_r}{\mathrm{d}\theta} = \mathbf{e}_\theta, \qquad \frac{\mathrm{d}\mathbf{e}_r}{\mathrm{d}\varphi} = \sin\theta\,\mathbf{e}_\varphi, \qquad \frac{\mathrm{d}\mathbf{e}_\theta}{\mathrm{d}\theta} - \mathbf{e}_r, \qquad \frac{\mathrm{d}\mathbf{e}_\theta}{\mathrm{d}\varphi} = \cos\theta\,\mathbf{e}_\varphi,$$

$$\frac{\mathrm{d}\mathbf{e}_\varphi}{\mathrm{d}\varphi} = -\cos\theta\,\mathbf{e}_\theta - \sin\theta\,\mathbf{e}_r, \tag{1.2.22}$$

where Greek variables stand for r, θ, or φ. Using the representations (1.2.20), we derive the dyadic expansion

$$\mathbf{L} \equiv \nabla\mathbf{u} = L_{\alpha\beta}\,\mathbf{e}_\alpha \otimes \mathbf{e}_\beta, \tag{1.2.23}$$

where summation is implied over the repeated indices, α and β. The spherical components of the velocity gradient tensor, $L_{\alpha\beta}$, are given in Table 1.2.1(b). The individual matrices, $\mathbf{e}_\alpha \otimes \mathbf{e}_\beta$, can be formulated using expressions (1.2.21).

Rate of expansion and vorticity

The rate of expansion is the trace of the velocity gradient tensor,

$$\nabla \cdot \mathbf{u} = L_{rr} + L_{\theta\theta} + L_{\varphi\varphi} = \frac{\partial u_r}{\partial r} + 2\frac{u_r}{r} + \frac{1}{r}\frac{\partial u_\theta}{\partial \theta} + \frac{u_\theta}{r}\cot\theta + \frac{1}{r\sin\theta}\frac{\partial u_\varphi}{\partial \varphi}. \tag{1.2.24}$$

The spherical polar components of the vorticity are

$$\omega_r = L_{\theta\varphi} - L_{\varphi\theta} = \frac{1}{r\sin\theta}\left(\frac{\partial(\sin\theta\,u_\varphi)}{\partial\theta} - \frac{\partial u_\theta}{\partial\varphi}\right), \qquad \omega_\theta = L_{\varphi r} - L_{r\varphi} = \frac{1}{r}\left(\frac{1}{\sin\theta}\frac{\partial u_r}{\partial\varphi} - \frac{\partial(r u_\varphi)}{\partial r}\right),$$

$$\omega_\varphi = L_{r\theta} - L_{\theta r} = \frac{1}{r}\left(\frac{\partial(r u_\theta)}{\partial r} - \frac{\partial u_r}{\partial\theta}\right). \tag{1.2.25}$$

1.2.4   Plane polar coordinates

Plane polar coordinates, (r, θ), are defined in Figure 1.2.3 (see also Section A.11, Appendix A). Using elementary trigonometry, we derive relations between the plane polar and corresponding Cartesian coordinates,

$$x = r\cos\theta, \qquad y = r\sin\theta, \tag{1.2.26}$$

and the inverse relations,

$$r = \sqrt{x^2 + y^2}, \qquad \theta = \arccos\frac{y}{r}. \tag{1.2.27}$$

Velocity gradient

To derive the components of the velocity gradient in plane polar coordinates, we express the gradient operator and velocity field as

$$\nabla = \mathbf{e}_r\frac{\partial}{\partial r} + \mathbf{e}_\theta\frac{1}{r}\frac{\partial}{\partial\theta}, \qquad \mathbf{u} = u_r\mathbf{e}_r + u_\theta\mathbf{e}_\theta, \tag{1.2.28}$$

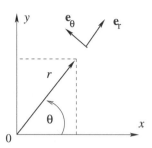

FIGURE 1.2.3   Illustration of plane polar coordinates, (r, θ), defined with respect to companion Cartesian coordinates, (x, y).

where $\mathbf{e}_r$ and $\mathbf{e}_\theta$ are unit vectors in the r and θ directions, and u_r and u_θ are the corresponding velocity components.

The plane polar unit vectors are related to the Cartesian unit vectors by

$$\mathbf{e}_r = \cos\theta\,\mathbf{e}_x + \sin\theta\,\mathbf{e}_y, \qquad \mathbf{e}_\theta = -\sin\theta\,\mathbf{e}_x + \cos\theta\,\mathbf{e}_y. \tag{1.2.29}$$

The inverse relations are

$$\mathbf{e}_x = \cos\theta\,\mathbf{e}_r - \sin\theta\,\mathbf{e}_\theta, \qquad \mathbf{e}_y = \sin\theta\,\mathbf{e}_r + \cos\theta\,\mathbf{e}_\theta. \tag{1.2.30}$$

All derivatives $\partial\mathbf{e}_\alpha/\partial\beta$ are zero, except for two derivatives,

$$\frac{\partial\mathbf{e}_r}{\partial\theta} = \mathbf{e}_\theta, \qquad \frac{\mathrm{d}\mathbf{e}_\theta}{\mathrm{d}\theta} = -\mathbf{e}_r, \tag{1.2.31}$$

where Greek variables stand for r or θ. Using the representations (1.2.28), we derive the dyadic expansion

$$\mathbf{L} \equiv \nabla\mathbf{u} = L_{\alpha\beta}\,\mathbf{e}_\alpha \otimes \mathbf{e}_\beta, \tag{1.2.32}$$

where summation is implied over the repeated indices α and β. The plane polar components of the velocity gradient tensor, $L_{\alpha\beta}$, are given in Table 1.2.1(c). The individual matrices, $\mathbf{e}_\alpha \otimes \mathbf{e}_\beta$, can be formulated using expressions (1.2.29).

Rate of expansion and vorticity

The rate of expansion is the trace of the velocity gradient tensor,

$$\nabla\cdot\mathbf{u} = L_{rr} + L_{\theta\theta} = \frac{1}{r}\Big(\frac{\partial(ru_r)}{\partial r} + \frac{\partial u_\theta}{\partial\theta}\Big). \tag{1.2.33}$$

The nonzero component of the vorticity pointing along the z axis is

$$\omega_z = L_{r\theta} - L_{\theta r} = \frac{1}{r}\Big(\frac{\partial(ru_\theta)}{\partial r} - \frac{\partial u_r}{\partial\theta}\Big). \tag{1.2.34}$$

For a fluid that rotates as a rigid body around the origin of the xy plane with angular velocity Ω, $u_r = 0$, $u_\theta = \Omega r$, and $\omega_z = 2\Omega$.

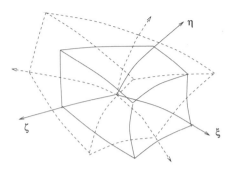

FIGURE 1.2.4   Illustration of nonorthogonal curvilinear coordinates. The solid lines represent covariant coordinates, (ξ, η, ζ), and the dashed lines represent the associated contravariant coordinates.

1.2.5   Axisymmetric flow

The velocity of an axisymmetric flow lies in an azimuthal plane of constant azimuthal angle, φ, measured around the axis of revolution. An example is laminar streaming (uniform) flow past a stationary sphere. Referring to cylindrical or spherical polar coordinates, we set $u_\varphi = 0$ and obtain

$$\mathbf{u} = u_x \mathbf{e}_x + u_\sigma \mathbf{e}_\sigma = u_r \mathbf{e}_r + u_\theta \mathbf{e}_\theta, \tag{1.2.35}$$

where the velocity components (u_x, u_σ) and (u_r, u_θ) depend on (x, σ) in cylindrical polar coordinates or (r, θ) in spherical polar coordinates, but are independent of the azimuthal angle, φ. The vorticity of an axisymmetric flow points in the azimuthal direction,

$$\boldsymbol{\omega} = \omega_\varphi \, \mathbf{e}_\varphi. \tag{1.2.36}$$

The azimuthal vorticity component, ω_φ, depends on (x, σ) or (r, θ), but is independent of φ.

1.2.6   Swirling flow

The velocity of a swirling flow points in the azimuthal direction,

$$\mathbf{u} = u_\varphi \, \mathbf{e}_\varphi. \tag{1.2.37}$$

An example of a swirling flow is the laminar flow due to the slow rotation of a sphere in a fluid of infinite expanse. The azimuthal velocity component, u_φ, depends on (x, σ) in cylindrical polar coordinates or (r, θ) in spherical polar coordinates, but is independent of the azimuthal angle, φ. A swirling flow can be superposed on an axisymmetric flow. The velocity of an axisymmetric flow in the presence of swirling motion is independent of the azimuthal angle, φ. The vorticity is oriented in any arbitrary direction.

1.2.7   Nonorthogonal curvilinear coordinates

A system of nonorthogonal curvilinear coordinates, (ξ, η, ζ), is illustrated in Figure 1.2.4. To describe a flow, we introduce covariant and contravariant base vectors and corresponding coordinates and vector components, as discussed in Sections A.12–A.17, Appendix A. The gradient operator and velocity vector can be expressed in contravariant or covariant component form. Correspondingly, the velocity gradient tensor can be expressed in its pure contravariant, pure covariant, or mixed component form, in four combinations.

The physical meaning of these representations stems from the geometrical interpretation of the covariant and contravariant base vectors, combined with the definition of the directional derivative of the velocity as the projection from the left of the velocity gradient tensor onto a unit vector pointing in a desired direction.

Problems

1.2.1 *Rigid-body rotation*

Compute the velocity gradient tensor, rate-of-deformation tensor, and vorticity of a two-dimensional flow expressing rigid-body rotation with angular velocity Ω. The plane polar velocity components are $u_r = 0$ and $u_\theta = \Omega r$.

1.2.2 *Two-dimensional tensor base in plane polar coordinates*

Formulate the four dyadic matrices $\mathbf{e}_\alpha \otimes \mathbf{e}_\beta$ in plane polar coordinates, where α and β are r or θ.

1.2.3 *Components of the velocity gradient tensor*

Derive the components of the velocity gradient tensor shown in (a) Table 1.2.1(a), (b) Table 1.2.1(b), and (c) Table 1.2.1(c).

1.3 Lagrangian labels of point particles

In Section 1.1, we introduced the fluid velocity, $\mathbf{u}$, in terms of the average velocity of the molecules constituting a fluid parcel, by taking the outer limit as the size of the parcel tends to zero. This point of view led us to regarding $\mathbf{u}$ a field function of position, $\mathbf{x}$, and time, t, writing $\mathbf{u}(\mathbf{x}, t)$. Now we compute the ratio between the mass and volume of a parcel and take the outer limit as the size of the parcel becomes infinitesimal to obtain the fluid density, ρ, as a function of position, $\mathbf{x}$, and time, t, writing $\rho(\mathbf{x}, t)$.

Eulerian framework

We can repeat the process for any appropriate kinematic or intensive thermodynamic variable, f, such as a spatial or temporal derivative of the velocity, the kinetic energy, the thermal energy, the enthalpy, or the entropy per unit mass of a fluid, and thus regard that variable as a function of $\mathbf{x}$ and t, writing

$$f(\mathbf{x}, t). \tag{1.3.1}$$

This point of view establishes an Eulerian framework for describing the kinematic structure of a flow and the physical or thermodynamic properties of a fluid.

Point particles

It is physically appealing and often mathematically convenient to describe the state of a fluid and structure of a flow in terms of the state and motion of the constituent point particles. As a first step, we identify the point particles by assigning to each one of them an identification vector, $\boldsymbol{\alpha}$, consisting of three dimensional or dimensionless scalar variables called Lagrangian labels that take values in a subset of an appropriate set of real vectors. In practice, the triplet

$$\boldsymbol{\alpha} = (\alpha_1, \alpha_2, \alpha_3) \tag{1.3.2}$$

can be identified with the Cartesian or some other curvilinear coordinates of the point particles at a specified instant in time. At any instant, the lines of constant α_1, α_2, and α_3 define a right-handed system of curvilinear coordinates in physical space, as discussed in Sections A.8–A.17, Appendix A.

Lagrangian framework

The value of any kinematic, physical, or intensive thermodynamic variable at a particular location, $\mathbf{x}$, and at a certain time instant, t, can be regarded as a property of the point particle that happens to be at that location at that particular instant. Thus, for any appropriate scalar, vector, or matrix function f that can be attributed to a point particle in a meaningful fashion, we may write

$$f(\mathbf{x}, t) = f[\mathbf{X}(\boldsymbol{\alpha}, t), t] = \beta \, \mathcal{F}(\boldsymbol{\alpha}, t), \tag{1.3.3}$$

where $\mathbf{X}(\boldsymbol{\alpha}, t)$ is the position of the point particle labeled $\boldsymbol{\alpha}$ at time instant t, β is a constant, and $\mathcal{F}$ is an appropriate variable expressing a suitable property of the point particles. Equation (1.3.3) suggests that, in order to obtain the value of the function f at a point $\mathbf{x}$ at time t, we may identify the point particle located at $\mathbf{x}$ at time t, read its label $\boldsymbol{\alpha}$, measure the variable $\mathcal{F}$, and multiply the value of $\mathcal{F}$ thus obtained by the coefficient β to produce f. Often f and $\mathcal{F}$ represent the same physical variable and their distinction is based solely upon the choice of independent variables used to describe a flow.

Velocity, vorticity, and velocity gradient

Applying (1.3.3) for the velocity, we obtain

$$\mathbf{u}(\mathbf{x}, t) = \mathbf{u}(\mathbf{X}(\boldsymbol{\alpha}, t), t) = \mathbf{U}(\boldsymbol{\alpha}, t). \tag{1.3.4}$$

In this case, the variable f is the fluid velocity, $\mathbf{u}$, the variable $\mathcal{F}$ is the velocity of the point particle labeled $\boldsymbol{\alpha}$ at time instant t, denoted by $\mathbf{U}(\boldsymbol{\alpha}, t)$, and the coefficient β is equal to unity. Equation (1.3.4) suggests that, to obtain the fluid velocity $\mathbf{u}$ at a point $\mathbf{x}$ at time t, we may identify the point particle that resides at $\mathbf{x}$ at time t, look up its label, $\boldsymbol{\alpha}$, and measure its velocity, $\mathbf{U}$.

Applying (1.3.3) for the vorticity, we obtain

$$\boldsymbol{\omega}(\mathbf{x}, t) = \boldsymbol{\omega}(\mathbf{X}(\boldsymbol{\alpha}, t), t) = 2\,\boldsymbol{\Omega}(\boldsymbol{\alpha}, t). \tag{1.3.5}$$

In this case, the variable f is the vorticity, $\boldsymbol{\omega}$, the variable $\mathcal{F}$ is the angular velocity of the point particle labeled $\boldsymbol{\alpha}$ at time instant t, denoted by $\boldsymbol{\Omega}(\mathbf{a}, t)$, and the coefficient β is equal to 2.

Applying (1.3.3) for the velocity gradient tensor, we obtain

$$L_{ij}(\mathbf{x}, t) = L_{ij}\big(\mathbf{X}(\boldsymbol{\alpha}, t), t\big) = \frac{\partial U_j}{\partial X_i}(\boldsymbol{\alpha}, t). \tag{1.3.6}$$

The expression between the two equal signs in (1.3.6) is the velocity gradient at the location of the point particle labeled $\boldsymbol{\alpha}$ at time t, computed in terms of the relative velocity of neighboring point particles, where $\mathbf{X}$ is the point-particle position. In this case, the coefficient β is equal to unity.

1.3.1 The material derivative

Since the velocity of a point particle is equal to the rate of change of its position, $\mathbf{X}$, we may express the point-particle velocity as

$$\mathbf{U}(\boldsymbol{\alpha}, t) = \left(\frac{\partial \mathbf{X}}{\partial t}\right)_{\boldsymbol{\alpha}}(\boldsymbol{\alpha}, t). \tag{1.3.7}$$

The partial derivative with respect to time keeping $\boldsymbol{\alpha}$ constant is known as the substantial, substantive, or material derivative, and is denoted by $\mathrm{D}/\mathrm{D}t$. Accordingly, equation (1.3.7) can be expressed in the compact form

$$\mathbf{U}(\boldsymbol{\alpha}, t) = \frac{\mathrm{D}\mathbf{X}}{\mathrm{D}t}(\boldsymbol{\alpha}, t). \tag{1.3.8}$$

In classical mechanics, the material derivative, $\mathrm{D}/\mathrm{D}t$, is identical to the time derivative of a body or particle, $\mathrm{d}/\mathrm{d}t$, as discussed in Section 1.5.

Chain rule

We have at our disposal two sets of independent variables that can be used to describe a flow, including the Eulerian set, $(\mathbf{x}, t)$, and the Lagrangian set, $(\boldsymbol{\alpha}, t)$. A relationship between the partial derivatives of a function f with respect to these two sets can be established by applying the chain rule, obtaining

$$\frac{\mathrm{D}f}{\mathrm{D}t} = \left(\frac{\partial f(\mathbf{X}(\boldsymbol{\alpha}, t), t)}{\partial t}\right)_{\boldsymbol{\alpha}} = \left(\frac{\partial f}{\partial t}\right)_{\mathbf{x}} + \left(\frac{\partial f}{\partial x_i}\right)_t \frac{\mathrm{D}X_i}{\mathrm{D}t}. \tag{1.3.9}$$

For simplicity, we drop the parentheses around the Eulerian partial derivatives on the right-hand side and use (1.3.8) and (1.3.4) to obtain

$$\frac{\mathrm{D}f}{\mathrm{D}t} = \frac{\partial f}{\partial t} + u_i \frac{\partial f}{\partial x_i} = \frac{\partial f}{\partial t} + \mathbf{u} \cdot \nabla f, \tag{1.3.10}$$

which relates the material derivative to temporal and spatial derivatives with respect to Eulerian variables. Summation is implied over the repeated index i in the central expression.

If all point particles retain their value of f as they move about the domain of a flow, the material derivative vanishes, $\mathrm{D}f/\mathrm{D}t = 0$, and we say that the field represented by f is convected by the flow.

Convective derivative

The term $\mathbf{u} \cdot \nabla f$ in (1.3.10) can be written as $|\mathbf{u}| \, \partial f / \partial l_u$, where $\partial f / \partial l_u$ is the rate of change of f with respect to arc length measured in the direction of the velocity, l_u. If the field f is steady, $\partial f / \partial t = 0$, and if the point point particles retain their value of f as they move in the domain of flow, $\mathrm{D}f / \mathrm{D}t = 0$, then $\partial f / \partial l_u$ must be zero. In that case, the value of f does not change in the direction of the velocity and is therefore constant along paths traveled by point particles, identified as streamlines. However, the value of f is generally different along different streamlines.

1.3.2   Point-particle acceleration

Applying (1.3.10) for the velocity, we derive an expression for the acceleration of a point particle in the Eulerian form

$$a_j \equiv \frac{\mathrm{D}u_j}{\mathrm{D}t} = \frac{\partial u_j}{\partial t} + u_i \frac{\partial u_j}{\partial x_i}, \tag{1.3.11}$$

where summation is implied over the repeated index, i. In vector notation,

$$\mathbf{a} \equiv \frac{\mathrm{D}\mathbf{u}}{\mathrm{D}t} = \frac{\partial \mathbf{u}}{\partial t} + \mathbf{u} \cdot \nabla \mathbf{u} = \frac{\partial \mathbf{u}}{\partial t} + \mathbf{u} \cdot \mathbf{L}, \tag{1.3.12}$$

where $\mathbf{L}$ is the velocity gradient tensor. Explicitly, the components of the point-particle acceleration are given in Table 1.3.1(*a*). If a point particle neither accelerates nor decelerates as it moves about the domain of a flow, then $\mathrm{D}\mathbf{u}/\mathrm{D}t = \mathbf{0}$.

The velocity distribution in a fluid that rotates steadily as a rigid body around the origin with angular velocity $\boldsymbol{\Omega}$ is $\mathbf{u} = \boldsymbol{\Omega} \times \mathbf{x}$, and the associated vorticity is $\boldsymbol{\omega} = 2\boldsymbol{\Omega}$. Setting $\partial \mathbf{u}/\partial t = \mathbf{0}$, noting that the velocity gradient tensor is equal to the vorticity tensor, $\boldsymbol{\Xi}$, and using (1.1.18), we obtain the jth component of the acceleration,

$$a_j = u_i \, \Xi_{ij} = \frac{1}{2} \, u_i \epsilon_{ijk} \, \omega_k = \epsilon_{jki} \, \Omega_k u_i, \tag{1.3.13}$$

which shows that

$$\mathbf{a} = \boldsymbol{\Omega} \times \mathbf{u} = \boldsymbol{\Omega} \times (\boldsymbol{\Omega} \times \mathbf{x}). \tag{1.3.14}$$

Invoking the geometrical interpretation of the cross product, we find that the acceleration points toward the axis of rotation and its magnitude is proportional to the distance from the axis of rotation.

Cylindrical polar coordinates

To derive the components of the point-particle acceleration in cylindrical polar coordinates, we express the velocity in these coordinates and the velocity gradient tensor in the dyadic form shown in (1.2.13). Formulating the product $\mathbf{u} \cdot \mathbf{L}$, we derive the expressions shown in Table 1.3.1(*b*). Alternative forms in terms of the material derivative of the cylindrical polar velocity components are

$$a_x = \frac{\partial u_x}{\partial t} + \mathbf{u} \cdot \nabla u_x = \frac{\mathrm{D}u_x}{\mathrm{D}t}, \qquad a_\sigma = \frac{\partial u_\sigma}{\partial t} + \mathbf{u} \cdot \nabla u_\sigma - \frac{u_\varphi^2}{\sigma} = \frac{\mathrm{D}u_\sigma}{\mathrm{D}t} - \frac{u_\varphi^2}{\sigma},$$

$$a_\varphi = \frac{\partial u_\varphi}{\partial t} + \mathbf{u} \cdot \nabla u_\varphi + \frac{u_\sigma u_\varphi}{\sigma} = \frac{\mathrm{D}u_\varphi}{\mathrm{D}t} + \frac{u_\sigma u_\varphi}{\sigma}. \tag{1.3.15}$$

(a)

$$a_x \equiv \frac{\mathrm{D}u_x}{\mathrm{D}t} = \frac{\partial u_x}{\partial t} + \mathbf{u} \cdot \nabla u_x = \frac{\partial u_x}{\partial t} + u_x \frac{\partial u_x}{\partial x} + u_y \frac{\partial u_x}{\partial y} + u_z \frac{\partial u_x}{\partial z}$$

$$a_y \equiv \frac{\mathrm{D}u_y}{\mathrm{D}t} = \frac{\partial u_y}{\partial t} + \mathbf{u} \cdot \nabla u_y = \frac{\partial u_y}{\partial t} + u_x \frac{\partial u_y}{\partial x} + u_y \frac{\partial u_y}{\partial y} + u_z \frac{\partial u_y}{\partial z}$$

$$a_z \equiv \frac{\mathrm{D}u_z}{\mathrm{D}t} = \frac{\partial u_z}{\partial t} + \mathbf{u} \cdot \nabla u_z = \frac{\partial u_z}{\partial t} + u_x \frac{\partial u_z}{\partial x} + u_y \frac{\partial u_z}{\partial y} + u_z \frac{\partial u_z}{\partial z}$$

(b)

$$a_x = \frac{\partial u_x}{\partial t} + u_x \frac{\partial u_x}{\partial x} + u_\sigma \frac{\partial u_x}{\partial \sigma} + \frac{u_\varphi}{\sigma} \frac{\partial u_x}{\partial \varphi}$$

$$a_\sigma = \frac{\partial u_\sigma}{\partial t} + u_x \frac{\partial u_\sigma}{\partial x} + u_\sigma \frac{\partial u_\sigma}{\partial \sigma} + \frac{u_\varphi}{\sigma} \frac{\partial u_\sigma}{\partial \varphi} - \frac{u_\varphi^2}{\sigma}$$

$$a_\varphi = \frac{\partial u_\varphi}{\partial t} + u_x \frac{\partial u_\varphi}{\partial x} + u_\sigma \frac{\partial u_\varphi}{\partial \sigma} + \frac{u_\varphi}{\sigma} \frac{\partial u_\varphi}{\partial \varphi} + \frac{u_\sigma u_\varphi}{\sigma}$$

(c)

$$a_r = \frac{\partial u_r}{\partial t} + u_r \frac{\partial u_r}{\partial r} + \frac{u_\theta}{r} \frac{\partial u_r}{\partial \theta} + \frac{u_\varphi}{r \sin\theta} \frac{\partial u_r}{\partial \varphi} - \frac{u_\theta^2 + u_\varphi^2}{r}$$

$$a_\theta = \frac{\partial u_\theta}{\partial t} + u_r \frac{\partial u_\theta}{\partial r} + \frac{u_\theta}{r} \frac{\partial u_\theta}{\partial \theta} + \frac{u_\varphi}{r \sin\theta} \frac{\partial u_\theta}{\partial \varphi} + \frac{u_r u_\theta}{r} - \frac{\cot\theta}{r} u_\varphi^2$$

$$a_\varphi = \frac{\partial u_\varphi}{\partial t} + u_r \frac{\partial u_\varphi}{\partial r} + \frac{u_\theta}{r} \frac{\partial u_\varphi}{\partial \theta} + \frac{u_\varphi}{r \sin\theta} \frac{\partial u_\varphi}{\partial \varphi} + \frac{u_\varphi}{r} \left(u_r + u_\theta \cot\theta \right)$$

(d)

$$a_r = \frac{\partial u_r}{\partial t} + u_r \frac{\partial u_r}{\partial \theta} + \frac{u_\theta}{r} \frac{\partial u_r}{\partial \theta} - \frac{u_\theta^2}{r}$$

$$a_\theta = \frac{\partial u_\theta}{\partial t} + u_r \frac{\partial u_\theta}{\partial r} + \frac{u_\theta}{r} \frac{\partial u_\theta}{\partial \theta} + \frac{u_r u_\theta}{r}$$

TABLE 1.3.1   Components of the point-particle acceleration in (a) Cartesian, (b) cylindrical polar, (c) spherical polar, and (d) plane polar coordinates.

These expressions illustrate that the cylindrical polar components of the acceleration are *not* simply equal to the material derivative of the corresponding polar components of the velocity. In the case of a fluid that rotates steadily as a rigid body around the x axis with angular velocity Ω, so that $u_x = 0$, $u_\sigma = 0$, and $u_\varphi = \Omega\sigma$, we find that $a_x = 0$, $a_\sigma = -u_\varphi^2/\sigma$, and $a_\varphi = 0$. These expressions are consistent with the corresponding Cartesian form (1.3.14).

Spherical polar coordinates

Working as in the case of cylindrical polar coordinates, we derive the spherical polar components of the point-particle acceleration shown in Table 1.3.1(*c*). Alternative forms in terms of the material derivative of the spherical polar velocity components are

$$a_r = \frac{Du_r}{Dt} - \frac{u_\theta^2 + u_\varphi^2}{r}, \qquad a_\theta = \frac{Du_\theta}{Dt} + \frac{u_r u_\theta}{r} - \frac{\cot\theta}{r}u_\varphi^2,$$

$$a_\varphi = \frac{Du_\varphi}{Dt} + \frac{u_\varphi}{r}\left(u_r + u_\theta\cot\theta\right). \tag{1.3.16}$$

We observe that the spherical polar components of the acceleration are *not* simply equal to the material derivative of the corresponding polar components of the velocity.

Plane polar coordinates

Working as in the case of cylindrical polar coordinates, we derive the plane polar components of the point-particle acceleration shown in Table 1.3.1(*d*). Alternative forms in terms of the material derivative of the plane polar velocity components are

$$a_r = \frac{Du_r}{Dt} - \frac{u_\theta^2}{r}, \qquad a_\theta = \frac{Du_\theta}{Dt} + \frac{u_r u_\theta}{r}. \tag{1.3.17}$$

We observe once again that the plane polar components of the acceleration are *not* simply equal to the material derivative of the corresponding polar components of the velocity.

1.3.3   Lagrangian mapping

We have regarded a fluid as a particulate medium consisting of an infinite collection of point particles identified by a vector label, $\boldsymbol{\alpha}$. The instantaneous position of a point particle, $\mathbf{X}$, is a function of $\boldsymbol{\alpha}$ and time, t. This functional dependence can be formalized in terms of a generally time-dependent mapping of the Cartesian labeling space of $\boldsymbol{\alpha}$ to the physical space, $\boldsymbol{\alpha} \to \mathbf{X}$, written in the symbolic form

$$\mathbf{X}(t) = \mathcal{C}_t(\boldsymbol{\alpha}), \tag{1.3.18}$$

as illustrated in Figure 1.3.1. The subscript t emphasizes that the mapping function $\mathcal{C}_t$ may change in time. Some authors unfortunately call the mapping function $\mathcal{C}_t(\boldsymbol{\alpha})$ a motion. We will assume that $\mathcal{C}_t$ is a differentiable function of the three scalar components of $\boldsymbol{\alpha}$.

It is important to realize that $\mathcal{C}_t$ is time-dependent even when the velocity field is steady, and is constant only if a point particle labeled $\boldsymbol{\alpha}$ is stationary. If we identify the label $\boldsymbol{\alpha}$ with the

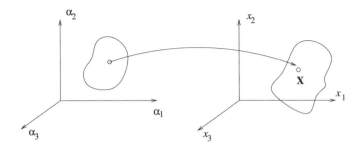

FIGURE 1.3.1   Lagrangian mapping of the parametric space, $\boldsymbol{\alpha}$, to the physical space, x. The position of a point particle in physical space is denoted by **X**.

Cartesian coordinates of the point particles at the origin of time, $t = 0$, then $\boldsymbol{\alpha} = \mathbf{X}(t = 0)$ and $\mathcal{C}_0(\boldsymbol{\alpha}) = \mathbf{X}(t = 0)$.

Lagrange Jacobian tensor

A differential vector in labeling space, $\mathrm{d}\boldsymbol{\alpha}$, is related to a differential vector in physical space, $\mathrm{d}\mathbf{X}$, by the equation

$$\mathrm{d}\mathbf{X} = \mathrm{d}\boldsymbol{\alpha} \cdot \mathcal{J} = \mathcal{J}^T \cdot \mathrm{d}\boldsymbol{\alpha}, \tag{1.3.19}$$

where the superscript T denotes the matrix transpose. We have introduced the Jacobian matrix of the mapping function, $\mathcal{C}_t$, with elements

$$\mathcal{J}_{ij} = \frac{\partial \mathcal{C}_{t_j}}{\partial \alpha_i}. \tag{1.3.20}$$

In index notation, equation (1.3.19) takes the from

$$\mathrm{d}X_i = \alpha_j \mathcal{J}_{ji} = \mathcal{J}_{ij}^T \mathrm{d}\alpha_j, \tag{1.3.21}$$

where summation is implied over the repeated index, j. Explicitly,

$$\mathcal{J} \equiv \widetilde{\nabla}\mathbf{X} = \begin{bmatrix} \dfrac{\partial X_1}{\partial \alpha_1} & \dfrac{\partial X_2}{\partial \alpha_1} & \dfrac{\partial X_3}{\partial \alpha_1} \\[2mm] \dfrac{\partial X_1}{\partial \alpha_2} & \dfrac{\partial X_2}{\partial \alpha_2} & \dfrac{\partial X_3}{\partial \alpha_2} \\[2mm] \dfrac{\partial X_1}{\partial \alpha_3} & \dfrac{\partial X_2}{\partial \alpha_3} & \dfrac{\partial X_3}{\partial \alpha_3} \end{bmatrix}, \tag{1.3.22}$$

where $\widetilde{\nabla} = (\partial/\partial\alpha_1, \partial/\partial\alpha_2, \partial/\partial\alpha_3)$ is the gradient in the labeling space. For convenience, we have defined $X_1 \equiv X$, $X_2 \equiv Y$, and $X_3 \equiv Z$. By definition, the components of the Lagrange Jacobian tensor satisfy the relations

$$\frac{\partial \mathcal{J}_{ij}}{\partial \alpha_k} = \frac{\partial \mathcal{J}_{kj}}{\partial \alpha_i} \tag{1.3.23}$$

for any i, j, and k. When these conditions are fulfilled, the Lagrangian field $\mathbf{X}(\boldsymbol{\alpha})$ can be constructed up to an arbitrary constant by integrating the differential equations (1.3.20).

Lagrangian metric

The differential vector $(\mathrm{d}\alpha_1, 0, 0)$ in labeling space is mapped to the following differential vector in physical space,

$$\mathrm{d}\mathbf{X}^{(1)} = \frac{\partial \mathcal{C}_t}{\partial \alpha_1}\, \mathrm{d}\alpha_1. \tag{1.3.24}$$

Similar equations can be written for $\mathrm{d}\mathbf{X}^{(2)}$ and $\mathrm{d}\mathbf{X}^{(3)}$,

$$\mathrm{d}\mathbf{X}^{(2)} = \frac{\partial \mathcal{C}_t}{\partial \alpha_2}\, \mathrm{d}\alpha_2, \qquad \mathrm{d}\mathbf{X}^{(3)} = \frac{\partial \mathcal{C}_t}{\partial \alpha_3}\, \mathrm{d}\alpha_3. \tag{1.3.25}$$

The three vectors, $\mathrm{d}\mathbf{X}^{(1)}$, $\mathrm{d}\mathbf{X}^{(2)}$, and $\mathrm{d}\mathbf{X}^{(3)}$, are not necessarily orthogonal.

A differential volume in labeling space, $\mathrm{d}V(\boldsymbol{\alpha})$, is mapped to a corresponding differential volume in physical space, $\mathrm{d}V(\mathbf{X})$. To derive a relationship between the magnitudes of these two volumes, we formulate the triple mixed product defining the volume,

$$(\mathrm{d}\mathbf{X}^{(1)} \times \mathrm{d}\mathbf{X}^{(2)}) \cdot \mathrm{d}\mathbf{X}^{(3)} = \left(\frac{\partial \mathcal{C}_t}{\partial \alpha_1} \times \frac{\partial \mathcal{C}_t}{\partial \alpha_2} \right) \cdot \frac{\partial \mathcal{C}_t}{\partial \alpha_3}\, (\mathrm{d}\alpha_1\, \mathrm{d}\alpha_2\, \mathrm{d}\alpha_3). \tag{1.3.26}$$

Now we assume that $\mathrm{d}\mathbf{X}^{(i)}$ are arranged according to the right-handed rule, identify the left-hand side with $\mathrm{d}V(\mathbf{X})$, and set $\mathrm{d}\alpha_1\, \mathrm{d}\alpha_2\, \mathrm{d}\alpha_3 = \mathrm{d}V(\boldsymbol{\alpha})$ to obtain

$$\mathrm{d}V(\mathbf{X}) = \mathcal{J}\, \mathrm{d}V(\boldsymbol{\alpha}), \tag{1.3.27}$$

where

$$\mathcal{J} \equiv \left(\frac{\partial \mathcal{C}_t}{\partial \alpha_1} \times \frac{\partial \mathcal{C}_t}{\partial \alpha_2} \right) \cdot \frac{\partial \mathcal{C}_t}{\partial \alpha_3} = \det(\boldsymbol{\mathcal{J}}) = \det(\boldsymbol{\mathcal{J}}^T) \tag{1.3.28}$$

is the Lagrange metric. Equation (1.3.27) identifies the determinant, $\mathcal{J}$, with the ratio of two corresponding infinitesimal volumes in physical and labeling space. For simplicity, we will denote $\mathrm{d}V(\mathbf{X})$ as $\mathrm{d}V$.

Dyadic expansion

Let $\widetilde{\mathbf{e}}_i$ be Cartesian unit vectors in the labeling space, $\boldsymbol{\alpha}$, and $\mathbf{e}_i$ be Cartesian unit vectors in physical space, $\mathbf{x}$, for $i = 1, 2, 3$. The Lagrange Jacobian tensor admits the bichromatic dyadic expansion

$$\boldsymbol{\mathcal{J}} = J_{ij}\, \widetilde{\mathbf{e}}_i \otimes \mathbf{e}_j, \tag{1.3.29}$$

where summation is implied over the repeated indices, i and j. Working as in Section 1.2 for the velocity gradient, we may expand the Lagrange Jacobian tensor in a different base comprised of Cartesian or curvilinear base vectors denoted by Greek indices,

$$\boldsymbol{\mathcal{J}} = J_{\alpha\beta}\, \widetilde{\mathbf{e}}_\beta \otimes \mathbf{e}_\gamma. \tag{1.3.30}$$

For example, $\widetilde{\mathbf{e}}_\beta$ can be cylindrical polar unit vectors in labeling space and $\mathbf{e}_\beta$ can be spherical polar unit vectors in physical space.

1.3.4   Deformation gradient

In the special case where the vector label $\boldsymbol{\alpha}$ is identified with the Cartesian coordinates of a point particle at a specified time, we set $\widetilde{\mathbf{e}}_i = \mathbf{e}_i$ and obtain the monochromatic dyadic expansion

$$\boldsymbol{\mathcal{J}} = J_{ij}\,\mathbf{e}_i \otimes \mathbf{e}_j = J_{\alpha\beta}\,\mathbf{e}_\beta \otimes \mathbf{e}_\gamma, \tag{1.3.31}$$

where summation is implied over the repeated indices, i and j corresponding to Cartesian coordinates, as well as over the repeated indices β and γ corresponding to general curvilinear coordinates. In this case, the transpose of the Lagrange Jacobian tensor is called the deformation gradient. It can be shown that the relative deformation gradient obeys transformation rules that render it a second-order Cartesian tensor (e.g., [365]). A distinguishing feature of the deformation gradient is that it is dimensionless.

Relative deformation gradient

Let us identify $\boldsymbol{\alpha}$ with the Cartesian coordinates of a point particle at an early time, t, and denote the corresponding coordinates of the point particles at the current time τ by $\boldsymbol{\xi}$, so that $\boldsymbol{\xi} = \boldsymbol{\mathcal{C}}_\tau(\boldsymbol{\alpha})$. Having made this choice, we recast equation (1.3.19) into the form

$$d\boldsymbol{\xi} = \mathbf{F}_t(\tau) \cdot d\boldsymbol{\alpha}, \tag{1.3.32}$$

where $\mathbf{F}_t(\tau)$ is the relative deformation gradient, also simply called the deformation gradient and denoted by $\mathbf{F}$, with components

$$F_{ij} = \frac{\partial \xi_i}{\partial \alpha_j}. \tag{1.3.33}$$

Rearranging (1.3.32) for an invertible deformation gradient, we obtain

$$d\boldsymbol{\alpha} = \mathbf{F}^{-1} \cdot d\boldsymbol{\xi}. \tag{1.3.34}$$

It is sometimes useful to introduce the displacement field, $\mathbf{v} \equiv \boldsymbol{\xi} - \boldsymbol{\alpha}$. Solving for $\boldsymbol{\xi}$ and substituting the result into (1.3.33), we obtain

$$F_{ij} = \delta_{ij} + \mathcal{D}_{ij}, \tag{1.3.35}$$

where δ_{ij} is Kronecker's delta and

$$\mathcal{D}_{ij} \equiv \frac{\partial v_i}{\partial \alpha_j} \tag{1.3.36}$$

is the displacement gradient tensor.

Notation in solid and continuum mechanics

In solid and continuum mechanics, equation (1.3.32) holds true, subject to two changes in notation: $\boldsymbol{\xi} \to \mathbf{x}$, and $\boldsymbol{\alpha} \to \mathbf{X}$. Thus, by convention, in the theory of deformable solids, $\mathbf{X}$ represents the Cartesian coordinates of a particle in a reference state and $\mathbf{x}$ are the corresponding coordinates in the current state. Different notation is employed in different texts.

Polar decomposition

The polar decomposition theorem guarantees that the deformation gradient tensor can be resolved into the product of an orthogonal tensor representing rotation, $\mathbf{R}$, and a symmetric and positive-definite tensor representing deformation, $\mathbf{U}$ or $\mathbf{V}$, so that

$$\mathbf{F} = \mathbf{R} \cdot \mathbf{U} = \mathbf{V} \cdot \mathbf{R}, \qquad (1.3.37)$$

where $\mathbf{U}$ is the right stretch tensor and $\mathbf{V}$ is the left stretch tensor. A small material vector $\mathrm{d}\boldsymbol{\alpha}$ deforms under the action of $\mathbf{U}$ and then rotates under the action of $\mathbf{R}$, or rotates under the action of $\mathbf{R}$ and then deforms under the action of $\mathbf{U}$. A practical way of carrying out the decomposition is suggested by the forthcoming equations (1.3.39) and (1.3.48).

Right Green–Lagrange strain tensor

Using (1.3.32), we find that the square of the length of a material vector, $\mathrm{d}\boldsymbol{\xi}$, corresponding to $\mathrm{d}\boldsymbol{\alpha}$, is given by

$$|\mathrm{d}\boldsymbol{\xi}|^2 = \mathrm{d}\boldsymbol{\xi} \cdot \mathrm{d}\boldsymbol{\xi} = (\mathbf{F} \cdot \mathrm{d}\boldsymbol{\alpha}) \cdot (\mathbf{F} \cdot \mathrm{d}\boldsymbol{\alpha}) = \mathrm{d}\boldsymbol{\alpha} \cdot (\mathbf{F}^T \cdot \mathbf{F}) \cdot \mathrm{d}\boldsymbol{\alpha}, \qquad (1.3.38)$$

where the superscript T denotes the matrix transpose. This expression motivates introducing the right Cauchy–Green strain tensor, $\mathbf{C}$, defined in terms of the deformation gradient as

$$\mathbf{C} = \mathbf{F}^T \cdot \mathbf{F} = \mathbf{U}^2. \qquad (1.3.39)$$

By definition,

$$|\mathrm{d}\boldsymbol{\xi}|^2 = \mathrm{d}\boldsymbol{\alpha} \cdot \mathbf{C} \cdot \mathrm{d}\boldsymbol{\alpha}. \qquad (1.3.40)$$

In terms of the displacement gradient tensor,

$$\mathbf{C} = \mathbf{I} + \boldsymbol{\mathcal{D}} + \boldsymbol{\mathcal{D}}^T + \boldsymbol{\mathcal{D}}^T \cdot \boldsymbol{\mathcal{D}}. \qquad (1.3.41)$$

The last quadratic term can be neglected in the case of small (linear) deformation.

Because $\mathbf{C}$ is symmetric and positive definite, it has three real and positive eigenvalues and three corresponding orthogonal eigenvectors (Problem 1.3.6). If $\mathrm{d}\boldsymbol{\phi}_j$ is the jth eigenvector of $\mathbf{C}$ with corresponding eigenvalue s_j^2, then

$$|\mathrm{d}\boldsymbol{\xi}_j| = s_j |\mathrm{d}\boldsymbol{\phi}_j|, \qquad (1.3.42)$$

where $\mathrm{d}\boldsymbol{\xi}_j$ is the image of the eigenvector and s_j is the stretch ratio, for $j = 1, 2, 3$.

The right Green–Lagrange strain tensor is defined as

$$\boldsymbol{\mathcal{E}} \equiv \frac{1}{2}(\mathbf{C} - \mathbf{I}) = \boldsymbol{\epsilon} + \frac{1}{2}\boldsymbol{\mathcal{D}}^T \cdot \boldsymbol{\mathcal{D}}, \qquad (1.3.43)$$

where

$$\boldsymbol{\epsilon} \equiv \frac{1}{2}\left(\boldsymbol{\mathcal{D}} + \boldsymbol{\mathcal{D}}^T\right) \qquad (1.3.44)$$

is the infinitesimal strain tensor. Substituting into (1.3.40) the expression

$$\mathbf{C} = \mathbf{I} + 2\boldsymbol{\mathcal{E}} \tag{1.3.45}$$

and rearranging, we obtain

$$|\mathrm{d}\boldsymbol{\xi}|^2 - |\mathrm{d}\boldsymbol{\alpha}|^2 = 2\,\mathrm{d}\boldsymbol{\alpha}\cdot\boldsymbol{\mathcal{E}}\cdot\mathrm{d}\boldsymbol{\alpha}. \tag{1.3.46}$$

We see that the right Green–Lagrange strain tensor determines the change in the squared length of a material vector due to deformation.

Left Green–Lagrange strain tensor

Using (1.3.34) and working in a similar fashion, we obtain

$$|\mathrm{d}\boldsymbol{\alpha}|^2 = \mathrm{d}\boldsymbol{\alpha}\cdot\mathrm{d}\boldsymbol{\alpha} = (\mathbf{F}^{-1}\cdot\mathrm{d}\boldsymbol{\xi})\cdot(\mathbf{F}^{-1}\cdot\mathrm{d}\boldsymbol{\xi}) = \mathrm{d}\boldsymbol{\xi}\cdot(\mathbf{F}\cdot\mathbf{F}^T)^{-1}\cdot\mathrm{d}\boldsymbol{\xi}. \tag{1.3.47}$$

This expression motivates introducing the left Cauchy–Green strain tensor

$$\mathbf{B} \equiv \mathbf{F}\cdot\mathbf{F}^T = \mathbf{V}^2. \tag{1.3.48}$$

By definition,

$$|\mathrm{d}\boldsymbol{\alpha}|^2 = \mathrm{d}\boldsymbol{\xi}\cdot\mathbf{B}^{-1}\cdot\mathrm{d}\boldsymbol{\xi}. \tag{1.3.49}$$

In terms of the displacement gradient tensor,

$$\mathbf{B} = \mathbf{I} + \boldsymbol{\mathcal{D}} + \boldsymbol{\mathcal{D}}^T + \boldsymbol{\mathcal{D}}\cdot\boldsymbol{\mathcal{D}}^T. \tag{1.3.50}$$

The last quadratic term can be neglected in the case of small (linear) deformation.

Because $\mathbf{B}$ is symmetric and positive definite, it has three real and positive eigenvalues and three corresponding orthogonal eigenvectors (Problem 1.3.6). The eigenvalues of $\mathbf{B}$ are the same as those of $\mathbf{C}$, but the corresponding eigenvectors are generally different. If $\mathrm{d}\boldsymbol{\phi}_j$ is the jth eigenvector of $\mathbf{C}$ corresponding to the eigenvalue s_j^2, then

$$\mathrm{d}\boldsymbol{\psi}_j = (\mathbf{F}^T)^{-1}\cdot\mathrm{d}\boldsymbol{\phi}_j \tag{1.3.51}$$

is an eigenvector of $\mathbf{B}$ corresponding to the same eigenvalue. Applying equation (1.3.49), we obtain

$$|\mathrm{d}\boldsymbol{\alpha}_j| = \frac{1}{s_j}\,|\mathrm{d}\boldsymbol{\psi}_j| \tag{1.3.52}$$

for $j = 1, 2, 3$, where s_j is the stretch ratio.

Constitutive equations

The Cauchy–Green strain tensors find important applications in developing constitutive equations that relate the stresses developing in a fluid to the deformation of fluid parcels, as discussed in Section 3.3.

Problems

1.3.1 *Lagrangian labeling*

Discuss whether it is possible to label all constituent point particles of a three-dimensional parcel using a single scalar variable, or even two scalar variables.

1.3.2 *Material derivative of the acceleration*

Derive an expression for the material derivative of the point-particle acceleration, $\mathrm{D}a/\mathrm{D}t$, in Cartesian coordinates in terms of derivatives of the velocity with respect to Eulerian variables.

1.3.3 *Flow due to the motion of a rigid body*

Consider the flow due to the steady motion of a rigid body translating with velocity $\mathbf{V}$ and rotating about a point $\mathbf{x}_0$ with angular velocity $\boldsymbol{\Omega}$ in an otherwise quiescent fluid of infinite expanse. In a frame of reference moving with the body, the flow is steady. Explain why the velocity field must satisfy the equation

$$\frac{\partial \mathbf{u}}{\partial t} = -\big[\mathbf{V} + \boldsymbol{\Omega} \times (\mathbf{x} - \mathbf{x}_0)\big] \cdot \nabla \mathbf{u}. \tag{1.3.53}$$

Does this equation also apply for a semi-infinite domain of flow bounded by an infinite plane wall?

1.3.4 *Temperature recording of a moving probe*

A temperature probe is moving with velocity $\mathbf{v}(t)$ in a temperature field, $T(\mathbf{x}, t)$. Develop an expression for the rate of the change of the temperature recorded by the probe in terms of $T(\mathbf{x}, t)$ and $\mathbf{v}(t)$.

1.3.5 *Relative deformation gradient*

(*a*) Explain why $\mathbf{F}_{(t)}(t) = \mathbf{I}$, where $\mathbf{I}$ is identity matrix.

(*b*) If a fluid is incompressible, the volume of all fluid parcels remains constant in time, as discussed in Section 1.5.4. Show that, for an incompressible fluid, $\det(\mathbf{F}_{(t)}(\tau)) = 1$.

(*c*) A fluid is undergoing simple shear flow along the x with velocity $u_x = \xi y$, $u_y = 0$, $u_z = 0$, where ξ is a constant shear rate with dimensionless of inverse time. Compute the relative deformation gradient.

1.3.6 *Cauchy–Green strain tensors*

Show that the tensors $\mathbf{B}$ and $\mathbf{C}$ are symmetric and positive definite. A tensor, $\mathbf{A}$, is positive definite if $\mathbf{x} \cdot \mathbf{A} \cdot \mathbf{x} > 0$ for any vector with appropriate length, $\mathbf{x}$.

1.4   Properties of fluid parcels and mass conservation

We have seen that a fluid parcel can be identified by labeling its constituent point particles using a Lagrangian vector field, $\boldsymbol{\alpha}$, that takes values inside a subset of a three-dimensional labeling space, $\mathcal{A}$. Using (1.3.27), we find that the parcel volume is

$$V_p = \iiint_{Parcel} \mathrm{d}V = \iiint_{\mathcal{A}} \mathcal{J} \, \mathrm{d}V(\boldsymbol{\alpha}), \tag{1.4.1}$$

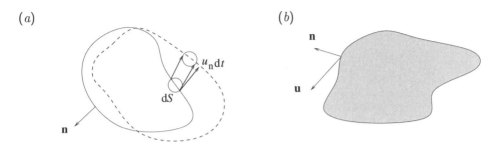

FIGURE 1.4.1   (*a*) Illustration of a convected fluid parcel or stationary control volume in a flow. Point particles over an infinitesimal patch of a material surface, dS, move during a small period of time, dt, spanning a cylindrical volume, d$V = u_n$dt dS, where u_n is the normal velocity. The dashed line outlines the new parcel shape. (*b*) The surface integral $\iint \mathbf{u} \cdot \mathbf{n}$ dS is the rate of change of the parcel volume. In contrast, the surface integral $\iint \rho\,\mathbf{u} \cdot \mathbf{n}$ dS is the rate of convective mass transport outward from a control volume.

where d$V(\boldsymbol{\alpha})$ is an infinitesimal volume in labeling space. The second integral shows that $\mathcal{J}$ plays the role of a volume density distribution function.

1.4.1   Rate of change of parcel volume and Euler's theorem in kinematics

Differentiating (1.4.1) with respect to time and noting that the domain of integration in labeling space is fixed, we find that the rate of change of volume of the parcel is given by

$$\frac{\mathrm{d}V_p}{\mathrm{d}t} = \frac{\mathrm{d}}{\mathrm{d}t}\iiint_{Parcel} \mathrm{d}V = \iiint_{\mathcal{A}} \frac{\mathrm{D}\mathcal{J}}{\mathrm{D}t}\,\mathrm{d}V(\boldsymbol{\alpha}). \qquad (1.4.2)$$

Note that a time derivative of an integral with respect to $\boldsymbol{\alpha}$ over a fixed integration domain, $\mathcal{A}$, is transferred as a material derivative inside the integral.

Since point particles move with the fluid velocity and the parcel shape changes only because of normal motion across the instantaneous parcel configuration, we may also write

$$\frac{\mathrm{d}V_p}{\mathrm{d}t} = \iint_{Parcel} \mathbf{u} \cdot \mathbf{n}\,\mathrm{d}S, \qquad (1.4.3)$$

where $\mathbf{n}$ is the unit normal vector pointing outward from the parcel and dS is a differential surface area, as illustrated in Figure 1.4.1. Both sides of (1.4.3) have units of length cubed divided by time. The thin closed line in Figure 1.4.1(*a*) describes the boundary of a fluid parcel at a certain time, and the bold closed line describes the boundary after time dt. The volume between the two boundaries is the change in the parcel volume after time dt. Further justification for (1.4.3) will be given in Section 1.10.

Using the divergence theorem to convert the surface to a volume integral, and introducing the rate of expansion, $\alpha \equiv \nabla \cdot \mathbf{u}$, we obtain from (1.4.3)

$$\frac{\mathrm{d}V_p}{\mathrm{d}t} = \iiint_{Parcel} \alpha\,\mathrm{d}V = \iiint_{\mathcal{A}} \alpha\,\mathcal{J}\,\mathrm{d}V(\boldsymbol{\alpha}). \qquad (1.4.4)$$

Comparing (1.4.2) with (1.4.4) and noting that the subset $\mathcal{A}$ is arbitrary to eliminate the integral signs, we obtain Euler's theorem of kinematics stating that

$$\frac{1}{\mathcal{J}}\frac{D\mathcal{J}}{Dt} = \alpha. \tag{1.4.5}$$

We have found that the rate of change of the Jacobian, $\mathcal{J}$, following a point particle is proportional to the rate of expansion of the fluid.

A formal but more tedious method of deriving (1.4.5) involves taking the material derivative of the determinant of the Jacobian matrix stated in (1.3.22), finding

$$\frac{D\mathcal{J}}{Dt} = \frac{D}{Dt}\det\left(\begin{bmatrix} \dfrac{\partial X_1}{\partial \alpha_1} & \dfrac{\partial X_2}{\partial \alpha_1} & \dfrac{\partial X_3}{\partial \alpha_1} \\[2mm] \dfrac{\partial X_1}{\partial \alpha_2} & \dfrac{\partial X_2}{\partial \alpha_2} & \dfrac{\partial X_3}{\partial \alpha_2} \\[2mm] \dfrac{\partial X_1}{\partial \alpha_3} & \dfrac{\partial X_2}{\partial \alpha_3} & \dfrac{\partial X_3}{\partial \alpha_3} \end{bmatrix}\right) = \mathcal{J}_1 + \mathcal{J}_2 + \mathcal{J}_3, \tag{1.4.6}$$

where

$$\mathcal{J}_1 = \det\left(\begin{bmatrix} \dfrac{\partial U_1}{\partial \alpha_1} & \dfrac{\partial X_2}{\partial \alpha_1} & \dfrac{\partial X_3}{\partial \alpha_1} \\[2mm] \dfrac{\partial U_1}{\partial \alpha_2} & \dfrac{\partial X_2}{\partial \alpha_2} & \dfrac{\partial X_3}{\partial \alpha_2} \\[2mm] \dfrac{\partial U_1}{\partial \alpha_3} & \dfrac{\partial X_2}{\partial \alpha_3} & \dfrac{\partial X_3}{\partial \alpha_3} \end{bmatrix}\right) = \det\left(\begin{bmatrix} \dfrac{\partial U_1}{\partial X_j}\dfrac{\partial X_j}{\partial \alpha_1} & \dfrac{\partial X_2}{\partial \alpha_1} & \dfrac{\partial X_3}{\partial \alpha_1} \\[2mm] \dfrac{\partial U_1}{\partial X_j}\dfrac{\partial X_j}{\partial \alpha_2} & \dfrac{\partial X_2}{\partial \alpha_2} & \dfrac{\partial X_3}{\partial \alpha_2} \\[2mm] \dfrac{\partial U_1}{\partial X_j}\dfrac{\partial X_j}{\partial \alpha_3} & \dfrac{\partial X_2}{\partial \alpha_3} & \dfrac{\partial X_3}{\partial \alpha_3} \end{bmatrix}\right) = \dfrac{\partial U_1}{\partial X_1}\mathcal{J}. \tag{1.4.7}$$

The determinants $\mathcal{J}_2$ and $\mathcal{J}_3$ are computed in a similar fashion. The corresponding matrices arise from $\mathcal{J}$ by replacing X_2 with U_2 in the second column or X_3 with U_3 in the third column. The results are then added to produce (1.4.5).

Expressing the material derivative in (1.4.5) in terms of Eulerian derivatives, we find that

$$\frac{\partial \mathcal{J}}{\partial t} + \nabla \cdot (\mathcal{J}\mathbf{u}) = 2\,\alpha\mathcal{J}, \tag{1.4.8}$$

which can be regarded as a transport equation for $\mathcal{J}$.

1.4.2   Reynolds transport theorem

The rate of change of a general scalar, vector, or tensor property field, $\mathcal{P}$, integrated over the volume of a fluid parcel is

$$\frac{d}{dt}\iiint_{Parcel} \mathcal{P}\,dV = \frac{d}{dt}\iiint_{\mathcal{A}} \mathcal{P}\mathcal{J}\,dV(\boldsymbol{\alpha}) = \iiint_{\mathcal{A}} \frac{D(\mathcal{P}\mathcal{J})}{Dt}\,dV(\boldsymbol{\alpha}). \tag{1.4.9}$$

For example, $\mathcal{P}$ can be identified with the density or specific thermal energy of the fluid. To derive the last expression in (1.4.9), we have observed once again that the volume of integration in labeling space is independent of time.

Expanding the derivative inside the last integral and using (1.4.5), we find that

$$\frac{\mathrm{d}}{\mathrm{d}t} \iiint_{Parcel} \mathcal{P} \, \mathrm{d}V = \iiint_{Parcel} \left(\frac{\mathrm{D}\mathcal{P}}{\mathrm{D}t} + \mathcal{P} \nabla \cdot \mathbf{u} \right) \mathrm{d}V. \tag{1.4.10}$$

Expressing the material derivative inside the integrand in terms of Eulerian derivatives and rearranging, we obtain

$$\frac{\mathrm{d}}{\mathrm{d}t} \iiint_{Parcel} \mathcal{P} \, \mathrm{d}V = \iiint_{Parcel} \left[\frac{\partial \mathcal{P}}{\partial t} + \nabla \cdot (\mathcal{P}\mathbf{u}) \right] \mathrm{d}V. \tag{1.4.11}$$

Finally, we apply the divergence theorem to the second term inside the last integral and derive the mathematical statement of the Reynolds transport theorem,

$$\frac{\mathrm{d}}{\mathrm{d}t} \iiint_{Parcel} \mathcal{P} \, \mathrm{d}V = \iiint_{Parcel} \frac{\partial \mathcal{P}}{\partial t} \, \mathrm{d}V + \iint_{Parcel} \mathcal{P} \mathbf{u} \cdot \mathbf{n} \, \mathrm{d}S, \tag{1.4.12}$$

where $\mathbf{n}$ is the normal unit vector pointing outward from the parcel and the last integral is computed over the parcel surface. Applying (1.4.12) with $\mathcal{P}$ set to a constant, we recover (1.4.3).

Balance of a transported field over a control volume

The volume integral on the right-hand side of (1.4.12) expresses the rate of accumulation of the physical or mathematical entity represented by $\mathcal{P}$ inside a fixed control volume that coincides with the instantaneous parcel shape. The surface integral on the right-hand side of (1.4.12), $\iint \mathcal{P}\mathbf{u} \cdot \mathbf{n} \, \mathrm{d}S$, expresses the rate of convective transport of the entity $\mathcal{P}$ outward from the control volume. For example, the integral $\iint \rho \mathbf{u} \cdot \mathbf{n} \, \mathrm{d}S$, represents the convective transport of mass outward from a control volume, where ρ is the fluid density. It is interesting to contrast this interpretation with our earlier discovery that the surface integral $\iint \mathbf{u} \cdot \mathbf{n} \, \mathrm{d}S$ represents the rate of change of volume of a fluid whose instantaneous boundary defines a control volume.

The left-hand side of (1.4.12) expresses the rate of accumulation of $\mathcal{P}$ inside the parcel due, for example, to diffusion across the parcel surface. Introducing a corresponding physical law, such as Fick's law of diffusion or Newton's second law of motion, transforms (1.4.12) into a transport equation expressing a balance over a fixed control volume. This interpretation is the cornerstone of transport phenomena (e.g., [36]).

1.4.3 Mass conservation and the continuity equation

In terms of the fluid density, ρ, the mass of a fluid parcel is given by

$$m_p = \iiint_{Parcel} \rho(\mathbf{X}, t) \, \mathrm{d}V(\mathbf{X}) = \iiint_{\mathcal{A}} \rho(\boldsymbol{\alpha}, t) \, \mathcal{J} \, \mathrm{d}V(\boldsymbol{\alpha}), \tag{1.4.13}$$

where $\mathbf{X}$ is the point-particle position, Applying (1.4.9) and (1.4.11) with $\mathcal{P} = \rho$ and requiring that mass neither disappears nor is produced in the flow–which is tantamount to stipulating that fluid parcels retain their mass as they move about the domain of flow–we set $\mathrm{d}m_p/\mathrm{d}t = 0$ and require that the integrands are identically zero. The result is the continuity equation expressing mass conservation for a compressible or incompressible fluid,

$$\frac{\mathrm{D}}{\mathrm{D}t}\big(\rho\,\mathrm{d}V(\mathbf{X})\big) = 0, \qquad \frac{\mathrm{D}}{\mathrm{D}t}\big(\rho\,\mathcal{J}\big) = 0, \qquad \frac{\mathrm{D}\rho}{\mathrm{D}t} + \rho\,\nabla\cdot\mathbf{u} = 0, \qquad \frac{\partial\rho}{\partial t} + \nabla\cdot(\rho\mathbf{u}) = 0. \qquad (1.4.14)$$

Mass conservation imposes a kinematic constraint, demanding that the structure of the velocity field be such that fluid parcels do not tend to occupy the same volume in space, leaving behind empty holes.

Integral mass balance

Now we apply the Reynolds transport theorem (1.4.12) with $\mathcal{P} = \rho$ and set the left-hand side to zero to obtain an integral statement of the continuity equation in Eulerian integral form,

$$\iiint_{V_c} \frac{\partial\rho}{\partial t}\,\mathrm{d}V = -\iint_{B_c} \rho\,\mathbf{u}\cdot\mathbf{n}\,\mathrm{d}S, \qquad (1.4.15)$$

where V_c is a fixed control volume that coincides with the instantaneous volume of a fluid parcel, and B_c is the boundary of the control volume. Equation (1.4.15) states that the rate of accumulation of mass inside a stationary control volume is equal to the rate of mass transport into the control volume through the boundaries.

Equation (1.4.15) can be produced by integrating the last equation in (1.4.14) over the control volume and using the divergence theorem to obtain a surface integral. Working similarly with (1.4.8), we obtain

$$\iiint_{V_c} \frac{\partial\mathcal{J}}{\partial t}\,\mathrm{d}V = -\iint_{B_c} \mathcal{J}\,\mathbf{u}\cdot\mathbf{n}\,\mathrm{d}S + 2\iiint_{V_c} \alpha\mathcal{J}\,\mathrm{d}V, \qquad (1.4.16)$$

which can be regarded as a transport equation for $\mathcal{J}$ in Eulerian integral form, where $\alpha \equiv \nabla\cdot\mathbf{u}$ is the rate of expansion.

1.4.4 Incompressible fluids and solenoidal velocity fields

Since point particles of an incompressible fluid retain their initial density as they move in the domain of flow, the material derivative of the density is zero,

$$\frac{\mathrm{D}\rho}{\mathrm{D}t} = 0. \qquad (1.4.17)$$

The continuity equation (1.4.14) simplifies into

$$\nabla\cdot\mathbf{u} \equiv \frac{\partial u_x}{\partial x} + \frac{\partial u_y}{\partial y} + \frac{\partial u_z}{\partial z} = 0. \qquad (1.4.18)$$

A vector field with vanishing divergence, satisfying (1.4.18), is called solenoidal. According to our discussion in Section 1.1, fluid parcels of an incompressible fluid translate, rotate, and deform while retaining their volume. The terms incompressible fluid, incompressible flow, and incompressible velocity field are sometimes used interchangeably. However, strictly speaking, compressibility is neither a kinematic property of the flow nor a structural property of the velocity field, but rather a physical property of the fluid. Combining (1.4.18) with (1.4.5), we find that

$$\frac{D\mathcal{J}}{Dt} = 0, \tag{1.4.19}$$

which states that the Jacobian is convected with the flow. The mapping function $\mathcal{C}_t$ introduced in (1.3.18) is then called isochoric, from the Greek word $\iota\sigma o\varsigma$, which means "equal," and the word $\chi\omega\rho o\varsigma$, which means "space."

Following are three examples of solenoidal velocity fields describing the motion of an incompressible fluid:

1. Since the divergence of the curl of any continuous vector field is identically zero, a velocity field that derives from a differentiable vector field, $\mathbf{A}$, as $\mathbf{u} = \nabla \times \mathbf{A}$ is solenoidal. The function $\mathbf{A}$ is called the vector potential, as discussed in Section 2.6.

2. Consider a velocity field that derives from the cross product of two arbitrary vector fields, $\mathbf{A}$ and $\mathbf{B}$, as $\mathbf{u} = \mathbf{A} \times \mathbf{B}$. Straightforward differentiation shows that $\nabla \cdot \mathbf{u} = \mathbf{B} \cdot \nabla \times \mathbf{A} - \mathbf{A} \cdot \nabla \times \mathbf{B}$. We observe that, if $\mathbf{A}$ and $\mathbf{B}$ are both irrotational, $\mathbf{u}$ will be solenoidal. Since any irrotational vector field can be expressed as the gradient of a scalar function, as discussed in Section 2.1, the velocity field $\mathbf{u} = \nabla \psi \times \nabla \chi$ is solenoidal for any pair of differentiable functions, ψ and χ.

3. Consider a velocity field that derives from the gradient of a scalar function, ϕ, called the potential function, $\mathbf{u} = \nabla \phi$. This velocity field is solenoidal, provided that ϕ satisfies Laplace's equation, $\nabla^2 \phi = 0$. In that case, ϕ is called a harmonic potential.

Polar coordinates

In cylindrical polar coordinates, (x, σ, φ), the continuity equation for an incompressible fluid takes the form

$$\nabla \cdot \mathbf{u} = \frac{\partial u_x}{\partial x} + \frac{1}{\sigma}\frac{\partial(\sigma u_\sigma)}{\partial \sigma} + \frac{1}{\sigma}\frac{\partial u_\varphi}{\partial \varphi} = 0. \tag{1.4.20}$$

In spherical polar coordinates, (r, θ, φ), the continuity equation for an incompressible fluid takes the form

$$\nabla \cdot \mathbf{u} = \frac{\partial u_r}{\partial r} + 2\frac{u_r}{r} + \frac{1}{r}\frac{\partial u_\theta}{\partial \theta} + \frac{u_\theta}{r}\cot\theta + \frac{1}{r\sin\theta}\frac{\partial u_\varphi}{\partial \varphi} = 0. \tag{1.4.21}$$

In plane polar coordinates, (r, θ), the continuity equation for an incompressible fluid takes the form

$$\nabla \cdot \mathbf{u} = \frac{1}{r}\frac{\partial(ru_r)}{\partial r} + \frac{1}{r}\frac{\partial u_\theta}{\partial \theta} = 0. \tag{1.4.22}$$

Expressions for the divergence of the velocity in more general orthogonal or nonorthogonal coordinates are given in Sections A.8–A.17, Appendix A.

Kinematic reciprocity of incompressible flows

Let $\mathbf{u}$ and $\mathbf{u}'$ be two unrelated solenoidal velocity fields, $\nabla \cdot \mathbf{u} = 0$ and $\nabla \cdot \mathbf{u}' = 0$. Assuming that neither velocity field contains singular points and using (1.4.18), we derive the identity

$$\frac{\partial}{\partial x_j} \left(u_i' \frac{\partial u_j}{\partial x_i} - u_i \frac{\partial u_j'}{\partial x_i} \right) = 0. \tag{1.4.23}$$

Integrating (1.4.23) over an arbitrary control volume that is bounded by a surface, D, and using the divergence theorem, we obtain

$$\iint_D \left(u_i' \frac{\partial u_j}{\partial x_i} - u_i \frac{\partial u_j'}{\partial x_i} \right) n_j \, \mathrm{d}S = 0, \tag{1.4.24}$$

where $\mathbf{n}$ is the unit vector normal to D. Equation (1.4.24) places an integral constraint on the mutual structure of any two incompressible flows over a common surface.

1.4.5 Rate of change of parcel properties

We turn to discussing the computation of the rate of change of an extensive kinematic, physical, or thermodynamic variable of a generally compressible fluid parcel. By definition, an extensive variable is proportional to the parcel's volume. For each extensive variable, there is a corresponding intensive variable so that when the latter is multiplied by the parcel volume and perhaps by a physical constant, it produces the extensive variable.

Pairs of extensive–intensive variables include momentum and velocity, thermal energy and temperature, kinetic energy and square of the magnitude of the velocity. In developing dynamical laws governing the behavior of fluid parcels, it is useful to have expressions for the rate of change of an extensive variable in terms of the rate of change of the corresponding intensive variable. Such expressions can be derived using the continuity equation (1.4.14).

Momentum

An important extensive variable is the linear momentum of a fluid parcel, defined as

$$\mathbf{M}_p \equiv \iiint_{Parcel} \rho(\mathbf{X},t)\, \mathbf{U}(\mathbf{X},t)\, \mathrm{d}V(\mathbf{X}) = \iiint_{\mathcal{A}} \rho(\boldsymbol{\alpha},t)\, \mathbf{U}(\boldsymbol{\alpha},t)\, \mathcal{J}\, \mathrm{d}V(\boldsymbol{\alpha}). \tag{1.4.25}$$

Using the continuity equation (1.4.14), we express the rate of the change of the linear momentum in terms of the acceleration of the point particles as

$$\frac{\mathrm{d}\mathbf{M}_p}{\mathrm{d}t} = \iiint_{\mathcal{A}} \left(\frac{\mathrm{D}\mathbf{U}}{\mathrm{D}t} \rho(\boldsymbol{\alpha},t)\, \mathcal{J} + \mathbf{U}(\boldsymbol{\alpha},t) \frac{\mathrm{D}(\rho J)}{\mathrm{D}t} \right) \mathrm{d}V(\boldsymbol{\alpha}), \tag{1.4.26}$$

and obtain

$$\frac{\mathrm{d}\mathbf{M}_p}{\mathrm{d}t} = \iiint_{Parcel} \rho \frac{\mathrm{D}\mathbf{U}}{\mathrm{D}t} \, \mathrm{d}V(\mathbf{X}). \tag{1.4.27}$$

Equation (1.4.27) is true for incompressible as well as compressible fluids.

Angular momentum

The angular momentum of a fluid parcel computed with respect to the origin is

$$\mathbf{A}_p \equiv \iiint_{Parcel} \rho(\mathbf{X}, t) \left[\mathbf{X} \times \mathbf{U}(\mathbf{X}, t) \right] dV(\mathbf{X}) = \iiint_{\mathcal{A}} \rho(\boldsymbol{\alpha}, t) \left[\mathbf{X} \times \mathbf{U}(\boldsymbol{\alpha}, t) \right] \mathcal{J} \, dV(\boldsymbol{\alpha}). \qquad (1.4.28)$$

Working as in the case of the linear momentum and recalling that $D\mathbf{X}/Dt = \mathbf{U}$, we find that

$$\frac{d\mathbf{A}_p}{dt} = \iiint_{Parcel} \rho \frac{D(\mathbf{X} \times \mathbf{U})}{Dt} \, dV(\mathbf{X}), \qquad (1.4.29)$$

and then

$$\frac{d\mathbf{A}_p}{dt} = \iiint_{Parcel} \rho \, \mathbf{X} \times \frac{D\mathbf{U}}{Dt} \, dV(\mathbf{X}), \qquad (1.4.30)$$

for compressible or incompressible fluids.

Generalization

For an arbitrary scalar, vector, or tensor intensive field, $\mathcal{F}$, we find that

$$\frac{d}{dt} \iiint_{Parcel} \mathcal{F}\rho \, dV = \frac{d}{dt} \iiint_{\mathcal{A}} \mathcal{F}\rho \mathcal{J} \, dV(\boldsymbol{\alpha}), \qquad (1.4.31)$$

yielding

$$\frac{d}{dt} \iiint_{Parcel} \mathcal{F}\rho \, dV = \iiint_{Parcel} \rho \frac{D\mathcal{F}}{Dt} \, dV, \qquad (1.4.32)$$

for compressible or incompressible fluids. Equation (1.4.27) emerges by setting $\mathcal{F} = \mathbf{U}$.

Problems

1.4.1 *Rate of change of an extensive variable*

Derive the right-hand sides of (1.4.30) and (1.4.32).

1.4.2 *Vector potential*

Show that the vector potentials $\mathbf{A}$ and $\mathbf{A} + \nabla f$ generate the same flow, where $\mathbf{A}$ and f are two arbitrary functions. Is this an incompressible flow?

1.4.3 *Reynolds transport theorem*

Show that (1.4.30) is consistent with (1.4.12).

1.5   Point-particle motion

In classical mechanics, a traveling point particle is identified by its Cartesian coordinates,

$$x = X(t), \qquad y = Y(t), \qquad z = Z(t). \qquad (1.5.1)$$

The position of the point particle is

$$\mathbf{X} = X(t)\,\mathbf{e}_x + Y(t)\,\mathbf{e}_y + Z(t)\,\mathbf{e}_z, \tag{1.5.2}$$

where $\mathbf{e}_x$, $\mathbf{e}_y$, and $\mathbf{e}_z$ are Cartesian unit vectors.

Velocity

By definition, the velocity of a point particle, $\mathbf{U}$, is equal to the rate of change of its position in space. If the x coordinate of a point particle has changed by an infinitesimal displacement, dX, over an infinitesimal time period, dt, then, by definition, $U_x = dX/dt$. Writing the counterparts of this equation for the y and z coordinates, we obtain

$$U_x = \frac{dX}{dt}, \qquad U_y = \frac{dY}{dt}, \qquad U_z = \frac{dZ}{dt}. \tag{1.5.3}$$

In vector notation,

$$\mathbf{U} = \frac{d\mathbf{X}}{dt} = U_x\,\mathbf{e}_x + U_y\,\mathbf{e}_y + U_z\,\mathbf{e}_z. \tag{1.5.4}$$

In the present context of isolated point-particle motion, the total derivative, d/dt, is the same as the material derivative, D/Dt.

Acceleration

The acceleration vector, $\mathbf{a}$, is defined as the rate of change of the velocity,

$$\mathbf{a} = a_x\,\mathbf{e}_x + a_y\,\mathbf{e}_y + a_z\,\mathbf{e}_z = \frac{d\mathbf{U}}{dt} = \frac{d^2\mathbf{X}}{dt^2}. \tag{1.5.5}$$

Accordingly, the Cartesian components of the point-particle acceleration are

$$a_x = \frac{d^2X}{dt^2}, \qquad a_y = \frac{d^2Y}{dt^2}, \qquad a_z = \frac{d^2Z}{dt^2}. \tag{1.5.6}$$

If the Cartesian coordinates of a point-particle are constant or change linearly in time, the acceleration is zero.

1.5.1   Cylindrical polar coordinates

The cylindrical polar coordinates of a point particle are determined by the functions

$$x = X(t), \qquad \sigma = \Sigma(t), \qquad \varphi = \Phi(t). \tag{1.5.7}$$

The position of a point particle is given by

$$\mathbf{X} = X(t)\,\mathbf{e}_x + \Sigma(t)\,\mathbf{e}_\sigma, \tag{1.5.8}$$

where $\mathbf{e}_\alpha$ are cylindrical polar unit vectors for $\alpha = x, \sigma, \varphi$. The rates of change of the cylindrical polar unit vectors following the motion of a point particle are given by the relations

$$\frac{d\mathbf{e}_x}{dt} = \mathbf{0}, \qquad \frac{d\mathbf{e}_\sigma}{dt} = \frac{d\Phi}{dt}\,\mathbf{e}_\varphi, \qquad \frac{d\mathbf{e}_\varphi}{dt} = -\frac{d\Phi}{dt}\,\mathbf{e}_\sigma. \tag{1.5.9}$$

Note that the first unit vector, $\mathbf{e}_x$, is fixed, while the second and third unit vectors, $\mathbf{e}_\sigma$ and $\mathbf{e}_\varphi$, change with position in space.

Velocity

Taking the time derivative of (1.5.8) and using relations (1.5.9), we derive an expression for the point particle velocity,

$$\mathbf{U} \equiv \frac{d\mathbf{X}}{dt} = \frac{d}{dt}(X\,\mathbf{e}_x + \Sigma\,\mathbf{e}_\sigma) = \frac{dX}{dt}\,\mathbf{e}_x + X\,\frac{d\mathbf{e}_x}{dt} + \frac{d\Sigma}{dt}\,\mathbf{e}_\sigma + \Sigma\,\frac{d\mathbf{e}_\sigma}{dt}, \tag{1.5.10}$$

and then

$$\mathbf{U} \equiv \frac{d\mathbf{X}}{dt} = \frac{dX}{dt}\,\mathbf{e}_x + \frac{d\Sigma}{dt}\,\mathbf{e}_\sigma + \Sigma\,\frac{d\Phi}{dt}\,\mathbf{e}_\varphi. \tag{1.5.11}$$

The cylindrical polar components of the velocity are then

$$U_x = \frac{dX}{dt}, \qquad U_\sigma = \frac{d\Sigma}{dt}, \qquad U_\varphi = \Sigma\,\frac{d\Phi}{dt}. \tag{1.5.12}$$

Since Φ is a dimensionless function, all three right-hand sides have units of length divided by time.

Acceleration

Differentiating expression (1.5.11) with respect to time and expanding the derivatives, we derive an expression for the acceleration,

$$\begin{aligned}
\mathbf{a} = \frac{d^2\mathbf{X}}{dt^2} &= \frac{d}{dt}\left(\frac{dX}{dt}\,\mathbf{e}_x + \frac{d\Sigma}{dt}\,\mathbf{e}_\sigma + \Sigma\,\frac{d\Phi}{dt}\,\mathbf{e}_\varphi\right) \\
&= \frac{d^2X}{dt^2}\,\mathbf{e}_x + \frac{d^2\Sigma}{dt^2}\,\mathbf{e}_\sigma + \frac{d\Sigma}{dt}\frac{d\mathbf{e}_\sigma}{dt} + \frac{d\Sigma}{dt}\frac{d\Phi}{dt}\,\mathbf{e}_\varphi + \Sigma\,\frac{d^2\Phi}{dt^2}\,\mathbf{e}_\varphi + \Sigma\,\frac{d\Phi}{dt}\frac{d\mathbf{e}_\varphi}{dt}.
\end{aligned} \tag{1.5.13}$$

Now we substitute relations (1.5.9) and find that

$$\mathbf{a} = \frac{d^2\mathbf{X}}{dt^2} = \frac{d^2X}{dt^2}\,\mathbf{e}_x + \frac{d^2\Sigma}{dt^2}\,\mathbf{e}_\sigma + \frac{d\Sigma}{dt}\frac{d\Phi}{dt}\,\mathbf{e}_\varphi + \frac{d\Sigma}{dt}\frac{d\Phi}{dt}\,\mathbf{e}_\varphi + \Sigma\,\frac{d^2\Phi}{dt^2}\,\mathbf{e}_\varphi - \Sigma\,\frac{d\Phi}{dt}\frac{d\Phi}{dt}\,\mathbf{e}_\sigma. \tag{1.5.14}$$

Finally, we consolidate the terms on the right-hand side and obtain the cylindrical polar components of the acceleration,

$$a_x = \frac{d^2X}{dt^2}, \qquad a_\sigma = \frac{d^2\Sigma}{dt^2} - \Sigma\left(\frac{d\Phi}{dt}\right)^2, \qquad a_\varphi = \Sigma\,\frac{d^2\Phi}{dt^2} + 2\,\frac{d\Sigma}{dt}\frac{d\Phi}{dt} = \frac{1}{\Sigma}\frac{d}{dt}\left(\Sigma^2\,\frac{d\Phi}{dt}\right). \tag{1.5.15}$$

Note that a change in the azimuthal angle, Φ, is accompanied by radial acceleration, a_σ.

1.5.2   Spherical polar coordinates

The spherical polar coordinates of a moving point particle are determined by the functions

$$r = R(t), \qquad \theta = \Theta(t), \qquad \varphi = \Phi(t). \tag{1.5.16}$$

The position of a point particle is described as

$$\mathbf{X} = R\,\mathbf{e}_r, \tag{1.5.17}$$

where $\mathbf{e}_\alpha$ are spherical polar unit vectors for $\alpha = r, \theta, \varphi$. The rates of change of the unit vectors following the motion of a point particle are given by the relations

$$\frac{d\mathbf{e}_r}{dt} = \frac{d\Phi}{dt}\sin\Theta\,\mathbf{e}_\varphi + \frac{d\Theta}{dt}\,\mathbf{e}_\theta, \qquad \frac{d\mathbf{e}_\theta}{dt} = \frac{d\Phi}{dt}\cos\Theta\,\mathbf{e}_\varphi - \frac{d\Theta}{dt}\,\mathbf{e}_r,$$

$$\frac{d\mathbf{e}_\varphi}{dt} = -\frac{d\Phi}{dt}\cos\Theta\,\mathbf{e}_\theta - \frac{d\Phi}{dt}\sin\Theta\,\mathbf{e}_r. \tag{1.5.18}$$

All three unit vectors change with position in space.

Velocity

Taking the time derivative of (1.5.17) and using relations (1.5.18), we derive an expression for the point-particle velocity,

$$\mathbf{U} = \frac{d\mathbf{X}}{dt} = \frac{dR}{dt}\,\mathbf{e}_r + R\frac{d\mathbf{e}_r}{dt} = \frac{dR}{dt}\,\mathbf{e}_r + R\frac{d\Phi}{dt}\sin\Theta\,\mathbf{e}_\varphi + R\frac{d\Theta}{dt}\,\mathbf{e}_\theta. \tag{1.5.19}$$

The spherical polar components of the velocity are

$$U_r = \frac{dR}{dt}, \qquad U_\theta = R\frac{d\Theta}{dt}, \qquad U_\varphi = R\sin\Theta\frac{d\Phi}{dt}. \tag{1.5.20}$$

Since the functions Θ and Φ are dimensionless, all three right-hand sides have units of length divided by time.

Acceleration

Differentiating expression (1.5.19) with respect to time, we obtain the point-particle acceleration, $\mathbf{a}$. Expanding the derivatives, we find that

$$\mathbf{a} \equiv \frac{d^2\mathbf{X}}{dt^2} = \mathbf{A} + \mathbf{B} + \mathbf{C}, \tag{1.5.21}$$

where the term

$$\mathbf{A} \equiv \frac{d}{dt}\left(\frac{dR}{dt}\,\mathbf{e}_r\right) = \frac{d^2R}{dt^2}\,\mathbf{e}_r + \frac{dR}{dt}\frac{d\mathbf{e}_r}{dt} = \frac{d^2R}{dt^2}\,\mathbf{e}_r + \frac{dR}{dt}\frac{d\Phi}{dt}\sin\Theta\,\mathbf{e}_\varphi + \frac{dR}{dt}\frac{d\Theta}{dt}\,\mathbf{e}_\theta \tag{1.5.22}$$

corresponds to the first term on the right-hand side of (1.5.19), the term

$$\mathbf{B} \equiv \frac{d}{dt}\left(R\frac{d\Phi}{dt}\sin\Theta\,\mathbf{e}_\varphi\right) = \frac{dR}{dt}\frac{d\Phi}{dt}\sin\Theta\,\mathbf{e}_\varphi + R\frac{d^2\Phi}{dt^2}\sin\Theta\,\mathbf{e}_\varphi + R\frac{d\Phi}{dt}\cos\Theta\frac{d\Theta}{dt}\,\mathbf{e}_\varphi \tag{1.5.23}$$

$$+ R\frac{d\Phi}{dt}\sin\Theta\frac{d\mathbf{e}_\varphi}{dt} = \frac{dR}{dt}\frac{d\Phi}{dt}\sin\Theta\,\mathbf{e}_\varphi + R\frac{d^2\Phi}{dt^2}\sin\Theta\,\mathbf{e}_\varphi + R\frac{d\Phi}{dt}\frac{d\Theta}{dt}\cos\Theta\,\mathbf{e}_\varphi$$

$$- R\left(\frac{d\Phi}{dt}\right)^2\sin\Theta\cos\Theta\,\mathbf{e}_\theta - R\left(\frac{d\Phi}{dt}\right)^2\sin^2\Theta\,\mathbf{e}_r \tag{1.5.24}$$

corresponds to the second term on the right-hand side of (1.5.19), and the term

$$\mathbf{C} \equiv \frac{\mathrm{d}}{\mathrm{d}t}\left(R\frac{\mathrm{d}\Theta}{\mathrm{d}t}\,\mathbf{e}_\theta\right) = \frac{\mathrm{d}R}{\mathrm{d}t}\frac{\mathrm{d}\Theta}{\mathrm{d}t}\,\mathbf{e}_\theta + R\frac{\mathrm{d}^2\Theta}{\mathrm{d}t^2}\,\mathbf{e}_\theta + R\frac{\mathrm{d}\Theta}{\mathrm{d}t}\frac{\mathrm{d}\mathbf{e}_\theta}{\mathrm{d}t}$$

$$= \frac{\mathrm{d}R}{\mathrm{d}t}\frac{\mathrm{d}\Theta}{\mathrm{d}t}\,\mathbf{e}_\theta + R\frac{\mathrm{d}^2\Theta}{\mathrm{d}t^2}\,\mathbf{e}_\theta + R\frac{\mathrm{d}\Phi}{\mathrm{d}t}\frac{\mathrm{d}\Theta}{\mathrm{d}t}\cos\Theta\,\mathbf{e}_\varphi - R\left(\frac{\mathrm{d}\Theta}{\mathrm{d}t}\right)^2\mathbf{e}_r \qquad (1.5.25)$$

corresponds to the third term on the right-hand side of (1.5.19). Consolidating the various terms, we derive the cylindrical polar components of the acceleration,

$$a_r = \frac{\mathrm{d}^2R}{\mathrm{d}t^2} - R\left(\frac{\mathrm{d}\Phi}{\mathrm{d}t}\right)^2\sin^2\Theta - R\left(\frac{\mathrm{d}\Theta}{\mathrm{d}t}\right)^2,$$

$$a_\theta = R\frac{\mathrm{d}^2\Theta}{\mathrm{d}t^2} + 2\frac{\mathrm{d}R}{\mathrm{d}t}\frac{\mathrm{d}\Theta}{\mathrm{d}t} - R\left(\frac{\mathrm{d}\Phi}{\mathrm{d}t}\right)^2\sin\Theta\cos\Theta, \qquad (1.5.26)$$

$$a_\varphi = R\frac{\mathrm{d}^2\Phi}{\mathrm{d}t^2}\sin\Theta + 2\frac{\mathrm{d}R}{\mathrm{d}t}\frac{\mathrm{d}\Phi}{\mathrm{d}t}\sin\Theta + 2R\frac{\mathrm{d}\Theta}{\mathrm{d}t}\frac{\mathrm{d}\Phi}{\mathrm{d}t}\cos\Theta.$$

Note that a change in the meridional angle, Θ, or azimuthal angle, Φ, is accompanied by radial acceleration.

1.5.3 Plane polar coordinates

The plane polar coordinates of a point particle are described by the functions

$$r = R(t), \qquad \theta = \Theta(t). \qquad (1.5.27)$$

The position of a point particle is described as

$$\mathbf{X} = R\,\mathbf{e}_r, \qquad (1.5.28)$$

where $\mathbf{e}_\alpha$ are plane polar unit vectors for $\alpha = r, \theta$. The rates of change of the unit vectors following the motion of a particle are given by

$$\frac{\mathrm{d}\mathbf{e}_r}{\mathrm{d}t} = \frac{\mathrm{d}\Theta}{\mathrm{d}t}\,\mathbf{e}_\theta, \qquad \frac{\mathrm{d}\mathbf{e}_\theta}{\mathrm{d}t} = -\frac{\mathrm{d}\Theta}{\mathrm{d}t}\,\mathbf{e}_r. \qquad (1.5.29)$$

Note that both unit vectors change with position in space.

Velocity

To derive the velocity components, we work as previously with cylindrical and spherical polar coordinates, and find that

$$U_r = \frac{\mathrm{d}R}{\mathrm{d}t}, \qquad U_\theta = R\frac{\mathrm{d}\Theta}{\mathrm{d}t}. \qquad (1.5.30)$$

Note that the right-hand sides have units of length divided by time.

Acceleration

The plane polar components of the particle acceleration are

$$a_r = \frac{\mathrm{d}^2 R}{\mathrm{d}t^2} - R\left(\frac{\mathrm{d}\Theta}{\mathrm{d}t}\right)^2, \qquad a_\theta = R\frac{\mathrm{d}^2\Theta}{\mathrm{d}t^2} + 2\frac{\mathrm{d}R}{\mathrm{d}t}\frac{\mathrm{d}\Theta}{\mathrm{d}t} = \frac{1}{R}\frac{\mathrm{d}}{\mathrm{d}t}\left(R^2\frac{\mathrm{d}\Theta}{\mathrm{d}t}\right). \qquad (1.5.31)$$

Note that a change in the polar angle, Θ, is accompanied by radial acceleration. If a particle moves along a circular path of constant radius R centered at the origin, $\mathrm{d}R/\mathrm{d}t = 0$, the acceleration components are

$$a_r = R\left(\frac{\mathrm{d}\Theta}{\mathrm{d}t}\right)^2 = \frac{U_\theta^2}{R}, \qquad a_\theta = R\frac{\mathrm{d}^2\Theta}{\mathrm{d}t^2}. \qquad (1.5.32)$$

A radial acceleration is necessary to follow the circular path.

1.5.4 Particle rotation around an axis

The new position of a point particle that has rotated around the x axis by angle φ_x, around the y axis by angle φ_y, or around the z axis by angle φ_z, is given by

$$\mathbf{x}^{new} = \mathcal{R}^{(x)}(\varphi_x) \cdot \mathbf{x}, \qquad \mathbf{x}^{new} = \mathcal{R}^{(y)}(\varphi_y) \cdot \mathbf{x}, \qquad \mathbf{x}^{new} = \mathcal{R}^{(z)}(\varphi_z) \cdot \mathbf{x}, \qquad (1.5.33)$$

where $\mathbf{x}$ is the old position and

$$\mathcal{R}^{(x)}(\varphi) = \begin{bmatrix} 1 & 0 & 0 \\ 0 & \cos\varphi & -\sin\varphi \\ 0 & \sin\varphi & \cos\varphi \end{bmatrix}, \qquad \mathcal{R}^{(y)}(\varphi) = \begin{bmatrix} \cos\varphi & 0 & \sin\varphi \\ 0 & 1 & 0 \\ -\sin\varphi & 0 & \cos\varphi \end{bmatrix},$$

$$\mathcal{R}^{(z)}(\varphi) = \begin{bmatrix} \cos\varphi & -\sin\varphi & 0 \\ \sin\varphi & \cos\varphi & 0 \\ 0 & 0 & 1 \end{bmatrix} \qquad (1.5.34)$$

are orthogonal rotation matrices, meaning that their transpose is equal to their inverse.

Rotation around an arbitrary axis

To rotate a point by a specified angle φ about an axis x_1' that passes through the origin, as shown in Figure 1.5.1, it is convenient to introduce a rotated Cartesian coordinate system comprised of primed axes, as discussed in Section 1.1.8. We then apply a forward transformation, $\mathbf{x} \to \mathbf{x}'$, followed by a rotation and then a backward transformation, $\mathbf{x}' \to \mathbf{x}$, to obtain the overall transformation

$$\mathbf{x}^{new} = \mathbf{P} \cdot \mathbf{x}, \qquad (1.5.35)$$

where

$$\mathbf{P} = \mathbf{A}^T \cdot \mathcal{R}^{(x)}(\varphi) \cdot \mathbf{A} \qquad (1.5.36)$$

is a projection matrix. The first row of the rotation matrix $\mathbf{A}$ contains the direction cosines of the x_1' axis,

$$a \equiv A_{11} = \cos\alpha_1, \qquad b \equiv A_{12} = \cos\alpha_2, \qquad c \equiv A_{13} = \cos\alpha_3, \qquad (1.5.37)$$

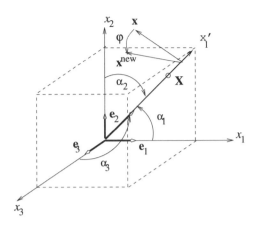

FIGURE 1.5.1   Rotation of a point **x** around the x_1' axis by angle φ. The new coordinates are found using the projection matrix (1.5.36).

where $a^2 + b^2 + c^2 = 1$. Carrying out the multiplications and simplifying, we derive the projection matrix

$$\mathbf{P} = \cos\varphi\,\mathbf{I} + (1 - \cos\varphi)\begin{bmatrix} a^2 & ab & ac \\ ab & b^2 & bc \\ ac & bc & c^2 \end{bmatrix} + \sin\varphi\begin{bmatrix} 0 & -c & b \\ c & 0 & -a \\ -b & a & 0 \end{bmatrix}, \qquad (1.5.38)$$

where **I** is the identity matrix. The matrix in the second term on the right-hand side of (1.5.38) is symmetric, whereas the matrix in the third term is skew-symmetric. This means that a rotation vector, **c**, with the property that $\mathbf{x}^{new} = \mathbf{c} \times \mathbf{x}$, cannot be found.

In practice, we may specify a point **X** on the x_1' axis, and compute the direction cosines

$$a = \frac{X}{|\mathbf{X}|}, \qquad b = \frac{Y}{|\mathbf{X}|}, \qquad c = \frac{Z}{|\mathbf{X}|}, \qquad (1.5.39)$$

as shown in Figure 1.5.1. Note that $\mathbf{P} \cdot \mathbf{X} = \mathbf{X}$, in agreement with physical intuition.

Problems

1.5.1 *Particle paths*

(*a*) A particle moves over a sphere of radius a in a path described by the equation $\varphi = 2\theta$, where $\theta = \Theta(t)$ is a given function of time. Illustrate the particle trajectory and compute the spherical polar components of the particle acceleration in terms of the function $\Theta(t)$.

(*b*) A particle moves in the xy plane on a spiral path described in plane polar coordinates (r, θ) by the equation $r = ae^\theta$, where a is a constant and $\theta = \Theta(t)$ is a given monotonic function of time. Compute the plane polar components of the particle acceleration in terms of the function $\Theta(t)$.

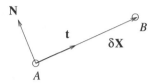

FIGURE 1.6.1   Illustration of a material vector with infinitesimal length and designated beginning, A, and end, B. The unit vector $\mathbf{N}$ is normal to the material vector.

1.5.2 *Rotation around an axis*

(a) Derive the projection matrix (1.5.38) following the procedure described in the text.

(b) Derive the projection matrix (1.5.38) by integrating the differential equation $d\mathbf{x}/dt = \mathbf{\Omega} \times \mathbf{x}$, where $\mathbf{\Omega} = \Omega(t)[a, b, c]$ is an angular velocity vector with constant orientation but possibly variable strength.

1.6   Material vectors and material lines

To lay the foundation for developing dynamical laws governing the motion and physical properties of fluid parcels, we study the evolution of material vectors, material lines, and material surfaces in a specified flow. The theoretical framework will be employed to describe the motion of interfaces between two immiscible fluids in a two-phase flow. In this section, we consider material vectors and material lines consisting of a fixed collection of point particles with permanent identity. In Section 1.7, we generalize the discussion to material surfaces. Elements of differential geometry of lines and surfaces will be introduced in the discourse, as required.

1.6.1   Material vectors

A material vector, $\delta\mathbf{X}$, is a small material line with infinitesimal length and a designated beginning and end, as shown Figure 1.6.1. Applying the definition of the point-particle velocity, $D\mathbf{X}/Dt = \mathbf{U} = \mathbf{u}(\mathbf{X})$, at the position of the two end point particles labeled A and B, expressing the velocity at the last point in a Taylor series about the first point, and keeping only the linear terms, we obtain

$$\frac{D\,\delta\mathbf{X}}{Dt} = \mathbf{U}^B - \mathbf{U}^A = \delta X_i \frac{\partial\mathbf{u}}{\partial x_i} = \delta\mathbf{X} \cdot \mathbf{L}, \tag{1.6.1}$$

where $\mathbf{L}$ is the velocity gradient tensor introduced in (1.1.2) evaluated at the position of the material vector, and summation is implied over the repeated index i.

Stretching

The rate of change of the length of the material vector, $\delta l \equiv |\delta\mathbf{X}|$, is given by

$$\frac{D\,\delta l}{Dt} = \frac{D\,(\delta\mathbf{X} \cdot \delta\mathbf{X})^{1/2}}{Dt} = \frac{1}{2\,(\delta\mathbf{X} \cdot \delta\mathbf{X})^{1/2}} \frac{D\,(\delta\mathbf{X} \cdot \delta\mathbf{X})}{Dt} = \frac{\delta\mathbf{X}}{\delta l} \cdot \frac{D\,\delta\mathbf{X}}{Dt}. \tag{1.6.2}$$

Using (1.6.1), we find that

$$\frac{\mathrm{D}\,\delta l}{\mathrm{D}t} = \frac{\delta \mathbf{X}}{\delta l} \cdot (\delta \mathbf{X} \cdot \mathbf{L}). \tag{1.6.3}$$

Rearranging, we obtain the evolution equation

$$\frac{1}{\delta l}\frac{\mathrm{D}\,\delta l}{\mathrm{D}t} = \mathbf{t} \cdot \mathbf{L} \cdot \mathbf{t}, \tag{1.6.4}$$

where $\mathbf{t} = \delta \mathbf{X}/\delta l$ is the unit vector in the direction of the material vector $\delta \mathbf{X}$. The right-hand side of (1.6.4) expresses the rate of extension of the fluid in the direction of the material vector.

Reorientation

To compute the rate of change of the unit vector, $\mathbf{t}$, that is parallel to a material vector, $\delta \mathbf{X}$, at all times, we write

$$\frac{\mathrm{D}\,\delta \mathbf{X}}{\mathrm{D}t} = \frac{\mathrm{D}(\mathbf{t}\,\delta l)}{\mathrm{D}t} = \delta l\,\frac{\mathrm{D}\mathbf{t}}{\mathrm{D}t} + \mathbf{t}\,\frac{\mathrm{D}\,\delta l}{\mathrm{D}t}, \tag{1.6.5}$$

and use (1.6.1) and (1.6.4) to obtain

$$\frac{\mathrm{D}\mathbf{t}}{\mathrm{D}t} = (\mathbf{t} \cdot \mathbf{L}) \cdot (\mathbf{I} - \mathbf{t} \otimes \mathbf{t}), \tag{1.6.6}$$

where $\mathbf{I}$ is the identity matrix. The operator $\mathbf{I} - \mathbf{t} \otimes \mathbf{t}$ on the right-hand side removes the component of a vector in the direction of the unit vector, $\mathbf{t}$, leaving only the normal component. Accordingly, the right-hand side of (1.6.6) involves derivatives of the velocity components in a plane that is normal to the material vector with respect to arc length measured in the direction of the material vector.

For example, if a material vector is aligned with the x axis, $\mathbf{t} = \mathbf{e}_x = [1, 0, 0]$, we find that $\mathrm{D}\mathbf{t}/\mathrm{D}t = [0, \partial u_y/\partial x, \partial u_z/\partial x]$, which shows that $\mathbf{t}$ tends to acquire components along the y and z axes.

Mutual reorientation

The cosine of the angle α subtended between two material unit vectors, $\mathbf{t}_1$ and $\mathbf{t}_2$, is given by an inner product, $\cos \alpha = \mathbf{t}_1 \cdot \mathbf{t}_2$. Taking the material derivative, we obtain

$$\frac{\mathrm{D}\cos \alpha}{\mathrm{D}t} = \frac{\mathrm{D}\mathbf{t}_1}{\mathrm{D}t} \cdot \mathbf{t}_2 + \frac{\mathrm{D}\mathbf{t}_2}{\mathrm{D}t} \cdot \mathbf{t}_1. \tag{1.6.7}$$

Substituting (1.6.6) and rearranging, we find that

$$\frac{\mathrm{D}\cos \alpha}{\mathrm{D}t} = (\mathbf{t}_1 \otimes \mathbf{t}_2 + \mathbf{t}_2 \otimes \mathbf{t}_1) : \mathbf{L} - \cos \alpha\, (\mathbf{t}_1 \otimes \mathbf{t}_1 + \mathbf{t}_2 \otimes \mathbf{t}_2) : \mathbf{L}. \tag{1.6.8}$$

Because the tensors $\mathbf{t}_1 \otimes \mathbf{t}_2 + \mathbf{t}_2 \otimes \mathbf{t}_1$ and $\mathbf{t}_1 \otimes \mathbf{t}_1 + \mathbf{t}_2 \otimes \mathbf{t}_2$ are symmetric, the velocity gradient $\mathbf{L}$ can be replaced by its symmetric constituent expressed by the vorticity tensor on the right-hand side. Accordingly, the first term on the right-hand side of (1.6.8) expresses the symmetric component of

(a)   (b)

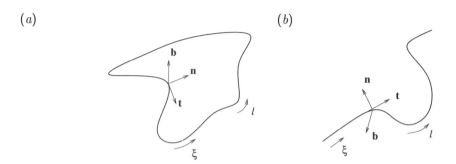

FIGURE 1.6.2 (a) A closed or open (b) material line is parametrized by a Lagrangian label, ξ. Shown are the tangent unit vector, $\mathbf{t}$, the normal unit vector, $\mathbf{n}$, and the binormal unit vector, $\mathbf{b}$.

the velocity gradient tensor in a dyadic base defined by the unit vectors $\mathbf{t}_1$ and $\mathbf{t}_2$. If the two vectors are initially perpendicular, $\cos\alpha = 0$ and the second term on the right-hand side of (1.6.8) is zero.

Normal vector

It is of interest to consider the evolution of a unit vector, $\mathbf{N}$, that is and remains normal to a material vector, as shown in Figure 1.6.1. Taking the material derivative of the constraint $\mathbf{N} \cdot \mathbf{N} = 1$, we find that $\mathbf{N} \cdot \mathrm{D}\mathbf{N}/\mathrm{D}t = 0$, which shows that the vector $\mathrm{D}\mathbf{N}/\mathrm{D}t$ lacks a component in the direction of $\mathbf{N}$. Taking the material derivative of the constraint $\mathbf{N} \cdot \delta\mathbf{X} = 0$ and using (1.6.1), we find that $\mathbf{t} \cdot \mathrm{D}\mathbf{N}/\mathrm{D}t = -\mathbf{t} \cdot \mathbf{L} \cdot \mathbf{N}$, where $\mathbf{t} = \delta\mathbf{X}/\delta l$ is the tangent unit vector. Combining these expressions, we derive the evolution equation

$$\frac{\mathrm{D}\mathbf{N}}{\mathrm{D}t} = -(\mathbf{t} \otimes \mathbf{t}) \cdot \mathbf{L} \cdot \mathbf{N} + \Omega\, \mathbf{t} \times \mathbf{N}, \tag{1.6.9}$$

where Ω is an unspecified rate of rotation of $\mathbf{N}$ about $\mathbf{t}$. In the case of two-dimensional or axisymmetric flow where the vectors $\mathbf{t}$ and $\mathbf{N}$ are restricted to remain in the xy or an azimuthal plane, $\Omega = 0$.

1.6.2 Material lines

Next, we consider a material line consisting of a fixed collection of point particles forming an open or closed loop, as shown in Figure 1.6.2. To identify the point particles, we introduce a scalar label, ξ, taking values inside an appropriate set of real numbers, Ξ, and regard the position of the point particles, $\mathbf{X}$, as a function of ξ and time, t, writing $\mathbf{X}(\xi, t)$. The unit vector

$$\mathbf{t} = \frac{1}{h}\frac{\partial \mathbf{X}}{\partial \xi} \tag{1.6.10}$$

is tangential to the material line, where $h = |\partial\mathbf{X}/\partial\xi|$. Because $\mathbf{t}$ is a unit vector, $\mathbf{t} \cdot \mathbf{t} = 1$, we have $\mathbf{t} \cdot \partial\mathbf{t}/\partial\xi = \frac{1}{2}\partial(\mathbf{t} \cdot \mathbf{t})/\partial\xi = 0$, which shows that the vector $\partial\mathbf{t}/\partial\xi$ is normal to $\mathbf{t}$.

The arc length of an infinitesimal section of the material line is $dl = h\,d\xi$ and the total arc length of the material line is

$$L = \int_\Xi h\,d\xi. \tag{1.6.11}$$

This expression shows that h is a scalar metric coefficient for the arc length similar to the Jacobian metric for the volume of a fluid parcel, $\mathcal{J}$. If we identify the label ξ with the instantaneous arc length along the material line, l, then $h = 1$.

1.6.3   Frenet–Serret relations

It is useful to introduce a system of orthogonal curvilinear coordinates constructed with reference to a material line. The principal unit vector normal to the material line, $\mathbf{n}$, is defined by the relation

$$\frac{\partial \mathbf{t}}{\partial l} = -\kappa \mathbf{n}, \tag{1.6.12}$$

where l is the arc length along the material line and κ is the signed curvature of the material line. The binormal unit vector is defined by the equation

$$\mathbf{b} = \mathbf{t} \times \mathbf{n}. \tag{1.6.13}$$

The three unit vectors, $\mathbf{t}$, $\mathbf{n}$, and $\mathbf{b}$, define three mutually orthogonal directions that can be used to construct a right-handed, orthogonal, curvilinear system of axes so that

$$\mathbf{t} = \mathbf{n} \times \mathbf{b}, \qquad \mathbf{n} = \mathbf{b} \times \mathbf{t}, \tag{1.6.14}$$

as shown in Figure 1.6.2. The plane containing the pair $(\mathbf{t}, \mathbf{n})$ is called the osculating plane, the plane containing the pair $(\mathbf{n}, \mathbf{b})$ is called the normal plane, and the plane containing the pair $(\mathbf{b}, \mathbf{t})$, is called the rectifying plane.

Differentiating (1.6.13) with respect to arc length, l, expanding the derivative on the right-hand side, and using (1.6.12), we find that the vector $d\mathbf{b}/dl$ is perpendicular to $\mathbf{t}$. Since $\mathbf{b}$ is a unit vector, $d(\mathbf{b} \cdot \mathbf{b})/dl = 2\mathbf{b} \cdot (d\mathbf{b}/dl) = 0$, which shows that $d\mathbf{b}/dl$ is perpendicular to $\mathbf{b}$. We conclude that $\mathbf{b}$ must be parallel to $\mathbf{n}$ and write

$$\frac{\partial \mathbf{b}}{\partial l} = -\tau \mathbf{n}, \tag{1.6.15}$$

where τ is the torsion of the material line.

Rewriting (1.6.13) as $\mathbf{n} = \mathbf{b} \times \mathbf{t}$, differentiating this expression with respect to l, and using (1.6.12) and (1.6.15), we obtain

$$\frac{\partial \mathbf{n}}{\partial l} = \kappa \mathbf{t} + \tau \mathbf{b}, \tag{1.6.16}$$

involving the curvature and the torsion of the line.

Equations (1.6.12), (1.6.15), and (1.6.16) comprise the Frenet–Serret relations. In the literature, the curvature, torsion, or both, may appear with opposite signs as a matter of convention.

Matrix formulation

The Frenet–Serret relations can be collected into the matrix form

$$\frac{d}{dl} \begin{bmatrix} \mathbf{t} \\ \mathbf{n} \\ \mathbf{b} \end{bmatrix} = \begin{bmatrix} 0 & -\kappa & 0 \\ \kappa & 0 & \tau \\ 0 & -\tau & 0 \end{bmatrix} \cdot \begin{bmatrix} \mathbf{t} \\ \mathbf{n} \\ \mathbf{b} \end{bmatrix}, \tag{1.6.17}$$

involving a singular skew-symmetric matrix on the right-hand side. One eigenvalue of this matrix is zero and the other two eigenvalues are complex conjugates, given by $\lambda = \pm i(\kappa^2 + \tau^2)^{1/2}$.

Complex variable formulation

It is sometimes useful to introduce a complex surface vector field, $\mathbf{q} \equiv \mathbf{n} + i\,\mathbf{b}$, where i is the imaginary unit, $i^2 = -1$. Combining the second with the third Frenet–Serret relations (1.6.15) and (1.6.16), we obtain

$$\frac{d\mathbf{q}}{dl} + i\tau\mathbf{q} = \kappa\mathbf{t}. \tag{1.6.18}$$

Multiplying both sides by the integrating factor

$$\Phi(l) \equiv \exp\left(i \int_0^l \tau(\ell)\,d\ell\right) \tag{1.6.19}$$

and rearranging, we obtain the compact form

$$\frac{d\mathbf{Q}}{dl} = -\psi\,\mathbf{t}, \tag{1.6.20}$$

where we have defined

$$\mathbf{Q}(l) \equiv \Phi(l)\,\mathbf{q}(l), \qquad \psi(l) \equiv -\kappa(l)\,\Phi(l). \tag{1.6.21}$$

Note that $\Phi\Phi^* = |\Phi|^2 = 1$, $\mathbf{Q}\cdot\mathbf{Q} = 0$, and $\mathbf{Q}\cdot\mathbf{Q}^* = 2$, where an asterisk denotes the complex conjugate. The first Frenet–Serret relation (1.6.12) yields

$$\frac{d\mathbf{t}}{dl} = -\kappa\mathbf{n} = -\mathrm{Real}(\kappa\mathbf{q}^*) = -\mathrm{Real}(\kappa\Phi\Phi^*\mathbf{q}^*) = \mathrm{Real}(\psi\mathbf{Q}^*) = \frac{1}{2}\,(\psi\mathbf{Q}^* + \psi^*\mathbf{Q}). \tag{1.6.22}$$

Darboux rotation vector formulation

An alternative representation of the Frenet–Serret relations is

$$\frac{d\mathbf{t}}{dl} = \boldsymbol{\chi}\times\mathbf{t}, \qquad \frac{d\mathbf{n}}{dl} = \boldsymbol{\chi}\times\mathbf{n}, \qquad \frac{d\mathbf{b}}{dl} = \boldsymbol{\chi}\times\mathbf{b}, \tag{1.6.23}$$

where $\boldsymbol{\chi} = \tau\mathbf{t} - \kappa\mathbf{b}$ is the Darboux rotation vector lying in the rectifying plane.

1.6.4   Evolution equations for a material line

Having established the Frenet–Serret framework, we proceed to compute the rate of change of the total length of a material line. Differentiating (1.6.11) with respect to time and noting that the limits of integration are fixed, we transfer the time derivative inside the integral as a material derivative and obtain

$$\frac{dL}{dt} = \int_{\Xi} \frac{Dh}{Dt}\, d\xi = \int_{Line} \frac{1}{h} \frac{Dh}{Dt}\, dl. \tag{1.6.24}$$

The last integrand expresses the local rate of extension of the material line.

Evolution of the metric

Taking the material derivative of the definition $h = |\partial \mathbf{X}/\partial \xi|$ and working as in (1.6.2), we obtain

$$\frac{Dh}{Dt} = \mathbf{t} \cdot \frac{\partial \mathbf{U}}{\partial \xi} = \frac{\partial (\mathbf{t} \cdot \mathbf{U})}{\partial \xi} - \mathbf{U} \cdot \frac{\partial \mathbf{t}}{\partial \xi}. \tag{1.6.25}$$

Using the Frenet–Serret relation (1.6.12) and setting the point-particle velocity equal to the fluid velocity, $\mathbf{U} = \mathbf{u}$, we obtain

$$\frac{Dh}{Dt} = \frac{\partial (\mathbf{u} \cdot \mathbf{t})}{\partial \xi} + \kappa \frac{\partial l}{\partial \xi} \mathbf{u} \cdot \mathbf{n}. \tag{1.6.26}$$

The two terms on the right-hand side express the change in length of an infinitesimal section of the material line due to stretching along the line, and extension due to motion normal to the line. These interpretations become more clear by considering the behavior of a circular material line that exhibits tangential motion with vanishing normal velocity, or is expanding in the radial direction while remaining in its plane (Problem 1.6.1).

Now substituting (1.6.26) into (1.6.24) and performing the integration, we obtain a revealing expression for the rate of the change of the length of a material line,

$$\frac{dL}{dt} = (\mathbf{u} \cdot \mathbf{t})_{start}^{end} + \int_{Line} \kappa\, \mathbf{u} \cdot \mathbf{n}\, dl. \tag{1.6.27}$$

If a material line forms a closed loop, the first term on the right-hand side vanishes and the tangential motion does not contribute to the rate of change of the total arc length of the line.

Evolution of the tangent unit vector

The rate of change of the tangent unit vector, $\mathbf{t}$, follows from (1.6.6), repeated for convenience,

$$\frac{D\mathbf{t}}{Dt} = (\mathbf{t} \cdot \mathbf{L}) \cdot (\mathbf{I} - \mathbf{t} \otimes \mathbf{t}) = \frac{\partial \mathbf{u}}{\partial l} \cdot (\mathbf{I} - \mathbf{t} \otimes \mathbf{t}) \tag{1.6.28}$$

or

$$\frac{D\mathbf{t}}{Dt} = \left(\frac{\partial \mathbf{u}}{\partial l} \cdot \mathbf{n}\right) \mathbf{n} + \left(\frac{\partial \mathbf{u}}{\partial l} \cdot \mathbf{b}\right) \mathbf{b}, \tag{1.6.29}$$

where $\mathbf{I}$ is the identity matrix. The dyadic decomposition $\mathbf{I} = \mathbf{t} \otimes \mathbf{t} + \mathbf{n} \otimes \mathbf{n} + \mathbf{b} \otimes \mathbf{b}$ was employed to derive the two expressions in (1.6.28). The right-hand side of (1.6.28) involves derivatives of the velocity components in the normal plane with respect to arc length along the material line. We see that the vector $\mathrm{D}\mathbf{t}/\mathrm{D}t$ lacks a component in the direction of $\mathbf{t}$.

Evolution of the curvature

The first Frenet–Serret relation (1.6.12) can be recast into the form

$$\kappa \mathbf{n} = -\frac{\partial \mathbf{t}}{\partial l} = -\frac{1}{h}\frac{\partial \mathbf{t}}{\partial \xi}. \tag{1.6.30}$$

Taking the material derivative, we find that

$$\kappa \frac{\mathrm{D}\mathbf{n}}{\mathrm{D}t} + \mathbf{n}\frac{\mathrm{D}\kappa}{\mathrm{D}t} = \frac{1}{h^2}\frac{\mathrm{D}h}{\mathrm{D}t}\frac{\partial \mathbf{t}}{\partial \xi} - \frac{1}{h}\frac{\partial}{\partial \xi}\left(\frac{\mathrm{D}\mathbf{t}}{\mathrm{D}t}\right) = -\frac{\kappa}{h}\frac{\mathrm{D}h}{\mathrm{D}t}\mathbf{n} - \frac{\partial}{\partial l}\left(\frac{\mathrm{D}\mathbf{t}}{\mathrm{D}t}\right). \tag{1.6.31}$$

Next, we project this equation onto $\mathbf{n}$, recall that $\mathbf{n}\cdot\mathbf{n} = 1$ and thus $\mathbf{n}\cdot\mathrm{D}\mathbf{n}/\mathrm{D}t = 0$, and rearrange to obtain an evolution equation for the curvature,

$$\frac{\mathrm{D}\kappa}{\mathrm{D}t} = -\frac{\kappa}{h}\frac{\mathrm{D}h}{\mathrm{D}t} - \mathbf{n}\cdot\frac{\partial}{\partial l}\left(\frac{\mathrm{D}\mathbf{t}}{\mathrm{D}t}\right) = -\kappa\,\mathbf{t}\cdot\frac{\partial \mathbf{u}}{\partial l} - \mathbf{n}\cdot\frac{\partial}{\partial l}\left(\frac{\mathrm{D}\mathbf{t}}{\mathrm{D}t}\right). \tag{1.6.32}$$

Taking the derivative of the rate of change $\mathrm{D}\mathbf{t}/\mathrm{D}t$ given in (1.6.28), we obtain

$$\frac{\partial}{\partial l}\left(\frac{\mathrm{D}\mathbf{t}}{\mathrm{D}t}\right) = \frac{\partial^2 \mathbf{u}}{\partial l^2}\cdot(\mathbf{I} - \mathbf{t}\otimes\mathbf{t}) + \kappa\frac{\partial \mathbf{u}}{\partial l}\cdot(\mathbf{t}\otimes\mathbf{n} + \mathbf{n}\otimes\mathbf{t}). \tag{1.6.33}$$

Substituting this expression into (1.6.32) and simplifying, we derive the final form

$$\frac{\mathrm{D}\kappa}{\mathrm{D}t} = -2\kappa\frac{\partial \mathbf{u}}{\partial l}\cdot\mathbf{t} - \frac{\partial^2 \mathbf{u}}{\partial l^2}\cdot\mathbf{n}. \tag{1.6.34}$$

In the case of an expanding circle of radius $a(t)$ in the xy plane, we set $\mathbf{u} = U\left[\cos\theta, \sin\theta\right]$, $\mathbf{t} = \left[-\sin\theta, \cos\theta\right]$, and $\mathbf{n} = \left[\cos\theta, \sin\theta\right]$, where U is the velocity of expansion and θ is the polar angle. Substituting $l = a\theta$ and $\kappa = 1/a$, we obtain an expected equation, $\mathrm{d}a/\mathrm{d}t = U$.

Evolution of the normal vector

To obtain an evolution equation for the normal unit vector, we project (1.6.31) onto the binormal vector and rearrange to find that

$$\mathbf{b}\cdot\frac{\mathrm{D}\mathbf{n}}{\mathrm{D}t} = -\frac{1}{\kappa}\mathbf{b}\cdot\frac{\partial}{\partial l}\left(\frac{\mathrm{D}\mathbf{t}}{\mathrm{D}t}\right) = -\frac{1}{\kappa}\frac{\partial^2 \mathbf{u}}{\partial l^2}\cdot\mathbf{b}. \tag{1.6.35}$$

Combining this expression with (1.6.9), we obtain

$$\frac{\mathrm{D}\mathbf{n}}{\mathrm{D}t} = -\left(\frac{\partial \mathbf{u}}{\partial l}\cdot\mathbf{n}\right)\mathbf{t} - \frac{1}{\kappa}\left(\frac{\partial^2 \mathbf{u}}{\partial l^2}\cdot\mathbf{b}\right)\mathbf{b}. \tag{1.6.36}$$

If a line is and remains in the osculating plane containing $\mathbf{t}$ and $\mathbf{n}$, the second term on the right-hand side does not appear. Expression (1.6.36) can also be derived by substituting (1.6.34) into (1.6.31).

Evolution of the binormal vector

An evolution equation of the binormal vector can be derived by combining the definition (1.6.13) with the evolution equations (1.6.36) and (1.6.6), finding

$$\frac{D\mathbf{b}}{Dt} = \frac{D\mathbf{t}}{Dt} \times \mathbf{n} - \frac{D\mathbf{n}}{Dt} \times \mathbf{t} = -\left(\frac{\partial \mathbf{u}}{\partial l} \cdot \mathbf{b}\right)\mathbf{t} + \frac{1}{\kappa}\left(\frac{\partial^2 \mathbf{u}}{\partial l^2} \cdot \mathbf{b}\right)\mathbf{n}. \tag{1.6.37}$$

We see that $D\mathbf{b}/Dt$ is perpendicular to $\mathbf{b}$, as required for the length of $\mathbf{b}$ to remain constant and equal to unity at any time.

Evolution of the torsion

An evolution equation for the torsion can be derived by taking the material derivative of (1.6.15), finding

$$\frac{D\tau}{Dt}\mathbf{n} + \tau\frac{D\mathbf{n}}{Dt} = -\frac{1}{h}\frac{Dh}{Dt}\tau\mathbf{n} - \frac{\partial}{\partial l}\left(\frac{D\mathbf{b}}{Dt}\right). \tag{1.6.38}$$

Projecting this equation onto $\mathbf{n}$, we obtain

$$\frac{D\tau}{Dt} = -\frac{\tau}{h}\frac{Dh}{Dt} - \mathbf{n}\cdot\frac{\partial}{\partial l}\left(\frac{D\mathbf{b}}{Dt}\right) = -\tau\,\mathbf{t}\cdot\frac{\partial \mathbf{u}}{\partial l} - \mathbf{n}\cdot\frac{\partial}{\partial l}\left(\frac{D\mathbf{b}}{Dt}\right). \tag{1.6.39}$$

Substituting (1.6.37), using the Frenet–Serret relations, and simplifying, we obtain

$$\frac{D\tau}{Dt} = -\frac{\partial \mathbf{u}}{\partial l}\cdot(\tau\mathbf{t} + \kappa\mathbf{b}) + \frac{1}{\kappa}\frac{\partial^2 \mathbf{u}}{\partial l^2}\cdot\left(\tau\mathbf{n} + \frac{1}{\kappa}\frac{\partial\kappa}{\partial l}\mathbf{b}\right) - \frac{1}{\kappa}\frac{\partial^3 \mathbf{u}}{\partial l^3}\cdot\mathbf{b}. \tag{1.6.40}$$

The presence of a third derivative with respect to arc length is an interesting feature of this equation.

Spin vector

The evolution equations for the tangent, normal, and binormal unit vectors derived previously in this section can be collected into the unified forms

$$\frac{D\mathbf{t}}{Dt} = \mathbf{s}\times\mathbf{t}, \qquad \frac{D\mathbf{n}}{Dt} = \mathbf{s}\times\mathbf{n}, \qquad \frac{D\mathbf{b}}{Dt} = \mathbf{s}\times\mathbf{b}, \tag{1.6.41}$$

where

$$\mathbf{s} = -\frac{1}{\kappa}\left(\frac{\partial^2 \mathbf{u}}{\partial l^2}\cdot\mathbf{b}\right)\mathbf{t} - \left(\frac{\partial \mathbf{u}}{\partial l}\cdot\mathbf{b}\right)\mathbf{n} + \left(\frac{\partial \mathbf{u}}{\partial l}\cdot\mathbf{n}\right)\mathbf{b} \tag{1.6.42}$$

is the spin vector. Formulas (1.6.41) complement the Darboux relations (1.6.23).

Consistency between (1.6.41) and (1.6.23) requires that

$$\frac{D(\boldsymbol{\chi}\times\mathbf{t})}{Dt} = \frac{d(\mathbf{s}\times\mathbf{t})}{dl}, \qquad \frac{D(\boldsymbol{\chi}\times\mathbf{n})}{Dt} = \frac{d(\mathbf{s}\times\mathbf{n})}{dl}, \qquad \frac{D(\boldsymbol{\chi}\times\mathbf{b})}{Dt} = \frac{d(\mathbf{s}\times\mathbf{b})}{dl}. \tag{1.6.43}$$

Expanding the derivatives in the first equations, we obtain

$$\frac{D\boldsymbol{\chi}}{Dt}\times\mathbf{t} + \boldsymbol{\chi}\times\frac{D\mathbf{t}}{Dt} = \frac{D\boldsymbol{\chi}}{Dt}\times\mathbf{t} + \boldsymbol{\chi}\times\mathbf{s}\times\mathbf{t} = \frac{d\mathbf{s}}{dl}\times\mathbf{t} + \mathbf{s}\times\frac{d\mathbf{t}}{dl} = \frac{d\mathbf{s}}{dl}\times\mathbf{t} + \mathbf{s}\times\boldsymbol{\chi}\times\mathbf{t}, \tag{1.6.44}$$

yielding

$$\left(\frac{\mathrm{D}\boldsymbol{\chi}}{\mathrm{D}t} - \frac{\mathrm{d}\mathbf{s}}{\mathrm{d}l}\right) \times \mathbf{t} = 2\,\mathbf{s} \times \boldsymbol{\chi} \times \mathbf{t}. \tag{1.6.45}$$

Working similarly with the second and third equations in (1.6.43), we obtain the compatibility condition

$$\frac{\mathrm{D}\boldsymbol{\chi}}{\mathrm{D}t} = \frac{\mathrm{d}\mathbf{s}}{\mathrm{d}l} - 2\,\mathbf{s} \times \boldsymbol{\chi}, \tag{1.6.46}$$

which can be regarded as an evolution equation for the Darboux vector.

Point particle velocity and acceleration

The velocity of a point particle that belongs to a material line can be resolved into tangential, normal, and binormal components,

$$\mathbf{U} = u_t\,\mathbf{t} + u_n\,\mathbf{n} + u_b\,\mathbf{b}. \tag{1.6.47}$$

The point-particle acceleration is

$$\mathbf{a} \equiv \frac{\mathrm{D}\mathbf{U}}{\mathrm{D}t} = \frac{\mathrm{D}u_t}{\mathrm{D}t}\,\mathbf{t} + \frac{\mathrm{D}u_n}{\mathrm{D}t}\,\mathbf{n} + \frac{\mathrm{D}u_b}{\mathrm{D}t}\,\mathbf{b} + u_t\,\frac{\mathrm{D}\mathbf{t}}{\mathrm{D}t} + u_n\,\frac{\mathrm{D}\mathbf{n}}{\mathrm{D}t} + u_b\,\frac{\mathrm{D}\mathbf{b}}{\mathrm{D}t}, \tag{1.6.48}$$

where the time derivatives of the tangent, normal, and binormal vectors are computed using the evolution equations derived previously in this section.

Generalized Frenet–Serret triad

It is sometimes convenient to replace the Frenet–Serret triad, $(\mathbf{t}, \mathbf{n}, \mathbf{b})$, with a rotated triad, $(\mathbf{t}, \mathbf{d}_1, \mathbf{d}_2)$. The orthonormal unit vectors $\mathbf{d}_1$ and $\mathbf{d}_2$ lie in a normal plane and are rotated with respect to $\mathbf{n}$ and $\mathbf{b}$ about the tangent vector $\mathbf{t}$ by angle α, as shown in Figure 1.6.3, so that

$$\mathbf{d}_1 = \cos\alpha\,\mathbf{n} + \sin\alpha\,\mathbf{b}. \qquad \mathbf{d}_2 = -\sin\alpha\,\mathbf{n} + \cos\alpha\,\mathbf{b}. \tag{1.6.49}$$

The rotated triad satisfies the modified Frenet–Serret relations

$$\frac{\mathrm{d}\mathbf{t}}{\mathrm{d}l} = \boldsymbol{\chi} \times \mathbf{t}, \qquad \frac{\mathrm{d}\,\mathbf{d}_1}{\mathrm{d}l} = \boldsymbol{\chi} \times \mathbf{d}_1, \qquad \frac{\mathrm{d}\,\mathbf{d}_2}{\mathrm{d}l} = \boldsymbol{\chi} \times \mathbf{d}_2, \tag{1.6.50}$$

where

$$\boldsymbol{\chi} = \chi_t\mathbf{t} + \chi_{d_1}\mathbf{d}_1 + \chi_{d_2}\mathbf{d}_2 \tag{1.6.51}$$

is a Darboux rotation vector with components

$$\chi_t = \tau, \qquad \chi_{d_1} = -\kappa\sin\alpha, \qquad \chi_{d_2} = -\kappa\cos\alpha. \tag{1.6.52}$$

The evolution equations for the rotated triad take the form

$$\frac{\mathrm{D}\mathbf{t}}{\mathrm{D}t} = \mathbf{s} \times \mathbf{t}, \qquad \frac{\mathrm{D}\,\mathbf{d}_1}{\mathrm{D}t} = \mathbf{s} \times \mathbf{d}_1, \qquad \frac{\mathrm{D}\,\mathbf{d}_2}{\mathrm{D}t} = \mathbf{s} \times \mathbf{d}_2, \tag{1.6.53}$$

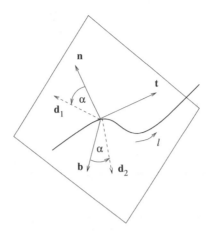

FIGURE 1.6.3   The orthonormal triad, $(\mathbf{t}, \mathbf{d}_1, \mathbf{d}_2)$, arises from the Frenet–Seret triad, $(\mathbf{t}, \mathbf{n}, \mathbf{b})$, by rotating the normal and binormal vectors, $\mathbf{n}$ and $\mathbf{b}$, about the tangent vector, $\mathbf{t}$, by angle α.

where

$$\mathbf{s} = s_t \mathbf{t} - \left(\frac{\partial \mathbf{u}}{\partial l} \cdot \mathbf{b}\right) \mathbf{n} + \left(\frac{\partial \mathbf{u}}{\partial l} \cdot \mathbf{n}\right) \mathbf{b} \qquad (1.6.54)$$

is the spin vector with an unspecified tangential component, s_t. Formulas (1.6.53) complement the Darboux equations (1.6.23). Consistency between these formulas requires the compatibility condition (1.6.46).

Problems

1.6.1 *Expanding and stretching circle*

Consider a circle of radius a, identify the label ξ with the polar angle measured around the center, θ, and evaluate the right-hand side of (1.6.26) for $u_r = U$ and $u_\theta = V \cos\theta$, where U and V are two constants.

1.6.2 *Helical line*

A helical line revolving around the x axis is described in the parametric form by the equations

$$x = b\frac{\varphi}{2\pi}, \qquad y = a\cos\varphi, \qquad z = a\sin\varphi, \qquad (1.6.55)$$

where a is the radius of the circumscribed cylinder, φ is the azimuthal angle, and b is the helical pitch. Show that the curvature and torsion of the helical line are constant and equal to $\kappa = a/(a^2 + b^2)$ and $\tau = b/(a^2 + b^2)$.

1.6.3 *Rigid-body motion*

Derive an expression for the spin vector defined in (1.6.54) for a material line in rigid-body motion.

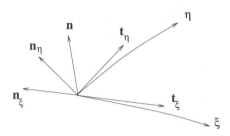

FIGURE 1.7.1   Illustration of a system of two curvilinear axes, (ξ, η), in a three-dimensional material surface. The unit vectors $\mathbf{n}$, $\mathbf{n}_\xi$, and $\mathbf{n}_\eta$ are normal to the surface, the ξ axis, and the η axis.

1.7   Material surfaces

A material surface is an open or closed, infinite or finite surface consisting of a fixed collection of point particles with permanent identity. Physically, a material surface can be identified with the boundary of a fluid parcel or with the interface between two immiscible viscous fluids. To describe the shape and motion of a material surface, we identify the point particles comprising the surface by two scalar labels, ξ and η, called surface curvilinear coordinates, taking values in a specified region of the (ξ, η) parametric plane. We will assume that the pair (ξ, η) forms a right-handed orthogonal or nonorthogonal coordinate system, as illustrated in Figure 1.7.1.

Using the label ξ and η, we effectively establish a mapping of the curved material surface in the three-dimensional physical space to a certain area in the parametric (ξ, η) plane. We may say that (ξ, η) are convected coordinates, meaning that the constituent point particles retain the values of (ξ, η) as they move in the domain of flow.

1.7.1   Tangential vectors and metric coefficients

To establish relations between the geometrical properties of a material surface and the position of point particles in the material surface, $\mathbf{X}(\xi, \eta, t)$, we introduce the tangential unit vectors

$$\mathbf{t}_\xi = \frac{1}{h_\xi} \frac{\partial \mathbf{X}}{\partial \xi}, \qquad \mathbf{t}_\eta = \frac{1}{h_\eta} \frac{\partial \mathbf{X}}{\partial \eta}, \qquad (1.7.1)$$

where

$$h_\xi = \left| \frac{\partial \mathbf{X}}{\partial \xi} \right|, \qquad h_\eta = \left| \frac{\partial \mathbf{X}}{\partial \eta} \right| \qquad (1.7.2)$$

are metrics associated with the curvilinear coordinates. The arc length of an infinitesimal section of the ξ or η axis is

$$dl_\xi = h_\xi \, d\xi, \qquad dl_\eta = h_\eta \, d\eta. \qquad (1.7.3)$$

Any linear combination of the tangential vectors (1.7.1) is also a tangential vector. If $\mathbf{t}_\eta$ is perpendicular to $\mathbf{t}_\xi$, the system of surface coordinates (ξ, η) is orthogonal, $\mathbf{t}_\xi \cdot \mathbf{t}_\eta = 0$.

Since the lengths of the vectors $\mathbf{t}_\xi$ and $\mathbf{t}_\eta$ are equal to unity, $\mathbf{t}_\xi \cdot \mathbf{t}_\xi = 1$ and $\mathbf{t}_\eta \cdot \mathbf{t}_\eta = 1$. Taking the ξ or η derivatives of these equations, we find that

$$\mathbf{t}_\xi \cdot \frac{\partial \mathbf{t}_\xi}{\partial \xi} = 0, \qquad \mathbf{t}_\xi \cdot \frac{\partial \mathbf{t}_\xi}{\partial \eta} = 0, \qquad \mathbf{t}_\eta \cdot \frac{\partial \mathbf{t}_\eta}{\partial \xi} = 0, \qquad \mathbf{t}_\eta \cdot \frac{\partial \mathbf{t}_\eta}{\partial \eta} = 0. \tag{1.7.4}$$

It is evident from the definitions (1.7.1) that the unit tangent vectors satisfy the relation

$$\frac{\partial(h_\xi \mathbf{t}_\xi)}{\partial \eta} = \frac{\partial(h_\eta \mathbf{t}_\eta)}{\partial \xi} = \frac{\partial^2 \mathbf{X}}{\partial \xi \partial \eta}. \tag{1.7.5}$$

Expanding the first two derivatives and projecting the resulting equation onto $\mathbf{t}_\xi$ or $\mathbf{t}_\eta$, we obtain

$$\frac{\partial h_\xi}{\partial \eta} = h_\eta \frac{\partial \mathbf{t}_\eta}{\partial \xi} \cdot \mathbf{t}_\xi + \frac{\partial h_\eta}{\partial \xi} \mathbf{t}_\xi \cdot \mathbf{t}_\eta, \qquad \frac{\partial h_\eta}{\partial \xi} = h_\xi \frac{\partial \mathbf{t}_\xi}{\partial \eta} \cdot \mathbf{t}_\eta + \frac{\partial h_\xi}{\partial \eta} \mathbf{t}_\eta \cdot \mathbf{t}_\xi. \tag{1.7.6}$$

Evolution of the coordinate metric coefficients

To compute the rate of change of the scaling factor h_ξ expressing the rate of extension of the ξ lines, we take the material derivative of the first equation in (1.7.2). Noting that, by definition, $\mathbf{U} = (\partial \mathbf{X}/\partial t)_{\xi,\eta}$, and working as in (1.6.26), we find that

$$\frac{Dh_\xi}{Dt} = \mathbf{t}_\xi \cdot \frac{\partial \mathbf{U}}{\partial \xi} = \frac{\partial(\mathbf{t}_\xi \cdot \mathbf{U})}{\partial \xi} + \mathbf{U} \cdot \mathbf{n}_\xi \kappa_\xi \frac{\partial l_\xi}{\partial \xi}, \tag{1.7.7}$$

where κ_ξ is the curvature of the ξ line and $\mathbf{n}_\xi$ is the principal unit vector normal to the ξ line defined by the equation $\partial \mathbf{t}_\xi / \partial l_\xi = -\kappa_\xi \mathbf{n}_\xi$, as shown in Figure 1.7.1. The two terms on the right-hand side of (1.7.7) express, respectively, changes in the length of an infinitesimal section of a ξ line due to stretching along the ξ line, and expansion due to motion normal to the ξ line. The rate of change of h_η is given by equation (1.7.7) with ξ replaced by η in each place.

Orthogonal coordinates

In the case of orthogonal curvilinear coordinates, $\mathbf{t}_\xi \cdot \mathbf{t}_\eta = 0$, the second term on the right-hand side of each equation in (1.7.6) does not appear. Since the vector $\partial \mathbf{t}_\eta / \partial \xi$ lacks a component in the direction of $\mathbf{t}_\eta$ and the vector $\partial \mathbf{t}_\xi / \partial \eta$ lacks a component in the direction of $\mathbf{t}_\xi$, we may write

$$\frac{\partial \mathbf{t}_\eta}{\partial \xi} = \frac{1}{h_\eta} \frac{\partial h_\xi}{\partial \eta} \mathbf{t}_\xi, \qquad \frac{\partial \mathbf{t}_\xi}{\partial \eta} = \frac{1}{h_\xi} \frac{\partial h_\eta}{\partial \xi} \mathbf{t}_\eta. \tag{1.7.8}$$

1.7.2 Normal vector and surface metric

The unit vector normal to a material surface at a point is

$$\mathbf{n} = \frac{1}{h_s} \frac{\partial \mathbf{X}}{\partial \xi} \times \frac{\partial \mathbf{X}}{\partial \eta}, \tag{1.7.9}$$

where

$$h_s \equiv \left| \frac{\partial \mathbf{X}}{\partial \xi} \times \frac{\partial \mathbf{X}}{\partial \eta} \right| \tag{1.7.10}$$

is the surface metric, as shown in Figure 1.7.1. Combining (1.7.1) with (1.7.9), we find that

$$\mathbf{n} = \frac{h_\xi h_\eta}{h_s} \mathbf{t}_\xi \times \mathbf{t}_\eta. \qquad (1.7.11)$$

Since any tangential vector is perpendicular to the normal vector, we can write

$$\mathbf{n} \cdot \frac{\partial \mathbf{X}}{\partial \xi} = 0, \qquad \mathbf{n} \cdot \frac{\partial \mathbf{X}}{\partial \eta} = 0. \qquad (1.7.12)$$

Differentiating the first equation with respect to η and the second equation with respect to ξ, expanding the derivatives and combining the results to eliminate the common term $\mathbf{n} \cdot \partial^2 \mathbf{X}/\partial \xi \partial \eta$, we obtain the useful relation

$$\frac{\partial \mathbf{n}}{\partial l_\xi} \cdot \mathbf{t}_\eta = \frac{\partial \mathbf{n}}{\partial l_\eta} \cdot \mathbf{t}_\xi. \qquad (1.7.13)$$

Surface area

The area of a differential element of the material surface is $dS = h_s \, d\xi \, d\eta$. The total area of the material surface is

$$S = \iint_\Omega h_s \, d\xi \, d\eta, \qquad (1.7.14)$$

where Ω is the range of variation of ξ and η over the surface. Equation (1.7.14) confirms that h_s is a metric coefficient associated with the surface coordinates, analogous to the Jacobian, $\mathcal{J}$, associated with the Lagrangian labels of three-dimensional fluid parcels. To compute the rate of change of the surface area of a material surface, we differentiate (1.7.14) with respect to time and note that the limits of integration are fixed to obtain

$$\frac{dS}{dt} = \iint_\Omega \frac{Dh_s}{Dt} \, d\xi \, d\eta = \iint_S \frac{1}{h_s} \frac{Dh_s}{Dt} \, dS, \qquad (1.7.15)$$

where S denotes the surface. The last integrand, expressing the rate of expansion of an infinitesimal material patch, is identified with the rate of dilatation of the surface, as discussed in Section 1.7.3.

Orthogonal coordinates

If the surface coordinates (ξ, η) are orthogonal, $\mathbf{t}_\xi \cdot \mathbf{t}_\eta = 0$ and $|\mathbf{t}_\xi \times \mathbf{t}_\eta| = 1$, the surface metric is the product of the two surface coordinate metrics, $h_s = h_\xi h_\eta$. Projecting equations (1.7.8) onto $\mathbf{n}$, we find that

$$\mathbf{n} \cdot \frac{\partial \mathbf{t}_\eta}{\partial \xi} = 0, \qquad \mathbf{n} \cdot \frac{\partial \mathbf{t}_\xi}{\partial \eta} = 0, \qquad (1.7.16)$$

yielding

$$\mathbf{t}_\eta \cdot \frac{\partial \mathbf{n}}{\partial \xi} = 0, \qquad \mathbf{t}_\xi \cdot \frac{\partial \mathbf{n}}{\partial \eta} = 0. \qquad (1.7.17)$$

1.7.3   Evolution equations

It will be convenient to introduce the material normal vector,

$$\mathbf{N} \equiv h_s \mathbf{n}. \tag{1.7.18}$$

Using the definition of the normal unit vector, $\mathbf{n}$, stated in (1.7.9), and the dynamical law (1.6.1) for the rate of change of a material vector, we compute

$$\frac{\mathrm{D}\mathbf{N}}{\mathrm{D}t} = \frac{\mathrm{D}}{\mathrm{D}t}\left(\frac{\partial \mathbf{X}}{\partial \xi}\right) \times \frac{\partial \mathbf{X}}{\partial \eta} + \frac{\partial \mathbf{X}}{\partial \xi} \times \frac{\mathrm{D}}{\mathrm{D}t}\left(\frac{\partial \mathbf{X}}{\partial \eta}\right) \tag{1.7.19}$$

and then

$$\frac{\mathrm{D}\mathbf{N}}{\mathrm{D}t} = \left(\frac{\partial \mathbf{X}}{\partial \xi} \cdot \mathbf{L}\right) \times \frac{\partial \mathbf{X}}{\partial \eta} + \frac{\partial \mathbf{X}}{\partial \xi} \times \left(\frac{\partial \mathbf{X}}{\partial \eta} \cdot \mathbf{L}\right). \tag{1.7.20}$$

In index notation,

$$\frac{\mathrm{D}N_i}{\mathrm{D}t} = \epsilon_{ijk}\frac{\partial X_l}{\partial \xi}L_{lj}\frac{\partial X_k}{\partial \eta} + \epsilon_{ikj}\frac{\partial X_k}{\partial \xi}\frac{\partial X_l}{\partial \eta}L_{lj}, \tag{1.7.21}$$

which can be rearranged into

$$\frac{\mathrm{D}N_i}{\mathrm{D}t} = L_{lj}\epsilon_{ijk}\left(\frac{\partial X_l}{\partial \xi}\frac{\partial X_k}{\partial \eta} - \frac{\partial X_k}{\partial \xi}\frac{\partial X_l}{\partial \eta}\right), \tag{1.7.22}$$

and then restated as

$$\frac{\mathrm{D}N_i}{\mathrm{D}t} = L_{lj}\epsilon_{ijk}\,\epsilon_{plk}\,\epsilon_{pmn}\frac{\partial X_m}{\partial \xi}\frac{\partial X_n}{\partial \eta}. \tag{1.7.23}$$

Now using the rules of repeated multiplication of the alternating matrix discussed in Section A.4, Appendix A, we obtain

$$\frac{\mathrm{D}N_i}{\mathrm{D}t} = L_{lj}(\delta_{ip}\delta_{jl} - \delta_{il}\delta_{jp})\,\epsilon_{pmn}\frac{\partial X_m}{\partial \xi}\frac{\partial X_n}{\partial \eta} \tag{1.7.24}$$

or

$$\frac{\mathrm{D}N_i}{\mathrm{D}t} = L_{jj}\,\epsilon_{imn}\frac{\partial X_m}{\partial \xi}\frac{\partial X_n}{\partial \eta} - L_{lj}\,\epsilon_{jmn}\frac{\partial X_m}{\partial \xi}\frac{\partial X_n}{\partial \eta}. \tag{1.7.25}$$

Switching back to vector notation, we express the final result in the form

$$\frac{1}{h_s}\frac{\mathrm{D}\mathbf{N}}{\mathrm{D}t} = (\alpha\mathbf{I} - \mathbf{L})\cdot\mathbf{n} = -(\mathbf{n}\times\nabla)\times\mathbf{u}, \tag{1.7.26}$$

where $\alpha = \nabla \cdot \mathbf{u}$ is the rate of expansion [432]. Because the operator $\mathbf{n} \times \nabla$ involves tangential derivatives, only the surface distribution of the velocity is required to evaluate the rate of change of the material normal vector, in agreement with physical intuition. Consequently, the rate of change of the material normal vector can be expressed solely in terms of the known instantaneous geometry and motion of the surface and is independent of the flow off the surface [433].

Metric coefficient

An evolution equation for the surface metric coefficient, h_s, can be derived by expressing (1.7.10) in the form

$$h_s^2 = \Big(\frac{\partial \mathbf{X}}{\partial \xi} \times \frac{\partial \mathbf{X}}{\partial \eta}\Big) \cdot \Big(\frac{\partial \mathbf{X}}{\partial \xi} \times \frac{\partial \mathbf{X}}{\partial \eta}\Big). \tag{1.7.27}$$

Taking the material derivative, we obtain

$$2\, h_s \frac{\mathrm{D}h_s}{\mathrm{D}t} = 2\,\Big(\frac{\partial \mathbf{X}}{\partial \xi} \times \frac{\partial \mathbf{X}}{\partial \eta}\Big) \cdot \frac{\mathrm{D}}{\mathrm{D}t}\Big(\frac{\partial \mathbf{X}}{\partial \xi} \times \frac{\partial \mathbf{X}}{\partial \eta}\Big), \tag{1.7.28}$$

which can be restated as

$$\frac{\mathrm{D}h_s}{\mathrm{D}t} = \mathbf{n} \cdot \frac{\mathrm{D}\mathbf{N}}{\mathrm{D}t}. \tag{1.7.29}$$

Using expression (1.7.26), we obtain the evolution equation

$$\frac{1}{h_s}\frac{\mathrm{D}h_s}{\mathrm{D}t} = \alpha - \mathbf{n}\cdot\mathbf{L}\cdot\mathbf{n}, \tag{1.7.30}$$

where $\alpha = \nabla \cdot \mathbf{u}$ is the rate of expansion

Surface divergence

Since the scalar $\mathbf{n}\cdot\mathbf{L}\cdot\mathbf{n}$ represents the normal derivative of the normal component of the velocity, the right-hand side of (1.7.30) represents the divergence of the velocity in the tangential plane,

$$\nabla_s \cdot \mathbf{u} \equiv \alpha - \mathbf{n}\cdot\mathbf{L}\cdot\mathbf{n} = (\mathbf{P}\cdot\nabla)\cdot\mathbf{u} = \mathrm{trace}(\mathbf{P}\cdot\mathbf{L}\cdot\mathbf{P}), \tag{1.7.31}$$

called the surface divergence of the velocity, where $\mathbf{P} = \mathbf{I} - \mathbf{n}\otimes\mathbf{n}$ is a tangential projection operator and $\mathbf{I}$ is the identity matrix. Equation (1.7.30) becomes

$$\frac{1}{h_s}\frac{\mathrm{D}h_s}{\mathrm{D}t} = \nabla_s \cdot \mathbf{u}. \tag{1.7.32}$$

If a material surface is inextensible, the surface divergence of the velocity is zero. For example, if $\mathbf{n} = [1,0,0]$, we obtain $\nabla_s \cdot \mathbf{u} = \partial u_y/\partial y + \partial u_z/\partial z$.

Taking the material derivative of (1.7.10) and carrying out straightforward differentiations, we find that

$$\frac{\mathrm{D}h_s}{\mathrm{D}t} = \Big(h_\eta \frac{\partial \mathbf{U}}{\partial \xi} \times \mathbf{t}_\eta - h_\xi \frac{\partial \mathbf{U}}{\partial \eta} \times \mathbf{t}_\xi\Big)\cdot \mathbf{n}. \tag{1.7.33}$$

Rearranging the triple scalar products, we obtain

$$\frac{\mathrm{D}h_s}{\mathrm{D}t} = \Big(h_\eta \frac{\partial \mathbf{U}}{\partial \xi} \cdot (\mathbf{t}_\eta \times \mathbf{n}) - h_\xi \frac{\partial \mathbf{U}}{\partial \eta} \cdot (\mathbf{t}_\xi \times \mathbf{n})\Big). \tag{1.7.34}$$

Comparing this equation with (1.7.32), we derive an alternative expression for the surface divergence of the velocity,

$$\nabla_s \cdot \mathbf{u} = \frac{h_\xi h_\eta}{h_s} \left(\frac{\partial \mathbf{U}}{\partial l_\xi} \cdot (\mathbf{t}_\eta \times \mathbf{n}) + \frac{\partial \mathbf{U}}{\partial l_\eta} \cdot (\mathbf{n} \times \mathbf{t}_\xi) \right). \tag{1.7.35}$$

If the surface coordinates are orthogonal, $\mathbf{t}_\eta \times \mathbf{n} = \mathbf{t}_\xi$ and $\mathbf{n} \times \mathbf{t}_\xi = \mathbf{t}_\eta$.

Normal vector

Combining (1.7.26) with (1.7.30), we derive an evolution equation for the unit vector normal to the surface,

$$\frac{D\mathbf{n}}{Dt} = \frac{1}{h_s} \left(\frac{D\mathbf{N}}{Dt} - \mathbf{n} \frac{Dh_s}{Dt} \right) = \mathbf{n}(\alpha - \nabla_s \cdot \mathbf{u}) - \mathbf{L} \cdot \mathbf{n}. \tag{1.7.36}$$

Rearranging the expression inside the last parentheses, we obtain

$$\frac{D\mathbf{n}}{Dt} = (\mathbf{n} \otimes \mathbf{n}) \cdot \mathbf{L} \cdot \mathbf{n} - \mathbf{L} \cdot \mathbf{n} = -\mathbf{P} \cdot \mathbf{L} \cdot \mathbf{n}, \tag{1.7.37}$$

where $\mathbf{P} = \mathbf{I} - \mathbf{n} \otimes \mathbf{n}$ is a tangential projection operator and $\mathbf{I}$ is the identity matrix (Problem 1.7.1). We observe that $\mathbf{n} \cdot D\mathbf{n}/Dt = 0$, as required. Expression (1.7.37) is consistent with the first term on the right-hand side of (1.6.9). Rearranging the last expression in (1.7.37), we confirm that the rate of change of the normal vector, $D\mathbf{n}/Dt$, has only a tangential component,

$$\frac{D\mathbf{n}}{Dt} = -\left[\mathbf{n} \times (\mathbf{L} \cdot \mathbf{n}) \right] \times \mathbf{n} = \mathbf{w} \times \mathbf{n}, \tag{1.7.38}$$

where $\mathbf{w} \equiv -\mathbf{n} \times (\mathbf{L} \cdot \mathbf{n})$.

Density-weighted metric

Combining (1.7.26) and (1.7.30) with the continuity equation (1.2.6), we obtain the compact forms

$$\frac{1}{\rho h_s} \frac{D(\rho \mathbf{N})}{Dt} = -\mathbf{L} \cdot \mathbf{n}, \qquad \frac{1}{\rho h_s} \frac{D(\rho h_s)}{Dt} = -\mathbf{n} \cdot \mathbf{L} \cdot \mathbf{n}. \tag{1.7.39}$$

These equations find useful applications in developing evolution equations for physical quantities defined over a material surface.

Significance of tangential and normal motions

The role of tangential and normal fluid motions on the dilatation of a surface can be demonstrated by resolving the velocity into tangential and normal components,

$$\mathbf{U} = U_\xi \mathbf{t}_\xi + U_\eta \mathbf{t}_\eta + U_n \mathbf{n}. \tag{1.7.40}$$

Substituting this expression into the right-hand side of (1.7.33), we obtain several terms, including the term

$$h_\eta \left(\frac{\partial(U_\xi \mathbf{t}_\xi)}{\partial \xi} \times \mathbf{t}_\eta \right) \cdot \mathbf{n} = h_\eta \left(\frac{\partial}{\partial \xi} \left(h_\eta U_\xi \frac{\mathbf{t}_\xi}{h_\eta} \right) \times \mathbf{t}_\eta \right) \cdot \mathbf{n}$$

$$= \left(\frac{\partial(h_\eta U_\xi)}{\partial \xi} \mathbf{t}_\xi \times \mathbf{t}_\eta \right) \cdot \mathbf{n} + h_\eta^2 U_\xi \left[\frac{\partial}{\partial \xi} \left(\frac{\mathbf{t}_\xi}{h_\eta} \right) \times \mathbf{t}_\eta \right] \cdot \mathbf{n}. \tag{1.7.41}$$

Simplifying and rearranging the last triple mixed product, we obtain

$$h_\eta \left(\frac{\partial(U_\xi \mathbf{t}_\xi)}{\partial \xi} \times \mathbf{t}_\eta \right) \cdot \mathbf{n} = \frac{h_s}{h_\xi h_\eta} \frac{\partial(h_\eta U_\xi)}{\partial \xi} + h_\eta^2 U_\xi \frac{\partial}{\partial \xi} \left(\frac{\mathbf{t}_\xi}{h_\eta} \right) \cdot (\mathbf{t}_\eta \times \mathbf{n}). \qquad (1.7.42)$$

Working in a similar fashion with another term, we find that

$$h_\xi \left(\frac{\partial(U_\xi \mathbf{t}_\xi)}{\partial \eta} \times \mathbf{t}_\xi \right) \cdot \mathbf{n} = h_\xi U_\xi \left(\frac{\partial \mathbf{t}_\xi}{\partial \eta} \times \mathbf{t}_\xi \right) \cdot \mathbf{n} = h_\xi U_\xi \frac{\partial \mathbf{t}_\xi}{\partial \eta} \cdot (\mathbf{t}_\xi \times \mathbf{n}). \qquad (1.7.43)$$

Combining these results, we obtain

$$h_\eta \, \mathbf{n} \cdot \left(\frac{\partial(U_\xi \mathbf{t}_\xi)}{\partial \xi} \times \mathbf{t}_\eta \right) - h_\xi \, \mathbf{n} \cdot \left(\frac{\partial(U_\xi \mathbf{t}_\xi)}{\partial \eta} \times \mathbf{t}_\xi \right)$$

$$= \frac{h_s}{h_\xi h_\eta} \frac{\partial(h_\eta U_\xi)}{\partial \xi} + U_\xi \left[h_\eta^2 \frac{\partial}{\partial \xi} \left(\frac{\mathbf{t}_\xi}{h_\eta} \right) \cdot (\mathbf{t}_\eta \times \mathbf{n}) - h_\xi \frac{\partial \mathbf{t}_\xi}{\partial \eta} \cdot (\mathbf{t}_\xi \times \mathbf{n}) \right]. \qquad (1.7.44)$$

Working in a similar fashion with the second tangential velocity, we obtain

$$h_\eta \, \mathbf{n} \cdot \left(\frac{\partial(U_\eta \mathbf{t}_\eta)}{\partial \xi} \times \mathbf{t}_\eta \right) - h_\xi \, \mathbf{n} \cdot \left(\frac{\partial(U_\eta \mathbf{t}_\eta)}{\partial \eta} \times \mathbf{t}_\xi \right)$$

$$= \frac{h_s}{h_\xi h_\eta} \frac{\partial(h_\xi U_\eta)}{\partial \eta} + U_\eta \left[h_\xi^2 \frac{\partial}{\partial \eta} \left(\frac{\mathbf{t}_\eta}{h_\xi} \right) \cdot (\mathbf{t}_\xi \times \mathbf{n}) - h_\eta \frac{\partial \mathbf{t}_\eta}{\partial \xi} \cdot (\mathbf{t}_\eta \times \mathbf{n}) \right]. \qquad (1.7.45)$$

The term on the right-hand side of (1.7.33) involving the normal component of the velocity takes the form

$$h_\eta \mathbf{n} \cdot \left(\frac{\partial(U_n \mathbf{n})}{\partial \xi} \times \mathbf{t}_\eta \right) - h_\xi \mathbf{n} \cdot \left(\frac{\partial(U_n \mathbf{n})}{\partial \eta} \times \mathbf{t}_\xi \right) = U_n \mathbf{n} \cdot \left(h_\eta \frac{\partial \mathbf{n}}{\partial \xi} \times \mathbf{t}_\eta - h_\xi \frac{\partial \mathbf{n}}{\partial \eta} \times \mathbf{t}_\xi \right). \qquad (1.7.46)$$

Substituting (1.7.44), (1.7.45), and (1.7.46) into (1.7.33), we obtain four terms,

$$\frac{1}{h_s} \frac{Dh_s}{Dt} = \frac{1}{h_\xi h_\eta} \left(\frac{\partial(h_\eta U_\xi)}{\partial \xi} + \frac{\partial(h_\xi U_\eta)}{\partial \eta} \right)$$

$$+ U_\xi \frac{1}{h_s} \left[h_\eta^2 \frac{\partial}{\partial \xi} \left(\frac{\mathbf{t}_\xi}{h_\eta} \right) \cdot (\mathbf{t}_\eta \times \mathbf{n}) - h_\xi \frac{\partial \mathbf{t}_\xi}{\partial \eta} \cdot (\mathbf{t}_\xi \times \mathbf{n}) \right]$$

$$+ U_\eta \frac{1}{h_s} \left[h_\xi^2 \frac{\partial}{\partial \eta} \left(\frac{\mathbf{t}_\eta}{h_\xi} \right) \cdot (\mathbf{t}_\xi \times \mathbf{n}) - h_\eta \frac{\partial \mathbf{t}_\eta}{\partial \xi} \cdot (\mathbf{t}_\eta \times \mathbf{n}) \right]$$

$$+ U_n \left(\frac{h_\eta}{h_s} \frac{\partial \mathbf{n}}{\partial \xi} \cdot (\mathbf{t}_\eta \times \mathbf{n}) - \frac{h_\xi}{h_s} \frac{\partial \mathbf{n}}{\partial \eta} \cdot (\mathbf{t}_\xi \times \mathbf{n}) \right), \qquad (1.7.47)$$

with the understanding that the point-particle velocity is the fluid velocity, $\mathbf{U} = \mathbf{u}$.

Surface divergence of the surface velocity

The sum of the first three terms on the right-hand side of (1.7.47) involving the tangential velocities is equal to the surface divergence of the surface velocity, $\mathbf{u}_s = \mathbf{u} \cdot \mathbf{P}$, defined as

$$\nabla_s \cdot \mathbf{u}_s \equiv \text{trace}(\mathbf{P} \cdot \nabla \mathbf{u}_s), \qquad (1.7.48)$$

where $\mathbf{P} = \mathbf{I} - \mathbf{n} \otimes \mathbf{n}$ is a tangential projection operator and $\mathbf{I}$ is the identity matrix. Physically, the surface divergence of the surface velocity expresses dilatation due to expansion in the plane the surface, which may occur even in the case of an incompressible fluid.

Dilatation due to expansion

The last term on the right-hand side of (1.7.47) expresses dilatation due to normal motion associated with expansion or contraction. In Section 1.8, we will see that the expression enclosed by the last large parentheses is equal to twice the mean curvature of the surface, κ_m, as shown in (1.8.19),

$$2\kappa_m = \frac{h_\xi h_\eta}{h_s}\left(\frac{\partial \mathbf{n}}{\partial l_\xi}\cdot(\mathbf{t}_\eta \times \mathbf{n}) + \frac{\partial \mathbf{n}}{\partial l_\eta}\cdot(\mathbf{n} \times \mathbf{t}_\xi)\right). \tag{1.7.49}$$

The expression on the right-hand side can be computed readily from a grid of surface marker points by numerical differentiation.

Evolution of the surface metric

In summary, we have derived an expression that delineates the significance of tangential and normal motions,

$$\frac{1}{h_s}\frac{Dh_s}{Dt} = \nabla_s \cdot \mathbf{u}_s + 2\kappa_m \mathbf{u}\cdot\mathbf{n}. \tag{1.7.50}$$

Only the first term on the right-hand side appears over a stationary surface where the normal velocity is zero. The evolution equation (1.7.50) for a material surface is a generalization of the evolution equation (1.6.26) for a material line.

Orthogonal surface curvilinear coordinates

The preceding interpretations become more evident by assuming that the surface coordinates are orthogonal, $\mathbf{t}_\xi \cdot \mathbf{t}_\eta = 0$. The first term on the right-hand side of (1.7.47) is the standard expression for the surface divergence of the velocity in orthogonal curvilinear coordinates,

$$\nabla_s \cdot \mathbf{u}_s = \frac{1}{h_\xi h_\eta}\left(\frac{\partial(h_\eta U_\xi)}{\partial \xi} + \frac{\partial(h_\xi U_\eta)}{\partial \eta}\right). \tag{1.7.51}$$

The term enclosed by the first square brackets on the right-hand side of (1.7.47) is zero. To show this, we note that $\mathbf{t}_\xi \times \mathbf{n} = -\mathbf{t}_\eta$ and $\mathbf{t}_\eta \times \mathbf{n} = \mathbf{t}_\xi$, recall that $\mathbf{t}_\xi \cdot \partial\mathbf{t}_\xi/\partial\xi = 0$ because $\mathbf{t}_\xi$ is a unit vector, and find that the terms enclosed by the first pair of square brackets simplify into

$$h_\eta^2 \frac{\partial}{\partial \xi}\left(\frac{\mathbf{t}_\xi}{h_\eta}\right)\cdot\mathbf{t}_\xi + h_\xi\frac{\partial\mathbf{t}_\xi}{\partial\eta}\cdot\mathbf{t}_\eta = h_\eta^2\frac{\partial}{\partial\xi}\left(\frac{1}{h_\eta}\right) + h_\xi\frac{\partial\mathbf{t}_\xi}{\partial\eta}\cdot\mathbf{t}_\eta = -\frac{\partial h_\eta}{\partial\xi} + h_\xi\frac{\partial\mathbf{t}_\xi}{\partial\eta}\cdot\mathbf{t}_\eta, \tag{1.7.52}$$

which is zero in light of (1.7.6). Working in a similar fashion, we find that the term enclosed by the second square brackets on the right-hand side of (1.7.47) is also zero.

1.7.4 Flow rate of a vector field through a material surface

The flow rate of a vector field, $\mathbf{q}$, through a material surface, $\mathcal{S}$, is defined as

$$Q = \iint_{\mathcal{S}} \mathbf{q} \cdot \mathbf{n}\, dS = \iint_{\Omega} \mathbf{q} \cdot \mathbf{N}\, d\xi\, d\eta, \qquad (1.7.53)$$

where ξ and η vary over the domain Ω in parametric space, and $\mathbf{N} \equiv h_s \mathbf{n}$ is the material normal vector. In various applications of fluid mechanics and transport phenomena, $\mathbf{q}$ can be identified with the velocity, the vorticity, the temperature gradient, or the gradient of the concentration of a chemical species. When $\mathbf{q}$ is the velocity, Q is the volumetric flow rate. When $\mathbf{q}$ is the vorticity, Q is the circulation around a loop bounding an open surface, as discussed in Section 1.12.2.

Taking the time derivative of (1.7.53), using (1.7.26), and introducing the rate of expansion, $\alpha = \nabla \cdot \mathbf{u}$, we obtain

$$\frac{dQ}{dt} = \iint_{\mathcal{S}} \left(\frac{D\mathbf{q}}{Dt} + \alpha\, \mathbf{q} - \mathbf{q} \cdot \mathbf{L} \right) \cdot \mathbf{n}\, dS. \qquad (1.7.54)$$

The second and third terms inside the integrand express the effect of surface dilatation. Expressing the material derivative, $D\mathbf{q}/Dt$, in terms of Eulerian derivatives and using the vector identity (A.6.11) in Appendix A, to write

$$\nabla \times (\mathbf{q} \times \mathbf{u}) = \alpha\, \mathbf{q} - \mathbf{u}\,(\nabla \cdot \mathbf{q}) + \mathbf{u} \cdot \nabla \mathbf{q} - \mathbf{q} \cdot \mathbf{L}, \qquad (1.7.55)$$

we recast (1.7.54) into the form

$$\frac{dQ}{dt} = \iint_{\mathcal{S}} \left(\frac{\partial \mathbf{q}}{\partial t} + \nabla \times (\mathbf{q} \times \mathbf{u}) + \mathbf{u}\,(\nabla \cdot \mathbf{q}) \right) \cdot \mathbf{n}\, dS. \qquad (1.7.56)$$

If the vector field $\mathbf{q}$ is solenoidal, $\nabla \cdot \mathbf{q} = 0$, the third term inside the integral does not appear. The Zorawski condition states that, for the flow rate Q to remain constant in time, the expression inside the tall parentheses on the right-hand side of (1.7.56) must be zero.

Problems

1.7.1 *Evolution of the unit vector normal to a material surface*

Explain how expression (1.7.37) arises from (1.7.36).

1.7.2 *Expanding and stretching sphere*

Consider a spherical surface of radius a and identify ξ with the meridional angle, θ, and η with the azimuthal angle, φ. Evaluate the right-hand side of (1.7.47) for $u_r = U$, $u_\theta = V \cos\theta$, and $u_\varphi = W \cos\varphi$, where U, V, and W are three constant velocities.

1.7.3 *Change of volume of a parcel resting on a material surface*

Consider a small flattened fluid parcel with a flat side of area dS resting on a material surface. The volume of the parcel is $dV = \mathbf{n} \cdot \delta\mathbf{X}\, dS$, where $\delta\mathbf{X}$ is a material vector across the thickness of the fluid parcel. Using (1.6.1) and (1.7.30), confirm that $D\,\delta V/Dt = \alpha\,\delta V$, where $\alpha = \nabla \cdot \mathbf{u}$ is the rate of expansion.

1.8   Differential geometry of surfaces

Studying the differential geometry of surfaces is prerequisite for describing the shape and motion of material surfaces and interfaces between two immiscible fluids in a specified flow. Necessary concepts and useful relationships are discussed in this section. Further information can be found in specialized monographs (e.g., [389]).

1.8.1   Metric tensor and the first fundamental form of a surface

Material point particles in a material surface can be identified by two surface curvilinear coordinates, (ξ, η), as discussed in Section 1.7. A material vector embedded in a material surface can be described in parametric form as

$$\mathrm{d}\mathbf{X} = \frac{\partial \mathbf{X}}{\partial \xi}\,\mathrm{d}\xi + \frac{\partial \mathbf{X}}{\partial \eta}\,\mathrm{d}\eta = h_\xi\,\mathrm{d}\xi\,\mathbf{t}_\xi + h_\eta\,\mathrm{d}\eta\,\mathbf{t}_\eta. \tag{1.8.1}$$

The square of the length of the vector is

$$\mathrm{d}\mathbf{X} \cdot \mathrm{d}\mathbf{X} = \frac{\partial \mathbf{X}}{\partial \xi} \cdot \frac{\partial \mathbf{X}}{\partial \xi}\,(\mathrm{d}\xi)^2 + 2\,\frac{\partial \mathbf{X}}{\partial \xi} \cdot \frac{\partial \mathbf{X}}{\partial \eta}\,\mathrm{d}\xi\,\mathrm{d}\eta + \frac{\partial \mathbf{X}}{\partial \eta} \cdot \frac{\partial \mathbf{X}}{\partial \eta}\,(\mathrm{d}\eta)^2. \tag{1.8.2}$$

Introducing the surface metric tensor, $g_{\alpha\beta}$, with components

$$g_{\xi\xi} \equiv \frac{\partial \mathbf{X}}{\partial \xi} \cdot \frac{\partial \mathbf{X}}{\partial \xi} = h_\xi^2, \qquad g_{\xi\eta} = g_{\eta\xi} = \frac{\partial \mathbf{X}}{\partial \xi} \cdot \frac{\partial \mathbf{X}}{\partial \eta}, \qquad g_{\eta\eta} \equiv \frac{\partial \mathbf{X}}{\partial \eta} \cdot \frac{\partial \mathbf{X}}{\partial \eta} = h_\eta^2, \tag{1.8.3}$$

we recast (1.8.2) into the compact form

$$\mathrm{d}\mathbf{X} \cdot \mathrm{d}\mathbf{X} \equiv g_{\xi\xi}\,(\mathrm{d}\xi)^2 + 2\,g_{\xi\eta}\,\mathrm{d}\xi\,\mathrm{d}\eta + g_{\eta\eta}\,(\mathrm{d}\eta)^2. \tag{1.8.4}$$

In the nomenclature of differential geometry, equation (1.8.4) is called the first fundamental form of the surface.

An alternative expression of the first fundamental form is

$$\mathrm{d}\mathbf{X} \cdot \mathrm{d}\mathbf{X} = (g_{\xi\xi} + 2\lambda\,g_{\xi\eta} + \lambda^2\,g_{\eta\eta})\,(\mathrm{d}\xi)^2, \tag{1.8.5}$$

where $\lambda \equiv \mathrm{d}\eta/\mathrm{d}\xi$. Since the binomial with respect to λ on the right-hand side of (1.8.5) is real and positive for any value of λ, the roots of the binomial must be complex, the discriminant must be negative, and the determinant of the metric tensor must be positive,

$$\det(\mathbf{g}) = g_{\xi\xi}\,g_{\eta\eta} - g_{\xi\eta}^2 > 0. \tag{1.8.6}$$

Using the definition (1.7.10), we find that

$$h_s^2 = \det(\mathbf{g}), \tag{1.8.7}$$

which is consistent with inequality (1.8.6).

Orthogonal coordinates

If the coordinates ξ and η are orthogonal, the metric tensor is diagonal, $g_{\xi\eta} = g_{\eta\xi} = 0$, $h_s^2 = g_{\xi\xi}g_{\eta\eta}$, and $h_s = h_\xi h_\eta$.

1.8.2   Second fundamental form of a surface

The normal vector changes across the length of an infinitesimal material vector by the differential amount

$$ d\mathbf{n} = \frac{\partial \mathbf{n}}{\partial \xi} \, d\xi + \frac{\partial \mathbf{n}}{\partial \eta} \, d\eta. \tag{1.8.8} $$

Projecting (1.8.8) on (1.8.1), we obtain the second fundamental form of the surface,

$$ -d\mathbf{X} \cdot d\mathbf{n} \equiv f_{\xi\xi} \, d\xi^2 + 2\, f_{\xi\eta} \, d\xi \, d\eta + f_{\eta\eta} \, d\eta^2, \tag{1.8.9} $$

where

$$ f_{\xi\xi} = -\frac{\partial \mathbf{X}}{\partial \xi} \cdot \frac{\partial \mathbf{n}}{\partial \xi} = \frac{\partial^2 \mathbf{X}}{\partial \xi^2} \cdot \mathbf{n}, \qquad f_{\eta\eta} = -\frac{\partial \mathbf{X}}{\partial \eta} \cdot \frac{\partial \mathbf{n}}{\partial \eta} = \frac{\partial^2 \mathbf{X}}{\partial \eta^2} \cdot \mathbf{n}, \tag{1.8.10} $$

and

$$ f_{\xi\eta} = -\frac{1}{2}\left(\frac{\partial \mathbf{X}}{\partial \xi} \cdot \frac{\partial \mathbf{n}}{\partial \eta} + \frac{\partial \mathbf{X}}{\partial \eta} \cdot \frac{\partial \mathbf{n}}{\partial \xi} \right) = \frac{\partial^2 \mathbf{X}}{\partial \xi \partial \eta} \cdot \mathbf{n} = -\frac{\partial \mathbf{X}}{\partial \xi} \cdot \frac{\partial \mathbf{n}}{\partial \eta} = -\frac{\partial \mathbf{X}}{\partial \eta} \cdot \frac{\partial \mathbf{n}}{\partial \xi}. \tag{1.8.11} $$

Equations (1.7.13). were used to derive the last two expressions for $f_{\xi\eta}$. The rate of change of the tensors $\mathbf{g}$ and $\mathbf{f}$ following a point particle in a surface can be computed from their definition using the evolution equations discussed in Section 1.7 (Problem 1.8.1).

Orthogonal coordinates

Using equations (1.7.17), we find that, if the coordinates ξ and η are orthogonal, the off-diagonal components of $f_{\alpha\beta}$ vanish, $f_{\xi\eta} = f_{\eta\xi} = 0$.

1.8.3   Curvatures

The normal curvature of a surface at a point in the direction of the ξ axis, denoted by $\mathcal{K}_\xi$, is defined as the curvature of the trace of the surface in a plane that contains the normal vector, $\mathbf{n}$, and the tangential unit vector, $\mathbf{t}_\xi$, drawn with the heavy line in Figure 1.8.1. In practice, the trace of the surface on the normal plane may not be available and the normal curvature must be extracted from the curvature of another surface line that is tangential to $\mathbf{t}_\xi$ at a point, denoted by κ_ξ, such as the dashed line in Figure 1.8.1. If l_ξ is the arc length along such a line, then, by definition,

$$ \frac{\partial \mathbf{t}_\xi}{\partial l_\xi} \equiv -\kappa_\xi \mathbf{n}_\xi, \tag{1.8.12} $$

where $\mathbf{n}_\xi$ is the unit vector normal to the dashed line in Figure 1.8.1.

Meusnier's theorem

Meusnier's theorem states the the normal curvature in the direction of the ξ axis is given by

$$ \mathcal{K}_\xi \equiv \frac{\partial \mathbf{n}}{\partial l_\xi} \cdot \mathbf{t}_\xi = -\frac{\partial \mathbf{t}_\xi}{\partial l_\xi} \cdot \mathbf{n} = \kappa_\xi \, \mathbf{n}_\xi \cdot \mathbf{n} = -\frac{f_{\xi\xi}}{g_{\xi\xi}} = -\frac{f_{\xi\xi}}{h_\xi^2}. \tag{1.8.13} $$

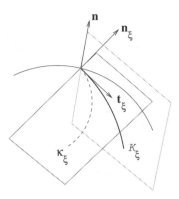

FIGURE 1.8.1   The normal curvature of a surface at a point in the direction of the ξ axis, denoted
by $\mathcal{K}_\xi$, is defined as the curvature of the trace of the surface in a plane that contains the normal
vector, $\mathbf{n}$, and tangential vector, $\mathbf{t}_\xi$.

A similar equation can be written for the η axis. To generalize these expressions, we consider a
tangent vector constructed as a linear combination of the two tangent vectors corresponding to the
surface coordinates, ξ and η,

$$\boldsymbol{\tau}_\lambda = \frac{\partial \mathbf{X}}{\partial \xi} + \lambda \frac{\partial \mathbf{X}}{\partial \eta}, \tag{1.8.14}$$

where λ is a free, positive or negative parameter. The normal curvature in the direction of $\boldsymbol{\tau}_\lambda$ is
given by

$$\mathcal{K}_\lambda = \kappa_\lambda \, \mathbf{n}_\lambda \cdot \mathbf{n} = -\frac{\partial \mathbf{t}_\lambda}{\partial l_\lambda} \cdot \mathbf{n} = \frac{\partial \mathbf{n}}{\partial l_\lambda} \cdot \mathbf{t}_\lambda = -\frac{f_{\xi\xi} + 2f_{\xi\eta}\lambda + f_{\eta\eta}\lambda^2}{g_{\xi\xi} + 2g_{\xi\eta}\lambda + g_{\eta\eta}\lambda^2}, \tag{1.8.15}$$

where $\mathbf{t}_\lambda = \boldsymbol{\tau}_\lambda / |\boldsymbol{\tau}_\lambda|$ is a tangent unit vector and l_λ is the corresponding arc length. Equation (1.8.13)
arises for $\lambda = 0$, and its counterpart for the η axis arises in the limit as $\lambda \to \pm\infty$.

Mean curvature

Using (1.8.14), we find that two tangent vectors corresponding to λ_1 and λ_2 are perpendicular if
they satisfy the relation

$$g_{\xi\xi} + (\lambda_1 + \lambda_2)\, g_{\xi\eta} + \lambda_1 \lambda_2 \, g_{\eta\eta} = 0. \tag{1.8.16}$$

Solving for λ_2 in terms of λ_1, we obtain

$$\lambda_2 = -\frac{g_{\xi\xi} + \lambda_1 g_{\xi\eta}}{g_{\xi\eta} + \lambda_1 g_{\eta\eta}}. \tag{1.8.17}$$

Now using (1.8.15), we find that the mean value of the directional normal curvatures in any two
perpendicular planes is independent of the plane orientation. Motivated by this observation, we

introduce the mean curvature of the surface at a point,

$$\kappa_m \equiv \frac{1}{2}\left(\mathcal{K}_{\lambda_1} + \mathcal{K}_{\lambda_2}\right) = -\frac{1}{2}\frac{g_{\xi\xi}f_{\xi\xi} - 2g_{\xi\eta}f_{\xi\eta} + g_{\eta\eta}f_{\eta\eta}}{g_{\xi\xi}g_{\eta\eta} - g_{\xi\eta}^2}. \tag{1.8.18}$$

An alternative expression arising from (1.7.47) is

$$\kappa_m = \frac{1}{2}\frac{h_\xi h_\eta}{h_s}\left(\frac{\partial \mathbf{n}}{\partial l_\xi}\cdot(\mathbf{t}_\eta \times \mathbf{n}) - \frac{\partial \mathbf{n}}{\partial l_\eta}\cdot(\mathbf{t}_\xi \times \mathbf{n})\right). \tag{1.8.19}$$

It can be shown by straightforward algebraic manipulations that (1.8.19) is equivalent to (1.8.18) (Problem 1.8.2).

Principal curvatures

The maximum and minimum directional normal curvatures, $\mathcal{K}_\lambda$, over all possible values of λ, are called the principal curvatures. The corresponding values of λ are found by setting $\partial \mathcal{K}_\lambda/\partial \lambda = 0$ and using the last expression in (1.8.15) to obtain a quadratic equation,

$$(f_{\eta\eta}g_{\xi\eta} - g_{\eta\eta}f_{\xi\eta})\lambda^2 + (f_{\eta\eta}g_{\xi\xi} - g_{\eta\eta}f_{\xi\xi})\lambda + f_{\xi\eta}g_{\xi\xi} - g_{\xi\eta}f_{\xi\xi} = 0. \tag{1.8.20}$$

The sum and the product of the two roots are given by

$$\lambda_1 + \lambda_2 = -\frac{f_{\eta\eta}g_{\xi\xi} - g_{\eta\eta}f_{\xi\xi}}{f_{\eta\eta}g_{\xi\eta} - g_{\eta\eta}f_{\xi\eta}}, \qquad \lambda_1\lambda_2 = \frac{f_{\xi\eta}g_{\xi\xi} - g_{\xi\eta}f_{\xi\xi}}{f_{\eta\eta}g_{\xi\eta} - g_{\eta\eta}f_{\xi\eta}}. \tag{1.8.21}$$

Direct substitution shows that these equations satisfy (1.8.16), and this demonstrates that, if $\mathcal{K}_{max}$ is the maximum principal curvature corresponding to a particular orientation, then $\mathcal{K}_{min}$ will be the minimum principal curvature corresponding to the perpendicular orientation. The mean curvature of the surface is

$$\kappa_m = \frac{1}{2}\left(\mathcal{K}_{max} + \mathcal{K}_{min}\right). \tag{1.8.22}$$

Euler's theorem for the curvature

We may assume that without loss of generality, that the maximum principal curvature occurs along the ξ axis, corresponding to $\lambda = 0$, and the minimum principal curvature occurs along its orthogonal η axis, corresponding to $\lambda \to \infty$. In that case, $g_{\xi\eta} = 0$ and $f_{\xi\eta} = 0$, yielding

$$\mathcal{K}_{max} = -\frac{f_{\xi\xi}}{g_{\xi\xi}}, \qquad \mathcal{K}_{min} = -\frac{f_{\eta\eta}}{g_{\eta\eta}}, \qquad \mathcal{K}_\lambda = -\frac{f_{\xi\xi} + f_{\eta\eta}\lambda^2}{g_{\xi\xi} + g_{\eta\eta}\lambda^2}. \tag{1.8.23}$$

Now let α be the angle subtended between the direction corresponding to λ and the principal direction of the maximum curvature, varying in the range $[0, \frac{1}{2}\pi]$. Using the geometrical interpretation of the inner vector product, we find that

$$\left(\frac{\partial \mathbf{X}}{\partial \xi} + \lambda\frac{\partial \mathbf{X}}{\partial \eta}\right)\cdot\frac{\partial \mathbf{X}}{\partial \xi} = \left|\frac{\partial \mathbf{X}}{\partial \xi} + \lambda\frac{\partial \mathbf{X}}{\partial \eta}\right|\left|\frac{\partial \mathbf{X}}{\partial \xi}\right|\cos\alpha \tag{1.8.24}$$

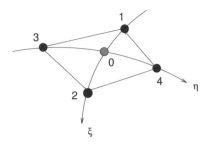

FIGURE 1.8.2   Five points are arranged along two curvilinear axes, (ξ, η) in a surface. If the axes are orthogonal, knowledge of the position of these points is sufficient for computing the mean curvature of the surface at the intersection.

and

$$\left(\frac{\partial \mathbf{X}}{\partial \xi} + \lambda \frac{\partial \mathbf{X}}{\partial \eta}\right) \cdot \frac{\partial \mathbf{X}}{\partial \eta} = \left|\frac{\partial \mathbf{X}}{\partial \xi} + \lambda \frac{\partial \mathbf{X}}{\partial \eta}\right| \left|\frac{\partial \mathbf{X}}{\partial \eta}\right| \sin \alpha, \tag{1.8.25}$$

yielding

$$g_{\xi\xi} = (g_{\xi\xi} + g_{\eta\eta} \lambda^2) \cos^2 \alpha, \qquad \lambda^2 g_{\eta\eta} = (g_{\xi\xi} + g_{\eta\eta} \lambda^2) \sin^2 \alpha. \tag{1.8.26}$$

Combining these expressions with (1.8.23), we derive Euler's theorem for the curvature, stating that the normal curvature in an arbitrary direction is related to the principal curvatures by

$$\mathcal{K}_\lambda = \cos^2 \alpha \, \mathcal{K}_{max} + \sin^2 \alpha \, \mathcal{K}_{min}. \tag{1.8.27}$$

When $\alpha = 0$ or $\frac{1}{2}\pi$, we obtain one or the other principal curvature.

Numerical methods

In numerical practice, the curvature of a surface can be computed by tracing two curvilinear axes with a set of marker points whose position is described as $\mathbf{X}(\xi, \eta)$, and then constructing parametric representations for the ξ and η coordinate lines using methods of curve fitting and function interpolation, as discussed in Section B.4, Appendix B. The partial derivatives of $\mathbf{X}$ with respect to ξ and η can be computed by numerical differentiation, as discussed in Section B.5, Appendix B.

Assume that the positions of five points along the ξ and η axes are given, $\mathbf{X}_i$ for $i = 0\text{--}4$, as shown in Figure 1.8.2. Using centered differences to approximate the tangential vectors at the location of the central point, $\mathbf{X}_0$, we obtain

$$\mathbf{t}_\xi \simeq \frac{\mathbf{X}_2 - \mathbf{X}_1}{|\mathbf{X}_2 - \mathbf{X}_1|}, \qquad \mathbf{t}_\eta \simeq \frac{\mathbf{X}_4 - \mathbf{X}_3}{|\mathbf{X}_4 - \mathbf{X}_3|}. \tag{1.8.28}$$

The numerical accuracy is of first order with respect to $\Delta\xi$ and $\Delta\eta$. If the points are evenly spaced with respect to ξ and η, the accuracy becomes of second order with respect to $\Delta\xi$ and $\Delta\eta$.

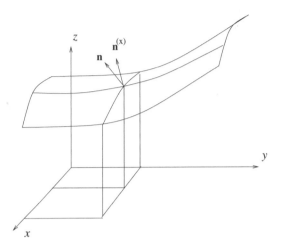

FIGURE 1.8.3 Explicit description of a surface as $z = q(x, y)$, showing the normal unit vector, $\mathbf{n}$, and the unit vector normal to the intersection of the surface with a plane that is normal to the y axis, denoted by $\mathbf{n}^{(x)}$.

Next, we compute an approximation to the normal vector at the central point,

$$\mathbf{n} \simeq \frac{\mathbf{t}_\xi \times \mathbf{t}_\eta}{|\mathbf{t}_\xi \times \mathbf{t}_\eta|}, \tag{1.8.29}$$

and use finite differences to approximate

$$\left(\frac{\partial \mathbf{t}_\xi}{\partial l_\xi}\right)_\eta \simeq \frac{1}{|\mathbf{X}_2 - \mathbf{X}_1|} \left(\frac{\mathbf{X}_2 - \mathbf{X}_0}{|\mathbf{X}_2 - \mathbf{X}_0|} - \frac{\mathbf{X}_0 - \mathbf{X}_1}{|\mathbf{X}_0 - \mathbf{X}_1|}\right). \tag{1.8.30}$$

The normal unit vector, $\mathbf{n}_\xi$, follows by dividing the right-hand side of (1.8.30) by its length, and the directional curvature is extracted from the formula

$$\kappa_\xi = -\mathbf{n}_\xi \cdot \frac{\partial \mathbf{t}_\xi}{\partial l_\xi}. \tag{1.8.31}$$

Similar equations are used to compute $\mathbf{n}_\eta$ and κ_η. If the curvilinear axes are orthogonal, the results can be substituted into (1.8.13) and then into (1.8.18) to obtain an approximation to the mean curvature at the central point, $\mathbf{X}_0$. In that case, but not more generally, knowledge of the position of five points is sufficient for estimating the mean curvature.

Description as $z = q(x, y)$

Assume that a surface is described explicitly by the function $z = q(x, y)$, as shown in Figure 1.8.3. The unit vector normal to the intersection of the surface with a plane that is perpendicular to the y axis is

$$\mathbf{n}^{(x)} = \frac{1}{(1 + q_x^2)^{1/2}} (-q_x \mathbf{e}_x + \mathbf{e}_z), \tag{1.8.32}$$

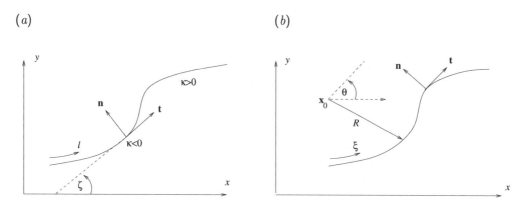

FIGURE 1.8.4   (*a*) Illustration of a line in the xy plane showing the tangential and normal unit vectors, **t** and **n**, and the sign of the curvature, κ. (*b*) Illustration of plane polar coordinates, (r, θ), in the xy plane centered at a point, $\mathbf{x}_0$.

and the unit vector normal to the surface is

$$\mathbf{n} = \frac{1}{(1 + q_x^2 + q_y^2)^{1/2}} \left(-q_x \mathbf{e}_x - q_y \mathbf{e}_y + \mathbf{e}_z \right), \tag{1.8.33}$$

where $q_x = \partial q / \partial x$, $q_y = \partial q / \partial y$, and $\mathbf{e}_x$, $\mathbf{e}_y$, and $\mathbf{e}_z$ are unit vectors along the x, y, and z axes. Applying Meusnier' formula (1.8.13), we find that the corresponding normal curvature is

$$\mathcal{K}_x = \kappa_x \, \mathbf{n}^{(x)} \cdot \mathbf{n} = \kappa_x \left(\frac{1 + q_x^2}{1 + q_x^2 + q_y^2} \right)^{1/2}, \tag{1.8.34}$$

where κ_x is the curvature of the planar intersection. We observe that $\mathcal{K}_x = \kappa_x$ only when $q_y = 0$, corresponding to a cylindrical surface.

A similar expression can be written for $\mathcal{K}_y$. The mean curvature of the surface is *not* necessarily equal to the average of $\mathcal{K}_x$ and $\mathcal{K}_y$.

1.8.4   Curvature of a line in a plane

The curvature of a line in the xy plane, κ, can be computed from the first Frenet–Serret relation in terms of the derivative of the unit tangent vector, **t**, with respect to the arc length, l,

$$\frac{\mathrm{d}\mathbf{t}}{\mathrm{d}l} = -\kappa \, \mathbf{n}. \tag{1.8.35}$$

By convention, **t** and the normal unit vector **n** form a right-handed system of axes, as shown in Figure 1.8.4(*a*). Projecting both sides of (1.8.35) onto **n**, we obtain

$$\kappa = -\mathbf{n} \cdot \frac{\mathrm{d}\mathbf{t}}{\mathrm{d}l}. \tag{1.8.36}$$

Since $\mathbf{t} \cdot \mathbf{n} = 0$ and $\mathrm{d}(\mathbf{t} \cdot \mathbf{n})/\mathrm{d}l = 0$, we also have

$$\kappa = \mathbf{t} \cdot \frac{\mathrm{d}\mathbf{n}}{\mathrm{d}l}. \tag{1.8.37}$$

If the line is tangential to the x axis at a point, $\mathbf{t} = \begin{bmatrix} 1, 0 \end{bmatrix}$ and $\kappa = \mathrm{d}n_x/\mathrm{d}x$.

Description in Cartesian coordinates as $y = q(x)$

Assume that a line in the xy plane does not turn upon itself but has a monotonic shape, as shown in Figure 1.8.4(a). The shape of the line can be described by a single-valued function, $y = q(x)$. Using elementary geometry, we find that the normal unit vector, tangent unit vector, and rate of change of the arc length with respect to x are given by

$$\mathbf{n} = \frac{1}{\sqrt{1+q'^2}} \left(-q' \, \mathbf{e}_x + \mathbf{e}_y \right), \qquad \mathbf{t} = \frac{1}{\sqrt{1+q'^2}} \left(\mathbf{e}_x + q' \, \mathbf{e}_y \right), \qquad \frac{\mathrm{d}l}{\mathrm{d}x} = \sqrt{1+q'^2}, \tag{1.8.38}$$

where $\mathbf{e}_x$ and $\mathbf{e}_y$ are unit vectors for the x and y axes, and a prime denotes a derivative with respect to x. The curvature of the line is

$$\kappa = -\mathbf{n} \cdot \frac{\mathrm{d}\mathbf{t}}{\mathrm{d}l} = -\frac{\mathrm{d}x}{\mathrm{d}l} \, \mathbf{n} \cdot \frac{\mathrm{d}\mathbf{t}}{\mathrm{d}x} = -\frac{1}{(1+q'^2)^{3/2}} \left(-q' \, \mathbf{e}_x + \mathbf{e}_y \right) \left(\mathbf{e}_x + q' \, \mathbf{e}_y \right)'. \tag{1.8.39}$$

Carrying out the differentiation and simplifying, we find that

$$\kappa = -\frac{q''}{(1+q'^2)^{3/2}} = \frac{1}{q'} \left(\frac{1}{\sqrt{1+q'^2}} \right)' = -\left(\frac{q'}{\sqrt{1+q'^2}} \right)'. \tag{1.8.40}$$

The slope angle, ζ, is defined by the equation $\tan \zeta = q'$, where $-\pi/2 < \zeta < \pi/2$, as shown in Figure 1.8.4(a). We note that $1 + q'^2 = 1/\cos^2 \zeta$, and obtain

$$\kappa = \frac{1}{q'} \frac{\mathrm{d}\cos\zeta}{\mathrm{d}x}. \tag{1.8.41}$$

Description in plane polar coordinates as $r = R(\theta)$

Consider a system of plane polar coordinates, (r, θ), in the xy plane, centered at a point, $\mathbf{x}_0$, as shown in Figure 1.8.4(b). The shape of a line can be described by a function $r = R(\theta)$. Using elementary geometry, we find that the normal unit vector, tangent unit vector, and rate of change of arc length with respect to θ are given by

$$\mathbf{n} = \frac{-R\,\mathbf{e}_r + R' \, \mathbf{e}_\theta}{\sqrt{R^2 + R'^2}} \qquad \mathbf{t} = \frac{R' \, \mathbf{e}_r + R\,\mathbf{e}_\theta}{\sqrt{R^2 + R'^2}} \qquad \frac{\mathrm{d}l}{\mathrm{d}\theta} = \sqrt{R^2 + R'^2}, \tag{1.8.42}$$

where $\mathbf{e}_r$ and $\mathbf{e}_\theta$ are unit vectors in the radial and polar directions, and a prime denotes a derivative with respect to θ. The curvature of the line is

$$\kappa = -\mathbf{n} \cdot \frac{\mathrm{d}\mathbf{t}}{\mathrm{d}l} = -\frac{\mathrm{d}\theta}{\mathrm{d}l} \, \mathbf{n} \cdot \frac{\mathrm{d}\mathbf{t}}{\mathrm{d}\theta} = -\frac{1}{(R^2 + R'^2)^{3/2}} \left(-R\,\mathbf{e}_r + R' \, \mathbf{e}_\theta \right) \cdot \left(R' \, \mathbf{e}_r + R\,\mathbf{e}_\theta \right)'. \tag{1.8.43}$$

Recalling that $\partial \mathbf{e}_r / \partial \theta = \mathbf{e}_\theta$ and $\partial \mathbf{e}_\theta / \partial \theta = -\mathbf{e}_r$, carrying out the differentiation and simplifying, we obtain

$$\kappa = \frac{RR'' - 2R'^2 - R^2}{(R^2 + R'^2)^{3/2}}. \tag{1.8.44}$$

In the case of a circular line of radius a, we set $R(\theta) = a$ and obtain $\kappa = -1/a$. The negative sign merely reflects our convention.

Parametric representation in Cartesian coordinates

A line in the xy plane can be described in parametric form in terms of a variable, ξ, that increases monotonically in the direction of the tangent unit vector, $\mathbf{t}$. Regarding the x and y coordinates of a point along the line as functions of ξ, we write $x = X(\xi)$ and $y = Y(\xi)$. Substituting these functions into the first expression for the curvature given in (1.8.40), writing

$$q' = \frac{dy}{dx} = \frac{Y_\xi}{X_\xi}, \qquad q'' = \frac{q'_\xi}{X_\xi}, \tag{1.8.45}$$

and carrying out the differentiations, we obtain the formula

$$\kappa = \frac{X_{\xi\xi} Y_\xi - Y_{\xi\xi} X_\xi}{(X_\xi^2 + Y_\xi^2)^{3/2}}, \tag{1.8.46}$$

where a subscript ξ denotes a derivative with respect to ξ. Formulas (1.8.40) arise by setting $\xi = x$. If ξ increases in a direction that is opposite to that of $\mathbf{t}$, a minus sign is introduced in front of the fraction on the right-hand side of (1.8.46).

Often in practice, the functions $X(\xi)$ and $Y(\xi)$ are reconstructed numerically from data describing the location of marker points along the line using, for example, cubic spline interpolation, as discussed in Appendix B. The derivatives of these functions are computed by numerical differentiation, as discussed in Sections B.4 and B.5, Appendix B. In the simplest implementation, the interpolating variable, ξ, is identified with the arc length of the polygonal line connecting successive marker points.

Shape of a line in terms of the curvature

In the convenient case where the parameter ξ is the arc length along the line, l, the denominator in (1.8.46) is equal to unity, yielding

$$\kappa = X_{ll} Y_l - Y_{ll} X_l. \tag{1.8.47}$$

Differentiating the expression $X_l^2 + Y_l^2 = 1$ with respect to l and using the resulting equation to simplify (1.8.47), we obtain

$$\kappa = \frac{X_{ll}}{Y_l} = -\frac{Y_{ll}}{X_l}. \tag{1.8.48}$$

These expressions allow us to reconstruct a curve in terms of the curvature, $\kappa(l)$, by integrating the ordinary differential equations $X_{ll} = \kappa Y_l$ and $Y_{ll} = -\kappa X_l$ subject to suitable boundary conditions.

Parametric description in plane polar coordinates

A line in the xy plane can be described in parametric form in plane polar coordinates, (r, θ), centered at a chosen point, $\mathbf{x}_0 = (x_0, y_0)$, in terms of a variable ξ that increases monotonically in the direction of the tangent unit vector, $\mathbf{t}$, as shown in Figure 1.8.4(b), so that $r = R(\xi)$ and $\theta = \Theta(\xi)$. Using the transformation rules

$$x(\xi) = x_0 + R(\xi) \cos \Theta(\xi), \qquad y(\xi) = y_0 + R(\xi) \sin \Theta(\xi), \qquad (1.8.49)$$

and applying the chain rule, we find that the expression for the curvature (1.8.46) becomes

$$\kappa = \frac{RR_{\xi\xi}\Theta_\xi - 2R_\xi^2\Theta_\xi - RR_\xi\Theta_{\xi\xi} - R^2\Theta_\xi^3}{(R_\xi^2 + R^2\Theta_\xi^2)^{3/2}}, \qquad (1.8.50)$$

where a subscript denotes a derivative with respect to ξ. Formula (1.8.44) arises be setting $\xi = \theta$.

1.8.5 Mean curvature of a surface as the divergence of the normal vector

The formulas derived in Sections 1.8.3 and 1.8.4 can be used to obtain useful expressions for the mean curvature of a three-dimensional surface. Assume that the x axis is normal, and the yz plane is tangential to a surface at a point. The mean curvature of the surface at that point is given by the surface divergence of the normal vector,

$$2\kappa_m = \nabla_s \cdot \mathbf{n} \equiv (\mathbf{P} \cdot \nabla) \cdot \mathbf{n} = \frac{\partial n_y}{\partial y} + \frac{\partial n_z}{\partial z}, \qquad (1.8.51)$$

where $\mathbf{P} = \mathbf{I} - \mathbf{n} \otimes \mathbf{n}$ is the surface projection operator and ∇ is the three-dimensional gradient. Requiring that

$$n_x^2 + n_y^2 + n_z^2 = 1, \qquad (1.8.52)$$

taking a derivative with respect to x, and noting that $n_x = 1$, $n_y = 0$, and $n_z = 0$ at the origin of the chosen coordinates, we find that $\partial n_x / \partial x = 0$, which shows that the normal derivative of the normal component of the normal vector is zero.

This property allows us to write a general expression for the mean curvature with reference to an arbitrary system of Cartesian coordinates whose axes are not necessarily tangential or normal to the surface at a point,

$$2\kappa_m \equiv \nabla \cdot \mathbf{n} = \frac{\partial n_x}{\partial x} + \frac{\partial n_y}{\partial y} + \frac{\partial n_z}{\partial z}. \qquad (1.8.53)$$

If a surface is described implicitly by the equation $F(x, y, z) = 0$, the unit vector normal to the surface is given by

$$\mathbf{n} = \frac{1}{|\nabla F|} \nabla F, \qquad (1.8.54)$$

and the mean curvature is given by

$$2\kappa_m = \nabla \cdot \left(\frac{1}{|\nabla F|} \nabla F \right) = \frac{1}{|\nabla F|} \nabla^2 F - \frac{1}{|\nabla F|^3} \nabla F \cdot (\nabla\nabla F) \cdot \nabla F, \qquad (1.8.55)$$

where $\nabla\nabla F$ is the matrix of second derivatives.

Explicit representation as $z = q(x, y)$

Assume that a surface is described explicitly as $z = q(x, y)$. The components of the normal unit vector are

$$n_x = -\frac{q_x}{(1 + q_x^2 + q_y^2)^{1/2}}, \qquad n_y = -\frac{q_y}{(1 + q_x^2 + q_y^2)^{1/2}}, \qquad n_z = \frac{1}{(1 + q_x^2 + q_y^2)^{1/2}}, \qquad (1.8.56)$$

where a subscript x or y denotes a derivative with respect to x or y. Applying (1.8.53), we obtain

$$2\kappa_m = -\frac{\partial}{\partial x}\left(\frac{q_x}{(1 + q_x^2 + q_y^2)^{1/2}}\right) - \frac{\partial}{\partial y}\left(\frac{q_y}{(1 + q_x^2 + q_y^2)^{1/2}}\right). \qquad (1.8.57)$$

Carrying out the differentiations, we obtain

$$2\kappa_m = -\frac{(1 + q_y^2)\, q_{xx} - 2q_x q_y q_{xy} + (1 + q_x^2)\, q_{yy}}{(1 + q_x^2 + q_y^2)^{3/2}}. \qquad (1.8.58)$$

This formula also arises from (1.8.55) by setting $F(x, y, z) = z - f(x, y)$. For a nearly flat surface, $2\kappa_m \simeq -q_{xx} - q_{yy}$.

Cylindrical polar coordinates

If a surface is described explicitly in cylindrical polar coordinates, (x, σ, φ), by the function $x = q(\sigma, \varphi)$, we set $F = x - q(\sigma, \varphi)$ and compute

$$\nabla F = \mathbf{e}_x - q_\sigma\, \mathbf{e}_\sigma - Q_\varphi\, \mathbf{e}_\varphi, \qquad (1.8.59)$$

where $Q_\varphi = q_\varphi/\sigma$ and a subscript σ or φ attached to q denotes a derivative with respect to σ or φ. The mean curvature computed using (1.8.55) is

$$2\kappa_m = -\frac{1}{(1 + q_\sigma^2 + Q_\varphi^2)^{3/2}}\left[(1 + Q_\varphi^2)\, q_{\sigma\sigma} + 2q_\sigma Q_\varphi\, \frac{Q_\varphi - q_{\sigma\varphi}}{\sigma} + (1 + q_\sigma^2)\left(Q_{\varphi\varphi} + \frac{q_\sigma}{\sigma}\right)\right], \qquad (1.8.60)$$

where $Q_{\varphi\varphi} = q_{\varphi\varphi}/\sigma^2$. In the case of an axisymmetric surface, $x = q(\varphi)$, the φ derivatives in (1.8.60) are set to zero. For a nearly flat surface,

$$2\kappa_m \simeq -\left(q_{\sigma\sigma} + Q_{\varphi\varphi} + \frac{q_\sigma}{\sigma}\right). \qquad (1.8.61)$$

The union of the three terms inside the parentheses is the Laplacian of the function $q(\sigma, \varphi)$.

Spherical polar coordinates

If a surface is described explicitly in spherical polar coordinates, (r, θ, φ), by the function $r = q(\theta, \varphi)$, we set $F = r - q(\theta, \varphi)$ and compute

$$\nabla F = \mathbf{e}_r - \frac{q_\theta}{r}\, \mathbf{e}_\theta - \frac{q_\varphi}{r \sin\theta}\, \mathbf{e}_\varphi, \qquad (1.8.62)$$

where a subscript θ or φ denotes a derivative with respect to θ or φ. The mean curvature is computed using (1.8.55) (Problem 1.8.4). For a nearly spherical surface of radius a, we find that

$$2\kappa_m \simeq \frac{2}{a} - \left(\frac{\cot\theta}{r^2}\, q_\theta + \frac{1}{r^2}\, q_{\theta\theta} + \frac{1}{r^2 \sin^2\theta}\, q_{\varphi\varphi}\right). \qquad (1.8.63)$$

The union of the three terms inside the parentheses is the Laplacian of the function $q(\theta, \varphi)$.

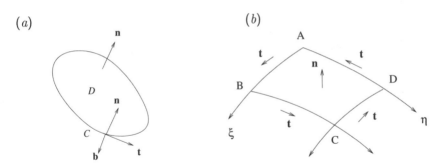

FIGURE 1.8.5 (*a*) A contour integral of the binormal vector, **b**, can be used to define the mean curvature. (*b*) The surface-average value of the mean curvature over an interfacial patch can be computed from the position of four marker points.

1.8.6   Mean curvature as a contour integral

The domain of definition of the normal unit vector can be extended from a surface into the whole three-dimensional space using expression (1.8.54). A version of Stokes' theorem discussed in Section A.7, Appendix A, states that

$$\oint_C \mathbf{F} \times \mathbf{t}\, dl = \iint_D \left[\mathbf{n} \nabla \cdot \mathbf{F} - (\nabla \mathbf{F}) \cdot \mathbf{n} \right] dS \qquad (1.8.64)$$

for any arbitrary differentiable vector function $\mathbf{F}$, where C is a closed contour bounding a surface, D, $\mathbf{n}$ is the unit vector normal to D pointing toward a designated side, $\mathbf{t}$ is the unit vector tangential to C, and $\mathbf{b} \equiv \mathbf{t} \times \mathbf{n}$ is the binormal vector, as shown in Figure 1.8.5(*a*). Applying this identity with $\mathbf{F} = \mathbf{n}$, we obtain

$$\oint_C \mathbf{n} \times \mathbf{t}\, dl = \iint_D \left[\mathbf{n} \nabla \cdot \mathbf{n} - (\nabla \mathbf{n}) \cdot \mathbf{n} \right] dS. \qquad (1.8.65)$$

Because $\mathbf{n}$ is a unit vector, $(\nabla \mathbf{n}) \cdot \mathbf{n} = \frac{1}{2}\nabla(\mathbf{n} \cdot \mathbf{n}) = 0$, yielding

$$\iint_D 2\kappa_m \mathbf{n}\, dS = \oint_C \mathbf{n} \times \mathbf{t}\, dl. \qquad (1.8.66)$$

If the surface is small, we may assume that the mean curvature and normal vector are constant and obtain an approximation for the mean curvature,

$$\kappa_m \simeq \frac{1}{2\Delta S}\, \mathbf{n} \cdot \oint_C \mathbf{n} \times \mathbf{t}\, dl, \qquad (1.8.67)$$

where ΔS is the surface area of D. The counterpart of this expression for a section of a two-dimensional surface with arc length Δl is

$$\kappa_m \simeq \frac{1}{2\Delta l}\, \mathbf{n} \cdot \Delta \mathbf{t}, \qquad (1.8.68)$$

where $\Delta \mathbf{t}$ is the difference in the unit tangent vectors between the last and first segment points.

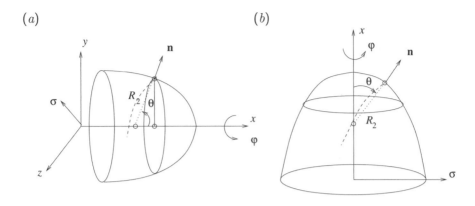

FIGURE 1.8.6 The second principal curvature of an axisymmetric surface, $\kappa_2 = 1/R_2$, is the curvature of the dashed line representing the trace of the surface in a plane that is normal to the surface and also normal to a plane of constant azimuthal angle, φ.

Numerical methods

To illustrate the practical application of the method, we consider a rectangular surface element defined by four points, A–D, that lie at the intersections of a pair of ξ lines and a pair of η lines representing surface curvilinear coordinates, as shown in Figure 1.8.5(b). Applying the trapezoidal rule to approximate the contour integral in (1.8.67), we obtain

$$
\begin{aligned}
\kappa_m \simeq \frac{1}{4\Delta S}\, \mathbf{n} \cdot \Big\{ &\Big[\Big(\mathbf{n} \times \frac{\partial \mathbf{X}}{\partial \xi}\Big)_A + \mathbf{n} \times \frac{\partial \mathbf{X}}{\partial \xi}\Big)_B\Big] (\xi_B - \xi_A) \\
&+ \Big[\Big(\mathbf{n} \times \frac{\partial \mathbf{X}}{\partial \eta}\Big)_B + \mathbf{n} \times \frac{\partial \mathbf{X}}{\partial \eta}\Big)_C\Big] (\eta_C - \eta_B) \\
&+ \Big[\Big(\mathbf{n} \times \frac{\partial \mathbf{X}}{\partial \xi}\Big)_C + \mathbf{n} \times \frac{\partial \mathbf{X}}{\partial \xi}\Big)_D\Big] (\xi_D - \xi_C) \\
&+ \Big[\Big(\mathbf{n} \times \frac{\partial \mathbf{X}}{\partial \eta}\Big)_D + \mathbf{n} \times \frac{\partial \mathbf{X}}{\partial \eta}\Big)_A\Big] (\eta_A - \eta_D) \Big\}.
\end{aligned}
\tag{1.8.69}
$$

The normal and tangent unit vectors on the right-hand side can be computed from the position of surface marker points using standard methods of numerical differentiation, as discussed in Section B.5, Appendix B.

1.8.7 Curvature of an axisymmetric surface

The mean curvature of an axisymmetric surface, illustrated in Figure 1.8.6, is the average of the two principal curvatures. The first principal curvature is the curvature of the trace of the surface in an azimuthal σx plane, denoted by κ_1, and the second principal curvature is the curvature of the trace of the surface in the orthogonal plane, denoted by κ_2. In the case of a sphere, the two principal curvatures are equal.

Description as $\sigma = w(x)$

The shape of an axisymmetric surface in a meridional plane of constant azimuthal angle φ can be described by the function $\sigma = w(x)$, as shown in Figure 1.8.6(a). The normal unit vector is

$$\mathbf{n} = \frac{1}{\sqrt{1+w'^2}}\,(\mathbf{e}_\sigma - w'\,\mathbf{e}_x), \qquad (1.8.70)$$

where a prime denotes a derivative with respect to x. The mean curvature is given by the divergence of the normal vector,

$$2\kappa_m = \frac{\partial n_x}{\partial x} + \frac{1}{\sigma}\frac{\partial(\sigma n_\sigma)}{\partial\sigma} = -\Big(\frac{w'}{\sqrt{1+w'^2}}\Big)' + \frac{1}{\sigma}\frac{\partial}{\partial\sigma}\Big(\frac{\sigma}{\sqrt{1+w'^2}}\Big). \qquad (1.8.71)$$

Carrying out the differentiations, we obtain

$$2\kappa_m = -\frac{w''}{(1+w'^2)^{3/2}} + \frac{1}{w}\frac{1}{\sqrt{1+w'^2}}, \qquad (1.8.72)$$

which can be rearranged into the expression

$$2\kappa_m = \frac{1}{w}\frac{1+w'^2 - ww''}{(1+w'^2)^{3/2}}. \qquad (1.8.73)$$

The first term on the right-hand side of (1.8.72) is the principal curvature in an azimuthal plane,

$$\kappa_1 = -\frac{w''}{(1+w'^2)^{3/2}}. \qquad (1.8.74)$$

The second term on the right-hand side of (1.8.72) is the second principal curvature,

$$\kappa_2 = \frac{1}{R_2}, \qquad R_2 = \sigma\sqrt{1+w'^2} = \frac{\sigma}{\sin\theta} = \frac{\sigma}{n_\sigma}, \qquad (1.8.75)$$

where R_2 is the second principal radius of curvature and the angle θ is defined in Figure 1.8.6(a). We have found that the second principal radius of curvature, R_2, is the signed distance of the point where the curvature is evaluated from the intersection of the extension of the normal vector with the x axis. If n_σ is negative, R_2 is also negative, and *vice versa*.

Description as $x = q(\sigma)$

The shape of an axisymmetric surface in an azimuthal plane can be described by the function $x = q(\sigma)$, as shown in Figure 1.8.6(b). The normal unit vector is

$$\mathbf{n} = \frac{1}{\sqrt{1+q'^2}}\,(\mathbf{e}_x - q'\,\mathbf{e}_\sigma), \qquad (1.8.76)$$

where a prime denotes a derivative with respect to σ. The mean curvature is the divergence of the normal vector,

$$2\kappa_m = \frac{\partial n_x}{\partial x} + \frac{1}{\sigma}\frac{\partial(\sigma n_\sigma)}{\partial\sigma} = -\frac{1}{\sigma}\Big(\frac{\sigma q'}{\sqrt{1+q'^2}}\Big)'. \qquad (1.8.77)$$

Carrying out the differentiations, we obtain

$$2\kappa_m = -\frac{q''}{(1+q'^2)^{3/2}} - \frac{1}{\sigma}\frac{q'}{\sqrt{1+q'^2}},$$ (1.8.78)

which is consistent with the more general expression (1.8.60).

The first term on the right-hand side of (1.8.78) is the principal curvature in an azimuthal plane,

$$\kappa_1 = -\frac{q''}{(1+q'^2)^{3/2}} = \frac{1}{q'}\left(\frac{1}{\sqrt{1+q'^2}}\right)'.$$ (1.8.79)

The second term is the second principal curvature,

$$\kappa_2 = \frac{1}{R_2}, \qquad R_2 = -\frac{\sigma}{q'}\sqrt{1+q'^2} = \frac{\sigma}{\sin\theta} = \frac{\sigma}{n_\sigma},$$ (1.8.80)

where R_2 is the signed second principal radius of curvature, and the angle θ is defined in Figure 1.8.6(b).

Problems

1.8.1 *Rate of change of the surface metric tensor*

Express the material derivative of the surface metric tensor, $g_{\alpha\beta}$, in terms of the velocity field.

1.8.2 *Mean curvature*

(a) Show that (1.8.19) can be reduced to the last expression in (1.8.18). *Hint:* Begin by expressing the normal vector in the cross products in terms of the tangent vectors using (1.7.11), and then expand the triple cross products.

(b) Derive an expression for the rate of change of the mean curvature following a point particle in terms of the velocity.

1.8.3 *Curvature of a line in a plane*

Derive the expression for the curvature shown in (1.8.50).

1.8.4 *Mean curvature in spherical polar coordinates*

Derive an expression for the mean curvature of a surface that is described explicitly in spherical polar coordinates, (r, θ, φ), by the function $r = q(\theta, \varphi)$.

Computer Problems

1.8.5 *Mean curvature of a spheroid*

Consider a spheroidal surface with one semi-axis equal to a and two semi-axes equal to b. In terms of the native orthogonal surface curvilinear coordinates, ξ and $\eta = \varphi$, the position of a point on the spheroid is specified in parametric form as $X = a \cos \xi$, $Y = b \sin \xi \cos \varphi$, and $Z = b \sin \xi \sin \varphi$, where $0 \leq \xi \leq \pi$ is a dimensionless parameter and φ is the azimuthal angle.

(a) Derive an expression for the mean curvature. Prepare a graph of the scaled mean curvature $a\kappa_m$ against φ for aspect ratios $b/a = 0.1$ (prolate spheroid), 1.0 (sphere), and 10 (oblate spheroid). Verify that, when $a = b$, the mean curvature takes the uniform value $1/a = 1/b$ for a sphere.

(b) Repeat (a) using (1.8.28)–(1.8.31) with sufficiently small increments $\Delta \xi$ and $\Delta \eta$ to approximate the curvature.

1.8.6 *Curvature of a line by parametric interpolation*

The following set of points trace a smooth closed loop in the xy plane:

x	1.1	2.2	3.9	4.1	3.0	2.0	1.3
y	1.0	0.1	1.1	3.0	3.9	4.1	3.0

Write a program that uses cubic spline interpolation with periodic end conditions to compute the Cartesian coordinates and curvature of a point along the loop as a function of a suitably chosen parameter ξ that increases monotonically from 0 to 1 along the loop, from start to finish (Section B.5, Appendix B). One plausible choice for ξ is the polygonal arc length, defined as the length of the polygonal line connecting successive nodes, divided by the value at the last node. The program should return the coordinates (x, y) corresponding to a specified value of ξ in the interval $[0, 1]$, where $\xi = 0$ and 1 correspond to $x = 1.1$ and $y = 1.0$. Generate a table of 32 values of the quadruplet (ξ, x, y, κ) at 32 evenly spaced intervals of ξ between 0 and 1.

1.9 Interfacial surfactant transport

An impure interface between two immiscible fluids is sometimes occupied by a molecular layer of a surfactant. Dividing the number of surfactant molecules inside an infinitesimal patch centered at a given point by the patch surface area, we obtain the surface concentration of the surfactant, Γ. The molecules of an insoluble surfactant are convected and diffuse over the interface, but do not enter the bulk of the fluid. Our objective is to derive an evolution equation for the surface concentration of an insoluble surfactant (e.g., [438]).

1.9.1 Two-dimensional interfaces

We begin by considering a chain of material point particles distributed along the inner or outer side of a two-dimensional interface and label the point particles using a Lagrangian parameter, ξ. The point particle position can be described in parametric form as $\mathbf{X}(\xi)$, as shown in Figure 1.9.1. Let l be the arc length along the interface measured from an arbitrary point particle labeled ξ_0. The number of surfactant molecules residing inside a test section of the interface confined between ξ_0 and ξ is

$$n(\xi, t) = \int_{l(\xi_0, t)}^{l(\xi, t)} \Gamma(\xi', t) \, dl(\xi') = \int_{\xi_0}^{\xi} \Gamma(\xi', t) \frac{\partial l}{\partial \xi'} \, d\xi', \tag{1.9.1}$$

where ξ' is an integration variable. Let q be the flux of surfactant molecules along the interface due to diffusion. Conservation of the total number of surfactant molecules inside the test section

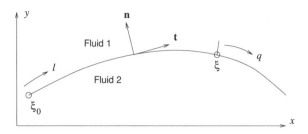

FIGURE 1.9.1   Point particles along a two-dimensional interface are identified by a parameter, ξ.

requires that

$$\frac{\partial n}{\partial t} = q(\xi_0) - q(\xi), \tag{1.9.2}$$

where the time derivative is taken keeping ξ fixed. Substituting the expression for n from the last integral of (1.9.1) and transferring the derivative inside the integral as a material derivative, D/Dt, we obtain

$$\int_{\xi_0}^{\xi} \frac{\mathrm{D}}{\mathrm{D}t} \left(\Gamma(\xi', t) \frac{\partial l}{\partial \xi'} \right) \mathrm{d}\xi' = q(\xi_0) - q(\xi). \tag{1.9.3}$$

Now we take the limit as ξ tends to ξ_0 and derive the differential equation

$$\frac{\mathrm{D}}{\mathrm{D}t} \left(\Gamma \frac{\partial l}{\partial \xi} \right) = -\frac{\partial q}{\partial \xi}. \tag{1.9.4}$$

Expanding the material derivative on the left-hand side, we obtain

$$\frac{\mathrm{D}\Gamma}{\mathrm{D}t} \frac{\partial l}{\partial \xi} + \Gamma \frac{\mathrm{D}}{\mathrm{D}t} \left(\frac{\partial l}{\partial \xi} \right) = -\frac{\partial q}{\partial \xi}. \tag{1.9.5}$$

To compute the second material derivative on the left-hand side, we use the Pythagorean theorem to write

$$\frac{\partial l}{\partial \xi} = \left[\left(\frac{\partial X}{\partial \xi} \right)^2 + \left(\frac{\partial Y}{\partial \xi} \right)^2 \right]^{1/2}, \tag{1.9.6}$$

and then compute

$$\frac{\mathrm{D}}{\mathrm{D}t} \left(\frac{\partial l}{\partial \xi} \right) = \frac{1}{2} \frac{1}{\partial l/\partial \xi} \left[2 \frac{\partial X}{\partial \xi} \frac{\mathrm{D}}{\mathrm{D}t} \left(\frac{\partial X}{\partial \xi} \right) + 2 \frac{\partial Y}{\partial \xi} \frac{\mathrm{D}}{\mathrm{D}t} \left(\frac{\partial Y}{\partial \xi} \right) \right]. \tag{1.9.7}$$

Interchanging the order of the material and the ξ derivative inside the square brackets on the right-hand side and setting $\mathrm{D}X/\mathrm{D}t = u_x$ and $\mathrm{D}Y/\mathrm{D}t = u_y$, we find that

$$\frac{\mathrm{D}}{\mathrm{D}t} \left(\frac{\partial l}{\partial \xi} \right) = \frac{1}{\partial l/\partial \xi} \left(\frac{\partial X}{\partial \xi} \frac{\partial u_x}{\partial \xi} + \frac{\partial Y}{\partial \xi} \frac{\partial u_y}{\partial \xi} \right) = \frac{1}{\partial l/\partial \xi} \frac{\partial \mathbf{X}}{\partial \xi} \cdot \frac{\partial \mathbf{u}}{\partial \xi}. \tag{1.9.8}$$

Rearranging, we obtain

$$\frac{\mathrm{D}}{\mathrm{D}t}\left(\frac{\partial l}{\partial \xi}\right) = \frac{\partial l}{\partial \xi}\frac{\partial \mathbf{X}}{\partial l}\cdot\frac{\partial \mathbf{u}}{\partial l} = \frac{\partial l}{\partial \xi}\,\mathbf{t}\cdot\frac{\partial \mathbf{u}}{\partial l}, \tag{1.9.9}$$

where $\mathbf{t} = \partial \mathbf{X}/\partial l$ is the tangent unit vector shown in Figure 1.9.1. Substituting the final expression into (1.9.5), we obtain

$$\frac{\mathrm{D}\Gamma}{\mathrm{D}t} + \Gamma\,\mathbf{t}\cdot\frac{\partial \mathbf{u}}{\partial l} = -\frac{\partial q}{\partial l}, \tag{1.9.10}$$

which is the targeted evolution equation for the surfactant concentration.

In certain applications, it is convenient to describe the surfactant surface concentration in Eulerian form in terms of x and t, as shown in Figure 1.9.1. The material derivative is then

$$\frac{\mathrm{D}\Gamma}{\mathrm{D}t} \equiv \left(\frac{\partial \Gamma}{\partial t}\right)_\xi = \frac{\partial \Gamma}{\partial t} + u_x\frac{\partial \Gamma}{\partial x}, \tag{1.9.11}$$

where the last expression involves Eulerian derivatives with respect to x and t. Similar expressions can be written with reference to curvilinear axes.

Fick's law

The diffusive flux of a surfactant along an interface can be described by Fick's law,

$$q = -D_s\frac{\partial \Gamma}{\partial l}, \tag{1.9.12}$$

where D_s is the surfactant surface diffusivity with unit of length squared divided by time. In practice, the surfactant diffusivity is typically small. Substituting this expression into (1.9.10), we derive a convection–diffusion equation,

$$\frac{\mathrm{D}\Gamma}{\mathrm{D}t} + \Gamma\,\mathbf{t}\cdot\frac{\partial \mathbf{u}}{\partial l} = \frac{\partial}{\partial l}\left(D_s\frac{\partial \Gamma}{\partial l}\right). \tag{1.9.13}$$

Stretching and expansion

It is illuminating to decompose the interfacial velocity into a tangential component and a normal component,

$$\mathbf{u} = u_t\,\mathbf{t} + u_n\,\mathbf{n}, \tag{1.9.14}$$

where $\mathbf{n}$ is the normal unit vector, as shown in Figure 1.9.1. The tangential and normal velocities are $u_t = \mathbf{u}\cdot\mathbf{t}$ and $u_n = \mathbf{u}\cdot\mathbf{n}$. Noting that $\mathbf{t}\cdot\mathbf{n} = 0$, $\mathbf{t}\cdot\mathbf{t} = 1$, and $\mathbf{n}\cdot\mathbf{n} = 1$, and using the Frenet-Serret relations $\mathrm{d}\mathbf{t}/\mathrm{d}l = -\kappa\,\mathbf{n}$ and $\mathrm{d}\mathbf{n}/\mathrm{d}l = \kappa\,\mathbf{t}$, we compute

$$\mathbf{t}\cdot\frac{\partial \mathbf{u}}{\partial l} = \frac{\partial u_t}{\partial l} + u_n\,\mathbf{t}\cdot\frac{\partial \mathbf{n}}{\partial l} = \frac{\partial u_t}{\partial l} + \kappa\,u_n, \tag{1.9.15}$$

where κ is the interfacial curvature. Substituting this expression into (1.9.13), we obtain

$$\frac{D\Gamma}{Dt} + \Gamma\left(\frac{\partial u_t}{\partial l} + \kappa u_n\right) = \frac{\partial}{\partial l}\left(D_s \frac{\partial \Gamma}{\partial l}\right). \tag{1.9.16}$$

The first term inside the parentheses on the left-hand side expresses the effect of interfacial stretching, and the second expresses the effect of interfacial expansion.

Stretching of a flat interface

As an application, we consider a flat interface along the x axis that is stretched uniformly under the influence of a tangential velocity field, u_x. Identifying the arc length l with x and setting $\kappa = 0$, we find that the transport equation (1.9.16) reduces to

$$\frac{D\Gamma}{Dt} + \Gamma \frac{\partial u_x}{\partial x} = \frac{\partial}{\partial x}\left(D_s \frac{\partial \Gamma}{\partial x}\right). \tag{1.9.17}$$

The material derivative can be resolved into Eulerian derivatives with respect to x and t, yielding

$$\frac{\partial \Gamma}{\partial t} + u_x \frac{\partial \Gamma}{\partial x} + \Gamma \frac{\partial u_x}{\partial x} = \frac{\partial}{\partial x}\left(D_s \frac{\partial \Gamma}{\partial x}\right) \quad \text{or} \quad \frac{\partial \Gamma}{\partial t} + \frac{\partial(u_x\Gamma)}{\partial x} = \frac{\partial}{\partial x}\left(D_s \frac{\partial \Gamma}{\partial x}\right). \tag{1.9.18}$$

This equation could have been derived directly by performing a surfactant molecular balance over a stationary differential control volume along the x axis, taking into consideration the convective and diffusive flux contributions.

In the case of a uniformly stretched interface, $u_x = \alpha x$, where α is a constant rate of extension. If the surfactant concentration is uniform at the initial instant, it will remain uniform at any time, governed by the linear equation

$$\frac{d\Gamma}{dt} + \alpha\,\Gamma = 0. \tag{1.9.19}$$

The solution reveals that, when α is positive, the surfactant concentration decreases exponentially due to dilution, $\Gamma(t) = \Gamma(t = 0)\,\exp(-\alpha t)$.

Expansion of a circular interface

As a second application, we consider a cylindrical interface with circular cross-section of radius $a(t)$ centered at the origin, expanding under the influence of a uniform radial velocity in the absence of circumferential motion. In corresponding plane polar coordinates, (r, θ), the transport equation (1.9.16) with $\kappa = 1/a$ and constant diffusivity becomes

$$\frac{D\Gamma}{Dt} + \Gamma \frac{u_r}{a} = \frac{D_s}{a^2} \frac{\partial^2 \Gamma}{\partial \theta^2}. \tag{1.9.20}$$

Resolving the material derivative into Eulerian derivatives with respect to r and t, we obtain

$$\frac{\partial \Gamma}{\partial t} + \Gamma \frac{u_r}{a} = \frac{D_s}{a^2} \frac{\partial^2 \Gamma}{\partial \theta^2}. \tag{1.9.21}$$

If the surfactant concentration is uniform at the initial instant, it will remain uniform at any time, governed by the linear equation

$$\frac{\mathrm{d}\Gamma}{\mathrm{d}t} + \Gamma \frac{u_r}{a} = 0. \tag{1.9.22}$$

In the case of expansion, $u_r > 0$, the surfactant concentration decreases due to dilution, $\mathrm{d}\Gamma/\mathrm{d}t < 0$. In the case of contraction, $u_r < 0$, the surfactant concentration increases due to compaction, $\mathrm{d}\Gamma/\mathrm{d}t > 0$.

Interfacial markers

The material derivative expresses the rate of change of the surfactant concentration following the motion of point particles residing inside an interface. In numerical practice, it may be expedient to follow the motion of interfacial marker points that move with the normal component of the fluid velocity and with an arbitrary tangential velocity, v_t. If $v_t = 0$, the marker points move normal to the interface at any instant. The velocity of a marker point is

$$\mathbf{v} = u_n \, \mathbf{n} + v_t \, \mathbf{t}. \tag{1.9.23}$$

By definition,

$$\frac{D\Gamma}{Dt} = \frac{\mathrm{d}\Gamma}{\mathrm{d}t} + (u_t - v_t) \frac{\partial \Gamma}{\partial l}, \tag{1.9.24}$$

where $\mathrm{d}/\mathrm{d}t$ is the rate of change of the surfactant concentration following a marker point. Substituting this expression into the transport equation (1.9.16), we obtain

$$\frac{\mathrm{d}\Gamma}{\mathrm{d}t} + (u_t - v_t) \frac{\partial \Gamma}{\partial l} + \Gamma \left(\frac{\partial u_t}{\partial l} + \kappa \, u_n \right) = \frac{\partial}{\partial l} \left(D_s \frac{\partial \Gamma}{\partial l} \right), \tag{1.9.25}$$

which can be restated as

$$\frac{\mathrm{d}\Gamma}{\mathrm{d}t} + \frac{\partial (u_t \Gamma)}{\partial l} - v_t \frac{\partial \Gamma}{\partial l} + \Gamma \kappa u_n = \frac{\partial}{\partial l} \left(D_s \frac{\partial \Gamma}{\partial l} \right). \tag{1.9.26}$$

The second term on the left-hand side represents the interfacial convective flux.

1.9.2   Axisymmetric interfaces

Next, we consider a chain of material point particles distributed along the inner or outer side of the trace of an axisymmetric interface in an azimuthal plane, and label the point particles using a parameter, ξ, so that their position in an azimuthal plane is described parametrically as $\mathbf{X}(\xi)$. Let l be the arc length along the trace of the interface measured from an arbitrary point particle labeled ξ_0, as illustrated in Figure 1.9.2.

To derive an evolution equation for the surface surfactant concentration, we introduce cylindrical polar coordinates, (x, σ, φ), and express the number of surfactant molecules inside a ring-like material section of the interface confined between ξ_0 and ξ as

$$n(\xi, t) = 2\pi \int_{l(\xi_0, t)}^{l(\xi, t)} \Gamma(\xi', t) \, \sigma(\xi') \, \mathrm{d}l(\xi') = 2\pi \int_{\xi_0}^{\xi} \Gamma(\xi', t) \frac{\partial l}{\partial \xi'} \sigma(\xi') \, \mathrm{d}\xi'. \tag{1.9.27}$$

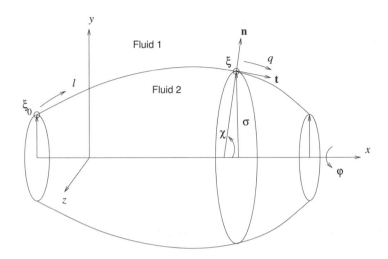

FIGURE 1.9.2   Point particles along the trace of an axisymmetric interface in an azimuthal plane are identified by a parameter ξ. The angle χ is subtended between the x axis and the straight line defined by the extension of the normal vector.

Conservation of the total number of surfactant molecules inside the test section requires that

$$\frac{\partial n}{\partial t} = 2\pi\sigma_0\, q(\xi_0) - 2\pi\sigma q(\xi),\qquad(1.9.28)$$

where q is the flux of surfactant molecules along the interface by diffusion, and the time derivative is taken keeping ξ fixed. The counterpart of the balance equation (1.9.4) is

$$\frac{\mathrm{D}}{\mathrm{D}t}\left(\Gamma(\xi,t)\,\sigma(\xi,t)\,\frac{\partial l}{\partial \xi}\right) = -\frac{\partial(\sigma q)}{\partial \xi},\qquad(1.9.29)$$

and the counterpart of equation (1.9.10) is

$$\frac{\mathrm{D}\Gamma}{\mathrm{D}t} + \Gamma\left(\mathbf{t}\cdot\frac{\partial \mathbf{u}}{\partial l} + \frac{u_\sigma}{\sigma}\right) = -\frac{1}{\sigma}\frac{\partial(\sigma q)}{\partial l}.\qquad(1.9.30)$$

In deriving this equation, we have set $\mathrm{D}\Sigma/\mathrm{D}t = u_\sigma$, where Σ is the distance of a point particle from the axis of revolution.

Stretching and expansion

In terms of the normal and tangential fluid velocities, u_n and u_t, the transport equation (1.9.30) takes the form

$$\frac{\mathrm{D}\Gamma}{\mathrm{D}t} + \Gamma\left(\frac{\partial u_t}{\partial l} + \kappa u_n + \frac{u_\sigma}{\sigma}\right) = -\frac{1}{\sigma}\frac{\partial(\sigma q)}{\partial l},\qquad(1.9.31)$$

where κ is the curvature of the interface in an azimuthal plane. Next, we introduce the angle χ subtended between the x axis and the normal vector to the interface at a point, as shown in Figure 1.9.2. Substituting $u_\sigma = u_n \sin \chi - u_t \cos \chi$, we obtain

$$\frac{D\Gamma}{Dt} + \Gamma \left(\frac{\partial u_t}{\partial l} - \frac{\cos \chi}{\sigma} u_t + (\kappa + \frac{\sin \chi}{\sigma}) u_n \right) = -\frac{1}{\sigma} \frac{\partial(\sigma q)}{\partial l}. \qquad (1.9.32)$$

The sum of the two terms inside the innermost parentheses on the left-hand side is twice the mean curvature of the interface, $2\kappa_m$. The first two terms inside the large parentheses can be consolidated to yield the final form

$$\frac{D\Gamma}{Dt} + \Gamma \left(\frac{1}{\sigma} \frac{\partial(\sigma u_t)}{\partial l} + 2 \kappa_m u_n \right) = -\frac{1}{\sigma} \frac{\partial(\sigma q)}{\partial l}. \qquad (1.9.33)$$

The first term inside the large parentheses on the left-hand side expresses the effect of axisymmetric interfacial stretching, and the second expresses the effect of interfacial expansion.

Marker points

An evolution equation for the rate of change of the surface surfactant concentration following interfacial marker points that are not necessarily point particles can be derived working as in Section 1.9.1 for two-dimensional flow. The result is

$$\frac{d\Gamma}{dt} + (u_t - v_t) \frac{\partial \Gamma}{\partial l} + \Gamma \left(\frac{1}{\sigma} \frac{\partial(\sigma u_t)}{\partial l} + 2 \kappa_m u_n \right) = -\frac{1}{\sigma} \frac{\partial(\sigma q)}{\partial l}, \qquad (1.9.34)$$

which can be restated as

$$\frac{d\Gamma}{dt} + \frac{1}{\sigma} \frac{\partial(\sigma u_t \Gamma)}{\partial l} - v_t \frac{\partial \Gamma}{\partial l} + \Gamma 2 \kappa_m u_n = -\frac{1}{\sigma} \frac{\partial(\sigma q)}{\partial l}. \qquad (1.9.35)$$

When $v_t = 0$, the marker points move normal to the interface and the third term on the right-hand side does not appear.

1.9.3   Three-dimensional interfaces

To derive an evolution equation for the surface surfactant concentration over a three-dimensional interface, we introduce convected surface curvilinear coordinates, (ξ, η), embedded in the interface, as discussed in Section 1.7. The total number of surfactant molecules residing inside a material interfacial patch consisting of a fixed collection of point particles is

$$n = \iint_{Patch} \Gamma \, dS = \iint_{\Xi} \Gamma \, h_s \, d\xi \, d\eta, \qquad (1.9.36)$$

where h_s is the surface metric and Ξ is the fixed support of the patch in the (ξ, η) plane. A mass balance requires that

$$\frac{dn}{dt} = -\oint_C \mathbf{b} \cdot \mathbf{q} \, dl, \qquad (1.9.37)$$

where C is the edge of the patch, $\mathbf{b} = \mathbf{t} \times \mathbf{n}$ is the binormal vector, $\mathbf{t}$ is the unit tangent vector, $\mathbf{n}$ is the unit vector normal to the patch, and $\mathbf{q}$ is the tangential diffusive surface flux. Substituting the last integral in (1.9.36) in place of n into the left-hand side of (1.9.37), and transferring the time derivative inside the integral as a material derivative, we obtain

$$\iint_{\Xi} \frac{\mathrm{D}(\Gamma h_s)}{\mathrm{D}t}\, \mathrm{d}\xi\, \mathrm{d}\eta = -\oint \mathbf{b}\cdot\mathbf{q}\, \mathrm{d}S. \tag{1.9.38}$$

Next, we expand the material derivative inside the integral, use (1.7.32) to evaluate the rate of change of the surface metric, and apply the divergence theorem to convert the line integral on the right-hand side into a surface integral, obtaining

$$\iint_{Patch} \Big(\frac{\mathrm{D}\Gamma}{\mathrm{D}t} + \Gamma\,\nabla_s\cdot\mathbf{u} \Big)\, \mathrm{d}S = -\iint_{Patch} \nabla_s\cdot\mathbf{q}\, \mathrm{d}S, \tag{1.9.39}$$

where $\nabla_s \equiv (\mathbf{I}-\mathbf{n}\otimes\mathbf{n})\cdot\nabla$ is the surface gradient and $\nabla_s\cdot\mathbf{u}$ is the surface divergence of the velocity. Because the size of the material patch is arbitrary, the integrands must balance to zero, yielding the transport equation

$$\frac{\mathrm{D}\Gamma}{\mathrm{D}t} + \Gamma\,\nabla_s\cdot\mathbf{u} = -\nabla_s\cdot\mathbf{q}. \tag{1.9.40}$$

Stretching and expansion

Physical insights can be obtained by resolving the surface velocity into its tangential and normal constituents, $\mathbf{u} = \mathbf{u}_s + u_n\,\mathbf{n}$, where $u_n = \mathbf{u}\cdot\mathbf{n}$ is the normal velocity and $\mathbf{u}_s$ is the tangential (surface) velocity. Substituting this expression into the surface divergence of the velocity on the left-hand side of (1.9.40), we obtain

$$\frac{\mathrm{D}\Gamma}{\mathrm{D}t} + \Gamma\left(\nabla_s\cdot\mathbf{u}_s + \mathbf{n}\cdot\nabla_s u_n + u_n\nabla_s\cdot\mathbf{n}\right) = -\nabla_s\cdot\mathbf{q}. \tag{1.9.41}$$

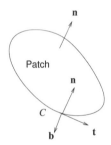

Illustration of an interfacial patch occupied by surfactant.

The inner product, $\mathbf{n}\cdot\nabla_s u_n$, is identically zero due to the orthogonality of the surface gradient and normal vector. Setting the surface divergence $\nabla_s\cdot\mathbf{n}$ equal to twice the mean curvature of the interface, $2\kappa_m$, we obtain the transport equation

$$\frac{\mathrm{D}\Gamma}{\mathrm{D}t} + \Gamma\left(\nabla_s\cdot\mathbf{u}_s + 2\kappa_m u_n\right) = -\nabla_s\cdot\mathbf{q}. \tag{1.9.42}$$

The first term inside the parentheses on the left-hand side expresses the effect of interfacial stretching, and the second expresses the effect of interfacial expansion. In the case of a stationary interface, $u_n = 0$, only the tangential velocity field affects the surfactant concentration.

Fick's law

The tangential diffusive flux vector can be described by Fick's law,

$$\mathbf{q} = -D_s\,\nabla_s\Gamma, \tag{1.9.43}$$

where D_s is the surfactant surface diffusivity. Substituting this expression into (1.9.40), we derive a convection–diffusion equation,

$$\frac{D\Gamma}{Dt} + \Gamma \nabla_s \cdot \mathbf{u} = \nabla_s \cdot (D_s \nabla_s \Gamma). \tag{1.9.44}$$

When the diffusivity is constant, the diffusion term becomes $D_s \nabla_s^2 \Gamma$, where $\nabla_s^2 \equiv \nabla_s \cdot \nabla_s$ is the surface Laplacian operator, sometimes also called the Laplace–Beltrami operator. Expression for this operator in orthogonal and nonorthogonal curvilinear coordinates can be derived using the formula presented in Appendix A.

Interfacial markers

In numerical practice, it may be expedient to follow the motion of interfacial marker points moving with the normal component of the fluid velocity, $u_n \mathbf{n}$, and with an arbitrary tangential velocity, $\mathbf{v}_s$. The marker-point velocity is

$$\mathbf{v} = u_n \mathbf{n} + \mathbf{v}_s. \tag{1.9.45}$$

When $\mathbf{v}_s = \mathbf{0}$, a marker point moves with the fluid velocity normal to the interface alone. Consequently, if the interface is stationary, the marker point is also stationary. When $\mathbf{v}_s = \mathbf{u} - u_n \mathbf{n}$, the marker points are material point particles moving with the fluid velocity. The rate of change of the surfactant concentration following the marker points is related to the material derivative by

$$\frac{d\Gamma}{dt} = \frac{D\Gamma}{Dt} - (\mathbf{u}_s - \mathbf{v}_s) \cdot \nabla \Gamma. \tag{1.9.46}$$

Combining this expression with (1.9.42), we obtain

$$\frac{d\Gamma}{dt} + (\mathbf{u}_s - \mathbf{v}_s) \cdot \nabla \Gamma + \Gamma (\nabla_s \cdot \mathbf{u}_s + 2\kappa_m u_n) = -\nabla_s \cdot \mathbf{q}. \tag{1.9.47}$$

Rearranging, we derive the alternative form

$$\frac{d\Gamma}{dt} + \nabla_s \cdot (\Gamma \mathbf{u}_s) - \mathbf{v}_s \cdot \nabla_s \Gamma + \Gamma 2\kappa_m u_n = -\nabla_s \cdot \mathbf{q}. \tag{1.9.48}$$

If the point particles move normal to the interface, the third term on the left-hand side expressing a convective contribution does not appear.

Problems

1.9.1 *Two-dimensional and axisymmetric transport*

Show that, in the case of two-dimensional flow depicted in Figure 1.9.1 or axisymmetric flow depicted in Figure 1.9.2, equation (1.9.48) reduces, respectively, to (1.9.26) or (1.9.35), by setting $\mathbf{v}_s = v_t \mathbf{t}$.

1.9.2 *Transport in a spherical surface*

Derive the specific form of (1.9.48) over a spherical surface of radius a in terms of the meridional angle, θ, and azimuthal angle, φ.

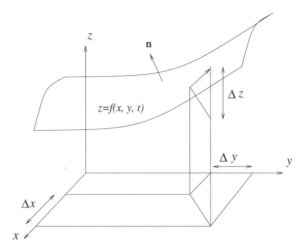

FIGURE 1.10.1   The shape of a material surface can be described in Eulerian form in global Cartesian coordinates as $z = f(x, y)$.

1.10   Eulerian description of material lines and surfaces

It is sometimes convenient, or even necessary, to describe a material line or surface in Eulerian parametric form using Cartesian or other global curvilinear coordinates instead of surface curvilinear coordinates discussed earlier in this chapter. The Eulerian description is particularly useful in studies of interfacial flow where a material line or surface is typically identified with an interface between two immiscible fluids or with a free surface separating a gas from a liquid.

A function that describes the shape of a material line or surface in Eulerian form satisfies an evolution equation that emerges by requiring that the motion of point particles on either side of the line or surface is consistent with the stationary or evolving shape of the line or surface described by the Eulerian form. This evolution equation can be regarded as a kinematic compatibility condition, analogous to a kinematic boundary condition specifying the normal or tangential component of the velocity on a rigid or deformable boundary.

1.10.1   Kinematic compatibility

With reference to Figure 1.10.1, we describe the shape of a material surface in Cartesian coordinates by the function $z = f(x, y, t)$. If the material surface is evolving, the function f changes in time, as indicated by the third of its arguments. To derive an evolution equation for f, we consider the position of a point particle in the material surface at times t and $t + \Delta t$, where Δt is a small time interval, recall that the point particle moves with the fluid velocity, and exercise geometrical reasoning to write

$$f(x + u_x \Delta t, y + u_y \Delta t, t + \Delta t) = f(x, y, t) + u_z \Delta t. \qquad (1.10.1)$$

Expanding the left-hand side of (1.10.1) in a Taylor series with respect to the first two of its arguments about the triplet (x, y, t), we find that

$$f(x, y, t + \Delta t) + \frac{\partial f}{\partial x} u_x \Delta t + \frac{\partial f}{\partial y} u_y \Delta t = f(x, y) + u_z \Delta t. \tag{1.10.2}$$

Now dividing each term by Δt, taking the limit as Δt becomes infinitesimal, and rearranging, we derive the targeted evolution equation

$$\frac{\partial f}{\partial t} + u_x \frac{\partial f}{\partial x} + u_y \frac{\partial f}{\partial y} - u_z = 0, \tag{1.10.3}$$

expressing kinematic compatibility. If the shape of the material surface is stationary, $\partial f / \partial t = 0$, the remaining three terms on the left-hand side of (1.10.3) must balance to zero.

Level-set formulation

A material surface can be described implicitly by the equation

$$F(x, y, z, t) \equiv f(x, y, t) - z = 0. \tag{1.10.4}$$

The function $F(x, y, z, t)$ is negative above the material surface, positive below the material surface, and zero over the material surface. Equation (1.10.3) can be restated in terms of the material derivative of F as

$$\frac{DF}{Dt} \equiv \frac{\partial F}{\partial t} + \mathbf{u} \cdot \nabla F = 0. \tag{1.10.5}$$

Introducing the upward normal unit vector, $\mathbf{n} = -\nabla F / |\nabla F|$, we obtain

$$\frac{1}{|\nabla F|} \frac{\partial F}{\partial t} = u_n, \tag{1.10.6}$$

where $u_n \equiv \mathbf{u} \cdot \mathbf{n}$ is the normal velocity component. This form shows that u_n must be continuous across a material surface. The tangential velocity component may undergo a discontinuity that is inconsequential in the context of kinematics.

The implicit function theorem allows us to generalize these results and state that, if a material surface is described in a certain parametric form as

$$F(\mathbf{x}, t) = c, \tag{1.10.7}$$

where c is a constant, then the evolution of the level-set function $F(\mathbf{x}, t)$ is governed by equation (1.10.5) or (1.10.6). This is another way of saying that the scalar field represented by the function $F(\mathbf{x}, t)$ is convected by the flow. Stated differently, a material surface is a convected level-set of the function $F(\mathbf{x}, t)$ determined by the constant c.

As an example, we describe a material line in a two-dimensional flow in plane polar coordinates, (r, θ), as $r = f(\theta, t)$, and introduce the level-set function

$$F(r, \theta, t) \equiv f(\theta, t) - r. \tag{1.10.8}$$

Kinematic compatibility requires that

$$\frac{\mathrm{D}f}{\mathrm{D}t} = \frac{\partial f}{\partial t} + \frac{u_\theta}{r}\frac{\partial f}{\partial \theta} - u_r = 0, \qquad (1.10.9)$$

where the velocity is evaluated on either side of the material line, subject to the condition that the normal velocity is continuous across the material line.

1.10.2   Generalized compatibility condition

Since the gradient of the level-set function, ∇F, is perpendicular to a material surface, an arbitrary tangential component, $\mathbf{v}_t$, can be added to the interfacial fluid velocity $\mathbf{u}$ in (1.10.5), yielding

$$\frac{\partial F}{\partial t} + (\,u_n\,\mathbf{n} + \mathbf{v}_t\,)\cdot\nabla F = 0. \qquad (1.10.10)$$

For example, we may set

$$\mathbf{v}_t = \beta\,(\mathbf{I} - \mathbf{n}\otimes\mathbf{n})\cdot\mathbf{u} = \beta\,\mathbf{n}\times\mathbf{u}\times\mathbf{n}, \qquad (1.10.11)$$

where β is an arbitrary time-dependent coefficient allowed to vary over the material surface. The projection operator $\mathbf{I} - \mathbf{n}\otimes\mathbf{n}$ extracts the tangential component of a vector that it multiplies. Equation (1.10.10) reveals that it is kinematically consistent to allow imaginary interfacial particles to move with their own velocity,

$$\mathbf{v} = u_n\,\mathbf{n} + \beta(\mathbf{I} - \mathbf{n}\otimes\mathbf{n})\cdot\mathbf{u}, \qquad (1.10.12)$$

that can be different than the fluid velocity. When $\beta \neq 1$, the interfacial particles represent marker points devoid of physical interpretation.

1.10.3   Line curvilinear coordinates

An evolving material line in the xy plane can be described parametrically by the equation

$$\mathbf{x}(\xi, t) = \mathbf{x}^R(\xi) + \zeta(\xi, t)\,\mathbf{n}^R(\xi), \qquad (1.10.13)$$

where $\mathbf{x}^R(\xi)$ describes a steady reference line, ξ is a parameter increasing in a specified direction along the reference line, $\mathbf{n}^R(\xi)$ is the unit vector normal to the reference line, and $\zeta(\xi, t)$ is the normal displacement, as illustrated in Figure 1.10.2. This representation is acceptable when the material line is sufficiently close to the reference line so that $\zeta(\xi, t)$ is a single-valued and continuous shape function. In practice, the material line can be the trace of a cylindrical surface whose generators are parallel to the z axis.

It will be necessary to introduce the arc length along the material line, l_ξ, increasing in the direction of the parameter ξ. A point particle in an evolving material line moves tangentially and normal to the reference line. If at time t the point particle is at a position corresponding to l_ξ, then at time $t + \Delta t$ the same point particle will be at a position corresponding to $l_\xi + \Delta l_\xi$, where

$$\Delta l_\xi = \frac{R}{R + \zeta}\,u_\xi \Delta t = \frac{1}{1 + \kappa^R \zeta}\,u_\xi \Delta t, \qquad (1.10.14)$$

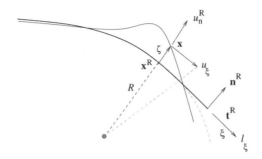

FIGURE 1.10.2   An evolving material line can be described in Eulerian form with respect to a stationary reference line in terms of the normal displacement, ζ. The radius of curvature of the reference line is denoted by R.

u_ξ is the velocity component tangential to the reference line evaluated at the position of the material line, R is the signed radius of curvature of the reference line, and $\kappa^R = 1/R$ is the curvature of the reference line. For the configuration illustrated in Figure 1.10.2, R and κ^R are positive. Expression (1.10.14) arises by approximating the line locally with a circular arc. Using the second Frenet–Serret relation, $\partial \mathbf{n}^R/\partial l_\xi = \kappa^R \mathbf{t}^R$, we find that

$$\kappa^R \zeta = \mathbf{t}^R \cdot \frac{\partial(\zeta \mathbf{n}^R)}{\partial l_\xi} = \frac{\partial \mathbf{x}^R}{\partial l_\xi} \cdot \frac{\partial(\zeta \mathbf{n}^R)}{\partial l_\xi}, \tag{1.10.15}$$

where $\mathbf{t}^R$ is the unit vector tangent to the reference line. Now exercising geometrical reasoning, we find that

$$\zeta(l_\xi + \Delta l_\xi, t + \Delta t) = \zeta(l_\xi, t) + u_n^R \, \Delta t, \tag{1.10.16}$$

where u_n^R is the velocity component normal to the reference line evaluated at the position of the material line, as illustrated in Figure 1.10.2. Expanding the left-hand side in a Taylor series and linearizing with respect to Δt, we derive the compatibility condition

$$\frac{\partial \zeta}{\partial t} + \frac{u_\xi}{1 + \kappa^R \zeta} \frac{\partial \zeta}{\partial l_\xi} - u_n^R = 0. \tag{1.10.17}$$

In the case of a rectilinear reference line, $\kappa^R = 0$, we recover the compatibility condition derived earlier in this section.

Differentiating expression (1.10.13) with respect to l_ξ and using the second Frenet–Serret relation, we obtain a vector that is tangential to the material line,

$$\frac{\partial \mathbf{x}}{\partial \xi} = (1 + \kappa^R \zeta) \, \mathbf{t}^R + \frac{\partial \zeta}{\partial \xi} \, \mathbf{n}^R. \tag{1.10.18}$$

This expression shows that an arbitrary tangential velocity with components

$$v_\xi = \beta \left(1 + \kappa^R \zeta\right), \qquad v_n^R = \beta \frac{\partial \zeta}{\partial \xi} \tag{1.10.19}$$

can be included on the left-hand side of (1.10.17), where $\beta(\xi)$ is an arbitrary coefficient.

1.10.4 Surface curvilinear coordinates

An evolving material surface can be described parametrically by the equation

$$\mathbf{x}(\xi, \eta, t) = \mathbf{x}^R(\xi, \eta) + \zeta(\xi, \eta, t)\, \mathbf{n}^R(\xi, \eta), \tag{1.10.20}$$

where the function $\mathbf{x}^R(\xi, \eta)$ describes a steady reference surface, ξ and η comprise a system of two surface curvilinear coordinates in the reference surface, $\mathbf{n}^R(\xi, \eta)$ is the unit vector normal to the reference surface, and $\zeta(\xi, \eta, t)$ is the normal displacement from the reference surface. This representation is appropriate when the material surface is sufficiently close to the reference surface so that $\zeta(\xi, \eta, t)$ is a single-valued and continuous shape function.

Working as in Section 1.10.3 for a material line, we find that, in the case of orthogonal surface curvilinear coordinates, (ξ, η), the shape function evolves according to the equation

$$\frac{\partial \zeta}{\partial t} + \frac{u_\xi}{1 + \mathcal{K}_\xi \zeta} \frac{\partial \zeta}{\partial l_\xi} + \frac{u_\eta}{1 + \mathcal{K}_\eta \zeta} \frac{\partial \zeta}{\partial l_\eta} - u_n^R = 0, \tag{1.10.21}$$

where l_ξ and l_η is the arc length in the reference surface along the two surface curvilinear coordinates, $\mathcal{K}_\xi$ and $\mathcal{K}_\eta$ are the corresponding principal curvatures, u_ξ and u_η are velocity components tangential to the reference surface evaluated at the material surface, and u_n^R is the velocity component normal to the reference surface. If the reference surface is a sphere of radius a, we may identify ξ with the meridional angle, θ, η with the azimuthal angle, φ, and set $l_\theta = a\theta$, $l_\varphi = (a \sin \theta)\, \varphi$, $\kappa_\xi = 1/a$, and $\kappa_\eta = 1/a$.

Equation (1.10.21) can be expressed in coordinate-free vector notation in terms of the surface gradient of the reference surface, $\nabla_s^R = \mathbf{P}^R \cdot \nabla$, and corresponding surface velocity, $\mathbf{u}_s^R = \mathbf{P}^R \cdot \mathbf{u}$, as

$$\frac{\partial \zeta}{\partial t} + \mathbf{u}_s^R \cdot \nabla_s^R \mathcal{F} - u_n^R = 0, \tag{1.10.22}$$

where $\mathbf{P}^R = \mathbf{I} - \mathbf{n}^R \otimes \mathbf{n}^R$ is the tangential projection operator and $\mathbf{I}$ is the identity matrix. The surface function, $\mathcal{F}(l, \alpha)$, satisfies the differential equation

$$\frac{\partial \mathcal{F}}{\partial l} = \frac{1}{1 + \mathcal{K}\zeta} \frac{\partial \zeta}{\partial l}, \tag{1.10.23}$$

where l is the arc length along the intersection of the material surface with a normal plane, corresponding to the tangent unit vector $\mathbf{t}$, $\mathcal{K}$ is the corresponding principal curvature, and α is the angle subtended between the intersection and a specified tangential direction. For example, $\alpha = 0$ may correspond to the direction of maximum principal curvature. The pair (l, α) constitutes a system of plane polar coordinates tangential to the material surface at a point of interest. By analogy with (1.10.15), we write

$$\mathcal{K}\zeta = \mathbf{t} \cdot \frac{\partial(\zeta \mathbf{n}^R)}{\partial l} = \frac{\partial \mathbf{x}^R}{\partial l} \cdot \frac{\partial(\zeta \mathbf{n}^R)}{\partial l}, \tag{1.10.24}$$

which demonstrates that the right-hand side of (1.10.23) is independent of α.

Problems

1.10.1 *Expanding sphere*

Consider a radially expanding sphere of radius $a(t)$ described in spherical polar coordinates by the equation $F(r, t) = r - a(t) = 0$. Use (1.10.5) to compute the radial velocity at the surface of a sphere in terms of F.

1.10.2 *Boundary condition at a propagating wavy material line*

Consider a material line in the xy plane described by the function $y = a \sin[k(x - ct)]$, where k is the wave number and c is the phase velocity. Use (1.10.5) to derive a boundary condition for the velocity.

1.10.3 *Line and surface curvilinear coordinates*

(a) Develop a compatibility condition when the normal vector on the right-hand side of (1.10.13) is replaced with an arbitrary unit vector that has a normal component.

(b) Repeat (a) for (1.10.20).

1.11   Streamlines, stream tubes, path lines, and streak lines

An instantaneous streamline in a flow is a line whose tangential vector at every point is parallel to the current velocity vector. A closed streamline forms a simple or twisted loop, whereas an open streamline crosses the boundaries of a flow or else extends to infinity. A streamline can meet another streamline or a multitude of other streamlines at a stagnation point. Stagnation points may occur in the interior of a flow or at the boundaries. Since the velocity is a single-valued function of position, the velocity at a stagnation point must necessarily vanish.

Autonomous differential equations

To describe a streamline, we introduce a variable τ that increases monotonically along the streamline in the direction of the velocity vector. One acceptable choice for τ is the time it takes for a point particle to move along the streamline from a specified initial position as it is convected by the frozen instantaneous velocity field. If the flow is steady, τ is the real time, t. The streamline is then described by an autonomous system of differential equations, with no time dependence on the right-hand side,

$$\frac{\mathrm{d}\mathbf{x}}{\mathrm{d}\tau} = \mathbf{u}(\mathbf{x}, t = t_0),$$

(1.11.1)

where t_0 is a specified time. Recasting (1.11.1) into the form

$$\frac{\mathrm{d}x}{u_x} = \frac{\mathrm{d}y}{u_y} = \frac{\mathrm{d}z}{u_z} = \mathrm{d}\tau,$$

(1.11.2)

confirms that an infinitesimal vector that is tangential to a streamline is parallel to the velocity.

1.11.1 Computation of streamlines

To compute a streamline passing through a specified point, $\mathbf{x}_0$, we integrate the differential equations (1.11.1) forward or backward with respect to τ, subject to the initial condition, $\mathbf{x}(\tau = 0) = \mathbf{x}_0$, using a standard numerical method, such as a Runge–Kutta method discussed in Section B.8, Appendix B. The integration terminates when the magnitude of the velocity becomes exceedingly small, signaling approach to a stagnation point.

Setting the time step

If the integration step, Δt, is kept constant during the integration, the travel distance along a streamline will be proportional to the magnitude of the local velocity at each step. A large number of steps will be required in regions of slow flow with a simple streamline pattern. To circumvent this difficulty, we may set the time step inversely proportional to the local magnitude of the velocity, thereby ensuring a nearly constant travel distance in each step. This method has the practical disadvantage that the computed streamline may artificially cross stagnation points where it is supposed to end. A remedy is to set the travel distance equal to, or less than, the finest length scale in the flow.

Ideally, the time step should be adjusted according to both the magnitude of the velocity and local curvature of the computed streamline, so that sharply turning streamlines are described with sufficient accuracy and the computation does not stall at regions of slow flow with simple structure. However, implementing these conditions increases the complexity of the computer code. Unless a high level of spatial resolution is required, the method of constant travel distance should be employed.

1.11.2 Stream surfaces and stream tubes

The collection of all streamlines passing through an open line in a flow forms a stream surface, and the collection of all streamlines passing through a closed loop forms a stream tube. Consider two closed loops wrapping once around a stream tube, and draw two surfaces, D_1 and D_2, that are bounded by each loop, as shown in Figure 1.11.1. The volumetric flow rate across each surface is

$$Q_i = \iint_{D_i} \mathbf{u} \cdot \mathbf{n} \, dS \tag{1.11.3}$$

for $i = 1, 2$, where $\mathbf{n}$ is the unit vector normal to D_1 or D_2. Using the divergence theorem, we compute

$$Q_2 = Q_1 - \iint_{S_t} \mathbf{u} \cdot \mathbf{n} \, dS + \iiint_{V_t} \nabla \cdot \mathbf{u} \, dV, \tag{1.11.4}$$

where S_t is the surface of the stream tube extending between the two loops, and V_t is the volume enclosed by the surfaces D_1, D_2, and S_t. Since the velocity is tangential to the stream tube and therefore perpendicular to the normal vector on S_t, the surface integral over S_t is zero and (1.11.4) simplifies into

$$Q_2 = Q_1 + \iiint_{V_t} \nabla \cdot \mathbf{u} \, dV. \tag{1.11.5}$$

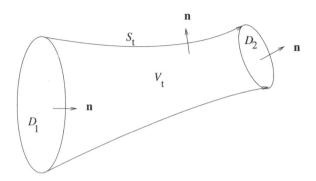

FIGURE 1.11.1   Illustration of a stream tube and two closed loops wrapping around the stream tube.

This equation suggests that the volumetric flow rate may increase or decrease along a stream tube according to whether the fluid inside the stream tube undergoes expansion or contraction.

Incompressible fluids

If the fluid is incompressible, $\nabla \cdot \mathbf{u} = 0$, the flow rate across any cross-section of a stream tube is constant, $Q_1 = Q_2$. Consequently, in the absence of singularities, a stream tube that carries a finite amount of fluid may not collapse into a nonsingular point where the fluid velocity is nonzero. If this occurred, the flow rate at the point of collapse would have to vanish, which contradicts the assumption that the stream tube carries a finite amount of fluid. A similar argument can be made to show that a streamline, approximated as a stream tube with infinitesimal cross-section, may not end suddenly in a flow, but must meet another streamline or multiple streamlines at a stagnation point, form a closed loop, extend to infinity, or cross the boundaries of the flow. A third consequence of (1.11.5) is that the distance between two adjacent streamlines in a two-dimensional incompressible flow is inversely proportional to the local magnitude of the fluid velocity. The faster the velocity, the closer the streamlines.

1.11.3   Streamline coordinates

Useful insights can be obtained by considering the motion of point particles with reference to the instantaneous structure of the velocity field near a streamline. The point particles translate tangentially to the streamlines and rotate around the local vorticity vector, which may point in an arbitrary direction with respect to the streamlines. (We note parenthetically that a flow where the vorticity vector is tangential to the velocity vector at every point, $\mathbf{u} \cdot \boldsymbol{\omega} = 0$, is called a Beltrami flow.) Our main goal is to establish a relation between the direction and magnitude of the vorticity vector and the structure of the streamline pattern in two- and three-dimensional flow.

Frenet–Serret framework

As a preliminary, we introduce a system of orthogonal curvilinear coordinates constructed with reference to a streamline. We begin by labeling point particles along the streamline by the arc length, l, and observe that the unit vector $\mathbf{t} = \mathrm{d}\mathbf{x}/\mathrm{d}l$ is tangential to the streamline and therefore

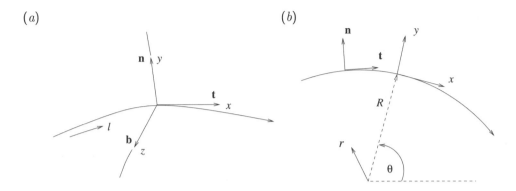

FIGURE 1.11.2   Illustration of a streamline and associated curvilinear axes constructed with reference to the streamline in (a) three-dimensional, and (b) two-dimensional flow.

parallel to the velocity, as shown in Figure 1.11.2(a). The principal unit vector normal to the streamline, $\mathbf{n}$, is defined by the first Frenet–Serret relation

$$\mathbf{n} = -\frac{1}{\kappa}\frac{\partial \mathbf{t}}{\partial l}, \tag{1.11.6}$$

where κ is the signed curvature of the streamline, as discussed in Section 1.6.3. The curvature of the streamline shown in Figure 1.11.2(a) is positive, $\kappa > 0$. Next, we introduce the binormal unit vector defined by the equation $\mathbf{b} = \mathbf{t} \times \mathbf{n}$. The three unit vectors, $\mathbf{t}$, $\mathbf{n}$, and $\mathbf{b}$, define three mutually orthogonal directions that can be used to construct a right-handed, orthogonal, curvilinear system of axes, so that $\mathbf{t} = \mathbf{n} \times \mathbf{b}$ and $\mathbf{n} = \mathbf{b} \times \mathbf{t}$, as discussed in Section 1.6.2. The second and third Frenet–Serret relations are

$$\frac{\partial \mathbf{n}}{\partial l} = \kappa\,\mathbf{t} + \tau\,\mathbf{b}, \qquad \frac{\partial \mathbf{b}}{\partial l} = -\tau\,\mathbf{n}, \tag{1.11.7}$$

where τ is the torsion of the streamline.

Local Cartesian coordinates

In the next step, we introduce a system of Cartesian coordinates with origin at a particular point on a chosen streamline. The x, y, and z axes point in the directions of $\mathbf{t}$, $\mathbf{n}$, and $\mathbf{b}$, as shown in Figure 1.11.2(a). Since the velocity is tangential to the streamline, $u_y = 0$ and $u_z = 0$ at the origin by definition at any instant. Moreover,

$$\frac{\partial u_y}{\partial x} = \frac{\partial \mathbf{u}}{\partial l} \cdot \mathbf{n} = \frac{\partial(\mathbf{u}\cdot\mathbf{n})}{\partial l} - \frac{\partial \mathbf{n}}{\partial l}\cdot\mathbf{u}, \qquad \frac{\partial u_z}{\partial x} = \frac{\partial \mathbf{u}}{\partial l} \cdot \mathbf{b} = \frac{\partial(\mathbf{u}\cdot\mathbf{b})}{\partial l} - \frac{\partial \mathbf{b}}{\partial l}\cdot\mathbf{u}. \tag{1.11.8}$$

Because the velocity is tangential to the streamline and thus perpendicular to the normal and binormal vectors along the entire streamline, $\mathbf{u}\cdot\mathbf{n} = 0$, $\mathbf{u}\cdot\mathbf{b} = 0$, and the first term after the second equal sign of each equation in (1.11.8) is identically zero. Using the Frenet–Serret formulas (1.11.7) to simplify the second term, we obtain

$$\frac{\partial u_y}{\partial x} = -\kappa u_x, \qquad \frac{\partial u_z}{\partial x} = 0. \tag{1.11.9}$$

In Section 1.11.4, these expressions will be used to relate the vorticity to the structure of the velocity field around a streamline.

Vorticity in two-dimensional flow

Consider a two-dimensional flow in the xy plane, as illustrated in Figure 1.11.2(b). Using relations (1.11.9), we find that the z vorticity component at the origin of the local Cartesian axes is given by

$$\omega_z = \frac{\partial u_y}{\partial x} - \frac{\partial u_x}{\partial y} = -\kappa u_x - \mathbf{n} \cdot (\nabla \mathbf{u}) \cdot \mathbf{t}. \tag{1.11.10}$$

Introducing plane polar coordinates with origin at the center of curvature of a streamline at a point, as shown in Figure 1.11.2(b), and noting that $u_x = -u_\theta$ along the y axis pointing in the radial direction, we obtain

$$\omega_z = \frac{u_\theta}{R} + \left(\frac{\partial u_\theta}{\partial r}\right)_{r=R}, \tag{1.11.11}$$

where $R = 1/\kappa$ is the radius of curvature of the streamline. This expression reveals that point particles spin about the z axis due to the global motion of the fluid associated with the curvature of the streamline, but also due to velocity variations in the normal direction.

Vorticity in streamline coordinates

The vorticity vector can be resolved into three components corresponding to the tangent, normal, and binormal directions at a point along a streamline. The corresponding vorticity components can be expressed in terms of the structure of the velocity field around a streamline. We will demonstrate that the streamline decomposition takes the form

$$\boldsymbol{\omega} = \left(\mathbf{n} \cdot \mathbf{L} \cdot \mathbf{b} - \mathbf{b} \cdot \mathbf{L} \cdot \mathbf{n}\right)\mathbf{t} + \frac{\partial u_t}{\partial l_b}\mathbf{n} - \left(\frac{\partial u_t}{\partial l_n} + \kappa u_t\right)\mathbf{b}, \tag{1.11.12}$$

where $\mathbf{L} = \nabla \mathbf{u}$ is the velocity gradient tensor, u_t is the tangential velocity component, l_n is the arc length in the normal direction, l_b is the arc length in the binormal direction, and κ is the curvature of the streamline. Expression (1.11.12) reveals the following:

- Point particles spin about the tangential vector due to the twisting of the streamline pattern, which is possible only in a genuine three-dimensional flow.

- Point particles spin about the normal vector due to velocity variations in the binormal direction, which is also possible only in a genuine three-dimensional flow.

- Point particles spin about the binormal vector due to velocity variations in the normal direction and also due to the global fluid motion of the fluid associated with the curvature of the streamline.

To derive the tangential vorticity component, we refer to Figure 1.11.2(a) and project the definition of the vorticity, $\boldsymbol{\omega} = \nabla \times \mathbf{u}$, onto the unit tangent vector, $\mathbf{t}$. Rearranging the resulting expression and introducing the normal and binormal vectors, we obtain

$$\omega_t \equiv \mathbf{t} \cdot \boldsymbol{\omega} = \mathbf{t} \cdot (\nabla \times \mathbf{u}) = (\mathbf{n} \times \mathbf{b}) \cdot (\nabla \times \mathbf{u}) = \mathbf{n} \cdot (\nabla \mathbf{u}) \cdot \mathbf{b} - \mathbf{b} \cdot (\nabla \mathbf{u}) \cdot \mathbf{n}. \qquad (1.11.13)$$

By definition, $\omega_t = 0$ in two-dimensional or axisymmetric flow.

To derive the normal vorticity component, we write $\mathbf{u} = u_t \mathbf{t}$ and invoke once again the definition of the vorticity, $\boldsymbol{\omega} = \nabla \times \mathbf{u}$, to obtain

$$\boldsymbol{\omega} = \nabla \times (u_t \mathbf{t}) = u_t \nabla \times \mathbf{t} + \nabla u_t \times \mathbf{t}. \qquad (1.11.14)$$

Projecting (1.11.14) onto the normal vector, $\mathbf{n}$, and rearranging, we find that

$$\omega_n \equiv \boldsymbol{\omega} \cdot \mathbf{n} = u_t \, (\nabla \times \mathbf{t}) \cdot \mathbf{n} + (\nabla u_t \times \mathbf{t}) \cdot \mathbf{n} = u_t \, (\nabla \times \mathbf{t}) \cdot (\mathbf{b} \times \mathbf{t}) + (\mathbf{t} \times \mathbf{n}) \cdot \nabla u_t, \qquad (1.11.15)$$

and then

$$\omega_n = u_t \, \mathbf{b} \cdot (\mathbf{t} \times \nabla \times \mathbf{t}) + \mathbf{b} \cdot \nabla u_t = u_t \, \mathbf{b} \cdot \big[(\nabla \mathbf{t}) \cdot \mathbf{t} - \mathbf{t} \cdot \nabla \mathbf{t} \big] + \frac{\partial u_t}{\partial l_b}. \qquad (1.11.16)$$

Because the length of the tangent unit vector $\mathbf{t}$ is constant, $(\nabla \mathbf{t}) \cdot \mathbf{t} = \frac{1}{2} \nabla (\mathbf{t} \cdot \mathbf{t}) = 0$. Using (1.11.6), we find that $\mathbf{b} \cdot (\mathbf{t} \cdot \nabla \mathbf{t}) = -\kappa \, \mathbf{b} \cdot \mathbf{n} = 0$, and conclude that $\omega_n = \partial u_t / \partial l_b$. By definition, $\omega_n = 0$ in two-dimensional or axisymmetric flow.

To derive the binormal vorticity component, we project (1.11.14) onto the binormal vector, $\mathbf{b}$, and work in a similar fashion to obtain the binormal vorticity components,

$$\omega_b \equiv \boldsymbol{\omega} \cdot \mathbf{b} = u_t \, (\nabla \times \mathbf{t}) \cdot \mathbf{b} + (\nabla u_t \times \mathbf{t}) \cdot \mathbf{b} = u_t \, (\nabla \times \mathbf{t}) \cdot (\mathbf{t} \times \mathbf{n}) + (\mathbf{t} \times \mathbf{b}) \cdot \nabla u_t, \qquad (1.11.17)$$

yielding

$$\omega_b = -u_t \, \mathbf{n} \cdot (\mathbf{t} \times \nabla \times \mathbf{t}) - \mathbf{n} \cdot \nabla u_t = u_t \, \mathbf{n} \cdot \frac{\partial \mathbf{t}}{\partial l} - \frac{\partial u_t}{\partial l_n} = -\kappa u_t - \frac{\partial u_t}{\partial l_n}. \qquad (1.11.18)$$

In the case of two-dimensional flow, the vorticity vector points in the binormal direction and (1.11.18) reduces to (1.11.10).

1.11.4 Path lines and streaklines

A path line represents the trajectory of a point particle that has been released from a certain position, $\mathbf{X}_0$, at some time instant, t_0. If the flow is steady, the path line coincides with the streamline passing through the point $\mathbf{X}_0$. The shape of a path line is described by the generally nonautonomous ordinary differential equation

$$\frac{d\mathbf{X}}{dt} = \mathbf{u}\big[\mathbf{X}(t), t\big], \qquad (1.11.19)$$

where $\mathbf{X}$ is the position of the point particle along its path. Formal integration yields the position of the point particle at time t,

$$\mathbf{X}(t; t_0) = \mathbf{X}_0(t_0) + \int_{t_0}^{t} \mathbf{u}(\mathbf{X}(\tau; t_0), \tau) \, d\tau = \mathbf{X}_0(t_0) + \int_{0}^{t-t_0} \mathbf{u}(\mathbf{X}(t_0 + \xi; t_0), t_0 + \xi) \, d\xi, \quad (1.11.20)$$

which can be regarded as a parametric representation of the path line in time. To compute a path line, we select an ejection location and time and integrate equation (1.11.19) in time using a standard numerical method, such as a Runge–Kutta method discussed in Section B.8, Appendix B.

Streaklines

A streakline is the instantaneous chain of point particles that have been released from the same or different locations at the same or different prior times in a flow. Streaklines can be produced in the laboratory by ejecting a dye from a stationary or moving needle. Regarding the injection time, t_0 as a Lagrangian marker variable, we describe the shape of a streakline at a particular time t by (1.11.20). When the point particles are injected at the same location, the term $\mathbf{X}_0(t_0)$ after each equal sign is constant.

Problems

1.11.1 *Beltrami and complex lamellar flows*

Explain why a two-dimensional or axisymmetric flow cannot be a Beltrami flow where the vorticity is parallel to the velocity, but is necessarily a complex lamellar flow where the velocity is perpendicular to its curl.

1.11.2 *Fluid in rigid-body rotation*

Use (1.11.12) to compute the vorticity of a fluid in rigid-body rotation. Show that the result is consistent with the definition of the vorticity, $\boldsymbol{\omega} = \nabla \times \mathbf{u}$.

1.11.3 *Linear flows*

Sketch and discuss the streamline pattern of the following linear flows: (a) purely rotational two-dimensional flow with $u_x = -\xi y$, $u_y = \xi x$, and $u_z = 0$, (b) two-dimensional extensional flow with $u_x = \xi x$, $u_y = -\xi y$, and $u_z = 0$, (c) axisymmetric extensional flow with $u_x = \xi x$, $u_y = -\frac{1}{2}\xi y$, and $u_z = -\frac{1}{2}\xi z$. What are the cylindrical polar velocity components in the third flow? In all cases, ξ is a constant rate of extension with dimensions of inverse time.

 Computer Problems

1.11.4 *Drawing a streamline*

Write a computer program that computes the streamline passing through a specified point in a given two-dimensional flow. The integration should be carried out using the modified Euler method discussed in Section B.8, Appendix B. The size of the step, Δt, should be selected so that the integration proceeds by a preset distance at every step.

1.11.5 *Drawing streamlines in a box*

Write a computer program that returns the velocity at a point inside a rectangular domain of flow in the xy plane confined between $a_x \leq x \leq b_x$ and $a_y \leq y \leq b_y$, where a_x, b_x, a_y, and b_y are specified constants. The input should include the two components of a two-dimensional velocity field at the nodes of an $N_x \times N_y$ Cartesian grid with nodes located at $x_i = a_x + (i-1)\Delta x$ and $y_j = a_y + (j-1)\Delta y$ for $i = 1, \ldots, N_x + 1$ and $j = 1, \ldots, N_y + 1$, where $\Delta x = (b_x - a_x)/N_x$ and $\Delta y = (b_y - a_y)/M$ are the grid spacings. The velocity between grid points should be computed by bilinear interpolation, as discussed in Section B.4, Appendix B.

1.11.6 *Drawing the streamline pattern in a box*

(*a*) Combine the programs of Problem 1.11.4 and 1.11.5 into a program that draws streamlines in a rectangular domain.

(*b*) Run the program to draw the streamline pattern in the box $0 \leq x \leq 1$ and $0 \leq y \leq 2$, with $N_x = 16$ and $N_y = 32$ divisions. The x and y velocity components at the grid points are

$$u_{ij} = \exp(\pi x_i) - \pi x_i \cos(\pi y_j), \qquad v_{ij} = \sin(\pi y_j) - \pi y_j \exp(\pi x_i). \qquad (1.11.21)$$

Is this velocity field solenoidal?

1.12   Vortex lines, vortex tubes, and circulation around loops

In Section 1.1, we saw that the vorticity vector at a point in a flow is parallel to the instantaneous angular velocity vector of a point particle that happens to be at that location. The magnitude of the vorticity vector is twice the magnitude of the angular velocity of the point particle. Using the definition $\boldsymbol{\omega} = \nabla \times \mathbf{u}$, and recalling that the divergence of the curl of any continuous vector field is identically zero, we find that the vorticity field is solenoidal,

$$\nabla \cdot \boldsymbol{\omega} = 0. \qquad (1.12.1)$$

This property imposes restrictions on the structure of the vorticity field, similar to those imposed on the structure of the velocity field for an incompressible fluid.

1.12.1   Vortex lines and tubes

An instantaneous vortex line is parallel to the vorticity vector and therefore to the point-particle angular velocity vector at each point. The collection of all vortex lines passing through a closed loop generates a surface called a vortex tube, as illustrated in Figure 1.12.1. Remembering that the vorticity field is solenoidal and repeating the arguments following equation (1.11.5), we find that a vortex line may not end in the interior of a flow. Instead, it must form a closed loop, meet other vortex lines at a stagnation point of the vorticity field, extend to infinity, or cross and exit the boundaries of the flow.

   The vortex tubes of a two-dimensional flow are cylindrical surfaces perpendicular to the plane of the flow. The vortex lines of an axisymmetric flow with no swirling motion are concentric circles and the vortex tubes form concentric axisymmetric surfaces. The vortex lines of an axisymmetric flow with swirling motion are spiral lines.

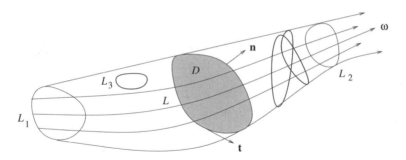

FIGURE 1.12.1   Illustration of a vortex tube. The circulation around a loop that wraps around the tube once is equal to the flow rate of the vorticity across a surface, D, bounded by the loop.

1.12.2   Circulation

The circulation around a closed loop residing in the domain of a flow, L, is defined as

$$C = \oint_L \mathbf{u} \cdot d\mathbf{X} = \oint_L \mathbf{u} \cdot \mathbf{t}\, dl, \qquad (1.12.2)$$

where $\mathbf{X}$ is the position of point particles along the loop, l is the arc length along the loop, and the tangent unit vector, $\mathbf{t} = d\mathbf{X}/dl$, is oriented in a specified direction. Using Stokes' theorem, we derive an alternative expression for the circulation,

$$C = \iint_D (\nabla \times \mathbf{u}) \cdot \mathbf{n}\, dS = \iint_D \boldsymbol{\omega} \cdot \mathbf{n}\, dS, \qquad (1.12.3)$$

where D is an arbitrary surface bounded by the loop, as illustrated in Figure 1.12.1. The direction of the normal unit vector, $\mathbf{n}$, is chosen so that $\mathbf{t}$ and $\mathbf{n}$ constitute a right-handed system of axes. Equation (1.12.3) states that the circulation around a loop is equal to the flow rate of the vorticity across any surface that is bounded by the loop.

Invariance of the circulation around a vortex tube

Consider two material loops, L_1 and L_2, wrapping once around a vortex tube, as shown in Figure 1.12.1. Using (1.12.3), we find that the difference in circulation around these loops is given by

$$C_2 - C_1 = \iint_{D_1} \boldsymbol{\omega} \cdot \mathbf{n}\, dS - \iint_{D_2} \boldsymbol{\omega} \cdot \mathbf{n}\, dS, \qquad (1.12.4)$$

where D_1 is an arbitrary surface bounded by L_1 and D_2 is an arbitrary surface bounded by L_2. Since the vorticity is tangential to the vortex tube, we may add to the right-hand side of (1.12.4) a corresponding integral over the surface of the tube extending between the loops L_1 and L_2. Using the divergence theorem to convert the surface integrals into a volume integral, we obtain

$$C_2 - C_1 = \iiint_{V_t} \nabla \cdot \boldsymbol{\omega}\, dV, \qquad (1.12.5)$$

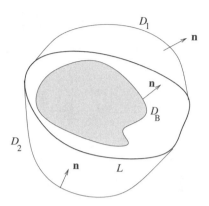

FIGURE 1.12.2 Illustration of the surface of an arbitrary body, D_B, a loop in the flow, L, and two surfaces bounded by the loop.

where the integration domain, V_t, is the volume of the vortex tube enclosed by D_1, D_2, and the surface of the vortex tube. Since the vorticity field is solenoidal, $\nabla \cdot \boldsymbol{\omega} = 0$, the right-hand side of (1.12.5) is identically zero. We conclude that the circulation around any loop that wraps a vortex tube once, denoted by κ and called the cyclic constant or strength of the vortex tube, is independent of the location and shape of the loop around the vortex tube at any instant.

Similar arguments can be made to show that the circulation around a loop that lies on a vortex tube but does not wrap around the vortex tube, such as the loop L_3 shown in Figure 1.12.1, is zero (Problem 1.12.1). The circulation around a closed loop that wraps m times around a vortex tube is equal to $m\kappa$, where κ is the strength of the vortex tube.

Flow of vorticity across a body

As an application, we consider an infinite flow that is bounded internally by a closed surface, D_B, which can be regarded as the surface of a body, and argue that the flow of vorticity across D_B must be identically zero. This becomes evident by introducing two surfaces, D_1 and D_2, that join at an arbitrary closed loop, L, subject to the condition that the union of the two surfaces completely encloses the body, as shown in Figure 1.12.2. Integrating (1.12.1) over the volume enclosed by D_1, D_2, and D_B, applying the divergence theorem to convert the volume integral to a surface integral, and recalling once again that the vorticity field is solenoidal, we obtain

$$\iint_{D_1} \boldsymbol{\omega} \cdot \mathbf{n} \, dS - \iint_{D_2} \boldsymbol{\omega} \cdot \mathbf{n} \, dS = \iint_{D_B} \boldsymbol{\omega} \cdot \mathbf{n} \, dS, \qquad (1.12.6)$$

where $\mathbf{n}$ is the normal unit vector oriented as shown in Figure 1.12.2. Because each integral on the left-hand side of (1.12.6) is equal to the circulation around L, the integral on the right-hand side, and therefore the flow of vorticity across D_B, must necessarily vanish,

$$\iint_{D_B} \boldsymbol{\omega} \cdot \mathbf{n} \, dS = 0. \qquad (1.12.7)$$

1.12.3   Rate of change of circulation around a material loop

Differentiating the definition of the circulation (1.12.2) with respect to time and expanding the derivative, we obtain the evolution equation

$$\frac{\mathrm{d}C}{\mathrm{d}t} = \frac{\mathrm{d}}{\mathrm{d}t} \oint_L \mathbf{u} \cdot \mathrm{d}\mathbf{X} = \oint_L \frac{\mathrm{D}\mathbf{u}}{\mathrm{D}t} \cdot \mathrm{d}\mathbf{X} + \oint_L \mathbf{u} \cdot \frac{\mathrm{D}\,\mathrm{d}\mathbf{X}}{\mathrm{D}t}. \tag{1.12.8}$$

Concentrating on the last integral, we use (1.6.1) to write

$$\oint_L \mathbf{u} \cdot \frac{\mathrm{D}\,\mathrm{d}\mathbf{X}}{\mathrm{D}t} = \oint_L (\mathrm{d}\mathbf{X} \cdot \mathbf{L}) \cdot \mathbf{u} = \frac{1}{2} \oint_L \mathrm{d}\mathbf{X} \cdot \nabla(\mathbf{u} \cdot \mathbf{u}) = 0, \tag{1.12.9}$$

and find that (1.12.8) takes the simplified form

$$\frac{\mathrm{d}C}{\mathrm{d}t} = \oint_L \frac{\mathrm{D}\mathbf{u}}{\mathrm{D}t} \cdot \mathrm{d}\mathbf{X}, \tag{1.12.10}$$

where $\mathbf{L} = \nabla\mathbf{u}$ is the velocity gradient tensor. Equation (1.12.10) identifies the rate of change of circulation around a material loop with the circulation of the acceleration field, $\mathrm{D}\mathbf{u}/\mathrm{D}t$, around the loop. If the acceleration field is irrotational, $\mathrm{D}\mathbf{u}/\mathrm{D}t$ can be expressed as the gradient of a potential function, as discussed in Section 2.1, and the closed line integral is zero. The rate of change of circulation then vanishes and the circulation around the loop is preserved during the motion.

An alternative evolution equation for the circulation can be derived in Eulerian form by applying the general evolution equation (1.7.56) for $\mathbf{q} = \boldsymbol{\omega}$. Recalling that the vorticity field is solenoidal, $\nabla \cdot \boldsymbol{\omega} = 0$, we find that

$$\frac{\mathrm{d}C}{\mathrm{d}t} = \iint_D \left(\frac{\partial\boldsymbol{\omega}}{\partial t} + \nabla \times (\boldsymbol{\omega} \times \mathbf{u}) \right) \cdot \mathbf{n}\,\mathrm{d}S, \tag{1.12.11}$$

where D is an arbitrary surface bounded by the loop.

Problems

1.12.1 *A loop on a vortex tube*

Show that the circulation around the loop L_3 illustrated in Figure 1.12.1 is zero.

1.12.2 *Flow inside a cylindrical container due to a rotating lid*

Sketch the vortex line pattern of a flow inside a cylindrical container that is closed at the bottom, driven by the rotation of the top lid.

1.12.3 *Solenoidality of the vorticity*

Show that $\boldsymbol{\omega} = \nabla\chi \times \nabla\psi$, is an acceptable vorticity field and $\mathbf{u} = \chi\nabla\psi$ is an acceptable associated velocity field, where χ and ψ are two arbitrary functions. Explain why this velocity field is complex lamellar, that is, the velocity is perpendicular to the vorticity at every point.

1.13   Line vortices and vortex sheets

The velocity field in a certain class of flows exhibits sharp variations across thin columns or layers of fluid. Examples include flows containing shear layers forming between two streams that merge at different velocities and around the edges of jets, turbulent flows, and flows due to tornadoes and whirls. A salient feature of these flows is that the support of the vorticity is compact, which means that the magnitude of the vorticity takes significant values only inside well-defined regions, concisely called vortices, while the flow outside the vortices is precisely or nearly irrotational.

1.13.1   Line vortex

Consider a flow where the vorticity vanishes everywhere except near a vortex tube with small cross-sectional area centered at a line, L. Taking the limit as the cross-sectional area of the vortex tube tends to zero while the circulation around the tube, κ, remains constant, we obtain a tubular vortex structure with infinitesimal cross-sectional area, infinite vorticity, and finite circulation, called a line vortex.

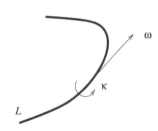

Illustration of a line vortex.

Since, by definition, the vorticity is tangential to a vortex line, the vorticity field associated with a line vortex can be described by the generalized distribution

$$\boldsymbol{\omega}(\mathbf{x}) = \kappa \int_L \mathbf{t}(\mathbf{x}')\, \delta_3(\mathbf{x} - \mathbf{x}')\, \mathrm{d}l(\mathbf{x}'), \tag{1.13.1}$$

where δ_3 is the three-dimensional delta function, $\mathbf{t}$ is the unit vector tangent to the line vortex, and l is the arc length along the line vortex. The velocity field induced by a line vortex will be discussed in Section 2.11.

1.13.2   Vortex sheet

Next, we consider a flow where the vorticity is zero everywhere, except inside a thin sheet centered at a surface, E. Taking the limit as thickness of the sheet tends to zero while the circulation around any loop that pierces the sheet through any two fixed points remains constant, we obtain a vortex sheet with infinitesimal cross-sectional area, infinite vorticity, and finite circulation, as shown in Figure 1.13.1(a).

Vorticity and circulation

The vorticity field associated with a vortex sheet, sometimes also called a sheet vortex, is described by the generalized distribution

$$\boldsymbol{\omega}(\mathbf{x}) = \iint_E \boldsymbol{\zeta}(\mathbf{x}')\, \delta_3(\mathbf{x} - \mathbf{x}')\, \mathrm{d}S(\mathbf{x}'), \tag{1.13.2}$$

where $\boldsymbol{\zeta}$ is a tangential vector field, called the strength of the vortex sheet. Since the vorticity field is solenoidal, $\nabla \cdot \boldsymbol{\omega} = 0$, the surface divergence of $\boldsymbol{\zeta}$ must vanish,

$$(\mathbf{I} - \mathbf{N} \otimes \mathbf{N}) \cdot \nabla \cdot \boldsymbol{\zeta} = 0, \tag{1.13.3}$$

(a)   (b)

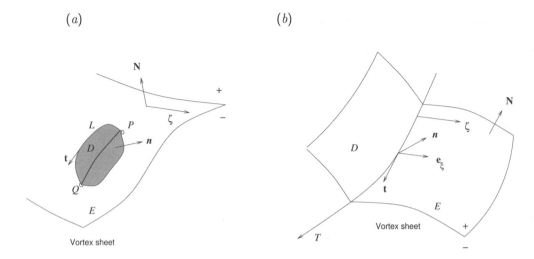

FIGURE 1.13.1   (a) Illustration of a three-dimensional vortex sheet. The loop L pierces the vortex sheet at the points P and Q. (b) Closeup of the intersection between a surface, D, and a vortex sheet; T is the intersection of D and E, and the unit vector $\mathbf{n}$ is normal to D.

where $\mathbf{I}$ is the identity matrix and $\mathbf{N}$ is the unit vector normal to the vortex sheet. To satisfy this constraint, a vortex sheet must be a closed surface, terminate at the boundaries of the flow, or extend to infinity.

Using Stokes' theorem, we find that the circulation around a loop L that pierces a vortex sheet at two points, P and Q, as shown in Figure 1.13.1(a), is given by

$$C = \oint_L \mathbf{u} \cdot \mathbf{t}\, dl = \iint_D \boldsymbol{\omega} \cdot \mathbf{n}\, dS, \tag{1.13.4}$$

where D is an arbitrary surface bounded by L and $\mathbf{n}$ is the unit vector normal to D. Substituting into the last integral the vorticity distribution (1.13.2), we obtain

$$C = \iint_D \left(\iint_E \boldsymbol{\zeta}(\mathbf{x}')\, \delta_3(\mathbf{x} - \mathbf{x}')\, dS(\mathbf{x}') \right) \cdot \mathbf{n}(\mathbf{x})\, dS(\mathbf{x}). \tag{1.13.5}$$

Switching the order of integration on the right-hand side, we obtain

$$C = \iint_E \boldsymbol{\zeta}(\mathbf{x}') \cdot \iint_D \left(\delta_3(\mathbf{x} - \mathbf{x}') \cdot \mathbf{n}(\mathbf{x})\, dS(\mathbf{x}) \right) dS(\mathbf{x}'). \tag{1.13.6}$$

The inner integral is over the arbitrary surface, D, and the outer integral is over the vortex sheet.

Velocity jump and strength of a vortex sheet

Next, we identify the intersection of the surface D with the vortex sheet, E, denoted by T, define the unit vector tangent to T, denoted by $\mathbf{t}$, and introduce a unit vector, $\mathbf{e}_\xi$, that lies in the vortex sheet, E, and is perpendicular to $\mathbf{t}$, as shown in Figure 1.13.1(b). If l is the arc length along $\mathbf{t}$ and l_ξ is the arc length along $\mathbf{e}_\xi$, then

$$dS(\mathbf{x}') = dl(\mathbf{x}')\,dl_\xi(\mathbf{x}') \tag{1.13.7}$$

is a differential surface element over the vortex sheet, E. The differential arc length dl_n in the direction of the normal unit vector $\mathbf{n}$ corresponding to a given dl_ξ is

$$dl_n = \mathbf{n}\cdot\mathbf{e}_\xi\,dl_\xi. \tag{1.13.8}$$

Substituting (1.13.7) and (1.13.8) into (1.13.6), we obtain

$$C = \int_T \boldsymbol{\zeta}(\mathbf{x}')\frac{1}{\mathbf{n}(\mathbf{x}')\cdot\mathbf{e}_\xi(\mathbf{x}')}\cdot\left(\int_{-\delta}^{\delta}\Big[\iint_D\Big(\delta_3(\mathbf{x}-\mathbf{x}')\cdot\mathbf{n}(\mathbf{x})\,dS(\mathbf{x})\Big]\,dl_n(\mathbf{x})\Big)\,dl(\mathbf{x}'), \tag{1.13.9}$$

where δ is a small length. Using the properties of the three-dimensional delta function to simplify the term inside the large parentheses, we find that

$$C = \int_T \frac{\mathbf{n}(\mathbf{x}')\cdot\boldsymbol{\zeta}(\mathbf{x}')}{\mathbf{n}(\mathbf{x}')\cdot\mathbf{e}_\xi(\mathbf{x}')}\,dl(\mathbf{x}'). \tag{1.13.10}$$

The strength of the vortex sheet, $\boldsymbol{\zeta}$, lies in the plane containing $\mathbf{t}$ and $\mathbf{e}_\xi$, which is perpendicular to the plane containing $\mathbf{e}_\xi$ and $\mathbf{n}$. Thus,

$$[(\mathbf{n}\times\mathbf{e}_\xi)\times\mathbf{e}_\xi]\cdot\boldsymbol{\zeta} = 0, \tag{1.13.11}$$

which is equivalent to

$$(\mathbf{n}\cdot\mathbf{e}_\xi)(\mathbf{e}_\xi\cdot\boldsymbol{\zeta}) = \mathbf{n}\cdot\boldsymbol{\zeta} \tag{1.13.12}$$

(Problem 1.13.2). Equation (1.13.9) simplifies into

$$C = \int_T \mathbf{e}_\xi\cdot\boldsymbol{\zeta}\,dl. \tag{1.13.13}$$

Taking the limit as the loop, L, collapses onto the vortex sheet on both sides, while the point Q tends to the point P, we find that

$$(\mathbf{u}^+ - \mathbf{u}^-)\cdot\mathbf{t} = \mathbf{e}_\xi\cdot\boldsymbol{\zeta}, \tag{1.13.14}$$

where $\mathbf{t}$ is the unit vector tangent to T. The superscripts plus and minus designate, respectively, the velocity just above and below the vortex sheet.

   Equation (1.13.14) reveals that the tangential component of the velocity undergoes a discontinuity across a vortex sheet. Mass conservation requires that the normal component of the velocity is continuous across the vortex sheet. Equation (1.13.14) then allows us to write

$$\mathbf{u}^+ - \mathbf{u}^- = \boldsymbol{\zeta}\times\mathbf{N}, \tag{1.13.15}$$

where $\mathbf{N}$ is the unit vector normal to the vortex sheet, as shown in Figure 1.13.1(a). We conclude that $\boldsymbol{\zeta}$, $\mathbf{N}$, and the difference $\mathbf{u}^+ - \mathbf{u}^-$ define three mutually perpendicular directions, so that

$$\boldsymbol{\zeta} = \mathbf{N} \times (\mathbf{u}^+ - \mathbf{u}^-). \qquad (1.13.16)$$

We have found that a vortex sheet represents a singular surface across which the tangential component of the velocity changes from one value above to another value below. The difference between these two values is given by the right-hand side of (1.13.15) in terms of the strength of the vortex sheet.

Principal velocity

The mean value of the velocity above and below a vortex sheet is called the principal velocity of the vortex sheet, or more precisely, the principal value of the velocity of the vortex sheet, denoted by

$$\mathbf{u}^{PV} \equiv \frac{1}{2}\left(\mathbf{u}^+ + \mathbf{u}^-\right). \qquad (1.13.17)$$

Combining (1.13.17) with (1.13.15), we obtain an expression for the velocity on either side of the vortex sheet in terms of the strength of the vortex sheet and the principal velocity,

$$\mathbf{u}^\pm = \mathbf{u}^{PV} + \frac{1}{2}\boldsymbol{\zeta} \times \mathbf{N}. \qquad (1.13.18)$$

Given the strength of the vortex sheet, the principal velocity is much easier to evaluate than the physical fluid velocity on either side of the vortex sheet.

1.13.3   Two-dimensional flow

Next, we describe the structure of line vortices and vortex sheets in two-dimensional flow in the xy plane. The vorticity is perpendicular to the xy plane, $\boldsymbol{\omega} = \omega_z \mathbf{e}_z$, and the vortex lines are infinite straight lines parallel to the z axis.

Point vortex

A rectilinear line vortex is called a point vortex. The location of a point vortex is identified by its trace in the xy plane, $\mathbf{x}_0 = (x_0, y_0)$. The vorticity in the xy plane is described by the singular distribution

$$\omega_z(\mathbf{x}) = \kappa \delta_2(\mathbf{x} - \mathbf{x}_0), \qquad (1.13.19)$$

where δ_2 is the two-dimensional delta function and κ is the strength of the point vortex. More will be said about point vortices in Chapter 11.

Vortex sheet

A two-dimensional vortex sheet is a cylindrical vortex sheet whose generators and strength, $\boldsymbol{\zeta}$, are oriented along the z axis, as illustrated in Figure 1.13.2 (a). We can set

$$\boldsymbol{\zeta} = \gamma \, \mathbf{e}_z, \qquad (1.13.20)$$

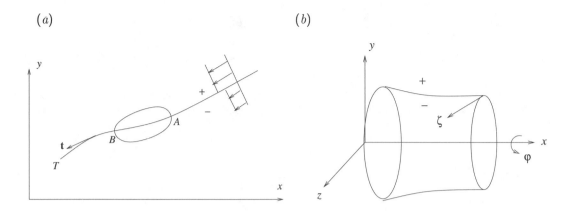

FIGURE 1.13.2   Illustration of (a) a two-dimensional vortex sheet with positive strength, γ, in the xy plane and (b) an axisymmetric vortex sheet.

where γ is the strength of the vortex sheet and $\mathbf{e}_z$ is the unit vector along the z axis. The vorticity field in the xy plane is represented by the singular distribution

$$\omega_z(\mathbf{x}) = \int_T \gamma(\mathbf{x}')\,\delta_2(\mathbf{x} - \mathbf{x}')\,\mathrm{d}l(\mathbf{x}'), \qquad (1.13.21)$$

where δ_2 is the two-dimensional Dirac delta function and T is the trace of the vortex sheet in the xy plane. Using (1.13.15), we find that the discontinuity in the velocity across a two-dimensional vortex sheet is given by

$$\mathbf{u}^+ - \mathbf{u}^- = \gamma\,\mathbf{t}, \qquad (1.13.22)$$

where $\mathbf{t}$ is a unit vector tangent to T, as shown in Figure 1.13.2. In terms of the principal velocity of the vortex sheet, $\mathbf{u}^{PV}$, the velocity at the upper and lower surface of the vortex sheet is given by

$$\mathbf{u}^+ = \mathbf{u}^{PV} + \frac{1}{2}\gamma\,\mathbf{t}, \qquad \mathbf{u}^- = \mathbf{u}^{PV} - \frac{1}{2}\gamma\,\mathbf{t}. \qquad (1.13.23)$$

When γ is positive, we obtain the local velocity profile shown in Figure 1.13.2(a).

The circulation around a loop that lies in the xy plane and pierces the vortex sheet at the points A and B is given by

$$C = \iint_D \omega_z(\mathbf{x})\,\mathrm{d}A(\mathbf{x}) = \iint_D \left[\int_T \gamma(\mathbf{x}')\delta_2(\mathbf{x} - \mathbf{x}')\,\mathrm{d}l(\mathbf{x}')\right]\mathrm{d}A(\mathbf{x}) = \int_{T_{AB}} \gamma(\mathbf{x}')\,\mathrm{d}l(\mathbf{x}'), \quad (1.13.24)$$

where D is the area in the xy plane enclosed by the loop and T_{AB} is the section of T between the points A and B, as illustrated in Figure 1.13.2(a). Fixing the point A and regarding the circulation, C, as a function of location of the point B along T, denoted by Γ, we obtain

$$\frac{\mathrm{d}\Gamma}{\mathrm{d}l} = \gamma. \qquad (1.13.25)$$

This definition allows us to express the vorticity distribution (1.13.21) in terms of Γ as

$$\omega_z(\mathbf{x}) = \int_T \delta_2(\mathbf{x} - \mathbf{x}') \, d\Gamma(\mathbf{x}').$$ (1.13.26)

Comparing (1.13.26) with (1.13.19) allows us to regard a cylindrical vortex sheet as a continuous distribution of point vortices. More will be said about vortex sheets and their self-induced motion in Chapter 11.

1.13.4 Axisymmetric flow

The vorticity vector in an axisymmetric flow without swirling motion points in the azimuthal direction and the vortex lines are concentric circles, as shown in Figure 1.13.2(b).

Line vortex ring

The position of an axisymmetric line vortex, also called a line vortex ring, is described by its trace in an azimuthal plane of constant azimuthal angle, φ, usually identified with the union of the first and second quadrants of the xy plane corresponding to $\varphi = 0$. The azimuthal component of the vorticity is given by the counterpart of equation (1.13.19),

$$\omega_\varphi(\mathbf{x}) = \kappa \, \delta_2(\mathbf{x} - \mathbf{x}_0),$$ (1.13.27)

where δ_2 is the two-dimensional delta function in an azimuthal plane. More will be said about line vortex rings and their self-induced motion in Chapter 11.

Vortex sheet

An axisymmetric vortex sheet can be identified by its trace in an azimuthal plane of constant azimuthal angle, φ. The strength of the vortex sheet, ζ, is oriented in the azimuthal direction, as shown in Figure 1.13.2(b). The vorticity distribution in an azimuthal plane is given by (1.13.21), where δ_2 is the two-dimensional delta function.

Problems

1.13.1 *A line vortex extending between two bodies*

Consider an infinite flow that contains two bodies and a single line vortex that begins on the surface of the first body and ends at the surface of the second body. Discuss whether this is an acceptable and experimentally realizable flow.

1.13.2 *Three-dimensional vortex sheet*

With reference to the discussion of the three-dimensional vortex sheet, show that the equation $[(\mathbf{n} \times \mathbf{e}_\xi) \times \mathbf{e}_\xi] \cdot \boldsymbol{\zeta} = 0$ is equivalent to $(\mathbf{n} \cdot \mathbf{e}_\xi)(\mathbf{e}_\xi \cdot \boldsymbol{\zeta}) = \mathbf{n} \cdot \boldsymbol{\zeta}$.

Kinematic analysis of a flow 2

In Chapter 1, we examined the behavior of fluid parcels, material vectors, material lines, and material surfaces in a specified flow field. The velocity field was assumed to be known as a function of Eulerian variables, including space and time, or Lagrangian variables, including point-particle labels and time. In this chapter, we discuss alternative methods of describing a flow in terms of auxiliary scalar or vector fields. By definition, the velocity field is related to an auxiliary field through a differential or integral relationship. Examples of auxiliary fields include the vorticity and the rate of expansion introduced in Chapter 1. Additional fields introduced in this chapter are the velocity potential for irrotational flow, the vector potential for incompressible fluids, the stream function for two-dimensional flow, the Stokes stream function for axisymmetric flow, and a pair of stream functions for a general three-dimensional incompressible flow. Some auxiliary fields, such as the rate of expansion, the vorticity, and the stream functions, have a clear physical interpretation. Other fields are mathematical devices motivated by analytical simplification.

Describing a flow in terms of an auxiliary field is motivated by two reasons. First, the number of scalar ancillary fields necessary to describe the flow of an incompressible fluid is less than the dimensionality of the flow by one unit, and this allows for analytical and computational simplifications. For example, we will see that a two-dimensional incompressible flow can be described in terms of a single scalar flow, called the stream function. A three-dimensional incompressible flow can be described in terms of a pair of scalar stream functions. The reduction in the number of scalar functions with respect to the number of nonvanishing velocity components is explained by observing that the latter may not be assigned independently, but must be coordinated so as to satisfy the continuity equation for incompressible fluids, $\nabla \cdot \mathbf{u} = 0$. Imposing additional constraints reduces further the number of required scalar functions. Thus, a three-dimensional incompressible and irrotational flow can be expressed in terms of a single scalar function, called the potential function.

In some cases, expressing the velocity field in terms of an auxiliary field allows us to gain physical insights into how the fluid parcel motion affects the global structure or evolution of a flow. For example, representing the velocity field in terms of the vorticity and rate of expansion illustrates the effect of spinning and expansion or contraction of small fluid parcels on the overall fluid motion.

The auxiliary fields discussed in this chapter are introduced with reference to the kinematic structure of a flow discussed in Chapter 1. Another class of auxiliary fields are defined with reference to the dynamics of a flow and, in particular, with respect to the stresses developing in the fluid as a result of the motion. Examples of these dynamical fields will be discussed in Chapter 6.

2.1 Irrotational flows and the velocity potential

If the vorticity vanishes at every point in a flow, $\boldsymbol{\omega} = \nabla \times \mathbf{u} = \mathbf{0}$, the flow is called irrotational. In Chapter 3, we will examine the mechanisms by which vorticity enters, is produced, and evolves in a flow. The analysis will show that, in real life, hardly any flow can be truly irrotational, except during an infinitesimal initial period of time where a fluid starts moving from the state of rest. It appears then that the concept of irrotational flow is merely a mathematical idealization.

However, a number of flows encountered in practical applications are nearly irrotational or consist of adjacent regions of nearly irrotational and nearly rotational flow. For example, high-speed streaming flow past an airfoil is irrotational everywhere except inside a thin boundary layer lining the airfoil and inside a slender wake. The flow produced by the propagation of waves at the surface of the ocean is irrotational everywhere except inside a thin boundary layer along the free surface. In most cases, the conditions under which a flow will be nearly or partially irrotational are not known *a priori*, but must be assessed by carrying out a detailed experimental or theoretical investigation.

Potential and irrotational flows

Since the curl of the gradient of any differentiable function is identically zero, any potential flow whose velocity derives as

$$\mathbf{u} = \nabla \phi \tag{2.1.1}$$

is irrotational, where ϕ is a scalar function called the potential function or the scalar velocity potential. We conclude that a potential flow is also an irrotational flow. Families of irrotational flows can be produced by making different selections for ϕ. Next, we inquire whether the inverse is also true, that is, whether an irrotational flow can be expressed as the gradient of a potential function, as shown in (2.1.1).

Simply and multiply connected domains

It is necessary to consider separately the cases of simply and multiply connected domains of flow. To make this distinction, we draw a closed loop inside the domain of a flow of interest. If the loop can be shrunk to a point that lies inside the domain of flow without crossing any boundaries, the loop is reducible. If the loop must cross one or more boundaries in order to shrink to a point that lies inside the domain of flow, the loop is irreducible. If any loop that can possibly be drawn inside a particular domain of flow is reducible, the domain is simply connected. If an irreducible loop can be found, the domain is multiply connected. Because a loop that wraps around a toroidal or cylindrical boundary of infinite extent, possibly with wavy corrugations, is irreducible, the domain in the exterior of the boundary is doubly connected. A domain containing two distinct toroidal or cylindrical boundaries of infinite extent is triply connected.

2.1.1 Simply connected domains

Using Stokes' theorem, we find that the circulation around a reducible loop, L, is given by

$$\oint_L \mathbf{u} \cdot \mathbf{t}\, dl = \iint_D \boldsymbol{\omega} \cdot \mathbf{n}\, dS = 0, \tag{2.1.2}$$

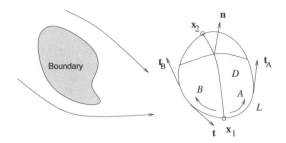

FIGURE 2.1.1   Illustration of a reducible loop, L, in a three-dimensional singly connected domain of flow, showing the decomposition of the circulation integral into two paths. The surface D is bounded by the loop.

where D is an arbitrary surface bounded by L, $\mathbf{t}$ is the unit vector tangential to L, and $\mathbf{n}$ is the unit vector normal to D oriented according to the counterclockwise convention with respect to $\mathbf{t}$, as illustrated in Figure 2.1.1. One consequence of equation (2.1.2) is that the circulation around any reducible loop that lies inside an irrotational flow is zero.

Next, we choose two points $\mathbf{x}_1$ and $\mathbf{x}_2$ on a reducible loop and decompose the line integral in (2.1.2) into two parts,

$$\int_{\mathbf{x}_1}^{\mathbf{x}_2} \mathbf{u} \cdot \mathbf{t}_A \, \mathrm{d}l = \int_{\mathbf{x}_1}^{\mathbf{x}_2} \mathbf{u} \cdot \mathbf{t}_B \, \mathrm{d}l. \tag{2.1.3}$$

The integral on the left-hand side is taken along path A with corresponding tangent vector $\mathbf{t}_A = \mathbf{t}$, while the integral on the right-hand side is taken along path B with corresponding tangent vector $\mathbf{t}_B = -\mathbf{t}$, as shown in Figure 2.1.1. Equation (2.1.3) states that the circulation around any path connecting two ordered points $\mathbf{x}_1$ and $\mathbf{x}_2$ on a reducible loop is the same. Consequently, a single-valued scalar function of position can be introduced, ϕ, such that

$$\int_{\mathbf{x}_1}^{\mathbf{x}_2} \mathbf{u} \cdot \mathbf{t} \, \mathrm{d}l = \phi(\mathbf{x}_2) - \phi(\mathbf{x}_1). \tag{2.1.4}$$

Taking the limit as the second point, $\mathbf{x}_2$, tends to the first point, $\mathbf{x}_1$, and using a linearized Taylor series expansion, we find that

$$\mathbf{u} \cdot \mathbf{t} = \mathbf{t} \cdot \nabla \phi. \tag{2.1.5}$$

Since the integration path, and thus the tangent unit vector $\mathbf{t}$, is arbitrary, we conclude that $\mathbf{u} = \nabla \phi$, as shown in (2.1.1).

We have demonstrated that an irrotational flow in a singly connected domain can be described by a single-valued potential function, ϕ. Stated differently, an irrotational flow in a singly connected domain is also a potential flow described in terms of a differentiable velocity potential.

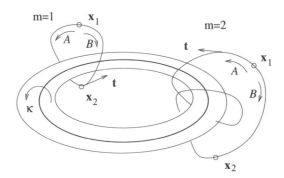

FIGURE 2.1.2   Illustration of two irreducible loops in a doubly connected domain of flow with number
of turns $m = 1$ and 2. The flow can be resolved into the flow due to a line vortex with circulation κ
inside the boundaries and a complementary irrotational flow described in terms by a single-valued
potential.

2.1.2   Multiply connected domains

A conceptual difficulty arises in the case of flow in a multiply connected domain. The reason is that
the circulation around an irreducible loop is not necessarily zero, but may depend on the number of
turns that the loop performs around a boundary, m. Two loops with $m = 1$ and 2 are illustrated in
Figure 2.1.2. The circulation around a loop that performs multiple turns is

$$\oint_L \mathbf{u} \cdot \mathbf{t}\, dl = m\kappa, \tag{2.1.6}$$

where κ is the lowest value of the circulation corresponding to a loop that performs a single turn,
called the cyclic constant of the flow around the boundary.

In the case of an irreducible loop, a surface D bounded by L must cross the boundaries of
the flow. Since D does not lie entirely inside the fluid, Stokes' theorem (2.1.2) cannot be applied.
Progress can be made by breaking up the circulation integral around the loop into two parts and
working as in (2.1.3) and (2.1.4), to find that

$$\Delta\phi_A - \Delta\phi_B = m\kappa, \tag{2.1.7}$$

where $\Delta\phi_A$ and $\Delta\phi_B$ denote the change in the potential function from the beginning to the end of
the paths A and B illustrated in Figure 2.1.2.

Regularization of a multi-valued potential

Equation (2.1.7) suggests that the potential function in a multiply connected domain can be a multi-
valued function of position. To avoid analytical and computational complications, we decompose
the velocity field of a flow of interest into two components,

$$\mathbf{u} = \mathbf{v} + \nabla\phi, \tag{2.1.8}$$

where **v** represents a known irrotational flow whose cyclic constants around the boundaries are the same as those of the flow of interest. For example, **v** could be identified with the flow due to a line vortex lying outside the domain of flow in the interior of a toroidal boundary. The strength of the line vortex is equal to the cyclic constant of the flow around the boundary, as illustrated in Figure 2.1.2. Because the velocity potential ϕ defined in (2.1.8) is a single-valued function of position, it can be computed by standard analytical and numerical methods without any added considerations.

Applications of the decomposition (2.1.8) will be discussed in Chapter 7 with reference to flow past a two-dimensional airfoil.

2.1.3   Jump in the potential across a vortex sheet

Consider a two-dimensional vortex sheet separating two regions of irrotational flow. The velocity on either side of the vortex sheet can be expressed in terms of two velocity potentials, ϕ^+ and ϕ^-. Substituting (2.1.1) into the equation defining the strength of the vortex sheet, $\mathbf{u}^+ - \mathbf{u}^- = \gamma\,\mathbf{t}$, and introducing the circulation along the vortex sheet, Γ, defined by the equation $d\Gamma/dl = \gamma$, we find that

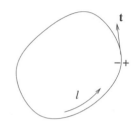

$$\nabla\phi^+ - \nabla\phi^- = \gamma\,\mathbf{t} = \frac{d\Gamma}{dl}\,\mathbf{t}, \qquad (2.1.9)$$

Illustration of a two-dimensional vortex sheet.

where γ is the strength of the vortex sheet and l is the arc length along the vortex sheet measured in the direction of the tangent unit vector, **t**.

Projecting (2.1.9) onto **t** and integrating with respect to arc length, l, we find that the jump in the velocity potential across the vortex sheet is given by

$$\phi^+ - \phi^- = \Gamma. \qquad (2.1.10)$$

We have assumed that ϕ^+ and ϕ^- have the same value at the designated origin of arc length along the vortex sheet, $l = 0$. Equation (2.1.10) finds useful applications in computing the self-induced motion of two-dimensional vortex sheets discussed in Section 11.5.

Three-dimensional vortex sheets

Next, we consider two points on a three-dimensional vortex sheet, A and B. Expressing the velocity on either side of the vortex sheet as the gradient of the corresponding potentials and integrating the equation $\mathbf{u}^+ - \mathbf{u}^- = \boldsymbol{\zeta} \times \mathbf{n}$ along a tangential path connecting the two points, we obtain

$$(\phi^+ - \phi^-)_B = (\phi^+ - \phi^-)_A + \int_A^B (\boldsymbol{\zeta} \times \mathbf{n}) \cdot \mathbf{t}\, dl, \qquad (2.1.11)$$

where $\boldsymbol{\zeta}$ is the strength of the vortex sheet and **t** is the tangent unit vector along the path. If **t** is tangential to $\boldsymbol{\zeta}$ at every point, the integration path coincides with a vortex line and the integrand in (2.1.11) is identically zero. We conclude that the jump of the velocity potential across a vortex sheet is constant along vortex lines.

2.1.4 The potential in terms of the rate of expansion

The velocity field of a potential flow is not necessarily solenoidal and the associated flow is not necessarily incompressible. Taking the divergence of (2.1.1), we find that

$$\nabla \cdot \mathbf{u} = \nabla^2 \phi, \tag{2.1.12}$$

which can be regarded as a Poisson equation for ϕ, forced by the rate of expansion, $\nabla \cdot \mathbf{u}$.

Using the Poisson inversion formula derived in Section 2.2, we obtain an expression for the potential of a three-dimensional flow in terms of the rate of expansion,

$$\phi(\mathbf{x}) = -\frac{1}{4\pi} \iiint_{Flow} \frac{1}{r} \alpha(\mathbf{x}') \, dV(\mathbf{x}') + H(\mathbf{x}), \tag{2.1.13}$$

where $r = |\mathbf{x} - \mathbf{x}'|$, $\alpha(\mathbf{x}') \equiv \nabla' \cdot \mathbf{u}(\mathbf{x}')$ is the rate of expansion, the gradient ∇' involves derivatives with respect to $\mathbf{x}'$, and $H(\mathbf{x})$ is a harmonic function determined by the boundary conditions, $\nabla^2 H = 0$. If the domain of flow extends to infinity, to ensure that the volume integral in (2.1.13) is finite, we require that the rate of expansion decays at a rate that is faster than $1/d^2$, where d is the distance from the origin.

Two-dimensional flow

The counterpart of (2.1.13) for two-dimensional flow in the xy plane is

$$\phi(\mathbf{x}) = \frac{1}{2\pi} \iint_{Flow} \ln\left(\frac{r}{a}\right) \alpha(\mathbf{x}') \, dA(\mathbf{x}') + H(\mathbf{x}), \tag{2.1.14}$$

where $dA = dx \, dy$ is an elementary area, a is a specified constant length, and H is a harmonic function in the xy plane determined by the boundary conditions.

Velocity field

To obtain an expression for the velocity in terms of the rate of expansion, we take the gradient of both sides of (2.1.13) and (2.1.14). Interchanging the gradient with the integral, we find that

$$\mathbf{u}(\mathbf{x}) = \nabla\phi(\mathbf{x}) = \frac{1}{4\pi} \iiint_{Flow} \frac{\hat{\mathbf{x}}}{r^3} \alpha(\mathbf{x}') \, dV(\mathbf{x}') + \nabla H(\mathbf{x}) \tag{2.1.15}$$

for three-dimensional flow, and

$$\mathbf{u}(\mathbf{x}) = \nabla\phi(\mathbf{x}) = \frac{1}{2\pi} \iint_{Flow} \frac{\hat{\mathbf{x}}}{r^2} \alpha(\mathbf{x}') \, dA(\mathbf{x}') + \nabla H(\mathbf{x}) \tag{2.1.16}$$

for two-dimensional flow, where $\hat{\mathbf{x}} = \mathbf{x} - \mathbf{x}'$. Later in this section, we will see that the integrals on the right-hand sides of the last two equations can be interpreted as volume or areal distributions of point sources. The densities of the distributions are equal to the local volumetric or areal rate of expansion.

2.1.5 Incompressible fluids and the harmonic potential

If the fluid is incompressible, the rate of expansion on the left-hand side of (2.1.12) vanishes and the potential ϕ satisfies Laplace's equation,

$$\nabla^2\phi = 0. \tag{2.1.17}$$

In this case, ϕ is a harmonic function called a harmonic potential.

Curvilinear coordinates

In cylindrical polar coordinates, (x, σ, φ), Laplace's equation reads

$$\nabla^2\phi = \frac{\partial^2\phi}{\partial x^2} + \frac{1}{\sigma}\frac{\partial}{\partial\sigma}\left(\sigma\frac{\partial\phi}{\partial\sigma}\right) + \frac{1}{\sigma^2}\frac{\partial^2\phi}{\partial\varphi^2}. \tag{2.1.18}$$

In spherical polar coordinates, (r, θ, φ), Laplace's equation reads

$$\nabla^2\phi = \frac{1}{r^2}\frac{\partial}{\partial r}\left(r^2\frac{\partial\phi}{\partial r}\right) + \frac{1}{r^2\sin\theta}\frac{\partial}{\partial\theta}\left(\sin\theta\frac{\partial\phi}{\partial\theta}\right) + \frac{1}{r^2\sin\theta}\frac{\partial^2\phi}{\partial\varphi^2}. \tag{2.1.19}$$

In plane polar coordinates, (r, θ), Laplace's equation reads

$$\nabla^2\phi = \frac{1}{r}\frac{\partial}{\partial r}\left(r\frac{\partial\phi}{\partial r}\right) + \frac{1}{r^2}\frac{\partial^2\phi}{\partial\theta^2}. \tag{2.1.20}$$

Expressions in more general orthogonal or nonorthogonal curvilinear coordinates are provided in Sections A.8–A.17, Appendix A.

Kinetic energy of the fluid and significance of normal boundary motion

The kinetic energy of an incompressible fluid with uniform density in a singly connected domain can be expressed as a boundary integral involving the boundary distribution of the potential. To derive this expression, we apply the rules of product differentiation to write

$$\mathcal{K} \equiv \frac{1}{2}\rho \iiint \mathbf{u}\cdot\mathbf{u}\,\mathrm{d}V = \frac{1}{2}\rho \iiint \mathbf{u}\cdot\nabla\phi\,\mathrm{d}V = \frac{1}{2}\rho \iiint \left(\nabla\cdot(\phi\mathbf{u}) - \phi\nabla\cdot\mathbf{u}\right)\mathrm{d}V, \tag{2.1.21}$$

where the integrals are computed over the volume of flow. Because the velocity field is solenoidal, $\nabla\cdot\mathbf{u} = 0$, the second term inside the integral on the right-hand side is zero. Applying the divergence theorem to convert the volume integral into a surface integral over the boundaries of the flow, B, we find that

$$\mathcal{K} = -\frac{1}{2}\rho \iint_B \phi\,\mathbf{u}\cdot\mathbf{n}\,\mathrm{d}S, \tag{2.1.22}$$

where $\mathbf{n}$ is the unit vector normal to the boundaries pointing into the flow.

Equation (2.1.22) shows that, if the normal velocity component is zero over all boundaries, the kinetic energy and therefore the velocity must be zero and the fluid must be quiescent. In turn,

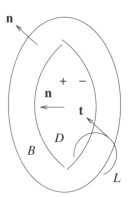

FIGURE 2.1.3   A doubly connected domain of flow containing a toroidal boundary is rendered singly connected by introducing a fictitious boundary surface ending at the boundary.

the velocity potential must be constant throughout the domain of flow. When the velocity potential has the same constant value over all boundaries, mass conservation requires that the right-hand side of (2.1.22) is identically zero. Consequently, the fluid must be quiescent and the potential must be constant and equal to its boundary value throughout the domain of flow.

Multiply connected domains

We can derive corresponding results for flow in a multiply connected domain where the velocity potential can be a multi-valued function of position. As an example, we consider a domain containing a toroidal boundary, as shown in Figure 2.1.3, and draw an arbitrary surface, D, that ends at the boundary, B. Regarding D as a fictitious boundary, we obtain a simply connected domain. Repeating the preceding analysis, we find that the kinetic energy of the fluid is

$$\mathcal{K} = -\frac{1}{2}\rho \iint_B \phi\, \mathbf{u} \cdot \mathbf{n}\, dS - \frac{1}{2}\rho \iint_D (\phi^+ - \phi^-)\, \mathbf{u} \cdot \mathbf{n}\, dS, \qquad (2.1.23)$$

where $\phi^{\pm}$ is the potential on either side of D. The circulation around the loop L shown in Figure 2.1.3 is equal to the cyclic constant of the flow around the toroidal boundary,

$$\kappa = \oint_L \mathbf{u} \cdot \mathbf{t}\, dl = \oint_L \mathbf{t} \cdot \nabla\phi\, dl = \oint_L \frac{\partial \phi}{\partial l}\, dl = \phi^+ - \phi^-. \qquad (2.1.24)$$

Consequently, the kinetic energy of the fluid is given by the expression

$$\mathcal{K} = -\frac{1}{2}\rho \iint_B \phi\, \mathbf{u} \cdot \mathbf{n}\, dS - \frac{1}{2}\rho\kappa Q, \qquad (2.1.25)$$

where Q is the flow rate across the artificial boundary, D.

Expression (2.1.25) shows that the kinetic energy is zero and the fluid is quiescent under two sets of conditions: (*a*) either the potential is constant or the normal component of the velocity is zero at the boundaries, and (*b*) either the cyclic constant, κ, or the flow rate, Q, is zero.

Uniqueness of solution in a singly connected domain

One important consequence of the integral representation of the kinetic energy is that, given boundary conditions for the normal component of the velocity, an incompressible irrotational flow in a singly connected domain is unique and the corresponding harmonic potential is determined uniquely up to an arbitrary constant.

To prove uniqueness of solution, we consider two harmonic potentials representing two distinct flows and note that their difference is also an acceptable harmonic potential, that is, the difference potential describes an acceptable incompressible irrotational flow. If the two flows satisfy the same boundary conditions for the normal component of the velocity, the difference flow must vanish and the corresponding harmonic potential must be constant. As a consequence, the two chosen flows must be identical and the corresponding harmonic potentials may differ at most by a scalar constant. Similar reasoning allows us to conclude that the boundary distribution of the potential uniquely defines an irrotational flow in a singly connected domain.

Uniqueness of solution in a multiply connected domain

In the case of flow in a doubly connected domain, we use (2.1.25) and find that specifying (*a*) boundary conditions either for the normal component of the velocity or for the potential, and (*b*) either the value of the cyclic constant κ, or the flow rate across a surface ending at a toroidal boundary Q, uniquely determines an irrotational flow. Similar conclusions can be drawn for a multiply connected domain.

Kelvin's minimum dissipation theorem

Kelvin demonstrated that, of all solenoidal velocity fields that satisfy prescribed boundary conditions for the normal component of the velocity, the velocity field corresponding to an irrotational flow has the least amount of kinetic energy. Stated differently, vorticity increases the kinetic energy of a fluid.

To prove Kelvin's theorem, we assume that $\mathbf{u}$ is an irrotational velocity field described by a velocity potential, ϕ, and $\mathbf{v}$ is another arbitrary solenoidal rotational velocity field. Moreover, we stipulate that the normal velocity component is the same over the boundaries, $\mathbf{u} \cdot \mathbf{n} = \mathbf{v} \cdot \mathbf{n}$. The difference in the kinetic energies of the two flows, $\Delta \mathcal{K} \equiv \mathcal{K}(\mathbf{v}) - \mathcal{K}(\mathbf{u})$, is

$$\Delta \mathcal{K} = \frac{1}{2}\rho \iiint (\mathbf{v} \cdot \mathbf{v} - \mathbf{u} \cdot \mathbf{u})\, \mathrm{d}V = \frac{1}{2}\rho \iiint (\mathbf{v} - \mathbf{u}) \cdot (\mathbf{v} - \mathbf{u})\, \mathrm{d}V + \rho \iiint (\mathbf{v} - \mathbf{u}) \cdot \mathbf{u}\, \mathrm{d}V. \qquad (2.1.26)$$

Manipulating the last integral on the right-hand side and using the divergence theorem, we obtain

$$\iiint (\mathbf{v} - \mathbf{u}) \cdot \nabla \phi\, \mathrm{d}V = \iint_B \phi\, (\mathbf{v} - \mathbf{u}) \cdot \mathbf{n}\, \mathrm{d}S, \qquad (2.1.27)$$

which is zero in light of the equality of the normal component of the boundary velocity. We conclude that the right-hand side of (2.1.26) is positive and thereby demonstrate that the kinetic energy of a rotational flow with velocity $\mathbf{v}$ is greater than the kinetic energy of the corresponding irrotational flow with velocity $\mathbf{u}$.

2.1.6 Singularities of incompressible and irrotational flow

Singular solutions of Laplace's equation for the harmonic potential constitute a fundamental set of flows used as building blocks for constructing and representing arbitrary irrotational flows.

Point source

Consider the solution of (2.1.17) in an infinite three- or two-dimensional domain in the absence interior boundaries, subject to a singular forcing term on the right-hand side,

$$\nabla^2 \phi^{3D} = m\, \delta_3(\mathbf{x} - \mathbf{x}_0), \qquad \nabla^2 \phi^{2D} = m\, \delta_2(\mathbf{x} - \mathbf{x}_0), \tag{2.1.28}$$

where δ_3 is the three-dimensional (3D) delta function, δ_2 is the two-dimensional (2D) delta function, m is a constant, and $\mathbf{x}_0$ is an arbitrary point in the domain of flow. Using the method of Fourier transforms, or else by employing a trial solution in the form of a power or logarithm of the distance from the singular point, $r = |\mathbf{x} - \mathbf{x}_0|$, we find that

$$\phi^{3D} = -\frac{m}{4\pi} \frac{1}{r}, \qquad \phi^{2D} = \frac{m}{2\pi} \ln \frac{r}{a}, \tag{2.1.29}$$

where a is an arbitrary length.

Since the potential ϕ satisfies Laplace's equation everywhere except at the point $\mathbf{x}_0$, it describes an acceptable incompressible and irrotational flow with velocity

$$\mathbf{u}^{3D} = \nabla \phi^{3D} = \frac{m}{4\pi} \frac{\mathbf{x} - \mathbf{x}_0}{r^3}, \qquad \mathbf{u}^{2D} = \nabla \phi^{2D} = \frac{m}{2\pi} \frac{\mathbf{x} - \mathbf{x}_0}{r^2}, \tag{2.1.30}$$

respectively, for three- or two-dimensional flow. The corresponding streamlines are straight radial lines emanating from the singular point, $\mathbf{x}_0$. It is a straightforward exercise to verify that the flow rate across a spherical surface or circular contour centered at the point $\mathbf{x}_0$ is equal to m. Consequently, the function ϕ^{3D} or ϕ^{2D} can be identified with the potential due to a point source with strength m in an infinite domain of flow. In Section 2.2.3, we will discuss the flow due to a point source in a bounded domain of flow.

In light of (2.1.30), the integrals on the right-hand sides of (2.1.15) and (2.1.16) can be interpreted as volume or areal distributions of point sources of mass forced by the rate of expansion. In Section 2.2.2, we will see that the potential due to a point sink provides us with the free-space Green's function of Laplace's equation

Point-source dipole

Now we consider two point sources with strengths of equal magnitude and opposite sign located at the points $\mathbf{x}_0$ and $\mathbf{x}_1$. Exploiting the linearity of Laplace's equation (2.1.17), we construct the associated potential by linear superposition,

$$\phi^{3D} = -\frac{m}{4\pi} \frac{1}{|\mathbf{x} - \mathbf{x}_0|} + \frac{m}{4\pi} \frac{1}{|\mathbf{x} - \mathbf{x}_1|},$$

$$\phi^{2D} = \frac{m}{2\pi} \ln \frac{|\mathbf{x} - \mathbf{x}_0|}{a} - \frac{m}{2\pi} \ln \frac{|\mathbf{x} - \mathbf{x}_1|}{a}, \tag{2.1.31}$$

respectively, for three- and two-dimensional flow. Placing the point $\mathbf{x}_1$ near $\mathbf{x}_0$, expanding the potential of the second point source in a Taylor series with respect to $\mathbf{x}_1$ about the location of the first point source, $\mathbf{x}_0$, and retaining only the linear terms, we find that

$$\phi^{3D} = \frac{m}{4\pi} (\mathbf{x}_1 - \mathbf{x}_0) \cdot \nabla_0 \frac{1}{|\mathbf{x} - \mathbf{x}_0|},$$

$$\phi^{2D} = -\frac{m}{2\pi} (\mathbf{x}_1 - \mathbf{x}_0) \cdot \nabla_0 \ln \frac{|\mathbf{x} - \mathbf{x}_0|}{a},$$

(2.1.32)

where the derivatives of the gradient ∇_0 are taken with respect to $\mathbf{x}_0$.

Now we take the limit as the distance $|\mathbf{x}_0 - \mathbf{x}_1|$ tends to zero while the product $\mathbf{d} \equiv m (\mathbf{x}_0 - \mathbf{x}_1)$ is held constant, and carry out the differentiations to derive the velocity potential associated with a three- or two-dimensional point-source dipole,

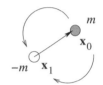

$$\phi^{3D} = -\frac{1}{4\pi} \frac{\mathbf{x} - \mathbf{x}_0}{r^3} \cdot \mathbf{d}, \qquad \phi^{2D} = -\frac{1}{2\pi} \frac{\mathbf{x} - \mathbf{x}_0}{r^2} \cdot \mathbf{d}, \qquad (2.1.33)$$

where $r = |\mathbf{x} - \mathbf{x}_0|$ is the distance of the field point, $\mathbf{x}$, from the singular point, $\mathbf{x}_0$. The corresponding velocity fields are given by

A point source with strength m and a point sink with strength $-m$ merge into a dipole.

$$\mathbf{u}^{3D} = \nabla\phi^{3D} = \frac{1}{4\pi} \left(-\frac{1}{r^3} \mathbf{I} + 3 \frac{\hat{\mathbf{x}} \otimes \hat{\mathbf{x}}}{r^5} \right) \cdot \mathbf{d},$$

$$\mathbf{u}^{2D} = \nabla\phi^{2D} = \frac{1}{2\pi} \left(-\frac{1}{r^2} \mathbf{I} + 2 \frac{\hat{\mathbf{x}} \otimes \hat{\mathbf{x}}}{r^4} \right) \cdot \mathbf{d},$$

(2.1.34)

where $\mathbf{I}$ is the identity matrix and $\hat{\mathbf{x}} = \mathbf{x} - \mathbf{x}_0$. The flow rate across a spherical surface, and therefore across any closed surface, enclosing a three-dimensional point-source dipole is zero. Similarly, the flow rate across any closed loop enclosing a two-dimensional point-source dipole is zero.

The streamline pattern in a meridional plane associated with a three-dimensional dipole is illustrated in Figure 2.1.4(a). The streamline pattern in the xy plane associated with a two-dimensional dipole is illustrated in Figure 2.1.4(b). In both cases, the dipole is oriented along the x axis, $\mathbf{d} = d\,\mathbf{e}_x$, where $d > 0$ and $\mathbf{e}_x$ is the unit vector along the x axis.

Problems

2.1.1 *A harmonic velocity field*

Consider a velocity field, $\mathbf{u}$, with the property that each Cartesian velocity component is a harmonic function, $\nabla^2 u_i = 0$, for $i = x, y, z$. Show that, if the rate of expansion is zero or constant, the corresponding vorticity field is irrotational, $\nabla \times \boldsymbol{\omega} = \mathbf{0}$.

2.1.2 *Flow between two surfaces*

Consider an incompressible potential flow in a domain that is bounded by two closed surfaces, as shown in Figure 2.1.1(a). The normal component of the velocity is zero over one surface, and the

(a)   (b)

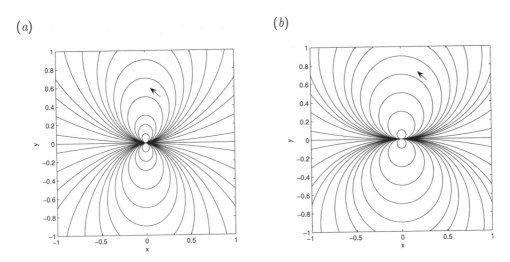

FIGURE 2.1.4   Streamline pattern due to (a) a three-dimensional and (b) a two-dimensional potential dipole pointing in the positive direction of the x axis.

tangential component of the velocity is zero over the other surface. Does this imply that the velocity field must vanish throughout the whole domain of flow?

2.1.3 *Infinite flow*

Consider a three-dimensional incompressible and irrotational flow in an infinite domain where the velocity vanishes at infinity. In Section 2.3, we will show that the velocity potential must tend to a constant value at infinity, as shown in equation (2.3.18). Based on this observation, show that, if the flow has no interior boundaries and no singular points, the velocity field must necessarily vanish throughout the whole domain of flow.

2.1.4 *Irrotational vorticity field*

(a) Show that an irrotational vorticity field, $\nabla \times \boldsymbol{\omega} = \mathbf{0}$, can be expressed as the gradient of a harmonic function.

(b) Consider an irrotational vorticity field of an infinite flow with no interior boundaries, where the vorticity vanishes at infinity. Use the results of Problem 2.1.3 to show that the vorticity must necessarily vanish throughout the whole domain.

2.1.5 *Point source*

Show that the flow rate across any surface that encloses a three-dimensional or two-dimensional point source is equal to strength of the point source, m, but the flow rate across any surface that does not enclose the point source is zero.

2.2   The reciprocal theorem and Green's functions of Laplace's equation

In Section 2.3, we will develop an integral representation for the velocity potential of an irrotational flow in terms of the rate of expansion of the fluid, the boundary values of the velocity potential, and the boundary distribution of the normal derivative of the potential, which is equal to the normal component of the velocity. To prepare the ground for these developments, in this section we introduce a reciprocal theorem for harmonic functions and discuss the Green's functions of Laplace's equation.

2.2.1   Green's identities and the reciprocal theorem

Green's first identity states that any two twice differentiable functions, f and g, satisfy the relation

$$f \, \nabla^2 g = \nabla \cdot (f \nabla g) - \nabla f \cdot \nabla g, \qquad (2.2.1)$$

which can be proven by straightforward differentiation working in index notation. Interchanging the roles of f and g, we obtain

$$g \, \nabla^2 f = \nabla \cdot (g \nabla f) - \nabla g \cdot \nabla f. \qquad (2.2.2)$$

Subtracting (2.2.2) from (2.2.1), we derive Green's second identity,

$$f \, \nabla^2 g - g \, \nabla^2 f = \nabla \cdot (f \nabla g - g \nabla f). \qquad (2.2.3)$$

If both functions f and g satisfy Laplace's equation, the left-hand side of (2.2.3) is zero, yielding a reciprocal relation for harmonic functions,

$$\nabla \cdot (f \nabla g - g \nabla f) = 0. \qquad (2.2.4)$$

Integrating (2.2.4) over a control volume that is bounded by a singly or multiply connected surface, D, and using the divergence theorem to convert the volume integral into a surface integral, we obtain the integral form of the reciprocal theorem,

$$\iint_D (f \, \nabla g - g \, \nabla f) \cdot \mathbf{n} \, \mathrm{d}S = 0, \qquad (2.2.5)$$

where $\mathbf{n}$ it the unit vector normal to D pointing either into the control volume or outward from the control volume. Equation (2.2.5) imposes an integral constraint on the boundary values and boundary distribution of the normal derivatives of any pair of nonsingular harmonic functions.

2.2.2   Green's functions in three dimensions

It is useful to introduce a special class of harmonic functions that are singular at a chosen point, $\mathbf{x}_0$. A three-dimensional Green's function satisfies the singularly forced Laplace's equation

$$\nabla^2 \mathcal{G}(\mathbf{x}, \mathbf{x}_0) + \delta_3(\mathbf{x} - \mathbf{x}_0) = 0, \qquad (2.2.6)$$

where δ_3 is the three-dimensional delta function, $\mathbf{x}$ is a field point, and $\mathbf{x}_0$ is the location of the Green's function, also called the singular point or the pole. When the domain of flow extends to infinity, the Green's function decays at least as fast as the inverse of the distance from the pole, $1/|\mathbf{x} - \mathbf{x}_0|$.

Green's functions of the first kind and Neumann functions

In addition to satisfying (2.2.6), a Green's function of the first kind is required to be zero over a specified surface, S_B, representing the boundary of a flow,

$$\mathcal{G}(\mathbf{x}, \mathbf{x}_0) = 0 \qquad (2.2.7)$$

when $\mathbf{x}$ lies on S_B. Unless qualified, a Green's function of the first kind is simply called a Green's function.

The normal derivative of a Green's function of the second kind, also called a Neumann function, is zero over a specified surface, S_B,

$$\mathbf{n}(\mathbf{x}) \cdot \nabla \mathcal{G}(\mathbf{x}, \mathbf{x}_0) = 0 \qquad (2.2.8)$$

when the point $\mathbf{x}$ lies on S_B, where $\mathbf{n}$ is the unit vector normal to S_B.

Physical interpretation

Comparing (2.2.6) with (2.1.28), we find that, physically, a Green's function represents the steady temperature field due to a *point source of heat* with unit strength located at the point $\mathbf{x}_0$, in the presence of an isothermal or insulated boundary, S_B. A Green's function of the first kind represents the temperature field due to a point source of heat subject to the condition of zero boundary temperature. A Green's function of the second kind represents the temperature field due to a point source of heat subject to the condition of zero boundary flux.

In an alternative interpretation, a Green's function is the harmonic potential due to a *point sink of mass* with unit strength located at the point $\mathbf{x}_0$ in a bounded or infinite domain of flow. A Green's function of the second kind represents the harmonic potential due to a point sink of mass, subject to the condition of zero boundary velocity implementing impermeability. This interpretation explains why a Green's function of the second kind cannot be found in a domain that is completely enclosed by a surface, S_B.

Free-space Green's function

The free-space Green's function corresponds to an infinite domain of flow in the absence of interior boundaries. Solving (2.2.6) by the method of Fourier transforms or simply by applying the first equation in (2.1.29) with $m = -1$, we obtain

$$\mathcal{G}(\mathbf{x}, \mathbf{x}_0) = \frac{1}{4\pi r}, \qquad (2.2.9)$$

where $r = |\mathbf{x} - \mathbf{x}_0|$ is the distance of the field point from the pole.

2.2.3 Green's functions in bounded domains

A Green's function consists of a singular part given by the free-space Green's function (2.2.9), and a complementary part that is nonsingular throughout and over the boundaries of the solution domain, represented by the function $\mathcal{H}(\mathbf{x}, \mathbf{x}_0)$, so that

$$\mathcal{G}(\mathbf{x}, \mathbf{x}_0) = \frac{1}{4\pi r} + \mathcal{H}(\mathbf{x}, \mathbf{x}_0). \qquad (2.2.10)$$

The decomposition (2.2.10) shows that, as the observation point, $\mathbf{x}$, approaches the singular point, $\mathbf{x}_0$, all Green's functions exhibit a common singular behavior. The precise form of $\mathcal{H}$ depends on the geometry of the boundary associated with the Green's function, S_B. In the absence of a boundary, $\mathcal{H}$ is identically zero. For a limited class of simple boundary geometries, the complementary part, $\mathcal{H}$, can be found by the method of images. The construction involves introducing Green's functions and their derivatives at strategically selected locations outside the domain of flow.

Semi-infinite domain bounded by a plane wall

The Green's function for a semi-infinite domain bounded by a plane wall located at $x = w$ is

$$\mathcal{G}(\mathbf{x}, \mathbf{x}_0) = \frac{1}{4\pi r} \pm \frac{1}{4\pi r_{im}}, \qquad (2.2.11)$$

where $r = |\mathbf{x} - \mathbf{x}_0|$, $r_{im} = |\mathbf{x} - \mathbf{x}_0^{im}|$, and

$$\mathbf{x}_0^{im} = (2w - x_0, y_0, z_0) \qquad (2.2.12)$$

is the image of the pole, $\mathbf{x}_0$, with respect to the wall. The minus and plus signs apply, respectively, for the Green's function of the first or second kind.

Interior and exterior of a sphere

The Green's functions of the first kind for a domain bounded internally or externally by a spherical surface of radius a centered at the point $\mathbf{x}_c$ is given by

$$\mathcal{G}(\mathbf{x}, \mathbf{x}_0) = \frac{1}{4\pi r} - \frac{1}{4\pi} \frac{a}{|\mathbf{x}_0 - \mathbf{x}_c|} \frac{1}{r_{inv}}, \qquad (2.2.13)$$

where $r = |\mathbf{x} - \mathbf{x}_0|$, $r_{inv} = |\mathbf{x} - \mathbf{x}_0^{inv}|$, and $\mathbf{x}_0^{inv}$ is the inverse of the singular point $\mathbf{x}_0$ with respect to the sphere located at

$$\mathbf{x}_0^{inv} = \mathbf{x}_c + \frac{a^2}{|\mathbf{x}_0 - \mathbf{x}_c|^2} (\mathbf{x}_0 - \mathbf{x}_c). \qquad (2.2.14)$$

By construction, $r_{inv}/r = a/|\mathbf{x}_0 - \mathbf{x}_c|$, which demonstrates that the Green's function is zero when the point $\mathbf{x}$ lies on the sphere.

The corresponding Green's function of the second kind representing the flow due to a point sink of mass in the presence of an interior impermeable sphere will be derived in Section 7.5.5. A Green's function of the second kind representing the flow inside a sphere cannot be found. The physical reason is that fluid cannot escape through the impermeable boundaries of an interior, completely enclosed domain.

(a) (b)

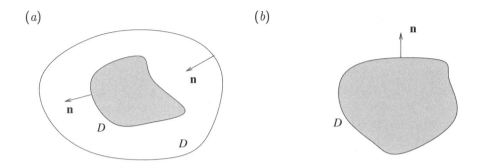

FIGURE 2.2.1 Illustration of (a) a finite control volume in a flow bounded by an interior and an exterior closed surface and (b) an infinite control volume bounded by an interior closed surface.

2.2.4 Integral properties of Green's functions

Consider a singly or multiply connected control volume, V_c, bounded by one surface or a collection of surfaces denoted by D, as illustrated in Figure 2.2.1(a). The boundary associated with the Green's function, S_B, is one of these surfaces. Integrating (2.2.6) over V_c and using the divergence theorem and the properties of the delta function, we find that $\mathcal{G}$ satisfies the integral constraint

$$\iint_D \mathbf{n}(\mathbf{x}) \cdot \nabla \mathcal{G}(\mathbf{x}, \mathbf{x}_0) \, dS(\mathbf{x}) = \begin{cases} 0 & \text{when } \mathbf{x}_0 \text{ is outside } V_c, \\ \frac{1}{2} & \text{when } \mathbf{x}_0 \text{ is on } D, \\ 1 & \text{when } \mathbf{x}_0 \text{ is inside } V_c, \end{cases} \qquad (2.2.15)$$

where the normal unit vector, $\mathbf{n}$, points into the control volume, V_c. When the point $\mathbf{x}_0$ is located precisely on the boundary, D, the improper but convergent integral on the left-hand side of (2.2.15) is called a principal value integral (PV). Using relations (2.2.15), we derive the identity

$$\iint_D \mathbf{n}(\mathbf{x}) \cdot \nabla \mathcal{G}(\mathbf{x}, \mathbf{x}_0) \, dS(\mathbf{x}) = \iint_D^{PV} \mathbf{n}(\mathbf{x}) \cdot \nabla \mathcal{G}(\mathbf{x}, \mathbf{x}_0) \, dS(\mathbf{x}) \pm \frac{1}{2}, \qquad (2.2.16)$$

where plus or minus sign on the left-hand side applies, respectively, when the point $\mathbf{x}_0$ lies inside or outside the control volume.

2.2.5 Symmetry of Green's functions

We return to Green's second identity (2.2.3) and identify the functions f and g with a Green's function, $\mathcal{G}$, whose singular point is located, respectively, at the points $\mathbf{x}_1$ and $\mathbf{x}_2$, so that $f = \mathcal{G}(\mathbf{x}, \mathbf{x}_1)$ and $g = \mathcal{G}(\mathbf{x}, \mathbf{x}_2)$. Using the definition (2.2.6), we obtain

$$-\mathcal{G}(\mathbf{x}, \mathbf{x}_1)\, \delta(\mathbf{x} - \mathbf{x}_2) + \mathcal{G}(\mathbf{x}, \mathbf{x}_2)\, \delta(\mathbf{x} - \mathbf{x}_1) = \nabla \cdot \left[\mathcal{G}(\mathbf{x}, \mathbf{x}_1)\, \nabla \mathcal{G}(\mathbf{x}, \mathbf{x}_2) - \mathcal{G}(\mathbf{x}, \mathbf{x}_2)\, \nabla \mathcal{G}(\mathbf{x}, \mathbf{x}_1) \right]. \quad (2.2.17)$$

Next, we integrate (2.2.17) over a control volume, V_c, that is bounded by a surface associated with the Green's function, S_B. In the case of infinite flow, V_c is also bounded by an outer surface with large dimensions, S_∞. Using the divergence theorem, we convert the volume integral on the right-hand side to a surface integral, invoke the distinguishing properties of the delta function to write, for example,

$$\iiint_{V_c} \mathcal{G}(\mathbf{x}, \mathbf{x}_1)\, \delta_3(\mathbf{x} - \mathbf{x}_2)\, \mathrm{d}V(\mathbf{x}) = \mathcal{G}(\mathbf{x}_2, \mathbf{x}_1), \tag{2.2.18}$$

and find that

$$\mathcal{G}(\mathbf{x}_2, \mathbf{x}_1) - \mathcal{G}(\mathbf{x}_1, \mathbf{x}_2) = \iint_{S_B, S_\infty} [\mathcal{G}(\mathbf{x}, \mathbf{x}_1)\, \nabla\mathcal{G}(\mathbf{x}, \mathbf{x}_2) - \mathcal{G}(\mathbf{x}, \mathbf{x}_2)\, \nabla\mathcal{G}(\mathbf{x}, \mathbf{x}_1)] \cdot \mathbf{n}(\mathbf{x})\, \mathrm{d}S(\mathbf{x}). \tag{2.2.19}$$

Since either the Green's function itself or its normal derivative is zero over S_B, the integral over S_B on the right-hand side disappears. In the case of infinite flow, we let the large surface S_∞ expand to infinity and find that, because the integrand decays at a rate that is faster than inverse quadratic, the corresponding integrals make vanishing contributions. We conclude that a Green's function of the first or second kind (Neumann function) satisfy the symmetry property

$$\mathcal{G}(\mathbf{x}_2, \mathbf{x}_1) = \mathcal{G}(\mathbf{x}_1, \mathbf{x}_2), \tag{2.2.20}$$

which allows us to switch the observation point and the pole.

Physical interpretation

In physical terms, equation (2.2.20) states that the temperature or velocity potential at a point, $\mathbf{x}_2$, due to a point source of heat or point sink of mass located at another point, $\mathbf{x}_1$, is equal to the temperature or velocity potential at the point $\mathbf{x}_1$ due to a corresponding singularity located at $\mathbf{x}_2$. In Chapter 10, we will see that Green's functions appear as kernels in integral equations for the boundary distribution of the harmonic potential or its normal derivatives, and the symmetry property (2.2.20) has important implications on the properties of the solution.

One noteworthy consequence of (2.2.20) is that, when the pole of a Green's function of the first kind is placed at the boundary S_B where the Green's function is required to vanish, the Green's function is identically zero, that is, $\mathcal{G}(\mathbf{x}, \mathbf{x}_0) = 0$ when $\mathbf{x}_0$ is on S_B for any $\mathbf{x}$. This behavior can be understood in physical terms by identifying the Green's function with the temperature field established when a point source of heat is placed on an isothermal body. Because the heat of the point source is immediately absorbed by the body, a temperature field is not established.

2.2.6 Green's functions with multiple poles

Adding N Green's functions with distinct poles, $\mathbf{x}_i$ for $i = 1, \ldots, N$, we obtain a Green's function with a multitude of poles,

$$\mathcal{G}(\mathbf{x}, \mathbf{x}_1, \ldots, \mathbf{x}_N) = \sum_{i=1}^{N} \mathcal{G}(\mathbf{x}, \mathbf{x}_i). \tag{2.2.21}$$

Physically, this Green's function can be identified with the temperature field due to a collection of point sources of heat or with the velocity potential due to a collection of point sinks of mass.

A periodic Green's function represents the temperature field or harmonic potential due to a simply, doubly, or triply periodic array of point sources of heat or point sinks of mass. In certain cases, a periodic Green's function cannot be computed simply by adding an infinite number of Green's functions with single poles, as the superposition produces divergent sums. Instead, the solution must be found by solving the defining equation

$$\nabla^2 \mathcal{G}(\mathbf{x}, \mathbf{x}_0) + \sum_i \delta_3(\mathbf{x} - \mathbf{x}_i) = 0, \tag{2.2.22}$$

where the sum is computed over the periodic array. The solution can be found using Fourier series expansions.

2.2.7 Multipoles of Green's functions

Differentiating a Green's function with respect to the position of the pole, $\mathbf{x}_0$, we obtain a vectorial singular solution called the Green's function dipole,

$$\mathcal{G}'(\mathbf{x}, \mathbf{x}_0) \equiv \nabla_0 \mathcal{G}(\mathbf{x}, \mathbf{x}_0), \tag{2.2.23}$$

where the zero subscript of the gradient indicates differentiation with respect to $\mathbf{x}_0$. Higher derivatives with respect to the singular point yield high-order singularities that are multipoles of the Green's function. The next three singularities are the quadruple,

$$\mathcal{G}''(\mathbf{x}, \mathbf{x}_0) = \nabla_0 \nabla_0 \mathcal{G}(\mathbf{x}, \mathbf{x}_0), \tag{2.2.24}$$

the sextuple,

$$\mathcal{G}'''(\mathbf{x}, \mathbf{x}_0) = \nabla_0 \nabla_0 \nabla_0 \mathcal{G}(\mathbf{x}, \mathbf{x}_0), \tag{2.2.25}$$

and the octuple,

$$\mathcal{G}''''(\mathbf{x}, \mathbf{x}_0) = \nabla_0 \nabla_0 \nabla_0 \nabla_0 \mathcal{G}(\mathbf{x}, \mathbf{x}_0). \tag{2.2.26}$$

Four indices are afforded by the octuple.

The free-space Green's function dipole and quadruple are given by

$$\mathcal{G}'(\mathbf{x}, \mathbf{x}_0) = \frac{1}{4\pi} \frac{\hat{\mathbf{x}}}{r^3}, \qquad \mathcal{G}''(\mathbf{x}, \mathbf{x}_0) = \frac{1}{4\pi} \left(-\frac{1}{r^3} \mathbf{I} + 3 \frac{\hat{\mathbf{x}} \otimes \hat{\mathbf{x}}}{r^5} \right), \tag{2.2.27}$$

where $\hat{\mathbf{x}} = \mathbf{x} - \mathbf{x}_0$, $r = |\hat{\mathbf{x}}|$, and $\mathbf{I}$ is the identity matrix. Comparing (2.2.27) with the first expressions in (2.1.33) and (2.1.34), we find that the velocity potential and velocity field due to a point-source dipole with strength $\mathbf{d}$ are given by

$$\phi = \mathcal{G}' \cdot \mathbf{d}, \qquad \mathbf{u} = \mathcal{G}'' \cdot \mathbf{d}, \tag{2.2.28}$$

where the constant $\mathbf{d}$ expresses the direction and strength of the dipole. Working in a similar fashion, we find that the velocity potential and velocity field due to a point-source quadruple are given by

$$\phi = \mathcal{G}'' : \mathbf{q}, \qquad \mathbf{u} = \mathcal{G}''' : \mathbf{q}, \tag{2.2.29}$$

where $\mathbf{q}$ is a constant two-index matrix expressing the strength and spatial structure of the quadruple.

2.2.8   Green's functions in two dimensions

Green's identities, the reciprocal theorem, and the apparatus of Green's functions can be extended in a straightforward manner to two-dimensional flow. The Green's functions satisfy the counterpart of equation (2.2.6) in the plane,

$$\nabla^2 \mathcal{G}(\mathbf{x}, \mathbf{x}_0) + \delta_2(\mathbf{x} - \mathbf{x}_0) = 0, \tag{2.2.30}$$

where ∇^2 is the two-dimensional Laplacian and δ_2 is the two-dimensional delta function.

Free-space Green's function

The free-space Green's function is given by

$$\mathcal{G}(\mathbf{x}, \mathbf{x}_0) = -\frac{1}{2\pi} \ln \frac{r}{a}, \tag{2.2.31}$$

where a is a specified constant length. It is important to note that the free-space Green's function increases at a logarithmic rate with respect to distance from the singular point, r. In contrast, its three-dimensional counterpart decays like $1/r$. The two-dimensional Green's function is dimensionless, whereas its three-dimensional counterpart has units of inverse length.

Semi-infinite domain bounded by a plane

Using the method of images, we find that the Green's function for a semi-infinite domain bounded by a plane wall located at $y = w$ is given by

$$\mathcal{G}(\mathbf{x}, \mathbf{x}_0) = -\frac{1}{2\pi} \left(\ln \frac{r}{a} \pm \ln \frac{r_{im}}{a} \right), \tag{2.2.32}$$

where $r = |\mathbf{x} - \mathbf{x}_0|$, $r_{im} = |\mathbf{x} - \mathbf{x}_0^{im}|$, and $\mathbf{x}_0^{im} = (x_0, 2w - y_0)$ is the image of the singular point, $\mathbf{x}_0$, with respect to the wall. The minus or plus sign apply, respectively, for the Green's function of the first or second kind (Neumann function).

Interior and exterior of a circle

The Green's function of the first kind in a domain that is bounded internally or externally by a circle of radius a centered at the point $\mathbf{x}_c$ is given by

$$\mathcal{G}(\mathbf{x}, \mathbf{x}_0) = -\frac{1}{2\pi} \left[\ln \frac{r}{a} + \ln \left(\frac{a}{|\mathbf{x}_0 - \mathbf{x}_c|} \frac{a}{r_{inv}} \right) \right], \tag{2.2.33}$$

where $r = |\mathbf{x} - \mathbf{x}_0|$, $r_{inv} = |\mathbf{x} - \mathbf{x}_0^{inv}|$, and $\mathbf{x}_0^{inv}$ is the inverse of the singular point $\mathbf{x}_0$ with respect to the circle located at

$$\mathbf{x}_0^{inv} = \mathbf{x}_c + \frac{a^2}{|\mathbf{x}_0 - \mathbf{x}_c|^2} (\mathbf{x}_0 - \mathbf{x}_c). \tag{2.2.34}$$

By construction, $r_{inv}/r = a/|\mathbf{x}_0 - \mathbf{x}_c|$, which shows that the Green's function is zero when the field point $\mathbf{x}$ lies on the circle.

The Green's function of the second kind representing the flow due to a point sink of mass in the presence of an interior circular boundary will be derived in Section 7.8. The Green's function of the second kind for flow in the interior of a circular boundary cannot be found.

Further properties of Green's functions

The reciprocal relation and identities (2.2.15) and (2.2.16) are also valid in two dimensions, provided that the control volume, V_c, is replaced by a control area, A_c, and the boundary, D, is replaced by a contour, C, enclosing A_c. The two-dimensional Green's functions satisfy the symmetry property (2.2.20). The proof is carried out working as in the case of three-dimensional flow.

Infinite flow

An apparent complication is encountered in the case of infinite two-dimensional flow. In the limit as the contour C_∞ expands to infinity, the Green's function may increase at a logarithmic rate, and the integrals of the two terms in (2.2.19) over the large contour, C_∞, which is the counterpart of the large surface S_∞ in the three-dimensional flow, may not vanish. However, expanding the Green's functions inside the integrand of (2.2.19) in a Taylor series with respect to $\mathbf{x}_0$ about the origin, we find that the sum of the two integrals makes a vanishing contribution and the combined integral over C_∞ cancels out.

Problems

2.2.1 *Free-space Green's function*

(*a*) Working in index notation, confirm that the free-space Green's function satisfies Laplace's equation everywhere except at the pole.

(*b*) Identify D with a spherical surface centered at the pole, $\mathbf{x}_0$, and show that the free-space Green's function satisfies the integral constraint (2.2.15).

2.2.2 *Solution of Poisson's equation*

Use the distinguishing properties of the delta function to show that the general solution of the Poisson's equation (2.1.12) is

$$\phi(\mathbf{x}) = - \iiint_{Flow} \mathcal{G}(\mathbf{x}, \mathbf{x}') \nabla' \cdot \mathbf{u}(\mathbf{x}') \, dV(\mathbf{x}') + H(\mathbf{x}), \tag{2.2.35}$$

where $\mathcal{G}$ is a Green's function and H is a nonsingular harmonic function.

2.2.3 *Symmetry of Green's functions*

(*a*) Verify that the Green's functions given in (2.2.11) and (2.2.13) satisfy the symmetry property (2.2.20).

(*b*) Discuss whether (2.2.20) implies that

$$\nabla \mathcal{G}(\mathbf{x}, \mathbf{x}_0) = \nabla \mathcal{G}(\mathbf{x}_0, \mathbf{x}) \quad \text{or} \quad \nabla \mathcal{G}(\mathbf{x}, \mathbf{x}_0) = \nabla_0 \mathcal{G}(\mathbf{x}, \mathbf{x}_0). \tag{2.2.36}$$

The gradient, ∇_0, involves derivatives with respect to $\mathbf{x}_0$.

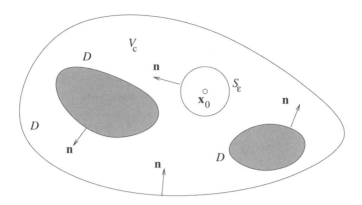

FIGURE 2.3.1 Illustration of a control volume used to derive a boundary-integral representation of the potential of an irrotational flow at a point, $\mathbf{x}_0$.

2.2.4 *Green's function sextuple*

(*a*) Derive the three-dimensional free-space Green's function sextuple and discuss its physical interpretation in terms of merged point sources and point sinks.

(*b*) Repeat (*a*) for the two-dimensional sextuple.

2.3   Integral representation of three-dimensional potential flow

Having introduced the reciprocal theorem and the Green's functions of Laplace's equation, we proceed to develop integral representations for the velocity potential of an irrotational, incompressible or compressible flow. We begin by considering a three-dimensional flow of an incompressible fluid in a simply connected domain. Applying the reciprocal relation (2.2.4) with a nonsingular single-valued harmonic potential ϕ in place of f and a Green's function of Laplace's equation, $\mathcal{G}(\mathbf{x}, \mathbf{x}_0)$, in place of g, we obtain

$$\nabla \cdot [\, \phi(\mathbf{x})\, \nabla \mathcal{G}(\mathbf{x}, \mathbf{x}_0) - \mathcal{G}(\mathbf{x}, \mathbf{x}_0)\, \nabla \phi(\mathbf{x}) \,] = 0. \tag{2.3.1}$$

Next, we select a control volume, V_c, that is bounded by a collection of surfaces, D, as illustrated in Figure 2.3.1 and consider two cases.

Evaluation outside a control volume

When the singular point of the Green's function, $\mathbf{x}_0$, is located outside V_c, the left-hand side of (2.3.1) is nonsingular throughout V_c. Repeating the procedure that led us from (2.2.17) to (2.2.19), we obtain

$$\iint_D \phi(\mathbf{x})\, \mathbf{n}(\mathbf{x}) \cdot \nabla \mathcal{G}(\mathbf{x}, \mathbf{x}_0)\, \mathrm{d}S(\mathbf{x}) = \iint_D \mathcal{G}(\mathbf{x}, \mathbf{x}_0)\, \mathbf{n}(\mathbf{x}) \cdot \nabla \phi(\mathbf{x})\, \mathrm{d}S(\mathbf{x}). \tag{2.3.2}$$

By convention, the normal unit vector, $\mathbf{n}$, points into the control volume.

Evaluation inside a control volume

When the singular point, $\mathbf{x}_0$, resides inside V_c, the left-hand side of (2.3.1) becomes infinite at the point $\mathbf{x}_0$. To apply the divergence theorem, we exclude from the control volume a small spherical volume of radius ϵ centered at $\mathbf{x}_0$, denoted by S_ϵ, as shown in Figure 2.3.1. The result is an equation that is identical to (2.3.2) except that the boundaries of the control volume include the spherical surface S_ϵ,

$$\iint_{D,S_\epsilon} \phi(\mathbf{x})\, \mathbf{n}(\mathbf{x}) \cdot \nabla \mathcal{G}(\mathbf{x}, \mathbf{x}_0)\, dS(\mathbf{x}) = \iint_{D,S_\epsilon} \mathcal{G}(\mathbf{x}, \mathbf{x}_0)\, \mathbf{n}(\mathbf{x}) \cdot \nabla \phi(\mathbf{x})\, dS(\mathbf{x}). \qquad (2.3.3)$$

By convention, the normal unit vector, $\mathbf{n}$, points into the control volume. Considering the integrals over S_ϵ, we write $dS = \epsilon^2 d\Omega$, where Ω is the solid angle defined as the area of a sphere of unit radius centered at $\mathbf{x}_0$. Using (2.2.8), we obtain

$$\nabla \mathcal{G}(\mathbf{x}, \mathbf{x}_0) = -\frac{1}{4\pi\epsilon^2}\, \mathbf{n}(\mathbf{x}) + \nabla \mathcal{H}(\mathbf{x}, \mathbf{x}_0), \qquad (2.3.4)$$

where $\mathbf{n} = \frac{1}{\epsilon}(\mathbf{x} - \mathbf{x}_0)$ is the normal unit vector pointing into the control volume, as shown in Figure 2.3.1. Taking the limit as ϵ tends to zero, using (2.2.8) and (2.3.4) and recalling that the complementary component $\mathcal{H}$ is nonsingular throughout the domain of flow, we find that

$$\iint_{S_\epsilon} \mathcal{G}(\mathbf{x}, \mathbf{x}_0)\, \mathbf{n}(\mathbf{x}) \cdot \nabla \phi(\mathbf{x})\, dS(\mathbf{x}) = \frac{1}{4\pi\epsilon} \iint_{S_\epsilon} \mathbf{n}(\mathbf{x}) \cdot \nabla \phi(\mathbf{x})\, \epsilon^2\, d\Omega(\mathbf{x}) \rightarrow 0 \qquad (2.3.5)$$

and

$$\iint_{S_\epsilon} \phi(\mathbf{x})\, \mathbf{n}(\mathbf{x}) \cdot \nabla \mathcal{G}(\mathbf{x}, \mathbf{x}_0)\, dS(\mathbf{x}) \rightarrow -\phi(\mathbf{x}_0)\, \frac{1}{4\pi\epsilon^2} \iint_{S_\epsilon} \epsilon^2\, d\Omega(\mathbf{x}) = -\phi(\mathbf{x}_0). \qquad (2.3.6)$$

Substituting (2.3.5) and (2.3.6) into (2.3.3), we obtain the final result

$$\phi(\mathbf{x}_0) = -\iint_D \mathcal{G}(\mathbf{x}, \mathbf{x}_0)\, \mathbf{n}(\mathbf{x}) \cdot \nabla \phi(\mathbf{x})\, dS(\mathbf{x}) + \iint_D \phi(\mathbf{x})\, \mathbf{n}(\mathbf{x}) \cdot \nabla \mathcal{G}(\mathbf{x}, \mathbf{x}_0)\, dS(\mathbf{x}), \qquad (2.3.7)$$

which provides us with a boundary-integral representation of a harmonic function in terms of the boundary values and the boundary distribution of its normal derivative, which is equal to the normal component of the velocity. To compute the value of ϕ at a chosen point, $\mathbf{x}_0$, inside a selected control volume, we simply compute the two boundary integrals on the right-hand side of (2.3.7) by analytical or numerical methods.

The integral representation (2.3.7) can be derived directly using the properties of the delta function, following the procedure that led us from (2.2.17) to (2.2.19). However, the present derivation based on the exclusion of a small sphere centered at the evaluation point bypasses the use of generalized functions.

Physical interpretation

The symmetry property (2.2.20) allows us to switch the order of the arguments of the Green's function in the integral representation (2.3.7), obtaining

$$\phi(\mathbf{x}_0) = -\iint_D \mathcal{G}(\mathbf{x}_0, \mathbf{x})\, \mathbf{n}(\mathbf{x}) \cdot \nabla \phi(\mathbf{x})\, dS(\mathbf{x}) + \iint_D \phi(\mathbf{x})\, \mathbf{n}(\mathbf{x}) \cdot \nabla \mathcal{G}(\mathbf{x}_0, \mathbf{x})\, dS(\mathbf{x}). \qquad (2.3.8)$$

The two integrals on the right-hand side represent boundary distributions of Green's functions and Green's function dipoles oriented normal to the boundaries of the control volume, amounting to boundary distributions of point sources and point-source dipoles. Making an analogy with corresponding results in the theory of electrostatics concerning the surface distribution of electric charges and charge dipoles, we call the first integral in (2.3.8) the single-layer harmonic potential and the second integral the double-layer harmonic potential. The distribution densities of these potentials are equal, respectively, to the normal derivative and to the boundary values of the harmonic potential.

Boundary-integral representation for the velocity

Taking the gradient of (2.3.8) with respect to the evaluation point, $\mathbf{x}_0$, we derive a boundary-integral representation of the velocity,

$$\mathbf{u}(\mathbf{x}_0) = - \iint_D \nabla_0 \mathcal{G}(\mathbf{x}_0, \mathbf{x}) \, \mathbf{n}(\mathbf{x}) \cdot \nabla \phi(\mathbf{x}) \, \mathrm{d}S(\mathbf{x}) + \iint_D \phi(\mathbf{x}) \, \mathbf{n}(\mathbf{x}) \cdot [\nabla \nabla_0 \mathcal{G}(\mathbf{x}_0, \mathbf{x})] \, \mathrm{d}S(\mathbf{x}). \quad (2.3.9)$$

The integrals on the right-hand side represent the velocity due to boundary distributions of Green's functions and Green's function dipoles.

Green's third identity

Applying the boundary-integral representation (2.3.8) with the free-space Green's function given in (2.2.7), we derive Green's third identity

$$\phi(\mathbf{x}_0) = -\frac{1}{4\pi} \iint_D \frac{1}{r} \mathbf{n}(\mathbf{x}) \cdot \nabla \phi(\mathbf{x}) \, \mathrm{d}S(\mathbf{x}) + \frac{1}{4\pi} \iint_D \frac{\phi(\mathbf{x})}{r^3} \mathbf{n}(\mathbf{x}) \cdot (\mathbf{x}_0 - \mathbf{x}) \, \mathrm{d}S(\mathbf{x}), \quad (2.3.10)$$

where $r = |\mathbf{x} - \mathbf{x}_0|$ is the distance of the evaluation point, $\mathbf{x}_0$, from the integration point, $\mathbf{x}$.

2.3.1   Unbounded flow decaying at infinity

Next, we consider a flow in an infinite domain enclosed by one interior boundary or a collection of closed interior boundaries, B, called a periphractic domain from the Greek words $\pi\epsilon\rho\iota$ which means "about" and $\varphi\rho\alpha\kappa\tau\eta\varsigma$ which means "fence," subject to the assumption that the velocity decays at infinity. We will show that the velocity potential tends to a constant value at infinity. We begin by selecting a point $\mathbf{x}_0$ inside the domain of flow and define a control volume that is enclosed by the interior boundaries, B, and a spherical surface of large radius R centered at the point $\mathbf{x}_0$, denoted by S_∞. Applying Green's third identity (2.3.10), we obtain

$$\phi(\mathbf{x}_0) = \frac{Q}{4\pi R} + \bar{\phi}_\infty(R, \mathbf{x}_0) - \frac{1}{4\pi} \iint_B \frac{1}{r} \mathbf{n}(\mathbf{x}) \cdot \nabla \phi(\mathbf{x}) \, \mathrm{d}S(\mathbf{x})$$

$$+ \frac{1}{4\pi} \iint_B \frac{\phi(\mathbf{x})}{r^3} \mathbf{n}(\mathbf{x}) \cdot (\mathbf{x}_0 - \mathbf{x}) \, \mathrm{d}S(\mathbf{x}), \quad (2.3.11)$$

where Q is the flow rate across S_∞ or any other closed surface enclosed by S_∞, defined as

$$Q = - \iint_{S_\infty} \mathbf{n}(\mathbf{x}) \cdot \nabla \phi(\mathbf{x}) \, \mathrm{d}S(\mathbf{x}) = \iint_B \mathbf{n}(\mathbf{x}) \cdot \nabla \phi(\mathbf{x}) \, \mathrm{d}S(\mathbf{x}), \quad (2.3.12)$$

and $\bar{\phi}_\infty$ is the average value of ϕ over S_∞,

$$\bar{\phi}_\infty(R,\mathbf{x}_0) \equiv \frac{1}{4\pi R^2} \iint_{S_\infty} \phi(\mathbf{x})\,dS(\mathbf{x}). \tag{2.3.13}$$

The normal unit vector, $\mathbf{n}$, points inward over S_∞. Concentrating on the flow rate, we write

$$Q = \iint_{S_\infty}\left(\frac{\partial\phi}{\partial r}\right)_{r=R} dS(\mathbf{x}) = \iint_{S_\infty}\left(\frac{\partial\phi}{\partial r}\right)_{r=R} R^2\,d\Omega(\mathbf{x}) = R^2\frac{d}{dR}\iint_{S_\infty}\phi\,d\Omega(\mathbf{x}), \tag{2.3.14}$$

where Ω is the solid angle defined as the surface area of a sphere of unit radius centered at the point $\mathbf{x}_0$. Substituting the definition of $\bar{\phi}_\infty$, we obtain

$$\frac{Q}{4\pi R^2} = \frac{\partial\bar{\phi}_\infty}{\partial R}. \tag{2.3.15}$$

Integrating this equation with respect to R treating Q as a constant, we obtain

$$\bar{\phi}_\infty(R,\mathbf{x}_0) = -\frac{Q}{4\pi R} + c(\mathbf{x}_0), \tag{2.3.16}$$

where c is independent of R. Taking derivatives of the last equation with respect to $\mathbf{x}_0$ keeping R fixed, we find that

$$\nabla_0 c = \nabla_0\bar{\phi}_\infty = \nabla_0 \iint_{S_\infty}\phi(\mathbf{x})\,d\Omega(\mathbf{x}) = \iint_{S_\infty}\nabla\phi(\mathbf{x})\,d\Omega(\mathbf{x}) = 0. \tag{2.3.17}$$

The value of zero emerges by letting R in the last integral tend to infinity and invoking the original assumption that the velocity decays at infinity. We have thus shown that c is an absolute constant. Substituting (2.3.15) into (2.3.11), we derive a simplified boundary integral representation lacking the integral over the large surface,

$$\phi(\mathbf{x}_0) = c - \frac{1}{4\pi}\iint_B \frac{1}{r}\mathbf{n}(\mathbf{x})\cdot\nabla\phi(\mathbf{x})\,dS(\mathbf{x}) + \frac{1}{4\pi}\iint_B \frac{\phi(\mathbf{x})}{r^3}\mathbf{n}(\mathbf{x})\cdot(\mathbf{x}_0-\mathbf{x})\,dS(\mathbf{x}). \tag{2.3.18}$$

This equation finds useful applications in computing an exterior flow.

Letting the point $\mathbf{x}_0$ in (2.3.18) tend to infinity and recalling that r is the distance of the evaluation point, $\mathbf{x}_0$, from the integration point, $\mathbf{x}$, over the interior boundary B, we deduce the asymptotic behavior

$$\phi(\mathbf{x}_0) = c - \frac{Q}{4\pi|\mathbf{x}_0|} + \cdots, \tag{2.3.19}$$

where the dots represent decaying terms. This expression demonstrates that the potential of an infinite three-dimensional flow that decays at infinity tends to a constant value far from the interior boundaries.

Based on (2.3.19), we find that the expression for the kinetic energy (2.1.22) becomes

$$\mathcal{K} = -\frac{1}{2}\rho\iint_B \phi\,\mathbf{u}\cdot\mathbf{n}\,dS + \frac{1}{2}\rho c Q, \tag{2.3.20}$$

where B is an interior boundary. Following the arguments of Section 2.1, we deduce that, given the boundary distribution of the normal component of the velocity, a potential flow that decays at infinity is unique and the corresponding harmonic potential is determined uniquely up to an arbitrary constant.

2.3.2 Simplified boundary-integral representations

The boundary-integral representation (2.3.8) can be simplified by judiciously reducing the domain of integration of the hydrodynamic potentials. This is accomplished by employing Green's functions that are designed to observe the geometry, symmetry, or periodicity of a flow. For example, if the Green's function or its normal derivative vanishes over a particular boundary, the corresponding single- or double-layer potential is identically zero. If the velocity and harmonic potential are periodic in one, two, or three directions, it is beneficial to use a Green's function that observes the periodicity of the flow so that the integrals over the periodic boundaries enclosing one period cancel each other and do not appear in the final boundary-integral representation.

2.3.3 Poisson integral for a spherical boundary

The Green's function of the first kind for flow inside or outside a sphere of radius a centered at the origin arises from (2.2.13) by setting $\mathbf{x}_c = \mathbf{0}$, finding

$$\mathcal{G}(\mathbf{x}, \mathbf{x}_0) = \frac{1}{4\pi} \left(\frac{1}{r} - \frac{a}{|\mathbf{x}_0|} \frac{1}{r_{inv}} \right), \tag{2.3.21}$$

where $r = |\mathbf{x} - \mathbf{x}_0|$, $r_{inv} = |\mathbf{x} - \mathbf{x}_0^{inv}|$, and $\mathbf{x}_0^{inv}$ is the inverse of the point $\mathbf{x}_0$ with respect to the sphere, located at $\mathbf{x}_0^{inv} = \mathbf{x}_0 \, a^2/|\mathbf{x}_0|^2$. The gradient of the Green's function is

$$\nabla \mathcal{G}(\mathbf{x}, \mathbf{x}_0) = -\frac{1}{4\pi} \left(\frac{\mathbf{x} - \mathbf{x}_0}{r^3} - \frac{a}{|\mathbf{x}_0|} \frac{\mathbf{x} - \mathbf{x}_0^{inv}}{r_{inv}^3} \right). \tag{2.3.22}$$

Evaluating this expression at a point $\mathbf{x}$ on the sphere and projecting the result onto the inward normal unit vector, $\mathbf{n} = -\frac{1}{a}\mathbf{x}$, we obtain

$$\mathbf{n}(\mathbf{x}) \cdot \nabla G(\mathbf{x}, \mathbf{x}_0) = \frac{1}{4\pi a} \left(\frac{1}{r^3}(a^2 - \mathbf{x}_0 \cdot \mathbf{x}) - \frac{a}{|\mathbf{x}_0|} \frac{1}{r_{inv}^3}(a^2 - \mathbf{x}_0^{inv} \cdot \mathbf{x}) \right). \tag{2.3.23}$$

Recalling that $r_{inv}/r = a/|\mathbf{x}_0|$ for a point on the sphere and simplifying, we find that

$$\mathbf{n}(\mathbf{x}) \cdot \nabla \mathcal{G}(\mathbf{x}, \mathbf{x}_0) = \frac{1}{4\pi a r^3} \left[(a^2 - \mathbf{x}_0 \cdot \mathbf{x}) - \frac{|\mathbf{x}_0|^2}{a^2} \left(a^2 - \frac{a^2}{|\mathbf{x}_0|^2} \mathbf{x}_0 \cdot \mathbf{x} \right) \right] = \frac{a^2 - |\mathbf{x}_0|^2}{4\pi a r^3}. \tag{2.3.24}$$

Substituting this expression into the first integral on the right-hand side of (2.5.1) and noting that the second integral vanishes because the Green's function is zero on the spherical boundary, we find that the potential at a point $\mathbf{x}_0$ located in the interior or exterior of a sphere of radius a centered at the origin is given by the Poisson integral

$$\phi(\mathbf{x}_0) = \pm \frac{a^2 - |\mathbf{x}_0|^2}{4\pi a} \iint_\circ \frac{\phi(\mathbf{x})}{|\mathbf{x} - \mathbf{x}_0|^3} \, dS(\mathbf{x}), \tag{2.3.25}$$

where $\circ$ denotes the sphere. The plus or minus sign applies when the evaluation point, $\mathbf{x}_0$, is located inside or outside the sphere (e.g., [204], p. 240). Applying this equation for constant interior ϕ provides us with an interesting identity.

2.3.4   Compressible fluids

To derive an integral representation for compressible flow, we apply Green's second identity (2.2.4) with a general potential ϕ in place of f and a Green's function $\mathcal{G}$ in place of g, finding

$$-\mathcal{G}(\mathbf{x}, \mathbf{x}_0)\,\nabla^2\phi(\mathbf{x}) = \nabla \cdot [\,\phi(\mathbf{x})\,\nabla\mathcal{G}(\mathbf{x}, \mathbf{x}_0) - \mathcal{G}(\mathbf{x}, \mathbf{x}_0)\,\nabla\phi(\mathbf{x})\,]. \tag{2.3.26}$$

Following the procedure discussed earlier in this section for incompressible flow, we derive the integral representation

$$\phi(\mathbf{x}_0) = -\iint_D \mathcal{G}(\mathbf{x}, \mathbf{x}_0)\,\mathbf{n}(\mathbf{x}) \cdot \nabla\phi(\mathbf{x})\,\mathrm{d}S(\mathbf{x})$$
$$+ \iint_D \phi(\mathbf{x})\,\mathbf{n}(\mathbf{x}) \cdot \nabla\mathcal{G}(\mathbf{x}, \mathbf{x}_0)\,\mathrm{d}S(\mathbf{x}) - \iiint_{V_c} \mathcal{G}(\mathbf{x}_0, \mathbf{x})\,\nabla^2\phi(\mathbf{x})\,\mathrm{d}V(\mathbf{x}), \tag{2.3.27}$$

which is identical to (2.3.8) except that the right-hand side includes a volume integral over the control volume V_c enclosed by the boundary D, called the volume potential. Physically, the volume potential represents a volume distribution of point sources.

Adopting the free-space Green's function, we obtain the counterpart of Green's third identity for compressible flow,

$$\phi(\mathbf{x}_0) = -\frac{1}{4\pi}\iint_D \frac{1}{r}\,\mathbf{n}(\mathbf{x}) \cdot \nabla\phi(\mathbf{x})\,\mathrm{d}S(\mathbf{x})$$
$$+ \frac{1}{4\pi}\iint_D \frac{\mathbf{x}_0 - \mathbf{x}}{r^3} \cdot \mathbf{n}(\mathbf{x})\,\phi(\mathbf{x})\,\mathrm{d}S(\mathbf{x}) - \frac{1}{4\pi}\iiint_{V_c} \frac{1}{r}\,\nabla^2\phi(\mathbf{x})\,\mathrm{d}S(\mathbf{x}). \tag{2.3.28}$$

The normal unit vector, $\mathbf{n}$, points into the control volume, V_c.

Problem

2.3.1 *Boundary-integral equation for a uniform potential*

Apply (2.3.2) and (2.3.7) with ϕ set to a constant to derive the first and third equations in (2.2.15).

2.4   Mean-value theorems in three dimensions

An important property of functions that satisfy Laplace's equation emerges by selecting a spherical control volume with radius a residing entirely in their domain of definition and centered at a chosen point, $\mathbf{x}_0$. Identifying the surface D in (2.3.10) with the spherical boundary and substituting $r = a$ and $\mathbf{n} = \frac{1}{a}(\mathbf{x}_0 - \mathbf{x})$, we obtain

$$\phi(\mathbf{x}_0) = -\frac{1}{4\pi a}\iint_\circ \mathbf{n}(\mathbf{x}) \cdot \nabla\phi(\mathbf{x})\,\mathrm{d}S(\mathbf{x}) + \frac{1}{4\pi a^2}\iint_\circ \phi(\mathbf{x})\,\mathrm{d}S(\mathbf{x}), \tag{2.4.1}$$

where $\circ$ denotes the sphere. Next, we use the divergence theorem and recall that $\nabla^2\phi = 0$ to find that the first integral on the right-hand side is zero. The result is a mean-value theorem expressed by the equation

$$\phi(\mathbf{x}_0) = \frac{1}{4\pi a^2}\iint_\circ \phi(\mathbf{x})\,\mathrm{d}S(\mathbf{x}), \tag{2.4.2}$$

stating that the mean value of a function that satisfies Laplace's equation over the surface of a sphere is equal to value of the function at the center of the sphere. One interesting consequence of the mean-value theorem is that, if the harmonic potential of an infinite flow with no interior boundaries decays at infinity, it must vanish throughout the whole space.

Using (2.4.2), we find that the mean value of a harmonic function over the volume of a sphere of radius a is equal to value of the function at the center of the sphere,

$$\frac{1}{V_\circ} \iiint_\circ \phi(\mathbf{x})\, dV(\mathbf{x}) = \frac{1}{V_\circ} \int_0^a \left[\iint_{S_r} \phi(\mathbf{x})\, dS(\mathbf{x}) \right] dr = \frac{1}{V_\circ} \int_0^a 4\pi r^2\, \phi(\mathbf{x}_0)\, dr = \phi(\mathbf{x}_0), \qquad (2.4.3)$$

where $V_\circ = \frac{4\pi}{3} a^3$ is the volume of the sphere and S_r denotes a sphere of radius r. Rearranging, we obtain

$$\phi(\mathbf{x}_0) = \frac{1}{\frac{4\pi}{3} a^3} \iiint_\circ \phi(\mathbf{x})\, dV(\mathbf{x}), \qquad (2.4.4)$$

stating that the mean value of a function that satisfies Laplace's equation over the volume of a sphere is equal to value of the function at the center of the sphere.

Extrema of harmonic functions

The mean-value theorem can be used to show that the minimum or maximum of a nonsingular harmonic function occurs only at the boundaries. To see this, we temporarily assume that an extremum occurs at a point inside the domain of a flow, $\mathbf{x}_0$, and apply the mean-value theorem to find that there must be at least one point on the surface of a sphere centered at the alleged point of extremum where the value of ϕ is higher or lower than $\phi(\mathbf{x}_0)$, so that the mean value of ϕ over the sphere is equal to $\phi(\mathbf{x}_0)$. However, this contradicts the original assumption.

Constant boundary potential

Now we consider a domain that is completely enclosed by a surface over which a harmonic potential ϕ is constant. According to the mean-value theorem, the constant value must be both a minimum and a maximum. Consequently, ϕ must have the same value at every point. This conclusion does not apply for an infinite domain bounded by interior surfaces.

Maximum of the magnitude of the velocity

Another consequence of the mean-value theorem is that, in the absence of singularities, the magnitude of the velocity, $\mathbf{u} = \nabla\phi$, reaches a maximum at the boundaries. To see this, we temporarily assume that the maximum occurs at a point inside the domain of a flow, $\mathbf{x}_0$, and introduce the unit vector in the direction of the local velocity, $\mathbf{e}_0 \equiv \mathbf{u}(\mathbf{x}_0)/|\mathbf{u}(\mathbf{x}_0)|$. The square of the magnitude of the velocity at a point, $\mathbf{x}$, inside the domain of flow is

$$\mathbf{u}(\mathbf{x}) \cdot \mathbf{u}(\mathbf{x}) = |\mathbf{u}(\mathbf{x}) \cdot \mathbf{e}_0|^2 + |(\mathbf{I} - \mathbf{e}_0 \otimes \mathbf{e}_0) \cdot \mathbf{u}(\mathbf{x})|^2. \qquad (2.4.5)$$

Accordingly,

$$\mathbf{u}(\mathbf{x}) \cdot \mathbf{u}(\mathbf{x}) \geq |\mathbf{u}(\mathbf{x}) \cdot \mathbf{e}_0|^2, \qquad (2.4.6)$$

with the equality holding when $\mathbf{x} = \mathbf{x}_0$. However, since the positive scalar function $|\mathbf{u}(\mathbf{x}) \cdot \mathbf{e}_0|$ satisfies Laplace's equation, a point $\mathbf{x}$ can be found such that

$$|\mathbf{u}(\mathbf{x}) \cdot \mathbf{e}_0|^2 > |\mathbf{u}(\mathbf{x}_0) \cdot \mathbf{e}_0|^2 = \mathbf{u}(\mathbf{x}_0) \cdot \mathbf{u}(\mathbf{x}_0). \tag{2.4.7}$$

Combining the last two inequalities, we find that

$$\mathbf{u}(\mathbf{x}) \cdot \mathbf{u}(\mathbf{x}) > \mathbf{u}(\mathbf{x}_0) \cdot \mathbf{u}(\mathbf{x}_0), \tag{2.4.8}$$

which contradicts the original assumption. One corollary of this result is that, if the velocity of an infinite irrotational flow with no interior boundaries vanishes at infinity, it must vanish throughout the whole space.

Mean-value theorem for singular functions

The mean-value theorem applies for a harmonic function that is free of singularities inside and over the surface of a sphere of radius a centered at a chosen point, $\mathbf{x}_0$. Consider the potential $\phi(\mathbf{x}) = \mathcal{G}(\mathbf{x}, \mathbf{x}_1)$, where $\mathcal{G}$ is a Green's function of Laplace's equation and the singular point $\mathbf{x}_1$ lies inside a sphere. The counterpart of equation (2.4.1) stemming from the reciprocal theorem is

$$\mathcal{G}(\mathbf{x}_0, \mathbf{x}_1) - \frac{1}{4\pi|\mathbf{x}_1 - \mathbf{x}_0|} = -\frac{1}{4\pi a} \iint_\circ \mathbf{n}(\mathbf{x}) \cdot \nabla\mathcal{G}(\mathbf{x}, \mathbf{x}_1) \, dS(\mathbf{x}) + \frac{1}{4\pi a^2} \iint_\circ \mathcal{G}(\mathbf{x}, \mathbf{x}_1) \, dS(\mathbf{x}), \tag{2.4.9}$$

where the normal unit vector, $\mathbf{n}$, points outward from the sphere. The first integral on the right-hand side is equal to -1. Rearranging, we obtain

$$\frac{1}{4\pi a^2} \iint_\circ \mathcal{G}(\mathbf{x}, \mathbf{x}_1) \, dS(\mathbf{x}) = \mathcal{G}(\mathbf{x}_0, \mathbf{x}_1) - \frac{1}{4\pi|\mathbf{x}_1 - \mathbf{x}_0|} + \frac{1}{4\pi a}. \tag{2.4.10}$$

If $\mathcal{G}$ is the free-space Green's function,

$$\frac{1}{4\pi a^2} \iint_\circ \mathcal{G}(\mathbf{x}, \mathbf{x}_1) \, dS(\mathbf{x}) = \frac{1}{4\pi a} \tag{2.4.11}$$

for any point $\mathbf{x}_1$ inside the sphere.

Biharmonic functions

Next, we derive a mean-value theorem for a function, Φ, that satisfies the biharmonic equation, $\nabla^4 \Phi = 0$. Working as in the case of harmonic functions, we consider a spherical control volume of radius a centered at a chosen point, $\mathbf{x}_0$. Identifying D with the spherical boundary of the control volume and substituting in (2.3.28) $r = a$, $\mathbf{n} = \frac{1}{a}(\mathbf{x}_0 - \mathbf{x})$, and $f = \Phi$, we obtain

$$\Phi(\mathbf{x}_0) = -\frac{1}{4\pi a} \iint_\circ \mathbf{n}(\mathbf{x}) \cdot \nabla\Phi(\mathbf{x}) \, dS(\mathbf{x}) + \frac{1}{4\pi a^2} \iint_\circ \Phi(\mathbf{x}) \, dS(\mathbf{x}) - \frac{1}{4\pi a} \iiint_\circ \nabla^2 \Phi(\mathbf{x}) \, dS(\mathbf{x}). \tag{2.4.12}$$

Applying the divergence theorem, we find that the first surface integral on the right-hand side is equal to the negative of the integral of the Laplacian $\nabla^2 \Phi$ over the volume of the sphere. We note that the function $\nabla^2\Phi$ satisfies Laplace's equation and use the mean-value theorem expressed by (2.4.4) to replace its volume integral over the sphere with $\frac{4\pi}{3}a^3\nabla^2\Phi(\mathbf{x}_0)$, obtaining

$$\Phi(\mathbf{x}_0) = -\frac{1}{3}a^2\nabla^2\Phi(\mathbf{x}_0) + \frac{1}{4\pi a^2}\iint_\circ \Phi(\mathbf{x})\,\mathrm{d}S(\mathbf{x}) - \frac{1}{4\pi a}\iiint_\circ \nabla^2\Phi(\mathbf{x})\,\mathrm{d}S(\mathbf{x}). \qquad (2.4.13)$$

Finally, we use the mean-value theorem expressed by (2.4.4) for the harmonic function $\nabla^2\Phi$ to simplify the last integral, and rearrange to derive the mean-value theorem expressed by

$$\frac{1}{4\pi a^2}\iint_\circ \Phi(\mathbf{x})\,\mathrm{d}S(\mathbf{x}) = \Phi(\mathbf{x}_0) + \frac{1}{6}a^2\nabla^2\Phi(\mathbf{x}_0), \qquad (2.4.14)$$

where $\circ$ denotes a sphere. This equation relates the mean value of a biharmonic function over the surface of a sphere to the value of the function and its Laplacian at the center of the sphere.

Working as in (2.4.4), we find that the mean value of a biharmonic function over the volume of a sphere is given by

$$\frac{1}{\frac{4\pi}{3}a^3}\iiint_\circ \Phi(\mathbf{x})\,\mathrm{d}V(\mathbf{x}) = \Phi(\mathbf{x}_0) + \frac{1}{10}a^2\nabla^2\Phi(\mathbf{x}_0). \qquad (2.4.15)$$

If Φ happens to satisfy Laplace's equation, $\nabla^2\Phi = 0$, which is inclusive of a biharmonic equation, equations (2.4.14) and (2.4.15) reduce to (2.4.2) and (2.4.4).

Problem

2.4.1 *Mean-value theorem for a singular biharmonic function*

A Green's function of the biharmonic equation in three dimensions satisfies the equation

$$\nabla^4\mathcal{G} + \delta_3(\mathbf{x} - \mathbf{x}_0) = 0, \qquad (2.4.16)$$

where δ_3 is the three-dimensional delta function. Confirm that the free-space Green's function is $\mathcal{G} = r/8\pi$, where $r = |\mathbf{x} - \mathbf{x}_0|$. Derive the counterpart of (2.4.10).

2.5   Two-dimensional potential flow

A boundary-integral representation and further properties of the single-valued harmonic potential of a two-dimensional flow can be derived working as in Section 2.3 for three-dimensional flow.

2.5.1   Boundary-integral representation

Equations (2.3.7) and (2.3.8) are also valid for two-dimensional flow, provided that the control volume is replaced by a control area, and the boundary, D, is replaced by a closed contour, C, enclosing the control area. The integral representation (2.3.7) takes the form

$$\phi(\mathbf{x}_0) = -\oint_C \mathcal{G}(\mathbf{x}, \mathbf{x}_0)\,\mathbf{n}(\mathbf{x}) \cdot \nabla\phi(\mathbf{x})\,\mathrm{d}l(\mathbf{x}) + \oint_C \phi(\mathbf{x})\,\mathbf{n}(\mathbf{x}) \cdot \nabla\mathcal{G}(\mathbf{x}, \mathbf{x}_0)\,\mathrm{d}l(\mathbf{x}), \qquad (2.5.1)$$

where the normal unit vector, $\mathbf{n}$, points into the control area enclosed by C. Green's third identity for a single-valued potential takes the form

$$\phi(\mathbf{x}_0) = \frac{1}{2\pi} \oint_C \ln\left(\frac{r}{a}\right) \mathbf{n}(\mathbf{x}) \cdot \nabla\phi(\mathbf{x})\, dl(\mathbf{x}) + \frac{1}{2\pi} \oint_C \frac{\phi(\mathbf{x})}{r^2} \mathbf{n}(\mathbf{x}) \cdot (\mathbf{x}_0 - \mathbf{x})\, dl(\mathbf{x}), \tag{2.5.2}$$

where $r = |\mathbf{x} - \mathbf{x}_0|$ and a is a chosen constant length.

Flow in an infinite domain

In the case of flow in an infinite domain, we stipulate that the velocity decays at infinity and derive the counterpart of (2.3.18) with a straightforward change in notation. Letting the point $\mathbf{x}_0$ tend to infinity, we find that the velocity potential behaves like

$$\phi(\mathbf{x}_0) = \frac{Q}{2\pi} \ln \frac{|\mathbf{x}_0|}{a} + c + \cdots, \tag{2.5.3}$$

where c is a constant, $|\mathbf{x}_0|$ is the distance from the origin assumed to be in the vicinity of the boundary C, a is a constant length, and

$$Q = \oint_C \mathbf{n} \cdot \nabla\phi\, dl \tag{2.5.4}$$

is the flow rate across C (Problem 2.5.3). Now we consider the kinetic energy of the flow expressed by the potential

$$\phi - \frac{Q}{2\pi} \ln \frac{|\mathbf{x}_0|}{a}. \tag{2.5.5}$$

Repeating the procedure of Section 2.3 for three-dimensional flow, and find that a two-dimensional flow in an infinite domain described by a single-valued harmonic potential is determined uniquely by specifying the normal component of the velocity along the interior boundaries.

Compressible flow

The integral representation of the potential of a compressible flow stated in (2.3.27) is also valid for two-dimensional flow, provided that the control volume is replaced by a control area, and the boundary, D, is replaced by a closed contour, C, enclosing the control area.

2.5.2 Mean-value theorems

The mean value of a harmonic potential, ϕ, along the perimeter a circle and over the area of a circular disk of radius a centered at a point, $\mathbf{x}_0$, is equal to the value of the potential at the centerpoint,

$$\frac{1}{2\pi a} \oint_\circ \phi(\mathbf{x})\, dl(\mathbf{x}) = \frac{1}{\pi a^2} \iint_\circ \phi(\mathbf{x})\, dA(\mathbf{x}) = \phi(\mathbf{x}_0). \tag{2.5.6}$$

where $\circ$ denotes the circle or the disk inside the circle, l is the arc length around the circle, and $dA = dx\, dy$ is the differential area inside the circle.

Working as in the case of three-dimensional flow, we derive mean-value theorems for functions that satisfy the biharmonic equation, $\nabla^4 \Phi = 0$, stating that

$$\frac{1}{2\pi a} \oint_{\circ} \Phi(\mathbf{x}) \, dl(\mathbf{x}) = \Phi(\mathbf{x}_0) + \frac{1}{4} a^2 \nabla^2 \Phi(\mathbf{x}_0) \tag{2.5.7}$$

where $\circ$ denotes the circle, and

$$\frac{1}{\pi a^2} \iint_{\circ} \Phi(\mathbf{x}) \, dA(\mathbf{x}) = \Phi(\mathbf{x}_0) + \frac{1}{8} a^2 \nabla^2 \Phi(\mathbf{x}_0), \tag{2.5.8}$$

where $\circ$ denotes the disk inside the circle (Problem 2.5.2). If Φ is a harmonic function, the last term on the right-hand side of (2.5.7) or (2.5.8) does not appear.

2.5.3   Poisson integral for a circular boundary

The Green's function of two-dimensional flow inside or outside a circle of radius a centered at the origin arises from (2.2.33) by setting $\mathbf{x}_c = \mathbf{0}$,

$$\mathcal{G}(\mathbf{x}, \mathbf{x}_0) = -\frac{1}{2\pi} \left[\ln \frac{r}{a} + \ln \left(\frac{a}{|\mathbf{x}_0|} \frac{a}{r_{inv}} \right) \right], \tag{2.5.9}$$

where $r = |\mathbf{x} - \mathbf{x}_0|$, $r_{inv} = |\mathbf{x} - \mathbf{x}_0^{inv}|$, and $\mathbf{x}_0^{inv}$ is the inverse point of the point $\mathbf{x}_0$ with respect to the circle located at $\mathbf{x}_0^{inv} = \mathbf{x}_0 \, a^2 / |\mathbf{x}_0|^2$. The gradient of the Green's function is

$$\nabla \mathcal{G}(\mathbf{x}, \mathbf{x}_0) = -\frac{1}{2\pi} \left(\frac{\mathbf{x} - \mathbf{x}_0}{r^2} - \frac{\mathbf{x} - \mathbf{x}_0^{inv}}{r_{inv}^2} \right). \tag{2.5.10}$$

Evaluating this expression at a point $\mathbf{x}$ on the circle and projecting the resulting equation onto the inward normal unit vector, $\mathbf{n} = -\mathbf{x}/a$, we obtain

$$\mathbf{n}(\mathbf{x}) \cdot \nabla \mathcal{G}(\mathbf{x}, \mathbf{x}_0) = \frac{1}{2\pi a} \left[\frac{1}{r^2} (a^2 - \mathbf{x}_0 \cdot \mathbf{x}) - \frac{1}{r_{inv}^2} (a^2 - \mathbf{x}_0^{inv} \cdot \mathbf{x}) \right]. \tag{2.5.11}$$

Recalling that for a point on the circle $r_{inv}/r = a/|\mathbf{x}_0|$ and simplifying, we find that

$$\mathbf{n}(\mathbf{x}) \cdot \nabla \mathcal{G}(\mathbf{x}, \mathbf{x}_0) = \frac{1}{2\pi a r^2} \left[(a^2 - \mathbf{x}_0 \cdot \mathbf{x}) - \frac{|\mathbf{x}_0|^2}{a^2} \left(a^2 - \frac{a^2}{|\mathbf{x}_0|^2} \mathbf{x}_0 \cdot \mathbf{x} \right) \right] = \frac{1}{2\pi a r^2} (a^2 - |\mathbf{x}_0|^2). \tag{2.5.12}$$

Substituting this expression into the first integral on the right-hand side of (2.5.1) and noting that the second integral disappears because the Green's function is zero on the circular boundary, we find that the potential at a point $\mathbf{x}_0$ located in the interior or exterior of a circle of radius a centered at the origin is given by the Poisson integral

$$\phi(\mathbf{x}_0) = \pm \frac{a^2 - |\mathbf{x}_0|^2}{2\pi a} \oint_{\circ} \frac{\phi(\mathbf{x})}{|\mathbf{x} - \mathbf{x}_0|^2} \, dl(\mathbf{x}), \tag{2.5.13}$$

where $\circ$ denotes the circle. The plus or minus sign corresponds, respectively, to interior or exterior flow (e.g., [106], p. 242). Applying this equation for constant interior ϕ provides us with an interesting identity.

Problems

2.5.1 *Two-dimensional infinite flow*
Derive the asymptotic expression (2.5.3).

2.5.2 *Mean-value theorems for biharmonic functions in two dimensions*
Prove the mean-value theorems expressed by (2.5.7) and (2.5.8).

2.5.3 *Singular harmonic functions*
Derive the counterpart of (2.4.10) in two dimensions.

2.6   The vector potential for incompressible fluids

In Section 1.5.4, we saw that a velocity field, $\mathbf{u}$, that arises as the curl of an arbitrary continuous vector field, $\mathbf{A}$, is solenoidal, $\nabla \cdot \mathbf{u} = 0$. Now we will demonstrate that the inverse is also true, that is, given a solenoidal velocity field, $\nabla \cdot \mathbf{u} = 0$, it is always possible to find a vector potential, $\mathbf{A}$, so that

$$\mathbf{u} = \nabla \times \mathbf{A}. \tag{2.6.1}$$

The three scalar components of this equation are

$$u_1 = \frac{\partial A_3}{\partial x_2} - \frac{\partial A_2}{\partial x_3}, \qquad u_2 = \frac{\partial A_1}{\partial x_3} - \frac{\partial A_3}{\partial x_1}, \qquad u_3 = \frac{\partial A_2}{\partial x_1} - \frac{\partial A_1}{\partial x_2}. \tag{2.6.2}$$

We begin by stipulating that the first component of $\mathbf{A}$, denoted by A_1, is a function of x_1 alone, and set

$$A_1 = f_1(x_1), \tag{2.6.3}$$

where f_1 is an arbitrary function. Integrating the second and third equations in (2.6.2) with respect to x_1 from an arbitrary position, $x_1 = a$, we find that

$$A_2(\mathbf{x}) = \int_a^{x_1} u_3(\mathbf{x}') \, dx_1' + f_2(x_2, x_3), \qquad A_3(\mathbf{x}) = -\int_a^{x_1} u_2(\mathbf{x}') \, dx_1' + f_3(x_2, x_3), \tag{2.6.4}$$

where f_2, f_3 are two arbitrary somewhat related functions of x_2 and x_3. Substituting expressions (2.6.4) into the first equation in (2.6.2), we obtain

$$u_1(\mathbf{x}) = -\int_a^{x_1} \left(\frac{\partial u_2}{\partial x_2} + \frac{\partial u_3}{\partial x_3}\right)(\mathbf{x}') \, dx_1' + \left(\frac{\partial f_3}{\partial x_2} - \frac{\partial f_2}{\partial x_3}\right)(x_2, x_3). \tag{2.6.5}$$

Since the velocity field $\mathbf{u}$ is solenoidal, we may simplify the integrand in (2.6.5) to obtain

$$u_1(\mathbf{x}) = \int_a^{x_1} \frac{\partial u_1}{\partial x_1}(\mathbf{x}') \, dx_1' + \left(\frac{\partial f_3}{\partial x_2} - \frac{\partial f_2}{\partial x_3}\right)(x_2, x_3). \tag{2.6.6}$$

Performing the integration using the fundamental theorem of calculus, we find that

$$\frac{\partial f_3}{\partial x_2} - \frac{\partial f_2}{\partial x_3} = u_1(a, x_2, x_3). \tag{2.6.7}$$

Equation (2.6.7) imposes a differential constraint on the otherwise arbitrary functions f_2 and f_3. The stipulation (2.6.3) and the restriction (2.6.7) can be used to derive the vector form

$$\nabla \times \mathbf{f} = u_1(a, x_2, x_3)\, \mathbf{e}_1, \tag{2.6.8}$$

where $\mathbf{e}_1$ is the unit vector along the first Cartesian axis.

In summary, we have constructed the components of a vector potential $\mathbf{A}$ explicitly in terms of the components of the solenoidal velocity field $\mathbf{u}$, as shown in (2.6.3) and (2.6.4), thereby demonstrating the existence of $\mathbf{A}$.

Solenoidal velocity potential

The vector potential corresponding to a particular incompressible flow is not unique. For example, since the curl of the gradient of any differentiable vector field is zero, the gradient of an arbitrary nonsingular scalar function can be added to a vector potential without altering the velocity field. This observation allows us to assert that it is always possible to find a solenoidal vector potential, $\mathbf{B}$, such that $\nabla \cdot \mathbf{B} = 0$. If $\mathbf{A}$ is a certain non-solenoidal potential, then $\mathbf{A} = \mathbf{B} - \nabla F$ will be a solenoidal potential, provided that the function F satisfies Poisson's equation, $\nabla^2 F = \nabla \cdot \mathbf{B}$.

Stream functions

In Section 2.9, we will show that the vector potential of a two-dimensional or axisymmetric flow can be described in terms of one scalar function, called, respectively, the Helmholtz two-dimensional stream function or the Stokes axisymmetric stream function. Both are defined uniquely up to an arbitrary scalar constant. The vector potential of a three-dimensional flow can be described in terms of two scalar functions, called the stream functions. For example, the vector potential of a three-dimensional flow can be expressed in the form $\mathbf{A} = \psi \nabla \chi$, where ψ and χ are two scalar stream functions. It is interesting to note that $\nabla \times \mathbf{A} = \nabla \psi \times \nabla \chi$, which shows that $\mathbf{A} \cdot (\nabla \times \mathbf{A}) = 0$, that is, the field $\mathbf{A}$ is perpendicular to its curl. A vector field that possesses this property is called a complex lamellar field ([14], p. 63). In contrast, a vector field that is parallel to its own curl, $\mathbf{A} \times (\nabla \times \mathbf{A}) = \mathbf{0}$, is called a Beltrami field.

Problems

2.6.1 *Explicit form of a vector potential*

Show that $\mathbf{A} = \chi \nabla \psi$ and $\mathbf{A} = -\psi \nabla \chi$ are two acceptable vector potentials for the velocity field $\mathbf{u} = \nabla \chi \times \nabla \psi$, where χ and ψ are two arbitrary functions.

2.6.2 *Deriving a vector potential*

Repeat the derivation of the vector potential discussed in the text, but this time assume that $A_1 = f_1(x_2)$ in place of (2.6.3).

2.6.3 *A property of the vector potential*

Show that the velocity and associated vector potential satisfy the symmetry property

$$\mathbf{u} \cdot \nabla \mathbf{A} = (\nabla \mathbf{A}) \cdot \mathbf{u}. \tag{2.6.9}$$

Hint: Begin by writing $\mathbf{u} \times \mathbf{u} = (\nabla \times \mathbf{A}) \times \mathbf{u} = \mathbf{0}$.

2.7 Representation of an incompressible flow in terms of the vorticity

Continuing the study of incompressible flow, we turn to examining the relationship between the vorticity distribution, $\boldsymbol{\omega}(\mathbf{x})$, and the structure of the velocity field, $\mathbf{u}(\mathbf{x})$. Specifically, we seek to develop a representation for the velocity field in terms of the associated vorticity distribution by inverting the equation defining the vorticity,

$$\boldsymbol{\omega} = \nabla \times \mathbf{u}. \tag{2.7.1}$$

To carry out this inversion, we may use (2.6.1), (2.6.3), and (2.6.4), where $\mathbf{u}$ and $\mathbf{A}$ are replaced, respectively, by $\boldsymbol{\omega}$ and $\mathbf{u}$. However, it is expedient to work in an alternative fashion by expressing the velocity in terms of a vector potential, as shown in (2.6.1).

Primary and complementary potentials

Let us assume that $\mathbf{A}$ is the most general vector potential capable of reproducing a velocity field of interest, $\mathbf{u}$. Taking the curl of both sides of the definition $\mathbf{u} = \nabla \times \mathbf{A}$ and manipulating the repeated curl on the right-hand side, we derive a differential equation for $\mathbf{A}$,

$$\boldsymbol{\omega} = \nabla \times \nabla \times \mathbf{A} = \nabla(\nabla \cdot \mathbf{A}) - \nabla^2 \mathbf{A}. \tag{2.7.2}$$

It is useful to resolve $\mathbf{A}$ into two additive components,

$$\mathbf{A} = \mathbf{B} + \mathbf{C}, \tag{2.7.3}$$

where the primary part $\mathbf{B}$ is a particular solution of Poisson's equation

$$\nabla^2 \mathbf{B} = -\boldsymbol{\omega}, \tag{2.7.4}$$

and the complementary part $\mathbf{C}$ is the general solution of the equation

$$\nabla \times \nabla \times \mathbf{C} = -\nabla(\nabla \cdot \mathbf{B}). \tag{2.7.5}$$

With these definitions, we can be sure that

$$\nabla \times \nabla \times (\mathbf{B} + \mathbf{C}) = \boldsymbol{\omega}, \tag{2.7.6}$$

as required. The computation and significance of the primary and complementary parts will be discussed later in this section.

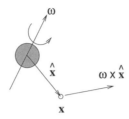

FIGURE 2.7.1 A rotating fluid particle induces a rotary flow expressed by the Biot–Savart integral.

2.7.1 Biot–Savart integral

Consider an interior or exterior flow in a three-dimensional domain. If the domain extends to infinity, we require that the velocity decays at least as fast as $1/d$ so that the corresponding vorticity decays at least as fast as $1/d^2$, where d is the distance from the origin. If the velocity does not decay at infinity, we subtract out the nondecaying far-field component and consider the vector potential associated with the remaining decaying component.

Under these stipulations, we obtain a particular solution of (2.7.4) using the Poisson inversion formula with the free-space Green's function,

$$\mathbf{B}(\mathbf{x}) = \frac{1}{4\pi} \iiint_{Flow} \frac{\boldsymbol{\omega}(\mathbf{x}')}{r}\, \mathrm{d}V(\mathbf{x}'), \tag{2.7.7}$$

where $r = |\mathbf{x} - \mathbf{x}'|$. Substituting (2.7.7) into the right-hand side of (2.7.3), taking the curl of the resulting expression, and switching the curl with the integral operator on the right-hand side, we derive an expression for the velocity,

$$u_i(\mathbf{x}) = \epsilon_{ijk} \Big[\frac{1}{4\pi} \iiint_{Flow} \frac{\partial}{\partial x_j}\Big(\frac{1}{r}\Big)\, \omega_k(\mathbf{x}')\, \mathrm{d}V(\mathbf{x}') + \frac{\partial C_k}{\partial x_j} \Big]. \tag{2.7.8}$$

Carrying out the differentiations under the integral sign and switching back to vector notation, we derive an integral representation for the velocity,

$$\mathbf{u}(\mathbf{x}) = \frac{1}{4\pi} \iiint_{Flow} \frac{1}{r^3} \big[\boldsymbol{\omega}(\mathbf{x}') \times \hat{\mathbf{x}} \big]\, \mathrm{d}V(\mathbf{x}') + \nabla \times \mathbf{C}, \tag{2.7.9}$$

where $\hat{\mathbf{x}} = \mathbf{x} - \mathbf{x}'$. The first two terms on the right-hand side express the primary flow, and the last term represents the complementary flow.

The volume integral on the right-hand side of (2.7.9) is similar to the Biot–Savart integral in electromagnetics expressing the magnetic field due to an electrical current. By analogy, the integral in (2.7.9) expresses the velocity field induced by the rotation of point particles distributed along the vortex lines, as shown in Figure 2.7.1. In formal mathematics, the Biot–Savart integral represents a volume distribution of singular fundamental solutions, called rotlets or vortons, with vector distribution density equal to the vorticity.

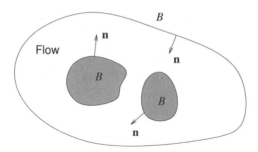

FIGURE 2.7.2 Illustration of a flow domain enclosed by a collection of interior and exterior boundaries, collectively denoted as B.

Representation in terms of the curl of the vorticity

Working in a slightly different manner, we define the primary velocity field $\mathbf{v} = \nabla \times \mathbf{B}$, take the curl of (2.7.7), interchange the curl with the integral sign on the right-hand side, and switch the variable of differentiation from $\mathbf{x}$ to $\mathbf{x}'$ while simultaneously introducing a minus sign to obtain

$$v_i(\mathbf{x}) = -\epsilon_{ijk} \frac{1}{4\pi} \iiint_{Flow} \frac{\partial}{\partial x'_j} \left(\frac{1}{r} \right) \omega_k(\mathbf{x}') \, dV(\mathbf{x}'). \qquad (2.7.10)$$

Manipulating the derivative inside the integral, we find that

$$v_i(\mathbf{x}) = \epsilon_{ijk} \frac{1}{4\pi} \iiint_{Flow} \left[-\frac{\partial}{\partial x'_j} \left(\frac{\omega_k(\mathbf{x}')}{r} \right) + \frac{1}{r} \frac{\partial \omega_k(\mathbf{x}')}{\partial x'_j} \right] dV(\mathbf{x}'). \qquad (2.7.11)$$

Finally, we apply Stokes' theorem to obtain

$$v_i(\mathbf{x}) = \epsilon_{ijk} \frac{1}{4\pi} \iint_B \frac{1}{r} \omega_k(\mathbf{x}') \, n_j(\mathbf{x}') \, dV(\mathbf{x}') + \epsilon_{ijk} \frac{1}{4\pi} \iiint_{Flow} \frac{1}{r} \frac{\partial \omega_k(\mathbf{x}')}{\partial x'_j} \, dV(\mathbf{x}'), \qquad (2.7.12)$$

where B represents the boundaries of the flows and the normal unit vector $\mathbf{n}$ points into the flow, as shown in Figure 2.7.2.

Now switching to vector notation and using the decomposition (2.7.3), we obtain the integral representation

$$\mathbf{u}(\mathbf{x}) = \frac{1}{4\pi} \iiint_{Flow} \frac{1}{r} \left[\nabla \times \boldsymbol{\omega}(\mathbf{x}') \right] dV(\mathbf{x}') + \frac{1}{4\pi} \iint_B \frac{1}{r} \left[\mathbf{n}(\mathbf{x}') \times \boldsymbol{\omega}(\mathbf{x}') \right] dS(\mathbf{x}') + \nabla \times \mathbf{C}. \qquad (2.7.13)$$

The second term on the right-hand side involving the boundary integral disappears when the vortex lines cross the boundaries at a right angle. The volume integral disappears when the curl of the vorticity field is irrotational, in which case all Cartesian components of the velocity satisfy Laplace's equation, $\nabla^2 \mathbf{u} = \mathbf{0}$. The curl of the vorticity of a two-dimensional or axisymmetric flow lies in the plane of the flow or in an azimuthal plane and the volume integral produces respective motion in these planes.

Infinite decaying flow

In the case of flow in an infinite domain with no interior boundaries decaying far from the origin, the representation (2.7.13) simplifies into

$$\mathbf{u}(\mathbf{x}) = \frac{1}{4\pi} \iiint_{Flow} \frac{1}{r} \nabla \times \boldsymbol{\omega}(\mathbf{x}')\, dV(\mathbf{x}') = -\frac{1}{4\pi} \iiint_{Flow} \frac{1}{r} \nabla^2 \mathbf{u}(\mathbf{x}')\, dV(\mathbf{x}'). \qquad (2.7.14)$$

To derive the last expression, we have noted that $\nabla^2 \mathbf{u} = -\nabla \times \boldsymbol{\omega}$ for any solenoidal velocity field, $\nabla \cdot \mathbf{u} = 0$.

2.7.2   The complementary potential

Before attempting to solve (2.7.5) for the complementary vector potential, $\mathbf{C}$, we must compute the divergence $\nabla \cdot \mathbf{B}$ on the right-hand side. Taking the divergence of (2.7.7) and interchanging the order of differentiation and integration on the right-hand side, we find that

$$\nabla \cdot \mathbf{B}(\mathbf{x}) = \frac{1}{4\pi} \iiint_{Flow} \frac{\partial}{\partial x_i} \left(\frac{1}{r} \right) \omega_i(\mathbf{x}')\, dV(\mathbf{x}'). \qquad (2.7.15)$$

Next, we change the variable of differentiation from $\mathbf{x}$ to $\mathbf{x}'$ while simultaneously introducing a minus sign, writing

$$\nabla \cdot \mathbf{B}(\mathbf{x}) = -\frac{1}{4\pi} \iiint_{Flow} \frac{\partial}{\partial x_i'} \left(\frac{1}{r} \right) \omega_i(\mathbf{x}')\, dV(\mathbf{x}'), \qquad (2.7.16)$$

and then

$$\nabla \cdot \mathbf{B}(\mathbf{x}) = \frac{1}{4\pi} \iiint_{Flow} \left[-\frac{\partial}{\partial x_i'} \left(\frac{\omega_i(\mathbf{x}')}{r} \right) + \frac{1}{r} \frac{\partial \omega_i(\mathbf{x}')}{\partial x_i'} \right] dV(\mathbf{x}'). \qquad (2.7.17)$$

Because the vorticity field is solenoidal, the second term inside the integral is zero. Using the divergence theorem, we find that

$$\nabla \cdot \mathbf{B}(\mathbf{x}) = \frac{1}{4\pi} \iint_B \frac{1}{r} \left[\boldsymbol{\omega}(\mathbf{x}') \cdot \mathbf{n}(\mathbf{x}') \right] dS(\mathbf{x}'), \qquad (2.7.18)$$

where the normal unit vector, $\mathbf{n}$, points into the domain of flow, as shown in Figure 2.7.2. The boundary integral vanishes when the vorticity vector is perpendicular to the normal vector or, equivalently, the vortex lines are tangential to the boundaries of the flow. This is always true in the case of two-dimensional or axisymmetric flow, but not necessarily true in the case of three-dimensional flow (Problem 2.7.1). In the case of an unconfined or partially unbounded flow, the boundaries of the flow include the whole or part of a spherical surface of large radius R that lies inside the fluid. As the radius of the large surface tends to infinity, the corresponding integral vanishes provided that the vorticity decays faster than $1/d$, where d is the distance from the origin.

Complementary velocity

When the boundary integral on the right-hand side of (2.7.18) vanishes, $\nabla \cdot \mathbf{B} = 0$ and (2.7.5) shows that the complementary velocity field, $\mathbf{w} \equiv \nabla \times \mathbf{C}$, is irrotational. In that case, $\mathbf{w}$ can be expressed as the gradient of a potential function, ϕ, so that $\mathbf{w} = \nabla\phi$. To ensure that the velocity field $\mathbf{u}$ is solenoidal, we require that ϕ satisfies Laplace's equation, $\nabla^2\phi = 0$. In the more general case where the boundary integral on the right-hand side of (2.7.18) does not vanish, we write

$$\mathbf{w} = \tilde{\mathbf{w}} + \nabla\phi, \qquad (2.7.19)$$

where $\tilde{\mathbf{w}}^S$ is a particular solenoidal solution of the equation $\nabla \times \mathbf{w} = -\nabla(\nabla \cdot \mathbf{B})$ and the function ϕ satisfies Laplace's equation, $\nabla^2\phi = 0$. If a particular solution is available, $\mathbf{w}^P$, a solenoidal solution can be constructed by setting $\tilde{\mathbf{w}} = \mathbf{w}^P + \nabla F$, where the function F satisfies the Poisson equation $\nabla^2 F = -\nabla \cdot \mathbf{w}^P$. In the case of infinite flow with no interior boundaries, $\tilde{\mathbf{w}}$, $\nabla\phi$, and $\mathbf{w}$ are all zero.

Interpretation of the complementary flow in terms of the extended vorticity

We will show that the rotational component of the complementary velocity field, $\mathbf{w}$, can be identified with the flow associated with the extension of the vortex lines outward from the boundaries of the flow in the sense of the Biot–Savart integral. We begin by considering an external flow that is bounded by a closed interior boundary, B, and introduce a nonsingular solenoidal vector field, $\boldsymbol{\zeta}$, inside the volume V_B enclosed by B, subject to the constraint

$$\boldsymbol{\zeta} \cdot \mathbf{n} = \boldsymbol{\omega} \cdot \mathbf{n} \qquad (2.7.20)$$

over B, where $\mathbf{n}$ is the unit vector normal to B pointing into the flow. A nonsingular solenoidal field $\boldsymbol{\zeta}$ that satisfies (2.7.20) exists only if the flow rate of $\boldsymbol{\zeta}$ across the boundary B is zero, which is always true: using (2.7.20) and recalling that the vorticity field is solenoidal, $\nabla \cdot \boldsymbol{\omega} = 0$, we find that

$$\oint\!\!\!\oint_B \boldsymbol{\zeta} \cdot \mathbf{n}\, \mathrm{d}S = \oint\!\!\!\oint_B \boldsymbol{\omega} \cdot \mathbf{n}\, \mathrm{d}S = -\iiint_{Flow} \nabla \cdot \boldsymbol{\omega}\, \mathrm{d}V = 0. \qquad (2.7.21)$$

Having established the existence of $\boldsymbol{\zeta}$, we recast (2.7.5) into the form

$$\nabla^2 \mathbf{C} = \nabla(\nabla \cdot \mathbf{B} + \nabla \cdot \mathbf{C}). \qquad (2.7.22)$$

A particular solution comprised of three scalar functions satisfying Laplace's equation, $\nabla^2 \mathbf{C} = \mathbf{0}$, and $\nabla \cdot \mathbf{C} = -\nabla \cdot \mathbf{B}$, is

$$\mathbf{C}(\mathbf{x}) = \frac{1}{4\pi} \iiint_{V_B} \frac{1}{r} \boldsymbol{\zeta}(\mathbf{x}')\, \mathrm{d}V(\mathbf{x}'), \qquad (2.7.23)$$

where $r = |\mathbf{x} - \mathbf{x}'|$ (Problem 2.7.4). Expression (2.7.23) relates the complementary vector potential, $\mathbf{C}$, to the extended field, $\boldsymbol{\zeta}$, through a Biot–Savart integral. We may then identify $\boldsymbol{\zeta}$ with the extension of the vorticity field outward from the domain of flow into the boundary. The extension can be implemented arbitrarily, subject to the constraint imposed by (2.7.20).

Taking the curl of (2.7.23) and repeating the manipulations that led us from (2.7.10) to (2.7.13), we derive two equivalent expressions for the complementary velocity field,

$$\mathbf{w}(\mathbf{x}) = \frac{1}{4\pi} \iiint_{V_B} \frac{1}{r^3} \left[\boldsymbol{\zeta}(\mathbf{x}') \times \hat{\mathbf{x}} \right] \mathrm{d}V(\mathbf{x}') \qquad (2.7.24)$$

and

$$\mathbf{w}(\mathbf{x}) = \frac{1}{4\pi} \iiint_{V_B} \frac{1}{r} \left[\nabla' \times \boldsymbol{\zeta}(\mathbf{x}') \right] \mathrm{d}V(\mathbf{x}') - \frac{1}{4\pi} \iint_B \frac{1}{r} \left[\mathbf{n}(\mathbf{x}') \times \boldsymbol{\zeta}(\mathbf{x}') \right] \mathrm{d}S(\mathbf{x}'), \qquad (2.7.25)$$

where the normal unit vector, $\mathbf{n}$, points into the flow. If the extended vorticity $\boldsymbol{\zeta}$ is irrotational, the volume integral on the right-hand side of (2.7.25) vanishes, leaving an expression for the complementary flow in terms of a surface integral over the boundary, B, involving the tangential component of $\boldsymbol{\zeta}$ alone. One way to ensure that $\boldsymbol{\zeta}$ is irrotational is to set $\boldsymbol{\zeta} = \nabla H$, where the function H satisfies Laplace's equation, $\nabla^2 H = 0$, and then compute H by solving Laplace's equation inside V_B subject to the Neumann boundary condition (2.7.20).

The complementary flow in terms of the boundary velocity

Switching to a different point of view, we apply Green's third identity (2.3.28) with $\mathbf{u}$ in place of f, $\mathbf{x}$ in place of $\mathbf{x}_0$, and $\mathbf{x}'$ in place of $\mathbf{x}$, finding

$$\mathbf{u}(\mathbf{x}) = -\frac{1}{4\pi} \iint_D \frac{1}{r} \left[\mathbf{n}(\mathbf{x}') \cdot \nabla \mathbf{u}(\mathbf{x}') \right] \mathrm{d}S(\mathbf{x}') \qquad (2.7.26)$$

$$+ \frac{1}{4\pi} \iint_D \frac{1}{r^3} \left[\hat{\mathbf{x}} \cdot \mathbf{n}(\mathbf{x}') \right] \mathbf{u}(\mathbf{x}') \, \mathrm{d}S(\mathbf{x}') - \frac{1}{4\pi} \iiint_{V_c} \frac{1}{r} \nabla^2 \mathbf{u}(\mathbf{x}') \, \mathrm{d}S(\mathbf{x}'),$$

where V_c is a control volume, D is the boundary of the control volume, $\hat{\mathbf{x}} = bx - \mathbf{x}'$, and $r = |\mathbf{x} - \mathbf{x}'|$. Using the vector identity $\nabla^2 \mathbf{u} = -\nabla \times \boldsymbol{\omega}$, applicable for a solenoidal velocity field, $\mathbf{u}$, and manipulating the integrand of the second integral, we obtain

$$\mathbf{u}(\mathbf{x}) = -\frac{1}{4\pi} \iint_B \frac{1}{r} \left[\mathbf{n}(\mathbf{x}') \cdot \nabla \mathbf{u}(\mathbf{x}') \right] \mathrm{d}S(\mathbf{x}') \qquad (2.7.27)$$

$$+ \frac{1}{4\pi} \iint_B \left[\mathbf{n}(\mathbf{x}') \cdot \nabla' \Big(\frac{1}{r} \Big) \right] \mathbf{u}(\mathbf{x}') \, \mathrm{d}S(\mathbf{x}') + \frac{1}{4\pi} \iiint_{V_c} \frac{1}{r} \nabla' \times \boldsymbol{\omega}(\mathbf{x}') \, \mathrm{d}S(\mathbf{x}').$$

Now comparing (2.7.27) with (2.7.13), we derive a boundary-integral representation for the complementary velocity field in terms of the boundary velocity and tangential component of the boundary vorticity,

$$\mathbf{w}(\mathbf{x}) = -\frac{1}{4\pi} \iint_B \left[\frac{1}{r} \big\{ \mathbf{n}(\mathbf{x}') \cdot \nabla \mathbf{u}(\mathbf{x}') + \mathbf{n}(\mathbf{x}') \times \boldsymbol{\omega}(\mathbf{x}') \big\} - \left[\mathbf{n}(\mathbf{x}') \cdot \nabla' \Big(\frac{1}{r} \Big) \right] \mathbf{u}(\mathbf{x}') \right] \mathrm{d}S(\mathbf{x}'). \quad (2.7.28)$$

Departing from the definition $\boldsymbol{\omega} = \nabla \times \mathbf{u}$ and working in index notation, we find that the expression enclosed by the curly brackets inside the integrand is equal to $[\nabla' \mathbf{u}(\mathbf{x}')] \cdot \mathbf{n}(\mathbf{x}')$, where the gradient ∇' involves derivatives with respect to $\mathbf{x}'$. Making this substitution, we obtain an expression for the complementary flow in terms of the boundary velocity and velocity gradient tensor,

$$\mathbf{w}(\mathbf{x}) = -\frac{1}{4\pi} \iint_B \left[\frac{1}{r} \nabla' \mathbf{u}(\mathbf{x}') - \mathbf{u}(\mathbf{x}') \, \nabla' \Big(\frac{1}{r} \Big) \right] \cdot \mathbf{n}(\mathbf{x}') \, \mathrm{d}S(\mathbf{x}'). \qquad (2.7.29)$$

In index notation,

$$w_i(\mathbf{x}) = -\frac{1}{4\pi} \iint_B \left[\frac{1}{r} \frac{\partial u_j(\mathbf{x}')}{\partial x'_i} - u_i(\mathbf{x}') \frac{\partial}{\partial x'_j} \Big(\frac{1}{r} \Big) \right] n_j(\mathbf{x}') \, \mathrm{d}S(\mathbf{x}'). \qquad (2.7.30)$$

We recall that the normal unit vector, $\mathbf{n}$, points into the flow.

The complementary flow in terms of a boundary vortex sheet

A more appealing representation of the complementary flow emerges by using the divergence theorem to convert the surface integral of the first term inside the integral in (2.7.30) into a volume integral, obtaining

$$w_i(\mathbf{x}) = \frac{1}{4\pi} \iiint_{Flow} \frac{\partial}{\partial x_j'} \left(\frac{1}{r} \frac{\partial u_j(\mathbf{x}')}{\partial x_i'} \right) dV(\mathbf{x}') + \frac{1}{4\pi} \iint_B u_i(\mathbf{x}') \frac{\partial}{\partial x_j'} \left(\frac{1}{r} \right) n_j(\mathbf{x}') dS(\mathbf{x}'). \quad (2.7.31)$$

Next, we recast the last integrand into the form

$$\frac{\partial}{\partial x_j'} \left(\frac{1}{r} \frac{\partial u_j(\mathbf{x}')}{\partial x_i'} \right) = \frac{\partial^2}{\partial x_i' \partial x_j'} \left(\frac{u_j(\mathbf{x}')}{r} \right) - \frac{\partial}{\partial x_j'} \left[u_j(\mathbf{x}') \frac{\partial}{\partial x_i'} \left(\frac{1}{r} \right) \right] \quad (2.7.32)$$

and use the continuity equation to simplify the first integral on the right-hand side, obtaining

$$\frac{\partial}{\partial x_j'} \left(\frac{1}{r} \frac{\partial u_j(\mathbf{x}')}{\partial x_i'} \right) = \frac{\partial}{\partial x_i'} \left[u_j(\mathbf{x}') \frac{\partial}{\partial x_j'} \left(\frac{1}{r} \right) \right] - \frac{\partial}{\partial x_j'} \left[u_j(\mathbf{x}') \frac{\partial}{\partial x_i'} \left(\frac{1}{r} \right) \right]. \quad (2.7.33)$$

Substituting this expression into the volume integral of (2.7.31) and using the divergence theorem to convert it to a surface integral, we obtain

$$w_i(\mathbf{x}) = -\frac{1}{4\pi} \iint_B u_j(\mathbf{x}') \frac{\partial}{\partial x_j'} \left(\frac{1}{r} \right) n_i(\mathbf{x}') dS(\mathbf{x}') \quad (2.7.34)$$

$$+ \iint_B u_j(\mathbf{x}') \frac{\partial}{\partial x_i'} \left(\frac{1}{r} \right) n_j(\mathbf{x}') dS(\mathbf{x}') + \frac{1}{4\pi} \iint_B u_i(\mathbf{x}') \frac{\partial}{\partial x_j'} \left(\frac{1}{r} \right) n_j(\mathbf{x}') dS(\mathbf{x}').$$

The first and third integrands can be combined to form a double cross product, resulting in an expression for the complementary flow in terms of the tangential and normal components of the boundary velocity,

$$\mathbf{w}(\mathbf{x}) = \frac{1}{4\pi} \iint_B \left[\mathbf{n}(\mathbf{x}') \times \mathbf{u}(\mathbf{x}') \right] \times \nabla' \left(\frac{1}{r} \right) dS(\mathbf{x}') + \frac{1}{4\pi} \iint_B \left[\mathbf{u}(\mathbf{x}') \cdot \mathbf{n}(\mathbf{x}') \right] \nabla' \left(\frac{1}{r} \right) dS(\mathbf{x}'). \quad (2.7.35)$$

The first boundary integral on the right-hand side of (2.7.35) expresses the velocity field due to a vortex sheet with strength $\mathbf{u} \times \mathbf{n}$ wrapping around the boundaries of the flow. Since the strength of the vortex sheet is equal to the tangential component of the velocity, the purpose of the vortex sheet is to annihilate the tangential component of the boundary velocity. The second integral expresses a boundary distribution of point sources whose strength is equal to the normal velocity .

2.7.3 Two-dimensional flow

The counterpart of expression (2.7.7) for two-dimensional flow in the xy plane is

$$\mathbf{B}(\mathbf{x}) = -\frac{1}{2\pi} \iint_{Flow} \ln \left(\frac{r}{a} \right) \omega_z(\mathbf{x}') \mathbf{e}_z \, dA(\mathbf{x}'), \quad (2.7.36)$$

where a is a constant length and $\mathbf{e}_z$ is the unit vector along the z axis. Note that both the vorticity, $\boldsymbol{\omega}$, and vector potential, $\mathbf{B}$, are oriented along the z axis. Since the vortex lines do not cross the boundaries of the flow, the divergence of $\mathbf{B}$ is zero and the velocity field is given by

$$\mathbf{u}(\mathbf{x}) = \frac{1}{2\pi} \iint_{Flow} \frac{1}{r^2} \left[\boldsymbol{\omega}(\mathbf{x}') \times \hat{\mathbf{x}} \right] \mathrm{d}A(\mathbf{x}') + \nabla H(\mathbf{x}), \qquad (2.7.37)$$

where $\hat{\mathbf{x}} = \mathbf{x} - \mathbf{x}'$ and H is a two-dimensional harmonic function, $\nabla^2 H = 0$. An alternative integral representation for the velocity is

$$\begin{aligned}
\mathbf{u}(\mathbf{x}) = &-\frac{1}{2\pi} \int_C \ln\left(\frac{r}{a}\right) \left[\mathbf{n}(\mathbf{x}') \times \boldsymbol{\omega}(\mathbf{x}') \right] \mathrm{d}l(\mathbf{x}') \\
&-\frac{1}{2\pi} \iint_{Flow} \ln\left(\frac{r}{a}\right) \nabla \times \boldsymbol{\omega}(\mathbf{x}') \, \mathrm{d}A(\mathbf{x}') + \nabla H(\mathbf{x}),
\end{aligned} \qquad (2.7.38)$$

where C is the boundary of the flow. In the case of infinite flow with no interior boundaries, both the boundary integral and the gradient ∇H on the right-hand side of (2.7.38) vanish.

Problems

2.7.1 *Vortex lines at boundaries*

Explain why the vortex lines may not cross a rigid boundary that is either stationary or translates in a viscous fluid, but must necessarily cross a boundary that rotates as a rigid body. The velocity of the fluid at the boundary is assumed to be equal to the velocity of the boundary, which means that the no-slip and no-penetration conditions apply.

2.7.2 *A vortex line that starts and ends on a body*

Consider an infinite incompressible flow containing a single line vortex that starts and ends at the surface of a body. There are many ways to extend the line vortex into the body subject to the constraint (2.7.20). Show that the flows induced by any two extended line vortices have identical rotational components; equivalently, the difference between these two flows expresses irrotational motion.

Hint: Consider the closed loop formed by the two extended vortex lines and use the Biot–Savart integral to show that the flow induced by this loop is irrotational everywhere except on the loop.

2.7.3 *Complementary flow*

(*a*) Confirm that (2.7.23) is a particular solution of (2.7.22).

(*b*) Derive (2.7.29) from (2.7.28).

2.7.4 *Reduction of the Biot–Savart integral from three to two dimensions*

Consider an infinite two-dimensional flow in the xy plane that decays at infinity. Substituting (1.1.23) into (2.7.9), setting $\mathrm{d}V = \mathrm{d}z\,\mathrm{d}A$, where $\mathrm{d}A = \mathrm{d}x\,\mathrm{d}y$, and performing the integration with respect to z, derive the integral representation (2.7.37). *Hint:* reference to standard tables of definite integrals (e.g., [150], p. 86) shows that

$$\int_{-\infty}^{\infty} \frac{\mathrm{d}z}{(x^2 + y^2 + z^2)^{3/2}} = \frac{2}{x^2 + y^2}. \qquad (2.7.39)$$

2.7.5 *Complementary flow for an impenetrable boundary*

Milne-Thomson ([268], p. 570) derives a simplified version of (2.7.35) for an incompressible flow that satisfies the no-penetration condition, $\mathbf{u} \cdot \mathbf{n} = 0$, over all boundaries of the flow, B. We begin by introducing a solenoidal velocity vector potential, $\mathbf{A}$, where $\nabla \cdot \mathbf{A} = 0$, according to our discussion in Section 2.6. Next, we set $\mathbf{A} = \nabla \times \mathbf{B}$, where $\mathbf{B}$ is a vector potential for $\mathbf{A}$, and obtain

$$\mathbf{u} = \nabla \times \nabla \times \mathbf{B} = \nabla(\nabla \cdot \mathbf{B}) - \nabla^2 \mathbf{B}. \tag{2.7.40}$$

Assuming that $\nabla \cdot \mathbf{B} = 0$, we find a particular solution in terms of the Poisson integral

$$\mathbf{B}(\mathbf{x}) = \frac{1}{4\pi} \iiint_{Flow} \frac{1}{r} \mathbf{u}(\mathbf{x}') \, dV(\mathbf{x}'), \tag{2.7.41}$$

which is analogous to (2.7.7). Differentiating, we find that

$$\mathbf{A}(\mathbf{x}) = \nabla \times \mathbf{B}(\mathbf{x}) = \frac{1}{4\pi} \nabla \times \iiint_{Flow} \frac{1}{r} \mathbf{u}(\mathbf{x}') \, dV(\mathbf{x}') = \frac{1}{4\pi} \iiint_{Flow} \mathbf{u}(\mathbf{x}') \times \nabla'\left(\frac{1}{r}\right) dV(\mathbf{x}'). \tag{2.7.42}$$

Manipulating the last integral, we obtain

$$\mathbf{A}(\mathbf{x}) = \frac{1}{4\pi} \iiint_{Flow} \frac{1}{r} \nabla' \times \mathbf{u}(\mathbf{x}') \, dV(\mathbf{x}') - \frac{1}{4\pi} \iiint_{Flow} \nabla' \times \left(\frac{\mathbf{u}(\mathbf{x}')}{r}\right) dV(\mathbf{x}'). \tag{2.7.43}$$

Finally, we invoke the definition of the vorticity and use the divergence theorem to find

$$\mathbf{A}(\mathbf{x}) = \frac{1}{4\pi} \iiint_{Flow} \frac{1}{r} \boldsymbol{\omega}(\mathbf{x}') \, dV(\mathbf{x}') + \frac{1}{4\pi} \iint_B \frac{1}{r} \left[\mathbf{n}(\mathbf{x}') \times \mathbf{u}(\mathbf{x}')\right] dS(\mathbf{x}'), \tag{2.7.44}$$

where the normal unit vector, $\mathbf{n}$, points into the flow. Based on (2.7.44), we obtain a general expression for the velocity field,

$$\mathbf{u}(\mathbf{x}) = \frac{1}{4\pi} \iiint_{Flow} \frac{1}{r^3} \left[\boldsymbol{\omega}(\mathbf{x}') \times \hat{\mathbf{x}}\right] dV(\mathbf{x}') + \frac{1}{4\pi} \iint_B \left[\mathbf{n}(\mathbf{x}') \times \mathbf{u}(\mathbf{x}')\right] \times \frac{\hat{\mathbf{x}}}{r^3} dS(\mathbf{x}'). \tag{2.7.45}$$

Note that the second integral on the right-hand side is a simplified version of the right-hand side of (2.7.35). Show that

$$\nabla \cdot \mathbf{B}(\mathbf{x}) = \frac{1}{4\pi} \iint_B \frac{1}{r} \left[\mathbf{u}(\mathbf{x}') \cdot \mathbf{n}(\mathbf{x}')\right] dS(\mathbf{x}') = 0, \tag{2.7.46}$$

so that the condition for the derivation (2.7.44) is fulfilled.

2.8 Representation of a flow in terms of the rate of expansion and vorticity

Previously in this chapter, we have shown that an irrotational velocity field can be expressed as the gradient of a potential function, whereas a solenoidal velocity field can be expressed as the curl of a vector potential. A velocity field that is both irrotational and solenoidal can be expressed in terms of either a potential function or a vector potential. In this section, we consider a flow that is neither incompressible nor irrotational.

Hodge–Helmholtz decomposition

If the domain of flow is infinite, we require that the velocity, $\mathbf{u}$, decays at a rate that is faster than $1/d$, where d is the distance from the origin. This constraint ensures that the rate of expansion and vorticity decay at a rate that is faster than $1/d^2$. If the velocity does not decay at infinity, we subtract out the nondecaying far-field component and consider the remaining decaying flow. Under these assumptions, the velocity field is subject to the fundamental theorem of vector analysis, also known as the Hodge or Helmholtz decomposition theorem, stating that $\mathbf{u}$ can be resolved into two constituents,

$$\mathbf{u} = \nabla\phi + \mathbf{w}, \tag{2.8.1}$$

where $\nabla\phi$ is an irrotational field expressed in terms of a velocity potential, ϕ, and $\mathbf{w}$ is a solenoidal vector field,

$$\nabla \cdot \mathbf{w} = 0. \tag{2.8.2}$$

This last constraint allows us to express $\mathbf{w}$ in terms of a vector potential, $\mathbf{A}$, and therefore recast (2.8.1) into the form

$$\mathbf{u} = \nabla\phi + \nabla \times \mathbf{A}. \tag{2.8.3}$$

To demonstrate the feasibility of the Hodge–Helmholtz decomposition, we take the curl of (2.8.1) and derive an expression for the vorticity,

$$\boldsymbol{\omega} = \nabla \times \mathbf{u} = \nabla \times \mathbf{w}. \tag{2.8.4}$$

Since the difference field $\mathbf{u} - \mathbf{w}$ is irrotational, it can be expressed as a gradient, $\nabla\phi$, as discussed in Section 2.1, and this completes the proof. Taking the divergence of (2.8.1), we find that

$$\nabla \cdot \mathbf{u} = \nabla^2\phi, \tag{2.8.5}$$

which, along with (2.8.4), shows that the rate of expansion of the flow with velocity $\nabla\phi$ and the vorticity of the flow with velocity $\mathbf{w}$ are identical to those of the flow $\mathbf{u}$.

Integral representation

Combining (2.8.1) with (2.1.15), (2.7.9), and (2.7.19), we derive a representation for the velocity in terms of the rate of expansion, the vorticity, and an unspecified irrotational and solenoidal velocity field described by a harmonic potential, H,

$$\mathbf{u}(\mathbf{x}) = \frac{1}{4\pi} \iiint_{Flow} \frac{\hat{\mathbf{x}}}{r^3} \, \alpha(\mathbf{x}') \, dV(\mathbf{x}') + \frac{1}{4\pi} \iiint_{Flow} \frac{1}{r^3} \left[\boldsymbol{\omega}(\mathbf{x}') \times \hat{\mathbf{x}} \right] dV(\mathbf{x}') + \mathbf{v}(\mathbf{x}) + \nabla H(\mathbf{x}), \tag{2.8.6}$$

where $\alpha(\mathbf{x}') = \nabla' \cdot \mathbf{u}(\mathbf{x}')$ is the rate of expansion. The complementary rotational field, $\mathbf{v}$, can be computed by extending the vortex lines outward from the domain of flow across the boundaries, as discussed in Section 2.7. In the case of unbounded flow that decays at infinity, the last two terms on the right-hand side of (2.8.6) do not appear.

Specifying the distribution of the rate of expansion, $\nabla \cdot \mathbf{u}$, and vorticity, $\boldsymbol{\omega}$, throughout the domain of flow defines the first three terms on the right-hand side of (2.8.6). Consequently, the gradient, ∇H, and thus the velocity, $\mathbf{u}$, are defined uniquely by prescribing the boundary distribution of the normal velocity component, $\mathbf{u} \cdot \mathbf{n}$. An important consequence is that it is not generally permissible to arbitrarily specify the distribution of both the rate of expansion and vorticity in a flow while requiring more than one scalar condition at the boundaries.

To illustrate the last point, we consider an incompressible flow due to a line vortex in an infinite domain containing a stationary body and compute ∇H by enforcing the no-penetration condition requiring that the velocity component normal to the body is zero. To satisfy an additional boundary condition, such as the no-slip condition requiring that the tangential component of the boundary velocity is zero, we complement the vorticity field with a vortex sheet situated over the surface of the body. The velocity induced by the vortex sheet annihilates the tangential velocity induced by the line vortex. The vorticity distribution associated with the vortex sheet must be taken into account when computing the second integral on the right-hand side of (2.8.6).

Problems

2.8.1 *Integral representation*

Discuss whether it is consistent to introduce a velocity field that satisfies the stipulated boundary conditions, compute the associated rate of expansion and vorticity, and then use (2.8.6) to deduce the complementary velocity field, $\mathbf{v}$, and harmonic potential, H.

2.8.2 *Poincaré decomposition*

Derive an expression for the velocity in terms of the vorticity by applying equations (2.6.1), (2.6.3), and (2.6.4) with $\boldsymbol{\omega}$ in place of $\mathbf{u}$ and $\mathbf{u}$ in place of $\mathbf{A}$.

2.9 Stream functions for incompressible fluids

Previously in this chapter, we developed integral representations for a velocity field in terms of the the rate of expansion, the vorticity, and the boundary velocity, and differential representations in terms of the potential function and the vector potential. The integral representations allowed us to obtain insights into the effect of the local fluid motion on the global structure of a flow. The differential representations allowed us to describe a flow using a reduced number of scalar functions. For example, in the case of irrotational flow, we describe the flow simply in terms of the potential function. In this section, we address the issue of the minimum number of scalar functions necessary to describe an arbitrary incompressible rotational flow.

2.9.1 Two-dimensional flow

Examining the streamline pattern of a two-dimensional incompressible flow, we find that the flow rate across an arbitrary line, L, that begins at a point, A, on a particular streamline and ends

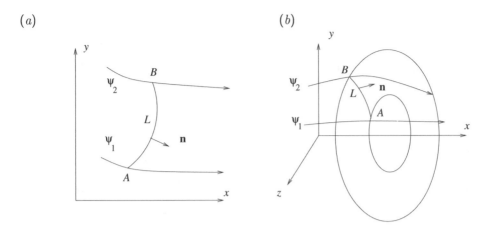

FIGURE 2.9.1   Illustration of two streamlines in (a) a two-dimensional or (b) axisymmetric flow, used to introduce the stream function.

at another point, B, on another streamline is constant, independent of the precise location of the points A and B along the two streamlines, as illustrated in Figure 2.9.1(a). Consequently, we may assign to every streamline a numerical value of a function, ψ, so that the difference in the values of ψ corresponding to two different streamlines is equal to the instantaneous flow rate across any line that begins at a point A on the first streamline and ends at another point B on the second streamline. Accordingly, we write

$$\psi_2 - \psi_1 = \int_A^B \mathbf{u} \cdot \mathbf{n}\, \mathrm{d}l, \tag{2.9.1}$$

where the integral is computed along the line L depicted in Figure 2.9.1(a). The right-hand side of (2.9.1) expresses the flow rate across L. Since the streamlines fill up the entire domain of a flow, we may regard ψ a field function of position, x and y, and time t, called the stream function.

Next, we consider two adjacent streamlines and apply the trapezoidal rule to approximate the integral in (2.9.1), obtaining the differential relation

$$\mathrm{d}\psi = u_x\, \mathrm{d}y - u_y\, \mathrm{d}x. \tag{2.9.2}$$

Since the right-hand side of (2.9.2) is a complete differential, we may write

$$u_x = \frac{\partial \psi}{\partial y}, \qquad u_y = -\frac{\partial \psi}{\partial x}, \tag{2.9.3}$$

which can be recast into the compact vector form

$$\mathbf{u} = \nabla \times (\psi\, \mathbf{e}_z), \tag{2.9.4}$$

where $\mathbf{e}_z$ is the unit vector along the z axis. Equation (2.9.4) suggests that a vector potential of a two-dimensional flow is

$$\mathbf{A} = \psi \mathbf{e}_z. \tag{2.9.5}$$

It is evident from the definitions (2.9.3) that $\nabla \cdot \mathbf{u} = 0$, which shows that the gradient $\nabla \psi$ is perpendicular to the streamlines. Accordingly, the stream function is constant along the streamlines of a steady or unsteady flow.

In summary, we have succeeded to express the velocity field and vector potential of a two-dimensional flow in terms of a single scalar function, the stream function, ψ. Using the definitions (2.9.3), we find that the stream function of a certain flow is determined uniquely up to an arbitrary constant.

Point source

The stream function associated with a two-dimensional point source of strength m located at a point, $\mathbf{x}_0$, introduced in (2.1.29), is given by

$$\psi(\mathbf{x}) = \frac{m}{2\pi}\, \theta, \tag{2.9.6}$$

where θ is the polar angle subtended between the vector $\mathbf{x} - \mathbf{x}_0$ and the x axis. This example makes it clear that, when the domain of flow contains point sources or point sinks or is multiply connected and the flow rate across a surface that encloses a boundary is nonzero, the stream function is a multi-valued function of position. An example is the flow due to the radial expansion of a two-dimensional bubble.

Poisson equation

Taking the curl of (2.9.4), we confirm that the vorticity is parallel to the z axis, $\boldsymbol{\omega} = \omega_z \mathbf{e}_z$. The nonzero component of the vorticity is the negative of the Laplacian of the stream function,

$$\omega_z = -\nabla^2 \psi. \tag{2.9.7}$$

Using the Poisson formula to invert (2.9.7), we derive an expression for the stream function in terms of the vorticity,

$$\psi(\mathbf{x}) = -\frac{1}{2\pi} \iint_{Flow} \ln\left(\frac{r}{a}\right) \omega_z(\mathbf{x}')\, dA(\mathbf{x}') + H(\mathbf{x}), \tag{2.9.8}$$

where $r = |\mathbf{x} - \mathbf{x}'|$, a is a specified length, and H is a harmonic function in the xy plane. In the case of infinite flow with no interior boundaries, H is a constant usually set to zero. It is worth observing that (2.9.5) and (2.9.8) are consistent with the more general form (2.7.36).

Plane polar coordinates

Returning to (2.9.4) and (2.9.7), we express the radial and angular components of the velocity and the nonzero component of the vorticity in plane polar coordinates, obtaining

$$u_r = \frac{1}{r}\frac{\partial \psi}{\partial \theta}, \qquad u_\theta = -\frac{\partial \psi}{\partial r}, \qquad \omega_z = -\frac{1}{r}\frac{\partial}{\partial r}\left(r\frac{\partial \psi}{\partial r}\right) - \frac{1}{r^2}\frac{\partial^2 \psi}{\partial \theta^2}. \tag{2.9.9}$$

Substituting expression (2.9.6), we confirm the radial direction of the velocity due to a point source.

2.9.2 Axisymmetric flow

To describe an axisymmetric flow without swirling motion, we introduce cylindrical coordinates, (x, σ, φ), as illustrated in Figure 2.9.1(b). Working as in the case of two-dimensional flow, we assign to every streamline a numerical value of a scalar function, $\psi(x, \sigma, t)$, called the Stokes stream function, so that the difference in the values ψ between any two streamlines is proportional to the instantaneous flow rate across an axisymmetric surface whose trace in a meridional plane begins at a point A on one streamline and ends at another point B on the second streamline. By definition,

$$\psi_2 - \psi_1 = \int_A^B \mathbf{u} \cdot \mathbf{n} \, \sigma \, \mathrm{d}l. \tag{2.9.10}$$

Multiplying the integral on the right-hand side by 2π produces the flow rate through an axisymmetric surface whose trace in an azimuthal plane is the line L shown in Figure 2.9.1(b).

Next, we consider two streamlines that lie in the same azimuthal plane and are separated by a small distance, and apply the trapezoidal rule to express (2.9.10) in the differential form

$$\mathrm{d}\psi = u_x \, \sigma \, \mathrm{d}\sigma - u_\sigma \, \sigma \, \mathrm{d}x, \tag{2.9.11}$$

which suggests the differential relations

$$u_x = \frac{1}{\sigma} \frac{\partial \psi}{\partial \sigma}, \qquad u_\sigma = -\frac{1}{\sigma} \frac{\partial \psi}{\partial x}. \tag{2.9.12}$$

Combining these equations, we derive the vector form

$$\mathbf{u} = \nabla \times \left(\frac{\psi}{\sigma} \mathbf{e}_\varphi \right), \tag{2.9.13}$$

where $\mathbf{e}_\varphi$ is the azimuthal unit vector. Thus,

$$\mathbf{A} = \frac{\psi}{\sigma} \mathbf{e}_\varphi \tag{2.9.14}$$

is the vector potential of an axisymmetric flow in the absence of swirling motion. Equations (2.9.12) show that, $\mathbf{u} \cdot \nabla \psi = 0$, which demonstrates that the gradient of the stream function, $\nabla \psi$, is perpendicular to the streamlines of the flow. Accordingly, the Stokes stream function, ψ, is constant along the streamlines of an axisymmetric flow.

In summary, we have managed to express the velocity field and vector potential of an axisymmetric flow in terms of a single scalar function, the Stokes stream function, ψ. It is clear from the definitions (2.9.12) that ψ is determined uniquely up to an arbitrary scalar constant. If the domain of flow is multiply connected, ψ can be a multi-valued function of position. An example is provided by the flow due to the expansion of a toroidal bubble.

It is worth noting that the stream function of two-dimensional flow has units of velocity multiplied by length, whereas the Stokes stream function of axisymmetric flow has units of velocity multiplied by length squared.

Point source

As an example, the Stokes stream function associated with a three-dimensional point source with strength m located at the point $\mathbf{x}_0$, introduced in (2.1.29), is given by

$$\psi(\mathbf{x}) = -\frac{m}{4\pi}\cos\theta = -\frac{m}{4\pi}\frac{x - x_0}{|\mathbf{x} - \mathbf{x}_0|}, \qquad (2.9.15)$$

where θ is the polar angle subtended between the vector $\mathbf{x} - \mathbf{x}_0$ and the x axis.

Vorticity

Taking the curl of (2.9.13), we find that the vorticity is oriented in the azimuthal direction, $\boldsymbol{\omega} = \omega_\varphi \mathbf{e}_\varphi$, where $\mathbf{e}_\varphi$ is the corresponding unit vector. The azimuthal component of the vorticity is given by

$$\omega_\varphi = -\frac{1}{\sigma}E^2\psi = -\frac{1}{\sigma}\left(\frac{\partial^2\psi}{\partial x^2} + \frac{\partial^2\psi}{\partial\sigma^2} - \frac{1}{\sigma}\frac{\partial\psi}{\partial\sigma}\right), \qquad (2.9.16)$$

where E^2 is a second-order linear differential operator defined as

$$E^2 \equiv \frac{\partial^2}{\partial x^2} + \frac{\partial^2}{\partial\sigma^2} - \frac{1}{\sigma}\frac{\partial}{\partial\sigma}. \qquad (2.9.17)$$

The inverse relation, providing us with the stream function in terms of the vorticity, is discussed in Section 2.12.

Spherical polar coordinates

In spherical polar coordinates, the radial and meridional velocity components of an axisymmetric flow are given by

$$u_r = \frac{1}{r^2\sin\theta}\frac{\partial\psi}{\partial\theta}, \qquad u_\theta = -\frac{1}{r\sin\theta}\frac{\partial\psi}{\partial r}. \qquad (2.9.18)$$

The azimuthal component of the vorticity is given by

$$\omega_\varphi = -\frac{1}{\sigma}E^2\psi = -\frac{1}{\sin\theta}\left(\frac{\partial^2\psi}{\partial r^2} + \frac{1}{r^2}\frac{\partial^2\psi}{\partial\theta^2} - \frac{\cot\theta}{r^2}\frac{\partial\psi}{\partial\theta}\right), \qquad (2.9.19)$$

where the second-order differential operator E^2 was defined in (2.9.17). In spherical polar coordinates,

$$E^2 \equiv \frac{\partial^2}{\partial r^2} + \frac{\sin\theta}{r^2}\frac{\partial}{\partial\theta}\left(\frac{1}{\sin\theta}\frac{\partial}{\partial\theta}\right) = \frac{\partial^2}{\partial r^2} + \frac{1}{r^2}\frac{\partial^2}{\partial\theta^2} - \frac{\cot\theta}{r^2}\frac{\partial}{\partial\theta}, \qquad (2.9.20)$$

involving derivatives with respect to the radial distance, r, and meridional angle, θ.

2.9.3   Three-dimensional flow

Finally, we consider the most general case of a genuine three-dimensional flow. Inspecting the streamline pattern, we identify two distinct families of stream surfaces, where each family fills the

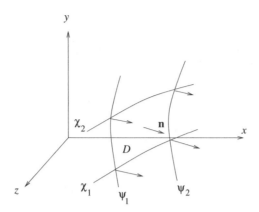

FIGURE 2.9.2   Illustration of two pairs of stream surfaces in a three-dimensional flow used to define the
stream functions.

entire domain of flow, as illustrated in Figure 2.9.2. A stream surface consists of all streamlines
passing through a specified line.

Focusing on a stream tube that is confined between two pairs of stream surfaces, one pair in
each family, we note that the flow rate, Q, across any open surface that is bounded by the stream
tube, D, is constant. We then assign to the four stream surfaces the labels ψ_1, ψ_2, χ_1, and χ_2, as
shown in Figure 2.9.2, so that

$$Q = \iint_D \mathbf{u} \cdot \mathbf{n}\, \mathrm{d}S = (\psi_2 - \psi_1)(\chi_2 - \chi_1). \tag{2.9.21}$$

In this light, ψ and χ emerge as field functions of space and time, called the stream functions. Now
using Stokes' theorem, we write

$$\iint_D \left[\nabla \times (\psi \nabla \chi)\right] \cdot \mathbf{n}\, \mathrm{d}S = \oint_C \psi\, \mathbf{t} \cdot \nabla \chi\, \mathrm{d}l = (\psi_2 - \psi_1)(\chi_2 - \chi_1) = Q, \tag{2.9.22}$$

where C is the contour enclosing D and $\mathbf{t}$ is the unit vector tangent to C pointing in the counterclockwise direction with respect to $\mathbf{n}$. Comparing (2.9.22) with (2.9.21) and remembering that the
surface D is arbitrary, we write

$$\mathbf{u} = \nabla \times (\psi \nabla \chi) = \nabla \psi \times \nabla \chi = -\nabla \times (\chi \nabla \psi). \tag{2.9.23}$$

Since the gradients $\nabla \psi$ and $\nabla \chi$ are normal to the corresponding stream tubes, the expression
$\mathbf{u} = \nabla \psi \times \nabla \chi$ underscores that the intersection between two stream tubes is a streamline. Our
analysis has revealed that

$$\mathbf{A} = \psi \nabla \chi, \qquad \mathbf{A} = -\chi \nabla \psi, \tag{2.9.24}$$

are two acceptable vector potentials of an incompressible flow.

In summary, we have succeeded to express a three-dimensional solenoidal velocity field in terms of two scalar functions, the stream functions ψ and χ. The spatial distribution of the stream functions depends on the choice of the families of stream surfaces chosen to derive (2.9.21). However, having made a choice, the stream functions are determined uniquely up to an arbitrary scalar constant. The stream function for two-dimensional flow in the xy plane and the Stokes stream function for axisymmetric flow derive from (2.9.23) by setting, respectively, $\chi = z$ and $\chi = \varphi$, where φ is the azimuthal angle (Problem 2.9.1).

Taking the curl of (2.9.23), we derive an expression for the vorticity,

$$\boldsymbol{\omega} = \mathbf{L}(\psi) \cdot \nabla \chi - \mathbf{L}(\chi) \cdot \nabla \psi, \tag{2.9.25}$$

where $\mathbf{L} = -\mathbf{I} \nabla^2 + \nabla \nabla$ and $\mathbf{I}$ is the identity matrix. Because each term on the right-hand side of (2.9.25) represents a solenoidal vector field, the vorticity field is solenoidal for any choice of ψ and χ (Problem 2.9.2).

Problems

2.9.1 *Stream functions*

Show that the two-dimensional and Stokes stream functions derive from (2.9.23) by setting, respectively, $\chi = z$ and $\chi = \varphi$.

2.9.2 *A linear velocity field*

Derive the stream function and sketch the streamlines of a two-dimensional flow with velocity components $u_x = \xi(x + y)$, $u_y = \xi(x - y)$, where ξ is a constant shear rate. Discuss the physical interpretation of this flow.

2.9.3 *Point-source dipoles*

Derive the stream functions of a two- or three-dimensional point-source dipole pointing along the x axis.

2.9.4 *Vorticity and stream functions*

(*a*) Derive expression (2.9.25).

(*b*) Show that $\mathbf{u} = \mathbf{K}(f) \cdot \mathbf{a}$ is a solenoidal velocity field, where f is an arbitrary function, $\mathbf{a}$ is an arbitrary constant, and $\mathbf{K} = -\mathbf{I} \nabla^2 + \nabla \nabla$ is a second-order operator.

2.9.5 *Stokes stream function*

Derive the Stokes stream function and sketch the streamlines of an axisymmetric flow with radial and meridional velocity components given by

$$u_r = -U \cos \theta \left(1 - \frac{a^3}{r^3} \right), \qquad u_\theta = \frac{1}{2} U \cos \theta \left(2 + \frac{a^3}{r^3} \right), \tag{2.9.26}$$

where U and a are two constants. Verify that the velocity field is solenoidal, compute the vorticity, and discuss the physical interpretation of this flow.

🖳 *Computer Problem*

2.9.6 *Streamlines*

Draw the streamlines of the flows described in (*a*) Problem 2.9.2 and (*b*) Problem 2.9.5.

2.10   Flow induced by vorticity

We return to discussing the structure and properties of an incompressible flow associated with a specified distribution of vorticity with compact support. For simplicity, we assume that the flow takes place in an infinite domain in the absence of interior boundaries, and the velocity field decays far from the region where the magnitude of the vorticity is significant. The presence of interior or exterior boundaries can be taken into account by introducing an appropriate complementary flow, as discussed in Section 2.7.

2.10.1   Biot–Savart integral

Our point of departure is the Biot–Savart law expressed by equations (2.7.3), (2.7.7), (2.7.9), and (2.7.13). In the case of infinite flow without interior boundaries, we obtain a simplified expression for the vector potential,

$$\mathbf{A}(\mathbf{x}) = \frac{1}{4\pi} \iiint_{Flow} \frac{1}{r} \boldsymbol{\omega}(\mathbf{x}') \, dV(\mathbf{x}'), \qquad (2.10.1)$$

where $\hat{\mathbf{x}} = \mathbf{x} - \mathbf{x}'$ and $r = |\mathbf{x} - \mathbf{x}'|$. The velocity field is given by the integral representation

$$\mathbf{u}(\mathbf{x}) = \frac{1}{4\pi} \iiint_{Flow} \frac{1}{r^3} \left[\boldsymbol{\omega}(\mathbf{x}') \times \hat{\mathbf{x}} \right] dV(\mathbf{x}') \qquad (2.10.2)$$

or

$$\mathbf{u}(\mathbf{x}) = \frac{1}{4\pi} \iiint_{Flow} \frac{1}{r} \nabla' \times \boldsymbol{\omega}(\mathbf{x}') \, dV(\mathbf{x}'), \qquad (2.10.3)$$

where the derivatives of the gradient ∇' operate with respect to $\mathbf{x}'$.

Structure of the far flow

Let us assume that the vorticity is concentrated inside a compact region in the neighborhood of a point, $\mathbf{x}_0$, and vanishes far from the vortex region. To study the structure of the far flow, we select a point $\mathbf{x}$ far from the vortex, expand the integral in (2.10.2) in a Taylor series with respect to the integration point $\mathbf{x}'$ about the point $\mathbf{x}_0$, and retain only the constant and linear terms to obtain

$$\mathbf{u}(\mathbf{x}) = \frac{1}{4\pi} \iiint_{Flow} \boldsymbol{\omega}(\mathbf{x}') \, dV(\mathbf{x}') \times \frac{\mathbf{x} - \mathbf{x}_0}{|\mathbf{x} - \mathbf{x}_0|^3}$$
$$+ \frac{1}{4\pi} \iiint_{Flow} \boldsymbol{\omega}(\mathbf{x}') \times \left[(\mathbf{x}' - \mathbf{x}_0) \cdot \nabla' \left(\frac{\mathbf{x} - \mathbf{x}'}{r^3} \right)_{\mathbf{x}'=\mathbf{x}_0} \right] dV(\mathbf{x}') + \cdots . \qquad (2.10.4)$$

Next, we examine the physical interpretation of the two leading terms on the right-hand side.

Rotlet or vorton

The first term on the right-hand side of (2.10.4) represents the flow due to a singularity called a rotlet or vorton, located at the point $\mathbf{x}_0$. The strength of this singularity is equal to the integral of the vorticity over the domain of flow. Recalling that the vorticity field is solenoidal, we write

$$\iiint_{Flow} \boldsymbol{\omega} \, dV = \iiint_{Flow} \nabla \cdot (\boldsymbol{\omega} \otimes \mathbf{x}) \, dV = \iint_{S_\infty} \mathbf{x} \, (\boldsymbol{\omega} \cdot \mathbf{n}) \, dS, \tag{2.10.5}$$

where S_∞ is a closed surface with large size enclosing the vortex region. Assuming that the vorticity decays fast enough for the last integral in (2.10.5) tends to zero as S_∞ expands to infinity, we find that the leading term on the right-hand side of (2.10.4) makes a zero contribution.

Point-source dipole

Concentrating on the second term on the right-hand side of (2.10.4), we change the variable of differentiation in the gradient inside the integrand from $\mathbf{x}'$ to $\mathbf{x}$, while simultaneously introducing a minus sign, and obtain

$$\mathbf{u}(\mathbf{x}) = \frac{1}{4\pi} \iiint_{Flow} [\,\hat{\mathbf{x}}' \cdot \nabla \Big(\frac{\hat{\mathbf{x}}}{|\hat{\mathbf{x}}|^3} \Big) \,] \times \boldsymbol{\omega}(\mathbf{x}') \, dV(\mathbf{x}'), \tag{2.10.6}$$

where $\hat{\mathbf{x}} = \mathbf{x} - \mathbf{x}_0$ and $\hat{\mathbf{x}}' = \mathbf{x}' - \mathbf{x}_0$. In index notation,

$$u_i(\mathbf{x}) = -\frac{1}{4\pi} \epsilon_{jlk} \frac{\partial^2}{\partial x_i \partial x_j} \Big(\frac{1}{|\hat{\mathbf{x}}|} \Big) \iiint_{Flow} \hat{x}'_i \, \omega_l(\mathbf{x}') \, dV(\mathbf{x}'). \tag{2.10.7}$$

To simplify the right-hand side of (2.10.7), we introduce the identity

$$\iiint_{Flow} (\hat{x}_i \omega_l + \hat{x}_i \omega_l) \, dV = \iiint_{Flow} \frac{\partial(\hat{x}_i \hat{x}_l \omega_k)}{\partial x_k} \, dV = \iint_{S_\infty} \hat{x}_i \, \hat{x}_l \, \omega_k n_k \, dS, \tag{2.10.8}$$

where $\mathbf{n}$ is the normal unit vector over the boundaries pointing outward from the flow. Assuming that the vorticity over the large surface S_∞ decays sufficiently fast so that the last integral in (2.10.8) vanishes, we find that the integral on the right-hand side of (2.10.7) is antisymmetric with respect to the indices i and l, and write

$$\iiint_{Flow} \hat{x}'_i \, \omega_l(\mathbf{x}') \, dV(\mathbf{x}') = \frac{1}{2} \, \epsilon_{mil} \epsilon_{mnk} \iiint_{Flow} \hat{x}'_n \, \omega_k(\mathbf{x}') \, dV(\mathbf{x}'). \tag{2.10.9}$$

Substituting the left-hand side of (2.10.9) for the integral on the right-hand side of (2.10.7), contracting the repeated multiplications of the alternating matrix, noting that the function $1/|\mathbf{x} - \mathbf{x}_0|$ satisfies Laplace's equation at every point except $\mathbf{x}_0$, and switching to vector notation, we finally obtain

$$\mathbf{u}(\mathbf{x}) = \frac{1}{4\pi} \mathbf{d} \cdot \nabla \nabla \Big(\frac{1}{|\mathbf{x} - \mathbf{x}_0|} \Big), \tag{2.10.10}$$

where

$$\mathbf{d} = \frac{1}{2} \iiint_{Flow} \hat{\mathbf{x}}' \times \boldsymbol{\omega}(\mathbf{x}') \, dV(\mathbf{x}'). \tag{2.10.11}$$

Expression (2.10.10) shows that, far from the vortex, the flow is similar to that due to a point-source dipole located at the point $\mathbf{x}_0$ whose strength is proportional to the moment of the vorticity. Cursory inspection of the volume and surface integrals in the preceding expressions reveals that (2.10.10) is valid, provided that the vorticity decays at least as fast as $1/|\mathbf{x} - \mathbf{x}_0|^4$ ([24], p. 520).

2.10.2   Kinetic energy

An expression for the kinetic energy of the fluid, $\mathcal{K}$, can be derived in terms of the vorticity distribution, $\boldsymbol{\omega} = \nabla \times \mathbf{u}$. Assuming that the density of the fluid is uniform throughout the domain of flow, we introduce the vector potential, $\mathbf{A}$, and write

$$\mathcal{K} = \frac{1}{2}\rho \iiint_{Flow} \mathbf{u} \cdot \mathbf{u}\, dV = \frac{1}{2}\rho \iiint_{Flow} \mathbf{u} \cdot \nabla \times \mathbf{A}\, dV. \tag{2.10.12}$$

Manipulating the integrand, we find that

$$\mathcal{K} = \frac{1}{2}\rho \iiint_{Flow} \left[\mathbf{A} \cdot \boldsymbol{\omega} - \nabla \cdot (\mathbf{u} \times \mathbf{A}) \right] dV. \tag{2.10.13}$$

Next, we use the divergence theorem to convert the volume integral of the second term inside the integral on the right-hand side into an integral over a large surface, S_∞. Assuming that the velocity decays sufficiently fast for the integral to vanish, we obtain

$$\mathcal{K} = \frac{1}{2}\rho \iiint_{Flow} \mathbf{A} \cdot \boldsymbol{\omega}\, dV. \tag{2.10.14}$$

The vector potential, $\mathbf{A}$, can be expressed in terms of the vorticity distribution using (2.10.1).

In the case of axisymmetric flow, we express the vector potential, $\mathbf{A}$, in terms of the Stokes stream function, ψ, and write $dV = 2\pi\sigma dx\, d\sigma$ to obtain

$$\mathcal{K} = \pi\rho \iint_{Flow} \psi(x, \sigma)\, \omega_\varphi(x, \sigma)\, dx\, d\sigma, \tag{2.10.15}$$

where ω_φ is the azimuthal component of the vorticity. An analogous expression for two-dimensional flow is discussed in Section 11.1.

A useful representation of the kinetic energy in terms of the velocity and vorticity emerges by using the identity

$$\nabla \cdot \left[\mathbf{u}\,(\mathbf{u} \cdot \mathbf{x}) - \frac{1}{2}\,(\mathbf{u} \cdot \mathbf{u})\,\mathbf{x} \right] + \frac{1}{2}\,\mathbf{u} \cdot \mathbf{u} = \mathbf{u} \cdot (\mathbf{x} \times \boldsymbol{\omega}) \tag{2.10.16}$$

([24], p. 520). Solving for the last term on the left-hand side, substituting the result into the first integral of (2.10.12), and using the divergence theorem to simplify the integral, we find that

$$\mathcal{K} = \rho \iiint_{Flow} \mathbf{u} \cdot (\mathbf{x} \times \boldsymbol{\omega})\, dV. \tag{2.10.17}$$

The velocity can be expressed in terms of the vorticity distribution using (2.10.2) or (2.10.3).

2.10.3   Flow due to a vortex sheet

To compute the velocity field due to a three-dimensional vortex sheet, we substitute the vorticity distribution (1.13.2) into the first integral of (2.10.2), finding

$$\mathbf{u}(\mathbf{x}) = \frac{1}{4\pi} \iint_{Sheet} \frac{1}{r^3} \left[\boldsymbol{\zeta}(\mathbf{x}') \times \hat{\mathbf{x}} \right] \mathrm{d}S(\mathbf{x}'). \qquad (2.10.18)$$

This expression provides us with the velocity field in terms of an integral over the the vortex sheet representing a surface distribution of rotlets or vortons. The coefficient of the dipole can be computed from (2.10.11) and is found to be

$$\mathbf{d} = \frac{1}{2} \iint_{Sheet} \hat{\mathbf{x}} \times \boldsymbol{\zeta}(\mathbf{x}) \, \mathrm{d}S(\mathbf{x}), \qquad (2.10.19)$$

where $\hat{\mathbf{x}} = \mathbf{x}' - \mathbf{x}_0$ and $\mathbf{x}_0$ is an arbitrary point.

The irrotational flow induced by the vortex sheet can be described in terms of a velocity potential, ϕ. In Section 10.6, we will see that ϕ can be represented in terms of a distribution of point-source dipoles oriented normal to the vortex sheet. Combining equations (10.6.1) written for the free-space Green's function and equation (10.6.8), we obtain

$$\phi(\mathbf{x}) = -\frac{1}{4\pi} \iint_{Sheet} \frac{1}{r^3} \left[\hat{\mathbf{x}} \cdot \mathbf{n}(\mathbf{x}') \right] (\phi^+ - \phi^-)(\mathbf{x}') \, \mathrm{d}S(\mathbf{x}'), \qquad (2.10.20)$$

where $\mathbf{n}$ is the unit vector normal to the vortex sheet pointing into the upper side corresponding to the plus sign. Taking the gradient of (2.10.20), integrating by parts, and using the definition of $\boldsymbol{\zeta}$, we recover (2.10.18).

Problems

2.10.1 *Impulse*

The impulse required to generate the motion of a fluid with uniform density is expressed by the momentum integral $\mathbf{P} = \rho \iiint \mathbf{u} \, \mathrm{d}V$. Show that two flows with different vorticity distributions but identical dipole strengths, $\mathbf{d}$, require the same impulse.

2.10.2 *Far flow*

Derive the second-order term in the asymptotic expansion (2.10.5) and discuss its physical interpretation. Comment on the asymptotic behavior of the flow when the coefficient of the dipole vanishes.

2.10.3 *Identities*

Prove identities (2.10.8) and (2.10.16).

2.11   Flow due to a line vortex

The flow due to a line vortex provides us with useful insights into the structure and dynamics of flows with concentrated vorticity, such as turbulent flows and flows due to tornadoes and whirls. Consider

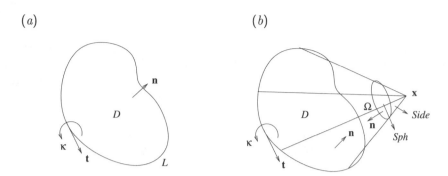

FIGURE 2.11.1   (*a*) Illustration of a closed line vortex, L, and a surface bounded by the line vortex, D. (*b*) The velocity potential at a point, **x**, is proportional to the solid angle subtended by the line vortex, Ω.

the flow induced by a line vortex, L, with strength κ, illustrated in Figure 2.11.1(*a*). To compute the vector potential, **A**, and associated velocity field, **u**, we substitute the vorticity distribution (1.13.1) into the Biot–Savart integrals (2.10.1) and (2.10.2), and find that

$$\mathbf{A}(\mathbf{x}) = \frac{\kappa}{4\pi} \int_L \frac{1}{r} \, \mathbf{t}(\mathbf{x}') \, \mathrm{d}l(\mathbf{x}'), \qquad \mathbf{u}(\mathbf{x}) = \frac{\kappa}{4\pi} \int_L \frac{1}{r^3} \big[\, \mathbf{t}(\mathbf{x}') \times \hat{\mathbf{x}} \,\big] \, \mathrm{d}l(\mathbf{x}'), \qquad (2.11.1)$$

where $\hat{\mathbf{x}} = \mathbf{x} - \mathbf{x}'$, $r = |\hat{\mathbf{x}}|$, and **t** is the unit tangential vector along the line vortex.

Using (2.10.11) and applying Stokes' theorem, we find that the associated coefficient of the dipole prevailing in the far field is given by

$$\mathbf{d} = \frac{\kappa}{2} \int_L (\mathbf{x} - \mathbf{x}_0) \times \mathbf{t}(\mathbf{x}) \, \mathrm{d}l(\mathbf{x}) = \kappa \iint_D \mathbf{n}(\mathbf{x}) \, \mathrm{d}S(\mathbf{x}), \qquad (2.11.2)$$

where $\mathbf{x}_0$ is an arbitrary point, D is an arbitrary closed surface bounded by the line vortex, and **n** is the unit vector normal to D, as shown in Figure 2.11.1(*a*).

2.11.1   Velocity potential

The irrotational flow induced by a line vortex can be expressed in terms of a harmonic potential, ϕ. At the outset, we acknowledge that, because the circulation around a loop that encloses the line vortex once is nonzero and equal to the cyclic constant of the flow around the line vortex, κ, the potential is a multi-valued function of position.

To derive an expression for the potential, we consider the closed line vortex depicted in Figure 2.11.1(*a*), express the velocity given in the second equation of (2.11.1) in index notation, and apply Stokes' theorem to convert the line integral along the line vortex into a surface integral over an arbitrary surface D that is bounded by the line vortex, obtaining

$$u_i(\mathbf{x}) = \frac{\kappa}{4\pi} \int_L \epsilon_{ijn} t_j(\mathbf{x}') \frac{\hat{x}_n}{r^3} \, \mathrm{d}l(\mathbf{x}') = \frac{\kappa}{4\pi} \iint_D \epsilon_{kmj} \frac{\partial}{\partial x'_m} \Big(\epsilon_{ijn} \frac{\hat{x}_n}{r^3} \Big) n_k(\mathbf{x}') \, \mathrm{d}S(\mathbf{x}'), \qquad (2.11.3)$$

which can be rearranged into

$$u_i(\mathbf{x}) = \frac{\kappa}{4\pi} \iint_D \epsilon_{kmj}\,\epsilon_{inj}\, \frac{\partial^2}{\partial x'_m \partial x_n}\left(\frac{1}{r}\right) n_k(\mathbf{x'})\,\mathrm{d}S(\mathbf{x'}). \tag{2.11.4}$$

Expanding the product of the alternating tensors and recalling that $1/r$ satisfies Laplace's equation in three dimensions, everywhere except at the point $\mathbf{x'}$, we obtain

$$u_i(\mathbf{x}) = -\frac{\kappa}{4\pi} \frac{\partial}{\partial x_i}\left(\iint_D \frac{\partial}{\partial x'_k}\left(\frac{1}{r}\right) n_k(\mathbf{x'})\,\mathrm{d}S(\mathbf{x'}) \right). \tag{2.11.5}$$

The right-hand side of (2.11.5) expresses the velocity as the gradient of the velocity potential

$$\phi(\mathbf{x}) = -\frac{\kappa}{4\pi} \iint_D \mathbf{n}(\mathbf{x'}) \cdot \nabla'\left(\frac{1}{r}\right) \mathrm{d}S(\mathbf{x'}) = -\frac{\kappa}{4\pi} \iint_D \frac{\mathbf{x}-\mathbf{x'}}{r^3} \cdot \mathbf{n}(\mathbf{x'})\,\mathrm{d}S(\mathbf{x'}), \tag{2.11.6}$$

which is consistent with the far-field expansion (2.10.10) for a more general flow.

Solid angle

A geometrical interpretation of the potential, ϕ, emerges by introducing a conical surface that contains all rays emanating from a specified field point, $\mathbf{x}$, and passing through the line vortex, as illustrated in Figure 2.11.1(b). We define a control volume that is bounded by the surface D, a section of a sphere with radius R centered at the point $\mathbf{x}$ and confined by the conical surface, denoted by Sph, and the section of the conical surface contained between the spherical surface and the line vortex, denoted by $Side$. Departing from the integral representation (2.11.6) and using the divergence theorem, we find that

$$\phi(\mathbf{x}) = \frac{\kappa}{4\pi} \iint_{Sph,Side} \frac{\mathbf{x}-\mathbf{x'}}{r^3} \cdot \mathbf{n}(\mathbf{x'})\,\mathrm{d}S(\mathbf{x'}). \tag{2.11.7}$$

Because the normal vector is perpendicular to the vector $\mathbf{x}-\mathbf{x'}$, the integral over the conical side surface $Side$ is identically zero. The normal vector over the spherical surface Sph is given by

$$\mathbf{n} = \frac{\alpha}{R}\,(\mathbf{x'}-\mathbf{x}), \tag{2.11.8}$$

where $\alpha = 1$ if the spherical surface is on the right or left side of D, and $\alpha = -1$ otherwise. For the configuration depicted in Figure 2.11.1(b), we select $\alpha = 1$. Equation (2.11.7) then yields

$$\phi(\mathbf{x}) = \alpha\,\frac{\kappa}{4\pi} \iint_{Sph} \frac{\mathbf{x}-\mathbf{x'}}{r^3} \cdot \frac{\mathbf{x'}-\mathbf{x}}{R}\,\mathrm{d}S(\mathbf{x'}) = -\frac{\kappa}{4\pi}\frac{\alpha}{R^2} \iint_{Sph} \mathrm{d}S(\mathbf{x'}) \tag{2.11.9}$$

or

$$\phi(\mathbf{x}) = -\frac{\kappa}{4\pi}\,\Omega(\mathbf{x}), \tag{2.11.10}$$

where $\Omega(\mathbf{x})$ is the solid angle subtended by the line vortex at the point $\mathbf{x}$. For the configuration depicted in Figure 2.11.1(b), the solid angle is positive.

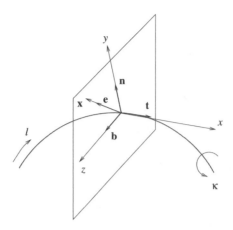

FIGURE 2.11.2   Closeup of a line vortex illustrating a local coordinate system with the x, y, and z axes parallel to the tangential, normal, and binormal vectors. For the configuration shown, the curvature of the line is positive by convention.

The solid angle, Ω, and therefore the scalar potential ϕ, is a multi-valued function of position. In the case of flow due to line vortex ring, Ω changes from -2π to 2π as the point $\mathbf{x}$ crosses the plane of the ring through the interior. This means that ϕ undergoes a corresponding discontinuity equal to κ, in agreement with our discussion in Section 2.1 regarding the behavior of the potential in multiply connected domains.

2.11.2   Self-induced velocity

If we attempt to compute the velocity at a point $\mathbf{x}$ located at a line vortex using the second integral representation in (2.11.1), we will encounter an essential difficulty due to the strong singularity of the integrand. To resolve the local structure of the flow, we introduce local Cartesian coordinates with origin at a chosen point on the line vortex, where the x, y, and z axes are parallel to the unit tangent, normal, and binormal vectors, $\mathbf{t}$, $\mathbf{n}$, and $\mathbf{b}$, respectively, as shown in Figure 2.11.2.

Next, we consider the velocity at a point $\mathbf{x}$ that lies in the normal plane containing $\mathbf{n}$ and $\mathbf{b}$, at the position $\mathbf{x} = \sigma \mathbf{e}$, where $\mathbf{e}$ is a unit vector that lies in the normal plane and σ is the distance from the x axis. Using the integral representation for the velocity (2.11.1), we obtain

$$\mathbf{u}(\mathbf{x}) = \frac{\kappa}{4\pi} \int_L \mathbf{t}(\mathbf{x}') \times \frac{\sigma \mathbf{e} - \mathbf{x}'}{r^3} \, dl(\mathbf{x}'), \tag{2.11.11}$$

which can be rearranged into

$$\mathbf{u}(\mathbf{x}) = \frac{\kappa}{4\pi} \left[-\sigma \mathbf{e} \times \int_L \frac{\mathbf{t}(\mathbf{x}')}{r^3} \, dl(\mathbf{x}') + \int_L \frac{\mathbf{x}'}{r^3} \times \mathbf{t}(\mathbf{x}') \, dl(\mathbf{x}') \right]. \tag{2.11.12}$$

To desingularize the last two integrals, we expand the position vector, $\mathbf{x}'$, and tangential vector along the line vortex, $\mathbf{t}(\mathbf{x}')$, in a Laurent series with respect to arc length, l, measured from the origin in the direction of $\mathbf{t}$, and retain two leading terms to obtain

$$\mathbf{x}' = \left(\frac{\partial \mathbf{x}'}{\partial l}\right)_0 l + \frac{1}{2}\left(\frac{\partial^2 \mathbf{x}'}{\partial l^2}\right)_0 l^2 + \cdots, \qquad \mathbf{t}(\mathbf{x}') = \mathbf{t}_0 + \left(\frac{\partial \mathbf{t}}{\partial l}\right)_0 l + \cdots. \qquad (2.11.13)$$

Using the Frenet-Serret relation

$$\frac{d\mathbf{t}}{dl} = -c\,\mathbf{n}, \qquad (2.11.14)$$

where c is the curvature of the line vortex, we find that

$$\mathbf{x}' = \mathbf{t}_0\, l - \frac{1}{2}\, c_0\, \mathbf{n}_0\, l^2 + \cdots, \qquad \mathbf{t}(\mathbf{x}') = \mathbf{t}_0 - c_0\, \mathbf{n}_0\, l + \cdots, \qquad (2.11.15)$$

where c_0 is the curvature of the line vortex at the origin. For the configuration depicted in Figure 2.11.2, the curvature is positive, $c_0 > 0$.

Considering the first integral in (2.11.12), we truncate the integration domain at $l = \pm a$ and use the expansion for the tangent vector shown in (2.11.15) to obtain

$$\mathcal{I}(\mathbf{x}) \equiv \mathbf{e} \times \int_{-a}^{a} \frac{\mathbf{t}(\mathbf{x}')}{r^3}\, dl(\mathbf{x}') = \mathbf{e} \times \mathbf{t}_0 \int_{-a}^{a} \frac{1}{r^3}\, dl(\mathbf{x}') - c_0\, \mathbf{e} \times \mathbf{n}_0 \int_{-a}^{a} \frac{l}{r^3}\, dl(\mathbf{x}') + \cdots, \qquad (2.11.16)$$

where $r = |\mathbf{x} - \mathbf{x}'|$ and a is a specified length. Next, we set $r^2 \simeq \sigma^2 + l^2$, and obtain

$$\mathcal{I}(\mathbf{x}) = \frac{\mathbf{e} \times \mathbf{t}_0}{\sigma^2} \int_{-a/\sigma}^{a/\sigma} \frac{d\eta}{(1+\eta^2)^{3/2}} - c_0\, \frac{\mathbf{e} \times \mathbf{n}_0}{\sigma} \int_{-a/\sigma}^{a/\sigma} \frac{\eta\, d\eta}{(1+\eta^2)^{3/2}} + \cdots, \qquad (2.11.17)$$

where $\eta \equiv l/\sigma$. Evaluating the integrals in the limit as σ tends to zero and correspondingly a/σ tends to infinity, we find that

$$\mathcal{I}(\mathbf{x}) = 2\, \frac{\mathbf{e} \times \mathbf{t}_0}{\sigma^2} + \cdots, \qquad (2.11.18)$$

where the three dots denote terms that increase slower that $1/\sigma^2$ as $\sigma \to 0$.

Working in a similar fashion with the second integral in (2.11.12), we find that

$$\mathcal{J}(\mathbf{x}) \equiv \int_{-a}^{a} \frac{\mathbf{x}'}{r^3} \times \mathbf{t}(\mathbf{x}')\, dl(\mathbf{x}') = \int_{-a}^{a} \frac{\mathbf{x}'}{r^3} \times \mathbf{t}_0\, dl(\mathbf{x}') - c_0 \int_{-a}^{a} \frac{\mathbf{x}'}{r^3} \times \mathbf{n}_0\, l\, dl(\mathbf{x}') + \cdots \qquad (2.11.19)$$

$$= \int_{-a}^{a} \frac{\mathbf{t}_0\, l - \frac{1}{2} c_0\, \mathbf{n}_0\, l^2}{r^3} \times \mathbf{t}_0\, dl(\mathbf{x}') - c_0 \int_{-a}^{a} \frac{\mathbf{t}_0\, l - \frac{1}{2} c_0\, \mathbf{n}_0\, l^2}{r^3} \times \mathbf{n}_0\, l\, dl(\mathbf{x}') + \cdots,$$

yielding

$$\mathcal{J}(\mathbf{x}) = -\frac{1}{2}\, c_0\, \mathbf{b}_0 \int_{-a}^{a} \frac{l^2}{r^3}\, dl(\mathbf{x}') + \cdots, \qquad (2.11.20)$$

where $\mathbf{b}_0 = \mathbf{t}_0 \times \mathbf{n}_0$. Setting $r^2 = \sigma^2 + l^2$ and defining $\eta \equiv l/\sigma$, we obtain

$$\boldsymbol{J}(\mathbf{x}) = -\frac{1}{2}\, c_0\, \mathbf{b}_0 \int_{-a/\sigma}^{a/\sigma} \frac{\eta^2 \mathrm{d}\eta}{(1+\eta^2)^{3/2}} + \cdots. \tag{2.11.21}$$

Evaluating the integrals in the limit as σ/a tends to zero, we find that

$$\boldsymbol{J}(\mathbf{x}) = -c_0\, \mathbf{b}_0 \ln\left(\frac{a}{\sigma}\right) + \cdots, \tag{2.11.22}$$

where the three dots indicate terms whose magnitude increases slower than $|\ln \sigma|$ as $\sigma \to 0$.

Finally, we substitute (2.11.18) and (2.11.22) into (2.11.12) and obtain an expression for the velocity near the line vortex, first derived by Luigi Sante Da Rios in 1906 (see [339]),

$$\mathbf{u}(\mathbf{x}) = \frac{\kappa}{4\pi\sigma}\, \mathbf{t}_0 \times \mathbf{e} - \frac{\kappa}{4\pi}\, c_0\, \mathbf{b}_0 \ln\left(\frac{a}{\sigma}\right) + \cdots. \tag{2.11.23}$$

The first term on the right-hand side of (2.11.23) describes the expected swirling motion around the line vortex, which is similar to that around a point vortex in two-dimensional flow discussed in Section 2.13.1. In the limit as σ/a tends to zero, the second term diverges, showing that the self-induced velocity of a curved line vortex ($c_0 \neq 0$ is infinite. This singular behavior reflects the severe approximation involved in the mathematical fabrication of singular vortex structures. In Section 11.10.1, we will discuss the regularization of (2.11.16) accounting for the finite size of the vortex core.

2.11.3   Local induction approximation (LIA)

The local induction approximation (LIA) amounts to computing the self-induced velocity of a line vortex by retaining only the second term on the right-hand side of (2.11.23), obtaining

$$\mathbf{u}(\mathbf{X}) = \frac{\mathrm{D}\mathbf{X}}{\mathrm{D}t} = -\frac{\kappa}{4\pi}\, c(\mathbf{X})\, \mathbf{b}(\mathbf{X}) \ln\frac{1}{\epsilon}, \tag{2.11.24}$$

where $\mathbf{X}$ is the position of a point particle along the centerline of the line vortex, ϵ is a small dimensionless parameter expressing the size of the vortex core, and $\mathrm{D}/\mathrm{D}t$ is the material derivative.

If the product of the curvature and the binormal vector, $c\mathbf{b}$, is constant along the line vortex, the line vortex translates as a rigid body. Examples include a circular vortex ring, an advancing helical vortex advancing rotating about its axis, and a planar nearly rectilinear line vortex with small amplitude sinusoidal undulations rotating as a rigid body about the centerline (Problems 2.11.2, 2.11.3). Numerical methods for computing the motion of line vortices based on the LIA are discussed in Section 11.10.1.

Da Rios Equations

Da Rios (1906, see [339]) used the LIA to derive a coupled nonlinear system of ordinary differential equations governing the evolution of the curvature and torsion of a line vortex. To simplify the notation, we introduce a scaled time, $\tilde{t} \equiv -t\kappa \ln \epsilon/(4\pi)$, and recast (2.11.24) into the form

$$\mathbf{V} \equiv \dot{\mathbf{X}} = -c\,\mathbf{b}, \tag{2.11.25}$$

where a dot denotes a derivative with respect to $\tilde{t}$. The Frenet–Serret relations derived in Section 1.6.2 state that

$$\mathbf{t}' = -c\,\mathbf{n}, \qquad \mathbf{b}' = -\tau\,\mathbf{n}, \qquad \mathbf{n}' = c\,\mathbf{t} + \tau\,\mathbf{b}, \qquad (2.11.26)$$

where τ is the torsion of the line vortex and a prime denotes a derivative with respect to arc length along the line vortex, l.

Equations (1.6.34) and (1.6.40) provide us with evolution equations for the curvature and torsion. In the present notation, these equations take the form

$$\dot{c} = -2c\,\mathbf{V}' \cdot \mathbf{t} - \mathbf{V}'' \cdot \mathbf{n},$$
$$\dot{\tau} = -\mathbf{V}' \cdot (\tau\mathbf{t} + c\mathbf{b}) + \frac{1}{c}\,\mathbf{V}'' \cdot \left(\tau\mathbf{n} + \frac{c'}{c}\,\mathbf{b}\right) - \frac{1}{c}\,\mathbf{V}''' \cdot \mathbf{b}. \qquad (2.11.27)$$

Differentiating (2.11.25) and using the Frenet-Serret relations, we find that

$$\mathbf{V}' = -c'\mathbf{b} + c\tau\,\mathbf{n}, \qquad \mathbf{V}'' = c^2\tau\mathbf{t} + (2c'\tau + c\tau')\,\mathbf{n} + \phi\,\mathbf{b},$$
$$\mathbf{V}''' \cdot \mathbf{b} = \tau(2c'\tau + c\tau') + \phi', \qquad (2.11.28)$$

where we have defined

$$\phi \equiv c\tau^2 - c''. \qquad (2.11.29)$$

Substituting these expressions into (2.11.27) and simplifying, we obtain the Da Rios equations

$$\dot{c} = -2c'\tau - c\tau' \qquad (2.11.30)$$

and

$$\dot{\tau} = c\,c' + \frac{c'}{c^2}\,\phi - \frac{1}{c}\,\phi' = \left(\frac{1}{2}c^2 - \frac{\phi}{c}\right)'. \qquad (2.11.31)$$

Equation (2.11.30) can expressed in the form

$$\frac{\mathrm{D}c^2}{\mathrm{D}\tilde{t}} = -2\,(c^2\tau)'. \qquad (2.11.32)$$

Integrating with respect to arc length along a closed line vortex provides us with a geometrical conservation law [33].

Schrödinger equation

Hasimoto [171] discovered that the local induction approximation can be reformulated as a nonlinear Schrödinger equation for a properly defined complex scalar function. To demonstrate this reduction, we recall the evolution equations for the tangent, normal, and binormal unit vectors stated in (1.6.28), (1.6.36), and (1.6.37). In the present notation, these equations read

$$\dot{\mathbf{t}} = \mathbf{V}' - (\mathbf{V}' \cdot \mathbf{t})\,\mathbf{t}, \qquad \dot{\mathbf{n}} = -(\mathbf{V}' \cdot \mathbf{n})\,\mathbf{t} - \frac{1}{c}\,(\mathbf{V}'' \cdot \mathbf{b})\,\mathbf{b},$$
$$\dot{\mathbf{b}} = -(\mathbf{V}' \cdot \mathbf{b})\,\mathbf{t} + \frac{1}{c}\,(\mathbf{V}'' \cdot \mathbf{b})\,\mathbf{n}. \qquad (2.11.33)$$

Substituting the velocity from (2.11.28), we find that

$$\dot{\mathbf{t}} = -c'\mathbf{b} + c\tau\,\mathbf{n}, \qquad \dot{\mathbf{n}} = -c\tau\mathbf{t} - \chi\mathbf{b}, \qquad \dot{\mathbf{b}} = c'\mathbf{t} + \chi\,\mathbf{n}, \qquad (2.11.34)$$

where

$$\chi \equiv \frac{\phi}{c} = \tau^2 - \frac{c''}{c}. \qquad (2.11.35)$$

Next, we introduce the complex vector field $\mathbf{q} \equiv \mathbf{n} + \mathrm{i}\,\mathbf{b}$, where i is the imaginary unit, $\mathrm{i}^2 = -1$, as discussed in Section 1.6.3. The second and third equations in (2.11.34) can be unified into the complex form

$$\dot{\mathbf{q}} = \mathrm{i}\,[\,(c' + \mathrm{i}\,c\tau)\,\mathbf{t} + \chi\,\mathbf{q}\,], \qquad (2.11.36)$$

and then rearranged into an evolution equation for the function $\mathbf{Q}$ defined in (1.6.21),

$$\dot{\mathbf{Q}} = \mathrm{i}\,(-\psi'\,\mathbf{t} + \chi\,\mathbf{Q}). \qquad (2.11.37)$$

Combining this expression with (1.6.20), we derive the evolution equation

$$\frac{\partial^2 \mathbf{Q}}{\partial \tilde{t}\,\partial l} = -\frac{\partial(\psi\,\mathbf{t})}{\partial \tilde{t}} = \mathrm{i}\,(-\psi\,\mathbf{t} + \chi\,\mathbf{Q})'. \qquad (2.11.38)$$

Expanding the time and arc length derivatives, we obtain

$$\dot{\psi}\,\mathbf{t} + \psi\,\dot{\mathbf{t}} = \mathrm{i}\,(\psi''\,\mathbf{t} + \psi'\,\mathbf{t}' - \chi'\,\mathbf{Q} - \chi\,\mathbf{Q}'). \qquad (2.11.39)$$

Separating the tangential from the normal components, we obtain

$$\dot{\psi} = \mathrm{i}\,(\psi'' + \psi\,\chi), \qquad \psi\,\dot{\mathbf{t}} = \mathrm{i}\,(\psi'\,\mathbf{t}' - \chi'\,\mathbf{Q}). \qquad (2.11.40)$$

Making substitutions in the second equation, we obtain

$$\psi\,(-c'\mathbf{b} + c\tau\mathbf{n}) = \mathrm{i}\,(-c\psi' + \chi'\,\frac{\psi}{c})\,\mathbf{n} - \chi'\,\frac{\psi}{c}\,\mathbf{b}, \qquad (2.11.41)$$

which requires that

$$\chi' = c\,c' = \frac{1}{2}\,(c^2)' = \frac{1}{2}\,(\psi\psi^*)', \qquad \mathrm{i}\chi'\,\psi = c^2(\psi\tau + \mathrm{i}\psi'). \qquad (2.11.42)$$

Integrating the first equation, we find that

$$\chi = \frac{1}{2}\,\big(\,\psi\psi^* + A(t)\,\big), \qquad (2.11.43)$$

where $A(t)$ is a specified function of time, Substituting this expression into the first equation of (2.11.40) we derive a nonlinear Schrödinger equation,

$$\frac{1}{\mathrm{i}}\frac{\partial\psi}{\partial\tilde{t}} = \frac{\partial^2\psi}{\partial l^2} + \frac{1}{2}\,\big(|\psi|^2 + A(t)\,\big)\,\psi, \qquad (2.11.44)$$

which is known to admit solutions in the form of nonlinear traveling waves called solitons.

Problems

2.11.1 *A rectilinear line vortex*

Use the expression for the vector potential given in (2.11.1) to compute the velocity field due to a rectilinear line vortex. Compare the results with the velocity field due to a point vortex discussed in Section 2.13.1.

2.11.2 *A helical line vortex*

A helical line vortex revolving around the x axis is described in the parametric form by the equations $x = b\varphi$, $y = a\cos\varphi$, $z = a\sin\varphi$, where a is the radius of the circumscribed cylinder, φ is the azimuthal angle, and $2\pi b$ is the helical pitch. Show that the velocity induced by a helical line vortex with strength κ is given by [167]

$$u_x = \frac{\kappa}{4\pi}\Big(a\frac{\partial I_2}{\partial y} + a\frac{\partial I_3}{\partial z}\Big), \qquad u_y = \frac{\kappa}{4\pi}\Big(b\frac{\partial I_1}{\partial z} - a\frac{\partial I_2}{\partial x}\Big), \qquad u_z = -\frac{\kappa}{4\pi}\Big(b\frac{\partial I_1}{\partial y} + a\frac{\partial I_3}{\partial x}\Big), \quad (2.11.45)$$

where

$$\begin{bmatrix} I_1 \\ I_2 \\ I_3 \end{bmatrix} = \int_{-\infty}^{\infty} \begin{bmatrix} 1 \\ \cos\theta \\ \sin\theta \end{bmatrix} \frac{\mathrm{d}\theta}{[\,(x - b\theta)^2 + (y - a\cos\theta)^2 + (z - a\sin\theta)^2]^{1/2}}. \qquad (2.11.46)$$

2.11.3 *A sinusoidal line vortex*

Show that a planar, nearly rectilinear line vortex with small amplitude sinusoidal undulations rotates about its axis as a rigid body.

2.12   Axisymmetric flow induced by vorticity

In the case of axisymmetric flow without swirling motion, we take advantage of the known orientation of the vorticity vector to simplify the Biot–Savart integral by performing the integration in the azimuthal direction by analytical or accurate numerical methods. To begin, we introduce cylindrical polar coordinates, (x, σ, φ), and substitute $\boldsymbol{\omega}(x,\sigma) = \omega_\varphi(x,\sigma)\,\mathbf{e}_\varphi$ into the Biot–Savart integral (2.10.2). Expressing the differential volume as $\mathrm{d}V = \sigma\,\mathrm{d}\varphi\,\mathrm{d}A$, where $\mathrm{d}A = \mathrm{d}x\,\mathrm{d}\sigma$, we obtain

$$\mathbf{u}(\mathbf{x}) = -\frac{1}{4\pi} \iint_{Flow} \Big(\int_0^{2\pi} \frac{\hat{\mathbf{x}} \times \mathbf{e}_\varphi'}{r^3}\,\mathrm{d}\varphi' \Big)\, \omega_\varphi(x',\sigma')\,\sigma'\,\mathrm{d}A(\mathbf{x}'), \qquad (2.12.1)$$

where $r = |\mathbf{x} - \mathbf{x}'|$. Next, we substitute

$$\hat{y} = \sigma\cos\varphi - \sigma'\cos\varphi', \qquad \hat{z} = \sigma\sin\varphi - \sigma'\sin\varphi', \qquad (2.12.2)$$

set $\mathbf{e}_\varphi' = (0, -\sin\varphi', \cos\varphi')$, and compute the outer vector product, finding

$$\mathbf{u}(\mathbf{x}) = \frac{1}{4\pi} \iint_{Flow} \Big(\int_0^{2\pi} \frac{1}{r^3} \begin{bmatrix} -\sigma\cos\hat{\varphi} + \sigma' \\ \hat{x}\cos\varphi' \\ \hat{x}\sin\varphi' \end{bmatrix}\,\mathrm{d}\varphi' \Big)\, \omega_\varphi(x',\sigma')\,\sigma'\,\mathrm{d}A(\mathbf{x}'), \qquad (2.12.3)$$

where $\hat{x} = x - x'$ and $\hat{\varphi} = \varphi - \varphi'$. The radial and azimuthal velocity components are

$$u_\sigma = \cos\varphi\, u_y + \sin\varphi\, u_z, \qquad u_\varphi = -\sin\varphi\, u_y + \cos\varphi\, u_z. \tag{2.12.4}$$

Making substitutions, we obtain

$$\begin{bmatrix} u_x \\ u_\sigma \\ u_\varphi \end{bmatrix}(\mathbf{x}) = \frac{1}{4\pi} \iint_{Flow} \left(\int_0^{2\pi} \frac{1}{r^3} \begin{bmatrix} -\sigma\cos\hat{\varphi} + \sigma' \\ \hat{x}\cos\varphi' \\ -\hat{x}\sin\varphi' \end{bmatrix} d\varphi' \right) \omega_\varphi(x',\sigma')\,\sigma'\,dA(\mathbf{x}'). \tag{2.12.5}$$

Finally, we substitute

$$r^2 = \hat{x}^2 + \sigma^2 + \sigma'^2 - 2\sigma\sigma'\cos\hat{\varphi} = \hat{x}^2 + (\sigma + \sigma')^2 - 4\sigma\sigma'\cos^2(\tfrac{1}{2}\hat{\varphi}), \tag{2.12.6}$$

and integrate to find that $u_\varphi = 0$, as expected, and

$$\begin{bmatrix} u_x \\ u_\sigma \end{bmatrix}(x,\sigma) = \frac{1}{4\pi} \iint_{Flow} \begin{bmatrix} -\sigma\, I_{31}(\hat{x},\sigma,\sigma') + \sigma'\, I_{30}(\hat{x},\sigma,\sigma') \\ \hat{x}\, I_{31}(\hat{x},\sigma,\sigma') \end{bmatrix} \omega_\varphi(x',\sigma')\,\sigma'\,dA(\mathbf{x}'), \tag{2.12.7}$$

where

$$I_{nm}(\hat{x},\sigma,\sigma') = \int_0^{2\pi} \frac{\cos^m\hat{\varphi}}{\left[\hat{x}^2 + (\sigma+\sigma')^2 - 4\sigma\sigma'\cos^2(\tfrac{1}{2}\hat{\varphi})\right]^{n/2}}\, d\hat{\varphi}. \tag{2.12.8}$$

Working in a similar fashion with the vector potential given in (2.10.1) in terms of the vorticity distribution, and recalling the $\mathbf{A} = (\psi/\sigma)\,\mathbf{e}_\varphi$, we derive a corresponding representation for the Stokes stream function,

$$\psi(x,\sigma) = \frac{\sigma}{4\pi} \iint_{Flow} I_{11}(\hat{x},\sigma,\sigma')\,\omega_\varphi(x',\sigma')\,\sigma'\,dA(\mathbf{x}'), \tag{2.12.9}$$

where $\hat{x} = x - x'$.

Computation of the integrals I_{nm}

To compute the integrals I_{nm} defined in (2.12.8), we write

$$I_{nm}(\hat{x},\sigma,\sigma') = \frac{4}{\left[\hat{x}^2 + (\sigma+\sigma')^2\right]^{n/2}}\, J_{nm}(k) = 4 \left(\frac{k}{\sqrt{4\sigma\sigma'}}\right)^n J_{nm}(k), \tag{2.12.10}$$

where

$$J_{nm}(k) \equiv \int_0^{\pi/2} \frac{(2\cos^2\eta - 1)^m}{(1 - k^2\cos^2\eta)^{n/2}}\, d\eta, \tag{2.12.11}$$

$\eta = \hat{\varphi}/2$, and

$$k^2 \equiv \frac{4\sigma\sigma'}{\hat{x}^2 + (\sigma+\sigma')^2} \tag{2.12.12}$$

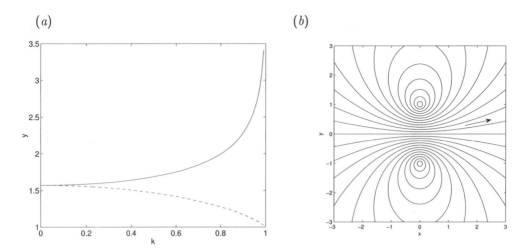

FIGURE 2.12.1 (a) Graphs of the complete elliptic integral of the first kind, $F(k)$ (solid line) and second kind, $E(k)$ (broken line), computed by an efficient iterative method. (b) Streamline pattern in a plane passing through the axis of revolution of a line vortex ring with positive circulation; lengths have been scaled by the ring radius.

is a dimensionless group varying the range $0 \leq k^2 \leq 1$. As $x \to x'$ and $\sigma \to \sigma'$, the composite variable k^2 tends to unity.

The integrals J_{nm} can be expressed in terms of the complete elliptic integrals of the first and second kind, F and E, defined as

$$F(k) \equiv \int_0^1 \frac{d\tau}{(1-\tau^2)^{1/2}(1-k^2\tau^2)^{1/2}} = \int_0^{\pi/2} \frac{du}{\sqrt{1-k^2\sin^2 u}} \qquad (2.12.13)$$

and

$$E(k) \equiv \int_0^{\pi/2} \sqrt{1-k^2\sin^2 u}\, du, \qquad (2.12.14)$$

where τ and u are dummy integration variables. Graphs of these integrals are shown in Figure 2.12.1(a). As k tends to unity, $F(k)$ diverges as $F(k) \simeq \ln(4/\sqrt{1-k^2})$, whereas $E(k)$ tends to unity.

Resorting to tables of indefinite integrals, we derive the expressions

$$J_{10}(k) = F(k), \qquad J_{11}(k) = \frac{2-k^2}{k^2}\, F(k) - \frac{2}{k^2}\, E(k),$$

$$J_{30}(k) = \frac{E(k)}{1-k^2}, \qquad J_{31}(k) = -\frac{2}{k^2}\, F(k) + \frac{2-k^2}{k^2(1-k^2)}\, E(k) \qquad (2.12.15)$$

(e.g., [150], p. 590). Substituting the expression for $J_{11}(k)$ into (2.12.10) and then into (2.12.9), we obtain the Stokes stream function

$$\psi(x,\sigma) = \frac{1}{2\pi} \iint_{Flow} \left(\frac{2-k^2}{k} F(k) - \frac{2}{k} E(k) \right) \sqrt{\sigma\sigma'}\, \omega_\varphi(x',\sigma')\, \mathrm{d}A(\mathbf{x}'). \tag{2.12.16}$$

Computation of the complete elliptic integrals

To evaluate the complete elliptic integrals of the first and second kind, we may compute the sequence

$$K_0 = k, \qquad K_p = \frac{1 - (1 - K_{p-1}^2)^{1/2}}{1 + (1 - K_{p-1}^2)^{1/2}} \tag{2.12.17}$$

for $p = 1, 2, \ldots$, and then set

$$F(k) = \frac{\pi}{2}\,(1 + K_1)(1 + K_2)(1 + K_3)\cdots, \qquad E(k) = F(k)\left(1 - \frac{1}{2}k^2 P\right), \tag{2.12.18}$$

where

$$P = 1 + \frac{1}{2} K_1 \left(1 + \frac{1}{2} K_2 (\cdots) \right). \tag{2.12.19}$$

Alternative polynomial approximations are available (e.g., [2], Chapter 17). In Matlab, complete elliptic integrals can be computed using the native function *ellipke*.

2.12.1   Line vortex rings

The azimuthal vorticity component associated with a circular line vortex ring of radius Σ located at the axial position X is $\omega_\varphi(\mathbf{x}) = \kappa\,\delta_2(\mathbf{x} - \mathbf{X})$, where δ_2 is the two-dimensional delta function in an azimuthal plane, and $\mathbf{x}_r = X\mathbf{e}_x + \Sigma\mathbf{e}_\sigma$ is the trace of the ring in that plane. Substituting this expression into (2.12.7) and performing the integration using the distinctive properties of the delta function, we obtain the axial and radial velocity components

$$\begin{bmatrix} u_x \\ u_\sigma \end{bmatrix}(x,\sigma) = \frac{\kappa}{4\pi}\, \Sigma \begin{bmatrix} -\sigma\, I_{31}(\hat{x},\sigma,\Sigma) + \Sigma\, I_{30}(\hat{x},\sigma,\Sigma) \\ \hat{x}\, I_{31}(\hat{x},\sigma,\Sigma) \end{bmatrix}, \tag{2.12.20}$$

where $\hat{x} = x - X$. The corresponding Stokes stream function is found from (2.12.9),

$$\psi(x,\sigma) = \frac{\kappa}{4\pi}\, \sigma\Sigma\, I_{11}(\hat{x},\sigma,\Sigma). \tag{2.12.21}$$

The streamline pattern in a plane passing through the axis of revolution is shown in Figure 2.12.1(*b*).

Velocity potential

The velocity potential due to a line vortex ring can be deduced from expressions (2.11.6) and (2.11.9) for line vortices. Identifying the surface D in (2.11.6) with a circular disk of radius Σ bounded by the vortex ring, we obtain

$$\phi(x,\sigma) = -\frac{\kappa}{4\pi}\, \Omega_{ring}(\mathbf{x}), \tag{2.12.22}$$

where Ω_{ring} is the angle subtended by the ring from the point $\mathbf{x}$, given by

$$\Omega(\mathbf{x}) = \hat{x} \int_0^{2\pi} \int_0^{\Sigma} \frac{\sigma' \, d\sigma'}{[\hat{x}^2 + (\sigma - \sigma' \cos\varphi')^2 + \sigma'^2 \sin^2\varphi']^{3/2}} \, d\varphi'. \tag{2.12.23}$$

Integrating with respect to the azimuthal angle, we obtain

$$\Omega(\hat{x}, \sigma, \Sigma) = \hat{x} \int_0^{\Sigma} I_{30}(\hat{x}, \sigma, \sigma') \sigma' \, d\sigma' = 4\hat{x} \int_0^{\sigma_R} \frac{J_{30}(k)}{[\hat{x}^2 + (\sigma + \sigma')^2]^{3/2}} \sigma' \, d\sigma', \tag{2.12.24}$$

where the functions I_{30} and J_{30} are defined in (2.12.8), (2.12.10), and (2.12.11), and k^2 is defined in (2.12.12). Finally, we obtain

$$\Omega(\hat{x}, \sigma, \Sigma) = 4\hat{x} \int_0^{\Sigma} \frac{E(k)}{1-k^2} \frac{\sigma'}{[\hat{x}^2 + (\sigma + \sigma')^2]^{3/2}} \, d\sigma', \tag{2.12.25}$$

where $E(k)$ is the complete elliptic integral of the second kind. An alternative method of computing Ω is available [302].

2.12.2   Axisymmetric vortices with linear vorticity distribution

Consider a flow containing an axisymmetric vortex embedded in an otherwise irrotational fluid. Inside the vortex, the azimuthal vorticity component varies linearly with distance from the axis of revolution, $\omega_\varphi = \Omega\sigma$, where Ω is a constant. Using (2.12.9) or (2.12.16), we find that the Stokes stream function of the induced flow is

$$\psi(x, \sigma) = \frac{\Omega}{4\pi} \sigma \iint_{A_V} I_{11}(\hat{x}, \sigma, \sigma') \sigma'^2 \, dA(\mathbf{x}'), \tag{2.12.26}$$

or

$$\psi(x, \sigma) = \frac{\Omega}{2\pi} \iint_{A_V} \left(\frac{2-k^2}{k} F(k) - \frac{2}{k} E(k) \right) \sqrt{\sigma\sigma'} \, \sigma' \, dA(\mathbf{x}'), \tag{2.12.27}$$

where A_V is the area occupied by the vortex in an azimuthal plane, $dA = dx \, d\sigma$ is a corresponding differential area, $\hat{x} = x - x'$, and the integrals I_{nm} are defined in (2.12.8).

Radial velocity

The radial velocity component is found by differentiating the stream function, finding

$$u_\sigma(x, \sigma) = -\frac{1}{\sigma} \frac{\partial\psi}{\partial x} = \frac{\Omega}{4\pi} \iint_{A_V} \frac{\partial}{\partial x'} \left(I_{11}(\hat{x}, \sigma, \sigma') \right) \sigma'^2 \, dA(\mathbf{x}'), \tag{2.12.28}$$

where $\hat{x} = x - x'$. Using the divergence theorem, we derive a simplified expression in terms of an integral along the trace of the vortex contour in an azimuthal plane, C,

$$u_\sigma(x, \sigma) = \frac{\Omega}{4\pi} \int_C I_{11}(\hat{x}, \sigma, \sigma') n_x(\mathbf{x}') \sigma'^2 \, dl(\mathbf{x}'), \tag{2.12.29}$$

where $\mathbf{n}$ is the normal unit vector pointing outward from the vortex contour.

Axial velocity

To develop a corresponding expression for the axial velocity component, we introduce the velocity potential function

$$\phi(x,\sigma) = -\frac{\Omega}{4\pi} \iint_{A_V} \Omega_{ring}(\hat{x},\sigma,\sigma')\,\sigma'\,dA(\mathbf{x}'), \qquad (2.12.30)$$

and work as in (2.12.28) to obtain

$$u_x(x,\sigma) = \frac{1}{\sigma}\frac{\partial\phi}{\partial x} = \frac{\Omega}{4\pi} \iint_{A_V} \frac{\partial}{\partial x'}\Big(\Omega_{ring}(\hat{x},\sigma,\sigma')\Big)\,\sigma'\,dA(\mathbf{x}'). \qquad (2.12.31)$$

Now using the divergence theorem, we find that

$$u_x(x,\sigma) = \frac{\Omega}{4\pi} \int_C \Omega_{ring}(\hat{x},\sigma,\sigma')\,\sigma'\,n_x(\mathbf{x}')\,dl(\mathbf{x}'). \qquad (2.12.32)$$

To compute the right-hand side of (2.12.32), we introduce a branch cut so that the solid angle becomes single-valued [302].

To circumvent introducing a branch cut, we resort to the integral representation of the Stokes stream function and write

$$u_x(x,\sigma) = \frac{1}{\sigma}\frac{\partial\psi}{\partial\sigma} = \frac{\Omega}{4\pi} \iint_{A_V} \frac{1}{\sigma}\frac{\partial}{\partial\sigma}\Big(\sigma\,I_{11}(\hat{x},\sigma,\sigma')\Big)\,\sigma'^2\,dA(\mathbf{x}'). \qquad (2.12.33)$$

If we were able to find two functions, $F(\hat{x},\sigma,\sigma')$ and $G(\hat{x},\sigma,\sigma')$, such that

$$\sigma'^2\frac{\partial}{\partial\sigma}\Big(\sigma\,I_{11}(\hat{x},\sigma,\sigma')\Big) = \sigma\Big(\frac{\partial F}{\partial\hat{x}} - \frac{\partial G}{\partial\sigma'}\Big), \qquad (2.12.34)$$

then we could write the contour integral representation

$$u_x(x,\sigma) = -\frac{\Omega}{4\pi} \int_C \big[\,F(\hat{x},\sigma,\sigma')\,n_x(x',\sigma') + G(\hat{x},\sigma,\sigma')\,n_\sigma(x',\sigma')\,\big]\,dl(x',\sigma'). \qquad (2.12.35)$$

Shariff, Leonard & Fertziger [368] considered the right-hand side of (2.12.1) and expressed the curl of the vorticity in terms of generalized functions. Their analysis implies that

$$F(\hat{x},\sigma,\sigma') = \hat{x}\,\sigma'I_{10}(\hat{x},\sigma,\sigma'), \qquad G(\hat{x},\sigma,\sigma') = \sigma\,\sigma'I_{11}(\hat{x},\sigma,\sigma'), \qquad (2.12.36)$$

yielding the computationally convenient expression

$$u_x(x,\sigma) = -\frac{\Omega}{4\pi} \int_C \big[\,\hat{x}\,I_{10}(\hat{x},\sigma,\sigma')\,n_x(x',\sigma') + \sigma\,I_{11}(\hat{x},\sigma,\sigma')\,n_\sigma(x',\sigma')\,\big]\,\sigma'\,dl(x',\sigma'). \qquad (2.12.37)$$

Taken together, equations (2.12.32) and (2.12.37) provide us with a basis for the contour dynamics formulation of axisymmetric vortex flow with linear vorticity distribution discussed in Section 11.9.3.

(a)   (b)

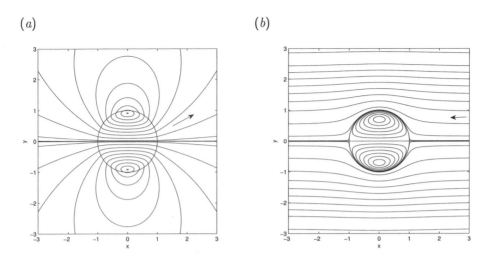

FIGURE 2.12.2 Streamline pattern associated with Hill's spherical vortex (a) in a stationary frame of reference and (b) in a frame of reference traveling with the vortex.

Hill's vortex

Hill's spherical vortex provides us with an important example of an axisymmetric vortex with linear vorticity distribution, $\omega_\varphi = \Omega\sigma$. To describe the flow, we introduce cylindrical polar coordinates with origin at the center of the vortex, (x, σ, φ), and associated spherical polar coordinates, (r, θ, φ). In a frame of reference moving with the vortex, the Stokes stream function inside Hill's vortex is given by

$$\psi_{int} = \frac{\Omega}{10}\,\sigma^2(a^2 - r^2),\tag{2.12.38}$$

where a is the vortex radius. Outside the vortex, the stream function is

$$\psi_{ext} = -\frac{\Omega}{15}\,a^2\sigma^2\left(1 - \frac{a^3}{r^3}\right),\tag{2.12.39}$$

where r is the distance from the origin, $r^2 = x^2 + \sigma^2$, $x = r\cos\theta$, and $\sigma = r\sin\theta$. We may readily verify that $\psi_{int} = \psi_{ext} = 0$ and $(\partial\psi_{int}/\partial r)_\theta = (\partial\psi_{ext}/\partial r)_\theta$ at $r = a$, ensuring that the velocity is continuous across the vortex contour. In fact, the exterior flow is potential flow past a sphere discussed in Section 7.5.2. Cursory inspection of the exterior flow reveals that the spherical vortex translates steadily along the x axis with velocity

$$V = \frac{2}{15}\,\Omega a^2,\tag{2.12.40}$$

while maintaining its spherical shape.

The streamline pattern in a stationary frame of reference and in a frame of reference translating with the vortex is shown in Figure 2.12.2. Comparing the pattern shown in Figure 2.12.2(a) with that shown in Figure 2.12.1(b) for a line vortex ring, we note a similarity in the structure of the exterior flow and conclude that the particular way in which the vorticity is distributed inside a vortex plays a secondary role in determining the structure of the flow far from the vortex.

Vortex rings finite core

Hill's vortex constitutes a limiting member of a family of steadily translating vortex rings parametrized by the cross-sectional area. The opposite extreme member in the family is a line vortex ring with infinitesimal cross-sectional area. The structure and stability of the rings have been studied by analytical and numerical methods, as discussed in Section 11.9 [282, 302].

Problem

2.12.1 *Hill's spherical vortex*

Departing from (2.12.38) and (2.12.39), confirm that (a) the azimuthal component of the vorticity is $\omega_\varphi = \Omega\sigma$ inside Hill's vortex and vanishes in the exterior of the vortex; (b) the velocity is continuous across the vortex boundary; (c) the velocity of translation of the vortex is given by (2.12.40).

 Computer Problems

2.12.2 *Complete elliptic integrals*

Write a computer program that computes the complete elliptic integrals of the first and second kind, F and E, according to (2.12.18). Confirm that your results are consistent with tabulated values.

2.12.3 *A line vortex ring*

(a) Write a program that computes the velocity field induced by a line vortex ring and reproduce the streamline pattern shown in Figure 2.12.2.

(b) Write a program that computes the velocity potential associated with a line vortex ring.

2.12.4 *Streamlines of Hill's spherical vortex*

Reproduce the streamlines pattern shown in Figure 2.12.3(a). The velocity could be computed by numerical differentiation setting, for example, $\partial\psi/\partial\sigma \simeq [\psi(\sigma+\epsilon) - \psi(\sigma-\epsilon)]/(2\epsilon)$, where ϵ is a small increment.

2.13   Two-dimensional flow induced by vorticity

Integral representations for the velocity of a two-dimensional flow in the xy plane in terms of the vorticity can be derived using the general formulas presented in Section 2.8 for three-dimensional flow. In this case, we stipulate that the vortex lines are rectilinear, parallel to the z axis. However, it is expedient to begin afresh from the integral representation (2.7.37) providing us with expressions for the x and y velocity components,

$$u_x(\mathbf{x}) = -\frac{1}{2\pi} \iint \frac{\hat{y}}{\hat{x}^2 + \hat{y}^2}\, \omega_z(\mathbf{x}')\, \mathrm{d}A(\mathbf{x}'), \qquad u_y(\mathbf{x}) = \frac{1}{2\pi} \iint \frac{\hat{x}}{\hat{x}^2 + \hat{y}^2}\, \omega_z(\mathbf{x}')\, \mathrm{d}A(\mathbf{x}'), \quad (2.13.1)$$

where $\mathbf{x}' = \mathbf{x} - \mathbf{x}'$ and the integration is performed over the domain of flow. Using (2.7.36), we find that the associated stream function is

$$\psi(\mathbf{x}) = -\frac{1}{4\pi} \iint \ln\left(\frac{\hat{x}^2 + \hat{y}^2}{a^2}\right) \omega_z(\mathbf{x}')\, dA(\mathbf{x}'), \qquad (2.13.2)$$

where a is a constant length. An alternative representation for the velocity originating from the integral representation (2.7.38) is

$$\mathbf{u}(\mathbf{x}) = -\frac{1}{4\pi} \iint \ln\left(\frac{\hat{x}^2 + \hat{y}^2}{a^2}\right) \nabla' \times \left[\omega_z(\mathbf{x}')\,\mathbf{e}_z\right] dA(\mathbf{x}'), \qquad (2.13.3)$$

where $\mathbf{e}_z$ is the unit vector along the z axis.

2.13.1 Point vortices

To derive the velocity field and stream function due to a point vortex located at a point, $\mathbf{x}_0$, we substitute the singular distribution $\omega_z(\mathbf{x}) = \kappa\, \delta_2(\mathbf{x} - \mathbf{x}_0)$ into (2.13.1) and (2.13.2), obtaining

$$u_x(\mathbf{x}) = -\frac{\kappa}{2\pi} \frac{\hat{y}}{\hat{x}^2 + \hat{y}^2}, \qquad u_y(\mathbf{x}) = \frac{\kappa}{2\pi} \frac{\hat{x}}{\hat{x}^2 + \hat{y}^2}, \qquad \psi(\mathbf{x}) = -\frac{\kappa}{4\pi} \ln\left(\frac{\hat{x}^2 + \hat{y}^2}{a^2}\right), \quad (2.13.4)$$

where δ_2 is the two-dimensional delta function, κ is the strength of the point vortex, and $\hat{\mathbf{x}} = \mathbf{x} - \mathbf{x}_0$. The polar velocity component is

$$u_\theta(\mathbf{x}) = \frac{\kappa}{2\pi} \frac{1}{r}, \qquad (2.13.5)$$

where $r = |\mathbf{x} - \mathbf{x}_0|$. The streamlines are concentric circles centered at $\mathbf{x}_0$, the velocity decays like $1/r$, and the cyclic constant of the motion around the point vortex is equal to κ. The velocity field is irrotational everywhere except at the singular point, $\mathbf{x}_0$. The associated multi-valued velocity potential is

$$\phi(\mathbf{x}) = \frac{\kappa}{2\pi}\, \theta, \qquad (2.13.6)$$

where θ is the polar angle subtended between the vector $\mathbf{x} - \mathbf{x}_0$ and the x axis. We observe that the harmonic potential increases by κ each time a complete turn is performed around the point vortex.

Complex-variable formulation

It is sometimes useful to introduce the complex variable $z = x + iy$, and the complex velocity $v = u_x + iu_y$, where i is the imaginary unit, $i^2 = -1$. The first two equations in (2.13.4) can be collected into the complex form

$$v^*(z) = u_x - i\, u_y = \frac{\kappa}{2\pi i} \frac{1}{z - z_0}, \qquad (2.13.7)$$

where an asterisk denotes the complex conjugate.

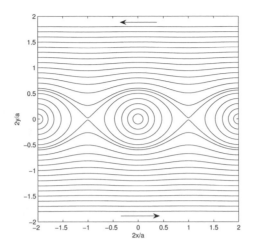

FIGURE 2.13.1   Mesmerizing streamline pattern due to an array of evenly spaced point vortices with positive (counterclockwise) circulation separated by distance a.

2.13.2   An infinite array of evenly spaced point vortices

Next, we consider the flow due to an infinite row of point vortices with uniform strength separated by distance a, as illustrated in Figure 2.13.1. The mth point vortex is located at $x_m = x_0 + ma$ and $y_m = y_0$, where (x_0, y_0) is the position of one arbitrary point vortex labeled 0, and m is an integer. If we attempt to compute the velocity induced by the array by summing the individual contributions, we will encounter unphysical divergent sums.

To overcome this difficulty, we express the stream function corresponding to the velocity field induced by the individual point vortices as

$$\psi_0(x, y) = -\frac{\kappa}{2\pi} \ln\left(\frac{r_0}{a}\right), \qquad \psi_m(x, y) = -\frac{\kappa}{2\pi} \ln\left(\frac{r_m}{|m|\,a}\right) \tag{2.13.8}$$

for $m = \pm 1, \pm 2, \ldots$, where

$$r_m \equiv \left[(x - x_m)^2 + (y - y_m)^2\right]^{1/2} \tag{2.13.9}$$

is the distance of the field point, $\mathbf{x} = (x, y)$, from the location of the mth point vortex. The denominators of the fractions in the arguments of the logarithms on the right-hand sides of (2.13.8) have been chosen judiciously to facilitate forthcoming algebraic manipulations.

It is important to observe that, as m tends to $\pm\infty$, the fraction on the right-hand side of the second equation in (2.13.8) tends to unity and its logarithm tends to vanish, thereby ensuring that remote point vortices make decreasingly small contributions. If the denominators were not included, remote point vortices would have made contributions that are proportional to the logarithm of the distance between a point vortex from the point (x, y) where the stream function is evaluated.

Next, we express the stream function due to the infinite array as the sum of a judiciously selected constant expressed by the term after the first equal sign in (2.13.10), and the individual stream functions stated in (2.13.8), obtaining

$$\psi(x, y) = -\frac{\kappa}{2\pi} \ln(\sqrt{2}\pi) + \sum_{m=-\infty}^{\infty} \psi_m(x, y). \tag{2.13.10}$$

Substituting the expressions for the individual Green's functions, we find that

$$\psi(x, y) = -\frac{\kappa}{2\pi} \ln \left(\frac{\sqrt{2}\,\pi\,r_0}{a} \right) - \frac{\kappa}{2\pi} \sum_{m=\pm 1, \pm 2, \ldots} \ln \left(\frac{r_m}{|m|\,a} \right), \tag{2.13.11}$$

which can be restated as

$$\psi(x, y) = -\frac{\kappa}{2\pi} \ln \left[\frac{\sqrt{2}\,\pi\,r_0}{a} \prod_{m=\pm 1, \pm 2, \ldots} \frac{r_m}{|m|\,a} \right], \tag{2.13.12}$$

where Π denotes the product. An identity allows us to compute the infinite product on the right-hand side of (2.13.12) in closed form, obtaining

$$\frac{\sqrt{2}\,\pi\,r_0}{a} \prod_{m=\pm 1, \pm 2, \ldots} \frac{r_m}{|m|\,a} = \{\cosh[k(y - y_0)] - \cos[k(x - x_0)]\}^{1/2} \tag{2.13.13}$$

(e.g., [4], p. 197). Substituting the right-hand side of (2.13.13) into (2.13.12), we derive the desired expression for the stream function

$$\psi(x, y) = -\frac{\kappa}{4\pi} \ln \left(\cosh[k(y - y_0)] - \cos[k(x - x_0)] \right), \tag{2.13.14}$$

where $k = 2\pi/a$ is the wave number. Differentiating the right-hand side of (2.13.14) with respect to x or y, we obtain the corresponding velocity components,

$$u_x(x, y) = -\frac{\kappa}{2a} \frac{\sinh[k(y - y_0)]}{\cosh[k(y - y_0)] - \cos[k(x - x_0)]},$$

$$u_y(x, y) = \frac{\kappa}{2a} \frac{\sin[k(x - x_0)]}{\cosh[k(y - y_0)] - \cos[k(x - x_0)]}. \tag{2.13.15}$$

As the wave number k tends to zero, expressions (2.13.14) and (2.13.15) reproduce the stream function and velocity field associated with a single point vortex. The streamline pattern due to the periodic array exhibits a cat's eye pattern, as illustrated in Figure 2.13.1.

Because of symmetry, the velocity at the location of one point vortex induced by all other point vortices is zero and the array is stationary. Far above or below the array, the x component of the velocity tends to the value $-\kappa/a$ or κ/a, while the y component decays at an exponential rate. This behavior renders the infinite array a useful model of the flow generated by the instability of a shear layer separating two parallel streams that merge at different velocities. The Kelvin–Helmholtz instability causes the shear layer to roll up into compact vortices represented by the point vortices of the periodic array.

2.13.3   Point-vortex dipole

Consider the flow due to two point vortices with strengths of equal magnitude and opposite sign located at the points $\mathbf{x}_0$ and $\mathbf{x}_1$. Taking advantage of the linearity of (2.13.1), we construct the associated stream function by superposing the stream functions due to the individual point vortices. Placing the second point, $\mathbf{x}_1$, near the first point, $\mathbf{x}_0$, and taking the limit as the distance $|\mathbf{x}_0 - \mathbf{x}_1|$ tends to zero while the product $\kappa(\mathbf{x}_0 - \mathbf{x}_1)$ remains constant and equal to $\boldsymbol{\lambda}$, we derive the stream function due to a point-vortex dipole,

$$\psi(\mathbf{x}) = -\frac{1}{2\pi}\,\boldsymbol{\lambda}\cdot\nabla_0 \ln\left(\frac{r}{a}\right) = \frac{1}{2\pi}\,\frac{\mathbf{x}-\mathbf{x}_0}{r^3}\cdot\boldsymbol{\lambda}, \tag{2.13.16}$$

where $r = |\mathbf{x} - \mathbf{x}_0|$ and the derivatives in ∇_0 operate with respect to $\mathbf{x}_0$. The associated velocity field follows readily by differentiation as

$$\begin{aligned} u_x(\mathbf{x}) &= \frac{1}{2\pi}\left(\frac{\lambda_y}{r^2} - 2\,\frac{y-y_0}{r^4}\,(\mathbf{x}-\mathbf{x}_0)\cdot\boldsymbol{\lambda}\right), \\[2mm] u_y(\mathbf{x}) &= -\frac{1}{2\pi}\left(\frac{\lambda_x}{r^2} - 2\,\frac{x-x_0}{r^4}\,(\mathbf{x}-\mathbf{x}_0)\cdot\boldsymbol{\lambda}\right). \end{aligned} \tag{2.13.17}$$

The streamline pattern of the flow due to a point-vortex dipole oriented along the y axis, with $\boldsymbol{\lambda} = \lambda\,\mathbf{e}_y$ and $\lambda > 0$ is identical to that due to a potential dipole oriented along the x axis shown in Figure 2.1.3(b). An alternative method of deriving the flow due to a point-vortex dipole employs the properties of generalized delta functions. We set

$$\boldsymbol{\omega}(\mathbf{x}) = \boldsymbol{\lambda}\cdot\nabla\delta(\mathbf{x}-\mathbf{x}_0), \tag{2.13.18}$$

and then use (2.13.1) and (2.13.2) to derive (2.13.16) and (2.13.17).

2.13.4   Vortex sheets

To derive the flow due to a two-dimensional vortex sheet, we substitute (1.13.26) into (2.13.1) and obtain

$$u_x(\mathbf{x}) = -\frac{1}{2\pi}\int_C \frac{\hat{y}}{\hat{x}^2+\hat{y}^2}\,\mathrm{d}\Gamma(\mathbf{x}'), \qquad u_y(\mathbf{x}) = \frac{1}{2\pi}\int_C \frac{\hat{x}}{\hat{x}^2+\hat{y}^2}\,\mathrm{d}\Gamma(\mathbf{x}'), \tag{2.13.19}$$

where $\mathbf{x}' = \mathbf{x}-\mathbf{x}'$, $\mathrm{d}\Gamma = \gamma\,\mathrm{d}l$ is the differential of the circulation along the vortex sheet, and C is the trace of the vortex sheet in the xy plane. Comparing (2.13.19) with (2.13.2), we interpret a vortex sheet as a continuous distribution of point vortices.

It is sometimes useful to introduce the complex variable $z = x + \mathrm{i}y$ and define the complex velocity $v = u_x + \mathrm{i}u_y$, where i is the imaginary unit, $\mathrm{i}^2 = -1$. Equations (2.13.19) may then be stated in the complex form

$$v^*(z) = \frac{1}{2\pi\mathrm{i}}\int_C \frac{\mathrm{d}\Gamma(z')}{z-z'}, \tag{2.13.20}$$

where an asterisk denotes the complex conjugate.

Periodic vortex sheet

Using identity (2.13.13), we find that the velocity due a periodic vortex sheet that is repeated along the x axis with period a is

$$u_x(x,y) = -\frac{1}{2a} \int_T \frac{\sinh[k(y-y')]}{\cosh[k(y-y')] - \cos[k(x-x')]}\, d\Gamma(\mathbf{x}'),$$

$$u_y(x,y) = \frac{1}{2a} \int_T \frac{\sin[k(x-x_0)]}{\cosh[k(y-y')] - \cos[k(x-x')]}\, d\Gamma(\mathbf{x}'),$$

(2.13.21)

where $k = 2\pi/a$ is the wave number and T is the trace of the vortex sheet in the xy plane inside one period. As the scaled wave number ka tends to zero, (2.13.21) reduces to (2.13.19).

2.13.5 Vortex patches

Consider a two-dimensional flow induced by a region of constant vorticity, $\omega_z = \Omega$, called a vortex patch, immersed in an otherwise stationary fluid. Introducing the two-dimensional vector potential expressed in terms of the stream function as $\mathbf{A} = (0, 0, \psi)$, and using the integral representation (2.13.2), we find that the induced velocity is

$$\mathbf{u}(\mathbf{x}) = \frac{\Omega}{4\pi} \iint_{Patch} \nabla' \times \left[\ln \frac{\hat{x}^2 + \hat{y}^2}{a^2}\, \mathbf{e}_z \right] dA(\mathbf{x}'),$$

(2.13.22)

where $\mathbf{e}_z$ is the unit vector along the z axis and the gradient ∇' involves derivatives with respect to the integration point, $\mathbf{x}'$. Using Stokes' theorem, we convert the area integral into a line integral along the closed contour of the patch, C, obtaining

$$\mathbf{u}(\mathbf{x}) = -\frac{\Omega}{4\pi} \oint_C \ln \frac{\hat{x}^2 + \hat{y}^2}{a^2}\, \mathbf{t}(\mathbf{x}')\, dl(\mathbf{x}'),$$

(2.13.23)

where $\mathbf{t}$ is the tangent unit vector pointing in the counterclockwise direction around C. An alternative way of deriving (2.13.23) departs from the identity

$$\nabla \times \boldsymbol{\omega} = \Omega \oint_C \delta_2(\mathbf{x} - \mathbf{x}')\, \mathbf{t}(\mathbf{x}')\, dl(\mathbf{x}'),$$

(2.13.24)

where δ_2 is the two-dimensional delta function in the xy plane. Substituting (2.13.24) into (2.13.3) and using the properties of the delta function, we recover (2.13.23). If a flow contains a number of disconnected vortex patches with different vorticity, the integral in (2.13.23) is computed along each vortex contour and then multiplied by the corresponding values of the constant vorticity, Ω.

Periodic vortex patches and vortex layers

Using identity (2.13.13), we find that the velocity due to a periodic array of vortex patches with constant vorticity Ω arranged along the x axis with period a is given by

$$\mathbf{u}(\mathbf{x}) = -\frac{\Omega}{4\pi} \oint_{A_V} \ln[\cos(k\hat{y}) - \cos(k\hat{x})]\, \mathbf{t}(\mathbf{x}')\, dl(\mathbf{x}'),$$

(2.13.25)

(a) (b)

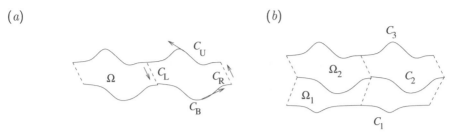

FIGURE 2.13.2   Illustration of (a) a periodic vortex layer with constant vorticity and (b) a periodic compound vortex layer consisting of two adjacent vortex layers with constant vorticity Ω_1 and Ω_2.

where $k = 2\pi/a$ is the wave number, C is the contour of one arbitrary patch in the array, and $\mathbf{t}$ is the tangent unit vector pointing in the counterclockwise direction.

As an application, we consider the velocity due to the periodic vortex layer illustrated in Figure 2.13.2(a). We select one period of the vortex layer and identify C with the union of the upper contour, C_U, lower contour, C_B, left contour, C_L, and right contour, C_R. The contributions from the contours C_L and C_R to the integral in (2.13.25) cancel because the corresponding tangent vectors point in opposite directions and the logarithmic function is periodic inside the integral. Consequently, the contour C reduces to the union of C_U and C_B.

Problems

2.13.1 *An array of point vortices*

Verify that, as the period a tends to infinity, (2.13.14) and (2.13.15) yield the velocity and stream function due to a single point vortex.

2.13.2 *Compound vortex layer*

Consider a periodic compound vortex layer consisting of two adjacent layers with constant vorticity Ω_1 and Ω_2, as illustrated in Figure 2.13.2(b). Derive an expression for the velocity in terms of contour integrals along C_1, C_2, and C_3.

3

Stresses, the equation of motion, and vorticity transport

In the first two chapters, we examined the kinematic structure of a flow and investigated possible ways of describing the velocity field in terms of secondary variables, such as the velocity potential, the vector potential, and the stream functions. However, in our discussion, we made no reference to the physical processes that are responsible for establishing a flow or to the conditions that are necessary for sustaining the motion of the fluid. To investigate these and related issues, in this chapter we introduce the fundamental ideas and physical variables needed to describe and compute the forces developing in a fluid at rest as the result of the motion. The theoretical framework will culminate in an equation of motion governing the structure of a steady flow and the evolution of an unsteady flow from a specified initial state. The point of departure for deriving the equation of motion is Newton's second law of motion for a material fluid parcel, stating that the rate of change of momentum of the parcel be equal to the sum of all forces exerted on the volume of the fluid occupying the parcel as well as on the parcel boundaries.

We will discuss constitutive equations relating the stresses developing on the surface of a material fluid parcel to the parcel deformation. In the discourse, we will concentrate on a special but common class of incompressible fluids, called Newtonian fluids, whose response is described by a linear constitutive equation. The equation of motion for an incompressible Newtonian fluid takes the form of a second-order partial differential equation in space for the velocity, called the Navier–Stokes equation. Supplementing the Navier–Stokes equation with the continuity equation to ensure mass conservation, and then introducing appropriate boundary and initial conditions, we obtain a complete set of governing equations. Analytical, asymptotic, and numerical methods for solving these equations under a broad range of conditions will be discussed in subsequent chapters.

The equation of motion can be regarded as a dynamical law for the evolution of the velocity field, providing us with an expression for the Eulerian time derivative, $\partial \mathbf{u}/\partial t$, or material derivative, $D\mathbf{u}/Dt$. To derive a corresponding law for the evolution of the vorticity field expressing the rate of rotation of fluid parcels, $\boldsymbol{\omega} = \nabla \times \mathbf{u}$, we take the curl of the Navier–Stokes equation and derive the vorticity transport equation. Inspecting the various terms in the vorticity transport equation allows us to develop insights into the dynamics of rotational flows. Vortex dynamics provides us with a natural framework for analyzing and computing flows dominated by the presence or motion of compact vortex structures, including line vortices, vortex patches, and vortex sheets.

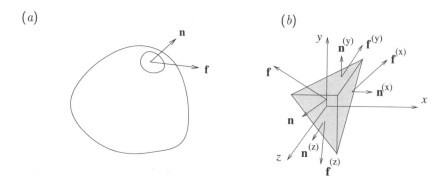

FIGURE 3.1.1   (*a*) Illustration of the traction vector exerted on a small surface on the boundary of a fluid parcel or control volume, **f**, and normal unit vector, **n**. The traction vector has a normal component and a tangential component; the normal component is the normal stress, and the tangential component is the shear stress. (*b*) Illustration of three small triangular surfaces perpendicular to the three Cartesian axes forming a tetrahedral control volume; **n** is the unit vector normal to the slanted face of the control volume, and **f** is the corresponding traction.

3.1   Forces acting in a fluid, traction, and the stress tensor

Consider a fluid parcel consisting of the same material, as shown in Figure 3.1.1(*a*). The adjacent material imparts to molecules distributed over the surface of the parcel a local force due to a short-range intermolecular force field, and thereby generate a normal and a tangential frictional force. Now consider a stationary control volume that is occupied entirely by a moving fluid. As the fluid flows, molecules enter and leave the control volume from all sides carrying momentum and thereby imparting to the control volume at a particular instant in time a normal force. Short-range intermolecular forces cause attraction between molecules on either side of the boundary of the control volume, and thereby generate an effective tangential frictional force.

Traction and surface force

The force exerted on an infinitesimal surface element located at the boundary of a fluid parcel or at the boundary of a control volume, d**F**, divided by the element surface area, dS, is called the traction and is denoted by

$$\mathbf{f} \equiv \frac{d\mathbf{F}}{dS}. \tag{3.1.1}$$

The traction depends on the location, orientation, and designated side of the infinitesimal surface element. The location is determined by the position vector, **x**, and the orientation and side are determined by the normal unit vector, **n**. The traction has units of force divided by squared length. In terms of the traction, the surface force exerted on a fluid parcel is

$$\mathbf{F} = \iint_{Parcel} \mathbf{f} \, dS. \tag{3.1.2}$$

The same expression provides us with the surface force exerted on a control volume occupied by fluid.

Normal and tangential components

It is useful to decompose the traction into a normal component, $\mathbf{f}^N$, pointing the direction of the normal unit vector, $\mathbf{n}$, and a complementary tangential or shear component, $\mathbf{f}^T$, given by

$$\mathbf{f}^N = (\mathbf{f} \cdot \mathbf{n})\,\mathbf{n}, \qquad \mathbf{f}^T = \mathbf{n} \times (\mathbf{f} \times \mathbf{n}) = \mathbf{f} \cdot (\mathbf{I} - \mathbf{n} \otimes \mathbf{n}), \tag{3.1.3}$$

where $\mathbf{I}$ is the identity matrix. The projection matrix $\mathbf{I} - \mathbf{n} \otimes \mathbf{n}$ extracts the tangential component of a vector that it multiplies.

Body force

A long-range ambient force field acting on the molecules of a fluid parcel imparts to the parcel a body force given by

$$\mathbf{F}^B = \iiint_{Parcel} \kappa\,\mathbf{b}\,dV, \tag{3.1.4}$$

where $\mathbf{b}$ is the strength of the body force field and κ is a companion physical constant that may depend on time as well on position in the domain of flow. Examples include the gravitational or an electromagnetic force field. In the case of the gravitational force field, $\mathbf{b}$ is the acceleration of gravity, $\mathbf{g}$, and κ is the fluid density, ρ. In the remainder of this book we will assume that the body force field is due to gravity alone.

3.1.1   Stress tensor

Next, we introduce a system of Cartesian coordinates and consider a small tetrahedral control volume with three sides perpendicular to the x, y, and z axes, as illustrated in Figure 3.1.1(b). The traction exerted on each of the three planar sides is denoted, respectively, by $\mathbf{f}^{(x)}$, $\mathbf{f}^{(y)}$, and $\mathbf{f}^{(z)}$. Stacking these tractions above one another in three rows, we formulate the matrix of stresses

$$\sigma_{ij} = f_j^{(i)}. \tag{3.1.5}$$

The first row of $\boldsymbol{\sigma}$ contains the components of $\mathbf{f}^{(x)}$, the second row contains the components of $\mathbf{f}^{(y)}$, and the third row contains the components of $\mathbf{f}^{(z)}$, so that

$$\boldsymbol{\sigma} = \begin{bmatrix} \sigma_{xx} & \sigma_{xy} & \sigma_{xz} \\ \sigma_{yx} & \sigma_{yy} & \sigma_{yz} \\ \sigma_{zx} & \sigma_{zy} & \sigma_{zz} \end{bmatrix} \equiv \begin{bmatrix} f_x^{(x)} & f_y^{(x)} & f_z^{(x)} \\ f_x^{(y)} & f_y^{(y)} & f_z^{(y)} \\ f_x^{(z)} & f_y^{(z)} & f_z^{(z)} \end{bmatrix}. \tag{3.1.6}$$

The diagonal elements of $\boldsymbol{\sigma}$ are the *normal stresses* exerted on the three mutually orthogonal sides of the control volume that are normal to the x, y, and z axes. The off-diagonal elements are the *tangential or shear stresses*. Later in this section, we will show that $\boldsymbol{\sigma}$ satisfies a transformation rule that qualifies it as a second-order Cartesian tensor.

Traction in terms of stress

We will demonstrate that the traction exerted on the slanted side of the infinitesimal tetrahedron illustrated in Figure 3.1.1(*b*) can be computed from knowledge of its orientation and the value of the stress tensor at the origin. Knowledge of the body force is not required.

Our point of departure is Newton's second law of motion for the fluid parcel enclosed by the tetrahedron, stating that the rate of change of momentum of the parcel is equal to the sum of the surface and body forces exerted on the parcel. Using (3.1.2) and (3.1.4), we obtain

$$\frac{\mathrm{d}}{\mathrm{d}t} \iiint_{Parcel} \rho \mathbf{u} \, \mathrm{d}V = \iint_{Parcel} \mathbf{f} \, \mathrm{d}S + \iiint_{Parcel} \rho \mathbf{g} \, \mathrm{d}V. \tag{3.1.7}$$

Because the size of the tetrahedron is infinitesimal, the traction exerted on each side is approximately constant. Neglecting variations in the momentum and body force over the parcel volume, we recast equation (3.1.7) into the algebraic form

$$\frac{\mathrm{D}(\rho \mathbf{u} \, \Delta V)}{\mathrm{D}t} = \mathbf{f}^{(x)} \Delta S_x + \mathbf{f}^{(y)} \Delta S_y + \mathbf{f}^{(z)} \Delta S_z + \mathbf{f} \Delta S + \rho \mathbf{g} \, \Delta V, \tag{3.1.8}$$

where $\mathrm{D}/\mathrm{D}t$ is the material derivative, ΔV is the volume of the tetrahedron, $\Delta S^{(x)}$, $\Delta S^{(y)}$, $\Delta S^{(z)}$, and ΔS are the surface areas of the four sides of the tetrahedron, and $\mathbf{f}$ is the traction exerted on the slanted side. Dividing each term in (3.1.8) by ΔS and rearranging, we obtain

$$\frac{1}{\Delta S} \left(\frac{\mathrm{D}(\rho \mathbf{u} \, \Delta V)}{\mathrm{D}t} - \rho \mathbf{g} \, \Delta V \right) = \mathbf{f}^{(x)} \frac{\Delta S^{(x)}}{\Delta S} + \mathbf{f}^{(y)} \frac{\Delta S^{(y)}}{\Delta S} + \mathbf{f}^{(z)} \frac{\Delta S^{(z)}}{\Delta S} + \mathbf{f}. \tag{3.1.9}$$

In the limit as the size of the parcel tends to zero, the ratio $\Delta V/\Delta S$ vanishes and the left-hand side disappears.

Now we introduce the unit vector normal to the slanted side pointing outward from the tetrahedron, $\mathbf{n}$, and use the geometrical relations

$$n_x = -\frac{\Delta S^{(x)}}{\Delta S}, \qquad n_y = -\frac{\Delta S^{(y)}}{\Delta S}, \qquad n_z = -\frac{\Delta S^{(z)}}{\Delta S}. \tag{3.1.10}$$

Substituting these expressions along with the definition of the stress matrix tensor (3.1.5) into (3.1.9) and rearranging, we find that

$$\mathbf{f} = \mathbf{n} \cdot \boldsymbol{\sigma}. \tag{3.1.11}$$

In index notation, the *j*th component of the traction is

$$f_j = n_i \, \sigma_{ij}, \tag{3.1.12}$$

where summation is implied over the repeated index, i. Equation (3.1.11) states that the traction, $\mathbf{f}$, is a linear function of the normal unit vector, $\mathbf{n}$, with a matrix of proportionality that is equal to the stress tensor, $\boldsymbol{\sigma}$.

Force on a fluid parcel in terms of the stress

Combining (3.1.2) with (3.1.11) and using the divergence theorem, we find that the surface force exerted on a fluid parcel is given by

$$\mathbf{F} = \iint_{Parcel} \mathbf{n} \cdot \boldsymbol{\sigma} \, \mathrm{d}S = \iiint_{Parcel} \nabla \cdot \boldsymbol{\sigma} \, \mathrm{d}V. \tag{3.1.13}$$

In index notation, the jth component of the surface force is

$$F_j = \iint_{Parcel} n_i \, \sigma_{ij} \, \mathrm{d}S = \iiint_{Parcel} \frac{\partial \sigma_{ij}}{\partial x_i} \, \mathrm{d}V. \tag{3.1.14}$$

Newton's third law requires that the parcel exerts on the ambient fluid a force of equal magnitude in the opposite direction.

The total force exerted on the parcel is the sum of the surface force given in (3.1.13) and the body force given in (3.1.4),

$$\mathbf{F}^{total} = \iiint_{Parcel} (\nabla \cdot \boldsymbol{\sigma} + \rho \mathbf{g}) \, \mathrm{d}V. \tag{3.1.15}$$

If the divergence of the stress tensor balances the body force, the total force exerted on the parcel is zero. Later in this chapter, we will see that this is true when the effect of fluid inertia is negligibly small.

Force acting on a fluid sheet

Consider a fluid parcel in the shape of an flattened sheet. Letting the thickness of the sheet tend to zero, we find that the surface force exerted on one side of the sheet is equal and opposite to that exerted on the other side of the sheet, so that the sum of the two forces balances to zero.

Force exerted on a boundary

To compute the force exerted on a boundary, we consider a flattened fluid parcel having the shape of a thin sheet attached to the boundary. As the thickness of the sheet tends to zero, the mass of the parcel becomes infinitesimal and the sum of the hydrodynamic forces exerted on either side of the parcel must balance to zero (Problem 3.1.1). Newton's third law requires that the force exerted on the side of the parcel adjacent to the boundary is equal in magnitude and opposite in direction to that exerted by the fluid on the boundary. Thus, the hydrodynamic force exerted on the boundary is given by

$$\mathbf{F} = \iint_{Boundary} \mathbf{n} \cdot \boldsymbol{\sigma} \, \mathrm{d}S, \tag{3.1.16}$$

where $\mathbf{n}$ is the unit vector normal to the boundary pointing into the fluid.

$\boldsymbol{\sigma}$ is a tensor

We will show that the matrix of stresses, $\boldsymbol{\sigma}$, is a Cartesian tensor, called the Cauchy stress tensor or simply the stress tensor. Following the standard procedure outlined in Section 1.1.7, we introduce two Cartesian systems of axes, x_i and x_i', related by the linear transformation $x_i' = A_{ij}x_j$, where $\mathbf{A}$ is an orthogonal matrix, meaning that its inverse is equal to its transpose. The stresses in the x_i' system are denoted by $\boldsymbol{\sigma}'$ and those in the x_i system are denoted by $\boldsymbol{\sigma}$. The traction exerted on a small surface in the x_i' system is denoted by $\mathbf{f}'$, and the same traction exerted on the same surface in the x_i system is denoted by $\mathbf{f}$. Next, we introduce the vector transformation

$$f_j' = A_{jl}f_l, \tag{3.1.17}$$

and use (3.1.13) to obtain

$$n_i'\sigma_{ij}' = A_{jl}n_k\,\sigma_{kl}. \tag{3.1.18}$$

Substituting $n_k = n_i'A_{ik}$, we find that

$$n_i'\sigma_{ij}' = A_{jl}A_{ik}n_i'\,\sigma_{kl}, \tag{3.1.19}$$

which demonstrates that $\boldsymbol{\sigma}$ satisfies the distinguishing property of second-order Cartesian tensors shown in (1.1.40) with $\boldsymbol{\sigma}$ in place of $\mathbf{T}$.

One important consequence of tensorial nature of $\boldsymbol{\sigma}$ is the existence of three scalar stress invariants. In particular, both the trace and the determinant of $\boldsymbol{\sigma}$ are independent of the choice of the working Cartesian axes.

3.1.2   Torque

The torque, $\mathbf{T}$, exerted on a fluid parcel, computed with respect to a chosen point, $\mathbf{x}_0$, consists of the torque due to the surface force, the torque due to the body force, and the torque due to an external torque field with intensity $\mathbf{c}$. Adding these contributions, we obtain the total torque

$$\mathbf{T}^{total} = \iint_{Parcel} \hat{\mathbf{x}} \times (\mathbf{n} \cdot \boldsymbol{\sigma})\,\mathrm{d}S + \iiint_{Parcel} \hat{\mathbf{x}} \times (\rho\,\mathbf{g})\,\mathrm{d}V + \iiint_{Parcel} \lambda\,\mathbf{c}\,\mathrm{d}V, \tag{3.1.20}$$

where $\hat{\mathbf{x}} = \mathbf{x} - \mathbf{x}_0$ and λ is an appropriate physical constant associated with the torque field, $\mathbf{c}$. For example, a torque field arises in a suspension of magnetized or bipolar particles by applying an electrical field.

Using the divergence theorem, we convert the surface integral on the right-hand side of (3.1.20) into a volume integral, obtaining in index notation

$$T_i^{total} = \iiint_{Parcel} \left(\epsilon_{ijk}\sigma_{jk} + \epsilon_{ijk}\hat{x}_j \frac{\partial \sigma_{lk}}{\partial x_l} + \rho\,\epsilon_{ijk}\hat{x}_j g_k + \lambda\,c_i \right) \mathrm{d}V, \tag{3.1.21}$$

where summation is implied over the repeated indices, j, k, and l.

In the absence of an external torque field, $\mathbf{c} = \mathbf{0}$, the torque exerted on a parcel with respect to a point $\mathbf{x}_1$, denoted by $\mathbf{T}^{total}(\mathbf{x}_1)$, is related to the torque with respect to another point, $\mathbf{x}_0$, by

$$\mathbf{T}^{total}(\mathbf{x}_1) = \mathbf{T}^{total}(\mathbf{x}_0) + (\mathbf{x}_1 - \mathbf{x}_0) \times \mathbf{F}^{total}. \tag{3.1.22}$$

When the rate of change of linear momentum of a fluid parcel is negligible, the total force exerted on the parcel vanishes and the torque is independent of the point with respect to which it is defined.

Torque exerted on a boundary

The torque exerted on a boundary with respect to a chosen point, $\mathbf{x}_0$, is given by the simplified version of (3.1.20),

$$\mathbf{T} = \iint_{Boundary} \hat{\mathbf{x}} \times (\mathbf{n} \cdot \boldsymbol{\sigma}) \, \mathrm{d}S, \tag{3.1.23}$$

where $\hat{\mathbf{x}} = \mathbf{x} - \mathbf{x}_0$ and $\mathbf{n}$ is the unit vector normal to the boundary pointing into the fluid. When the surface force exerted on the boundary is zero, the torque is independent of the location of the center of torque, $\mathbf{x}_0$.

3.1.3 Stresses in curvilinear coordinates

We have defined the components of the stress tensor in terms of the traction exerted on three mutually perpendicular infinitesimal planar surfaces that are normal to the x, y, and z axes, denoted by $\mathbf{f}^{(x)}$, $\mathbf{f}^{(y)}$, and $\mathbf{f}^{(z)}$. These definitions can be extended to general orthogonal or nonorthogonal curvilinear coordinates discussed in Sections A.8–A.17, Appendix A.

Orthogonal curvilinear coordinates

Working as in the case of Cartesian coordinates, we define the components of the stress tensor in general orthogonal curvilinear coordinates, (ξ, η, ζ), as shown in Figure 3.1.2(a). Examples are cylindrical, spherical, and plane polar coordinates shown in Figure 3.1.2(b–d).

Let $\mathbf{f}^{(\alpha)}$ be the traction exerted on an infinitesimal surface that is perpendicular to the α coordinate line at a point, where a Greek index stands for ξ, η, or ζ. The nine components of the stress tensor, $\sigma_{\alpha\beta}$, are defined by the equation

$$\mathbf{f}^{(\alpha)} = \sigma_{\alpha\beta} \, \mathbf{e}_\beta, \tag{3.1.24}$$

where $\mathbf{e}_\beta$ is the unit vector in the direction of the β coordinate line, and summation over the index β is implied on the right-hand side. The stress tensor itself is given by the dyadic decomposition

$$\boldsymbol{\sigma} = \sigma_{\alpha\beta} \, \mathbf{e}_\alpha \otimes \mathbf{e}_\beta, \tag{3.1.25}$$

where the matrices $\mathbf{e}_\alpha \otimes \mathbf{e}_\beta$ provide us with a base of the three-dimensional tensor space, as discussed in Section 1.1.

Conversely, the components of the stress tensor can be extracted from the stress tensor by double-dot projection,

$$\sigma_{\alpha\beta} = \boldsymbol{\sigma} : (\mathbf{e}_\alpha \otimes \mathbf{e}_\beta). \tag{3.1.26}$$

The double dot product of two matrices is defined in Section A.4, Appendix A.

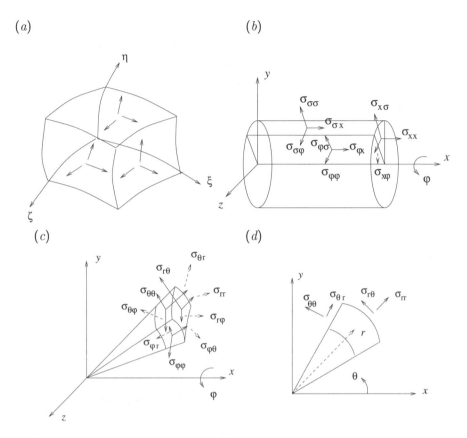

FIGURE 3.1.2   (*a*) Definition of the nine components of the stress tensor in orthogonal curvilinear coordinates. Components of the stress tensor in (*b*) cylindrical, (*c*) spherical, and (*d*) plane polar coordinates.

Cylindrical, spherical, and plane polar coordinates

The components of the stress tensor in cylindrical, spherical, and plane polar coordinates are defined in Figure 3.1.2(*b–d*). Plane polar coordinates arise from spherical polar coordinates by setting $\varphi = 0$, or from cylindrical polar coordinates by setting $x = 0$ and relabeling σ as r and φ as θ.

Nonorthogonal curvilinear coordinates

A system of nonorthogonal curvilinear coordinates, (ξ, η, ζ), is illustrated in Figure 3.1.3. Following standard practice, we introduce covariant and contravariant base vectors and corresponding coordinates, as discussed in Section A.12, Appendix A. The traction exerted on an arbitrary surface can be expressed in contravariant or covariant component form. We may consider the traction exerted on a small surface that is perpendicular to the contravariant or covariant coordinate lines. Accordingly, the stress tensor can be expressed in terms of its pure contravariant, pure covariant, or mixed com-

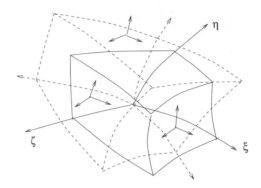

FIGURE 3.1.3   Definition of the nine components of the stress tensor in nonorthogonal curvilinear coordinates. The solid lines represent covariant coordinates and the dashed lines represent the associated contravariant coordinates.

ponents, in four combinations. The physical meaning of the pure and mixed representations stems from the geometrical interpretation of the covariant and contravariant base vectors, combined with the definition of the traction as the projection from the left of the stress tensor onto a unit vector pointing in a specified direction.

Problems

3.1.1 *Normal component of the traction*

Verify that, in terms of the stress tensor, the normal component of the traction is given by

$$\mathbf{f}^N = [\boldsymbol{\sigma} : (\mathbf{n} \otimes \mathbf{n})]\, \mathbf{n}. \tag{3.1.27}$$

The double dot product of two matrices is defined in Section A.4, Appendix A.

3.1.2 *Hydrodynamic torque exerted on a boundary*

Show that, if the force exerted on a boundary is zero, the torque is independent of the location of the point with respect to which the torque is evaluated.

3.1.3 *Mean value of the stress tensor over a parcel*

The mean value of the stress tensor over the volume of a fluid parcel is defined as

$$\bar{\boldsymbol{\sigma}} \equiv \frac{1}{V_P} \iiint_{Parcel} \boldsymbol{\sigma}\, \mathrm{d}V, \tag{3.1.28}$$

where V_p is the parcel volume. Show that

$$\bar{\sigma}_{ij} = -\frac{1}{V_P} \left(\iint_{Parcel} \sigma_{ik} n_k x_j\, \mathrm{d}S + \iiint_{Parcel} \frac{\partial \sigma_{ik}}{\partial x_k} x_j\, \mathrm{d}V \right), \tag{3.1.29}$$

where $\mathbf{n}$ is the normal unit vector pointing into the parcel.

3.1.4 *Computing the traction*

Assume that the stress tensor is given by

$$\boldsymbol{\sigma} = \begin{bmatrix} x & xy & xyz \\ xy & y & z \\ xyz & z & z \end{bmatrix} \tag{3.1.30}$$

in dyn/cm^2, where lengths are measured in cm. Evaluate the normal and shear component of the traction (*a*) over the surface of a sphere centered at the origin, and (*b*) over the surface of a cylinder that is coaxial with the x axis.

3.2 Cauchy equation of motion

To derive the counterpart of Newton's second law of motion for a fluid, we combine (3.1.7) with (3.1.15) and use expression (1.4.27) for the rate of change of momentum of a fluid parcel defined in (1.4.25). Noting that the shape and size of the parcel are arbitrarily chosen to discard the integral sign, we derive Cauchy's equation of motion

$$\rho \frac{D\mathbf{u}}{Dt} = \nabla \cdot \boldsymbol{\sigma} + \rho \mathbf{g}, \tag{3.2.1}$$

which is applicable for compressible or incompressible fluids. The effect of the fluid inertia is represented by the term on the left-hand side.

Hydrodynamic volume force

The divergence of the stress tensor is the hydrodynamic volume force,

$$\boldsymbol{\Sigma} \equiv \nabla \cdot \boldsymbol{\sigma}. \tag{3.2.2}$$

The equation of motion (3.2.1) then takes the compact form

$$\rho \frac{D\mathbf{u}}{Dt} = \boldsymbol{\Sigma} + \rho \mathbf{g}, \tag{3.2.3}$$

which illustrates that the hydrodynamic volume force complements the body force.

Eulerian form

Expressing the material derivative, D/Dt, in terms of Eulerian derivatives, we obtain the Eulerian form of the equation of motion,

$$\rho \left(\frac{\partial \mathbf{u}}{\partial t} + \mathbf{u} \cdot \nabla \mathbf{u} \right) = \boldsymbol{\Sigma} + \rho \mathbf{g}. \tag{3.2.4}$$

The second term on the left-hand side can be regarded as a fictitious nonlinear inertial force. Using the continuity equation,

$$\frac{\partial \rho}{\partial t} + \nabla \cdot (\rho \mathbf{u}) = 0, \tag{3.2.5}$$

we derive the alternative form

$$\frac{\partial(\rho\mathbf{u})}{\partial t} + \nabla\cdot(\rho\,\mathbf{u}\otimes\mathbf{u}) = \mathbf{\Sigma} + \rho\,\mathbf{g}. \tag{3.2.6}$$

Stress–momentum tensor

It is useful to introduce the stress–momentum tensor,

$$\boldsymbol{\tau} \equiv \boldsymbol{\sigma} - \rho\,\mathbf{u}\otimes\mathbf{u}, \tag{3.2.7}$$

and recast (3.2.6) into the compact form

$$\frac{\partial(\rho\mathbf{u})}{\partial t} = \nabla\cdot\boldsymbol{\tau} + \rho\,\mathbf{g}. \tag{3.2.8}$$

If the flow is steady, the left-hand side of (3.2.8) is zero and the divergence of the stress–momentum tensor balances the body force.

3.2.1 Integral momentum balance

Integrating (3.2.8) over a fixed control volume V_c that lies entirely inside the fluid, and using the divergence theorem to convert the volume integral of the divergence of the stress–momentum tensor into a surface integral over the boundary D of V_c, we obtain an integral or macroscopic momentum balance,

$$\iiint_{V_c} \frac{\partial(\rho\mathbf{u})}{\partial t}\,dV = -\iint_D \boldsymbol{\tau}\cdot\mathbf{n}\,dS + \iiint_{V_c} \rho\,\mathbf{g}\,dV, \tag{3.2.9}$$

where $\mathbf{n}$ is the normal unit vector pointing into the control volume. Decomposing the stress–momentum tensor into its constituents and rearranging, we obtain

$$\iiint_{V_c} \frac{\partial(\rho\mathbf{u})}{\partial t}\,dV + \iint_D (\rho\mathbf{u})\,(\mathbf{u}\cdot\mathbf{n})\,dS = -\iint_D \boldsymbol{\sigma}\cdot\mathbf{n}\,dS + \iiint_{V_c} \rho\,\mathbf{g}\,dV. \tag{3.2.10}$$

Equation (3.2.10) states that the rate of change of momentum of the fluid occupying a fixed control volume is balanced by the flow rate of momentum normal to the boundaries, the force exerted on the boundaries, and the body force exerted on the fluid residing inside the control volume. The integral momentum balance allows us to develop approximate relations between global properties of a steady or unsteady flow. In engineering analysis, we typically derive expressions for boundary forces in terms of boundary velocities, subject to rational simplifications for inlet and outlet velocity profiles. The formulation can be generalized to describe other transported fields, such as heat or the concentration of a chemical species (e.g., [36]).

3.2.2 Energy balance

A differential energy balance can be obtained by projecting the equation of motion onto the velocity vector at a chosen point. Projecting the Eulerian form (3.2.4) and rearranging, we obtain

$$\frac{1}{2}\rho\left(\frac{\partial|\mathbf{u}|^2}{\partial t} + \mathbf{u}\cdot\nabla|\mathbf{u}|^2\right) = \mathbf{\Sigma}\cdot\mathbf{u} + \rho\,\mathbf{g}\cdot\mathbf{u}. \tag{3.2.11}$$

Combining this equation with the continuity equation (3.2.5) and rearranging the right-hand side, we obtain

$$\frac{\partial}{\partial t}\left(\frac{1}{2}\rho|\mathbf{u}|^2\right) + \nabla\cdot\left(\frac{1}{2}\rho|\mathbf{u}|^2\mathbf{u}\right) = \nabla\cdot(\boldsymbol{\sigma}\cdot\mathbf{u}) - \boldsymbol{\sigma}:\nabla\mathbf{u} + \rho\mathbf{g}\cdot\mathbf{u}. \qquad (3.2.12)$$

The double dot product of two matrices is defined in Section A.4, Appendix A.

Integral energy balance

An integral or macroscopic balance arises by integrating the differential energy balance over a fixed control volume, V_c, bounded by a surface, D. Integrating (3.2.12) and using the divergence theorem, we obtain

$$\iiint_{V_c}\frac{\partial}{\partial t}\left(\frac{1}{2}\rho|\mathbf{u}|^2\right)\mathrm{d}V = \iint_D\left(\frac{1}{2}\rho|\mathbf{u}|^2\right)\mathbf{u}\cdot\mathbf{n}\,\mathrm{d}S$$
$$- \iint_D\mathbf{u}\cdot\mathbf{f}\,\mathrm{d}S - \iiint_{V_c}\boldsymbol{\sigma}:\nabla\mathbf{u}\,\mathrm{d}V + \iiint_{V_c}\rho\mathbf{g}\cdot\mathbf{u}\,\mathrm{d}V, \qquad (3.2.13)$$

where $\mathbf{f} = \mathbf{n}\cdot\boldsymbol{\sigma}$ is the traction and $\mathbf{n}$ is the normal unit vector pointing into the control volume. The four terms on the right-hand side of (3.2.13) represent, respectively, the rate of supply of kinetic energy into the control volume by convection, the rate of working of the traction at the boundary of the control volume, the rate of energy dissipation, and the rate of working against the body force. We have found that the rate of dissipation of internal energy per unit volume of fluid is given by the double dot product of the stress tensor and velocity gradient tensor, $\boldsymbol{\sigma}:\nabla\mathbf{u}$.

Rate of working against the body force

When the fluid is incompressible and the density is uniform throughout the domain of flow, the last volume integral in (3.2.13) expressing the rate of working against the body force can be transformed into a surface integral over the boundary, D, by writing

$$\mathcal{W}_B \equiv \iiint_{V_c}\rho\mathbf{g}\cdot\mathbf{u}\,\mathrm{d}V = \rho\iiint_{V_c}\nabla\cdot[(\mathbf{g}\cdot\mathbf{x})\,\mathbf{u}]\,\mathrm{d}V = -\rho\iint_D(\mathbf{g}\cdot\mathbf{x})(\mathbf{u}\cdot\mathbf{n})\,\mathrm{d}S, \qquad (3.2.14)$$

where $\mathbf{n}$ is the normal unit vector pointing into the control volume. If the normal component of the velocity obeys the no-penetration boundary condition for a translating body, $\mathbf{u}\cdot\mathbf{n} = \mathbf{V}\cdot\mathbf{n}$, we may apply the divergence theorem to find that $\mathcal{W}_B = \rho V_{cv}\mathbf{g}\cdot\mathbf{V}$, where V_{cv} is the volume of the control volume and $\mathbf{V}$ is a constant velocity. The last term in (3.2.13) may then be identified with the rate of working necessary to elevate the fluid inside the control volume with velocity $\mathbf{V}$.

3.2.3   Energy dissipation inside a fluid parcel

The total energy of the fluid residing inside a parcel is comprised of the kinetic energy due to the motion of the fluid, the potential energy due to an external body force field, and the internal thermodynamic energy. The instantaneous kinetic and potential energies are given by

$$\mathcal{K} = \frac{1}{2}\iiint_{Parcel}\rho\mathbf{u}\cdot\mathbf{u}\,\mathrm{d}V, \qquad \mathcal{P} = -\iiint_{Parcel}\rho\mathbf{X}\cdot\mathbf{g}\,\mathrm{d}V, \qquad (3.2.15)$$

where $\mathbf{X}$ is the position of the point particles occupying the parcel. Taking the material derivative of the kinetic energy integral, and recalling that, because of mass conservation, $\mathrm{D}(\rho\,dV)/\mathrm{D}t = 0$, we express the rate of change of the kinetic energy in terms of the point-particle acceleration,

$$\frac{d\mathcal{K}}{dt} = \frac{1}{2} \iiint_{Parcel} \rho \frac{\mathrm{D}(\mathbf{u}\cdot\mathbf{u})}{\mathrm{D}t}\, dV = \iiint_{Parcel} \rho\, \mathbf{u} \cdot \frac{\mathrm{D}\mathbf{u}}{\mathrm{D}t}\, dV. \tag{3.2.16}$$

Now we use the equation of motion to express the acceleration in terms of the stress tensor and body force, obtaining

$$\frac{d\mathcal{K}}{dt} = \iiint_{Parcel} \mathbf{u} \cdot (\nabla \cdot \boldsymbol{\sigma})\, dV + \iiint_{Parcel} \rho\, \mathbf{u} \cdot \mathbf{g}\, dV. \tag{3.2.17}$$

Further manipulation yields

$$\frac{d\mathcal{K}}{dt} = \iiint_{Parcel} \nabla \cdot (\mathbf{u} \cdot \boldsymbol{\sigma})\, dV - \iiint_{Parcel} \boldsymbol{\sigma} : \nabla \mathbf{u}\, dV + \iiint_{Parcel} \rho\, \mathbf{u} \cdot \mathbf{g}\, dV. \tag{3.2.18}$$

The last integral is equal to $-d\mathcal{P}_p/dT$. Using the divergence theorem to convert the first integral on the right-hand side into a surface integral and rearranging, we derive an energy balance expressed by the equation

$$\frac{d(\mathcal{K} + \mathcal{P})}{dt} = -\iint_{Parcel} \mathbf{u} \cdot \mathbf{f}\, dS - \iiint_{Parcel} \boldsymbol{\sigma} : \nabla \mathbf{u}\, dV, \tag{3.2.19}$$

where $\mathbf{f} = \mathbf{n} \cdot \boldsymbol{\sigma}$ is the traction and $\mathbf{n}$ is the normal unit vector pointing into the parcel.

The first integral on the right-hand side of (3.2.19) is the rate of working of the traction on the parcel surface. It then follows from the first thermodynamic principle that the second integral expresses the rate of change of internal energy, which is equal to the rate energy dissipation inside the parcel, $\mathcal{I}$, expended for increasing the temperature of the fluid,

$$\frac{d\mathcal{I}}{dt} = \iiint_{Parcel} \boldsymbol{\sigma} : \nabla \mathbf{u}\, dV, \tag{3.2.20}$$

which is consistent with the third term on the right-hand side of (3.2.13).

3.2.4   Symmetry of the stress tensor in the absence of a torque field

The angular momentum balance for a fluid parcel requires that the rate of change of the angular momentum of the fluid occupying the parcel computed with respect to a specified point, $\mathbf{x}_0$, be equal to the total torque exerted on the parcel given in (3.1.20) or (3.1.21),

$$\frac{d}{dt} \iiint_{Parcel} \rho\, \hat{\mathbf{x}} \times \mathbf{u}\, dV = \mathbf{T}^{total}, \tag{3.2.21}$$

where $\hat{\mathbf{x}} = \mathbf{x} - \mathbf{x}_0$. Transferring the derivative inside the integral as a material derivative and using the continuity equation, we obtain

$$\iiint_{Parcel} \rho \frac{\mathrm{D}\hat{\mathbf{x}}}{\mathrm{D}t} \times \mathbf{u}\, dV + \iiint_{Parcel} \rho\, \mathbf{x} \times \frac{\mathrm{D}\mathbf{u}}{\mathrm{D}t}\, dV = \mathbf{T}^{total}. \tag{3.2.22}$$

The first integrand is equal to $\rho \mathbf{u} \times \mathbf{u}$, which is identically zero. Switching to index notation, replacing the total torque with the right-hand side of (3.1.21) and rearranging, we obtain

$$\iiint_{Parcel} \left[\epsilon_{ijk} \hat{x}_j \left(\rho \frac{\mathrm{D} u_k}{\mathrm{D} t} - \frac{\partial \sigma_{jk}}{\partial x_j} - \rho g_k \right) - \epsilon_{ilk} \sigma_{lk} - \lambda c_i \right] \mathrm{d}V = 0. \qquad (3.2.23)$$

Since the volume of the parcel is arbitrary, we may discard the integral sign and use the equation of motion (3.2.1) to simplify the integrand, finding

$$\epsilon_{ilk} \sigma_{lk} + \lambda c_i = 0. \qquad (3.2.24)$$

Multiplying (3.2.24) by ϵ_{imn} and manipulating the product of the alternating tensors, we find that

$$\sigma_{mn} - \sigma_{nm} = -\lambda \epsilon_{mni} c_i, \qquad (3.2.25)$$

which shows that, in the absence of an external torque field, $\mathbf{c}$, the stress tensor must be symmetric, $\sigma_{ij} = \sigma_{ji}$ or $\boldsymbol{\sigma} = \boldsymbol{\sigma}^T$, where the superscript T designates the matrix transpose. In that case, only three of the six nondiagonal components of the stress tensor are independent, and the remaining three nondiagonal components are equal to their transpose counterparts. The traction may then be computed as

$$\mathbf{f} \equiv \mathbf{n} \cdot \boldsymbol{\sigma} = \boldsymbol{\sigma} \cdot \mathbf{n}. \qquad (3.2.26)$$

In the remainder of this book, we will tacitly assume that the conditions for the stress tensor to be symmetric are satisfied.

Principal directions

The symmetry of the stress tensor in the absence of a torque field guarantees the existence of three real eigenvalues and corresponding orthogonal eigenvectors. The traction exerted on an infinitesimal planar surface that is perpendicular to an eigenvector points in the normal direction, that is, it lacks a shearing component. Setting the Cartesian axes parallel to the eigenvectors renders the stress tensor diagonal. In the case of an isotropic fluid, defined as a fluid that has no favorable direction, the eigenvectors of the stress tensor must coincide with those of the rate-of-deformation tensor, as will be discussed in Section 3.3.

Orthogonal curvilinear coordinates

In the absence of a torque field, the components of the stress tensor in orthogonal curvilinear coordinates, (ξ, η, ζ), are symmetric, that is, $\sigma_{\alpha\beta} = \sigma_{\beta\alpha}$, where Greek indices stand for ξ, η, or ζ. For example, in cylindrical polar coordinates, $\sigma_{x\varphi} = \sigma_{\varphi x}$.

3.2.5   Hydrodynamic volume force in curvilinear coordinates

The three scalar components of the equation of motion (3.2.3) can be expressed in orthogonal or nonorthogonal curvilinear coordinates by straightforward yet tedious manipulations. The procedure will be illustrated for plane polar coordinates, and expressions will be given in cylindrical and spherical polar coordinates. Corresponding expressions for the point-particle acceleration on the left-hand side of the equation of motion, $\mathrm{D}\mathbf{u}/\mathrm{D}t$, are given in Table 1.5.1. Substituting these expressions into the equation of motion provides us with three scalar component equations corresponding to the chosen coordinates.

Plane polar coordinates

To derive the plane polar components of the hydrodynamic volume force, $\boldsymbol{\Sigma} = \nabla \cdot \boldsymbol{\sigma}$, we express the gradient and stress tensor in the corresponding forms

$$\nabla = \mathbf{e}_r \frac{\partial}{\partial r} + \mathbf{e}_\theta \frac{1}{r} \frac{\partial}{\partial \theta}, \qquad \boldsymbol{\sigma} = \sigma_{rr} \, \mathbf{e}_r \otimes \mathbf{e}_r + \sigma_{r\theta} \, \mathbf{e}_r \otimes \mathbf{e}_\theta + \sigma_{\theta r} \, \mathbf{e}_\theta \otimes \mathbf{e}_r + \sigma_{\theta\theta} \, \mathbf{e}_\theta \otimes \mathbf{e}_\theta, \quad (3.2.27)$$

and compute

$$\boldsymbol{\Sigma} = \nabla \cdot \boldsymbol{\sigma} = \mathbf{e}_r \cdot \frac{\partial \boldsymbol{\sigma}}{\partial r} + \frac{1}{r} \mathbf{e}_\theta \cdot \frac{\partial \boldsymbol{\sigma}}{\partial \theta}. \tag{3.2.28}$$

Expanding the derivatives, we obtain

$$\boldsymbol{\Sigma} = \frac{\partial \sigma_{rr}}{\partial r} \, \mathbf{e}_r + \frac{\partial \sigma_{r\theta}}{\partial r} \, \mathbf{e}_\theta + \frac{1}{r} \left(\frac{\partial \sigma_{\theta r}}{\partial \theta} \, \mathbf{e}_r + \frac{\partial \sigma_{\theta\theta}}{\partial \theta} \mathbf{e}_\theta \right)$$
$$+ \sigma_{\alpha\beta} \left(\mathbf{e}_r \cdot \frac{\partial (\mathbf{e}_\alpha \otimes \mathbf{e}_\beta)}{\partial r} + \frac{1}{r} \mathbf{e}_\theta \cdot \frac{\partial (\mathbf{e}_\alpha \otimes \mathbf{e}_\beta)}{\partial \theta} \right), \tag{3.2.29}$$

where summation is implied over the repeated indices, α and β, standing for r or θ. Expanding the derivatives of the products and grouping similar terms, we obtain

$$\boldsymbol{\Sigma} = \left(\frac{\partial \sigma_{rr}}{\partial r} + \frac{1}{r} \frac{\partial \sigma_{\theta r}}{\partial \theta} \right) \mathbf{e}_r + \left(\frac{\partial \sigma_{r\theta}}{\partial r} + \frac{1}{r} \frac{\partial \sigma_{\theta\theta}}{\partial \theta} \right) \mathbf{e}_\theta$$
$$+ \sigma_{\alpha\beta} \left(\mathbf{e}_r \cdot \frac{\partial \mathbf{e}_\alpha}{\partial r} \mathbf{e}_\beta + \mathbf{e}_r \cdot \mathbf{e}_\alpha \frac{\partial \mathbf{e}_\beta}{\partial r} + \frac{1}{r} \mathbf{e}_\theta \cdot \frac{\partial \mathbf{e}_\alpha}{\partial \theta} \mathbf{e}_\beta + \frac{1}{r} \mathbf{e}_\theta \cdot \mathbf{e}_\alpha \frac{\partial \mathbf{e}_\beta}{\partial \theta} \right). \tag{3.2.30}$$

Now we recall that all derivatives $\partial \mathbf{e}_\alpha / \partial \beta$ are zero, except for two derivatives,

$$\frac{\partial \mathbf{e}_r}{\partial \theta} = \mathbf{e}_\theta, \qquad \frac{\mathrm{d} \mathbf{e}_\theta}{\mathrm{d} \theta} = -\mathbf{e}_r, \tag{3.2.31}$$

and find that

$$\boldsymbol{\Sigma} = \left(\frac{\partial \sigma_{rr}}{\partial r} + \frac{1}{r} \frac{\partial \sigma_{\theta r}}{\partial \theta} \right) \mathbf{e}_r + \left(\frac{\partial \sigma_{r\theta}}{\partial r} + \frac{1}{r} \frac{\partial \sigma_{\theta\theta}}{\partial \theta} \right) \mathbf{e}_\theta + \sigma_{r\beta} \frac{1}{r} \mathbf{e}_\theta \cdot \frac{\partial \mathbf{e}_r}{\partial \theta} \mathbf{e}_\beta + \sigma_{\theta\beta} \frac{1}{r} \frac{\partial \mathbf{e}_\beta}{\partial \theta}, \tag{3.2.32}$$

where summation is implied over the repeated index, β. Simplifying, we derive the expressions given in Table 3.2.1(c).

Cylindrical and spherical polar coordinates

Expressions for the components of the hydrodynamic volume force $\boldsymbol{\Sigma}$ in cylindrical and spherical polar coordinates are collected in Table 3.2.1(a, b).

(a)

$$\Sigma_x = \frac{\partial \sigma_{xx}}{\partial x} + \frac{1}{\sigma}\frac{\partial(\sigma\sigma_{\sigma x})}{\partial \sigma} + \frac{1}{\sigma}\frac{\partial \sigma_{\varphi x}}{\partial \varphi}$$

$$\Sigma_\sigma = \frac{\partial \sigma_{x\sigma}}{\partial x} + \frac{1}{\sigma}\frac{\partial(\sigma\sigma_{\sigma\sigma})}{\partial \sigma} + \frac{1}{\sigma}\frac{\partial \sigma_{\varphi\sigma}}{\partial \varphi} - \frac{\sigma_{\varphi\varphi}}{\sigma}$$

$$\Sigma_\varphi = \frac{\partial \sigma_{x\varphi}}{\partial x} + \frac{1}{\sigma^2}\frac{\partial(\sigma^2 \sigma_{\varphi\sigma})}{\partial \sigma} + \frac{1}{\sigma}\frac{\partial \sigma_{\varphi\varphi}}{\partial \varphi}$$

(b)

$$\Sigma_r = \frac{1}{r^2}\frac{\partial(r^2\sigma_{rr})}{\partial r} + \frac{1}{r\sin\theta}\frac{\partial(\sigma_{r\theta}\sin\theta)}{\partial \theta} + \frac{1}{r\sin\theta}\frac{\partial \sigma_{\varphi r}}{\partial \varphi} - \frac{\sigma_{\theta\theta} + \sigma_{\varphi\varphi}}{r}$$

$$\Sigma_\theta = \frac{1}{r^2}\frac{\partial(r^2\sigma_{r\theta})}{\partial r} + \frac{1}{r\sin\theta}\frac{\partial(\sigma_{\theta\theta}\sin\theta)}{\partial \theta} + \frac{1}{r\sin\theta}\frac{\partial \sigma_{\varphi\theta}}{\partial \varphi} + \frac{\sigma_{r\theta} - \sigma_{\varphi\varphi}\cot\theta}{r}$$

$$\Sigma_\varphi = \frac{1}{r^2}\frac{\partial(r^2\sigma_{r\varphi})}{\partial r} + \frac{1}{r}\frac{\partial \sigma_{\theta\varphi}}{\partial \theta} + \frac{1}{r\sin\theta}\frac{\partial \sigma_{\varphi\varphi}}{\partial \varphi} + \frac{\sigma_{r\varphi} + 2\,\sigma_{\theta\varphi}\cot\theta}{r}$$

(c)

$$\Sigma_r = \frac{1}{r}\frac{\partial(r\sigma_{rr})}{\partial r} + \frac{1}{r}\frac{\partial \sigma_{\theta r}}{\partial \theta} - \frac{\sigma_{\theta\theta}}{r}, \qquad \Sigma_\theta = \frac{1}{r^2}\frac{\partial(r^2\sigma_{r\theta})}{\partial r} + \frac{1}{r}\frac{\partial \sigma_{\theta\theta}}{\partial \theta}$$

TABLE 3.2.1   (b) The x, σ, and φ components of the hydrodynamic volume force in cylindrical polar coordinates. (b) The r, θ, and φ components of the hydrodynamic volume force in spherical polar coordinates. (c) The r and θ components of the hydrodynamic volume force in plane polar coordinates.

3.2.6   Noninertial frames

Cauchy's equation of motion was derived under the assumption that the frame of reference is inertial, which means that the Cartesian axes are either stationary or translate in space with constant velocity, but neither accelerate nor rotate. It is sometimes convenient to work in a noninertial frame whose origin translates with respect to an inertial frame with time-dependent velocity, $\mathbf{U}(t)$, while its axes rotate about the instantaneous position of the origin with a time-dependent angular velocity, $\boldsymbol{\Omega}(t)$. Since Newton's law only applies in an inertial frame, modifications are necessary in order to account for the linear and angular acceleration.

Velocity

Our first task is to compute the velocity of a point particle in an inertial frame in terms of its coordinates in a noninertial frame. We begin by introducing three unit vectors, $\mathbf{e}_1$, $\mathbf{e}_2$, and $\mathbf{e}_3$, associated with the noninertial coordinates, y_1, y_2, and y_3, and describe the position of a point particle in the inertial frame as

$$\mathbf{X} = \mathbf{x}_0 + \mathbf{Y}, \tag{3.2.33}$$

where $\mathbf{x}_0$ is the instantaneous position of the origin of the noninertial frame, $\mathbf{Y} = Y_i \, \mathbf{e}_i$ is the point-particle position in the noninertial system, and Y_i are the corresponding coordinates; summation is implied over the repeated index, i. By definition, $\mathbf{x}_0$ and $\mathbf{e}_i$ evolve in time according to the linear equations

$$\frac{\mathrm{d}\mathbf{x}_0}{\mathrm{d}t} = \mathbf{U}, \qquad \frac{\mathrm{d}\mathbf{e}_i}{\mathrm{d}t} = \boldsymbol{\Omega} \times \mathbf{e}_i. \tag{3.2.34}$$

Taking the material derivative of (3.2.33), we obtain an expression for the velocity in the inertial system,

$$\mathbf{u} \equiv \frac{\mathrm{D}\mathbf{X}}{\mathrm{D}t} = \frac{\mathrm{d}\mathbf{x}_0}{\mathrm{d}t} + \frac{\mathrm{D}Y_i}{\mathrm{D}t} \, \mathbf{e}_i + Y_i \, \frac{\mathrm{d}\mathbf{e}_i}{\mathrm{d}t}. \tag{3.2.35}$$

Using (3.2.34), we find that

$$\mathbf{u} = \mathbf{U} + \frac{\mathrm{D}Y_i}{\mathrm{D}t} \, \mathbf{e}_i + \boldsymbol{\Omega} \times \mathbf{Y}. \tag{3.2.36}$$

The second term on the right-hand side is the velocity of a point particle in the noninertial system,

$$\mathbf{v} = \frac{\mathrm{D}Y_i}{\mathrm{D}t} \, \mathbf{e}_i. \tag{3.2.37}$$

The velocity components in the noninertial system are $v_i = \mathrm{D}Y_i/\mathrm{D}t$.

Acceleration

The acceleration of a point particle in the inertial system is

$$\mathbf{a}(\mathbf{X}) \equiv \frac{\mathrm{D}\mathbf{u}}{\mathrm{D}t} = \frac{\mathrm{D}^2\mathbf{X}}{\mathrm{D}t^2}. \tag{3.2.38}$$

Taking the material derivative of (3.2.36), we obtain

$$\mathbf{a}(\mathbf{X}) = \frac{\mathrm{d}\mathbf{U}}{\mathrm{d}t} + \frac{\mathrm{D}^2 Y_i}{\mathrm{D}t^2} \, \mathbf{e}_i + \frac{\mathrm{D}Y_i}{\mathrm{D}t} \frac{\mathrm{d}\mathbf{e}_i}{\mathrm{d}t} + \boldsymbol{\Omega} \times \left(\frac{\mathrm{D}Y_i}{\mathrm{D}t} \, \mathbf{e}_i \right) + \frac{\mathrm{d}\boldsymbol{\Omega}}{\mathrm{d}t} \times \mathbf{Y} + \boldsymbol{\Omega} \times \left(Y_i \, \frac{\mathrm{d}\mathbf{e}_i}{\mathrm{d}t} \right). \tag{3.2.39}$$

The second term on the right-hand side represents the acceleration of the point particles in the noninertial frame,

$$\mathbf{a}(\mathbf{Y}) = \frac{\mathrm{D}^2 Y_i}{\mathrm{D}t^2} \, \mathbf{e}_i. \tag{3.2.40}$$

Using the second relation in (3.2.34) to simplify the third and last terms on the right-hand side of (3.2.39), we find that

$$\mathbf{a}(\mathbf{X}) = \frac{\mathrm{d}\mathbf{U}}{\mathrm{d}t} + \mathbf{a}(\mathbf{Y}) + 2\,\boldsymbol{\Omega} \times \mathbf{v} + \frac{\mathrm{d}\boldsymbol{\Omega}}{\mathrm{d}t} \times \mathbf{Y} + \boldsymbol{\Omega} \times (\boldsymbol{\Omega} \times \mathbf{Y}). \tag{3.2.41}$$

The right-hand side involves position, velocity, and acceleration in the noninertial system alone.

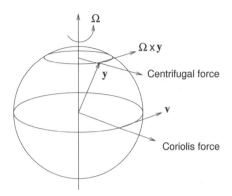

FIGURE 3.2.1   Illustration of the Coriolis and centrifugal forces on the globe due to the rotation of the earth.

Equation of motion

Substituting the right-hand side of (3.2.41) for the point-particle acceleration into the equation of motion (3.2.3), and rearranging, we derive a generalized equation of motion in the noninertial frame,

$$\rho \frac{D\mathbf{v}}{Dt} = \mathbf{\Sigma} + \rho\,\mathbf{g} + \mathbf{f}^{I}, \qquad (3.2.42)$$

where $\mathbf{v}$ is the velocity field in the noninertial frame,

$$\mathbf{f}^{I} \equiv -\rho \left(\frac{d\mathbf{U}}{dt} + 2\mathbf{\Omega} \times \mathbf{v} + \mathbf{\Omega} \times (\mathbf{\Omega} \times \mathbf{y}) + \frac{d\mathbf{\Omega}}{dt} \times \mathbf{y} \right) \qquad (3.2.43)$$

is a fictitious inertial force per unit volume of fluid, and $\mathbf{y}$ is the position in the noninertial frame. The fictitious inertial force consists of (a) the linear acceleration force, $-\rho\,d\mathbf{U}/dt$; (b) the Coriolis force, $-2\rho\,\mathbf{\Omega} \times \mathbf{v}$; (c) the centrifugal force, $-\rho\,\mathbf{\Omega} \times (\mathbf{\Omega} \times \mathbf{y})$; and (d) the angular-acceleration force, $-\rho\,(d\mathbf{\Omega}/dt) \times \mathbf{y}$.

Centrifugal and Coriolis forces

The Coriolis and centrifugal forces on the globe are illustrated in Figure 3.2.1.   A Coriolis force develops at the equator when a fluid moves along or normal to the equator with respect to the rotating surface of the earth. A centrifugal force develops everywhere except at the poles, independent of the fluid velocity. The negative of the last term on the right-hand side of the σ component of the point particle acceleration in cylindrical polar coordinates shown in Table 1.3.1, multiplied by the fluid density, $-\rho\,u_{\varphi}^{2}/\sigma$, is also called the centrifugal force. The negative of the last term on the right-hand side of the φ component of the point particle acceleration in cylindrical polar coordinates shown in Table 1.3.1, multiplied by the fluid density, $-\rho\,u_{\sigma}u_{\varphi}/\sigma$, is also called the Coriolis force. However, this terminology is not entirely consistent, for the centrifugal and the Coriolis forces are attributed to a noninertial system. Similar terms appear in the r and θ components of the point-particle acceleration in plane polar coordinates.

To understand the physical motivation for this nomenclature, we consider a fluid in rigid-body rotation. Describing the motion in a stationary frame of reference, we identify the term $\rho u_\varphi^2/\sigma$ with the centripetal point-particle acceleration force, which must be balanced by an opposing radial force. Describing the motion in a noninertial frame of reference that rotates with the fluid, we find that the Coriolis force is zero, and the centrifugal force is $-\rho\, u_\varphi^2/\sigma$. The temptation to interpret the centripetal acceleration force, in the inertial system as the centrifugal force is then apparent.

Problems

3.2.1 *Angular momentum balance with respect to an arbitrary point in space*

Write the counterpart of the balance (3.1.20) when the angular momentum and torque are computed with respect to an arbitrary point, $\mathbf{x}_0$. Then proceed as in the text to derive (3.2.24).

3.2.2 *Hydrodynamic volume force in polar coordinates*

Derive the components of the hydrodynamic volume force, $\mathbf{\Sigma} \equiv \nabla \cdot \boldsymbol{\sigma}$, in (a) cylindrical polar and (b) spherical polar coordinates.

3.3   Constitutive equations for the stress tensor

Molecular motions in a fluid that has been in a macroscopic state of rest for a sufficiently long period of time reach dynamical equilibrium whereupon the stress field assumes the isotropic form

$$\boldsymbol{\sigma} = -p_{th}\,\mathbf{I}, \tag{3.3.1}$$

where $\mathbf{I}$ is the identity matrix. The thermodynamic pressure, p_{th}, is a function of the density and temperature, and depends on the chemical composition of the fluid in a manner that is determined by an appropriate equation of state.

Ideal-gas law

In the case of an ideal gas, the thermodynamic pressure, p_{th}, is related to the density, ρ, by Clapeyron's ideal gas law,

$$p_{th} = \frac{RT}{M}\,\rho, \tag{3.3.2}$$

where $R = 8.314 \times 10^3$ kg m^2/(s^2 kmole K) is the ideal gas constant, T is Kelvin's absolute temperature, which is equal to the Celsius centigrade temperature reduced by 273 units, M is the molar mass, defined as the mass of one mole comprised of a collection of N_A molecules, and $N_A = 6.022 \times 10^{26}$ is the Avogadro number. The molar mass of an element is equal to the atomic weight of the element listed in the periodic table expressed in grams.

Effect of fluid motion

Physical intuition suggests that the instantaneous structure of the stress field inside a fluid that has been in a state of motion for some time depends not only on the current thermodynamic conditions, but also on the history of the motion of all point particles comprising the fluid, from inception of the motion, up to the present time. Leaving aside physiochemical interactions that are independent of the fluid motion, we argue that the stress field depends on the structure of the velocity field at all prior times. This reasoning leads us to introduce a constitutive equation for the stress tensor that relates the stress at a point at a particular instant $t = \tau$ to the structure of the velocity field at all prior times,

$$\boldsymbol{\sigma}(t = \tau) = \boldsymbol{\mathcal{G}}[\mathbf{u}(t \leq \tau)]. \tag{3.3.3}$$

The nonlinear functional operator $\boldsymbol{\mathcal{G}}$ may involve derivatives or integrals of the velocity with the respect to space and time. Coefficients appearing in the specific functional form of $\boldsymbol{\mathcal{G}}$ are regarded as rheological properties of the fluid.

Reaction pressure and deviatoric stress tensor

It is useful to recast equation (3.3.3) into the alternative form

$$\boldsymbol{\sigma} \equiv -p\,\mathbf{I} + \check{\boldsymbol{\sigma}}, \tag{3.3.4}$$

where p is the reaction pressure defined by the equation

$$p \equiv -\frac{1}{3}\,\mathrm{trace}(\boldsymbol{\sigma}), \tag{3.3.5}$$

and $\check{\boldsymbol{\sigma}}$ is the deviatoric part of the stress tensor. Since $\boldsymbol{\sigma}$ is a tensor, its trace and therefore the reaction pressure are invariant under changes of the axes of the Cartesian coordinates. Physically, the negative of the reaction pressure can be identified with the mean value of the normal component of the traction exerted on a small surface located at a certain point, averaged over all possible orientations. In a different interpretation, the reaction pressure is identified with the mean value of the normal component of the traction exerted on the surface of a small spherical parcel (Problem 3.3.1). In the remainder of this book, the reaction pressure will be called simply the pressure.

Coordinate invariance, objectivity, and fading memory

To be admissible, a constitutive equation must satisfy a number of conditions (e.g., [365, 396]). The condition of coordinate invariance requires that the constitutive equation should be valid independent of the coordinate system where the position vector, velocity, and stress are described. Thus, the functional form (3.3.3) must hold true independently of whether the stress and the velocity are expressed in Cartesian, cylindrical polar, spherical polar, or any other type of curvilinear coordinates. The condition of material objectivity requires that the instantaneous stress field should be independent of the motion of the observer. The condition of fading memory requires that the instantaneous structure of the stress field must depend on the recent motion of the fluid stronger than it does on the ancient history.

Significance of parcel deformation

Next, we argue that the history of deformation of a small fluid parcel rather than the fluid velocity itself is significant as far as determining deviatoric stresses on the parcel surface. This argument is supported by the observation that rigid-body motion does not generate a deviatoric stress field. Replacing the velocity in the arguments of the operator in (3.3.3) with the deformation gradient $\mathbf{F}$ introduced in (1.3.31), we obtain

$$\boldsymbol{\sigma}(t = \tau) = \mathcal{G}[\mathbf{F}(t \leq \tau)]. \tag{3.3.6}$$

If the deformation gradient is zero throughout the domain of flow, the fluid executes rigid-body motion.

3.3.1 Simple fluids

The stress tensor in a simple fluid at the position of a point particle labeled $\boldsymbol{\alpha}$ is a function of the history of the deformation gradient, $\mathbf{F}$, evaluated at all prior positions of the point particle over all past times up to the present time [408]. The constitutive equation for a simple fluid thus takes the form

$$\boldsymbol{\sigma}(\boldsymbol{\alpha}, t = \tau) = \mathcal{G}[\mathbf{F}(\boldsymbol{\alpha}, t \leq \tau)]. \tag{3.3.7}$$

It can be shown that a simple fluid is necessarily isotropic, that is, it has no favorable or unfavorable directions in space ([365], p. 67).

3.3.2 Purely viscous fluids

A purely viscous fluid is a simple fluid with the added property that the stress at the location of a point particle at a particular time instant is a function of the deformation gradient evaluated at the position of the point particle at that particular instant (e.g., [365], p. 134). Thus, the constitutive equation for a purely viscous fluid takes the simplified form

$$\boldsymbol{\sigma}(\boldsymbol{\alpha}, t = \tau) = \mathcal{G}[\mathbf{F}(\boldsymbol{\alpha}, t = \tau)]. \tag{3.3.8}$$

A purely viscous fluid lacks memory, that is, it is inelastic. Enforcing the principle of material objectivity, we find that the functional form of the operator $\mathcal{G}$ in (3.3.8) must be such that the antisymmetric part of $\mathbf{F}$ drops out and the stress tensor must be a function of the rate-of-deformation tensor, $\mathbf{E} \equiv \frac{1}{2}(\mathbf{L} + \mathbf{L}^T) - \frac{1}{3}\alpha\mathbf{I}$. We thus obtain a constitutive relation first proposed by Stokes in 1845,

$$\boldsymbol{\sigma}(\boldsymbol{\alpha}, t = \tau) = \mathcal{G}[\mathbf{E}(\boldsymbol{\alpha}, t = \tau)]. \tag{3.3.9}$$

Because $\boldsymbol{\sigma}$ as well as $\mathbf{E}$, $\mathbf{E}^2$, $\mathbf{E}^3$, ..., are all second-order tensors transforming in a similar fashion, as discussed in Section 1.1.7, the most general form of (3.3.9) is

$$\boldsymbol{\sigma}(\boldsymbol{\alpha}, t = T) = f_0(I_1, I_2, I_3)\,\mathbf{I} + f_1(I_1, I_2, I_3)\,\mathbf{E} + f_2(I_1, I_2, I_3)\,\mathbf{E}^2 + \cdots, \tag{3.3.10}$$

where f_i are functions of the three invariants of the rate-of-deformation tensor introduced in (1.1.46). Built in equation (3.3.10) is the assumption that the principal directions of the stress tensor are identical to those of the rate-of-deformation tensor, as required by the stipulation that the fluid is isotropic.

3.3.3   Reiner–Rivlin and generalized Newtonian fluids

Using the Caley–Hamilton theorem [317], we can express the power $\mathbf{E}^n$ for $n \geq 3$ as a linear combination of $\mathbf{I}$, $\mathbf{E}$, and $\mathbf{E}^2$, with coefficients that are functions of the three invariants. This manipulation allows us to retain only the three leading terms shown on the right-hand side of (3.3.10). The resulting constitutive equation describes a Reiner–Rivlin fluid. A generalized Newtonian fluid arises by eliminating the quadratic term, setting $f_2 = 0$. Examples of generalized Newtonian fluids are the power-law and the Bingham plastic fluids.

As an example, we consider unidirectional shear flow along the x axis with velocity varying along the y axis, $\mathbf{u} = [u_x(y), 0, 0]$. The shear stress is determined from the constitutive equation

$$\sigma_{xy} = \mu_0 \left| \frac{du_x}{dy} \right|^n \frac{du_x}{dy}, \tag{3.3.11}$$

where μ_0 and n are two physical constants. This scalar constitutive equation corresponds to a generalized Newtonian fluid, called the power-law fluid. Setting $n = 0$ we obtain a Newtonian fluid.

3.3.4   Newtonian fluids

A Newtonian fluid is a Reiner–Rivlin fluid whose stress depends linearly on the rate of deformation tensor. All functions f_n with $n > 1$ in (3.3.10) are zero, f_1 is constant, and f_0 is a linear function of the third invariant, $I_3 = \nabla \cdot \mathbf{u}$, which is equal to the rate of expansion, α. We thus set

$$f_0 = -p, \qquad f_1 = 2\mu, \tag{3.3.12}$$

where p is the reaction pressure, allowed to be a linear function of the rate of expansion, and the coefficient μ is a physical constant with dimensions of mass per time per length called the dynamic viscosity or simply the viscosity of the fluid. The viscosity is often measured in units of poise, which is equal to 1 gr/(cm sec). In general, the viscosity of a gas increases, whereas the viscosity of a liquid decreases as the temperature is raised. The viscosity of water and air at three temperatures is listed in the first column of Table 3.3.1. The constitutive equation for a Newtonian fluid thus takes the linear form

$$\boldsymbol{\sigma} = -p\,\mathbf{I} + 2\mu\,\mathbf{E}. \tag{3.3.13}$$

Because the trace of the rate-of-deformation tensor $\mathbf{E}$ is zero, the trace of $\boldsymbol{\sigma}$ is equal to $-3p$, as required.

Dilatational viscosity

The reaction pressure in a fluid that has been left alone in a macroscopic state of rest for a long time period of time reduces to the thermodynamic pressure, p_{th}, determined by the local density of the fluid and temperature according to an assumed equation of state. Under flow conditions, we note the assumed linearity of the stress tensor on the rate-of-deformation tensor and recall that the fluid is isotropic to write

$$p = p_{th} - \kappa\alpha, \tag{3.3.14}$$

Temperature $T(°C)$	Water μ(cp)	Air μ(cp)	Water ν (cm^2/s)	Air ν (cm^2/s)
20	1.002	0.0181	1.004×10^{-2}	15.05×10^{-2}
40	0.653	0.0191	0.658×10^{-2}	18.86×10^{-2}
80	0.355	0.0209	0.365×10^{-2}	20.88×10^{-2}

TABLE 3.3.1   The dynamic viscosity (μ) and kinematic viscosity (ν) of water and air at three temperatures; cp stands for centipoise, which is one hundredth of the viscosity unit poise: 1 cp = 0.01 g/(cm sec).

where $\alpha \equiv \nabla \cdot \mathbf{u}$ is the rate of expansion and κ is a physical constant with dimensions of mass per time and length, called the dilatational or expansion viscosity [351]. The Newtonian constitutive equation then takes the form

$$\boldsymbol{\sigma} = -p_{th}\,\mathbf{I} + \kappa\alpha\,\mathbf{I} + 2\mu\,\mathbf{E}. \tag{3.3.15}$$

The trace of the stress tensor is

$$\mathrm{trace}(\boldsymbol{\sigma}) = -3\,p_{th} + 3\kappa\alpha. \tag{3.3.16}$$

The coefficient 3κ is sometimes called the bulk viscosity. The constitutive equation (3.3.15) is known to describe with high accuracy the stress distribution in a broad range of fluids whose molecules are small compared to the macroscopic dimensions of the flow and whose spatial configuration is sufficiently simple. Substituting into (3.3.15) the definition of the rate-of-deformation tensor, $\mathbf{E} \equiv \frac{1}{2}\left(\mathbf{L} + \mathbf{L}^T\right) - \frac{1}{3}\alpha\,\mathbf{I}$, we obtain

$$\boldsymbol{\sigma} = -p_{th}\,\mathbf{I} + \lambda\,\alpha\,\mathbf{I} + \mu\left(\mathbf{L} + \mathbf{L}^T\right), \tag{3.3.17}$$

where $\mathbf{L}$ is the velocity gradient tensor, the superscript T denotes the matrix transpose, and

$$\lambda = \kappa - \frac{2}{3}\mu \tag{3.3.18}$$

is the second coefficient of viscosity.

Internal energy

Substituting the Newtonian constitutive equation (3.3.15) into (3.2.20), we find that the rate of production of internal energy inside a fluid parcel is

$$\frac{\mathrm{d}\mathcal{I}_p}{\mathrm{d}t} = -\iiint_{Parcel} p_{th}\alpha\,\mathrm{d}V + \iiint_{Parcel} \kappa\,\alpha^2\,\mathrm{d}V + 2\iiint_{Parcel} \mu\,\mathbf{E}:\mathbf{E}\,\mathrm{d}V. \tag{3.3.19}$$

The first term on the right-hand side expresses reversible production of energy in the usual sense of thermodynamics. The second term expresses irreversible dissipation of energy due to the expansion or compression of the fluid, which further justifies calling κ the expansion viscosity. The third term expresses irreversible dissipation of energy due to pure deformation.

3.3 *Constitutive equations for the stress tensor* 209

Incompressible Newtonian fluids

The rate of expansion vanishes in the case of an incompressible Newtonian fluid, $\alpha = 0$. Expressing the rate of deformation tensor in terms of the velocity gradient tensor, we recast the Newtonian constitutive equation (3.3.13) into the form

$$\boldsymbol{\sigma} = p\mathbf{I} + \mu\left[\nabla\mathbf{u} + (\nabla\mathbf{u})^T\right]. \tag{3.3.20}$$

Explicitly, the nine components of the stress tensor are given by the matrix equation

$$
\begin{bmatrix}
\sigma_{xx} & \sigma_{xy} & \sigma_{xz} \\
\sigma_{yx} & \sigma_{yy} & \sigma_{yz} \\
\sigma_{zx} & \sigma_{zy} & \sigma_{zz}
\end{bmatrix}
=
\begin{bmatrix}
-p + 2\mu\dfrac{\partial u_x}{\partial x} & \mu\left(\dfrac{\partial u_y}{\partial x} + \dfrac{\partial u_x}{\partial y}\right) & \mu\left(\dfrac{\partial u_z}{\partial x} + \dfrac{\partial u_x}{\partial z}\right) \\[2ex]
\mu\left(\dfrac{\partial u_x}{\partial y} + \dfrac{\partial u_y}{\partial x}\right) & -p + 2\mu\dfrac{\partial u_y}{\partial y} & \mu\left(\dfrac{\partial u_z}{\partial y} + \dfrac{\partial u_y}{\partial z}\right) \\[2ex]
\mu\left(\dfrac{\partial u_x}{\partial z} + \dfrac{\partial u_z}{\partial x}\right) & \mu\left(\dfrac{\partial u_y}{\partial z} + \dfrac{\partial u_z}{\partial y}\right) & -p + 2\mu\dfrac{\partial u_z}{\partial z}
\end{bmatrix}. \tag{3.3.21}
$$

It is important to recognize that, by requiring incompressibility, which is tantamount to assuming that fluid parcels maintain their original volume and density, we essentially discard the functional dependence of the hydrodynamic to the thermodynamic pressure, and (3.3.14) may no longer be applied.

Polar coordinates

Explicit expressions for the components of the stress tensor in terms of the velocity and pressure in cylindrical, spherical, and plane polar coordinates are given in Table 3.3.2. Note that the stress components remain symmetric in orthogonal curvilinear coordinates. In the case of two-dimensional flow expressing rigid-body rotation with angular velocity Ω around the origin in the xy plane, we substitute $u_r = 0$ and $u_\theta = \Omega r$ and find that $\sigma_{rr} = -p$, $\sigma_{r\theta} = 0$, $\sigma_{\theta r} = 0$, $\sigma_{\theta\theta} = -p$, where p is the pressure.

3.3.5 Inviscid fluids

Inviscid fluids, also called ideal fluids, are Newtonian fluids with vanishing viscosity. The constitutive equation for an incompressible inviscid fluid derives from (3.3.20) by setting $\mu = 0$, yielding

$$\boldsymbol{\sigma} = -p\,\mathbf{I}. \tag{3.3.22}$$

In real life, no fluid can be truly inviscid and equation (3.3.22) must be regarded as a mathematical idealization arising in the limit as the rate-of-deformation tensor, $\mathbf{E}$, tends to become vanishingly small. Under certain conditions, superfluid helium behaves like an inviscid fluid and may thus be used in the laboratory to visualize ideal flows.

It is instructive to note the similarity in functional form between (3.3.22) and (3.3.1). However, it is important to acknowledge that the pressure in (3.3.22) is a flow variable, whereas the pressure in (3.3.1) is a thermodynamic variable determined by an appropriate equation of state.

(a)

$$\sigma_{xx} = -p + 2\mu \frac{\partial u_x}{\partial x}, \qquad \sigma_{x\sigma} = \sigma_{\sigma x} = \mu \left(\frac{\partial u_x}{\partial \sigma} + \frac{\partial u_\sigma}{\partial x} \right), \qquad \sigma_{x\varphi} = \sigma_{\varphi x} = \mu \left(\frac{\partial u_\varphi}{\partial x} + \frac{1}{\sigma} \frac{\partial u_x}{\partial \varphi} \right)$$

$$\sigma_{\sigma\sigma} = -p + 2\mu \frac{\partial u_\sigma}{\partial \sigma}, \qquad \sigma_{\sigma\varphi} = \sigma_{\varphi\sigma} = \mu \left[\sigma \frac{\partial}{\partial \sigma} \left(\frac{u_\varphi}{\sigma} \right) + \frac{1}{\sigma} \frac{\partial u_\sigma}{\partial \varphi} \right],$$

$$\sigma_{\varphi\varphi} = -p + 2\mu \left(\frac{1}{\sigma} \frac{\partial u_\varphi}{\partial \varphi} + \frac{u_\sigma}{\sigma} \right)$$

(b)

$$\sigma_{rr} = -p + 2\mu \frac{\partial u_r}{\partial r}, \qquad \sigma_{r\theta} = \sigma_{\theta r} = \mu \left[r \frac{\partial}{\partial r} \left(\frac{u_\theta}{r} \right) + \frac{1}{r} \frac{\partial u_r}{\partial \theta} \right]$$

$$\sigma_{r\varphi} = \sigma_{\varphi r} = \mu \left[\frac{1}{r \sin\theta} \frac{\partial u_r}{\partial \varphi} + r \frac{\partial}{\partial r} \left(\frac{u_\varphi}{r} \right) \right], \qquad \sigma_{\theta\theta} = -p + 2\mu \left(\frac{1}{r} \frac{\partial u_\theta}{\partial \theta} + \frac{u_r}{r} \right)$$

$$\sigma_{\theta\varphi} = \sigma_{\varphi\theta} = \mu \left[\frac{\sin\theta}{r} \frac{\partial}{\partial \theta} \left(\frac{u_\varphi}{\sin\theta} \right) + \frac{1}{r \sin\theta} \frac{\partial u_\theta}{\partial \varphi} \right]$$

$$\sigma_{\varphi\varphi} = -p + \mu \frac{2}{r \sin\theta} \left(\frac{\partial u_\varphi}{\partial \varphi} + u_r \sin\theta + u_\theta \cos\theta \right)$$

(c)

$$\sigma_{rr} = -p + 2\mu \frac{\partial u_r}{\partial r}, \qquad \sigma_{r\theta} = \sigma_{\theta r} = \mu \left[r \frac{\partial}{\partial r} \left(\frac{u_\theta}{r} \right) + \frac{1}{r} \frac{\partial u_r}{\partial \theta} \right]$$

$$\sigma_{\theta\theta} = -p + 2\mu \left(\frac{1}{r} \frac{\partial u_\theta}{\partial \theta} + \frac{u_r}{r} \right)$$

TABLE 3.3.2 Constitutive relations for the components of the stress tensor for an incompressible Newtonian fluid in (a) cylindrical, (b) spherical, and (c) plane polar coordinates. Note that the stress components remain symmetric in orthogonal curvilinear coordinates.

Problems

3.3.1 *Molar mass of steam*

What is the molar mass of steam?

3.3.2 *Hydrodynamic pressure*

Demonstrate that $-p$ is the average value of the normal component of the traction exerted on the surface of a spherical fluid parcel with infinitesimal dimensions.

3.4   Force and energy dissipation in incompressible Newtonian fluids

In the remainder of this book, we concentrate on the motion of incompressible Newtonian fluids. In the present section, we use the constitutive equation (3.3.20) to derive specific expressions for the traction, force, and torque exerted on a fluid parcel and on the boundaries of the flow in terms of the velocity and the pressure, assess the rate of change of the internal energy due to viscous dissipation, and derive the specific form of the integral energy balance.

3.4.1   Traction and surface force

Substituting (3.3.20) into (3.1.11), we find that, the traction exerted on a fluid parcel is given by the following expressions in terms of the instantaneous pressure and velocity fields,

$$\mathbf{f} = -p\,\mathbf{n} + 2\mu\,\mathbf{E}\cdot\mathbf{n} = -p\,\mathbf{n} + \mu\left[\,(\nabla\mathbf{u})\cdot\mathbf{n} + \mathbf{n}\cdot\nabla\mathbf{u}\,\right], \tag{3.4.1}$$

where $\mathbf{E} = \frac{1}{2}(\mathbf{L} + \mathbf{L}^T)$ is the rate-of-deformation tensor for an incompressible fluid and $\mathbf{L} = \nabla\mathbf{u}$ is the velocity gradient tensor. In index notation,

$$f_i = -p\,n_i + \mu\left(\frac{\partial u_j}{\partial x_i}\,n_j + n_j\,\frac{\partial u_i}{\partial x_j}\right). \tag{3.4.2}$$

The third term on the right-hand side expresses the spatial derivative of the ith velocity component in the direction of the normal vector. When the fluid is inviscid, we obtain a simplified form involving the pressure alone, $\mathbf{f} = -p\,\mathbf{n}$.

In terms of the vorticity tensor, $\boldsymbol{\Xi} = \frac{1}{2}(\mathbf{L} - \mathbf{L}^T)$, we obtain

$$\mathbf{f} = -p\,\mathbf{n} + 2\mu\,(\mathbf{n}\cdot\nabla\mathbf{u} + \mathbf{n}\cdot\boldsymbol{\Xi}). \tag{3.4.3}$$

Expressing the vorticity tensor in terms of the vorticity vector, we find that

$$\mathbf{f} = -p\,\mathbf{n} + 2\mu\,\mathbf{n}\cdot\nabla\mathbf{u} + \mu\,\mathbf{n}\times\boldsymbol{\omega}. \tag{3.4.4}$$

The union of the second and third terms on the right-hand side expresses the traction due to the deviatoric component of the stress tensor. If the velocity field is irrotational, the last term on the right-hand side vanishes and the part of the traction corresponding to the deviatoric component of the stress tensor is proportional to the derivative of the velocity in the direction of the normal vector, $\mathbf{n}\cdot\nabla\mathbf{u}$.

Normal and tangential components

It is useful to decompose the traction into a normal component, $\mathbf{f}^N$, and a tangential or shear component, $\mathbf{f}^T$, so that $\mathbf{f} = \mathbf{f}^N + \mathbf{f}^T$. In general, both the normal and tangential components are nonzero. At a surface that cannot withstand a shear stress, defined as a free surface, the tangential component is zero. Projecting (3.4.4) onto the normal unit vector, we obtain the normal component of the traction,

$$\mathbf{f}^N = \left[-p + 2\mu\,\mathbf{n}\cdot(\nabla\mathbf{u})\cdot\mathbf{n}\right]\mathbf{n}. \tag{3.4.5}$$

Introducing a local Cartesian coordinate system with two axes perpendicular to the normal unit vector, $\mathbf{n}$, we find that the second term on the right-hand side of (3.4.5) expresses the viscous stress associated with the normal derivative of the normal velocity component. Multiplying (3.4.1) with the projection matrix $\mathbf{I} - \mathbf{n} \otimes \mathbf{n}$, we obtain the tangential component of the traction,

$$\mathbf{f}^T = 2\mu\,\mathbf{n} \cdot \mathbf{E} \cdot (\mathbf{I} - \mathbf{n} \otimes \mathbf{n}) = 2\mu\,\mathbf{n} \times (\mathbf{n} \cdot \mathbf{E}) \times \mathbf{n}. \tag{3.4.6}$$

The structure of the flow must be such that the right-hand side of (3.4.6) is identically zero at a free surface.

3.4.2 Force and torque exerted on a fluid parcel

Substituting (3.4.1) into (3.1.13) and adding the body force, we obtain the total force exerted on a parcel of an incompressible Newtonian fluid,

$$\mathbf{F}^{tot} = - \iint_{Parcel} p\,\mathbf{n}\,\mathrm{d}S + 2 \iint_{Parcel} \mu\,\mathbf{E} \cdot \mathbf{n}\,\mathrm{d}S + \iiint_{Parcel} \rho\,\mathbf{g}\,\mathrm{d}V, \tag{3.4.7}$$

where $\mathbf{n}$ is the normal unit vector pointing outward from the parcel. Substituting (3.4.1) into (3.1.20), we find that, in the absence of an external torque field, the torque with respect to a point, $\mathbf{x}_0$, exerted on the parcel is given by

$$\mathbf{T}^{tot} = - \iint_{Parcel} p\,(\mathbf{x} - \mathbf{x}_0) \times \mathbf{n}\,\mathrm{d}S + 2 \iint_{Parcel} \mu\,(\mathbf{x} - \mathbf{x}_0) \times (\mathbf{E} \cdot \mathbf{n})\,\mathrm{d}S + \iiint_{Parcel} \rho\,(\mathbf{x} - \mathbf{x}_0)\,\mathrm{d}V \times \mathbf{g}, \tag{3.4.8}$$

where $\mathbf{n}$ is the normal unit vector pointing outward from the parcel.

3.4.3 Hydrodynamic pressure and stress

When the density of the fluid and acceleration of gravity are uniform throughout the domain of flow, it is beneficial to eliminate the body force from expressions (3.4.7) and (3.4.8) by introducing the hydrodynamic pressure and corresponding hydrodynamic stress tensor, denoted by a tilde, defined as

$$\tilde{p} \equiv p - \rho\,\mathbf{g} \cdot \mathbf{x}, \qquad \tilde{\sigma} \equiv -\tilde{p}\mathbf{I} + 2\mu\mathbf{E} = \sigma + \rho\,(\mathbf{g} \cdot \mathbf{x})\,\mathbf{I}. \tag{3.4.9}$$

In hydrodynamic variables, the total force and torque with respect to a point $\mathbf{x}_0$ exerted on a fluid parcel are given by the surface integrals

$$\mathbf{F}^{tot} = - \iint_{Parcel} \tilde{p}\,\mathbf{n}\,\mathrm{d}S + 2 \iint_{Parcel} \mu\,\mathbf{E} \cdot \mathbf{n}\,\mathrm{d}S \tag{3.4.10}$$

and

$$\mathbf{T}^{tot} = - \iint_{Parcel} \tilde{p}\,(\mathbf{x} - \mathbf{x}_0) \times \mathbf{n}\,\mathrm{d}S + 2 \iint_{Parcel} \mu\,(\mathbf{x} - \mathbf{x}_0) \times (\mathbf{E} \cdot \mathbf{n})\,\mathrm{d}S. \tag{3.4.11}$$

3.4.4   Force and torque exerted on a boundary

To derive an expression for the hydrodynamic force exerted on a boundary, B, we substitute (3.4.1) into (3.1.16) and obtain

$$\mathbf{F} = - \iint_B p\, \mathbf{n}\, \mathrm{d}S + 2 \iint_B \mu\, \mathbf{E} \cdot \mathbf{n}\, \mathrm{d}S, \qquad (3.4.12)$$

where $\mathbf{n}$ is the normal unit vector pointing into the fluid. The first term on the right-hand side represents the form drag, and the second term represents the skin friction.

To obtain an expression for the torque with respect to a point, $\mathbf{x}_0$, we substitute (3.4.1) into (3.1.23) and obtain

$$\mathbf{T} = - \iint_B p\, (\mathbf{x} - \mathbf{x}_0) \times \mathbf{n}\, \mathrm{d}S + 2 \iint_B \mu\, (\mathbf{x} - \mathbf{x}_0) \times (\mathbf{E} \cdot \mathbf{n})\, \mathrm{d}S. \qquad (3.4.13)$$

In hydrostatics, or when viscous stresses are insignificant, the force and torque can be computed from knowledge of the pressure distribution over the boundary.

3.4.5   Energy dissipation inside a parcel

Substituting the Newtonian constitutive equation (3.3.20) into (3.2.20) and using the continuity equation, $\nabla \cdot \mathbf{u} = 0$, we find that the rate of viscous dissipation inside a fluid parcel is given by

$$\frac{\mathrm{d}\mathcal{I}_p}{\mathrm{d}t} = \iiint_{Parcel} \Phi\, \mathrm{d}V, \qquad (3.4.14)$$

where

$$\Phi \equiv 2\mu\, \mathbf{E} : \mathbf{E} \qquad (3.4.15)$$

expresses the rate of viscous dissipation per unit volume of fluid. Equation (3.4.14) is a special version of (3.3.19) applicable to compressible Newtonian fluids. Viscous forces dissipate energy, converting it into thermal energy and thereby raising the temperature of the fluid. When the rate of viscous dissipation is negligible, the sum of the kinetic and potential energies remains constant in time.

3.4.6   Rate of working of the traction

Integrating both sides of the identity

$$\mathbf{u} \cdot (\nabla \cdot \boldsymbol{\sigma}) = \nabla \cdot (\mathbf{u} \cdot \boldsymbol{\sigma}) - \boldsymbol{\sigma} : \nabla \mathbf{u} \qquad (3.4.16)$$

over the volume of a fluid parcel, and using the divergence theorem, we obtain

$$\iiint_{Parcel} \mathbf{u} \cdot (\nabla \cdot \boldsymbol{\sigma})\, \mathrm{d}V = - \iint_{Parcel} \mathbf{u} \cdot \mathbf{f}\, \mathrm{d}S - \iiint_{Parcel} \boldsymbol{\sigma} : \nabla \mathbf{u}\, \mathrm{d}V, \qquad (3.4.17)$$

where $\mathbf{f} = \mathbf{n} \cdot \boldsymbol{\sigma}$ is the boundary traction and $\mathbf{n}$ is the normal unit vector pointing into the parcel. Substituting the Newtonian constitutive equation (3.3.20), we obtain

$$\iiint_{Parcel} \mathbf{u} \cdot (\nabla \cdot \boldsymbol{\sigma}) \, \mathrm{d}V = - \iint_{Parcel} \mathbf{u} \cdot \mathbf{f} \, \mathrm{d}S - \iiint_{Parcel} \Phi \, \mathrm{d}V. \tag{3.4.18}$$

The two terms on the right-hand side of (3.4.18) represent, respectively, the rate of working of the surface traction and the rate of viscous dissipation.

Working in a similar fashion with the deviatoric part of the stress tensor $\check{\boldsymbol{\sigma}}$ defined in (3.3.4), we obtain

$$\iiint_{Parcel} \mathbf{u} \cdot (\nabla \cdot \check{\boldsymbol{\sigma}}) \, \mathrm{d}V = - \iint_{Parcel} \mathbf{u} \cdot \check{\mathbf{f}} \, \mathrm{d}S - \iiint_{Parcel} \Phi \, \mathrm{d}V, \tag{3.4.19}$$

where $\check{\mathbf{f}} = \mathbf{n} \cdot \check{\boldsymbol{\sigma}}$ is the deviatoric part of the traction.

In the case of a fluid with uniform viscosity, we use the Newtonian constitutive relation (3.3.20) and the continuity equation to obtain

$$\nabla \cdot \check{\boldsymbol{\sigma}} = \mu \nabla^2 \mathbf{u} = -\mu \nabla \times \boldsymbol{\omega}. \tag{3.4.20}$$

Using this relation, we find that, if the flow is irrotational or the vorticity is uniform throughout the domain of flow, the rate of working of the deviatoric viscous traction is balanced by the rate of viscous dissipation,

$$\iint_{Parcel} \mathbf{u} \cdot \check{\mathbf{f}} \, \mathrm{d}S = - \iiint_{Parcel} \Phi \, \mathrm{d}V. \tag{3.4.21}$$

3.4.7 Energy integral balance

Substituting the Newtonian constitutive equation (3.3.20) into (3.2.13), we derive the explicit form of the integral energy balance over a fixed control volume, V_c, bounded by a surface, D,

$$\iiint_{V_c} \frac{\partial}{\partial t} (\tfrac{1}{2} \rho |\mathbf{u}|^2) \, \mathrm{d}V = \iint_D (\tfrac{1}{2} \rho |\mathbf{u}|^2) \mathbf{u} \cdot \mathbf{n} \, \mathrm{d}S - \iint_D \mathbf{u} \cdot \mathbf{f} \, \mathrm{d}S - \iiint_{V_c} \Phi \, \mathrm{d}V + \iiint_{V_c} \rho \mathbf{g} \cdot \mathbf{u} \, \mathrm{d}V, \tag{3.4.22}$$

where the normal unit vector, $\mathbf{n}$, points into the control volume. The five integrals in (3.4.22) represent, respectively, the rate of accumulation of kinetic energy inside the control volume, the rate of convection of kinetic energy into the control volume, the rate of working of surface forces, the rate of viscous dissipation, change of potential energy associated with the body force.

When the flow is irrotational or the vorticity is uniform throughout the domain of flow, we use (3.4.21) and obtain the simplified form

$$\iiint_{V_c} \frac{\partial}{\partial t} \left(\tfrac{1}{2} \rho |\mathbf{u}|^2 \right) \, \mathrm{d}V = \iint_D (\tfrac{1}{2} \rho |\mathbf{u}|^2) \mathbf{u} \cdot \mathbf{n} \, \mathrm{d}S + \iint_D p \mathbf{u} \cdot \mathbf{n} \, \mathrm{d}S + \iiint_{V_c} \rho \mathbf{g} \cdot \mathbf{u} \, \mathrm{d}V. \tag{3.4.23}$$

This energy balance is also valid when the flow is not irrotational or has uniform vorticity, but the fluid can be considered to be inviscid.

Problem

3.4.1 *Energy dissipation*

Compute the energy dissipation inside a parcel of an Newtonian fluid in simple shear flow with velocity $u_x = \xi y$, $u_y = 0$, $u_z = 0$, where ξ is a constant shear rate.

3.5   The Navier–Stokes equation

Substituting the constitutive equation for an incompressible Newtonian fluid (3.3.20) into Cauchy's equation of motion (3.2.1), we obtain the Navier–Stokes equation

$$\rho \frac{Du}{Dt} = -\nabla p + 2\nabla \cdot (\mu E) + \rho g. \tag{3.5.1}$$

The density, ρ, and viscosity, μ, are allowed to vary in time and space in the domain of flow. Substituting into (3.5.1) the definition of the rate-of-deformation tensor for an incompressible fluid, $E = \frac{1}{2}[\nabla u + (\nabla u)^T]$, expanding the derivatives, using the continuity equation for an incompressible fluid, $\nabla \cdot u = 0$, and expressing the material derivative of the velocity in terms of Eulerian derivatives, we recast (3.5.1) into the explicit form

$$\rho\left(\frac{\partial u}{\partial t} + u \cdot \nabla u\right) = -\nabla p + \mu \nabla^2 u + 2\nabla \mu \cdot E + \rho g. \tag{3.5.2}$$

The four terms on the right-hand side express, respectively, the pressure force, the viscous force, a force due to viscosity variations, and the body force. The union of the first three terms comprises the hydrodynamic volume force,

$$\Sigma = -\nabla p + 2\nabla \cdot (\mu E) = -\nabla p + \mu \nabla^2 u + 2\nabla \mu \cdot E. \tag{3.5.3}$$

The three Cartesian components of the Navier–Stokes equation for a fluid with uniform viscosity are displayed in Table 3.5.1.

   When the fluid density and acceleration of gravity are uniform throughout the domain of flow, it is convenient to work with the hydrodynamic pressure, $\tilde{p} \equiv p - \rho g \cdot x$, and associated hydrodynamic stress tensor indicated by a tilde introduced in (3.4.9), obtaining the equation of motion

$$\rho \frac{Du}{Dt} = \nabla \cdot \tilde{\sigma} = -\nabla \tilde{p} + 2\nabla \cdot (\mu E), \tag{3.5.4}$$

which is distinguished by the absence of the body force. In solving the Navier–Stokes equation, the distinction between the regular and modified pressure becomes relevant only when boundary conditions for the pressure or traction are imposed.

Kinematic viscosity

Dividing both sides of (3.5.2) by the density, we obtain the new form

$$\frac{\partial u}{\partial t} + u \cdot \nabla u = -\frac{1}{\rho}\nabla p + \nu \nabla^2 u + \frac{2}{\rho}\nabla \mu \cdot E + g, \tag{3.5.5}$$

$$\rho\left(\frac{\partial u_x}{\partial t} + u_x\frac{\partial u_x}{\partial x} + u_y\frac{\partial u_x}{\partial y} + u_z\frac{\partial u_x}{\partial z}\right) = -\frac{\partial p}{\partial x} + \mu\left(\frac{\partial^2 u_x}{\partial x^2} + \frac{\partial^2 u_x}{\partial y^2} + \frac{\partial^2 u_x}{\partial z^2}\right) + \rho g_x$$

$$\rho\left(\frac{\partial u_y}{\partial t} + u_x\frac{\partial u_y}{\partial x} + u_y\frac{\partial u_y}{\partial y} + u_z\frac{\partial u_y}{\partial z}\right) = -\frac{\partial p}{\partial y} + \mu\left(\frac{\partial^2 u_y}{\partial x^2} + \frac{\partial^2 u_y}{\partial y^2} + \frac{\partial^2 u_y}{\partial z^2}\right) + \rho g_y$$

$$\rho\left(\frac{\partial u_z}{\partial t} + u_x\frac{\partial u_z}{\partial x} + u_y\frac{\partial u_z}{\partial y} + u_z\frac{\partial u_z}{\partial z}\right) = -\frac{\partial p}{\partial z} + \mu\left(\frac{\partial^2 u_z}{\partial x^2} + \frac{\partial^2 u_z}{\partial y^2} + \frac{\partial^2 u_z}{\partial z^2}\right) + \rho g_z$$

TABLE 3.5.1 Eulerian form of the three Cartesian components of the Navier–Stokes equation for an incompressible fluid with uniform viscosity.

where $\nu = \mu/\rho$ is a new physical constant with dimensions of length squared over time analogous to the molecular or thermal diffusivity, called the kinematic viscosity of the fluid. Sample values of the kinematic viscosity for water and air are given in Table 3.3.1.

Cylindrical, spherical, and plane polar coordinates

The polar components of the hydrodynamic volume force arise by substituting the constitutive equation for an incompressible Newtonian fluid in the general expressions for the hydrodynamic volume force in terms of the stresses shown in Table 3.2.2. After a fair amount of algebra, we derive the expressions shown in Table 3.5.2. The corresponding polar components of the Navier–Stokes equation arise by substituting these expressions in the right-hand side of (3.2.3).

3.5.1 Vorticity and viscous force

The identity $\nabla^2\mathbf{u} = -\nabla \times \boldsymbol{\omega}$, applicable for any solenoidal velocity field, $\nabla \cdot \mathbf{u} = 0$, allows us to express the Laplacian of the velocity on the right-hand side of (3.5.2) in terms of the vorticity, and thereby establish a relationship between the structure of the vorticity field and the intensity of the viscous force. Assuming, for simplicity, that the fluid viscosity is uniform throughout the domain of flow, we recast (3.5.2) into the form

$$\rho\frac{D\mathbf{u}}{Dt} = -\nabla p - \mu\nabla \times \boldsymbol{\omega} + \rho\mathbf{g}, \qquad (3.5.6)$$

which shows that viscous forces are important only in regions where the curl of the vorticity is significant.

The equation of motion (3.5.6) reveals that irrotational flows and flows whose vorticity field is irrotational behave like inviscid flows. The dynamics of these flows is determined by a balance between the inertial force due to the point-particle acceleration, the pressure force, and the body force. Viscosity is important only insofar as to establish the vorticity distribution. Once this has been achieved, viscosity plays no further role in the force balance. Flows with irrotational vorticity fields include two-dimensional flows with constant vorticity and axisymmetric flows without swirling

(a)

$$\Sigma_x = -\frac{\partial p}{\partial x} + \mu \nabla^2 u_x = -\frac{\partial p}{\partial x} + \mu \left(\frac{\partial^2 u_x}{\partial x^2} + \frac{1}{\sigma} \frac{\partial}{\partial \sigma} \left(\sigma \frac{\partial u_x}{\partial \sigma} \right) + \frac{1}{\sigma^2} \frac{\partial^2 u_x}{\partial \varphi^2} \right)$$

$$\Sigma_\sigma = -\frac{\partial p}{\partial \sigma} + \mu \left(\nabla^2 u_\sigma - \frac{u_\sigma}{\sigma^2} - \frac{2}{\sigma^2} \frac{\partial u_\varphi}{\partial \varphi} \right)$$

$$= -\frac{\partial p}{\partial \sigma} + \mu \left(\frac{\partial^2 u_\sigma}{\partial x^2} + \frac{\partial}{\partial \sigma} \left(\frac{1}{\sigma} \frac{\partial(\sigma u_\sigma)}{\partial \sigma} \right) + \frac{1}{\sigma^2} \frac{\partial^2 u_\sigma}{\partial \varphi^2} - \frac{2}{\sigma^2} \frac{\partial u_\varphi}{\partial \varphi} \right)$$

$$\Sigma_\varphi = -\frac{1}{\sigma} \frac{\partial p}{\partial \varphi} + \mu \left(\nabla^2 u_\varphi - \frac{u_\varphi}{\sigma^2} + \frac{2}{\sigma^2} \frac{\partial u_\sigma}{\partial \varphi} \right)$$

$$= -\frac{1}{\sigma} \frac{\partial p}{\partial \varphi} + \mu \left(\frac{\partial^2 u_\varphi}{\partial x^2} + \frac{\partial}{\partial \sigma} \left(\frac{1}{\sigma} \frac{\partial(\sigma u_\varphi)}{\partial \sigma} \right) + \frac{1}{\sigma^2} \frac{\partial^2 u_\varphi}{\partial \varphi^2} + \frac{2}{\sigma^2} \frac{\partial u_\sigma}{\partial \varphi} \right)$$

(b)

$$\Sigma_r = -\frac{\partial p}{\partial r} + \mu \left(\nabla^2 u_r - 2\frac{u_r}{r^2} - \frac{2}{r^2} \frac{\partial u_\theta}{\partial \theta} - \frac{2}{r^2} u_\theta \cot\theta - \frac{2}{r^2 \sin\theta} \frac{\partial u_\varphi}{\partial \varphi} \right)$$

$$\Sigma_\theta = -\frac{1}{r} \frac{\partial p}{\partial \theta} + \mu \left(\nabla^2 u_\theta + \frac{2}{r^2} \frac{\partial u_r}{\partial \theta} - \frac{u_\theta}{r^2 \sin^2\theta} - \frac{2}{r^2} \frac{\cot\theta}{\sin\theta} \frac{\partial u_\varphi}{\partial \varphi} \right)$$

$$\Sigma_\varphi = -\frac{1}{r \sin\theta} \frac{\partial p}{\partial \varphi} + \mu \left[\nabla^2 u_\varphi + \frac{1}{r^2 \sin^2\theta} \left(-u_\varphi + 2\frac{\partial u_r}{\partial \varphi} + 2\cos\theta \frac{\partial u_\theta}{\partial \varphi} \right) \right]$$

where: $$\nabla^2 f = \frac{1}{r^2} \frac{\partial}{\partial r} \left(r^2 \frac{\partial f}{\partial r} \right) + \frac{1}{r^2 \sin\theta} \frac{\partial}{\partial \theta} \left(\sin\theta \frac{\partial f}{\partial \theta} \right) + \frac{1}{r^2 \sin^2\theta} \frac{\partial^2 f}{\partial \varphi^2}$$

(c)

$$\Sigma_r = -\frac{\partial p}{\partial r} + \mu \left(\nabla^2 u_r - \frac{u_r}{r^2} - \frac{2}{r^2} \frac{\partial u_\theta}{\partial \theta} \right) = -\frac{\partial p}{\partial r} + \mu \left[\frac{\partial}{\partial r} \left(\frac{1}{r} \frac{\partial(r u_r)}{\partial r} \right) + \frac{1}{r^2} \frac{\partial^2 u_r}{\partial \theta^2} - \frac{2}{r^2} \frac{\partial u_\theta}{\partial \theta} \right]$$

$$\Sigma_\theta = -\frac{1}{r} \frac{\partial p}{\partial \theta} + \mu \left(\nabla^2 u_\theta - \frac{u_\theta}{r^2} + \frac{2}{r^2} \frac{\partial u_r}{\partial \theta} \right) = -\frac{1}{r} \frac{\partial p}{\partial \theta} + \mu \left[\frac{\partial}{\partial r} \left(\frac{1}{r} \frac{\partial(r u_\theta)}{\partial r} \right) + \frac{1}{r^2} \frac{\partial^2 u_\theta}{\partial \theta^2} + \frac{2}{r^2} \frac{\partial u_r}{\partial \theta} \right]$$

TABLE 3.5.2   Components of the hydrodynamic volume force for a Newtonian fluid in (a) cylindrical, (b) spherical, and (c) plane polar coordinates. The Laplacian operator ∇^2 in spherical polar coordinates is given in the fourth entry of (b).

motion where the azimuthal vorticity component increases linearly with distance from the axis of revolution, $\omega_\varphi = \Omega \sigma$, where Ω is a constant.

3.5.2   Uniform-density and barotropic fluids

We now focus our attention on fluids with uniform density and on barotropic fluids whose pressure is a function of the density alone, $p(\rho)$. Although the second stipulation is not entirely consistent with the assumptions underlying the derivation of the Navier–Stokes equation for an incompressible fluid, it is sometimes a reasonable approximation. We may then write

$$\frac{1}{\rho}\nabla p = \nabla \mathcal{F}, \tag{3.5.7}$$

where $\mathcal{F} = p/\rho$ in the case of uniform-density fluids. In the case of barotropic fluids, the function $\mathcal{F}$ is found by integrating the ordinary differential equation

$$\frac{d\mathcal{F}}{d\rho} = \frac{1}{\rho}\frac{dp}{d\rho}. \tag{3.5.8}$$

For simplicity, in our discussion we consider uniform-density fluids. Adaptations for barotropic fluids can be made by straightforward modifications.

3.5.3   Bernoulli function

Expressing the material derivative in (3.5.6) in terms of Eulerian derivatives and using the identity

$$\mathbf{u}\cdot\nabla\mathbf{u} = \frac{1}{2}\nabla(\mathbf{u}\cdot\mathbf{u}) - \mathbf{u}\times\boldsymbol{\omega}, \tag{3.5.9}$$

we derive a new form of the Navier–Stokes equation,

$$\frac{\partial\mathbf{u}}{\partial t} + \nabla\left(\frac{1}{2}\mathbf{u}\cdot\mathbf{u} + \frac{p}{\rho} - \mathbf{g}\cdot\mathbf{x}\right) = \mathbf{u}\times\boldsymbol{\omega} - \nu\nabla\times\boldsymbol{\omega}. \tag{3.5.10}$$

Nonlinear terms appear in the second term on the left-hand side involving the square of the velocity and in the first term on the right-hand side involving the velocity and the vorticity. The term inside the parentheses on the left-hand side of (3.5.10) is the Bernoulli function,

$$\mathcal{B}(\mathbf{x},t) \equiv \frac{1}{2}\mathbf{u}\cdot\mathbf{u} + \frac{p}{\rho} - \mathbf{g}\cdot\mathbf{x}. \tag{3.5.11}$$

Physically, the Bernoulli function expresses the mass distribution density of the total energy consisting of the kinetic energy, the internal energy due to the pressure, and the potential energy due to the body force. In terms of the Bernoulli function, the Navier–Stokes equation takes the compact form

$$\rho\left(\frac{\partial\mathbf{u}}{\partial t} + \nabla\mathcal{B}\right) = \rho\,\mathbf{u}\times\boldsymbol{\omega} - \mu\nabla\times\boldsymbol{\omega}. \tag{3.5.12}$$

Later in this section, we will see that, under certain conditions, the Bernoulli function is constant along streamlines or even throughout the domain of flow. When this occurs, $\mathcal{B}$ is called the Bernoulli constant.

Change of the Bernoulli function along a streamline in a steady flow

Now we project (3.5.12) at a point in a steady flow onto the unit vector that is tangent to the streamline passing through that point, $\mathbf{t} = \mathbf{u}/|\mathbf{u}|$. The projection of the first term on the right-hand side vanishes identically, yielding

$$\mathbf{t} \cdot \nabla \mathcal{B} = \frac{\partial \mathcal{B}}{\partial l} = -\nu\, \mathbf{t} \cdot (\nabla \times \boldsymbol{\omega}),$$

(3.5.13)

where $\nu = \mu/\rho$ is the kinematic viscosity and l is the arc length along the streamline measured in the direction of $\mathbf{t}$. Equation (3.5.13) states that the rate of change of the Bernoulli function with respect to arc length along a streamline is proportional to the component of the viscous force tangential to the streamline. When the viscous force opposes the motion of the fluid, $\mathcal{B}$ decreases along the streamline, $\partial \mathcal{B}/\partial l < 0$.

Integrating (3.5.13) along a closed streamline and using the fundamental theorem of calculus to set the integral of the left-hand side to zero, we obtain

$$\oint \mathbf{t} \cdot (\nabla \times \boldsymbol{\omega})\, \mathrm{d}l = 0,$$

(3.5.14)

which shows that the circulation of the curl of the vorticity along a closed streamline is zero in a steady flow.

3.5.4   Vortex force

The first term on the right-hand side of (3.5.12), $\rho\, \mathbf{u} \times \boldsymbol{\omega}$, is called the vortex force. In a Beltrami flow where, by definition, the velocity is parallel to the vorticity at every point, the vortex force is identically zero and the nonlinear term $\mathbf{u} \cdot \nabla \mathbf{u}$ is equal to the gradient of the kinetic energy per unit mass of the fluid.

Two-dimensional and axisymmetric flow

In the case of a two-dimensional or axisymmetric flow, we introduce the two-dimensional stream function or the axisymmetric Stokes stream function, ψ, and express the outer product of the velocity and vorticity, respectively, as

$$\mathbf{u} \times \boldsymbol{\omega} = -\omega_z \nabla \psi, \qquad \mathbf{u} \times \boldsymbol{\omega} = -\frac{\omega_\varphi}{\sigma} \nabla \psi.$$

(3.5.15)

In the second equation for axisymmetric flow, σ is the distance from the axis of revolution, ω_φ is the azimuthal component of the vorticity, and ∇ is the gradient in the $x\sigma$ azimuthal plane.

Later in this chapter, we will discuss a class of steady two-dimensional flows where the vorticity is constant along the streamlines and may thus be regarded as a function of the stream function, $\omega_z = f(\psi)$, and a class of axisymmetric flows where the ratio ω_φ/σ is constant along the streamlines and may thus be regarded as a function of the Stokes stream function, $\omega_\varphi/\sigma = f(\psi)$. Under these circumstances, the right-hand sides of equations (3.5.15) can be written as a gradient,

$$\mathbf{u} \times \boldsymbol{\omega} = -\nabla F(\psi),$$

(3.5.16)

where the function F is the indefinite integral of the function f,

$$\frac{dF}{d\psi} = f(\psi). \tag{3.5.17}$$

Equation (3.5.16) shows that, because the cross product $\mathbf{u} \times \boldsymbol{\omega}$ constitutes an irrotational vector field, the vortex force can be grouped with the gradient on the left-hand side of (3.5.13). In the case of two-dimensional flow with constant vorticity, or axisymmetric flow where $f(\psi) = \Omega$, with Ω being constant, we find that $F = \Omega\psi + c$, where c is constant throughout the domain of flow.

3.5.5 Reciprocity of Newtonian flows

Consider the flow an incompressible Newtonian fluid with density ρ, viscosity μ, velocity $\mathbf{u}$, and corresponding stress field $\boldsymbol{\sigma}$, and another unrelated flow of an incompressible Newtonian fluid with density ρ', viscosity μ', velocity $\mathbf{u}'$, and corresponding stress field $\boldsymbol{\sigma}'$. Projecting the velocity of the second flow onto the divergence of the stress tensor of the first flow at some point, we obtain

$$\mathbf{u}' \cdot (\nabla \cdot \boldsymbol{\sigma}) = u_i' \frac{\partial \sigma_{ij}}{\partial x_j} = \frac{\partial(u_i'\sigma_{ij})}{\partial x_j} - \sigma_{ij}\frac{\partial u_i'}{\partial x_j}. \tag{3.5.18}$$

Substituting the definition of the stress tensor, we obtain

$$\mathbf{u}' \cdot (\nabla \cdot \boldsymbol{\sigma}) = \frac{\partial(u_i'\sigma_{ij})}{\partial x_j} + \left(p\,\delta_{ij} - \mu\frac{\partial u_i}{\partial x_j} - \mu\frac{\partial u_j}{\partial x_i} \right)\frac{\partial u_i'}{\partial x_j}. \tag{3.5.19}$$

Using the continuity equation to eliminate the term involving the pressure, we find that

$$\mathbf{u}' \cdot (\nabla \cdot \boldsymbol{\sigma}) = \frac{\partial(u_i'\sigma_{ij})}{\partial x_j} - \mu\left(\frac{\partial u_i}{\partial x_j} + \frac{\partial u_j}{\partial x_i}\right)\frac{\partial u_i'}{\partial x_j}. \tag{3.5.20}$$

Setting $\mu' = \mu$ and $\mathbf{u} = \mathbf{u}'$, integrating over the volume of a fluid parcel, and using the divergence theorem, we recover (3.4.18).

Now interchanging the roles of the two flows, we obtain

$$\mathbf{u} \cdot \nabla \cdot \boldsymbol{\sigma}' = \frac{\partial(u_i\sigma_{ij}')}{\partial x_j} - \mu'\left(\frac{\partial u_i'}{\partial x_j} + \frac{\partial u_j'}{\partial x_i}\right)\frac{\partial u_i}{\partial x_j}. \tag{3.5.21}$$

Multiplying (3.5.20) by μ' and (3.5.21) by μ, and subtracting corresponding sides of the resulting equations, we obtain the differential statement of the generalized Lorentz reciprocal identity

$$\frac{\partial}{\partial x_j}\left(\mu'u_i'\sigma_{ij} - \mu u_i\sigma_{ij}'\right) = \mu'\mathbf{u}' \cdot \nabla \cdot \boldsymbol{\sigma} - \mu\,\mathbf{u} \cdot \nabla \cdot \boldsymbol{\sigma}', \tag{3.5.22}$$

which imposes a constraint on the mutual structure of the velocity and stress fields of two unrelated incompressible Newtonian flows. Equations (3.5.20), (3.5.21), and (3.5.22) also apply for the hydrodynamic stress tensor defined in (3.4.9) (Problem 3.5.1).

Expressing the divergence of the stress tensors on the right-hand side of (3.5.22) in terms of the point-particle acceleration and the body force using the equation of motion, we obtain

$$\frac{\partial}{\partial x_j}\left(\mu' u_i' \sigma_{ij} - \mu u_i \sigma_{ij}'\right) = \rho \mu' \mathbf{u}' \cdot \frac{D\mathbf{u}}{Dt} - \rho' \mu \mathbf{u} \cdot \frac{D\mathbf{u}'}{Dt} - (\rho \mu' \mathbf{u}' - \rho' \mu \mathbf{u}) \cdot \mathbf{g}. \qquad (3.5.23)$$

In terms of the hydrodynamic stress tensor indicated by a tilde, we obtain the simpler form

$$\frac{\partial}{\partial x_j}\left(\mu' u_i' \tilde{\sigma}_{ij} - \mu u_i \tilde{\sigma}_{ij}\right) = \rho \mu' \mathbf{u}' \cdot \frac{D\mathbf{u}}{Dt} - \rho' \mu \mathbf{u} \cdot \frac{D\mathbf{u}'}{Dt}. \qquad (3.5.24)$$

Integrating (3.5.24) over a chosen control volume, V_c, that is bounded by a surface, D, and using the divergence theorem to convert the volume integral of the left-hand side into a surface integral over D, we derive the corresponding integral form

$$\iint_D \left(\mu' \mathbf{u}' \cdot \tilde{\mathbf{f}} - \mu \mathbf{u} \cdot \tilde{\mathbf{f}}'\right) dV = -\iiint_{V_c} \left(\rho \mu' \mathbf{u}' \cdot \frac{D\mathbf{u}}{Dt} - \rho' \mu \mathbf{u} \cdot \frac{D\mathbf{u}'}{Dt}\right) dV, \qquad (3.5.25)$$

where the normal unit vector, $\mathbf{n}$, points into the control volume. In Section 6.8, we will see that the reciprocal identities (3.5.24) and (3.5.25) find extensive applications in the study of Stokes flow where the effect of fluid inertia is negligibly small.

Problems

3.5.1 *Flow past a body*

Consider a steady streaming flow of a viscous fluid in the horizontal direction past a stationary body. Discuss whether the Bernoulli function $\mathcal{B}$ increases or decreases in the streamwise direction.

3.5.2 *Hydrostatics*

Consider a body of fluid with uniform density that is either stationary or translates as a rigid body. Show that the general solution of the equation of motion is $p = \rho \mathbf{g} \cdot \mathbf{x} + c(t)$ or $\tilde{p} = c(t)$, where $c(t)$ is an arbitrary function of time, and $\tilde{p}$ is the hydrodynamic pressure. Discuss the computation of $c(t)$ for a flow of your choice.

3.5.3 *Oseen flow*

(*a*) Consider a steady streaming (uniform) flow with velocity $\mathbf{V}$ past a stationary body. Far from the body, the flow can be decomposed into the incident flow and a disturbance flow $\mathbf{v}$ due to the body, $\mathbf{u} = \mathbf{V} + \mathbf{v}$. Introduce a similar decomposition for the pressure, substitute the decompositions into the Navier–Stokes equation, and neglect quadratic terms in $\mathbf{v}$ to derive a linear equation.

(*b*) Repeat (*a*) for uniform flow past a semi-infinite plate aligned with the flow. Specifically, derive a linear equation for the disturbance stream function applicable far from the plate.

3.5.4 *Reciprocal identity*

(*a*) Show that equations (3.5.20), (3.5.21), and (3.5.22) are also applicable for the hydrodynamic stress tensor defined in (3.4.9).

(*b*) Derive (3.5.25) from (3.5.23).

3.6 Euler and Bernoulli equations

When the viscous force on the right-hand side of (3.5.6) is negligible, the Navier–Stokes equation reduces to Euler's equation,

$$\rho \frac{D\mathbf{u}}{Dt} = -\nabla p + \rho \mathbf{g}, \tag{3.6.1}$$

which is strictly applicable for inviscid fluids. One important difference between the Euler equation and the Navier–Stokes equation is that the former is a first-order partial differential equation whereas the latter is a second-order partial differential equation in space with respect to the velocity. In Section 3.7, we will see that this difference has important implications on the number of boundary conditions required to compute a solution.

When the density of the fluid and acceleration of gravity are uniform throughout the domain of a flow, it is convenient to introduce the hydrodynamic pressure incorporating the effect of the body force, $\tilde{p} \equiv p - \rho \mathbf{g} \cdot \mathbf{x}$, and recast Euler's equation (3.6.1) into the form

$$\rho \frac{D\mathbf{u}}{Dt} = -\nabla \tilde{p}. \tag{3.6.2}$$

This expression reveals that the point-particle acceleration field, $D\mathbf{u}/Dt$, is irrotational. As a consequence, if a flow governed by the Euler equation is irrotational at the initial instant, it will remain irrotational at all times. The permanence of irrotational flow for a fluid with uniform density and negligible viscous forces will be discussed in the context of vorticity dynamics in Section 3.11.

In terms of the Bernoulli function,

$$\mathcal{B}(\mathbf{x}, t) \equiv \frac{1}{2} \mathbf{u} \cdot \mathbf{u} + \frac{p}{\rho} - \mathbf{g} \cdot \mathbf{x}, \tag{3.6.3}$$

Euler's equation reads

$$\rho \left(\frac{\partial \mathbf{u}}{\partial t} + \nabla \mathcal{B} \right) = \rho \, \mathbf{u} \times \boldsymbol{\omega}. \tag{3.6.4}$$

Nonlinear terms appear in the definition of the Bernoulli function as well as in the cross product of the velocity and the vorticity on the right-hand side.

3.6.1 Bernoulli function in steady rotational flow

In the case of steady rotational flow, Euler's equation (3.6.4) takes the simple form

$$\nabla \mathcal{B} = \mathbf{u} \times \boldsymbol{\omega}. \tag{3.6.5}$$

In a Beltrami flow where the velocity is parallel to the vorticity at every point, the right-hand side of (3.6.5) is zero and the Bernoulli function is constant throughout the domain of flow.

Considering the more general case of a flow where velocity and vorticity are not necessarily aligned, we apply (3.6.5) at a certain point on a chosen streamline and project both sides onto the tangent unit vector. Note that the right-hand side of (3.6.5) is normal to the velocity and therefore also to the streamline, and integrating the resulting expression with respect to arc length along the streamline, we find that $\mathcal{B}$ is constant along the streamline. Symbolically, we write

$$\mathcal{B} = \mathcal{F}(\text{streamline}). \tag{3.6.6}$$

The value of $\mathcal{B}$ at a particular streamline is usually computed by applying (3.6.5) at an entrance or exit point and then requiring appropriate boundary conditions for the boundary velocity or traction.

Spatial distribution of the Bernoulli function

To investigate the variation of the Bernoulli function across streamlines, we introduce the unit vector tangent to a streamline, $\mathbf{t} = \mathbf{u}/|\mathbf{u}|$, the unit vector normal to the streamline, $\mathbf{n}$, and the associated binormal vector, $\mathbf{b} = \mathbf{t} \times \mathbf{n}$. The triplet $(\mathbf{t}, \mathbf{n}, \mathbf{b})$ defines a system of orthogonal right-handed coordinates. Projecting (3.6.5) onto the normal unit vector, $\mathbf{n}$, and rearranging the triple scalar product on the right-hand side, we find that

$$\mathbf{n} \cdot \nabla \mathcal{B} = (\mathbf{u} \times \boldsymbol{\omega}) \cdot \mathbf{n} = -(\mathbf{u} \times \mathbf{n}) \cdot \boldsymbol{\omega} = -u\,(\mathbf{t} \times \mathbf{n}) \cdot \boldsymbol{\omega}, \tag{3.6.7}$$

which can be rearranging into

A local coordinate system attached to a point on a chosen streamline.

$$\frac{1}{u}\frac{\partial \mathcal{B}}{\partial l_n} = -\boldsymbol{\omega} \cdot \mathbf{b}, \tag{3.6.8}$$

where $u = |\mathbf{u}|$ and l_n is the arc length measured in the direction of $\mathbf{n}$. Projecting (3.6.5) onto the binormal unit vector, $\mathbf{b}$, and working in a similar fashion, we find that

$$\frac{1}{u}\frac{\partial \mathcal{B}}{\partial l_b} = \boldsymbol{\omega} \cdot \mathbf{n}, \tag{3.6.9}$$

where l_b is the arc length measured in the direction of $\mathbf{b}$. If the vorticity is locally parallel to the velocity at the chosen point, the right-hand sides of (3.6.8) and (3.6.9) are zero and $\mathcal{B}$ reaches a local extremum at that point.

Two-dimensional and axisymmetric flow

In the case of two-dimensional or axisymmetric flow that fulfills the prerequisites for (3.5.16), we obtain the explicit expressions

$$\mathcal{B} = -F(\psi) + c, \tag{3.6.10}$$

where c is constant throughout the domain of flow. For example, in the case of two-dimensional flow with uniform vorticity, $\omega_z = \Omega$, or axisymmetric flow with azimuthal vorticity $\omega_\varphi = \Omega\sigma$, where Ω is a constant, we find that $\mathcal{B} = -\Omega\psi + c$. Since the curl of the vorticity and thus the magnitude of viscous forces are identically zero, these flows represent exact solutions of the Navier–Stokes equation. The velocity field can be computed working in the context of kinematics, and the pressure follows from knowledge of the stream function using the derived expression for $\mathcal{B}$. One example is the flow inside Hill's spherical vortex discussed in Section 2.12.2.

3.6.2   Bernoulli equations

Bernoulli's equations are integrated forms of simplified versions of the Navier–Stokes equation for certain special classes of flows.

Irrotational flow

The right-hand side of (3.6.4) is zero in a steady or unsteady irrotational flow. Introducing the velocity potential, ϕ, substituting $\mathbf{u} = \nabla\phi$ in the temporal derivative on the left-hand side of (3.6.4), and integrating the resulting equation in space, we derive Bernoulli's equation

$$\frac{\partial\phi}{\partial t} + \mathcal{B} = c(t),$$ (3.6.11)

where $c(t)$ is an unspecified time-dependent function. In practice, the value of $c(t)$ is determined by applying (3.6.11) at a point on a selected boundary of the flow and then introducing an appropriate boundary condition for the velocity or pressure. Equation (3.6.11) can be interpreted as an evolution equation for ϕ.

In applications involving unsteady flow with free surfaces, it is convenient to work with an alternative form of (3.6.11) involving the material derivative of the velocity potential, $\mathrm{D}\phi/\mathrm{D}t = \partial\phi/\partial t + \mathbf{u}\cdot\nabla\phi$, which is equal to the rate of change of the potential following a point particle. Adding $\mathbf{u}\cdot\nabla\phi$ and subtracting $\mathbf{u}\cdot\mathbf{u}$ from the left-hand side of (3.6.11), we obtain

$$\frac{\mathrm{D}\phi}{\mathrm{D}t} - \frac{1}{2}\mathbf{u}\cdot\mathbf{u} + \frac{p}{\rho} - \mathbf{g}\cdot\mathbf{x} = c(t).$$ (3.6.12)

Working in a similar manner, we find that the rate of change of the potential as seen by an observer who travels with arbitrary velocity $\mathbf{v}$ is given by

$$\frac{\mathrm{d}\phi}{\mathrm{d}t} + (\frac{1}{2}\mathbf{u} - \mathbf{v})\cdot\mathbf{u} + \frac{p}{\rho} - \mathbf{g}\cdot\mathbf{x} = c(t).$$ (3.6.13)

When $\mathbf{V} = \mathbf{u}$, we recover (3.6.12). Equations (3.6.12) and (3.6.13) find applications in the numerical computation of free-surface flow using boundary-integral methods discussed in Chapter 10.

Steady irrotational flow

The Bernoulli equation for steady irrotational flow takes the simple form

$$\mathcal{B} = c,$$ (3.6.14)

where c is a constant. Since the velocity attains its maximum at the boundaries, the dynamic pressure, $\tilde{p} \equiv p - \rho\mathbf{g}\cdot\mathbf{x}$ attains a minimum at the boundaries.

Two-dimensional flow with constant vorticity

A judicious decomposition of the velocity field of a generally unsteady two-dimensional flow whose vorticity is and remains uniform and constant in time, equal to Ω, allows us to describe the flow in terms of a velocity potential and also derive a Bernoulli's equation for the pressure. In Section 3.13, we will see that the physical conditions for the vorticity to be uniform are that it is uniform at the initial instant and the effect of viscosity is negligibly small. The key idea is to introduce the stream function, ψ, and write

$$\mathbf{u}(\mathbf{x}, t) = \nabla \times (\psi \, \mathbf{e}_z) = \mathbf{v}(\mathbf{x}) + \nabla \phi(\mathbf{x}, t), \tag{3.6.15}$$

where $\mathbf{v}$ is a steady two-dimensional flow with uniform vorticity, $\omega_z = \Omega$. The term $\nabla \phi$ represents a complementary, generally unsteady irrotational flow, where ϕ is a suitable harmonic potential. For example, $\mathbf{v}$ can be the velocity of simple shear flow along the x axis varying along the y axis, $\mathbf{v} = (-\Omega y, 0)$. Substituting (3.6.15) into (3.5.10), and noting that the viscous force is identically zero, we obtain

$$\nabla \left(\frac{\partial \phi}{\partial t} \right) + \nabla \mathcal{B} = [\nabla \times (\psi \, \mathbf{e}_z)] \times (\Omega \, \mathbf{e}_z) = -\Omega \, \nabla \psi. \tag{3.6.16}$$

Integrating, we derive Bernoulli's equation involving the stream function,

$$\frac{\partial \phi}{\partial t} + \mathcal{B} + \Omega \, \psi = c(t), \tag{3.6.17}$$

where $c(t)$ is a time-dependent function.

3.6.3 Bernoulli's equation in a noninertial frame

The equation of motion in an accelerating and rotating frame of reference contains three types of inertial forces that must be taken into account when integrating the Navier–Stokes equation to derive Bernoulli's equations, as discussed in Section 3.2.6. Euler's equation in a noninertial frame reads

$$\frac{\partial \mathbf{v}}{\partial t} + \nabla \mathcal{B} = \mathbf{v} \times \boldsymbol{\varrho} - \frac{d\mathbf{U}}{dt} - 2\,\boldsymbol{\Omega} \times \mathbf{v} - \boldsymbol{\Omega} \times (\boldsymbol{\Omega} \times \mathbf{y}) - \frac{d\boldsymbol{\Omega}}{dt} \times \mathbf{y}, \tag{3.6.18}$$

where $\boldsymbol{\varrho}$ is the vorticity in the noninertial frame,

$$\mathcal{B}(\mathbf{x}, t) \equiv \frac{1}{2} \mathbf{v} \cdot \mathbf{v} + \frac{p}{\rho} - \mathbf{g} \cdot \mathbf{y} \tag{3.6.19}$$

is the corresponding Bernoulli function, $\mathbf{y}$ is the position vector in the noninertial frame, and the rest of the symbols are defined in Section 3.2.6.

Consider a noninertial frame that translates with linear velocity $\mathbf{U}(t)$ with respect to an inertial frame, and assume that the velocity field in the noninertial frame, $\mathbf{v}$, is irrotational. Expressing $\mathbf{v}$ in terms of an unsteady velocity potential, Φ, defined so that $\mathbf{v} = \nabla \Phi$, and taking into account the acceleration-reaction body force, we obtain the modified Bernoulli's equation

$$\left(\frac{\partial \Phi}{\partial t} \right)_{\mathbf{y}} + \frac{1}{2} \mathbf{v} \cdot \mathbf{v} + \frac{p}{\rho} + \left(\mathbf{g} - \frac{d\mathbf{U}}{dt} \right) \cdot \mathbf{y} = c(t), \tag{3.6.20}$$

where $c(t)$ is an unspecified function of time. When $\mathbf{U}$ is constant or zero, we recover the standard Bernoulli equation for unsteady irrotational flow.

In another application, we consider a noninertial frame whose axes rotate with constant angular velocity $\mathbf{\Omega}$ about the origin of an inertial frame. We will assume that the flow is irrotational in the inertial frame and appears to be steady in the noninertial frame. Euler's equation in the noninertial frame (3.6.18) simplifies into

$$\nabla \mathcal{B} = \mathbf{v} \times (\varrho + 2\,\mathbf{\Omega}) - \mathbf{\Omega} \times (\mathbf{\Omega} \times \mathbf{y}). \qquad (3.6.21)$$

Integrating in space, we derive a modified Bernoulli equation,

$$\mathcal{B} - \frac{1}{2}\,|\mathbf{\Omega} \times \mathbf{y}|^2 = \mathcal{F}(\text{streamline}), \qquad (3.6.22)$$

which is inclusive of the Bernoulli equation (3.6.6) for an inertial frame. In the literature of geophysical fluid dynamics, the term $\rho(\mathbf{g} \cdot \mathbf{y} + \frac{1}{2}\,|\mathbf{\Omega} \times \mathbf{y}|^2)$ is sometimes called the geopotential.

Problems

3.6.1 *Fluid sloshing in a tank*

A fluid is sloshing inside a tank executing rotational oscillations about the z axis with angular velocity $\Omega(t)$. In a stationary frame of reference (x, y, z), the induced irrotational flow can be described in terms of a velocity potential $\mathbf{u} = \nabla \phi$. Derive an expression for the rate of change of the potential ϕ in a frame of reference where the tanks appears to be stationary.

3.6.2 *Bernoulli's equations in a noninertial frame*

Derive (3.6.20) and (3.6.22).

3.7   Equations and boundary conditions governing the motion of an incompressible Newtonian fluid

The flow of an incompressible Newtonian fluid is governed by the Navier–Stokes equation (3.5.1) expressing Newton's second law for the motion of a small fluid parcel, and the continuity equation, $\nabla \cdot \mathbf{u} = 0$, expressing mass conservation. The union of these two equations provides us with four scalar partial differential equations for four scalar functions, including the three components of the velocity, $\mathbf{u}$, and the pressure, p.

To complete the mathematical statement of a fluid flow problem, we must supply an appropriate number of boundary conditions for certain flow variables including the velocity, the pressure, or the traction. In the case of unsteady flow, we must also supply a suitable initial condition for the velocity. The initial pressure field can be found by requiring that the velocity is and remains solenoidal at all times, as will be discussed in Sections 9.1 and 13.2. The initial and boundary conditions arise from considerations that are independent of those that led us to the continuity equation and equation of motion.

Counting the boundary conditions

Since the Navier–Stokes equation is a second-order partial differential equation for the velocity, we must supply three scalar boundary conditions over each boundary of the flow. In practice, we specify either the three components of the velocity, or the three components of the traction, or a combination of the velocity and the traction. If we specify the normal component of the velocity over all boundaries of a domain whose volume remains constant in time, we must ensure that the total volumetric flow rate into the domain of flow is zero, otherwise the continuity equation cannot be satisfied. In certain computational procedures for solving the equations of incompressible Newtonian flow, the boundary velocity is specified at discrete points and the total flow rate is not precisely zero due to numerical error. This imperfection may provide an entry point for numerical instability.

Irrotational and inviscid flow

In the case of irrotational flow or flow of an inviscid fluid governed by the Euler equation, the order of the Navier–Stokes equation is reduced from two to one by one count. Consequently, only one scalar boundary condition over each boundary is required. It is then evident that, with some fortuitous exceptions, the assumption of irrotational flow cannot be made at the outset.

3.7.1 The no-penetration condition over an impermeable boundary

Since the molecules of a fluid cannot penetrate an impermeable surface, the component of the fluid velocity normal to an impermeable boundary, $\mathbf{n} \cdot \mathbf{u}$, must be equal to the corresponding normal component of the boundary velocity, $\mathbf{n} \cdot \mathbf{V}$. The no-penetration condition requires that

$$\mathbf{n} \cdot \mathbf{u} = \mathbf{n} \cdot \mathbf{V}, \tag{3.7.1}$$

where the boundary velocity $\mathbf{V}$ may be constant or vary over the boundary. An impermeable boundary is not necessarily a rigid boundary, as it may represent, for instance, a fluid interface or the surface of a flexible body.

Consider a steady or evolving impenetrable boundary whose position is described in Eulerian form by the equation $F(\mathbf{x}, t) = c$, where c is a constant. The no-penetration condition can be stated as $DF/Dt = 0$, evaluated at the boundary, where D/Dt is the material derivative. The fluid velocity in the material derivative can be amended with the addition of an arbitrary tangential component.

Rigid-body motion

If a boundary moves as a rigid body, translating with velocity $\mathbf{U}$ and rotating about a point $\mathbf{x}_0$ with angular velocity $\mathbf{\Omega}$, the boundary velocity is $\mathbf{V} = \mathbf{U} + \mathbf{\Omega} \times (\mathbf{x} - \mathbf{x}_0)$ and the no-penetration condition requires that

$$\mathbf{n} \cdot \mathbf{u} = \mathbf{n} \cdot [\, \mathbf{U} + \mathbf{\Omega} \times (\mathbf{x} - \mathbf{x}_0) \,] \tag{3.7.2}$$

evaluated at the boundary.

Two-dimensional flow

In the case of two-dimensional flow in the xy plane past an impermeable boundary executing rigid-body motion, translating while rotating about an axis that is parallel to the z axis and passes through the point $\mathbf{x}_0$, equation (3.7.2) takes the form

$$\mathbf{V} = \mathbf{U} + \Omega_z \, \mathbf{e}_z \times (\mathbf{x} - \mathbf{x}_0), \tag{3.7.3}$$

where $\mathbf{e}_z$ is the unit vector along the z axis and Ω_z is the angular velocity of rotation around the z axis. Next, we write $\mathbf{n} = (dy/dl, -dx/dl)$, where dx and dy are differential increments around the boundary corresponding to the differential arc length dl measured in the counterclockwise direction. Expressing the velocity in terms of the stream function, ψ, we obtain

$$\mathbf{u} \cdot \mathbf{n} = \frac{\partial \psi}{\partial y}\frac{dy}{dl} + \frac{\partial \psi}{\partial x}\frac{dx}{dl} = \frac{\partial \psi}{\partial l} = \mathbf{U} \cdot \mathbf{n} + \Omega_z \left[\mathbf{e}_z \times (\mathbf{x} - \mathbf{x}_0) \right] \cdot \mathbf{n} \qquad (3.7.4)$$

or

$$\frac{\partial \psi}{\partial l} = U_x \frac{dy}{dl} - U_y \frac{dx}{dl} - \Omega_z \left((x - x_0)\frac{dx}{dl} + (y - y_0)\frac{dy}{dl} \right). \qquad (3.7.5)$$

Integrating with respect to arc length, we obtain the scalar boundary condition

$$\psi = U_x\, y - U_y\, x - \frac{1}{2}\, \Omega_z\, |\mathbf{x} - \mathbf{x}_0|^2 + c(t), \qquad (3.7.6)$$

where $c(t)$ is a time-dependent constant.

Equation (3.7.6) reveals that the stream function is constant over a stationary impermeable surface in a steady or unsteady flow. The value of the stream function is determined by the flow rate between the surface and the rest of the boundaries. In general, the value of the stream function over different disconnected stationary boundaries may not be specified *a priori*, but must be computed as part of the solution.

Axisymmetric flow

In the case of axisymmetric flow past a body that translates along its axis of revolution with velocity $\mathbf{U} = U\mathbf{e}_x$, we introduce the Stokes stream function, ψ, and derive the scalar boundary condition

$$\psi = \frac{1}{2} U\sigma^2 + c(t), \qquad (3.7.7)$$

where σ is the distance from the axis of revolution and $c(t)$ is a time-dependent constant. This expressions shows that the Stokes stream function takes a constant value over a stationary impermeable axisymmetric surface.

3.7.2   No-slip at a fluid-fluid or fluid-solid interface

Experimental observations of a broad class of fluids under a wide range of conditions have shown that the tangential component of the fluid velocity is continuous across a fluid-solid or fluid-fluid interface, which means that the slip velocity is zero and the no-slip condition applies. It is important to emphasize that the no-slip condition has been demonstrated independent of the mechanical or chemical constitution of the boundaries and properties of the fluid. One exception arises in the case of rarefied gas flow where the mean free path of the molecules is comparable to the size of the boundaries and a description of the flow in the context of continuum mechanics is no longer appropriate.

The no-slip boundary condition requires that the tangential component of the fluid velocity is zero over a stationary solid boundary. Combining this condition with the no-penetration condition, we find that the fluid velocity over a stationary impermeable solid boundary is equal to the boundary velocity. If a boundary executes rigid body motion, translating with linear velocity $\mathbf{U}$ and rotating about a point $\mathbf{x}_0$ with angular velocity $\mathbf{\Omega}$, then $\mathbf{u} = \mathbf{U} + \mathbf{\Omega} \times (\mathbf{x} - \mathbf{x}_0)$.

Physical origin

In the case of gases, the molecules of the fluid are adsorbed on the solid surface over a time period that is long enough for thermal equilibrium to be established, yielding a macroscopic no-slip condition on a solid surface. An analogous explanation is possible in the case of liquids based on the formation of short-lived bonds between liquid and solid molecules due to weak intermolecular forces. However, because these physical mechanisms depend on the properties of the solid and fluid molecules, the no-slip condition may occasionally break down.

An alternative explanation is that the proper boundary condition on a solid surface is the condition of vanishing tangential traction, and an apparent no-slip condition arises on a macroscopic level due to inherent boundary irregularities. Detailed computations with model geometries lend support to this explanation [340].

3.7.3   Slip at a fluid-solid interface

An alternative boundary condition used on occasion in place of the no-slip condition prescribes that the tangential component of the fluid velocity, $\mathbf{u}$, relative to the boundary velocity, $\mathbf{V}$, is proportional to the tangential component of the surface traction. According to the Navier–Maxwell–Basset formula, the tangential slip velocity is given by

$$\mathbf{u}_{slip} \equiv (\mathbf{u} - \mathbf{V}) \cdot (\mathbf{I} - \mathbf{n} \otimes \mathbf{n}) = \frac{1}{\beta} \frac{a}{\mu} \mathbf{f} \cdot (\mathbf{I} - \mathbf{n} \otimes \mathbf{n}) = \frac{1}{\beta} \frac{a}{\mu} \mathbf{n} \times \mathbf{f} \times \mathbf{n}, \qquad (3.7.8)$$

where $\mathbf{f} \equiv \mathbf{n} \cdot \boldsymbol{\sigma}$ is the traction, $\boldsymbol{\sigma}$ is the stress tensor, $\mathbf{n}$ is the unit normal vector pointing into the fluid, $\mathbf{I} - \mathbf{n} \otimes \mathbf{n}$ is the tangential projection operator, and a is a specified length scale. The dimensionless Basset coefficient, β, ranges from zero in the case of perfect slip and vanishing shear stress, to infinity in the case of no-slip and finite shear stress. The slip length is defined as

$$\lambda \equiv \frac{a}{\beta}. \qquad (3.7.9)$$

In the case of perfect slip, $\beta = 0$, the drag force is due exclusively to the form drag due to the pressure.

In rarefied gases, the slip coefficient, β, can be rigorously related to the mean free path, λ_f, defining the Knudsen number, $\mathrm{Kn} \equiv \lambda_f / a$ by the Maxwell relation

$$\beta = \frac{1}{\mathrm{Kn}} \frac{\sigma}{2 - \sigma}, \qquad (3.7.10)$$

where σ is the tangential momentum accommodation coefficient (TMAC) expressing the fraction of molecules that undergo diffusive rather than specular reflection (e.g., [72, 362]). In the limit $\sigma \to 2$, we obtain the no-slip boundary condition, $\beta \to \infty$. In the limit $\sigma \to 0$, we obtain the perfect-slip boundary condition, $\beta \to 0$.

The slip boundary condition has been used with success to describe flow over the surface of a porous medium where the slip velocity accounts for the presence of pores, and flow in the neighborhood of a moving three-phase contact line where the slip velocity removes the singular behavior of the traction in the immediate neighborhood of the contact line [114].

3.7.4 Derivative boundary conditions

We have discussed several types of boundary conditions with physical origin. Derivative boundary conditions emerge by combining the physical boundary conditions with the equations governing the motion of the fluid, including the continuity equation and the equation of motion. For example, invoking the continuity equation, we find that the normal derivative of the normal component of the velocity over an impermeable solid surface where the no-slip condition applies must be zero, as discussed in Section 3.9.1. A derivative boundary condition for the pressure over a solid boundary will be discussed in Chapter 13.

3.7.5 Conditions at infinity

Mathematical idealization of a flow in a domain with large dimensions produces flow in a totally or partially infinite domain. Examples include infinite flow past a stationary object, and semi-infinite shear flow over a plane wall with a cavity or a protrusions. To complete the definition of the mathematical problem, we require a condition for the asymptotic behavior of the flow at infinity.

To implement the far-field condition, we decompose the velocity field into an unperturbed component and a disturbance component, and require that the ratio of the magnitudes of the disturbance velocity and the unperturbed velocity decays at infinity. This condition does not necessarily imply that the disturbance velocity vanishes at infinity. For example, in the case of uniform flow past a sphere, the disturbance velocity due to the sphere decays at infinity; whereas in the case of parabolic flow past a sphere, the disturbance velocity grows at a rate that is less than quadratic. In the case of simple shear flow over an infinite plane wall with periodic corrugations or protrusions, we require that the disturbance flow due to the protrusions grows at a less-than-a-linear rate. The solution reveals that the disturbance velocity tends to a constant value far from the wall, thereby introducing an *a priori* unknown slip velocity.

3.7.6 Truncated domains of flow

In computing an external or infinite internal flow by numerical methods, it is a standard practice to truncate the physical domain of flow at a certain level that allows for an affordable computational cost. In the case of flow past a body in a wind tunnel, we introduce a computational domain confined by an in-flow plane, an out-flow plane, and side-flow boundaries. Ideally, the boundary conditions at the computational boundaries should be derived by performing a far-field asymptotic analysis. Unfortunately, this is possible only for a limited number of flows [155]. The assumption of fully developed flow for interior viscous flow is often used in practice. The choice of far-field boundary condition must be exercised with a great deal of caution in order to ensure mass conservation and prevent the violation of momentum and energy integral balances. The effect of domain truncation must be assessed carefully before a numerical solution can be claimed to have any degree of physical relevance.

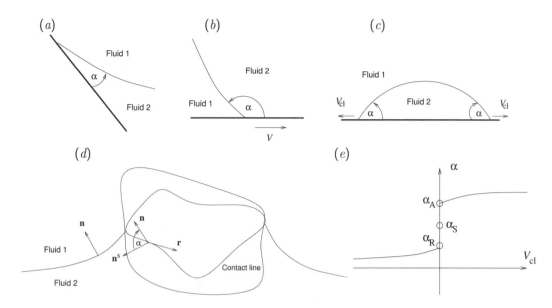

FIGURE 3.7.1   Illustration of (a) a stationary interface meeting a stationary solid surface at a static
contact line, (b) a stationary interface meeting a moving solid surface at a dynamic contact line, (c)
a contact line moving over a stationary solid surface due to a spreading liquid drop. The contact
angle, α, is measure on the side of the fluid labeled 2. (d) Contact line around the surface of a
floating particle. (e) Typical dependence of the dynamic contact angle, α, on the velocity of the
contact line on a stationary surface, V_{cl}.

3.7.7   Three-phase contact lines

In applications involving liquid films and small droplets and bubbles attached to solid surfaces, we
encounter stationary or moving interfaces between two fluids ending at stationary or moving solid
boundaries, as shown in Figure 3.7.1. The line where an interface meets a solid surface is called a
three-phase contact line or simply a contact line. A contact line can be stationary or move over a
surface spontaneously or in response to an imposed fluid flow. For example, a contact line is moving
down an inclined plane due to a developing film or rolling liquid drop.

The angle subtended between (a) the vector that is normal to the contact line and tangential
to an interface, and (b) the vector that is normal to the contact line and lies on the solid boundary
is called the contact angle, α. The contact angle is measured from the side of a designated fluid, as
shown in Figure 3.7.1. In the case of a liquid-gas interface, α is measured by convention from the
side of the liquid. With reference to Figure 3.7.1(d), the contact angle measured on the side of the
fluid labeled 2 is given by

$$\cos \alpha = \mathbf{n}^s \cdot \mathbf{n}, \qquad\qquad (3.7.11)$$

where $\mathbf{n}^s$ is the unit vector normal to the surface and $\mathbf{n}$ is the unit vector normal to the interface pointing into the fluid labeled 1. The plane containing $\mathbf{n}^s$ and $\mathbf{n}$ at a point is normal to the contact line at that point.

Shape of a contact line

A contact line may have an arbitrary shape determined by the degree of surface roughness and can have any desired shape. On a perfectly smooth surface, the shape of the contact line is determined by the contact angle. This means that the contact line may not be assigned *a priori* but must be found as part of the solution to satisfy a constraint on the contact angle.

Static and dynamic contact angle

A static contact angle is established at a stationary contact line that is pinned on a stationary solid surface. A dynamic contact angle is established at a stationary contact line that lies on a moving solid surface, or at a moving contact line that lies on a stationary solid surface, as illustrated in Figure 3.7.1(b, c). The static contact angle is a physical constant determined by the molecular properties of the solid and fluids. The dynamic contact angle depends not only on the constitution of the solid and fluids, but also on the relative velocity of the surface and contact line [114].

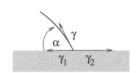

Three tensions balance at a contact line according to Young's equation.

A rational framework for predicting the static contact angle employs three interfacial tensions applicable to each fluid-fluid or fluid-solid pair at the contact line, denoted by γ_1, γ_2, and γ. A tangential force balance yields the Young equation,

$$\gamma_2 = \gamma_1 + \gamma \cos \alpha. \tag{3.7.12}$$

Similar equations can be written for a contact line that lies at a corner or cusp where two tensions are not necessarily aligned.

Dependence of the dynamic contact angle on the contact angle velocity

A typical graph of the dynamic contact angle plotted against the contact line velocity over a stationary surface, V_{cl}, is shown in Figure 3.7.1(e). With reference to Figure 3.7.1(c), positive velocity V_{cl} corresponds to a spreading droplet. Measurements show that $\partial\alpha/\partial V_{cl} > 0$ is independent of the fluids involved. The extrapolated values of α in the limit as V_{sl} tends to zero from positive or negative values are called the advancing or receding contact angle, α_A and α_R. Contact angle hysteresis is evidenced by the dependence of the estimated values of α_A and α_R on the laboratory procedures. Further support is provided by the observation that, when $V_{cl} = 0$, the contact angle may take any value between α_A and α_R.

Conversely, the velocity of a contact line can be regarded as a function of the difference, $\alpha - \alpha_S$, $\alpha - \alpha_A$, or $\alpha - \alpha_R$. From this viewpoint, contact angle motion is driven by deviations of the contact angle from an appropriate threshold. In the case of a spreading drop illustrated in Figure 3.7.1(c), $\alpha > \alpha_A$ resulting in a positive (outward) contact line velocity. In the case of a heavy drop moving down an inclined plane, the contact angle at the front of the drop is higher than α_A, while the contact angle at the back of the drop is less than α_R, resulting in a forward bulging shape.

Singularity at a moving contact angle

A local analysis of the equation of motion in the immediate neighborhood of a contact line moving on a stationary solid surface reveals that the tangential component of the surface force becomes singular at the contact line. More important, the singularity is nonintegrable, meaning that the drag force exerted on the surface assumes an infinite value, as discussed in Section 6.2. This unphysical behavior suggests that, either the equation of motion breaks down in the immediate vicinity of the dynamic contact line, or else the no-slip boundary condition ceases to be valid. The second explanation motivates replacing the no-slip boundary condition with the slip-condition shown in (3.7.8). Computations have shown that this approximation does not have a profound effect on the global structure of the flow [114]. A precursor liquid film of fluid 1 on the solid surface is sometimes introduced to regularize the motion near a moving contact line.

Problems

3.7.1 *No-penetration boundary condition*

Discuss possible ways of implementing the no-penetration boundary condition in terms of the stream functions for three-dimensional flow.

3.7.2 *A drop sliding down an inclined plane*

Sketch the shape of heavy three-dimensional drop sliding down an inclined plane and discuss the shape of the contact line.

3.8   Interfacial conditions

The components of the stress tensor on either side of an interface between two immiscible fluids, labeled 1 and 2, are not necessarily equal. Consequently, the traction may undergo a discontinuity defined as

$$\Delta \mathbf{f} \equiv \mathbf{f}^{(1)} - \mathbf{f}^{(2)} = \left(\boldsymbol{\sigma}^{(1)} - \boldsymbol{\sigma}^{(2)} \right) \cdot \mathbf{n}, \qquad (3.8.1)$$

where $\mathbf{n}$ is the normal unit vector pointing into fluid 1 by convention, and $\boldsymbol{\sigma}^{(1)}$, $\boldsymbol{\sigma}^{(2)}$ are the stress tensors in the two fluids evaluated on either side of the interface, as shown in Figure 3.8.1(a). The discontinuity in the interfacial traction, $\Delta \mathbf{f}$, is determined by the physiochemical properties of the fluids and molecular constitution of the interface, and is affected by the local temperature and local concentration of surface-active substances, commonly called surfactants, populating the interface. A laundry detergent or dishwashing liquid is a familiar household surfactant.

Jump in the hydrodynamic traction

When it is beneficial to work with the hydrodynamic pressure and stress defined in (3.4.9), indicated by a tilde, we introduce the associated jump in the hydrodynamic interfacial traction, also denoted by a tilde, defined as

$$\Delta \tilde{\mathbf{f}} \equiv \tilde{\mathbf{f}}^{(1)} - \tilde{\mathbf{f}}^{(2)} = \left(\tilde{\boldsymbol{\sigma}}^{(1)} - \tilde{\boldsymbol{\sigma}}^{(2)} \right) \cdot \mathbf{n}. \qquad (3.8.2)$$

(a)   (b)

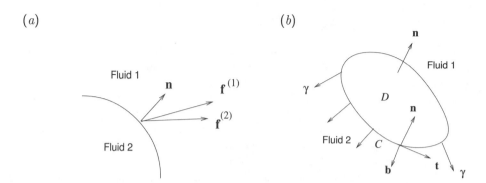

FIGURE 3.8.1 (a) Illustration of an interface between two immiscible fluids labeled 1 and 2, showing the traction on either side of the interface. The normal unit vector, $\mathbf{n}$, points into fluid 1 by convention. (b) Surface tension pulls the edges of an interfacial patch of a three-dimensional interface in the tangential plane.

The jump in the hydrodynamic traction is related to the jump in the physical traction by

$$\Delta \tilde{\mathbf{f}} = \Delta \mathbf{f} + (\rho_1 - \rho_2) \, (\mathbf{g} \cdot \mathbf{x}) \, \mathbf{n}, \tag{3.8.3}$$

where ρ_1 and ρ_2 are the fluid densities.

3.8.1   Isotropic surface tension

An uncontaminated interface between two immiscible fluids exhibiting isotropic surface tension, γ, playing the role of surface pressure. In a macroscopic interpretation, the surface tension pulls the edges of an interfacial patch in the tangential plane, as shown in Figure 3.8.1(b). The surface tension of a clean interface between water and air at $20°$C is $\gamma = 73$ dyn/cm. The surface tension of a clean interface between glycerin and air at the same temperature is $\gamma = 63$ dyn/cm.

Effect of temperature and surfactant

The surface tension generally decreases as the temperature is raised and becomes zero at a critical threshold. If an interface is populated by a surfactant, the surface tension is determined by the local surface surfactant concentration, Γ. The negative of the derivative of the surface tension, $-\mathrm{d}\gamma/\mathrm{d}\Gamma$, is sometimes called the Gibbs surface elasticity. Typically, the surface tension decreases as Γ increases and reaches a plateau at a saturation concentration where the interface is covered by a monolayer of surfactant molecules, as discussed in Section 3.8.2.

Interfacial force balance

To derive an expression for the jump in the interfacial traction, $\Delta \mathbf{f}$, we neglect the mass of the interfacial stratum and write a force balance over a small interfacial patch, D, that is bounded by the contour C,

$$\iint_D \Delta \mathbf{f}\, \mathrm{d}S + \oint_C \gamma\, \mathbf{t} \times \mathbf{n}\, \mathrm{d}l = \mathbf{0}, \tag{3.8.4}$$

where $\mathbf{n}$ is the unit vector normal to D pointing into fluid 1, $\mathbf{t}$ is the unit vector tangent to C, and $\mathbf{b} = \mathbf{t} \times \mathbf{n}$ is the unit binormal vector, as shown in Figure 3.8.1(b). Next, we extend smoothly the domain of definition of the surface tension and normal vector from the interface into the whole three-dimensional space. The extension of the surface tension can be implemented without any constraints. In practice, this can be done by solving a partial differential equation with the Dirichlet boundary condition over the interface. The extension of the normal vector can be implemented by setting $\mathbf{n} = \nabla F / |\nabla F|$, where the equation $F(x, y, z, t) = 0$ describes the location of the interface.

A variation of Stokes' theorem expressed by equation (A.7.8), Appendix A, states that, for any arbitrary vector function of position, $\mathbf{q}$,

$$\oint_C \mathbf{q} \times \mathbf{t}\, \mathrm{d}l = \iint_D \left(\mathbf{n} \nabla \cdot \mathbf{q} - (\nabla \mathbf{q}) \cdot \mathbf{n} \right) \mathrm{d}S. \tag{3.8.5}$$

Applying this identity for the extended vector field, $\mathbf{g} = \gamma \mathbf{n}$, and combining the resulting expression with (3.8.4), we obtain

$$\iint_D \Delta \mathbf{f}\, \mathrm{d}S = \iint_D \left(\mathbf{n} \nabla \cdot (\gamma \mathbf{n}) - [\nabla (\gamma \mathbf{n})] \cdot \mathbf{n} \right) \mathrm{d}S. \tag{3.8.6}$$

Expanding the derivatives inside the integral and noting that $(\nabla \mathbf{n}) \cdot \mathbf{n} = \frac{1}{2} \nabla (\mathbf{n} \cdot \mathbf{n}) = 0$, we obtain

$$\iint_D \Delta \mathbf{f}\, \mathrm{d}S = \iint_D \left(\mathbf{n} \left[\gamma (\nabla \cdot \mathbf{n}) - \mathbf{n} \nabla \gamma \right] - \nabla \gamma \right) \mathrm{d}S. \tag{3.8.7}$$

Now we take the limit as the surface patch shrinks to a point, rearrange the integrand on the right-hand side, and discard the integral sign on account of the arbitrary integration domain to obtain the desired expression

$$\Delta \mathbf{f} = \gamma \mathbf{n} \nabla \cdot \mathbf{n} - (\mathbf{I} - \mathbf{n} \otimes \mathbf{n}) \cdot \nabla \gamma, \tag{3.8.8}$$

which can be restated as

$$\Delta \mathbf{f} = \gamma \mathbf{n} \nabla \cdot \mathbf{n} - (\mathbf{n} \times \nabla \gamma) \times \mathbf{n}. \tag{3.8.9}$$

The divergence of the normal vector on the right-hand side of (3.8.8) is equal to twice the mean curvature of the interface, $\nabla \cdot \mathbf{n} = 2\,\kappa_m$. Thus,

$$\Delta \mathbf{f} = \gamma\, 2\kappa_m\, \mathbf{n} - (\mathbf{n} \times \nabla \gamma) \times \mathbf{n}. \tag{3.8.10}$$

By definition, the mean curvature is positive when the interface has a spherical shape with fluid 2 residing in the interior and the normal vector pointing outward into the exterior fluid labeled 1. The computation of the mean curvature is discussed in Section 1.8.

Laplace pressure and Marangoni traction

The two terms on the right-hand side of (3.8.10) express, respectively, a discontinuity in the normal direction and a discontinuity in the tangential plane. The normal discontinuity is called the Laplace or capillary pressure, and the tangential discontinuity is called the Marangoni traction. The term "capillary" derives from the Latin word "capilla" meaning hair, on the justification that the capillary pressure is important inside small tubes whose size is comparable to that of a hair. When the surface tension is constant, a tangential component does not appear and the jump in the interfacial traction is oriented in the normal direction.

Capillary force and torque

Surface tension pulls the boundary of a flow around a three-phase contact line in a direction that is tangential to the interface and lies in a plane that is normal to the contact line. With reference to Figure 3.7.1(d), the resultant capillary force and torque with respect to an arbitrary point, $\mathbf{x}_0$, are given by

$$\mathbf{F}^{cap} = \oint \gamma \, \mathbf{r} \times \mathbf{n} \, dl, \qquad \mathbf{T}^{cap}(\mathbf{x}_0) = \oint \gamma \, (\mathbf{x} - \mathbf{x}_0) \times (\mathbf{r} \times \mathbf{n}) \, dl, \qquad (3.8.11)$$

where $\mathbf{n}$ is the unit vector tangent to the interface, $\mathbf{r} = d\mathbf{x}/dl$ is the unit vector tangent to the contact line, the integration is performed around the contact line, and l is the arc length around the contact line. Using a vector identity, we obtain an alternative expression for the capillary torque,

$$\mathbf{T}^{cap}(\mathbf{x}_0) = \oint \gamma \left([(\mathbf{x} - \mathbf{x}_0) \cdot \mathbf{n}] \, \mathbf{r} - [(\mathbf{x} - \mathbf{x}_0) \cdot \mathbf{r}] \, \mathbf{n} \right) dl. \qquad (3.8.12)$$

The surface tension inside the integrals on the right-hand sides of (3.8.11) and (3.8.12) is allowed to be a function of position.

Capillary torque on a spherical particle

As an application, we consider the capillary torque due to a contact line over a spherical particle of radius a for constant surface tension, γ. The unit vector normal to the particle surface at a point, $\mathbf{x}$, is $\mathbf{n}^s = \frac{1}{a} (\mathbf{x} - \mathbf{x}_c)$, where $\mathbf{x}_c$ is the particle center. Consequently, $\cos\alpha = \frac{1}{a} (\mathbf{x} - \mathbf{x}_c) \cdot \mathbf{n}$, where α is the contact angle computed from (3.7.11). Identifying the center of torque $\mathbf{x}_0$ with the the particle center, $\mathbf{x}_c$, and noting that $(\mathbf{x} - \mathbf{x}_c) \cdot \mathbf{r} = 0$, we obtain

$$\mathbf{T}^{cap}(\mathbf{x}_c) = \gamma a \oint \cos\alpha \, \mathbf{r} \, dl. \qquad (3.8.13)$$

We observe that, if the contact angle is constant along the contact line, the torque with respect to the center of the spherical particle is identically zero irrespective of the shape of the contact line [373]. This result underscores the importance of contact angle hysteresis in imparting a non-zero capillary torque for any contact line shape.

3.8.2 Surfactant equation of state

We have mentioned that variations in the surface concentration of a surfactant, Γ, cause corresponding variations in the surface tension, γ. The relation between Γ and γ is determined by an appropriate surface equation of state.

Linear surface equation of state

When the surfactant concentration is below the saturation threshold, a linear relationship can be assumed between Γ and γ according to Gibbs' surface equation of state,

$$\gamma = \gamma_c - RT\Gamma, \tag{3.8.14}$$

where R is the ideal gas constant, T is the absolute temperature, and γ_c is the surface tension of a clean interface that is devoid of surfactants (e.g., [3]). It is convenient to introduce a reference surfactant concentration, Γ_0, corresponding to surface tension γ_0. Rearranging (3.8.14), we obtain

$$\gamma = \gamma_c \left(1 - \beta \frac{\Gamma}{\Gamma_0}\right), \tag{3.8.15}$$

where

$$\beta \equiv \frac{RT}{\gamma_c} \Gamma_0 \tag{3.8.16}$$

is a dimensionless parameter expressing the significance of the surfactant concentration. By definition, $\gamma_0 = \gamma_c (1 - \beta)$. The parameter β is related to the Gibbs surface elasticity employed in the surfactant literature by

$$E \equiv -\frac{\partial \gamma}{\partial \Gamma} = \beta \frac{\gamma_c}{\Gamma_0} = \frac{\beta}{1 - \beta} \frac{\gamma_0}{\Gamma_0}. \tag{3.8.17}$$

The dimensionless Marangoni number is defined as

$$\mathrm{Ma} \equiv -\frac{\partial \gamma}{\partial \Gamma} \frac{\Gamma_0}{\gamma_0} = E \frac{\Gamma_0}{\gamma_0} = \frac{\beta}{1 - \beta}. \tag{3.8.18}$$

Using this definition, we find that $\beta = \mathrm{Ma}/(1 + \mathrm{Ma})$, $\gamma_c = (1 + \mathrm{Ma})\, \gamma_0$, and

$$\gamma = \gamma_c - \mathrm{Ma} \frac{\gamma_0}{\Gamma_0} \Gamma. \tag{3.8.19}$$

Nonlinear surface equation of state

More generally, because the surface elasticity depends on the local surfactant concentration, the relationship between the surface tension and the surface surfactant concentration is nonlinear. For small variations in the surfactant concentration around a reference value, Γ_0, we may use Henry's equation of state expressed by $\gamma = \gamma_0 - E\,(\Gamma - \Gamma_0)$, where $E \equiv -(\partial \gamma / \partial \Gamma)_{\Gamma_0}$.

For large variations in the surfactant surface concentration, we may use Langmuir's equation of state derived on the assumption of second-order adsorption/desorption kinetics,

$$\gamma = \gamma_c + RT\Gamma_\infty \ln\left(1 - \frac{\Gamma}{\Gamma_\infty}\right) = \gamma_c \left[1 + \frac{\beta}{\eta} \ln\left(1 - \eta \frac{\Gamma}{\Gamma_0}\right)\right], \tag{3.8.20}$$

where Γ_∞ is the surfactant concentration at maximum packing and $\eta = \Gamma_0/\Gamma_\infty$ is the surface coverage at the reference state (e.g., [48]). Applying this equation for $\Gamma = \Gamma_0$, we obtain

$$\gamma_c = \frac{\gamma_0}{1 + \frac{\beta}{\eta} \ln(1 - \eta)}. \tag{3.8.21}$$

In deriving Langmuir's equation of state, ideal behavior is assumed to neglect cohesive and repulsive interactions between surfactant molecules in the interfacial monolayer.

3.8.3   Interfaces with involved mechanical properties

The jump in traction across an interface with a complex molecular structure can be described rigorously or phenomenologically in terms of interfacial shear and dilatational surface viscosities, moduli of elasticity, and flexural stiffness (e.g., [118, 315, 366]). For example, the membrane enclosing a red blood cell is a viscoelastic sheet whose mechanical response is characterized by a high dilatational modulus that ensures surface incompressibility, a moderate shear elastic modulus that allows for significant deformation, and a substantial energy-dissipating surface viscosity. Mathematical frameworks for describing direction dependent interfacial tensions can be developed working under the auspices of shell theory for the molecular interfacial stratum or by employing models of three-dimensional molecular networks.

3.8.4   Jump in pressure across an interface

The normal component of the traction, and therefore the pressure, undergoes a discontinuity across an interface. To compute the jump in the pressure in terms of the velocity, we decompose the jump in the traction, $\Delta \mathbf{f}$, into its normal and tangential components, and identify the normal component in each fluid with the term inside the square bracket on the right-hand side of (3.4.5), finding

$$\Delta f^N \equiv \Delta \mathbf{f} \cdot \mathbf{n} = -p_1 + p_2 + 2\,\mathbf{n} \cdot \left(\mu_1 \nabla \mathbf{u}^{(1)} - \mu_2 \nabla \mathbf{u}^{(2)} \right) \mathbf{n}, \tag{3.8.22}$$

where $\mathbf{n}$ is the normal unit vector pointing into fluid 1. Rearranging, we obtain an expression for the jump in the interfacial pressure in terms of the normal component of $\Delta \mathbf{f}$ and the viscous normal stress,

$$\Delta p \equiv p_1 - p_2 = -\Delta \mathbf{f} \cdot \mathbf{n} + 2\,\mathbf{n} \cdot \left(\mu_1 \nabla \mathbf{u}^{(1)} - \mu_2 \nabla \mathbf{u}^{(2)} \right) \mathbf{n}, \tag{3.8.23}$$

where $\Delta \mathbf{f}$ is given by an appropriate interfacial constitutive equation. When the fluids are either stationary or inviscid, only the first term on the right-hand side of (3.8.23) survives. When the interface exhibits isotropic tension, we use (3.8.8) and derive Laplace's law, $\Delta p = -2\kappa_m \gamma$.

3.8.5   Boundary condition for the velocity at an interface

Certain numerical methods for solving problems of interfacial flow require removing the pressure from the dynamic boundary condition for the jump in the interfacial traction, $\Delta \mathbf{f}$. The tangential component of $\Delta \mathbf{f}$ involves derivatives of the velocity alone and does not require further manipulation. To eliminate the pressure difference from the normal component of $\Delta \mathbf{f}$, we apply the equation of motion on either side of the interface, formulate the difference of the resulting expressions, and project the difference onto a tangent unit vector, $\mathbf{t}$, to obtain

$$\mathbf{t} \cdot \nabla(p_2 - p_1) = \left(-\rho_2 \frac{D\mathbf{u}^{(2)}}{Dt} + \rho_1 \frac{D\mathbf{u}^{(1)}}{Dt} + \mu_2 \nabla^2 \mathbf{u}^{(2)} - \mu_1 \nabla^2 \mathbf{u}^{(1)} + (\rho_2 - \rho_1)\mathbf{g} \right) \cdot \mathbf{t}. \tag{3.8.24}$$

Substituting expression (3.8.23) for the pressure difference into the left-hand side of (3.8.24) provides us with the desired boundary condition in terms of the velocity.

If the fluids are inviscid, the velocity is allowed to be discontinuous across the interface, yielding the simplified form

$$\mathbf{t} \cdot \nabla \Delta f^N = \mathbf{t} \cdot \nabla (p_2 - p_1) = \left(-\rho_2 \frac{D\mathbf{u}^{(2)}}{Dt} + \rho_1 \frac{D\mathbf{u}^{(1)}}{Dt} + (\rho_2 - \rho_1)\mathbf{g} \right) \cdot \mathbf{t}, \qquad (3.8.25)$$

where $\Delta f^N = \Delta \mathbf{f} \cdot \mathbf{n}$ is the normal component of the jump in the traction. Equation (3.8.25) imposes a constraint on the tangential components of the point-particle acceleration on either side of the interface between two immiscible fluids.

3.8.6 Generalized equation of motion for flow in the presence of an interface

To compute the flow of two adjacent immiscible fluids with different physical properties, we may separately solve the governing equations in each fluid subject to the interfacial kinematic and dynamic matching conditions discussed earlier in this chapter. As an alternative, the interfacial boundary conditions can be incorporated into a generalized equation of motion that applies inside as well as at the interface between the two fluids. The generalized equation of motion arises by regarding an interface as a singular surface of concentrated body force, obtaining

$$\rho \frac{D\mathbf{u}}{Dt} = \nabla \cdot \boldsymbol{\sigma} + \rho \mathbf{g} + \mathbf{q}, \qquad (3.8.26)$$

where $\mathbf{q}$ is a singular forcing function expressing an interfacial distribution of point forces,

$$\mathbf{q}(\mathbf{x}) = - \iint_I \delta_3(\mathbf{x} - \mathbf{x}')\, \Delta \mathbf{f}(\mathbf{x}')\, dS(\mathbf{x}'), \qquad (3.8.27)$$

I stands for the interface, δ_3 is the three-dimensional delta function, and $\Delta \mathbf{f}$ is the jump in the traction across the interface. The velocity is required to be continuous, but the physical properties of the fluids and stress tensor are allowed to undergo discontinuities across the interface.

To demonstrate that (3.8.27) provides us with a consistent representation of the flow by incorporating the precise form of the dynamic boundary condition (3.8.1), we compute the volume integral over a thin layer of fluid of thickness ϵ centered at the interface, I. The upper side of the interface corresponding to fluid 1 is denoted by I_+, and the lower side corresponding to fluid 2 is denoted by I_-. Using the divergence theorem to manipulate the integral of the divergence of the stress and the properties of the delta function to manipulate the integral of $\mathbf{q}$, and then taking the limit as the layer collapses onto the interface, we obtain

$$\iiint_L \rho \frac{D\mathbf{u}}{Dt}\, dV = \iint_{I_+, I_-} \mathbf{n} \cdot \boldsymbol{\sigma}\, dS + \iiint_L \rho \mathbf{g}\, dV - \iint_I \Delta \mathbf{f}\, dS, \qquad (3.8.28)$$

where $\mathbf{n}$ is the normal unit vector pointing into fluid 1, and L denotes the layer. In the limit as ϵ tends to zero, the volume integrals of the point-particle acceleration and gravitational force vanish. The surface integrals can be rearranged to yield

$$\mathbf{0} = \iint_I \mathbf{n}^+ \cdot \boldsymbol{\sigma}^+\, dS - \iint_I \mathbf{n}^+ \cdot \boldsymbol{\sigma}^-\, dS - \iint_I \Delta \mathbf{f}\, dS, \qquad (3.8.29)$$

where $\mathbf{n}^+$ is the normal unit vector pointing into fluid 1, $\boldsymbol{\sigma}^+$ is the stress on the upper side of the interface, and $\boldsymbol{\sigma}^-$ is the stress on the lower side of the interface. It is now evident that (3.8.29), and therefore (3.8.26), is consistent with the interfacial condition (3.8.1).

A formal way of deriving (3.8.26) involves replacing the step functions, inherent in the representation of the physical properties of the fluids, and the delta functions, inherent in the distribution of the interfacial force, with smooth functions that change gradually over a thin interfacial layer of thickness ϵ. As long as ϵ is noninfinitesimal, the regular form of the equation of motion applies in the bulk of the fluids as well as inside the interfacial layer. Taking the limit as ϵ tends to zero, we derive (3.8.26).

Two-dimensional flow

In the case of two-dimensional flow, the interfacial distribution (3.8.27) takes the form of a line integral,

$$\mathbf{q}(\mathbf{x}) = - \int_I \Delta\mathbf{f}(\mathbf{x}')\, \delta_2(\mathbf{x} - \mathbf{x}')\, \mathrm{d}l(\mathbf{x}'), \tag{3.8.30}$$

where l is the arc length along the interface and δ_2 is the two-dimensional delta function in the plane of the flow.

Problems

3.8.1 *Interfaces with isotropic tension*

Derive the counterpart of equation (3.8.8) for two-dimensional flow and discuss its physical interpretation.

3.8.2 *Generalized equation of motion for two-dimensional flow*

Write the counterparts of equations (3.8.28) and (3.8.29) for two-dimensional flow.

3.9 Traction, vorticity, and kinematics at boundaries and interfaces

The boundary conditions at impermeable rigid boundaries, free surfaces, and fluid interfaces discussed in Sections 3.7 and 3.8 allow us to simplify the expressions for the corresponding traction, vorticity, force and torque, and thereby obtain useful insights in the kinematics and dynamics of a viscous flow.

3.9.1 Rigid boundaries

Consider flow past an impermeable rigid boundary that is either stationary or executes rigid-body motion, including translation and rotation. In the case of rigid-body motion, we describe the flow in a frame of reference where the boundary appears to be stationary. Concentrating at a particular point on the boundary, we introduce a local Cartesian system with origin at that point, two axes z and x tangential to the boundary, and the y axis normal to the boundary pointing into the fluid.

Expanding the velocity in a Taylor series with respect to x, y, and z, and enforcing the no-slip and no-penetration boundary conditions, we find that all velocity components and their first partial derivatives with respect to x and z at the origin are zero,

$$\mathbf{u}(0) = \mathbf{0}, \qquad \left(\frac{\partial \mathbf{u}}{\partial x}\right)_0 = \mathbf{0}, \qquad \left(\frac{\partial \mathbf{u}}{\partial z}\right)_0 = \mathbf{0}. \tag{3.9.1}$$

The continuity equation, $\nabla \cdot \mathbf{u} = 0$, requires that the first partial derivative of the normal component of the velocity with respect to y is also zero at the origin,

$$\left(\frac{\partial u_y}{\partial y}\right)_0 = 0. \tag{3.9.2}$$

The corresponding normal derivatives of the tangential velocity, $\partial u_x/\partial y$ and $\partial u_z/\partial y$, are not necessarily zero. With reference to arbitrary Cartesian coordinates that are not necessarily tangential to the boundary, equations (3.9.1) and (3.9.2) take the form

$$\mathbf{u} = \mathbf{0}, \qquad (\mathbf{I} - \mathbf{n} \otimes \mathbf{n}) \cdot \nabla \mathbf{u} = (\mathbf{n} \times \nabla \mathbf{u}) \times \mathbf{n} = 0, \qquad \mathbf{n} \cdot (\nabla \mathbf{u}) \cdot \mathbf{n} = 0, \tag{3.9.3}$$

evaluated at the boundary.

Vorticity and traction

Using conditions (3.9.3) to simplify expressions (3.4.3) and (3.4.5), we find that the components of the traction at the origin of the local coordinate system are given by

$$f_x = \mu \frac{\partial u_x}{\partial y}, \qquad f_y = -p, \qquad f_z = \mu \frac{\partial u_z}{\partial y}. \tag{3.9.4}$$

With reference to a general coordinate system, the traction is

$$\mathbf{f} = -p\,\mathbf{n} + \mu\,\mathbf{n} \cdot (\nabla \mathbf{u}) \cdot (\mathbf{I} - \mathbf{n} \otimes \mathbf{n}). \tag{3.9.5}$$

A local Cartesian system at a point on a boundary.

The second term on the right-hand side involves derivatives of the tangential components of the velocity in a direction normal to the boundary.

Using the definition of the vorticity in conjunction with the kinematic constraint expressed by the second equation in (3.9.3), we find that the component of the vorticity normal to the boundary is zero. At the origin of the local coordinate system, the tangential components of the vorticity are

$$\omega_x = \frac{\partial u_z}{\partial y} = \frac{1}{\mu} f_z, \qquad \omega_z = -\frac{\partial u_x}{\partial y} = -\frac{1}{\mu} f_x, \tag{3.9.6}$$

These expressions along with (3.9.3) suggest that, with reference to a general coordinate system,

$$\mathbf{n} \cdot \nabla \mathbf{u} = \mathbf{n} \cdot \nabla \mathbf{u} \cdot (\mathbf{I} - \mathbf{n} \otimes \mathbf{n}) = \boldsymbol{\omega} \times \mathbf{n}. \tag{3.9.7}$$

The general expression for the traction given in (3.4.4), $\mathbf{f} = -p\,\mathbf{n} + 2\mu\,\mathbf{n} \cdot \nabla \mathbf{u} + \mu\,\mathbf{n} \times \boldsymbol{\omega}$, simplifies into

$$\mathbf{f} = -p\,\mathbf{n} + \mu\,\boldsymbol{\omega} \times \mathbf{n}. \tag{3.9.8}$$

Substituting (3.9.5) and (3.9.8) into (3.4.11) and (3.4.12), we derive simplified expressions for the hydrodynamic force and torque exerted on a rigid boundary in terms of the vorticity and normal derivatives of the tangential components of the velocity.

Velocity gradient tensor

Combining the second equation in (3.9.3) with (3.9.7) we derive an expression for the velocity gradient tensor in terms of the vorticity,

$$\nabla \mathbf{u} = \mathbf{n} \otimes (\boldsymbol{\omega} \times \mathbf{n}). \tag{3.9.9}$$

The symmetric part of the velocity gradient tensor is the rate-of-deformation tensor and the anti-symmetric part is the vorticity tensor. In the case of a compressible fluid, the term $\alpha \mathbf{n} \otimes \mathbf{n}$ is added to the left-hand side of (3.9.9), where $\alpha \equiv \nabla \cdot \mathbf{u}$ is the rate of expansion.

Skin-friction and surface vortex lines

Equation (3.9.7) shows that the vorticity vector is perpendicular to the skin-friction vector defined as the tangential component of the traction vector,

$$\mathbf{f} \cdot (\mathbf{I} - \mathbf{n} \otimes \mathbf{n}) = (\mathbf{n} \times \mathbf{f}) \times \mathbf{n} = \mu \boldsymbol{\omega} \times \mathbf{n}. \tag{3.9.10}$$

Equation (3.9.7) shows that the vorticity vector is perpendicular to the skin-friction vector,

$$\boldsymbol{\omega} \cdot [\mathbf{f} \cdot (\mathbf{I} - \mathbf{n} \otimes \mathbf{n})] = 0, \qquad \boldsymbol{\omega} \cdot [\mathbf{n} \times \mathbf{f} \times \mathbf{n}] = 0. \tag{3.9.11}$$

A line over a boundary whose tangent vector is parallel to the skin friction vector at each point is a skin-friction line. A line over the boundary whose tangent vector is parallel to the vorticity at each point is a boundary or surface vortex line. Equations (3.9.11) show that the skin friction lines are orthogonal to the surface vorticity lines. Equation (3.9.5) suggests that a skin-friction line can be described in parametric form by an autonomous ordinary differential equation,

$$\frac{d\mathbf{x}}{d\tau} = \mathbf{n} \cdot (\nabla \mathbf{u}) \cdot (\mathbf{I} - \mathbf{n} \otimes \mathbf{n}) = \boldsymbol{\omega} \times \mathbf{n}, \tag{3.9.12}$$

where τ is an arbitrary time-like parameter.

Topology of skin-friction lines

Singular points around which fluid particles move normal to the boundary faster than they move along the boundary occur when the right-hand side of (3.9.12) vanishes and therefore the surface vorticity becomes zero. Singular points are classified as nodal points of attachment or separation, foci of attachment or separation, and saddle points. Examples of skin-friction and vorticity lines in each case are shown, respectively, with dashed and solid lines in Figure 3.9.1 [238, 405].

A nodal point belongs to an infinite number of skin-friction lines all of which except for one, the one labeled AA in Figure 3.9.1(a), are tangential to a single line labeled BB. A focal point belongs to an infinite set of skin-friction lines that spiral away from or into the focal point, as shown in Figure 3.9.1(b). A saddle point is the point of intersection of two skin-friction lines, as shown

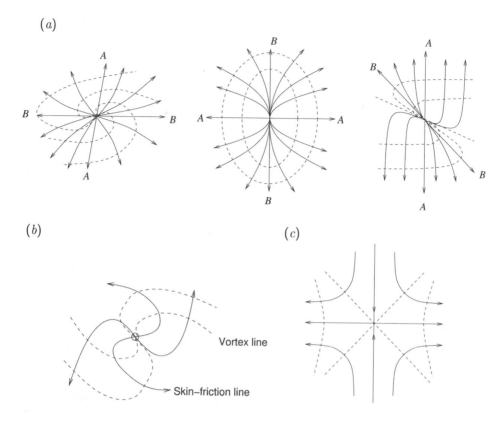

FIGURE 3.9.1   Illustration of singular points of the skin-friction pattern on a solid boundary [238]: (a) nodal points, (b) a focus, and (c) a saddle point.

in Figure 3.9.1(c). Topological constraints require that the number of nodal points and foci on a boundary exceed the number of saddle points by two. An in-depth discussion of the topography and topology of skin-friction lines is presented by Lighthill [238] and Tobak & Peake [405].

Motion of point particles near boundaries

The relation between the skin-friction lines and the trajectories of point particles in the vicinity of a boundary becomes evident by introducing a system of orthogonal surface curvilinear coordinates over the boundary, (ξ, η). The point particles move with velocity that is nearly tangential to the boundary, except when they find themselves in the neighborhood of a singular point where the streamlines turn away from the boundary. Denoting the instantaneous distance of a point particle away from the boundary by ζ, we expand the velocity in a Taylor series with respect to distance in the normal direction and find that the rate of change of distance traveled by a point particle along the curvilinear axes is

$$\frac{\mathrm{d}l_\xi}{\mathrm{d}t} = \mathbf{U}(\xi, \eta, t) \cdot \mathbf{t}_\xi = \zeta \, (\mathbf{n} \cdot \nabla \mathbf{u}) \cdot \mathbf{t}_\xi = \zeta \, \boldsymbol{\omega} \cdot \mathbf{t}_\eta,$$

$$\frac{\mathrm{d}l_\eta}{\mathrm{d}t} = \mathbf{U}(\xi, \eta, t) \cdot \mathbf{t}_\eta = \zeta \, (\mathbf{n} \cdot \nabla \mathbf{u}) \cdot \mathbf{t}_\eta = -\zeta \, \boldsymbol{\omega} \cdot \mathbf{t}_\xi, \qquad (3.9.13)$$

where ω is the vorticity and the velocity gradient and vorticity are evaluated at the surface, $\zeta = 0$. Dividing these equations side by side, we derive an alternative expression of (3.9.12),

$$\frac{\mathrm{d}l_\xi}{\mathbf{n} \cdot (\nabla \mathbf{u}) \cdot \mathbf{t}_\xi} = \frac{\mathrm{d}l_\eta}{\mathbf{n} \cdot (\nabla \mathbf{u}) \cdot \mathbf{t}_\eta}. \qquad (3.9.14)$$

3.9.2 Free surfaces

Equations (3.9.3) do not apply at a stationary, moving, or evolving free surface where the tangential velocity component is not necessarily zero. However, because a free surface cannot support shear stress, the tangential component of the traction must be zero,

$$\mathbf{f} \times \mathbf{n} = (\mathbf{n} \cdot \boldsymbol{\sigma}) \times \mathbf{n} = \mathbf{0}. \qquad (3.9.15)$$

Expressing the stress tensor in terms of the pressure and the rate-of-deformation tensor using the Newtonian constitutive equation, $\boldsymbol{\sigma} = -p\mathbf{I} + \mu \nabla \mathbf{u} + \mu (\nabla \mathbf{u})^T$, we find that

$$\mathbf{f} \times \mathbf{n} = \mu \left((\mathbf{n} \cdot \nabla \mathbf{u}) \times \mathbf{n} - \mathbf{n} \times [(\nabla \mathbf{u}) \cdot \mathbf{n}] \right) = \mathbf{0}. \qquad (3.9.16)$$

Rearranging, we derive the condition of zero shear stress

$$(\mathbf{n} \cdot \nabla \mathbf{u}) \times \mathbf{n} = \mathbf{n} \times [(\nabla \mathbf{u}) \cdot \mathbf{n}]. \qquad (3.9.17)$$

The tangential component of the vorticity at an arbitrary surface is

$$\boldsymbol{\omega} \cdot (\mathbf{I} - \mathbf{n} \otimes \mathbf{n}) = (\mathbf{n} \times \boldsymbol{\omega}) \times \mathbf{n} = [(\nabla \mathbf{u}) \cdot \mathbf{n} - \mathbf{n} \cdot \nabla \mathbf{u}] \times \mathbf{n}. \qquad (3.9.18)$$

Implementing the condition of zero shear stress stated in (3.9.17), we obtain

$$\boldsymbol{\omega}_{fs} \cdot (\mathbf{I} - \mathbf{n} \otimes \mathbf{n}) = -2 \, (\mathbf{n} \cdot \nabla \mathbf{u}) \times \mathbf{n} = 2 \, [(\nabla \mathbf{u}) \cdot \mathbf{n}] \times \mathbf{n} \qquad (3.9.19)$$

or

$$\boldsymbol{\omega}_{fs} \cdot (\mathbf{I} - \mathbf{n} \otimes \mathbf{n}) = 2 \left(\nabla(\mathbf{u} \cdot \mathbf{n}) - (\nabla \mathbf{n}) \cdot \mathbf{u}] \right) \times \mathbf{n}. \qquad (3.9.20)$$

The first term on the right-hand side of (3.9.20) contains derivatives of the normal component of the velocity tangential to the free surface. The derivatives of the normal vector in the second term are related to the normal curvatures of the free surface in the normal plane that contains the velocity and its conjugate plane, as discussed in the next section for steady flow.

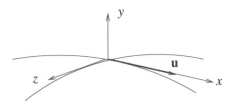

FIGURE 3.9.2 A local Cartesian system attached to a free surface with the x axis pointing in the direction of the velocity vector.

Steady flow

The first term on the right-hand side of (3.9.19) vanishes over a stationary free surface where the normal velocity is zero, $\mathbf{u} \cdot \mathbf{n} = 0$. It is useful to introduce a local Cartesian system with origin at a point on a free surface, the x axis pointing in the direction of the velocity vector, and the y axis pointing normal to the free surface, as shown in Figure 3.9.2. Setting $\mathbf{n} = [0, 1, 0]$, we find that the tangential component of the vorticity is given by

$$\boldsymbol{\omega}_t = 2\,[0,1,0] \times \big((\nabla\mathbf{n}) \cdot [u_x, 0, 0]\big) = 2u_x\,(\frac{\partial n_x}{\partial z}\,\mathbf{e}_x - \frac{\partial n_x}{\partial x}\,\mathbf{e}_z\,), \qquad (3.9.21)$$

where $\mathbf{e}_x$ is the unit vector along the x axis and $\mathbf{e}_z$ is the unit vector along the z axis. The derivative, $\kappa_x = \partial n_x/\partial x$, is the principal curvature of the free surface in the xy plane.

3.9.3 Two-dimensional flow

Simplified expressions for the boundary traction and vorticity can be derived in the case of two-dimensional flow in the xy plane.

Rigid boundaries

First, we consider the structure of a flow near a stationary rigid boundary with reference to the local coordinate system illustrated in Figure 3.9.3(a). Using (3.9.3), we find that the z component of the vorticity and tangential component of the wall shear stress at the origin are given by

$$\omega_z = -\frac{\partial u_x}{\partial y}, \qquad \sigma_{xy} = -\mu\,\omega_z = \mu\,\frac{\partial u_x}{\partial y}. \qquad (3.9.22)$$

Thus, $\sigma_{xy} = -\mu\omega_z$, revealing an intimate connection between the wall vorticity and the shear stress at a solid boundary.

Stagnation point at a solid boundary

Next, we concentrate on the structure of the flow in the vicinity of a stagnation point illustrated in Figure 3.9.3(b). Near the wall, the fluid moves in opposite directions on either side of the stagnation point. Because the wall shear stress and vorticity have opposite signs on either side of the stagnation

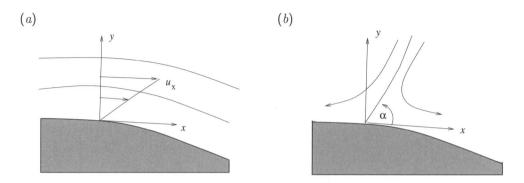

FIGURE 3.9.3 Illustration of (a) a two-dimensional flow near a solid wall and (b) a stagnation point on a solid wall.

point, both must be zero at the stagnation point. Conversely, the vanishing of the shear stress or vorticity provides us with a criterion for the occurrence of a stagnation point.

To demonstrate further that the shear stress is zero at a stagnation point by considering a point that lies on the dividing streamline near the stagnation point, at the position $(\mathrm{d}x, \mathrm{d}y)$. Setting the position vector, $(\mathrm{d}x, \mathrm{d}y)$, parallel to the velocity at that point, and expressing the velocity in a Taylor series about the stagnation point, we find that

$$\tan\alpha = \frac{\mathrm{d}y}{\mathrm{d}x} = \frac{u_y}{u_x} = \frac{\dfrac{\partial u_y}{\partial x}\,\mathrm{d}x + \dfrac{\partial u_y}{\partial y}\,\mathrm{d}y}{\dfrac{\partial u_x}{\partial x}\,\mathrm{d}x + \dfrac{\partial u_x}{\partial y}\,\mathrm{d}y}, \tag{3.9.23}$$

where α is the angle that the dividing streamline forms with the x axis, and all partial derivatives are evaluated at the stagnation point. Using the continuity equation to write $\partial u_y/\partial y = -\partial u_x/\partial x$ and enforcing the no-slip boundary condition, we find that all partial derivatives $\partial u_y/\partial y$, $\partial u_x/\partial x$, $\partial u_y/\partial x$, at the origin are zero. Since the slope of the dividing streamline is finite, the partial derivative $\partial u_x/\partial y$ must also be zero, yielding the condition of vanishing vorticity and wall shear stress at a stagnation point.

Now assuming that the wall is flat, we seek to predict the slope of the dividing streamline in terms of the structure of the velocity. Since the first-order terms in the fraction in (3.9.23) vanish, we must retain the second-order contributions, obtaining

$$\tan\alpha = \frac{\mathrm{d}y}{\mathrm{d}x} = \frac{u_y}{u_x} = \frac{\dfrac{\partial^2 u_y}{\partial x^2}\,(\mathrm{d}x)^2 + 2\dfrac{\partial^2 u_y}{\partial x \partial y}\,(\mathrm{d}x)(\mathrm{d}y) + \dfrac{\partial^2 u_y}{\partial y^2}\,(\mathrm{d}y)^2}{\dfrac{\partial^2 u_x}{\partial x^2}\,(\mathrm{d}x)^2 + 2\dfrac{\partial^2 u_x}{\partial x \partial y}\,(\mathrm{d}x)(\mathrm{d}y) + \dfrac{\partial^2 u_x}{\partial y^2}\,(\mathrm{d}y)^2} \tag{3.9.24}$$

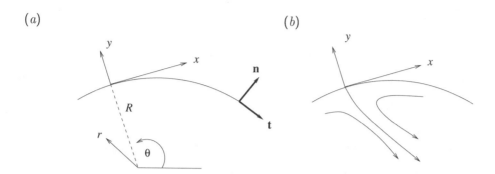

FIGURE 3.9.4   (*a*) Illustration of a two-dimensional flow under a free surface. The origin of the plane polar coordinates, (r, θ), is set at the center of curvature at a point on the surface. (*b*) Illustration of a stagnation point at a stationary free surface; the dividing streamline crosses the free surface at a right angle.

or

$$\tan \alpha = \frac{\mathrm{d}y}{\mathrm{d}x} = \frac{u_y}{u_x} = \frac{\dfrac{\partial^2 u_y}{\partial x^2} + 2 \dfrac{\partial^2 u_y}{\partial x \partial y} \tan \alpha + \dfrac{\partial^2 u_y}{\partial y^2} \tan^2 \alpha}{\dfrac{\partial^2 u_x}{\partial x^2} + 2 \dfrac{\partial^2 u_x}{\partial x \partial y} \tan \alpha + \dfrac{\partial^2 u_x}{\partial y^2} \tan^2 \alpha}. \tag{3.9.25}$$

Enforcing the no-slip boundary condition and using the continuity equation, we find that

$$\frac{\partial^2 u_y}{\partial x^2} = 0, \qquad \frac{\partial^2 u_y}{\partial x \partial y} = -\frac{\partial^2 u_x}{\partial x^2} = 0. \tag{3.9.26}$$

Writing $\partial^2 u_y / \partial y^2 = -\partial^2 u_x / \partial x \partial y$ and rearranging (3.9.25), we obtain

$$\tan \alpha = -3 \left(\frac{\partial^2 u_x}{\partial x \partial y} \right) \bigg/ \left(\frac{\partial^2 u_x}{\partial y^2} \right) = -3 \left(\frac{\partial f_x}{\partial x} \right) \bigg/ \left(\frac{\partial p}{\partial x} \right), \tag{3.9.27}$$

where f_x is the boundary traction, p is the hydrodynamic pressure, and all variables are evaluated at the stagnation point [289]. To derive the expression in the denominator on the right-hand side of (3.9.27), we have applied the Navier–Stokes equation at the stagnation point and used the no-slip boundary condition to set $\mu \, \partial^2 u_x / \partial y^2 = \partial p / \partial x$, where p is the hydrodynamic pressure. Equation (3.9.27) applies also at a stagnation point on curved wall, provided that the derivatives with respect to x are replaced by derivatives with respect to arc length, l [248].

Vorticity at an evolving free surface

Next, we consider the flow below an evolving two-dimensional free surface and introduce Cartesian coordinates where the x axis is tangential and the y axis is perpendicular to the free surface at a point, as shown in Figure 3.9.4(*a*). The z vorticity component and shear stress at the free surface are

$$\omega_z = \frac{\partial u_y}{\partial x} - \frac{\partial u_x}{\partial y}, \qquad \sigma_{xy} = \mu \left(\frac{\partial u_y}{\partial x} + \frac{\partial u_x}{\partial y} \right). \tag{3.9.28}$$

Setting the shear stress to zero, as required by definition for a free surface, we obtain

$$\omega_z = 2 \frac{\partial u_y}{\partial x} = 2\mathbf{t} \cdot (\nabla \mathbf{u}) \cdot \mathbf{n} = 2\mathbf{t} \cdot \nabla(\mathbf{u} \cdot \mathbf{n}) - 2\mathbf{t} \cdot (\nabla \mathbf{n}) \cdot \mathbf{u}, \tag{3.9.29}$$

where $\mathbf{t}$ is the tangent unit vector. Using the Frenet–Serret relation $\mathbf{t} \cdot \nabla \mathbf{n} = \partial \mathbf{n}/\partial l = \kappa \, \mathbf{t}$, we write

$$\omega_z = 2\mathbf{t} \cdot \nabla(\mathbf{u} \cdot \mathbf{n}) - 2\kappa \mathbf{u} \cdot \mathbf{t}, \tag{3.9.30}$$

where l is the arc length along the free surface and κ is the curvature of the free surface. In terms of the stream function, ψ, we obtain

$$\omega_z = -2 \frac{\partial^2 \psi}{\partial l^2} - 2\kappa \frac{\partial \psi}{\partial l_n}, \tag{3.9.31}$$

where l_n is the arc length in the direction of the normal unit vector, $\mathbf{n}$.

Vorticity at a stationary free surface

Since the free surface of a steady flow is also a streamline, we may use (1.11.9) to simplify further the right-hand side of the expression for the shear stress given in (3.9.28), obtaining

$$\sigma_{xy} = -\mu \kappa \mathbf{u} \cdot \mathbf{t} + \mu \frac{\partial u_x}{\partial y}. \tag{3.9.32}$$

Introducing plane polar coordinates with origin at the center of curvature of the free surface at an arbitrary point, (r, θ), as shown in Figure 3.9.4(a), we find that

$$\sigma_{xy} = \mu \frac{u_\theta}{R} - \mu \frac{\partial u_\theta}{\partial r} = -\mu R \left[\frac{\partial}{\partial r} \left(\frac{u_\theta}{r} \right) \right]_{r=R}, \tag{3.9.33}$$

where $R = 1/\kappa$ is the radius of curvature and all expressions are evaluated at $r = R$. Setting the shear stress to zero and combining (3.9.32) with (1.11.10), we derive a simple expression for the vorticity over a stationary free surface in terms of the tangential velocity and curvature of the free surface,

$$\omega_z = -2\kappa \, \mathbf{u} \cdot \mathbf{t}. \tag{3.9.34}$$

This expression demonstrates that the vorticity is zero at a stagnation point along a free-surface where the velocity is zero, at an inflection point on a free-surface where the curvature is zero, and over a planar free surface.

It is reassuring to observe that (3.9.34) also arises from the general relation (3.9.20) for three-dimensional flow by setting $\partial n_x/\partial z = 0$ to obtain $\boldsymbol{\omega}_T = -2u_x \kappa_x \mathbf{e}_z$.

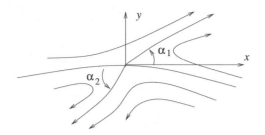

FIGURE 3.9.5   Illustration of a stagnation point at a stationary fluid interface. When the shear stress is continuous across the interface, the slopes of the dividing streamlines obey a refraction law.

Stagnation point at a stationary free surface

A stagnation-point flow at a free surface is illustrated in Figure 3.9.4(b). In terms of the local coordinates x and y, the slope of the dividing streamline at the stagnation point is given by (3.9.23), where all partial derivatives are evaluated at the stagnation point. Taking into account the first equation in (1.11.9) and (3.9.32), we find that $\partial u_x/\partial y$ and $\partial u_y/\partial x$ are zero at the stagnation point. The continuity equation requires that $\partial u_x/\partial x = -\partial u_y/\partial y$, which shows that $\tan \alpha$ must either be zero, in which case we obtain a degenerate stagnation point, or negative infinity. We conclude that the dividing streamline must meet a free surface at a right angle.

Stagnation point at a stationary interface

A stagnation point at the interface between two viscous fluids labeled 1 and 2 is illustrated in Figure 3.9.5. Applying (3.9.23) at a point at the dividing streamline in the upper fluid and using (1.11.9) to set $\partial u_y/\partial x = 0$ and the continuity equation to write $\partial u_y/\partial y = -\partial u_x/\partial x$, we obtain

$$\tan \alpha_1 = -2 \frac{\partial u_x^{(1)}/\partial x}{\partial u_x^{(1)}/\partial y}. \tag{3.9.35}$$

Dividing corresponding sides of (3.9.35) with their counterparts for the second fluid and noting that, because the velocity is continuous across the interface, the derivative $\partial u_x/\partial x$ is shared by the two fluids, we obtain

$$\frac{\tan \alpha_1}{\tan \alpha_2} = \frac{\partial u_x^{(2)}/\partial y}{\partial u_x^{(1)}/\partial y}. \tag{3.9.36}$$

Now using (3.9.32) and requiring that the velocity is continuous across the interface, we find that the jump in the tangential component of the traction across the interface at the stagnation point is given by

$$(\mathbf{f} \cdot \mathbf{t})^{(1)} - (\mathbf{f} \cdot \mathbf{t})^{(2)} = \mu_1 \frac{\partial u_x^{(1)}}{\partial y} - \mu_2 \frac{\partial u_x^{(2)}}{\partial y}. \tag{3.9.37}$$

Assuming that the shear stress is continuous across the interface, we set the left-hand side of (3.9.37) to zero and combine the resulting equation with (3.9.36) to derive the remarkably simple formula

$$\frac{\tan \alpha_1}{\tan \alpha_2} = \frac{\mu_1}{\mu_2}, \tag{3.9.38}$$

which can be regarded as a refraction law for the dividing streamline [248]. When the viscosities of the two fluids are matched, the dividing streamlines join smoothly at the stagnation point.

Problems

3.9.1 *Force on a boundary*

Draw a sequence of surfaces enclosing a stationary rigid body in a flow and uniformly tending to the surface of the body. Denoting the typical separation between a surface and the body by ϵ, we compute the force $\mathbf{F}(\epsilon)$ exerted on the surface and plot the magnitude of $\mathbf{F}$ against ϵ. Do you expect that the graph will show a sharp variation as ϵ tends to zero?

3.9.2 *Force on a boundary*

Derive (3.9.33) working in the plane polar coordinates illustrated in Figure 3.9.4(*a*).

3.9.3 *Traction normal to a line in a two-dimensional flow*

(*a*) Show that the component of the traction normal to an arbitrary line in a two-dimensional flow is given by

$$\mathbf{f} \cdot \mathbf{n} = -p + 2\mu \frac{\partial(\mathbf{u} \cdot \mathbf{t})}{\partial l} - 2\mu \kappa \mathbf{u} \cdot \mathbf{n} = -p - 2\mu \frac{\partial^2 \psi}{\partial l \partial l_n} + 2\mu \kappa \frac{\partial \psi}{\partial l}, \tag{3.9.39}$$

where l is the arc length along the line, $\mathbf{t}$ is the tangent unit vector, κ is the curvature of the line, l_n is the arc length measured in the direction of the normal vector, $\mathbf{n}$, and ψ is the stream function.

(*b*) Use equation (3.9.39) to derive a boundary condition for the pressure at a free surface in the presence of surface tension.

(*c*) Use equation (3.9.39) to derive a boundary condition for the jump in pressure across a fluid interface in the presence of surface tension.

3.10   Scaling of the Navier–Stokes equation and dynamic similitude

Evaluating the various terms of the Navier–Stokes equation at a point in a flow, we may find a broad range of magnitudes. Depending on the structure of the flow and location in the flow, some terms may make dominant contributions and should be retained, while other terms may make minor contributions and could be neglected without compromising the integrity of the simplified description. To identify this opportunity and benefit from concomitant simplifications, we inspect the structure the a flow and find that the magnitude of the velocity typically changes by an amount U over a distance L. This means that the magnitudes of the first or second spatial derivatives of the velocity are comparable, respectively, to the magnitudes of the ratios U/L and U/L^2. Furthermore, we typically find that the magnitude of the velocity at a particular point in the flow changes by U

over a time period T, which means that the magnitude of the first partial derivative of the velocity with respect to time is comparable to the ratio U/T.

Characteristic scales

Typically, but not always, the characteristic length L is related to the size of the boundaries, the characteristic velocity U is determined by the particular mechanism driving the flow, and the characteristic time T is either imposed by external means or simply defined as $T = L/U$. In the case of unidirectional flow through a channel or tube, U can be identified with the maximum velocity across the channel or tube. In the case of uniform flow past a stationary body, U can be identified with the velocity of the incident flow. In the case of forced oscillatory flow, T can be identified with the period of oscillation.

Nondimensionalization

Next, we scale all terms in the Navier–Stokes equation using the aforementioned scales and compare their relative magnitudes. To accomplish the first task, we introduce the dimensionless variables

$$\hat{\mathbf{u}} \equiv \frac{\mathbf{u}}{U}, \qquad \hat{\mathbf{x}} \equiv \frac{\mathbf{x}}{L}, \qquad \hat{t} \equiv \frac{t}{T}, \qquad \hat{p} \equiv \frac{pL}{\mu U}. \qquad (3.10.1)$$

Expressing the dimensional variables in terms of corresponding dimensionless variables and substituting the result into the Navier–Stokes equation, we obtain the dimensionless form

$$\beta \frac{\partial \hat{\mathbf{u}}}{\partial \hat{t}} + \mathrm{Re}\, \hat{\mathbf{u}} \cdot \widehat{\nabla} \hat{\mathbf{u}} = -\widehat{\nabla} \hat{p} + \widehat{\nabla}^2 \hat{\mathbf{u}} + \frac{\mathrm{Re}}{\mathrm{Fr}^2} \frac{\mathbf{g}}{g}, \qquad (3.10.2)$$

where the gradient $\widehat{\nabla}$ involves derivatives with respect to the dimensionless position vector, $\hat{\mathbf{x}}$, and $g = |\mathbf{g}|$. Since the magnitudes of the dimensionless variables and their derivatives in (3.10.2) are of order unity, and since $\mathbf{g}/g$ is a unit vector expressing the direction of the body force, the relative importance of the various terms is determined by the magnitude of their multiplicative factors, which are the frequency parameter, β, the Reynolds number, Re, and the Froude number, Fr, defined as

$$\beta \equiv \frac{L^2}{\nu T}, \qquad \mathrm{Re} \equiv \frac{UL}{\nu}, \qquad \mathrm{Fr} \equiv \frac{U}{\sqrt{gL}}, \qquad (3.10.3)$$

where ν is the kinematic viscosity of the fluid with dimensions of length squared divided by time.

The frequency parameter, β, expresses the relative magnitudes of the inertial acceleration force and the viscous force, or equivalently, the ratio of the characteristic diffusion time, L^2/ν, to the time scale of the flow, T. The Reynolds number expresses the relative magnitudes of the inertia convective force and the viscous force, or equivalently, the ratio of the characteristic diffusion time, L^2/ν, to the convective time, L/U. The Froude number expresses the relative magnitudes of the inertial convective force and the body force. The group

$$\frac{\mathrm{Re}}{\mathrm{Fr}} = \frac{gL^2}{\nu U} \qquad (3.10.4)$$

expresses the relative magnitudes of the body force and the viscous force.

In the absence of external forcing, T, is identified with the convective time scale, L/U, β reduces to Re, and the dimensionless Navier–Stokes equation (3.10.2) involves only two independent parameters, Re and Fr.

3.10.1 Steady, quasi-steady, and unsteady Stokes flow

When Re $\ll 1$ and $\beta \ll 1$, both terms on the left-hand side of (3.10.2) are small compared to the terms on the right-hand side and can be neglected. Reverting to dimensional variables, we find that the rate of change of momentum of fluid parcels $\rho D\mathbf{u}/Dt$ is negligible and the flow is governed by the Stokes equation

$$-\nabla p + \mu \nabla^2 \mathbf{u} + \rho \mathbf{g} = \mathbf{0}, \qquad (3.10.5)$$

stating that pressure, viscous, and body forces balance at every instant. A flow that is governed by the Stokes equation is called a Stokes or creeping flow.

The absence of a temporal derivative in the equation of motion does not necessarily mean that that the flow is steady, but only implies that the forces exerted on fluid parcels are at equilibrium. Consequently, the instantaneous structure of the flow depends only on the present boundary configuration and boundary conditions, which means that the flow is in a quasi-steady state. Stated differently, the history of motion enters the physical description only insofar as to determine the current boundary configuration.

Steady and quasi-steady Stokes flows are encountered in a variety of natural and engineering applications. Examples include slurry transport, blood flow in capillary vessels, flow due to the motion of ciliated micro-organisms, flow past microscopic aerosol particles, flow due to the coalescence of liquid drops, and flow in the mantle of the earth due to natural convection. In certain cases, the Reynolds number is low due to small boundary dimensions as in the case of flow past a red blood cell, which has an average diameter of 8μm. In other applications, the Reynolds number is low due to the small magnitude of the velocity or high kinematic viscosity. An example is the flow due to the motion of an air bubble in a very viscous liquid such as honey or glycerin.

The linearity of the Stokes equation allows us to conduct extensive theoretical studies, analyze the properties of the flow, and generate desired solutions by linear superposition using a variety of analytical and computational methods. An extensive discussion of the properties and methods of computing Stokes flow will be presented in Chapter 6.

Unsteady Stokes flow

When Re $\ll 1$ but $\beta \sim 1$, the nonlinear inertial convective term on the left-hand side of (3.10.2) is negligible compared to the rest of the terms and can be discarded. Reverting to dimensional variables, we find that the nonlinear component of the point-particle acceleration, $\mathbf{u} \cdot \nabla \mathbf{u}$, is insignificant, and the rate of change of momentum of a point particle can be approximated with the Eulerian inertial acceleration force, $\rho \, \partial \mathbf{u}/\partial t$. The motion of the fluid is governed by the unsteady Stokes equation, also called the linearized Navier–Stokes equation,

$$\rho \frac{\partial \mathbf{u}}{\partial t} = -\nabla p + \mu \nabla^2 \mathbf{u} + \rho \mathbf{g}. \qquad (3.10.6)$$

Because of the presence of the acceleration term involving a time derivative on the left-hand side, the instantaneous structure of the flow depends not only on the instantaneous boundary configuration and boundary conditions, but also on the history of fluid motion.

Physically, the unsteady Stokes equation describes flows characterized by sudden acceleration or deceleration. Three examples are the flow occurring in hydrodynamic braking, the flow occurring during the impact of a particle on a solid surface, and the initial stages of the flow due to a particle settling from rest in an ambient fluid. The linearity of the unsteady Stokes equation allows us to compute solutions for oscillatory and general time-dependent motion using a variety of methods, including Fourier and Laplace transforms in time and space, as well as construct solutions by linear superposition. The properties and methods of computing unsteady Stokes flow are discussed in Sections 6.15–6.18.

3.10.2   Flow at high Reynolds numbers

It may appear that, when $\mathrm{Re} \gg 1$ and $\beta \gg 1$, both the pressure and viscous terms on the right-hand side of the equation of motion (3.10.2) are negligible compared to the terms on the left-hand side. While this may be true in certain cases, because of the subjective scaling of the pressure, the magnitude of the dimensionless pressure gradient may not remain of order unity in the limit of vanishing Re and β.

To allow for this possibility, we rescale the pressure by introducing a new dimensionless pressure defined as

$$\hat{\pi} \equiv \frac{p}{\rho U^2} = \frac{1}{\mathrm{Re}}\,\hat{p}. \tag{3.10.7}$$

Working as previously, we obtain a new dimensionless form of the Navier–Stokes equation,

$$\beta \frac{\partial \hat{\mathbf{u}}}{\partial \hat{t}} + \mathrm{Re}\,\hat{\mathbf{u}} \cdot \hat{\nabla}\hat{\mathbf{u}} = -\mathrm{Re}\,\hat{\nabla}\hat{\pi} + \hat{\nabla}^2 \hat{\mathbf{u}} + \frac{\mathrm{Re}}{\mathrm{Fr}^2}\frac{\mathbf{g}}{|\mathbf{g}|}. \tag{3.10.8}$$

Now considering the limits $\mathrm{Re} \gg 1$ and $\beta \gg 1$, we find that the viscous term becomes small compared to the rest of the terms and can be neglected, yielding Euler's equation,

$$\frac{\beta}{\mathrm{Re}}\frac{\partial \hat{\mathbf{u}}}{\partial \hat{t}} + \hat{\mathbf{u}} \cdot \hat{\nabla}\hat{\mathbf{u}} = -\hat{\nabla}\hat{\pi} + \frac{1}{\mathrm{Fr}^2}\frac{\mathbf{g}}{|\mathbf{g}|}, \tag{3.10.9}$$

where $\beta/\mathrm{Re} = L/(TU)$. We have found that, at high Reynolds numbers, a laminar flow behaves like an inviscid flow in the absence of small-scale turbulent motion. By neglecting the viscous term, we have lowered the order of the Navier–Stokes equation from two to one, by one unit. This reduction has important implications on the number of required boundary conditions, spatial structure of the flow, and properties of the solution.

3.10.3   Dynamic similitude

Let us consider a steady or unsteady flow in an domain of finite or infinite extent, specify suitable boundary conditions, and introduce the dimensionless variables shown in (3.10.1) to obtain the dimensionless Navier–Stokes equation (3.10.2). In terms of the dimensionless variables, the continuity equation becomes

$$\hat{\nabla} \cdot \hat{\mathbf{u}} = 0, \tag{3.10.10}$$

and the boundary conditions can be expressed in the symbolic form

$$\mathcal{F}(\hat{\mathbf{u}}, \hat{\sigma}) = \mathbf{0}, \tag{3.10.11}$$

where $\hat{\sigma}$ is the nondimensional stress tensor. The function $\mathcal{F}$ may contain dimensionless constants defined with respect to the physical properties of the fluid, the characteristic scales U, L, and T, the body force, and other physical constants that depend on the nature of the boundary conditions, such as the surface tension, γ, and the slip coefficient. Three dimensionless numbers pertinent to an interface between two immiscible fluids are the Bond number (Bo), the Weber number (We), and the capillary number (Ca), defined as

$$\mathrm{Bo} = \frac{\rho g L^2}{\gamma}, \qquad \mathrm{We} = \frac{\rho L U^2}{\gamma}. \qquad \mathrm{Ca} = \frac{\mu U}{\gamma}. \tag{3.10.12}$$

The Bond number expresses the significance of the gravitational force relative to surface tension. The Weber number expresses the significance of the inertial force relative to surface tension. The capillary number expresses the significance of the viscous stresses relative to surface tension.

In the space of dimensionless variables, the flow is governed by the equation of motion (3.10.2) and the continuity equation (3.10.10), and the solution satisfies the conditions stated in (3.10.11). It is evident then that the structure of a steady flow and the evolution of an unsteady flow depend on (a) the values of the dimensionless numbers β, Re, and Fr, (b) the functional form of the boundary conditions stated in (3.10.11), and (c) the values of the dimensionless numbers involved in (3.10.11). Two flows in two different physical domains will have a similar structure provided that, in the space of corresponding dimensionless variables, the boundary geometry, initial state, and boundary conditions are the same. Similarity of structure means that the velocity or pressure field of the first flow can be deduced from those of the second flow, and *vice versa*, by multiplication with an appropriate factor.

Dynamic similitude can be exploited to study the flow of a particular fluid in a certain domain by studying the flow of another fluid in a similar, larger or smaller, domain. This is achieved by adjusting the properties of the second fluid, flow domain, or both, to match the values of β, Re, Fr, and any other dimensionless numbers entering the boundary conditions. Miniaturization or scale-up of a domain of flow is important in the study of large-scale flows, such as flow past aircraft, and small-scale flows, such as flow past microorganisms and small biological cells and flow over surfaces with small-scale roughness and imperfections.

Problems

3.10.1 *Scaling for a translating body in a moving fluid*

A rigid body translates steadily with velocity $\mathbf{V}$ in an infinite fluid that moves with uniform velocity $\mathbf{U}$. Discuss the scaling of the various terms in the Navier–Stokes equation and derive the appropriate Reynolds number.

3.10.2 *Dimensionless form of interfacial boundary conditions*

(*a*) Express the interfacial boundary conditions (3.8.8) in dimensionless form and identify an appropriate dimensionless number.

(*b*) Develop a relation between the Weber number, the Reynolds number, and the capillary number.

3.10.3 *Walking and running*

Compute the Reynolds number of a walking or running adult person. Repeat for an ant and an elephant.

3.10.4 *Flow past a sphere*

We want to study the structure of uniform (streaming) flow of water with velocity $U = 40$ km/hr past a stationary sphere with diameter $D = 0.5$ cm. For this purpose, we propose to study uniform flow of air past another sphere with larger diameter, $D = 10$ cm. What is the appropriate air speed?

3.11 Evolution of circulation around material loops and dynamics of the vorticity field

Previously in this chapter, we discussed the structure and dynamics of an incompressible Newtonian flow with reference to the equation of motion. Useful insights into the physical mechanisms governing the motion of the fluid can be obtained by studying the evolution of the circulation around material loops and the dynamics of the vorticity field. The latter illustrates the physical processes contributing to the rate of change of the angular velocity of small fluid parcels. We will see that the rate of change of the vorticity obeys a set of rules that are amenable to appealing and illuminating physical interpretations.

3.11.1 Evolution of circulation around material loops

To compute the rate of change of the circulation around a closed reducible or irreducible material loop of a Newtonian fluid, L, we combine (1.12.10) with the Navier–Stokes equation (3.5.5), assume that the viscosity of the fluid and the acceleration of gravity are uniform throughout the domain of flow, and obtain

$$\frac{\mathrm{d}C}{\mathrm{d}t} = \oint_L \left(-\frac{1}{\rho}\nabla p + \nu\nabla^2\mathbf{u} + \mathbf{g} \right) \cdot \mathrm{d}\mathbf{X}, \qquad (3.11.1)$$

where $\mathbf{X}$ is the position of a point particle along the loop. Since the integral of the derivative of a function is equal to the function itself, the integral of the last term on the right-hand side of (3.11.1) makes a vanishing contribution. A similar reasoning reveals that, when the density of the fluid is uniform or the fluid is barotropic, the integral of the pressure term is also zero. Setting the Laplacian of the velocity equal to the negative of the curl of the vorticity, we obtain the simplified form

$$\frac{\mathrm{d}C}{\mathrm{d}t} = \nu \oint_L (\nabla^2\mathbf{u}) \cdot \mathrm{d}\mathbf{X} = -\nu \oint_L (\nabla \times \boldsymbol{\omega}) \cdot \mathrm{d}\mathbf{X}. \qquad (3.11.2)$$

If the vorticity field is irrotational, $\nabla \times \boldsymbol{\omega} = \mathbf{0}$, the integral on the right-hand vanishes and the rate of change of the circulation is zero.

3.11.2 Kelvin circulation theorem

When viscous forces are insignificant, we obtain

$$\frac{\mathrm{d}C}{\mathrm{d}t} = \frac{\mathrm{d}}{\mathrm{d}t} \iint_D \boldsymbol{\omega} \cdot \mathbf{n}\,\mathrm{d}S = 0, \qquad (3.11.3)$$

where D is an arbitrary surface bounded by the loop and $\mathbf{n}$ is the unit vector normal to D. Equation (3.11.3) expresses Kelvin's circulation theorem, stating that the circulation around a closed material loop in a flow with negligible viscous forces is preserved during the motion. As a consequence, the circulation around a loop that wraps around a toroidal boundary, and thus the cyclic constant of the motion around the boundary, remain constant in time.

Now we consider a reducible loop, take the limit as the loop shrinks to a point, and find that, when the conditions for Kelvin's circulation theorem to apply are fulfilled, point particles maintain their initial vorticity, which means that they keep spinning at a constant angular velocity as they translate and deform in the domain of flow. Physically, the absence of a viscous torque exerted on a fluid parcel guarantees that the angular momentum of the parcel is preserved during the motion.

3.11.3 Helmholtz theorems

One consequence of Kelvin's circulation theorem is Helmholtz' first theorem, stating that, if the vorticity vanishes at the initial instant, it must vanish at all subsequent times. This behavior is sometimes described as permanence of irrotational flow. Another consequence of Kelvin's circulation theorem is Helmholtz' second theorem, stating that vortex tubes behave like material surfaces maintaining their initial circulation. Thus, line vortices are material lines consisting of a permanent collection of point particles.

3.11.4 Dynamics of the vorticity field

The Lagrangian form of the equation of motion provides us with the acceleration of point particles, while the Eulerian form of the equation of motion provides us with the rate of change of the velocity at a point in a flow in terms of the instantaneous velocity and pressure fields. To derive corresponding expressions for the angular velocity of point particles and rate of change of the vorticity at a point in a flow, we take the curl of the Navier–Stokes equation (3.5.5). Using the identity

$$\nabla \times (\mathbf{u} \cdot \nabla \mathbf{u}) = \mathbf{u} \cdot \nabla \boldsymbol{\omega} - \boldsymbol{\omega} \cdot \nabla \mathbf{u}, \qquad (3.11.4)$$

and assuming that the acceleration of gravity is constant throughout the domain of flow, we derive the vorticity transport equation,

$$\frac{D\boldsymbol{\omega}}{Dt} \equiv \frac{\partial \boldsymbol{\omega}}{\partial t} + \mathbf{u} \cdot \nabla \boldsymbol{\omega} = \boldsymbol{\omega} \cdot \nabla \mathbf{u} + \frac{1}{\rho^2}\nabla \rho \times \nabla p + \nu \nabla^2 \boldsymbol{\omega} + \nabla \nu \times \nabla^2 \mathbf{u} + 2\nabla \times \left(\frac{1}{\rho}\nabla \mu \cdot \mathbf{E}\right), \quad (3.11.5)$$

where $\mathbf{E}$ is the rate-of-deformation tensor. Projecting both sides of (3.11.5) onto half the moment of inertia of a small fluid parcel provides us with an angular momentum balance.

Now restricting our attention to fluids whose viscosity is uniform throughout the domain of flow, we obtain the simplified form

$$\frac{D\boldsymbol{\omega}}{Dt} = \frac{\partial \boldsymbol{\omega}}{\partial t} + \mathbf{u} \cdot \nabla \boldsymbol{\omega} = \boldsymbol{\omega} \cdot \nabla \mathbf{u} + \frac{1}{\rho^2} \nabla \rho \times \nabla p + \nu \nabla^2 \boldsymbol{\omega}. \tag{3.11.6}$$

We could solve the Navier–Stokes for the pressure gradient and substitute the result into the right-hand side of (3.11.6) to derive an expression in terms of the velocity and vorticity, but this is necessary neither in theoretical analysis nor in numerical computation. The left-hand side of (3.11.6) expresses the material derivative of the vorticity, which is equal to the rate of change of the vorticity following a point particle, or half the rate of change of the angular velocity of the point particle.

Reorientation and vortex stretching

The first term on the right-hand side of (3.11.6), $\boldsymbol{\omega} \cdot \nabla \mathbf{u}$, expresses generation of vorticity due to the interaction between the vorticity field and the velocity gradient tensor, $\mathbf{L} = \nabla \mathbf{u}$. Comparing (3.11.6) with (1.6.1), we find that, under the action of this term, the vorticity vector evolves as a material vector, rotating with the fluid while being stretched under the action of the flow. Thus, generation of a Cartesian vorticity component in a particular direction is due to both reorientation of the vortex lines under the action of the local flow, and compression or stretching of the vorticity vector due to the straining component of the local flow. The second mechanism is known as vortex stretching.

To illustrate further the nature of the nonlinear term, $\boldsymbol{\omega} \cdot \nabla \mathbf{u}$, we begin with the statement $\boldsymbol{\omega} \times \boldsymbol{\omega} = \mathbf{0}$ and use a vector identity to write

$$\boldsymbol{\omega} \times \boldsymbol{\omega} = (\nabla \times \mathbf{u}) \times \boldsymbol{\omega} = \boldsymbol{\omega} \cdot \nabla \mathbf{u} - \boldsymbol{\omega} \cdot (\nabla \mathbf{u})^T = \mathbf{0}, \tag{3.11.7}$$

where the superscript T denotes the matrix transpose. Consequently,

$$\boldsymbol{\omega} \cdot \nabla \mathbf{u} = \boldsymbol{\omega} \cdot (\nabla \mathbf{u})^T = (\nabla \mathbf{u}) \cdot \boldsymbol{\omega}. \tag{3.11.8}$$

This identity allows us to write

$$\omega_j \frac{\partial u_i}{\partial x_j} = \omega_j \frac{\partial u_j}{\partial x_i} = \omega_j \, E_{ji} = \omega_j \left(\beta \, \frac{\partial u_i}{\partial x_j} + (1 - \beta) \, \frac{\partial u_j}{\partial x_i} \right), \tag{3.11.9}$$

where β is an arbitrary constant. If the vorticity vector happens to be an eigenvector of the rate-of-deformation tensor, $\mathbf{E}$, it will amplify or shrink in its direction, behaving like a material vector.

Baroclinic production

The second term on the right-hand side of (3.11.6), $\frac{1}{\rho^2} \nabla \rho \times \nabla p$, expresses baroclinic generation of vorticity due to the interaction between the pressure and density fields. To illustrate the physical process underlying this coupling, we consider a vertically stratified fluid that is set in motion by the application of a horizontal pressure gradient. The heavier point particles at the top accelerate slower than the lighter fluid particles at the bottom. Consequently, a vertical fluid column buckles in the counterclockwise direction, thus generating vorticity with positive sign. When the density of the fluid is uniform or the pressure is a function of the density alone, the baroclinic production term vanishes. In the second case, the gradients of the pressure and density are parallel and their cross product is identically zero.

Diffusion

The third term on the right-hand side of (3.11.6), $\nu\nabla^2\boldsymbol{\omega}$, expresses diffusion of vorticity with diffusivity that is equal to the kinematic viscosity of the fluid, ν. Under the action of this term, the Cartesian components of the vorticity vector diffuse like passive scalars.

Eulerian form

An alternative form of the vorticity transport equation (3.11.6) emerges by restating the nonlinear term in the equation of motion in terms of the right-hand side of (3.5.9) before taking the curl. Noting that the curl of the gradient of a continuous function is identically zero, we derive the Eulerian form of the vorticity transport equation,

$$\frac{\partial\boldsymbol{\omega}}{\partial t} + \nabla\times(\boldsymbol{\omega}\times\mathbf{u}) = \frac{1}{\rho^2}\nabla\rho\times\nabla p + \nu\nabla^2\boldsymbol{\omega}. \tag{3.11.10}$$

The second term on the left-hand side incorporates the effects of convection, reorientation, and vortex stretching.

3.11.5   Production of vorticity at an interface

In Section 3.8.6, we interpreted an interface between two immiscible fluids as a singular surface of distributed force. Dividing (3.8.26) by the density and taking the curl of the resulting equation, we find that the associated rate of production of vorticity is

$$\nabla\times\left(\frac{1}{\rho}\mathbf{q}\right) = -\frac{1}{\rho^2}\nabla\rho\times\mathbf{q} + \frac{1}{\rho}\nabla\times\mathbf{q}, \tag{3.11.11}$$

where $\mathbf{q}$ is the singular forcing function defined in (3.8.27). The first term of the right-hand side of (3.11.11) can be combined with the baroclinic production term in the vorticity transport equation. If the discontinuity in the surface force is normal to an interface, the vectors $\nabla\rho$ and $\mathbf{q}$ are aligned and the interface does not make a contribution to the baroclinic production.

Concentrating on the second term on the right-hand side of (3.11.11), we define $\mathbf{w} = \nabla\times\mathbf{q}$ and take the curl of (3.8.27) to find

$$w_i = -\epsilon_{ijk}\iint_I \frac{\partial\delta_3(\mathbf{x}-\mathbf{x}')}{\partial x_j}\,\Delta f_k(\mathbf{x}')\,\mathrm{d}S(\mathbf{x}'), \tag{3.11.12}$$

where I denotes the interface. In vector notation,

$$\mathbf{w} = -\iint_I \nabla\delta_3(\mathbf{x}-\mathbf{x}')\times\Delta\mathbf{f}(\mathbf{x}')\,\mathrm{d}S(\mathbf{x}') = \iint_I \nabla'\delta_3(\mathbf{x}-\mathbf{x}')\times\Delta\mathbf{f}(\mathbf{x}')\,\mathrm{d}S(\mathbf{x}'), \tag{3.11.13}$$

where a prime over the gradient designates differentiation with respect to the integration point, $\mathbf{x}'$. The last expression shows that the second term on the right-hand side of (3.11.11) generates a sheet of vortex dipoles causing a discontinuity in the velocity gradient across the interface.

Two-dimensional flow

The nature of the last term on the right-hand side of (3.11.11) can be illustrated best by considering a two-dimensional flow in the xy plane and computing

$$w_z = \int_I \nabla' \delta_2(\mathbf{x} - \mathbf{x}') \cdot [\Delta \mathbf{f}(\mathbf{x}') \times \mathbf{e}_z] \, dl(\mathbf{x}'), \qquad (3.11.14)$$

where δ_2 is the two-dimensional delta function in the plane of the flow, l is the arc length along the interface, and $\mathbf{e}_z$ is the unit vector along the z axis [261]. Decomposing the jump in the interfacial traction into its normal and tangential components designated by the superscripts N and T, we obtain

$$w_z = \int_I \nabla' \delta_2(\mathbf{x} - \mathbf{x}') \cdot [-\Delta f^T \mathbf{n} + \Delta f^N \mathbf{t}](\mathbf{x}') \, dl(\mathbf{x}'). \qquad (3.11.15)$$

We have assumed that the triplet $(\mathbf{t}, \mathbf{n}, \mathbf{e}_z)$ forms a right-handed coordinates so that $\mathbf{t} \times \mathbf{e}_z = -\mathbf{n}$ and $\mathbf{n} \times \mathbf{e}_z = \mathbf{t}$.

The tangential component inside the integrand can be manipulated further by writing

$$\int_I \Delta f^N(\mathbf{x}') \, \mathbf{t}(\mathbf{x}') \cdot \nabla' \delta_2(\mathbf{x} - \mathbf{x}') \, dl(\mathbf{x}') = \int_I \mathbf{t}(\mathbf{x}') \cdot \nabla'[\Delta f^N(\mathbf{x}') \, \delta_2(\mathbf{x} - \mathbf{x}')] \, dl(\mathbf{x}')$$

$$- \int_I \delta_2(\mathbf{x} - \mathbf{x}') \, \mathbf{t}(\mathbf{x}') \cdot \nabla'[\Delta f^N(\mathbf{x}')] \, dl(\mathbf{x}'). \qquad (3.11.16)$$

When the interface is closed or periodic, the first integral on the right-hand side vanishes and (3.11.15) simplifies into

$$w_z = -\int_I \Delta f^T(\mathbf{x}') \, \mathbf{n}(\mathbf{x}') \cdot \nabla' \delta_2(\mathbf{x} - \mathbf{x}') \, dl(\mathbf{x}') + \int_I \delta_2(\mathbf{x} - \mathbf{x}') \, d\,\Delta f^N(\mathbf{x}'). \qquad (3.11.17)$$

The first term on the right-hand side expresses the generation of a sheet of vortex dipoles due to the tangential component of the discontinuity in the interfacial traction. The second term expresses the generation of a vortex sheet due to the normal component of the discontinuity in the interfacial traction.

As an application, we consider an infinite flat interface parallel to the x axis and assume that the normal component Δf^N is zero along the entire interface. Expression (3.11.17) becomes

$$w_z = -\Delta f^T \int_I \frac{\partial \delta_2(\mathbf{x} - \mathbf{x}')}{\partial y'} \, dx'. \qquad (3.11.18)$$

Problem

3.11.1 *Two-dimensional flow*

Write the counterpart of equation (3.11.5) for the z vorticity component in two-dimensional flow in the xy plane.

3.12 Vorticity transport in a homogeneous or barotropic fluid

In the case of uniform-density fluids or barotropic fluids whose pressure is a function of the density alone, the vorticity transport equation (3.11.6) simplifies into

$$\frac{D\boldsymbol{\omega}}{Dt} \equiv \frac{\partial \boldsymbol{\omega}}{\partial t} + \mathbf{u} \cdot \nabla \boldsymbol{\omega} = \boldsymbol{\omega} \cdot \nabla \mathbf{u} + \nu \nabla^2 \boldsymbol{\omega}. \tag{3.12.1}$$

The associated Eulerian form given in (3.11.10) becomes

$$\frac{\partial \boldsymbol{\omega}}{\partial t} + \nabla \times (\boldsymbol{\omega} \times \mathbf{u}) = \nu \nabla^2 \boldsymbol{\omega}. \tag{3.12.2}$$

Equation (3.12.1) shows that the vorticity at a particular point evolves due to convection, vortex stretching, and viscous diffusion. The vorticity field, and thus the velocity field, is steady only when the combined action balances to zero. Combining (3.12.2) with (1.12.11), we derive Kelvin's circulation theorem for a flow with negligible viscous forces.

Since the vorticity field is solenoidal, $\nabla \cdot \boldsymbol{\omega} = 0$, we may write $\nabla^2 \boldsymbol{\omega} = -\nabla \times (\nabla \times \boldsymbol{\omega})$ and restate (3.12.2) as

$$\nabla \times \left(\frac{\partial \mathbf{u}}{\partial t} + \boldsymbol{\omega} \times \mathbf{u} + \nu \nabla \times \boldsymbol{\omega} \right) = \mathbf{0}, \tag{3.12.3}$$

which ensures that the vector field inside the parentheses is and remains irrotational.

Generalized Beltrami flows

By definition, the nonlinear term on the left-hand side of (3.12.2) vanishes in a generalized Beltrami flow,

$$\nabla \times (\boldsymbol{\omega} \times \mathbf{u}) = \mathbf{0}. \tag{3.12.4}$$

Equation (3.12.2) then reduces into the unsteady heat conduction equation for the Cartesian components of the vorticity,

$$\frac{\partial \boldsymbol{\omega}}{\partial t} = \nu \nabla^2 \boldsymbol{\omega}. \tag{3.12.5}$$

The class of generalized Beltrami flows includes, as a subset, Beltrami flows whose vorticity is aligned with the velocity at every point, $\boldsymbol{\omega} \times \mathbf{u} = \mathbf{0}$.

If $\mathbf{u}$ and $\boldsymbol{\omega}$ are the velocity and vorticity of a generalized Beltrami flow, their negative pair, $-\mathbf{u}$ and $-\boldsymbol{\omega}$, also satisfies (3.12.2), and the reversed flow is also a generalized Beltrami flow (Problem 3.12.9). Since the direction of the velocity along the streamlines is reversed when we change the sign of the velocity, a generalized Beltrami flow is sometimes called a two-way flow.

Flow with insignificant viscous forces

When viscous forces are insignificant, the vorticity transport equation (3.12.1) simplifies into

$$\frac{D\boldsymbol{\omega}}{Dt} = \boldsymbol{\omega} \cdot \nabla \mathbf{u}, \tag{3.12.6}$$

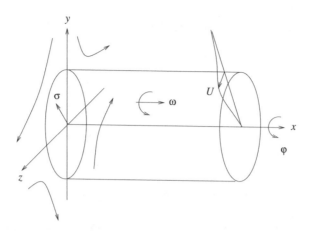

FIGURE 3.12.1   Illustration of the Burgers columnar vortex surviving in the presence of an axisymmetric straining flow.

which is the foundation of Helmholtz' theorems discussed in Section 3.11.3. Referring to (1.6.1), we find that the vorticity vector behaves like a material vector in the following sense: if dl is a small material vector aligned with the vorticity vector at an instant, then $\omega/|\omega| = dl/|dl|$ at all times, where ω is the vorticity at the location of the material vector.

Diffusion of vorticity through boundaries

According to equation (3.12.1), the rate of change of the vorticity vanishes throughout an irrotational flow, and this seemingly suggests that the flow will remain irrotational at all times. However, this erroneous deduction ignores the possibility that vorticity may enter the flow through the boundaries. The responsible physical mechanism is analogous to that by which heat enters an initially isothermal domain across the boundaries of a conductive medium due to a sudden change in the boundary temperature.

The process by which vorticity enters a flow can be illustrated by considering the flow around an impermeable boundary that is placed suddenly in an incident flow. To satisfy the no-penetration boundary condition, we introduce an irrotational disturbance flow. However, we are still left with a finite slip-velocity amounting to a boundary vortex sheet. Viscosity causes the singular vorticity distribution associated with the vortex sheet to diffuse into the fluid, complementing the disturbance irrotational flow, while the boundary vorticity is continuously adjusting in response to the developing flow. If the fluid is inviscid, vorticity cannot diffuse into the flow, the vortex sheet adheres to the boundary, and the flow remains irrotational.

3.12.1   Burgers columnar vortex

The Burgers columnar vortex provides us with an example of a steady flow where diffusion of vorticity is counterbalanced by convection and stretching [62, 354]. The velocity field arises by superimposing a swirling flow describing a columnar vortex and an irrotational, axisymmetric, uniaxial extensional flow, as shown in Figure 3.12.1. In cylindrical polar coordinates, (x, σ, φ), with the x axis pointing in the direction of the vortex, the velocity components are

$$u_x = \xi x, \qquad u_\sigma = -\frac{1}{2}\xi\sigma, \qquad u_\varphi = U(\sigma), \qquad (3.12.7)$$

where ξ is the rate of extension and $U(\sigma)$ is the *a priori* unknown azimuthal velocity profile. The radial and azimuthal vorticity components are identically zero, $\omega_\sigma = 0$ and $\omega_\varphi = 0$, and the axial vorticity component is given by

$$\omega_x(\sigma) = \frac{1}{\sigma}\frac{\mathrm{d}(\sigma U)}{\mathrm{d}\sigma}. \qquad (3.12.8)$$

Substituting the expressions for the velocity into the axial component of the vorticity transport equation (3.12.2), setting the time derivative to zero to ensure steady state and rearranging, we obtain a linear, homogeneous, second-order ordinary differential equation,

$$\xi\frac{\mathrm{d}(\sigma^2\omega_x)}{\mathrm{d}\sigma} + 2\nu\frac{\mathrm{d}}{\mathrm{d}\sigma}\left(\sigma\frac{\mathrm{d}\omega_x}{\mathrm{d}\sigma}\right) = 0. \qquad (3.12.9)$$

A nontrivial solution that is finite at the axis of revolution is

$$\omega_x = A\xi\exp\left(-\frac{\xi}{4\nu}\sigma^2\right), \qquad (3.12.10)$$

where A is a dimensionless constant. Inspecting the argument of the exponential term, we find that the diameter of the vortex is constant along the axis of revolution, and the size of the vortex is comparable to the viscous length scale $(4\nu/\xi)^{1/2}$.

The azimuthal velocity profile is computed by integrating equation (3.12.8), finding the Gaussian distribution

$$U(\sigma) = 2A\nu\frac{1}{\sigma}\left[1 - \exp\left(-\frac{\xi}{4\nu}\sigma^2\right)\right]. \qquad (3.12.11)$$

The azimuthal velocity is zero at the axis of revolution and decays like that due to a point vortex far from the axis of revolution. Accordingly $A = C/(4\pi\nu)$, where C is the circulation around a loop with infinite radius. The rate of extension, ξ, determines the radius of the Burgers vortex; as ξ tends to infinity, we obtain the flow due to a point vortex (see also Problem 3.12.1).

3.12.2   A vortex sheet diffusing in the presence of stretching

An example of a flow that continues to evolve until diffusion of vorticity is balanced by convection is provided by a diffusing vortex sheet separating two uniform streams that merge along the x axis with velocities $\pm U_0$, as illustrated in Figure 3.12.2. The vortex sheet is subjected to a two-dimensional extensional velocity field in the yz plane stretching the vortex lines along the z axis. Stipulating that the vortex lines are and remain oriented along the z axis, we express the three component of the velocity and the z component of the vorticity as

$$u_y = -\xi y, \qquad u_z = -\xi z, \qquad u_x = U(y, t) \qquad \omega_z = -\frac{\mathrm{d}U}{\mathrm{d}y}, \qquad (3.12.12)$$

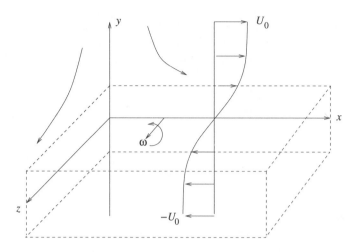

<small>FIGURE 3.12.2</small>   A diffusing vortex sheet is subjected to a two-dimensional straining flow that stretches
the vortex lines and allows for steady state to be established. For the velocity profile shown, $\omega_z < 0$.

where $U(y)$ is the velocity profile across the vortex layer and ξ is the rate of elongation of the
extensional flow.

Similarity solution

The absence of an intrinsic characteristic length scale suggests that the vortex layer may develop in
a self-similar fashion so that

$$\omega_z = \frac{\xi}{\delta(t)} f(\eta), \tag{3.12.13}$$

where $\delta(t)$ is the nominal thickness of the vortex layer to be computed as part of the solution,

$$\eta \equiv \frac{y}{\delta(t)} \tag{3.12.14}$$

is a dimensionless similarity variable, and f is a function with dimensions of length. Substituting
expressions (3.12.12) and (3.12.13) into the z component of the vorticity transport equation and
rearranging, we obtain

$$\delta \left(\frac{\mathrm{d}\delta}{\mathrm{d}t} + \xi\delta \right) = -\nu \frac{f''}{f + \eta f'}, \tag{3.12.15}$$

where a prime denotes a derivative with respect to η. We observe that the left-hand side of (3.12.15)
is a function of t alone, whereas the right-hand side is a function of both t and y. Each side
must be equal to the same constant, set for convenience equal to $2\nu A^2$, where A is a dimensionless
coefficient. Solving the resulting ordinary differential equations for δ and f subject to the initial
condition $\delta(0) = 0$ and the stipulation that f is an even function of η, we obtain

$$\delta^2(t) = A^2 \frac{2\nu}{\xi} \left(1 - \mathrm{e}^{-2\xi t} \right), \qquad f(\eta) = f(0) \exp(-A^2\eta^2). \tag{3.12.16}$$

Substituting expressions (3.12.16) into (3.12.13), we derive the vorticity distribution

$$\omega_z = B \frac{\xi}{(1 - e^{-2\xi t})^{1/2}} \exp\left[-\frac{\xi y^2}{2\nu(1 - e^{-2\xi t})}\right], \tag{3.12.17}$$

where

$$B = \frac{1}{A}\left(\frac{\xi}{2\nu}\right)^{1/2} f(0) \tag{3.12.18}$$

is a new dimensionless constant. To compute the velocity profile, $U(y)$, we substitute (3.12.17) into the left-hand side of the last equation in (3.12.12). Integrating with respect to y, we obtain an expression in terms of the error function,

$$\mathrm{erf}\, w \equiv \frac{2}{\sqrt{\pi}} \int_0^w e^{-v^2}\, dv. \tag{3.12.19}$$

The constant B then follows as

$$B = -U_0 \left(\frac{2}{\pi \xi \nu}\right)^{1/2} \tag{3.12.20}$$

(Problem 3.12.2). At the initial instant, the velocity profile $U(y, 0)$ undergoes a discontinuity and the thickness of the vortex layer is zero. As time progresses, the thickness of the vortex layer tends to the asymptotic value $\delta_\infty = A(2\nu/\xi)^{1/2}$ where diffusion of vorticity away from the zx plane is balanced by convection.

Burgers vortex layer

At steady steady, we obtain the Burgers vortex layer with vorticity and velocity profiles given by

$$\omega_z(y) = -U_0 \left(\frac{2\xi}{\pi \nu}\right)^{1/2} \exp\left(-\frac{\xi y^2}{2\nu}\right), \qquad U(y) = U_0\, \mathrm{erf}\left[\left(\frac{\xi}{2\nu}\right)^{1/2} y\right]. \tag{3.12.21}$$

The pressure field can be computed using Bernoulli's equation for the yz plane [62].

Unstretched layer

In the absence of straining flow, $\xi = 0$, the vortex layer diffuses to occupy the entire plane. To describe the developing vorticity field, we linearize the exponential terms in (3.12.17) with respect to their arguments and obtain

$$\omega_z(t) = -U_0 \left(\frac{1}{\pi \nu t}\right)^{1/2} \exp\left(-\frac{y^2}{4\nu t}\right). \tag{3.12.22}$$

Integrating with respect to y, we derive the associated velocity profile

$$U(t) = U_0\, \mathrm{erfc}\left(\frac{y}{2\sqrt{\nu t}}\right), \tag{3.12.23}$$

where $\mathrm{erfc} = 1 - \mathrm{erc}$ is the complementary error function.

3.12.3 Axisymmetric flow

In the case of axisymmetric flow without swirling motion, the axial and radial components of the vorticity transport equation (3.12.2) are trivially satisfied. The azimuthal component of the vorticity transport equation provides us with a scalar evolution equation for the azimuthal vorticity component,

$$\frac{D}{Dt}\left(\frac{\omega_\varphi}{\sigma}\right) = \frac{\nu}{\sigma^2}\,E^2(\sigma\,\omega_\varphi),\qquad(3.12.24)$$

where E^2 is a second-order operator defined in (2.9.17) and (2.9.20). The vortex stretching term is inherent in the material derivative of ω_φ/σ on the right-hand side of the ratio (3.12.24). If the azimuthal vorticity is proportional to the radial distance at the initial instant, $\omega_\varphi = \Omega\sigma$, where Ω is a constant, the right-hand side of (3.12.24) vanishes and the vorticity remains proportional to the radial distance at all times.

Flow with negligible viscous forces

Equation (3.12.24) shows that, if viscous effects are insignificant, point particles move in an azimuthal plane while their vorticity is continuously adjusted so that the ratio ω_φ/σ remains constant in time, equal to that at the initial instant,

$$\frac{D}{Dt}\left(\frac{\omega_\varphi}{\sigma}\right) = 0.\qquad(3.12.25)$$

Accordingly, the strength of a line vortex ring is proportional to its radius. The underlying physical mechanism can be traced to preservation of circulation around vortex tubes.

3.12.4 Enstrophy and intensification of the vorticity field

The enstrophy of a flow is defined as the integral of the square of the magnitude of the vorticity over the flow domain,

$$\mathcal{E} \equiv \iiint_{Flow} \boldsymbol{\omega} \cdot \boldsymbol{\omega}\, dV.\qquad(3.12.26)$$

Projecting the vorticity transport equation onto the vorticity vector, we find that the enstrophy of a flow with uniform physical properties evolves according to the equation

$$\frac{d\mathcal{E}}{dt} = 2\iiint_{Flow}(\boldsymbol{\omega}\otimes\boldsymbol{\omega}):\mathbf{E}\,dV - 2\nu\iiint_{Flow}\nabla\boldsymbol{\omega}:\nabla\boldsymbol{\omega}\,dV + \nu\iint_{B}\mathbf{n}\cdot\nabla(\boldsymbol{\omega}\cdot\boldsymbol{\omega})\,dS,\qquad(3.12.27)$$

where B stands for the boundaries of the flow. The three terms on the right-hand side represent, respectively, intensification of vorticity due to vortex stretching, the counterpart of viscous dissipation for the vorticity, and surface diffusion across the boundaries.

3.12.5 Vorticity transport equation in a noninertial frame

The equation of motion in an accelerating frame of reference that translates with time-dependent velocity $\mathbf{U}(t)$ while rotating about the origin with angular velocity $\boldsymbol{\Omega}(t)$ includes the inertial forces shown in (3.2.43). Let be $\mathbf{v}$ the velocity in the noninertial frame defined in (3.2.36), and ϱ be the corresponding vorticity. Including the inertial forces in the inertial form of the Navier–Stokes equation, taking the curl of the resulting equation, and rearranging, we find that the modified or intrinsic vorticity, $\boldsymbol{\omega} \equiv \varrho + 2\boldsymbol{\Omega}$, satisfies the inertial form of the vorticity transport equation,

$$\frac{\mathrm{D}\boldsymbol{\omega}}{\mathrm{D}t} = \frac{\partial \boldsymbol{\omega}}{\partial t} + \mathbf{v} \cdot \nabla \boldsymbol{\omega} = \boldsymbol{\omega} \cdot \nabla \mathbf{v} + \nu \nabla^2 \boldsymbol{\omega} \tag{3.12.28}$$

(Problem 3.12.6). The simple form of (3.12.28) is exploited for the efficient computation of an incompressible Newtonian flow in terms of the velocity and intrinsic vorticity, as discussed in Section 13.2 (e.g., [383]).

Problems

3.12.1 *Burgers columnar vortex*

Evaluate the constant A in terms of the maximum value of the meridional velocity, $U(\sigma)$.

3.12.2 *Vortex layer*

Derive the velocity profile associated with (3.12.17) and the value of the constant B shown in (3.12.20).

3.12.3 *Vorticity due to the motion of a body*

Discuss the physical process by which vorticity enters the flow due to a body that is suddenly set in motion in a viscous fluid.

3.12.4 *Ertel's theorem*

Show that, for any scalar function of position and time, $f(x, t)$,

$$\frac{\mathrm{D}}{\mathrm{D}t}(\boldsymbol{\omega} \cdot \nabla f) = \boldsymbol{\omega} \cdot \nabla\left(\frac{\mathrm{D}f}{\mathrm{D}t}\right). \tag{3.12.29}$$

Based on this equation, explain why, if the field represented by f is convected by the flow, $\mathrm{D}f/\mathrm{D}t = 0$, the scalar $\boldsymbol{\omega} \cdot \nabla f$ will also be convected by the flow.

3.12.5 *Generalized Beltrami flow*

If we switch the sign of the velocity of a generalized Beltrami flow, should we also switch the sign of the modified pressure gradient to satisfy the equation of motion?

3.12.6 *Vorticity transport equation in a noninertial frame*

Derive the vorticity transport equation (3.12.28).

3.13   Vorticity transport in two-dimensional flow

The vorticity transport equation for two-dimensional flow in the xy plane simplifies considerably due to the absence of vortex stretching. We find that the x and y components of the general vorticity transport equation (3.12.2) are trivially satisfied, while the z component yields a scalar convection–diffusion equation for the z vorticity component,

$$\frac{D\omega_z}{Dt} \equiv \frac{\partial \omega_z}{\partial t} + \mathbf{u} \cdot \nabla \omega_z = \nu \nabla^2 \omega_z, \qquad (3.13.1)$$

where ∇^2 is the two-dimensional Laplacian operator in the xy plane.

It is tempting to surmise that ω_z behaves like a passive scalar. However, we must remember that vorticity distribution has a direct influence on the development of the velocity field mediated by the Biot–Savart integral.

Flow with negligible viscous forces

When the viscous force is insignificant, the right-hand side of (3.13.1) is zero, yielding the Lagrangian conservation law

$$\frac{D\omega_z}{Dt} = 0. \qquad (3.13.2)$$

Physically, point particles move while maintaining their initial vorticity, that is, they spin at a constant angular velocity. Two consequences of this result are that point vortices maintain their strength and patches of constant vorticity preserve their initial vorticity.

Prandtl–Batchelor theorem

In Section 3.11.1, we showed that the circulation of the curl of the vorticity along a closed streamline is zero in a steady flow,

$$\oint \mathbf{t} \cdot (\nabla \times \boldsymbol{\omega}) \, \mathrm{d}l = 0. \qquad (3.13.3)$$

In the case of two-dimensional flow, this equation states that the contribution of the tangential component of the viscous force to the total force exerted on the volume of a fluid enclosed by a streamline is zero. Since the z vorticity component is constant along a streamline in the absence of viscous forces, it can be regarded as a function of the stream function, ψ. The line integral in (3.5.14) then becomes

$$\oint \mathbf{t} \cdot (\nabla \times \boldsymbol{\omega}) \, \mathrm{d}l = \oint \mathbf{t} \cdot (\mathbf{e}_x \frac{\partial \omega_z}{\partial y} - \mathbf{e}_y \frac{\partial \omega_z}{\partial x}) \, \mathrm{d}l = \frac{\mathrm{d}\omega_z}{\mathrm{d}\psi} \oint \mathbf{t} \cdot (\mathbf{e}_x \frac{\partial \psi}{\partial y} - \mathbf{e}_y \frac{\partial \psi}{\partial x}) \, \mathrm{d}l, \qquad (3.13.4)$$

where $\mathbf{e}_x$ and $\mathbf{e}_y$ are unit vectors along the x and y axis. Invoking the definition of the stream function, we find that

$$\oint \mathbf{t} \cdot (\nabla \times \boldsymbol{\omega}) \, \mathrm{d}l = \frac{\mathrm{d}\omega_z}{\mathrm{d}\psi} \oint \mathbf{t} \cdot (\mathbf{e}_x \, u_x + \mathbf{e}_y \, u_y) \, \mathrm{d}l = \frac{\mathrm{d}\omega_z}{\mathrm{d}\psi} \oint \mathbf{t} \cdot \mathbf{u} \, \mathrm{d}l. \qquad (3.13.5)$$

Setting the last expression to zero to satisfy (3.13.3), we find that either the circulation around the streamline is zero or the vorticity is constant in regions of recirculating flow.

Preservation of the total circulation in the absence of boundaries

The total circulation of a two-dimensional flow is defined as the integral of the vorticity over the area of the flow,

$$\Omega \equiv \iint_{Flow} \omega_z \, dA. \tag{3.13.6}$$

Consider an infinite flow that decays at infinity in the absence of interior boundaries. Integrating (3.13.1) over the whole domain of flow and using the continuity equation, we find that

$$\frac{d\Omega}{dt} = \iint_{Flow} (-\mathbf{u} \cdot \nabla \omega_z + \nu \nabla^2 \omega_z) \, dA = \iint_{Flow} \nabla \cdot (-\omega_z \mathbf{u} + \nu \nabla \omega_z) \, dA. \tag{3.13.7}$$

Using the divergence theorem, we convert the last areal integral into a line integral over a large loop enclosing the flow. Since the velocity is assumed to vanish at infinity, the line integral is zero and the total circulation of the flow remains constant in time. Physically, vorticity neither is produced nor can escape in the absence of boundaries.

3.13.1 Diffusing point vortex

To illustrate the action of viscous diffusion, we consider the decay of a point vortex with strength κ placed at the origin. The initial vorticity distribution can be expressed in terms of a two-dimensional delta function in the xy plane, as $\omega_z(\mathbf{x}, t = 0) = \kappa \, \delta_2(\mathbf{x})$. Assuming that the flow remains axisymmetric with respect to the z axis at all times, we introduce plane polar coordinates centered at the point vortex, (r, θ), and simplify (3.13.1) into the scalar unsteady diffusion equation

$$\frac{\partial \omega_z}{\partial t} = \frac{\nu}{r} \frac{\partial}{\partial r} \left(r \frac{\partial \omega_z}{\partial r} \right). \tag{3.13.8}$$

Since the nonlinear term vanishes, the flow due to a diffusing point vortex is a generalized Beltrami flow.

The absence of an intrinsic length scale suggests that the solution is a function of the dimensionless similarity variable $\eta \equiv r/(\nu t)^{1/2}$, so that

$$\omega_z = \frac{\kappa}{\nu t} f(\eta), \tag{3.13.9}$$

where f is an *a priori* unknown function. Substituting this functional form into (3.13.8), we derive a second-order linear ordinary differential equation,

$$2 \, (\eta f')' + \eta^2 f' + 2 f \eta = 0, \tag{3.13.10}$$

where a prime denotes a derivative with respect to η. Requiring that the derivatives of f are finite at the origin and the total circulation of the flow is equal to κ at any time, we obtain the solution

$$\omega_z = \frac{\kappa}{4 \pi \nu t} \exp \left(-\frac{r^2}{4 \nu t} \right). \tag{3.13.11}$$

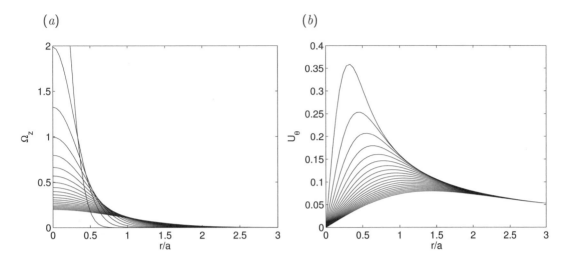

FIGURE 3.13.1   Profiles of (a) the dimensionless vorticity, $\Omega_z \equiv \omega_z a^2/\kappa$, and ($b$) dimensionless velocity, $U_\theta \equiv a u_\theta/\kappa$, around a diffusing point vortex at a sequence of dimensionless times, $\tau \equiv \nu t/a^2 = 0.02, 0.04, \ldots$, where a is a reference length.

Integrating the definition $\omega_z = (1/r)\partial(r u_\theta)/\partial r$, we derive an expression for the polar velocity component,

$$u_\theta = \frac{\kappa}{2\pi r}\left[1 - \exp\left(-\frac{r^2}{4\nu t}\right)\right], \tag{3.13.12}$$

describing the flow due to a diffusing point vortex known as the Oseen vortex. Radial profiles of the dimensionless vorticity, $\Omega_z \equiv \omega_z a^2/\kappa$, and reduced velocity, $U_\theta \equiv a u_\theta/\kappa$, are shown in Figure 3.13.1 at a sequence of dimensionless times, $\tau \equiv \nu t/a^2$, where a is a reference length. As time progresses, the vorticity diffuses away from the point vortex and tends to occupy the whole plane. Viscosity smears the vorticity distribution and reduces the point vortex into a vortex blob.

3.13.2   Generalized Beltrami flows

The simple form of the vorticity transport equation for two-dimensional flow can be exploited to derive exact solutions representing steady and unsteady, viscous and inviscid flows. When the gradient of the vorticity is and remains perpendicular to the velocity vector, the nonlinear convective term in the vorticity transport equation (3.13.1) is identically zero, yielding a generalized Beltrami flow whose vorticity evolves according to the unsteady diffusion equation,

$$\frac{\partial \omega_z}{\partial t} = \nu \nabla^2 \omega_z. \tag{3.13.13}$$

Since the gradient of the vorticity is perpendicular to the streamlines, and thus the vorticity is constant along the streamlines, the vorticity can be regarded as a function of the stream function,

$$\omega_z = -\nabla^2 \psi = f(\psi). \tag{3.13.14}$$

Substituting (3.13.14) into (3.13.13) and rearranging, we obtain

$$\nabla^2 \left(\frac{\partial \psi}{\partial t} + \nu \, f(\psi) \right) = 0, \tag{3.13.15}$$

which is fulfilled if the stream function evolves according to the equation

$$\frac{\partial \psi}{\partial t} = -\nu \, f(\psi) = \nu \, \nabla^2 \psi. \tag{3.13.16}$$

Specifying the function f and solving for ψ provides us with families of two-dimensional generalized Beltrami flows. Setting f equal to zero or a constant value, we obtain irrotational flows and flows with constant vorticity. Other families of steady and unsteady generalized Beltrami flows have been discovered [419, 420].

Differentiating (3.13.16) with respect to x and y, we find that the Cartesian components of the velocity satisfy the unsteady heat conduction equation,

$$\frac{\partial \mathbf{u}}{\partial t} = \nu \, \nabla^2 \mathbf{u}. \tag{3.13.17}$$

To derive the associated pressure field, we substitute the first equation in (3.5.15) into the Navier–Stokes equation (3.5.10), replace $\partial \mathbf{u}/\partial t$ with the right-hand side of (3.13.17), and express the curl of the vorticity on the right-hand as the negative of the Laplacian of the velocity. Simplifying and integrating the resulting equation with respect to the spatial variables, we obtain

$$p = -\frac{1}{2} \rho \, \mathbf{u} \cdot \mathbf{u} + \rho \, \mathbf{g} \cdot \mathbf{x} - F(\psi) + c, \tag{3.13.18}$$

where c is a constant and $F(\psi)$ is the indefinite integral of the function $f(\psi)$, as shown in (3.5.17).

Taylor cellular flow

An interesting unsteady generalized Beltrami flow arises by stipulating that the function $f(\psi)$ introduced in (3.13.14) is linear in ψ. Setting $f(\psi) = \alpha^2 \psi$, where α is a constant with dimensions of inverse length, and integrating in time the differential equation comprised of the first two terms in (3.13.16), we obtain

$$\psi(x, y, t) = \chi(x, y) \exp(-\alpha^2 \nu t). \tag{3.13.19}$$

Equation (3.13.16) requires that the function χ satisfies Helmholtz' equation

$$\nabla^2 \chi = -\alpha^2 \, \chi. \tag{3.13.20}$$

The particular solution

$$\chi = A \, \cosh(\beta x) \, \cos(\gamma y) \tag{3.13.21}$$

describes an exponentially decaying cellular periodic flow with wave numbers in the x and y directions equal to β and γ, where $\beta^2 + \gamma^2 = \alpha^2$ and A is an arbitrary constant [397]. The corresponding

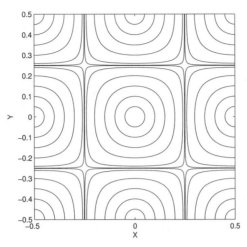

FIGURE 3.13.2   Streamline pattern of the doubly periodic decaying Taylor cellular flow with square cells.

pressure distribution is found from (3.13.18). The streamline pattern for $\beta = \gamma$ in the plane of dimensionless axes $X = \beta x$ and $Y = \gamma y$ is shown in Figure 3.13.2.

Inviscid flow

When the effect of viscosity is insignificant, the terms multiplied by the kinematic viscosity in (3.13.16) vanish. Accordingly, any time-independent solution of (3.13.14) represents an acceptable steady inviscid flow. Examples include the flow due to a point vortex, the flow due to an infinite array of point vortices, and any unidirectional shear flow with arbitrary velocity profile. The corresponding pressure distributions are found from (3.13.18).

3.13.3   Extended Beltrami flows

Another opportunity for linearizing the vorticity transport equation arises by stipulating that the vorticity distribution takes the particular form

$$\omega_z = -\nabla^2 \psi = -\chi(x, y, t) - c\psi, \tag{3.13.22}$$

where χ is a specified function and c is a specified constant with dimensions of inverse squared length. The vorticity transport equation reduces to a linear partial differential equation for ψ,

$$\frac{\partial \nabla^2 \psi}{\partial t} + \frac{\partial \chi}{\partial x} \frac{\partial \psi}{\partial y} - \frac{\partial \chi}{\partial y} \frac{\partial \psi}{\partial x} - \nu c^2 \psi = \nu \left(\nabla^2 \chi + c \chi \right), \tag{3.13.23}$$

whose solution can be found for a number of cases in analytical form (e.g., [419, 420]).

Kovasznay flow

As an example, consider a steady flow with $\chi = -cUy$, where U is a constant [212]. Expression (3.13.22) becomes

$$\omega_z = -\nabla^2\psi = c\,(Uy - \psi).$$ (3.13.24)

The steady version of the vorticity transport equation (3.13.23) simplifies into

$$\frac{U}{\nu c}\frac{\partial\psi}{\partial x} - \psi = -Uy.$$ (3.13.25)

A solution is

$$\psi = U\left[y - \alpha\,\frac{2\pi}{k}\,\sin(ky)\,\exp(\beta kx)\right],$$ (3.13.26)

where α is an arbitrary dimensionless constant, k is an arbitrary wave number in the y direction, and $\beta = c\nu/(kU)$ is a dimensionless number.

It remains to verify that (3.13.26) is consistent with the assumed functional form (3.13.24). Substituting (3.13.26) into (3.13.24), we obtain a quadratic equation for β,

$$\beta^2 - \frac{\mathrm{Re}}{2\pi}\beta - 1 = 0,$$ (3.13.27)

where $\mathrm{Re} = 2\pi U/k\nu$ is a Reynolds number. Retaining the root with the negative value, we obtain

$$\beta = -\frac{1}{2}\left[\left(\frac{\mathrm{Re}^2}{4\pi^2} + 4\right)^{1/2} - \frac{\mathrm{Re}}{2\pi}\right].$$ (3.13.28)

Streamline patterns in the plane of dimensionless axes $X = x/a$ and $Y = y/a$, are illustrated in Figure 3.13.3 for $\alpha = 1.0$ and $\mathrm{Re} = 10$ and 50, where $a = 2\pi/k$ is the separation of two successive cells along the y axis. In both cases, a stagnation point is present at the origin. The flow can be envisioned as being established in the wake of an infinite array of cylinders arranged along the y axis, subject to a uniform incident flow along the x axis. As the Reynolds number is raised, the regions of recirculating flow become increasingly slender.

Problems

3.13.1 *Prandtl–Batchelor theorem for axisymmetric flow*

Derive the Prandtl–Batchelor theorem for axisymmetric flow with negligible viscous forces where the ratio ω_φ/σ is constant along a streamline.

3.13.2 *Flow over a porous plate with suction*

Consider uniform flow with velocity U far above and parallel to a porous plate. Fluid is withdrawn with a uniform velocity V through the plate. Show that the velocity field is given by

$$u_x = U\,(1 - \mathrm{e}^{-yV/\nu}), \qquad u_y = V.$$ (3.13.29)

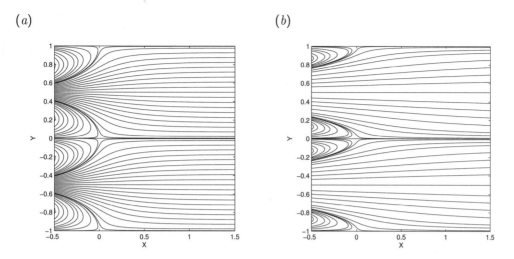

FIGURE 3.13.3   Streamline pattern of the Kovasznay flow for $\alpha = 1.0$ and Reynolds number (a) Re = 10 or (b) 50.

Discuss the interpretation of the solution in terms of vorticity diffusion toward and convection away from the plate.

3.13.3 *Extended Beltrami flow*

Show that the flow described by the stream function

$$\psi = -Uy\left(1 - e^{-xU/\nu}\right) \tag{3.13.30}$$

can be derived as an extended Beltrami flow. Discuss the physical interpretation of this flow.

Hydrostatics

4

When a fluid is stationary or translates as a rigid body, the continuity equation is satisfied in a trivial manner and the Navier–Stokes equation reduces into a first-order differential equation expressing a balance between the pressure gradient and the body force. This simplified equation of hydrostatics can be integrated by elementary methods, subject to appropriate boundary conditions, to yield the pressure distribution inside the fluid. The integration produces a scalar constant that is determined by specifying the level of the pressure at an appropriate point over a boundary. The product of the normal unit vector and the pressure may then be integrated over the surface of an immersed or submerged body to produce the buoyancy force. The torque with respect to a chosen point can be computed in a similar way. Details of this procedure and applications will be discussed in this chapter for fluids that are stationary or undergo steady or unsteady rigid-body motion.

In the case of two immiscible stationary fluids in contact, the interfacial boundary conditions discussed in Section 3.8 require that the normal component of the traction undergoes a discontinuity that is balanced by surface tension. In hydrostatics, the normal component of the traction is equal to the negative of the pressure and the tangential component is identically zero. Accordingly, the interface must assume a shape that is compatible with the pressure distribution on either side and conforms with boundary conditions for the contact angle at a three-phase contact line. Conversely, the pressure distribution in the two fluids cannot be computed independently, but must be found simultaneously with the interfacial shape so that all boundary conditions are fulfilled. Depending on the shape of the contact line, the magnitude of the contact angle, and the density difference between the two fluids, a stationary interface may take a variety of interesting and sometimes unexpected shapes (e.g., [52, 197]). For example, an interface inside a dihedral angle confined between two planes may climb to infinity, as discussed in Section 4.2.1. The computation of interfacial shapes presents us with a challenging mathematical problem involving highly nonlinear ordinary and partial differential equations [126].

Our first task in this chapter is to derive the pressure distribution in stationary, translating, and rotating fluids. The equations governing interfacial hydrostatics are then discussed, and numerical procedures for computing the shape of interfaces with two-dimensional (cylindrical) and axisymmetric shapes are outlined. The computation of three-dimensional interfacial shapes and available software are reviewed briefly at the conclusion of this chapter as specialized topics worthy of further investigation.

4.1 Pressure distribution in rigid-body motion

We begin by considering the pressure distribution in a stationary fluid. However, since any fluid that moves as a rigid body appears to be stationary in a suitable frame of reference, we also include fluids that execute steady or unsteady translation and rotation.

4.1.1 Stationary and translating fluids

The Navier–Stokes equation for a fluid that is either stationary or translates with uniform velocity $\mathbf{u} = \mathbf{U}(t)$ simplifies into a linear equation,

$$\rho \frac{\mathrm{d}\mathbf{U}}{\mathrm{d}t} = -\nabla p + \rho \mathbf{g}. \tag{4.1.1}$$

Assuming that the density of the fluid is uniform and solving for the pressure, we obtain

$$p = \rho \left(\mathbf{g} - \frac{\mathrm{d}\mathbf{U}}{\mathrm{d}t} \right) \cdot \mathbf{x} + P(t), \tag{4.1.2}$$

where the function $P(t)$ is found by enforcing an appropriate boundary condition reflecting the physics of the problem under consideration.

Force on an immersed body

The force exerted on a body immersed in a stationary or translating fluid can be computed in terms of the stress tensor using the general formula (3.1.16). Setting $\boldsymbol{\sigma} = -p\mathbf{I}$, we find that

$$\mathbf{F} = - \iint_{Body} p \, \mathbf{n} \, \mathrm{d}S, \tag{4.1.3}$$

where $\mathbf{n}$ is the normal unit vector pointing into the fluid. Substituting expression (4.1.2) and using the divergence theorem to convert the surface integral into a volume integral over the volume of the body, we derive Archimedes' buoyancy force,

$$\mathbf{F} = -\rho V_B \left(\mathbf{g} - \frac{\mathrm{d}\mathbf{U}}{\mathrm{d}t} \right), \tag{4.1.4}$$

where V_B is the volume occupied by the body.

Torque on an immersed body

The torque exerted on a body with respect to a chosen point, $\mathbf{x}_0$, is found working in a similar manner using (3.1.23), finding

$$\mathbf{T} = -\rho V_B \left(\mathbf{x}_c - \mathbf{x}_0 \right) \times \left(\mathbf{g} - \frac{\mathrm{d}\mathbf{U}}{\mathrm{d}t} \right), \tag{4.1.5}$$

where

$$\mathbf{x}_c \equiv \frac{1}{V_B} \iiint_{Body} \mathbf{x} \, \mathrm{d}V(\mathbf{x}) \tag{4.1.6}$$

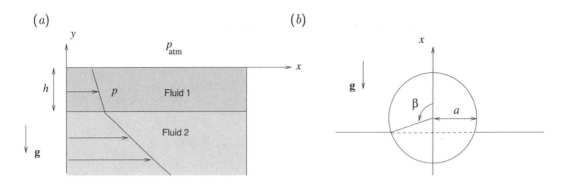

FIGURE 4.1.1   (a) Depiction of a liquid layer resting between an underlying heavier liquid and overlying air, illustrating the hydrostatic pressure profile. (b) A spherical particle floats at the flat surface of a liquid at floating angle β; the dashed line represents the horizontal circular contact line.

is the volume centroid of the body. Placing the pivot point $\mathbf{x}_0$ at the volume centroid, $\mathbf{x}_c$, makes the torque vanish. Using the divergence theorem, we derive a convenient representation for the volume centroid in terms of a surface integral,

$$\mathbf{x}_c = \frac{1}{2V_B} \iint_{Body} \begin{bmatrix} x^2 & 0 & 0 \\ 0 & y^2 & 0 \\ 0 & 0 & z^2 \end{bmatrix} \cdot \mathbf{n} \, dS(\mathbf{x}), \qquad (4.1.7)$$

where $\mathbf{n}$ is the normal unit vector pointing into the fluid.

A floating liquid layer

As an application, we consider the pressure distribution in a stationary liquid layer labeled 1 with thickness h, resting between air and a pool of a heavier liquid labeled 2, as shown in Figure 4.1.1(a). Assuming that both the free surface and the liquid interface are perfectly flat, we introduce Cartesian coordinates with origin at the free surface and the y axis pointing in the vertical direction upward so that

$$g_x = 0, \qquad g_y = -g, \qquad (4.1.8)$$

where $g = |\mathbf{g}|$ is the magnitude of the acceleration of gravity. Applying the general equation (4.1.2) with $\mathbf{U} = \mathbf{0}$ for each fluid, we obtain the pressure distributions

$$p_1 = -\rho_1 g y + P_1, \qquad p_2 = -\rho_2 g y + P_2, \qquad (4.1.9)$$

where P_1 and P_2 are two constant pressures.

To compute the constant P_1, we require that the pressure at the free surface of fluid 1 is equal to the atmospheric pressure, p_{atm}, and find that $P_1 = p_{atm}$. To compute the constant P_2, we apply both equations in (4.1.9) at the interface located at $y = -h$, and subtract the resulting expressions to find that $p_2 - p_1 = \Delta\rho g h + P_2 - p_{atm}$, where $\Delta\rho = \rho_2 - \rho_1$. Requiring continuity of the normal component of the traction or pressure across the interface, we obtain $P_2 = p_{atm} - \Delta \rho g h$. Substituting the values of P_1 and P_2 into (4.1.9), we obtain the explicit pressure distributions

$$p_1 = -\rho_1 g y + p_{atm}, \qquad p_2 = -\rho_2 g y + p_{atm} - \Delta \rho g h. \tag{4.1.10}$$

The satisfaction of the dynamic condition $p_1(-h) = p_2(-h)$ can be readily verified.

Pressure in a vibrating container

As a second application, we consider the pressure distribution in a liquid inside a container that vibrates harmonically in the vertical direction with angular frequency Ω. Assuming that the free surface remains flat, we introduce Cartesian axes with origin at the mean position of the free surface and the y axis pointing against the direction of gravity, so that $g_y = -g$. The position of the free surface is described as $y = a \cos(\Omega t)$, where a is the amplitude of the oscillation. The velocity of the fluid is $\mathbf{U}(t) = -a\Omega \sin \Omega t \, \mathbf{e}_y$, where $\mathbf{e}_y$ is the unit vector along the y axis. Using (4.1.2), we find that the pressure distribution in the liquid is given by

$$p = -\rho y \left[g - a\Omega^2 \cos(\Omega t) \right] + P(t). \tag{4.1.11}$$

The function $P(t)$ is evaluated by requiring that the pressure at the free surface is equal to the atmospheric pressure, p_{atm}, at any instant, obtaining

$$P(t) = p_{atm} + \rho a \cos(\Omega t) \left[g - a\Omega^2 \cos(\Omega t) \right]. \tag{4.1.12}$$

Substituting this expression into (4.1.11) and rearranging, we obtain

$$p = p_{atm} + \rho \left[y - a \cos(\Omega t) \right] \left[-g + a\Omega^2 \cos(\Omega t) \right]. \tag{4.1.13}$$

The term inside the first square brackets on the right-hand side is the vertical component of the position vector in a frame of reference moving with the free surface. The term inside the second square brackets is the sum of the vertical components of the gravitational acceleration and inertial acceleration $-\mathrm{d}\mathbf{U}(t)/\mathrm{d}t$. It is then evident that expression (4.1.13) is consistent with the pressure distribution that would have arisen if we worked in a frame of reference where the free surface appears to be stationary, accounting for the fictitious inertial force due to the vertical vibration.

4.1.2 Rotating fluids

The velocity distribution in a fluid in steady rigid-body rotation around the origin with constant angular velocity $\mathbf{\Omega}$ is $\mathbf{u} = \mathbf{\Omega} \times \mathbf{x}$, and the associated vorticity is $\boldsymbol{\omega} = 2\mathbf{\Omega}$. Remembering that the Laplacian of the velocity is equal to the negative of the curl of the vorticity, we find that the viscous force is identically zero. Evaluating the acceleration from (1.3.14), we obtain the simplified equation of motion

$$\rho \, \mathbf{\Omega} \times (\mathbf{\Omega} \times \mathbf{x}) = -\nabla p + \rho \, \mathbf{g}. \tag{4.1.14}$$

In cylindrical polar coordinates, (x, σ, φ), with the x axis pointing in the direction of the angular velocity vector, $\mathbf{\Omega}$, equation (4.1.14) takes the form

$$-\rho \, \Omega^2 \sigma \, \mathbf{e}_\sigma = -\nabla p + \rho \, \mathbf{g}, \tag{4.1.15}$$

where Ω is the magnitude of $\boldsymbol{\Omega}$ and $\mathbf{e}_\sigma$ is the unit vector in the radial (σ) direction. Expressing the gradient of the pressure in cylindrical coordinates and integrating with respect to σ, we find that

$$p = \rho\,\mathbf{g} \cdot \mathbf{x} + \frac{1}{2}\,\rho\,\Omega^2\sigma^2 + P(t), \qquad (4.1.16)$$

which can be restated as

$$p = \rho\,\mathbf{g} \cdot \mathbf{x} + \frac{1}{2}\,\rho\,|\boldsymbol{\Omega} \times \mathbf{x}|^2 + P(t), \qquad (4.1.17)$$

where $P(t)$ is an inconsequential function of time. The additional pressure expressed by the second term on the right-hand side is necessary in order to balance the radial centrifugal force due to the rotation.

Rotating container

As an application, we consider the pressure distribution inside a container that rotates steadily around the horizontal x axis. Setting the y axis in the direction of gravity pointing upward so that the component of the acceleration of gravity are $g_x = 0$, $g_y = -g$, and $g_z = 0$, we find the pressure distribution

$$p = -\rho g y + \frac{1}{2}\,\rho\,\Omega^2(y^2 + z^2) + P, \qquad (4.1.18)$$

which can be restated as

$$p = \frac{1}{2}\,\rho\,\Omega^2 \left[\left(y - \frac{g}{\Omega^2} \right)^2 + z^2 \right] - \frac{1}{2}\,\rho\,\frac{g^2}{\Omega^2} + P. \qquad (4.1.19)$$

The expression shows that surfaces of constant pressure are concentric horizontal cylinders with axis passing through the point $z = 0$ and $y = g/\Omega^2$. The minimum pressure occurs at the common axis.

Free surface of a rotating fluid

In another application, we consider the shape of the free surface of a liquid inside a horizontal cylindrical beaker that rotates about its axis of revolution with angular velocity Ω. Setting the x axis in the direction of gravity pointing upward so that $g_x = -g$, we find that the pressure distribution in the fluid is given by

$$p = -\rho g x + \frac{1}{2}\,\rho\,\Omega^2\sigma^2 + P, \qquad (4.1.20)$$

where σ is the distance from the x axis. Evaluating the pressure at the free surface, neglecting the effect of surface tension, and requiring that the pressure at the free surface is equal to the ambient atmospheric pressure, we obtain an algebraic equation for the shape of the free surface describing a paraboloid.

Rotary oscillations

Now we assume that the fluid rotates around the origin as a rigid body with time-dependent angular velocity $\boldsymbol{\Omega}(t)$. Substituting the associated velocity field $\mathbf{u} = \boldsymbol{\Omega}(t) \times \mathbf{x}$ in the Navier–Stokes equation, we obtain the pressure distribution $p = p_1 + p_2$, where p_1 is given by the right-hand side of (4.1.17) and p_2 satisfies the equation

$$\rho \frac{d\mathbf{\Omega}}{dt} \times \mathbf{x} = -\nabla p_2. \tag{4.1.21}$$

Taking the curl of both sides of (4.1.21) and noting the the curl of the gradient of any twice differentiable function is identically zero, we arrive at the solvability condition $d\mathbf{\Omega}/dt = \mathbf{0}$, which contradicts the original assumption of unsteady rotation. The physical implication is that rotary oscillation requires a velocity field other than rigid-body motion. For example, subjecting a glass of water to rotary oscillation does not mean that the fluid in the glass will engage in rigid-body oscillatory rotation. In fact, at high frequencies, the motion of the fluid will be confined inside a thin boundary layer around the glass surface, and the main body of the fluid outside the boundary layer will be stationary.

4.1.3 Compressible fluids

The pressure distribution (4.1.2) was derived under the assumption that the density is uniform throughout the fluid. When the fluid is compressible, p is the thermodynamic pressure related to the fluid density, ρ, and to the Kelvin absolute temperature, T, by an appropriate equation of state.

Ideal gas

In the case of an ideal gas, the pressure is given by the ideal gas law, $p = RT\rho/M$, as discussed in Section 3.3. Solving for the density and substituting the resulting expression into (4.1.1) for a stationary fluid, $\mathbf{U} = \mathbf{0}$, we obtain a first-order differential equation,

$$\frac{1}{p} \nabla p = \nabla \ln\left(\frac{p}{p_0}\right) = \frac{M}{RT}\,\mathbf{g}, \tag{4.1.22}$$

where p_0 is an unspecified reference pressure. When the temperature is constant, the solution is

$$\ln \frac{p}{p_0} = \frac{M}{RT}\,\mathbf{g} \cdot \mathbf{x}. \tag{4.1.23}$$

We have found that the pressure exhibits an exponential dependence on distance instead of the linear dependence observed in an incompressible liquid.

As an application, we consider the pressure distribution in the atmosphere regarded as an ideal gas with molar mass $M = 28.97$ kg/kmole, at temperature 25°C, corresponding to absolute temperature $T = 298$ K. In Cartesian coordinates with origin at sea level and the y axis pointing upward, the components of the acceleration of gravity vector are given by $g_x = 0$, $g_y = -g$, and $g_z = 0$, where $g = 9.80665$ m/s². Equation (4.1.23) provides us with an exponentially decaying profile,

$$p = p_0 \exp(-\frac{Mg}{RT}\,y), \tag{4.1.24}$$

where p_0 is the pressure at sea level. Substituting $R = 8.314 \times 10^3$ kg m²/(s² kmole K) and taking $p_0 = 1.0\,\text{atm} = 1.0133 \times 10^5$ Pascal $= 1.0133 \times 10^5 \text{kg m}^{-1}\,\text{s}^{-2}$, we find that the pressure at the elevation of $y = 1$ km $= 1000$ m is

$$p = 1.0 \exp \left(- \frac{28.97 \times 9.80665}{8.314 \times 10^3 \times 298} 1000 \right) \text{atm} = 0.892 \text{ atm}. \qquad (4.1.25)$$

The corresponding density distribution is found by substituting the pressure distribution (4.1.24) in the ideal gas law.

Problems

4.1.1 *Two layers resting on a pool*

A liquid layer labeled 2 with thickness h_2 is resting at the surface of a pool of a heavier fluid labeled 3, underneath a layer of another liquid layer labeled 1 with thickness h_1. The pressure above layer 1 is atmospheric. Compute the pressure distribution in the two layers and in the pool.

4.1.2 *Rotating drop*

Describe the shape of a suspended axisymmetric drop rotating as a rigid body about the vertical axis in the absence of interfacial tension.

4.1.3 *Pressure distribution in the atmosphere*

Derive the pressure distribution in the atmosphere, approximated as an ideal gas, when the temperature varies linearly with elevation, $T = T_0 - \beta y$, where T_0 is the temperature at sea level, β is a constant called the lapse rate, and the y axis points against the direction of gravity.

4.1.4 *Compressible gas in rigid-body motion*

Generalize (4.1.23) to the case of an ideal gas translating as a rigid body with arbitrary time-dependent velocity, while rotating with constant angular velocity.

4.1.5 *A floating sphere*

A spherical particle of radius a is floating at the surface of a liquid underneath a zero-density gas, as shown in Figure 4.1.1(*b*). Assuming that the interface is flat, show that the floating angle, β, satisfies the cubic equation

$$\varrho^3 - 3\varrho + 2(2s - 1) = 0, \qquad (4.1.26)$$

where $\varrho \equiv \cos \beta$, $s \equiv W_s/(\rho g V_s)$ is a dimensionless constant, W_s is the weight of the sphere, and V_s is the volume of the sphere (e.g., [318]). If the sphere is made of a homogeneous material with density ρ_B, then $s = \rho_B/\rho$ is the density ratio.

🖳 Computer Problem

4.1.6 *A floating sphere*

Solve equation (4.1.26) and prepare a graph of the floating angle β against s in the range $[0, 1]$.

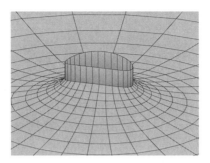

FIGURE 4.2.1   A hydrostatic meniscus forming outside a vertical elliptical cylinder with aspect ratio 0.6, for contact angle $\alpha = \pi/4$ and scaled capillary length $\ell/c = 4.5036$, where c is the major semi-axis of the ellipse. The computer code that generated this shape is included in Directory *men_ell* inside Directory *03_hydrostat* of the software library FDLIB (Appendix C).

4.2   The Laplace–Young equation

An important class of problems in interfacial hydrodynamics concerns the shape of a curved interface with uniform surface tension, γ, separating two stationary fluids labeled 1 and 2. An example is shown in Figure 4.2.1 [321]. Substituting in the interfacial force balance (3.8.8) the pressure distribution (4.1.2) with vanishing acceleration, we find that the jump in the interfacial traction defined in (3.8.1) is given by

$$\Delta \mathbf{f} = (\boldsymbol{\sigma}^{(1)} - \boldsymbol{\sigma}^{(2)}) \cdot \mathbf{n} = (p_2 - p_1)\,\mathbf{n} = [\,\Delta\rho\,\mathbf{g} \cdot \mathbf{x} + (P_2 - P_1)\,]\,\mathbf{n} = \gamma\,2\kappa_m\,\mathbf{n}, \qquad (4.2.1)$$

where $\Delta\rho = \rho_2 - \rho_1$ is the density difference, P_1 and P_2 are two pressure functions of time attributed to each fluid, $\mathbf{n}$ is the normal unit vector pointing into fluid 1 by convention, and κ_m is the mean curvature of the interface. Rearranging, we obtain the Laplace–Young equation

$$2\kappa_m = \frac{\Delta\rho}{\gamma}\,\mathbf{g} \cdot \mathbf{x} + \lambda, \qquad (4.2.2)$$

where $\lambda \equiv (P_2 - P_1)/\gamma$ is a new function of time with dimensions of inverse length.

Capillary length and Bond number

The capillary length is a physical constant with dimensions of length defined as

$$\ell \equiv \left(\frac{\gamma}{|\Delta\rho|g}\right)^{1/2}, \qquad (4.2.3)$$

where g is the magnitude of the acceleration of gravity. For water at 20°C, the capillary length is approximately 2.5 mm. A dimensionless Bond number can be defined in the terms of the capillary length as

$$\mathrm{Bo} \equiv \frac{|\Delta\rho|ga^2}{\gamma} = \left(\frac{a}{\ell}\right)^2, \qquad (4.2.4)$$

Two-dimensional interface described as $y = f(x)$

$$\kappa = -\frac{f''}{(1 + f'^2)^{3/2}} = \frac{1}{f'}\left(\frac{1}{\sqrt{1 + f'^2}}\right)' = -\left(\frac{f'}{\sqrt{1 + f'^2}}\right)'$$

A prime denotes a derivative with respect to x

Two-dimensional interface described in plane polar coordinates as $r = R(\theta)$

$$\kappa = \frac{RR'' - 2R'^2 - R^2}{(R^2 + R'^2)^{3/2}}$$

A prime denotes a derivative with respect to θ

Two-dimensional interface described parametrically as $x = X(\xi)$ and $y = Y(\xi)$

$$\kappa = \frac{X_{\xi\xi}Y_\xi - Y_{\xi\xi}X_\xi}{(X_\xi^2 + Y_\xi^2)^{3/2}}$$

A subscript ξ denotes a derivative with respect to ξ

Two-dimensional interface described parametrically as $r = R(\xi)$ and $\theta = \Theta(\xi)$

$$\kappa = \frac{RR_{\xi\xi}\Theta_\xi - 2R_\xi^2\Theta_\xi - RR_\xi\Theta_{\xi\xi} - R^2\Theta_\xi^3}{(R_\xi^2 + R^2\Theta_\xi^2)^{3/2}}$$

A subscript ξ denotes a derivative with respect to ξ

TABLE 4.2.1   Curvature of a two-dimensional interface in several parametric forms. Fluid 1 lies above
fluid 2.

where a is a properly chosen length. For example, a can be the radius of a circular tube surrounded by an interface. In terms of the capillary length or Bond number, the Laplace–Young equation (4.2.2) takes the form

$$2\kappa_m = \pm\frac{1}{\ell^2}\,\mathbf{e}_g \cdot \mathbf{x} + \lambda \qquad \text{or} \qquad 2\kappa_m = \pm\frac{\text{Bo}}{a^2}\,\mathbf{e}_g \cdot \mathbf{x} + \lambda, \tag{4.2.5}$$

where $\mathbf{e}_g \equiv \frac{1}{g}\mathbf{g}$ is the unit vector pointing in the direction of gravity, the plus sign applies when $\Delta\rho > 0$, and the minus sign applies when $\Delta\rho < 0$.

Mean curvature and differential equations

Expressing the mean curvature in an appropriate parametric form reduces (4.2.2) into a nonlinear ordinary or partial differential equation describing the shape of the interface. Ordinary differential equations arise for two-dimensional or axisymmetric shapes, and partial differential equations arise for three-dimensional shapes. The computation of the mean curvature was discussed in Section 1.8. Expressions for the directional and mean curvature of two-dimensional, axisymmetric, and three-dimensional surfaces are summarized in Tables 4.2.1–4.2.3 in several direct or parametric forms. The independent variables in the differential equations that arise from the Laplace-Young equation are determined by the chosen parametrization.

Axisymmetric interface described in cylindrical polar coordinates as $x = f(\sigma)$

$$\kappa_1 = -\frac{f''}{(1+f'^2)^{3/2}}, \quad \kappa_2 = -\frac{1}{\sigma}\frac{f'}{\sqrt{1+f'^2}}, \quad 2\kappa_m = \kappa_1 + \kappa_2 = -\frac{1}{\sigma}\left(\frac{\sigma f'}{\sqrt{1+f'^2}}\right)'$$

A prime denotes a derivative with respect to σ

Axisymmetric interface described in cylindrical polar coordinates as $\sigma = w(x)$

$$\kappa_1 = -\frac{w''}{(1+w'^2)^{3/2}}, \quad \kappa_2 = \frac{1}{w}\frac{1}{\sqrt{1+w'^2}}, \quad 2\kappa_m = \kappa_1 + \kappa_2 = \frac{1}{w}\frac{1+w'^2 - ww''}{(1+w'^2)^{3/2}}$$

A prime denotes a derivative with respect to x

TABLE 4.2.2   Principal and mean curvatures of an axisymmetric interface in several parametric forms with fluid 1 lying in the outer space; κ_1 is the principal curvature in an azimuthal plane, and κ_2 is the principal curvature in the conjugate plane.

Three-dimensional interface described as $z = f(x, y)$

$$2\kappa_m = -\frac{(1+f_y^2)f_{xx} - 2f_x f_y f_{xy} + (1+f_x^2)f_{yy}}{(1+f_x^2+f_y^2)^{3/2}} \simeq -f_{xx} - f_{yy} \quad \text{(nearly flat)}$$

Three-dimensional interface described in cylindrical polar coordinates as $x = q(\sigma, \varphi)$

$$2\kappa_m = -\frac{1}{(1+q_\sigma^2+Q_\varphi^2)^{3/2}}\left[(1+Q_\varphi^2)q_{\sigma\sigma} + 2q_\sigma Q_\varphi \frac{Q_\varphi - q_{\sigma\varphi}}{\sigma} + (1+q_\sigma^2)\left(Q_{\varphi\varphi} + \frac{q_\sigma}{\sigma}\right)\right]$$

$$Q_\varphi = q_\varphi/\sigma, \qquad Q_{\varphi\varphi} = q_{\varphi\varphi}/\sigma^2$$

Three-dimensional interface described in spherical polar coordinates as $r = f(\theta, \varphi)$

$$\mathbf{n} = \frac{\nabla F}{|\nabla F|}, \qquad \nabla F = \mathbf{e}_r - \frac{f_\theta}{r}\mathbf{e}_\theta - \frac{f_\varphi}{r\sin\theta}\mathbf{e}_\varphi, \qquad 2\kappa_m = \nabla \cdot \mathbf{n}$$

$$2\kappa_m \simeq \frac{2}{r} - \frac{\cot\theta}{r^2}f_\theta - \frac{f_{\theta\theta}}{r^2} - \frac{f_{\varphi\varphi}}{r^2\sin^2\theta} \quad \text{(nearly spherical)}$$

TABLE 4.2.3   Mean curvature of a three-dimensional interface in several parametric forms with fluid 1 lying in the upper half-space. A subscript denotes a partial derivative.

Boundary conditions

The solution of the differential equations that arise from the Laplace–Young equation is subject to a boundary condition that specifies either the contact angle at a three-phase contact line where the interface meets a solid boundary or a third fluid, or the shape of the contact line. The first Neumann-like boundary condition is employed when a solid boundary is perfectly smooth, whereas the second Dirichlet-like boundary condition is employed when a solid surface exhibits appreciable roughness (e.g., [115]). Problems where the shape of the contact line is specified are much easier to solve than those where the contact angle is specified.

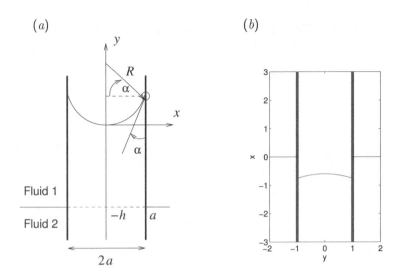

FIGURE 4.2.2 (*a*) Illustration of a meniscus developing between two parallel vertical plates separated by distance 2*a* for contact angle $\alpha < \pi/2$. (*b*) When α is greater than $\pi/2$, the meniscus submerges and the capillary rise h is negative. The meniscus shape depicted in (*b*) was produced by the FDLIB code *men_2d* (Appendix C) [318].

Interfaces with constant mean curvature

Under certain conditions, the right-hand side of (4.2.2) is nearly constant and the interface takes a shape with constant mean curvature. Constant mean-curvature shapes in three dimensions include (*a*) a sphere or a section of a sphere, (*b*) a circular cylinder or a section of a circular cylinder, (*c*) an unduloid defined as an axisymmetric surface whose trace in an azimuthal plane coincides with the focus of an ellipse rolling over the axis of revolution, (*d*) a catenoid discussed in Problem 4.2.3, and (*e*) a nodoid. Lines of constant curvature in two dimensions include a circle and a section of a circle.

Meniscus between two vertical plates

As an example, we consider the shape of a two-dimensional meniscus subtended between two vertical flat plates separated by distance 2*a*, as illustrated in Figure 4.2.2. The height of the meniscus midway between the plates is the capillary rise, h. Assuming that the meniscus takes the shape of a circular arc of radius R, and using elementary trigonometry, we find that $a = R\cos\alpha$, where α is the contact angle. The curvature of the interface is reckoned as negative when the interface is concave, as shown in Figure 4.2.2(*a*), and positive when the interface is convex, as shown in Figure 4.2.2(*b*).

The pressure profiles in the upper or lower fluids are described by (4.1.9). Equating the pressures on either side of the flat interface outside and far from the plates, located at $y = -h$, we obtain

$$\lambda \equiv \frac{P_2 - P_1}{\gamma} = -\frac{\Delta \rho g h}{\gamma}, \qquad (4.2.6)$$

where $\Delta\rho = \rho_2 - \rho_1 > 0$. Next, we substitute this expression for λ into the Laplace–Young equation (4.2.2), compute the radius of curvature from the expression $R = a/\cos\alpha$, introduce the approximation

$$2\kappa_m \simeq -\frac{1}{R} = -\frac{\cos\alpha}{a}, \tag{4.2.7}$$

evaluate the resulting equation at the midplane, $x = 0$, and find that

$$h \simeq \frac{\gamma}{\Delta\rho g a}\cos\alpha = \frac{\ell^2}{a}\cos\alpha. \tag{4.2.8}$$

The sign of the capillary rise in (4.2.8) is determined by the contact angle, α. When $\alpha < \pi/2$, the meniscus rises; when $\alpha > \pi/2$, the meniscus submerges; when $\alpha = \pi/2$ the meniscus remains flat at the level of the free surface outside the plates. The maximum possible elevation or submersion height is ℓ^2/a.

   Equation (4.2.8) can be derived directly by performing a force balance on the liquid column raised capillary action. Setting the weight of the column reduced by the buoyancy force, $2ah\,\Delta\rho g$, equal to the vertical component of the capillary force exerted at the two contact lines, $2\gamma\cos\alpha$, and rearranging, we obtain precisely (4.2.8). A formal justification for this calculation will be given in Section 4.2.3. A numerical procedure for computing the meniscus shape without any approximations will be discussed in Section 4.3.2.

   We may now return to (4.2.2) and establish the conditions under which the assumption that the right-hand side is nearly constant is valid. Requiring that the variation in the elevation of the interface, $|R|\,(1-\sin\alpha)$, is smaller than $|h|$, we find that

$$\left(\frac{a}{\ell}\right)^2 \ll 1 + \sin\alpha, \tag{4.2.9}$$

which shows that the plate separation must be sufficiently smaller than the capillary length, otherwise gravitational effects are important along the length of the meniscus.

   The predictions of equation (4.2.8) also apply when the plate separation, $2a$, changes slowly in the z direction normal to the xy plane. An example is provided by the meniscus forming between two vertical plates with small-amplitude sinusoidal corrugations.

Axisymmetric meniscus inside a circular tube

In a related application, we consider the axisymmetric meniscus inside a circular capillary tube of radius a, as shown in Figure 4.2.3. Assuming that the meniscus has constant mean curvature, taking the shape of a sphere with signed radius R, and working as in the case of the two-dimensional meniscus between two parallel plates, we find an approximate expression for the capillary rise,

$$h \simeq 2\frac{\gamma}{\Delta\rho g a}\cos\alpha = 2\frac{\ell^2}{a}\cos\alpha \tag{4.2.10}$$

which differs from (4.2.8) only by a fact of two on the right-hand side (Problem 4.2.1). The condition for this prediction to be accurate is

$$\left(\frac{a}{\ell}\right)^2 \ll 2\,(1 + \sin\alpha). \tag{4.2.11}$$

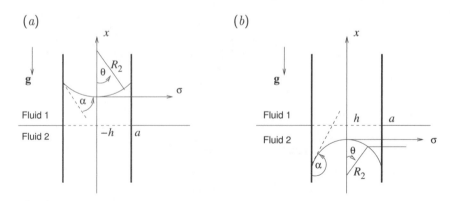

FIGURE 4.2.3 Illustration of an axisymmetric meniscus developing inside a vertical circular tube of radius
a for (a) contact angle $\alpha < \pi/2$ and (b) $\alpha > \pi/2$. The signed length R_2 is the second principal
radius of curvature.

Alternatively, expression (4.2.10) can be derived by setting the weight of the raised column reduced
by the buoyancy force, $\pi a^2 h \Delta \rho g$, equal to the vertical component of the capillary force exerted
around the contact line, $2\pi a \gamma \cos \alpha$. A numerical procedure for computing the meniscus shape
without any approximations will be discussed in Section 4.4.1.

4.2.1   Meniscus inside a dihedral corner

Consider the shape of a meniscus between two vertical intersecting plates forming a dihedral corner
with semi-angle β, as shown in Figure 4.2.4. Deep inside the corner, the meniscus may rise all the
way up to infinity under the action of surface tension. Assuming that the curvature of the trace of
the interface in a vertical plane is much smaller than the curvature of the trace of the interface in a
horizontal plane, we approximate $2\kappa_m \simeq -1/R$, where R is the radius of curvature of the interface
in a horizontal plane defined in Figure 4.2.4.

Setting the x axis upward against the direction of gravity with origin at the level of the
undeformed interface far from the walls, we find that the Laplace–Young equation (4.2.2) reduces
into

$$\frac{1}{R} \simeq \frac{\Delta \rho\, g}{\gamma}\, x, \tag{4.2.12}$$

where $\Delta \rho = \rho_2 - \rho_1$. Using elementary trigonometry, we write the geometrical condition

$$c = (R + d)\sin\beta = R\cos\alpha, \tag{4.2.13}$$

where α is the contact angle, d is the meniscus thickness at the vertical bisecting plane, as shown in
Figure 4.2.4, and the length c is defined in Figure 4.2.4. Combining equations (4.2.12) and (4.2.13),
we obtain

$$d = \frac{1-k}{k}\, R = \frac{1-k}{k}\, \frac{\gamma}{\Delta \rho\, x}, \tag{4.2.14}$$

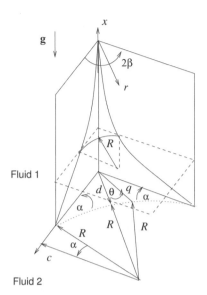

FIGURE 4.2.4   Illustration of a meniscus inside a dihedral corner confined between two vertical plates
   intersecting at angle 2β.

where $k \equiv \sin \beta / \cos \alpha$. The meniscus rises if $d > 0$, which is possible only when $\alpha + \beta < \frac{1}{2}\pi$. When
this condition is met, the meniscus takes a hyperbolic shape that diverges at the sharp corner. In
practice, the sharp corner has a nonzero curvature that causes the meniscus to rise up to a finite
capillary height. Expressions (4.2.13) and (4.2.14) validate *a posteriori* our assumption that the
curvature of the meniscus in a vertical plane is much smaller than the curvature of the interface in
a horizontal plane.

The area occupied by the liquid in a plane perpendicular to the x axis, denoted by $A(x)$, scales
as $A(x) \sim d^2$. The volume of liquid residing inside the tapering liquid tongue, denoted by V, scales
as $V \sim \int_{x_0}^{\infty} d(x)^2 \, \mathrm{d}x$, where x_0 is a positive elevation marking the beginning of the local solution.
Since $d \sim 1/x$, we find that $V \sim 1/x_0 \sim d_0$, where d_0 is the meniscus thickness corresponding to x_0.
We conclude that a finite amount of liquid resides inside the tapering liquid tongue.

The shape of the meniscus can be described in plane polar coordinates with origin at the
corner, (r, θ), by the function $r = q(x, \theta)$, as shown in Figure 4.2.4. Using the law of cosines, we find
that

$$R^2 = (d + R)^2 + q^2 - 2(d + R)\, q \, \cos\theta. \tag{4.2.15}$$

Solving this quadratic equation for q and eliminating R and d in favor of x using (4.2.12) and
(4.2.14), we find that the meniscus is described by the equation

$$x = \frac{\gamma}{\Delta \rho g} \frac{\cos\theta - \sqrt{k^2 - \sin^2 \theta}}{kr}, \tag{4.2.16}$$

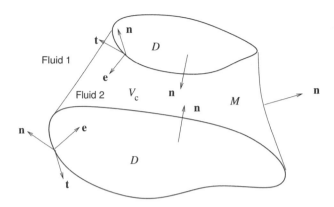

FIGURE 4.2.5   Illustration of a meniscus bounded by two closed contact angles drawn as heavy lines.

where $-\beta \leq \theta \leq \beta$. The contact lines corresponding to $\theta = \pm\beta$ are described by the equation

$$x_{cl} = \frac{\gamma}{\Delta\rho g} \frac{\cos(\alpha + \beta)}{\sin\beta} \frac{1}{r}. \tag{4.2.17}$$

We have found that the meniscus climbs up to infinity when $\alpha + \beta < \pi/2$. For a right-angled corner where $\beta = \pi/4$, such as that occurring inside a square tube, the contact line must be less than $\pi/4$. When $\alpha + \beta > \pi/2$, the meniscus does not rise to infinity but takes a bounded shape [93]. In the absence of gravity, a solution does not exist when $\alpha + \beta > \pi/2$. This discontinuous behavior underscores the important effect of boundary geometry in capillary hydrostatics.

4.2.2   Capillary force

Consider a meniscus confined by one closed contact line or a multitude of contact lines with arbitrary shape, as shown in Figure 4.2.5. Multiplying both sides of the Laplace–Young equation (4.2.2) by the normal unit vector, $\mathbf{n}$, and integrating the resulting equation over the entire surface of the meniscus, M, we obtain

$$\iint_M 2\kappa_m \, \mathbf{n} \, \mathrm{d}S = \frac{\Delta\rho}{\gamma} \iint_M (\mathbf{g} \cdot \mathbf{x}) \, \mathbf{n} \, \mathrm{d}S + \lambda \iint_M \mathbf{n} \, \mathrm{d}S. \tag{4.2.18}$$

Next, we identify a surface or a collection of surfaces, D, whose union with the meniscus, M, encloses a finite control volume, V_c, as depicted in Figure 4.2.5. Using the divergence theorem to manipulate the first integral on the right-hand side, we obtain

$$\iint_M 2\kappa_m \, \mathbf{n} \, \mathrm{d}S = \frac{\Delta\rho}{\gamma} \iint_D (\mathbf{g} \cdot \mathbf{x}) \, \mathbf{n} \, \mathrm{d}S + \lambda \iint_D \mathbf{n} \, \mathrm{d}S + \frac{\Delta\rho}{\gamma} V_c \, \mathbf{g}, \tag{4.2.19}$$

where the normal unit vector $\mathbf{n}$ over D points into the control volume, as shown in Figure 4.2.5. Applying Stokes' theorem expressed by equation (A.7.8), Appendix A, for $\mathbf{G} = \mathbf{n}$, and recalling that

$2\kappa_m = \nabla \cdot \mathbf{n}$, we obtain

$$\iint_M 2\kappa_m \, \mathbf{n} \, \mathrm{d}S = \oint_C \mathbf{n} \times \mathbf{t} \, \mathrm{d}l, \tag{4.2.20}$$

where the line integral is computed around each contact line, C, $\mathbf{t}$ is the tangent unit vector, and l is the arc length along the contact line. Substituting this expression into (4.2.19) and setting $\mathbf{e} \equiv \mathbf{n} \times \mathbf{t}$, we obtain the force balance

$$\gamma \oint_C \mathbf{e} \, \mathrm{d}l = \iint_D \left[\Delta\rho \left(\mathbf{g} \cdot \mathbf{x} \right) + \gamma\lambda \right] \mathbf{n} \, \mathrm{d}S + \Delta\rho V_c \mathbf{g}. \tag{4.2.21}$$

The left-hand side expresses the capillary force exerted around the contact lines. The last term on the right-hand side expresses the weight of the fluid residing inside the control volume, reduced by the buoyancy force. The first term on the right-hand side expresses a surface pressure force.

As an example, we consider the meniscus subtended between two vertical circular cylinders bounded by a horizontal flat bottom, and identify D with union of the surface of the cylinders below the contact lines and the the the flat bottom. If the contact lines are horizontal circles, the horizontal component of the right-hand side of (4.2.21) is zero and the horizontal component of the capillary force exerted on the outer cylinder is equal in magnitude and opposite in direction with that exerted on the outer cylinder.

4.2.3   Small deformation

When the deformation of an interface from a known equilibrium shape with constant mean curvature is small compared to its overall size, the Laplace–Young equation can be simplified by linearizing the expression for the mean curvature about the known equilibrium position corresponding, for example, to a flat or spherical shape.

As an example, we assume that a three-dimensional interface is described in Cartesian coordinates as $x = f(y, z)$, where the magnitude of f is small compared to the global dimensions of the interface and $f = 0$ yields a flat equilibrium shape consistent with the boundary conditions. Substituting into (4.2.2) the linearized expression for the mean curvature with respect to f shown in Table 4.2.1, we obtain a Helmholtz equation for f,

$$\frac{\partial^2 f}{\partial y^2} + \frac{\partial^2 f}{\partial z^2} = -\frac{\Delta\rho}{\gamma} \mathbf{g} \cdot \mathbf{x} - \lambda, \tag{4.2.22}$$

where $\Delta\rho = \rho_2 - \rho_1$. The solution must be found subject to an appropriate boundary condition at the contact line.

Problems

4.2.1 *Meniscus between a planar and a wavy plate*

Consider the meniscus established between a flat vertical plate and another vertical plane with small-amplitude vertical sinusoidal corrugations. Develop an expression for the capillary height with respect to the horizontal coordinate, z.

4.2.2 *Meniscus inside a cylindrical capillary*

Derive (4.2.10) and discuss the physical implications of (4.2.11).

4.2.3 *Catenoid*

(*a*) Verify that the equation $\sigma = f(x) = c_1 [1 + f'(x)^2]^{1/2}$, where c_1 is a constant, describes an axisymmetric interface with zero mean curvature. The surface is called a catenoid and its trace in an azimuthal plane is called a catenary.

(*b*) Show that the catenoid is described by the equation $\sigma = c_1 \cosh[(x - c_2)/c_2]$, where c_2 is a new constant.

🖥 Computer Problem

4.2.4 *A film between two rings*

A thin liquid film is subtended between two coaxial circular rings of equal radius, a, separated by distance b. Assuming that gravitational effects are insignificant and requiring that the pressures on either side of the film are equal, we find that the film must take a shape with zero mean curvature. Assuming that the film takes the shape of a catenoid discussed in Problem 4.2.3, compute the constants c_1 and c_2 by requiring that the catenoid passes through the rings, and prepare graphs of the dimensionless ratios c_1/a and c_2/a against the aspect ratio b/a. Plot the film profile in an azimuthal plane for $b/a = 0.20, 0.50, 1.00$, and 1.33.

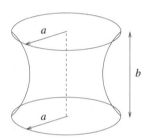

A liquid film supported by two coaxial rings.

Note: When $0 < b/a < 1.33$, two real solutions for c_1/a arise. The physically relevant solution is the one with the larger value. For $b/a > 1.33$, the solution is complex, indicating that a catenoid cannot be established.

4.3   Two-dimensional interfaces

We proceed to discuss specific methods of computing the shape of two-dimensional interfaces with uniform surface tension governed by the Laplace–Young equation (4.2.2). Our analysis will be carried out in Cartesian coordinates where the y axis points against the acceleration of gravity vector, so that $\mathbf{g} = (0, -g, 0)$. Using the first entry of Table 4.2.1, we find that the shape of an interface that is described by the equation $y = f(x)$ is governed by a second-order nonlinear ordinary differential equation,

$$\kappa = -\frac{f''}{(1 + f'^2)^{3/2}} = \frac{1}{f'}\left(\frac{1}{(1 + f'^2)^{1/2}}\right)' = -\left(\frac{f'}{(1 + f'^2)^{1/2}}\right)' = -\frac{f}{\ell^2} + \lambda, \qquad (4.3.1)$$

where $\ell = (g/\Delta\rho g)^{1/2}$ is the capillary length, $\Delta\rho = \rho_2 - \rho_1$ is the difference in the densities of the fluids on either side of the interface, and λ is a constant with dimensions of inverse length to be determined as part of the solution; fluid 2 is assumed to lie underneath fluid 1. Integrating (4.3.1)

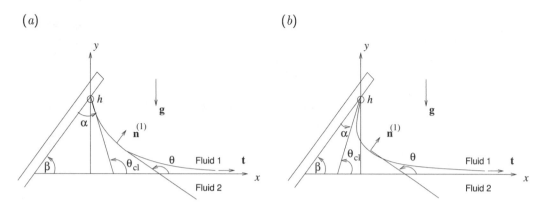

FIGURE 4.3.1   Illustration of a semi-infinite interface attached to an inclined plate with (a) a monotonic or (b) reentrant shape. Far from the plate, the interface becomes horizontal.

once with respect to x, we obtain a first-order equation,

$$\frac{1}{(1 + f'^2)^{1/2}} = -\frac{1}{2} \frac{f^2}{\ell^2} + \lambda f + \delta, \tag{4.3.2}$$

where δ is a new dimensionless constant. Expressing the slope of the interface in terms of the slope angle θ subtended between the tangent to the interface and the x axis, defined such that $f' = \tan\theta$, as shown in Figure 4.3.1, we recast (4.3.2) into the form

$$|\cos\theta| = -\frac{1}{2} \frac{f^2}{\ell^2} + \lambda f + \delta. \tag{4.3.3}$$

Rearranging (4.3.2), we derive a first-order ordinary differential equation,

$$\frac{df}{dx} = \pm \left[\left(\frac{2}{2\lambda f + 2\delta - f^2/\ell^2} \right)^2 - 1 \right]^{1/2}. \tag{4.3.4}$$

The plus or minus sign must be selected according to the expected interfacial shape. Performing a second integration subject to a stipulated boundary condition provides us with specific shapes.

4.3.1   A semi-infinite meniscus attached to an inclined plate

In the first application, we consider the shape of a semi-infinite meniscus attached to a flat plate that is inclined at an angle β with respect to a horizontal plane, as shown in Figure 4.3.1. The origin of the Cartesian axes has been set at the level of the undeformed interface underneath the contact line. Far from the plate, as x tends to infinity, the interface tends to become flat. The contact angle subtended between the inclined plate and the tangent to the interface at the contact point has a prescribed value, α. The angle subtended between the tangent to the interface at the contact line and the x axis is

$$\theta_{cl} = \alpha + \beta, \tag{4.3.5}$$

as shown in Figure 4.3.1. Since both the inclination angle β and the contact angle α vary between 0 and π, the angle θ_{cl} ranges between 0 and 2π. When $\theta_{cl} = \pi$, the meniscus is perfectly flat. This observation provides us with a practical method of measuring the contact angle in the laboratory, by varying the plate inclination angle, β, until the interface appears to be entirely flat; at that point $\alpha = \pi - \beta$. Requiring that the interface becomes flat as x tends to infinity, and therefore the curvature and slope both tend to zero, and using expressions (4.3.1) and (4.3.3), we obtain

$$\lambda = 0, \qquad \delta = 1. \tag{4.3.6}$$

The constants P_1 and P_2 defined in (4.1.6) are equal to the pressure at the horizontal interface far from the inclined plate.

Small deformation

If the slope of the interface is uniformly small along the entire meniscus, $f' \ll 1$ for $0 \leq x < \infty$, equation (4.3.1) simplifies into a linear differential equation, $f'' \simeq f/\ell^2$. A solution that decays as x tends to infinity is

$$f \simeq h\,e^{-x/\ell}, \tag{4.3.7}$$

where $h \equiv f(0)$ is the elevation of the interface at the contact line. Enforcing the contact angle boundary condition $f'(0) = \tan\theta_{cl}$, we obtain

$$\frac{h}{\ell} = -\tan\theta_{cl} \simeq \pi - \alpha - \beta, \tag{4.3.8}$$

which is valid when $\theta_{cl} \simeq \pi$.

Monotonic shapes

When the interface takes a monotonic shape, as illustrated in Figure 4.3.1(a), the slope angle θ_{cl} lies in the second or third quadrant, $\frac{1}{2}\pi \leq \theta_{cl} \leq \frac{3}{2}\pi$. Applying (4.3.3) at the contact line with $\lambda = 0$ and $\delta = 1$, we find that the capillary rise $h \equiv f(0)$ is given by

$$\frac{h^2}{\ell^2} = 2\left(1 - |\cos\theta_{cl}|\right). \tag{4.3.9}$$

Using standard trigonometric identities, we obtain

$$\frac{h}{\ell} = 2\cos\frac{\theta_{cl}}{2}, \tag{4.3.10}$$

Since $\frac{1}{2}\pi \leq \theta_{cl} \leq \frac{3}{2}\pi$, the maximum possible capillary height, occurring when $\theta_{cl} = \frac{1}{2}\pi$ or $\frac{3}{2}\pi$, is $|h|_{max} = \sqrt{2}\ell$. In the limit as θ_{cl} tends to π, we find that $h/\ell \to \pi - \theta_{cl}$, in agreement with the asymptotic solution (4.3.8).

Having obtained the elevation of the interface at the plate, we proceed to compute the interfacial shape. Equation (4.3.4) with $\lambda = 0$ and $\delta = 1$ becomes

$$\frac{df}{dx} = \pm\left[\left(\frac{2}{2 - f^2/\ell^2}\right)^2 - 1\right]^{1/2}, \tag{4.3.11}$$

where the plus or minus sign is selected according to the expected interfacial shape.

It is convenient to introduce the nondimensional variables $F \equiv f/\ell$ and $X \equiv x/\ell$. Simplifying (4.3.11), we obtain

$$\frac{dF}{dX} = \pm F \frac{\sqrt{4 - F^2}}{2 - F^2}. \tag{4.3.12}$$

Note that the denominator on the right-hand side becomes zero when $F = \pm\sqrt{2}$. At that point, the slope of the meniscus becomes infinite, signaling a transition to a reentrant shape. Equation (4.3.12) is accompanied by the contact line boundary condition

$$F(0) = \frac{h}{\ell} \equiv H = 2\cos\frac{\theta_{cl}}{2}. \tag{4.3.13}$$

Solving (4.3.12) using a standard numerical method, such as a Runge–Kutta method, provides us with a family of shapes parametrized by the angle θ_{cl}.

In practice, it is preferable to regard X as a function of F and consider the inverse of the differential equation (4.3.12),

$$\frac{dX}{dF} = \pm \frac{2 - F^2}{F\sqrt{4 - F^2}}. \tag{4.3.14}$$

Integrating with respect to F, we obtain

$$X = \pm \int_F^H \frac{2 - \omega^2}{\omega\sqrt{4 - \omega^2}} \, d\omega, \tag{4.3.15}$$

where the maximum possible value of $|H|$ is $\sqrt{2}$. Evaluating the integral with the help of mathematical tables, we find that

$$X = \pm[\,\Phi(F) - \Phi(H)\,], \tag{4.3.16}$$

where

$$\Phi(F) = \ln \frac{2 + \sqrt{4 - F^2}}{|F|} - \sqrt{4 - F^2} \tag{4.3.17}$$

(e.g., [150], pp. 81–85). To obtain the shape of the meniscus, we plot X against F in the range $0 \leq F \leq H$ if $H > 0$, or in the range $H \leq F \leq 0$ if $H < 0$.

Reentrant shapes

The preceding analysis assumes that the interface has a monotonic shape, which is true if θ_{cl} lies in the second and third quadrants, $\frac{1}{2}\pi \leq \theta_{cl} \leq \frac{3}{2}\pi$. Outside this range, the interface turns upon itself, as shown in Figure 4.3.1(*b*). Since fluid 2 lies above fluid 1 beyond the turning point where the meniscus is vertical, the sign of the curvature must be switched. The capillary rise is given by equation (4.3.9) with the minus sign replaced by the plus sign on the right-hand side,

$$\frac{h^2}{\ell^2} = 2\left(1 + \cos\theta_{cl}\right). \tag{4.3.18}$$

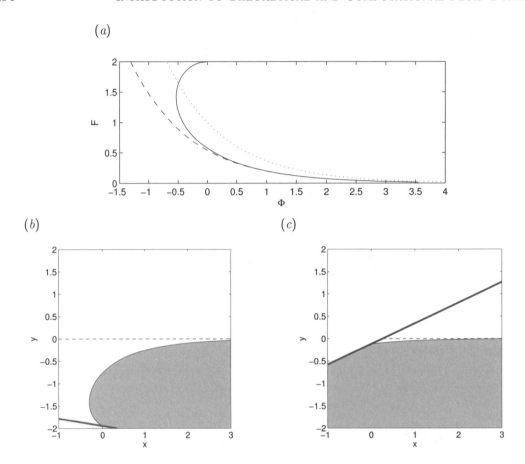

FIGURE 4.3.2   (a) The shape of the meniscus for any value of the contact angle, α, and arbitrary plate inclination angle, β, can be deduced from the master curve drawn with the solid line, representing the function Φ. The dashed and dotted lines represent approximate solutions for small interfacial deformation. (b, c) Shape of a semi-infinite meniscus attached to an inclined plate produced by the FDLIB code *men_2d_plate* (Appendix C). The plate inclination angles are different, but the contact angle is the same in both cases.

Using standard trigonometric identities, we recover (4.3.10), which shows that the maximum possible capillary height, achieved for a horizontal plate, $\theta_{cl} = 0$ or π, is $|h|_{max} = 2\ell$. The solution (4.3.16) is valid in the range $0 \leq F \leq H$ if $H > 0$, or in the range $H \leq F \leq 0$ if $H < 0$, with the maximum value of the dimensionless capillary height $|H|$ being 2.

Master curve

The shape of the interface in the complete range of θ_{cl} can be deduced from the master curve drawn with the solid line in Figure 4.3.2(a), representing the function $\Phi(F)$ defined in (4.3.17). To obtain the shape of the meniscus for a particular plate inclination angle, β, we identify the intersection of the inclined plate with the master curve consistent with the specified contact angle, α. When the capillary height is negative, we work with the mirror image of the master curve. The dashed line in Figure 4.3.2(a) represents the leading-order approximation for small deformation,

$$\Phi(F) \simeq -\ln|F| + 2(\ln 2 - 1). \tag{4.3.19}$$

Dividing the small-deformation solution (4.3.7) by ℓ and taking the logarithm of the resulting expression, we find $X \simeq -\ln|F| + \ln|H|$, which shows that $\Phi(F) \simeq -\ln|F|$, represented by the dotted line in Figure 4.3.2(a).

Numerical methods

As a practical alternative, the shape of the interface can be constructed numerically according to the following steps:

1. Compute the slope angle θ_{cl} from equation (4.3.5).
2. Compute the capillary rise h using the formulas

$$\frac{1}{\sqrt{2}}\frac{h}{\ell} = \begin{cases} (1 + |\cos(\theta_{cl})|)^{1/2} & \text{if} \quad 0 < \theta_{cl} < \frac{1}{2}\pi, \\ (1 - |\cos(\theta_{cl})|)^{1/2} & \text{if} \quad \frac{1}{2}\pi < \theta_{cl} < \pi, \\ -(1 - |\cos(\theta_{cl})|)^{1/2} & \text{if} \quad \pi < \theta_{cl} < \frac{3}{2}\pi, \\ -(1 + |\cos(\theta_{cl})|)^{1/2} & \text{if} \quad \frac{3}{2}\pi < \theta_{cl} < 2\pi. \end{cases} \tag{4.3.20}$$

3. Integrate the differential equation (4.3.11) from $f = h$ to 0 with initial condition $x(f = h) = 0$. If h is negative, we use a negative spatial step.

The method is implemented in the Matlab code *men_2d_plate* included in the software library FDLIB discussed in Appendix C. The graphics generated by the code for two plate inclination angles and fixed contact angle is shown in Figure 4.3.2(b, c).

4.3.2 Meniscus between two vertical plates

In another application, we consider the shape of the meniscus between two vertical flat plates separated by distance $2a$, as shown in Figure 4.2.2(a). We begin the analysis by requiring that the pressure at the level of the undeformed interface outside the plates, located at $y = -h$, is equal to a reference pressure P_0, and use (4.1.9) to find that $P_1 = P_0 - \rho_1 gh$ and $P_2 = P_0 - \rho_2 gh$. Based on these values, we compute the constant

$$\lambda \equiv \frac{P_2 - P_1}{g} = -\frac{\Delta\rho gh}{g} = -\frac{h}{\ell^2}, \tag{4.3.21}$$

where $\ell = (g/\Delta\rho g)^{1/2}$ is the capillary length.

Because of symmetry, the slope of the interface is zero midway between the plates. Using (4.3.1) and (4.3.2), obtain $\lambda = -f''(0)$ and $\delta = 1$. Combing the expression for λ with (4.3.21), we find that

$$\frac{h}{\ell^2} = f''(0), \tag{4.3.22}$$

which provides us with an expression for the capillary rise in terms of the *a priori* unknown curvature of the meniscus at the centerline. Substituting the derived values of λ and δ into (4.3.3), we obtain a relation between the slope of the interface and the elevation of the meniscus,

$$f(f + 2h) = 2\ell^2(1 - |\cos\theta|), \tag{4.3.23}$$

where the angle θ is defined such that $f' = \tan\theta$.

Differential equations

Equation (4.3.4) provides us with a first-order differential equation describing the shape of the interface in terms of the *a priori* unknown capillary rise, h,

$$\frac{df}{dx} = \pm\left[\left(\frac{2\ell^2}{2\ell^2 - f(f + 2h)}\right)^2 - 1\right]^{1/2}. \tag{4.3.24}$$

The boundary conditions require that $f(0) = 0$ and $f'(a) = \cot\alpha$, where α is the contact angle.

In practice, it is convenient to work with the second-order equation (4.3.1). Substituting $\lambda = -h/\ell^2$, we obtain

$$f'' = \frac{1}{\ell^2}(f + h)(1 + f'^2)^{3/2}. \tag{4.3.25}$$

The boundary conditions require that $f(0) = 0$, $f'(0) = 0$, and $f'(a) = \cot\alpha$. If the pair (f, h) is a solution for a specified contact angle, α, then the pair $(-f, -h)$ is a solution for the reflected contact angle, $\pi - \alpha$. Near the midplane, $f' \ll 1$ and the differential equation simplifies into $\tilde{f}'' \simeq \tilde{f}/\ell^2$, where $\tilde{f} = f + h$. The solution reveals a local exponential shape, $f \simeq h(1 - e^{-x/\ell})$.

Bond number

To recast equation (4.3.25) in dimensionless form, we introduce the dimensionless variables $F = f/a$, $X = x/a$, and $H = h/a$. Substituting these expressions into (4.3.25), we obtain

$$F'' = \mathrm{Bo}\,(F + H)(1 + F'^2)^{3/2}, \tag{4.3.26}$$

where $\mathrm{Bo} \equiv (a/\ell)^2 = \Delta\rho g a^2/\gamma$ is a Bond number and a prime denotes a derivative with respect to X. The boundary conditions require that $F(0) = 0$, $F'(0) = 0$, and $F'(1) = \cot\alpha$. It is now evident that the shape of the meniscus is determined by the Bond number and contact angle, α.

Numerical method

A standard procedure for solving equation (4.3.26) involves recasting it as a system of two first-order nonlinear ordinary differential equations,

$$\frac{dF}{dX} = G, \qquad \frac{dG}{dX} = \mathrm{Bo}\,(F + H)(1 + F'^2)^{3/2}, \tag{4.3.27}$$

with boundary conditions $F(0) = 0$, $G(0) = 0$, and $G(1) = \cot\alpha$. Having specified values for Bo and α, we compute the solution by iteration using a shooting method according to the following steps:

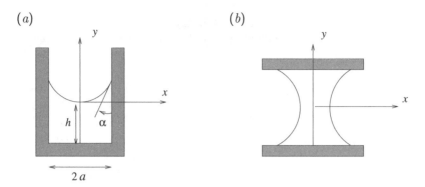

FIGURE 4.3.3   Illustration of (*a*) the free surface of a liquid in a rectangular container and (*b*) a two-dimensional liquid bridge supported by two horizontal plates.

1. Guess the capillary rise, H. A suitable guess provided by equation (4.2.8) is $H = \cos\alpha/\mathrm{Bo}$.

2. Integrate (4.3.27) from $X = 0$ to 1 with initial conditions $F(0) = 0$ and $G(0) = 0$.

3. Check whether the numerical solution satisfies the third boundary condition, $G(1) = \cot\alpha$. If not, repeat the computation with a new and improved value for H.

The improvement in the third step can be made using a method for solving nonlinear algebraic equations discussed in Section B.3, Appendix B. A meniscus shape computed by this method is shown in Figure 4.2.2(*b*).

4.3.3   Meniscus inside a closed container

A related problem addresses the shape of the interface of a fixed volume of liquid inside a rectangular container that is closed at the bottom, as shown in Figure 4.3.3(*a*). Applying (4.3.1) and (4.3.2) at the free surface midway between the vertical walls, we find that $\lambda = -f''(0)$ and $\delta = 1$. However, we may no longer relate the constant λ to the central elevation, h.

Working as in the case of a meniscus between two parallel plates, we derive the counterpart of equations (4.3.27),

$$\frac{\mathrm{d}F}{\mathrm{d}X} = G, \qquad \frac{\mathrm{d}G}{\mathrm{d}X} = \mathrm{Bo}\,(F + \Lambda)\,(1 + F'^2)^{3/2}, \tag{4.3.28}$$

where

$$\Lambda \equiv -\frac{\ell^2}{a}\lambda = \frac{\ell^2}{a}f''(0) \tag{4.3.29}$$

is an *a priori* unknown dimensionless constant. The boundary conditions require that $F(0) = 0$, $G(0) = 0$, and $G(1) = \cot\alpha$. The solution can be found using the shooting method described in Section 4.3.2, where guesses and corrections are made with respect to Λ. Assuming that the meniscus has a circular shape provides us with an educated guess, $\Lambda = \cos\alpha/\mathrm{Bo}$.

Once the interfacial shape has been found, the dimensionless midplane elevation, $H = h/a$, can be computed by requiring that the volume per unit width of the liquid has a specified value v, yielding the integral constraint

$$H = \frac{v}{2a^2} - \int_0^1 F(X)\,\mathrm{d}X. \qquad (4.3.30)$$

If H turns out to be zero or negative, the meniscus will touch or cross the bottom of the container and the solution will become devoid of physical relevance. In that case, the fluid will arrange itself inside each corner of the container according to the specified value of the contact angle. The shape of the meniscus must then be found using a different type of parametrization (Problem 4.3.1).

Problems

4.3.1 *Meniscus inside a container*

Assuming that the meniscus crosses the bottom of the container shown in Figure 4.3.3(*a*), develop an alternative appropriate parametric representation, derive the governing differential equations, and state the accompanying boundary conditions.

4.3.2 *A liquid bridge between two horizontal plates*

Derive a differential equation describing the shape of the free surface of a two-dimensional liquid bridge subtended between two horizontal parallel plates shown in Figure 4.3.3(*b*) in a convenient parametric form.

 Computer Problems

4.3.3 *Meniscus attached to a wall*

Integrate equation (4.3.12) to generate profiles of the meniscus for a eight evenly spaced values of $\alpha + \beta$ between 0 and 2π, separated by $\pi/4$.

4.3.4 *Meniscus between two plates*

Write a program that computes the shape of a two-dimensional meniscus subtended between two vertical plates for a specified Bond number and contact angle. The integration of the ordinary differential equations should be carried out using a method of your choice. Run the program to compute the profile of a meniscus of water between two plates separated by a distance $2b = 5$ mm, for contact angle $\alpha = \pi/4$ and surface tension $\gamma = 70$ dyn/cm.

4.4   Axisymmetric interfaces

To compute the shape of axisymmetric interfaces, we work as in the case of two-dimensional interfaces discussed in Section 4.3. The problem is reduced to solving a second-order ordinary differential equation involving an unspecified constant which is typically evaluated by requiring an appropriate boundary condition. Some minor complications arise when the differential equation is applied at the axis of symmetry.

4.4.1 Meniscus inside a capillary tube

As a first case study, we consider the shape of a meniscus inside an open vertical cylindrical tube of radius a immersed in a quiescent pool, as shown in Figure 4.2.3 [92]. Describing the interface as $x = f(\sigma)$ and carrying out a preliminary analysis as in Section 4.3.2 for the two-dimensional meniscus between two vertical plates, we derive the value of λ shown in (4.3.21), repeated here for convenience, $\lambda = -\Delta\rho g h/g = -h/\ell^2$.

Using the expression for the mean curvature shown in the second row of Table 4.2.2, we derive a second-order nonlinear ordinary differential equation,

$$f'' = \frac{1}{\ell^2}\,(f + h)\,(1 + f'^2)^{3/2} - \frac{1}{\sigma}\,f'\,(1 + f'^2). \tag{4.4.1}$$

The boundary conditions require that $f(0) = 0$, $f'(0) = 0$, and $f'(a) = \cot\alpha$, where α is contact line. Note that, if the pair (f, h) is a solution for a specified contact angle, α, then the pair $(-f, -h)$ is a solution for the reflected contact angle, $\pi - \alpha$. Equation (4.4.1) differs from its two-dimensional counterpart (4.3.25) by one term expressing the second principal curvature.

An apparent difficulty arises when we attempt to evaluate (4.4.1) at the tube centerline, $\sigma = 0$, as the second term on the right-hand side becomes indeterminate. However, using the regularity condition $f'(0) = 0$, we find that the mean curvature of the interface at the centerline is equal to $-f''(0)$, which can be substituted into the Laplace–Young equation to give

$$f''(0) = \frac{h}{2\ell}. \tag{4.4.2}$$

This expression can also be derived by applying the l'Hôpital rule to evaluate the right-hand side of (4.4.1).

Near the center of the cylinder, the interfacial slope is small, $f' \ll 1$. Linearizing (4.4.1), we obtain the zeroth-order modified Bessel equation

$$f'' = \frac{1}{\ell^2}\,(f + h) - \frac{f'}{\sigma}. \tag{4.4.3}$$

An acceptable solution that remains finite at the origin is

$$f(\sigma) \simeq h\,[\,\mathrm{I}_0(\sigma/\ell) - 1\,], \tag{4.4.4}$$

where I_0 is the zeroth-order modified Bessel function.

Parametric description

Following standard practice, we resolve (4.4.1) into a system of two first-order differential equations, which we then solve using a shooting method. To ensure good performance, we introduce a parametric representation in terms of the slope angle θ defined by the equation $\tan\theta = f'$, where $0 \le \theta \le -\alpha + \pi/2$. The contact angle boundary condition is automatically satisfied by this parametrization. To derive the equations governing the shape of the interface in parametric form, we write

$$f'' = \frac{\mathrm{d}f'}{\mathrm{d}\sigma} = \frac{\mathrm{d}f'}{\mathrm{d}\theta}\frac{\mathrm{d}\theta}{\mathrm{d}\sigma} = \frac{\mathrm{d}\tan\theta}{\mathrm{d}\theta}\frac{\mathrm{d}\theta}{\mathrm{d}\sigma} = \frac{1}{\cos^2\theta}\frac{\mathrm{d}\theta}{\mathrm{d}\sigma} \tag{4.4.5}$$

and

$$\frac{\mathrm{d}x}{\mathrm{d}\theta} = \frac{\mathrm{d}x}{\mathrm{d}\sigma}\frac{\mathrm{d}\sigma}{\mathrm{d}\theta} = \tan\theta\,\frac{\mathrm{d}\sigma}{\mathrm{d}\theta}. \tag{4.4.6}$$

Substituting (4.4.5) into the left-hand side of (4.4.1), replacing f' on the right-hand side by $\tan\theta$, and simplifying, we obtain

$$\frac{\mathrm{d}\sigma}{\mathrm{d}\theta} = \frac{\cos\theta}{Q}, \tag{4.4.7}$$

where

$$Q = \frac{x+h}{\ell^2} - \frac{\sin\theta}{\sigma}. \tag{4.4.8}$$

Substituting (4.4.7) into the right-hand side of (4.4.6), we find that

$$\frac{\mathrm{d}x}{\mathrm{d}\theta} = \frac{\sin\theta}{Q}. \tag{4.4.9}$$

Equations (4.4.7) and (4.4.9) provide us with the desired system of ordinary differential equations describing the shape of the meniscus in parametric form. To complete the definition of the problem, we must supply three boundary conditions: two because we have a system of two first-order ordinary differential equations, and one more because of the unspecified capillary height h. Fixing the origin and the radial location of the contact line, we stipulate that

$$\sigma(0) = 0, \qquad x(0) = 0, \qquad \sigma(\theta_{cl}) = a, \tag{4.4.10}$$

where $\theta_{cl} = \frac{1}{2}\pi - \alpha$.

The denominator Q defined in (4.4.8) becomes unspecified at the origin where $\sigma = 0$ and $\theta = 0$. Combining (4.4.2) with (4.4.5) and (4.4.6), we find that, at the origin,

$$\left(\frac{\mathrm{d}\sigma}{\mathrm{d}\theta}\right)_0 = 2\frac{\ell^2}{h}, \qquad \left(\frac{\mathrm{d}x}{\mathrm{d}\theta}\right)_0 = 0, \tag{4.4.11}$$

which is used to initialize the computation. The first equation in (4.4.11) can be derived by applying the l'Hôpital rule to evaluate the right-hand side of the first equation in (4.4.7), and then solving for $\mathrm{d}\sigma/\mathrm{d}\theta$ (Problem 4.4.1).

In terms of the dimensionless variables $\Sigma = \sigma/a$ and $X = x/a$, equations (4.4.7)–(4.4.10) become

$$\frac{\mathrm{d}\Sigma}{\mathrm{d}\theta} = \frac{\cos\theta}{W}, \qquad \frac{\mathrm{d}X}{\mathrm{d}\theta} = \frac{\sin\theta}{W}, \tag{4.4.12}$$

where

$$W = \mathrm{Bo}\,(X + H) - \frac{\sin\theta}{Q}, \qquad (4.4.13)$$

$H = h/a$, and $\mathrm{Bo} = (a/\ell)^2 = \Delta\rho g a^2/\gamma$ is the Bond number. The boundary conditions require that

$$\Sigma(0) = 0, \qquad X(0) = 0, \qquad \Sigma(\theta_{cl}) = 1. \qquad (4.4.14)$$

At the origin, $\theta = 0$,

$$\left(\frac{\mathrm{d}\Sigma}{\mathrm{d}\theta}\right)_0 = \frac{2}{\mathrm{Bo}}\frac{1}{H}, \qquad \left(\frac{\mathrm{d}X}{\mathrm{d}\theta}\right)_0 = 0. \qquad (4.4.15)$$

Numerical method

Having specified the Bond number, Bo, and contact angle, α, we compute the solution using a shooting method as described in Section 4.3.2 for the analogous problem of a meniscus between two plates according to the following steps:

1. Guess the reduced capillary height, H. An educated guess provided by equation (4.2.10) is $H = 2\cos\alpha/\mathrm{Bo}$.

2. Integrate (4.4.12) from $\theta = 0$ to θ_{cl}. To initialize the computation, we use (4.4.15).

3. Check whether the numerical solution satisfies the third boundary condition in (4.4.14). If not, the computation is repeated with a new and improved value for H.

The improvement in the third step can be done using the secant method discussed in Section B.3, Appendix B.

4.4.2   Meniscus outside a circular tube

In the second case study, we consider an infinite axisymmetric meniscus developing around a vertical circular cylinder of radius a. The circular horizontal contact line is located at $x = h$. The shape of the interface is described by a function, $x = f(\sigma)$, where σ is the distance from the cylinder centerline and the origin of the x axis is defined such that $f(\sigma)$ decays to zero far from the cylinder, $\sigma \to \infty$, as shown in Figure 4.4.1(a). The Laplace–Young equation reduces into a second-order ordinary differential equation,

$$f'' = (1 + f'^2)\left(-\frac{f'}{\sigma} + \frac{f}{\ell^2}\sqrt{1 + f'^2}\right), \qquad (4.4.16)$$

where a prime denotes a derivative with respect to σ. Prescribing the contact angle provides us with the boundary condition $f' = -\cot\alpha$ at $\sigma = a$.

Far from the cylinder, the interfacial slope is small, $f' \ll 1$, and the Laplace–Young equation (4.4.16) reduces to the zeroth-order Bessel equation,

$$f'' = -\frac{f'}{\sigma} + \frac{f}{\ell^2}. \qquad (4.4.17)$$

(a)   (b)

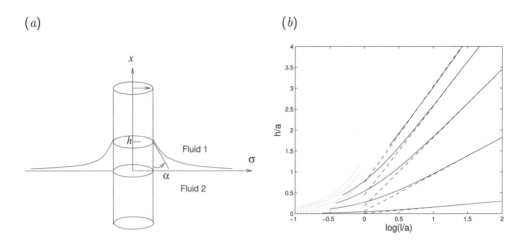

FIGURE 4.4.1   (a) Illustration of an axisymmetric meniscus developing around a vertical circular cylinder. (b) Dependence of the capillary rise, h, on the capillary length, ℓ, for contact angle $\alpha/\pi = 0.02$
(highest line), 0.125, 0.250, 0.375, and 0.480 (lowest line). The dotted lines show the capillary
height of a two-dimensional meniscus attached to a vertical flat plate. The dashed lines represent
the predictions of an approximate asymptotic solution for high capillary lengths.

An acceptable solution that decays at infinity is proportional to the modified Bessel function of zero
order, K_0,

$$f(\sigma) \simeq \xi a \, K_0(\sigma/\ell), \qquad (4.4.18)$$

where ξ is a dimensionless constant. It is beneficial to eliminate ξ by formulating the ratio between
the shape function f and its derivative, finding

$$f(\sigma) + \ell \, \frac{K_0(\sigma/\ell)}{K_1(\sigma/\ell)} \, f'(\sigma) \simeq 0, \qquad (4.4.19)$$

where K_1 is the first-order modified Bessel function.

The boundary-value problem can be solved numerically using a standard shooting method
combined with secant updates, as discussed in Section B.3, Appendix B. In the numerical method, the
solution domain is truncated at a large radial distance, σ_{max}, where the Robin boundary condition
(4.4.19) is applied [320]. The computer code is available in the software library FDLIB [318] (see
Appendix C).

Graphs of the capillary rise around the contact line are shown in Figure 4.4.1(b) together with
the predictions of an asymptotic analysis discussed later in this section, represented by the dashed
lines. As ℓ/a becomes smaller, the curvature of the cylinder becomes decreasingly important and
the results reduce to those for a two-dimensional meniscus attached to a vertical flat plate. Families
of interfacial shapes are presented in Figure 4.4.2 for two contact angles.

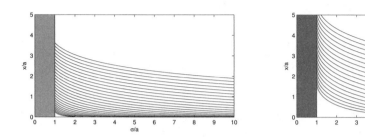

FIGURE 4.4.2 Interfacial shapes of an axisymmetric meniscus outside a circular cylinder parametrized
by the capillary length for contact angle (*a*) $\alpha = \pi/4$ and (*b*) $\pi/50$. The higher the capillary length,
the higher the interfacial profile.

Asymptotics

When the interfacial elevation is sufficiently smaller than the capillary length, $f/\ell \ll 1$, while the
slope f' is not necessarily small, equation (4.4.16) takes the gravity-free form

$$f'' = -(1 + f'^2)\frac{f'}{\sigma}, \tag{4.4.20}$$

describing a zero-mean-curvature interface. A solution of this nonlinear equation that satisfies the
prescribed contact angle boundary condition at the cylinder surface, $f'(\sigma = a) = -\cot\alpha$, is

$$f(\sigma) \sim a\cos\alpha\,\ln\frac{\delta}{\hat{\sigma} + \sqrt{\hat{\sigma}^2 - \cos^2\alpha}}, \tag{4.4.21}$$

where $\hat{\sigma} \equiv \sigma/a$ and δ is a dimensionless constant. As σ/ℓ tends to zero, the outer asymptotic
solution (4.4.18) yields

$$f(\sigma) \simeq -\xi a\left(\ln\frac{\sigma}{\ell} + \mathrm{E}\right), \tag{4.4.22}$$

where $\mathrm{E} = 0.577215665\ldots$ is Euler's constant. Matching this expression with the functional form
of (4.4.21) in the limit as $\hat{\sigma}$ tends to infinity, we recover the coefficient of the outer field, λ, the
dimensionless coefficient δ, and then an expression for the capillary rise, $h \equiv f(\sigma = a)$,

$$\xi = \cos\alpha, \qquad \frac{h}{\ell} \simeq \cos\alpha\left(\ln\frac{4\ell/a}{1 + \sin\alpha} - \mathrm{E}\right) \tag{4.4.23}$$

(Derjaguin (1946) *Dokl. Akad. Nauk USSR* **51**, 517). The predictions of this equation, represented
by the dashed lines in Figure 4.4.1(*b*), are in excellent agreement with the numerical solution. A
formal asymptotic expansion with respect to the small parameter a/ℓ is available [242].

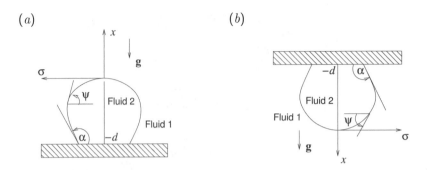

FIGURE 4.4.3   Illustration of (a) an axisymmetric sessile liquid drop resting on a horizontal plane, and (b) an axisymmetric pendant liquid drop hanging under a horizontal plate.

4.4.3   A sessile drop resting on a flat plate

In the third case study, we consider the shape of an axisymmetric bubble or drop with a specified volume resting on a horizontal plane wall, as shown in Figure 4.4.3(a). Because x may not be a single-valued function of σ, the functional form $x = \mathcal{F}(\sigma)$ is no longer appropriate and we work with the alternative representation $\sigma = f(x)$. Substituting the expression for the mean curvature from the third row of Table 4.2.1 into the Laplace–Young equation (4.3.1), we obtain the second-order ordinary differential equation

$$f'' = \left(\frac{x}{\ell^2} - \lambda\right)(1 + f'^2)^{3/2} + \frac{1 + f'^2}{f}. \tag{4.4.24}$$

Evaluating (4.4.24) at the origin, we find that $\lambda = -f''(0)$, which shows that the constant λ is equal to twice the mean curvature at the centerline.

Parametric description

To formulate the numerical problem, we introduce the slope angle ψ defined by the equation $\cot \psi = -f'$, ranging from zero at the centerline to the contact angle α at the contact line, and regard x and σ along the interface as functions of ψ, as shown in Figure 4.4.3(a). The contact angle boundary condition is satisfied automatically by this parametrization. Following the procedure that led us from equation (4.4.1) to equations (4.4.7) and (4.4.9), we resolve (4.4.24) into a system of two first-order equations,

$$\frac{\mathrm{d}x}{\mathrm{d}\psi} = \frac{\sin\theta}{Q}, \qquad \frac{\mathrm{d}\sigma}{\mathrm{d}\psi} = -\frac{\cos\theta}{Q}, \tag{4.4.25}$$

where

$$Q = \frac{\sin\psi}{\sigma} + \frac{x}{\ell^2} - \lambda. \tag{4.4.26}$$

The boundary conditions require that $\sigma(0) = 0$ and $x(0) = 0$. One more condition emerges by requiring that the drop volume drop has a specified value, V,

$$\pi \int_{-d}^{0} \sigma^2 \, \mathrm{d}x = V, \tag{4.4.27}$$

where d is the height of the drop defined in Figure 4.4.3(a).

The denominator, Q, becomes indeterminate at the centerline, $\psi = 0$. Using the l'Hôpital rule to evaluate the right-hand side of the second equation in (4.4.25), we find that

$$\left(\frac{\mathrm{d}\sigma}{\mathrm{d}\psi}\right)_0 = -\frac{1}{\left(\frac{\mathrm{d}\psi}{\mathrm{d}\sigma}\right)_0 - \lambda}. \tag{4.4.28}$$

Rearranging we find that, at the origin,

$$\left(\frac{\mathrm{d}x}{\mathrm{d}\psi}\right)_0 = 0, \qquad \left(\frac{\mathrm{d}\sigma}{\mathrm{d}\psi}\right)_0 = \frac{2}{\lambda}. \tag{4.4.29}$$

In terms of the non-dimensional variables $X = x/\ell$ and $\Sigma = \sigma/\ell$, equations (4.4.25) become

$$\frac{\mathrm{d}X}{\mathrm{d}\psi} = \frac{\sin\theta}{W}, \qquad \frac{\mathrm{d}\Sigma}{\mathrm{d}\psi} = -\frac{\cos\theta}{W}, \tag{4.4.30}$$

with boundary conditions $\Sigma(0) = 0$ and $X(0) = 0$, where

$$W = \frac{\sin\psi}{\Sigma} + X - \eta, \tag{4.4.31}$$

and $\eta \equiv \lambda\ell$. The volume constraint (4.4.27) becomes

$$\pi \int_{-D}^{0} \Sigma^2 \, \mathrm{d}X = \widehat{V}, \tag{4.4.32}$$

where $D = d/\ell$ and $\widehat{V} = V/\ell^3$. Equations (4.4.29) state that, at the origin,

$$\left(\frac{\mathrm{d}X}{\mathrm{d}\psi}\right)_0 = 0, \qquad \left(\frac{\mathrm{d}\Sigma}{\mathrm{d}\psi}\right)_0 = \frac{2}{\eta}. \tag{4.4.33}$$

The shape of the drop depends on the reduced volume, $\widehat{V}$, and contact angle, α. Given the values of $\widehat{V}$ and α, the drop shape can be found using a shooting method according to the following steps:

1. Guess the value of the constant η. A good estimate can be obtained by assuming that the drop has a spherical shape with radius $a = [3V/(4\pi)]^{1/3}$. Since the constant λ is equal to the mean curvature at the centerline, we may set $\lambda = 1/a$ and compute $\eta = \ell/a$.

2. Solve an initial-value problem from by integrating (4.4.30) from $\psi = 0$ to α. To initialize the computation, we use (4.4.33).

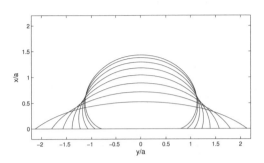 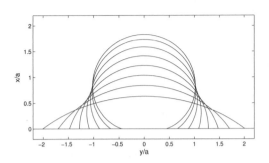

FIGURE 4.4.4   (Left) Shapes of a sessile drop for dimensionless volume $\widehat{V} \equiv V/\ell^3 = 4\pi/3$ and contact angle $\alpha/\pi = 0.1, 0.2, \ldots, 0.9$, and (right) corresponding shapes for a smaller volume, $\widehat{V} = 4\pi/3^4$. The axes are scaled by the equivalent drop radius, $a = (3V/4\pi)^{1/3}$.

3. Check whether (4.4.32) is fulfilled. If not, repeat the computation with a new and improved value for η.

Drop profiles for volume $\widehat{V} = 4\pi/3$ and $4\pi/3^4$ and contact angle $\alpha/\pi = 0.1, 0.2, \ldots, 0.9$ are shown in Figure 4.4.4. When the reduced volume $\hat{V}$ is small, the drops take the shape of sections of sphere. As the drop volume increases, the interface tends to flatten at the top due to the action of gravity.

Problems

4.4.1 *A sessile drop*

Derive the first equation in (4.4.11) by applying the l'Hôpital rule evaluate the right-hand side of the first equation in (4.4.7) and then solving for $d\sigma/d\psi$.

4.4.2 *A pendant drop*

Derive in dimensionless form the differential equations and boundary conditions governing the shape of an axisymmetric drop pending underneath a horizontal flat plate, as shown in Figure 4.4.2(*b*).

4.4.3 *A rotating drop*

Derive in dimensionless form the differential equations and boundary conditions governing the shape of an axisymmetric drop resting on, or pending underneath a horizontal flat plate that rotates steadily about the axis of the drop with angular velocity Ω.

Computer Problems

4.4.4 *Meniscus inside a capillary tube*

Compute the meniscus of water inside a capillary tube of radius $a = 2.5$ mm, for contact angle $\alpha = \pi/4$ and surface tension $\gamma = 70$ dyn/cm. *Hint:* the solution yields $h = 0.359$ mm.

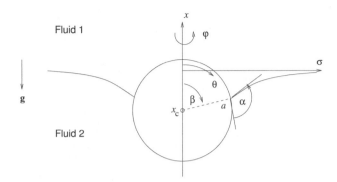

FIGURE 4.4.5   Illustration of a spherical particle straddling the interface between two immiscible fluids.

4.4.5 *A sessile drop*

Compute the shape of a sessile water drop with volume $V = 2$ ml, for contact angle $\alpha = 3\pi/4$ and surface tension $\gamma = 70$ dyn/cm.

4.4.6 *A straddling particle*

A spherical particle of radius a is floating at the interface between two fluids, as shown in Figure 4.4.5. Develop a numerical procedure for computing the particle center position, x_c, and floating angle, β, in terms of the contact angle, α, the densities of the two fluids, and the density of the sphere [317, 318].

4.5   Three-dimensional interfaces

A standard method of computing the shape of a three-dimensional interface involves describing the interface in parametric form in terms of two surface variables, ξ and η, and then computing the position of a collection of marker points located at intersections of grid lines. The successful parametrization is determined by the nature of the expected interfacial shape. The position of the marker points is computed by solving the Laplace–Young equation, which in this case reduces to a second-order nonlinear partial differential equation with respect to ξ and η, using finite-difference, finite-element, spectral or variational methods. The accurate computation of the mean curvature is a major concern. Only a few numerical methods are able to produce satisfactory accuracy due to the geometrical nonlinearity of corresponding expressions.

Numerical implementations and variational formulations are discussed by Brown [58], Milinazzo & Shinbrot [266], Hornung & Mittelmann [191], and Li & Pozrikidis [237]. Finite-difference calculations of liquid bridges connecting three equal spheres are presented by Rynhart *et al.* [356]. Finite-element calculations of wetting modes of superhydrophobic surfaces are presented by Zheng, Yu & Zhao [441]. Finite-difference methods combined with conformal mapping have been implemented in recent work [320, 319, 321].

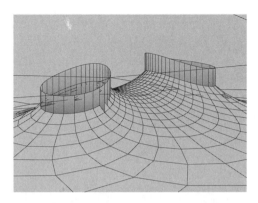

FIGURE 4.6.1   A meniscus forming between two vertical circular cylinders whose centers are separated by distance equal to the cylinder diameter, for contact angle $\alpha = \pi/4$, and capillary length $\ell/a = 4.5036$, where a is the cylinder radius.

4.6   Software

Directory *03_hydrostat* of the fluid mechanics library FDLIB contains a collection of programs that compute two-dimensional, axisymmetric, and three-dimensional hydrostatic shapes, as discussed in Appendix C.

Meniscus around an elliptical cylinder

The meniscus forming around a vertical elliptical cylinder is shown in Figure 4.2.1 for contact angle $\alpha = \pi/4$. The interfacial shape was computed by solving the Young–Laplace equation using a finite-difference method implemented in boundary-fitted elliptic coordinates generated by conformal mapping [320]. The computer code resides in Directory *men_ell* inside Directory *03_hydrostat* of FDLIB. For clarity, the height of the cylinder shown in Figure 4.2.1 reflects the elevation of the contact line. We observe that the elliptical cross-section causes significant oscillations in the capillary elevation around the value corresponding to a circle. High elevation occurs on the broad side of the cylinder, and low elevation occurs at the tip of the cylinder.

We conclude that high boundary curvature causes a local decline in the capillary height below the value corresponding to a flat plate. In the case of a meniscus attached to a wavy wall, the contact line elevation is higher in the troughs than in the crests of the corrugations [179]. This behavior is consistent with our analysis in Section 4.2.1 on the shape of a meniscus inside a dihedral corner.

Meniscus between two vertical cylinders

The meniscus developing between two vertical circular cylinders with the same radius is shown in Figure 4.6.1. The interfacial shape was generated using the computer code *men_cc* in Directory *03_hydrostat* of FDLIB. The Young–Laplace equation is solved by a finite-difference method in boundary-fitted bipolar coordinates. In the configuration shown in Figure 4.6.1, the cylinder centers are separated by distance equal to the common cylinder diameters. The contact angle has a specified value, $\alpha = \pi/4$, around both contact lines. We observe that the rise of the meniscus is most pronounced inside the gap between the cylinders.

Surface evolver

The `surface evolver` program is able to describe triangulated surfaces whose shape is determined by surface tension and other types of surface energy, subject to various constraints. Further information is given in the Internet site: http://www.susqu.edu/brakke/evolver/evolver.html.

 Computer Problem

4.6.1 *Meniscus outside an elliptical cylinder*

Use the FDLIB code *men_ell* to compute the meniscus around a vertical elliptical cylinder with constant perimeter and aspect ratio 1.0 (circle), 0.5, 0.25, and 0.125. Discuss the effect of the aspect ratio on the minimum and maximum interface elevation around the contact line.

Exact solutions

5

Having established the equations governing the motion of an incompressible Newtonian fluid and associated boundary conditions, we proceed to derive specific solutions. Not surprisingly, we find that computing analytical solutions is hindered by the presence of the nonlinear convection in the equation of motion due to the point-particle acceleration, $\mathbf{u} \cdot \nabla \mathbf{u}$, rendering the system of governing equations quadratic with respect to the velocity. Consequently, analytical solutions can be found only for a limited class of flows where the nonlinear term either happens to vanish or is assumed to make an insignificant contribution. Under more general circumstances, solutions must be found by approximate, asymptotic, and numerical methods for solving ordinary and partial differential equations. Fortunately, the availability of an extensive arsenal of classical and modern methods allows us to successfully tackle a broad range of problems under a wide range of physical conditions.

In this chapter, we discuss a family of flows whose solution can be found either analytically or numerically by solving ordinary and elementary one-dimensional partial differential equations. We begin by considering unidirectional flows in channels and tubes where the nonlinear convection term in the equation of motion is identically zero. The reason is that fluid particles travel along straight paths with constant velocity and vanishing acceleration. Swirling flows inside or outside a circular tube and between concentric cylinders provides us with a family of flows with circular streamlines where the nonlinear term due to the particle acceleration assumes a simple tractable form. Solutions for unsteady flows will be derived by separation of variables and similarity solutions, when possible.

Following the discussion of unidirectional flows, we address a rare class of flows in entirely or partially infinite domains that can be computed by solving ordinary differential equations without any approximations. Reviews and discussion of further exact solutions can be found in articles by Berker [31], Whitham [428], Lagestrom [218], Rott [355], and Wang [419, 420, 421]. These exact solutions are complemented by approximate, asymptotic, and numerical solutions for low- or high-Reynolds number flows discussed later in this book.

In Chapter 6, we discuss the properties and computation of low-Reynolds-number flows where the nonlinear convection term, $\mathbf{u} \cdot \nabla \mathbf{u}$, makes a negligible contribution to the equation of motion. These flows are governed by the continuity equation and the linearized Navier–Stokes equation describing steady, quasi-steady, or unsteady Stokes flow. A quasi-steady flow is an unsteady flow forced parametrically through time-dependent boundary conditions. Linearization allows us to build an extensive theoretical framework and derive exact solutions for a broad range of problems by analytical and efficient numerical methods.

310

In Chapters 7, 8, and 10, we discuss precisely and nearly irrotational flows where the vorticity is confined inside slender wakes and boundary layers. The velocity field in the bulk of the flow is obtained by solving Laplace's equation for a harmonic velocity potential. Once the outer irrotational flow is available, the flow inside boundary layers and wakes is computed by solving simplified versions of the equation of motion originating from the boundary-layer approximation. In Chapter 11, we discuss numerical methods for computing inviscid or nearly inviscid flows containing regions or islands of concentrated vorticity, including point vortices, vortex rings, vortex filaments, vortex patches, and vortex sheets. In the final Chapters 12 and 13, we discuss finite-difference methods for solving the complete system of governing equations for Navier–Stokes flow without any approximations apart from those involved in the implementation of the numerical method.

5.1   Steady unidirectional flows

A distinguishing feature of unidirectional flow is that all but one velocity components are identically zero in a properly defined Cartesian or cylindrical polar system of coordinates, and the non-vanishing velocity component is constant in the streamwise direction. These features allow us to considerably simplify the equation of motion and derive analytical solutions in readily computable form. The study of unidirectional flows can be traced to the pioneering works of Rayleigh, Navier, Stokes, Couette, Poiseuille, and Nusselt, for channel, tube, and film flow.

5.1.1   Rectilinear flows

One important class of unidirectional flows includes steady rectilinear flows of a fluid with uniform physical properties through a straight channel or tube due to an externally imposed pressure gradient, gravity, or longitudinal boundary motion. The streamlines are straight lines, the fluid particles travel with constant velocity and vanishing acceleration along straight paths, and the pressure gradient is uniform throughout the flow. Neglecting entrance effects and assuming that the flow occurs along the x axis, we set the negative of the streamwise pressure gradient equal to a constant χ,

$$\frac{\partial p}{\partial x} \equiv -\chi. \tag{5.1.1}$$

When the pressure at the entrance of a channel or tube is equal to that at the exit, $\partial p/\partial x = 0$ and $\chi = 0$.

Setting in the Navier–Stokes equation $\partial u_x/\partial t = 0$ for steady flow, $\partial u_x/\partial x = 0$ for fully developed flow, and $u_y = u_z = 0$ for unidirectional flow, we obtain three linear scalar equations corresponding to the x, y, and z directions,

$$\mu\left(\frac{\partial^2 u_x}{\partial y^2} + \frac{\partial^2 u_x}{\partial z^2}\right) = -(\chi + \rho g_x), \qquad \frac{\partial p}{\partial y} = \rho g_y, \qquad \frac{\partial p}{\partial z} = \rho g_z. \tag{5.1.2}$$

When the x axis is horizontal, $g_x = 0$. Nonlinear inertial terms are absent because the magnitude of the velocity is constant along the streamlines.

The pressure distribution is recovered by integrating (5.1.1) and the last two equations in (5.1.2), yielding

$$p = -\chi x + \rho\,(g_y y + g_z z) + p_0, \tag{5.1.3}$$

where p_0 is a reference constant pressure found by enforcing an appropriate boundary condition. When $\chi = -\rho g_x$, the pressure assumes the hydrostatic profile, $p = -\rho \mathbf{g} \cdot \mathbf{x} + p_0$.

Poisson equation

The x component of the equation of motion (5.1.2) provides us with a Poisson equation for the streamwise velocity component with a constant forcing term on the right-hand side,

$$\frac{\partial^2 u_x}{\partial y^2} + \frac{\partial^2 u_x}{\partial z^2} = -\frac{\chi + \rho g_x}{\mu}. \tag{5.1.4}$$

The boundary conditions specify the distribution of u_x along a channel or tube wall or the shear stress $\mu \mathbf{n} \cdot \nabla u_x$ along a free surface, where ∇ is the two-dimensional gradient operating in the yz plane and $\mathbf{n}$ is the unit vector normal to the free surface. In the case of multilayer flow, the shear stress, $\mu \mathbf{n} \cdot \nabla u_x$, is required to be continuous across the layer interface. Accordingly, the normal derivative of the velocity $\mathbf{n} \cdot \nabla u_x$ undergoes a jump determined by the viscosities of the two fluids.

Pressure-, gravity-, and boundary-driven flow

Three important cases can be recognized corresponding to pressure-, gravity-, and boundary-driven flow:

- When a channel or tube is horizontal and the walls are stationary, we obtain pressure-driven flow due to an externally imposed pressure gradient, $\partial p / \partial x = -\chi$.

- When the walls are stationary and the pressure does not change in the direction of the flow, $\chi = 0$, we obtain gravity-driven flow.

- When $\chi = -\rho g_x$, the pressure assumes the hydrostatic distribution and the flow is driven by the longitudinal translation of the whole wall or a section of a wall.

In the third case, the Poisson equation (5.1.4) reduces to Laplace's equation, $\nabla^2 u_x = 0$, whose solution is independent of the viscosity, μ. Mixed cases of pressure-, gravity-, and boundary-driven flow can be constructed by linear superposition.

5.1.2 Flow through a channel with parallel walls

In the first application, we consider two-dimensional flow in a channel confined between two parallel walls separated by distance h, as shown in Figure 5.1.1. The bottom and top walls translate parallel to themselves with constant velocities V_1 and V_2. In the framework of steady unidirectional flow, the streamwise velocity, u_x, depends on y alone. The Poisson equation (5.1.4) simplifies into a second-order ordinary differential equation,

$$\frac{\mathrm{d}^2 u_x}{\mathrm{d}y^2} = -\frac{\chi + \rho g_x}{\mu}. \tag{5.1.5}$$

Setting the origin of the Cartesian axes at the lower wall, integrating twice with respect to y, and enforcing the no-slip boundary conditions $u = V_1$ at $y = 0$ and $u = V_2$ at $y = h$, we find that the

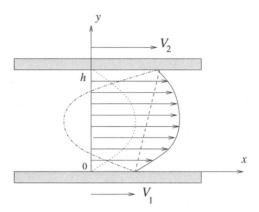

FIGURE 5.1.1   Velocity profiles of flow through a two-dimensional channel with parallel walls separated by distance h. The linear profile (dashed line) corresponds to Couette flow, the parabolic profile (dotted line) corresponds to plane Hagen–Poiseuille flow, and the intermediate profile (dot-dashed line) corresponds to flow with vanishing flow rate.

velocity profile takes the parabolic distribution

$$u_x(y) = V_1 + (V_2 - V_1)\frac{y}{h} + \frac{\chi + \rho g_x}{2\mu}\, y\,(h - y). \tag{5.1.6}$$

The first two terms on the right-hand side represent a boundary-driven flow. The third term represents pressure- and gravity-driven flow. When $\chi = -\rho g_x$, we obtain Couette flow with a linear velocity profile, also called simple shear flow [94]. When $V_1 = 0$ and $V_2 = 0$, we obtain plane Hagen–Poiseuille pressure- or gravity-driven flow (e.g., [392]). Velocity profiles corresponding to the extreme cases of Couette flow, Hagen–Poiseuille flow, and a flow with vanishing flow rate are shown in Figure 5.1.1.

The flow rate per unit width of the channel is found be integrating the velocity profile,

$$Q = \int_0^h u_x \, \mathrm{d}y = \frac{1}{2}\,(V_1 + V_2)\,h + \frac{1}{12}\frac{\chi + \rho g_x}{\mu}\, h^3. \tag{5.1.7}$$

Note the dependence on h or h^3, respectively, in the first or second term on the right-hand side. The mean fluid velocity is

$$u_x^{mean} \equiv \frac{Q}{h} = \frac{1}{2}\,(V_1 + V_2) + \frac{1}{12}\frac{\chi + \rho g_x}{\mu}\, h^2. \tag{5.1.8}$$

In the case of pressure- or gravity-driven flow, the maximum velocity occurring at the midplane, $u_x^{max} = u_x(y = \frac{1}{2}h)$ is related to the mean velocity by $u_x^{max} = \frac{3}{2}u_x^{mean}$.

5.1.3 Nearly unidirectional flows

The exact solutions for steady unidirectional flow can be used to construct approximate solutions for problems involving nearly unidirectional flow. Two examples based on unidirectional channel flow are presented in this section. The methodology will be illustrated further and formalized in Section 6.4 under the auspices of low-Reynolds-number flow.

Flow in a closed channel

Consider flow inside an elongated, closed, horizontal channel confined between two parallel belts that translate parallel to themselves with arbitrary velocities V_1 and V_2. Since fluid cannot escape through side walls, the flow rate through any cross-section of the channel must be zero, $Q = 0$, and the fluid must recirculate inside the channel. Near the side walls, we obtain a reversing flow. Far from the side walls, we obtain

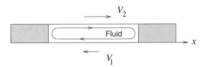

Flow in a long horizontal channel that is closed at both ends.

nearly unidirectional flow. Equation (5.1.7) reveals the spontaneous onset of a pressure field with pressure gradient

$$-\chi \equiv \frac{\partial p}{\partial x} = 6 \frac{\mu}{h^2} (V_1 + V_2),\qquad(5.1.9)$$

inducing a back flow that counteracts the primary flow due to the belt motion. The velocity profile is illustrated by the dot-dashed line in Figure 5.1.1.

Settling of a slab down a channel with parallel walls

In the second application, we consider the gravitational settling of a rectangular slab with width $2b$ and length L along the midplane of a two-dimensional container confined between two vertical plates separated by distance $2a$. The container is closed at the bottom and open to the atmosphere at the top. Our objective is to estimate the settling velocity, V, and the pressure difference established between the bottom and top due to the motion of the slab.

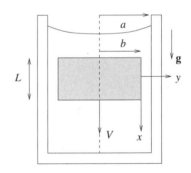

A rectangular slab settling down the centerline of a container.

To simplify the analysis, we observe that the flow between each side of the slab and the adjacent wall is nearly unidirectional flow with an *a priori* unknown negative pressure gradient, χ. Thus, the velocity profile across the right gap is given by (5.1.6) with $V_1 = V$, $V_2 = 0$, channel width $h = a - b$, and $g_x = g$. If p_{top} is the pressure at the top of the slab, then $p_{bot} = p_{top} - \chi L$ is the pressure at the bottom. The negative of the pressure gradient, χ, and velocity of settling, V, must be computed by enforcing two scalar constraints.

One constraint arises by stipulating that the rate of displacement of the liquid by the slab is equal to the upward flow rate through the gaps,

$$Q = -Vb,\qquad(5.1.10)$$

where Q is given in (5.1.7) with $g_x = g$. Simplifying, we obtain

$$\mu V + \frac{1}{6} (\chi + \rho g) \frac{(a-b)^3}{a+b} = 0. \tag{5.1.11}$$

Note that Q is negative, as required for the fluid to escape upward. A second constraint arises by balancing the x components of the surface and body forces exerted on the slab. Requiring that the pressure force at the top and bottom counterbalances the weight of the slab and the force due to the shear stress along the sides, we find that

$$(p_{top} - p_{bot}) b + \rho_s g L b + \mu \left(\frac{du_x}{dy} \right)_{y=0} L = 0 \tag{5.1.12}$$

or

$$(\chi + \rho g) b + (\rho_s - \rho) g b + \left(-\mu \frac{V}{h} + \frac{\chi + \rho g}{2} h \right) = 0, \tag{5.1.13}$$

where ρ_s is the density of the slab material. After simplification, we obtain

$$\mu V - \frac{1}{2} (\chi + \rho g) (a^2 - b^2) = \Delta\rho g b (a - b), \tag{5.1.14}$$

where $\Delta\rho = \rho_s - \rho$. Solving (5.1.11) and (5.1.15) for V and $\chi + \rho g$, we obtain

$$V = \frac{1}{4} \frac{\Delta\rho\, g a^2}{\mu} \frac{\delta (1 - \delta)^3}{1 + \delta + \delta^2}, \qquad \chi + \rho g = -\frac{3}{2} \Delta\rho\, g \frac{\delta (1 + \delta)}{1 + \delta + \delta^2}, \tag{5.1.15}$$

where $\delta = b/a$, independent of the slab length, L. We observe that the settling velocity, V, vanishes when $\delta = 0$ because the weight of the slab is infinitesimal, and also when $\delta = 1$ because the sides of the slab stick to the side plates. The maximum settling velocity occurs at the intermediate value $\delta \simeq 0.204$. In the case of a neutrally buoyant slab, $\Delta\rho = 0$, the fluid is quiescent and the pressure assumes the hydrostatic distribution, $\chi = -\rho g$.

5.1.4 Flow of a liquid film down an inclined plane

Gravity-driven flow of a liquid film down an inclined surface is encountered in a variety of engineering applications including the manufacturing of household items and magnetic recording media. Under certain conditions discussed in Section 9.12 in the context of hydrodynamic stability, the free surface is flat and the liquid film has uniform thickness, h, as shown in Figure 5.1.2.

In the inclined Cartesian coordinates depicted in Figure 5.1.2, the x velocity component, u_x, depends on the distance from the inclined plane alone, y. The components of the acceleration of the gravity vector are $\mathbf{g} = g\,(\sin\beta, -\cos\beta, 0)$, where β is the plane inclination angle with respect to the horizontal, defined such that $\beta = 0$ corresponds to a horizontal plane and $\beta = \frac{1}{2}\pi$ corresponds to a vertical plane. Since the pressure outside and therefore inside the film is independent of x, we set $\chi = 0$. The pressure distribution across the film is

$$p = \rho g \cos\beta\, (h - y) + p_{atm}, \tag{5.1.16}$$

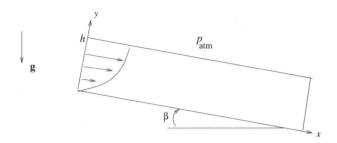

FIGURE 5.1.2   Illustration of a flat liquid film with thickness h flowing down an inclined plane due to gravity.

where p_{atm} is the ambient atmospheric pressure. Integrating the differential equation (5.1.5) twice subject to the no-slip boundary condition at the wall, $u_x = 0$ at $y = 0$, and the condition of vanishing shear stress at the free surface, $\partial u_x/\partial y = 0$ at $y = h$, we derive a semi-parabolic velocity profile first deduced independently by Hopf in 1910 and Nusselt in 1916,

$$u_x = \frac{g}{2\nu} \sin \beta \, y \, (2h - y), \qquad (5.1.17)$$

where $\nu = \mu/\rho$ is the kinematic viscosity of the fluid (e.g., [139]). The velocity at the free surface and flow rate per unit width of the film are

$$u_x^{max} = \frac{g}{2\nu} h^2 \sin \beta, \qquad Q = \int_0^h u_x \, \mathrm{d}y = \frac{g}{3\nu} h^3 \sin \beta. \qquad (5.1.18)$$

The mean velocity of the liquid is

$$u_x^{mean} \equiv \frac{Q}{h} = \frac{g}{3\nu} h^2 \sin \beta = \frac{2}{3} u_x^{max}. \qquad (5.1.19)$$

In Section 6.4.4, these equations will be used to describe unsteady, nearly unidirectional film flow down an inclined plane with a mildly sloped free surface profile.

Problem

5.1.1 *Two-layer flow*

(a) Consider the flow of two superposed layers of two different fluids in a channel with parallel walls separated by distance h. Derive the velocity profile across the two fluids in terms of the physical properties of the fluids and flow rates for the mixed case of boundary- and pressure-driven flow.

(b) Repeat (a) for gravity driven flow of two layers flowing down an inclined plate.

5.2   Steady tube flows

We proceed to derive analytical solutions for unidirectional pressure-, gravity-, and boundary-driven flows along straight tubes with various cross-sectional shapes. Unlike in channel flow, the velocity

(a)   (b)

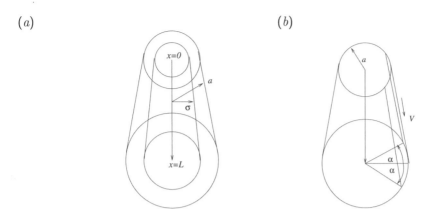

FIGURE 5.2.1   Illustration of (a) pressure- and gravity-driven flows and (b) boundary-driven flow through a tube with circular cross-section of radius a.

profile in tube flow depends on two spatial coordinates determining the position over the tube cross-section in the yz plane.

5.2.1   Circular tube

Pressure-driven flow through a circular tube, illustrated in Figure 5.2.1(a), was first considered by Hagen and Poiseuille in his treatise of blood flow (e.g., [392]).

Pressure- and gravity-driven flows

To derive the solution for axisymmetric pressure- and gravity-driven flows, we express (5.1.4) in cylindrical polar coordinates, (x, σ, φ), and derive a second-order ordinary differential equation,

$$\frac{1}{\sigma} \frac{\mathrm{d}}{\mathrm{d}\sigma} \left(\sigma \frac{\mathrm{d}u_x}{\mathrm{d}\sigma} \right) = -\frac{\chi + \rho g_x}{\mu}, \tag{5.2.1}$$

accompanied by a regularity condition at the centerline, $\mathrm{d}u_x/\mathrm{d}\sigma = 0$ at $\sigma = 0$, and the no-slip boundary condition at the tube wall, $u_x = 0$ at $\sigma = a$. Two straightforward integrations of (5.2.1) subject to these conditions yield the parabolic velocity profile

$$u_x = \frac{\chi + \rho g_x}{4\mu} \, (a^2 - \sigma^2). \tag{5.2.2}$$

The maximum velocity occurs at the tube centerline, $\sigma = 0$, and is given by

$$u_x^{max} = \frac{\chi + \rho g_x}{4\mu} \, a^2. \tag{5.2.3}$$

The flow rate through a tube cross-section is found by integrating the velocity,

$$Q \equiv \iint u_x \mathrm{d}y \, \mathrm{d}z = 2\pi \int_0^a u_x \, \sigma \, \mathrm{d}\sigma = \frac{\chi + \rho g_x}{8\mu} \, \pi a^4. \tag{5.2.4}$$

The mean fluid velocity is

$$u_x^{mean} \equiv \frac{Q}{\pi a^2} = \frac{\chi + \rho g_x}{8\mu} a^2 = \frac{1}{2} u_x^{max}. \tag{5.2.5}$$

Conversely, the centerline velocity is twice the mean velocity,

$$u_x^{max} = 2 \, u_x^{mean} = 2 \frac{Q}{\pi a^2}. \tag{5.2.6}$$

The dependence of the flow rate on the fourth power of the tube radius shown in (5.2.4) is sometimes called Poiseuille's law. This functional relationship was first inferred by Poiseuille based on experimental observations at a time when neither the no-slip boundary condition nor the parabolic velocity profile had been established.

Integral momentum balance

It is instructive to rederive the parabolic velocity profile (5.2.2) by performing an integral momentum balance over a cylindrical control volume of length L and radius σ, as shown in Figure 5.2.1(a). Requiring that the sum of the x component of the surface force exerted on the control volume balances the body force along the x axis, we obtain

$$\iint_{bottom,top} \left[(\sigma_{xx})_{x=L} - (\sigma_{xx})_{x=0} \right] \mathrm{d}S + \left(\sigma_{x\sigma} \right)_\sigma 2\pi\sigma L + \rho g_x \pi \sigma^2 L = 0. \tag{5.2.7}$$

Expressing the stresses in terms of the velocity and pressure, we obtain

$$-\iint_{bottom,top} \left[(p)_{x=L} - (p)_{x=0} \right] \mathrm{d}S + \mu \left(\frac{\mathrm{d}u_x}{\mathrm{d}\sigma} \right)_\sigma 2\pi\sigma L + \rho g_x \pi \sigma^2 L = 0. \tag{5.2.8}$$

Treating the pressure difference inside the first integral as a constant, integrating over the cross-section, and solving for the first derivative of the velocity, we obtain

$$\frac{\mathrm{d}u_x}{\mathrm{d}\sigma} = -\frac{\chi + \rho g_x}{2\mu} \sigma, \tag{5.2.9}$$

where $\chi = -\big((p)_{x=L} - (p)_{x=0}\big)/L$ is the negative of the pressure gradient. Equation (5.2.9) is the first integral of (5.2.1). Integrating (5.2.9) once subject to the no-slip boundary condition $u_x = 0$ at $\sigma = a$, we recover the parabolic profile (5.2.2).

Reynolds number and stability

The Reynolds number of the flow through a circular tube based on the tube diameter, $2a$, and the maximum velocity, u_x^{max}, is

$$\mathrm{Re} = \frac{2a\rho u_x^{max}}{\mu} = \frac{4\rho Q}{\pi \mu a} = \frac{\chi + \rho g_x}{2\mu^2} a^3. \tag{5.2.10}$$

Note the dependence on the third power of the tube radius. In practice, the rectilinear flow described by (5.2.2) is established when the Reynolds number is below a critical threshold that depends on

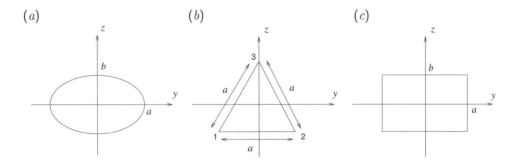

FIGURE 5.2.2   Illustration of pressure- and gravity-driven flows through a tube with (a) elliptical, (b) equilateral triangular, and (c) rectangular cross-section.

the wall roughness, as discussed in Section 9.8.6 in the context of hydrodynamic stability. As the Reynolds number increases above this threshold, the streamlines becomes wavy, random motion appears, and a turbulent flow is ultimately established.

Boundary-driven flow

Next, we consider flow in a horizontal circular tube driven by the translation of a sector of the tube wall with aperture angle 2α confined between the azimuthal planes $-\alpha < \varphi < \alpha$ in the absence of a pressure drop, as shown in Figure 5.2.1(b). To compute the velocity profile, we solve Laplace's equation, $\nabla^2 u_x = 0$, subject to the no-slip boundary condition $u_x = 0$ at $\sigma = a$, except that $u_x = V$ at $\sigma = a$ and $-\alpha < \varphi < \alpha$, where V is the boundary velocity. The solution can be found using the Poisson integral formula (2.5.13), and is given by

$$u_x(\mathbf{x}_0) = V \, \frac{a^2 - \sigma^2}{2\pi} \int_{-\alpha}^{\alpha} \frac{d\varphi}{a^2 + \sigma^2 - 2a\sigma \cos(\varphi_0 - \varphi)}. \tag{5.2.11}$$

Performing the integration, we obtain

$$u_x(\sigma, \varphi) = \frac{V}{\pi} \left[\arctan\left(\frac{a+\sigma}{a-\sigma} \tan \frac{\alpha - \varphi}{2}\right) + \arctan\left(\frac{a+\sigma}{a-\sigma} \tan \frac{\alpha + \varphi}{2}\right) \right]. \tag{5.2.12}$$

In the limit $\alpha \to \pi$ the fluid translates as a rigid body with the wall velocity V in a plug-flow mode. The flow rate must be found by integrating the velocity profile using numerical methods [317].

5.2.2   Elliptical tube

Now we consider flow through a tube an elliptical cross-section, as shown in Figure 5.2.2(a), where a and b are the tube semi-axes along the y and z axes. The velocity profiles for pressure- and gravity-driven flows follows from the observation that the equation of the ellipse, $\mathcal{F}(y, z) = 0$, involves a quadratic shape function of the cross-sectional coordinates y and z satisfying Poisson's equation with a constant forcing term on the right-hand side,

$$\mathcal{F}(y, z) = 1 - \frac{y^2}{a^2} - \frac{z^2}{b^2}. \tag{5.2.13}$$

The velocity profile is found by inspection,

$$u_x(y, z) = \frac{\chi + \rho g_x}{2\mu} \frac{a^2 b^2}{a^2 + b^2} \mathcal{F}(y, z). \tag{5.2.14}$$

The maximum velocity occurs at the center of the elliptical cross-section. The flow rate is found by integrating the velocity over the elliptical cross-section [311], yielding

$$Q = \pi \frac{\chi + \rho g_x}{4\mu} \frac{a^3 b^3}{a^2 + b^2}. \tag{5.2.15}$$

When $a = b$, we recover our results in Section 5.2.1 for a circular tube.

5.2.3   Triangular tube

Next, we consider flow through a tube whose cross-section is an equilateral triangle with side length a, as shown in Figure 5.2.2(b). The three vertices of the triangular contour are located at the (y, z) doublets

$$\mathbf{v}_1 = a \frac{1}{2} \left(-1, -\frac{1}{\sqrt{3}}\right), \qquad \mathbf{v}_2 = a \frac{1}{2} \left(1, -\frac{1}{\sqrt{3}}\right), \qquad \mathbf{v}_3 = a \left(0, \frac{1}{\sqrt{3}}\right), \tag{5.2.16}$$

where the origin has been set at the centroid of the triangle. The tube contour is described by the cubic equation $\mathcal{F}(y, z) = 0$, where

$$\mathcal{F}(y, z) = \frac{1}{a^3} (2\sqrt{3}\, z + a)(\sqrt{3}\, z + 3y - a)(\sqrt{3}\, z - 3y - a) \tag{5.2.17}$$

is a dimensionless shape function. The term inside the first parentheses is zero on the lower side, the term inside the second parentheses is zero on the left side, and the term inside the third parentheses is zero on the right side. Carrying out the multiplications, we obtain

$$\mathcal{F}(y, z) = 1 - 9(\widehat{y}^2 + \widehat{z}^2) - 6\sqrt{3}(3\widehat{z}\widehat{y}^2 - \widehat{z}^3), \tag{5.2.18}$$

where $\widehat{y} = y/a$ and $\widehat{z} = z/a$. We note that

$$\nabla^2 \mathcal{F}(y, z) = -\frac{36}{a^2}, \tag{5.2.19}$$

and conclude that the velocity profile in pressure- and gravity-driven flows is a cubic polynomial in y and z, given by

$$u_x(y, z) = \frac{\chi + \rho g_x}{36\mu} a^2 \mathcal{F}(y, z) \tag{5.2.20}$$

(Clairborne 1952, see [382]). Integrating the velocity over the tube cross-section, we derive an expression for the flow rate,

$$Q = \frac{\sqrt{3}}{320} \frac{\chi + \rho g_x}{\mu} a^4. \tag{5.2.21}$$

The dependence of the flow rate on the fourth power of a linear tube dimension, in this case the side length, a, is typical of pressure-driven tube flow.

Sparrow [382] derives the velocity profile in a isosceles triangular tube in terms of an infinite series. Other triangular profiles will be discussed in Section 5.2.9.

5.2.4 Rectangular tube

In the fourth configuration, we consider flow through a tube with rectangular cross-section whose sides along the y and z axes are equal to $2a$ and $2b$, as shown in Figure 5.2.2(c). Setting the origin at the centerpoint, expressing the velocity as the sum of a parabolic component with respect to z that satisfies the Poisson equation (5.1.4) and a homogeneous component that satisfies Laplace's equation, and then using Fourier expansions in the y direction to compute the homogeneous component subject to appropriate boundary conditions, we obtain the velocity profile

$$u_x(y,z) = \frac{\chi + \rho g_x}{2\mu} \, b^2 \mathcal{F}(y,z), \tag{5.2.22}$$

where

$$\mathcal{F}(y,z) = 1 - \frac{z^2}{b^2} + 4 \sum_{n=1}^{\infty} \frac{(-1)^n}{\alpha_n^3} \frac{\cosh(\alpha_n \frac{y}{b})}{\cosh(\alpha_n \frac{a}{b})} \, \cos(\alpha_n \frac{z}{b}) \tag{5.2.23}$$

is a dimensionless function and $\alpha_n = (n - \frac{1}{2})\pi$. Integrating the velocity profile over the tube cross-section, we derive the flow rate

$$Q = \frac{4}{3} \frac{\chi + \rho g_x}{\mu} \, ab^3 \, \phi(\frac{a}{b}), \tag{5.2.24}$$

where

$$\phi(\xi) = 1 - \frac{6}{\xi} \sum_{n=1}^{\infty} \frac{1}{\alpha_n^5} \, \tanh(\alpha_n \xi). \tag{5.2.25}$$

As ξ tends to infinity, the dimensionless function $\phi(\xi)$ tends to unity. As ξ tends to zero, $\phi(\xi)$ behaves like ξ^2. Both cases correspond to flow through a two-dimensional channel with parallel walls described by equation (5.1.6).

5.2.5 Annular tube

Next, we consider unidirectional flow through the annular space confined between two concentric circular cylinders with radii R_1 and $R_2 > R_1$, as shown in Figure 5.2.3. The cylinders are allowed to translate with respective velocities equal to V_1 and V_2 along their length, in the presence of an independent axial pressure gradient. Expressing the Poisson equation (5.1.4) in cylindrical polar coordinates, integrating twice with respect to the distance from the centerline, σ, and enforcing the boundary conditions $u_x = V_1$ at $\sigma = R_1$ and $u_x = V_2$ at $\sigma = R_2$, we derive the velocity profile

$$u_x(\sigma) = V_2 + (V_1 - V_2) \frac{\ln(R_2/\sigma)}{\ln \delta} + \frac{\chi + \rho g_x}{4\mu} \left(R_2^2 - \sigma^2 - (R_2^2 - R_1^2) \frac{\ln(R_2/\sigma)}{\ln \delta} \right), \tag{5.2.26}$$

where $\delta = R_2/R_1 > 1$ is the tube radius ratio. The wall shear stress on each cylinder arises by straightforward differentiation. The flow rate tube is found by straightforward integration, yielding

$$Q \equiv 2\pi \int_{R_1}^{R_2} u_x(\sigma) \, \sigma \, d\sigma = \pi \left[V_2 R_2^2 - V_1 R_1^2 - \frac{1}{2}(V_2 - V_1) \frac{R_2^2 - R_1^2}{\ln \delta} \right]$$

$$+ \pi \frac{\chi + \rho g_x}{8\mu} (R_2^2 - R_1^2)\left(R_2^2 + R_1^2 - \frac{R_2^2 - R_1^2}{\ln \delta} \right). \tag{5.2.27}$$

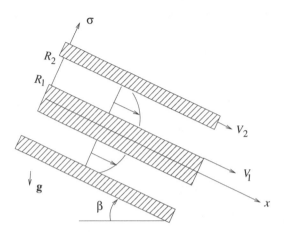

FIGURE 5.2.3   Illustration of flow in the annular space confined two inclined infinite concentric cylinders.

If the tube is closed at both ends, the pressure drop is such that the flow rate is zero. When the clearance of the annular channel is small compared to the inner cylinder radius, $h \equiv R_2 - R_1 \ll R_1$, the curvature of the walls becomes insignificant and expressions (5.2.26) and (5.2.27) reduce to (5.1.6) and (5.1.7) with $y = \sigma - R_1$ describing flow in a channel confined between two parallel walls (Problem 5.2.8).

A cylindrical object moving inside a circular tube

As an application, we consider flow past or due to the motion of a cylindrical object with length $2b$ and radius c inside a cylindrical tube of radius a, as shown in Figure 5.2.4. If the length of the object, $2b$, is large compared to the annular gap, $a - c$, the flow inside the annular gap is approximately unidirectional. The velocity profile inside the gap arises by applying (5.2.26) with $R_1 = c$, $R_2 = a$, $V_1 = V$, $V_2 = 0$, and $g_x = 0$, yielding

$$u_x(\sigma) = V \frac{\ln(a/\sigma)}{\ln \delta} + \frac{\chi}{4\mu} \left(a^2 - \sigma^2 - (a^2 - c^2) \frac{\ln(a/\sigma)}{\ln \delta} \right), \tag{5.2.28}$$

where $\delta = a/c > 1$. The corresponding flow rate is

$$Q_{gap} = \pi V \left(\frac{1}{2} \frac{a^2 - c^2}{\ln \delta} - c^2 \right) + \pi \frac{\chi}{8\mu} (a^2 - c^2)(a^2 + c^2 - \frac{a^2 - c^2}{\ln \delta}). \tag{5.2.29}$$

The axial force exerted on the object can be computed by summing the pressure force normal to the two discoidal sides and the shear force exerted on the cylindrical side with surface area $4\pi bc$, yielding

$$F_x = \pi c^2 \left((p)_{x=-b} - (p)_{x=b} \right) + 4\pi \mu bc \left(\frac{\partial u_x}{\partial \sigma} \right)_{\sigma=c}, \tag{5.2.30}$$

where

$$\left(\frac{\partial u_x}{\partial \sigma} \right)_{\sigma=c} = \frac{c}{\ln \delta} \left[-\frac{V}{c^2} + \frac{\chi}{4\mu} \left(-2\ln \delta + \delta^2 - 1 \right) \right]. \tag{5.2.31}$$

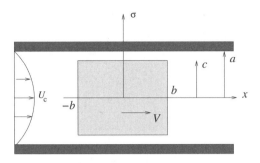

FIGURE 5.2.4   Flow past or due to the motion of a tightly fitting cylinder of radius c inside a cylindrical tube of radius $a > c$.

Setting $(p)_{x=-b} - (p)_{x=b} = 2b\chi$ and simplifying, we obtain

$$F_x = ba^2 \frac{\pi}{\ln \delta} \Big(-\frac{4\mu V}{a^2} + \chi \frac{\delta^2 - 1}{\delta^2} \Big). \tag{5.2.32}$$

Two modular cases are of interest. In the first case, the cylinder moves with velocity V inside a tube that is closed at both ends. Mass conservation requires that the flow rate through the narrow annular gap balances the rate of displacement of the fluid on either side of the cylinder, $Q_{gap} = -V\pi c^2$. Evaluating the flow rate using (5.2.27) with $R_1 = c$, $R_2 = a$, $V_1 = V$, $V_2 = 0$, and $g_x = 0$, and rearranging, we obtain the negative of the pressure gradient,

$$\chi_V = -\frac{4\mu V}{a^2} \frac{\delta^2}{-\delta^2 + 1 + (\delta^2 + 1)\ln \delta}. \tag{5.2.33}$$

The second fraction on the right-hand side is identified with a positive dimensionless pressure gradient coefficient, c_p^V. The velocity profile arises by setting in (5.2.28) $\chi = \chi_V$. Substituting (5.2.33) into (5.2.32), we obtain the drag force

$$F_x = -4\pi b\mu V \frac{\delta^2 + 1}{-\delta^2 + 1 + (\delta^2 + 1)\ln \delta}. \tag{5.2.34}$$

The associated drag coefficient is

$$c_V \equiv -\frac{F_x}{6\pi\mu cV} = \frac{2}{3}\varepsilon \frac{\delta^2 + 1}{-\delta^2 + 1 + (\delta^2 + 1)\ln \delta}, \tag{5.2.35}$$

where $\varepsilon \equiv b/c$ is the object aspect ratio.

In the second modular case, the cylinder is stationary and the fluid is forced to move at a constant flow rate, $Q_{gap} = \frac{1}{2}U_c\pi a^2$, where U_c is the centerline velocity of an equivalent Poiseuille flow. Evaluating the flow rate using (5.2.27) with $R_1 = c$, $R_2 = a$, $V_1 = 0$, $V_2 = 0$, and $g_x = 0$, and rearranging, we obtain the negative of the pressure gradient,

$$\chi_U = \frac{4\mu U_c}{a^2} \frac{\delta^4}{\delta^4 - 1 - (\delta^2 - 1)^2/\ln \delta}. \tag{5.2.36}$$

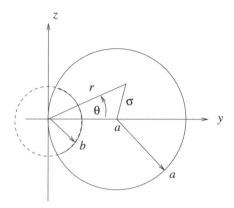

FIGURE 5.2.5   Illustration of a circular tube of radius a indented with a circular sector of radius b.

The second fraction on the right-hand side can be identified with a positive dimensionless pressure gradient coefficient, c_p^U. The velocity profile arises by setting in (5.2.28) $V = 0$ and $\chi = \chi_U$. Substituting (5.2.36) into (5.2.32), we obtain the drag force

$$F_x = 4\pi\mu b\, U_c \frac{\delta^2}{-\delta^2 + 1 + (\delta^2 + 1)\ln\delta}. \tag{5.2.37}$$

The associated drag coefficient is

$$c_U \equiv \frac{F_x}{6\pi\mu c U_c} = \frac{2}{3}\,\varepsilon\,\frac{\delta^2}{-\delta^2 + 1 + (\delta^2 + 1)\ln\delta}. \tag{5.2.38}$$

The negative of the pressure drop and drag force coefficient in the general case arise by superposition,

$$\chi = \chi_V + \chi_U = -\frac{4\mu}{a^2}\left(V c_p^V - U_c c_p^U\right), \qquad F_x = F_x^V + F_x^U = -6\pi\mu c\left(V c_D^V - U_c c_D^U\right). \tag{5.2.39}$$

In the case of a freely suspended particle, we set $F_x = 0$ and find that $V/U_c = c_U/c_V$.

5.2.6   Further shapes

Berker [31] discusses solutions for pressure- and gravity-driven flows through tubes with a variety of cross-sectional shapes, including the half moon, the circular sector, the limaçon, and the eccentric annulus. The velocity distribution and flow rate are typically expressed in terms of an infinite series similar to those shown in (5.2.22) and (5.2.25).

Indented circular tube

As an example, we consider pressure- and gravity-driven flows through a circular tube of radius a indented with a circular sector of radius b, as shown in Figure 5.2.5. The origin of the Cartesian axes has been placed at the leftmost point of the unperturbed circular contour. The axial velocity, u_x, satisfies the Poisson equation

$$\nabla^2 u_x = \frac{1}{r}\frac{\partial}{\partial r}\left(r\frac{\partial u_x}{\partial r}\right) + \frac{1}{r^2}\frac{\partial^2 u_x}{\partial \theta^2} = -\frac{\chi + \rho g_x}{\mu}, \tag{5.2.40}$$

where r is the distance from the origin and θ is the corresponding polar angle. Using the law of cosines, we find that

$$a^2 - \sigma^2 = 2ar\cos\theta - r^2, \tag{5.2.41}$$

where σ is the distance from the center of the outer circle. The boundary conditions specify that $u_x = 0$ at $\sigma = a$ and $r = 2a\cos\theta$ (outer circle), and $u_x = 0$ at $r = b$ (inner circle). The velocity profile can be constructed by modifying the parabolic profile of the Poiseuille flow, obtaining

$$u_x = \frac{\chi + \rho g_x}{4\mu}(a^2 - \sigma^2)(1 - \frac{b^2}{r^2}) = \frac{\chi + \rho g_x}{4\mu}(2\frac{a}{r}\cos\theta - 1)(r^2 - b^2). \tag{5.2.42}$$

When $b = 0$, we recover Poiseuille flow through a circular tube.

5.2.7 General solution in complex variables

It is possible to derive the general solution for the velocity field in pressure-, gravity-, and boundary-driven flow through a straight tube with arbitrary cross-section based on a formulation in complex variables. In the case of pressure- or gravity-driven flows, we decompose the velocity field into a particular component and a homogeneous component,

$$u_x(y, z) = \frac{\chi + \rho g_x}{\mu}\left[v(y, z) + f(y, z)\right], \tag{5.2.43}$$

where v is a particular solution satisfying the Poisson equation $\nabla^2 v + 1 = 0$, and the function f satisfies the Laplace equation, $\nabla^2 f = 0$. A convenient choice for v is the quadratic function

$$v(x, y) = a^2 - \frac{1}{2}\left[\alpha(y - y_R)^2 + (1 - \alpha)(z - z_R)^2\right], \tag{5.2.44}$$

where a is a chosen length, α is an arbitrary dimensionless parameter, and (y_R, z_R) are the coordinates of an arbitrary reference point in the yz plane. Straightforward differentiation confirms that the function v satisfies Poisson's equation $\nabla^2 v + 1 = 0$ for any value of α, as required.

The no-slip boundary condition requires that $f = -v$ around the tube contour. The problem has been reduced to computing the harmonic function f subject to the Dirichlet boundary condition. The theory of functions of a complex variable guarantees that f can be regarded as the real part of an analytic function, $G(w)$, where $w = y + iz$ is the complex variable in the yz plane and i is the imaginary unit, $f = \text{Real}\{G(w)\}$. To compute the solution, we introduce a function $w = \mathcal{F}(\zeta)$ that maps a disk of radius ϱ centered at the origin of the ζ complex plane to the cross-section of the tube in the physical w plane, as discussed in Section 7.11. The tube contour is described parametrically by the functions

$$y_c(\theta) = \text{Real}\{\mathcal{F}(\varrho e^{i\theta})\}, \qquad z_c(\theta) = \text{Imag}\{\mathcal{F}(\varrho e^{i\theta})\}, \tag{5.2.45}$$

where θ is the polar angle in the ζ plane, $0 \leq \theta < 2\pi$. Writing $\zeta = \sigma e^{i\varphi}$ and using the Poisson integral (2.5.13), we obtain

$$f(y,z) = \frac{\varrho^2 - \sigma^2}{2\pi} \int_0^{2\pi} \frac{v[\,y_c(\theta'), z_c(\theta')\,]}{\varrho^2 + \sigma^2 - 2\varrho\sigma \cos(\theta - \theta')} \, d\theta',$$

(5.2.46)

where

$$y(\sigma, \theta) = \mathrm{Real}\{\mathcal{F}(\sigma e^{i\theta})\}, \qquad z(\sigma, \theta) = \mathrm{Imag}\{\mathcal{F}(\sigma e^{i\theta})\},$$

(5.2.47)

and $0 \leq \sigma \leq \varrho$. In general, the integral in (5.2.46) must be evaluated by numerical methods. The computation of the required mapping function $\mathcal{F}(\zeta)$ is discussed in Section 7.11.

5.2.8   Boundary-integral formulation

The axial velocity profile satisfying the Poisson equation can be computed by solving an integral equation for a properly defined homogeneous component based on the decomposition (5.2.43). Pertinent computer codes and a discussion of the numerical method can be found in the boundary-element library BEMLIB [313] (Appendix C).

5.2.9   Triangles and polygons with rounded edges

Motivated by the solution for flow through an elliptic tube given in (5.2.14), we consider the velocity profile

$$u_x(y,z) = \frac{\chi + \rho g_x}{2\mu} \frac{a^2 b^2}{a^2 + b^2} \mathcal{F}(y,z),$$

(5.2.48)

where a and b are two specified lengths,

$$\mathcal{F}(y,z) = 1 - \frac{y^2}{a^2} - \frac{z^2}{b^2} - \mathcal{H}(y,z),$$

(5.2.49)

is a suitable function, and $\mathcal{H}(y,z)$ is a harmonic function in the yz plane satisfying

$$\frac{\partial^2 \mathcal{H}}{\partial y^2} + \frac{\partial^2 \mathcal{H}}{\partial z^2} = 0.$$

(5.2.50)

The tube contour where $u_x = 0$ is described implicitly by the equation $\mathcal{F}(y,z) = 0$. When $\mathcal{H} = 0$, we obtain an elliptical tube with semiaxes a and b. More generally, the tube shape must be found by numerical methods.

To obtain tubes with polygonal profiles, we define the complex variable $w = y + iz$ and introduce the function

$$\mathcal{H}(y,z) = \alpha \, \mathrm{Imag}\Big\{ e^{i\phi} \Big(\frac{w}{c}\Big)^m \Big\},$$

(5.2.51)

where α is a dimensionless coefficient, i is the imaginary unit, $i^2 = -1$, c is a specified length, m is a specified integer determining the polygonal shape, and ϕ is a specified phase angle.

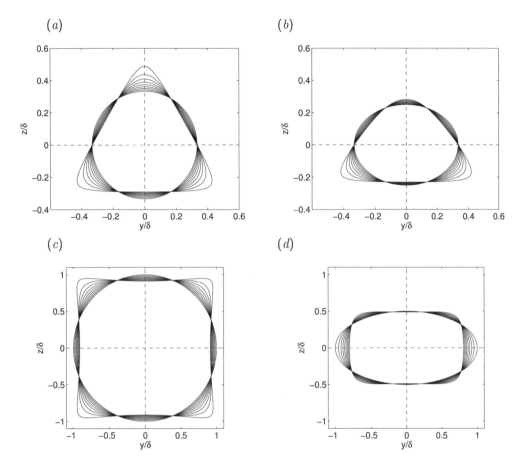

FIGURE 5.2.6   (*a, b*) Tube contours resembling triangles with rounded corners ranging between a circle and a triangle. (*c, d*) Tube contours resembling squares or rectangles with rounded corners.

Rounded triangles

For $a = b = c = \frac{1}{3}\,\delta$, $m = 3$, and $\phi = 0$, we obtain

$$\mathcal{F}(y, z) = 1 - 9\,(\hat{y}^2 + \hat{z}^2) - 27\alpha\,(3\hat{z}\hat{y}^2 - \hat{z}^3), \qquad (5.2.52)$$

where $\hat{y} = y/\delta$, $\hat{z} = z/\delta$, and δ is a specified length [422]. When $\alpha = 0$, the tube contour is a circle of radius $\frac{1}{3}\,\delta$ centered at the origin. Comparison with (5.2.18) reveals that, when $\alpha = 2/3^{3/2}$, the tube contour is an equilateral triangle with side length equal to δ. Figure 5.2.6(*a*) shows contours describing equilateral triangles with rounded corners for intermediate values of α.

For $a = \frac{1}{3}\,\delta$, $b = \frac{1}{4}\,\delta$, $c = \frac{1}{3}\,\delta$, $m = 3$, and $\phi = 0$, we obtain a family of isosceles triangles with rounded corners parametrized by α, as shown in Figure 5.2.6(*b*).

Rounded squares and rectangles

A variety of other shapes resembling m-sided polygons with rounded corners can be produced working in a similar fashion. For $a = b = c = \delta$, $m = 4$, and $\phi = 0$, we obtain

$$\mathcal{F}(y, z) = 1 - \hat{y}^2 + \hat{z}^2 - \alpha\,(\hat{y}^4 - 6\,\hat{y}^2\hat{z}^2 + \hat{z}^4). \tag{5.2.53}$$

When $\alpha = 0$, the tube contour is a circle of radius δ. As α increases, squares with rounded corners appear, as shown in Figure 5.2.6(c).

For $a = \delta$, $b = \frac{1}{2}\delta$, $c = \delta$, $m = 4$, and $\phi = 0$, we obtain a family of rectangles with rounded corners parametrized by α, as shown in Figure 5.2.6(d).

Problems

5.2.1 *Flow through a triangular tube with partitions*

Consider pressure-driven flow through a channel whose cross-section in the yz plane is parametrized by an index n as follows: $n = 1$ corresponds to a channel whose cross-section is an equilateral triangle with side length equal to a; $n = 2$ corresponds to a partitioned channel that derives from the channel for $n = 1$ by introducing three straight segments connecting the midpoints of the sides of the original triangle; each time n is increased by one unit, each triangle is partitioned into four smaller triangles by connecting midpoints of its sides. The number of triangles at the nth level is equal to 4^{n-1}. Show that the flow rate through the nth member of the family is given by

$$Q = \frac{1}{8^{n-1}} \frac{\sqrt{3}}{320} \frac{\chi + \rho g_x}{\mu}\, a^4. \tag{5.2.54}$$

Discuss the behavior of the flow rate in the limit as n tends to infinity.

5.2.2 *A two-dimensional paint brush*

Taylor developed a simple model for estimating the amount of paint deposited onto a plane wall during brushing [400]. The brush is modeled as an infinite array of semi-infinite parallel plates separated by distance $2a$, sliding at a right angle with velocity V over a flat painted surface. The space between the plates is filled with a liquid.

(a) Show that the velocity distribution inside a brush channel is

$$u_x(y, z) = V\left[1 + 2\sum_{n=1}^{\infty} \frac{(-1)^n}{\alpha_n} \exp(-\alpha_n \frac{z}{a})\, \cos(\alpha_n \frac{y}{a})\right], \tag{5.2.55}$$

where $\alpha_n = (n - \frac{1}{2})\pi$ (e.g., [318]). The origin has been set on the painted surface midway between two plates; the z axis is perpendicular to the painted surface.

(b) The amount of liquid deposited on the surface per channel is

$$Q = \int_0^\infty\!\!\int_{-a}^{a} [V - u_x(y, z)]\,\mathrm{d}y\,\mathrm{d}z. \tag{5.2.56}$$

Show that

$$Q = -2V \sum_{n=1}^{\infty} \frac{(-1)^n}{\alpha_n} \int_0^{\infty} \int_{-a}^{a} \exp(-\alpha_n \frac{z}{a}) \cos(\alpha_n \frac{y}{a}) \, dy \, dz = 4Va^2 \sum_{n=1}^{\infty} \frac{1}{\alpha_n^3}, \qquad (5.2.57)$$

yielding $Q \simeq 1.085 \, Va^2$. The dependence of the flow rate on the second power of a linear boundary dimension, in this case, a, is typical of boundary-driven flow. Explain why the thickness of the film left behind the brush is $h = Q/(2Va)$.

(*c*) Repeat (*a*) and (*b*) assuming that the brush planes are inclined at an angle α with respect to the painted surface. Discuss the relation between Q and α.

5.2.3 *Flow through a tapered tube*

Consider a conical tube with a slowly varying radius, $a(x)$. Derive an expression for the pressure drop necessary to drive a flow with a given flow rate.

5.2.4 *Free-surface flow in a square duct*

Consider unidirectional gravity-driven flow along a tilted square duct with side length $2a$, inclined at angle β with respect to the horizontal. The top of the duct is open to the atmosphere and the free surface is assumed to be flat. The fluid velocity at the bottom, left, and right walls, and the shear stress at the top free surface are required to be zero. It is convenient to introduce Cartesian coordinates where the left and right walls are located at $y = \pm a$ and the bottom wall is located at $z = -a$. Show that the velocity field and flow rate are given by

$$u_x = \frac{g}{2\nu} \sin \beta \left[a^2 - y^2 + 4a^2 \sum_{n=1}^{\infty} \frac{(-1)^n}{\alpha_n^3} \frac{\cosh[\alpha_n(1+z/a)]}{\cosh(2\alpha_n)} \cos(\alpha_n \frac{y}{a}) \right] \qquad (5.2.58)$$

and

$$Q = 4 \frac{g}{\nu} \sin \beta \, a^4 \left(1 - \sum_{n=1}^{\infty} \frac{\tanh(2\alpha_n)}{\alpha_n^5} \right), \qquad (5.2.59)$$

where $\alpha_n = \frac{\pi}{2}(2n-1)$ and ν is the kinematic viscosity of the fluid.

5.2.5 *Coating a rod*

A cylindrical rod of radius a is pulled upward with velocity V from a liquid pool after it has been coated with a liquid film of thickness h. The coated liquid is draining downward due to the action of gravity. Show that, in cylindrical polar coordinates where the x axis is coaxial with the rod and points upward against the acceleration of gravity, the velocity profile across the film is

$$u_x = V + \frac{g}{4\nu} \left(\sigma^2 - a^2 - 2(a+h)^2 \right) \ln \frac{\sigma}{a}. \qquad (5.2.60)$$

What is the value of V for the film thickness film to remain constant in time?

5.2.6 *Flow through a rectangular channel*

Prepare a plot the function $\phi(\xi)$ defined in (5.2.25). Discuss the behavior as ξ tends to zero or infinity.

5.2.7 *Settling of a cylindrical slab inside a vertical tube*

Consider a solid cylindrical slab with radius b and density ρ_s settling with velocity V under the action of gravity along the centerline of a vertical circular tube with radius $a > b$, where $a \simeq b$. The tube is closed at the bottom and exposed to the atmosphere at the top. Derive an expression for the scaled settling velocity, $\mu V/(\rho_s - \rho)ga^2$, in terms of the radii ratio, $\delta \equiv b/a$, where ρ is the density of the fluid inside the cylinder.

5.2.8 *Flow between two concentric cylinders*

Show that as the channel width $h = R_2 - R_1$ becomes decreasingly small compared to the inner radius, R_1, expressions (5.2.26) and (5.2.27) reduce to (5.1.6) and (5.1.7) with $y = \sigma - R_1$ [318].

[X] **Computer Problem**

5.2.9 *Flow through a flexible hose*

A gardener delivers water through a circular hose made of a flexible material. By pinching the end of the hose, she is able to obtain elliptical cross-sectional shapes with variable aspect ratio, while the perimeter of the hose remains constant. Compute the delivered flow rate as a function of the aspect ratio of the cross-section for a given pressure gradient.

5.3 Steadily rotating flows

Previously in this chapter, we discussed rectilinear unidirectional flow where the nonvanishing velocity component is directed along the x axis. Now we turn our attention to swirling flow with circular streamlines generated by boundary rotation.

5.3.1 Circular Couette flow

In the first application, we consider swirling flow in the annular space between two concentric cylinders with radii R_1 and R_2 rotating about their common axis with angular velocities Ω_1 and Ω_2. Assuming that the axial and radial velocity components are zero and the azimuthal component is independent of the axial and azimuthal position, we set $u_\varphi = u(\sigma)$, $u_x = 0$, and $u_\sigma = 0$. The axial component of the equation of motion is satisfied by the pressure distribution, $p = \rho \mathbf{g} \cdot \mathbf{x} + \tilde{p}(\sigma)$, where $\tilde{p}(\sigma) = p - \rho \mathbf{g} \cdot \mathbf{x}$ is the hydrodynamic pressure excluding hydrostatic variations. The radial and azimuthal components of the equation of motion yield the ordinary differential equations

$$\frac{d\tilde{p}}{d\sigma} = \rho \frac{u_\varphi^2}{\sigma}, \qquad \frac{d}{d\sigma}\left(\frac{1}{\sigma}\frac{d(\sigma u_\varphi)}{d\sigma}\right) = 0. \tag{5.3.1}$$

The first equation states that the centrifugal force is balanced by a spontaneously developing radial pressure gradient. Integrating the second equation subject to the boundary conditions $u_\varphi = \Omega_1 R_1$ at $\sigma = R_1$ and $u_\varphi = \Omega_2 R_2$ at $\sigma = R_2$, we obtain the velocity profile

$$u_\varphi(\sigma) = \frac{\Omega_2 - \alpha\Omega_1}{1 - \alpha}\sigma - \frac{\Omega_2 - \Omega_1}{1 - \alpha}\frac{R_1^2}{\sigma}, \tag{5.3.2}$$

where $\alpha \equiv \left(R_1/R_2\right)^2 < 1$ is a geometrical dimensionless parameter. When $\Omega_2 = \Omega_1 \equiv \Omega$, the fluid rotates like a rigid body with azimuthal velocity $u_\varphi(\sigma) = \Omega\sigma$. When $\Omega_2 = \alpha\,\Omega_1$, we obtain the flow due to a point vortex with strength $\kappa = 2\pi\Omega_1 R_1^2 = 2\pi\Omega_2 R_2^2$ located at the axis, generating irrotational flow. As the clearance of the channel becomes decreasingly small compared to the radii of the cylinders, $\alpha \to 1$, the flow in the gap resembles plane Couette flow in a channel with parallel walls (Problem 5.3.1).

The pressure distribution is found by substituting the velocity profile (5.3.2) into the first equation of (5.3.1). When the fluid rotates as a rigid body, $\Omega_2 = \Omega_1 \equiv \Omega$, we obtain

$$p = \frac{1}{2}\,\rho\,\Omega^2\sigma^2 + \rho\,\mathbf{g}\cdot\mathbf{x} + p_0, \tag{5.3.3}$$

where p_0 is a constant. The first term on the right-hand side is the pressure field established in response to the centrifugal force. When the flow resembles that due to a point vortex, we obtain

$$p = -\frac{1}{2}\,\rho\,U_1^2\,\frac{R_1^2}{\sigma^2} + \rho\,\mathbf{g}\cdot\mathbf{x} + p_0, \tag{5.3.4}$$

where $U_1 = \Omega_1 R_1$.

The circular Couette flow device provides us with a simple method for computing the viscosity of a fluid in terms of the torque exerted on the inner or outer cylinder, given by

$$T \equiv -2\pi\sigma^2\,\sigma_{\sigma\varphi} = 4\pi\mu\,R_1^2\,\frac{\Omega_2 - \Omega_1}{1-\alpha}. \tag{5.3.5}$$

Note that the shear stress, $\sigma_{\sigma\varphi}$, and thus the torque is zero in the case of rigid-body rotation.

The stability of the circular Couette flow will be discussed in Section 9.9. Linear stability analysis shows that the laminar flow discussed in this section is established only when the pair (Ω_1, Ω_2) falls inside a certain range. Wavy motion leading to cellular flow is established outside this range.

5.3.2   Ekman flow

Consider a semi-infinite body of fluid residing in the lower half-space from $z = 0$ to $-\infty$ and rotating around the z axis with constant angular velocity Ω. We refer to a noninertial frame of reference that rotates with the fluid with angular velocity, $\mathbf{\Omega} = \Omega\,\mathbf{e}_z$, and assume that the vertical velocity component, u_z, is zero while the horizontal velocity components, u_x and u_y, depend only on z and are independent of x and y. Simplifying the equation of motion in the rotating noninertial frame by setting the left-hand side of (3.2.43) to zero, we obtain

$$\mathbf{0} = -\nabla p + \mu\,\nabla^2\mathbf{u} + \rho\,\mathbf{g} + \rho\left[-2\,\mathbf{\Omega}\times\mathbf{u} - \mathbf{\Omega}\times(\mathbf{\Omega}\times\mathbf{x})\right]. \tag{5.3.6}$$

The first term inside the square brackets on the right-hand side is the Coriolis force, and the second term is the centrifugal force. The x, y, and z components of (5.3.6) are

$$0 = -\frac{\partial p}{\partial x} + \mu\,\frac{\mathrm{d}^2 u_x}{\mathrm{d}z^2} + \rho g_x + \rho\,\Omega\left(2u_y - \Omega x\right),$$

$$0 = -\frac{\partial p}{\partial y} + \mu\,\frac{\mathrm{d}^2 u_y}{\mathrm{d}z^2} + \rho g_y + \rho\,\Omega\left(-2u_x - \Omega y\right), \qquad 0 = -\frac{\partial p}{\partial z} + \rho g_z. \tag{5.3.7}$$

The solution is

$$p = \rho\, \mathbf{g} \cdot \mathbf{x} + \frac{1}{2}\, \rho\, \Omega^2 (x^2 + y^2) + p_0, \qquad (5.3.8)$$

where p_0 is a constant reference pressure, and the velocity components u_x and u_y satisfy the coupled ordinary differential equations

$$\mu\, \frac{\mathrm{d}^2 u_x}{\mathrm{d}z^2} = -2\rho\, \Omega u_y, \qquad \mu\, \frac{\mathrm{d}^2 u_y}{\mathrm{d}z^2} = 2\rho\, \Omega u_x. \qquad (5.3.9)$$

It is convenient to collect these equations into the compact complex form

$$\frac{\mathrm{d}^2 (u_x + \mathrm{i}\, u_y)}{\mathrm{d}z^2} = 2\,\mathrm{i}\, \frac{\Omega}{\nu} (u_x + \mathrm{i}u_y), \qquad (5.3.10)$$

where i is the imaginary unit, $\mathrm{i}^2 = -1$, and $\nu = \mu/\rho$ is the kinematic viscosity of the fluid. Stipulating that $u_x = U_x$ and $u_y = U_y$ at $z = 0$, and that u_x and u_y decay as z tends to negative infinity, we obtain

$$u_x + \mathrm{i}\, u_y = (U_x + \mathrm{i}\, U_y) \exp\Big[-(1 + \mathrm{i})\, \frac{|z|}{\delta} \Big], \qquad (5.3.11)$$

where $\delta \equiv (2\nu/\Omega)^{1/2}$ is the Ekman layer thickness. We observe that the velocity components in a plane that is normal to the axis of rotation decay exponentially with downward distance from the surface of the liquid.

Problem

5.3.1 *Flow between concentric cylinders*

Show that, as the cylinder radii R_1 and R_2 tend to become equal, the velocity profile in the gap between the two cylinders resembles the linear profile of plane-Couette flow in a channel with clearance $h = R_2 - R_1$ [318].

5.4   Unsteady unidirectional flows

Unsteady unidirectional flows share many of the simplifying features of steady unidirectional flows discussed earlier in this chapter. An unsteady flow can be driven by an imposed time-dependent pressure gradient or by time-dependent boundary motion. In the case of rectilinear flow along the x axis, the y and z components of the equation of motion are satisfied by the pressure distribution

$$p = -\chi(t)\, x + \rho\, (g_y y + g_z z) + p_0, \qquad (5.4.1)$$

where $\chi(t) \equiv -\partial p / \partial x$ is the negative of the streamwise pressure gradient. The x component of the equation of motion provides us with an unsteady conduction equation in the presence of a time-dependent source term for the streamwise velocity component, u_x,

$$\rho\, \frac{\partial u_x}{\partial t} = \chi(t) + \mu\, \Big(\frac{\partial^2 u_x}{\partial y^2} + \frac{\partial^2 u_x}{\partial z^2} \Big) + \rho g_x. \qquad (5.4.2)$$

The linearity of this equation allows us to derive exact solutions for two-dimensional, axisymmetric, and more general boundary configurations.

5.4.1   Flow due to the in-plane vibrations of a plate

In 1845, Stokes studied the flow due to the in-plane vibrations of a horizontal flat plate along the x axis in an otherwise quiescent semi-infinite fluid located above the plate, $y > 0$. Assuming that u_x is a function of time, t, and distance from the plate, y, and setting $\partial p / \partial x = \rho g_x$, which is equivalent to $\chi = -\rho g_x$, we find that the pressure distribution assumes the hydrostatic profile and the velocity satisfies a simplified version of the general governing equation (5.4.2),

$$\frac{\partial u_x}{\partial t} = \nu \, \frac{\partial^2 u_x}{\partial y^2}, \tag{5.4.3}$$

known as the one-dimensional unsteady heat conduction equation, where $\nu = \mu / \rho$ is the kinematic viscosity of the fluid.

The velocity is required to decay far from the plate, as $y \to \infty$. The no-slip boundary condition at the plate requires that

$$u_x(y = 0, t) = V \cos(\Omega t), \tag{5.4.4}$$

where Ω is the angular frequency of the oscillations and V is the amplitude of the vibrations. It is convenient to express this boundary condition in the form

$$u_x(y = 0, t) = V \operatorname{Real}\{ e^{-\mathrm{i}\Omega t} \}, \tag{5.4.5}$$

where i is the imaginary unit, $\mathrm{i}^2 = -1$. Motivated by the linearity of (5.4.3), we also set

$$u_x(y, t) = V \operatorname{Real}\{ f(y) \, e^{-\mathrm{i}\Omega t} \}, \tag{5.4.6}$$

where $f(y)$ is a dimensionless complex function satisfying the boundary condition $f(0) = 1$ and the far-field condition $f(\infty) = 0$. Substituting (5.4.6) into (5.4.3), we derive a linear ordinary differential equation

$$-\mathrm{i}\Omega f = \nu f'', \tag{5.4.7}$$

where a prime denotes a derivative with respect to y. The solution is readily found to be

$$f(y) = \exp[-(1 - \mathrm{i}) \, \hat{y}], \tag{5.4.8}$$

where $\hat{y} \equiv y / \delta$ and

$$\delta = \sqrt{\frac{2\nu}{\Omega}} \tag{5.4.9}$$

is the Stokes boundary-layer thickness. Substituting (5.4.8) into (5.4.6), we obtain

$$u_x(y, t) = V \operatorname{Real}\{ \exp[-\mathrm{i}\Omega t - (1 - \mathrm{i}) \, \hat{y}] \}, \tag{5.4.10}$$

yielding the velocity profile

$$u_x(y, t) = V \cos(\Omega t - \hat{y}) \, e^{-\hat{y}} \tag{5.4.11}$$

(a) (b)

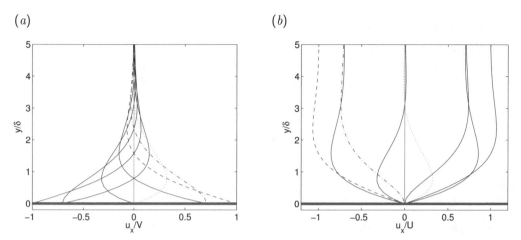

FIGURE 5.4.1 Velocity profiles (a) due to the in-plane oscillation of a plate in a semi-infinite viscous fluid and (b) in oscillatory flow over a stationary plate. Profiles are shown at phase angles $\Omega t/\pi = 0$ (dashed line), 0.25 (dash-dotted line), 0.50 (dotted line), 0.75, 1.0, 1.25, 1.50, 1.75, and 2.0. The y coordinate is scaled by the Stokes boundary layer thickness δ defined in (5.4.9).

for $\hat{y} \geq 0$. We have found that the velocity profile is an exponentially damped wave with wave number $1/\delta$ and associated wavelength $2\pi\delta$, propagating in the y direction with phase velocity $c_p = \Omega\delta = (2\nu\Omega)^{1/2}$. Since the amplitude of the velocity decays exponentially with distance from the plate, the motion of the fluid is negligible outside a wall layer of thickness δ. Velocity profiles at a sequence of time instants over one period of the oscillations are shown in Figure 5.4.1(a).

Wall shear stress

Differentiating (5.4.10) with respect to y and evaluating the derivative at the plate, $y = 0$, we obtain an expression for the shear stress over the plate,

$$\left(\sigma_{xy}\right)_{y=0} = \mu \left(\frac{\partial u_x}{\partial y}\right)_{y=0} = V(\mu\rho\Omega)^{1/2} \cos(\Omega t - \frac{3\pi}{4}) = \frac{\mu V}{\delta} \sqrt{2} \cos(\Omega t - \frac{3\pi}{4}). \qquad (5.4.12)$$

It is interesting to note that the phase shift between the wall shear stress and the velocity of the plate is $3\pi/4$, independent of the angular frequency, Ω. This is a unique feature of oscillatory flow driven by the motion by a flat surface in the absence of other boundaries. We will see later in this section that, for other geometries, the phase shift depends on the frequency of oscillation.

5.4.2 Flow due to an oscillatory pressure gradient above a plate

Next, we study the complementary problem of unidirectional flow above a stationary plate due to an oscillatory pressure gradient, setting

$$\chi \equiv -\frac{\partial p}{\partial x} = -\rho g_x + \zeta \sin(\Omega t), \qquad (5.4.13)$$

where ζ is a constant amplitude. The velocity profile is governed by the following simplified version of the general equation (5.4.2),

$$\rho \frac{\partial u_x}{\partial t} = \zeta \sin(\Omega t) + \mu \frac{\partial^2 u_x}{\partial y^2}. \tag{5.4.14}$$

The solution is

$$u_x(y,t) = v(y,t) - \frac{\zeta}{\rho \Omega} \cos(\Omega t), \tag{5.4.15}$$

where the complementary velocity v is given by the right-hand side of (5.4.11) with $V = \zeta/(\rho\Omega)$, yielding

$$u_x(y,t) = \frac{\zeta}{\rho \Omega} \left[\cos(\Omega t - \hat{y}) \, \mathrm{e}^{-\hat{y}} - \cos(\Omega t) \right]. \tag{5.4.16}$$

The velocity is zero over the plate located at $y = 0$, and describes uniform oscillatory flow with velocity $u_x(\infty, t) = -U \cos(\Omega t)$ far from the plate, where $U = \zeta/(\rho \Omega)$. Physically, the flow consists of an outer potential core and a Stokes boundary layer with thickness $\delta = (2\nu/\Omega)^{1/2}$ adhering to the plate. Velocity profiles at a sequence of time instants over one period are shown in Figure 5.4.1(b).

5.4.3   Flow due to the sudden translation of a plate

Consider a semi-infinite flow above a flat plate that is suddenly set in motion parallel to itself with constant velocity, V, underneath an otherwise quiescent fluid. The evolving velocity profile is found by solving (5.4.3) subject to the initial condition $u_x(y, t = 0) = 0$ and the boundary and far-field conditions

$$u_x(y = 0, t > 0) = V, \qquad u_x(y \to \infty, t) = 0. \tag{5.4.17}$$

Because of the linearity of the governing equation and boundary conditions, we expect that the fluid velocity will be proportional to the plate velocity, V, and set

$$u_x = V f(y, t, \nu), \tag{5.4.18}$$

where f is a dimensionless function. The arguments of $f(y, t, \nu)$ must combine into dimensionless groups. In the absence of external length and time scales, we formulate a combination in terms of the similarity variable

$$\eta = \frac{y}{\sqrt{\nu t}}, \tag{5.4.19}$$

so that $f(y, t, \nu) = F(\eta)$, where $F(\eta)$, is an unknown dimensionless function. This means that the velocity as seen by an observer who finds herself at the position $y = \sqrt{\nu t}$, and is thus traveling along the y axis with velocity $v = \mathrm{d}y/\mathrm{d}t = \sqrt{\nu/4t}$, remains constant in time. The boundary and far-field conditions require that $F(0) = 1$ and $F(\infty) = 0$. Substituting into (5.4.3) the self-similar velocity profile $u_x = V F(\eta)$, we obtain

$$\frac{\mathrm{d}F}{\mathrm{d}\eta} \frac{\partial \eta}{\partial t} = \nu \frac{\partial}{\partial y} \left(\frac{\mathrm{d}F}{\mathrm{d}\eta} \frac{\partial \eta}{\partial y} \right). \tag{5.4.20}$$

(a)   (b)

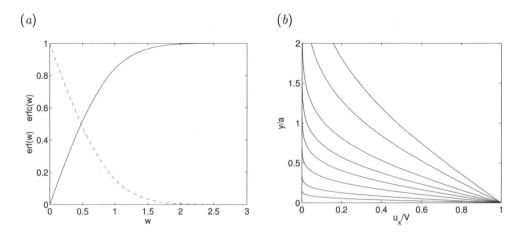

FIGURE 5.4.2   (a) Graphs of the error function (solid line) and complementary error function (dashed line) computed using algebraic approximations. (b) Velocity profiles of the flow due to the sudden motion of a plate in a semi-infinite fluid at dimensionless times $\nu t/a^2 = 0.001, 0.005, 0.02, 0.05, 0.1, 0.2, 0.5,$ and 1.0, where a is an arbitrary length scale,

Carrying out the differentiations, we derive a second-order nonlinear ordinary differential equation

$$-\frac{1}{2}\,\eta\,\frac{\mathrm{d}F}{\mathrm{d}\eta} = \frac{\mathrm{d}^2 F}{\mathrm{d}\eta^2}. \tag{5.4.21}$$

Integrating twice subject to the aforementioned boundary and far-field conditions, we obtain the velocity profile

$$\frac{u_x}{V} = F(\eta) = \mathrm{erfc}\left(\frac{\eta}{2}\right) \equiv 1 - \mathrm{erf}\left(\frac{\eta}{2}\right) \equiv 1 - \frac{2}{\sqrt{\pi}} \int_0^{\eta/2} \mathrm{e}^{-\xi^2}\,\mathrm{d}\xi. \tag{5.4.22}$$

The complementary error function, $\mathrm{erfc}(w)$, and the error function, $\mathrm{erf}(w)$, can be computed by accurate polynomial approximations. Graphs are presented in Figure 5.4.2(a). As w tends to infinity, $\mathrm{erf}(w)$ tends to unity and correspondingly $\mathrm{erfc}(w)$ tends to zero. The converse is true as w tends to zero.

A sequence of evolving velocity profiles plotted with respect to y/a are shown in Figure 5.4.2(b) at a sequence of dimensionless times $\nu t/a^2$, where a is an arbitrary length. At the initial instant, the velocity profile is proportional to the discontinuous Heaviside function, reflecting the sudden onset of a vortex sheet attached to the plate,

$$u_x(y, t = 0) = V\mathcal{H}(y). \tag{5.4.23}$$

The dimensionless Heaviside function, $\mathcal{H}(t)$, is defined such that $\mathcal{H}(t) = 0$ for $t < 0$ and $\mathcal{H}(t) = 1$ for $t > 0$. As time progresses, the vortex sheet diffuses into the fluid, transforming into a vortex layer of growing thickness.

Wall shear stress

The wall shear stress is given by

$$
\left(\sigma_{xy}\right)_{y=0} = \mu \left(\frac{\partial u_x}{\partial y}\right)_{y=0} = \mu V \left(\frac{\mathrm{d}F}{\mathrm{d}\eta}\right)_{\eta=0} \left(\frac{\partial \eta}{\partial y}\right)_{y=0} = -\mu \frac{V}{\sqrt{\pi \nu t}}. \tag{5.4.24}
$$

A singularity appears at the initial instant, $t = 0$, when the plate starts moving. Physically, the plate cannot be pushed impulsively with constant velocity, but must be accelerated gradually from the initial to the final value over a finite period of time. Setting the shear stress equal to $-\mu V/\delta$, where δ is the effective thickness of the vorticity layer, we find that δ increases in time like the square root of time, $\delta \sim \sqrt{\nu t}$.

Diffusing vortex sheet

The velocity field given in (5.4.22) describes the flow due to the sudden introduction of an infinite plane parallel to a uniform stream flowing along the x axis, as well as the flow associated with a diffusing vortex sheet separating two uniform streams flowing with velocities V and $-V$ above and below the x axis, as discussed in Section 3.12; see equation (3.12.23).

5.4.4   Flow due to the arbitrary translation of a plate

Consider the flow due to the sudden translation of a plate discussed in Section 5.4.3. Since the plate velocity is a step function in time, we may set $u_x(0,t) = V\,\mathcal{H}(t)$, where $\mathcal{H}(t)$ is the dimensionless Heaviside function defined such that $\mathcal{H}(t) = 0$ for $t < 0$ and $\mathcal{H}(t) = 1$ for $t > 0$. A discontinuity occurs at the origin of time, $t = 0$. Since the derivative of the Heaviside function is the one-dimensional Dirac delta function, $\delta_1(t)$, we may write

$$
\left(\frac{\partial u_x}{\partial t}\right)_{y=0} = V \left(\frac{\partial F}{\partial t}\right)_{y=0}(t) = V\,\delta_1(t), \tag{5.4.25}
$$

which reveals that

$$
\left(\frac{\partial F}{\partial t}\right)_{y=0}(t) = \delta_1(t). \tag{5.4.26}
$$

In the case of arbitrary plate translation with arbitrary time-dependent velocity $V(t)$ and $V(0) = 0$, we obtain

$$
u_x(0,t) = V(t) = \int_0^{t+\delta t} V(\tau)\,\delta_1(t-\tau)\,\mathrm{d}\tau = \int_0^{t+\delta t} V(\tau)\left(\frac{\partial F}{\partial t}\right)_{y=0}(t-\tau)\,\mathrm{d}\tau, \tag{5.4.27}
$$

where δt is an infinitesimal time interval. This expression allows us to construct the flow due to the translation of the plate by generalization to arbitrary y. Assuming that the fluid is quiescent for $t < 0$, we obtain

$$
u_x(y,t) = \int_0^t V(\tau)\left(\frac{\partial F}{\partial t}\right)(y,t-\tau)\,\mathrm{d}\tau. \tag{5.4.28}
$$

Substituting expression (5.4.22) and taking the time derivative, we find that

$$u_x(y,t) = \frac{y}{2\sqrt{\pi\nu}} \int_0^t \frac{V(\tau)}{(t-\tau)^{3/2}} \exp\left[-\frac{y^2}{4\nu(t-\tau)}\right] d\tau. \tag{5.4.29}$$

This solution also describes the temperature distribution in a semi-infinite conductive medium due to an arbitrary boundary temperature (e.g., ([66]). The right-hand side of (5.4.29) expresses a distribution of Green's functions of the unsteady diffusion equation (6.17.4) given in (6.17.5).

5.4.5   Flow in a channel with parallel walls

Continuing the investigation of unsteady unidirectional flow, we consider time-dependent boundary-driven (Couette) or pressure-driven (Poiseuille) flow in a channel confined between two parallel walls located at $y = 0$ and h. The Couette flow is governed by equation (5.4.3) and the Poiseuille flow is governed by equation (5.4.14).

Oscillatory Couette flow

Assume that the upper wall is stationary while the lower wall oscillates in its plane along the x axis with angular frequency Ω. The lower wall velocity is $u_x(y = 0, t) = V\cos(\Omega t)$, where V is a constant amplitude. Working as in Section 5.4.1 for flow in a semi-infinite fluid due to an oscillating plate, we derive the velocity profile

$$u_x(y,t) = V\,\mathrm{Real}\{\,f(y)\,\mathrm{e}^{-\mathrm{i}\Omega t}\,\}, \tag{5.4.30}$$

where

$$f(y) = \frac{\exp[-(1-\mathrm{i})\hat{y}] - \exp[-(1-\mathrm{i})(2\hat{h}-\hat{y})]}{1 - \exp[-(1-\mathrm{i})\,2\hat{h}]}, \tag{5.4.31}$$

$\hat{y} \equiv y/\delta$, $\hat{h} \equiv h/\delta$, and $\delta \equiv (2\nu/\Omega)^{1/2}$ is the Stokes boundary-layer thickness. The structure of the flow can be regarded as a function of the dimensionless Womersley number defined with respect to half the channel width,

$$\mathrm{Wo} \equiv \frac{h}{2}\sqrt{\frac{\Omega}{\nu}} = \frac{1}{\sqrt{2}}\frac{h}{\delta} = \frac{1}{\sqrt{2}}\hat{h}. \tag{5.4.32}$$

When the Womersley number is small, the flow evolves in quasi-steady fashion and the velocity profile is nearly linear with respect to y at any time. As the Womersley number tends to infinity, we recover the results of Section 5.4.1 for flow due to an oscillating plate in a semi-infinite fluid. Comparing the present flow with that due to the oscillating plate, we find that the phase shift between the velocity and shear stress at the lower wall depends on the angular frequency, Ω. This difference underscores the hydrodynamic importance of the second plate or another nearby boundary.

Transient Couette flow

Next, we assume that the upper wall is stationary while the lower wall is suddenly set in motion parallel to itself along the x axis with constant velocity V. Initially, the flow resembles that due to the motion of a flat plate immersed in a semi-infinite fluid. As time progresses, we obtain Couette flow with a linear velocity profile. To expedite the solution, we decompose the flow into the steady Couette flow and a transient flow that decays at long times. Applying the method of separation of variables in y and t, we find that the velocity is given by a Fourier series [318],

$$u_x(y, t) = V \left(1 - \frac{y}{h} - \frac{2}{\pi} \sum_{n=1}^{\infty} \frac{1}{n} \exp\left(-\frac{n^2 \pi^2 \nu}{h^2} t\right) \sin \frac{n \pi y}{h} \right). \tag{5.4.33}$$

Steady state is established approximately when $t \simeq h^2/\nu$.

Working in an alternative fashion, we apply the Laplace transform method and obtain

$$u_x(y, t) = V \left[\mathrm{erfc}\left(\frac{y}{2\sqrt{\nu t}}\right) - \mathrm{erfc}\left(\frac{2h - y}{2\sqrt{\nu t}}\right) + \mathrm{erfc}\left(\frac{2h + y}{2\sqrt{\nu t}}\right) \right.$$
$$\left. - \mathrm{erfc}\left(\frac{4h - y}{2\sqrt{\nu t}}\right) + \mathrm{erfc}\left(\frac{4h + y}{2\sqrt{\nu t}}\right) + \cdots \right], \tag{5.4.34}$$

which is more appropriate for computing the flow at short times.

Oscillatory plane Poiseuille flow

In the next configuration, we consider flow due to an oscillatory streamwise pressure gradient described by

$$\chi \equiv -\frac{\partial p}{\partial x} = -\rho g_x + \zeta \sin(\Omega t), \tag{5.4.35}$$

where ζ is a constant amplitude and Ω is the angular frequency of the oscillations. Straightforward computation shows that the velocity profile is given by

$$u_x(y, t) = \frac{\zeta}{\rho \Omega} \mathrm{Real}\left\{ f(y) \, \mathrm{e}^{-\mathrm{i}\Omega t} \right\}, \tag{5.4.36}$$

where

$$f(y) = \frac{\cosh\left[(-1 + \mathrm{i})(\hat{y} - \frac{1}{2}\,\hat{h})\right]}{\cosh\left[(-1 + \mathrm{i})\frac{1}{2}\,\hat{h}\right]} - 1, \tag{5.4.37}$$

is a complex function, $\hat{y} \equiv y/\delta$, $\hat{h} \equiv h/\delta$, and $\delta \equiv (2\nu/\Omega)^{1/2}$ is the Stokes boundary-layer thickness (e.g., [318]). It is sometimes useful to regard the structure of the flow as a function of the dimensionless Womersley number defined in (5.4.32). When the Womersley number is small, we obtain quasi-steady plane Hagen–Poiseuille flow with parabolic velocity profile.

Considering the limit of high frequencies, we replace the hyperbolic cosine in the denominator on the right-hand side of (5.4.37) with half the exponential of its argument, resolve the hyperbolic cosine in the numerator into its two exponential constituents, and derive the velocity profile

$$u_x(y, t) \simeq \frac{\zeta}{\rho \Omega} \mathrm{Real}\left\{ \left[\mathrm{e}^{-(1-\mathrm{i})\,\hat{y}} + \mathrm{e}^{-(1-\mathrm{i})\,(\hat{h} - \hat{y})} - 1 \right] \exp(-\mathrm{i}\Omega t) \right\}. \tag{5.4.38}$$

Comparing (5.4.38) with (5.4.16), we find that the flow consists of an irrotational core executing rigid-body motion with velocity $u_x = -\frac{\zeta}{\rho\Omega}\cos(\Omega t)$, and two Stokes boundary layers, one attached to each wall. Inspecting the precise form of the velocity profile at high frequencies, we find that the amplitude of the velocity may significantly exceed that of the plug flow in the central core.

Transient plane Poiseuille flow

In the last application, we consider unidirectional flow due to sudden tilting or application of a constant pressure gradient $\partial p/\partial x = -\chi$. Applying the method of separation of variables and using Fourier expansions, we obtain the velocity profile

$$u_x(y,t) = \frac{\chi + \rho g_x}{2\mu}\left[y(h-y) - \frac{8}{\pi^3}h^2\sum_{n=1,3,\dots}^{\infty}\frac{1}{n^3}\exp(-\frac{n^2\pi^2\nu}{h^2}t)\sin\frac{n\pi y}{h}\right]. \qquad (5.4.39)$$

The velocity vanishes at the initial instant, $t = 0$. At long times, we recover the Hagen–Poiseuille parabolic profile.

5.4.6 Finite-difference methods

The analytical solutions derived in this section can be approximated with numerical solutions obtained by standard methods for solving parabolic partial differential equations, as discussed in Chapter 12. To illustrate the methodology, we consider transient flow in a channel with clearance h due to the application of a constant pressure gradient, $\partial p/\partial x = -\chi$. It is convenient to introduce the dimensionless variables

$$f = \frac{\mu u_x}{\chi h^2}, \qquad \tau = \frac{\nu t}{h^2}, \qquad \xi = \frac{y}{h}, \qquad (5.4.40)$$

where $\tau \geq 0$ and $0 \leq \xi \leq 1$. The governing equation takes the dimensionless form

$$\frac{\partial f}{\partial \tau} = 1 + \frac{\partial^2 f}{\partial \xi^2}. \qquad (5.4.41)$$

The boundary conditions require that $f = 0$ at $\xi = 0$ and 1 at any time, τ. The initial condition requires that $f = 0$ at $\eta = 0$ for $0 \leq \xi \leq 1$.

Discretization and difference equations

Following standard practice, we divide the solution domain in ξ into N evenly spaced intervals with uniform size $\Delta\xi$ defined by $N + 1$ grid points, $\xi_i = (i-1)/N$, where $i = 1, \dots, N + 1$. Our objective is to compute the values of the function f at the grid points at a sequence of dimensionless times separated by the time interval $\Delta\tau$. Applying equation (5.4.41) at the ith grid point and approximating the time derivative using a backward difference and the spatial derivatives using a central difference, we obtain the backward-time/centered-space (BTCS) finite-difference equation

$$\frac{f_i^{n+1} - f_i^n}{\Delta\tau} = 1 + \frac{f_{i-1}^{n+1} - 2f_i^{n+1} + f_{i+1}^{n+1}}{\Delta\xi^2}, \qquad (5.4.42)$$

for $i = 2, \ldots, N$, where f_i^n denotes the value of f at the ith grid point at time $\tau = n\Delta\tau$ (Chapter 12 and Section B.4, Appendix B). Rearranging, we obtain the linear equation

$$-\alpha f_{i-1}^{n+1} + (1 + 2\alpha) f_i^{n+1} - \alpha f_{i+1}^{n+1} = f_i^n + \Delta\tau \qquad (5.4.43)$$

for $i = 2, \ldots, N$, where $\alpha \equiv \Delta\tau/\Delta\xi^2$. The no-slip boundary condition at the lower and upper walls requires that $f_1 = 0$ and $f_{N+1} = 0$. Collecting the nodal values f_i^n, into a vector, $\mathbf{f}^n$, for $i = 2, \ldots, N$, we derive a system of linear algebraic equations,

$$\mathbf{A} \cdot \mathbf{f}^{n+1} = \mathbf{f}^n + \Delta\tau\,\mathbf{e}, \qquad (5.4.44)$$

where $\mathbf{A}$ is a tridiagonal matrix originating from (5.4.43) and all entries of the vector $\mathbf{e}$ are equal to unity. The algorithm involves solving the system (5.4.44) at successive time instants using the Thomas algorithm discussed in Section B.1.4, Appendix B.

Problem

5.4.1 *Flow due to the application of a constant shear stress on a planar surface*

Show that the velocity field due to the sudden application of a constant shear stress τ along the planar boundary of a semi-infinite fluid residing in the upper half-plane, $y \geq 0$, is

$$u_x(y, t) = \frac{\tau}{\mu}\sqrt{\nu t}\,\Big[\,\frac{2}{\sqrt{\pi}}\,\exp(-\tfrac{1}{4}\eta^2) - \eta\,\mathrm{erfc}\frac{\eta}{2}\,\Big], \qquad (5.4.45)$$

where $\eta = y/(\nu t)^{1/2}$ is a similarity variable. Discuss the asymptotic behavior at long times.

 ### Computer Problems

5.4.2 *Flow in a two-dimensional channel due to an oscillatory pressure gradient*

Plot the profile of the amplitude of the velocity at a sequence of Womersley numbers. Identify and discuss the occurrence of overshooting at high frequencies.

5.4.3 *Transient Couette flow in a channel with parallel walls*

Consider transient Couette flow in a two-dimensional channel confined between two parallel walls located at $y = 0$ and d, as discussed in Section 5.4.5. To isolate the singular behavior at short times, we write

$$\frac{u_x}{V} = F(\eta) + G(y, t), \qquad (5.4.46)$$

where the function $F(\eta)$ is given in (5.4.22). The nonsingular dimensionless function $G(y, t)$ satisfies the one-dimensional unsteady diffusion equation (5.4.3) with boundary conditions $G(0, t) = 0$ and $G(d, t) = -F[d/(\nu t)^{1/2}]$, and initial condition $G(y, 0) = 0$. Write a program that advances the function G in time using a finite-difference method with the BTCS discretization discussed in the text. Plot and discuss velocity profiles at a sequence of dimensionless times, $\xi = t\nu/d^2$.

5.5   Unsteady flows inside tubes and outside cylinders

Continuing our study of unsteady flow, we address a family of unidirectional and swirling flows inside or outside a circular tube of radius a that is either filled with a viscous fluid or is immersed in an infinite ambient fluid. It is natural to express the governing equation (5.4.2) in cylindrical polar coordinates, (x, σ, φ), where the x axis coincides with the tube centerline.

Rectilinear flow

In the case of rectilinear unidirectional flow along the tube axis, the axial velocity component satisfies the linear equation

$$\rho \frac{\partial u_x}{\partial t} = -\frac{\partial p}{\partial x} + \mu \frac{1}{\sigma} \frac{\partial}{\partial \sigma}\Big(\sigma \frac{\partial u_x}{\partial \sigma}\Big) + \rho g_x. \tag{5.5.1}$$

Expanding the derivative on the right-hand side, we obtain

$$\rho \frac{\partial u_x}{\partial t} = -\frac{\partial p}{\partial x} + \mu \Big(\frac{\partial^2 u_x}{\partial \sigma^2} + \frac{1}{\sigma} \frac{\partial u_x}{\partial \sigma}\Big) + \rho g_x. \tag{5.5.2}$$

If the flow is due to the cylinder translation along its length, we obtain the counterpart of Couette flow (Problem 5.5.1).

Swirling flow

In the case of swirling flow with circular streamlines, the pressure assumes the hydrostatic distribution and the azimuthal velocity component satisfies the linear equation

$$\rho \frac{\partial u_\varphi}{\partial t} = \mu \frac{\partial}{\partial \sigma}\Big(\frac{1}{\sigma} \frac{\partial(\sigma u_\varphi)}{\partial \sigma}\Big) \tag{5.5.3}$$

or

$$\rho \frac{\partial u_\varphi}{\partial t} = \mu \Big(\frac{\partial^2 u_\varphi}{\partial \sigma^2} + \frac{1}{\sigma} \frac{\partial u_\varphi}{\partial \sigma} - \frac{u_\varphi}{\sigma^2}\Big). \tag{5.5.4}$$

The last term inside the parentheses on the right-hand side distinguishes this equation from its counterpart for rectilinear flow.

5.5.1   Oscillatory Poiseuille flow

Pulsating flow inside a tube due to an oscillatory pressure gradient is a useful model of the flow through large blood vessels (e.g., [140]). Setting

$$\chi \equiv -\frac{\partial p}{\partial x} = -\rho g_x + \zeta \sin(\Omega t), \tag{5.5.5}$$

where ζ is the amplitude of the pressure gradient and Ω is the angular frequency of the oscillations, we derive the solution

$$u_x(\sigma, t) = \frac{\zeta}{\rho \Omega} \operatorname{Real}\Big\{ \Big(\frac{\mathrm{J}_0^*[-(1-\mathrm{i})\,\hat{\sigma}]}{\mathrm{J}_0^*[-(1-\mathrm{i})\,\hat{a}]} - 1\Big) \exp(-\mathrm{i}\Omega t) \Big\}, \tag{5.5.6}$$

where J_0 is the zeroth-order Bessel function, $\hat{\sigma} = \sigma/\delta$, $\hat{a} = a/\delta$, a is the tube radius, $\delta = (2\nu/\Omega)^{1/2}$ is the Stokes boundary-layer thickness, and an asterisk denotes the complex conjugate (e.g., [178], p. 226; [318]). The zeroth-order Bessel function of complex argument can be evaluated using the definition

$$J_0(\varrho\, e^{3\pi i/4}) \equiv \text{ber}_0(\varrho) + i\,\text{ker}_0(\varrho), \qquad (5.5.7)$$

where ϱ is a real positive number. The Kelvin functions of zeroth order, $\text{ber}_0(\varrho)$ and $\text{ker}_0(\varrho)$, can be computed by polynomial approximations (e.g., [2], pp. 379, 384).

The structure of the flow can be regarded as a function of the dimensionless Womersley number defined with respect to the tube radius,

$$\text{Wo} \equiv a\,\sqrt{\frac{\Omega}{\nu}} = \sqrt{2}\,\frac{a}{\delta} = \sqrt{2}\,\hat{a}. \qquad (5.5.8)$$

When the Womersley number is small, we obtain quasi-steady Poiseuille flow with a nearly parabolic velocity profile. As the Womersley number tends to infinity, we obtain a compound flow consisting of a plug-flow core and an axisymmetric boundary layer wrapping around the tube wall. At high frequencies, the amplitude of the velocity may exhibit an overshooting, sometimes mysteriously called the annular effect, similar to that discussed in Section 5.4.5 for oscillatory plane Poiseuille flow.

5.5.2 Transient Poiseuille flow

Next, we consider the transient flow inside a circular tube of radius a generated by the sudden application of a constant pressure gradient along the tube axis, $\partial p/\partial x = -\chi$. The solution satisfies the boundary condition $u_x = 0$ at $\sigma = a$ at any time, and the initial condition $u_x = 0$ at $t = 0$ for $0 \leq \sigma \leq a$. To expedite the analysis, we resolve the flow into the steady parabolic Poiseuille flow prevailing at long times with velocity

$$u_x^P(\sigma) = \frac{1}{4\mu}\,(\chi + \rho g_x)(a^2 - \sigma^2), \qquad (5.5.9)$$

and a transient flow with velocity $v_x(\sigma, t)$, so that $u_x = u_x^P + v_x$. The transient flow satisfies equation (5.5.2) with $\chi + \rho g_x = 0$, observes the boundary condition $v_x = 0$ at $\sigma = a$ at any time, satisfies the initial condition $v_x = -u_x^P$ at $t = 0$ for $0 \leq \sigma \leq a$, and decays to zero at long times.

Applying the method of separation of variables with respect to radial distance and time for the transient flow, we derive the velocity profile

$$u_x(\sigma, t) = \frac{\chi + \rho g_x}{4\mu}\left[a^2 - \sigma^2 - 8\,a^2 \sum_{n=1}^{\infty} \frac{1}{\alpha_n^3}\,\frac{J_0(\alpha_n\,\sigma/a)}{J_1(\alpha_n)}\,\exp\left(-\alpha_n^2\,\frac{\nu t}{a^2}\right)\right], \qquad (5.5.10)$$

where J_0 and J_1 are the Bessel functions of zeroth and first order, and α_n are the real positive zeros of J_0 (e.g., [318]). The first five zeros of J_0 are

$$2.40482, \quad 5.52007, \quad 8.65372, \quad 11.79153, \quad 14.93091, \qquad (5.5.11)$$

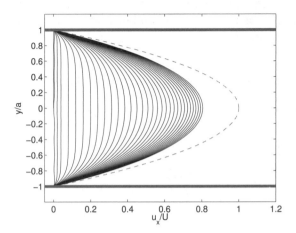

FIGURE 5.5.1   Transient flow in a circular tube of radius a due to the sudden application of a constant
pressure gradient or sudden tilting. Velocity profiles are shown at dimensionless times $\nu t/a^2 =$
$0.001, 0.005, 0.010, 0.020, 0.030, \ldots$. The dashed parabolic line corresponds to the Poiseuille flow
with centerline velocity U established at long times.

accurate to the fifth decimal place (e.g., [2]). Velocity profiles at a sequence of dimensionless times
$t\nu/a^2$ are shown in Figure 5.5.1. Note the presence of boundary layers near the wall and the
occurrence of plug flow at the core at short times. The steady parabolic velocity profile is established
when $t \simeq a^2/\nu$.

Finite-difference methods

To compute a numerical solution, it is convenient to introduce the dimensionless variables

$$f = \frac{\mu u_x}{(\chi + \rho g_x)a^2}, \qquad \tau = \frac{\nu t}{a^2}, \qquad \xi = \frac{\sigma}{a}, \qquad (5.5.12)$$

where $\tau \geq 0$ and $0 \leq \xi \leq 1$. The governing equation takes the dimensionless form

$$\frac{\partial f}{\partial \tau} = 1 + \frac{\partial^2 f}{\partial \xi^2} + \frac{1}{\xi}\frac{\partial f}{\partial \xi}. \qquad (5.5.13)$$

The initial condition requires that $f = 0$ at $\eta = 0$ for $0 \leq \xi \leq 1$ and the no-slip boundary condition
requires that $f = 0$ at $\xi = 1$ at any time.

Working as in Section 5.4.6, we derive the backward-time centered-space (BTCS) finite-
difference equation

$$\frac{f_i^{n+1} - f_i^n}{\Delta \tau} = 1 + \frac{f_{i+1}^{n+1} - 2f_i^n + f_{i-1}^n}{\Delta \xi^2} + \frac{1}{\xi_i}\frac{f_{i+1}^{n+1} - f_{i-1}^n}{2\Delta \xi}, \qquad (5.5.14)$$

for $i = 2, \ldots, N$, where f_i^n denotes the value of f at the ith grid point at dimensionless time $\tau = n\Delta\tau$ (Chapter 12 and Section B.5, Appendix B). Rearranging, we obtain a linear equation,

$$(\beta_i - \alpha) f_{i-1}^{n+1} + (1 + 2\alpha) f_i^{n+1} - (\beta_i + \alpha) f_{i+1}^{n+1} = f_i^n + \Delta\xi \tag{5.5.15}$$

for $i = 2, \ldots, N$, where $\alpha \equiv \Delta\tau/\Delta\xi^2$ and $\beta_i \equiv \Delta\tau/(2\xi_i\Delta\xi)$. The no-slip boundary condition at the tube wall requires that $f_{N+1} = 0$.

One more equation is needed to formulate a system of N equations for the N unknowns, f_i^n for $i = 1, \ldots, N$. This last equation arises by applying the governing differential equation (5.5.13) at the tube centerline. Since the third term on the right-hand side becomes indeterminate as $\xi \to 0$, we use l'Hôpital's rule and find that, at $\xi = 0$,

$$\frac{\partial f}{\partial \tau} = 1 + 2 \frac{\partial^2 f}{\partial \xi^2}. \tag{5.5.16}$$

Applying the finite-difference approximation provides us with the desired difference equation

$$(1 + 4\alpha)f_1^{n+1} - 4\alpha f_2^{n+1} = f_1^n + \Delta\xi. \tag{5.5.17}$$

Now collecting the nodal values f_i^n into a vector, $\mathbf{f}^n$, for $i = 1, \ldots, N$, and appending to (5.5.17) equations (5.4.43), we obtain a linear system of algebraic equations,

$$\mathbf{A} \cdot \mathbf{f}^{n+1} = \mathbf{f}^n + \Delta\tau\, \mathbf{e}, \tag{5.5.18}$$

where $\mathbf{A}$ is a tridiagonal matrix and the vector $\mathbf{e}$ is filled with ones. The algorithm involves solving system (5.5.18) at successive time instants using the Thomas algorithm discussed in Section B.1.4, Appendix B.

5.5.3 Transient Poiseuille flow subject to constant flow rate

Transient flow in a pipe subject to a constant flow rate, Q, is complementary to the starting flow due to the sudden application of a constant pressure gradient studied in Section 5.5.2. At the initial instant, the velocity profile is flat, $u_x(t = 0) = Q/(\pi a^2)$, and the pressure gradient takes an unphysical infinite value. As time progresses, the velocity profile tends to the become parabolic while the pressure gradient tends to the value corresponding to the Poiseuille law. The transient velocity profile and axial pressure gradient can be found by the method of separation of variables in the form of a Fourier–Bessel series,

$$u_x = 2 \frac{Q}{\pi a^2} \left[1 - \frac{\sigma^2}{a^2} + 2 \sum_{n=1}^{\infty} \frac{1}{\beta_n^2} \left(\frac{J_0(\beta_n \sigma/a)}{J_0(\beta_n)} \right) \exp(-\beta_n^2 \frac{\nu t}{a^2}) \right], \tag{5.5.19}$$

and

$$\chi \equiv -\frac{\partial p}{\partial x} = 8\mu \frac{Q}{\pi a^4} \left[1 + \frac{1}{2} \sum_{n=1}^{\infty} \exp\left(-\beta_n^2 \frac{\nu t}{a^2} \right) \right], \tag{5.5.20}$$

where β_n, for $n = 1, 2, \ldots$, are the positive zeros of the second-order Bessel function, J_2. The first five zeros of J_2 are

$$5.13562, \quad 8.41724, \quad 11.61984, \quad 14.79595, \quad 17.95982, \tag{5.5.21}$$

accurate to the fifth decimal place (e.g., [2]). The singular behavior of the pressure gradient at $t = 0$ is evident by the divergence of the sum in (5.5.20).

5.5.4   Transient swirling flow inside a hollow cylinder

A hollow circular cylinder of radius a is filled with a quiescent viscous fluid. Suddenly, the cylinder starts rotating around its axis with constant angular velocity, Ω, generating a swirling flow. After a sufficiently long time has elapsed, the fluid executes rigid-body rotation with angular velocity Ω. Solving the governing equation (5.5.4) by separation of variables, we derive the transient azimuthal velocity profile

$$u_\varphi(\sigma t) = \Omega \left[\sigma + 2a \sum_{n=1}^{\infty} \frac{1}{\alpha_n} \frac{J_1(\alpha_n \hat{\sigma})}{J_0(\alpha_n)} \exp\left(-\alpha_n^2 \frac{\nu t}{a^2} \right) \right], \tag{5.5.22}$$

where $\hat{\sigma} = \sigma/a$ and α_n are the positive zeros of the Bessel function of the first kind, J_1. The first five zeros of J_1 are

$$3.83171, \quad 7.01559, \quad 10.17347, \quad 13.32369, \quad 16.47062, \tag{5.5.23}$$

accurate to the fifth decimal place (e.g., [2]). At short times, the circumferential flow near the cylinder surface resembles that due to the impulsive translation of a flat plate.

5.5.5   Transient swirling flow outside a cylinder

A solid circular cylinder of radius a is immersed in a quiescent infinite fluid. Suddenly, the cylinder starts rotating around its axis with constant angular velocity, Ω, generating a swirling flow. After a sufficiently long time has elapsed, the flow resembles that due to a rectilinear line vortex with strength $\kappa = 2\pi\Omega a^2$ situated at the centerline. An expression for the transient velocity profile can be derived in terms of Bessel and modified Bessel functions using of Laplace transform methods ([218], p.72; [437], p. 316, [371], p. 143).

A practical alternative is to compute the solution computed by numerical methods. To remove the singular behavior due to the impulsive motion at the initial instant, we set

$$u_\varphi = \frac{\Omega a^2}{\sigma} f(\eta, \xi), \tag{5.5.24}$$

where $\eta = (\sigma - a)/(\nu t)^{1/2}$ and $\xi = (\nu t)^{1/2}/a$, are dimensionless variables, and f is a dimensionless function ([371], p. 199). Considering (5.5.3), we compute the derivatives

$$\frac{\partial u_\varphi}{\partial t} = \frac{\Omega a^2}{2\sigma t} (-\eta f_\eta + \xi f_\xi), \qquad \frac{\partial(\sigma u_\varphi)}{\partial \sigma} = \frac{\Omega a}{\xi} f_\eta, \tag{5.5.25}$$

and

$$\frac{\partial}{\partial \sigma} \left(\frac{1}{\sigma} \frac{\partial (\sigma u_\varphi)}{\partial \sigma} \right) = \frac{\Omega}{\xi \sigma} \left(-\frac{a}{\sigma} f_\eta + \frac{1}{\xi} f_{\eta\eta} \right), \tag{5.5.26}$$

where a subscript η denotes a derivative with respect to η and a subscript ξ denotes a derivative with respect to ξ. Substituting these expressions into (5.5.3) and simplifying, we obtain

$$\frac{1}{2\xi} \left(-\eta f_\eta + \xi f_\xi \right) = -\frac{a}{\sigma} f_\eta + \frac{1}{\xi} f_{\eta\eta}. \tag{5.5.27}$$

Setting $\sigma/a = 1 + \eta\xi$ and rearranging, we find that f satisfies a convection–diffusion equation,

$$f_\xi + \left(\frac{2}{1+\eta\xi} - \frac{\eta}{\xi} \right) f_\eta = \frac{2}{\xi} f_{\eta\eta}, \tag{5.5.28}$$

with boundary conditions $f(0,\xi) = 1$ and $f(\infty,\xi) = 0$, and initial condition $f(\eta,0) = \mathrm{erfc}(\eta/2)$. The solution can be found by numerical methods (Problem 5.5.7).

Problems

5.5.1 *Axial flow due to the motion of a circular cylinder*

(*a*) Derive the velocity profile inside a circular cylinder executing axial translational vibrations.

(*b*) Repeat (*a*) for the flow outside a cylinder immersed in an infinite fluid.

(*c*) Consider the flow outside a cylinder that suddenly starts moving parallel to its length with constant velocity in an otherwise quiescent fluid. Discuss the development of the flow and derive an expression for the drag force at short times ([218], p.73; [23]).

5.5.2 *Axial flow between concentric annular tubes*

Derive the velocity profile of the axial flow between two concentric cylinders due to sudden application of a constant pressure gradient (e.g., [437], p. 318).

5.5.3 *Oscillatory swirling flow outside and inside a circular cylinder*

(*a*) A solid circular cylinder of radius a executes rotational oscillations around the centerline with angular frequency Ω in an otherwise quiescent infinite fluid, thus generating swirling flow. Compute the velocity in terms of Bessel functions ([371], p. 141).

(*b*) Repeat (*a*) for interior flow.

5.5.4 *Transient swirling flow in an annulus*

An annular channel containing a viscous fluid is confined between two concentric cylinders. Suddenly, the cylinders start rotating around their axis with different but constant angular velocities. Derive an expression for the developing velocity profile in terms of Bessel and related functions (e.g., [437], p. 316). Discuss the structure of the flow when the cylinders rotate with the same angular velocity.

(a)   (b)

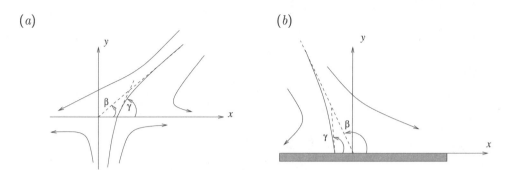

FIGURE 5.6.1   Illustration of stagnation-point flow (a) in the interior of a fluid and (b) over a solid
wall. In both cases, the origin of the y axis is set at the stagnation point of the outer flow. The
angle β defines the slope of the straight dividing streamline of the outer inviscid flow. The angle
γ defines the actual slope of the generally curved dividing streamline of the viscous flow.

Computer Problems

5.5.5 *Axial pulsating flow in a circular pipe*

Plot the velocity profiles of pulsating pressure-driven flow inside a circular pipe at a sequence of
times and for several frequencies. Discuss the structure of the flow with reference to the occurrence
of boundary layers.

5.5.6 *Axial starting flow in a pipe*

Write a program that computes the velocity profiles of starting flow in a circular pipe due to the
sudden application of a constant pressure gradient using the finite-difference method discussed in
the text. Present profiles of the dimensionless velocity at a sequence of dimensionless times.

5.5.7 *Transient swirling flow outside a spinning cylinder*

Write a program that solves equation (5.5.28) using a finite-difference method with the BTCS
discretization discussed in the text. Plot and discuss velocity profiles at a sequence of times.

5.6   Stagnation-point and related flows

A family of two-dimensional, axisymmetric, and three-dimensional flows involving stagnation points
can be computed by solving systems of ordinary differential equations for properly defined functions
and independent variables. The stagnation points may occur in the interior of the fluid or at the
boundaries, as illustrated in Figure 5.6.1(a, b). The precise location of the stagnation points and the
slope of the dividing streamlines or stream surfaces are determined by the the outer flow far from the
stagnation point. Although the prescribed outer flow represents an exact solution of the equation of
motion, it does not satisfy the required boundary conditions along the dividing streamlines, stream
surfaces, or solid boundaries. Our goal is to compute a local solution that satisfies the boundary
conditions and agrees with the outer solution far from the stagnation points.

    Reviews of exact solutions for stagnation-point flows obtained by solving ordinary differential equations can be found in References [42, 421]. In this section, we discuss several fundamental configurations.

5.6.1   Two-dimensional stagnation-point flow inside a fluid

Jeffery derived a class of exact solutions of the equations of two-dimensional incompressible flow [199]. Peregrine [292] pointed out that one of these solutions represents oblique stagnation-point flow in the interior of a fluid, as shown in Figure 5.6.1(*a*).

Outer flow

We begin constructing the solution by introducing the outer flow prevailing far from the stagnation point, denoted by the superscript ∞. The far-flow stream function, velocity, and vorticity in the upper half-space, $y > 0$, are given by

$$\psi^\infty = \xi\left(xy\sin\alpha + \frac{1}{2}\,y^2\,\cos\alpha\right),$$

$$u_x^\infty = \xi\left(x\sin\alpha + y\cos\alpha\right), \qquad u_y^\infty = -\xi\,y\sin\alpha, \qquad \omega_z^\infty = -\xi\cos\alpha, \qquad (5.6.1)$$

where ξ is a constant rate of elongation and α is a free parameter. Setting the stream function to zero, we find that α is related to the angle β subtended between the straight dividing streamline and the x axis by

$$\tan\beta = -2\tan\alpha, \qquad\qquad (5.6.2)$$

as shown in Figure 5.6.1(*a*). Equations (5.6.1) describe oblique stagnation-point flow with constant vorticity against a planar surface in the upper half space. The values $\alpha = 0$ and π correspond, respectively, to unidirectional simple shear flow toward the positive or negative direction of the x axis with shear rate ξ. The value $\alpha = \pi/2$ corresponds to irrotational orthogonal stagnation-point flow in the upper half-space with rate of extension ξ.

    The far-flow stream function, velocity components, and vorticity in the lower half-space, $y < 0$, are given by

$$\psi^\infty = \xi xy\sin\alpha, \qquad u_x^\infty = \xi x\sin\alpha, \qquad u_y^\infty = -\xi y\sin\alpha, \qquad \omega_z^\infty = 0, \qquad (5.6.3)$$

These equations describe irrotational orthogonal stagnation-point flow with shear rate equal to $\xi\sin\alpha$ in the lower half space.

    The outer flow in the upper and lower half-space constitutes an exact solution of the Navier–Stokes equation for a fluid with uniform physical properties and zero or constant vorticity. The velocity is continuous across the horizontal dividing streamline at $y = 0$. However, the derivatives of the velocity undergo a discontinuity that renders the outer flow admissible only in the context of ideal flow. When viscosity is present, a vortex layer develops around the dividing streamline along the x axis rendering the derivatives of the velocity and therefore the stresses continuous throughout the domain of flow.

Viscous flow

Motivated by the functional form of the stream function of the outer flow in the upper and lower half-space, shown in (5.6.1) and (5.6.3), we express the stream function of the viscous flow in the form

$$\psi = x\mathcal{F}(y) + \mathcal{G}(y), \tag{5.6.4}$$

where $\mathcal{F}(y)$ and $\mathcal{G}(y)$ are two unknown functions satisfying the far-field conditions

$$\mathcal{F}(\pm\infty) \simeq \xi y \sin\alpha, \qquad \mathcal{G}(+\infty) \simeq \frac{1}{2}\xi y^2 \sin\alpha, \qquad \mathcal{G}(-\infty) \simeq 0, \tag{5.6.5}$$

and the symbol $\simeq$ denotes the leading-order contributions. Straightforward differentiation yields the components of the velocity and the vorticity,

$$u_x = x\mathcal{F}'(y) + \mathcal{G}'(y), \qquad u_y = -\mathcal{F}(y), \qquad \omega_z = -x\mathcal{F}''(y) - \mathcal{G}''(y). \tag{5.6.6}$$

Substituting expressions (5.6.6) into the steady version of the two-dimensional vorticity transport equation (3.13.1),

$$u_x \frac{\partial \omega_z}{\partial x} + u_y \frac{\partial \omega_z}{\partial y} = \nu \left(\frac{\partial^2 \omega_z}{\partial x^2} + \frac{\partial^2 \omega_z}{\partial y^2} \right), \tag{5.6.7}$$

and factoring out the x variable, we derive two fourth-order ordinary differential equations,

$$\nu \mathcal{F}'''' + \mathcal{F}\mathcal{F}''' - \mathcal{F}'\mathcal{F}'' = 0, \qquad \nu \mathcal{G}'''' + \mathcal{F}\mathcal{G}''' - \mathcal{F}''\mathcal{G}' = 0. \tag{5.6.8}$$

The first equation involves the function $\mathcal{F}$ alone. A solution that satisfies the first far-field condition (5.6.5) is the far flow itself,

$$\mathcal{F}(y) = \xi y \sin\alpha. \tag{5.6.9}$$

Equation (5.6.6) then yields $\omega_z = -\mathcal{F}''$. Substituting (5.6.9) into the second equation of (5.6.8), we obtain an equation for g,

$$\mathcal{G}'''' + \frac{\xi}{\nu} \sin\alpha \, y \, \mathcal{G}''' = 0. \tag{5.6.10}$$

Integrating (5.6.10) twice subject to the aforementioned far-field (5.6.5) and requiring that the vorticity is continuous across the dividing streamline located at $y = 0$, we obtain an expression for the negative of the vorticity,

$$\mathcal{G}'' = -\omega_z = \begin{cases} \xi\cos\alpha + A\operatorname{erfc}(\lambda y) & \text{for } y > 0, \\ (A + \xi\cos\alpha)\operatorname{erfc}(\lambda y) & \text{for } y < 0, \end{cases} \tag{5.6.11}$$

where $\lambda = (\xi \sin\alpha/\nu)^{1/2}$ and A is a constant that must be found by specifying the rate of decay of vorticity far from the x axis. Further integrations of (5.6.11) require two additional stipulations on the behavior of the flow far from the stagnation point.

5.6.2   Two-dimensional stagnation-point flow against a flat plate

Next, we consider oblique two-dimensional stagnation-point flow in the upper half space against a flat plate, as illustrated in Figure 5.6.1(*b*). Far from the plate, the magnitude of the vorticity and stream function assume the far-field distribution given in (5.6.1). Since the outer flow satisfies neither the no-slip nor the no-penetration boundary condition at the wall, it is not an acceptable solution of the equations of viscous flow. A numerical solution that satisfies the full Navier–Stokes equation and boundary conditions was derived by Hiemenz [176] for orthogonal flow and by Stuart [390] for the more general case of oblique flow.

Working as in the case of the free stagnation-point flow, we express the stream function as shown in (5.6.4), and find that the functions $\mathcal{F}$ and $\mathcal{G}$ satisfy the differential equations (5.6.8). To satisfy the no-slip and no-penetration conditions on the wall, we require that

$$\mathcal{F}(0) = 0, \qquad \mathcal{F}'(0) = 0, \qquad \mathcal{F}'(0) = 0. \tag{5.6.12}$$

Setting without loss of generality the stream function to zero at the the plate located at $y = 0$, we obtain

$$\mathcal{G}(0) = 0. \tag{5.6.13}$$

As y tends to infinity, the function $\mathcal{F}$ must behave to leading order like $\mathcal{F} \simeq \xi y \sin \alpha$, and the function $\mathcal{G}$ must behave like $\mathcal{G} \simeq \frac{1}{2} \xi y^2 \cos \alpha$.

Computation of the function $\mathcal{F}$

Integrating the first equation in (5.6.8) once subject to the aforementioned far-field condition, we obtain a third-order equation,

$$\nu \mathcal{F}''' + \mathcal{F}\mathcal{F}'' - \mathcal{F}'^2 = -\xi^2 \sin^2 \alpha. \tag{5.6.14}$$

It is convenient to introduce a dimensionless independent variable, η, and a dimensionless dependent variable scaled with respect to the forcing function on the right-hand side, $f(\eta)$, defined such that

$$\eta = \left(\frac{\xi}{\nu} \sin \alpha \right)^{1/2} y, \qquad \mathcal{F}(y) = (\xi \nu \sin \alpha)^{1/2} f(\eta), \tag{5.6.15}$$

and thereby obtain the dimensionless equation

$$f''' + f f'' - f'^2 + 1 = 0, \tag{5.6.16}$$

where a prime denotes a derivative with respect to η. The boundary conditions require that $f(0) = 0$ and $f'(0) = 0$, and the far-field condition requires that $f(\infty) \simeq \eta$ or $f'(\infty) = 1$.

To solve the third-order equation (5.6.16), we denote $f = x_1$, introduce the auxiliary functions $f' = x_2$ and $f'' = x_3$, and resolve (5.6.16) in a system of three first-order ordinary differential equations,

$$\frac{\mathrm{d}}{\mathrm{d}\eta} \begin{bmatrix} x_1 \\ x_2 \\ x_3 \end{bmatrix} = \begin{bmatrix} x_2 \\ x_3 \\ x_2^2 - x_1 x_3 - 1 \end{bmatrix}, \tag{5.6.17}$$

with boundary conditions $x_1(0) = 0$, $x_2(0) = 0$, and $x_2(\infty) = 1$. The solution can be computed using a shooting method for boundary-value problems according to the following steps:

1. Guess the value of $x_3(0)$.

2. Integrate (5.6.17) as an initial-value problem from $\eta = 0$ to η_{max}, where η_{max} is a sufficiently large truncation level. In practice, setting η_{max} as low as 3.0 yields satisfactory accuracy.

3. Check whether $x_2(\eta_{max}) = 1$ within a preset tolerance. If not, return to Step 1 and repeat the computations with a new and improved value for $x_3(0)$.

The new value of $x_3(0)$ in Step 3 can be found using a method for solving the nonlinear algebraic equation $Q[x_3(0)] \equiv x_2(\eta_{max}) - 1.0 = 0$, as discussed in Section B.3, Appendix B [317].

In practice, it is expedient to carry out the iterations using Newton's method based on variational equations. First, we define the derivatives

$$x_4 \equiv \frac{\mathrm{d}x_1}{\mathrm{d}x_3(0)}, \qquad x_5 \equiv \frac{\mathrm{d}x_2}{\mathrm{d}x_3(0)}, \qquad x_6 \equiv \frac{\mathrm{d}x_3}{\mathrm{d}x_3(0)}, \tag{5.6.18}$$

expressing the sensitivity of the solution to the initial guess. Differentiating equations (5.6.17) with respect to $x_3(0)$, we obtain

$$\frac{\mathrm{d}x_4}{\mathrm{d}\eta} = x_5, \qquad \frac{\mathrm{d}x_5}{\mathrm{d}\eta} = x_6, \qquad \frac{\mathrm{d}x_6}{\mathrm{d}\eta} = 2\,x_2 x_5 - x_4 x_3 - x_1 x_6, \tag{5.6.19}$$

subject to the initial conditions $x_4(\eta = 0) = 0$, $x_5(\eta = 0) = 0$, and $x_6(\eta = 0) = 1$. The shooting method involves the following steps:

1. Guess a value for $x_3(t = 0)$.
2. Integrate equations (5.6.17) and (5.6.19) with the aforementioned initial conditions.
3. Replace the current initial value, $x_3(t = 0)$, with the updated value

$$x_3(t = 0) \leftarrow x_3(t = 0) - \frac{x_2(\eta = \infty) - 1}{x_5(\eta = \infty)}. \tag{5.6.20}$$

The iterations converge quadratically for a reasonable initial guess. Graphs of the functions f, f', and f'' computed by this method are shown in Figure 5.6.2(a). The numerical solution reveals that $x_3(0) \simeq 1.232588$. At large values of η, the function $f(\eta)$ behaves like

$$f(\eta) = \eta - b + \cdots, \tag{5.6.21}$$

where $b \simeq 0.647900$, to shown accuracy.

Computation of the function $\mathcal{G}$

Having computed the function $\mathcal{F}(y)$ in terms of $f(\eta)$, we substitute the result into the first equation of (5.6.8) and obtain a linear homogeneous equation for the function $\mathcal{G}(y)$. Integrating once, we derive a third-order equation,

$$\nu \mathcal{G}''' + \mathcal{F}\mathcal{G}'' - \mathcal{F}'\mathcal{G}' = c, \tag{5.6.22}$$

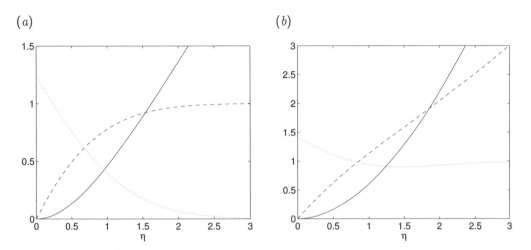

FIGURE 5.6.2   Two-dimensional oblique stagnation-point flow against a flat plate. (a) Graphs of the functions f (solid line), f' (dashed line), and f'' (dotted line). (b) Graphs of the functions g (solid line), g' (dashed line), and g'' (dotted line).

where c is a constant. In terms of the dimensionless function $f(\eta)$, we obtain

$$\nu\,\mathcal{G}''' + (\xi\nu\sin\alpha)^{1/2} f(\eta)\,\mathcal{G}'' - \xi\sin\alpha f'(\eta)\,\mathcal{G}' = c. \tag{5.6.23}$$

Letting y or η tend to infinity and using (5.6.21) and the far-field condition $\mathcal{G}(\infty) \simeq \frac{1}{2}\xi y^2 \cos\alpha$, we obtain

$$c = -b\,\xi\cos\alpha(\xi\nu\sin\alpha)^{1/2}, \tag{5.6.24}$$

where the constant b is defined in (5.6.21).

In the case of orthogonal stagnation-point flow corresponding to $\alpha = \frac{1}{2}\pi$, or simple shear flow corresponding to $\alpha = 0$ or π, we find that $c = 0$, which means that equation (5.6.23) admits the trivial solution $\mathcal{G} = 0$. In the more general case where $\alpha \neq 0, \frac{1}{2}\pi, \pi$, we express g in terms of a dimensionless function $g(\eta)$ defined by

$$\mathcal{G}(y) = \nu\cot\alpha\, g(\eta), \tag{5.6.25}$$

where η has been defined previously in (5.6.15). Substituting (5.6.25) and (5.6.24) into (5.6.23), we obtain the dimensionless equation

$$g''' + fg'' - f'g' = -b, \tag{5.6.26}$$

where a prime denotes a derivative with respect to η. The solution must satisfy the boundary conditions $g(0) = 0$ and $g'(0) = 0$, and a far-field condition dictating that, as η tends to infinity, g behaves like $g \simeq \frac{1}{2}\eta^2$. Equation (5.6.16) suggests that $g = bf$ is a particular solution of (5.6.26), and this allows us to eliminate the forcing function on the right-hand side and obtain a homogeneous equation.

In practice, it is expedient to compute the solution directly by numerical methods. Denoting $g = x_7$ and introducing the auxiliary functions $g' = x_8$ and $g'' = x_9$, we rewrite (5.6.26) in the standard form of a first-order system of ordinary differential equations,

$$\frac{\mathrm{d}}{\mathrm{d}\eta} \begin{bmatrix} x_7 \\ x_8 \\ x_9 \end{bmatrix} = \begin{bmatrix} x_8 \\ x_9 \\ -x_1 x_9 + x_2 x_8 - b, \end{bmatrix}. \qquad (5.6.27)$$

The solution must satisfy the boundary conditions $x_7(0) = 0$ and $x_8(0) = 0$, and the far-field condition $x_9(\infty) = 1$. The solution can be found using the shooting method described previously for the function F, where shooting is now done with respect to $x_9(0)$. The functions x_1 and x_2 are known at discrete points from the numerical solution of (5.6.16). Because equation (5.6.26) is linear, two shootings followed by linear interpolation are sufficient. The numerical results, plotted in Figure 5.6.2(b), show that $x_9(0) \simeq 1.406544$, to shown accuracy [108].

Vorticity

The vorticity can be computed in terms of the functions f'' and g'' as

$$\omega_z = \omega_z^\infty - \xi x \sin\alpha \left(\frac{\xi \sin\alpha}{\nu}\right)^{1/2} f''(\eta) + \xi \left[1 - g''(\eta)\right] \cos\alpha. \qquad (5.6.28)$$

The first term on the right-hand side represents the constant vorticity of the incident flow. The third term represents a vortex layer with uniform thickness attached to the wall. Far from the origin, the second term on the right-hand side dominates and the ratio between the vorticity and the x component of the incident flow is

$$\left|\frac{\omega_z}{u_x^\infty}\right| \simeq \left(\frac{\xi \sin\alpha}{\nu}\right)^{1/2} f''(\eta), \qquad (5.6.29)$$

which reveals the presence of a vortex layer of constant thickness lining the wall. In physical terms, diffusion of vorticity from the wall is balanced by convection, thereby confining the vorticity gradient near the wall.

The thickness of the vortex layer, δ, can be defined as the point where $x_3(\eta) = 0.01$. The numerical solution shows that $f''(2.4) \simeq 0.01$, and thus

$$\delta = 2.4 \left(\frac{\nu}{\xi \sin\alpha}\right)^{1/2}. \qquad (5.6.30)$$

As α tends to zero, the thickness of the vortex layer understandably diverges to infinity.

Pressure field

Having computed the velocity field, we substitute the result into the Navier–Stokes equation and obtain a partial differential equation for the pressure. Straightforward integration yields the pressure

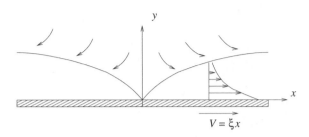

FIGURE 5.6.3   Illustration of two-dimensional flow in an semi-infinite fluid due a symmetrically stretching sheet.

field

$$p = c\rho x - \frac{1}{2}\rho\xi^2\sin^2\alpha\,x^2 - \mu\xi\sin\alpha\left[f'(\eta) + \frac{1}{2}f^2(\eta)\right] + \rho\,\mathbf{g}\cdot\mathbf{x} + p_0, \tag{5.6.31}$$

where the constant c was defined in (5.6.24) and p_0 is a reference pressure. When $\alpha = 0$ or π, the
pressure assumes the hydrostatic distribution, $p = \rho\,\mathbf{g}\cdot\mathbf{x} + p_0$.

Stability

The behavior of perturbations introduced in the base flow derived in this section can be assessed by
performing a linear stability analysis, as discussed in Chapter 9. Conditions for stability subject to
suitable perturbations have been established [224].

5.6.3   Flow due to a stretching sheet

Consider a steady two-dimensional flow produced by the symmetric stretching of an elastic sheet in
an otherwise quiescent semi-infinite fluid, as shown in Figure 5.6.3. The x component of the velocity
at the surface of the sheet, located at $y = 0$, is $u_x = \xi x$, where ξ is a constant rate of extension with
dimensions of inverse time.

Following our analysis in Section 5.6.2 for orthogonal stagnation-point flow against a flat plate,
$\alpha = \frac{1}{2}\pi$, we introduce the dimensionless variable $\eta = (\xi/\nu)^{1/2}y$, and express the stream function in
the form $\psi(x,y) = (\xi\nu)^{1/2}x\,f(\eta)$. The boundary conditions require that $f(0) = 0$ and $f'(0) = 1$, and
the far-field condition of vanishing x velocity requires that $f'(\infty) = 0$. Making substitutions into the
vorticity transport equation, we find that the dimensionless function $f(\eta)$ satisfies the homogeneous
Hiemenz equation

$$f''' + ff'' - f'^2 = 0, \tag{5.6.32}$$

where a prime denotes a derivative with respect to η. The solution is given by

$$f(\eta) = 1 - \exp(-\eta), \tag{5.6.33}$$

yielding Crane's [96] stream function

$$\psi(x,y) = (\xi\nu)^{1/2}x\,(1 - \mathrm{e}^{-\eta}), \tag{5.6.34}$$

(a)

(b)

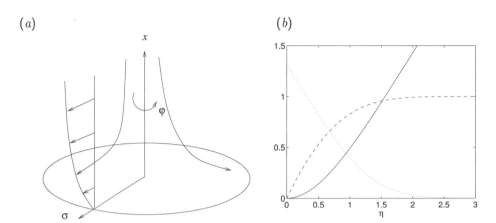

FIGURE 5.6.4 (a) Illustration of axisymmetric orthogonal stagnation-point flow. (b) Graphs of the functions f (solid line), f' (dashed line), and f'' (dotted line). The radial velocity profile is represented by the dashed line.

which provides us with an exact solution of the Navier–Stokes equation with corresponding pressure field

$$p = -\mu\xi \left[f'(\eta) + \frac{1}{2} f^2(\eta) \right] + \rho\, \mathbf{g} \cdot \mathbf{x} + p_0, \tag{5.6.35}$$

where p_0 is a reference pressure. The y velocity component tends to a constant value, $v = -(\xi\nu)^{1/2}$, far from the stretching sheet. Thus, the stretching of the sheet causes a uniform flow against the sheet to satisfy the mass balance.

The effect of wall suction or blowing at a constant velocity can be incorporated into the similarity solution (e.g., [423]). In Section 8.3.2, Crane's solution will be generalized to situations where the in-plane wall velocity exhibits an arbitrary power-law dependence on the x position, subject to the boundary-layer approximation.

5.6.4 Axisymmetric orthogonal stagnation-point flow

Next, we consider irrotational, axisymmetric, orthogonal stagnation-point flow against a flat plate located at $x = 0$, and introduce cylindrical polar coordinates, (x, σ, φ), as shown in Figure 5.6.4(a). Far from the plate, as x tends to infinity, the Stokes stream function, axial velocity component, and radial velocity component assume the far-field distributions

$$\psi^\infty = -\xi x \sigma^2, \qquad u_x^\infty = -2\xi x, \qquad u_\sigma^\infty = \xi\sigma, \tag{5.6.36}$$

where ξ is a constant rate of extension with dimensions of inverse length.

Working as in Section 5.6.2 for two-dimensional orthogonal stagnation-point flow, we express the Stokes stream function, ψ, in terms of a dimensionless variable, $\eta = (\xi/\nu)^{1/2}x$, and set

$$\psi = -(\xi\nu)^{1/2}\sigma^2 f(\eta), \tag{5.6.37}$$

where $f(\eta)$ is a dimensionless function. The axial and radial velocity components and the azimuthal vorticity component are given by

$$u_x = -2\,(\xi\nu)^{1/2} f(\eta), \qquad u_\sigma = \xi\sigma f'(\eta), \qquad \omega_\varphi = \xi\sigma(\xi\nu)^{1/2} f''(\eta), \qquad (5.6.38)$$

where a prime denotes a derivative with respect to η, The no-penetration and no-slip boundary conditions require that $f(0) = 0$ and $f'(0) = 0$, and the far-field condition requires that $f'(\infty) = 1$.

Substituting expressions (5.6.38) into the vorticity transport equation for axisymmetric flow, and integrating once subject to the far-field condition, we derive Homann's [188] ordinary differential equation

$$f''' + 2ff'' - f'^2 + 1 = 0, \qquad (5.6.39)$$

which differs from the Hiemenz equation (5.6.16) only by the value of one coefficient. The solution can be found using a shooting method similar to that used to solve (5.6.17), yielding $f''(0) = 1.3120$, as shown in Figure 5.6.4(b). Having computed the velocity field, we substitute the result into the Navier–Stokes equation and integrate in space to obtain the pressure field

$$p = -\frac{1}{2}\,\rho\,\xi^2\sigma^2 - 2\mu\xi\,\big[\,f'(\eta) + f^2(\eta)\,\big] + \rho\,\mathbf{g}\cdot\mathbf{x} + p_0, \qquad (5.6.40)$$

where p_0 is a reference pressure.

As in the case of two-dimensional stagnation-point flow, the vorticity is confined inside a vortex layer of constant thickness lining the wall. Orthogonal axisymmetric stagnation-point flow provides us with an example of a flow where intensification of the vorticity due to vortex stretching is balanced by diffusion and convection under the influence of the incident flow.

5.6.5 Flow due to a radially stretching disk

Consider a steady axisymmetric flow due to the radial stretching of an elastic sheet in an otherwise quiescent fluid residing in the upper half space, $x > 0$. The radial component of the velocity at the surface of the sheet, located at $x = 0$, is $u_\sigma = \xi\sigma$, where ξ is the rate of extension with dimensions of inverse time. Repeating the analysis of Section 5.6.4, we find that the function $f(\eta)$ satisfies the homogeneous Homann equation

$$f''' + 2ff'' - f'^2 = 0, \qquad (5.6.41)$$

where a prime denotes a derivative with respect to the similarity variable $\eta = (\xi/\nu)^{1/2} x$. The no-penetration and no-slip boundary conditions require that $f(0) = 0$ and $f'(0) = 1$, and the far-field condition requires that $f'(\infty) = 0$. It will be noted that (5.6.41) differs from (5.6.32) only by a numerical factor of two. The corresponding pressure distribution is given by

$$p = -2\mu\xi\,\big[\,f'(\eta) + f^2(\eta)\,\big] + \rho\,\mathbf{g}\cdot\mathbf{x} + p_0, \qquad (5.6.42)$$

where p_0 is a reference pressure. Although an analytical solution of (5.6.41) is not available, a numerical solution can be obtained readily by elementary numerical methods [418].

In Section 8.5.3, the formulation will be generalized to situations where the radial disk velocity exhibits an arbitrary power-law dependence on the radial position, subject to the boundary-layer approximation.

5.6.6   Three-dimensional orthogonal stagnation-point flow

Generalizing the flow configurations considered previously in this section, now we consider three-dimensional orthogonal stagnation-point flow against a flat plate located at $x_3 = 0$. Far from the plate, as x_3 tends to infinity, the velocity components assume the far-field distributions

$$u_1^\infty = \xi_1 x_1, \qquad u_2^\infty = \xi_2 x_2, \qquad u_3^\infty = -(\xi_1 + \xi_2) x_3, \qquad (5.6.43)$$

where ξ_1 and ξ_2 are two arbitrary shear rates with dimensions of inverse length. When $\xi_1 = 0$ or $\xi_2 = 0$, we obtain two-dimensional orthogonal stagnation-point flow in the $x_1 x_3$ or $x_2 x_3$ plane. When $\xi_1 = \xi_2$, we obtain axisymmetric orthogonal stagnation-point flow.

Our experience with two-dimensional and axisymmetric flows suggests expressing the solution in the form

$$u_1 = \xi_1 x_1 q'(\eta), \qquad u_2 = \xi_2 x_2 w'(\eta), \qquad u_3 = -\sqrt{\xi_1 \nu} \left[q(\eta) + \lambda w(\eta) \right], \qquad (5.6.44)$$

where $\lambda = \xi_2/\xi_1$ is a dimensionless parameter,

$$\eta = x_3 \left(\frac{\xi_1}{\nu} \right)^{1/2}, \qquad (5.6.45)$$

is a dimensionless coordinate, and a prime denotes a derivative with respect to η [193]. The unknown functions q and w are required to satisfy the boundary conditions $q(0) = 0$, $w(0) = 0$, $q'(0) = 0$, and $w'(0) = 0$, and the far-field conditions $q'(\infty) = 1$ and $w'(\infty) = 1$. The functional forms (5.6.44) automatically satisfy the continuity equation.

It is convenient to work with the hydrodynamic pressure excluding hydrostatic variations, $\tilde{p} \equiv p - \rho \mathbf{g} \cdot \mathbf{x}$. Substituting expressions (5.6.44) into the x_3 component of the equation of motion, we find that the derivatives of the hydrodynamic pressure $\partial \tilde{p}/\partial x_1$ and $\partial \tilde{p}/\partial x_2$ are independent of the vertical coordinate x_3, and thus identical to those of the incident irrotational flow. Using Bernoulli's equation, we find that

$$\frac{\partial \tilde{p}}{\partial x_1} = -\rho \xi_1^2 x_1, \qquad \frac{\partial \tilde{p}}{\partial x_2} = -\rho \xi_2^2 x_2. \qquad (5.6.46)$$

Substituting these expressions along with (5.6.44) into the x_1 and x_2 components of the equation of motion, we derive two coupled nonlinear ordinary differential equations for the functions q and w,

$$q''' + (q + \lambda w) q'' - q'^2 + 1 = 0, \qquad w''' + (w + \lambda q) w'' - \lambda (w'^2 - 1) = 0. \qquad (5.6.47)$$

In the case of axisymmetric flow where $\lambda = 1$, both equations reduce to (5.6.39) with $q = w = f$. In the case of two-dimensional flow with $\lambda = 0$, the first equation reduces to (5.6.16) with $q = f$.

The boundary value problem expressed by (5.6.47) can be solved using a shooting method where guesses are made for $q''(0)$ and $w''(0)$. Numerical results show that $q''(0) = 1.233$ and $w''(0) = 0.570$ when $\lambda = 0.0$; $q''(0) = 1.247$ and $w''(0) = 0.805$ when $\lambda = 0.25$; $q''(0) = 1.267$ and $w''(0) = 0.998$ when $\lambda = 0.50$; $q''(0) = 1.288$ and $w''(0) = 1.164$ when $\lambda = 0.75$; $q''(0) = 1.312$ and $w''(0) = 1.312$ when $\lambda = 1.00$.

Unsteady flow

The scaling of the velocity shown in (5.6.44) is useful for computing unsteady flow due to the start-up of an outer stagnation-point flow. Substituting into the continuity equation and equation of motion the functional forms

$$u_1 = \xi_1 \, x_1 K_1(x_3, t), \qquad u_2 = \xi_2 \, x_2 K_2(x_3, t), \qquad u_3 = -(\xi_1 \nu)^{1/2} \, K_3(x_3, t),$$

$$p = -\frac{1}{2} \, \rho \, (\xi_1^2 x_1^2 + \xi_2^2 x_2^2) + \mu \, (\xi_1 + \xi_2) \, K_4(x_3, t) + \rho \, \mathbf{g} \cdot \mathbf{x} + p_0, \tag{5.6.48}$$

we obtain a system of four differential equations for the functions K_i with respect to x_3 and t, where $i = 1$–4. The solution can be found by standard numerical methods (e.g., [371] p. 207). There is a particular protocol of startup where the flow admits a similarity solution and the problem is reduced to solving an ordinary differential equation for a carefully crafted similarity variable [419].

Problems

5.6.1 *Two-dimensional stagnation-point flow against an oscillating plate*

Consider two-dimensional orthogonal stagnation-point flow with rate of extension ξ against a flat plate that oscillates in its plane along the x axis with angular frequency Ω and amplitude V. The flow can be described by the stream function

$$\psi = (\nu \xi)^{1/2} \, \mathrm{Real} \big\{ \, x f(\eta) + V \left(\nu/\xi \right)^{1/2} q(\eta) \, \exp(-\mathrm{i}\Omega t) \, \big\}, \tag{5.6.49}$$

where $\eta = (\xi/\nu)^{1/2} y$ is a scaled coordinate, i is the imaginary unit, and f, q are two unknown complex functions (e.g., [353]; [428] p. 402). The no-slip and no-penetration conditions at the plate require that $f(0) = 0$, $f'(0) = 0$, $q(0) = 0$, and $q'(0) = 1$. The far-field condition requires that $f(\infty) \simeq \eta$ or $f'(\infty) = 1$, and $q'(\infty) = 0$.

(*a*) Show that the function f satisfies equation (5.6.16), and is therefore identical to that for steady stagnation-point flow, whereas the function q satisfies the linear equation

$$q''' + f q'' - (f' + \mathrm{i}\,\hat{\Omega})\, q' = 0, \tag{5.6.50}$$

where $\hat{\Omega} \equiv \Omega/\xi$ is a dimensionless parameter. Demonstrate that the pressure field is unaffected by the oscillation.

(*b*) Explain why the problem is equivalent to that of stagnation-point flow against a stationary flat plate, where the stagnation point oscillates about a mean position on the plate.

(*c*) Develop the mathematical formulation when the plate oscillates along the z axis ([428] p. 405).

5.6.2 *Stagnation-point flow against a moving plate*

Discuss the computation of two-dimensional orthogonal stagnation-point flow against a plate that translates in its plane with a general time-dependent velocity [425].

5.6.3 *Flow due to a stretching plane*

Develop a similarity solution for the flow in a semi-infinite fluid due to the stretching of a surface located at $x_3 = 0$. The tangential velocity at the surface is given by $u_1 = \xi_1 x_1$ and $u_2 = \xi_2 x_2$ [418].

Computer Problems

5.6.4 *Stagnation-point flow*

(*a*) Write a program that uses the shooting method to solve the system (5.6.17) and (5.6.27) for two-dimensional stagnation-point flow and reproduce the graphs in Figure 5.6.2(*a, b*).

(*b*) Write a program that uses the shooting method to solve the system (5.6.39) for axisymmetric orthogonal stagnation-point flow and reproduce the graphs in Figure 5.6.4(*b*).

(*c*) Write a computer program that uses the shooting method to solve equations (5.6.41) subject to the stated boundary conditions. Compare the results with the Crane solution for the corresponding two-dimensional flow.

(*d*) Write a computer program that uses the shooting method to solve equations (5.6.47) and prepare graphs of the results for $\lambda = 0.30, 0.60$, and 0.90.

5.6.5 *Orthogonal stagnation-point flow against an oscillating plane*

Solve the differential equation (5.6.50) using a method of your choice and prepare a graph of the solution.

5.7 Flow due to a rotating disk

Consider an infinite horizontal disk rotating in its plane around the x axis and about the centerpoint with constant angular velocity Ω in a semi-infinite fluid, as shown in Figure 5.7.1(*a*). The rotation of the disk generates a swirling flow that drives a secondary axisymmetric stagnation-point flow against the disk along the axis of rotation. The secondary flow is significant inside a boundary layer attached to the disk, and decays to zero far from the disk. The thickness of the boundary layer depends on the angular velocity of rotation.

Von Kármán [415] observed that the equation of motion can be reduced into a system of ordinary differential by introducing the transformations

$$u_x = (\nu\Omega)^{1/2}\, H(\eta), \qquad u_\sigma = \Omega\,\sigma\, F(\eta), \qquad u_\varphi = \Omega\,\sigma\, G(\eta), \qquad (5.7.1)$$

where $\eta = x(\Omega/\nu)^{1/2}$ is a scaled x coordinate and H, F, and G are three dimensionless functions. The no-slip boundary condition at the disk surface requires that $H(0) = 0$, $F(0) = 0$, and $G(0) = 1$. The continuity equation requires that $H'(0) = 0$. The far-field condition requires that $F(\infty) = 0$, $G(\infty) = 0$, and $H'(\infty) = 0$. Note that by demanding that $H'(\infty) = 0$ or $H(\infty)$ is a constant, we allow for the onset of uniform axial flow toward the disk. Substituting expressions (5.7.1) into the Navier–Stokes equation, we find that the pressure must assume the functional form

$$p = -\mu\Omega\, Q(\eta) + \rho\, \mathbf{g} \cdot \mathbf{x} + p_0, \qquad (5.7.2)$$

where Q is a dimensionless function and p_0 is a constant. Note that the hydrodynamic pressure does not exhibit the usual radial dependence due to the centripetal acceleration.

Combining the equation of motion with the continuity equation, we obtain a system of four ordinary differential equations for the functions H, F, G, and Q. Combining these equations further

(a) $\qquad\qquad\qquad\qquad\qquad\qquad\qquad$ (b)

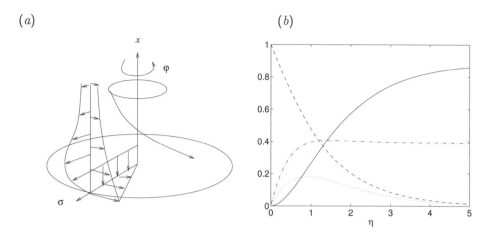

FIGURE 5.7.1   (a) Illustration of flow due to a rotating disk. (b) Graphs of the dimensionless functions $-H$ (solid line), G (dashed line), F (dotted line), and Q (dot-dashed line).

to eliminate the functions Q and F, we derive two third-order coupled ordinary differential equations for the functions H and G,

$$H''' - H''H + \frac{1}{2}H'^2 - 2G^2 = 0, \qquad G'' - G'H + GH' = 0, \qquad (5.7.3)$$

subject to the aforementioned boundary conditions. Once the solution has been found, the functions F and Q arise from the expressions

$$F = -\frac{1}{2}H', \qquad Q = \frac{1}{2}H^2 - H'. \qquad (5.7.4)$$

An approximate solution of equations (5.7.3) is available [90]. To compute a numerical solution, we recast these equations as a system of five first-order equations for H, H', H'', G, and G'. Denoting $H = x_1$ and $G = x_4$, we obtain

$$\frac{\mathrm{d}}{\mathrm{d}t} \begin{bmatrix} x_1 \\ x_2 \\ x_3 \\ x_4 \\ x_5 \end{bmatrix} = \begin{bmatrix} x_2 \\ x_3 \\ x_1 x_3 - \frac{1}{2}x_2^2 + 2x_4^2 \\ x_5 \\ x_1 x_5 - x_4 x_2 \end{bmatrix}, \qquad (5.7.5)$$

with boundary and far-field conditions

$$x_1(0) = 0, \qquad x_2(0) = 0, \qquad x_4(0) = 1, \qquad x_2(\infty) = 0, \qquad x_4(\infty) = 0. \qquad (5.7.6)$$

Since the initial values $x_3(0)$ and $x_5(0)$ are *a priori* unknown, we must solve a boundary-value problem with two unknown end conditions. The solution can be found using a shooting method in two variables according to the following steps:

1. Guess the values of $H''(0) = x_3(0)$ and $G'(0) = x_5(0)$.

2. Integrate the system by solving an initial value problem from $\eta = 0$ up to a sufficiently large value. In practice, $\eta = 10$ yields satisfactory accuracy.

3. Check whether the far-field conditions $H'(\infty) = x_2(\infty) = 0$ and $G(\infty) = x_4(\infty) = 0$ are fulfilled. If not, return to Step 2 and repeat the procedure with improved guesses for $x_3(0)$ and $x_5(0)$.

Results computed using this method are plotted in Figure 5.7.1(b). The numerical solution shows that $H''(0) = -1.0204$ and $G'(0) = -0.6159$, to shown accuracy.

The thickness of the boundary layer on the disk, δ, can be identified with the elevation of the point where the function $F(\eta)$ drops to a small value. The numerical solution shows that $F(\eta) = 0.01$ when

$$\delta \simeq 5.4 \, (\nu/\Omega)^{1/2}. \tag{5.7.7}$$

At the edge of the boundary layer, the axial velocity takes the value $u_x = -0.89 \, (\nu\Omega)^{1/2}$. The associated flow feeds the boundary layer in order to sustain the radial motion of the fluid away from the centerline.

Torque

Neglecting end effects, we find that the axial component of the torque exerted on a rotating disk of radius a is given by

$$T_z = \frac{\pi}{2} \, G'(0) \, \rho a^4 (\nu\Omega^3)^{1/2}. \tag{5.7.8}$$

Stability

Observation reveals and stability analysis confirms that the steady flow over the disk is stable as long as the radius of the disk, a, expressed by the Reynolds number, $\mathrm{Re} = a(\Omega/\nu)^{1/2}$, is lower than about 285 [251]. Above this threshold, the flow develops non-axisymmetric waves yielding a spiral vortex pattern.

Unsteady flow

The von Kármán scaling of the velocity with respect to radial distance σ shown in (5.7.1) is also useful in the case of unsteady flow due, for example, to the sudden rotation of the disk with constant or variable angular velocity. Substituting into the continuity equation and equation of motion the functional forms

$$u_x = (\nu\Omega)^{1/2} \, H(x, t), \qquad u_\sigma = \sigma\Omega \, F(x, t), \qquad u_\varphi = \sigma\Omega \, G(x, t),$$
$$p = -\mu\Omega \, Q(x, t) + \rho \mathbf{g} \cdot \mathbf{x} + p_0, \tag{5.7.9}$$

we obtain a system of four partial differential equations for the functions H, F, G, and Q with respect to x and t whose solution can be found by numerical methods ([371] p. 204; [419]).

Problems

5.7.1 *Rotary oscillations*

Consider the flow due to a disk executing rotary oscillations. Derive four differential equations for the functions H, F, G, and Q with respect to x and t. Confirm that, in the limit of zero frequency, these equations reduce to (5.7.3) and (5.7.4).

5.7.2 *Flow due to two coaxial rotating disks*

Consider the flow between two parallel disks of infinite extent separated by distance b, rotating around a common axis with angular velocity Ω_1 and Ω_2. Introduce expressions (5.7.1), where Ω is replaced by the angular velocity of the lower disk Ω_1, and show that the functions H, F, G, and Q satisfy equations (5.7.3) and (5.7.4) with boundary conditions

$$F(0) = 0, \quad G(0) = 1, \quad H(0) = 0, \quad F(\eta_1) = 0, \quad F(\eta_1) = \frac{\Omega_2}{\Omega_1}, \quad H(\eta_1) = 0, \qquad (5.7.10)$$

where $\eta_1 = (\Omega_1/\nu)^{1/2}b$.

5.8   Flow inside a corner due to a point source

Jeffery and Hamel independently considered two-dimensional flow between two semi-infinite planes intersecting at an angle 2α, as illustrated in Figure 5.8.1 [161, 199]. The motion of the fluid is driven by a point source located at the apex. This configuration can be regarded as a model of the local two-dimensional flow in a rapidly converging or diverging channel.

We begin the analysis by introducing plane polar coordinates with origin at the apex, (r, θ), as shown in Figure 5.8.1. Assuming that the flow is purely radial, we set $u_\theta = 0$ and use the continuity equation to find that the radial velocity component takes the functional form

$$u_r = \frac{A}{r}\,F(\eta), \qquad (5.8.1)$$

where $\eta = \theta/\alpha$, $F(\eta)$ is a dimensionless function, and A is a dimensional constant related to flow rate, Q, by the equation

$$Q = A\,\alpha \int_{-1}^{1} F(\eta)\,\mathrm{d}\eta. \qquad (5.8.2)$$

To satisfy the no-slip boundary condition at the two walls located at $\theta = \pm\alpha$, we require that $F(\pm1) = 0$. Since the velocity field is purely radial, closed streamlines do not arise. However, it is possible that the direction of the flow may reverse inside a certain portion of the channel. When this occurs, the fluid moves against the direction of the primary flow driven by the point source.

We are free to select the relative magnitudes of A and F. To remove this degree of freedom, we specify that the maximum value of F is equal to unity, which is equivalent to requiring that $F' = 0$ when $F = 1$, where a prime denotes a derivative with respect to η. Substituting (5.8.1) into the two-dimensional vorticity transport equation, we obtain a third-order nonlinear equation for F,

$$F''' + 4\alpha^2 F' + 2\alpha\,\mathrm{Re}\,FF' = 0, \qquad (5.8.3)$$

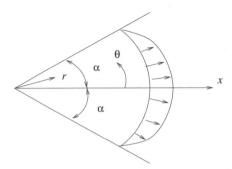

FIGURE 5.8.1 Illustration of the Jeffery–Hamel flow between two semi-infinite planes intersecting at angle 2α, driven by a point source located at the apex.

where $\mathrm{Re} \equiv \alpha A/\nu$ is the Reynolds number. To simplify the notation, we have allowed Re to be negative in the case of net inward flow. Integrating (5.8.3) once with respect to η, we derive the Jeffery–Hamel equation

$$F'' + 4\alpha^2 F + \alpha \, \mathrm{Re} \, F^2 = \delta, \tag{5.8.4}$$

where δ is a constant. Substituting (5.8.1) into the radial component of the Navier–Stokes equation, using (5.8.4), and integrating once with respect to r, we obtain the hydrodynamic pressure distribution,

$$\tilde{p} \equiv p - \rho \, \mathbf{g} \cdot \mathbf{x} = \frac{2\mu}{r^2} \, A \left(F(\eta) - \frac{1}{4}\delta \right). \tag{5.8.5}$$

Stokes flow

The third term on the left-hand side of (5.8.4) represents the effect of fluid inertia. Neglecting this term yields a linear equation whose solution is readily found by elementary analytical methods. Solving for F and computing A in terms of Q from (5.8.2), we obtain

$$u_r = \frac{Q}{r} \, \frac{\cos 2\theta - \cos 2\alpha}{\sin 2\alpha - 2\alpha \cos 2\alpha}. \tag{5.8.6}$$

A different way of deriving this solution based on the stream function is discussed in Section 6.2.5.

Inertial flow

To compute flow at nonzero Reynolds numbers, we multiply (5.8.4) by F' and rearrange to obtain

$$\frac{1}{2} \, (F'^2)' + 2\alpha^2 (F^2)' + \frac{1}{3} \, \alpha \, \mathrm{Re} \, (F^3)' = \delta \, F'. \tag{5.8.7}$$

One more integration subject to the condition $F'(\eta) = 0$ when $F = 1$ yields the first-order equation

$$F' = \pm(1-F)^{1/2} \left[\frac{2}{3} \alpha \, \mathrm{Re} \, F(1+F) + 4\alpha^2 F + c \right]^{1/2}, \tag{5.8.8}$$

where c is a new constant. Applying (5.8.8) at $\eta = \pm 1$ reveals that c is related to the magnitude of the shear stress at the walls by

$$c = |F'(\pm 1)|. \tag{5.8.9}$$

When $AF'(1) < 0$ or $AF'(-1) < 0$, we obtain a region of reversed flow adjacent to the walls. Given the values of α and Re, the problem is reduced to solving (5.8.8) and computing the value of the constant c subject to the boundary conditions $F(\pm 1) = 0$. Once this is done, the results can be substituted into (5.8.2) to yield the corresponding value of Q. Rosenhead [350] derived an exact solution in terms of elliptic functions.

Numerical methods

To compute a numerical solution, we use a shooting method involving the following steps [190]: Guess a value for c; integrate (5.8.8) by solving an initial value problem with the plus or minus sign and initial condition $F(-1) = 0$ from $\eta = -1$ to $= 1$; examine whether $F(1) = 0$; if not, repeat the computation with an improved estimate for c. The results show that the flow exhibits a variety of features, especially at high Reynolds numbers, discussed in detail by a number of authors [24, 136, 137, 428, 190]. Linear stability analysis shows that the flow becomes unstable above a critical Reynolds number that decreases rapidly as the angle α becomes wider [160].

 ***Computer Problem***

5.8.1 *Computing the Jeffery–Hamel flow by a shooting method*

Solve (5.8.8) for $\alpha = \pi/8$ at a sequence of Reynolds numbers, Re $= 1, 10, 20$, and 50. Compute the corresponding dimensionless flow rate, Q/ν.

5.9   Flow due to a point force

Consider the flow due to a narrow jet of a fluid discharging into a large tank that is filled with the same fluid. As the diameter of the jet decreases while the flow rate is kept constant, we obtain the flow due to a point source of momentum, also called a point force, located at the point of discharge. A different realization of the flow due to a point force is provided by the motion of a small particle settling under the action of gravity in an otherwise quiescent ambient fluid.

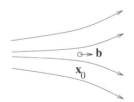

Flow due to a point source of momentum.

   The velocity and pressure fields due to a steady point force applied at a point $\mathbf{x}_0$ satisfy the continuity equation and the steady version of Navier–Stokes equation with a singular forcing term on the right-hand side,

$$\rho\,\mathbf{u} \cdot \nabla\mathbf{u} = -\nabla\tilde{p} + \mu\nabla^2\mathbf{u} + \mathbf{b}\,\delta(\mathbf{x} - \mathbf{x}_0), \tag{5.9.1}$$

where $\mathbf{b}$ is a constant vector determining the magnitude and direction of the point force. The effect of the distributed body force due to gravity has been absorbed into the hydrodynamic pressure,

$\tilde{p} \equiv p - \rho \mathbf{g} \cdot \mathbf{x}$. Physically, equation (5.9.1) describes the flow due to the steady gravitational settling of a particle with infinitesimal dimensions but finite weight in an ambient fluid of infinite expanse, viewed in a frame of reference translating with the particle.

Assuming that the flow is axisymmetric, we introduce spherical polar coordinates, (r, θ, φ), where the x axis points in the direction of the point force, $\mathbf{b}$, and describe the flow in terms of the Stokes stream function, ψ. The absence of an intrinsic length scale suggests the functional form

$$\psi(r, \theta) = \nu r \, \mathcal{F}(\cos \theta), \tag{5.9.2}$$

where $\mathcal{F}$ is a dimensionless function and $\nu = \mu/\rho$ is the kinematic viscosity of the fluid. For future convenience, $\cos \theta$ has been used instead of the polar angle θ in the argument. The spherical polar components of the velocity are

$$u_r = \frac{1}{r^2 \sin \theta} \frac{\partial \psi}{\partial \theta} = -\frac{\nu}{r} \mathcal{F}', \qquad u_\theta = -\frac{1}{r \sin \theta} \frac{\partial \psi}{\partial r} = -\frac{\nu}{r \sin \theta} \mathcal{F}, \tag{5.9.3}$$

and the azimuthal component of the vorticity is

$$\omega_\varphi = -\frac{\nu}{r} \sin \theta \, \mathcal{F}'', \tag{5.9.4}$$

where a prime denotes a derivative with respect to $\cos \theta$. It is clear from these functional forms that the streamlines are self-similar, that is, one streamline derives from another by a proper adjustment of the radial scale.

Progress can be made by substituting (5.9.3) into the vorticity transport equation to derive a fourth-order ordinary differential equation for $\mathcal{F}$. However, it is more expedient to work directly with the generalized equation of motion (5.9.1), thus obtaining a simultaneous solution for the pressure. Motivated again by the absence of an intrinsic length scale, we express the hydrodynamic pressure in the functional form

$$\tilde{p}(r, \theta) = p_0 + \mu \frac{\nu}{r^2} \mathcal{Q}(\cos \theta), \tag{5.9.5}$$

where $\mathcal{Q}$ is a dimensionless function and p_0 is an unspecified constant. Substituting (5.9.3) and (5.9.5) into the radial and meridional components of the Navier–Stokes equation for steady axisymmetric flow, given by

$$\rho \left(u_r \frac{\partial u_r}{\partial r} + \frac{u_\theta}{r} \frac{\partial u_r}{\partial \theta} - \frac{u_\theta^2}{r} \right) = -\frac{\partial \tilde{p}}{\partial r} + \mu \left(\nabla^2 u_r - 2 \frac{u_r}{r^2} - \frac{2}{r^2} \frac{\partial u_\theta}{\partial \theta} - \frac{2}{r^2} u_\theta \cot \theta \right),$$

$$\tag{5.9.6}$$

$$\rho \left(u_r \frac{\partial u_\theta}{\partial r} + \frac{u_\theta}{r} \frac{\partial u_\theta}{\partial \theta} + \frac{u_r u_\theta}{r} \right) = -\frac{1}{r} \frac{\partial \tilde{p}}{\partial \theta} + \mu \left(\nabla^2 u_\theta + \frac{2}{r^2} \frac{\partial u_r}{\partial \theta} - \frac{u_\theta}{r^2 \sin^2 \theta} \right),$$

where

$$\nabla^2 f = \frac{1}{r^2} \frac{\partial}{\partial r} \left(r^2 \frac{\partial f}{\partial r} \right) + \frac{1}{r^2 \sin \theta} \frac{\partial}{\partial \theta} \left(\sin \theta \frac{\partial f}{\partial \theta} \right) + \frac{1}{r^2 \sin^2 \theta} \frac{\partial^2 f}{\partial \varphi^2}, \tag{5.9.7}$$

we obtain two coupled, third-order, nonlinear ordinary differential equations,

$$(\mathcal{F}\mathcal{F}' - \sin^2\theta\,\mathcal{F}'')' + \frac{\mathcal{F}^2}{\sin^2\theta} + 2\mathcal{Q} = 0, \qquad \mathcal{F}'' + \frac{1}{2}\left(\frac{\mathcal{F}^2}{\sin^2\theta}\right)' + \mathcal{Q}' = 0. \tag{5.9.8}$$

Integrating the second equation once and combining the result with the first equation to eliminate $\mathcal{Q}$, we obtain an equation for $\mathcal{F}$ alone,

$$\left[(1-\eta^2)\,\mathcal{F}'' - \mathcal{F}\mathcal{F}'\right]' + 2\mathcal{F}' = A, \tag{5.9.9}$$

where $\eta = \cos\theta$, and A is an integration constant. Integrating once more, we derive a second-order equation,

$$(1-\eta^2)\,\mathcal{F}'' - \mathcal{F}\mathcal{F}' + 2\mathcal{F}' - A\eta = c, \tag{5.9.10}$$

where c is a new integration constant.

To evaluate the constants A and c, we note that both the positive and negative parts of the x axis are streamlines and set $\mathcal{F}(\pm 1) = 0$. Next, we require that the radial component of the velocity and the vorticity are finite along the x axis and use (5.9.8) to find that, as η tends to ± 1, both $\mathcal{F}'(\pm 1)$ and the limit of $(1-\eta^2)^{1/2}\mathcal{F}''(\eta)$ must be bounded. Equation (5.9.10) then yields $A = 0$ and $c = 0$. Integrating (5.9.10), we find

$$(1-\eta^2)\,\mathcal{F}' + 2\eta\mathcal{F} - \frac{1}{2}\mathcal{F}^2 = \alpha, \tag{5.9.11}$$

where α is a new integration constant. Applying (5.9.11) at the x axis corresponding to $\eta = \pm 1$ and requiring a nonsingular behavior we obtain $\alpha = 0$. Integrating (5.9.11), we finally obtain the general solution

$$\mathcal{F} = 2\,\frac{\sin^2\theta}{d - \cos\theta} \tag{5.9.12}$$

where d is a new integration constant whose value must be adjusted so that the flow field satisfies (5.9.1) at the singular point, $\mathbf{x}_0$ [221, 385]. Substituting (5.9.12) into the first equation of (5.9.8) yields the function $\mathcal{Q}$,

$$\mathcal{Q} = 4\,\frac{d\cos\theta - 1}{(d - \cos\theta)^2}. \tag{5.9.13}$$

To evaluate the constant d, we apply the macroscopic momentum balance (3.2.9) over a spherical volume of fluid of radius R centered at the point force. We note that the flow is steady and use the properties of the delta function to find

$$\iint_{Sphere} (\boldsymbol{\sigma} - \rho\,\mathbf{u}\otimes\mathbf{u})\cdot\mathbf{n}\,dS = -\mathbf{b}, \tag{5.9.14}$$

where $\mathbf{n}$ is the normal vector pointing outward from the spherical surface. Using (5.9.12) and (5.9.13) to evaluate the stress tensor and substituting the result in the x component of (5.9.14), we obtain an implicit equation for d,

$$\mathrm{Re}^2 \equiv \frac{|\mathbf{b}|}{8\pi\mu\nu} = \frac{8}{3}\,\frac{d}{d^2-1} + d^2\,\ln\frac{d-1}{d+1} + 2d, \tag{5.9.15}$$

(a)   (b)

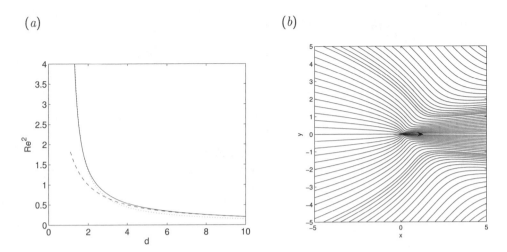

FIGURE 5.9.1   (a) The relationship between the effective Reynolds number, Re, defined in (5.9.15)
and the constant d. The dashed line represents the low-Re asymptotic given in (5.9.17), and the
dotted line represents the high-Re asymptotic given in (5.9.17). (b) Streamline pattern of the flow
due to a point source of momentum or point force for $d = 1.1$.

when Re is an effective Reynolds number of the flow. It is now evident that the structure of the
flow depends on the value of Re determining the intensity of the point force. In fact, this conclusion
could have been drawn on the basis of dimensional analysis at the outset. A graph of Re against
the constant d in its range of definition, $d > 1$, is shown with the solid line in Figure 5.9.1(a). The
streamline pattern for $d = 1.1$ is shown in Figure 5.9.1(b).

Low-Reynolds-number flow

To study the structure of flow for large values of d, we expand the right-hand side of (5.9.15) in
a Taylor series with respect to $1/d$ and find that $\mathrm{Re}^2 \simeq 2/d$, which confirms that Re is small,
as shown in Figure 5.9.1(a). Making appropriate approximations, we find that the corresponding
stream function and modified pressure are given by

$$\psi = \frac{|\mathbf{b}|}{8\pi\mu}\, r \sin^2 \theta, \qquad \tilde{p} = \frac{|\mathbf{b}|}{4\pi}\, \frac{1}{r^2}\, \cos\theta. \qquad (5.9.16)$$

Evaluating the various terms of the Navier–Stokes equation shows that the nonlinear inertial terms
are smaller than the pressure and viscous terms by a factor of $1/d$ and the solution (5.9.15) satisfies
the equations of Stokes flow. The streamline pattern is symmetric about the midplane $x = 0$, as
illustrated in Figure 6.4.1(a).

High-Reynolds-number flow

To study the opposite limit as d tends to unity, we expand the right-hand side of (5.9.15) in a Taylor series with respect to $d - 1$, and find that

$$\mathrm{Re}^2 = \frac{4}{3(d-1)}, \tag{5.9.17}$$

which confirms that Re is large. Making appropriate approximations, we find that, as long as $\cos\theta$ is not too close to unity, that is, sufficiently far from the positive part of the x axis, the stream function and hydrodynamic pressure are given by

$$\psi = 2\nu r\,(1 + \cos\theta), \qquad \tilde{p} = -\mu\,\frac{\nu}{r^2}\,\frac{1}{1 - \cos\theta}, \tag{5.9.18}$$

independent of the magnitude of Re.

Further flows due to a point force

The flow due to a point force in an infinite domain discussed in this section has been generalized to account for a simultaneous swirling motion and for the presence of a flat or conical boundaries [146, 372, 428]. Other configurations are discussed in Chapter 6 in the context of Stokes flow.

Problem

5.9.1 *Edge of the Landau jet*

The edge of the jet can be defined as the surface where the associated streamlines are at minimum distance from the x axis. Show that this is a conical surface with $\theta = \beta$, where $\cos\beta = 1/d$.

Flow at low Reynolds numbers

6

The distinguishing and most salient feature of low-Reynolds-number flow is that the nonlinear convective force, $\rho\, \mathbf{u} \cdot \nabla \mathbf{u}$, makes a small or negligible contribution to Cauchy's equation of motion expressing Newton's law of motion for a small fluid parcel. The motion of the fluid is governed by the continuity equation and the linearized Navier–Stokes equation describing steady, quasi-steady, or unsteady Stokes flow, as discussed in Section 3.10. The linearity of the governing equations allows us to study in detail the mathematical properties of the solution and the physical structure of the flow. Solutions can be obtained by a variety of analytical and numerical methods for a broad range of configurations involving single- and multi-fluid flow. Examples include methods based on separation of variables, singularity representations, and boundary-integral formulations. In this chapter, we discuss the general properties of steady, quasi-steady, and unsteady flow at low Reynolds numbers, and then proceed to derive exact and approximate solutions for flow near boundary corners, flow of liquid films, lubrication flow in confined geometries, and flow past or due to the motion of rigid particles and liquid drops.

The study of uniform flow past a stationary particle or flow due to the translation of a particle in an infinite quiescent fluid will lead us to examine the structure of the far field where the Stokes-flow approximation is no longer appropriate. In the far flow, the neglected inertial term $\rho\, \mathbf{u} \cdot \nabla \mathbf{u}$ can be approximated with the linear form $\rho\, \mathbf{U} \cdot \nabla \mathbf{u}$, where $\mathbf{U}$ is the uniform velocity of the incident flow, yielding Oseen's equation and corresponding Oseen flow. Computing solutions of the Oseen equation is complicated by the quadratic coupling of $\mathbf{U}$ and $\mathbf{u}$. Nevertheless, the preserved linearity of the equation of motion with respect to $\mathbf{u}$ allows us to conduct some analytical studies and develop efficient methods of numerical computation. In concluding this chapter, we discuss oscillatory and more general time-dependent flow governed by the unsteady Stokes equation (3.10.6). Although the new feature of time-dependent motion renders the analysis somewhat more involved, we are still able to build a firm theoretical foundation and derive exact and approximate solutions using efficient analytical and numerical methods.

6.1 Equations and fundamental properties of Stokes flow

The scaling analysis of Section 3.10 revealed that the flow of an incompressible Newtonian fluid with uniform physical properties at low values of the frequency parameter, $\beta \equiv L^2/(\nu T)$, and Reynolds number, $\mathrm{Re} \equiv UL/\nu$, is governed by the continuity equation expressing mass conservation, $\nabla \cdot \mathbf{u} = 0$, and the Stokes equation expressing a force balance,

$$\nabla \cdot \sigma + \rho \mathbf{g} = \mathbf{0}, \qquad (6.1.1)$$

or

$$-\nabla p + \mu \nabla^2 \mathbf{u} + \rho \mathbf{g} = \mathbf{0}, \qquad (6.1.2)$$

where $\mathbf{u}$ is the velocity, p is pressure, μ is the fluid viscosity, ρ is the fluid density, and $\mathbf{g}$ is the acceleration of gravity.

In terms of the hydrodynamic pressure excluding the effect of the body force, $\tilde{p} \equiv p - \rho\,\mathbf{g}\cdot\mathbf{x}$, and associated stress tensor, $\tilde{\boldsymbol{\sigma}} = -\tilde{p}\mathbf{I} + 2\mu\mathbf{E}$, the Stokes equation becomes

$$\nabla \cdot \tilde{\boldsymbol{\sigma}} = \mathbf{0}, \qquad (6.1.3)$$

or

$$-\nabla \tilde{p} + \mu \nabla^2 \mathbf{u} = \mathbf{0}, \qquad (6.1.4)$$

where $\mathbf{E} = \frac{1}{2}\left[\nabla\mathbf{u} + (\nabla\mathbf{u})^T\right]$ is the rate-of-deformation tensor for an incompressible fluid.

A flow that is governed by one of the four equivalent equations (6.1.1)–(6.1.4) is called a Stokes or creeping flow.

Force and torque on a fluid parcel

The total force exerted on an arbitrary fluid parcel, including the hydrodynamic and the body force, is given in (3.1.15), repeated below for convenience,

$$\mathbf{F}^{tot} = \iiint_{Parcel} (\nabla \cdot \boldsymbol{\sigma} + \rho\,\mathbf{g})\,\mathrm{d}V. \qquad (6.1.5)$$

Referring to (6.1.1), we see that the hydrodynamic force balances the body force and the total force exerted on the parcel is zero. The assumption of creeping flow guarantees that the rate of change of momentum of a fluid parcel is negligibly small.

The total torque with respect to a point, $\mathbf{x}_0$, exerted on an arbitrary fluid parcel was given in (3.1.20) and (3.1.21), repeated below for convenience,

$$\mathbf{T}^{tot} = \iint_{Parcel} \hat{\mathbf{x}} \times (\mathbf{n} \cdot \boldsymbol{\sigma})\,\mathrm{d}S + \iiint_{Parcel} \hat{\mathbf{x}} \times (\rho\,\mathbf{g})\,\mathrm{d}V + \iiint_{Parcel} \lambda\,\mathbf{c}\,\mathrm{d}V, \qquad (6.1.6)$$

where $\hat{\mathbf{x}} = \mathbf{x} - \mathbf{x}_0$ and λ is an appropriate physical constant associated with the torque field $\mathbf{c}$. Converting the surface integral to a volume integral and switching to index notation, we obtain

$$T_i^{tot} = \iiint_{Parcel} \left(\epsilon_{ijk}\,\sigma_{jk} + \epsilon_{ijk}\,\hat{x}_j\,\frac{\partial \sigma_{lk}}{\partial x_l} + \rho\,\epsilon_{ijk}\hat{x}_j g_k + \lambda\,c_i\right)\mathrm{d}V. \qquad (6.1.7)$$

In the absence of an external torque field, the stress tensor is symmetric and the total torque exerted on any fluid parcel is zero.

6.1.1   The pressure satisfies Laplace's equation

Taking the divergence of the Stokes equation (6.1.2) and using the continuity equation for an incompressible fluid, $\nabla \cdot \mathbf{u} = 0$, we find that the pressure satisfies Laplace's equation,

$$\nabla^2 p = 0. \tag{6.1.8}$$

The general properties of harmonic functions discussed in Section 2.1.5 ensure that, in the absence of singular points, the pressure must attain extreme values at the boundaries of the flow or else grow unbounded at infinity. The hydrodynamic pressure, $\tilde{p}$, excluding hydrostatic variations, also satisfies Laplace's equation, $\nabla^2 \tilde{p} = 0$.

Flow in an infinite domain

Consider a flow in a domain of infinite expanse, possibly in the presence of interior boundaries, and assume that the velocity decays and the pressure tends to a constant at infinity. Using the results of Section 2.3, we find that the hydrodynamic pressure decays like $1/d^m$, where d is the distance from the origin and m is an integer. Substituting in the Stokes equation $p = 1/d$, we obtain an equation for the velocity that does not admit a solenoidal solution. Consequently, the pressure must decay at least as fast as $1/d^2$. Balancing the orders of the pressure gradient and of the Laplacian of the velocity, we find that the velocity must decay at least as fast as $1/d$. Repeating these arguments for two-dimensional Stokes flow, we find that the velocity must increase or decay at a rate that is less than logarithmic, $\ln(d/a)$, where a is a constant length scale.

6.1.2   The velocity satisfies the biharmonic equation

Taking the Laplacian of (6.1.2) and using (6.1.8), we find that the Cartesian components of the velocity satisfy the biharmonic equation,

$$\nabla^4 \mathbf{u} = \mathbf{0}. \tag{6.1.9}$$

The mean-value theorems stated in (2.4.14) and (2.4.14) provide us with the identities

$$\frac{1}{4\pi a^2} \iint_{Sphere} \mathbf{u}(\mathbf{x}) \, dS(\mathbf{x}) = \mathbf{u}(\mathbf{x}_0) + \frac{1}{6} a^2 \nabla^2 \mathbf{u}(\mathbf{x}_0) \tag{6.1.10}$$

and

$$\frac{1}{\frac{4\pi}{3} a^3} \iiint_{Sphere} \mathbf{u}(\mathbf{x}) \, dV(\mathbf{x}) = \mathbf{u}(\mathbf{x}_0) + \frac{1}{10} a^2 \nabla^2 \mathbf{u}(\mathbf{x}_0), \tag{6.1.11}$$

relating the mean value of the velocity over the surface or volume of a sphere of radius a centered at a point, $\mathbf{x}_0$, to the value of the velocity and its Laplacian at the center of the sphere.

Irrotational flow

Equation (6.1.9) shows that any solenoidal velocity field, $\mathbf{u} = \nabla \phi$, satisfies the Stokes equation with a corresponding constant hydrodynamic pressure. However, since an irrotational flow alone cannot be made to satisfy more than one scalar boundary condition, it is generally unable by itself to represent an externally or internally bounded flow.

6.1.3   The vorticity satisfies Laplace's equation

Taking the curl of the Stokes equation and noting that the curl of the gradient of any twice differentiable vector field is zero, we find that the Cartesian components of the vorticity satisfy Laplace's equation,

$$\nabla^2 \boldsymbol{\omega} = \mathbf{0}. \tag{6.1.12}$$

Consequently, each Cartesian component of the vorticity attains an extreme value at the boundaries of the flow. Another important consequence of (6.1.12) is the absence of localized concentration of vorticity in Stokes flow. The onset of regions of recirculating fluid does not imply the presence of compact vortices, as it typically does in the case of high-Reynolds-number flow discussed in Chapters 10 and 11.

Any linear flow with constant vorticity satisfies the equations of Stokes flow with a corresponding constant hydrodynamic pressure, including the simple shear flow $\mathbf{u} = [\xi y, 0, 0]$, where ξ is a constant shear rate.

Mean-value theorems

The mean-value theorems for functions that satisfy Laplace's equation stated in (2.4.2) and (2.4.14) provide us with the identities

$$\boldsymbol{\omega}(\mathbf{x}_0) = \frac{1}{4\pi a^2} \iint_{Sphere} \boldsymbol{\omega}(\mathbf{x}) \, \mathrm{d}S(\mathbf{x}) = \frac{1}{\frac{4\pi}{3} a^3} \iiint_{Sphere} \boldsymbol{\omega}(\mathbf{x}) \, \mathrm{d}V(\mathbf{x}), \tag{6.1.13}$$

stating that the mean value of the vorticity over the surface or volume of a sphere of radius a centered at a point, $\mathbf{x}_0$, is equal to the value of the vorticity at the center of the sphere.

We note that the outward normal vector over the surface of the sphere is $\mathbf{n} = \frac{1}{a}(\mathbf{x} - \mathbf{x}_0)$ and use the divergence theorem to write

$$\iint_{Sphere} (\mathbf{x} - \mathbf{x}_0) \times \mathbf{u}(\mathbf{x}) \, \mathrm{d}S(\mathbf{x}) = a \iint_{Sphere} \mathbf{n}(\mathbf{x}) \times \mathbf{u}(\mathbf{x}) \, \mathrm{d}S(\mathbf{x}) = a \iiint_{Sphere} \boldsymbol{\omega}(\mathbf{x}) \, \mathrm{d}V(\mathbf{x}). \tag{6.1.14}$$

Combining this equation with (6.1.13), we obtain

$$\boldsymbol{\omega}(\mathbf{x}_0) = \frac{1}{\frac{4\pi}{3} a^4} \iint_{Sphere} (\mathbf{x} - \mathbf{x}_0) \times \mathbf{u}(\mathbf{x}) \, \mathrm{d}S(\mathbf{x}). \tag{6.1.15}$$

This equation relates the mean value of the angular moment of the velocity over the surface of a sphere to the angular velocity of the fluid at the center of the sphere.

6.1.4   Two-dimensional flow

The vorticity of a two-dimensional flow in the xy plane arises from the stream function, ψ, as $\omega_z = -\nabla^2 \psi$. Since the vorticity is a harmonic function, $\nabla^2 \omega_z = 0$, the stream function satisfies the biharmonic equation,

$$\nabla^4 \psi = 0. \tag{6.1.16}$$

The mean value of the stream function along a circle or over a circular disk are given by the mean-value theorem stated in (2.5.7) and (2.5.8).

6.1.5 Axisymmetric and swirling flows

To describe an axisymmetric Stokes flow, we introduce cylindrical polar coordinates, (x, σ, φ), and accompanying spherical polar coordinates, (r, θ, φ), and express the velocity field in terms of the Stokes stream function, ψ. The azimuthal component of the vorticity vector was given in equation (2.9.19), repeated for convenience

$$\omega_\varphi = -\frac{1}{\sigma} E^2 \psi, \tag{6.1.17}$$

where E^2 is a second-order differential operation defined in (2.9.17) and (2.9.20) as

$$E^2 \equiv \frac{\partial^2}{\partial x^2} + \frac{\partial^2}{\partial \sigma^2} - \frac{1}{\sigma}\frac{\partial}{\partial \sigma} = \frac{\partial^2}{\partial r^2} + \frac{\sin\theta}{r^2}\frac{\partial}{\partial \theta}\left(\frac{1}{\sin\theta}\frac{\partial}{\partial \theta}\right) = \frac{\partial^2}{\partial r^2} + \frac{1}{r^2}\frac{\partial^2}{\partial \theta^2} - \frac{\cot\theta}{r^2}\frac{\partial}{\partial \theta}. \tag{6.1.18}$$

The steady vorticity transport equation for axisymmetric Stokes flow requires that the Stokes stream function satisfies a fourth-order linear partial differential equation,

$$E^2(E^2\psi) = E^4\psi = 0. \tag{6.1.19}$$

Far from the axis of revolution, this equation reduces to the biharmonic equation corresponding to two-dimensional flow in an azimuthal plane.

Swirling flow

To describe a swirling Stokes flow, such as that produced by the slow rotation of a prolate spheroid around its axis of revolution, x, we introduce the swirl function, $\Theta(x, \sigma)$, defined by the equation $u_\varphi = \Theta/\sigma$, where σ is the distance from the axis of revolution. The vorticity points along the x axis, indicated by the unit vector $\mathbf{e}_x$,

$$\omega = \frac{1}{\sigma}\frac{\partial \Theta}{\partial \sigma}\,\mathbf{e}_x. \tag{6.1.20}$$

Swirling motion does not generate a pressure field in Stokes flow. Resorting to (6.1.9), we find that the swirl function satisfies a second-order partial differential equation,

$$E^2\Theta = 0, \tag{6.1.21}$$

where the operator E^2 is defined (6.1.18).

6.1.6 Reversibility of Stokes flow

Let us assume that $\mathbf{u}$ and p are a pair of velocity and pressure fields satisfying the equations of Stokes flow with a body force, $\rho\mathbf{g}$. It is evident that $-\mathbf{u}$, $-p$, and $-\rho\mathbf{g}$ will also satisfy the equations of Stokes flow. We conclude that the reversed flow is a mathematically acceptable and physically viable solution. It is important to note that the direction of the hydrodynamic force and torque acting on any surface are reversed when the signs $\mathbf{u}$ and p are switched. Reversibility is not shared by inertial flow at nonzero Reynolds numbers, for the sign of the nonlinear acceleration term $\mathbf{u} \cdot \nabla \mathbf{u}$ does not change when the sign of the velocity is reversed.

Reversibility of Stokes flow can be invoked to deduce the structure of a flow without actually computing the solution. Consider a solid sphere convected under the action of simple shear flow in the vicinity of a plane wall. In principle, the hydrodynamic force acting on the sphere may have a component perpendicular to the wall and a component parallel to the wall. Let us assume temporarily that the component of the force perpendicular to the wall pushes the sphere away from the wall. Reversing the direction of the shear flow must reverse the direction of the force and thus push the sphere towards the wall. However, this anisotropy is physically unacceptable in light of the fore-and-aft symmetry of the domain of flow. We conclude that the normal component of the force on the sphere is zero and the sphere must keep moving parallel to the wall.

Using the concept of reversibility, we infer that the streamline pattern around an axisymmetric body with fore-and-aft symmetry moving along its axis must also be axisymmetric and fore-and-aft symmetric. The streamline pattern over a two-dimensional rectangular cavity must be symmetric with respect to the midplane. A neutrally buoyant spherical particle convected in a parabolic flow inside a cylindrical tube may not move toward the center of the tube or the wall, but must retain the initial radial position.

6.1.7   Energy integral balance

In the absence of inertial forces, the rate of accumulation and rate of convection of kinetic energy into a fixed control volume, V_c, bounded by a surface, D, are zero. Combining (3.4.22) with (3.4.15), we derive a simplified energy integral balance,

$$\iint_D \mathbf{u} \cdot \tilde{\mathbf{f}} \, dS = -2\mu \iiint_{V_c} \mathbf{E} : \mathbf{E} \, dV, \qquad (6.1.22)$$

where $\tilde{\mathbf{f}} = \mathbf{n} \cdot \tilde{\boldsymbol{\sigma}}$ is the hydrodynamic traction excluding the effect of the body force, and $\mathbf{n}$ is the unit normal vector pointing into the control volume. Equation (6.1.22) states that the rate of working of the hydrodynamic traction on the boundary is balanced by the rate of viscous dissipation inside the control volume.

Flow due to rigid-body motion

As an application, we consider the flow produced by the motion of a rigid body that translates with velocity $\mathbf{V}$ while rotating about the point $\mathbf{x}_0$ with angular velocity $\boldsymbol{\Omega}$ in an ambient fluid of infinite expanse. We select a control volume V_c that is confined by the surface of the body, B, and a large surface, S_∞, and apply (6.1.22) with D representing the union of B and S_∞. Letting S_∞ tend to infinity and noting that the velocity decays like $1/d$, where d is the distance from the origin, we find that the corresponding surface integral on the left-hand side makes a vanishing contribution. Requiring the boundary condition $\mathbf{u} = \mathbf{V} + \boldsymbol{\Omega} \times (\mathbf{x} - \mathbf{x}_0)$ over B, we obtain

$$\mathbf{V} \cdot \mathbf{F} + \boldsymbol{\Omega} \cdot \mathbf{T} = -2\mu \iiint_{V_c} \mathbf{E} : \mathbf{E} \, dV, \qquad (6.1.23)$$

where $\mathbf{F}$ and $\mathbf{T}$ are the hydrodynamic force and torque with respect to $\mathbf{x}_0$ exerted on the body, excluding contributions from the body force.

Equation (6.1.23) expresses a balance between the rate of supply of mechanical energy necessary to sustain the motion of the body and the rate of viscous dissipation in the fluid. Since the right-hand side is nonpositive for any combination of $\mathbf{V}$ and $\boldsymbol{\Omega}$, each projection $\mathbf{V} \cdot \mathbf{F}$ and $\boldsymbol{\Omega} \cdot \mathbf{T}$ must be negative. Physically, these inequalities assert that the drag force exerted on a translating body and the torque exerted on a rotating body always resist the motion of the body.

6.1.8   Uniqueness of Stokes flow

Let us assume now that either the velocity or the boundary traction or the projection of the velocity onto the boundary traction, $\mathbf{u} \cdot \tilde{\mathbf{f}}$, is zero over the boundaries of the flow. The left-hand side of (6.1.22) is zero and the rate of deformation tensor vanishes throughout the domain of flow. As a result, the velocity must express rigid-body motion including translation and rotation (Problem 1.1.3). An important consequence of this conclusion is uniqueness of solution of the equations of Stokes flow noted by Helmholtz in 1868.

To prove uniqueness of solution, we temporarily consider two solutions corresponding to a given set of boundary conditions for the velocity or traction. The flow expressing the difference between these two solutions satisfies homogeneous boundary conditions and must represent rigid-body motion. If the boundary conditions specify the velocity on a portion of the boundary, rigid-body motion is not permissible, the difference flow must vanish, and the solution of the stated problem is unique. However, if the boundary conditions specify the traction over all boundaries, an arbitrary rigid-body motion can be added to any particular solution.

6.1.9   Minimum dissipation principle

Helmholtz showed that the rate of viscous dissipation in a Stokes flow with velocity $\mathbf{u}$ is lower than that in any other incompressible flow with velocity $\mathbf{u}'$. The boundary velocity is assumed to be the same in both flows. One consequence of the minimum energy dissipation principle is that the drag force and torque exerted on rigid body moving steadily in a fluid under conditions of Stokes flow are lower than those for flow at nonzero Reynolds numbers.

To demonstrate the minimum dissipation principle, we introduce the rate-of-deformation tensor, $\mathbf{E}$, and consider the identity

$$\iiint_{Flow} \mathbf{E}' : \mathbf{E}' \, dV - \iiint_{Flow} \mathbf{E} : \mathbf{E} \, dV = \iiint_{Flow} (\mathbf{E}' - \mathbf{E}) : (\mathbf{E}' - \mathbf{E}) \, dV + \iiint_{Flow} (\mathbf{E}' - \mathbf{E}) : \mathbf{E} \, dV. \quad (6.1.24)$$

The first integral on the right-hand side is nonnegative. We will show that the second integral on the right-hand side is zero. Working toward that goal, we write

$$\iiint_{Flow} (\mathbf{E}' - \mathbf{E}) : \mathbf{E} \, dV = \iiint_{Flow} \frac{\partial (u_i' - u_i)}{\partial x_j} E_{ij} \, dV, \quad (6.1.25)$$

and use the divergence theorem to obtain

$$\iiint_{Flow} (\mathbf{E}' - \mathbf{E}) : \mathbf{E} \, dV = - \iint_{B} (u_i' - u_i) E_{ij} \, n_j \, dS - \iiint_{Flow} (u_i' - u_i) \frac{\partial E_{ij}}{\partial x_j} \, dV, \quad (6.1.26)$$

where $\mathbf{n}$ is the unit vector normal to the boundaries, B, pointing into the fluid. Since $\mathbf{u}$ and $\mathbf{u}'$ are identical over the boundary, the boundary integral on the right-hand side is zero. In fact, the volume integral on the right-hand side is also zero. To show this, we use the continuity equation and the Stokes equation to obtain

$$\iiint_{Flow} (u_i' - u_i) \frac{\partial E_{ij}}{\partial x_j} \, \mathrm{d}V = \frac{1}{2} \iiint_{Flow} (u_i' - u_i) \nabla^2 u_i \, \mathrm{d}V = \frac{1}{2\mu} \iiint_{Flow} (u_i' - u_i) \frac{\partial p}{\partial x_i} \, \mathrm{d}V. \qquad (6.1.27)$$

Using the continuity equation once more, we write

$$\iiint_{Flow} (\mathbf{u}' - \mathbf{u}) \cdot \nabla p \, \mathrm{d}V = \iiint_{Flow} \nabla \cdot [p(\mathbf{u}' - \mathbf{u})] \, \mathrm{d}V = -\frac{1}{\mu} \iint_B p(\mathbf{u}' - \mathbf{u}) \cdot \mathbf{n} \, \mathrm{d}S. \qquad (6.1.28)$$

Since $\mathbf{u}$ and $\mathbf{u}'$ have the same values over the boundaries, the last integral is zero.

6.1.10   Complex-variable formulation of two-dimensional Stokes flow

Two-dimensional Stokes flow is amenable to a formulation in complex variables that is analogous to, but somewhat more involved than that, of two-dimensional potential flow discussed in Section 7.10. In the case of potential flow, the complex-variable formulation is based on the observation that the harmonic potential, ϕ, and stream function, ψ, comprise a pair of conjugate harmonic functions. In the case of Stokes flow, the complex-variable formulation can be developed in two different ways based on the pressure and vorticity or with reference to the stress components.

Pressure–vorticity formulation

Consider a two-dimensional Stokes flow in the xy plane. Using the identity $\nabla^2 \mathbf{u} = -\nabla \times \boldsymbol{\omega}$ along with the Stokes equation, $\mu \nabla^2 \mathbf{u} = \nabla p$, we obtain

$$\mu \nabla \times \boldsymbol{\omega} = -\nabla p, \qquad (6.1.29)$$

where p is the hydrodynamic pressure excluding hydrostatic variations. Writing the two scalar components of (6.1.29) in the x and y directions, we find that the z vorticity component, denoted by ω, and the hydrodynamic pressure, p, satisfy the Cauchy–Riemann equations,

$$\frac{\partial \omega}{\partial x} = \frac{1}{\mu} \frac{\partial p}{\partial y}, \qquad \frac{\partial \omega}{\partial y} = -\frac{1}{\mu} \frac{\partial p}{\partial x}. \qquad (6.1.30)$$

As a consequence, the complex function

$$f(z) = \omega + \mathrm{i} \frac{p}{\mu} \qquad (6.1.31)$$

is an analytic function of the complex variable $z = x + \mathrm{i}y$, where i is the imaginary unit. Assigning to the complex function $f(z)$ various forms provides us with different families of two-dimensional Stokes flows.

To describe the structure of a flow in terms of the stream function, it is convenient to work with the first integral of $f(z)$ defined by the equation $\mathrm{d}F/\mathrm{d}z = f$. The vorticity and pressure fields are given by

$$\omega = \frac{\partial F_R}{\partial x} = \frac{\partial F_I}{\partial y}, \qquad \frac{p}{\mu} = -\frac{\partial F_R}{\partial y} = \frac{\partial F_I}{\partial x}, \qquad (6.1.32)$$

where the subscripts R and I denote the real and imaginary parts. To derive an expression for the stream function in terms of F, we note that, if H_0, H_1, H_2, and H_3 are four harmonic functions and c_0, c_1, c_2, and c_3 are four arbitrary constants, then

$$\psi = c_0 H_0 + c_1 x H_1 + c_2 y H_2 + c_3 \left(x^2 + y^2\right) H_3 \tag{6.1.33}$$

satisfies the biharmonic equation, $\nabla^4 \psi = 0$, and it is therefore an acceptable stream function. The corresponding vorticity is

$$\omega = -\nabla^2 \psi = -2c_1 \frac{\partial H_1}{\partial x} - 2c_2 \frac{\partial H_2}{\partial y} - 4c_3 \left(H_3 + x \frac{\partial H_3}{\partial x} + y \frac{\partial H_3}{\partial y}\right). \tag{6.1.34}$$

Comparing this expression with the first expression in (6.1.32), we set $H_1 = F_R$, $H_2 = F_I$, $H_3 = 0$, and $c_1 + c_2 = -1/2$, and restate (6.1.33) as

$$\psi = c_0 H + c_1 (x - x_0) F_R - \left(\frac{1}{2} + c_1\right)(y - y_0) F_I, \tag{6.1.35}$$

where c_0 and y_0 are two arbitrary constants and H is a harmonic function of x and y.

Equations (6.1.32) and (6.1.35) provide us with the vorticity, pressure, and stream function of a two-dimensional Stokes flow in terms of a generating complex function, F, and a real function, H. Having specified these functions, we obtain a family of flows with identical vorticity and pressure fields parametrized by the constants c_0 and c_1. For example, setting F equal to a constant produces irrotational flows with vanishing vorticity and constant pressure.

In the particular case where $c_1 = -1/4$, expression (6.1.35) takes the compact form

$$\psi = c_0 H - \frac{1}{4}(x - x_0) F_R - \frac{1}{4}(y - y_0) F_I = -\frac{1}{4} \text{real}\left[(z - z_0)^* F(z) + G(z)\right], \tag{6.1.36}$$

where $G(z)$ is an arbitrary analytic function and an asterisk denotes the complex conjugate. Straightforward differentiation yields the associated velocity field,

$$u_x + \mathrm{i}\, u_y = \frac{\mathrm{i}}{4}\left[F + (z - z_0) F'^* + G'^*\right], \tag{6.1.37}$$

where a prime denotes a derivative with respect to z.

An extensive discussion and further applications of the complex variable formulation of two-dimensional Stokes flow is available elsewhere ([223], Chapter 7). The complex-variable formulation is amenable to a boundary-integral representation that can be used to derive and numerically solve boundary-integral equations [306].

Formulation in terms of the Airy stress function

A two-dimensional Stokes flow can be described by the Airy stress function, Φ, defined in terms of the hydrodynamic stress that excludes hydrostatic pressure variations, $\tilde{\sigma}$, by the equations

$$\tilde{\sigma}_{xx} = \mu \frac{\partial^2 \Phi}{\partial y^2}, \qquad \tilde{\sigma}_{xy} = \tilde{\sigma}_{yx} = -\mu \frac{\partial^2 \Phi}{\partial x \partial y}, \qquad \tilde{\sigma}_{yy} = \mu \frac{\partial^2 \Phi}{\partial x^2}. \tag{6.1.38}$$

The Stokes equation is satisfied for any differentiable function Φ. Using the continuity equation and recalling that the pressure is a harmonic function, we find that Φ satisfies the biharmonic equation, $\nabla^4 \Phi = 0$. Expressing the three independent components of the stress tensor in terms of the pressure, p, and stream function, ψ, and using the Stokes equation to eliminate the pressure, we obtain

$$\frac{\partial^2 \Phi}{\partial x^2} - \frac{\partial^2 \Phi}{\partial y^2} = -4 \frac{\partial^2 \psi}{\partial x \partial y}, \qquad \frac{\partial^2 \psi}{\partial x^2} - \frac{\partial^2 \psi}{\partial y^2} = \frac{\partial^2 \Phi}{\partial x \partial y}. \tag{6.1.39}$$

Next, we write $x = \frac{1}{2}(z + z^*)$ and $y = -\frac{i}{2}(z - z^*)$, and regard Φ and ψ as functions of z and z^*, where an asterisk denotes the complex conjugate. Applying the chain rule, we find that

$$\left(\frac{\partial \Phi}{\partial z^*} \right)_z = \frac{\partial \Phi}{\partial x} \frac{\partial x}{\partial z^*} + \frac{\partial \Phi}{\partial y} \frac{\partial y}{\partial z^*} = \frac{1}{2} \left(\frac{\partial \Phi}{\partial x} + i \frac{\partial \Phi}{\partial y} \right) \tag{6.1.40}$$

and

$$\left(\frac{\partial^2 \Phi}{\partial z^{*2}} \right)_z = \frac{1}{4} \left(\frac{\partial^2 \Phi}{\partial x^2} - \frac{\partial^2 \Phi}{\partial y^2} + 2i \frac{\partial^2 \Phi}{\partial x \partial y} \right). \tag{6.1.41}$$

Similar expressions can be derived for ψ. Combining these results, we find that the complex function $\chi = \Phi - 2i\psi$ satisfies the differential equation

$$\left(\frac{\partial^2 \chi}{\partial z^{*2}} \right)_z = 0. \tag{6.1.42}$$

Integrating (6.1.42) twice, we obtain

$$\chi(z) = x^* \chi_1(z) + \chi_2(z), \tag{6.1.43}$$

where χ_1 and χ_2 are two arbitrary analytic functions of z. Different selections for χ_1 and χ_2 produce various types of two-dimensional Stokes flow. For example, setting

$$\chi = \frac{1}{2} \left[(\zeta^* - \frac{1}{3}\zeta)\zeta - 4 \right] \zeta \tag{6.1.44}$$

produces two-dimensional Hagen–Poiseuille flow in a two-dimensional channel confined between two parallel walls located at $y = \pm a$, where $\zeta = z/a$.

The formulation in terms of the Airy stress function and stream function is amenable to a boundary-integral representation that can be used to derive integral equations [306].

6.1.11   Swirling flow

A swirling flow can be described by a single scalar function, $\chi(x, \sigma)$, defined in cylindrical polar coordinates, (x, σ, φ), by the equations

$$\tilde{\sigma}_{x\varphi} = \mu \frac{1}{\sigma^2} \frac{\partial \chi}{\partial \sigma}, \qquad \tilde{\sigma}_{\sigma\varphi} = -\mu \frac{1}{\sigma^2} \frac{\partial \chi}{\partial x} \tag{6.1.45}$$

([247], p. 325). All other stress components are identically zero. The Stokes equation is satisfied for any function χ. For an incompressible Newtonian fluid,

$$\tilde{\sigma}_{x\varphi} = \mu \frac{\partial u_\varphi}{\partial x}, \qquad \tilde{\sigma}_{\sigma\varphi} = \mu\sigma \frac{\partial}{\partial\sigma}\left(\frac{u_\varphi}{\sigma}\right), \tag{6.1.46}$$

requiring that

$$\sigma \frac{\partial}{\partial\tilde{\sigma}}\left(\frac{\tilde{\sigma}_{x\varphi}}{\sigma}\right) = \frac{\partial\tilde{\sigma}_{\sigma\varphi}}{\partial x}. \tag{6.1.47}$$

Substituting expressions (6.1.45), we find that χ satisfies the equation

$$\frac{\partial^2\chi}{\partial x^2} + \sigma^3 \frac{\partial}{\partial\sigma}\left(\frac{1}{\sigma^3}\frac{\partial\chi}{\partial\sigma}\right) = 0. \tag{6.1.48}$$

The function $\phi \equiv \chi\cos 2\varphi/\sigma^2$ satisfies Laplace's equation, $\nabla^2\phi = 0$.

Problems

6.1.1 *Reversibility of Stokes flow*

Explain why the drag force exerted on a solid sphere rotating in the vicinity an infinite plane wall may not have a component perpendicular to the wall.

6.1.2 *Mean-value theorems*

(*a*) Derive (6.1.15) by expanding the velocity field in a Taylor series about the center of the sphere.

(*b*) Derive the counterpart of (6.1.15) for two-dimensional flow.

6.1.3 *Complex variable formulation*

With reference to the representation (6.1.35), find the values of c_0 and c_1 and the functions H and $F(z)$ that produce (*a*) unidirectional simple shear flow, (*b*) orthogonal stagnation-point flow, (*c*) oblique stagnation-point flow.

6.1.4 *Airy stress function*

Find the functions χ_1 and χ_2 corresponding to the flows described in Problem 6.1.3.

6.1.5 *Swirling flow*

(*a*) Confirm that the Stokes equation is satisfied for any choice of the function χ defined in (6.1.45).

(*b*) Show that χ is constant along a line of vanishing traction.

(*c*) Show that the derivative of χ normal to a line C rotating as a rigid body is zero. The azimuthal velocity component is $u_\varphi = \Omega\sigma$ over C, where Ω is the angular velocity of rotation.

(*d*) Confirm that the function $\phi \equiv \chi\cos 2\varphi/\sigma^2$ satisfies Laplace's equation, $\nabla^2\phi = 0$.

6.1.6 *Shear flow over a wavy wall*

Consider a two-dimensional semi-infinite simple shear flow with shear rate ξ past an infinite sinusoidal wall with wavelength L and amplitude ϵL, where ϵ is a small dimensionless coefficient. The wall profile in the xy plane is described by the function $y = y_w(x) = \epsilon L \cos(kx)$, where $k = 2\pi/L$ is the wave number. Show that the stream function is given by the asymptotic expansion

$$\psi = \frac{1}{2} \xi y^2 - \epsilon \xi y L e^{-ky} \cos(kx) + \cdots, \qquad (6.1.49)$$

where the dots indicate quadratic and higher-order terms in ϵ. Compute the drag force exerted on the wall over one period accurate to first order in ϵ.

6.2   Stokes flow in a wedge-shaped domain

Because the velocity of a viscous fluid is zero over a stationary solid boundary, the Reynolds number of the flow in the vicinity of the boundary is necessarily small and the local structure of the flow is governed by the equations of Stokes flow. The role of the outer flow is to provide a driving mechanism that determines the asymptotic behavior of the flow far from the boundary. In certain applications, the flow is due to boundary motion but the Reynolds number of the flow near the boundary is low due to the small length scales involved. Under these circumstances, the structure of the inner Stokes flow is determined by the local boundary geometry and nature of boundary conditions driving the flow.

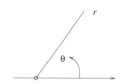

Illustration of a wedge-shaped domain.

In this section, we derive a general solution and study two-dimensional Stokes flow in a wedge-shaped domain bounded by two intersecting stationary, translating, or rotating planar walls representing rigid boundaries or free surfaces of vanishing shear stress. In Section 6.3, we will consider in detail the structure of two-dimensional flow inside and around a corner with stationary walls driven by an ambient flow.

6.2.1   General solution for the stream function

To circumvent the computation of the pressure and thus reduce the number of unknowns, it is convenient to describe the flow in terms of the stream function, ψ. Since the boundary conditions are applied at intersecting plane walls, it is appropriate to introduce plane polar coordinates with origin at the vertex, (r, θ), and separate the radial from the angular dependence by writing

$$\psi(r, \theta) = q(r) f(\theta), \qquad (6.2.1)$$

where $q(r)$ and $f(\theta)$ are two unknown functions.

Working in hindsight, we stipulate the power-law functional dependence

$$q(r) = A r^\lambda. \qquad (6.2.2)$$

The complex constant A is a measure of the intensity of the outer flow, and the complex exponent λ determines the structure of the inner Stokes flow. It is understood that either the real or the imaginary part of the complex stream function ψ provides us with an acceptable solution.

The nonzero vorticity component directed along the z axis is

$$\omega_z = -\nabla^2 \psi = -\frac{1}{r}\frac{\partial}{\partial r}\left(r\frac{\partial \psi}{\partial r}\right) - \frac{1}{r^2}\frac{\partial^2 \psi}{\partial \theta^2} = -\frac{f}{r}\frac{\mathrm{d}}{\mathrm{d}r}\left(r\frac{\mathrm{d}q}{\mathrm{d}r}\right) - \frac{q}{r^2}\frac{\mathrm{d}^2 f}{\mathrm{d}\theta^2}. \tag{6.2.3}$$

Substituting (6.2.1) and (6.2.2), we find that

$$\omega_z = -A\,r^{\lambda-2}\left(\lambda^2 f + \frac{\mathrm{d}^2 f}{\mathrm{d}\theta^2}\right). \tag{6.2.4}$$

Requiring that the vorticity is a harmonic function, $\nabla^2 \omega_z = 0$, provides us with a fourth-order homogeneous ordinary differential equation for f,

$$f'''' + 2(\lambda^2 - 2\lambda + 2)\,f'' + \lambda^2(\lambda-2)^2 f = 0, \tag{6.2.5}$$

where a prime denotes a derivative with respect to θ. Substituting an eigensolution of the form

$$f(\theta) = \exp(\kappa\theta), \tag{6.2.6}$$

we obtain a quadratic equation for the generally complex constant κ,

$$\kappa^4 + 2(\lambda^2 - 2\lambda + 2)\,\kappa^2 + \lambda^2(\lambda-2)^2 = 0. \tag{6.2.7}$$

When κ has an imaginary component, the eigensolutions can be expressed in terms of trigonometric functions. Solving for κ and substituting the result into (6.2.6), we obtain the general solution

$$f(\theta) = \begin{cases} B\sin(\lambda\theta - \beta) + C\sin[(\lambda-2)\theta - \gamma] & \text{if } \lambda \neq 0, 1, 2, \\ B\sin(2\theta - \beta) + C\theta + D & \text{if } \lambda = 0, 2, \\ B\sin(\theta - \beta) + C\theta\sin(\theta - \gamma) & \text{if } \lambda = 1, \end{cases} \tag{6.2.8}$$

where B, C, and D are complex constants and β and γ are real constants. A solution with $\lambda = 0$ was presented in (5.8.6) in the context of the Jeffery–Hamel flow.

6.2.2 Two-dimensional stagnation-point flow on a plane wall

In the first application, we consider Stokes flow near a stagnation point on a plane wall, as illustrated in Figure 6.2.1. The no-slip and no-penetration boundary conditions require that $f = 0$ and $f' = 0$ at $\theta = 0$ and π. In addition, we require that $f = 0$ at the dividing streamline located at $\theta = \alpha$. Making the judicious choice $\lambda = 3$ and using the general solution (6.2.8), we obtain

$$f(\theta) = B\sin(3\theta - \beta) + C\sin(\theta - \gamma). \tag{6.2.9}$$

Enforcing the boundary conditions, we derive the specific form

$$f(\theta) = F\sin^2\theta\,\sin(\theta - \alpha), \tag{6.2.10}$$

which can be substituted into (6.2.1) to yield the stream function in plane polar or Cartesian coordinates,

$$\psi = Gr^3\sin^2\theta\,\sin(\theta - \alpha) = Gy^2(y\cos\alpha - x\sin\alpha), \tag{6.2.11}$$

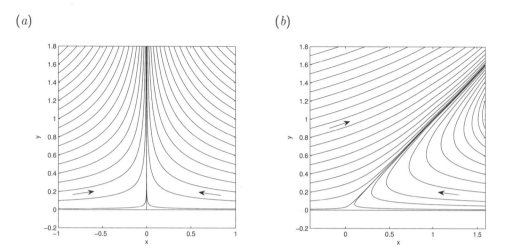

(a) (b)

FIGURE 6.2.1 Stokes flow near a stagnation point on a plane wall. The angle subtended between the wall and the dividing streamline, measured from the positive x semiaxis, is (a) $\alpha = \pi/2$ or (b) $\pi/4$. As α tends to zero, we obtain unidirectional shear flow with parabolic velocity profile.

where F and G are two real constants determined by the intensity of the outer flow that is responsible for the onset of the stagnation point. The Cartesian components of the velocity and pressure gradient are

$$u_x = Gy\,(3y\cos\alpha - 2x\sin\alpha), \qquad u_y = Gy^2\sin\alpha, \qquad \nabla p = 2\mu G \begin{bmatrix} 3\cos\alpha \\ \sin\alpha \end{bmatrix}. \qquad (6.2.12)$$

It is worth noting that the pressure gradient is constant and points into the same quadrant as the velocity along the dividing streamline. Streamline patterns for $\alpha = \pi/2$ and $\pi/4$ are shown in Figure 6.2.1. As the angle α tends to zero, we obtain unidirectional shear flow with parabolic velocity profile.

6.2.3 Taylor's scraper

Taylor [400] studied the flow near the edge of a flat plate moving along the x axis with velocity V, scraping fluid off a plane wall at an angle α. In a frame of reference moving with the scraper, the wall appears to move along the x axis with velocity $-V$, against a stationary scraper, as shown in Figure 6.2.2. The no-slip boundary condition requires that $(1/r)(\partial\psi/\partial\theta) = -V$ at the wall positioned at $\theta = 0$. To satisfy this condition, we set $\lambda = 1$ and find that the radial component of the velocity depends on the polar angle θ alone, and is independent of r. Enforcing the boundary conditions $f = 0$, $f' = -V$ at $\theta = 0$ and $f = 0$, $f' = 0$ at $\theta = \alpha$, and using the general solution (6.2.8), we obtain the stream function

$$\psi(\theta) = \frac{V}{\alpha^2 - \sin^2\alpha}\, r\,[\alpha(\theta - \alpha)\sin\theta - \theta\sin(\theta - \alpha)\sin\alpha]. \qquad (6.2.13)$$

Streamline patterns for $V > 0$ are shown in Figure 6.2.2 for three scraping angles.

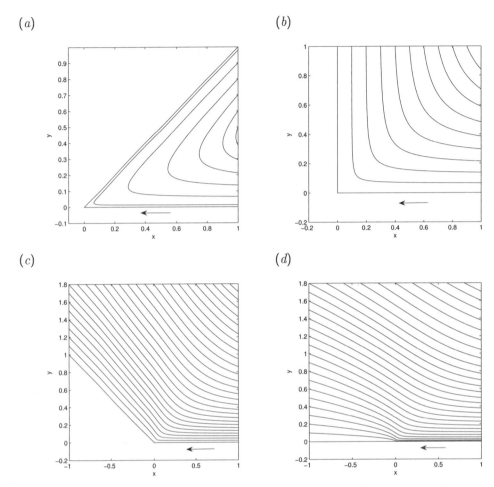

FIGURE 6.2.2 Streamline pattern of the flow due a scraper moving over a plane wall in a frame of reference moving with the scraper. The inclined wall on the left side of each frame is stationary. The scraping angle is (a) $\alpha = \pi/4$, (b) $\pi/2$, (c) $3\pi/4$, and (d) π.

The wall shear stress is given by

$$\sigma_{r\theta}(\theta = 0) = \frac{\mu}{r}\left(\frac{\partial u_r}{\partial \theta}\right)_{\theta=0} = \frac{\mu}{r^2}\left(\frac{\partial^2 \psi}{\partial \theta^2}\right)_{\theta=0} = -\mu\,\frac{V}{r}\,\frac{2\alpha - \sin(2\alpha)}{\alpha^2 - \sin\alpha^2}. \tag{6.2.14}$$

The nonintegrable $1/r$ functional form suggests that an infinite force is required to maintain the motion of the scraper. In real life, the corner between the scraper and the wall has a small but nonzero curvature that alters the local flow and removes the singular behavior. If the edge of the scraper is perfectly sharp, a small gap will be present between the scraper and the wall to allow for a small amount of leakage. This situation occurs at the top corners of a fluid-filled cavity, where the motion is driven by a translating lid.

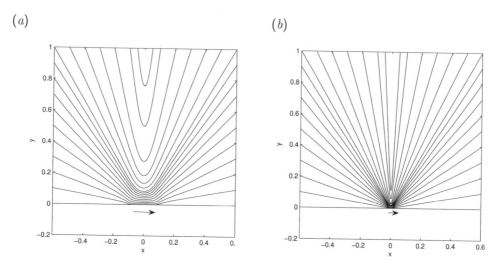

FIGURE 6.2.3   Streamline pattern of the flow due the translation of a section of a plane wall (a) close to the moving section and (b) far from the moving section.

The case $\alpha = \pi$ describes flow driven by the motion of a semi-infinite belt extending from the origin to infinity, as shown in Figure 6.2.2(d). A slight rearrangement of the general solution (6.2.13) yields the simple form

$$\psi(\theta) = Vr\,\frac{\theta - \pi}{\pi}\,\sin\theta = Vy\,\frac{\theta - \pi}{\pi}. \qquad (6.2.15)$$

In Section 6.2.3, this elementary flow will be used to construct a general solution for arbitrary in-plane wall motion.

6.2.4   Flow due to the in-plane motion of a wall

Using the solution (6.2.15) and exploiting the feasibility of linear superposition, we find that the flow due to the in-plane motion of a section of the wall extending between $x = -a$ and a with uniform velocity V is described by the stream function

$$\psi = -Vy\left(\frac{\theta^- - \pi}{\pi} - \frac{\theta^+ - \pi}{\pi}\right) = -Vy\,\frac{\theta^+ - \theta^-}{\pi}, \qquad (6.2.16)$$

where $\tan\theta^+ = y/(x - a)$ and $\tan\theta^- = y/(x + a)$. As a/r tends to zero, we obtain the asymptotic solution

$$\psi \simeq -\frac{V}{\pi}\left(\frac{\partial\theta}{\partial x}\right)_0 2a = \frac{2aV}{\pi}\frac{y^2}{x^2 + y^2}, \qquad (6.2.17)$$

which describes the flow far from the moving section. Streamline patterns near and far from the moving section are shown in Figure 6.2.3. The far flow in the first quadrant resembles that due to a point source with strength $m = 2aV/\pi$ located at the origin, in the presence of two intersecting walls along the x and y axes where the no-slip and no-penetration conditions apply. The flow in the second quadrant admits a similar interpretation.

Exploiting further the feasibility of superposition, we use the far-flow solution (6.2.17) to derive the stream function of the flow above a plane wall subject to an arbitrary distribution of a tangential in-plane velocity, $\mathcal{V}(x)$,

$$\psi(x, y) = \frac{1}{\pi} \int_{-\infty}^{\infty} \frac{y^2}{(x - x')^2 + y^2} \mathcal{V}(x') \, dx'. \tag{6.2.18}$$

Expression (6.2.16) arises by setting $\mathcal{V}$ equal to a constant V for $|x| < a$, and to zero otherwise. When $\mathcal{V}$ is equal to a constant V over the entire x axis, we obtain $\psi = Vy$, describing uniform flow.

Differentiating (6.2.18), we compute the velocity components

$$u_x(x, y) = \frac{1}{\pi} \int_{-\infty}^{\infty} \frac{\partial}{\partial y} \left(\frac{y^2}{(x - x')^2 + y^2} \right) \mathcal{V}(x') \, dx',$$

$$u_y(x, y) = -\frac{1}{\pi} \int_{-\infty}^{\infty} \frac{\partial}{\partial x} \left(\frac{y^2}{(x - x')^2 + y^2} \right) \mathcal{V}(x') \, dx' = \frac{2}{\pi} \int_{-\infty}^{\infty} \frac{(x - x')}{[(x - x')^2 + y^2]^2} y^2 \, \mathcal{V}(x') \, dx'. \tag{6.2.19}$$

In the limit $y \to 0$, the derivative inside the integral for the x velocity component tends to $\pi \delta(x - x')$, where δ is the one-dimensional delta function. When $\mathcal{V}(x) = \xi x$, we obtain orthogonal stagnation-point flow with constant rate of elongation, ξ.

6.2.5 Flow near a belt plunging into a pool or a plate withdrawn from a pool

Consider the flow due a belt plunging into a liquid pool or a plate withdrawn from a liquid pool with velocity V at an angle α, as shown in Figure 6.2.4. Over the belt or plate, located at $\theta = -\alpha$, we require the no-slip boundary condition $\partial \psi / \partial \theta = Vr$. Motivated by this expression, we set the exponent $\lambda = 1$. At the free surface of the pool, we require the no-penetration condition and the condition of vanishing shear stress.

Using the general solution (6.2.8) and implementing the four boundary conditions $f = 0$, $f' = V$ at $\theta = -\alpha$ and $f = 0$, $f'' = 0$ at $\theta = 0$, we derive the stream function [271]

$$\psi = -\frac{2}{2\alpha - \sin(2\alpha)} Vr \left(\theta \cos \theta \sin \alpha - \alpha \sin \theta \cos \alpha \right). \tag{6.2.20}$$

Streamline patterns for $\alpha = \pi/4$ and $\pi/2$ are shown in Figure 6.2.4. As in the case of the scraper discussed in Section 6.2.2, the radial velocity component depends only on the polar angle, θ, and is independent of radial distance, r. Because the shear stress diverges like $1/r$ near the corner, an infinite force is required to maintain the motion of the belt. In reality, the free surface will distort near the contact point to prevent this unphysical singular behavior.

(a)   (b)

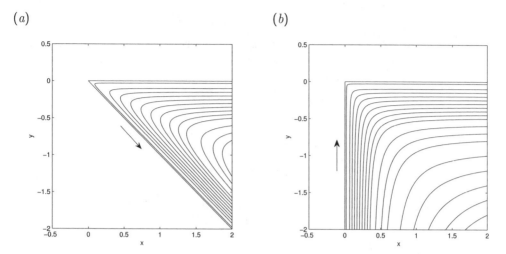

FIGURE 6.2.4   Flow due to (a) a belt plunging into a liquid pool at angle $\alpha = \pi/4$ and (b) a plate withdrawn from a liquid pool at angle $\alpha = \pi/2$. In both cases, the shear stress is zero at the horizontal free surface.

6.2.6   Jeffery–Hamel flow

Consider Stokes flow between two intersecting planes located at $\theta = \pm\alpha$, due to a point source with strength m located at the apex, as discussed in Section 5.8 for Navier–Stokes flow. Using the present formulation, we derive the stream function

$$\psi(\theta) = \frac{m}{2} \frac{\sin 2\theta - 2\theta \cos 2\alpha}{\sin 2\alpha - 2\alpha \cos 2\alpha}, \qquad (6.2.21)$$

satisfying $\psi(\alpha) - \psi(-\alpha) = m$. This expression is consistent with the radial velocity given in (5.8.6). Since the shear stress is zero at the midplane, $\theta = 0$, the solution also describes the flow inside the corner bounded by a stationary wall located at $\theta = \pm\alpha$ and a free surface located at the midplane $\theta = 0$, where the flow is driven by a point source with strength $\frac{1}{2}m$ located at the apex.

The denominator in (6.2.21) vanishes and the stream function diverges when $\alpha = 257.45°$ due to the neglected inertial forces [272]. This failure underscores the possible limitations of Stokes flow (see also Problem 6.2.3).

Problems

6.2.1 *Belt plunging into a pool*

With reference to Figure 6.2.4, derive an expression for the radial velocity at the free surface and compare it with the velocity of the belt.

6.2.2 *Flow due to an imposed shear stress near a static contact line*

Consider a flow driven by the application of a uniform shear stress, τ, along a flat interface located at $\theta = \alpha$, in the presence of a stationary plane wall located at $\theta = 0$. Derive the stream function

$$\psi = \frac{\tau}{8\mu} \frac{r^2}{\cos 2\alpha - 1 + \alpha \sin 2\alpha} \big[(1 - \cos 2\alpha - 2\alpha \sin 2\alpha) \sin 2\theta$$
$$+ (2\alpha \cos 2\alpha - \sin 2\alpha)(\cos 2\theta - 1) + 2\theta(1 - \cos 2\alpha) \big]. \tag{6.2.22}$$

Discuss the structure of the flow in the limit as $\alpha \to 0$ [271].

6.2.3 *Flow due to the counter-rotation of two hinged plates*

(a) Consider the flow between two hinged plates located at $\theta = \pm\alpha(t)$, rotating against each other with angular velocity $\pm\Omega$. Derive the stream function [271]

$$\psi = \frac{\Omega}{2} r^2 \frac{\sin 2\theta - 2\theta \cos 2\alpha}{\sin 2\alpha - 2\alpha \cos 2\alpha}. \tag{6.2.23}$$

Show that the denominator vanishes and the stream function diverges when $\alpha = 257.45°$. Discuss the physical implication of this failure with reference to the neglected inertial forces [272].

(b) Generalize the results to arbitrary angular velocities of rotation.

6.3 Stokes flow inside and around a corner

Continuing our discussion of Stokes flow in a wedge-shaped domain, we study the structure of the flow near or around a corner that is confined by two stationary intersecting walls, as illustrated in Figures 6.3.1 and 6.3.2 [103, 271]. The local flow near the corner is driven by the motion of the fluid far from the walls. Because all boundary conditions for the stream function are homogeneous, the solution must be found by solving an eigenvalue problem for the exponent λ introduced in (6.2.2). Stated differently, the exponent λ cannot be assigned *a priori* but must be found as part of the solution. We will consider separately the cases of flow with antisymmetric or symmetric streamline pattern, as shown in Figures 6.3.1 and 6.3.2. Antisymmetric flow occurs when the outer flow exhibits a corresponding symmetry, or else the midplane of symmetry represents a free surface with vanishing shear stress.

6.3.1 Antisymmetric flow

Concentrating first on the case of antisymmetric flow, illustrated in Figure 6.3.1, we require that the velocity is zero at the walls and derive the boundary conditions $\psi = 0$ and $\partial\psi/\partial\theta = 0$ at $\theta = \pm\alpha$. Assuming that $\lambda \neq 0, 1, 2$, using (6.2.8), and stipulating that the velocity is antisymmetric with respect to the midplane, $\partial\psi/\partial\theta = 0$ at $\theta = 0$, we find that

$$f(\theta) = A \cos \lambda\theta + B \cos[(\lambda - 2)\theta], \tag{6.3.1}$$

where A and B are two complex constants. When λ is complex, the complex conjugate of the right-hand side of (6.3.1) is also added to extract the real part. Enforcing the boundary conditions at the

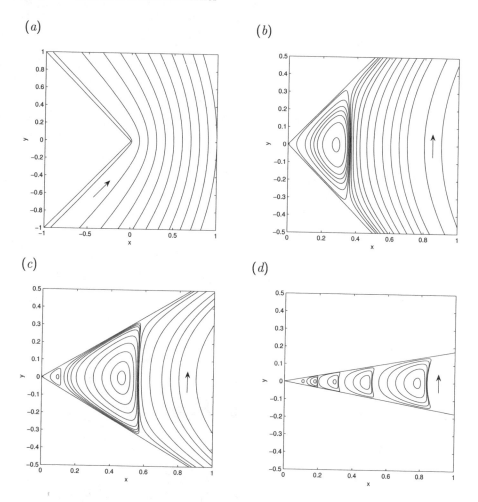

FIGURE 6.3.1 Antisymmetric Stokes flow near two intersecting stationary plane walls for corner semi-angle (a) $\alpha = 135°$, (b) $45°$, (c) $30°$, and (d) $10°$.

walls, we obtain two homogeneous equations,

$$
\begin{bmatrix} \cos \lambda\alpha & \cos[(\lambda-2)\alpha] \\ \lambda \sin \lambda\alpha & (\lambda-2)\sin[(\lambda-2)\alpha] \end{bmatrix} \cdot \begin{bmatrix} A \\ B \end{bmatrix} = \begin{bmatrix} 0 \\ 0 \end{bmatrix}.
\tag{6.3.2}
$$

A nontrivial solution for A and B is possible only when the determinant of the coefficient matrix is zero, providing us with the secular equation

$$
(\lambda - 2) \cos \lambda\alpha \sin[(\lambda - 2)\alpha] - \lambda \sin \lambda\alpha \cos[(\lambda - 2)\alpha] = 0,
\tag{6.3.3}
$$

which can be recast into the compact form of a nonlinear algebraic equation for λ,

$$
\sin[2\alpha(\lambda - 1)] = (1 - \lambda) \sin 2\alpha.
\tag{6.3.4}
$$

(*a*)   (*b*)

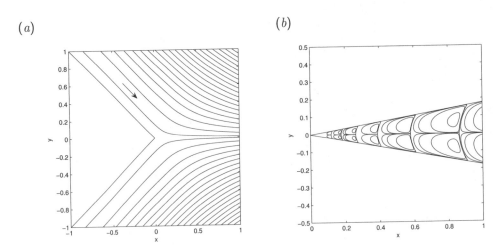

FIGURE 6.3.2 Symmetric Stokes flow near two intersecting stationary planes for corner semi-angle
(*a*) $\alpha = 135°$ and (*b*) $10°$.

An obvious solution is $\lambda = 1$. However, since for this value the third rather than the first expression in (6.2.8) should have been chosen, this choice is disqualified. Combining (6.3.2) with (6.3.1) and (6.2.1), we find that the stream function is given by

$$\psi = Cr^\lambda \Big(\cos\lambda\theta \cos[(\lambda - 2)\alpha] - \cos\lambda\alpha \cos[(\lambda - 2)\theta] \Big), \qquad (6.3.5)$$

where C is a real constant.

Non-reversing flow

Dean & Montagnon [103] pointed out that real and positive solutions other than the trivial solution exist in the range $73° < \alpha < 180°$ or $0.41\pi < a < \pi$; the lower limit is a numerical approximation. In practice, the nontrivial solutions can be computed by solving equation (6.3.4) using Newton's method (Problem 6.3.1). We are mainly interested in the smallest solution plotted with the solid line in the left column of Figure 6.3.3(*a*), describing the strongest possible velocity field that is expected to prevail in practice. When $\alpha = 90°$, we find that $\lambda = 2$ corresponding to simple shear flow. When $\alpha = 180°$, we find that $\lambda = 1.5$ corresponding to flow around a flat plate. The streamline pattern for $\alpha = 135°$ is shown in Figure 6.3.1(*a*).

Eddies

Moffatt pointed out that, approximately when $\alpha < 73°$, equation (6.3.4) has complex solutions [271]. To study the structure of the flow, we write $\lambda = \lambda_R + i\lambda_I$ and decompose (6.3.4) into its real and imaginary parts indicated by the subscripts R and I, obtaining

$$\sin[2\alpha(\lambda_R - 1)] \cosh(2\alpha\lambda_I) = (1 - \lambda_R)\sin 2\alpha,$$
$$\cos[2\alpha(\lambda_R - 1)] \sinh(2\alpha\lambda_I) = -\lambda_I \sin 2\alpha. \qquad (6.3.6)$$

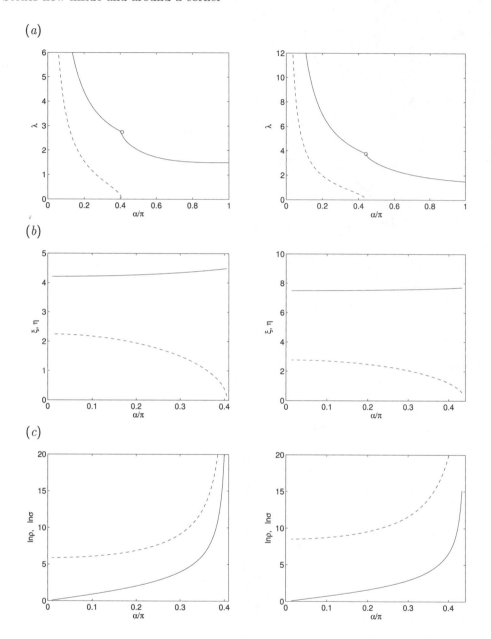

FIGURE 6.3.3   The left column describes antisymmetric flow and the right column describes symmetric flow inside a corner confined between two intersecting planes. (a) Dependence of the real (solid lines) and imaginary (broken lines) part of the exponent λ with the smallest real part on the corner semi-angle, α. The eigenvalue is real when $0.41\pi < \alpha < \pi$ in antisymmetric flow or when $0.43\pi < \alpha < \pi$ in symmetric flow. (b) Graphs of ξ (solid lines) and η (broken lines) defined in (6.3.6) in the regime where eddies develop. (c) Graphs of the geometric ratio, ρ (solid lines), and decay ratio, σ (broken lines), defined in equations (6.3.14) and (6.3.15).

To simplify the notation, we introduce the auxiliary variables

$$\xi = 2\alpha(\lambda_R - 1), \qquad \eta = 2\alpha\lambda_I, \qquad \kappa = \frac{\sin 2\alpha}{2\alpha}, \qquad (6.3.7)$$

note that $\lambda = 1 + (\xi + i\eta)/(2\alpha)$, and recast (6.3.6) into the form

$$\sin\xi \cosh\eta = -\kappa\xi, \qquad \cos\xi \sinh\eta = -\kappa\eta. \qquad (6.3.8)$$

Since both the sine and cosine of ξ must be negative, ξ must lie in the range $(2n-1)\pi < \xi < (2n-\frac{1}{2})\pi$, where n is an integer. The solution of the two nonlinear equations (6.3.8) can be computed using Newton's method with suitable initial guesses for ξ and η. The solution branch with the smallest real part, λ_R, is shown in the left column of Figure 6.3.3(a, b). Combining (6.3.2) with (6.3.1) and (6.2.1), we find that the stream function for $\alpha < 73°$ is given by

$$\psi = C\, r^{\lambda_R} \operatorname{Real}\{\, e^{i\lambda_I \ln r} Q \,\}, \qquad (6.3.9)$$

where C is a real constant and

$$Q = \cos(\lambda\theta) \cos[(\lambda - 2)\alpha] - \cos[(\lambda - 2)\theta] \cos(\lambda\alpha). \qquad (6.3.10)$$

Thus,

$$\psi = C\, r^{\lambda_R} \,[\operatorname{Real}\{Q\} \cos(\lambda_I \ln r) - \operatorname{Imag}\{Q\} \sin(\lambda_I \ln r)]. \qquad (6.3.11)$$

Streamline patterns for $\alpha = 45°$, $30°$, and $10°$ are shown in Figure 6.3.1(b–d).

Moffatt observed that, as r tends to zero, the stream function changes sign an infinite number of times [271]. Physically, an infinite sequence of self-similar eddies develop inside the corner. To identify these eddies, we express the stream function in the form

$$\psi = C\, r^{\lambda_R} \,|Q| \, \cos[\,\lambda_I \ln r + \operatorname{Arg}\{Q\}\,], \qquad (6.3.12)$$

where Arg denotes the argument of a complex number. Setting $\psi = 0$ or $\lambda_I \ln r + \arg(Q) = (n+\frac{1}{2})\pi$, we find that the shape of the dividing streamlines is described by the equation

$$r_n = \exp\left(\frac{1}{\lambda_I}\left[(n + \frac{1}{2})\pi - \operatorname{Arg}\{Q\}\right]\right), \qquad (6.3.13)$$

where $n = \ldots, -1, 0, 1, \ldots$. Applying (6.3.13) at a certain value of θ, we find that the ratio of the radial positions of two successive dividing streamlines is

$$\rho \equiv \frac{r_{n+1}}{r_n} = e^{\pi/\lambda_I}. \qquad (6.3.14)$$

Using (6.3.12), we find that the ratio of the magnitude of the corresponding polar component of the velocity is

$$\sigma \equiv \frac{u_{\theta_{n+1}}}{r_{\theta_n}} = e^{(\lambda_R - 1)\pi/\lambda_I}. \qquad (6.3.15)$$

The geometric ratio, ρ, and amplification ratio, σ, are plotted on a linear-logarithmic scale with a solid or broken line in the left column of Figure 6.3.3(c).

As the angle α tends to zero, the geometrical ratio ρ tends to unity, which means that the eddies tend to become evenly spaced between two nearly parallel walls. As α increases from zero, ρ takes values on the order of unity for small and moderate angles, revealing a slow spatial modulation of the eddy size. As α tends to the critical angle $\alpha \simeq 73°$, ρ increases rapidly. Physically, the eddy size decays quickly into the corner flow. The amplification ratio σ takes large values over an extended range of angles, α, and this reveals that the intensity of the eddy flow decays rapidly inside the corner. The minimum value of σ achieved in the limit $\alpha \to 0$ is approximately 360. When $\alpha = 45°$, σ is approximately 2000. Consequently, it takes one eddy for the magnitude of the velocity to decrease by three orders of magnitude!

Channel flow

As α tends to zero, the intersecting planes tend to become parallel yielding a two-dimensional channel with parallel walls. The emerging flow can be identified with the flow far inside a deep cavity driven by an overpassing fluid. To derive an appropriate similarity solution, we may consider the limit as α tends to zero. However, it is more expedient to begin afresh working in Cartesian coordinates where the walls are located at $y = \pm a$.

Setting $\psi = Af(y)\exp(-k|x|)$, requiring that ψ satisfies the biharmonic equation, and enforcing the no-penetration condition $\psi = 0$ at $y = \pm a$, we obtain

$$\psi = B\left(a\cos ky - y\cot ka\sin ky\right)\exp(-k|x|), \tag{6.3.16}$$

where A and B are complex constants and k is the complex wave number,. Requiring the no-slip condition $\partial\psi/\partial y = 0$ at $y = \pm a$ provides us with an algebraic equation, $\sin 2ka = -2ka$, which admits only imaginary solutions. The solution with the smallest positive real part, $2ka = 4.21 + 2.26\mathrm{i}$, describes a periodic eddy pattern with approximate wavelength $L \simeq 4\pi a/2.26 = 5.56a$.

6.3.2   Symmetric flow

Next, we address the case of symmetric flow illustrated in Figure 6.3.2. Requiring that the velocity vanishes at the walls located at $\theta = \pm\alpha$ and the shear stress vanishes at the midplane $\theta = 0$, we derive the boundary conditions $\psi = 0$ and $\partial\psi/\partial\theta = 0$ at $\theta = \pm\alpha$, and $\psi = 0$ and $\partial^2\psi/\partial\theta^2 = 0$ at $\theta = 0$. Working as in the case of the antisymmetric flow, we set

$$f = A\sin\lambda\theta + B\sin[(\lambda - 2)\theta], \tag{6.3.17}$$

where A and B are two complex constants. Enforcing the boundary conditions, we derive an eigenvalue problem expressed by

$$\sin[2\alpha(\lambda - 1)] = (\lambda - 1)\sin 2\alpha. \tag{6.3.18}$$

The obvious solutions $\lambda = 1, 2$ are disqualified for the reason that the second and third rather than the first expression in (6.2.8) should have been chosen.

Moffatt pointed out that real and positive solutions exist in the range $78° < \alpha < 180°$ [271]. The solution branch with the smaller magnitude is plotted with the solid line in the right column of Figure 6.3.3(a). When $\alpha = 90°$, we find that $\lambda = 2$ corresponding to simple shear flow. When $\alpha = 180°$, we find that $\lambda = 1.5$ corresponding to flow along a semi-infinite flat plate. When approximately $\alpha < 78° = 0.43\pi$, equation (6.3.18) has complex solutions that can be computed working in terms of ξ and η as in the case of antisymmetric flow. The solution branch with the smallest real part λ_R is shown in the right column of Figure 6.3.3(a, b). The corresponding geometric ratio, ρ, and amplification ratio, σ, are plotted on a linear-logarithmic scale in the right column of Figure 6.3.3(c). Comparing these results with those shown in the left column of Figure 6.3.3 for antisymmetric flow reveals that the size of the eddies decays less rapidly. However, the intensity of the eddy motion decays at an appreciably higher rate.

Channel flow

As α tends to zero, the intersecting planes tend to become parallel and the asymptotic solution describes Stokes flow inside a liquid layer resting on a plane wall. To derive a similarity solution, it is expedient to introduce Cartesian coordinates where the wall is located at $y = 0$ and the midplane or free surface is located at $y = a$. Working as in the case of antisymmetric flow, we find that the wavelength of the periodic eddy pattern is $L \simeq 2\pi a/2.76 = 2.28\,a$ [271].

6.3.3   Flow due to a disturbance at a corner

We have considered solutions corresponding to λ with a positive real part, yielding a rapidly decaying sequence of eddies. If λ_1 is a solution of (6.3.4) or (6.3.18) with positive real part, then $\lambda_2 = 2 - \lambda_1$ is also a solution with positive or negative real part corresponding to a flow whose velocity decays like $r^{-\lambda_1+1}$ as $r \to \infty$. In this light, the streamline patterns shown in Figures 6.3.1 and 6.3.2 can be reinterpreted as those due to a disturbance near a corner, where the velocity decays far from the corner.

Problem

6.3.1 *Flow inside the corner between two free surfaces*

Consider flow between two planar intersecting free surfaces with vanishing shear stress [271]. The flow is driven by fluid motion far from the corner An example is the flow near the cusped interface of a bubble subjected to straining flow.

(a) Consider the case of antisymmetric flow described by (6.3.1). Show that the exponent λ satisfies the equation

$$(\lambda - 1) \cos \lambda\alpha \cos[(\lambda - 2)\alpha] = 0, \qquad (6.3.19)$$

which has only real solutions, precluding the occurrence of eddies. Confirm that, when $\alpha > \pi/2$, the smallest positive solution is $\lambda = 2 - \pi/(2\alpha)$, corresponding to rotational flow. Confirm that, when $\alpha < \pi/2$, the smallest positive solution is $\lambda = \pi/(2\alpha)$, corresponding to irrotational flow.

(b) Repeat (a) for symmetric flow.

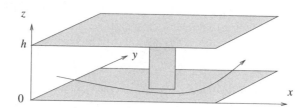

FIGURE 6.4.1   Illustration of flow past an obstacle in a Hele–Shaw cell confined between two parallel plates separated by distance h.

Computer Problem

6.3.2 *Solution of the eigenvalue problem*

(*a*) Write a program that solves equations (6.3.4) and (6.3.8) and reproduce the plots shown in Figure 6.3.3.

(*b*) Repeat (*a*) for symmetric flow.

6.4   Nearly unidirectional flows

When the curvature of the streamlines is small, the flow can be regarded as locally unidirectional, which means that the velocity profile can be assumed to depend on the local pressure gradient and direction of the body force. The union of these assumptions comprises the framework of lubrication flow. To formally study the structure of a nearly unidirectional flow, we may introduce asymptotic expansions for flow variables of interest in terms of the curvature of the streamlines. The analysis confirms that the approximate solution obtained in the framework of lubrication flow is the leading-order approximation of the exact solution (e.g., [223]).

6.4.1   Flow in the Hele–Shaw cell

The Hele–Shaw cell is a narrow channel with parallel walls separated by distance h. The clearance of the channel can be blocked by stationary or moving objects, such as circular disks, flattened air bubbles and liquid drops, as depicted in Figure 6.4.1. The flow can be driven by an imposed pressure gradient, gravity, or the motion of the objects.

When the channel height, h, is small compared to the global dimensions of the channel and size of the objects, pressure variations normal to the channel walls are negligibly small. The flow may then be assumed to be locally unidirectional with a parabolic velocity profile described by (5.1.6). If the walls are located at $z = 0$ and h, as shown in Figure 6.4.1, the nonzero velocity components are

$$u_x(x,y,z) = \frac{1}{2\mu}\left(-\frac{\partial p}{\partial x} + \rho g_x\right) z(h-z), \qquad u_y(x,y,z) = \frac{1}{2\mu}\left(-\frac{\partial p}{\partial y} + \rho g_y\right) z(h-z) \quad (6.4.1)$$

to leading-order approximation. Integrating these profiles across the clearance of the channel, we obtain the mean fluid velocity

$$\bar{\mathbf{u}}(x,y) \equiv \frac{1}{h}\int_0^h \mathbf{u}\,\mathrm{d}z = \frac{h^2}{12\mu}\,\nabla(-p + \rho\,\mathbf{g}\cdot\mathbf{x}), \tag{6.4.2}$$

where the gradient ∇ operates with respect to x and y.

It is evident from (6.4.2) that the negative of the hydrodynamic pressure, $-\tilde{p} \equiv -p + \rho\,\mathbf{g}\cdot\mathbf{x}$, plays the role of a potential function for the mean velocity, $\bar{\mathbf{u}}$. Consequently, the two-dimensional vector field $\bar{\mathbf{u}}(x,y,t)$ is irrotational. Mass conservation requires that the two-dimensional vector field $\bar{\mathbf{u}}$ is solenoidal, $\nabla\cdot\bar{\mathbf{u}} = 0$. Substituting expression (6.4.2) for the velocity, we find that the hydrodynamic or physical pressure is a harmonic function of x and y,

$$\nabla^2 p = 0. \tag{6.4.3}$$

These results suggest that flow in a Hele–Shaw cell with uniform gap past an obstacle is identical to two-dimensional irrotational flow of an incompressible fluid past a body with corresponding geometry. Interestingly, highly viscous flow inside the Hele–Shaw cell provides us with a method of visualizing irrotational flow normally occurring when the effect of viscosity is negligibly small.

Computing the flow in the Hele–Shaw is thus reduced to solving Laplace's equation (6.4.3). Over an impermeable boundary, we require the no-penetration condition expressed as a Neumann boundary condition for the pressure. Over the inlet and outlet, we typically impose a Dirichlet boundary condition specifying the level of the pressure. Over an interface between two immiscible fluids, we require continuity of the normal velocity and specify that the pressure undergoes a discontinuity determined by the interfacial curvature and surface tension.

The assumptions that lead us to the Laplace equation (6.4.3) cease to be valid in the vicinity of a boundary where the flow becomes three-dimensional and the assumption of local unidirectional flow is no longer appropriate. However, boundary effects in the Hele–Shaw cell are confined inside thin fluid layers whose thickness is of order of the plate separation, h, and are thus negligible to a leading-order approximation. Neglecting inertial nonlinear effects due to the curvature of the streamlines introduces an additional error that is small as long as the fluid velocity is not exceedingly high.

Flow in a coating die

A coating die is an industrial device used to generate a thin sheet of a liquid for subsequent coating onto a moving substratum, as illustrated in Figure 6.4.2(a). The die assembly consists of a circular inlet tube that feeds a liquid into a Hele–Shaw cell. A schematic illustration of the cross-section of the inlet tube is shown in Figure 6.4.2(b). The die should be designed to deliver a uniform flow rate at the outlet of the Hele-Shaw cell so that a layer of uniform thickness can be coated onto a moving support.

To develop a model for the flow inside the die, we neglect the the effect of gravity and assume that the average velocity of the liquid through the channel is related to the pressure gradient by equation (6.4.2) and the pressure satisfies Laplace's equation (6.4.3). The boundary conditions

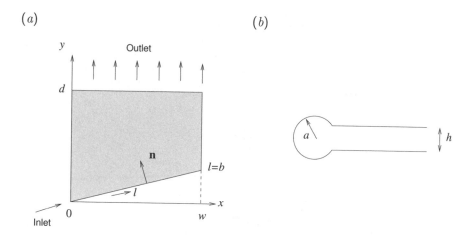

FIGURE 6.4.2   Flow in a coating die consisting of an inlet tube and a Hele-Shaw cell; (a) top view of the die, and (b) cross-section of the inlet tube.

require that the pressure at the outlet is equal to the atmospheric pressure, $p = p_a$ at $y = d$, and the fluid cannot penetrate the side walls, $\partial p/\partial y = 0$ at $x = 0$ and w.

To complete the definition of the problem, we must provide a boundary condition for the pressure at the inlet. A mass balance for the flow along the inlet tube requires that

$$\frac{\mathrm{d}Q}{\mathrm{d}l} = -h\,\mathbf{n}\cdot\bar{\mathbf{u}} = \frac{h^3}{12\mu}\,\mathbf{n}\cdot\nabla p, \tag{6.4.4}$$

where Q is the flow rate and l is the arc length along the centerline of the inlet tube, $0 < l < b$, and $\mathbf{n}$ is the normal vector in the xy plane pointing into the Hele–Shaw cell. Evaluating the flow rate along the inlet using Poiseuille's law, we set

$$Q = -\frac{\pi a^4}{8\mu}\frac{\partial p}{\partial l}. \tag{6.4.5}$$

The inlet tube is allowed to be tapered, that is, the tube radius a can be a function of arc length along the centerline, l. Combining (6.4.4) with (6.4.5), we derive the desired inlet boundary condition in the form of a second-order partial differential equation,

$$\frac{\partial^2 p}{\partial l^2} + \frac{4}{a}\frac{\mathrm{d}a}{\mathrm{d}l}\frac{\partial p}{\partial l} + \frac{2h^3}{3\pi a^4}\,\mathbf{n}\cdot\nabla p = 0, \tag{6.4.6}$$

supplemented by the value of the pressure at the inlet, $l = 0$, and the condition that the flow rate vanishes at the end of the inlet tube, $\partial p/\partial l = 0$ at $l = b$.

Hele–Shaw cell with uneven walls

The analysis of flow in a Hele–Shaw cell with perfectly parallel walls discussed previously in this section can be extended to situations where the clearance of the channel exhibits slow modulations

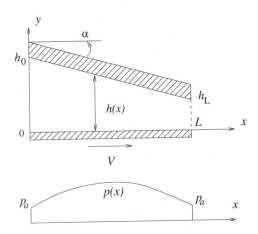

FIGURE 6.4.3   Lubrication flow between a horizontal and an inclined flat surface (top), and the developing pressure field (bottom). In this configuration, the pressure is the same on either end of the lubrication zone.

so that the plate separation h is a function of x and y. Equations (6.4.1) and (6.4.2) are still valid provided that the origin is set at the lower wall and the upper wall is located at $z = h(x, y)$. Mass conservation requires that $\nabla \cdot (h\bar{\mathbf{u}}) = 0$, yielding a Helmholtz equation,

$$\nabla \cdot [h^3(\nabla p - \rho \mathbf{g})] = 0. \tag{6.4.7}$$

Physically, this equation describes the steady distribution of temperature in a plate whose thermal conductivity is proportional to h^3.

6.4.2   Hydrodynamic lubrication

In another application of the lubrication approximation, we consider viscous flow between a horizontal flat plate that moves parallel to itself with velocity V underneath a stationary mildly sloped surface representing, for example, the assembly of a rocker bearing, as illustrated in Figure 6.4.3. If the slope of the inclined surface is sufficiently small, the flow at any x station across the gap between the planar and the inclined surface can be approximated with plane Couette–Poiseuille flow between two parallel plates separated by a variable distance, $h(x)$.

To leading-order approximation, the hydrodynamic pressure is independent of position across the channel, y, so that $p(x)$. According to (5.1.7), the flow rate through the channel is given by

$$Q = \frac{1}{2} V h - \frac{\mathrm{d}p}{\mathrm{d}x} \frac{h^3}{12\mu}. \tag{6.4.8}$$

Mass conservation requires that Q is a constant, independent of x. To compute the pressure distribution along the gap, we solve (6.4.8) for the unknown pressure gradient, $\mathrm{d}p/\mathrm{d}x$, and obtain the ordinary differential equation

$$\frac{\mathrm{d}p}{\mathrm{d}x} = 6\mu \frac{V}{h^2} - 12\mu \frac{Q}{h^3}. \tag{6.4.9}$$

The boundary conditions specify that $p = p_0$ at $x = 0$ and $p = p_L$ at $x = L$, where p_0 and p_L are specified pressures at the beginning and end of the lubrication zone confined in the interval $0 \le x \le L$. One boundary condition is required because (6.4.9) is a first-order ordinary differential equation, and another boundary condition is required because the flow rate Q is an unknown that must be found as part of the solution.

Linear profile

To make the analysis more specific, we consider an inclined upper wall with linear profile, as shown in Figure 6.4.3, and set $h = h_0 - \alpha x$, where α is the constant slope and h_0 the clearance of the channel at the beginning of the lubrication zone. Equation (6.4.9) becomes

$$\frac{\mathrm{d}p}{\mathrm{d}x} = 6\mu \frac{V}{(h_0 - \alpha x)^2} - 12\mu \frac{Q}{(h_0 - \alpha x)^3}. \tag{6.4.10}$$

To compute the flow rate, Q, we require that the integral of the right-hand side of (6.4.10) with respect to x across the lubrication zone is $p_L - p_0$.

For illustration, we assume that the pressure is equal to the atmospheric pressure p_a at both ends of the lubrication zone, $p_0 = p_L = p_a$. Setting the integral of the right-hand side of (6.4.10) from $x = 0$ to L to zero, we obtain

$$Q = V \frac{h_0 h_L}{h_0 + h_L}, \tag{6.4.11}$$

where $h_L = h_0 - \alpha L$ is the clearance of the channel at the end of the lubrication zone. Substituting this expression for Q into (6.4.10) and integrating with respect to x, and recover the pressure distribution

$$p = p_a + 6\mu V \frac{\alpha}{h_0 + h_L} \frac{x(L - x)}{(h_0 - \alpha x)^2}. \tag{6.4.12}$$

It is convenient to introduce the dimensionless geometric parameter

$$\kappa \equiv \frac{h_0}{\alpha L} = \frac{h_0}{h_0 - h_L}, \tag{6.4.13}$$

taking values in the range $(1, \infty)$ or $(-\infty, 0)$. In the first range, the inclined wall slopes downward, as shown in Figure 6.4.3. In the second range, the inclined wall slopes upward, leaning away from the direction of translation. Values of κ in the range $[0, 1]$ are prohibited by the requirement that the upper wall does not slope downward so much as to touch the lower wall before the end of the lubrication zone. When $\kappa = 1$, the two walls meet at $x = L$, and the channel is closed at the right end. Equations (6.4.11) and (6.4.12) take the form

$$Q = V h_0 \frac{\kappa - 1}{2\kappa - 1}, \qquad p = p_a + 6\mu V \frac{L}{h_0^2} \frac{\kappa^2}{2\kappa - 1} \frac{\hat{x}(1 - \hat{x})}{(\kappa - \hat{x})^2}, \tag{6.4.14}$$

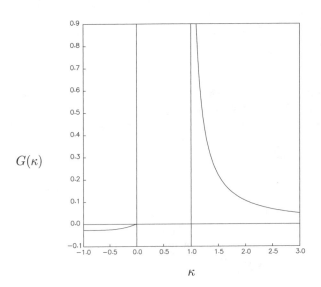

$G(\kappa)$

κ

FIGURE 6.4.4   Graph of the function $G(\kappa)$ expressing the hydrodynamic lift force exerted on the inclined plane shown in Figure 6.4.3.

where $\hat{x} = x/L$ is the dimensionless position varying in the interval $[0, 1]$. The maximum pressure occurs at the location

$$x_{max} = \frac{\kappa}{2\kappa - 1} L \tag{6.4.15}$$

and is given by

$$p_{max} = p_a + 6\mu V \frac{\kappa}{(\kappa - 1)(2\kappa - 1)} \frac{L}{h_0^2}. \tag{6.4.16}$$

As κ increases from unity to infinity, the location of the maximum pressure is shifted from the left end point, $x = L$, to the midpoint, $x = \frac{1}{2}L$.

The y component of the hydrodynamic force exerted on the sloped surface can be approximated with the negative of the normal force exerted on the planar surface. The latter is found by integrating the pressure over the planar surface, finding

$$F_y \simeq \int_0^L p \, dx = p_a L + 6\mu V \frac{L^2}{h_0^2} \kappa^2 \left(\ln \frac{h_0}{h_L} - 2 \frac{h_0 - h_L}{h_0 + h_L} \right). \tag{6.4.17}$$

It is useful to recast this expression into the form

$$F_y = p_a L + 6\mu V \frac{L^2}{h_0^2} G(\kappa), \tag{6.4.18}$$

where

$$G(\kappa) = \kappa^2 \left(\ln \frac{\kappa}{\kappa - 1} - \frac{2}{2\kappa - 1} \right). \tag{6.4.19}$$

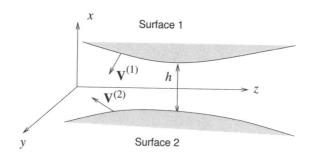

FIGURE 6.4.5 Lubrication flow between two solid but not necessarily rigid surfaces in close contact moving with different velocities.

The second term on the right-hand side of (6.4.19) is the lubrication lifting or load force. A graph of the function $G(\kappa)$ is shown in Figure 6.4.4.

When $\kappa > 1$, the plane wall moves toward the minimum gap and the lubrication force is positive. Under these conditions, given V, the lift force will be able to balance the weight of an overlying object whose lower surface is represented by the inclined plane, provided that κ is sufficiently close to unity. Alternatively, given κ, the lift force will be able to balance the weight of an overlying object provided that the velocity V is sufficiently large. When $\kappa < 0$, the wall moves toward the maximum gap and the lubrication force pulls the object toward the moving plane, closing the gap and choking the flow.

6.4.3 Reynolds lubrication equation

Next, we consider the motion of a fluid between two solid, but possibly deformable, surfaces in close contact, consisting of a rigid or elastic material. In the Cartesian coordinates defined in Figure 6.4.5, the distance between the surfaces, h, is a function of the lateral coordinates y, z, and time, t. The velocities of the upper and lower surfaces, $\mathbf{V}^{(1)}$ and $\mathbf{V}^{(2)}$, are allowed to vary with position over the surfaces and time, reflecting rigid-body motion and boundary deformation. Applying the kinematic boundary condition (1.10.5) at the upper and lower surfaces, and subtracting corresponding sides of the resulting expressions, we derive an evolution equation for the film thickness, $h(y, z, t)$,

$$\frac{\partial h}{\partial t} = (\mathbf{V}^{(2)} - \mathbf{V}^{(1)}) \cdot \nabla h - V_x^{(2)} + V_x^{(1)}, \qquad (6.4.20)$$

where $\nabla = (\partial/\partial y, \partial/\partial z)$ is the two-dimensional gradient in the yz plane.

Mass conservation requires that

$$\frac{\partial h}{\partial t} = -\nabla \cdot (h\overline{\mathbf{u}}), \qquad (6.4.21)$$

where

$$\overline{u}_y = \frac{1}{h} \int u_y \, \mathrm{d}x, \qquad \overline{u}_z = \frac{1}{h} \int u_z \, \mathrm{d}x \qquad (6.4.22)$$

are the mean velocities of the fluid along the gap, and the integration with respect to x is performed over the film thickness. Now adopting the approximations of lubrication flow, we use the unidirectional channel flow solution (5.1.8) to write

$$\bar{u}_y(y, z, t) = \frac{1}{2} \left(V_y^{(1)} + V_y^{(2)} \right) - \frac{h^2}{12\mu} \left(\frac{\partial p}{\partial y} - \rho g_y \right),$$

$$\bar{u}_z(y, z, t) = \frac{1}{2} \left(V_z^{(1)} + V_z^{(2)} \right) - \frac{h^2}{12\mu} \left(\frac{\partial p}{\partial z} - \rho g_z \right). \qquad (6.4.23)$$

Finally, we combine (6.4.20), (6.4.21), and (6.4.23) to obtain the Reynolds lubrication equation governing the distribution of the pressure inside the gap,

$$\frac{1}{12\mu} \nabla \cdot \left[h^3 \left(\nabla p - \rho \mathbf{g} \right) \right] = \frac{1}{2} \nabla \cdot \left[h \left(\mathbf{V}^{(1)} + \mathbf{V}^{(2)} \right) \right] + \left(\mathbf{V}^{(2)} - \mathbf{V}^{(1)} \right) \cdot \nabla h - V_x^{(2)} + V_x^{(1)}. \qquad (6.4.24)$$

The left-hand side of (6.4.24) expresses the effects of pressure- and gravity-driven flow. The right-hand side represents the effects of Couette and squeezing flow.

Squeezing flow between two disks

As an application of the Reynolds lubrication equation, we consider the axisymmetric flow between two parallel, horizontal, coaxial planar disks moving against each other under the action of an imposed force. The disks may represent, for example, the flattened surfaces of two colliding bodies. Without loss of generality, we assume that the upper disk moves with velocity $V(t)$ against a stationary lower disk. Introducing cylindrical polar coordinates with the x axis pointing toward the moving disk, we write $\mathbf{V}^{(1)} = -V(t)\mathbf{e}_x$ and $\mathbf{V}^{(2)} = \mathbf{0}$, where $\mathbf{e}_x$ is the unit vector along the x axis, and the two surfaces are located at $x = 0$ and $h(t)$.

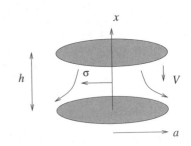

Illustration of squeezing flow between two disks.

All but the last term on the right-hand side of (6.4.24) vanish, yielding

$$\frac{1}{\sigma} \frac{\partial}{\partial \sigma} \left(\sigma \frac{\partial p}{\partial \sigma} \right) = -\frac{12\mu V}{h^3}, \qquad (6.4.25)$$

where σ is the distance from the axis of revolution. Integrating once with respect to σ subject to the condition that the pressure gradient is finite at the centerline, we obtain

$$\frac{\partial p}{\partial \sigma} = -\frac{6\mu V}{h^3} \sigma, \qquad (6.4.26)$$

which shows that the radial pressure gradient increases linearly with distance from the centerline. Integrating (6.4.26) once with respect to σ, we derive an expression for the hydrodynamic pressure due to the squeezing flow,

$$p = -\frac{3\mu V}{h^3} \sigma^2 + \mathcal{P}(x), \qquad (6.4.27)$$

where the function $\mathcal{P}(x)$ incorporates the effect of the ambient pressure and body force. The radial velocity profile is found by integrating the lubrication equation of channel flow

$$\frac{\partial p}{\partial \sigma} = \mu \frac{\partial^2 u_\sigma}{\partial x^2}, \qquad (6.4.28)$$

with boundary condition $u_\sigma = 0$ at $x = 0$ and h, yielding

$$u_\sigma = 3V \frac{\sigma}{h^3} x(h - x). \qquad (6.4.29)$$

The axial velocity, u_x, arises by integrating the continuity equation using (6.4.29) and accounting for the boundary conditions at the lower surface, yielding

$$u_x = -\frac{6V}{h^3} \int_0^x \varrho \, (h - \varrho) \, d\varrho = -\frac{V}{h^3} x^2 \left(3h - 2x\right), \qquad (6.4.30)$$

independent of σ (Problem 6.4.3). We observe that $u_x(h) = -V$, as required.

   The vertical component of the force exerted on the upper disk is found by integrating the hydrodynamic normal stress over the surface of the upper disk from the origin up to the disk radius, $\sigma = a$,

$$F_x = 2\pi \int_0^a p\sigma \, d\sigma = 2\pi \int_0^a \left(-\frac{3\mu V}{h^3} \sigma^2 + \mathcal{P}(h) \right) \sigma \, d\sigma, \qquad (6.4.31)$$

yielding

$$F_x = \frac{3}{2} \frac{\pi \mu V a^4}{h^3} + \pi a^2 \, \mathcal{P}(h). \qquad (6.4.32)$$

This force is equal in magnitude and opposite in sign to the weight of a body whose lower surface is represented by the upper disk, pushing against a flat surface represented by the lower disk. The magnitude of the force shown in (6.4.32) is equal to the force that one must apply in order to remove a piece of adhesive tape from a solid surface, pulling it normal to the surface with velocity V. A thin layer of an adhesive viscous liquid is assumed to separate the surface from the tape. A generalization of (6.4.32) to non-Newtonian fluids is available [231].

Further applications

Further applications of hydrodynamic lubrication are discussed in a comprehensive monograph by Hamrock [164]. Lubrication theory finds important applications in describing the flow inside the narrow gaps between two colliding particles or liquid drops where a squeezing flow is established.

6.4.4   Film flow down an inclined plane

The lubrication approximation can be used to describe the flow of a nearly uniform liquid film down an inclined plane, as illustrated in Figure 6.4.6. The film thickness, $h(x,t)$, depends on the downstream position, x, and time, t. We will assume that the spatial variation of the film thickness

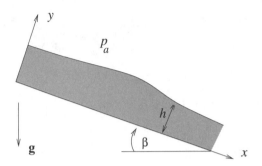

FIGURE 6.4.6   Evolution of a thin liquid film flowing under the action of gravity down an inclined plane.

is small, $\partial h/\partial x < 1$, so that the flow is nearly unidirectional at every x station. Adapting the first equation in (5.1.17) for the velocity profile and equation (5.1.18) for the flow rate, we obtain

$$u_x(x, y, t) = \frac{1}{2\mu} \left(\rho g \sin \beta - \frac{\partial p}{\partial x} \right) y (2h - y), \qquad Q(x, t) = \frac{h^3}{3\mu} \left(\rho g \sin \beta - \frac{\partial p}{\partial x} \right), \qquad (6.4.33)$$

where β is the plane inclination angle. As part of the lubrication approximation, we assume that the hydrodynamic pressure developing inside the film due to the flow is independent of the transverse position, y. Accounting for the pressure drop across the free surface due to surface tension and adding the hydrostatic variation, we derive the pressure distribution

$$p(x, y, t) = \rho g \cos \beta (h - y) + \gamma \kappa + p_a, \qquad (6.4.34)$$

where κ is the curvature of the free surface in the xy plane and p_a is the atmospheric pressure prevailing above the film. Substituting this equation in the expression for the flow rate and introducing the approximation $\kappa \simeq -h_{xx}$, we obtain

$$Q(x, t) = \frac{h^3}{3\mu} (\rho g \sin \beta - \rho g \cos \beta h_x + \gamma h_{xxx}), \qquad (6.4.35)$$

where the subscript x denotes a partial derivative with respect to x. Mass conservation requires that the film thickness and flow rate satisfy the equation

$$h_t = -Q_x, \qquad (6.4.36)$$

where the subscript t denotes a partial derivative with respect to t. It will be noted that equation (6.4.21) is a generalization of equation (6.4.36). Substituting (6.4.35) into (6.4.36), we derive a fourth-order nonlinear partial differential equation describing the evolution of the film thickness,

$$h_t + \frac{1}{3\mu} \left[h^3 \left(\rho g \sin \beta - \rho g \cos \beta h_x + \gamma h_{xxx} \right) \right]_x = 0. \qquad (6.4.37)$$

To complete the definition of the problem, an initial film profile and four periodicity or boundary conditions must be provided.

If the film surface travels without change in shape with phase velocity c, the film thickness is a function of the translating coordinate $\xi \equiv x - ct$ alone. Substituting $h(x,t) = \eta(\xi)$ in the evolution equation (6.4.37) and integrating once with respect to ξ, we obtain

$$c = \frac{1}{3\mu\eta} \left[\eta^3 \left(\rho g \sin \beta - \rho g \cos \beta \, \eta_\xi + \gamma \, \eta_{\xi\xi\xi} \right) \right] + \frac{d}{\eta}, \tag{6.4.38}$$

where d is a constant and η is a shape function. This equation can be used to describe a propagating gravity current down an inclined plane (e.g., [263]).

Dimensionless form

It is useful to introduce the dimensionless variables $\hat{h} = h/\bar{h}$, $\hat{x} = x/\bar{h}$, and $\hat{t} = g\bar{h} \sin \beta t/(2\nu)$, where $\bar{h}$ is a reference length identified with a mean film thickness, and $\nu = \mu/\rho$ is the kinematic viscosity of the fluid. Equation (6.4.37) becomes

$$\hat{h}_{\hat{t}} + \frac{2}{3} \left[\hat{h}^3 (1 - \hat{h}_{\hat{x}} \cot \beta + \Gamma \, \hat{h}_{\hat{x}\hat{x}\hat{x}}) \right]_{\hat{x}} = 0, \tag{6.4.39}$$

where $\Gamma = \gamma/(\rho g \bar{h}^2 \sin \beta)$ is an inverse Bond number determined by the physical properties of the fluid, the mean film thickness, and the plane inclination angle.

6.4.5   Film leveling on a horizontal plane

In the particular case where the film rests on a horizontal plane, $\beta = 0$, the evolution equation for the film thickness undergoes significant simplifications. As a physical application, we consider the leveling of an uneven liquid layer coated on a flat horizontal surface and evolving under the simultaneous action of gravity and surface tension. Assuming that the flow inside the film is locally unidirectional and the velocity profile is parabolic at every x position along the wall, we find that the evolution of the film thickness is governed by the following simplified version of (6.4.37),

$$h_t + \frac{g}{3\nu} \left[h^3 \left(-h_x + \frac{\gamma}{\rho g} h_{xxx} \right) \right]_x = 0. \tag{6.4.40}$$

In terms of the dimensionless variables $\hat{h} \equiv h/\bar{h}$, $\hat{x} \equiv x/\bar{h}$, and $\hat{t} = g\bar{h}t/(2\nu)$, we obtain the dimensionless equation

$$\hat{h}_{\hat{t}} + \frac{2}{3} \left[\hat{h}^3 (-\hat{h}_{\hat{x}} + \Gamma \hat{h}_{\hat{x}\hat{x}\hat{x}}) \right]_{\hat{x}} = 0, \tag{6.4.41}$$

where $\Gamma = g/(\rho g \bar{h}^2)$ is an inverse Bond number and $\bar{h}$ is the mean film thickness.

Numerical methods

To compute the evolution of a periodic film from a specified initial state, we divide one period of the film with length L into N evenly spaced intervals separated by the grid points $x_i = (i-1)L/N$, where $i = 1, \ldots, N + 1$. Our objective is to compute the film thickness at the grid points at a sequence of time instants separated by a time interval, Δt.

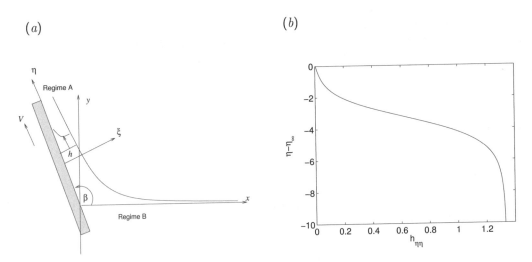

FIGURE 6.4.7 (a) Illustration of a flat plate withdrawn with constant velocity, V, from a liquid pool. (b) Distribution of the curvature along the film surface far from the pool.

The value of h at the ith grid point at the time instant $t = n\Delta t$ is denoted as h_i^n, where $n = 0, 1, \ldots$, and h_i^0 corresponds to the prescribed initial shape. Applying equation (6.4.40) at the ith grid point and approximating the temporal derivative using a forward difference and the spatial derivatives using central differences, we obtain

$$h_i^{n+1} = h_i^n - \frac{g}{3\nu} \frac{\Delta t}{\Delta x} \left(\mathcal{F}_{i+1}^n - \mathcal{F}_i^n \right) \tag{6.4.42}$$

for $i = 1, \ldots, N$, where $\mathcal{F} \equiv h^3 \left(-h_x + \frac{\gamma}{\rho g} h_{xxx} \right)$. Values of the function $\mathcal{F}$ at the grid points can be computed using the finite-difference approximations discussed in Section B.5, Appendix B, subject to the periodicity condition on both sides. The difference equation (6.4.42) provides us with an explicit method of updating the film thickness at the grid points. In practice, the ratio $\Delta t/\Delta x$ must be kept sufficiently small to prevent the onset of numerical instability, as discussed in Chapter 12. An implicit method involving the solution of a system of nonlinear algebraic equations at each time step can be implemented to circumvent this restriction.

6.4.6 Coating a plate

Consider a flat plate withdrawn with constant velocity V at an angle β from a liquid pool, as shown in Figure 6.4.7(a). We are interested in computing the thickness of the liquid film coated on the plate far from the pool, denoted by h_∞. The flow can be divided naturally into two regimes: (a) Regime A prevails inside the film far from the pool and (b) Regime B prevails near the neck of the interface, as shown in Figure 6.4.7(a). The effect of gravity is insignificant in Regime A due to the assumed small film thickness. Hydrostatics dominates in Regime B due to the slow flow, and the shape of the interface is determined by a balance between the gravitational and the capillary force.

Our plan is to separately consider the two regimes and then sensibly match the corresponding free surface profiles.

Film flow

Concentrating on Regime A, we introduce inclined Cartesian coordinates, (ξ, η), as shown in Figure 6.4.7(a). Adapting the first equation in (5.1.17) for the velocity profile and equation (5.1.18) for the flow rate of film flow, we obtain

$$u_\eta(\eta) = -\frac{1}{2\mu}\frac{\partial p}{\partial \eta}\,\xi\,(2h - \xi) + V, \qquad Q = -\frac{h^3}{3\mu}\frac{\partial p}{\partial \eta} + Vh, \tag{6.4.43}$$

where $h(\eta)$ is the film thickness and $Q = Vh_\infty$ is the constant flow rate. The pressure inside the film is determined by the normal interfacial force balance. Neglecting the effect of gravity and approximating the curvature with the linear form $\kappa \simeq -h_{\eta\eta}$, we obtain

$$p(\eta) \simeq -\gamma h_{\eta\eta} + p_a, \tag{6.4.44}$$

where p_a is the ambient pressure and the subscript η denotes a derivative with respect to η. Substituting this pressure distribution into the expression for the flow rate given in (6.4.43), setting $Q = Vh_\infty$, and rearranging, we obtain a third-order nonlinear differential equation,

$$h_{\eta\eta\eta} = -\frac{\mathrm{Ca}}{3}\frac{h - h_\infty}{h^3}. \tag{6.4.45}$$

where $\mathrm{Ca} \equiv \mu V/\gamma$ is a capillary number determining the relative magnitude of viscous and capillary forces over the interface. The far-field condition requires that $h \to h_\infty$ as $y \to \infty$. Linearizing the right-hand side of (6.4.45) in the limit as $\eta \to \infty$ and $h \to h_\infty$, we obtain

$$(h - h_\infty)_{\eta\eta\eta} \simeq -\frac{\mathrm{Ca}}{3}\frac{h - h_\infty}{h_\infty^3}. \tag{6.4.46}$$

The only acceptable solution of this third-order linear ordinary differential equation that tends to h_∞ as $\eta \to \infty$ is

$$h - h_\infty \simeq c\,\exp\left[-\left(\frac{\mathrm{Ca}}{3}\right)^{1/3}\frac{\eta}{h_\infty}\right], \tag{6.4.47}$$

where c is a constant. Differentiating this expression and eliminating the undesirable constant c, we obtain

$$h_\eta \simeq -\left(\frac{\mathrm{Ca}}{3}\right)^{1/3}\frac{h - h_\infty}{h_\infty}, \qquad h_{\eta\eta} \simeq \left(\frac{\mathrm{Ca}}{3}\right)^{2/3}\frac{h - h_\infty}{h_\infty}. \tag{6.4.48}$$

Free-surface deformation

Next, we consider Regime B and introduce the solution for the semi-infinite hydrostatic meniscus attached to an inclined plate derived in Section 4.3.1 with zero apparent contact angle, $\alpha = 0$, and arbitrary plate inclination angle, β. Our analysis has shown that the curvature of the interface at the apparent contact line is

$$\kappa_{cl} = -\frac{h_{cl}}{\ell^2}, \tag{6.4.49}$$

where h_{cl} is the capillary height at the apparent contact line, $\ell \equiv (\gamma/\rho g)^{1/2}$ is the capillary length, and ρ is the density of the coated liquid. In the case of a vertical plate, $\beta = \pi/2$ and $h_{cl}/\ell = \sqrt{2}$.

Matching

To prevent a discontinuity in the pressure, we require that the curvature of the interface at the beginning of Regime A is the same as that at the end of Regime B, yielding

$$h_{\eta\eta} \simeq \frac{h_{cl}}{\ell^2} \tag{6.4.50}$$

at the matching zone. The validity of this condition can be justified by formal matched asymptotic expansions [430].

Landau–Levich formula

It is convenient to introduce the dimensionless variables

$$\hat{h} = \frac{h}{h_\infty}, \qquad \hat{\eta} = \frac{\eta}{h_\infty}\, \mathrm{Ca}^{1/3}. \tag{6.4.51}$$

Equation (6.4.45) governing the film shape in Regime A becomes

$$\hat{h}_{\hat{\eta}\hat{\eta}\hat{\eta}} = -\frac{3}{\hat{h}^3}\,(\hat{h} - 1). \tag{6.4.52}$$

The primary and derived far-field conditions (6.4.48) take the form

$$\hat{h} \to 1, \qquad \hat{h}_{\hat{\eta}} \to -\frac{1}{3^{1/3}}\,(\hat{h}-1), \qquad \hat{h}_{\hat{\eta}\hat{\eta}} \to \frac{1}{3^{2/3}}\,(\hat{h}-1). \tag{6.4.53}$$

The matching condition (6.4.50) requires that

$$\hat{h}_{\hat{\eta}\hat{\eta}} \simeq h_{cl}\,\frac{h_\infty}{\ell^2}\,\frac{1}{\mathrm{Ca}^{2/3}} \tag{6.4.54}$$

at the beginning of the film flow zone.

Equation (6.4.52) can be integrated backward with respect to $\hat{\eta}$ using a standard numerical method, starting from an arbitrary and inconsequential initial position, $\hat{\eta}_\infty$, where $\hat{h} \simeq 1$. The necessary initial conditions for the numerical integration are provided by (6.4.53). A numerical solution obtained for $\hat{h}_\infty = 1.01$ is shown in Figure 6.4.7(*b*), where dimensionless variables indicated by a hat are implied in the labels of the horizontal and vertical axes.

The numerical computations reveal that the second derivative tends to $\hat{h}_{\hat{\eta}\hat{\eta}} \simeq 1.337578$ as $\hat{\eta} - \hat{\eta}_\infty \to -\infty$. Substituting this result into (6.4.54) and rearranging, we obtain the Landau–Levich formula [222]

$$\frac{h_\infty}{\ell} \simeq 1.337578 \, \frac{\ell}{h_{cl}} \, \text{Ca}^{2/3}. \tag{6.4.55}$$

The predictions of this equation are in good agreement with laboratory observations at small capillary numbers.

Flow in the pool

Once the film thickness has been determined, the flow in the pool far from the meniscus can be reconstructed by superposing (*a*) the flow due to a plate withdrawn from the pool studied in Section 6.2.4, and (*b*) the Jeffery–Hamel flow due to a point sink with strength $m = 2Vh_\infty$ studied in Section 6.2.5. The point sink emulates the liquid leaving the pool to be coated on the upward moving plate. The factor of two in the strength of the point source is present because the horizontal free surface represents that midplane of the dihedral corner of the Jeffery–Hamel flow.

In plane polar coordinates, (r, θ), defined with respect to the xy axes such that $x = r \cos\theta$ and $y = r \sin\theta$, as shown in Figure 6.4.7(*a*), the stream function is

$$\psi = \frac{2}{2(\pi - \beta) - \sin(2\beta)} \, Vr \, (\theta \cos\theta \sin\beta + \alpha \, \sin\theta \, \cos\beta) + Vh_\infty \frac{\sin 2\theta - 2\theta \cos 2\beta}{\sin 2\beta + 2\alpha \cos 2\beta}. \tag{6.4.56}$$

Typical streamline patterns are shown in Figure 6.4.8 for two plate withdrawal angles.

Problems

6.4.1 *Lubrication flow in a channel*

With reference to Figure 6.4.3, compute the maximum lift force subject to the constraint that the mean clearance of the channel, $\frac{1}{2}(h_0 + h_L)$, is held constant.

6.4.2 *Lubrication flow between a flat and a curved surface*

Consider the arrangement in Figure 6.4.3, but replace the inclined plane with a section of a circular arc of radius a whose center lies in the plane $x = L$. Repeat the lubrication analysis discussed in the text and recompute the lift force in terms of h_L, L, and a.

6.4.3 *Squeezing flow between two disks*

Compute the axial velocity profile across the gap, $u_x(x, \sigma)$.

Computer Problem

6.4.4 *Film leveling*

Write a program that computes the evolution of a periodic film using the finite-difference method discussed in the text for $L = 2\pi\bar{h}$ with initial condition $h = \bar{h}[1 + \epsilon \sin(2\pi x/L)]$, where ϵ is a

(a)   (b)

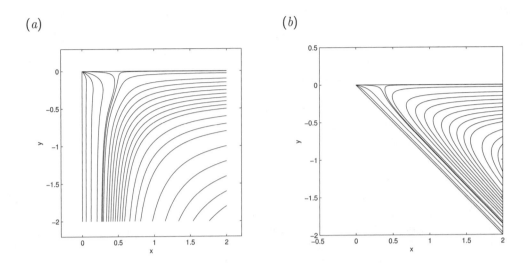

FIGURE 6.4.8   Streamline pattern in a liquid pool far from a plate withdrawn at inclination angle (a)
      $\beta = \pi/2$ or (b) $\pi/4$.

dimensionless amplitude. Run the program for $\epsilon = 0.50$ and $\Gamma = 0, 0.5$, and 1.0. Plot transient film profiles at a sequence of times and discuss the behavior of the solution in each case.

6.5   Flow due to a point force

The velocity field due to a point force applied at a certain point in an a particular domain of a flow plays an important role in the analysis and computation of Stokes flow in a variety of applications. Examples include the flow due to the motion of small particles in a suspension and the propulsion of a microscopic organism. Physically, the flow due to a point force can be identified with the flow generated by the slow motion of a small particle in an otherwise quiescent fluid, as discussed in Section 5.9 in the more general context of Navier–Stokes flow. Later in this chapter, we will see that surface distributions of point forces provide us with boundary-integral representations that can be used to derive boundary-integral equations.

The velocity and hydrodynamic pressure fields due to a point force are found by solving the continuity equation, $\nabla \cdot \mathbf{u} = 0$, and the singularly forced Stokes equation,

$$-\nabla p + \mu \nabla^2 \mathbf{u} + \mathbf{b}\, \delta_3(\mathbf{x} - \mathbf{x}_0) = \mathbf{0}, \qquad (6.5.1)$$

where $\mathbf{x}_0$ is the location of the point force, the constant $\mathbf{b}$ represents the direction and magnitude of the point force, and δ_3 is the three-dimensional delta function. Equation (6.5.1) is the linearized version of (5.9.1) in the absence of fluid inertia. The solution of (6.5.1) must be found subject to a boundary condition requiring that the velocity is zero over a stationary solid boundary of the flow denoted by S_B, that is, $\mathbf{u} = \mathbf{0}$ when $\mathbf{x}$ is on S_B.

6.5.1   Green's functions of Stokes flow

To expedite the solution of (6.5.1), we introduce the Green's function tensor of three-dimensional Stokes flow, $\mathcal{G}_{ij}$, with dimensions of inverse length, defined by the equation

$$u_i(\mathbf{x}) = \frac{1}{8\pi\mu}\,\mathcal{G}_{ij}(\mathbf{x}, \mathbf{x}_0)\,b_j, \tag{6.5.2}$$

where summation is implied over the repeated index, j, and the factor $1/(8\pi\mu)$ has been introduced to facilitate further manipulations. To satisfy the condition of zero velocity on a solid boundary, S_B, we require that $\mathcal{G}_{ij}(\mathbf{x}, \mathbf{x}_0) = 0$ when $\mathbf{x}$ lies on S_B. Other boundary conditions are imposed on free surfaces or interfaces between two immiscible fluids.

The vorticity, pressure, and stress fields due to the point force can be expressed in terms of a vorticity tensor, $\boldsymbol{\Omega}$, pressure vector, $\boldsymbol{\mathcal{P}}$, and stress tensor, $\boldsymbol{\mathcal{T}}$, as

$$\omega_i(\mathbf{x}) = \frac{1}{8\pi\mu}\,\Omega_{ij}(\mathbf{x}, \mathbf{x}_0)\,b_j, \qquad p(\mathbf{x}) = \frac{1}{8\pi}\,\mathcal{P}_j(\mathbf{x}, \mathbf{x}_0)\,b_j, \qquad \sigma_{ik}(\mathbf{x}) = \frac{1}{8\pi}\,\mathcal{T}_{ijk}(\mathbf{x}, \mathbf{x}_0)\,b_j, \tag{6.5.3}$$

where summation is implied over the repeated index, j. The stress tensor is given by

$$\mathcal{T}_{ijk}(\mathbf{x}, \mathbf{x}_0) = -\delta_{ik}\,\mathcal{P}_j(\mathbf{x}, \mathbf{x}_0) + \frac{\partial \mathcal{G}_{ij}(\mathbf{x}, \mathbf{x}_0)}{\partial x_k} + \frac{\partial \mathcal{G}_{kj}(\mathbf{x}, \mathbf{x}_0)}{\partial x_i}. \tag{6.5.4}$$

We note that $\mathcal{T}_{ijk} = \mathcal{T}_{kji}$, as required by the symmetry of the stress tensor, $\boldsymbol{\sigma}$. If the domain of flow is infinite, $\boldsymbol{\Omega}$, $\boldsymbol{\mathcal{P}}$, and $\boldsymbol{\mathcal{T}}$ are required to decay far from the point force.

6.5.2   Differential properties of Green's functions

Taking the divergence of (6.5.2) and using the continuity equation we obtain the identity

$$\frac{\partial \mathcal{G}_{ij}(\mathbf{x}, \mathbf{x}_0)}{\partial x_i} = 0, \tag{6.5.5}$$

where summation is implied over the repeated index, i. Substituting expressions (6.5.3) into (6.5.1) and discarding the arbitrary constant $\mathbf{b}$, we obtain

$$-\frac{\partial \mathcal{P}_j(\mathbf{x}, \mathbf{x}_0)}{\partial x_k} + \nabla^2 \mathcal{G}_{kj}(\mathbf{x}, \mathbf{x}_0) + 8\pi\,\delta_{kj}\,\delta_3(\mathbf{x} - \mathbf{x}_0) = 0 \tag{6.5.6}$$

and

$$\frac{\partial \mathcal{T}_{ijk}(\mathbf{x}, \mathbf{x}_0)}{\partial x_i} = \frac{\partial \mathcal{T}_{kji}(\mathbf{x}, \mathbf{x}_0)}{\partial x_i} = -8\pi\,\delta_{kj}\,\delta_3(\mathbf{x} - \mathbf{x}_0), \tag{6.5.7}$$

which are equivalent statements of the Stokes equation written for the Green's function. Working in index notation, we can show that

$$\frac{\partial}{\partial x_k}\left[\epsilon_{ilm}\,x_l\,\mathcal{T}_{mjk}(\mathbf{x}, \mathbf{x}_0)\right] = -8\pi\epsilon_{ilj}x_l\,\delta(\mathbf{x} - \mathbf{x}_0), \tag{6.5.8}$$

where ϵ_{ilj} is the alternating tensor.

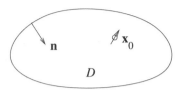

FIGURE 6.5.1   Illustration of a point force and a closed surface, D, in the available domain of flow. The point force may be located inside (as shown), outside, or precisely on D.

6.5.3   Integral properties of Green's functions

Integrating (6.5.5) over a volume of fluid that is enclosed by a surface, D, and using the divergence theorem, we derive the identity

$$\iint_D \mathcal{G}_{ij}(\mathbf{x}, \mathbf{x}_0)\, n_i(\mathbf{x})\, \mathrm{d}S(\mathbf{x}) = 0, \tag{6.5.9}$$

where $\mathbf{n}$ is the unit vector normal to D. This identity holds true independent of whether the singular point $\mathbf{x}_0$ is located inside, precisely on, or outside D. Working in a similar fashion, we derive the identities

$$\iint_D \mathcal{T}_{ijk}(\mathbf{x}, \mathbf{x}_0)\, n_i(\mathbf{x})\, \mathrm{d}S(\mathbf{x}) = \iint_D \mathcal{T}_{kji}(\mathbf{x}, \mathbf{x}_0)\, n_i(\mathbf{x})\, \mathrm{d}S(\mathbf{x}) = \beta\, \delta_{jk} \tag{6.5.10}$$

and

$$\iint_D \epsilon_{ilm} \mathcal{T}_{mjk}(\mathbf{x}, \mathbf{x}_0)\, n_k(\mathbf{x})\, \mathrm{d}S(\mathbf{x}) = \beta\, \epsilon_{ilj}\, x_{0_l}, \tag{6.5.11}$$

where $\mathbf{n}$ is unit normal vector pointing into the control volume, as shown in Figure 6.5.1, and x_{0_l} on the right-hand side of (6.5.11) denotes the lth component of $\mathbf{x}_0$. The coefficient β on the right-hand sides is equal to 8π, 4π, or 0, depending on whether the point $\mathbf{x}_0$ is located inside, precisely on, or outside a smooth surface, D. When the point $\mathbf{x}_0$ is on D, the surface integrals are improper but convergent principal-value integrals.

6.5.4   Symmetry of Green's functions

In Section 6.8.3, we will show that the Green's functions satisfy the symmetry property

$$\mathcal{G}_{ij}(\mathbf{x}, \mathbf{x}_0) = \mathcal{G}_{ij}(\mathbf{x}, \mathbf{x}_0), \tag{6.5.12}$$

which provides us with a relation between the velocity at the point $\mathbf{x}$ due to a point force placed at the point $\mathbf{x}_0$, and the velocity at the point $\mathbf{x}_0$ due to a point force placed at $\mathbf{x}$. Among other purposes, identity (6.5.12) serves as a useful check of accuracy in the computation of a Green's function.

6.5.5   Point force in free space

To compute the flow due to a point force in an infinite domain of flow with no interior boundaries (free space), we express the delta function on the right-hand side of (6.5.1) in terms of the Green's function of Laplace's equation,

$$\delta_3(\mathbf{x} - \mathbf{x}_0) = -\frac{1}{4\pi} \nabla^2\left(\frac{1}{r}\right), \tag{6.5.13}$$

where $r = |\mathbf{x} - \mathbf{x}_0|$. Recalling that the hydrodynamic pressure satisfies Laplace's equation and balancing the dimensions of the pressure gradient with those of the delta function in (6.5.1), we set

$$p = -\frac{1}{4\pi} \nabla\left(\frac{1}{r}\right) \cdot \mathbf{b}, \tag{6.5.14}$$

and find that (6.5.1) becomes

$$\mu\nabla^2\mathbf{u} = \frac{1}{4\pi} \mathbf{b} \cdot (\mathbf{I}\nabla^2 - \nabla\nabla)\left(\frac{1}{r}\right), \tag{6.5.15}$$

where $\mathbf{I}$ is the identity matrix. This form motivates expressing the velocity in terms of a generating scalar function, $\mathcal{H}$, as

$$\mathbf{u} = \frac{1}{\mu}(\mathbf{I}\nabla^2\mathcal{H} - \nabla\nabla\,\mathcal{H}) \cdot \mathbf{b}. \tag{6.5.16}$$

The continuity equation is satisfied for any twice differentiable function $\mathcal{H}$. Substituting (6.5.16) into (6.5.15) and discarding the arbitrary constant $\mathbf{b}$, we obtain

$$(\mathbf{I}\nabla^2 - \nabla\nabla)\left(\nabla^2\mathcal{H} - \frac{1}{4\pi r}\right) = 0, \tag{6.5.17}$$

which is satisfied by any solution of the Poisson equation

$$\nabla^2\mathcal{H} - \frac{1}{4\pi r} = 0. \tag{6.5.18}$$

Taking the Laplacian of this equation, we find that $\mathcal{H}$ is the Green's function of the biharmonic equation,

$$\nabla^4\mathcal{H} + \delta_3(\mathbf{x} - \mathbf{x}_0) = 0, \tag{6.5.19}$$

which is given by $\mathcal{H} = r/(8\pi)$.

Stokeslet

Substituting $\mathcal{H} = r/(8\pi)$ into (6.5.16) and carrying out the differentiations, we derive the flow due to a point force in the standard form (6.5.2). The associated free-space Green's function, $\mathcal{G}_{ij}$, also called the Stokeslet or the Oseen–Burgers tensor and denoted by $\mathcal{S}_{ij}$, is given by

$$\mathcal{S}_{ij}(\hat{\mathbf{x}}) = \frac{\delta_{ij}}{r} + \frac{\hat{x}_i\hat{x}_j}{r^3}, \tag{6.5.20}$$

(a)   (b)

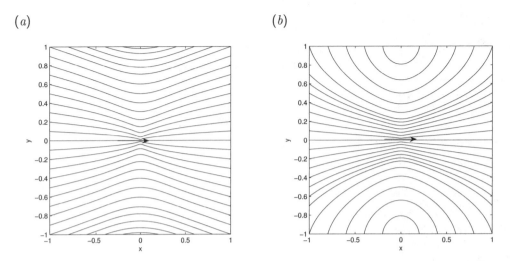

FIGURE 6.5.2   Streamline pattern due to (a) a thee-dimensional and (b) a two-dimensional point force pointing along the x axis.

where $\hat{\mathbf{x}} = \mathbf{x} - \mathbf{x}_0$ and $r = |\hat{\mathbf{x}}|$. The corresponding vorticity, pressure, and stress fields are given by (6.5.6) with

$$\Omega_{ij}(\hat{\mathbf{x}}) = 2\,\epsilon_{ijl}\,\frac{\hat{x}_l}{r^3}, \qquad \mathcal{P}_j(\hat{\mathbf{x}}) = 2\,\frac{\hat{x}_j}{r^3}, \qquad \mathcal{T}_{ijk}(\hat{\mathbf{x}}) = -6\,\frac{\hat{x}_i\hat{x}_j\hat{x}_k}{r^5}. \tag{6.5.21}$$

The expression for $\mathcal{T}_{ijk}$ is found by substituting the expressions for $\mathcal{S}_{ij}$ and $\mathcal{P}_j$ into (6.5.4).

Stream function

The Stokes stream function due to a point force located at the origin and pointing along the x axis is given by

$$\psi = \frac{b_x}{8\pi\mu}\,r\sin^2\theta, \tag{6.5.22}$$

where b_x is the strength of the point force, r is the distance from the origin, and θ is the meridional angle. The streamline pattern is shown in Figure 6.5.2(a). We observe that the point force causes a forward motion of the fluid throughout the domain of flow.

Force on a spherical surface enclosing a point force

It is instructive to compute the hydrodynamic traction and force exerted on a fluid sphere of radius a centered at a point force. Using (6.5.3) and (6.5.21) and setting $n_k = \hat{x}_k/a$, we obtain

$$f_i(\mathbf{x}) = \sigma_{ik}(\mathbf{x})\,n_k(\mathbf{x}) = \frac{1}{8\pi}\,\mathcal{T}_{ijk}(\mathbf{x},\mathbf{x}_0)\,n_k(\mathbf{x})\,b_j = -\frac{3}{4\pi}\,\frac{\hat{x}_i\hat{x}_j}{a^5}\,b_j. \tag{6.5.23}$$

The hydrodynamic force exerted on the spherical surface is

$$F_i = \iint_{Sphere} f_i(\mathbf{x})\, dS(\mathbf{x}) = -\frac{3}{4\pi a^4} \iint_{Sphere} \hat{x}_i \hat{x}_j\, dS(\mathbf{x})\, b_j. \tag{6.5.24}$$

Using the divergence theorem, we compute

$$\iint_{Sphere} \hat{x}_i \hat{x}_j\, dS(\mathbf{x}) = a \iint_{Sphere} \hat{x}_i n_j\, dS(\mathbf{x}) = a \iiint_{Sphere} \frac{\partial \hat{x}_i}{\partial \hat{x}_j}\, dS(\mathbf{x}) = \frac{4\pi}{3} a^4 \delta_{ij}. \tag{6.5.25}$$

Substituting this expression into (6.5.24), we find that

$$\mathbf{F} = -\mathbf{b}, \tag{6.5.26}$$

independent of the radius of the sphere, a. Physically, in the absence of inertia, the net force exerted on a volume of fluid enclosed by two concentric spheres must vanish.

Torque on a spherical surface enclosing a point force

The torque with respect to the location of the point force, $\mathbf{x}_0$, on any surface enclosing the point force is zero. The torque with respect to the location of any other point is nonzero (Problem 6.5.2).

Fourier transforms

It is instructive to rederive the Stokeslet using the method of Fourier transforms. The three-dimensional complex Fourier transform of a rapidly decaying scalar function $f(\mathbf{x})$ defined in the whole three-dimensional space is

$$\hat{f}(\mathbf{k}) = \frac{1}{(2\pi)^{3/2}} \iiint f(\mathbf{x})\, \exp(i\,\mathbf{k}\cdot\mathbf{x})\, dV(\mathbf{x}), \tag{6.5.27}$$

where i is the imaginary unit and $\mathbf{k}$ is a real wave number vector. The inverse transform is

$$f(\mathbf{x}) = \frac{1}{(2\pi)^{3/2}} \iiint \hat{f}(\mathbf{k})\, \exp(-i\,\mathbf{k}\cdot\mathbf{x})\, dV(\mathbf{k}). \tag{6.5.28}$$

The Fourier transforms of the partial derivatives are

$$\left(\widehat{\frac{\partial f}{\partial x_i}}\right)(\mathbf{k}) = -i\, k_i \hat{f}(\mathbf{k}). \tag{6.5.29}$$

Using this expression, we obtain the Fourier transform of the gradient, $(\widehat{\nabla f})(\mathbf{k}) = -i\mathbf{k}\hat{f}(\mathbf{k})$.

Taking the Fourier transform of the continuity equation (6.5.5) for the Stokeslet, we obtain

$$k_i\, \hat{S}_{ij}(\mathbf{k}) = 0. \tag{6.5.30}$$

Taking the Fourier transform of (6.5.6) for a Stokeslet placed at the origin, $\mathbf{x}_0 = \mathbf{0}$, and using the properties of the delta function, we find that

$$i\, k_l \hat{\mathcal{P}}_j(\mathbf{k}) - \mathbf{k}\cdot\mathbf{k}\, \hat{S}_{lj}(\mathbf{k}) + \frac{4}{\sqrt{2\pi}} \delta_{lj} = 0. \tag{6.5.31}$$

Multiplying this equation by k_l, summing over l, and using (6.5.30) to eliminate the Stokeslet, we obtain

$$i\,\mathbf{k}\cdot\mathbf{k}\,\hat{\mathcal{P}}_j(\mathbf{k}) + \frac{4}{\sqrt{2\pi}}\,k_j = 0. \tag{6.5.32}$$

Rearranging, we obtain the Fourier transform of the pressure,

$$\hat{\mathcal{P}}_j(\mathbf{k}) = i\,\frac{4}{\sqrt{2\pi}}\,\frac{k_j}{\mathbf{k}\cdot\mathbf{k}}. \tag{6.5.33}$$

Substituting this expression into (6.5.31), we derive the Fourier transform of the Stokeslet,

$$\hat{\mathcal{S}}_{ij}(\mathbf{k}) = \frac{4}{\sqrt{2\pi}}\,\frac{1}{\mathbf{k}\cdot\mathbf{k}}\left(\delta_{ij} - \frac{k_i k_j}{\mathbf{k}\cdot\mathbf{k}}\right). \tag{6.5.34}$$

Now taking the inverse Fourier transform of the pressure, we find that

$$\mathcal{P}_j(\mathbf{x}) = \frac{i}{\pi^2}\iiint \frac{k_j}{\mathbf{k}\cdot\mathbf{k}}\exp(-i\,\mathbf{k}\cdot\mathbf{x})\,dV(\mathbf{k}) = -\frac{1}{\pi^2}\frac{\partial}{\partial x_j}\left(\iiint \frac{1}{\mathbf{k}\cdot\mathbf{k}}\exp(-i\,\mathbf{k}\cdot\mathbf{x})\,dV(\mathbf{k})\right). \tag{6.5.35}$$

Expressing the wave number vector, $\mathbf{k}$, in spherical polar coordinates, (r,θ,φ), where the meridional angle θ is measured from the orientation of $\mathbf{k}$, we compute the integral

$$\mathcal{I} \equiv \iiint \frac{1}{\mathbf{k}\cdot\mathbf{k}}\exp(-i\,\mathbf{k}\cdot\mathbf{x})\,dV(\mathbf{k}) = 2\pi\int_0^\infty \left(\int_0^\pi \exp(-i\,kr\cos\theta)\,\sin\theta\,d\theta\right)dk, \tag{6.5.36}$$

finding

$$\mathcal{I} = 4\pi\int_0^\infty \frac{\sin(kr)}{kr}\,dk = \frac{2\pi^2}{r}, \tag{6.5.37}$$

where $r = |\mathbf{x}|$. Substituting this expression into (6.5.35) reproduces the expression for the pressure Green's function. The Stokeslet is derived working in a similar fashion (Problem 6.5.1(a)).

6.5.6   Point force above an infinite plane wall

In 1907, Lorentz derived an expression for the flow due to a point force in a semi-infinite fluid above an infinite plane wall [245]. If the wall is located at $x = w$, as shown in Figure 6.5.3, the Green's function satisfies the boundary condition $\mathcal{G}(x = w, y, z; \mathbf{x}_0) = \mathbf{0}$, where $\mathbf{x}_0$ is the location of the point force.

Blake [37] demonstrated that the Green's function can be constructed from a Stokeslet, $\mathcal{S}$, and a few image singularities, including a Stokeslet, a potential dipole, and a point-force dipole,

$$\mathcal{G}(\mathbf{x},\mathbf{x}_0) = \mathcal{S}(\mathbf{x} - \mathbf{x}_0) - \mathcal{S}(\mathbf{x} - \mathbf{x}_0^{im}) - 2\,h_0^2\,\Phi(\mathbf{x} - \mathbf{x}_0^{im}) + 2\,h_0\,\Psi(\mathbf{x} - \mathbf{x}_0^{im}), \tag{6.5.38}$$

where $h_0 = x_0 - w$ is the distance of the point force from the wall, and

$$\mathbf{x}_0^{im} = \begin{bmatrix} 2w - x_0, y_0, z_0 \end{bmatrix} \tag{6.5.39}$$

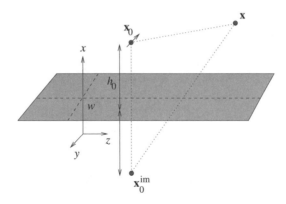

FIGURE 6.5.3   Illustration of a point force located at a point, $\mathbf{x}_0$, above an infinite wall positioned at
$x = w$. The point $\mathbf{x}_0^{im}$ is the image of the point force with respect to the wall.

is the image of the singular point, $\mathbf{x}_0$, with respect to the wall. The tensors $\boldsymbol{\Phi}$ and $\boldsymbol{\Psi}$ contain,
respectively, potential dipoles and Stokeslet dipoles,

$$\Phi_{ij}(\mathbf{x}) = \pm \frac{\partial}{\partial x_j}\left(-\frac{x_i}{r^3}\right) = \pm\left(-\frac{\delta_{ij}}{r^3} + 3\frac{x_i x_j}{r^5}\right) = \pm \mathcal{D}_{ij} \tag{6.5.40}$$

and

$$\Psi_{ij}(\mathbf{x}) = \pm\left(-\frac{\partial S_{i1}}{\partial x_j}\right) = x_1 \Phi_{ij}(\mathbf{x}) \pm \frac{-\delta_{j1}\, x_i + \delta_{i1}\, x_j}{r^3}, \tag{6.5.41}$$

where $r = |\mathbf{x}|$ and $\mathcal{D}_{ij}$ is the potential dipole discussed in Section 6.6.1. The minus sign of $\pm$ applies
for $j = 1$, corresponding to the x direction, and the plus sign applies for $j = 2, 3$, corresponding to
the y and z directions [313].

Pressure field

By analogy with (6.5.38), we express the pressure vector in the form

$$\mathcal{P}(\mathbf{x}, \mathbf{x}_0) = \mathcal{P}^S(\mathbf{x} - \mathbf{x}_0) - \mathcal{P}^S(\mathbf{x} - \mathbf{x}_0^{im}) + 2h_0\, \mathcal{P}^{\Psi}(\mathbf{x} - \mathbf{x}_0^{im}), \tag{6.5.42}$$

where $\mathcal{P}^S$ is the pressure vector associated with the Stokeslet given in (6.5.24), and

$$\mathcal{P}_i^{\Psi}(\mathbf{x}) = \pm 2\mathcal{D}_{i1}. \tag{6.5.43}$$

The minus sign applies for $i = 1$, corresponding to the x direction, and the plus sign applies for
$i = 2$ and 3, corresponding to the y and z directions. Because they are irrotational singularities, the
potential dipoles do not make a contribution to the pressure.

6.5.7 Two-dimensional point force

The flow due to a two-dimensional point force is found by solving the continuity equation, $\nabla \cdot \mathbf{u} = 0$, and the singularly forced Stokes equation

$$-\nabla p + \mu \nabla^2 \mathbf{u} + \mathbf{b}\, \delta_2(\mathbf{x} - \mathbf{x}_0) = \mathbf{0}, \tag{6.5.44}$$

where $\mathbf{b}$ is the strength of the point force, δ_2 is the two-dimensional delta function and ∇^2 is the two-dimensional Laplacian operator in the xy plane. Working as in the case of three-dimensional flow, we express the velocity, vorticity, pressure, and stress fields in the standard forms

$$u_i(\mathbf{x}) = \frac{1}{4\pi\mu}\, \mathcal{G}_{ij}(\mathbf{x}, \mathbf{x}_0)\, b_j, \qquad \omega_i(\mathbf{x}) = \frac{1}{4\pi\mu}\, \Omega_{ij}(\mathbf{x}, \mathbf{x}_0)\, b_j, \qquad p(\mathbf{x}) = \frac{1}{4\pi}\, \mathcal{P}_j(\mathbf{x}, \mathbf{x}_0)\, b_j,$$

$$\sigma_{ik}(\mathbf{x}) = \frac{1}{4\pi}\, \mathcal{T}_{ijk}(\mathbf{x}, \mathbf{x}_0)\, b_j, \tag{6.5.45}$$

involving the velocity, vorticity, pressure, and stress Green's functions, $\mathcal{G}$, Ω, $\mathcal{P}$, and $\mathcal{T}$, where summation is implied over the repeated index, j.

Two-dimensional point force in free space

To compute the flow due to a two-dimensional point force in free space, we express the two-dimensional delta function on the right-hand side of (6.5.1) in terms of the Green's function of Laplace's equation,

$$\delta_2(\mathbf{x} - \mathbf{x}_0) = \frac{1}{2\pi}\, \nabla^2 \ln \frac{r}{a}, \tag{6.5.46}$$

where a is a constant length. Noting that the pressure is a harmonic function and balancing the dimensions of the pressure gradient and the delta function, we set

$$p = \frac{1}{2\pi} \nabla (\ln \frac{r}{a}) \cdot \mathbf{b}. \tag{6.5.47}$$

Next, we introduce the generating function $\mathcal{H}$ defined in (6.5.16) and derive the Poisson equation

$$\nabla^2 \mathcal{H} + \frac{1}{2\pi} \ln \frac{r}{a} = 0, \tag{6.5.48}$$

which shows that $\mathcal{H}$ is the Green's function of the biharmonic equation,

$$\nabla^4 \mathcal{H} + \delta_2(\mathbf{x} - \mathbf{x}_0) = 0, \tag{6.5.49}$$

given by

$$\mathcal{H} = \frac{1}{8\pi} \left(-\ln \frac{r}{a} + 1 \right) r^2. \tag{6.5.50}$$

Substituting this expression into (6.5.16), we derive the velocity field in the standard form (6.5.45), where $\mathcal{G}_{ij}$ is the free-space Green's function,

$$\mathcal{S}_{ij} = -\delta_{ij} \ln \frac{r}{a} + \frac{\hat{x}_i \hat{x}_j}{r^2}, \tag{6.5.51}$$

also called the two-dimensional Stokeslet, $\hat{\mathbf{x}} = \mathbf{x} - \mathbf{x}_0$ and $r = |\hat{\mathbf{x}}|$. The associated vorticity, pressure, and stress fields are given by (6.5.45) with

$$\Omega_{ij}(\hat{\mathbf{x}}) = 2\,\epsilon_{ijl}\,\frac{\hat{x}_l}{r^2}, \qquad \mathcal{P}_j(\hat{\mathbf{x}}) = 2\,\frac{\hat{x}_j}{r^2}, \qquad \mathcal{T}_{ijk}(\hat{\mathbf{x}}) = -4\,\frac{\hat{x}_i\hat{x}_j\hat{x}_k}{r^4}. \qquad (6.5.52)$$

All these fields decay like $1/r$ with respect to the distance from the singular point, $\mathbf{x}_0$. In contrast, the corresponding field for three-dimensional flow exhibit a faster, $1/r^2$ decay.

Stream function

The stream function associated with a point force is found by integrating the velocity,

$$\psi = \frac{1}{4\pi\mu}\left(1 - \ln\frac{r}{a}\right)(\hat{y}\,\mathbf{e}_x - \hat{x}\,\mathbf{e}_y) \cdot \mathbf{b}, \qquad (6.5.53)$$

where $\mathbf{e}_x$ and $\mathbf{e}_y$ are unit vectors along the x and y axes. The streamline pattern due to a point force pointing along the x axis is illustrated in Figure 6.5.2(b). Lengths have been normalized by a.

In Section 6.1.11, we discussed the complex variable formulation of two-dimensional Stokes flow. To obtain the stream function due to a point force with unit strength located at the point z_0 and oriented along the x axis, we set in (6.1.35)

$$c_0 = \frac{1}{4\pi}, \qquad H = y - y_0, \qquad c_1 = 0, \qquad F(z) = \frac{\mathrm{i}}{2\pi}\ln\frac{|z - z_0|}{a}, \qquad (6.5.54)$$

and obtain

$$\psi = \frac{1}{4\pi}\,(y - y_0)\left(-\ln\frac{|z - z_0|}{a} + 1\right). \qquad (6.5.55)$$

The alternative choice

$$c_0 = -\frac{1}{4\pi}, \qquad H = x - x_0, \qquad c_1 = -\frac{1}{2}, \qquad F(z) = -\frac{1}{2\pi}\ln\frac{|z - z_0|}{a}, \qquad (6.5.56)$$

yields the stream function due to a point force of unit strength oriented along the y axis,

$$\psi = \frac{1}{4\pi}\,(x - x_0)\left(-\ln\frac{|z - z_0|}{a} + 1\right). \qquad (6.5.57)$$

These expressions are consistent with the compact form (6.5.53).

Properties of Green functions

Two-dimensional Green's functions have been derived for a variety of boundary geometries (e.g., [306, 313]). All Green's functions exhibit a common singular logarithmic behavior at the location of the point force, identical to that of the free-space Green's function. The Green's functions of infinite flow are required to decay at infinity or increase, at most, at a logarithmic rate. Using the continuity equation and the Stokes equation, we find that the two-dimensional Green's functions satisfy (6.5.5), (6.5.6), and (6.5.7) provided that the factor 8π on the right-hand side of the second and third equations is replaced with 4π. Equations (6.5.9)–(6.5.11) are also valid for two-dimensional Green's functions provided that the surface integral over a closed surface D is replaced by a line integral along a smooth closed contour, C. The coefficient β on the right-hand sides of (6.5.10) and (6.5.11) is equal to 4π, 2π, or 0, depending on whether the point $\mathbf{x}_0$ is located inside, on, or outside a smooth contour C. When $\mathbf{x}_0$ is on C, the line integrals are improper principal-value integrals.

6.5.8 Classification, computation, and further properties of Green's functions

The Green's functions of Stokes flow can be classified in several categories according to the topology of the domain of flow. First, we have the free-space Green's function for infinite unbounded flow. Second, we have the Green's functions for infinite or semi-infinite domains of flow bounded by solid surfaces. Third, we have the Green's functions of interior flow in a completely confined domain. Singly, doubly, and triply periodic Green's functions express the flow due to periodic arrays of point forces. As the observation point, $\mathbf{x}$, approaches the location of the point force, $\mathbf{x}_0$, all Green's functions exhibit a common singular behavior that is identical to that of the free-space Green's function. The Green's functions for infinite unbounded flow are required to decay at infinity at a rate that matches, or is lower than that of, the free-space Green's function.

Interior flow

The Green's functions of interior flow may carry a degree of freedom determining the flow rate or pressure drop in an appropriate direction. For example, in the case of flow inside an infinite tube, the flow due to a periodic array of point forces oriented along the tube may generate either zero axial flow rate or zero pressure drop. If a Green's function for one condition is available, the Green's function for the other condition can be produced readily by adding an appropriate pressure-driven flow, such as a Poiseuille flow.

Library of Green's functions

A detailed discussion and explicit expressions of Green's functions for a variety of boundary geometries can be found in References [306, 313]. Computer codes are available in the boundary-element software library BEMLIB [313] (see also Appendix C).

Problems

6.5.1 *Stokeslet via Fourier transforms*

(a) Invert the three-dimensional Fourier transform to derive the Stokeslet.

(b) Derive the two-dimensional Stokeslet and associated pressure using the two-dimensional complex Fourier transform.

6.5.2 *Torque on a surface enclosing a point force*

Using (6.5.23), show that the torque with respect to the location of a point force exerted on any surface that encloses the point force is zero. What is the torque with respect to another point in space?

6.5.3 *Properties of Green's functions*

(a) Prove identity (6.5.8) for three-dimensional flow.

(b) Show that (6.5.8) is also valid for two-dimensional flow, provided that the factor 8π on the right-hand side is replaced by 4π.

6.5.4 *Two-dimensional semi-infinite flow*

Derive the velocity induced by a two-dimensional point force above a plane wall located at $y = w$.

6.6   Fundamental solutions of Stokes flow

The linearity of the equations of Stokes flow allows us to generate solutions by superposing fundamental solutions that satisfy the governing equations but not necessarily the required boundary conditions. The superposition is designed so that the flow expressed by discrete collections or continuous distributions of properly selected fundamental solutions satisfies the boundary conditions in an exact or approximate sense. The fundamental solutions may become infinite at a singular point or else diverge at infinity. In this section, we develop three classes of fundamental solutions originating from the point source, the point force, and a quadratic flow. In Section 6.7, we will use these fundamental solutions to study flow past particles and liquid drops.

6.6.1   Point source and derivative singularities

In Section 6.1, we saw that any irrotational velocity field satisfies the equations of Stokes flow with an associated constant pressure field set to zero. One example is the flow due to a point source located at the point $\mathbf{x}_0$, given in (2.1.30). Differentiating the point source with respect to the position of the singular point, $\mathbf{x}_0$, we obtain a sequence of derivative vectorial or tensorial singularities expressing irrotational flow. The first three members of this chain are the potential dipole, the potential quadruple, and the potential sextuple. The velocity and stress fields due to a point source with scalar strength m, a potential dipole with vectorial strength $\mathbf{d}$, and a potential quadruple with tensorial strength $\mathbf{q}$ are shown in Table 6.6.1. The associated pressure fields are uniformly zero, and the vorticity fields vanish throughout the domain of flow.

6.6.2   Point force and derivative singularities

A second family of fundamental solutions originates from the point force. Differentiating the Stokeslet with respect to the singular point, $\mathbf{x}_0$, we obtain derivative singularities representing multipoles of the point force. The velocity, pressure, and stress fields due to a point force with vectorial strength $\mathbf{b}$ and a point-force dipole with tensorial strength $\mathbf{p}$ located at the point $\mathbf{x}_0$ in free space are shown in Table 6.6.2. The point-force dipole in free space is also called the Stokeslet dipole.

Couplet or rotlet

The coefficient of the Stokeslet dipole, $\mathbf{p}$, can be resolved into a symmetric component, $\mathbf{s} = \frac{1}{2}(\mathbf{p}+\mathbf{p}^T)$, and an antisymmetric component, $\mathbf{r} = \frac{1}{2}(\mathbf{p} - \mathbf{p}^T)$, where the superscript T indicates the matrix

Point source with strength m

$$u_i = \frac{1}{4\pi} \, m \, \mathcal{M}_i \qquad \sigma_{ik} = \frac{\mu}{4\pi} \, m \, \mathcal{T}_{ik}^{\mathcal{M}}$$

$$\mathcal{M}_i = \frac{\hat{x}_i}{r^3} \qquad \mathcal{T}_{ik}^{\mathcal{M}} = 2\frac{\delta_{ij}}{r^3} - 6\frac{\hat{x}_i\hat{x}_k}{r^5} = -2\mathcal{D}_{ik}$$

Potential dipole with strength d

$$u_i = \frac{1}{4\pi} \, \mathcal{D}_{ij} d_j \qquad \sigma_{ik} = \frac{\mu}{4\pi} \, \mathcal{T}_{ijk}^{\mathcal{D}} \, d_j$$

$$\mathcal{D}_{ij} = \frac{\partial \mathcal{M}_i}{\partial x_{0_j}} = -\frac{\delta_{ij}}{r^3} + 3\frac{\hat{x}_i\hat{x}_j}{r^5}$$

$$\mathcal{T}_{ijk}^{\mathcal{D}} = \frac{\partial \mathcal{T}_{ik}^{\mathcal{M}}}{\partial x_{0_j}} = 6\frac{\delta_{jk}\hat{x}_i + \delta_{ki}\hat{x}_j + \delta_{ij}\hat{x}_k}{r^5} - 30\frac{\hat{x}_i\hat{x}_j\hat{x}_k}{r^7} = -2\mathcal{Q}_{ijk}$$

Potential quadruple with strength q

$$u_i = \frac{1}{4\pi} \, \mathcal{Q}_{ijl} \, q_{jl} \qquad \sigma_{ik} = \frac{\mu}{4\pi} \, \mathcal{T}_{ijlk}^{\mathcal{Q}} \, q_{jl}$$

$$\mathcal{Q}_{ijl} = \frac{\partial \mathcal{D}_{ij}}{\partial x_{0_l}} = -3\frac{\delta_{ij}\hat{x}_l + \delta_{jl}\hat{x}_i + \delta_{li}\hat{x}_j}{r^5} + 15\frac{\hat{x}_i\hat{x}_j\hat{x}_l}{r^7}$$

$$\mathcal{T}_{ijlk}^{\mathcal{Q}} = \frac{\partial \mathcal{T}_{ijk}^{\mathcal{D}}}{\partial x_{0_l}} = -6\frac{\delta_{ij}\delta_{lk} + \delta_{il}\delta_{jk} + \delta_{ik}\delta_{jl}}{r^5} + 30\frac{\delta_{jk}\hat{x}_k + \delta_{ik}\hat{x}_j + \delta_{jk}\hat{x}_i}{r^7}\hat{x}_l$$

$$+30\frac{\delta_{il}\hat{x}_j\hat{x}_k + \delta_{jl}\hat{x}_i\hat{x}_k + \delta_{kl}\hat{x}_i\hat{x}_j}{r^7} - 210\frac{\hat{x}_i\hat{x}_j\hat{x}_l\hat{x}_k}{r^9}$$

TABLE 6.6.1   Velocity and stress fields at a point, $\mathbf{x}$, due to irrotational singularities of three-dimensional Stokes flow located at the point $\mathbf{x}_0$ in free space; $r = |\hat{\mathbf{x}}|$ and $\hat{\mathbf{x}} = \mathbf{x} - \mathbf{x}_0$. The associated hydrodynamic pressure fields are constant.

transpose. The velocity due to the dipole can be expressed as

$$u_i = \frac{1}{8\pi\mu} \left([\mathcal{S}_{ijl}^{\mathcal{D}}]^+ \, s_{jl} + [\mathcal{S}_{ijl}^{\mathcal{D}}]^- \, r_{jl} \right), \tag{6.6.1}$$

where the square brackets $[\ ]^+$ denote the symmetric part and the square brackets $[\ ]^-$ denote the antisymmetric part with respect to the indices j and l. Cursory inspection yields the specific

Point force with strength **b**

$$u_i = \frac{1}{8\pi\mu}\, \mathcal{S}_{ij}\, b_j \qquad p = \frac{1}{8\pi}\, \mathcal{P}_j^{\mathcal{S}}\, b_j \qquad \sigma_{ik} = \frac{1}{8\pi}\, \mathcal{T}_{ijk}^{\mathcal{S}}\, b_j$$

$$\mathcal{S}_{ij} = \frac{\delta_{ij}}{r} + \frac{\hat{x}_i \hat{x}_j}{r^3} \qquad \mathcal{P}_j^{\mathcal{S}} = 2\frac{\hat{x}_j}{r^3} = 2\mathcal{M}_j \qquad \mathcal{T}_{ijk}^{\mathcal{S}} = -6\frac{\hat{x}_i \hat{x}_j \hat{x}_k}{r^5}$$

Point-force dipole with strength **p**

$$u_i = \frac{1}{8\pi\mu}\, \mathcal{S}_{ijl}^{\mathcal{D}}\, p_{jl} \qquad p = \frac{1}{8\pi}\, \mathcal{P}_{jl}^{\mathcal{SD}}\, p_{jl} \qquad \sigma_{ik} = \frac{1}{8\pi}\, \mathcal{T}_{ijlk}^{\mathcal{SD}}\, p_{jl}$$

$$\mathcal{S}_{ijl}^{\mathcal{D}} = \frac{\partial \mathcal{S}_{ij}}{\partial x_{0_l}} = \frac{1}{r^3}\left(\delta_{ij}\hat{x}_l - \delta_{il}\hat{x}_j - \delta_{jl}\hat{x}_i\right) + 3\frac{\hat{x}_i \hat{x}_j \hat{x}_l}{r^5}$$

$$\mathcal{P}_{jl}^{\mathcal{SD}} = \frac{\partial \mathcal{P}_j^{\mathcal{S}}}{\partial x_{0_l}} = 2\frac{\partial}{\partial x_{0_l}}\left(\frac{\hat{x}_j}{r^3}\right) = -2\frac{\delta_{jl}}{r^3} + 6\frac{\hat{x}_j \hat{x}_l}{r^5} = 2\mathcal{D}_{jl}$$

$$\mathcal{T}_{ijlk}^{\mathcal{SD}} = \frac{\partial \mathcal{T}_{ijk}^{\mathcal{S}}}{\partial x_{0_l}} = \frac{6}{r^5}\left(\delta_{il}\hat{x}_j \hat{x}_k + \delta_{jl}\hat{x}_k \hat{x}_i + \delta_{kl}\hat{x}_i \hat{x}_j\right) - 30\frac{\hat{x}_i \hat{x}_j \hat{x}_k \hat{x}_l}{r^7}$$

Couplet or rotlet with strength **c**

$$u_i = \frac{1}{8\pi\mu}\, \mathcal{C}_{im}\, c_m \qquad p = 0 \qquad \sigma_{ik} = \frac{1}{8\pi}\, \mathcal{T}_{imk}^{\mathcal{C}}\, c_m$$

$$\mathcal{C}_{im} = -\frac{1}{2}\,\epsilon_{jlm}\, \mathcal{S}_{ijl}^{\mathcal{D}} = \epsilon_{iml}\frac{\hat{x}_l}{r^3} \qquad \mathcal{T}_{imk}^{\mathcal{C}} = -\frac{1}{2}\,\epsilon_{jlm}\, \mathcal{T}_{ijlk}^{\mathcal{SD}} = 3\frac{\epsilon_{ijm}\hat{x}_k + \epsilon_{kjm}\hat{x}_i}{r^5}\,\hat{x}_j$$

TABLE 6.6.2   Velocity, pressure, and stress fields at the point **x** due to a three-dimensional point force or point-force dipole located at the point $\mathbf{x}_0$ in free space; $r = |\hat{\mathbf{x}}|$, $\hat{\mathbf{x}} = \mathbf{x} - \mathbf{x}_0$, $\mathcal{M}$ is the point source, and $\mathcal{D}$ is the point-source dipole.

expressions

$$[\mathcal{S}_{ijl}^{\mathcal{D}}]^+ = -\delta_{jl}\frac{\hat{x}_i}{r^3} + 3\frac{\hat{x}_i \hat{x}_j \hat{x}_l}{r^5}, \qquad [\mathcal{S}_{ijl}^{\mathcal{D}}]^- = \frac{1}{r^3}\left(\delta_{ij}\hat{x}_l - \delta_{il}\hat{x}_j\right), \tag{6.6.2}$$

so that $[\mathcal{S}_{ijl}^{D}]^+ + [\mathcal{S}_{ijl}^{D}]^- = [\mathcal{S}_{ijl}^{D}]$. Exploiting the antisymmetry of **r**, we write

$$r_{jl} = -\frac{1}{2}\,\epsilon_{jlm}\, c_m, \tag{6.6.3}$$

where the vector **c** is defined as

$$c_m = -\epsilon_{mjl}\, r_{jl} = -\epsilon_{mjl}\, p_{jl}. \tag{6.6.4}$$

Expression (6.6.1) for the velocity due to the point-force dipole takes the form

$$u_i = \frac{1}{8\pi\mu}\left([\mathcal{S}^{\mathcal{D}}_{ijl}]^+ s_{jl} - \frac{1}{2}[\mathcal{S}^{\mathcal{D}}_{ijl}]^- \epsilon_{jlm}\, c_m\right) = \frac{1}{8\pi\mu}\left([\mathcal{S}^{\mathcal{D}}_{ijl}]^+ s_{jl} + \mathcal{C}_{im}\, c_m\right), \tag{6.6.5}$$

where

$$\mathcal{C}_{im} \equiv -\frac{1}{2}\epsilon_{jlm}[\mathcal{S}^{\mathcal{D}}_{ijl}]^- = -\frac{1}{2}\epsilon_{jlm}\, S^{\mathcal{D}}_{ijl} = \epsilon_{iml}\frac{\hat{x}_l}{r^3} \tag{6.6.6}$$

is a new fundamental solution called the couplet or rotlet. The pressure field associated with the couplet is constant, and the stress field is given in the third entry of Table 6.6.2.

Stresslet

Inspecting the symmetric component of the Stokeslet doublet given in (6.6.2), we recognize a point source and a complementary fundamental solution called the stresslet, defined as

$$\Sigma_{ijl} \equiv [\mathcal{S}^{\mathcal{D}}_{ijl}]^+ + \delta_{jl}\mathcal{M}_i = \frac{1}{2}(S^{\mathcal{D}}_{ijl} + S^{\mathcal{D}}_{ilj}) + \delta_{jl}\mathcal{M}_i = 3\frac{\hat{x}_i\hat{x}_j\hat{x}_l}{r^5}, \tag{6.6.7}$$

where $\mathcal{M}_i$ is the point source. The velocity, pressure, and stress field due to a stresslet with strength **s** is given in the first entry of Table 6.6.3.

Point-force quadruple

Differentiating the Stokeslet doublet, we obtain the Stokeslet quadruple shown in the second entry of Table 6.6.3. Contracting the last two indices of the quadruple, we obtain the singularity

$$\mathcal{S}^{\mathcal{SQ}}_{ijll} = -2\left(-\frac{\delta_{ij}}{r^3} + 3\frac{\hat{x}_i\hat{x}_j}{r^5}\right). \tag{6.6.8}$$

The term inside the parentheses on the right-hand side is recognized as the potential dipole,

$$\mathcal{D}_{ij} = -\frac{1}{2}\mathcal{S}^{\mathcal{SQ}}_{ijll} = -\frac{1}{2}\nabla_0^2\mathcal{S}_{ij}, \tag{6.6.9}$$

where $\mathcal{S}$ is the Stokeslet. This expression allows to restate all irrotational singularities, with the exception of the point source, in terms of derivatives of the Laplacian of the Green's function.

6.6.3 Contribution of singularities to the global properties of a flow

Inspecting the functional form of the fundamental solutions discussed in this section, we find that the flow rate Q through any closed surface that encloses a singularity is zero, except for the point source where $Q = m$, and the stresslet where $Q = \frac{1}{2}s_{ii}$. The force exerted on any surface enclosing a singularity is zero, except for the Stokeslet where $\mathbf{F} = -\mathbf{b}$. The torque with respect to a point

Stresslet with strength s

$$u_i = \frac{1}{8\pi\mu} \Sigma_{ijl}\, s_{jl} \qquad p = \frac{1}{8\pi} \mathcal{P}^\Sigma_{jl}\, s_{jl} \qquad \sigma_{ik} = \frac{1}{8\pi} \mathcal{T}^\Sigma_{ijlk}\, s_{jl}$$

$$\Sigma_{ijl} = 3\frac{\hat{x}_i \hat{x}_j \hat{x}_l}{r^5} \qquad \mathcal{P}^\Sigma_{jl} = \mathcal{P}^{S\mathcal{D}}_{jl} = \frac{\partial \mathcal{P}^S_j}{\partial x_{0_l}} = 2\frac{\partial}{\partial x_{0_l}}\left(\frac{\hat{x}_j}{r^3}\right) = -2\frac{\delta_{jl}}{r^3} + 6\frac{\hat{x}_j \hat{x}_l}{r^5} = 2\mathcal{D}_{jl}$$

$$\mathcal{T}^\Sigma_{ijlk} = -\delta_{ik}\mathcal{P}^\Sigma_{jl} + \frac{\partial \Sigma_{ijl}}{\partial x_k} + \frac{\partial \Sigma_{kjl}}{\partial x_i} = \frac{1}{2}\left(\mathcal{T}^{S\mathcal{D}}_{ijlk} + \mathcal{T}^{S\mathcal{D}}_{iljk}\right) + \delta_{jl}\mathcal{T}^M_{ik}$$

$$= \delta_{ik}\delta_{jl}\frac{2}{r^3} + \frac{3}{r^5}\left(\delta_{ij}\hat{x}_k\hat{x}_l + \delta_{il}\hat{x}_j\hat{x}_k + \delta_{kj}\hat{x}_i\hat{x}_l + \delta_{kl}\hat{x}_i\hat{x}_j\right) - 30\frac{\hat{x}_i\hat{x}_j\hat{x}_k\hat{x}_l}{r^7}$$

Point-force quadruple with strength t

$$u_i = \frac{1}{8\pi\mu} \mathcal{S}^\mathcal{Q}_{ijlm}\, t_{jlm} \qquad p = \frac{1}{8\pi} \mathcal{P}^{S\mathcal{Q}}_{jlm}\, t_{jlm} \qquad \sigma_{ik} = \frac{1}{8\pi} \mathcal{T}^{S\mathcal{Q}}_{ijlmk}\, t_{jlm}$$

$$\mathcal{S}^\mathcal{Q}_{ijlm} = \frac{\partial \mathcal{S}^\mathcal{D}_{ijl}}{\partial x_{0_m}} = \frac{\partial^2 \mathcal{S}_{ij}}{\partial x_{0_l}\partial x_{0_m}} = \frac{1}{r^3}\left(-\delta_{ij}\delta_{lm} + \delta_{il}\delta_{jm} + \delta_{jl}\delta_{im}\right)$$

$$-\frac{3}{r^5}\left[\left(-\delta_{ij}\hat{x}_l + \delta_{il}\hat{x}_j + \delta_{jl}\hat{x}_i\right)\hat{x}_m + \delta_{im}\hat{x}_j\hat{x}_l + \delta_{jm}\hat{x}_i\hat{x}_l + \delta_{lm}\hat{x}_i\hat{x}_j\right] + 15\frac{\hat{x}_i\hat{x}_j\hat{x}_l\hat{x}_m}{r^7}$$

$$\mathcal{P}^{S\mathcal{Q}}_{jlm} = \frac{\partial \mathcal{P}^{S\mathcal{D}}_{jl}}{\partial x_{0_m}} = \frac{\partial^2 \mathcal{P}^S_j}{\partial x_{0_l}\partial x_{0_m}} = -\frac{6}{r^5}\left(\delta_{jl}\hat{x}_m + \delta_{jm}\hat{x}_l + \delta_{lm}\hat{x}_j\right) + 30\frac{\hat{x}_j\hat{x}_l\hat{x}_m}{r^7}$$

$$\mathcal{T}^{S\mathcal{Q}}_{ijlmk} = \frac{\partial \mathcal{T}^{S\mathcal{D}}_{ijlk}}{\partial x_{0_m}} = \frac{\partial^2 \mathcal{T}^S_{ijk}}{\partial x_{0_l}\partial x_{0_m}}$$

TABLE 6.6.3 Velocity, pressure, and stress fields at the point **x** due to a three-dimensional rotlet, stresslet, or point-force quadruple located at the point **x**$_0$ in free space; $r = |\hat{\mathbf{x}}|$ and $\hat{\mathbf{x}} = \mathbf{x} - \mathbf{x}_0$, $\mathcal{D}$ is the point-source dipole, and $\mathcal{T}^M$ is the stress tensor due to a point source.

x$_1$ exerted on a spherical surface that encloses a singularity is zero, except for the point force, the point force dipole, and the couplet where

$$\mathbf{T} = (\mathbf{x}_1 - \mathbf{x}_0) \times \mathbf{b}, \qquad T_i = \epsilon_{ijl}\, p_{jl}, \qquad \mathbf{T} = -\mathbf{L}. \tag{6.6.10}$$

These expressions will find applications in Section 6.7 where we discuss exact and approximate singularity representations for several flows.

Stokeson with strength w

$$u_i = \mathcal{N}_{ij}\, w_j \qquad p = \mu\, \mathcal{P}_j^{\mathcal{N}}\, w_j \qquad \sigma_{ik} = \mu\, \mathcal{T}_{ijk}^{\mathcal{N}}\, w_j$$

$$\mathcal{N}_{ij} = 2r^2 \delta_{ij} - \hat{x}_i \hat{x}_j \qquad \mathcal{P}_j^{\mathcal{N}} = 10\,\hat{x}_j \qquad \mathcal{T}_{ijk}^{\mathcal{N}} = 3\left(-4\,\delta_{ik}\hat{x}_j + \delta_{ij}\hat{x}_k + \delta_{kj}\hat{x}_i\right)$$

Stokeson dipole with strength h

$$u_i = \mathcal{S}_{ijl}^{\mathcal{ND}}\, h_{jl} \qquad p = \mu\, \mathcal{P}_{jl}^{\mathcal{ND}}\, h_{jl} \qquad \sigma_{ik} = \mu\, \mathcal{T}_{ijlk}^{\mathcal{ND}}\, h_{jl}$$

$$\mathcal{S}_{ijl}^{\mathcal{ND}} = \frac{\partial \mathcal{N}_{ij}}{\partial x_{0_l}} = -\delta_{ij}4\hat{x}_l + \delta_{il}\hat{x}_j + \delta_{jl}\hat{x}_i \qquad \mathcal{P}_{jl}^{\mathcal{ND}} = \frac{\partial \mathcal{P}_j^{\mathcal{N}}}{\partial x_{0_l}} = -10\,\delta_{jl}$$

$$\mathcal{T}_{ijlk}^{\mathcal{ND}} = \frac{\partial \mathcal{T}_{ijk}^{\mathcal{N}}}{\partial x_{0_l}} = 3\left(-4\delta_{ik}\delta_{jl} + \delta_{ij}\delta_{kl} + \delta_{kj}\delta_{il}\right)$$

TABLE 6.6.4   Velocity, pressure, and stress field at a point $\mathbf{x}$ due to three-dimensional interior rotational singularities of Stokes flow located at the point $\mathbf{x}_0$ in free space; $r = |\hat{\mathbf{x}}|$ and $\hat{\mathbf{x}} = \mathbf{x} - \mathbf{x}_0$.

6.6.4   Interior flow

Families of fundamental solutions that have no singular points but diverge at infinity can be constructed for interior flow [87]. To derive these singularities, we recall that the pressure is a harmonic function and set

$$p = 10\mu\,(\mathbf{x} - \mathbf{x}_0) \cdot \mathbf{w}, \tag{6.6.11}$$

where $\mathbf{w}$ is a constant vector and the factor of ten has been introduced for future convenience. Substituting this expression into the Stokes equation, we obtain a Poisson equation for the velocity, $\nabla^2 \mathbf{u} = 10\,\mathbf{w}$, to be solved subject to the constraint $\nabla \cdot \mathbf{u} = 0$ imposed by the continuity equation. The solution is

$$\mathbf{u} = \mathcal{N} \cdot \mathbf{w}, \tag{6.6.12}$$

where $\mathcal{N}$ is a new singularity called the Stokeson, shown in the first entry of Table 6.6.4.

The derivatives of the Stokeson with respect to the singular point, $\mathbf{x}_0$, are legitimate fundamental solutions of interior Stokes flow. Differentiating the Stokeson once, we obtain the Stokeson dipole shown in the second entry of Table 6.6.4. The symmetric part of the Stokeson dipole is the stresson, and the antisymmetric part is the roton. The stresson represents linear, purely straining flow with vanishing vorticity and pressure, and the roton represents rigid-body rotation (Problem 6.6.3).

6.6.5   Flow bounded by solid surfaces

The apparatus of fundamental solutions can be extended in a straightforward fashion to flows that are bounded by solid or other surfaces [306]. The Green's functions discussed in Section 6.5, representing the flow due to a point force, provide us with one class of fundamental solutions. Other families can be constructed by differentiating the point force with respect to the singular point. For example, the point-source dipole arises from the Laplacian of the point force.

Point source

The velocity due to a point source located at a point $\mathbf{x}_0$ in a totally or partially infinite domain of flow arises from the pressure vector due to the point force at the same point in the same domain, denoted by $\mathcal{P}(\mathbf{x}, \mathbf{x}_0)$, as

$$\mathbf{u}(\mathbf{x}) = \frac{1}{4\pi}\, m\, \mathcal{M}(\mathbf{x}, \mathbf{x}_0), \tag{6.6.13}$$

where m is the strength of the point source and

$$\mathcal{M}(\mathbf{x}, \mathbf{x}_0) = -\frac{1}{2}\, \mathcal{P}(\mathbf{x}_0, \mathbf{x}). \tag{6.6.14}$$

Note that the arguments $\mathbf{x}$ and $\mathbf{x}_0$ are switched on the left- and right-hand sides of (6.6.14). For example, in the case of the free-space Green's function,

$$\mathcal{P}(\mathbf{x}_0, \mathbf{x}) = \frac{2}{r^3}\, (\mathbf{x}_0 - \mathbf{x}), \qquad \mathcal{M}(\mathbf{x}, \mathbf{x}_0) = \frac{1}{r^3}\, (\mathbf{x} - \mathbf{x}_0), \tag{6.6.15}$$

where $r = |\mathbf{x} - \mathbf{x}_0|$.

Point source above an infinite wall

In the case of flow in a semi-infinite domain bounded by a plane wall located at $x = w$, we use expression (6.5.42) for the pressure vector and obtain

$$\mathcal{P}(\mathbf{x}_0, \mathbf{x}) = \mathcal{P}^S(\mathbf{x}_0 - \mathbf{x}) - \mathcal{P}^S(\mathbf{x}_0 - \mathbf{x}^{im}) + 2h\, \mathcal{P}^{\Psi}(\mathbf{x}_0 - \mathbf{x}^{im}), \tag{6.6.16}$$

where $h = x - w$ and

$$\mathbf{x}^{im} = (2w - x, y, z) \tag{6.6.17}$$

is the image of the point $\mathbf{x}$ with respect to the wall. In index notation,

$$\mathcal{P}_i(\mathbf{x}_0, \mathbf{x}) = 2\, \mathcal{M}_i^{FS}(\mathbf{x}_0 - \mathbf{x}) - 2\, \mathcal{M}_i^{FS}(\mathbf{x}_0 - \mathbf{x}^{im}) \pm 4h\, \mathcal{D}_{i1}^{FS}(\mathbf{x}_0 - \mathbf{x}^{im}), \tag{6.6.18}$$

where $\mathcal{M}^{FS}$ is the point source in free space $\mathcal{D}^{FS}$ is the point-source dipole in free space. Using (6.6.14) and substituting the specific expressions for the singularities, we obtain

$$\mathcal{M}(\mathbf{x}, \mathbf{x}_0) = \frac{\hat{\mathbf{x}}}{r^3} + \frac{1}{r_{im}^3}\begin{bmatrix} x_0 - 2w + x \\ y_0 - y \\ z_0 - z \end{bmatrix} + 2\frac{x - w}{r_{im}^5}\begin{bmatrix} -r_{im}^2 + 3\,(x_0 - x^{im})^2 \\ -3\,(x_0 - x^{im})(y_0 - y) \\ -3\,(x_0 - x^{im})(z_0 - z) \end{bmatrix}, \tag{6.6.19}$$

where

$$\hat{\mathbf{x}} = \mathbf{x} - \mathbf{x}_0, \qquad r = |\mathbf{x} - \mathbf{x}_0|, \qquad r_{im} = |\mathbf{x}_0 - \mathbf{x}^{im}|. \qquad (6.6.20)$$

Next, we introduce the image of the point $\mathbf{x}_0$ with respect to the wall,

$$\mathbf{x}_0^{im} = (2w - x_0, y_0, z_0), \qquad (6.6.21)$$

define the distance $\hat{\mathbf{x}}_{im} = \mathbf{x} - \mathbf{x}_0^{im}$, and write

$$x - w = (x + x_0 - 2w) - (x_0 - w) = \hat{x}_{im} - h_0, \qquad (6.6.22)$$

where $h_0 = x_0 - w$, we derive the expression

$$\mathcal{M}(\mathbf{x}, \mathbf{x}_0) = \mathcal{M}^{FS}(\hat{\mathbf{x}}) + \frac{1}{r_{im}^3} \begin{bmatrix} \hat{x}_{im} \\ -\hat{y}_{im} \\ -\hat{z}_{im} \end{bmatrix} + 2\frac{\hat{x}_{im} - h_0}{r_{im}^5} \begin{bmatrix} -r_{im}^2 + 3\,\hat{x}_{im}^2 \\ 3\,\hat{x}_{im}\hat{y}_{im} \\ 3\,\hat{x}_{im}\hat{z}_{im} \end{bmatrix}, \qquad (6.6.23)$$

which can be rearranged into the expression

$$\mathcal{M}(\mathbf{x}, \mathbf{x}_0) = \mathcal{M}^{FS}(\hat{\mathbf{x}}) + \mathcal{M}^{FS}(\hat{\mathbf{x}}_{im}) - \frac{2}{r_{im}^3} \begin{bmatrix} 0 \\ \hat{y}^{im} \\ \hat{z}^{im} \end{bmatrix} + 2\frac{\hat{x}_{im}}{r_{im}^5} \begin{bmatrix} -r_{im}^2 + 3\,\hat{x}_{im}^2 \\ 3\,\hat{x}_{im}\hat{y}_{im} \\ 3\,\hat{x}_{im}\hat{z}_{im} \end{bmatrix} + 2h_0\,\mathcal{D}_{i1}. \;(6.6.24)$$

Combining the third with the fourth term on the right-hand side, we derive the final expression

$$\mathcal{M}(\mathbf{x}, \mathbf{x}_0) = \mathcal{M}^{FS}(\hat{\mathbf{x}}) + \mathcal{M}^{FS}(\hat{\mathbf{x}}_{im}) + 2\left(-\frac{1}{r_{im}^3} + 3\frac{\hat{x}_{im}^2}{r_{im}^5}\right) \begin{bmatrix} \hat{x}_{im} \\ \hat{y}_{im} \\ \hat{z}_{im} \end{bmatrix} + 2h_0\,\mathcal{D}_{i1}. \qquad (6.6.25)$$

The first two terms on the right-hand side represent point sources located, respectively, at $\mathbf{x}_0$ and $\mathbf{x}_0^{im}$. The third term is a Stokeslet doublet, $\mathcal{S}_{ixx}^D$, and the fourth term is a potential dipole placed at $\mathbf{x}_0^{im}$.

6.6.6 Two-dimensional flow

Fundamental solutions for two-dimensional flow are derived working as previously in this section for three-dimensional flow. Examples of two-dimensional fundamental solutions are the Green's functions discussed in Section 6.5, and the point source and its derivatives discussed in Section 2.1.6. Families of irrotational and rotational singularities are presented in Table 6.6.5–6.6.7. Some of these singularities will be employed in Section 6.14 to compute streaming flow past a circular cylinder.

Problems

6.6.1 *Vorticity due to singularities*

Derive the vorticity fields due to the singularities listed in Table 6.6.2.

Two-dimensional point source with strength m

$$u_i = \frac{1}{2\pi} m\, \mathcal{M}_i \qquad \sigma_{ik} = \frac{\mu}{2\pi} m\, \mathcal{T}_{ik}^{\mathcal{M}}$$

$$\mathcal{M}_i = \frac{\hat{x}_i}{r^2} \qquad \mathcal{T}_{ik}^{\mathcal{M}} = 2\frac{\delta_{ij}}{r^2} - 4\frac{\hat{x}_i \hat{x}_k}{r^4} = -2\mathcal{D}_{ik}$$

Two-dimensional potential dipole with strength $\mathbf{d}$

$$u_i = \frac{1}{2\pi} \mathcal{D}_{ij}\, d_j \qquad \sigma_{ik} = \frac{\mu}{2\pi} \mathcal{T}_{ijk}^{\mathcal{D}}\, d_j$$

$$\mathcal{D}_{ij} = \frac{\partial \mathcal{M}_i}{\partial x_{0_j}} = -\frac{\delta_{ij}}{r^2} + 2\frac{\hat{x}_i \hat{x}_j}{r^4} \qquad \mathcal{T}_{ijk}^{\mathcal{D}} = \frac{\partial \mathcal{T}_{ik}^{\mathcal{M}}}{\partial x_{0_j}} = 4\frac{\delta_{jk}\hat{x}_i + \delta_{ki}\hat{x}_j + \delta_{ij}\hat{x}_k}{r^4} - 16\frac{\hat{x}_i \hat{x}_j \hat{x}_k}{r^6} = -2\mathcal{Q}_{ijk}$$

Two-dimensional potential quadruple with strength $\mathbf{q}$

$$u_i = \frac{1}{2\pi} \mathcal{Q}_{ijl}\, q_{jl} \qquad \sigma_{ik} = \frac{\mu}{2\pi} \mathcal{T}_{ijlk}^{\mathcal{Q}}\, q_{jl}$$

$$\mathcal{Q}_{ijl} = \frac{\partial \mathcal{D}_{ij}}{\partial x_{0_l}} = -2\frac{\delta_{ij}\hat{x}_l + \delta_{jl}\hat{x}_i + \delta_{li}\hat{x}_j}{r^4} + 8\frac{\hat{x}_i \hat{x}_j \hat{x}_l}{r^6}$$

$$\mathcal{T}_{ijlk}^{\mathcal{Q}} = \frac{\partial \mathcal{T}_{ijk}^{\mathcal{D}}}{\partial x_{0_l}} = -4\frac{\delta_{ij}\delta_{lk} + \delta_{il}\delta_{jk} + \delta_{ik}\delta_{jl}}{r^4} + 16\frac{\delta_{jk}\hat{x}_k + \delta_{ik}\hat{x}_j + \delta_{jk}\hat{x}_i}{r^6}\hat{x}_l$$

$$+16\frac{\delta_{il}\hat{x}_j \hat{x}_k + \delta_{jl}\hat{x}_i \hat{x}_k + \delta_{kl}\hat{x}_i \hat{x}_j}{r^6} - 96\frac{\hat{x}_i \hat{x}_j \hat{x}_l \hat{x}_k}{r^8}$$

TABLE 6.6.5 Velocity and stress field at the point $\mathbf{x}$ due to irrotational singularities of two-dimensional Stokes flow located at the point $\mathbf{x}_0$ in free space; $r = |\hat{\mathbf{x}}|$ and $\hat{\mathbf{x}} = \mathbf{x} - \mathbf{x}_0$. The associated hydrodynamic pressure fields are constant. Corresponding singularities of three-dimensional flow are shown in Table 6.6.1.

6.6.2 *Two-dimensional point-force quadruple*

Derive the expressions for the two-dimensional point-force quadruple shown in Table 6.6.7.

6.6.3 *Stresson and roton*

(*a*) Show that the stresson represents purely straining linear flow with vanishing vorticity and pressure, while the roton represents rigid-body rotation.

(*b*) Derive the two-dimensional stresson and roton.

Two-dimensional point force with strength **b**

$$u_i = \frac{1}{4\pi\mu}\, \mathcal{S}_{ij}\, b_j \qquad p = \frac{1}{4\pi}\, \mathcal{P}_j^{\mathcal{S}}\, b_j \qquad \sigma_{ik} = \frac{1}{4\pi}\, \mathcal{T}_{ijk}^{\mathcal{S}}\, b_j$$

$$\mathcal{S}_{ij} = -\delta_{ij}\ln\frac{r}{a} + \frac{\hat{x}_i \hat{x}_j}{r^2} \qquad \mathcal{P}_j^{\mathcal{S}} = 2\frac{\hat{x}_j}{r^2} = 2\mathcal{M}_j \qquad \mathcal{T}_{ijk}^{\mathcal{S}} = -4\frac{\hat{x}_i \hat{x}_j \hat{x}_k}{r^4}$$

Two-dimensional point-force dipole with strength **p**

$$u_i = \frac{1}{4\pi\mu}\, \mathcal{S}_{ijl}^{\mathcal{D}}\, p_{jl} \qquad p = \frac{1}{4\pi}\, \mathcal{P}_{jl}^{\mathcal{SD}}\, p_{jl} \qquad \sigma_{ik} = \frac{1}{4\pi}\, \mathcal{T}_{ijlk}^{\mathcal{SD}}\, p_{jl}$$

$$\mathcal{S}_{ijl}^{\mathcal{D}} = \frac{\partial \mathcal{S}_{ij}}{\partial x_{0_l}} = \frac{1}{r^2}\left(\delta_{ij}\hat{x}_l - \delta_{il}\hat{x}_j - \delta_{jl}\hat{x}_i\right) + 2\frac{\hat{x}_i \hat{x}_j \hat{x}_l}{r^5}$$

$$\mathcal{P}_{jl}^{\mathcal{SD}} = \frac{\partial \mathcal{P}_j^{\mathcal{S}}}{\partial x_{0_l}} = 2\frac{\partial}{\partial x_{0_l}}\left(\frac{\hat{x}_j}{r^2}\right) = -2\frac{\delta_{jl}}{r^2} + 4\frac{\hat{x}_j \hat{x}_l}{r^4} = 2\mathcal{D}_{ij}$$

$$\mathcal{T}_{ijlk}^{\mathcal{SD}} = \frac{\partial \mathcal{T}_{ijk}^{\mathcal{S}}}{\partial x_{0_l}} = \frac{4}{r^4}\left(\delta_{il}\hat{x}_j \hat{x}_k + \delta_{jl}\hat{x}_k \hat{x}_i + \delta_{kl}\hat{x}_i \hat{x}_j\right) - 16\frac{\hat{x}_i \hat{x}_j \hat{x}_k \hat{x}_l}{r^6}$$

Two-dimensional rotlet with strength **c**

$$u_i = \frac{1}{4\pi\mu}\, \mathcal{C}_{im}\, c_m \qquad p = 0 \qquad \sigma_{ik} = \frac{1}{4\pi}\, \mathcal{T}_{imk}^{\mathcal{C}}\, c_m$$

$$\mathcal{C}_{im} = \epsilon_{iml}\frac{\hat{x}_l}{r^2} \qquad \mathcal{T}_{imk}^{\mathcal{C}} = -\frac{1}{2}\,\epsilon_{jml}\,\mathcal{T}_{ijlk}^{\mathcal{SD}} = 2\frac{\epsilon_{ijm}\hat{x}_k + \epsilon_{kjm}\hat{x}_i}{r^4}\,\hat{x}_j$$

TABLE 6.6.6   Velocity, pressure, and stress field at the point **x** due to two-dimensional rotational singularities of Stokes flow located at the point $\mathbf{x}_0$ in free space; $r = |\hat{\mathbf{x}}|$ and $\hat{\mathbf{x}} = \mathbf{x} - \mathbf{x}_0$, a is a chosen length, $\mathcal{M}$ is the two-dimensional point source, and $\mathcal{D}$ is the two-dimensional point-source dipole.

6.6.4 *Point source above a wall*

Draw the streamline pattern of the axisymmetric flow due to a three-dimensional point source above an infinite plane wall.

6.6.5 *Two-dimensional point source above a wall*

Derive the velocity field due to a two-dimensional point source above an infinite wall located at $y = w$ [306].

Two-dimensional stresslet with strength **s**

$$u_i = \frac{1}{4\pi\mu} \Sigma_{ijl}\, s_{jl} \qquad p = \frac{1}{4\pi} \mathcal{P}^{\Sigma}_{jl}\, s_{jl} \qquad \sigma_{ik} = \frac{1}{4\pi} \mathcal{T}^{\Sigma}_{ijlk}\, s_{jl}$$

$$\Sigma_{ijl} = 2\,\frac{\hat{x}_i \hat{x}_j \hat{x}_l}{r^5} \qquad \mathcal{P}^{\Sigma}_{jl} = \mathcal{P}^{SD}_{jl} = \frac{\partial \mathcal{P}^{S}_j}{\partial x_{0_l}} = 2\frac{\partial}{\partial x_{0_l}}\left(\frac{\hat{x}_j}{r^2}\right) = -2\frac{\delta_{jl}}{r^2} + 4\frac{\hat{x}_j \hat{x}_l}{r^4} = 2\mathcal{D}_{jl}$$

$$T^{\Sigma}_{ijlk} = -\delta_{ik}\,\mathcal{P}^{\Sigma}_{jl} + \frac{\partial \Sigma_{ijl}}{\partial x_k} + \frac{\partial \Sigma_{kjl}}{\partial x_i} = \frac{1}{2}\left(T^{SD}_{ijlk} + T^{SD}_{ilkk}\right) + \delta_{jl}\,\mathcal{T}^{\mathcal{M}}_{ik}$$

$$= \delta_{ik}\delta_{jl}\,\frac{1}{r^2} + \frac{2}{r^4}\left(\delta_{ij}\hat{x}_k\hat{x}_l + \delta_{il}\hat{x}_j\hat{x}_k + \delta_{kj}\hat{x}_i\hat{x}_l + \delta_{kl}\hat{x}_i\hat{x}_j\right) - 16\,\frac{\hat{x}_i\hat{x}_j\hat{x}_k\hat{x}_l}{r^6}$$

Two-dimensional point-force quadruple with strength **t**

$$u_i = \frac{1}{4\pi\mu}\mathcal{S}^{Q}_{ijlm}\, t_{jlm} \qquad p = \frac{1}{4\pi}\mathcal{P}^{SQ}_{jlm}\, t_{jlm} \qquad \sigma_{ik} = \frac{1}{4\pi}\mathcal{T}^{SQ}_{ijlmk}\, t_{jlm}$$

$$\mathcal{S}^{Q}_{ijlm} = \frac{\partial \mathcal{S}^{D}_{ijl}}{\partial x_{0_m}} = \frac{\partial^2 \mathcal{S}_{ij}}{\partial x_{0_l}\partial x_{0_m}} = \frac{1}{r^2}\left(-\delta_{ij}\delta_{lm} + \delta_{il}\delta_{jm} + \delta_{jl}\delta_{im}\right)$$

$$-\frac{2}{r^4}\left[(-\delta_{ij}\hat{x}_l + \delta_{il}\hat{x}_j + \delta_{jl}\hat{x}_i)\hat{x}_m + \delta_{im}\hat{x}_j\hat{x}_l + \delta_{jm}\hat{x}_i\hat{x}_l + \delta_{lm}\hat{x}_i\hat{x}_j\right] + 8\,\frac{\hat{x}_i\hat{x}_j\hat{x}_l\hat{x}_m}{r^6}$$

$$\mathcal{P}^{SQ}_{jlm} = \frac{\partial \mathcal{P}^{SD}_{jl}}{\partial x_{0_m}} = \frac{\partial^2 \mathcal{P}^{S}_j}{\partial x_{0_l}\partial x_{0_m}} = -\frac{2}{r^4}\left(\delta_{jl}\hat{x}_m + \delta_{jm}\hat{x}_l + \delta_{lm}\hat{x}_j\right) + 8\,\frac{\hat{x}_j\hat{x}_l\hat{x}_m}{r^6}$$

$$\mathcal{T}^{SQ}_{ijlmk} = \frac{\partial \mathcal{T}^{SD}_{ijlk}}{\partial x_{0_m}} = \frac{\partial^2 \mathcal{T}^{S}_{ijk}}{\partial x_{0_l}\partial x_{0_m}}$$

TABLE 6.6.7 Velocity, pressure, and stress field at the point **x** due to two-dimensional rotational singularities of Stokes flow located at the point $\mathbf{x}_0$ in free space; $r = |\hat{\mathbf{x}}|$ and $\hat{\mathbf{x}} = \mathbf{x} - \mathbf{x}_0$, $\mathcal{D}$ is the point-source dipole, and $\mathcal{T}^{\mathcal{M}}$ is the stress tensor due to a point source.

6.7 Stokes flow past or due to the motion of particles and liquid drops

The fundamental solutions derived in Section 6.6 can be used as building blocks to produce the flow past or due to the motion of solid particles and liquid drops. In Section 6.12, we will see that the singularity representations allow us to develop generalized Faxén laws providing us with the force, the torque, and higher moments of the traction exerted on a particle immersed in an arbitrary flow in terms of the velocity and derivatives of the velocity of the ambient flow at selected locations. Exact singularity representations will be developed in this section for flow due to the translation or rotation of a sphere, arbitrary linear flow past a stationary sphere, flow due to the translation or rotation of a prolate spheroid, flow due to the translation of a spherical drop, and linear flow past a stationary sphere with the slip boundary condition over the surface of the sphere. Approximate singularity representations can be constructed by numerical methods.

6.7.1 Translating sphere

In the first case study, we consider the flow generated by a solid sphere of radius a translating with velocity $\mathbf{V}$ in an otherwise quiescent fluid of infinite expanse. The functional forms of the singularities listed in Tables 6.6.1 and 6.6.2 suggest representing the flow in terms of a Stokeslet with strength $\mathbf{b}$ and a potential dipole with strength $\mathbf{d}$, both situated at the center of the sphere, $\mathbf{x}_0$. The induced velocity field is given by

$$u_i(\mathbf{x}) = \frac{1}{8\pi\mu}\, \mathcal{S}_{ij}(\mathbf{x}, \mathbf{x}_0)\, b_j + \frac{1}{4\pi}\, \mathcal{D}_{ij}(\mathbf{x}, \mathbf{x}_0)\, d_j. \tag{6.7.1}$$

Introducing the explicit forms of the singularities, we obtain

$$u_i(\mathbf{x}) = \frac{1}{8\pi\mu}\left(\frac{\delta_{ij}}{r} + \frac{\hat{x}_i \hat{x}_j}{r^3}\right) b_j + \frac{1}{4\pi}\left(-\frac{\delta_{ij}}{r^3} + 3\frac{\hat{x}_i \hat{x}_j}{r^5}\right) d_j. \tag{6.7.2}$$

Enforcing the no-slip and no-penetration boundary conditions, $\mathbf{u} = \mathbf{V}$ at $r = a$, we obtain two algebraic equations for the coefficients of the singularities,

$$a^2 \mathbf{b} - 2\mu \mathbf{d} = 8\pi\mu a^3\, \mathbf{V}, \qquad a^2 \mathbf{b} + 6\mu \mathbf{d} = \mathbf{0}. \tag{6.7.3}$$

The solution is

$$\mathbf{b} = 6\pi\mu a \mathbf{V}, \qquad \mathbf{d} = -\pi a^3 \mathbf{V}. \tag{6.7.4}$$

Substituting these expressions into (6.7.2) and grouping similar terms, we derive an explicit representation of the velocity field,

$$u_i(\mathbf{x}) = \frac{1}{4}\frac{a}{r}\left(3 + \frac{a^2}{r^2}\right) V_i + \frac{3}{4}\frac{a}{r}\left(1 - \frac{a^2}{r^2}\right)\frac{\hat{x}_i \hat{x}_j}{r^2} V_j. \tag{6.7.5}$$

Because of the point force singularity, the flow generated by the sphere decays like $1/r$ far from the sphere.

Stream function

In spherical polar coordinates centered at the sphere with the x axis pointing in the direction of translation indicated by the unit vector $\mathbf{e}_x$, the Stokes stream function is

$$\psi(r, \theta) = \frac{1}{4}\, ar\left(3 - \frac{a^2}{r^2}\right)\sin^2\theta\, \mathbf{V} \cdot \mathbf{e}_x. \tag{6.7.6}$$

Since the stream function diverges at infinity, an unphysical infinite amount of fluid is convected along the x due to the motion of the sphere. The streamline pattern in a stationary frame of reference and in a frame of reference moving with the sphere are depicted in Figure 6.7.1.

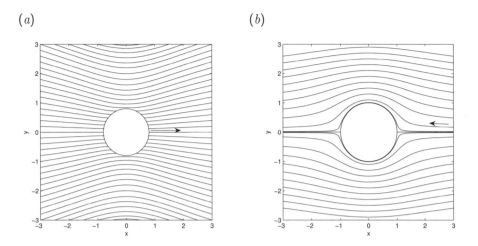

FIGURE 6.7.1   Streamline pattern of the flow due to a sphere translating along the x axis under conditions of Stokes flow in (a) a stationary frame of reference and (b) a frame of reference moving with the sphere.

Surface traction

The hydrodynamic traction exerted on the sphere is readily computed from the strength of the singularities using Tables 6.6.1 and 6.6.2, and is found to be

$$f_i = \frac{1}{8\pi} T^{\mathcal{S}}_{ijk} b_j n_k + \mu \frac{1}{4\pi} T^{\mathcal{D}}_{ijk} d_j n_k. \tag{6.7.7}$$

Substituting the expressions for the singularities, we find that

$$f_i = -\frac{3}{4\pi} \frac{\hat{x}_i \hat{x}_j \hat{x}_k}{a^5} b_j n_k + \mu \frac{3}{2\pi} \left(\frac{\delta_{jk} \hat{x}_i + \delta_{ki} \hat{x}_j + \delta_{ij} \hat{x}_k}{a^5} - 5 \frac{\hat{x}_i \hat{x}_j \hat{x}_k}{a^7} \right) d_j n_k. \tag{6.7.8}$$

Setting $n_k = \hat{x}_k/a$, observing that $\hat{x}_k \hat{x}_k = a^2$, and simplifying, we obtain

$$f_i = -\frac{3}{4\pi} \frac{\hat{x}_i \hat{x}_j}{a^4} b_j + \mu \frac{3}{2\pi} \left(\frac{2\hat{x}_i \hat{x}_j + \delta_{ij} a^2}{a^6} - 5 \frac{\hat{x}_i \hat{x}_j}{a^6} \right) d_j. \tag{6.7.9}$$

Substituting expressions (6.7.4) and simplifying, we find that

$$f_i = -\frac{9}{2} \frac{\hat{x}_i \hat{x}_j}{a^3} V_j - \mu \frac{3}{2} \left(\frac{-3\,\hat{x}_i \hat{x}_j + \delta_{ij} a^2}{a^3} \right) V_j, \tag{6.7.10}$$

yielding the remarkably simple expression

$$f_i = -\frac{3}{2} \frac{\mu}{a} V_i. \tag{6.7.11}$$

We have found that the traction is a constant vector oriented in the direction of translation. This is a unique, remarkable yet fortuitous feature of the spherical geometry.

Stokes law

To compute the hydrodynamic force exerted on the sphere, we can either integrate the traction over the surface of the sphere or use the properties of the singularities discussed in Section 6.6. The result is the Stokes law,

$$\mathbf{F} = \iint_{Sphere} \mathbf{f} \, dS = -6\pi\mu a \, \mathbf{V}. \tag{6.7.12}$$

The torque exerted on the sphere with respect to the center of the sphere is zero. The torque with respect to any other point is nonzero.

Settling or rising velocity

As an application, we compute the velocity of a sphere that is settling or rising under the action of gravity in an infinite fluid. Requiring that the hydrodynamic drag force exerted on the sphere given in (6.7.12) cancels the buoyancy force and the weight of the sphere, we obtain

$$-6\pi\mu a \, \mathbf{V} + \frac{4}{3}\pi a^3 \left(\rho_s - \rho\right) \mathbf{g} = \mathbf{0}, \tag{6.7.13}$$

where ρ_s is the density of the sphere. Rearranging, we obtain

$$\mathbf{V} = \frac{2}{9} \frac{a^2(\rho_s - \rho)}{\mu} \, \mathbf{g}. \tag{6.7.14}$$

A heavy sphere falls in the direction of gravity and a light sphere rises against the direction of gravity.

6.7.2 Sphere in linear flow

In the second case study, we consider an infinite linear flow with velocity $\mathbf{u}^\infty = \mathbf{A} \cdot (\mathbf{x} - \mathbf{x}_0)$ past a stationary sphere centered at the point $\mathbf{x}_0$, where $\mathbf{A}$ is a constant matrix with zero trace representing the transpose of the velocity gradient, $\mathbf{A}^T = \nabla \mathbf{u}^\infty$. Inspecting the functional forms of the fundamental solutions derived in Section 6.6 suggests representing the disturbance flow due to the sphere by a Stokeslet doublet, $\boldsymbol{\mathcal{S}}^{\mathcal{D}}$, and a potential quadruple, $\boldsymbol{\mathcal{Q}}$, placed at the center of the sphere,

$$u_i(\mathbf{x}) = u_i^\infty(\mathbf{x}) + \frac{1}{8\pi\mu} \mathcal{S}_{ijl}^{\mathcal{D}}(\mathbf{x}, \mathbf{x}_0) \, p_{jl} + \frac{1}{4\pi} \mathcal{Q}_{ijl}(\mathbf{x}, \mathbf{x}_0) \, q_{jl}. \tag{6.7.15}$$

Substituting the specific expressions for the singularities and defining $\hat{\mathbf{x}} = \mathbf{x} - \mathbf{x}_0$, we find that

$$u_i(\mathbf{x}) = A_{ij}\,\hat{x}_j + \frac{1}{8\pi\mu}\left(\frac{\delta_{ij}\hat{x}_l - \delta_{il}\hat{x}_j - \delta_{jl}\hat{x}_i}{r^3} + 3\frac{\hat{x}_i\hat{x}_j\hat{x}_l}{r^5}\right)p_{jl}$$

$$+ \frac{3}{4\pi}\left(-\frac{\delta_{ij}\hat{x}_l + \delta_{jl}\hat{x}_i + \delta_{li}\hat{x}_j}{r^5} + 5\frac{\hat{x}_i\hat{x}_j\hat{x}_l}{r^7}\right)q_{jl}. \tag{6.7.16}$$

Rearranging, we obtain

$$u_i(\mathbf{x}) = A_{ij}\,\hat{x}_j + \frac{1}{8\pi\mu}\frac{\delta_{ij}\hat{x}_l - \delta_{il}\hat{x}_j}{r^3}\,p_{jl} - \frac{3}{4\pi}\frac{\delta_{ij}\hat{x}_l + \delta_{li}\hat{x}_j}{r^5}\,q_{jl}$$

$$-\frac{1}{8\pi\mu}\frac{\hat{x}_i}{r^3}\,p_{ll} - \frac{3}{4\pi}\frac{\hat{x}_i}{r^5}\,q_{ll} + \frac{3}{8\pi\mu}\frac{\hat{x}_i\hat{x}_j\hat{x}_l}{r^5}\,p_{jl} + \frac{15}{4\pi}\frac{\hat{x}_i\hat{x}_j\hat{x}_l}{r^7}\,q_{jl}. \qquad (6.7.17)$$

Requiring that the velocity at the surface of the sphere is zero, we obtain

$$\mathbf{A} + \frac{1}{8\pi\mu}\frac{1}{a^3}(\mathbf{p} - \mathbf{p}^T) - \frac{3}{4\pi}\frac{1}{a^5}(\mathbf{q} + \mathbf{q}^T) = \mathbf{0}, \qquad p_{ll} = q_{ll} = 0, \qquad \mathbf{p} = -\frac{10\mu}{a^2}\,\mathbf{q}, \qquad (6.7.18)$$

where p_{ll} is the trace of $\mathbf{p}$ and q_{ll} is the trace of $\mathbf{q}$. Splitting the matrix $\mathbf{A}$ into a symmetric and an antisymmetric part in the first equation, we obtain

$$\frac{1}{2}(\mathbf{A} - \mathbf{A}^T) + \frac{1}{8\pi\mu}\frac{1}{a^3}(\mathbf{p} - \mathbf{p}^T) = \mathbf{0}, \qquad \frac{1}{2}(\mathbf{A} + \mathbf{A}^T) - \frac{3}{4\pi}\frac{1}{a^5}(\mathbf{q} + \mathbf{q}^T) = \mathbf{0}. \qquad (6.7.19)$$

Eliminating $\mathbf{p}$ in favor of $\mathbf{q}$ from the first equation in (6.7.19) using the third equation in (6.7.18), and simplifying the second equation, we find that

$$\mathbf{q} - \mathbf{q}^T = \frac{2}{5}\pi a^5(\mathbf{A} - \mathbf{A}^T), \qquad \mathbf{q} + \mathbf{q}^T = \frac{2}{3}\pi a^5(\mathbf{A} + \mathbf{A}^T). \qquad (6.7.20)$$

Adding these equations and rearranging, we obtain

$$\mathbf{q} = \frac{2}{15}\pi a^5(4\mathbf{A} + \mathbf{A}^T), \qquad \mathbf{p} = -\frac{4}{3}\pi\mu a^3(4\mathbf{A} + \mathbf{A}^T). \qquad (6.7.21)$$

The symmetric and antisymmetric components of the coefficient of the point-force dipole , satisfying $\mathbf{p} = \mathbf{s} + \mathbf{r}$, are

$$\mathbf{s} = -\frac{20}{3}\pi\mu a^3\,\mathbf{E}, \qquad \mathbf{r} = -4\pi\mu a^3\,\mathbf{\Xi}, \qquad (6.7.22)$$

where

$$\mathbf{E} = \frac{1}{2}(\mathbf{A} + \mathbf{A}^T), \qquad \mathbf{\Xi} = \frac{1}{2}(\mathbf{A} - \mathbf{A}^T) \qquad (6.7.23)$$

are the rate-of-strain tensor and vorticity tensor of the linear flow.

Surface traction

The disturbance traction exerted on the sphere is given by

$$f_i^D(\mathbf{x}) = \frac{1}{8\mu}T_{ijlk}^{SD}(\mathbf{x},\mathbf{x}_0)\,p_{jl}\,n_k + \frac{\mu}{4\pi}T_{ijlk}^{Q}(\mathbf{x},\mathbf{x}_0)\,q_{jl}\,n_k = 3\frac{\mu}{a}A_{ij}\hat{x}_j. \qquad (6.7.24)$$

It is remarkable that the disturbance traction is proportional to the local velocity of the incident flow.

Force and torque

The force exerted on the sphere is zero due to the absence of a point force. To compute the coefficient of the couplet inherent in the Stokeslet dipole, we use expression (6.6.4), obtaining

$$c_m = 4\pi\mu a^3 \epsilon_{mjl} A_{jl}. \tag{6.7.25}$$

Using the third equation in (6.6.10), we find that the torque with respect to the center of the sphere exerted on the sphere is

$$T_m = -c_m = -4\pi\mu a^3 \epsilon_{mjl} A_{jl}. \tag{6.7.26}$$

In terms of the vorticity of the ambient flow, $\omega_i^\infty = -\epsilon_{ijk} A_{jk}$,

$$\mathbf{T} = 4\pi a^3 \omega^\infty. \tag{6.7.27}$$

In Section 6.7.12, we will see that this is a more general result applicable to any incident unbounded flow.

6.7.3   Rotating sphere

The velocity over the surface of a rigid sphere that rotates with angular velocity $\mathbf{\Omega}$ is identical to the disturbance velocity over the surface of a stationary sphere that is immersed in a linear flow with transpose velocity gradient tensor

$$A_{ik} = -\epsilon_{ijk} \Omega_j. \tag{6.7.28}$$

Noting that the matrix $\mathbf{A}$ is antisymmetric, $\mathbf{A} = -\mathbf{A}^T$, and using the solution for a sphere in linear flow discussed in Section 6.7.2, we obtain $\mathbf{q} = \frac{2\pi}{5}a^5\mathbf{A}$ and $\mathbf{p} = -4\pi\mu a^3\mathbf{A}$. Because $\mathbf{q}$ is antisymmetric and Q_{ijl} is symmetric with respect to the last two indices, the quadruple makes a vanishing contribution. Using (6.7.16), we obtain the velocity field

$$u_i = -a^3\frac{1}{2}\mathcal{S}^{\mathcal{D}}_{ijl} A_{jl} = a^3\frac{1}{2}\mathcal{S}^{\mathcal{D}}_{ijl} \epsilon_{jml} \Omega_m = a^3 \mathcal{C}_{im}\Omega_m = a^3\epsilon_{iml} \Omega_m \frac{\hat{x}_l}{r^3}, \tag{6.7.29}$$

which shows that the flow is represented by a couplet with strength

$$\mathbf{c} = 8\pi\mu a^3 \mathbf{\Omega} \tag{6.7.30}$$

situated at the center of the rotating sphere. The torque exerted on the sphere then follows as

$$\mathbf{T} = -\mathbf{c} = -8\pi\mu a^3 \mathbf{\Omega}. \tag{6.7.31}$$

Combining (6.7.27) with (6.7.31), we find that a freely suspended sphere bearing zero torque rotates at an angular velocity that is equal to half the vorticity of the ambient flow.

Combining (6.7.24) with (6.7.28), we find that the traction exerted on the sphere is

$$f_i(\mathbf{x}) = -3\frac{\mu}{a}\epsilon_{ijk}\Omega_j \hat{x}_k \qquad \text{or} \qquad \mathbf{f}(\mathbf{x}) = -3\frac{\mu}{a}\mathbf{\Omega}\times\hat{\mathbf{x}}. \tag{6.7.32}$$

It is remarkable that the traction is proportional to the velocity over the surface of the rotating sphere.

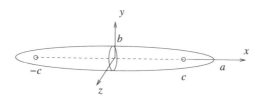

FIGURE 6.7.1 The flow due to the rotation or translation of a prolate spheroid can be represented by a singularity distribution over the focal length of the spheroid.

6.7.4 Rotation and translation of a prolate spheroid

Consider a prolate spheroid with major semiaxis a aligned with the x axis and minor semiaxis b in the yz plane, as illustrated in Figure 6.7.1. The focal semiaxis, c, is defined by the equation $c^2 = a^2 - b^2$. The eccentricity of the spheroid, $e \equiv c/a$, ranges in the interval $[0,1)$, where $e = 0$ corresponds to a sphere and $e \simeq 1$ corresponds to a slender needle.

Swirling flow due to rotation

The swirling flow due to a prolate spheroid rotating with angular velocity Ω_x around its major axis can be represented by a distribution of couplets oriented along the x axis over the focal length of the spheroid,

$$u_i(\mathbf{x}) = \Omega_x c^2 \beta \int_{-c}^{c} \left(1 - \frac{\xi^2}{c^2}\right) C_{ix}(\mathbf{x}, \mathbf{X}(\xi)) \, d\xi, \tag{6.7.33}$$

where $\mathbf{X}(\xi) = [\xi, 0, 0]$ is a point along the centerline and

$$\beta = \frac{1}{\dfrac{2e}{1 - e^2} - \ln \dfrac{1+e}{1-e}} \tag{6.7.34}$$

is a dimensionless coefficient [86]. Integrating the strength density of the couplets over the focal length, we find that the torque exerted on the spheroid is

$$\mathbf{T} = -\frac{32}{3} \pi \mu \Omega_x \beta c^3 \mathbf{e}_x, \tag{6.7.35}$$

where $\mathbf{e}_x$ is the unit vector along the x axis. In the limit $e \to 0$, we find that $\beta \to 3/(4e^3)$, reproducing our earlier results for the sphere.

Translation

The flow due to a prolate spheroid translating with velocity $\mathbf{V}$ in an infinite fluid can be represented by distributions of Stokeslets and potential dipoles over the focal length of the spheroid pointing in the direction of translation, with constant or parabolic distribution densities, respectively [87]. The induced velocity field is

$$u_i(\mathbf{x}) = \int_{-c}^{c} \left(S_{ij}(\mathbf{x}, \mathbf{X}(\xi)) - \frac{1}{2}\left(1 - \frac{\xi^2}{c^2}\right) b^2 \mathcal{D}_{ij}(\mathbf{x}, \mathbf{X}(\xi)) \right) d\xi \, \alpha_{jk} V_k, \tag{6.7.36}$$

where $\mathbf{X}(\xi) = [\xi, 0, 0]$ is a point along the centerline and α_{jk} is a dimensionless diagonal matrix with nonzero components

$$\alpha_{xx} = \frac{e^2}{-2e + (1 + e^2) \ln \dfrac{1+e}{1-e}}, \qquad \alpha_{yy} = \alpha_{zz} = \frac{2e^2}{2e - (1 - 3e^2) \ln \dfrac{1+e}{1-e}}. \qquad (6.7.37)$$

The force exerted on the spheroid is found by integrating the strength density of the point forces over the focal length,

$$\mathbf{F} = -16\pi\mu c\, \boldsymbol{\alpha} \cdot \mathbf{V}. \qquad (6.7.38)$$

In the limit $e \to 0$, all three coefficients a_{11}, a_{22}, and a_{33} tend to $3/(8e)$, in agreement with our earlier results for a sphere.

6.7.5 Translating spherical liquid drop

Next, we consider the flow due to a spherical liquid drop with viscosity $\lambda\mu$ translating with velocity $\mathbf{V}$ in an infinite ambient fluid with viscosity μ. Cursory inspection of the menu of available singularities suggests the following representation for the exterior flow,

$$u_i^{ext}(\mathbf{x}) = \frac{1}{8\pi\mu} \mathcal{S}_{ij}(\mathbf{x}, \mathbf{x}_0)\, b_j + \frac{1}{4\pi} \mathcal{D}_{ij}(\mathbf{x}, \mathbf{x}_0)\, d_j, \qquad (6.7.39)$$

and the following representation for the interior flow,

$$u_i^{int}(\mathbf{x}) = c_i + \mathcal{N}_{ij}(\mathbf{x}, \mathbf{x}_0)\, w_j, \qquad (6.7.40)$$

where $\mathcal{S}$ is the Stokeslet, $\mathcal{D}$ is the potential dipole, $\mathcal{N}$ is the Stokeson, and $\mathbf{c}$ is a constant representing a uniform flow. Substituting the expressions for the singularities and requiring that the velocity is continuous across the drop surface, we find that

$$a^2\mathbf{b} - 2\mu a\mathbf{d} - 8\pi\mu a^3\mathbf{c} - 16\pi\mu a^5\mathbf{w} = \mathbf{0}, \qquad a^2\mathbf{b} + 6\mu a\mathbf{d} + 8\pi\mu a^5\mathbf{w} = \mathbf{0}. \qquad (6.7.41)$$

Requiring that the component of the velocity normal to the drop surface is zero in a frame of reference moving with the drop, $(\mathbf{u}^{int} - \mathbf{V}) \cdot \mathbf{n} = 0$, we obtain an additional condition,

$$\mathbf{c} + a^2\mathbf{w} = \mathbf{V}. \qquad (6.7.42)$$

A fourth equation emerges by requiring that the shear stress is continuous across the interface. Using the formulas given in the tables of Section 6.6, we find that the shear stress exerted on the exterior and interior sides of the interface are given by

$$f_k^{shear,ext} = \frac{1}{4\pi} \Big[-3 \frac{\hat{x}_i \hat{x}_j}{a^4} b_j + 6\mu \Big(\frac{\delta_{ij}}{a^4} - 3 \frac{\hat{x}_i \hat{x}_j}{a^6} \Big) d_j \Big] (\delta_{ik} - n_i n_k) \qquad (6.7.43)$$

or

$$f_k^{shear,int} = 3\lambda\mu \Big(a\, \delta_{ij} - 3 \frac{\hat{x}_i \hat{x}_j}{a} \Big) w_j\, (\delta_{ik} - n_i n_k). \qquad (6.7.44)$$

Note that the projection operator $\mathbf{I} - \mathbf{n} \otimes \mathbf{n}$ represented by the last term on the right-hand side extracts the tangential component of the surface traction. Equating the right-hand sides of (6.7.43) and (6.7.44), we obtain

$$\mathbf{d} = 2\pi\lambda a^5 \, \mathbf{w}. \tag{6.7.45}$$

Solving the system of equations (6.7.41), (6.7.42), and (6.7.45) for the coefficients of the fundamental solutions, we obtain

$$\mathbf{b} = 2\pi\mu a \, \frac{3\lambda+2}{\lambda+1} \, \mathbf{V}, \quad \mathbf{d} = -\pi a^3 \, \frac{\lambda}{\lambda+1} \, \mathbf{V}, \quad \mathbf{c} = \frac{1}{2} \, \frac{2\lambda+3}{\lambda+1} \, \mathbf{V}, \quad \mathbf{w} = -\frac{1}{2a^2} \, \frac{1}{\lambda+1} \, \mathbf{V}. \tag{6.7.46}$$

In the limit of high viscosity ratio, $\lambda \to \infty$, the coefficients $\mathbf{b}$ and $\mathbf{d}$ tend to corresponding values for the solid sphere given in (6.7.4), $\mathbf{c}$ tends to $\mathbf{V}$, and $\mathbf{w}$ vanishes, yielding a uniform interior flow.

Having computed the coefficients of the singularities, we derive exact representations for the velocity field inside and outside the drop in a stationary frame of reference,

$$u_i^{ext} = \frac{1}{4} \frac{1}{\lambda+1} \left[\frac{a}{r} \left(3\lambda+2+\lambda\frac{a^2}{r^2} \right) V_i + \frac{a^3}{r^3} \left(3\lambda+2-3\lambda\frac{a^2}{r^2} \right) \frac{\hat{x}_i \hat{x}_j}{a^2} V_i \right] \tag{6.7.47}$$

and

$$u_i^{int} = \frac{1}{2} \frac{1}{\lambda+1} \left[\left(2\lambda+3-2\frac{a^2}{r^2} \right) V_i + \frac{\hat{x}_i \hat{x}_j}{a^2} V_j \right]. \tag{6.7.48}$$

The streamline patterns for $\lambda = 1$ in a stationary frame of reference and in a frame of reference moving with the drop is depicted in Figure 6.7.2.

Inside the drop, the azimuthal component of the vorticity, ω_φ, increases linearly with distance from the axis, σ, and the flow is identical to that inside Hill's spherical vortex. As a consequence, the interior flow satisfies the unsimplified Navier–Stokes equation with the nonlinear terms included on the left-hand side. However, this is not true for the flow outside the drop.

Drag force

The hydrodynamic drag force exerted on the drop is found from the coefficient of the point force,

$$\mathbf{F} = -\mathbf{b} = -2\pi\mu a \, \frac{3\lambda+2}{\lambda+1} \, \mathbf{V}. \tag{6.7.49}$$

This expression was derived independently by Hadamard and Rybczynski in 1911 by solving a partial-differential equation for the Stokes stream function. Equation (6.7.49) can be expressed in terms of a drag coefficient as

$$c_D \equiv \frac{F}{\pi\rho V^2 a^2} = \frac{4}{\text{Re}} \frac{3\lambda+2}{\lambda+1}, \tag{6.7.50}$$

where ρ is the fluid density, V is the magnitude of the drop velocity, and $\text{Re} = 2a\rho V/\mu$ is the Reynolds number based on the drop diameter, $2a$. Inertial corrections to this formula are available, as reviewed by Clift, Grace, & Weber ([89], Chapter 5).

(a)   (b)

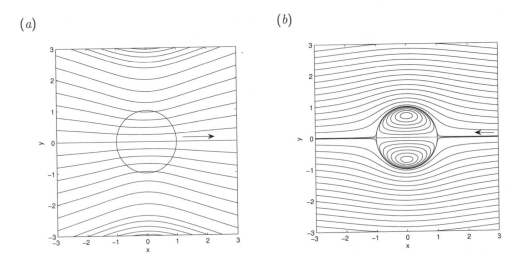

FIGURE 6.7.2 Streamline pattern due to the translation of a spherical liquid drop in an infinite fluid
with the same viscosity, $\lambda = 1$, (a) in a stationary frame of reference and (b) in a frame of reference
moving with the drop.

Settling or rising velocity

Balancing the hydrodynamic force given in (6.7.49), the buoyancy force, and the weight of the drop,
we find that the velocity of a drop that is rising or settling under the action of gravity is

$$\mathbf{V} = \frac{\rho_d - \rho}{\mu}\, a^2\, \frac{2}{3}\, \frac{\lambda+1}{3\lambda+2}\, \mathbf{g}, \tag{6.7.51}$$

where ρ_d is the drop density. In the limit $\lambda \to \infty$, we recover the settling velocity of a solid sphere.

Normal force balance

The singularity solution was derived without reference to the component of the traction normal
to the drop interface. A tacit assumption appears to be that surface tension is high enough for
the drop to maintain a spherical shape even though the difference in the normal component of the
traction across the interface may not be uniform so that it can be balanced by the product of the
surface tension and twice the mean curvature, $2\gamma/a$. In fact, we will show that this assumption is
not necessary.

Using the expressions given in tables of Section 6.6, we find that the normal component of the
traction on either side of the interface is

$$(\mathbf{f} \cdot \mathbf{n})^{ext} = \frac{3}{4\pi a^5}\, (a^2\mathbf{b} + 4\mu\mathbf{d}) \cdot \hat{\mathbf{x}} - \rho\, \mathbf{g} \cdot \hat{\mathbf{x}} + c_1, \quad (\mathbf{f} \cdot \mathbf{n})^{int} = -6\lambda\mu\, \mathbf{w} \cdot \hat{\mathbf{x}} - \rho_d\, \mathbf{g} \cdot \hat{\mathbf{x}} + c_2, \tag{6.7.52}$$

where $\mathbf{n}$ is the outward unit normal vector and c_1, c_2 are two undefined constants. Subtracting these expressions, we obtain

$$(\mathbf{f} \cdot \mathbf{n})^{ext} - (\mathbf{f} \cdot \mathbf{n})^{int} = \frac{3}{4\pi a^5} \left(a^2 \mathbf{b} + 4\mu \mathbf{d} - 8\pi a^2 \lambda \mu \, \mathbf{w} \right) \cdot \hat{\mathbf{x}} - (\rho - \rho_d) \mathbf{g} \cdot \hat{\mathbf{x}} + c_1 - c_2. \qquad (6.7.53)$$

Substituting (6.7.51) into (6.7.46) and then into (6.7.53), we find that the jump in the normal component of the traction is constant and equal to $c_1 - c_2 = -2\gamma/a$. Since the singularity solution satisfies all required boundary conditions, surface tension is not necessary for a settling or rising drop to maintain the spherical shape. However, surface tension does affect the stability of the spherical shape.

Stability

Linear stability analysis for small perturbations shows that the spherical drop is stable, except in the complete absence of surface tension. In practice, we find that, if the surface tension is sufficiently high or a perturbation is sufficiently small, the drop is able to restore its spherical shape. Otherwise, the interface either elongates into an oblate shape or obtains a toroidal shape. Physically, the back side of the drop is convected outward or inward under the influence of the local stagnation-point flow. Stages in the evolution of a settling drop with a prolate or oblate initial shape are shown in Figure 6.7.3 for $\lambda = 1$ in the absence or presence of surface tension. The evolution of the interface was computed using the boundary-integral method for Stokes flow [305].

6.7.6 Flow past a spherical particle with the slip boundary condition

Consider an infinite linear flow with velocity $\mathbf{u}^\infty = \mathbf{U} + \mathbf{A} \cdot \mathbf{x}$ past a translating and rotating spherical particle of radius a, where $\mathbf{U}$ is a constant velocity and $\mathbf{A}$ is the transpose of the velocity gradient tensor, $\mathbf{A}^T = \nabla \mathbf{u}^\infty$ [249]. The continuity equation requires the trace of $\mathbf{A}$ to be zero. The no-penetration condition and a slip boundary condition are enforced at the particle surface,

$$\mathbf{u} = \mathbf{V} + \boldsymbol{\Omega} \times (\mathbf{x} - \mathbf{x}_c) + \mathbf{u}^S, \qquad (6.7.54)$$

where $\mathbf{V}$ is the velocity of translation of the particle center, $\mathbf{x}_c$, and $\boldsymbol{\Omega}$ is the angular velocity of rotation about $\mathbf{x}_c$. The first two terms on the right-hand side of (6.7.54) describe rigid-body motion, while the third term expresses a slip velocity given by the Navier–Maxwell–Basset slip formula

$$\mathbf{u}^S = \frac{1}{\beta} \frac{a}{\mu} \mathbf{f} \cdot (\mathbf{I} - \mathbf{n} \otimes \mathbf{n}) = \frac{\lambda}{\mu} \mathbf{n} \times \mathbf{f} \times \mathbf{n}, \qquad (6.7.55)$$

where $\mathbf{f} \equiv \boldsymbol{\sigma} \cdot \mathbf{n}$ is the traction, $\boldsymbol{\sigma}$ is the stress tensor, $\mathbf{n}$ is the unit normal vector pointing into the fluid, $\mathbf{I} - \mathbf{n} \otimes \mathbf{n}$ is the tangential projection operator, β is the dimensionless Basset particle slip coefficient ranging from zero for vanishing shear stress and perfect slip to infinity for nonzero shear stress and no-slip, and $\lambda = a/\beta$ is the slip length, as discussed in Section 3.7.4.

For convenience, we set the particle center at the origin, $\mathbf{x}_c = \mathbf{0}$, and express the velocity field in terms of five singularities,

$$u_i = U_i + A_{ij} x_j + a \, \mathcal{S}_{ij} \, b_j + a^3 \, \mathcal{D}_{ij} \, d_j + a^3 \, \mathcal{C}_{ij} \, c_j + a^3 \, \mathcal{S}_{ijl}^{\mathcal{D}} \, p_{jl} + a^5 \, \mathcal{Q}_{ijl} \, q_{jl}, \qquad (6.7.56)$$

where $\mathcal{S}_{ij}$ is the Stokeslet, $\mathcal{D}_{ij}$ is the potential dipole, $\mathcal{C}_{ij}$ is the rotlet, $\mathcal{S}_{ijl}^{\mathcal{D}}$ is the Stokeslet dipole, and $\mathcal{Q}_{ijl}$ is the potential quadrapole.

(a)   (b)

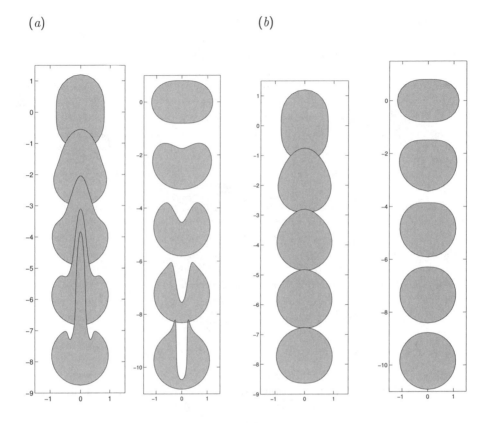

FIGURE 6.7.3   Stages in the evolution of a settling drop with an oblate or prolate initial shape (a) in
the absence and (b) in the presence of surface tension for unit viscosity ratio, $\lambda = 1$. Snapshots
are shown at equal time intervals. When the surface tension is sufficiently strong, the drop restores
its spherical shape.

The no-penetration boundary condition requires that $u_i x_i = V_i x_i$ at the particle surface,
$r = a$. Substituting the expressions for the singularities and simplifying, we find that

$$u_j x_j = U_j x_j + A_{jl}\, x_j\, x_l + 2\,(b_j + d_j)\, x_j + (-\delta_{jl}\, a^2 + 3\, x_j x_l)\, (p_{jl} + 3\, q_{jl}) = V_j x_j, \qquad (6.7.57)$$

and thus

$$\mathbf{b} + \mathbf{d} = \frac{1}{2}\,(\mathbf{V} - \mathbf{U}), \qquad 3\,[\mathbf{p}]^+ + 9\,\mathbf{q} = -\mathbf{E}, \qquad (6.7.58)$$

where the brackets $[\ ]^+$ denote the symmetric part of the enclosed tensor, and $\mathbf{E}$ is the rate-of-
deformation tensor, $\mathbf{E} = [\mathbf{A}]^+$.

The tangential velocity component over the particle surface is given by

$$u_i^| = u_i - u_m \frac{x_i x_m}{a^2} = U_i + A_{ij} \, x_j + a \, \mathcal{S}_{ij} \, b_j + a^3 \, \mathcal{D}_{ij} \, d_j + a^3 \, \mathcal{C}_{ij} \, c_j + a^3 \, \mathcal{S}_{ijl}^D \, p_{jl} + a^5 \, Q_{ijl} \, q_{jl}$$
$$- \frac{x_i x_m}{a^2} \left(U_m + A_{mj} \, x_j + a \, \mathcal{S}_{mj} \, g_j + a^3 \, \mathcal{D}_{mj} \, d_j + a^3 \, \mathcal{C}_{mj} \, c_j + a^3 \, \mathcal{S}_{mjl}^D \, p_{jl} + a^5 \, Q_{mjl} \, q_{jl} \right). \quad (6.7.59)$$

Making substitutions and simplifying, we find that

$$u_i^| = (U_j + b_j - d_j)(\delta_{ij} - \frac{x_i x_j}{a^2}) + \epsilon_{ilj} \, c_l \, x_j$$
$$+ (A_{ij} + p_{ij} - p_{ji} - 3 \, q_{ij} - 3 \, q_{ji}) \, x_j + \frac{1}{a^2} \left(-A_{lj} + 6 \, q_{jl} \right) x_i x_j x_l. \quad (6.7.60)$$

The stress field is given by

$$\sigma_{ik} = \mu \Big(A_{ik} + A_{ki} + a \, \mathcal{T}_{ijk}^S \, b_j + a^3 \, \mathcal{T}_{ijk}^S \, d_j + a^3 \, \mathcal{T}_{ijk}^C \, c_j + a^3 \, \mathcal{T}_{ijlk}^{SD} \, p_{jl} + a^5 \, \mathcal{T}_{ijlk}^Q \, q_{jl} \Big). \quad (6.7.61)$$

The traction exerted on the surface of the sphere is $f_i = \sigma_{ik} x_k / a$, and its tangential component is

$$f_i^| = f_i - f_m \frac{x_i x_m}{a^2} = \mu \frac{x_k}{a} (A_{ik} + A_{ki} + a \, \mathcal{T}_{ijk}^S \, b_j + a^3 \, \mathcal{T}_{ijk}^S \, d_j + a^3 \, \mathcal{T}_{ijk}^C \, c_j + a^3 \, \mathcal{T}_{ijlk}^{SD} \, p_{jl} + a^5 \, \mathcal{T}_{ijlk}^Q \, q_{jl})$$
$$- \mu \frac{x_i x_m x_k}{a^3} \Big(A_{mk} + A_{km} + a \, \mathcal{T}_{mjk}^S \, g_j + a^3 \, \mathcal{T}_{mjk}^D \, d_j + a^3 \, \mathcal{T}_{mjk}^C \, c_j$$
$$+ a^3 \, \mathcal{T}_{mjlk}^{SD} \, p_{jl} + a^5 \, \mathcal{T}_{mjlk}^Q \, q_{jl} \Big). \quad (6.7.62)$$

Making substitutions and simplifying, we obtain

$$f_i^| = 6\mu \, d_j \, (\delta_{ij} - \frac{x_j x_i}{a^2}) + \frac{3\mu}{a} \, c_l \, \epsilon_{ijl} \, x_j + \frac{\mu}{a} \, (A_{ij} + A_{ji} + 6 \, p_{ji} + 24 \, q_{ij} + 24 \, q_{ji} - 6 \, q_{kk} \delta_{ij}) \, x_j$$
$$+ \frac{\mu}{a^3} (6 \, q_{kk} - 2 A_{jl} - 6 p_{jl} - 48 \, q_{jl}) \, x_i x_j x_l. \quad (6.7.63)$$

The slip boundary condition at the particle surface requires that

$$u_i^| - V_i + V_m \frac{x_i x_m}{a^2} - \epsilon_{ijk} \Omega_j x_k = \frac{a}{\mu \beta} \, f_i^|. \quad (6.7.64)$$

Substituting the expressions for the tangential components of the velocity and traction, and grouping similar terms, we find that

$$\mathbf{U} - \mathbf{V} + \mathbf{b} - \mathbf{d} = \frac{6}{\beta} \, \mathbf{d}, \qquad \mathbf{c} = \frac{\beta}{\beta + 3} \, \mathbf{\Omega}, \qquad (6.7.65)$$

and also

$$\mathbf{A} + 2 \, [\mathbf{p}]^- - 6 \, \mathbf{q} = \frac{1}{\beta} \Big(2 \, \mathbf{E} + 6 \, \mathbf{p}^T + 48 \, \mathbf{q} - 6 \, \text{trace}(\mathbf{q}) \, \mathbf{I} \Big) \qquad (6.7.66)$$

and

$$(\beta - 2) \, \mathbf{E} - 6 \, [\mathbf{p}]^+ - 6 \, (8 + \beta) \, \mathbf{q} + 6 \, \text{trace}(\mathbf{q}) \, \mathbf{I} = \mathbf{0}, \qquad (6.7.67)$$

where the square brackets $[\ ]^-$ denote the antisymmetric part of the enclosed tensor and the superscript T denotes the matrix transpose. Combining the first equation in (6.7.58) with the first equation in (6.7.65), we obtain

$$\mathbf{b} = \frac{3}{4}\frac{\beta+2}{\beta+3}(\mathbf{V}-\mathbf{U}), \qquad \mathbf{d} = \frac{1}{4}\frac{\beta}{\beta+3}(\mathbf{U}-\mathbf{V}). \tag{6.7.68}$$

The force exerted on the sphere is

$$\mathbf{F} = -8\pi\mu a\,\mathbf{b} = -6\pi\mu a\,\frac{\beta+2}{\beta+3}(\mathbf{V}-\mathbf{U}). \tag{6.7.69}$$

As β tends to zero, the Stokes-law coefficient of six tends to four, indicating a substantial reduction in the drag force. Combining the second equation in (6.7.58) with equations (6.7.66) and (6.7.67), we find that

$$[\mathbf{p}]^+ = -\frac{5}{6}\frac{\beta+2}{\beta+5}\mathbf{E}, \qquad [\mathbf{p}]^- = -\frac{1}{2}\frac{\beta}{\beta+3}\Xi, \qquad \mathbf{q} = \frac{1}{6}\frac{\beta}{\beta+5}\mathbf{E}, \tag{6.7.70}$$

where Ξ is the vorticity tensor, $\Xi = [\mathbf{A}]^-$. Thus,

$$\mathbf{p} = [\mathbf{p}]^+ + [\mathbf{p}]^- = -\frac{(4\beta^2+20\,\beta+15)\,\mathbf{A} + (\beta^2+5\,\beta+15)\mathbf{A}^T}{6\,(\beta+3)(\beta+5)}. \tag{6.7.71}$$

The coefficient of the couplet inherent in the antisymmetric part of the Stokeslet dipole is

$$c_m^{\mathcal{SD}} = -\epsilon_{mjl}\,a^3\,[p_{jl}]^- = \frac{1}{2}\,\epsilon_{mjl}\,a^3\,\frac{\beta}{\beta+3}\,A_{jl} = -\frac{a^3}{2}\,\frac{\beta}{\beta+3}\,\omega_m^\infty, \tag{6.7.72}$$

where $\boldsymbol{\omega}^\infty$ is the vorticity of the ambient linear flow. The torque experienced by the sphere is

$$\mathbf{T} = -8\pi\mu\,(a^3\,\mathbf{c} + \mathbf{c}^{\mathcal{SD}}) = -4\pi\mu a^3\,\frac{\beta}{\beta+3}\,(2\boldsymbol{\Omega} - \boldsymbol{\omega}^\infty). \tag{6.7.73}$$

A torque-free particle rotates with angular velocity that is equal to half the vorticity of the incident flow. When $\beta = 0$, a rotating particle does not generate a flow and the torque vanishes.

6.7.7   Approximate singularity representations

Exact singularity representations are known only for a limited class of flows bounded by spherical or spheroidal surfaces [86, 87, 208]. To derive approximate representations for more general flows and arbitrary boundary shapes, we resort to asymptotic and numerical methods. The general strategy is to express the flow in terms of discrete or continuous singularity distributions, and then compute the coefficients of the singularities or densities of the distributions, and possibly their location, so as to satisfy the required boundary conditions in some approximate sense.

Flow past a sphere

Burgers represented the flow due to a translating sphere in terms of a point force situated at the center of the sphere and computed the strength of the point force by requiring that the mean velocity vanishes over the surface of the sphere ([61], p. 120). The approximate solution reproduces precisely Stokes' law (Problem 6.7.4(a)). The perfect agreement is explained by recalling that the exact solution is composed by a Stokeslet and a potential dipole, and noting that average velocity of the dipole is zero over the surface of a sphere. Burgers performed a similar computation for the disturbance flow due to a sphere held stationary in a paraboloidal flow and found that the drag force is still in perfect agreement with the exact solution which can be found readily using Faxén relations discussed in Section 6.12 (Problem 6.7.4(b)).

Numerical methods

To develop a formal procedure for computing singularity representations, we introduce a positive functional expressing the difference between the specified boundary conditions and the corresponding boundary values computed using the singularity representation. When the problem requires that $\mathbf{u} = \mathbf{V}$ over a boundary, D, we introduce a positive functional $\mathcal{F}(\mathbf{u} - \mathbf{V})$ that vanishes when $\mathbf{u} = \mathbf{V}$. Other boundary conditions are implemented in a straightforward fashion.

The general strategy is to represent the velocity $\mathbf{u}$ in terms of a certain singularity distribution, and then minimize the functional, $\mathcal{F}$, with respect to the strength of the singularities and possibly the location of their poles (e.g., [442]). When velocity boundary conditions are specified, a wise choice is the least-squares collocation functional

$$\mathcal{F}(\mathbf{u} - \mathbf{V}) = \sum_{i=1}^{M} |\mathbf{u}(\mathbf{x}_i) - \mathbf{V}(\mathbf{x}_i)|^2, \tag{6.7.74}$$

where $\mathbf{x}_i$, is a collection of M collocation points over D. Minimizing (6.7.74) with respect to the strength of the singularities yields a system of linear equations. When the force or torque acting on the boundary D are specified, the strengths of the singularities must satisfy additional linear constraints originating from (6.6.10).

The least-squares collocation method has found extensive applications in a variety of numerical studies of Stokes flow involving suspended particles. In certain cases, instead of using singularity expansions, it is more expedient to use alternative expansions in terms of spherical harmonic functions ([220], §335; [208], Chapter 13). A generalized implementation allows the singularities to relocate as part of the optimization (e.g., [442]).

Problems

6.7.1 *Flow due to a translating sphere*

Derive the Stokes stream function (6.7.6) by solving the partial differential equation (6.1.19) subject to appropriate boundary conditions.

6.7.2 *Minimum resistance shapes*

(a) Consider a family of spheroids with constant volume and variable aspect ratio, b/a. Compute the aspect ratio of the spheroid with the minimum drag force in axial or transverse translation.

(b) Repeat (a) for the torque in the case of rotation around the spheroid major axis.

6.7.3 *Pressure on a sphere in irrotational or Stokes flow*

Discuss the differences in the pressure distribution over the surface a rigid translating sphere under conditions of potential or Stokes flow.

🖳 *Computer Problem*

6.7.4 *Burgers solution*

(*a*) Following Burgers [61], we represent the flow due to the translation of a sphere in terms of a point force alone as

$$\mathbf{u}(x) = \frac{1}{8\pi\mu} \, \boldsymbol{\mathcal{S}}(\mathbf{x}, \mathbf{x}_0) \cdot \mathbf{b}, \tag{6.7.75}$$

where $\boldsymbol{\mathcal{S}}$ is the Stokeslet located at the center of the sphere, $\mathbf{x}_0$. Compute the coefficient $\mathbf{b}$ by requiring that the average fluid velocity on the surface of the sphere is equal to the velocity of translation. Calculate the drag force exerted on the sphere and compare it with the exact value given by Stokes' law.

(*b*) Perform a similar computation for paraboloidal flow past a stationary sphere of radius a, with velocity

$$\mathbf{u}^P = \frac{U}{a^2} \left(y^2 + z^2, \quad 0, \quad 0 \right), \tag{6.7.76}$$

where U is a constant. Compare your result for the force exerted on the sphere with the exact value given by the Faxén law, $\mathbf{F} = 4\pi\mu a \, [U, 0, 0]$ (Section 6.12).

6.8   The Lorentz reciprocal theorem and its applications

Consider two unrelated Navier–Stokes flows with velocity fields $\mathbf{u}$ and $\mathbf{u}'$ and associated hydrodynamic stress tensors, excluding hydrostatic pressure variations, $\tilde{\boldsymbol{\sigma}}$ and $\tilde{\boldsymbol{\sigma}}'$. A generalized reciprocal theorem relating the two velocity and stress fields was stated in equation (3.5.22),

$$\frac{\partial}{\partial x_j} \left(\mu' u_i' \tilde{\sigma}_{ij} - \mu \, u_i \tilde{\sigma}_{ij}' \right) = \mu' \mathbf{u}' \cdot \nabla \cdot \tilde{\boldsymbol{\sigma}} - \mu \mathbf{u} \cdot \nabla \cdot \tilde{\boldsymbol{\sigma}}', \tag{6.8.1}$$

where μ and μ' are the fluid viscosities. For simplicity, we omit the tilde above the hydrodynamic stress, pressure, and traction. If both flows satisfy the Stokes equation, the right-hand side of (6.8.1) is zero, yielding the Lorentz reciprocal identity [245]

$$\frac{\partial}{\partial x_j} (\mu' u_i' \sigma_{ij} - \mu u_i \sigma_{ij}') = 0, \tag{6.8.2}$$

which is the counterpart of Green's second identity for two scalar harmonic potentials.

Assuming that the two Stokes flows of interest are free of singular points inside a control volume, V_c, bounded by a surface, D, we integrate (6.8.2) over the control volume and use the divergence theorem to obtain

$$\iint_D (\mu' u_i' f_i - \mu u_i f_i')\, \mathrm{d}S = 0, \tag{6.8.3}$$

where $\mathbf{f} = \mathbf{n} \cdot \boldsymbol{\sigma}$ is the boundary traction and the unit normal vector $\mathbf{n}$ may point either into or out of the control volume.

The reciprocal identities (6.8.2) and (6.8.3) allow us to obtain information about a certain flow without explicitly solving the equations of Stokes flow, but merely by using information about another flow. An alternative form of the reciprocal theorem involving the velocity and the pressure is stated in Problem 6.8.1.

6.8.1   Flow past a stationary particle

As an application of the reciprocal theorem, we consider an arbitrary incident flow with velocity $\mathbf{u}^\infty$ past a stationary particle. The particle causes a disturbance velocity, $\mathbf{u}^D$, which can be added to the incident velocity to produce the total velocity, $\mathbf{u} = \mathbf{u}^\infty + \mathbf{u}^D$.

Force

To compute the force exerted on the particle, we turn to (6.8.3) and identify $\mathbf{u}'$ with the velocity field generated when the particle translates with velocity $\boldsymbol{\mathcal{V}}$. Exploiting the linearity of the Stokes equation, we express the corresponding traction exerted at the particle surface as

$$\mathbf{f}^t = -\mu\, \mathbf{R}^t \cdot \boldsymbol{\mathcal{V}}, \tag{6.8.4}$$

where $\mathbf{R}^t$ is a traction resistance matrix for translation indicated by the superscript t, not to be confused with the matrix transpose.

Next, we select a control volume enclosed by the particle surface, denoted by P, and a surface with large size enclosing the particle, denoted by S_∞. Applying the reciprocal relation (6.8.3) for the pair of flows $\mathbf{u}^t$ and $\mathbf{u}^D$ with equal fluid viscosities, we obtain

$$\iint_{S_\infty, P} \mathbf{u}^t \cdot \mathbf{f}^D\, \mathrm{d}S = \iint_{S_\infty, P} \mathbf{u}^D \cdot \mathbf{f}^t\, \mathrm{d}S. \tag{6.8.5}$$

Letting the surface S_∞ tend to infinity and noting that the velocity decays at least as fast as $1/d$ and the traction decays at least as fast as $1/d^2$, where d is the distance from a designated particle center, we find that the contribution of the surface integrals over S_∞ disappears. Equation (6.8.5) then yields

$$\boldsymbol{\mathcal{V}} \cdot \mathbf{F}^D = \iint_P \mathbf{u}^D \cdot \mathbf{f}^t\, \mathrm{d}S, \tag{6.8.6}$$

where $\mathbf{F}^D$ is the disturbance force exerted on the particle. In the absence of inertia, the force exerted on any fluid volume by the ambient flow is zero and the disturbance force, $\mathbf{F}^D$, is equal to the total force, $\mathbf{F}$. Enforcing the boundary condition $\mathbf{u} = \mathbf{0}$ or $\mathbf{u}^D = -\mathbf{u}^\infty$ on P and using (6.8.4), we finally obtain

$$\mathbf{F} = \mu \iint_P \mathbf{u}^\infty \cdot \mathbf{R}^t \, \mathrm{d}S. \tag{6.8.7}$$

This expression allows us to compute the hydrodynamic force exerted on a stationary particle in terms of the incident velocity over the particle surface and the traction resistance matrix for translation, $\mathbf{R}^t$ [56].

Torque

Next, we consider the torque exerted on a rigid particle that is held stationary in an ambient flow. Repeating the previous steps for the force, we introduce a rotary traction resistance matrix , $\mathbf{R}^r$, defined by the equation

$$\mathbf{f}^r = -\mu \, \mathbf{R}^r \cdot \mathcal{O}, \tag{6.8.8}$$

where $\mathbf{f}^r$ is the traction established when the particle rotates about a specified point, $\mathbf{x}_0$, with angular velocity $\mathcal{O}$. The analysis provides us with the counterpart of (6.8.7) for the torque with respect to $\mathbf{x}_0$,

$$\mathbf{T} = \mu \iint_P \mathbf{u}^\infty \cdot \mathbf{R}^r \, \mathrm{d}S. \tag{6.8.9}$$

This expression allows us to evaluate the hydrodynamic torque exerted on a stationary particle in terms the incident velocity over the particle surface and the traction resistance matrix for rotation, $\mathbf{R}^r$ [56].

Spherical particle

As an application, we use expressions (6.7.11) and (6.7.32) for the traction on a spherical particle of radius a, and deduce the translational and rotary traction resistance matrices

$$R_{ij}^t = \frac{3}{2}\frac{1}{a}\,\delta_{ij}, \qquad R_{ij}^r = \frac{3}{a}\,\epsilon_{ijk}\,\hat{x}_k, \tag{6.8.10}$$

where $\hat{\mathbf{x}} = \mathbf{x} - \mathbf{x}_0$ is the distance from the particle center, $\mathbf{x}_0$. Substituting these expressions into (6.8.7) and (6.8.9), we obtain

$$\mathbf{F} = \frac{3}{2}\frac{\mu}{a}\iint_P \mathbf{u}^\infty \, \mathrm{d}S, \qquad \mathbf{T} = 3\frac{\mu}{a}\iint_P \hat{\mathbf{x}} \times \mathbf{u}^\infty \, \mathrm{d}S. \tag{6.8.11}$$

Because the matrix $\mathbf{R}^t$ is constant and diagonal, the force is proportional to the surface average of the incident velocity, $\mathbf{u}^\infty$.

Flow in a bounded domain

Similar results are obtained for an incident flow past a particle that is held stationary in a bounded domain of flow. For example, the force and torque exerted on the particle that is held stationary inside a tube can be computed from knowledge of the traction exerted on the particle when it translates or rotates in an otherwise stationary fluid inside the tube.

Generalized Faxén laws

Equations (6.8.7) and (6.8.9) constitute one version of the generalized Faxén laws for the force and torque. Other versions of the Faxén law s producing the force, torque, and higher moments of the traction in terms of the incident velocity and its spatial derivatives evaluated at specific points inside a particle are discussed in Section 6.12.

6.8.2   Force and torque on a moving particle

Next, we consider a flow with velocity $\mathbf{u}^\infty$ past a suspended rigid particle that translates with linear velocity $\mathbf{V}$ while rotating about a point $\mathbf{x}_0$ with angular velocity $\mathbf{\Omega}$. The presence or motion of the particle generates a disturbance velocity, $\mathbf{u}^D$, which can be added to the incident velocity to yield the total velocity, $\mathbf{u} = \mathbf{u}^D + \mathbf{u}^\infty$.

Force and torque from the resistance problem

Applying the reciprocal identity for the disturbance flow and the flow produced when the particle translates in an otherwise quiescent fluid with velocity $\mathcal{V}$, we derive equation (6.8.6). Enforcing the boundary condition of rigid-body motion, $\mathbf{u} = \mathbf{V} + \mathbf{\Omega} \times (\mathbf{x} - \mathbf{x}_0)$, restated as

$$\mathbf{u}^D = \mathbf{V} + \mathbf{\Omega} \times (\mathbf{x} - \mathbf{x}_0) - \mathbf{u}^\infty, \tag{6.8.12}$$

and recalling that the disturbance force is equal to the total force, we obtain

$$\mathbf{V} \cdot \mathbf{F}^t + \mathbf{\Omega} \cdot \mathbf{T}^t = \iint_P \mathbf{u}^\infty \cdot \mathbf{f}^t \, \mathrm{d}S + \mathcal{V} \cdot \mathbf{F}, \tag{6.8.13}$$

where $\mathbf{F}^t$ and $\mathbf{T}^t$ are the force and torque with respect to the point $\mathbf{x}_0$ exerted on the particle when it translates with velocity $\mathcal{V}$. Working in a similar fashion in the case of particle rotation, we obtain the corresponding equation

$$\mathbf{V} \cdot \mathbf{F}^r + \mathbf{\Omega} \cdot \mathbf{T}^r = \iint_P \mathbf{u}^\infty \cdot \mathbf{f}^R \, \mathrm{d}S + \mathcal{O} \cdot \mathbf{F}, \tag{6.8.14}$$

where $\mathbf{F}^r$ and $\mathbf{T}^r$ are the force and torque with respect to the point $\mathbf{x}_0$ exerted on the particle when it rotates around the point $\mathbf{x}_0$ with angular velocity $\mathcal{O}$.

Now we take advantage of the linearity of the equations of Stokes flow with respect to the boundary conditions and write

$$\mathbf{F}^t = -\mu \mathbf{X} \cdot \mathcal{V}, \qquad \mathbf{T}^t = -\mu \mathbf{Z} \cdot \mathcal{V}, \qquad \mathbf{F}^r = -\mu \mathbf{W} \cdot \mathcal{O}, \qquad \mathbf{T}^r = -\mu \mathbf{Y} \cdot \mathcal{O}, \tag{6.8.15}$$

where $\mathbf{X}$, $\mathbf{Z}$ are resistance matrices for translation and $\mathbf{W}$, $\mathbf{Y}$ are resistance matrices for rotation. Using the reciprocal theorem, we may show that the resistance matrices $\mathbf{X}$ and $\mathbf{Y}$ are symmetric and $\mathbf{Z}$ is the transpose of $\mathbf{W}$, $\mathbf{W} = \mathbf{Z}^T$ (Problem 6.8.3). Using the definitions (6.8.15), we recast (6.8.13) and (6.8.14) into the forms

$$
\mathbf{V} \cdot \mathbf{X} + \mathbf{\Omega} \cdot \mathbf{Z} = \iint_P \mathbf{u}^\infty \cdot \mathbf{R}^t \, dS - \frac{1}{\mu} \mathbf{F},
$$
$$
\mathbf{V} \cdot \mathbf{Z}^T + \mathbf{\Omega} \cdot \mathbf{Y} = \iint_P \mathbf{u}^\infty \cdot \mathbf{R}^r \, dS - \frac{1}{\mu} \mathbf{T}.
$$

(6.8.16)

Given the force, $\mathbf{F}$, and torque, $\mathbf{T}$, exerted on the particle and the resistance matrices for translation and rotation, we obtain a linear system of equations for the particle translational and rotational velocities, $\mathbf{V}$ and $\mathbf{\Omega}$.

Force and torque from the mobility problem

Consider a force-free and torque-free rigid particle convected in an ambient flow with velocity $\mathbf{u}^\infty$. Applying the reciprocal theorem, we find that the particle velocity of translation, $\mathbf{V}$, and angular velocity of rotation, $\mathbf{\Omega}$, satisfy the equations

$$
\mathbf{V} \cdot \mathbf{F} = \iint_P \mathbf{u}^\infty \cdot \mathbf{f}^F \, dS, \qquad \mathbf{\Omega} \cdot \mathbf{T} = \iint_P \mathbf{u}^\infty \cdot \mathbf{f}^T \, dS,
$$

(6.8.17)

where $\mathbf{f}^F$ and $\mathbf{f}^T$ are the tractions exerted on the particle when it moves under the influence of an external force, $\mathbf{F}$, or an external torque, $\mathbf{T}$ (Problem 6.8.4). An important consequence of (6.8.17) is that the translational and angular velocities of a force-free and torque-free particle immersed in an arbitrary ambient flow can be computed from the distribution of the incident velocity over the particle surface and the traction exerted on the particle surface when it moves under the influence of an external force or torque.

6.8.3   Symmetry of Green's functions

An important consequence of the reciprocal theorem is the symmetry of the Green's functions of Stokes flow,

$$
\mathcal{G}_{ij}(\mathbf{x}, \mathbf{x}_0) = \mathcal{G}_{ij}(\mathbf{x}, \mathbf{x}_0).
$$

(6.8.18)

Analogous symmetry properties exist for the Green's functions of potential flow, linear elastostatics and, more generally, for the Green's function of linear, elliptic partial differential equations involving self-adjoint differential operators. Physically, the symmetry property states that the velocity at a point, $\mathbf{x}$, due to a point force at another point, $\mathbf{x}_0$, can be deduced from the velocity at the point $\mathbf{x}_0$ due to a point force located at $\mathbf{x}$.

To prove (6.8.18), we introduce the velocity fields induced by two point forces located at $\mathbf{x}_1$ and $\mathbf{x}_2$ with strengths $\mathbf{b}^{(1)}$ and $\mathbf{b}^{(2)}$, given by

$$
u_i^{(1)} = \frac{1}{8\pi\mu} \mathcal{G}_{ij}(\mathbf{x}, \mathbf{x}_1) \, b_j^{(1)}, \qquad u_i^{(2)} = \frac{1}{8\pi\mu} \mathcal{G}_{ij}(\mathbf{x}, \mathbf{x}_2) \, b_j^{(2)},
$$

(6.8.19)

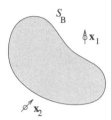

FIGURE 6.8.1   Illustration of two point forces near a boundary, S_B, introduced to prove the symmetry of the Green's functions of Stokes flow.

as illustrated in Figure 6.8.1. The corresponding stress fields satisfy the equations

$$\frac{\partial \sigma_{ij}^{(1)}}{\partial x_j} = -b_i^{(1)} \delta_3(\mathbf{x} - \mathbf{x}^{(1)}), \qquad \frac{\partial \sigma_{ij}^{(2)}}{\partial x_j} = -b_i^{(2)} \delta_3(\mathbf{x} - \mathbf{x}^{(2)}), \qquad (6.8.20)$$

where δ_3 is the three-dimensional delta function. Substituting these expressions into the right-hand side of (6.8.2) with equal fluid viscosities, $\mu' = \mu$, we obtain

$$\frac{\partial}{\partial x_j}(u_i^{(2)} \sigma_{ij}^{(1)} - u_i^{(1)} \sigma_{ij}^{(2)})$$

$$= -\frac{1}{8\pi\mu} \left[\mathcal{G}_{ik}(\mathbf{x}, \mathbf{x}^{(2)}) \, b_k^{(2)} b_i^{(1)} \, \delta_3(\mathbf{x} - \mathbf{x}^{(1)}) - \mathcal{G}_{ik}(\mathbf{x}, \mathbf{x}^{(1)}) \, b_k^{(2)} b_i^{(2)} \, \delta_3(\mathbf{x} - \mathbf{x}^{(2)}) \right]. \quad (6.8.21)$$

Next, we integrate (6.8.21) over a control volume that is confined by a solid boundary where the Green's functions are required to be zero, S_B, two spherical surfaces of infinitesimal radii enclosing $\mathbf{x}_1$ and $\mathbf{x}_2$, and, in the case of infinite flow, a large surface enclosing S_B as well as the points $\mathbf{x}_1$ and $\mathbf{x}_2$. Using the divergence theorem, we convert the volume integral on the left-hand side of the resulting equation into a surface integral over all surfaces enclosing the control volume.

   The surface integral over S_B vanishes because the Green's function and hence the velocities $\mathbf{u}^{(1)}$ and $\mathbf{u}^{(2)}$ are zero over S_B. Letting the radius of the large surface tend to infinity and the radii of the small spheres enclosing $\mathbf{x}_1$ and $\mathbf{x}_2$ to zero, we find that the corresponding surface integrals make insignificant contributions. As a result, the whole integral of the left-hand side of (6.8.21) vanishes. Using the properties of the delta function to manipulate the right-hand side, we finally arrive at the equation

$$b_k^{(1)} \, b_i^{(2)} \left[\mathcal{G}_{ki}(\mathbf{x}_1, \mathbf{x}_2) - \mathcal{G}_{ik}(\mathbf{x}_2, \mathbf{x}_1) \right] = 0. \qquad (6.8.22)$$

Property (6.8.18) follows by observing that the constants $\mathbf{b}^{(1)}$ and $\mathbf{b}^{(2)}$ are arbitrary.

   Repeating this procedure with some modifications, we can show that the two-dimensional Green's functions also satisfy the symmetry property (6.8.18) (Problem 6.8.5).

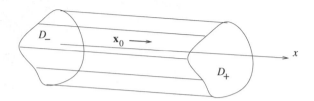

FIGURE 6.8.2 Illustration of a point force inside cylindrical tube with arbitrary cross-section. The corresponding flow affords one degree of freedom determining the axial flow rate or pressure drop.

6.8.4 Pressure drop due to a point force in a tube

Consider a point force located at the point $\mathbf{x}_0$ inside an infinite cylindrical tube with arbitrary cross-section whose generators are parallel to the x axis, as shown in Figure 6.8.2. Applying the reciprocal theorem for (a) the flow induced by the point force expressed by a corresponding Green's function, and (b) unidirectional pressure-driven flow indicated by the superscript P, we find that

$$\frac{\partial}{\partial x_j}\Big[u_i^P(\mathbf{x})\,\frac{1}{8\pi}\,\mathcal{T}_{ikj}(\mathbf{x},\mathbf{x}_0) - \frac{1}{8\pi\mu}\,\mathcal{G}_{ik}(\mathbf{x},\mathbf{x}_0)\,\sigma_{ij}^P(\mathbf{x})\Big] = -u_i^p(\mathbf{x})\,\delta_3(\mathbf{x}-\mathbf{x}_0), \qquad (6.8.23)$$

where k is a free index indicating the direction of the point force. We will assume that the velocity generated by the point force decays far upstream and downstream, and the pressure tends to a constant value far from the point force, so that

$$\mathcal{P}_j(x \to \pm\infty, \mathbf{x}_0) \to \alpha_{\pm\infty}\,\frac{8\pi}{a^2}\,\delta_{jx}, \qquad \mathcal{T}_{ijk}(x \to \pm\infty, \mathbf{x}_0) \to -\alpha_{\pm\infty}\,\frac{8\pi}{a^2}\,\delta_{jx}\,\delta_{ik}, \qquad (6.8.24)$$

where $\alpha_{\pm\infty}$ are dimensionless constants and a is a specified length. These circumstances arise if the tube is closed at both ends.

Next, we integrate (6.8.23) over a volume of fluid inside the tube bounded by a far upstream cross-sectional surface, D_+, and a far downstream cross-sectional surface, D_-. Using the divergence theorem, we convert the volume integral into a surface integral. Recalling that $\mathbf{u}^P$ and $\mathcal{G}$ are both zero when the point $\mathbf{x}$ lies at the tube surface, and observing that $\mathcal{G}$ is virtually zero on $D_{\pm\infty}$ due to the fast decay of the flow due to a point force, we find that

$$\iint_{D_\pm} u_i^P(\mathbf{x})\,\mathcal{T}_{ikj}(\mathbf{x},\mathbf{x}_0)\,n_j(\mathbf{x})\,\mathrm{d}S(\mathbf{x}) = 8\pi\,u_i^p(\mathbf{x}_0), \qquad (6.8.25)$$

where $\mathbf{n}$ is the unit normal vector pointing into the control volume. Evaluating the integral using (6.8.24), we obtain $(\alpha_{+\infty} - \alpha_{-\infty})\,Q_p = a^2 u_i^p(\mathbf{x}_0)$, where

$$Q_p = \iint_{D_{-\infty}} \mathbf{u}^P \cdot \mathbf{n}\,\mathrm{d}S \qquad (6.8.26)$$

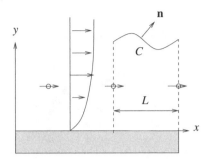

FIGURE 6.8.3   Illustration of a periodic array of two-dimensional point forces above a plane wall.

is the axial flow rate of the pressure-driven flow. Rearranging, we derive an expression for the pressure drop due to a point force inside a closed tube,

$$\alpha_{+\infty} - \alpha_{-\infty} = \frac{a^2}{Q_p}\, u_i^p(\mathbf{x}_0). \tag{6.8.27}$$

This pressure drop induces a back flow that cancels the forward flow induced by the point force. The functional dependence of the pressure drop on the position of the point force is precisely that exhibited by the velocity profile of the corresponding pressure-driven flow.

Flow in a circular tube

In the case of flow through a circular tube of radius a, we use the parabolic Poiseuille flow profile and find that

$$\alpha_{+\infty} - \alpha_{-\infty} = 2\,\frac{a^2 - \sigma_0^2}{a^2}, \tag{6.8.28}$$

where σ_0 is the distance of the point force from the centerline.

6.8.5   Streaming flow due to an array of point forces above a wall

Consider a periodic array of two-dimensional point forces located at a sequence of points $\mathbf{x}_l$ parallel to an infinite plane wall positioned at $y = 0$, as illustrated in Figure 6.8.3, where $l = -\infty, \ldots, 0, \ldots, \infty$. Applying the reciprocal theorem for (a) the flow induced by the periodic array expressed in terms of the corresponding periodic Green's function, and (b) unidirectional simple shear flow with velocity $u_x^{ssf} = \xi y$, $u_y^{ssf} = 0$, we obtain

$$\frac{\partial}{\partial x_j}\left[\, u_i^{ssf}(\mathbf{x})\,\frac{1}{4\pi}\,\mathcal{T}_{ikj}(\mathbf{x}, \mathbf{x}_0) - \frac{1}{4\pi\mu}\,\mathcal{G}_{ik}(\mathbf{x}, \mathbf{x}_0)\,\sigma_{ij}^{ssf}(\mathbf{x})\,\right] = -u_i^{ssf}(\mathbf{x})\sum_{l=-\infty}^{\infty}\delta_2(\mathbf{x} - \mathbf{x}_l), \tag{6.8.29}$$

where k is a free index indicating the direction of the point forces and δ_2 is the two-dimensional delta function.

When the point forces are perpendicular to the wall, the induced velocity decays far from the wall. When the point forces are parallel to the wall, the induced velocity tends to a constant value far from the wall,

$$\mathcal{G}_{ij}(\mathbf{x}, \mathbf{x}_0) \to 4\pi\alpha \, \delta_{ix} \, \delta_{jx}, \tag{6.8.30}$$

as $y \to \infty$, where α is a dimensionless constant. The corresponding pressure and stress fields decay at an exponential rate.

Next, we integrate (6.8.29) over one period of the flow confined between the wall and a periodic line far above the wall, C, and use the divergence theorem to convert the areal integral into a line integral along the boundary of one period of the flow. Noting that the net contribution of the periodic segments is zero and recalling that $\mathbf{u}^{ssf}$ and $\mathcal{G}$ are zero when $\mathbf{x}$ lies on the wall, we obtain

$$\int_C \mathcal{G}_{ik}(\mathbf{x}, \mathbf{x}_0) \, \sigma_{ij}^{ssf}(\mathbf{x}) \, n_j(\mathbf{x}) \, \mathrm{d}l(\mathbf{x}) = 4\pi\mu \, u_i^{ssf}(\mathbf{x}_0), \tag{6.8.31}$$

where $\mathbf{n}$ is the unit normal vector pointing outward from the control volume. Substituting (6.8.30) and evaluating the stress field of the simple shear flow, we find that

$$\alpha\xi L = u_i^{ssf}(\mathbf{x}_0), \tag{6.8.32}$$

where L is the point force separation. Rearranging, we find that

$$\alpha = \frac{y_0}{L}. \tag{6.8.33}$$

We have shown that the velocity of the streaming flow induced by an array of point forces parallel to a wall is proportional to the distance of the point forces from the wall.

The results are readily generalized to the flow induced by a doubly periodic array of three-dimensional point forces arranged in a Bravais lattice parallel to an infinite plane wall located at $x = 0$. The counterpart of (6.8.30) is

$$\mathcal{G}_{ij}(\mathbf{x}, \mathbf{x}_0) \to 8\pi \frac{x_0}{\sqrt{A}} \, J_{ij}, \tag{6.8.34}$$

where x_0 is the distance of the point forces from the wall, $\mathbf{J}$ is the identity matrix except that the first diagonal element is set to zero, and A is the area occupied by one periodic cell in the yx plane parallel to the wall.

Problems

6.8.1 *Alternative statement of the reciprocal theorem*

Show that two Stokes flows of a certain fluid with velocity $\mathbf{u}$ and $\mathbf{u}'$ and corresponding hydrodynamic pressure fields p and p' satisfy the alternative reciprocal relation [166]

$$\frac{\partial}{\partial x_k}\left[u_i'\left(-\delta_{ik}p + \mu\frac{\partial u_i}{\partial x_k}\right) - u_i\left(-\delta_{ik}p' + \mu\frac{\partial u_i'}{\partial x_k}\right)\right] = 0. \tag{6.8.35}$$

6.8.2 *Faxén laws*

(a) Prove equation (6.8.9).

(b) Departing from (6.8.11), show that the force exerted on a spherical particle of radius a placed at the axis of a paraboloidal flow with velocity $\mathbf{u}^\infty = U[(y/a)^2 + (z/a)^2, 0, 0]$ is $\mathbf{F} = 4\pi\mu U a[1, 0, 0]$.

6.8.3 *Symmetry of the resistance tensors*

Use the reciprocal theorem to show that the resistance matrices $\mathbf{X}$ and $\mathbf{Y}$ introduced in (6.8.15) are symmetric, and $\mathbf{W}$ is the transpose of $\mathbf{Z}$ [54, 55].

6.8.4 *Force and torque from the mobility problem*

Prove equations (6.8.17).

6.8.5 *Symmetry of the two-dimensional Green's functions*

Prove that the two-dimensional Green's functions satisfy the symmetry property (6.5.12). The proof presents an apparent but not essential difficulty due to the possible divergence of the Green's functions at infinity [306].

6.9   Boundary-integral representation of Stokes flow

The solution of linear, elliptic, and homogeneous partial differential equations can be represented in terms of boundary integrals involving the unknown function and its derivatives (e.g., [386, 387]). One example is the boundary-integral representation of harmonic functions discussed in Section 2.3. Another example is Somigliana's identity for the displacement field in linear elastostatics (e.g., [247], p. 245; [295]). In the case of Stokes flow, we obtain a boundary-integral representation involving the boundary values of the velocity and traction.

Reciprocal theorem with a point force

A convenient starting point for deriving the boundary-integral representation is the Lorentz reciprocal identity (6.8.2) applied for a particular flow of interest with velocity $\mathbf{u}$ and stress $\boldsymbol{\sigma}$, and the flow due to a point force with strength $\mathbf{b}$ located a point, $\mathbf{x}_0$, with velocity and stress fields

$$u_i'(\mathbf{x}) = \frac{1}{8\pi\mu}\, \mathcal{G}_{ij}(\mathbf{x}, \mathbf{x}_0)\, b_j, \qquad \sigma_{ik}'(\mathbf{x}) = \frac{1}{8\pi}\, \mathcal{T}_{ijk}(\mathbf{x}, \mathbf{x}_0)\, b_j. \qquad (6.9.1)$$

Substituting these expressions into (6.8.2) with $\mu = \mu'$ and discarding the arbitrary constant $\mathbf{b}$, we obtain the equation

$$\frac{\partial}{\partial x_k}\Big(\mathcal{G}_{ij}(\mathbf{x}, \mathbf{x}_0)\, \sigma_{ik}(\mathbf{x}) - \mu\, u_i(\mathbf{x})\, \mathcal{T}_{ijk}(\mathbf{x}, \mathbf{x}_0)\Big) = 0, \qquad (6.9.2)$$

which is valid everywhere except at the singular point, $\mathbf{x}_0$.

Reciprocal identity with a point force

Next, we select a control volume, V_c, bounded by a closed, singly or multiply connected surface, D, consisting of fluid surfaces, fluid interfaces, or solid surfaces, as illustrated in Figure 6.9.1, and place

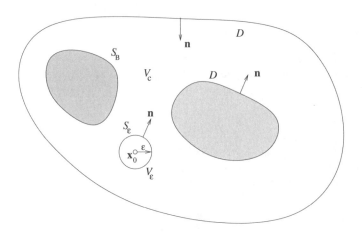

FIGURE 6.9.1   A control volume used to derive the boundary-integral representation of Stokes flow and associated integral equations.

the point force outside the control volume. Noting that the function inside the large parentheses in (6.9.2) is nonsingular throughout V_c, we integrate (6.9.2) over V_c and use the divergence theorem to convert the volume integral over V_c into a surface integral over D, obtaining

$$\iint_D \left[\mathcal{G}_{ij}(\mathbf{x}, \mathbf{x}_0)\, \sigma_{ik}(\mathbf{x}) - \mu\, u_i(\mathbf{x})\, \mathcal{T}_{ijk}(\mathbf{x}, \mathbf{x}_0) \right] n_k(\mathbf{x})\, \mathrm{d}S(\mathbf{x}) = 0. \qquad (6.9.3)$$

In equation (6.9.3) as well as in all subsequent equations, the unit normal vector, $\mathbf{n}$, points into the control volume, V_c.

Integral representation

Now we place the point $\mathbf{x}_0$ inside the control volume, V_c, and introduce a small spherical volume V_ϵ of radius ϵ centered at $\mathbf{x}_0$. The function inside the square brackets in (6.9.2) is nonsingular throughout the reduced volume, $V_c - V_\epsilon$. Integrating (6.9.2) over $V_c - V_\epsilon$ and using the divergence theorem to convert the volume integral into a surface integral, we obtain

$$\iint_{D,S_\epsilon} \left[\mathcal{G}_{ij}(\mathbf{x}, \mathbf{x}_0)\, \sigma_{ik}(\mathbf{x}) - \mu\, u_i(\mathbf{x})\, \mathcal{T}_{ijk}(\mathbf{x}, \mathbf{x}_0) \right] n_k(\mathbf{x})\, \mathrm{d}S(\mathbf{x}) = 0, \qquad (6.9.4)$$

where S_ϵ is the spherical surface enclosing the volume V_ϵ, as illustrated in Figure 6.9.1. Letting the radius ϵ shrink to zero, we find that, to leading order in ϵ, the tensors $\mathcal{G}$ and $\mathcal{T}$ reduce to the Stokeslet and its associated stress tensor,

$$\mathcal{G}_{ij} \simeq \frac{\delta_{ij}}{\epsilon} + \frac{\hat{x}_i \hat{x}_j}{\epsilon^3}, \qquad\qquad \mathcal{T}_{ijk} \simeq -6\, \frac{\hat{x}_i \hat{x}_j \hat{x}_k}{\epsilon^5}, \qquad (6.9.5)$$

where $\hat{\mathbf{x}} = \mathbf{x} - \mathbf{x}_0$. Over S_ϵ, the normal vector is $\mathbf{n} = \frac{1}{\epsilon}\hat{\mathbf{x}}$ and the differential surface area is $\mathrm{d}S = \epsilon^2 \mathrm{d}\Omega$, where Ω is the differential solid angle. Substituting these expressions along with (6.9.5) into (6.9.4), we obtain

$$\iint_D \left[\mathcal{G}_{ij}(\mathbf{x}, \mathbf{x}_0)\, \sigma_{ik}(\mathbf{x}) - \mu\, u_i(\mathbf{x})\, \mathcal{T}_{ijk}(\mathbf{x}, \mathbf{x}_0) \right] n_k(\mathbf{x})\, \mathrm{d}S(\mathbf{x})$$

$$= -\iint_{S_\epsilon} \left[\left(\frac{\delta_{ij}}{\epsilon} + \frac{\hat{x}_i \hat{x}_j}{\epsilon^3} \right) \sigma_{ik}(\mathbf{x}) + 6\mu\, u_i(\mathbf{x})\, \frac{\hat{x}_i \hat{x}_j \hat{x}_k}{\epsilon^5} \right] \hat{x}_k\, \mathrm{d}S(\mathbf{x}) = 0. \qquad (6.9.6)$$

As $\epsilon \to 0$, $\mathbf{u}$ and $\boldsymbol{\sigma}$ over S_ϵ tend to their respective values at the center of the small volume V_ϵ, which are $\mathbf{u}(\mathbf{x}_0)$ and $\boldsymbol{\sigma}(\mathbf{x}_0)$. Since $\hat{\mathbf{x}}$ decreases linearly with ϵ as $\epsilon \to 0$, the contribution of the stress term inside the integral on the right-hand side of (6.9.6) decreases linearly with ϵ whereas the contribution of the velocity term tends to a constant value. Thus, in the limit $\epsilon \to 0$, equation (6.9.6) becomes

$$\iint_D [\mathcal{G}_{ij}(\mathbf{x}, \mathbf{x}_0)\, \sigma_{ik}(\mathbf{x}) - \mu\, u_i(\mathbf{x})\, \mathcal{T}_{ijk}(\mathbf{x}, \mathbf{x}_0)]\, n_k(\mathbf{x})\, \mathrm{d}S(\mathbf{x}) = -6\mu u_i(\mathbf{x}_0)\, \frac{1}{\epsilon^4} \iint_{S_\epsilon} \hat{x}_i \hat{x}_j\, \mathrm{d}S(\mathbf{x}). \quad (6.9.7)$$

Using the divergence theorem, we compute

$$\iint_{S_\epsilon} \hat{x}_i \hat{x}_j\, \mathrm{d}S(\mathbf{x}) = \epsilon \iint_{S_\epsilon} \hat{x}_i n_j\, \mathrm{d}S(\mathbf{x}) = \epsilon \iiint_{V_\epsilon} \frac{\partial \hat{x}_i}{\partial \hat{x}_j}\, \mathrm{d}S(\mathbf{x}) = \delta_{ij}\, \frac{4\pi}{3}\, \epsilon^4. \qquad (6.9.8)$$

Substituting this expression into (6.9.7), we obtain

$$u_j(\mathbf{x}_0) = -\frac{1}{8\pi\mu} \iint_D \mathcal{G}_{ij}(\mathbf{x}, \mathbf{x}_0)\, f_i(\mathbf{x})\, \mathrm{d}S(\mathbf{x}) + \frac{1}{8\pi} \iint_D u_i(\mathbf{x})\, \mathcal{T}_{ijk}(\mathbf{x}, \mathbf{x}_0)\, n_k(\mathbf{x})\, \mathrm{d}S(\mathbf{x}), \qquad (6.9.9)$$

where $\mathbf{f} = \boldsymbol{\sigma} \cdot \mathbf{n}$ is the boundary traction.

Equation (6.9.9) provides us with an integral representation of the velocity field in terms of boundary distributions of the Green's function, $\mathcal{G}$, and associated stress tensor, $\mathcal{T}$. The densities of these distributions are proportional, respectively, to the boundary values of the traction and velocity. Making an analogy with corresponding results in the theory of electrostatics, we term the first distribution involving the boundary traction the single-layer potential, and the second distribution involving the boundary velocity the double-layer potential.

Physical interpretation

The symmetry property (6.5.12) allows us to switch the order of the indices of the Green's function in the single-layer potential, provided that we also switch the order of the arguments, $\mathbf{x}$ and $\mathbf{x}_0$, obtaining

$$u_j(\mathbf{x}_0) = -\frac{1}{8\pi\mu} \iint_D \mathcal{G}_{ji}(\mathbf{x}_0, \mathbf{x})\, f_i(\mathbf{x})\, \mathrm{d}S(\mathbf{x}) + \frac{1}{8\pi} \iint_D u_i(\mathbf{x})\, \mathcal{T}_{ijk}(\mathbf{x}, \mathbf{x}_0)\, n_k(\mathbf{x})\, \mathrm{d}S(\mathbf{x}). \quad (6.9.10)$$

It is now evident that the single-layer potential represents a boundary distribution of point forces with surface strength equal to $-\mathbf{f}$. The physical interpretation of the double-layer potential will be discussed later in this section.

Flow in an infinite domain

Mathematical idealization produces entirely or partially unbounded domains of flow. Two examples are the flow due the motion of a small particle in an infinitely dilute suspension, and semi-infinite shear flow over a wall containing a solitary depression or projection. To apply the boundary integral equation, we select a control volume that is confined by a solid or fluid boundary, B, and a large surface extending to infinity, S_∞. If the fluid is quiescent at infinity, the velocity must decay at least as fast as $1/d$ and the pressure and stress must decay at least as fast as $1/d^2$, where d is a typical distance from B, as discussed in Section 6.1. Since the Green's function decays at least as fast as $1/d$ and the associated stress tensor decays at least as fast as $1/d^2$, as S_∞ expands to infinity, the single- and double-layer potentials make vanishing contributions. As a result, the domain D of the boundary-integral equation is conveniently reduced to B.

Simplification with the use of proper Green's functions

The domain of integration of the single- and double-layer potentials consists of all fluid and solid surfaces enclosing a selected volume of flow. If the velocity happens to vanish over a portion of the boundary, the corresponding double-layer integral does not make a contribution to the double-layer potential. Similarly, if the surface force happens to vanish over another portion of the boundary, the corresponding single-layer integral does not make a contribution to the single-layer potential. A further reduction in the domain of integration is possible by using a Green's function that observes the symmetry, periodicity, or other special features of the flow.

Consider a flow that is bounded by an internal or external solid stationary surface, B. Since the velocity is zero over B, the corresponding double-layer integral does not appear. To also eliminate the single-layer potential, we use a Green's function that vanishes over B, that is, $\mathcal{G}(\mathbf{x}, \mathbf{x}_0) = \mathbf{0}$ when $\mathbf{x}$ is on B. In this way, B is conveniently excluded from the domain of the boundary-integral equation.

Integral identities

Useful identities can be derived by applying the boundary-integral equation for simple flows that are known to be exact solutions of the equations of Stokes flow. In the case of rigid-body motion with translational velocity $\mathbf{V}$ and angular velocity $\mathbf{\Omega}$, the fluid velocity is $\mathbf{u} = \mathbf{V} + \mathbf{\Omega} \times \hat{\mathbf{x}}$, where $\hat{\mathbf{x}} = \mathbf{x} - \mathbf{x}_c$ and $\mathbf{x}_c$ is the center of rotation. Setting $\mathbf{f} = -p\mathbf{n}$, where p is the constant hydrodynamic pressure, we derive the identities

$$\iint_D T_{ijk}(\mathbf{x}, \mathbf{x}_0)\, n_k(\mathbf{x})\, dS(\mathbf{x}) = 8\pi\alpha\, \delta_{ij} \tag{6.9.11}$$

and

$$\epsilon_{ilm} \iint_D \hat{x}_m\, T_{ijk}(\mathbf{x}, \mathbf{x}_0) n_k(\mathbf{x})\, dS(\mathbf{x}) = 8\pi\alpha\, \epsilon_{jlm}\, \hat{x}_{0_m}, \tag{6.9.12}$$

where D is the smooth boundary of an arbitrary control volume, $\hat{x}_{0_m}$ denotes the mth component of $\mathbf{x}_0$, and $\alpha = 1$ or 0 depending on whether the point $\mathbf{x}_0$ is located inside or outside D. These equations reiterate identities (6.5.10) and (6.5.11). Similar identities can be written for linear or parabolic flow.

6.9.1 Flow past a translating and rotating rigid particle

In the particular case of flow past a stationary or moving rigid particle, the boundary-integral representation is amenable to an important simplification. We start by decomposing the velocity, $\mathbf{u}$, into an undisturbed component prevailing in the absence of the particle, $\mathbf{u}^\infty$, and a disturbance component due to the particle, $\mathbf{u}^D$, so that $\mathbf{u} = \mathbf{u}^\infty + \mathbf{u}^D$. On the surface of the particle, the fluid velocity is $\mathbf{u} = \mathbf{V} + \mathbf{\Omega} \times (\mathbf{x} - \mathbf{x}_c)$ and the disturbance velocity is $\mathbf{u}^D = -\mathbf{u}^\infty + \mathbf{V} + \mathbf{\Omega} \times (\mathbf{x} - \mathbf{x}_c)$, where $\mathbf{V}$ is the velocity of translation and $\mathbf{\Omega}$ is the angular velocity of rotation about the point $\mathbf{x}_c$. Applying the integral representation (6.9.10) for the disturbance flow, neglecting the integrals at infinity due to their decay and simplifying, we find that, for a point $\mathbf{x}_0$ in the flow,

$$u_j^D(\mathbf{x}_0) = -\frac{1}{8\pi\mu} \iint_P f_i^D(\mathbf{x})\, \mathcal{G}_{ij}(\mathbf{x}, \mathbf{x}_0)\, \mathrm{d}S(\mathbf{x}) + \frac{1}{4\pi} \iint_P u_i^\infty(\mathbf{x})\, \mathcal{T}_{ijk}(\mathbf{x}, \mathbf{x}_0)\, n_k(\mathbf{x})\, \mathrm{d}S(\mathbf{x}), \quad (6.9.13)$$

where P is the particle surface. Since the point $\mathbf{x}_0$ lies in the exterior of the control volume occupied by the particle, we may use the reciprocal identity (6.9.1) for the incident flow $\mathbf{u}^\infty$ to write

$$\mu \iint_P u_i^\infty(\mathbf{x})\, \mathcal{T}_{ijk}(\mathbf{x}, \mathbf{x}_0)\, n_k(\mathbf{x})\, \mathrm{d}S(\mathbf{x}) = \iint_P f_i^\infty(\mathbf{x})\, \mathcal{G}_{ij}(\mathbf{x}, \mathbf{x}_0)\, \mathrm{d}S(\mathbf{x}). \quad (6.9.14)$$

Now combining the two representations to eliminate the double-layer integral from the right-hand side of (6.9.13) and adding the incident velocity field $\mathbf{u}^\infty$ to both sides of the resulting equation, we derive a simplified single-layer representation,

$$u_j(\mathbf{x}_0) = u_j^\infty(\mathbf{x}_0) - \frac{1}{8\pi\mu} \iint_P f_i(\mathbf{x})\, \mathcal{G}_{ij}(\mathbf{x}, \mathbf{x}_0)\, \mathrm{d}S(\mathbf{x}). \quad (6.9.15)$$

The disturbance velocity field is represented by a surface distribution of point forces whose strength density is the boundary traction.

Behavior of the flow far from a particle

To study the behavior of the flow far from a particle, we expand the Green's function in a Taylor series with respect to $\mathbf{x}$ about a chosen point, $\mathbf{x}_c$, that is located somewhere in the vicinity or inside the body, and derive a multipole expansion for the velocity,

$$\begin{aligned}
u_j(\mathbf{x}_0) = u_j^\infty(\mathbf{x}_0) - \frac{1}{8\pi\mu} \bigg(&\mathcal{G}_{ji}(\mathbf{x}_0, \mathbf{x}_c) \iint_P f_i(\mathbf{x})\, \mathrm{d}S(\mathbf{x}) \\
&+ \frac{\partial \mathcal{G}_{ij}}{\partial x_{c_k}}(\mathbf{x}_0, \mathbf{x}_c) \iint_P (x_k - x_{c,k})\, f_i(\mathbf{x})\, \mathrm{d}S(\mathbf{x}) + \cdots \bigg).
\end{aligned} \quad (6.9.16)$$

The first integral on the right-hand side of (6.9.16) is the force exerted on the particle, $\mathbf{F}$. The second integral is the first moment of the boundary traction. Subsequent integrals express higher moments of the boundary traction.

The first term on the right-hand side of (6.9.16) represents the flow due to a point force. The second term represents the flow due to a point-force dipole. Subsequent terms represent the flow due to point-force quadruples and higher-order multipoles. The coefficient of the point-force dipole, $\mathbf{q}$, can be decomposed into the coefficient of the stresslet, $\mathbf{s}$, and an antisymmetric component, $\mathbf{r}$, amounting to a torque mediating couplet, as discussed in Section 6.6.2. Equation (6.9.16) shows that, far from a particle, the disturbance flow due to the particle is similar to that due to a point force with strength $-\mathbf{F}$. When $\mathbf{F} = \mathbf{0}$, the far flow is similar to that produced by a point-force dipole. Consequently, in the case of flow in an infinite domain, the disturbance flow decays as $1/d$ or $1/d^2$, respectively, for $\mathbf{F} \neq \mathbf{0}$ or $\mathbf{F} = \mathbf{0}$, where d is the distance from the particle center.

Translating sphere

Substituting into (6.9.15) the expression for the traction exerted on a sphere translating with velocity $\mathbf{V}$ in an infinite quiescent fluid, given in (6.7.11), we obtain

$$u_j(\mathbf{x}_0) = \frac{3}{16\pi a} V_i \iint_{Sphere} \mathcal{S}_{ij}(\mathbf{x}, \mathbf{x}_0) \, \mathrm{d}S(\mathbf{x}), \tag{6.9.17}$$

where $\mathcal{S}_{ij}$ is the Stokeslet. Substituting into the left-hand side the singularity representation (6.7.1) expressed in the form

$$u_j(\mathbf{x}_0) = V_i \left(\frac{3}{4} a + \frac{1}{8} a^3 \nabla_0^2 \right) \mathcal{S}_{ij}(\mathbf{x}_0, \mathbf{x}_c), \tag{6.9.18}$$

discarding the arbitrary velocity V_i, and rearranging, we obtain the identity

$$\frac{1}{4\pi a^2} \iint_{Sphere} \mathcal{S}_{ij}(\mathbf{x}, \mathbf{x}_0) \, \mathrm{d}S(\mathbf{x}) = \left(1 + \frac{1}{6} a^2 \nabla_c^2 \right) \mathcal{S}_{ij}(\mathbf{x}_0, \mathbf{x}_c), \tag{6.9.19}$$

where $\mathbf{x}_c$ is the center of the sphere, the gradient ∇_c involves derivatives with respect to $\mathbf{x}_c$, and the point $\mathbf{x}_0$ lies outside the sphere. This expression is consistent with the mean-value theorem for biharmonic functions stated in (2.4.14).

For a point $\mathbf{x}_0$ that lies inside the sphere, the representation (6.9.17) yields $u_j(\mathbf{x}_0) = V_j$, and we find that

$$\iint_{Sphere} \mathcal{S}_{ij}(\mathbf{x}, \mathbf{x}_0) \, \mathrm{d}S(\mathbf{x}) = \frac{16}{3} \pi a \, \delta_{ij}, \tag{6.9.20}$$

independent of the precise position of $\mathbf{x}_0$ inside the sphere. This identity can be confirmed readily when the point $\mathbf{x}_0$ is the center of the sphere.

6.9.2 Significance of the double-layer potential

We return to discussing the physical significance of the double-layer potential involving the stress tensor of the Green's function, $\mathcal{T}_{ijk}$. Decomposing the stress tensor into its pressure and viscous constituents using (6.5.4), we obtain

$$\iint_D u_i(\mathbf{x}) \, \mathcal{T}_{ijk}(\mathbf{x}, \mathbf{x}_0) \, n_k(\mathbf{x}) \, \mathrm{d}S(\mathbf{x}) = - \iint_D \mathcal{P}_j(\mathbf{x}, \mathbf{x}_0) \, (\mathbf{u} \cdot \mathbf{n})(\mathbf{x}) \, \mathrm{d}S(\mathbf{x})$$

$$+ \iint_D \frac{\partial \mathcal{G}_{ji}(\mathbf{x}, \mathbf{x}_0)}{\partial x_k} (u_i n_k + u_k n_i)(\mathbf{x}) \, \mathrm{d}S(\mathbf{x}). \tag{6.9.21}$$

The second integral on the right-hand side represents a distribution of symmetric point-force dipoles.

Interpretation of the pressure vector as a point source

Consider the first integral on the right-hand side of (6.9.21) in the case of an entirely or partially infinite domain of flow. Since the boundary velocity distribution is arbitrary, the pressure vector, $\mathcal{P}(\mathbf{x}, \mathbf{x}_0)$, must represent the velocity field of a Stokes flow. In fact, inspecting the behavior of the pressure near the point force reveals that $\mathcal{P}(\mathbf{x}, \mathbf{x}_0)$ represents the velocity at the point $\mathbf{x}_0$ due to a point source located at the point $\mathbf{x}$. It is now evident that the first integral on the right-hand side of (6.9.21) represents a distribution of point sources whose density is proportional to the normal component of the fluid velocity, $\mathbf{u} \cdot \mathbf{n}$. This contribution vanishes over a solid boundary where $\mathbf{u} = \mathbf{0}$ and over a stationary fluid interface where $\mathbf{u} \cdot \mathbf{n} = 0$.

Using the definition of the Green's function and the symmetry property of the velocity Green's function tensor, we write

$$\frac{\partial \mathcal{P}_j(\mathbf{x}, \mathbf{x}_0)}{\partial x_k} = \nabla^2 \mathcal{G}_{jk}(\mathbf{x}_0, \mathbf{x}). \tag{6.9.22}$$

This equation reveals that all derivatives of the pressure can be expressed in terms of the Laplacian of the Green's function.

Interior and triply periodic flow

The interpretation of the pressure Green's function discussed earlier in this section does not apply in the case of a completely enclosed or triply periodic domain of flow. The reason is that the velocity distribution is subject to the constraint of zero flow rate across the boundary. Consequently, $\mathcal{P}(\mathbf{x}, \mathbf{x}_0)$ may no longer be interpreted as the flow due to a point source.

6.9.3   Flow due to a stresslet

Our discussion has shown that the stress tensor associated with a point force in an entirely or partially infinite, but not completely enclosed, domain of flow represents a fundamental solution of Stokes flow. Specifically, the pair

$$u_j(\mathbf{x}_0) = \frac{1}{16\pi\mu} T_{ijk}(\mathbf{x}, \mathbf{x}_0)\, s_{ik}, \qquad p(\mathbf{x}_0) = \frac{1}{8\pi} \mathcal{P}_{ik}^{\Sigma}(\mathbf{x}, \mathbf{x}_0)\, s_{ik} \tag{6.9.23}$$

represents the velocity and pressure field at the point $\mathbf{x}_0$ due to a stresslet with strength $\mathbf{s}$ placed at the point $\mathbf{x}$, where $\mathcal{P}^{\Sigma}$ is the corresponding pressure tensor. In the case of the free-space Green's function, the pressure tensor due to the stresslet is

$$\mathcal{P}_{ik}^{STR}(\mathbf{x}, \mathbf{x}_0) = -2\frac{\delta_{ik}}{r^3} + 6\frac{\hat{x}_i \hat{x}_k}{r^5}, \tag{6.9.24}$$

as shown in the first entry of Table 6.6.3.

6.9.4   Boundary-integral representation for the pressure

The interpretation of the double-layer potential in terms of point sources and point-force dipoles motivates a boundary-integral representation of the pressure in terms of two distributions corresponding to the single- and double-layer potentials,

$$p(\mathbf{x}_0) = -\frac{1}{8\pi} \iint_D \mathcal{P}_i(\mathbf{x}, \mathbf{x}_0) f_i(\mathbf{x}) \, dS(\mathbf{x}) + \frac{\mu}{4\pi} \iint_D u_i(\mathbf{x}) \mathcal{P}_{ik}^{\Sigma}(\mathbf{x}, \mathbf{x}_0) n_k(\mathbf{x}) \, dS(\mathbf{x}). \quad (6.9.25)$$

The vector $\mathcal{P}$ and tensor $\mathcal{P}^{\Sigma}$ express the pressure corresponding to the Green's function and associated stress tensor expressing a stresslet.

6.9.5 Flow past an interface

Consider a flow with velocity $\mathbf{u}^{\infty}$ past a closed fluid interface representing a bubble, drop, capsule, or cell. The boundary-integral representation (6.9.10) at a point $\mathbf{x}_0$ outside the interface for the disturbance flow indicated by the superscript D becomes

$$u_j^D(\mathbf{x}_0) = -\frac{1}{8\pi\mu} \iint_{\mathcal{I}} \mathcal{G}_{ji}(\mathbf{x}_0, \mathbf{x}) f_i^D(\mathbf{x}) \, dS(\mathbf{x}) + \frac{1}{8\pi} \iint_{\mathcal{I}} u_i^D(\mathbf{x}) \mathcal{T}_{ijk}(\mathbf{x}, \mathbf{x}_0) n_k(\mathbf{x}) \, dS(\mathbf{x}), \quad (6.9.26)$$

where $\mathcal{I}$ denotes the interface. Since the point $\mathbf{x}_0$ is located in the exterior of the interface, we may use the reciprocal identity (6.9.1) for the incident flow $\mathbf{u}^{\infty}$ to write

$$\iint_{\mathcal{I}} \mathcal{G}_{ji}(\mathbf{x}_0, \mathbf{x}) f_i^{\infty}(\mathbf{x}) \, dS(\mathbf{x}) = \mu \iint_{\mathcal{I}} u_i^{\infty}(\mathbf{x}) \mathcal{T}_{ijk}(\mathbf{x}, \mathbf{x}_0) n_k(\mathbf{x}) \, dS(\mathbf{x}). \quad (6.9.27)$$

Combining the last two representations to formulate the total velocity and traction, $\mathbf{u} = \mathbf{u}^{\infty} + \mathbf{u}^D$ and $\mathbf{f} = \mathbf{f}^{\infty} + \mathbf{f}^D$, and adding the incident velocity field $\mathbf{u}^{\infty}$ to both sides of the resulting equation, we obtain

$$u_j(\mathbf{x}_0) = u_j^{\infty}(\mathbf{x}_0) - \frac{1}{8\pi\mu} \iint_{\mathcal{I}^+} \mathcal{G}_{ji}(\mathbf{x}_0, \mathbf{x}) f_i(\mathbf{x}) \, dS(\mathbf{x}) + \frac{1}{8\pi} \iint_{\mathcal{I}^+} u_i(\mathbf{x}) \mathcal{T}_{ijk}(\mathbf{x}, \mathbf{x}_0) n_k(\mathbf{x}) \, dS(\mathbf{x}). \quad (6.9.28)$$

The notation $\mathcal{I}^+$ emphasizes that the velocity and traction are evaluated on the outer side of the interface indicated by the normal vector, $\mathbf{n}$.

Behavior of the flow far from an interface

To study the behavior of the flow far from a closed interface, we expand the Green's function in a Taylor series with respect to $\mathbf{x}$ about a chosen point, $\mathbf{x}_c$, located somewhere in the vicinity or inside the interface, and derive a multipole expansion for the velocity,

$$\begin{aligned} u_j(\mathbf{x}_0) = u_j^{\infty}(\mathbf{x}_0) &- \frac{1}{8\pi\mu} \left(\mathcal{G}_{ji}(\mathbf{x}_0, \mathbf{x}_c) \iint_{\mathcal{I}^+} f_i(\mathbf{x}) \, dS(\mathbf{x}) \right. \\ &+ \left. \frac{\partial \mathcal{G}_{ij}}{\partial x_{c_k}}(\mathbf{x}_0, \mathbf{x}_c) \iint_{\mathcal{I}^+} [(x_k - x_{c,k}) f_i(\mathbf{x}) - \mu(u_k n_i + u_i n_k)(\mathbf{x})] \, dS(\mathbf{x}) + \cdots \right) \\ &- \frac{1}{8\pi} \mathcal{P}_j(\mathbf{x}_c, \mathbf{x}_0) \iint_{\mathcal{I}} \mathbf{u} \cdot \mathbf{n} \, dS + \cdots. \end{aligned} \quad (6.9.29)$$

The first series on the right-hand side contains multipoles of the point force, while the second series contains multipoles of the pressure. It is evident now that the pressure term $\mathcal{P}(\mathbf{x}_c, \mathbf{x}_0)$ represents the velocity field at the point $\mathbf{x}_0$ due to a point source at the point $\mathbf{x}_c$, in agreement with our earlier observations. Formula (6.9.22) suggests that all terms after the first term in the second series on the right-hand side of (6.9.29) can be incorporated into the first series. Consequently, the disturbance flow far from the interface can be represented in terms of an expansion of multipoles of the point force supplemented by a point source.

6.9.6   Flow past a liquid drop or capsule

Consider a flow with velocity $\mathbf{u}^\infty$ and viscosity μ past a liquid drop, capsule, or cell that contains a fluid with viscosity $\lambda\mu$. The velocity at a point in the ambient fluid, $\mathbf{x}_0$, is given by the integral representation (6.9.28). Applying the reciprocal theorem at that point for the interior flow, we obtain

$$\frac{1}{8\pi\lambda\mu} \iint_{\mathcal{I}^-} \mathcal{G}_{ji}(\mathbf{x}_0, \mathbf{x})\, f_i(\mathbf{x})\, \mathrm{d}S(\mathbf{x}) = \frac{1}{8\pi} \iint_{\mathcal{I}^-} u_i(\mathbf{x})\, \mathcal{T}_{ijk}(\mathbf{x}, \mathbf{x}_0)\, n_k(\mathbf{x})\, \mathrm{d}S(\mathbf{x}), \qquad (6.9.30)$$

where $\mathcal{I}^-$ denotes the interior side of the interface. Combining (6.9.28) with (6.9.30) to formulate the jump in the traction across the interface, $\Delta\mathbf{f} \equiv \mathbf{f}^+ - \mathbf{f}^-$, and assuming that the velocity is continuous across the interface, we derive the integral representation

$$u_j(\mathbf{x}_0) = u_j^\infty(\mathbf{x}_0) - \frac{1}{8\pi\mu} \iint_{\mathcal{I}} \mathcal{G}_{ji}(\mathbf{x}_0, \mathbf{x})\, \Delta f_i(\mathbf{x})\, \mathrm{d}S(\mathbf{x}) + \frac{1-\lambda}{8\pi} \iint_{\mathcal{I}} u_i(\mathbf{x})\, \mathcal{T}_{ijk}(\mathbf{x}, \mathbf{x}_0)\, n_k(\mathbf{x})\, \mathrm{d}S(\mathbf{x}),$$

$$(6.9.31)$$

where $\mathbf{n}$ is the unit normal vector pointing into the ambient fluid. In the case of equal viscosities, $\lambda = 1$, the double-layer integral does not appear.

Since the force on the exterior side of the interface is equal to the force on the interior side of the interface to satisfy global equilibrium, and since the particle volume is fixed, the multipole expansion (6.9.29) simplifies into

$$u_j(\mathbf{x}_0) = u_j^\infty(\mathbf{x}_0) + \frac{\partial \mathcal{G}_{ij}}{\partial x_{c_k}}(\mathbf{x}_0, \mathbf{x}_c)\, p_{ik} + \cdots, \qquad (6.9.32)$$

where

$$p_{ik} = \iint_{\mathcal{I}^+} \left[(x_k - x_{c,k})\, \Delta f_i(\mathbf{x}) - \mu(1 - \lambda)\, (u_k n_i + u_i n_k)(\mathbf{x}) \right] \mathrm{d}S(\mathbf{x}). \qquad (6.9.33)$$

In the case of equal fluid viscosities, $\lambda = 1$, the second term inside the integral does not appear.

Working in a similar fashion, we find that the velocity field in the interior of a drop, capsule, or cell is given by the right-hand side of (6.9.31), provided that all terms, including the first term representing the ambient flow, are divided by the viscosity ratio, λ ([306], Chapter 5).

6.9.7   Axisymmetric flow

In the case of axisymmetric flow, we express the Cartesian velocity and traction components in terms of their radial and meridional counterparts, and perform the integration in the azimuthal direction,

φ, to reduce the surface integrals into line integrals along the boundary contour in an azimuthal plane. The corresponding kernels of the single- and double-layer potential involve complete elliptical integrals arising from integrals with respect to the azimuthal angle, φ. Further discussion and specific expressions can be found elsewhere (e.g., [306, 313]).

6.9.8 Two-dimensional flow

To derive the boundary-integral representation of two-dimensional Stokes flow, we repeat our earlier steps for three-dimensional flow. Applying the reciprocal theorem, we find that

$$\oint_C \mathcal{G}_{ij}(\mathbf{x}, \mathbf{x}_0)\, f_i(\mathbf{x})\, \mathrm{d}l(\mathbf{x}) = \mu \oint_C u_i(\mathbf{x})\, \mathcal{T}_{ijk}(\mathbf{x}, \mathbf{x}_0)\, n_k(\mathbf{x})\, \mathrm{d}l(\mathbf{x}), \tag{6.9.34}$$

where the point $\mathbf{x}_0$ lies outside a selected domain of flow enclosed by the contour C, and l is the arc length along C. For a point $\mathbf{x}_0$ located inside the selected domain of flow, we obtain the boundary-integral representation

$$u_j(\mathbf{x}_0) = -\frac{1}{4\pi\mu} \oint_C \mathcal{G}_{ij}(\mathbf{x}, \mathbf{x}_0)\, f_i(\mathbf{x})\, \mathrm{d}l(\mathbf{x}) + \frac{1}{4\pi} \oint_C u_i(\mathbf{x})\, \mathcal{T}_{ijk}(\mathbf{x}, \mathbf{x}_0)\, n_k(\mathbf{x})\, \mathrm{d}l(\mathbf{x}), \tag{6.9.35}$$

where the unit normal vector, $\mathbf{n}$, points into the domain of flow. All properties, interpretations, and simplifications of the boundary-integral equation discussed earlier for three-dimensional flow also apply to two-dimensional flow. Exceptions arise in the case of infinite flow past a body and flow produced by the motion of a body in an infinite fluid. In both cases, if the force exerted on the body is nonzero, the disturbance velocity far from the body increases at a logarithmic rate and the boundary integrals at infinity may not be overlooked.

The pressure vector, $\mathcal{P}$, and stress tensor, $\mathcal{T}$, associated with a Green's function for an entirely or partially infinite domain of flow constitute two fundamental solutions of Stokes flow. Specifically, $\mathcal{P}(\mathbf{x}, \mathbf{x}_0)$ represents the velocity at the point $\mathbf{x}_0$ due to a two-dimensional point source located at $\mathbf{x}$. Similarly, the flow given in (6.9.23) represents the velocity at the point $\mathbf{x}_0$ due to a two-dimensional stresslet located at $\mathbf{x}$. The corresponding pressure field may be expressed in terms of a pressure tensor $\mathcal{P}^{STR}$, as shown in (6.9.23). In the case of flow in infinite space,

$$\mathcal{P}_{ik}^{STR}(\mathbf{x}, \mathbf{x}_0) = -2\frac{\delta_{ik}}{r^2} + 4\frac{\hat{x}_i \hat{x}_k}{r^4}. \tag{6.9.36}$$

The pressure field of a two-dimensional Stokes flow can be expressed in terms of two distributions corresponding to the single- and double-layer potentials,

$$p(\mathbf{x}_0) = -\frac{1}{4\pi} \oint_C \mathcal{P}_i(\mathbf{x}, \mathbf{x}_0)\, f_i(\mathbf{x})\, \mathrm{d}l(\mathbf{x}) + \frac{\mu}{4\pi} \oint_C u_i(\mathbf{x})\, \mathcal{P}_{ik}^{STR}(\mathbf{x}, \mathbf{x}_0)\, n_k(\mathbf{x})\, \mathrm{d}l(\mathbf{x}). \tag{6.9.37}$$

The vector $\mathcal{P}$ and tensor $\mathcal{P}^{STR}$ express the pressure field corresponding to the Green's function and associated stress tensor representing a stresslet.

Problems

6.9.1 *Reciprocal theorem with a point force*

Derive the boundary integral representation (6.9.10) on the basis of (6.5.7) using the properties of the delta function without reference to the small volume V_ϵ.

6.9.2 *Alternative boundary integral representation*

Integrate (6.8.35) over a selected control volume and use the divergence theorem to derive the alternative boundary integral representation

$$u_j(\mathbf{x}_0) = -\frac{1}{8\pi\mu} \iint_D \mathcal{G}_{ji}(\mathbf{x}_0, \mathbf{x}) \left(-p\, n_i + \frac{\partial u_i}{\partial x_k}\, n_k \right)(\mathbf{x})\, \mathrm{d}S(\mathbf{x}) \qquad (6.9.38)$$

$$+\frac{1}{8\pi} \iint_D u_i(\mathbf{x}) \left(-\mathcal{P}_j(\mathbf{x}, \mathbf{x}_0)\, n_i(\mathbf{x}) + \frac{\partial \mathcal{G}_{ji}(\mathbf{x}_0, \mathbf{x})}{\partial x_k}\, n_k(\mathbf{x}) \right) n_k(\mathbf{x})\, \mathrm{d}S(\mathbf{x})$$

involving the velocity and the pressure ([166], p. 81).

6.9.3 *Flow due to the motion of a thin sheet*

Consider the flow produced by the motion of a piece of paper or fabric with infinitesimal thickness in a viscous fluid. Show that the double-layer integral in the boundary integral equation can be eliminated naturally and the flow can be represented in terms of a single-layer integral whose distribution density admits a simple physical interpretation.

6.10 Boundary-integral equation methods

The boundary-integral representation developed in Section 6.9 provides us with a powerful tool for computing Stokes flow by solving integral equations over the boundaries of a selected control volume of flow. The main benefit of this approach is that the dimensionality of the computational problem is reduced by one unit with respect to that of the physical problem. Thus, computing a three-dimensional flow is reduced to solving an integral equation over a two-dimensional domain consisting of all surfaces enclosing a selected volume of flow. Arbitrary boundary geometries can be easily accommodated by surface approximation and interpolation [317].

To derive the boundary-integral equation, we examine the behavior of the boundary integrals as the field point, $\mathbf{x}_0$, approaches the boundary D of a selected domain of flow. Due to the weak singularity of the Stokeslet, the single-layer integral remains continuous as the point $\mathbf{x}_0$ approaches and then crosses D from either side. If the velocity and normal vector vary continuously over D, the double-layer potential undergoes a discontinuity of magnitude $8\pi\mathbf{u}(\mathbf{x}_0)$ as the point $\mathbf{x}_0$ crosses D,

$$\lim_{\mathbf{x}_0 \to D} \iint_D u_i(\mathbf{x})\, \mathcal{T}_{ijk}(\mathbf{x}, \mathbf{x}_0)\, n_k(\mathbf{x})\, \mathrm{d}S(\mathbf{x}) = \iint_D^{PV} u_i(\mathbf{x})\, \mathcal{T}_{ijk}(\mathbf{x}, \mathbf{x}_0)\, n_k(\mathbf{x})\, \mathrm{d}S(\mathbf{x}) \pm 4\pi\, u_j(\mathbf{x}_0), \quad (6.10.1)$$

where PV designates the principal value of the double-layer potential defined as the improper double-layer integral computed when the point $\mathbf{x}_0$ is located precisely on D. The plus sign applies when the point $\mathbf{x}_0$ approaches D from the side of the selected volume of flow indicated by the normal vector,

n, and the minus sign otherwise. To confirm the discontinuity, we write

$$\iint_D u_i(\mathbf{x})\, \mathcal{T}_{ijk}(\mathbf{x},\mathbf{x}_0)\, n_k(\mathbf{x})\, dS(\mathbf{x}) = \iint_D [u_i(\mathbf{x}) - u_i(\mathbf{x}_0)]\, \mathcal{T}_{ijk}(\mathbf{x},\mathbf{x}_0)\, n_k(\mathbf{x})\, dS(\mathbf{x})$$

$$+ u_i(\mathbf{x}_0) \iint_D \mathcal{T}_{ijk}(\mathbf{x},\mathbf{x}_0) n_k(\mathbf{x})\, dS(\mathbf{x}). \tag{6.10.2}$$

The first integral on the right-hand side is continuous due to the reduced singularity of the kernel. A discontinuity is evident from identity (6.9.11).

Substituting (6.10.1) with the plus sign into (6.9.9) or with the minus sign into (6.9.3), we find that, when the field point $\mathbf{x}_0$ is located on D,

$$u_j(\mathbf{x}_0) = -\frac{1}{4\pi\mu} \iint_D \mathcal{G}_{ij}(\mathbf{x},\mathbf{x}_0)\, f_i(\mathbf{x})\, dS(\mathbf{x}) + \frac{1}{4\pi} \iint_D^{PV} u_i(\mathbf{x})\, \mathcal{T}_{ijk}(\mathbf{x},\mathbf{x}_0)\, n_k(\mathbf{x})\, dS(\mathbf{x}). \tag{6.10.3}$$

This integral equation imposes an integral constraint on the mutual distributions of the boundary velocity and traction. Conversely, the integral equation allows us to compute the boundary velocity from the boundary traction, and *vice versa*, as required.

In summary, equations (6.9.2), (6.9.10), and (6.10.3) apply, respectively, when the point $\mathbf{x}_0$ is located outside, inside, and precisely on a smooth boundary of a selected control volume, D.

6.10.1 Integral equations

Prescribing the boundary velocity **u** over D provides us with a Fredholm integral equation of the first kind for the boundary traction originating from (6.10.3),

$$\iint_D \mathcal{G}_{ij}(\mathbf{x},\mathbf{x}_0)\, f_i(\mathbf{x})\, dS(\mathbf{x}) = -4\pi\mu u_j(\mathbf{x}_0) + \mu I_j^D(\mathbf{x}_0), \tag{6.10.4}$$

where

$$I_j^D(\mathbf{x}_0) \equiv \iint_D^{PV} u_i(\mathbf{x})\, \mathcal{T}_{ijk}(\mathbf{x},\mathbf{x}_0)\, n_k(\mathbf{x})\, dS(\mathbf{x}) \tag{6.10.5}$$

is the known double-layer potential.

Alternatively, prescribing the boundary traction **f** over D provides us with a Fredholm integral equation of the second kind for the boundary velocity originating from (6.10.3),

$$u_i(\mathbf{x}_0) = \frac{1}{4\pi} \iint_D^{PV} u_i(\mathbf{x})\, \mathcal{T}_{ijk}(\mathbf{x},\mathbf{x}_0)\, n_k(\mathbf{x})\, dS(\mathbf{x}) - \frac{1}{4\pi\mu} I_j^S(\mathbf{x}_0), \tag{6.10.6}$$

where

$$I_j^D(\mathbf{x}_0) \equiv \iint_D \mathcal{G}_{ij}(\mathbf{x},\mathbf{x}_0)\, f_i(\mathbf{x})\, dS(\mathbf{x}) \tag{6.10.7}$$

is the known single-layer potential.

Prescribing the velocity over a portion of D and the traction over the remaining portion of D provides us with a Fredholm integral equation of mixed kind for the unknown boundary distributions. Applications of the mixed formulation can be found in flows bounded by solid boundaries and free surfaces or interfaces. Once the integral equations have been solved, the velocity, pressure, and stress fields can be computed using the boundary-integral representations (6.9.10) and (6.9.25).

Weakly singular integral equations

Inspecting the integral equations (6.10.4) and (6.10.6), we observe that the kernels $\mathcal{G}_{ij}(\mathbf{x}, \mathbf{x}_0)$ and $\mathcal{T}_{ijk}(\mathbf{x}, \mathbf{x}_0)\, n_k(\mathbf{x})$ become singular as the observation point, $\mathbf{x}$, approaches the pole, $\mathbf{x}_0$. In fact, careful inspection reveals that the singularities of the kernels are not square integrable. This behavior prevents us from studying the properties of the solution using the Fredholm and Hilbert–Schmidt theories for integral equations with square integrable kernels. However, when the boundary D is a Lyapunov surface, which means that it has a continuous normal vector, the kernels are weakly singular, the single- and double-layer potentials are compact linear operators, and the theoretical framework still applies (e.g., [208, 306]).

6.10.2   Two-dimensional flow

Repeating the preceding derivations with minor modifications, we derive corresponding results for two-dimensional flow. As the field point $\mathbf{x}_0$ approaches and then crosses a boundary contour C from either side, the single-layer potential remains continuous. If the velocity and normal vector vary continuously over a smooth contour C, the double-layer potential undergoes a discontinuity of magnitude $4\pi\mathbf{u}(\mathbf{x}_0)$,

$$\lim_{\mathbf{x}_0 \to C} \int_C u_i(\mathbf{x})\, \mathcal{T}_{ijk}(\mathbf{x}, \mathbf{x}_0)\, n_k(\mathbf{x})\, \mathrm{d}l(\mathbf{x}) = \int_C^{PV} u_i(\mathbf{x})\, \mathcal{T}_{ijk}(\mathbf{x}, \mathbf{x}_0)\, n_k(\mathbf{x})\, \mathrm{d}l(\mathbf{x}) \pm 2\pi\, u_j(\mathbf{x}_0), \qquad (6.10.8)$$

where the plus sign applies when $\mathbf{x}_0$ approaches C from the side of the flow indicated by the normal vector, $\mathbf{n}$, and the minus sign otherwise (Problem 6.10.1). The principal value of the double-layer integral is the value of the improper double-layer integral computed when the evaluation point $\mathbf{x}_0$ is located precisely on C.

Combining (6.10.8) with (6.9.35), we find that, for a point $\mathbf{x}_0$ that lies on C,

$$u_j(\mathbf{x}_0) = -\frac{1}{2\pi\mu} \oint_D \mathcal{G}_{ij}(\mathbf{x}, \mathbf{x}_0)\, f_i(\mathbf{x})\, \mathrm{d}l(\mathbf{x}) + \frac{1}{2\pi} \int_C^{PV} u_i(\mathbf{x})\, \mathcal{T}_{ijk}(\mathbf{x}, \mathbf{x}_0)\, n_k(\mathbf{x})\, \mathrm{d}l(\mathbf{x}), \qquad (6.10.9)$$

where PV denotes the principal value of the double-layer potential. This integral representation serves as a point of departure for deriving integral equations of the first or second kind for the boundary velocity or traction involving the principal value of the double-layer potential.

Weakly singular integral equations

The kernel of the single-layer potential exhibits a weak logarithmic singularity associated with the two-dimensional Stokeslet. The kernel of the principal value of the double-layer potential undergoes a discontinuity as the integration point, $\mathbf{x}$, crosses over the evaluation point, $\mathbf{x}_0$, along a smooth contour. This behavior can be easily accommodated in numerical methods for solving the integral equations.

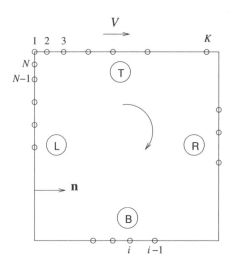

FIGURE 6.10.1   Application of the boundary-integral equation method to flow in a rectangular cavity driven by a moving lid.

6.10.3   Flow in a rectangular cavity

As an application, we consider flow in a rectangular cavity driven by a moving lid, as shown in Figure 6.10.1. The top, left, bottom, and right walls are denoted by T, L, B, and R. Using the boundary-integral representation (6.9.35), and enforcing the velocity boundary conditions, and noting the unit normal vector points against the y axis over the lid, we obtain

$$u_j(\mathbf{x}_0) = -\frac{1}{4\pi\mu} \int_{T,L,B,R} \mathcal{S}_{ij}(\mathbf{x},\mathbf{x}_0)\, f_i(\mathbf{x})\, \mathrm{d}l(\mathbf{x}) - \frac{V}{4\pi} \int_T \mathcal{T}_{xjy}(\mathbf{x},\mathbf{x}_0)\, \mathrm{d}l(\mathbf{x}), \qquad (6.10.10)$$

where $\mathcal{S}$ is the Stokeslet, $\mathcal{T}$ is the corresponding stress tensor, V is the lid velocity, and the point $\mathbf{x}_0$ lies inside the cavity. The integral equation (6.10.9) takes the corresponding form

$$\int_{T,L,B,R} \mathcal{S}_{ij}(\mathbf{x},\mathbf{x}_0)\, f_i(\mathbf{x})\, \mathrm{d}l(\mathbf{x}) = -2\pi\mu u_j(\mathbf{x}_0) - \mu V \int_T^{PV} \mathcal{T}_{xjk}(\mathbf{x},\mathbf{x}_0)\, \mathrm{d}l(\mathbf{x}), \qquad (6.10.11)$$

where the point $\mathbf{x}_0$ is located on T, L, B, or R. The problem has been reduced to solving an integral equation of the first kind for the boundary traction $\mathbf{f}$ inside the single-layer integral on the left-hand side.

Boundary-element method

To solve the integral equation (6.10.11), we may discretize the boundaries of the flow into straight elements and approximate the traction components with constant functions constant over each element, as shown in Figure 6.10.1. This approximation allows us to recast (6.10.11) into the discretized form

$$\sum_{l=1}^{N} f_i^{(l)} A_{ij}^{(l)}(\mathbf{x}_0) = -2\pi\mu u_j(\mathbf{x}_0) - \mu V \sum_{l=1}^{K} B_j^{(l)}(\mathbf{x}_0), \qquad (6.10.12)$$

where $\mathbf{f}^{(l)}$ is the constant value of the traction over the lth element, N is the total number of segments, and the sum on the right-hand side runs over the K lid elements. The influence coefficient, $\mathbf{A}^{(l)}$, and forcing vector, $\mathbf{B}^{(l)}$, are defined as

$$A_{ij}^{(l)} \equiv \int_{E_l} \mathcal{G}_{ij}(\mathbf{x}, \mathbf{x}_0)\, dl(\mathbf{x}), \qquad B_j^{(l)} \equiv \int_{E_l}^{PV} \mathcal{T}_{xjy}(\mathbf{x}, \mathbf{x}_0)\, dl(\mathbf{x}), \qquad (6.10.13)$$

where E_l denotes the lth boundary element. Applying (6.10.12) at the midpoint of the mth element, denoted by $\mathbf{X}^{(m)}$, we obtain a system of N linear equations for $\mathbf{f}^{(l)}$,

$$\sum_{l=1}^{N} f_i^{(l)} A_{ij}^{(l)}(\mathbf{X}^{(m)}) = -2\pi\mu u_j(\mathbf{X}^{(m)}) - \mu V \sum_{l=1}^{K} B_j^{(l)}(\mathbf{X}^{(m)}), \qquad (6.10.14)$$

where $m = 1, \ldots, N$. Most of the influence coefficients defined in (6.10.13) can be computed by numerical integration.

When the mth element midpoint, $\mathbf{X}^{(m)}$, is located over the lid, the kernel $\mathcal{T}_{xjy}$ is identically zero and $B_j^{(l)}(\mathbf{X}^{(m)}) = 0$ for $l, m = 1, \ldots, K$. The associated influence coefficient $\mathbf{A}^{(l)}(\mathbf{X}^{(m)})$ corresponding to the lid segments can be computed exactly by noting that the off-diagonal components are zero, $A_{xy}^{(l)}(\mathbf{X}^{(m)}) = 0$ and $A_{yx}^{(l)}(\mathbf{X}^{(m)}) = 0$, and evaluating the integrals

$$A_{xx}^{(l)} = \int_{E_l} \left(-\ln\frac{|X^{(m)} - x|}{a} + 1\right) dx, \qquad A_{yy}^{(l)} = -\int_{E_l} \ln\frac{|X^{(m)} - x|}{a}\, dx \qquad (6.10.15)$$

by elementary analytical methods, where a is an arbitrary length and $l, m = 1, \ldots, K$. Similar expressions can be derived for the left, bottom, and right wall segments at midpoints $\mathbf{X}^{(m)}$ located at corresponding sides.

Formulation of a linear system

Proceeding with the logistics of the boundary-element implementation, we collect the two components of the element tractions $\mathbf{f}^{(l)}$ into a long vector

$$\mathbf{w} = \left[f_x^{(1)}, \quad f_x^{(2)}, \quad \cdots, \quad f_x^{(N)}, \quad f_y^{(1)}, \quad f_y^{(2)}, \quad \cdots, \quad f_y^{(N)} \right], \qquad (6.10.16)$$

and compile equations (6.10.14) to formulate a system of linear equations,

$$\mathbf{C} \cdot \mathbf{w} = \mathbf{d}, \qquad (6.10.17)$$

where $\mathbf{C}$ is an influence coefficient matrix and $\mathbf{d}$ is a constant vector. The first set of N equations in (6.10.17) correspond to the x components of (6.10.14), and the second set of N equations correspond to the y components of (6.10.14) for $\mathbf{X}^{(1)}, \mathbf{X}^{(2)}, \ldots, \mathbf{X}^{(N)}$.

Regularization

The linear system (6.10.17) is nearly singular due to the unspecified level of the pressure inside the flow. To prevent numerical difficulties, we may specify the normal component of the traction at a chosen element, discard one equation, and then solve a system of $2N - 1$ equations for the remaining unknowns. More reliable methods of removing the singularity by eigenvalue deflation are available (e.g., [313]). In the present case, we may exploit the anticipated symmetry of the flow setting, for example,

$$f_x^{(1)} = f_x^{(K)}, \qquad f_y^{(1)} = -f_y^{(K)}. \tag{6.10.18}$$

When an even number of elements is employed, implementing the symmetry condition reduces the size of the final system of equations by a factor of two. The simplified linear system contains equations corresponding to $N/2$ collocation points distributed over the left or right half of the cavity.

6.10.4 Boundary-element methods

A variety of boundary-element methods have been developed for computing Stokes flow past or due to the motion of particles and interfaces representing drops, capsules, and biological cells (e.g., [306, 313]). A numerical simulation illustrating stages in the axisymmetric settling or rise of a liquid drop normal to a plane wall is shown in Figure 6.10.2. The drop contour in an azimuthal plane is described by a set of marker points, and the shape of the interface is reconstructed by cubic-spline interpolation (e.g., [317]). The marker-point velocity is computed from the boundary-integral representation, and the position of the marker points is advanced in time using the second-order Runge–Kutta method discussed in Section B.8, Appendix B. Regridding is performed after each time step to ensure adequate spatial resolution.

Problem

6.10.1 *Boundary-integral equation of three-dimensional flow*

(*a*) Follow the limiting procedure outlined in Section 6.9 to derive the boundary-integral equation for a point $\mathbf{x}_0$ on a smooth boundary, D. (*Hint*: consider a point $\mathbf{x}_0$ on D, define a control volume that is enclosed by D and excludes a hemisphere of radius ϵ centered at $\mathbf{x}_0$, and let ϵ tend to zero.)

(*b*) Show that equation (6.10.3) is valid at a corner point, $\mathbf{x}_0$, provided that $u_j(\mathbf{x}_0)$ on the left-hand side is replaced by $u_i(\mathbf{x}_0) \, c_{ij}(\mathbf{x}_0)$, where the matrix $c_{ij}(\mathbf{x}_0)$ is determined by the geometry of the corner. Evaluate $c_{ij}(\mathbf{x}_0)$ for a point at the corner of a two-dimensional wedge [306].

6.10.2 *Boundary-integral equation of two-dimensional flow*

(*a*) Follow the limiting procedure outlined in Section 6.9 to derive the boundary-integral equation for a point $\mathbf{x}_0$ on a smooth boundary contour, C, of a two-dimensional flow.

(*b*) Show that the boundary-integral equation (6.10.9) is valid at a corner point, $\mathbf{x}_0$, provided that the coefficient $1/2\pi$ in front of the single- and double-layer potential is replaced by $1/(2\alpha)$, where α is the interior corner angle.

(a)   (b)

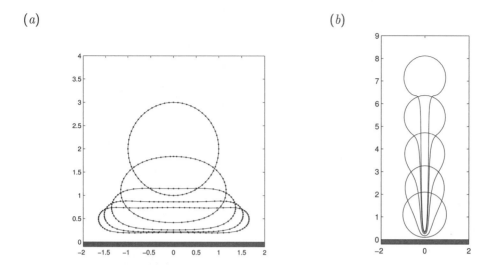

FIGURE 6.10.2   Stages in (a) the gravitational spreading of a liquid drop over a plane wall and (b)
the gravitational rise of a liquid drop away from a wall, in the absence of surface tension. The
viscosity of the drop is equal to that of the ambient fluid. Interfacial profiles are shown at equal
time intervals.

Computer Problem

6.10.3 *Flow in a cavity*

(a) Consider flow in a rectangular cavity driven by a moving lid, as shown in Figure 6.10.1. Write a
program that computes the boundary traction using the boundary-element method described in the
text. Prepare graphs of the boundary shear stress for cavities with depth to width ratio 0.10, 0.50,
1.0, 2.0, and 10. Discuss your results with reference to the occurrence of stagnation points along the
boundaries of the flow.

(b) Draw streamline patterns for the cases studied in (a) and discuss the structure of the flow.

6.11   Slender-body theory

Flow past or due to the motion of slender bodies is encountered in studies of fiber suspensions and
biological flagella or cilia motion, and in applications involving flow over surfaces with attached
rods, such as nanomats consisting of carbon nanotubes. Hancock [165] and Gray & Hanckock [151]
represented the flow due to a moving flagellum in terms of point forces and companion potential
dipoles distributed along the centerline. In the simplest version of their theory, known as the
resistance-coefficient theory, the strength of the singularities is assumed to be proportional to the
local velocity of translation. Improved theories represent the flow due to the motion of a cilium in
terms of *a priori* unknown singularities distributed along the centerline. Enforcing the boundary
conditions provides us with Fredholm integral equations for the densities of the distributions.

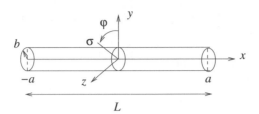

FIGURE 6.11.1   In slender-body theory, the flow due to the axial or transverse translation of an elongated circular cylinder is described by distributions of point forces and potential dipoles over the centerline.

In this section, we review the basic concepts of the slender-body theory and present a theoretical framework that supports the basic premise of the slender-body approximation. Our analysis will confirm that the centerline singularity distributions are rational approximations of surface distributions arising from the boundary-integral representation, thus validating an otherwise heuristic approach.

6.11.1 Axial motion of a cylindrical rod

Consider the flow due to the translation of a slender cylindrical rod of radius b and length $L = 2a$ situated between $x = -a$ and a, and introduce cylindrical polar coordinates, (x, σ, φ), as shown in Figure 6.11.1. When the rod translates along the centerline with velocity V_x, the induced axisymmetric flow is represented by a uniform distribution of point forces oriented along the x axis with linear strength density $\alpha \mu V_x$, where μ is the fluid viscosity and α is an *a priori* unknown dimensionless coefficient to be found as part of the solution. The axial component of the velocity at a point on the rod located at the radial position $\sigma = b$ is given by

$$u_x(x, b) = V_x \frac{\alpha}{8\pi} \int_{-a}^{a} \left(\frac{1}{[(x - \xi)^2 + b^2]^{1/2}} + \frac{(x - \xi)^2}{[(x - \xi)^2 + b^2]^{3/2}} \right) d\xi. \qquad (6.11.1)$$

Introducing the dimensionless variable $w = (x - \xi)/b$ and rearranging the integrand, we obtain

$$u_x(x, b) = V_x \frac{\alpha}{8\pi} \int_{-A}^{B} \left(\frac{2}{(w^2 + 1)^{1/2}} - \frac{1}{(w^2 + 1)^{3/2}} \right) dw, \qquad (6.11.2)$$

where

$$A = \frac{a + x}{b}, \qquad B = \frac{a - x}{b}. \qquad (6.11.3)$$

Both A and B are positive when $-a < x < a$. Performing the integration, we find that

$$u_x(x, b) = V_x \frac{\alpha}{8\pi} \left(2 \ln \left[(B + \sqrt{B^2 + 1})(A + \sqrt{A^2 + 1}) \right] - \frac{B}{\sqrt{B^2 + 1}} - \frac{A}{\sqrt{A^2 + 1}} \right). \qquad (6.11.4)$$

The radial component of the velocity at the surface of the cylinder is

$$u_\sigma(x,b) = b\,V_x\,\frac{\alpha}{8\pi}\int_{-a}^{a}\frac{x-\xi}{[(x-\xi)^2+b^2]^{3/2}}\,\mathrm{d}\xi = V_x\,\frac{\alpha}{8\pi}\int_{-A}^{B}\frac{w}{(1+w^2)^{3/2}}\,\mathrm{d}w. \qquad (6.11.5)$$

Performing the integration, we find that

$$u_\sigma(x,b) = V_x\,\frac{\alpha}{8\pi}\Big(\frac{1}{\sqrt{A^2-1}}-\frac{1}{\sqrt{B^2-1}}\Big). \qquad (6.11.6)$$

Asymptotics

As the aspect ratio b/a tends to zero, the linear functions $A(x)$ and $B(x)$ tend to infinity and (6.11.4) takes the asymptotic form

$$u_x(x,b) \simeq V_x\,\frac{\alpha}{4\pi}\big[\ln(4AB)-1\big]. \qquad (6.11.7)$$

The product $AB = (a^2-x^2)/b^2$ describes a parabolic distribution that vanishes at both cylinder ends, $x = \pm a$. Applying (6.11.7) at the midpoint of the cylinder and at a point located at a quarter of the length of the rod away from the tips, we obtain

$$u_x(0,b) = V_x\,\frac{\alpha}{4\pi}\Big(2\ln\frac{L}{b}-1\Big), \qquad u_x\Big(\frac{a}{2},b\Big) = V_x\,\frac{\alpha}{4\pi}\Big(2\ln\frac{L}{b}+\ln\frac{3}{4}-1\Big). \qquad (6.11.8)$$

To satisfy the no-slip boundary condition, we require that $u_x(x,b) = V_x$. When the aspect ratio L/b is large, the two velocities in (6.11.8) are not too far apart, and the choice

$$\alpha = \frac{2\pi}{\ln\frac{L}{b}-\frac{1}{2}} \qquad (6.11.9)$$

provides us with a reasonable singularity representation. The drag force exerted on the cylinder is given by

$$\mathbf{F} = -\alpha\mu V_x L\,\mathbf{e}_x, \qquad (6.11.10)$$

where $\mathbf{e}_x$ is the unit vector along the x axis. The radial surface velocity given in (6.11.6) is of order b/L, which is small as long the aspect ratio is large, corroborating the consistency of the approximate representation.

Surface distribution

To justify the flow representation in terms of a line distribution of point forces along the centerline, we consider the underlying dimensionless velocity field

$$v_i(\mathbf{x}) = \frac{1}{8\pi}\int_{-a}^{a}S_{ix}\big(\mathbf{x},\mathbf{X}(\xi)\big)\,\mathrm{d}\xi, \qquad (6.11.11)$$

where $\mathbf{X}(\xi) = [\xi,0,0]$, and compare it with that induced by an equivalent uniform distribution of point forces over the cylindrical surface,

$$v_i(\mathbf{x}) = \frac{1}{8\pi}\,\frac{1}{2\pi b}\iint S_{ix}\big(\mathbf{x},\mathbf{X}(\xi)\big)\,\mathrm{d}S, \qquad (6.11.12)$$

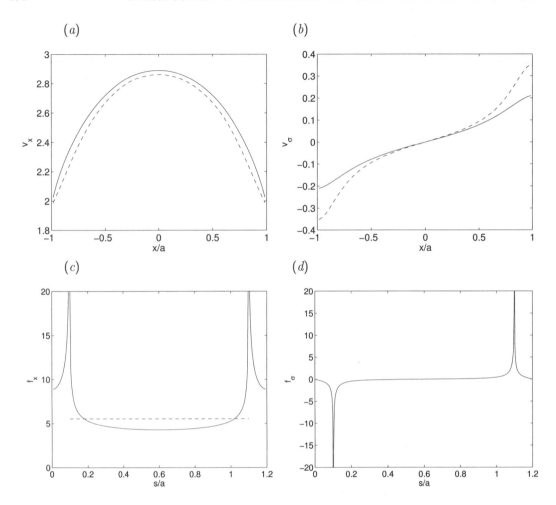

FIGURE 6.11.2   (*a*) Axial and (*b*) radial surface velocity due to a centerline distribution (solid line) and an equivalent surface distribution (dashed line) of point forces, for a cylinder with aspect ratio $L/a = 10$. (*c*) Axial and (*d*) radial components of the traction on a cylinder with aspect ratio $L/b = 10$ computed by a boundary-element method (solid lines). The dashed line in (*c*) represents the prediction of slender-body theory.

where $\mathbf{X}(\xi) = [\xi, b\cos\varphi, b\sin\varphi]$ and $dS = b\,d\xi\,d\varphi$ is a differential surface area. The surface integral in (6.11.12) was computed by numerical integration in terms of the axisymmetric Green's function of Stokes flow [318]. Graphs of the axial and radial velocities, v_x and v_σ, computed from (6.11.11) and (6.11.12) are shown in Figure 6.11.2(*a*, *b*) with the solid and broken line, respectively, for a cylinder with aspect ratio $L/a = 10$. The good agreement justifies replacing the surface distribution with a line distribution along the centerline. Even better agreement is obtained for higher aspect ratios, L/a.

Comparison with an exact solution

To assess the accuracy of the slender-body theory, in Figure 6.11.2(*c, d*) we plot with solid lines the axial and radial components of the traction over the surface of a cylinder with aspect ratio $L/b = 10$, computed by an accurate boundary-element method. The traction has been scaled by $\mu V_x/L$ and plotted with respect to arc length along the the trace of the cylinder in a meridional plane, measured from the axis of revolution, denoted by s. Benign singularities are observed at the sharp corners on either end of the cylinder. The horizontal dashed line in Figure 6.11.2(*c*) represents the predictions of the slender-body theory, $f_x = \alpha \mu V_x/(2\pi b)$. The good agreement over the main body of the cylinder lends credence to the slender-body representation.

6.11.2   Transverse motion of a cylindrical rod

Next, we consider the flow due to transverse translation of a cylindrical rod along the y axis with velocity V_y. Working in hindsight, we introduce a representation in terms of a distribution of point forces with constant density $\beta \mu V_y$ and a distribution of potential dipoles with constant density $\gamma b^2 V_y$ along the rod centerline, where β and γ are *a priori* unknown dimensionless coefficients. Both singularities are oriented in the direction of translation along the y axis.

The y velocity component at the surface of the cylinder, $\sigma = b$, is

$$u_y(x,b,\varphi) = V_y \frac{\beta}{8\pi} \int_{-a}^{a} \left(\frac{1}{[(x-\xi)^2 + b^2]^{1/2}} + \frac{y^2}{[(x-\xi)^2 + b^2]^{3/2}} \right) d\xi$$

$$+ V_y\, b^2\, \frac{\gamma}{4\pi} \int_{-a}^{a} \left(-\frac{1}{[(x-\xi)^2 + b^2]^{3/2}} + 3\frac{y^2}{[(x-\xi)^2 + b^2]^{5/2}} \right) d\xi, \qquad (6.11.13)$$

where $y = b\cos\varphi$ and φ is the azimuthal angle. Introducing the dimensionless variable $w = (x-\xi)/b$ and rearranging the integrand, we obtain

$$u_y(x,b,\varphi) = V_y \frac{\beta}{8\pi} \int_{-A}^{B} \left(\frac{1}{(w^2+1)^{1/2}} + \frac{\cos^2\varphi}{(w^2+1)^{3/2}} \right) dw$$

$$+ V_y \frac{\gamma}{4\pi} \int_{-A}^{B} \left(-\frac{1}{(w^2+1)^{3/2}} + 3\frac{\cos^2\varphi}{(w^2+1)^{5/2}} \right) dw. \qquad (6.11.14)$$

Performing the integration, we find that

$$u_y(x,b,\varphi) = V_y \left(\frac{\beta}{8\pi}\mathcal{F} + \frac{\gamma}{4\pi}\mathcal{G} \right), \qquad (6.11.15)$$

where

$$\mathcal{F} = \ln \left[(B + \sqrt{B^2+1})(A + \sqrt{A^2+1}) \right] + \cos^2\varphi \left(\frac{B}{\sqrt{B^2+1}} + \frac{A}{\sqrt{A^2+1}} \right) \qquad (6.11.16)$$

and

$$\mathcal{G} = -\frac{B}{\sqrt{B^2+1}} - \frac{A}{\sqrt{A^2+1}} + \cos^2\varphi \left(B\frac{2B^2+3}{(B^2+1)^{3/2}} + A\frac{2A^2+3}{(A^2+1)^{3/2}} \right). \qquad (6.11.17)$$

The z velocity component is given by

$$u_z(x, b, \varphi) = V_y \frac{\beta}{8\pi} \int_{-a}^{a} \frac{yz}{[b^2 + (x - \xi)^2]^{3/2}} \, \mathrm{d}\xi + V_y \, b^2 \frac{\gamma}{2\pi} 3 \int_{-a}^{a} \frac{yz}{[b^2 + (x - \xi)^2]^{5/2}} \, \mathrm{d}\xi. \quad (6.11.18)$$

Asymptotics

In the limit as the aspect ratio a/b and thus A and B tend to infinity, we obtain

$$u_y(x, b, \varphi) = V_y \frac{\beta}{8\pi} [\ln(4AB) + 2 \cos^2 \varphi] + V_y \frac{\gamma}{2\pi} (-1 + 2 \cos^2 \varphi). \quad (6.11.19)$$

To eliminate the φ dependence, we set

$$\gamma = -\frac{1}{4} \beta \quad (6.11.20)$$

and obtain

$$u_y(x, b) \simeq V_y \frac{\beta}{8\pi} [\ln(4AB) + 1] \simeq V_y \frac{\beta}{8\pi} (2 \ln \frac{L}{b} + 1). \quad (6.11.21)$$

Working as in the case of axial motion, we require that $u_y = V_y$ at $x = 0$ and find that

$$\beta = \frac{4\pi}{\ln \frac{L}{b} + \frac{1}{2}}. \quad (6.11.22)$$

Integrating the distributed point forces, we find that the drag force exerted on the cylinder is given by

$$\mathbf{F} = -\beta \mu V_y L \, \mathbf{e}_y, \quad (6.11.23)$$

where $\mathbf{e}_y$ is the unit vector along the y axis. It is worth noting that the ratio of the magnitude of the drag force in transverse and axial motions is approximately equal to two.

It remains to confirm that the x and z velocity components are zero on the surface of the cylinder. Working as for the y velocity component, we find that

$$u_z(x, b, \varphi) = V_y \frac{\beta}{8\pi} \sin 2\varphi + V_y \frac{\gamma}{4\pi} \sin 2\varphi, \quad (6.11.24)$$

which is zero in light of (6.11.22). The magnitude of the x velocity component at the surface of the cylinder is of order b/L, which is small as long the aspect ratio is large.

Surface distribution

To justify the flow representation in terms of a line distribution of point forces and potential dipoles along the centerline, we consider the underlying dimensionless velocity field

$$v_i(\mathbf{x}) = \frac{1}{8\pi} \int_{-a}^{a} \left(S_{iy}(\mathbf{x}, \mathbf{X}(\xi)) - \frac{1}{4} b^2 \, \mathcal{D}_{iy}(\mathbf{x}, \mathbf{X}(\xi)) \right) \mathrm{d}\xi, \quad (6.11.25)$$

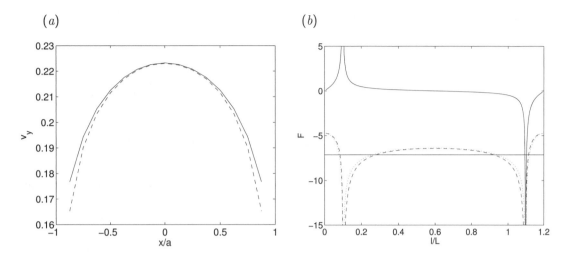

FIGURE 6.11.3   (a) Velocity due to a centerline distribution of point forces and dipoles (solid line), and an equivalent surface distribution of point forces (dashed line), for cylinder aspect ratio $L/a = 10$. (b) Graphs of the dimensionless cylindrical polar components of the traction around the cylinder plotted with respect to arc length; $\mathcal{F}_x$ (solid lines), $\mathcal{F}_\sigma$ (dashed lines), and $\mathcal{F}_\varphi$ (dotted lines).

where $\mathbf{X}(\xi) = [\xi, 0, 0]$. An approximation of the y surface velocity corresponding to (6.11.21) is

$$v_y(x) = \frac{1}{8\pi}[\ln(4AB) + 1]. \tag{6.11.26}$$

This velocity field is now compared with that induced by an equivalent uniform distribution of point forces pointing along the y axis over the cylindrical surface,

$$v_i(\mathbf{x}) = \frac{1}{8\pi}\frac{1}{2\pi b}\iint S_{iy}(\mathbf{x}, \mathbf{X}(\xi))\, \mathrm{d}S, \tag{6.11.27}$$

where $\mathbf{X}(\xi) = [\xi, b\cos\varphi, b\sin\varphi]$ and $\mathrm{d}S = b\,\mathrm{d}\xi\,\mathrm{d}\varphi$ is a differential surface area. The surface integral in (6.11.27) was computed by numerical integration.

Shown in Figure 6.11.3(a) is the y component of the surface velocity induced by the centerline point force and potential dipole distributions (solid line) and by the corresponding surface distribution of point forces (dashed line) for a cylinder with aspect ratio $L/a = 10$. The good agreement provides support for the basic premise of the slender-body representation.

Comparison with an exact solution

To assess the accuracy of the slender-body theory, we express the cylindrical polar components of the traction over the surface of the circular cylinder, $\mathbf{f} = \boldsymbol{\sigma}\cdot\mathbf{n}$, in the form

$$f_x = \frac{\mu V_y}{L}\,\mathcal{F}_x(l)\,\cos\varphi, \qquad f_\sigma = \frac{\mu V_y}{L}\,\mathcal{F}_\sigma(l)\,\cos\varphi, \qquad f_\varphi = -\frac{\mu V_y}{L}\,\mathcal{F}_\varphi(l)\,\sin\varphi, \tag{6.11.28}$$

where l is the arc length along the trace of the cylinder contour in a meridional plane, $\mathcal{F}_\alpha(l)$ are dimensionless functions, and $\alpha = x, \sigma, \varphi$. The y and z components of the traction are

$$f_y = f_\sigma \cos\varphi - f_\varphi \sin\varphi = \frac{\mu V_y}{L} \left(\mathcal{F}_\sigma(l) \cos^2\varphi + \mathcal{F}_\varphi(l) \sin^2\varphi \right) \qquad (6.11.29)$$

and

$$f_y = f_\sigma \sin\varphi - f_\varphi \cos\varphi = \frac{\mu V_y}{L} \left(\mathcal{F}_\sigma(l) - \mathcal{F}_\varphi(l) \right) \sin\varphi \cos\varphi. \qquad (6.11.30)$$

The traction coefficient functions $\mathcal{F}_\alpha$ were computed by solving an integral equation using a highly accurate boundary-element method. Graphs of the dimensionless traction Fourier coefficients are plotted in Figure 6.11.3(b) with respect to arc length for a cylinder with moderate aspect ratio, $L/b = 10$. Singularities are observed at the sharp corners at either end. The coefficients $\mathcal{F}_\sigma(l)$ and $\mathcal{F}_\varphi(l)$ are nearly identical over the main body of the cylinder, meaning that the streamwise traction coefficient f_y is nearly uniform around the cylinder cross-section and the traction f_z is nearly zero. The horizontal line in Figure 6.11.3(b) represents the predictions of the slender-body theory, $f_y = \beta \mu V_y / (2\pi b)$. The successful comparison provides further validation for the slender-body approximation.

6.11.3 Arbitrary translation of a cylindrical rod

The slender-body representations for axial or transverse motion of a circular cylinder can be unified by expressing the velocity field in arbitrary translation in the form

$$u_i(\mathbf{x}) = \frac{1}{8\pi\mu} \int_{-a}^{a} \mathcal{S}_{ij}\big(\mathbf{x}, \mathbf{X}(\xi)\big) b_j(\xi) \, d\xi + \frac{1}{4\pi} \int_{-a}^{a} \mathcal{D}_{ij}\big(\mathbf{x}, \mathbf{X}(\xi)\big) d_j(\xi) \, d\xi, \qquad (6.11.31)$$

where $\mathcal{S}$ is the Stokeslet, $\mathcal{D}$ is the potential dipole, $\mathbf{X}(\xi) = [\xi, 0, 0]$ is a point along the centerline, and

$$\mathbf{b} = \mu \begin{bmatrix} \alpha V_x \\ \beta V_y \\ \beta V_z \end{bmatrix}, \qquad \mathbf{d} = -\frac{1}{4} b^2 \begin{bmatrix} 0 \\ \beta V_y \\ \beta V_z \end{bmatrix} \qquad (6.11.32)$$

are the densities of point force and potential dipole distributions. The force exerted on the cylinder is given by

$$\mathbf{F} = -\mu L \left(\alpha V_x \mathbf{e}_x + \beta V_y \mathbf{e}_y + \beta V_z \mathbf{e}_z \right), \qquad (6.11.33)$$

where $\mathbf{e}_x$, $\mathbf{e}_y$, and $\mathbf{e}_z$ are unit vectors in the subscripted directions.

A settling needle

As an application, we consider a slender needle settling under the action of gravity at an angle θ with respect to the vertical direction, as shown in Figure 6.11.4, where $\theta = 0$ corresponds to a horizontal needle and $\theta = \pi/2$ corresponds to a vertical needle. Reversibility of Stokes flow requires that the

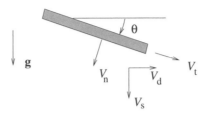

FIGURE 6.11.4 Illustration of a slender needle settling under the action of gravity in an infinite fluid.

needle maintain its initial inclination. Balancing the drag force with the weight of the needle, W, we find that

$$W \sin \theta = \mu L \alpha V_t, \qquad W \cos \theta = \mu L \beta V_n, \qquad (6.11.34)$$

where V_t is the velocity of the needle tangential to the centerline and V_n is the velocity of the needle normal to the centerline. Setting $\beta \simeq 2\alpha$, we find that the ratio between the horizontal drift velocity, $V_d = V_t \cos \theta - V_n \sin \theta$, and the vertical settling velocity, $V_s = V_t \sin \theta + V_n \cos \theta$, is

$$\frac{V_d}{V_s} = \frac{2 \cos \theta \sin \theta - \sin \theta \cos \theta}{2 \sin^2 \theta + \cos^2 \theta} = \frac{\sin \theta \cos \theta}{1 + \cos^2 \theta}. \qquad (6.11.35)$$

The maximum value of this ratio is attained when $\theta = \pi/4$.

6.11.4 Slender-body theory for arbitrary centerline shapes

Generalizing (6.11.31) for a slender body with a curved centerline, $\mathcal{C}$, we introduce a representation in terms of point forces with linear density $\mathbf{b}$ and potential dipoles with linear density $\mathbf{d}$,

$$u_i(\mathbf{x}) = u_i^\infty(\mathbf{x}) + \frac{1}{8\pi\mu} \int_{\mathcal{C}} \mathcal{S}_{ij}\big(\mathbf{x}, \mathbf{X}(l)\big) \, b_j(l) \, \mathrm{d}l + \frac{1}{4\pi} \int_{\mathcal{C}} \mathcal{D}_{ij}\big(\mathbf{x}, \mathbf{X}(l)\big) \, d_j(l) \, \mathrm{d}l, \qquad (6.11.36)$$

where $\mathbf{u}^\infty$ is the velocity of an imposed flow, l is the arc length along the centerline, and $\mathbf{X}(l)$ is the position along the centerline [177, 239, 240]. Motivated by the second expression in (6.11.32), we extract the normal component of the dipole and set the tangential component to zero to obtain

$$\mathbf{d} = -\frac{1}{4} b^2 \, \mathbf{b} \cdot (\mathbf{I} - \mathbf{t} \otimes \mathbf{t}), \qquad (6.11.37)$$

where $\mathbf{t}$ is the unit tangent vector along the centerline and $\mathbf{I} - \mathbf{n} \otimes \mathbf{t}$ is a normal projection operator.

Numerical methods

Given the velocity along the rod centerline, equation (6.11.36) provides us with an integral equation of the first kind for the point force strength density, $\mathbf{b}$. A numerical solution can be computed by dividing the centerline into straight segments and assuming that the force density $\mathbf{b}$ takes the constant value $\mathbf{b}^{(k)}$ over the kth segment (e.g., [177]). Enforcing (6.11.36) at the midpoint of each segment provides us with a system of linear equations for the segment values, $\mathbf{b}^{(k)}$.

The influence coefficients are computed by integrating along the centerline numerically or analytically, as discussed previously in this section for a circular cylinder. The element length must be sufficiently larger than the rod radius, b, otherwise the basic premise of the slender-body approximation is violated and numerical instabilities arise. To improve the performance of the method, the formulation can be recast in terms of an integral equation of the second kind for the Stokeslet density distribution [200].

Integral equations

The numerical procedure can be formalized in terms of the integral equation

$$\mathbf{u}(l_0) = \mathbf{u}^{\infty}(l_0) + \frac{1}{4\pi\mu} (\mathbf{b} \cdot \mathbf{t} \otimes \mathbf{t})_0 (2\ln\frac{2q}{b} - 1) + \frac{1}{8\pi\mu} [\mathbf{b} \cdot (\mathbf{I} - \mathbf{t} \otimes \mathbf{t})]_0 (2\ln\frac{2q}{b} + 1)$$

$$+ \frac{1}{8\pi\mu} \int_{|l-l_0|>q} \boldsymbol{\mathcal{S}}(l_0, l) \cdot \mathbf{b}(l)\, \mathrm{d}l + \frac{1}{4\pi} \int_{|l-l_0|>q} \boldsymbol{\mathcal{D}}(l_0, l) \cdot \mathbf{d}(l)\, \mathrm{d}l, \qquad (6.11.38)$$

where q is a cut-off length allowed to depend on l_0, and the subscript 0 indicates evaluation at l_0 [156]. The second term on the right-hand side originates from the first equation in (6.11.8) and the third term on the right-hand side originates from (6.11.21) with $L = 2q$. The second term disappears when $q = \frac{b}{2}\sqrt{e} \simeq 0.82b$. The last integral on the right-hand side of (6.11.38) can be neglected due to the fast decay far from the evaluation point at l_0.

6.11.5   Motion near boundaries

In the case of flow in a domain that is bounded by a solid surface, we introduce the complement of the Stokeslet ensuring the satisfaction of the no-slip and no-penetration condition on the surface, given by $\boldsymbol{\mathcal{S}}^c = \boldsymbol{\mathcal{G}} - \boldsymbol{\mathcal{S}}$, where $\boldsymbol{\mathcal{G}}$ is the Stokes flow Green's function for the chosen domain of flow. In the case of flow in a semi-infinite domain bounded an infinite plane wall, the complementary flow can be computed by the method of images, as discussed in Section 6.5. The complement of the dipole can be computed in terms of the Laplacian of the complement of the point force [156]. The following term is then added to the right-hand side of (6.11.38),

$$\mathbf{v}(l_0) \equiv \frac{1}{8\pi\mu} \int_C \boldsymbol{\mathcal{S}}^c(l_0, l) \cdot \mathbf{b}(l)\, \mathrm{d}l + \frac{1}{4\pi} \int_C \boldsymbol{\mathcal{D}}^c(l_0, l) \cdot \mathbf{d}(l)\, \mathrm{d}l. \qquad (6.11.39)$$

Note that the integrals are performed along the entire length of the centerline. The last integral on the right-hand side of (6.11.39) can be neglected due to the fast decay away from the evaluation point at l_0 [177].

Shear flow past a rod attached to a wall

As an application, we consider simple shear flow along the y axis past a straight rod of length d and radius b attached normal to a plane wall. The velocity components of the incident flow are $u_x^{\infty} = 0$, $u_y^{\infty} = \xi x$, and $u_z^{\infty} = 0$, where ξ is the shear rate. The density distribution of the point forces is

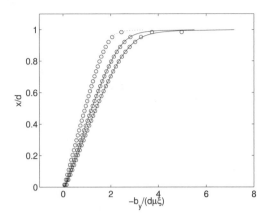

FIGURE 6.11.5   Critical evaluation of slender-body theory for shear flow past a rod attached to a plane wall. The circles represent numerical results for the dimensionless centerline force density, $-b_y/(\mu\xi L)$, obtained by an accurate boundary-element method for a cylinder with aspect ratio $b/d = 0.001, 0.005,$ and 0.01. The lines represent the results of a highly accurate boundary-element method for $b/d = 0.001$ (dotted line), 0.005, and 0.01.

found by solving the integral equation

$$\mathbf{u}(l_0) = \mathbf{u}^\infty(l_0) + \frac{1}{4\pi\mu}(\mathbf{b}\cdot\mathbf{t}\otimes\mathbf{t})_0\Big(2\ln\frac{2q}{b} - 1\Big) + \frac{1}{8\pi\mu}\big[\mathbf{b}\cdot(\mathbf{I} - \mathbf{t}\otimes\mathbf{t})\big]_0\Big(2\ln\frac{2q}{b} + 1\Big)$$

$$+\frac{1}{8\pi\mu}\int_{|l-l_0|<q}(\boldsymbol{\mathcal{G}} - \boldsymbol{\mathcal{S}})(l_0,l)\cdot\mathbf{b}(l)\,\mathrm{d}l + \frac{1}{8\pi\mu}\int_{|l-l_0|>q}\boldsymbol{\mathcal{G}}(l_0,l)\cdot\mathbf{b}(l)\,\mathrm{d}l. \qquad (6.11.40)$$

In the numerical method, the rod centerline is divided into N equal elements of length $\Delta l = d/N$. Applying (6.11.40) at the midpoint of each element, setting $2q = \Delta l$, and requiring $\mathbf{u}(l_0) = \mathbf{0}$, we obtain a system of linear equations for the linear density, $\mathbf{b}$. The integrals on the right-hand side of (6.11.40) can be computed over the union of the elements using the Gauss–Legendre quadrature (Section B.6, Appendix B). Numerical results obtained with 32 elements for three rod aspect ratios, b/d, are represented by the circles in Figure 6.11.5. The predictions of the slender-body theory shown in this figure are compared with highly accurate results obtained by a boundary-element method for $b/d = 0.001$ (dotted line), 0.005, and 0.01 [309]. The excellent agreement over the entire length of the rod except near the tip where the slender-body theory is expected to fail corroborates the fundamental premises of the slender-body approximation.

6.11.6   Extensions and generalizations

Asymptotic analyses for slender bodies with arbitrary cross-sectional shapes, flexible bodies, bodies whose volume changes in time, and bodies executing general types of motion have shown that, in general, distributions of point forces, potential dipoles, couplets, and point sources are required (e.g., [25]).

Problem

6.11.1 *Exact and approximate representations for prolate spheroids*

Prepare a graph of the linear distribution density of the point forces given by the exact solution (6.7.36), and compare it with the constant distributions given in (6.11.9) and (6.11.22) for a family of prolate spheroids with increasing aspect ratio. Compare the exact with the approximate drag force and estimate the aspect ratio where the slender-body solution is sufficiently accurate.

6.12   Generalized Faxén laws

In order to compute the velocity of translation and angular velocity of rotation of a particle that is suspended in an ambient fluid, we must have available the force, $\mathbf{F}$, and torque, $\mathbf{T}$, exerted on the particle when it is held stationary in an incident ambient flow. The generalized Faxén laws provide us with such expressions in terms of the incident velocity field or its derivatives at particular locations in the interior or over the surface of the particle. Faxén first developed relations for a spherical particle [122].

6.12.1   Force on a stationary particle

First, we consider the force exerted on a solid particle that is held stationary in an ambient flow with velocity $\mathbf{u}^\infty$ [207, 209]. We begin by assuming that the flow generated by a particle translating with velocity $\mathbf{V}$ can be represented in terms of a discrete or continuous distribution of point forces in the form

$$u_i(\mathbf{x}) = V_k \mathcal{M}_{kj}[\mathcal{G}_{ij}(\mathbf{x}, \mathbf{x}_0)], \tag{6.12.1}$$

where $\mathcal{G}_{ij}$ is a Stokes flow Green's function for the velocity and $\mathcal{M}_{kj}$ is an integral, differential, or integro-differential operator acting with respect to the point $\mathbf{x}_0$ to generate point forces and their derivatives expressing multipoles of the point force inside or over the surface of the particle. To satisfy the boundary condition $\mathbf{u} = \mathbf{V}$ at the particle surface, we require that

$$\mathcal{M}_{kj}[G_{ij}(\mathbf{x}, \mathbf{x}_0)] = \delta_{ki}, \tag{6.12.2}$$

when the point $\mathbf{x}$ lies at the particle surface. The traction on the particle surface is given by the corresponding expression

$$f_i^t(\mathbf{x}) = \mu V_k \mathcal{M}_{kj}[\mathcal{T}_{ijl}(\mathbf{x} - \mathbf{x}_0)] n_l(\mathbf{x}), \tag{6.12.3}$$

where the superscript t stands for translation. Combining (6.12.3) with (6.8.4) and (6.8.7), we obtain an expression for the force exerted on the particle,

$$F_k = -\mu \iint_P u_i^\infty(\mathbf{x}) \mathcal{M}_{kj}[\mathcal{T}_{ijl}(\mathbf{x} - \mathbf{x}_0)] n_l(\mathbf{x}) \, dS(\mathbf{x}), \tag{6.12.4}$$

where P denotes the particle surface.

The boundary integral representation (6.9.9) for the ambient flow, $\mathbf{u}^\infty$, at a point $\mathbf{x}_0$ located inside the volume occupied by the particle takes the form

$$u_j^\infty(\mathbf{x}_0) = \frac{1}{8\pi\mu} \iint_P f_i^\infty(\mathbf{x})\, \mathcal{G}_{ij}(\mathbf{x}, \mathbf{x}_0)\, \mathrm{d}S(\mathbf{x}) - \frac{1}{8\pi} \iint_P u_i^\infty(\mathbf{x})\, \mathcal{T}_{ijk}(\mathbf{x}, \mathbf{x}_0)\, n_k(\mathbf{x})\, \mathrm{d}S(\mathbf{x}), \quad (6.12.5)$$

where the unit normal vector $\mathbf{n}$ points outward from the particle. Operating on both sides by $\mathcal{M}_{kj}$, we obtain

$$\mathcal{M}_{kj}[u_j^\infty(\mathbf{x}_0)] = \frac{1}{8\pi\mu} \iint_P f_i^\infty(\mathbf{x})\, \mathcal{M}_{kj}[\mathcal{G}_{ij}(\mathbf{x}, \mathbf{x}_0)]\, \mathrm{d}S(\mathbf{x})$$
$$- \frac{1}{8\pi} \iint_P u_i^\infty(\mathbf{x})\, \mathcal{M}_{kj}[\mathcal{T}_{ijk}(\mathbf{x}, \mathbf{x}_0)]\, n_k(\mathbf{x})\, \mathrm{d}S(\mathbf{x}). \quad (6.12.6)$$

Incorporating the boundary condition (6.12.2) in the first integral on the right-hand side of (6.12.6), and noting that the force due to the ambient flow over a fluid parcel occupied by the particle is zero, we find that the first integral is identically zero. Combining the resulting equation with (6.12.4), we derive the generalized Faxén relation

$$F_k = 8\pi\mu\mathcal{M}_{kj}[\mathbf{u}_j(\mathbf{x}_0)], \quad (6.12.7)$$

providing us with the hydrodynamic force exerted on a stationary solid particle. This expression allows us to compute the force exerted on the particle in terms of the velocity of the incident flow alone.

Spherical particle

In Section 6.7.1, we found that the flow due to the translation of a spherical particle in an infinite fluid can be presented in terms of a point force and a potential dipole placed at the particle center, $\mathbf{x}_0$. Introducing the representation (6.7.1) with coefficients given in (6.7.4) and using (6.6.9) to express the potential dipole, $\mathcal{D}$, in terms of the Stokeslet, $\mathcal{S}$, we obtain

$$u_i(\mathbf{x}) = V_k\, \delta_{kj} \left(\frac{3}{4}\, a + \frac{1}{8}\, a^3\, \nabla_0^2 \right) \mathcal{S}_{ij}(\mathbf{x}, \mathbf{x}_0), \quad (6.12.8)$$

where a is the particle radius, $\mathbf{x}_0$ is the particle center, and the Laplacian ∇_0^2 operates with respect to $\mathbf{x}_0$. Faxén's law for the force then follows from (6.12.7) as

$$\mathbf{F} = 6\pi\mu a\, \mathbf{u}^\infty(\mathbf{x}_0) + \pi\mu a^3 \left(\nabla^2 \mathbf{u}^\infty \right)(\mathbf{x}_0). \quad (6.12.9)$$

Thus, the force exerted on a spherical particle can be found from the incident velocity and its Laplacian at the particle center. Comparing the Faxén relation (6.12.9) with the first relation in (6.8.11), we confirm the mean-value theorems stated in (6.1.10).

Prolate spheroid

Using the singularity representation shown in (6.7.36), we derive Faxén relation for the force on a prolate spheroid,

$$F_k = 8\pi\mu\alpha_{kj} \int_{-c}^{c} \left(u_j^\infty(\mathbf{X}(\xi)) + \frac{1}{4}\, b^2 \left(1 - \frac{\xi^2}{c^2} \right) (\nabla^2 u_j^\infty)(\mathbf{X}(\xi)) \right) \mathrm{d}\xi, \quad (6.12.10)$$

where $\mathbf{X}(\xi) = [\xi, 0, 0]$ is a point along the centerline. As the eccentricity tends to zero, $c \to 0$, we recover the Faxén law for a sphere.

6.12.2 Torque on a stationary particle

Next, we consider the torque exerted on a particle that is held stationary in an ambient flow with velocity $\mathbf{u}^\infty$. Assuming that the flow generated when the particle rotates about the origin with angular velocity $\mathbf{\Omega}$ can be represented in terms of a discrete or continuous distribution of Green's functions as

$$u_i = \Omega_k \mathcal{N}_{kj}[\mathcal{G}_{ij}(\mathbf{x}, \mathbf{x}_0)], \qquad (6.12.11)$$

and working as previously for the force, we find that the torque exerted on the particle is given by the generalized Faxén relation

$$T_k = 8\pi\mu \mathcal{N}_{kj}[u_j(\mathbf{x}_0)]. \qquad (6.12.12)$$

This expression allows us to compute the torque exerted on the particle in terms of the velocity of the incident flow alone.

Spherical particle

In Section 6.7.3, we found that the flow due to the rotation of a spherical particle in an infinite fluid can be represented in terms of a couplet alone. Combining (6.7.29) with (6.6.6), we obtain

$$u_i(\mathbf{x}) = \frac{1}{2}\, \Omega_k a^3 \epsilon_{klj} \frac{\partial \mathcal{S}_{ij}}{\partial x_{0_j}}(\mathbf{x}, \mathbf{x}_0), \qquad (6.12.13)$$

where $\mathcal{S}$ is the Stokeslet and $\mathbf{x}_0$ is the particle center. Faxén's law then follows from (6.12.12) as

$$\mathbf{T} = 4\pi\mu a^3 \left(\nabla \times \mathbf{u}^\infty\right)(\mathbf{x}_0). \qquad (6.12.14)$$

We have found that the torque exerted on a spherical particle can be computed from the curl of the velocity representing the local vorticity of the incident flow at the particle center. Comparing the Faxén relation (6.12.14) with that shown in (6.8.11), we confirm the mean-value theorem expressed by identity (6.1.11). Combining (6.12.14) with (6.7.31), we find that a freely suspended sphere bearing zero torque rotates at an angular velocity that is equal to half the vorticity of the ambient flow.

Prolate spheroid

Based on the singularity representation (6.7.33), we find that the Faxén law for the axial component of the torque on a prolate spheroid is expressed by

$$T_x = 4\pi\mu\beta c^2 \int_{-c}^{c} \left(1 - \frac{\xi^2}{c^2}\right) \left(\nabla \times \mathbf{u}^\infty\right)(\mathbf{X}(\xi))\, \mathrm{d}\xi, \qquad (6.12.15)$$

where $\mathbf{X}(\xi) = [\xi, 0, 0]$. As the eccentricity tends to zero, $c \to 0$, we recover the Faxén law for a spherical particle.

6.12.3   Traction moments on a stationary particle

Generalized Faxén relations can be developed for higher moments of the traction on a stationary particle. The zeroth-order moment is the force and the antisymmetric part of the first-order moment is the torque.

Coefficient of the stresslet

The symmetric part of the first-order moment is the coefficient of the stresslet expressing the symmetric part of the point force dipole in the multipole expansion (6.9.16). To compute the coefficient of the stresslet for a particle that is held stationary in an arbitrary flow, we assume that the disturbance flow generated when the particle is held stationary in a purely straining flow with velocity $\mathbf{u}^\infty = \mathbf{E} \cdot \mathbf{x}$ admits the singularity representation

$$u_i^D = E_{lk} \mathcal{L}_{klj}[\mathcal{G}_{ij}(\mathbf{x}, \mathbf{x}_0)], \qquad (6.12.16)$$

where $\mathbf{E}$ is a symmetric matrix with zero trace. The coefficient of the stresslet arising when the particle is held stationary in an arbitrary ambient flow with velocity $\mathbf{u}^\infty$ is

$$s_{lk} = -4\pi\mu \left(\mathcal{L}_{klj}^0[u_j(\mathbf{x}_0)] + \mathcal{L}_{lkj}^0[u_j(\mathbf{x}_0)] \right), \qquad (6.12.17)$$

where $\mathcal{L}^0$ operates with respect to $\mathbf{x}_0$ [306].

Spherical particle

The singularity representation (6.7.15) with coefficients given in (6.7.21), combined with the definitions of the singularities discussed in Section 6.6, show that, in the case of a spherical particle of radius a,

$$u_i^D = -\frac{5}{6} E_{jl}\, a^3 \frac{\partial}{\partial x_{0_l}} \left(1 + \frac{1}{10} a^2 \nabla_0^2\right) \mathcal{S}_{ij}(\mathbf{x}, \mathbf{x}_0), \qquad (6.12.18)$$

where $\mathcal{S}$ is the Stokeslet. The Faxén relation for the coefficient of the stresslet then follows from (6.12.17) as

$$s_{ij} = \frac{10}{3}\, \pi\mu a^3 \left(1 + \frac{1}{10} a^2 \nabla^2\right)\left(\frac{\partial u_j^\infty}{\partial x_i} + \frac{\partial u_i^\infty}{\partial x_j}\right), \qquad (6.12.19)$$

where the right-hand side is evaluated at the center of the sphere.

6.12.4   Arbitrary particle shapes

To derive the Faxén relations for a particular particle shape, we require singularity representations for the flow due to the particle translation or rotation, as well as for the disturbance flow produced when the particle is held stationary in a straining ambient flow. The singularity representations can be computed using numerical and approximate methods, as discussed in Section 6.7.7.

6.12.5 Faxén laws for a fluid particle

Our next objective is to derive Faxén relations for a fluid particle that is held stationary in an ambient flow with velocity $\mathbf{u}^\infty$. Working as in the case of a solid particle, we introduce the disturbance flow due to the particle, $\mathbf{u}^D$, and decompose the velocity field as $\mathbf{u} = \mathbf{u}^\infty + \mathbf{u}^D$. Since the particle is stationary, the normal component of the velocity $\mathbf{u}$ is zero at the interface. The tangential component of the velocity and interfacial traction are required to be continuous across the interface, but the normal component of the traction is allowed to undergo a discontinuity balanced by the capillary pressure due to surface tension or by a gravitational pressure due to differences in the density of the internal and external fluid.

Force on a stationary drop or bubble

First, we consider the Faxén relation for the force on a stationary drop or bubble. Applying the reciprocal theorem for the disturbance flow with velocity $\mathbf{u}^D$ due to the presence of the drop, and the flow with velocity $\mathbf{u}^t$ produced when the drop translates with velocity $\mathbf{V}$, we obtain

$$\iint_{P_{ext}} \mathbf{u}^t \cdot \mathbf{f}^D \, \mathrm{d}S = \iint_{P_{ext}} \mathbf{u}^D \cdot \mathbf{f}^t \, \mathrm{d}S, \tag{6.12.20}$$

where the subscript *ext* indicates the exterior particle surface. Now we substitute $\mathbf{u}^D = \mathbf{u} - \mathbf{u}^\infty$ into the right-hand side, slightly modify the left-hand side, and rearrange to obtain

$$\iint_{P_{ext}} (\mathbf{u}^t - \mathbf{V}) \cdot \mathbf{f}^D \, \mathrm{d}S + \mathbf{V} \cdot \mathbf{F}^D = \iint_{P_{ext}} \mathbf{u} \cdot \mathbf{f}^t \, \mathrm{d}S - \iint_{P_{ext}} \mathbf{u}^\infty \cdot \mathbf{f}^t \, \mathrm{d}S, \tag{6.12.21}$$

where $\mathbf{F}^D$ is the disturbance force exerted on the particle.

Next, we describe the velocity of the exterior flow $\mathbf{u}^t$ by the singularity representation (6.12.1) and project the boundary-integral representation (6.12.6) onto $\mathbf{V}$ and obtain

$$V_k \mathcal{M}_{kj}[u_j(\mathbf{x}_0)] = \frac{1}{8\pi\mu} \iint_{P_{ext}} f_i^\infty(\mathbf{x}) \, V_k \, \mathcal{M}_{kj}\big[\,\mathcal{G}_{ij}(\mathbf{x},\mathbf{x}_0)\,\big] \, \mathrm{d}S(\mathbf{x})$$
$$- \frac{1}{8\pi} \iint_{P_{ext}} u_i^\infty(\mathbf{x}) \, V_k \, \mathcal{M}_{kj}\big[\,\mathcal{T}_{ijk}(\mathbf{x},\mathbf{x}_0)\,\big] \, n_k(\mathbf{x}) \, \mathrm{d}S(\mathbf{x}), \tag{6.12.22}$$

which can be restated as

$$V_k \mathcal{M}_{kj}\big[\,\mathbf{u}_j(\mathbf{x}_0)\,\big] = \frac{1}{8\pi\mu} \iint_{P_{ext}} f_i^\infty(\mathbf{x}) \, u_i^t(\mathbf{x}) \, \mathrm{d}S(\mathbf{x}) - \frac{1}{8\pi\mu} \iint_{P_{ext}} u_i^\infty(\mathbf{x}) \, f_i^t(\mathbf{x}) \, \mathrm{d}S(\mathbf{x}), \tag{6.12.23}$$

where μ is the viscosity of the ambient fluid and $\mathbf{n}$ is the unit normal vector pointing outward from the particle.

Combining (6.12.21) with (6.12.23), we find that

$$\iint_{P_{ext}} (\mathbf{u}^t - \mathbf{V}) \cdot \mathbf{f}^D \, \mathrm{d}S + \mathbf{V} \cdot \mathbf{F}^D = \iint_{P_{ext}} \mathbf{u} \cdot \mathbf{f}^t \, \mathrm{d}S - \iint_{P_{ext}} \mathbf{u}^t \cdot \mathbf{f}^\infty \, \mathrm{d}S + 8\pi\mu \mathbf{V} \cdot \mathcal{M}[\mathbf{u}(\mathbf{x}_0)] \tag{6.12.24}$$

or

$$\iint_{P_{ext}} (\mathbf{u}^t - \mathbf{V}) \cdot \mathbf{f} \, \mathrm{d}S + \mathbf{V} \cdot \mathbf{F} = \iint_{P_{ext}} \mathbf{u} \cdot \mathbf{f}^t \, \mathrm{d}S + 8\pi\mu \mathbf{V} \cdot \mathcal{M}[\mathbf{u}(\mathbf{x}_0)]. \qquad (6.12.25)$$

Since the velocities $\mathbf{u}^t - \mathbf{V}$ and $\mathbf{u}$ are tangential to the interface and the tangential components of $\mathbf{f}$ and $\mathbf{f}^t$ are continuous across the interface, the domain of integration of both integrals in (6.12.25) can be switched from the exterior to the interior side of the interface. Applying the reciprocal theorem for the disturbance flow $\mathbf{u}^t - \mathbf{V}$ and for the interior flow $\mathbf{u}$, we then find that the first two integrals in (6.12.25) are exactly the same. Rearranging, we derive a generalized Faxén relation for the force expressed by (6.12.7).

Spherical drop or bubble

In Section 6.7.6, we found that the exterior flow produced by the translation of a spherical drop admits the singularity representation (6.7.39) with coefficients given in (6.7.46),

$$u_i^{ext} = V_j \frac{1}{8} a \left(2 \frac{3\lambda + 2}{\lambda + 1} + a^2 \frac{\lambda}{\lambda + 1} \nabla_0^2 \right) S_{ij}(\mathbf{x}, \mathbf{x}_0), \qquad (6.12.26)$$

where λ is the ratio of the drop to the ambient fluid viscosity, $\mathbf{x}_0$ is the drop center, and $\mathcal{S}$ is the Stokeslet. The Faxén law for the force takes the form

$$\mathbf{F} = \pi\mu a \left(2 \frac{3\lambda + 2}{\lambda + 1} + a^2 \frac{\lambda}{\lambda + 1} \nabla_0^2 \right) \mathbf{u}^\infty(\mathbf{x}_0). \qquad (6.12.27)$$

As λ tends to infinity, we recover expression (6.12.9) for a spherical particle.

Torque and high-order moments

Following a similar procedure, we find that the generalized Faxén relations for the torque and coefficient of the stresslet for a solid particles discussed earlier in this section also apply for gas bubbles and liquid drops [208, 209, 306].

Problems

6.12.1 *Faxén laws for a prolate spheroid*

Confirm that, as the spheroid aspect ratio tends to unity, the Faxén laws for the force and torque reduce to those for a sphere.

6.12.2 *Approximate Faxén relation for a sphere*

Derive an approximate Faxén law for the force on a solid sphere based on the approximate singularity representation discussed in Problem 6.7.1(a). Compare the approximate with the exact law.

6.13 Effect of inertia and Oseen flow

The solution of the equations of Stokes flow provides us with the leading-order approximation to the structure of a flow at low Reynolds numbers. A fundamental assumption is that inertial forces are uniformly negligible throughout the domain of flow. In the case of interior flow, this assumption can be validated by making the Reynolds number, defined with respect to the size of the boundaries, sufficiently small. Having obtained the solution of the equations of Stokes flow, we may proceed to study the effect of fluid inertia by expanding the velocity and pressure in regular perturbation series with respect to the Reynolds number [412].

Exterior flow

In the case of exterior flow in an infinite domain, the assumption that inertial forces are negligible throughout the domain of flow may not be valid. As an example, we consider uniform (streaming) flow with velocity U past a stationary three-dimensional body. The Stokes-flow solution reveals that, when the force exerted on the body is nonzero, the perturbation flow decays like $1/d$, which means that the magnitude of the left-hand side of the Navier–Stokes equation decays like $\rho U/d^2$, whereas the magnitude of the right-hand side decays like μ/d^3, where d is the distance from the body. This means that inertial forces dominate and the Stokes-flow approximation ceases to be valid at distances greater than $\mu/(\rho U)$ from the body.

Failure of the Stokes-flow approximation

An important consequence of the failure of the Stokes-flow approximation at large distances from a body is that the Stokes-flow solution does not necessarily satisfy the far-field condition. In the case of three-dimensional flow, this difficulty is shielded by the $1/d$ decay of the flow due to a point force, where d is the distance from the point force. However, a paradoxical behavior is encountered in the case of two-dimensional flow due to the logarithmic divergence of the flow induced by a point force, as discussed in Section 6.14.

These difficulties can be resolved by admitting that the flow near the body is governed by the equations of Stokes flow, whereas the flow far from the body is governed by the equations of Navier–Stokes or Oseen flow incorporating a linearized convective contribution, as discussed in Section 6.13.1. The method of matched asymptotic expansions may then be applied to derive a uniformly valid solution.

6.13.1 Oseen flow

In the case of steady streaming (uniform) flow with velocity $\mathbf{U}$ past a stationary body, Oseen proposed replacing the Stokes equation with the linearized Navier–Stokes equation

$$\rho\,\mathbf{U}\cdot\nabla\mathbf{u} = -\nabla p + \mu\nabla^2\mathbf{u}, \qquad (6.13.1)$$

subject to the condition that $\mathbf{u}$ tends to $\mathbf{U}$ far from the body [287]. The left-hand side captures the dominant contribution of the inertial forces far from the body. When the Reynolds number defined with respect to the distance from the body is sufficiently low, the left-hand side of (6.13.1) is small compared to the right-hand side and the Oseen equation (6.13.1) describes Stokes flow.

Flow due to a moving body

Oseen's equation for flow due to a body that translates steadily with velocity $\mathbf{V}$ in an otherwise quiescent fluid takes the form of the unsteady Stokes equation

$$\rho \frac{\partial \mathbf{u}}{\partial t} = -\nabla p + \mu \nabla^2 \mathbf{u}, \tag{6.13.2}$$

subject to the condition that $\mathbf{u}$ decays to zero far from the body. Since in a frame of reference moving with the body the velocity field is steady,

$$\frac{\partial \mathbf{u}}{\partial t} + \mathbf{V} \cdot \nabla \mathbf{u} = \mathbf{0} \qquad \text{or} \qquad \frac{\partial \mathbf{u}}{\partial t} = -\mathbf{V} \cdot \nabla \mathbf{u}, \tag{6.13.3}$$

which shows that (6.13.2) is identical to (6.13.1) with $\mathbf{V} = -\mathbf{U}$. We have found that the flow due to the steady translation of a body governed by the unsteady Stokes equation is equivalent to Oseen flow past a stationary body.

Axisymmetric flow

In the case of axisymmetric flow, it is convenient to align the x axis with the velocity of the outer streaming flow so that $\mathbf{U} = U\mathbf{e}_x$, introduce cylindrical polar coordinates, (x, σ, φ), and describe the flow in terms of the Stokes stream function, ψ. A linear vorticity transport equation corresponding to the Oseen equation (6.13.1) arises by simplifying the general vorticity transport equation (3.12.24), obtaining

$$\frac{U}{\sigma} \frac{\partial \omega_\varphi}{\partial x} = \frac{\nu}{\sigma^2} E^2(\sigma \, \omega_\varphi), \tag{6.13.4}$$

where $\nu = \mu/\rho$ is the kinematic viscosity of the fluid,

$$\omega_\varphi = -\frac{1}{\sigma} E^2 \psi = -\frac{1}{\sigma} \left(\frac{\partial^2 \psi}{\partial x^2} + \frac{\partial^2 \psi}{\partial \sigma^2} - \frac{1}{\sigma} \frac{\partial \psi}{\partial \sigma} \right), \tag{6.13.5}$$

is the azimuthal vorticity, and E^2 is a second-order linear differential operator defined in (2.9.17), repeated below for convenience,

$$E^2 \equiv \frac{\partial^2}{\partial x^2} + \frac{\partial^2}{\partial \sigma^2} - \frac{1}{\sigma} \frac{\partial}{\partial \sigma}. \tag{6.13.6}$$

Substituting into (6.13.4) expression (6.13.5) and rearranging, we obtain

$$\left(E^2 - \frac{U}{\nu} \frac{\partial}{\partial x} \right) (E^2 \psi) = 0. \tag{6.13.7}$$

When $U = 0$, this equation describes axisymmetric Stokes flow.

In spherical polar coordinates, (r, θ, φ), equation (6.13.7) becomes

$$\left[E^2 - \frac{U}{\nu} \left(\cos\theta \frac{\partial}{\partial r} - \sin\theta \frac{1}{r} \frac{\partial}{\partial \theta} \right) \right] (E^2 \psi) = 0, \tag{6.13.8}$$

where

$$E^2 \equiv \frac{\partial^2}{\partial r^2} + \frac{\sin\theta}{r^2} \frac{\partial}{\partial \theta} \left(\frac{1}{\sin\theta} \frac{\partial}{\partial \theta} \right) = \frac{\partial^2}{\partial r^2} + \frac{1}{r^2} \frac{\partial^2}{\partial \theta^2} - \frac{\cot\theta}{r^2} \frac{\partial}{\partial \theta}. \tag{6.13.9}$$

6.13.2   Flow due to a point force

The linearity of the governing equations with respect to velocity and pressure allows us to construct solutions for Oseen flow by superposing fundamental solutions represented by Green's functions. The flow due to a point force satisfies (6.13.1) with a singular forcing function added to the right-hand side,

$$\rho\, \mathbf{U} \cdot \nabla \mathbf{u} = -\nabla p + \mu \nabla^2 \mathbf{u} + \mathbf{b}\, \delta_3(\mathbf{x} - \mathbf{x}_0), \tag{6.13.10}$$

where $\mathbf{b}$ is a constant vector and δ_3 is the three-dimensional delta function. The fundamental solution of the Oseen equation, called the Oseenlet and denoted by $\mathcal{O}$, can be derived working as in Section 6.5 for the Stokeslet [166]. The pressure due to an Oseenlet is identical to that of the Stokeslet. The velocity is given by the counterpart of expression (6.5.16),

$$\mathbf{u} = \frac{1}{8\pi\mu}\, \mathcal{O} \cdot \mathbf{b} = \frac{1}{\mu} \left(\mathbf{I} \nabla^2 \mathcal{H} - \nabla \nabla \mathcal{H} \right) \cdot \mathbf{b}, \tag{6.13.11}$$

which ensures the satisfaction of the continuity equation. The generating function $\mathcal{H}$ satisfies the linear differential equation

$$-\frac{1}{\nu}\, \mathbf{U} \cdot \nabla \mathcal{H} + \nabla^2 \mathcal{H} - \frac{1}{4\pi r} = 0, \tag{6.13.12}$$

where $\nu = \mu/\rho$ is the kinematic viscosity of the fluid. The Laplacian, $\mathcal{Q} \equiv \nabla^2 \mathcal{H}$, is the Green's function of the linear convection–diffusion equation,

$$\frac{1}{\nu}\, \mathbf{U} \cdot \nabla \mathcal{Q} = \nabla^2 \mathcal{Q} + \delta_3(\mathbf{x} - \mathbf{x}_0), \tag{6.13.13}$$

given by

$$\mathcal{Q}(\mathbf{x}, \mathbf{x}_0) = \Phi(\mathbf{x}, \mathbf{x}_0)\, \exp\left[\frac{1}{2\nu}\, \mathbf{U} \cdot (\mathbf{x} - \mathbf{x}_0)\right], \tag{6.13.14}$$

where $\Phi(\mathbf{x}, \mathbf{x}_0)$ is the Green's function of the Helmholtz equation

$$\nabla^2 \Phi(\mathbf{x}, \mathbf{x}_0) - c^2\, \Phi(\mathbf{x}, \mathbf{x}_0) + \delta_3(\mathbf{x} - \mathbf{x}_0) = 0 \tag{6.13.15}$$

with $c = |\mathbf{U}|/(2\nu)$, given by

$$\Phi(\mathbf{x}, \mathbf{x}_0) = \frac{1}{4\pi r}\, e^{-cr}. \tag{6.13.16}$$

Thus,

$$\mathcal{Q} \equiv \nabla^2 \mathcal{H} = \frac{1}{4\pi r}\, e^{-\eta}, \tag{6.13.17}$$

where

$$\eta = \frac{|\mathbf{U}|}{2\nu}\, (r - \mathbf{e} \cdot \hat{\mathbf{x}}), \tag{6.13.18}$$

$\mathbf{e} = \mathbf{U}/|\mathbf{U}|$ is the unit vector in the direction of the streaming flow, $\hat{\mathbf{x}} = \mathbf{x} - \mathbf{x}_0$, and $r = |\hat{\mathbf{x}}|$.

To solve equation (6.13.17) for $\mathcal{H}$, we work in hindsight and assume that $\mathcal{H}$ is a function of the dimensionless variable η alone. The gradient is

$$\nabla\mathcal{H} = \frac{|\mathbf{U}|}{2\nu}\frac{d\mathcal{H}}{d\eta}\left(\frac{\hat{\mathbf{x}}}{r} - \mathbf{e}\right), \tag{6.13.19}$$

and the Laplacian is

$$\nabla^2\mathcal{H} = \nabla\cdot\nabla\mathcal{H} = \left(\frac{|\mathbf{U}|}{2\nu}\right)^2 \frac{d^2\mathcal{H}}{d\eta^2}\left(\frac{\hat{\mathbf{x}}}{r} - \mathbf{e}\right)\cdot\left(\frac{\hat{\mathbf{x}}}{r} - \mathbf{e}\right) + \frac{|\mathbf{U}|}{2\nu}\frac{d\mathcal{H}}{d\eta}\nabla\cdot\left(\frac{\hat{\mathbf{x}}}{r}\right). \tag{6.13.20}$$

Simplifying the expression for the Laplacian, we obtain

$$\nabla^2\mathcal{H} = \left(\frac{|\mathbf{U}|}{2\nu}\right)^2 \frac{d^2\mathcal{H}}{d\eta^2}\frac{2}{r}(r - \mathbf{e}\cdot\hat{\mathbf{x}}) + \frac{|\mathbf{U}|}{2\nu}\frac{d\mathcal{H}}{d\eta}\frac{2}{r} = \frac{|\mathbf{U}|}{\nu}\frac{1}{r}\left(\eta\frac{d^2\mathcal{H}}{d\eta^2} + \frac{d\mathcal{H}}{d\eta}\right). \tag{6.13.21}$$

Substituting the last expression into (6.13.17) and rearranging, we derive the ordinary differential equation

$$\frac{d}{d\eta}\left(\eta\frac{d\mathcal{H}}{d\eta}\right) = \frac{1}{4\pi}\frac{\nu}{|\mathbf{U}|}e^{-\eta}, \tag{6.13.22}$$

whose solution is

$$\frac{d\mathcal{H}}{d\eta} = \frac{1}{4\pi}\frac{\nu}{|\mathbf{U}|}\frac{1-e^{-\eta}}{\eta}, \qquad \mathcal{H}(\eta) = \frac{1}{4\pi}\frac{\nu}{|\mathbf{U}|}\int_0^\eta \frac{1-e^{-\xi}}{\xi}\,d\xi. \tag{6.13.23}$$

Expression (6.13.19) for the gradient now becomes

$$\nabla\mathcal{H} = \frac{1}{8\pi}\frac{1-e^{-\eta}}{\eta}\left(\frac{\hat{\mathbf{x}}}{r} - \mathbf{e}\right). \tag{6.13.24}$$

The Oseenlet is computed from the expression

$$\mathcal{O} = 8\pi\left(\mathcal{Q}\,\mathbf{I} - \nabla\nabla\mathcal{H}\right). \tag{6.13.25}$$

Making substitutions, we find that

$$\mathcal{O}_{ij}(\mathbf{x},\mathbf{x}_0) = 2\frac{1}{r}e^{-\eta}\delta_{ij} + \frac{\partial}{\partial x_i}\left[\frac{1-e^{-\eta}}{\eta}\left(\frac{\hat{x}_j}{r} - e_j\right)\right]. \tag{6.13.26}$$

Carrying out the differentiations on the right-hand side, we derive lengthy expressions.

Behavior near the point force

To assess the behavior of the Oseenlet near the point force, we let η tend to zero, expand the integrand in (6.13.23) in a Taylor series, and perform the integration to find that

$$\mathcal{H}(\eta) = \frac{1}{4\pi}\frac{\nu}{|\mathbf{U}|}\left(\eta - \frac{1}{4}\eta^2 + \cdots\right) = \frac{1}{8\pi}\left[r - \mathbf{e}\cdot\hat{\mathbf{x}} - \frac{1}{4}\frac{|\mathbf{U}|}{2\nu}(r - \mathbf{e}\cdot\hat{\mathbf{x}})^2 + \cdots\right]. \tag{6.13.27}$$

Keeping only the first two terms and discarding the irrelevant linear function $\mathbf{e}\cdot\mathbf{x}$, we recover the expression for the Stokeslet, $\mathcal{H} = r/(8\pi)$.

6.13.3   Drag force on a sphere

Consider streaming flow along the x axis with velocity $U > 0$, and introduce an Oseenlet with strength b_x parallel to the streaming flow. The induced x velocity component is

$$u_x = \frac{1}{\mu} \left(\nabla^2 \mathcal{H} - \frac{\partial^2 \mathcal{H}}{\partial x^2} \right) b_x = \frac{1}{\mu} \left(\frac{\partial^2 \mathcal{H}}{\partial y^2} + \frac{\partial^2 \mathcal{H}}{\partial z^2} \right) b_x. \tag{6.13.28}$$

Substituting the asymptotic form (6.13.27), we find that, near the singular point, in limit as r tends to zero, the streamwise velocity component behaves as

$$u_x \simeq \frac{1}{8\pi\mu} \left(S_{xx} - \frac{U}{2\nu} \right) b_x, \tag{6.13.29}$$

where S_{xx} is the Stokeslet. The second term on the right-hand side expresses an induced uniform flow with velocity

$$\tilde{u}_x = -\frac{1}{8\pi\mu} \frac{U}{2\nu} b_x. \tag{6.13.30}$$

In the case of streaming flow past a sphere of radius a, we set $b_x = -6\pi\mu a U$ from the Stokes flow solution, and compute the induced velocity

$$\tilde{u}_x \simeq \frac{3}{8} \frac{aU}{\nu} = \frac{3}{16} \mathrm{Re}, \tag{6.13.31}$$

where $\mathrm{Re} = 2aU/\nu$ is the Reynolds number based on the sphere diameter, $2a$. Using Stokes's law, we find that the force exerted on the sphere is given by

$$F_x = 6\pi\mu a \left(U + \tilde{u}_x \right) = 6\pi\mu a U \left(1 + \frac{3}{16} \mathrm{Re} \right), \tag{6.13.32}$$

yielding the drag coefficient

$$c_D \equiv \frac{F}{\pi\rho U^2 a^2} = \frac{12}{\mathrm{Re}} \left(1 + \frac{3}{16} \mathrm{Re} \right). \tag{6.13.33}$$

The second term on the right-hand side expresses the first inertial correction to Stokes's law. The procedure can be extended using the method of matched asymptotic expansions to derive further corrections to the drag force [412].

6.13.4   Stream function due to a point force in the direction of the flow

When the point force is parallel to the uniform velocity of the far flow, $\mathbf{U}$, the flow induced by the point force is axisymmetric. To derive an expression for the Stokes stream function, ψ, we introduce spherical polar coordinates with the x pointing in the direction of the far flow so that $\mathbf{U} = U_x \mathbf{e}_x$. Substituting into (6.13.7) the educated guess

$$E^2 \psi = \phi \, e^{\frac{1}{2}X}, \tag{6.13.34}$$

we obtain

$$\Big(E^2 - \frac{1}{2\,\ell^2} \Big)\,\phi = 0, \tag{6.13.35}$$

where $X = U_x x/\nu$ is a dimensionless variable and $\ell = \nu/|U_x|$ is a length scale. Working in hindsight, we introduce spherical polar coordinates, (r, θ, φ), set

$$\phi(r, \theta) = \sin^2\theta\,\chi(R), \tag{6.13.36}$$

and derive the ordinary differential equation

$$\chi'' - \Big(\frac{2}{R} + \frac{1}{4}\Big)\,\chi = 0, \tag{6.13.37}$$

where $R = r/\ell$ is a dimensionless variable and a prime denotes a derivative with respect to R. A decaying solution is

$$\chi(R) = c\,\Big(1 + \frac{2}{R}\Big)\,\mathrm{e}^{-\frac{1}{2}R}, \tag{6.13.38}$$

where c is a constant. Substituting these expressions into (6.13.36), we obtain a second-order equation for the stream function,

$$E^2\psi = c\,\sin^2\theta\,\Big(1 + \frac{2}{R}\Big)\,\mathrm{e}^{-\frac{1}{2}(R-X)}, \tag{6.13.39}$$

where $X = \pm R\cos\theta$, the plus sign applies if $U_x > 0$ and the minus sign applies if $U_x < 0$. Solving this equation, we find that the Stokes stream function associated with a point force with strength b_x is given by

$$\psi^{pfx} = \frac{b_x}{4\pi\rho U_x}\,\frac{1}{R}\,(R + X)\,\Big(1 - \mathrm{e}^{-\frac{1}{2}(R-X)}\Big). \tag{6.13.40}$$

In the limit $R \to 0$, we recover the stream function for the Stokeslet given in (6.5.22).

6.13.5 Oseen–Lamb flow past a sphere

Exact solutions to the Oseen equation for flow past particles are not available. An approximate solution describing streaming flow with velocity U_x along the x axis past a stationary sphere of radius a situated at the origin was proposed by Lamb [219]. To derive this solution, we introduce spherical polar coordinates with the x axis in the direction of the far flow, so that $U_x > 0$, and represent the disturbance flow due to the sphere in terms of a potential dipole and a point force located at the center of the sphere and pointing along the x axis. The Stokes stream function is constructed by linear superposition,

$$\psi = \psi^\infty + \psi^{pfx} + \frac{d_x}{4\pi r}\,\sin^2\theta, \tag{6.13.41}$$

where $\psi^\infty = \frac{1}{2}\,U_x r^2 \sin^2\theta$ is the stream function of the incident streaming flow, ψ^{pfx} is the stream function due to a point force point along the axis given in (6.13.40), and d_x is the strength of the dipole.

(a)(b)

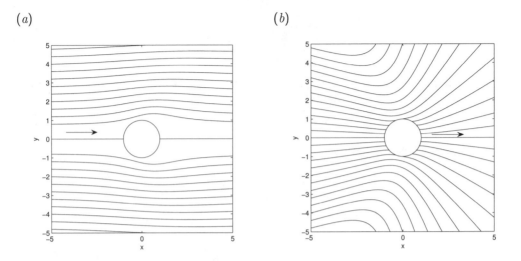

FIGURE 6.13.1 Approximate streamline pattern of Oseen flow (a) past a sphere and (b) due to the translation of a sphere at $\mathrm{Re} = 2$.

Resorting to the singularity representation of Stokes flow past a stationary sphere, we set

$$b_x = -6\pi\mu a U_x, \qquad d_x = \pi a^3 U_x, \tag{6.13.42}$$

and obtain

$$\psi = \psi^\infty + a^2 U_x \left[-\frac{3}{\mathrm{Re}} (1 + \cos\theta) \left(1 - \exp[-\frac{\mathrm{Re}}{4}\frac{r}{a}(1 - \cos\theta)] \right) + \frac{1}{4}\frac{a}{r}\sin^2\theta \right], \tag{6.13.43}$$

where $\mathrm{Re} = 2a\rho U_x/\mu$ is the Reynolds number defined with respect to the sphere diameter, $2a$. Near the sphere, the ratio r/a is of order unity and the argument of the exponential term is small. Expanding the exponential term in a Taylor series, we obtain the stream function for Stokes flow, supplemented by a small correction whose magnitude is proportional to the Reynolds number. Unlike in Stokes flow, the streamline pattern in Oseen flow is no longer symmetric with respect to the midplane of the sphere, as shown in Figure 6.13.1(a) for $\mathrm{Re} = 2$.

Flow due to a translating sphere

The stream function of the flow due to a sphere translating in the positive direction of the x axis with velocity V arises from (6.13.43) as

$$\psi = a^2 V \left[\frac{3}{\mathrm{Re}} (1 - \cos\theta) \left(1 - \exp[-\frac{\mathrm{Re}}{4}\frac{r}{a}(1 + \cos\theta)] \right) - \frac{1}{4}\frac{a}{r}\sin^2\theta \right]. \tag{6.13.44}$$

Far from the sphere, the streamlines are radial lines everywhere except inside a wake, as shown in Figure 6.13.1(b) for $\mathrm{Re} \equiv 2a\rho V/\mu = 2$.

6.13.6   Two-dimensional Oseen flow

The pressure field due to a two-dimensional point force in Oseen flow is identical to that in Stokes flow. The velocity field is given by

$$\mathbf{u} = \frac{1}{4\pi\mu}\,\mathcal{O}\cdot\mathbf{b} = \frac{1}{\mu}\,(\mathbf{I}\nabla^2\mathcal{H} - \nabla\nabla\,\mathcal{H})\cdot\mathbf{b}, \tag{6.13.45}$$

where $\mathcal{O}$ is the two-dimensional Oseenlet, regarded as the counterpart of the two-dimensional Stokeslet for Stokes flow. The generating function, $\mathcal{H}$, satisfies the linear differential equation

$$-\frac{1}{\nu}\,\mathbf{U}\cdot\nabla\mathcal{H} + \nabla^2\mathcal{H} + \frac{1}{4\pi}\ln\frac{r}{a} = 0, \tag{6.13.46}$$

where $r = |\mathbf{x} - \mathbf{x}_0|$ and a is a chosen length. The Laplacian, $\mathcal{Q} \equiv \nabla^2\mathcal{H}$, is the Green's function of the linear convection–diffusion equation

$$\frac{1}{\nu}\,\mathbf{U}\cdot\nabla\mathcal{Q} = \nabla^2\mathcal{Q} + \delta_2(\mathbf{x} - \mathbf{x}_0), \tag{6.13.47}$$

given by

$$\mathcal{Q}(\mathbf{x},\mathbf{x}_0) = \Phi(\mathbf{x},\mathbf{x}_0)\,\exp\Big[\frac{1}{2\nu}\mathbf{U}\cdot(\mathbf{x}-\mathbf{x}_0)\Big], \tag{6.13.48}$$

where $\Phi(\mathbf{x},\mathbf{x}_0)$ is the Green's function of the Helmholtz equation

$$\nabla^2\Phi(\mathbf{x},\mathbf{x}_0) - c^2\,\Phi(\mathbf{x},\mathbf{x}_0) + \delta_2(\mathbf{x}-\mathbf{x}_0) = 0 \tag{6.13.49}$$

with $c = |\mathbf{U}|/(2\nu)$, given by

$$\Phi(\mathbf{x},\mathbf{x}_0) = \frac{1}{2\pi}\,\mathrm{K}_0(cr), \tag{6.13.50}$$

and K_0 is the zeroth-order modified Bessel function of the second kind. Thus,

$$\mathcal{Q} \equiv \nabla^2\mathcal{H} = \frac{1}{2\pi}\,\mathrm{K}_0(cr)\,\exp(c\,\mathbf{e}\cdot\hat{\mathbf{x}}), \tag{6.13.51}$$

where $\mathbf{e} = \mathbf{U}/|\mathbf{U}|$ is the unit vector in the direction of the streaming flow, $\hat{\mathbf{x}} = \mathbf{x} - \mathbf{x}_0$, and $r = |\hat{\mathbf{x}}|$.

   To simplify the analysis, we may assume that the streaming flow occurs along the x axis and set $\mathbf{U} = U\mathbf{e}_x$, $\mathbf{e} = \mathbf{e}_x$, and $\mathbf{e}\cdot\hat{\mathbf{x}} = \hat{x} \equiv x - x_0$. The solution of (6.13.51) is given by the indefinite integral

$$\mathcal{H} = \frac{1}{4\pi c}\int[\mathrm{K}_0(cr)\,\mathrm{e}^{c\hat{x}} + \ln(cr)]\,\mathrm{d}x. \tag{6.13.52}$$

The two-dimensional Oseenlet may now be computed from the expression

$$\mathcal{O} = 4\pi\,(\mathbf{I}\nabla^2\mathcal{H} - \nabla\nabla\mathcal{H}) = 4\pi\,(\mathcal{Q}\mathbf{I} - \nabla\nabla\mathcal{H}). \tag{6.13.53}$$

For a point force pointing in the direction of the outer streaming flow along the x axis, we obtain

$$\mathcal{O}_{xx} = \mathrm{K}_0(cr)\,\mathrm{e}^{c\hat{x}} + \Big(\mathrm{K}_1(cr)\,\mathrm{e}^{c\hat{x}} - \frac{1}{cr}\Big)\frac{\hat{x}}{r}, \qquad \mathcal{O}_{yx} = \Big(\mathrm{K}_1(cr)\,\mathrm{e}^{c\hat{x}} - \frac{1}{cr}\Big)\frac{\hat{y}}{r}. \qquad (6.13.54)$$

For a point force pointing normal to the outer streaming flow, we obtain

$$\mathcal{O}_{yx} = \Big(\mathrm{K}_1(cr)\,\mathrm{e}^{c\hat{x}} - \frac{1}{cr}\Big)\frac{\hat{y}}{r}, \qquad \mathcal{O}_{yy} = \mathrm{K}_0(cr)\,\mathrm{e}^{c\hat{x}} - \Big(\mathrm{K}_1(cr)\,\mathrm{e}^{c\hat{x}} - \frac{1}{cr}\Big)\frac{\hat{x}}{r}. \qquad (6.13.55)$$

We note that $\mathcal{O}_{xx} + \mathcal{O}_{yy} = 4\pi Q$, as required. These expressions can also be derived by applying Fourier transforms, as discussed in Section 6.5 for the Stokeslet ([230], Problem 6.13.4).

Behavior near the point force

Using the asymptotic expansions $\mathrm{K}_0(cr) \simeq -\ln(cr) + \ln 2 - \gamma$ and $\mathrm{K}_1(cr) \simeq 1/(cr)$ in the limit $cr \to 0$, we obtain

$$\mathcal{O}_{xx} = \mathcal{S}_{xx}(c\hat{\mathbf{x}}) + \ln 2 - \gamma + \cdots, \qquad \mathcal{O}_{yx} = \mathcal{S}_{yx}(c\hat{\mathbf{x}}) + \cdots,$$
$$\mathcal{O}_{xy} = \mathcal{S}_{yx}(c\hat{\mathbf{x}}) + \cdots, \qquad \mathcal{O}_{yy} = \mathcal{S}_{xx}(c\hat{\mathbf{x}}) + \ln 2 - \mathrm{E} + \cdots, \qquad (6.13.56)$$

where $\mathrm{E} = 0.577215665\cdots$ is Euler's constant, $\mathcal{S}$ is the two-dimensional Stokeslet, and the dots denote decaying terms. We observe a leading-order Stokes flow solution complemented by streaming flow in the direction of the point force. In Section 6.14, these expressions will be used to describe streaming flow past a circular cylinder.

6.13.7   Flow past particles, drops, and bubbles

Inertial effects on the flow past and due to the motion of rigid particles and liquid drops at nonzero Reynolds numbers have been studied extensively using asymptotic and numerical methods. A wealth of information is compiled by Clift, Grace, & Weber [89].

Problems

6.13.1 *Effects of inertia*

(a) Explain why the Stokes flow approximation is valid throughout the domain of shear flow past a stationary body.

(b) Consider a particle translating above an infinite plane wall. Estimate the distance from the particle where inertial effects become important.

6.13.2 *Vorticity in Oseen flow*

Consider Oseen flow due to the translation of a three-dimensional body. Show that the vorticity field satisfies the unsteady heat conduction equation

$$\rho\frac{\partial\boldsymbol{\omega}}{\partial t} = -\mathbf{U}\cdot\nabla\boldsymbol{\omega} = \nu\nabla^2\boldsymbol{\omega}. \qquad (6.13.57)$$

Explain in physical terms why the body acts like an effective source of vorticity.

6.13.3 *Two-dimensional Oseenlet*

Derive the two-dimensional Oseenlet using the method of Fourier transforms.

6.14   Flow past a circular cylinder

A natural set of singularities for representing Stokes flow due to the translation of a circular cylinder in an infinite fluid is the two-dimensional Stokeslet, $\mathcal{S}$, and the two-dimensional potential dipole, $\mathcal{D}$, both placed at the center of the cylinder, $\mathbf{x}_c$. However, since the velocity due to a two-dimensional point force increases logarithmically with distance from the singular point, the far-flow condition of decaying velocity cannot be satisfied. Physically, the Stokes flow approximation ceases to be valid at a certain distance from the cylinder determined by the Reynolds number. Bearing in mind this essential difficulty, we proceed to derive the solution in three stages: first, we consider the Stokes flow in the vicinity of the cylinder; second, we describe the far flow using Oseen's equation; third we match the functional forms of the two solutions.

6.14.1   Inner Stokes flow

Inspecting the singularities of two-dimensional Stokes flow derived in Section 6.6 suggests describing the flow by a point force with strength $\mathbf{b}$ and a potential dipole with strength $\mathbf{d}$ placed at the center of the cylinder. Since the velocity due to a two-dimensional point force in Stokes flow diverges at infinity, it is permissible to complement the singularity representation with a uniform flow expressed by a constant, $\mathbf{c}$, setting

$$u_i(\mathbf{x}) = \frac{1}{4\pi\mu}\,\mathcal{S}_{ij}(\mathbf{x},\mathbf{x}_0)\,b_j + \frac{1}{2\pi}\,\mathcal{D}_{ij}(\mathbf{x},\mathbf{x}_0)\,d_j + c_i. \qquad (6.14.1)$$

Introducing the explicit forms of the singularities, we obtain

$$u_i(\mathbf{x}) = \frac{1}{4\pi\mu}\,(-\delta_{ij}\ln\frac{r}{\delta} + \frac{\hat{x}_i\hat{x}_j}{r^2})\,b_j + \frac{1}{2\pi}\,(-\frac{\delta_{ij}}{r^2} + 2\,\frac{\hat{x}_i\hat{x}_j}{r^4})\,d_j + c_i, \qquad (6.14.2)$$

where $\hat{\mathbf{x}} = \mathbf{x} - \mathbf{x}_c$ is the distance from the center of the cylinder and δ is a specified length to be determined as part of the solution.

Now we enforce the boundary condition $\mathbf{u} = \mathbf{V}$ at the surface of the cylinder located at $r = a$, and obtain a system of two equations for three unknowns, $\mathbf{b}$, $\mathbf{d}$, and $\mathbf{c}$,

$$-\frac{1}{4\pi\mu}\ln\frac{a}{\delta}\,\mathbf{b} - \frac{1}{2\pi a^2}\,\mathbf{d} + \mathbf{c} = \mathbf{V}, \qquad a^2\,\mathbf{b} + 4\mu\,\mathbf{d} = \mathbf{0}, \qquad (6.14.3)$$

where $\mathbf{V}$ is the cylinder velocity. The solution is

$$\mathbf{b} = 2\pi\mu\,\alpha\,\mathbf{V}, \qquad \mathbf{d} = -\frac{1}{2}\,\pi a^2\,\alpha\,\mathbf{V}, \qquad \mathbf{c} = [\,1 + \frac{1}{2}\,\alpha\,(\ln\frac{a}{\delta} - \frac{1}{2})\,]\,\mathbf{V}, \qquad (6.14.4)$$

where α is an indeterminate dimensionless constant. Substituting these expressions into (6.14.2), we derive an explicit expression for the velocity field,

$$u_i(\mathbf{x}) = \frac{\alpha}{2}\,(-\ln\frac{r}{a} - \frac{1}{2} + \frac{1}{2}\frac{a^2}{r^2}) + \frac{\alpha}{2}\,(1 - \frac{a^2}{r^2})\,\frac{\hat{x}_i\hat{x}_j}{r^2}\,V_j + V_i. \qquad (6.14.5)$$

Note that the length δ has disappeared after these substitutions. The force per unit length exerted on the cylinder is

$$\mathbf{F} = -\mathbf{s} = -2\pi\mu\,\alpha\,\mathbf{V}. \tag{6.14.6}$$

The torque with respect to the center of the cylinder is zero.

Surface traction

The hydrodynamic traction exerted on the cylinder can be computed from the strength of the singularities using the expressions given in Tables 6.5.3 and 6.5.4,

$$f_i = \frac{1}{4\pi}\,T_{ijk}^S b_j n_k + \mu\,\frac{1}{2\pi}\,T_{ijk}^D d_j n_k. \tag{6.14.7}$$

We find that

$$f_i = -\frac{1}{\pi}\frac{\hat{x}_i\hat{x}_j\hat{x}_k}{a^4}\,b_j\,n_k + \mu\,\frac{2}{\pi}\,\Big(\frac{\delta_{jk}\hat{x}_i + \delta_{ki}\hat{x}_j + \delta_{ij}\hat{x}_k}{a^4} - 4\,\frac{\hat{x}_i\hat{x}_j\hat{x}_k}{a^6}\,\Big)\,d_j\,n_k. \tag{6.14.8}$$

Setting $n_k = \hat{x}_k/a$, $\hat{x}_k\hat{x}_k = a^2$, and simplifying, we obtain

$$f_i = -\frac{1}{\pi}\frac{\hat{x}_i\hat{x}_j}{a^3}\,b_j + \mu\,\frac{2}{\pi}\,\Big(\frac{2\hat{x}_i\hat{x}_j + \delta_{ij}a^2}{a^5} - 4\,\frac{\hat{x}_i\hat{x}_j}{a^5}\,\Big)\,d_j. \tag{6.14.9}$$

Substituting the expressions for $\mathbf{b}$ and $\mathbf{d}$ given in (6.14.4) and simplifying, we find that

$$f_i = -2\,\alpha\,\frac{\hat{x}_i\hat{x}_j}{a^3}\,V_j - \alpha\,\mu\,\Big(\frac{-2\,\hat{x}_i\hat{x}_j + \delta_{ij}a^2}{a^3}\Big)\,V_j, \tag{6.14.10}$$

yielding

$$f_i = -\alpha\,\frac{\mu}{a}\,V_i. \tag{6.14.11}$$

We have found that, as in the case of a translating sphere, the traction is a constant vector oriented in the direction of translation. This is a unique and rather fortuitous feature of the circular geometry.

Integral representation

We have seen that the potential dipole can be expressed in terms of the Laplacian of the Stokeslet,

$$\boldsymbol{\mathcal{D}}(\mathbf{x}, \mathbf{x}_0) = -\frac{1}{2}\nabla_0^2\,\boldsymbol{\mathcal{S}}(\mathbf{x}, \mathbf{x}_0). \tag{6.14.12}$$

Using the second equation in (6.14.3) to write

$$\mathbf{d} = -\frac{a^2}{4\mu}\,\mathbf{b}, \tag{6.14.13}$$

we recast the singularity representation (6.14.1) into the form

$$u_i(\mathbf{x}) = \frac{1}{4\pi\mu}\,\mathcal{S}_{ij}(\mathbf{x}, \mathbf{x}_0)\,s_j + \frac{a^2}{16\pi\mu}\,\nabla_0^2\mathcal{S}_{ij}(\mathbf{x}, \mathbf{x}_0)\,s_j + c_i. \tag{6.14.14}$$

Applying the mean-value theorem for biharmonic functions (2.5.7) for the Stokeslet, we obtain

$$\frac{1}{2\pi a}\oint_{Cylinder} \mathcal{S}_{ij}(\mathbf{x},\boldsymbol{\xi})\,dl(\boldsymbol{\xi}) = S_{ij}(\mathbf{x},\mathbf{x}_0) + \frac{1}{4}a^2\nabla_0^2 S_{ij}(\mathbf{x},\mathbf{x}_0),\qquad(6.14.15)$$

where the point $\mathbf{x}$ lies outside the cylinder and the point $\boldsymbol{\xi}$ lies at the surface of the cylinder. Combining the last two equations, we find that

$$u_i(\mathbf{x}) = \frac{\alpha}{4\pi a}\oint_{Cylinder} \mathcal{S}_{ij}(\mathbf{x},\boldsymbol{\xi})\,dl(\boldsymbol{\xi})\,V_j + c_i.\qquad(6.14.16)$$

The integral on the right-hand side represents a uniform distribution of two-dimensional point forces with strength density $f_j = \alpha\mu V_j/a$ around the cylinder contour. This interpretation is an essential aspect of the slender-body theory discussed in Section 6.11.

6.14.2   Oseen flow far from the cylinder

Next, we describe the flow far from the cylinder (outer solution) by an Oseenlet with strength b_x parallel to the x axis located at the center of the cylinder, $\mathbf{x}_c$. The asymptotic expressions (6.13.56) show that, as r tends to zero, the inner limit of the outer solution is described by a Stokeslet complemented by uniform flow,

$$u_i^{outer}(\mathbf{x}) \simeq -\frac{1}{4\pi\mu}\Big(-\delta_{ij}\ln\frac{r}{\delta}+\frac{\hat{x}_i\hat{x}_j}{r^2}\Big)b_j + \frac{1}{4\pi\mu}\big(2\ln 2 - \gamma\big)b_i,\qquad(6.14.17)$$

where $\delta = \nu/|\mathbf{V}|$ and $\mathbf{b} = [b_x,0]$. To satisfy the force balance, the strength of the point force, $\mathbf{b}$, must be the same as that occurring in the inner Stokes flow solution.

6.14.3   Matching

In Section 6.14.1, we saw that the inner Stokes flow consists of a point force, a potential dipole, and a uniform flow. The outer limit of the inner flow computed from (6.14.2) by neglecting the fast-decaying dipole is

$$u_i^{inner}(\mathbf{x}) \simeq \frac{1}{4\pi\mu}\big(-\delta_{ij}\ln\frac{r}{\delta}+\frac{\hat{x}_i\hat{x}_j}{r^2}\big)b_j + c_i.\qquad(6.14.18)$$

Comparing (6.14.17) with (6.14.18) and requiring functional consistency between the inner and outer limits, we set

$$\mathbf{c} = \frac{1}{4\pi\mu}\big(2\ln 2 - \gamma\big)\mathbf{b}.\qquad(6.14.19)$$

Substituting into this equation the expressions for $\mathbf{b}$ and $\mathbf{c}$ given in (6.14.4), we obtain

$$1 + \frac{\alpha}{2}\Big(\ln\frac{\mathrm{Re}}{2}-\frac{1}{2}\Big) = \frac{\alpha}{2}\big(2\ln 2 - \gamma\big),\qquad(6.14.20)$$

where $\mathrm{Re} = D\mu|\mathbf{V}|/\mu$ is the Reynolds number defined with respect to the cylinder diameter, $D = 2a$. Solving for the coefficient α, we obtain [219]

$$\alpha = \frac{2}{\ln\dfrac{8}{\mathrm{Re}} - \gamma + \dfrac{1}{2}} \simeq \frac{2}{\ln\dfrac{7.4}{\mathrm{Re}}}. \tag{6.14.21}$$

Substituting (6.14.21) into (6.14.6), we obtain an expression for the drag force,

$$\mathbf{F} \simeq -2\pi\mu\, \frac{2}{\ln\dfrac{7.4}{\mathrm{Re}}}\, \mathbf{V}. \tag{6.14.22}$$

As Re tends to zero, the drag force diverges, and this underscores the ill-posedness of two-dimensional streaming Stokes flow.

The analysis can formally extended using the method of matched asymptotic expansions to derive further terms in the expression for the drag force [412].

Problems

6.14.1 *Flow past an elliptical cylinder*

Outline a procedure for computing the drag force on an infinite elliptical cylinder immersed in streaming flow along the major or minor axis.

6.14.2 *Flow due to a circular drop*

Develop a singularity representation of the flow due to a moving circular liquid drop.

6.15   Unsteady Stokes flow

In the remainder of this chapter, we turn our attention to a class of flows occurring at low Reynolds numbers, $\mathrm{Re} \equiv UL/\nu \ll 1$, but not necessarily at low values of the dimensionless frequency parameter, $\beta \equiv L^2/\nu T$, as discussed in Section 3.10.2, where U is a characteristic velocity, L is a characteristic length, T is a characteristic time, and ν is the kinematic viscosity of the fluid. More generally, we consider flows where $\mathrm{Re} \ll \beta$, requiring that that characteristic time scale is much smaller than the convective time scale, $T \ll L/U$. The motion of the fluid is governed by the continuity equation for an incompressible fluid, $\nabla \cdot \mathbf{u} = 0$, and the unsteady Stokes equation (3.10.6), repeated below for convenience,

$$\rho\, \frac{\partial \mathbf{u}}{\partial t} = -\nabla p + \mu \nabla^2 \mathbf{u}, \tag{6.15.1}$$

where ρ is the fluid density, μ is the fluid viscosity, and p is the hydrodynamic pressure incorporating the effect of the body force.

Flows governed by the unsteady Stokes equation include those generated by hydrodynamic braking and by the translational or rotational vibrations of rigid and liquid particles suspended in a viscous fluid, provided that the amplitude of the oscillation is small compared to the particle size. Other examples include oscillatory flow past a suspended particle due to the passage of a sound wave, and the flow due to the transient motion of a particle in a quiescent, accelerating, or decelerating ambient fluid. Unsteady Stokes flow occurs in the inner ear due to the small-amplitude vibrations of the eardrum transmitted through the ossicular bone chain.

6.15.1   Properties of unsteady Stokes flow

Taking the divergence of the unsteady Stokes equation (6.15.1) and using the continuity equation, we find that the pressure satisfies Laplace's equation,

$$\nabla^2 p = 0, \tag{6.15.2}$$

as in the case of steady Stokes flow. Taking the curl of (6.15.1), we find that the Cartesian components of the vorticity evolve according to the unsteady heat conduction equation,

$$\frac{\partial \boldsymbol{\omega}}{\partial t} = \nu \, \nabla^2 \boldsymbol{\omega}, \tag{6.15.3}$$

where $\nu = \mu/\rho$ is the kinematic viscosity of the fluid.

Any irrotational velocity field described by a harmonic potential, $\mathbf{u} = \nabla \phi$, satisfies the equations of unsteady Stokes flow. The associated hydrodynamic pressure is given by the simplified Bernoulli equation

$$p = -\rho \, \frac{\partial \phi}{\partial t} + c(t), \tag{6.15.4}$$

where $c(t)$ is a time-dependent function. However, because a general potential flow is not generally able to satisfy both the no-penetration and the no-slip conditions over a solid boundary, it must be complemented by a rotational flow.

Unsteady Stokes flow shares many of the properties of steady Stokes flow discussed previously in this chapter, including reversibility, the reciprocal identity, Faxén's laws, and uniqueness subject to specified boundary conditions for the velocity (e.g., [217, 306]).

6.15.2   Laplace transform

The Laplace transform allows us to eliminate the temporal dependence from the governing equations, replacing it with an algebraic dependence. Assuming that the motion has started at time $t = 0^-$ to allow for the possibility of an impulse causing a discontinuity at $t = 0$, we specify that all flow variables grow, at most, at an exponential rate in time and introduce the one-sided Laplace transform of the velocity, pressure, and vorticity, denoted by a caret,

$$\begin{bmatrix} \hat{\mathbf{u}} \\ \hat{p} \\ \hat{\boldsymbol{\omega}} \end{bmatrix} (\mathbf{x}, s) = \int_{0^-}^{\infty} \begin{bmatrix} \mathbf{u} \\ p \\ \boldsymbol{\omega} \end{bmatrix} (\mathbf{x}, t) \, \mathrm{e}^{-st} \, \mathrm{d}t, \tag{6.15.5}$$

where s is a complex variable. Taking the Laplace transform of the unsteady Stokes equation (6.15.1) and the continuity equation, $\nabla \cdot \mathbf{u} = 0$, and using the properties of the Laplace transform, we obtain Brinkman's equation and the continuity equation in the Laplace domain,

$$s\rho\,\hat{\mathbf{u}} = -\nabla\hat{p} + \mu\nabla^2\hat{\mathbf{u}}, \qquad \nabla \cdot \hat{\mathbf{u}} = 0. \tag{6.15.6}$$

An additional term appears on the right-hand side of the Stokes equation in the presence of a distributed or localized body force.

Having computed the solution subject to appropriate boundary conditions, we recover the physical variables in the time domain in terms of the Bromwich integral in the complex s plane,

$$\begin{bmatrix} \mathbf{u} \\ p \\ \omega \end{bmatrix} (\mathbf{x}, t) = \frac{1}{2\pi\mathrm{i}} \int_{\gamma-\mathrm{i}\infty}^{\gamma+\mathrm{i}\infty} \begin{bmatrix} \hat{\mathbf{u}} \\ \hat{p} \\ \hat{\omega} \end{bmatrix} (\mathbf{x}, s)\, \mathrm{e}^{st}\, \mathrm{d}s, \tag{6.15.7}$$

where i is the imaginary unit and γ is a sufficiently large real positive number chosen so that, as we move upward, all singularities of the Laplace transformed functions lie on the left of the integration path.

6.15.3   Oscillatory flow

The linearity of the equations of unsteady Stokes flow requires that the velocity, vorticity, hydrodynamic pressure, and hydrodynamic stress of an oscillatory flow with angular frequency Ω exhibit identical harmonic dependencies. We may thus write

$$\begin{bmatrix} \mathbf{u} \\ \omega \\ p \\ \sigma \end{bmatrix} = \begin{bmatrix} \bar{\mathbf{u}} \\ \bar{\omega} \\ \bar{p} \\ \bar{\sigma} \end{bmatrix} \exp(-\mathrm{i}\Omega t), \tag{6.15.8}$$

where i is the imaginary unit and a bar denotes the amplitude of the underlying variable. Substituting these expressions into (6.15.1) and into the continuity equation, $\nabla \cdot \mathbf{u} = 0$, we obtain the equations of oscillatory Stokes flow,

$$-\mathrm{i}\Omega\rho\,\mathbf{u} = -\nabla p + \mu\nabla^2\mathbf{u}, \qquad \nabla \cdot \mathbf{u} = 0. \tag{6.15.9}$$

It is worth noting the similarity with the Laplace transformed equations (6.15.6), subject to the substitution $s \to -\mathrm{i}\Omega$.

Solutions of the equation of oscillatory flow can be produced by superposing fundamental solutions or by developing boundary integral representations, as in the case of Stokes flow. A fundamental building block is the Green's function, representing the flow due to a point force whose strength oscillates harmonically in time.

Oscillatory point force

Consider the flow due to an oscillatory point force located at a point, $\mathbf{x}_0$. The strength of the point force is given by the real or imaginary part of

$$\mathbf{b} = \bar{\mathbf{b}}\exp(-\mathrm{i}\Omega t), \tag{6.15.10}$$

where $\bar{\mathbf{b}}$ is a constant amplitude. The induced velocity and pressure fields are computed by solving the equation

$$-i\Omega\rho\,\mathbf{u} = -\nabla p + \mu\nabla^2\mathbf{u} + \mathbf{b}\,\delta_3(\mathbf{x} - \mathbf{x}_0), \qquad (6.15.11)$$

subject to the continuity equation, $\nabla \cdot \mathbf{u} = 0$, where δ_3 is the three-dimensional delta function. To compute the solution, we work as in Section 6.5.5 for Stokes flow, and set

$$\mathbf{u} = \frac{1}{\mu}\left(\mathbf{I}\,\nabla^2\mathcal{H} - \nabla\nabla\mathcal{H}\right)\cdot\mathbf{b}. \qquad (6.15.12)$$

The generating function $\mathcal{H}$ satisfies the inhomogeneous Helmholtz equation,

$$\nabla^2\mathcal{H} + i\,\frac{\Omega}{\nu}\,\mathcal{H} - \frac{1}{4\pi r} = 0, \qquad (6.15.13)$$

where $r = |\mathbf{x} - \mathbf{x}_0|$ is the distance from the point force. Taking the Laplacian of (6.15.13), we find that the Laplacian of the generating function, $\mathcal{Q} \equiv \nabla^2\mathcal{H}$, is the Green's function of the Helmholtz equation

$$\nabla^2\mathcal{Q} + i\,\frac{\Omega}{\nu}\,\mathcal{Q} + \delta_3(\mathbf{x} - \mathbf{x}_0) = 0, \qquad (6.15.14)$$

which is given by

$$\mathcal{Q} = \frac{1}{4\pi r}\,\mathrm{e}^{-R}, \qquad (6.15.15)$$

where

$$R \equiv r\left(-i\,\frac{\Omega}{\nu}\right)^{1/2} = \frac{1-i}{\sqrt{2}}\,\frac{r}{\ell} \qquad (6.15.16)$$

is a dimensionless complex distance and

$$\ell \equiv \sqrt{\nu/\Omega} \qquad (6.15.17)$$

is a kinematic diffusion length. Note that the real part of scaled distance R is positive to ensure exponential decay.

To compute the function $\mathcal{H}$, we introduce spherical polar coordinates with origin at the point force, set

$$\mathcal{Q} \equiv \nabla^2\mathcal{H} = \frac{1}{r^2}\frac{\mathrm{d}}{\mathrm{d}r}\left(r^2\frac{\mathrm{d}\mathcal{H}}{\mathrm{d}r}\right) = \frac{1}{4\pi r}\,\mathrm{e}^{-R} \qquad (6.15.18)$$

and find that

$$\frac{\mathrm{d}\mathcal{H}}{\mathrm{d}r} = \frac{1}{4\pi R^2}\left[1 - (1+R)\,\mathrm{e}^{-R}\right]. \qquad (6.15.19)$$

Integrating, we obtain

$$\mathcal{H} = \frac{1}{4\pi} \left(i \frac{\nu}{\Omega} \right)^{1/2} \frac{e^{-R} - 1}{R} = \frac{1}{4\pi} \frac{1+i}{\sqrt{2}} \ell \frac{e^{-R} - 1}{R}. \tag{6.15.20}$$

Oscillatory Stokeslet

Substituting (6.15.20) into (6.15.12), we derive the velocity field due to an oscillating point force in the standard form

$$u_i(\mathbf{x}, \mathbf{x}_0) = \frac{1}{8\pi\mu} S_{ij}(\hat{\mathbf{x}}) \, b_j, \tag{6.15.21}$$

where $\hat{\mathbf{x}} = \mathbf{x} - \mathbf{x}_0$. The oscillatory Stokeslet is given by

$$\mathcal{S} = 8\pi \left(\mathbf{I} \nabla^2 \mathcal{H} - \nabla\nabla\mathcal{H} \right) = 8\pi \left[\mathcal{Q}\mathbf{I} - \nabla \left(\frac{d\mathcal{H}}{dr} \frac{\hat{\mathbf{x}}}{r} \right) \right], \tag{6.15.22}$$

yielding

$$\mathcal{S} = 8\pi \left[\left(\mathcal{Q} - \frac{1}{r} \frac{d\mathcal{H}}{dr} \right) \mathbf{I} + \frac{1}{r^2} \left(\frac{1}{r} \frac{d\mathcal{H}}{dr} - \frac{d^2\mathcal{H}}{dr^2} \right) \hat{\mathbf{x}} \otimes \hat{\mathbf{x}} \right], \tag{6.15.23}$$

where $\mathbf{I}$ is the identity matrix. Making substitutions, we obtain

$$S_{ij}(\hat{\mathbf{x}}) = \frac{\delta_{ij}}{r} \mathcal{A}(R) + \frac{\hat{x}_i \hat{x}_j}{r^3} \mathcal{C}(R), \tag{6.15.24}$$

where

$$\mathcal{A}(R) = 2 \left(1 + \frac{1}{R} + \frac{1}{R^2} \right) e^{-R} - \frac{2}{R^2}, \qquad \mathcal{C}(R) = -2 \left(1 + \frac{3}{R} + \frac{3}{R^2} \right) e^{-R} + \frac{6}{R^2}, \tag{6.15.25}$$

are frequency-dependent functions. One may confirm that $\mathcal{A}(0) = \mathcal{C}(0) = 1$, which shows that, at small frequencies or close to the point force, the unsteady Stokeslet reduces to the regular Stokeslet for Stokes flow.

The vorticity, pressure, and stress fields associated with the oscillatory Stokeslet are given by (6.15.8) with

$$\omega_i = \frac{1}{8\pi\mu} \Omega_{ij} \, b_j, \qquad p = \frac{1}{8\pi} \mathcal{P}_j \, b_j, \qquad \sigma_{ik} = \frac{1}{8\pi} \mathcal{T}_{ijk} \, b_j, \tag{6.15.26}$$

where

$$\Omega_{ij} = 2 \, \epsilon_{ijk} \frac{\hat{x}_l}{r^3} (R+1) e^{-R}, \qquad \mathcal{P}_j = 2 \frac{\hat{x}_j}{r^3}, \tag{6.15.27}$$

$$\mathcal{T}_{ijk} = -\frac{2}{r^3} \left(\delta_{ij}\hat{x}_k + \delta_{kj}\hat{x}_i \right) \left[(R+1) e^{-R} - \mathcal{C} \right] - \frac{2}{r^3} \delta_{ik}\hat{x}_j (1 - \mathcal{C}) - 2 \frac{\hat{x}_i \hat{x}_j \hat{x}_k}{r^5} \left[5\mathcal{C} - 2(R+1)e^{-R} \right],$$

and the function $\mathcal{C}$ is given in (6.15.25). Note that $\mathcal{T}_{ijk}$ is symmetric with respect to the indices i and k.

Traction and force on a spherical surface

It is instructive to compute the traction on a spherical surface of radius r centered at the oscillating point force, given by $f_i = \sigma_{ij} n_j$, where $\mathbf{n} = \frac{1}{r}(\mathbf{x} - \mathbf{x}_0)$ is the outward normal vector. Making substitutions, we obtain

$$f_i = \frac{1}{8\pi} \left(\frac{\delta_{ij}}{r} \mathcal{K}(R) + \frac{\hat{x}_i \hat{x}_j}{r^4} \mathcal{L}(R) \right) b_j, \tag{6.15.28}$$

where

$$\mathcal{K}(R) = 2\left[\mathcal{C} - (R+1) e^{-R} \right], \qquad \mathcal{L}(R) = 2\left[(R+1) e^{-R} - 1 - 3\mathcal{C} \right]. \tag{6.15.29}$$

Performing a Taylor series expansion, we find that $\mathcal{K}(0) = 0$ and $\mathcal{L}(0) = -6$, which is consistent with equation (6.5.23) for Stokes flow. The force exerted on the spherical surface is found by integrating the traction,

$$\mathbf{F} = \frac{1}{6} (3\mathcal{K} + \mathcal{L}) \mathbf{b} = -\frac{1}{3} \left[2(R+1) e^{-R} + 1 \right] \mathbf{b}. \tag{6.15.30}$$

We note that the force exerted on a small sphere with infinitesimal radius ($R = 0$) is equal to $-\mathbf{b}$, whereas the force exerted on a sphere with infinite radius ($R \to \infty$) is equal to $-\frac{1}{3} \mathbf{b}$. The difference between these two values is equal to the rate of change of momentum of the fluid surrounding the point force.

Behavior at low frequencies

To examine the asymptotic behavior of the flow due to an oscillating point force at low frequencies, we expand the generating function $\mathcal{H}$ given in (6.15.20) in a Taylor series for small R, obtaining

$$\mathcal{H} = \frac{1}{4\pi} \frac{1+\mathrm{i}}{\sqrt{2}} \ell \left(-1 + \frac{1}{2} R - \frac{1}{6} R^2 - \frac{1}{24} R^3 + \cdots \right). \tag{6.15.31}$$

Simplifying, we obtain

$$\mathcal{H} = -\frac{1}{4\pi} \frac{1+\mathrm{i}}{\sqrt{2}} \ell + \frac{r}{8\pi} \left(1 - \frac{1}{3} R + \frac{1}{12} R^2 + \cdots \right). \tag{6.15.32}$$

The constant term does not generate a flow; the linear term generates a steady Stokeslet, the quadratic term generates a uniform flow; further terms generate a sequence of unclassified singularities. It is interesting to recall that the Oseenlet also generates a uniform flow in addition to the Stokeslet near the point force. The corresponding expansion for the oscillating Stokeslet is

$$\mathcal{S}(\hat{\mathbf{x}}) = \mathcal{S}^{(0)}(\hat{\mathbf{x}}) + \mathcal{S}^{(1)}(\hat{\mathbf{x}}) + R^2 \mathcal{S}^{(2)}(\hat{\mathbf{x}}) + R^3 \mathcal{S}^{(3)}(\hat{\mathbf{x}}) + \cdots, \tag{6.15.33}$$

where $\mathcal{S}^{(0)}$ is the steady Stokeslet and

$$\mathcal{S}^{(1)}_{ij} = -\frac{4}{3} \frac{1-\mathrm{i}}{\sqrt{2}} \frac{1}{\ell} \delta_{ij}, \qquad \mathcal{S}^{(2)}_{ij} = \frac{1}{4} \left(3 \frac{\delta_{ij}}{r} - \frac{\hat{x}_i \hat{x}_j}{r^3} \right), \qquad \mathcal{S}^{(3)}_{ij} = \frac{2}{15} \left(2 \frac{\delta_{ij}}{r} - \frac{\hat{x}_i \hat{x}_j}{r^3} \right). \tag{6.15.34}$$

All constituent tensors, $\boldsymbol{\mathcal{S}}^{(n)}$, have dimensions of inverse length. As noted, the tensor $\boldsymbol{\mathcal{S}}^{(1)}$ represents uniform streaming flow.

Behavior at high frequencies

To examine the behavior of the flow at high frequencies or far from the point force, we expand $\boldsymbol{\mathcal{S}}$ in an asymptotic series for large R, obtaining

$$\mathcal{S}_{ij}(\hat{\mathbf{x}}) = \frac{2}{R^2}\left(-\frac{\delta_{ij}}{r} + 3\frac{\hat{x}_i\hat{x}_j}{r^3}\right) + 2\,\mathrm{e}^{-R}\left(\frac{\delta_{ij}}{r} - \frac{\hat{x}_i\hat{x}_j}{r^3}\right) + \cdots. \tag{6.15.35}$$

The first term on the right-hand side is the steady potential dipole. We conclude that, at high frequencies or far from the point force, the unsteady Stokeslet produces irrotational flow.

Green's functions of oscillatory flow

A Green's function of oscillatory Stokes flow represents the flow due to an oscillatory point force in an infinite or bounded domain of flow. When the domain extends to infinity, the velocity, vorticity, pressure and stress tensors are required to decay as the observation point moves far from the point force. The Green's function for semi-infinite flow bounded by a plane wall is discussed in References [303, 306]. It can be shown using the reciprocal theorem that the Green's functions for the velocity, $\boldsymbol{\mathcal{G}}$, satisfy the symmetry property (6.5.12). The pressure vector, $\boldsymbol{\mathcal{P}}$, and stress tensor, $\boldsymbol{\mathcal{T}}$, are acceptable unsteady Stokes flows representing the flow due to an oscillatory point source and the flow due to an oscillatory stresslet.

Boundary-integral formulation of oscillatory flow

Working as in Section 6.9 for Stokes flow, we derive an identical boundary-integral representation,

$$\alpha\,u_j(\mathbf{x}_0) = -\frac{1}{8\pi\mu}\iint_D f_i(\mathbf{x})\,\mathcal{S}_{ij}(\mathbf{x},\mathbf{x}_0)\,\mathrm{d}S(\mathbf{x}) + \frac{1}{8\pi}\iint_D u_i(\mathbf{x})\,\mathcal{T}_{ijk}(\mathbf{x},\mathbf{x}_0)\,n_k(\mathbf{x})\,\mathrm{d}S(\mathbf{x}), \tag{6.15.36}$$

where D is the smooth boundary of a selected control volume, $\mathbf{n}$ is the unit normal vector pointing into the control volume, and $\alpha = 1, \frac{1}{2}, 0$, respectively, when the the point $\mathbf{x}_0$ lies inside, at the boundary of, or outside the control volume. In the second case, the principal value of the double-layer potential represented by the integral on the right-hand side is implied.

For example, applying the boundary-integral representation (6.15.36) for streaming oscillatory flow with uniform velocity $\mathbf{u} = \mathbf{U}$ and pressure $p = \mathrm{i}\Omega\rho\,\mathbf{U}\cdot\mathbf{x}$, and setting $\mathbf{f} = -p\mathbf{n}$, we derive the identity

$$\alpha\,\delta_{ij} = \mathrm{i}\,\frac{\Omega}{8\pi\nu}\iint_D x_i\,n_k(\mathbf{x})\,\mathcal{S}_{kj}(\mathbf{x},\mathbf{x}_0)\,\mathrm{d}S(\mathbf{x}) + \frac{1}{8\pi}\iint_D \mathcal{T}_{ijk}(\mathbf{x},\mathbf{x}_0)\,n_k(\mathbf{x})\,\mathrm{d}S(\mathbf{x}), \tag{6.15.37}$$

where $\nu = \mu/\rho$ is the kinematic viscosity of the fluid.

6.15.4 Flow due to a vibrating particle

Consider the flow due to the translational or rotational vibrations of a particle in an otherwise quiescent infinite ambient fluid. The characteristic time scale of the flow is $T = 1/\Omega$, the characteristic velocity is $U = \Omega \bar{d}$, and the characteristic length is the particle size, L, where Ω is the angular frequency and $\bar{d}$ is the amplitude of the vibration. Requiring that $UL/\nu \equiv \mathrm{Re} \ll \beta \equiv L^2/\nu T$, we obtain $T \ll L/U$ or $\bar{d} \ll L$, which shows that the equations of unsteady Stokes flow apply when the amplitude of vibration is much smaller than the particle size. Enforcing the boundary condition of rigid-body motion at the mean position of the particle instead of the instantaneous position introduces an error that is comparable to $\bar{d}/L$, which is negligible as long as the amplitude of the oscillation is small compared to the particle size. The boundary-integral formulation provides us with the integral representation (6.15.36), where the integration domain, D, is the mean particle surface.

Translational vibrations

In the case of translational vibrations with velocity $\mathbf{V} = \bar{\mathbf{V}} \exp(-\mathrm{i}\Omega t)$, we set $\mathbf{u} = \mathbf{V}$ at the mean position of the particle surface, P, and obtain the integral representation

$$u_j(\mathbf{x}_0) = -\frac{1}{8\pi\mu} \iint_P f_i(\mathbf{x})\, S_{ij}(\mathbf{x}, \mathbf{x}_0)\, \mathrm{d}S(\mathbf{x}) + \frac{1}{8\pi} V_i \iint_P \mathcal{T}_{ijk}(\mathbf{x}, \mathbf{x}_0)\, n_k(\mathbf{x})\, \mathrm{d}S(\mathbf{x}), \qquad (6.15.38)$$

for a point $\mathbf{x}_0$ lying in the ambient fluid. Combining this representation with identity (6.15.37) written with $\alpha = 0$ to eliminate the double-layer potential, we derive a single-layer representation,

$$u_j(\mathbf{x}_0) = -\frac{1}{8\pi\mu} \iint_P \left[f_i(\mathbf{x}) + \mathrm{i}\,\Omega\rho \left(\mathbf{V} \cdot \mathbf{x} \right) n_i \right] S_{ij}(\mathbf{x}, \mathbf{x}_0)\, \mathrm{d}S(\mathbf{x}). \qquad (6.15.39)$$

Now we move the point $\mathbf{x}_0$ to the particle surface and derive an integral equation that can be solved by numerical methods (e.g., [304, 243]). A similar reduction of the boundary-integral representation to a single-layer potential is not possible in the case of rotational oscillations.

Low-frequency asymptotics

To study the behavior of the flow at low angular velocities in the general case of translational and rotational vibrations, we introduce the dimensionless complex frequency

$$\lambda \equiv \left(-\mathrm{i}\,\frac{\Omega}{\nu} \right)^{1/2} L = \frac{1 - \mathrm{i}}{\sqrt{2}}\,\frac{L}{\ell}, \qquad (6.15.40)$$

where L is a particle length scale. Next, we expand the velocity and traction in asymptotic series with respect to λ,

$$\mathbf{u} = \mathbf{u}^{(0)} + \lambda \mathbf{u}^{(1)} + \lambda^2\, \mathbf{u}^{(2)} + \cdots, \qquad \mathbf{f} = \mathbf{t}^{(0)} + \lambda \mathbf{f}^{(1)} + \lambda^2\, \mathbf{f}^{(2)} + \cdots. \qquad (6.15.41)$$

The boundary conditions of rigid-body motion require that

$$\mathbf{u}^{(0)} = \mathbf{V} + \mathbf{\Omega} \times (\mathbf{x} - \mathbf{x}_c), \qquad \mathbf{u}^{(1)} = \mathbf{u}^{(2)} = \cdots = \mathbf{0} \qquad (6.15.42)$$

at the mean particle surface, where

$$\mathbf{V} = \bar{\mathbf{V}} \exp(-\mathrm{i}\Omega t), \qquad \mathbf{\Omega} = \bar{\mathbf{\Omega}} \exp(-\mathrm{i}\Omega t), \qquad (6.15.43)$$

are the translational and rotational velocities about the center of rotation, $\mathbf{x}_c$. Substituting these expansions along with (6.15.33) into (6.15.36), and collecting terms of zeroth and first order in λ, we obtain the integral equations

$$u_j^{(0)}(\mathbf{x}_0) = -\frac{1}{8\pi\mu} \iint_P f_i^{(0)}(\mathbf{x})\, \mathcal{S}_{ij}^{(0)}(\mathbf{x}, \mathbf{x}_0)\, \mathrm{d}S(\mathbf{x})\, n_k(\mathbf{x})\, \mathrm{d}S(\mathbf{x}) \qquad (6.15.44)$$

and

$$-\frac{1}{6\pi\mu a} F_j^{(0)} = -\frac{1}{8\pi\mu} \iint_P f_i^{(1)}(\mathbf{x})\, \mathcal{S}_{ij}^{(0)}(\mathbf{x}, \mathbf{x}_0)\, \mathrm{d}S(\mathbf{x}). \qquad (6.15.45)$$

The double-layer potential has disappeared from (6.15.44) due to the boundary condition or rigid-body motion.

The solution of the zeroth-order integral equation (6.15.44) can be expressed in the form

$$\mathbf{f}^{(0)} = -\mu\,(\mathbf{R}^t \cdot \mathbf{V} + \mathbf{R}^r \cdot \mathbf{\Omega}), \qquad (6.15.46)$$

where $\mathbf{R}^t$ and $\mathbf{R}^r$ are, respectively, the steady translational and rotary surface traction resistance matrices introduced in (6.8.4) and (6.8.8). The steady force and torque exerted on the particle can be expressed in terms of the resistance matrices introduced in (6.8.15),

$$\mathbf{F}^{(0)} = -\mu\,(\mathbf{X} \cdot \mathbf{V} + \mathbf{Z} \cdot \mathbf{\Omega}), \qquad \mathbf{T}^{(0)} = -\mu\,(\mathbf{Z}^T \cdot \mathbf{V} + \mathbf{Y} \cdot \mathbf{\Omega}). \qquad (6.15.47)$$

Comparing the last four equations, we conclude that the first correction to the traction, force, and torque exerted on the particle are given by

$$\begin{bmatrix} \mathbf{f} \\ \mathbf{F} \\ \mathbf{T} \end{bmatrix}^{(1)} = \frac{1}{6\pi a} \begin{bmatrix} \mathbf{R}^t \\ \mathbf{X} \\ \mathbf{Z}^T \end{bmatrix} \cdot \mathbf{F}^{(0)} = -\frac{\mu}{6\pi a} \begin{bmatrix} \mathbf{R}^t \\ \mathbf{X} \\ \mathbf{Z}^T \end{bmatrix} \cdot (\mathbf{X} \cdot \mathbf{V} + \mathbf{P} \cdot \mathbf{\Omega}). \qquad (6.15.48)$$

We have found that the first-order correction can be computed from the resistance matrices for steady Stokes flow.

6.15.5 Two-dimensional oscillatory point force

The flow due to a two-dimensional oscillatory point force can be derived working as in Section 6.15.3 for the three-dimensional point force. We find that the generating function $\mathcal{H}$ satisfies an inhomogeneous Helmholtz equation in the xy plane,

$$\nabla^2 \mathcal{H} + \mathrm{i}\frac{\Omega}{\nu}\mathcal{H} + \frac{1}{2\pi}\ln\frac{r}{a} = 0, \qquad (6.15.49)$$

where $r = |\mathbf{x} - \mathbf{x}_0|$, ∇^2 is the Laplacian operator in the xy plane, and a is a constant length. Taking the Laplacian of (6.15.49), we find that the Laplacian of the generating function, $\mathcal{Q} \equiv \nabla^2\mathcal{H}$, is the fundamental solution of a Helmholtz equation in the xy plane,

$$\nabla^2 \mathcal{Q} + \mathrm{i}\frac{\Omega}{\nu}\mathcal{Q} + \delta_2(\mathbf{x} - \mathbf{x}_0) = 0. \qquad (6.15.50)$$

The free-space Green's function is given by

$$Q = \frac{1}{2\pi} \left(\ker_0 \varrho - i \kei_0 \varrho \right), \qquad (6.15.51)$$

where $r = |\mathbf{x} - \mathbf{x}_0|$, $\ell = \sqrt{\nu/\Omega}$, $\varrho = r/\ell$, and $\ker_0$, $\kei_0$ are Kelvin functions (e.g., [2], p. 379). To compute the solution, we express the Laplacian in plane polar coordinates,

$$\nabla^2 \mathcal{H} = \frac{1}{r} \frac{d}{dr} \left(r \frac{d\mathcal{H}}{dr} \right) = Q, \qquad (6.15.52)$$

and integrate once to find that

$$\frac{d\mathcal{H}}{dr} = \frac{1}{2\pi r} \int_0^r \xi \left(\ker_0 \varrho - i \kei_0 \varrho \right) d\xi. \qquad (6.15.53)$$

Abramowitz & Stegun ([2], p. 380) provide us with the indefinite integrals

$$\int \xi \ker_0(\xi) \, d\xi = -\frac{\xi}{\sqrt{2}} \left(\ker_1 \xi - \kei_1 \xi \right) = \xi \kei_0' \xi \qquad (6.15.54)$$

and

$$\int \xi \kei_0(\xi) \, d\xi = -\frac{\xi}{\sqrt{2}} \left(\kei_1 \xi + \ker_1 \xi \right) = -\xi \ker_0' \xi. \qquad (6.15.55)$$

Substituting these integrals into (6.15.53), we obtain

$$\frac{d\mathcal{H}}{dr} = \frac{\ell}{2\pi} \left(\kei_0' \varrho + i \ker_0' \varrho \right). \qquad (6.15.56)$$

The oscillatory Stokeslet is given by

$$\boldsymbol{S} = 4\pi (\mathbf{I} \nabla^2 \mathcal{H} - \nabla\nabla\mathcal{H}) = 4\pi \left[\mathbf{I} \, Q - \nabla \left(\frac{d\mathcal{H}}{dr} \frac{\hat{\mathbf{x}}}{r} \right) \right], \qquad (6.15.57)$$

which can be recast into the form

$$\boldsymbol{S} = \mathcal{A}(\varrho) \, \mathbf{I} + \frac{\mathcal{C}(\varrho)}{r^2} \, \hat{\mathbf{x}} \otimes \hat{\mathbf{x}}, \qquad (6.15.58)$$

where

$$\mathcal{A}(\varrho) = 4\pi \left(\mathcal{G} - \frac{1}{r} \frac{d\mathcal{H}}{dr} \right) = 2 \left(\ker_0 \varrho - \frac{1}{\varrho} \kei_0' \varrho \right) - 2 i \left(\kei_0 \varrho + \frac{1}{\varrho} \ker_0' \varrho \right),$$

$$\mathcal{C}(\varrho) = 4\pi \left(\frac{1}{r} \frac{d\mathcal{H}}{dr} - \frac{d^2\mathcal{H}}{dr^2} \right) = 2 \left(-\ker_0 \varrho + \frac{2}{\varrho} \kei_0' \varrho \right) + 2 i \left(\kei_0 \varrho + \frac{2}{\varrho} \ker_0' \varrho \right)$$

$$(6.15.59)$$

[307]. As R tends to zero, we recover the steady two-dimensional Stokeslet.

6.15.6   Inertial effects and steady streaming

The equations of unsteady Stokes flow describe the structure of a flow when the Reynolds number, Re, is much smaller than the frequency parameter β, so that the nonlinear acceleration term, $\rho\, \mathbf{u}\cdot\nabla\mathbf{u}$, is insignificant compared to the Eulerian term, $\rho\,\partial\mathbf{u}/\partial t$. The leading-order solution can be regarded as a base state for studying nonlinear inertial effects using regular perturbation expansions.

Oscillatory flow

The Stokes flow solution for oscillatory flow, is given by the real or imaginary part of the right-hand side of (6.15.8). Selecting the real part, we introduce a regular asymptotic expansion for the velocity with respect to Re,

$$\mathbf{u}(\mathbf{x},t) = \frac{1}{2}\left(\bar{\mathbf{u}}^{(0)}\, e^{-i\Omega t} + \bar{\mathbf{u}}^{(0)*} e^{i\Omega t}\right) + \mathrm{Re}\, \mathbf{u}^{(1)}(\mathbf{x},t) + \cdots, \tag{6.15.60}$$

where the superscript (0) denotes the Stokes flow solution, the superscript (1) denotes the first inertial correction, an asterisk designates the complex conjugate, and the three dots denote high-order corrections with respect to Re, The nonlinear convection term on the left-hand side of the equation of motion becomes

$$\mathbf{u}\cdot\nabla\mathbf{u} = \frac{1}{4}\,\mathcal{F}(\mathbf{u}^{(0)}) + O(\mathrm{Re}), \tag{6.15.61}$$

where

$$\mathcal{F}(\mathbf{u}^{(0)}) \equiv \left[\bar{\mathbf{u}}^{(0)}\cdot\nabla\bar{\mathbf{u}}^{(0)}\right] e^{-2i\Omega t} + \left[\bar{\mathbf{u}}^{(0)*}\cdot\nabla\bar{\mathbf{u}}^{(0)*}\right] e^{2i\Omega t} + \nabla\cdot\left(\bar{\mathbf{u}}^{(0)}\otimes\bar{\mathbf{u}}^{(0)*} + \bar{\mathbf{u}}^{(0)*}\otimes\bar{\mathbf{u}}^{(0)}\right). \tag{6.15.62}$$

Substituting expressions (6.15.60) and (6.15.62) into the Navier–Stokes equation, nondimensionalizing all terms as discussed in Section 3.10, retaining terms of first order in Re, and reverting to physical variables, we obtain a forced unsteady Stokes equation

$$\rho\,\frac{\partial\mathbf{u}^{(1)}}{\partial t} = -\nabla p^{(1)} + \mu\,\nabla^2\mathbf{u}^{(1)} - \frac{1}{4}\,\rho\,\mathcal{F}, \tag{6.15.63}$$

where $p^{(1)}$ is the first inertial correction to the pressure. We observe that the nonlinear acceleration term produces second harmonics of the fundamental frequency, Ω, represented by the terms inside the square brackets on the right-hand side of (6.15.62). The time-independent term enclosed by the parentheses on the right-hand side of (6.15.62) is responsible for the onset of inertial steady streaming superimposed on the underlying oscillatory flow.

Steady streaming

To describe the steady streaming flow, we integrate equation (6.15.63) over one period to eliminate the time dependence, and derive a forced Stokes equation,

$$-\nabla\bar{p}^{(1)} + \mu\,\nabla^2\bar{\mathbf{u}}^{(1)} - \frac{1}{4}\,\rho\,\nabla\cdot\left(\bar{\mathbf{u}}^{(0)}\otimes\bar{\mathbf{u}}^{(0)*} + \bar{\mathbf{u}}^{(0)*}\otimes\bar{\mathbf{u}}^{(0)}\right) = \mathbf{0}, \tag{6.15.64}$$

where a bar designates a time average. The problem is reduced to solving the forced Stokes flow equation (6.15.64) together with the continuity equation, subject to appropriate boundary conditions.

At high frequencies, the forcing term is confined inside a Stokes layer surrounding the boundaries of the flow, and a solution can be found using the boundary-layer approximation (e.g., [24], p. 358).

The mathematical properties and computation of steady streaming due to an oscillatory flow over a curved boundary have been studied extensively with reference to the flow induced by the propagation of sound waves (e.g., [203, 210, 290, 345]).

Problems

6.15.1 *Oscillatory point force*

(*a*) Confirm that, as the scaled distance R tends to zero, the vorticity and stress tensors $\boldsymbol{\Omega}$ and $\mathbf{T}$ defined in (6.15.27) reduce to those for steady Stokes flow.

(*b*) Repeat (*a*) for the two-dimensional oscillatory point force.

6.15.2 *Oscillating sphere*

Derive the specific form of (6.15.48) for an oscillating sphere.

6.16 Singularity methods for oscillatory flow

The apparatus of the singularity method for Stokes flow can be extended to unsteady Stokes flow. The generalized Faxén relations based on singularity representations discussed in Section 6.12 are also valid for unsteady Stokes flow [303]. While the general principles of the singularity method remain the same, the specific expressions for the singularities become more involved and the computation of singularity representations becomes more cumbersome. In this section, we derive of families of singularities of oscillatory Stokes flow originating from the point source or point force, and then employ them to generate exact solutions.

6.16.1 Singularities of oscillatory flow

Working as in the case of Stokes flow, we generate two families of singularities consisting of irrotational and rotational flows. The former originate from the point source, and the latter originate from the point force.

Oscillatory point source and derivative singularities

A point source with oscillatory strength, $m = \bar{m}\exp(-\mathrm{i}\Omega t)$, located at the point $\mathbf{x}_0$, produces irrotational flow. The induced velocity and pressure fields are

$$\mathbf{u}(\mathbf{x}) = \frac{m}{4\pi}\frac{\mathbf{x}-\mathbf{x}_0}{r^3}, \qquad p(\mathbf{x}) = -\mathrm{i}\Omega\rho\,\frac{m}{4\pi}\frac{1}{r}, \tag{6.16.1}$$

where $\hat{\mathbf{x}} = \mathbf{x} - \mathbf{x}_0$ and $r = |\hat{\mathbf{x}}|$. As the frequency Ω decreases, the pressure field vanishes throughout the domain of flow.

Differentiating the point source with respect to the singular point, $\mathbf{x}_0$, we obtain a sequence of derivative singularities expressing irrotational flow. The first two are the potential dipole and the potential quadruple. The velocity fields are shown in Table 6.6.1 and the corresponding pressure fields are found by differentiating the pressure of the point source.

Multipoles of the oscillatory point force

Differentiating a Green's function representing an oscillatory point force with respect to the singular point, $\mathbf{x}_0$, we obtain a sequence of singularities expressing multipoles of the oscillatory point force. The associated pressure fields are identical to those of the corresponding singularities of steady Stokes flow.

Symmetric Stokeslet quadruple

The Laplacian of the free-space Green's function with respect to the singular point provides us with symmetric Stokeslet quadruple,

$$\mathcal{S}_{ij}^{SQ} = -\frac{1}{2}\nabla_0^2 \mathcal{S}_{ij} = \left[-\frac{\delta_{ij}}{r^3}\left(1 + R + R^2\right) + \frac{\hat{x}_i\hat{x}_j}{r^5}\left(3 + 3R + R^2\right) \right] \mathrm{e}^{-R}, \tag{6.16.2}$$

where $R \equiv r(-\mathrm{i}\Omega/\nu)^{1/2}$ is a dimensionless distance. One may verify by straightforward substitution that

$$\mathcal{S}^{SQ} = \mathcal{D} - \frac{1}{2}\frac{R^2}{r^2}\,\mathcal{S}, \tag{6.16.3}$$

where $\mathcal{D}$ is the potential dipole given in Table 6.6.1 and $\mathcal{S}$ is the unsteady Stokeslet given in (6.15.24). As $R \to 0$, the symmetric Stokeslet quadruple reduces to the potential dipole, $\mathcal{D}$.

The velocity and stress fields due to a symmetric Stokeslet quadruple with strength $\mathbf{q}$ are given by

$$u_i = \frac{1}{4\pi}\mathcal{S}_{ij}^{SQ}\,q_j, \qquad \sigma_{ik} = \frac{\mu}{4\pi}\mathcal{T}_{ijk}^{SQ}\,q_j, \tag{6.16.4}$$

where

$$\mathcal{T}_{ijk}^{SQ} = \left(\frac{\delta_{ij}\hat{x}_k + \delta_{jk}\hat{x}_i}{r^5}\left(6 + 6R + 3R^2 + R^3\right) + 2\delta_{ik}\frac{\hat{x}_j}{r^5}\left(3 + 3R + R^2\right) \right.$$
$$\left. - 2\frac{\hat{x}_i\hat{x}_j\hat{x}_k}{r^7}\left(15 + 15R + 6R^2 + R^3\right) \right)\mathrm{e}^{-R}. \tag{6.16.5}$$

The traction exerted on a spherical surface of radius r centered at the singular point is

$$f_i = \sigma_{ij}n_j = \frac{\mu}{4\pi}\left(\frac{\delta_{ij}}{r^4}\,\mathcal{E}(R) - \frac{\hat{x}_i\hat{x}_j}{r^6}\,\mathcal{F}(R) \right)\mathrm{e}^{-R}\,q_j, \tag{6.16.6}$$

where

$$\mathcal{E}(R) = 6 + 6R + 3R^2 + R^3, \qquad \mathcal{F}(R) = 18 + 18R + 7R^2 + R^3. \tag{6.16.7}$$

Substituting $\mathcal{E}(0) = 6$ and $\mathcal{F}(0) = 18$, we confirm agreement with corresponding results for Stokes flow. The associated force is

$$\mathbf{F} = \frac{\mu}{r^2}\left[\mathcal{E}(R) - \frac{1}{3}\mathcal{F}(R)\right]\mathrm{e}^{-R}\,\mathbf{q} = \frac{2}{3}\frac{1}{r^2}R^2\left(1 + R\right)\mathrm{e}^{-R}\,\mathbf{q}. \tag{6.16.8}$$

We observe that the force is zero only in the case of Stokes flow, $R \to 0$.

Oscillatory rotlet or couplet

The velocity, pressure, and stress fields due to an oscillatory couplet or rotlet with strength $\mathbf{c}$ are given by

$$u_i = \frac{1}{8\pi\mu}\,\mathcal{C}_{im}\,c_m, \qquad p = 0, \qquad \sigma_{ik} = \frac{1}{8\pi}\,\mathcal{T}^C_{imk}\,c_m, \tag{6.16.9}$$

where

$$\mathcal{C}_{im} = \epsilon_{iml}\,\frac{\hat{x}_l}{r^3}\,(1+R)\,\mathrm{e}^{-R}. \tag{6.16.10}$$

The torque exerted on a spherical surface of radius r centered at the couplet is

$$\mathbf{T} = -\left(1 + R + \frac{1}{3}\,R^2\right)\mathrm{e}^{-R}\,\mathbf{c}. \tag{6.16.11}$$

As $R \to 0$, we recover our previous results for Stokes flow $\mathbf{T} = -\mathbf{c}$.

Interior flow

The velocity, pressure, and stress fields due to an oscillating Stokeson with strength $\mathbf{s} = \bar{\mathbf{s}}\exp(-\mathrm{i}\Omega t)$ is given by [303]

$$u_i = \mathcal{S}^N_{ij}\,s_j, \qquad p = \mu\,\mathcal{P}^N_j\,s_j, \qquad \sigma_{ik} = \mu\,\mathcal{T}^N_{ijk}\,s_j. \tag{6.16.12}$$

The pressure vector is the same as that of the steady Stokeson, $\mathcal{P}^N_j = 10\,\hat{x}_j$. The velocity tensor is given by

$$\mathcal{S}^N_{ij} = \delta_{ij}2r^2\mathcal{Q}(R) - \hat{x}_i\hat{x}_j\mathcal{W}(R), \tag{6.16.13}$$

and traction exerted on a spherical surface of radius r centered at the Stokeson is

$$f_i = \sigma_{ij}n_j = \mu\left(\delta_{ij}r\,\mathcal{I}(R) + \frac{\hat{x}_i\hat{x}_j}{r}\,\mathcal{J}(R)\right)s_j, \tag{6.16.14}$$

where the functions $\mathcal{Q}$, $\mathcal{W}$, $\mathcal{I}$, and $\mathcal{J}$, are given in Table 6.16.1. In the limit of steady Stokes flow, we find that $\mathcal{I}(0) = 3$ and $\mathcal{J}(0) = -9$, yielding the Stokeson in Stokes flow.

6.16.2   Singularity representations

The analytical and numerical computations of singularity representations for unsteady Stokes flow are similar to those of Stokes flow discussed in Section 6.7. If the exterior flow due to the small-amplitude oscillations of a rigid or fluid particle can be expressed in terms of a collection of singularities, including the oscillating Stokeslet and its derivatives expressing multipoles of the point force, then the force exerted on the particle can be computed directly from the total strength of the Stokeslets, $\mathbf{b}$, as

$$\mathbf{F} = -\mathbf{b} - \mathrm{i}\Omega\rho V_p\mathbf{V}, \tag{6.16.15}$$

where V_p is the particle volume and $\mathbf{V}$ is the particle velocity of translation [303].

$$Q(R) = -\frac{1}{4R^5}\left(2 - R^3 - 30(1 + R^2)\sinh R + 30R\cosh R\right)$$

$$W(R) = \frac{15}{R^5}\left((3 + R^3)\sinh R - 3R\cosh R\right)$$

$$\mathcal{I}(R) = \frac{15}{R^5}\left(R(6 + R^2)\cosh R - 3(2 + R^2)\sinh R\right)$$

$$\mathcal{J}(R) = -10 - \frac{15}{R^5}\left(R(18 + R^2)\cosh R - (7R^2 + 18)\sinh R\right)$$

TABLE 6.16.1   Functions related to the velocity field and force exerted on a spherical surface due to an oscillating Stokeson in interior flow.

Rotational vibration of a sphere

The flow due to the rotational vibration of a sphere of radius a can be represented in terms of a rotlet with strength

$$\mathbf{c} = 8\pi\mu a^3 \frac{e^\lambda}{1 + \lambda}\,\boldsymbol{\Omega}, \tag{6.16.16}$$

where $\lambda^2 = -i\Omega a^2/\nu$ is a dimensionless frequency. Using (6.16.11), we find that the torque exerted on the sphere is

$$\mathbf{T} = -(1 + \lambda + \tfrac{1}{3}\lambda^2)e^{-\lambda}\mathbf{c} = -\frac{8}{3}\pi\mu a^3\left(3 + \frac{\lambda^2}{1 + \lambda}\right)\boldsymbol{\Omega}. \tag{6.16.17}$$

As $\lambda \to 0$, we recover our earlier results for Stokes flow.

Translational vibration of a sphere

The flow due to the translational vibration of a sphere of radius a whose center is oscillating about a mean position, $\mathbf{x}_0$, can be represented in terms of an unsteady Stokeslet with strength $\mathbf{b}$ and a symmetric Stokeslet quadruple with strength $\mathbf{q}$,

$$u_i(\mathbf{x}) = \frac{1}{8\pi\mu}S_{ij}(\mathbf{x}, \mathbf{x}_0)\,b_j + \frac{1}{4\pi}S_{ij}^{SQ}(\mathbf{x}, \mathbf{x}_0)\,q_j. \tag{6.16.18}$$

Substituting the expressions for the singularities and enforcing the boundary condition $\bar{\mathbf{u}} = \bar{\mathbf{V}}$ at $r = a$, we find that

$$\mathbf{b} = 6\pi\mu a(1 + \lambda + \tfrac{1}{2}\lambda^2)\,\mathbf{V}, \qquad \mathbf{q} = -6\pi a^3\frac{1}{\lambda}(e^\lambda - 1 - \lambda - \tfrac{1}{3}\lambda^2)\,\mathbf{V}, \tag{6.16.19}$$

where $\lambda^2 = -i\Omega a^2/\nu$ is a dimensionless frequency. As $\lambda \to 0$, we obtain expressions (6.7.4) for steady Stokes flow.

Using (6.15.30) and (6.16.8) or (6.16.15) along with the coefficients (6.16.19), we obtain the force exerted on the sphere,

$$\mathbf{F} = -\frac{1}{3}\left[2e^{-\lambda}(\lambda+1)+1\right]\mathbf{b} - \frac{i\Omega}{\rho}\frac{2}{3}e^{-\lambda}(1+\lambda)\,\mathbf{q} = -\mathbf{b} - i\Omega\rho V_s\mathbf{V}, \qquad (6.16.20)$$

where $V_s = \frac{4\pi}{3}a^3$ is the volume of the sphere. Making substituting, we recover the Boussinesq–Basset formula (e.g., [413])

$$\mathbf{F} = -6\pi\mu a\left(1+\lambda+\frac{1}{9}\lambda^2\right)\mathbf{V}. \qquad (6.16.21)$$

It is remarkable that the expression for the force is a binomial in the complex frequency parameter λ. The three terms in the parentheses on the right-hand side of (6.16.21) represent, respectively, the steady Stokes drag force, the unsteady Boussinesq–Basset force, and an acceleration reaction associated with the virtual mass in potential flow. The quadratic dependence is true only for the spherical shape [303].

Further singularity representations

Exact singularity representations for unsteady flow are limited to those for flow produced by the translational or rotational oscillations of a solid or liquid sphere, and for the disturbance flow due to a sphere that is held stationary in an oscillating linear ambient flow ([303]; [208], Chapter 5; [308]). It should be noted that, since a general oscillating linear flow is not an exact solution of the equations of unsteady Stokes flow, the solution for linear ambient flow must be regarded as the response of the sphere to the linear term in the Taylor series expansion about the center of the sphere of a general unsteady incident flow. Approximate singularity representations of the flow due to the translational oscillations of a prolate spheroid can be computed in terms of oscillatory point forces and quadruples distributed over the focal length of the spheroid [303]. The densities of the distributions range from those for steady flow at low frequencies to those for inviscid potential flow at high frequencies.

Problems

6.16.1 *Oscillating liquid drop*

Derive the singularity representation of an oscillating liquid drop [303].

6.17   Unsteady Stokes flow due to a point force

Having investigated the properties and computation of oscillatory Stokes flow, now we turn our attention to a more general time-dependent motion.

6.17.1   Impulsive point force

The velocity and pressure fields of an unsteady Stokes flow due to an impulsive point force with strength $\mathbf{b}$ applied at a point $\mathbf{x}_0$ at time t_0 satisfy the continuity equation, $\nabla \cdot \mathbf{u} = 0$, and the singularly forced equation

$$\rho \frac{\partial \mathbf{u}}{\partial t} = -\nabla p + \mu \nabla^2 \mathbf{u} + \mathbf{b}\, \delta_3(\mathbf{x} - \mathbf{x}_0)\, \delta_1(t - t_0), \tag{6.17.1}$$

where δ_3 is the three-dimensional delta function and δ_1 is the one-dimensional delta function. The solution for the velocity and pressure can be expressed in the forms

$$\mathbf{u}(\hat{\mathbf{x}}, \hat{t}) = \frac{1}{\mu}\left(\nabla^2 \mathcal{H} - \nabla\nabla\mathcal{H}\right) \cdot \mathbf{b}, \qquad p(\hat{\mathbf{x}}, \hat{t}) = -\frac{1}{4\pi} \nabla\left(\frac{1}{r}\right) \cdot \mathbf{b} = \frac{1}{4\pi} \frac{\mathbf{x} - \mathbf{x}_0}{r^3} \cdot \mathbf{b}, \tag{6.17.2}$$

where $r = |\mathbf{x} - \mathbf{x}_0|$ is the distance of the field point, $\mathbf{x}$, from the point force, and $\hat{t} = t - t_0$ is the time elapsed since the introduction of the point force. Working as in Section 6.5.5, we find that the generating function, $\mathcal{H}$, with dimensions of length over time, satisfies the equation

$$\frac{1}{\nu}\frac{\partial \mathcal{H}}{\partial t} = \nabla^2 \mathcal{H} - \frac{1}{4\pi r}\delta(t - t_0), \tag{6.17.3}$$

where $\nu = \mu/\rho$ is the kinematic viscosity of the fluid. The continuity equation, $\nabla \cdot \mathbf{u} = 0$, is satisfied for any nonsingular function $\mathcal{H}$.

The Laplacian of the generating function, $\mathcal{Q} \equiv \nabla^2 \mathcal{H}$, is the Green's function of the unsteady diffusion equation, satisfying

$$\frac{1}{\nu}\frac{\partial \mathcal{Q}}{\partial t} = \nabla^2 \mathcal{Q} + \delta_3(\mathbf{x} - \mathbf{x}_0)\, \delta(t - t_0), \tag{6.17.4}$$

given by

$$\mathcal{Q}(r, \hat{t}) = \frac{\nu}{(4\pi\nu\hat{t})^{3/2}} \exp\left(-\frac{r^2}{4\nu\hat{t}}\right) \tag{6.17.5}$$

(see also Section 12.2.1). Solving the Poisson equation $\nabla^2 \mathcal{H} = \mathcal{Q}$ in spherical polar coordinates for a radial field,

$$\nabla^2 \mathcal{H} = \frac{1}{r^2}\frac{\partial}{\partial r}\left(r^2 \frac{\partial \mathcal{H}}{\partial r}\right) = \mathcal{Q}, \tag{6.17.6}$$

we find that

$$\mathcal{H}(r, \hat{t}) = \frac{1}{4\pi r}\left(\frac{\nu}{\pi\hat{t}}\right)^{1/2} \int_0^r \left[1 - \exp\left(-\frac{\xi^2}{4\nu\hat{t}}\right)\right] d\xi. \tag{6.17.7}$$

In terms of the error function,

$$\mathrm{erf}(w) \equiv \frac{2}{\sqrt{\pi}} \int_0^w e^{-\varrho^2} d\varrho, \tag{6.17.8}$$

we obtain

$$\mathcal{H}(r, \hat{t}) = \frac{1}{4\pi r}\left(\frac{\nu}{\pi\hat{t}}\right)^{1/2}\left[r - (\pi\nu\hat{t})^{1/2}\,\mathrm{erf}\left(\frac{r}{(4\nu\hat{t})^{1/2}}\right)\right]. \tag{6.17.9}$$

Straightforward differentiation according to (6.17.2) produces the evolving velocity field.

Long-time asymptotics

The behavior of the flow at long times can be found by expanding the integrand in (6.17.7) in a Taylor series with respect to the argument of the exponential, finding

$$\mathcal{H}(r,\hat{t}) = \frac{1}{4\pi r} \left(\frac{\nu}{\pi \hat{t}} \right)^{1/2} \int_0^r \left[\frac{\xi^2}{4\nu\hat{t}} - \frac{1}{2} \left(\frac{\xi^2}{4\nu\hat{t}} \right)^2 + \cdots \right] d\xi. \qquad (6.17.10)$$

Performing the integration, we obtain

$$\mathcal{H}(r,\hat{t}) = \frac{1}{4\pi} \left(\frac{\nu}{\pi \hat{t}} \right)^{1/2} \left[\frac{1}{3} \frac{r^2}{4\nu\hat{t}} - \frac{1}{10} \frac{r^4}{(4\nu\hat{t})^2} + \cdots \right]. \qquad (6.17.11)$$

Substituting this expansion into (6.17.2), we obtain

$$\mathbf{u}(\hat{\mathbf{x}},\hat{t}) = \frac{1}{4\pi\mu} \left(\frac{\nu}{\pi \hat{t}} \right)^{1/2} \left[\frac{4}{3} \frac{1}{4\nu\hat{t}} \mathbf{I} - 6 \frac{r^3}{(4\nu\hat{t})^2} \boldsymbol{\mathcal{S}}^{(3)} + \cdots \right] \cdot \mathbf{b}, \qquad (6.17.12)$$

where $\mathbf{I}$ is the identity matrix and the tensor $\boldsymbol{\mathcal{S}}^{(3)}$ is defined in (6.15.34). This expression finds applications in the computation of the long-time decay of the angular velocity autocorrelation function of small particles executing Brownian motion [183].

Laplace transform

The flow due to a localized impulse can also be derived by applying the Laplace transform to obtain equations (6.15.6) with a forcing term in the equation of motion due to the impulse,

$$s\rho \,\hat{\mathbf{u}} = \nabla \hat{p} + \mu \nabla^2 \hat{\mathbf{u}} + \mathbf{b}\,\delta_3(\mathbf{x} - \mathbf{x}_0), \qquad \nabla \cdot \hat{\mathbf{u}} = 0. \qquad (6.17.13)$$

The solution is

$$\hat{u}_i(\hat{\mathbf{x}},s) = \frac{1}{8\pi\mu} \mathcal{S}_{ij}(\hat{\mathbf{x}},s)\, b_j, \qquad (6.17.14)$$

where $\hat{\mathbf{x}} = \mathbf{x} - \mathbf{x}_0$ and $\boldsymbol{\mathcal{S}}$ is the unsteady Stokeslet tensor given in (6.15.24) with $R = r\sqrt{s/\nu}$. Inversion yields the results produced earlier in this section using th method of Green's functions.

To compute the long time behavior of the flow, we use expansion (6.15.33) for the oscillatory Stokeslet at small s, finding

$$\hat{\mathbf{u}} = \frac{1}{8\pi\mu} \left(\boldsymbol{\mathcal{S}}^{(0)}(\hat{\mathbf{x}}) - \frac{4}{3} R \frac{1}{r} \mathbf{I} + R^2 \boldsymbol{\mathcal{S}}^{(2)}(\hat{\mathbf{x}}) + R^3 \boldsymbol{\mathcal{S}}^{(3)}(\hat{\mathbf{x}}) + \cdots \right) \cdot \mathbf{b}. \qquad (6.17.15)$$

Inverting this expansion reproduces expression (6.17.12) (Problem 6.17.2).

6.17.2 Persistent point force

The generating function for a stationary point force with constant strength introduced at time t_{start} at the fixed point $\mathbf{x}_0$ can be found by integrating the generating function due to an impulsive point force, given in (6.17.7), with respect to t_0 from t_{start} until the present time t, obtaining

$$\tilde{\mathcal{H}}(r, t - t_{start}) \equiv \int_{t_{start}}^t \mathcal{H}(r, t - t_0)\, dt_0 = \int_0^{t-t_{start}} \mathcal{H}(r,\tau)\, d\tau, \qquad (6.17.16)$$

where $\tau = t - t_0$ (e.g., [61]). Making substitutions, we find that

$$\tilde{\mathcal{H}}(r, t - t_{start}) = \frac{1}{4\pi^{3/2} r} \int_0^r \xi \, \mathcal{I}(\eta) \, d\xi, \tag{6.17.17}$$

where $\eta \equiv \xi / [\nu (t - t_{start})]^{1/2}$ is a dimensionless similarity variable,

$$\mathcal{I}(\eta) \equiv \int_0^{t - t_{start}} \frac{1}{\xi} \left(\frac{\nu}{\tau} \right)^{1/2} \left[1 - \exp \left(-\frac{\xi^2}{4\nu\tau} \right) \right] d\tau = 2 \int_\eta^\infty \frac{1 - e^{-\varrho^2/4}}{\varrho^2} \, d\varrho, \tag{6.17.18}$$

and $\rho \equiv \xi / (\nu\tau)^{1/2}$ is a dimensionless integration variable. Evaluating the integral for small values of η, we obtain

$$\mathcal{I} = \sqrt{\pi} - \frac{1}{2}\eta + \frac{1}{48}\eta^3 + \cdots. \tag{6.17.19}$$

Substituting this expansion into (6.17.17) and performing the integration, we obtain

$$\tilde{\mathcal{H}} = \frac{r}{8\pi^{3/2}} \left(\sqrt{\pi} - \frac{1}{3}R + \frac{1}{120}R^3 + \cdots \right), \tag{6.17.20}$$

where $R \equiv r / [\nu (t - t_{start})]^{1/2}$, which is consistent with the expression derived in Section 6.5.5 for the Stokeslet.

6.17.3   Wandering point force with constant strength

The generating function for a moving point force with constant strength introduced at time t_{start} can be found by specifying the trajectory of the singular point, $\mathbf{x}_0(t)$, and integrating (6.17.7) with respect to t_0 from t_{start} up to the present time, t. The Laplacian of the generating function, denoted by $\tilde{\mathcal{Q}}(\mathbf{x}, t)$, is given by

$$\tilde{\mathcal{Q}}(\mathbf{x}, t) \equiv \int_{t_{start}}^t \mathcal{Q}(\mathbf{x}, \mathbf{x}_0(t_0), t - t_0) \, dt_0 = \int_0^{t - t_{start}} \mathcal{Q}(\mathbf{x}, \mathbf{x}_0(t - \tau), \tau) \, d\tau, \tag{6.17.21}$$

where $\tau = t - t_0$ is the elapsed time and $\mathcal{Q}$ is given in (6.17.5).

In the case of a point force moving with constant velocity, $\mathbf{V}$, the position of the point force is $\mathbf{x}_0(t - \tau) = \mathbf{x}_0(t) - \mathbf{V}\tau$. Making substitutions, we obtain

$$\tilde{\mathcal{Q}}(\mathbf{x}, t) = \int_0^{t - t_{start}} \frac{\nu}{(4\pi\nu\tau)^{3/2}} \exp \left(-\frac{(\mathbf{x} - \mathbf{x}_0(t) - \mathbf{V}\tau)^2}{4\nu\tau} \right) d\hat{t}. \tag{6.17.22}$$

Letting the upper integration limit tend to infinity and performing the integration, we obtain

$$\tilde{\mathcal{Q}} = \frac{1}{4\pi r} \exp \left[-\frac{|\mathbf{V}|}{2\nu} (r - \mathbf{e} \cdot \hat{\mathbf{x}}) \right], \tag{6.17.23}$$

where $\hat{\mathbf{x}} = \mathbf{x} - \mathbf{x}_0(t)$, $r = |\hat{\mathbf{x}}|$, and $\mathbf{e} = \mathbf{V}/|\mathbf{V}|$ is the unit vector in the direction of translation. Expression (6.17.23) is consistent with the Laplacian of the generating function associated with the Oseenlet, given in (6.13.17).

Problems

6.17.1 *Green's function*

Derive the solution of (6.17.5) in two dimensions.

6.17.2 *Decay of an impulsive point force*

Invert (6.17.15) using tables of the Laplace transform to reproduce expansion (6.17.12).

6.17.3 *Two-dimensional impulsive point force*

Derive the counterpart of (6.17.9) in two-dimensional flow.

6.18 Unsteady particle motion

In Sections 6.15 and 6.16, we studied oscillatory Stokes flow past a stationary particle and flow due to a vibrating particle. To describe an arbitrary time-dependent flow, we may solve the unsteady Stokes equation using the method of Laplace transform discussed in Section 6.15.2. Solutions in the Laplace transformed domain can be produced by boundary-integral and singularity methods, as discussed previously for oscillatory flow. The inversion of the Laplace transform can be carried out by analytical or numerical methods.

6.18.1 Force on a translating sphere

The Laplace transform of the force exerted on a solid sphere of radius a moving with a specified time-dependent velocity $\mathbf{V}(t)$ from rest, $\mathbf{V}(0) = \mathbf{0}$, follows from (6.16.21) as

$$\widehat{\mathbf{F}} = -6\pi\mu a \left(1 + \lambda + \frac{1}{9}\lambda^2\right)\widehat{\mathbf{V}}, \tag{6.18.1}$$

where $\lambda^2 = sa^2/\nu$. Taking the inverse Laplace transform, we obtain

$$\mathbf{F}(t) = -6\pi\mu a \mathbf{V}(t) - 6a^2\sqrt{\pi\rho\mu} \int_0^t \left(\frac{\mathrm{d}\mathbf{V}}{\mathrm{d}t}\right)_\tau \frac{\mathrm{d}\tau}{\sqrt{t-\tau}} - \frac{1}{2}\rho V_s \frac{\mathrm{d}\mathbf{V}}{\mathrm{d}t}, \tag{6.18.2}$$

where $V_s = \frac{4\pi}{3}a^3$ is the volume of the sphere. The three terms on the right-hand side of (6.18.2) represent the Stokes drag force, the Boussinesq–Basset viscous memory force, and an acceleration reaction. The inverse square root kernel of the Boussinesq–Basset memory integral reflects the diffusion of vorticity away from the particle surface.

Equation (6.18.2) can be used to derive an integro-differential equation describing the gravitational settling of a spherical particle released from rest. Using Newton's second law of motion incorporating the hydrodynamic drag force, the buoyancy force, and the particle weight, we obtain

$$\left(\rho_s + \frac{1}{2}\rho\right) V_s \frac{\mathrm{d}\mathbf{V}}{\mathrm{d}t} = -6\pi\mu a \mathbf{V}(t) - 6a^2\sqrt{\pi\rho\mu} \int_0^t \left(\frac{\mathrm{d}\mathbf{V}}{\mathrm{d}t}\right)_\tau \frac{\mathrm{d}\tau}{\sqrt{t-\tau}} + (\rho_s - \rho) V_s \mathbf{g}, \tag{6.18.3}$$

where ρ_s is the density of the sphere. Equation (6.18.3) can be recast as a second-order ordinary differential equation in time for $\mathbf{V}(t)$, which can be integrated by standard numerical methods ([89],

p. 288; [437], p. 376). It should be pointed out that, because the boundary condition is applied at a fixed position, equations (6.18.1)–(6.18.3) are strictly valid only for small travel distances, much less than the particle size.

Equations similar to (6.18.2) and (6.18.3) have been developed for spherical bubbles and drops and nonspherical rigid particles [89].

Torque on a rotating sphere

Consider a sphere of radius a rotating about its center with time-depended angular velocity, $\mathbf{\Omega}(t)$. The counterpart of expression (6.16.17) for the Laplace transform of the angular velocity and torque exerted on the sphere is

$$\widehat{\mathbf{T}} = -\frac{8}{3}\pi\mu a^3 \left(3 + \frac{\lambda^2}{1+\lambda}\right)\widehat{\mathbf{\Omega}} = -\frac{8}{3}\pi\mu a^3 \left(3 + \frac{1}{\sqrt{\nu/a^2}} \frac{s}{\sqrt{\nu/a^2}+\sqrt{s}}\right)\widehat{\mathbf{\Omega}}, \qquad (6.18.4)$$

where $\lambda^2 = sa^2/\nu$. Assuming that the sphere has started rotating at $t = 0$ in a quiescent fluid, we obtain the torque by taking the inverse Laplace transform denoted by $\mathcal{L}^{-1}$. Resorting to tables of the Laplace transform, we read

$$\mathcal{L}^{-1}\left(\frac{1}{s+\sqrt{\alpha s}}\right) = \mathcal{L}^{-1}\left(\frac{1}{\sqrt{s}}\frac{1}{\sqrt{\alpha}+\sqrt{s}}\right) = e^{\alpha t}\operatorname{erfc}(\sqrt{\alpha t}), \qquad (6.18.5)$$

where α is a constant. Now we write

$$\widehat{\mathbf{T}} = -\frac{8}{3}\pi\mu a^3 \left(3 + \left[\frac{1}{\sqrt{\nu/a^2}}\frac{1}{\sqrt{s}} - \frac{1}{\sqrt{s}}\frac{1}{\sqrt{\nu/a^2}+\sqrt{s}}\right]s\right)\widehat{\mathbf{\Omega}}, \qquad (6.18.6)$$

and find that

$$\mathbf{T}(t) = -\frac{8}{3}\pi\mu a^3 \left[3\,\mathbf{\Omega}(t) + \int_0^t \left(\frac{d\mathbf{\Omega}}{dt}\right)_\tau \left(\frac{1}{\sqrt{\pi\alpha(t-\tau)}} - e^{\alpha(t-\tau)}\operatorname{erc}\sqrt{\alpha(t-\tau)}\right)d\tau\right], \qquad (6.18.7)$$

where $\alpha = \nu/a^2$ is the inverse of the viscous time scale [303]. We note the presence of an instantaneous viscous torque together with a viscous history torque. There is no added mass term.

6.18.2   Particle motion in an ambient flow

To compute the force exerted on a translating particle, it is expedient to refer to a frame of reference moving with the particle. If the flow in a stationary frame is described by the unsteady Stokes equation, the flow in the moving frame is described by the nonlinear Navier–Stokes equation. Geometrical nonlinearities arise in the case of general unsteady motion when the particle position is not fixed. Equations describing particle motion in an ambient flow accounting for the effect of fluid inertia have been developed under various approximations [258]. Comprehensive reviews can be found in References [89, 308, 413].

Problem

6.18.1 *Force on a sphere in time-dependent motion*
Derive (6.18.2) from (6.16.21) using the method of Laplace transform.

 Computer Problem

6.18.2 *Gravitational settling of a sphere*
Develop and implement a computational procedure for solving the integro-differential equation (6.18.3). Compute and plot the velocity of the settling sphere as a function of time. To evaluate the singular memory integral, subtract out and then integrate analytically the singularity.

Irrotational flow

7

The vorticity transport equation (3.12.1) for fluids whose density is uniform throughout the domain of flow, or for barotropic fluids whose pressure is a function of the density alone, shows that the rate of vorticity production is zero throughout an irrotational flow. Consequently, vorticity may enter the flow only by diffusion across the boundaries. Once vorticity has entered the flow, it is convected by the existing velocity field while diffusing with a diffusivity that is equal to the kinematic viscosity of the fluid, and intensifies or attenuates due to vortex stretching. An unbounded flow does not have any vorticity entrance ports. Accordingly, if an unbounded flow is irrotational at the initial instant, it will remain irrotational at all times.

Under certain conditions, the distance across which the vorticity penetrates the fluid from the boundaries is small compared to the overall size of the boundaries, and the bulk of the flow is nearly irrotational. This occurs when the rate of diffusion of vorticity into the flow is comparable to the rate of convection of vorticity by tangential fluid motion along the boundary. Balancing the orders of magnitude of the rate of convection and diffusion of vorticity yields the formal requirement that the Reynolds number, Re, defined with respect to the typical size of the boundaries, should be sufficiently high, Re $\gg$ 1. When this condition is met, vorticity is convected along thin layers wrapping around the boundaries, and is then channeled into slender wakes or deposited into regions of rotational flow, commonly called vortices or regions of separated flow. The point where a vortex layer detaches from a boundary and enters the bulk of the flow defines the position where a boundary layer is said to separate. The relevant physical mechanisms and the mathematical description of these processes will be discussed in Chapter 8.

Examples of nearly irrotational flows include high-Reynolds-number flows past an aircraft wing or streamlined ground vehicles. Under normal operating conditions, the vorticity is confined inside thin boundary layers lining the wing or the surface of a vehicle, as well as inside wakes developing behind these bodies. Another example is the flow generated by the propagation of surface waves in the ocean, which is irrotational everywhere except inside a thin boundary layer containing a small amount of vorticity along the free surface.

Whether the vorticity of a high-Reynolds-number flow will remain confined within slender wakes or generate steady or unsteady regions of recirculating flow is not known *a priori*, but must be assessed by observation, analysis, or numerical computation. In practice, as the Reynolds number increases, the onset of hydrodynamic instability causes unsteady or oscillatory motion that renders the structure of the flow hard to analyze and difficult to predict. Turbulent motion characterized by temporal and spatial fluctuations appears beyond a critical threshold.

Prandtl proposed decoupling the irrotational flow far from the boundary layers from the flow inside the boundary layers. In the first step, a potential-flow problem is solved assuming that the domain of irrotational flow extends all the way up to the boundaries and neglecting the presence of wakes and regions of recirculating fluid. In the second step, the potential-flow solution is used to study the development and instability of boundary layers, subject to appropriate simplifying assumptions. In this chapter and in Chapter 10, we discuss the properties and computation of the outer potential flow. In Chapters 8 and 9, we discuss the structure and stability of the boundary-layer flow.

7.1   Equations and computation of irrotational flow

The description and computation of an irrotational flow is considerably simplified by introducing the velocity potential, ϕ, defined by the equation

$$\mathbf{u} = \nabla\phi, \tag{7.1.1}$$

where $\mathbf{u}$ is the velocity. Taking the curl of (7.1.1) and remembering that the curl of the gradient of a twice differentiable function is identically zero, we find that the vorticity vanishes and confirm that a potential flow is also an irrotational flow. In Section 2.1, we saw that the converse is also true, that is, any irrotational flow can be described in terms of a single- or multi-valued velocity potential, ϕ. The advantages of introducing the potential become evident by observing that, instead of computing the three components of the velocity, we only need to compute the scalar potential function of an irrotational flow.

Incompressibility condition

The continuity equation for an incompressible fluid requires that the velocity field be solenoidal, $\nabla \cdot \mathbf{u} = 0$, and the potential ϕ be a harmonic function,

$$\nabla^2\phi = 0. \tag{7.1.2}$$

Laplace's equation (7.1.2) can be solved by a variety of analytical and numerical methods subject to one scalar boundary condition over each boundary. Since the normal component of the velocity at the edge of a boundary layer over an impermeable stationary body is small, we must enforce the no-penetration condition, $\mathbf{n} \cdot \nabla\phi = 0$. The irrotational flow computed subject to this condition exhibits a nonzero tangential velocity, which amounts to a surface of discontinuity identified as a vortex sheet. This apparent jump is a macroscopic representation of a vortex layer or boundary layer whose thickness is small compared to the macroscopic length scale of the flow.

Bernoulli equation

Since the velocity field of an irrotational flow can be computed without reference to the equation of motion, kinematics and dynamics are decoupled. The equation of motion provides us with a first-order linear partial differential equation for the pressure, which can be integrated to yield Bernoulli's equation (3.9.22), repeated for convenience,

$$\frac{\partial\phi}{\partial t} + \mathcal{B} = c(t), \tag{7.1.3}$$

where

$$\mathcal{B} \equiv \frac{1}{2}\mathbf{u} \cdot \mathbf{u} + \frac{p}{\rho} - \mathbf{g} \cdot \mathbf{x} \tag{7.1.4}$$

is the Bernoulli function and $c(t)$ is a specified function of time. We have assumed that the density of the fluid is uniform throughout the domain of flow.

It is worth emphasizing that an irrotational flow and the associated pressure field computed using Bernoulli's equation constitute an exact solution of the full Navier–Stokes equation with the viscous force included. However, with some fortuitous exceptions, the solution cannot satisfy more than one scalar boundary condition. One exception is the two-dimensional flow generated by the steady rotation of a circular cylinder around its center, which is identical to the irrotational circulatory flow due to a point vortex. Another exception is the flow generated by the radial expansion or contraction of a spherical bubble which is identical to the irrotational flow due to a point source, as discussed in Section 7.1.3.

7.1.1 Force and torque exerted on a boundary

If the pressure is available, the force and torque with respect to a point, $\mathbf{x}_0$, exerted on a boundary, B, can be computed using simplified versions of the general expressions (3.4.12) and (3.4.13),

$$\mathbf{F} = -\iint_B p\,\mathbf{n}\,\mathrm{d}S, \qquad \mathbf{T} = -\iint_B p\,(\mathbf{x} - \mathbf{x}_0) \times \mathbf{n}\,\mathrm{d}S, \tag{7.1.5}$$

where $\mathbf{n}$ is the normal unit vector pointing into the flow. Strictly speaking, these equations provide us with the force and torque exerted on a fluid surface around the edge of boundary layers, subject to the assumption that viscous stresses make negligible contributions. These predictions must be corrected taking into consideration the viscous stresses near the boundaries and the pressure drop occurring across boundary layers, as discussed in Chapter 8.

7.1.2 Viscous dissipation in irrotational flow

Although the viscous force in the equation of motion is zero in an irrotational flow, the viscous stresses and rate of viscous dissipation are nonzero and depend on the structure of the flow and the fluid viscosity, as discussed in Section 3.4. This realization makes an important distinction between irrotational and inviscid flow.

Departing from (3.4.14) and (3.4.15), noting that the rate-of-deformation tensor is the same as the velocity gradient tensor in an irrotational flow, using the continuity equation, and applying the divergence theorem, we find that the rate of viscous dissipation in an irrotational flow with uniform viscosity is given by

$$\iiint_{Flow} \Phi\,\mathrm{d}V = 2\mu \iiint_{Flow} (\nabla\nabla\phi) : (\nabla\nabla\phi) = -\mu \iint_B \mathbf{n} \cdot \nabla(\mathbf{u} \cdot \mathbf{u})\,\mathrm{d}S, \tag{7.1.6}$$

where $\mathbf{n}$ is the normal unit vector pointing into the flow and B denotes the boundaries. The right-hand side of (7.1.6) expresses the rate of viscous dissipation in terms of the derivatives of the square of the magnitude of the velocity normal to the boundaries.

Using equation (3.4.21), we find that the rate of viscous dissipation inside a fluid parcel balances the rate of working of the deviatoric viscous traction $\check{\mathbf{f}} = \check{\boldsymbol{\sigma}} \cdot \mathbf{n}$,

$$\iiint_{Parcel} \Phi \, \mathrm{d}V = - \iint_{Parcel} \mathbf{u} \cdot \check{\mathbf{f}} \, \mathrm{d}S, \tag{7.1.7}$$

where $\check{\boldsymbol{\sigma}}$ is the deviatoric component of the stress tensor and $\mathbf{n}$ is the normal unit vector pointing into the parcel. Further discussion of the relation between the drag force exerted on a boundary and the rate of viscous dissipation is available [201].

7.1.3   A radially expanding or contracting spherical bubble

An interesting example of viscous irrotational flow is provided by the radial expansion or contraction of a spherical bubble with time-dependent radius $a(t)$ in an ambient fluid of infinite expanse. The velocity field can be represented by a three-dimensional point source with time-dependent strength $m(t)$ placed at the center of the bubble. In spherical polar coordinates (r, θ, φ) with origin at the bubble center, the velocity potential and radial component of the velocity are given by

An expanding bubble produces irrotational flow represented by a point source.

$$\phi = -\frac{m(t)}{4\pi}\frac{1}{r}, \qquad u_r = \frac{\partial \phi}{\partial r} = \frac{m(t)}{4\pi}\frac{1}{r^2}, \tag{7.1.8}$$

where r is the distance from the center of the bubble. The no-penetration condition at the surface of the bubble requires that $\mathrm{d}a/\mathrm{d}t = u_r(a)$, which can be rearranged into an expression for the strength of the point source,

$$m(t) = 4\pi a^2 \frac{\mathrm{d}a}{\mathrm{d}t}. \tag{7.1.9}$$

Far from the bubble, the pressure assumes the hydrostatic distribution

$$p \simeq \rho\, \mathbf{g} \cdot (\mathbf{x} - \mathbf{x}_c) + p_\infty(t), \tag{7.1.10}$$

where $\mathbf{x}_c$ is the bubble center and $p_\infty(t)$ is the far-field hydrodynamic pressure. Substituting (7.1.8) and (7.1.9) into Bernoulli's equation (7.1.3) and rearranging, we obtain

$$p - \rho\, \mathbf{g} \cdot (\mathbf{x} - \mathbf{x}_c) - p_\infty(t) = \rho\, \frac{1}{3} \Big[\, \frac{1}{r}\frac{\mathrm{d}^2 a^3}{\mathrm{d}t^2} - \frac{1}{6}\frac{1}{r^4}\Big(\frac{\mathrm{d}a^3}{\mathrm{d}t}\Big)^2 \Big]. \tag{7.1.11}$$

To derive an evolution equation for the bubble radius in terms of the internal bubble pressure, $p_B(t)$, we enforce the radial component of the interfacial boundary condition (3.8.8) with constant surface tension, γ,

$$\sigma_{rr} + p_B(t) = \gamma\, 2\kappa_m, \tag{7.1.12}$$

set $\kappa_m = 1/a$, and obtain

$$p_B(t) = p(r = a, t) - 2\mu \Big(\frac{\partial u_r}{\partial r}\Big)_{r=a} + 2\,\frac{\gamma}{a}. \tag{7.1.13}$$

Note that the normal viscous stress expressed by the second term on the right-hand side makes a nonzero contribution to the interfacial force balance. Rearranging, we obtain

$$p(r = a, t) = p_B(t) - 4\frac{\mu}{a}\frac{da}{dt} - 2\frac{\gamma}{a}. \tag{7.1.14}$$

We will assume that the bubble is so small that hydrostatic variations can be neglected over the bubble diameter, so that $\rho\mathbf{g}\cdot\mathbf{x} \simeq \rho\mathbf{g}\cdot\mathbf{x}_c$ over the bubble surface. Using (7.1.11) the evaluate the pressure at the bubble surface, we derive a second-order nonlinear evolution equation governing the bubble radius,

$$a\frac{d^2a}{dt^2} + \frac{3}{2}\left(\frac{da}{dt}\right)^2 + \frac{4\nu}{a}\frac{da}{dt} = \frac{1}{\rho}\left(p_B(t) - p_\infty(t) - 2\frac{\gamma}{a}\right), \tag{7.1.15}$$

known as the generalized Rayleigh equation. In general, the solution of (7.1.15) subject to a certain initial condition and a specified ambient pressure field or pressure inside the bubble must be found by numerical methods [299].

When $p_B(t) = p_\infty(t)$, time does not appear explicitly on the right-hand side of (7.1.15). In that case, it is possible to reduce the order of the equation by regarding da/dt a function of a, and writing $da/dt = f(a)$. Substituting this expression into (7.1.15), we obtain a first-order differential equation,

$$a\frac{df}{da} + \frac{3}{2}f + \frac{4\nu}{a} = -2\frac{\gamma}{\rho}\frac{1}{af}. \tag{7.1.16}$$

If viscous effects and surface tension are negligible, we obtain the exact solution $f = da/dt = c/a^{3/2}$, where c is a constant. Using the definition of f and integrating once with respect to time we obtain the exact solution

$$a = a_0\left[1 + \frac{5}{2}\frac{1}{a_0}\left(\frac{da}{dt}\right)_{t=0}t\right]^{2/5}, \tag{7.1.17}$$

where $a_0 = a(t = 0)$ is the initial bubble radius.

The collapse of a bubble near a wall has important consequences on mechanical damage due to cavitation [38]. Observations and numerical simulations have revealed that the bubble no longer has a spherical shape, but develops a dimple that drives a strong damaging jet of ambient fluid toward the wall.

7.1.4 Velocity variation around a streamline

The condition of vanishing vorticity imposes a constraint on the structure of the velocity field around a streamline in an irrotational flow, with interesting consequences. Setting the components of the vorticity in (1.11.12) to zero, we obtain

$$\mathbf{n}\cdot(\nabla\mathbf{u})\cdot\mathbf{b} = \mathbf{b}\cdot(\nabla\mathbf{u})\cdot\mathbf{n}, \qquad \frac{\partial u}{\partial l_b} = 0, \qquad \frac{\partial u}{\partial l_n} = -\kappa u, \tag{7.1.18}$$

where u is the magnitude of the velocity. The third equation shows that, if a streamline has a circular shape and the fluid moves in the clockwise direction so that the curvature κ is positive, $\partial u/\partial l_n$ is negative and the velocity on the inner side of the circle is greater than the velocity on the outer side. The associated counterclockwise rotation of fluid parcels balances the clockwise rotation due to the global motion of the fluid along the circular streamline. Generalizing this result, we find that the velocity of an irrotational flow around a sharp corner reaches a maximum at the salient edge.

7.1.5   Minimum of the pressure

In Section 2.4, we used the mean-value theorem to show that the velocity potential and magnitude of the velocity reach extreme values at the boundaries. Now we will demonstrate that the hydrodynamic pressure, $\tilde{p} \equiv p - \rho \mathbf{g} \cdot \mathbf{x}$, reaches a minimum at the boundaries. Taking the Laplacian of Bernoulli's equation (7.1.3), we find that $\tilde{p}$ satisfies Poisson's equation with a negative forcing function,

$$\nabla^2 \tilde{p} = -\frac{1}{2} \rho \nabla^2 (\mathbf{u} \cdot \mathbf{u}) = -\rho \nabla \mathbf{u} : \nabla \mathbf{u}. \tag{7.1.19}$$

To derive the last expression, we have noted that the individual Cartesian velocity components satisfy Laplace's equation in an irrotational flow, $\nabla^2 u_x = \nabla^2 u_y = \nabla^2 u_z = 0$. Integrating (7.1.19) over a control volume, V_c, bounded by a closed surface, D, and using the divergence theorem, we obtain

$$\iint_D \mathbf{n} \cdot \nabla \tilde{p} \, dS = -\rho \iiint_{V_c} \nabla \mathbf{u} : \nabla \mathbf{u} \, dV < 0, \tag{7.1.20}$$

where the normal unit vector $\mathbf{n}$ points outward from the control volume. Next, we identify D with a small surface that encloses a point where the hydrodynamic pressure allegedly reaches a minimum, so that $\tilde{p}$ is constant over this surface. By construction, $\mathbf{n} \cdot \nabla \tilde{p}$ is positive over the surface, which contradicts inequality (7.1.20).

Problems

7.1.1 *Expanding or contracting bubble*

(a) Verify that the velocity field (7.1.8) satisfies (7.1.7).

(b) Derive (7.1.17) subject to the underlying assumptions.

7.1.2 *Pressure at a bend*

Consider flow along a straight pipe with a sudden bend. Assuming that the velocity profile along the straight section of the pipe is uniform and the flow around the bend is irrotational, show that the maximum velocity and thus the minimum pressure occurs at the wall of the pipe on the inside of the bend.

7.1.3 *Minimum of the pressure*

Discuss whether the pressure, p, must reach a minimum at the boundaries.

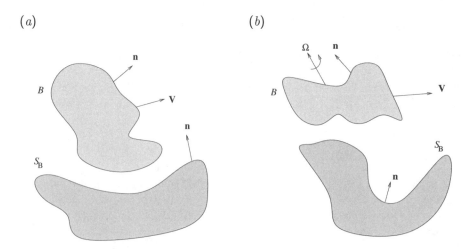

FIGURE 7.2.1   Illustration of (a) flow past a translating rigid body and (b) flow due to the motion or deformation of a body in the presence of a stationary boundary, S_B.

7.2   Flow past or due to the motion of a three-dimensional body

An important area of study in the field of incompressible potential flow concerns flow past a stationary or moving body and the flow due to the motion or deformation of a body in an effectively inviscid fluid. In this section, we discuss the general properties of these flows concentrating on the behavior of the flow far from the body and kinetic energy of the fluid, and in Section 7.3, we discuss the developing forces and torques. Interestingly, we will find that the behavior of the far flow, the kinetic energy of the fluid, and force exerted on the body are related by simple functional forms.

7.2.1   Flow past a translating rigid body

We begin by considering an ambient potential flow past a three-dimensional rigid body that translates with velocity $\mathbf{V}$ in the possible presence of a stationary boundary, as illustrated in Figure 7.2.1(a). It is beneficial to decompose the harmonic potential into two constituents, $\phi = \phi_\infty + \phi_d$, where ϕ_∞ describes the undisturbed ambient flow prevailing in the absence of the body and ϕ_d is the disturbance potential due to the body.

Boundary-integral representation

Next, we write the boundary integral representation (2.3.8) for the disturbance potential ϕ_d at a point in the flow, $\mathbf{x}_0$. The boundary, D, consists of the surface of the body, B, the surface of an impenetrable stationary boundary, S_B, and a large surface enclosing B and S_B, denoted by S_∞. The double-layer integral over S_B vanishes because of the zero normal velocity. The single-layer integral over S_B can be made to vanish by using an appropriate Green's function. As the large surface expands to infinity, the integrals over S_∞ decay to zero. Enforcing the boundary no-penetration condition $\mathbf{u} \cdot \mathbf{n} = 0$ over B, expressed in the form

$$\mathbf{n} \cdot \nabla \phi_d = \mathbf{V} \cdot \mathbf{n} - \mathbf{n} \cdot \nabla \phi_\infty, \qquad (7.2.1)$$

we obtain the integral representation

$$\phi_d(\mathbf{x}_0) = -\iint_B \mathcal{G}(\mathbf{x}_0, \mathbf{x}) \, \mathbf{n}(\mathbf{x}) \cdot [\mathbf{V} - \nabla \phi_\infty(\mathbf{x})] \, \mathrm{d}S(\mathbf{x}) + \iint_B \phi_d(\mathbf{x}) \, \mathbf{n}(\mathbf{x}) \cdot \nabla \mathcal{G}(\mathbf{x}_0, \mathbf{x}) \, \mathrm{d}S(\mathbf{x}). \quad (7.2.2)$$

In the case of flow in an infinite domain with no interior boundaries, $\mathcal{G}(\mathbf{x}, \mathbf{x}_0)$ is the free-space Green's function, $\mathcal{G}(\mathbf{x}, \mathbf{x}_0) = 1/(4\pi|\mathbf{x} - \mathbf{x}_0|)$.

Now we apply the integral relation (2.3.2) for a test flow with velocity $\mathbf{V} - \nabla \phi^\infty$ and corresponding potential $\mathbf{V} \cdot \mathbf{x} - \phi_\infty$. Identifying the control volume with the volume occupied by the body and the boundary of the control volume with the surface of the body, we obtain

$$\iint_B [\mathbf{V} \cdot \mathbf{x} - \phi_\infty(\mathbf{x})] \, \mathbf{n}(\mathbf{x}) \cdot \nabla \mathcal{G}(\mathbf{x}_0, \mathbf{x}) \, \mathrm{d}S(\mathbf{x}) = \iint_B \mathcal{G}(\mathbf{x}_0, \mathbf{x}) \, \mathbf{n}(\mathbf{x}) \cdot [\mathbf{V} - \nabla \phi_\infty(\mathbf{x})] \, \mathrm{d}S(\mathbf{x}). \quad (7.2.3)$$

Use of this identity is permissible because the point $\mathbf{x}_0$ lies in the exterior of the body. Combining (7.2.2) with (7.2.3) to eliminate the single-layer potential, we derive a simplified representation for the disturbance potential in terms of a double-layer potential. Adding to both sides of the representation the potential of the incident flow, we obtain

$$\phi(\mathbf{x}_0) = \phi_\infty(\mathbf{x}_0) - \iint_B [\mathbf{V} \cdot \mathbf{x} - \phi(\mathbf{x})] \, \mathbf{n}(\mathbf{x}) \cdot \nabla \mathcal{G}(\mathbf{x}_0, \mathbf{x}) \, \mathrm{d}S(\mathbf{x}). \qquad (7.2.4)$$

The modular cases of flow past a stationary body and flow due to the translation of a body arise by setting, respectively, $\mathbf{V}$ equal to zero or ϕ_∞ equal to a constant.

Far-field expansion

The representation (7.2.4) provides us with a convenient starting point for assessing the asymptotic behavior of the flow from the body. Assuming that the point $\mathbf{x}_0$ lies far from the body, we select a point $\mathbf{x}_c$ inside or in the vicinity of the body, expand the Green's function dipole in a Taylor series with respect to the point $\mathbf{x}$ about the point $\mathbf{x}_c$, and retain only the leading term to obtain

$$\phi(\mathbf{x}_0) = \phi_\infty(\mathbf{x}_0) - \left(\iint_B [\mathbf{V} \cdot \mathbf{x} - \phi(\mathbf{x})] \, \mathbf{n}(\mathbf{x}) \, \mathrm{d}S(\mathbf{x}) \right) \cdot \nabla_c \mathcal{G}(\mathbf{x}_0, \mathbf{x}_c) + \cdots, \qquad (7.2.5)$$

where the subscript c signifies differentiation with respect to $\mathbf{x}_c$. Comparing equation (7.2.5) with (2.2.23), we find that, far from the body, the disturbance flow is similar to that due to a point-source dipole located at the point $\mathbf{x}_c$ with strength

$$\mathbf{d} = V_B \mathbf{V} - \iint_B \phi(\mathbf{x}) \, \mathbf{n}(\mathbf{x}) \, \mathrm{d}S(\mathbf{x}), \qquad (7.2.6)$$

where V_B is the volume occupied by the body. The first term on the right-hand side of (7.2.6) arises by applying the divergence theorem to manipulate the surface integral in (7.2.5) involving $\mathbf{V}$.

We have found that the coefficient of the dipole can be computed from knowledge of the velocity of the body and distribution of the velocity potential over the surface of the body. In Section 7.3, we will see that the coefficient of the dipole can be used to compute the force exerted on a rigid body in accelerating motion.

7.2.2 Flow due to the motion and deformation of a body

Next, we consider the flow due to the translation, rotation, or deformation of a body in an otherwise quiescent fluid, as illustrated in Figure 7.2.1(b). The body may represent a swimming fish or a moving underwater vehicle. To assess the far field behavior of the flow, we resort to the boundary-integral representation for the harmonic potential (2.3.8), assume that the point $\mathbf{x}_0$ is located far from the body, select a point $\mathbf{x}_c$ inside or in the vicinity of the body, and expand the Green's function and its gradient in a Taylor series with respect to $\mathbf{x}$ about a point, $\mathbf{x}_c$. The result is an asymptotic expansion for the far flow in terms of a Green's function and its multipoles,

$$\phi(\mathbf{x}_0) = \mathcal{G}(\mathbf{x}_0, \mathbf{x}_c) \iint_B \mathbf{n}(\mathbf{x}) \cdot \nabla\phi(\mathbf{x}) \, \mathrm{d}S(\mathbf{x})$$

$$- \iint_B [-\phi(\mathbf{x})\,\mathbf{n}(\mathbf{x}) + (\mathbf{x} - \mathbf{x}_c)\,\mathbf{n}(\mathbf{x}) \cdot \nabla\phi(\mathbf{x})] \, \mathrm{d}S(\mathbf{x}) \cdot \nabla_c \mathcal{G}(\mathbf{x}_0, \mathbf{x}_c) + \cdots. \qquad (7.2.7)$$

The leading term on the right-hand side of (7.2.7) represents the flow due to a point source whose strength is equal to the flow rate across the surface of the body. The second integral is the coefficient of the point-source dipole,

$$\mathbf{d} \equiv \iint_B \left[-\phi(\mathbf{x})\,\mathbf{n}(\mathbf{x}) + (\mathbf{x} - \mathbf{x}_c)\,\mathbf{n}(\mathbf{x}) \cdot \nabla\phi(\mathbf{x}) \right] \mathrm{d}S(\mathbf{x}). \qquad (7.2.8)$$

Using the divergence theorem, we find that the integral on the right-hand side of (7.2.8) remains unchanged when the domain of integration, B, is replaced by any other closed surface enclosing B. When the flow rate across a surface enclosing the body is zero, the integral is independent of the location of the pivot point, $\mathbf{x}_c$.

Motion of a rigid body

As a specific application of (7.2.8), we consider the flow due to a rigid body translating with velocity $\mathbf{V}$ and rotating with angular velocity $\mathbf{\Omega}$ about a point, $\mathbf{x}_c$. The no-penetration condition at the surface of the body requires that

$$\mathbf{n}(\mathbf{x}) \cdot \nabla\phi(\mathbf{x}) = [\mathbf{V} + \mathbf{\Omega} \times (\mathbf{x} - \mathbf{x}_c)] \cdot \mathbf{n}(\mathbf{x}). \qquad (7.2.9)$$

Substituting (7.2.9) into (7.2.8) and rearranging the triple mixed product, we obtain

$$\mathbf{d} = V_B \mathbf{V} - \iint_B \phi(\mathbf{x})\,\mathbf{n}(\mathbf{x}) \, \mathrm{d}S(\mathbf{x}) + \iint_B (\mathbf{x} - \mathbf{x}_c) \big([(\mathbf{x} - \mathbf{x}_c) \times \mathbf{n}(\mathbf{x})] \cdot \mathbf{\Omega} \big) \, \mathrm{d}S(\mathbf{x}). \qquad (7.2.10)$$

As in the case of flow past a translating body discussed earlier in this section, the strength of the dipole can be computed from the distribution of the velocity potential over the surface of the body.

Applying the divergence theorem, we find that the second integral on the right-hand side of (7.2.10) can be made to disappear by identifying the point $\mathbf{x}_c$ with the volume centroid of the body defined as

$$\mathbf{X}_c \equiv \frac{1}{V_B} \iiint_{Body} \mathbf{x} \, \mathrm{d}V(\mathbf{x}). \qquad (7.2.11)$$

A representation of the centroid in terms of a surface integral is given in (4.1.7).

Decomposition into fundamental modes of rigid-body motion

The linearity of the governing equations and boundary conditions allows us to express the velocity potential of the flow due to the motion of a rigid body as a linear combination of the linear velocity and angular velocity of rotation about a point, $\mathbf{x}_c$, in the form

$$\phi(\mathbf{x}) = V_i(t)\,\Phi_i\big(\mathbf{x}, \mathbf{x}_c(t), \mathbf{e}(t)\big) + \Omega_i(t)\,\Phi_{i+3}\big(\mathbf{x}, \mathbf{x}_c(t), \mathbf{e}(t)\big). \tag{7.2.12}$$

The six velocity potentials Φ_i for $i = 1, \ldots, 6$, represent three fundamental modes of translation and three fundamental modes of rotation; the tall parentheses contain the arguments of Φ_i. The vector $\mathbf{e}$ describes the instantaneous orientation of the body. The no-penetration boundary condition requires that

$$\mathbf{n} \cdot \nabla \Phi_i = \begin{cases} n_i & \text{for } i = 1, 2, 3, \\ \big[(\mathbf{x} - \mathbf{x}_c) \times \mathbf{n}\big]_{i-3} & \text{for } i = 4, 5, 6, \end{cases} \tag{7.2.13}$$

where the subscript on the square brackets in the second expression denotes the Cartesian components.

It is convenient to introduce a six-dimensional vector, $\mathbf{N}$, whose first and second three-entry blocks contain, respectively, the normal unit vector, $\mathbf{n}$, and the vector $(\mathbf{x} - \mathbf{x}_c) \times \mathbf{n}$,

$$\mathbf{N} \equiv \big[\mathbf{n},\ (\mathbf{x} - \mathbf{x}_c) \times \mathbf{n}\big]. \tag{7.2.14}$$

The no-penetration boundary condition (7.2.13) takes the form

$$\mathbf{n} \cdot \nabla \Phi_i = \mathbf{N}_i \tag{7.2.15}$$

for $i = 1, \ldots, 6$.

Coefficient of the dipole and added mass

Substituting (7.2.13) in the integral on the right-hand side of (7.2.10), we obtain a compact expression for the coefficient of the dipole,

$$\mathbf{d} = V_B(\mathbf{V} + \boldsymbol{\alpha} \cdot \mathbf{V} + \boldsymbol{\beta} \cdot \boldsymbol{\Omega} + \boldsymbol{\gamma} \cdot \boldsymbol{\Omega}), \tag{7.2.16}$$

where

$$\alpha_{ij} = -\frac{1}{V_B} \iint_B n_i \Phi_j \, dS, \qquad \beta_{ij} = -\frac{1}{V_B} \iint_B n_i \Phi_{j+3} \, dS,$$

$$\gamma_{ij} = \frac{1}{V_B} \epsilon_{lmj} \iint_B \hat{x}_i \hat{x}_l n_m \, dS = \frac{1}{V_B} \epsilon_{ijl} \iiint_B \hat{x}_l \, dV \tag{7.2.17}$$

for $i, j = 1, 2, 3$ are added-mass tensors, and $\hat{\mathbf{x}} = \mathbf{x} - \mathbf{x}_c$. This terminology is explained in Section 7.3.3 where the force and torque exerted on a rigid body in arbitrary unsteady motion is discussed.

If the point $\mathbf{x}_c$ is placed at centroid of the body, $\mathbf{X}_c$, defined in (7.2.11), the tensor $\boldsymbol{\gamma}$ is identically zero and expression (7.2.16) simplifies into

$$\mathbf{d} = V_B(\mathbf{V} + \boldsymbol{\alpha} \cdot \mathbf{V} + \boldsymbol{\beta} \cdot \boldsymbol{\Omega}), \tag{7.2.18}$$

which provides us with a method of computing the elements of the added-mass tensors $\boldsymbol{\alpha}$ and $\boldsymbol{\beta}$ from the coefficient of the dipole corresponding to six independent modes of rigid-body motion.

Symmetry of the added-mass tensor, α

In the case of flow due to the motion of a rigid body in an domain of infinite expanse in the absence of exterior or interior boundaries, the 3×3 added-mass tensor α_{ij} is symmetric, $\alpha_{ij} = \alpha_{ji}$. To demonstrate this property, we use the first set of boundary conditions in (7.2.13), recall that Φ_i are harmonic functions, $\nabla^2 \Phi_i = 0$, and apply the divergence theorem to write

$$\iint_B n_i \Phi_j \, \mathrm{d}S = \iint_B (\mathbf{n} \cdot \nabla \Phi_i) \, \Phi_j \, \mathrm{d}S = - \iiint_B (\nabla \Phi_i) \cdot (\nabla \Phi_j) \, \mathrm{d}V. \qquad (7.2.19)$$

We have noted that Φ_i decays fast enough so that the integral on the left-hand side of (7.2.19) over a large closed surface is infinitesimal. Interchanging the roles of Φ_i and Φ_j, we obtain an expression with identical right-hand side, completing the proof.

The symmetry of α_{ij} implies that the component of the coefficient of the dipole in a particular direction due to translation in another direction is equal to the component of the dipole in the second direction due to translation in the first direction.

Kinetic energy of the flow due to the motion of a rigid body and grand added mass

Equation (2.1.22) provides us with an expression for the instantaneous kinetic energy of a potential flow in terms of a boundary integral involving the boundary distribution of ϕ and the boundary distribution of the normal velocity component. In the case of flow due to the motion of a rigid body, possibly in the presence of stationary boundaries, we use the boundary conditions (7.2.9) to obtain

$$\mathcal{K} = -\frac{1}{2} \rho \iint_B \mathbf{u} \cdot \mathbf{n} \phi \, \mathrm{d}S = -\frac{1}{2} \rho \iint_B \mathbf{V} \cdot \mathbf{n} \phi \, \mathrm{d}S - \frac{1}{2} \rho \iint_B \left[\boldsymbol{\Omega} \times (\mathbf{x} - \mathbf{x}_c) \right] \cdot \mathbf{n} \phi \, \mathrm{d}S. \qquad (7.2.20)$$

Rearranging, we find that

$$\mathcal{K} = -\frac{1}{2} \rho \mathbf{V} \cdot \iint_B \phi \mathbf{n} \, \mathrm{d}S - \frac{1}{2} \rho \boldsymbol{\Omega} \cdot \iint_B \phi (\mathbf{x} - \mathbf{x}_c) \times \mathbf{n} \, \mathrm{d}S. \qquad (7.2.21)$$

To expedite the analysis, we introduce the six-dimensional vector

$$\mathbf{W} = \left[V_x, V_y, V_z, \Omega_x, \Omega_y, \Omega_z \right], \qquad (7.2.22)$$

and substitute the decomposition (7.2.12) into the right-hand side of (7.2.21) to obtain the compact quadratic form

$$\mathcal{K} = \frac{1}{2} \rho V_B W_i A_{ij} W_j, \qquad (7.2.23)$$

where $\mathbf{A}$ is the 6×6 grand added-mass tensor defined as

$$A_{ij} = -\frac{1}{V_B} \iint_B N_i \Phi_j \, \mathrm{d}S, \qquad (7.2.24)$$

and the vector $\mathbf{N}$ is defined in (7.2.15). Using (7.2.15), we obtain

$$A_{ij} = -\frac{1}{V_B} \iint_B (\mathbf{n} \cdot \nabla \Phi_i) \, \Phi_j \, \mathrm{d}S. \tag{7.2.25}$$

The tensors $\boldsymbol{\alpha}$ and $\boldsymbol{\beta}$ defined in (7.2.17) comprise, respectively, the top and bottom 3×3 diagonal blocks of $\mathbf{A}$.

It is evident from the right-hand side of (7.2.25) that $\mathbf{A}$ depends on the instantaneous body shape, orientation, and presence of other stationary objects, but is independent of the body's linear or angular acceleration. Physically, $\mathbf{A}$ is a measure of the sensitivity of the kinetic energy of the fluid to the velocity of translation and angular velocity of rotation, and may thus be regarded as an influence matrix for the kinetic energy.

Symmetry of the grand added-mass tensor

We will show that the grand added-mass tensor $\mathbf{A}$ is symmetric, that is, $A_{ij} = A_{ji}$. Consequently, the kinetic energy of the fluid produced when a body translates in a particular direction and rotates about another direction with unit linear and angular velocities is the same as that produced when the body translates in the second direction and rotates about the first direction with unit linear and angular velocities. The symmetry property follows immediately by using the divergence theorem to write

$$\iint_B (\mathbf{n} \cdot \nabla \Phi_i) \, \Phi_j \, \mathrm{d}S = -\iiint_{Flow} \nabla \cdot (\Phi_j \nabla \Phi_i) \, \mathrm{d}S = -\iiint_{Flow} \nabla \Phi_j \cdot \nabla \Phi_j \, \mathrm{d}S. \tag{7.2.26}$$

Interchanging the role of the two potentials on the left-hand side produces the same right-hand side. Since $\mathbf{A}$ is symmetric, it has only 21 independent components. Six of these components can be made to vanish by making appropriate choices for the center of rotation $\mathbf{x}_c$ and directions of the Cartesian axes.

7.2.3   Kinetic energy of the flow due to the translation of a rigid body

In the case of the flow due to a translating but nonrotating body, we use (7.2.23) and (7.2.18) and obtain an expression for the kinetic energy in terms of the coefficient of the dipole,

$$\mathcal{K} = \frac{1}{2} \rho V_B \mathbf{V} \cdot \boldsymbol{\alpha} \cdot \mathbf{V} = \frac{1}{2} \rho \, (\mathbf{d} - V_B \mathbf{V}) \cdot \mathbf{V}. \tag{7.2.27}$$

Since the matrix $\boldsymbol{\alpha}$ is symmetric, we can find three mutually perpendicular principal directions where $\boldsymbol{\alpha}$ is diagonal and the quadratic cross terms of the velocity, $V_i V_j$ for $i \neq j$, do not appear in (7.2.27). In the case of an axisymmetric body, one principal direction coincides with the axis of the body and the other two principal directions lie in the perpendicular plane.

Later in this chapter, we will discuss a method of computing a potential flow in terms of discrete or continuous distributions of dipoles inside the volume or over the surface of a body. Equations (7.2.18) and (7.2.27) will provide us with the added-mass tensor $\boldsymbol{\alpha}$ and kinetic energy of fluid in terms of the total strength of the dipoles, thus circumventing the need for detailed computation.

Coefficient of the dipole

As a final note, we relate the sensitivity of the kinetic energy of the fluid with respect to the velocity of translation to the coefficient of the dipole. Beginning with (7.2.27) and using (7.2.18), we obtain

$$\frac{\partial \mathcal{K}}{\partial V_k} = \frac{1}{2}\rho V_B \frac{\partial}{\partial V_k}(\mathbf{V}\cdot\boldsymbol{\alpha}\cdot\mathbf{V}) = \rho V_B \alpha_{kj}V_j = \rho(d_k - V_B V_k). \qquad (7.2.28)$$

Rearranging, we find that

$$d_k = \frac{1}{\rho}\frac{\partial \mathcal{K}}{\partial V_k} + V_B V_k, \qquad (7.2.29)$$

which relates the coefficient of the dipole to the rate of change of kinetic energy with respect to the velocity of the body.

Problems

7.2.1 *Flow past a rotating body*

Discuss whether it is possible to derive a double-layer representation similar to (7.2.4) for irrotational flow past a rotating rigid body.

7.2.2 *Tensorial nature of the grand added-mass matrix*

Show that **A** is a second-order tensor of the sixth-dimensional Cartesian space with coordinates $(x_1, x_2, x_3, x_1, x_2, x_3)$ ([437], p. 101).

7.2.3 *Elemental potentials*

Derive the boundary conditions (7.2.13).

7.3 Force and torque exerted on a three-dimensional body

We proceed to study in more detail the force and torque exerted on a three-dimensional moving or stationary body in potential flow, given in equations (7.1.5) in terms of the pressure. Our main goal is to develop simplified expressions in terms of the velocity potential and velocity field using Bernoulli's equation (7.1.3). The density of the fluid is assumed to be uniform throughout the domain of flow.

7.3.1 Steady flow past a stationary body

We begin by considering steady potential flow past a stationary rigid body, as illustrated in Figure 7.3.1. The no-penetration boundary condition applies at the surface of the body.

Force

Substituting Bernoulli's equation (7.1.3) into the first equation of (7.1.5), we obtain the force exerted on the body,

$$\mathbf{F} = \frac{1}{2}\rho \iint_B (\mathbf{u}\cdot\mathbf{u})\,\mathbf{n}\,dS - \rho V_B \mathbf{g}, \qquad (7.3.1)$$

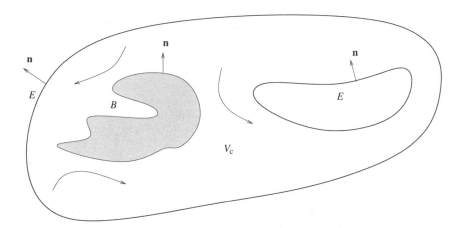

FIGURE 7.3.1 The force exerted on a stationary rigid body, B, immersed in a potential flow can be computed in terms of surface integrals over the boundary E of a control volume that is partly confined by the body.

where B stands for the surface of the body, $\mathbf{n}$ is the normal unit vector pointing into the fluid, and V_B is the volume of the body.

To simplify the computation of the integral on the right-hand side of (7.3.1), we introduce a control volume, V_c, that is enclosed by the surface of the body, B, and a closed boundary or a collection of closed boundaries, denoted by E, as shown in Figure 7.3.1. Using the divergence theorem, we write

$$
\iiint_{V_c} \nabla (\mathbf{u} \cdot \mathbf{u}) \, dV = - \iint_B (\mathbf{u} \cdot \mathbf{u}) \, \mathbf{n} \, dS + \iint_E (\mathbf{u} \cdot \mathbf{u}) \, \mathbf{n} \, dS. \tag{7.3.2}
$$

Since the flow is irrotational, the velocity gradient tensor is symmetric, $\nabla \mathbf{u} = (\nabla \mathbf{u})^T$, and

$$
\iiint_{V_c} \nabla (\mathbf{u} \cdot \mathbf{u}) \, dV = 2 \iiint_{V_c} (\nabla \mathbf{u}) \cdot \mathbf{u} \, dV = 2 \iiint_{V_c} \mathbf{u} \cdot \nabla \mathbf{u} \, dV = 2 \iiint_{V_c} \nabla \cdot (\mathbf{u}\mathbf{u}) \, dV. \tag{7.3.3}
$$

The continuity equation, $\nabla \cdot \mathbf{u} = 0$, was used to derive the last expression. Applying once again the divergence theorem and using the no-penetration boundary condition, $\mathbf{u} \cdot \mathbf{n} = 0$ over B, we obtain

$$
\iiint_{V_c} \nabla (\mathbf{u} \cdot \mathbf{u}) \, dV = \iint_E \mathbf{u} \, (\mathbf{u} \cdot \mathbf{u}) \, dA, \tag{7.3.4}
$$

where the normal unit vector $\mathbf{n}$ over E points outward from the control volume, as shown in Figure 7.3.1. Combining (7.3.2), (7.3.4), and (7.3.1), we derive an expression for the force as a surface integral over E,

$$
\mathbf{F} = \frac{1}{2} \rho \iint_E \left[(\mathbf{u} \cdot \mathbf{u}) \, \mathbf{n} - 2(\mathbf{u} \cdot \mathbf{n}) \, \mathbf{u} \right] dS - \rho V_B \mathbf{g}. \tag{7.3.5}
$$

The integral on the right-hand side of (7.3.5) is independent of the shape of E. Identifying E with B and using the no-penetration condition, $\mathbf{u} \cdot \mathbf{n} = 0$, we recover (7.3.1). The usefulness of the generalized expression (7.3.5) will become evident when we discuss particular applications.

Torque

Substituting Bernoulli's equation (7.1.3) into the second equation of (7.1.5) and rearranging, we find that the torque with respect to a point, $\mathbf{x}_c$, exerted on the body is given by

$$\mathbf{T} = \frac{1}{2} \rho \iint_B (\mathbf{x} - \mathbf{x}_c) \times \mathbf{n} \, (\mathbf{u} \cdot \mathbf{u}) \, dS - \rho V_B (\mathbf{X}_c - \mathbf{x}_c) \times \mathbf{g}, \qquad (7.3.6)$$

where $\mathbf{X}_c$ is the center of volume of the body defined in (7.2.12). When the center of the torque, $\mathbf{x}_c$, is identified with the volume centroid, the torque due to the body force disappears. Working as previously for the force, we derive the generalized expression

$$\mathbf{T} = \frac{1}{2} \rho \iint_E (\mathbf{x} - \mathbf{x}_c) \times [(\mathbf{u} \cdot \mathbf{u}) \, \mathbf{n} - 2(\mathbf{u} \cdot \mathbf{n}) \, \mathbf{u}] \, dS - \rho V_B (\mathbf{X}_c - \mathbf{x}_c) \times \mathbf{g}, \qquad (7.3.7)$$

where E is an arbitrary surface which, along with B, encloses a control volume, and the normal unit vector $\mathbf{n}$ over E points outward from the control volume. Using the divergence theorem, we can show that the integral on the right-hand side of (7.3.7) is independent of the shape of E. Identifying E with B and using the no-penetration condition, $\mathbf{u} \cdot \mathbf{n} = 0$, we recover (7.3.6).

7.3.2 Force and torque on a body near a point source or point-source dipole

As an application, we consider the flow due to a point source with strength m located at a point, $\mathbf{x}_s$, in the vicinity of a body, in the absence of any other interior or exterior boundaries. Decomposing the velocity field into a singular component due to the point source and a nonsingular complementary component due to the presence of the body, denoted by $\mathbf{v}$, we obtain

$$\mathbf{u} = \frac{m}{4\pi} \frac{\mathbf{x} - \mathbf{x}_s}{r^3} + \mathbf{v}, \qquad (7.3.8)$$

where $r = |\mathbf{x} - \mathbf{x}_s|$ is the distance from the point source.

Next, we identify E with a spherical surface of small radius ϵ centered at the point source, denoted by S_ϵ, and a large surface enclosing the body and the point source. As the large surface expands to infinity, the corresponding integrals in (7.3.5) make infinitesimal contributions. Over the surface of the small sphere enclosing the point force, the inward unit normal vector is given by $\mathbf{n} = -\frac{1}{\epsilon} (\mathbf{x} - \mathbf{x}_s)$. Consequently,

$$\mathbf{u} = -\frac{m}{4\pi\epsilon^2} \mathbf{n} + \mathbf{v}, \qquad \mathbf{u} \cdot \mathbf{u} = \left(\frac{m}{4\pi\epsilon^2}\right)^2 - \frac{m}{2\pi\epsilon^2} \mathbf{n} \cdot \mathbf{v} + \mathbf{v} \cdot \mathbf{v}, \qquad \mathbf{u} \cdot \mathbf{n} = -\frac{m}{4\pi\epsilon^2} + \mathbf{v} \cdot \mathbf{n}. \quad (7.3.9)$$

Substituting these expressions into (7.3.5), evaluating the velocity $\mathbf{v}$ at the location of the point source, and simplifying, we obtain

$$\mathbf{F} = \frac{1}{2} \rho \iint_{S_\epsilon} [-\frac{m}{2\pi\epsilon^2} \mathbf{n} \, (\mathbf{v}_s \cdot \mathbf{n}) - 2(-\frac{m}{4\pi\epsilon^2} + \mathbf{v}_s \cdot \mathbf{n})(-\frac{m}{4\pi\epsilon^2} \mathbf{n} + \mathbf{v}_s)] \, dS - \rho V_B \mathbf{g}, \qquad (7.3.10)$$

where $\mathbf{v}_s \equiv \mathbf{v}(\mathbf{x}_s)$. Simplifying and performing the integration in the limit as $\epsilon \to 0$, we derive the remarkably simple expression

$$\mathbf{F} = \rho m\,\mathbf{v}(\mathbf{x}_s) - \rho V_B \mathbf{g}. \tag{7.3.11}$$

Substituting (7.3.9) into (7.3.7) and working in a similar fashion, we obtain an expression for the torque,

$$\mathbf{T} = \rho m\,(\mathbf{x}_s - \mathbf{x}_c) \times \mathbf{v}(\mathbf{x}_s) - \rho V_B(\mathbf{X}_c - \mathbf{x}_c) \times \mathbf{g}. \tag{7.3.12}$$

We have found that the force and torque can be computed in terms of the reflected velocity, $\mathbf{v}$, at the position of the point source.

Force and torque on a body near a point-source dipole

Corresponding results can be derived for the flow due to a point-source dipole with strength $\mathbf{d}$ located at a point, $\mathbf{x}_d$. In this case, we the integrals over two small spherical surfaces enclosing a point source and a point sink separated by a small distance. The force exerted on the body is

$$\mathbf{F} = \rho\,\mathbf{d} \cdot \nabla\mathbf{v}(\mathbf{x}_d) - \rho V_B \mathbf{g}, \tag{7.3.13}$$

and the corresponding torque is

$$\mathbf{T} = \rho\,\mathbf{d} \times \mathbf{v}(\mathbf{x}_d) - \rho V_B(\mathbf{X}_c - \mathbf{x}_c) \times \mathbf{g}, \tag{7.3.14}$$

where $\mathbf{v}$ is the reflected velocity due to the body ([268], p. 498).

Multiple singularities

If the flow is due to a collection of singularities including point sources and point-source dipoles, the right-hand sides of (7.3.11), (7.3.12), (7.3.13), and (7.3.14) are summed over all singularities.

7.3.3   Force and torque on a moving rigid body

Now we concentrate on the force and torque exerted on a rigid body that translates with time-dependent velocity, $\mathbf{V}(t)$, while rotating with time-dependent angular velocity, $\mathbf{\Omega}(t)$, about a moving material point, $\mathbf{x}_c(t)$, in an otherwise quiescent fluid.

Force

Substituting Bernoulli's equation (7.1.3) into the expression for the force (7.1.5), we find that

$$\mathbf{F} = -\iint_B p\,\mathbf{n}\,\mathrm{d}S = \rho \iint_B \left(\frac{\partial \phi}{\partial t} + \frac{1}{2}\,\mathbf{u} \cdot \mathbf{u} \right) \mathbf{n}\,\mathrm{d}S - \rho V_B \mathbf{g}. \tag{7.3.15}$$

To assess the contribution of the six fundamental modes of rigid body motion, we express ϕ in terms of the six fundamental potentials introduced in (7.2.12), Expanding the time derivative, we obtain

$$\frac{\partial \phi}{\partial t} = \Phi_i\,\frac{\mathrm{d}V_i}{\mathrm{d}t} + \Phi_{i+3}\,\frac{\mathrm{d}\Omega_i}{\mathrm{d}t} + \frac{\partial \Phi_i}{\partial t}\,V_i + \frac{\partial \Phi_{i+3}}{\partial t}\,\Omega_i, \tag{7.3.16}$$

where summation is implied over $i = 1, 2, 3$, and

$$\Phi_j \big(\mathbf{x}, \mathbf{x}_c(t), \mathbf{e}(t) \big) \tag{7.3.17}$$

for $j = 1$–6. The material point $\mathbf{x}_c(t)$ defines the instantaneous position of the body, and the unit vector $\mathbf{e}(t)$ defines the instantaneous orientation. Substituting (7.3.16) into (7.3.15), we find that

$$\mathbf{F} = \rho \left(\iint_B \mathbf{n}\, \Phi_i \, \mathrm{d}S \right) \frac{\mathrm{d}V_i}{\mathrm{d}t} + \rho \left(\iint_B \mathbf{n}\, \Phi_{i+3} \, \mathrm{d}S \right) \frac{\mathrm{d}\Omega_i}{\mathrm{d}t} + \mathbf{F}^S(\mathbf{V}, \boldsymbol{\Omega}), \tag{7.3.18}$$

where all terms that depend on the instantaneous position, orientation, linear and angular velocities of the body, but are independent of the linear or angular acceleration, have been collected into the term $\mathbf{F}^S$ expressing the force exerted on a body in steady motion. The first two terms on the right-hand side of (7.3.18) express the acceleration reaction. In terms of the 3×3 added-mass tensors $\boldsymbol{\alpha}$ and $\boldsymbol{\beta}$ introduced in (7.2.17), we obtain

$$\mathbf{F} = -\rho V_B \Big(\boldsymbol{\alpha} \cdot \frac{\mathrm{d}\mathbf{V}}{\mathrm{d}t} + \boldsymbol{\beta} \cdot \frac{\mathrm{d}\boldsymbol{\Omega}}{\mathrm{d}t} \Big) + \mathbf{F}^S(\mathbf{V}, \boldsymbol{\Omega}). \tag{7.3.19}$$

The first term on the right-hand side of (7.3.19) is a generalized version of Newton's second law of motion, where the coefficient matrix $\rho V_B \boldsymbol{\alpha}$ plays the role of an effective or virtual mass. Since the product ρV_B is the mass of fluid displaced by the body, it is natural to call $\boldsymbol{\alpha}$ the coefficient of virtual inertia or added mass.

Torque

Working in a similar fashion, we derive an expression for the torque with respect to a point, $\mathbf{x}_c(t)$, exerted on a moving rigid body,

$$\mathbf{T} = \rho \left(\iint_B (\mathbf{x} - \mathbf{x}_c) \times \mathbf{n}\, \Phi_i \, \mathrm{d}S \right) \frac{\mathrm{d}V_i}{\mathrm{d}t} + \rho \left(\iint_B (\mathbf{x} - \mathbf{x}_c) \times \mathbf{n}\, \Phi_{i+3} \, \mathrm{d}S \right) \frac{\mathrm{d}\Omega_i}{\mathrm{d}t} + \mathbf{T}^S(\mathbf{V}, \boldsymbol{\Omega}). \tag{7.3.20}$$

The first two terms on the right-hand side express the torque due to unsteady translation and rotation. The third term is analogous to that on the right-hand side of (7.3.18).

Grand coefficient of virtual inertia

Equations (7.3.18) and (7.3.20) can be collected into a unified form in terms of a six-dimensional force–torque vector comprised of the three components of the force and three components of the torque, $\mathbf{R} = \big[\mathbf{F}, \mathbf{T} \big]$. Introducing the grand added-mass tensor $\mathbf{A}$ defined in (7.2.25) and the extended velocity vector $\mathbf{W}$ defined in (7.2.22), we obtain

$$\mathbf{R} = -\rho V_B \frac{\mathrm{d}\mathbf{W}}{\mathrm{d}t} \cdot \mathbf{A} + \mathbf{R}^S, \tag{7.3.21}$$

where the superscript S denotes the value of $\mathbf{R}$ in steady motion. Since the matrix $\mathbf{A}$ is symmetric, the order of the vector-matrix multiplication in the first term on the right-hand side of (7.3.21) is inconsequential. The first term on the right-hand side of (7.3.21) expresses a variation of Newton's second law of motion where the matrix coefficient $\rho V_B \mathbf{A}$ plays the role of an effective or virtual mass

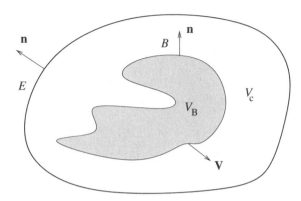

FIGURE 7.4.1 Illustration of an arbitrary body, B, translating with velocity $\mathbf{V}$ in an infinite ambient
fluid. To compute the force and torque exerted on the body, we introduce a closed surface, E,
enclosing the body.

for the extended velocity vector. This observation provides justification for calling $\mathbf{A}$ the grand
coefficient of virtual inertia or grand added mass.

Problems

7.3.1 *Force and torque on a body due to a dipole or a quadruple*

(*a*) Derive expressions (7.3.13) and (7.3.14).

(*b*) Derive corresponding expressions for a point source quadruple.

7.3.2 *Torque on a moving rigid body*

Derive expression (7.3.19).

7.4   Force and torque on a translating rigid body in streaming flow

Of particular interest is the flow due to a rigid body translating in a fluid of infinite expanse in the
absence of exterior or interior boundaries, and uniform (streaming) flow past a stationary body. Our
objective is to derive specific expressions for the force and torque.

7.4.1   Force on a translating body

First, we consider a body translating with time-dependent velocity $\mathbf{V}(t)$ in an otherwise quiescent
fluid of infinite expanse, as shown in Figure 7.4.1. Expression (7.3.19) for the forced exerted on the
body simplifies into

$$\mathbf{F} = -\rho V_B\, \boldsymbol{\alpha} \cdot \frac{\mathrm{d}\mathbf{V}}{\mathrm{d}t} + \rho \iint_B \left(\frac{1}{2}\mathbf{u}\cdot\mathbf{u} + V_i \frac{\partial\, \Phi_i[\mathbf{x},\mathbf{x}_c(t),\mathbf{e}]}{\partial t} \right) \mathbf{n}\, \mathrm{d}S - \rho V_B \mathbf{g}, \qquad (7.4.1)$$

where the material vector $\mathbf{e}$ is independent of time. The three terms on the right-hand side of (7.4.1) represent, respectively, the acceleration reaction, the steady force, and the buoyancy force. Since the body translates in a fluid of infinite expanse in the absence of exterior or interior boundaries, the fundamental potentials Φ_i depend on the distance from the designated center of the body, $\mathbf{x}_c$,

$$\Phi_i\big(\mathbf{x}, \mathbf{x}_c(t), \mathbf{e}\big) = \Phi_i\big(\mathbf{x} - \mathbf{x}_c(t), \mathbf{e}\big), \tag{7.4.2}$$

where, by definition, $\mathrm{d}\mathbf{x}_c/\mathrm{d}t = \mathbf{V}$. Concentrating on the steady force, we write

$$\iint_B V_i \frac{\partial\,\Phi_i[\mathbf{x} - \mathbf{x}_c(t), \mathbf{e}]}{\partial t}\,\mathbf{n}\,\mathrm{d}S = -\iint_B V_i \frac{\partial\,\Phi_i[\mathbf{x} - \mathbf{x}_c(t), \mathbf{e}]}{\partial x_j} \frac{\mathrm{d}x_{c_j}}{\mathrm{d}t}\,\mathbf{n}\,\mathrm{d}S = -\iint_B (\mathbf{u} \cdot \mathbf{V})\,\mathbf{n}\,\mathrm{d}S. \tag{7.4.3}$$

Substituting this expression into (7.4.1), we find that

$$\mathbf{F} = -\rho V_B\,\boldsymbol{\alpha}\cdot\frac{\mathrm{d}\mathbf{V}}{\mathrm{d}t} + \rho\iint_B \Big(\frac{1}{2}\mathbf{u}\cdot\mathbf{u} - \mathbf{u}\cdot\mathbf{V}\Big)\,\mathbf{n}\,\mathrm{d}S - \rho V_B\mathbf{g}. \tag{7.4.4}$$

We will show that the integral on the right-hand side of (7.4.4) is identically zero and the expression for the force simplifies into

$$\mathbf{F} = -\rho V_B\,\boldsymbol{\alpha}\cdot\frac{\mathrm{d}\mathbf{V}}{\mathrm{d}t} - \rho V_B\mathbf{g}. \tag{7.4.5}$$

Vanishing of the steady force

To simplify the computation of the integral of the quadratic term involving $\mathbf{u}\cdot\mathbf{u}$ on the right-hand side of (7.4.4), we introduce a control volume, V_c, confined between the body, B, and a closed exterior surface, E, as shown in Figure 7.4.1. Using the divergence theorem, we restate the integral of the quadratic term as

$$\iint_B (\mathbf{u}\cdot\mathbf{u})\,\mathbf{n}\,\mathrm{d}S = \iint_E (\mathbf{u}\cdot\mathbf{u})\,\mathbf{n}\,\mathrm{d}S - \iiint_{V_c} \nabla(\mathbf{u}\cdot\mathbf{u})\,\mathrm{d}V, \tag{7.4.6}$$

where the normal unit vector $\mathbf{n}$ over E points outward from the control volume. Since the flow is irrotational, $\nabla\mathbf{u} = (\nabla\mathbf{u})^T$ and

$$\iiint_{V_c} \nabla(\mathbf{u}\cdot\mathbf{u})\,\mathrm{d}V = 2\iiint_{V_c} (\nabla\mathbf{u})\cdot\mathbf{u}\,\mathrm{d}V = 2\iiint_{V_c} \mathbf{u}\cdot\nabla\mathbf{u}\,\mathrm{d}V = 2\iiint_{V_c} \nabla\cdot(\mathbf{u}\mathbf{u})\,\mathrm{d}V. \tag{7.4.7}$$

To derive the last expression, we have used the continuity equation, $\nabla\cdot\mathbf{u} = 0$. Applying the divergence theorem once again and enforcing the no-penetration boundary condition, we obtain

$$\iiint_{V_c} \nabla\cdot(\mathbf{u}\mathbf{u})\,\mathrm{d}V = -\iint_B (\mathbf{V}\cdot\mathbf{n})\,\mathbf{u}\,\mathrm{d}S + \iint_E (\mathbf{u}\cdot\mathbf{n})\,\mathbf{u}\,\mathrm{d}S. \tag{7.4.8}$$

Now combining (7.4.6) with (7.4.8), we obtain

$$\iint_B (\mathbf{u}\cdot\mathbf{u})\,\mathbf{n}\,\mathrm{d}S = 2\iint_B (\mathbf{V}\cdot\mathbf{n})\,\mathbf{u}\,\mathrm{d}S + \iint_E [(\mathbf{u}\cdot\mathbf{u})\,\mathbf{n} - 2\,(\mathbf{u}\cdot\mathbf{n})\,\mathbf{u}]\,\mathrm{d}S. \tag{7.4.9}$$

Substituting this expression in the right-hand side of (7.4.4) we find that

$$\mathbf{F} = -\rho V_B\, \boldsymbol{\alpha} \cdot \frac{d\mathbf{V}}{dt} + \rho \iint_B \left[\, \mathbf{u}\,(\mathbf{V}\cdot\mathbf{n}) - (\mathbf{u}\cdot\mathbf{V})\,\mathbf{n}\,\right] dS$$
$$+ \frac{1}{2}\,\rho \iint_E \left[(\mathbf{u}\cdot\mathbf{u})\,\mathbf{n} - 2\,\mathbf{u}\,(\mathbf{u}\cdot\mathbf{n})\right] dS - \rho V_B \mathbf{g}. \tag{7.4.10}$$

The first integrand on the right-hand side can be written as $\mathbf{V}\times(\mathbf{u}\times\mathbf{n})$.

Using the irrotationality condition, $\nabla\mathbf{u} = (\nabla\mathbf{u})^T$, we establish that the integral over B on the right-hand side of (7.4.10) remains unchanged when the domain of integration is replaced by an arbitrary surface enclosing B (Problem 7.4.2(a)). Replacing B with E and combining the two integrals on the right-hand side of (7.4.10), we obtain the compact form

$$\mathbf{F} = -\rho V_B\, \boldsymbol{\alpha} \cdot \frac{d\mathbf{V}}{dt} + \rho \iint_E \left(\, \mathbf{u}\,[(\mathbf{V}-\mathbf{u})\cdot\mathbf{n}] + [\mathbf{u}\cdot(\tfrac{1}{2}\mathbf{u}-\mathbf{V})]\,\mathbf{n}\,\right) dS - \rho V_B \mathbf{g}. \tag{7.4.11}$$

Identifying E with B reproduces (7.4.4). Now we identify E with a large spherical surface and note that, because in the far flow the velocity behaves like that due to a point-source dipole and thus decays like $1/r^3$, the integral on the right-hand side (7.4.11) is identically zero, yielding (7.4.5).

We have thus arrived at the remarkable conclusion that the hydrodynamic force exerted on a three-dimensional body translating steadily in a fluid of infinite expanse with no exterior or interior boundaries is zero. It is important to emphasize that the presence of boundaries may cause the development of a drag force, a lift force force, or both (Problem 7.4.1). Other exceptions arise in the case of an infinite cylindrical body with a smooth or corrugated surface.

D'Alembert's paradox

The component of the steady force in the direction of translation can be computed in an alternative fashion using an energy argument. Concentrating on the integral energy balance (3.4.23), repeated here for convenience,

$$\iiint_{V_c} \frac{\partial}{\partial t}\left(\frac{1}{2}\,\rho\,|\mathbf{u}|^2\right) dV = \iint_D \left(\frac{1}{2}\,\rho\,|\mathbf{u}|^2\right)\mathbf{u}\cdot\mathbf{n}\, dS + \iint_D p\,\mathbf{u}\cdot\mathbf{n}\, dS + \iiint_{V_c} \rho\,\mathbf{g}\cdot\mathbf{u}\, dV, \tag{7.4.12}$$

we use the divergence theorem to convert the first boundary integral on the right-hand side into a volume integral. Combining the resulting expression with the integral on the left-hand side and expressing the body-force term as shown in (3.2.14), we obtain

$$\iiint_{V_c} \frac{D}{Dt}\left(\frac{1}{2}\,\rho\,|\mathbf{u}|^2\right) dV = \iint_D p\,\mathbf{u}\cdot\mathbf{n}\, dS - \rho \iint_D (\mathbf{g}\cdot\mathbf{x})(\mathbf{u}\cdot\mathbf{n})\, dS, \tag{7.4.13}$$

where D/Dt is the material derivative. Now we identify the control volume with the whole domain of flow and recognize the left-hand side of (7.4.13) as the rate of change of the kinetic energy, $d\mathcal{K}/dt$. The boundary, D, consists of the surface of the body and a surface of large size extending to infinity. We note that, far from the body, the flow behaves like that due to a point-source dipole and eliminate the corresponding integrals over the large surface. Enforcing the no-penetration condition over the surface of the body and using the divergence theorem to manipulate the surface integral involving the body force, we obtain

$$\frac{d\mathcal{K}}{dt} = -\mathbf{V} \cdot (\mathbf{F} + \rho V_B \mathbf{g}). \tag{7.4.14}$$

The parentheses on the right-hand side enclose the hydrodynamic force, excluding the buoyancy force.

Now working independently, we differentiate (7.2.27) with respect to time and recall that the virtual mass matrix $\boldsymbol{\alpha}$ is symmetric to express the rate of change of kinetic energy in the form

$$\frac{d\mathcal{K}}{dt} = \rho V_B \alpha_{ij} V_i \frac{dV_j}{dt}. \tag{7.4.15}$$

Combining (7.4.14) with (7.4.15) and using (7.2.18) to express the added mass in terms of the coefficient of the dipole, we obtain

$$\frac{d\mathcal{K}}{dt} = -\mathbf{V} \cdot (\mathbf{F} + \rho V_B \mathbf{g}) = \rho V_B \alpha_{ij} V_i \frac{dV_j}{dt} = -\rho (\mathbf{d} - V_B \mathbf{V}) \cdot \frac{d\mathbf{V}}{dt}, \tag{7.4.16}$$

which shows that the component of the hydrodynamic component of the force in the direction of translation is nonzero only when the body accelerates, and vanishes when the body executes steady motion (d'Alembert's paradox). Physically, when a body translates with constant velocity in the absence of exterior and interior boundaries, the whole flow pattern is convected with the body and the kinetic energy remains constant in time, $d\mathcal{K}/dt = 0$. Because the kinetic energy of the fluid remains constant in the absence of dissipation, work is not required to sustain the motion of the fluid.

7.4.2 Force on a stationary body in uniform unsteady flow

The force exerted on a rigid body that is held stationary in an incident streaming (uniform) flow with time-dependent velocity $\mathbf{U}(t)$ can be found using the results for flow due to the body translation. In a frame of reference translating with velocity $\mathbf{U}(t)$, the far flow vanishes and the body appears to translate with velocity $\mathbf{V}(t) = -\mathbf{U}(t)$ in an otherwise quiescent fluid. Subtracting from the acceleration reaction given in (7.4.5) the fictitious inertial force $-\rho V_B \, d\mathbf{U}(t)/dt$, we obtain

$$\mathbf{F} = -\rho V_B \left(\boldsymbol{\alpha} \cdot \frac{d\mathbf{V}}{dt} + \mathbf{g} \right) + \rho V_B \frac{d\mathbf{U}}{dt}. \tag{7.4.17}$$

Rearranging, we find that

$$\mathbf{F} = \rho V_B (\mathbf{I} + \boldsymbol{\alpha}) \cdot \frac{d\mathbf{U}}{dt} - \rho V_B \mathbf{g}. \tag{7.4.18}$$

The term associated with the identity matrix $\mathbf{I}$ on the right-hand side expresses the force exerted on the fluid displaced by the body in accelerating streaming flow.

7.4.3   Force on a body translating in a uniform unsteady flow

To compute the force exerted on a body that translates with velocity $\mathbf{V}(t)$ in a uniform unsteady flow with velocity $\mathbf{U}(t)$, we work in a frame of reference that translates with velocity $\mathbf{U}(t) - \mathbf{V}(t)$. The far flow vanishes and the body appears to translate with velocity $\mathbf{W}(t) = \mathbf{V}(t) - \mathbf{U}(t)$ in an otherwise quiescent fluid. Subtracting from the acceleration reaction given in (7.4.5) the fictitious inertial force $-\rho V_B \, \mathrm{d}[\mathbf{U}(t) - \mathbf{V}(t)]/\mathrm{d}t$, we obtain

$$\mathbf{F} = -\rho V_B \Big(\boldsymbol{\alpha} \cdot \frac{\mathrm{d}\mathbf{W}}{\mathrm{d}t} + \mathbf{g} \Big) + \rho V_B \, \frac{\mathrm{d}(\mathbf{U} - \mathbf{V})}{\mathrm{d}t}. \tag{7.4.19}$$

Rearranging, we find that

$$\mathbf{F} = \rho V_B (\mathbf{I} + \boldsymbol{\alpha}) \cdot \frac{\mathrm{d}\mathbf{U}}{\mathrm{d}t} - \rho V_B \, \boldsymbol{\alpha} \cdot \frac{\mathrm{d}\mathbf{V}}{\mathrm{d}t} - \rho V_B \mathbf{g}, \tag{7.4.20}$$

which is inclusive of (7.4.5) and (7.4.18).

Acceleration of a body translating in uniform unsteady flow

To compute the acceleration of a body that translates in an infinite fluid in the absence of exterior or interior boundaries, we apply Newton's second law of motion stating that the rate of change of the linear momentum of the body balances the hydrodynamic and gravitational force. Using expression (7.4.20) for the hydrodynamic force, we obtain

$$\rho_B V_B \, \frac{\mathrm{d}\mathbf{V}}{\mathrm{d}t} = \rho V_B (\mathbf{I} + \boldsymbol{\alpha}) \cdot \frac{\mathrm{d}\mathbf{U}}{\mathrm{d}t} - \rho V_B \, \boldsymbol{\alpha} \cdot \frac{\mathrm{d}\mathbf{V}}{\mathrm{d}t} - \rho V_B \mathbf{g} + \rho_B V_B \mathbf{g}, \tag{7.4.21}$$

where ρ_B is the density of the body. Rearranging, we derive a linear ordinary differential equation,

$$(\kappa \mathbf{I} + \boldsymbol{\alpha}) \cdot \frac{\mathrm{d}\mathbf{V}}{\mathrm{d}t} = (\mathbf{I} + \boldsymbol{\alpha}) \cdot \frac{\mathrm{d}\mathbf{U}}{\mathrm{d}t} + (\kappa - 1) \, \mathbf{g}, \tag{7.4.22}$$

where $\kappa = \rho_B / \rho$ is the density ratio. In Section 7.5 we will discuss applications of (7.4.22) to the motion of a sphere.

7.4.4   Torque on a rigid body translating in an infinite fluid

Next, we consider the torque exerted on a body that translates with velocity $\mathbf{V}$ in an otherwise quiescent fluid. Departing from (7.3.20) and working as previously for the force, we find that the torque with respect to a point, $\mathbf{x}_c$, exerted on the body is

$$\mathbf{T} = \rho \Big(\iint_B (\hat{\mathbf{x}} \times \mathbf{n}) \, \Phi_i \, \mathrm{d}S \Big) \frac{\mathrm{d}V_i}{\mathrm{d}t} + \rho \iint_B \Big(\tfrac{1}{2} \mathbf{u} \cdot \mathbf{u} - \mathbf{u} \cdot \mathbf{V} \Big) \hat{\mathbf{x}} \times \mathbf{n} \, \mathrm{d}S - \rho V_B (\mathbf{X}_c - \mathbf{x}_c) \times \mathbf{g}, \tag{7.4.23}$$

where $\mathbf{X}_c$ is the volume centroid, $\hat{\mathbf{x}} = \mathbf{x} - \mathbf{X}_c$, and summation is implied over the repeated index i over $1, 2, 3$. This expression is the counterpart of (7.4.4) for the force.

To simplify the computation of the integral of the quadratic term involving $\mathbf{u} \cdot \mathbf{u}$ on the right-hand side of (7.4.23), we introduce a control volume, V_c, confined between the body, B, and a closed exterior surface , E, as shown in Figure 7.4.1. Using the divergence theorem stated in equation (A.7.3), Appendix A, we restate the integral of the quadratic term as

$$\iint_B (\mathbf{u}\cdot\mathbf{u})\,\hat{\mathbf{x}}\times\mathbf{n}\,dS = \iint_E (\mathbf{u}\cdot\mathbf{u})\,\hat{\mathbf{x}}\times\mathbf{n}\,dS + \iiint_{V_c} \nabla\times[(\mathbf{u}\cdot\mathbf{u})\,\hat{\mathbf{x}}]\,dV, \tag{7.4.24}$$

where the normal unit vector $\mathbf{n}$ over E points outward from the control volume. Focusing on the volume integral, we recall that, because the flow is irrotational, the velocity gradient tensor is symmetric, $\nabla\mathbf{u}=(\nabla\mathbf{u})^T$, yielding

$$\iiint_{V_c} \nabla\times[(\mathbf{u}\cdot\mathbf{u})\,\hat{\mathbf{x}}]\,dV = 2\iiint_{V_c}[(\nabla\mathbf{u})\cdot\mathbf{u}]\times\hat{\mathbf{x}}\,dV = 2\iiint_{V_c}[\mathbf{u}\cdot\nabla\mathbf{u}]\times\hat{\mathbf{x}}\,dV \tag{7.4.25}$$

and then

$$\iiint_{V_c} \nabla\times[(\mathbf{u}\cdot\mathbf{u})\,\hat{\mathbf{x}}]\,dV = 2\iiint_{V_c}\nabla\cdot(\mathbf{uu})\times\hat{\mathbf{x}}\,dV = 2\iiint_{V_c}\nabla\cdot[\mathbf{u}(\mathbf{u}\times\hat{\mathbf{x}})]\,dV. \tag{7.4.26}$$

To derive the penultimate integral, we have used the continuity equation, $\nabla\cdot\mathbf{u}=0$. Applying the divergence theorem once more and enforcing the no-penetration boundary condition, we obtain

$$\iiint_{V_c} \nabla\times[(\mathbf{u}\cdot\mathbf{u})\,\hat{\mathbf{x}}]\,dV = 2\iint_E(\mathbf{u}\cdot\mathbf{n})\,\mathbf{u}\times\hat{\mathbf{x}}\,dS - 2\iint_B(\mathbf{V}\cdot\mathbf{n})\,\mathbf{u}\times\hat{\mathbf{x}}\,dS. \tag{7.4.27}$$

Combining (7.4.24) with (7.4.27), we find that

$$\iint_B(\mathbf{u}\cdot\mathbf{u})\,\hat{\mathbf{x}}\times\mathbf{n}\,dS = -2\iint_B(\mathbf{V}\cdot\mathbf{n})\,\mathbf{u}\times\hat{\mathbf{x}}\,dS + \iint_E\big[(\mathbf{u}\cdot\mathbf{u})\,\hat{\mathbf{x}}\times\mathbf{n}+2(\mathbf{u}\cdot\mathbf{n})\,\mathbf{u}\times\hat{\mathbf{x}}\big]\,dS. \tag{7.4.28}$$

Substituting this expression in (7.4.23), we obtain

$$\mathbf{T} = \rho\Big(\iint_B(\hat{\mathbf{x}}\times\mathbf{n})\,\Phi_i\,dS\Big)\frac{dV_i}{dt} - \rho\iint_B\hat{\mathbf{x}}\times\big[(\mathbf{u}\cdot\mathbf{V})\mathbf{n}-\mathbf{u}(\mathbf{V}\cdot\mathbf{n})\big]\,dS$$
$$+\rho\iint_E\hat{\mathbf{x}}\times\big[\tfrac{1}{2}(\mathbf{u}\cdot\mathbf{u})\mathbf{n}-(\mathbf{u}\cdot\mathbf{n})\mathbf{u}\big]\,dS - \rho V_B(\mathbf{X}_c-\mathbf{x}_c)\times\mathbf{g}. \tag{7.4.29}$$

Identifying E with B reproduces (7.4.23). The integrand in the second integral on the right-hand side of (7.4.29) can be simplified by writing $(\mathbf{u}\cdot\mathbf{V})\mathbf{n}-\mathbf{u}(\mathbf{V}\cdot\mathbf{n})=\mathbf{V}\times(\mathbf{u}\times\mathbf{n})$.

Next, we add and subtract one term from the second integral on the right-hand side of (7.4.29), obtaining

$$\mathbf{T} = \rho\Big(\iint_B\hat{\mathbf{x}}\times\mathbf{n}\,\Phi_i\,dS\Big)\frac{dV_i}{dt} - \rho\iint_B\big[\hat{\mathbf{x}}\times[\mathbf{V}\times(\mathbf{u}\times\mathbf{n})]+\phi\,\mathbf{V}\times\mathbf{n}\big]\,dS$$
$$+\iint_B\phi\,\mathbf{V}\times\mathbf{n}\,dS + \rho\iint_E\hat{\mathbf{x}}\times\big[\tfrac{1}{2}(\mathbf{u}\cdot\mathbf{u})\mathbf{n}-(\mathbf{u}\cdot\mathbf{n})\mathbf{u}\big]\,dS - \rho V_B(\mathbf{X}_c-\mathbf{x}_c)\times\mathbf{g}. \tag{7.4.30}$$

The second integral on the right-hand side of (7.4.30) remains unchanged when the surface of the body, B, is replaced by a closed surface enclosing B (Problem 7.4.2(b)). Replacing B with a large spherical surface and noting that, far from the body, the flow is similar to that due to a point-source dipole and the velocity decays like $1/d^3$, where d is the distance from the origin, we find that the second and fourth integrals on the right-hand side of (7.4.30) are identically zero.

Finally, we express the third integral on the right-hand side of (7.4.30) in terms of the coefficient of the dipole using (7.2.10), we find that

$$\mathbf{T} = \rho \left(\iint_B (\hat{\mathbf{x}} \times \mathbf{n}) \, \Phi_i \, dS \right) \frac{dV_i}{dt} - \rho \mathbf{V} \times \mathbf{d} - \rho V_B (\mathbf{X}_c - \mathbf{x}_c) \times \mathbf{g}, \qquad (7.4.31)$$

where summation of the repeated index i is implied over $1, 2, 3$.

Steady flow

In the case of steady flow and in the absence of significant gravitational effects, the hydrodynamic torque is

$$\mathbf{T} = -\rho \mathbf{V} \times \mathbf{d}, \qquad (7.4.32)$$

which suggests that the torque exerted on a steadily translating body vanishes only when the coefficient of the dipole is oriented in the direction of translation so that the cross product is zero.

Using (7.2.18) to express the coefficient of the dipole, $\mathbf{d}$, in terms of the added mass tensor $\boldsymbol{\alpha}$, and remembering that $\boldsymbol{\alpha}$ is symmetric, we find

$$\mathbf{T} = -V_B \rho \mathbf{V} \times (\boldsymbol{\alpha} \cdot \mathbf{V}). \qquad (7.4.33)$$

we deduce that it is possible to find three mutually perpendicular directions of translation, called the directions of permanent translation, where coefficient of the dipole is parallel to the direction of the translation. A body translating in these directions does not show a tendency for rotation.

Problems

7.4.1 *A body translating in a bounded domain*

(a) Explain why equation (7.4.2) is valid for the two potentials Φ_y and Φ_z corresponding to translation parallel to a plane wall that is perpendicular to the x axis in a semi-infinite domain of flow.

(b) Consider a body translating parallel to a plane wall in a semi-infinite fluid. Show that equation (7.4.11) is valid provided that D is identified with the wall, and simplify the integrand.

(c) Consider a body translating inside or outside a cylindrical boundary with arbitrary cross-section. Discuss whether (7.4.2) is valid for any of the fundamental potentials.

7.4.2 *Invariant integrals*

(a) Show that the integral over the surface of the body, B, on the right-hand side of (7.4.10) remains unchanged when B is replaced by a closed surface enclosing B.

(b) Repeat (a) for the second integral on the right-hand side of (7.4.30).

7.5   Flow past or due to the motion of a sphere

Having discussed the general features of flow past or due to the motion of a three-dimensional body, we proceed to consider in detail the particular case of flow past or due to the motion of a sphere.

7.5.1   Flow due to a translating sphere

The flow due to a sphere of radius a translating with velocity $\mathbf{V}$ in a fluid of infinite expanse in the absence of boundaries can be represented by a point-source dipole with strength

$$\mathbf{d} = 2\pi a^3 \mathbf{V} \tag{7.5.1}$$

placed at the instantaneous center of the sphere, $\mathbf{x}_0$. In this case, not only the far flow, but also the entire flow is described by a potential dipole. Using (2.1.33) and (2.1.34), we find that the harmonic potential and velocity field are given by

$$\phi = -\frac{1}{2}\frac{a^3}{r^3}\mathbf{V}\cdot\hat{\mathbf{x}}, \qquad \mathbf{u} = \frac{1}{2}\frac{a^3}{r^3}\big[-\mathbf{V} + \frac{3}{r^2}(\mathbf{V}\cdot\hat{\mathbf{x}})\hat{\mathbf{x}}\big], \tag{7.5.2}$$

where $r = |\hat{\mathbf{x}}|$ and $\hat{\mathbf{x}} = \mathbf{x} - \mathbf{x}_0$. The streamline pattern shown in Figure 7.5.1(a) is identical to that outside Hill's spherical vortex illustrated in Figure 2.13.3(a).

In spherical polar coordinates, (r, θ, φ), where the x axis corresponding to $\theta = 0, \pi$ is parallel to the velocity of the sphere, $\mathbf{V} = V\mathbf{e}_x$, the Stokes stream function is

$$\psi = \frac{1}{2}V\frac{a^3}{r}\sin^2\theta. \tag{7.5.3}$$

The surface velocity is equal to V at the centerline located at $\theta = 0$ and π, and $-\frac{1}{2}V$ at the midplane located at $\theta = \frac{1}{2}\pi$.

Added mass, kinetic energy, and force

Substituting into expressions (7.2.18), (7.2.27), and (7.4.5) the coefficient of the dipole $\mathbf{d} = 2\pi a^3 \mathbf{V}$, we compute the added mass tensor, kinetic energy of the fluid, and force exerted on the sphere,

$$\alpha = \frac{1}{2}\mathbf{I}, \qquad \mathcal{K} = \frac{1}{4}\rho V_B |\mathbf{V}|^2, \qquad \mathbf{F} = -\rho V_B\big(\frac{1}{2}\frac{d\mathbf{V}}{dt} + \mathbf{g}\big), \tag{7.5.4}$$

where $V_B = \frac{4\pi}{3}a^3$ is the volume of the sphere. The last expression shows that the external force necessary to overcome the hydrodynamic force and thus accelerate a sphere is equal to the force necessary to accelerate half the volume of fluid displaced by the sphere.

Fundamental potentials

Using (7.5.2) and noting that a sphere rotating about its center does not generate a flow, we find that the six fundamental potentials defined in (7.2.13) are given by

$$\Phi_i = -\frac{1}{2}\frac{a^3}{r}x_i, \qquad \Phi_{i+3} = 0 \tag{7.5.5}$$

(a)   (b)

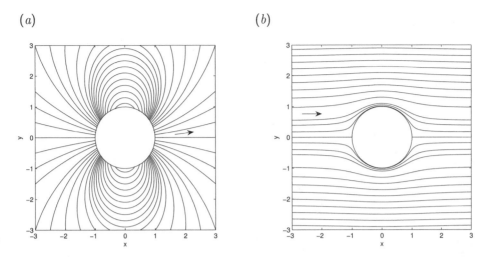

FIGURE 7.5.1   Streamline pattern of irrotational flow (a) due to a translating sphere and (b) past a stationary sphere.

for $i = 1, 2, 3$, where the point $\mathbf{x}_c$ has been identified with the center of the sphere, $\mathbf{x}_0$. If we set the point $\mathbf{x}_c$ at a location other than the center of the sphere, the first three potentials associated with translation will be modified and the last three potentials associated with rotation will express a nontrivial flow.

7.5.2   Uniform flow past a sphere

The solution for uniform (streaming) flow with velocity $\mathbf{U}(t)$ past a stationary sphere of radius a arises from that for flow due to a translating sphere working in a frame of reference where the fluid far from the sphere appears to be stationary. Substituting in (7.5.2) $\mathbf{V} = -\mathbf{U}$ and adding the potential or velocity of the far flow, we obtain

$$\phi = \left(1 + \frac{1}{2}\frac{a^3}{r^3}\right)\mathbf{U} \cdot \hat{\mathbf{x}}, \qquad \mathbf{u} = \mathbf{U} + \frac{1}{2}\frac{a^3}{r^3}\left[\mathbf{U} - \frac{3}{r^2}(\mathbf{U} \cdot \hat{\mathbf{x}})\hat{\mathbf{x}}\right], \qquad (7.5.6)$$

where $r = |\hat{\mathbf{x}}|$, $\hat{\mathbf{x}} = \mathbf{x} - \mathbf{x}_0$, and $\mathbf{x}_0$ is the center of the sphere. The streamline pattern in the xy plane, shown in Figure 7.5.1(b), is identical to that outside Hill's spherical vortex in a frame of reference traveling with the vortex, shown in Figure 2.13.3(b).

In spherical polar coordinates where the x axis is parallel to the far-field velocity, $\mathbf{U} = U\mathbf{e}_x$, the Stokes stream function is given by

$$\psi = \frac{1}{2}Ur^2\sin^2\theta\left(1 - \frac{a^3}{r^3}\right), \qquad (7.5.7)$$

where $\sigma = r\sin\theta$ is the distance from the axis of revolution. The surface velocity is zero at the front and back of the sphere corresponding to $\theta = 0$ and π, and is equal to $\frac{3}{2}U$ at the midplane corresponding to $\theta = \frac{1}{2}\pi$.

7.5.3   Unsteady motion of a sphere

Consider a sphere translating with generally time-dependent velocity $\mathbf{V}(t)$ in a uniform unsteady flow with velocity $\mathbf{U}(t)$ due, for example, to the passage of a sound wave. Substituting the expression for the added mass given in (7.5.4) into (7.4.20) and (7.4.22), we find that the force experienced by the sphere is

$$\mathbf{F} = \frac{3}{2}\rho V_B \frac{d\mathbf{U}}{dt} - \frac{1}{2}\rho V_B \frac{d\mathbf{V}}{dt} - \rho V_B \mathbf{g}, \qquad (7.5.8)$$

and the acceleration of the sphere is

$$\frac{d\mathbf{V}}{dt} = \frac{3}{2\kappa + 1}\frac{d\mathbf{U}}{dt} + 2\frac{\kappa - 1}{2\kappa + 1}\mathbf{g}, \qquad (7.5.9)$$

where $\kappa = \rho_s/\rho$ and ρ_s is the density of the sphere. In the absence of the gravitational force, the acceleration of a sphere that is heavier than the fluid, $\kappa > 1$, is lower than that of the fluid, while the acceleration of a sphere that is lighter than the fluid, $\kappa < 1$, is higher than that of the fluid.

Gravitational settling or rise of a sphere

As an application, we consider the gravitational settling or rise of a sphere in a quiescent fluid of infinite expanse. Setting in (7.5.9) $\mathbf{U} = \mathbf{0}$, we find that

$$\frac{d\mathbf{V}}{dt} = 2\frac{\kappa - 1}{2\kappa + 1}\mathbf{g}. \qquad (7.5.10)$$

One interesting consequence of this expression is that the acceleration of a sphere with negligible mass, $\kappa = 0$, such as an air bubble, is twice the acceleration of gravity.

Terminal velocity of a settling or rising sphere

When a sphere starts rising or settling in a viscous fluid, the induced flow is irrotational during an infinitesimal period of time at the beginning of the motion, and expression (7.5.9) predicts the exact initial acceleration. As time progresses, the viscous drag force makes a significant contribution to the force balance and vorticity enters the flow by diffusion across boundary layers, as discussed in Chapter 8. At long times, the sphere reaches a terminal velocity where the viscous drag force balances the weight of the sphere and the buoyancy force. At that stage, the unsteady force given in (7.5.8) does not contribute to the force balance.

7.5.4   A spherical bubble rising at high Reynolds numbers

Observations have shown that, when surface tension is sufficiently high, a spherical air bubble rising in an infinite fluid maintains a nearly spherical shape. At high Reynolds numbers, the vorticity is confined inside a thin boundary layer that wraps around the surface of the bubble and inside a narrow wake behind the bubble, while the main part of the flow is nearly irrotational, as discussed Section 8.1. To compute the terminal velocity of the bubble, we turn to the energy balance (7.4.14), include on the right-hand side the viscous dissipation term given by (3.4.13), note that the kinetic energy of the fluid remains constant in time, and thus obtain the Levich equation

$$\mathbf{V} \cdot (\mathbf{F} + \rho V_B \mathbf{g}) = - \iiint_{Flow} \Phi \, dV \tag{7.5.11}$$

([235] p. 444; [259]). In Section 8.1.2, we will see that, at high Reynolds numbers, the rate of dissipation inside the boundary layer around the surface of the bubble is negligible compared to the viscous dissipation in the bulk of the fluid. Consequently, the overall rate of dissipation can be computed using (7.1.6). According to (7.1.7), the right-hand side of (7.5.11) is equal to the rate of working of the deviatoric part of the boundary traction computed using the solution for irrotational flow.

Substituting (7.5.2) into the last integral of (7.1.6) and carrying out the integration produces the value $12\pi a \mu |\mathbf{V}|^2$, which shows that the hydrodynamic drag force exerted on the sphere is

$$\mathbf{D} \equiv \mathbf{F} + \rho V_B \mathbf{g} = -12\pi a \mu \, \mathbf{V}. \tag{7.5.12}$$

It is interesting that the magnitude of the drag force is twice that predicted by Stokes's law for a solid sphere moving at low Reynolds numbers given in (6.7.12). The drag coefficient for an air bubble moving with velocity V in a viscous fluid at high Reynolds numbers is then

$$c_D \equiv \frac{D}{\pi \rho V^2 a^2} = \frac{24}{\mathrm{Re}}, \tag{7.5.13}$$

where $\mathrm{Re} = 2aV/\nu$ is the Reynolds number and ν is the kinematic viscosity of the fluid.

Newton's second law of motion requires that the force $\mathbf{F}$ is balanced by the weight of the bubble, $\mathbf{F} + \rho_B V_B \mathbf{g} = \mathbf{0}$. Neglecting the density of the bubble, $\rho_B \simeq 0$, we set $\mathbf{F} = \mathbf{0}$ and use (7.5.12) to find the terminal velocity

$$V = \frac{1}{9} \frac{g a^2}{\nu}. \tag{7.5.14}$$

It is worth emphasizing that this prediction arises from a relatively simple calculation using the potential flow solution alone.

7.5.5 Weiss's theorem for arbitrary flow past a stationary sphere

Consider an arbitrary incident irrotational flow past a stationary sphere of radius a centered at the point $\mathbf{x}_0$ in a domain of infinite expanse. The incident flow is required to be free of singularities inside the volume occupied by the sphere. Weiss's theorem provides us with the disturbance potential due to the sphere, denoted by the subscript d, in the form

$$\phi_d(\mathbf{x}) = \frac{a}{r} \phi_\infty \left[\mathbf{x}^{inv}(a) \right] - \frac{2}{ar} \int_0^a \phi_\infty \left[\mathbf{x}^{inv}(\eta) \right] \eta \, d\eta, \tag{7.5.15}$$

where $\phi_\infty(\mathbf{x})$ is the harmonic potential of the incident flow,

$$\mathbf{x}^{inv}(\eta) = \mathbf{x}_0 + \frac{\eta^2}{r^2} (\mathbf{x} - \mathbf{x}_0) \tag{7.5.16}$$

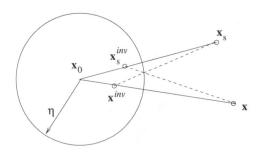

FIGURE 7.5.2   Illustration of the inverse of a field point, $\mathbf{x}$, and inverse of a point source, $\mathbf{x}_s$, with respect to a sphere of radius η centered at $\mathbf{x}_0$.

is the inverse of the field point, $\mathbf{x}$, with respect to a sphere of radius η centered at $\mathbf{x}_0$, and $r = |\mathbf{x}-\mathbf{x}_0|$ is the distance of $\mathbf{x}$ from the center of the sphere [426]. The right-hand side of (7.5.15) involves the potential of the incident flow along a line starting at the center of the sphere and ending at the inverse of the evaluation point, $\mathbf{x}$, with respect to the sphere, $\mathbf{x}^{inv}(a)$, as shown in Figure 7.5.2.

Point source outside a sphere

As an application, we consider the flow due to a point source with strength m placed at a point, $\mathbf{x}_s$, outside a stationary sphere. Straightforward application of (7.5.15) with $\phi_\infty(\mathbf{x}) = -m/(4\pi|\mathbf{x} - \mathbf{x}_s|)$ yields

$$\phi_d(\mathbf{x}) = -\frac{m}{4\pi}\frac{a}{r}\frac{1}{|\mathbf{x}^{inv}(a) - \mathbf{x}_s|} + \frac{m}{2\pi}\frac{1}{ar}\int_0^a \frac{\eta}{|\mathbf{x}^{inv}(\eta) - \mathbf{x}_s|}\,\mathrm{d}\eta, \qquad (7.5.17)$$

which can be restated as

$$\phi_d(\mathbf{x}) = -\frac{ma}{4\pi r}\frac{1}{\left|(\mathbf{x} - \mathbf{x}_0)\dfrac{a^2}{r^2} - (\mathbf{x}_s - \mathbf{x}_0)\right|} + \frac{m}{2\pi ar}\int_0^a \frac{\eta}{\left|(\mathbf{x} - \mathbf{x}_0)\dfrac{\eta^2}{r^2} - (\mathbf{x}_s - \mathbf{x}_0)\right|}\,\mathrm{d}\eta. \qquad (7.5.18)$$

The inverse point of the point source with respect to a sphere of radius η centered at $\mathbf{x}_0$ is

$$\mathbf{x}_s^{inv}(\eta) = \mathbf{x}_0 + \frac{\eta^2}{r_s^2}(\mathbf{x}_s - \mathbf{x}_0), \qquad (7.5.19)$$

where $r_s = |\mathbf{x}_s - \mathbf{x}_0|$ is the distance of the point source from the center of the sphere. By definition of the inverse points,

$$r_s\, r_s^{inv} = r\, r^{inv} = \eta^2, \qquad (7.5.20)$$

where

$$r_s^{inv} = |\mathbf{x}_s^{inv}(\eta) - \mathbf{x}_0|, \qquad r^{inv} = |\mathbf{x}^{inv}(\eta) - \mathbf{x}_0| \qquad (7.5.21)$$

are the distances of the inverse points from the center of the sphere, $\mathbf{x}_0$, as illustrated in Figure 7.5.2. Rearranging (7.5.20), we find that

$$\frac{r}{r_s} = \frac{r_s^{inv}}{r^{inv}}, \tag{7.5.22}$$

which demonstrates that the triangles $(\mathbf{x}_0, \mathbf{x}^{inv}, \mathbf{x}_s)$ and $(\mathbf{x}_0, \mathbf{x}_s^{inv}, \mathbf{x})$ depicted in Figure 7.5.2 are similar. Accordingly,

$$\frac{r}{r_s} = \frac{|\mathbf{x}_s^{inv}(\eta) - \mathbf{x}|}{|\mathbf{x}^{inv}(\eta) - \mathbf{x}_s|} \tag{7.5.23}$$

and expression (7.5.17) can be recast into the form

$$\phi_d(\mathbf{x}) = -\frac{m}{4\pi}\frac{a}{r_s}\frac{1}{|\mathbf{x} - \mathbf{x}_s^{inv}(a)|} + \frac{m}{2\pi}\frac{1}{ar_s}\int_0^a \frac{1}{|\mathbf{x} - \mathbf{x}_s^{inv}(\eta)|}\,\eta\,\mathrm{d}\eta. \tag{7.5.24}$$

It is convenient to write

$$\mathbf{x}_s^{inv}(\eta) = \mathbf{x}_0 + \xi\mathbf{e}_s, \tag{7.5.25}$$

where $\mathbf{e}_s = (\mathbf{x}_s - \mathbf{x}_0)/r_s$ is the unit vector pointing from the center of the sphere to the point source, and $\xi = \eta^2/r_s$. Straightforward manipulation of the integral in (7.5.24) yields the more useful form

$$\phi_d(\mathbf{x}) = -\frac{m}{4\pi}\frac{a}{r_s}\frac{1}{|\mathbf{x} - (\mathbf{x}_0 + a^2\mathbf{e}_s/r_s)|} + \frac{m}{4\pi a}\int_0^{a^2/r_s} \frac{\mathrm{d}\xi}{|\mathbf{x} - (\mathbf{x}_0 + \xi\mathbf{e}_s)|}. \tag{7.5.26}$$

The first term on the right-hand side of (7.5.26) represents a point source with strength ma/r_s placed at the inverse point of the original point source with respect to the sphere. The second term represents a continuous distribution of point sinks extending from the center of the sphere to the inverse point of the point source. The net discharge of the point source and point sinks balances to zero.

The force exerted on the sphere can be deduced immediately from the exact solution using expression (7.3.11), where $\mathbf{v}(\mathbf{x}_s) = \nabla\phi_d(\mathbf{x}_s)$. Taking the gradient of (7.5.24), we find that

$$\mathbf{v}(\mathbf{x}_s) = \left(\frac{m}{4\pi}\frac{a}{r_s}\frac{1}{|\mathbf{x}_s - \mathbf{x}_s^{inv}(a)|^2} - \frac{m}{2\pi}\frac{1}{ar_s}\int_0^a \frac{1}{|\mathbf{x}_s - \mathbf{x}_s^{inv}(\eta)|^2}\,\eta\,\mathrm{d}\eta\right)\mathbf{e}_s. \tag{7.5.27}$$

Next, we substitute

$$\mathbf{x}_s - \mathbf{x}_s^{inv}(\eta) = \left(1 - \frac{\eta^2}{r_s^2}\right)(\mathbf{x}_s - \mathbf{x}_0), \qquad |\mathbf{x}_s - \mathbf{x}_s^{inv}(\eta)| = \left(r_s - \frac{\eta^2}{r_s}\right), \tag{7.5.28}$$

and obtain

$$\mathbf{v}(\mathbf{x}_s) = \frac{m}{4\pi}\left(\frac{ar_s}{(r_s^2 - a^2)^2} - \frac{r_s}{a}\int_0^{a^2} \frac{\mathrm{d}\eta^2}{(r_s^2 - \eta^2)^2}\right)\mathbf{e}_s. \tag{7.5.29}$$

The integral on the right-hand side is equal to $a^2/[r_s^2(r_s^2 - a^2)]$. Making substitutions and simplifying, we find that

$$\mathbf{v}(\mathbf{x}_s) = \frac{m}{4\pi} \frac{a^3}{r_s (r_s^2 - a^2)^2} \, \mathbf{e}_s. \tag{7.5.30}$$

The hydrodynamic component of the force is thus

$$\mathbf{F} = \rho m^2 \frac{1}{4\pi} \frac{a^3}{r_s(r_s^2 - a^2)^2} \, \mathbf{e}_s \tag{7.5.31}$$

([268], p. 496). We observe that the force is oriented from the center of the sphere toward the point source. The direction of the force can be explained by noting that the magnitude of the velocity at the region of the sphere near the point source is higher than that elsewhere. Bernoulli's equation shows that the inverse is true for the pressure, and the net result is that the sphere is attracted to the point source.

7.5.6   Butler's theorems for axisymmetric flow past a stationary sphere

Next, we consider an axisymmetric flow past a stationary sphere in a domain of infinite expanse, in the absence of exterior or interior boundaries. The incident flow is described in spherical polar coordinates centered at the sphere by a Stokes stream function, $\psi_\infty(r, \theta)$, assumed to be free of singularities inside the volume occupied by the sphere. The level of ψ_∞ is adjusted so that $\psi_\infty = 0$ at the center of the sphere, $\mathbf{x}_c$. In the absence of singularities inside the sphere, ψ_∞ is of order r^2 near the center of the sphere, where r is the distance from the center of the sphere.

Butler's sphere theorem [63] states that the disturbance stream function due to the presence of the sphere, denoted by the subscript d, is given by

$$\psi_d = -\frac{r}{a} \, \psi_\infty(\frac{a^2}{r}, \theta). \tag{7.5.32}$$

It is reassuring to confirm that the stream function given in (7.5.7) for uniform flow past a sphere is in agreement with (7.5.32).

A second theorem due to Butler states that, if all singularities of $\psi_\infty(r, \theta)$ lie inside the sphere, and if ψ_∞ decays like $1/r$ far from the sphere, then (7.5.32) provides us with the disturbance flow in the interior of the sphere.

Orthogonal stagnation-point flow past a sphere

As an application of Butler's theorem, we consider irrotational orthogonal stagnation-point flow past a sphere with

$$\psi_\infty = \xi r^3 \cos\theta \sin^2\theta, \tag{7.5.33}$$

where ξ is a constant rate of extension. The corresponding velocity components are shown in (5.6.36). Using Butler's theorem, we find that the disturbance stream function due to the sphere is

$$\psi_d = -\xi \frac{a^5}{r^2} \cos\theta \sin^2\theta. \tag{7.5.34}$$

We can verify that $\psi_\infty + \psi_d = 0$ at the surface of the sphere, ensuring that the no-penetration condition is satisfied.

Point source outside a sphere

We return to discussing the flow due to a point source in the presence of a sphere considered in Section 7.5.5 in terms of the velocity potential. Assuming that the point source is located on the x axis at the point $\mathbf{x}_s = (x_s, 0, 0)$, we use Butler's theorem to find that the incident Stokes stream function to be used with (7.5.32) is

$$\psi_\infty = -\frac{m}{4\pi} \left(\frac{x - x_s}{|\mathbf{x} - \mathbf{x}_s|} + \frac{x_s}{|x_s|} \right) = -\frac{m}{4\pi} \left(\frac{r\cos\theta - x_s}{(r^2 + x_s^2 - 2rx_s\cos\theta)^{1/2}} + \frac{x_s}{|x_s|} \right). \tag{7.5.35}$$

The disturbance stream function due to the sphere then follows as

$$\psi_d = \frac{m}{4\pi} \frac{r}{a} \left(\frac{a^2\cos\theta - rx_s}{(a^4 + r^2x_s^2 - 2a^2rx_s\cos\theta)^{1/2}} + \frac{x_s}{|x_s|} \right). \tag{7.5.36}$$

It can be shown that the flow expressed by (7.5.36) is identical to that described by the potential given in (7.5.18) [63].

Line vortex ring outside a sphere

In a third application, we consider the axisymmetric flow due to a line vortex ring with radius σ_R and strength κ oriented perpendicular to the x axis at the plane $x = x_R$, in the presence of a sphere of radius $a < |x_R|$ centered at the origin. The incident Stokes stream function to be used with Butler's theorem is found from (2.12.21),

$$\psi_\infty(r,\theta) = \frac{\kappa}{4\pi} \sigma_R r \sin\theta \int_0^{2\pi} \frac{\cos\varphi}{\mathcal{R}} \, d\varphi, \tag{7.5.37}$$

where

$$\mathcal{R} \equiv \left[(r\cos\theta - x_R)^2 + (r\sin\theta + \sigma_R)^2 - 4\sigma_R r\sin\theta\cos^2(\tfrac{1}{2}\varphi) \right]^{1/2}. \tag{7.5.38}$$

Applying Butler's theorem, we find that the disturbance stream function is

$$\psi_d(r,\theta) = -\frac{\kappa}{4\pi} a\,\sigma_R r \sin\theta \int_0^{2\pi} \frac{\cos\varphi}{\mathcal{P}} \, d\varphi, \tag{7.5.39}$$

where

$$\mathcal{P} \equiv \left[(a^2\cos\theta - rx_R)^2 + (a^2\sin\theta + r\sigma_R)^2 - 4a^2\sigma_R r\sin\theta\cos^2(\tfrac{1}{2}\varphi) \right]^{1/2}. \tag{7.5.40}$$

Now we introduce the square of the distance from the origin, $r_R^2 = x_R^2 + \sigma_R^2$, and note that

$$(a^2\cos\theta - rx_R)^2 + (a^2\sin\theta + r\sigma_R)^2 = r_R^2 \left[\left(r\cos\theta - x_R\frac{a^2}{r_R^2}\right)^2 + \left(r\sin\theta + \sigma_R\frac{a^2}{r_R^2}\right)^2 \right]. \tag{7.5.41}$$

Using this expression, we find that the disturbance flow is generated by an image ring whose strength and location are given by

$$\kappa_{im} = -\kappa \frac{a}{r_R}, \qquad x_R^{im} = x_R \frac{a^2}{r_R^2}, \qquad \sigma_R^{im} = \sigma_R \frac{a^2}{r_R^2}. \qquad (7.5.42)$$

The image ring passes through the inverse point of the trace of the primary ring with respect to the sphere in any azimuthal plane.

 Computer Problems

7.5.1 *Weiss's theorem for streaming flow*

Use Weiss's theorem to derive the velocity potential corresponding to uniform flow past a sphere given in (7.5.2).

7.5.2 *Weiss's theorem for a rectilinear vortex*

(a) Consider a rectilinear line vortex with strength κ parallel to the z axis, located at $x = b$ and $y = 0$, in the presence of a sphere of radius a centered at the origin. Show that the disturbance potential is given by

$$\phi^d = \frac{\kappa a}{2\pi r} \arctan(\frac{y}{x - r^2 b/a^2}) - \frac{\kappa}{\pi a r} \int_0^a \arctan(\frac{y}{x - r^2 b/\eta^2}) \, \eta \, d\eta. \qquad (7.5.43)$$

(b) Demonstrate that the disturbance potential can be described by an image system consisting of a line vortex ring of radius $a^2/(2b)$ lying in the xz plane and centered at $x = a^2/(2b)$ and $z = 0$, and a vortex sheet subtended by the image ring [426].

7.5.3 *Settling and rising spheres*

Explain in physical terms why the acceleration of a settling solid sphere is lower than the acceleration of gravity, whereas the acceleration of a rising fluid sphere can be as high as twice the acceleration of gravity.

Computer Problem

7.5.4 *Flow due to a point source in the presence of a sphere*

Draw the streamline patterns of flow due to a point source located in the exterior of a sphere at radial position $r_s/a = 1.2$, 2.0, and 4.0.

7.6   Flow past or due to the motion of a nonspherical body

In Section 7.5, we saw that computing the flow past or due to the motion of a spherical body can be expedited with the application of certain powerful theorems which, unfortunately, are specific to the spherical shape. A standard method of computing the flow past or due to the motion of a nonspherical body involves solving Laplace's equation for the velocity potential in appropriate curvilinear coordinates that conform with the body geometry, subject to the no-penetration boundary condition. The procedure is explained in detail in classical texts of fluid mechanics (e.g., [220] Chapter 5; [268]). Two examples concerning the motion of a prolate or oblate spheroid will be discussed in Sections 7.6.1 and 7.6.2.

Singularity methods

An interesting method of computing a potential flow past or due to the motion of a body involves representing the harmonic potential in terms of discrete or continuous distributions of singular fundamental solutions, including the point source and the point-source dipole, and then computing the strengths of the singularities, and possibly their optimal location, by analytical or numerical methods in order to satisfy stipulated boundary conditions. One advantage of the singularity representation is that it provides us with the kinetic energy of the fluid, the added-mass tensor, and the acceleration reaction directly from the strength of the singularities, thereby bypassing the computation of the velocity field and pressure distribution over the body and inside the flow. Applications will be discussed in Sections 7.6.1 and 7.6.3.

Boundary-integral methods

Another powerful class of numerical methods for computing potential flow originates from the boundary-integral representation discussed in Chapter 2, as well as from generalized single- or double-layer representations discussed in Chapter 10. The numerical procedure involves applying the boundary-integral representation at the boundaries of the flow to derive Fredholm integral equations of the first or second kind for the boundary distributions of unknown functions, including the harmonic potential and its normal derivative. The integral equations can be solved by highly accurate boundary-element or panel methods. The engineering importance and popularity of the boundary-integral methods justifies their separate discussion in Chapter 10.

7.6.1   Translation of a prolate spheroid

Chwang & Wu [86] demonstrated that the irrotational flow due to the translation of a prolate spheroid with major semi-axis a and minor semi-axis b can be represented in terms of point-source dipoles distributed over the focal length of the spheroid, extending between the two foci of the generating ellipse, $-c \leq x \leq c$, where $c = ea$,

$$e \equiv \left[1 - \left(\frac{b}{a} \right)^2 \right]^{1/2} \tag{7.6.1}$$

is the eccentricity of the spheroid, and the origin has been set at the center of the spheroid, as shown in Figure 7.6.1. The dipoles point in the direction of translation and have a parabolic linear density distribution that vanishes at the two focal points. The induced velocity potential is

$$\phi(\mathbf{x}) = -\frac{1}{2} \int_{-c}^{c} \left(\frac{V_x}{\delta_x} (x - \xi) + \frac{V_y}{\delta_y} y + \frac{V_z}{\delta_z} z \right) \frac{c^2 - \xi^2}{|\mathbf{x} - \boldsymbol{\xi}|^3} \, d\xi, \tag{7.6.2}$$

where the point $\boldsymbol{\xi} = (\xi, 0, 0)$ lies on the x axis, $\mathbf{V}$ is the velocity of translation, and δ_x, δ_y, δ_z are dimensionless coefficients given by

$$\delta_x = \frac{2e}{1 - e^2} - \ln \frac{1 + e}{1 - e}, \qquad \delta_y = \delta_z = e \frac{2e^2 - 1}{1 - e^2} + \frac{1}{2} \ln \frac{1 + e}{1 - e}. \tag{7.6.3}$$

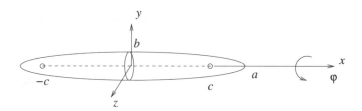

FIGURE 7.6.1   The potential flow due to the translation of a prolate spheroid can be represented by potential dipoles distributed over the focal length of the spheroid pointing in the direction of translation.

Using Maclaurin series expansions, we find that, as the eccentricity tends to zero and the prolate spheroid tends to a sphere, $\delta_x, \delta_y, \delta_z \simeq \frac{4}{3} e^3$.

The total coefficient of the dipole is found by integrating the density of the parabolic distribution, yielding

$$d_i = 2\pi \frac{V_i}{\delta_i} \int_{-c}^{c} (c^2 - \xi^2) \, d\xi = \frac{8\pi}{3} c^3 \frac{V_i}{\delta_i}, \tag{7.6.4}$$

where summation is *not* implied over the repeated index i. Substituting expression (7.6.4) into (7.2.27) and noting that the volume of the spheroid is $V_B = \frac{4\pi}{3} ab^2$, we obtain the kinetic energy of the fluid

$$\mathcal{K} = \frac{4\pi}{3} \rho c^3 \left(\frac{V_x^2}{\delta_x} + \frac{V_y^2}{\delta_y} + \frac{V_z^2}{\delta_z} \right) - \frac{2\pi}{3} \rho ab^2 \left(V_x^2 + V_y^2 + V_z^2 \right). \tag{7.6.5}$$

Using (7.2.18) we find that the nonzero elements of the diagonal added-mass tensor are

$$\alpha_{xx} = 2 \left(\frac{a}{b} \right)^2 \frac{e^3}{\delta_x} - 1, \qquad \alpha_{yy} = \alpha_{zz} = 2 \left(\frac{a}{b} \right)^2 \frac{e^3}{\delta_y} - 1. \tag{7.6.6}$$

As the eccentricity of the spheroid tends to zero, we recover the results of Section 7.5.1 for a sphere, $\alpha_{xx} = \alpha_{yy} = \alpha_{zz} = 1/2$.

Solution in prolate spheroidal coordinates

A more explicit but somewhat indirect description of the flow is possible in terms of prolate spheroidal coordinates, (ζ, μ, φ), related to Cartesian coordinates, (x, y, z), and associated cylindrical polar coordinates, (x, σ, φ), by the equations

$$x = c\mu\zeta, \qquad y = \sigma \cos\varphi, \qquad z = \sigma \sin\varphi, \qquad \sigma = c \sqrt{\zeta^2 - 1} \sqrt{1 - \mu^2}, \tag{7.6.7}$$

where the dimensionless parameter ζ ranges from unity to infinity and the dimensionless μ varies between -1 and 1 (e.g., [220], p. 141). It is convenient to set $\mu = \cos\chi$, where the parameter χ varies in the range $[0, \pi]$ along the contour of the generating ellipse. The surfaces of constant

ζ or μ are confocal prolate spheroids and hyperboloids of two sheets. The surface of the spheroid corresponds to $\zeta = \zeta_0 = 1/e$. As $\zeta_0 \to 1$, the spheroid reduces to a slender needle.

The velocity potential and Stokes stream function of the axisymmetric flow due to a prolate spheroid translating along the x axis are given by

$$\phi = 2c \frac{V_x}{\delta_x} \left(1 - \frac{1}{2} \zeta \ln \frac{\zeta+1}{\zeta-1}\right) \mu, \qquad \psi = \frac{V_x}{\delta_x} \sigma^2 \left(\frac{\zeta}{\zeta^2-1} - \frac{1}{2} \zeta \ln \frac{\zeta+1}{\zeta-1}\right), \qquad (7.6.8)$$

where the dimensionless coefficient δ_x is defined in (7.6.3).

The velocity potential of the three-dimensional flow due to a prolate spheroid translating normal to the x axis is given by

$$\phi = \frac{1}{c} \sigma^2 \left[\frac{1}{2} \ln\left(\frac{\zeta+1}{\zeta-1}\right) - \frac{\zeta}{\zeta^2-1}\right] \left(\frac{V_y}{\delta_y} \cos\varphi + \frac{V_z}{\delta_z} \sin\varphi\right), \qquad (7.6.9)$$

where the dimensionless coefficients δ_y and δ_z are defined in (7.6.3).

7.6.2   Translation of an oblate spheroid

In the case of an oblate spheroid, we introduce oblate spheroidal coordinates, (ζ, μ, φ), related to Cartesian coordinates, (x, y, z), and associated cylindrical polar coordinates, (x, σ, φ), by the equations

$$x = c\mu\zeta, \qquad y = \sigma\cos\varphi, \qquad z = \sigma\sin\varphi, \qquad \sigma = c\sqrt{\zeta^2+1}\sqrt{1-\mu^2} \qquad (7.6.10)$$

where the dimensionless parameter ζ varies from zero to infinity and the dimensionless parameter μ varies in the interval $[-1, 1]$. We may set $\mu \equiv \cos\chi$, where the angle χ varies from 0 to π along the contour of the generating ellipse (e.g., [220], p. 144).

Consider an oblate spheroid with minor semiaxis along the x axis equal to a and major semiaxis normal to the x axis equal to b, where $a < b$, as shown in Figure 7.6.2. The focal length of the spheroid is $c = eb$, where

$$c^2 = b^2 - a^2, \qquad e \equiv \left[1 - \left(\frac{a}{b}\right)^2\right]^{1/2} \qquad (7.6.11)$$

The dimensionless eccentricity, e, varies in the interval $(0, 1]$ whose lower or upper limit corresponds to a circular disk or a sphere. The surface of the spheroid corresponds to $\zeta = \zeta_0 = a/c$. The surfaces of constant z or μ are, respectively, planetary ellipsoids and hyperboloids of revolution of one sheet with common focal circle at $x = 0$ and $\sigma = c$.

The velocity potential and Stokes stream function of the axisymmetric flow due to axial translation along the x axis are given by

$$\phi = b \frac{V_x}{\delta_x} (1 - \zeta \operatorname{arccot}\zeta) \mu, \qquad \psi = \frac{1}{2e} \frac{V_x}{\delta_x} \sigma^2 \left(\frac{\zeta}{\zeta^2+1} - \operatorname{arccot}\zeta\right), \qquad (7.6.12)$$

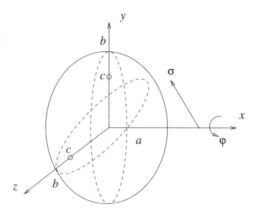

FIGURE 7.6.2   Illustration of the potential flow due to the translation of an oblate spheroid. Solutions
for the axisymmetric flow due to axial translation and three-dimensional flow due to transverse
translation are available.

where

$$\delta_x = \sqrt{1 - e^2} - \frac{1}{e}\, \text{arcsine}. \tag{7.6.13}$$

The velocity potential of the three-dimensional flow due to transverse translation normal to the x
axis is given by

$$\phi = \frac{1}{c}\sigma^2 \left(\frac{\zeta}{\zeta^2 + 1} - \text{arccot}\zeta \right) \left(\frac{V_y}{\delta_y}\, \cos\varphi + \frac{V_z}{\delta_z}\, \sin\varphi \right), \tag{7.6.14}$$

where

$$\delta_y = \delta_z = \frac{1}{\zeta_0}\, \frac{\zeta_0^2 + 2}{\zeta_0^2 + 1} - \text{arccot}\zeta_0. \tag{7.6.15}$$

As the aspect ratio a/b tends to zero and the eccentricity tends to unity, we obtain the flow due
to the translation of a circular disk of infinitesimal thickness (Problem 7.6.1). In the opposite limit
where a/b tends to unity and the eccentricity tends to zero, we recover the results of Section 7.5.1
for a sphere.

7.6.3   Singularity representations

An efficient numerical method of computing axisymmetric or three-dimensional potential flow due to
the axial or transverse translation of an axisymmetric body involves representing the flow in terms
of a collection of singularities deployed at selected locations along the centerline or in the interior of
the body, and then computing the strengths of the singularities so as to satisfy the no-penetration
boundary condition in some approximate fashion.

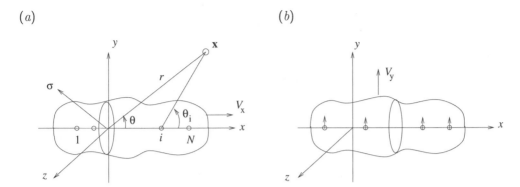

FIGURE 7.6.3   Representation of the flow due to the translation of an axisymmetric body in terms of (*a*) point sources and sinks for axial motion and (*b*) point-source dipoles for transverse motion.

Axial translation of an axisymmetric body

In the case of flow due to the axial translation of an axisymmetric body, we introduce a representation in terms of N point sources or sinks distributed along at centerline, as shown in Figure 7.6.3(a). For a body with fore-and-aft symmetry, the singularities are arranged symmetrically with respect to the midplane. Since the flow rate across any surface enclosing the body is zero, the sum of the strengths of the singularities must balance to zero.

It is convenient to work in a frame of reference translating with the body along with the x axis with velocity V_x. In this frame, the body appears to be stationary and the fluid approaches the body along the x axis with velocity $-V_x$. Referring to spherical polar coordinates centered at each singularity, and using the expression (2.9.15) for each singularity, we derive the Stokes stream function

$$\psi(x,\sigma) = -\frac{1}{2}\,V_x\,\sigma^2 - \frac{1}{4\pi}\sum_{i=1}^{N} m_i \cos\theta_i, \qquad (7.6.16)$$

where σ is the distance from the axis of symmetry, m_i is the strength of the ith singularity, and θ_i is the angle subtended between the field point $\mathbf{x}$ and the ith singularity located at the point x_i on the x axis, satisfying

$$\theta_i = \arctan[\sigma/(x - x_i)], \qquad (7.6.17)$$

as shown in Figure 7.6.3(a). The problem has been reduced to computing the strengths of the singularities so that $\psi = 0$ over the contour of the body in an azimuthal plane. Having obtained the strengths of the singularities, we recover the coefficient of the dipole, $\mathbf{d} = (\mathrm{d}_x, 0, 0)$, as

$$d_x = \sum_{i=1}^{N} x_i\,m_i. \qquad (7.6.18)$$

(a) (b)

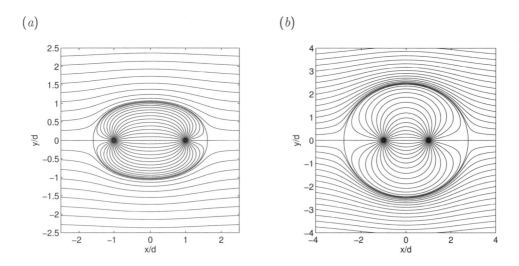

FIGURE 7.6.4 Rankine ovoids generated by introducing a point and a point sink separated by distance $2d$ in streaming flow for (a) $\lambda = 0.2$ and (b) 0.05. As λ decreases, the dividing stream surface becomes a sphere.

Since the strengths of the singularities add up to zero, the result is independent of the chosen origin of the x axis.

Rankine ovoids

Introducing one point source and one point sink with strengths of equal magnitude and opposite sign, $\pm m$, located at $x = \pm d$, we find that $\psi = 0$ over a family of axisymmetric bodies parametrized by the dimensionless number $\lambda = V_x d^2/m$, known as Rankine ovoids. Two examples for $\lambda = 0.5$ and 0.05 ar shown in Figure 7.6.4. As λ tends to zero, the two singularities merge to yield a point-source dipole with strength

$$d_x = 2md = \frac{2}{\lambda} V_x d^3. \qquad (7.6.19)$$

Referring to (7.5.1), we find that, in this limit, the ovoids reduce to a sphere of radius a satisfying the equation $d_x = 2\pi a^3 V_x$. Solving for the radius of the sphere, we find that $(a/d)^3 = 1/(\pi\lambda)$.

Numerical methods

Given a body of a certain shape, the strengths of the point sources or sinks, m_i, can be computed by pointwise collocation. The procedure involves requiring that $\psi = 0$ at N_c selected points over the contour of the body in an azimuthal plane, where $N_c \geq N$, and then solving a linear system of equations for the strength of the singularities, m_i. Selecting $N_c > N$ and then solving an overdetermined system of linear equations by a least-squares or minimization method helps reduce the sensitivity of the solution to the position of the collocation points.

Continuous singularity distribution

In a generalization approach, the flow is represented by a continuous distribution of point sources with linear density $\varrho(x)$ over a portion of the centerline inside the body between two selected limits, $x = -a$ and b. The discrete singularity representation emerges by setting $\varrho(x)$ equal to a weighted sum of one-dimensional delta functions with different singular points. In a frame of reference translating with a body, the Stokes stream function is

$$\psi(x,\sigma) = -\frac{1}{2} V_x \, \sigma^2 - \frac{1}{4\pi} \int_{-a}^{b} \varrho(\xi) \cos \theta(\xi) \, d\xi, \tag{7.6.20}$$

where $\theta(\xi) = \arctan[\sigma/(x - \xi)]$. Requiring that $\psi = 0$ over the body contour in an azimuthal plane provides us with an integral equation of the first kind for the linear density $\varrho(\xi)$. The solution can be found by standard numerical methods, including the collocation method and the method of weighted residuals (e.g., [104]).

Using the expression for the potential due to a point source given in Section 2.1.6, we find that the harmonic potential corresponding to the distribution (7.6.20) is given by

$$\phi(\mathbf{x}) = -V_x x - \frac{1}{4\pi} \int_{-a}^{b} \varrho(\xi) \frac{1}{|\mathbf{x} - \boldsymbol{\xi}|} \, d\xi, \tag{7.6.21}$$

where the point $\boldsymbol{\xi} = (\xi, 0, 0)$ lies along the x axis. In fact, it can be shown that the continuous distribution (7.6.20) is equivalent to a series expansion in terms of Legendre polynomials with argument $\cos \theta$, scaled over the interval $[-a, b]$. The coefficients of the series are related to the moments of the distribution density function, $\varrho(\xi)$ ([24], p. 460).

Transverse translation of axisymmetric bodies

To describe the flow due to the transverse translation of an axisymmetric body, we introduce a representation in terms of point-source dipoles with strength $\mathbf{d}_i$ pointing in the direction of translation. The singularities are placed at strategic location along the centerline inside the body, as shown in Figure 7.6.3(b). The velocity field due to translation along the y axis is

$$\mathbf{u}(\mathbf{x}) = \frac{1}{4\pi} V_y \sum_{i=1}^{N} d_i \left(-\frac{1}{|\mathbf{x} - \mathbf{x}_i|^3} \mathbf{e}_y + 3 \frac{y}{|\mathbf{x} - \mathbf{x}_i|^5} (\mathbf{x} - \mathbf{x}_i) \right), \tag{7.6.22}$$

where $\mathbf{x}_i = (x_i, 0, 0)$ is the location of the ith singularity and $\mathbf{e}_y$ is the unit vector along the y axis. Evaluating (7.6.22) at the surface of the body, enforcing the no-penetration boundary condition, $\mathbf{u} \cdot \mathbf{n} = \mathbf{V} \cdot \mathbf{n} = V_y n_y$, and writing $n_y = (y/\sigma) \, n_\sigma$ and $n_z = (z/\sigma) \, n_\sigma$, we obtain

$$\sum_{i=1}^{N} d_i \left(-\frac{1}{|\mathbf{x} - \mathbf{x}_i|^3} + 3 \frac{\sigma}{n_\sigma} \frac{(x - x_i) n_x + \sigma n_\sigma}{|\mathbf{x} - \mathbf{x}_i|^5} \right) = 4\pi, \tag{7.6.23}$$

where $\sigma^2 = y^2 + z^2$. To compute the strengths of the dipoles, $\mathbf{d}_i$, we apply (7.6.23) at N_c selected collocation points along the contour of the body in an azimuthal plane and work as in the case of axial translation. In the case of a sphere, we recover the exact solution with one dipole placed at the center of the sphere.

Continuous singularity distribution for transverse flow

In a generalized implementation of the singularity method, the flow is represented by a continuous distribution of point-source dipoles pointing in the direction of translation. The density of the distribution is computed by a collocation or weighted residual method. In the case of a prolate spheroid, a parabolic distribution over the focal length of the spheroid produces the exact solution, as discussed in Section 7.6.1. For more general shapes, the optimal distribution domain that provides us with the highest accuracy can be found by numerical optimization or experimentation.

Singularity methods for three-dimensional flow

Singularity methods for three-dimensional flow arise as straightforward extensions of those for axisymmetric flow. The types of singularities employed and their location inside the body must be selected by exercising physical intuition. The strengths of the singularities are typically computed by pointwise collocation. When the flow is represented in terms of a collection of N_s point sources with strengths m_i and N_d point-source dipoles with strengths $\mathbf{d}_i$, the coefficient of the dipole is computed as

$$\mathbf{d} = \sum_{i=1}^{N_s} m_i \mathbf{x}_i + \sum_{i=1}^{N_d} \mathbf{d}_i. \qquad (7.6.24)$$

In an advanced implementation of the singularity method, the locations of the singularities are not specified at the outset but computed instead as part of the solution to minimize an appropriate positive functional enforcing the boundary conditions (e.g., [442]). This modification allows us to achieve a high degree of accuracy with a small number of singularities.

Problem

7.6.1 *Flow due to the axial translation of a circular disk*

Consider the flow due to the axial translation of an oblate spheroid. Take the limit as the eccentricity e tends to unity to derive the velocity potential due to the translation of a circular disk with infinitesimal thickness (e.g., [220], p. 144).

 Computer Problems

7.6.2 *Rankine ovoids*

Compute and plot the contours of Rankine ovoids in an azimuthal plane for a sequence of values of the slenderness parameter $\lambda = a^2 V_x / m$.

7.6.3 *Flow due to translation of an axisymmetric body*

Write a program that uses a collocation method to compute the strengths of the point sources or point-source dipoles according to (7.6.20) or (7.6.23). Run the program for a prolate spheroid with aspect ratio $a/b = 2$ and compare the computed coefficient of the dipole with the exact value given in (7.6.4). The singularities should be distributed evenly over the focal length of the spheroid. Discuss the convergence of the method with respect to the number of singularities employed.

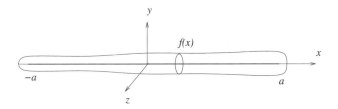

FIGURE 7.7.1 Illustration of a slender axisymmetric body. The flow due to the translation of the body can be represented in terms of singularities distributed over an interval inside the body.

7.7 Slender-body theory

When the length of an axisymmetric body is much larger than its cross-sectional size, it is possible to represent the flow in terms of singularities distributed along the centerline and then compute the densities of the distributions by asymptotic methods. The approximate solution obtained in this fashion provides us with a reasonable alternative to full numerical computation.

7.7.1 Axial motion

Consider a slender axisymmetric body whose surface is described in cylindrical polar coordinates, (x, σ, φ), as $\sigma = f(x)$, where $-a < x < a$. For simplicity, and without loss of generality, we set the origin of the x axis midway between the two ends, as shown in Figure 7.7.1. Assuming that the body translates parallel to the x axis with velocity V_x, we introduce a representation in terms of a distribution of point sources along the centerline with linear density ϱ, and obtain the velocity potential

$$\phi(\mathbf{x}) = -\frac{1}{4\pi} \int_{-a}^{a} \frac{\varrho(\xi)}{|\mathbf{x} - \boldsymbol{\xi}|} \, d\xi, \qquad (7.7.1)$$

where the point $\boldsymbol{\xi} = (\xi, 0, 0)$ lies on the x axis.

The density of the distribution, ϱ, must be adjusted to satisfy the no-penetration condition at the surface of the body. To achieve this, we write a mass balance over a fixed control volume that is confined between the instantaneous surface of the body and two planes that are perpendicular to the x axis. Let $A = \pi f^2$ be the cross-sectional area of the body and $Q(x)$ be the instantaneous flow rate in the exterior of the body through a plane that is perpendicular to the x axis. In the limit as the distance between the two planes becomes infinitesimal, we obtain

$$\frac{\partial A}{\partial t} = \frac{\partial Q}{\partial x} \qquad \text{or} \qquad 2\pi f \frac{\partial f}{\partial t} = \frac{\partial Q}{\partial x}, \qquad (7.7.2)$$

Next, we introduce the instantaneous flow rate $q(x)$ across the whole area of a plane that is perpendicular to the x axis and use the point-source representation to write

$$\frac{\partial q}{\partial x} = \varrho. \qquad (7.7.3)$$

Since, in a frame of reference moving with the body, the radius of the body is constant,

$$\frac{\partial f}{\partial t} + V_x \frac{\partial f}{\partial x} = 0. \tag{7.7.4}$$

Combining this equation with (7.7.2) and approximating Q with q, we find that the density distribution of the point sources is given by

$$\varrho = \frac{\partial A}{\partial t} = -V_x \frac{\partial A}{\partial x}. \tag{7.7.5}$$

Although the assumptions underlying these derivations fail near the end of the body where (7.7.2) ceases to be valid, expression (7.7.5) provides us with a reasonable approximation of the flow over the main portion of the body.

The coefficient of the dipole, $\mathbf{d}$, is parallel to the centerline. Using (7.7.5) and integrating by parts under the stipulation $f(\pm a) = 0$, we find that

$$d_x = \int_{-a}^{a} \xi \, \varrho(\xi) \, \mathrm{d}\xi = -V_x \int_{-a}^{a} \xi \, \frac{\mathrm{d}(\pi f^2)}{\mathrm{d}\xi} \, \mathrm{d}\xi = V_x V_B, \tag{7.7.6}$$

where V_B is the volume of the body. Substituting (7.7.6) into (7.2.18) and (7.2.23), we find that, at this level of approximation, the added mass and kinetic energy of the fluid are both zero. A higher-order approximation is required to account for the presence of the body.

Prolate spheroids

To assess the accuracy of the slender-body theory, we compare the approximate solution (7.7.6) with the exact solution for a prolate spheroid given in (7.6.4). As the aspect ratio b/a tends to zero, the exact solution yields

$$d_x = \frac{4\pi}{3} V_x a^3 \left(\frac{b}{a}\right)^2 \left[1 - \left(\frac{b}{a}\right)^2 \ln \frac{b}{a}\right] + \cdots. \tag{7.7.7}$$

Neglecting the second term inside the square bracket yields precisely the right-hand side of (7.7.6). The slender-body approximation incurs a relative error on the order $(b/a)^2 \ln(b/a)$, which is small for bodies with moderate and high aspect ratio. Consequently, slender-body theory provides us with an efficient method for approximating the coefficient of the dipole with the product of the volume of the body and velocity of translation in axial motion.

Asymptotic expansions

Equation (7.7.6) can be derived formally working within the framework of asymptotic expansions. We begin by considering the no-penetration condition, $u_x n_x + u_\sigma n_\sigma = V_x n_x$, and introduce the approximations $n_x \simeq -\partial f / \partial x$ and $n_\sigma \simeq 1$. Since u_x and u_σ have comparable magnitudes while n_x is small compared to n_σ, we may approximate

$$u_\sigma \simeq -V_x \frac{\partial f}{\partial x}. \tag{7.7.8}$$

Differentiating (7.7.1) with respect to σ, we express the radial velocity as

$$u_\sigma(\mathbf{x}) = \frac{\partial \phi}{\partial \sigma} = \frac{1}{4\pi} \sigma \int_{-a}^{a} \frac{\varrho(\xi)}{[(x-\xi)^2 + \sigma^2]^{3/2}} \, d\xi. \tag{7.7.9}$$

Rearranging, we obtain

$$u_\sigma(\mathbf{x}) = \frac{\partial \phi}{\partial \sigma} = \frac{1}{4\pi\sigma} \int_{-(a+x)/\sigma}^{(a-x)/\sigma} \frac{\varrho(\xi)}{(\eta^2 + 1)^{3/2}} \, d\eta, \tag{7.7.10}$$

where $\eta = (\xi - x)/\sigma$ is an auxiliary variable. Next, we evaluate (7.7.9) at a point on the surface of the body, $\sigma = f(x)$, expand $\varrho(\xi)$ in a Taylor series about x and retain only the leading constant term. As f/a tends to zero, the limits of integration with respect to η become infinite yielding the approximation

$$u_\sigma(\mathbf{x}) = \frac{1}{4\pi f(x)} \varrho(x) \int_{-\infty}^{\infty} \frac{d\eta}{(\eta^2 + 1)^{3/2}} = \frac{1}{2\pi} \frac{\varrho(x)}{f(x)}. \tag{7.7.11}$$

Substituting the approximate boundary condition (7.7.8) reproduces (7.7.5).

Working in a similar fashion, we find that the axial component of the velocity at a point on the surface of the body is given by

$$u_x(\mathbf{x}) = \frac{\partial \phi}{\partial x} = \frac{1}{4\pi} \int_{-a}^{a} \frac{x-\xi}{[(x-\xi)^2 + \sigma^2]^{3/2}} \varrho(\xi) \, d\xi. \tag{7.7.12}$$

Expanding $\varrho(\xi)$ in a Taylor series about x and retaining only the constant and linear terms, we obtain

$$u_x(\mathbf{x}) = -\frac{1}{4\pi} \frac{d\varrho}{dx} \ln \frac{4(a^2 - x^2)}{f^2(x)}, \tag{7.7.13}$$

which can be used along with (7.7.11) and Bernoulli's equation to evaluate the surface pressure.

7.7.2   Transverse motion

In the case of a slender axisymmetric body translating along the y axis that is normal to the centerline with velocity V_y, we approximate the flow near the body with that due to translating circular cylinder whose radius is equal to the local radius of the body, $f(x)$.

In Section 7.8, we will see that streaming flow past a circular cylinder can be represented in terms of a two-dimensional dipole oriented along the y axis with strength $2\pi V_y f^2 = 2V_y A$, where A is the cross-sectional area of the body. Since the two-dimensional dipole emerges by integrating the three-dimensional dipole along the x axis, it is appropriate to introduce a representation in terms of a distribution of three-dimensional dipoles with linear distribution density $2\pi V_y f^2(x)$, obtaining the potential

$$\phi(\mathbf{x}) = -\frac{1}{2} V_y \int_{-a}^{a} \frac{y}{|\mathbf{x} - \boldsymbol{\xi}|^3} f^2(\xi) \, d\xi, \tag{7.7.14}$$

where the point $\boldsymbol{\xi} = (\xi, 0, 0)$ lies on the x axis. This representation can be derived more rigorously working as in the case of axial motion in the context of asymptotic expansions (Problem 7.7.1(b)). The coefficient of the dipole,

$$d_y \equiv 2\pi V_y \int_{-a}^{a} f^2(x) \, \mathrm{d}x = 2 V_y V_B, \tag{7.7.15}$$

is equal to twice that for axial motion shown in (7.7.7). A comparison with the exact solution for prolate spheroids confirms the consistency and accuracy of the slender-body representation (7.7.14) for bodies with moderate and large aspect ratio (Problem 7.7.1(c)). The nonzero component of the added-mass matrix and kinetic energy of the fluid are $\alpha_{yy} = 1$ and $\mathcal{K} = \frac{1}{2}\rho V_B V_y^2$.

Problems

7.7.1 *Slender-body theory*

(a) Compute the coefficient of the dipole for a sphere according to (7.7.6). Compare your result with the exact solution.

(b) Derive the representation (7.7.14) working in the context of asymptotic expansions as in the case of axial motion discussed in the text.

(c) Compute the asymptotic limit of the coefficient of the dipole for a prolate spheroid given in (7.6.4) in the limit as b/a tends to zero, and thus verify that the leading-order term is given by $d_y = 2V_y V_B$, in agreement with the slender-body approximation.

7.7.2 *Application of slender-body theory*

Derive the distribution density of the singularity representation corresponding to the axial translation of an axisymmetric body whose surface is described by $\sigma = f(x) = b\,(1 - x^2/a^2)$, where $-a < x < a$ and b is a given length. Evaluate the pressure at the surface of the body.

7.8 Flow past or due to the motion of a two-dimensional body

Because the domain of flow past a two-dimensional body is inevitably doubly connected, the velocity potential is a single valued function of position only when the circulation around a body is zero. The nonuniqueness of the potential requires special attention in studying the properties and computing the structure of a two-dimensional flow.

7.8.1 Flow past a stationary or translating rigid body

Consider a potential flow past a two-dimensional body translating with velocity $\mathbf{V}$, as illustrated in Figure 7.8.1(a). The velocity field can be resolved into three components, $\mathbf{u} = \mathbf{u}_\infty + \mathbf{v} + \mathbf{u}_d$, where $\mathbf{u}_\infty = \nabla \phi_\infty$ is the velocity of a specified incident irrotational flow with potential ϕ_∞, $\mathbf{v}$ is the velocity due to a point vortex located at a chosen point inside the body, and $\mathbf{u}_d$ is a disturbance velocity expressed in terms of a single-valued harmonic potential ϕ_d, so that $\mathbf{u}_d = \nabla \phi_d$. The strength of the point vortex is equal to the circulation around the body.

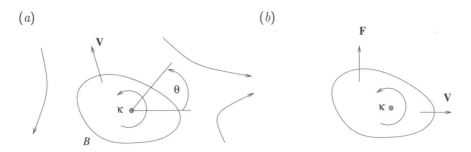

FIGURE 7.8.1   (*a*) Illustration of irrotational flow past a translating two-dimensional rigid body with nonzero circulation around the body.   (*b*) Direction of the lift force, **F**, on a translating two-dimensional body with positive circulation around the body.

The no-penetration boundary condition requires that $\mathbf{u} \cdot \mathbf{n} = \mathbf{V} \cdot \mathbf{n}$ over the contour of the body, B. Making substitutions and rearranging, we obtain

$$\mathbf{n} \cdot \nabla \phi_d = \mathbf{n} \cdot \mathbf{V} - \mathbf{n} \cdot \nabla \phi_\infty - \mathbf{n} \cdot \mathbf{v}. \tag{7.8.1}$$

The counterpart of the boundary-integral representation (7.2.2) for the single-valued disturbance potential, ϕ_d, is

$$\phi_d(\mathbf{x}_0) = - \oint_B \mathcal{G}(\mathbf{x}_0, \mathbf{x})\, \mathbf{n}(\mathbf{x}) \cdot \left[\mathbf{V} - \nabla \phi_\infty(\mathbf{x}) - \mathbf{v}(\mathbf{x}) \right] \mathrm{d}l(\mathbf{x})$$
$$+ \oint_B \phi_d(\mathbf{x})\, \mathbf{n}(\mathbf{x}) \cdot \nabla \mathcal{G}(\mathbf{x}_0, \mathbf{x})\, \mathrm{d}l(\mathbf{x}), \tag{7.8.2}$$

where l is the arc length around the body contour, B.

Using the counterpart of (7.2.3) for two-dimensional flow to simplify the single-layer potential on the right-hand side of (7.8.2), and adding to both sides of the resulting equation the potential of the incident flow, ϕ_∞, we obtain

$$\Phi(\mathbf{x}_0) = \phi_\infty(\mathbf{x}_0) + \oint_B \mathcal{G}(\mathbf{x}_0, \mathbf{x})\, \mathbf{n}(\mathbf{x}) \cdot \mathbf{v}(\mathbf{x})\, \mathrm{d}l(\mathbf{x})$$
$$+ \oint_B \left[-\mathbf{V} \cdot \mathbf{x} + \Phi(\mathbf{x}) \right] \mathbf{n}(\mathbf{x}) \cdot \nabla \mathcal{G}(\mathbf{x}_0, \mathbf{x})\, \mathrm{d}l(\mathbf{x}), \tag{7.8.3}$$

where $\Phi \equiv \phi_\infty + \phi_d$ is a single-valued potential. The total velocity potential is

$$\phi = \Phi + \frac{\kappa}{2\pi}\, \theta, \tag{7.8.4}$$

where θ is the polar angle measured around the point vortex. In the absence of circulation around the body, $\kappa = 0$, $\mathbf{v}$ vanishes and Φ reduces to ϕ. The modular cases of flow past a stationary body and flow due to the translation of a body in an otherwise quiescent fluid emerge by setting $\mathbf{V}$ or ϕ_∞ to zero, respectively.

Far-field expansion

To assess the behavior of the flow far from the body, we follow a procedure similar to that outlined in Section 7.2 for three-dimensional flow. The result is the asymptotic expansion

$$\phi(\mathbf{x}_0) = \phi_\infty(\mathbf{x}_0) + \mathbf{d} \cdot \nabla_c \,\mathcal{G}(\mathbf{x}_0, \mathbf{x}_c) + \frac{\kappa}{2\pi}\, \theta + \cdots, \tag{7.8.5}$$

where the gradient ∇_c involves derivatives with respect to an arbitrary point inside or in the vicinity of the body, $\mathbf{x}_c$,

$$\mathbf{d} = A_B \mathbf{V} - \oint_B \left[\mathbf{n}(\mathbf{x})\,\Phi(\mathbf{x}) - (\mathbf{x} - \mathbf{x}_c)\,\mathbf{v}(\mathbf{x}) \cdot \mathbf{n}(\mathbf{x}) \right] \mathrm{d}l(\mathbf{x}) \tag{7.8.6}$$

is the coefficient of the dipole, A_B is the cross-sectional area of the body in the xy plane. It will be noted that $\mathbf{d}$ is defined in terms of the single-valued part of the velocity potential Φ along the body contour.

7.8.2 Flow due to the motion and deformation of a body

Next, we consider the flow due to the translation, rotation, and deformation of a body in an otherwise quiescent fluid. We begin by resolving the velocity into two components, $\mathbf{u} = \mathbf{u}_d + \mathbf{v}$, where $\mathbf{u}_d$ represents a disturbance flow described by a single-valued harmonic potential, ϕ_d, so that $\mathbf{u}_d = \nabla \phi_d$, and $\mathbf{v}$ is the velocity field due to a point vortex located inside the body. The integral formulation provides us with the boundary-integral representation

$$\phi_d(\mathbf{x}_0) = -\oint_B \mathcal{G}(\mathbf{x}_0, \mathbf{x})\,\mathbf{n}(\mathbf{x}) \cdot [\mathbf{u} - \mathbf{v}](\mathbf{x})\,\mathrm{d}l(\mathbf{x}) + \oint_B \phi_d(\mathbf{x})\,\mathbf{n}(\mathbf{x}) \cdot \nabla\mathcal{G}(\mathbf{x}_0, \mathbf{x})\,\mathrm{d}l(\mathbf{x}), \tag{7.8.7}$$

where the point $\mathbf{x}_0$ lies in the domain of flow.

Far-field expansion

To assess the behavior of the flow far from the body, we expand the Green's function and its dipole in a Taylor series with respect to $\mathbf{x}$ about a chosen point, $\mathbf{x}_c$, note that

$$\oint_B \mathbf{v} \cdot \mathbf{n}\,\mathrm{d}l = 0, \tag{7.8.8}$$

due to mass conservation, and obtain

$$\phi_d(\mathbf{x}_0) = -\mathcal{G}(\mathbf{x}_0, \mathbf{x}_c) \oint_B \mathbf{n}(\mathbf{x}) \cdot \mathbf{u}(\mathbf{x})\,\mathrm{d}l(\mathbf{x}) + \mathbf{d} \cdot \nabla_c \mathcal{G}(\mathbf{x}_0, \mathbf{x}_c) + \cdots, \tag{7.8.9}$$

where

$$\mathbf{d} = -\oint_B \left[\mathbf{n}(\mathbf{x})\,\phi_d(\mathbf{x}) - (\mathbf{x} - \mathbf{x}_c)\,\mathbf{n}(\mathbf{x}) \cdot \mathbf{u}_d(\mathbf{x}) \right] \mathrm{d}l(\mathbf{x}), \tag{7.8.10}$$

is the coefficient of the dipole. We can demonstrate using the divergence theorem that the integral on the right-hand side of (7.8.10) remains unchanged when the contour of integration, B, is replaced by any other contour enclosing B.

The first term on the right-hand side of (7.8.9) represents the flow due to a point source associated with the change in the area occupied by the body. To leading order, far from the body, the flow behaves like that due to a point source and a point vortex whose strength is equal to the circulation around the body. In the absence of circulation and when the area occupied by the body does not change in time, the far flow resembles that due to a two-dimensional potential dipole.

7.8.3 Motion of a rigid body

The single-valued disturbance velocity $\mathbf{u}_d$ due to a rigid body that translates with velocity $\mathbf{V}$ while rotating with angular velocity $\mathbf{\Omega} = \Omega_z \mathbf{e}_z$ around the z axis about a point , $\mathbf{x}_c$, satisfies the boundary condition

$$\mathbf{n}(\mathbf{x}) \cdot \mathbf{u}_d(\mathbf{x}) = \left[\mathbf{V} + \mathbf{\Omega} \times (\mathbf{x} - \mathbf{x}_c) - \mathbf{v}\right] \cdot \mathbf{n}(\mathbf{x}). \tag{7.8.11}$$

Substituting this condition into (7.8.10), we obtain an expression for the coefficient of the dipole in terms of the disturbance velocity potential around the body,

$$\mathbf{d} = A_B \mathbf{V} - \oint_B \left[\mathbf{n}(\mathbf{x})\,\phi_d(\mathbf{x}) + (\mathbf{x} - \mathbf{x}_c)\,\mathbf{n}(\mathbf{x}) \cdot \mathbf{v}(\mathbf{x})\right] \mathrm{d}l(\mathbf{x})$$

$$+ \oint_B \mathbf{\Omega} \cdot \left[(\mathbf{x} - \mathbf{x}_c) \times \mathbf{n}(\mathbf{x})\right] (\mathbf{x} - \mathbf{x}_c)\,\mathrm{d}l(\mathbf{x}). \tag{7.8.12}$$

The last integral on the right-hand side can be made to vanish by identifying the point $\mathbf{x}_c$ with the areal centroid of the body,

$$\mathbf{X}_c \equiv \frac{1}{A_B} \iint_{Body} \mathbf{x}\,\mathrm{d}A(\mathbf{x}). \tag{7.8.13}$$

We may use the divergence theorem in two dimensions to derive a representation of the areal centroid in terms of a contour integral analogous to that shown in (4.1.7).

Decomposition into fundamental modes of rigid-body motion and circulation

The linearity of the equations and boundary conditions governing potential flow due to the motion of a rigid body allows us to express the potential as a linear combination of the velocity of translation, $\mathbf{V}$, angular velocity rotation about the z axis, $\mathbf{\Omega} = \Omega_z \mathbf{e}_z$, and strength of a point vortex κ located at a certain location inside the body and producing a desired degree of circulation. Consequently, we can write

$$\phi(\mathbf{x}) = V_x\,\Phi_1[\mathbf{x}, \mathbf{x}_c, \mathbf{e}(t)] + V_y\,\Phi_2[\mathbf{x}, \mathbf{x}_c, \mathbf{e}(t)] + \Omega_z\,\Phi_3[\mathbf{x}, \mathbf{x}_c, \mathbf{e}(t)] + \kappa\left(\frac{\theta}{2\pi} + \Phi_4[\mathbf{x}, \mathbf{x}_c, \mathbf{e}(t)]\right), \tag{7.8.14}$$

where Φ_i are four fundamental single-valued velocity potentials, $\mathbf{x}_c$ is the chosen center of rotation, the unit vector $\mathbf{e}$ describes the instantaneous body orientation, and θ is the polar angle around a point vortex located inside the body. The square brackets contain the arguments of the fundamental potentials, Φ_i.

Substituting (7.8.14) into (7.8.12) and identifying for convenience $\mathbf{x}_c$ with the areal centroid $\mathbf{X}_c$ to discard the last term on the right-hand side, we obtain

$$\mathbf{d} = A_B\left[(\mathbf{I} + \boldsymbol{\alpha}) \cdot \mathbf{V} + \Omega_z\,\boldsymbol{\zeta} + \kappa\,\boldsymbol{\eta}\right] + \oint_B (\mathbf{x} - \mathbf{X}_c)\left[\mathbf{n}(\mathbf{x}) \cdot \mathbf{v}(\mathbf{x})\right] \mathrm{d}l(\mathbf{x}), \tag{7.8.15}$$

where

$$\alpha_{ij} = -\frac{1}{A_B} \oint_B n_i \Phi_j \, dl, \qquad \zeta_i = -\frac{1}{A_B} \oint_B n_i \Phi_3 \, dl, \qquad \eta_i = -\frac{1}{A_B} \oint_B n_i \Phi_4 \, dl, \qquad (7.8.16)$$

for $i = 1, 2$. Working as in Section 7.2, we can show that the 2×2 added-mass matrix $\boldsymbol{\alpha}$ is symmetric.

Kinetic energy of the flow due to the motion of a rigid body

The $1/d$ decay of the velocity in the case of nonzero circulation is responsible for infinite kinetic energy in a two-dimensional flow, where d is the distance from the origin. This unphysical behavior can be resolved by observing that the onset of circulation during startup is accompanied by the generation of an equal amount of vorticity of opposite sign deposited at infinity, causing a faster decay. Useful information can be obtained by considering the kinetic energy of the fluid inside an area that is enclosed by the body and a large circle of radius R. Decomposing ϕ into a single-valued disturbance component, ϕ_d, and a component associated with the point vortex, and using the counterpart of expression (2.1.22) for two-dimensional flow, we obtain

$$\mathcal{K} = -\frac{1}{2} \rho \oint_B \phi_d \left[\mathbf{n} \cdot \nabla \phi_d \right] dl + \cdots. \qquad (7.8.17)$$

If the circulation around the body is nonzero, the omitted term represented by the dots is infinite; otherwise, it is zero. The finite part of the kinetic energy represented by the first term on the right-hand side of (7.8.17) is amenable to the analysis of Section 7.2 for three-dimensional flow, subject to straightforward changes in notation.

Force on a translating rigid body

To compute the force exerted on a translating two-dimensional body, we resort to Bernoulli's equation. Working as in Section 7.3 for three-dimensional flow, we derive the expression

$$\mathbf{F} = -\rho A_B \, \boldsymbol{\alpha} \cdot \frac{d\mathbf{V}}{dt} + \rho \oint_B \left[\mathbf{u} \left(\mathbf{V} \cdot \mathbf{n} \right) - \left(\mathbf{V} \cdot \mathbf{u} \right) \mathbf{n} \right] dl - \rho A_B \mathbf{g}. \qquad (7.8.18)$$

The three terms on the right-hand side represent, respectively, the acceleration reaction, a steady force, and the buoyancy force. Using the divergence theorem, we find that the integral on the right-hand side of (7.8.18) remains unchanged when the contour of integration is replaced by any other closed contour enclosing B. When the circulation around the body is nonzero, the velocity decays like $1/d$ and the integral over a circle with large radius takes the Kutta–Joukowski value $\kappa \mathbf{e}_z \times \mathbf{V}$, where d is the distance from the origin and $\mathbf{e}_z$ is the unit vector along the z axis. Substituting this expression into (7.8.18), we obtain the final expression

$$\mathbf{F} = -\rho A_B \, \boldsymbol{\alpha} \cdot \frac{d\mathbf{V}}{dt} + \rho \kappa \, \mathbf{e}_z \times \mathbf{V} - \rho A_B \mathbf{g}, \qquad (7.8.19)$$

which shows that the steady drag force is zero and the lift force is proportional to the circulation around a translating two-dimensional body, independent of the body shape.

The direction of the lift force for positive circulation is illustrated in Figure 7.8.1(b). Physically, the circulation accelerates the fluid above the body and decelerates the fluid below the body, thus causing a pressure difference that results in a lift force. The lifting action of an airfoil hinges on its ability to generate a sufficient amount of positive circulation, so that the lift force counterbalances the aircraft weight. The generation of circulation relies on viscous effects inside boundary layers developing around the airfoil, as discussed in Chapter 8.

The torque exerted on a two-dimensional body can be computed working as in Section 7.3 for three-dimensional flow.

Problems

7.8.1 *Boundary conditions for the fundamental potentials*

State the boundary conditions satisfied by the four fundamental potentials introduced in (7.8.14).

7.8.2 *Force on a translating body*

Derive (*a*) expression (7.8.18) and (*b*) the Kutta–Joukowski value for the lift force from (7.1.3) and (7.1.5).

7.9   Computation of two-dimensional flow past a body

Having established general properties of two-dimensional flow past or due to the motion of a rigid body, we proceed to discussing specific solutions obtained by analytical or numerical methods. We will discuss flow due to the translation of a circular cylinder, streaming flow past a circular cylinder, flow representation in terms of a boundary vortex sheet, and then overview numerical and asymptotic methods.

7.9.1   Flow due to the translation of a circular cylinder with circulation

In the first application, we consider the flow due to a circular cylinder of radius a translating with generally time-dependent velocity $\mathbf{V}(t)$ in the xy plane, while rotating about its center around the z axis with angular velocity $\Omega_z(t)$ in an infinite fluid. Since the component of the velocity normal to the surface of the cylinder due to the cylinder rotation is zero, rotation does not generate fluid motion.

Singularity representation

The flow due to translation with a specified circulation around the cylinder, C, can be represented by a point-source dipole with strength

$$\mathbf{d} = 2\pi a^2 \mathbf{V}, \tag{7.9.1}$$

and a point vortex with strength $\kappa = C$, both placed at the center of the cylinder. The harmonic potential and velocity field are given by

$$\phi(\mathbf{x}) = -\frac{a^2}{r^2}\,\mathbf{V}\cdot\hat{\mathbf{x}} + \frac{\kappa}{2\pi}\,\theta, \qquad \mathbf{u}(\mathbf{x}) = a^2\left[-\frac{1}{r^2}\,\mathbf{V} + \frac{2}{r^4}\,(\mathbf{V}\cdot\hat{\mathbf{x}})\,\hat{\mathbf{x}}\right] + \frac{\kappa}{2\pi}\frac{1}{r}\,\mathbf{e}_\theta, \tag{7.9.2}$$

where $\hat{\mathbf{x}} = \mathbf{x} - \mathbf{x}_c$, $r = |\hat{\mathbf{x}}_c|$ is the distance of the field point, $\mathbf{x}$, from the instantaneous center of the cylinder, $\mathbf{x}_c$, θ is the polar angle around the center of the cylinder, and $\mathbf{e}_\theta$ is the corresponding unit vector.

Fundamental potentials

Comparing the potential given in (7.9.2) with (7.8.14), we deduce that the four fundamental potentials for translation, rotation, and circulation, are given by

$$\Phi_1 = -\frac{a^2}{r^2}\,\hat{x}, \qquad \Phi_2 = -\frac{a^2}{r^2}\,\hat{y}, \qquad \Phi_3 = 0, \qquad \Phi_4 = 0. \tag{7.9.3}$$

Since the velocity due to the point vortex is parallel to the surface of the cylinder, the last term on the right-hand side of (7.8.15) is zero.

Coefficient of the dipole, added mass, and force

Using (7.9.3) and the definitions (7.8.16), we find that the coefficients ζ and β vanish, and the coefficient of the dipole is given by

$$\mathbf{d} = \pi a^2 (\mathbf{I} + \boldsymbol{\alpha}) \cdot \mathbf{V}. \tag{7.9.4}$$

Remembering that $\mathbf{d} = 2\pi a^2 \mathbf{V}$ according to (7.9.1), we obtain $\boldsymbol{\alpha} = \mathbf{I}$, which can be substituted into (7.8.19) to yield the force exerted on the cylinder,

$$\mathbf{F} = -\rho \pi a^2 \frac{\mathrm{d}\mathbf{V}}{\mathrm{d}t} + \rho \kappa \mathbf{e}_z \times \mathbf{V} - \rho \pi a^2 \mathbf{g}, \tag{7.9.5}$$

where the last term represents the buoyancy force.

Stream function and streamlines

In plane polar coordinates where the x axis is parallel to the velocity of translation, $\mathbf{V}$, the stream function is given by

$$\psi(\mathbf{x}) = \frac{a^2}{r} \sin\theta\, \mathbf{V} \cdot \mathbf{e}_x - \frac{\kappa}{2\pi} \ln\frac{r}{a}. \tag{7.9.6}$$

where $\mathbf{V} = V\mathbf{e}_x$. This expression reveals that the structure of the flow is determined by the dimensionless circulation parameter $\lambda \equiv \kappa/(4\pi a V)$. Streamline patterns for several values of λ with $\kappa > 0$ and $V > 0$ are shown in Figure 7.9.1. In all cases, the lift force points in the positive direction of the y axis due to the high velocity, and thus low pressure, on the upper side of the cylinder compared to that on the lower side.

7.9.2   Uniform flow past a stationary circular cylinder with circulation

The solution for unsteady uniform flow with velocity $\mathbf{U}(t)$ past a stationary or rotating circular cylinder with circulation C around the cylinder can be derived from that due to a translating

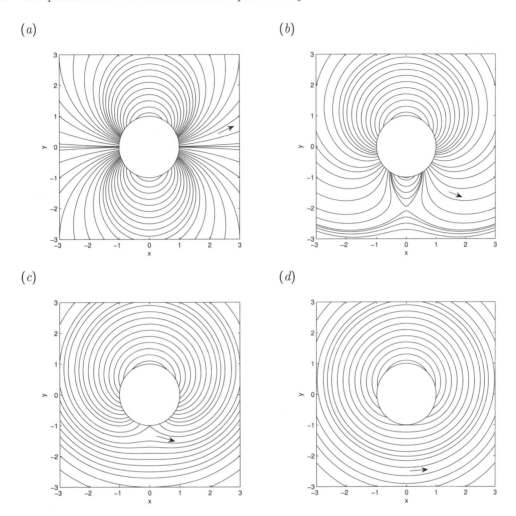

FIGURE 7.9.1   Streamline patterns of the flow due to the translation of a circular cylinder in the positive
direction of the x axis with counterclockwise circulation corresponding to circulation parameter (a)
$\lambda \equiv \kappa/(4\pi a V) = 0$, ($b$) 0.25, ($c$) 0.5, and ($d$) 1.0.

cylinder discussed in Section 7.9.1. Working in a frame of reference where the flow far from the
cylinder appears to be stationary, we obtain

$$\phi(\mathbf{x}) = \left(1 + \frac{a^2}{r^2}\right) \mathbf{U} \cdot \hat{\mathbf{x}} + \frac{\kappa}{2\pi}\,\theta, \qquad \mathbf{u}(\mathbf{x}) = \mathbf{U} + a^2 \left[\frac{1}{r^2}\mathbf{U} - \frac{2}{r^4}(\mathbf{U}\cdot\hat{\mathbf{x}})\,\hat{\mathbf{x}}\right] + \frac{\kappa}{2\pi}\frac{1}{r}\,\mathbf{e}_\theta, \quad (7.9.7)$$

where $\kappa = C$ is the strength of a point vortex, $\hat{\mathbf{x}} = \mathbf{x} - \mathbf{x}_c$, $r = |\hat{\mathbf{x}}_c|$ is the distance of the field
point $\mathbf{x}$ from the center of the cylinder, $\mathbf{x}_c$, θ is the polar angle measured around $\mathbf{x}_c$, and $\mathbf{e}_\theta$ is the
corresponding unit vector. In plane polar coordinates where the x axis is parallel to $\mathbf{U}$, the stream

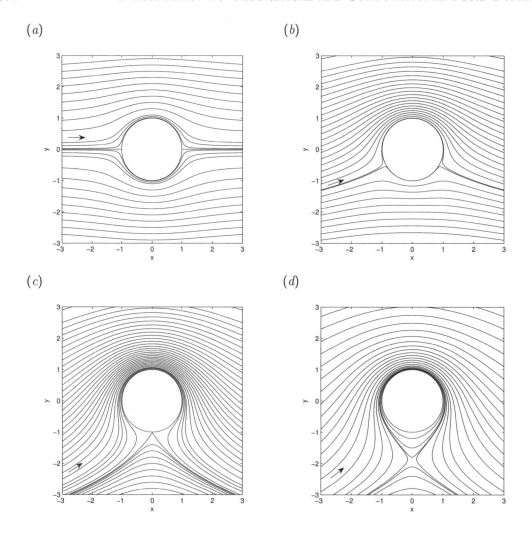

FIGURE 7.9.2 Streamline patterns of uniform flow past a stationary or rotating circular cylinder for values of the circulation parameter (a) $\beta \equiv -\kappa/(4\pi aU) = 0$, (b) 0.5, (c) 1.0, and (d) 1.2.

function is

$$\psi(r,\theta) = U\left(r - \frac{a^2}{r}\right)\sin\theta - \frac{\kappa}{2\pi}\ln\frac{r}{a}, \tag{7.9.8}$$

where $\mathbf{U} = U\mathbf{e}_x$. Note that the stream function is zero when $r = a$, as required. Equation (7.9.8) reveals that the structure of the flow depends on the dimensionless circulation parameter $\beta \equiv -\kappa/(4\pi aU)$. Streamline patterns for positive values of β with $\kappa < 0$ and $U > 0$ are shown in Figure 7.9.2.

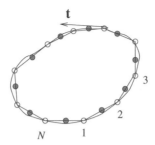

FIGURE 7.9.3   The flow past a stationary two-dimensional body can be represented by a vortex sheet.

The tangential component of the velocity around the surface of the cylinder is readily found by differentiating the stream function (7.9.8) with respect to the radial distance, r, yielding

$$u_\theta(r = a) = -\left(\frac{\partial \psi}{\partial r}\right)_{r=a} = -2U \sin \theta + \frac{\kappa}{2\pi a}. \tag{7.9.9}$$

We observe that u_θ becomes zero when $\sin \theta = -\beta$, and this reveals the onset of a symmetric pair of stagnation points on the surface of the cylinder when $0 < \beta < 1$. For larger values of β, the stagnation points move off the surface of the cylinder and then merge to yield a free stagnation point inside the flow, as shown in Figure 7.9.2. It is evident from (7.9.9) that, when $\beta > 0$, the magnitude of the velocity at the top of the cylinder is higher than that at the bottom of the cylinder. Bernoulli's equation then shows that the surface pressure at the top is lower than that at the bottom, which provides us with a physical explanation for the occurrence of a lift force toward the positive direction of the y axis noted by Magnus in 1853.

7.9.3   Representation in terms of a boundary vortex sheet

An interesting representation of a two-dimensional potential flow past a stationary body with vanishing or nonzero circulation around the body emerges by pretending that the interior of the body is occupied by a stationary fluid. We may then regard the disturbance flow due to the body as though it were induced by a two-dimensional vortex sheet situated at the surface of the body, as illustrated in Figure 7.9.3.

The strength of the vortex sheet is equal to the tangential component of the fluid velocity, $u_t = \mathbf{u} \cdot \mathbf{t}$, where $\mathbf{t}$ is the tangential unit vector pointing in the counterclockwise direction. This point of view suggests expressing the stream function at an arbitrary point $\mathbf{x}_0$ in the flow as

$$\psi(\mathbf{x}_0) = \psi_\infty(\mathbf{x}_0) - \frac{1}{2\pi} \int_0^L \ln \frac{|\mathbf{x}_0 - \mathbf{x}|}{a} \, u_t(\mathbf{x}) \, dl(\mathbf{x}) + c, \tag{7.9.10}$$

where ψ_∞ is the stream function of an incident flow, L is the total arc length of the contour of the body in the xy plane, c is an arbitrary constant, and a is a chosen length.

Applying (7.9.10) at a point at the surface of the body and enforcing the no-penetration boundary condition, $\psi = 0$, we obtain an integral equation of the first kind for the tangential boundary velocity,

$$\frac{1}{2\pi} \int_0^L \ln \frac{|\mathbf{x}_0 - \mathbf{x}|}{a} u_t(\mathbf{x}) \, dl(\mathbf{x}) = \psi_\infty(\mathbf{x}_0) + c. \qquad (7.9.11)$$

The constant c determines implicitly the circulation around the body.

Numerical methods

A simple numerical method for solving the integral equation (7.9.11) involves tracing the contour of the body with N marker points arranged in the counterclockwise direction, and then approximating the contour of the body with a polygonal line connecting successive marker points. Assuming that the value of u_t is constant and equal to u_i over the ith segment, we obtain the discrete version of (7.9.11),

$$A_i(\mathbf{x}_0) \, u_i = \psi_\infty(\mathbf{x}_0) + c, \qquad (7.9.12)$$

where summation is implied over the repeated index $i = 1, \ldots, N$, and

$$A_i(\mathbf{x}_0) = \frac{1}{2\pi} \int_{\mathbf{x}_i}^{\mathbf{x}_{i+1}} \ln \frac{|\mathbf{x}_0 - \mathbf{x}|}{a} \, dl(\mathbf{x}) \qquad (7.9.13)$$

are influence coefficients. The tangential velocities u_i can be computed by pointwise collocation. The methodology involves applying (7.9.12) at the midpoint of each segment, $\mathbf{y}_j = \frac{1}{2}(\mathbf{x}_j + \mathbf{x}_{j+1})$, to derive a system of N linear equations,

$$A_i(\mathbf{y}_j) \, u_i = \psi_\infty(\mathbf{y}_j) + c \qquad (7.9.14)$$

for $j = 1, \ldots, N$, where summation is implied over the repeated index, i. To ensure that the circulation around the body has a prescribed value, κ, we introduce the additional constraint

$$\sum_{i=1}^N u_i \, |\mathbf{x}_i - \mathbf{x}_{i+1}| = \kappa, \qquad (7.9.15)$$

and solve a system of the $N + 1$ linear equations comprised of (7.9.14) and (7.9.15) for $N + 1$ unknowns, u_i and c.

The off-diagonal influence coefficients of the influence matrix, $A_i(\mathbf{y}_j)$ for $i \neq j$, can be computed using a standard integration method, such as the trapezoidal rule. or a Gaussian quadrature (Section B.6, Appendix B). The diagonal influence coefficients corresponding to the host collocation point, $A_j(\mathbf{y}_j)$, can be computed analytically as

$$A_j(\mathbf{y}_j) = \frac{1}{\pi} |\mathbf{y}_j - \mathbf{x}_j| \, (\ln \frac{|\mathbf{y}_j - \mathbf{x}_j|}{a} - 1) \qquad (7.9.16)$$

for $j = 1, \ldots, N$. An integral identity ensures that the choice of the length, a, is immaterial on the accuracy of the computations.

Flow past a plate with infinitesimal thickness

In the case of uniform flow past a flat plate with infinitesimal thickness positioned along the x axis between the points $x = \pm a$, the upper and lower surfaces of the body coincide and the integral representation (7.9.10) reduces to

$$\psi(\mathbf{x}_0) = \psi_\infty(\mathbf{x}_0) - \frac{1}{2\pi} \int_{-a}^{a} \ln \frac{|\mathbf{x}_0 - x\, \mathbf{e}_x|}{a} \Delta u(x) \, \mathrm{d}x + c, \tag{7.9.17}$$

where $\Delta u \equiv u_t^- - u_t^+$ is the discontinuity in the tangential velocity across the plate, the plus superscript indicates evaluation at the upper side of the plate, and the minus superscript indicates evaluation at the lower side of the plate. Equation (7.9.11) provides us with an integral equation,

$$\frac{1}{2\pi} \int_{-a}^{a} \ln \frac{|x_0 - x|}{a} \Delta u(x) \, \mathrm{d}x = \psi_\infty(x_0) + c, \tag{7.9.18}$$

where the constant c determines the circulation around the plate, κ.

To establish the relation between κ and c, we note that

$$\frac{1}{2\pi} \int_{-a}^{a} \ln \frac{|x_0 - x|}{a} \frac{\mathrm{d}x}{\sqrt{a^2 - x^2}} = -\frac{1}{\sqrt{2}}, \tag{7.9.19}$$

independent of x_0, which demonstrates that

$$\Delta u(x) = -\frac{\sqrt{2}}{\sqrt{a^2 - x^2}} c \tag{7.9.20}$$

is a solution of (7.9.18) with $\psi_\infty = 0$. The circulation around the plate is

$$\kappa = \int_{-a}^{a} \Delta u(x) \, \mathrm{d}x = -\sqrt{2}\pi c. \tag{7.9.21}$$

Combining the last two equations, we obtain

$$\Delta u(x) = \frac{\kappa}{\pi} \frac{1}{\sqrt{a^2 - x^2}}. \tag{7.9.22}$$

Note that singularities occur at both ends of the plate, $x = \pm a$.

In the case of uniform flow with velocity $\mathbf{U} = (U_x, U_y)$, the stream function of the incident flow is $\psi_\infty = U_x y - U_y x$ and the solution of the integral equation (7.9.18) is given by

$$\Delta u(x) = \left(2U_y x + \frac{\kappa}{\pi} \right) \frac{1}{\sqrt{a^2 - x^2}} \tag{7.9.23}$$

(Section 7.12.2). When the circulation assumes the Kutta value $\kappa = -2\pi a U_y$, the singularity at the trailing edge, $x = a$, disappears and fluid particles on either side of the plate leave the plate in the tangential direction without making a sharp turn.

7.9.4   Computation of flow past bodies with arbitrary geometry

The singularity methods described in Section 7.6.3 for three-dimensional flow can be easily adapted to flow past or due to the motion of a two-dimensional body. However, the efficiency of these methods competes with that of another class of methods based on a complex-variable formulation combined with the conformal mapping techniques discussed in Sections 7.10–7.13. A third class of powerful methods relies on the boundary-integral formulation discussed in Chapter 10.

Slender-body theory

Slender-body theory provides us with useful approximate solutions for flow past an elongated body with high aspect ratio. Consider the flow due to the axial translation of a slender two-dimensional body whose contour is symmetric with respect to the x axis, in the absence of circulation around the body. The upper and lower surfaces of the body are described by the function $y = \pm f(x)$ for $-a < x < a$. Working as in the case of flow due to the motion of an axisymmetric body discussed in Section 7.7, we represent the flow due to translation with velocity V_x along the x axis in terms of two-dimensional point sources distributed along the centerline. The induced harmonic potential is

$$\phi(\mathbf{x}) = \frac{V_x}{2\pi} \int_{-a}^{a} \ln \frac{|\mathbf{x} - \boldsymbol{\xi}|}{a} \, \varrho(\xi) \, \mathrm{d}\xi, \tag{7.9.24}$$

where ϱ is the density of the distribution, a is a chosen length, and the point $\boldsymbol{\xi} = (\xi, 0)$ lies at the x axis. The no-penetration condition requires that

$$u_x n_x + u_y n_y = V_x n_x \tag{7.9.25}$$

around the body contour. Introducing the approximations $n_x \simeq -\mathrm{d}f/\mathrm{d}x$ and $n_y \simeq 1$ on the upper surface of the body, and noting that u_x and u_y have comparable magnitudes while n_x is small compared to n_y, we derive the approximation

$$u_y(x, f) \simeq -V_x \frac{\mathrm{d}f}{\mathrm{d}x}. \tag{7.9.26}$$

Next, we differentiate (7.9.24) with respect to y to produce u_y, obtaining

$$u_y(\mathbf{x}) = \frac{V_x}{2\pi} y \int_{-a}^{a} \frac{\varrho(\xi)}{(x-\xi)^2 + y^2} \, \mathrm{d}\xi = \frac{V_x}{2\pi} \int_{-(a+x)/y}^{(a-x)/y} \frac{\varrho(\xi)}{\eta^2 + 1} \, \mathrm{d}\eta, \tag{7.9.27}$$

where $\eta = (\xi - x)/y$ is an auxiliary integration variable. Next, we evaluate (7.9.27) at a point $\mathbf{x} = (x, f)$ on the upper surface of the body, expand $\varrho(\xi)$ in a Taylor series about x, and retain only the leading constant term. As the body aspect ratio f/a tends to zero, the limits of integration with respect to η tens to infinity, yielding the approximation

$$u_y(x, f) \simeq \frac{V_x}{2\pi} \varrho(x) \int_{-\infty}^{\infty} \frac{\mathrm{d}\eta}{\eta^2 + 1} = \frac{1}{2} V_x \, \varrho(x). \tag{7.9.28}$$

Substituting (7.9.28) into (7.9.26), we find that

$$\varrho = -2 \frac{\mathrm{d}f}{\mathrm{d}x}. \tag{7.9.29}$$

The coefficient of the dipole is

$$d_x = V_x \int_{-a}^{a} \xi \, \varrho(\xi) \, \mathrm{d}\xi = -2V_x \int_{-a}^{a} \xi \, \frac{\mathrm{d}f}{\mathrm{d}\xi} \, \mathrm{d}\xi = 2A_B V_x, \tag{7.9.30}$$

where A_B is the area occupied by the body. Remarkably but fortuitously, this formula produces the exact answer for a circular cylinder derived in Section 7.9.1.

Problems

7.9.1 *Lift on a circular cylinder*

Substitute (7.9.2) into Bernoulli's equation to obtain the pressure distribution over the cylinder and then use (7.1.5) to compute the force exerted on the cylinder.

7.9.2 *Formulation in terms of a boundary vortex sheet*

(a) Derive the counterparts of (7.9.10) and (7.9.11) for axisymmetric flow past a stationary body.

(b) Discuss the implementation of the method for three-dimensional flow, that is, derive an integral equation for the two tangential components of the boundary vortex sheet. *Hint:*   recall that the vorticity field is solenoidal.

 Computer Problem

7.9.3 *Boundary vortex sheet*

(a) Write a program that solves the system of equations (7.9.14) and (7.9.15) for a specified body geometry described in terms of a collection of marker points.

(b) Run the program for the case of uniform flow along the x axis past a circular cylinder. Plot the distribution of tangential velocity along the surface of the cylinder for several values of the circulation and compare the numerical results with the exact solution.

(c) Run the program for the case of uniform flow along the x axis past an elliptic cylinder with aspect ratio equal to 4. Plot the distribution of the tangential velocity for several values of the circulation.

(d) Run the program for uniform flow with velocity $U_x = U_y$ past a flat plate aligned with the x axis. Plot the distribution of tangential velocity for several values of the circulation. Compare the numerical results with the exact solution given in (7.9.23).

7.10   Complex-variable formulation of two-dimensional flow

A powerful method of analyzing and computing the two-dimensional irrotational flow of an incompressible fluid is based on a complex variable formulation that employs the harmonic potential, ϕ, and the stream function, ψ. The theory of analytic functions of a complex variable allows us to derive solutions for a variety of flows in domains with arbitrary geometry using efficient and elegant analytical and numerical methods.

7.10.1   The complex potential

Consider a two-dimensional irrotational flow of an incompressible fluid in the xy plane, and introduce the corresponding two-dimensional harmonic potential, ϕ, satisfying Laplace's equation in the plane due to the continuity equation, $\nabla^2 \phi = 0$, and associated stream function, ψ. Since the vorticity is identically zero, the stream function is a harmonic function in the xy plane, $\omega_z = -\nabla^2 \psi = 0$. The velocity vector field is tangential to the instantaneous streamlines of constant ψ and perpendicular to the instantaneous equipotential lines of constant ϕ. A rectilinear or curvilinear grid consisting of lines of constant ψ and ϕ is sometimes called a flow net.

Cauchy–Riemann equations

The companion harmonic functions ϕ and ψ are related through their definition by the Cauchy–Riemann equations,

$$u_x = \frac{\partial \phi}{\partial x} = \frac{\partial \psi}{\partial y}, \qquad u_y = \frac{\partial \phi}{\partial y} = -\frac{\partial \psi}{\partial x}. \qquad (7.10.1)$$

Since ϕ and ψ is a pair of conjugate harmonic functions, they can be identified with the real or imaginary part of an analytic function of a complex variable, $z = x + iy$,

$$w(z) \equiv \phi + i\,\psi, \qquad (7.10.2)$$

called the complex potential, where i is the imaginary unit, $i^2 = -1$. A function of a complex variable, z, is analytic at a point, z_0, if its first derivative with respect to z is finite and independent of the direction of differentiation at every point in a neighborhood of z_0. An entire function is analytic at each point in the complex plane. In the remainder of this chapter, $z = x + iy$ will denote the complex variable rather than the Cartesian coordinate normal to the xy plane.

Velocity

Differentiating (7.10.2) and using the Cauchy–Reimann equations (7.10.1), we obtain the two velocity components in the complex form

$$\frac{dw}{dz} = u_x - i\,u_y \equiv u^*, \qquad (7.10.3)$$

where $u \equiv u_x + i\,u_y$ is the complex velocity, and an asterisk denotes the complex conjugate. If the velocity field is continuous, the complex potential must be an analytic function of z everywhere except at isolated singular points. Selecting different analytic functions for the complex potential, $w(z)$, allows us to construct diverse families of incompressible irrotational flows.

Uniform flow

The complex potential of uniform (streaming) flow with velocity $\mathbf{U} = (U_x, U_y)$ in the xy plane is given by

$$w = (U_x - i\,U_y)\,z = U^* z, \qquad (7.10.4)$$

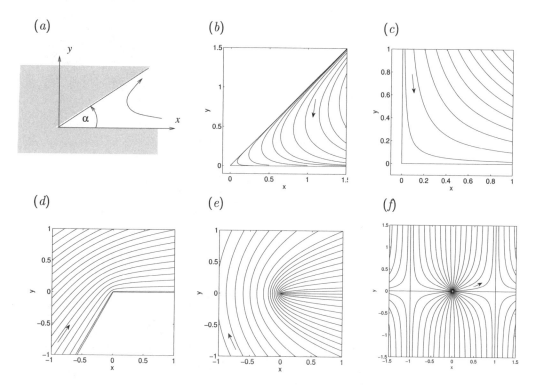

FIGURE 7.10.1   (*a*) Illustration of potential flow inside or around a corner with aperture angle α. (*b*–*e*) Streamline patterns of flow in a corner with aperture angle (*b*) $\alpha = \pi/4$, (*c*) $\pi/2$, (*d*) $3\pi/4$, and (*e*) 2π. (*f*) Streamline pattern of the flow due to a periodic array of point sources arranged along the x axis; the x and y axes have been scaled with half the point source separation.

where $U = U_x + iU_y$ is a complex velocity. The potential function, stream function, and complex velocity are given by

$$\phi = U_x\, x + U_y\, y, \qquad \psi = U_x y - U_y x, \qquad \frac{\mathrm{d}w}{\mathrm{d}z} = u^* = U_x - i\, U_y. \tag{7.10.5}$$

Flow inside and around a corner

The complex potential of irrotational flow inside or around a corner with aperture angle $\alpha = \pi/m$, as shown in Figure 7.10.1(*a*), is given by

$$w = Az^m, \tag{7.10.6}$$

where A is a real constant with appropriate dimensions. In plane polar coordinates, (r, θ), defined such that $x = r\cos\theta$, $y = r\sin\theta$, and $z = r\exp(i\theta)$, one wall is located at $\theta = 0$ and the second wall is located at $\theta = \alpha$. The potential function, stream function, and complex velocity are given by

$$\phi = Ar^m \cos(m\theta), \qquad \psi = Ar^m \sin(m\theta), \qquad \frac{\mathrm{d}w}{\mathrm{d}z} = u^* = Amz^{m-1}. \tag{7.10.7}$$

Point source

$$w = \frac{m}{2\pi} \ln \frac{z - z_0}{a}, \quad \phi = \frac{m}{2\pi} \ln \frac{|z - z_0|}{a}, \quad \psi = \frac{m}{2\pi} \arg \frac{|z - z_0|}{a}, \quad \frac{\mathrm{d}w}{\mathrm{d}x} = \frac{m}{2\pi} \frac{z^* - z_0^*}{|z - z_0|^2}$$

---

m is the strength of the point source and a is a specified length

---

Point-source dipole

$$w = -\frac{d_x + \mathrm{i}\,d_y}{2\pi} \frac{1}{z - z_0}, \qquad \phi = -\frac{1}{2\pi} \frac{d_x(x - x_0) + d_y(y - y_0)}{|z - z_0|^2}$$

$$\psi = \frac{1}{2\pi} \frac{d_x(y - y_0) - d_y(x - x_0)}{|z - z_0|^2}, \qquad \frac{\mathrm{d}w}{\mathrm{d}x} = \frac{d_x + \mathrm{i}d_y}{2\pi} \frac{1}{(z - z_0)^2}$$

$\mathbf{d} = (d_x, d_y)$ is the strength of the point source dipole

---

Point vortex

$$w = \frac{\kappa}{2\pi\mathrm{i}} \ln \frac{z - z_0}{a}, \quad \phi = \frac{\kappa}{2\pi} \arg \frac{|z - z_0|}{a}, \quad \psi = -\frac{\kappa}{2\pi} \ln \frac{|z - z_0|}{a}, \quad \frac{\mathrm{d}w}{\mathrm{d}x} = \frac{\kappa}{2\pi\mathrm{i}} \frac{z^* - z_0^*}{|z - z_0|^2}$$

κ is the strength of the point vortex, a is a specified length

---

TABLE 7.10.1   The complex potential, w, the corresponding real harmonic potential, ϕ, the stream function, ψ, and the complex velocity field, $\mathrm{d}w/\mathrm{d}z = u_x - \mathrm{i}u_y$ of elementary potential flows due to singularities located at z_0. An asterisk designates the complex conjugate.

Streamline patterns for several angles aperture angles are shown in Figure 7.10.1(b–e). When $\alpha = \pi$, corresponding to $m = 1$, we obtain uniform streaming flow along the x axis. When $\alpha = 2\pi$, corresponding to $m = 1/2$, we obtain flow around a semi-infinite flat plate represented by the positive part of the x axis.

Point sources and point vortices

The complex potential and velocity due to a point source, a point-source dipole, and a point vortex are given in Tables 7.10.1. The complex potential and velocity due to a periodic array of point sources and a periodic array point vortices are given in 7.10.2. The streamline pattern of the flow due to a periodic array of point sources is shown Figure 7.10.1(f). The streamline pattern due to a periodic array of point vortices is shown in Figure 2.10.1.

Periodic array of point sources along the x axis

$$w = \frac{m}{2\pi} \ln \left[\sin \left(\frac{k}{2} (z - z_0) \right) \right]$$

$$\phi = \frac{m}{4\pi} \ln \left[\cosh \left(k(y - y_0) \right) - \cos \left(k(x - x_0) \right) \right]$$

$$\psi = \frac{m}{2\pi} \arg \left[\sin \left(\frac{k}{2} (x - x_0) \right) \cosh \left(\frac{k}{2} (y - y_0) \right) + \mathrm{i} \cos \left(\frac{k}{2} (x - x_0) \right) \sinh \left(\frac{k}{2} (y - y_0) \right) \right]$$

$$\frac{dw}{dz} = \frac{m}{2a} \cot \left[\frac{k}{2} (z - z_0) \right]$$

Periodic array of point vortices along the x axis

$$w = \frac{\kappa}{2\pi \mathrm{i}} \ln \left[\sin \left(\frac{k}{2} (z - z_0) \right) \right]$$

$$\phi = \frac{\kappa}{2\pi} \arg \left[\sin \left(\frac{k}{2} (x - x_0) \right) \cosh \left(\frac{k}{2} (y - y_0) \right) + \mathrm{i} \cos \left(\frac{k}{2} (x - x_0) \right) \sinh \left(\frac{k}{2} (y - y_0) \right) \right]$$

$$\psi = -\frac{\kappa}{4\pi} \ln \left[\cosh \left(k(y - y_0) \right) - \cos \left(k(x - x_0) \right) \right]$$

$$\frac{dw}{dz} = \frac{\kappa}{2a\mathrm{i}} \cot \left[\frac{k}{2} (z - z_0) \right]$$

TABLE 7.10.2   The complex potential and velocity field due to a periodic arrangement of point sources or point vortices along the x axis; a is the period, $k = 2\pi/a$ is the wave number, and z_0 is the location of one point source or point vortex in the array.

Point source between two parallel walls

The velocity field due to a point source located at a point, $\mathbf{x}_0$, between two parallel walls separated by distance h can be represented by two infinite arrays of point sources with period $L = 2h$ running perpendicular to the walls. The first array contains the point source, and the second array contains the image of the point source with respect to the upper or lower wall. The strengths of the point sources in both arrays are equal. The wavenumber of each periodic array is $2\pi/L = \pi/h$. Using Table 7.10.2, we find that, If the walls are parallel to the x axis and thus perpendicular to the y axis, the corresponding harmonic potential is given by

$$\phi(\mathbf{x}) = \frac{m}{4\pi} \ln \left[\left(\cosh \left[\frac{\pi}{h} (x - x_0) \right] - \cos \left[\frac{\pi}{h} (y - y_0) \right] \right) \right.$$
$$\left. \left(\cosh \left[\frac{\pi}{h} (x - x_0) \right] - \cos \left[\frac{\pi}{h} (y - y_0^{im}) \right] \right) \right], \tag{7.10.8}$$

where y_0^{im} is the image of the point source with respect to the upper or lower wall. Since the two images are separated by distance $2h$, the choice is inconsequential. For example, of the lower wall is located at $y = y_w$, then the image point with respect to the lower wall is $y_0^{im} = 2y_w - y_0$ and the corresponding image point with respect to the upper wall is $y_0^{im} = 2y_w + 2h - y_0$. Setting $m = -1$ we obtain the Green's function of the second kind (Neumann function) of Laplace's equation whose normal derivative vanishes over the lower and upper walls.

Green's function in a domain between two parallel walls

The Green's function of Laplace's equation in a infinite strip confined between two parallel plane walls separated by distance h can be represented by an infinite array of point sinks complemented by an infinite array of point sources. If the walls are parallel to the x axis and thus perpendicular to the y axis, the Green's function is given by

$$\mathcal{G}(\mathbf{x}, \mathbf{x}_0) = -\frac{1}{4\pi} \ln \frac{\cosh[\frac{\pi}{h}(x - x_0)] - \cos[\frac{\pi}{h}(y - y_0)]}{\cosh[\frac{\pi}{h}(x - x_0)] - \cos[\frac{\pi}{h}(y - y_0^{im})]}, \quad (7.10.9)$$

where y_0^{im} is the image of the singular point with respect to the upper or lower wall. As $\mathbf{x} \to \mathbf{x}_0$, we recover the free-space Green's function, $\mathcal{G}(\mathbf{x}, \mathbf{x}_0) = -\frac{1}{2\pi} \ln(|\mathbf{x} - \mathbf{x}_0|/h)$.

7.10.2 Computation of the complex potential from the potential or stream function

The theory of harmonic functions provides us with a method of computing the stream function from the potential function, and *vice versa*, yielding the complex potential in terms of one its harmonic components. The procedure involves integrating the Cauchy–Riemann equations along a path in the complex plane, obtaining

$$\phi(x, y) = \phi(x_0, y_0) - \int_{y_0}^{y} \frac{\partial \psi}{\partial x}(x_0, y') \, dy' + \int_{x_0}^{x} \frac{\partial \psi}{\partial y}(x', y_0) \, dx',$$

$$\psi(x, y) = \psi(x_0, y_0) + \int_{y_0}^{y} \frac{\partial \phi}{\partial x}(x_0, y') \, dy' - \int_{x_0}^{x} \frac{\partial \phi}{\partial y}(x', y_0) \, dx', \quad (7.10.10)$$

where (x_0, y_0) is an arbitrary point where the potential function and stream function are arbitrarily assigned [106]. The integrals can be computed by analytical or numerical methods.

Far-field expansion of the complex potential for infinite flow

Insights on the structure of infinite flow past or due to the motion of a two-dimensional body can be gained by studying the behavior of the complex potential far from the body. To assess the most general form of the complex potential corresponding to a velocity field that either decays or tends to a constant at infinity, we expand the analytic function dw/dz into a Laurent series about a point z_0 outside a circle of sufficiently large radius centered at a chosen point, z_0. The series begins with a constant term and contains only negative powers of the complex distance $z - z_0$.

Integrating the series with respect to z and setting for convenience the integration constant to zero, we obtain the expansion

$$w(z) = U^*(z - z_0) + \frac{m - i\kappa}{2\pi} \ln \frac{z - z_0}{a} + b_0 + \sum_{n=1}^{\infty} \frac{b_n}{(z - z_0)^n}, \quad (7.10.11)$$

where a is a constant length and b_n are constant complex coefficients. Referring to Table 7.10.1, we recognize the second term on the right-hand side of (7.10.11) as a point source and a point vortex located at the point z_0. The strength of the point source, m, is equal to the flow rate across a closed contour enclosing the body. The strength of the point vortex, κ, is equal to the cyclic constant of the motion around the body. For example, in the case of flow due to the isotropic expansion of a circular bubble centered at the point z_0, κ and all coefficients b_n in (7.10.11) are zero.

7.10.3   Blasius theorems

Blasius developed an elegant method of computing the force and torque exerted on a body that is held stationary in a steady ambient flow [40]. As a first step, we introduce the complex force, $F = F_x + iF_y$, and use Bernoulli's equation for steady irrotational flow to express the hydrodynamic force exerted on the body as

$$F^* = -\mathrm{i} \oint_{Body} p \,\mathrm{d}z^* = \frac{1}{2} \mathrm{i}\rho \oint_{Body} \left(\frac{\mathrm{d}w}{\mathrm{d}z}\right) \left(\frac{\mathrm{d}w}{\mathrm{d}z}\right)^* \mathrm{d}z^* = \frac{1}{2} \mathrm{i}\rho \oint_{Body} \left(\frac{\mathrm{d}w}{\mathrm{d}z}\right) \mathrm{d}w^*, \qquad (7.10.12)$$

where p is the real hydrodynamic pressure, ρ is the fluid density, the path of integration is taken in the counterclockwise direction, and an asterisk denotes the complex conjugate. Since the stream function is constant and the complex potential is real around the body contour, we can replace $\mathrm{d}w^*$ in the last integral with $\mathrm{d}w$, obtaining

$$F^* = \frac{1}{2} \mathrm{i}\rho \oint_{Body} \left(\frac{\mathrm{d}w}{\mathrm{d}z}\right)^2 \mathrm{d}z. \qquad (7.10.13)$$

Working in a similar fashion with the real torque, T, we find that

$$T = \oint_{Body} p\,(x\,\mathrm{d}x + y\,\mathrm{d}y) = -\frac{1}{2}\rho\, \mathrm{Real}\left\{ \oint_{Body} \left(\frac{\mathrm{d}w}{\mathrm{d}z}\right)^2 z\,\mathrm{d}z \right\}. \qquad (7.10.14)$$

Assuming that the functions inside the last integrals in (7.10.13) and (7.10.14) are analytic throughout the domain of flow, we use Cauchy's integral theorem to replace the integrals along the contour of the body with corresponding integrals along another closed contour enclosing the body. Introducing the Laurent series of the complex potential stated in (7.10.11) allows us to compute the integrals in terms of the coefficients of the far-field expansion.

Uniform flow

As an application, we consider uniform flow with velocity $\mathbf{U}$ past a stationary body. Substituting into (7.10.13) the Laurent series (7.10.11) with $m = 0$, we obtain

$$F^* = \frac{1}{2} \mathrm{i}\rho \oint_{Body} \left(U^* + \frac{\kappa}{2\pi \mathrm{i}} \frac{1}{z - z_0} - \sum_{i=1}^{\infty} \frac{n b_n}{(z - z_0)^{n+1}} \right)^2 \mathrm{d}z. \qquad (7.10.15)$$

Expanding the square, we find that

$$F^* = \frac{1}{2} \mathrm{i}\rho \oint_{Body} \left(U^{*2} + U^* \frac{\kappa}{\pi \mathrm{i}} \frac{1}{z - z_0} + \cdots \right) \mathrm{d}z, \qquad (7.10.16)$$

where the three dots represent terms that decay at a quadratic or higher rate. Evaluating the integral by the method of residues, we obtain the real components of the force,

$$F_x = \rho \kappa U_y, \qquad F_y = -\rho \kappa U_x, \qquad (7.10.17)$$

which shows that uniform incident past a body combined with circulation around the body generates a lift force that is independent of the body shape. These expressions are in agreement with our earlier results in Section 7.8 obtained by different methods.

Working in a similar fashion, we find that the hydrodynamic torque with respect to a point, z_0, is given by

$$T = -2\pi\rho \, \mathrm{Imag}\{ U^* b_1 \}. \qquad (7.10.18)$$

Because the complex constant b_1 depends on the body shape, the result for the torque is less general than that for the force.

Force on a translating body

The hydrodynamic force exerted on a body that translates with velocity $\mathbf{V}$ follows from (7.10.17) by setting $\mathbf{U} = -\mathbf{V}$, yielding

$$F_x = -\rho \kappa V_y, \qquad F_y = \rho \kappa V_x, \qquad (7.10.19)$$

in agreement with our previous results in Section 7.8. A lift force is established when $V_x > 0$ in the presence of positive circulation, $\kappa > 0$.

7.10.4   Flow due to the translation of a circular cylinder

Combining the first equation in (7.9.2) with (7.9.6) and using Table 7.10.1, we find that the complex potential of the flow due to a circular cylinder of radius a translating along the x axis with velocity V_x and along the y axis with velocity V_y is given by

$$w(z) = -V \frac{a^2}{z - z_c} + \frac{\kappa}{2\pi \mathrm{i}} \ln \frac{z - z_c}{a}, \qquad (7.10.20)$$

where κ is the circulation around the cylinder, z_c is the instantaneous center of the cylinder, and $V = V_x + \mathrm{i} V_y$ is the complex velocity of translation. The first term on the right-hand side of (7.10.20) represents a point-source dipole and the second term represents a point vortex.

7.10.5   Uniform flow past a circular cylinder

Combining the first equation in (7.9.7) with (7.9.8) and using Table 7.10.1, we find that the complex potential of uniform flow past a stationary circular cylinder of radius a is given by

$$w(z) = U^* z + U \frac{a^2}{z - z_c} + \frac{\kappa}{2\pi \mathrm{i}} \ln \frac{z - z_c}{a}, \qquad (7.10.21)$$

where $U = U_x + \mathrm{i} U_y$ is the complex velocity far from the cylinder. The three terms on the right-hand side of (7.10.21) represent, respectively, uniform flow, the flow due to a point-source dipole, and the flow due to a point vortex. The complex potential (7.10.21) can be used to derive the complex potential of uniform flow past a body with arbitrary cross-section using the method of conformal mapping, as discussed in Section 7.11.

7.10.6 Arbitrary flow past a circular cylinder

Milne–Thomson's circle theorem allows us to derive an exact expression for the complex potential of an arbitrary incident flow past a stationary circular cylinder [267]. A prerequisite is that the incident flow prevailing in the absence of the cylinder is free of singularities in the interior of the cylinder. As a preliminary, we regard the incident complex potential, w_∞, as a function of $z - z_c$, where z_c is the center of the cylinder. The complex potential in the presence of the cylinder is

$$w(z - z_c) = w_\infty(z - z_c) + w_\infty^\dagger \left(\frac{a^2}{z - z_c} \right), \tag{7.10.22}$$

where $w_\infty^\dagger(z) \equiv w_\infty^*(z^*)$ and a is the cylinder radius.

Uniform flow

As an example, we consider uniform flow past a circular cylinder with vanishing circulation around the cylinder, and set $w_\infty = U^* z$. The restriction of vanishing circulation stems from the required absence of singularities inside the cylinder. To apply the circle theorem, we write

$$w_\infty(z - z_c) = U^*(z - z_c) + U^* z_c \tag{7.10.23}$$

and use (7.10.22) to obtain

$$w(z) = U^* z + U \frac{a^2}{z - z_c} + U z_c^*, \tag{7.10.24}$$

which is consistent with (7.10.21). The inconsequential constant $U z_c^*$ on the right-hand side plays no role on the velocity field.

Point source outside a circular cylinder

In the case of an incident flow due to a point source located at the point z_s in the exterior of a circular cylinder, we set

$$w_\infty(z) = \frac{m}{2\pi} \ln \frac{z - z_s}{a}. \tag{7.10.25}$$

To apply the circle theorem, we write

$$w_\infty(z - z_c) = \frac{m}{2\pi} \left(\ln[z - z_c - (z_s - z_c)] - \ln a \right) \tag{7.10.26}$$

and use (7.10.22) to find that the complex potential in the presence of the cylinder is

$$w(z) = \frac{m}{2\pi} \left(\ln \frac{z - z_s}{a} + \ln \left[\frac{a}{z - z_c} - \frac{(z_s - z_c)^*}{a} \right] \right). \tag{7.10.27}$$

Rearranging, we obtain

$$w(z) = \frac{m}{2\pi} \left(\ln \frac{z - z_s}{a} - \ln \frac{z - z_c}{a} + \ln \left[1 - \frac{(z - z_c)(z_s - z_c)^*}{a^2} \right] \right) \tag{7.10.28}$$

and then

$$w(z) = \frac{m}{2\pi} \left(\ln \frac{z - z_s}{z - z_c} + \ln \left[\frac{z - z_c}{a} - \frac{a}{(z_s - z_c)^*} \right] + \ln \frac{(z_c - z_s)^*}{a} \right). \tag{7.10.29}$$

The last term on the right-hand side of (7.10.29) can be discarded as inconsequential. The resulting expression shows that the disturbance flow due to the cylinder can be represented in terms of a point sink with strength $-m$ located at the center of the cylinder and a point source with strength m located at the inverse point of the primary point source with respect to the cylinder,

$$z_s^{inv} = z_c + \frac{a^2}{|z_s - z_c|^2} (z_s - z_c) = z_c + \frac{a^2}{(z_s - z_c)^*}. \tag{7.10.30}$$

The real part of (7.10.29) with $m = -1$ is the two-dimensional Green's function of the second kind or Neumann function, $\mathcal{N}(z, z_0)$, in the exterior of a circular cylinder.

Point vortex outside or inside a cylinder

In the case of an incident flow due to a point vortex located at the point z_v in the exterior of a cylinder, we set

$$w_\infty(z) = \frac{\kappa}{2\pi i} \ln \frac{z - z_v}{a} = \frac{\kappa}{2\pi i} \left(\ln \frac{z - z_c - (z_v - z_c)}{a} \right). \tag{7.10.31}$$

The complex potential of the flow in the presence of the cylinder is

$$w(z) = \frac{\kappa}{2\pi i} \left(\ln \frac{z - z_v}{a} - \ln \left[\frac{a}{z - z_c} - \frac{(z_v - z_c)^*}{a} \right] \right). \tag{7.10.32}$$

Working as in the case of the point source, we find that the disturbance flow due to the cylinder can be represented in terms of a point vortex with strength κ located at the center of the cylinder and an image point vortex with strength $-\kappa$ located at the inverse point of the primary point vortex with respect to the cylinder, $z_v^{inv} = z_c + a^2/(z_v - z_c)^*$. The velocity induced by the first point vortex is tangential to the cylinder and can be disregarded without affecting the no-penetration condition. We may consider the exterior point vortex as the image of the interior point vortex and use the solution derived in this section to describe the flow due to a point vortex inside a cylinder.

Problems

7.10.1 *Linear irrotational flow past a circular cylinder*

Consider a linear flow with velocity $\mathbf{u}^\infty = \mathbf{A} \cdot \mathbf{x}$ past a circular cylinder of radius a centered at the origin, where $\mathbf{A}$ is a constant symmetric matrix with zero trace. (*a*) Derive the corresponding complex potential. (*b*) Compute the force and torque exerted on a cylinder centered at the origin using the Blasius formulas (7.10.13) and (7.10.14).

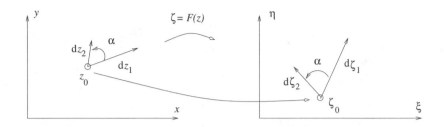

FIGURE 7.11.1   Conformal mapping of the z plane to the ζ plane using the mapping function $\zeta = F(z)$. The angle α subtended between two infinitesimal vectors is preserved after mapping.

7.10.2 *Flow due to a point source*

Explain in physical terms why it is not possible to find a complex potential associated with a point source inside a cylinder or any other closed surface. What will happen if a small perforation is introduced at the contour of the cylinder?

7.10.3 *Flow due to a point-source dipole in the presence of a cylinder*

Derive the complex potential due to a point-source dipole located (a) outside and (b) inside a circular cylinder.

7.11   Conformal mapping

Conformal mapping allows us to compute potential flow in a two-dimensional domain with arbitrary geometry from knowledge of a corresponding elementary flow in a domain with simple geometry. To develop the method, we introduce a complex variable, $\zeta = \xi + i\eta$, where ξ and η are two real variables, and a complex function $F(z)$ that is analytic in some region of the physical complex plane, where $z = x + iy$, and map a point in the physical plane to another point in the ζ plane so that

$$\zeta = F(z), \tag{7.11.1}$$

as shown in Figure 7.11.1. When the function $F(z)$ is multivalued, we introduce an appropriate branch cut in the z plane so as to render the mapping unique. The image of an open or closed line in the z plane is another open or closed line in the ζ plane with different orientation and shape.

Inverse mapping function

It will be necessary to also introduce the inverse function mapping a point in the ζ plane back to a point in the z plane,

$$z = f(\zeta). \tag{7.11.2}$$

When the function $f(\zeta)$ is multi-valued, we introduce an appropriate branch cut in the ζ plane so as to render the inverse mapping unique.

Differential vectors

Differentiating $F(z)$ and $f(\zeta)$ with respect to z or ζ, we find that the ratio of two corresponding infinitesimal differential vectors in the z and ζ planes, dz and $d\zeta$, are related by

$$\frac{d\zeta}{dz} = F'(z) = |F'(z)| \exp\big(i\operatorname{Arg}\{F'(z)\}\big), \qquad \frac{dz}{d\zeta} = f'(\zeta) = |f'(\zeta)| \exp\big(i\operatorname{Arg}\{f'(\zeta)\}\big), \quad (7.11.3)$$

where Arg denotes the argument of a complex number, defined as the polar angle around the origin of the complex plane. Since the functions $F(z)$ and $f(\zeta)$ are analytic, the derivatives in equations (7.11.3) depend on z or ζ but are independent of the orientation of corresponding differential pairs, dz and $d\zeta$.

The first equation in (7.11.3) states that the length of the differential vector $d\zeta$ is equal to the length of dz multiplied by the scalar factor $|F'(z)|$. The direction of $d\zeta$ is rotated with respect to that of dz by an angle that is equal to the argument of $F'(z)$. A singular behavior is expected at a critical point of the mapping (7.11.1) where $F'(z)$ becomes infinite or, equivalently, $f'(z)$ is zero. A similar interpretation of the differentials pertains to the second equation in (7.11.3).

A horse is a horse, of course, of course

Let us select a point z_0 in the z plane and draw two infinitesimal vectors, dz_1 and dz_2, that start at the chosen point, z_0, as shown in Figure 7.11.1. The images of these vectors form a pair of corresponding vectors in the ζ plane, $d\zeta_1$ and $d\zeta_2$, starting at ζ_0. Using the first or second equation in (7.11.3), we find that the angle subtended between the vectors dz_1 and dz_2 is the same as the angle subtended between the vectors $d\zeta_1$ and $d\zeta_2$, except if z_0 happens to be a critical point. The equality of the angles can be traced back to the analyticity of the complex mapping function $F(z)$. As a consequence, the image of a tiny loop having the shape of a tiny little donkey in the z plane will look like a tiny little donkey in the ζ plane, possibly rotated and amplified or shrunk, but definitely looking like a donkey. This property justifies calling the mapping (7.11.1) and its inverse conformal.

7.11.1   Cauchy–Riemann equations

Since the real and imaginary parts of the the mapping function $F(z)$ satisfy the Cauchy–Riemann equations for an analytical complex function, we can write $\partial F_R/\partial x = \partial F_I/\partial y$ and $\partial F_R/\partial y = -\partial F_I/\partial x$, where the subscripts R and I denote the real and imaginary parts. Equivalent statements are

$$\frac{\partial \xi}{\partial x} = \frac{\partial \eta}{\partial y}, \qquad \frac{\partial \xi}{\partial y} = -\frac{\partial \eta}{\partial x}. \qquad (7.11.4)$$

As a consequence, ξ and η are harmonic functions of x and y,

$$\frac{\partial^2 \xi}{\partial x^2} + \frac{\partial^2 \xi}{\partial y^2} = 0, \qquad \frac{\partial^2 \eta}{\partial x^2} + \frac{\partial^2 \eta}{\partial y^2} = 0. \qquad (7.11.5)$$

Using equations (7.11.4), we derive the orthogonality condition

$$\frac{\partial \boldsymbol{\chi}}{\partial x} \cdot \frac{\partial \boldsymbol{\chi}}{\partial y} = 0, \qquad (7.11.6)$$

where $\boldsymbol{\chi}$ is the real position vector in the ζ plane.

The Cauchy–Riemann equations associated with the inverse mapping function, $f(\zeta)$, take the corresponding forms

$$\frac{\partial x}{\partial \xi} = \frac{\partial y}{\partial \eta}, \qquad \frac{\partial x}{\partial \eta} = -\frac{\partial y}{\partial \xi}. \tag{7.11.7}$$

As a consequence, x and y are harmonic functions of ξ and η,

$$\frac{\partial^2 x}{\partial \xi^2} + \frac{\partial^2 x}{\partial \eta^2} = 0, \qquad \frac{\partial^2 y}{\partial \xi^2} + \frac{\partial^2 y}{\partial \eta^2} = 0. \tag{7.11.8}$$

Using equations (7.11.7), we derive the orthogonality condition

$$\frac{\partial \mathbf{x}}{\partial \xi} \cdot \frac{\partial \mathbf{x}}{\partial \eta} = 0, \tag{7.11.9}$$

where $\mathbf{x}$ is the real position vector in the z plane. This condition is the cornerstone of orthogonal grid generation by conformal mapping, which involves producing the images of lines of constant ξ and constant η and identifying their intersections as computational nodes (e.g., [317]).

7.11.2   Elementary conformal mapping functions

It is instructive to consider elementary conformal mapping functions that find frequent application in the theory of potential flow. An extensive discussion can be found in texts on complex analysis (e.g., [4]).

Linear function with shift

The linear function with shift,

$$\zeta = F(z) = az + b, \tag{7.11.10}$$

multiplies the distance from the origin, $|z|$, by the scalar factor $|a|$, rotates a line connecting the origin to a point z by an angle that is equal to the argument of a, and then shifts every point by the complex number b. Circles in the z plane remain circles in the ζ plane. The inverse mapping function,

$$z = f(\zeta) = \frac{\zeta - b}{a}, \tag{7.11.11}$$

performs a similar transformation.

Möbius transformation

The Möbius transformation, also called the partial fractional transformation,

$$\zeta = F(z) = \frac{Az + B}{Cz + D}, \tag{7.11.12}$$

maps the whole complex plane onto itself, where A–D are four complex constants. Circles and straight lines in the z plane are mapped to circles or straight lines in the ζ plane, and *vice versa*. When $C = 0$, we obtain a linear transformation with shift. The inverse transformation,

$$z = f(\zeta) = \frac{D\zeta - B}{-C\zeta + A},$$
(7.11.13)

performs a similar function.

Exponential function

The exponential function,

$$\zeta = F(z) = \lambda \exp \frac{2\pi z}{b},$$
(7.11.14)

where b and λ are two real and positive constants, maps the semi-infinite horizontal strip $x > 0$ and $0 < y < b$ to the exterior a circle of radius λ centered at the origin, as shown in Figure 7.11.2(a). The vertical side of the strip is mapped to the circle, and the two semi-infinite horizontal sides are mapped to a semi-infinite section of the ξ axis, $\xi > \lambda$. The inverse mapping function,

$$z = f(\zeta) = \frac{b}{2\pi} \ln \frac{\zeta}{\lambda},$$
(7.11.15)

becomes unique by introducing a branch cut along the positive ξ axis, $\xi > \lambda$, and specifying that the argument of ζ takes values in the range $[0, 2\pi)$.

The same exponential function (7.11.14) maps the rectangle confined between $0 < x < a$ and $0 < y < b$ to an annulus with inner radius λ and outer radius

$$\mu = \lambda \exp \frac{2\pi a}{b},$$
(7.11.16)

as shown in Figure 7.11.2(b). As the aspect ratio a/b tends to infinity, the annulus reduces to the exterior of a disk of radius λ.

The exponential function,

$$\zeta = F(z) = \lambda \exp\left(-\frac{2\pi z}{b}\right),$$
(7.11.17)

maps the semi-infinite horizontal strip $x > 0$ and $0 < y < b$ to a disk of radius λ centered at the origin of the ζ plane. The left infinity of the strip is mapped to the center of the disk. The inverse mapping function,

$$z = f(\zeta) = -\frac{b}{2\pi} \ln \frac{\zeta}{\lambda},$$
(7.11.18)

becomes unique by introducing a branch cut along the positive ξ axis, $0 < \xi < \lambda$, and specifying that the argument of ζ takes values in the range $[0, 2\pi)$.

(a)

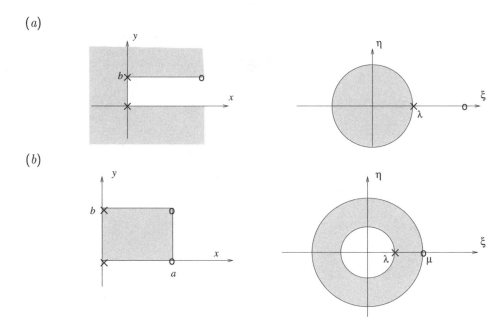

(b)

FIGURE 7.11.2   The exponential function maps (a) a semi-infinite strip to the exterior of a circular disk, and (b) a rectangle to an annulus.

Logarithmic function

The logarithmic function,

$$\zeta = F(z) = \frac{b}{2\pi} \ln \frac{z}{a}, \tag{7.11.19}$$

where a and b are two real and positive constants, maps the z plane to the infinite horizontal strip $-\infty < \xi < \infty$, and $0 < \eta < b$. The mapping becomes unique by introducing a branch cut along the positive x axis. The inverse mapping function is

$$z = a \exp \frac{2\pi\zeta}{b}. \tag{7.11.20}$$

The same function (7.11.19) maps the upper half z plane to the infinite horizontal strip $-\infty < \xi < \infty$, and $0 < \eta < \frac{1}{2}b$.

7.11.3   Gradient of a scalar

The gradient of an arbitrary function $g(x,y)$ in the z plane or $g(\xi, \eta)$ in the ζ plane is a vector with components

$$\nabla g \equiv \left(\frac{\partial g}{\partial x}, \ \frac{\partial g}{\partial y} \right), \qquad \widehat{\nabla} g \equiv \left(\frac{\partial g}{\partial \xi}, \ \frac{\partial g}{\partial \eta} \right). \tag{7.11.21}$$

Using the chain rule, we find that

$$\frac{\partial g}{\partial x} = \frac{\partial g}{\partial \xi}\frac{\partial \xi}{\partial x} + \frac{\partial g}{\partial \eta}\frac{\partial \eta}{\partial x}, \qquad \frac{\partial g}{\partial y} = \frac{\partial g}{\partial \xi}\frac{\partial \xi}{\partial y} + \frac{\partial g}{\partial \eta}\frac{\partial \eta}{\partial y}, \tag{7.11.22}$$

which can be collected into the vector form

$$\nabla g = \boldsymbol{\mathcal{H}} \cdot \widehat{\nabla} g, \tag{7.11.23}$$

where $\boldsymbol{\mathcal{H}}$ is the Jacobian matrix,

$$\boldsymbol{\mathcal{H}} \equiv \left[\begin{array}{cc} \partial\xi/\partial x & \partial\eta/\partial x \\ \partial\xi/\partial y & \partial\eta/\partial y \end{array} \right]. \tag{7.11.24}$$

Transformation metric

Using the Cauchy–Riemann equations (7.11.4), we derive an expression for the determinant of the Jacobian matrix,

$$h^2 \equiv \det(\boldsymbol{\mathcal{H}}) = \left(\frac{\partial\xi}{\partial x}\right)^2 + \left(\frac{\partial\xi}{\partial y}\right)^2 = \left(\frac{\partial\eta}{\partial x}\right)^2 + \left(\frac{\partial\eta}{\partial y}\right)^2 = |F'(z)|^2, \tag{7.11.25}$$

where $h = |F'(z)|$ is the metric of the transformation. The inverse of the Jacobian matrix is

$$\boldsymbol{\mathcal{H}}^{-1} = \frac{1}{h^2}\,\boldsymbol{\mathcal{H}}^T, \tag{7.11.26}$$

where the superscript T denotes the matrix transpose. We conclude that

$$\boldsymbol{\mathcal{H}} \cdot \boldsymbol{\mathcal{H}}^T = h^2 \mathbf{I}, \tag{7.11.27}$$

where $\mathbf{I}$ is the identity matrix.

Gradient projection

Using (7.11.23), we find that the projection of the gradients of two arbitrary functions, g_1 and g_2, at corresponding points is given by

$$(\nabla g_1) \cdot (\nabla g_2) = (\boldsymbol{\mathcal{H}} \cdot \widehat{\nabla} g_1) \cdot (\boldsymbol{\mathcal{H}} \cdot \widehat{\nabla} g_2) = \widehat{\nabla} g_1 \cdot (\boldsymbol{\mathcal{H}}^T \cdot \boldsymbol{\mathcal{H}}) \cdot \widehat{\nabla} g_2 = h^2\,(\widehat{\nabla} g_1) \cdot (\widehat{\nabla} g_2). \tag{7.11.28}$$

Setting $g_1 = g_2 = g$, we find that

$$|\nabla g| = h\,|\widehat{\nabla} g|. \tag{7.11.29}$$

Expression (7.11.28) reveals that, if the vectors ∇g_1 and ∇g_2 are orthogonal, $(\nabla g_1) \cdot (\nabla g_2) = 0$, the vectors $\widehat{\nabla} g_1$ and $\widehat{\nabla} g_2$ are also orthogonal, $(\widehat{\nabla} g_1) \cdot (\widehat{\nabla} g_2) = 0$.

7.11.4 Laplacian of a scalar

The Laplacian of an arbitrary function g in the z or ζ plane is defined as

$$\nabla^2 g \equiv \nabla \cdot \nabla g = \frac{\partial^2 g}{\partial x^2} + \frac{\partial^2 g}{\partial y^2}, \qquad \widehat{\nabla}^2 g \equiv \widehat{\nabla} \cdot \widehat{\nabla} g = \frac{\partial^2 g}{\partial \xi^2} + \frac{\partial^2 g}{\partial \eta^2}. \tag{7.11.30}$$

Working as for the gradient using the chain rule, we find that

$$\nabla^2 g = h^2 \widehat{\nabla}^2 g \tag{7.11.31}$$

(see also Section A.9, Appendix A). A function g that satisfies Laplace's equation in the xy plane, $\nabla^2 g = 0$, will also satisfy Laplace's equation in the $\xi\eta$ plane, $\widehat{\nabla}^2 g = 0$.

Working in a similar fashion, we find that

$$\nabla^4 g = h^4 \widehat{\nabla}^4 g + 4 |F''(z)|^2 \widehat{\nabla}^2 g. \tag{7.11.32}$$

A function g that satisfies the biharmonic equation in the xy plane, $\nabla^4 g = 0$, will not necessarily satisfy the biharmonic equation in the $\xi\eta$ plane, $\widehat{\nabla}^4 g \neq 0$.

Convection–diffusion in irrotational flow

As an application, we consider a steady temperature field, $T(x, y)$, governed by the steady convection–diffusion equation in a two-dimensional irrotational flow with velocity $\mathbf{u} = \nabla\phi$,

$$u_x \frac{\partial T}{\partial x} + u_y \frac{\partial T}{\partial y} = \mathbf{u} \cdot \nabla T = (\nabla\phi) \cdot (\nabla T) = \kappa \nabla^2 T, \tag{7.11.33}$$

where κ is the thermal diffusivity. Using the preceding relations for the gradient and Laplacian, we find that $T(\xi, \eta)$ satisfies the same equation in the $\xi\eta$ plane,

$$(\widehat{\nabla}\phi) \cdot (\widehat{\nabla} T) = \kappa \, \widehat{\nabla}^2 T. \tag{7.11.34}$$

Accordingly, a solution in the transformed plane can be used to produce a solution in the physical plane, and *vice versa*, subject to appropriate boundary conditions [26].

7.11.5 Flow in the ζ plane

The complex potential of an irrotational flow in the physical xy plane, $w(z) = \phi + i\psi$, is a function of the corresponding complex variable, $z = x + iy$, where ϕ is the potential function and ψ is the stream function. Since to every point in the z plane we may assign a corresponding point in the ζ plane according to (7.11.1), we may also regard w a function of ζ. Combining (7.10.2) with (7.11.2), we write

$$w(z) = w[f(\zeta)] \equiv W(\zeta) = \Phi(\xi, \eta) + i\,\Psi(\xi, \eta), \tag{7.11.35}$$

where Φ and Ψ are two real functions of ξ and η. Because an analytic function of another analytic function is also analytic, the functions Φ and Ψ are harmonic with respect to their arguments, that is, they satisfy Laplace's equation

$$\frac{\partial^2 \Phi}{\partial \xi^2} + \frac{\partial^2 \Phi}{\partial \eta^2} = 0, \qquad \frac{\partial^2 \Psi}{\partial \xi^2} + \frac{\partial^2 \Psi}{\partial \eta^2} = 0. \tag{7.11.36}$$

Moreover, Ψ and Ψ constitute a pair of conjugate harmonic functions.

These properties allow us to identify Φ with the potential function and Ψ with the stream function of a flow in the ζ plane, so that

$$\frac{dW}{d\zeta} = U_\xi - iU_\eta,$$

(7.11.37)

where $\mathbf{U} = (U_\xi, U_\eta)$ is the velocity in the ζ plane. To find the relation between the magnitude and direction of the velocity in the z plane, $\mathbf{u}$, and the corresponding velocity in the ζ plane, $\mathbf{U}$, we use the chain rule to write

$$\frac{dW}{d\zeta} = \frac{dw}{dz}\frac{dz}{d\zeta} = \frac{dw}{dz}f'(\zeta),$$

(7.11.38)

which shows that the ratio of the two complex velocities is given by

$$\frac{U_\xi - iU_\eta}{u_x - iu_y} = f'(\zeta) = |f'(\zeta)| \exp\left(i\arg[f'(\zeta)]\right).$$

(7.11.39)

Equation (7.11.39) allows us to compute $\mathbf{u}$ from $\mathbf{U}$, and *vice versa*, in terms of the mapping function $f(\zeta)$. Comparing (7.11.39) with the second equation in (7.11.3), we find that differential vectors and velocities are amplified in inverse proportion so that the flow rates across corresponding differential line elements are the same in both planes.

7.11.6 Flow due to singularities

It is interesting to investigate the nature of the flow in the ζ plane due to a singularity in the z plane, such as a point source or a point vortex. This can be done by expanding the harmonic potential $W(\zeta)$ in a Taylor series about a point ζ_0, which is the image of the pole of the singularity in the z plane, z_0.

In the case of a point source in the z plane, we choose a reference length, a, and write

$$W(\zeta) = w[f(\zeta)] = w(z) = \frac{m}{2\pi}\ln\frac{z-z_0}{a} = \frac{m}{2\pi}\ln\frac{f(\zeta)-f(\zeta_0)}{a}.$$

(7.11.40)

Expanding $f(\zeta)$ in a Taylor series about the singular point ζ_0, we find we find that

$$W(\zeta) \simeq \frac{m}{2\pi}\ln[f'(\zeta_0)\frac{\zeta-\zeta_0}{a}] \simeq \frac{m}{2\pi}\ln\frac{\zeta-\zeta_0}{a},$$

(7.11.41)

which shows that the flow in the ζ plane contains a point source with identical strength placed at the image point, ζ_0. An exception occurs when the point z_0 happens to be a critical point of the conformal mapping. Replacing m with $-i\kappa$, we obtain a corresponding result for a point vortex with strength κ.

In the case of a point-source dipole with complex strength $d = d_x + id_y$, we find that

$$W(\zeta) \simeq -\frac{d}{2\pi}\frac{1}{f'(\zeta_0)}\frac{1}{\zeta-\zeta_0},$$

(7.11.42)

which shows that the flow in the ζ plane contains a point-source dipole with modified strength and orientation. Replacing d with $-i\lambda$ we obtain a corresponding result for a point-vortex dipole.

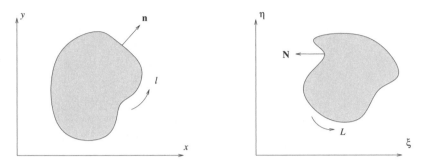

FIGURE 7.11.3   Mapping of a flow domain in the z plane to a corresponding domain in the ζ plane. An impermeable boundary in the z plane remains impermeable in the ζ plane.

7.11.7   Transformation of boundary conditions

Consider the boundary of an irrotational flow in the z plane and the corresponding boundary of the flow in the ζ plane computed using the conformal mapping (7.11.1). It is clear from (7.11.35) that corresponding values of $w(z)$ and $W(\zeta)$ are identical. Consequently, a Dirichlet boundary condition that specifies the real or imaginary part of $w(z)$ or $W(z)$ is preserved after mapping.

The Neumann boundary condition remains a Neumann boundary condition but undergoes a quantitative transformation. To see this, we express the normal unit vector in the z plane as $\mathbf{n} = (dy/dl, -dx/dl)$ and apply the chain rule to obtain

$$\mathbf{n} \cdot \nabla \phi = \frac{\partial \phi}{\partial x} \frac{dy}{dl} - \frac{\partial \phi}{\partial y} \frac{dx}{dl} = \frac{\partial \phi}{\partial \xi} \left(\frac{\partial \xi}{\partial x} \frac{dy}{dl} - \frac{\partial \xi}{\partial y} \frac{dx}{dl} \right) + \frac{\partial \phi}{\partial \eta} \left(\frac{\partial \eta}{\partial x} \frac{dy}{dl} - \frac{\partial \eta}{\partial y} \frac{dx}{dl} \right), \qquad (7.11.43)$$

where l is the arc length around the boundary, as shown in Figure 7.11.3. Now using the Cauchy–Riemann equations, we find that

$$\mathbf{n} \cdot \nabla \phi = \frac{\partial \phi}{\partial \xi} \left(\frac{\partial \eta}{\partial y} \frac{dy}{dl} + \frac{\partial \eta}{\partial x} \frac{dx}{dl} \right) + \frac{\partial \phi}{\partial \eta} \left(-\frac{\partial \xi}{\partial y} \frac{dy}{dl} - \frac{\partial \xi}{\partial y} \frac{dx}{dl} \right) = \frac{\partial \phi}{\partial \xi} \frac{d\eta}{dl} - \frac{\partial \phi}{\partial \eta} \frac{d\xi}{dl} \qquad (7.11.44)$$

and then

$$\mathbf{n} \cdot \nabla \phi = \frac{dL}{dl} \mathbf{N} \cdot \widehat{\nabla} \Phi = |F'(z)| \, \mathbf{N} \cdot \widehat{\nabla} \Phi, \qquad (7.11.45)$$

where $\mathbf{N}$ is the normal unit vector pointing into the flow in the ζ plane and L is the arc length along the transformed boundary, as shown in Figure 7.11.3. Equation (7.11.45) ensures that an impermeable boundary in the z plane where $\mathbf{n} \cdot \nabla \phi = 0$ remains impermeable in the ζ plane, that is, $\mathbf{N} \cdot \widehat{\nabla} \Phi = 0$.

7.11.8   Streaming flow past a circular cylinder

As an application, we consider uniform flow in the z plane past a circular cylinder of radius a centered at the origin, with arbitrary circulation around the cylinder. The corresponding complex

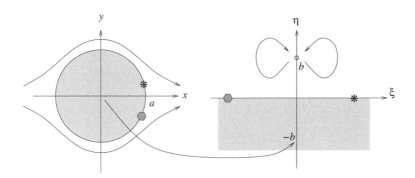

FIGURE 7.11.4 The exterior of a circle of radius a in the z plane is mapped to the upper half-plane using the linear fractional transformation (7.11.47). Corresponding points are shown with the same symbol. Uniform flow past a cylinder with nonzero circulation around the cylinder in the z plane transforms to flow due to a point-source dipole and a point vortex supplemented by their images in the ζ plane. Streamlines for zero circulation are shown.

potential is given by (7.10.21) with $z_c = 0$,

$$w(z) = U^*z + Ua\,\frac{a}{z} + \frac{\kappa}{2\pi i}\,\ln\frac{z}{a}. \tag{7.11.46}$$

The following linear fractional transformation and its inverse map the exterior or interior of a circle of radius a in the z plane to the upper or lower half ζ plane, and *vice versa*,

$$\zeta = F(z) = ib\,\frac{z+a}{z-a}, \qquad z = f(\zeta) = a\,\frac{\zeta+ib}{\zeta-ib}, \tag{7.11.47}$$

where b is a real positive constant, as shown in Figure 7.11.4. Infinity is mapped to the first singular point $\zeta = ib$, and the center of the circle is mapped to the second singular point $\zeta = -ib$.

Substituting into (7.11.46) the inverse transformation given by the second expression in (7.11.47), we derive the complex potential of the flow in the ζ plane,

$$W(\zeta) = U^*a\,\frac{\zeta+ib}{\zeta-ib} + Ua\,\frac{\zeta-ib}{\zeta+ib} + \frac{\kappa}{2\pi i}\,\ln\frac{\zeta+ib}{\zeta-ib}. \tag{7.11.48}$$

After a straightforward manipulation, we obtain

$$W(\zeta) = (U^* + U)a + 2iab\left(\frac{U^*}{\zeta-ib} - \frac{U}{\zeta+ib}\right) + \frac{\kappa}{2\pi i}\,\ln\frac{\zeta+ib}{\zeta-ib}. \tag{7.11.49}$$

Referring to Table 7.10.1, we find that the flow in the ζ plane is represented by a point-source dipole with strength $d = -4\pi iabU^*$ placed at the first singular point, $\zeta = ib$, and a point vortex with strength κ placed at the second singular point, $\zeta = -ib$. Both singularities are accompanied by their images with respect to the wall to satisfy the no-penetration condition.

7.11.9   A point vortex inside or outside a circular cylinder

Consider a point vortex with strength κ located at the point ζ_v above a plane wall identified with the ξ axis in the ζ plane. The corresponding complex potential consists of the potential due to the point vortex and the potential due to its image with respect to the wall,

$$W(\zeta) = \frac{\kappa}{2\pi \mathrm{i}} \ln \frac{\zeta - \zeta_v}{\zeta - \zeta_v^*}.\tag{7.11.50}$$

Substituting the conformal mapping function given by the first expression in (7.11.47) and rearranging, we obtain the corresponding complex potential of the flow in the z plane,

$$w(z) = \frac{\kappa}{2\pi \mathrm{i}} \left(\ln \frac{(z+a)(z_v-a) - (z-a)(z_v+a)}{(z+a)(z_v^*-a) + (z-a)(z_v^*+a)} + \ln \frac{z_v^* - a}{z_v - a} \right).\tag{7.11.51}$$

Simplifying the first fraction and discarding the inconsequential constant last fraction, we obtain

$$w(z) = \frac{\kappa}{2\pi \mathrm{i}} \ln \frac{z_v - z}{zz_v^* - a^2} = \frac{\kappa}{2\pi \mathrm{i}} \left(\ln \frac{z_v - z}{z - a^2/z_v^*} - \ln(-z_v^*) \right).\tag{7.11.52}$$

Discarding the constant last term, we find that the flow consists of a primary point vortex in the z plane and a reflected point vortex with opposite strength located at the image of the point vortex with respect to the cylinder, in agreement with (7.10.32).

7.11.10   Applications of conformal mapping

To apply the theory of conformal mapping theory, we may consider a certain flow in the z plane and study the corresponding flow in the ζ plane subject to a chosen mapping function $\zeta = F(z)$. However, in a typical application, we are faced with the inverse problem where a function $z = f(\zeta)$ that maps a domain of flow with simple geometry in the ζ plane to a physical domain of a flow with a prescribed geometry in the z plane is required. Simple domains include the semi-infinite plane and the interior or exterior of a circular disk, as discussed in Section 7.12. The corresponding flow in the ζ plane should be available readily and preferably in closed form. The issue of existence and uniqueness of the forward mapping function is addressed by the Riemann mapping theorem.

7.11.11   Riemann's mapping theorem

Riemann's mapping theorem guarantees that any two simply connected domains can be conformally mapped onto one another. The reason is that any simply connected domain in the z plane can be mapped to a unit disk centered at the origin of the ζ plane. An indirect correspondence may then be established between the two regions in terms of the individual conformal mappings. Exceptions are the whole plane and the whole plane minus a single point.

A three-parameter family of functions $\zeta = F(z)$ can be found that map a simply connected domain, D, to another simply connected domain D'. To remove the three degrees of freedom, we generalize the definition of a point in the complex plane to include infinity and choose one of the following three sets of conditions:

1. Stipulate that an arbitrary point, z_0, inside D is mapped to the origin, $\zeta = 0$, and specify the direction of an infinitesimal vector starting at z_0 whose image lies on the ξ axis.

2. Stipulate that an arbitrary point, z_0, inside D is mapped to the origin $\zeta = 0$, and another arbitrary point ζ_1 at the boundary of D is mapped to an arbitrary point ζ_1 at the boundary of D'.

3. Stipulate that three arbitrary points, z_0, z_1, and z_2, at the boundary of D are mapped in the same order to three arbitrary corresponding points ζ_0, ζ_1, and ζ_2 at the boundary of D'.

As an example, the Möbius transformation (7.11.12) is defined by specifying the three ratios A/D, B/D, and C/D.

Problems

7.11.1 *Conformal mapping*

Discuss the properties of the conformal mapping $\zeta = F(z) = az^n$, where a is a complex constant and n is a real constant.

7.11.2 *Properties of conjugate flows*

Consider an open contour in the z plane and the corresponding contour in the ζ plane. Show that the flow rate across and circulation along these contours corresponding to the complex potentials $w(z)$ and $W(\zeta)$ are identical.

7.11.3 *A point vortex inside or outside a cylindrical surface*

Assume that the function $\zeta = F(z)$ maps the interior or exterior of a simply connected domain D enclosed by an impenetrable boundary to the interior or exterior of disk of radius a centered at the origin. Show that the complex potential associated with a point vortex with strength κ located at z_0 is

$$w(z) = \frac{\kappa}{2\pi i} \ln \Big(\frac{F(z) - F(z_0)}{a - F(z)F(z_0^*)/a} \Big). \tag{7.11.53}$$

7.11.4 *Green's function in the presence of a cylinder with arbitrary cross-section*

Let the function $\zeta = F(z)$ map the interior of a simply-connected domain D to a disk of radius a centered at the origin. Show that the complex potential corresponding to the Green's function of Laplace's equation in the exterior of D is

$$w(z) = -\frac{1}{2\pi} \ln \Big(\frac{F(z) - F(z_0)}{a - F(z)F(z_0^*)/a} \Big). \tag{7.11.54}$$

7.11.5 *Laplacian and biharmonic operators*

Prove equations (7.11.31) and (7.11.32).

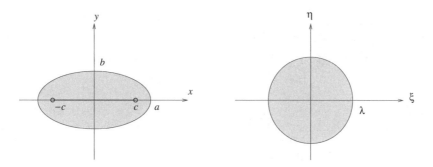

FIGURE 7.12.1   Mapping the exterior of an ellipse with major and minor semiaxes a and b in the z plane to the exterior of a circle of radius λ in the ζ plane.

7.12   Applications of conformal mapping to flow past a two-dimensional body

To compute a streaming flow past a two-dimensional body with arbitrary cross-section, we map the exterior of the body in the z plane to the exterior of a disk of radius λ centered at the origin of the ζ plane. The flow in the physical z plane is recovered from the flow in the ζ plane using the exact solution (7.10.21),

$$W(\zeta) = U^*\zeta + U\frac{\lambda^2}{\zeta} + \frac{\kappa}{2\pi i}\ln\frac{\zeta}{\lambda}. \tag{7.12.1}$$

To ensure that the far flows in the two complex planes behave in a similar manner so that uniform flow far from the disk in the ζ plane is uniform flow far from the body in the z plane, we require that the mapping function, $\zeta = F(z)$, and its inverse, $z = f(\zeta)$, are linear functions far from the body, as $|z|$ or $|\zeta|$ tends to infinity, so that their first derivatives tend to a constant.

7.12.1   Flow past an elliptical cylinder

Consider uniform flow past an elliptical cylinder centered at the origin of the z plane with major semiaxis equal to a and minor semiaxis equal to b, where $b < a$, as shown in Figure 7.12.1. The focal length of the cylinder, c, and the dimensionless eccentricity, e, are defined as

$$c = \sqrt{a^2 - b^2} = ae, \qquad e = \left[1 - \left(\frac{b}{a}\right)^2\right]^{1/2}. \tag{7.12.2}$$

The inverse mapping function,

$$z = f(\zeta) = \zeta + \frac{1}{4}\frac{c^2}{\zeta}, \tag{7.12.3}$$

projects the exterior of a disk of radius $\lambda = \frac{1}{2}(a + b)$ centered at the origin of the ζ plane to the exterior of the ellipse in the z plane. Since $f(\zeta)$ exhibits the required linear behavior at infinity, it is acceptable for the study of uniform flow.

Decomposing (7.12.3) into its real and imaginary parts, we obtain the explicit coordinate transformations

$$x = \xi \left(1 + \frac{1}{4} \frac{c^2}{|\zeta|^2}\right), \qquad y = \eta \left(1 - \frac{1}{4} \frac{c^2}{|\zeta|^2}\right), \qquad (7.12.4)$$

where c is the focal length and $|\zeta|^2 = \zeta \zeta^* = \xi^2 + \eta^2$. The forward transformation mapping the exterior of the ellipse to the exterior of the disk is found by solving the quadratic equation (7.12.3) for ζ. We note that the root with the negative sign corresponds to a point inside the ellipse and choose the root with the positive sign to obtain

$$\zeta = F(z) = \frac{1}{2} \left(z + \sqrt{z^2 - c^2}\right). \qquad (7.12.5)$$

The square root on the right-hand side becomes unique by introducing a branch cut along the x axis between the two focal points, extending from $-c$ to c.

The complex potential of the flow in the z plane is found by substituting (7.12.5) into (7.12.1) and setting $\lambda = \frac{1}{2}(a+b)$ to obtain

$$w(z) = \frac{1}{2} U^* \left(z + \sqrt{z^2 - c^2}\right) + \frac{1}{2} U \frac{(a+b)^2}{z + \sqrt{z^2 - c^2}} + \frac{\kappa}{2\pi i} \ln \frac{z + \sqrt{z^2 - c^2}}{a+b}. \qquad (7.12.6)$$

When $a = b$, we recover the complex potential for flow past a circular cylinder of radius a.

7.12.2 Flow past a flat plate

As the aspect ratio of the ellipse decreases, $b/a \to 0$, we obtain a plate with length $2a$ aligned with the x axis. The transformation (7.12.3) reduces to

$$z = f(\zeta) = \zeta + \frac{1}{4} \frac{a^2}{\zeta}, \qquad (7.12.7)$$

mapping the exterior of a disk of radius $\lambda = \frac{1}{2} a$ centered at the origin of the ζ plane to the whole complex z plane. The circular contour of the disk is mapped to the flat plate. The inverse transformation (7.12.5) with $c = a$ becomes

$$\zeta = F(z) = \frac{1}{2} \left(z + \sqrt{z^2 - a^2}\right), \qquad (7.12.8)$$

where the branch cut of the square root coincides with the length of the plate. A different method of deriving (7.12.7) is discussed in Problem 7.13.1.

Substituting (7.12.8) into (7.12.1) and setting $\lambda = \frac{1}{2}a$, or else setting in (7.12.6) $b = 0$, provides us with the complex potential of the flow in the z plane,

$$w(z) = \frac{1}{2} U^* \left(z + \sqrt{z^2 - a^2}\right) + \frac{1}{2} U \frac{a^2}{z + \sqrt{z^2 - a^2}} + \frac{\kappa}{2\pi i} \ln \frac{z + \sqrt{z^2 - a^2}}{a}. \qquad (7.12.9)$$

Simplifying the first two terms on the right-hand side, we obtain

$$w(z) = U_x z - \mathrm{i}\, U_y \sqrt{z^2 - a^2} + \frac{\kappa}{2\pi \mathrm{i}} \ln \frac{z + \sqrt{z^2 - a^2}}{a}. \qquad (7.12.10)$$

The velocity field is given by

$$u_x - \mathrm{i}\, u_y = \frac{\mathrm{d}w}{\mathrm{d}z} = U_x - \mathrm{i}\, U_y\, \frac{z}{\sqrt{z^2 - a^2}} + \frac{\kappa}{2\pi \mathrm{i}}\, \frac{1}{\sqrt{z^2 - a^2}}. \qquad (7.12.11)$$

The tangential velocity on the upper or lower surfaces of the plate, indicated by a plus or minus superscript, is

$$u_x^{\pm}(x, y = 0)) = U_x \mp \left(U_y\, x + \frac{\kappa}{2\pi}\right) \frac{1}{\sqrt{a^2 - x^2}} \qquad (7.12.12)$$

for $-a \le x \le a$. We observe that the velocity diverges at both ends of the plate, $x = \pm a$. The force exerted on the plate follows from (7.10.17). Streamline patterns for several flow configurations are shown in Figure 7.12.2.

Kutta–Joukowski condition

The Kutta–Joukowski condition requires that the circulation established around the plate is such that the velocity is finite at the trailing edge and the two fluid streams merge smoothly above and below the airfoil (e.g., [99]). Physically, the action of viscosity during the unsteady startup process is such that, when the final potential flow state is established, viscous effects are significant only insofar as to ensure that the Kutta–Joukowski condition is satisfied. Equation (7.12.12) shows that, if the trailing edge is located at $x = -a$, the circulation around a flat plate necessary to satisfy the Kutta–Joukowski condition is

$$\kappa = 2\pi a U_y. \qquad (7.12.13)$$

Accordingly, when $\beta \equiv \kappa/(2\pi a U_y) = 1$, the streamlines merge smoothly at the trailing edge, as shown in Figure 7.12.2(*b*).

7.12.3   Joukowski's transformation

The study of potential flow past a two-dimensional body has been motivated to a large extent by applications in aircraft design. To investigate the performance of an aircraft, we must have available the structure of the nearly two-dimensional potential flow around an airfoil with a specified degree of circulation around the airfoil, including the circulation corresponding to the Kutta–Joukowski condition. Knowledge of the potential flow allows us to compute the viscous drag force exerted on the airfoil using the boundary-layer theory discussed in Chapter 8.

The analytical computation of potential flow past an airfoil with a specified shape is generally intractable. Early work in aerodynamics before the advent of high-speed computers concentrated on families of airfoil shapes generated by mapping a circle using carefully crafted transformations. One important family of shapes is generated by the Joukowski transformation,

$$z = f(\zeta) = \zeta + \frac{\sigma^2}{\zeta}, \qquad (7.12.14)$$

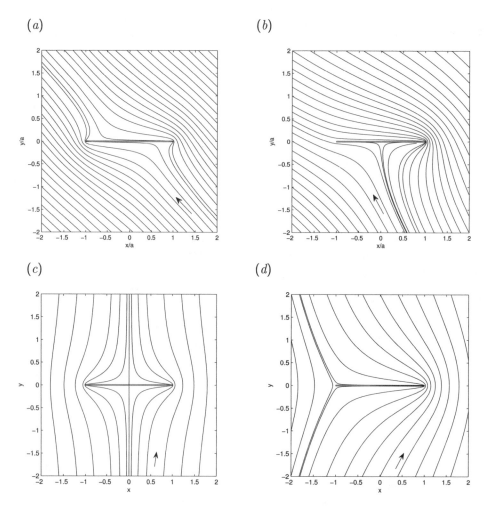

FIGURE 7.12.2   (a, b) Streaming flow past a flat plate with incident velocity $U_x = -U_y$ and circulation parameter (a) $\beta \equiv \kappa/(2\pi a U_y) = 0$ or (b) 1 (Kutta–Joukowski value). (c, d) Streamline patterns for incident velocity $U_x = 0$ and (c) $\beta = 0$ or (d) 1 (Kutta–Joukowski value).

where σ is a specified length, as shown in Figure 7.12.3. It will be noted that (7.12.14) includes as special cases (7.12.7) and (7.12.3), corresponding to the flat plate and the ellipse. An alternative version of (7.12.14) is

$$\frac{z - 2\sigma}{z + 2\sigma} = \left(\frac{\zeta - \sigma}{\zeta + \sigma}\right)^2. \qquad (7.12.15)$$

The critical points of the Joukowski transformation where $df/d\zeta = 0$ are located at $\zeta = \pm\sigma$, corresponding to $z = \pm 2\sigma$. A circle centered at the origin of the ζ plane and passing through both singular points is mapped to a flat plate with half length equal to $a = 2\sigma$ in the z plane. A smooth

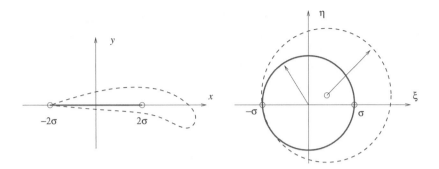

FIGURE 7.12.3   Mapping with the Joukowski transformation. The bold circle in the $\xi\eta$ plane transforms into a plate in the xy plane, and the dashed circle transforms into an airfoil.

curve in the ζ plane passing through the first singular point, $\zeta = -\sigma$, and enclosing the second singular point, $\zeta = \sigma$, is mapped to a cusped curve in the z plane. The cusp is located at the image of the first singular point, as shown in Figure 7.12.3.

In engineering practice, Joukowski's transformation is used to generate airfoils with a rounded leading edge and a cusped trailing edge. The airfoils are the images of circles in the ζ plane passing through the first singular point, $\zeta = -\sigma$, and enclosing the second singular point, $\zeta = \sigma$, as shown in Figure 7.12.3. When the center of the circle lies on the ξ axis, the airfoil is symmetric about the x axis. When the center of the circle is located in the first quadrant, the airfoil is downward cambered. A generalization of the Joukowski transformation is discussed in Problem 7.12.1. Other generalizations describing multi-parameter airfoils are available (e.g., [293], pp. 72–79).

7.12.4   Arbitrary shapes

Next, we address the more challenging problem of computing a function $z = f(\zeta)$ that maps the exterior of a circle with a specified radius λ in the ζ plane to the exterior of a body with arbitrary shape in the z plane. The mapping function for the ellipse shown in (7.12.3) and the mapping function of the Joukowski airfoil shown in (7.12.14) appear as truncated Laurent series. For a body with general shape, we expect the full infinite expansion,

$$z = f(\zeta) = \zeta + a^{(0)} + \sum_{n=1}^{\infty} a^{(n)} \frac{\lambda^n}{\zeta^n}, \tag{7.12.16}$$

where $a^{(n)}$ are complex coefficients determined by the body shape and orientation. The linear term on the right-hand side of (7.12.16) satisfies the requirement that $\mathrm{d}f/\mathrm{d}\zeta$ tends to unity as ζ tends to infinity. The constant $a^{(0)}$ determines the relative position of the body in the z plane and can be set to zero. A finite set of subsequent coefficients can be computed by stipulating that a collection of points around the body contour are mapped to a corresponding collection of points on the circular contour in the ζ plane. However, the precise location of the image points must be computed as a part of the solution.

Alternatively, nonstandard body shapes can be generated by keeping selected terms in the sum on the right-hand side of (7.12.16). For example, to obtain rounded squares, we set all coefficients to zero except for the third coefficient, $a^{(3)} < 0$, according to the Lewis transformation. An infinite sum that yields a perfect square is available (e.g., [210]).

Numerical methods

To formalize a numerical method, we map the exterior of a smooth body to the exterior of the disk of radius λ centered at the origin of the ζ plane. The position around the circular contour is $\zeta = \lambda \exp(i\varphi)$, where φ is the polar angle and i is the imaginary unit. Equation (7.12.16) provides us with a Fourier expansion,

$$z(\varphi) = \lambda e^{i\varphi} + a^{(0)} + \sum_{n=1}^{\infty} a^{(n)} e^{-ni\varphi}. \tag{7.12.17}$$

Separating the real part (R) from the imaginary part (I), we obtain

$$x(\varphi) = a_R^{(0)} + (\lambda + a_R^{(1)}) \cos\varphi + a_I^{(1)} \sin\varphi + \sum_{n=2}^{\infty} \left(a_R^{(n)} \cos(n\varphi) + a_I^{(n)} \sin(n\varphi) \right) \tag{7.12.18}$$

and

$$y(\varphi) = a_I^{(0)} + a_I^{(1)} \cos\varphi + (\lambda - a_R^{(1)}) \sin\varphi + \sum_{n=2}^{\infty} \left(a_I^{(n)} \cos(n\varphi) - a_R^{(n)} \sin(n\varphi) \right). \tag{7.12.19}$$

Using the Fourier orthogonality properties, we find that

$$\lambda = \frac{1}{2\pi} \int_0^{2\pi} \left(x(\varphi) \cos\varphi + y(\varphi) \sin\varphi \right) d\varphi, \tag{7.12.20}$$

and

$$a_R^{(n)} = \frac{1}{2\pi} \int_0^{2\pi} \left[x(\varphi) \cos(n\varphi) - y(\varphi) \sin(n\varphi) \right] d\varphi,$$

$$a_I^{(n)} = \frac{1}{2\pi} \int_0^{2\pi} \left[x(\varphi) \sin(n\varphi) + y(\varphi) \cos(n\varphi) \right] d\varphi \tag{7.12.21}$$

for $n \geq 0$. The numerical method involves the following steps:

1. Trace the contour of the body with N points distributed in the counterclockwise direction, located at (x_i, y_i), for $i = 1, \ldots, N$.

2. Assign values for φ to the marker points, φ_i.

3. Reconstruct the functions $x(\varphi)$ and $y(\varphi)$ by cubic-spline interpolation.

4. Compute λ and the coefficients $a^{(n)}$ for $n = 0, 1, \ldots, n_{max}$ by performing the integration in (7.12.20) and (7.12.21) analytically or using the trapezoidal rule implemented by a fast Fourier transform, where n_{max} is a specified truncation level (e.g., [317]).

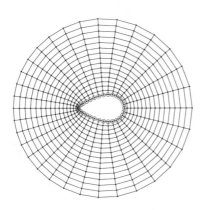 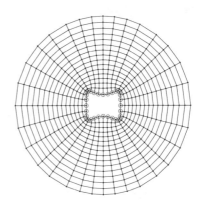

FIGURE 7.12.4 Grids of nodes generated by mapping the exterior of a body described by a chain a marker points, indicated by circles, to the exterior of a disk.

5. Compute the right-hand side of expressions (7.12.18) and (7.12.19), and formulate the differences between the specified and computed coordinates, $\Delta x_i \equiv x_i - x(\varphi_i)$ and $\Delta y_i \equiv y_i - y(\varphi_i)$.

6. Adjust the values φ_i to either drive the squares of the distances $|\Delta z_i|^2$ to zero or minimize the positive functional $\sum_{i=1}^{N} |\Delta z_i|^2$, using, for example, Newton's method.

Grids of points generated using the minimization method for truncation level $n_{max} = N/2$ are shown in Figure 7.12.4. The radial lines pass through a chain of marker points distributed around the body. The difference between the specified and reconstructed body contour diminishes rapidly as more terms are retained in the expansion.

Schwarz–Christoffel and other transformations

The Schwarz–Christoffel transformation allows us to map the exterior of an arbitrary body with polygonal shape in the z plane to the upper half ζ plane or exterior of a circle in the ζ plane. The mapping function is known in differential or integral flow in terms of a set of scalar coefficients. The particular form of the mapping function and the computation of the pertinent coefficients is discussed in Section 7.13. Generalizations of the Schwarz–Christoffel transformation to polygons with rounded edges, polygons consisting of circular arcs, and bodies with smooth shapes are available.

Another way of obtaining the mapping function is by mapping the exterior of a body of interest to the interior of an auxiliary body using the inverse transformation $\zeta = 1/z$, where the origin lies inside the body. At the second stage, we map the interior of the auxiliary body to the upper half-plane or interior of a disk. However, even if the original body has a polygonal shape, the auxiliary body will have a curved shape, which precludes the application of specialized methods for polygons. Other methods of computing a function that maps the interior of a body to a disk involve solving an integral equation over the body contour and developing a variational formulation (e.g., [65]).

Orthogonal grid generation

In Section 7.1, we mentioned that the mapping function, $z = f(\zeta)$, is useful for orthogonal grid generation. The function (7.12.16) maps the exterior of a circular disk with radius λ to the exterior of a body with arbitrary geometry. The exterior of the disk can be mapped further to a semi-infinite horizontal strip in the σ plane, $\sigma_R > \sigma_0$ and $0 < \sigma_I < 2\pi$, using the mapping function

$$\zeta = \lambda \exp(\sigma - \sigma_0), \tag{7.12.22}$$

where σ_0 is a specified threshold. The left end of the strip, $\sigma_R = \sigma_0$, is mapped to the contour of the body. An orthogonal grid can be generated by mapping lines of constant σ_R or constant σ_I inside the strip to the physical z plane.

For example, substituting (7.12.22) into the inverse mapping function (7.12.3) for an ellipse with major semiaxis a and minor semiaxis b, we obtain

$$z = a \cosh(\sigma - \sigma_0) + b \sinh(\sigma - \sigma_0). \tag{7.12.23}$$

It is convenient to set $\sigma_0 = \operatorname{atanh}(b/a)$ and obtain the simplified expression $z = A \cosh \sigma$, where

$$A = a \cosh \sigma_0 - b \sinh \sigma_0 = \frac{c^2}{a} \cosh \sigma_0, \tag{7.12.24}$$

and c is the focal length of the ellipse. The real and imaginary parts of σ provide us with elliptic coordinates.

Problems

7.12.1 *Bipolar coordinates*

Plane bipolar coordinates, (ξ, η), are defined by the conformal mapping function

$$x + \mathrm{i}\, y = A\mathrm{i} \, \cot \frac{\pi - \eta + \mathrm{i}\xi}{2}, \tag{7.12.25}$$

where A is a real constant. Decomposing the mapping function into its real and imaginary parts, we obtain

$$x = A \frac{\sinh \xi}{\cosh \xi + \cos \eta}, \qquad y = A \frac{\sin \eta}{\cosh \xi + \cos \eta}. \tag{7.12.26}$$

Assume that the dimensionless variable ξ varies in the range $\xi_1 \le \xi \le \xi_2$, and the dimensionless variable η varies in the range $[0, 2\pi)$.

(a) Show that the lower limit, $\xi = \xi_1 < 0$, corresponds to the surface of a left circle, the upper limit, $\xi = \xi_2 > 0$, corresponds to the surface of a right circle, and the value $\xi = 0$ corresponds to the y axis.

(b) Show that the center of the left circle is located at $x = x_{c_1} \equiv A \coth \xi_1$ and $y = 0$, the center of the right circle is located at $x = x_{c_2} \equiv A \coth \xi_2$ and $y = 0$, the radius of the left circle is $a_1 = -A/\sinh \xi_1$, and the radius of the right circle is $a_2 = A/\sinh \xi_2$.

(c) Confirm that

$$A^2 = y_{c_1}^2 - a_1^2 = y_{c_2}^2 - a_2^2 = (y_{c_2} - 2d)^2 - a_1^2 = (y_{c_1} + 2d)^2 - a_2^2, \qquad (7.12.27)$$

where $2d \equiv x_{c_2} - x_{c_1} = A\,(\coth \xi_2 - \coth \xi_1)$.

7.12.2 *Generalized Joukowski transformation*

The following generalized Joukowski transformation prevents the formation of a cuspidal trailing edge,

$$\frac{z - 2\sigma}{z + 2\sigma} = \left(\frac{\zeta - \sigma}{\zeta + \sigma}\right)^q, \qquad (7.12.28)$$

where $1 < q < 2$. Demonstrate that a circle centered at the origin of the ζ plane and passing through both singular points, $\zeta = \pm\sigma$ transforms into two circular arcs passing through the points $z = \pm q\sigma$.

 Computer Problem

7.12.3 *Joukowski airfoils*

(a) Generate a family of airfoils using the Joukowski transformation. The airfoils should be the images of circles passing through the first singular point, $\zeta = -\sigma$, and enclosing the second singular point $\zeta = \sigma$. The center of the circle should be located at the ξ axis at $\xi/\sigma = 1.0$, 1.2, 1.5, and 2.0.

(b) Repeat (a) with the center of the circle located at $\xi/\sigma = 1.0$, 1.2, 1.5, 2.0, and $\eta/\sigma = 0.2$.

7.13   The Schwarz–Christoffel transformation

The Schwarz–Christoffel transformation maps an interior or exterior domain bounded by straight segments, semi-infinite straight lines, or infinite straight lines, to a half-plane, or to the interior or exterior of a circular disk. The pertinent mapping functions are available in integral form in terms of a set of *a priori* unknown scalar coefficients. Computing the transformation is reduced to calculating these coefficients, which can be done analytically for simple shapes or numerically for more complicated shapes. Generalized Schwarz–Christoffel transformations that map domains bounded by smooth contours are available. In this section, we present the various forms of the Schwarz–Christoffel transformation for polygonal and smooth domains and illustrate the computation of the mapping parameters by analytical and numerical methods.

7.13.1   Mapping a polygon to a semi-infinite plane

One version of the Schwarz–Christoffel transformation maps the interior of an N-sided closed polygon in the z plane to the upper half ζ plane, as shown in Figure 7.13.1. To generate the transformation, we number the vertices around the polygon sequentially in the counterclockwise direction and compute the exterior angles subtended between the extension of each side and the next side, γ_i, as shown in Figure 7.13.1. By convention, $-\pi < \gamma_i < \pi$, where γ_i is positive when the ith corner is projecting into the exterior of the polygon, and negative otherwise. All angles γ_i are positive for a convex

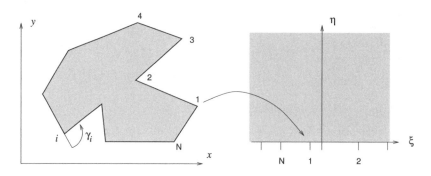

FIGURE 7.13.1   One version of the Schwarz–Christoffel transformation maps a polygon in the z plane to the upper half ζ plane.

polygon and all angles γ_i are equal for a regular polygon. In all cases, the sum of the angles satisfies the geometrical constraint

$$\sum_{i=1}^{N} \gamma_i = 2\pi, \tag{7.13.1}$$

where positive and negative angles may appear inside the sum. For example, in the case of a square, $N = 4$ and $\gamma_i = \frac{1}{2}\pi$ for $i = 1$–4. The Schwarz–Christoffel transformation provides us with the inverse mapping function in the differential form

$$\frac{\mathrm{d}z}{\mathrm{d}\zeta} = c \prod_{i=1}^{N} \frac{1}{(\zeta - \xi_i)^{\gamma_i/\pi}}, \tag{7.13.2}$$

where c is a complex constant and ξ_n are the images of the vertices of the polygon along the real ξ axis of the ζ plane. The associated integral form is

$$z = f(\zeta) = z_0 + c \int_{\zeta_0}^{\zeta} \left[\prod_{i=1}^{N} \frac{1}{(\varrho - \xi_i)^{\gamma_i/\pi}} \right] \mathrm{d}\varrho, \tag{7.13.3}$$

where z_0 and ζ_0 are a pair of corresponding points and ϱ is an integration variable. The complex powers in (7.13.2) and (7.13.3) are computed best by setting, for example,

$$(\varrho - \xi_i)^{\gamma_i/\pi} = \exp\left[\frac{\gamma_i}{\pi} \ln(\varrho - \xi_i) \right], \tag{7.13.4}$$

where the branch cut of the logarithmic function is the negative real axis. Given the polygon shape, the problem is reduced to specifying or computing the images of the vertices, ξ_i, and the values of the three constants, c, z_0, and ζ_0. Riemann's mapping theorem discussed in Section 7.11.10 allows us to arbitrarily choose the images of three vertices, ξ_i, and compute the rest of the vertices so as to map the polygon to the upper half-plane. It is permissible to set $\xi_1 = -\infty$ or $\xi_N = \infty$, in which case the corresponding factors in the product in (7.13.2) or (7.13.3) do not appear.

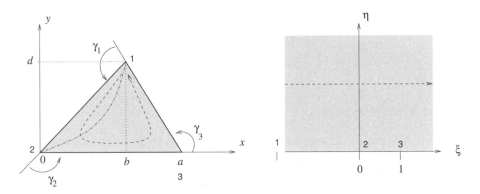

FIGURE 7.13.2   Mapping a triangle in the z plane to the upper half ζ plane. The dot-dashed line is the image of the η axis. Corresponding streamlines are shown as dashed lines.

Triangle

In the simplest application, we map the interior of a triangle with vertices at the points $z_1 = b + \mathrm{i}\,d$, $z_2 = 0$, and $z_3 = a$, to the upper half ζ plane, where a, b, and d are three real positive constants, as illustrated in Figure 7.13.2. The Riemann mapping theorem allows us to arbitrarily select the images of all three vertices. It is convenient to set $\xi_1 = -\infty$, $\xi_2 = 0$, and $\xi_3 = 1$. Applying (7.13.3) with $z_0 = 0$ and $\zeta_0 = 0$, we find that

$$z = f(\zeta) = c \int_0^\zeta \frac{\mathrm{d}\varrho}{\varrho^{\gamma_2/\pi}(\varrho - 1)^{\gamma_3/\pi}}. \tag{7.13.5}$$

Unfortunately, it is not possible to perform the integration analytically for an arbitrary triangular shape. To evaluate the constant c, we require that $f(1) = a$ and obtain

$$a = c \int_0^1 \frac{\mathrm{d}\varrho}{\varrho^{\gamma_2/\pi}(\varrho - 1)^{\gamma_3/\pi}} = \frac{c}{(-1)^{\gamma_3/\pi}} \frac{\Gamma(1 - \gamma_2/\pi)\,\Gamma(1 - \gamma_3/\pi)}{\Gamma(2 - \gamma_2/\pi - \gamma_3/\pi)}. \tag{7.13.6}$$

The values of the gamma function, Γ, can be read from tables or approximated by series expansions (e.g., [2], p. 255). Substituting (7.13.6) into (7.13.5), we obtain

$$z = f(\zeta) = a\, \frac{\Gamma(2 - \gamma_2/\pi - \gamma_3/\pi)}{\Gamma(1 - \gamma_2/\pi)\,\Gamma(1 - \gamma_3/\pi)} \int_0^\zeta \frac{\mathrm{d}\varrho}{\varrho^{\gamma_2/\pi}(1 - \varrho)^{\gamma_3/\pi}}. \tag{7.13.7}$$

The η axis is mapped to the dot-dashed line connecting the second to the first vertex in the xy plane, as shown in Figure 7.13.2.

To compute the singular integral on the right-hand side of (7.13.7), denoted by $\mathcal{I}$, we subtract out the singularity at the origin by restating the integral as

$$\mathcal{I} = \mathcal{J} + \int_0^\zeta \frac{\mathrm{d}\varrho}{\varrho^{\gamma_2/\pi}} = \mathcal{J} + \frac{1}{1 - \gamma_2/\pi}\, \zeta^{1 - \gamma_2/\pi}, \tag{7.13.8}$$

where

$$\mathcal{J} \equiv \int_0^\zeta \frac{1}{\varrho^{\gamma_2/\pi}} \left[\frac{1}{(1-\varrho)^{\gamma_3/\pi}} - 1 \right] d\varrho. \tag{7.13.9}$$

The integral on the right-hand side of (7.13.9) can be computed by standard numerical methods, as discussed in Section B.6, Appendix B.

Uniform flow along the ξ axis in the ζ plane corresponds to flow due to a point-source dipole placed at the vertex z_1 in the z plane, and oriented perpendicular to the bisector of the corresponding triangle angle. A streamline is illustrated with the dashed lines in Figure 7.13.2.

Semi-infinite strip

As the elevation of the first vertex, d, tends to infinity while b lies between 0 and a, the triangle becomes a vertical semi-infinite strip with width a. Setting $\gamma_2 = \gamma_3 = \frac{1}{2}\pi$, we obtain

$$z = f(\zeta) = c \int_0^\zeta \frac{d\varrho}{\varrho^{1/2}(\varrho - 1)^{1/2}} = c \int_{-1}^{2\zeta-1} \frac{d\lambda}{(\lambda^2 - 1)^{1/2}} = -i c \left[\arcsin(2\zeta - 1) + \frac{\pi}{2} \right], \tag{7.13.10}$$

where $\lambda = 2\varrho - 1$ and $\varrho = \frac{1}{2}(\lambda + 1)$. Requiring that $f(1) = a$, we find that $c = ia/\pi$ and obtain he mapping function

$$z = f(\zeta) = \frac{a}{\pi} \left[\arcsin(2\zeta - 1) + \frac{\pi}{2} \right]. \tag{7.13.11}$$

The forward mapping function follows by inversion,

$$\zeta = F(z) = \frac{1}{2} \left(1 - \cos \frac{\pi z}{a} \right). \tag{7.13.12}$$

Uniform flow along the ξ axis in the ζ plane corresponds to flow descending along the left vertical side of the strip and leaving up along the right side of the strip in the z plane (Problem 7.13.4).

Rectangle

Next, we map the interior of a rectangle with vertices at the points

$$z_1 = -a + ib, \qquad z_2 = -a, \qquad z_3 = a, \qquad z_4 = a + ib, \tag{7.13.13}$$

to the upper half ζ plane, as shown in Figure 7.13.3. Motivated by the symmetry of the rectangle, we specify that $\xi_2 = -1$ and $\xi_3 = 1$. Requiring that the origin of the z plane is mapped to the origin of the ζ plane, we set $z_0 = 0$ and $\zeta_0 = 0$. Anticipating the symmetry of the solution, we set $\xi_1 = -\lambda$ and $\xi_4 = \lambda$, where $\lambda > 1$ is a real constant to be computed as part of the solution. Substituting these values along with $\gamma_i = \pi/2$ into (7.13.3) for $i = 1$–4, we obtain the transformation

$$z = f(\zeta) = c \int_0^\zeta \frac{d\varrho}{(\varrho^2 - 1)^{1/2}(\varrho^2 - \lambda^2)^{1/2}}. \tag{7.13.14}$$

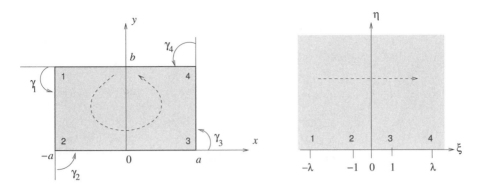

FIGURE 7.13.3   Mapping a rectangle in the z plane to the upper half ζ plane. Corresponding labels of the four vertices are shown.

To compute the values of the two constants c and λ, we require that, as we move along the ξ axis from the origin up to ξ_3 and then up to ξ_4, we find ourselves at the corresponding vertices z_3 and z_4. The first condition yields the real equation

$$c \int_0^1 \frac{\mathrm{d}\xi}{(1 - \varrho^2)^{1/2}(\lambda^2 - \xi^2)^{1/2}} = a, \tag{7.13.15}$$

which shows that c is real. The second condition yields the real equation

$$c \int_1^\lambda \frac{\mathrm{d}\xi}{(\xi^2 - 1)^{1/2}(\lambda^2 - \xi^2)^{1/2}} = -b. \tag{7.13.16}$$

Combining these equations to eliminate c, we obtain a nonlinear equation,

$$\mathcal{Q}(\lambda) \equiv \int_1^\lambda \frac{\mathrm{d}\xi}{(\xi^2 - 1)^{1/2}(\lambda^2 - \xi^2)^{1/2}} + \frac{b}{a} \int_0^1 \frac{\mathrm{d}\xi}{(1 - \xi^2)^{1/2}(\lambda^2 - \xi^2)^{1/2}} = 0. \tag{7.13.17}$$

The solution can be found using standard numerical methods, such as Newton's method discussed in Section B.3, Appendix B.

Certain aspects of the computation require special attention. The integral in (7.13.15) can be written as

$$\int_0^1 \frac{\mathrm{d}\xi}{(1 - \xi^2)^{1/2}(\lambda^2 - \xi^2)^{1/2}} = \frac{1}{\lambda} F\left(\frac{1}{\lambda}\right), \tag{7.13.18}$$

where

$$F(k) \equiv \int_0^1 \frac{\mathrm{d}\xi}{(1 - \xi^2)^{1/2}(1 - k^2\xi^2)^{1/2}} = \int_0^{\pi/2} \frac{\mathrm{d}u}{\sqrt{1 - k^2 \sin^2 u}} \tag{7.13.19}$$

is the complete elliptic integral of the first kind discussed in Section 2.12, $0 < k < 1$, and $\xi = \sin u$.

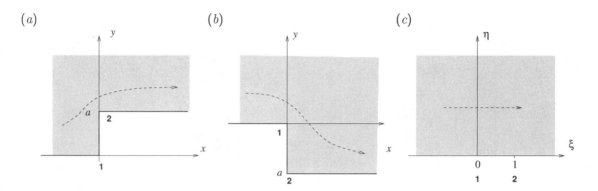

FIGURE 7.13.4   Mapping of a semi-infinite domain above a wall with (a) a forward or (b) backward facing step to the upper half ζ plane shown in (c).

To compute the singular integral on the left-hand side of (7.13.16), denoted by $\mathcal{I}$, we subtract out the singularities on either end of the integration domain by restating the integral into the form

$$\mathcal{I} = \mathcal{J} + \frac{1}{(\lambda^2 - 1)^{1/2}} \Big[\int_1^\lambda \frac{\mathrm{d}\xi}{(\xi^2 - 1)^{1/2}} + \int_1^\lambda \frac{\mathrm{d}\xi}{(\lambda^2 - \xi^2)^{1/2}} \Big], \qquad (7.13.20)$$

where

$$\mathcal{J} \equiv \int_1^\lambda \Big(\frac{1}{(\xi^2 - 1)^{1/2}(\lambda^2 - \xi^2)^{1/2}} - \frac{1}{(\lambda^2 - 1)^{1/2}} \Big[\frac{1}{(\xi^2 - 1)^{1/2}} + \frac{1}{(\lambda^2 - \xi^2)^{1/2}} \Big] \Big) \, \mathrm{d}\xi. \quad (7.13.21)$$

The $\mathcal{J}$ integral can be computed by standard numerical methods, as discussed in Section B.6, Appendix B. The two improper integrals enclosed by the square brackets on the right-hand side of (7.13.20) can be computed by elementary analytical methods.

Uniform flow along the ξ axis in the ζ plane corresponds to flow due to a horizontal point-source dipole located at the midpoint of the upper side of the rectangle in the z plane, $x = 0$, $y = 0$, as shown in Figure 7.13.3.

Semi-infinite region above a step

The Schwarz–Christoffel transformation can be extended to include domains that are bounded by generalized polygons with one or two vertices at infinity. One example is the semi-infinite strip discussed earlier in this section. A second example is the corner-like domain shown in Figure 7.10.1(a), where $\zeta = Az^{\alpha/\pi}$, $\alpha = \pi - \gamma$ is the internal turning angle, γ is the external turning angle according to the conventions of the Schwarz-Christoffel transformation, and A is a constant.

A third example is the semi-infinite region above a wall with a forward facing step whose boundary can be regarded as a generalized polygon with one vertex at infinity, as shown in Figure 7.13.4(a). Mapping the vertex at infinity in the z plane to the positive infinity of the ξ axis, specifying that $\xi_1 = 0$ and $\xi_2 = 1$, and setting $\gamma_1 = \frac{1}{2}\pi$ and $\gamma_2 = -\frac{1}{2}\pi$, we obtain the transformation

$$z = f(\zeta) = c \int_0^\zeta \left(\frac{\varrho - 1}{\varrho}\right)^{1/2} d\varrho = 2c \int_0^{\sqrt{\zeta}} \sqrt{\omega^2 - 1} \, d\omega, \tag{7.13.22}$$

where $\omega = \sqrt{\varrho}$. Performing the integration, we find that

$$z = f(\zeta) = c \left[\omega \sqrt{\omega^2 - 1} - \ln(\omega + \sqrt{\omega^2 - 1})\right]_0^{\sqrt{\zeta}}, \tag{7.13.23}$$

and then

$$z = f(\zeta) = c \left[\sqrt{\zeta(\zeta - 1)} - \ln(\sqrt{\zeta} + \sqrt{\zeta - 1}) + i\frac{\pi}{2}\right]. \tag{7.13.24}$$

The constant c is determined by requiring that $f(1) = ia$, yielding $c = 2a/\pi$.

Backward facing step

In the case of a backward facing step shown in Figure 7.13.4(b), we specify that $\xi_1 = 0$ and $\xi_2 = 1$, and set $\gamma_1 = -\frac{1}{2}\pi$ and $\gamma_2 = \frac{1}{2}\pi$, to obtain the transformation

$$z = f(\zeta) = c \int_0^\zeta \left(\frac{\varrho}{\varrho - 1}\right)^{1/2} d\varrho = 2c \int_0^{\sqrt{\zeta - 1}} \sqrt{\omega^2 + 1} \, d\omega, \tag{7.13.25}$$

where $\omega = \sqrt{\varrho - 1}$. Performing the integration, we find that

$$z = f(\zeta) = c \left[\omega \sqrt{\omega^2 + 1} + \ln(\omega + \sqrt{\omega^2 + 1})\right]_i^{\sqrt{\zeta - 1}}, \tag{7.13.26}$$

and then

$$z = f(\zeta) = c \left[\sqrt{\zeta(\zeta - 1)} + \ln(\sqrt{\zeta} + \sqrt{\zeta - 1}) - i\frac{\pi}{2}\right]. \tag{7.13.27}$$

The constant c is determined by requiring that $f(1) = -ia$, yielding $c = 2a/\pi$.

Angles of generalized polygons

The angle γ_i corresponding to the vertex of a generalized polygon located at infinity is equal to $2\pi - \theta$, where θ is the external angle formed by the intersection of the two corresponding sides when they are extended back from infinity.

For example, in the case of the generalized polygon with two collapsed sides and one vertex at infinity illustrated in Figure 7.13.5,

$$\gamma_1 = \frac{\pi}{2}, \qquad \gamma_2 = 2\pi - \gamma_3 = \frac{4\pi}{3}, \qquad \gamma_3 = \frac{2\pi}{3}, \qquad \gamma_4 = \frac{\pi}{2}, \qquad \gamma_5 = -\pi. \tag{7.13.28}$$

Note that the angle constraint (7.13.1) is satisfied.

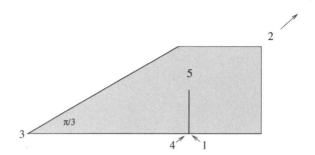

FIGURE 7.13.5   Illustration of a generalized polygon with two sides collapsed and one vertex at infinity.

Channel-like domain

Consider an infinite channel-like domain defined by N vertices, as shown in Figure 7.13.6. The left aperture angle defined in this figure, γ_L, is related to the vertex angles α_1 and α_2 by

$$\alpha_1 + \alpha_2 = \pi + \gamma_L. \tag{7.13.29}$$

For example, when $\alpha_1 = \frac{1}{2}\pi$ and $\alpha_2 = \frac{1}{2}\pi$, $\gamma_L = 0$. The right aperture angle, γ_R, is related to the corresponding vertex angles β_1 and β_2 by

$$\beta_1 + \beta_2 = \pi + \gamma_R. \tag{7.13.30}$$

For example, when $\beta_1 = \frac{1}{2}\pi$ and $\alpha_2 = 0$, $\gamma_R = \frac{1}{2}\pi$. The identity

$$\alpha_1 + \alpha_2 + \beta_1 + \beta_2 + \sum_{i=1}^{N} \gamma_i = 2\pi \tag{7.13.31}$$

requires that

$$\gamma_L + \gamma_R + \sum_{i=1}^{N} \gamma_i = 0, \tag{7.13.32}$$

where the angles γ_i are defined in Figure 7.13.6.

The Schwartz–Christoffel transformation mapping the channel to the upper half ζ plane is given by the differential form

$$\frac{\mathrm{d}z}{\mathrm{d}\zeta} = \frac{c}{\zeta^{1+\gamma_L/\pi}} \prod_{i=1}^{N} \frac{1}{(\zeta - \xi_i)^{\gamma_i/\pi}}, \tag{7.13.33}$$

where the left infinity is mapped to the origin of the ζ plane and the right infinity is mapped to the infinity of the ζ plane. For example, in the case of a channel confined between two parallel horizontal walls, we set $N = 2$, $\gamma_L = 0$, $\gamma_1 = 0$, and $\gamma_2 = 0$, yielding $\mathrm{d}z/\mathrm{d}\zeta = 1/\zeta$. Integrating, we obtain $z = c \ln(\zeta/\zeta_0)$, where c and ζ_0 are two constants.

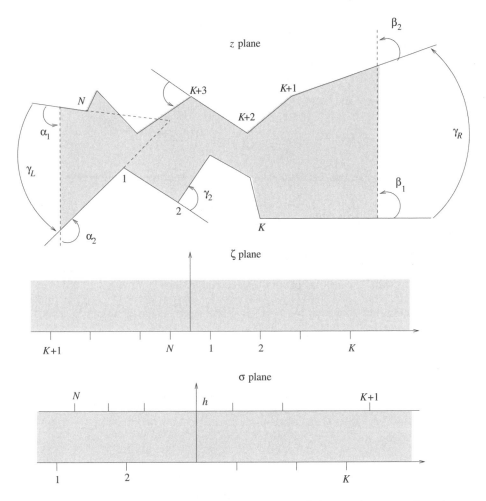

FIGURE 7.13.6   A channel-like domain with two vertices at infinity can be mapped to the upper half ζ plane and then to an infinite strip in the σ plane.

The upper half ζ plane can be subsequently mapped to a horizontal strip in the complex, $\sigma = \sigma_R + i\sigma_I$ plane, $-\infty < \sigma_R < \infty$, $0 < \sigma_I < h$, as shown in Figure 7.13.6. The mapping is mediated by the logarithmic function and its inverse,

$$\sigma = \frac{h}{\pi} \ln \frac{\zeta}{a}, \qquad \zeta = a \exp \frac{\pi\sigma}{h}, \qquad (7.13.34)$$

where a is a positive constant [129]. Making substitutions into (7.13.33), we find that

$$\frac{dz}{d\sigma} = \ell \exp\left(-\frac{\gamma_L \sigma}{h}\right) \prod_{i=1}^{N} \left[\exp \frac{\pi(\sigma - \sigma_i)}{h} - 1\right]^{-\gamma_i/\pi}, \qquad (7.13.35)$$

where ℓ is a new constant. Rearranging the right-hand side, we obtain

$$\frac{dz}{d\sigma} = \ell \exp\left(-\frac{\gamma_L \sigma}{h}\right) \prod_{i=1}^{N} \left[\exp\frac{-\gamma_i(\sigma - \sigma_i)}{2h}\right] \times \prod_{i=1}^{N} \left[\sinh\frac{\pi(\sigma - \sigma_i)}{2h}\right]^{-\gamma_i/\pi}. \tag{7.13.36}$$

Using (7.13.32), we finally obtain

$$\frac{dz}{d\sigma} = q \exp\left[(\gamma_L - \gamma_R)\frac{\sigma}{h}\right] \prod_{i=1}^{N} \left[\sinh\frac{\pi(\sigma - \sigma_i)}{2h}\right]^{-\gamma_i/\pi}, \tag{7.13.37}$$

where q is a new constant. We can separate the lower from the upper wall vertices to obtain

$$\frac{dz}{d\sigma} = p \exp\left[(\gamma_L - \gamma_R)\frac{\sigma}{h}\right] \prod_{i=1}^{K} \left[\sinh\frac{\pi(\sigma - \sigma_i)}{2h}\right]^{-\gamma_i/\pi} \prod_{i=K+1}^{N} \left[\cosh\frac{\pi(\sigma - \sigma_i + ih)}{2h}\right]^{-\gamma_i/\pi}, \tag{7.13.38}$$

where p is a new constant. In the case of a channel with symmetric extensions, $\gamma_L = \gamma_R$, the exponential factor in front of the products on the right-hand side does not appear. In the case of a channel with a flat top wall, the second product on the right-hand side of (7.13.38) does not appear.

In the case of a channel whose upper and lower walls are repeated along the x axis with period L, we obtain [130]

$$\frac{dz}{d\sigma} = p \prod_{m=-\infty}^{\infty} \left(\prod_{i=1}^{K} \left[\sinh\frac{\pi(\sigma - \sigma_i - mL)}{2h}\right]^{-\gamma_i/\pi} \prod_{i=K+1}^{N} \left[\cosh\frac{\pi(\sigma - \sigma_i + ih - mL)}{2h}\right]^{-\gamma_i/\pi}\right). \tag{7.13.39}$$

In practice, accurate results can be obtained even when the product is truncated at small values of m. In the case of a channel with a flat top wall, the second product on the right-hand side of (7.13.39) does not appear.

7.13.2 Mapping a polygon to the unit disk

In certain applications, it is desirable to map the interior of a polygon in the z plane to a unit disk centered at the origin of the complex τ plane. This can be done by first mapping the unit disk in the τ plane to the upper half ζ plane, and *vice versa*, using a linear fractional transformation and its inverse,

$$\zeta = \mathcal{W}(\tau) = i\mu\frac{1 + \tau}{1 - \tau}, \qquad \tau = \mathcal{W}^{-1}(\zeta) = \frac{\zeta - i\mu}{\zeta + i\mu}, \tag{7.13.40}$$

where μ is a real positive constant, as illustrated in Figure 7.13.7. The center of the disk is mapped to the point $\zeta = i\mu$. The origin of the ζ plane is mapped to the leftmost point of the unit circle, $\tau = -1$. In the second stage, the upper half-plane is mapped to the interior of the polygon.

Substituting the first expression in (7.13.40) into the right-hand side of (7.13.3), we obtain the required mapping function

$$z = f(\zeta) = f[\mathcal{W}(\tau)] \equiv q(\tau) = z_0 + 2i\mu c \int_{\tau_0}^{\tau} \prod_{i=1}^{N} \left[\left(i\mu\frac{1 + \vartheta}{1 - \vartheta} - \xi_i\right)^{-\gamma_i/\pi}\right] \frac{d\vartheta}{(1 - \vartheta)^2}. \tag{7.13.41}$$

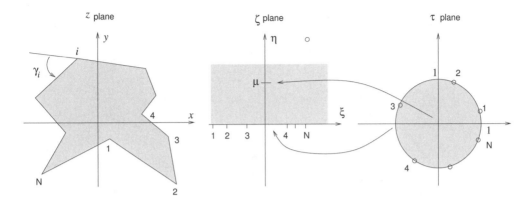

FIGURE 7.13.7   Mapping a polygon in the z plane to a unit disk centered at the origin of the τ plane.

Rearranging and taking into account (7.13.1) to simplify the integrand, we obtain

$$z = q(\tau) = z_0 + d \int_{\tau_0}^{\tau} \prod_{i=1}^{N} \left[\frac{1}{(\vartheta - \tau_i)^{\gamma_i/\pi}} \right] d\vartheta, \tag{7.13.42}$$

where d is a new constant. The image points, $\tau_i = (\xi_i - i\mu)/(\xi_i + i\mu)$ for $i = 1, \dots, N$, are distributed around the unit circle in the τ plane. We can place ξ_1 or ξ_N at infinity to obtain, respectively, $\tau_1 = 1$ or $\tau_N = 1$. Although the corresponding multiplier does not appear in (7.13.3), it must be included inside the product in (7.13.42). It will be noted that the disk transformation (7.13.42) is identical in form to the semi-infinite plane transformation (7.13.3). The only difference is that the images, τ_i, are distributed around the unit circle in the counterclockwise fashion instead of the real ξ axis.

Regular polygons

In the case of a regular N-sided polygon centered at the origin of the z plane, $z_0 = 0$, we set $\tau_0 = 0$, $\gamma_i = 2\pi/N$, and choose

$$\tau_i = \exp \left(2\pi i \, \frac{i-1}{N} \right) \tag{7.13.43}$$

for $i = 1, \dots, N$, so that the vertices are distributed uniformly around the unit circle with $\tau_1 = 1$. The transformation (7.13.42) with $z_0 = 0$ and $\tau_0 = 0$ becomes

$$z = q(\tau) = d \int_0^{\tau} \prod_{i=1}^{N} \left[\vartheta - \exp \left(2\pi i \, \frac{i-1}{N} \right) \right]^{-2/N} d\vartheta. \tag{7.13.44}$$

The product inside the integral is equivalent to a simple polynomial, τ_i,

$$\prod_{i=1}^{N} \left[\vartheta - \exp \left(2\pi i \, \frac{i-1}{N} \right) \right] = \vartheta^N - 1, \tag{7.13.45}$$

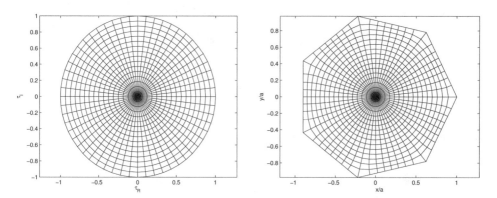

FIGURE 7.13.8   Corresponding grids generated by mapping a disk of unit radius in the τ plane to a heptagon of radius a in the z plane.

yielding

$$z = q(\tau) = A \int_0^\tau \frac{\mathrm{d}\vartheta}{(1 - \vartheta^N)^{2/N}},$$ (7.13.46)

where A is a new constant computed by requiring $q(1) = a$, and a is the radius of the polygon. Performing the integration with respect to ϑ, we obtain an inconvenient expression in terms of the Gauss hypergeometric function. Although an approximate Taylor series expansion is available [225], in practice, it is expedient to compute the mapping function by numerical integration. Corresponding grids generated by mapping the disk to a heptagon are shown in Figure 7.13.8.

Fractal shapes

Polygonal shapes can be generated by distributing a specified number of images τ_i around the unit disk and specifying the corresponding angles γ_i in the physical z plane, subject to the restriction imposed by (7.13.1). Fractal contours can be generated in the limit as the number of vertices tends to infinity [344].

7.13.3   Mapping the exterior of a polygon

Thus far, we have discussed mapping the interior of a polygon to the upper half-plane or unit disk. Corresponding transformations are available for mapping the exterior of a polygon.

Mapping to the unit disk

To map the exterior of a polygon in the z plane to the unit disk in the τ plane, we regard the exterior of the polygon as a doubly connected domain bounded by the polygon of interest and another auxiliary polygon of large size extending to infinity. The two polygons are connected by a branch cut, as shown in Figure 7.13.9. We number the vertices of the inner polygon of interest

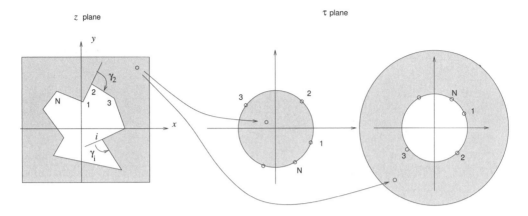

FIGURE 7.13.9   Mapping the exterior of a polygon to the interior or exterior of a unit disk in the τ complex plane.

sequentially in the *clockwise* direction, and compute the vertex angles, γ_i, where

$$\sum_{i=1}^{N} \gamma_i = -2\pi. \tag{7.13.47}$$

For the configuration shown in Figure 7.13.9, $\gamma_1 > 0$ and $\gamma_2 < 0$. The vertices of the exterior auxiliary polygon are mapped to the origin of the τ plane, so that the corresponding images are zero, $\tau_j = 0$, and the sum of the corresponding angles satisfies $\sum_j \gamma_j = 2\pi$.

Applying (7.13.2) for the union of the vertices of the outer and inner polygons with the aforementioned conventions, we obtain the transformation

$$z = q(\tau) = z_0 + d \int_{\tau_0}^{\tau} \prod_{i=1}^{N} \left[\frac{1}{(\vartheta - \tau_i)^{\gamma_i/\pi}} \right] \frac{d\vartheta}{\vartheta^2}, \tag{7.13.48}$$

where z_0 and τ_0 is an arbitrary pair of corresponding points and d is a complex constant. The image points, τ_i, are distributed in the counterclockwise direction around the unit disk centered at the origin of the τ complex plane, as shown in Figure 7.13.10 ([276], p. 193).

Regular polygons

In the case of a regular N-sided polygon centered at the origin of the z plane, we set $\gamma_i = -2\pi/N$ for $i = 1, \ldots, N$, and select the images shown in (7.13.43) so that the vertices are distributed uniformly around the unit circle with $\tau_1 = 1$. The transformation (7.13.48) reduces to

$$z = q(\tau) = z_0 + d \int_{\tau_0}^{\tau} (\vartheta^N - 1)^{2/N} \frac{d\vartheta}{\vartheta^2}. \tag{7.13.49}$$

The power $(\vartheta^N - 1)^{2/N}$ is computed best after the polynomial $\vartheta^N - 1$ has been factorized into a product of monomials involving the Nth roots of unity.

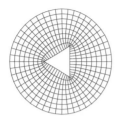 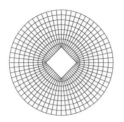 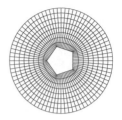

FIGURE 7.13.10   Grids generated by mapping the exterior of the unit circle to the exterior of an equilateral triangle, square, or regular pentagon.

Mapping to the exterior of the unit circle

To map the exterior of a polygon in the z plane to the exterior of a unit disk in the ζ plane, we substitute into (7.13.48) $\zeta \to 1/\zeta$ and obtain an identical expression. The image points, τ_i, are distributed in the clockwise direction around the unit disk centered at the origin of the τ complex plane, as shown in Figure 7.13.9. Grids generated by mapping the exterior of the unit circle truncated at a certain radial threshold to the exterior of a triangle, square, or pentagon are shown in Figure 7.13.10.

Mapping to the upper half-plane

Substituting into (7.13.48) the second transformation in (7.13.40), $\tau = \mathcal{W}^{-1}(\zeta)$, we obtain a function that maps the exterior of a polygon in the z plane to the upper half ζ plane,

$$z = f(\zeta) = z_0 + c \int_{\zeta_0}^{\zeta} \prod_{i=1}^{N} \left[\frac{1}{(\varrho - \xi_i)^{\gamma_i/\pi}} \right] \frac{\mathrm{d}\varrho}{(\varrho + \mathrm{i}\mu)^2 (\varrho - \mathrm{i}\mu)^2} \qquad (7.13.50)$$

([65], p. 153). The computation of the unknown parameters is analogous to that for the corresponding mapping of the interior of a polygon previously discussed.

7.13.4   Periodic domains

Consider a semi-infinite domain bounded by a periodic polygonal line consisting of N periodically repeated vertices, as shown in Figure 7.13.11. We will map each period of the domain into a vertical semi-infinite strip with width λ bounded by the ξ axis in the ζ plane. As a preliminary, we define the turning angles, γ_i, satisfying

$$\sum_{i=1}^{N} \gamma_i = 0, \qquad (7.13.51)$$

where $-\pi < \gamma_i < \pi$. Positive angles correspond to counterclockwise rotation and negative angles correspond to clockwise rotation.

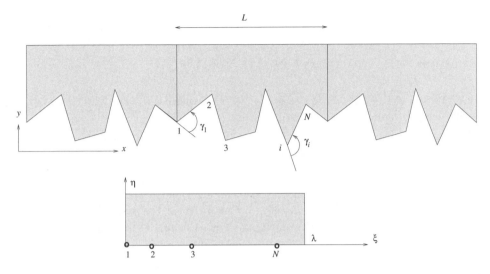

FIGURE 7.13.11   Mapping a semi-infinite domain bounded by a periodic polygonal line to the upper half-plane.

Modifying the Schwartz–Christoffel transformation (7.13.2) to account for all vertex images, we obtain the inverse mapping function

$$\frac{dz}{d\zeta} = c \prod_{i=1}^{N} \Big(\prod_{k=-\infty}^{\infty} \frac{1}{(\zeta + k\lambda - \xi_i)} \Big)^{\gamma_i/\pi} = c \prod_{i=1}^{N} \Big[(\zeta - \xi_i) \prod_{k=1}^{\infty} [(\zeta - \xi_i)^2 - k^2\lambda^2)] \Big]^{-\gamma_i/\pi}. \quad (7.13.52)$$

Further rearrangement yields

$$\frac{dz}{d\zeta} = c\lambda \prod_{i=1}^{N} \Big[\frac{\zeta - \xi_i}{\lambda} \prod_{k=1}^{\infty} \big[1 - \frac{(\zeta - \xi_i)^2}{k^2\lambda^2} \big] \Big]^{-\gamma_i/\pi}. \quad (7.13.53)$$

Next, we introduce the Euler product expansion of the sine function,

$$\sin(\pi w) = \pi w \prod_{k=1}^{\infty} \Big(1 - \frac{w^2}{k^2} \Big) \quad (7.13.54)$$

and substitute $w = (\zeta - \xi_i)/\lambda$ to obtain

$$\frac{dz}{d\zeta} = d \prod_{i=1}^{N} \Big[\sin \big(\pi \frac{\zeta - \xi_i}{\lambda} \big) \Big]^{-\gamma_i/\pi}, \quad (7.13.55)$$

where $d = c\lambda/\pi$. Requiring that the first point, z_1, be mapped to the origin of the ζ plane, $\xi_1 = 0$, and integrating, we find that

$$z = z_1 + d \int_0^{\zeta} \prod_{i=1}^{N} \Big[\sin \big(\pi \frac{\varrho - \xi_i}{\lambda} \big) \Big]^{-\gamma_i/\pi} d\varrho. \quad (7.13.56)$$

Far from the wall, as $\eta \to \infty$, we obtain a constant mapping slope,

$$\frac{dz}{d\zeta} \to d \, \exp\Big(-\mathrm{i} \sum_{i=1}^{N} \gamma_i \, \frac{\xi_i}{\lambda} \Big). \tag{7.13.57}$$

Because of (7.13.51), the argument of the exponent is insensitive to the definition of the origin of the ξ axis. Integrating along a periodic, we find that

$$d = \frac{L}{\lambda} \, \exp\Big(\mathrm{i} \sum_{i=1}^{N} \gamma_i \, \frac{\xi_i}{\lambda} \Big), \tag{7.13.58}$$

which can be used to evaluate the constant d once the positions ξ_i are available. Substituting this expression into (7.13.57), we find that $dz/d\zeta \to L/\lambda$ far from the wall. To compute the $N-1$ unknown images, $\xi_2, \ldots, \xi_N$, we integrate along the ξ axis from the first up to the jth vertex, we obtain

$$z_j = z_1 + d \int_0^{\xi_j} \prod_{i=1}^{N} \Big[\sin\Big(\pi \frac{\xi - \xi_i}{\lambda} \Big) \Big]^{-\gamma_i/\pi} d\xi \tag{7.13.59}$$

for $j = 2, \ldots, N$. Requiring that this expression reproduces the vertices of the polygonal boundary in the physical plane provides us with a system of nonlinear equations.

Triangular asperities

In the case of a wall with an infinite sequence of triangular asperities, $N = 2$ and $\gamma_2 = -\gamma_1$, as shown in Figure 7.13.12(a). Without loss of generality, we may assume that $\gamma_1 < 0$ so that the integrand tends to zero at the origin. The transformation (7.13.56) with $\xi_1 = 0$ becomes

$$z = z_1 + \frac{L}{\lambda} e^{\mathrm{i}\gamma_1(1 - \xi_2/\lambda)} \int_0^{\zeta} \Big[\frac{\sin(\pi\varrho/\lambda)}{\sin(\pi(\xi_2 - \varrho)/\lambda)} \Big]^{-\gamma_1/\pi} d\varrho. \tag{7.13.60}$$

Since $\gamma_1 < 0$, the exponent inside the integral is positive and a singularity does not appear at the origin, $\rho = 0$. The integral can be computed accurately using the numerical methods discussed in Section B.6, Appendix B. Applying equation (7.13.59) for $j = 2$ provides us with an algebraic equation for ξ_2, which can be recast into the form

$$\mathcal{F}(\xi_2) \equiv z_1 - z_2 + \frac{L}{\lambda} e^{\mathrm{i}(1 - \gamma_1\xi_2/\lambda)} \int_0^{\xi_2} \Big[\frac{\sin(\pi\xi/\lambda)}{\sin(\pi(\xi_2 - \xi)/\lambda)} \Big]^{-\gamma_1/\pi} d\xi = 0. \tag{7.13.61}$$

To compute the constant ξ_2, we solve the real equation $\mathcal{F}\mathcal{F}^* = 0$ using, for example, Newton's method discussed in Section B.3, Appendix B. To evaluate the integral on the right-hand side, denoted by $\mathcal{I}$, we remove the singularity of the integrand at $\xi = \xi_2$ by writing

$$\mathcal{I} = \int_0^{\xi_2} \Big(\Big[\frac{\sin(\pi\xi/\lambda)}{\sin(\pi(\xi_2 - \xi)/\lambda)} \Big]^{-\gamma_1/\pi} - \Big[\frac{\sin(\pi\xi_2/\lambda)}{\pi(\xi_2 - \xi)/\lambda} \Big]^{-\gamma_1/\pi} \Big) d\xi$$
$$+ \Big[\frac{\sin(\pi\xi_2/\lambda)}{\pi/\lambda} \Big]^{-\gamma_1/\pi} \int_0^{\xi_2} \Big[\frac{1}{\xi_2 - \xi} \Big]^{-\gamma_1/\pi} d\xi. \tag{7.13.62}$$

(a)

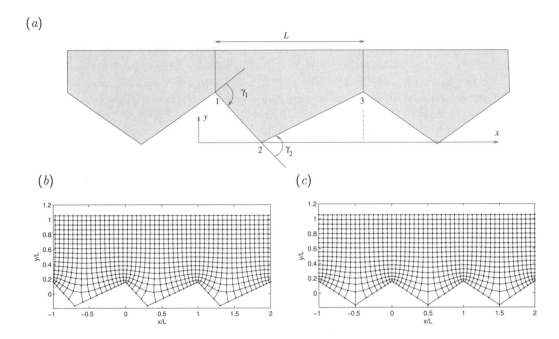

(b)   (c)

FIGURE 7.13.12   (a) Mapping a semi-infinite domain bounded by a periodic sequence of asperities to the upper half-plane. (b, c) Grids generated by mapping a uniform Cartesian grid in the ζ plane to the z plane.

The first integral on the right-hand side can be computed accurately using the numerical methods discussed in Section B.6, Appendix B. The second integral on the right-hand side is equal to ξ_2^m/m, where $m = 1 + \gamma_1/\pi > 0$. A grid of points generated by mapping a uniform Cartesian grid from the ζ to the z plane is shown in Figure 7.13.12(c). In the case of symmetric asperities where $x_2 = x_1 + \frac{1}{2}L$, while y_1 and y_2 are arbitrary, we set $\xi_2 = \frac{1}{2}\lambda$ and obtain

$$z = z_1 + \frac{L}{\lambda}\,\mathrm{e}^{\mathrm{i}\gamma_1/2} \int_0^\zeta \left[\tan\left(\frac{\pi\varrho}{\lambda}\right)\right]^{-\gamma_1/\pi} \mathrm{d}\varrho. \tag{7.13.63}$$

A grid generated by mapping a uniform Cartesian grid is shown in Figure 7.13.12(b).

Polygonal strip

Transformations mapping a periodic channel with polygonal walls to an infinite strip are available [129, 130, 132, 133]. The mapping functions are useful in developing orthogonal grids for use with finite-difference methods, or studying scalar conduction in domains with involved and fractal geometry [53].

7.13.5 Numerical methods

We have discussed transformations mapping the interior or exterior of a generalized polygon to the upper half-plane, to the unit disk or the exterior of a unit disk. For polygons with a small number of vertices, the constants involved in the mapping function can be computed analytically or by standard numerical methods. Unfortunately, as the number of vertices increases, high accuracy in evaluating improper integrals in the complex plane is required due to vertex crowding. To address these concerns, efficient iterative numerical methods for computing Schwarz–Christoffel and related transformations for polygonal domains with complicated geometry have been developed [53, 129, 133, 406].

The usefulness of the standard Schwarz–Christoffel transformation appears to be limited by the requirement of polygonal boundary geometry. However, any curved boundary can be approximated with a polygonal line connecting a chain of vertices. Thus, our ability to map a polygonal domain with a large number of vertices to a disk or to the half-plane can be exploited to derive approximate mapping functions for arbitrary domains. An alternative is to use generalized Schwarz–Christoffel transformations for smooth domains.

7.13.6 Generalized Schwarz–Christoffel transformations

The Schwarz–Christoffel transformation discussed in Section 7.13.1 maps the interior of a polygon to the semi-infinite plane, as shown in Figure 7.13.1. The differential form of the relevant inverse mapping function given in (7.13.2) can be recast into the judiciously designed form

$$\frac{dz}{d\zeta} = c \prod_{i=1}^{N} \exp\left(-\frac{\gamma_i}{\pi} \log(\zeta - \xi_i)\right).$$ (7.13.64)

An equivalent form is

$$\frac{dz}{d\zeta} = c \prod_{i=1}^{N} \exp\left(-\frac{1}{\pi} \int_{-\infty}^{\infty} \frac{d\varphi}{d\xi} \log(\zeta - \xi)\, d\xi\right),$$ (7.13.65)

where φ is the slope angle defined in Figure 7.13.13. In the case of an N-sided polygonal boundary, φ is a discontinuous function,

$$\frac{d\varphi}{d\xi} = \sum_{i=1}^{N} \gamma_i\, \delta(\xi - \xi_i),$$ (7.13.66)

and (7.13.65) reduces to (7.13.64), where δ is the one-dimensional delta function.

Smooth contours

In fact, the transformation (7.13.65) applies for polygonal as well as smooth shapes with nonsingular slope functions $\varphi(\xi)$. A slight rearrangement yields

$$\frac{dz}{d\zeta} = c \prod_{i=1}^{N} \exp\left(-\frac{1}{\pi} \int_{-\infty}^{\infty} \frac{d\varphi}{d\ell} \log(\zeta - \xi) \frac{d\ell}{d\xi}\, d\xi\right),$$ (7.13.67)

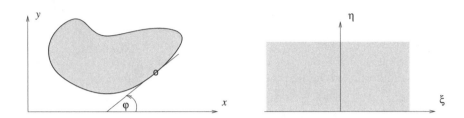

FIGURE 7.13.13   A generalized Schwarz–Christoffel transformation maps a smooth domain in the z plane to the upper half ζ plane.

involving the known function $\mathrm{d}\varphi/\mathrm{d}\ell$ and the unknown function $\mathrm{d}\ell/\mathrm{d}\xi$, where ℓ is an arc-length like parameter describing the physical contour [129, 133]. To compute the mapping function, we may trace a smooth contour enclosing a domain of interest in the z plane with a number of marker points, introduce the corresponding images, ξ_i, construct the function $\mathrm{d}\ell/\mathrm{d}\xi$ by interpolation, and work as in the case of a polygon to generate a system of nonlinear equations.

To map a smooth region to the unit disk in the τ plane, we use the transformation

$$\frac{\mathrm{d}z}{\mathrm{d}\tau} = d \prod_{i=1}^{N} \exp\left(-\frac{1}{\pi} \oint \frac{\mathrm{d}\varphi}{\mathrm{d}\ell} \log(\tau - v) \frac{\mathrm{d}\ell}{\mathrm{d}v} \, \mathrm{d}v \right), \tag{7.13.68}$$

corresponding to the polygon transformation (7.13.42), where d is a complex constant and the integral is performed in the counterclockwise direction around the unit circle. Writing $v = \exp(i\chi)$, we obtain

$$\frac{\mathrm{d}z}{\mathrm{d}\tau} = d \prod_{i=1}^{N} \exp\left(-\frac{1}{\pi} \int_{0}^{2\pi} \frac{\mathrm{d}\varphi}{\mathrm{d}\ell} \log(\tau - e^{i\chi}) \frac{\mathrm{d}\ell}{\mathrm{d}\chi} \, \mathrm{d}\chi \right). \tag{7.13.69}$$

A weak logarithmic singularity occurs when the point τ is located at the unit circle. Working in a similar fashion, we derive transformations that map the exterior of a smooth contour to the upper half-plane or exterior of the unit disk.

Rounded corners and circular arcs

Modified transformations for polygons with rounded corners [7] and polygons of circular arcs [194] are available.

Problem

7.13.1 *Mapping the exterior of a flat plate to the exterior of a disk*

Derive (7.12.7) from (7.13.48), thus obtaining a transformation that maps the exterior of a disk with radius $\lambda = \frac{1}{2}\,a$ centered at the origin of the ζ plane to the exterior of a flat plate subtended between the points $x = \pm a$ in the z plane.

Computer Problems

7.13.2 *Flow in a strip*

Compute and plot the streamlines of the flow inside the semi-infinite strip using the transformation (7.13.12), subject to uniform flow along the ξ axis in the ζ plane.

7.13.3 *Semi-infinite flow above a step*

Compute and plot the streamlines of potential flow in the semi-infinite region above a step corresponding to the transformation (7.13.41), subject to uniform flow along the ξ axis in the ζ plane.

7.13.4 *Mapping the interior of a rectangle to the upper half-plane*

Write a program that computes the constants involved in the transformation that maps the rectangle shown in Figure 7.13.3 to the upper half-plane. The algorithm should employ the iterative method discussed in the text. Plot the streamlines of the flow inside the rectangle corresponding to uniform flow along the ξ axis in the ζ plane.

7.13.5 *Plane with flat plates*

Consider a semi-infinite domain above a horizontal plane located at $y = 0$ hosting an infinite sequence of flat plates with length a separated by distance L. Substituting into (7.13.56) $N = 3$, $\gamma_1 = -\pi$, $\gamma_2 = \pi/2$, and $\gamma_3 = \pi/2$, along with $\xi_1 = 0$ and $\xi_3 = \lambda - \xi_2$, we obtain

$$z = z_1 + \frac{L}{\lambda} \int_0^\zeta \sin\left(\frac{\pi \varrho}{\lambda}\right) \left[\sin\left(\pi \frac{\varrho - \xi_2}{\lambda}\right) \sin\left(\pi \frac{\varrho + \xi_2}{\lambda}\right) \right]^{-1/2} d\varrho. \tag{7.13.70}$$

The plate aspect ratio, a/L, is determined implicitly by the dimensionless parameter ξ_2/λ ranging in the interval $(0, 1)$. Generate a grid of points similar to that shown in Figure 7.13.12(b) for several values of ξ_2/λ.

Boundary-layer analysis

8

There is an important class of flows where the gradient of the vorticity or the vorticity itself vanishes virtually everywhere, except inside thin layers or narrow columns of fluid wrapping around or trailing behind solid boundaries, free surfaces, and fluid interfaces, loosely identified as boundary layers. Inspecting the equation of motion, we find that viscous forces are small and can be neglected outside the boundary layers, but make important contributions and must be retained inside the boundary layers.

An important consequence of dropping the viscous force in the Navier–Stokes equation to obtain the Euler equation outside boundary layers is that the order of the governing equations with respect to the spatial partial derivatives is reduced from two to one. This makes it impossible to satisfy, in general, more than one scalar boundary condition over each boundary, as required for viscous fluids. The presence of boundary layers inside which the motion of the fluid is governed by a second-order partial differential equation due to the presence of the viscous force is thus imperative. An important conclusion is that, with some exceptions, viscous forces cannot be neglected uniformly throughout the domain of a flow.

Prandtl boundary layers occur in high-Reynolds-number flow past a streamlined body, such as an airfoil, or flow past a round but not so bluff body, such as a cylinder or a sphere. At the turn of the twentieth century, Prandtl argued that, at sufficiently high Reynolds numbers, but not so high that the flow becomes turbulent, the bulk of the flow is nearly irrotational and vorticity gradients are confined inside boundary layers lining the surface of a body as well as inside narrow wakes and possibly compact regions of recirculating flow [327]. The flow may then be analyzed in two stages: first, we consider the outer irrotational flow subject to the no-penetration boundary condition; second, we compute the flow inside the boundary layers subject to appropriate simplifications based on the realization that the flow is nearly unidirectional, driven by the tangential component of the outer flow. The boundary-layer flow satisfies the no-slip boundary condition at a wall and agrees with the outer-flow solution far from the wall.

Following Prandtl's original idea, a large body of theoretical and laboratory work has confirmed the physical relevance and efficiency of boundary-layer theory, opening a new era in the study of viscous flow (e.g., [395]). Prandtl boundary layers have been identified over surfaces translating at high speed or stretching in an otherwise quiescent fluid in a broad range of applications. The importance of boundary-layer flows in practical aerodynamics has spawned a large body of literature discussed in comprehensive monographs and reviews [41, 70, 71, 260, 352, 363, 376].

The physical processes determining the behavior of boundary layers developing over solid surfaces in unsteady flow become evident by considering the structure of a boundary layer developing along a rigid body that is held stationary in an impulsively started incident flow. At the initial instant, the velocity field is irrotational everywhere, except inside a thin vortex layer lining the body. Viewed at a distance, the vortex layer resembles a vortex sheet whose strength is equal to the negative of the tangential velocity of the potential flow. As soon as the motion has been initiated, the vortex sheet diffuses into the flow. Locally around the boundary, the flow resembles that developing over an infinite flat plate subject to an impulsively started uniform flow described by a similarity solution, as discussed in Section 5.4.3. When the thickness of the vortex layer becomes significant, convection of vorticity in the tangential and normal directions becomes important and the leading-order similarity solution is modified with the addition of a second-order term that is proportional to time since startup [391]. Examination of the second-order term shows that the boundary shear stress vanishes and back flow occurs at the point where the irrotational flow decelerates along the boundary. The time of separation predicted by the second-order solution is in remarkably good agreement with experimental observation.

More generally, changes in the conditions of an incident flow affect the distribution of the tangential velocity and pressure gradient that drive the flow inside the boundary layer, and therefore cause the vorticity to diffuse across the boundaries and enter or exit the flow. The precise description of the evolution of an unsteady boundary-layer flow presents us with significant analytical and computational challenges that have been tackled only for a limited number of flows [352].

A different class of boundary layers occurs at free surfaces and interfaces between two streams of the same or different fluids. In the case of interfacial flow, an irrotational incident flow is not capable of satisfying both the kinematic boundary condition of continuity of velocity and the dynamic boundary condition specifying the discontinuity in the interfacial traction, and the role of the boundary layer is to make an appropriate adjustment. An example is the boundary layer established around the surface of a gas bubble rising through a liquid. Another example is the boundary layer established around a liquid jet discharging into an ambient fluid.

In this chapter, we introduce the fundamental concepts involved in the formulation of boundary-layer theory and discuss analytical, approximate, and numerical methods for solving the governing equations in various types of steady and unsteady flow. We will consider Prandtl boundary layers developing over stationary surfaces in nonaccelerating and accelerating flow, and boundary layers developing over moving or stretching surfaces in an otherwise quiescent fluid. By way of application, the theory will be used to predict the expansion of a two-dimensional or axisymmetric oil slick floating on a pool of an otherwise quiescent fluid.

8.1 Boundary-layer theory

A fundamental assumption in developing a boundary-layer theory is that the flow consists of an outer region where vorticity gradients virtually vanish and the flow is described by the equations of inviscid flow, including Euler's equation and the continuity equation, and a boundary layer attached to a surface. Wakes and regions of recirculating flow are allowed, but are significant only insofar as to modify the structure of the outer flow.

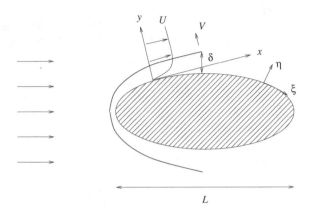

FIGURE 8.1.1   Illustration of the Prandtl boundary layer developing around a two-dimensional body. Shown are the typical velocity, U, the typical length, L, and the boundary-layer thickness, δ.

8.1.1   Prandtl boundary layer on a solid surface

To illustrate the physical arguments involved in the boundary-layer formulation and demonstrate the resulting mathematical simplifications, we consider the boundary layer developing around a mildly curved two-dimensional rigid body that is held stationary in an incident irrotational flow, as illustrated in Figure 8.1.1.

Continuity equation

We begin the boundary-layer analysis by introducing Cartesian coordinates where the x axis is tangential and the y axis is perpendicular to the surface of the body at a point, as shown in Figure 8.1.1. Next, we apply the continuity equation at a point in the vicinity of the origin,

$$\frac{\partial u}{\partial x} + \frac{\partial v}{\partial y} = 0, \tag{8.1.1}$$

where $u = u_x$ and $v = u_y$ are the components of the velocity along the x and y axes.

Let L be the typical dimension of the body, U be the typical magnitude of the velocity of the outer irrotational flow, δ be the typical thickness of the boundary layer, defined as the region around the body across which the velocity undergoes a rapid variation and the magnitude of the viscous force is significant, and V be the typical magnitude of the velocity component normal to the body at the edge of the boundary layer. We expect that the magnitude of the derivative $\partial u/\partial x$ inside the boundary layer will be comparable to the ratio U/L, and the magnitude of the derivative $\partial v/\partial y$ will be comparable to the ratio V/δ. The continuity equation (8.1.1) requires that

$$\frac{U}{L} \sim \frac{V}{\delta} \qquad \text{or} \qquad V \sim U\frac{\delta}{L}, \tag{8.1.2}$$

stating that V scales with the boundary-layer thickness, δ. In fact, V is roughly proportional to the boundary-layer thickness, δ.

Next, we turn to examining the two components of the equation of motion inside the boundary layer in the vicinity of the origin written for the hydrodynamic pressure that excludes the effect of the body force. For simplicity, the hydrodynamic pressure is denoted without a tilde, as p.

x component of the equation of motion

Considering the x component of the Navier–Stokes equation,

$$\frac{\partial u}{\partial t} + u\frac{\partial u}{\partial x} + v\frac{\partial u}{\partial y} = -\frac{1}{\rho}\frac{\partial p}{\partial x} + \nu\frac{\partial^2 u}{\partial x^2} + \nu\frac{\partial^2 u}{\partial y^2}, \qquad (8.1.3)$$

$$U^2/L \quad U^2/L \qquad\qquad \nu U/L^2 \quad \nu U/\delta^2$$

we introduce the scalings

$$u \sim U, \qquad \frac{\partial u}{\partial x} \sim \frac{U}{L}, \qquad v \sim V, \qquad \frac{\partial u}{\partial y} \sim \frac{U}{\delta}, \qquad \frac{\partial^2 u}{\partial x^2} \sim \frac{U}{L^2}, \qquad \frac{\partial^2 u}{\partial y^2} \sim \frac{U}{\delta^2}, \qquad (8.1.4)$$

where ν is the kinematic viscosity of the fluid. Moreover, we use the scaling (8.1.2) to eliminate V in favor of U, and find that the magnitudes of the various terms are as shown underneath equation (8.1.3). The scaling of the first term involving the time derivative on the left-hand side is determined by the temporal variation of the outer flow, which is left unspecified. At this point, there is no obvious way of scaling the x derivative of the dynamic pressure gradient on the right-hand side of (8.1.3) in the context of kinematics alone.

The scalings shown underneath equation (8.1.3), combined with the fundamental assumption that $\delta \ll L$, have two important consequences. First, the penultimate viscous term on the right-hand side is small compared to the last term and can be neglected, yielding the boundary-layer equation

$$\frac{\partial u}{\partial t} + u\frac{\partial u}{\partial x} + v\frac{\partial u}{\partial y} = -\frac{\partial p}{\partial x} + \nu\frac{\partial^2 u}{\partial y^2}. \qquad (8.1.5)$$

Second, the magnitude of the last viscous term must be comparable to the magnitude of the inertial terms on the left-hand side, requiring that

$$\frac{U^2}{L} \sim \nu\frac{U}{\delta^2}. \qquad (8.1.6)$$

Rearranging, we find that

$$\delta \sim \left(\frac{\nu L}{U}\right)^{1/2} = \frac{L}{\sqrt{\mathrm{Re}}}, \qquad (8.1.7)$$

where $\mathrm{Re} = UL/\nu$ is the Reynolds number.

y component of the equation of motion

Next, we consider the individual terms in the y component of the Navier–Stokes equation,

$$\frac{\partial v}{\partial t} + u\frac{\partial v}{\partial x} + v\frac{\partial v}{\partial y} = -\frac{1}{\rho}\frac{\partial p}{\partial y} + \nu\frac{\partial^2 v}{\partial x^2} + \nu\frac{\partial^2 v}{\partial y^2},$$

$$U^2\delta/L^2 \quad U^2\delta/L^2 \qquad\qquad U^2\delta/L^2 \quad U^2\delta/L^2 \qquad (8.1.8)$$

and introduce the scalings

$$v \sim V, \qquad \frac{\partial v}{\partial x} \sim \frac{V}{L}, \qquad u \sim U, \qquad \frac{\partial v}{\partial y} \sim \frac{V}{\delta}, \qquad \frac{\partial^2 v}{\partial x^2} \sim \frac{V}{L^2}, \qquad \frac{\partial^2 v}{\partial y^2} \sim \frac{V}{\delta^2}. \qquad (8.1.9)$$

Moreover, we express the kinematic viscosity ν in terms of δ using (8.1.6), setting $\nu \sim U\delta^2/L$, and find that the magnitudes of the various terms are as shown underneath equation (8.1.8).

The magnitudes of all nonlinear convective and viscous terms are of order δ. Unless the magnitude of the time derivative on the left-hand side is of order unity, the dynamic pressure gradient across the boundary layer must also be of order δ, $\partial p/\partial y \simeq \delta$. This deduction allows us to write to leading-order approximation

$$\frac{\partial p}{\partial y} \simeq 0. \qquad (8.1.10)$$

We conclude that nonhydrostatic pressure variations across the boundary layer are negligible, and the dynamic pressure inside the boundary layer is primarily a function of position along the wall.

Boundary-layer equations

To compute the streamwise pressure gradient, we evaluate the x component of the Euler equation at the edge of the boundary layer, obtaining

$$-\frac{1}{\rho}\frac{\partial p}{\partial x} = \frac{\partial U}{\partial t} + U\frac{\partial U}{\partial x}, \qquad (8.1.11)$$

where U is the tangential component of the velocity of the outer flow. The boundary-layer equation (8.1.5) then becomes

$$\frac{\partial u}{\partial t} + u\frac{\partial u}{\partial x} + v\frac{\partial u}{\partial y} = \frac{\partial U}{\partial t} + U\frac{\partial U}{\partial x} + \nu\frac{\partial^2 u}{\partial y^2}, \qquad (8.1.12)$$

where U is regarded as a known forcing function.

The continuity equation (8.1.1) and the boundary-layer equation (8.1.12) provide us with a system of two second-order nonlinear partial-differential equations for the x and y velocity components, u and v, to be solved subject to the no-slip and no-penetration boundary conditions requiring that u and v are zero along the boundary, and a far-field condition requiring that, as y/δ tends to infinity, u tends to the tangential component of the outer velocity, U. Because the boundary-layer equations do not involve a second partial derivative of v with respect to y, a far-field condition for v is not required. The pressure plays the role of a forcing function computed by solving the equations governing the outer irrotational flow.

Favorable and adverse pressure gradient

Evaluating (8.1.12) at the origin and enforcing the no-slip and no-penetration conditions in a steady flow, we obtain

$$\left(\frac{\partial^2 u}{\partial y^2}\right)_{y=0} = -\frac{U}{\nu}\frac{dU}{dx}, \qquad (8.1.13)$$

which shows that the sign of the curvature of the velocity profile at the boundary is opposite to that of the streamwise acceleration of the outer flow, dU/dx. Thus, the flow inside the boundary layer in a decelerating outer flow, corresponding to $dU/dx < 0$, reverses direction, causing convection of vorticity away from the boundary and the consequent formation of wakes and vortices inside the bulk of the flow.

When $dU/dx > 0$, the pressure gradient is negative, $dp/dx < 0$, and the boundary layer is subjected to a favorable pressure gradient. In the opposite case where $dU/dx < 0$, the pressure gradient is positive, $dp/dx > 0$, and the boundary layer is subjected to an adverse pressure gradient. Equation (8.1.13) shows that an adverse pressure gradient promotes flow separation.

Boundary-layer equations in curvilinear coordinates

The Prandtl boundary-layer equation (8.1.12) was developed with reference to the local Cartesian axes shown in Figure 8.1.1, and is strictly valid near the origin. To avoid redefining the Cartesian axes at every point along the boundary, we introduce a curvilinear coordinate system where the ξ axis is tangential to the boundary and the η axis is perpendicular to the boundary, as shown in Figure 8.1.1. The corresponding velocity components are denoted by u_ξ and u_η.

Performing the boundary-layer analysis, we find that the boundary-layer equations (8.1.1), (8.1.10), and (8.1.12) remain valid to leading-order approximation, provided that the Cartesian x and y coordinates are replaced by corresponding arc lengths in the ξ and η directions denoted, respectively, by l and ℓ. Equation (8.1.10) becomes

$$\frac{\partial p}{\partial \ell} = \frac{\rho U_\xi^2}{R}, \tag{8.1.14}$$

where R is the radius of curvature of the boundary. We conclude that the dynamic pressure drop across the boundary layer is of order δ, provided that R is not too small, that is, provided that the boundary is not too sharply curved. For simplicity, in the remainder of this chapter we denote l and ℓ, respectively, by x and y.

Parabolization of the equation of motion

The absence of a second partial derivative with respect to x renders the boundary-layer equation (8.1.12) a parabolic partial differential equation with respect to the streamwise position, x. This classification has important consequences on the nature of the solution and on the choice of numerical methods for computing the solution. Specifically, the system of equations (8.1.1) and (8.1.12) can be solved using a marching method with respect to x, beginning from a particular x station where the structure of the boundary layer is known. In contrast, the Navier–Stokes equation is an elliptic partial differential equation with respect to x and y, and the solution must by found simultaneously at every point in the flow, even when the velocity and pressure are specified at the inlet.

The parabolic nature of the boundary-layer equation (8.1.12) with respect to x implies that a perturbation introduced at some point along the boundary layer modifies the structure of the flow downstream but leaves the upstream flow unchanged. The absence of the second partial derivative with respect to x due to the boundary-layer approximation precludes a mechanism for upstream signal propagation.

Dimensionless form of the equation of motion

The estimates of the various terms in the equation of motion discussed earlier in this section suggest a particular way of defining dimensionless variables to be used in nondimensionalizing the equation of motion, given by

$$
\hat{x} = \frac{x}{L}, \qquad \hat{y} = \frac{y}{\delta} = \frac{y}{L}\,\mathrm{Re}^{1/2}, \qquad \hat{t} = \frac{tU}{L},
$$

$$
\hat{u} = \frac{u}{U}, \qquad \hat{v} = \frac{v}{V} = \frac{v}{U}\,\mathrm{Re}^{1/2}, \qquad \hat{p} = \frac{p}{\rho U^2}. \tag{8.1.15}
$$

The corresponding dimensionless form of the continuity equation and two components of the unsimplified Navier–Stokes equation are

$$
\frac{\partial \hat{u}}{\partial \hat{x}} + \frac{\partial \hat{v}}{\partial \hat{y}} = 0,
$$

$$
\frac{\partial \hat{u}}{\partial \hat{t}} + \hat{u}\frac{\partial \hat{u}}{\partial \hat{x}} + \hat{v}\frac{\partial \hat{u}}{\partial \hat{y}} = -\frac{\partial \hat{p}}{\partial \hat{x}} + \frac{1}{\mathrm{Re}}\frac{\partial^2 \hat{u}}{\partial \hat{x}^2} + \frac{\partial^2 \hat{u}}{\partial \hat{y}^2},
$$

$$
\frac{1}{\mathrm{Re}}\left(\frac{\partial \hat{v}}{\partial \hat{t}} + \hat{u}\frac{\partial \hat{v}}{\partial \hat{x}} + \hat{v}\frac{\partial \hat{v}}{\partial \hat{y}}\right) = -\frac{\partial \hat{p}}{\partial \hat{y}} + \frac{1}{\mathrm{Re}^2}\frac{\partial^2 \hat{v}}{\partial \hat{x}^2} + \frac{1}{\mathrm{Re}}\frac{\partial^2 \hat{v}}{\partial \hat{y}^2}. \tag{8.1.16}
$$

The magnitude of each partial derivative in equations (8.1.16) is of order unity. Taking the limit as Re tends to infinity and retaining only the leading-order terms, we obtain a simplified system of governing equations,

$$
\frac{\partial \hat{u}}{\partial \hat{x}} + \frac{\partial \hat{v}}{\partial \hat{y}} = 0, \qquad \frac{\partial \hat{u}}{\partial \hat{t}} + \hat{u}\frac{\partial \hat{u}}{\partial \hat{x}} + \hat{v}\frac{\partial \hat{u}}{\partial \hat{y}} = -\frac{\partial \hat{p}}{\partial \hat{x}} + \frac{\partial^2 \hat{u}}{\partial \hat{y}^2}, \qquad \frac{\partial \hat{p}}{\partial \hat{y}} = 0. \tag{8.1.17}
$$

Reverting to dimensional variables and using Bernoulli's equation, we recover equations (8.1.1), (8.1.10), and (8.1.12).

The absence of the Reynolds number in the governing equations (8.1.17) suggests that the Reynolds number is significant only insofar as to determine the physical thickness of the boundary layer and magnitude of the y velocity component according to (8.1.12), and does not have an influence on the structure of the flow inside the boundary layer. However, the assumption that the Reynolds number is sufficiently high is necessary for this asymptotic behavior to prevail.

Energy dissipation inside a boundary layer

Viewed at a distance, a boundary layer resembles a vortex sheet whose strength is equal to the tangential component of the velocity of the outer flow. Since the velocity gradient diverges at the location of a vortex sheet, the associated rate of viscous dissipation is infinite. Viewed from a point in its proximity, the boundary layer resembles a vortex layer with noninfinitesimal thickness incurring a finite rate of viscous dissipation. Specifically, the rate of viscous dissipation per unit mass of fluid inside the boundary layer is

$$
\nu\left(\frac{\partial u}{\partial y}\right)^2 \sim \nu\,\frac{U^2}{\delta^2}, \tag{8.1.18}
$$

which, according to (8.1.7), scales as U^3/L and therefore remains finite as the viscosity tends to vanish or the Reynolds number tends to infinity. The rate of viscous dissipation inside the boundary layer per unit surface area of the boundary is

$$\int_0^\delta \mu \left(\frac{\partial u}{\partial y}\right)^2 \mathrm{d}y \sim \mu \frac{U^2}{\delta}, \tag{8.1.19}$$

which, according to (8.1.7), scales as $\rho U^3/\mathrm{Re}^{1/2}$. Since the rate of viscous dissipation per unit volume of fluid outside the boundary layer scales with $\rho U^3/(L\mathrm{Re})$, the rate of viscous dissipation inside the boundary layer makes a dominant contribution.

Flow separation

Boundary-layer analysis for laminar flow is based on two key assumptions: the Reynolds number is sufficiently high, but not so high that the flow becomes turbulent, and the vorticity remains confined inside boundary layers wrapping around the boundaries. The physical relevance of the second assumption depends on the structure of the incident flow and boundary geometry. Streamlined bodies allow laminar boundary layers to develop over a large surface area, whereas bluff bodies cause vorticity to concentrate inside compact regions forming steady or unsteady wakes. For example, the alternating ejection of vortices of opposite sign into a wake is responsible for the von Kármán vortex street illustrated in Figure 8.1.2(a) (e.g., [187]). These limitations should be born in mind in carrying out a boundary-layer analysis.

Drag force on a body

The changes in the structure of a flow as a function of the Reynolds number are reflected on global flow diagnostics, such as the drag force exerted on a body and the angular frequency of the vortices developing in the wake of a body, Ω, expressed in terms of the Strouhal number, $\mathrm{St} = \Omega L/U$. The dimensionless drag coefficient for a circular cylinder with diameter D is defined as

$$c_D \equiv \frac{F}{\frac{1}{2}\rho U^2 D}, \tag{8.1.20}$$

where F is the drag force per unit length exerted on the cylinder. In Figure 8.1.2(b), the drag coefficient is plotted against the Reynolds number, $\mathrm{Re} = UD/\nu$, on a log-log scale. Using (6.14.22), we find that, in the limit of vanishing Reynolds number, the drag force is given by the modified Stokes law

$$F \simeq \frac{4\pi\mu U}{\ln\dfrac{7.4}{\mathrm{Re}}}. \tag{8.1.21}$$

Correspondingly, the drag coefficient is given by

$$c_D \simeq \frac{8\pi}{\mathrm{Re}\,\ln\dfrac{7.4}{\mathrm{Re}}}. \tag{8.1.22}$$

The predictions of this formula are represented by the dashed line in Figure 8.1.2(b). The apparent change in the functional form of the drag coefficient at a critical Reynolds number on the order of

(a)   (b)

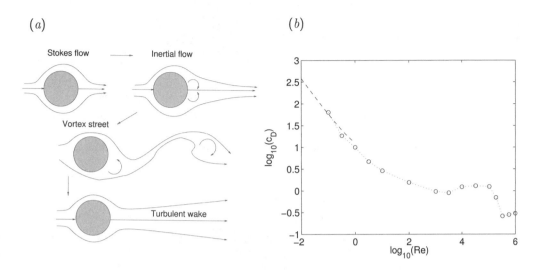

FIGURE 8.1.2 (*a*) Changes in the structure of streaming flow past a circular cylinder with increasing
Reynolds number showing boundary-layer separation and the development of a wake consisting of
a vortex street. (*b*) Laboratory data for the drag coefficient on a circular cylinder that is held
stationary in streaming flow plotted against the Reynolds number defined with respect to the
cylinder diameter. The dashed line represents an analytical prediction for low-Reynolds-number
flow.

10^3, shown in Figure 8.1.2(*b*), is due to the detachment of the boundary layer from the surface of
the cylinder at a certain point at the rear surface of the cylinder, in a process that is described as
flow separation, as shown in Figure 8.1.2(*a*). When the Reynolds number becomes on the order of
10^5, the flow becomes fully turbulent and the boundary layer reattaches, causing a sudden decline
in the drag coefficient.

Three-dimensional boundary layers

To analyze the structure of a boundary layer over a three-dimensional curved surface, we introduce
curvilinear coordinates with two axes lying in the surface and the third axis perpendicular to the
surface, and implement the boundary-layer approximation with respect to the normal coordinate.
The analysis reveals that the pressure drop across the boundary layer can be neglected, and the
tangential components of the Navier–Stokes equation can be parabolized by eliminating the second
partial derivative of the tangential velocity with respect to distance in a tangential direction to
leading-order approximation. The precise form of the boundary-layer equations for three-dimensional
flow are discussed in Section 8.5.2.

8.1.2   Boundary layers at free surfaces and fluid interfaces

Consider the flow near a free surface where the tangential component of the traction is required to
vanish. An irrotational ambient flow is not capable of satisfying the condition of vanishing shear

stress, and a boundary layer must be established. An example is the boundary layer developing over the surface of a spherical bubble rising through a liquid at high Reynolds numbers, as discussed in Section 7.5.4 [259]. An important difference between this boundary layer and the Prandtl boundary layer developing over a solid surface is that, in the case of a free surface, the velocity does not have to undergo a jump whose magnitude is comparable to the velocity of the incident flow.

Free surfaces

The tangential component of the vorticity at an arbitrary surface was given in (3.9.18) as

$$\boldsymbol{\omega} \cdot (\mathbf{I} - \mathbf{n} \otimes \mathbf{n}) = (\mathbf{n} \times \boldsymbol{\omega}) \times \mathbf{n} = [(\nabla \mathbf{u}) \cdot \mathbf{n} - \mathbf{n} \cdot \nabla \mathbf{u}] \times \mathbf{n}, \tag{8.1.23}$$

and the rotated shear stress was given in (3.9.16) as

$$\mathbf{f} \times \mathbf{n} = \mu \Big[(\mathbf{n} \cdot \nabla \mathbf{u}) \times \mathbf{n} - \mathbf{n} \times [(\nabla \mathbf{u}) \cdot \mathbf{n}] \Big], \tag{8.1.24}$$

where $\mathbf{f}$ is the traction and $\mathbf{n}$ is the normal unit vector. Combining these expressions, we obtain

$$\boldsymbol{\omega} \cdot (\mathbf{I} - \mathbf{n} \otimes \mathbf{n}) = \frac{1}{\mu} \mathbf{f} \times \mathbf{n} - 2 (\mathbf{n} \cdot \nabla \mathbf{u}) \times \mathbf{n} = -\frac{1}{\mu} \mathbf{f} \times \mathbf{n} + 2 [(\nabla \mathbf{u}) \cdot \mathbf{n}] \times \mathbf{n}. \tag{8.1.25}$$

Requiring that the tangential component of the traction vanishes at the free surface, we find that the tangential component of the vorticity is given by equation (3.9.19), repeated here for convenience,

$$\boldsymbol{\omega}_{fs} \cdot (\mathbf{I} - \mathbf{n} \otimes \mathbf{n}) = -2 (\mathbf{n} \cdot \nabla \mathbf{u}) \times \mathbf{n} = 2 [(\nabla \mathbf{u}) \cdot \mathbf{n}] \times \mathbf{n}, \tag{8.1.26}$$

where the subscript fs denotes the free surface.

It is useful to decompose the velocity field into the outer component, $\mathbf{u}_\infty$, and a complementary velocity field associated with the boundary layer, $\mathbf{u}_{bl}$, so that $\mathbf{u} = \mathbf{u}_\infty + \mathbf{u}_{bl}$. Assume that the outer velocity field $\mathbf{u}_\infty$ is irrotational. Applying (8.1.25) at the edge of the boundary layer where the vorticity is zero, we find that

$$\boldsymbol{\omega}_\infty \cdot (\mathbf{I} - \mathbf{n} \otimes \mathbf{n}) = \frac{1}{\mu} \mathbf{f}_\infty \times \mathbf{n} - 2 (\mathbf{n} \cdot \nabla \mathbf{u}_\infty) \times \mathbf{n} = -\frac{1}{\mu} \mathbf{f}_\infty \times \mathbf{n} + 2 [(\nabla \mathbf{u}_\infty) \cdot \mathbf{n}] \times \mathbf{n} = \mathbf{0}. \tag{8.1.27}$$

To leading order, equation (8.1.26) yields

$$\boldsymbol{\omega}_{fs} \cdot (\mathbf{I} - \mathbf{n} \otimes \mathbf{n}) \simeq -2 (\mathbf{n} \cdot \nabla \mathbf{u}_\infty) \times \mathbf{n} \simeq 2 [(\nabla \mathbf{u}_\infty) \cdot \mathbf{n}] \times \mathbf{n}. \tag{8.1.28}$$

Combining the last two equations, we find that

$$\boldsymbol{\omega}_{fs} \cdot (\mathbf{I} - \mathbf{n} \otimes \mathbf{n}) \simeq \frac{1}{\mu} \mathbf{f}_\infty \times \mathbf{n}. \tag{8.1.29}$$

For the vorticity inside the boundary layer to be of order U/L, the jump in the complementary velocity $\mathbf{u}_{bl}$ across the boundary layer must be of the same order of magnitude as the boundary layer thickness, δ, which scales with $\mathrm{Re}^{-1/2}$, and is therefore small as long as the Reynolds number is sufficiently high. The smallness of $\mathbf{u}_{bl}$ allows us to linearize the equation of motion and vorticity

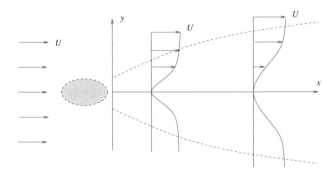

FIGURE 8.1.3   Illustration of a boundary layer developing in the wake of a two-dimensional body.

transport equation with respect to $\mathbf{u}$ about $\mathbf{u}_\infty$. Since the magnitude of $\mathbf{u}_{bl}$ is much smaller than that of $\mathbf{u}_\infty$, the tendency for back flow in a decelerating flow along a free surface is much weaker than along a solid wall, and the rate of viscous dissipation inside the boundary layer is comparable to that inside the incident flow.

Fluid interfaces

To describe the flow around a fluid interface, we introduce a boundary layer on either side of the interface, demand that the velocity is continuous across the interface, and require that the two boundary layers develop so as to satisfy the tangential force balance. The mathematical formulation is discussed by Harper & Moore [169] with reference to the boundary layer developing around the surface of a viscous drop rising through an otherwise quiescent ambient liquid at high Reynolds numbers.

8.1.3   Boundary layers in a homogeneous fluid

Boundary-layer theory assumes that the magnitude of the derivative of the tangential velocity normal to a surface is much larger than the magnitude of the corresponding tangential derivative. In the case of flow of a homogeneous fluid with uniform physical properties, the boundary-layer approximation can be applied in cases where the velocity exhibits sharp variations across thin layers of fluid, called free boundary layers or shear layers. An example is the shear layer developing during the viscous spreading of a vortex sheet discussed in Section 5.4.3.

Spreading of a two-dimensional wake

Another example is the spreading of a symmetric two-dimensional wake behind a streamlined object that is held stationary in a uniform (streaming) incident flow along the x axis with velocity U, as illustrated in Figure 8.1.3. Assuming that the rate of spreading with respect to downstream distance, x, is small, we implement the boundary-layer approximation and derive a linear partial differential equation for the streamwise velocity, $u(x, y)$,

$$U \frac{\partial u}{\partial x} = \nu \frac{\partial^2 u}{\partial y^2}. \tag{8.1.30}$$

Requiring that u tends to U as y tends to infinity, we obtain the Gaussian profile

$$u(x,y) = U \left[1 - \frac{c}{\sqrt{x}} \exp\left(- \frac{U}{4\nu x} y^2 \right) \right], \tag{8.1.31}$$

where c is a dimensional constant determined by the particular form of the velocity distribution at the beginning of the wake.

Writing a momentum integral balance over a rectangular control area with two parallel sides perpendicular to the x axis containing the object, we obtain an expression for the drag force per unit width exerted on the object generating the wake,

$$F = \rho \int_{-\infty}^{\infty} (U^2 - u^2) \, \mathrm{d}y \simeq \rho \int_{-\infty}^{\infty} U \, (U - u) \, \mathrm{d}y. \tag{8.1.32}$$

Substituting the velocity profile, performing the integration, and rearranging, we find that

$$c = \frac{1}{2\sqrt{\pi}} \frac{F}{\rho U^{3/2} \nu^{1/2}} \tag{8.1.33}$$

(Problem 8.1.4). Other free boundary layers are discussed in Problems 8.2.3, 8.5.5, and 8.5.6.

8.1.4   Numerical and approximate methods

The boundary-layer equations for two-dimensional flow provide us with a parabolic system of two nonlinear partial differential equations with respect to tangential arc length, l. The solution can be found using standard space-marching, weighted-residual, or finite-difference methods, as discussed in Chapter 12. Expressing the velocity in terms of the stream function allows us to discard the continuity equation and concentrate on solving one third-order nonlinear parabolic partial-differential equation for the stream function. Reviews of early and more recent implementations are presented by Blottner [41] and Cebeci & Bradshaw [71].

A variety of other specialized methods have been developed for computing the flow inside boundary layers developing over arbitrarily shaped bodies that either move or are held stationary in an arbitrary irrotational flow. A review of early work can be found in a comprehensive monograph edited by Rosenhead [352], a concise and illuminating discussion is given in a textbook by White ([427], Chapter 4), and more recent developments are discussed by Cebeci & Bradshaw [70, 71].

Problems

8.1.1 *Boundary conditions*

Discuss the number of boundary conditions necessary for computing the tangential and normal velocity components inside a two-dimensional Prandtl boundary layer.

8.1.2 *Lubrication flow and boundary layers*

Discuss the relation between the boundary-layer equations and the lubrication equations for nearly unidirectional viscous flow in a narrow channel discussed in Section 6.4.

8.1.3 *Prandtl's transposition theorem*

Show that, if the velocities $u(x, y, t)$ and $v(x, y, t)$ satisfy the Prandtl boundary-layer equations, then the velocities $\tilde{u}(\tilde{x}, \tilde{y}, t)$ and $\tilde{v}(\tilde{x}, \tilde{y}, t)$ will also satisfy the boundary-layer equations, where $\tilde{x} = x$, $\tilde{y} = y + f(x)$, $\tilde{u}(\tilde{x}, \tilde{y}, t) = u(x, y, t)$, $\tilde{v}(\tilde{x}, \tilde{y}, t) = v(x, y, t) + f'(x) u(x, y, t)$, and $f(x)$ is an arbitrary function ([352], p. 211).

8.1.4 *Spreading of a two-dimensional wake*

Derive (8.1.32) and the relation between the coefficient c and F given in (8.1.33).

8.1.5 *Von Mises' transformation*

Von Mises' transformation regards the velocity as a function of the independent variables x and ψ instead of x and y, where ψ is the stream function with zero value over an impermeable boundary.

(*a*) Show that, in the von Mises variables, and subject to the boundary conditions $u = 0$ and $v = 0$ at $y = 0$ and far-field condition that u tends to U as y/δ tends to infinity, the equation of motion (8.1.12) becomes

$$\frac{\partial u}{\partial t} + u\frac{\partial u}{\partial x} = \frac{\partial U}{\partial t} + U\frac{\partial U}{\partial x} + \nu u \frac{\partial}{\partial \psi}\left(u\frac{\partial u}{\partial \psi}\right), \tag{8.1.34}$$

with boundary conditions $u = 0$ at $\psi = 0$ and far-field condition that u tends to U as ψ tends to infinity.

(*b*) Show that, in the case of steady flow, the equation of motion (8.1.12) can be restated as a nonlinear parabolic equation,

$$\frac{\partial \chi}{\partial x} = \nu \left(U^2 - \chi\right)^{1/2} \frac{\partial^2 \chi}{\partial \psi^2}, \tag{8.1.35}$$

where $\chi \equiv U^2 - u^2$, with boundary condition $\chi = U^2$ at $\psi = 0$ and far-field condition requiring that χ decays to zero as ψ tends to infinity.

8.2   Boundary layers in nonaccelerating flow

When the outer tangential velocity, U, is constant, $\mathrm{d}U/\mathrm{d}x = 0$, the boundary-layer equation (8.1.12) simplifies into a nonlinear homogeneous convection–diffusion equation,

$$u\frac{\partial u}{\partial x} + v\frac{\partial u}{\partial y} = \nu \frac{\partial^2 u}{\partial y^2}, \tag{8.2.1}$$

to be solved for the x and y velocity components, $u = u_x$ and $v = u_y$, subject to the continuity equation (8.1.1).

8.2.1   Boundary-layer on a flat plate

The simplest possible boundary layer is encountered in the case of uniform (streaming) flow along a stationary semi-infinite plate that is held parallel to the incident stream, as shown in Figure 8.2.1(*a*). Because the length of the plate is infinite, our only choice for a characteristic length scale introduced

(a)   (b)

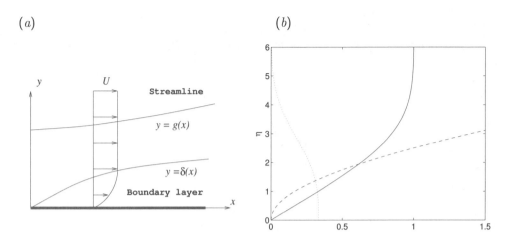

FIGURE 8.2.1   (a) Illustration of a boundary layer developing along a semi-infinite flat plate that is held parallel to a uniform incident stream. (b) Graphs of the Blasius self-similar streamwise velocity profile $u/U = f'$ (solid line), its integral f (dashed line), and derivative f'' (dotted line).

in Section 8.2, L, is the streamwise distance, x. Equation (8.1.7) provides us with an expression for the boundary-layer thickness in terms of the local Reynolds number, $\mathrm{Re}_x \equiv Ux/\nu$,

$$\delta(x) \sim \left(\frac{\nu x}{U}\right)^{1/2} = \frac{x}{\mathrm{Re}_x^{1/2}}. \tag{8.2.2}$$

Recall that this scaling arose by balancing the magnitudes of the inertial and viscous forces inside the boundary layer.

Self-similarity and the Blasius equation

Blasius [39] discovered that computing the solution of the governing partial differential equations (8.2.1) and (8.1.1) can be reduced to solving a single ordinary differential equation. To carry out this reduction, we assume that the flow develops in a self-similar fashion so that the streamwise velocity profile across the boundary layer is a function of a scaled dimensionless transverse position expressed by the similarity variable

$$\eta \equiv \frac{y}{\delta(x)} = \left(\frac{U}{\nu x}\right)^{1/2} y, \tag{8.2.3}$$

according to the functional form

$$u(x, y) = U F(\eta), \tag{8.2.4}$$

where $F(\eta)$ is an *a priori* unknown function to be computed as part of the solution. A key observation is that the self-similar streamwise profile derives from the stream function

$$\psi(x, y) = (U\nu x)^{1/2} f(\eta), \tag{8.2.5}$$

where the function $f(\eta)$ is the indefinite integral or antiderivative of the function $F(\eta)$, satisfying $\mathrm{d}f/\mathrm{d}\eta = F(\eta)$. The principal advantage of using the stream function is that the continuity equation is satisfied automatically and does not need to be considered in further analysis.

Differentiating (8.2.5) with respect to y or x, and invoking the definition of the stream function, we obtain expressions for the x velocity component,

$$u(x,y) = \frac{\partial \psi}{\partial y} = U\,(U\nu x)^{1/2}\, f'\, \frac{\mathrm{d}\eta}{\mathrm{d}y} = U\,(U\nu x)^{1/2}\, f'\, \left(\frac{U}{\nu x}\right)^{1/2}, \tag{8.2.6}$$

and y velocity component,

$$v(x,y) = -\frac{\partial \psi}{\partial x} = -(U\nu)^{1/2}\, \frac{\partial[\sqrt{x}\, f(\eta)]}{\partial x} = -\frac{1}{2}\left(\frac{U\nu}{x}\right)^{1/2} f - (U\nu x)^{1/2}\, f'\, \frac{\partial \eta}{\partial x}, \tag{8.2.7}$$

where a prime denotes a derivative with respect to η. Simplifying, we obtain

$$u(x,y) = U f', \qquad v(x,y) = -\frac{1}{2}\left(\frac{U\nu}{x}\right)^{1/2} (f - \eta f'). \tag{8.2.8}$$

Substituting these expressions into the boundary-layer equation (8.2.1), we derive the Blasius equation

$$f''' + \frac{1}{2}\, f f'' = 0. \tag{8.2.9}$$

An alternative form is

$$\frac{\mathrm{d}\ln f''}{\mathrm{d}\eta} = -\frac{1}{2}\, f. \tag{8.2.10}$$

Enforcing the no-slip and no-penetration boundary conditions, we obtain

$$f(0) = 0, \qquad f'(0) = 0. \tag{8.2.11}$$

Since the flow in the boundary layer reduces to the outer uniform flow far from the plate,

$$f'(\infty) = 1. \tag{8.2.12}$$

Equations (8.2.11) and (8.2.12) provide us with boundary and far-field conditions to be used in solving the Blasius equation (8.2.9). Since boundary conditions are specified at both ends of the computational domain, $[0, \infty)$, we must solve a boundary-value problem involving a third-order, nonlinear, ordinary differential equation.

Rescaling

If $f(\eta) = g(\eta)$ is a solution of (8.2.9), then $f(\eta) = \alpha g(\alpha\eta)$ is also a solution of (8.2.9) for any value of the constant α. Requiring that the function $g(\eta)$ satisfies the boundary conditions $g(0) = 0$, $g'(0) = 0$, and $g''(0) = \beta$, where β is an arbitrary value, and setting $\alpha = 1/[g'(\infty)]^{1/2}$, ensures that the function $\alpha g(\alpha\eta)$ satisfies the boundary conditions (8.2.11) and (8.2.12), and therefore represents the desired solution, $f(\eta)$. However, in practice, it is more expedient to compute the solution by a shooting method, as discussed later in this section.

Weyl integral equation

Integrating by parts and using the boundary conditions (8.2.11), we find that

$$\int_0^\eta (\xi - \eta)^2 f''(\xi)\, \mathrm{d}\xi = -2 \int_0^\eta (\xi - \eta) f'(\xi)\, \mathrm{d}\xi = 2 \int_0^\eta f(\xi)\, \mathrm{d}\xi, \qquad (8.2.13)$$

where a prime denotes a derivative with respect to the integration variable, ξ. Integrating (8.2.10) with respect to η and using (8.2.13), we derive Weyl's integral equation for f'',

$$\ln \frac{f''}{f_0''} = -\frac{1}{4} \int_0^\eta (\xi - \eta)^2 f''(\xi)\, \mathrm{d}\xi, \qquad (8.2.14)$$

where $f_0'' = f''(0)$ ([352], p. 233).

Curvature of the velocity profile at the wall

Applying the Blasius equation (8.2.9) at the plate, $\eta = 0$, and using the first boundary condition in (8.2.11), we find that

$$f'''(0) = 0, \qquad (8.2.15)$$

which shows that the curvature of the streamwise velocity profile vanishes at the wall, in agreement with equation (8.1.13).

Numerical solution

To solve the Blasius equation (8.2.9), we rename $x_1 = f$, denote the first and second derivatives of the function f as $x_2 \equiv \mathrm{d}f/\mathrm{d}\eta$ and $x_3 \equiv \mathrm{d}x_2/\mathrm{d}\eta = \mathrm{d}^2 f/\mathrm{d}\eta^2$, and decompose the third-order Blasius equation into a system of three first-order nonlinear equations,

$$\frac{\mathrm{d}x_1}{\mathrm{d}\eta} = x_2, \qquad \frac{\mathrm{d}x_2}{\mathrm{d}\eta} = x_3, \qquad \frac{\mathrm{d}x_3}{\mathrm{d}\eta} = -\frac{1}{2} x_1 x_3. \qquad (8.2.16)$$

The accompanying boundary and far-field conditions are

$$x_1(0) = 0, \qquad x_2(0) = 0, \qquad x_2(\infty) = 1, \qquad (8.2.17)$$

originating from (8.2.11) and (8.2.12). The solution can be computed using a shooting method according to the following steps:

1. Guess the value of $x_3(0) \equiv f''(\eta = 0)$.

2. Integrate equations (8.2.16) from $\eta = 0$ to $\eta = \infty$ subject to the initial conditions (8.2.11) using the guessed value of $x_3(0)$.

3. Check whether the far-field condition $x_2(\infty) = 1$ is satisfied; if not, improve the guess for $x_3(0)$ and return to Step 2.

Numerical experimentation shows that integrating up to $\eta = 10$ in the second step yields satisfactory accuracy. The improvement in the third step can be made using a method for solving nonlinear algebraic equations, as discussed in Section B.3, Appendix B.

Variational equations

In practice, it is expedient to carry out the iterations using Newton's method based on derived variational equations (e.g., [317]). First, we define the derivatives

$$x_4 \equiv \frac{\mathrm{d}x_1}{\mathrm{d}x_3(0)}, \qquad x_5 \equiv \frac{\mathrm{d}x_2}{\mathrm{d}x_3(0)}, \qquad x_6 \equiv \frac{\mathrm{d}x_3}{\mathrm{d}x_3(0)}, \qquad (8.2.18)$$

expressing the sensitivity of the solution to the initial guess. Differentiating equations (8.2.16) with respect to $x_3(0)$, we obtain

$$\frac{\mathrm{d}x_4}{\mathrm{d}\eta} = x_5, \qquad \frac{\mathrm{d}x_5}{\mathrm{d}\eta} = x_6, \qquad \frac{\mathrm{d}x_6}{\mathrm{d}\eta} = -\frac{1}{2}\left(x_4 x_3 + x_1 x_6\right), \qquad (8.2.19)$$

subject to the initial conditions

$$x_4(\eta = 0) = 0, \qquad x_5(\eta = 0) = 0, \qquad x_6(\eta = 0) = 1. \qquad (8.2.20)$$

The shooting method involves the following steps: guess a value for $x_3(0)$; integrate equations (8.2.17) and (8.2.19) with the aforementioned initial conditions; replace the current initial value, $x_3(0)$, with an updated value,

$$x_3(0) \leftarrow x_3(0) - \frac{x_2(\infty) - 1}{x_5(\infty)}. \qquad (8.2.21)$$

The iterations converge quadratically for a reasonable initial guess.

Self-similar velocity profile

Numerical computations show that the far-field boundary condition is satisfied when $f''(0) \simeq 0.332057$. The streamwise velocity profile, $u/U = f' \equiv \mathrm{d}f/\mathrm{d}\eta$, is drawn with the solid line in Figure 8.2.1(b), along with the profiles of f (dashed line) and $f'' \equiv \mathrm{d}^2 f/\mathrm{d}\eta^2$ (dotted line). Close inspection reveals that $u/U = 0.99$ when $\eta \simeq 4.9$. Based on this result, we define the 99% boundary-layer thickness,

$$\delta_{99} = 4.9 \left(\frac{\nu x}{U}\right)^{1/2} \qquad \text{or} \qquad \frac{\delta_{99}}{x} = \frac{4.9}{\mathrm{Re}_x}, \qquad (8.2.22)$$

where $\mathrm{Re}_x \equiv U x/\nu$ is the local Reynolds number. The 99.5% boundary-layer thickness is defined in a similar fashion. The numerical solution shows that the corresponding numerical coefficient on the right-hand side of equations (8.2.22) is approximately equal to 5.3.

Wall shear stress

The wall shear stress and drag force exerted on a boundary are of particular interest in the engineering design of equipment for high-speed flow. According to the Blasius similarity solution, the wall shear stress is given by

$$\sigma_{xy}^{w}(x) = \mu \left(\frac{\partial u}{\partial y} \right)_{y=0} = \frac{f''(0)}{\sqrt{\mathrm{Re}_x}} \rho U^2 = \frac{0.332}{\sqrt{\mathrm{Re}_x}} \rho U^2. \tag{8.2.23}$$

We observe that σ_{xy}^{w} takes an infinite value at the leading edge, and decreases as the inverse square root of the streamwise distance or local Reynolds number, Re_x, along the plate. However, the physical significance of the singular behavior at the origin is undermined by the breakdown of the assumptions that led us to the boundary-layer equations at the leading edge.

Drag force

Even though the shear stress is infinite at the leading edge, the inverse-square-root singularity is integrable and the drag force exerted on any finite section of the plate extending from the leading edge up to an arbitrary point is finite. Using the similarity solution, we find that the drag force exerted on the plate over a length extending from the leading edge up to a certain distance x, is given by

$$D(x) \equiv \int_0^x \sigma_{xy}^{w}(\xi)\, \mathrm{d}\xi = 0.332\, \rho\, U^{3/2} \nu^{1/2} \int_0^x \frac{\mathrm{d}\xi}{\sqrt{\xi}}. \tag{8.2.24}$$

Performing the integration, we find that

$$D(x) = \frac{0.664}{\sqrt{\mathrm{Re}_x}} \rho\, U^2 x. \tag{8.2.25}$$

Based on this expression, we define the dimensionless drag coefficient

$$c_D(x) \equiv \frac{D(x)}{\frac{1}{2}\rho U^2 x} = \frac{1.328}{\sqrt{\mathrm{Re}_x}}. \tag{8.2.26}$$

The predictions of the last two equations for the drag force and drag coefficient agree well with laboratory measurements up to $\mathrm{Re}_x \simeq 1.2 \times 10^5$. At that point, the flow inside the boundary layer develops a wavy pattern and ultimately becomes turbulent. Above the critical value of Re_x, the function $c_D(\mathrm{Re}_x)$ jumps to a different branch with significantly higher values. The transition from steady laminar flow, presently considered, to unsteady turbulent flow will be discussed in Section 9.8.2 in the context of hydrodynamic stability.

Vorticity transport

Neglecting the velocity component along the y axis, we find that the z component of the vorticity inside the boundary layer is given by

$$\omega_z(x, y) \simeq -\frac{\partial u}{\partial y} = -\frac{f''(\eta)}{\sqrt{\mathrm{Re}_x}} \frac{U^2}{\nu} = -f''(\eta) \frac{U}{\delta(x)}. \tag{8.2.27}$$

We observe that the magnitude of the vorticity at a particular location, η, decreases like the inverse of the local boundary-layer thickness due to the broadening of the velocity profile.

The streamwise rate of convection of vorticity across a plane that is perpendicular to the plate is given by

$$\int_0^\infty u(x,y)\,\omega_z(x,y)\,\mathrm{d}y \simeq -\int_0^\infty u(x,y)\,\frac{\partial u}{\partial y}\,\mathrm{d}y = -\frac{1}{2}U^2,\qquad(8.2.28)$$

independent of the downstream position, x. Thus, the flux of vorticity across the plate is zero and viscous diffusion of vorticity does not occur at the wall, in agreement with our earlier observation that the gradient of the vorticity vanishes at the wall. Consequently, all convected vorticity is generated at the leading edge where the boundary-layer approximation ceases to be valid. Viscous stresses at the leading edge somehow generate the proper amount of vorticity necessary for the establishment of the Blasius self-similar flow.

Displacement thickness

Because of the widening of the streamwise velocity profile, the streamlines inside the Blasius boundary layer are deflected upward, away from the plate, as shown in Figure 8.2.1(a). Consider a streamline outside the boundary layer, described by the equation $y = g(x)$, and write a mass balance over a control area that is enclosed by (a) the streamline, (b) a vertical plane at $x = 0$, (c) a vertical plane located at a certain distance x, and (d) the flat plate. Since the streamwise velocity profile is flat at the leading edge located at $x = 0$, we obtain

$$\int_0^{g(0)} U\,\mathrm{d}y = \int_0^{g(x)} u(x,y)\,\mathrm{d}y.\qquad(8.2.29)$$

Straightforward rearrangement yields

$$U\left[\,g(x) - g(0)\,\right] = \int_0^{g(x)} \left[\,U - u(x,y)\,\right]\mathrm{d}y.\qquad(8.2.30)$$

Taking the limit as the streamline under consideration moves farther from the plate, we find that

$$\lim_{x\to\infty}\left[\,g(x) - g(0)\,\right] = \delta^*(x),\qquad(8.2.31)$$

where

$$\delta^*(x) \equiv \int_0^\infty \left(1 - \frac{u}{U}\right)\mathrm{d}y\qquad(8.2.32)$$

is the displacement thickness. Using the numerical solution of the Blasius equation to evaluate the integral on the right-hand side, we derive the exact expression

$$\delta^*(x) = \left(\frac{\nu x}{U}\right)^{1/2}\int_0^\infty\left(1 - \frac{\mathrm{d}f}{\mathrm{d}\eta}\right)\mathrm{d}\eta = 1.721\left(\frac{\nu x}{U}\right)^{1/2},\qquad(8.2.33)$$

which shows that the displacement thickness, like the 99% boundary layer thickness, increases like the square root of the streamwise position, x.

Physically, the displacement thickness represents the vertical displacement of the streamlines far from from the plate with respect to their elevation at the leading edge. Laboratory experiments have shown that the laminar boundary layer undergoes a transition from the laminar to the turbulent state when the Reynolds number based on the displacement thickness reaches the approximate value 600, that is, when $\delta^* \sim 600\,\nu/U$. At that point, turbulent shear stresses become significant and the present analysis based on the assumption of laminar flow ceases to be valid.

Iterative improvement

The displacement thickness describes the surface of a fictitious impenetrable but slippery body that is held stationary in the incident irrotational flow. An improved boundary-layer theory can be developed by replacing the tangential velocity of the outer flow along the plate, U, with the corresponding tangential velocity of the irrotational flow past the fictitious body. The irrotational flow past the fictitious body must be computed after the displacement thickness has been established, as discussed in this section. This iterative improvement provides us with a venue for describing the flow in the framework of matched asymptotics (e.g., [277]).

Momentum thickness

It is illuminating to perform a momentum integral balance over the control area employed to define the displacement thickness [415]. Since the upper boundary of the control volume is a streamline, the associated tangential flow does not contribute to the rate of momentum input. Assuming that the normal stresses on the vertical sides are equal in magnitude and opposite in sign, which is justified by the assumption that the pressure drop across the boundary layer is negligibly small, and neglecting the traction along the top streamline, we obtain

$$\int_0^{g(0)} U\,(\rho U)\,\mathrm{d}y - \int_0^{g(x)} u\,(\rho u)\,\mathrm{d}y - D(x) = 0, \tag{8.2.34}$$

where $D(x)$ is the drag force exerted on the plate defined in (8.2.24). Now we make the two upper limits of integration equal by recasting (8.2.34) into the form

$$-\rho U^2 \left[g(x) - g(0) \right] - \rho \int_0^{g(x)} (U^2 - u^2)\,\mathrm{d}y - D(x) = 0. \tag{8.2.35}$$

Finally, we take the limit as the streamline defining the top of the control area is moved far from the plate, and use (8.2.30) to obtain the relation

$$D(x) = \rho\,U^2 \Theta(x), \tag{8.2.36}$$

where $\Theta(x)$ is the momentum thickness defined as

$$\Theta(x) \equiv \int_0^\infty \frac{u}{U}\left(1 - \frac{u}{U}\right)\mathrm{d}y. \tag{8.2.37}$$

Using the numerical solution of the Blasius equation, we find that

$$\Theta(x) = \left(\frac{\nu x}{U}\right)^{1/2} \int_0^\infty f'(\eta)\left[1 - f'(\eta)\right]\mathrm{d}\eta = 0.664 \left(\frac{\nu x}{U}\right)^{1/2}, \tag{8.2.38}$$

where $f'(\eta) = \mathrm{d}f/\mathrm{d}\eta$. The momentum thickness, like the displacement thickness and the 99% boundary-layer thickness, increases as the square root of the streamwise position, x. Differentiating (8.2.36) with respect to x and using (8.2.24), we derive the relation

$$\frac{\mathrm{d}\Theta}{\mathrm{d}x} = \frac{\sigma_{xy}^w}{\rho U^2},\tag{8.2.39}$$

which provides us with a method of extracting the wall shear stress from the momentum thickness, and *vice versa*, based on available information.

Shape factor

The ratio of the displacement to the momentum thicknesses is called the shape factor,

$$H \equiv \frac{\delta^*}{\Theta}.\tag{8.2.40}$$

Substituting the right-hand sides of expressions (8.2.33) and (8.2.38) into (8.2.40), we find that, for the Blasius boundary layer over a flat plate, $H = 2.591$. Inspecting the definitions of δ^* and Θ in equations (8.2.32) and (8.2.37), we find that the shape factor is greater than unity as long as the streamwise velocity u is less than U within a substantial portion of the boundary layer. The satisfaction of this restriction is consistent with physical intuition. The smaller the value of H, the more blunt the velocity profile across the boundary layer.

Relation between the wall shear stress and the momentum thickness

The momentum thickness is related to the wall shear stress, and *vice versa*, by the integral momentum balance expressed by equation (8.2.36). Differentiating (8.2.24) with respect to x, we find that

$$\frac{\mathrm{d}D(x)}{\mathrm{d}x} = \sigma_{xy}^w(x).\tag{8.2.41}$$

Expressing the drag force in terms of the momentum thickness using (8.2.36), we obtain

$$\sigma_{xy}^w(x) = \rho U^2 \frac{\mathrm{d}\Theta(x)}{\mathrm{d}x}.\tag{8.2.42}$$

If the shear stress is known, the momentum thickness can be computed by integrating with respect to x. If the momentum thickness is known, the shear stress can be computed by differentiating with respect to x.

Given the velocity profile across a boundary layer, the wall shear stress can be computed in two ways: directly by differentiation, and indirectly by evaluating the momentum thickness and then differentiating it with respect to streamwise position x to obtain the wall shear stress according to equation (8.2.42). The indirect method is less sensitive to the structure of the velocity profile near the wall. For the velocity profile that arises by solving the Blasius equation, the two aforementioned methods are equivalent. To see this, we substitute (8.2.38) into (8.2.42) and obtain (8.2.23). We conclude that the approximations that led us to the boundary-layer equations are identical to those that led us to the simplified integral momentum balance (8.2.32).

Solution by trial profiles

Nothing can stop us from introducing a self-similar velocity profile with some reasonable form that is either stipulated by physical intuition or suggested by laboratory observation. Our goal is to adjust an unspecified function involved in this form so that the two methods of computing the wall shear stress discussed previously in this section are equivalent. A reasonable velocity profile is

$$\frac{u}{U} = \frac{\mathrm{d}f(\eta)}{\mathrm{d}\eta} = \begin{cases} \sin \frac{\pi y}{2\Delta(x)} & \text{for } 0 \leq y \leq \Delta(x), \\ 1 & \text{for } y \geq \Delta(x), \end{cases} \tag{8.2.43}$$

where $\eta \equiv y/\Delta(x)$ and $\Delta(x)$ is an unspecified function playing the role of a boundary-layer thickness, similar to the δ_{99} thickness introduced in (8.2.22). Note that the velocity distribution (8.2.43) conforms with the required boundary and far-field conditions, $f'(0) = 0$, $f'''(0) = 0$, and $f'(\infty) = 1$, but does not satisfy the Blasius equation; a prime denotes a derivative with respect to η. Differentiating the profile (8.2.43) with respect to y, we obtain the wall shear stress

$$\sigma_{xy}^w(x) \equiv \mu \left(\frac{\mathrm{d}u}{\mathrm{d}y}\right)_{y=0} = \frac{\pi}{2} \mu \frac{U}{\Delta(x)}. \tag{8.2.44}$$

The displacement and momentum thicknesses defined in (8.2.32) and (8.2.37) are found to be

$$\delta^*(x) = (1 - \frac{2}{\pi}) \Delta(x) = 0.363 \, \Delta(x), \qquad \Theta(x) = (\frac{2}{\pi} - \frac{1}{2}) \Delta(x) = 0.137 \, \Delta(x), \tag{8.2.45}$$

and the shape factor is $H = 2.660$. It is reassuring to observe that the shape factor is remarkably close to that arising from the exact solution of the Blasius equation, $H = 2.591$.

Substituting the expressions for the momentum thickness and wall shear stress into (8.2.42), we derive an ordinary differential equation for $\Delta(x)$,

$$\frac{\pi}{2} \mu \frac{U}{\Delta(x)} = 0.137 \, \rho U^2 \frac{\mathrm{d}\Delta(x)}{\mathrm{d}x}. \tag{8.2.46}$$

Rearranging and integrating with respect to x subject to the initial condition $\Delta = 0$ at $x = 0$, we find that

$$\Delta(x) = 4.80 \left(\frac{\nu x}{U}\right)^{1/2}. \tag{8.2.47}$$

Substituting this expression into (8.2.44) and (8.2.45), we find that

$$\sigma_{xy}^w(x) = \frac{0.327}{\sqrt{\mathrm{Re}_x}} \rho U^2, \qquad \delta^*(x) = 1.4043 \left(\frac{\nu x}{U}\right)^{1/2}, \qquad \Theta(x) = 0.8544 \left(\frac{\nu x}{U}\right)^{1/2}. \tag{8.2.48}$$

These expressions are in excellent agreement with their exact counterparts shown in (8.2.23), (8.2.33), and (8.2.38). However, the agreement is fortuitous and thus atypical of the accuracy of the approximate method (Problem 8.2.2).

(a)   (b)

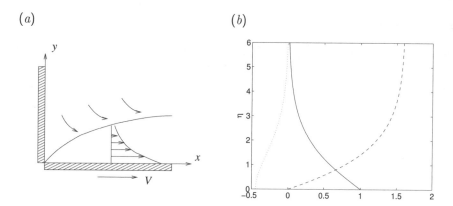

FIGURE 8.2.2 (*a*) Illustration of the Sakiadis boundary layer developing over a semi-infinite belt translating in an otherwise quiescent fluid. (*b*) Graphs of the self-similar streamwise velocity profile $u/V = f'$ (solid line), its integral f (dashed line), and derivative f'' (dotted line).

8.2.2 Sakiadis boundary layer on a moving flat surface

Sakiadis investigated the flow due to the translation of a semi-infinite belt with velocity V along the x axis normal to a vertical stationary wall, as illustrated in Figure 8.2.2(*a*) [360, 361]. At high velocities, a boundary layer is established over the belt, while the fluid is quiescent far from the belt. The Sakiadis boundary layer is the mirror image of the Blasius boundary layer established in streaming flow over a stationary plate discussed in Section 8.2.1, in that the wall and far-field velocities play opposite roles.

Similarity solution

Carrying out the boundary-layer analysis, we find that the flow in the Sakiadis boundary layer is governed by the Blasius ordinary differential equation (8.2.9), where the belt velocity, V, is used in place of the free-stream velocity, U, in the similarity variable

$$\eta \equiv \frac{y}{\delta(x)} = \left(\frac{V}{\nu x}\right)^{1/2} y. \qquad (8.2.49)$$

The function $f(\eta)$ satisfies the boundary conditions $f(0) = 0$ and $f'(0) = 1$ and the far-field condition $f'(\infty) = 0$. The solution can be found using the shooting method discussed in Section 8.2.1 for the Blasius boundary layer. Numerical computations show that the far-field condition is satisfied when $f''(0) = -0.443748$.

The profile of the streamwise velocity, $u/V = f'(\eta)$, is drawn with the solid line in Figure 8.2.2(*b*) along with the profiles of $f(\eta)$ and $f''(\eta)$, drawn with the dashed and dotted lines. As in the case of the Blasius boundary layer, the curvature of the velocity profile is zero at the wall. The 0.01% boundary-layer thickness, $\delta_{0.01}$, is defined as the elevation where $u/V = 0.01$. Using the numerical, solution, we find that

$$\delta_{0.01} = 6.37 \left(\frac{\nu x}{V} \right)^{1/2}.$$

(8.2.50)

Displacement thickness, momentum thickness, and wall shear stress

The displacement and momentum thicknesses of the Sakiadis boundary layer are defined as

$$\delta^* \equiv \int_0^\infty \frac{u}{V} \, \mathrm{d}y, \qquad \Theta \equiv \int_0^\infty \left(\frac{u}{V} \right)^2 \mathrm{d}y.$$

(8.2.51)

Using the numerical solution, we find that

$$\delta^* = 1.616 \left(\frac{\nu x}{V} \right)^{1/2}, \qquad \Theta = 0.888 \left(\frac{\nu x}{V} \right)^{1/2}.$$

(8.2.52)

Performing a momentum integral balance, we find that the drag force exerted on the moving plate from the origin up to a certain position, x, is given by

$$D(x) \equiv \int_0^x \sigma_{xy}^w(\xi) \, \mathrm{d}\xi = \rho V^2 \Theta(x).$$

(8.2.53)

The wall shear stress is given by

$$\sigma_{xy}^w = \rho V^2 \frac{\mathrm{d}\Theta}{\mathrm{d}x} = \frac{0.444}{\sqrt{\mathrm{Re}_x}} \rho V^2,$$

(8.2.54)

where $\mathrm{Re}_x = V x / \nu$ is the Reynolds number defined with respect to the belt velocity, V.

In fact, the velocity profile is reasonably well approximated with the complementary error function, $u/V \simeq \mathrm{erfc}(\eta/2)$. The corresponding numerical coefficients on the right-hand sides of (8.2.52) are 1.128 and 0.856 [361].

8.2.3   Spreading of a two-dimensional jet

To illustrate further the application of boundary-layer theory, we consider the spreading of a symmetric two-dimensional jet discharging from a slit along the x axis into an infinite pool of quiescent fluid, as shown in Figure 8.2.3. A free boundary layer whose thickness increases with streamwise position, x, is established along the midplane, $y = 0$.

Writing a momentum integral balance over an infinite control area with two parallel sides perpendicular to the x axis, we find that the momentum integral,

$$M = \rho \int_{-\infty}^\infty u^2 \, \mathrm{d}y,$$

(8.2.55)

is constant, independent of x. Motivated by this observation and carrying out a dimensional analysis, we express the stream function in the form

$$\psi = \left(\frac{M\nu}{\rho} \right)^{1/3} x^{1/3} f(\eta),$$

(8.2.56)

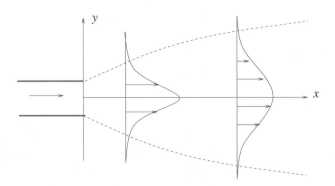

FIGURE 8.2.3   Illustration of a boundary layer developing due to the spreading of a two-dimensional or axisymmetric jet discharging along the x axis into a quiescent ambient fluid.

where f is a dimensionless function and

$$\eta \equiv \left(\frac{M}{\nu\mu}\right)^{1/3}\frac{y}{x^{2/3}} \tag{8.2.57}$$

is a dimensionless similarity variable. These expressions reveal that the boundary-layer thickness is proportional to $x^{2/3}$, while the centerline velocity decreases like $x^{-1/3}$. In contrast, the Blasius and Sakiadis boundary layer thicknesses are proportional to $x^{1/2}$.

Making substitutions in the steady version of the Prandtl boundary-layer equation (8.1.12) with U set to zero, we obtain a nonlinear homogeneous ordinary differential equation,

$$f''' + \frac{1}{3}ff'' + \frac{1}{3}f'^2 = f''' + \frac{1}{3}(ff')' = 0, \tag{8.2.58}$$

accompanied by the boundary conditions $f(0) = 0$, $f''(0) = 0$, and the far-field condition $f'(\infty) = 0$, where a prime denotes a derivative with respect to η. The integral constraint (8.2.55) requires that

$$\int_{-\infty}^{\infty} f'^2 \, \mathrm{d}\eta = 1. \tag{8.2.59}$$

The solution can be found by analytical methods and is given by

$$f(\eta) = 6\alpha \tanh(\alpha\eta), \tag{8.2.60}$$

where $\alpha = 1/48^{1/3}$. The axial volumetric flow rate per unit width is given by

$$Q(x) = \int_{-\infty}^{\infty} u \, \mathrm{d}y = 12\alpha \left(\frac{M\nu x}{\rho}\right)^{1/3}. \tag{8.2.61}$$

We observe that $Q(x)$ increases in the streamwise direction due to the entrainment of ambient fluid.

Problems

8.2.1 *Blasius boundary layer*

Show that the streamwise velocity component, u, is constant inside the Blasius boundary layer along a parabola described by $x = \alpha y^2$, where α is a constant.

8.2.2 *Von Kármán approximate method*

Assume that the velocity profile inside a boundary layer is given by $u/U = \tanh\big(y/\Delta(x)\big)$ instead of that shown in (8.2.43). Show that the effective boundary-layer thickness, wall shear stress, displacement thickness, and momentum thickness are given by the right-hand sides of equations (8.2.47) and (8.2.48), except that the numerical coefficients on the right-hand sides are equal to 2.553, 0.392, 1.770, and 0.664, respectively, ([427], p. 247). Discuss the accuracy of these results with reference to the exact solution.

8.2.3 *Shear layer*

Consider the flow between two parallel streams that merge along the x axis with velocities U_1 and U_2. The stream function can be expressed in the form $\psi = (\nu x)^{1/2} f(\eta)$, where f is a dimensionless function, $\eta = y/\delta$, and $\delta = (\nu x/U_1)^{1/2}$. Show that, in the boundary-layer approximation, the function f satisfies the Blasius equation and derive appropriate boundary conditions.

8.2.4 *Boundary layer on a moving flat surface*

Perform the boundary-layer analysis of the combined Blasius–Sakiadis boundary layer with an arbitrary streaming incident velocity, U, and arbitrary wall velocity, V.

 Computer Problem

8.2.5 *Blasius and Sakiadis boundary layers*

(*a*) Solve the Blasius problem using a method of your choice. Compute and display the velocity profile shown in Figure 8.2.1(*b*).

(*b*) Solve the Sakiadis problem using a method of your choice. Compute and display the velocity profile shown in Figure 8.2.2(*b*).

8.3 Boundary layers in accelerating or decelerating flows

Having examined the boundary layer on a semi-infinite flat plate that is aligned with a streaming (uniform) incident flow or translates in an otherwise quiescent fluid, distinguished by a constant far-field or tangential velocity, we proceed to consider situations where the incident or boundary velocity exhibits acceleration or deceleration with an accompanying favorable or adverse pressure gradient. Examples of physical circumstances where this occurs are illustrated in Figure 8.3.1. In related applications, the boundary layer is due to the stretching of a planar sheet with a specified in-plane velocity. In this section, we will discuss the classical Falkner–Skan boundary layer developing over a plate in accelerating or decelerating flow, and then address an analogous boundary layer developing over a stretching surface in steady or unsteady flow.

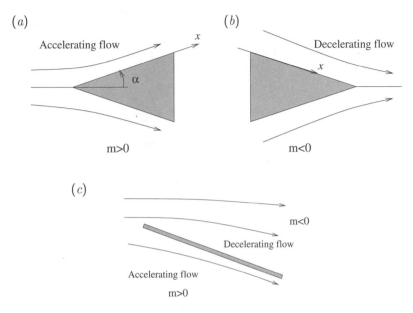

FIGURE 8.3.1   Examples of boundary layers in accelerating or decelerating flow. (a, b) Flow past a wedge-shaped body and (c) streaming flow past a flat plate at a finite angle of attack.

8.3.1   Falkner–Skan boundary layer

The tangential component of the velocity of the outer flow responsible for the Falkner–Skan boundary layer exhibits a power-law dependence on the streamwise position along a flat boundary, x,

$$U(x) = \xi x^m, \tag{8.3.1}$$

where ξ is a positive dimensional coefficient and m is a positive or negative exponent [121]. When $m = 0$, we recover the Blasius flow over a semi-infinite flat plate at zero angle of attack. When $m = 1$, we obtain orthogonal stagnation-point flow against a flat plate. Intermediate values of m correspond to accelerating symmetric flow past a wedge with semi-angle $\alpha = \pi m/(m+1)$, as illustrated in Figure 8.3.1(a).

Differentiating the outer velocity described by (8.3.1), we obtain an expression the streamwise acceleration of the outer flow

$$\frac{\mathrm{d}U}{\mathrm{d}x} = m\xi x^{m-1}, \tag{8.3.2}$$

which shows that the outer flow accelerates when $m > 0$ and decelerates when $m < 0$. In the first case, the continuity equation requires that the normal derivative of the velocity component of the outer flow normal to the boundary, V, is negative, $\partial V/\partial y < 0$. Since V is zero at the wall, it must be negative at the edge of the boundary layer. The motion of the fluid normal to the boundary contains the vorticity toward the boundary, thereby restricting the growth of the boundary layer.

Substituting (8.3.2) into the boundary-layer equation (8.1.12), we obtain the specific form

$$u\frac{\partial u}{\partial x} + v\frac{\partial u}{\partial y} = \xi^2 m x^{2m-1} + \nu\frac{\partial^2 u}{\partial y^2}. \tag{8.3.3}$$

Working as in Section 8.2 for the Blasius boundary layer, we identify the characteristic length L with the current streamwise position, x, and use (8.1.7) to write

$$\delta(x) \sim \left(\frac{\nu x}{U(x)}\right)^{1/2} = \left(\frac{\nu}{\xi}\right)^{1/2} x^{(1-m)/2}. \tag{8.3.4}$$

The exponent $(1-m)/2$ determines the rate of spatial growth of the boundary layer in accelerating or decelerating flow.

Similarity solution

Next, we postulate that the velocity profile across the boundary is self-similar, that is, u is a function of the similarity variable

$$\eta \equiv \frac{y}{\delta(x)} = y\left(\frac{U(x)}{\nu x}\right)^{1/2} = y\left(\frac{\xi}{\nu x^{1-m}}\right)^{1/2}, \tag{8.3.5}$$

and write

$$u(x,y) = U(x)\,F(\eta), \tag{8.3.6}$$

where $F(\eta)$ is a dimensionless function. This self-similar velocity profile can be derived from the stream function shown in (8.2.5), repeated here for convenience,

$$\psi(x,y) = (U\nu x)^{1/2}\, f(\eta), \tag{8.3.7}$$

where $U(x)$ is given in (8.3.1) and $F(\eta) = f'(\eta)$. Differentiating (8.3.7) and using the relations

$$\frac{\partial \eta}{\partial x} = \frac{m-1}{2}\frac{y}{x}\left(\frac{\xi}{\nu x^{1-m}}\right)^{1/2}, \qquad \frac{\partial \eta}{\partial y} = \left(\frac{\xi}{\nu x^{1-m}}\right)^{1/2}, \tag{8.3.8}$$

we obtain the velocity components

$$u(x,y) = \frac{\partial \psi}{\partial y} = U(x)\frac{\mathrm{d}f}{\mathrm{d}\eta} = \xi x^m \frac{\mathrm{d}f}{\mathrm{d}\eta} \tag{8.3.9}$$

and

$$v(x,y) = -\frac{\partial \psi}{\partial x} = \frac{1}{2}(\nu\xi x^{m-1})^{1/2}\left[(1-m)\,\eta\,\frac{\mathrm{d}f}{\mathrm{d}\eta} - (1+m)\,f\right]. \tag{8.3.10}$$

Further differentiation provides us with expressions for the derivatives of the velocity involved in the boundary-layer equation,

$$\frac{\partial u}{\partial x} = \xi m x^{m-1}\frac{\mathrm{d}f}{\mathrm{d}\eta} + U\frac{\mathrm{d}^2 f}{\mathrm{d}\eta^2}\frac{\partial \eta}{\partial x}, \qquad \frac{\partial u}{\partial y} = U\frac{\mathrm{d}^2 f}{\mathrm{d}\eta^2}\frac{\partial \eta}{\partial y}, \qquad \frac{\partial^2 u}{\partial y^2} = U\frac{\mathrm{d}^3 f}{\mathrm{d}\eta^3}\left(\frac{\partial \eta}{\partial y}\right)^2. \tag{8.3.11}$$

Substituting these expressions into (8.3.3), we derive a third-order nonlinear inhomogeneous ordinary differential equation for the function f,

$$f''' + \frac{1}{2}(m+1)ff'' - mf'^2 + m = 0. \qquad (8.3.12)$$

The boundary and far-field conditions require that $f(0) = 0$, $f'(0) = 0$, and $f'(\infty) = 1$. When $m = 0$, equations (8.3.9), (8.3.10), and (8.3.12) reduce to the Blasius equations (8.2.8) and (8.2.9). When $m = 1$, we obtain equation (5.6.16), providing us with an exact solution to the equation of motion describing orthogonal stagnation-point flow.

Velocity profile

Applying the Falker–Skan boundary-layer equation (8.3.12) at the wall, $y = 0$, and enforcing the aforementioned boundary conditions, we find that $f'''(0) = -m$, which is positive when $m < 0$, corresponding to decelerating flow. Thus, the curvature of the velocity profile at the wall is positive in a decelerating flow. Noting that f''' must become negative at a sufficiently high value of η for the streamwise velocity f' to tend to a constant value, reveals the presence of an inflection point. In Chapter 9, we will see that the inflection point renders the boundary layer susceptible to hydrodynamic instability mediated by the growth of small perturbations.

Numerical solution

Since boundary conditions are specified at both ends of the solution domain, $[0, \infty)$, we encounter a two-point boundary-value problem involving a third-order differential equation. To solve the Falkner–Skan equation, we rename $x_1 = f$, denote the first and second derivatives of the function f as $x_2 \equiv df/d\eta$ and $x_3 \equiv dx_2/d\eta = d^2f/d\eta^2$, and decompose the third-order equation into a system of three first-order nonlinear equations,

$$\frac{dx_1}{d\eta} = x_2, \qquad \frac{dx_2}{d\eta} = x_3, \qquad \frac{dx_3}{d\eta} = -\frac{1}{2}(m+1)x_1x_3 + mx_2^2 - m. \qquad (8.3.13)$$

The accompanying boundary and far-field conditions are shown in (8.2.17). Working as in the case of the Blasius equation, we implement Newton's method by defining the derivatives (8.2.18) satisfying the initial conditions (8.2.20). Differentiating equations (8.3.13) with respect to $x_3(0)$, we obtain

$$\frac{dx_4}{d\eta} = x_5, \qquad \frac{dx_5}{d\eta} = x_6, \qquad \frac{dx_6}{d\eta} = -\frac{1}{2}(m+1)(x_4x_3 + x_1x_6) + 2m\,x_2x_5. \qquad (8.3.14)$$

The shooting method is implemented by guessing a value for $x_3(\eta = 0)$, integrating equations (8.3.13) and (8.3.14) with the aforementioned initial conditions, and replacing the guessed value, $x_3(\eta = 0)$, with an updated value computed using formula (8.2.21).

The solution for $m = 1$, corresponding to orthogonal stagnation-point flow, was obtained in Section 5.6.1. Numerical solutions for other values of m have been presented by a number of authors following the original contributions of Hartree [170] and Stewartson [388]. A unique solution branch is found when $m > 0$, corresponding to accelerating flow, and multiple solution branches are found when $m < 0$, corresponding to decelerating flow [120, 427]. In the case of decelerating flow, the

(*a*)   (*b*)

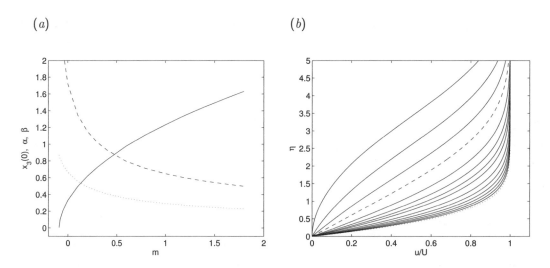

FIGURE 8.3.2   (*a*) Graphs of the wall shear stress coefficient $x_3(0) = f''(0)$ (solid line), coefficient α
pertaining to the displacement thickness (dashed line), and coefficient β pertaining to the momentum thickness (dotted line). (*b*) Velocity profiles across the Falkner–Skan boundary layer: from
bottom to top, $m = 1.8$ (dotted line), $1.6, 1.4, \ldots 0.4$, $0.3, 0.2, 0.1$, 0.0 (dashed line representing
the Blasius boundary layer), -0.05, -0.08, and -0.0904.

solution that appears to be most physical relevant is retained. Results for $x_3(0)$ obtained using the
shooting method discussed previously in this chapter are shown with the solid line in Figure 8.3.2(*a*).
Velocity profiles expressed by the derivative $f'(\eta)$ are shown in Figure 8.3.2(*b*) for several values of
m. The profiles for $m < 0$, corresponding to decelerating flow, exhibit an inflection point near the
wall.

Wall shear stress, vorticity transport, displacement and momentum thicknesses

The wall shear stress is found by differentiating the streamwise velocity profile and evaluating the
resulting expression at the wall, finding

$$\sigma_{xy}^w(x) = \mu \left(\frac{\partial u}{\partial y}\right)_{y=0} = \rho \sqrt{\nu}\, f''(0)\, \xi^{3/2}\, x^{(3m-1)/2}. \tag{8.3.15}$$

We observe that, when $m = 1/3$, the shear stress is independent of the streamwise position, x;
when $m > 1/3$, the shear stress increases with the streamwise position; when $m < 1/3$, the shear
stress decreases with the streamwise position; when $m = 1$, corresponding to orthogonal stagnation-
point flow, the shear stress increases linearly with the streamwise position; when $m = -0.0904$,
corresponding to decelerating flow, $f''(0) = 0$ and the shear stress, and thus the skin friction, vanish
uniformly along the wall. The rate of convection of vorticity across a plane that is perpendicular to
the wall is given by the right-hand side of (8.2.28). Because U depends on the streamwise position,
vorticity must diffuse across the wall in order to satisfy the vorticity balance at every x station.

The displacement and momentum thicknesses defined in equations (8.2.32) and (8.2.37) are given by

$$\delta^*(x) \equiv \int_0^\infty \left(1 - \frac{u}{U}\right) \mathrm{d}y = \alpha \left(\frac{\nu x}{U}\right)^{1/2} = \alpha \left(\frac{\nu}{\xi}\right)^{1/2} x^{(1-m)/2} \tag{8.3.16}$$

and

$$\Theta(x) \equiv \int_0^\infty \frac{u}{U}\left(1 - \frac{u}{U}\right) \mathrm{d}y = \beta \left(\frac{\nu x}{U}\right)^{1/2} = \beta \left(\frac{\nu}{\xi}\right)^{1/2} x^{(1-m)/2}, \tag{8.3.17}$$

where

$$\alpha = \int_0^\infty (1 - f') \, \mathrm{d}\eta, \qquad \beta = \int_0^\infty f'(1 - f') \, \mathrm{d}\eta \tag{8.3.18}$$

are dimensionless coefficients dependent on m, plotted with the dashed and dotted lines in Figure 8.3.2(a). When $m > 1$, both δ^* and Θ decrease in the streamwise direction due to the acceleration of the incident flow. When $m < 1$, both δ^* and Θ increase in the streamwise direction due to the deceleration of the incident flow. When $m = 1$, corresponding to orthogonal stagnation-point flow, the displacement and momentum thicknesses are constant along the wall. Physically, viscous diffusion of vorticity is balanced by convection against the flat plate.

8.3.2   Steady flow due to a stretching sheet

In Section 8.2.2, we introduced the Sakiadis boundary layer as the mirror image of the Blasius boundary layer satisfying complementary boundary conditions for the plate and far-flow velocities. A similar analogy can be made between the Falkner–Skan boundary layer and the boundary layer developing over a semi-infinite stretching elastic sheet emerging from a vertical slit into an otherwise quiescent fluid along the x axis, as illustrated in Figure 8.2.2(a). In the generalized Sakiadis boundary layer, the x component of the velocity at the surface of the sheet, located at $y = 0$, is described by the power law

$$V(x) = \xi x^m, \tag{8.3.19}$$

where ξ is the constant rate of extension and m is a positive or negative exponent.

Similarity solution

As a preliminary, we introduce the similarity variable η defined in (8.3.5), where the sheet velocity, $V(x)$, is used in place of the far-field velocity, $U(x)$,

$$\eta \equiv \frac{y}{\delta(x)} = y \left(\frac{V}{\nu x}\right)^{1/2} = y \left(\frac{\xi}{\nu x^{1-m}}\right)^{1/2}. \tag{8.3.20}$$

Performing the boundary-layer analysis in the absence of outer flow, $U = 0$, we find that the function $f(\eta)$ satisfies the homogeneous Falkner–Skan equation

$$f''' + \frac{1}{2}(m+1) f f'' - m f'^2 = 0, \tag{8.3.21}$$

(a)   (b)

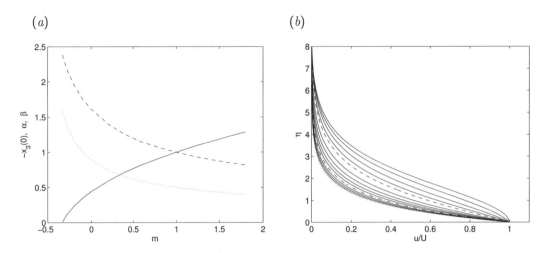

FIGURE 8.3.3   (a) Graphs of the negative of the wall shear stress coefficient $-x_3(0) = -f''(0)$ (solid line), coefficient α pertaining to the displacement thickness (dashed line), and coefficient β pertaining to the momentum thickness (dotted line). (b) Velocity profiles across the boundary layer above a stretching sheet: from bottom to top, $m = 1.8$ (dotted line), $1.6, 1.4, 1, 2, 1.0$ (dot-dashed line), $0.8, \ldots, 0.0$ (dashed line representing the Sakiadis boundary layer), $-0.10, -0.20, -0.28$, and -0.33.

where a prime denotes a derivative with respect to η. The boundary and far-field conditions require that $f(0) = 0$, $f'(0) = 1$, and $f'(\infty) = 0$. When $m = 0$, we obtain the Blasius equation describing the Sakiadis boundary layer over an inextensible belt. When $m = 1$, we obtain the homogeneous Hiemenz equation (5.6.32), which is satisfied by Crane's exact solution of the unsimplified Navier–Stokes equation, $f(\eta) = 1 - e^{-\eta}$, as discussed in Section 5.6.3.

Solutions for other values of m computed by a shooting method are shown in Figure 8.3.3. The wall shear stress tends to zero at a critical yet unphysical negative value of m. Applying (8.3.21) at the wall, $y = 0$, and enforcing the aforementioned boundary conditions, we find that $f'''(0) = m$, which is negative when $m < 0$, corresponding to decelerating flow. Thus, the curvature of the velocity profile at the wall is negative in a decelerating flow. Since f''' must become positive as η tends to infinity for the streamwise velocity f' to decay to zero, we infer the presence of an inflection point in the velocity profile. In Chapter 9, we will see that the inflection point renders the boundary layer susceptible to hydrodynamic instability mediated by the growth of small perturbations.

Displacement and momentum thicknesses

The displacement and momentum thicknesses defined in equations (8.2.51) are given by

$$\delta^*(x) \equiv \int_0^\infty \frac{u}{V}\,\mathrm{d}y = \alpha \left(\frac{\nu x}{V}\right)^{1/2} = \alpha \left(\frac{\nu}{\xi}\right)^{1/2} x^{(1-m)/2} \tag{8.3.22}$$

and

$$\Theta(x) \equiv \int_0^\infty \left(\frac{u}{V}\right)^2 dy = \beta \left(\frac{\nu x}{V}\right)^{1/2} = \beta \left(\frac{\nu}{\xi}\right)^{1/2} x^{(1-m)/2}, \tag{8.3.23}$$

where

$$\alpha = \int_0^\infty f' \, d\eta = f(\infty), \qquad \beta = \int_0^\infty f'^2 \, d\eta \tag{8.3.24}$$

are dimensionless coefficients plotted with the dashed and dotted lines in Figure 8.3.3(a) against the exponent, m. When $m > 1$, both δ^* and Θ decrease in the streamwise direction due to the severe stretching of the sheet. When $m < 1$, both δ^* and Θ increase in the streamwise direction due to the deceleration of the sheet. When $m = 1$, corresponding to the Crane boundary layer, the displacement and momentum thicknesses are constant over the sheet.

8.3.3   Unsteady flow due to a stretching sheet

Next, we consider a class of unsteady boundary layers generated by the time-dependent in-plane stretching of an elastic sheet occupying the positive part of the x axis in an otherwise quiescent fluid, as shown in Figure 8.2.2(a). The flow inside the boundary layer attached to the sheet is governed by the continuity equation (8.1.1) and the unsteady boundary-layer equation (8.1.12) in the absence of an outer potential flow,

$$\frac{\partial u}{\partial t} + u \frac{\partial u}{\partial x} + v \frac{\partial u}{\partial y} = \nu \frac{\partial^2 u}{\partial y^2}. \tag{8.3.25}$$

To simplify the analysis, we describe the flow in terms of an unsteady stream function, $\psi(x, y, t)$, defined in the usual way such that $u \equiv u_x = \partial\psi/\partial y$ and $v \equiv u_y = -\partial\psi/\partial x$.

Similarity solution

We are interested in situations where the sheet velocity, $V(x,t) = u(x, y = 0, t)$, allows for a similarity solution such that

$$\psi(x, y, t) = t^\kappa \tilde{\psi}(\tilde{x}, \tilde{y}), \qquad \tilde{x} = \frac{x}{t^\alpha}, \qquad \tilde{y} = \frac{y}{t^\beta}, \tag{8.3.26}$$

where the exponents κ, α, and β are determined as part of the solution and a tilde indicates a similarity variable. The two velocity components are given by

$$u = \frac{\partial\psi}{\partial y} = t^{\kappa-\beta} \frac{\partial\tilde{\psi}}{\partial\tilde{y}}, \qquad v = -\frac{\partial\psi}{\partial x} = -t^{\kappa-\alpha} \frac{\partial\tilde{\psi}}{\partial\tilde{x}}. \tag{8.3.27}$$

Substituting these expressions along with the expressions

$$\frac{\partial u}{\partial t} = t^{\delta-\beta-1} \left((\delta - \beta) \frac{\partial\tilde{\psi}}{\partial\tilde{y}} - \alpha\tilde{x} \frac{\partial^2\tilde{\psi}}{\partial\tilde{x}\partial\tilde{y}} - \beta\tilde{y} \frac{\partial^2\tilde{\psi}}{\partial\tilde{y}^2} \right), \qquad \frac{\partial u}{\partial x} = t^{\delta-\alpha-\beta} \frac{\partial^2\tilde{\psi}}{\partial\tilde{x}\partial\tilde{y}}, \tag{8.3.28}$$

and

$$\frac{\partial u}{\partial y} = t^{\kappa-2\beta} \frac{\partial^2\tilde{\psi}}{\partial\tilde{y}^2}, \qquad \frac{\partial^2 u}{\partial y^2} = t^{\kappa-3\beta} \frac{\partial^3\tilde{\psi}}{\partial\tilde{y}^3}, \tag{8.3.29}$$

in the boundary-layer equation (8.3.25), we obtain

$$(\kappa - \beta) \frac{\partial \tilde{\psi}}{\partial \tilde{y}} - \alpha \tilde{x} \frac{\partial^2 \tilde{\psi}}{\partial \tilde{x} \partial \tilde{y}} - \beta \tilde{y} \frac{\partial^2 \tilde{\psi}}{\partial \tilde{y}^2} + t^{\kappa - \alpha - \beta + 1} \left(\frac{\partial \tilde{\psi}}{\partial \tilde{y}} \frac{\partial^2 \tilde{\psi}}{\partial \tilde{x} \partial \tilde{y}} - \frac{\partial \tilde{\psi}}{\partial \tilde{x}} \frac{\partial^2 \tilde{\psi}}{\partial \tilde{y}^2} \right) = \nu t^{-2\beta + 1} \frac{\partial^3 \tilde{\psi}}{\partial \tilde{y}^3}. \quad (8.3.30)$$

To eliminate the explicit time dependence on both sides of this equation, we require that

$$\beta = \frac{1}{2}, \qquad \kappa = \alpha - \frac{1}{2}, \quad (8.3.31)$$

and derive a nonlinear partial differential equation for $\tilde{x}$ and $\tilde{y}$,

$$(\alpha - 1) \frac{\partial \tilde{\psi}}{\partial \tilde{y}} - \left(\alpha \tilde{x} - \frac{\partial \tilde{\psi}}{\partial \tilde{y}} \right) \frac{\partial^2 \tilde{\psi}}{\partial \tilde{x} \partial \tilde{y}} - \left(\frac{1}{2} \tilde{y} + \frac{\partial \tilde{\psi}}{\partial \tilde{x}} \right) \frac{\partial^2 \tilde{\psi}}{\partial \tilde{y}^2} = \nu \frac{\partial^3 \tilde{\psi}}{\partial \tilde{y}^3}. \quad (8.3.32)$$

The no-penetration boundary condition requires that $\tilde{\psi}(\tilde{x}, 0) = 0$, and the far-field condition requires that $\partial \tilde{\psi} / \partial \tilde{x} \to 0$ and $\partial \tilde{\psi} / \partial \tilde{y} \to 0$ as $\tilde{y} \to \infty$. The value of α and the choice of further boundary conditions depend on the specifics of the problem under consideration, as discussed in Sections 8.3.4 and 8.3.5.

Equation (8.3.32) admits the solution $\tilde{\psi}(\tilde{x}, \tilde{y}) = \tilde{x} f(\tilde{y}) + \tilde{\psi}_\infty$, resulting in an ordinary differential equation,

$$\nu f''' + \left(\frac{1}{2} \tilde{y} + f \right) f'' + f'(1 - f') = 0, \quad (8.3.33)$$

where $\tilde{\psi}_\infty$ is a constant and a prime denotes a derivative with respect to $\tilde{y}$. The no-penetration condition requires that $f(0) = 0$, and the far-field condition requires that $f(\tilde{y})$ decays to zero as $\tilde{y}$ tends to infinity.

Solution near the origin

We may assume that the general solution near the origin, $\tilde{x} \to 0$, takes the asymptotic form

$$\tilde{\psi}(\tilde{x}, \tilde{y}) \simeq \tilde{x}^q f(\eta), \quad (8.3.34)$$

where q is a positive exponent, $\eta = \tilde{y} / \tilde{x}^\varrho$ is a similarity variable, and the positive exponent ϱ is determined as part of the solution. Substituting (8.3.34) into (8.3.32) and simplifying, we find that

$$[\alpha(1 - q + \varrho) - 1] f' + \left(\alpha \varrho - \frac{1}{2} \right) \eta f'' + \tilde{x}^{q - \varrho - 1} [(q - \varrho) f'^2 - q f f''] = \nu \tilde{x}^{-2\varrho} f'''. \quad (8.3.35)$$

This equation is physically acceptable under two complementary conditions. First, setting the power of $\tilde{x}$ on the right-hand side of (8.3.35) to zero, $\varrho = 0$, we obtain

$$\nu f''' + \frac{1}{2} \tilde{y} f'' - [\alpha(1 - q) - 1] f' = q \tilde{x}^{q-1} (f'^2 - f f''). \quad (8.3.36)$$

When $q > 1$, the right-hand side vanishes as $\tilde{x} \to 0$, yielding a homogeneous integral equation,

$$\nu f''' + \frac{1}{2} \tilde{y} f'' - [\alpha(1 - q) - 1] f' = 0. \quad (8.3.37)$$

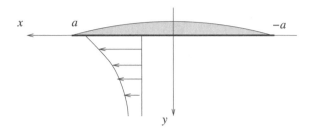

FIGURE 8.3.4 Illustration of an unsteady boundary layer extending between $x = -a(t)$ and $a(t)$ due to the spreading of a two-dimensional oil slick at the surface of the ocean.

Alternatively, we require that the power of $\tilde{x}$ on the right-hand side of (8.3.35) matches that of the last term on the left-hand side, $q - \varrho - 1 = -2\varrho$ or $\varrho = 1 - q$, and obtain

$$\nu f''' + q\,f f'' - (2q-1)\,f'^2 = \tilde{x}^{2\varrho}\,(2\alpha\varrho - 1)\left(f' + \frac{1}{2}\,\eta f''\right). \tag{8.3.38}$$

The right-hand side vanishes as $\tilde{x} \to 0$, yielding the homogeneous Falkner–Skan equation (8.3.21) with $m = 2q-1$. The properties of these equations have been investigated with respect to uniqueness of solution [135, 296, 297].

8.3.4 Gravitational spreading of a two-dimensional oil slick

As an application of the similarity solution derived in Section 8.3.3, we consider the flow due to the gravitational spreading of a two-dimensional oil slick over the flat surface of an otherwise quiescent pool of water, as shown in Figure 8.3.4. A thinning film of oil with time-dependent thickness $h(y, t)$ is confined inside the spilling zone, $|x| < a(t)$. At high Reynolds numbers, an unsteady boundary layer is established next to the film, while the fluid tends to become quiescent far from the film [60].

Oil flow

The oil phase will be denoted by the superscript or subscript o. To leading order, the x and y components of the equation of motion inside the oil film take the form

$$\frac{\partial \sigma^o_{xy}}{\partial y} = \frac{\partial p_o}{\partial x}, \qquad \frac{\partial p_o}{\partial y} = \rho_o g, \tag{8.3.39}$$

where σ^o_{xy} is the shear stress. Integrating the second equation and requiring that the pressure at the oil–air interface, located at $y = -h(x,t)$, is equal to the atmospheric pressure, p_{atm}, we obtain the oil pressure distribution

$$p_o = \rho_o g\,(y + h) + p_{atm}. \tag{8.3.40}$$

Substituting this expression into the first equation in (8.3.39), and integrating with respect to y subject to the condition of zero shear stress at the oil–air interface, we obtain the interfacial shear stress on the side of the oil,

$$\sigma_{xy}^o(y = 0) = \rho_o g\, h\, \frac{\partial h}{\partial x}. \tag{8.3.41}$$

A mass balance for the oil provides us with an evolution equation for the film thickness,

$$\frac{\partial h}{\partial t} + \frac{\partial(\bar{u}_o h)}{\partial x} = 0, \tag{8.3.42}$$

where $\bar{u}_o(x)$ is the mean velocity across the oil film. The total amount of floating oil per unit width in the transverse z direction is

$$A_o(t) = \int_0^{a(t)} h(x, t)\, \mathrm{d}x. \tag{8.3.43}$$

Continuity of velocity and shear stress at the oil–water interface requires that

$$u(y = 0) = u_o(y = 0) \equiv V(x), \qquad \mu \left(\frac{\partial u}{\partial y}\right)_{y=0} = \sigma_{xy}^o(y = 0) = \rho_o g h\, \frac{\partial h}{\partial x}, \tag{8.3.44}$$

where u is the x component of the water velocity.

Similarity solution

Progress can be made by assuming that the boundary layer develops according to the similarity solution discussed in Section 8.3.3. The edges of the spreading film are located at

$$x = \pm a(t) = \pm \tilde{x}^* t^\alpha, \tag{8.3.45}$$

where $\tilde{x}^*$ is a constant to be determined as part of the solution.

Substituting the similarity solution into (8.3.43), stipulating that $A_o(t) = q_o t^n$, and rearranging, we obtain

$$A_o(t) = q_o t^n = t^\alpha \int_0^{\tilde{x}^*} h(x, t)\, \mathrm{d}\tilde{x}, \tag{8.3.46}$$

where q_o is a constant coefficient and n is a specified exponent. This equation is satisfied only if

$$h(x, t) = t^{n-\alpha}\, \tilde{h}(\tilde{x}), \tag{8.3.47}$$

where $\tilde{h}(\tilde{x})$ is an unknown function to be determined as part of the solution. Substituting this expression along with the similarity solution into the interfacial shear stress condition given in the second equation in (8.3.44), we obtain

$$\mu\, t^{\kappa+3\alpha-2\beta-2n} \left(\frac{\partial^2 \tilde{\psi}}{\partial \tilde{y}^2}\right)_{\tilde{y}=0} = \rho_o g\, \tilde{h}\, \frac{\mathrm{d}\tilde{h}}{\mathrm{d}\tilde{x}}. \tag{8.3.48}$$

To eliminate the explicit time dependence on the left-hand side, we require that the exponent of time, t, on the left-hand side is zero. Combining this condition with (8.5.36), we obtain

$$\kappa = \frac{4n - 1}{8}, \qquad \alpha = \frac{4n + 3}{8}, \qquad \beta = \frac{1}{2}. \tag{8.3.49}$$

The interfacial condition (8.3.48) then becomes

$$\frac{\mathrm{d}\tilde{h}^2}{\mathrm{d}\tilde{x}} = 2\frac{\mu}{\rho_o g}\left(\frac{\partial^2\tilde{\psi}}{\partial\tilde{y}^2}\right)_{\tilde{y}=0}, \tag{8.3.50}$$

coupling the oil with the water flow.

The similarity solution predicts that the x component of the water velocity is

$$u = t^{\kappa-\beta}\frac{\partial\tilde{\psi}}{\partial\tilde{y}}, \tag{8.3.51}$$

and this motivates expressing the mean oil velocity in the corresponding form

$$\bar{u}_o = t^{\kappa-\beta}\,\tilde{x}\,\mathcal{G}(\tilde{x}) = t^{\kappa-\beta-\alpha}\,x\,\mathcal{G}(\tilde{x}) = \frac{x}{t}\,\mathcal{G}(\tilde{x}), \tag{8.3.52}$$

where $\mathcal{G}(\tilde{x})$ is an unknown function. Combining this expression with (8.3.47), we find that

$$\bar{u}_o\,h = t^{n-1}\,\tilde{x}\,\tilde{h}(\tilde{x})\,\mathcal{G}(\tilde{x}). \tag{8.3.53}$$

Substituting this expression along with (8.3.47) into the mass balance equation (8.3.42), we obtain an ordinary differential equation,

$$\frac{\mathrm{d}}{\mathrm{d}\tilde{x}}\big[\,\tilde{x}(\mathcal{G}-\alpha)\tilde{h}\,\big] + n\tilde{h} = 0. \tag{8.3.54}$$

When $n = 0$, this equation is satisfied by the constant function $\mathcal{G}(\tilde{x}) = \alpha$. Assuming with good reason that $\tilde{h}$ is nearly constant except at the leading edge, we derive the more general, albeit approximate, result

$$\mathcal{G}(\tilde{x}) \simeq \alpha - n. \tag{8.3.55}$$

A numerical solution of the boundary-layer equations using an approximate method based on an integral momentum balance will be presented in Section 8.4.4.

8.3.5   Molecular spreading of a two-dimensional oil slick

When the thickness of the oil film has reached molecular dimensions, the oil behaves like a surfactant, affecting the surface tension according to an assumed surface equation of state, $\gamma(h)$ [135, 297]. Balancing the induced Marangoni traction with the shear stress, we derive the interfacial condition

$$\mu\left(\frac{\partial u}{\partial y}\right)_{y=0} = -\frac{\partial\gamma}{\partial x}. \tag{8.3.56}$$

A mass balance over the oil film provides us with an evolution equation for the film thickness

$$\frac{\partial h}{\partial t} + \frac{\partial(u_s h)}{\partial x} = 0, \tag{8.3.57}$$

where u_s is the surface velocity parallel to the interface. This equation also describes the evolution of the surface concentration of an insoluble surfactant, Γ. In the preceding and forthcoming equations, Γ is substituted for h.

In the similarity solution, the film thickness and surface tension are assumed to depend exclusively on the scaled coordinate $\tilde{x}$, so that

$$h(x,t) = \tilde{h}(\tilde{x}), \qquad \gamma(x,t) = \tilde{\gamma}(\tilde{x}). \tag{8.3.58}$$

The interfacial condition (8.3.56) requires that

$$\mu\, t^{\kappa+\alpha-2\beta} \left(\frac{\partial^2 \tilde{\psi}}{\partial \tilde{y}^2} \right)_{\tilde{y}=0} = -\frac{\mathrm{d}\tilde{\gamma}}{\mathrm{d}\tilde{x}}. \tag{8.3.59}$$

To eliminate the explicit time dependence on the left-hand side, we set $\kappa + \alpha - 2\beta = 0$. Combining this equation with (8.3.31), we obtain

$$\kappa = \frac{1}{4}, \qquad \alpha = \frac{3}{4}, \qquad \beta = \frac{1}{2}. \tag{8.3.60}$$

The evolution equation (8.3.57) for the film thickness becomes

$$-\frac{3}{4}\,\tilde{x}\,\frac{\mathrm{d}\tilde{h}}{\mathrm{d}\tilde{x}} + \frac{\mathrm{d}}{\mathrm{d}\tilde{x}}\left[\tilde{\psi}_{\tilde{y}}^0\, \tilde{h} \right] = 0, \tag{8.3.61}$$

which can be restated as

$$\frac{\mathrm{d}}{\mathrm{d}\tilde{x}}\left[\left(\tilde{\psi}_{\tilde{y}}^0 - \frac{3}{4}\,\tilde{x} \right) \tilde{h} \right] + \frac{3}{4}\,\tilde{h} = 0, \tag{8.3.62}$$

where $\tilde{\psi}_{\tilde{y}}^0 \equiv (\partial \tilde{\psi}/\partial \tilde{y})_{\tilde{y}=0}$. Numerical solutions of the boundary-layer equations have been obtained by finite-difference methods [135, 297].

8.3.6   Arbitrary two-dimensional flow

We return to examining the boundary layer developing over a curved surface in accelerating or decelerating flow beyond the confines of the Falkner–Skan power law. Smith & Clutter [375] considered the boundary layer developing along a curved body and regarded the stream function, ψ, as a function of the arc length along the boundary, l, and scaled transverse position

$$\eta = \left(\frac{U}{\nu l} \right)^{1/2} \ell, \tag{8.3.63}$$

where ℓ is the arc length in the normal direction. Following Görtler [148], we write

$$\psi = (U\nu l)^{1/2}\, f(l, \eta), \tag{8.3.64}$$

where f is an unknown function. Substituting this functional form into the boundary-layer equation, we obtain a third-order partial differential equation,

$$f''' + \frac{1}{2}\,(m+1)\,ff'' - mf'^2 + m = l\left(f'\,\frac{\partial f'}{\partial l} - f''\,\frac{\partial f}{\partial l} \right), \tag{8.3.65}$$

where $m = (l/U)\,dU/dl$ and a prime denotes a partial derivative with respect to η. The boundary and far-field conditions require that $f(l,0) = 0$, $f'(l,0) = 0$, and $f'(l,\infty) = 1$. In the case of the Falkner–Skan boundary layer, the velocity profile is described by (8.3.1) with $l = x$ and $\ell = y$, the coefficient m is constant, the function f is independent of x and only depends on η, and the right-hand side of (8.3.65) is identically zero yielding the Falkner–Skan ordinary differential equation (8.3.12).

Approximating the partial derivatives with respect to l in (8.3.65) with first- or second-order backward differences, we obtain an ordinary differential equation for f with respect to η that can be solved subject to the aforementioned boundary conditions using a standard shooting method. In another implementation of the method, equation (8.3.65) is recast as a system of three first-order equations with respect to η, which is then integrated using a standard finite-difference method.

Problem

8.3.1 *Similarity patching*

Develop an algorithm that allows us to integrate the boundary-layer equations by pretending that the boundary layer consists of a sequence of finite patches of Falkner–Skan boundary layers, where the exponent m is constant aver each patch [374].

 ### Computer Problems

8.3.2 *Falkner–Skan boundary layer*

Write a computer program that solves the Falkner–Skan boundary-layer equation using a method of your choice. Reproduce Figure 8.3.2 and tabulate the corresponding dimensionless coefficients α and β defined in (8.3.18).

8.3.3 *Boundary layer due to a stretching sheet*

Write a computer program that solves the homogeneous Falkner–Skan equation for the generalized Sakiadis boundary layer using a method of your choice. Reproduce the counterpart of Figure 8.3.3.

8.3.4 *Method of Smith & Clutter*

Derive an ordinary differential equation for the function f with respect to η by approximating the partial derivatives of f with respect to l in (8.3.65) using second-order backward finite differences. Discuss the implementation of the method in a computer code.

8.4   Integral momentum balance analysis

Von Kármán [415] developed a powerful and elegant method of computing the flow inside a boundary layer along a two-dimensional body that is held stationary in an incident flow, based on an integral momentum balance. The analysis culminates in a partial differential equation with respect to time and position along the body for a properly defined boundary-layer thickness, such as the displacement or momentum thickness. In fact, Kármán's integral formulation is more general and can be applied to

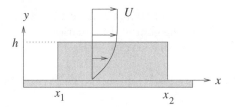

FIGURE 8.4.1   A rectangular control volume introduce to perform an integral momentum balance. The balance provides us with a relation between the displacement thickness, the momentum thickness, and the wall shear stress.

other types boundary-layer flows. In this section, we discuss the development and implementation of the method to flow past a solid surface or due to in-plane boundary motion in an otherwise quiescent fluid. When appropriate, the method will be validated by comparison with exact or numerical solutions of the boundary-layer equations derived by different methods.

8.4.1   Flow past a solid surface

Consider the boundary layer developing over a flat plate, as shown in Figure 8.4.1. The tangential component of the outer flow along the surface is denoted by $U(x)$. For simplicity, we assume that the physical properties of the fluid are uniform throughout the domain of flow. Our point of departure is the integral momentum balance stated in equation (3.2.9), applied over a control area, A_c, that is bounded by a contour, C,

$$\iint_{A_c} \frac{\partial(\rho \mathbf{u})}{\partial t}\, dA = - \oint_C (\boldsymbol{\sigma} - \rho\, \mathbf{u} \otimes \mathbf{u}) \cdot \mathbf{n}\, dl + \iint_{A_c} \rho \mathbf{g}\, dA, \tag{8.4.1}$$

where $\boldsymbol{\sigma}$ is the stress tensor and l is the arc length along C.

Next, we consider a control area confined between two vertical planes located at x_1 and x_2, the flat surface located at $y = 0$, and a horizontal plane located at the elevation $y = h$, as shown in Figure 8.4.1. Introducing the Newtonian constitutive equation for the stress tensor, neglecting the normal viscous stresses expressed by the term $2\mu\, \partial u_x/\partial x$ or $2\mu\, \partial u_y/\partial y$ over the vertical and top planes, and also neglecting gravitational effects, we find that the x component of the integral momentum balance becomes

$$\int_{x_1}^{x_2} \int_0^h \rho\, \frac{\partial u}{\partial t}\, dy\, dx - \int_0^h [u(\rho u)]_{x=x_1,y}\, dy + \int_0^h [u(\rho u)]_{x=x_2,y}\, dy$$

$$+ \int_{x_1}^{x_2} [v(\rho u)]_{x,y=h}\, dx = \int_0^h (-p)_{x=x_1,y}\, dy - \int_0^h (-p)_{x=x_2,y}\, dy - \int_{x_1}^{x_2} \sigma_{xy}^w(x)\, dx. \tag{8.4.2}$$

Taking the limit as x_1 tends to x_2, recalling that the pressure is constant across the boundary layer, setting $u(x, y = h) = U(x)$, and rearranging, we obtain an integro-differential relation,

$$\rho \int_o^h \frac{\partial u}{\partial t}\, dy = h \left(\frac{\partial p}{\partial x} \right)_{y=h} - \rho \frac{\partial}{\partial x} \int_0^h u^2\, dy - \rho\, U(x)\, (v)_{x,y=h} - \sigma_{xy}^w(x). \tag{8.4.3}$$

To reduce the number of unknowns, we eliminate the y velocity at the top plane, $v(x, y = h)$, in favor of the velocity profile, u, using the continuity equation, setting

$$(v)_{x,y=h} = -\int_0^h \frac{\partial u}{\partial x}(x, y')\, dy'. \tag{8.4.4}$$

Moreover, we use the x component of Euler's equation (6.5.3) to evaluate the streamwise pressure gradient, finding

$$\frac{\partial p}{\partial x} = -\rho\left(\frac{\partial U}{\partial t} + U\frac{\partial U}{\partial x}\right). \tag{8.4.5}$$

Substituting expressions (8.4.4) and (8.4.5) into (8.4.3) and rearranging, we obtain

$$\rho\int_o^h \frac{\partial(U-u)}{\partial t}\, dy = -\rho\frac{\partial}{\partial x}\int_0^h u(U-u)\, dy$$
$$-\rho\frac{\partial}{\partial x}\int_0^h U(U-u)\, dy + \rho U\frac{\partial}{\partial x}\int_0^h (U-u)\, dy + \sigma_{xy}^w(x), \tag{8.4.6}$$

which can be interpreted as an evolution law for the momentum deficit expressed by the term $\rho(U-u)$.

Von Kármán integral momentum balance equation

Now we let the scaled height h/δ tend to infinity and use the definitions of the displacement and momentum thicknesses stated in equations (8.2.32) and (8.2.37) to derive the von Kármán integral momentum balance

$$\rho\frac{\partial(U\delta^*)}{\partial t} + \rho\frac{\partial(U^2\Theta)}{\partial x} + \rho\frac{\partial(U^2\delta^*)}{\partial x} - \rho U\frac{\partial(U\delta^*)}{\partial x} - \sigma_{xy}^w = 0. \tag{8.4.7}$$

Rearranging, we derive a relation between the displacement thickness, the momentum thickness, and the wall shear stress in the dimensionless form

$$\frac{1}{U^2}\frac{\partial(U\delta^*)}{\partial t} + \frac{\partial\Theta}{\partial x} + (2\Theta + \delta^*)\frac{1}{U}\frac{\partial U}{\partial x} = \frac{\sigma_{xy}^w}{\rho U^2}. \tag{8.4.8}$$

If the flow is steady, the first term on the left-hand side does not appear. If, in addition, U is constant, equation (8.4.8) reduces to (8.2.39) describing the Blasius boundary layer developing over a semi-infinite flat plate that is held stationary in an incident stream at zero angle of attack.

Equation (8.4.8) can be derived from the continuity equation (8.1.1) and the boundary-layer equation (8.1.12) in a way that bypasses the integral momentum balance [301]. Multiplying the left-hand side of (8.1.1) by $U - u$, adding the product to the right-hand side of (8.1.12), using the continuity equation, integrating the result from $y = 0$ to ∞, and enforcing the boundary conditions, we obtain (8.4.8). This derivation ensures that no error is introduced by replacing the boundary-layer equations with the integral momentum balance.

Impulsive flow past a semi-infinite plate

As an example, we consider an impulsively started streaming flow with constant velocity U past a semi-infinite flat plate. Equation (8.4.8) takes the simple form

$$\frac{\partial \delta^*}{\partial t} + U \frac{\partial \Theta}{\partial x} = \frac{\sigma_{xy}^w}{\rho U}. \tag{8.4.9}$$

At short times, we obtain impulsive flow over a stationary infinite plate, which is the complement of the flow due to the sudden translation of an infinite plane discussed in Section 5.4.3. A steady boundary layer described by the Blasius solution is established at long times. A numerical solution describing the evolution of the flow can be found using the Kármán–Pohlhausen method discussed in Section 8.4.2, combined with the numerical methods discussed in Chapter 12.

Steady orthogonal stagnation-point flow

In the case of steady orthogonal stagnation-point flow, $U = \xi x$, where ξ is a constant rate of stretching. In Section 8.3, we found that the boundary-layer thickness is constant, independent of x. Setting in the steady version of equation (8.4.8) $\partial \Theta / \partial x = 0$ and simplifying, we obtain

$$\sigma_{xy}^w = \rho \left(2\Theta + \delta^*\right) \xi^2 x. \tag{8.4.10}$$

Substituting the expressions for the displacement and momentum thickness given in (8.3.16) and (8.3.17) for $m = 1$, we find that

$$\sigma_{xy}^w = \rho \sqrt{\nu} \left(2\beta + \alpha\right) \xi^{3/2} x, \tag{8.4.11}$$

which is consistent with (8.3.15) for $m = 1$, provided that $f''(0) = 2\beta + \alpha$. Substituting expressions (8.3.18), we obtain

$$f''(0) = \int_0^\infty \left(-2f'^2 + f' + 1\right) \mathrm{d}\eta. \tag{8.4.12}$$

To confirm this identity, we recast the Falkner–Skan equation (8.3.12) for $m = 1$ into the form

$$f''' + \left[\, f(f' - 1)\,\right]' - 2f'^2 + f' + 1 = 0. \tag{8.4.13}$$

Integrating and using the boundary and far-field conditions $f(0) = 0$, $f'(0) = 0$, and $f'(\infty) = 1$, reproduces (8.4.12).

Injection and suction

If fluid is injected into the flow or withdrawn from the flow through a porous wall with normal velocity V, the right-hand side of (8.4.8) contains the additional term $-V/U$, where V is positive in the case of injection and negative in the case of suction (problem 8.4.1).

8.4.2   The Kármán–Pohlhausen method

Kármán and Pohlhausen implemented an approximate method for computing the boundary-layer thickness and associated structure of the flow based on the momentum integral balance (8.4.8). The main idea is to introduce a meaningful boundary-layer thickness, $\Delta(x)$, similar to the δ_{99} boundary-layer thickness, and then employ a sensible velocity profile across the boundary layer, $u/U = F(\eta)$, where $\eta \equiv y/\Delta(x)$ is a dimensionless similarity variable. At the second stage, we compute $\Delta(x)$ to satisfy the integral momentum balance (8.4.8) for solving a differential equation.

The implementation of the method in the case of flow over a flat plate at zero angle of attack where $F(\eta)$ is a quarter of a period of a sinusoidal function, as shown in (8.2.43), was discussed in Section 8.2. In this section, we illustrate the implementation for arbitrary steady flows.

Pohlhausen polynomials

Pohlhausen described the velocity profile across the boundary layer by a fourth-order polynomial,

$$\frac{u}{U} = F(\eta) = \begin{cases} a(x)\,\eta + b(x)\,\eta^2 + c(x)\,\eta^3 + d(x)\,\eta^4 & \text{for } 0 < \eta < 1, \\ 1 & \text{for } \eta > 1, \end{cases} \tag{8.4.14}$$

where $a(x)$–$d(x)$ are four position-dependent coefficients to be computed as part of the solution. The functional form (8.4.14) satisfies the no-slip boundary condition at the wall corresponding to $\eta = 0$. To compute the four coefficients, we require four equations. First, we demand that the overall velocity profile is continuous with smooth first and second derivatives at the edge of the boundary layer corresponding to $\eta = 1$, and thus obtain three conditions,

$$F(\eta = 1) = 1, \qquad F'(\eta = 1) = 0, \qquad F''(\eta = 1) = 0, \tag{8.4.15}$$

where a prime denotes a derivative with respect to η. A fourth condition arises by applying the boundary-layer equation (8.1.13) at the wall located at $y = 0$, and then using the no-slip and no-penetration conditions to set the left-hand side to zero. Evaluating the streamwise pressure gradient using (8.4.5) with the time derivative on the right-hand side set to zero, we obtain

$$\nu \left(\frac{\partial^2 u}{\partial y^2} \right)_{y=0} = -UU', \tag{8.4.16}$$

where $U' \equiv \mathrm{d}U/\mathrm{d}x$. Next, we express the velocity in terms of the function $F(\eta)$ introduced in (8.4.14), and obtain

$$F''(\eta = 0) = -\Lambda, \tag{8.4.17}$$

where

$$\Lambda(x) \equiv \frac{\Delta^2(x)}{\nu} U' \tag{8.4.18}$$

is a dimensionless function expressing the ratio of the magnitude of the inertial acceleration forces in the outer irrotational flow to the magnitude of the viscous forces developing inside the boundary layer; if $\mathrm{d}U/\mathrm{d}x = 0$, then $\Lambda = 0$. By definition, the effective boundary-layer thickness $\Delta(x)$ is related to $\Lambda(x)$ by

$$\Delta(x) \equiv \left(\frac{\nu\Lambda}{U'} \right)^{1/2}. \tag{8.4.19}$$

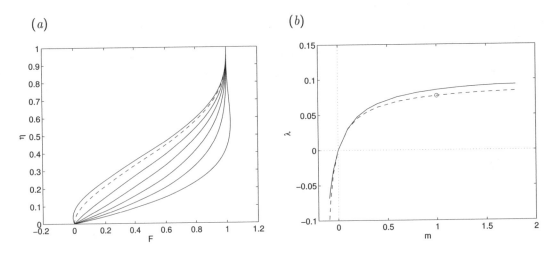

FIGURE 8.4.2   (a) Profiles of the Pohlhausen polynomials for $\Lambda = 20, 12$ (dashed), $6, 0, -6, -12$, and -15. (b) Exact (solid line) and approximate (dashed line) values of the Holstein–Bohlen function λ for the Falkner–Skan boundary layer in accelerating or decelerating flow.

Requiring that the Pohlhausen profile (8.4.14) satisfies equations (8.4.15) and (8.4.17), we obtain

$$a = 2 + \frac{\Lambda}{6}, \qquad b = -\frac{\Lambda}{2}, \qquad c = -2 + \frac{\Lambda}{2}, \qquad d = 1 - \frac{\Lambda}{6}. \tag{8.4.20}$$

Substituting these expressions into (8.4.14) and rearranging, we obtain the velocity profile in terms of the parameter Λ,

$$\frac{u}{U} = F(\eta) = \begin{cases} \eta\,(2 - 2\,\eta^2 + \eta^3) + \Lambda\,\frac{1}{6}\,\eta\,(1 - \eta)^3 & \text{for } 0 < \eta < 1, \\ 1 & \text{for } \eta > 1. \end{cases} \tag{8.4.21}$$

A family of profiles for $\Lambda = 20, 12$ (dashed line), $6, 0, -6, -12$, and -15, is shown in Figure 8.4.2(a). When $\Lambda > 12$, corresponding to a strongly accelerating external flow according to (10.5.14), the profile exhibits an overshooting, which undermines the physical relevance of the fourth-order polynomial expansion. When $\Lambda = -12$, the slope of the velocity profile is zero at the wall and the flow is on the verge of reversal. At that point, the approximations that led us to the boundary-layer equations cease to be valid, and the boundary layer is expected to separate from the wall producing a region of recirculating fluid attached to the wall.

The displacement thickness, momentum thickness, and wall shear stress can be computed in terms of $\Delta(x)$ and $\Lambda(x)$ using the profiles (8.4.21), and are found to be

$$\delta^* = \int_0^\infty \left[1 - F(\eta)\right] \mathrm{d}y \quad = \quad \Delta \frac{1}{10} (3 - \frac{1}{12}\Lambda),$$

$$\Theta = \int_0^\infty F(\eta) \left[1 - F(\eta)\right] \mathrm{d}y \quad = \quad \Delta \frac{1}{315} (37 - \frac{1}{3}\Lambda - \frac{5}{144}\Lambda^2), \qquad (8.4.22)$$

$$\sigma_{xy}^w = \frac{\mu U}{\Delta} \left(\frac{\partial F(\eta)}{\partial \eta}\right)_{\eta=0} \quad = \quad \frac{\mu U}{\Delta} (2 + \frac{1}{6}\Lambda).$$

Expressing $\Delta(x)$ in terms of $\Lambda(x)$ using the definition (8.4.19), we obtain corresponding expressions in terms of $\Lambda(x)$ and the known distribution $U'(x)$. Substituting expressions (8.4.22) into the momentum integral balance (8.4.8) provides us with a first-order nonlinear ordinary differential equation for $\Lambda(x)$ with respect to x. Having solved this equation, we recover the boundary-layer thickness, $\Delta(x)$, using the definition (8.4.19).

Blasius boundary layer

In the case of uniform flow past a semi-infinite flat plate that is held stationary at zero angle of attack, studied in Section 8.2, U is constant, $\Lambda = 0$, and the right-hand side of (8.4.19) is indeterminate. However, this is only an apparent difficulty. Substituting expressions (8.4.22) into (8.4.8) provides us with an ordinary differential equation for $\Delta(x)$,

$$\Delta \frac{\mathrm{d}\Delta}{\mathrm{d}x} = \frac{630}{37} \frac{\nu}{U}, \qquad (8.4.23)$$

which can be integrated from the leading edge up to a certain point, x, subject to the initial condition $\Delta(x = 0) = 0$, to give

$$\Delta(x) = 5.836 \left(\frac{\nu x}{U}\right)^{1/2}. \qquad (8.4.24)$$

It is instructive to compare this expression with that shown in (8.2.47), corresponding to a sinusoidal velocity profile, and note a significant difference in the numerical coefficient on the right-hand side. Substituting (8.4.24) into (8.4.22), we find that

$$\sigma_{xy}^w = \frac{0.343}{\sqrt{\mathrm{Re}_x}} \rho U^2, \qquad \delta^* = 1.751 \left(\frac{\nu x}{U}\right)^{1/2}, \qquad \Theta = 0.685 \left(\frac{\nu x}{U}\right)^{1/2}, \qquad (8.4.25)$$

which are reasonably close to the exact expressions given in (8.2.23), (8.2.33), and (8.2.38). The predicted shape factor is $H = 2.556$.

Holstein–Bohlen function

In the case of the Blasius boundary layer, we were able to derive an analytical solution for $\Delta(x)$. Under more general circumstances, the solution must be obtained by numerical methods. It is convenient to introduce the dimensionless Holstein–Bohlen function

$$\lambda(x) \equiv \frac{\Theta^2(x)}{\Delta^2(x)} \Lambda(x) = \frac{\Theta^2(x)}{\nu} U'(x), \qquad (8.4.26)$$

whose physical interpretation is similar to that of Λ defined in (8.4.18). Using the expression for the momentum thickness given in (8.4.22), we obtain a relationship between λ and Λ,

$$\lambda = \frac{1}{315^2}\,\Lambda\,(37 - \frac{1}{3}\Lambda - \frac{5}{144}\Lambda^2)^2. \tag{8.4.27}$$

The value $\Lambda = -12$ corresponds to $\lambda = -0.15673$ where the boundary layer is expected to separate, as shown in Figure 8.4.2(a). In the case of the Falkner–Skan boundary layer, we substitute into (8.4.26) the velocity distribution $U = \xi x^m$ and the analytical solution for the momentum thickness given in (8.3.17), and find that $\lambda = m\beta^2$, which is plotted with the solid line in Figure 8.4.2(b). As the exponent m increases from zero, yielding a strongly accelerating flow, λ increases and tends to a constant.

Numerical solution

To expedite the numerical solution, we multiply both sides of the momentum integral balance (8.4.8) at steady state by Θ, and rearrange to obtain

$$\frac{d}{dx}\left(\frac{\lambda}{U'}\right) \equiv \frac{1}{\nu}\frac{d\Theta^2}{dx} = 2\,\frac{T(\lambda)}{U}, \tag{8.4.28}$$

where

$$T(\lambda) \equiv S(\lambda) - \left[2 + H(\lambda)\right]\lambda, \tag{8.4.29}$$

$H \equiv \delta^*/\Theta$ is the dimensionless shape factor, and $S(\lambda)$ is a dimensionless shear function defined as

$$S(\lambda) \equiv \frac{\Theta}{\mu U}\,\sigma_{xy}^w. \tag{8.4.30}$$

Physically, the shear function expresses the ratio between the wall shear stress and the average value of the shear stress across the boundary layer, and is thus another measure of the bluntness of the boundary-layer velocity profile. Using expressions (8.4.25), we find that

$$H = \frac{315}{10}\,\frac{3 - \frac{1}{12}\Lambda}{37 - \frac{1}{3}\Lambda - \frac{5}{144}\Lambda^2}, \qquad S = \frac{1}{315}\,(2 + \frac{\Lambda}{6})\,(37 - \frac{1}{3}\Lambda - \frac{5}{144}\Lambda^2), \tag{8.4.31}$$

where Λ can be expressed in terms of λ using equation (8.4.27).

In the case of the Blasius or Falkner–Skan boundary layer, equation (8.4.28) is satisfied if λ is a zero of the nonlinear algebraic equation

$$2m\,T(\lambda) + (m-1)\lambda = 0. \tag{8.4.32}$$

A physically acceptable solution branch is shown with the dashed line in Figure 8.4.2(b). When $m = 1$, corresponding to orthogonal stagnation-point flow, we find that $\lambda = 0.0770356$ corresponding to $\Lambda = 7.0523231$, indicated by the reclusive circle in Figure 8.4.2(b).

More generally, equation (8.4.28) must be solved by numerical methods. The numerical procedure involves the following steps:

1. Given the value of λ at a particular position, x, compute the corresponding value of Λ by solving the nonlinear algebraic equation (8.4.27).

2. Evaluate the functions S and H using expressions (8.4.31).

3. Compute the right-hand side of (8.4.28) to obtain the rate of change of the ratio on the left-hand side with respect to x.

4. Advance the value of λ over a small step, Δx.

5. Return to Step 1 and repeat for another cycle.

The numerical integration typically begins at a stagnation point where the tangential velocity U vanishes and the right-hand side of (8.4.28) is undefined. The initial value of λ is found by solving equation (8.4.32) with a proper value for the exponent m pertinent to the flow near the stagnation point. Although the assumptions that led us to the boundary-layer equations cease to be valid at the stagnation point where the local flow is not nearly unidirectional, starting the integration at a stagnation point does not undermine the validity of the solution.

Evaluation at a stagnation point

To evaluate the right-hand side of (8.4.28) at a stagnation point located at $x = 0$, we apply the l'Hôpital rule to evaluate the fraction and obtain

$$\frac{\mathrm{d}}{\mathrm{d}x}\left(\frac{\lambda}{U'}\right) = 2\,\frac{\mathrm{d}T}{\mathrm{d}x}\frac{1}{U'} = 2\,\frac{\mathrm{d}T}{\mathrm{d}x}\frac{1}{U'}\frac{\mathrm{d}\lambda}{\mathrm{d}x}, \tag{8.4.33}$$

where all terms are evaluated at $x = 0$. Rearranging the last term, we obtain

$$\frac{\mathrm{d}}{\mathrm{d}x}\left(\frac{\lambda}{U'}\right) = 2\,\frac{\mathrm{d}T}{\mathrm{d}\lambda}\left[\frac{\mathrm{d}}{\mathrm{d}\lambda}\left(\frac{\lambda}{U'}\right) + \lambda\,\frac{U''}{(U')^2}\right]. \tag{8.4.34}$$

Combining the left-hand side with the first term inside the square brackets on the right-hand side, we find that

$$\frac{\mathrm{d}}{\mathrm{d}x}\left(\frac{\lambda}{U'}\right) = \frac{2\lambda}{1 - 2T'}\,T'\,\frac{U''}{(U')^2}, \tag{8.4.35}$$

where $T' = \mathrm{d}T/\mathrm{d}\lambda$. Evaluating the expression on the right-hand side using the definition of $T(\lambda)$, we extract the required initial value

$$\frac{\mathrm{d}}{\mathrm{d}x}\left(\frac{\lambda}{U'}\right)_{x=0} = -0.0652\left(\frac{U''}{(U')^2}\right)_{x=0}, \tag{8.4.36}$$

where we have assumed that U' is nonzero and finite at the origin.

Boundary layer around a curved body

The Kármán–Pohlhausen method was developed with reference to a planar boundary, with the x coordinate increasing along the boundary in the direction of the velocity of the outer flow. To tackle the more general case of a curved boundary, we simply replace x with the arc length along the boundary, l, measured in the direction of the tangential velocity of the incident flow, and begin the integration at a stagnation point. A difficulty arises at the critical point where the acceleration dU/dl becomes zero or infinite. However, the ambiguity can be removed by carrying out the integration at that point using the Falkner–Skan similarity solution with a proper value for the exponent m (Problem 8.4.4).

Boundary layer around a cylinder in streaming flow

As an application, we consider streaming flow past a stationary circular cylinder of radius a with vanishing circulation around the cylinder, as shown in Figure 8.4.3(a). Far from the cylinder, the flow occurs toward the negative direction of the x axis and the velocity tends to the uniform value $-U_\infty \, \mathbf{e}_x$, where $U_\infty > 0$ is the magnitude of the streaming flow and $\mathbf{e}_x$ is the unit vector along the x axis. Using the velocity potential for irrotational flow past a circular cylinder with zero circulation, given in (7.9.9), we find that the tangential component of velocity of the outer flow over the surface of the cylinder is

$$U_\theta(\theta) = \left(\frac{\partial \phi}{\partial \theta} \right)_{r=a} = 2U_\infty \sin\theta, \tag{8.4.37}$$

where θ is the polar angle measured around the center of the cylinder in the counterclockwise direction, as shown in Figure 8.4.3(a). The arc length around the cylinder measured from the front stagnation point is $l = a\theta$. The required derivatives of the velocity with respect to arc length are

$$\frac{dU_\theta}{dl} = \frac{1}{a} \frac{dU_\theta}{d\theta} = 2\frac{U_\infty}{a} \cos\theta, \qquad \frac{d^2U_\theta}{dl^2} = \frac{1}{a^2} \frac{d^2U_\theta}{d\theta^2} = -2\frac{U_\infty}{a^2} \sin\theta. \tag{8.4.38}$$

Equation (8.4.36) yields

$$\frac{d}{dl} \left(\frac{\lambda}{dU_\theta/dl} \right)_{l=0} = 0, \tag{8.4.39}$$

which is used to start up the computations.

Graphs of the solution are shown in Figure 8.4.3(b–e). The velocity profile across the boundary layer at different stations around the cylinder can be inferred from the scaled Pohlhausen profiles shown in Figure 8.4.2 using the local value of Λ. The shear function becomes negative when separation occurs, even though the average value of the shear stress across the boundary layer is positive. At the front stagnation point, the values of all functions shown in Figure 8.4.3 are close to those predicted by the Falkner–Skan similarity solution for orthogonal stagnation-point flow with $m = 1$ and shear rate $\xi = 2U_\infty/a$.

The numerical solution reveals that $\Lambda = -12$ at the meridional angle $\theta = 109.5°$. At that point, the shear stress becomes zero and the boundary layer is expected to separate. Comparing this result with the experimentally observed value $\theta = 80.5°$, we find a serious disagreement attributed to the deviation of the actual outer flow from the idealized potential flow described by (7.9.9), due to the presence of a wake. To improve the solution, we may describe the tangential velocity distribution U_θ by interpolation based on data collected in the laboratory. When this is done, the predictions of the boundary-layer analysis are in excellent agreement with laboratory observation ([427], p. 323).

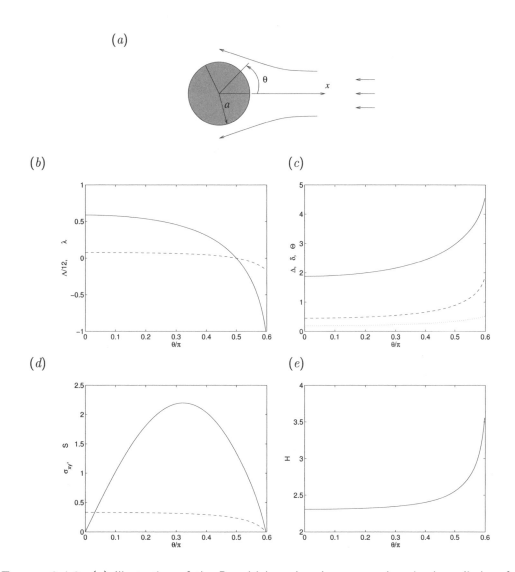

FIGURE 8.4.3   (a) Illustration of the Prandtl boundary layer around a circular cylinder of radius a held stationary in an incident streaming flow with velocity V. (b) Distribution of the dimensionless parameters $\frac{1}{12}\Lambda$ (solid line) and λ (dashed line) computed by the von Kármán-Pohlhausen method. (c) Boundary-layer thickness Δ (solid line), displacement thickness δ^* (dashed line), and momentum thickness Θ (dotted line); all are reduced by $(\nu a/V)^{1/2}$. (d) Distribution of the wall shear stress reduced by $\mu V/a$ (solid line) and shape factor S (dashed line). (e) Distribution of the shear function, H.

Extensions and improvements

The Kármán–Pohlhausen method described in this section can been improved in several ways. In one extension, the fourth-order Pohlhausen polynomial is replaced by another function based on theoretical arguments or laboratory observations. For example, a high-degree polynomial that satisfies additional boundary conditions at the wall and at the edge of the boundary layer can be employed (Problem 8.4.2). In another extension, the shear and shape functions $S(\lambda)$ and $H(\lambda)$ are described by empirical correlations.

Thwaites function

Thwaites proposed replacing the function $T(\lambda)$ defined in (8.4.29) by the linear function $T(\lambda) = 0.45\lambda - 6.0$, motivated by laboratory observations [402]. An analytical solution can be obtained,

$$\Theta^2(x) = \Theta^2(x_0) + \frac{0.45\nu}{U^6} \int_{x_0}^{x} U^5(\xi) \, d\xi, \qquad (8.4.40)$$

where x_0 is an arbitrary point (Problem 8.4.3). The predictions of this equation are in excellent agreement with experimental observations.

8.4.3   Flow due to in-plane boundary motion

A momentum integral balance can be written for the boundary layer developing over a flat surface due to tangential boundary motion with velocity $V(x,t)$ in an otherwise quiescent fluid. An example is the boundary layer due to a stretching sheet discussed in Sections 8.3.2 and 8.3.3. The counterpart of the von Kármán equation (8.4.8) is

$$\frac{1}{V^2} \frac{\partial(V\delta^*)}{\partial t} + \frac{\partial\Theta}{\partial x} + \frac{2}{V}\frac{\partial V}{\partial x}\Theta = -\frac{\sigma_{xy}^w}{\rho V^2}, \qquad (8.4.41)$$

where δ^* and Θ are the displacement and momentum thicknesses defined in (8.2.51). Equation (8.4.41) can be derived by multiplying the left-hand side of the continuity equation (8.1.1) by the streamwise velocity, u, adding the product to the right-hand side of (8.1.12), using the continuity equation, integrating the result from $y = 0$ to ∞, and enforcing the boundary conditions.

Kármán–Pohlhausen method

To implement the Kármán– Pohlhausen method for steady flow, we introduce the similarity variable $\eta = y/\Delta(x)$, where $\Delta(x)$ is an appropriate boundary-layer thickness, and define the dimensionless function

$$\Lambda \equiv \frac{\Delta^2}{\nu} V' \qquad (8.4.42)$$

expressing the ratio of the magnitude of the inertial acceleration forces to the magnitude of the viscous forces developing inside the boundary layer, where $V' \equiv dV/dx$. If $V' = 0$, then $\Lambda = 0$. Conversely, the boundary-layer thickness $\Delta(x)$ is related to the dimensionless parameter $\Lambda(x)$ by

$$\Delta \equiv \left(\frac{\nu\Lambda}{V'}\right)^{1/2}. \qquad (8.4.43)$$

The velocity profile across the boundary layer can be represented by the complementary Pohlhausen polynomial,

$$\frac{u}{V} = F_c(\eta) \equiv 1 - F(\eta), \tag{8.4.44}$$

satisfying the boundary conditions

$$F_c(\eta = 1) = 0, \qquad F_c'(\eta = 1) = 0, \qquad F_c''(\eta = 1) = 0, \qquad F_c''(\eta = 0) = \Lambda, \tag{8.4.45}$$

where a prime denotes a derivative with respect to η. The last condition arises by applying the boundary-layer equation at the wall, located at $y = 0$, and then using the no-slip and no-penetration boundary conditions in the absence of a pressure gradient to obtain

$$\nu \left(\frac{\partial^2 u}{\partial y^2} \right)_{y=0} = V V'. \tag{8.4.46}$$

Explicitly, the velocity profile is given by

$$\frac{u}{V} = F_c(\eta) = \begin{cases} 1 - 2\,\eta + 2\,\eta^3 - \eta^4 - \Lambda \frac{1}{6}\,\eta\,(1-\eta)^3 & \text{for } 0 < \eta < 1, \\ 0 & \text{for } \eta > 1. \end{cases} \tag{8.4.47}$$

The displacement thickness, momentum thickness, and wall shear stress can be computed in terms of $\Delta(x)$ and $\Lambda(x)$ using the profiles (8.4.21), and are found to be

$$\delta^* = \frac{\Delta}{10}\left(3 - \frac{\Lambda}{12}\right), \qquad \Theta = \frac{\Delta}{126}\left(23 - \frac{11}{12}\Lambda + \frac{1}{72}\Lambda^2\right), \qquad \sigma_{xy}^w = -\frac{\mu V}{\Delta}\left(2 + \frac{1}{6}\Lambda\right). \tag{8.4.48}$$

Expressing $\Delta(x)$ in terms of $\Lambda(x)$ using the definition (8.4.43), we obtain corresponding expressions in terms of $\Lambda(x)$ and the boundary distribution of the acceleration, $V'(x)$. Substituting expressions (8.4.48) into the momentum integral balance (8.4.41) provides us with a first-order nonlinear ordinary differential equation for $\Lambda(x)$ with respect to x. Having solved this equation, we recover the boundary-layer thickness, $\Delta(x)$, using the definition (8.4.43).

Sakiadis boundary layer

In the case of the Sakiadis boundary layer discussed in Section 8.3, the boundary velocity V is constant and $\Lambda = 0$. Substituting expressions (8.4.48) into the steady version of (8.4.41) provides us with an ordinary differential equation for $\Delta(x)$,

$$\Delta \frac{d\Delta}{dx} = 2\,\frac{126}{23}\,\frac{\nu}{V}, \tag{8.4.49}$$

which can be integrated subject to the initial condition $\Delta(x = 0) = 0$ to give

$$\Delta(x) = 4.6811 \left(\frac{\nu x}{V}\right)^{1/2}. \tag{8.4.50}$$

Substituting (8.4.50) into (8.4.48), we obtain

$$\delta^* = 1.751 \left(\frac{\nu x}{U}\right)^{1/2}, \qquad \Theta = 0.685 \left(\frac{\nu x}{U}\right)^{1/2}, \qquad \sigma_{xy}^w = \frac{0.42725}{\sqrt{\mathrm{Re}_x}}\,\rho V^2, \tag{8.4.51}$$

which are close to the exact expressions given in (8.2.52) and (8.2.54).

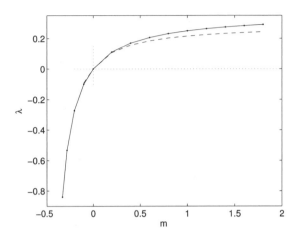

FIGURE 8.4.4   Exact (solid line) and approximate (dashed line) values of the Holstein–Bohlen parameter λ for the boundary layer due to a stretching sheet.

Holstein–Bohlen parameter

In the case of flow due to boundary stretching, the dimensionless Holstein–Bohlen function is defined as

$$\lambda(x) \equiv \frac{\Theta^2(x)}{\Delta^2(x)} \Lambda(x) = \frac{\Theta^2(x)}{\nu} V', \tag{8.4.52}$$

where $V' = \mathrm{d}V/\mathrm{d}x$. Using the expression for the momentum thickness given in (8.4.48), we obtain a relationship between λ and Λ,

$$\lambda = \frac{\Lambda}{126^2} \left(23 - \frac{11}{12} \Lambda + \frac{1}{72} \Lambda^2\right)^2. \tag{8.4.53}$$

In the case of the generalized Crane boundary layer, we substitute the velocity distribution $V = \xi x^m$ and the analytical solution for the momentum thickness given in (8.3.23) into the definition (8.4.52), and find that $\lambda = m\beta^2$, which is plotted with the solid line in Figure 8.4.4. When $m = 0$, we obtain $\lambda = 0$ due to the absence of acceleration. As the exponent m increases from zero, λ increases and tends to a constant value in a strongly accelerating flow.

Numerical solution

To expedite the numerical solution, we multiply both sides of the momentum integral balance (8.4.41) at steady state by Θ, and rearrange to obtain

$$\frac{\mathrm{d}}{\mathrm{d}x} \left(\frac{\lambda}{V'}\right) \equiv \frac{1}{\nu} \frac{\mathrm{d}\Theta^2}{\mathrm{d}x} = 2 \frac{S(\lambda) - 2\lambda}{V}, \tag{8.4.54}$$

where

$$S(\lambda) \equiv -\frac{\Theta}{\mu V}\, \sigma_{xy}^w \qquad (8.4.55)$$

is a dimensionless shear function. Using expressions (8.4.48) based on the Pohlhausen polynomials, we find that

$$S = \frac{1}{126}\left(2 + \frac{\Lambda}{6}\right)\left(23 - \frac{11}{12}\Lambda + \frac{1}{72}\Lambda^2\right). \qquad (8.4.56)$$

Equation (8.4.54) can be solved using the numerical methods discussed in Section 8.4.2.

In the case of the Sakiadis or generalized Crane boundary layer, equation (8.4.54) is satisfied for a constant λ satisfying the nonlinear algebraic equation

$$2m\left(S(\lambda) - 2\lambda\right) + (m-1)\lambda = 0. \qquad (8.4.57)$$

A physically acceptable solution branch is shown with the dashed line in Figure 8.4.4. The approximate solution is reasonably close to the exact solution plotted with the solid line.

8.4.4 Similarity solution of the flow due to stretching sheet

In Section 8.3.3 we discussed a similarity solution for the flow due to stretching sheet. In the terminology introduced in that section,

$$x = t^\alpha\, \tilde{x}, \qquad y = t^\beta\, \tilde{y}, \qquad u = t^{\kappa - \beta}\, \tilde{u}(\tilde{x}, \tilde{y}), \qquad V = t^{\kappa - \beta}\, \tilde{V}(\tilde{x}),$$

$$\delta^* = t^\beta\, \tilde{\delta}^*(\tilde{x}), \qquad \Theta = t^\beta\, \tilde{\Theta}(\tilde{x}), \qquad \sigma_{xy}^w = \mu\left(\frac{\partial u}{\partial y}\right)_{y=0} = \mu\, t^{\kappa - 2\beta}\left(\frac{\partial \tilde{u}}{\partial \tilde{y}}\right)_{\tilde{y}=0}, \qquad (8.4.58)$$

where a tilde denotes a similarity variable. Substituting these expressions into the integral momentum balance expressed by (8.4.41), and recalling that $\kappa = \alpha - \beta$, we obtain an ordinary differential equation,

$$\kappa\, \frac{\tilde{\delta}^*}{\tilde{V}} - \alpha\, \frac{\tilde{x}}{\tilde{V}^2}\, \frac{d(\tilde{V}\tilde{\delta}^*)}{d\tilde{x}} + \frac{d\tilde{\Theta}}{d\tilde{x}} + \frac{2}{\tilde{V}}\, \frac{d\tilde{V}}{d\tilde{x}}\, \tilde{\Theta} = -\frac{\nu}{\tilde{V}^2}\left(\frac{\partial \tilde{u}}{\partial \tilde{y}}\right)_{\tilde{y}=0}. \qquad (8.4.59)$$

A similarity solution in the $\tilde{x}\tilde{y}$ domain can be sought in terms of the complementary Pohlhausen polynomials defined in (8.4.47),

$$\tilde{u}(\tilde{x}, \tilde{y}) = \tilde{V}(\tilde{x})\, F_c(\eta), \qquad (8.4.60)$$

where $\eta = \tilde{x}/\tilde{\Delta}(\tilde{x})$ is a similarity variable and $\tilde{\Delta}(\tilde{x})$ is the boundary-layer thickness. To evaluate the parameter Λ inherent in the Pohlhausen polynomials, we apply the boundary-layer equation at the surface of the stretching sheet and find that

$$\nu\left(\frac{\partial^2 u}{\partial y^2}\right)_{y=0} = \frac{\partial V}{\partial t} + V\, \frac{\partial V}{\partial x} = t^{\kappa - 3/2}\left[(\kappa - \beta)\, \tilde{V} - \alpha\tilde{x}\, \tilde{V}' + \tilde{V}\tilde{V}'\right]. \qquad (8.4.61)$$

After simplifications, we obtain

$$\nu \left(\frac{\partial^2 \tilde{u}}{\partial \tilde{y}^2}\right)_{\tilde{y}=0} = \tilde{V}\,\mathcal{V}, \qquad \text{where} \qquad \mathcal{V} \equiv \kappa - \beta + \left(1 - \alpha\,\frac{\tilde{x}}{\tilde{V}}\right)\tilde{V}', \qquad (8.4.62)$$

and define

$$\Lambda(\tilde{x}) \equiv \frac{\tilde{\Delta}^2(\tilde{x})}{\nu}\,\mathcal{V} \qquad \text{or} \qquad \tilde{\Delta}(\tilde{x}) \equiv \left(\frac{\nu\Lambda}{\mathcal{V}}\right)^{1/2}. \qquad (8.4.63)$$

The last two equations can be used to compute Λ from $\tilde{\Delta}$, and *vice versa*.

Substituting into the right-hand side of (8.4.59) the expression

$$\left(\frac{\partial \tilde{u}}{\partial \tilde{x}}\right)_{\tilde{x}=0} = -\frac{\tilde{V}}{\tilde{\Delta}}\left(2 + \frac{\Lambda}{6}\right), \qquad (8.4.64)$$

originating from the third equation in (8.4.48), and rearranging, we obtain the ordinary differential equation

$$\frac{\mathrm{d}\tilde{\Theta}}{\mathrm{d}\tilde{x}} - \frac{\alpha}{\tilde{V}}\,\tilde{x}\,\frac{\mathrm{d}\tilde{\delta}^*}{\mathrm{d}\tilde{x}} = \frac{\alpha}{\tilde{V}}\left[\left(\frac{\tilde{x}}{\tilde{V}}\frac{\mathrm{d}\tilde{V}}{\mathrm{d}\tilde{x}} - \frac{\kappa}{\alpha}\right)\tilde{\delta}^* - \frac{2}{\alpha}\frac{\mathrm{d}\tilde{V}}{\mathrm{d}\tilde{x}}\tilde{\Theta} + \frac{\nu}{\alpha}\frac{1}{\tilde{\Delta}}\left(2 + \frac{\Lambda}{6}\right)\right]. \qquad (8.4.65)$$

The solution must be found by numerical methods according to the physics of the problem under consideration.

8.4.5 Gravitational spreading of an oil slick

In the case of gravitational spreading of an oil slick discussed in Section 8.3.4, equation (8.4.65) is integrated from the leading edge, $\tilde{x} = \tilde{x}^*$, where the boundary-layer thickness is zero, $\Lambda = 0$, $\tilde{\delta}^* = 0$, and $\tilde{\Theta} = 0$, toward the centerpoint. The interfacial condition (8.3.50) provides us with a differential equation for the square of the film thickness,

$$\frac{\mathrm{d}\tilde{h}^2}{\mathrm{d}\tilde{x}} = 2\,\frac{\mu}{\rho_o g}\left(\frac{\partial \tilde{u}}{\partial \tilde{y}}\right)_{\tilde{y}=0} = -\frac{2\mu}{\rho_o g}\,\frac{\tilde{V}}{\tilde{\Delta}}\left(2 + \frac{\Lambda}{6}\right), \qquad (8.4.66)$$

where ρ_o is the oil density. To circumvent the integrable singularity at the leading edge, $\tilde{x} = \tilde{x}^*$, we introduce the variable $\omega = (\tilde{x}^* - \tilde{x})^{1/2}$ and recast (8.4.66) into the form

$$\frac{\mathrm{d}\tilde{h}^2}{\mathrm{d}\omega} = \omega\,\frac{4\mu}{\rho_o g}\,\frac{\tilde{V}}{\tilde{\Delta}}\left(2 + \frac{\Lambda}{6}\right). \qquad (8.4.67)$$

Integrating from the leading edge where $\omega = 0$ up to the centerline, we obtain the film thickness.

Now we set $\tilde{V} \simeq \alpha\tilde{x}$ for $n = 0$ or approximate $\tilde{V} \simeq (\alpha - n)\tilde{x}$ for general n, and obtain $\mathcal{V} \simeq \kappa - \beta - n$. Equation (8.4.65) becomes

$$\frac{\mathrm{d}(\tilde{\Theta} - c\,\tilde{\delta}^*)}{\mathrm{d}\tilde{x}} = \frac{8}{3 - 4n}\,\frac{1}{\tilde{x}}\left[\frac{1}{2}\tilde{\delta}^* - 2\tilde{\Theta} + \nu\,\frac{1}{\tilde{\Delta}}\left(2 + \frac{\Lambda}{6}\right)\right], \qquad (8.4.68)$$

where

$$c = \frac{\alpha}{\alpha - n} = \frac{3 + 4n}{3 - 4n}. \qquad (8.4.69)$$

To integrate equation (8.4.68), we multiply both sides by $\tilde{\Theta} - c\,\tilde{\delta}^*$ and rearrange to obtain

$$\frac{\mathrm{d}(\tilde{\Theta} - c\,\tilde{\delta}^*)^2}{\mathrm{d}\tilde{x}} = \frac{2}{\tilde{x}} \frac{8}{3 - 4n} \left(\left[\frac{1}{2}\tilde{\delta}^* - 2\tilde{\Theta} \right] (\tilde{\Theta} - c\,\tilde{\delta}^*) + \nu\,\mathcal{H}(\Lambda)\,c\left(2 + \frac{\Lambda}{6}\right) \right), \qquad (8.4.70)$$

where

$$\mathcal{H}(\Lambda) \equiv \frac{\tilde{\Theta} - c\,\tilde{\delta}^*}{\tilde{\Delta}} = \frac{1}{126}\left(23 - \frac{11}{12}\Lambda + \frac{1}{72}\Lambda^2\right) - \frac{c}{10}\left(3 - \frac{\Lambda}{12}\right). \qquad (8.4.71)$$

Using relations (8.4.48) and the expression for the boundary-layer thickness in terms of Λ from (8.4.63), we find that

$$(\tilde{\Theta} - c\,\tilde{\delta}^*)^2 = \nu\,\frac{\Lambda}{\mathcal{V}}\left[\frac{1}{126}\left(23 - \frac{11}{12}\Lambda + \frac{1}{72}\Lambda^2\right) - \frac{c}{10}\left(3 - \frac{\Lambda}{12}\right)\right]^2, \qquad (8.4.72)$$

which can be used to compute the value of Λ, given the the left-hand side, by numerically solving an algebraic equation. Equation (8.4.70) accompanied by (8.4.72) can be integrated from the leading edge toward the centerpoint using elementary numerical methods.

The unknown value $\tilde{x}^*$ corresponding to the leading edge is determined by the oil slick volume per unit width, $A_o = q_o\,t^n$,

$$\tilde{x}^* = \zeta \left(\frac{g^2 q_o^4}{\nu}\right)^{1/8}, \qquad (8.4.73)$$

where ζ is a numerical coefficient. In practice, we may guess a value for $\tilde{x}^*$, solve the preceding equations, compute the volume of oil, and then rescale $\tilde{x}^*$ to ensure a specified volume. When $n = 0$, the numerical computations show that $\zeta = 1.74$. The distribution of the boundary-layer thickness for $n = 0$ computed using the numerical method discussed in this section is shown in Figure 8.4.5. We see that the film thickness $\tilde{h}$ is nearly constant, except near the leading edge. The results are in excellent agreement with those obtained by other numerical methods [60, 75].

Problems

8.4.1 *Boundary layer with suction*

Show that, if a fluid is injected into a flow or withdrawn from a flow through a porous wall with normal velocity V, the right-hand side of (8.4.8) contains the additional term $-V/U$.

8.4.2 *Boundary conditions at a wall*

Assume that the velocity profile across a boundary layer is described by the function $u/U = F(\eta)$. Show that, in addition to satisfying the no-slip condition $F(0) = 0$ and the far-field conditions (8.4.15), the velocity profile is subject to the boundary condition $F'''(0) = 0$.

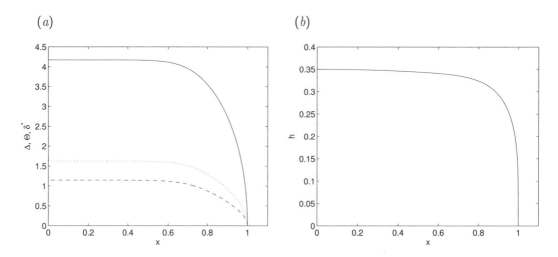

FIGURE 8.4.5   (a) Boundary layer thickness Δ (solid line), momentum thickness Θ (dashed line), and displacement thickness δ^* (dotted line), along a two-dimensional oil slick expanding due to gravity, reduced by $\tilde{x}^*$ and plotted against the dimensionless similarity distance $\tilde{x}/\tilde{x}^*$. (b) Corresponding distribution of the film thickness for a fixed amount of oil, $n = 0$.

8.4.3 *Thwaites boundary layer*

Replace the function $T(\lambda)$ given in (8.4.29) with the general linear function $T(\lambda) = a\lambda - b$ and derive a generalized form of (8.4.40), where a and b are two constants.

▣ *Computer Problem*

8.4.4 *Boundary layer around a circular cylinder*

(a) Write a program that uses the Kármán–Pohlhausen method to compute the boundary layer around a circular cylinder with vanishing circulation around the cylinder, subjected to an incident uniform flow. Run the code and reproduce the graphs shown in Figure 8.4.4.

(b) Repeat (a) with nonzero circulation of your choice and discuss the behavior of the flow.

8.5   Axisymmetric and three-dimensional flows

The formulation of the boundary-layer theory and the exact or approximate solution of the boundary-layer equations for axisymmetric and three-dimensional flows are carried out according to the practices and ideas discussed earlier in this chapter for two-dimensional flow, with some minor modifications.

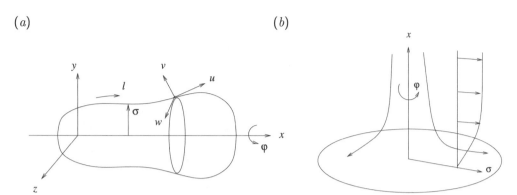

FIGURE 8.5.1   Illustration of (*a*) a boundary layer in axisymmetric flow past a body and (*b*) axisymmetric orthogonal stagnation-point flow against a flat surface.

8.5.1   Axisymmetric boundary layers

To carry out the boundary-layer analysis for axisymmetric flow in the presence or absence of swirling motion, we introduce the arc length along the trace of the boundary in a plane of constant azimuthal angle φ, denoted by l, and the arc length in the normal direction, denoted by ℓ, as shown in Figure 8.5.1(*a*). The corresponding tangential and normal velocity components are denoted by u and v, and the azimuthal velocity component responsible for the swirling motion is denoted by w.

Continuity equation

The continuity equation for axisymmetric flow in the absence or presence of swirling motion takes the form

$$\frac{1}{\sigma}\frac{\partial(\sigma u)}{\partial l} + \frac{\partial v}{\partial \ell} = 0. \tag{8.5.1}$$

When the distance from the axis of symmetry, σ, is much larger than the typical boundary-layer thickness around the body contour, the continuity equation (8.5.1) can be recast into the approximate convenient form

$$\frac{\partial(\sigma u)}{\partial l} + \frac{\partial(\sigma v)}{\partial \ell} = 0. \tag{8.5.2}$$

To satisfy this equation, we may introduce a stream function, $\psi(l, \ell)$, defined by the equations

$$u = \frac{1}{\sigma}\frac{\partial \psi}{\partial \ell}, \qquad v = -\frac{1}{\sigma}\frac{\partial \psi}{\partial l}. \tag{8.5.3}$$

In the case of a cylindrical body, $l = x$, $\ell = \sigma$, $u = u_x$, and $v = u_\sigma$. In the case of flow over a plane that is perpendicular to the x axis, as shown in Figure 8.5.1(*b*), $l = \sigma$ and $\ell = x$, $u = u_\sigma$ and $v = u_x$.

Boundary-layer equations

The normal component of the equation of motion states that the pressure drop across the boundary layer can be neglected to leading-order approximation. Consequently, the pressure distribution, $p(l, \varphi)$, can be computed using Euler's equation for the incident flow,

$$-\frac{1}{\rho}\frac{\partial p}{\partial l} = \frac{\partial U}{\partial t} + U\frac{\partial U}{\partial l} - \frac{W^2}{\sigma}\frac{\partial \sigma}{\partial l}, \qquad -\frac{1}{\rho}\frac{1}{\sigma}\frac{\partial p}{\partial \varphi} = \frac{\partial W}{\partial t} + U\frac{\partial W}{\partial l} + \frac{UW}{\sigma}\frac{\partial \sigma}{\partial l}, \qquad (8.5.4)$$

where U is the meridional tangential velocity and W is the azimuthal swirling velocity of the outer irrotational flow. The corresponding components of the equation of motion simplify into the boundary-layer equations

$$\frac{\partial u}{\partial t} + u\frac{\partial u}{\partial l} + v\frac{\partial u}{\partial \ell} - \frac{w^2}{\sigma}\frac{\partial \sigma}{\partial l} = -\frac{1}{\rho}\frac{\partial p}{\partial l} + \nu\frac{\partial^2 u}{\partial \ell^2} \qquad (8.5.5)$$

and

$$\frac{\partial w}{\partial t} + u\frac{\partial w}{\partial l} + v\frac{\partial w}{\partial \ell} + \frac{uw}{\sigma}\frac{\partial \sigma}{\partial l} = -\frac{1}{\rho}\frac{1}{\sigma}\frac{\partial p}{\partial \varphi} + \nu\frac{\partial^2 w}{\partial \ell^2}. \qquad (8.5.6)$$

In the absence of swirling motion, equation (8.5.6) is trivially satisfied and equation (8.5.5) reduces into the boundary-layer equation (8.1.12) for two-dimensional flow with a straightforward change in notation,

$$\frac{\partial u}{\partial t} + u\frac{\partial u}{\partial l} + v\frac{\partial u}{\partial \ell} = \frac{\partial U}{\partial t} + U\frac{\partial U}{\partial l} - \frac{1}{\rho}\frac{\partial p}{\partial l} + \nu\frac{\partial^2 u}{\partial \ell^2}. \qquad (8.5.7)$$

The effect of axisymmetry is manifested in the continuity equation (8.5.1) alone.

Mangler's transformation

Mangler invented a remarkable transformation that reduces the boundary-layer equations (8.5.1) and (8.5.7) for axisymmetric flow without swirling motion to the simpler boundary-layer equations (8.1.1) and (8.1.12) for two-dimensional flow written for the judiciously chosen modified variables

$$\tilde{x} = \frac{1}{L^2}\int_0^l \sigma^2 \, \mathrm{d}l, \qquad \tilde{y} = \frac{\sigma}{L}\ell, \qquad \tilde{u} = u, \qquad \tilde{v} = \frac{L}{\sigma}\left(v + \frac{\ell}{\sigma}\frac{\mathrm{d}\sigma}{\mathrm{d}l}u\right), \qquad (8.5.8)$$

where L is a specified length [254]. To demonstrate the reduction of the continuity equation, we write

$$\frac{\partial}{\partial l} = \frac{\partial \tilde{x}}{\partial l}\frac{\partial}{\partial \tilde{x}} + \frac{\partial \tilde{y}}{\partial l}\frac{\partial}{\partial \tilde{y}} \simeq \frac{\sigma^2}{L^2}\frac{\partial}{\partial \tilde{x}} + \frac{\ell}{L}\frac{\mathrm{d}\sigma}{\mathrm{d}l}\frac{\partial}{\partial \tilde{y}} = \frac{\sigma^2}{L^2}\frac{\partial}{\partial \tilde{x}} + \frac{\tilde{y}}{\sigma}\frac{\mathrm{d}\sigma}{\mathrm{d}l}\frac{\partial}{\partial \tilde{y}} \qquad (8.5.9)$$

and

$$\frac{\partial}{\partial \ell} = \frac{\partial \tilde{x}}{\partial \ell}\frac{\partial}{\partial \tilde{x}} + \frac{\partial \tilde{y}}{\partial \ell}\frac{\partial}{\partial \tilde{y}} \simeq \frac{\sigma}{L}\frac{\partial}{\partial \tilde{y}} \qquad . \qquad (8.5.10)$$

The simplified continuity equation (8.5.2) requires that

$$\frac{\partial(\sigma u)}{\partial \hat{x}} + \frac{L}{\sigma}\frac{\partial(\sigma v)}{\partial \hat{y}} + \frac{L^2}{\sigma^2}\frac{\hat{y}}{\sigma}\frac{d\sigma}{dl}\frac{\partial(\sigma u)}{\partial \hat{y}} = 0, \tag{8.5.11}$$

which can be approximated as

$$\frac{\partial u}{\partial \hat{x}} + \frac{\partial}{\partial \hat{y}}\left(\frac{L}{\sigma} v + \frac{L^2}{\sigma^2}\frac{\hat{y}}{\sigma}\frac{d\sigma}{dl} u\right) = 0. \tag{8.5.12}$$

Now invoking the last definition in (8.5.8), we obtain the continuity equation for the transformed variables in the two-dimensional $\tilde{x}\tilde{y}$ domain.

Mangler's transformation reduces the problem of computing a boundary layer over an axisymmetric body to the problem of computing a boundary layer over a body with modified geometry in a fictitious two-dimensional flow. A practical drawback is that the pressure distribution of the fictitious flow can be considerably more involved than that of the physical flow.

Momentum integral balance

Working as in the case of two-dimensional flow, we derive an integral momentum balance for axisymmetric flow in the absence of swirling motion,

$$\frac{1}{U^2}\frac{\partial(U\delta^*)}{\partial t} + \frac{1}{\sigma}\frac{\partial(\sigma\Theta)}{\partial l} + (2\Theta + \delta^*)\frac{1}{U}\frac{\partial U}{\partial l} = \frac{\sigma^w_{sh}}{\rho U^2}, \tag{8.5.13}$$

which is a slight modification of (8.4.8) (Problem 8.5.1). The numerator of the fraction on the right-hand side, σ^w_{sh}, is the wall shear stress. The displacement and momentum thicknesses are defined as in the case of two-dimensional flow in terms of integrals with respect to normal arc length, ℓ.

Using the fourth-degree Pohlhausen polynomial and the Holstein–Bohlen dimensionless parameter introduced in (8.4.26), we find that the momentum integral balance for steady flow takes the form

$$\frac{1}{\sigma^2}\frac{d}{dl}\left(\frac{\sigma^2\lambda}{U'}\right) = 2\frac{T(\lambda)}{U}, \tag{8.5.14}$$

where the function $T(\lambda)$ defined in (8.4.29), and a prime denotes a derivative with respect to tangential arc length, l (Problem 8.5.1). At the front stagnation point of an axisymmetric body, σ reduces to l and the tangential velocity U behaves like ξl, where ξ is the local shear rate. Making substitutions in the left-hand side of (8.5.14) and requiring that $d\lambda/dl$ is finite, we obtain the starting value $\lambda = 0.0571$. Equation (8.5.6) is integrated using the methods discussed earlier for two-dimensional flow. Boundary-layer separation occurs when $\lambda = -0.15673$, corresponding to $\Lambda = -12$.

In the case of uniform flow past a stationary sphere, numerical integration of (8.5.14) shows that the boundary layer separates at the meridional angle $\theta = 110°$, which is substantially higher than the measured value $83°$ (Problem 8.5.6). The discrepancy is attributed to the poor representation of the outer irrotational flow due to the presence of a wake.

Görtler, Smith & Clutter formulation

Following the formulation of Smith & Clutter [375], we regard the Stokes stream function, ψ, as a function of the arc length along the trace of the boundary in an azimuthal plane, l, and of the scaled transverse position, $\eta = (U/\nu l)^{1/2}\ell$, where ℓ is the arc length in the normal direction. Following Görtler [148], we write

$$\psi = (U\nu l)^{1/2}\,\sigma f(l,\eta), \tag{8.5.15}$$

and introduce the dimensionless parameter $m = (l/U)\mathrm{d}U/\mathrm{d}l$. The counterpart of equation (8.3.65) for axisymmetric flow is

$$f'' + \left(\frac{m+1}{2} + \frac{l}{\sigma}\frac{\mathrm{d}\sigma}{\mathrm{d}l}\right)ff'' - mf'^2 + m = l\left(f'\frac{\partial f'}{\partial l} - f''\frac{\partial f}{\partial l}\right), \tag{8.5.16}$$

where a prime denotes a partial derivative with respect to η. The solution can be computed using a shooting method or a direct finite-difference method as in the case of two-dimensional flow.

Integral momentum balance of flow due to boundary motion

The integral momentum balance of the flow due to in-plane boundary motion with tangential boundary velocity $V(l)$ in an otherwise quiescent fluid is expressed by the dimensionless equation

$$\frac{1}{V^2}\frac{\partial(V\delta^*)}{\partial t} + \frac{1}{\sigma V^2}\frac{\partial(\sigma V^2\Theta)}{\partial l} = -\frac{\sigma_{\sigma x}^w}{\rho V^2}, \tag{8.5.17}$$

where δ^* and Θ are the displacement and momentum thicknesses defined in (8.2.51) with ℓ in place of the normal distance, y.

8.5.2   Steady flow against or due to a radially stretching surface

As an application, we consider axisymmetric orthogonal stagnation-point flow against an infinite flat surface located at $x = 0$, as illustrated in Figure 8.5.1(b). The radial velocity of the incident flow is described by the power-law

$$U(\sigma) = \xi\sigma^m, \tag{8.5.18}$$

where ξ is a positive constant and m is a positive exponent. The flow inside the boundary layer is governed by the continuity equation,

$$\frac{1}{\sigma}\frac{\partial(\sigma u_\sigma)}{\partial\sigma} + \frac{\partial u_x}{\partial x} = 0, \tag{8.5.19}$$

and the boundary-layer equation,

$$u_\sigma\frac{\partial u_\sigma}{\partial\sigma} + u_x\frac{\partial u_\sigma}{\partial x} = \nu\frac{\partial^2 u_\sigma}{\partial x^2}. \tag{8.5.20}$$

To carry out the boundary-layer analysis, we introduce the dimensionless similarity variable

$$\eta = \left(\frac{U}{\nu\sigma}\right)^{1/2}x, \tag{8.5.21}$$

and express the Stokes stream function in the form

$$\psi = -(U\nu\sigma)^{1/2}\,\sigma\,f(\eta). \tag{8.5.22}$$

The radial velocity component is

$$u_\sigma = -\frac{1}{\sigma}\frac{\partial\psi}{\partial x} = U f'(\eta). \tag{8.5.23}$$

The wall shear stress is found by differentiating the radial velocity profile and evaluating the resulting expression at the wall, yielding

$$\sigma_{\sigma x}^w(\sigma) = \mu\left(\frac{\partial u}{\partial x}\right)_{x=0} = \rho\sqrt{\nu}\,f''(0)\,\xi^{3/2}\,\sigma^{(3m-1)/2}. \tag{8.5.24}$$

Performing the boundary-layer analysis, we find that the function $f(\eta)$ satisfies a third-order ordinary differential equation,

$$f''' + \left[\frac{1}{2}(m+1)+1\right]ff'' - mf'^2 + m = 0. \tag{8.5.25}$$

The boundary conditions require that $f(0) = 0$ and $f'(0) = 0$, and the far-field condition requires that $f'(\infty) = 1$. When $m = 1$, we recover the Homann equation (5.6.39), providing us with an exact solution of the unsimplified Navier–Stokes equation.

Radially stretching surface

A complementary problem addresses the flow due to the radial stretching of an elastic sheet in its plane with radial velocity $V(\sigma) = \xi\sigma^m$ in an otherwise quiescent fluid. The axisymmetric boundary layer developing over the sheet is described by the Stokes stream function

$$\psi = -(V\nu\sigma)^{1/2}\,\sigma\,f(\eta), \tag{8.5.26}$$

where

$$\eta = \left(\frac{V}{\nu\sigma}\right)^{1/2} x \tag{8.5.27}$$

is a dimensionless similarity variable, the function $f(\eta)$ satisfies the homogeneous equation

$$f''' + \left[\frac{1}{2}(m+1)+1\right]ff'' - mf'^2 = 0, \tag{8.5.28}$$

and a prime denotes a derivative with respect to η. The boundary conditions require that $f(0) = 0$ and $f'(0) = 1$, and the far-field condition requires that $f'(\infty) = 0$.

8.5.3   Unsteady flow due to a radially stretching surface

Next, we consider a class of unsteady boundary layers generated by the in-plane radial stretching of a surface located at $x = 0$, as shown in Figure 8.5.1(b). The corresponding two-dimensional boundary layers were discussed in Section 8.3.3. The flow inside the axisymmetric boundary layers is governed by the continuity equation (8.5.1) and the counterpart of equation (8.3.25) for axisymmetric flow,

$$\frac{\partial u}{\partial t} + u \frac{\partial u}{\partial \sigma} + v \frac{\partial u}{\partial x} = \nu \frac{\partial^2 u}{\partial x^2}, \tag{8.5.29}$$

where $u = u_\sigma$ is the radial velocity and $v = u_x$ is the axial velocity. The flow will be described in terms of a time-dependent Stokes stream function, $\psi(\sigma, x, t)$, defined such that

$$u \equiv u_\sigma = -\frac{1}{\sigma} \frac{\partial \psi}{\partial x}, \qquad v \equiv u_x = \frac{1}{\sigma} \frac{\partial \psi}{\partial \sigma}. \tag{8.5.30}$$

The ordered pair, (σ, x), defines a system of two Cartesian coordinates similar to the (x, y) coordinates in two-dimensional flow.

Similarity solution

We are interested in situations where the radial sheet velocity, $u_\sigma(\sigma, x = 0, t) \equiv V(\sigma, t)$, allows for a similarity solution such that

$$\psi(\sigma, x, t) = -t^\kappa \, \tilde{\psi}(\tilde{\sigma}, \tilde{x}), \qquad \tilde{\sigma} = \frac{\sigma}{t^\alpha}, \qquad \tilde{x} = \frac{x}{t^\beta}, \tag{8.5.31}$$

where the exponents κ, α, and β will be determined as part of the solution. The radial and axial velocity components are given by

$$u = -\frac{1}{\sigma} \frac{\partial \psi}{\partial x} = t^{\kappa - \beta} \frac{1}{\sigma} \frac{\partial \tilde{\psi}}{\partial \tilde{x}}, \qquad v = \frac{1}{\sigma} \frac{\partial \psi}{\partial \sigma} = -t^{\kappa - \alpha} \frac{1}{\sigma} \frac{\partial \tilde{\psi}}{\partial \tilde{\sigma}}. \tag{8.5.32}$$

Substituting these expressions along with the expressions

$$\frac{\partial u}{\partial t} = t^{\kappa - \beta - 1} \frac{1}{\sigma} \left((\kappa - \beta) \frac{\partial \tilde{\psi}}{\partial \tilde{x}} - \alpha \tilde{\sigma} \frac{\partial^2 \tilde{\psi}}{\partial \tilde{x} \partial \tilde{\sigma}} - \beta \tilde{x} \frac{\partial^2 \tilde{\psi}}{\partial \tilde{x}^2} \right) \tag{8.5.33}$$

and

$$\frac{\partial u}{\partial \sigma} = t^{\kappa - \alpha - \beta} \frac{1}{\sigma} \left(\frac{\partial^2 \tilde{\psi}}{\partial \tilde{\sigma} \partial \tilde{x}} - \frac{1}{\tilde{\sigma}} \frac{\partial \tilde{\psi}}{\partial \tilde{x}} \right), \qquad \frac{\partial u}{\partial x} = t^{\kappa - 2\beta} \frac{1}{\sigma} \frac{\partial^2 \tilde{\psi}}{\partial \tilde{x}^2}, \qquad \frac{\partial^2 u}{\partial x^2} = t^{\kappa - 3\beta} \frac{1}{\sigma} \frac{\partial^3 \tilde{\psi}}{\partial \tilde{x}^3}, \tag{8.5.34}$$

into the boundary-layer equation (8.5.29), we obtain

$$(\kappa - \beta) \frac{\partial \tilde{\psi}}{\partial \tilde{x}} - \alpha \tilde{\sigma} \frac{\partial^2 \tilde{\psi}}{\partial \tilde{\sigma} \partial \tilde{x}} - \beta \tilde{x} \frac{\partial^2 \tilde{\psi}}{\partial \tilde{x}^2}$$

$$+ t^{\kappa - 2\alpha - \beta + 1} \frac{1}{\tilde{\sigma}} \left(\frac{\partial \tilde{\psi}}{\partial \tilde{x}} \left[\frac{\partial^2 \tilde{\psi}}{\partial \tilde{\sigma} \partial \tilde{x}} - \frac{1}{\tilde{\sigma}} \frac{\partial \tilde{\psi}}{\partial \tilde{x}} \right] - \frac{\partial \tilde{\psi}}{\partial \tilde{\sigma}} \frac{\partial^2 \tilde{\psi}}{\partial \tilde{x}^2} \right) = \nu t^{-2\beta + 1} \frac{\partial^3 \tilde{\psi}}{\partial \tilde{x}^3}. \tag{8.5.35}$$

To eliminate the explicit time dependence on both sides of this equation, we set

$$\beta = \frac{1}{2}, \qquad \kappa = 2\alpha - \frac{1}{2}, \tag{8.5.36}$$

and derive a nonlinear partial differential equation for $\tilde{\sigma}$ and $\tilde{x}$,

$$\left(2\alpha - 1 - \frac{1}{\tilde{\sigma}^2}\frac{\partial\tilde{\psi}}{\partial\tilde{x}}\right)\frac{\partial\tilde{\psi}}{\partial\tilde{x}} - \left(\alpha\tilde{\sigma} - \frac{1}{\tilde{\sigma}}\frac{\partial\tilde{\psi}}{\partial\tilde{x}}\right)\frac{\partial^2\tilde{\psi}}{\partial\tilde{\sigma}\partial\tilde{x}} - \left(\frac{1}{2}\tilde{x} + \frac{1}{\tilde{\sigma}}\frac{\partial\tilde{\psi}}{\partial\tilde{\sigma}}\right)\frac{\partial^2\tilde{\psi}}{\partial\tilde{x}^2} = \nu\frac{\partial^3\tilde{\psi}}{\partial\tilde{x}^3}. \qquad (8.5.37)$$

The no-penetration boundary condition requires that $\tilde{\psi}(\tilde{\sigma},0) = 0$, and the far-field condition requires that $\partial\tilde{\psi}/\partial\tilde{\sigma} \to 0$ and $\partial\tilde{\psi}/\partial\tilde{x} \to 0$ as $\tilde{x} \to \infty$.

Equation (8.5.37) admits the solution $\tilde{\psi}(\tilde{\sigma},\tilde{x}) = \tilde{\sigma}^2 f(\tilde{x}) + \tilde{\psi}_\infty$, resulting in an ordinary differential equation,

$$\nu f''' + \left(\frac{1}{2}\tilde{x} + 2f\right)f'' + f'(1 - f') = 0, \qquad (8.5.38)$$

where $\tilde{\psi}_\infty$ is a constant and a prime denotes a derivative with respect to $\tilde{x}$. The no-penetration condition requires that $f(0) = 0$ and the far-field condition requires that $f(\tilde{x})$ decays to zero as $\tilde{x}$ tends to infinity.

Solution near the centerpoint

We may assume that the general solution near the centerline, $\tilde{\sigma} \to 0$, takes the asymptotic form

$$\tilde{\psi}(\tilde{\sigma},\tilde{x}) \simeq \tilde{\sigma}^q f(\eta), \qquad (8.5.39)$$

where q is a positive exponent, $\eta = \tilde{x}/\tilde{\sigma}^\varrho$ is a similarity variable, and the positive exponent ϱ will be determined as part of the solution. Substituting (8.5.39) into (8.5.37) and simplifying, we obtain

$$[\alpha(2 - q + \varrho) - 1]\,f' + \left(\alpha\varrho - \frac{1}{2}\right)\eta f'' + \tilde{\sigma}^{q-\varrho-2}\left[(q - \varrho - 1)\,f'^2 - q\,ff''\right] = \nu\tilde{\sigma}^{-2\varrho}\,f'''. \qquad (8.5.40)$$

This equation is physically acceptable under two complementary conditions. First, setting the power of $\tilde{\sigma}$ on the right-hand side of (8.5.40) to zero, $\varrho = 0$, we obtain

$$\nu f''' + \frac{1}{2}\tilde{x}\,f'' - [\alpha(2 - q) - 1]\,f' = \tilde{\sigma}^{q-2}\left[(q - 1)f'^2 - qff''\right]. \qquad (8.5.41)$$

When $q > 2$, the right-hand side vanishes as $\tilde{\sigma} \to 0$, yielding a homogeneous integral equation,

$$\nu f''' + \frac{1}{2}\tilde{x}\,f'' - [\alpha(2 - q) - 1]\,f' = 0. \qquad (8.5.42)$$

Alternatively, we require that the power of $\tilde{\sigma}$ on the right-hand side of (8.5.40) matches that of the last term on the left-hand side, $q - \varrho - 2 = -2\varrho$ or $\varrho = 2 - q$, and thus obtain

$$\nu f''' + q\,ff'' - (2q - 3)\,f'^2 = \tilde{\sigma}^{2\varrho}\,(2\alpha\varrho - 1)\left(f' + \frac{1}{2}\eta f''\right). \qquad (8.5.43)$$

The right-hand side vanishes as $\tilde{\sigma} \to 0$, yielding the homogeneous Falkner–Skan equation (8.5.28) with $m = 2q - 3$. The properties of the equations presented in this section have been investigated with respect to uniqueness of solution [296].

Integral momentum balance

The integral momentum balance (8.5.17) provides us with the equation

$$\frac{1}{V^2}\frac{\partial(V\delta^*)}{\partial t} + \frac{\partial\Theta}{\partial\sigma} + \left(\frac{2}{V}\frac{\partial V}{\partial\sigma} + \frac{1}{\sigma}\right)\Theta = -\frac{\sigma_{\sigma x}^w}{\rho V^2}. \tag{8.5.44}$$

To compute the similarity solution, we introduce further similarity variables denoted by a tilde,

$$u = t^{\kappa-\alpha-\beta}\,\tilde{u}(\tilde{\sigma},\tilde{x}), \qquad V = t^{\kappa-\alpha-\beta}\,\tilde{V}(\tilde{\sigma}), \qquad \delta^* = t^\beta\,\tilde{\delta}^*(\tilde{\sigma}),$$

$$\Theta = t^\beta\,\tilde{\Theta}(\tilde{\sigma}), \qquad \sigma_{\sigma x}^w = \mu\left(\frac{\partial u}{\partial x}\right)_{x=0} = \mu\,t^{\kappa-\alpha-2\beta}\left(\frac{\partial\tilde{u}}{\partial\tilde{x}}\right)_{\tilde{x}=0}. \tag{8.5.45}$$

Substituting these expressions into the integral momentum balance (8.5.44) and recalling that $\kappa = 2\alpha - \beta$, we obtain an ordinary differential equation,

$$(\kappa-\alpha)\frac{\tilde{\delta}^*}{\tilde{V}} - \tilde{\sigma}\frac{\alpha}{\tilde{V}^2}\frac{d(\tilde{V}\tilde{\delta}^*)}{d\tilde{\sigma}} + \frac{d\tilde{\Theta}}{d\tilde{\sigma}} + \left(\frac{2}{\tilde{V}}\frac{d\tilde{V}}{d\tilde{\sigma}} + \frac{1}{\tilde{\sigma}}\right)\tilde{\Theta} = -\frac{\nu}{\tilde{V}^2}\left(\frac{\partial\tilde{u}}{\partial\tilde{x}}\right)_{\tilde{x}=0}. \tag{8.5.46}$$

Given the radial velocity, $\tilde{V}(\tilde{\sigma})$, the solution can be found by sensible approximate methods.

Complementary Pohlhausen polynomials

A similarity solution in the $\tilde{\sigma}\tilde{x}$ plane can be sought in terms of the complementary Pohlhausen polynomials defined in (8.4.47),

$$\tilde{u}(\tilde{x},\tilde{y}) = \tilde{V}(\tilde{x})\,F_c(\eta), \tag{8.5.47}$$

where $\eta = \tilde{x}/\tilde{\Delta}(\tilde{\sigma})$ is a similarity variable and $\tilde{\Delta}(\tilde{\sigma})$ is the boundary-layer thickness. To evaluate the parameter Λ inherent in the Pohlhausen polynomials, we apply the boundary-layer equation at the stretching sheet and find that

$$\nu\left(\frac{\partial^2 u}{\partial x^2}\right)_{x=0} = \frac{\partial V}{\partial t} + V\frac{\partial V}{\partial\sigma} = t^{\kappa-\alpha-3/2}\left[(\kappa-\alpha-\beta)\,\tilde{V} - \alpha\tilde{\sigma}\,\tilde{V}' + \tilde{V}\tilde{V}'\right], \tag{8.5.48}$$

where a prime denotes a derivative with respect to $\tilde{\sigma}$. After simplifications, we obtain

$$\nu\left(\frac{\partial^2\tilde{u}}{\partial\tilde{x}^2}\right)_{\tilde{x}=0} = \tilde{V}\,\mathcal{V}, \qquad \text{where} \qquad \mathcal{V} \equiv \kappa-\alpha-\beta+\left(1-\alpha\frac{\tilde{\sigma}}{\tilde{V}}\right)\tilde{V}', \tag{8.5.49}$$

and define

$$\Lambda(\tilde{\sigma}) \equiv \frac{\tilde{\Delta}^2(\tilde{\sigma})}{\nu}\,\mathcal{V} \qquad \text{or} \qquad \tilde{\Delta}(\tilde{\sigma}) \equiv \left(\frac{\nu\Lambda}{\mathcal{V}}\right)^{1/2}. \tag{8.5.50}$$

The last two equations can be used to compute Λ from $\tilde{\Delta}$ and *vice versa*.

Substituting into the right-hand side of (8.5.46) the expression

$$\left(\frac{\partial\tilde{u}}{\partial\tilde{y}}\right)_{\tilde{y}=0} = -\frac{\tilde{V}}{\tilde{\Delta}}\left(2+\frac{\Lambda}{6}\right), \tag{8.5.51}$$

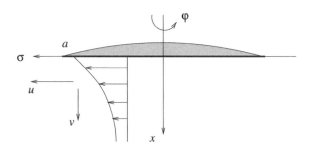

FIGURE 8.5.2 Illustration of an unsteady boundary layer due to the axisymmetric spreading of an oil slick of radius $a(t)$ at the surface of the ocean.

originating from the third equation in (8.4.48), and rearranging, we obtain the ordinary differential equation

$$\frac{d\tilde{\Theta}}{d\tilde{\sigma}} - \tilde{\sigma}\frac{\alpha}{\tilde{V}}\frac{d\tilde{\delta}^*}{d\tilde{\sigma}} = \frac{\alpha}{\tilde{V}}\left[\left(\frac{\tilde{\sigma}}{\tilde{V}}\frac{d\tilde{V}}{d\tilde{\sigma}} - \frac{\kappa - \alpha}{\alpha}\right)\tilde{\delta}^* - \frac{2}{\alpha}\frac{d\tilde{V}}{d\tilde{\sigma}}\tilde{\Theta} + \frac{\nu}{\alpha}\frac{1}{\tilde{\Delta}}\left(2 + \frac{\Lambda}{6}\right)\right] - \frac{\tilde{\Theta}}{\tilde{\sigma}}. \qquad (8.5.52)$$

The solution must be found by numerical methods according to the physics of the problem under consideration.

8.5.4   Gravitational spreading of an axisymmetric oil slick

As an application of the similarity solution derived in Section 8.5.3, we consider the gravitational spreading of an axisymmetric oil slick over the flat surface of an otherwise quiescent pool of water, as shown in Figure 8.5.2. A thinning film of oil with time-dependent thickness $h(\sigma, t)$ is confined inside the spilling zone, $\sigma < a(t)$. At high Reynolds numbers, an unsteady boundary layer develops next to the film, while the fluid tends to become quiescent far from the film. The two-dimensional version of this problem was discussed in Sections 8.3.4 and 8.4.5.

Oil flow

To leading order, the σ and x components of the equation of motion inside the thin oil film, denoted by the subscript or subscript o, take the form

$$\frac{\partial \sigma_{\sigma x}^o}{\partial x} = \frac{\partial p_o}{\partial \sigma}, \qquad \frac{\partial p_o}{\partial x} = \rho_o g, \qquad (8.5.53)$$

where $\sigma_{\sigma x}^o$ is the shear stress. Integrating the second equation and requiring that the pressure at the oil–air interface, located at $x = -h(\sigma, t)$, is equal to the atmospheric pressure, p_{atm}, we obtain the oil pressure distribution

$$p_o = \rho_o g\,(x + h) + p_{atm}. \qquad (8.5.54)$$

Substituting this expression into the first equation of (8.5.53) and integrating with respect to x subject to the condition of zero shear stress at the oil-air interface, we obtain

$$\sigma_{\sigma x}^o(x = 0) = \rho_o g\,h\,\frac{\partial h}{\partial \sigma}. \qquad (8.5.55)$$

A mass balance for the oil provides us with an evolution equation for the film thickness,

$$\frac{\partial h}{\partial t} + \frac{1}{\sigma}\frac{\partial(\sigma \bar{u}_o h)}{\partial \sigma} = 0, \tag{8.5.56}$$

where $\bar{u}_o(\sigma)$ is the mean radial velocity across the oil film. The total volume of floating oil is

$$V_o(t) = 2\pi \int_0^{a(t)} h(\sigma, t)\,\sigma\,\mathrm{d}\sigma. \tag{8.5.57}$$

Continuity of velocity and shear stress at the oil–water interface requires that

$$u(x=0) = u_o(x=0) \equiv V(\sigma), \qquad \mu\left(\frac{\partial u}{\partial x}\right)_{x=0} = \sigma^o_{\sigma x}(x=0) = \rho_o g h \frac{\partial h}{\partial \sigma}, \tag{8.5.58}$$

where u is the radial (σ) component of the water velocity.

Similarity solution

We assume that the boundary layer develops according to the similarity solution derived in Section 8.5.3. The radius of the spreading film is

$$\sigma = a(t) = \tilde{\sigma}^*\,t^\alpha, \tag{8.5.59}$$

where $\tilde{\sigma}^*$ is a constant to be determined as part of the solution.

Substituting the similarity solution into (8.5.57), stipulating that $V_o(t) = q_o t^n$, and rearranging, we obtain

$$V_o(t) = q_o t^n = 2\pi\, t^{2\alpha} \int_0^{\tilde{\sigma}} h(\tilde{\sigma})\,\tilde{\sigma}\,\mathrm{d}\tilde{\sigma}, \tag{8.5.60}$$

where q_o is a constant coefficient and n is a specified exponent. This equation is satisfied only if

$$h(\sigma, t) = t^{n-2\alpha}\,\tilde{h}(\tilde{\sigma}), \tag{8.5.61}$$

where $\tilde{h}(\tilde{\sigma})$ is an unknown function to be determined as part of the solution. Substituting this expression along with the similarity solution into the interfacial shear stress condition given by the second equation in (8.5.58), we obtain

$$\mu\, t^{\kappa + 4\alpha - 2\beta - 2n}\frac{1}{\tilde{\sigma}}\left(\frac{\partial^2 \tilde{\psi}}{\partial \tilde{x}^2}\right)_{\tilde{x}=0} = \rho_o g\, \tilde{h}\,\frac{\partial \tilde{h}}{\partial \tilde{\sigma}}. \tag{8.5.62}$$

To eliminate the explicit time dependence on the left-hand side, we require that the exponent of time, t, on the left-hand side is zero. Combining this condition with (8.5.36), we obtain

$$\kappa = \frac{2}{3}n, \qquad \alpha = \frac{3+4n}{12}, \qquad \beta = \frac{1}{2}. \tag{8.5.63}$$

The interfacial condition (8.5.62) then becomes

$$\frac{\partial \tilde{h}^2}{\partial \tilde{\sigma}} = 2\frac{\mu}{\rho_o g}\frac{1}{\tilde{\sigma}}\left(\frac{\partial^2 \tilde{\psi}}{\partial \tilde{x}^2}\right)_{\tilde{x}=0}, \tag{8.5.64}$$

coupling the oil with the water flow.

The similarity solution predicts that

$$u = t^{\kappa - \alpha - \beta} \frac{1}{\tilde{\sigma}} \frac{\partial \tilde{\psi}}{\partial \tilde{x}}, \tag{8.5.65}$$

and this motivates setting

$$\bar{u}_{\mathrm{o}} = t^{\kappa - \alpha - \beta} \, \tilde{\sigma} \, \mathcal{G}(\tilde{\sigma}) = t^{\kappa - 2\alpha - \beta} \, \sigma \, \mathcal{G}(\tilde{x}) = \frac{\sigma}{t} \, \mathcal{G}(\tilde{\sigma}), \tag{8.5.66}$$

where $\mathcal{G}(\tilde{\sigma})$ is an unknown function. Combining this expression with (8.5.61), we find that

$$\sigma \bar{u}_{\mathrm{o}} h = t^{n-1} \, \tilde{\sigma}^2 \, \tilde{h}(\tilde{\sigma}) \, \mathcal{G}(\tilde{\sigma}). \tag{8.5.67}$$

Substituting this expression along with (8.5.61) into the balance equation (8.5.56), we derive the ordinary differential equation

$$\frac{1}{\tilde{\sigma}} \frac{\mathrm{d}}{\mathrm{d}\tilde{\sigma}} \left[\tilde{\sigma}^2 \tilde{h} (\mathcal{G} - \alpha) \right] + n \tilde{h} = 0. \tag{8.5.68}$$

When $n = 0$, this equation is satisfied by the constant function $\mathcal{G}(\tilde{\sigma}) = \alpha$. Assuming with good reason that $\tilde{h}$ is nearly constant except at the leading edge, we derive the more general albeit approximate result $\mathcal{G}(\tilde{x}) \simeq \alpha - n/2$.

Integral momentum balance

To compute the boundary layer, we may integrate the integral momentum balance (8.5.52) from the leading edge located at $\tilde{\sigma} = \tilde{\sigma}^*$ where the thickness of the boundary-layer is zero, $\Lambda = 0$, $\tilde{\delta}^* = 0$, and $\tilde{\Theta} = 0$, toward the centerpoint. The interfacial condition (8.5.64) provides us with a differential equation for the square of the film thickness,

$$\frac{\mathrm{d}\tilde{h}^2}{\mathrm{d}\tilde{\sigma}} = 2 \, \frac{\mu}{\rho_{\mathrm{o}} g} \left(\frac{\partial \tilde{u}}{\partial \tilde{x}} \right)_{\tilde{x}=0} = -\frac{2\mu}{\rho_{\mathrm{o}} g} \, \tilde{V} \, \frac{1}{\tilde{\Delta}} \left(2 + \frac{\Lambda}{6} \right). \tag{8.5.69}$$

To circumvent the integrable singularity at the leading edge located at $\tilde{\sigma} = \tilde{\sigma}^*$, we recast this equation into the form

$$\frac{\mathrm{d}\tilde{h}^2}{\mathrm{d}\omega} = \frac{4\mu}{\rho_{\mathrm{o}} g} \, \omega \, \tilde{V} \, \frac{1}{\tilde{\Delta}} \left(2 + \frac{\Lambda}{6} \right), \tag{8.5.70}$$

where $\omega = (\tilde{\sigma}^* - \tilde{\sigma})^{1/2}$. Integrating from the leading edge where $\omega = 0$ up to the centerline, we obtain the film thickness.

Spreading of a fixed amount of oil

In the case of gravitational spreading of a fixed amount of oil, $n = 0$, we have found that $\tilde{V} = \alpha \tilde{x}$, $\kappa = 0$, $\alpha = \frac{1}{4}$, and $\mathcal{V} = \kappa - \alpha - \beta = -\frac{3}{4}$. Equation (8.5.52) simplifies into

$$\frac{\mathrm{d}(\tilde{\Theta} - \tilde{\delta}^*)}{\mathrm{d}\tilde{\sigma}} = \frac{1}{\tilde{\sigma}} \left[\left(2 - \frac{\kappa}{\alpha} \right) \tilde{\delta}^* - 2 \tilde{\Theta} + \frac{\nu}{\alpha} \frac{1}{\tilde{\Delta}} \left(2 + \frac{\Lambda}{6} \right) \right] - 2 \frac{\tilde{\Theta}}{\tilde{\sigma}}. \tag{8.5.71}$$

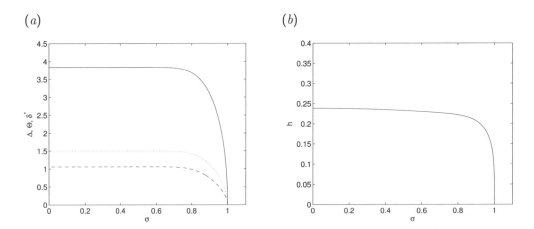

FIGURE 8.5.3   (a) Boundary-layer thickness Δ (solid line), momentum thickness Θ (dashed line), and displacement thickness δ^* (dotted line), over an axisymmetric oil slick expanding due to gravity, reduced by $\tilde{\sigma}^*$ and plotted against the dimensionless similarity distance $\tilde{\sigma}/\tilde{\sigma}^*$. (b) Corresponding distribution of the film thickness for a fixed amount of oil, $n = 0$.

To integrate this equation, we multiply both sides by $\tilde{\Theta} - \tilde{\delta}^*$ and rearrange to obtain

$$\frac{\mathrm{d}(\tilde{\Theta} - \tilde{\delta}^*)^2}{\mathrm{d}\tilde{\sigma}} = \frac{2}{\tilde{\sigma}} \left(\left[(2 - \frac{\kappa}{\alpha})\,\tilde{\delta}^* - 2\,\tilde{\Theta} \right] (\tilde{\Theta} - \tilde{\delta}^*) + \frac{\nu}{\alpha}\,\mathcal{H}(\Lambda)\,(2 + \frac{\Lambda}{6}) \right) - \frac{\tilde{\Theta}}{\tilde{\sigma}}\,(\tilde{\Theta} - \tilde{\delta}^*), \qquad (8.5.72)$$

where the function $\mathcal{H}$ is given in (8.4.71). Using relations (8.4.48) and the expression for the boundary-layer thickness in terms of Λ given in (8.5.50), we obtain equation (8.4.72), which can be used to compute the value of Λ given the left-hand side, by numerically solving an algebraic equation. Equation (8.5.72) accompanied by (8.4.72) can be integrated from the leading edge up to the centerline of the oil slick using elementary numerical methods.

The unknown value $\tilde{\sigma}^*$ corresponding to the radius of the leading edge is determined from the oil slick volume, $V_o = q_o t^n$,

$$\tilde{\sigma}^* = \zeta \left(\frac{g^2 q_o^4}{\nu} \right)^{1/12}, \qquad (8.5.73)$$

where ζ is a numerical coefficient. In practice, we may guess a value for $\tilde{\sigma}^*$, solve the boundary-layer equations, compute the volume of oil, and then rescale $\tilde{\sigma}^*$ to meet a specified volume. When $n = 0$, the numerical computations show that $\zeta = 1.14$. The distribution of the boundary-layer thickness computed using the numerical method discussed in this section is shown in Figure 8.5.3. The results are in excellent agreement with those obtained by different methods [75].

8.5.5   Molecular spreading of an axisymmetric oil slick

When the thickness of the oil film discussed in Section 8.5.4 has reached molecular dimensions, the oil behaves like a surfactant, affecting the surface tension according to an assumed surface equation of state, $\gamma(h)$ [297]. The two-dimensional version of this problem was discussed in Section 8.3.5. Balancing the induced Marangoni traction with the shear stress, we obtain

$$\mu \left(\frac{\partial u}{\partial x} \right)_{x=0} = -\frac{\partial \gamma}{\partial \sigma}. \tag{8.5.74}$$

A mass balance over the oil film provides us with the evolution equation

$$\frac{\partial h}{\partial t} + \frac{1}{\sigma} \frac{\partial (\sigma u_s h)}{\partial \sigma} = 0, \tag{8.5.75}$$

where $u_s = u_\sigma(\sigma, x = 0, t)$ is the radial surface velocity.

In the similarity solution, the film thickness and the surface tension are assumed to depend exclusively on the scaled radial coordinate $\tilde{\sigma}$, so that

$$h(\sigma, t) = \tilde{h}(\tilde{\sigma}), \qquad \gamma(\sigma, t) = \tilde{\gamma}(\tilde{\sigma}). \tag{8.5.76}$$

The interfacial condition (8.5.74) requires that

$$\mu \, t^{\kappa - 2\beta} \frac{1}{\tilde{\sigma}} \left(\frac{\partial^2 \tilde{\psi}}{\partial \tilde{x}^2} \right)_{\tilde{x}=0} = -\frac{d\tilde{\gamma}}{d\tilde{\sigma}}. \tag{8.5.77}$$

To eliminate the explicit time dependence on the left-hand side, we require that $\kappa - 2\beta = 0$. Combining this condition with (8.5.36), we obtain

$$\delta = 1, \qquad \alpha = \frac{3}{4}, \qquad \beta = \frac{1}{2}. \tag{8.5.78}$$

The evolution equation (8.5.75) for the film thickness becomes

$$-\frac{3}{4} \tilde{\sigma} \frac{d\tilde{h}}{d\tilde{\sigma}} + \frac{1}{\tilde{\sigma}} \frac{d}{d\tilde{\sigma}} \left[\tilde{\psi}_{\tilde{x}}^0 \, \tilde{h} \right] = 0, \tag{8.5.79}$$

where $\tilde{\psi}_{\tilde{x}}^0 \equiv (\partial \tilde{\psi} / \partial \tilde{x})_{\tilde{x}=0}$. Numerical solutions of the boundary-layer equations can be obtained by finite-difference methods [297].

8.5.6 Three-dimensional flow

The computation of boundary layers in three-dimensional flow is complicated by the arbitrary orientation of the tangential component of the velocity of the incident flow. The boundary-layer equations arise from two tangential components of the equation of motion by discarding the second partial derivatives of the tangential velocity components with respect to arc length in the tangential directions. The normal component of the equation of motion ensures that the pressure drop is negligible across the boundary layer. Accordingly, the tangential pressure gradient can be set equal to that of the incident irrotational flow.

In global laboratory Cartesian coordinates, the boundary-layer equations for an irrotational incident flow with tangential boundary velocity $\mathbf{U}$ take the form

$$\frac{\partial \mathbf{u}}{\partial t} \cdot \mathbf{P} + \mathbf{u} \cdot (\nabla \mathbf{u}) \cdot \mathbf{P} = \frac{\partial \mathbf{U}}{\partial t} \cdot \mathbf{P} + \frac{1}{2} \mathbf{P} \cdot \nabla (\mathbf{U} \cdot \mathbf{U}) + \nu (\mathbf{n} \otimes \mathbf{n}) : (\nabla \nabla \mathbf{u}) \cdot \mathbf{P}, \tag{8.5.80}$$

where $\mathbf{P} = \mathbf{I} - \mathbf{n} \otimes \mathbf{n}$ is the tangential projection operator. The continuity equation retains its full form for three-dimensional flow.

Surface curvilinear coordinates

In practice, equation (8.5.80) is decomposed into two tangential components corresponding to a pair of orthogonal or nonorthogonal surface curvilinear coordinates, ξ_1 and ξ_2, wrapping around the boundary, and a third coordinate, η, varying in the normal direction. An appealing choice for lines of constant ξ_1 and ξ_2 are the surface streamlines of the incident potential flow and the curves intersecting them at a right angle at each point, called the intrinsic coordinates. The two component boundary-layer equations involve derivatives of the velocity with respect to ξ_1, ξ_2, and η multiplied by the metrics of the three curvilinear coordinates. The wall shear stress, also called the skin friction, and momentum thickness may point in arbitrary tangential directions. In surface curvilinear coordinates, the momentum thickness is a rank-two tensor. The momentum integral balance and the Kármán–Pohlhausen method and its variations are developed in a straightforward fashion but assume more involved vectorial forms (e.g., [253]). A discussion of approximate and numerical methods for computing boundary layers in three-dimensional flow can be found in monographs by Rosenhead [352], Moore [260], and Cebeci & Bradshaw [70].

Problems

8.5.1 *Momentum integral balance*

(*a*) Derive the integral momentum balance equation (8.5.13).

(*b*) Derive the Kármán–Pohlhausen differential equation (8.5.14).

8.5.2 *Mangler's transformation for a sphere*

Compute and discuss the transformed variables according to Mangler for streaming (uniform) flow past (*a*) a cone and (*b*) a sphere.

8.5.3 *Spreading of an axisymmetric jet*

Consider the spreading of an axisymmetric jet that emerges through a circular orifice along the x axis into an infinite quiescent ambient fluid ([363], p. 230). A momentum integral balance over a control volume bounded by two parallel planes that are perpendicular to the x axis requires that the momentum integral,

$$M = 2\pi\rho \int_0^\infty u^2 \sigma \, \mathrm{d}\sigma, \qquad (8.5.81)$$

is independent of x. Implementing the boundary-layer approximation in the axial component of the equation of motion written in cylindrical polar coordinates, (x, σ, φ), and noting that the ambient pressure is constant, we obtain the boundary-layer equation

$$u \frac{\partial u}{\partial x} + v \frac{\partial u}{\partial \sigma} = \frac{\nu}{\sigma} \frac{\partial}{\partial \sigma} \left(\sigma \frac{\partial u}{\partial \sigma} \right), \qquad (8.5.82)$$

where $u = u_x$ is the axial velocity and $v = u_\sigma$ is the radial velocity. The corresponding Stokes stream function can be expressed in the form

$$\psi(x, \sigma) = \nu x \, f(\eta), \tag{8.5.83}$$

where f is a dimensionless function and $\eta \equiv \sigma/x$ is a dimensionless similarity variable. Substituting (8.5.83) into (8.5.82), we obtain a third-order homogeneous ordinary differential equation,

$$\left(f'' - \frac{f'}{\eta}\right)' + \frac{1}{\eta}\left(ff'' + f'^2\right) - \frac{ff'}{\eta^2} = 0. \tag{8.5.84}$$

We observe that, if $f(\eta)$ is a solution of (8.5.84), then $f(c\eta)$ is also a solution for any constant c. The boundary conditions specify that $f(0) = 0$, $f'(0) = 0$, and $f'(\infty) = 0$.

Solve (8.5.84) subject to the integral constraint (8.5.81), to obtain the solution

$$f(\eta) = \frac{4\alpha\eta^2}{4 + \alpha\eta^2}, \tag{8.5.85}$$

where $\alpha = 3M/(16\pi\mu\nu)$ is a dimensionless constant. Show that the axial volumetric flow rate is given by $Q = 8\pi\nu x$, independent of M.

8.5.4 *Spreading of an axisymmetric wake*

Consider the widening of an axisymmetric wake behind a streamlined object that is held stationary in uniform (streaming) flow along the x axis with velocity U (e.g., [363], p. 234). Show that, if the rate of widening is sufficiently small, boundary-layer predicts the axial velocity profile

$$u(x, \sigma) = U\left(1 - \frac{c}{x} \exp\left[-\frac{U\sigma^2}{4\nu x}\right]\right), \tag{8.5.86}$$

where $c = F/(4\pi\mu U)$ is a dimensional constant and F is the drag force exerted on the object. Compare this solution with that derived in Section 8.1.4 for a two-dimensional wake.

8.5.5 *Three-dimensional boundary layers*

Present the explicit forms of the tangential components of (8.5.80) corresponding to a system of orthogonal surface curvilinear coordinates, ξ_1 and ξ_2, wrapping around the body, and a third coordinate, η, that is normal to the body.

 **Computer Problem**

8.5.6 *Boundary layer around a sphere*

Write a program that uses the Kármán–Pohlhausen method to compute the boundary layer around a sphere that is held stationary in an incident uniform flow. Plot the distribution of the wall shear stress and estimate the meridional angle, θ, where the boundary layer is expected to separate.

8.6   Oscillatory boundary layers

An important class of boundary layers arise in oscillatory flow past a stationary body or suspended particle, pulsating internal flow inside a channel or tube, and in the flow caused by natural or forced vibrations of rigid or deformable objects. For example, an oscillating boundary layer develops at the bottom of the ocean due to the periodic flow induced by the propagation of free-surface or internal gravity waves.

Role of vorticity

To illustrate the physical mechanism governing the structure of an oscillatory boundary layer, we consider uniform pulsating flow with angular velocity Ω past a stationary rigid body. Reversal of the velocity of the outer flow over one period causes the sign of the boundary vorticity to alternate in time. As a result, vorticity of positive and negative sign diffuses across the boundary and enters the flow, and the boundary layer consists of successive vortex layers that penetrate and tend to cancel each other as they are convected by the incident flow. An example is the Stokes boundary layer arising in oscillatory unidirectional streaming flow over an infinite plate, as discussed in Section 5.4.1. The Stokes boundary layer consists of traveling bands of vorticity with alternating sign penetrating the flow by a distance that is comparable to the Stokes boundary-layer thickness, $\delta = (2\nu/\Omega)^{1/2}$, where Ω is the angular frequency of the oscillations and ν is the kinematic viscosity of the fluid. Similar boundary layers occur in oscillatory flow through tubes, as discussed in Section 5.5.

Driven Stokes boundary layer

When the Reynolds number of an incident oscillatory flow is high, $\mathrm{Re} = \Omega L^2/\nu \gg 1$, and the geometry of the boundary is smooth and sufficiently simple, the thickness of an oscillatory boundary layer is small compared to the boundary size, L. Under these circumstances, the outer flow is irrotational and the boundary layer is driven by the tangential component of the outer velocity, $\mathbf{u}^\infty$, given by

$$\mathbf{u}_t^\infty \equiv (\mathbf{n} \times \mathbf{u}^\infty) \times \mathbf{n} = (\mathbf{I} - \mathbf{n} \otimes \mathbf{n}) \cdot \mathbf{u}^\infty \equiv \bar{U}_t \cos(\Omega t)\,\mathbf{t}, \qquad (8.6.1)$$

where $\mathbf{n}$ is the normal unit vector pointing into the flow, $\mathbf{t}$ is a tangent unit vector, and $\bar{U}_t$ is the amplitude of the tangential velocity. Neglecting the curvature of the boundary, we approximate the flow at a particular point on the boundary with the flow over a flat plate due to an overpassing oscillatory streaming flow with velocity $\bar{U}_t \cos(\Omega t)\mathbf{t}$, as discussed in Section 5.4.2. Introducing a local coordinate system with the y axis perpendicular to the boundary at a point and origin at that point, and using (5.4.16), we find that the velocity profile across the oscillatory boundary layer is given by

$$\mathbf{u} = \bar{U}_t \left[\, \cos(\Omega t) - \cos(\Omega t - \hat{y})\,\mathrm{e}^{-\hat{y}} \,\right]\mathbf{t}, \qquad (8.6.2)$$

where $\hat{y} = y/\delta$ and $\delta = (2\nu/\Omega)^{1/2}$ is the Stokes boundary-layer thickness.

Drag force

The force exerted on the boundary consists of the acceleration reaction associated with the outer unsteady irrotational flow and a viscous drag due to the Stokes boundary layer. It might appear that the local solution (8.6.2) alone is sufficient for computing the viscous drag. However, this approach leads us to the erroneous conclusion that the phase shift between the drag force and the velocity at the edge of the boundary layer is equal to $3\pi/4$, independent of the boundary geometry ([24], p. 335). The viscous normal stresses due to the boundary curvature must be taken into consideration.

Computation of the damping force

Consider a streaming oscillatory flow with velocity $\mathbf{u}^\infty = \mathbf{U}\cos(\Omega t)$ past a stationary body, where $\mathbf{U}$ is a constant velocity. The component of the drag force in the direction of $\mathbf{U}$ and in phase with $\mathbf{U}$, is called the damping force. Batchelor proposed that the damping force can be computed by setting the rate of energy dissipation inside the Stokes boundary layer equal to the rate of working necessary to hold the body at a fixed position, both averaged over one cycle ([24], p. 336).

To perform this calculation, we work in a frame of reference where the body appears to execute translational oscillations with velocity $\mathbf{v} = \mathbf{V}\cos(\Omega t)$ and the velocity decays far from the body, where $\mathbf{V} = -\mathbf{U}$. Noting that the average kinetic energy of the fluid is constant over each period of the oscillation, $T = 2\pi/\Omega$, and integrating the energy balance over one period, we obtain

$$\int_0^T \left(\iint_{Body} \mathbf{u} \cdot \mathbf{f} \, \mathrm{d}S \right) \mathrm{d}t = - \int_0^T \left(\iiint_{Flow} \Phi \, \mathrm{d}V \right) \mathrm{d}t, \tag{8.6.3}$$

where the normal unit vector, $\mathbf{n}$, points into the flow.

Substituting (8.6.2) into (3.4.15), we find that the rate of dissipation inside a small volume within the boundary layer is given by

$$\Phi = \mu \left(\frac{\partial u_t}{\partial y} \right)^2 = \rho \,\Omega\, \bar{U}_t^2 \cos^2\!\left(\Omega t - \frac{3\pi}{4} - \frac{y}{\delta}\right) \mathrm{e}^{-2y/\delta}, \tag{8.6.4}$$

where u_t is the tangential velocity. The average rate of viscous dissipation over one period of the oscillation is thus given by

$$\bar{\mathcal{D}} \equiv \frac{1}{T} \int_0^T \!\!\int_0^\infty \left(\iint_{Body} \Phi \, \mathrm{d}S \right) \mathrm{d}y \, \mathrm{d}t = \frac{\mu}{2\delta} \iint_{Body} \bar{U}_t^2 \, \mathrm{d}S. \tag{8.6.5}$$

Next, we express the hydrodynamic drag force as

$$\mathbf{D} = \bar{\mathbf{D}} \cos(\Omega t - \alpha), \tag{8.6.6}$$

where $\bar{\mathbf{D}}$ is the amplitude and α is the phase shift between the drag force and the velocity. The average rate of working of the body due to its motion against the fluid, expressed by the integral on the left-hand side of (8.6.3), is

$$\bar{\mathcal{W}} \equiv \frac{1}{T} \int_0^T \mathbf{V} \cdot \mathbf{D} \cos(\Omega t) \cos(\Omega t - \alpha) \, \mathrm{d}t = \frac{1}{2} \cos\alpha \, \mathbf{V} \cdot \mathbf{D}. \tag{8.6.7}$$

Substituting (8.6.5) and (8.6.7) into (8.6.3), we obtain

$$\mathbf{U} \cdot \mathbf{D} = \frac{1}{\cos\alpha} \frac{\mu}{\delta} \iint_B U_t^2 \, \mathrm{d}S. \tag{8.6.8}$$

Our analysis in Sections 6.15 and 6.16 has indicated that, unless the boundary has sharp edges, in which case the linearized equation of motion ceases to be valid, $\mathbf{D}$ is an analytic function of $(-\mathrm{i}/\delta)^{1/2}$

and $\alpha = \pi/4$. To compute the damping force, we must evaluate the surface integral of the square of the amplitude of the tangential velocity of the outer flow.

As an example, we use the solution for streaming irrotational flow past a sphere of radius a to evaluate the integral in (8.6.8) and find that

$$\mathbf{U} \cdot \mathbf{D} = \left(\frac{\Omega a^2}{\nu}\right)^{1/2} 6\pi\mu a |\mathbf{V}|^2. \qquad (8.6.9)$$

In the case of streaming irrotational flow past a cylinder of radius a and length L, we find that

$$\mathbf{U} \cdot \mathbf{D} = \left(\frac{\Omega a^2}{\nu}\right)^{1/2} 4\pi\mu L |\mathbf{V}|^2. \qquad (8.6.10)$$

Problem

8.6.1 *Damping force on a sphere and a cylinder*

Integrate the potential flow solution to compute the dissipation integral (8.6.8) for (a) a sphere and (b) a cylinder.

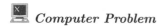 **Computer Problem**

8.6.2 *Damping force on a spheroid*

Integrate the potential flow solution to compute the dissipation integral (8.6.8) for axial flow past a prolate spheroid. Plot your results against the spheroid aspect ratio and confirm agreement for the sphere.

Hydrodynamic stability

9

In previous chapters, we discussed analytical and numerical methods for computing the structure of a steady flow and the evolution of an unsteady flow under a broad range of conditions. The spectrum of flows considered includes creeping flows at vanishing Reynolds numbers, irrotational flows at high Reynolds numbers, and flows dominated by vortex motion. In this chapter, we address the issue of whether the steady flows considered can be physically realized in practice.

In nature and technology, a flow is always established through a transient process, beginning from a certain initial state. It is important to realize that the evolution of a physical flow from the initial state will not necessarily lead to a steady state that can be captured by analytical or numerical methods. For example, imposing a pressure drop across the length of a circular tube does not guarantee the onset of unidirectional Poiseuille flow with a parabolic velocity profile discussed in Section 5.2.1. In fact, in 1883 Reynolds observed that, at sufficiently high Reynolds numbers, Re, the tube flow develops wavy motions and the assumption of unidirectional motion ceases to be valid.

Flows in industrial, laboratory, and natural settings are subjected to disturbances with small or large amplitude due to equipment and building vibration, Brownian motion of microscopic suspended particles, and other reasons. In some technological applications, perturbations with a suitable amplitude and form are purposely introduced in order to initiate a certain desirable action, such as enhance mixing or delay boundary-layer separation. It is then possible that natural or artificial disturbances may amplify in time or space leading to unsteady motion or to a new steady state.

The behavior of a disturbance depends on its specific form as well as on the structure of the unperturbed flow. In the context of hydrodynamic stability, the unperturbed flow is a base flow. Perturbations may exhibit different types of behavior depending on the Reynolds number and other dimensionless numbers characterizing the base flow. In some cases, perturbations may grow while being convected with the base flow, causing a convective instability. In other cases, perturbations may spread out to contaminate the whole domain of flow, causing an absolute instability. More involved types of behavior are possible [195]. For example, disturbances in a globally unstable flow may initiate self-excited modes. Establishing criteria for the resilience and thus physical relevance of a particular steady or unsteady base flow is the main directive of hydrodynamic stability analysis.

To assess the stability of a base flow, we may subject it to a broad range of perturbations and observe their subsequent evolution. If all perturbations decay, the flow is stable. If certain perturbations amplify, the flow is unstable and may not be realized in a physical setting unless an external mechanism that controls the growth of the unstable perturbations is provided. In certain cases, assessing the behavior of perturbations can be done by simple physical arguments. More generally, the study of perturbations requires detailed analysis and numerical computation.

In principle, the behavior of perturbations can be studied by solving the equation of motion together with the continuity equation subject to imposed boundary conditions and a suitable initial condition. However, since the number of admissible disturbances is innumerable, it is futile to attempt to exhaust all possibilities. One way to make progress is to assume that the magnitude of the perturbation is small and remain small during an initial period of time, linearize the governing equations with respect to all flow variables around the base state, and solve the governing equations for a wide range of initial conditions using, for example, the method of Laplace transform. This approach encapsulates the essence of linear stability analysis. However, even after linearization, a general solution can be obtained by analytical methods only for a limited family of flows by performing a normal-mode analysis that considers perturbations with exponential growth or decay in time.

If linear stability analysis indicates that perturbations grow in time, the base flow hosting the perturbations is unstable. However, since neglected nonlinear effects may promote the growth or perturbations, the converse is true only if the magnitude of perturbations is and remains sufficiently small in time. In certain cases, nonlinear effects slow down or even suppress the growth of unstable perturbations and lead to a new steady or periodic state. Assessing the effect of nonlinearities can be challenging. Progress can be made working under the auspices of weakly nonlinear stability theory where a disturbance is described by a perturbation series and the analysis is carried out up to the second or higher order with respect to a perturbation parameter. A full assessment of the nonlinear effects requires the use of numerical methods, such as the finite-difference methods discussed in Chapter 13.

In this chapter, we introduce the basic concepts underlying the formulation of the linear stability problem for internal, external, interfacial, and free-surface flows. In the case of interfacial flows, we employ the method of domain perturbation where a fluid is artificially extended into another fluid or a boundary by analytical continuation. Having laid the theoretical foundation and derived the governing equations, we present the normal-mode stability analysis and study the properties of a fundamental class of viscous and inviscid flows. Solutions of the linearized equations will be obtained by analytical and numerical methods for solving ordinary and linear partial differential equations involving an eigenvalue.

Further discussion of linear and nonlinear stability analysis can be found in classical texts by Lin [241], Chandrasekhar [73], and Betchov & Criminale [34], and in other relevant reviews and monographs [95, 112, 195, 257, 364].

9.1 Evolution equations and formulation of the linear stability problem

To prepare the ground for computing the evolution of perturbations in a nearly steady flow, we summarize the equations governing the evolution of the velocity, vorticity, and pressure fields in a homogeneous incompressible fluid with uniform physical properties, in the presence of a uniform body force.

Evolution of the velocity

For the purpose of assessing stability, it is useful to regard the Navier–Stokes equation (3.5.5) as an evolution equation for the velocity,

$$\frac{\partial \mathbf{u}}{\partial t} = \mathcal{F}_{\mathbf{u}}(\mathbf{u}, p), \tag{9.1.1}$$

where

$$\mathcal{F}_{\mathbf{u}}(\mathbf{u}, p) \equiv -\mathbf{u} \cdot \nabla \mathbf{u} - \frac{1}{\rho} \nabla p + \nu \nabla^2 \mathbf{u} + \mathbf{g} \tag{9.1.2}$$

is a forcing function, ρ is the fluid density, and ν is the kinematic viscosity.

Evolution of the pressure

An explicit evolution equation for the pressure is not available. Instead, the condition of fluid incompressibility imposes a local constraint, requiring that the instantaneous pressure field is such that the rate of expansion remains zero at all times, $\nabla \cdot \mathbf{u} = 0$. To illustrate the way in which this condition provides us with an implicit evolution equation for the pressure, we take the divergence of (9.1.1) and derive an evolution equation for the rate of expansion, $\alpha \equiv \nabla \cdot \mathbf{u}$,

$$\frac{\partial \alpha}{\partial t} = \nabla \cdot \mathcal{F}_{\mathbf{u}} = -\nabla \cdot (\mathbf{u} \cdot \nabla \mathbf{u}) - \frac{1}{\rho} \nabla^2 p + \nu \nabla^2 \alpha. \tag{9.1.3}$$

Requiring that the left-hand side and the last term on the right-hand side are zero, we derive a Poisson equation for the pressure,

$$\nabla^2 p = -\rho \nabla \cdot (\mathbf{u} \cdot \nabla \mathbf{u}) = -\rho \nabla \nabla : (\mathbf{u}\mathbf{u}), \tag{9.1.4}$$

subject to boundary conditions discussed in Section 13.3.3. Conversely, if the pressure field develops so that (9.1.4) is satisfied at all times, the gradient of the pressure on the right-hand side of (9.1.2) will be such that $\nabla \cdot \mathcal{F}_{\mathbf{u}} = 0$ at any time (Problem 9.1.1).

Now we take the partial derivative of (9.1.4) with respect to time, expand the derivatives on the right-hand side, and use (9.1.1) to derive a Poisson equation for the rate of change of the pressure,

$$\nabla^2 \left(\frac{\partial p}{\partial t} \right) = -\rho \nabla \cdot (\mathcal{F}_{\mathbf{u}} \cdot \nabla \mathbf{u}) - \rho \nabla \cdot (\mathbf{u} \cdot \nabla \mathcal{F}_{\mathbf{u}}) = -2\rho \nabla \nabla : (\mathbf{u}\mathcal{F}_{\mathbf{u}}). \tag{9.1.5}$$

The solution can be found using the Poisson inversion formula, yielding a standard evolution equation written in the symbolic form

$$\frac{\partial p}{\partial t} = \mathcal{F}_p(\mathbf{u}, \mathcal{F}_{\mathbf{u}}(\mathbf{u}, p)). \tag{9.1.6}$$

The functional form of $\mathcal{F}_p$ depends on the instantaneous structure of the flow and required boundary conditions.

Solenoidal projection

To illustrate further the role of the pressure in maintaining the velocity field solenoidal, we consider the right-hand side of (9.1.2) with the pressure term omitted, and define the pressureless forcing function

$$\mathcal{F}_{\mathbf{u}}(\mathbf{u},0) \equiv -\mathbf{u}\cdot\nabla\mathbf{u} + \nu\nabla^2\mathbf{u} + \mathbf{g}. \qquad (9.1.7)$$

The Helmholtz decomposition theorem discussed in Section 2.8 allows us to express the function $\mathcal{F}_{\mathbf{u}}(\mathbf{u}, p=0)$ as the sum of a solenoidal and an irrotational field,

$$\mathcal{F}_{\mathbf{u}}(\mathbf{u},0) = \nabla\times\mathbf{A} + \nabla\phi, \qquad (9.1.8)$$

where $\mathbf{A}$ is a vector potential and ϕ is a potential function. By definition,

$$\mathcal{F}_{\mathbf{u}}(\mathbf{u},p) = \mathcal{F}_{\mathbf{u}}(\mathbf{u},0) - \frac{1}{\rho}\nabla p = \nabla\times\mathbf{A} + \nabla\phi - \frac{1}{\rho}\nabla p. \qquad (9.1.9)$$

If the fluid is incompressible, $\mathcal{F}_{\mathbf{u}}(\mathbf{u},p)$ is solenoidal and

$$\nabla p = \rho\,\nabla\phi, \qquad (9.1.10)$$

which shows that the pressure gradient projects $\mathcal{F}_{\mathbf{u}}(\mathbf{u},0)$ into the space of solenoidal functions, thereby transforming it into a solenoidal field, $\mathcal{F}_{\mathbf{u}}(\mathbf{u},p)$. In Section 13.5, we will see that this interpretation is the foundation of a class of numerical methods for integrating the equation of motion in time, called projection or pressure-correction methods.

Evolution of the vorticity

An evolution equation for the vorticity emerges by recasting the vorticity transport equation (3.12.1) for an incompressible fluid with uniform physical properties into the form

$$\frac{\partial\omega}{\partial t} = \mathcal{F}_{\omega}(\omega,\mathbf{u}), \qquad (9.1.11)$$

where

$$\mathcal{F}_{\omega}(\omega,\mathbf{u}) \equiv -\nabla\times(\omega\times\mathbf{u}) + \nu\,\nabla^2\omega = -\mathbf{u}\cdot\nabla\omega + \omega\cdot\nabla\mathbf{u} + \nu\,\nabla^2\omega. \qquad (9.1.12)$$

One noteworthy feature of (9.1.12) is the absence of the pressure. This feature is exploited for the expedient analytical or numerical computation of flows based on the vorticity transport equation, as discussed in Chapters 11 and 13.

Summary of evolution equations

Compiling equations (9.1.2), (9.1.6), and (9.1.11), we obtain a complete system of evolution equations for the velocity, pressure, and vorticity in vector form,

$$\frac{\partial\mathbf{u}}{\partial t} = \mathcal{F}_{\mathbf{u}}(\mathbf{u},p), \qquad \frac{\partial p}{\partial t} = \mathcal{F}_p(p,\mathbf{u}), \qquad \frac{\partial\omega}{\partial t} = \mathcal{F}_{\omega}(\omega,\mathbf{u}). \qquad (9.1.13)$$

The left-hand side of each equation in (9.1.13) is zero in a steady flow, and the structure of the velocity, pressure, and vorticity fields is governed by the equations of steady flow,

$$\mathcal{F}_{\mathbf{u}}(\mathbf{u}^S, p^S) = 0, \qquad \mathcal{F}_p(p^S, \mathbf{u}^S) = 0, \qquad \mathcal{F}_{\boldsymbol{\omega}}(\boldsymbol{\omega}^S, \mathbf{u}^S) = 0, \tag{9.1.14}$$

where the superscript S indicates a steady state.

9.1.1 Linear evolution from a steady state

Next, we consider a nearly steady flow that deviates only slightly from a certain steady state. This physical condition can be implemented by expressing the velocity, pressure, and vorticity fields as

$$\mathbf{u} = \mathbf{u}^S + \epsilon \mathbf{u}^U, \qquad p = p^S + \epsilon p^U, \qquad \boldsymbol{\omega} = \boldsymbol{\omega}^S + \epsilon \boldsymbol{\omega}^U, \tag{9.1.15}$$

where $\epsilon \ll 1$ is a small dimensionless coefficient and the superscript U denotes the unsteady component. Both the steady and unsteady components are required to satisfy the continuity equation for an incompressible fluid,

$$\nabla \cdot \mathbf{u}^S = 0, \qquad \nabla \cdot \mathbf{u}^U = 0. \tag{9.1.16}$$

The steady variables depend only on position, $\mathbf{x}$, whereas the unsteady variables depend on position and time, t.

Substituting expressions (9.1.15) into the right-hand sides of (9.1.2) and (9.1.12) and discarding terms with quadratic dependence on ϵ, we obtain the linear forms

$$\mathcal{F}_{\mathbf{u}}(\mathbf{u}, p) \simeq \epsilon \left[-\mathbf{u}^S \cdot \nabla \mathbf{u}^U - \mathbf{u}^U \cdot \nabla \mathbf{u}^S - \frac{1}{\rho} \nabla p^U + \nu \nabla^2 \mathbf{u}^U \right] \tag{9.1.17}$$

and

$$\mathcal{F}_{\boldsymbol{\omega}}(\boldsymbol{\omega}, \mathbf{u}) \simeq \epsilon \left[-\mathbf{u}^S \cdot \nabla \boldsymbol{\omega}^U - \mathbf{u}^U \cdot \nabla \boldsymbol{\omega}^S + \boldsymbol{\omega}^U \cdot \nabla \mathbf{u}^S + \boldsymbol{\omega}^S \cdot \nabla \mathbf{u}^U + \nu \nabla^2 \boldsymbol{\omega}^U \right]. \tag{9.1.18}$$

To derive these equations, we have used (9.1.14) to eliminate the terms involving the steady field alone on the right-hand sides.

Next, we substitute expressions (9.1.15), (9.1.17), and (9.1.18) into the evolution equations (9.1.1) and (9.1.11), and thus derive a linear evolution equation for the unsteady components of the velocity and vorticity,

$$\frac{\partial \mathbf{u}^U}{\partial t} = -\mathbf{u}^S \cdot \nabla \mathbf{u}^U - \mathbf{u}^U \cdot \nabla \mathbf{u}^S - \frac{1}{\rho} \nabla p^U + \nu \nabla^2 \mathbf{u}^U \tag{9.1.19}$$

and

$$\frac{\partial \boldsymbol{\omega}^U}{\partial t} = -\mathbf{u}^S \cdot \nabla \boldsymbol{\omega}^U - \mathbf{u}^U \cdot \nabla \boldsymbol{\omega}^S + \boldsymbol{\omega}^U \cdot \nabla \mathbf{u}^S + \boldsymbol{\omega}^S \cdot \nabla \mathbf{u}^U + \nu \nabla^2 \boldsymbol{\omega}^U. \tag{9.1.20}$$

These two equations can be collected into the compact form

$$\frac{\partial}{\partial t} \begin{bmatrix} \mathbf{u}^U \\ \boldsymbol{\omega}^U \end{bmatrix} = \begin{bmatrix} \mathbf{A} & \mathbf{B} & \mathbf{0} \\ \mathbf{C} & \mathbf{0} & \mathbf{D} \end{bmatrix} \cdot \begin{bmatrix} \mathbf{u}^U \\ p^U \\ \boldsymbol{\omega}^U \end{bmatrix}, \tag{9.1.21}$$

where $\mathbf{A}$, $\mathbf{B}$, $\mathbf{C}$, and $\mathbf{D}$ are four differential operators given by

$$\mathbf{A} = -\mathbf{u}^S \cdot \nabla - (\nabla \mathbf{u}^S)^T \cdot + \nu \nabla^2, \qquad \mathbf{B} = -\frac{1}{\rho} \nabla,$$

$$\mathbf{C} = -(\nabla \omega^S)^T \cdot + \omega^S \cdot \nabla, \qquad \mathbf{D} = -\mathbf{u}^S \cdot \nabla + (\nabla \mathbf{u}^S)^T \cdot + \nu \nabla^2 \ . \tag{9.1.22}$$

We observe that the precise form of $\mathbf{A}$, $\mathbf{B}$, $\mathbf{C}$, and $\mathbf{D}$ depends on the structure of the base flow.

Substituting the right-hand side of (9.1.15) for the velocity and pressure into (9.1.4), and linearizing the right-hand side of the resulting equation, we obtain a Poisson equation for the unsteady component of the pressure,

$$\nabla^2 p^U = -\rho \nabla \cdot (\mathbf{u}^S \cdot \nabla \mathbf{u}^U + \mathbf{u}^U \cdot \nabla \mathbf{u}^S) = -2\rho \nabla\nabla : (\mathbf{u}^U \mathbf{u}^S). \tag{9.1.23}$$

We can work in a similar fashion with the evolution equation (9.1.6), recasting it into a form that is similar to that shown in (9.1.21). However, this is not necessary for the purposes of our discourse.

Linear stability analysis

Equations (9.1.21) and (9.1.23) provide us with a system of linear homogeneous partial differential equations for the evolution of the unsteady component of a nearly steady flow. Solving this system allows us to study the departure of a flow from a steady state during an initial period of time where the magnitude of the unsteady component is small compared to that of the base flow.

Now we consider a base flow at steady state and identify the unsteady component with a physical perturbation. Depending on the structure of the base flow and form of the perturbation, a disturbance may behave in different ways. If the magnitude of the disturbance, defined in some local or global sense, grows, remains constant, or decays asymptotically in time, the disturbance is unstable, marginally stable, or stable. If every possible disturbance decays in time, the base flow is linearly stable. If certain disturbances grow, the base flow is linearly unstable. An unstable flow can be physically realized only if unstable disturbances are screened out by a control mechanism that detects and suppresses the growth of perturbations.

To this end, we underline the limitations and advisory role of linear stability theory by pointing out that a flow that is stable according to linear stability theory will not necessarily appear in practice. The reason is that nonlinear interactions and small deviations from an assumed perfect geometry of the domain of flow can be responsible for unstable behavior. For example, Poiseuille flow in a tube is always stable according to linear theory but unstable in real life when the Reynolds number is sufficiently large, as discussed in Section 9.8.6.

9.1.2   Evolution of stream surfaces and interfaces

A perturbation causes a stream surface or interface between two immiscible fluids to deviate from the steady shape corresponding to the base flow. In the case of two-dimensional flow in the xy plane, the shape of a streamline or fluid interface can be described parametrically as

$$\mathbf{x}(l, t) = \mathbf{X}(l) + \epsilon q(l, t) \, \mathbf{n}(l), \tag{9.1.24}$$

(a)   (b)

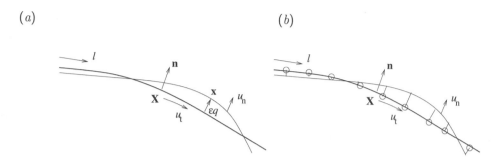

FIGURE 9.1.1   (a) Illustration of a material line or interface deviating from a reference shape corresponding to a two-dimensional base flow. (b) A perturbed material line or interface can be described by a chain of marker particles parametrized by the arc length along the unperturbed interface, l.

where $\mathbf{X}(l)$ is a steady streamline or interface regarded as a reference line, l is the arc length along the reference line, $\mathbf{n}(l)$ is the unit vector normal to the reference line, and $\epsilon q(l,t)$ is the normal displacement, as shown in Figure 9.1.1(a). Correspondingly, the velocity component normal to the reference line evaluated at the position of the perturbed stream surface or interface can be expressed in the form

$$u_n(l,t) = \epsilon v_n(l,t),   \tag{9.1.25}$$

where v_n is the leading-order normal velocity component. Substituting the expressions $\zeta = \epsilon q$ and $u_n = \epsilon v_n$ into the general kinematic compatibility condition (1.10.17), and linearizing with respect to ϵ, we obtain the evolution equation

$$\frac{\partial q}{\partial t} + u_t(l)\,\frac{\partial q}{\partial l} - v_n = 0,   \tag{9.1.26}$$

where u_t is the velocity component of the base flow tangential to the reference line. Given the tangential and normal velocities, the shape of an evolving interface can be computed by integrating in time the convection equation (9.1.26) by analytical or numerical methods. Equation (9.1.26) also describes the evolution of the trace of an axisymmetric interface in an azimuthal plane in axisymmetric flow. Similar equations can be written for three-dimensional flow.

Numerical methods

In a numerical implementation, a two-dimensional or axisymmetric interface is typically described by a chain of N marked points or nodes, as shown in Figure 9.1.1(b). The position of the ith marker point corresponding to arc length l_i is

$$\mathbf{x}_i = \mathbf{X}_i + \epsilon q_i(t)\,\mathbf{n},   \tag{9.1.27}$$

where $q_i(t)$ is the nodal value of the shape function $q(t)$. Since tangential motion does not cause a change in shape, we have stipulated that the marker points move normal to the unperturbed stationary interface.

The shape function, $q(l)$, and its derivative, $q'(l) \equiv dq/dl$, can be constructed in terms of the nodal values, q_i, by cubic spline or another type interpolation, as discussed in Section B.4, Appendix B. The numerical method provides us with the nodal derivatives,

$$q'(l_i) = \left(\frac{\partial q'(l_i)}{\partial q_j} \right)_{q=0} q_j, \tag{9.1.28}$$

where summation is implied over the repeated index, j. Note the linear dependence of $q'(l_i)$ on the nodal displacements, q_j. Linearizing the normal velocity at the ith node with respect to the position of all nodes, and working in a similar fashion, we obtain an expression for the nodal velocities

$$v_n(l_i) = \left(\frac{\partial v_n(l_i)}{\partial q_j} \right)_{q=0} q_j. \tag{9.1.29}$$

Applying equation (9.1.26) at the ith node and substituting expressions (9.1.28) and (9.1.29), we derive a system of linear ordinary differential equations,

$$\frac{dq_i}{dt} = M_{ij} q_j, \tag{9.1.30}$$

where summation is implied over the repeated index, j, and

$$M_{ij} \equiv \left(\frac{\partial v_n(l_i)}{\partial q_j} \right)_{q=0} - u_t(l_i) \left(\frac{\partial q'(l_i)}{\partial q_j} \right)_{q=0} \tag{9.1.31}$$

is a constant matrix. The general solution of the linear system (9.1.30) can be found in terms of the eigenvalues, eigenvectors, and possibly generalized eigenvectors of the matrix $\mathbf{M}$ (e.g., [317]). If the matrix $\mathbf{M}$ is diagonalizable, we obtain exponential solutions described as regular normal modes. As an alternative, equation (9.1.30) can be integrated in time by numerical methods.

Discrete perturbations

In practice, the jth column of the matrix $\mathbf{M}$ introduced in (9.1.30) can be generated by a simple numerical procedure, described as the method of discrete perturbations, according to the following steps:

1. All marker points are positioned along the unperturbed interface at chosen arc length positions l_i, where $i = 1, \ldots, N$.

2. The jth marker point is displaced normal to the interface by a small distance, δ.

3. Cubic spline or another interpolation with appropriate boundary conditions is performed for a function $\mathcal{G}(l)$ using the data set

$$\mathcal{G}(l_1) = 0, \quad \ldots, \quad \mathcal{G}(l_{j-1}) = 0, \quad \mathcal{G}(l_j) = \delta, \quad \mathcal{G}(l_{j+1}) = 0, \quad \ldots, \quad \mathcal{G}(l_N) = 0, \tag{9.1.32}$$

and the nodal derivatives, $\mathcal{G}'(l_i)$, are evaluated. For arbitrary δ,

$$\left(\frac{\partial q'(l_i)}{\partial q_j} \right)_{q=0} = \frac{1}{\delta} \mathcal{G}'(l_i). \tag{9.1.33}$$

The value of δ must be small if nonlinear interpolation is employed.

4. The velocity component normal to the unperturbed contour is evaluated at all nodes. For small δ, we compute

$$\left(\frac{\partial v_n(l_i)}{\partial q_j}\right)_{q=0} \simeq \frac{1}{\delta} v_n(l_i). \tag{9.1.34}$$

The value of δ must be such that the differentiation error is larger than the numerical error in evaluating the velocity.

5. The results of Steps 3 and 4 are used to compute M_{ij} according to (9.1.31)

The computation is repeated for $j = 1, \ldots, N$ to generate all columns of $\mathbf{M}$ and the eigenvalues are computed by standard numerical methods. This formulation is especially advantageous when the normal velocity can be computed from a contour- or boundary-integral representation, as in the case of Stokes flow, potential flow, or inviscid flow containing vortex patches. The main advantage is that solving the equations of inviscid or viscous flow is entirely bypassed and the analysis is conducted using an essentially geometrical approach [43]. Applications are discussed in Section 9.3.5.

9.1.3 Simplified evolution equations

In previous chapters, we discussed approximate solutions for low- and high-Reynolds number flow. For example, in Section 6.4.4, we found that the thickness of a liquid film coated on a horizontal surface, $h(x, t)$, is governed by the partial differential equation (6.4.40), repeated here for convenience,

$$h_t + \frac{\rho g}{3\mu}\left[h^3\left(-h_x + \frac{\gamma}{\rho g}h_{xxx}\right)\right]_x = 0, \tag{9.1.35}$$

where ρ is the film density, μ is the film viscosity, γ is the surface tension, g is the acceleration of gravity, a subscript t denotes a derivative with respect to time, and a subscript x denotes a derivative with respect to x. To derive a linearized evolution equation for the film thickness, we set

$$h(x, t) = \bar{h} + \epsilon \eta(x, t), \tag{9.1.36}$$

where $\bar{h}$ is the mean film thickness and $\eta(x, t)$ is a perturbation shape function. Substituting this expression into (9.1.35) and linearizing with respect to ϵ, we obtain a fourth-order linear partial differential equation with respect to η,

$$\eta_t + \frac{\rho g}{3\mu}\bar{h}^3\left(-\eta_x + \frac{\gamma}{\rho g}\eta_{xxx}\right)_x = 0. \tag{9.1.37}$$

An initial condition and a boundary or periodicity condition for η must be provided. The solution can be computed by finite-difference, finite-element, spectral, or other numerical methods.

Problems

9.1.1 *Pressure Poisson equation*

Consider an initially solenoidal velocity field. Show that if the pressure is computed from (9.1.4), the velocity field will remain solenoidal at all times. What will happen if the initial velocity field is not initially solenoidal?

9.1.2 *Nonlinear evolution equation in operator form*

Recast (9.1.13) in operator form similar to that shown in (9.1.21).

9.1.3 *Linearized vorticity transport for two-dimensional flow*

Discuss the physical significance of the various terms in (9.1.20) for two-dimensional flow.

9.2 Initial-value problems

We proceed to discussing possible ways of solving the initial-value problem governed by the linearized equation of motion and vorticity transport equation (9.1.21), subject to appropriate boundary conditions and a specified initial condition describing a perturbation. The analysis will illustrate the difficulty in computing the evolution of an unsteady flow by analytical methods even after linearization.

9.2.1 Inviscid Couette flow

In the case of inviscid Couette flow along the x axis in a channel confined between two plane walls located at $y = \pm a$, an analytical solution of the linearized initial-value problem can be found in integral form [283]. The velocity of the base flow is $\mathbf{u}^S = (\xi y, 0)$, where ξ is a constant shear rate with units of inverse time.

We focus our attention on two-dimensional perturbations in the plane of the flow and introduce the disturbance stream function, $\psi(x, y, t)$. The no-penetration boundary condition at the two walls requires that $\psi = 0$ at $y = \pm a$. The z component of the linearized vorticity transport equation (9.1.20) shows that, in the absence of viscous forces, the vorticity associated with the disturbance flow is simply convected by the base flow,

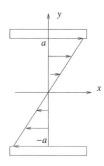

Couette flow in a channel.

$$\frac{\partial \omega_z^U}{\partial t} + \xi y \frac{\partial \omega_z^U}{\partial x} = 0, \tag{9.2.1}$$

where $\omega_z^U = -\nabla^2 \psi$ and the superscript U denotes the unsteady component. Inverting the hyperbolic operator on the left-hand side, we obtain the general solution

$$\nabla^2 \psi = -\omega_z^U(x, y, t) = -\Omega_z(x - \xi t y, y), \tag{9.2.2}$$

where $\Omega_z(x, y) \equiv \omega_z^U(x, y, t = 0)$ is the disturbance vorticity at the initial instant.

Fourier expansions

For simplicity, we consider perturbations that are initially symmetric with respect to the origin of the x axis, $x = 0$, and express the stream function as a sine Fourier series in the y direction and a cosine Fourier integral in the x direction. Antisymmetric perturbations can be treated in a similar manner and a general perturbation can be expressed in terms of symmetric and antisymmetric modes

(Problem 9.2.1). Taking into account the no-penetration boundary condition $\psi = 0$ at $y = \pm a$, we write

$$\psi(x, y, t = 0) = \sum_{n=1}^{\infty} \sin\left(\frac{n\pi}{2a}(y + a)\right) \int_{-\infty}^{\infty} b_n(\lambda) \cos \lambda x \, d\lambda, \tag{9.2.3}$$

where $b_n(\lambda)$ are real Fourier coefficients with dimensions of velocity multiplied by squared length. Without loss of generality, we may assume that b_n are even functions, $b_n(\lambda) = b_n(-\lambda)$; odd functions are annihilated upon integration. Straightforward differentiation yields the initial vorticity field,

$$\Omega_z(x, y) \equiv -\nabla^2 \psi(x, y, t = 0) = \sum_{n=1}^{\infty} \sin\left(\frac{n\pi}{2a}(y + a)\right) \int_{-\infty}^{\infty} \left[\left(\frac{n\pi}{2a}\right)^2 + \lambda^2\right] b_n(\lambda) \cos \lambda x \, d\lambda. \tag{9.2.4}$$

Fourier transform

The Fourier transform of a rapidly decaying function, $f(x)$, is defined by the integral

$$\hat{f}(k) \equiv \int_{-\infty}^{\infty} f(x) \, e^{-ikx} \, dx, \tag{9.2.5}$$

where i is the imaginary unit. The Fourier transform of the derivatives are $\hat{f}'(k) = ik \, \hat{f}(k)$ and $\hat{f}''(k) = -k^2 \, \hat{f}(k)$. The inverse transform is

$$f(x) = \frac{1}{2\pi} \int_{-\infty}^{\infty} \hat{f}(k) \, e^{ikx} \, d\kappa. \tag{9.2.6}$$

Derivation of a one-dimensional Poisson equation

Substituting (9.2.4) into (9.2.2) and taking the x Fourier transform of the emerging equation, we find that

$$\left(-k^2 + \frac{\partial^2}{\partial y^2}\right) \hat{\psi}(k, y, t) = \mathcal{F}(k, y, t), \tag{9.2.7}$$

where

$$\mathcal{F}(k, y, t) = -\sum_{n=1}^{\infty} \sin\left(\frac{n\pi}{2a}(y + a)\right) \int_{-\infty}^{\infty} \left[\left(\frac{n\pi}{2a}\right)^2 + \lambda^2\right] b_n(\lambda) \left[\int_{-\infty}^{\infty} \cos[\lambda(x - \xi ty)] \, e^{-ikx} \, dx\right] d\lambda. \tag{9.2.8}$$

Next, we compute the inner integral

$$\int_{-\infty}^{\infty} \cos[\lambda(x - \xi ty)] \, e^{-ikx} \, dx = \frac{1}{2} \int_{-\infty}^{\infty} \left(\exp[i\lambda(x - \xi ty)] + \exp[-i\lambda(x - \xi ty)]\right) e^{-ikx} \, dx$$

$$= \frac{1}{2} \left(e^{-i\lambda \xi ty} \int_{-\infty}^{\infty} \exp[i(\lambda - \kappa)x] \, dx + e^{i\lambda \xi ty} \int_{-\infty}^{\infty} \exp[-i(\lambda + \kappa)x] \, dx\right), \tag{9.2.9}$$

yielding

$$\int_{-\infty}^{\infty} \cos[\lambda(x - \xi t y)] \, e^{-ikx} \, dx = \pi \, e^{-i\lambda\xi t y} \left(\delta(\lambda - \kappa) + \delta(\lambda + \kappa) \right), \qquad (9.2.10)$$

where δ is the one-dimensional Dirac delta function. Substituting this expression into (9.2.8), using the distinctive properties of the delta function, and recalling that $b_n(k) = b_n(-k)$, we find that (9.2.7) becomes

$$\left(-k^2 + \frac{\partial^2}{\partial y^2} \right) \hat{\psi}(k, y, t) = -2\pi \sum_{n=1}^{\infty} \left[\left(\frac{n\pi}{2a} \right)^2 + k^2 \right] \sin \left(\frac{n\pi}{2a} (y + a) \right) b_n(k) \, e^{-ik\xi t y}. \qquad (9.2.11)$$

Restating the sine as the sum of two complex exponentials, we find that

$$\left(-k^2 + \frac{\partial^2}{\partial y^2} \right) \hat{\psi}(k, y, t) = \pi i \sum_{n=1}^{\infty} \left[\left(\frac{n\pi}{2a} \right)^2 + k^2 \right] b_n(k) \left(A_n - A_{-n} \right), \qquad (9.2.12)$$

where

$$A_n = \exp \left(i \left[\frac{n\pi}{2a} (y + a) - k\xi t y \right] \right). \qquad (9.2.13)$$

Particular and general solution

A particular solution of (9.2.12) can be expressed in the form

$$\hat{\psi}^p(k, y, t) = -\pi i \sum_{n=1}^{\infty} \left[\left(\frac{n\pi}{2a} \right)^2 + k^2 \right] b_n(k) \left(\alpha_n A_n - \alpha_{-n} A_{-n} \right), \qquad (9.2.14)$$

where α_n, κ_n, and λ_n are functions of k and t. Substituting (9.2.14) into (9.2.12) and matching the functional forms on the left- and right-hand sides, we find that

$$\alpha_n = \frac{1}{\left(\frac{n\pi}{2a} - k\xi t \right)^2 + k^2}. \qquad (9.2.15)$$

To simplify the notation, we define the negative-order coefficients, $b_{-n}(k) \equiv -b_n(k)$, and obtain

$$\hat{\psi}^p(k, y, t) = -\pi i \sum_{n=-\infty}^{\infty} {}' \left[\left(\frac{n\pi}{2a} \right)^2 + k^2 \right] b_n(k) \, \alpha_n A_n, \qquad (9.2.16)$$

where the prime indicates that the term $n = 0$ is excluded from the sum. Substituting the expression for α_n, adding a judiciously selected homogeneous solution, and taking the inverse Fourier transform, we derive the general solution,

$$\psi(x, y, t) = -\frac{i}{2} \sum_{n=-\infty}^{\infty} {}' \int_{-\infty}^{\infty} \frac{\left(\frac{n\pi}{2a} \right)^2 + k^2}{\left(\frac{n\pi}{2a} - k\xi t \right)^2 + k^2} \, b_n(k) \qquad (9.2.17)$$

$$\times \left[B_n - e^{ik\xi t y} \left(\kappa_n \, \sinh[k(a - y)] + \lambda_n \, \sinh[k(a + y)] \right) \right] e^{ik(x - \xi t y)} \, dk,$$

where the functions κ_n and λ_n depend on k and t but not on y, and

$$B_n = \exp\left(\frac{n\pi i}{2a}(y+a)\right). \tag{9.2.18}$$

To satisfy the no-penetration boundary condition, we require that the term enclosed by the large square brackets is zero when $y = \pm a$, yielding

$$\kappa_n e^{-ik\xi ta}\sinh(2ka) = B_n(-a) = 1,$$
$$\lambda_n e^{ik\xi ta}\sinh(2ka) = B_n(a) = e^{n\pi i} = (-1)^n. \tag{9.2.19}$$

Selecting the real part on the right-hand side of (9.2.16), we obtain ([437], p. 482)

$$\psi(x,y,t) = \frac{1}{2}\sum_{n=-\infty}^{\infty}{}'\int_{-\infty}^{\infty}\frac{\left(\frac{n\pi}{2a}\right)^2 + k^2}{\left(\frac{n\pi}{2a} - k\xi t\right)^2 + k^2}\, b_n(k)\,\Phi_n(x,y,t,k)\,\mathrm{d}k, \tag{9.2.20}$$

where

$$\Phi_n(x,y,t,k) = \sin\left(\frac{n\pi}{2a}(y+a) + k(x - \xi t y)\right) \tag{9.2.21}$$
$$-\frac{1}{\sinh(2ka)}\Big(\sinh[k(a-y)]\,\sin[k(x+\xi ta)] + \sinh[k(a+y)]\,\sin[k(x-\xi ta) + n\pi]\Big).$$

Evaluating the right-hand side of (9.2.20) at long times shows that the disturbance decays like $1/t^2$, and this means that the base flow is stable (Eliassen, Høilland & Riis [119]; see [257, 112]). Later in this chapter, we will see that viscous forces render certain types of perturbations unstable in a certain range of Reynolds numbers.

9.2.2 Laplace transform

In the case of inviscid plane Couette flow, we were able to solve the initial-value problem governed by (9.1.21) and (9.1.23) exactly by analytical methods. To compute the solution for more general base flows, we resort to approximate and numerical methods for linear partial differential equations. The Laplace transform allows us to eliminate the temporal dependence from the governing equations, replacing it with an algebraic dependence. Assume that a perturbation has been introduced at the origin of time and then grows, at most, at an exponential rate in time. The one-sided Laplace transform of the velocity, pressure, and vorticity is defined as

$$\begin{bmatrix} \hat{\mathbf{u}} \\ \hat{p} \\ \hat{\omega} \end{bmatrix}^U(\mathbf{x},s) = \int_{0+}^{\infty}\begin{bmatrix} \mathbf{u} \\ p \\ \omega \end{bmatrix}^U(\mathbf{x},t)\,\mathrm{e}^{-st}\,\mathrm{d}t, \tag{9.2.22}$$

where s is a complex variable. Taking the Laplace transform of (9.1.21) provides us with a system of linear, inhomogeneous, second-order system partial differential equations in the spatial variables for the Laplace transformed functions,

$$s \begin{bmatrix} \hat{\mathbf{u}} \\ \hat{\omega} \end{bmatrix}^U = \begin{bmatrix} \mathbf{A} & \mathbf{B} & \mathbf{0} \\ \mathbf{C} & \mathbf{0} & \mathbf{D} \end{bmatrix} \cdot \begin{bmatrix} \hat{\mathbf{u}} \\ \hat{p} \\ \hat{\omega} \end{bmatrix}^U + \begin{bmatrix} \hat{\mathbf{u}} \\ \hat{\omega} \end{bmatrix}^U (\mathbf{x}, t = 0^+), \qquad (9.2.23)$$

where 0^+ indicates the limit $t \to 0$ from positive times when the disturbance is present. Having computed the solution subject to appropriate boundary conditions, we recover the physical variables in the time domain in terms of the Bromwich integral in the complex s plane,

$$\begin{bmatrix} \mathbf{u} \\ p \\ \omega \end{bmatrix}^U (\mathbf{x}, t) = \frac{1}{2\pi \mathrm{i}} \int_{\gamma - \mathrm{i}\infty}^{\gamma + \mathrm{i}\infty} \begin{bmatrix} \hat{\mathbf{u}} \\ \hat{p} \\ \hat{\omega} \end{bmatrix}^U (\mathbf{x}, s) \, \mathrm{e}^{st} \, \mathrm{d}s, \qquad (9.2.24)$$

where i is the imaginary unit and γ is a sufficiently large real positive number chosen so that all singularities of the Laplace transformed functions lie on the left of the integration path. The integral in (9.2.24) can be evaluated using the method of residues by introducing a closed contour that encloses all singularities of the integrand. In the absence of branch points of the Laplace-transformed variables, the contour can be identified with the union of the vertical line $s = \gamma$ and a semi-circular contour of large radius.

9.2.3   Green's functions

Green's functions are solutions of the linearized equation of motion and associated vorticity transport equation for the velocity, vorticity, and pressure, subject to a localized forcing. To compute a Green's function, we add to the right-hand side of the vorticity transport equation the singular function

$$\delta(x - x_0) \, \delta(y - y_0) \, \delta(z - z_0) \, \delta(t) \, \mathbf{b}, \qquad (9.2.25)$$

where $\mathbf{x}_0$ is an arbitrary point in the domain of flow, $\mathbf{b}$ is a constant vector, and δ is the one-dimensional delta function. The solution for the velocity is $\mathcal{G}(\mathbf{x}, \mathbf{x}_0, t) \cdot \mathbf{b}$, where $\mathcal{G}(\mathbf{x}, \mathbf{x}_0, t)$ is the Green's function tensor for the velocity. The solution of the initial-value problem can be expressed as (a) a volume integral of the Green's function over the domain of flow multiplied by an appropriate density distribution function, $\mathbf{q}$, determined by the initial condition, and (b) a convolution integral in time, in the form

$$\mathbf{u}(\mathbf{x}, t) = \int_0^t \left[\iiint_{Flow} \mathcal{G}(\mathbf{x}, \mathbf{x}_0, t - \tau) \cdot \mathbf{q}(\mathbf{x}_0) \, \mathrm{d}V(\mathbf{x}_0) \right] \mathrm{d}\tau. \qquad (9.2.26)$$

Unfortunately, because Green's functions are hard to compute, this approach finds limited applications in practice [195].

Problem

9.2.1 *Inviscid Couette flow*

(a) Verify that (9.2.20) reduces to (9.2.3) at the initial instant, $t = 0$.

(*b*) Derive the counterpart of (9.2.20) for antisymmetric perturbations where the stream function at the initial instant is given by a sine Fourier integral with respect to x.

 Computer Problem

9.2.2 *Perturbation flow*

Plot the streamline patterns according to (9.2.3) and (9.2.20) at the initial instant and at a later time of your choice for $b_n(k) = 0$ and $b_1(k) = \xi a^3 e^{-ka}$. Discuss the behavior of the perturbation.

9.3   Normal-mode analysis

To study the evolution of every possible type of perturbation on a given base flow is practically impossible. Progress can be made by expressing an initial disturbance as a combination of linearly independent fundamental modes, and then examining the evolution of each individual mode. This approach assumes that a complete set of fundamental modes is available.

A convenient set of fundamental modes that are analogous to the eigenvectors of a diagonalizable matrix are normal modes with exponential dependence on time. The unsteady component of the flow takes the form

$$
\begin{bmatrix} \mathbf{u} \\ p \\ \omega \end{bmatrix}^{NM} (\mathbf{x}, t) = \begin{bmatrix} \mathbf{V} \\ \Pi \\ \Omega \end{bmatrix} (\mathbf{x}, \varrho)\, e^{-i\varrho t},
\tag{9.3.1}
$$

where the superscript NM stands for "normal mode", ϱ is a complex constant called the complex growth rate, complex cyclic frequency, or complex angular velocity, i is the imaginary unit satisfying $i^2 = -1$, and $\mathbf{V}$, Π, and Ω are complex functions of $\mathbf{x}$ and ϱ. All dependent variables are assumed to be complex, with the understanding that both the real and imaginary parts represent admissible solutions.

Substituting (9.3.1) into the linearized equation of motion and vorticity transport equation stated in (9.1.21), we obtain a system of linear homogeneous equations governing the spatial structure of the normal modes,

$$
(i\varrho - \mathbf{u}^S \cdot \nabla + \nu \nabla^2)\, \mathbf{V} - \mathbf{V} \cdot \nabla \mathbf{u}^S - \frac{1}{\rho} \nabla \Pi = 0,
\tag{9.3.2}
$$

and

$$
(i\varrho - \mathbf{u}^S \cdot \nabla + \nu \nabla^2)\, \Omega - \mathbf{V} \cdot \nabla \omega^S + \Omega \cdot \nabla \mathbf{u}^S + \omega^S \cdot \nabla \mathbf{V} = 0.
\tag{9.3.3}
$$

The continuity equation requires that

$$
\nabla \cdot \mathbf{V} = 0.
\tag{9.3.4}
$$

By introducing the normal modes, we have factored out the time dependence and derived a system of partial differential equations in space involving the *a priori* unknown complex growth rate, ϱ.

Discrete and continuous spectra

Nontrivial solutions of the system of differential equations (9.3.2), (9.3.3), and (9.3.4) exist only when the complex growth rate, ϱ, takes values in a set of complex numbers called the spectrum of eigenvalues of the base flow. The spectrum consists of a discrete part containing separated eigenvalues, and a continuous part containing families of eigenvalues that vary in a continuous fashion along a curve with respect to some parameter (e.g., [95], p. 52). The discrete part of the spectrum may contain a finite number of eigenvalues or no eigenvalues at all. The continuous part of the spectrum may be null.

For example, in the case of viscous unidirectional flow in a channel, only a discrete spectrum consisting of an infinite number of discrete eigenvalues that form a complete set can be found. In the case of inviscid unidirectional channel flow, only a continuous spectrum consisting of stable normal modes can be found. The nature of the two components of the spectrum will be illustrated in Section 9.5.2 for inviscid Couette flow.

Completeness

Whether or not the normal modes provide us with a complete base of eigenfunctions capable of describing an arbitrary disturbance by linear superposition depends on the topology of the domain of flow and presence of singularities [217]. If the normal modes provide us with a complete base, an arbitrary disturbance can be expressed as a linear combination of (*a*) the discrete normal modes multiplied by appropriate complex coefficients, and (*b*) distributions of the continuous normal modes weighed by appropriate complex distribution density functions. It is important to keep in mind that, even though the individual normal modes may grow or decay at an exponential rate, their super-position may exhibit a different type of temporal behavior (e.g., [68]). Generalized normal modes with nonexponential dependence on time, corresponding to the eigenvectors of nondiagonalizable matrices, may arise in some flows. These generalized modes behave as $t^m \exp(-i\varrho t)$, with an *a priori* unknown algebraic exponent, m.

Stable and unstable normal modes

Decomposing the complex growth rate ϱ into its real and imaginary parts, $\varrho = \varrho_R + i\varrho_I$, we recast (9.3.1) into the form

$$
\begin{bmatrix} \mathbf{u} \\ p \\ \omega \end{bmatrix}^{NM} (\mathbf{x}, t) = \begin{bmatrix} \mathbf{V} \\ \Pi \\ \Omega \end{bmatrix} (\mathbf{x}, \varrho)\, e^{\varrho_I t}\, e^{-i\varrho_R t}, \tag{9.3.5}
$$

which shows that ϱ_I is the temporal growth rate of the normal mode and ϱ_R is the cyclic frequency or angular velocity of the perturbation. If ϱ_I is positive, the disturbance grows exponentially in time and the base flow is linearly unstable; if ϱ_I is negative, the disturbance decays and the normal mode is linearly stable; if ϱ_I is zero, the disturbance is neutrally stable.

If a perturbation consists of a number of superimposed normal modes, the mode corresponding to the eigenvalue with the maximum growth rate, ϱ_I, called the most unstable or most dangerous normal mode, will dominate the rest of the modes. The computation of this growth rate and associated eigensolution is a prime objective of the linear stability analysis. To establish the critical conditions under which a flow is expected to be unstable, we examine the behavior of neutrally stable perturbations with $\varrho_I = 0$, with respect to the dimensionless numbers that characterize the base flow, such as the Reynolds number or the Weber number.

Laplace transform

It is instructive to apply the method of Laplace transform for computing the evolution of normal modes. Combining the normal-mode form (9.3.1) with the definition of the Laplace transform shown in (9.2.22), we find that the transformed variables, indicated by a caret, are given by

$$
\begin{bmatrix} \hat{\mathbf{u}} \\ \hat{p} \\ \hat{\omega} \end{bmatrix}^{NM} (\mathbf{x}, s, \varrho) = \frac{1}{s + i\varrho} \begin{bmatrix} \mathbf{V} \\ \Pi \\ \Omega \end{bmatrix} (\mathbf{x}, \varrho). \tag{9.3.6}
$$

Substituting the right-hand side of (9.3.6) into (9.2.23) and simplifying, we recover (9.3.2) and (9.3.3), and thereby reconcile the inhomogeneous problem expressed by (9.2.23) with the eigenvalue problem expressed by (9.3.2) and (9.3.3).

Equation (9.3.6) shows that $\varrho = -is$ is a simple pole of the flow variables in the Laplace-transformed domain. The associated normal modes may then be used to evaluate the Bromwich integral in (9.2.24) using the method of residues. Conversely, the poles of the Laplace-transformed variables correspond to normal modes that fall within the discrete part of the spectrum. The continuous part of the spectrum is associated with branch cuts of branch points in the complex s plane.

9.3.1 Interfacial flow

In the case of interfacial flow, we substitute into the linearized kinematic compatibility condition (9.1.26) the normal-mode expansions

$$
q(l, t) = Q(l) \, e^{-i\varrho t}, \qquad v_n(l, t) = V_n(l) \, e^{-i\varrho t}, \tag{9.3.7}
$$

and obtain an ordinary differential equation,

$$
-i\varrho \, Q(l) + u_t(l) \, \frac{dQ}{dl} - V_n(l) = 0, \tag{9.3.8}
$$

where $Q(l)$ is a shape function and $V_n(l)$ is the corresponding normal velocity. Rearranging and observing that $V_n(l)$ is an implicit function of the normal displacement, $Q(l)$, we obtain an eigenvalue problem governed by an ordinary differential equation,

$$
V_n(l) - u_t(l) \, \frac{dQ}{dl} = -i\varrho \, Q(l). \tag{9.3.9}
$$

The second term on the left-hand side does not appear in the absence of base flow. The form of the shape function $Q(l)$ can be deduced by physical intuition subject to volume and possibly contact angle constraints, or else computed as part of the solution.

Linear stability analysis by discrete perturbations

In the numerical implementation, a two-dimensional or axisymmetric interface is typically described by a chain of N marker points or nodes, as shown in Figure 9.1.1(b). In the case of a normal-mode perturbation, the position of the ith marker point corresponding to arc length l_i is

$$\mathbf{x}_i = \mathbf{X}_i + \epsilon\, Q_i\, e^{-i\varrho t}\, \mathbf{n}_i \tag{9.3.10}$$

for $i = 1, \ldots, N$, where Q_i is the nodal value of the shape function Q, and $\mathbf{n}_i$ is the corresponding unperturbed normal unit vector. Because tangential motion does not cause a change in shape, we have stipulated that the marker points move normal to the unperturbed stationary interface.

The shape function, $Q(l)$, and its derivative, $Q'(l)$, can be constructed in terms of the nodal values, Q_i, by cubic spline or some other type of interpolation, yielding an expression for the derivative at the ith node,

$$Q'_i = \left(\frac{\partial Q'_i}{\partial Q_j}\right)_{Q=0} Q_j, \tag{9.3.11}$$

where summation is implied over the repeated index, j (e.g., [317]). Similarly, the normal velocity at the ith node can be linearized with respect to the position of all nodes, yielding the corresponding expression

$$V_n(l_i) = \left(\frac{\partial V_n(l_i)}{\partial Q_j}\right)_{Q=0} Q_j. \tag{9.3.12}$$

Applying equation (9.3.9) at the ith node and substituting these expressions, we obtain an algebraic eigenvalue problem expressed by the linear system

$$M_{ij} Q_j = -i\varrho\, Q_i, \tag{9.3.13}$$

where

$$M_{ij} \equiv \left(\frac{\partial V_n(l_i)}{\partial Q_j}\right)_{Q=0} - u_t(l_i) \left(\frac{\partial Q'_i}{\partial Q_j}\right)_{Q=0}. \tag{9.3.14}$$

The matrix $\mathbf{M}$ can be generated by the discrete perturbation method discussed in Section 9.1.7. The method has been applied successfully to study the capillary instability of an infinite quiescent viscous cylindrical thread containing a periodic array of particles arranged along the centerline, and the gravitational instability of a pendant drop [43, 322]. Two additional applications are discussed in this section.

Relaxation of a viscous drop

In the first application, we study the relaxation of a slightly deformed neutrally buoyant spherical liquid drop suspended in an infinite fluid with the same viscosity, μ. Restricting our attention to axisymmetric perturbations with respect to an arbitrarily chosen x axis that passes through the center of the unperturbed drop, we distribute interfacial nodes around the semi-circular interfacial contour in an azimuthal plane, as shown in Figure 9.3.1(a). Because of the absence of a base flow, the

(a)   (b)

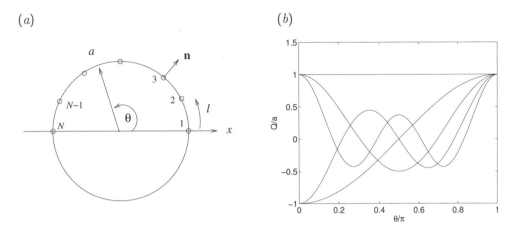

FIGURE 9.3.1   (*a*) Illustration of a stationary spherical viscous liquid drop suspended in an infinite quiescent ambient fluid. (*b*) Graphs of the first five eigenfunctions.

unperturbed tangential velocity is identically zero, $u_t = 0$, and the second term on the right-hand side of (9.3.14) does not appear. Referring to (9.3.9), we set $l = a\theta$ and identify the normal velocity with the radial velocity, $u_n = u_r$, where a is the drop radius and θ is the polar angle.

The discrete perturbation method described in this section is implemented in the computer code *drop_ax* placed in directory *08_stab* of the software library FDLIB (Appendix C). The interfacial velocity is computed using the boundary-integral method for Stokes flow discussed in Section 6.9. Solving the eigenvalue problem provides us with the dimensionless growth rate $\hat{\rho}_I \equiv \varrho_I \mu a / \gamma$, where γ is the surface tension. The computations reveal a double zero eigenvalue followed by a sequence of negative eigenvalues, $\hat{\rho}_I = -0.46, -0.76, -1.04, -1.31, -1.57, -1.83, \ldots$, corresponding to decaying modes. The first five eigenfunctions, $Q(\theta)$, normalized so that $Q(\pi) = 1$, are shown in Figure 9.3.1(*b*). The first two eigenfunctions, corresponding to the zero eigenvalues, express inadmissible radial expansion, $Q = 1$, and inconsequential translation, $Q = -\cos\theta$. Further eigenfunctions are precisely proportional to the Legendre polynomials, $L_n(\cos\theta)$, for $n > 1$, ensuring that the deformation conserves the drop volume to first order with respect to the dimensionless deformation parameter, ϵ.

Stability of a settling viscous drop

In the second application, we consider the instability of a spherical viscous liquid drop with radius a and viscosity $\lambda\mu$, settling with velocity V in the direction of gravity along the x axis in an infinite ambient fluid with viscosity μ under conditions of Stokes flow, as illustrated in Figure 9.3.2(*a*). Using the singularity solution given in (6.7.47) and (6.7.48), we find that, in a frame of reference translating with the drop, the interfacial velocity over the unperturbed spherical interface is

$$u_i(r = a) = \frac{1}{2} \frac{1}{\lambda + 1} \left[(2\lambda + 1) V_i + \frac{\hat{x}_i \hat{x}_j}{a^2} V_j \right] - V_i = \frac{1}{2} \frac{1}{\lambda + 1} \left(-V_i + \frac{\hat{x}_i \hat{x}_j}{a^2} V_j \right), \qquad (9.3.15)$$

(a)   (b)

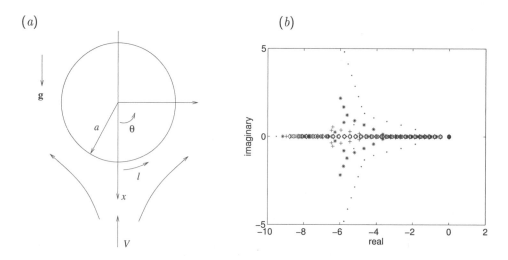

FIGURE 9.3.2 (a) Illustration of a settling viscous liquid drop observed in a frame of reference translating with the drop velocity, V. (b) Spectrum of dimensionless complex growth rates $-i\varrho\mu a/\gamma$ for
Ca $= 0$ (circles), 0.5 (squares), 1.0 (diamonds), 2.0 (plus signs), 3.0 (asterisks), and 5.0 (dots).

where $\hat{\mathbf{x}} = \mathbf{x} - \mathbf{x}_0$ is the distance from the drop center, $\mathbf{x}_0$, λ is the viscosity ratio, and $V_x = V$,
$V_y = 0$, $V_z = 0$ are the Cartesian components of the drop velocity. We can easily verify that the
normal velocity component is zero, as required, $\mathbf{u} \cdot \mathbf{n} = 0$. In a frame of reference moving with the
drop, the tangential surface velocity is

$$u_t(l) = u_\theta(r = a) = V\,\frac{1}{2}\,\frac{\sin\theta}{\lambda + 1}, \tag{9.3.16}$$

where $l = a\theta$ and θ is the meridional angle. The maximum surface velocity occurs at the equator,
$\theta = \frac{1}{2}\pi$. Using expression (6.7.51) for the velocity of settling, we formulate the capillary number

$$\text{Ca} \equiv \frac{\mu V}{\gamma}\,\frac{3}{2}\,\frac{3\lambda + 2}{\lambda + 1} = \frac{a^2(\rho_d - \rho)g}{\gamma} \tag{9.3.17}$$

determining the significance of the capillary pressure due to surface tension, γ, relative to the viscous
stresses due to the motion, where ρ_d is the drop fluid density.

Referring to (9.3.9), we identify the normal velocity with the radial velocity, $u_n = u_r$. The discrete perturbation method is implemented in the computer code *drop_ax* discussed in the preceding
section on drop relaxation. The spectrum of dimensionless complex growth rates, $-i\varrho\mu a/\gamma$, is shown
in Figure 9.3.2(b) for several capillary numbers, Ca. Zero eigenvalues representing inconsequential
translation appear for any capillary number. The eigenvalues are real at small capillary numbers but
become complex as the capillary number increases beyond a critical threshold. Since the real part
of $-i\varrho\mu a/\gamma$ is zero or negative independent of the capillary number, the interface is stable except in
the complete absence of surface tension corresponding to infinite capillary number.

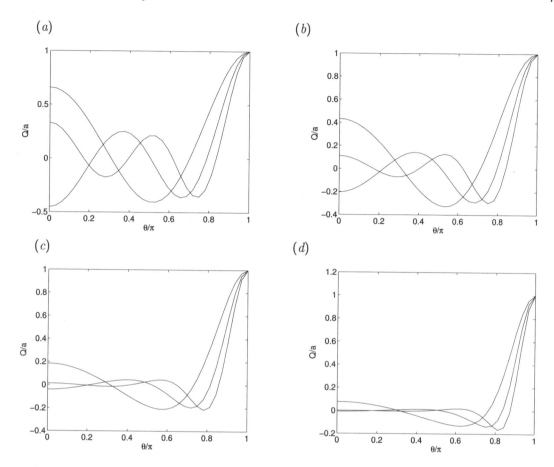

FIGURE 9.3.3 The first three eigenfunctions representing the interfacial deformation of a moving viscous drop for capillary number (a) Ca = 0.5, (b) 1.0, (c) 2.0, and (d) 3.0.

The first three eigenfunctions corresponding to nonzero eigenvalues are shown in Figure 9.3.3 for Ca = 0.5, 1.0, 2.0, and 3.0. The corresponding eigenfunctions for zero capillary number, corresponding to a stationary drop relaxing after a normal-mode perturbation has been imposed, are shown in Figure 9.3.1(b). In the case of a moving drop, the interfacial velocity field causes the normal-mode perturbations to concentrate at the back of the drop where a sharp peak arises at high capillary numbers. The peak becomes infinitely tall in the absence of surface tension due to the uninhibited action of the local stagnation-point flow.

9.3.2 Simplified evolution equations

The availability of simplified evolution equations considerably expedites the linear stability analysis in terms of normal modes. As an example, we investigate the instability or leveling of a film coated on a horizontal surface in the framework of the lubrication approximation. To study spatially periodic normal-mode perturbations with wavelength L, we substitute into the linearized equation (9.1.37) the normal-mode form

$$\eta(x,t) = \bar{h}\cos(kx)\,\exp(-i\varrho t), \tag{9.3.18}$$

where $k = 2\pi/L$ is the wave number and $\bar{h}$ is the mean film thickness. Simplifying and rearranging, we derive a purely imaginary growth rate,

$$\varrho_I = -\frac{\bar{h}^3}{3\mu}\,k^2(\rho g + \gamma k^2). \tag{9.3.19}$$

Since $\rho_I < 0$ under any conditions, normal-mode perturbations decay exponentially due to the combined action of gravity and surface tension.

Problems

9.3.1 *Normal modes in operator form*

Recast equations (9.3.2) and (9.3.3) in operator form similar to that shown in equation (9.1.21).

9.3.2 *Eigenvalues and eigenvectors of a matrix*

Discuss the number of eigenvalues and the number of possible distinct eigenvectors of an $N \times N$ square matrix. State the conditions under which the eigenvectors form a complete base of the Nth-dimensional vector space (e.g., [317]).

9.3.3 *Ordinary differential equations*

Consider a system of first-order linear ordinary differential equations, $\mathrm{d}\mathbf{x}/\mathrm{d}t = \mathbf{A} \cdot \mathbf{x}$, where $\mathbf{x}$ is the unknown solution vector and $\mathbf{A}$ is a square matrix. Express the general solution in terms of exponential functions in time involving the eigenvalues of $\mathbf{A}$. Discuss the possible occurrence of non-exponential solutions (e.g., [317]).

9.4 Unidirectional flows

Unidirectional flows are encountered in a broad range of natural and engineering applications. Examples include shear-layer flows forming between two merging streams, atmospheric boundary-layer flows, pressure-driven flows in channels and tubes, and the flow of liquid films down inclined planes. The velocity, pressure, and vorticity of a steady unidirectional base flow along the x axis varying along the y axis are described as

$$\mathbf{u}^S = U(y)\,\mathbf{e}_x, \qquad p^S(x), \qquad \boldsymbol{\omega}^S = -\frac{\mathrm{d}U}{\mathrm{d}y}\,\mathbf{e}_z, \tag{9.4.1}$$

where the function $U(y)$ describes the velocity profile, $\mathbf{e}_x$ is the unit vector along the x axis, and $\mathbf{e}_z$ is the unit vector along the z axis. In certain applications, the pressure p^S is constant; an example is Couette flow between two plates. In other applications, the pressure varies linearly with respect to x and the streamwise pressure gradient is constant; an example is Hagen–Poiseuille flow through a channel.

Linearized equation of motion

To carry out the normal-mode stability analysis, we substitute expressions (9.4.1) into the governing equations (9.3.2) and (9.3.3). Simplifying, we obtain the linearized equation of motion

$$\left(i\varrho - U\frac{\partial}{\partial x} + \nu \nabla^2\right)\mathbf{V} - U' V_y\, \mathbf{e}_x - \frac{1}{\rho}\nabla \Pi = \mathbf{0}, \tag{9.4.2}$$

and linearized vorticity transport equation

$$\left(i\varrho - U\frac{\partial}{\partial x} + \nu \nabla^2\right)\mathbf{\Omega} + U'' V_y\, \mathbf{e}_x + U'\left(\Omega_y\, \mathbf{e}_x - \frac{\partial \mathbf{V}}{\partial z}\right) = \mathbf{0}, \tag{9.4.3}$$

where a prime denotes a derivative with respect to y. The physical properties of the fluid are assumed to be uniform throughout the domain of flow.

Fourier decomposition

A normal-mode disturbance can be expressed as a double complex Fourier integral with respect to x and z in the horizontal plane where the velocity of the unperturbed base flow is constant. Because the governing equations (9.4.2) and (9.4.3) are linear, each Fourier mode evolves independently and can be studied in isolation. Motivated by this observation, we consider perturbations of the form

$$\begin{bmatrix} \mathbf{V} \\ \Pi \\ \mathbf{\Omega} \end{bmatrix}(\mathbf{x}, \varrho) = \begin{bmatrix} \mathbf{F} \\ G \\ \mathbf{Q} \end{bmatrix}(y, \varrho) \times \exp[i(k_x x + k_z z)], \tag{9.4.4}$$

where $\mathbf{F}$, G, and $\mathbf{Q}$ are complex functions of y to be determined as part of the solution, and k_x, k_z are complex wave numbers in the x and z directions comprising the two-dimensional wave number vector

$$\mathbf{k} = (k_x, 0, k_z). \tag{9.4.5}$$

Substituting (9.4.4) into the continuity equation (9.3.4) and carrying out the differentiations, we obtain the equation

$$k_x F_x + k_z F_z - i F_y' = 0, \tag{9.4.6}$$

where a prime denotes a derivative with respect to y. Recalling that the vorticity is defined as the curl of the velocity, we find that the functions $\mathbf{F}$ and $\mathbf{Q}$ are related by

$$Q_x = F_z' - i k_z F_y, \qquad Q_y = i(k_z F_x - k_x F_z), \qquad Q_z = i k_x F_y - F_x'. \tag{9.4.7}$$

Substituting (9.4.4) into (9.4.3) and using (9.4.6) and (9.4.7), we obtain the vector equation

$$\left[i(\varrho - U k_x) - \nu(k_x^2 + k_z^2)\right]\mathbf{Q} + \nu \mathbf{Q}'' - i U' \begin{bmatrix} k_x F_z \\ k_z F_y \\ k_z F_z \end{bmatrix} + U'' F_y \begin{bmatrix} 0 \\ 0 \\ 1 \end{bmatrix} = \mathbf{0}. \tag{9.4.8}$$

Two scalar boundary conditions must be specified over each boundary.

Temporal and spatial instability

When the wave number $\mathbf{k}$ is real and the angular velocity ϱ is complex, we obtain a spatially periodic disturbance evolving in time, corresponding to temporal instability. When $\mathbf{k}$ is complex and ϱ is real, we obtain a disturbance evolving in space while its amplitude at a particular point in the flow oscillates in time, corresponding to spatial instability. In the more general case where both $\mathbf{k}$ and ϱ are complex, we obtain a disturbance that evolves both in time and space, corresponding to a hybrid instability. The solution of the general problem can be constructed by superposing temporal and spatial modes.

9.4.1   Squire's theorems

Before tackling the solution of the linearized equations for normal-mode perturbations, we digress to discuss Squire's theorems relating the behavior of three-dimensional spatially periodic perturbations evolving in time to the behavior of equivalent two-dimensional perturbations evolving in the xy plane that is perpendicular to the vorticity of the base flow [384].

In the case of two-dimensional disturbances in the xy plane, $k_z = 0$, the x and y components of (9.4.8) are trivially satisfied and the z component becomes

$$\left[\mathrm{i}\left(\varrho - U k_x\right) - \nu\, k_x^2 \right] Q_z + \nu\, Q_z'' + U'' F_y = 0, \tag{9.4.9}$$

where Q_z is given by the third equation in (9.4.7).

Returning to the three-dimensional problem, we introduce the unit wave number vector, $\hat{\mathbf{k}}$, and its reciprocal vector, $\hat{\mathbf{l}}$, given by

$$\hat{\mathbf{k}} = \frac{1}{k}\left(k_z, 0, k_z\right), \qquad \hat{\mathbf{l}} = \frac{1}{k}\left(-k_z, 0, k_z\right), \tag{9.4.10}$$

defined such that $\hat{\mathbf{k}} \cdot \hat{\mathbf{l}} = 0$. where $k \equiv |\mathbf{k}|$. Assuming that the wave number is real and projecting equation (9.4.8) onto the reciprocal unit vector, $\hat{\mathbf{l}}$, we obtain

$$\left[\mathrm{i}\left(\varrho\frac{k}{k_x} - Uk\right) - \nu\frac{k}{k_x} k^2 \right] \mathcal{J} + \frac{k}{k_x}\nu\, \mathcal{J}'' + U'' F_y = 0. \tag{9.4.11}$$

We have introduced the component of the vorticity vector in the direction of the reciprocal vector,

$$\mathcal{J} \equiv \mathbf{Q} \cdot \hat{\mathbf{l}} = \mathrm{i}\,k\,F_y - \frac{1}{k}\left(k_x F_x' + k_z F_z'\right). \tag{9.4.12}$$

Now we make the substitutions

$$\tilde{\varrho} = \frac{k}{k_x}\,\varrho, \qquad \tilde{k}_x = k, \qquad \tilde{k}_z = 0, \qquad \tilde{\nu} = \nu\,\frac{k}{k_x}, \tag{9.4.13}$$

and

$$\tilde{F}_x = \frac{1}{k}\left(k_x F_x + k_z F_z\right), \qquad \tilde{F}_y = F_y, \qquad \tilde{F}_z = 0, \qquad \tilde{Q}_z = \mathcal{J}, \tag{9.4.14}$$

and observe that equations (9.4.11) and (9.4.12) reduce to equation (9.4.9) and the third equation in (9.4.7) written for the tilde variables and physical parameters. This means that the study of three-dimensional perturbations can be reduced to the study of two-dimensional perturbations with a suitable change of the wavelength of the perturbation and kinematic viscosity of the fluid determining the Reynolds number of the base flow.

We conclude that, to assess the stability of a unidirectional flow with uniform physical properties, it is sufficient to consider two-dimensional disturbances. Once the tilded variables corresponding to an equivalent two-dimensional problem are available, the non-tilded variables corresponding to the three-dimensional physical problem are recovered using relations (9.4.13) and (9.4.14). It should be emphasized that Squire's theorem may not be valid when the boundaries of the flow deform in response to a perturbation, and are applicable only for spatially periodic modes corresponding to temporal stability.

Critical Reynolds number

The third definition in (9.4.13) shows that the kinematic viscosity of the fluid for the tilde variables corresponding to the equivalent two-dimensional problem is higher than for the non-tilde variables corresponding to the three-dimensional problem. Accordingly, the Reynolds number of the flow in the former variables is lower than that in the latter variables. This observation provides us with a basis for Squire's theorem for viscous flow: to compute the maximum Reynolds number for stability, it is sufficient to consider two-dimensional disturbances whose wave number vector points in the direction of the base flow.

Inviscid flow

The first definition in (9.4.13) shows that the growth rate in the equivalent two-dimensional problem is higher than that in the physical three-dimensional problem. When the fluid is inviscid, the two problems occur at the same infinite Reynolds number, yielding Squire's theorem for inviscid flow: for every unstable three-dimensional perturbation, we can find another two-dimensional perturbation with different wavelength that is more unstable.

9.4.2 The Orr–Sommerfeld equation

Motivated by Squire's theorem, we consider a unidirectional flow along the x axis, restrict our attention to two-dimensional disturbances in the xy plane, and seek to compute the eigenvalues and eigenfunctions associated with the normal modes. To reduce the number of unknown functions, we introduce a vector potential for the velocity, $\mathbf{F}$, writing

$$\mathbf{F}(\mathbf{x}) = \nabla \times (q\,\mathbf{e}_z), \qquad Q_z(\mathbf{x}) = -\nabla^2 q, \qquad (9.4.15)$$

where $q(\mathbf{x})$ is a complex function playing the role of a stream function and $\mathbf{e}_z$ is the unit vector along the z axis. To conform with the periodicity of the flow, we set

$$q(x, y) = f(y)\,e^{ikx}, \qquad (9.4.16)$$

where $f(y)$ is a complex function with dimensions of velocity multiplied by length. To simplify the notation, we have set $k = k_x$. Substituting expression (9.4.16) into (9.4.15) and then into (9.4.9), we obtain a second-order linear ordinary differential equation,

$$w'' - \left[k^2 + i\frac{k}{\nu}(U - c) \right] w = i\frac{k}{\nu} U'' f, \qquad (9.4.17)$$

where a prime denotes a derivative with respect to y. The auxiliary function

$$w \equiv -f'' + k^2 \qquad (9.4.18)$$

provides us with the amplitude of the vorticity of the disturbance flow.

Substituting (9.4.18) into (9.4.17), we obtain a fourth-order linear homogeneous ordinary differential equation,

$$f'''' - 2k^2 f'' + k^4 f = i\frac{k}{\nu} \left[(U - c)(f'' - k^2 f) - U'' f \right], \qquad (9.4.19)$$

derived independently by Orr [283] in 1907 and by Sommerfeld [378] in 1908, known as the Orr–Sommerfeld equation, where $c \equiv \varrho/k$ is the complex phase velocity.

Dimensionless form

Nondimensionalizing all variables using as characteristic length L, velocity V, and time L/V, as discussed in Section 3.10, we recast the Orr–Sommerfeld equation into the dimensionless form

$$\hat{f}'''' - 2\hat{k}^2 \hat{f}'' + \alpha^4 \hat{f} = i \operatorname{Re} \hat{k} \left[(\hat{U} - \hat{c})(f'' - \alpha^2 f) - \hat{U}'' f \right]. \qquad (9.4.20)$$

We have introduced the following dimensionless variables and constants,

$$\hat{f} = \frac{f}{VL}, \qquad \hat{U} = \frac{U}{L}, \qquad \hat{c} = \frac{c}{V}, \qquad \hat{\mathbf{x}} = \frac{\mathbf{x}}{L}, \qquad \hat{k} = kL, \qquad \operatorname{Re} = \frac{VL}{\nu}. \qquad (9.4.21)$$

A prime in (9.4.20) indicates a derivative with respect to the dimensionless transverse position, $\hat{y}$.

9.4.3 Temporal instability

In the temporal instability problem, a real wave number, k, is specified, and the Orr–Sommerfeld equation is regarded as a linear ordinary differential equation involving an unspecified complex eigenvalue, c. The real part of c is the phase velocity of the perturbation. Having obtained c, we compute the complex growth rate, $\varrho = kc$, identify its real and imaginary parts, $\varrho = \varrho_R + i\varrho_I$, and extract the growth rate of the perturbation, $\varrho_I = kc_I$. At neutral stability, $\varrho_I = 0$, both c and ϱ are real. In dimensionless variables, the solution of the temporal stability problem depends on the Reynolds number, Re, and dimensionless wave number, $\hat{k}$.

If the velocity U in (9.4.19) is replaced by $U - U_0$, where U_0 is an arbitrary constant, the eigenvalues of the Orr–Sommerfeld equation will be shifted from c to $c - U_0$. Physically, referring to a frame of reference that translates steadily in the direction of the flow with velocity U_0 changes the phase velocity of the perturbation but leaves the growth rate unaffected, in agreement with physical intuition.

(a)   (b)

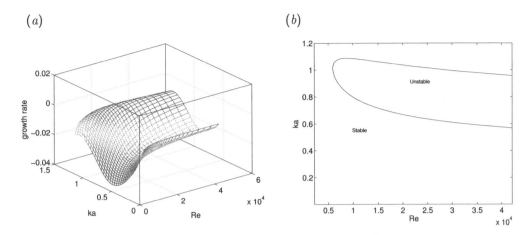

FIGURE 9.4.1   (a) Dependence of the highest growth rate scaled by V/a on the Reynolds number, Re, and scaled wave number, ka, for plane Hagen–Poiseuille flow. (b) Stability phase diagram separating stable from unstable normal modes.

Plane Poiseuille flow

As an example, we consider plane Hagen–Poiseuille flow along the x axis in a channel confined between two parallel plates located at $y = \pm a$ (e.g., [369]). The velocity of the undisturbed flow is described by the parabolic profile

$$U(y) = U_0 \left(1 - \frac{y^2}{a^2} \right), \tag{9.4.22}$$

where U_0 is the centerline velocity. The Reynolds number is defined as $\mathrm{Re} = U_0 a/\nu$ where ν is the kinematic viscosity of the fluid. A graph of the highest growth rate representing the most unstable mode, scaled by V/a, is shown in Figure 9.4.1(a), and a stability phase diagram in the (Re, ka) plane is shown Figure 9.4.1(b). The complex phase velocity was computed using a finite-difference method discussed in Section 9.7. The neutral stability curve where $c_I = 0$ separates stable from unstable modes in Figure 9.4.1(b). The flow is stable below a critical Reynolds, $\mathrm{Re}_c \simeq 5772$ for any wave number, as discussed in Section 9.8.5.

9.4.4   Spatial instability

In the spatial instability problem, we set $k = k_R + \mathrm{i}k_I$ and $\varrho = \varrho_R$, where k_R is the real wave number of the perturbation, $-k_I$ is the corresponding spatial growth rate, and ϱ_R is the real cyclic frequency or angular velocity of the perturbation. Specifying ϱ_R provides us with an eigenvalue problem for the complex wave number, k. At neutral stability, $k_I = 0$ and k is real. In dimensionless variables, the solution of the spatial stability problem depends on the Reynolds number, Re, and dimensionless real cyclic frequency $L\varrho_R/V$, where L and V are characteristic length and velocity scales.

9.4.5   Relation between the temporal and the spatial instability

The temporal and spatial stability problems are based on different expectations for the growth or decay of perturbations–evolution in time, versus evolution in space. At neutral stability, the wavenumber, k, and growth rate, ϱ, are both real and the solutions of the two problems are identical. This means that the threshold wave number, cyclic frequency, and Reynolds number that separated regions or stable and unstable flow are all the same.

Considering the general case where k and ϱ are both complex, we introduce dimensionless variables and express the solution of the eigenvalue problem associated with the Orr–Sommerfeld equation in the functional form

$$\mathcal{F}(\hat{\varrho}, \hat{k}, \mathrm{Re}) = 0, \qquad (9.4.23)$$

where $\hat{\varrho} = L\varrho/V$ is a dimensionless complex growth rate and $\hat{k} = kL$ is the dimensionless complex wave number. Assuming that $\mathcal{F}$ is an analytic function of its arguments, we follow a solution branch and write

$$\mathrm{d}\mathcal{F} = \frac{\partial \mathcal{F}}{\partial \hat{\varrho}}\, \mathrm{d}\hat{\varrho} + \frac{\partial \mathcal{F}}{\partial \hat{k}}\, \mathrm{d}\hat{k} + \frac{\partial \mathcal{F}}{\partial \mathrm{Re}}\, \mathrm{d}\mathrm{Re} = 0. \qquad (9.4.24)$$

Next, we denote the real wave number of the temporal stability problem for neutral stability at a particular Reynolds number by $\hat{k}_0$, and the corresponding dimensionless real cyclic frequency of the spatial stability problem for neutral stability at the same Reynolds number by $\hat{\varrho}_0$, where, by definition, $\mathcal{F}(\hat{k}_0, \hat{\varrho}_0, \mathrm{Re}) = 0$. If the Reynolds number is changed by a differential amount, dRe, while $\hat{k}_0$ is held constant, $\hat{\varrho}_0$ will change by $\mathrm{d}\hat{\varrho}$ so that

$$\frac{\partial \mathcal{F}}{\partial \hat{\varrho}}\, (\mathrm{d}\hat{\varrho})_{\hat{k}} + \frac{\partial \mathcal{F}}{\partial \mathrm{Re}}\, \mathrm{d}\mathrm{Re} = 0. \qquad (9.4.25)$$

If the Reynolds number is changed by a differential amount, dRe, while ϱ_0 is held constant, $\hat{k}_0$ will change by $\mathrm{d}\hat{k}$ so that

$$\frac{\partial \mathcal{F}}{\partial \hat{k}}\, (\mathrm{d}\hat{k})_{\hat{\varrho}} + \frac{\partial \mathcal{F}}{\partial \mathrm{Re}}\, \mathrm{d}\mathrm{Re} = 0. \qquad (9.4.26)$$

Combining equations (9.4.25) and (9.4.26) to eliminate dRe, we obtain

$$(\mathrm{d}\hat{\varrho})_{\hat{k}} = \frac{\partial \mathcal{F}/\partial \hat{k}}{\partial \mathcal{F}/\partial \hat{\varrho}}\, (\mathrm{d}\hat{k})_{\hat{\varrho}} = -\left(\frac{\partial \hat{\varrho}}{\partial \hat{k}}\right)_{\mathrm{Re}} (\mathrm{d}\hat{k})_{\hat{\varrho}} \equiv c_G\, \mathrm{d}\hat{k}, \qquad (9.4.27)$$

where $c_G = (\partial \hat{\varrho}/\partial \hat{k})_{\mathrm{Re}}$ is the complex group velocity of the spatially growing waves under conditions of neutral stability [141]. Equation (9.4.27) can be used to extract information on the temporal stability problem using results from the spatial stability problem, and *vice versa*.

9.4.6   Base flow with linear velocity profile

Consider a shear flow with linear velocity profile, $U(y) = \xi y + U_0$, where ξ is a constant shear rate and U_0 is a constant reference velocity. The second derivative of the velocity profile U'' vanishes throughout the domain of flow and the right-hand side of the vorticity transport equation (9.4.17) is identically zero. To simplify the functional form of (9.4.17), we introduce the new variable

$$\eta \equiv \left(\frac{\nu}{k\xi} \right)^{2/3} \left[k^2 + \mathrm{i}\, \frac{k}{\nu}\, (\xi y + U_0 - c) \right], \tag{9.4.28}$$

which is a linear function of y. Substituting (9.4.28) into (9.4.17), we obtain the Stokes equation

$$\frac{\mathrm{d}^2 w}{\mathrm{d}\eta^2} + \eta w = 0, \tag{9.4.29}$$

whose solution can be found in terms of contour integrals in the complex plane (e.g., [123]).

9.4.7   Computation of the disturbance flow

Having solved the eigenvalue problem, we combine expressions (9.3.5), (9.4.4), (9.4.15), and (9.4.16), and derive the stream function of the disturbance flow associated with a normal mode,

$$\psi^{NM}(x, y, t) = f(y)\, \exp[\mathrm{i}(kx - \varrho t)] = f(y)\, \exp[\mathrm{i}k(x - ct)]. \tag{9.4.30}$$

To expedite the study of the linear stability problem, the functional form (9.4.30) is sometimes assumed at the outset. Differentiating the stream function with respect to y or x, we derive the two normal-mode disturbance velocity components,

$$\begin{bmatrix} u_x \\ u_y \end{bmatrix}^{NM} = \begin{bmatrix} f'(y) \\ -\mathrm{i}\,kf(y) \end{bmatrix} \exp[\mathrm{i}(kx - \varrho t)], \tag{9.4.31}$$

and the disturbance vorticity,

$$\omega_z^{NM}(x, y, t) = \left[f''(y) + k^2 f(y) \right] \exp[\mathrm{i}(kx - \varrho t)]. \tag{9.4.32}$$

To obtain the corresponding normal-mode pressure, we substitute these expressions into the linearized Navier–Stokes equation (9.4.2) and integrate the resulting first-order differential equation.

Problem

9.4.1 *Squire's theorem*

Discuss whether Squire's theorem applies to the case of film flow down an inclined plane where the free surface may deform in response to a perturbation.

9.5   The Rayleigh equation for inviscid fluids

When the Reynolds number on the right-hand side of (9.4.20) is large or the fluid is effectively inviscid, viscous forces are negligible and the Orr–Sommerfeld equation reduces to the Rayleigh equation corresponding to the Euler equation,

$$(U - c)(f'' - k^2 f) - U'' f = 0, \tag{9.5.1}$$

where a prime denotes a derivative with respect to y. Rearranging, we derive the standard form of a second-order linear ordinary differential equation with variable coefficients,

$$f'' - \left(k^2 + \frac{U''}{U-c}\right)f = 0,$$ (9.5.2)

where k, or c, or both are allowed to be complex. For future reference, we introduce the variable $q \equiv f/(U-c)$, and recast (9.5.2) into the form

$$[(U-c)^2 q']' - k^2 (U-c)^2 q = 0.$$ (9.5.3)

The first term on the left-hand side involves a linear self-adjoint operator, $\mathcal{L}(q) \equiv [(U-c)^2 q']'$.

Riccati equation

To expedite the solution, it is sometimes useful to introduce the variable $F \equiv f'/f$ and rearrange (9.5.2) into the Riccati equation,

$$F' = -F^2 + k^2 + \frac{U''}{U-c}.$$ (9.5.4)

The reduction in the order of Rayleigh's equation from two to one is offset by the occurrence of a nonlinearity due to the quadratic term on the right-hand side.

Critical layer

The second term inside the parentheses on the left-hand side of the Rayleigh equation (9.5.2) becomes singular at the critical layer where the the unperturbed fluid velocity is equal to the phase velocity of the perturbation, $U = c$. For this to occur, c must be real, which means that the flow is neutrally stable. The associated disturbances fall within the continuous part of the spectrum (e.g., [67]). We will see that the occurrence of this singularity has important implications on the efficiency of numerical methods used to compute normal modes.

9.5.1 Base flow with linear velocity profile

In the case of a base flow with linear velocity profile, $U(y) = \xi y + U_0$, where ξ is a constant shear rate and U_0 is a constant reference velocity. Rayleigh's equation (9.5.1) simplifies into

$$(U-c)(f'' - k^2 f) = 0,$$ (9.5.5)

which is satisfied by any solution of the one-dimensional Helmholtz equation, $f'' - k^2 f = 0$. The associated eigenvalues provide us with the discrete part of the spectrum. Equation (9.4.32) shows that the corresponding eigenfunctions represent irrotational flow. Using (9.4.31), we find that the associated normal-mode velocity potential is given by

$$\phi^{NM}(x,y,t) = -\frac{i}{k} f'(y) \exp[ik(x-ct)].$$ (9.5.6)

These eigensolutions will be used in Section 9.6 to determine the stability of a vortex layer with constant vorticity separating two uniform streams.

Continuous spectrum

A second family of solutions of (9.5.5) corresponding to the continuous part of the spectrum emerges by allowing the expression inside the second set of parentheses on the left-hand side to be singular when $U = c$. The singularity occurs at the position $y = \eta \equiv (U - U_0)/\xi$, which varies between the lower and the upper boundary of the flow. Since c is real, the corresponding normal modes are neutrally stable.

The eigensolutions can be expressed as $f(y) = \xi a \mathcal{G}(y)$, where a is a reference length. The function $\mathcal{G}$, with dimensions of length, satisfies the differential equation

$$\mathcal{G}'' - k^2 \mathcal{G} + \delta(y - \eta) = 0, \tag{9.5.7}$$

where δ is the one-dimensional delta function with dimensions of inverse length. The solution of (9.5.7) is the Green's function of the one-dimensional Helmholtz equation forced at the position where the phase velocity becomes equal to the fluid velocity of the base linear flow. Referring to (9.4.32), we find that the disturbance represents the flow due to a flat vortex sheet with sinusoidal strength situated along the x axis at the position $y = \eta$. An acceptable solution of (9.5.7) must be continuous with discontinuous derivatives at $y = \eta$, satisfy the homogeneous equation $\mathcal{G}'' - k^2 \mathcal{G} = 0$ everywhere except at $y = \eta$, respect the specified kinematic conditions at the boundaries of the flow, and conform with the jump condition

$$\mathcal{G}'(\eta_+) - \mathcal{G}'(\eta_-) = -1, \tag{9.5.8}$$

where the plus and minus subscripts indicate evaluation immediately above and below the critical level $y = \eta$ (e.g., [67]).

9.5.2 Inviscid Couette flow

As an application, we investigate the stability of inviscid plane Couette flow in a channel confined between two parallel walls located at $y = \pm a$, discussed earlier in Section 9.2. The velocity profile is $U(y) = \xi y$, where ξ is a constant shear rate, and the origin of the y axis has been set midway between the walls so that $-a \leq y \leq a$. Because a solution of the Helmholtz equation $f'' - k^2 f = 0$ consistent with the no-penetration condition $f = 0$ at $y = \pm a$ cannot be found, the discrete part of the spectrum is null. Physically, when the normal component of the boundary velocity is zero, an irrotational velocity field cannot be established in a confined domain.

Continuous spectrum

The general solution of (9.5.7), corresponding to the continuous part of spectrum , is parametrized by the elevation $\eta = U/\xi$ where $c = U$, ranging from $-a$ to a. We find that

$$\mathcal{G}(k, y, \eta) = \frac{1}{k \sinh(2ka)} \begin{cases} \sinh[k(a + \eta)] \sinh[k(a - y)] & \text{for} \quad \eta < y < a, \\ \sinh[k(a - \eta)] \sinh[k(a + y)] & \text{for} \quad -a < y < \eta. \end{cases} \tag{9.5.9}$$

It is a straightforward exercise to verify that (9.5.9) satisfies the four conditions stated at the end of Section 9.5.1 (Problem 9.5.2).

Since $c = \xi\eta$, the complex stream function corresponding to an arbitrary perturbation can be expressed in terms of the eigenfunctions (9.5.9) by superposition as

$$\psi(x, y, t) = -\frac{1}{2\pi} \int_{-\infty}^{\infty} \int_{-a}^{a} q(k, \eta)\, \mathcal{G}(k, y, \eta)\, \exp[\mathrm{i}k(x - \xi\eta t)]\, \mathrm{d}\eta\, \mathrm{d}k, \tag{9.5.10}$$

where the dimensionless function q is determined by the form of the perturbation at the initial instant. To demonstrate this dependence explicitly, we express the initial stream function as a Fourier integral with respect to x in the form

$$\psi(x, y, t = 0) = \frac{1}{2\pi} \int_{-\infty}^{\infty} \hat{\psi}_0(k, y)\, \mathrm{e}^{\mathrm{i}kx}\, \mathrm{d}k, \tag{9.5.11}$$

where $\hat{\psi}_0(k, y)$ is the one-dimensional Fourier transform of the initial stream function with respect to x. Comparing (9.5.11) with (9.5.10), we write

$$\hat{\psi}_0(k, y) = -\int_{-a}^{a} q(k, \eta)\, \mathcal{G}(k, y, \eta)\, \mathrm{d}\eta. \tag{9.5.12}$$

Operating on (9.5.12) with $\partial^2/\partial y^2 - k^2$, switching the order of the second derivative and the integral sign on the right-hand side, and using (9.5.7) and the distinguishing properties of the delta function, we find that

$$q(k, y) = \left(\frac{\partial^2}{\partial y^2} - k^2\right) \hat{\psi}_0(k, y), \tag{9.5.13}$$

which provides us with a relation between q and the Fourier transform, $\hat{\psi}_0$. Remembering that $\omega_z = -\nabla^2\psi$ and taking the Laplacian of (9.5.11), we find that $\xi a q(k, y)$ is the Fourier transform of the initial vorticity distribution with respect to x.

Fourier expansion

Further progress can be made by expanding $\hat{\psi}_0(k, y)$ in a sine Fourier series with respect to y, similar to that shown in (9.4.2),

$$\hat{\psi}_0(k, y) = 2\pi \sum_{n=1}^{\infty} b_n(k) \sin\left(\frac{n\pi}{2a}(y + a)\right), \tag{9.5.14}$$

where b_n are complex coefficients. Substituting (9.5.14) into the right-hand side of (9.5.13), we obtain a sine Fourier series with respect to y for $q(k, y)$,

$$q(k, y) = -2\pi \sum_{n=1}^{\infty} b_n(k) \left[\left(\frac{n\pi}{2a}\right)^2 + k^2\right] \sin\left(\frac{n\pi}{2a}(y + a)\right). \tag{9.5.15}$$

Finally, we substitute (9.5.15) into the integrand of (9.5.10) and carry out the integration with respect to η to derive the general solution in terms of b_n,

$$\psi(x, y, t) = \sum_{n=1}^{\infty} \int_{-\infty}^{\infty} b_n(k) \left[\left(\frac{n\pi}{2a}\right)^2 + k^2\right] \mathrm{e}^{\mathrm{i}kx}\, \Psi_n(k, y, t)\, \mathrm{d}k, \tag{9.5.16}$$

where

$$\Psi_n(k,y,t) \equiv \int_{-a}^{a} \mathcal{G}(k,y,\eta) \sin\left(\frac{n\pi}{2a}(\eta+a)\right) e^{-ik\xi\eta t} \, d\eta \qquad (9.5.17)$$

is a kernel defined in terms of a Green's function transform.

Green's function transform

To compute the integral on the right-hand side of (9.5.17), we note that $\mathcal{G}'' = \mathcal{G}_{\eta\eta}$, where a subscript η denotes a derivative with respect to η. Multiplying (9.5.7) by $\exp[i(\alpha\eta+\beta)]$ and integrating with respect to η from $-a$ to a, we find that

$$\int_{-a}^{a} \mathcal{G}_{\eta\eta}(k,y,\eta) e^{i(\alpha\eta+\beta)} \, d\eta - k^2 \int_{-a}^{a} \mathcal{G}(k,y,\eta) e^{i(\alpha\eta+\beta)} \, d\eta + e^{i(\alpha y+\beta)} = 0, \qquad (9.5.18)$$

where α and β are two constants. Integrating by parts, we obtain

$$\left[\mathcal{G}_\eta(k,y,\eta) e^{i(\alpha\eta+\beta)}\right]_{\eta=-a}^{\eta=a} - (\alpha^2+k^2) \int_{-a}^{a} \mathcal{G}(k,y,\eta) e^{i(\alpha\eta+\beta)} \, d\eta + e^{i(\alpha y+\beta)} = 0. \qquad (9.5.19)$$

Rearranging, we obtain

$$\mathcal{F}(k,y,\alpha,\beta) \equiv \int_{-a}^{a} \mathcal{G}(k,y,\eta) e^{i(\alpha\eta+\beta)} \, d\eta = \frac{1}{\alpha^2+k^2}\left(e^{i(\alpha y+\beta)} + \left[\mathcal{G}_\eta(k,y,\eta) e^{i(\alpha\eta+\beta)}\right]_{\eta=-a}^{\eta=a}\right). \qquad (9.5.20)$$

Carrying out the differentiation on the right-hand side, we obtain

$$\mathcal{F}(k,y,\alpha,\beta) = \frac{1}{\alpha^2+k^2}\left(e^{i(\alpha y+\beta)} - e^{i\beta}\frac{1}{\sinh(2ka)}\left[\sinh[k(a+y)]e^{i\alpha a} + \sinh[k(a-y)]e^{-i\alpha a}\right]\right). \qquad (9.5.21)$$

Stream function

We return to (9.5.17), express the sine term in terms of complex exponentials, and compute

$$\Psi_n(k,y,t) = \frac{1}{2i}\Big(\mathcal{F}(k,y,\alpha_n,\beta_n) - \mathcal{F}(k,y,\alpha_{-n},\beta_{-n})\Big), \qquad (9.5.22)$$

where

$$\alpha_n = \frac{n\pi}{2a} - k\xi t, \qquad \beta_n = \frac{n\pi}{2}. \qquad (9.5.23)$$

Substituting the expression for $\mathcal{F}$ and simplifying, we obtain

$$\Psi_n(k,y,t) = \frac{1}{2i}\big[\mathcal{H}_n(k,y,t) - \mathcal{H}_{-n}(k,y,r)\big], \qquad (9.5.24)$$

where

$$\mathcal{H}_n(k,y) = \frac{1}{\left(\frac{n\pi}{2a} - k\xi t\right)^2 + k^2} \left[\exp[\mathrm{i}\left(\frac{n\pi}{2a}(y+a) - k\xi ty\right)] \right. \tag{9.5.25}$$

$$\left. - \frac{1}{\sinh(2ka)} \left(\sinh[k(a+y)]\exp[\mathrm{i}(n\pi - k\xi ta)] + \sinh[k(a-y)]\exp(\mathrm{i}k\xi ta)\right)\right].$$

Substituting (9.5.24) into (9.5.16) and defining the negative coefficients $b_{-n}(k) \equiv -b_n(k)$, we obtain

$$\psi(x,y,t) = \frac{1}{2\mathrm{i}} \sum_{n=-\infty}^{\infty}{}' \int_{-\infty}^{\infty} \frac{\left(\frac{n\pi}{2a}\right)^2 + k^2}{\left(\frac{n\pi}{2a} - k\xi t\right)^2 + k^2} b_n(k)\, \mathcal{J}_n(k,y,t)\, \mathrm{d}k, \tag{9.5.26}$$

where a prime indicates that the term $n = 0$ is excluded from the sum, and

$$\mathcal{J}_n(k,y) = \exp\left[\mathrm{i}\left(\frac{n\pi}{2a}(y+a) + k(x - \xi ty)\right)\right] \tag{9.5.27}$$

$$- \frac{1}{\sinh(2ka)}\left(\sinh[k(a+y)]\exp[\mathrm{i}(k(x - \xi ta) + n\pi)] + \sinh[k(a-y)]\exp[\mathrm{i}k(x + \xi ta)]\right).$$

The imaginary part of $\mathcal{J}_n$ is precisely the function Φ_n given in (9.2.21).

We have confirmed that the set of the normal modes falling within the continuous part of the spectrum comprises a complete set.

9.5.3   General theorems on the temporal instability of shear flows

Several general theorems allow us to assess the temporal stability of inviscid unidirectional flows and obtain estimates for the location of the phase velocity in the complex plane by mere inspection.

Rayleigh criterion on the significance of an inflection point

Multiplying both sides of (9.5.2) by f^* and rearranging, we obtain

$$(f'f^*)' - |f'|^2 = \left(k^2 + \frac{U''}{U-c}\right)|f|^2, \tag{9.5.28}$$

where an asterisk denotes the complex conjugate and a prime denotes a derivative with respect to y. Assuming that the flow is confined between two impenetrable planar boundaries located at $y = a$ and b, we integrate (9.5.28) with respect to y from a to b and enforce the no-penetration condition $f(a) = f(b) = 0$ to obtain

$$-\int_a^b |f'|^2\, \mathrm{d}y = \int_a^b \left(k^2 + \frac{U''}{|U-c|^2}(U-c^*)\right)|f|^2\, \mathrm{d}y. \tag{9.5.29}$$

Equating the real and imaginary parts of the left-hand and right-hand sides, we find that

$$-\int_a^b |f'|^2\, \mathrm{d}y = \int_a^b \left(k^2 + \frac{U''}{|U-c|^2}(U-c_R)\right)|f|^2\, \mathrm{d}y \tag{9.5.30}$$

and

$$c_I \int_a^b \left(k^2 + \frac{U''}{|U - c|^2} \right) |f|^2 \, \mathrm{d}y = 0. \qquad (9.5.31)$$

The last equation requires that either $c_I = 0$, corresponding to neutral stability, or the integral on the left-hand side is zero, requiring that U'' changes sign at least once between $y = a$ and b. We conclude that, for a normal mode of a unidirectional shear flow to be unstable, the velocity profile must exhibit at least one inflection point [335]. This is a necessary but not sufficient condition, that is, a normal-mode disturbance of a unidirectional shear flow whose velocity profile has an inflection point is not necessarily unstable.

An important consequence of Rayleigh's criterion is that the normal modes of infinite simple shear flow, Couette or Poiseuille channel flow, are stable. Curious though it may seem, viscous forces are required to render perturbations unstable. Physically, viscous stresses sustain a perturbation for a longer period of time, thereby giving them a better chance to grow.

Fjørtoft's condition for instability

We can go beyond Rayleigh's criterion by combining (9.5.30) and (9.5.31) to find that, unless a perturbation is neutrally stable, we must have

$$\int_a^b (U - U_0) \frac{U''}{|U - c|^2} (U - c^*) |f|^2 \, \mathrm{d}y = - \int_a^b (k^2 |f|^2 + |f'|^2) \, \mathrm{d}y < 0, \qquad (9.5.32)$$

where U_0 is an arbitrary constant [128]. Consider a flow whose velocity profile has a single inflection point, and identify U_0 with the velocity at the inflection point. The sign of the product $U''(U - U_0)$ is constant throughout the integration domain. For the integral on the left-hand side of (9.5.32) to be negative, the sign of $U''(U - U_0)$ must also be negative, otherwise the disturbance will be neutrally stable. Consequently, for a normal mode of a unidirectional shear flow to be unstable, the maximum of the absolute value of the vorticity of the base flow must occur at an inflection point.

Combining Rayleigh's and Fjørtoft's theorems, we find that the normal modes of the flow shown in Figure 9.5.1(a) are stable, whereas those of the flow shown in Figure 9.5.1(b) may be stable or unstable.

Sufficient condition for instability

Rayleigh's and Fjørtoft's theorems provide us with necessary but not sufficient conditions for instability. Tollmien indicated that, in the case of shear flow in a channel with a symmetric and monotonically varying profile $U(y)$ similar to that observed in boundary layers, these conditions are also sufficient. The proof and further extensions of the theory are discussed in detail by Drazin & Reid [112] and Yih ([437], p. 473).

9.5.4   Howard's semi-circle theorem

Numerical methods for computing the complex phase velocity c may require an accurate initial guess. In some cases, this can be found by using Gershgorin's theorem applicable to standard algebraic

(a)   (b)

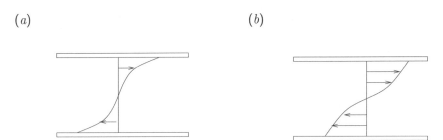

FIGURE 9.5.1   Applications of Rayleigh's theorem on the significance of an inflection point and Fjørtoft's theorem. Normal modes in flow (a) are stable, but those in flow (b) may be unstable.

eigenvalue problems (Problem 9.5.4) [317]. A more general but less accurate method of locating eigenvalues is provided by Howard's semi-circle theorem stating that c must fall inside a half-disk in the upper half complex plane [192]. The center of the disk lies at the point $\frac{1}{2}(U_{max} + U_{min})$ on the real axis, and the disk radius is equal to $\frac{1}{2}(U_{max} - U_{min})$, where U_{max} and U_{min} are the maximum and minimum values of U. One consequence of Howard's semi-circle theorem is that there is always a point where the real part of c is equal to U and the disturbance travels with the local fluid velocity. The region around that point is the critical layer.

To prove the semi-circle theorem, we multiply (9.5.3) by q^* and integrate the resulting equation from $y = a$ to b, where an asterisk indicates the complex conjugate. Enforcing the no-penetration condition $q(a) = 0$ and $q(b) = 0$, we obtain

$$\int_a^b (U - c)^2 \Phi \, dy = 0, \tag{9.5.33}$$

where $\Phi \equiv |q'|^2 + k^2|q^2|$ is a non-negative function. Decomposing the integral into its real and imaginary parts, we obtain

$$\int_a^b (U^2 - 2c_R U + c_R^2 - c_I^2) \Phi \, dy = 0, \qquad c_I \int_a^b (U - c_R) \Phi \, dy = 0. \tag{9.5.34}$$

The second equation is satisfied for neutrally stable perturbations, $c_I = 0$. Leaving this case aside, we set the second integral to zero and find that the phase velocity c_R must take values between U_{max} and U_{min}, which demonstrates that a disturbance cannot move faster than the fluid [335].

Next, we work in two stages. First, we use the second equation in (9.5.34) to simplify the first equation, obtaining

$$\int_a^b (U^2 - c_R^2 - c_I^2) \Phi \, dy = 0. \tag{9.5.35}$$

Second, we make the independent observation that, since Φ is non-negative,

$$\int_a^b (U - U_{min})(U - U_{max}) \Phi \, dy < 0, \tag{9.5.36}$$

and use the second equation in (9.5.34) to restate (9.5.36) as

$$\int_a^b \left[U^2 - c_R(U_{max} + U_{min}) + U_{max}U_{min} \right] \Phi \, dy < 0. \tag{9.5.37}$$

Finally, we combine (9.5.35) with (9.5.37) and obtain

$$\int_a^b \left[c_R^2 + c_I^2 - c_R(U_{max} + U_{min}) + U_{max}U_{min} \right] \Phi \, dy < 0 \tag{9.5.38}$$

or

$$\int_a^b \left[\left(c_R - \frac{U_{max} + U_{min}}{2} \right)^2 + c_I^2 - \left(\frac{U_{max} - U_{min}}{2} \right)^2 \right] \Phi \, dy < 0. \tag{9.5.39}$$

Since Φ is non-negative, the term inside the square brackets in the integrand must be nonpositive, and c must be located inside a disk in the complex plane centered at the real axis at the point $\frac{1}{2}(U_{max} + U_{min})$. The disk radius is $\frac{1}{2}(U_{max} - U_{min})$. The upper half of the disk contains unstable normal modes with positive values, c_I.

Problems

9.5.1 *Neutral stability of a shear layer*

Confirm that $c = 0$ is an eigenvalue and $f(y) = A \, \mathrm{sech}(y/b)$ is the corresponding neutrally stable eigenfunction of Rayleigh's equation for infinite shear flow with velocity profile $U(y) = U_0 \tanh(y/b)$, where A and U_0 are two constants. What are the dimensions of A?

9.5.2 *Green's function of the Helmholtz equation*

Verify that (9.5.9) satisfies the four conditions stated at the end of Section 9.5.1.

9.5.3 *Stability of inviscid flow with sinusoidal velocity profile*

Consider a unidirectional inviscid shear flow with velocity profile $U(y) = U_0 \sin(y/d)$ in a channel confined between two parallel plane walls located at $y = a$ and b, where d is a specified constant length and U_0 is a specified constant velocity. Show that, if there are no values of $d/(n\pi)$ between a and b, where n is an integer, the flow is stable.

9.5.4 *Gerschgorin's theorem*

Show that the eigenvalues of a square $N \times N$ matrix, **A**, must be located inside the union of N disks in the complex plane. The disks are centered at the diagonal elements of **A** and the radii of the disks are equal to the sum of the magnitudes of the off-diagonal elements of the corresponding rows [317].

 Computer Problem

9.5.5 *Inviscid plane Couette flow*

Draw the streamline pattern corresponding to the eigenfunctions (9.5.9) for $b_n = 0$ when $n > 1$ and a function $b_1(k)$ of your choice, at several time instants. Discuss the structure of the evolving flow.

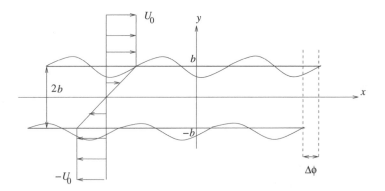

FIGURE 9.6.1   Illustration of a vortex layer with thickness $2b$ separating two uniform streams flowing along the x axis with uniform velocities $\pm U_0$. The horizontal lines represent the upper and lower vortex contours in the unperturbed configuration. The sinusoidal lines represent the upper and lower vortex in the perturbed configuration.

9.6   Instability of a uniform vortex layer

Having established general theorems regarding the temporal instability of unidirectional inviscid shear flows based on the Rayleigh equation, we proceed to consider a case where an exact solution can be found. The results will provide us with insights into the structure of normal modes and reveal important details on the stability properties of general free shear flows with monotonically varying velocity profiles.

Base flow

Consider a homogeneous unbounded shear flow consisting of an infinite a vortex layer with thickness $2b$ and uniform vorticity Ω, as illustrated in Figure 9.6.1, considered by Rayleigh ([337], Vol. II, p. 393). The vortex layer separates two uniform semi-infinite streams translating along the x axis with velocities $\pm U_0$, where $\Omega = -U_0/b$. In the unperturbed state, the velocity profile is flat above and below the vortex layer and varies linearly across the vortex layer. This piecewise linear velocity profile can be regarded as an approximation of a continuous profile described, for example, by the hyperbolic tangent function, $U(y) = U_0 \tanh(y/b)$.

Rayleigh equation

Since $U'' = 0$ everywhere in the flow except at the edges of the vortex layer, the normal modes are governed by the simplified Rayleigh equation (9.5.5). Leaving aside neutrally stable modes with $c = U$ corresponding to the continuous part of the spectrum, we obtain the Helmholtz equation, $f'' - k^2 f = 0$, describing irrotational perturbations, where k is the real wave number. The general solution is

$$f_n(y) = a_n\, e^{ky} + b_n\, e^{-ky}, \tag{9.6.1}$$

where a_n and b_n are constant coefficients, and $n = -1, 0, 1$ indicate, respectively, the region below, inside, and above the vortex layer. To ensure that the disturbance decays far from the vortex layer, we set

$$a_1 = 0, \qquad b_{-1} = 0. \qquad (9.6.2)$$

The disturbance stream function and velocity of the normal modes are given by expressions (9.4.30) and (9.4.31). Since the disturbance flow is irrotational throughout the domain of flow, we may introduce the velocity potential given in (9.5.6), if required.

Deformation of vortex boundaries

The perturbation causes the upper and lower vortex boundaries to deform in response to the flow. The location of the boundaries in the deformed state can be described by the real or imaginary parts of the functions

$$y = \eta_1(x,t) = b + \epsilon A_1 \exp[ik(x - ct)], \qquad y = \eta_{-1}(x,t) = -b + \epsilon A_{-1} \exp[ik(x - ct)], \quad (9.6.3)$$

where ϵ is a small dimensionless number, $A_{\pm 1}$ are two complex constants, the subscript -1 denotes the lower vortex contour, and the subscript 1 denotes the upper vortex contour.

Continuity of the y velocity component

To compute the eight unknowns comprised of $A_{\pm 1}$ and a_n, b_n, for $n = -1, 0, 1$, we require six equations in addition to (9.6.2). Two equations emerge by requiring that the y component of the velocity is continuous across the boundaries of the perturbed vortex layer. Equation (9.4.31) shows that this will be true provided that

$$f_{-1}(\eta_{-1}) = f_0(\eta_{-1}), \qquad f_0(\eta_{-1}) = f_1(\eta_1). \qquad (9.6.4)$$

To first order in ϵ, we obtain

$$f_{-1}(-b) = f_0(-b), \qquad f_0(b) = f_1(b). \qquad (9.6.5)$$

Using expressions (9.6.1), we find that

$$a_0 + \beta\, b_0 = a_{-1}, \qquad \beta\, a_0 + b_0 = b_1, \qquad (9.6.6)$$

where $\beta \equiv \exp(2kb)$ is a dimensionless coefficient.

Continuity of the x velocity component

Next, we require that the x component of the disturbance velocity is continuous across the boundaries of the vortex layer. To implement this condition, we evaluate the normal-mode velocity at $y = \eta_{\pm 1}$ using equations (9.4.31), expand the resulting expressions in a Taylor series about $y = \pm b$, and retain only linear terms with respect to ϵ. For the upper boundary, we find that

$$u_x(y = \eta_1) = U(y = \eta_1) + \epsilon u_x^{NM}(y = \eta_1) \simeq U(y = b) + \left(\frac{dU}{dy}\right)_{y=b}(\eta_1 - b) + \epsilon u_x^{NM}(y = b). \quad (9.6.7)$$

Noting that $U(y = b) = -\Omega b$ and making substitutions, we obtain

$$u_x(y = \eta_1) = -\Omega b + \epsilon \left(A_1 \frac{dU}{dy} + f' \right)_{y=b} \exp[ik(x - ct)], \tag{9.6.8}$$

where $dU/dx = -\Omega$ inside the vortex layer and $dU/dx = 0$ outside the vortex layer. Continuity of velocity then requires that

$$f_1'(b) = f_0'(b) - \Omega A_1. \tag{9.6.9}$$

Working in a similar manner with the lower vortex boundary, we obtain that

$$f_{-1}'(-b) = f_0'(-b) - \Omega A_{-1}. \tag{9.6.10}$$

Substituting (9.6.1) into (9.6.9) and (9.6.10), we derive the relations

$$V A_1 - \beta a_0 + b_0 = b_1, \qquad V A_{-1} - a_0 + \beta b_0 = -a_{-1}, \tag{9.6.11}$$

where $V \equiv \Omega e^{kb}/k$ is a convenient intermediate velocity.

Kinematic compatibility

Since the boundaries of the vortex layer are material lines convected by the flow,

$$\frac{D}{Dt} \left(\eta_{\pm 1}(x, t) - y \right) = 0, \tag{9.6.12}$$

where D/Dt is the material derivative. Expanding all terms in Taylor series with respect to ϵ and retaining only the linear contributions, we find that

$$\frac{\partial \eta_{\pm 1}}{\partial t} + U(y = \pm b) \frac{\partial \eta_{\pm 1}}{\partial x} - \epsilon u_y^{NM}(y = \pm b) = 0. \tag{9.6.13}$$

Substituting (9.6.3) and (9.4.31) into (9.6.13), we obtain

$$A_1 = \frac{f_1(b)}{c - U_0} = b_1 \frac{e^{-kb}}{c - U_0}, \qquad A_{-1} = \frac{f_{-1}(-b)}{c + U_0} = a_{-1} \frac{e^{-kb}}{c + U_0}. \tag{9.6.14}$$

Now we are in a position to compute the complex phase velocity and associated growth rate.

Computation of the growth rate

Equations (9.6.2), (9.6.6), (9.6.11), and (9.6.14) provide us with a system of eight linear homogeneous equations for eight unknown coefficients. Setting the determinant of the coefficient matrix to zero to ensure the existence of a nontrivial solution, provides us with a venue for the computation of the complex phase velocity, c. To solve the linear system, we substitute (9.6.14) into (9.6.11) and obtain

$$\beta a_0 - b_0 = b_1 \left(-1 + \frac{\Omega}{k(c - U_0)} \right), \qquad a_0 - \beta b_0 = a_{-1} \left(1 + \frac{\Omega}{k(c + U_0)} \right). \tag{9.6.15}$$

Finally we substitute (9.6.6) into (9.6.15) and derive a homogeneous system of two linear equations,

$$
\begin{bmatrix}
\Omega & \beta(2kc + 2kU_0 + \Omega) \\
\beta(2kc - 2kU_0 - \Omega) & -\Omega
\end{bmatrix}
\cdot
\begin{bmatrix}
a_0 \\
b_0
\end{bmatrix}
= \mathbf{0}.
\tag{9.6.16}
$$

Setting the determinant of the matrix on the left-hand side to zero provides us with a quadratic equation for the complex phase velocity whose solution is

$$
c = \pm \frac{U_0}{2kb} \left[(1 - 2kb)^2 - e^{-4kb} \right]^{1/2}.
\tag{9.6.17}
$$

The quantity under the radical is negative when $kb < 0.639 \cdots$, and positive otherwise. In the first case, c is purely imaginary and the plus sign in (9.6.17) yields an unstable mode with growth rate $\varrho_I = kc$ and vanishing phase velocity. The minus sign produces a companion stable normal mode. When $kb > 0.639 \cdots$, c is real and the normal modes translate upstream or downstream with phase velocity c, while maintaining their initial amplitude. The lack of dissipation in an inviscid fluid prevents the decay of the kinetic energy of a perturbation.

Flow instability

We have found that a normal-mode disturbance can be unstable only when the scaled wave number, kb, is less than approximately 0.639, which means that the ratio between the wave length and the layer thickness, $L/(2b)$, is larger than approximately 4.92. Normal-mode disturbances with shorter wavelengths travel along the vortex layer with phase velocity that depends on the scaled wave number, kb. The significance of these results on the behavior of general spatially periodic perturbations that do not necessarily represent normal modes is discussed in Section 9.6.3.

Growth rate and phase velocity

The imaginary part of the phase velocity and growth rate of unstable normal modes are plotted with the dashed and solid lines in Figure 9.6.2(a) against the scaled wave number, kb. The scaled imaginary part of the phase velocity, c_I/U_0, decreases from unity when $kb = 0$ to zero at the critical threshold for instability, $kb \simeq 0.639$. The scaled growth rate, $4b\varrho_I/U_0 = 4kbc_I/U_0$, reaches a maximum at the most unstable or most dangerous normal mode corresponding to $kb \simeq 0.398$. The dotted line in Figure 9.6.2(a) displays the dimensionless phase velocity of stable normal modes, c_R/U_0, corresponding to the plus sign in equation (9.6.17). As kb increases beyond the stability threshold, the phase velocity tends to the velocity of the upper stream so that the ratio c_R/U_0 tends to unity. If the minus sign were chosen instead in equation (9.6.17), the phase velocity would have opposite sign.

9.6.1   Waves on vortex boundaries

It is illuminating to examine the deformation of the upper and lower vortex contours due to a normal-mode perturbation. To compute the ratio of the complex amplitudes of the waves on the upper and lower vortex boundaries, we divide equations (9.6.14), and use equations (9.6.6) and (9.6.16) to obtain

$$
\frac{A_{-1}}{A_1} = \frac{a_{-1}}{a_1} \frac{c - U_0}{c + U_0} = \frac{(1 - 2kb)\beta^2 - 1 + 2kb\beta^2 \hat{c}}{2kb\beta(1 + \hat{c})},
\tag{9.6.18}
$$

(a)

(b)

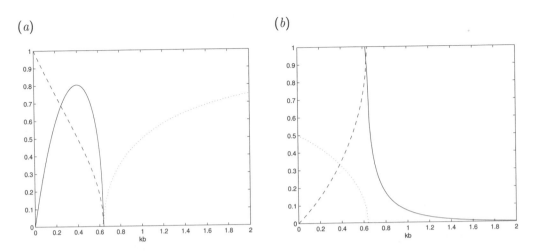

FIGURE 9.6.2   (a) Graphs of the dimensionless growth rate, $4b\varrho_I/U_0$ (solid line), and scaled imaginary part of the complex phase velocity, c_I/U_0 (dashed line), in the unstable regime of wave numbers. The dotted line in the stable regime of wave numbers represents the scaled phase velocity, c_R/U_0. (b) The solid line represents the ratio of amplitudes of traveling waves along the upper and lower vortex contours, $|A_{-1}/A_1|$, in the regime of stable wave numbers. The dashed line represents the phase shift $\Delta\phi/\pi$ between growing waves on the upper and lower vortex contours in the regime of unstable numbers. The dotted line represents the phase shift between the disturbance in circulation and the wave along the upper vortex contour, $\Delta\phi_\gamma/\pi$, in the regime of unstable wave numbers.

where $\hat{c} \equiv c/U_0$ is the scaled complex phase velocity computed from (9.6.17). The ratio $|A_{-1}/A_1|$ is equal to unity in the regime of unstable wave numbers. The solid in Figure 9.6.2(b) displays the amplitude ratio $|A_{-1}/A_1|$ in the regime of stable wave numbers, corresponding to the plus sign in equation (9.6.17). At high scaled wave numbers, kb, the lower contour is only slightly deformed with respect to the upper contour. If the minus sign were chosen instead in equation (9.6.17), the ratio $|A_{-1}/A_1|$ would be the inverse of that plotted in Figure 9.6.2(b), which means that the upper contour would be only slightly deformed with respect to the lower contour. These observations demonstrate that the two normal modes act in complementary ways to prevent spatial bias in a unidirectional flow.

The dashed line in Figure 9.6.2(b) represents the phase shift of the sinusoidal waves on the upper and lower vortex contours for the unstable normal mode in the regime of unstable wave numbers, $\Delta\phi = \arg(A_{-1}/A_1)$. As kb tends to zero, $\Delta\phi$ vanishes, indicating that the boundary waves tend to grow in phase. As kb increases toward the threshold of neutral stability, $\Delta\phi$ increases monotonically toward the maximum value of π, indicating that boundary waves tend to grow out of phase. The phase shift of the stable normal mode is the negative of that of the unstable mode shown with the dashed line in Figure 9.6.2(b). The profiles shown in Figure 9.6.3 illustrate the structure of unstable and stable normal modes.

(a)

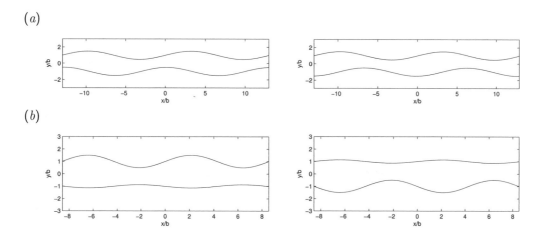

(b)

FIGURE 9.6.3   (a) Structure of an unstable (left) or stable (right) normal mode for $kb = 0.48438$. (b) Illustration of two stable modes for $kb = 0.73438$.

9.6.2   Disturbance in circulation

The occurrence of phase shift between the waves on the upper and lower vortex contours suggests that rotational fluid is redistributed inside the vortex layer due to a normal-mode perturbation. The resulting periodic accumulation of vorticity can be regarded as a physical mechanism that amplifies or dampens the perturbations. To examine the redistribution of rotational fluid in quantitative terms, we consider the strength of the perturbed vortex layer defined as

$$\gamma(x) \equiv \Omega\left(\eta_1 - \eta_{-1}\right) = \Omega\left[2b - \epsilon(A_{-1} - A_1)\right]\exp[ik(x - ct)], \qquad (9.6.19)$$

and confirm that the perturbation causes a disturbance in the strength of the vortex layer. The associated phase shift with respect to the displacement of the upper vortex contour is

$$\Delta\phi_\gamma = \arg\left(\frac{A_{-1} - A_1}{A_1}\right). \qquad (9.6.20)$$

Using (9.6.18), we compute

$$\frac{A_{-1} - A_1}{A_1} = \frac{2kb\beta(1 + \beta) + 1 - \beta^2 + 2kb\beta(1 - \beta)\hat{c}}{2kb\beta(1 + \hat{c})}. \qquad (9.6.21)$$

The phase shift $\Delta\phi_\gamma$ is zero in the regime of stable wave numbers where the waves on the upper and lower vortex contours are out of phase, $\Delta\phi = \pi$.

A graph of $\Delta\phi_\gamma/\pi$ corresponding to the plus sign in equation (9.6.17) is shown with the dotted line in Figure 9.6.2(b) in the regime of unstable wave numbers. As kb tends to zero, $\Delta\phi_\gamma$ tends to $\frac{1}{2}\pi$. Physically, rotational fluid tends to accumulate midway between the crests and troughs of growing waves.

Vortex sheet

A surface of discontinuity, regarded as a vortex sheet, arises in the limit as kb tends to zero while the upper and lower stream velocity, $U_0 = -\Omega b$, is held constant. Expanding (9.6.17) in a Taylor series with respect to kb yields the complex growth rate

$$\varrho \equiv kc \to \pm i\, kU_0, \tag{9.6.22}$$

which shows that a vortex sheet is unstable for all wave numbers. More important, the growth rate becomes infinite as the wave length of a disturbance becomes infinitesimally small. This singular behavior undermines the physical relevance of a vortex sheet and imposes essential difficulties in computing its self-induced motion, as discussed in Section 11.5.

9.6.3 Behavior of general periodic disturbances

The behavior of general periodic disturbances that deform the boundaries of the vortex layer in a sinusoidal fashion with scaled wave number kb can be studied by decomposing the disturbance into two normal modes with the same wave number. The procedure involves solving a system of three linear equations for the amplitudes and phase shifts of the normal modes (Problem 9.6.1(b)). When $kb < 0.639\cdots$, one of the two normal modes is unstable and the original disturbance is unstable, provided that it does not coincide precisely with a stable normal mode. When $kb > 0.639\cdots$, the disturbance is stable. Now we consider a periodic disturbance with wavelength L and wave number $k = 2\pi/L$ deforming the vortex boundaries in a nonsinusoidal fashion. The initial location of the boundaries can be described by a Fourier series in x, and each term in the series can be decomposed into a pair of normal modes. The disturbance is unstable only if $kb < 0.639\cdots$.

9.6.4 Nonlinear instability

The profiles shown in Figure 9.6.4 illustrate the evolution of a periodic disturbance computed by the method of contour dynamics discussed in Section 11.8.3. The computer code is available in Directory *11_vortex* of the software library FDLIB discussed in Appendix C. We observe an initial linear growth followed by a nonlinear evolution manifested by the development of nonsinusoidal contour shapes. The amplification of the disturbance produces a periodic array of compact vortices connected by thin filaments of rotational fluid. The dynamics of this flow provides us with an example of an instability leading to a new nearly steady state after nonlinear saturation.

9.6.5 Generalized Rayleigh equation

The normal-mode solution derived in this section satisfies the Rayleigh equation (9.5.1), provided that the second derivative of the unperturbed velocity profile is expressed in terms of the one-dimensional delta function, $\delta(x)$, as

$$U'' = \Omega\,[\,\delta(x+b) - \delta(x-b)\,]. \tag{9.6.23}$$

The resulting generalized Rayleigh equation is

$$(U-c)[f'' - k^2 f] - \Omega\,[\,\delta(x+b) - \delta(x-b)\,]\,f = 0. \tag{9.6.24}$$

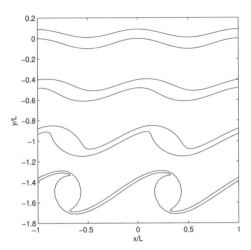

FIGURE 9.6.4   Linear amplification and nonlinear evolution of a periodic disturbance on a vortex layer
computed by the method of contour dynamics. Profiles of the vortex contours are shown at equal
time intervals from top to bottom.

The solution is required to be a continuous function, $f(y)$. Integrating (9.6.24) with respect to y
over a small distance centered at the upper or lower vortex contour, located at $y = \pm b$, using the
properties of the delta function, and rearranging, we find that

$$f'(b^+) - f'(b^-) = \Omega\, \frac{f(b)}{U_0 - c}, \qquad f'(-b^+) - f'(-b^-) = \Omega\, \frac{f(b)}{U_0 + c}, \tag{9.6.25}$$

where the superscript $\pm$ indicate evaluation on the upper $(+)$ or lower $(-)$ side. These expressions
are consistent with conditions (9.6.9) and (9.6.10), subject to expressions (9.6.14).

Problems

9.6.1 *Vortex layer*

(a) Superimpose on the vortex layer shown in Figure 9.6.1 a uniform flow in the x direction with
velocity U_1. Compute the complex phase velocity and discuss the effect of U_1 on the behavior of
normal modes.

(b) Consider a disturbance that deforms the upper and lower vortex boundaries in a sinusoidal
fashion with the same wavelength but different amplitudes and a specified phase shift. Express the
disturbance in terms of two normal modes.

9.6.2 *Vortex layer attached to a wall*

Consider a vortex layer with uniform vorticity attached to an impermeable wall on the lower side
and exposed to a uniform (streaming) flow on the upper side. Compute the complex phase velocity
of the normal modes [329].

9.6.3 *Compound vortex layer*

Formulate the linear stability problem of an unbounded inviscid flow containing two attached vortex layers with arbitrary uniform vorticity and thickness separating two parallel uniform streams [324].

 Computer Problems

9.6.4 *Normal modes*

(*a*) Plot the shape of the vortex contours corresponding to stable and unstable normal modes for $kb = 0.2$ and 1.0. Discuss the results in terms of redistribution of rotational fluid.

(*b*) Plot and discuss the corresponding streamlines of the perturbed flow.

9.6.5 *Compound vortex layer*

With reference to Problem 9.6.3, write a program that computes the complex phase velocity of the normal modes as a function of kb_1 when the vorticity of the upper layer is equal in magnitude and opposite in sign to that of the lower layer, where b_1 is the upper-layer thickness. Prepare plots of the growth rate of unstable disturbances against kb_1 for different values of the lower-to-upper layer thickness ratio b_2/b_1. Discuss the physical significance of your results. *Hint:* two distinct bands of unstable wave numbers may appear.

9.7 Numerical solution of the Rayleigh and Orr–Sommerfeld equations

Analytical solutions to the Orr–Sommerfeld and Rayleigh equations are available only for a limited class of creeping or inviscid base flows. To study the stability of general unsteady unidirectional flows at nonzero and noninfinite Reynolds numbers, we resort to approximate, asymptotic, and numerical methods. Analytical and numerical techniques are discussed in articles by Gersting & Jankowski [143] and Davey [101] and in a monograph by Drazin & Reid [112]. In this section, we overview a fundamental class of numerical methods and discuss their implementation for the temporal stability problem where a real wave number is specified and the corresponding complex phase velocity is to be found as part of the solution.

9.7.1 Finite-difference methods for inviscid flow

We begin by developing finite-difference methods for solving the Rayleigh equation (9.5.1) for the case inviscid shear flow inside a channel confined between two parallel impermeable walls located at $y = -A$ and B. Moving the complex phase velocity c to the right-hand side, we obtain

$$Uf'' - (k^2U + U'')f = c(f'' - k^2f), \qquad (9.7.1)$$

where a prime denotes a derivative with respect to y. Note that the right-hand side is independent of the velocity profile of the base flow.

In the first step of the numerical implementation, we introduce a one-dimensional uniform grid of nodes located at y_i for $i = 0, \ldots, N+1$, where $y_0 = -A$ and $y_{N+1} = B$, as shown in Figure 9.7.1 To satisfy the no-penetration boundary condition, we require that $f_0 = 0$ and $f_{N+1} = 0$. Because

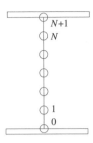

Figure 9.7.1   A finite-difference grid for solving the Rayleigh or Orr–Sommerfeld equation determining the linear stability of unidirectional flow.

the stream function takes the same value at the two walls, the disturbance flow does not generate a net flow rate. Next, we approximate the second derivative, f'', at the ith node with a central difference, and thereby replace the differential equation (9.7.1) with the finite-difference equation

$$U(y_i)\frac{f_{i-1}-2f_i+f_{i+1}}{\Delta y^2}-\left[k^2U(y_i)+U''(y_i)\right]f_i=c\left(\frac{f_{i-1}-2f_i+f_{i+1}}{\Delta y^2}-k^2f_i\right),\qquad(9.7.2)$$

where $f_i=f(y_i)$ is the value of the eigenfunction $f(y)$ at the ith node. Denoting

$$U_i\equiv U(y_i),\qquad U_i''\equiv U''(y_i),\qquad h\equiv\Delta y,\qquad(9.7.3)$$

and rearranging, we obtain

$$U_i\,f_{i-1}-\left[2\,U_i+h^2\,(k^2U_i+U_i'')\right]f_i+U_i\,f_{i+1}=c\left[f_{i-1}-(2+k^2h^2)\,f_i+f_{i+1}\right].\qquad(9.7.4)$$

Collecting the unknown values f_i into an N-dimensional vector, $\mathbf{f}=[f_1,f_2,\ldots,f_{N-1},f_N]$, we formulate a linear system of equations,

$$\mathbf{A}\cdot\mathbf{f}=c\,\mathbf{B}\cdot\mathbf{f},\qquad(9.7.5)$$

where

$$\mathbf{A}\equiv\begin{bmatrix}-2U_1-h^2(k^2U_1+U_1'') & U_1 & 0 & \cdots & & 0\\ U_2 & -2U_2-h^2(k^2U_2+U_2'') & U_2 & \cdots & & 0\\ \vdots & & \ddots & & \ddots & \vdots\\ 0 & & 0 & & \cdots & U_N & -2\,U_N-h^2(k^2U_N+U_N'')\end{bmatrix}$$
$$(9.7.6)$$

is an $N\times N$ tridiagonal matrix,

$$\mathbf{B}\equiv\begin{bmatrix}-2-(kh)^2 & 1 & 0 & 0 & \cdots & 0\\ 1 & -2-(kh)^2 & 1 & 0 & \cdots & 0\\ \vdots & \vdots & \vdots & \vdots & \vdots & \vdots\\ 0 & 0 & 0 & \cdots & 1 & -2-(kh)^2\end{bmatrix}\qquad(9.7.7)$$

is another $N \times N$ tridiagonal matrix. The structures of the matrices $\mathbf{A}$ and $\mathbf{B}$ depend on the finite-difference method chosen to approximate the derivatives. The structure of the matrix $\mathbf{B}$ is independent of the velocity profile, $U(y)$.

Solving the generalized eigenvalue problem

We have reduced the problem of solving the Rayleigh equation to a generalized algebraic eigenvalue problem expressed by the linear algebraic system (9.7.5). The solution provides us with N complex eigenvalues and corresponding eigenfunctions. We are especially interested in the eigenvalue with the maximum growth rate, $\varrho_I = kc_I$, corresponding to the most unstable normal mode. As the discretization level N increases, the maximum growth rate obtained by solving the generalized eigenvalue problem tends to that arising from the exact solution of the continuous problem expressed by (9.7.1). Powerful numerical methods for solving the generalized algebraic eigenvalue problem are available (e.g., [205]). Reliable functions methods based on sophisticated algorithms are implemented in Matlab.

Reduction to a standard eigenvalue problem

In one approach, the generalized eigenvalue problem is recast into a standard algebraic eigenvalue problem expressed by the equation

$$\mathbf{D} \cdot \mathbf{f} = c \mathbf{f}. \tag{9.7.8}$$

The complex growth rate, c, is an eigenvalue of the new matrix $\mathbf{D} \equiv \mathbf{B}^{-1} \cdot \mathbf{A}$. The eigenvalue with the largest magnitude can be computed by the power method, and other eigenvalues can be obtained by successive deflation, as discussed in Section B.2, Appendix B. A good initial guess for c is provided by Gerschgorin's theorem discussed in Section B.2, Appendix B. A practical concern is that the computation of the inverse matrix, $\mathbf{B}^{-1}$, necessary to obtain $\mathbf{D}$, can be prohibitively expensive or introduce significant numerical round-off error.

In an alternative approach, equation (9.7.5) is restated as a homogeneous system of linear algebraic equations,

$$\mathbf{E}(c) \cdot \mathbf{f} = \mathbf{0}, \tag{9.7.9}$$

and the eigenvalues, c, are identified with the roots of the characteristic polynomial, $\det[\mathbf{E}(c)] = 0$. Rearranging (9.7.5), we find that

$$\mathbf{E} \equiv \begin{bmatrix} -2 - h^2(k^2 + \frac{U_1''}{U_1 - c}) & 1 & 0 & 0 & \cdots & 0 \\ 1 & -2 - h^2(k^2 + \frac{U_2''}{U_2 - c}) & 1 & 0 & \cdots & 0 \\ \vdots & & \ddots & \ddots & \ddots & \vdots \\ 0 & 0 & 0 & \cdots & 1 & -2 - h^2(k^2 + \frac{U_N''}{U_N - c}) \end{bmatrix}. \tag{9.7.10}$$

Because the matrix $\mathbf{E}$ is tridiagonal, its determinant can be computed using an efficient algorithm discussed in Section B.2.1, Appendix B.

Having specified the wave number, k, we compute the eigenvalues in three repetitive steps: guess a complex value c; compute $\det(\mathbf{E})$ using a method for tridiagonal matrices; correct the value of c to ensure that both the real and imaginary parts of $\det(\mathbf{E})$ are zero. The correction in the third step can be made using Newton's method discussed in Section B.3, Appendix B, setting

$$c^{new} = c^{old} - \det[\mathbf{E}(c^{old})] / \left(\frac{\mathrm{d}\det[\mathbf{E}(c)]}{\mathrm{d}c}\right)_{c=c^{old}}. \tag{9.7.11}$$

Since $\det[\mathbf{E}(c)]$ is an analytic function of c, we may select a real or complex number with small magnitude, ϵ, and approximate the derivative $\mathrm{d}\det[\mathbf{E}(c)]/\mathrm{d}c$ with a finite difference. Using, for example, a forward difference, we find that

$$\frac{\mathrm{d}\det[\mathbf{E}(c)]}{\mathrm{d}c} \simeq \frac{\det[\mathbf{E}(c+\epsilon)] - \det[\mathbf{E}(c)]}{\epsilon}. \tag{9.7.12}$$

High accuracy in the computation of this derivative is not required.

Shear flow with hyperbolic tangent velocity profile

As an application, we consider the instability of a shear flow whose velocity profile is described by the hyperbolic tangent function,

$$U(y) = U_0 \tanh \frac{y}{b}, \tag{9.7.13}$$

where U_0 is a constant velocity and b is a constant length expressing the width of the shear flow. The flow occurs inside a channel confined between two parallel walls located at $y = \pm a$. The broken and solid lines in Figure 9.7.2 show the dimensionless imaginary part of the scaled phase velocity, c_I/U_0, and the scaled growth rate, $4b\varrho_I/U_0$, of unstable normal modes computed by the finite-difference method discussed in this section. The depicted family of curves correspond to a sequence of scaled channel widths, a/b. As expected, the presence of the walls reduces the growth rate of the perturbations. As a/b tends to infinity, we obtain infinite shear flow. In this limit, the critical wave number for neutral stability is $kb = 1.0$, and the value of c_I/U_0 in the limit $kb \to 0$ is equal to unity [264].

It is instructive to compare the graphs presented in Figure 9.7.1 with corresponding graphs presented in Figure 9.6.2(a) for a vortex layer. The comparison reveals that the detailed structure of the velocity profile–piecewise linear versus hyperbolic tangent–affects only mildly the stability of a shear flow.

9.7.2 Finite-difference methods for viscous flow

Finite-difference methods for the Orr–Sommerfeld equation determining the stability of viscous unidirectional flow can be developed working as in Section 9.7.1 for inviscid flow. The first numerical implementation of a finite-difference method can be traced to Thomas' seminal work on the stability of plane Hagen–Poiseuille flow [403]. To implement the finite-difference method, we rearrange the Orr–Sommerfeld equation into

$$\mathrm{i}\frac{\nu}{k}f'''' + (U - \mathrm{i}\,2k\nu)f'' - (k^2 U + U'' - \mathrm{i}k^3\nu)f = c\,(f'' - k^2 f). \tag{9.7.14}$$

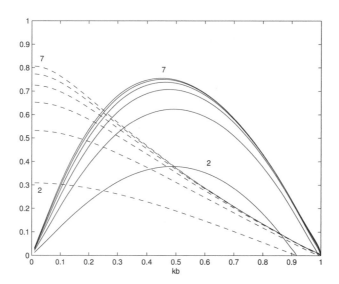

FIGURE 9.7.2   Stability graph for inviscid shear flow with hyperbolic tangent velocity profile, $U(y) = U_0 \tanh(y/b)$. Graphs of the scaled imaginary part of the complex phase velocity, c_I/U_0 (dashed lines), and dimensionless growth rate, $4b\varrho_I/U_0$ (solid lines), in the unstable regime of wave numbers for several channel half-width $a/b = 2.0, 3.0, 4.0, 5.0, 6.0,$ and 7.0.

The no-penetration and no-slip boundary conditions require that $f = 0$ and $f' = 0$ over a stationary plane wall confining the flow. Applying (9.7.14) at the ith interior node and approximating the second and fourth derivatives with central differences, we obtain the difference equation

$$\mathrm{i}\, \frac{\nu}{k} \frac{f_{i-2} - 4f_{i-1} + 6f_i - 4f_{i+1} + f_{i+2}}{\Delta y^4} + \left[\, U(y_i) - \mathrm{i}\,2k\nu \,\right] \frac{f_{i-1} - 2f_i + f_{i+1}}{\Delta y^2}$$
$$- \left[\, k^2 U(y_i) + U''(y_i) - \mathrm{i}\,\nu k^3 \,\right] f_i = c \left(\frac{f_{i-1} - 2f_i + f_{i+1}}{\Delta y^2} - k^2 f_i \right), \qquad (9.7.15)$$

where $f_i \equiv f(y_i)$. In the absence of viscous effects, we set $\nu = 0$ and obtain the corresponding discretized Rayleigh equation (9.7.2).

Compiling the difference equations at the nodes provides us with a system of linear equations that is similar to (9.7.5), except that $\mathbf{A}$ is a complex pentadiagonal matrix due to the presence of the fourth derivative. The matrix $\mathbf{B}$ is the same as that in the case of inviscid flow. The final algebraic system can be recast into the form $\mathbf{E}(c) \cdot \mathbf{f} = \mathbf{0}$, where $\mathbf{E}$ is a pentadiagonal matrix. Unfortunately, the determinant of $\mathbf{E}$ may no longer be computed by a specialized algorithm, and must be found using a general-purpose method such as Gauss elimination or LU decomposition discussed in Section B.1, Appendix B. Results obtained by solving the generalized eigenvalue problem using a Matlab function will be presented in Section 9.8.1. The computer code is available in program orr in Directory 08_stab of FDLIB (Appendix C).

9.7.3   Weighted-residual methods

The complex eigenfunction $f(y)$ can be expanded in a truncated sum of a complete set of preferably orthogonal basis functions $g_k(y)$ that satisfy the prescribed boundary conditions,

$$f(y, c) = \sum_{k=1}^{N} a_k(c) \, g_k(y), \qquad (9.7.16)$$

where $a_k(c)$ are unknown constants, c it the complex phase velocity, and N is a specified truncation level. Substituting (9.7.16) into the Rayleigh or Orr–Sommerfeld equation, multiplying the resulting equation by a chosen set of weight functions, $w_k(y)$, and integrating the product with respect to y over the solution domain, we obtain a homogeneous system of algebraic equations for the coefficients $a_k(c)$. In Galerkin's method, the weighting functions, w_k, are identified with the expansion functions, g_k (e.g., [318]). Requiring that the final system of equations has a nontrivial solution provides us with a generalized eigenvalue problem for c.

The choice of basis functions plays an important role in the accuracy and efficiency of the numerical method. In the case of plane Poiseuille flow between two plane walls separated by the distance $2a$, an appropriate choice for symmetric disturbances is $g_k = (1 - \hat{y}^2) \, \hat{y}^{2(k-1)}$, where $\hat{y} = y/a$ is a scaled position and the origin of the y axis has been placed midway between the two walls [143]. Analysis shows that expansions in Chebyshev polynomials yield higher accuracy than expansions in other seemingly more relevant sets of orthogonal functions [284]. In the case of infinite shear flow, the arguments of the basis functions should exhibit exponential decay [274].

9.7.4   Solving differential equations

In a different approach, the Orr–Sommerfeld or Rayleigh equation is integrated with respect to y using a standard numerical method, such as the Runge–Kutta method discussed in Section B.8, Appendix B. Shooting with respect to the complex growth rate is then performed to satisfy the boundary conditions. This approach bypasses the computation of eigenvalues of large matrices with the benefit of conceptual and analytical simplicity. However, certain difficulties associated with numerical instability may arise.

Reduction to a first-order system

To integrate the Orr–Sommerfeld equation, we decompose it into a system of four first-order differential equations,

$$\frac{\mathrm{d}\mathbf{q}}{\mathrm{d}t} = \mathbf{M} \cdot \mathbf{q}, \qquad (9.7.17)$$

where $\mathbf{q} = \begin{bmatrix} f, & f', & f'' - k^2 f, & f''' - k^2 f' \end{bmatrix}$,

$$\mathbf{M} = \begin{bmatrix} 0 & 1 & 0 & 0 \\ k^2 & 0 & 1 & 0 \\ 0 & 0 & 0 & 1 \\ -\hat{k} U'' & 0 & k^2 + \hat{k}\,(U - c) & 0 \end{bmatrix}, \qquad (9.7.18)$$

and $\hat{k} \equiv ik/\nu$.

To illustrate the implementation of the method, we consider plane Poiseuille flow in a channel confined between two parallel walls located at $y = \pm a$. The no-slip and no-penetration boundary conditions require that $f(\pm a)$ and $f'(\pm a)$ are all zero. The method proceeds according to the following steps:

1. Guess a complex value for c.

2. Integrate (9.7.17) from $y = -a$ to a with initial condition $\mathbf{q} = [0,0,1,0]$ to generate a first solution, called $\mathbf{q}_1$.

3. Integrate (9.7.17) from $y = -a$ to a with initial condition $\mathbf{q} = [0,0,0,1]$ to generate a second solution, called $\mathbf{q}_2$.

4. The linear combination, $\mathbf{q}_3 = \mathbf{q}_1 + \beta \mathbf{q}_2$, satisfies (9.7.17) with

$$f_3(a) = f_1(a) + \beta f_2(a), \qquad f_3'(a) = f_1'(a) + \beta f_2'(a), \qquad (9.7.19)$$

where β is a constant. Requiring that $f_3(a) = 0$ and $f_3'(a) = 0$, we obtain the compatibility condition

$$G(c) \equiv f_1(a)f_2'(a) - f_2(a)f_1'(a) = 0. \qquad (9.7.20)$$

In general, this condition will not be satisfied for the guessed value of c.

5. Improve the value of c using, for example, Newton's method, and return to Step 2.

In searching for symmetric eigenmodes, we require that $f'(0)$, $f'''(0)$, $f(a)$, and $f'(a)$ are zero and carry out the integrations in Steps 2 and 3 from the centerline of the channel toward the upper wall with initial condition $[1,0,0,0]$ in Step 2, and $[0,0,1,0]$ in Step 3.

Although seemingly innocuous, the procedure may suffer from numerical instability leading to unreliable results when the system (9.7.17) is integrated in the vicinity of walls, especially at high Reynolds numbers. To bypass this difficulty, we may perform forward and backward integration and combine the solutions to eliminate spurious oscillations. Alternatives are filtering out numerical instabilities, orthonormalizing the solution during the numerical integration, and performing parallel shooting (e.g., [143]). Similar difficulties are encountered when integrating the Rayleigh equation for inviscid flow. The numerical computations proceed smoothly when integrating from a region where the velocity profile of the base flow is nearly uniform to the main core of the shear flow, but numerical instability is encountered when the integration continues into the region of uniform flow. Physically, the perturbation decays exponentially into the region of vanishing shear rate.

The practical difficulties associated with integrating the Orr–Sommerfeld or Rayleigh's equation can be avoided by working with a modified nonlinear set of equations constructed according to Riccati's method, or by using the compound matrix method briefly discussed in the remainder of this section.

9.7.5 Riccati equation

An alternative approach is based on Riccati's equation (9.5.4), repeated here for convenience,

$$F' = -F^2 + k^2 + \frac{U''}{U - c}, \qquad (9.7.21)$$

where $F \equiv f'/f$. To illustrate the method, we consider unidirectional inviscid shear flow in a channel confined between two parallel walls located at $y = \pm a$. The algorithm is implemented according to the following steps:

- Guess a complex value for c.

- Integrate Riccati's equation from $y = -a$ to a using as initial condition the specified boundary condition $F(-a) = 0$.

- Adjust the value of c to achieve $F(a) = 0$. The correction can be made using Newton's method.

Approximations may cause numerical instability that degrades the accuracy of the computations in the case of unbounded flow. A remedy is to map the infinite domain of flow onto a finite strip using the transformation

$$\eta = \tanh \frac{y}{b}, \tag{9.7.22}$$

where b is a characteristic length scale comparable to the effective thickness of the shear flow and z is a new dimensionless variable [264]. As y varies from $-\infty$ to $+\infty$, η varies from -1 to 1. Substituting (9.7.22) into Riccati's equation (9.5.4), we obtain

$$\frac{\mathrm{d}F}{\mathrm{d}\eta} = \frac{b}{1 - \eta^2} \left(-F^2 + k^2 + \frac{U''}{U - c} \right), \tag{9.7.23}$$

where a prime indicates a derivative with respect to y.

To develop appropriate boundary conditions, we resort to Rayleigh's equation and find that, as y tends to $\pm\infty$, the function $f(y)$ behaves like $\exp(\mp ky)$. Using the definition $F = f'/f$, we then obtain $F(\eta = -1) = k$ and $F(\eta = 1) = -k$. The numerical procedure involves three repetitive steps: guess a complex value for c; integrate (9.7.23) from $\eta = -1$ to 1 with initial condition $F(\eta = -1) = k$; adjust the value of c to achieve $F(\eta = 1) = -k$ using, for example, Newton's method.

The right-hand side of (9.7.23) is indeterminate at $\eta = -1$. To avoid this apparent difficulty, we may begin the integration from $\eta = -1 + \epsilon$ using as initial condition $F(\eta = -1 + \epsilon) = k$, where ϵ is a small positive number. Alternatively, we compute the right-hand side at $\eta = -1$ using the l'Hôpital rule, as illustrated in the following example.

Hyperbolic-tangent shear flow

As an application, we consider an unbounded shear flow with hyperbolic-tangent velocity profile, $U(y) = U_0 \tanh(y/b)$, where U_0 is the far-field velocity and b is half the shear layer thickness [264]. In this case, $U(y) = U_0 \eta$, and $U'' = -2\,(U_0/b^2)\,\eta\,(1 - \eta^2)$, where a prime denotes a derivative with respect to y. Substituting these expressions into (9.7.23), we obtain a nonlinear differential equation,

$$\frac{\mathrm{d}F}{\mathrm{d}\eta} = b\,\frac{k^2 - F^2}{1 - \eta^2} - \frac{2}{b}\,\frac{\eta}{\eta - c/U_0}. \tag{9.7.24}$$

To compute $\mathrm{d}F/\mathrm{d}\eta$ at $\eta = \pm 1$, we use the l'Hôpital rule to evaluate the first term on the right-hand side, finding

$$\left(\frac{\mathrm{d}F}{\mathrm{d}\eta}\right)_{\eta=\pm 1} = \mp kb \left(\frac{\mathrm{d}F}{\mathrm{d}\eta}\right)_{\eta=\pm 1} - \frac{2}{b}\frac{1}{1 \mp c/U_0}, \tag{9.7.25}$$

which can be rearranged into

$$\left(\frac{\mathrm{d}F}{\mathrm{d}\eta}\right)_{\eta=\pm 1} = -\frac{2}{b(1 \pm kb)(1 \mp c/U_0)}. \tag{9.7.26}$$

Viscous flow

Riccati's method reduces the Orr–Sommerfeld equation into a quadratic system of four first-order differential equations for the four entries of the 2×2 Riccati matrix $\mathbf{R}$ defined by the equation $\mathbf{u} = \mathbf{R} \cdot \mathbf{v}$. The two-dimensional vectors $\mathbf{u}$ and $\mathbf{v}$ contain, respectively, the second and fourth, and first and third entries of the vector $\mathbf{q}$ defined in (9.7.17). The implementation of this powerful method is discussed by Davey [101].

9.7.6   Compound matrix method

The compound matrix method involves developing differential equations for the four minors of a 4×2 solution matrix [112, 279]. The two columns of the solution matrix contain the values and the first three derivatives of two independent solutions of the Orr–Sommerfeld equation subject to two distinct sets of initial conditions.

🖥 Computer Problems

9.7.1 *Finite-difference method for the Rayleigh equation*

(a) Consider spatially periodic perturbations in an inviscid shear flow with velocity profile

$$U(y) = U_0\big(\delta \tanh \hat{y} + (\delta - 1)\exp(-\hat{y}^2)\big), \tag{9.7.27}$$

where $\hat{y} = y/b$, b is a specified length, and the parameter δ takes values in the interval $[0, 1]$. The limiting values $\delta = 1$ and 0 yield, respectively, a shear layer with hyperbolic tangent velocity profile and a symmetric wake with Gaussian velocity profile. Assuming that the flow occurs in a bounded domain, $-a \leq y \leq a$, prepare and discuss a graph of the maximum growth rate against the wave number kb for $\delta = 0, 0.5, 1.0$, and $a/b = 2.0, 3.0, 4.0$.

(b) Repeat (a) for the velocity profile

$$U(y) = U_0\big(\delta \tanh \hat{y} + (\delta - 1)\operatorname{sech}^2\hat{y}\big). \tag{9.7.28}$$

The limiting values $\delta = 1$ and 0 yield, respectively, a shear layer with hyperbolic tangent velocity profile and the Bickley jet [416].

9.7.2 *Riccati's method*

Repeat Problem 9.7.1 using Riccati's equation.

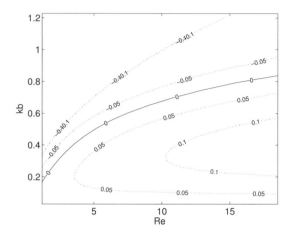

FIGURE 9.8.1   Stability graph for a viscous shear layer with hyperbolic tangent velocity profile, $U = U_0 \tanh(y/b)$. Contour plot of the dimensionless growth rate, $\varrho_I/(bU_0)$, against the Reynolds number, $\text{Re} = U_0 b/\nu$, and wave number, kb.

9.8   Instability of viscous unidirectional flows

The fundamental significance and practical importance of unidirectional and nearly unidirectional shear flows has motivated numerous studies of their stability by analytical and numerical methods. Drazin & Reid review channel, free shear layer, boundary layer, and jet flows ([112], p. 211). Huerre & Monkewitz summarize in tabular form the stability characteristics of several families of external shear flows and provide a comprehensive list of references ([195], Table 3). In this section, we review the stability characteristics of a few representative flows.

9.8.1   Free shear layers

A free shear layer is the diffuse interface between two parallel fluid streams merging with different velocities. Consider a symmetric free shear layer with hyperbolic tangent velocity profile, $U = U_0 \tanh(y/b)$, where U_0 is the magnitude of the velocity far above and below the shear layer and b is half the shear layer thickness. A contour plot of the dimensionless growth rate, $\varrho_I/(bU_0)$, with respect to the Reynolds number, $\text{Re} = U_0 b/\nu$, and wave number, kb. is shown in Figure 9.8.1. The graphs were produced using the finite-difference method for the Orr–Sommerfeld equation discussed in Section 9.7.2. The neutral stability curve corresponds to vanishing dimensionless growth rate (zero contour line). In Section 9.7.1, we mentioned that, in the limit of infinite Reynolds number, the flow is unstable for disturbances with wave number $kb < 1$. The results in Figure 9.8.1 show that, in fact, the flow is unstable at any finite Reynolds number, except in the theoretical limit of Stokes flow. The range of unstable wave numbers broadens and the maximum growth rate increases as the Reynolds number becomes higher [35]. We conclude the viscosity has a stabilizing influence on free shear flows.

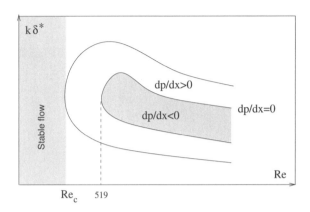

FIGURE 9.8.2 Stability graph of boundary layers with self-similar velocity profiles. Schematic illustration of regions of stable and unstable flow plotted with respect to the Reynolds number, $\text{Re} = U\delta^*/\nu$, and wave number, $k\delta^*$, where δ^* is the displacement thickness.

The velocity profile of a free shear layer represents an exact solution of the equations of steady inviscid flow. However, because the velocity profile does not satisfy the steady version of the Navier–Stokes equation with the viscous force present, it is not an acceptable representation of a steady unidirectional viscous flow. Including viscous effects causes the shear layer to spread out and the vorticity to diffuse away from the central region, as discussed in Chapter 5. Neglecting the temporal or spatial evolution of a nearly steady or nearly unidirectional base flow may introduce significant error [195].

9.8.2   Boundary layers

The normal-mode analysis of a self-similar velocity profile associated with a boundary layer is conducted under the approximation of locally unidirectional flow. The results show that the base flow is stable when the Reynolds number is below a critical threshold, Re_c, that depends on the streamwise pressure gradient of the outer irrotational flow, dp/dx, as shown in Figure 9.8.2. Boundary layers with zero or favorable pressure gradient, $dp/dx < 0$, are stable in the limit of inviscid flow, as required by Rayleigh's theorem on the significance of inflection points. However, since the velocity profile of a boundary layer with adverse pressure gradient, $dp/dx > 0$, exhibits an inflection point, the boundary layer is susceptible to instability in the limit of inviscid flow. Solving the Orr–Sommerfeld equation reveals that instability occurs at sufficiently high Reynolds numbers.

Schematic contour plots of the growth rate as a function of the Reynolds number and scaled wave number are shown in Figure 9.8.2. We observe a family of loops separating a regime of stable flow on the left from a regime of unstable flow on the right. The unstable normal modes are called Tollmien–Schlichting waves. For the Blasius boundary layer corresponding to $dp/dx = 0$, the critical Reynolds number for instability is $\text{Re}_c = 519$, where $\text{Re} = U\delta^*/\nu$ and δ^* is the displacement thickness. When instability first occurs at Re_c, the wave number of the marginally stable mode is $k\delta^* = 0.3$, corresponding to wavelength $L = 18\,\delta^*$. In the opposite extreme case of a boundary

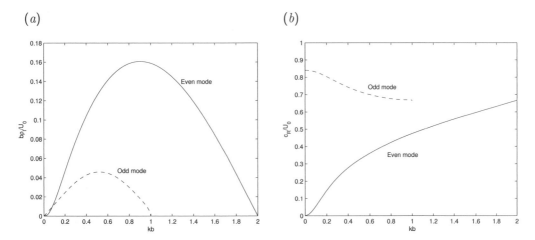

FIGURE 9.8.3   Normal modes of the inviscid Bickley jet with velocity profile $U(y) = U_0 \operatorname{sech}^2(y/b)$; (*a*) growth rate and (*b*) phase velocity of even (symmetric) or odd (antisymmetric) modes.

layer in orthogonal stagnation-point flow, $\mathrm{Re}_c \simeq 1.4 \times 10^4$. The Blasius boundary layer and other boundary layers with favorable pressure gradient are stable in the limit of inviscid flow but unstable at finite and sufficiently high Reynolds numbers. This behavior demonstrates once again that viscous stresses may have a destabilizing influence by sustaining the perturbations.

9.8.3   Jets

The stability properties of two-dimensional jets are generally similar to those of free shear layers discussed in Section 9.8.1. Physically, the edges of a jet constitute shear layers susceptible to the generic instability of free shear flows. The interaction of the two edges leads to multiple normal modes corresponding to symmetric (sinuous) or antisymmetric (varicose) perturbations. In the case of a circular jet, we obtain axisymmetric or spiral deformations.

Bickley jet

The Bickley jet is a prototypical two-dimensional jet with a symmetric velocity profile described by

$$U(y) = U_0 \operatorname{sech}^2 \hat{y},\tag{9.8.1}$$

where $\hat{y} = y/b$ and b is half the jet thickness. In the theoretical limit of inviscid flow, the jet exhibits the sinuous and varicose unstable normal modes described in Figure 9.8.3. The results presented in this figure were produced using the finite-difference method discussed in Section 9.7.1. The wave number, phase velocity, and eigenfunctions of neutrally stable waves are $kb = 2$, $c = \frac{2}{3} U_0$, and $f(y) = bU_0 \operatorname{sech}^2 \hat{y}$ for the symmetric mode, and $kb = 1$, $c = \frac{2}{3} U_0$, and $f(y) = bU_0 \operatorname{sech}\hat{y} \tanh \hat{y}$ for the antisymmetric mode ([112], p. 233). Because the growth rate of the symmetric mode is higher than that of the antisymmetric mode, the amplification of an arbitrary disturbance leads to the formation of a staggered array of vortices in the arrangement of the von Kármán vortex street, as discussed in Section 11.3. Solving the Orr–Sommerfeld equation, we find that the flow is unstable when the Reynolds number, $\mathrm{Re} = bU_0/\nu$, exceeds the critical threshold $\mathrm{Re}_c = 4$. The wave number at neutral stability corresponding to Re_c is $kb = 0.20$.

9.8.4   Couette and Poiseuille flow

Normal-mode analysis reveals that the plane Couette flow is stable at any Reynolds number, $\mathrm{Re} \equiv \xi a^2/\nu$, and for any wave number; ξ is the wall shear rate or slope of the velocity profile and a is half the channel width. Unstable behavior observed in practice at Reynolds numbers as low as $\mathrm{Re} = 350$ is attributed to nonlinear effects due to the finite amplitude of the disturbances, wall roughness, and deviation from unidirectional motion due to entrance effects. Transient growth may introduce nonlinearity that is responsible for a secondary instability (e.g., [364]).

Normal-mode analysis reveals that the plane Poiseuille flow becomes unstable when the Reynolds number, $\mathrm{Re} \equiv U_{cl}a/\nu$, exceeds the critical value $\mathrm{Re}_c = 5772$, where a is half the channel width, and U_{cl} is the centerline velocity. At lower Reynolds numbers, the flow is stable. The wave number of the normal mode that first becomes unstable at the critical Reynolds number is $ka \simeq 1.020$. In practice, the flow becomes unstable when the Reynolds number exceeds the approximate threshold 1500.

Poiseuille flow in a circular tube is stable against axisymmetric as well as more general three-dimensional perturbations. In practice, the flow becomes unstable when the Reynolds number, $\mathrm{Re} \equiv U_{cl}a/\nu$, exceeds a threshold as low as 1100, where a is the tube radius and U_{cl} is the centerline velocity. The unstable behavior is attributed to the reasons stated previously in this section for Couette flow.

Problem

9.8.1 *Bickley jet*

Confirm the properties of the neutrally stable waves given in the text for the symmetric and anti-symmetric mode.

 Computer Problems

9.8.2 *Boundary layers*

Prepare a graph of the temporal growth rate against the wave number for two self-similar boundary-layer profiles of your choice with adverse pressure gradient in the limit of inviscid flow. Show that, as $\mathrm{d}p/\mathrm{d}x$ tends to zero, the band of unstable wave numbers shrinks to zero.

9.8.3 *Bickley jet*

Reproduce the graphs of the growth rate and phase velocity shown in Figure 9.8.3.

9.9 Inertial instability of rotating flows

A new type of instability arises when the base flow rotates so that the unperturbed streamlines are not rectilinear. Centrifugal forces give rise to an effective distributed body force causing the onset of a pressure gradient across the streamlines to satisfy the force balance, and this may destabilize the flow.

9.9.1 Rayleigh criterion for inviscid flow

The simplest manifestation of the inertial instability due to rotation occurs in the case of purely swirling flow. Consider a system of cylindrical polar coordinates, (x, σ, φ), concentric with the axis of revolution of a swirling flow. Rayleigh [338] offered an energy argument to show that, in the case of inviscid flow, a necessary and sufficient condition for the flow to be stable against axisymmetric disturbances is that the distribution of circulation in the radial direction, $C = 2\pi\sigma u_\varphi$, satisfies the inequality

$$\frac{\mathrm{d}C^2}{\mathrm{d}\sigma} > 0. \tag{9.9.1}$$

To derive Rayleigh's criterion, we consider two fluid rings with radii σ_1 and σ_2 and equal volumes, δV, compute their respective kinetic energies and add them to obtain the combined kinetic energy in the unperturbed state in terms of the ring radii and circulations,

$$\mathcal{K}_U = \frac{1}{2}\left[(u_\varphi^2)_1 + (u_\varphi^2)_2\right]\rho\,\delta V = \frac{1}{8\pi^2}\left[\left(\frac{C_1}{\sigma_1}\right)^2 + \left(\frac{C_2}{\sigma_2}\right)^2\right]\rho\,\delta V, \tag{9.9.2}$$

where ρ is the fluid density. Next, we assume that the rings interchange radial positions and compute the combined kinetic energy in the perturbed state, $\mathcal{K}_P$, in terms of the radii and circulations, subject to the restriction of constant circulation imposed by Kelvin's circulation theorem,

$$\mathcal{K}_P = \frac{1}{8\pi^2}\left[\left(\frac{C_1}{\sigma_2}\right)^2 + \left(\frac{C_2}{\sigma_1}\right)^2\right]\rho\,\delta V. \tag{9.9.3}$$

Requiring that the perturbation is supplied with additional energy, $\mathcal{K}_P > \mathcal{K}_U$, we find that

$$(C_2^2 - C_1^2)\left(\frac{1}{\sigma_1^2} - \frac{1}{\sigma_2^2}\right) > 0. \tag{9.9.4}$$

In the limit as σ_1 tends to σ_2, we find that $\delta C^2 \delta\sigma > 0$, which proves Rayleigh's criterion stated in (9.9.1).

Von Kármán provided an appealing physical interpretation of Rayleigh's criterion [414]. Consider a fluid ring that is displaced in a way that preserves axisymmetry from an initial radial position, σ_1, to a new radial position, σ_2, due to the instability. Kelvin's circulation theorem requires that the circulation around the ring, C, is preserved after the perturbation. The old and new azimuthal components of the velocity of the ring, $(u_\varphi)_1$ and $(u_\varphi)_2$, are related by $C_1 = 2\pi\sigma_1(u_\varphi)_1 = 2\pi\sigma_2(u_\varphi)_2$. The centrifugal force per unit volume due to the rotation of the fluid at the unperturbed and perturbed states are

$$F_1 = \rho\,\frac{(u_\varphi^2)_1}{\sigma_1}, \qquad F_2 = \rho\,\frac{(u_\varphi^2)_2}{\sigma_2} = \frac{\rho}{4\pi^2}\frac{C_1^2}{\sigma_2^3}. \tag{9.9.5}$$

(a)   (b)

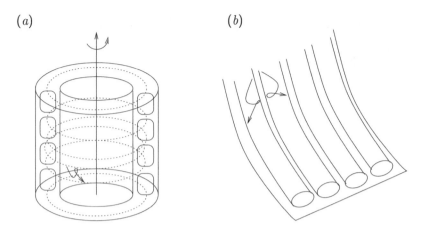

FIGURE 9.9.1   Instability of rotating flows: (a) Taylor instability of a fluid between two rotating concentric cylinders and (b) Görtler instability of a high-Reynolds-number flow along a curved wall.

If the pressure field remains unchanged, the radial pressure gradient at the new position has the undisturbed value

$$\left(\frac{\mathrm{d}p}{\mathrm{d}\sigma}\right)_2 = \frac{\rho}{4\pi^2}\frac{C_2^2}{\sigma_2^3}. \tag{9.9.6}$$

Rayleigh's criterion (9.9.1) affirms that the flow is stable as long as the pressure gradient is able to overcome the centrifugal acceleration, and thus push the fluid ring back to its original position.

9.9.2   Taylor instability

Taylor studied the instability of the circular Couette flow generated by the rotation of two coaxial cylinders, as discussed in Section 5.3.1 [398]. Observation shows and linear stability analysis confirms that the flow is unstable for combinations of the inner and cylinder angular velocities Ω_1 and Ω_2 that fall inside a certain regime determined by the radii of the inner and outer cylinders, R_1 and R_2. When the inner cylinder is fixed and the outer cylinder rotates, the flow is stable; in the opposite case, the flow can be unstable. For fixed radii, R_1 and R_2, there is a region in the (Ω_1, Ω_2) plane where unstable modes grow and saturate, leading to a new steady state involving axisymmetric rolling patterns with coiled streamlines, known as Taylor vortex flow, as illustrated in Figure 9.9.1(a). More complicated wavy and turbulent states of the Taylor vortices occur for other combinations of the cylinder angular velocities. A normal-mode stability analysis reveals that, if Rayleigh's circulation criterion is fulfilled, the base flow is stable at any Reynolds number.

9.9.3   Görtler instability

A more complex manifestation of the inertia instability of rotating fluids occurs in boundary-layer flow over a concave wall at high Reynolds numbers [147]. Under certain conditions, the boundary layer develops an alternating sequence of rolling structures illustrated in Figure 9.9.1(b). Regarding the wall as the stationary outer cylinder of a circular Couette flow allows us to make an analogy between the instability of this flow and that of the flow between two concentric cylinders.

Problem

9.9.1 *Dean instability*

Consider pressure-driven flow through a curved cylindrical tube having, for example, a toroidal or spiral shape. Discuss the possibility of inertial instability due to the centrifugal force.

9.10　Instability of a planar interface in potential flow

Having investigated the instability of homogeneous unidirectional shear flows, in the remainder of this chapter we turn to examining the instability of the interface between two fluids across which the density and viscosity may undergo a step discontinuity and the surface tension may enter the interfacial force balance.

　　We begin in this section by consider the instability of the interface between two inviscid fluids that merge with different velocities forming a unidirectional base flow. Since the tangential component of the fluid velocity is allowed to undergo a discontinuity across the interface between two inviscid fluids, the interface can be regarded as a flat vortex sheet whose self-induced motion causes the growth or decay of perturbations. Conversely, the disturbance flow can be attributed to the instability of an interfacial vortex sheet. The second interpretation allows us to study the nonlinear stages of the motion using the vortex methods discussed in Chapter 11. In Sections 9.11–9.13, we discuss the instability of the interface between viscous fluids due to the fluid inertia, gravity, viscosity stratification, or interfacial tension.

9.10.1　Base flow and linear stability analysis

Consider two inviscid fluids with uniform densities, ρ_1 and ρ_2, moving in the horizontal direction with uniform velocities U_1 and U_2, as shown in Figure 9.10.1. For simplicity, we assume that the interface exhibits a constant surface tension denoted by T (the standard symbol γ is reserved for the strength of an interfacial vortex sheet). In the unperturbed state, the interface is a flat vortex sheet with uniform strength $\gamma = U_2 - U_1$. In the configuration shown in Figure 9.10.1 where $U_1 > U_2$, the strength of the vortex sheet is negative.

　　A two-dimensional spatially periodic normal-mode disturbance causes the interface to deform in a sinusoidal fashion. The profile of the perturbed interface can be described by the real or imaginary part of the function $y = \epsilon \eta(x, t)$, where ϵ is a small dimensionless coefficient,

$$\eta(x, t) = A \exp[\mathrm{i}k(x - ct)] \qquad (9.10.1)$$

is the normal-mode wave form of the perturbation, A is the complex amplitude of the interfacial wave, $k = 2\pi/L$ is the wave number, L is the wavelength, and c is the complex phase velocity. To formulate the linear stability problem, we describe the perturbations on either side of the interface separately using the linearized equation of motion, and then match the two disturbance flows by requiring appropriate kinematic and dynamic conditions.

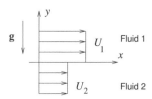

FIGURE 9.10.1   Illustration of a periodic vortex sheet representing the interface between two inviscid fluids in uniform motion along the x axis.

Rayleigh equation

Since $U'' = 0$ above and below the interface, Rayleigh's equation (9.5.2) simplifies into the Helmholtz equation, $f'' - k^2 f = 0$, inside each fluid. Because the vorticity of the base flow vanishes above and below the interface, the disturbance flow is irrotational. The general solution of the Helmholtz equation is

$$f_n(y) = A_n \, e^{ky} + B_n \, e^{-ky}, \tag{9.10.2}$$

where $n = 1, 2$ for the upper or lower fluid, and A_n, B_n are four constants. Demanding that the disturbance decays far from the interface, we set

$$A_1 = 0, \qquad B_2 = 0. \tag{9.10.3}$$

Using (9.5.6), (9.10.2), and (9.10.3), we derive expressions for the normal-mode potential,

$$\phi_1^{NM} = i \, B_1 \, e^{-ky} \, \exp[ik(x - ct)], \qquad \phi_2^{NM} = -i \, A_2 \, e^{ky} \, \exp[ik(x - ct)]. \tag{9.10.4}$$

To compute the remaining three unknowns, B_1, A_2, and c, we require three equations.

Kinematic compatibility

The velocity on either side of the interface must be such that the motion of point particles adjacent to the interface is consistent with the evolving shape of the interface expressed by (9.10.1), as discussed in Section 1.10. Introducing the kinematic constraint $\mathrm{D}(y - \epsilon\eta)/\mathrm{D}t = 0$ on either side of the interface, expanding all terms in Taylor series with respect to ϵ, and retaining only the linear contributions, we find that

$$\frac{\partial \eta}{\partial t} + U_n \frac{\partial \eta}{\partial x} - u_y^{NM}(y = 0) = 0, \tag{9.10.5}$$

where $\mathrm{D}/\mathrm{D}t$ is the material derivative and the subscript n of U_n indicates that the velocity is computed on the upper ($n = 1$) or lower ($n = 2$) side of the interface. Setting $u_y = \partial\phi/\partial y$, and substituting into (9.10.5) expressions (9.10.1) and (9.10.4), we derive two equations,

$$A = \frac{A_2}{c - U_2} = \frac{B_1}{c - U_1}. \tag{9.10.6}$$

It is clear that the phase velocity cannot be the upper or lower stream velocity.

Interfacial force balance

Next, we require that the normal stress undergoes a discontinuity across the interface that is balanced by the surface tension, τ. Recalling that the normal stress is equal to the negative of the pressure in an inviscid fluid, we obtain

$$p_2^{NM} - p_1^{NM} = -\tau \frac{\partial^2 \eta}{\partial x^2}, \qquad (9.10.7)$$

evaluated at the interface. Consistent with the linear analysis, we have approximated the curvature of the disturbed interface with the negative of the second derivative of the shape function η with respect to x (see Table 4.2.1).

To express the pressure in terms of the velocity, we use Bernoulli's equation. Linearizing the quadratic terms and substituting the result into (9.10.7), we obtain

$$-\rho_2 \left(\frac{\partial \phi_2^{NM}}{\partial t} + U_2 \frac{\partial \phi_2^{NM}}{\partial x} + g\eta \right) + \rho_1 \left(\frac{\partial \phi_1^{NM}}{\partial t} + U_1 \frac{\partial \phi_1^{NM}}{\partial x} + g\eta \right) = -\tau \frac{\partial^2 \eta}{\partial x^2}, \qquad (9.10.8)$$

evaluated at $y = 0$. Substituting the expression for the potential given in (9.10.4), we obtain

$$-\rho_2 \left[k(U_2 - c) A_2 + gA \right] + \rho_1 \left[-k(U_1 - c) B_1 + gA \right] = k^2 A \tau. \qquad (9.10.9)$$

Growth rates

To compute the complex phase velocity, we solve equations (9.10.6) for A_2 and B_1 in terms of A, substitute the result into (9.10.9), and eliminate the arbitrary constant A to derive a quadratic equation for c whose solution is

$$c = \frac{1}{\rho_1 + \rho_2} \left(\rho_1 U_1 + \rho_2 U_2 \pm \sqrt{\mathcal{D}} \right) \qquad (9.10.10)$$

where

$$\mathcal{D} = \frac{\Delta \rho \, g}{k} (\rho_1 + \rho_2) + k\tau(\rho_1 + \rho_2) - \rho_1 \rho_2 \, \Delta U^2 \qquad (9.10.11)$$

is the discriminant, $\Delta U = U_2 - U_1$, and $\Delta \rho = \rho_2 - \rho_1$. If $\mathcal{D}$ is positive, c is real and the disturbance travels with constant amplitude and phase velocity $c_R = c$.

If $\mathcal{D}$ is negative, c is complex, one of the two solutions corresponding to the $\pm$ sign has a positive imaginary part, and some disturbances are unstable. The phase velocity and growth rate of unstable waves are

$$c_R = \frac{\rho_1 U_1 + \rho_2 U_2}{\rho_1 + \rho_2}, \qquad \varrho_I = k \frac{\sqrt{|\mathcal{D}|}}{\rho_1 + \rho_2}. \qquad (9.10.12)$$

The critical wave numbers for neutral stability, k_c, are found by setting the discriminant $\mathcal{D}$ to zero, yielding a quadratic equation,

$$\tau k_c^2 - \Delta U^2 \frac{\rho_1 \rho_2}{\rho_1 + \rho_2} k_c + \Delta \rho \, g = 0. \qquad (9.10.13)$$

The two real roots enclose a finite band of unstable wave numbers.

Critical velocity difference

To ensure that a given wave number is stable, we require that $\mathcal{D} > 0$ and derive a constraint on the magnitude of the velocity difference,

$$\Delta U^2 < \frac{\Delta \rho \, g}{k} \frac{\rho_1 + \rho_2}{\rho_1 \rho_2} \Big(1 + \frac{\tau k^2}{\Delta \rho \, g} \Big). \tag{9.10.14}$$

The critical velocity difference under which the flow is stable to all disturbances, $|\Delta U_c|$, corresponds to the minimum of the right-hand side of (9.10.14), regarded as a function of k, which occurs when $k = (\Delta \rho \, g / \tau)^{1/2}$, yielding

$$|\Delta U_c|^4 = 4 \, \Delta \rho \, g \, \tau \Big(\frac{\rho_1 + \rho_2}{\rho_1 \rho_2} \Big)^2. \tag{9.10.15}$$

When the upper fluid is heavier than the lower fluid, $\Delta \rho < 0$, the flow is unstable even in the absence of flow.

Interfacial vortex sheet

It is interesting to consider the perturbation in the strength of the vortex sheet, γ, representing the interface. Recalling that the circulation along the vortex sheet is equal to the jump in the velocity potential across the vortex sheet and using (9.10.4) and (9.10.6), we find that

$$\gamma = U_2 - U_1 - \epsilon \Big(\frac{\partial \phi_1^{NM}}{\partial x} - \frac{\partial \phi_2^{NM}}{\partial x} \Big)_{y=0} = U_2 - U_1 + \epsilon k \, (B_1 + A_2) \exp[ik(x - ct)]. \tag{9.10.16}$$

Making substitutions, we obtain

$$\gamma = U_2 - U_1 + 2\epsilon A k \, [\, c - \frac{1}{2} \, (U_1 + U_2) \,] \exp[ik(x - ct)]. \tag{9.10.17}$$

Comparing (9.10.17) with (9.10.1), we deduce that the perturbation in the strength of the vortex sheet has a phase shift with respect to the perturbation of the interface equal to the argument of the shifted complex velocity, $c - \frac{1}{2} (U_1 + U_2)$.

9.10.2   Kelvin–Helmholtz instability

When the densities of the fluids above and below the vortex sheet are matched, $\rho_1 = \rho_2 = \rho$, the right-hand side of (9.10.15) is zero. Consequently, some perturbations are unstable for any velocity difference, ΔU. The critical wave numbers are found from (9.10.13),

$$k_{c_1} = 0, \qquad k_{c_2} = \frac{1}{2} \frac{\rho \Delta U^2}{\tau}. \tag{9.10.18}$$

Intermediate wave numbers yield unstable normal modes with growth rate

$$\varrho_I = \frac{1}{2} k \, |\Delta U| \, \Big(1 - 2 \, \frac{k \tau}{\rho \Delta U^2} \Big)^{1/2}, \tag{9.10.19}$$

and phase velocity equal to the mean velocity of the unperturbed streams, $c_R = \frac{1}{2}(U_1 + U_2)$. The phase shift between the disturbance in the strength and shape of the vortex sheet is equal to $\frac{1}{2}\pi$ for unstable wave numbers and zero for neutrally stable wave numbers. The maximum growth rate corresponding to the most dangerous normal mode occurs when $k = \frac{2}{3}k_{c_2}$. Large wave numbers with small wavelengths are stabilized by the restraining action of surface tension.

Absence of surface tension

In the absence of surface tension, k_{c_2} is shifted to infinity, all wave numbers are unstable, and (9.10.19) with $U_1 = U_0$ and $U_2 = -U_0$ reduces to (9.6.22), yielding a linear relationship between the growth rate and the wave number.

Nonlinear evolution

The initial linear growth and subsequent nonlinear development of the Kelvin–Helmholtz instability are illustrated in Figure 9.10.2(a) in the absence of surface tension. The evolution of the interface depicted in this figure was computed using the point-vortex method for vortex sheets discussed in Section 11.5. We observe that the growth of sinusoidal waves leads to the formation of a periodic array of spiral structures. Numerical evidence strongly suggests that the curvature of the interface becomes discontinuous at a point immediately before the spiral structures develop (e.g., [215]). Since a geometrical singularity arises spontaneously from a smooth initial condition, the problem of vortex-sheet motion is classified as ill-posed. Comparing the instability of the vortex sheet illustrated in Figure 9.10.2(a) with the instability of a vortex layer illustrated in Figure 9.6.5 demonstrates the strong effect of the velocity discontinuity due to mathematical idealization involved in the construction of a vortex sheet.

9.10.3 Rayleigh–Taylor instability

The Rayleigh–Taylor instability occurs when the unperturbed fluids are quiescent, $U_1 = 0$ and $U_2 = 0$, and the upper fluid is heavier than the lower fluid, $\rho_1 > \rho_2$ or $\Delta\rho < 0$, in which case the fluids are said to be unstably or inversely stratified [336]. The strength of the unperturbed vortex sheet representing the flat interface is zero. Setting into (9.10.13) $\Delta U = 0$, we obtain the critical wave number

$$k_c = \left(\frac{|\Delta\rho|\,g}{T}\right)^{1/2}.$$
(9.10.20)

For smaller wave numbers, one normal mode is stable and the second normal mode is unstable with associated growth rate

$$\varrho_I = kc_I = \left(k\,\frac{|\Delta\rho|\,g - k^2 T}{\rho_1 + \rho_2}\right)^{1/2}$$
(9.10.21)

and zero phase velocity. The maximum growth rate occurs when $k = k_c/\sqrt{3}$. Normal modes with wave numbers larger than k_c are stabilized by surface tension.

In the absence of surface tension, all wave numbers are unstable and the growth rate is given by the simplified expression

$$\varrho_I = \sqrt{Akg},$$
(9.10.22)

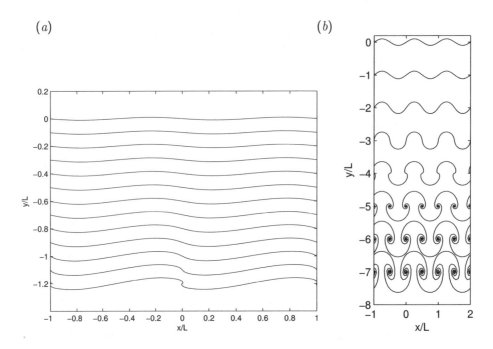

FIGURE 9.10.2   Initial linear growth and subsequent nonlinear evolution of (a) the Kelvin–Helmholtz instability and (b) the Rayleigh–Taylor instability of a vortex sheet in the absence of surface tension. Profiles of the interface are shown at equal time intervals from top to bottom.

where

$$\mathcal{A} = \left| \frac{\rho_2 - \rho_1}{\rho_2 + \rho_1} \right| \tag{9.10.23}$$

is the Atwood ratio. Because, the higher the wave number, the higher the growth rate of the perturbation, a singularity may occur after a finite evolution time due to the amplification of small-scale irregularities.

Nonlinear evolution

The initial growth and long-time evolution of the Rayleigh–Taylor instability is illustrated in Figure 9.10.2(b) for $\mathcal{A} = 0.5$ in the absence of surface tension. In these simulations, the evolution of the interface was computed using a variation of the point-vortex method discussed in Section 11.5, incorporating regularization. The development of a convoluted pattern with a secondary Kelvin–Helmholtz instability appearing along the sides of plunging sections of the interface is a striking feature of the nonlinear motion.

Accelerating interfaces

Consider two adjacent fluids accelerating with velocity $\mathbf{V}(t)$ normal to a flat interface. The analysis of this section remains valid in a noninertial frame translating with the interface, provided that the inertial acceleration force, $-\rho \, d\mathbf{V}/dt$, is added to the gravitational force, $\rho \mathbf{g}$ [399]. In this light, the Rayleigh–Taylor instability emerges as the instability of an accelerating interface between two fluids in the possible presence of a body force directed normal to the interface.

Wall effects

Container walls may have a strong stabilizing influence on the interfacial instability by placing limits on the maximum wave length (or minimum wave number) of normal modes that are allowed to enter the physical system. As an example, we consider the instability of a horizontal interface inside a vertical circular container of radius a that is partially filled with a liquid labeled 2 lying underneath another liquid labeled 1. In the unperturbed state, the interface is flat and the pressure distribution in each fluid is determined by hydrostatics.

Introducing cylindrical coordinates, (x, σ, φ), with the x axis pointing upward against the direction of gravity, we describe the position of the interface by the real or imaginary part of the function $x = \epsilon \eta(\sigma, \varphi, t)$, where

$$\eta(\sigma, \varphi, t) = A \, \mathrm{J}_m(k\sigma) \, \exp[\mathrm{i}(m\varphi - \varrho t)], \qquad (9.10.24)$$

is a normal-mode shape function, A is a complex constant determined by the initial perturbation, k is a real coefficient playing the role of a radial wave number, m is an integer expressing the azimuthal structure of the normal mode, and ϱ is the complex growth rate. The associated velocity potential is

$$\phi_j^{NM}(\sigma, \varphi, t) = B_{m,j} \, \mathrm{e}^{\pm kx} \, \mathrm{J}_m(k\sigma) \, \exp[\mathrm{i}(m\varphi - \varrho t)] \qquad (9.10.25)$$

for $j = 1$ (upper fluid) or 2 (lower fluid), where $B_{m,n}$ are two complex constants. The plus or minus sign in the first exponential apply for the lower or upper fluid, so that the perturbation decays far from the interface (Problem 9.10.4).

To satisfy the no-penetration condition at the cylinder surface, we require that $\partial \phi_n^{NM}/\partial \sigma = 0$ at $\sigma = a$, and find that an acceptable radial wave number k must satisfy the equation

$$\left(\frac{\mathrm{J}_m(k\sigma)}{\mathrm{d}\sigma} \right)_{\sigma=a} = 0. \qquad (9.10.26)$$

Referring to standard mathematical tables, we find that, for $m = 0$, corresponding to an axisymmetric perturbation, $ka = 3.83, 7.02, 10.17, \ldots$; for $m = 1$, $ka = 1.84, 5.33, 8.53, \ldots$; for $m = 2$, $ka = 3.05, 6.70, 9.97 \ldots$. We note that the smallest value of ka is 1.84 corresponding to $m = 1$, and invoke criterion (9.10.20) to find that half of the interface plunges and the other half rises when $\rho_1 > \rho_2$ and the cylinder radius exceeds the critical value

$$a_c = 1.84 \left(\frac{\tau}{|\Delta \rho| \, g} \right)^{1/2}. \qquad (9.10.27)$$

When the cylinder radius is smaller than this critical value, the interface is stable in a configuration where heavy fluid lies above a light fluid supported by interfacial tension.

9.10.4 · Gravity–capillary waves

Our analysis can be adapted to study the propagation of waves on the free surface of an otherwise quiescent liquid, such as water in the ocean. Setting $U_1 = 0$ and $U_2 = 0$, neglecting the density of the upper fluid, $\rho_1 = 0$, and denoting for convenience $\rho_2 = \rho$, we obtain from (9.10.10) the phase velocity

$$c = \pm \left(\frac{g}{k} \left(1 + \frac{\tau k^2}{\rho g} \right) \right)^{1/2}. \tag{9.10.28}$$

The normal modes represent neutrally stable waves of constant amplitude traveling to the left or right with phase velocity that depends on the wave number, k. The initial shape of an interface consisting of a packet of waves with different wave numbers evolves as each wave travels with a different phase velocity. The dispersion of the packet is determined by the functional relationship between the phase velocity and the wave number $c(k)$ shown in (9.10.28). This interpretation justifies calling (9.10.28) a dispersion relation.

Equipartition of energy in progressive gravity waves

The evolving profile of a progressive gravity wave is described by $y(x,t) = \epsilon \eta(x,t)$, where $\eta = A \cos[k(x - ct)]$ and A is a real constant. The potential energy of the fluid inside one period is

$$\mathcal{P} = \rho g \int_0^L \left(\int_0^\eta y \, \mathrm{d}y \right) \mathrm{d}x = \frac{1}{2} \rho g \, \epsilon^2 A^2 \int_0^L \cos^2[k(x - ct)] \, \mathrm{d}x = \frac{1}{2} \frac{\rho g}{k} \epsilon^2 A^2, \tag{9.10.29}$$

where $L = 2\pi/k$ is the wavelength. Resorting to (2.1.22), we find that the linearized kinetic energy of the fluid inside one period can be expressed as an integral along the free surface,

$$\mathcal{K} \simeq -\frac{1}{2} \rho \int_0^L \phi \, \mathbf{u} \cdot \mathbf{n} \, \mathrm{d}x, \tag{9.10.30}$$

where $\mathbf{n}$ is the unit vector normal to the free surface pointing into the flow. Using (9.10.4), we evaluate the free-surface potential $\phi = \epsilon A c \sin[k(x - ct)]$. Substituting this expression along with the expression

$$\mathbf{u} \cdot \mathbf{n} = c \, n_x \simeq c \, \frac{\partial \eta}{\partial x} = -c \, \epsilon A k \sin[k(x - ct)] \tag{9.10.31}$$

into (9.10.30), we obtain

$$\mathcal{K} \simeq \frac{1}{2} \rho \, c^2 \epsilon^2 A^2 \int_0^L \sin^2[k(x - ct)] \, \mathrm{d}x = \frac{1}{2k} \rho \, c^2 \epsilon^2 A^2. \tag{9.10.32}$$

In the absence of surface tension, $c^2 = g/k$, and the potential and kinetic energies are equal.

Standing waves

Superposing two waves with the same amplitude traveling in opposite directions, corresponding to the two signs in (9.10.28), we obtain a standing gravity–capillary wave oscillating with angular velocity kc. Whether a progressive or a standing wave arises in practice depends on the physical mechanism that is responsible for initiating the disturbance and on the boundary conditions at the point where the interface meets a container.

9.10.5   Displacement of immiscible fluids in a porous medium

A liquid labeled 2 displaces another liquid labeled 1 through a homogeneous and isotropic porous medium. The flow of each fluid is described by Darcy's law, $\mathbf{u} = \nabla\phi$, where $\mathbf{u}$ is the macroscopic fluid velocity,

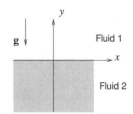

$$\phi = \frac{\kappa}{\mu}\left(-p + \rho\,\mathbf{g}\cdot\mathbf{x}\right) \qquad (9.10.33)$$

is the driving potential, p is a macroscopic pressure, and κ is the permeability of the fluid through the porous medium. The continuity equation requires that the potential, ϕ, and thus the macroscopic pressure, p, satisfy Laplace's equation, $\nabla^2\phi = 0$ and $\nabla^2 p = 0$.

Displacement of immiscible fluids in a porous medium

In the unperturbed state, the interface is flat and the two fluids translate along the y axis pointing upward against the direction of gravity with the same uniform seepage velocity, U. The unperturbed velocity potentials, indicated by the superscript (0), are given by $\phi_1^{(0)} = Uy$ and $\phi_2^{(0)} = Uy$. The interfacial velocity is $V = U/n$, where $n < 1$ is the porosity.

The horizontal interface is susceptible to the Saffman–Taylor instability [359]. To carry out the linear stability analysis for two-dimensional perturbations in the xy plane, we describe the flow in a frame of reference moving with the unperturbed interface. Working as in Section 9.10.1, we describe the profile of the perturbed interface by the real or imaginary part of the function $y = \epsilon\eta(x,t)$, where ϵ is a small dimensionless coefficient,

$$\eta(x,t) = A\exp[ik(x - ct)] \qquad (9.10.34)$$

is the normal-mode wave form of the perturbation, A is the complex amplitude of the interfacial wave, $k = 2\pi/L$ is the wave number, L is the wavelength, and c is the complex phase velocity. We will describe the perturbations on either side of the interface separately, and then match the two disturbance flows by requiring appropriate kinematic and dynamic conditions.

The normal-mode potentials are given in (9.10.4) in terms of the unknown constants B_1 and A_2. To compute the three unknowns, B_1, A_2, and c, we require three equations. Introducing the kinematic constraint $D(y - \epsilon\eta)/Dt = 0$ on either side of the interface, expanding all terms in Taylor series with respect to ϵ, and retaining only the linear contributions, we find that

$$n\frac{\partial\eta}{\partial t} - u_y^{NM}(y = 0) = 0, \qquad (9.10.35)$$

where D/Dt is the material derivative. Making substitutions, we find that $B_1 = A_2 = ncA$.

The driving potential in the ith fluid on either side of the perturbed interface is $\phi_i = \phi_i^{(0)} + \epsilon\phi_i^{NM}$ for $i = 1, 2$, evaluated at $y = \epsilon\eta$. Linearizing with respect to ϵ, we find that

$$\phi_1(y = \epsilon\eta) \simeq \epsilon\left[U\eta + \phi_1^{NM}(y = 0)\right] = \epsilon(U + i\,nc)\,\eta,$$
$$\phi_2(y = \epsilon\eta) \simeq \epsilon\left[U\eta + \phi_2^{NM}(y = 0)\right] = \epsilon(U - i\,nc)\,\eta. \qquad (9.10.36)$$

From Darcy's law, the pressure is related to the potential by $p_i = -(\lambda_i \phi_i + \rho_i g y)$ for $i = 1, 2$, where $\lambda_2 \equiv \mu_2/\kappa_2$ and $\lambda_1 \equiv \mu_1/\kappa_1$. Substituting (9.10.36), we obtain an expression for the interfacial pressure jump,

$$p_2 - p_1 = -\epsilon \left[\Delta\lambda\, U - i\,(\lambda_1 + \lambda_2)\, nc + \Delta\rho\, g \right] \eta, \qquad (9.10.37)$$

where $\Delta\rho = \rho_2 - \rho_1$ and $\Delta\lambda = \lambda_2 - \lambda_1$. Balancing the pressure difference with an effective surface tension, T, we set $p_2 - p_1 = -\epsilon T\, \partial^2\eta/\partial x^2 = \epsilon T k^2 \eta$. A note should be made that the physical origin of this equation is unclear in light of the particulate medium. Equation (9.10.37) then provides us with an expression for the complex phase velocity,

$$\varrho_I \equiv kc = -\frac{k}{n} \frac{\Delta\rho\, g + \Delta\lambda\, U + T k^2}{\lambda_1 + \lambda_2}. \qquad (9.10.38)$$

The flow is stable if $\varrho_I < 0$, which requires that

$$\Delta\rho\, g > (\lambda_1 - \lambda_2)\, U - T\kappa^2 \qquad \text{or} \qquad U < \frac{\Delta\rho\, g + T\kappa^2}{\lambda_1 - \lambda_2}, \qquad (9.10.39)$$

and unstable otherwise. The second inequality shows that, when $\kappa_1 = \kappa_2$ and in the absence of surface tension, the interface between a low-viscosity fluid displacing a high-viscosity fluid with the same density is unstable.

Problems

9.10.1 *Oblique waves*

Derive the counterpart of (9.10.10) for interfacial waves directed obliquely with respect to the direction of the flow. The wave numbers in the x and z directions are, respectively, k_x and k_z [95].

9.10.2 *Instability of a flat accelerating interface*

Discuss the stability of the interface between two liquids inside a freely falling capsule.

9.10.3 *Rayleigh–Taylor instability*

(a) Explain why an appropriate Reynolds number of the flow due to the Rayleigh–Taylor instability in the early stages of the motion is $\mathrm{Re} = g^{1/2}/(k^{3/2}\nu)$. The flow is virtually irrotational when $\mathrm{Re} \gg 1$. What is the condition for the flow to remain virtually irrotational at long times?

(b) Assume that an infinite horizontal interface located at $y = 0$ suffers a three-dimensional deformation described by the real imaginary part of the function $y(z, x, t) = \epsilon\eta(z, x, t)$, where $\eta(z, x, t) = A\, S(z, x)\exp(-i\varrho t)$ is a shape function, A is a complex constant, and ϱ is the complex growth rate. Show that the function $S(x, z)$ satisfies Helmholtz' equation

$$\frac{\partial^2 S}{\partial z^2} + \frac{\partial^2 S}{\partial x^2} = -k^2 S, \qquad (9.10.40)$$

subject to periodicity or no-penetration conditions. Show that the growth rate satisfies (9.10.21).

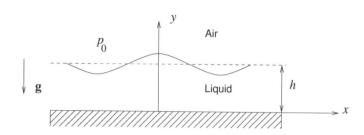

FIGURE 9.11.1   Gravitational leveling of a liquid film resting on a horizontal plane in a constant-pressure environment.

9.10.4 *Rayleigh–Taylor instability inside a circular cylinder*

(*a*) Verify that the potential given in (9.10.25) satisfies Laplace's equation.

(*b*) Show that the interfacial position given in (9.10.24) is consistent with the kinematic boundary condition at the interface.

(*c*) Compute the constants $B_{m,n}$ in terms of the constant A.

9.10.5 *Rayleigh–Taylor instability between two vertical plates*

Discuss the Rayleigh–Taylor instability of a flat two-dimensional interface between two parallel vertical plates separated by distance $2a$. The contact line is free to move under the action of the flow due to the instability.

9.10.6 *Displacement of immiscible fluids in the Hele–Shaw cell*

Consider two immiscible fluids displacing each other in the Hele–Shaw cell as discussed in Section 6.4.1. According to (6.4.2), the flow is described by Darcy's law with permeability $\kappa = h^2/12$ and unit porosity, $n = 1$. Reformulate the Saffman–Taylor linear stability problem assuming that the pressure jump across the interface is given by $p_2 - p_1 = T\left(2/h - \epsilon\, \partial^2\eta/\partial x^2\right)$. Derive criteria for stable displacement and discuss the physical relevance of the interfacial pressure boundary condition.

9.11   Film leveling on a horizontal wall

In Section 9.10, we considered the instability of an inviscid interfacial flow where the stress tensor is determined by the pressure alone. A more general framework is necessary in the case of viscous interfacial flow where normal and tangential interfacial force balances arise. To demonstrate the methodology, in this section we perform the stability analysis of a liquid film resting on a horizontal wall under a constant-pressure medium in the absence of a mean flow, as illustrated in Figure 9.11.1. The interface is occupied by an insoluble surfactant that is convected and diffuses over the interface, but not into the bulk of the fluids, altering the surface tension according to a specified equation of state.

Base state

In the unperturbed state, the film is quiescent and the surfactant is distributed uniformly over the flat interface. The pressure distribution assumes the hydrostatic profile

$$p^{(0)}(y) = -\rho g(y - h) + p_0, \tag{9.11.1}$$

where ρ is the film density, g is the acceleration of gravity, p_0 is the ambient pressure, and the superscript (0) denotes the unperturbed base state.

Normal-mode analysis

To carry out the normal-mode stability analysis for two-dimensional perturbations, we describe the position of the interface by the real or imaginary part of the function

$$y = f(x, t) = h + \epsilon \eta(x, t), \tag{9.11.2}$$

where h is the unperturbed film thickness, ϵ is a dimensionless coefficient whose magnitude is much less than unity,

$$\eta(x, t) = A \exp[ik(x - ct)] \tag{9.11.3}$$

is the normal-mode wave form of the perturbation, A is the complex amplitude of the interfacial wave, $k = 2\pi/L$ is the wave number, L is the wavelength, and c is the complex phase velocity.

Stream function

The stream function, ψ, is defined by the equations $u = \partial\psi/\partial y$ and $v = -\partial\psi/\partial x$. In the linear analysis, we set $\psi = \epsilon\psi^{(1)}$ and introduce the normal-mode form

$$\psi^{(1)}(x, y, t) = \phi(\hat{y}) \exp[ik(x - ct)], \tag{9.11.4}$$

where $\hat{y} \equiv ky$ and the superscript (1) denotes the disturbance. The perturbation velocity components are $u_x = \epsilon u_x^{(1)}$ and $u_y = \epsilon u_y^{(1)}$, where $u_x^{(1)} = \partial\psi^{(1)}/\partial y$ and $u_y^{(1)} = -\partial\psi^{(1)}/\partial x$.

The Reynolds number of the flow is so small that the motion of the fluid is governed by the equations of Stokes flow. Requiring that the stream function satisfies the biharmonic equation, $\nabla^4\psi^{(1)} = 0$, we obtain

$$\phi(\hat{y}) = a_1 e^{\hat{y}} + a_2 \hat{y} e^{\hat{y}} + a_3 e^{-\hat{y}} + a_4 \hat{y} e^{-\hat{y}}, \tag{9.11.5}$$

where a_i are four dimensionless complex coefficients for $i = 1$–4.

Wall conditions

The no-penetration and no-slip boundary conditions over the wall require that

$$\phi(\hat{y} = 0) = 0, \qquad \phi'(\hat{y} = 0) = 0, \tag{9.11.6}$$

where a prime denotes a derivative with respect to $\hat{y}$. Making substitutions, we obtain

$$a_3 = -a_1, \qquad 2a_1 + a_2 + a_4 = 0. \tag{9.11.7}$$

Kinematic compatibility

Kinematic compatibility requires that $\mathrm{D}[f(x,t) - y]/\mathrm{D}t = 0$, where $\mathrm{D}/\mathrm{D}t$ is the material derivative, yielding

$$\frac{\partial f}{\partial t} + u_x \frac{\partial f}{\partial x} - u_y = 0, \tag{9.11.8}$$

where the velocity is evaluated at the film surface. Linearizing in the absence of a mean flow, we obtain

$$\frac{\partial \eta}{\partial t} + \frac{\partial \psi^{(1)}}{\partial x} = 0 \tag{9.11.9}$$

evaluated at the undisturbed film thickness, $y = h$. Substituting the normal-mode forms, we obtain $-\mathrm{i}Akc + \mathrm{i}k\phi(\hat{k}) = 0$, yielding

$$\phi(\hat{k}) = cA, \tag{9.11.10}$$

where $\hat{k} = kh$ is a dimensionless wave number.

Surfactant concentration and surface tension

Surfactant concentration inhomogeneities over the film surface cause corresponding variations in surface tension. The distribution of the interfacial surfactant concentration, Γ, and surface tension, γ, are described by the companion functions

$$\Gamma(x,t) = \Gamma_0 + \epsilon\,\Gamma_1 \exp[\mathrm{i}k(x - ct)], \qquad \gamma(x,t) = \gamma_0 + \epsilon\,\gamma_1 \exp[\mathrm{i}k(x - ct)], \tag{9.11.11}$$

where Γ_0, γ_0 are uniform values corresponding to the flat film, and Γ_1, γ_1 are complex amplitudes. Since the perturbations are small, we can write

$$\frac{\gamma_1}{\gamma_0} = -\mathrm{Ma}\,\frac{\Gamma_1}{\Gamma_0}, \tag{9.11.12}$$

where Ma is the Marangoni number defined in Section 3.8.2.

Surfactant transport

The linearized form of the surfactant transport equation (1.9.13) is

$$\frac{\partial \Gamma^{(1)}}{\partial t} + \Gamma_0 \frac{\partial u_x^{(1)}}{\partial x} = D_s \frac{\partial^2 \Gamma^{(1)}}{\partial x^2}, \tag{9.11.13}$$

where

$$\Gamma^{(1)}(x,t) = \Gamma_1 \exp[\mathrm{i}k(x - ct)], \tag{9.11.14}$$

D_s is the surface surfactant diffusivity, and the velocity is evaluated at the unperturbed film surface, $y = h$. Expressing the velocity in terms of the stream function, we obtain

$$\frac{\partial \Gamma^{(1)}}{\partial t} + \Gamma_0 \frac{\partial^2 \psi^{(1)}}{\partial x \partial y} = D_s \frac{\partial^2 \Gamma^{(1)}}{\partial x^2}. \tag{9.11.15}$$

Substituting the normal-mode forms and rearranging, we find that the complex amplitude of the surfactant concentration is given by

$$\frac{\Gamma_1}{\Gamma_0} = k \, \frac{\phi'(\hat{k})}{c + ikD_s}. \tag{9.11.16}$$

Correspondingly, the complex amplitude of the surface tension is given by

$$\frac{\gamma_1}{\gamma_0} = -k \, \frac{\text{Ma}}{c + ikD_s} \, \phi'(\hat{k}). \tag{9.11.17}$$

Evaluating the derivative and remembering that $a_3 = -a_1$, we obtain

$$\frac{\gamma_1}{\gamma_0} = -k \, e^{\hat{k}} \, \frac{\text{Ma}}{c + ikD_s} \left[a_1(1 + q) + a_2(1 + \hat{k}) + a_4(1 - \hat{k}) \, q \right], \tag{9.11.18}$$

where $q = \exp(-2\hat{k})$.

Tangential component of the interfacial force balance

The linearized tangential component of the interfacial force balance requires that

$$\mu \left(\frac{\partial u_x^{(1)}}{\partial y} + \frac{\partial u_y^{(1)}}{\partial x} \right)_{y=h} = \frac{\partial \gamma^{(1)}}{\partial x}, \tag{9.11.19}$$

where

$$\gamma^{(1)}(x, t) = \gamma_1 \exp[ik(x - ct)]. \tag{9.11.20}$$

The term on the right-hand side of (9.11.19) represents the Marangoni traction. Expressing the velocity in terms of the stream function and rearranging, we obtain

$$\left(\frac{\partial^2 \psi^{(1)}}{\partial y^2} - \frac{\partial \psi^{(1)}}{\partial x^2} \right)_{y=h} = \frac{1}{\mu} \frac{\partial \gamma^{(1)}}{\partial x}, \tag{9.11.21}$$

Substituting the normal-mode forms, we find that

$$\left(\frac{d^2\phi}{d\hat{y}^2} + \phi \right)_{\hat{y}=\hat{k}} = i \, \frac{\gamma_1}{\mu k}. \tag{9.11.22}$$

Computing the derivatives, recalling that $a_3 = -a_1$, and rearranging, we find that

$$(1 - q) \, a_1 + (\hat{k} + 1) \, a_2 + (\hat{k} - 1) \, q \, a_4 = i \, \frac{\gamma_1}{2\mu k} \, e^{-\hat{k}}. \tag{9.11.23}$$

Normal component of the interfacial force balance

The linearized pressure field is $p = p^{(0)} + \epsilon p^{(1)}$. The linearized normal component of the interfacial stress balance provides us with an expression for the perturbation pressure,

$$p^{(1)} = 2\mu \, \frac{\partial u_y^{(1)}}{\partial y} + \rho g \, \eta - \gamma_0 \, \frac{\partial^2 \eta}{\partial x^2} + \gamma_1 \, \kappa_0, \tag{9.11.24}$$

where κ_0 is the interfacial curvature in the unperturbed configuration. All terms in (9.11.24) are evaluated at the unperturbed film surface, $y = h$. The first term on the right-hand side expresses the contribution of the viscous normal stress at the interface. Because in the unperturbed configuration the film surface is flat, $\kappa_0 = 0$, the last term on the right-hand side is identically zero. Consequently, surface tension variations do not affect the normal force balance.

Differentiating (9.11.24) with respect to x and using the tangential projection of the Stokes equation to evaluate the pressure derivative on the left-hand side, we obtain the preferred pressure-free form

$$\frac{\partial p^{(1)}}{\partial x} = \mu \, \nabla^2 u_x^{(1)} = 2\mu \, \frac{\partial^2 u_y^{(1)}}{\partial x \partial y} + \rho g \, \frac{\partial \eta}{\partial x} - \gamma_0 \, \frac{\partial^3 \eta}{\partial x^3}, \qquad (9.11.25)$$

where all terms are evaluated at the unperturbed film surface, $y = h$. Introducing the stream function, we obtain

$$\mu \left(3 \, \frac{\partial^3 \psi^{(1)}}{\partial x^2 \partial y} + \frac{\partial^3 \psi^{(1)}}{\partial y^3} \right) = \rho g \, \frac{\partial \eta}{\partial x} - \gamma_0 \, \frac{\partial^3 \eta}{\partial x^3}. \qquad (9.11.26)$$

Substituting the normal-mode forms and using (9.11.10), we obtain

$$\mu c \left[-3 \, \phi'(\hat{k}) + \phi'''(\hat{k}) \right] = \mathrm{i} \left(\frac{\rho g}{k^2} + \gamma_0 \right) \phi(\hat{k}), \qquad (9.11.27)$$

where a prime denotes a derivative with respect to $\hat{y}$. Evaluating the derivatives and remembering that $a_3 = -a_1$, we find

$$c \left[-(1+q)\, a_1 - a_2 \hat{k} + a_4 \hat{k} \, q \right] = \mathrm{i} \, \frac{1}{2\mu} \left(\frac{\rho g}{k^2} + \gamma_0 \right) \left(a_1 (1-q) + a_2 \, \hat{k} + a_4 \, \hat{k} \, q \right). \qquad (9.11.28)$$

Formulation of an eigenvalue problem

Collecting the second equation in (9.11.7), the shear stress balance (9.11.23), the normal stress balance (9.11.28), and equation (9.11.18), we derive a linear system of homogeneous equations, $\mathbf{M} \cdot \mathbf{w} = \mathbf{0}$, where $\mathbf{w} = [a_1, a_2, a_4, \gamma_1/(\mu k)]$ is an unknown vector. The coefficient matrix is given by

$$\mathbf{M} = \begin{bmatrix} 2 & 1 & 1 & 0 \\ 1-q & \hat{k}+1 & (\hat{k}-1)\,q & -\mathrm{i}\,\frac{1}{2}\,\mathrm{e}^{-\hat{k}} \\ c\,(1+q) + \mathrm{i}\,\Pi\,(1-q) & \hat{k}\,(c+\mathrm{i}\,\Pi) & \hat{k}\,(-c+\mathrm{i}\,\Pi)\,q & 0 \\ \mathrm{Ma}\,(1+q) & \mathrm{Ma}\,(1+\hat{k}) & \mathrm{Ma}\,(1-\hat{k})\,q & \mu\,(c+\mathrm{i}k D_s)\,\mathrm{e}^{-\hat{k}} \end{bmatrix}, \qquad (9.11.29)$$

where

$$\Pi \equiv \frac{1}{2\mu} \left(\frac{\rho g}{k^2} + \gamma_0 \right). \qquad (9.11.30)$$

Setting the determinant of the matrix $\mathbf{M}$ to zero provides us with a quadratic equation with two purely imaginary roots for c, yielding negative growth rates, $\varrho_I \equiv -\mathrm{i}k c_I$.

Dimensionless numbers

A Bond number can be defined with respect to either the wavelength of the perturbation or the layer thickness,

$$\mathrm{Bo} \equiv \frac{\rho g L^2}{\gamma_0}, \qquad \mathrm{Bo}' \equiv \frac{\rho g h^2}{\gamma_0}, \tag{9.11.31}$$

Corresponding dimensionless property groups can be introduced expressing the importance of the surfactant diffusivity,

$$\alpha = \frac{\gamma_0 L}{\mu D_s}, \qquad \alpha' = \frac{\gamma_0 h}{\mu D_s} = \alpha \, \frac{h}{L}. \tag{9.11.32}$$

Multiplying the third row of (9.11.29) by $\mu \hat{k}/\gamma_0$, we obtain the new matrix

$$\mathbf{M}' = \begin{bmatrix} 2 & 1 & 1 & 0 \\ 1-q & \hat{k}+1 & (\hat{k}-1)\,q & -\mathrm{i}\,\frac{1}{2}\,\mathrm{e}^{-\hat{k}} \\ \hat{c}\,\hat{k}(1+q)+\mathrm{i}\,\Lambda(1-q)/\hat{k} & \hat{c}\,\hat{k}^2+\mathrm{i}\,\Lambda & (-\hat{c}\,\hat{k}^2+\mathrm{i}\,\Lambda)\,q & 0 \\ \mathrm{Ma}\,(1+q) & \mathrm{Ma}\,(1+\hat{k}) & \mathrm{Ma}\,(1-\hat{k})\,q & (\hat{c}+\mathrm{i}\hat{k}/\alpha')\,\mathrm{e}^{-\hat{k}} \end{bmatrix}, \tag{9.11.33}$$

where $\hat{c} \equiv c\mu/\gamma_0$ is a dimensionless complex phase velocity and

$$\Lambda \equiv \hat{k}^2 \, \frac{\mu}{\gamma_0} \, \Pi = \frac{\hat{k}^2}{8\pi^2} \, (\mathrm{Bo} + 4\pi^2) = \frac{1}{2} \, (\mathrm{Bo}' + \hat{k}^2). \tag{9.11.34}$$

Setting the determinant of the matrix $\mathbf{M}'$ to zero provides us with a quadratic equation with two purely imaginary roots for $\hat{c}$, yielding negative growth rates, $\hat{\varrho}_I \equiv -\mathrm{i}\hat{k}\hat{c}_I$.

Absence of surfactant

In the absence of surfactant, $\mathrm{Ma} = 0$, we set $\gamma_0 = \gamma$ and $\gamma_1 = 0$ and retain the first three equations in (9.11.29) involving the dimensionless group Λ. After some algebra, we find that the real part of the phase velocity is zero, as expected by symmetry, and the dimensionless growth rate is given by

$$\hat{\varrho}_I \equiv \frac{\mu h \varrho_I}{\gamma} = -\Lambda \, \frac{1}{\hat{k}} \, \frac{\frac{1}{2}\,\sinh(2\hat{k}) - \hat{k}}{\hat{k}^2 + \cosh^2 \hat{k}}. \tag{9.11.35}$$

Since the fraction on the right-hand side is positive for any $\hat{k}$, the growth rate is negative and the film is stable.

In Section 9.3.6, we derived an expression for the growth rate subject to the lubrication approximation. In the present notation, equation (9.3.19) becomes

$$\varrho_I = -\mathrm{i}\,\frac{h^3}{3\mu} \, k^2 (\rho g + \gamma k^2) = -2 \, \frac{\gamma}{3\mu h} \, \Lambda \, \hat{k}^2. \tag{9.11.36}$$

This expression is the asymptotic form of (9.11.35) in the limit is $\hat{k}$ tends to zero (Problem 9.11.1(b)).

Computer code

Directory *coat0_s* of the software library FDLIB contains a code that computes the growth rates based on the equations derived in this section (Section D.8, Appendix D.)

Layer underneath a semi-infinite fluid

The linear stability analysis of a liquid film resting on a horizontal wall underneath another semi-infinite viscous fluid is presented in Section D.14, Appendix D. The corresponding growth rate is given in (D.14.28) in terms of the viscosity ratio, λ; in the present configuration, $\lambda = 0$.

Problem

9.11.1 *Inertial effects and asymptotic behavior*

(*a*) State the conditions for the effect of fluid inertia to be negligible.

(*b*) Confirm that expression (9.11.36) arises from (9.11.35) in the limit is $\hat{k}$ tends to zero.

 **Computer Problem**

9.11.2 *Growth rate*

Prepare graphs of the dimensionless growth rate against the dimensionless wave number $\hat{k}$ in the presence of surfactant and discuss the effect of the Marangoni number.

9.12   Film flow down an inclined plane

Gravity-driven flow of a liquid film down an inclined surface is observed in everyday life and is encountered in numerous applications. Examples include the flow of rain water down the windshield of an automobile, the flow of a cooling film down a heated surface, and the flow of a layer of a photographic gelatin emulsion down an inclined plane for subsequent deposition onto a moving substrate. The mathematical formulation of the linear stability problem requires careful attention to ensure the consistent implementation of the dynamic boundary condition at the location of the evolving free surface. Apart from the shear-flow instability and instability due to density stratification identified in previous sections, we will encounter a new mode of instability associated with differences in the viscosity between the film and ambient fluid.

9.12.1   Base flow

The base flow whose stability we wish to investigate is described by the Nusselt flat-film solution discussed in Section 5.1.4, as illustrated in Figure 9.12.1. To simplify the notation, it is helpful to nondimensionalize all variables using as characteristic velocity the unperturbed velocity at the free surface, u_x^{max}, characteristic length the unperturbed film thickness, h, characteristic time the ratio h/u_x^{max}, and characteristic pressure and stress the combination $\mu u_x^{max}/h$. In dimensionless

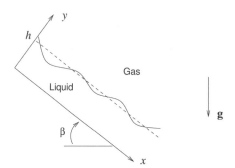

FIGURE 9.12.1   Periodic disturbances on a liquid film flowing down an inclined plane. The dashed line represents the unperturbed film surface.

variables, the base flow velocity and pressure are given by

$$\hat{u}_x^{(0)} \equiv \frac{u_x^{(0)}}{u_x^{max}} = \frac{2\nu}{gh^2 \sin \beta} u_x^{(0)} = \hat{y}\left(1 - \hat{y}\right), \qquad \hat{p}^{(0)} \equiv \frac{2}{\rho gh \sin \beta} p^{(0)} = 2\cot\beta\left(1 - \hat{y}\right), \quad (9.12.1)$$

where β is the plane inclination angle, h is the unperturbed film thickness, ν is the kinematic viscosity of the fluid, $\hat{y} = y/h$ is a dimensionless distance from the plane, and the superscript (0) denotes the steady state. The dimensionless volumetric flow rate per unit width of the film and mean velocity of the fluid are

$$\hat{Q}^{(0)} = \frac{2\nu}{gh^3 \sin \beta} Q^{(0)} = \frac{2}{3}, \qquad \hat{u}_x^{mean} = \frac{2\nu}{gh^2 \sin \beta} u_x^{mean} = \frac{2}{3}. \qquad (9.12.2)$$

For future reference, we introduce the dimensionless stress tensor in the xy plane,

$$\boldsymbol{\sigma}^{(0)} = \frac{2}{\rho gh \sin \beta} \sigma^{(0)} = 2\left(1 - \hat{y}\right)\begin{bmatrix} \cot\beta & 1 \\ 1 & -\cot\beta \end{bmatrix}. \qquad (9.12.3)$$

For convenience, we have assumed that the pressure takes the reference value of zero at the free surface so that $\boldsymbol{\sigma}^{(0)} = \mathbf{0}$ at $\hat{y} = 1$.

9.12.2   Linear stability analysis

In the inclined Cartesian coordinates depicted in Figure 9.12.1, the Navier–Stokes equation assumes the dimensionless form

$$\mathrm{Re}\left(\frac{\partial \hat{\mathbf{u}}}{\partial \hat{t}} + \hat{\mathbf{u}} \cdot \hat{\nabla}\hat{\mathbf{u}}\right) = -\hat{\nabla}p + \hat{\nabla}^2\hat{\mathbf{u}} + 2\begin{bmatrix} 1 \\ -\cot\beta \end{bmatrix}, \qquad (9.12.4)$$

where

$$\hat{t} = \frac{tu_x^{max}}{h}, \qquad \hat{\nabla} = \left(\frac{\partial}{\partial \hat{x}}, \frac{\partial}{\partial \hat{y}}\right), \qquad \hat{\mathbf{x}} = \frac{1}{h}\mathbf{x} \qquad (9.12.5)$$

are the dimensionless time, gradient operator, and position vector. The Reynolds number is

$$\text{Re} = \frac{h u_x^{max}}{\nu} = \frac{g h^3}{2\nu^2} \sin\beta = \frac{2}{\sin\beta} \text{Fr}^2, \tag{9.12.6}$$

where $\text{Fr} \equiv u_x^{max}/\sqrt{gh}$ is the Froude number.

Orr–Sommerfeld equation

A normal-mode disturbance alters the velocity field from the steady field, $\mathbf{u}^{(0)}(\mathbf{x})$, to an unsteady field,

$$\mathbf{u}(\mathbf{x}, t) = \mathbf{u}^{(0)}(\mathbf{x}) + \epsilon \mathbf{u}^{(1)}(\mathbf{x}, t), \tag{9.12.7}$$

where ϵ is a small dimensionless coefficient and the superscript (1) denotes the perturbation. Nondimensionalizing the normal-mode velocities given in (9.4.31) using the scales discussed in Section 9.12.1, we derive the dimensionless disturbance velocity

$$\hat{u}_x^{(1)} = \hat{f}'(\hat{y}) \exp[i\hat{k}(\hat{x} - \hat{c}\hat{t})], \qquad \hat{u}_y^{(1)} = -i\hat{k}\hat{f}(\hat{y}) \exp[i\hat{k}(\hat{x} - \hat{c}\hat{t})], \tag{9.12.8}$$

where $\hat{k} = kh$ is the dimensionless wave number, $\hat{c} = c/u_x^{max}$ is a dimensionless complex phase velocity, and $\hat{f} = f/(h u_x^{max})$ is a scaled eigenfunction. Substituting the velocity profile (9.12.1) into the dimensionless version of the Orr–Sommerfeld equation (9.4.20), we obtain

$$\hat{f}'''' - 2\hat{k}^2 \hat{f}'' + \hat{k}^4 \hat{f} = i \, \text{Re} \, \hat{k} \left[(2\hat{y} - \hat{y}^2 - \hat{c})(\hat{f}'' - \hat{k}^2 \hat{f}) + 2\hat{f} \right], \tag{9.12.9}$$

where a prime denotes a derivative with respect to $\hat{y}$.

To complete the definition of the linear stability problem, we require four boundary conditions for the eigenfunction $\hat{f}$. Two boundary conditions arise by requiring that both velocity components are zero at the plane, yielding

$$\hat{f}(0) = \hat{f}'(0) = 0. \tag{9.12.10}$$

Two additional boundary conditions emerge by demanding that the shear stress is zero and the normal stress is consistent with the jump in interfacial traction due to surface tension at the free surface. To prepare the ground for implementing these conditions, we digress to describe the kinematics of the perturbed flow at the free surface.

Kinematic compatibility

In the normal-mode analysis, the scaled elevation of the deformed free surface is described by the real or imaginary part of the function

$$\hat{y} = \hat{\zeta}(\hat{x}, \hat{t}) = 1 + \epsilon \hat{\eta}(\hat{x}, \hat{t}), \tag{9.12.11}$$

where ϵ is a dimensionless coefficient whose magnitude is much less than unity,

$$\hat{\eta}(\hat{x}, \hat{t}) = A \exp[i\hat{k}(\hat{x} - \hat{c}\hat{t})] \tag{9.12.12}$$

is the normal-mode wave form of the perturbation, and A is a dimensionless complex constant. Substituting (9.12.11) into the kinematic boundary condition

$$\frac{\partial \hat{\zeta}}{\partial t} + \hat{u}_x \frac{\partial \hat{\zeta}}{\partial x} - \hat{u}_y = 0, \tag{9.12.13}$$

we obtain

$$i \epsilon \hat{k} A \left(-\hat{c} + \hat{u}_x \right) \exp[i\hat{k}(\hat{x} - \hat{c}\hat{t})] - \hat{u}_y = 0, \tag{9.12.14}$$

where the velocity components $\hat{u}_x$ and $\hat{u}_y$ are evaluated at the location of the perturbed free surface. To compute the free surface velocity, we expand the velocity in a Taylor series around the undeformed position,

$$\hat{\mathbf{u}}(\hat{y} = \hat{\zeta}) = \hat{\mathbf{u}}(\hat{y} = 1) + \left(\frac{\partial \hat{\mathbf{u}}}{\partial \hat{y}} \right)_{\hat{y}=1} \epsilon \hat{\eta} + \cdots. \tag{9.12.15}$$

Expressing the velocity in terms of the base component and normal-mode perturbation, we obtain

$$\hat{\mathbf{u}}(\hat{y} = \hat{\zeta}) = \hat{\mathbf{u}}^{(0)}(\hat{y} = 1) + \epsilon \left[\left(\frac{\partial \hat{\mathbf{u}}^{(0)}}{\partial \hat{y}} \right)_{\hat{y}=1} \hat{\eta} + \hat{\mathbf{u}}^{(1)}(\hat{y} = 1) \right] + \cdots, \tag{9.12.16}$$

where the dots represent terms with quadratic or higher-order dependence on ϵ. Substituting (9.12.11) into (9.12.16) and the result into (9.12.14), and retaining terms of first order with respect to ϵ, we find that $A = \hat{f}(1)/(\hat{c} - 1)$. Consequently, the deformed free surface is described by the real or imaginary part of the function

$$\hat{\eta}(\mathbf{x}, \hat{t}) = \frac{\hat{f}(1)}{\hat{c} - 1} \exp[i\hat{k}(\hat{x} - \hat{c}\hat{t})]. \tag{9.12.17}$$

Free-surface conditions

At the free surface, the hydrodynamic traction undergoes a jump that is balanced by surface tension or a more general interfacial force field. For a clean interface with constant surface tension γ,

$$\hat{\sigma} \cdot \mathbf{n} = -\frac{1}{\text{Bo}} \hat{\kappa} \, \mathbf{n}, \tag{9.12.18}$$

where $\hat{\sigma}$ is the dimensionless stress tensor, $\mathbf{n}$ is the normal unit vector pointing into the ambient gas, $\hat{\kappa}$ is the dimensionless curvature of the free surface in the xy plane, and

$$\text{Bo} = \frac{\rho g h^2}{2\gamma} \sin \beta \tag{9.12.19}$$

is a Bond number. It is sometimes convenient to introduce the inverse Bond number, Γ, Weber number, We, and a property group, S, that depends exclusively on the physical properties of the fluid, defined as

$$\Gamma = \frac{1}{\text{Bo}} = \frac{\text{Re}}{\text{We}} = \frac{S}{\text{Re}^{2/3}} \frac{1}{\sin^{1/3} \beta}, \qquad \text{We} = \frac{\rho h (u_x^{max})^2}{\gamma}, \qquad S = \gamma \left(\frac{2\rho}{g\mu^4} \right)^{1/3}. \tag{9.12.20}$$

For water at room temperature, S is approximately equal to 4,280.

Next, we expand both sides of (9.12.18) in perturbation series with respect to ϵ. For this purpose, we decompose the stress tensor and normal unit vector into their steady and unsteady or disturbance components,

$$\hat{\boldsymbol{\sigma}} = \hat{\boldsymbol{\sigma}}^{(0)} + \epsilon\,\hat{\boldsymbol{\sigma}}^{(1)}, \qquad \mathbf{n} = \mathbf{n}^{(0)} + \epsilon\,\mathbf{n}^{(1)}, \tag{9.12.21}$$

where

$$\mathbf{n}^{(0)} = \begin{bmatrix} 0, & 1 \end{bmatrix}, \qquad \mathbf{n}^{(1)} = \begin{bmatrix} -\dfrac{\partial\hat{\eta}}{\partial\hat{x}}, & 0 \end{bmatrix}. \tag{9.12.22}$$

To compute the left-hand side of (9.12.18) in terms of flow variables at the undisturbed position of the free surface, we apply once again the method of domain perturbation, writing

$$
\begin{aligned}
(\hat{\boldsymbol{\sigma}} \cdot \mathbf{n})_{\hat{y}=\hat{\zeta}} &= \left[\hat{\boldsymbol{\sigma}}(\hat{y}=1) + \epsilon\hat{\eta}\left(\frac{\partial\hat{\boldsymbol{\sigma}}}{\partial\hat{y}}\right)_{\hat{y}=1} + \cdots \right] \cdot \mathbf{n} \\
&= \left[\hat{\boldsymbol{\sigma}}^{(0)}(\hat{y}=1) + \epsilon\,\hat{\boldsymbol{\sigma}}^{(1)}(\hat{y}=1) + \epsilon\hat{\eta}\left(\frac{\partial\hat{\boldsymbol{\sigma}}^{(0)}}{\partial\hat{y}}\right)_{\hat{y}=1} + \cdots \right] \cdot (\mathbf{n}^{(0)} + \epsilon\mathbf{n}^{(1)}) \\
&= \epsilon\left[\hat{\boldsymbol{\sigma}}^{(1)}(\hat{y}=1) + \hat{\eta}\left(\frac{\partial\hat{\boldsymbol{\sigma}}^{(0)}}{\partial\hat{y}}\right)_{\hat{y}=1} \right] \cdot \mathbf{n}^{(0)} + \cdots.
\end{aligned}
\tag{9.12.23}
$$

Substituting the last expression into the left-hand side of (9.12.18), approximating $\hat{\kappa} = -\epsilon\,\partial^2\hat{\eta}/\partial\hat{x}^2$ on the right-hand side, and equating terms of same order in ϵ, we obtain

$$\hat{\boldsymbol{\sigma}}^{(1)}(\hat{y}=1) \cdot \begin{bmatrix} 0 \\ 1 \end{bmatrix} = 2\hat{\eta}\begin{bmatrix} 1 \\ -\cot\beta \end{bmatrix} + \Gamma\frac{\partial^2\hat{\eta}}{\partial\hat{x}^2}\begin{bmatrix} 0 \\ 1 \end{bmatrix}, \tag{9.12.24}$$

where Γ is the inverse Bond number. Next, we decompose (9.12.24) into its two scalar constituents expressing tangential and normal force balances,

$$\hat{\sigma}^{(1)}_{xy}(\hat{y}=1) = 2\hat{\eta}, \qquad \hat{\sigma}^{(1)}_{yy}(\hat{y}=1) = -2\hat{\eta}\cot\beta + \Gamma\frac{\partial^2\hat{\eta}}{\partial\hat{x}^2}. \tag{9.12.25}$$

Expressing the stress tensor in terms of its pressure and the viscous constituents, we obtain two scalar linearized boundary conditions,

$$\frac{\partial\hat{u}^{(1)}_x}{\partial\hat{y}} + \frac{\partial\hat{u}^{(1)}_y}{\partial\hat{x}} = 2\hat{\eta}, \qquad \hat{p}^{(1)} = 2\frac{\partial\hat{u}^{(1)}_y}{\partial\hat{y}} + 2\hat{\eta}\cot\beta - \Gamma\frac{\partial^2\hat{\eta}}{\partial\hat{x}^2}, \tag{9.12.26}$$

where all terms are evaluated at the unperturbed free surface position, $\hat{y}=1$.

Tangential force balance

Substituting (9.12.8) along with (9.12.17) into the first equation in (9.12.26) and rearranging, we obtain

$$\hat{f}''(1) + \left(\hat{k}^2 - \frac{2}{\hat{c}-1}\right)\hat{f}(1) = 0, \tag{9.12.27}$$

expressing a linearized tangential force balance.

Normal force balance

To derive the free surface condition corresponding to the second equation in (9.12.26), we require an expression for the disturbance pressure in terms of the velocity. This expression emerges by applying the dimensionless form of the linearized Navier–Stokes equation at the free surface, projecting it onto the tangent unit vector, $\mathbf{t}$, and rearranging to find

$$\mathbf{t} \cdot \hat{\nabla} p^{(1)} = -\operatorname{Re}\left(\frac{\partial \hat{\mathbf{u}}^{(1)}}{\partial \hat{t}} + \hat{\mathbf{u}}^{(0)} \cdot \hat{\nabla} \hat{\mathbf{u}}^{(1)} + \hat{\mathbf{u}}^{(1)} \cdot \hat{\nabla} \hat{\mathbf{u}}^{(0)} \right) + (\hat{\nabla}^2 \hat{\mathbf{u}}^{(1)}) \cdot \mathbf{t}. \tag{9.12.28}$$

Now differentiating the second equation in (9.12.26) with respect to $\hat{x}$, we obtain

$$\frac{\partial \hat{p}^{(1)}}{\partial \hat{x}} = 2 \frac{\partial^2 \hat{u}_y^{(1)}}{\partial \hat{x} \partial \hat{y}} + 2 \frac{\partial \hat{\eta}}{\partial \hat{x}} \cot \beta - \Gamma \frac{\partial^3 \hat{\eta}}{\partial \hat{x}^3}, \tag{9.12.29}$$

evaluated at $\hat{y} = 1$. Equating the right-hand sides of (9.12.28) and (9.12.29) and retaining only terms of order ϵ, we obtain

$$-\operatorname{Re}\left(\frac{\partial \hat{u}_x^{(1)}}{\partial \hat{t}} + \frac{\partial \hat{u}_x^{(1)}}{\partial \hat{x}} \right) + \hat{\nabla}^2 \hat{u}_x^{(1)} = 2 \frac{\partial^2 \hat{u}_y^{(1)}}{\partial \hat{x} \partial \hat{y}} + 2 \frac{\partial \hat{\eta}}{\partial \hat{x}} \cot \beta - \Gamma \frac{\partial^3 \hat{\eta}}{\partial \hat{x}^3}, \tag{9.12.30}$$

evaluated at $\hat{y} = 1$. Finally, we substitute (9.12.8) along with (9.12.17) into (9.12.30) and obtain the desired boundary condition

$$\hat{f}'''(1) + \hat{k}\left[\mathrm{i}\operatorname{Re}(\hat{c} - 1) - 3\hat{k} \right] \hat{f}'(1) - \mathrm{i}\frac{\hat{k}}{\hat{c}-1}\left(2\cot\beta + \Gamma\hat{k}^2 \right) \hat{f}(1) = 0, \tag{9.12.31}$$

involving the Reynolds number, Re, and the inverse Bond number, Γ.

Formulation of an eigenvalue problem

The problem has been reduced to solving an eigenvalue problem expressed by the Orr–Sommerfeld equation (9.12.9), subject to four boundary conditions (9.12.10), (9.12.27), and (9.12.31). Analytical solutions are not available and we must resort to approximate, asymptotic, and numerical methods (e.g., [298, 74, 77]). Anshus & Goren [10] noted that a main difficulty in solving the Orr–Sommerfeld equation (9.12.9) stems from the presence of the non-constant coefficient on the right-hand side involving the unperturbed velocity profile, and proposed replacing the velocity distribution with the maximum velocity at the free surface. At low and moderate Reynolds numbers, approximate growth rates computed in this fashion are close to the exact growth rates computed by numerical methods.

Long-wave solution

Useful insights can be obtained by considering disturbances with long wavelength or small scaled wave number, $\hat{k}$ [435]. It turns out from the numerical solution of the exact eigenvalue problem that the fastest growing mode does have a long wavelength, and the approximate analysis for small scaled wave numbers provides us with accurate predictions. Expanding the eigenfunction $\hat{f}$ and the complex phase velocity $\hat{c}$ in Taylor series with respect to $\hat{k}$, we obtain

$$\hat{f} = \hat{f}_0 + \hat{k}\hat{f}_1 + \hat{k}^2 \hat{f}_2 + \cdots, \qquad \hat{c} = \hat{c}_0 + \hat{k}\hat{c}_1 + \hat{k}^2 \hat{c}_2 + \cdots. \tag{9.12.32}$$

Substituting these expansions into the Orr–Sommerfeld equation and associated boundary conditions, assuming that both Re and Γ are of order unity, and collecting terms of same order in $\hat{k}$, we derive a sequence of eigenvalue problems.

The zeroth-order problem is governed by the differential equation

$$\hat{f}_0'''' = 0, \tag{9.12.33}$$

with boundary conditions

$$\hat{f}_0(0) = 0, \qquad \hat{f}_0'(0) = 0, \qquad \hat{f}_0''(1) - \frac{2}{\hat{c}_0 - 1}\,\hat{f}_0(1) = 0, \qquad \hat{f}_0'''(1) = 0. \tag{9.12.34}$$

The solution is readily found to be

$$\hat{f}_0 = B\,\hat{y}^2, \qquad \hat{c}_0 = 2, \tag{9.12.35}$$

where B is an arbitrary constant expressing the initial magnitude of the perturbation. Physically, this solution tells us that long waves translate with a phase velocity that is equal to twice the velocity of the liquid film at the free surface. It is interesting to note that this behavior contradicts Rayleigh's theorem for inviscid flow discussed in Section 9.5.

The first-order problem is governed by the differential equation

$$\hat{f}_1'''' = -\mathrm{i}\,\mathrm{Re}\left[\,(\hat{y}^2 - 2\hat{y} + \hat{c}_0)\,\hat{f}_0'' - 2\hat{f}_0\,\right], \tag{9.12.36}$$

with boundary conditions

$$\hat{f}_1(0) = 0, \qquad \hat{f}_1'(0) = 0, \qquad \hat{f}_1''(1) - \frac{2}{\hat{c}_0 - 1}\,\hat{f}_1(1) = -\frac{2\hat{c}_1}{(\hat{c}_0 - 1)^2}\,\hat{f}_0(1),$$

$$\hat{f}_0'''(1) = -\mathrm{i}\,\mathrm{Re}\,(\hat{c}_0 - 1)\,\hat{f}_0'(1) + 2\mathrm{i}\,\frac{\cot\beta}{\hat{c}_0 - 1}\,\hat{f}_0(1). \tag{9.12.37}$$

Since $\hat{k}$ has been used as a perturbation variable, it appears neither in the zeroth-order equation (9.12.33) nor in the first-order equation (9.12.36). Substituting (9.12.35) into (9.12.36), we obtain the simple form

$$\hat{f}_1'''' = 4\mathrm{i}\,B\,\mathrm{Re}\,(\hat{y} - 1). \tag{9.12.38}$$

A solution that satisfies the two boundary conditions at the wall is given by the fifth-degree polynomial

$$\hat{f}_1 = \mathrm{i}\,\frac{1}{30}\,B\,\mathrm{Re}\,\hat{y}^4(\hat{y} - 5) + C\hat{y}^3, \tag{9.12.39}$$

where C is a new constant. Substituting (9.12.39) into the third and fourth boundary conditions in (9.12.37), we derive two linear homogeneous equations for B and C,

$$\begin{bmatrix} \hat{c}_1 - \frac{8}{15}\,\mathrm{iRe} & 2 \\ -2\mathrm{i}\cot\beta & g \end{bmatrix} \cdot \begin{bmatrix} B \\ C \end{bmatrix} = \mathbf{0}. \tag{9.12.40}$$

Setting the determinant of the coefficient matrix to zero to ensure a nontrivial solution, we obtain $C = \frac{1}{3} \mathrm{i} B \cos \beta$ and

$$\hat{c}_1 = \mathrm{i} \frac{8}{15} \left(\mathrm{Re} - \frac{5}{4} \cot \beta \right). \tag{9.12.41}$$

Now collecting the zeroth- and first-order solutions, we obtain

$$\hat{c} \simeq 2 + \mathrm{i} \hat{k} \frac{8}{15} \left(\mathrm{Re} - \frac{5}{4} \cot \beta \right). \tag{9.12.42}$$

Recalling that the dimensionless growth rate of the disturbance is equal to the imaginary part of $-\mathrm{i} \hat{k} \hat{c}$, we obtain Benjamin's [28] criterion for long waves,

$$\mathrm{Re} > \frac{5}{4} \cot \beta. \tag{9.12.43}$$

Substituting the definition of the Reynolds number from (9.12.6), we find that long waves will grow when

$$h > \left(\frac{10}{4} \frac{\nu^2}{g} \frac{\cot \beta}{\sin \beta} \right)^{1/3}. \tag{9.12.44}$$

This inequality places a limit on the maximum film thickness for stable flat-film flow. A vertical film corresponding to $\beta = \pi/2$ is unstable at all Reynolds numbers.

Approximate methods

Benjamin [28] developed an approximate method for solving the linear stability problem based on an infinite series expansion,

$$\hat{f}(\hat{y}) = \sum_{n=0}^{\infty} A_n \hat{y}^n. \tag{9.12.45}$$

Substituting (9.12.9) into the Orr–Sommerfeld equation, we obtain a recurrence relationship among four groups of coefficients, A_n, A_{n-2}, A_{n-4}, and A_{n-6}. Enforcing the boundary conditions provides us with an algebraic eigenvalue problem with infinite dimensions. In practice, the series (9.12.45) and associated linear algebraic eigenvalue problem are truncated at a finite level. Benjamin noted that truncating amounts to expressing the solution in a Taylor series with respect to $\hat{k}$ and Re, and then truncating that series. For example, keeping 16 terms produces results that are accurate to third order in $\hat{k}$ and Re. Benjamin solved the corresponding eigenvalue problem analytically and derived an involved relationship between the critical wave number for neutral stability, the Reynolds number, Re, and the Weber number, We.

Instead of implementing the long wave approximation in the final linearized system and boundary conditions, Benney [29] worked with an approximate form of the Navier–Stokes equation that assumes long waves of finite amplitude at the outset. The governing equations are developed as discussed in Section 6.4. By carrying out a linear stability analysis of the resulting evolution equation for the film thickness, Benney produced the second and third coefficients of the complex growth rate $\hat{c}$ shown in (9.12.32),

$$\hat{c}_2 = -2 - \frac{32}{63} \operatorname{Re} (\operatorname{Re} - \frac{5}{4} \cot \beta), \tag{9.12.46}$$

and

$$\hat{c}_3 = i \left(-\frac{1}{3} \Gamma - \frac{157}{56} \operatorname{Re} + \frac{6}{5} \cot \beta - \frac{8}{45} \operatorname{Re} \cot^2 \beta + \frac{138904}{155925} \operatorname{Re}^2 \cot \beta - \frac{1213952}{2027025} \operatorname{Re}^3 \right). \tag{9.12.47}$$

The neutral stability curves where the imaginary part of $\hat{c}$ becomes zero, are given by $\hat{k}_c = 0$ and $(-\hat{c}_0/\hat{c}_3)^{1/2}$.

Problems

9.12.1 *Benney's method*

Show that, according to Benney's analysis, the neutral stability curves along which the imaginary part of c is zero, are given by

$$\operatorname{Re} = \frac{5}{4} \cot \beta + \hat{k}_c^2 (0.625 \, \Gamma + 4.320870 \cot \beta - 0.000006 \cot^2 \beta) + O(\hat{c}_c^4), \tag{9.12.48}$$

where $\hat{c}_c$ is the critical dimensionless wave number.

9.12.2 *Two-layer film flow*

Consider the flow of two superposed liquid films down an inclined plane. Formulate and discuss the equations and boundary conditions governing the temporal linear stability problem.

9.12.3 *Two-layer channel flow*

Consider the flow of two superposed layers in plane Couette or Poiseuille flow inside a channel confined between two parallel walls. Formulate and discuss the equations and boundary conditions governing the temporal linear stability problem [189, 436].

9.13   Capillary instability of a curved interface

A curved interface between two immiscible fluids is susceptible to a capillary instability due to surface tension. Rayleigh observed that distortions of the interface cause normal stress differences that may promote the growth of perturbations [334]. Manifestations of the capillary instability are found in a broad range of applications. A capillary instability occurs when a cylindrical jet of a fluid penetrates another immiscible fluid, causing it to develop corrugations and finally to break up into an array of droplets. Stirred liquid drops develop viscous filaments that disintegrate into smaller drops to form an emulsion. In nature, the capillary instability of a quiescent liquid column is responsible for the formation of spider's web. In biomechanics, the capillary instability of an annular layer coated on the exterior or interior side of a cylindrical surface, such as a pulmonary airway, is responsible for the formation of annular rings. The capillary instability is exploited in ink-jet printer technology where an ejected column of ink breaks up into an array of droplets that are subsequently guided onto a printed surface by an electrical field [46].

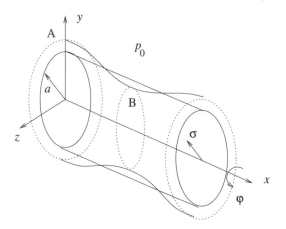

FIGURE 9.13.1   Illustration of a liquid thread suspended in an infinite constant-pressure medium.

9.13.1   Rayleigh instability of an inviscid liquid column

Consider a cylindrical jet of an inviscid liquid with circular cross-section of radius a discharging with uniform velocity into an ambient gas with negligible density and constant pressure, p_0, as shown in Figure 9.13.1. Assuming that gravitational forces are negligible and recalling that the mean curvature of the unperturbed interface is $\kappa_m = 1/(2a)$, we find that the pressure inside the jet is $p_0 + a\gamma$, where γ is the surface tension.

Normal-mode analysis

It is convenient to work in a frame of reference where the jet appears to be a stationary column of fluid and the ambient gas executes an inconsequential backward motion. The flow due to a perturbation will be described in cylindrical coordinates coaxial with the undisturbed cylindrical interface, (x, σ, φ). In the case of axisymmetric perturbations, the jet radius is described by the real part of the function

$$\sigma = f(x,t) = a + \epsilon\eta(x,t), \qquad (9.13.1)$$

where ϵ is a dimensionless coefficient whose magnitude is much less than unity,

$$\eta(x,t) = A\exp[ik(x - ct)] \qquad (9.13.2)$$

is the normal-mode wave form of the perturbation, A is the complex amplitude of the interfacial wave, $k = 2\pi/L$ is the wave number, L is the wavelength, and c is the complex phase velocity.

Confining our attention to irrotational and spatially periodic perturbations, we introduce a velocity potential for the flow inside the liquid, $\phi = \epsilon\phi^{(1)}$, where

$$\phi^{(1)}(x, \sigma, t) = B\,\Phi(\sigma)\,\exp[ik(x - ct)] \qquad (9.13.3)$$

is a perturbation potential, B is a complex constant, and Φ is a complex eigenfunction. Requiring that the perturbation potential satisfies Laplace's equation written in cylindrical polar coordinates, and demanding a nonsingular behavior at the centerline, $\sigma = 0$, we find that

$$\Phi(\sigma) = J_0(ik\sigma) = I_0(k\sigma), \qquad (9.13.4)$$

where J_0 is the zeroth-order Bessel function and I_0 is the zeroth-order modified Bessel function. To derive the last expression, we used the property of Bessel functions $J_n(ix) = i^n I_n(x)$, where n is a positive integer and x is a real argument. To compute the ratio B/A and complex phase velocity, c, we require two constraints.

Kinematic compatibility

The first constraint arises by substituting the expressions given into (9.13.2) and (9.13.3) in the linearized kinematic condition at the interface stated in (1.10.3), $\partial\eta/\partial t - \partial\phi^{(1)}/\partial\sigma = 0$, obtaining

$$-ikcA = B\,\frac{d\Phi}{d\sigma} = B\,\frac{dJ_0(ik\sigma)}{d\sigma} = -i\,kB\,J_1(ik\sigma), \qquad (9.13.5)$$

evaluated at the unperturbed interface, $\sigma = a$, where J_1 is the first-order Bessel function. Simplifying, we obtain

$$cA = B\,J_1(ika) = iB\,I_1(ka), \qquad (9.13.6)$$

where I_1 is the first-order modified Bessel function.

Interfacial normal force balance

The second constraint arises by requiring that the pressure drop across the interface is balanced by surface tension, $p - p_0 = 2\kappa_m\gamma$, where κ_m is the mean curvature of the perturbed interface. Linearizing the expression for the mean curvature given in Table 4.2.1, we obtain

$$\frac{p - p_0}{\gamma} = \frac{1}{a^2}\,(a - \epsilon\eta) - \epsilon\,\frac{\partial^2\eta}{\partial x^2}. \qquad (9.13.7)$$

The second term on the right-hand side is the curvature in a meridional plane and the second term is the curvature in its conjugate orthogonal plane. The pressure in the liquid can be evaluated using Bernoulli's equation (3.6.11) for unsteady irrotational flow. Discarding the square of the velocity and substituting into (9.13.7) the resulting linearized form, we obtain

$$\rho\,\frac{\partial\phi^{(1)}}{\partial t} = \gamma\left(\frac{\eta}{a^2} + \frac{\partial^2\eta}{\partial x^2}\right). \qquad (9.13.8)$$

Substituting into (9.13.8) expressions (9.13.2) and (9.13.3), and evaluating the left-hand side at the location of the unperturbed interface, $\sigma = a$, we find that

$$-ikc\rho B\,I_0(ka) = \frac{\gamma}{a^2}\,A\left[1 - (ka)^2\right]. \qquad (9.13.9)$$

Growth rate and phase velocity

Dividing corresponding sides of (9.13.6) and (9.13.9) to eliminate the constants A and B, and rearranging, we obtain the square of the complex phase velocity

$$c^2 = -\frac{\gamma}{\rho k a^2} \frac{\mathrm{I}_1(ka)}{\mathrm{I}_0(ka)} \left[1 - (ka)^2\right], \tag{9.13.10}$$

The modified Bessel functions, $\mathrm{I}_0(ka)$ and $\mathrm{I}_1(ka)$, are positive for any scaled wavenumber, ka. Accordingly, the sign of the right-hand side of (9.13.10) is determined by the magnitude of ka.

When $ka > 1$, the right-hand side of (9.13.10) is positive and the complex phase velocity, c, takes two real values corresponding to stable normal modes. Axisymmetric waves with $ka > 1$ travel undamped to the left or right with phase velocity

$$c = \pm \left(\frac{\gamma}{\rho k a^2} \frac{\mathrm{I}_1(ka)}{\mathrm{I}_0(ka)} \left[1 - (ka)^2\right]\right)^{1/2}. \tag{9.13.11}$$

The permanence of the amplitude of the interfacial wave is a consequence of the absence of viscous dissipation in an inviscid flow.

When $ka < 1$, the right-hand side of (9.13.10) is negative, and the complex phase velocity, c, takes two complex conjugate values with zero real part. The negative imaginary part corresponds to a stable mode, while the positive imaginary part corresponds to an unstable mode with growth rate

$$\varrho_I = k c_I = \left(\frac{\gamma k}{\rho a^2} \frac{\mathrm{I}_1(ka)}{\mathrm{I}_0(ka)} \left[1 - (ka)^2\right]\right)^{1/2}. \tag{9.13.12}$$

The fastest growth occurs approximately when $ka = 0.679$, and the corresponding maximum growth rate is

$$(\varrho_I)_{max} = 0.34 \left(\frac{\gamma k}{\rho a^2}\right)^{1/2}. \tag{9.13.13}$$

Physically, perturbations with large wavelength amplify due to a surface pressure generated by surface tension.

We have found that, when the Plateau–Rayleigh criterion $ka < 1$ is fulfilled, the jet falls prey to the capillary instability [334]. The growth of perturbations in the unstable regime can be explained by considering the pressure distribution in the liquid column at the initial instant, neglecting pressure variations due to the fluid acceleration. Considering station A in Figure 9.13.1, we note that the curvature of the interface in a meridional plane increases due to the interfacial corrugation, whereas the curvature in a plane that is perpendicular to the x axis decreases due to the increased jet radius; the inverse is true at station B. The linearized form of the expression for the mean curvature on the right-hand side of (9.13.9) then shows that the pressure at station A will be lower than the pressure at station B and the fluid will be driven towards the crest when $0 < ka < 1$, thereby amplifying the interfacial corrugations.

Nonlinear evolution

The initial amplification of interfacial corrugations due to the Rayleigh instability is followed by the formation of liquid drops connected by axisymmetric bridges, eventually breaking up into smaller satellite drops. Mansour & Lundgren [255] present numerical simulations based on the boundary-integral method discussed in Chapter 10.

Three-dimensional corrugations

To study the behavior of non-axisymmetric disturbances, we describe the shape of the interface by the real or imaginary part of the shape function

$$\eta(x,t) = A \exp[\mathrm{i}(kx + m\varphi - ct)], \tag{9.13.14}$$

where m an integer representing the azimuthal wave number. Repeating the preceding analysis, we find that the growth rate is given by the generalized expression [334]

$$c^2 = -\frac{\gamma}{\rho k a^2} \frac{\mathrm{I}'_m(ka)}{\mathrm{I}_m(ka)} \left[1 - (ka)^2 - m^2 \right], \tag{9.13.15}$$

where I_m are mth-order modified Bessel functions. When $m = 0$, we recover (9.13.10). Since the ratio of the modified Bessel functions on the right-hand side of (9.13.15) is positive for any integer $m > 0$, the right-hand side is positive for any $m \geq 1$ and the interface is stable to non-axisymmetric perturbations. Thus, a cylindrical column of fluid is expected to break up by developing axisymmetric corrugations, in agreement with observation.

9.13.2   Capillary instability of a viscous thread

Next, we consider the instability of a viscous thread suspended in an infinite ambient viscous fluid, as illustrated in Figure 9.13.2(a). The inner fluid is labeled 1 and the outer fluid is labeled 2. The viscosity of the inner fluid is μ_1 and the viscosity of outer fluid is $\mu_2 = \lambda \mu_1$, where λ is the viscosity ratio. In the absence of surfactants, the interface exhibits uniform surface tension, γ. In the unperturbed configuration, the fluids are quiescent. We will assume that the motion of the fluid on either side of the interface due to a perturbation is governed by the continuity equation and the Stokes equation with appropriate physical constants for each fluid.

Normal-mode analysis

To carry out the normal-mode stability analysis for axisymmetric perturbations, we introduce cylindrical polar coordinates, (x, σ, φ), where the x axis coincides with the axis of revolution of the unperturbed interface. The radial position of the interface is described by the real or imaginary part of the function

$$\sigma = f(x,t) = a + \epsilon \eta(x,t), \tag{9.13.16}$$

where a is the unperturbed thread radius, ϵ is a dimensionless coefficient whose magnitude is much less than unity,

$$\eta(x,t) = A \exp[\mathrm{i}k(x - ct)] \tag{9.13.17}$$

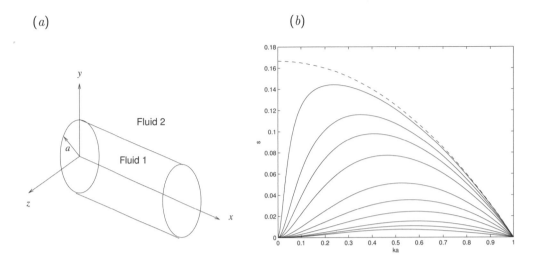

FIGURE 9.13.2 (a) Illustration of an infinite viscous thread suspended in a quiescent infinite viscous fluid. (b) Dependence of the dimensionless growth rate, $s \equiv \varrho_I a \mu_1/\gamma$, on the scaled wave number, ka, for viscosity ratio $\lambda = 0$ (dashed line) corresponding to a thread suspended in vacuum, 0.01, 0.05, 0.1, 0.2, 0.5, 1.0, 2.0, 5.0, 10.0, and 20.0 (lowest line).

is the normal-mode wave form of the perturbation, A is the complex amplitude of the interfacial wave, $k = 2\pi/L$ is the wave number, L is the wavelength, and c is the complex phase velocity.

Stream function, velocity and vorticity

Taking advantage of the axial symmetry of the flow, we describe the perturbation flow in terms of a perturbation Stokes stream function ψ_1 for the inner fluid and ψ_2 for the outer fluid. The axial and radial components of the perturbation velocity are given by

$$u_{x_j} = \epsilon \, \frac{1}{\sigma} \, \frac{\partial \psi_j^{1)}}{\partial \sigma}, \qquad u_{\sigma_j} = -\epsilon \, \frac{1}{\sigma} \, \frac{\partial \psi_j^{(1)}}{\partial x} \tag{9.13.18}$$

for $j =$1, 2, where the superscript (1) denotes the perturbation. The azimuthal component of the vorticity is

$$\omega_{\varphi_j} = -\epsilon \, \frac{1}{\sigma} \, E^2 \psi_j^{(1)}, \tag{9.13.19}$$

where

$$E^2 \equiv \frac{\partial^2}{\partial x^2} + \frac{\partial^2}{\partial \sigma^2} - \frac{1}{\sigma} \frac{\partial}{\partial \sigma} \tag{9.13.20}$$

is a second-order differential operator. The vorticity transport equation for axisymmetric Stokes flow requires that

$$E^4 \psi_j^{(1)} \equiv E^2 \left(E^2 \psi_j^{(1)} \right) = 0, \tag{9.13.21}$$

which can be decomposed into two second-order constitutive equations

$$E^2 \psi_j^{(1)} = \tilde{\psi}_j^{(1)}, \qquad E^2 \tilde{\psi}_j^{(1)} = 0, \qquad (9.13.22)$$

where a tilde denotes an intermediate solution.

Normal-mode expansion

The stream function is expressed in the normal-mode form

$$\psi_j^{(1)}(x, \sigma, t) = \phi_j(\sigma) \, \exp[ik(x - ct)], \qquad (9.13.23)$$

where $\phi_j(\sigma)$ are eigenfunctions. Substituting this expression into (9.13.22), we obtain

$$\left(\frac{d^2}{d\sigma^2} - \frac{1}{\sigma} \frac{d}{d\sigma} - k^2 \right) \phi_j = \tilde{\phi}_j, \qquad (9.13.24)$$

where

$$\left(\frac{d^2}{d\sigma^2} - \frac{1}{\sigma} \frac{d}{d\sigma} - k^2 \right) \tilde{\phi}_j = 0. \qquad (9.13.25)$$

The general solution is

$$\phi_1(\sigma) = \sigma \big[A_{1,1} \, \mathrm{I}_1(\hat{\sigma}) + A_{2,1} \, \sigma \, \mathrm{I}_0(\hat{\sigma}) \big] \qquad (9.13.26)$$

for the thread fluid labeled 1, and

$$\phi_2(\sigma) = \sigma \big[B_{1,2} \, \mathrm{K}_1(\hat{\sigma}) + B_{2,2} \, \sigma \, \mathrm{K}_0(\hat{\sigma}) \big] \qquad (9.13.27)$$

for the ambient fluid labeled 2, where I_0, I_1, K_0, and K_1 are Bessel functions, $A_{i,j}$, $B_{i,j}$ are complex constants, and $\hat{\sigma} \equiv k\sigma$ is the dimensionless radial position. The first term on the right-hand side of (9.13.26) or (9.13.27) satisfies (9.13.24) with $\tilde{\phi}_j = 0$. The second term satisfies (9.13.24) with $\tilde{\phi}_j$ being a nontrivial solution of (9.13.25).

Substituting the preceding expressions into (9.13.18) and using the properties of the Bessel functions,

$$\mathrm{I}_0'(z) = \mathrm{I}_1(z), \qquad \mathrm{I}_1'(z) = \mathrm{I}_0(z) - \frac{1}{z} \mathrm{I}_1(z), \qquad \mathrm{K}_0'(z) = -\mathrm{K}_1(z),$$

$$\mathrm{K}_1'(z) = -\mathrm{K}_0(z) - \frac{1}{z} \mathrm{K}_1(z), \qquad (9.13.28)$$

where a prime denotes a derivative with respect to z, we find that

$$u_{x_j} = \epsilon \, U_{x_j}(\sigma) \, \exp[ik(x - ct)], \qquad u_{\sigma_j} = \epsilon \, U_{\sigma_j}(\sigma) \, \exp[ik(x - ct)], \qquad (9.13.29)$$

where

$$U_{x_1}(\sigma) = \frac{1}{\sigma} \frac{d\phi_1}{d\sigma} = A_{1,1} k \, \mathrm{I}_0(\hat{\sigma}) + A_{2,1} \big[2 \, \mathrm{I}_0(\hat{\sigma}) + \hat{\sigma} \, \mathrm{I}_1(\hat{\sigma}) \big],$$

$$U_{\sigma_1}(\sigma) = -\frac{ik}{\sigma} \phi_1 = -ik \big[A_{1,1} \mathrm{I}_1(\hat{\sigma}) + A_{2,1} \sigma \, \mathrm{I}_0(\hat{\sigma}) \big] \qquad (9.13.30)$$

for the thread fluid labeled 1, and

$$U_{x_2}(\sigma) = \frac{1}{\sigma}\frac{d\phi_2}{d\sigma} = -B_{1,2}\, k\, K_0(\hat{\sigma}) + B_{2,2}\left[\, 2\, K_0(\hat{\sigma}) - \hat{\sigma}\, K_1(\hat{\sigma})\,\right],$$

$$U_{\sigma_2}(\sigma) = -\frac{ik}{\sigma}\,\phi_2 = -ik\left[\, B_{1,2}K_1(\hat{\sigma}) + B_{2,2}\,\sigma\, K_0(\hat{\sigma})\,\right] \qquad (9.13.31)$$

for the ambient fluid labeled 2.

Pressure field

The normal-mode pressure field is

$$p_j = P_j + \epsilon\,\chi_j(\sigma)\,\exp[ik(x - ct)], \qquad (9.13.32)$$

where P_1 and P_2 are the uniform unperturbed pressures, $P_1 - P_2 = \gamma/a$ is the unperturbed capillary pressure, and γ is the surface tension. To compute the disturbance pressure field, we consider the x component of the Stokes equation for axisymmetric flow,

$$\frac{\partial p_j}{\partial x} = \mu_j\left[\frac{\partial^2 u_{x_j}}{\partial x^2} + \frac{1}{\sigma}\frac{\partial}{\partial\sigma}\Big(\sigma\frac{\partial u_{x_j}}{\partial\sigma}\Big)\right]. \qquad (9.13.33)$$

Substituting (9.13.32) along with the first equation into (9.13.29) and linearizing, we obtain

$$\chi_j(\sigma) = i\mu_j kU_{x_j} - \frac{i\mu_j}{k}\frac{1}{\sigma}\frac{d}{d\sigma}\Big(\sigma\frac{dU_{x_j}}{d\sigma}\Big) = -\frac{i\mu_j}{k}\Big[\frac{1}{\sigma}\frac{d}{d\sigma}\big(\sigma\frac{dU_{x_j}}{d\sigma}\big) - k^2\, U_{x_j}\Big] \qquad (9.13.34)$$

or

$$\chi_j(\sigma) = -\frac{i\mu_j}{k}\Big(\frac{d^2 U_{x_j}}{d\sigma^2} + \frac{1}{\sigma}\frac{dU_{x_j}}{d\sigma} - k^2\, U_{x_j}\Big). \qquad (9.13.35)$$

Making substitutions, we find that

$$\chi_1(\sigma) = -i\, 2\mu_1 kA_{2,1}I_0(\hat{\sigma}), \qquad \chi_2(\sigma) = -i\, 2\mu_2 kB_{2,2}K_0(\hat{\sigma}). \qquad (9.13.36)$$

Note that only one term of the general solution in each fluid contributes to the pressure field.

Continuity of velocity at the interface

Requiring that the x and σ velocity components are continuous at the interface and using expressions (9.13.30) and (9.13.31), we derive the linearized kinematic interfacial conditions

$$A_{1,1}\, k\, I_0(\hat{k}) + A_{2,1}\left[\, 2\, I_0(\hat{k}) + \hat{k}\, I_1(\hat{k})\,\right] = -B_{1,2}\, k\, K_0(\hat{k}) + B_{2,2}\left[\, 2\, K_0(\hat{k}) - \hat{k}\, K_1(\hat{k})\,\right],$$

$$A_{1,1}\, I_1(\hat{k}) + A_{2,1}\, a\, I_0(\hat{k}) = B_{1,2}\, K_1(\hat{k}) + B_{2,2}\, a\, K_0(\hat{k}), \qquad (9.13.37)$$

where $\hat{k} = ka$ is a scaled wave number.

Kinematic compatibility

Kinematic compatibility requires that $D[f(x,t) - \sigma]/Dt = 0$ at the interface, where D/Dt is the material derivative and the function f describes the position of the interface according to (9.13.16). Carrying out the differentiation, we obtain

$$\frac{\partial f}{\partial t} + u_x \frac{\partial f}{\partial x} - u_\sigma = 0, \qquad (9.13.38)$$

where the velocity is evaluated at the interface. Substituting the preceding expressions and linearizing, we obtain an expression for the interfacial amplitude,

$$A = -\frac{U_{\sigma_1}(\sigma = a)}{ick} = \frac{1}{c}\left[A_{1,1}I_1(\hat{k}) + A_{2,1}a\,I_0(\hat{k})\right]. \qquad (9.13.39)$$

Because of the continuity condition (9.13.37), the outer fluid velocity, $U_{\sigma_2}(\sigma = a)$, could have been used in this expression.

Tangential interfacial force balance

Balancing the tangential hydrodynamic forces exerted on an infinitesimal portion of the interface, we derive the linearized shear stress condition

$$\left(\sigma_{x\sigma_1} - \sigma_{x\sigma_2}\right)_{\sigma=a} = \mu_1\left(\frac{\partial u_{x_1}}{\partial \sigma} + \frac{\partial u_{\sigma_1}}{\partial x}\right)_{\sigma=a} - \mu_2\left(\frac{\partial u_{x_2}}{\partial \sigma} + \frac{\partial u_{\sigma_2}}{\partial x}\right)_{\sigma=a} = 0. \qquad (9.13.40)$$

Substituting the normal-mode expansions, we obtain

$$\mu_1\left(\frac{dU_{x_1}}{d\sigma} + ikU_{\sigma_1}\right)_{\sigma=a} - \mu_2\left(\frac{dU_{x_2}}{d\sigma} + ikU_{\sigma_2}\right)_{\sigma=a} = 0. \qquad (9.13.41)$$

Using the expressions

$$\frac{dU_{x_1}}{d\sigma} = A_{1,1}\,k^2 I_1(\hat{\sigma}) + A_{2,1}k\left[2\,I_1(\hat{\sigma}) + \hat{\sigma}\,I_0(\hat{\sigma})\right],$$

$$\frac{dU_{x_2}}{d\sigma} = B_{1,2}\,k^2 K_1(\hat{\sigma}) - B_{2,2}\,k\left[2\,K_1(\hat{\sigma}) - \hat{\sigma}\,K_0(\hat{\sigma})\right], \qquad (9.13.42)$$

along with expressions (9.13.30) and (9.13.31), we obtain

$$2\mu_1 k\left(A_{1,1}k\,I_1(\hat{k}) + A_{2,1}[I_1(\hat{k}) + \hat{k}I_0(\hat{k})]\right)$$
$$-2\mu_2 k\left(B_{1,2}k\,K_1(\hat{k}) - B_{2,2}[K_1(\hat{k}) - \hat{k}K_0(\hat{k})]\right) = 0. \qquad (9.13.43)$$

Consolidating the various terms, we derive a linear algebraic equation,

$$F_1 A_{1,1} + F_2 A_{2,1} + F_7 B_{1,2} + F_8 B_{2,2} = 0, \qquad (9.13.44)$$

where

$$F_1 = kI_1(\hat{k}), \quad F_2 = I_1(\hat{k}) + \hat{k}\,I_0(\hat{k}), \quad F_7 = -\lambda k\,K_1(\hat{k}), \quad F_8 = \lambda\left[K_1(\hat{k}) - \hat{k}\,K_0(\hat{k})\right]. \qquad (9.13.45)$$

Normal stress interfacial condition

Balancing the normal force exerted on an infinitesimal portion of the interface with the interfacial tension, γ, and linearizing, we derive the normal stress balance

$$\left(\sigma_{\sigma\sigma_1} - \sigma_{\sigma\sigma_2}\right)_{\sigma=a} = \left(-p + 2\mu_1 \frac{\partial u_{\sigma_1}}{\partial\sigma}\right)_{\sigma=a} - \left(-p + 2\mu_2 \frac{\partial u_{\sigma_2}}{\partial\sigma}\right)_{\sigma=a} = \gamma\, 2\kappa_m, \quad (9.13.46)$$

where κ_m is the mean curvature of the interface. In the linearized approximation,

$$2\kappa_m = -\frac{1}{a} + \epsilon\left(\frac{\eta}{a^2} + \eta_{xx}\right) = -\frac{1}{a} + \epsilon\frac{A}{a^2}\left(1 - \hat{k}^2\right)\exp[ik(x - ct)]. \quad (9.13.47)$$

Using (9.13.39) to eliminate the interface amplitude, A, we obtain

$$2\kappa_m = -\frac{1}{a} + \epsilon\, 2\kappa_m^{(1)}\exp[ik(x - ct)], \quad (9.13.48)$$

where

$$2\kappa_m^{(1)} = \frac{1 - \hat{k}^2}{ca^2}\left[A_{1,1}\, \mathrm{I}_1(\hat{k}) + A_{2,1}\, a\, \mathrm{I}_0(\hat{k})\right]. \quad (9.13.49)$$

Substituting the normal-mode expansions into (9.13.46), we obtain

$$\left(-\chi_1 + 2\mu_1 \frac{\partial U_{\sigma_1}}{\partial\sigma}\right)_{\sigma=a} - \left(-\chi_2 + 2\mu_2 \frac{\partial U_{\sigma_2}}{\partial\sigma}\right)_{\sigma=a} = \gamma\frac{1 - \hat{k}^2}{ca^2}\left[A_{1,1}\, \mathrm{I}_1(\hat{k}) + A_{2,1}\, a\, \mathrm{I}_0(\hat{k})\right]. \quad (9.13.50)$$

Using the expressions

$$\frac{dU_{\sigma_1}}{d\sigma} = -i\,k\left(A_{1,1}\, k\left[\mathrm{I}_0(\hat{\sigma}) - \frac{1}{\hat{\sigma}}\mathrm{I}_1(\hat{\sigma})\right] + A_{2,1}\left[\mathrm{I}_0(\hat{\sigma}) + \hat{\sigma}\,\mathrm{I}_1(\hat{\sigma})\right]\right) \quad (9.13.51)$$

and

$$\frac{dU_{\sigma_2}}{d\sigma} = -i\,k\left(-B_{1,2}\, k\left[\mathrm{K}_0(\hat{\sigma}) + \frac{1}{\hat{\sigma}}\mathrm{K}_1(\hat{\sigma})\right] + B_{2,2}\left[\mathrm{K}_0(\hat{\sigma}) - \hat{\sigma}\,\mathrm{K}_1(\hat{\sigma})\right]\right), \quad (9.13.52)$$

along with the last expression in (9.13.36) for the pressure, we obtain

$$\alpha_{1,1}A_{1,1} + \alpha_{2,1}aA_{2,1} + \beta_{1,2}B_{1,2} + \beta_{2,2}aB_{2,2} = 0, \quad (9.13.53)$$

where

$$\alpha_{1,1} = \hat{k}\,\mathrm{I}_0(\hat{k}) - \mathrm{I}_1(\hat{k}) - \Pi\,\mathrm{I}_1(\hat{k}), \qquad \alpha_{2,1} = \hat{k}\,\mathrm{I}_1(\hat{k}) - \Pi\,\mathrm{I}_0(\hat{k}),$$
$$\beta_{1,2} = \lambda\left[\hat{k}\,\mathrm{K}_0(\hat{k}) + \mathrm{K}_1(\hat{k})\right], \qquad \beta_{2,2} = \lambda\,\hat{k}\,\mathrm{K}_1(\hat{k}). \quad (9.13.54)$$

We have introduced the dimensionless group

$$\Pi = i\,\frac{\gamma}{2\mu_1 c}\frac{1 - \hat{k}^2}{\hat{k}}. \quad (9.13.55)$$

If c is imaginary, as expected due to the absence of a mean base flow, the parameter Π is real.

(a)

$$
\mathbf{M} = \begin{bmatrix}
\mathrm{I}_1(\hat{k}) & \mathrm{I}_0(\hat{k}) & -\mathrm{K}_1(\hat{k}) & -\mathrm{K}_0(\hat{k}) \\
\hat{k}\,\mathrm{I}_0(\hat{k}) & 2\,\mathrm{I}_0(\hat{k}) + \hat{k}\mathrm{I}_1(\hat{k}) & \hat{k}\,\mathrm{K}_0(\hat{k}) & -2\,\mathrm{K}_0(\hat{k}) + \hat{k}\mathrm{K}_1(\hat{k}) \\
\hat{k}\,\mathrm{I}_1(\hat{k}) & \mathrm{I}_1(\hat{k}) + \hat{k}\,\mathrm{I}_0(\hat{k}) & -\lambda\,\hat{k}\,\mathrm{K}_1(\hat{k}) & \lambda\,[\mathrm{K}_1(\hat{k}) - \hat{k}\mathrm{K}_0(\hat{k})] \\
\hat{k}\mathrm{I}_0(\hat{k}) - \mathrm{I}_1(\hat{k}) - \Pi\,\mathrm{I}_1(\hat{k}) & \hat{k}\,\mathrm{I}_1(\hat{k}) - \Pi\,\mathrm{I}_0(\hat{k}) & \lambda\,[\hat{k}\,\mathrm{K}_0(\hat{k}) + \mathrm{K}_1(\hat{k})] & \lambda\,\hat{k}\,\mathrm{K}_1(\hat{k})
\end{bmatrix}
$$

(b)

$$
\mathbf{M}' = \begin{bmatrix}
\mathrm{I}_1(\hat{k}) & \hat{k}\,\mathrm{I}_0(\hat{k}) - \mathrm{I}_1(\hat{k}) \\
\hat{k}\,\mathrm{I}_0(\hat{k}) & \hat{k}\,[\mathrm{I}_0(\hat{k}) + \hat{k}\,\mathrm{I}_1(\hat{k})] \\
\hat{k}\,\mathrm{I}_1(\hat{k}) & \hat{a}^2\mathrm{I}_0(\hat{k}) \\
\hat{k}\,\mathrm{I}_0(\hat{k}) - \mathrm{I}_1(\hat{k}) - \Pi\,\mathrm{I}_1(\hat{k}) & (1 + \hat{k}^2)\mathrm{I}_1(\hat{k}) - \hat{k}\mathrm{I}_0(\hat{k}) - \Pi\,[\hat{k}\mathrm{I}_0(\hat{k}) - \mathrm{I}_1(\hat{k})]
\end{bmatrix}
$$

$$
\begin{bmatrix}
-\mathrm{K}_1(\hat{k}) & -\hat{k}\,\mathrm{K}_0(\hat{k}) - \mathrm{K}_1(\hat{k}) \\
\hat{k}\,\mathrm{K}_0(\hat{k}) & \hat{k}\,[-\mathrm{K}_0(\hat{k}) + \hat{k}\mathrm{K}_1(\hat{k})] \\
-\lambda\,\hat{k}\,\mathrm{K}_1(\hat{k}) & -\lambda\,\hat{k}^2\mathrm{K}_0(\hat{k}) \\
\lambda\,[\hat{k}\,\mathrm{K}_0(\hat{k}) + \mathrm{K}_1(\hat{k})] & \lambda\,[(1 + \hat{k}^2)\,\mathrm{K}_1(\hat{k}) + \hat{k}\,\mathrm{K}_0(\hat{k})]
\end{bmatrix}
$$

(c)

$$
\mathbf{M}'' = \begin{bmatrix}
\mathrm{I}_1(\hat{k}) & \hat{k}\,\mathrm{I}_0(\hat{k}) - \mathrm{I}_1(\hat{k}) \\
\mathrm{I}_0(\hat{k}) & \mathrm{I}_0(\hat{k}) + \hat{k}\,\mathrm{I}_1(\hat{k}) \\
\mathrm{I}_1(\hat{k}) & \hat{k}\,\mathrm{I}_0(\hat{k}) \\
\hat{k}\,\mathrm{I}_0(\hat{k}) - \mathrm{I}_1(\hat{k}) - \Pi\,\mathrm{I}_1(\hat{k}) & (\hat{k}^2 + 1)\,\mathrm{I}_1(\hat{k}) - \hat{k}\,\mathrm{I}_0(\hat{k}) - \Pi\,[\hat{k}\,\mathrm{I}_0(\hat{k}) - \mathrm{I}_1(\hat{k})]
\end{bmatrix}
$$

$$
\begin{bmatrix}
\mathrm{K}_1(\hat{k}) & -\hat{k}\,\mathrm{K}_0(\hat{k}) - \mathrm{K}_1(\hat{k}) \\
-\mathrm{K}_0(\hat{k}) & -\mathrm{K}_0(\hat{k}) + \hat{k}\,\mathrm{K}_1(\hat{k}) \\
\lambda\,\mathrm{K}_1(\hat{k}) & -\lambda\,\hat{k}\,\mathrm{K}_0(\hat{k}) \\
-\lambda\,[\hat{k}\,\mathrm{K}_0(\hat{k}) + \mathrm{K}_1(\hat{k})] & \lambda\,[(1 + \hat{k}^2)\,\mathrm{K}_1(\hat{k}) + \hat{k}\,\mathrm{K}_0(\hat{k})]
\end{bmatrix}
$$

TABLE 9.13.1 Matrices arising from the linear stability analysis of a viscous liquid thread. Setting the determinant of any one of these matrices to zero provides us with an algebraic equation for the complex phase velocity.

Formulation of an eigenvalue problem

To formulate an algebraic eigenvalue problem, we collect equations (9.13.37) (9.13.44), and (9.13.53) into a homogeneous linear system, $\mathbf{M} \cdot \mathbf{q} = \mathbf{0}$, where $\mathbf{q} = \left(A_{1,1},\ aA_{2,1},\ B_{1,2},\ aB_{2,2} \right)$, and the 4×4 complex coefficient matrix $\mathbf{M}$ is given in Table 9.13.1(a). Multiplying the second column of $\mathbf{M}$ by $\hat{k}$, subtracting from the product the first column, and also multiplying the fourth column by $\hat{k}$ and adding to the product the third column, we obtain the modified matrix $\mathbf{M}'$ shown in Table 9.13.1(b).

$$D_1 = \det\left(\begin{bmatrix} \hat{k}\,I_0(\hat{k}) - I_1(\hat{k}) & K_1(\hat{k}) & -\hat{k}\,K_0(\hat{k}) - K_1(\hat{k}) \\ I_0(\hat{k}) + \hat{k}\,I_1(\hat{k}) & -K_0(\hat{k}) & -K_0(\hat{k}) + \hat{k}\,K_1(\hat{k}) \\ \hat{k}\,I_0(\hat{k}) & \lambda\,K_1(\hat{k}) & -\lambda\,\hat{k}\,K_0(\hat{k}) \end{bmatrix} \right)$$

$$D_2 = \det\left(\begin{bmatrix} I_1(\hat{k}) & K_1(\hat{k}) & -\hat{k}\,K_0(\hat{k}) - K_1(\hat{k}) \\ I_0(\hat{k}) & -K_0(\hat{k}) & -K_0(\hat{k}) + \hat{k}\,K_1(\hat{k}) \\ I_1(\hat{k}) & \lambda\,K_1(\hat{k}) & -\lambda\,\hat{k}\,K_0(\hat{k}) \end{bmatrix} \right)$$

$$D_3 = \det\left(\begin{bmatrix} I_1(\hat{k}) & \hat{k}\,I_0(\hat{k}) - I_1(\hat{k}) & -\hat{k}\,K_0(\hat{k}) - K_1(\hat{k}) \\ I_0(\hat{k}) & I_0(\hat{k}) + \hat{k}\,I_1(\hat{k}) & -K_0(\hat{k}) + \hat{k}\,K_1(\hat{k}) \\ I_1(\hat{k}) & \hat{k}\,I_0(\hat{k}) & -\lambda\,\hat{k}\,K_0(\hat{k}) \end{bmatrix} \right)$$

$$D_4 = \det\left(\begin{bmatrix} I_1(\hat{k}) & \hat{k}\,I_0(\hat{k}) - I_1(\hat{k}) & K_1(\hat{k}) \\ I_0(\hat{k}) & I_0(\hat{k}) + \hat{k}\,I_1(\hat{k}) & -K_0(\hat{k}) \\ I_1(\hat{k}) & \hat{k}\,I_0(\hat{k}) & \lambda\,K_1(\hat{k}) \end{bmatrix} \right)$$

TABLE 9.13.2   Determinants of matrices arising from the linear stability analysis of a viscous liquid thread.

Switching the sign in the third column and eliminating the factor $\hat{k}$ in each term in the second row and third rows, we obtain the new matrix $\mathbf{M}''$ shown in Table 9.13.1(c). Expanding the determinant of this matrix with respect to the last row and setting the resulting expression to zero, we derive the equation

$$\Pi = \frac{1}{\Phi}\left(\left[\hat{k}\,I_0(\hat{k}) - I_1(\hat{k})\right] D_1 - \left[(\hat{k}^2 + 1)\,I_1(\hat{k}) - \hat{k}\,I_0(\hat{k})\right] D_2 \right.$$
$$\left. -\lambda\left[\hat{k}\,K_0(\hat{k}) + K_1(\hat{k})\right] D_3 + \lambda\left[(1 + \hat{k}^2)\,K_1(\hat{k}) + \hat{k}\,K_0(\hat{k})\right] D_4 \right), \qquad (9.13.56)$$

where $\Phi = I_1(\hat{k})\,D_1 - \left[\hat{k}\,I_0(\hat{k}) - I_1(\hat{k})\right]D_2$, and D_1–D_4 are four determinants defined with respect to the first three rows of $\mathbf{M}''$ and given in Table 9.13.2. Invoking the definition of Π from (9.13.55), we finally derive a purely imaginary growth rate,

$$\varrho \equiv kc = i\,\frac{\gamma}{2\mu_1 a\Pi}\,(1 - \hat{k}^2), \qquad (9.13.57)$$

where Π is computed from (9.13.56) [410]. A code that produces the growth rate is available in the software library FDLIB (Section D.17, Appendix D, Directory thread0).

Graphs of the dimensionless growth rate $s \equiv a\mu_1\varrho_I/\gamma$ plotted against the scaled wave number, ka, are shown in Figure 9.13.2(b) for several viscosity ratios, $\lambda = \mu_2/\mu_1$. The results demonstrate that the thread is unstable for scaled wave numbers that are lower than the Rayleigh–Plateau threshold, $ka < 1$. When the ambient fluid is inviscid, $\lambda = 0$, the growth rate tends to a finite limit

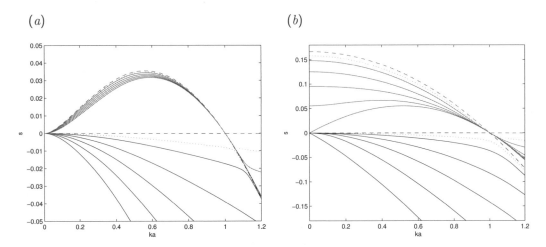

FIGURE 9.13.3   Effect of an insoluble surfactant on the instability of a viscous thread suspended in an infinite fluid. The graphs demonstrate the dependence of the dimensionless growth rate $s \equiv \varrho_I a \mu_1 / \gamma$ on the scaled wave number for $\beta = 0.001$ (dashed lines), 0.05 (dotted lines), $0.1, 0.2, \ldots, 0.5$, and negligible surfactant diffusivity, $\alpha = \infty$. The viscosity ratio is (a) $\lambda = 1$ and (b) 0.

as ka tends to zero. In contrast, when the ambient is viscous, the growth rate vanishes as ka tends to zero. We conclude that the ambient fluid viscosity has a profound effect on the evolution of long waves.

Effect of an insoluble surfactant

The presence of an insoluble surfactant is incorporated into the linear stability analysis presented in Directory ann210, Section D.2, Appendix D. The effect of the surfactant is determined by the dimensionless parameter β defined in (3.8.16) or Marangoni number Ma defined in (3.8.18), and a dimensionless number expressing the significance of the surfactant diffusivity, D_s,

$$\alpha \equiv \frac{a \gamma_0}{\mu_1 D_s}, \qquad (9.13.58)$$

where γ_0 is the surface tension of the unperturbed thread [214]. In practice, the surfactant surface diffusivity is typically negligible and α is large. When $\beta = 0$ or Ma $= 0$, the surface tension is not affected by the presence of the passively advected surfactant. A code that produces the growth rates of the two normal modes is available in the software library FDLIB, as discussed in Section D.2, Appendix D, Directory ann210.

Graphs of the dimensionless growth rate $s \equiv a \mu_1 \varrho_I / \gamma_0$ plotted against the scaled wave number, ka, are shown in Figure 9.13.3 for viscosity ratio $\lambda = 1$ and 0. Two normal modes arise in the presence of an insoluble surfactant. The first normal mode is unstable for scaled wave numbers that are lower than the Rayleigh–Plateau threshold, $ka < 1$. The second normal mode is stable for all wave numbers. The corresponding growth rates arise from the splitting of the contiguous lobe in

(a)   (b)

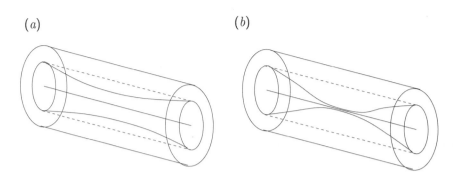

FIGURE 9.13.4 Evolution of an annular layer coating the interior side of a circular tube subject to an axisymmetric perturbation. The straight dotted lines show the initial position of the interface. Frame (a) illustrates the initial amplification of sinusoidal interfacial corrugations and the formation of primary drops connected with axisymmetric bridges, as shown in frame (b). The bridges eventually breakup to yield two alternating sequences of drops.

the absence of surfactant, $\beta = 0$. Because the surfactant reduces the growth rate of the unstable mode, it has a dampening action on the capillary instability. The effect is small when $\lambda = 1$, but significant when $\lambda = 0$.

9.13.3 Annular layers

The capillary instability of an annular layer coated on the external or internal surface of a hollow cylinder is of interest in pulmonary physiology with reference to the spontaneous occlusion of bronchioles (e.g., [281]). A code that produces the growth rate is available in the software library FDLIB for Navier–Stokes and Stokes flow, as discussed in Sections D.1 and D.2, Appendix D, Directories ann21 and ann210. The mathematical formulation takes into account the effect of an insoluble surfactant. A schematic illustration of the evolution of an annular layer coated on the interior of a circular tube, subject to an axisymmetric perturbation is shown in Figure 9.13.4. We observe the initial amplification of interfacial corrugations and the formation of primary drops connected with axisymmetric bridges. The bridges eventually break up to yield two alternating sequences of drops. If the thickness of the annular layers is small, the instability leads to the developing of lenticular structures occluding the tube or a periodic array of axisymmetric toroidal lobes attached to the cylinder.

Problems

9.13.1 *Coalescence of an array of drops*

Consider a one-dimensional array of touching spherical drops arranged in a straight line, representing a chain of melted glass beads. Explain why one should expect the drops to coalesce forming a liquid column, and the column to break up into a new array of drops. Compute the initial-to-final drop radius ratio.

🖥 *Computer Problem*

9.13.2 *Capillary instability of an inviscid jet*

Prepare a plot of the growth rate against the scaled wave number ka based on equations (9.13.12). To compute the modified Bessel functions, use computer library functions or polynomial approximations.

9.13.3 *Capillary instability of a viscous thread*

Reproduce the graphs of the growth rate shown in Figure 9.13.3(a).

9.14   FDLIB Software

Appendix C introduces the fluid mechanics software library FDLIB. Directory *08_stab* of FDLIB contains a collection of programs that perform the linear stability analysis. The User Guide of this directory is given in Appendix D.

Boundary-integral methods for potential flow

10

Potential flow is encountered in a variety of natural contexts and engineering applications. A familiar example is high-Reynolds-number flow past a streamlined body discussed in Chapters 7 and 8. Since the vorticity is confined inside thin boundary layers and narrow wakes, the main part of the flow is irrotational and the associated velocity field can described in terms of a scalar potential, $\mathbf{u} = \nabla \phi$. Using the continuity equation for incompressible fluids, $\nabla \cdot \mathbf{u} = 0$, we find that ϕ is a harmonic function, $\nabla^2 \phi = 0$. The computation of the flow is thus reduced to solving Laplace's equation subject to the no-penetration condition over the impermeable boundaries of the flow or to a dynamic boundary condition specifying the pressure at a free surface.

Another example of potential flow from a different physical context is provided by the flow of a viscous fluid through a channel confined between two parallel plates separated by a small distance h, called the Hele–Shaw cell, as discussed in Section 6.4.1. We have found that the fluid velocity averaged across the width of the channel, $\bar{\mathbf{u}}$, is proportional the gradient of the hydrodynamic pressure, $\tilde{p} = p - \rho \mathbf{g} \cdot \mathbf{x}$. Accordingly, $\tilde{p}$ plays the role of a velocity potential, $\bar{\mathbf{u}} = \nabla \phi$, where $\phi = -\tilde{p} h^2/(12\mu)$ and μ is the fluid viscosity. Mass conservation requires that the hydrodynamic pressure is a harmonic function, $\nabla^2 \tilde{p} = 0$. The computation of the Hele–Shaw flow is thus reduced to solving Laplace's equation subject to the Dirichlet, Neumann, or another boundary condition over different boundaries of the flow.

A related application concerns the flow of a viscous fluid through an isotropic porous medium, such as a fibrous matrix or ground rock. According to Darcy's law, the macroscopic velocity of the fluid, $\mathbf{U}$, defined as the average velocity of the fluid over a volume that is small compared to the global dimensions of the flow but large compared to the size of the fibers or grains, is related to the corresponding macroscopic hydrodynamic pressure, $\tilde{p}$, through the equation $\mathbf{U} = -(\kappa/\mu)\nabla \tilde{p}$, where κ is a physical constant called the permeability of the fluid through the porous medium. It is evident that the macroscopic pressure plays the role of a potential function, $\phi = (\kappa/\mu)\tilde{p}$. The continuity equation requires that $\tilde{p}$ is a harmonic function, $\nabla^2 \tilde{p} = 0$. The computation of a porous-medium flow is then reduced to solving Laplace's equation subject to the Neumann boundary conditions over the impermeable boundaries of the flow or to the Dirichlet boundary condition over boundaries that are exposed to the atmosphere.

Motivated by the pervasiveness of Laplace's equation in the various branches of fluid mechanics and other physical and engineering sciences, we devote this chapter to discussing a powerful class of numerical methods for computing potential flow in domains with arbitrary geometry, known as

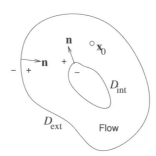

FIGURE 10.1.1   Illustration of a flow domain bounded by an internal and an external surface whose
union comprises the boundary, D. In the case of exterior flow decaying at infinity, D_{ext} does not
appear.

boundary-integral, boundary-element (BEM), boundary-integral-equation (BIE), or panel methods.
The formulation and basic implementation of the method will be discussed in Section 10.1–10.4, and
subsequent sections are devoted to extensions and advanced topics.

10.1   The boundary-integral equation

In Section 2.3, we developed an integral representation of a function ϕ that satisfies Laplace's
equation, $\nabla^2\phi = 0$, in terms of two boundary integrals representing boundary distributions of
point sources and point-source dipoles. The densities of these distributions are proportional to the
boundary values of the function and its normal derivative. For a point $\mathbf{x}_0$ located inside a selected
control volume bounded by a surface, D, as shown in Figure 10.1.1, we derived the representation
(2.3.8), repeated below for convenience,

$$\phi(\mathbf{x}_0) = -\iint_D \mathcal{G}(\mathbf{x}_0, \mathbf{x})\, \mathbf{n}(\mathbf{x}) \cdot \nabla\phi(\mathbf{x})\, \mathrm{d}S(\mathbf{x}) + \iint_D \phi(\mathbf{x})\, \mathbf{n}(\mathbf{x}) \cdot \nabla\mathcal{G}(\mathbf{x}_0, \mathbf{x})\, \mathrm{d}S(\mathbf{x}), \qquad (10.1.1)$$

where $\mathcal{G}(\mathbf{x}_0, \mathbf{x})$ is a Green's function of Laplace's equation and $\mathbf{n}$ is the unit vector normal to D
pointing into the control volume. The Green's function satisfies the symmetry property $\mathcal{G}(\mathbf{x}_0, \mathbf{x}) = \mathcal{G}(\mathbf{x}, \mathbf{x}_0)$. The first integral on the right-hand side of (10.1.1), representing a distribution of point
sources, is the single-layer potential. The second integral, representing a distribution of point-source
dipoles oriented normal to D, is the double-layer potential.

Equation (10.1.1) is also valid at a point, $\mathbf{x}_0$, located inside a partially or entirely infinite
domain of flow that is bounded by an interior boundary D, provided that the velocity decays and
therefore the potential ϕ tends to the reference value of zero at infinity.

Integral equations

The representation (10.1.1) allows us to compute the value of $\phi(\mathbf{x}_0)$ at any point $\mathbf{x}_0$ inside a selected
control volume in terms of the distribution of ϕ and its normal derivative, $\mathbf{n}\cdot\nabla\phi$, over the boundaries
of the control volume. In practice, physical arguments provide us with one scalar boundary condition

for the distribution of ϕ, $\mathbf{n} \cdot \nabla \phi$, or their linear combination, but not for both. It would appear then that (10.1.1) is useful only when both boundary distributions are available. The basic idea of the boundary-integral method is to let the point $\mathbf{x}_0$ approach the boundary D, and thereby reduce (10.1.1) into a Fredholm integral equation for the unknown distribution.

10.1.1   Behavior of the hydrodynamic potentials

Before taking the limit of (10.1.1) as the point $\mathbf{x}_0$ approaches the boundary, D, we must examine the behavior of the single- and double-layer potentials. To avoid mathematical complications, we assume that the boundary D is a Lyapunov surface, which means that it has a continuously varying normal unit vector in the absence of corners or cusps.

Single-layer potential

Examining the singularity of the single-layer integrand in (10.1.1), we find that, as the point $\mathbf{x}_0$ approaches and then crosses D, the single-layer potential varies in a continuous fashion.

Double-layer potential

The double-layer potential exhibits a jump across the integration domain, D. The discontinuity is seen most clearly by writing

$$\lim_{\mathbf{x}_0 \to D^{\pm}} \iint_D \phi(\mathbf{x})\, \mathbf{n}(\mathbf{x}) \cdot \nabla \mathcal{G}(\mathbf{x}_0, \mathbf{x})\, \mathrm{d}S(\mathbf{x}) = \mathcal{J}_0 + \phi_0 \lim_{\mathbf{x}_0 \to D^{\pm}} \iint_D \mathbf{n}(\mathbf{x}) \cdot \nabla \mathcal{G}(\mathbf{x}_0, \mathbf{x})\, \mathrm{d}S(\mathbf{x}), \qquad (10.1.2)$$

where $\phi_0 \equiv \phi(\mathbf{x}_0)$,

$$\mathcal{J}_0(\mathbf{x}_0) \equiv \lim_{\mathbf{x}_0 \to D^{\pm}} \iint_D \left[\phi(\mathbf{x}) - \phi(\mathbf{x}_0) \right] \mathbf{n}(\mathbf{x}) \cdot \nabla \mathcal{G}(\mathbf{x}_0, \mathbf{x})\, \mathrm{d}S(\mathbf{x}), \qquad (10.1.3)$$

and the superscripts $+$ or $-$ designate that the point $\mathbf{x}_0$ approaches D from within the control volume, indicated by the normal vector, or from the exterior side, as shown in Figure 10.1.1. Examining the singularity of the integrand, we find that the function $\mathcal{J}_0(\mathbf{x}_0)$ varies continuously as $\mathbf{x}_0$ approaches and then crosses D. To assess the behavior of the integral in the right-hand side of (10.1.2), we recall that D represents the collection of all boundaries of an enclosed domain of flow and use identities (2.2.15) to find that

$$\lim_{\mathbf{x}_0 \to D_{ext}^{\pm}} \iint_D \phi(\mathbf{x})\, \mathbf{n}(\mathbf{x}) \cdot \nabla \mathcal{G}(\mathbf{x}_0, \mathbf{x})\, \mathrm{d}S(\mathbf{x}) = \mathcal{J}_0(\mathbf{x}_0) + \beta_{\pm} \phi(\mathbf{x}_0) \qquad (10.1.4)$$

for an exterior boundary, where $\beta_+ = 1$ and $\beta_- = 0$, and

$$\lim_{\mathbf{x}_0 \to D_{int}^{\pm}} \iint_D \phi(\mathbf{x})\, \mathbf{n}(\mathbf{x}) \cdot \nabla \mathcal{G}(\mathbf{x}_0, \mathbf{x})\, \mathrm{d}S(\mathbf{x}) = \mathcal{J}_0(\mathbf{x}_0) + \gamma_{\pm} \phi(\mathbf{x}_0) \qquad (10.1.5)$$

for an interior boundary, where $\gamma_+ = 0$ and $\gamma_- = -1$. These equations demonstrate that, as the point $\mathbf{x}_0$ crosses an exterior or interior boundary to enter the domain of flow, the double-layer integral undergoes a discontinuity equal to $\phi(\mathbf{x}_0)$.

Principal-value of the double-layer integral

When the evaluation point $\mathbf{x}_0$ is located precisely on D, the double-layer integral is an improper but convergent integral, called the principal-value integral and denoted by PV. Subtracting the singularity, as shown in (10.1.2), and using identities (2.2.15), we find that

$$\iint_{D_{ext}}^{PV} \phi(\mathbf{x})\,\mathbf{n}(\mathbf{x}) \cdot \nabla \mathcal{G}(\mathbf{x}_0,\mathbf{x})\,\mathrm{d}S(\mathbf{x}) = \mathcal{J}_0(\mathbf{x}_0) + \frac{1}{2}\,\phi(\mathbf{x}_0) \tag{10.1.6}$$

for an exterior boundary, and

$$\iint_{D_{int}}^{PV} \phi(\mathbf{x})\,\mathbf{n}(\mathbf{x}) \cdot \nabla \mathcal{G}(\mathbf{x}_0,\mathbf{x})\,\mathrm{d}S(\mathbf{x}) = \mathcal{J}_0(\mathbf{x}_0) - \frac{1}{2}\,\phi(\mathbf{x}_0) \tag{10.1.7}$$

for an interior boundary.

Combining (10.1.4) and (10.1.5) with (10.1.6) and (10.1.7), we obtain a unified relationship between the limits of the double-layer potential and its principal value for an exterior or interior boundary,

$$\lim_{\mathbf{x}_0 \to D^{\pm}} \iint_D \phi(\mathbf{x})\,\mathbf{n}(\mathbf{x}) \cdot \nabla \mathcal{G}(\mathbf{x}_0,\mathbf{x})\,\mathrm{d}S(\mathbf{x}) = \iint_D^{PV} \phi(\mathbf{x})\,\mathbf{n}(\mathbf{x}) \cdot \nabla \mathcal{G}(\mathbf{x}_0,\mathbf{x})\,\mathrm{d}S(\mathbf{x}) \pm \frac{1}{2}\,\phi(\mathbf{x}_0). \tag{10.1.8}$$

The significance of this equation hinges on the realization that the principal value of the double-layer potential is much easier to compute than the limit of the double-layer potential as the point $\mathbf{x}_0$ approaches the boundary from either side.

10.1.2   Boundary-integral equation

Having assessed the behavior of the single-layer and double-layer potentials, we now take the limit of the integral representation (10.1.1) as the field point $\mathbf{x}_0$ approaches the boundary D, and use (10.1.8) to derive the boundary-integral equation

$$\phi(\mathbf{x}_0) = -2 \iint_D \mathcal{G}(\mathbf{x}_0,\mathbf{x})\,\mathbf{n}(\mathbf{x}) \cdot \nabla \phi(\mathbf{x})\,\mathrm{d}S(\mathbf{x}) + 2 \iint_D^{PV} \phi(\mathbf{x})\,\mathbf{n}(\mathbf{x}) \cdot \nabla \mathcal{G}(\mathbf{x}_0,\mathbf{x})\,\mathrm{d}S(\mathbf{x}), \tag{10.1.9}$$

where the point $\mathbf{x}_0$ lies on D. It will be noted that (10.1.9) is identical to (10.1.1), except that the right-hand side is multiplied by a factor of two and the principal value of the double-layer integral is employed. Because of the symmetry property $\mathcal{G}(\mathbf{x}_2,\mathbf{x}_1) = \mathcal{G}(\mathbf{x}_1,\mathbf{x}_2)$, the arguments of the Green's function in the single-layer and double-layer potentials can be freely switched.

An alternative method of deriving (10.1.9) involves assuming that the point $\mathbf{x}_0$ lies on a smooth boundary, D, and then repeating the analysis of Section 2.3, excluding from the control volume a hemisphere centered at the point $\mathbf{x}_0$. Generalizing this method, we find that (10.1.9) is also valid at a point $\mathbf{x}_0$ located at a boundary wedge or corner, provided that $\phi(\mathbf{x}_0)$ on the left-hand side is multiplied by the factor $\alpha/2\pi$, where α is the solid angle subtended by the corner on the side of the control volume (Problem 10.1.1). In the case of a smooth boundary, $\alpha = 2\pi$.

10.1.3   Integral equations of the second kind

Specifying the distribution of the normal derivative $\mathbf{n} \cdot \nabla \phi$ over D reduces (10.1.9) into a Fredholm integral equation of the second kind for the boundary distribution of ϕ,

$$\phi(\mathbf{x}_0) = 2 \iint_D^{PV} \phi(\mathbf{x}) \, \mathbf{n}(\mathbf{x}) \cdot \nabla \mathcal{G}(\mathbf{x}_0, \mathbf{x}) \, \mathrm{d}S(\mathbf{x}) - 2\,\mathcal{I}^S(\mathbf{x}_0), \qquad (10.1.10)$$

where

$$\mathcal{I}^S(\mathbf{x}_0) \equiv \iint_D \mathcal{G}(\mathbf{x}_0, \mathbf{x}) \, \mathbf{n}(\mathbf{x}) \cdot \nabla \phi(\mathbf{x}) \, \mathrm{d}S(\mathbf{x}) \qquad (10.1.11)$$

is the known single-layer potential. Recasting (10.1.10) into the standard form of an integral equation, we obtain

$$\phi(\mathbf{x}_0) = \iint_D^{PV} \phi(\mathbf{x}) \, \mathcal{K}(\mathbf{x}_0, \mathbf{x}) \, \mathrm{d}S(\mathbf{x}) - 2\,\mathcal{I}^S(\mathbf{x}_0), \qquad (10.1.12)$$

where

$$\mathcal{K}(\mathbf{x}_0, \mathbf{x}) = 2\,\mathbf{n}(\mathbf{x}) \cdot \nabla \mathcal{G}(\mathbf{x}_0, \mathbf{x}) \qquad (10.1.13)$$

is the kernel of the double-layer potential.

Regularity

To assess the singular nature of the kernel of the double-layer potential, we note the asymptotic behavior (2.2.10) and find that, as $\mathbf{x} \to \mathbf{x}_0$,

$$\mathcal{K}(\mathbf{x}_0, \mathbf{x}) \simeq 2\,\mathbf{n}(\mathbf{x}) \cdot \nabla \left(\frac{1}{r} \right) = -\frac{1}{2\pi}\,\mathbf{n}(\mathbf{x}) \cdot \frac{\mathbf{x} - \mathbf{x}_0}{r^3}. \qquad (10.1.14)$$

When the normal unit vector varies continuously over D, as $\mathbf{x} \to \mathbf{x}_0$, the vectors $\mathbf{x}_0 - \mathbf{x}$ and $\mathbf{n}$ tend to become perpendicular and their dot product tends to vanish. Consequently, the singularity of the kernel $\mathcal{K}(\mathbf{x}_0, \mathbf{x})$ is of order $1/r^{2-\epsilon}$, where ϵ is a positive number determined by the smoothness of the boundary. If D is a smooth surface with finite curvature, $\epsilon = 0$ and the singularity of the kernel is of order $1/r$. Since the order of the singularity is less than that dimensionality of the domain of integration, which is equal to two, the kernel $\mathcal{K}(\mathbf{x}_0, \mathbf{x})$ and the integral equation (10.1.12) are weakly singular. Later in this chapter, we will see that this property has important implications on the existence and uniqueness of solution, as well as on the feasibility of computing the solution by iterative methods.

Uniqueness of solution

In Section 10.7, we will study the properties of the integral equation (10.1.12) and will find that, when D represents the interior boundary of an infinite external flow, the integral equation has a unique solution. However, when D is the exterior boundary of an internal flow in the absence of interior boundaries, the integral equation has either no solution or an infinite number of solutions that differ by an arbitrary constant. The physical relevance of these results becomes evident by identifying ϕ with a temperature field and noting that a steady temperature distribution in a confined domain will exist only if the flow rate of heat through the boundaries is zero. When this condition is met, the temperature can be set to an arbitrary level.

Flow past a stationary rigid body

As an application, we consider an ambient irrotational flow past a stationary impermeable body. In Section 7.2, we showed that the velocity potential is given by the simplified representation (7.2.4) with $\mathbf{V} = \mathbf{0}$ involving the double-layer potential alone,

$$\phi(\mathbf{x}_0) = \phi_\infty(\mathbf{x}_0) + \iint_B \phi(\mathbf{x})\, \mathbf{n}(\mathbf{x}) \cdot \nabla \mathcal{G}(\mathbf{x}_0, \mathbf{x})\, \mathrm{d}S(\mathbf{x}), \qquad (10.1.15)$$

where B is the surface of the body. Taking the limit as the point $\mathbf{x}_0$ approaches the surface of the body and using (10.1.8), we derive an integral equation of the second kind for ϕ,

$$\phi(\mathbf{x}_0) = 2\,\phi_\infty(\mathbf{x}_0) + 2 \iint_B^{PV} \phi(\mathbf{x})\, \mathbf{n}(\mathbf{x}) \cdot \nabla \mathcal{G}(\mathbf{x}_0, \mathbf{x})\, \mathrm{d}S(\mathbf{x}). \qquad (10.1.16)$$

Substituting the free-space Green's function, $\mathcal{G}(\mathbf{x}_0, \mathbf{x}) = 1/(4\pi r)$, where $r = |\mathbf{x} - \mathbf{x}_0|$, we obtain the specific form

$$\phi(\mathbf{x}_0) = 2\,\phi_\infty(\mathbf{x}_0) - \frac{1}{2\pi} \iint_B^{PV} \phi(\mathbf{x})\, \frac{\mathbf{x} - \mathbf{x}_0}{r^3} \cdot \mathbf{n}(\mathbf{x})\, \mathrm{d}S(\mathbf{x}). \qquad (10.1.17)$$

In Section 10.7, we will show that equation (10.1.17) has a unique solution for any ambient potential ϕ_∞ representing an arbitrary irrotational incident flow.

Flow past a sphere

As a specific application, we consider streaming flow with velocity $\mathbf{U}$ past a stationary sphere with radius a centered at the origin of the chosen Cartesian coordinates. The exact solution for the velocity potential was given in (7.5.6) as

$$\phi(\mathbf{x}) = \left(1 + \frac{1}{2}\frac{a^3}{|\mathbf{x}|^3}\right) \mathbf{U} \cdot \mathbf{x}. \qquad (10.1.18)$$

Substituting into (10.1.17) $\phi_\infty(\mathbf{x}) = \mathbf{U} \cdot \mathbf{x}$ and the boundary distribution $\phi(\mathbf{x}) = \frac{3}{2}\mathbf{U} \cdot \mathbf{x}$, and simplifying, we obtain the identity

$$\mathbf{x}_0 = \frac{3}{2\pi} \iint_{Sphere}^{PV} \mathbf{x}\, \frac{\mathbf{x} - \mathbf{x}_0}{r^3} \cdot \mathbf{n}(\mathbf{x})\, \mathrm{d}S(\mathbf{x}), \qquad (10.1.19)$$

where $r = |\mathbf{x} - \mathbf{x}_0|$, $\mathbf{n}(\mathbf{x}) = \frac{1}{a}\mathbf{x}$ is the normal unit vector, the integration is performed over the surface of the sphere, and the point $\mathbf{x}_0$ lies at the surface of the sphere (Problem 10.1.2).

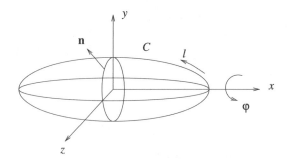

FIGURE 10.1.2   In the case of axisymmetric flow, the boundary integrals can be reduced to line integrals along the trace of the boundaries in an azimuthal plane.

10.1.4   Integral equations of the first kind

Specifying the distribution of ϕ over the boundary, D, reduces (10.1.9) into a Fredholm integral equation of the first kind for the boundary distribution of $\mathbf{n} \cdot \nabla \phi$,

$$\iint_D \mathcal{G}(\mathbf{x}_0, \mathbf{x})\, \mathbf{n}(\mathbf{x}) \cdot \nabla \phi(\mathbf{x})\, \mathrm{d}S(\mathbf{x}) = \mathcal{F}(\mathbf{x}_0), \tag{10.1.20}$$

with forcing function $\mathcal{F} = \mathcal{I}^D - \frac{1}{2}\,\phi$, where

$$\mathcal{I}^D(\mathbf{x}_0) \equiv \iint_D^{PV} \phi(\mathbf{x})\, \mathbf{n}(\mathbf{x}) \cdot \nabla \mathcal{G}(\mathbf{x}_0, \mathbf{x})\, \mathrm{d}S(\mathbf{x}) \tag{10.1.21}$$

is the double-layer integral. The kernel of the integral equation (10.1.20) is the Green's function, $\mathcal{G}(\mathbf{x}_0, \mathbf{x})$. Inspecting the singular behavior shown in (2.2.10), we find that, if the normal unit vector varies continuously over the boundary, D, the integral equation is weakly singular. Further properties of the solution will be discussed in Section 10.7.

10.1.5   Flow in an axisymmetric domain

In the case of axisymmetric flow, the boundary integrals of the single- or double-layer potentials can be reduced to line integrals along the contour of the boundaries in an azimuthal plane, denoted by C, as shown in Figure 10.1.2. This reduction considerably simplifies the solution of the integral equations and the computation of the velocity potential inside the domain of flow.

To implement these simplifications, we introduce cylindrical polar coordinates, (x, σ, φ), defined with respect to Cartesian coordinates such that $y = \sigma \cos \varphi$ and $z = \sigma \sin \varphi$. The azimuthal velocity component is zero, $u_\varphi = 0$, and the potential ϕ is independent of the azimuthal angle, φ. Next, we express the Cartesian components of the velocity and normal unit vector in cylindrical polar coordinates,

$$u_y = u_\sigma \cos \varphi, \qquad u_z = u_\sigma \sin \varphi, \qquad n_y = n_\sigma \cos \varphi, \qquad n_z = n_\sigma \sin \varphi, \tag{10.1.22}$$

and compute the normal derivative

$$\mathbf{n} \cdot \nabla \phi = \mathbf{n} \cdot \mathbf{u} = u_x n_x + u_\sigma n_\sigma. \tag{10.1.23}$$

Applying the boundary-integral equation (10.1.9) with the free-space Green's function given by $\mathcal{G}(\mathbf{x}_0, \mathbf{x}) = 1/(4\pi|\mathbf{x} - \mathbf{x}_0|)$, and expressing the differential surface area as $\mathrm{d}S = \sigma \,\mathrm{d}\varphi \,\mathrm{d}l$, we obtain

$$\phi(\mathbf{x}_0) = -\frac{1}{2\pi} \int_C \left(\int_0^{2\pi} \frac{1}{|\mathbf{x}_0 - \mathbf{x}|} \,\mathrm{d}\varphi \right) [\mathbf{n} \cdot \mathbf{u}](l) \,\sigma(l) \,\mathrm{d}l$$

$$+ \frac{1}{2\pi} \int_C^{PV} \left(\int_0^{2\pi} \frac{\mathbf{x}_0 - \mathbf{x}}{|\mathbf{x} - \mathbf{x}|^3} \cdot \mathbf{n}(\mathbf{x}) \,\mathrm{d}\varphi \right) \phi(l) \,\sigma(l) \,\mathrm{d}l, \tag{10.1.24}$$

where l is the arc length along the trace of the boundary in the xy plane, C.

It is convenient to introduce the single- and double-layer axisymmetric kernels defined as

$$\mathcal{I}(\mathbf{x}_0, l) \equiv \int_0^{2\pi} \frac{\mathrm{d}\varphi}{|\mathbf{x}_0 - \mathbf{x}|}, \qquad \mathcal{K}(\mathbf{x}_0, l) \equiv \int_0^{2\pi} \frac{\mathbf{x}_0 - \mathbf{x}}{|\mathbf{x} - \mathbf{x}|^3} \cdot \mathbf{n}(\mathbf{x}) \,\mathrm{d}\varphi. \tag{10.1.25}$$

Substituting into the double-layer kernel the expression

$$(\mathbf{x}_0 - \mathbf{x}) \cdot \mathbf{n} = (x_0 - x)n_x + (\sigma_0 \cos\varphi_0 - \sigma \cos\varphi)n_\sigma \cos\varphi + (\sigma_0 \sin\varphi_0 - \sigma \sin\varphi)n_\sigma \sin\varphi$$

$$= (x_0 - x)n_x + \sigma_0 n_\sigma \cos(\varphi_0 - \varphi) - \sigma n_\sigma, \tag{10.1.26}$$

we obtain

$$\mathcal{I}(\mathbf{x}_0, l) = I_{10}(\hat{x}, \sigma, \sigma_0),$$

$$\mathcal{K}(\mathbf{x}_0, l) = (\hat{x} n_x - \sigma n_\sigma) I_{10}(\hat{x}, \sigma, \sigma_0) + \sigma_0 \, n_\sigma I_{30}(\hat{x}, \sigma, \sigma_0), \tag{10.1.27}$$

where $\hat{x} = x_0 - x$ and the integrals I_{nm} are defined in (2.12.10) and (2.12.11). The integral equation (10.1.24) takes the one-dimensional form

$$\phi(\mathbf{x}_0) = -\frac{1}{2\pi} \int_C \mathcal{I}(\mathbf{x}_0, l)[\mathbf{n} \cdot \mathbf{u}](l) \,\sigma(l) \,\mathrm{d}l + \frac{1}{2\pi} \int_C^{PV} \mathcal{K}(\mathbf{x}_0, l) \,\phi(l) \,\sigma(l) \,\mathrm{d}l. \tag{10.1.28}$$

Integral equations of the first or second kind with respect to arc length, l, can be derived depending on the imposed boundary conditions as discussed previously in this section.

Three-dimensional flow in an axisymmetric domain

In the case of three-dimensional flow in an axisymmetric domain, the velocity potential and Cartesian components of the velocity can be expanded in Fourier series with respect to the azimuthal angle, φ. Substituting these series into the boundary-integral equation, we derive corresponding Fourier series for the single- and double-layer potentials. Separating similar terms provides us with a system of integral equations for the Fourier coefficients. The procedure has been used successfully to derive analogous integral equations for Stokes flow [313].

Problems

10.1.1 *General form of the boundary integral equation*

Show that the boundary-integral representation and boundary-integral equation can be recast into
the generalized form

$$c\,\phi(\mathbf{x}_0) = -2 \iint_D \mathcal{G}(\mathbf{x}_0,\mathbf{x})\,\mathbf{n}(\mathbf{x}) \cdot \nabla\phi(\mathbf{x})\,\mathrm{d}S(\mathbf{x}) + 2 \iint_D^{PV} \phi(\mathbf{x})\,\mathbf{n}(\mathbf{x}) \cdot \nabla\mathcal{G}(\mathbf{x}_0,\mathbf{x})\,\mathrm{d}S(\mathbf{x}), \quad (10.1.29)$$

where $c = 1$ or 0 when the point $\mathbf{x}_0$ is located inside or outside selected control volume, $c = \frac{1}{2}$ when
$\mathbf{x}_0$ lies on a smooth patch of D, and $c = \alpha/(2\pi)$ when $\mathbf{x}_0$ lies at corner or wedge, where α is the
solid angle subtended by the corner facing the control volume. In all cases, the normal unit vector,
$\mathbf{n}$, points into the control volume.

10.1.2 *Flow past a sphere*

To confirm identity (10.1.19), we write

$$\iint_{Sphere}^{PV} \mathbf{x}\,\frac{\mathbf{x}-\mathbf{x}_0}{r^3} \cdot \mathbf{n}(\mathbf{x})\,\mathrm{d}S(\mathbf{x}) = \iint_{Sphere} (\mathbf{x}-\mathbf{x}_0)\,\frac{\mathbf{x}-\mathbf{x}_0}{r^3} \cdot \mathbf{n}(\mathbf{x})\,\mathrm{d}S(\mathbf{x}) - \mathbf{x}_0 \iint_{Sphere}^{PV} \mathbf{n}(\mathbf{x}) \cdot \nabla\Big(\frac{1}{r}\Big)\,\mathrm{d}S(\mathbf{x}).$$

$$(10.1.30)$$

To compute the first integral on the right-hand side, we introduce Cartesian coordinates with origin
at the center of the sphere and the x axis passing through the point $\mathbf{x}_0$. Owing to symmetry, the y
and z components of the integral are identically zero. Carrying out the integration in terms of the
azimuthal angle, θ, show that the x component of the integral is equal to $-\frac{4\pi}{3}\,\mathbf{x}_0$. Use (2.2.15) to
show that the second integral on the right-hand side of (10.1.30) is equal to -2π. Combining these
results, confirm identity (10.1.19).

10.2   Two-dimensional flow

The results of Section 10.1 for three-dimensional flow can be adapted readily to two-dimensional
interior or exterior flow. The counterpart of the boundary-integral equation (10.1.9) is

$$\phi(\mathbf{x}_0) = -2 \int_C \mathcal{G}(\mathbf{x}_0,\mathbf{x})\,\mathbf{n}(\mathbf{x}) \cdot \nabla\phi(\mathbf{x})\,\mathrm{d}l(\mathbf{x}) + 2 \int_C^{PV} \phi(\mathbf{x})\,\mathbf{n}(\mathbf{x}) \cdot \nabla\mathcal{G}(\mathbf{x}_0,\mathbf{x})\,\mathrm{d}l(\mathbf{x}), \quad (10.2.1)$$

where $\mathcal{G}(\mathbf{x}_{,0}\,,\mathbf{x})$ is a two-dimensional Green's function, C is the boundary of a selected control area,
l is the arc length along C, and $\mathbf{n}$ is the unit vector pointing into the control area. In the remainder
of this section, we illustrate the application of (10.2.1) to several types of flow.

10.2.1   Flow past a rigid body with circulation

In the first application, we consider irrotational flow past a stationary two-dimensional body with
nonzero circulation around the body. Setting $\mathbf{V} = \mathbf{0}$ in the integral representation (7.8.3) and using

the free-space Green's function, $\mathcal{G}(\mathbf{x}_0, \mathbf{x}) = -\frac{1}{2\pi} \ln(|\mathbf{x} - \mathbf{x}_0|/a)$, we obtain an integral representation for the single-valued potential Φ,

$$\Phi(\mathbf{x}_0) = \phi_\infty(\mathbf{x}_0) - \frac{1}{2\pi} \oint_B \ln\left(\frac{r}{a}\right) \mathbf{n}(\mathbf{x}) \cdot \mathbf{v}(\mathbf{x})\, \mathrm{d}l(\mathbf{x}) - \frac{1}{2\pi} \oint_B \Phi(\mathbf{x})\, \mathbf{n}(\mathbf{x}) \cdot \frac{\mathbf{x} - \mathbf{x}_0}{r^2}\, \mathrm{d}l(\mathbf{x}), \quad (10.2.2)$$

where $r = |\mathbf{x} - \mathbf{x}_0|$, a is a chosen length, and $\mathbf{v}$ is the velocity due to a point vortex whose strength is equal to the circulation of the flow around the body, located at an arbitrary point inside the body. Letting the point $\mathbf{x}_0$ approach the body, we derive an integral equation of the second kind for Φ,

$$\Phi(\mathbf{x}_0) = -\frac{1}{\pi} \oint_B^{PV} \Phi(\mathbf{x})\, \mathbf{n}(\mathbf{x}) \cdot \frac{\mathbf{x} - \mathbf{x}_0}{r^2}\, \mathrm{d}l(\mathbf{x}) + 2\phi_\infty(\mathbf{x}_0) - \frac{1}{\pi} \oint_B \ln\left(\frac{r}{a}\right) \mathbf{n}(\mathbf{x}) \cdot \mathbf{v}(\mathbf{x})\, \mathrm{d}l(\mathbf{x}). \quad (10.2.3)$$

In Section 10.7, we will show that this integral equation admits a unique solution.

10.2.2   Fluid sloshing in a tank

An interesting formulation arises in the case of irrotational flow due to the sloshing of a liquid inside a tank, as illustrated in Figure 10.2.1(a). Applying the integral equation (10.2.1) with the free-space Green's function and using the no-penetration condition over the walls, we obtain

$$\phi(\mathbf{x}_0) = -\frac{1}{\pi} \oint_{W,S}^{PV} \phi(\mathbf{x})\, \mathbf{n}(\mathbf{x}) \cdot \frac{\mathbf{x} - \mathbf{x}_0}{r^2}\, \mathrm{d}l(\mathbf{x}) + \frac{1}{\pi} \int_S \ln\left(\frac{r}{a}\right) \mathbf{n}(\mathbf{x}) \cdot \mathbf{u}(\mathbf{x})\, \mathrm{d}l(\mathbf{x}), \quad (10.2.4)$$

where W represents the four side and bottom walls, S represents the free surface, and a is a chosen length. An efficient computational procedure for describing the evolution of the free surface subject to an initial condition involves the following steps (e.g., [275]):

1. Trace the free surface with a collection of marker points and assign to the marker points initial values for ϕ.

2. Solve the integral equation (10.2.4) for the normal velocity component over the free surface, S, and for the potential over the walls, W. In this case, (10.2.4) is an integral equation of mixed kind.

3. Differentiate the potential along the free surface to compute the tangential component of the fluid velocity.

4. Advance the position of the marker points over a small time interval with the normal and tangential fluid velocities. To prevent clustering, it is preferable to move the marker points with the normal velocity alone. This modification provides us with a kinematically consistent description of the free surface.

5. At the new position, update the marker-point potential using Bernoulli's equation. Applying equation (7.1.3) at the free surface, we find that, when the marker points move with the fluid velocity, the rate of change of the potential following their motion is given by

$$\frac{\mathrm{D}\phi}{\mathrm{D}t} = \frac{\partial \phi}{\partial t} + \mathbf{u} \cdot \nabla\phi = \frac{\partial \phi}{\partial t} + |\mathbf{u}|^2 = \frac{1}{2}|\mathbf{u}|^2 - \frac{p}{\rho} + \mathbf{g} \cdot \mathbf{x} + c(t). \quad (10.2.5)$$

(a)   (b)

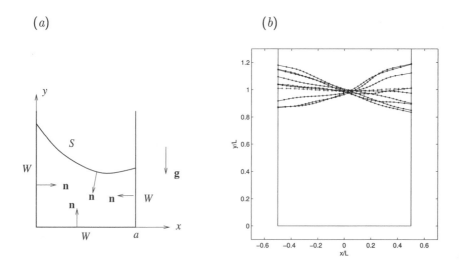

FIGURE 10.2.1 (a) Illustration of liquid sloshing inside a two-dimensional tank. The evolution of the free surface is computed by a boundary-integral method while the free-surface potential is updated using Bernoulli's equation. (b) Results of numerical simulations when the tank starts moving along the x axis to the left and then it stops. The initial position of the free surface is shown as a dashed line, and the interfacial marker points are shown as dots. Profiles are shown at equal time intervals. Note the deformation of the free surface and subsequent sloshing of the liquid.

If the marker points move with the normal velocity, we use the alternative evolution equation

$$\frac{\mathrm{d}\phi}{\mathrm{d}t} = \frac{\partial\phi}{\partial t} + (\mathbf{u}\cdot\mathbf{n})\,\mathbf{n}\cdot\nabla\phi + |\mathbf{u}|^2 = (\mathbf{u}\cdot\mathbf{n})^2 - \frac{1}{2}|\mathbf{u}|^2 - \frac{p}{\rho} + \mathbf{g}\cdot\mathbf{x} + c(t). \qquad (10.2.6)$$

The time-dependent integration constant, $c(t)$, in (10.2.5) and (10.2.6) can be set to zero without any consequences on the motion of the fluid. The pressure on the right-hand sides derives from the dynamic boundary condition at the free surface. For example, in the case of a free surface with surface tension γ, we set $p = p_{atm} + \gamma\kappa$, where κ is the curvature of the free surface in the xy plane and p_{atm} is the ambient atmospheric pressure.

6. Return to Step 2 and repeat the cycle.

If the tank undergoes vertical or horizontal acceleration with velocity $\mathbf{V}(t)$, we work in a frame of reference where the tank appears to be stationary. The acceleration of gravity, $\mathbf{g}$, on the right-hand side of (10.2.5) and (10.2.6) is then replaced by $\mathbf{g} - \mathrm{d}\mathbf{V}/\mathrm{d}t$ to account for the fictitious inertial acceleration force.

Green's function for a semi-infinite strip

The boundary-integral equation (10.2.4) can be simplified by using a Green's function of the second kind (Neumann function) whose normal derivative is zero over the bottom, left, and right walls,

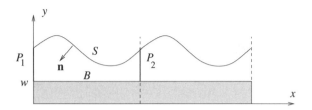

FIGURE 10.2.2   Flow due to the propagation of free-surface gravity waves over a flat bottom.

constructed by the method of images. The image system consists of four periodic arrays of point sinks with equal strengths. Using the expression for the potential due to a periodic array of point sources shown in Table 7.8.1, we obtain

$$\mathcal{G}(\mathbf{x}, \mathbf{x}_0) = -\frac{1}{4\pi} \sum_{m=1}^{4} \ln \Big(\cosh[k(y - y_m)] - \cos[k(x - x_m)] \Big), \tag{10.2.7}$$

where $k = \pi/a$ is the wave number, and

$$\mathbf{x}_1 = (x_0, y_0), \qquad \mathbf{x}_2 = (-x_0, y_0), \qquad \mathbf{x}_3 = (x_0, -y_0), \qquad \mathbf{x}_4 = (-x_0, -y_0). \tag{10.2.8}$$

Using the Green's function (10.2.7) allows us to eliminate the double-layer potential over the walls and provides us with a Fredholm integral equation of the first kind for the normal component of the velocity over the free surface alone.

Stages in the deformation of the free surface computed using the numerical method described in this section in conjunction with the boundary-element method described in Section 10.3.1 are shown in Figure 10.2.1(b) in the absence of surface tension. The tank starts moving along the x axis with a constant acceleration for a certain time period, and then it stops. The numerical simulation illustrates the initial deformation of the free surface and subsequent sloshing of the fluid inside the tank.

10.2.3   Propagation of gravity waves

In the third application, we consider two-dimensional potential flow due to the propagation of periodic gravity waves on the free surface of a liquid above a flat bottom wall located at $y = w$, as shown in Figure 10.2.2. The free surface and velocity field are repeated along the x axis with period a. Because the no-slip condition is not enforced at the bottom wall, the fluid can translate as a rigid-body along the x axis with arbitrary velocity. To remove this ambiguity, we stipulate without loss of generality that the potential ϕ observes the periodicity of the flow, $\phi(x + a) = \phi(x)$.

To simplify the boundary-integral formulation, we use a Green's function of the second kind (Neumann function) that is repeated in the x direction with period a. The normal derivative of the Green's function vanishes at the bottom, $y = w$. Using the method of images, we find that

$$\mathcal{G}(\mathbf{x}, \mathbf{x}_0) = -\frac{1}{4\pi} \ln\left(\cosh[k(y-y_0)] - \cos[k(x-x_0)] \right)$$

$$-\frac{1}{4\pi} \ln\left(\cosh[k(y+y_0-2w)] - \cos[k(x-x_0)] \right), \qquad (10.2.9)$$

where $k = 2\pi/a$ is the wave number. Next, we identify our control area with one period of the flow enclosed by the contours S, P_1, P_2, and B, as shown in Figure 10.2.2, and derive the integral equation (10.2.1). The double-layer integral is identically zero over B because the normal derivative of the Green's function vanishes when the point $\mathbf{x}$ is on B. The single-layer integral is also identically zero over B because of the no-penetration boundary condition. The integrals over the vertical segments P_1 and P_2 cancel each other because ϕ is periodic and the corresponding normal unit vectors point into opposite directions. These simplifications allow us to reduce the contour, C, to one period of the free surface alone. The evolution of the flow can be computed using the numerical procedure for the problem of fluid sloshing discussed in Section 10.2.2.

Problems

10.2.1 *Fluid sloshing in a tank*

Consider fluid sloshing inside a two-dimensional tank that undergoes rotational oscillations around the z axis about a fixed point. Indicate the necessary modifications of the computational procedure discussed in the text. *Hint:* consider Bernoulli's equation for a flow with constant vorticity.

10.2.2 *Gravity waves in a semi-infinite fluid*

Develop a boundary-integral formulation for the flow due to the propagation of a two-dimensional periodic gravity wave in a fluid with infinite depth.

10.3 Numerical methods

Analytical solutions of the boundary-integral equations are feasible only for a limited number of boundary geometries and types of flow. To compute the solution under general circumstances, we resort to approximate, asymptotic, and numerical methods. Reviewing the available strategies for solving Fredholm integral equations of mathematical physics, we find a variety of methods with varying degrees of accuracy and sophistication. A discussion of general and specialized methods can be found in monographs and reviews by Atkinson [16, 17, 19] and other authors [104, 174, 306, 313].

10.3.1 Boundary-element method

In a popular class of methods, known as boundary-element or panel methods, the boundary is discretized into a set of N_E boundary elements defined by element nodes. The boundary elements may have a variety of shapes, including flat and curved triangles or rectangles described by parametric interpolation (e.g., [306, 313]). In two dimensions, the boundary elements can be straight segments, circular arcs, or other interpolated segments. In the next step, the unknown boundary function is approximated with a truncated polynomial expansion in terms of properly defined surface coordinates over each element. The union of the local expansions provides us with an approximate solution that is not necessarily continuous across the element edges or nodes.

The union of all element expansions contains M unknown coefficients. When the unknown function is approximated with a constant function over each element, M is equal to the number of elements, N_E. When higher-order expansions are employed, M is greater than N_E. If the order of the expansion of the unknown function is identical to that of the Cartesian coordinates of a point over an element, we obtain an isoparametric representation. If it is higher, we obtain a superparametric representation.

Global expansions

To facilitate the logistics of the computation, it is convenient to identify the coefficients of the expansion over each element with the values of the unknown function at corresponding element nodes. The functions multiplying these coefficients are local or element basis functions. Compiling the local expansions and collecting the coefficients corresponding to shared element nodes, we obtain an expansion in terms of N global basis functions. It should be noted that, although superficially innocuous, certain types of local expansions may introduce instabilities in the numerical solution, as discussed later in this section.

Influence coefficients

The union of the local expansions of the unknown function is now substituted into the integral equation, and the N coefficients are extracted from the single- and double-layer potentials to yield an algebraic equation,

$$\mathcal{F}(c_1, c_2, \ldots, c_N) = 0, \tag{10.3.1}$$

where c_i are the coefficients of the global expansion. Since $\mathcal{F}$ is a linear function of c_i, we can write

$$\mathcal{F}(c_1, c_2, \ldots, c_N) = A_1 c_1 + A_2 c_2 + \cdots + A_N c_N + B, \tag{10.3.2}$$

where

$$B = \mathcal{F}(c_1 = 0, \ldots, c_N = 0). \tag{10.3.3}$$

The influence coefficients, A_i, are integrals of the single- or double-layer potential over selected boundary elements. In certain cases, it is expedient to compute the influence coefficients by numerical differentiation,

$$A_m = \mathcal{F}(c_1 = 0, \ldots, c_{m-1} = 0, c_m = 1, c_{m+1} = 0, \ldots, c_N = 0) - B, \tag{10.3.4}$$

implementing an exact finite-difference approximation.

Collocation and weighted residual methods

The coefficients of the local expansions can be computed by the collocation method or by the method of weighted residuals. In the collocation method, the discretized integral equation is applied at N

selected collocation points over the boundary. In the method of weighted residuals, the discretized integral equation is multiplied by a set of N chosen weighting functions and the product is integrated over the boundary. The goal in both cases is to produce a system of N linear algebraic equations for the N coefficients of the global expansion.

Different choices for the weighting functions in the method of weighted residuals produce different schemes. Identifying the weighting functions with the global basis functions, we obtain Galerkin's method. Identifying the weighting functions with delta functions whose singular points are placed at selected locations over the boundary, we recover the collocation method. Because of increased computational requirements, Galerkin's method is suitable only for problems with notable geometrical simplicity [17].

Formulation of a linear system

In the last step, the derived system of linear algebraic equations, symbolically written in the form $\mathbf{A} \cdot \mathbf{x} = \mathbf{b}$, is solved for the N coefficients of the global expansion contained in the unknown vector $\mathbf{x}$, where $\mathbf{A}$ is an influence matrix. Since, in general, the matrix $\mathbf{A}$ is dense and nonsymmetric, the solution must be found using a general-purpose numerical method. Two possible choices are Gauss elimination and an iterative method such as Jacobi's method or the method of conjugate gradients and its variations discussed in Section B.1, Appendix B. Iterative methods are preferred for large systems due to computational savings. A direct method requires $O(K^3)$ multiplications, whereas an iterative method requires $O(IK^2)$ multiplications, where K is the size of linear system and I is the number of necessary iterations. The former is much larger than the latter when $K > I$.

In solving the final system of linear equations for the coefficients of the local expansions, issues of well-posedness, existence, and uniqueness of the solution of the integral equation arise. Ill-posed integral equations and integral equations with no solution or multiple solutions lead to ill-conditioned linear systems with sensitive, oscillatory, or nonconvergent solutions. Integral equations of the first kind may be susceptible to oscillations due to numerical error (e.g., [104]). Fortunately, experience has shown that integral equations of the first kind that arise from boundary-integral representations are well-behaved.

10.3.2   Flow past a two-dimensional airfoil

To illustrate the application of the boundary-element method, we consider flow past a two-dimensional Joukowski airfoil with nonzero circulation around the airfoil, as illustrated in Figure 10.3.1. Our point of departure is the integral equation of the second kind (10.2.3) for the single-valued potential, Φ, repeated below for convenience,

$$\Phi(\mathbf{x}_0) = -\frac{1}{\pi} \oint_C^{PV} \Phi(\mathbf{x})\, \mathbf{n}(\mathbf{x}) \cdot \frac{\mathbf{x} - \mathbf{x}_0}{r^2}\, dl(\mathbf{x}) + 2\phi_\infty(\mathbf{x}_0) - \frac{1}{\pi} \oint_C \ln\left(\frac{r}{a}\right) \mathbf{n}(\mathbf{x}) \cdot \mathbf{v}(\mathbf{x})\, dl(\mathbf{x}), \quad (10.3.5)$$

where C is the airfoil contour, $r = |\mathbf{x} - \mathbf{x}_0|$, and a is a chosen length. We begin by discretizing the contour of the airfoil into N boundary elements, E_j, where $j = 1, \ldots, N$. Approximating Φ and $\mathbf{v} \cdot \mathbf{n}$ with constant functions over the jth element, respectively equal to Φ_j and v_j, we recast (10.3.5) into the discrete form

$$\Phi(\mathbf{x}_0) = -F_j(\mathbf{x}_0)\, \Phi_j + 2\,\phi_\infty(\mathbf{x}_0) + G_j(\mathbf{x}_0)\, v_j, \quad (10.3.6)$$

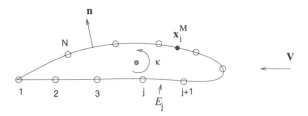

FIGURE 10.3.1   Illustration of potential flow past a two-dimensional Joukowski airfoil computed by a
boundary-element method. The jth element is defined by the jth and $j + 1$ nodes.

where summation is implied over the repeated index, j. The influence coefficients $F_j(\mathbf{x}_0)$ and $G_j(\mathbf{x}_0)$
are defined as

$$F_j(\mathbf{x}_0) = \frac{1}{\pi} \int_{E_j}^{PV} \frac{\mathbf{x} - \mathbf{x}_0}{r^2} \cdot \mathbf{n}(\mathbf{x}) \, dl(\mathbf{x}), \qquad G_j(\mathbf{x}_0) = -\frac{1}{\pi} \int_{E_j} \ln\left(\frac{r}{a}\right) dl(\mathbf{x}). \qquad (10.3.7)$$

The principal-value integral is relevant only when the evaluation point, $\mathbf{x}_0$, lies on the element E_j.

Applying equation (10.3.6) at the midpoint of each element, $\mathbf{x}_i^M$, we obtain a system of linear
equations for the collection of all Φ_j,

$$\Phi_i = -F_j(\mathbf{x}_i^M)\, \Phi_j + 2\, \phi_\infty(\mathbf{x}_i^M) + G_j(\mathbf{x}_i^M)\, v_j, \qquad (10.3.8)$$

which can be rewritten in the standard form of a linear system,

$$\left[\delta_{ij} + F_j(\mathbf{x}_i^M) \right] \Phi_j = 2\, \phi_\infty(\mathbf{x}_i^M) + G_j(\mathbf{x}_i^M)\, v_j, \qquad (10.3.9)$$

where δ_{ij} is Kronecker's delta. The solution can be found by the method of Gauss elimination
discussed in Section B.1.2, Appendix B.

Computation of the influence coefficients

The accurate computation of the influence coefficients, $F_j(\mathbf{x}_i^M)$ and $G_j(\mathbf{x}_i^M)$, is an important aspect
of the numerical method. When $i \neq j$, the integrals involved in the definition of these coefficients
can be computed by standard methods, including the trapezoidal rule and the Gauss–Legendre
quadrature discussed in Section B.6, Appendix B.

Consider the diagonal coefficients, $F_i(\mathbf{x}_i^M)$, where summation is *not* implied over i. When the
elements are straight segments and the integration point, $\mathbf{x}$, lies at the ith element, the distance $\mathbf{x} -$
$\mathbf{x}_i^M$ is perpendicular to the normal unit vector, the corresponding integrand in (10.3.7) is identically
zero, and $F_i(\mathbf{x}_i^M) = 0$. When the elements are curved segments such as circular arcs, we apply a
Taylor series expansion about the midpoint to find that $F_i(\mathbf{x}_i^M) \simeq 0$ to leading-order approximation.
These simplifications explain why the principal value of the double-layer integral is used instead of
the limit of the double-layer integral as the evaluation point approaches a boundary.

When $i = j$, the integral $G_i(\mathbf{x}_i^M)$ exhibits a logarithmic singularity, which can be integrated analytically over a straight segment to yield

$$G_i(\mathbf{x}_i^M) = -\frac{1}{\pi} \Delta l_i \Big[\ln \Big(\frac{\Delta l_i}{2a}\Big) - 1 \Big]. \qquad (10.3.10)$$

where $\Delta l_i = |\mathbf{x}_{i+1} - \mathbf{x}_i|$ is the length of the ith element. A similar analytical integration is possible when the boundary elements are circular arcs (Problem 10.3.1).

Velocity and pressure

Having solved the linear system (10.3.8) or (10.3.9), we may differentiate the boundary distribution of the velocity potential with respect to arc length using a standard numerical method to obtain the boundary distribution of the tangential velocity, and then apply Bernoulli's equation to compute the boundary distribution of the pressure and pressure drag coefficient. The drag and lift forces exerted on the airfoil can be computed from the pressure by numerically integrating along the airfoil contour, as discussed in Section 7.1. The results show that the drag force vanishes and the lift force has the Kutta–Joukowski value (7.8.19).

Problems

10.3.1 *Flow past a two-dimensional airfoil*

Discuss the significance of the location of the point vortex with regard to the integral equation (10.3.5).

10.3.2 *Boundary-element integrals*

(*a*) Derive (10.3.10) and investigate whether it is possible to compute $G_j(\mathbf{x}_i^M) = 0$ analytically over a straight segment when $i \neq j$.

(*b*) Derive the counterpart of (10.3.10) when the boundary elements are circular arcs.

Computer Problems

10.3.3 *Flow past a two-dimensional airfoil*

Compute the pressure distribution, drag force, and lift force on a Joukowski airfoil of your choice. Compare your results for the lift and drag force with theoretical predictions (Section 7.8).

10.3.4 *Flow in a coating die*

Compute the velocity distribution at the exit of the coating die shown in Figure 6.3.2 for $h = 3$ mm, $L = 20$ cm, $W = 20$ cm, sloping angle $\alpha = 20°$, and constant tube radius $R = 6$ mm. The pressure at the inlet point is 3 atm, and the working fluid is glycerin.

10.3.5 *Unidirectional flow in a pipe*

Compute the distribution of shear stress around the boundary of a cylindrical pipe with elliptical cross-section and aspect ratio of your choice in unidirectional pressure-driven flow. Compare your results with the exact solution derived in Section 5.2.2.

10.4   Generalized boundary-integral representations

In Section 7.2, we developed the boundary-integral representation (7.2.4) for the harmonic potential of an irrotational flow past a stationary or translating rigid body involving the double-layer potential alone. Two important advantages of this representation are conceptual simplicity and computational convenience in solving the associated integral equation. Motivated by these simplifications, we seek to eliminate one of the two hydrodynamic potentials from the general boundary-integral representation and associated integral equation, and thereby construct generalized boundary-integral representations and associated integral equations.

Exterior flow

Considering an exterior flow that decays far from an interior boundary, D, we introduce the harmonic potential χ in the interior of D so that $\chi = \phi$ over D. Applying the reciprocal relation (2.3.2) for χ at a point $\mathbf{x}_0$ in the exterior of D and subtracting (2.3.2) from (2.3.8), we obtain a generalized representation in terms of a single-layer potential alone,

$$\phi(\mathbf{x}_0) = -\iint_D \mathcal{G}(\mathbf{x}_0, \mathbf{x})\, q(\mathbf{x})\, \mathrm{d}S(\mathbf{x}), \tag{10.4.1}$$

where $q = \mathbf{n} \cdot \nabla(\phi - \chi)$ is the distribution density and $\mathbf{n}$ is the normal unit vector pointing into the exterior of D. Working in a similar fashion, we introduce a harmonic potential ϱ in the interior of D so that $\mathbf{n} \cdot \nabla\varrho = \mathbf{n} \cdot \nabla\phi$ over D, and derive the generalized double-layer representation

$$\phi(\mathbf{x}_0) = \iint_D \vartheta(\mathbf{x})\, \mathbf{n}(\mathbf{x}) \cdot \nabla \mathcal{G}(\mathbf{x}_0, \mathbf{x})\, \mathrm{d}S(\mathbf{x}). \tag{10.4.2}$$

where $\vartheta = \phi - \varrho$ is the distribution density and $\mathbf{n}$ is the normal unit vector pointing into the exterior of D. The derivation of (10.4.1) and (10.4.2) hinges on the existence of the complementary potentials χ and ϱ. In Section 10.7, we will see that it is always possible to find the internal potential, χ, and thus represent any exterior flow in terms of a single-layer potential alone. In contrast, a complementary potential ϱ exists only when the flow rate of ϕ across the boundary D is zero,

$$\iint_D \mathbf{n} \cdot \nabla\phi\, \mathrm{d}S = 0. \tag{10.4.3}$$

Consequently, only a limited class of exterior flows that satisfy (10.4.3) can be represented in terms of a double-layer potential alone.

Interior flow

In the case of interior flow, we introduce a harmonic potential χ in the exterior of D so that $\chi = \phi$ over D, and χ and decays at infinity. Repeating the preceding analysis, we derive the single-layer representation (10.4.1). Working in a similar fashion, we introduce the harmonic potential ϱ in the exterior of D so that $\mathbf{n} \cdot \nabla\varrho = \mathbf{n} \cdot \nabla\phi$ over D and ϱ decays at infinity, and derive the double-layer representation (10.4.2). Analysis undertaken in Section 10.7 guarantees the existence of both χ and ϱ, and thus ensures that any nonsingular interior flow can be represented in terms of a single-layer or double-layer potential alone.

Investigation of generalized representations

The simplicity of the generalized boundary-integral representations motivates us to investigate of the properties of the single- and double-layer potentials and associated integral equations, as well as assess the extent to which these representations are capable of describing an arbitrary interior or exterior flow. These topics will be the theme of the following three sections.

Problems

10.4.1 *Temperature field between bodies held at a constant temperature*

Consider the temperature field in the region between a number of closed surfaces that are held at different constant temperatures. Identify ϕ with the temperature and show that the double-layer integral on the right-hand side of (2.3.8) vanishes, yielding a representation in terms of a single-layer potential alone.

10.4.2 *Interior flow with an interior boundary*

Consider a domain of flow enclosed by an external surface in the presence of a closed interior boundary. Discuss whether it is possible to develop a double-layer representation defined over both the interior and exterior surfaces.

10.5   The single-layer potential

The single-layer potential is the harmonic function due to a distribution of Green's functions over a chosen surface, D,

$$\phi(\mathbf{x}_0) = \iint_D \mathcal{G}(\mathbf{x}_0, \mathbf{x})\, q(\mathbf{x})\, \mathrm{d}S(\mathbf{x}), \tag{10.5.1}$$

where q is the surface density of the distribution, as shown in Figure 10.5.1. Physically, the single-layer potential can be identified with the temperature field due to a distribution of point sources of heat or with the velocity potential due to a distribution of point sinks of mass with density q over the surface D.

The single-layer potential is continuous throughout the domain of flow as well as across the distribution domain, D. When $\mathcal{G}$ is a Green's function of the first kind, ϕ vanishes over the associated boundary S_B. When $\mathcal{G}$ is a Green's function of the second kind (Neumann function), the normal derivative of ϕ vanishes over S_B. Using the general properties of the Green's functions we find that, far from D, ϕ decays at a rate that is equal to, or faster than $1/d$, where d is the distance from the origin. The precise rate of decay depends on the geometry of S_B.

Integral identity

For simplicity, we assume that the distribution domain D is a closed surface with a continuously varying normal unit vector, $\mathbf{N}$, pointing outward from D, as shown in Figure 10.5.1. The integral

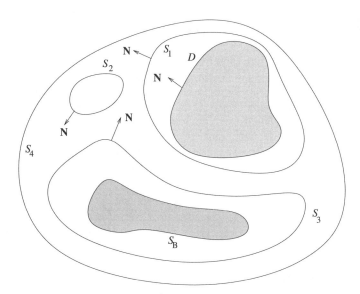

FIGURE 10.5.1   Distribution domain of a hydrodynamic potential and illustration of several closed
surfaces in a flow.

identities (2.2.15) allow us to write

$$\iint_D \mathbf{N}(\mathbf{x}) \cdot \nabla \mathcal{G}(\mathbf{x}, \mathbf{x}_0) \, \mathrm{d}S(\mathbf{x}) = \begin{cases} 0 & \text{when } \mathbf{x}_0 \text{ is outside } D, \\ -\frac{1}{2} & \text{when } \mathbf{x}_0 \text{ is on } D, \\ -1 & \text{when } \mathbf{x}_0 \text{ is inside } D. \end{cases} \qquad (10.5.2)$$

This identities are also true for any closed surface that does not enclose any boundaries, such as
the surface S_1 or S_2 illustrated in Figure 10.5.1, provided that the normal unit vector, $\mathbf{N}$, points
outward.

Flow rates

Integrating the normal derivative of (10.5.1) over a surface that encloses the distribution domain,
D, but no other boundaries, such as the surface S_1 illustrated in Figure 10.5.1, we find that the
volumetric flow rate through the surface is

$$Q_1 = \iint_{S_1} \mathbf{N}(\mathbf{x}_0) \cdot \nabla_0 \phi(\mathbf{x}_0) \, \mathrm{d}S(\mathbf{x}_0) = \iint_D \left(\iint_{S_1} \mathbf{N}(\mathbf{x}_0) \cdot \nabla_0 \mathcal{G}(\mathbf{x}_0) \, \mathrm{d}S(\mathbf{x}_0) \right) q(\mathbf{x}) \, \mathrm{d}S(\mathbf{x}). \quad (10.5.3)$$

Using (10.5.2) with S_1 in place of D, we obtain

$$Q_1 = - \iint_D q(\mathbf{x}) \, \mathrm{d}S(\mathbf{x}), \qquad (10.5.4)$$

which shows that Q_1 vanishes only when the total strength of the distribution defined as the integral
in (10.5.4) is zero. Working in a similar manner, we find that the volumetric flow rate across a closed

fluid surface that does not enclose any boundaries, such as the surface S_2 illustrated in Figure 10.5.1, is zero (Problem 10.5.1).

10.5.1 Derivatives of the single-layer potential

The derivative of the single-layer potential normal to a small surface element located at a point $\mathbf{x}_0$ with normal unit vector $\mathbf{n}(\mathbf{x}_0)$ is given by

$$\mathbf{n}(\mathbf{x}_0) \cdot \nabla_0 \phi(\mathbf{x}_0) = \mathbf{n}(\mathbf{x}_0) \cdot \iint_D \nabla_0 \mathcal{G}(\mathbf{x}_0, \mathbf{x}) \, q(\mathbf{x}) \, \mathrm{d}S(\mathbf{x}), \tag{10.5.5}$$

where the subscript 0 of the del operator designates differentiation with respect to $\mathbf{x}_0$. By definition, when $\mathcal{G}$ is a Green's function of the second kind, the normal derivative of ϕ vanishes over the associated boundary, S_B.

Normal derivative

To investigate the behavior of the normal derivative across the distribution domain, D, we take the limit of (10.5.5) as the point $\mathbf{x}_0$ approaches D, introduce the free-space Green's function $\mathcal{G}^{FS} = 1/(4\pi|\mathbf{x} - \mathbf{x}_0|)$, note that $\nabla_0 \mathcal{G}^{FS} = -\nabla \mathcal{G}^{FS}$, and write the identity

$$\mathbf{n}(\mathbf{x}_0) \cdot \nabla_0 \phi(\mathbf{x}_0) = J(\mathbf{x}_0) + \iint_D \mathbf{N}(\mathbf{x}) \cdot \nabla_0 \mathcal{G}^{FS}(\mathbf{x}_0, \mathbf{x}) \, q(\mathbf{x}_0) \, \mathrm{d}S(\mathbf{x}), \tag{10.5.6}$$

where

$$J(\mathbf{x}_0) \equiv \iint_D \left[\mathbf{N}(\mathbf{x}_0) \cdot \nabla_0 \mathcal{G}(\mathbf{x}_0, \mathbf{x}) \, q(\mathbf{x}) - \mathbf{N}(\mathbf{x}) \cdot \nabla_0 \mathcal{G}^{FS}(\mathbf{x}_0, \mathbf{x}) \, q(\mathbf{x}_0) \right] \mathrm{d}S(\mathbf{x}). \tag{10.5.7}$$

Because of the weak singularity of the integrand in (10.5.7), the integral is continuous across D. Considering the second term on the right-hand side of (10.5.6), we use (10.5.2) to obtain

$$\left[\mathbf{N}(\mathbf{x}_0) \cdot \nabla_0 \phi(\mathbf{x}_0) \right]^+ = J(\mathbf{x}_0), \qquad \left[\mathbf{N}(\mathbf{x}_0) \cdot \nabla_0 \phi_-(\mathbf{x}_0) \right]^- = J(\mathbf{x}_0) + q(\mathbf{x}_0), \tag{10.5.8}$$

where the plus superscript denotes the exterior side of D indicated by the direction of $\mathbf{N}$, and the minus superscript denotes the interior side. Equations (10.5.8) demonstrate that the normal derivative of the single-layer potential suffers a discontinuity of magnitude $-q$ across the distribution domain D,

$$\left[\mathbf{N}(\mathbf{x}_0) \cdot \nabla_0 \phi(\mathbf{x}_0) \right]^+ - \left[\mathbf{N}(\mathbf{x}_0) \cdot \nabla_0 \phi_-(\mathbf{x}_0) \right]^- = -q(\mathbf{x}_0). \tag{10.5.9}$$

Principal value of the normal derivative

The principal value (PV) of the normal derivative of the single-layer potential is defined as the value of the right-hand side of (10.5.6) computed when the point $\mathbf{x}_0$ is located precisely on D. Since D has been assumed to be a smooth surface, we use (10.5.2) and find that

$$\left[\mathbf{N}(\mathbf{x}_0) \cdot \nabla_0 \phi(\mathbf{x}_0) \right]^{PV} = J(\mathbf{x}_0) + \frac{1}{2} q(\mathbf{x}_0). \tag{10.5.10}$$

Combining equations (10.5.8) and (10.5.10), we obtain

$$\left[\, \mathbf{N}(\mathbf{x}_0) \cdot \nabla_0 \phi(\mathbf{x}_0)\, \right]^+ = \left[\, \mathbf{N}(\mathbf{x}_0) \cdot \nabla_0 \phi(\mathbf{x}_0)\, \right]^{PV} - \frac{1}{2}\, q(\mathbf{x}_0),$$

$$\left[\, \mathbf{N}(\mathbf{x}_0) \cdot \nabla_0 \phi(\mathbf{x}_0)\, \right]^- = \left[\, \mathbf{N}(\mathbf{x}_0) \cdot \nabla_0 \phi(\mathbf{x}_0)\, \right]^{PV} + \frac{1}{2}\, q(\mathbf{x}_0). \tag{10.5.11}$$

Tangential derivative

Since the single-layer potential is a continuous function of position, its tangential derivatives vary continuously over the distribution domain, D. Combining this observation with the jump condition (10.5.11), we express the gradient of the single-layer potential on either side of D as

$$\left[\, \nabla_0 \phi(\mathbf{x}_0)\, \right]^+ = \left[\, \nabla_0 \phi(\mathbf{x}_0)\, \right]^{PV} - \frac{1}{2}\, q(\mathbf{x}_0)\, \mathbf{N}(\mathbf{x}_0),$$

$$\left[\, \nabla_0 \phi(\mathbf{x}_0)\, \right]^- = \left[\, \nabla_0 \phi(\mathbf{x}_0)\, \right]^{PV} + \frac{1}{2}\, q(\mathbf{x}_0)\, \mathbf{N}(\mathbf{x}_0). \tag{10.5.12}$$

10.5.2   Integral equations

We have investigated in sufficient detail the properties of the single-layer potential. Next, we proceed to derive integral equations originating from the single-layer representation (10.5.1).

Dirichlet problem

In the Dirichlet problem, the distribution of the single-layer potential ϕ is specified over D, and the density of the distribution, q, is required. Applying (10.5.1) at a point $\mathbf{x}_0$ on D provides us with a Fredholm integral equation of the first kind for q. Numerical evidence suggests that, in general, the solution of this equation is unique and the computational problem is tractable by standard numerical methods, including the panel method discussed in Section 10.3.

Neumann problem

In the Neumann problem, the normal derivative of ϕ is specified over the external or internal side of D, and the distribution density, q, is required. Using (10.5.11) to express the normal derivative in terms of the principal value, we obtain a Fredholm integral equation of the second kind for q,

$$q(\mathbf{x}_0) = \pm 2 \iint_D^{PV} \mathbf{N}(\mathbf{x}_0) \cdot \nabla_0 \mathcal{G}(\mathbf{x}_0, \mathbf{x})\, q(\mathbf{x})\, \mathrm{d}S(\mathbf{x}) \mp 2 \left[\, \mathbf{N}(\mathbf{x}_0) \cdot \nabla_0 \phi(\mathbf{x}_0)\, \right]^{\pm}, \tag{10.5.13}$$

where the plus superscript corresponds to the exterior side of D and the minus superscript corresponds to the interior side of D. It is convenient to introduce the linear integral operator

$$\mathcal{O}_{\pm}\big[q(\mathbf{x}_0)\big] \equiv q(\mathbf{x}_0) \mp 2 \iint_D^{PV} \mathbf{N}(\mathbf{x}_0) \cdot \nabla_0 \mathcal{G}(\mathbf{x}_0, \mathbf{x})\, q(\mathbf{x})\, \mathrm{d}S(\mathbf{x}), \tag{10.5.14}$$

and recast (10.5.13) into the compact form

$$\mathcal{O}_{\pm}[q] = \mp 2 \left[\, \mathbf{N} \cdot \nabla_0 \phi_{\pm}\, \right]^{\pm}, \tag{10.5.15}$$

where $\mathcal{O}_+$ corresponds to the exterior problem and $\mathcal{O}_-$ corresponds to the interior problem.

In Section 10.7, we will find that the integral equation (10.5.13) has a unique solution in the case of exterior flow, but either no solution or an infinite number of solutions in the case of interior flow. In the case of interior flow, an infinite number of solutions exist only when the constraint (10.7.25) is fulfilled. Fortunately, the satisfaction of this constraint is guaranteed by the continuity equation for any flow that does not contain point sources and point sinks with a nonzero net discharge. We conclude that the single-layer representation is capable of representing any external flow and any nonsingular internal flow.

Flow past a rigid body

As a specific application, we consider an irrotational flow with velocity $\mathbf{u}_\infty$ past a stationary three-dimensional rigid body, and decompose the velocity potential into the potential of the incident flow, ϕ_∞, and a disturbance potential due to the body, ϕ_D, so that $\phi = \phi_\infty + \phi_D$. Introducing a single-layer representation for ϕ_D, we obtain

$$\phi(\mathbf{x}_0) = \phi_\infty(\mathbf{x}_0) + \iint_B \mathcal{G}(\mathbf{x}_0, \mathbf{x})\, q(\mathbf{x})\, \mathrm{d}S(\mathbf{x}), \tag{10.5.16}$$

where B denotes the surface of the body. Enforcing the no-penetration boundary condition $\mathbf{N} \cdot \nabla \phi = 0$ over B, we derive the integral equation

$$q(\mathbf{x}_0) = 2 \iint_B^{PV} \mathbf{N}(\mathbf{x}_0) \cdot \nabla_0 \mathcal{G}(\mathbf{x}_0, \mathbf{x})\, q(\mathbf{x})\, \mathrm{d}S(\mathbf{x}) + 2\,\mathbf{N}(\mathbf{x}_0) \cdot \mathbf{u}_\infty(\mathbf{x}_0). \tag{10.5.17}$$

According to our previous discussion, the integral equation (10.5.17) has a unique solution for any incident flow.

Flow past a two-dimensional airfoil

In a second application, we consider irrotational flow with velocity $\mathbf{u}_\infty$ past the two-dimensional airfoil with nonzero circulation, as illustrated in Figure 10.3.1. It is convenient to decompose the velocity field into the incident component, $\mathbf{u}_\infty$, the velocity due a point vortex whose strength, κ, is equal to the circulation around the airfoil $\mathbf{v}$, and a disturbance velocity described by a single-valued potential, $\mathbf{u}_D$, so that $\mathbf{u} = \mathbf{u}_\infty + \mathbf{v} + \mathbf{u}_D$. The point vortex is located at a chosen point inside the airfoil. Expressing $\mathbf{u}_D$ in terms of a single-layer potential with density q and working as in the previous case of three-dimensional flow, we derive the integral equation

$$q(\mathbf{x}_0) = 2 \oint_C^{PV} \mathbf{N}(\mathbf{x}_0) \cdot \nabla_0 \mathcal{G}(\mathbf{x}_0, \mathbf{x})\, q(\mathbf{x})\, \mathrm{d}l(\mathbf{x}) + 2\,\mathbf{N}(\mathbf{x}_0) \cdot \big[\mathbf{u}_\infty(\mathbf{x}_0) + \mathbf{v}(\mathbf{x}_0)\big], \tag{10.5.18}$$

where C represents the airfoil contour. A numerical solution can be found by the boundary-element method discussed in Section 10.3 (Problem 10.5.4).

Problems

10.5.1 *Integral properties of the single-layer potential*

Show that, when $\mathcal{G}$ is a Green's function of the second kind, the flow rate across the surface S_3 illustrated in Figure 10.5.1 is zero. Explain why that is not generally true when $\mathcal{G}$ is a Green's function of the first kind.

10.5.2 *Solution of a homogeneous integral equation*

Show that, when D is the surface of a sphere, a constant q satisfies the homogeneous equation $\mathcal{O}_- = 0$, where the operator $\mathcal{O}_-$ is defined in (10.5.15).

Hint: Consider the distribution of ϕ outside a sphere over which ϕ is constant, and use the standard boundary-integral representation to obtain (10.5.1), where $q = \mathbf{N} \cdot \nabla \phi$. Note that the distribution of ϕ inside the sphere must be constant and the flux over the interior side of the sphere must be zero.

10.5.3 *Flow past a body computed inside the body*

Explain why the potential (10.5.16) is uniform inside the body.

 Computer Problem

10.5.4 *Flow past a two-dimensional airfoil with circulation*

Use the single-layer representation combined with the boundary-element method discussed in Section 10.3 to compute the pressure distribution around a Joukowski airfoil discussed in Problem 10.2.3. Compute the lift and drag force.

10.6   The double-layer potential

The double-layer potential is defined as the harmonic field due to a surface distribution of Green's function dipoles oriented perpendicular to a distribution domain, D,

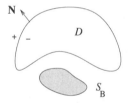

$$\phi(\mathbf{x}_0) = \iint_D q(\mathbf{x}) \, \mathbf{N}(\mathbf{x}) \cdot \nabla \mathcal{G}(\mathbf{x}_0, \mathbf{x}) \, dS(\mathbf{x}), \qquad (10.6.1)$$

Distribution domain of the double-layer potential.

where q is the distribution density and $\mathbf{N}$ is the normal unit vector pointing outward from the volume enclosed by D. Far from D, the double-layer potential decays at least as fast as $1/d^2$, where d is a typical distance from D. The precise rate of decay depends on the shape of the boundary S_B associated with the Green's function, $\mathcal{G}$. The flow rate across any closed surface that does not enclose S_B is zero (Problem 10.6.1).

Jump condition

The double-layer potential is continuous throughout the domain of flow, but suffers a discontinuity across the distribution domain, D. To assess the discontinuity, we write

$$\lim_{\mathbf{x}_0 \to D} \phi(\mathbf{x}_0) = \lim_{\mathbf{x}_0 \to D} \mathcal{I}(\mathbf{x}_0) + q(\mathbf{x}_0) \iint_D \mathbf{N}(\mathbf{x}) \cdot \nabla \mathcal{G}(\mathbf{x}_0, \mathbf{x}) \, dS(\mathbf{x}), \qquad (10.6.2)$$

where

$$\mathcal{I}(\mathbf{x}_0) \equiv \iint_D [\, q(\mathbf{x}) - q(\mathbf{x}_0)] \, \mathbf{N}(\mathbf{x}) \cdot \nabla \mathcal{G}(\mathbf{x}_0, \mathbf{x}) \, dS(\mathbf{x}). \qquad (10.6.3)$$

We note that $\mathcal{I}(\mathbf{x}_0)$ is continuous across D and use identity (10.5.2) to evaluate the second integral on the right-hand side of (10.6.2), finding

$$\lim_{\mathbf{x}_0 \to D_+} \phi(\mathbf{x}_0) \equiv \phi_+(\mathbf{x}_0) = \mathcal{I}(\mathbf{x}_0), \qquad \lim_{\mathbf{x}_0 \to D_-} \phi(\mathbf{x}_0) \equiv \phi_-(\mathbf{x}_0) = \mathcal{I}(\mathbf{x}_0) - q(\mathbf{x}_0), \qquad (10.6.4)$$

where D_+ and D_- signify that $\mathbf{x}_0$ approaches D from the external side, indicated by the direction of the normal unit vector, $\mathbf{N}$, or from the internal side. When the point $\mathbf{x}_0$ is located precisely on D, we obtain the principal value (PV) of the double-layer potential,

$$\phi_{PV}(\mathbf{x}_0) = \mathcal{I}(\mathbf{x}_0) - \frac{1}{2} q(\mathbf{x}_0). \qquad (10.6.5)$$

Comparing the last three equations, we find that

$$\phi_+(\mathbf{x}_0) = \phi_{PV}(\mathbf{x}_0) + \frac{1}{2} q(\mathbf{x}_0), \qquad \phi_-(\mathbf{x}_0) = \phi_{PV}(\mathbf{x}_0) - \frac{1}{2} q(\mathbf{x}_0), \qquad (10.6.6)$$

and

$$\phi_{PV}(\mathbf{x}_0) = \frac{1}{2} \left[\phi_+(\mathbf{x}_0) + \phi_-(\mathbf{x}_0) \right], \qquad (10.6.7)$$

which shows that the double-layer potential suffers a discontinuity across the distribution domain,

$$\phi_+(\mathbf{x}_0) - \phi_-(\mathbf{x}_0) = q(\mathbf{x}_0). \qquad (10.6.8)$$

It is instructive to observe the opposite signs on the right-hand sides of (10.6.6) for the double-layer potential and (10.5.11) for the normal derivative of the single-layer potential.

10.6.1   Restatement as a single-layer potential

The double-layer potential can be restated as a single-layer potential defined over D. To show this, we introduce the extensions of the distribution density function, q, into the interior and exterior of D, denoted by q_- and q_+, respectively, defined such that

$$\nabla^2 q_- = 0 \quad \text{inside } D, \qquad \nabla^2 q_+ = 0 \quad \text{outside } D, \qquad q_- = q_+ = q \quad \text{on } D. \qquad (10.6.9)$$

In addition, we require that q_+ decays to zero at infinity. The general properties of harmonic functions discussed in Section 2.1 guarantee the existence and uniqueness of both q_- and q_+.

Next, we select a point $\mathbf{x}_0$ in the exterior of D and note that both $\mathcal{G}(\mathbf{x}_0, \mathbf{x})$ and $q_-(\mathbf{x})$ are nonsingular harmonic functions throughout the interior of D. Applying the reciprocal identity (2.2.1) with q_- in place of f and $\mathcal{G}(\mathbf{x}_0, \mathbf{x})$ in place of g, we restate (10.6.1) in terms of a single-layer potential with distribution density $\mathbf{N} \cdot \nabla q_-$,

$$\phi_{ext}(\mathbf{x}_0) = \iint_D \mathcal{G}(\mathbf{x}, \mathbf{x}_0) \, \mathbf{N}(\mathbf{x}) \cdot \nabla q_-(\mathbf{x}) \, \mathrm{d}S(\mathbf{x}). \qquad (10.6.10)$$

Using the divergence theorem, we write

$$\iint_D \mathbf{N}(\mathbf{x}) \cdot \nabla q_-(\mathbf{x}) \, \mathrm{d}S(\mathbf{x}) = \iiint_{V_-} \nabla^2 q_-(\mathbf{x}) \, \mathrm{d}S(\mathbf{x}) = 0, \qquad (10.6.11)$$

where V_- is the interior of D. This identity demonstrates that the total strength of the single-layer potential is zero and confirms that, far from D, ϕ_{ext} decays like a Green's function dipole.

Working in a similar manner, we find that the double-layer potential at a point $\mathbf{x}_0$ in the interior of D can be restated as a single-layer potential with distribution density $\mathbf{N} \cdot \nabla q_+$,

$$\phi_{int}(\mathbf{x}_0) = \iint_D \mathcal{G}(\mathbf{x}, \mathbf{x}_0)\, \mathbf{N}(\mathbf{x}) \cdot \nabla q_+(\mathbf{x})\, dS(\mathbf{x}). \qquad (10.6.12)$$

The total strength of the single-layer potential is zero.

Taking the limit as the point $\mathbf{x}_0$ approaches D from either side and using (10.6.10) and (10.6.12) in conjunction with (10.6.8), we obtain

$$\phi_+(\mathbf{x}_0) - \phi_-(\mathbf{x}_0) = \iint_D \mathcal{G}(\mathbf{x}, \mathbf{x}_0)\, \mathbf{N}(\mathbf{x}) \cdot \nabla[q_- - q_+](\mathbf{x})\, dS(\mathbf{x}) = q(\mathbf{x}_0). \qquad (10.6.13)$$

To confirm this equation, we write the boundary-integral equation (10.1.9) for q_+ with $\mathbf{n} = \mathbf{N}$ and for q_- with $\mathbf{n} = -\mathbf{N}$. Adding the two equations, we obtain

$$q_+(\mathbf{x}_0) + q_-(\mathbf{x}_0) = -2 \iint_D \mathcal{G}(\mathbf{x}_0, \mathbf{x})\, \mathbf{N}(\mathbf{x}) \cdot \nabla[q_+ - q_-](\mathbf{x})\, dS(\mathbf{x})$$

$$+ 2 \iint_D^{PV} [q_+ - q_-](\mathbf{x})\, \mathbf{N}(\mathbf{x}) \cdot \nabla \mathcal{G}(\mathbf{x}_0, \mathbf{x})\, dS(\mathbf{x}), \qquad (10.6.14)$$

where the evaluation point, $\mathbf{x}_0$, lies on D. Identity (10.6.13) arises by substituting into this equation the third equation in (10.6.9).

10.6.2 Derivatives of the double-layer potential and the velocity field

The gradient of the double-layer potential is continuous throughout the domain of flow but undergoes a discontinuity across the distribution domain D. To study the behavior of the normal derivative across D, we write the boundary-integral equation (10.1.9) for the exterior flow with $\mathbf{n} = \mathbf{N}$ and for the interior flow with $\mathbf{n} = -\mathbf{N}$. Adding the two equations, we obtain

$$\phi_+(\mathbf{x}_0) + \phi_-(\mathbf{x}_0) = -2 \iint_D \mathcal{G}(\mathbf{x}_0, \mathbf{x})\, \mathbf{N}(\mathbf{x}) \cdot \nabla[\phi_+ - \phi_-](\mathbf{x})\, dS(\mathbf{x})$$

$$+ 2 \iint_D^{PV} [\phi_+ - \phi_-](\mathbf{x})\, \mathbf{N}(\mathbf{x}) \cdot \nabla \mathcal{G}(\mathbf{x}_0, \mathbf{x})\, dS(\mathbf{x}), \qquad (10.6.15)$$

where $\mathbf{x}_0$ is on D. Using (10.6.8) to simplify the integrand of the double-layer potential on the right-hand side, invoking the definition (10.6.1) to identify the double-layer potential with the principal value of ϕ, and taking into account (10.6.7), we finally obtain

$$\iint_D \mathcal{G}(\mathbf{x}_0, \mathbf{x})\, \mathbf{N}(\mathbf{x}) \cdot \nabla[\phi_+ - \phi_-](\mathbf{x})\, dS(\mathbf{x}) = 0, \qquad (10.6.16)$$

where $\mathbf{x}_0$ is on D. Since this equation holds true for any arbitrary double-layer potential, the integrand must be identically zero and the normal derivative of the double-layer potential must be continuous across D.

Representation as a vortex sheet

It is evident from (10.6.6) and (10.6.8) that the tangential derivative of the double-layer potential exhibits a discontinuity across D,

$$\left[\nabla\phi\right]^{\pm} = \left[\nabla\phi\right]^{PV} \pm \frac{1}{2}\left(\mathbf{I} - \mathbf{N} \otimes \mathbf{N}\right) \cdot \nabla q = \left[\nabla\phi\right]^{PV} \pm \frac{1}{2}\left(\mathbf{N} \times \nabla q\right) \times \mathbf{N}, \qquad (10.6.17)$$

and

$$\mathbf{P} \cdot \nabla\left[\phi_{+} - \phi_{-}\right] = \mathbf{P} \cdot \nabla q = \left(\mathbf{N} \times \nabla q\right) \times \mathbf{N}, \qquad (10.6.18)$$

where $\mathbf{P} = \mathbf{I} - \mathbf{N} \otimes \mathbf{N}$ is the tangential projection operator. Equation (10.6.18) states that the tangential component of the velocity undergoes a discontinuity across D. Referring to (1.13.15), we find that the flow can be regarded as though it were induced by a vortex sheet with strength

$$\zeta = \mathbf{N} \times \nabla q \qquad (10.6.19)$$

situated over D, in the possible presence of an external or internal boundary, S_B. We thus arrive at the interesting conclusion that the flow due to a double-layer potential can be regarded as the flow due to a vortex sheet whose strength derives from the tangential derivatives of the double-layer distribution density, q.

Vector potential and velocity field

In the case of flow in a unbounded domain, we use the free-space Green's function, $\mathcal{G} = 1/(4\pi r)$, where $r = |\mathbf{x} - \mathbf{x}_0|$, note the property $\nabla\mathcal{G}(r) = -\nabla_0\mathcal{G}(r)$, and find that a vector potential for the velocity is

$$\mathbf{A}(\mathbf{x}_0) = -\iint_{D} \mathbf{N}(\mathbf{x}) \times \nabla\mathcal{G}(\mathbf{x}_0, \mathbf{x})\, q(\mathbf{x})\, \mathrm{d}S(\mathbf{x}), \qquad (10.6.20)$$

where the subscript 0 indicates differentiation with respect to $\mathbf{x}_0$. One may readily verify by straightforward differentiation that $\mathbf{u} = \nabla \times \mathbf{A} = \nabla\phi$.

To derive the velocity field, we use the Biot–Savart integral in (2.10.18), and find that

$$\mathbf{u}(\mathbf{x}_0) = \nabla_0\phi(\mathbf{x}_0) = \iint_{D} \nabla\mathcal{G}(r) \times \left[\nabla q(\mathbf{x}) \times \mathbf{N}(\mathbf{x})\right] \mathrm{d}S(\mathbf{x}). \qquad (10.6.21)$$

An alternative derivation of (10.6.21) is discussed in Section 10.6.5.

10.6.3   Integral equations for the Dirichlet problem

Specifying the distribution of ϕ on either side of D reduces (10.6.6) into a Fredholm integral equation of the second kind for the distribution density, q,

$$q(\mathbf{x}_0) = -2\left[\phi_{PV}(\mathbf{x}_0) - \phi_{+}(\mathbf{x}_0)\right], \qquad q(\mathbf{x}_0) = 2\left[\phi_{PV}(\mathbf{x}_0) - \phi_{-}(\mathbf{x}_0)\right], \qquad (10.6.22)$$

where the plus sign indicates the outer side, the minus sign indicates the inner side, and

$$\phi_{PV}(\mathbf{x}_0) = \int\!\!\!\int_D^{PV} q(\mathbf{x})\,\mathbf{N}(\mathbf{x})\cdot\nabla\mathcal{G}(\mathbf{x},\mathbf{x}_0)\,\mathrm{d}S(\mathbf{x}) \tag{10.6.23}$$

is a principal-value integral. It is useful to introduce the operator

$$\mathcal{P}_{\pm}[q(\mathbf{x}_0)] \equiv q(\mathbf{x}_0) \pm 2 \int\!\!\!\int_D^{PV} q(\mathbf{x})\,\mathbf{N}(\mathbf{x})\cdot\nabla\mathcal{G}(\mathbf{x},\mathbf{x}_0)\,\mathrm{d}S(\mathbf{x}), \tag{10.6.24}$$

and rewrite (10.6.22) in the compact form

$$\mathcal{P}_{\pm}[q] = \pm 2\,\phi_{\pm}. \tag{10.6.25}$$

Identity (10.5.2) shows that the homogeneous equation, $\mathcal{P}^+[q] = 0$, has a nontrivial solution with constant q for any shape D. Accordingly, equation (10.6.22) for the exterior problem has either no solution or an infinite number of solutions differing by an arbitrary constant.

The properties of the integral equations (10.6.25) will be discussed extensively in Section 10.7. The results will show that the exterior flow problem has a solution only when the flow rate of the forcing function ϕ_+ across D is zero. When this condition is met, the integral equation has a multiplicity of solutions that differ by an arbitrary constant. In contrast, the integral equation for interior flow has a unique solution for any forcing function ϕ_-. Thus, the double-layer representation is capable of representing any internal flow and a limited class of external flows.

Flow past a rigid body

As an application, we consider irrotational flow with velocity $\mathbf{u}_\infty$ past a three-dimensional stationary rigid body bounded by the surface B. Following standard practice, we decompose the velocity potential into an undisturbed component, ϕ_∞, and a disturbance component due to the body, ϕ_D, so that $\phi = \phi_\infty + \phi_D$. Introducing a double-layer representation for ϕ_D, we write

$$\phi(\mathbf{x}_0) = \phi_\infty(\mathbf{x}_0) + \int\!\!\!\int_B q(\mathbf{x})\,\mathbf{N}(\mathbf{x})\cdot\nabla\mathcal{G}(\mathbf{x},\mathbf{x}_0)\,\mathrm{d}S(\mathbf{x}). \tag{10.6.26}$$

The no-penetration boundary condition requires that $\mathbf{N}\cdot\nabla\phi = 0$ over the exterior side of B.

To derive an integral equation, we recall that the normal derivative of the double-layer potential is continuous across B. Accordingly, the normal component of the velocity vanishes on the interior side of B, the internal velocity field computed from (10.6.26) is zero, and ϕ takes a constant value, c, in the interior of B. Applying (10.6.26) at a point on the internal side of B and using (10.6.6) provides us with an integral equation of the second kind for q,

$$q(\mathbf{x}_0) = 2 \int\!\!\!\int_B^{PV} q(\mathbf{x})\,\mathbf{N}(\mathbf{x})\cdot\nabla\mathcal{G}(\mathbf{x},\mathbf{x}_0)\,\mathrm{d}S(\mathbf{x}) + 2\left[\phi_\infty(\mathbf{x}_0) - c\right]. \tag{10.6.27}$$

Using identities (10.5.2), we obtain the alternative integral equation

$$q(\mathbf{x}_0) = 2 \int\!\!\!\int_B^{PV} \left[q(\mathbf{x}) + 2\,c\right]\mathbf{N}(\mathbf{x})\cdot\nabla\mathcal{G}(\mathbf{x},\mathbf{x}_0)\,\mathrm{d}S(\mathbf{x}) + 2\,\phi_\infty(\mathbf{x}_0). \tag{10.6.28}$$

In Section 10.7, we will show that (10.6.27) has a unique solution for any incident flow.

Oscillations of a liquid drop

In a second application, we consider the oscillations of an inviscid liquid drop suspended in ambient gas. An efficient numerical procedure for computing the motion of the drop surface involves the following steps [22]:

1. Describe the surface in terms of a collection of point particles moving with the fluid velocity.

2. Assign to the point particles initial values for the potential, ϕ.

3. Differentiate ϕ along the free surface to compute the tangential component of the fluid velocity.

4. Solve the integral equation (10.6.22) for q for the interior problem using the free-space Green's function.

5. Compute the vector potential over the free surface using (10.6.20) and then obtain the normal component of the fluid velocity using the equation $\mathbf{N} \cdot \mathbf{u} = \mathbf{N} \cdot (\nabla \times \mathbf{A})$. The right-hand side involves tangential derivatives of $\mathbf{A}$.

6. Advect the marker points with the fluid velocity. The tangential and normal components of the velocity are available from Steps 3 and 5.

7. At the new position, update the potential using (10.2.5).

8. Return to Step 3 and repeat the computation for another step.

Other free-surface problems involving irrotational flow can be solved by similar methods.

10.6.4   Integral equations for the Neumann problem

Next, we consider the Dirichlet problem where the distribution of the normal derivative $\mathbf{N} \cdot \nabla \phi$ is specified on the internal or external side of D, and the distribution density q is sought. Projecting equation (10.6.17) onto the normal unit vector and specifying the normal derivative on one side of D provides us with a Fredholm integral equation of the second kind for the tangential component of ∇q. Once this equation has been solved, q can be reconstructed from tangential derivatives over D.

Flow past a stationary body

As a specific application, we consider irrotational flow with velocity $\mathbf{u}_\infty$ past a stationary rigid body enclosed by a surface, B. We begin by decomposing the velocity potential into the incident component, ϕ_∞, and a disturbance component due to the body, ϕ_D, so that $\phi = \phi_\infty + \phi_D$. Next, we introduce a double-layer representation for ϕ_D to obtain the integral representation (10.6.26). According to (10.6.17) and (10.6.21), the velocity on either side of B is given by

$$
\nabla_0 \phi^\pm(\mathbf{x}_0) = \nabla_0 \phi_\infty(\mathbf{x}_0) + \iint_B^{PV} \nabla \mathcal{G}(r) \times \left[\nabla q(\mathbf{x}) \times \mathbf{N}(\mathbf{x}) \right] \, dS(\mathbf{x})
$$
$$
\pm \frac{1}{2} \left[\mathbf{N}(\mathbf{x}_0) \times \nabla q(\mathbf{x}_0) \right] \times \mathbf{N}(\mathbf{x}_0).
$$

(10.6.29)

The no-penetration condition requires that $\mathbf{N} \cdot \nabla \phi = 0$ on the external side of B. Since the normal derivative of the double-layer potential is continuous across B, the internal flow must vanish and $\nabla \phi_- = \mathbf{0}$. Equation (10.6.17) then yields

$$
\nabla \phi^+ = (\mathbf{N} \times \nabla q) \times \mathbf{N},
$$

(10.6.30)

which relates the tangential velocity to the strength of the vortex sheet, $\boldsymbol{\zeta}$, given in (10.6.19). Substituting $\nabla\phi_- = \mathbf{0}$ into the left-hand side of (10.6.29) provides us with a vectorial integral equation for the strength of the vortex sheet [213],

$$\frac{1}{2}\,\boldsymbol{\zeta}(\mathbf{x}_0) \times \mathbf{N}(\mathbf{x}_0) = \nabla_0\phi_\infty(\mathbf{x}_0) + \int\!\!\!\int_B^{PV} \boldsymbol{\zeta}(\mathbf{x}) \times \nabla\mathcal{G}(r)\,\mathrm{d}S(\mathbf{x}). \tag{10.6.31}$$

In the case of two-dimensional or axisymmetric flow, we obtain an integral equation for the tangential velocity (see Section 10.7).

10.6.5 Free-space flow

As a last topic of this section, we examine in more detail the case of the free-space Green's function, $\mathcal{G} = 1/(4\pi r)$, where $r = |\mathbf{x} - \mathbf{x}_0|$. Exploiting the property $\nabla\mathcal{G}(r) = -\nabla_0\mathcal{G}(r)$, we restate the double-layer potential (10.6.1) as

$$\phi(\mathbf{x}_0) = -\nabla_0 \cdot \int\!\!\!\int_D q(\mathbf{x})\,\mathbf{N}(\mathbf{x})\,\mathcal{G}(\mathbf{x}_0, \mathbf{x})\,\mathrm{d}S(\mathbf{x}), \tag{10.6.32}$$

where the subscript 0 indicates differentiation with respect to $\mathbf{x}_0$. Taking the gradient of (10.6.32) and using the identity

$$\nabla(\nabla \cdot \mathbf{A}) = \nabla \times \nabla \times \mathbf{A} + \nabla^2\mathbf{A}, \tag{10.6.33}$$

applicable for any twice differentiable function $\mathbf{A}$, we find that ([91], p. 57)

$$\nabla_0\phi(\mathbf{x}_0) = -\nabla_0 \times \nabla_0 \times \int\!\!\!\int_D q(\mathbf{x})\,\mathbf{N}(\mathbf{x})\,\mathcal{G}(\mathbf{x}_0, \mathbf{x})\,\mathrm{d}S(\mathbf{x}). \tag{10.6.34}$$

Now we extend the domain of definition of q inward and outward from D according to (10.6.9), select a point $\mathbf{x}_0$ in the exterior of D, and use the divergence theorem to write

$$\int\!\!\!\int_D q(\mathbf{x})\,\mathbf{N}(\mathbf{x})\,\mathcal{G}(\mathbf{x}_0, \mathbf{x})\,\mathrm{d}S(\mathbf{x}) = \int\!\!\!\int\!\!\!\int_{V_-} \nabla[\,q_-(\mathbf{x})\,\mathcal{G}(\mathbf{x}_0, \mathbf{x})\,]\,\mathrm{d}V(\mathbf{x}). \tag{10.6.35}$$

Equation (10.6.34) yields

$$\nabla_0\phi_{ext}(\mathbf{x}_0) = -\nabla_0 \times \int\!\!\!\int\!\!\!\int_{V_-} \nabla \times [\,q_-(\mathbf{x})\,\nabla\mathcal{G}(\mathbf{x}_0, \mathbf{x})\,]\,\mathrm{d}V(\mathbf{x}) \tag{10.6.36}$$

and then

$$\nabla_0\phi_{ext}(\mathbf{x}_0) = \nabla_0 \times \int\!\!\!\int\!\!\!\int_{V_-} \nabla \times [\,\nabla q_-(\mathbf{x})\,\mathcal{G}(\mathbf{x}_0, \mathbf{x})\,]\,\mathrm{d}V(\mathbf{x}). \tag{10.6.37}$$

A variation of Stokes' theorem for a twice differentiable vector function, $\mathbf{F}$, states that

$$\int\!\!\!\int\!\!\!\int_{V_-} \nabla \times \mathbf{F}\,\mathrm{d}V = \int\!\!\!\int_D \mathbf{N} \times \mathbf{F}\,\mathrm{d}S, \tag{10.6.38}$$

as discussed in Section A.7, Appendix A. Applying (10.6.38) for the volume integral in (10.6.38), we obtain

$$\nabla_0 \phi_{ext}(\mathbf{x}_0) = \nabla_0 \times \iint_D \mathbf{N}(\mathbf{x}) \times \nabla q_-(\mathbf{x})\, \mathcal{G}(\mathbf{x}_0, \mathbf{x})\, \mathrm{d}S(\mathbf{x}), \qquad (10.6.39)$$

which can be rearranged into

$$\nabla_0 \phi_{ext}(\mathbf{x}_0) = \iint_D \left[\mathbf{N}(\mathbf{x}) \times \nabla q_-(\mathbf{x}) \right] \times \nabla \mathcal{G}(\mathbf{x}_0, \mathbf{x})\, \mathrm{d}S(\mathbf{x}). \qquad (10.6.40)$$

Working in a similar manner, we find that the gradient of the double-layer potential at a point $\mathbf{x}_0$ in the interior of D is given by the corresponding expression

$$\nabla_0 \phi_{int}(\mathbf{x}_0) = \iint_D \left[\mathbf{N}(\mathbf{x}) \times \nabla q_+(\mathbf{x}) \right] \times \nabla \mathcal{G}(\mathbf{x}_0, \mathbf{x})\, \mathrm{d}S(\mathbf{x}). \qquad (10.6.41)$$

Equations (10.6.40) and (10.6.41) provide us with the gradient of the double-layer potential in terms of distributions of the Green's function dipoles. Because the corresponding distribution densities, $-\mathbf{N}(\mathbf{x}) \times \nabla q_+(\mathbf{x})$ and $-\mathbf{N}(\mathbf{x}) \times \nabla q_-(\mathbf{x})$, are tangential to D, they can be determined from the distribution of q over D independent of the extensions, q_- and q_+. To recover (10.6.21), we write

$$\mathbf{N} \times \nabla q_- = \mathbf{N} \times \nabla q_+ = \mathbf{N} \times \nabla q. \qquad (10.6.42)$$

Problems

10.6.1 *Flow rate due to the double-layer potential*

Show that the flow rate of the double-layer potential across any closed surface that does not enclose S_B is zero.

10.6.2 *Restatement as a single-layer potential*

Explain why the total strength of the interior single-layer representation (10.6.12) is zero.

10.6.3 *Potential of doublets*

The potential due to a surface distribution of Green's function doublets is defined as

$$\phi(\mathbf{x}_0) = \iint_D \mathbf{q}(\mathbf{x}) \cdot \nabla \mathcal{G}(\mathbf{x}_0, \mathbf{x})\, \mathrm{d}S(\mathbf{x}), \qquad (10.6.43)$$

where $\mathbf{q}$ is the vectorial density distribution. When $\mathbf{q}$ is normal to D, we obtain a double-layer potential. Show that, as $\mathbf{x}_0$ approaches D, the limiting values of ϕ are given by

$$\phi_\pm(\mathbf{x}_0) = \phi^{PV}(\mathbf{x}_0) \pm \frac{1}{2} \mathbf{N}(\mathbf{x}_0) \cdot \mathbf{q}(\mathbf{x}_0), \qquad (10.6.44)$$

where the plus sign applies for the side indicated by the direction of the normal unit vector, $\mathbf{N}$, and the minus sign for the other side.

10.6.4 *Velocity due to the double-layer potential*

Evaluate expressions (10.6.40) and (10.6.41) on either side of D and subtract out the dominant singularity of the integrands to obtain

$$\nabla_0 \phi^{\pm}(\mathbf{x}_0) = \iint_{D^{\pm}} [\mathbf{N}(\mathbf{x}) \times \nabla q(\mathbf{x}) - \mathbf{N}(\mathbf{x}_0) \times \nabla q(\mathbf{x}_0)] \times \nabla \mathcal{G}(\mathbf{x}_0, \mathbf{x}) \, dS(\mathbf{x})$$

$$+ \mathbf{N}(\mathbf{x}_0) \times \nabla q(\mathbf{x}_0) \times \iint_{D_{\pm}} \nabla \mathcal{G}(\mathbf{x}_0, \mathbf{x}) \, dS(\mathbf{x}). \tag{10.6.45}$$

Explain why the first integral on the right-hand side is continuous across D, whereas the second integral undergoes a discontinuity of magnitude $\mathbf{N}$ across D.

 Computer Problem

10.6.5 *Flow past a two-dimensional airfoil*

Solve the integral equation (10.6.28) to obtain the pressure distribution, drag force, and lift force on a Joukowski airfoil discussed in Problem 10.3.3, using the boundary-element method discussed in Section 10.3.

10.7   Investigation of integral equations of the second kind

In previous sections, we made occasional reference to the existence and uniqueness of solution of the integral equations that arise from standard and generalized boundary-integral representations. In this section, we examine in detail the properties of the integral equations. Since the integral equations of the first kind are amenable only to rudimentary theoretical investigations (e.g., [300], Chapter 6), we focus our attention to integral equations of the second kind and study the existence and uniqueness of solution as well as the feasibility of computing the solution by iterative methods.

Summary of integral equations

For convenience, we assume that the flow is bounded externally or internally by a single closed surface, D, and denote the outward normal unit vector by $\mathbf{N}$. The integral equation (10.1.9) arising from the boundary-integral representation with the Dirichlet boundary condition then becomes

$$\phi_+(\mathbf{x}_0) = 2 \iint_D^{PV} \phi_+(\mathbf{x}) \, \mathbf{N}(\mathbf{x}) \cdot \nabla \mathcal{G}(\mathbf{x}_0, \mathbf{x}) \, dS(\mathbf{x}) - 2 \mathcal{I}_+(\mathbf{x}_0) \tag{10.7.1}$$

for the exterior problem, and

$$\phi_-(\mathbf{x}_0) = -2 \iint_D^{PV} \phi_-(\mathbf{x}) \, \mathbf{N}(\mathbf{x}) \cdot \nabla \mathcal{G}(\mathbf{x}_0, \mathbf{x}) \, dS(\mathbf{x}) + 2 \mathcal{I}_-(\mathbf{x}_0) \tag{10.7.2}$$

for the interior problem, where

$$\mathcal{I}_{\pm}(\mathbf{x}_0) \equiv \iint_D \mathcal{G}(\mathbf{x}_0, \mathbf{x}) \, \mathbf{N}(\mathbf{x}) \cdot \nabla \phi_{\pm}(\mathbf{x}) \, dS(\mathbf{x}) \tag{10.7.3}$$

is the known single-layer potential computed on the exterior $(+)$ or interior $(-)$ side of D. For convenience, we repeat the integral equation (10.5.13) arising from the single-layer representation with the Neumann boundary condition,

$$q(\mathbf{x}_0) = \pm 2 \iint_D^{PV} q(\mathbf{x}) \, \mathbf{N}(\mathbf{x}_0) \cdot \nabla_0 \mathcal{G}(\mathbf{x}_0, \mathbf{x}) \, \mathrm{d}S(\mathbf{x}) \mp 2 \, \mathbf{N}(\mathbf{x}_0) \cdot \nabla_0 \phi_\pm(\mathbf{x}_0), \tag{10.7.4}$$

and the integral equation (10.6.22) arising from the double-layer representation with the Dirichlet boundary condition,

$$q(\mathbf{x}_0) = \mp 2 \iint_D^{PV} q(\mathbf{x}) \, \mathbf{N}(\mathbf{x}) \cdot \nabla \mathcal{G}(\mathbf{x}, \mathbf{x}_0) \, \mathrm{d}S(\mathbf{x}) \pm 2 \, \phi_\pm(\mathbf{x}_0), \tag{10.7.5}$$

where the plus sign corresponds to the exterior problem and the minus sign corresponds to the interior problem.

10.7.1   Generalized homogeneous equations

To investigate the properties of the integral equations, we consider the corresponding generalized homogeneous equations that arise by discarding the forcing function on the right-hand side and multiplying the principal-value integral by a complex constant, β, or its complex conjugate, β^*, obtaining

$$\phi(\mathbf{x}_0) = 2\beta \iint_D^{PV} \psi(\mathbf{x}) \, \mathbf{N}(\mathbf{x}) \cdot \nabla \mathcal{G}(\mathbf{x}, \mathbf{x}_0) \, \mathrm{d}S(\mathbf{x}) \tag{10.7.6}$$

and

$$\chi(\mathbf{x}_0) = 2\beta^* \iint_D^{PV} \chi(\mathbf{x}) \, \mathbf{N}(\mathbf{x}_0) \cdot \nabla_0 \mathcal{G}(\mathbf{x}_0, \mathbf{x}) \, \mathrm{d}S(\mathbf{x}), \tag{10.7.7}$$

where ψ and χ are eigenfunctions and an asterisk denotes the complex conjugate. In the standard terminology of integral equation theory, β is a complex eigenvalue of the double-layer operator

$$\mathcal{L}[\psi(\mathbf{x}_0)] \equiv 2 \iint_D^{PV} \psi(\mathbf{x}) \, \mathbf{N}(\mathbf{x}) \cdot \nabla \mathcal{G}(\mathbf{x}, \mathbf{x}_0) \, \mathrm{d}S(\mathbf{x}), \tag{10.7.8}$$

and β^* is a complex eigenvalue of the adjoint of the double-layer operator

$$\mathcal{L}^\dagger[\chi(\mathbf{x}_0)] \equiv 2 \iint_D^{PV} \chi(\mathbf{x}) \, \mathbf{N}(\mathbf{x}_0) \cdot \nabla_0 \mathcal{G}(\mathbf{x}_0, \mathbf{x}) \, \mathrm{d}S(\mathbf{x}). \tag{10.7.9}$$

Note that the adjoint operator $\mathcal{L}^\dagger$ arises from $\mathcal{L}$, and *vice versa*, by switching the arguments $\mathbf{x}_0$ and $\mathbf{x}$ of the kernel. Using these definitions, equations (10.7.6) and (10.7.7) are written concisely as

$$\psi = \beta \, \mathcal{L}[\psi], \qquad \chi = \beta^* \mathcal{L}^\dagger[\chi]. \tag{10.7.10}$$

Self-adjoint operators

It can be shown by direct substitution that any two complex functions, q_1 and q_2, satisfy the distinguishing property of adjoint operators,

$$(\mathcal{L}[q_1],\, q_2) = (q_1,\, \mathcal{L}^\dagger[q_2]), \tag{10.7.11}$$

where the projection operator $(\cdot, \cdot)$ is defined as

$$(q_1,\, q_2) \equiv \iint_D q_1(\mathbf{x})\, q_2^*(\mathbf{x})\, \mathrm{d}S(\mathbf{x}), \tag{10.7.12}$$

and an asterisk indicates the complex conjugate (Problem 10.7.1).

10.7.2   Riesz–Fredholm theory

The significance of the generalized homogeneous equations (10.7.10) relies on the Riesz–Fredholm theory of compact linear operators (e.g., [91], Chapter 1). The applicability of this theory in the present context is justified by the weak singularity of the kernel of the double-layer operator and its adjoint in the absence of boundary corners. This property guarantees that $\mathcal{L}$ and $\mathcal{L}^\dagger$ are completely continuous or compact operators. Certain important results of the Riesz–Fredholm theory are summarized in this section.

Spectrum

The homogeneous equations (10.7.10) have complex conjugate eigenvalues and the same number of eigensolutions. The set of all eigenvalues is the spectrum of the double-layer operator or its adjoint. The eigenvalues can be numerable or innumerable. However, an infinite number of eigenvalues may not accumulate at any point in the complex plane, except at infinity. The spectral radius of $\mathcal{L}$ or $\mathcal{L}^\dagger$ is defined as the magnitude of the eigenvalue, β, with the minimum norm.

Fredholm alternative

Fredholm's alternative ensures that, if the coefficient α does not belong to the spectrum of $\mathcal{L}$, then the inhomogeneous equation

$$q = \alpha \mathcal{L}[q] + F \tag{10.7.13}$$

has a unique solution, where F is a forcing function. If α belongs to the spectrum of $\mathcal{L}$, the inhomogeneous equation has either no solution or an infinite number of solutions. An infinite number of solutions exist only if all eigenfunctions of the adjoint double-layer operator, χ, corresponding to α, are orthogonal to the forcing function F, that is, only when $(F, \chi) = 0$, where the projection $(\cdot, \cdot)$ is defined in (10.7.12). Similar results apply to the inhomogeneous equation involving the adjoint operator defined in (10.7.9),

$$q = \alpha \mathcal{L}^\dagger[q] + F, \tag{10.7.14}$$

where F is a specified forcing function.

Successive substitutions

The inhomogeneous equations (10.7.13) and (10.7.14) can be solved by the method of successive substitutions, also called the method of fixed-point iterations or successive approximations, only if α is less than the spectral radius of the double-layer operator. The method is implemented by guessing a solution, computing the right-hand side of the integral equation, and then replacing the guessed with the computed solution. In the extreme case when $\alpha = 0$, the solution is obtained after one iteration. When the initial guess is set equal to zero, $q = 0$, we obtain a series of approximations called the Neumann series of the integral equations.

10.7.3 Spectrum of the double-layer operator

Next, we proceed to investigate the spectrum of the double-layer operator and its adjoint. As a preliminary, we integrate equation (10.7.7) over D and use equation (10.5.2) to derive a constraint on an eigenvalue and the corresponding eigenfunction,

$$(1 + \beta^*) \iint_D \chi \, dS = 0. \tag{10.7.15}$$

We begin by introducing the single-layer potential

$$\phi(\mathbf{x}_0) = \iint_D \mathcal{G}(\mathbf{x}_0, \mathbf{x}) \, \chi(\mathbf{x}) \, dS(\mathbf{x}), \tag{10.7.16}$$

and use (10.5.11) and (10.7.7) to obtain

$$\mathbf{N}(\mathbf{x}) \cdot \nabla \phi_+(\mathbf{x}) = \frac{1}{2}\left(\frac{1}{\beta^*} - 1\right) \chi(\mathbf{x}), \qquad \mathbf{N}(\mathbf{x}) \cdot \nabla \phi_-(\mathbf{x}) = \frac{1}{2}\left(\frac{1}{\beta^*} + 1\right) \chi(\mathbf{x}). \tag{10.7.17}$$

Working as in (2.1.21), we find that the kinetic energy of the flow in the interior and exterior of D are given by

$$\mathcal{K}_+ = -\frac{1}{2}\rho \iint_{D_+} \phi \, \mathbf{N} \cdot \nabla \phi^* \, dS, \qquad \mathcal{K}_- = \frac{1}{2}\rho \iint_{D_-} \phi \, \mathbf{N} \cdot \nabla \phi^* \, dS. \tag{10.7.18}$$

Substituting (10.7.17) into (10.7.18), we find that

$$\mathcal{K}_+ = \frac{1}{4}\rho \left(1 - \frac{1}{\beta}\right) \iint_{D_+} \phi \chi^* \, dS, \qquad \mathcal{K}_- = \frac{1}{4}\rho \left(1 + \frac{1}{\beta}\right) \iint_{D_-} \phi \chi^* \, dS. \tag{10.7.19}$$

Combining these expressions to eliminate the integral on the right-hand side, we obtain

$$\beta = \frac{1 + \delta}{1 - \delta}, \tag{10.7.20}$$

where

$$\delta \equiv \mathcal{K}_+ / \mathcal{K}_-. \tag{10.7.21}$$

Since δ is real and positive, β is real and lies outside the closed range $[-1, 1]$. However, this deduction may fail in the following three cases.

Case $\beta = 1$

When $\beta = 1$, equation (10.7.19) shows that $\mathcal{K}_+ = 0$, which requires that ϕ is constant throughout V_+. Since ϕ must vanish at infinity, the value of this constant is necessarily zero. Since ϕ is continuous across D, ϕ must also vanish throughout V_- and the normal derivative on the internal side of D must be zero. Applying (10.7.17) on the internal side of D shows that $\chi = 0$, which demonstrates that $\beta = 1$ cannot be an eigenvalue of the double-layer potential.

Case $\beta = -1$

When $\beta = -1$, equation (10.7.19) shows that $\mathcal{K}_- = 0$, which requires that ϕ is constant throughout V_- as well as on the interior side of D. Equation (10.7.17) yields the eigenfunction

$$\chi = \mathbf{N} \cdot \nabla \phi_+. \qquad (10.7.22)$$

The constraint (10.7.15) is satisfied and (10.7.22) constitutes a perfectly acceptable eigenfunction. Writing the boundary-integral representation for ϕ defined in (10.7.16) allows us to identify χ with the negative of the normal gradient of a potential function on the external side of D established when ϕ is constant over D.

Although we are not be able to compute the eigensolution χ explicitly, except for one particular case discussed in Problem 10.7.2, we can show that it must satisfy the integral constraint

$$\iint_D \chi \, \mathrm{d}S \neq 0. \qquad (10.7.23)$$

Since ϕ is constant over D, equation (10.7.19) shows that, if (10.7.23) were not true, $\mathcal{K}_+$ would have to vanish, which means that ϕ and thus χ would have to be zero. Equation (10.5.2) shows that the corresponding eigensolution of (10.7.6) is constant, $\psi = c$. One important consequence of the last two equations is that the geometric multiplicity of the eigenvalue $\beta = -1$ is exactly equal to one. If this were not true, we would be able to find a generalized or principal eigensolution η satisfying the inhomogeneous equation $\eta = -\mathcal{L}^A[\eta] + \chi$. According to Fredholm's alternative, for a solution of this equation to exist, χ must be orthogonal to ψ. However, since ψ is constant, we obtain a contradiction with (10.7.23). We conclude that the algebraic and geometric multiplicities of the eigenvalue $\beta = -1$ are equal to one.

Case of zero kinetic energy

A third exception occurs when one or both of $\mathcal{K}_\pm$ are zero. However, this takes us back to the two cases cases previously discussed.

Summary

We have found that (a) the eigenvalues of the double-layer potential are real and lie outside the semi-closed range $(-1, 1]$, and (b) $\beta = -1$ is a marginal eigenvalue for any shape, D. The spectral radius of the double-layer operator or its adjoint is equal to unity. Based on these results, we make the following deductions regarding the properties of the integral equations for interior or exterior flow:

- Equations (10.7.2) and (10.7.4) for the exterior problem and equation (10.7.5) for the interior problem have a unique solution for any boundary shape, D.

- Equation (10.7.1) for the interior problem has either no solution or a multiplicity of solutions that differ by an arbitrary constant. According to Fredholm's alternative, multiple solutions exist only if the projection of the forcing function onto the adjoint eigensolution χ, given in (10.7.22), is zero. However, this is true for any flow that does not contain point sources and point sinks. To see this, we denote the single-layer potential as $\mathcal{I}^S$ and formulate its projection on χ to obtain

$$\left(\chi, \mathcal{I}^S \right) = \iint_D \chi(\mathbf{x}_0) \left(\iint_D \mathcal{G}(\mathbf{x}_0, \mathbf{x}) \, \mathbf{N}(\mathbf{x}) \cdot \nabla \phi_-(\mathbf{x}) \, \mathrm{d}S(\mathbf{x}) \right) \mathrm{d}S(\mathbf{x}_0) \tag{10.7.24}$$

$$= \iint_D \mathbf{N}(\mathbf{x}) \cdot \nabla \phi_-(\mathbf{x}) \left(\iint_D \chi(\mathbf{x}_0) \, \mathcal{G}(\mathbf{x}_0, \mathbf{x}) \, \mathrm{d}S(\mathbf{x}) \right) \mathrm{d}S(\mathbf{x}_0) = \iint_D \mathbf{N}(\mathbf{x}) \cdot \nabla \phi_-(\mathbf{x}) \, c \, \mathrm{d}S(\mathbf{x}_0) = 0,$$

where c is the constant value of the potential over D corresponding to (10.7.22).

- Equation (10.7.4) for the interior problem has either no solution or a multiplicity of solutions. According to Fredholm's alternative, multiple solutions exist only if the projection of the forcing function onto the constant eigensolution of the double-layer potential corresponding to $\beta = -1$ is zero,

$$\iint_D \mathbf{N} \cdot \nabla \phi_- \, \mathrm{d}S = 0. \tag{10.7.25}$$

This requires that the flow rate of the field represented by the forcing function across D is zero.

- Equation (10.7.5) for the exterior problem has either no solution or a multiplicity of solutions that differ by an arbitrary constant According to Fredholm's alternative, an infinity of solutions exist only if the projection of the forcing function onto the adjoint eigensolution χ corresponding to $\beta = -1$ is zero,

$$\iint_D \phi_+ \chi \, \mathrm{d}S = -c \iint_D \mathbf{N} \cdot \nabla \phi_+ \, \mathrm{d}S = 0, \tag{10.7.26}$$

where c is the constant value of the potential over D defined in (10.7.16). The second expression in (10.7.26) arises from the reciprocal theorem for harmonic functions discussed in Section 2.2.

- None of the equations discussed can be solved by the method of successive substitutions.

The spectrum of the sphere

It is instructive to compute the spectrum of the double-layer operator when D is the surface of a sphere of radius a. Noting that the single-layer potential (10.7.16) is continuous across D, and using (10.7.17), we derive two matching conditions at $r = a$,

$$\phi_+ = \phi_-, \qquad \frac{\partial \phi_+}{\partial r} = \kappa \frac{\partial \phi_-}{\partial r}, \tag{10.7.27}$$

where $\kappa = (1 - \beta^*)/(1 + \beta^*)$ and r is the distance from the center of the sphere. The problem is reduced to computing nontrivial interior and exterior harmonic potentials ϕ_+ and ϕ_- subject to (10.7.27).

Expanding ϕ_+ and ϕ_- in solid spherical harmonics, we obtain

$$\phi_+ = \sum_{n=0}^{\infty} b_n \left(\frac{a}{r}\right)^{n+1} \mathcal{S}_n(\theta, \varphi), \qquad \phi_- = \sum_{n=0}^{\infty} c_n \left(\frac{r}{a}\right)^{n+1} \mathcal{S}_n(\theta, \varphi), \tag{10.7.28}$$

where $\mathcal{S}_n$ are surface spherical harmonics and b_n, c_n are constants (e.g., [220]). Enforcing the matching conditions (10.7.28), we find that

$$b_n = c_n, \qquad -b_n(n+1) = \kappa n c_n, \tag{10.7.29}$$

which requires that $\kappa = -(n+1)/n$ and $b = -2n - 1$. Thus, the spectrum of the double-layer operator for the sphere is the set of all negative integers.

Problems

10.7.1 *Adjoint of the double-layer operator*

Show that any two complex functions, q_1 and q_2, satisfy (10.7.11), where the projection $(\cdot, \cdot)$ is defined in (10.7.12).

10.7.2 *A known eigensolution*

Verify that, when D is the surface of a sphere, (10.7.7) has a constant eigensolution with corresponding eigenvalue $\beta = -1$.

10.7.3 *Null space of operators*

Find the null space of the operators $\mathcal{O}_\pm$ and $\mathcal{P}_\pm$ introduced in (10.5.14) and (10.6.24), that is, find all functions q that satisfy $\mathcal{O}_\pm[q] = 0$ or $\mathcal{P}_\pm[q] = 0$.

10.7.4 *Multiple boundaries*

Show that, when D consists of a number of closed surfaces, D_i, where $i = 1, \ldots, N$, the most general eigensolution of (10.7.6) with $\beta = -1$ is $\psi = c_k$ over D_k, where c_k are arbitrary constants. What is the most general adjoint eigensolution χ?

10.8   Regularization of integral equations of the second kind

We have seen that the integral equations (10.7.2) and (10.7.4) for interior flow and the integral equation (10.7.5) for exterior flow have either no solution or an infinite number of solutions when appropriate restrictions on the forcing functions are satisfied. The integral equations (10.7.2) and (10.7.4) for exterior flow and the integral equation (10.7.5) for interior flow have a unique solution which, however, cannot be computed by the method of successive substitutions.

The occurrence of multiple solutions and our inability to compute a solution by the method of successive substitutions can be circumvented by replacing the original integral equations with regularized integral equations that have a unique solution amenable to the method of successive substitutions. The solution of the regularized equations or a modification of the solution also satisfy the original integral equations.

10.8.1 Wielandt deflation

To set up the ground for developing the regularized integral equations, we establish a connection between linear integral equations and linear calculus of matrices. Consider a real square matrix, $\mathbf{A}$, and introduce its transpose, $\mathbf{B} = \mathbf{A}^T$. The matrices $\mathbf{A}$ and $\mathbf{B}$ satisfy the distinguishing relation of adjoint operators,

$$\big(\mathbf{y}, \mathbf{A} \cdot \mathbf{x}\big) = \big(\mathbf{B} \cdot \mathbf{y}, \mathbf{x}\big), \tag{10.8.1}$$

where $\mathbf{y}$ and $\mathbf{x}$ are two arbitrary vectors and the parentheses denote the inner vector product. Consequently, $\mathbf{A}$ and $\mathbf{B}$ have complex conjugate eigenvalues and the same number of eigenvectors and generalized eigenvectors or principal vectors (e.g., [45, 318]). Each eigenvector of $\mathbf{A}$ corresponding to a particular eigenvalue, λ_1, is orthogonal to each eigenvector of $\mathbf{B}$ corresponding to a different eigenvalue, λ_2.

Let us assume that $\mathbf{u}$ is a real eigenvector of $\mathbf{A}$ corresponding to an eigenvalue, λ_1, with adjoint eigenvector $\mathbf{v}$,

$$\mathbf{A} \cdot \mathbf{u} = \mathbf{u} \cdot \mathbf{B} = \lambda_1 \mathbf{u}, \qquad \mathbf{B} \cdot \mathbf{v} = \mathbf{v} \cdot \mathbf{A} = \lambda_1 \mathbf{v}. \tag{10.8.2}$$

Wielandt's theorem states that the modified matrix

$$A_{ij}^{(1)} \equiv A_{ij} - \lambda_1 u_i w_j \tag{10.8.3}$$

shares its eigenvalues with $\mathbf{A}$, except that λ_1 has been replaced by zero, where $\mathbf{w}$ is an arbitrary vector satisfying $\mathbf{w} \cdot \mathbf{u} = 1$ (e.g., [45], p. 361; [429], p. 596). Wielandt's theorem also states that the modified matrix

$$B_{ij}^{(1)} \equiv B_{ij} - \lambda_1 v_i z_j \tag{10.8.4}$$

shares its eigenvalues with $\mathbf{B}$, except that λ_1 has been replaced by zero, where $\mathbf{z}$ is an arbitrary vector satisfying $\mathbf{z} \cdot \mathbf{v} = 1$. The adjoint of $\mathbf{B}^{(1)}$, which is equal to its transpose,

$$A_{ij}^{(2)} \equiv A_{ij} - \lambda_1 z_i v_j, \tag{10.8.5}$$

shares its eigenvalues with the adjoint of $\mathbf{B}$, which is the matrix $\mathbf{A}$, except that λ_1 has been replaced by zero.

The process of modifying a matrix to reduce the number of nonzero eigenvalues is called eigenvalue spectrum deflation. Single deflation eliminates one eigenvalue and leaves all other eigenvalues and adjoint eigenvectors unaffected, but alters the corresponding eigenvectors (Problem 10.8.1). To implement eigenvalue deflation, we must have available one eigenvalue and either the corresponding eigenvector, $\mathbf{u}$, or the adjoint eigenvector, $\mathbf{v}$.

10.8.2 Regularization of (10.7.4) for exterior flow

To demonstrate the application of eigenvalue deflation to integral equations, we consider equation (10.7.4) for exterior flow, repeated below for convenience,

$$q(\mathbf{x}_0) = 2 \iint_D^{PV} q(\mathbf{x}) \, \mathbf{N}(\mathbf{x}_0) \cdot \nabla_0 \mathcal{G}(\mathbf{x}_0, \mathbf{x}) \, \mathrm{d}S(\mathbf{x}) - 2 \, \mathbf{N}(\mathbf{x}_0) \cdot \nabla_0 \phi_+(\mathbf{x}_0), \qquad (10.8.6)$$

which was shown to have a unique solution. Unfortunately, because the spectral radius of the adjoint double-layer operator is equal to unity, the solution cannot be computed by the method of successive substitutions.

Let us discretize the domain D into N boundary elements and approximate the unknown function $q(\mathbf{x})$ with a constant function over each element. In the numerical representation, equation (10.8.6) takes the discrete form

$$\mathbf{x} = \mathbf{A} \cdot \mathbf{x} + \mathbf{b}, \qquad (10.8.7)$$

where $\mathbf{A}$ is an influence matrix, the vector $\mathbf{x}$ contains the constant values of q over the boundary elements, and the vector $\mathbf{b}$ contains the values of the forcing function at the midpoints of the boundary elements. Instead of solving (10.8.7), we propose to solve the regularized equation

$$\mathbf{x} = \mathbf{A}^{(2)} \cdot \mathbf{x} + \mathbf{c}, \qquad (10.8.8)$$

where $\mathbf{A}^{(2)}$ is defined in (10.8.5) and $\mathbf{c}$ is a constant vector.

To ensure that the solution of (10.8.8) also satisfies (10.8.7), we require that

$$\mathbf{c} = \mathbf{b} + \lambda_1 \, \mathbf{z} \, (\mathbf{v} \cdot \mathbf{x}). \qquad (10.8.9)$$

Projecting (10.8.7) onto $\mathbf{v}$ and recalling that, by definition, $\mathbf{v} \cdot \mathbf{A} = \lambda_1 \mathbf{v}$, we obtain

$$\mathbf{v} \cdot \mathbf{x} = \frac{1}{1 - \lambda_1} \, \mathbf{v} \cdot \mathbf{b}. \qquad (10.8.10)$$

Substituting this expression into the last term on the right-hand side of (10.8.9), we obtain

$$\mathbf{c} = \mathbf{b} + \frac{\lambda_1}{1 - \lambda_1} \mathbf{z} \, (\mathbf{v} \cdot \mathbf{b}). \qquad (10.8.11)$$

When $\mathbf{v} \cdot \mathbf{b} = 0$, we obtain $\mathbf{c} = \mathbf{b}$. Equation (10.8.8) then takes the explicit form

$$\mathbf{x} = \mathbf{A}^{(2)} \cdot \mathbf{x} + \left(\mathbf{I} + \frac{\lambda_1}{1 - \lambda_1} \mathbf{z} \otimes \mathbf{v} \right) \cdot \mathbf{b}, \qquad (10.8.12)$$

where $\mathbf{I}$ is the identity matrix. The solution of (10.8.12) is identical to that of (10.8.7).

Identifying $\lambda_1 = -1$ with the eigenvalue of $\mathbf{A}$ with the maximum norm corresponding to the eigenvalue of the adjoint double-layer operator with the minimum norm, $\beta = -1$, and noting that

λ_1 is not a eigenvalue of $\mathbf{A}^{(2)}$, we find that the spectral radius of $\mathbf{A}^{(2)}$ is less than unity. Equation (10.8.12) may then be solved by the method of successive substitutions.

Deflated integral equation

Now we return to (10.8.6), recall that $\psi = c$ is the adjoint eigenfunction corresponding to the eigenvalue with minimum norm, $\beta = -1$, note that $\mathbf{v}$ plays the role of ψ, and introduce a function $\eta(\mathbf{x})$ corresponding to $\mathbf{z}$. To satisfy the condition $\mathbf{z} \cdot \mathbf{v} = 1$, we set $\eta = 1/(cS_D)$, so that the integral of the product $\psi\eta$ over D is equal to unity, where S_D is the surface area of D. Motivated by (10.8.9), we replace (10.8.6) with the regularized integral equation

$$
\begin{aligned}
q(\mathbf{x}_0) = 2 \iint_D^{PV} q(\mathbf{x})\, \mathbf{N}(\mathbf{x}_0) \cdot \nabla_0 \mathcal{G}(\mathbf{x}_0, \mathbf{x})\, \mathrm{d}S(\mathbf{x}) + \frac{1}{S_D} \iint_D q\, \mathrm{d}S \\
- 2\,\mathbf{N}(\mathbf{x}_0) \cdot \nabla_0 \phi_+(\mathbf{x}_0) + \frac{1}{S_D} \iint_D \mathbf{N} \cdot \nabla\phi_+ \mathrm{d}S.
\end{aligned}
\tag{10.8.13}
$$

To demonstrate that a solution of (10.8.13) also satisfies (10.8.6), we integrate (10.8.13) over D and use identity (10.5.2) to find

$$
\iint_D q\, \mathrm{d}S = - \iint_D \mathbf{N} \cdot \nabla\phi^+ \mathrm{d}S,
\tag{10.8.14}
$$

which can be substituted into (10.8.13) to produce (10.8.6).

The generalized homogeneous equation of (10.8.13) arises by discarding the forcing function expressed by the last two terms on the right-hand side, and multiplying the remaining terms on the right-hand side by the eigenvalue β^*, yielding

$$
\chi(\mathbf{x}_0) = 2\beta^* \left(\iint_D^{PV} \chi(\mathbf{x})\, \mathbf{N}(\mathbf{x}_0) \cdot \nabla_0 \mathcal{G}(\mathbf{x}_0, \mathbf{x})\, \mathrm{d}S(\mathbf{x}) + \frac{1}{2S_D} \iint_D \chi\, \mathrm{d}S \right).
\tag{10.8.15}
$$

Integrating (10.8.15) over D and using (10.5.2), we find that the last integral on the right-hand side is zero, which allows us to identify the eigenvalues and eigenfunctions of (10.8.15) with those of (10.7.7). However, since criterion (10.7.23) is violated, $\beta = -1$ is no longer an eigenvalue and the spectral radius of the integral operator on the right-hand side of (10.8.13) is higher than unity. This guarantees that equation (10.8.13) can be solved by the desirable method of successive approximations.

When the boundary D consists of a collection of closed boundaries, D_k, where $k = 1, \dots, N_k$, the counterpart of equation (10.8.13) at a point $\mathbf{x}_0$ that lies on the kth boundary is

$$
\begin{aligned}
q(\mathbf{x}_0) = 2 \iint_D^{PV} \mathbf{N}(\mathbf{x}_0) \cdot \nabla_0 \mathcal{G}(\mathbf{x}_0, \mathbf{x})\, q(\mathbf{x})\, \mathrm{d}S(\mathbf{x}) + \frac{1}{S_{D_k}} \iint_{D_k} q\, \mathrm{d}S \\
- 2\,\mathbf{N}(\mathbf{x}_0) \cdot \nabla_0 \phi_+(\mathbf{x}_0) + \frac{1}{S_{D_k}} \iint_{D_k} \mathbf{N} \cdot \nabla\phi_+ \mathrm{d}S,
\end{aligned}
\tag{10.8.16}
$$

where S_{D_k} is the surface area of the kth boundary. We observe that each surface undergoes an individual deflation.

10.8.3   Regularization of (10.7.5) for interior flow

Next, we apply the method of eigenvalue deflation to the integral equation (10.7.5) arising from the double-layer representation with Dirichlet boundary conditions for interior flow,

$$q(\mathbf{x}_0) = 2 \int\!\!\!\!\int_D^{PV} q(\mathbf{x})\,\mathbf{N}(\mathbf{x}) \cdot \nabla \mathcal{G}(\mathbf{x}, \mathbf{x}_0)\, dS(\mathbf{x}) - 2\,\phi_-(\mathbf{x}_0). \tag{10.8.17}$$

In Section 10.7, we demonstrated that this integral equation admits a unique solution. However, because the spectral radius of the adjoint double-layer operator is equal to unity, the solution cannot be found by the method of successive substitutions.

Working as in Section 10.8.2, we discretize the domain D into N boundary elements and assume that q is constant over each element to obtain the linear algebraic system (10.8.7) with a different influence matrix, $\mathbf{A}$, and known vector, $\mathbf{b}$. Instead of solving this system, we propose to solve the alternative system

$$\mathbf{x} = \mathbf{A}^{(1)} \cdot \mathbf{x} + \mathbf{c}, \tag{10.8.18}$$

where the matrix $\mathbf{A}^{(1)}$ is defined in (10.8.3). To ensure that the solution of the altered system (10.8.18) also satisfies the original system $\mathbf{x} = \mathbf{A} \cdot \mathbf{x} + \mathbf{b}$, we require that

$$\mathbf{c} = \mathbf{b} + \lambda_1\,\mathbf{u}\,(\mathbf{w} \cdot \mathbf{x}). \tag{10.8.19}$$

Since $\mathbf{A}^{(1)} \cdot \mathbf{u} = \mathbf{0}$, we can rewrite (10.8.18) as

$$\mathbf{y} = \mathbf{A}^{(1)} \cdot \mathbf{y} + \mathbf{b}, \tag{10.8.20}$$

where

$$\mathbf{y} \equiv \left(\mathbf{I} - \lambda_1\,\mathbf{u} \otimes \mathbf{w} \right) \cdot \mathbf{x}. \tag{10.8.21}$$

Projecting (10.8.21) onto $\mathbf{w}$ and recalling that $\mathbf{w} \cdot \mathbf{u} = 1$, we obtain $\mathbf{w} \cdot \mathbf{y} = (1 - \lambda_1)\,\mathbf{w} \cdot \mathbf{x}$. Substituting this expression into (10.8.21) and rearranging, we obtain

$$\mathbf{x} = \left(\mathbf{I} + \frac{\lambda_1}{1 - \lambda_1}\,\mathbf{u} \otimes \mathbf{w} \right) \cdot \mathbf{y}. \tag{10.8.22}$$

Instead of solving the original equation $\mathbf{x} = \mathbf{A} \cdot \mathbf{x} + \mathbf{b}$ for $\mathbf{x}$, we prefer to solve equation (10.8.20) for $\mathbf{y}$ and then use (10.8.22) to recover $\mathbf{x}$.

Identifying $\lambda_1 = -1$ with the eigenvalue of $\mathbf{A}$ with the maximum norm corresponding to the eigenvalue $\beta = -1$ of the double-layer operator with the minimum norm, and noting that λ_1 is not a eigenvalue of $\mathbf{A}^{(1)}$, we find that the spectral radius of $\mathbf{A}^{(1)}$ is less than unity. Equation (10.8.20) may then be solved by the method of successive substitutions.

Deflated integral equation

Now we return to (10.8.17), recall that $\beta = -1$ is an eigenvalue of the generalized homogeneous equation (10.7.6) with constant eigenfunction, $\psi = c$, note that $\mathbf{u}$ plays the role of $\mathbf{y}$, and introduce a function η corresponding to the vector $\mathbf{w}$. To satisfy the constraint $\mathbf{w} \cdot \mathbf{u} = 1$, we set $\eta = 1/(cS_D)$, where S_D is the surface area of D. Finally, we write the counterparts of equations (10.8.20) and (10.8.22),

$$Q(\mathbf{x}_0) = 2 \iint_D^{PV} Q(\mathbf{x}) \, \mathbf{N}(\mathbf{x}) \cdot \nabla \mathcal{G}(\mathbf{x}, \mathbf{x}_0) \, \mathrm{d}S(\mathbf{x}) + \frac{1}{S_D} \iint_D Q \, \mathrm{d}S - 2 \, \phi_-(\mathbf{x}_0), \qquad (10.8.23)$$

and

$$q = Q - \frac{1}{2S_D} \iint_D Q \, \mathrm{d}S. \qquad (10.8.24)$$

The numerical procedure involves solving (10.8.23) for Q and then using (10.8.24) to recover q.

The generalized adjoint homogeneous equation corresponding to (10.8.23) is given in (10.8.15). Our earlier analysis has shown that (10.8.23) has a unique solution that can be computed by the method of successive substitutions. Straightforward substitution of (10.8.24) into (10.8.23) yields the original integral equation (10.8.17) (Problem 10.8.2).

10.8.4 Regularization of (10.7.1) for exterior flow

The deflation of equation (10.7.1) for exterior flow is similar to that of equation (10.7.5) for interior flow, as discussed in Section 10.8.3. The counterparts of equations (10.8.23) and (10.8.24) are

$$\Phi(\mathbf{x}_0) = 2 \iint_D^{PV} \Phi(\mathbf{x}) \, \mathbf{N}(\mathbf{x}) \cdot \nabla \mathcal{G}(\mathbf{x}, \mathbf{x}_0) \, \mathrm{d}S(\mathbf{x}) + \frac{1}{S_D} \iint_D \Phi \, \mathrm{d}S - 2 \, \mathcal{I}_+^S(\mathbf{x}_0), \qquad (10.8.25)$$

and

$$\phi_+ = \Phi - \frac{1}{2S_D} \iint_D \Phi \, \mathrm{d}S, \qquad (10.8.26)$$

where $\mathcal{I}_+^S$ is the single-layer potential on the right-hand side of (10.7.1).

If D consists of a number of disconnected closed boundaries, we obtain a modified set of equations. For a point $\mathbf{x}_0$ that lies at the kth boundary, D_k, we obtain

$$\Phi(\mathbf{x}_0) = 2 \iint_D^{PV} \Phi(\mathbf{x}) \, \mathbf{N}(\mathbf{x}) \cdot \nabla \mathcal{G}(\mathbf{x}, \mathbf{x}_0) \, \mathrm{d}S(\mathbf{x}) + \frac{1}{S_{D_k}} \iint_{D_k} \Phi \, \mathrm{d}S - 2 \, \mathcal{I}_+^S(\mathbf{x}_0), \qquad (10.8.27)$$

and

$$\phi_+ = \Phi - \frac{1}{2S_{D_k}} \iint_{D_k} \Phi \, \mathrm{d}S, \qquad (10.8.28)$$

where S_{D_k} is the surface area of the kth boundary. We observe that each surface undergoes an individual deflation.

10.8.5   Regularization of (10.7.5) for exterior flow

Next, we consider the integral equation (10.7.5) that arises from the double-layer representation with the Dirichlet boundary condition for exterior flow. The counterpart of (10.8.18) is

$$\mathbf{x} = -\mathbf{A}^{(1)} \cdot \mathbf{x} + \mathbf{c}, \tag{10.8.29}$$

where the influence matrix $\mathbf{A}$ is the same as that appearing in (10.8.18) but the vector $\mathbf{c}$ is different. Working in the familiar way, we replace (10.8.29) with

$$\mathbf{y} = -\mathbf{A}^{(1)} \cdot \mathbf{y} + \mathbf{b}, \tag{10.8.30}$$

where

$$\mathbf{y} = (\mathbf{I} + \lambda_1 \mathbf{u} \otimes \mathbf{w}) \cdot \mathbf{x}. \tag{10.8.31}$$

To perform eigenvalue deflation, we must apply (10.8.31) with $\lambda_1 = -1$. However, inverting this equation to recover $\mathbf{x}$ is prohibited by the singular nature of the matrix $\mathbf{I} - \mathbf{uw}$ (Problem 10.8.3). We conclude that (10.7.5) cannot be regularized for exterior flow. Similar difficulties are encountered when we attempt to deflate equations (10.7.1) and (10.7.5) for interior flow.

10.8.6   Completed double-layer representation for exterior flow

Our inability to represent an arbitrary exterior flow with a double-layer potential alone stems from the fast decay of the point-force dipole, requiring that the flow rate of the velocity $\nabla \phi$ across the distribution domain D or any other surface enclosing D vanish. This restriction is fulfilled in the case of flow past or due to the motion of a rigid body, but not in the case of flow due to the expansion of a bubble.

To demonstrate the limitations of the double-layer representation, we regard the integral equation (10.6.25) for exterior flow, $\mathcal{P}_+[q] = 2\phi_+$, as a mapping of the space of density functions, q, to the space of the exterior boundary distributions, ϕ_+. Functional analysis of compact operators shows that the image of the mapping $\mathcal{P}_+$ is orthogonal to the null space of the adjoint operator $\mathcal{O}_-$ defined in (10.5.15), which contains the eigenfunction χ given in (10.7.22). A solution of (10.6.25) for exterior flow will exist only when the projection of ϕ^+ on χ is zero, as required by (10.7.26).

Range completion

To complete the deficient range of the double-layer operator in the case of exterior flow, we introduce a composite representation,

$$\phi(\mathbf{x}) = \iint_D q(\mathbf{x})\, \mathbf{N}(\mathbf{x}) \cdot \nabla \mathcal{G}(\mathbf{x}_0, \mathbf{x})\, dS(\mathbf{x}) + \Phi(\mathbf{x}_0), \tag{10.8.32}$$

where the harmonic function Φ contributes a finite flow rate across D. Applying (10.8.32) at the exterior side of D and using (10.6.6), we obtain an integral equation of the second kind for q,

$$q(\mathbf{x}_0) = -2 \iint_D^{PV} q(\mathbf{x})\, \mathbf{N}(\mathbf{x}) \cdot \nabla \mathcal{G}(\mathbf{x}_0, \mathbf{x})\, dS(\mathbf{x}) - 2\,\Phi(\mathbf{x}_0) + 2\,\phi_+(\mathbf{x}_0). \tag{10.8.33}$$

Equation (10.8.32) is the completed double-layer representation, and equation (10.8.33) is the associated integral equation.

Completion with a point source

One possible choice for Φ is the potential associated with a Green's function located at a point, $\mathbf{x}_s$, inside D,

$$\Phi(\mathbf{x}_0) = c\,\mathcal{G}(\mathbf{x}_0, \mathbf{x}_s), \tag{10.8.34}$$

where c is a constant determined by the distribution density, q. To remove this dependency while preserving linearity, we set

$$c = A \iint_D q(\mathbf{x})\,\mathrm{d}S(\mathbf{x}), \tag{10.8.35}$$

where A is a new constant independent of q. With these choices, the integral equation (10.8.33) becomes

$$q(\mathbf{x}_0) = -2 \iint_D^{PV} q(\mathbf{x})\,\mathbf{N}(\mathbf{x}) \cdot \nabla\mathcal{G}(\mathbf{x}_0, \mathbf{x})\,\mathrm{d}S(\mathbf{x}) - 2A\,\mathcal{G}(\mathbf{x}_0, \mathbf{x}_s) \iint_D q(\mathbf{x})\,\mathrm{d}S(\mathbf{x}) + 2\,\phi_+(\mathbf{x}_0). \tag{10.8.36}$$

To show that this equation has a unique solution, we consider the corresponding adjoint homogeneous equation

$$\chi(\mathbf{x}_0) = -2 \iint_D^{PV} \chi(\mathbf{x})\,\mathbf{N}(\mathbf{x}_0) \cdot \nabla_0\mathcal{G}(\mathbf{x}_0, \mathbf{x})\,\mathrm{d}S(\mathbf{x}) - 2A\,\mathcal{G}(\mathbf{x}_0, \mathbf{x}_s) \iint_D \chi(\mathbf{x})\,\mathrm{d}S(\mathbf{x}), \tag{10.8.37}$$

where χ is an eigenfunction. Integrating (10.8.37) over D and using (10.5.2), we find that the last integral is identically zero. Equation (10.8.37) then reduces to (10.7.7) with $\beta^* = -1$, in violation of (10.7.23). Consequently, the integral equation (10.8.37) has only the trivial null solution for any value of A.

Problems

10.8.1 *Wielandt's deflation*

Show that the Wielandt deflation eliminates one eigenvalue, leaves all other eigenvalues and adjoint eigenvectors unaffected, but alters the eigenvectors.

10.8.2 *Deflated equation*

Show that (10.8.23) and (10.8.24) produce the solution of (10.8.17).

10.8.3 *A singular matrix*

Show that, if two vectors, $\mathbf{u}$ and $\mathbf{w}$, satisfy $\mathbf{u} \cdot \mathbf{w} = 1$, the matrix $\mathbf{I} - \mathbf{u} \otimes \mathbf{w}$ is singular. *Hint:* demonstrate that one eigenvalue is zero.

10.8.4 *Completion with a single-layer potential*

One choice for the function Φ defined in (10.8.32) is a single-layer potential with density ζ,

$$\Phi(\mathbf{x}_0) = \iint_D \mathcal{G}(\mathbf{x}_0, \mathbf{x})\, \zeta(\mathbf{x})\, \mathrm{d}S(\mathbf{x}). \tag{10.8.38}$$

To ensure the linearity of the integral equation, we require that ζ is a linear function of q, setting $\zeta = Aq$, where A is a constant. Show that, with this choice, the solution of (10.8.33) is unique for any positive A.

10.8.5 *Completion with a collection of point sources*

Identify the function Φ introduced in (10.8.32) with the potential due to a collection of point sources located in the interior of D. Discuss the uniqueness of solution of the resulting integral equation.

10.9 Iterative solution of integral equations of the second kind

Compared to direct methods, iterative methods for solving integral equations of the second kind have two significant advantages: reduced algebraic manipulations associated with grouping unknown coefficients of the local expansions in the boundary-element implementation, and affordability in terms of computer memory requirements, central processor (CPU) time, and programming effort. To illustrate the implementation of iterative methods, we consider a prototypical integral equation of the second kind,

$$q = \alpha\, \mathcal{O}[q] + F, \tag{10.9.1}$$

where $\mathcal{O}$ is a compact linear integral operator defined over a specified line or surface, D, q is an unknown density distribution, and F is a given forcing function. We proceed by discretizing D into N boundary elements and approximate the integral operator $\mathcal{O}[q]$ with an integration quadrature over each element to obtain

$$q(\mathbf{x}) = \sum_{\substack{\text{Elements} \\ n=1,\ldots,N}} \left(\sum_{\substack{\text{Quadrature points} \\ k=1,\ldots,K}} A_{nk}(\mathbf{x}, \mathbf{x}_{n,k})\, q(\mathbf{x}_{n,k}) \right) + F(\mathbf{x}), \tag{10.9.2}$$

where $\{\mathbf{x}_{n,k}\}$ is the union of the K quadrature base points over each element and $\mathbf{A}$ is an influence matrix (see Section B.6, Appendix B).

Nÿstrom method

In Nÿstrom's method, equation (10.9.2) is enforced at the integration quadrature points over each element to yield a system of NK linear algebraic equations for the unknown values, $q(\mathbf{x}_{n,k})$ [15]. Jacobi's method for linear systems is the discrete implementation of the method of successive substitutions involving three steps: guess the values $q(\mathbf{x}_{nk})$, compute the right-hand side of (10.9.2) at all quadrature base points, and replace the assumed with the computed values, $q^{new}(\mathbf{x}_{nk})$. Other iterative schemes based on generalized conjugate gradient methods are discussed in Section B.1, Appendix B.

To reduce the computational cost, it is helpful to compute the influence matrix $A_{n,k}$ once at the outset and then store and recall its elements during the iterations. When N is on the order of a few hundred and the unknown function is assumed to be a constant or linear function over each element, the computer memory requirements are modest. When the number of boundary elements is large or high-order approximations are employed, the size of the matrix $A_{n,k}$ may exceed the available memory, and this necessitates recomputing some or all of the elements, $A_{n,k}$, before each iteration.

Element expansions

In a more general approach capturing the spirit of the spectral-element method, the unknown density, q, is expanded into a polynomial or trigonometric series with respect to properly defined surface variables over each element, and the coefficients of the expansion are identified with the values of q at corresponding element nodes. Initial values of q are assigned, the polynomial approximations are evaluated over all elements, and the integral equation is enforced at the nodes. Finally, the double-layer integral is computed and added to the forcing function F, thereby producing new values for q at the nodes. The procedure is repeated until the values of q at all nodes change by less than a preset minimum after one iteration. This method has the advantage of directly producing the value of q at the nodes, and thus circumventing the need for further interpolation. This feature is especially desirable when q represents a primary variable, such as the velocity.

Convergence

It is important to emphasize that the iterative procedures described in this section converge only when the magnitude of the coefficient α is less than the spectral radius of the operator $\mathcal{O}$, and the integration domain, D, is a smooth Lyapunov surface. If D contains sharp edges or corners, the iterations may diverge. In order to compute a solution, equation (10.9.2) must be restated in the standard form of a linear system of equations and the solution must be computed by direct inversion.

10.9.1   Computer implementations with parallel processor architecture

The availability of computers with parallel processor architecture allows us to tackle problems with pronounced geometric complexity. The standard computational problem requires solving a linear system of equations, such as that shown in (10.9.2), by a direct or iterative method. Parallelization of the computational tasks in a direct method ranges from hard to unfeasible [32]. For example, parallelization in Gaussian elimination is prohibited by pivoting. Iterative methods are ideally suited for parallel computation, the main consideration being the efficient distribution of tasks among the available processors to minimize communication time.

In the context of parallel computation, we may distinguish two classes of problems according to the topology of the flow domain. The first class includes problems in domains that are bounded by distinct closed surfaces representing, for example, the boundaries of a collection of suspended particles. The second class includes problems in domains with contiguous but complex boundaries representing, for example, the surface of a vehicle or aircraft.

Assuming that the boundary of a flow consists of a collection of M smooth closed surfaces, we decompose the integral operator in (10.9.1) into the sum of M operators, where each operator is supported by a distinct boundary, and write

$$q(\mathbf{x}) = \alpha \sum_{m=1}^{M} \mathcal{O}_m[q] + F(\mathbf{x}). \tag{10.9.3}$$

We then guess the distribution of q over each boundary, call it $q^{(0)}$, and iterate individually on each surface using the method of successive substitutions, where each boundary is assigned to a different processor. Over the ith surface, we iterate based on the equation ˙

$$q(\mathbf{x}) = \alpha \mathcal{O}_i[q] + G(\mathbf{x}) \qquad \text{where} \qquad G(\mathbf{x}) = \alpha \sum_{m=1}^{M} {}' \mathcal{O}_m[q] + F(\mathbf{x}), \tag{10.9.4}$$

and the prime after the sum indicates that the term $m = i$ is excluded from the summation. These local iterations are guaranteed to converge as long as $|\alpha|$ is less than the spectral radius of the corresponding operators. After a number of local iterations have been carried out, the initial guess $q^{(0)}$ is updated across all processors, a second global iteration is carried out, and the procedure is repeated until a converged solution has been found. The frequency and protocol of communication between the processors play an important role in determining the overall efficiency of the method (e.g., [102, 138]).

Computer Problem

10.9.1 *Nÿstrom method*
Solve the integral equation

$$q(x) = \frac{1}{2} \int_0^1 q(\xi) \, |x - \xi| \, \mathrm{d}\xi + \mathrm{e}^x \tag{10.9.5}$$

using Nÿstrom's method implemented by the trapezoidal rule with 32 divisions.

11

Vortex motion

An important class of flows at high Reynolds numbers are characterized by the presence or motion of compact regions of concentrated vorticity, concisely called vortices, including vortex filaments and vortex layers. The analysis of these flows and the development of pertinent numerical methods is conducted under the assumption that viscous forces are confined inside thin boundary layers and the fluid in the main part of the flow is effectively inviscid. In nature and technology, vortices are generated by a variety of mechanisms, including the deposition of vorticity inside compact wakes behind bluff bodies, the rollup of separated boundary layers and vortex sheets ejected from sharp corners, and the instability of shear layers between two fluids that merge at different velocities. In everyday life, vortices emerge from the rollup of vortex sheets on either side of a blade that moves parallel to itself broadside along a free surface. Vortices have been described poetically as the muscles of a turbulent flow.

To compute the evolution of a flow that is dominated by vortex motion, it is expedient to compute the evolution of the vorticity field using the vorticity transport equation, while simultaneously describing the evolution of the velocity field by inverting the definition of the vorticity, $\omega = \nabla \times \mathbf{u}$. This method of computation, concisely described as a vortex method, is preferable to directly integrating in time the equation of motion in primary variables, including the velocity and the pressure. The direct approach will be discussed in Chapter 13 in the context of the finite-difference method.

The computational advantages of vortex methods stem from substantial simplifications in the vorticity transport equation that are not necessarily reflected in the equation of motion. In the case of two-dimensional flow, because of the absence of vortex stretching, the vorticity evolves under the action of convection and viscous diffusion alone. When viscous effects are insignificant, a region of concentrated vorticity remains compact at all times and the vortices are simply convected by the flow. These features allow us to reduce the computational domain by considering only those regions of the flow where the vorticity takes substantial values, above a preset threshold. Another important advantage of vortex methods is that computing the pressure is not required. The significance of this simplification will become evident in Chapter 13 where we discuss the subtleties involved in deriving and implementing boundary conditions for the pressure from specified conditions for the velocity.

A typical computational algorithm of a vortex method for computing the evolution of an unsteady flow involves the following four main steps:

1. Given an initial velocity field, compute the associated initial vorticity field. If the initial vorticity field is specified, this step is not necessary.
2. Use the vorticity transport equation to advance the vorticity field by one time step.

3. Invert the definition of the vorticity, $\boldsymbol{\omega} = \nabla \times \mathbf{u}$, to obtain the velocity field at the new time instant.

4. Return to Step 2 and repeat the computation for another time step.

To obtain the velocity field from the vorticity field in Step 2, we may use the Biot–Savart integral discussed in Sections 2.10–2.13. This approach is particularly effective in the case of inviscid or slightly viscous flows considered in this chapter. Other methods discussed in Chapter 13 in the context of the finite-difference formulation involve expressing the velocity in terms of a solenoidal vector potential, $\mathbf{A}$, so that $\mathbf{u} = \nabla \times \mathbf{A}$ and $\nabla \cdot \mathbf{A} = 0$, and then solving a Poisson equation for $\mathbf{A}$, $\nabla^2 \mathbf{A} = -\boldsymbol{\omega}$, or inverting Laplace's equation for the velocity $\nabla^2 \mathbf{u} = -\nabla \times \boldsymbol{\omega}$. These methods are appropriate for viscous flows the vorticity is spread all over the domain of flow.

In this chapter, we outline and discuss the fundamental principles, governing equations, and computational algorithms underlying a general class of vortex methods for two-dimensional, axisymmetric, and three-dimensional flow. Further information on specific topics can be found in review articles by Clements & Maull [88], Saffman & Baker [358], Leonard [233, 234], Pullin [330], Puckett [328], and in monographs by Newton [278] and Saffman [357].

11.1 Invariants of vortex motion

In computing vortex motion of an inviscid fluid in a domain of infinite expanse in the absence of interior boundaries and subject to the condition that the velocity decays at infinity, it is wise to check the accuracy of the numerical results by monitoring the evolution of known invariants of the motion.

11.1.1 Three-dimensional flow

In Section 2.10, we saw that, if that the vorticity of a three-dimensional flow decays faster than $1/d^3$, where d is the distance from the origin, the integral of the vorticity over the domain of flow is zero. Thus, the total vorticity,

$$\iint_{Flow} \boldsymbol{\omega}\, dV = \mathbf{0} \tag{11.1.1}$$

is a first invariant of the motion.

Linear and angular impulse

Two additional invariants are the linear impulse, $\mathcal{P}$, and the angular impulse, $\mathcal{A}$, required to generate the motion and thereby impart to the fluid the necessary linear and angular momentum, defined as

$$\mathcal{P} = \frac{1}{2}\rho \iint_{Flow} \mathbf{x} \times \boldsymbol{\omega}\, dV = \mathbf{0}, \qquad \mathcal{A} = \frac{1}{3}\rho \iint_{Flow} \mathbf{x} \times (\mathbf{x} \times \boldsymbol{\omega})\, dV = \mathbf{0} \tag{11.1.2}$$

(e.g., [24], Section 7.2). The angular impulse is proportional to the first moment of the distributed linear impulse expressed by the integrand defining $\mathcal{P}$. To show that $\mathcal{P}$ and $\mathcal{A}$ remain constant in time, we take the time derivative of expressions (11.1.2) and invoke the vorticity transport equation for a homogeneous fluid (Problem 11.1.1(a)).

Kinetic energy

In the absence of viscous dissipation, the total kinetic energy of the fluid is conserved. Thus,

$$\mathcal{K} = \frac{1}{2}\rho \iint_{Flow} \mathbf{u}\cdot\mathbf{u}\,dV = \frac{1}{2}\rho\iint_{Flow}\mathbf{A}\cdot\boldsymbol{\omega}\,dV = \rho\iint_{Flow}\mathbf{u}\cdot(\mathbf{x}\times\boldsymbol{\omega})\,dV \tag{11.1.3}$$

is a fourth invariant of the motion, where $\mathbf{A}$ is a solenoidal vector potential for the velocity defined by the equation $\mathbf{u} = \nabla\times\mathbf{A}$ (Section 2.6).

Helicity

Projecting Euler's equation onto the vorticity vector and then integrating the product over the volume of the flow, we find that the helicity,

$$\mathcal{H} \equiv \iint_{Flow}\mathbf{u}\cdot\boldsymbol{\omega}\,dV, \tag{11.1.4}$$

is a fifth invariant of the motion. Physically, the helicity is a measure of the net linkage of the vortex lines [273].

Viscous effects

In the presence of viscous forces, the total vorticity defined in (11.1.1) remains zero and the linear and angular impulse defined in (11.1.2) remain invariant during the motion. However, the kinetic energy of the fluid and the helicity change during the evolution of an unsteady viscous flow. It is not surprising that the kinetic energy decreases due to viscous dissipation. Using the energy balance (3.4.22) for infinite flow, we find that

$$\frac{d\mathcal{K}}{dt} = -2\mu\iiint_{Flow}\mathbf{E}:\mathbf{E}\,dV = -\mu\iiint_{Flow}[\boldsymbol{\omega}\cdot\boldsymbol{\omega} + 2\nabla\cdot(\mathbf{u}\cdot\mathbf{L})]\,dV, \tag{11.1.5}$$

where $\mathbf{L}$ is the velocity gradient tensor and $\mathbf{E}$ is the rate-of-deformation tensor. Applying the divergence theorem and simplifying, we find that

$$\frac{d\mathcal{K}}{dt} = -\mu\iiint_{Flow}\boldsymbol{\omega}\cdot\boldsymbol{\omega}\,dV \tag{11.1.6}$$

(Problem 11.1.1(b)). The integral on the right-hand side of (11.1.6) is the enstrophy of the flow.

Projecting the vorticity transport equation onto the vorticity vector, we derive an evolution equation for the enstrophy,

$$\frac{d}{dt}\iiint_{Flow}\boldsymbol{\omega}\cdot\boldsymbol{\omega}\,dV = 2\iiint_{Flow}(\boldsymbol{\omega}\otimes\boldsymbol{\omega}):\mathbf{E}\,dV - 2\nu\iiint_{Flow}\nabla\boldsymbol{\omega}:\nabla\boldsymbol{\omega}\,dV, \tag{11.1.7}$$

where ν is the kinematic viscosity of the fluid (Problem 11.1.1(b)). A more general evolution equation for flow in a bounded domain is given in (3.12.27).

11.1.2 Axisymmetric flow

In the case of axisymmetric flow with no swirling motion, the vorticity points in the azimuthal direction, $\boldsymbol{\omega} = \omega_\varphi \mathbf{e}_\varphi$. In cylindrical polar coordinates, (x, σ, φ), the expressions for the linear momentum, angular momentum, kinetic energy, and helicity simplify into

$$\mathcal{P} = \pi\rho \iint_{Flow} \omega_\varphi\, \sigma^2\, \mathrm{d}A\, \mathbf{e}_x, \qquad \boldsymbol{\mathcal{A}} = \mathbf{0}, \qquad \mathcal{K} = \pi\rho \iint_{Flow} \psi\, \omega_\varphi\, \mathrm{d}A, \qquad \mathcal{H} = 0, \qquad (11.1.8)$$

where σ is the distance from the axis of revolution, $\mathrm{d}A = \mathrm{d}x\, \mathrm{d}\sigma$ is the differential area in a plane of constant azimuthal angle, φ, $\mathbf{e}_x$ and $\mathbf{e}_\varphi$ are coordinate unit vectors, and ψ is the Stokes stream function for axisymmetric flow; see equation (2.10.15). The helicity is identically zero because the velocity is normal to the vorticity in an axisymmetric flow.

11.1.3 Two-dimensional flow

In the case of two-dimensional flow, we obtain a new set of invariants, some but not all of which are related to the invariants of three-dimensional flow. In Section 3.13, we saw that the total circulation of a two-dimensional flow is conserved. Accordingly,

$$\mathcal{C} = \iint_{Flow} \omega_z\, \mathrm{d}A \qquad (11.1.9)$$

is an invariant of the motion for inviscid or viscous fluids, where $\mathrm{d}A = \mathrm{d}x\, \mathrm{d}y$ is a differential area in the xy plane of the flow and ω_z is the nonvanishing vorticity component.

Vorticity centroid and dispersion length

Two additional invariants are the centroid of vorticity and square of the dispersion length, defined, respectively, as

$$\boldsymbol{\mathcal{X}} = \frac{1}{\mathcal{C}} \iint_{Flow} \mathbf{x}\, \omega_z\, \mathrm{d}A, \qquad \mathcal{D}^2 = \frac{1}{\mathcal{C}} \iint_{Flow} (\mathbf{x} \cdot \mathbf{x})\, \omega_z\, \mathrm{d}A. \qquad (11.1.10)$$

These invariants are related to the linear and angular impulse that must be expended in order to generate the motion of the fluid (e.g., [24], Section 7.2). Due to the constancy of $\mathcal{C}$, the location of the origin of the Cartesian axes is immaterial.

Kinetic energy

The kinetic energy of an infinite two-dimensional flow with nonvanishing total circulation, $\mathcal{C}$, is infinite. However, the quantity

$$\mathcal{W} = \frac{1}{2}\rho \iint_{Flow} \psi\, \omega_z\, \mathrm{d}A \qquad (11.1.11)$$

is finite and remains constant during the motion, where ψ is the stream function, assumed to decay to zero at infinity. Physically, $\mathcal{W}$ expresses the part of the kinetic energy of the fluid that depends on the particular way in which the vorticity is distributed in the flow (e.g., [24], Section 7.2).

Viscous effects

In the case of two-dimensional viscous flow with no interior boundaries, the total circulation, $\mathcal{C}$, and vorticity centroid, $\mathcal{X}$, remain constant in time. The square of the dispersion length increases at a constant rate according to the evolution equation

$$\frac{d\mathcal{D}^2}{dt} = 4\nu \tag{11.1.12}$$

(e.g., [24], p. 536). This increase reflects the tendency of the vorticity to diffuse away from the initial distribution and occupy all available space. Viscous dissipation causes $\mathcal{W}$ to monotonically decrease during the evolution of an unsteady flow.

Problem

11.1.1 *Invariants of the motion*

(a) Show that $\mathcal{P}$ and $\mathcal{A}$ are conserved in a three-dimensional flow.

(b) Derive the evolution equations (11.1.6) and (11.1.7).

(c) Derive the evolution equation (11.1.12).

11.2   Point vortices

We begin the discussion of vortex motion by considering two-dimensional flow of an effectively inviscid fluid containing a collection of N point vortices defined discussed in Sections 1.13.3 and 2.13. The vorticity transport equation guarantees that the point vortices maintain their strength as they move about the domain of flow. Using the Biot–Savart integral, we find that, in the absence of boundaries, the velocity of each point vortex is equal to the sum of the velocities induced by all other point vortices.

Point-vortex motion

The fundamental steps of vortex methods outlined in the Introduction of this chapter, combined with the expressions for the induced velocity given in Section 2.13.1, provide us with a system of coupled nonlinear ordinary differential equations for the coordinates of the point vortices, $\mathbf{X}_i$, in the xy plane,

$$\frac{dX_i}{dt} = -\frac{1}{2\pi} \sum_{j=1}^{N}{}' \kappa_j \frac{Y_i - Y_j}{|\mathbf{X}_i - \mathbf{X}_j|^2}, \qquad \frac{dY_i}{dt} = \frac{1}{2\pi} \sum_{j=1}^{N}{}' \kappa_j \frac{X_i - X_j}{|\mathbf{X}_i - \mathbf{X}_j|^2}, \tag{11.2.1}$$

where κ_j is the strength of the jth point vortex and the prime indicates that the term $i = j$ is excluded from the sum. An implicit supplement to this system is the condition of strength invariance, $d\kappa_i/dt = 0$, for $i = 1, \dots, N$.

To compute the motion of a collection of point vortices from a given initial configuration, we integrate equations (11.2.1) in time using a standard numerical method, such as the Runge–Kutta method, as discussed in Section B.8, Appendix B.

To position of the point vortices in the complex plane is governed by the coupled differential equations

$$\frac{\mathrm{d}Z_i^*}{\mathrm{d}t} = \frac{1}{2\pi\mathrm{i}} \sum_{j=1}^{N}{}' \frac{1}{Z_i - Z_j}, \qquad (11.2.2)$$

where $Z_i = X_i + \mathrm{i}Y_i$ and i is the imaginary unit. Equations (11.2.1) are the real and imaginary parts of the unifying complex form (11.2.2).

Stream function and invariants of the motion

The stream function due to a collection of N point vortices arises by summing the individual stream functions of each point vortex,

$$\psi(\mathbf{x}) = -\frac{1}{2\pi} \sum_{i=1}^{N} \kappa_i \ln \frac{|\mathbf{x} - \mathbf{X}_i|}{a}, \qquad (11.2.3)$$

where a is a chosen length. Expressing the vorticity in terms of the two-dimensional delta function and using (11.2.3), we find that the scalar invariants of the motion (11.1.10)–(11.1.11) simplify into

$$\mathcal{C} = \sum_{i=1}^{N} \kappa_i, \qquad \mathcal{X} = \frac{1}{\mathcal{C}} \sum_{i=1}^{N} \kappa_i \, \mathbf{X}_i, \qquad \mathcal{D}^2 = \frac{1}{\mathcal{C}} \sum_{i=1}^{N} \kappa_i \, \mathbf{X}_i \cdot \mathbf{X}_i,$$

$$\mathcal{W} = -\frac{\rho}{4\pi} \sum_{i=1}^{N} \sum_{j=1}^{N}{}' \kappa_i \kappa_j \ln \frac{|\mathbf{X}_i - \mathbf{X}_j|}{a}. \qquad (11.2.4)$$

The availability of these invariants allows us to predict the trajectories of two point vortices, as discussed in Section 11.2.1.

Hamiltonian formulation

Kirchhoff noted that the system of ordinary differential equations (11.2.1) can be recast into the Hamiltonian form

$$\kappa_i \frac{\mathrm{d}X_i}{\mathrm{d}t} = \frac{\partial \mathcal{W}}{\partial Y_i}, \qquad \kappa_i \frac{\mathrm{d}Y_i}{\mathrm{d}t} = -\frac{\partial \mathcal{W}}{\partial X_i}, \qquad (11.2.5)$$

where summation over the repeated index i is *not* implied on the left-hand sides. This formulation allows us to study the motion of point vortices in the context of Hamiltonian dynamics [278].

11.2.1 Two point vortices

The motion of two point vortices with strength κ_1 and κ_2 separated by distance d can be predicted from the requirement that $\mathcal{X}$, $\mathcal{D}^2$, and $\mathcal{W}$ remain constant in time. We find that the point vortices move along concentric circles centered at the centroid of vorticity $\mathcal{X}$ with angular velocity

$$\Omega = \frac{\kappa_1 + \kappa_2}{2\pi d^2}, \qquad (11.2.6)$$

(*a*)   (*b*)

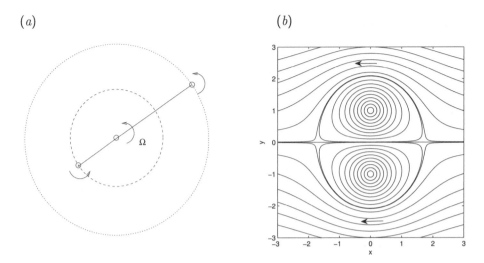

FIGURE 11.2.1   (*a*) Two point vortices move along concentric circles centered at the centroid of vorticity. (*b*) Streamline pattern due to a pair of point vortices with strengths of equal magnitude
and opposite sign, drawn in a frame of reference moving with the vortex pair. The x axis may be
regarded as an impenetrable wall.

while maintaining their initial separation, as illustrated in Figure 11.2.1(*a*). If $\kappa_1 = \kappa_2$, the point
vortices rotate about the midpoint. If $\kappa_1 \gg \kappa_2$, the second point vortex nearly orbits around the
nearly stationary first point vortex.

Translating vortex pair

If the strengths of the two point vortices have equal magnitude and opposite sign, $\pm\kappa$, the total
circulation of the flow is zero, the centroid of vorticity is shifted to infinity, and the point vortices
move in parallel straight lines that are perpendicular to their separation with velocity

$$V = \frac{\kappa}{2\pi d}. \tag{11.2.7}$$

The associated streamline pattern in a frame of reference moving with the point vortices is shown in
Figure 11.2.1(*b*). The dividing streamline takes the shape of an oblate oval with major and minor
axes equal to 2.09*a* and 1.73*a*, respectively, where $a = d/2$ is the half the point vortex separation.
It is instructive to observe the similarity in the streamline pattern shown in Figure 11.2.1(*b*) with
that shown in Figure 2.12.2 for axisymmetric flow due to a line vortex ring.

11.2.2   More than two point vortices

The motion of three or more point vortices in an infinite domain is known to follow periodic orbits
but also exhibit complex and chaotic behavior [12]. Steadily rotating point vortex arrangements
include concentric polygons, rectilinear distributions, and star shapes [13]. Examples will be given
in this section.

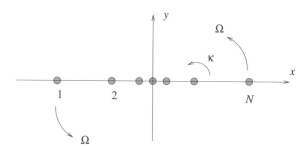

FIGURE 11.2.2   A collection of N point vortices with identical strengths, κ, distributed at positions corresponding to the zeros of the N-degree Hermite polynomial rotates as a whole with angular velocity Ω.

Vortex polygons

Havelock [172] considered the motion of N identical point vortices arranged at the vertices of a regular polygon that is inscribed inside a circle of radius a. Assume that at an instant the point vortices are located at the complex positions

$$z_i = x_i + \mathrm{i}y_i = a \exp\left(2\pi\mathrm{i}\,\frac{i-1}{N-1}\right) \tag{11.2.8}$$

for $i = 1, \ldots, N$, where i is the imaginary unit, so that $z_1 = a$. Using Table 7.10.1, we find that the complex potential of the induced flow is

$$w(z) = \frac{\kappa}{2\pi\mathrm{i}} \ln \frac{z^N - a^N}{a^N}. \tag{11.2.9}$$

The complex velocity induced at the location of the first point vortex, $z_1 = a$, by all other point vortices is

$$\lim_{z\to a} \frac{\mathrm{d}}{\mathrm{d}z}\left[w - \frac{\kappa}{2\pi\mathrm{i}} \ln \frac{z-a}{a}\right] = \frac{\kappa}{2\pi\mathrm{i}} \lim_{z\to a}\left[N\frac{z^{N-1}}{z^N - a^N} - \frac{1}{z-a}\right] = \frac{\kappa}{2\pi\mathrm{i}}\,\frac{N-1}{2a}. \tag{11.2.10}$$

Accordingly, the vortex polygon rotates intact with angular velocity

$$\Omega = (N-1)\frac{\kappa}{4\pi a^2}. \tag{11.2.11}$$

For $N = 2$, we recover formula (11.2.6) with $\kappa_1 = \kappa_2 = \kappa$ and $d = 2a$. Linear stability analysis reveals that the rotation is stable when $N < 7$, marginally stable when $N = 7$, and unstable when $N > 7$ [172].

11.2.3   Rotating collinear arrays of point vortices

Next, we consider the mutually induced motion of a collection of N point vortices with equal strengths, κ, initially situated along the x axis at positions, x_i, where $i = 1, \ldots, N$, as shown in Figure 11.2.2. As a preliminary, we introduce the Nth-degree generating polynomial

$$\Phi(x) = (x - x_1)(x - x_2) \cdots (x - x_N). \tag{11.2.12}$$

Using elementary calculus, we find that

$$\frac{\Phi''(x_i)}{2\,\Phi'(x_i)} = \frac{1}{x_i - x_1} + \cdots + \frac{1}{x_i - x_{i-1}} + \frac{1}{x_i - x_{i+1}} + \cdots + \frac{1}{x_i - x_N} \qquad (11.2.13)$$

(e.g., [317]). Note that one singular term is missing from the sum. The y velocity component of the ith point vortex is given by

$$v_i = \frac{\kappa}{2\pi} \frac{\Phi''(x_i)}{2\Phi'(x_i)}. \qquad (11.2.14)$$

If the point vortex assembly rotates as a whole around the origin with angular velocity Ω, then

$$v_i = \Omega\, x_i. \qquad (11.2.15)$$

Combining the last two expressions, we obtain

$$\frac{\kappa}{2\pi} \frac{\Phi''(x_i)}{2\,\Phi'(x_i)} = \Omega\, x_i \qquad (11.2.16)$$

for $i = 1, 2\ldots, N$, which shows that x_i are the roots of the Nth-degree polynomial

$$\Pi(x) \equiv \Phi''(x) - \frac{4\pi\Omega}{\kappa}\, x\, \Phi'(x). \qquad (11.2.17)$$

Accordingly, we can write

$$\Pi(x) = c\, \Phi(x), \qquad (11.2.18)$$

where c is a constant. To ensure that the coefficients of the highest power of x on the right-hand sides of the last two expressions are the same, we set $c = -4\pi\Omega N/\kappa$ and find that Φ satisfies the differential equation

$$\Phi''(x) - \frac{4\pi\Omega}{\kappa}\, x\Phi'(x) + \frac{4\pi\Omega}{\kappa}\, N\Phi(x) = 0. \qquad (11.2.19)$$

In terms of the dimensionless position,

$$\widehat{x} \equiv x\, \sqrt{\frac{2\pi\Omega}{\kappa}}, \qquad (11.2.20)$$

equation (11.2.19) reduces to the Hermite equation

$$\ddot{\Phi}(\widehat{x}) - 2\,\widehat{x}\,\dot{\Phi}(\widehat{x}) + 2N\Phi(\widehat{x}) = 0, \qquad (11.2.21)$$

where a dot denotes a derivative with respect to $\widehat{x}$. The solution of the Hermite equation is the Nth-degree Hermite polynomial whose roots, σ_i, are available in analytical or tabular form (e.g., [317]). The position of the point vortices in physical space is

$$x_i \equiv \sigma_i \left(\frac{\kappa}{2\pi\Omega}\right)^{1/2} \qquad (11.2.22)$$

for $i = 1, \ldots, N$, where Ω is a free parameter whose sign matches the sign of κ.

11.2.4 Rotating clusters of point vortices

Next, we consider the mutually induced motion of a cluster of N point vortices with equal strengths, κ, situated at arbitrary positions in the complex z plane. Working as in the case of the linear array, we introduce the Nth-degree generating complex polynomial

$$\Phi(z) = (z - z_1)(z - z_2) \cdots (z - z_N), \tag{11.2.23}$$

where z_i is the position of the ith point vortex. Using elementary calculus, we find that

$$\frac{\Phi''(z_i)}{2\,\Phi'(z_i)} = \frac{1}{z_i - z_1} + \cdots + \frac{1}{z_i - z_{i-1}} + \frac{1}{z_i - z_{i+1}} + \cdots + \frac{1}{z_i - z_N}, \tag{11.2.24}$$

where a prime denotes a derivative with respect to z (e.g., [317]). Note that one detrimental singular term is missing from the sum. The complex velocity of the ith point vortex is given by

$$u_i - \mathrm{i}\,v_i = \frac{\kappa}{2\pi\mathrm{i}} \frac{\Phi''(z_i)}{2\Phi'(z_i)}, \tag{11.2.25}$$

where $u_i = u_x$ is the x velocity component and $v_i = u_y$ is the y velocity component of the ith point vortex.

If the point vortex assembly rotates intact as a cluster around the origin with angular velocity Ω, the velocity of the ith point vortex satisfies $u_i + \mathrm{i}\,v_i = \mathrm{i}\Omega\,z_i$ or

$$u_i - \mathrm{i}\,v_i = -\mathrm{i}\Omega\,z_i^* \tag{11.2.26}$$

for $i = 1, \ldots, N$, where an asterisk denotes the complex conjugate. Substituting expression (11.2.25) for the left-hand side, we find that

$$\frac{\kappa}{2\pi\mathrm{i}} \frac{\Phi''(z_i)}{2\,\Phi'(z_i)} = -\mathrm{i}\Omega\,z_i^*. \tag{11.2.27}$$

This expression shows that the positions z_i are zeros of the generally nonanalytic complex function

$$\Pi(z) \equiv \Phi''(z) - \frac{4\pi\Omega}{\kappa}\,z^*\,\Phi'(z), \tag{11.2.28}$$

constituting a polynomial in z and z^*.

A rotating polygon

In the case of a regular polygon of N point vortices with radius a, we substitute $\Phi(z) = z^N - a^N$ and find that

$$\Pi(z) \equiv N(N-1)\,z^{N-2} - \frac{4\pi\Omega}{\kappa}\,z^* N z^{N-1} = N z^{N-2} \left(N - 1 - \frac{4\pi\Omega}{\kappa}\,|z|^2 \right). \tag{11.2.29}$$

Requiring that $\Pi(z) = 0$ for $z^N = a^N$ where $|z| = a$, we recover precisely Havelock's expression (11.2.11).

Two nested polygons

Now we consider two nested concentric regular polygons of M point vortices with radii $|a_1|$ and $|a_2|$, so that $N = 2M$ and

$$\Phi(z) = (z^M - a_1^M)(z^M - a_2^M), \tag{11.2.30}$$

where a_1 and a_2 are two complex constants. Straightforward differentiation yields

$$\Phi'(z) = M z^{M-1} (2z^M - a_1^M - a_2^M) \tag{11.2.31}$$

and

$$\Phi''(z) = M z^{M-2} \left[2(2M-1) z^M - (M-1)(a_1^M + a_2^M) \right]. \tag{11.2.32}$$

Substituting these expressions into (11.2.28), we find that

$$\Pi(z) \equiv M z^{M-2} \left[2(2M-1) z^M - (M-1)(a_1^M + a_2^M) - \frac{4\pi\,\Omega}{\kappa} |z|^2 (2z^M - a_1^M - a_2^M) \right]. \tag{11.2.33}$$

Requiring that $\Pi(z) = 0$ when $z^M = a_1^M$ or $z^M = a_2^M$, we obtain two algebraic equations,

$$(3M-1)\, a_1^M - (M-1)\, a_2^M = \frac{4\pi\Omega}{\kappa} |a_1|^2 (a_1^M - a_2^M),$$

$$(3M-1)\, a_2^M - (M-1)\, a_1^M = \frac{4\pi\Omega}{\kappa} |a_2|^2 (a_2^M - a_1^M). \tag{11.2.34}$$

Dividing side by side to eliminate Ω and rearranging, we obtain

$$\alpha^M |\alpha|^2 - \frac{3M-1}{M-1} (\alpha^M + |\alpha|^2) + 1 = 0, \tag{11.2.35}$$

where $\alpha = a_2/a_1$ is the radii ratio. Further rearrangement yields

$$\beta^2 \alpha^{M+2} - \frac{3M-1}{M-1} (\alpha^M + \beta^2 \alpha^2) + 1 = 0, \tag{11.2.36}$$

where $\beta \equiv |\alpha|/\alpha$. If α is real, in which case the polygons are in-phase, $\beta^2 = 1$. If $\alpha = |\alpha| \exp(\pi i/M)$, in which the polygons are out of phase, $\beta^2 = \exp(2\pi i/M)$.

For $M = 2$ and $\beta^2 = 1$, we obtain the algebraic equation $\alpha^4 - 10\alpha^2 + 1 = 0$. The roots, $\alpha = \pm\varrho$ and $\alpha = \pm 1/\varrho$, describe a collinear array of four point vortices discussed in Section 11.2.3, where $\varrho = 3.14626436994197$ is the ratio of the first two positive roots of the fourth-order Hermite polynomial. For $M = 2$ and $\beta^2 = -1$, we obtain the algebraic equation $\alpha^4 = 1$ whose roots produce a rotating square. For $M = 3$ and $\beta^2 = 1$, we obtain the algebraic equation $\alpha^5 - 4\alpha^3 - 4\alpha^2 + 1 = 0$. The real roots, $\alpha = \varrho$ and $\alpha = 1/\varrho$, yield two nested triangles, as shown in the first panel of Figure 11.2.3, where $\varrho = 2.36920540709247$.

Similar structures are obtained for higher values of M, as shown in Figure 11.2.3. The point vortex clusters shown in this figure were generated using the program *pvpoly2* in directory *09_vortex* of the software library FDLIB (Appendix C).

FIGURE 11.2.3   Equilibrium structures of two nested polygons of point vortices with identical strengths rotating steadily in an infinite fluid. In-phase and out-of-phase polygons are observed for $M > 4$.

Rotating clusters with a central point vortex

In the presence of a central point vortex with strength κ_0 located at the origin, we obtain the counterpart of (11.2.25)

$$u_i - \mathrm{i}\, v_i = \frac{\kappa}{2\pi \mathrm{i}} \frac{\Phi''(z_i)}{2\Phi'(z_i)} + \frac{\kappa_0}{2\pi \mathrm{i}} \frac{1}{z_i}, \tag{11.2.37}$$

where $u = u_x$ and $v = u_y$. The counterpart of (11.2.27) for a rotating cluster is

$$\frac{\kappa}{2\pi \mathrm{i}} \frac{\Phi''(z_i)}{2\,\Phi'(z_i)} = -\mathrm{i}\,\Omega\, z_i^* - \frac{\kappa_0}{2\pi \mathrm{i}} \frac{1}{z_i}. \tag{11.2.38}$$

This expression shows that the positions z_i are zeros of the generally nonanalytic complex function

$$\Lambda(z) \equiv \Phi''(z) - \frac{4\pi\Omega}{\kappa}\, z^*\, \Phi'(z) + \frac{2}{z}\, \lambda_0\, \Phi'(z), \tag{11.2.39}$$

where $\lambda_0 \equiv \kappa_0/\kappa$.

In the case of a regular polygon of N point vortices with radius a, we substitute $\Phi(z) = z^N - a^N$ and find that

$$\Lambda(z) \equiv N(N-1)\, z^{N-2} - \frac{4\pi\Omega}{\kappa}\, z^* N z^{N-1} + 2N\lambda_0 z^{N-2}. \tag{11.2.40}$$

Simplifying, we obtain

$$\Lambda(z) = N z^{N-2} \left(N - 1 + 2\lambda_0 - \frac{4\pi\Omega}{\kappa} |z|^2 \right). \tag{11.2.41}$$

Requiring that $\Lambda(z) = 0$ when $z^N = a^N$, we obtain

$$\Omega = (N - 1 + 2\lambda_0) \frac{\kappa}{4\pi a^2}, \tag{11.2.42}$$

which is a generalization of (11.2.11).

In the case of two nested concentric M-sided regular polygons with radii $|a_1|$ and $|a_2|$, we have $N = 2M$ and $\Phi(z) = (z^M - a_1^M)(z^M - a_2^M)$. Working as previously , we obtain

$$\Lambda(z) \equiv M z^{M-2} \left[2(2M - 1) z^M - (M - 1) (a_1^M + a_2^M) \right.$$
$$\left. - \left(\frac{4\pi\Omega}{\kappa} |z|^2 - 2\lambda_0 \right) (2z^M - a_1^M - a_2^M) \right]. \tag{11.2.43}$$

The counterparts of equations (11.2.44) are

$$(3M - 1 + 2\lambda_0) a_1^M - (M - 1 + 2\lambda_0) a_2^M = \frac{4\pi\Omega}{\kappa} |a_1|^2 (a_1^M - a_2^M),$$
$$(3M - 1 + 2\lambda_0) a_2^M - (M - 1 + 2\lambda_0) a_1^M = \frac{4\pi\Omega}{\kappa} |a_2|^2 (b^M - a^M). \tag{11.2.44}$$

Dividing side by side to eliminate Ω and rearranging, we obtain

$$\beta^2 \alpha^{M+2} - \frac{3M - 1 + 2\lambda_0}{M - 1 + 2\lambda_0} (\alpha^M + \beta^2 \alpha^2) + 1 = 0, \tag{11.2.45}$$

where $\alpha = b/a$ is the radii ratio and $\beta \equiv |\alpha|/\alpha$. If α is real, the polygons are in-phase; if $\alpha = |\alpha| \exp(\pi i/M)$, the polygons are out of phase.

Generalization

To generalize the preceding results, we consider a rotating assembly of point vortex consisting of an M_1-sided polygon of point vortices of radius $|a_1|$ and strength κ_1, and a concentric M_2-sided polygon of point vortices of radius $|a_2|$ and strength κ_2. It is convenient to introduce the polynomials $\Phi_1(z) = z^{M_1} - a_1^{M_1}$ and $\Phi_2(z) = z^{M_2} - a_2^{M_2}$, and express the velocity of the ith point vortex in the first array as

$$u_i - \mathrm{i} v_i = \frac{\kappa_1}{2\pi\mathrm{i}} \frac{\Phi_1''(z_i)}{2\Phi_1'(z_i)} + \frac{\kappa_2}{2\pi\mathrm{i}} \frac{\Phi_2'(z_i)}{\Phi_2(z_i)} = -\mathrm{i}\Omega z_i^*, \tag{11.2.46}$$

and the velocity of the ith point vortex in the second array as

$$u_i - \mathrm{i} v_i = \frac{\kappa_2}{2\pi\mathrm{i}} \frac{\Phi_2''(z_i)}{2\Phi_2'(z_i)} + \frac{\kappa_1}{2\pi\mathrm{i}} \frac{\Phi_1'(z_i)}{\Phi_1(z_i)} = -\mathrm{i}\Omega z_i^*. \tag{11.2.47}$$

These expressions show that the position of the point vortices in the first polygon are the zeros of the function

$$\Pi_1(z) = \Phi_1''(z)\, \Phi_2(z) + \Phi_1'(z)\Big(2\lambda\, \Phi_2'(z) - \frac{4\pi\Omega}{\kappa_1} z^*\, \Phi_2(z) \Big), \tag{11.2.48}$$

and the position of the point vortices in the second polygon are the zeros of the function

$$\Pi_2(z) = \Phi_2''(z)\, \Phi_1(z) + \Phi_2'(z)\frac{1}{\lambda}\Big(2\,\Phi_1'(z) - \frac{4\pi\Omega}{\kappa_1} z^*\, \Phi_1(z) \Big), \tag{11.2.49}$$

where $\lambda = \kappa_2/\kappa_1$. Substituting the expressions for Φ_1 and Φ_2, we find that

$$\Pi_1(z) = M_1 z^{M_1-2}\Big[(M_1-1)(z^{M_2} - a_2^{M_2}) + 2\lambda\, M_2 z^{M_2} - \frac{4\pi\Omega}{\kappa_1}|z|^2\,(z^{M_2} - a_2^{M_2}) \Big] \tag{11.2.50}$$

and

$$\Pi_2(z) = M_2 z^{M_2-2}\Big[(M_2-1)(z^{M_1} - a_1^{M_1}) + \frac{1}{\lambda}\Big(2M_1 z^{M_1} - \frac{4\pi\Omega}{\kappa_1}|z|^2\,(z^{M_1} - a_1^{M_1}) \Big) \Big]. \tag{11.2.51}$$

Requiring that $\Pi_1(z) = 0$ when $z^{M_1} = a_1^{M_1}$ and $\Pi_2(z) = 0$ when $z^{M_2} = a_2^{M_2}$, we obtain two equations,

$$(M_1-1)(a_1^{M_2} - a_2^{M_2}) + 2\lambda\, M_2 a_1^{pM_1} = \frac{4\pi\Omega}{\kappa_1}|a_1|^2\,(a_1^{M_2} - a_2^{M_2}),$$

$$\lambda\,(M_2-1)(a_2^{M_1} - a_1^{M_1}) + 2\,M_1 a_2^{M_2/p} = \frac{4\pi\Omega}{\kappa_1}|a_2|^2\,(a_2^{M_1} - a_1^{M_1}). \tag{11.2.52}$$

When $M_1 = M_2 = M$ and $\lambda = 1$, we recover equations (11.2.34). Dividing side by side to eliminate Ω, we obtain an equation for the radii ratio, a_2/a_1. When $M_1 = M_2$ and $\lambda \neq 1$, we obtain a generalization of (11.2.53),

$$\beta^2\alpha^{M+2} - \frac{(2+\lambda)M - \lambda}{M-1}\alpha^M - \frac{(2\lambda+1)M - 1}{M-1}\beta^2\alpha^2 + \lambda = 0, \tag{11.2.53}$$

where $\alpha \equiv a_2/a_1$ and $\beta \equiv |\alpha|/\alpha$.

The analysis presented in this section can be extended to a higher number of nested point vortex polygons with the same or different strengths [13].

11.2.5   Point vortices in bounded domains

The velocity induced by a point vortex residing in an internally or externally bounded domain of flow must be complemented with that due to a nonsingular reflected irrotational flow ensuring that the no-penetration boundary condition is satisfied. The point vortex itself moves under the influence of the complementary flow. In some cases, the complementary flow can expressed in terms of image point vortices with appropriate strengths located outside the domain of flow. The motion of the image point vortices is deduced from the instantaneous position of the primary vortex.

The stream function of the flow due to a point vortex located at the point $\mathbf{X}$ in the presence of an impermeable boundary is the Green's function of the first kind of the two-dimensional Laplace equation multiplied by the strength of the point vortex, κ, that is, $\psi(\mathbf{x}, \mathbf{X}) = \kappa \, \mathcal{G}(\mathbf{x}, \mathbf{X})$. In the case of unbounded flow, we employ the free-space Green's function, $\mathcal{G}(\mathbf{x}, \mathbf{X}) = -1/(2\pi) \ln |\mathbf{x} - \mathbf{X}|$. Specific expressions for several Green's functions generated by the method of images are given in Section 2.2.8.

Point vortex above a plane wall

The image of a point vortex with respect to a plane wall is another point vortex with strength of equal magnitude and opposite sign placed in the instantaneous mirror image of the original point vortex with respect to the wall. Examining the velocity field induced by the image point vortex, we find that the original point vortex moves parallel to the wall with velocity $V = \kappa/(4\pi b)$, where b is the distance of the point vortex from the wall. The streamline pattern in a frame of reference moving with the point vortex is identical to that shown in Figure 11.2.1(b), provided that the x axis is identified with the wall.

Point vortex inside a circular cylinder

The image of a point vortex located at a point $\mathbf{X}$ inside a circular cylinder of radius a centered at the point $\mathbf{x}_c$ is another point vortex with strength of equal magnitude and opposite sign placed at the instantaneous inverse point of the point vortex with respect to the cylinder,

$$\mathbf{X}^{inv} = \mathbf{x}_c + \frac{a^2}{|\mathbf{X} - \mathbf{x}_c|^2} (\mathbf{X} - \mathbf{x}_c). \tag{11.2.54}$$

This construction shows that a single point vortex describes a circular path that is concentric with the cylinder, moving around the center of the cylinder with polar velocity

$$u_\theta = \frac{\kappa}{2\pi} \frac{|\mathbf{X} - \mathbf{x}_c|}{a^2 - |\mathbf{X} - \mathbf{x}_c|^2}. \tag{11.2.55}$$

The streamline pattern of the induced flow is shown in Figure 11.2.4(a).

Point vortex outside a circular cylinder

The image of a point vortex located at a point $\mathbf{X}$ outside a circular cylinder of radius a centered at the point $\mathbf{x}_c$ is another point vortex with strength of equal magnitude and opposite sign placed at the instantaneous inverse point of the point vortex with respect to the cylinder. The image system can be supplemented with a third point vortex located at the center of the cylinder whose strength, κ_c, determines the circulation around the cylinder. Consequently, the primary point vortex of interest describes a circular path, moving around the center of the cylinder with polar velocity

$$u_\theta = -\frac{\kappa}{2\pi} \frac{|\mathbf{X} - \mathbf{x}_c|}{a^2 - |\mathbf{X} - \mathbf{x}_c|^2} + \frac{\kappa_c}{2\pi} \frac{1}{|\mathbf{X} - \mathbf{x}_c|}. \tag{11.2.56}$$

The streamline pattern for $\kappa_c = 0$ is shown in Figure 11.2.4(b).

(a)

(b)

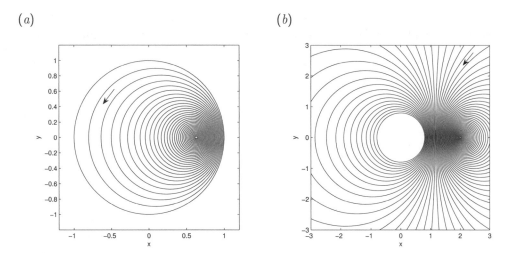

FIGURE 11.2.4   Streamline pattern due to a point vortex (a) inside and (b) outside a circular cylinder.

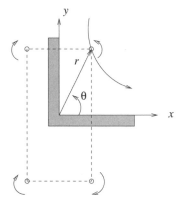

FIGURE 11.2.5   Trajectory of a point vortex in the first quadrant near a right-angle corner, and illustration of the three images.

Point vortex near a corner

The image system of a point vortex in a quarterly infinite domain bounded by a right-angle corner is the set of three point vortices shown in Figure 11.2.5. In plane polar coordinates with origin at the apex where the walls located at $\theta = 0$ and $\pi/2$, the point vortex moved along a curve described by $r = c/\sin(2\theta)$, where the constant c is determined by the initial position. A generalization of this configuration to flow inside a corner with arbitrary aperture angle is discussed in Problem 11.2.2.

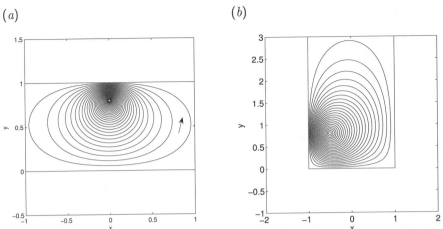

FIGURE 11.2.6   Streamline pattern due to a point vortex (a) between two parallel walls and (b) in a semi-infinite strip.

Point vortex between two parallel walls

Next, we consider the motion of a point vortex between two parallel walls separated by distance h. The image system consists of two periodic arrays of point vortices with separation $2h$ arranged perpendicular to the walls. The first array contains the original point vortex, and the second array contains the image of the point vortex with respect to the upper or lower wall. The strength of the point vortices in the second array is equal in magnitude and opposite in sign to that of the point vortices in the first array.

Assume that the point vortex is located at (X, Y) and the walls are parallel to the x axis. Using (2.13.14), we find that the stream function of the induced flow is

$$\psi(x) = -\frac{\kappa}{4\pi} \ln \frac{\cosh[k(x-X)] - \cos[k(y-Y)]}{\cosh[k(x-X)] - \cos[k(y-Y_{im})]}, \qquad (11.2.57)$$

where $k = \pi/h$ is a wave number and Y_{im} is the y position of the image. Using (2.13.15), we find that the point vortex moves along the x axis parallel to the walls under the influence of the flow induced by the second array containing its image. The velocity of the point vortex is

$$V_x = \frac{\kappa}{4h} \frac{\sin(2kb)}{1 - \cos(2kb)}, \qquad (11.2.58)$$

where b is the distance of the point vortex from the lower wall. The streamline pattern in shown in Figure 11.2.6(a). In the limit as kb tends to zero, we recover the earlier results for a point vortex in a semi-infinite fluid above an infinite plane wall.

Point vortex in a semi-infinite strip

Consider a point vortex placed between two parallel walls separated by distance h and intersecting at a right angle a third wall to form a semi-infinite rectangular strip, as illustrated in Figure 11.2.6(b).

The image system consists of the image system associated with the two parallel walls, and the reflection of the image system with respect to the third intersecting wall. The strengths of the reflected point vortices are the negatives of those of their images. The motion of the point vortex can be computed by numerical methods.

Complex boundary shapes

The complementary irrotational flow due to a boundary with arbitrary geometry can be computed using the conformal mapping methods discussed in Section 7.12 or the boundary-integral methods discussed in Chapter 10.

11.2.6 Periodic arrangement of point vortices

Using equation (2.13.15), we find that the motion of an infinite periodic collection of N point vortices repeated in the x direction with period a is governed by the differential equations

$$
\begin{aligned}
\frac{\mathrm{d}X_i}{\mathrm{d}t} &= -\frac{1}{2a} \sum_{j=1}^{N}{}' \kappa_j \frac{\sinh[k(Y_i - Y_j)]}{\cosh[k(Y_i - Y_j)] - \cos[k(X_i - X_j)]}, \\
\frac{\mathrm{d}Y_i}{\mathrm{d}t} &= \frac{1}{2a} \sum_{j=1}^{N}{}' \kappa_j \frac{\sin[k(X_i - X_j)]}{\cosh[k(Y_i - Y_j)] - \cos[k(X_i - X_j)]},
\end{aligned}
\tag{11.2.59}
$$

where $k = 2\pi/a$ is the wave number and the prime indicates that the term $j = i$ is excluded from the sum.

Problems

11.2.1 *Equation of motion of point vortices in complex variables*

Introduce the complex variable $Z = X + \mathrm{i}Y$ and confirm that equations (11.2.1) and (11.2.59) can be expressed in the compact forms

$$
\frac{\mathrm{d}Z_i^*}{\mathrm{d}t} = \frac{1}{2\pi\mathrm{i}} \sum_{j=1}^{N}{}' \frac{\kappa_j}{Z_i - Z_j}, \qquad \frac{\mathrm{d}Z_i^*}{\mathrm{d}t} = \frac{1}{2a\mathrm{i}} \sum_{j=1}^{N}{}' \kappa_j \cot\left[\frac{1}{2} k\left(Z_i - Z_j\right)\right],
\tag{11.2.60}
$$

where i is the imaginary unit, $\mathrm{i}^2 = -1$, an asterisk denotes the complex conjugate, and the prime indicates that the term $j = i$ is excluded from the sum.

11.2.2 *Motion of a point vortex near a corner*

Consider the motion of a point vortex in the vicinity of a corner with internal angle π/n. Using conformal mapping show, that in plane polar coordinates with origin at the apex and the walls located at $\theta = 0$ and π/n, the trajectory of the point vortex is described by the equation $r = c/\sin(n\theta)$, where c is a constant determined by the initial position of the point vortex ([24], p. 536).

11.2.3 *Two point vortices behind a cylinder*

Föppl considered uniform potential flow along the x axis past a stationary cylinder of radius a centered at the origin, with vanishing circulation around the cylinder, as shown in Figure 11.2.7

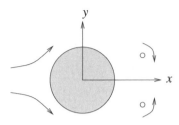

FIGURE 11.2.7   Two point vortices behind a cylinder in streaming flow emulate a stationary wake.

([220], p. 223). A pair of point vortices with opposite strength is symmetrically located above and below the x axis. Verify that the point vortices will remain stationary provided that they lie on the curve described by the equation $2ry = r^2 - a^2$, and the strengths of the point vortices are $\kappa = \pm Ur(1 - a^2/r^2)(1 - a^4/r^4)$, where r is the distance from the origin.

11.2.4 *Point vortex polygons*

Derive the counterpart of (11.2.53) in the presence of a central point vortex with strength κ_0.

🖥 *Computer Problems*

11.2.5 *A periodic array of point vortices between two walls*

Consider of an infinite array of point vortices separated by distance a between two parallel plane walls separated by distance h. Compute and plot the velocity of the array against the ratios b/h and a/h, where b is the distance of the point vortices from the lower wall. Discuss the results of your computations.

11.2.6 *Stability of a polygonal arrangement of point vortices*

(*a*) Write a program that computes the motion of point vortices initially arranged at the vertices of an N-sided regular polygon of radius a. Perform simulations for $N = 6, 7$, and 8, and discuss the differences in behavior due to the amplification of round-off error.

(*b*) Assume that the motion occurs in the presence of a circular cylinder of radius b concentric with the polygon. Compute and plot the angular velocity of the point vortices for $N = 2, 4$, and 8, and $b/a = 0.10, 0.50, 0.90, 1.1, 1.50$, and 2.0. Compare your results with the analytical solution for the case of unbounded flow.

11.3   Rows of point vortices

An infinite row of identical point vortices is a useful model of the flow developing from the instability of a vortex layer, as discussed in Section 9.6. The associated streamline pattern is shown in Figure 2.13.1. When the point vortices are precisely collinear and equidistant, they remain stationary in a steady flow. To assess the stability of the array, we displace the point vortices in the horizontal and vertical directions and compute their subsequent motion. If the initial displacement is small compared to the point vortex separation, a, the motion can be described analytically in the context of linear stability theory discussed in Chapter 9.

11.3.1   Stability of a row or point vortices

Consider an infinite array of point vortices distributed along the x axis. Following the general formalism of the normal mode analysis discussed in Section 9.3, we assume that the position of the perturbed point vortices is given by the real part of the right-hand sides of the equations

$$X_m = a\,(m + \xi_m), \qquad Y_m = a\,\eta_m \qquad (11.3.1)$$

for $m = 0, \pm 1, \pm 2, \ldots$, where

$$\xi_m = \epsilon_1 \exp[\mathrm{i}(mka - \varrho t)], \qquad \eta_m = \epsilon_2 \exp[\mathrm{i}(mka - \varrho t)] \qquad (11.3.2)$$

are dimensionless complex displacements, ϵ_1 and ϵ_2 are dimensionless complex coefficients with small but comparable magnitudes, $k = 2\pi/L$ is the wave number of the perturbation, L is the corresponding wavelength, and ϱ is the complex growth rate. If the imaginary part of ϱ is positive, the periodic array is unstable.

The physical requirement that the separation between two consecutive point vortices is smaller than the wavelength of the perturbation, $L = k/2\pi$, imposes the geometrical constraint $a/L = ka/(2\pi) < 1$. When the ratio L/a is an integer, the array is perturbed in a commensurate and spatially periodic fashion.

The motion of the point vortices is governed by the differential equations (11.2.1). Linearizing the terms inside the sums with respect to ϵ_1 and ϵ_2, we obtain

$$\frac{Y_m - Y_n}{|\mathbf{X}_m - \mathbf{X}_n|^2} \simeq \frac{1}{a}\,\frac{\eta_m - \eta_n}{(m - n)^2}, \qquad \frac{X_m - X_n}{|\mathbf{X}_m - \mathbf{X}_n|^2} \simeq \frac{1}{a}\,\frac{m - n + \xi_m - \xi_n + \cdots}{(m - n)^2 + 2(m - n)(x_m - x_n) + \cdots} \qquad (11.3.3)$$

for any arbitrary integer pair, (m, n). Simplifying the second expansion, we obtain

$$\frac{X_m - X_n}{|\mathbf{X}_m - \mathbf{X}_n|^2} \simeq \frac{1}{a}\left(\frac{1}{m - n} - \frac{\xi_m - \xi_n}{(m - n)^2} + \cdots\right). \qquad (11.3.4)$$

The x position of each point vortex evolves according to the linearized ordinary differential equation

$$\frac{\mathrm{d}X_m}{\mathrm{d}t} = -\frac{\kappa}{2\pi a} \sum_{n=-\infty}^{\infty}{}' \frac{\eta_m - \eta_n}{(m - n)^2} = -\epsilon_2 \frac{\kappa}{2\pi a} \exp[\mathrm{i}(mka - \varrho t)] \sum_{n=-\infty}^{\infty}{}' \frac{1 - \mathrm{e}^{\mathrm{i}(n-m)ka}}{(m - n)^2}, \qquad (11.3.5)$$

where the prime indicates that the term $m = n$ is excluded from the sum. Substituting into the left-hand side the first expressions in (11.3.1) and (11.3.2) and simplifying, we obtain

$$\epsilon_1\, a\, \mathrm{i}\varrho = \epsilon_2 \frac{\kappa}{2\pi a}\, \alpha, \qquad (11.3.6)$$

where

$$\alpha = \sum_{l=-\infty}^{\infty} {}' \frac{1 - e^{ilka}}{l^2} = \sum_{l=-\infty}^{\infty} {}' \frac{1 - \cos(lka)}{l^2} = ka \left(1 - \frac{ka}{2\pi}\right), \tag{11.3.7}$$

and the prime indicates that the singular term $l = 0$ is excluded from the sum.

Unfortunately, applying formula (11.2.1) for the y velocity component produces a divergent sum. To circumvent this apparent but not essential difficulty, we retain a large but finite number of $2N + 1$ point vortices in the array and obtain

$$\frac{dY_m}{dt} = \frac{\kappa}{2\pi a} \left(\sum_{n=-N}^{N} {}' \frac{1}{m - n} - \epsilon_1 \exp[i(mka - \varrho t)] \sum_{n=-\infty}^{\infty} {}' \frac{1 - e^{i(n-m)ka}}{(m - n)^2} \right). \tag{11.3.8}$$

Because the first sum on the right-hand side of (11.3.8) is independent of the position of the point vortices, it can be discarded with no consequences on the flow. This manipulation is an example of a mathematical renormalization. Substituting into the left-hand side the second expressions in (11.3.1) and (11.3.2), and simplifying, we obtain

$$\epsilon_2 \, a \, i\varrho = \epsilon_1 \frac{\kappa}{2\pi a} \alpha. \tag{11.3.9}$$

Combining equations (11.3.6) and (11.3.9), we find that

$$\varrho = \pm i\pi \frac{\kappa}{a^2} \frac{a}{L} \left(1 - \frac{a}{L}\right), \qquad \frac{\epsilon_2}{\epsilon_1} = \mp 1. \tag{11.3.10}$$

Since $a/L < 1$, the plus or minus sign can be selected according to the sign of the strength of the point vortices so that the imaginary part of ϱ is positive, corresponding to instability. The corresponding normal mode displacement causes the point vortices to move away from their original location at an exponential rate (see also Problem 11.3.1).

An unstable and a stable normal mode displacement are illustrated in Figure 11.3.1 for negative point vortex strength, $\kappa < 0$, corresponding, respectively, to $\epsilon_2/\epsilon_1 = 1$ and -1. In the case of the unstable mode, the point vortices are displaced so that they concentrate near regions where the interface slopes downward, causing a local counterclockwise rotation that promotes the instability. The opposite is true in the case of the stable normal mode with $\epsilon_2/\epsilon_1 = -1$.

Vortex pairing

When the wavelength of the perturbation, L, is a multiple of the point vortex separation, a, the ratio $n \equiv L/a$ is an integer expressing the number of point vortices residing inside each period. The expression for the complex growth rate in (11.3.10) becomes

$$\varrho = \pm i\pi \frac{\kappa}{a^2} \frac{n - 1}{n^2}. \tag{11.3.11}$$

For fixed point vortex strength, κ, the highest value of ϱ_I corresponding to the most unstable perturbation is obtained for the smallest possible integer, $n = 2$, associated with a pairing interaction

FIGURE 11.3.1   An unstable and a stable normal mode displacement of an infinite array of point vortices with negative circulation.

between two adjacent point vortices. An unstable normal mode perturbation displaces the point vortices upward and downward in succession, yielding a triangular wave.

Vortex sheet

An interesting interpretation of expression (11.3.11) for the growth rate arises by expressing the strength of the point vortices in terms of the difference in the velocity of the streams far above and below the point-vortex array, ΔU, the wavelength of the perturbation, L, and the number of point vortices inside each period, $n = L/a$, finding $\kappa = -a\,\Delta U = -L\,\Delta U/n$. The growth rate is

$$\varrho = \pm i\pi\,\frac{\Delta U}{L}\,\frac{n-1}{n} = \pm i\,\frac{1}{2}\,k\Delta U\,\frac{n-1}{n}, \tag{11.3.12}$$

where $k = 2\pi/L$ is the wave number. Increasing the number of point vortices inside each wavelength, n, from two to infinity, while keeping the velocity difference constant and the wavelength of the perturbation fixed, doubles the growth rate. The limit $n \to \infty$ corresponds to a continuous distribution of point vortices representing a vortex sheet.

11.3.2   Von Kármán vortex street

The von Kármán vortex street consists of two parallel rows of point vortices arranged along the x axis with separation a and displaced along the y axis by distance b, as shown in Figure 11.3.2. The strengths of the point vortices in the upper row are equal in magnitude and opposite in sign to the strengths of the point vortices in the lower row. The von Kármán vortex street is a useful model of the wake developing behind a bluff object.

Symmetric street

Using expressions (11.2.59), we find that, when the point vortices are located above each other in the arrangement of the symmetric vortex street shown in Figure 11.3.2(*a*), they translate parallel to the x axis with velocity

$$V_x = \frac{\kappa}{2a}\,\frac{\sinh(kb)}{\cosh(kb)-1} = \frac{\kappa}{2a}\,\coth(\tfrac{1}{2}kb), \tag{11.3.13}$$

where $k = 2\pi/a$ is the wave number. Linear stability analysis shows that the symmetric vortex street is unstable and should not be expected to occur in practice (Problem 11.3.2).

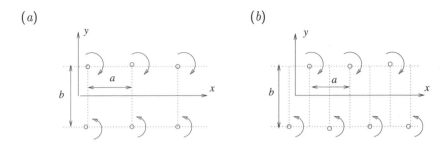

FIGURE 11.3.2 Illustration of (a) a symmetric and (b) an antisymmetric street of point vortices. The strengths of the point vortices in the top row are equal in magnitude and opposite in sign that those of the point vortices in the bottom row.

Antisymmetric street

When the vortices in the top row are displaced with respect to the point vortices in the bottom row along the x axis by a distance that is equal to half their separation, forming the antisymmetric vortex street shown in Figure 11.3.2(b), they translate along the x axis with velocity

$$V_x = \frac{\kappa}{2a} \frac{\sinh(kb)}{\cosh(kb) + 1} = \frac{\kappa}{2a} \tanh(\frac{1}{2}kb). \qquad (11.3.14)$$

Linear stability analysis reveals that the antisymmetric vortex street is unstable, except in the special case where $\cosh(kb) = 2$ corresponding to $b/a = 0.281$, where the flow is marginally stable (Problem 11.3.2). However, nonlinearities promote the instability and the antisymmetric vortex street is considered unstable.

Problem

11.3.1 *Stability of a row of point vortices*

Use expressions (11.2.59) to study the stability of a row of point vortices.

11.3.2 *Stability of a vortex street*

Carry out the linear stability analysis of the (a) symmetric and (b) antisymmetric vortex street ([220], pp. 225, 228).

Problem

11.3.3 *Kelvin–Helmholtz instability of a row of point vortices*

An infinite array of point vortices separated by distance a induces an upper streaming flow with velocity U and a lower streaming flow with velocity $-U$. The strength of each point vortex is related to U by $\kappa = -2U/a$. Assume that the array is subjected to a sinusoidal perturbation with fixed wave length $L = na$ and amplitude b_0, where n is an integer.

(a) Write a program that computes the motion of the point vortices and plot profiles of the evolving array. The ordinary differential equations governing the motion of the point vortices should be integrated in time using the modified Euler method discussed in Section A.8, Appendix B. The program should compute the ratio of the maximum displacement of the point vortices from the x axis, b, to the initial amplitude, b_0.

(b) Consider an unstable normal mode perturbation where the initial position of the mth point vortex is given by

$$X_m = c_1 \, a \, (m-1) + b_0 \, \sin \varphi_m, \qquad Y_m = c_2 \, b_0 \, \sin \varphi_m, \qquad (11.3.15)$$

where m is an integer, $c_1 = 1$, $c_2 = 1$, $\varphi_m = 2\pi(m-1)/n$, $b_0 > 0$ is a perturbation amplitude, and $\kappa < 0$. Compute the motion of the point vortices for $b_0/L = 0.10$ and $n = 4, 8, 16$, and 32. Plot the ratio b/b_0 against time on a linear-log scale in each case and compare your results with the predictions of the linear stability analysis discussed in the text.

(c) Repeat (b) for $c_1 = 1$ and $c_2 = -1$, corresponding to a stable normal mode.

(d) Repeat (b) for $c_1 = 0$ and $c_2 = 1$ and discuss the difference in behavior from case (b).

11.4   Vortex blobs

A point vortex is a mathematical idealization emerging in the limit as the cross-section of a two-dimensional rectilinear vortex filament tends to zero, while the circulation around the filament is held fixed. Unfortunately, the infinitesimal cross-section of the point vortices causes an erratic mutually induced motion, especially when two point vortices are in close proximity, due to the divergent velocity. To suppress the erratic motion, we may replace the singular delta function representing the vorticity distribution with a regularized distribution supported by a finite cross-sectional area, and thereby obtain a vortex blob [81]. The instantaneous vorticity field due to a vortex blob with strength κ located at the point $\mathbf{X}$ is given by

$$\omega_z = \kappa \, \zeta(\mathbf{x} - \mathbf{X}), \qquad (11.4.1)$$

where the distribution function, ζ, is a regularization of the two-dimensional delta function, δ_2, normalized so that its integral over the entire xy plane is equal to unity,

$$\iint \zeta(\mathbf{x}) \, \mathrm{d}x \, \mathrm{d}y = 1, \qquad (11.4.2)$$

as discussed in Section A.3, Appendix A. The distribution function ζ has dimensions of inverse squared length.

11.4.1   Radially symmetric blobs

Assuming that the vorticity of a blob is symmetric around the centerline, we introduce the specific form

$$\zeta(\mathbf{x} - \mathbf{X}) = \frac{\pi}{\varepsilon^2} \, f(r/\epsilon), \qquad (11.4.3)$$

where $r = |\mathbf{x} - \mathbf{X}|$ is the distance from the center of the blob, ε is the blob radius, and f is a dimensionless distribution function decaying far from the blob. As ε tends to zero, ζ reduces to the two-dimensional delta function. For the integral of ζ over the entire xy plane to be equal to unity, the function f must satisfy the integral constraint

$$\int_0^\infty w f(w)\, \mathrm{d}w = \frac{1}{2\pi^2}. \tag{11.4.4}$$

A possible choice for the core distribution function is the piecewise constant top-hat distribution

$$f(w) = \begin{cases} 1/\pi^2 & \text{for } w \leq 1, \\ 0 & \text{for } w \geq 1, \end{cases} \tag{11.4.5}$$

describing the circular Rankine vortex with constant vorticity inside the core. Another choice is the Gaussian distribution

$$f(w) = \frac{1}{\pi^2}\, \mathrm{e}^{-w^2}. \tag{11.4.6}$$

Both choices satisfy the integral constraint (11.4.4).

Velocity field

In plane polar coordinates centered at an axisymmetric vortex blob, the radial velocity is zero. To compute the polar velocity, u_θ, we resort to the definition of vorticity and derive the ordinary differential equation

$$\frac{1}{r}\frac{\mathrm{d}}{\mathrm{d}r}(r u_\theta) = \kappa\, \frac{\pi}{\varepsilon^2}\, f(r/\epsilon). \tag{11.4.7}$$

Straightforward integration subject to the condition that the velocity is zero at the centerline yields the velocity profile

$$u_\theta = \frac{\kappa}{2\pi r}\, \mathcal{F}(r/\epsilon), \tag{11.4.8}$$

where

$$\mathcal{F}(r/\epsilon) = 2\pi^2 \int_0^{r/\epsilon} w'\, f(w')\, \mathrm{d}w' \tag{11.4.9}$$

is a dimensionless function. Equation (11.4.4) shows that, as the blob radius ϵ tends to zero for fixed r, the function $\mathcal{F}(r/\epsilon)$ tends to unity. The Cartesian components of the velocity at a point $\mathbf{x}$ due to a blob located at $\mathbf{X}$ are

$$u_x(\mathbf{x}) = -\frac{\kappa}{2\pi}\, \frac{\hat{y}}{\hat{x}^2 + \hat{y}^2}\, \mathcal{F}(r/\epsilon), \qquad u_y(\mathbf{x}) = \frac{\kappa}{2\pi}\, \frac{\hat{x}}{\hat{x}^2 + \hat{y}^2}\, \mathcal{F}(r/\epsilon), \tag{11.4.10}$$

where $\hat{\mathbf{x}} = \mathbf{x} - \mathbf{X}$ is the distance from the centerpoint. This velocity differs from that due to a point vortex only by the presence of the radial distribution function, $\mathcal{F}(r/\epsilon)$. Far from the center of the blob, $\mathcal{F}(r/\epsilon)$ tends to unity and the blob induces a velocity field similar to that due to a point vortex.

For the top-hat distribution shown in (11.4.5), straightforward computation of the integral in (11.4.8) yields

$$\mathcal{F}(w) = \begin{cases} w^2 & \text{for } w \leq 1, \\ 1 & \text{for } w \geq 1, \end{cases} \tag{11.4.11}$$

which shows that the fluid inside a Rankine blob rotates like a rigid body, while the flow outside the Rankine blob is identical to that due to a point vortex.

For the Gaussian distribution shown in (11.4.6), we obtain

$$\mathcal{F}(w) = 1 - e^{-w^2}. \tag{11.4.12}$$

Far small w, the function $\mathcal{F}(w)$ is quadratic in w and the velocity depends linearly on the distance from the center of the blob. Far from the blob, the function $\mathcal{F}(w)$ tends to unity and the flow resembles that due to a point vortex.

Biot–Savart integral

It is instructive to rederive the velocity field due to a vortex blob based on the Biot–Savart integral (2.1.26). Setting the origin at the center of a blob, we find that the y component of the velocity at a point x_0 located on the x axis is given by

$$u_y(x_0) = \frac{\kappa}{2\pi} \int_0^{2\pi} \int_0^\infty \frac{x_0 - r\cos\theta}{x_0^2 + r^2 - 2x_0 r\cos\theta} \, \zeta(\mathbf{x}) \, r \, \mathrm{d}r \, \mathrm{d}\theta. \tag{11.4.13}$$

Substituting (11.4.3) into the integrand of (11.4.13), we obtain

$$u_y(x_0) = \frac{\kappa}{2} \int_0^\infty \left(\int_0^{2\pi} \frac{x_0 - r\cos\theta}{x_0^2 + r^2 - 2x_0 r\cos\theta} \, \mathrm{d}\theta \right) f(w) \, w \, \mathrm{d}w, \tag{11.4.14}$$

where $w = r/\epsilon$. Finally, we note that

$$\int_0^{2\pi} \frac{1 - w\cos\theta}{1 + w^2 - 2w\cos\theta} \, \mathrm{d}\theta = \begin{cases} 2\pi & \text{if } w < 1, \\ 0 & \text{if } w > 1, \end{cases} \tag{11.4.15}$$

and set $w = r/x_0$ to obtain the right-hand side of (11.4.8) with x_0 in place of r.

11.4.2   Motion of a collection of vortex blobs

The motion of the center of a vortex blob that belongs to a collection of N axisymmetric blobs with identical structure is described by the following modified version of (11.2.1),

$$\frac{\mathrm{d}X_i}{\mathrm{d}t} = -\frac{1}{2\pi} \sum_{j=1}^N {}' \kappa_j \frac{Y_i - Y_j}{|\mathbf{X}_i - \mathbf{X}_j|^2} \, \mathcal{F}\left(\frac{\mathbf{X}_i - \mathbf{X}_j}{\varepsilon_j} \right),$$

$$\frac{\mathrm{d}Y_i}{\mathrm{d}t} = \frac{1}{2\pi} \sum_{j=1}^N {}' \kappa_j \frac{X_i - X_j}{|\mathbf{X}_i - \mathbf{X}_j|^2} \, \mathcal{F}\left(\frac{\mathbf{X}_i - \mathbf{X}_j}{\varepsilon_j} \right), \tag{11.4.16}$$

where κ_j is the strength of the jth blob and the prime indicates that the term $i = j$ is excluded from the sum. In order to preserve the total circulation, the strength of each blob remains constant in time.

11.4.3   Periodic arrays of vortex blobs

Combining (11.2.59) with (11.4.16), we find that the motion of a collection of N vortex blobs that are repeated periodically in the x direction with period a is described by the system

$$\frac{dX_i}{dt} = -\frac{1}{2a} \sum_{j=1}^{N}{}' \kappa_j \frac{\sinh[k(Y_i - Y_j)]}{\cosh[k(Y_i - Y_j)] - \cos[k(X_i - X_j)]} + U_i,$$

$$\frac{dY_i}{dt} = \frac{1}{2a} \sum_{j=1}^{N}{}' \kappa_j \frac{\sin[k(X_i - X_j)]}{\cosh[k(Y_i - Y_j)] - \cos[k(X_i - X_j)]} + V_i, \qquad (11.4.17)$$

where

$$U_i = -\frac{1}{2\pi} \sum_{j=1}^{N} \left[\kappa_j \sum_{m=-\infty}^{\infty} \frac{Y_i - Y_j}{R_{ijm}^2} \mathcal{G}_{ijm} \right], \qquad V_i = \frac{1}{2\pi} \sum_{j=1}^{N} \left[\kappa_j \sum_{m=-\infty}^{\infty} \frac{X_i - X_j - ma}{R_{ijm}^2} \mathcal{G}_{ijm} \right],$$

$$(11.4.18)$$

$$\mathcal{G}_{ijm} \equiv \mathcal{F}\left(\frac{R_{ijm}}{\varepsilon_j}\right) - 1, \qquad R_{ijm}^2 = (X_i - X_j - ma)^2 + (Y_i - Y_j)^2.$$

The sums on the right-hand side of (11.4.17) express the velocity due to a periodic array of point vortices. The sums in (11.4.18) express corrections due to the finite size of the blobs. The inner infinite sums over m converge rapidly and can be truncated after a sufficient number of terms.

11.4.4   Diffusing blobs

In real life, the radius of each vortex blob increases due to viscous diffusion. Consider a single blob with Gaussian vorticity distribution, $f(w) = \exp(-w^2)$. Substituting this form into the vorticity transport equation written in plane polar coordinates, we find that the blob radius, ε, increases in time according to the equation

$$\frac{d\varepsilon}{dt} = 2\frac{\nu}{\varepsilon} \qquad \text{or} \qquad \frac{d\varepsilon^2}{dt} = \nu, \qquad (11.4.19)$$

where ν is the kinematic viscosity of the fluid. We have found that the square of the radius or a Gaussian blob, ε^2, increases at a linear rate in time. Equations (11.4.16) and (11.4.19) provide us with a system of ordinary differential equations for the position of the centers and radii of a collection of vortex blobs. To preserve the total circulation, which is an invariant of the motion, we require that the strength of each blob remains constant during the motion. However, in practice, the vorticity distribution of each blob becomes non-axisymmetric due to mutual interactions.

Problem

11.4.1 *Vortex blob Biot–Savart integral*

Derive the definite integral (11.4.15) and then compute the velocity field induced by a vortex blob.

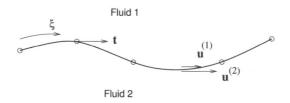

FIGURE 11.5.1   Illustration of a two-dimensional vortex sheet described by a string of marker points moving with the fluid velocity normal to the vortex sheet, while executing an arbitrary tangential motion.

Computer Problems

11.4.2 *Stability of a polygonal arrangement of vortex blobs*

(*a*) Write a program that computes the motion of vortex blobs with Gaussian vorticity distribution initially arranged at the vertices of an N-sided regular polygon of radius a. Perform simulations with $N = 6$ and blob radius $\varepsilon/a = 0.01, 0.05, 0.10, 0.20$, and 0.50. Plot the angular velocity of rotation of the vortex arrangement against b/a and discuss the effect of the blob radius.

(*b*) Repeat (*a*) with diffusing blobs whose radii increase according to (11.4.19).

11.4.3 *Vortex street with vortex blobs*

Compute and plot the velocity of translation of vortex blobs with Gaussian vorticity distribution in a symmetric or antisymmetric vortex street as a function of the blob radius ε/a, for street aspect ratio $b/a = 0.10, 0.25$, and 0.50.

11.5   Two-dimensional vortex sheets

A two-dimensional vortex sheet is a cylindrical surface across which the tangential component of the fluid velocity, residing in a plane that is perpendicular to the generators of the vortex sheet, undergoes a discontinuity called the strength of the vortex sheet, γ, as shown in Figure 11.5.1. By definition,

$$\Delta \mathbf{u} \equiv \mathbf{u}^{(2)} - \mathbf{u}^{(1)} = \gamma \mathbf{t}, \qquad (11.5.1)$$

where $\mathbf{u}^{(i)}$ is the fluid velocity on the ith side of the vortex sheet, $i = 1, 2$, and $\mathbf{t}$ is the tangent unit vector. Since nonzero viscosity causes the unphysical onset of infinite shear stress and viscous dissipation, the notion of a vortex sheet whose thickness is and remains infinitesimal in time is acceptable only in the context of inviscid flow.

Principal velocity

The principal velocity of the vortex sheet is defined as the mean value of the velocity of the fluid on either side of the vortex sheet,

$$\mathbf{u}^{PV} \equiv \frac{1}{2}\left(\mathbf{u}^{(1)} + \mathbf{u}^{(2)}\right). \tag{11.5.2}$$

Combining this definition with (11.5.1), we obtain

$$\mathbf{u}^{(1)} = \mathbf{u}^{PV} - \frac{1}{2}\gamma\,\mathbf{t}, \qquad \mathbf{u}^{(2)} = \mathbf{u}^{PV} + \frac{1}{2}\gamma\,\mathbf{t}, \tag{11.5.3}$$

subject to the conventions defined in Figure 11.5.1.

Marker points

To describe the motion of a vortex sheet, we mark its trace in the xy plane with a string of marker points identified by a permanent label, ξ, as depicted in Figure 11.5.1, and follow the evolution of the vortex sheet by computing the trajectories of the marker points. Kinematic compatibility requires that the component of the velocity of the marker points normal to the vortex sheet be equal to the normal component of the fluid velocity on either side of the vortex sheet. The tangential component of the marker-point velocity can be arbitrary. If $\mathbf{X}(\xi, t)$ is the position and $\mathbf{V}(\xi, t)$ is the velocity of a marker point,

$$\frac{\partial \mathbf{X}}{\partial t} = \mathbf{V}(\xi, t) = (\mathbf{u}^{(i)} \cdot \mathbf{n})\,\mathbf{n} + V_t(\xi)\,\mathbf{t} \tag{11.5.4}$$

for $i = 1, 2$, where $V_t(\xi)$ is an arbitrary tangential velocity allowed to vary in time and across the marker points.

Using equations (11.5.3), we express the marker-point velocity in terms of the velocity of the fluid on either side of the vortex sheet, the principal velocity of the vortex sheet (PV), and the strength of the vortex sheet, as

$$\mathbf{V}(a, t) = (1 - \alpha)\,\mathbf{u}^{(1)} + \alpha\,\mathbf{u}^{(2)} = \mathbf{u}^{PV} + \left(\alpha - \frac{1}{2}\right)\gamma\,\mathbf{t}, \tag{11.5.5}$$

where α is a dimensionless parameter related to the tangential marker-point velocity by

$$V_t = \mathbf{u}^{PV} \cdot \mathbf{t} + \left(\alpha - \frac{1}{2}\right)\gamma. \tag{11.5.6}$$

When $\alpha = 0, \frac{1}{2}, 1$ the marker points move, respectively, with the velocity of the fluid on the first side of the vortex sheet, the principal velocity of the vortex sheet, and the velocity of the fluid on the second side of the vortex sheet. In the remainder of our discussion, α will be assumed to be a specified constant.

It will be useful to express the fluid velocity on either side of the vortex sheet in terms of the marker-point velocity, $\mathbf{V}$, as

$$\mathbf{u}^{(1)} = \mathbf{V} + \alpha_1\gamma\,\mathbf{t}, \qquad \mathbf{u}^{(2)} = \mathbf{V} + \alpha_2\gamma\,\mathbf{t}, \tag{11.5.7}$$

where $\alpha_1 = -\alpha$ and $\alpha_2 = 1 - \alpha$. Equations (11.5.7) can be collected into the compact form

$$\mathbf{u}^{(i)} = \mathbf{V} + \alpha_i \gamma \, \mathbf{t} \tag{11.5.8}$$

for $i = 1, 2$.

Biot–Savart integral

Substituting into (11.5.5) the Biot–Savart integral (2.13.19), we obtain an integro-differential equation for the motion of the marker points,

$$\frac{\partial \mathbf{X}}{\partial t} = \mathbf{V}(\xi) = \frac{1}{2\pi} \int_C^{PV} \begin{bmatrix} -Y(\xi) + Y(\xi') \\ X(\xi) - X(\xi') \end{bmatrix} \frac{\gamma(\xi')}{|\mathbf{X}(\xi) - \mathbf{X}(\xi')|^2} \frac{\partial l}{\partial \xi'} \, d\xi' + \left(\alpha - \frac{1}{2}\right) \gamma \, \mathbf{t} + \mathbf{v}, \tag{11.5.9}$$

where C is the trace of the vortex sheet in the xy plane, PV denotes the principal-value integral, and l is the arc length along the vortex sheet measured in the direction of the tangent unit vector, $\mathbf{t}$. The first term on the right-hand side of (11.5.9) expresses the self-induced velocity of the vortex sheet. The third term expresses a generally time-dependent external rotational or irrotational flow with velocity $\mathbf{v}$.

11.5.1   Evolution of the strength of the vortex sheet

To compute the trajectories of the marker points, we require an evolution equation for the strength of the vortex sheet following the marker points, expressed by the derivative $(\partial \gamma / \partial t)_\xi$. Our point of departure is Euler's equation on either side of the vortex sheet,

$$\frac{\partial \mathbf{u}^{(i)}}{\partial t} + \mathbf{u}^{(i)} \cdot \nabla \mathbf{u}^{(i)} = -\frac{1}{\rho_i} \nabla p_i + \mathbf{g} \tag{11.5.10}$$

for $i = 1, 2$, where p_i is the pressure and ρ_i is the fluid density. Slightly rearranging the left-hand side, we obtain

$$\frac{\partial \mathbf{u}^{(i)}}{\partial t} + \mathbf{V} \cdot \nabla \mathbf{u}^{(i)} + (\mathbf{u}^{(i)} - \mathbf{V}) \cdot \nabla \mathbf{u}^{(i)} = -\frac{1}{\rho_i} \nabla p_i + \mathbf{g}. \tag{11.5.11}$$

The first two terms on the left-hand side express the derivative $(\partial \mathbf{u}^{(i)} / \partial t)_\xi$, representing the rate of change of the fluid velocity on either side of the vortex sheet following the marker points.

Differentiating equation (11.5.8), relating the fluid velocity to the marker-point velocity, we obtain

$$\left(\frac{\partial \mathbf{u}^{(i)}}{\partial t}\right)_\xi = \left(\frac{\partial \mathbf{V}}{\partial t}\right)_\xi + \alpha_i \left(\frac{\partial (\gamma \mathbf{t})}{\partial t}\right)_\xi. \tag{11.5.12}$$

Substituting the right-hand side of (11.5.12) in place of the first two terms in (11.5.11), and setting $\mathbf{u}^{(i)} - \mathbf{V} = \alpha_i \gamma \, \mathbf{t}$, we find that

$$\rho_i \left(\frac{\partial \mathbf{V}}{\partial t}\right)_\xi + \rho_i \alpha_i \left[\left(\frac{\partial (\gamma \mathbf{t})}{\partial t}\right)_\xi + \gamma \frac{\partial (\mathbf{V} + \alpha_i \gamma \mathbf{t})}{\partial l}\right] = -\nabla p_i + \rho_i \, \mathbf{g}, \tag{11.5.13}$$

where l is the arc length along the vortex sheet. Next, we take the difference of the equations that arise by writing (11.5.13) for $i = 1$ and 2, finding

$$\Delta\rho \left(\frac{\partial \mathbf{V}}{\partial t}\right)_\xi + \tilde{\rho}\left[\left(\frac{\partial(\gamma \mathbf{t})}{\partial t}\right)_\xi + \gamma \frac{\partial \mathbf{V}}{\partial l}\right] + \widehat{\Delta\rho}\,\gamma\, \frac{\partial(\gamma \mathbf{t})}{\partial l} = -\nabla\Delta p + \Delta\rho\,\mathbf{g}, \qquad (11.5.14)$$

where $\Delta\rho = \rho_2 - \rho_1$ is the density difference, $\Delta p = p_2 - p_1$ is the pressure difference, and

$$\tilde{\rho} = \alpha_2\rho_2 - \alpha_1\rho_1 = (1-\alpha)\,\rho_2 + \alpha\rho_1, \qquad \widehat{\Delta\rho} = \alpha_2^2\rho_2 - \alpha_1^2\rho_1 = (1-\alpha)^2\rho_2 - \alpha^2\rho_1. \quad (11.5.15)$$

Projecting (11.5.14) onto the tangent unit vector, $\mathbf{t}$, to formulate the tangential derivatives of the pressure difference, and noting that, because $\mathbf{t}$ is a unit vector $\mathbf{t} \cdot \partial\mathbf{t}/\partial l = 0$ and $\mathbf{t} \cdot \partial\mathbf{t}/\partial t = 0$, we find that

$$\Delta\rho\,\mathbf{t}\cdot\left(\frac{\partial \mathbf{V}}{\partial t}\right)_\xi + \tilde{\rho}\left[\left(\frac{\partial\gamma}{\partial t}\right)_\xi + \gamma\mathbf{t}\cdot\frac{\partial \mathbf{V}}{\partial l}\right] + \widehat{\Delta\rho}\,\gamma\,\frac{\partial\gamma}{\partial l} = -\frac{\partial \Delta p}{\partial l} + \Delta\rho\,\mathbf{g}\cdot\mathbf{t}. \qquad (11.5.16)$$

Rearranging, we obtain the evolution equation

$$\tilde{\rho}\left(\frac{\partial\gamma}{\partial t}\right)_\xi + \Delta\rho\,\mathbf{t}\cdot\left(\frac{\partial \mathbf{V}}{\partial t}\right)_\xi = -\tilde{\rho}\,\gamma\,\mathbf{t}\cdot\frac{\partial \mathbf{V}}{\partial l} - \widehat{\Delta\rho}\,\gamma\,\frac{\partial\gamma}{\partial l} - \frac{\partial \Delta p}{\partial l} + \Delta\rho\,\mathbf{g}\cdot\mathbf{t}. \qquad (11.5.17)$$

The jump in the pressure across the vortex sheet on the right-hand side of (11.5.17) depends on the properties of the interface separating the two fluids, as discussed in Section 3.8. For example, if the vortex sheet represents the interface between two immiscible fluids with surface tension τ,

$$\Delta p = \tau\kappa, \qquad (11.5.18)$$

where κ is the curvature of the vortex sheet in the xy plane, reckoned as positive when the interface is downward convex. In the absence of surface tension, $\Delta p = 0$. Combining equation (11.5.17) with the interfacial condition (11.5.18), we obtain the desired evolution equation for the strength of the vortex sheet following the marker points involving the marker-point velocity distribution, $\mathbf{V}(\xi)$, and the tangential component of the marker-point acceleration, $\mathbf{t} \cdot (\partial\mathbf{V}/\partial t)_\xi$.

Fredholm integral equation

The tangential acceleration in the second term on the left-hand side of (11.5.17) is an implicit function of the time derivative $(\partial\gamma/\partial t)_\xi$. To establish this relationship, we differentiate the right-hand side of (11.5.9) with respect to time keeping ξ constant. Projecting the resulting expression onto the tangent unit vector and recalling once again that $\mathbf{t}$ is a unit vector, we find that

$$\mathbf{t}(\xi)\cdot\left(\frac{\partial\mathbf{V}}{\partial t}\right)_\xi = \frac{1}{2\pi}\,\mathbf{t}(\xi)\cdot\int_C^{PV}\frac{\partial}{\partial t}\left(\begin{bmatrix} -Y(\xi) + Y(\xi') \\ X(\xi) - X(\xi') \end{bmatrix}\frac{\gamma(\xi')}{|\mathbf{X}(\xi)-\mathbf{X}(\xi')|^2}\,\frac{\partial l}{\partial\xi'}\right)\,\mathrm{d}\xi' + \mathcal{G}(\xi), \quad (11.5.19)$$

where

$$\mathcal{G}(\xi) \equiv \left(\alpha - \frac{1}{2}\right)\left(\frac{\partial\gamma}{\partial t}\right)_\xi + \mathbf{t}(\xi)\cdot\left(\frac{\partial\mathbf{v}}{\partial t}\right)_\xi. \qquad (11.5.20)$$

The time derivative inside the integral is taken keeping both ξ and ξ' constant. Expanding the derivative under the integral sign, we obtain

$$\mathbf{t}(\xi) \cdot \left(\frac{\partial \mathbf{V}}{\partial t} \right)_\xi = \frac{1}{2\pi} \mathbf{t}(\xi) \cdot \int_C^{PV} \begin{bmatrix} -Y(\xi) + Y(\xi') \\ X(\xi) - X(\xi') \end{bmatrix} \frac{1}{|\mathbf{X}(\xi) - \mathbf{X}(\xi')|^2} \left(\frac{\partial \gamma}{\partial t} \right)_{\xi'} dl(\xi') + \mathcal{J}(\xi) + \mathcal{G}(\xi).$$

$$(11.5.21)$$

The function

$$\mathcal{J}(\xi) \equiv \frac{1}{2\pi} \mathbf{t}(\xi) \cdot \int_C^{PV} \frac{\partial}{\partial t} \left(\begin{bmatrix} -Y(\xi) + Y(\xi') \\ X(\xi) - X(\xi') \end{bmatrix} \frac{1}{|\mathbf{X}(\xi) - \mathbf{X}(\xi')|^2} \frac{\partial l}{\partial \xi'} \right) \gamma(\xi') \, d\xi' \qquad (11.5.22)$$

expresses the tangential component of the acceleration of the point particles if they were moving with the principal velocity of the vortex sheet while preserving the current circulation. In practice, this function can be computed by numerical time differentiation. Substituting the right-hand side of (11.5.21) into the second term on the right-hand side of (11.5.17), rearranging, and observing that $\tilde{\rho} + \Delta\rho \left(\alpha - \frac{1}{2} \right) = \frac{1}{2}(\rho_1 + \rho_2)$, we obtain a linear Fredholm integral equation of the second kind for $(\partial\gamma/\partial t)_\xi$,

$$\left(\frac{\partial \gamma}{\partial t} \right)_\xi + \frac{\mathcal{A}}{\pi} \mathbf{t}(\xi) \cdot \int_C^{PV} \begin{bmatrix} -Y(\xi) + Y(\xi') \\ X(\xi) - X(\xi') \end{bmatrix} \frac{1}{|\mathbf{X}(\xi) - \mathbf{X}(\xi')|^2} \left(\frac{\partial \gamma}{\partial t} \right)_{\xi'} dl(\xi') = 2\mathcal{F}(\xi), \quad (11.5.23)$$

where

$$\mathcal{A} = \frac{\rho_2 - \rho_1}{\rho_1 + \rho_2} \qquad (11.5.24)$$

is the Atwood ratio. The forcing function on the right-hand side of (11.5.23) is given by

$$\mathcal{F}(\xi) = -\mathcal{A} \left[\mathcal{J} + \mathbf{t} \cdot \left[\left(\frac{\partial \mathbf{v}}{\partial t} \right)_\xi - \mathbf{g} \right] \right] - \gamma \left(\mathcal{B} \, \mathbf{t} \cdot \frac{\partial \mathbf{V}}{\partial l} + \widehat{\mathcal{A}} \frac{\partial \gamma}{\partial l} \right) - \frac{1}{\rho_1 + \rho_2} \frac{\partial \Delta p}{\partial l}, \qquad (11.5.25)$$

where

$$\mathcal{B} = \frac{\tilde{\rho}}{\rho_1 + \rho_2} = \frac{(1-\alpha)\rho_2 + \alpha\rho_1}{\rho_1 + \rho_2}, \qquad \widehat{\mathcal{A}} = \frac{\widehat{\Delta\rho}}{\rho_1 + \rho_2} = \frac{(1-\alpha)^2\rho_2 - \alpha^2\rho_1}{\rho_1 + \rho_2} \qquad (11.5.26)$$

are dimensionless coefficients defined in terms of the fluid densities and the marker-point velocities.

Evolution of circulation along the vortex sheet

The circulation along the vortex sheet, Γ, is defined by the differential relation $d\Gamma = \gamma \, dl$, where $\Gamma = 0$ at a designated marker point. Straightforward differentiation yields

$$\left(\frac{\partial \, d\Gamma}{\partial t} \right)_\xi = \left(\frac{\partial \, (\gamma dl)}{\partial t} \right)_\xi = \left(\frac{\partial \gamma}{\partial t} \right)_\xi dl + \gamma \, \mathbf{t} \cdot \frac{\partial \mathbf{V}}{\partial l} \, dl, \qquad (11.5.27)$$

which can be restated as

$$\frac{\partial^2 \Gamma}{\partial t \partial \xi} = \left[\left(\frac{\partial \gamma}{\partial t} \right)_\xi + \gamma \, \mathbf{t} \cdot \frac{\partial \mathbf{V}}{\partial l} \right] \frac{\partial l}{\partial \xi}. \qquad (11.5.28)$$

Solving (11.5.28) for $(\partial\gamma/\partial t)_\xi$, substituting the result into the first term on the left-hand side of (11.5.17), multiplying the emerging equation by $\partial l/\partial\xi$, and simplifying, we derive an integro-differential evolution equation for the circulation along the vortex sheet,

$$\tilde{\rho}\,\frac{\partial^2\Gamma}{\partial t\partial\xi} + \Delta\rho\,\frac{\partial\mathbf{x}}{\partial\xi}\cdot\left(\frac{\partial\mathbf{V}}{\partial t}\right)_\xi = \frac{\partial}{\partial\xi}\left[-\frac{1}{2}\,\widehat{\Delta\rho}\,\gamma^2 - \Delta p + \Delta\rho\,\mathbf{g}\cdot\mathbf{x}\right]. \tag{11.5.29}$$

The tangential acceleration in the second term on the left-hand side is computed from (11.5.21). The right-hand side of (11.5.29) expresses the rate of production of vorticity along the vortex sheet.

11.5.2   Motion with the principal velocity of the vortex sheet

When the marker points move with the principal velocity of the vortex sheet, $\alpha = \frac{1}{2}$, we obtain $\alpha_1 = -\frac{1}{2}$, $\alpha_2 = \frac{1}{2}$, and

$$\tilde{\rho} = \frac{1}{2}\,(\rho_1 + \rho_2), \qquad \widehat{\Delta\rho} = \frac{1}{4}\,\Delta\rho, \qquad \mathcal{B} = \frac{1}{2}, \qquad \widehat{\mathcal{A}} = \frac{1}{4}\,\mathcal{A}. \tag{11.5.30}$$

The forcing function defined in (11.5.25) takes the simpler form

$$\mathcal{F}(\xi) = -\mathcal{A}\left[\mathcal{J} + \mathbf{t}\cdot\left[\left(\frac{\partial\mathbf{v}}{\partial t}\right)_\xi - \mathbf{g}\right] + \frac{1}{4}\,\gamma\,\frac{\partial\gamma}{\partial l}\right] - \frac{1}{2}\,\gamma\,\mathbf{t}\cdot\frac{\partial\mathbf{V}}{\partial l} - \frac{1}{\rho_1+\rho_2}\,\frac{\partial\Delta p}{\partial l}. \tag{11.5.31}$$

The motion of the vortex sheet is determined exclusively by the Atwood ratio, $\mathcal{A}$. The evolution equation (11.5.17) becomes

$$\left(\frac{\partial\gamma}{\partial t}\right)_\xi + \gamma\,\mathbf{t}\cdot\frac{\partial\mathbf{V}}{\partial l} = -2\,\mathcal{A}\left[\mathbf{t}\cdot\left(\frac{\partial\mathbf{V}}{\partial t}\right)_\xi + \frac{1}{4}\,\gamma\,\frac{\partial\gamma}{\partial l} - \mathbf{t}\cdot\mathbf{g}\right] - \frac{2}{\rho_1+\rho_2}\,\frac{\partial\Delta p}{\partial l}. \tag{11.5.32}$$

The first two terms inside the square brackets on the right-hand side can be written in the compact form

$$\mathbf{t}\cdot\left(\frac{\partial\mathbf{V}}{\partial t}\right)_\xi + \frac{1}{4}\,\gamma\,\frac{\partial\gamma}{\partial l} = \mathbf{t}\cdot\bar{\mathbf{a}}, \tag{11.5.33}$$

where

$$\bar{\mathbf{a}} \equiv \frac{1}{2}\left(\frac{D\mathbf{u}^{(1)}}{Dt} + \frac{D\mathbf{u}^{(2)}}{Dt}\right) \tag{11.5.34}$$

is the mean value of the fluid acceleration on either side of the vortex sheet and D/Dt is the material derivative. To show this, we write

$$\frac{D\mathbf{u}^{(i)}}{Dt} = \frac{\partial\mathbf{u}^{(i)}}{\partial t} + \mathbf{u}^{(i)}\cdot\nabla\mathbf{u}^{(i)} = \frac{\partial\mathbf{u}^{(i)}}{\partial t} + (\mathbf{V}+\alpha_i\gamma\mathbf{t})\cdot\nabla\mathbf{u}^{(i)} \tag{11.5.35}$$

for $i = 1, 2$. Applying this equation for $i = 1$ and 2 with $\alpha_1 = -\frac{1}{2}$ and $\alpha_2 = \frac{1}{2}$, and summing the resulting equations, we obtain

$$2\,\bar{\mathbf{a}} = 2\,\frac{\partial\mathbf{V}}{\partial t} + (\mathbf{V}-\frac{1}{2}\gamma\mathbf{t})\cdot\nabla\mathbf{u}^{(1)} + (\mathbf{V}+\frac{1}{2}\gamma\mathbf{t})\cdot\nabla\mathbf{u}^{(2)}. \tag{11.5.36}$$

Rearranging the right-hand side, we find that

$$2\,\bar{\mathbf{a}} = 2\,\frac{\partial \mathbf{V}}{\partial t} + \mathbf{V} \cdot \nabla(\mathbf{u}^{(1)} + \mathbf{u}^{(2)}) + \frac{1}{2}\gamma\,\mathbf{t} \cdot \nabla(\mathbf{u}^{(2)} - \mathbf{u}^{(1)}). \tag{11.5.37}$$

The union of the first two terms on the right-hand side of (11.5.37) is equal to $2\,(\partial\mathbf{V}/\partial t)_\xi$. Substituting $\mathbf{u}^{(2)} - \mathbf{u}^{(1)} = \gamma\,\mathbf{t}$ and projecting the resulting equation on $\mathbf{t}$, we obtain (11.5.33).

Matched fluid densities

When the vortex sheet separates two fluids with equal densities, $\rho_1 = \rho_2 = \rho$, the Atwood ratio is zero, $\mathcal{A} = 0$, yielding an explicit expression for the rate of change of the strength of the vortex sheet following the marker points,

$$\left(\frac{\partial\gamma}{\partial t}\right)_\xi = -\gamma\,\mathbf{t} \cdot \frac{\partial\mathbf{V}}{\partial l} - \frac{1}{\rho}\frac{\partial\Delta p}{\partial l}. \tag{11.5.38}$$

The evolution equation (11.5.29) for the circulation along the vortex sheet can be integrated with respect to arc length to give

$$\left(\frac{\partial\Gamma}{\partial t}\right)_\xi = -\frac{\Delta p}{\rho} + c(t), \tag{11.5.39}$$

where $c(t)$ is an arbitrary function of time. In the absence of surface tension, the pressure is continuous across the vortex sheet. Setting without loss of generality $c(t) = 0$, we obtain the remarkably simple evolution equation $(\partial\Gamma/\partial t)_\xi = 0$, stating that the marker points preserve their circulation. To simplify the formulation, we may label the marker points by their initial circulation, setting $\xi = \Gamma$.

11.5.3   The Boussinesq approximation

In certain applications involving gravity-driven flow, density variations play an important role in determining the magnitude of the gravitational force, but make a minor contribution to the magnitude of the inertial force. To simplify the computation of the flow, we may adopt the Boussinesq approximation, replacing the individual fluid densities ρ_i with the mean value, $\rho = \frac{1}{2}\,(\rho_1 + \rho_2)$, in all but the gravity term in the equation of motion. The Boussinesq approximation is strictly valid in the limit as the Atwood ratio $\mathcal{A}$ tends to zero while the scaled acceleration of gravity gT^2/L tends to infinity, so that the product $\mathcal{A}g$ remains finite, where L and T are characteristic length and time scales.

Under the Boussinesq approximation, the factors and coefficients defined in (11.5.15) and (11.5.26) take the values

$$\tilde{\rho} = \rho, \qquad \widehat{\Delta\rho} = (1 - 2\alpha)\,\rho, \qquad \mathcal{B} = \frac{1}{2}, \qquad \widehat{\mathcal{A}} = \frac{1}{2} - \alpha. \tag{11.5.40}$$

The evolution equation for the strength of the vortex sheet reduces to

$$\left(\frac{\partial\gamma}{\partial t}\right)_\xi = -\gamma\left(\mathbf{t} \cdot \frac{\partial\mathbf{V}}{\partial l} + (1 - 2\alpha)\,\frac{\partial\gamma}{\partial l}\right) - \frac{1}{\rho}\frac{\partial\Delta p}{\partial l} + 2\,\mathcal{A}\mathbf{g} \cdot \mathbf{t}, \tag{11.5.41}$$

and the evolution equation for the circulation along the vortex sheet reduces to

$$\left(\frac{\partial \Gamma}{\partial t}\right)_{\xi} = -(\frac{1}{2} - \alpha)\gamma^2 - \frac{\Delta p}{\rho} + 2\mathcal{A}\mathbf{g}\cdot\mathbf{x} + c(t). \tag{11.5.42}$$

The simplifications resulting from the Boussinesq approximation are reflected in the absence of the tangential marker-point acceleration on the right-hand sides of (11.5.41) and (11.5.42).

11.5.4   A vortex sheet embedded in irrotational flow

In the particular but common case where a vortex sheet separates two regions of irrotational flow, the circulation along the vortex sheet is equal to the jump in the velocity potential across the vortex sheet, $\Gamma = \phi_2 - \phi_1$, as discussed in Section 2.1.3. Using the unsteady Bernoulli equation, we find that the change in the potential on either side of a vortex sheet as seen by an observer who is moving with the marker points is

$$\left(\frac{\partial \phi_i}{\partial t}\right)_{\xi} - \mathbf{V}\cdot\mathbf{u}^{(i)} + \frac{1}{2}\mathbf{u}^{(i)}\cdot\mathbf{u}^{(i)} + \frac{p_i}{\rho_i} - \mathbf{g}\cdot\mathbf{x} = c_i(t) \tag{11.5.43}$$

for $i = 1, 2$, where $c_i(t)$ two are unrelated functions. Substituting $\mathbf{u}^{(i)} = \mathbf{V} + \alpha_i \gamma\mathbf{t}$ and simplifying, we find that

$$\left(\frac{\partial \phi_i}{\partial t}\right)_{\xi} - \frac{1}{2}\mathbf{V}\cdot\mathbf{V} + \frac{1}{2}\alpha_i^2\gamma^2 + \frac{p_i}{\rho_i} - \mathbf{g}\cdot\mathbf{x} = c_i(t). \tag{11.5.44}$$

Next, we apply this equation for $i = 1, 2$ and take the difference of the resulting expressions. In the case of equal fluid densities, $\rho_1 = \rho_2 = \rho$, we obtain

$$\left(\frac{\partial \Gamma}{\partial t}\right)_{\xi} + \frac{1}{2}(\alpha_2^2 - \alpha_1^2)\gamma^2 + \frac{\Delta p}{\rho} = c_2(t) - c_1(t), \tag{11.5.45}$$

which is consistent with equation (11.5.29) derived by a different method.

Problems

11.5.1 *A vortex sheet embedded in potential flow*

Discuss how equation (11.5.43) can be used to derive (11.5.29) in the general case of different fluid densities.

11.5.2 *A fluid interface in oscillatory motion*

Consider a fluid resting on another fluid with different density inside a container that executes vertical oscillatory motion normal to the interface along the x axis with velocity $V_x = V_0 \sin(\Omega t)$, where V_0 is a constant amplitude and Ω is the angular frequency of the oscillations. In the absence of viscous forces, the motion of the interface can be described in terms of an vortex sheet situated over the interface. Show that, subject to the Boussinesq approximation, the evolution of the circulation along the vortex sheet for $\alpha = \frac{1}{2}$ is described by

$$\left(\frac{\partial \Gamma}{\partial t}\right)_{\xi} = \frac{\tau}{\rho}\kappa + 2\mathcal{A}\left[V_0 \cos(\Omega t) - g_x\right]x, \tag{11.5.46}$$

where τ is the surface tension, κ is the curvature, and $\mathcal{A}$ is the Atwood ratio.

FIGURE 11.6.1 Discretization of a vortex sheet into a collection of elements defined by end points and midpoints. In the point-vortex method, the vortex sheet is replaced by a collection of point vortices located at the midpoints.

11.6   The point-vortex method for vortex sheets

In Section 11.5, we derived an integro-differential equation governing the motion of marker points distributed along a two-dimensional vortex sheet, and an integral equation governing the evolution of the strength of the vortex sheet or circulation along the vortex sheet. These equations can be solved by standard numerical methods for differential and integral equations (e.g., [409]). The simplest approach is based on the point-vortex discretization, first implemented by Rosenhead at the dawn of scientific computing [349].

Trapezoidal rule

In the point vortex method, the principal value of the Biot–Savart integral (11.5.9) along a vortex sheet is approximated with the trapezoidal rule. To implement the method, we divide the vortex sheet into N elements whose end points are defined in terms of a chain of marker points, $\mathbf{X}_{i-1/2}$, where $i = 1, \ldots, N + 1$, as shown in Figure 11.6.1. The midpoint of the ith segment extending between $\mathbf{X}_{i-1/2}$ and $\mathbf{X}_{i+1/2}$ is denoted by $\mathbf{X}_i$, the element length is denoted by Δl_i, the difference in the circulation across the end points is denoted by $\Delta\Gamma_i$, and the unit vector tangent to the segment at the midpoint is denoted by $\mathbf{t}_i$. Using central differences, we approximate the tangent unit vector as

$$\mathbf{t}_i \simeq \frac{1}{\Delta l_i}\left(\mathbf{X}_{i+1/2} - \mathbf{X}_{i-1/2}\right). \tag{11.6.1}$$

More advanced differentiation methods based on cubic spline or trigonometric interpolation of the coordinates of the marker points with respect to the arc length of the polygonal line connecting the marker points offer higher accuracy.

Point vortex motion

The element end points and midpoints are marker points whose velocity is determined by equation (11.5.9). Applying (11.5.9) at the midpoints, $\mathbf{X}_i$, we derive a system of nonlinear differential equations for $\mathbf{X}_i$,

$$\frac{d\mathbf{X}_i}{dt} = \frac{1}{2\pi}\sum_{j=1}^{N}{}' \left[\begin{array}{c} -Y_i + Y_j \\ X_i - X_j \end{array}\right] \frac{\Delta\Gamma_j}{|\mathbf{X}_i - \mathbf{X}_j|^2} + \left(\alpha - \frac{1}{2}\right)\frac{\Delta\Gamma_i}{\Delta l_i}\,\mathbf{t}_i + \mathbf{v}(\mathbf{X}_i), \tag{11.6.2}$$

where the prime indicates that the term $i = j$ is excluded from the sum. Apart from the last two terms, equation (11.6.2) is identical to (11.2.1) describing the motion of a collection of N point vortices located at the element midpoints, $\mathbf{X}_i$. The element circulation, $\Delta\Gamma_j$, plays the role of the point vortex strength, κ_j. Computing the evolution of the vortex sheet is thus reduced to computing the motion of point vortices with generally evolving strength.

Having advanced the position of the midpoints, $\mathbf{X}_i$, using (11.6.2), we reset the position of the end points, $\mathbf{X}_{i+1/2}$, by linear or high-order interpolation. In the simplest approach, we set $\mathbf{X}_{i+1/2} = \frac{1}{2}(\mathbf{X}_i + \mathbf{X}_{i+1})$. As an alternative, we advance the position of the end points according to the velocity field induced by the point vortices, setting

$$\frac{d\mathbf{X}_{i+1/2}}{dt} = \frac{1}{2\pi}\sum_{j=1}^{N}\left[\begin{array}{c}-Y_{i+1/2}+Y_j\\X_{i+1/2}-X_j\end{array}\right]\frac{\Delta\Gamma_j}{|\mathbf{X}_{i+1/2}-\mathbf{X}_j|^2} + (\alpha - \frac{1}{2})\gamma_{i+1/2}\,\mathbf{t}(\mathbf{X}_{i+1/2}) + \mathbf{v}(\mathbf{X}_{i+1/2}).$$

$$(11.6.3)$$

The tangential unit vector in the last term of (11.6.3) can be computed by numerical differentiation setting, for example, $\mathbf{t}(\mathbf{X}_{i+1/2}) = (\mathbf{X}_{i+1} - \mathbf{X}_i)/|\mathbf{X}_{i+1} - \mathbf{X}_i|$.

Evolution of the strength of the point vortices

To derive an evolution equation for the element circulation, $\Delta\Gamma_i$, we integrate equation (11.5.29) with respect to the parameter ξ over the length of a segment. Applying the trapezoidal rule, we derive the approximate form

$$\tilde{\rho}\frac{d\Delta\Gamma_i}{dt} + \Delta\rho\left[\mathbf{t}\cdot\left(\frac{\partial\mathbf{V}}{\partial t}\right)_\xi\right]\Delta l_i = \left[-\frac{1}{2}\widehat{\Delta\rho}\gamma^2 - \Delta p + \Delta\rho\,\mathbf{g}\cdot\mathbf{x}\right]_{\mathbf{X}_{i-1/2}}^{\mathbf{X}_{i+1/2}}. \qquad (11.6.4)$$

The acceleration of the point vortices involved in the first term on the right-hand side is computed by differentiating the right-hand side of (11.6.2) with respect to time. To simplify the expressions, we set $\alpha = \frac{1}{2}$ and obtain

$$\left(\frac{\partial\mathbf{V}}{\partial t}\right)_a \equiv \frac{d\mathbf{V}_i}{dt} = \frac{1}{2\pi}\sum_{j=1}^{N}{}'\left[\begin{array}{c}-Y_i+Y_j\\X_i-X_j\end{array}\right]\frac{1}{|\mathbf{X}_i-\mathbf{X}_j|^2}\frac{d\Delta\Gamma_j}{dt}$$

$$+\frac{1}{2\pi}\sum_{j=1}^{N}{}'\frac{d}{dt}\left(\left[\begin{array}{c}-Y_i+Y_j\\X_i-X_j\end{array}\right]\frac{1}{|\mathbf{X}_i-\mathbf{X}_j|^2}\right)\Delta\Gamma_j + \left(\frac{\partial\mathbf{v}}{\partial t}\right)(\mathbf{X}_i). \qquad (11.6.5)$$

Carrying out the differentiation in the second sum on the right-hand side and simplifying, we find that

$$\frac{d}{dt}\left(\left[\begin{array}{c}-Y_i+Y_j\\X_i-X_j\end{array}\right]\frac{1}{|\mathbf{X}_i-\mathbf{X}_j|^2}\right) = \frac{1}{\hat{r}^4}\left[\begin{array}{c}2\hat{u}\hat{x}\hat{y}+\hat{v}\,(\hat{y}^2-\hat{x}^2)\\\hat{u}\,(\hat{y}^2-\hat{x}^2)-2\hat{v}\hat{x}\hat{y}\end{array}\right], \qquad (11.6.6)$$

where

$$\hat{x} = X_i - X_j, \qquad \hat{y} = Y_i - Y_j, \qquad \hat{r}^2 = \hat{x}^2 + \hat{y}^2, \qquad \hat{u} = V_{x_i} - V_{x_j}, \qquad \hat{v} = V_{y_i} - V_{y_j}. \quad (11.6.7)$$

Substituting (11.6.6) into (11.6.5) and then into (11.6.4), we obtain a system of linear algebraic equations for $d\Delta\Gamma_i/dt$. The solution can be computed by Jacobi iterations, which involves assuming values for $d\Delta\Gamma_i/dt$, substituting these values into the right-hand side of (11.6.5), and then computing the right-hand side of (11.6.4) to produce new and improved values for $d\Delta\Gamma_i/dt$, to be used in subsequent iterations.

Boussinesq approximation

Adopting the Boussinesq approximation and setting in (11.6.4) $\alpha = \frac{1}{2}$, we obtain an explicit evolution equation for the strength of the point vortices,

$$\frac{d\,\Delta\Gamma_i}{dt} = \left[-\frac{\Delta p}{\tilde{\rho}} + 2\,\mathcal{A}\,\mathbf{g}\cdot\mathbf{x} \right]_{\mathbf{x}_{i-1/2}}^{\mathbf{x}_{i+1/2}}, \tag{11.6.8}$$

where $\mathcal{A}$ is the Atwood ratio defined in (11.5.24). In the absence of surface tension, $\Delta p = 0$. We see that the Boussinesq approximation bypasses solving a system of equations for the rate of the change of the strength of the point vortices.

11.6.1   Regularization

In practice, unless a vortex sheet is sufficiently stretched during the evolution, the motion of the point vortices representing the vortex sheet becomes disorganized and eventually chaotic. The origin of this pathology can be traced back to the exponential amplification of the round-off error due to the instability of an array of point vortices discussed in Section 11.3.

Repositioning

One way to regularize the motion is to redistribute the point vortices along the vortex sheet after each time step or every few time steps, so that the point vortices become evenly spaced with respect to the instantaneous arc length [127]. The strength of the point vortices at the new positions is found from the strength of the point vortices at the old positions by interpolation with respect to ξ or with respect to arc length of the polygonal line connecting successive marker points.

Smoothing the position

In another approach, irregularities are filtered out by expressing the Cartesian coordinates of the point vortices in a global Fourier series or a local polynomial expansion in terms of an appropriate label, such as the polygonal arc length. The position of the point vortices is then recomputed by truncating the Fourier series or discarding odd terms in the polynomial expansion [215]. A simple yet effective method of smoothing is provided by the five-point formula

$$\mathbf{x}_i^s = \frac{1}{16}\left(-\mathbf{x}_{i-2} + 4\,\mathbf{x}_{i-1} + 10\,\mathbf{x}_i + 4\,\mathbf{x}_{i+1} - \mathbf{x}_{i+2} \right), \tag{11.6.9}$$

where the superscript s denotes a smoothed position [244].

Smoothing the kernel

In a third approach, the motion is regularized by replacing the point vortices with vortex blobs, as discussed in Section 11.4 [82]. Alternatively, the singular Biot–Savart integrand in (11.5.9) is replaced with a nearly singular integrand, and equation (11.5.7) is correspondingly modified. For $\alpha = \frac{1}{2}$, the motion of the point vortices is governed by the equations

$$\frac{d\mathbf{X}_i}{dt} = \frac{1}{2\pi} \sum_{j=1}^{N}{}' \begin{bmatrix} -Y_i + Y_j \\ X_i - X_j \end{bmatrix} \frac{\Delta\Gamma_j}{|\mathbf{X}_i - \mathbf{X}_j|^2 + \delta^2} + \mathbf{v}(\mathbf{X}_i), \qquad (11.6.10)$$

where δ is a small numerical parameter with dimensions of length, included to eliminate the unstable behavior due to the singular Biot–Savart integral [216].

11.6.2 Periodic vortex sheets

The equations derived in this section for a finite or closed vortex sheet can be modified in a straight-forward fashion to describe the motion of a vortex sheet that is repeated periodically in the x direction with period a. Using (2.13.23), we find that the motion of marker points deployed along the vortex sheet is governed by a modified version of (11.5.7),

$$\frac{\partial \mathbf{X}}{\partial t} = \frac{1}{2a} \int_C^{PV} \begin{bmatrix} -\sinh(k[Y(\xi) - y(\xi')]) \\ \sin(k[X(\xi) - x(\xi')]) \end{bmatrix} \frac{\gamma(\xi')}{\cosh(k[Y(\xi) - y(\xi')]) - \cos(k[X(\xi) - x(\xi')])} \frac{\partial l}{\partial \xi'} d\xi'$$

$$+ \left(\alpha - \frac{1}{2}\right) \gamma \mathbf{t} + \mathbf{v}, \qquad (11.6.11)$$

where $k = 2\pi/a$ is the wave number and C is the trace of the vortex sheet in the xy plane over one period. The evolution equation for the strength of the vortex sheet remains unchanged.

The counterpart of the regularized system (11.6.10) is

$$\frac{d\mathbf{X}_i}{dt} = \frac{1}{2a} \sum_{j=1}^{N}{}' \begin{bmatrix} -\sinh[k(Y_i - Y_j)] \\ \sin[k(X_i - X_j)] \end{bmatrix} \frac{\Delta\Gamma_j}{\cosh[k(Y_i - Y_j)] - \cos[k(X_i - X_j)] + \varepsilon^2} + \mathbf{v}(\mathbf{X}_i), \quad (11.6.12)$$

where N is the number of point vortices inside one period and ε is a small dimensionless regularization parameter. The evolution equation for the strength of the point vortices remain unchanged.

Stages in the evolution of a sinusoidally perturbed vortex sheet with negative strength immersed in a homogeneous fluid computed using (11.6.12) with $N = 96$ and $\varepsilon = 0.125$ or 0.250 are shown in Figure 11.6.2. The vortex sheet rolls up into a periodic sequence of spirals whose turns keep increasing in time. A higher number of tightly wound spiral turns develop as ε becomes smaller. The numerical results suggest that, in the limit as ε tends to zero, a cusp develops spontaneously at the center of the spiral after a finite period of time, immediately before the spiral starts forming its turns [216].

Problems

11.6.1 *Point-vortex method for an accelerating vortex sheet*

Write the counterpart of the evolution equation (11.6.8) in a frame of reference undergoing translational acceleration with time-dependent velocity $\mathbf{V}(t)$.

(a)   (b)

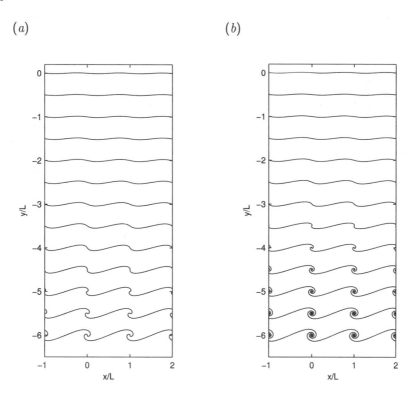

FIGURE 11.6.2   Stages in the evolution of a sinusoidally perturbed vortex sheet with negative strength computed using the regularized equation (11.6.12) with (a) $\varepsilon = 0.250$ or (b) 0.125. Profiles are shown at equal time intervals.

11.6.2 *Stability of a row of regularized point vortices*

Carry out the linear stability analysis of an infinite row of point vortices whose motion is described by the regularized evolution equation (11.6.12). Verify that, as ε tends to zero, the results are consistent with those for the point-vortex array discussed in Section 11.3.

11.6.3 *Smoothing*

Discuss the mathematical action of the smoothing formula (11.6.9) with reference to a triangular wave.

 Computer Problem

11.6.4 *Rayleigh–Taylor instability of a vortex sheet*

Consider the instability of the interface separating two immiscible quiescent fluids with different densities in the absence of surface tension. At the initial instant, the interface is subjected to a periodic sinusoidal perturbation in shape with wavelength L and amplitude b_0. When viscous effects are insignificant, the interface can be identified with a vortex sheet with vanishing initial strength. Write a program that computes the motion of the vortex sheet using the point-vortex method, subject to the Boussinesq approximation. The integration in time should be performed using the modified Euler method discussed in Section B.8, Appendix B. The motion should be regularized using a method of your choice. At the initial instant, assume that each point vortex is displaced along the y axis in a sinusoidal fashion, while its strength remains zero. Run the program for $b_0/L = 0.10$ and $N = 8$, 16, and 32 point vortices inside each period separated by distance $a = L/N$. Discuss the observed differences in behavior.

11.7   Two-dimensional flow with distributed vorticity

In Section 11.6, we discussed a method of computing the motion of two-dimensional vortex sheets discretized into a collection of point vortices or vortex blobs. The procedure involves computing the motion of the point vortices or vortex blobs while simultaneously updating their strengths by solving an integral equation. To extend this method to the more general case of two-dimensional flow with continuous vorticity distribution, we express the initial vorticity field in terms of a collection of N point vortices with strength κ_i located at $\mathbf{x}_i$,

$$\omega_z = \sum_{i=1}^{N} \kappa_i(t)\, \delta_2(\mathbf{x} - \mathbf{x}_i), \tag{11.7.1}$$

where δ_2 is the two-dimensional delta function. The motion of the point vortices is then computed using the methods discussed in Section 11.2 [85]. A regularized version of method discretizes the vorticity field into vortex blobs, as discussed in Section 11.4 (e.g., [328]).

Discretization of the vorticity field

At the beginning of the computation, we must distribute the point vortices in a way that is consistent with the initial state. In one implementation, a two-dimensional rectilinear or curvilinear grid forming a two-dimensional array of cells is introduced and a point vortex is placed in the middle of each cell. To ensure that the sum of the strengths of all point vortices is equal to the circulation around a large loop enclosing the flow, the strength of each point vortex is set equal to the integral of the vorticity or circulation of the fluid inside the corresponding cell. Other methods of discretizing the initial vorticity field are available [78].

11.7.1   Vortex-in-cell method (VIC)

The simulations become prohibitively expensive when a large number of point vortices are employed. Computing the velocity of N point vortices requires calculating N^2 mutual interactions, which can be unaffordable for N on the order of a few hundred or thousand. To circumvent this difficulty, a vortex-in-cell (VIC) method was proposed as a special implementation of the cloud-in-cell method (CIC) [85]. The numerical scheme helps reduce the computational cost by circumventing the direct computation of the mutual interactions.

Given the instantaneous position of the point vortices, the stream function is computed by inverting the Poisson equation

$$\nabla^2 \psi = -\omega_z. \qquad (11.7.2)$$

The velocity at the position of the point vortices arises by numerically differentiating the stream function.

Poisson equation on a grid

The Poisson equation for the stream function can be solved by finite-difference or fast Fourier transform methods based on spectral expansions. Both methods require computing the vorticity at grid points from the position and strength of the point vortices according to an algorithm that preserves the circulation around each cell.

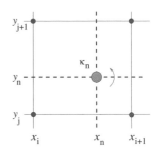

Illustration of a point vortex inside a host cell.

Assume that the nth point vortex lies inside a rectangular cell that is bounded by the vertical grid lines $x = x_i$ and x_{i+1}, and by the horizontal grid lines $y = y_j$ and y_{j+1}. The contribution of that point vortex to the vorticity of adjacent grid points can be computed according to the formulas

$$\omega_{i,j}^{(n)} = \beta\,(x_{i+1} - x_n)\,(y_{j+1} - y_n), \qquad \omega_{i+1,j}^{(n)} = \beta\,(x_n - x_i)\,(y_{j+1} - y_n),$$

$$\omega_{i,j+1}^{(n)} = \beta\,(x_{i+1} - x_n)\,(y_n - y_j), \qquad \omega_{i+1,j+1}^{(n)} = \beta\,(x_n - x_i)\,(y_n - y_j), \qquad (11.7.3)$$

where $\beta = 4\kappa_n/A$, A is the cell area, $\mathbf{x}_n = (x_n, y_n)$ is the position of the nth point vortex, and κ_n is the strength of the nth point vortex [85]. The contribution of the point vortex to all other grid points is zero. Implementations of the vortex-in-cell method for vortex blobs are available [6].

Fast algorithms

A variety of computational techniques have been developed for reducing the $O(N^2)$ operations of the point-vortex method to $O(N \ln N)$ or even $O(N)$ operations. Examples are tree codes and multipole expansions [328]. One particular method involves successively discretizing the domain of flow into smaller rectangles by subdividing a parent rectangle into four smaller rectangles until N rectangles have been formed. Near-neighbors and well-separated rectangles are identified and, for the purpose of computing velocities, point vortices that lie inside well-separated boxes are condensed into central point vortices whose strength is equal to the sum of the strengths of the condensed point vortices.

11.7.2   Viscous effects

A physically appealing method of accounting for viscous or diffusive effects is based on the observation that point particles in a macroscopically stationary fluid execute random motions due to thermal fluctuations. The stochastic displacement of a point particle in a particular direction over a time period, Δt, is a Gaussian distribution with zero mean value and variance equal to $2\kappa\Delta t$, where κ is the particle diffusivity.

Moore (1969) (see [265]) and Chorin [81] observed that, since point vortices and vortex blobs in an inviscid fluid maintain their original strength, viscous diffusion can be emulated by allowing the point vortices or vortex blobs to execute random motion in addition to the deterministic motion due to mutual interactions. The diffusion step displaces the position of the ith point vortex over one time step by $(\Delta x_i, \Delta y_i)$, where $\Delta x_1, \Delta y_1, \ldots, \Delta x_N, \Delta y_N$ is a set of independent random displacement whose probability density function forms a Gaussian distribution with zero mean and variance equal to $2\nu\Delta t$, where ν is the kinematic viscosity. Chorin [81] applied the random vortex method to simulate flow at high Reynolds number past a circular cylinder. Milinazzo & Saffman [265] and Roberts [347] showed that the casual application of the method may lead to significant numerical error due to slow convergence with respect to N. A large number of point vortices is required to ensure statistical equilibrium.

Another way of emulating the effects of viscosity is to employ vortex blobs whose radius increases in time due to diffusion, as discussed in Section 11.4. The lack of random motion justifies calling these vortex methods deterministic (e.g., [328]).

11.7.3 Vorticity generation at a solid boundary

Consider a viscous flow past an impermeable solid boundary where the tangential and normal velocity components are required to vanish. In general, the velocity field induced by the point vortices that arise by discretizing the vorticity field satisfies neither the no-penetration nor the no-slip condition. To annihilate the penetration velocity, we may introduce an appropriate complementary potential flow. However, the sum of the flow induced by the point vortices and the complementary potential flow will still have a finite tangential component, u_t, amounting to a boundary vortex sheet whose strength is equal to $-u_t$. Physically, the boundary vortex sheet represents a viscous boundary layer.

In the case of steady unseparated flow past a streamlined body, the thickness of the boundary layer depends on the Reynolds number as $1/\sqrt{\text{Re}}$. In the case of unsteady flow, vorticity diffuses away from the vortex sheet, it is convected by the ambient flow, and finally enters the bulk of the flow. Chorin [81] proposed modeling these physical processes according to the following steps:

1. Discretize the boundary into a collection of segments with arc length Δl.

2. Compute the tangential velocity, u_t at the midpoint of each segment.

3. Introduce a vortex blob with top-hat vorticity distribution, circulation equal to $u_t\Delta l$, and radius $r_b = \Delta l/(2\pi)$ at the midpoint of each segment.

4. Move the blobs with a velocity that is equal to the sum of (a) the velocity induced by the vorticity of the flow, (b) the velocity due to a potential flow that accounts for the the no-penetration condition, and (c) a random velocity that emulates viscous diffusion. If a vortex blob crosses the boundary, it is removed from the flow.

5. Return to Step 2 and repeat the computation for another time step.

Subtleties arise when we attempt to determine the motion of newly created blobs. An improved version of the method involves discretizing the vortex sheet into elemental vortex patches using the Prandtl boundary layer equations [83, 84]. Refinements and improvements are available [328].

Computer Problems

11.7.1 *Kirchhoff's vortex*

Write a program that computes the motion of an elliptical vortex patch with constant vorticity Ω immersed in an infinite otherwise quiescent fluid using the point-vortex method. The initial strength of the point vortices should be computed by projection onto a rectilinear grid. The algorithm should employ direct summation for computing the mutually induced velocities. The modified Euler method discussed in Section A.8, Appendix B, should be employed for integrating the differential equations. Perform computations for ellipse aspect ratios $a/b = 1.1, 1.5$, and 2, and an increasing number of point vortices, N. Assess the stage where the computational requirements become prohibitive for the available resources.

11.7.2 *Diffusing Kirchhoff's vortex*

Repeat Problem 11.7.1 including viscous effects implemented by random walks. The random displacements should be computed with the help of a random-number generator. Study the motion for ellipse aspect ratios $a/b = 1.0, 1.5$, and 2, and several values of the effective Reynolds number $\mathrm{Re} = \Omega a^2/\nu$. Carry out simulations with an increasing number of point vortices, N, and discuss the results.

11.8   Two-dimensional vortex patches

Two-dimensional vortex patches with uniform vorticity are amenable to analytical and computational studies that reveal the structure and dynamics of regions of concentrated vorticity in an otherwise irrotational flow [357]. In Section 2.13.5, we saw that the flow due to a vortex patch with uniform vorticity Ω can be expressed as a line integral around the patch contour, C. Equation (2.13.23), provides us with the contour integral representation

$$\mathbf{u}(\mathbf{x}) = -\frac{\Omega}{4\pi} \oint_C \ln \frac{\hat{x}^2 + \hat{y}^2}{a^2} \, \mathbf{t}(\mathbf{x}') \, \mathrm{d}l(\mathbf{x}'), \qquad (11.8.1)$$

where $\hat{\mathbf{x}} = \mathbf{x} - \mathbf{x}'$, a is a chosen length, and l is the arc length around the contour measured in the direction of the tangent unit vector, $\mathbf{t}$.

The vorticity transport equation for two-dimensional flow in the absence of viscous forces guarantees that the vorticity of the patch remains constant in time. Combining the contour integral representation with the vorticity transport equation, we find that the dynamics of the flow can be described on geometrical grounds by following the evolution of the vortex contour alone.

11.8.1   The Rankine vortex

The Rankine vortex is a circular vortex patch immersed in an otherwise quiescent infinite fluid, as shown in Figure 11.8.1(a). The fluid inside the vortex rotates like a rigid body about the center of the vortex with polar velocity

$$u_\theta = \frac{1}{2} \Omega r, \qquad (11.8.2)$$

(*a*)   (*b*)

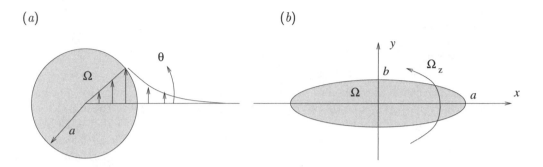

FIGURE 11.8.1   (*a*) Illustration of a Rankine vortex of radius a. The fluid executes rigid-body motion inside the vortex and the velocity decays like that due to a point vortex outside the circular core. (*b*) Illustration of Kirchhoff's elliptical vortex rotating with angular velocity Ω_z.

where r is the distance from the centerpoint. Outside a Rankine vortex of radius a, the flow is identical to that due to a point vortex with strength $\kappa = \Omega \pi a^2$ situated at the center of the vortex, and the polar velocity is

$$u_\theta = \frac{\Omega}{2} \frac{a^2}{r}. \tag{11.8.3}$$

Note that the composite velocity is continuous, but the derivative of the velocity suffers a discontinuity across the vortex contour located at $r = a$.

Small-amplitude waves around the vortex contour

To study the propagation of small-amplitude waves around the Rankine vortex, we displace the contour to the radial position

$$r = a \left[1 + \epsilon \cos(m\theta) \right], \tag{11.8.4}$$

where $\epsilon \ll 1$ is a small dimensionless number, θ is the polar angle measured around the center of the unperturbed vortex, and $m > 1$ is an integer determining the circumferential wave number. In Cartesian coordinates, the vortex contour is described by the equations

$$x = x_c(\theta) = a \left[1 + \epsilon \cos(m\theta) \right] \cos\theta, \qquad y = y_c(\theta) = a \left[1 + \epsilon \cos(m\theta) \right] \sin\theta. \tag{11.8.5}$$

Substituting these expressions into the contour integral representation (11.8.1), we derive an expression for the velocity induced by the deformed patch,

$$\begin{bmatrix} u_x \\ u_y \end{bmatrix}(\mathbf{x}) = -\frac{\Omega a}{4\pi} \int_0^{2\pi} \ln \frac{\left[x - x_c(\theta)\right]^2 + \left[y - y_c(\theta)\right]^2}{a^2} \begin{bmatrix} -\sin\theta + \epsilon \left[\cos(m\theta)\cos\theta\right]' \\ \cos\theta + \epsilon \left[\cos(m\theta)\sin\theta\right]' \end{bmatrix} d\theta, \tag{11.8.6}$$

where a prime denotes a derivative with respect to θ.

Linearization

Next, we evaluate the velocity at a point on the vortex contour located at $\theta = \theta_0$. As a preliminary, we compute the squares of the x and y distances

$$(\Delta x)^2 \equiv \left[x_c(\theta_0) - x_c(\theta)\right]^2 = a^2 \left[\cos\theta_0 - \cos\theta + \epsilon\big(\cos(m\theta_0)\cos\theta_0 - \cos(m\theta)\cos\theta\big)\right]^2 \quad (11.8.7)$$

and

$$(\Delta y)^2 \equiv \left[y_c(\theta_0) - y_c(\theta)\right]^2 = a^2 \left[\sin\theta_0 - \sin\theta + \epsilon\big(\cos(m\theta_0)\sin\theta_0 - \cos(m\theta)\sin\theta\big)\right]^2. \quad (11.8.8)$$

Adding these expressions, linearizing with respect to ϵ, and simplifying, we find that

$$(\Delta x)^2 + (\Delta y)^2 \simeq 2a^2 \left[1 - \cos(\theta_0 - \theta)\right]\big(1 + \epsilon\left[\cos(m\theta_0) + \cos(m\theta)\right]\big). \quad (11.8.9)$$

Substituting this expression into the argument of the logarithm in the contour integral representation (11.8.6) and linearizing, we obtain

$$\begin{bmatrix} u_x \\ u_y \end{bmatrix}(\theta_0) \simeq -\frac{\Omega a}{4\pi} \int_0^{2\pi} \ln[1 - \cos(\theta_0 - \theta)] \begin{bmatrix} -\sin\theta \\ \cos\theta \end{bmatrix} d\theta - \epsilon \frac{\Omega a}{4\pi} \begin{bmatrix} w_x \\ w_y \end{bmatrix}(\theta_0), \quad (11.8.10)$$

where

$$\begin{bmatrix} w_x \\ w_y \end{bmatrix}(\theta_0) = \int_0^{2\pi} \ln[1 - \cos(\theta - \theta_0)] \begin{bmatrix} \cos(m\theta)\cos\theta \\ \cos(m\theta)\sin\theta \end{bmatrix}' d\theta$$

$$+ \int_0^{2\pi} \left[\cos(m\theta_0) + \cos(m\theta)\right] \begin{bmatrix} -\sin\theta \\ \cos\theta \end{bmatrix} d\theta. \quad (11.8.11)$$

The first integral integral on the right-hand side of (11.8.10) expresses the circulatory flow of the perfectly circular vortex with velocity components $u_x = -\frac{1}{2}\Omega a \sin\theta_0$ and $u_y = \frac{1}{2}\Omega a \cos\theta_0$. The second integral on the right-hand side of (11.8.11) is identically zero for any $m > 1$. Accordingly,

$$\begin{bmatrix} u_x \\ u_y \end{bmatrix}(\theta_0) \simeq \frac{\Omega a}{2} \begin{bmatrix} -\sin\theta_0 \\ \cos\theta_0 \end{bmatrix} - \epsilon \frac{\Omega a}{4\pi} \int_0^{2\pi} \ln[1 - \cos w] \begin{bmatrix} \cos[m(w + \theta_0)]\cos(w + \theta_0) \\ \cos[m(w + \theta_0)]\sin(w + \theta_0) \end{bmatrix}' d\theta, \quad (11.8.12)$$

where $w \equiv \theta - \theta_0$ and a prime denotes a derivative with respect to w.

Normal velocity

The velocity component normal to the unperturbed circular contour arises by projecting the velocity given in (11.8.12) onto the radial unit vector, $\mathbf{n} = [\cos\theta_0, \sin\theta_0]$, finding

$$u_n(\theta_0) = -\epsilon \frac{\Omega a}{4\pi} \int_0^{2\pi} \ln[1 - \cos w] \begin{bmatrix} \cos[m(w + \theta_0)]\cos(w + \theta_0) \\ \cos[m(w + \theta_0)]\sin(w + \theta_0) \end{bmatrix}' \cdot \begin{bmatrix} \cos\theta_0 \\ \sin\theta_0 \end{bmatrix} d\theta. \quad (11.8.13)$$

Simplifying the integrand, we obtain

$$u_n(\theta_0) = \epsilon \frac{\Omega a}{4\pi} \int_0^{2\pi} \ln[1 - \cos w] \Big[\cos[m(w + \theta_0)]\sin w + m \sin[m(w + \theta_0)]\cos w\Big] dw, \quad (11.8.14)$$

which can be restated as

$$u_n(\theta_0) = -\epsilon \frac{\Omega a}{4\pi} \int_0^{2\pi} \ln[1 - \cos w] \Big[\cos[m(w + \theta_0)] \cos w \Big]' \, dw \tag{11.8.15}$$

and then

$$u_n(\theta_0) = \epsilon \frac{\Omega a}{4\pi} \sin(m\theta_0) \int_0^{2\pi} \ln[1 - \cos w] \Big[\sin(mw) \cos w \Big]' \, dw. \tag{11.8.16}$$

Integrating by parts, we find that the integral on the right-hand side is equal to 2π when $m > 0$, and -2π when $m < 0$. Thus,

$$u_n(\theta_0) = \epsilon \frac{m}{|m|} \frac{\Omega a}{2} \sin(m\theta_0). \tag{11.8.17}$$

The calculations can be repeated with a sine instead of a cosine on the right-hand side of (11.8.4), resulting in (11.8.17) with a negative cosine instead of a sine on the right-hand side.

Stability

Our analysis has indicated that, if

$$\eta(\theta, t) = a\epsilon \cos[m(\theta - ct)] \tag{11.8.18}$$

is the normal displacement of the boundary velocity, then

$$u_n(\theta, t) = \epsilon \frac{m}{|m|} \frac{\Omega a}{2} \sin[m(\theta - ct)] \tag{11.8.19}$$

is the velocity normal to the circular contour, where t stands for time and c is the phase velocity of the contour wave. Kinematic compatibility requires that

$$\frac{\partial \eta}{\partial t} + \frac{u_\theta}{a} \frac{\partial \eta}{\partial \theta} - u_n = 0, \tag{11.8.20}$$

where $u_\theta = \frac{1}{2}\Omega a$ is the tangential velocity of the unperturbed vortex. Substituting (11.8.18) and (11.8.19) into (11.8.20) and simplifying, we obtain

$$-c + \frac{\Omega}{2} - \frac{1}{|m|} \frac{\Omega}{2} = 0. \tag{11.8.21}$$

Rearranging, we derive an expression for the phase velocity first deduced by Kelvin in 1880 [404],

$$c = \frac{1}{2} \Omega \left(1 - \frac{1}{|m|}\right). \tag{11.8.22}$$

Since c is real, boundary waves rotate with constant amplitude and the Rankine vortex is linearly stable. The phase velocity increases from $c = \frac{1}{4}\Omega$ when $m = 2$ for long waves to $c = \frac{1}{2}\Omega$ as $m = \to \infty$ for short waves.

     We have analyzed the evolution of Kelvin waves around the Rankine vortex based on the contour representation in a way that bypasses the computation of the pressure based on the Euler or Bernoulli equation.

11.8.2   The Kirchhoff elliptical vortex

Kirchhoff discovered that the contour of an elliptical vortex patch, but not the fluid itself, rotates steadily with angular velocity that depends on the patch aspect ratio (e.g., [270, 357]). Consider a horizontal elliptical vortex patch whose contour is described by the equations

$$x_c(\eta) = a \cos \eta, \qquad y_c(\eta) = b \sin \eta, \tag{11.8.23}$$

where a and b are the semiaxes and η is the natural parameter of the ellipse ranging in the interval $[0, 2\pi)$. Without loss of generality, we assume that $a \geq b$, as shown in Figure 11.8.1(b). Substituting (11.8.23) into (2.13.23), we derive an integral representation for the induced velocity,

$$\begin{bmatrix} u_x \\ u_y \end{bmatrix} (\mathbf{x}) = -\frac{\Omega}{4\pi} \int_0^{2\pi} \ln \frac{(x - a\cos\eta)^2 + (y - b\sin\eta)^2}{a^2} \begin{bmatrix} -a\sin\eta \\ b\cos\eta \end{bmatrix} d\eta. \tag{11.8.24}$$

The parameter η should not be confused with the polar angle, θ.

Contour velocity

The velocity at a point on the vortex contour corresponding to $\eta = \eta_0$, located at $x = a\cos\eta_0$ and $y = b\sin\eta_0$, is

$$\begin{bmatrix} u_x \\ u_y \end{bmatrix} (\eta_0) = -\frac{\Omega}{4\pi} \int_0^{2\pi} \ln \left(\frac{\mathcal{R}^2}{a^2}\right) \begin{bmatrix} -a\sin\eta \\ b\cos\eta \end{bmatrix} d\eta, \tag{11.8.25}$$

where

$$\mathcal{R}^2 \equiv a^2 \left(\cos\eta_0 - \cos\eta\right)^2 + b^2 \left(\sin\eta_0 - \sin\eta\right)^2. \tag{11.8.26}$$

Simplifying, we find that

$$\mathcal{R}^2 = \left(a^2 + b^2\right)\left(1 - \kappa \cos\eta_+\right)\left(1 - \cos\eta_-\right), \tag{11.8.27}$$

where

$$\eta_+ = \eta + \eta_0, \qquad \eta_- = \eta - \eta_0, \qquad \kappa = \frac{a^2 - b^2}{a^2 + b^2}. \tag{11.8.28}$$

Substituting (11.8.27) into (11.8.25), we obtain

$$\begin{bmatrix} u_x \\ u_y \end{bmatrix} (\eta_0) = -\frac{\Omega}{4\pi} \left[\int_0^{2\pi} \ln(1 - \kappa \cos\eta_+) \begin{bmatrix} -a\sin(\eta_+ - \eta_0) \\ b\cos(\eta_+ - \eta_0) \end{bmatrix} d\eta_+ \right.$$
$$\left. + \int_0^{2\pi} \ln(1 - \cos\eta_-) \begin{bmatrix} -a\sin(\eta_- + \eta_0) \\ b\cos(\eta_- + \eta_0) \end{bmatrix} d\eta_- \right]. \tag{11.8.29}$$

Simplifying, we obtain

$$\begin{bmatrix} u_x \\ u_y \end{bmatrix} (\eta_0) = -\frac{\Omega}{4\pi} \begin{bmatrix} a\left(A - A_1\right)\sin\eta_0 \\ b\left(A + A_1\right)\cos\eta_0 \end{bmatrix}, \tag{11.8.30}$$

where

$$A(\kappa) = \int_0^{2\pi} \ln(1 - \kappa \cos \eta_+) \, \cos \eta_+ \, d\eta_+ = -2\pi \frac{1 - \sqrt{1 - \kappa^2}}{\kappa} = -2\pi \frac{a^2 - b^2}{(a+b)^2}, \qquad (11.8.31)$$

and $A_1 \equiv A(1) = -2\pi$. Substituting the expressions for A and A_1 into (11.8.30), we obtain

$$\begin{bmatrix} u_x \\ u_y \end{bmatrix} (\eta_0) = \Omega \frac{ab}{a+b} \begin{bmatrix} -\sin \eta_0 \\ \cos \eta_0 \end{bmatrix}. \qquad (11.8.32)$$

It is interesting that the tangential velocity at the tip of the vortex, $\eta_0 = 0$, is the same as that at the midpoint of the flat side, $\eta_0 = \frac{1}{2}\pi$.

Angular velocity of rotation

In a frame of reference rotating around the vortex center with angular velocity Ω_z, the velocity around the vortex contour is

$$\begin{bmatrix} v_x \\ v_y \end{bmatrix} (\eta_0) = \Omega \frac{ab}{a+b} \begin{bmatrix} -\sin \eta_0 \\ \cos \eta_0 \end{bmatrix} - \Omega_z \begin{bmatrix} -b \sin \eta_0 \\ a \cos \eta_0 \end{bmatrix}. \qquad (11.8.33)$$

We note that the vector $[b \cos \eta_0, a \sin \eta_0]$ is normal to the vortex contour at the position η_0, and conclude that the normal velocity, $\mathbf{v} \cdot \mathbf{n}$, will be zero provided that

$$\Omega \frac{ab}{a+b} (a - b) - \Omega_z (a^2 - b^2) = 0, \qquad (11.8.34)$$

yielding

$$\Omega_z = \Omega \frac{ab}{(a+b)^2} \qquad \text{or} \qquad \Omega_z = \frac{\kappa}{\pi} \frac{1}{(a+b)^2}, \qquad (11.8.35)$$

where $\kappa = \Omega \pi ab$ is the circulation around the vortex. We have shown that an elliptical vortex patch rotates intact with angular velocity Ω_z given in (11.8.35). As the aspect ratio a/b increases while the circulation around the vortex is held fixed, the rotating elliptical vortex reduces to a rotating vortex layer with elliptical distribution of circulation along the major axis. The limit $a/b \to \infty$ leads us to a rotating vortex sheet (Problem 11.8.1).

Tangential velocity

Substituting (11.8.35) into (11.8.33), we obtain the velocity around the vortex contour in a frame of reference rotating with the vortex,

$$\begin{bmatrix} v_x \\ v_y \end{bmatrix} (\eta_0) = \Omega_z \begin{bmatrix} -a \sin \eta_0 \\ b \cos \eta_0 \end{bmatrix}. \qquad (11.8.36)$$

The corresponding tangential velocity is

$$v_t(\eta_0) \equiv \frac{1}{J(\eta_0)} \mathbf{v} \cdot \begin{bmatrix} -a \sin \eta_0 \\ b \cos \eta_0 \end{bmatrix} = \Omega_z J(\eta_0), \qquad (11.8.37)$$

where

$$J(\eta) = \left(a^2 \sin^2 \eta + b^2 \cos^2 \eta\right)^{1/2} \tag{11.8.38}$$

is the metric of the arc length with respect to the native parameter η around the elliptical shape.

Stability of the Kirchhoff elliptical vortex

To study the evolution of the vortex contour for small deviations from the elliptical shape, we refer to a frame of reference rotating with the unperturbed vortex and describe the vortex contour as

$$\begin{bmatrix} x_c \\ y_c \end{bmatrix}(\eta) = \begin{bmatrix} a \cos \eta \\ b \sin \eta \end{bmatrix} + \epsilon a q(\eta, t)\, \mathbf{n}, \tag{11.8.39}$$

where ϵ is a small dimensionless number, $q(\eta)$ is a dimensionless deformation function, and $\mathbf{n}$ is the unit vector normal to a horizontal stationary elliptical patch, given by

$$\mathbf{n} = \frac{1}{\sqrt{\mathrm{d}x_c^2 + \mathrm{d}y_c^2}} \begin{bmatrix} \mathrm{d}y_c \\ -\mathrm{d}x_c \end{bmatrix} = \frac{1}{J(\eta)} \begin{bmatrix} b \cos \eta \\ a \sin \eta \end{bmatrix}. \tag{11.8.40}$$

To ensure that the deformation preserves the area of the elliptical patch, we require that

$$\oint q\, \mathrm{d}l = \int_0^{2\pi} q\, J\, \mathrm{d}\eta = 0. \tag{11.8.41}$$

Kinematic consistency requires the linearized condition

$$\epsilon a\, \frac{\partial q}{\partial t} + \epsilon a\, \frac{v_t}{J}\, \frac{\partial q}{\partial \eta} - v_n = 0, \tag{11.8.42}$$

where v_t is the velocity component tangential to the unperturbed contour given in (11.8.37), and v_n is the velocity component normal to the unperturbed contour. Substituting expression (11.8.37) and rearranging, we find that

$$\frac{\partial q}{\partial t} + \Omega_z\, \frac{\partial q}{\partial \eta} - \frac{v_n}{\epsilon a} = 0. \tag{11.8.43}$$

Given a relation between v_n and q, this linear partial differential equation can be integrated in time by numerical methods.

Normal modes

To study normal mode perturbations with exponential dependence in time, we assume that the shape function and normal velocity in the rotating frame behave as

$$q(\eta, t) = Q(\eta)\, \mathrm{e}^{-\mathrm{i}\varrho t}, \qquad v_n(\eta, t) = \epsilon \Omega a\, V_n(\eta)\, \mathrm{e}^{-\mathrm{i}\varrho t}, \tag{11.8.44}$$

where ϱ is the complex growth rate, $Q(\eta)$ is a complex dimensionless eigenfunction, and $V_n(\eta)$ is an accompanying dimensionless complex function. Substituting these expressions into (11.8.43), we obtain

$$-\mathrm{i}\varrho\, Q(\eta) + \Omega_z\, \frac{\mathrm{d}Q}{\mathrm{d}\eta} - \Omega\, V_n(\eta) = 0. \tag{11.8.45}$$

Given a linear relation between V_n and Q, we obtain an eigenvalue problem for the complex phase velocity, ϱ. To ensure the satisfaction of the area-preserving condition (11.8.41), we now set

$$Q = \frac{a}{J} \left(\cos m\eta + \gamma \sin m\eta \right), \qquad (11.8.46)$$

where $m > 1$ is the integer circumferential wave number and γ is an *a priori* unknown coefficient determining the phase of the normal mode with respect to the x axis. Substituting this expression into (11.8.45), we find that

$$-i\varrho\, Q_c(\eta) + \Omega_z \frac{dQ_c}{d\eta} - \Omega \left[V_n(\eta) \right]_{Q=Q_c} = -\gamma \left(-i\varrho\, Q_s(\eta) + \Omega_z \frac{dQ_s}{d\eta} - \Omega \left[V_n(\eta) \right]_{Q=Q_s} \right), \quad (11.8.47)$$

where we have defined

$$Q_c \equiv \frac{a}{J} \cos m\eta, \qquad Q_s \equiv \frac{a}{J} \sin m\eta. \qquad (11.8.48)$$

If V_n is available, equation (11.8.47) can be applied for two different values of η, and the resulting equations can be divided side by side to produce a quadratic algebraic equation for ϱ. Once the roots have been found, the coefficient γ determining the phase shift can be computed by applying (11.8.47) for an arbitrary value η.

Computation of the normal velocity

The contour integral representation produces us with an expression for the velocity field induced by a slightly deformed elliptical vortex in a stationary frame,

$$\begin{bmatrix} u_x \\ u_y \end{bmatrix} (\mathbf{x}) = -\frac{\Omega}{4\pi} \int_0^{2\pi} \ln \frac{\mathcal{R}_\epsilon^2}{a^2} \begin{bmatrix} -a \sin \eta + b\,\epsilon(\psi \cos \eta)' \\ b \cos \eta + a\,\epsilon(\psi \sin \eta)' \end{bmatrix} d\eta, \qquad (11.8.49)$$

where $\psi \equiv aq/J$, a prime denotes a derivative with respect to η, and

$$\mathcal{R}_\epsilon^2 \equiv (x - a \cos \eta - \epsilon b \psi \cos \eta)^2 + (y - b \sin \eta - \epsilon a \psi \sin \eta)^2. \qquad (11.8.50)$$

For a point located at the perturbed vortex contour, $\eta = \eta_0$,

$$\mathcal{R}_\epsilon^2 = \Big(a\,(\cos \eta_0 - \cos \eta) + \epsilon\, b\, [\psi(\eta_0) \cos \eta_0 - \psi(\eta) \cos \eta] \Big)^2$$
$$+ \Big(b\,(\sin \eta_0 - \sin \eta) + \epsilon\, a\, [\psi(\eta_0) \sin \eta_0 - \psi(\eta) \sin \eta] \Big)^2. \qquad (11.8.51)$$

Linearizing with respect to ϵ, we find that

$$\mathcal{R}_\epsilon^2 \simeq \mathcal{R}^2 + \epsilon\, 2ab\, \mathcal{R}_1, \qquad (11.8.52)$$

where $\mathcal{R}$ is given in (11.8.27) and

$$\mathcal{R}_1 = (\cos \eta_0 - \cos \eta)[\psi(\eta_0) \cos \eta_0 - \psi(\eta) \cos \eta]$$
$$+ (\sin \eta_0 - \sin \eta)[\psi(\eta_0) \sin \eta_0 - \psi(\eta) \sin \eta]. \qquad (11.8.53)$$

Substituting (11.8.52) into (11.8.49) and linearizing the logarithmic term, we obtain

$$\begin{bmatrix} u_x \\ u_y \end{bmatrix}(\eta_0) \simeq -\frac{\Omega}{4\pi} \int_0^{2\pi} \left(\ln \frac{\mathcal{R}^2}{a^2} + \epsilon\, 2ab \frac{\mathcal{R}_1}{\mathcal{R}^2} \right) \begin{bmatrix} -a\sin\eta + b\,\epsilon(\psi\cos\eta)' \\ b\cos\eta + a\,\epsilon(\psi\sin\eta)' \end{bmatrix} d\eta. \qquad (11.8.54)$$

Linearizing further, we find that the velocity in a frame of reference rotating with the elliptical vortex is given by

$$\begin{bmatrix} v_x \\ v_y \end{bmatrix}(\eta_0) \simeq \Omega_z \begin{bmatrix} -b\sin\eta_0 \\ a\cos\eta_0 \end{bmatrix} - \frac{\Omega}{4\pi}\int_0^{2\pi} \ln\frac{\mathcal{R}^2}{a^2} \begin{bmatrix} -a\sin\eta \\ b\cos\eta \end{bmatrix} d\eta$$

$$-\epsilon\left(\frac{\Omega}{4\pi}\, 2ab \int_0^{2\pi} \frac{\mathcal{R}_1}{\mathcal{R}^2} \begin{bmatrix} -a\sin\eta \\ b\cos\eta \end{bmatrix} d\eta + \frac{\Omega}{4\pi}\int_0^{2\pi}\ln\frac{\mathcal{R}^2}{a^2} \begin{bmatrix} b\,(\psi\cos\eta)' \\ a\,(\psi\sin\eta)' \end{bmatrix} d\eta + \Omega_z \begin{bmatrix} -a\psi\sin\eta_0 \\ b\psi\cos\eta_0 \end{bmatrix} \right).$$

$$(11.8.55)$$

The velocity component normal to the unperturbed elliptical contour is

$$v_n(\eta_0) \simeq -\epsilon\, \frac{1}{J(\eta_0)}\frac{\Omega}{4\pi} \left(2a^2b^2 \int_0^{2\pi} \frac{\mathcal{R}_1}{\mathcal{R}^2}\sin(\eta_0 - \eta)\, d\eta + \int_0^{2\pi} L(\eta, \eta_0)\ln\frac{\mathcal{R}^2}{a^2}\, d\eta \right), \qquad (11.8.56)$$

where

$$L(\eta, \eta_0) = \begin{bmatrix} b\,(\psi\cos\eta)' \\ a\,(\psi\sin\eta)' \end{bmatrix} \cdot \begin{bmatrix} b\cos\eta_0 \\ a\sin\eta_0 \end{bmatrix}. \qquad (11.8.57)$$

The integrable logarithmic singularity in the second integral on the right-hand side of (11.8.56) can be removed by expressing the integral in the form

$$\int_0^{2\pi} \left(L(\eta, \eta_0)\ln\frac{\mathcal{R}^2}{a^2} - L(\eta_0, \eta_0)\ln[1 - \cos(\eta_0 - \eta_0)] \right) d\eta - 2\pi L(\eta_0, \eta_0)\log 2. \qquad (11.8.58)$$

Because of the periodic integrands, the regularized integral can be computed efficiently by the trapezoidal rule. Machine accuracy can be achieved by using a large number of integration base points.

Comparing (11.8.56) with the second expression in (11.8.44), we deduce the requisite dimensionless function $V_n(\eta)$ in relation to the dimensionless shape function $Q(\eta)$ given in (11.8.46),

$$V_n(\eta_0) \simeq -\frac{1}{4\pi a}\frac{1}{J(\eta_0)} \left(2a^2b^2 \int_0^{2\pi} \frac{\mathcal{R}_1}{\mathcal{R}^2}\sin(\eta_0 - \eta)\, d\eta + \int_0^{2\pi} L(\eta, \eta_0)\ln\frac{\mathcal{R}^2}{a^2}\, d\eta \right). \qquad (11.8.59)$$

The procedure described in the paragraph following equation (11.8.48) may then be applied.

Love waves

Love [246] carried out a classical stability analysis and obtained the complex growth rate

$$\varrho = \pm\frac{\Omega}{2}\left[\left(2m\frac{\Omega_z}{\Omega} - 1 \right)^2 - \left(\frac{a-b}{a+b} \right)^{2m} \right]^{1/2}. \qquad (11.8.60)$$

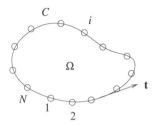

FIGURE 11.8.2   Illustration of a vortex patch with constant vorticity, Ω. The vortex contour, C, is described by N marker points.

Perturbations with $m = 2$ are stable for any aspect ratio a/b; perturbations with $m = 3$ are stable when $a/b < 3$; perturbations with $m > 3$ are stable below higher aspect ratio thresholds [270]. The solution of the quadratic equation for the complex growth rate discussed in the paragraph following equation (11.8.47) is in perfect agreement with Love's analytical predictions.

Numerical linear stability analysis

An algebraic eigenvalue problem originating from (11.8.45) can be formulated based on the contour integral representation, as discussed in Section 9.3.5 in the general context of interfacial flow. First, the elliptical vortex contour is traced with a string of N marker points corresponding to a sequence of parameter values, η_j, where $j = 1, \ldots, N$. The jth point is displaced by a small distance δ normal to the elliptical contour, so that $\epsilon a q_i = 0$ for $i = 1, \ldots, N$, except that $\epsilon a q_j = \delta$, and the deformed contour is reconstructed by cubic-splines interpolation. Next, the derivative $d_{ij} \equiv (\partial q / \partial \eta)_i$ is computed numerically at all points, and the normal velocity $v_{n_i}^{(j)}$ is evaluated from the contour integral representation using the numerical methods discussed in Section 11.8.3, for $i = 1, \ldots, N$. The results are now collected into a matrix

$$A_{ij} = \frac{1}{\delta}\left(v_{n_i}^{(j)} - \Omega_z\, d_{ij}\right), \tag{11.8.61}$$

whose eigenvalues are equal to $-i\varrho$. The corresponding eigenvectors are discrete representations of the shape function $Q(\eta)$. The solution of the eigenvalue problem reproduces Love's eigenvalues and corresponding eigenfunctions. Computer programs are available in directory *kirchhoff* inside directory *09_vortex* of the software library FDLIB (Appendix C).

11.8.3   Contour dynamics

In the method of contour dynamics invented by Zabusky, Hughes, and Roberts [440], the evolution of a vortex patch is followed numerically by computing the motion of the vortex contour. This is done by tracing the vortex contour with N point particles arranged in the counterclockwise direction, as shown in Figure 11.8.2, and then computing the motion of the ith point particle, $\mathbf{X}_i$, using the equation

$$\frac{d\mathbf{X}_i}{dt} = -\frac{\Omega}{4\pi} \oint_C \ln \frac{(X_i - x)^2 + (Y_i - y)^2}{a^2}\, \mathbf{t}(\mathbf{x})\, dl(\mathbf{x}) \tag{11.8.62}$$

for $i = 1, \ldots, N$, where a is a constant length and $\mathbf{t}$ is the tangent unit vector pointing in the counterclockwise direction around the vortex contour [113, 330]. To compute the integral in (11.8.62), we interpolate for the shape of the contour C from the position of the marker points, as discussed in Section B.4, Appendix B.

In the simplest implementation of the method, the vortex contour is approximated with a polygonal line connecting a chain of marker points. To compute the contour integral on the right-hand side of (11.8.62) over a segment that does not contain the ith marker point, $\mathbf{X}_i$, we use a standard numerical method, such as the trapezoidal rule, Simpson's rule, or a Gaussian quadrature, as discussed in Section B.6, Appendix B. However, since the integrand in (11.8.62) exhibits a logarithmic singularity, these methods cannot be applied for the two segments that contain $\mathbf{X}_i$ as an end point. The integration in these two cases can be done by elementary analytical methods [318].

Combining the trapezoidal rule for integration over the nonsingular segments with the analytical integration over the singular segments, we obtain

$$\frac{d\mathbf{X}_i}{dt} = -\frac{\Omega}{8\pi} \sum_{j=1}^{N}{}' (\mathbf{X}_{j+1} - \mathbf{X}_j) \left(\ln \frac{|\mathbf{X}_i - \mathbf{X}_j|^2}{a^2} + \ln \frac{|\mathbf{X}_i - \mathbf{X}_{j+1}|^2}{a^2} \right)$$

$$-\frac{\Omega}{4\pi} \sum_{j=i-1,i} (\mathbf{X}_{j+1} - \mathbf{X}_j) \left(\ln \frac{|\mathbf{X}_{j+1} - \mathbf{X}_j|^2}{a^2} - 2 \right), \qquad (11.8.63)$$

where the prime denotes that the terms $j = i - 1$ and $j = i$ are excluded from the sum. Equation (11.8.63) provides us with a system of N nonlinear, coupled ordinary differential equations for the position of the marker points. The system can be integrated in time using Euler's method, a Runge–Kutta method, or another method, as discussed in Section B.8, Appendix B.

Multiple vortices

When the flow contains a collection of vortex patches with the same or different vorticity, the right-hand side of (11.8.62) contains the sum of integrals over the contours of all patches multiplied by the corresponding values of the vorticity, and equation (11.8.63) undergoes corresponding modifications. Figure 11.8.3 shows the evolution of two initially circular vortex patches with identical vorticity in close proximity computed using a code encapsulated in the software library FDLIB (Appendix C). We observe that the patches coalesce into a larger patch under the action of their mutually induced velocity. Numerical investigation shows that the patches merge only if they are initially closer than a critical distance.

Periodic arrangements

Using the results of Section 2.13.5, we find that the motion of marker points distributed around the contour of a vortex patch that is repeated periodically along the x axis with period a is governed by the modified version of (11.8.62)

$$\frac{d\mathbf{X}_i}{dt} = -\frac{\Omega}{4\pi} \oint_C \ln \left(\cosh[k(Y_i - y)] - \cos[k(X_i - x)] \right) \mathbf{t}(\mathbf{x}) \, dl(\mathbf{x}), \qquad (11.8.64)$$

where C is the contour of one patch and $k = 2\pi/a$ is the wave number.

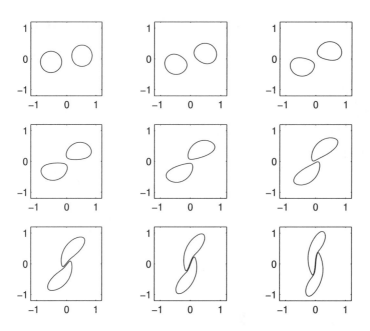

FIGURE 11.8.3   Evolution of two circular vortex patches with equal vorticity. Vortex contours are shown at equal time intervals. The vortex patches are attracted and gradually coalesce into a larger rotating elliptical vortex.

The contour of a patch can be approximated with a polygonal line connecting point particles. To compute the improper integral over the adjacent segments of the ith point particle, S_{i-1} and S_i, we write

$$\int_{S_j} \ln\left(\cosh[k(Y_i - y)] - \cos[k(X_i - x)]\right) \mathbf{t}(\mathbf{x}) \, dl(\mathbf{x}) = \mathcal{I} + \mathcal{J} \tag{11.8.65}$$

for $j = i - 1$ and i, where

$$\mathcal{I} \equiv \int_{S_j} \ln \frac{\cosh[k(Y_i - y)] - \cos[k(X_i - x)]}{k^2(X_i - x)^2 + k^2(Y_i - y)} \mathbf{t}(\mathbf{x}) \, dl(\mathbf{x}) \tag{11.8.66}$$

and

$$\mathcal{J} \equiv \int_{S_j} \ln\left[k^2(X_i - x)^2 + k^2(Y_i - y)^2\right] \mathbf{t}(\mathbf{x}) \, dl(\mathbf{x}). \tag{11.8.67}$$

The nonsingular integral in (11.8.66) can be computed with adequate accuracy using the trapezoidal rule or a Gaussian quadrature. The singularity has been shifted to the integral in (11.8.67), which can be computed by elementary methods.

Vortex layers

An example of an infinite vortex patch is the infinite vortex layer with constant vorticity discussed in Section 9.6. When the boundaries of the vortex layer are parallel, the flow is unidirectional and the velocity varies linearly with distance across the vortex layer and has two different constant values above and below the vortex layer. Figure 9.6.5 shows stages in the instability of a periodically perturbed vortex layer computed using the method of contour dynamics for periodic flow [323, 324].

Problem

11.8.1 *Rotating vortex sheet*

Show that a flat vortex sheet with strength $\gamma = \gamma_0 (1 - x^2/a^2)^{1/2}$ situated in the interval $-a \leq x \leq a$ rotates undeformed. Derive an expression for the angular velocity of rotation of the vortex sheet in terms of the constant γ_0.

 Computer Problems

11.8.2 *Kirchhoff's vortex*

Write a program that computes the motion of an elliptical vortex patch using the contour integral representation. Perform computations for axis ratio $a/b = 1.1, 1.5, 2.0, 3.0$, and 4.0 and verify that the patch rotates as a rigid body. Plot the computed angular velocity of rotation against the aspect ratio and compare the numerical results with the exact solution.

11.8.3 *Interaction of two vortex patches*

Write a program that computes the interaction of two identical circular vortex patches with equal vorticity in close proximity. Carry out computations for three cases where the initial separation of the vortex centers is 2.5, 3.0, and 4.0 the initial vortex radius. Discuss the motion of the patches with reference to vortex merger and contour filamentation. To improve the efficiency of your program, you may wish to exploit the symmetry of the two vortex contours with respect of the midpoint of the line connecting the two vortex centers.

11.9 Axisymmetric flow

Vortex methods for axisymmetric vortex flow are extensions of those for two-dimensional flow discussed earlier in this chapter. However, the curvature of the vortex lines and the occurrence of vortex stretching necessitates essential modifications that demonstrate the subtlety of three-dimensional vortex flows.

11.9.1 Coaxial line vortex rings

The motion of a collection of coaxial line vortex rings is analogous to that of point vortices. The axial velocity of each ring is the sum of the self-induced velocity and the velocities induced by all other rings, as discussed in Section 2.12.1. In the absence of other rings and flow boundaries, an axisymmetric vortex ring translates along the axis of revolution with its own self-induced velocity.

In contrast, an isolated rectilinear line vortex (point vortex) remains stationary. The formation, motion, and stability of vortex rings has attracted a great deal of attention in the scientific and popular literature [367].

If V_i is the self-induced velocity of the ith vortex ring that belongs to a collection of N coaxial rings, the axial position, X_i, and radius, Σ_i, of the ring evolve according to the equations

$$
\begin{aligned}
\frac{\mathrm{d}X_i}{\mathrm{d}t} &= V_i(\Sigma_i) + \sum_{j=1}^{N}{}' \kappa_j\, U_x(X_i - X_j, \Sigma_i, \Sigma_j), \\
\frac{\mathrm{d}\Sigma_i}{\mathrm{d}t} &= \sum_{j=1}^{N}{}' \kappa_j\, U_\sigma(X_i - X_j, \Sigma_i, \Sigma_j),
\end{aligned}
\tag{11.9.1}
$$

where the prime indicates that the term $i = j$ is excluded from the sum,

$$
\left[\begin{array}{c} U_x \\ U_\sigma \end{array}\right](x, \sigma, \sigma') = \frac{1}{4\pi} \left[\begin{array}{c} -\sigma\, I_{31}(x, \sigma, \sigma') + \sigma'\, I_{30}(x, \sigma, \sigma') \\ x\, I_{31}(x, \sigma, \sigma') \end{array}\right],
\tag{11.9.2}
$$

and the integrals I_{nm} are defined in (2.13.14). The vorticity transport equation requires that the strength of each ring, κ_i, remains constant in time.

Self-induced velocity

The divergence of the self-induced velocity of a line vortex ring with infinitesimal core underscores the importance of the core size. To compute the motion of a vortex ring, we may assume a sensible vorticity distribution over the core, and thus transform the vortex ring into an axisymmetric vortex blob. Kelvin's formula for the self-induced velocity of a vortex ring whose vorticity is distributed uniformly over a circular vortex core of radius a is

$$
V = \frac{\kappa}{4\pi\Sigma}\left(\ln\frac{8\Sigma}{a} - \frac{1}{4}\right),
\tag{11.9.3}
$$

where Σ is the ring centerline radius, $\kappa = \Omega\pi a^2$ is the vortex ring strength, and $\omega_\varphi = \Omega$ is the uniform core vorticity ([220], p. 241). Hicks's formula for the self-induced velocity of a vortex ring whose vorticity is concentrated around a vortex sheet of radius a is identical to (11.9.3), except that the factor $1/4$ is replaced by $1/2$ on the right-hand side [175]. For a ring with arbitrary core vorticity distribution, we expect that the self-induced velocity is

$$
V = \frac{\kappa}{4\pi\Sigma}\left(\ln\frac{8\Sigma}{a} - \frac{1}{2} + \alpha\right),
\tag{11.9.4}
$$

where α is a dimensionless constant taking the value of $1/4$ for a Kelvin ring and the value of zero for a Hicks ring (e.g., [125]).

As the radius of a ring, Σ, changes during the motion, the azimuthal vorticity, ω_φ, increases or decreases by the same proportion. In response to this change, the radius of the core, a, adjusts to preserve the ring strength and core volume, so that $\mathrm{d}(\Sigma a^2)/\mathrm{d}t = 0$. Expanding the derivative, we obtain an expression for the rate of change of the radius of the vortex core,

$$
\frac{\mathrm{d}a}{\mathrm{d}t} = -\frac{1}{2}\frac{a}{\Sigma}\frac{\mathrm{d}\Sigma}{\mathrm{d}t}.
\tag{11.9.5}
$$

The motion of the vortex rings is computed by simultaneously integrating in time equations (11.9.1) and (11.9.5) for the axial position, radial position, and radius of the vortex core.

Bounded domains

When a vortex ring resides in a domain that is bounded by an impermeable axisymmetric surface, the induced velocity must be complemented with that due to a potential flow that accounts for the no-penetration boundary condition. When the boundary geometry is sufficiently simple, the complementary flow can be expressed in terms of images of the vortex rings. The image of a vortex ring with respect to a plane wall is another ring with opposite strength placed at the instantaneous mirror-image position. The image of a ring with respect to a sphere was discussed in Section 7.5.

11.9.2   Vortex sheets

Methods for computing the motion of axisymmetric vortex sheets are similar to those for two-dimensional vortex sheets discussed in Section 11.6. The expressions for the self-induced velocity resulting from the Biot–Savart integral involve complete elliptic integrals of the first and second kind that must be evaluated by numerical methods. The evolution of the strength of the vortex sheet is governed by equation (11.5.17). However, the tangential component of the marker-point acceleration in an azimuthal plane is given by an expression that is more complicated than that shown on the right-hand side of (11.5.21). When a vortex sheet separates two fluids with identical densities and the interfacial marker points move with the principal velocity of the vortex sheet, the evolution of the strength of the vortex sheet and circulation along the trace of the vortex sheet in an azimuthal plane are governed by the simplified equations (11.5.38) and (11.5.39).

11.9.3   Vortex patches in axisymmetric flow

The vorticity transport equation for axisymmetric flow in the absence of viscous forces takes the simple form $\mathrm{D}(\omega_\varphi/\sigma)/\mathrm{D}t = 0$, where $\mathrm{D}/\mathrm{D}t$ is the material derivative. This equation shows that, if the azimuthal vorticity component, ω_φ, is equal to $\Omega\sigma$ at a particular instant, it will remain so at all times, where Ω is an arbitrary constant. In this case, computing the evolution of an axisymmetric vortex patch with linear vorticity distribution is reduced to describing the motion of the vortex contour.

Contour dynamics

The numerical implementation of the contour dynamics method for axisymmetric flow is similar to that for two-dimensional flow discussed in Section 11.8.3 (e.g., [302, 330]). Using (2.12.29) and (2.12.37), we find that the axial position, X, and radial position, Σ, of point particles distributed along the vortex contour evolve according to the equation

$$\frac{\mathrm{d}}{\mathrm{d}t}\left[\begin{array}{c} X \\ \Sigma \end{array}\right] = -\frac{\Omega}{4\pi}\int_C \left[\begin{array}{c} \hat{x}\,I_{10}(\hat{x},\Sigma,\sigma)\,n_x(x,\sigma) + \Sigma\,I_{11}(\hat{x},\Sigma,\sigma)\,n_\sigma(x,\sigma)] \\ -\sigma\,I_{11}(\hat{x},\Sigma,\sigma)\,n_x(\mathbf{x}) \end{array}\right]\sigma\,\mathrm{d}l(\mathbf{x}), \qquad (11.9.6)$$

where the integration is performed along the vortex contour, C, $\mathbf{n}$ is the normal unit vector pointing outward from the vortex patch, $\hat{x} = X - x$, and the integrals I_{nm} are defined in (2.13.14). Near the evaluation point, $\mathbf{X}$, the integrand in (11.9.6) behaves as shown in (11.8.62). The computation of the singular integrals over the adjacent segments of a marker point is done by subtracting out the singularity according to (11.8.66) (Problem 11.9.2).

Problems

11.9.1 *Self-induced velocity of a vortex ring with constant vorticity*

Carry out an asymptotic analysis of (2.12.7) to derive (11.9.3).

11.9.2 *Desingularization of contour integrals*

(*a*) Carry out an asymptotic analysis to show that the integrand in (11.9.6) behaves like that shown in (11.8.62).

(*b*) Explain how the singularity can be subtracted out from the integrand according to (11.9.5).

 Computer Problems

11.9.3 *Motion of coaxial line vortex rings*

Write a program that computes the motion of two line vortex rings whose vorticity is distributed uniformly over the core. Run the program to compute the interaction of two initially identical vortex rings for several ratios of the initial core-to-ring radius. Discuss the behavior of the ring doublet.

11.9.4 *A line vortex ring approaching head-on a plane wall or a sphere*

(*a*) Compute the trajectory of a line vortex ring with uniform vorticity distribution approaching a plane wall with its axis perpendicular to the wall.

(*b*) Repeat (*a*) for a vortex ring with initial radius Σ_0 approaching a sphere of radius a centered at the ring axis. Perform a series of computations for $\Sigma_0/a = 0.10$, 0.50, 1.0, and 2.0, and discuss differences in behavior. The location of the image ring is given in equation (7.4.10).

11.10 Three-dimensional flow

Vortex methods for three-dimensional flow are generalizations of those for two-dimensional and axisymmetric flow discussed earlier in this chapter. However, the extensions are not always straightforward and the numerical implementation may lead to conceptual difficulties and significant numerical error.

11.10.1 Line vortices

In Section 2.11.2, we saw that a curved line vortex with infinitesimal cross-section moves with an unphysical infinite self-induced velocity. This singular behavior emphasizes that the centerline curvature and the size of the vortex core plays a critical role in determining the motion of vortex filaments. Computing the self-induced motion of a line vortex with finite vortex core can be done by several approximate methods. The point of departure in all cases is the Biot–Savart integral

(2.11.1), repeated here for convenience,

$$\mathbf{u}(\mathbf{x}) = \frac{\kappa}{4\pi} \int_L \frac{\mathbf{t}(\mathbf{x}') \times \hat{\mathbf{x}}}{r^3} \, dl(\mathbf{x}'),$$ (11.10.1)

where $\hat{\mathbf{x}} = \mathbf{x} - \mathbf{x}'$, $r = |\hat{\mathbf{x}}|$, κ is the strength of the line vortex, and $\mathbf{t}$ is the tangent unit vector along a line vortex, L. Since a line vortex moves with the fluid velocity in the absence of viscous diffusion, the problem is reduced to solving an integro-differential equation for the position of point particles distributed along the line vortex, $d\mathbf{X}/dt = \mathbf{u}(\mathbf{X})$.

Desingularization of the Biot–Savart integral

Rosenhead [348] proposed replacing the Biot–Savart integral (11.10.1) with a modified integral involving a smooth kernel, yielding the velocity field

$$\mathbf{u}(\mathbf{x}) = \frac{\kappa}{4\pi} \int_L \mathbf{t}(\mathbf{x}') \times \frac{\hat{\mathbf{x}}}{(r^2 + \delta^2)^{3/2}} \, dl(\mathbf{x}').$$ (11.10.2)

The magnitude of the size parameter, δ, is small comparable to the size of the vortex core, which is assumed to be constant along the line vortex. A similar desingularization was discussed in Section 11.6 in the context of the point-vortex method for vortex sheets.

Truncation of the Biot–Savart integral

Hama [162, 163] retained the exact form of the Biot–Savart integral (11.10.1) but truncated the line integral to a small distance on either side of the point where the velocity is evaluated. Partial justification for this approximation is provided by the observation that setting the cut off length equal to $b\frac{1}{2} \exp \frac{1}{4}$ reproduces the velocity of a circular vortex ring whose vorticity is distributed uniformly over a core of radius b.

Local-induction approximation (LIA)

In the Da Rios local induction approximation (LIA) discussed in Section 2.11.3, we retain the last term in the asymptotic form of the Biot–Savart (2.11.12) but truncate the limits of integration on either side of the origin at a lower limit that is comparable to the size of the vortex core, b. To leading-order approximation, we obtain

$$\mathbf{u}(\mathbf{X}) = -\frac{\kappa}{4\pi} \, c(\mathbf{X}) \, \mathbf{b}(\mathbf{X}) \ln \frac{b}{a},$$ (11.10.3)

where c is the curvature of the line vortex and a is the truncation length (Da Rios 1906, see [339]; [162, 163]). According to the LIA, the velocity vector at a point on a line vortex is parallel to the local binormal vector, $\mathbf{b}$. The numerical procedure is similar to that used to compute the motion of two-dimensional vortex sheets in terms of point vortices.

11.10.2 Three-dimensional vortex sheets

To compute the motion of a three-dimensional vortex sheet, we introduce two surface curvilinear coordinates, ξ and η, and follow the motion of marker points distributed at the nodes of an interfacial grid labeled by discrete values of ξ and η. The strength of the vortex sheet, $\boldsymbol{\zeta} = \mathbf{n} \times (\mathbf{u}^{(1)} - \mathbf{u}^{(2)})$, is tangential to the instantaneous position of the vortex sheet, where the superscripts (1) and (2) designate evaluation at the upper or lower side of the vortex sheet and the normal unit vector $\mathbf{n}$ points toward the upper side.

The marker points move normal to the vortex sheet with the common normal fluid velocity on either side of the vortex sheet, while executing an arbitrary tangential motion. The velocity of a marker point can be expressed in terms of the principal velocity of the vortex sheet defined as the principal value of the Biot–Savart integral, as discussed in Section 11.5 for two-dimensional flow. To compute the evolution of the strength of the vortex sheet, $\boldsymbol{\zeta}$, we work as in Section 11.5, beginning with Euler's equation on either side of vortex sheet written in the form of (11.5.12). The complexity of the algebraic manipulations in the general case discourages a detailed derivation.

Vortex sheet immersed in a homogeneous fluid

Assume that a vortex sheet separates two fluids with equal densities, ρ, the flow on either side of the vortex sheet is irrotational, and the interfacial marker points move with the principal velocity of the vortex sheet. Introducing the velocity potential on either side of the vortex sheet and using the unsteady Bernoulli equation, we obtain

$$\left(\frac{\partial \phi_i}{\partial t}\right)_{\xi,\eta} - \mathbf{V} \cdot \mathbf{u}^{(i)} + \frac{1}{2} \mathbf{u}^{(i)} \cdot \mathbf{u}^{(i)} + \frac{p_i}{\rho} - \mathbf{g} \cdot \mathbf{x} = c_i(t) \tag{11.10.4}$$

for $i = 1, 2$, where $c_i(t)$ are time-dependent functions. Substituting

$$\mathbf{V} = \frac{1}{2} \left(\mathbf{u}^{(1)} + \mathbf{u}^{(2)}\right), \tag{11.10.5}$$

we obtain

$$\left(\frac{\partial \phi_i}{\partial t}\right)_{\xi,\eta} + \frac{1}{2} \mathbf{u}^{(1)} \cdot \mathbf{u}^{(2)} + \frac{p_i}{\rho} - \mathbf{g} \cdot \mathbf{x} = c_i(t). \tag{11.10.6}$$

Evaluating equation (11.10.6) on either side of the vortex sheet and subtracting the resulting expressions, we obtain

$$\left(\frac{\partial (\phi_1 - \phi_2)}{\partial t}\right)_{\xi,\eta} = -\frac{p_1 - p_2}{\rho} + c_1(t) - c_2(t). \tag{11.10.7}$$

The jump in pressure across the vortex sheet is related to surface tension of the interface represented by the vortex sheet, τ, by $p_1 - p_2 = \tau 2 \kappa_m$, where κ_m is the mean curvature. Equation (11.10.7) shows that, in the absence of surface tension and when $c_1(t) = c_2(t)$, the marker points maintain the difference in the potential across the vortex sheet, $\phi_1 - \phi_2$ [202].

Equation (11.10.7) provides us with a basis for a numerical procedure according to the following steps: advance the position of the marker points with the principal velocity of the vortex sheet; update the jump in the potential $\phi^+ - \phi^-$ according to (11.10.7); compute the surface gradient $\nabla(\phi^+ - \phi^-)$ over the vortex sheet; update the strength of the vortex sheet using the definition $\boldsymbol{\zeta} = \mathbf{n} \times \nabla(\phi^+ - \phi^-)$ [64, 311].

11.10.3   Particle methods

Generalized vortex-particle methods for three-dimensional flow with isolated line vortices or distributed vorticity include the standard vortex-particle method, which is the counterpart of the point-vortex method, the regularized vortex-particle method, which is the counterpart of the vortex-blob method, and the vortex-in-cell method. In the standard vortex particle method, the vorticity field is discretized into a form that is analogous to that shown in (11.7.1),

$$\boldsymbol{\omega}(\mathbf{x}) = \sum_{i=1}^{N} \boldsymbol{\alpha}_i(t)\, \delta_3(\mathbf{x} - \mathbf{x}_i), \qquad (11.10.8)$$

where δ_3 is the three-dimensional delta function. The ith term on the right-hand side of (11.10.8) represents a vortex particle with strength $\boldsymbol{\alpha}_i$ located at the point $\mathbf{x}_i$. The velocity field arises by taking the curl of the vector potential

$$\mathbf{A}(\mathbf{x}) = \frac{1}{4\pi} \sum_{i=1}^{N} \frac{\boldsymbol{\alpha}_i(t)}{\mathbf{x} - \mathbf{x}_i(t)}, \qquad (11.10.9)$$

computed according to (2.10.1). The vortex particles move with the fluid velocity while their strength evolves due to vortex stretching. Using equations (3.11.9) and (3.12.1), we obtain

$$\frac{\mathrm{d}\mathbf{x}_i}{\mathrm{d}t} = \mathbf{u}[\mathbf{x}_i(t), t], \qquad \frac{\mathrm{d}\boldsymbol{\alpha}_i}{\mathrm{d}t} = \boldsymbol{\alpha}_i(t) \cdot [\beta \nabla \mathbf{u} + (1 - \beta)(\nabla \mathbf{u})^T], \qquad (11.10.10)$$

where β is a free parameter. The choice $\beta = 0$ helps preserve the total vorticity of the flow, which is an invariant of the motion [431]. The alternative choice $\beta = \frac{1}{2}$ helps reduce the computational cost.

One difficulty with the representation (11.10.8) is that, in general, the discretized vorticity field and vector potential are not solenoidal. Improvements can be made to rectify this deficiency. Vortex methods for three-dimensional flow are reviewed by Leonard [234], Kino & Ghonien [211], Winckelmans & Leonard [431], and Puckett [328].

 Computer Problem

11.10.1 *Self-induced velocity of a vortex ring*

Write a program that computes the self-induced velocity of a circular vortex ring of radius a based on the regularized Biot–Savart integral (11.10.2). Prepare and discuss a plot of the ring velocity against δ/a.

Finite-difference methods for convection–diffusion

12

Finite-difference methods provide us with a powerful tool for generating numerical solutions to a broad range of partial differential equations. Before we can apply these methods to solving the equations governing fluid flow, we must have available reliable and accurate schemes for solving the convection–diffusion equation for a scalar or vector field. The development of such schemes and the investigation of their performance is the theme of the present chapter. Since the subject of finite-difference methods is broad and diverse, we confine our attention to outlining the fundamental principles and procedures and presenting a selected class of methods that either illustrate the methodology or find extensive applications in numerical practice. Further discussion of methods employed in computational fluid dynamics (CFD) can be found in specialized monographs and texts on numerical methods for partial-differential equations [5, 124, 184, 185, 269, 317, 343, 377].

It is helpful to keep in mind throughout this chapter that the particular way in which the convection–diffusion equation enters a numerical procedure for computing the structure of a steady flow or the evolution of an unsteady incompressible flow depends on the chosen computational strategy. In certain cases, the convection–diffusion equation is integrated with reference to the equation of motion. In other cases, the convection–diffusion equation is integrated with reference to the vorticity transport equation. Examples in each category will be discussed in Chapter 13.

12.1 Definitions and procedures

The most general problem addressed in this chapter is the computation of a generally vector function, $\mathbf{f}$, that satisfies the nonlinear convection–diffusion equation

$$\frac{\partial \mathbf{f}}{\partial t} + \mathbf{u}(\mathbf{x}, t, \mathbf{f}) \cdot \nabla \mathbf{f} = \kappa \nabla^2 \mathbf{f} \qquad (12.1.1)$$

in a specified one-, two-, or three-dimensional solution domain, subject to a given initial condition, $\mathbf{f}(\mathbf{x}, t = 0) = \mathbf{F}(\mathbf{x})$, where $\mathbf{F}(\mathbf{x})$ is a prescribed function. The convection velocity, $\mathbf{u}$, is a known function of position, $\mathbf{x}$, and time t, explicitly, as well as implicitly through its dependence on the solution $\mathbf{f}$. If $\mathbf{u}$ is constant, we obtain a linear convection–diffusion equation. If $\mathbf{u}$ depends on $\mathbf{x}$ and t but is independent of $\mathbf{f}$, we obtain a quasi-linear convection–diffusion equation.

The diffusivity, κ, is assumed to be a uniform throughout the solution domain. When κ is nonzero, equation (12.1.1) is a parabolic differential equation in time. When κ is zero, equation (12.1.1) is a hyperbolic differential equation in time. Later in this chapter, we will see that this seemingly academic classification has important consequences on the effectiveness of the various finite-difference schemes.

To complete the statement of the computational problem, we must introduce a proper number of boundary conditions. When the diffusivity κ is nonzero, the convection–diffusion equation is a second-order partial differential equation and we must supply a number of boundary conditions that matches the dimensionality of the unknown function $\mathbf{f}$. Thus, if $\mathbf{f}$ is a three-dimensional vector, we require three scalar conditions, one for each component of $\mathbf{f}$ over each boundary. Fewer boundary conditions are needed when κ is zero.

12.1.1 Finite-difference grids

The central goal of the finite-difference method is to produce the values of an unknown function at the nodes of a coordinate grid that covers the solution domain at a sequence of discrete time levels separated by a chosen time step, Δt. The finite-difference grid can be defined in Cartesian coordinates, (x, y, z), or any other orthogonal or nonorthogonal curvilinear coordinates, (ξ, η, ζ), as discussed in Sections A.8–A.17, Appendix A. The choice of coordinates is dictated by the geometry of the solution domain and is made with the prime objective of facilitating the implementation of the boundary conditions. For example, when the domain of solution is the exterior or interior of a sphere, the boundary conditions are naturally described in spherical polar coordinates with origin at the center of the sphere and the governing equations are best solved in these coordinates. Orthogonal coordinates are desirable for analytical simplicity and improved numerical stability.

In Cartesian coordinates, the finite-difference grid is comprised of an array of straight lines that run parallel to the x, y, and z axes, with grid spacings Δx, Δy, and Δz. The grid spacings may vary across the solution domain to allow for enhanced spatial resolution at regions where the solution is expected to exhibit sharp variations. The grid becomes finer as the grid spacings become smaller.

Interpolation

Once a discrete finite-difference solution is available, the values of the computed function between grid points and time levels can be obtained by applying standard methods of function interpolation, extrapolation, or approximation, as discussed in Sections B.4 and B.7, Appendix B.

12.1.2 Finite-difference discretization

A distinguishing feature of the finite-difference method is that the temporal and spatial partial derivatives in the governing equations are approximated with finite differences that relate the values of the unknown functions at a group of neighboring grid points at various time levels. This approximation replaces the governing partial differential equation (PDE) with a finite-difference equation (FDE). A compilation of finite-difference approximations to total and partial derivatives can be found in Section B.5, Appendix B. The process of replacing partial derivatives with algebraic differences is called finite-difference approximation or discretization of the differential equation.

Applying the finite-difference equation sequentially at internal grid nodes provides us with a system of linear or nonlinear algebraic equations that relate the values of the unknown functions at the nodes. In certain cases, the solution domain is extended beyond the natural boundaries of the physical problem and the finite-difference equation is applied at boundary nodes for the purpose of accurately implementing boundary conditions involving derivatives.

12.1.3 Consistency

The accuracy of a numerical computation based on a finite-difference method depends on the control parameters of the numerical method, including the grid spacing and time step. Assume that both are reduced simultaneously but arbitrarily so that they may have different orders of magnitude. If the finite-difference equation approximates the partial-differential equation with increasing accuracy in some sensible fashion, then the finite-difference method is consistent. However, if the grid spacing and time step must be reduced simultaneously is special ways for the finite-difference equation to approximate the partial-differential equation with increasing accuracy, then the finite-difference method is only conditionally consistent.

The consistency of a finite-difference equation that arises by applying well established finite-difference formulas to approximate temporal and spatial derivatives in the partial differential equation is guaranteed and does not need to be examined. By contrast, the consistency of a finite-difference equation that arises by heuristic or *ad-hoc* modifications of well-established finite-difference approximations is subject to confirmation.

Modified differential equation

The consistency of a finite-difference method can be assessed by pretending that all variables in the finite-difference equation are continuous functions of space and time, and then expanding them in a Taylor series around a selected grid point at a certain time instant to obtain a new differential equation, called the modified differential equation (MDE) [424]. If, in the limit as the size of the time step and grid spacings are reduced simultaneously but independently the modified differential equation reduces to the original partial differential equation, the finite-difference method is consistent. Phrased differently, if a finite-difference method is consistent, the difference between the MDE and the PDE involves terms that are proportional to powers, but not ratios, of the grid sizes and time step. The exponents of these powers define the order of the numerical error. Examples will be presented in this chapter.

Undetermined coefficients

Certain finite-difference equations emerge by applying the differential equation at a particular grid point, and then replacing it with a combination of values of the unknown function at a group of neighboring grid points at different times. The coefficients that multiply the values of the function are computed by imposing certain restrictions, including consistency with the differential equation and a desired degree of accuracy in approximating the partial derivatives.

12.1.4 Numerical stability

Let us assume that the exact solution of the convection–diffusion equation, or some other partial-differential equation, subject to a given initial condition and a proper number of boundary conditions, does not grow unbounded in time, but either stays constant or decays at every point. It is not unreasonable to demand that the finite-difference solution reproduces this behavior, that is, it provides us with a numerical approximation that is free of oscillations. If it does, the finite-difference method is stable; otherwise the method is unstable. If the exact solution of the differential equation grows in time, the finite-difference method is stable when it produces a numerical solution that grows at a rate that not greater than that of the exact solution.

The stability of relatively simple finite-difference methods for linear partial-differential equations can be assessed by several methods, including the von Neumann stability method, the projection matrix method, and the discrete-perturbation method. The stability of involved finite-difference methods is harder to investigate. In practice, stability is often warranted by the absence of noticeable spatial or temporal oscillations in the results of a computation.

The stability of finite-difference methods for nonlinear differential equations is typically examined by linearizing the differential equation about a particular grid point and then studying the performance of the finite-difference method with reference to the linearized equation, as discussed in Chapter 9 in the context of hydrodynamic stability. The local stability criteria obtained in this manner provide us with an accurate characterization of the overall performance of the numerical scheme.

12.1.5 Convergence

Stability imposes a modest restriction on the numerical method. Before the numerical results can be claimed to bear any degree of physical relevance with respect to the physical problem governed by the original partial-differential equation, we must ensure that, as the size of the grid and time step are made finer, the numerical solution converges to the exact solution.

Lax's equivalence theorem guarantees that, if a numerical solution of a linear partial-differential equation obtained using a consistent finite-difference approximation is stable, then, in the limit as the grid spacings and time step tend to zero, the numerical solution converges to the exact solution [227, 343]. Thus, consistency and stability ensure convergence and *vice versa*. The convergence of finite-difference methods for nonlinear differential equations is harder to assess. Experience has shown that, if the numerical method is consistent and locally stable, the finite-difference solution converges to the exact solution in the limit as the grid spacings and time step are refined.

12.1.6 Conservative form

If the convection velocity field is solenoidal, $\nabla \cdot \mathbf{u} = 0$, equation (12.1.1) can be recast into the equivalent conservative form

$$\frac{\partial \mathbf{f}}{\partial t} + \nabla \cdot \left[\mathbf{u}(\mathbf{f}, \mathbf{x}, t) \otimes \mathbf{f} \right] = \kappa \nabla^2 \mathbf{f}. \tag{12.1.2}$$

By contrast, the primary equation (12.1.1) expresses the nonconservative form. This terminology emphasizes that the components of the matrix $\mathbf{u} \otimes \mathbf{f}$ at the grid points telescope up to the boundaries, and thus conserve possible invariants of the solution in a finite-difference discretization. The conservative form is preferred over the nonconservative form due to improved accuracy and superior numerical stability. In the case of incompressible fluid flow, both the Navier–Stokes equation and the vorticity transport equation can be recast into a conservative form similar to that shown in (12.1.2).

Problems

12.1.1 *Developing finite-difference approximations*

(*a*) The first and second derivatives of a function, $f(x)$, can be approximated by the central difference formulas

$$f'(x) = af(x - h) + bf(x) + cf(x + h), \qquad f''(x) = Af(x - h) + Bf(x) + Cf(x + h), \quad (12.1.3)$$

where h is a small interval. Derive relations among the coefficients a–c and A–C so that the finite-difference approximations are consistent. This means that, in the limit as h tends to zero, the approximations reproduce the exact values of the first and second derivatives at x. Derive additional relations among the coefficients so that the discretization error is of second order in h and solve for the coefficients a–c and A–C.

(*b*) Compute the coefficients a–c and A–C so that the error of the backward or forward difference approximations

$$f'(x) = af(x - 2h) + bf(x - h) + cf(x), \quad f''(x) = Af(x) + Bf(x + h) + Cf(x + 2h), \quad (12.1.4)$$

is of second order in h.

(*c*) Compute the coefficients a–e so that the error of the central difference approximation

$$f'(x) = af(x - 2h) + bf(x - h) + cf(x) + df(x + h) + ef(x + 2h) \qquad (12.1.5)$$

is of fourth order in h.

12.1.2 *Finite-difference formulation for a linear ODE*

Consider the linear ordinary differential equation $f'' + 4f = 0$ in a domain extending between $x = 0$ and $\pi/2$, with boundary conditions $f'(0) = -2$ and $f(\pi/2) = -1$, where a prime indicates a derivative with respect to x.

(*a*) Compute the solution analytically.

(*b*) Discretize the solution domain into N evenly spaced intervals separated by the nodes $x_i = (i - 1)\Delta x$, where $\Delta x = \pi/(2N)$ and $i = 1, \ldots, N + 1$. Apply the differential equation at the ith node and approximate the second derivative using central differences to derive the finite-difference equation

$$f_{i-1} - 2(1 - 2\Delta x^2)f_i + f_{i+1} = 0 \qquad (12.1.6)$$

for $i = 2, \ldots, N$, where $f_i \equiv f(x_i)$. The boundary condition at $x = \pi/2$ requires that $f_{N+1} = -1$.

(*c*) To confirm the consistency of (12.1.6), we regard the grid values f_j as discrete realizations of a twice-differentiable function and evaluate them using a Taylor series expansion about the point x_i, thereby obtaining the corresponding modified differential equation (MDE). Show that, as Δx tends to zero, the MDE reduces to the original ordinary differential equation (ODE).

(*d*) Approximating the first derivative at $x = 0$ with a forward difference, we obtain the discrete boundary condition $f_2 - f_1 = -2\Delta x$. Collect all finite-difference equations into a linear system, $\mathbf{A} \cdot \mathbf{f} = \mathbf{b}$, where $\mathbf{A}$ is a tridiagonal matrix,

$$\mathbf{f} = [\,f_1, f_2, \ldots, f_N\,] \tag{12.1.7}$$

is the N-dimensional solution vector, and $\mathbf{b}$ is a known vector. Present the explicit forms of the matrix $\mathbf{A}$ and vector $\mathbf{b}$.

(*e*) The finite-difference equation (12.1.6), is second-order accurate in Δx, whereas the corresponding finite-difference equation for the boundary condition at $x = 0$ derived in (*d*) is first-order accurate in Δx. This discrepancy somewhat compromises the overall accuracy of the finite-difference method. To obtain a fully consistent second-order method, we extend the solution domain beyond the natural boundary at $x = 0$, introduce a fictitious exterior node, $x_0 = -\Delta x$, apply equation (12.1.6) for $i = 1$, and approximate the derivative with a central difference to obtain the discrete boundary condition $f_2 - f_0 = -4\Delta x$. Derive the corresponding linear system, $\mathbf{B} \cdot \mathbf{f} = \mathbf{c}$, and present the explicit forms of the tridiagonal coefficient matrix $\mathbf{B}$ and known vector $\mathbf{c}$.

12.1.3 *Finite-difference formulation for quasi-linear ODEs*

Repeat Problem 12.1.2(*c*–*e*) for the differential equation $f'' + 4(\cos x + \sin x)f = 0$ in the same domain and with the same boundary conditions.

 Computer Problems

12.1.4 *Finite-difference solution of a linear ODE*

Solve the tridiagonal systems of equations developed in Problem 12.1.2(*d, e*) using the Thomas algorithm discussed in Section B.1.4, Appendix B. Compare the numerical results for discretization levels $N = 2, 4, 8$, and 16 with the exact solution.

12.1.5 *Finite-difference solution of a quasi-linear ODE*

Solve the tridiagonal systems of equations developed in Problem 12.1.3 using the Thomas algorithm for discretization levels $N = 2, 4, 8$ and 16. Discuss the accuracy of the numerical solution.

12.2   One-dimensional diffusion

We begin by developing finite-difference methods for the one-dimensional scalar unsteady diffusion equation

$$\frac{\partial f}{\partial t} = \kappa \frac{\partial^2 f}{\partial x^2}, \tag{12.2.1}$$

which is a special case of the more general convection–diffusion equation (12.1.1), subject to the initial condition $f(x, 0) = F(x)$, where $F(x)$ is a known function. For simplicity, we assume that the solution domain extends over the entire x axis and stipulate the homogeneous far-field conditions $f(x = \pm\infty, t) = 0$. If the domain of solution were bounded in the interval $a \leq x \leq b$, we would have to require one boundary condition for f or $\partial f/\partial x$ at both ends, $x = a, b$, or boundary conditions for both f and $\partial f/\partial x$ at one end.

12.2.1   Exact solution in an infinite domain

When the solution domain extends over the entire x axis, the exact solution can be found in integral form in terms of the Green's function of the unsteady diffusion equation in one dimension, denoted by $\mathcal{G}(x, x_0, t, t_0)$, satisfying

$$\frac{\partial \mathcal{G}}{\partial t} = \kappa \frac{\partial^2 \mathcal{G}}{\partial x^2} + \delta_1(x - x_0)\,\delta(t - t_0), \tag{12.2.2}$$

where δ_1 is the one-dimensional delta function (see also Section 6.17). Physically, the Green's function represents the diffusing field due to a impulsive point source with unit strength applied at a point, x_0, at time t_0. Using Fourier transforms, we obtain the evolving Gaussian distribution

$$\mathcal{G}(x - x_0, t - t_0) = \frac{1}{\left[4\pi\kappa(t - t_0)\right]^{1/2}} \exp\left[-\frac{1}{4\kappa}\frac{(x - x_0)^2}{t - t_0}\right] \tag{12.2.3}$$

for $-\infty < x < \infty$ and $t > t_0$ (e.g., [66], p. 53). At the initiation instant, $t = t_0$, the Green's function is the one-dimensional delta function.

In terms of the Green's function, the solution of the unsteady diffusion equation (12.2.1) is constructed by superposition as

$$f(x, t) = \int_{-\infty}^{\infty} F(x + \varrho)\,\mathcal{G}(\varrho, t)\,\mathrm{d}\varrho, \tag{12.2.4}$$

where the integration variable, ϱ, is the distance from the evaluation point, x. Substituting the Green's function, we obtain the exact solution

$$f(x, t) = \frac{1}{\sqrt{4\pi\kappa t}} \int_{-\infty}^{\infty} F(x + \varrho) \exp\left(-\frac{\varrho^2}{4\kappa t}\right) \mathrm{d}\varrho \tag{12.2.5}$$

for $t > 0$. The integral on the right-hand side can be computed numerically using the Gauss-Hermite quadrature [317].

12.2.2   Finite-difference grid

Our objective is to generate the discrete version of the exact solution given in (12.2.5) using a finite-difference method. We will assume that the function $f(x, t)$ is and remains infinitesimal outside a sufficiently wide computational domain, $a \leq x \leq b$, during a certain initial period of evolution. As a first step toward implementing the finite-difference method, we introduce a two-dimensional grid that covers a semi-infinite strip in the space-time domain, $a \leq x \leq b$ and $0 \leq t < \infty$, as illustrated in Figure 12.2.1(a).

The goal of the finite-difference method is to provide us with the values of the function, f_i^n, at the grid points, x_i, where $i = 1, \ldots, N + 1$, at a sequence of successive time levels, t_n, beginning at the initial time level, $t_0 = 0$, subject to the boundary conditions $f_1^n = 0$ and $f_{N+1}^n = 0$. By definition,

$$f_i^n \equiv f(x_i, t_n). \tag{12.2.6}$$

The initial condition specifies that $f_i^0 = F(x_i)$.

(a) $\qquad\qquad\qquad\qquad\qquad\qquad\qquad\qquad$ (b)

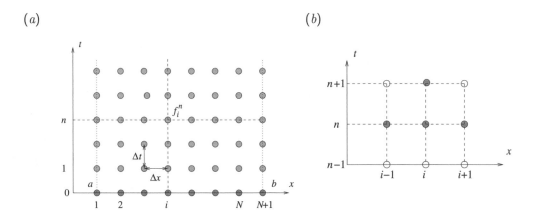

FIGURE 12.2.1   (a) Discretization of the space-time domain for solving the one-dimensional diffusion equation. The solution is advanced forward in time from the initial level, $t = 0$. (b) Computational stencil of the forward-time, centered-space (FTCS) discretization.

12.2.3   Explicit FTCS method

Applying (12.2.1) at the x_i grid point at the time instant t_n, and approximating the time derivative using a forward difference (FT) and the space derivative using a central difference (CS), we derive the FTCS finite-difference equation with first-order accuracy in time and second-order accuracy in space,

$$\frac{f_i^n - f_i^{n+1}}{\Delta t} + O(\Delta t) = \kappa \frac{f_{i-1}^n - 2f_i^n + f_{i+1}^n}{\Delta x^2} + O(\Delta x^2). \qquad (12.2.7)$$

The FTCS differentiation stencil is indicated with filled circles in Figure 12.2.1(b). Solving (12.2.7) for f_i^{n+1}, we obtain

$$f_i^{n+1} = \alpha f_{i-1}^n + (1 - 2\alpha) f_i^n + \alpha f_{i+1}^n, \qquad (12.2.8)$$

where

$$\alpha \equiv \frac{\kappa \Delta t}{\Delta x^2} \qquad (12.2.9)$$

is a positive dimensionless constant called the diffusion number. Equations (12.2.7) and (12.2.8) can be applied at the internal grid points, $i = 2, \dots, N$, but not at the boundary points, $i = 1, N + 1$.

Equation (12.2.8) provides us with a straightforward algorithm for computing the value of f at a grid point at the time level $n + 1$ in terms of the values of f at three grid points at the previous time level, n. Since the algorithm does not require solving a system of algebraic equations, it is classified as explicit. In summary, the FTCS discretization provides us with an explicit two-level method with first-order accuracy in time and second-order accuracy in space.

Consistency

To confirm the consistency of the difference equation (12.2.8), we regard all discrete variables as continuous functions of space and time, expand them in Taylor series about the point (x_i, t_n), and simplify to obtain the associated modified differential equation

$$f_t + \frac{1}{2} f_{tt} \Delta t + O(\Delta t^2) = \kappa f_{xx} + \frac{1}{12} \kappa f_{xxxx} \Delta x^2 + O(\Delta x^4), \qquad (12.2.10)$$

where subscripts denote partial derivatives with respect to corresponding variables, and all variables are evaluated at x_i and t_n. Since, in the limit as Δt and Δx tend to zero independently, the modified differential equation (12.2.10) reduces to the partial differential equation (12.2.1), the FTCS discretization is consistent.

Differentiating the governing equation (12.2.1) once with respect to t and twice with respect to x, and combining the resulting equations, we derive a fourth-order differential equation,

$$f_{tt} = \kappa^2 f_{xxxx}. \qquad (12.2.11)$$

Eliminating f_{xxxx} on the right-hand side of (12.2.10) in favor of f_{tt} and combining the resulting expression with the second term on the left-hand side, we find that, when $\alpha = 1/6$, the accuracy of the FTCS discretization becomes of second order in both space and time.

Successive mapping

To formalize the action of the FTCS method, we collect the grid values of f_i^n at the grid points $i = 2, \ldots, N$ into a vector, $\mathbf{f}^n$, and use (12.2.8) in conjunction with the Dirichlet boundary conditions $f_1^n = 0$ and $f_{N+1}^n = 0$ to obtain a linear system,

$$\mathbf{f}^{n+1} = \mathbf{B} \cdot \mathbf{f}^n, \qquad (12.2.12)$$

where $\mathbf{B}$ is a $(N-1) \times (N-1)$ tridiagonal matrix with superdiagonal, diagonal, and subdiagonal elements equal to α, $1-2\alpha$, and α. This matrix form shows that the solution vector, $\mathbf{f}^{n+1}$, arises by projecting the vector $\mathbf{f}^n$ onto the matrix $\mathbf{B}$, and thereby establishes a relation between time stepping and successive mapping in vector space.

Whether the vector $\mathbf{f}^n$ will amplify or shrink during the successive mappings depends on the spectral radius of the projection matrix, $\mathbf{B}$, denoted by $\varrho(\mathbf{B})$ and defined as the maximum magnitudes of the eigenvalues of $\mathbf{B}$. The theory of matrix calculus instructs us that, if $\varrho(\mathbf{B})$ is equal to unity, less than unity, or larger than unity, the length of $\mathbf{f}^n$ will stay constant, decrease, or increase during the successive mappings (e.g., [317]). Since, according to the exact solution, the magnitude of f decays due to diffusion at long times, we tolerate the first behavior, welcome the second behavior, and dismiss the third behavior as numerically unstable.

To compute the eigenvalues of the matrix $\mathbf{B}$, we write $\mathbf{B} = \alpha \mathbf{C} + \mathbf{I}$, where $\mathbf{C}$ is a tridiagonal matrix with super-diagonal, diagonal, and subdiagonal elements equal to 1, -2, and 1, and $\mathbf{I}$ is the identity matrix. The eigenvalues of $\mathbf{B}$ and $\mathbf{C}$, denoted by $\lambda(\mathbf{B})$ and $\lambda(\mathbf{C})$, are related by $\lambda(\mathbf{B}) = \alpha\lambda(\mathbf{C}) + 1$. A detailed computation shows that

$$\lambda_m(\mathbf{B}) = 1 - 4\alpha \cos^2\left(\frac{m\pi}{2N}\right) \qquad (12.2.13)$$

for $m = 1, \ldots, N - 1$ (e.g., [5] p. 57; [317]). This expression demonstrates that the spectral radius of $\mathbf{B}$ is less than unity only when $\alpha < \frac{1}{2}$, and the FTCS method is conditionally stable.

Effect of the boundary conditions

If boundary conditions other than the homogeneous Dirichlet conditions $f_1^n = 0$ and $f_{N+1}^n = 0$ are specified at one end or both ends of the computational domain, the mapping matrix $\mathbf{B}$ will be slightly altered. However, the performance of the finite-difference method will still be determined by the spectral radius, $\varrho(\mathbf{B})$. As an example, we impose the Neumann boundary condition at the left end, $\partial f/\partial x = q$ at $x = a$, and retain the homogeneous Dirichlet condition at the right end, $x = b$. To implement the left-end boundary condition with second-order accuracy, we extend the solution domain beyond the physical boundary, $x = a$, introduce a fictitious node, $x_0 = x_1 - \Delta x$, and use a central difference to obtain $f_2^n - f_0^n = 2q\Delta x$. Having extended the solution domain, we may apply the differential equation at the first node, x_1, and write (12.2.8) for $i = 1$.

To obtain the corresponding projection matrix, $\mathbf{B}$, we collect the values of f_i^n at the grid points $i = 1, \ldots, N$ into a vector, $\mathbf{f}^n$, and combine (12.2.8) with the discretized Neumann boundary condition to obtain a linear system, $\mathbf{f}^{n+1} = \mathbf{B} \cdot \mathbf{f}^n + \mathbf{b}$, where $\mathbf{B}$ is an $N \times N$ tridiagonal matrix with superdiagonal, diagonal, and subdiagonal elements respectively equal to α, $1 - 2\alpha$, and α, except that the second entry of the first row is equal to 2α. All entries of the vector $\mathbf{b}$ are zero, except for the first entry that is equal to $-2q\Delta x$. The presence of the vector $\mathbf{b}$ does not affect the significance of the projection matrix $\mathbf{B}$ regarding the behavior of the numerical solution. When $\varrho(\mathbf{B})$ is equal to unity, less than unity, or larger than unity, the length of $\mathbf{f}^n$ will stay constant, decrease, or increase during the successive mappings. Unfortunately, the eigenvalues of $\mathbf{B}$ are not known in analytical form.

Von Neumann stability analysis

The simple structure of the projection matrix associated with the FTCS method subject to Dirichlet boundary condition at both ends allowed us to compute its eigenvalues and spectral radius exactly, and thereby assess the stability of the numerical method without any approximations. Analytical expressions for the eigenvalues of the projection matrix may not be available for more advanced finite-difference discretizations and more general boundary conditions. We may certainly compute the eigenvalues using a numerical method, but this can be an arduous computational task.

Another way to assess the stability of the numerical method is by performing the von Neumann stability analysis of the finite-difference equation, neglecting the boundary conditions. The basic idea is to examine the behavior of the numerical solution subject to a sinusoidal initial condition with a prescribed wavelength, L. Motivated by the linearity of the governing equation, we separate the temporal from the spatial dependencies by writing

$$f_i^n = A^n \exp(\mathrm{i}\, i\theta), \tag{12.2.14}$$

where i is the imaginary unit, $\theta = 2\pi\Delta x/L$ is the phase angle, and A^n is a constant coefficient dependent on the time level, n. Note that the superscript n is a time index, not an exponent.

Substituting (12.2.14) into (12.2.8) and simplifying, we find that

$$\frac{A^{n+1}}{A^n} \equiv G = 1 - 2\alpha\,(1 - \cos\theta) = 1 - 4\alpha\sin^2\frac{\theta}{2}, \qquad (12.2.15)$$

where G is the growth factor, also called the gain or amplification factor. When $\alpha > \frac{1}{2}$, the magnitude of the right-hand side of (12.2.15) is greater than unity in a certain range of θ and the numerical method is unstable. When $\alpha < \frac{1}{2}$, the magnitude of the right-hand side of (12.2.15) is less than unity for any value of θ and the numerical method is stable. These results are consistent with our previous conclusions based on the spectral radius of the projection matrix, $\mathbf{B}$.

The efficiency of the von Neumann stability analysis is now apparent. One limitation of the method is that, in its simple form described in this section, it does not incorporate the effect of general boundary conditions, which may have a destabilizing influence on the finite-difference method.

Assessment of the FTCS method

Since the diffusion number α is proportional to the temporal step, Δt, and inversely proportional to the square of the spatial step, Δx^2, the stability constraint $\alpha < \frac{1}{2}$ of the FTCS method requires a time step that is excessively small and may require a prohibitive computational cost. The low-order accuracy combined with the conditional stability renders the FTCS method less attractive compared to its alternatives.

12.2.4   Explicit CTCS or Leap-Frog method

To achieve second-order accuracy in both time and space, we may use central differences in time and space. Applying (12.2.1) at the point x_i at the time instant t_n, we obtain the CTCS difference equation

$$\frac{f_i^{n+1} - f_i^{n-1}}{2\Delta t} + O(\Delta t^2) = \kappa\,\frac{f_{i-1}^n - 2f_i^n + f_{i+1}^n}{\Delta x^2} + O(\Delta x^2). \qquad (12.2.16)$$

Rearranging, we derive the three-time-level explicit algorithm

$$f_i^{n+1} = f_i^{n-1} + 2\alpha f_{i-1}^n - 4\alpha f_i^n + 2\alpha f_{i+1}^n. \qquad (12.2.17)$$

The solution at the first time level, $n = 1$, must be computed using a two-level method, such as the FTCS method discussed in Section 12.2.1, with a time step that is small enough to prevent the onset of deleterious oscillations.

To examine the stability of the method, we substitute (12.2.14) into (12.2.17), set

$$\frac{A^{n+1}}{A^n} = \frac{A^n}{A^{n-1}} = G, \qquad (12.2.18)$$

and obtain a quadratic equation for the gain, $G^2 + \beta G - 1 = 0$, where $\beta = 2\alpha(1 - \cos\theta) = 4\alpha\sin^2(\theta/2)$ is a real non-negative parameter. The solution is

$$G = \frac{1}{2}\left[-\beta \pm (\beta^2 + 4)^{1/2}\right]. \qquad (12.2.19)$$

Since the magnitude of the root corresponding to the minus sign is higher than unity, the CTCS method is unconditionally unstable and thus of no practical value.

12.2.5   Explicit Du Fort–Frankel method

Du Fort and Frankel [116] proposed a modification of the CTCS discretization to ensure second-order accuracy while improving numerical stability. The idea is to replace the middle term in the numerator on the right-hand side of (12.2.16) with an average value, yielding

$$\frac{f_i^{n+1} - f_i^{n-1}}{2\Delta t} + O(\Delta t^2) = \kappa \, \frac{f_{i-1}^n - (f_i^{n+1} + f_i^{n-1}) + f_{i+1}^n}{\Delta x^2} + O(\Delta x^2). \tag{12.2.20}$$

Rearranging, we obtain a three-level explicit algorithm expressed by the difference equation

$$f_i^{n+1} = \frac{1 - 2\alpha}{1 + 2\alpha} \, f_i^{n-1} + \frac{2\alpha}{1 + 2\alpha} \, (f_{i-1}^n + f_{i-1}^n). \tag{12.2.21}$$

The computations must be started using a two-level method.

Performing the von Neumann stability analysis, we find that the amplification factor satisfies the quadratic equation

$$(1 + 2\alpha) \, G^2 - 4\alpha \cos\theta \, G - 1 + 2\alpha = 0. \tag{12.2.22}$$

Examining the roots, we find that $|G| < 1$ for any value of α, which ensures that the Du Fort–Frankel method is unconditionally stable.

Consistency

Since (12.2.20) was derived by an *ad-hoc* modification of the well-founded CTCS discretization, its consistency must be examined by comparing the associated modified differential equation (MDE) with the given differential equation (12.2.1). To derive the MDE, we regard all discrete variables in (12.2.20) as continuous functions of space and time, expand them in Taylor series about (x_i, t_n), and simplify to obtain

$$f_t = \kappa \Big[f_{xx} - \Big(\frac{\Delta t}{\Delta x}\Big)^2 f_{tt} \Big]. \tag{12.2.23}$$

We note that equation (12.2.23) is an accurate approximation of (12.2.1) only if the ratio $(\Delta t/\Delta x)^2$ is sufficiently small, which requires that Δt is sufficiently smaller than Δx in appropriate units. This restriction is milder than that arising from a stability constraint on the diffusion number, α. We observe that (12.2.23) is classified as a hyperbolic differential equation due to the presence of the second derivative with respect to time, whereas (12.2.1) is classified as a parabolic differential equation, and conclude that adding a term with a wave-like character has a stabilizing influence on the numerical method.

Because of its advantages regarding numerical stability, the Du Fort–Frankel method enjoyed widespread popularity in early numerical implementations. When using this method, care must be taken so that the ratio $(\Delta t/\Delta x)^2$ is sufficiently small, otherwise the results will not be physically meaningful.

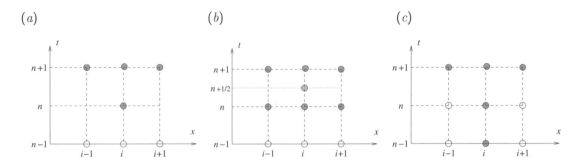

FIGURE 12.2.2   (*a*) Computational stencil of the BTCS discretization; the solution at three grid points at the new time level is computed in terms of one current value. (*b*) Computational stencil of the Crank–Nicolson discretization; the solution at three grid points at the new time level is computed in terms of three current values. (*c*) Computational stencil of a three-level implicit method; the solution at the new time level is computed in terms of values at two previous time levels.

12.2.6   Implicit BTCS or Laasonen method

Thus far, we have considered explicit methods where the solution at a particular time level is computed directly from the solution at one or two previous time levels without solving a system of equations. Now we turn to considering implicit discretizations that require solving a system of equations in the hope of gaining unconditional stability while ensuring consistency and thus relaxing the restriction on the time step.

Applying (12.2.1) at the point x_i at the time instant t_{n+1}, and approximating the time derivative with a backward difference and the space derivative with the central difference, we obtain the BTCS difference equation

$$\frac{f_i^{n+1} - f_i^n}{\Delta t} + O(\Delta t) = \kappa \frac{f_{i-1}^{n+1} - 2f_i^{n+1} + f_{i+1}^{n+1}}{\Delta x^2} + O(\Delta x^2). \tag{12.2.24}$$

The corresponding finite-difference stencil is shown with filled circles in Figure 12.2.2(*a*). Rearranging (12.2.24), we derive the two-level implicit algorithm expressed by the equation

$$-\alpha f_{i-1}^{n+1} + (1 + 2\alpha)f_i^{n+1} - \alpha f_{i+1}^{n+1} = f_i^n, \tag{12.2.25}$$

where α is the diffusion number defined in (12.2.9). The presence of three unknown nodal values on the left-hand side renders the BTCS method implicit.

Recasting (12.2.25) into a matrix form and implementing homogeneous Dirichlet boundary conditions, we obtain a system of linear equations, $\mathbf{A} \cdot \mathbf{f}^{n+1} = \mathbf{f}^n$, where $\mathbf{A}$ is an $(N-1) \times (N-1)$ tridiagonal matrix with superdiagonal, diagonal, and subdiagonal elements equal to $-\alpha$, $1 + 2\alpha$, and $-\alpha$. Solving for $\mathbf{f}^{n+1}$ yields $\mathbf{f}^{n+1} = \mathbf{A}^{-1} \cdot \mathbf{f}^n$, where $\mathbf{A}^{-1}$ is the inverse of $\mathbf{A}$. This expression shows that stepping in time is tantamount to successively mapping the initial vector $\mathbf{f}^0$ with the projection matrix $\mathbf{A}^{-1}$. In practice, in order to compute the solution at the $n+1$ time level, we solve a system of linear algebraic equations, $\mathbf{A} \cdot \mathbf{f}^{n+1} = \mathbf{f}^n$, which renders the BTCS method implicit. Since the coefficient matrix $\mathbf{A}$ is tridiagonal and diagonally dominant, the linear system can be solved efficiently using the Thomas algorithm discussed in Section B.1.4, Appendix B.

To study the stability of the BTCS method, we consider the eigenvalues of the projection matrix $\mathbf{A}^{-1} = (-\alpha \mathbf{C} + \mathbf{I})^{-1}$, where $\mathbf{C}$ is a tridiagonal matrix with superdiagonal, diagonal, and subdiagonal elements equal to 1, -2, and 1. Recalling that the eigenvalues of the inverse of a matrix are equal to the inverse of the eigenvalues, and using (12.2.13), we obtain

$$\lambda_m = \frac{1}{1 + 4\alpha \sin^2 \frac{m\pi}{2N}}, \tag{12.2.26}$$

where $m = 1, \ldots, N - 1$. Since the spectral radius of the projection matrix is less than unity, the method is unconditionally stable. An independent way of arriving at this result is by performing the von Neumann stability analysis, obtaining the amplification factor

$$G = \frac{1}{1 + 2\alpha(1 - \cos\theta)} = \frac{1}{1 + 4\alpha \sin^2 \frac{\theta}{2}}, \tag{12.2.27}$$

whose magnitude is less than unity for any value of α or θ.

The low-order temporal accuracy of the BTCS method places a restriction on the size of the time step for an accurate solution.

12.2.7 Implicit Crank–Nicolson method

Continuing our search for an efficient method, we target an algorithm that is unconditionally stable and second-order accurate in both time and space. The explicit FTCS method emerged by applying the diffusion equation (12.2.1) at the point x_i at time level t_n, whereas the implicit BTCS method emerged by applying (12.2.1) at the point x_i at time level t_{n+1}. Applying (12.2.1) at the intermediate grid point, $(x_i, t_{n+1/2})$, located halfway between the grid points (x_i, t_n) and (x_i, t_{n+1}), and setting the spatial derivative at the $t_{n+1/2}$ level equal to the average of the spatial derivatives at the t_n and t_{n+1} levels, we derive the Crank–Nicolson [97] finite-difference equation

$$\frac{f_i^{n+1} - f_i^n}{\Delta t} + O(\Delta t^2) = \frac{\kappa}{2} \left(\frac{f_{i-1}^{n+1} - 2f_i^{n+1} + f_{i+1}^{n+1}}{\Delta x^2} + \frac{f_{i-1}^n - 2f_i^n + f_{i+1}^n}{\Delta x^2} \right) + O(\Delta x^2). \tag{12.2.28}$$

The corresponding finite-difference stencil is shown with filled circles in Figure 12.2.2(*b*). Rearranging (12.2.28), we obtain a tridiagonal system of equations,

$$-\alpha f_{i-1}^{n+1} + 2\,(1 + \alpha) f_i^{n+1} - \alpha f_{i+1}^{n+1} = \alpha f_{i-1}^n + 2\,(1 - \alpha) f_i^n - \alpha f_{i+1}^n. \tag{12.2.29}$$

Deriving and examining the corresponding modified differential equation shows that the Crank–Nicolson method is consistent and second-order accurate in both time and space.

Stability

Recasting (12.2.29) into a matrix form, we obtain a system of linear equations, $\mathbf{A} \cdot \mathbf{f}^{n+1} = \mathbf{B} \cdot \mathbf{f}^n$, where $\mathbf{A}$ is a tridiagonal matrix with superdiagonal, diagonal, and subdiagonal elements equal to

$-\alpha$, $2(1 + \alpha)$, and $-\alpha$, and $\mathbf{B}$ is another tridiagonal matrix with superdiagonal, diagonal, and subdiagonal elements equal to α, $2(1 - \alpha)$, and α. Solving for $\mathbf{f}^{n+1}$, we find that

$$\mathbf{f}^{n+1} = \mathbf{A}^{-1} \cdot \mathbf{B} \cdot \mathbf{f}^n, \qquad (12.2.30)$$

which shows that stepping in time is equivalent to mapping successively with the projection matrix $\mathbf{P} = \mathbf{A}^{-1} \cdot \mathbf{B}$. It can be shown that the spectral radius of the projection matrix is always less than unity, thus ensuring that the Crank–Nicolson method is unconditionally stable (Problem 12.2.2). Carrying out the von Neumann stability analysis, we derive the amplification factor

$$G = \frac{1 - \alpha\,(1 - \cos\theta)}{1 + \alpha\,(1 - \cos\theta)} = \frac{1 - 2\alpha\sin^2(\theta/2)}{1 + 2\alpha\sin^2(\theta/2)}, \qquad (12.2.31)$$

which is always less than unity, confirming that the method is unconditionally stable.

Because of its remarkable properties regarding accuracy and stability, the Crank–Nicolson method has become a standard choice in practice.

Multiple substeps

It is instructive to regard the Crank–Nicolson method as the implementation of two substeps, where each substep lasts for time interval $\frac{1}{2}\Delta t$. The first substep is carried out using the explicit FTCS method, while the second substep is carried out using the implicit BTCS method according to the finite-difference equations

$$\frac{f_i^{n+1/2} - f_i^n}{\frac{1}{2}\Delta t} = \kappa\,\frac{f_{i-1}^n - 2f_i^n + f_{i+1}^n}{\Delta x^2}, \qquad \frac{f_i^{n+1} - f_i^{n+1/2}}{\frac{1}{2}\Delta t} = \kappa\,\frac{f_{i-1}^{n+1} - 2f_i^{n+1} + f_{i+1}^{n+1}}{\Delta x^2}. \quad (12.2.32)$$

Adding these equations to eliminate the intermediate variables at time $t_{n+1/2}$ reproduces (12.2.28). The unconditional stability of the second substep prevails over the conditional stability of the first substep and renders the method overall unconditionally stable.

12.2.8 Implicit three-level methods

Richtmyer and Morton survey three-level implicit methods ([343], p. 189). The general form of a five-point method with a T-shaped finite-difference stencil shown in Figure 12.2.2(c) is

$$(1 + \beta)\,\frac{f_i^{n+1} - f_i^n}{\Delta t} - \beta\,\frac{f_i^n - f_i^{n-1}}{\Delta t} = \kappa\,\frac{f_{i-1}^{n+1} - 2f_i^{n+1} + f_{i+1}^{n+1}}{\Delta x^2}, \qquad (12.2.33)$$

where β is an arbitrary positive constant. It can be shown that the numerical algorithm is unconditionally stable for any value of β.

The particular choice $\beta = \frac{1}{2}$ provides us with a method that is second-order accurate in time and capable of suppressing small-scale oscillations. These features render the difference equation (12.2.33) with $\beta = \frac{1}{2}$ preferable over the Crank–Nicolson method when the solution exhibits sharp spatial variations. The choice $\beta = \frac{1}{2}\left(1 + \frac{1}{6\alpha}\right)$ provides us with a method that is second-order accurate in time and fourth-order accurate in space.

Problems

12.2.1 *CTCS and the Du Fort–Frankel method*

(a) Express (12.2.17) in vector notation in terms of the solution vectors $\mathbf{f}^{n+1}$, $\mathbf{f}^{n}$, and $\mathbf{f}^{n-1}$.

(b) Perform a consistency analysis of the Du Fort–Frankel method and derive the modified differential equation (12.2.23).

12.2.2 *Spectral radius of the projection matrix of the Crank–Nicolson method*

Verify that the eigenvalues of the projection matrix for the Crank–Nicolson method are

$$\lambda_m(\mathbf{A}^{-1} \cdot \mathbf{B}) = \frac{1 - 2\alpha \sin^2(m\pi/2N)}{1 + 2\alpha \sin^2(m\pi/2N)} \tag{12.2.34}$$

for $m = 1, \dots, N - 1$. Show that the spectral radius of the projection matrix is less than unity and therefore the method is unconditionally stable.

12.2.3 *Generalized Crank–Nicolson method*

A general form of the Crank–Nicolson equation (12.2.28) arises by using a weighted average to approximate the spatial derivative, obtaining

$$\frac{f_i^{n+1} - f_i^n}{\Delta t} = \kappa \left(\gamma \frac{f_{i-1}^{n+1} - 2f_i^{n+1} + f_{i+1}^{n+1}}{\Delta x^2} + (1 - \gamma) \frac{f_{i-1}^n - 2f_i^n + f_{i+1}^n}{\Delta x^2} \right), \tag{12.2.35}$$

where γ is a positive parameter ranging in the interval $[0, 1]$. When $\gamma = 0$, we obtain the explicit FTCS method; when $\beta = 1/2$, we obtain the implicit Crank–Nicolson method; when $\beta = 1$, we obtain the implicit BTCS method.

(a) Show that the method is generally first-order accurate in t and second-order accurate in x. When $\gamma = \frac{1}{2}\left(1 - \frac{1}{6\alpha}\right)$, the accuracy of the method increases to second order in t and fourth order in x.

(b) Provide an interpretation of the method in terms of a sequence of two elementary substeps.

(c) Perform the von Neumann stability analysis to derive the amplification factor

$$G = \frac{1 - 4(1 - \gamma)\alpha \sin^2 \frac{\theta}{2}}{1 + 4\gamma\alpha \sin^2 \frac{\theta}{2}}. \tag{12.2.36}$$

Show that the method is unconditionally stable when $\frac{1}{2} < \beta < 1$, and conditionally stable when $0 < \beta < \frac{1}{2}$. Derive the stability restriction $2\alpha < 1/(1 - 2\gamma)$ ([343], p. 189). When $\gamma = \frac{1}{2}$, the critical value of α is shifted to infinity and the method becomes unconditionally stable.

12.2.4 *Dispersion*

The dispersion equation in one dimension reads

$$\frac{\partial f}{\partial t} = \gamma \frac{\partial^3 f}{\partial x^3}, \tag{12.2.37}$$

where the constant γ is the dispersion coefficient. A solution describing the propagation of a sinusoidal wave is $f(x, t) = \exp[-\mathrm{i}\, k(x - ct)]$, where k is a real wave number, $c = k^2\gamma$ is the phase velocity of the traveling wave, and i is the imaginary unit. Since the phase velocity depends on the wavelength, a wave packet disperses as each constituent wave travels with its own phase velocity. Physically, the dispersion equation acts to propagate an arbitrary initial distribution while modifying its shape.

(a) The FTCS discretization leads to an explicit five-point two-level scheme described by the finite-difference equation

$$f_i^{n+1} = -\frac{1}{2}\zeta f_{i-2}^n + \zeta f_{i-1}^n + f_i^n - \zeta f_{i+1}^n + \frac{1}{2}\zeta f_{i+2}^n, \tag{12.2.38}$$

where $\zeta = \gamma \Delta t / \Delta x^3$. The numerical error is of order Δt and Δx^2. Carry out the von Neumann stability analysis to derive the amplification factor

$$G = 1 + i\, 2\zeta \sin\theta (1 - \cos\theta) \tag{12.2.39}$$

and explain why the method is unconditionally unstable.

(b) The implicit five-point two-level BTCS discretization leads to the finite-difference equation

$$\frac{1}{2}\zeta f_{i-2}^{n+1} - \zeta f_{i-1}^{n+1} + f_i^{n+1} + \zeta f_{i+1}^{n+1} - \frac{1}{2}\gamma f_{i+2}^{n+1} = f_i^n. \tag{12.2.40}$$

Derive the associated amplification factor

$$G = \frac{1}{1 + i\, 2\zeta \sin\theta \,(1 - \cos\theta)} \tag{12.2.41}$$

and explain why the method is unconditionally stable.

(c) Repeat (b) for the counterpart of the Crank–Nicolson method.

12.2.5 *Fourth-order diffusion*

The fourth-order diffusion equation in one dimension reads

$$\frac{\partial f}{\partial t} = -\lambda \frac{\partial^4 f}{\partial x^4}, \tag{12.2.42}$$

where λ is a positive constant called the hyperdiffusivity or fourth-order diffusivity.

(a) The FTCS discretization leads to the explicit five-point difference equation

$$f_i^{n+1} = -\epsilon f_{i-2}^n + 4\epsilon f_{i-1}^n + (1 - 6\,\epsilon)\, f_i^n + 4\epsilon f_{i+1}^n - \epsilon f_{i+2}^n, \tag{12.2.43}$$

where $\epsilon = \lambda \Delta t / \Delta x^4$, with first-order accuracy in time and second-order accuracy in space. Carry out the von Neumann stability analysis, compute the amplification factor, and assess the stability of the method.

(b) Repeat (a) for the implicit five-point two-level BTCS method.

(c) Develop the counterpart of the Crank–Nicolson method.

12.2.6 *Eigenvalues and eigenvectors of a tridiagonal Toeplitz matrix*

A matrix with constant diagonal elements is called a Toeplitz matrix. Consider a $K \times K$ tridiagonal Toeplitz matrix whose subdiagonal, diagonal, and superdiagonal elements are respectively equal to a, b, and c [317].

(*a*) Verify that the eigenvalues are given by

$$\lambda_i = a - 2(bc)^{1/2} \cos\left(\frac{i\pi}{K+1}\right) \tag{12.2.44}$$

for $i = 1, \ldots, K$. Note that, when the product bc is negative, the eigenvalues are complex.

(*b*) Show that, when b and c are real and negative, the eigenvector of the ith eigenvalue is

$$u_j^{(i)} = r^j \sin\left(\frac{ij\pi}{K+1}\right) \tag{12.2.45}$$

for $j = 1, \ldots, K$, where $r = (c/b)^{1/2}$ is a real and positive constant.

(*b*) Deduce the eigenvectors when b and c are real and positive.

🖳 Computer Problem

12.2.7 *Transient Couette flow*

A two-dimensional channel confined between two parallel plates separated by distance 2 cm is filled with water. Suddenly, the upper plate starts moving parallel to itself with velocity $V = 10\left(1 - e^{-\xi t}\right)$ cm/s while the lower plate is stationary, where ξ is a constant with dimensions of inverse time. Compute and plot the developing velocity profiles for $\xi = 0.1$ or 1 s^{-1} using (*a*) the FTCS method, (*b*) the Du Fort–Frankel method, (*c*) the CTCS method, and (*d*) the Crank–Nicolson method. Discuss the selection of the temporal and spatial steps.

12.3   Diffusion in two and three dimensions

Having addressed the unsteady diffusion equation in one dimension, now we consider unsteady diffusion in two and three dimensions governed, respectively, by the equations

$$\frac{\partial f}{\partial t} = \kappa\left(\frac{\partial^2 f}{\partial x^2} + \frac{\partial^2 f}{\partial y^2}\right), \qquad \frac{\partial f}{\partial t} = \kappa\left(\frac{\partial^2 f}{\partial x^2} + \frac{\partial^2 f}{\partial y^2} + \frac{\partial^2 f}{\partial z^2}\right), \tag{12.3.1}$$

subject to a specified initial condition, $f(\mathbf{x}, t = 0) = F(\mathbf{x})$, where $F(\mathbf{x})$ is a known function. For simplicity, we assume that the solution domain extends over the whole two-dimensional plane or three-dimensional space, and the function f decays to zero far from the origin.

The majority of finite-difference methods discussed in Section 12.2 can be extended in a straightforward manner to handle a second or third dimension. Examples will be discussed in this section. However, we will see that the need for computationally efficient algorithms necessitates the development of new procedures.

12.3.1 Iterative solution of Laplace and Poisson equations

Before proceeding to discuss specific discretizations, we remark that the finite-difference methods for solving equations (12.3.1) provide us with iterative procedures for solving the Laplace and Poisson equations in two and three dimensions. As an example, we consider the Poisson equation in two dimensions,

$$\frac{\partial^2 f}{\partial x^2} + \frac{\partial^2 f}{\partial y^2} + g(x,y) = 0, \tag{12.3.2}$$

where g is a given source term. The idea is to introduce a fictitious unsteady diffusion problem in the presence of a source term, governed by the unsteady reaction–diffusion equation

$$\frac{\partial f}{\partial t} = \kappa \left(\frac{\partial^2 f}{\partial x^2} + \frac{\partial^2 f}{\partial y^2} + g \right), \tag{12.3.3}$$

subject to a certain initial condition. Since the asymptotic solution of (12.3.3) at long times is identical to the solution of (12.3.2), advancing in time the solution of the diffusion–reaction equation amounts to iterating on the solution of the Poisson equation with a projection matrix that arises from the finite-difference scheme used to integrate (12.3.3).

12.3.2 Finite-difference procedures

In three dimensions, the goal of the finite-difference method is to generate the values of the unknown function, f, at the nodes of a three-dimensional grid parametrized by three indices, i, j, and k. In two dimensions, the grid is parametrized by two indices i and j. For simplicity, we assume that the grid spacings, Δx, Δy, and Δz, are uniform but not necessarily identical throughout the solution domain.

von Neumann stability analysis

To carry out the von Neumann stability analysis in three dimensions, we consider the evolution of a sinusoidal wave and set

$$f_{ijk}^n = A^n \exp[\,\mathrm{i}\,(i\theta_x + j\theta_y + k\theta_z)\,], \tag{12.3.4}$$

where i is the imaginary unit and $\theta_x = 2\pi \Delta x / L_x$, $\theta_y = 2\pi \Delta y / L_y$, and $\theta_z = 2\pi \Delta z / L_z$, are phase angles with corresponding wavelengths L_x, L_y, and L_z. The objective is to study the magnitude of the amplification factor, $G \equiv A^{n+1}/A^n$. Because the stability criteria for two- or three-dimensional diffusion are more restrictive than those for one-dimensional diffusion, conditionally stable methods are prohibitively expensive.

12.3.3 Explicit FTCS method

To implement the FTCS method in two dimensions, we apply the diffusion equation at the (i,j) node of a two-dimensional grid at the time instant, t_n, and approximate the second derivatives in the x and y directions using central differences to obtain the difference equation

$$\frac{f_{i,j}^{n+1} - f_{i,j}^n}{\Delta t} = \kappa \left(\frac{f_{i-1,j}^n - 2\,f_{i,j}^n + f_{i+1,j}^n}{\Delta x^2} + \frac{f_{i,j-1}^n - 2\,f_{i,j}^n + f_{i,j+1}^n}{\Delta y^2} \right). \tag{12.3.5}$$

Rearranging, we derive the explicit form

$$f_{i,j}^{n+1} = \alpha_x(f_{i+1,j}^n + f_{i-1,j}^n) + (1 - 2\alpha_x - 2\alpha_y) f_{i,j}^n + \alpha_y(f_{i,j+1}^n + f_{i,j-1}^n), \qquad (12.3.6)$$

where

$$\alpha_x = \frac{\kappa\Delta t}{\Delta x^2}, \qquad \alpha_y = \frac{\kappa\Delta t}{\Delta y^2} \qquad (12.3.7)$$

are the diffusion numbers in the x and y directions. An alternative form of (12.3.6) is

$$f_{i,j}^{n+1} = \alpha_x\left[(f_{i+1,j}^n + f_{i-1,j}^n) + 2\left(\frac{1}{2\alpha_x} - 1 - \beta\right)f_{i,j}^n + \beta\left(f_{i,j+1}^n + f_{i,j-1}^n\right)\right], \qquad (12.3.8)$$

where $\beta = (\Delta x/\Delta y)^2$. As β tends to zero or infinity, we recover the FTCS formula for one-dimensional diffusion along the x or y axis.

Carrying out the von Neumann stability analysis, we derive the amplification factor

$$G = 1 - 4\left(\alpha_x \sin^2 \frac{\theta_x}{2} + \alpha_y \sin^2 \frac{\theta_y}{2}\right), \qquad (12.3.9)$$

which shows that the FTCS method in two dimensions is stable provided that $\alpha_x + \alpha_y < \frac{1}{2}$, or

$$\alpha_x < \frac{1}{2(1+\beta)}. \qquad (12.3.10)$$

This requirement imposes a strong constraint on the size of the time step and renders the explicit discretization inefficient. Similar difficulties are encountered in three dimensions.

12.3.4 Implicit BTCS method

To achieve unconditional stability, we resort to an implicit method. The fully implicit BTCS discretization in two dimensions yields the difference equation

$$\frac{f_{i,j}^{n+1} - f_{i,j}^n}{\Delta t} = \kappa\left(\frac{f_{i-1,j}^{n+1} - 2f_{i,j}^{n+1} + f_{i+1,j}^{n+1}}{\Delta x^2} + \frac{f_{i,j-1}^{n+1} - 2f_{i,j}^{n+1} + f_{i,j+1}^{n+1}}{\Delta y^2}\right), \qquad (12.3.11)$$

with first-order accuracy in time and second-order accuracy in space. Rearranging, we derive the implicit formula

$$-\alpha_x(f_{i-1,j}^{n+1} + f_{i+1,j}^{n+1}) + (1 + 2\alpha_x + 2\alpha_y) f_{i,j}^{n+1} - \alpha_y(f_{i,j-1}^{n+1} + f_{i,j+1}^{n+1}) = f_{i,j}^n. \qquad (12.3.12)$$

A stability analysis reveals that the method is unconditionally stable.

Unfortunately, the numerical implementation of (12.3.12) results in a pentadiagonal system of algebraic equations whose solution cannot be carried out using a specialized method, such as the Thomas algorithm. This complication renders the BTCS method prohibitively expensive.

12.3.5   ADI method in two dimensions

To reduce the computational burden of the implicit BTCS method for two-dimensional diffusion, Peaceman & Rachford [291] and Douglas [109] proposed splitting each time step into two substeps of equal duration $\frac{1}{2}\Delta t$, and approximating the spatial derivatives in a partially implicit fashion while working sequentially and in the x and y directions in an alternating fashion. The computations proceed according to the finite-difference equations

$$\frac{f_{i,j}^{n+1/2} - f_{i,j}^{n}}{\frac{1}{2}\Delta t} = \kappa \left[\frac{f_{i-1,j}^{n+1/2} - 2f_{i,j}^{n+1/2} + f_{i+1,j}^{n+1/2}}{\Delta x^2} \right] + \kappa \, \frac{f_{i,j-1}^{n} - 2f_{i,j}^{n} + f_{i,j+1}^{n}}{\Delta y^2} \qquad (12.3.13)$$

and

$$\frac{f_{i,j}^{n+1} - f_{i,j}^{n+1/2}}{\frac{1}{2}\Delta t} = \kappa \, \frac{f_{i-1,j}^{n+1/2} - 2f_{i,j}^{n+1/2} + f_{i+1,j}^{n+1/2}}{\Delta x^2} + \kappa \left[\frac{f_{i,j-1}^{n+1} - 2f_{i,j}^{n+1} + f_{i,j+1}^{n+1}}{\Delta y^2} \right], \qquad (12.3.14)$$

where $n + \frac{1}{2}$ is an intermediate time level and the square brackets on the right-hand sides enclose implicit discretizations. The first substep is carried out using the implicit BTCS method for the x direction, while the second substep is carried out using the implicit BTCS method for the y direction. To eliminate the bias associated with this particular arrangement, the order is alternated after the completion of each step. The overall method is second-order accurate in time and space.

     Rearranging the difference equations, we obtain a two-step implicit algorithm representing the x and y sweeps,

$$-\alpha_x f_{i-1,j}^{n+1/2} + 2(1+\alpha_x) f_{i,j}^{n+1/2} - \alpha_x f_{i+1,j}^{n+1/2} = \alpha_y f_{i,j-1}^{n} + 2(1-\alpha_y) f_{i,j}^{n} + \alpha_y f_{i,j+1}^{n} \qquad (12.3.15)$$

and

$$-\alpha_y f_{i,j-1}^{n+1} + 2(1+\alpha_y) f_{i,j}^{n+1} - \alpha_y f_{i,j+1}^{n+1} = \alpha_x f_{i-1,j}^{n+1/2} + 2(1-\alpha_x) f_{i,j}^{n+1/2} + \alpha_x f_{i+1,j}^{n+1/2}. \qquad (12.3.16)$$

Completing each time step requires solving two systems of tridiagonal equations, which can be done efficiently using the Thomas algorithm described in Section B.1.4, Appendix B.

     Carrying out the von Neumann stability analysis, we obtain the amplification factor

$$G = \frac{(1 - 2\alpha_x \sin^2 \frac{\theta_x}{2})(1 - 2\alpha_y \sin^2 \frac{\theta_y}{2})}{(1 + 2\alpha_x \sin^2 \frac{\theta_x}{2})(1 + 2\alpha_y \sin^2 \frac{\theta_y}{2})}. \qquad (12.3.17)$$

Cursory examination reveals that $|G| < 1$ under any conditions, which ensures that the ADI method is unconditionally stable. The second-order accuracy combined with the unconditional stability are the reasons that the ADI method is a standard choice in practice.

12.3.6   Crank–Nicolson method and approximate factorization

The ADI method in two dimensions allows us to advance the solution over one time step by solving two pseudo one-dimensional problems, which amounts to decoupling the diffusion processes in the two spatial directions. To demonstrate this clearly, we introduce the central-difference operators

$$\Delta_x^2 f_{i,j}^n \equiv f_{i-1,j}^n - 2 f_{i,j}^n + f_{i+1,j}^n, \qquad \Delta_y^2 f_{i,j}^n \equiv f_{i,j-1}^n - 2 f_{i,j}^n + f_{i,j+1}^n, \qquad (12.3.18)$$

and restate the ADI equations as

$$(2 - \alpha_x \Delta_x^2) f_{i,j}^{n+1/2} = (2 + \alpha_y \Delta_y^2) f_{i,j}^n, \qquad (2 - \alpha_y \Delta_y^2) f_{i,j}^{n+1} = (2 + \alpha_x \Delta_x^2) f_{i,j}^{n+1/2}. \qquad (12.3.19)$$

Combining these equations to eliminate the intermediate solution at the $n + \frac{1}{2}$ level, we obtain

$$(2 - \alpha_x \Delta_x^2)(2 - \alpha_y \Delta_y^2) f_{i,j}^{n+1} = (2 + \alpha_x \Delta_x^2)(2 + \alpha_y \Delta_y^2) f_{i,j}^n. \qquad (12.3.20)$$

The spatial decoupling is reflected in the factorized nature of the difference operators on either side.

The fully implicit Crank–Nicolson discretization of the two-dimensional diffusion equation in both spatial directions can be written in the symbolic form

$$(2 - \alpha_x \Delta_x^2 - \alpha_y \Delta_y^2) f_{i,j}^{n+1} = (2 + \alpha_x \Delta_x^2 + \alpha_y \Delta_y^2) f_{i,j}^n, \qquad (12.3.21)$$

which can be restated as

$$(2 - \alpha_x \Delta_x^2)(2 - \alpha_y \Delta_y^2) f_{i,j}^{n+1} = (2 + \alpha_x \Delta_x^2)(2 + \alpha_y \Delta_y^2) f_{i,j}^n + \alpha_x \alpha_y \Delta_x^2 \Delta_y^2 (f_{i,j}^{n+1} - f_{i,j}^n). \qquad (12.3.22)$$

The Crank–Nicolson algorithm can be shown to be unconditionally stable (problem 12.3.2).

The ADI form (12.3.20) derives from the Crank–Nicolson form (12.3.22) by neglecting the last term on the right-hand side. This simplification is permissible, for the neglected term is of fourth-order in the spatial variables, whereas (12.3.20) is meant to be accurate only up to second order in the spatial steps. We may say that the ADI method results from the approximate factorization of the difference operators on either side of the Crank–Nicolson equation (12.3.21).

12.3.7 Poisson equation in two dimensions

The explicit and semi-implicit methods for two-dimensional diffusion provide us with efficient iterative procedures for solving the Poisson's equation $\nabla^2 f + g = 0$, as discussed in Section 12.3.1. Algorithms arising from the explicit FTCS discretization on a uniform grid and its modified versions corresponding to the Gauss–Seidel and successive over-relaxation methods, are collected in Table 12.3.1. The explicit point Gauss–Seidel scheme (PGS) given in the second entry arises from the FTCS scheme given in the first entry by setting $\alpha_x = 1/[2(1 + \beta)]$, which marginally satisfies the stability criterion $\alpha_x + \alpha_y < 1/2$. Introducing a relaxation parameter, ω, yields the point successive over-relaxation (PSOR) method shown in the third entry of Table 12.3.1. All methods are second-order accurate in Δx and Δy.

The line Gauss–Seidel (LGS) scheme shown in the fourth entry of Table 12.3.1 arises from the semi-implicit discretization of the Poisson equation. The relaxation parameter ω of the line successive over-relaxation (LSOR) scheme shown in the last entry of Table 12.3.1 varies in the range $[0, 2]$. The implicit LGS and LSOR schemes require solving tridiagonal systems of equations for each grid line parallel to the x axis, which can be done efficiently using the Thomas algorithm. Similar equations can be written for the y axis.

To develop the ADI method, we introduce the parameter $\varrho = 2/\alpha_x$, and recast the ADI equations (12.3.15) as shown in the first entry of Table 12.3.2. Since the ADI method is unconditionally

Explicit FTCS

$$f_{i,j}^{n+1} = f_{i,j}^n + \alpha_x \left[f_{i+1,j}^n + f_{i-1,j}^n - 2\left(1 + \beta\right) f_{i,j}^n + \beta \left(f_{i,j+1}^n + f_{i,j-1}^n\right) + \Delta x^2 g_{i,j} \right]$$

Explicit point Gauss–Seidel (PGS)

$$f_{i,j}^{n+1} = \frac{1}{2\left(1 + \beta\right)} \left[f_{i+1,j}^n + f_{i-1,j}^n + \beta \left(f_{i,j+1}^n + f_{i,j-1}^n\right) + \Delta x^2 g_{i,j} \right]$$

Explicit point successive over-relaxation (PSOR)

$$f_{i,j}^{n+1} = \left(1 - \omega\right) f_{i,j}^n + \frac{\omega}{2\left(1 + \beta\right)} \left[f_{i+1,j}^n + f_{i-1,j}^n + \beta \left(f_{i,j+1}^n + f_{i,j-1}^n\right) + \Delta x^2 g_{i,j} \right]$$

Implicit line Gauss–Seidel for the x direction (LGS)

$$-f_{i-1,j}^{n+1} + 2\left(1 + \beta\right) f_{i,j}^{n+1} - f_{i+1,j}^{n+1} = \beta \left(f_{i,j-1}^n + f_{i,j+1}^n\right) + \Delta x^2 g_{i,j}$$

Implicit line successive over-relaxation for the x direction (LSOR)

$$-\omega f_{i-1,j}^{n+1} + 2\left(1 + \beta\right) f_{i,j}^{n+1} - \omega f_{i+1,j}^{n+1} = 2\left(1 - \omega\right)\left(1 + \beta\right) f_{i,j}^n + \omega \beta \left(f_{i,j-1}^n + f_{i,j+1}^n\right) + \Delta x^2 g_{i,j}$$

Table 12.3.1   Iterative methods for solving the Poisson equation, $\nabla^2 f + g = 0$, on a uniform rectangular grid, where $\beta = (\Delta x / \Delta y)^2$. All methods are second-order accurate in Δx and Δy. The relaxation parameter ω in the PSOR and LSOR schemes ranges between 0 and 2.

stable, it might appear that the fastest approach to the steady state will be achieved using a large value for Δt or a small value for ϱ. However, careful analysis shows that the minimum number of iterations for a specified level of accuracy is achieved when a certain repetitive sequence of values for ϱ is employed. In its generic implementation, the performance of the ADI method is comparable to that of the line successive over-relaxation method (LSOR) ([5]; [184], p. 446).

To develop the SOR-ADI method shown in the second entry of Table 12.3.2, we set $\varrho = 2\beta$ in the first ADI equation and $\varrho = 2$ in the second ADI equation so that the middle terms on the right-hand sides disappear. A relaxation parameter, ω, is then introduced to accelerate the convergence ([186], Vol. II, p. 9). To demonstrate the consistency of the SOR-ADI scheme, we assume that the solution depends only on x. The first SOR-ADI equation yields

$$-\omega f_{i-1,j}^{n+1/2} + 2(1 + \beta) f_{i,j}^{n+1/2} - \omega f_{i+1,j}^{n+1/2} = 2(1 + \beta - \omega) f_{i,j}^n + \frac{\omega}{2} \Delta x^2 g_{i,j}. \tag{12.3.23}$$

ADI

$$-f_{i-1,j}^{n+1/2} + (2 + \varrho)\, f_{i,j}^{n+1/2} - f_{i+1,j}^{n+1/2} = \beta\, f_{i,j-1}^{n} + (\varrho - 2\beta)\, f_{i,j}^{n} + \beta\, f_{i,j+1}^{n} + \Delta x^2\, g_{i,j},$$

$$-\beta\, f_{i,j-1}^{n+1} + (\varrho + 2\beta)\, f_{i,j}^{n+1} - \beta\, f_{i,j+1}^{n+1} = f_{i-1,j}^{n+1/2} + (\varrho - 2)\, f_{i,j}^{n+1/2} + f_{i+1,j}^{n+1/2} + \Delta x^2\, g_{i,j}$$

SOR-ADI

$$-\omega\, f_{i-1,j}^{n+1/2} + 2\,(1 + \beta)\, f_{i,j}^{n+1/2} - \omega\, f_{i+1,j}^{n+1/2}$$
$$= \omega\,\beta\, f_{i,j-1}^{n} + 2\,(1 + \beta)\,(1 - \omega)\, f_{i,j}^{n} + \omega\,\beta\, f_{i,j+1}^{n} + \omega\, \Delta x^2\, g_{i,j},$$

$$-\omega\,\beta\, f_{i,j-1}^{n+1} + 2\,(1 + \beta)\, f_{i,j}^{n+1} - \omega\,\beta\, f_{i,j+1}^{n+1}$$
$$= \omega\, f_{i-1,j}^{n+1/2} + 2\,(1 + \beta)\,(1 - \omega)\, f_{i,j}^{n+1/2} + \omega\, f_{i+1,j}^{n+1/2} + \omega\, \Delta x^2\, g_{i,j}$$

TABLE 12.3.2 ADI methods for solving the Poisson equation, $\nabla^2 f + g = 0$, on a uniform Cartesian grid, where $\beta = (\Delta x/\Delta y)^2$. All methods are second-order accurate in Δx and Δy.

When $\omega = 1 + \beta$, we obtain

$$-f_{i-1,j}^{n+1/2} + 2f_{i,j}^{n+1/2} - f_{i+1,j}^{n+1/2} = \Delta x^2 g_{i,j}, \tag{12.3.24}$$

which is the central difference discretization of the ordinary differential equation $d^2 f/dx^2 + g = 0$.

12.3.8   ADI method in three dimensions

The standard ADI method in three dimensions is carried out in three substeps of equal duration, $\frac{1}{3}\Delta t$. One spatial dimension is treated implicitly while the other two dimensions are treated explicitly in each substep, in the spirit of (12.3.13) and (12.3.14). The method is first-order accurate in time and second-order accurate in space. Unfortunately, the partial BTCS discretizations result in an algorithm that is stable only when $\alpha_x + \alpha_y + \alpha_z < \frac{3}{2}$, which places an unaffordable restriction on the size of the time step [317].

Douglas [111] developed an ADI method that is second-order accurate in both time and space and unconditionally stable. The method proceeds in a predictor–correction sense in three substeps. The first substep produces a predicted solution using the Crank–Nicolson method for the x direction, while treating the y and z directions explicitly according to the difference equation

$$(3 - \tfrac{1}{2}\alpha_x \Delta_x^2) f_{i,j,k}^* = (3 + \tfrac{1}{2}\alpha_x \Delta_x^2 + \alpha_y \Delta_y^2 + \alpha_z \Delta_z^2) f_{i,j,k}^n, \tag{12.3.25}$$

where an asterisk denotes the first provisional solution. The second substep produces a predicted solution using the x discretization of the first substep and the Crank–Nicolson discretization for the y direction, while treating the z direction explicitly according to the difference equation

$$(3 - \frac{1}{2}\alpha_y\Delta_y^2)f_{i,j,k}^{**} = (3 + \frac{1}{2}\alpha_x\Delta_x^2 + \frac{1}{2}\alpha_y\Delta_y^2 + \alpha_z\,\Delta_z^2)f_{i,j,k}^n + \frac{1}{2}\alpha_x\Delta_x^2 f_{i,j,k}^*, \qquad (12.3.26)$$

where a double asterisk denotes the second provisional solution. The third substep advances the solution using the x and y discretizations of the second substep, while using the Crank–Nicolson method for the z direction according to the difference equation

$$(3 - \frac{1}{2}\alpha_z\Delta_z^2)f_{i,j,k}^{n+1} = (3 + \frac{1}{2}\alpha_x\Delta_x^2 + \frac{1}{2}\alpha_y\Delta_y^2 + \frac{1}{2}\alpha_z\Delta_z^2)\,f_{i,j,k}^n + \frac{1}{2}\alpha_x\Delta_x^2 f_{i,j,k}^* + \frac{1}{2}\alpha_y\Delta_y^2 f_{i,j,k}^{**}.$$
$$(12.3.27)$$

Combining (12.3.26) with (12.3.25) and simplifying, we obtain

$$(3 - \frac{1}{2}\alpha_y\Delta_y^2)f_{i,j,k}^{**} = 3\,f_{i,j,k}^* - \frac{1}{2}\alpha_y\Delta_y^2 f_{i,j,k}^n. \qquad (12.3.28)$$

Combining (12.3.27) with (12.3.26) and simplifying, we obtain

$$(3 - \frac{1}{2}\alpha_z\Delta_z^2)f_{i,j,k}^{n+1} = 3\,f_{i,j,k}^{**} - \frac{1}{2}\alpha_z\Delta_z^2 f_{i,j,k}^n. \qquad (12.3.29)$$

We can deduce the overall action of the method by combining (12.3.29) and (12.3.28) to eliminate the second intermediate solution, obtaining

$$(3 - \frac{1}{2}\alpha_y\Delta_y^2)(3 - \frac{1}{2}\alpha_z\Delta_z^2)f_{i,j,k}^{n+1} = 3\,f_{i,j,k}^* - \frac{1}{2}\alpha_y\Delta_y^2 f_{i,j,k}^n - \frac{1}{2}\alpha_z(3 - \frac{1}{2}\alpha_y\Delta_y^2)\Delta_z^2 f_{i,j,k}^n. \quad (12.3.30)$$

Now combining (12.3.30) and (12.3.25) to eliminate the first intermediate solution, we obtain

$$(3 - \frac{1}{2}\alpha_x\Delta_x^2)(3 - \frac{1}{2}\alpha_y\Delta_y^2)(3 - \frac{1}{2}\alpha_z\Delta_z^2)f_{i,j,k}^{n+1}$$
$$= (3 + \frac{1}{2}\alpha_x\Delta_x^2 + \alpha_y\Delta_y^2 + \alpha_z\Delta_z^2)f_{i,j,k}^n - \frac{1}{2}(3 - \frac{1}{2}\alpha_x\Delta_x^2)\alpha_y\Delta_y^2 f_{i,j,k}^n$$
$$- \frac{1}{2}\alpha_z(3 - \frac{1}{2}\alpha_x\Delta_x^2)(3 - \frac{1}{2}\alpha_y\Delta_y^2)\Delta_z^2 f_{i,j,k}^n. \qquad (12.3.31)$$

Simplifying the right-hand side, we find that

$$(3 - \frac{1}{2}\alpha_x\Delta_x^2)(3 - \frac{1}{2}\alpha_y\Delta_y^2)(3 - \frac{1}{2}\alpha_z\Delta_z^2)f_{i,j,k}^{n+1} = \Big(1 + \frac{1}{2}(\alpha_x\Delta_x^2 + \alpha_y\Delta_y^2 + \alpha_z\Delta_z^2)$$
$$+ \frac{1}{4}(\alpha_x\alpha_y\Delta_x^2\Delta_y^2 + \alpha_x\alpha_z\Delta_x^2\Delta_z^2 + \alpha_y\alpha_z\Delta_y^2\Delta_z^2) - \frac{1}{8}\alpha_x\alpha_y\alpha_z\Delta_x^2\Delta_y^2\Delta_z^2\Big)f_{i,j,k}^n. \qquad (12.3.32)$$

The right-hand side can be rearranged to give

$$(3 - \frac{1}{2}\alpha_x\Delta_x^2)(3 - \frac{1}{2}\alpha_y\Delta_y^2)(3 - \frac{1}{2}\alpha_z\Delta_z^2)f_{i,j,k}^{n+1}$$
$$= (3 + \frac{1}{2}\alpha_x\Delta_x^2)(3 + \frac{1}{2}\alpha_y\Delta_y^2)(3 + \frac{1}{2}\alpha_z\Delta_z^2)f_{i,j,k}^n - \frac{1}{4}\alpha_x\alpha_y\alpha_z\Delta_x^2\Delta_y^2\Delta_z^2 f_{i,j,k}^n,$$
$$(12.3.33)$$

which reveals a nearly factorized form. Performing the von Neumann stability analysis, we find that the method is unconditionally stable [317].

12.3.9 Operator splitting and fractional steps

Another way of preserving the tridiagonal nature of the one-dimensional implicit discretization is by replacing the three-dimensional equation (12.3.1), or its two-dimensional counterpart, with a set of three or two one-dimensional evolution equations that operate successively in fractional steps [434]. In two dimensions, we obtain

$$\frac{\partial f}{\partial t} = \kappa \frac{\partial^2 f}{\partial x^2}, \qquad \frac{\partial f}{\partial t} = \kappa \frac{\partial^2 f}{\partial y^2}, \tag{12.3.34}$$

which amounts to allowing diffusion to operate sequentially in the two directions, each time neglecting the other dimension. Each fractional step proceeds for the full time interval, Δt, and time is reset to the initial value at the end of the first fractional step. To preserve the spatial isotropy of the Laplacian operator, the order of the equations in (12.3.34) can be switched after completing a full time step. The fractional steps can be carried out using different methods for the one-dimensional component diffusion equations, as discussed in Section 12.2. The stability restrictions of the overall method is the union of the restrictions pertaining to the individual fractional steps.

Approximate factorization implicit (AFI) method

The approximate factorization implicit (AFI) method emerges by applying the implicit BTCS discretization in each fractional step of (12.3.34). It can be shown that the AFI scheme derives from the approximate factorization of the fully implicitly BTCS discretization of (12.3.1) (Problem 12.3.2). Each fractional step requires solving a tridiagonal system of equations using the Thomas algorithm.

Crank–Nicolson method

Using the Crank–Nicolson method for each fractional step in two dimensions, we obtain the difference equations

$$(1 - \frac{1}{2}\alpha_x \Delta_x^2) f_{i,j}^* = (1 + \frac{1}{2}\alpha_x \Delta_x^2) f_{i,j}^n, \qquad (1 - \frac{1}{2}\alpha_y \Delta_y^2) f_{i,j}^{n+1} = (1 + \frac{1}{2}\alpha_y \Delta_y^2) f_{i,j}^*, \tag{12.3.35}$$

where an asterisk denotes an intermediate solution. Eliminating the intermediate solution yields the ADI equation in two dimensions. We conclude that the algorithm is second-order accurate in time and space, while enjoying unconditional stability.

Problems

12.3.1 *Generalized fully implicit Crank–Nicolson method in two dimensions*

The Crank–Nicolson method can be generalized into a more comprehensive scheme expressed by the difference equation

$$\left[1 - \gamma \left(\alpha_x \Delta_x^2 + \alpha_y \Delta_y^2 \right) \right] f_{i,j}^{n+1} = \left[1 + (1 - \gamma) \left(\alpha_x \Delta_x^2 + \alpha_y \Delta_y^2 \right) \right] f_{i,j}^n, \tag{12.3.36}$$

where γ is a dimensionless numerical parameter. The explicit FTCS discretization corresponds to $\gamma = 0$, the standard Crank–Nicolson discretization corresponds to $\gamma = \frac{1}{2}$, and the implicit BTCS discretization corresponds to $\gamma = 1$. Perform the von Neumann stability analysis to derive the amplification factor

$$G = \frac{1 - 4(1 - \gamma)\left[\alpha_x \sin^2(\frac{1}{2}\theta_x) + \alpha_y \sin^2(\frac{1}{2}\theta_y)\right]}{1 + 4\gamma\left[\alpha_x \sin^2(\frac{1}{2}\theta_x) + \alpha_y \sin^2(\frac{1}{2}\theta_y)\right]}. \tag{12.3.37}$$

Show that, when $\gamma > \frac{1}{2}$, $|G| < 1$ for any values of α_x, α_y, θ_x, and θ_y, and the method is stable; whereas when $\gamma < \frac{1}{2}$, the method is stable only if

$$\alpha_x + \alpha_y < \frac{1}{2(1 - 2\gamma)}. \tag{12.3.38}$$

Explain why the Crank–Nicolson method corresponding to $\gamma = \frac{1}{2}$ is unconditionally stable.

12.3.2 *AFI method*

Show that the AFI method can be regarded as the result of the approximate factorization of the fully implicit BTCS discretization of (12.3.1).

12.3.3 *ADI method for Poisson's equation in three dimensions*

Develop an iterative method for solving Poisson's equation in three dimensions based on the ADI method of Douglas [111].

 Computer Problems

12.3.4 *ADI method for the Poisson equation in two dimensions*

Write a computer program that solves the two-dimensional Poisson equation in a rectangular domain covered by an $N_x \times N_y$ grid based on the ADI method given in Table 12.3.2, subject to the Dirichlet boundary condition. Run the program to compute the solution inside a square box with a forcing function and boundary values of your choice. Examine the rate of convergence of your results with respect to ϱ. Test the reliability of your solution by comparing the results against a known analytical solution of your choice.

12.3.5 *ADI method for the Poisson equation in three dimensions*

Repeat Problem 12.3.4 for the three-dimensional Poisson equation inside a rectangular box discretized into an $N_x \times N_y \times N_z$ grid, based on the ADI method developed in Problem 12.3.3.

12.4 Convection

Having discussed the extreme case of pure diffusion in one two and three dimensions, now we turn our attention to the opposite extreme case of pure convection.

12.4.1 One-dimensional convection

We begin by considering the one-dimensional linear convection equation,

$$\frac{\partial f}{\partial t} + U \frac{\partial f}{\partial x} = 0, \tag{12.4.1}$$

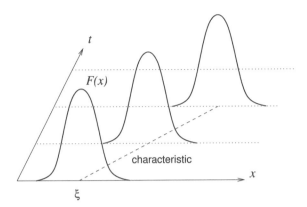

FIGURE 12.4.1   According to the linear convection equation, the initial distribution travels along the
   x axis with velocity U. The dashed line is a characteristic line along which the solution remains
   constant.

where U is a positive or negative constant convection velocity. For simplicity, we assume that the
solution domain extends over the whole x axis and stipulate the far-field conditions $f(\pm\infty, t) = 0$.
If the solution domain were finite, we would have to specify one boundary condition at one end. An
initial condition stipulating that $f(x, t = 0) = F(x)$ is provided, where $F(x)$ is a known function.

One may readily verify that the exact solution is given by $f(x,t) = F(x - Ut)$, which states
that the initial distribution, $F(x)$, travels along the x axis with constant velocity U, as illustrated in
Figure 12.4.1. If the velocity U is negative, the motion occurs toward the negative direction of the
x axis. The value of the function $f(x,t)$ remains constant along the characteristic lines $x = \xi + Ut$
in the xt plane, as shown with the dashed line in Figure 12.4.1, where ξ is an arbitrary point.

As a preliminary, we differentiate (12.4.1) twice, once with respect to t and the second time
with respect to x, and combine the resulting expressions to obtain the wave equation

$$\frac{\partial^2 f}{\partial t^2} = U^2 \frac{\partial^2 f}{\partial x^2}, \tag{12.4.2}$$

which is a prototypical second-order hyperbolic partial differential equation. The hyperbolic nature
of the first-order equation (12.4.1) thus becomes apparent. A function that satisfies the convection
equation (12.4.1) also satisfies the wave equation (12.4.2).

12.4.2   Explicit FTCS method

Applying the linear convection equation (12.4.1) at the x_i grid point at time t_n, and approximating
the time derivative, $\partial f/\partial t$, with a forward difference and the space derivative, $\partial f/\partial x$, with a central
difference, we obtain the FTCS difference equation

$$\frac{f_i^{n+1} - f_i^n}{\Delta t} + U \frac{f_{i+1}^n - f_{i-1}^n}{2\Delta x} = 0. \tag{12.4.3}$$

The numerical error due to the temporal and spatial discretizations is of order Δt and Δx^2. Since this difference equation emerged by applying standard finite-difference approximations, its consistency is guaranteed and need not be examined. Solving for f_i^{n+1}, we obtain the explicit formula

$$f_i^{n+1} = \frac{c}{2} f_{i-1}^n + f_i^n - \frac{c}{2} f_{i+1}^n, \tag{12.4.4}$$

where

$$c = \frac{U \Delta t}{\Delta x} \tag{12.4.5}$$

is the dimensionless convection number, also called the Courant number.

According to the difference formula (12.4.4), the new solution vector, $\mathbf{f}^{n+1}$, emerges by multiplying the previous solution vector, $\mathbf{f}^n$, with a tridiagonal projection matrix $\mathbf{B}$ whose subdiagonal, diagonal, and superdiagonal elements are, respectively, equal to $\frac{c}{2}$, 1, and $-\frac{c}{2}$. We know that, if the spectral radius of a projection matrix with a complete set of eigenvectors is equal to unity, less than unity, or larger than unity, the length of $\mathbf{f}^n$ will stay roughly constant, decrease, or increase during the successive projections. We remember that, according to the exact solution, the initial distribution is convected without change in shape, and accept the first behavior, tolerate the second behavior, and dismiss the third behavior as numerically unstable.

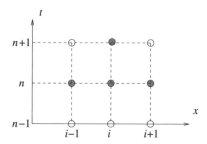

Computational stencil of the FTCS discretization.

The eigenvalues and thus the spectral radius of the $(N-1)\times(N-1)$ matrix $\mathbf{B}$ can be computed analytically and are given by

$$\lambda_m = 1 + \mathrm{i}\, c \cos\left(\frac{m\pi}{N}\right) \tag{12.4.6}$$

for $m = 1, \ldots, N-1$, where i is the imaginary unit [317]. Since the magnitude of all eigenvalues is greater than unity, the norm of the solution vector, $\mathbf{f}^n$, will amplify during the projections. Accordingly, the numerical method is unconditionally unstable.

Another way of assessing stability is by performing the von Neumann analysis, obtaining the amplification factor

$$G \equiv \frac{A^{n+1}}{A^n} = 1 + \mathrm{i}\, c \sin\theta, \tag{12.4.7}$$

where $\theta = 2\pi \Delta x / L$ is the phase angle. Since the magnitude of the right-hand side of (12.4.7) is greater than unity for any value of c and θ, the solution vector, $\mathbf{f}^n$, will amplify for any wavelength, L, and this confirms that the FTCS method is unconditionally unstable and must be abandoned. It is instructive to recall that the FTCS method for the diffusion equation was found to be conditionally stable.

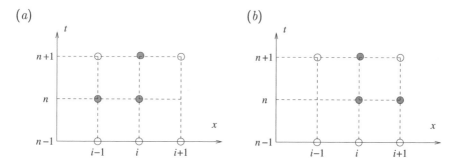

FIGURE 12.4.2 Computational stencil of the (a) FTBS and (b) FTFS discretization for solving the convection equation.

12.4.3   FTBS method

We proceed to exploring a different type of discretization in the hope of ensuring numerical stability. Now we apply (12.4.1) at the point x_i at the time instant t_n and approximate the time derivative, $\partial f / \partial t$, with a forward difference and the spatial derivative, $\partial f / \partial x$, with a backward difference to obtain the FTBS scheme

$$\frac{f_i^{n+1} - f_i^n}{\Delta t} + U \frac{f_i^n - f_{i-1}^n}{\Delta x} = 0. \tag{12.4.8}$$

The corresponding finite-difference stencil is shown in Figure 12.4.2(a). The numerical error is of order Δt and Δx. Rearranging (12.4.8), we derive the two-level explicit algorithm

$$f_i^{n+1} = c\, f_{i-1}^n + (1 - c)\, f_i^n. \tag{12.4.9}$$

The right-hand side is recognized as a weighted average.

The exact solution at the x_i grid point at the first step is

$$f_i^{exact}(\Delta t) = F(x_i - U\Delta t) = F(x_{i-1} + \Delta x - U\Delta t), \tag{12.4.10}$$

where $F(x)$ is the initial distribution. When U is positive and $\Delta x = U\Delta t$ or $c = 1$, we find that

$$f_i^{exact}(\Delta t) = F(x_{i-1}) = f_{i-1}^0, \tag{12.4.11}$$

which shows that the FTBS difference equation (12.4.9) reproduces the exact solution. It might appear that adjusting Δx and Δt so that $c = 1$ is a perfect choice. Unfortunately, we will see that stability constraints necessitate a compromise.

Performing the von Neumann stability analysis, we derive the amplification factor

$$G \equiv \frac{A^{n+1}}{A^n} = 1 - c + c\, \exp(\mathrm{i}\theta), \tag{12.4.12}$$

which reveals that G traces a circle of radius c centered at the point $(1 - c, 0)$ in the complex plane. To guarantee stability, we require that $|G| < 1$, and obtain the stability constraint $0 < c < 1$. When the convection velocity U is negative, this constraint cannot be fulfilled and the FTBS method is unstable.

12.4.4   FTFS method

Returning to (12.4.1), we apply forward differences for the time and space derivatives, and obtain the FTFS difference equation

$$\frac{f_i^{n+1} - f_i^n}{\Delta t} + U\, \frac{f_{i+1}^n - f_i^n}{\Delta x} = 0. \tag{12.4.13}$$

The corresponding finite-difference stencil is shown in Figure 12.4.2(b). The numerical error is of order Δt and Δx. Rearranging, we derive the two-level explicit formula

$$f_i^{n+1} = (1+c)\, f_i^n - c\, f_{i+1}^n. \tag{12.4.14}$$

When U is negative and $c = -1$, we find that $f_i^{n+1} = f_{i+1}^n$, which reproduces the exact solution.

Carrying out the von Neumann stability analysis, we derive the gain

$$G \equiv \frac{A^{n+1}}{A^n} = 1 + c - c\,\exp(-\mathrm{i}\,\theta), \tag{12.4.15}$$

which reveals that G is located on a circle of radius c centered at the point $(1+c, 0)$ in the complex plane. To ensure stability, we require that $|G| < 1$ and obtain the stability constraint $-1 < c < 0$. When the convection velocity U is positive, the method is unstable.

12.4.5   Upwind differencing, numerical diffusivity, and the CFL condition

The complementary successes of the FTBS and FTFS schemes, respectively, for positive and negative values of the convection velocity, U, suggests the method of upwind differencing: use FTBS when $U > 0$, use FTFS when $U < 0$, and always keep $|c| < 1$. The restriction $|c| < 1$ is known as the Courant–Friedrichs–Lewy (CFL) stability criterion. Upwind differencing is particularly effective when the convection velocity is not constant but varies in time and space over the solution domain. In physical terms, upwind differencing carries information on the structure of the solution forward from the direction of a traveling wave, and thus anticipates and suppresses the growth of unwanted perturbations.

12.4.6   Numerical diffusion

To explain further the conditional stability of the spatially biased FTBS and FTFS discretizations, respectively, for positive and negative values of the convection velocities, which should be contrasted with the unconditional instability of the FTCS discretization, we perform a consistency analysis.

FTBS discretization

Considering first the FTBS formula (12.4.9), we pretend that all discrete variables are continuous functions of space and time and expand them in Taylor series about the doublet (x_i, t_n) to obtain

$$f_i^n + (f_t)_i^n\, \Delta t + \frac{1}{2}\,(f_{tt})_i^n\, \Delta t^2 + O(\Delta t^3)$$

$$= c\left(f_i^n - (f_x)_i^n\, \Delta x + \frac{1}{2}\,(f_{xx})_i^n\, \Delta x^2 \right) + (1-c)\, f_i^n + c\, O(\Delta x^3), \tag{12.4.16}$$

where $f_t = \partial f/\partial t$, $f_x = \partial f/\partial x$, and $f_{xx} = \partial^2 f/\partial x^2$. Simplifying and rearranging, we find that

$$(f_t)_i^n \, \Delta t + c \, (f_x)_i^n \, \Delta x = \frac{1}{2} \left[c \, (f_{xx})_i^n \, \Delta x^2 - (f_{tt})_i^n \, \Delta t^2 \right] + O(\Delta t^3) + c \, O(\Delta x^3). \qquad (12.4.17)$$

Next, we divide both sides by Δt and recall the definition $c = U \Delta t / \Delta x$ to obtain

$$(f_t)_i^n + U \, (f_x)_i^n = \left[c \, \frac{\Delta x^2}{\Delta t} \, (f_{xx})_i^n \, \Delta x^2 - (f_{tt})_i^n \, \Delta t \right] + O(\Delta t^2) + \frac{c}{\Delta t} \, O(\Delta x^3). \qquad (12.4.18)$$

Finally, we substitute $f_{tt} = U^2 f_{xx}$ and obtain

$$(f_t)_i^n + U \, (f_x)_i^n = \frac{1}{2} U \Delta x \, (1 - c) \, (f_{xx})_i^n + O(\Delta t^2) + U \, O(\Delta x^2). \qquad (12.4.19)$$

Neglecting the quadratic terms, we derive the modified differential equation

$$f_t + U f_x = \frac{1}{2} U \Delta x \, (1 - c) \, f_{xx}. \qquad (12.4.20)$$

The right-hand side represents a small diffusive term with numerical diffusivity

$$\kappa_{num} = \frac{1}{2} U \Delta x \, (1 - c), \qquad (12.4.21)$$

which is positive when $0 < c < 1$. Negative numerical diffusivity is tantamount to numerical instability.

FTFS discretization

Working in a similar fashion with the FTFS method, we derive a modified differential equation involving a small diffusion term involving the numerical diffusivity

$$\kappa_{num} = -\frac{1}{2} U \Delta x \, (1 + c), \qquad (12.4.22)$$

which is nonnegative when $-1 < c < 0$. Combining this result with the von Neumann stability analysis, we confirm that positive numerical diffusivity helps ensure numerical stability.

FTCS discretization

What went wrong with the FTCS discretization? Carrying out a consistency analysis, we derive a modified differential equation involving a small diffusion term with numerical diffusivity

$$\kappa_{num} = -\frac{1}{2} c U \Delta x = -\frac{1}{2} U^2 \Delta t, \qquad (12.4.23)$$

which is negative under any conditions.

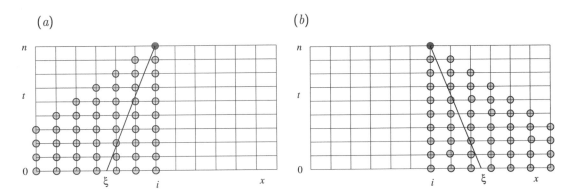

FIGURE 12.4.3   Cone of influence for the (a) FTBS and (b) FTFS method. The characteristic is drawn as a heavy line.

12.4.7   Numerical cone of influence

The exact solution of the linear convection equation at the x_i grid point at time t_n is found by traveling backward in time along the characteristic, as shown in Figure 12.4.3. When we have reached the initial time level, $t = 0$, where $x = \xi$, we stop and set

$$f_i^n = F(\xi). \tag{12.4.24}$$

According to the finite-difference equations derived in the section, to compute the value f_i^n, we use information at all grid points residing inside a planar angle with apex at the point (x_i, t_n), called the numerical cone of influence, as illustrated in Figure 12.4.3. The CFL condition, $|c| < 1$, requires that the characteristic lies inside the numerical cone of influence. If the numerical cone of influence does not include the characteristic that passes through a node, the numerical method will surely be unstable. The CFL condition is thus necessary but not sufficient for numerical stability.

12.4.8   Lax's modification of the FTCS method

Lax proposed a modification of the unconditionally unstable FTCS method meant to introduce a stabilizing diffusive action [226]. Replacing f_i^n in the finite-difference approximation of the time derivative with an average value, $\frac{1}{2}(f_{i+1}^n + f_{i-1}^n)$, we obtain the Lax difference equation

$$\frac{f_i^{n+1} - \frac{1}{2}\left(f_{i+1}^n + f_{i-1}^n\right)}{\Delta t} + U\,\frac{f_{i+1}^n - f_{i-1}^n}{2\,\Delta x} = 0. \tag{12.4.25}$$

Rearranging, we derive an explicit two-level formula,

$$f_i^{n+1} = \frac{1}{2}\left(1 + c\right) f_{i-1}^n + \frac{1}{2}\left(1 - c\right) f_{i+1}^n. \tag{12.4.26}$$

However, since this formula was obtained by an *ad hoc* modification of the well-founded FTCS discretization, its consistency must be examined. Expanding all variables in Taylor series about the

x_i grid point at time t_n, we derive the modified differential equation

$$f_t + U f_x = -\frac{1}{2} \Delta t \, f_{tt} + \frac{1}{2} \frac{\Delta x^2}{\Delta t} \, f_{xx}. \tag{12.4.27}$$

We conclude that the difference equation is consistent only when Δx and Δt are reduced simultaneously so that the ratio $\Delta x^2 / \Delta t$ tends to zero and the last term on the right-hand side becomes infinitesimal. Keeping the ratio $\Delta x / \Delta t$ constant yields a method that is first-order accurate in time and space.

Performing the von Neumann stability analysis, we derive the amplification factor

$$G = \cos\theta + \mathrm{i}\, c \, \sin\theta. \tag{12.4.28}$$

Requiring that $|G| < 1$, we find that $\cos^2\theta + c^2 \sin^2\theta < 1$, which reproduces the CFL condition, $|c| < 1$.

To explain the conditional stability of Lax's method, we substitute $f_{tt} = U^2 f_{xx}$ into the modified differential equation and group the first with the second term on the right-hand side to obtain the numerical diffusivity

$$\kappa_{num} = \frac{1}{2} U \Delta x \left(\frac{1}{c} - c \right), \tag{12.4.29}$$

which is positive when $|c| < 1$. As the time step and thus c tends to zero, the numerical diffusivity becomes excessive, undermining the physical relevance of the calculations.

12.4.9   The Lax–Wendroff method

Upwind (FTBS or FTFS) differencing offers numerical stability but suffers from low-order spatial accuracy. FTCS differencing offers second-order spatial accuracy but suffers from unconditional instability. We want to develop a two-level explicit method that combines accuracy and stability, that is, a method that is second-order accurate in time and space as well as conditionally stable. In the Lax–Wendroff method, the next value, f_i^{n+1}, is expressed as a linear combination of three previous values, f_{i-1}^n, f_i^n, and f_{i+1}^n,

$$f_i^{n+1} = a_{-1} f_{i-1}^n + a_0 f_i^n + a_1 f_{i+1}^n, \tag{12.4.30}$$

where a_{-1}, a_0, and a_1 are three constant coefficients [228]. This formula encompasses all two-level methods considered previously in this section.

To ensure consistency, we expand all variables in (12.4.30) in Taylor series about (x_i, t_n) and obtain

$$f_i^n + (f_t)_i^n \, \Delta t + \frac{1}{2} (f_{tt})_i^n \, \Delta t^2 = a_{-1} \left[f_i^n - (f_x)_i^n \, \Delta x + \frac{1}{2} (f_{xx})_i^n \, \Delta x^2 \right] + a_0 \, f_i^n$$

$$+ a_1 \left[f_i^n + (f_x)_i^n \, \Delta x + \frac{1}{2} (f_{xx})_i^n \, \Delta x^2 \right]. \tag{12.4.31}$$

Rearranging, we derive the modified differential equation

$$(1 - a_{-1} - a_0 - a_1) f_i^n + \Delta t \left[(f_t)_i^n + \frac{U}{c} (a_{-1} + a_1) (f_x)_i^n \right]$$
$$= \frac{1}{2} (a_{-1} + a_1) (f_{xx})_i^n \Delta x^2 - \frac{1}{2} (f_{tt})_i^n \Delta t^2. \tag{12.4.32}$$

Next, we substitute $f_{tt} = U^2 f_{xx}$ and require that the modified differential equation reduces to the convection equation as Δx and Δt tend to zero. Stipulating also that the second-order error due to the temporal discretization cancels the second-order error due to the spatial discretization, we obtain a system of algebraic equations for the three unknown coefficients,

$$\begin{bmatrix} 1 & 1 & 1 \\ 1 & 0 & -1 \\ 1 & 0 & 1 \end{bmatrix} \begin{bmatrix} a_{-1} \\ a_0 \\ a_1 \end{bmatrix} = \begin{bmatrix} 1 \\ c \\ c^2 \end{bmatrix}. \tag{12.4.33}$$

The solution is

$$a_{-1} = \frac{1}{2} c (c + 1), \qquad a_0 = 1 - c^2, \qquad a_1 = \frac{1}{2} c (c - 1). \tag{12.4.34}$$

Substituting these values into (12.4.30), we derive the Lax–Wendroff formula

$$f_i^{n+1} = \frac{1}{2} c (c + 1) f_{i-1}^n + (1 - c^2) f_i^n + \frac{1}{2} c (c - 1) f_{i+1}^n. \tag{12.4.35}$$

Rearranging, we find that

$$f_i^{n+1} = f_i^n + \frac{1}{2} c (f_{i-1}^n - f_{i+1}^n) + \frac{1}{2} c^2 (f_{i-1}^n - 2 f_{i+1}^n + f_{i+1}^n). \tag{12.4.36}$$

The first two terms on the right-hand side implement the FTCS discretization. The last term cancels a term originating from the second time derivative of the modified differential equation.

To assess the numerical stability of the Lax–Wendroff method, we carry out the von Neumann stability analysis and find that

$$G = 1 - c^2 + c^2 \cos\theta + i\, c \sin\theta, \tag{12.4.37}$$

which shows that the gain traces an ellipse with center at the point $(1 - c^2, 0)$ and semi-axes equal to c^2 and c in the complex plane. When $|c| = 1$, the ellipse reduces to the unit circle centered at the origin. Geometrical arguments reveal that $|G| < 1$ when $|c| < 1$, and this demonstrates that the method is stable when the CFL condition is fulfilled.

12.4.10 Explicit CTCS or leapfrog method

Another way to achieve second-order accuracy in time and space is by using central differences for both the time and space derivatives, obtaining the difference equation

$$\frac{f_i^{n+1} - f_i^{n-1}}{2\Delta t} + U \frac{f_{i+1}^n - f_{i-1}^n}{2\Delta x} = 0. \tag{12.4.38}$$

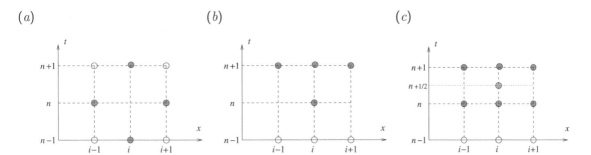

FIGURE 12.4.4   Computational stencil of (a) the explicit CTCS discretization involving three time levels, (b) the implicit BTCS discretization involving two time levels. and (c) the implicit Crank–Nicolson discretization involving three time levels.

The associated numerical error is on the order of Δt^2 and Δx^2. The computational stencil is shown in Figure 12.4.4(a). Rearranging, we derive the two-level explicit formula

$$f_i^{n+1} = f_i^{n-1} + c\,f_{i-1}^n - c\,f_{i+1}^n. \tag{12.4.39}$$

The solution at the first time level, $n = 1$, must be computed using a two-level method with a sufficiently small time step to prevent the onset of numerical instability.

To perform the von Neumann stability analysis, we substitute $f_i^n = A^n \exp(-\mathrm{i}\,i\theta)$ and find that

$$A^{n+1} = A^{n-1} + 2\,\mathrm{i}\,A_n\,c\,\sin\theta. \tag{12.4.40}$$

Dividing both sides by A^n, we obtain

$$\frac{A^{n+1}}{A^n} = \frac{A^{n-1}}{A^n} + 2\mathrm{i}\,c\,\sin\theta. \tag{12.4.41}$$

Next, we set $A^{n+1}/A^n = A^n/A^{n-1} = G$, and derive a quadratic equation, $G^2 - 2\mathrm{i}\,c\sin\theta\,G - 1 = 0$, whose solution is

$$G = \mathrm{i}\,\delta \pm \sqrt{1 - \delta^2}, \tag{12.4.42}$$

where $\delta = c\,\sin\theta$. To assess the magnitude of the amplification factor, we consider two complementary possibilities. If $\delta^2 > 1$, then $G = \mathrm{i}\,(\delta \pm \sqrt{\delta^2 - 1}\,)$ and the magnitude of one of the roots is greater than unity. If $\delta^2 < 1$, then $|G| = 1$, which shows that the magnitude of the solution stays constant during the successive mappings, in agreement with the exact solution. We note that $\delta^2 < 1$ when $|c| < 1$ for any value of θ and conclude that the CTCS method is stable provided that the CFL condition is fulfilled. Since the CTCS discretization for the unsteady diffusion equation was found to be unconditionally unstable, we infer that the presence of diffusivity does not necessarily promote numerical stability.

In practice, the efficiency of the CTCS method is undermined by increased memory requirements associated with storage of information at three time levels. In addition, numerical error that grows independently at every other time step, described as even-odd coupling, may arise.

12.4.11   Implicit BTCS method

Next, we turn to considering implicit discretizations in the hope of achieving unconditional stability and thus relaxing the restriction on the time step, Δt. The BTCS discretization yields the difference equation

$$\frac{f_i^{n+1} - f_i^n}{\Delta t} + U \frac{f_{i+1}^{n+1} - f_{i-1}^{n+1}}{2\,\Delta x} = 0. \tag{12.4.43}$$

The numerical error is of order Δt and Δx^2. The computational stencil is shown in Figure 12.4.4(b). Rearranging, we derive a two-level implicit formula,

$$-c\,f_{i-1}^{n+1} + 2f_i^{n+1} + c\,f_{i+1}^{n+1} = 2f_i^n. \tag{12.4.44}$$

In the case of periodic boundary conditions, time-stepping is carried out by solving a linear system of equations,

$$\mathbf{A} \cdot \mathbf{f}^{n+1} = \mathbf{f}^n, \tag{12.4.45}$$

with an $N \times N$ tridiagonal coefficient matrix,

$$\mathbf{A} = \begin{bmatrix} 2 & c & 0 & \cdots & 0 & -c \\ -c & 2 & c & \cdots & 0 & 0 \\ 0 & -c & 2 & c & \cdots & 0 \\ \cdots & \cdots & \cdots & \cdots & \cdots & \cdots \\ 0 & \cdots & 0 & -c & 2 & c \\ c & 0 & \cdots & 0 & -c & 2 \end{bmatrix}. \tag{12.4.46}$$

Note that this matrix is equal to twice the transpose of the projection matrix for the FTCS discretization. The projection matrix of the BTCS discretization is $\mathbf{P} = \mathbf{A}^{-1}$. Since the eigenvalues of $\mathbf{P}$ are the inverses of the eigenvalues of $\mathbf{A}$, the numerical method is unconditionally stable. Carrying out the von Neumann stability analysis, we derive the amplification factor

$$G = \frac{1}{1 - \mathrm{i}\,c\sin\theta}. \tag{12.4.47}$$

Since the magnitude of G is always less than unity, the BTCS method is confirmed to be unconditionally stable. The increased computational effort required for solving a linear system at each step is rewarded with unconditional stability. The reason is that the boundaries of the numerical cone of influence include all points at the $n+1$ time level. Consequently, the numerical cone of influence reduces to a rectangular strip, which is guaranteed to contain the characteristic line passing through any grid point.

12.4.12   Crank–Nicolson method

The Crank–Nicolson method arises by applying the differential equation midway between the present and next time levels, t_n and t_{n+1}, and approximating the spatial derivative with the average of two derivatives at two time levels according to the difference equation

$$\frac{f_i^{n+1} - f_i^n}{\Delta t} + U \frac{1}{2} \left(\frac{f_{i+1}^{n+1} - f_{i-1}^{n+1}}{2\,\Delta x} + \frac{f_{i+1}^n - f_{i-1}^n}{2\,\Delta x} \right) = 0. \tag{12.4.48}$$

The numerical error is of order Δt^2 and Δx^2. The computational stencil is shown in Figure 12.4.4(c). Rearranging, we derive the difference equation

$$-cf_{i-1}^{n+1} + 4f_i^{n+1} + cf_{i+1}^{n+1} = cf_{i-1}^n + 4f_i^n - cf_{i+1}^n, \qquad (12.4.49)$$

yielding a tridiagonal system of algebraic equations at each time step. Carrying out the von Neumann stability analysis, we find that the amplification factor is the ratio of two conjugate numbers,

$$G = \frac{2 + \mathrm{i}\,c\,\sin\theta}{2 - \mathrm{i}\,c\,\sin\theta}. \qquad (12.4.50)$$

Since $|G| = 1$, the method is unconditionally stable.

12.4.13   Summary

The restriction on the time step of conditionally stable explicit methods is not intolerable. The CFL condition requires that the time step, Δt, be roughly smaller than Δx in appropriate dimensionless units, which is not unreasonable. In contrast, the stability restriction for the diffusion equation discussed in Section 12.2 requires that the time step, Δt, be roughly smaller than Δx^2 in appropriate dimensionless units, which is typically intolerable. Although implicit methods allow us to use larger time steps, the associated numerical error may erode the accuracy and therefore the physical relevance of the solution. It is then not surprising that explicit methods are a standard choice in practice.

12.4.14   Modified dynamics and explicit numerical diffusion

We saw that finite-difference discretizations introduce some type of numerical error whose leading-order term is proportional to the second, third, or fourth spatial derivative of the solution. The presence of this term implements, respectively, numerical diffusion, dispersion, and fourth-order diffusion. A certain amount of numerical diffusion is necessary to ensure numerical stability and dampen small-scale irregularities. Monotone schemes produce solutions that are free of artificial oscillations. In practice, a prohibitively fine grid may be required to reduce the artificial smearing of physical sharp gradients. A remedy would be to use a higher-order method, but the associated modified differential equation typically contains a dispersive term that causes local oscillations.

The intensity of artificial oscillations can be reduced by adding to the original differential equation explicit diffusion or fourth-order diffusion. In one dimension, these are implemented, respectively, by the terms $\beta(x)\,\partial^2 f/\partial x^2$ and $-\gamma(x)\,\partial^4 f/\partial x^4$. The positive coefficients β and γ can be allowed to vary in space and time according to the structure of the solution. The optimal values of these coefficients depend on the particular problem under consideration and must be found by numerical experimentation (e.g., [185], p. 207). A weak justification for introducing explicit numerical diffusion is that the approximations that lead us to the convection equation have neglected physical diffusion-like processes. Artificially introducing these terms into the differential equation may not undermine the physical relevance of the solution.

12.4.15 Multistep methods

The performance of explicit methods can be improved by breaking up each time step into a number of elementary substeps executed using different methods. Some substeps predict the solution at the next time level using a crude method, while other substeps make corrections using a more sophisticated method. Accordingly, multistep methods are sometimes classified as predictor–corrector methods.

The overall action of multistep methods for linear convection can be reduced to that of single-step methods developed earlier in this section. Their discussion in the context of linear equations serves as a point of reference for developing methods for nonlinear convection.

Richtmyer's method

In Richtmyer's method [342], a complete step is carried out in two substeps of equal duration $\frac{1}{2}\Delta t$. The first substep is executed using Lax's method based on formula (12.4.26), and the second substep is executed using the CTCS method based on formula (12.4.39). The corresponding difference equations are

$$f_i^{n+1/2} = \frac{1}{2}\left(1+\frac{c}{2}\right)f_{i-1}^n + \frac{1}{2}\left(1-\frac{c}{2}\right)f_{i+1}^n, \qquad f_i^{n+1} = f_i^n + \frac{c}{2}f_{i-1}^{n+1/2} - \frac{c}{2}f_{i+1}^{n+1/2}, \qquad (12.4.51)$$

where $c = U\Delta t/\Delta x$. Eliminating from the second equation the solution at the intermediate $n+\frac{1}{2}$ level, we obtain the Lax–Wendroff formula (12.4.35) with grid spacing $2\Delta x$,

$$f_i^{n+1} = \frac{1}{2}\varrho(\varrho+1)f_{i-2}^n + (1-\varrho^2)f_i^n + \frac{1}{2}\varrho(\varrho-1)f_{i+2}^n, \qquad (12.4.52)$$

where $\varrho = \frac{1}{2}c = U\Delta t/(2\Delta x)$. Our analysis of the Lax–Wendroff formula guarantees that Richtmyer's method is second-order accurate in time and space, and stable as long as $|\varrho| < 1$ or $|c| < 2$.

Burstein's method

Burstein's method [47] differs from Richtmyer's method in that the convection equation is applied at the intermediate grid point $i+\frac{1}{2}$ at the first substep and at the regular grid point i at the second substep. In both cases, the spatial step is $\frac{1}{2}\Delta x$. The counterparts of equations (12.4.51) are

$$f_{i+1/2}^{n+1/2} = \frac{1}{2}(1+c)f_i^n + \frac{1}{2}(1-c)f_{i+1}^n, \qquad f_i^{n+1} = f_i^n + cf_{i-1/2}^{n+1/2} - cf_{i+1/2}^{n+1/2}, \qquad (12.4.53)$$

where $c = U\Delta t/\Delta x$. Eliminating the intermediate solution from the second equation, we obtain the Lax–Wendroff formula (12.4.35). We conclude that Burstein's method is second-order accurate in time and space, and stable as long as $|c| < 1$.

MacCormack method

In MacCormack's method [250], a predictor step is made using the explicit FTFS method to produce an approximation to f_i^{n+1} according to formula (12.4.14), denoted as f_i^*,

$$f_i^* = (1+c)f_i^n - cf_{i+1}^n. \qquad (12.4.54)$$

Rearranging, we obtain

$$f_i^* = f_i^n - c\,(f_{i+1}^n - f_i^n).\qquad(12.4.55)$$

The corrector step uses the explicit forward time discretization and a hybrid forward/backward space approximation involving the preliminary values according to the difference equation

$$\frac{f_i^{n+1} - f_i^n}{\Delta t} + \frac{U}{2}\,\Big(\frac{f_{i+1}^n - f_i^n}{\Delta x} + \frac{f_i^* - f_{i-1}^*}{\Delta x} \Big) = 0.\qquad(12.4.56)$$

Rearranging, we obtain

$$f_i^{n+1} = f_i^n - \frac{1}{2}\,c\,(f_{i+1}^n - f_i^n + f_i^* - f_{i-1}^*).\qquad(12.4.57)$$

To deduce the action of the method, we use equation (12.4.55) to eliminate f_{i+1}^n on the right-hand side of (12.4.57) in favor of f_i^*, we obtain

$$f_i^{n+1} = f_i^n - \frac{1}{2}\,(f_i^n - f_i^*) - \frac{1}{2}\,c\,(f_i^* - f_{i-1}^*),\qquad(12.4.58)$$

and then

$$f_i^{n+1} = \frac{1}{2}\,[\, f_i^n + (1 - c)\,f_i^* + c f_{i-1}^* \,].\qquad(12.4.59)$$

Eliminating the intermediate variables denoted by an asterisk using equation (12.4.54), we recover the Lax–Wendroff formula (12.4.35). We conclude that MacCormack's method is second-order accurate in time and space, and stable as long as the CFL condition is fulfilled, $|c| < 1$. In Section 12.4.16, we will see that MacCormack's method is particularly effective for problems of nonlinear convection.

Flux-Corrected-Transport

We have seen that explicit numerical diffusion may be necessary to enhance the performance of first-order methods. The idea behind flux-corrected transport is to use a predictor–corrector method where the predictor step involves an artificial dampening term [49, 50, 51]. The corrector step removes excessive dissipation by introducing diffusion with a negative diffusivity, termed anti-diffusion.

12.4.16 Nonlinear convection

Most methods for linear convection can be extended to nonlinear convection where the convection velocity, U, is no longer a constant. The main new feature is that U is evaluated at the grid point where the differential equation is applied to yield a difference equation. When a finite-difference method is conditionally stable, the time step must be sufficiently small to satisfy the stability criteria for linear convection with the local velocity over the entire solution domain. Implicit methods lead to a system of nonlinear algebraic equations that must be solved by iterative methods at a high computational cost. Accurate and efficient numerical solutions can be obtained using multistep methods, including and MacCormack's method discussed in this section for the inviscid Burgers equation.

Inviscid Burgers equation

The convective and conservative forms of Burgers equation are

$$\frac{\partial f}{\partial t} + f \frac{\partial f}{\partial x} = 0, \qquad \frac{\partial f}{\partial t} + \frac{\partial q}{\partial x} = 0, \tag{12.4.60}$$

where $q = \frac{1}{2} f^2$ is an effective local kinetic energy. Physically, Burgers's equation describes the propagation of an initial distribution with a convection velocity that is proportional to the local amplitude of the distribution. The dependence of the local propagation velocity on the local amplitude causes profile steepening and allows the formation of discontinuous fronts and shocks from smooth initial shapes.

Lax's method

Lax's modification of the FTCS discretization yields the explicit two-level formula

$$f_i^{n+1} = \frac{1}{2} \left(1 + c_{i-1}^n\right) f_{i-1}^n + \frac{1}{2} \left(1 - c_{i+1}^n\right) f_{i+1}^n, \tag{12.4.61}$$

where

$$c_{i-1}^n = r f_{i-1}^n, \qquad c_{i+1}^n = r f_{i+1}^n, \tag{12.4.62}$$

and $r = \Delta t / \Delta x$. The method is first-order accurate in time, second-order accurate in space, and stable provided that $|r f_{max}| < 1$.

Lax–Wendroff method

To achieve second-order accuracy, we discretize the conservative form as

$$f_i^{n+1} = f_i^n + \frac{1}{2} r \left(q_{i-1}^n - q_{i+1}^n\right) + \frac{1}{4} r^2 \left(A q_{i-1}^n - B q_i^n + C q_{i+1}^n\right), \tag{12.4.63}$$

where $A, B,$ and C are adjustable coefficients and $r = \Delta t / \Delta x$. The second term on the right-hand side of (12.4.63) arises from the second term on the right-hand side of (12.4.36) by setting $c = \frac{1}{2} \left(f_{i-1}^n + f_{i+1}^n\right)$. Working as in the case of the linear convection, we find that

$$A = f_{i-1}^n + f_i^n, \qquad B = f_{i-1}^n + 2 f_i^n + f_{i+1}^n, \qquad C = f_i^n + f_{i+1}^n \tag{12.4.64}$$

(e.g., [185], p. 207). The method is second-order accurate in space and time, and stable provided that $|r f_{max}| < 1$.

BTCS discretization

The implicit BTCS discretization results in a quadratic system of algebraic equations that must be solved by iteration,

$$-r \left(f_{i-1}^{n+1}\right)^2 + 4 f_i^{n+1} + r \left(f_{i+1}^{n+1}\right)^2 = 4 f_i^n, \tag{12.4.65}$$

where $r = \Delta t / \Delta x$. A suitable initial guess is provided by the converged solution at the previous time step. In practice, increased computational cost renders this method uneconomical.

MacCormack's method

Substituting into the difference formula (12.4.55) the expression $c = \frac{1}{2}\left(f_{i+1}^n + f_i^n\right)$, we obtain a tentative update denoted by an asterisk,

$$f_i^* = f_i^n - c\left(q_{i+1}^n - q_i^n\right). \tag{12.4.66}$$

Setting in the difference formula (12.4.59) $c = \frac{1}{2}\left(f_i^* + f_{i-1}^*\right)$, we obtain the solution at the next time level,

$$f_i^{n+1} = \frac{1}{2}\left[f_i^n + f_i^* - c\left(q_i^* - q_{i-1}^*\right)\right]. \tag{12.4.67}$$

The bias in the solution due to the forward or backward differencing is eliminated by alternating the order after the completion of a time step. The method is stable as long as $|rf_{max}| < 1$.

12.4.17   Convection in two and three dimensions

Generalizing our discussion of the one-dimensional convection equation, we consider linear convection in three dimensions governed by the equation

$$\frac{\partial f}{\partial t} + U\frac{\partial f}{\partial x} + V\frac{\partial f}{\partial y} + W\frac{\partial f}{\partial z} = 0. \tag{12.4.68}$$

The convection velocity, $\mathbf{U} = (U, V, W)$, is assumed to be constant in time and space. The case of two-dimensional convection arises by setting $W = 0$. An initial condition, $f(\mathbf{x}, t = 0) = F(\mathbf{x})$, and suitable boundary conditions are provided. Assuming, for simplicity, that the solution domain extends over the whole three-dimensional space or two-dimensional plane, we find that the exact solution is given by $F(\mathbf{x} - \mathbf{U}t)$, which states that the initial distribution, $F(\mathbf{x})$, travels with constant velocity $\mathbf{U}$. The characteristic lines along which the function f remains constant, are described by the equation $\mathbf{x} - \mathbf{U}t = \boldsymbol{\xi}$, where $\boldsymbol{\xi}$ is an arbitrary point in space.

Finite-difference methods for two-dimensional and three-dimensional convection arise by direct and straightforward extensions those for one-dimensional convection. In this section, we discuss certain characteristic examples.

Lax's method

Lax's method in two dimensions emerges by replacing (12.4.68) with the finite-difference equation

$$\frac{1}{\Delta t}\left[f_{i,j}^{n+1} - \frac{1}{4}\left(f_{i+1,j}^n + f_{i-1,j}^n + f_{i,j+1}^n + f_{i,j-1}^n\right)\right] + U\frac{f_{i+1,j}^n - f_{i-1,j}^n}{2\Delta x} + V\frac{f_{i,j+1}^n - f_{i,j-1}^n}{2\Delta y} = 0. \tag{12.4.69}$$

Rearranging, we obtain the explicit two-level algorithm

$$f_{i,j}^{n+1} = \frac{1}{2}\left(\frac{1}{2} + c_x\right)f_{i-1,j}^n + \frac{1}{2}\left(\frac{1}{2} - c_x\right)f_{i+1,j}^n + \frac{1}{2}\left(\frac{1}{2} + c_y\right)f_{i,j-1}^n + \frac{1}{2}\left(\frac{1}{2} - c_y\right)f_{i,j+1}^n, \tag{12.4.70}$$

where

$$c_x = \frac{U\Delta t}{\Delta x}, \qquad c_y = \frac{V\Delta t}{\Delta y} \qquad (12.4.71)$$

are convection numbers for the x and y directions. Performing the von Neumann stability analysis, we find that the method is stable provided that $c_x^2 + c_y^2 < \frac{1}{2}$. This condition imposes a strong restriction on the time step that renders Lax's method inefficient in practice. In three dimensions, the stability constraint is $c_x^2 + c_y^2 + c_z^2 < \frac{1}{3}$, which describes the interior of a sphere with radius $1/\sqrt{3}$ centered at the origin, where $c_z = W\Delta t/\Delta z$ and W is the convection velocity along the z axis.

Implicit methods

To achieve unconditional stability, we may resort to a fully implicit method. In three dimensions, we obtain the difference equation

$$\left[1 + \frac{1}{2}\left(c_x\Delta_x + c_y\Delta_y + c_z\Delta_z\right)\right]f_{i,j,k}^{n+1} = f_{i,j,k}^n, \qquad (12.4.72)$$

where

$$\Delta_x(f_{i,j,k}) \equiv f_{i+1,j,k} - f_{i-1,j,k}, \qquad \Delta_y(f_{i,j,k}) \equiv f_{i,j+1,k} - f_{i,j-1,k},$$
$$(12.4.73)$$
$$\Delta_z(f_{i,j,k}) \equiv f_{i,j,k+1} - f_{i,j,k-1}$$

are central-difference operators. Unfortunately, collecting the finite-difference equations results in a system of algebraic equations that are sparse but not tridiagonal. Numerical methods for such systems with large size require a prohibitive amount of computational effort.

As a remedy, we may use an ADI partially implicit discretization, which requires solving a one-dimensional equation in each substep (e.g., [110]; [182], Volume I, p. 442; [229]). The standard ADI algorithm is unconditionally stable in two dimensions and unconditionally unstable in three dimensions.

Operator splitting

Following the general blueprint of operator splitting and fractional step methods, we replace the three-dimensional convection equation (12.4.68) with a system of three one-dimensional sequential equations,

$$\frac{\partial f}{\partial t} + U\frac{\partial f}{\partial x} = 0, \qquad \frac{\partial f}{\partial t} + V\frac{\partial f}{\partial y} = 0, \qquad \frac{\partial f}{\partial t} + W\frac{\partial f}{\partial z} = 0. \qquad (12.4.74)$$

Each equation applies for the full time step, $t_n < t < t_n + \Delta t$.

Adopting the implicit BTCS discretization, we advance the solution in each step by solving tridiagonal systems of equations,

$$(1 + \frac{1}{2}c_x\Delta_x)f_{i,j,k}^* = f_{i,j,k}^n, \qquad (1 + \frac{1}{2}c_y\Delta_y)f_{i,j,k}^{**} = f_{i,j,k}^*, \qquad (1 + \frac{1}{2}c_z\Delta_x)f_{i,j,k}^{n+1} = f_{i,j,k}^{**}, \qquad (12.4.75)$$

where the starred and double starred variables denote the solution after the first and second fractional steps. Combining the three equations in (12.4.75), we derive an overall finite-difference scheme involving a factorized implicit operator in the left-hand side,

$$(1 + \frac{1}{2}\, c_x \Delta_x)(1 + \frac{1}{2}\, c_y \Delta_y)(1 + \frac{1}{2}\, c_z \Delta_z) f_{i,j,k}^{n+1} = f_{i,j,k}^n. \tag{12.4.76}$$

Comparing this formula with the fully implicit formula (12.4.72) shows that the fractional step method implements the approximate factorization of the explicit BTCS scheme.

Problems

12.4.1 *Numerical diffusivity*

Carry out a consistency analysis of the FTCS and Lax's methods and derive the corresponding numerical diffusivities.

12.4.2 *Lax–Wendroff method for the inviscid Burgers equation*

Derive equations (12.4.64) working as in the case of linear convection.

12.4.3 *Lax's method in two and three dimensions*

(*a*) Perform the von Neumann stability analysis to derive the stability criterion $c_x^2 + c_y^2 < \frac{1}{2}$. Derive the associated numerical diffusivity.

(*b*) Analyze the stability of the method in three dimensions and derive the stability criterion $c_x^2 + c_y^2 + c_z^2 < \frac{1}{3}$.

12.4.4 *ADI method in two dimensions*

Devise an ADI method for linear two-dimensional convection, and study its consistency and stability. Discuss how the method can be extended to three dimensions.

Computer Problems

12.4.5 *Burgers equation with MacCormack's method*

Use MacCormack's method to solve the inviscid Burgers equation in an infinite domain with initial condition (*a*) the inverted Heaviside step function, $F(x) = 1$ for $x < 0$ and $F(x) = 0$ for $x > 0$, and (*b*) the Gaussian function $F(x) = \exp(-x^2)$. Discuss the behavior of the numerical solution in each case.

12.4.6 *Burgers equation with an implicit method*

Repeat Problem 12.4.3 with the implicit BTCS discretization. Discuss the performance of the method and comment on the computational cost.

12.5 Convection–diffusion

The methods developed previously in this chapter for the modular cases of pure diffusion and pure convection can be extended in a straightforward fashion to handle combined convection–diffusion. In this section, we discuss methods for linear one-dimensional convection–diffusion governed by the parabolic differential equation

$$\frac{\partial f}{\partial t} + U \frac{\partial f}{\partial x} = \kappa \frac{\partial^2 f}{\partial x^2}, \tag{12.5.1}$$

subject to an initial condition, $f(x, t = 0) = F(x)$. Both the convection velocity, U, and diffusivity, κ, are assumed to be constant.

12.5.1 Explicit FTCS and Lax's method

The forward time–centered space (FTCS) discretization yields the difference equation

$$\frac{f_i^{n+1} - f_i^n}{\Delta t} + U \frac{f_{i+1}^n - f_{i-1}^n}{2\Delta x} = \kappa \frac{f_{i-1}^n - 2f_i^n + f_{i+1}^n}{\Delta x^2}. \tag{12.5.2}$$

Rearranging, we obtain the explicit two-level algorithm

$$f_i^{n+1} = \left(\frac{1}{2}c + \alpha\right) f_{i-1}^n + (1 - 2\alpha) f_i^n - \left(\frac{1}{2}c - \alpha\right) f_{i+1}^n, \tag{12.5.3}$$

where

$$c = \frac{U\Delta t}{\Delta x}, \qquad \alpha = \frac{\kappa \Delta t}{\Delta x^2} \tag{12.5.4}$$

are the dimensionless convection and diffusion numbers. The ratio of these two numbers is defined as the cell Reynolds number, sometimes also called the cell Péclet number,

$$\mathrm{Re}_c \equiv \frac{c}{\alpha} = \frac{U\Delta x}{\kappa}. \tag{12.5.5}$$

Physically, Re_c expresses the relative strengths of the convective and diffusive contributions to the difference equation underlying (12.5.1).

In previous sections, we found that the explicit FTCS scheme is conditionally stable for unsteady diffusion and unconditionally unstable for linear convection. Performing the von Neumann stability analysis for convection–diffusion, we derive the amplification factor

$$G = 1 - 2\alpha + 2\alpha \cos\theta + \mathrm{i}\,c \sin\theta, \tag{12.5.6}$$

which shows that G traces an ellipse passing through the point $(1, 0)$ in the complex plane. The center of the ellipse is located at the point $(1 - 2\alpha, 0)$, and the ellipse semi-axes are equal to 2α and c. To guarantee stability, we require that the ellipse lies inside the unit disk.

First, we demand that the length of each semi-axis is less than unity,

$$\alpha < \frac{1}{2}, \qquad c < 1. \tag{12.5.7}$$

A further restriction emerges by requiring that the curvature of the ellipse at the point $(1,0)$ is higher than the curvature of the unit circle. Combining these restrictions, we obtain

$$c^2 < 2\alpha < 1. \tag{12.5.8}$$

It is instructive to recall that the FTCS discretization for pure convection leads to an unconditionally unstable scheme. We conclude that the presence of diffusion has a stabilizing action on the numerical method.

The modified differential equation corresponding to the difference equation (12.5.10) is the convection–diffusion equation with effective diffusivity

$$\kappa_{eff} = \kappa - \frac{1}{2} U^2 \Delta t = \kappa \left(1 - \frac{1}{2}\frac{c^2}{\alpha}\right). \tag{12.5.9}$$

Inequality (12.5.8) states that a necessary condition for numerical stability is that the positive physical diffusivity overtakes the negative numerical diffusivity in absolute value.

The difference equation (12.5.3) can be recast into the form

$$f_i^{n+1} = \alpha\left(1 + \frac{1}{2}\operatorname{Re}_c\right)f_{i-1}^n + (1-2\alpha)f_i^n + \alpha\left(1 - \frac{1}{2}\operatorname{Re}_c\right)f_{i+1}^n. \tag{12.5.10}$$

Consider an initial condition where the value of the function f is positive at one grid point and zero at all other grid points. Since the initial distribution will be convected and diffuse, the value of f should be positive at all grid points at all subsequent time levels. For this to be true, all three coefficients on the right-hand side of equation (12.5.10) must be positive, suggesting the physical restriction

$$\operatorname{Re}_c < 2. \tag{12.5.11}$$

Violation of this inequality may lead to unphysical overshooting.

The stability restriction (12.5.8) requires an excessively small time step, and this renders the FTCS discretization uneconomical. Lax's modification leads to an unconditionally unstable method. Regrettably, the explicit FTCS scheme and its variations are of limited interest.

12.5.2 Upwind differencing

Based on our experience with the convection equation, we expect that the stability of the FTCS method will improve by applying upwind differencing for the convective derivative. Assuming that U is positive, we use a forward difference for the time derivative, a backward difference for the first spatial derivative, and a centered difference for the second spatial derivative, and thus obtain the FTBSCS scheme

$$\frac{f_i^{n+1} - f_i^n}{\Delta t} + U\frac{f_i^n - f_{i-1}^n}{\Delta x} = \kappa\frac{f_{i-1}^n - 2f_i^n + f_{i+1}^n}{\Delta x^2}. \tag{12.5.12}$$

The numerical error is of order Δt and Δx. Rearranging, we obtain

$$f_i^{n+1} = (c+\alpha)f_{i-1}^n + (1-c-2\alpha)f_i^n + \alpha f_{i+1}^n. \tag{12.5.13}$$

Carrying out a consistency analysis, we find that the corresponding modified differential equation is the convection–diffusion equation with effective diffusivity

$$\kappa_{eff} = \kappa \left[1 + \frac{1}{2} (1 - c) \, \mathrm{Re}_c \right].$$
(12.5.14)

Since Re_c vanishes as Δx tends to zero, the method is consistent.

Performing the von Neumann stability analysis, we derive the amplification factor

$$G = 1 - (c + 2\alpha) + (c + 2\alpha) \cos \theta + \mathrm{i} \, c \sin \theta,$$
(12.5.15)

which shows that G traces an ellipse passing through the point $(1, 0)$ in the complex plane. The center of the ellipse is located at $(1 - c - 2\,\alpha, 0)$, and the semi-axes of the ellipse are $c + 2\alpha$ and c. To guarantee stability, we demand that the ellipse lies inside the unit disk and derive the stability constraint

$$c^2 < c + 2\alpha < 1.$$
(12.5.16)

When U is negative, we use a forward difference for the convective spatial derivative and a centered difference for the diffusive derivative. Working in a similar fashion, we find that the method is stable if $c^2 < |c| + 2\alpha < 1$. In practice, the numerical diffusivity associated with upwind differencing can be significant. This undesirable feature combined with the first-order accuracy and the conditional stability renders the method inferior to its alternatives.

Generalized upwind differencing

Upwind differencing can be generalized into a more comprehensive scheme,

$$\frac{f_i^{n+1} - f_i^n}{\Delta t} + \frac{1}{2} U \left[(1 + \beta) \frac{f_i^n - f_{i-1}^n}{\Delta x} + (1 - \beta) \frac{f_{i+1}^n - f_i^n}{\Delta x} \right] = \kappa \frac{f_{i-1}^n - f_i^n + -f_{i+1}^n}{\Delta x^2},$$
(12.5.17)

where β is an arbitrary constant. Setting $\beta = 0$, we recover the explicit FTCS scheme. Setting $\beta = 1$ when $U > 0$ and $\beta = -1$ when $U < 0$, we recover the first-order upwind differencing scheme discussed in Section 12.5.2. Carrying out the von Neumann stability analysis, we find that the method is stable provided that

$$c^2 < \beta c + 2\alpha < 1,$$
(12.5.18)

which is consistent with the formulas derived earlier for $\beta = 0, \pm 1$.

Higher-order methods

To improve the accuracy and reduce the numerical diffusivity of the first-order upwind method, we may approximate the convective derivative using a third-order backward difference involving four points, while keeping the central difference for the second spatial derivative [232]. When U is positive, the finite-difference equation is

$$\frac{f_i^{n+1} - f_i^n}{\Delta t} + U \frac{f_{i-2}^n - 6 f_{i-1}^n + 3 f_i^n + 2 f_{i+1}^n}{6 \, \Delta x} = \kappa \frac{f_{i-1}^n - 2 f_i^n + f_{i+1}^n}{\Delta x^2}.$$
(12.5.19)

The discretization error is of order Δt and Δx^2. Rearranging, we derive the Leonard difference equation

$$f_i^{n+1} = -\frac{1}{6}\,c\,f_{i-2}^n + (c+\alpha)\,f_{i-1}^n + (1 - \frac{1}{2}c - 2\alpha)\,f_i^n + (-\frac{1}{3}c + \alpha)\,f_{i+1}^n. \tag{12.5.20}$$

The corresponding modified differential equation is the convection–diffusion equation with an effective diffusivity identical to that of the FTCS scheme given in (12.5.9). Carrying out the von Neumann stability analysis, we derive the amplification factor

$$G = -\frac{c}{6}\,\exp(2i\theta) + (c+\alpha)\,\exp(i\theta) + 1 - \frac{c}{2} - 2\alpha + (-\frac{c}{3} + \alpha)\,\exp(-i\theta). \tag{12.5.21}$$

The stability criteria are substantially milder than those of the first-order upwind method.

12.5.3   Explicit CTCS and the Du Fort–Frankel method

We have found that the CTCS discretization is unconditionally unstable for unsteady diffusion and conditionally stable for pure convection. Does adding convection to diffusion have a stabilizing influence? Surprisingly, the answer is negative. The CTCS discretization for the convection–diffusion equation leads to an unconditionally unstable scheme.

The Du Fort–Frankel method is based on a variation of the CTCS discretization implemented by the formula

$$\frac{f_i^{n+1} - f_i^{n-1}}{2\Delta t} + U\,\frac{f_{i+1}^n - f_{i-1}^n}{2\Delta x} = \kappa\,\frac{f_{i-1}^n - (f_i^{n+1} + f_i^{n-1}) + f_{i+1}^n}{\Delta x^2}. \tag{12.5.22}$$

This is the CTCS discretization, except that the middle term in the numerator on the right-hand side has been replaced by an average. Rearranging, we derive an explicit three-level formula,

$$f_i^{n+1} = \frac{c+2\alpha}{1+2\alpha}\,f_{i-1}^n + \frac{1-2\alpha}{1+2\alpha}\,f_i^{n-1} + \frac{-c+2\alpha}{1+2\alpha}\,f_{i+1}^n. \tag{12.5.23}$$

A consistency analysis shows that the Du Fort–Frankel method produces results that approximate the solution of the convection–diffusion equation only if the ratio $(\Delta t / \Delta x)^2$ is sufficiently small.

To carry out the von Neumann stability analysis, we substitute $f_i^n = A^n \exp(-i\,i\theta)$ and set $G = A^{n+1}/A^n = A^n/A^{n-1}$ to find

$$G = \frac{c+2\alpha}{1+2\alpha}\,\exp(i\theta) + \frac{1-2\alpha}{1+2\alpha}\,\frac{1}{G} + \frac{-c+2\alpha}{2\alpha+1}\,\exp(-i\theta). \tag{12.5.24}$$

Rearranging, we obtain the quadratic equation

$$(1+2\alpha)\,G^2 - 2\,(2\alpha\cos\theta + i\,c\sin\theta)\,G - 1 + 2\alpha = 0. \tag{12.5.25}$$

Detailed consideration reveals that $|G| < 1$ when $|c| < 1$, and this shows that the Du Fort–Frankel method is stable as long as the CFL condition is fulfilled.

The absence of a stability restriction on the diffusion number, α, allows us to use large time steps. However, a sufficiently small time step must be used to ensure that the solution is accurate and consistent with the convection–diffusion equation.

12.5.4 Implicit methods

Implicit methods are unconditionally stable for pure diffusion and convection, and remain unconditionally stable for mixed convection–diffusion.

BTCS

Implementing a backward difference for the time derivative and centered differences for the convective and diffusive spatial derivatives, we derive the fully implicit BTCS difference formula

$$\frac{f_i^{n+1} - f_i^n}{\Delta t} + U \frac{f_{i+1}^{n+1} - f_{i-1}^{n+1}}{2\Delta x} = \kappa \frac{f_{i-1}^{n+1} - 2f_i^{n+1} + f_{i+1}^{n+1}}{\Delta x^2}. \tag{12.5.26}$$

The numerical error is of order Δt and Δx^2. Rearranging, we derive a tridiagonal system of algebraic equations,

$$-(c+2\alpha)\, f_{i-1}^{n+1} + 2\,(1+2\alpha)\, f_i^{n+1} + (c-2\alpha)\, f_{i+1}^{n+1} = 2f_i^n. \tag{12.5.27}$$

The corresponding amplification factor is found to be

$$G = \frac{1}{1 + 2\alpha(1 - \cos\theta) - \mathrm{i}\,c\sin\theta}. \tag{12.5.28}$$

Since $|G| < 1$ for any values of α and c, the method is unconditionally stable. However, the physical restriction $\mathrm{Re}_c < 2$ must be satisfied for the results to be physically meaningful under any conditions.

Crank–Nicolson method

To improve the temporal accuracy of the BTCS method, we implement the fully implicit Crank–Nicolson method according to the difference equation

$$\frac{f_i^{n+1} - f_i^n}{\Delta t} + U \frac{1}{2} \Big(\frac{f_{i+1}^{n+1} - f_{i-1}^{n+1}}{2\Delta x} + \frac{f_{i+1}^n - f_{i-1}^n}{2\Delta x} \Big)$$
$$= \kappa \frac{1}{2} \Big(\frac{f_{i-1}^{n+1} - 2f_i^{n+1} + f_{i+1}^{n+1}}{\Delta x^2} + \frac{f_{i-1}^n - 2f_i^n + f_{i+1}^n}{\Delta x^2} \Big). \tag{12.5.29}$$

The numerical error is of order Δt^2 and Δx^2. Rearranging, we derive a tridiagonal system of algebraic equations,

$$-(c+2\alpha)f_{i-1}^{n+1} + 4\,(1+\alpha)f_i^{n+1} + (c-2\alpha)f_{i+1}^{n+1} = (c+2\alpha)f_{i-1}^n + 4\,(1-\alpha)f_i^n - (c-2\alpha)f_{i+1}^n. \tag{12.5.30}$$

The amplification factor is

$$G = \frac{2 - 2\alpha(1 - \cos\theta) + \mathrm{i}\,c\sin\theta}{2 + 2\alpha(1 - \cos\theta) - \mathrm{i}\,c\sin\theta}. \tag{12.5.31}$$

It can be shown that $|G| < 1$ for any value of α and c, and thus the method is unconditionally stable. The physical restriction $\mathrm{Re}_c < 2$ still applies.

Two-level fully implicit method

The general form of an implicit method involving three grid points and two time levels is

$$b_{-1} f_{i-1}^{n+1} + b_0 f_i^{n+1} + b_1 f_{i+1}^{n+1} = a_{-1} f_{i-1}^n + a_0 f_i^n + a_1 f_{i+1}^n, \tag{12.5.32}$$

where a_i and b_i are six constant coefficients. Consistency requires that

$$b_{-1} + b_0 + b_1 = a_{-1} + a_0 + a_1 = 1. \tag{12.5.33}$$

The last equality represents an arbitrary normalization that removes one degree of freedom. For example, in the case of the BTCS discretization, $a_{-1} = 0$, $a_0 = 1$, $a_1 = 0$, and

$$b_{-1} = -\frac{1}{2}(c + 2\alpha) \qquad b_0 = 1 + 2\alpha, \qquad b_1 = \frac{1}{2}(c - 2\alpha). \tag{12.5.34}$$

In the case of the Crank–Nicolson discretization,

$$a_{-1} = \frac{1}{4}(c + 2\alpha), \qquad a_0 = 1 - \alpha, \qquad a_1 = -\frac{1}{4}(c - 2\alpha),$$

$$b_{-1} = -\frac{1}{4}(c + 2\alpha), \qquad b_0 = 1 + \alpha, \qquad b_1 = \frac{1}{4}(c - 2\alpha). \tag{12.5.35}$$

Carrying out the von Neumann stability analysis for the general case, we derive the amplification factor

$$G = \frac{1 - (a_1 + a_{-1})(1 - \cos\theta) - \mathrm{i}\,(a_1 - a_{-1})\sin\theta}{1 - (b_1 + b_{-1})(1 - \cos\theta) - \mathrm{i}\,(b_1 - b_{-1})\sin\theta} \tag{12.5.36}$$

([294], p. 39). Formulas (12.5.28) and (12.5.31) are special cases of (12.5.36).

Three-level fully implicit method

We can achieve second-order temporal accuracy by using a scheme that combines three-level backward differencing in time and a fully implicit spatial discretization according to the difference equation

$$\frac{3 f_i^{n+1} - 4 f_i^n + f_i^{n-1}}{2\Delta t} + U \frac{f_{i+1}^{n+1} - f_{i-1}^{n+1}}{2\Delta x} = \kappa \frac{f_{i-1}^{n+1} - 2 f_i^{n+1} + f_{i+1}^{n+1}}{\Delta x^2}. \tag{12.5.37}$$

The numerical error is of order Δt^2 and Δx^2. Rearranging, we derive a tridiagonal system of algebraic equations for the solution vector at the next time level, t_{n+1}. A stability analysis reveals that the method is unconditionally stable. Comparing the amplification factor with that of the Crank–Nicolson method shows that small-amplitude oscillations are damped more effectively when the solution exhibits sharp variations. The physical restriction $\mathrm{Re}_c < 2$ must be fulfilled.

MacCormack explicit method

In this genuinely predictor–corrector method, the predictor step is implemented as an extension of formula (12.4.54),

$$f_i^* = (1 + c) f_i^n - c f_{i+1}^n + \alpha (f_{i-1}^n - 2 f_i^n + f_{i+1}^n), \tag{12.5.38}$$

and the corrector step is implemented as an extension of formula (12.4.59),

$$f_i^{n+1} = \frac{1}{2} \left(f_i^n + f_i^* - c \left(f_i^* - f_{i-1}^* \right) + \alpha \left(f_{i-1}^* - 2 f_i^* + f_{i+1}^* \right) \right). \tag{12.5.39}$$

The method is stable inside a square window described by $|c| < 0.90$ and $\alpha < 0.50$ (e.g., [184, 317]).

MacCormack implicit method

To develop the implicit MacCormack method, we repeat the steps that led us to equations (12.5.38) and (12.5.39), except that the diffusion term is treated implicitly in both the predictor and corrector steps. Details are provided by Hoffmann and Chiang ([185], Vol. I, p. 263).

12.5.5 Operator splitting and fractional steps

We have seen that certain discretizations work well for the convection equation, while other discretizations work well for the diffusion equation. This observation suggests the use of fractional steps where the convective and diffusive components are treated independently by different methods according to the constituent equations

$$\frac{\partial f}{\partial t} + U \frac{\partial f}{\partial x} = 0, \qquad \frac{\partial f}{\partial t} = \kappa \frac{\partial^2 f}{\partial x^2}. \tag{12.5.40}$$

Both equations apply for a full time step, $t_n < t < t_n + \Delta t$, with the understanding that the time is reset to the initial value, t_n, at the end of the first fractional step. The first step takes us from the current time level, f^n, to f^*, and the second step takes us from f^* to the new time level, f^{n+1}.

For example, applying the FTCS discretization, we obtain the difference equations

$$f_i^* = \frac{1}{2} c f_{i-1}^n + f_i^n - \frac{1}{2} c f_{i+1}^n, \qquad f_i^{n+1} = \alpha f_{i-1}^* + (1 - 2\alpha) f_i^* + \alpha f_{i+1}^*. \tag{12.5.41}$$

The stability restrictions are the same as those for the FTCS method applied to the undivided convection–diffusion equation.

Hopscotch method

The hopscotch method is named after a children's game ([269], p. 77; [149]). The explicit FTCS discretization is used to advance the solution at the odd-numbered grid points, x_{2i+1}, and then the implicit BTCS discretization is used to advance the solution at the even-numbered grid points, x_{2i}, where i is an integer. The order is reversed after the completion of each step. Since the solution at every other grid point at the new time level is known, the implicit step does not have to be done through matrix inversion and the method is effectively explicit. The hopscotch method is first-order accurate in time, second-order accurate in space, and stable as long as $|c| < 1$. There is no stability restriction on the diffusion number α.

The explicit FTCS update of the ith node from t_n to t_{n+1} yields

$$f_i^{n+1} = \left(\frac{1}{2} c + \alpha \right) f_{i-1}^n + (1 - 2\alpha) f_i^n - \left(\frac{1}{2} c - \alpha \right) f_{i+1}^n. \tag{12.5.42}$$

The implicit BTCS update of the same node from t_{n-1} to t_n yields

$$(\frac{1}{2}c + \alpha) f_{i-1}^n - (1 + 2\alpha) f_i^n - (\frac{1}{2}c - \alpha) f_{i+1}^n = -f_i^{n-1}. \qquad (12.5.43)$$

Combining these equations, we derive the difference equation

$$f_i^{n+1} = 2f_i^n - f_i^{n-1}, \qquad (12.5.44)$$

which implements linear extrapolation. This formula is applicable only in the FTCS step.

12.5.6   Nonlinear convection–diffusion

Our discussion of pure nonlinear convection in Section 12.4 also applies to the present case of mixed nonlinear convection–diffusion.

Burgers equation

A prototypical equation for studying the performance of numerical methods is the Burgers convection–diffusion equation whose convective and conservative forms are

$$\frac{\partial f}{\partial t} + f \frac{\partial f}{\partial x} = \kappa \frac{\partial^2 f}{\partial x^2}, \qquad \frac{\partial f}{\partial t} + \frac{\partial q}{\partial x} = \kappa \frac{\partial^2 f}{\partial x^2}, \qquad (12.5.45)$$

where $q = \frac{1}{2} f^2$ is the quadratic flux. The general solution in an unbounded domain extending over the entire x axis can be found analytically using the Cole–Hopf transformation [30],

$$f = -\frac{2\kappa}{u} \frac{du}{dx}. \qquad (12.5.46)$$

Direct substitution shows that the function u satisfies the linear unsteady diffusion

$$\frac{\partial u}{\partial t} = \kappa \frac{\partial^2 u}{\partial x^2}, \qquad (12.5.47)$$

whose exact solution is available in integral form. Note that the transformation fails as κ tends to zero, yielding the inviscid form. An exact solution is

$$f(x, t) = -\frac{\kappa}{L} \frac{\cosh(x/L)}{\sinh(x/L) + \exp(-\kappa t/L^2)}, \qquad (12.5.48)$$

where L is a specified length [30]. Note that a discontinuity occurs when the denominator becomes zero. As time progresses, the discontinuity progresses along the x axis and settles at the origin, $x = 0$.

The explicit MacCormack discretization arises by a straightforward modification of equations (12.4.66) and (12.4.67). The predictor step is

$$f_i^* = f_i^n - c\,(q_{i+1}^n - q_i^n) + \alpha\,(f_{i-1}^n - 2f_i^n + f_{i+1}^n), \qquad (12.5.49)$$

and the correction step is

$$f_i^{n+1} = \frac{1}{2} \left[f_i^n + f_i^* - c\,(q_i^* - q_{i-1}^*) \right] + \alpha\,(f_{i-1}^* - 2f_i^* + f_{i+1}^*). \qquad (12.5.50)$$

12.5.7 Convection–diffusion in two and three dimensions

Finite-difference methods for the three-dimensional linear convection–diffusion equation

$$\frac{\partial f}{\partial t} + U\frac{\partial f}{\partial x} + V\frac{\partial f}{\partial y} + W\frac{\partial f}{\partial z} = \kappa\left(\frac{\partial^2 f}{\partial x^2} + \frac{\partial^2 f}{\partial y^2} + \frac{\partial^2 f}{\partial z^2}\right), \tag{12.5.51}$$

and its simplified two-dimensional version, emerge by straightforward extensions of the methods for the one-dimensional equation discussed in Section 12.6. Selected examples are discussed in this section.

FTCS method

The fully explicit FTCS method is consistent, first-order accurate in time, and second-order accurate in space. Stability requires that

$$\alpha_x + \alpha_y + \alpha_z < \frac{1}{2}, \qquad \frac{c_x^2}{\alpha_x} + \frac{c_y^2}{\alpha_x} + \frac{c_z^2}{\alpha_z} < 2. \tag{12.5.52}$$

In two dimensions, the sums on the left-hand sides are over x and y [180].

Upwind differencing

Adopting a first-order upwind difference for the convective spatial derivative and the central difference for the diffusive derivative yields a conditionally stable method. The stability constraint in two dimensions is $4\alpha_x + |c_x| + |c_y| < 1$ ([294], p. 66).

Hopscotch method

The hopscotch method is implemented as discussed for one dimension in Section 12.6 [149]. In two dimensions, we first use the explicit FTCS method to advance the solution at the grid points $x_{i,j}$, where $i + j$ is an odd integer, and then use the implicit BTCS method to advance the solution at the grid points $x_{i,j}$, where $i + j$ is an even integer. The order is reversed after the completion of a time step. After the first time step, the FTCS difference equation is replaced with the equivalent equation

$$f_{i,j}^{n+1} = 2f_{i,j}^n - f_{i,j}^{n-1}. \tag{12.5.53}$$

The method is first-order accurate in time, second-order accurate in space, and stable as long as $|c_x| < 1$ and $|c_y| < 1$.

ADI method in two dimensions

Implicit methods are preferred due to their unconditional stability. The ADI method in two dimensions is implemented according to the finite-difference equations

$$\frac{f_{i,j}^{n+1/2} - f_{i,j}^n}{\frac{1}{2}\Delta t} + U\left[\frac{f_{i+1,j}^{n+1/2} - f_{i-1,j}^{n+1/2}}{2\Delta x}\right] + V\frac{f_{i,j+1}^n - f_{i,j-1}^n}{2\Delta y}$$
$$= \kappa\left[\frac{f_{i-1,j}^{n+1/2} - 2f_{i,j}^{n+1/2} + f_{i+1,j}^{n+1/2}}{\Delta x^2}\right] + \kappa\frac{f_{i,j-1}^n - 2f_{i,j}^n + f_{i,j+1}^n}{\Delta y^2} \tag{12.5.54}$$

and

$$\frac{f_{i,j}^{n+1} - f_{i,j}^{n+1/2}}{\frac{1}{2}\Delta t} + U \frac{f_{i+1,j}^{n+1/2} - f_{i-1,j}^{n+1/2}}{2\Delta x} + V \left[\frac{f_{i,j+1}^{n+1} - f_{i,j-1}^{n+1}}{2\Delta y} \right]$$

$$= \kappa \frac{f_{i-1,j}^{n+1/2} - 2f_{i,j}^{n+1/2} + f_{i+1,j}^{n+1/2}}{\Delta x^2} + \kappa \left[\frac{f_{i,j-1}^{n+1} - 2f_{i,j}^{n+1} + f_{i,j+1}^{n+1}}{\Delta y^2} \right], \qquad (12.5.55)$$

where the square brackets enclose implicit discretizations. The ADI method is unconditionally stable and second-order accurate in time and space ([294], p. 66).

ADI method in two dimensions with time-dependent velocities

When the velocity components U and V are not constant but vary in time, the ADI method becomes first-order accurate in Δt. To ensure second-order accuracy, we use the following averaged velocities in the first ADI step,

$$U = a_1 U^{n+1} + (1 - a_1 - a_2) U^n + a_2 U^{n-1},$$
$$V = b_1 V^{n+1} + (1 - b_1 - b_2) V^n + b_2 V^{n-1}, \qquad (12.5.56)$$

and the following averaged velocities in the second ADI step,

$$U = (1 - a_1 + a_2 + a_3) U^{n+1} + (a_1 - a_2 - 2a_3) u_x^n + a_3 U^{n-1},$$
$$V = (1 - b_1 + b_2 + b_3) V^{n+1} + (b_1 - b_2 - 2b_3) V^n + b_3 V^{n-1}, \qquad (12.5.57)$$

where a_i and b_i are six arbitrary constants ([294], p. 66).

Fractional steps

To implement the fractional step method in two or three dimensions, we consider convection–diffusion in each direction individually through a sequence of one-dimensional steps of equal duration, Δt. In three dimensions, the fractional steps are carried out according to the one-dimensional convection–diffusion equations

$$\frac{\partial f}{\partial t} + u_x \frac{\partial f}{\partial x} = \kappa \frac{\partial^2 f}{\partial x^2}, \qquad \frac{\partial f}{\partial t} + u_y \frac{\partial f}{\partial y} = \kappa \frac{\partial^2 f}{\partial y^2}, \qquad \frac{\partial f}{\partial t} + u_z \frac{\partial f}{\partial z} = \kappa \frac{\partial^2 f}{\partial z^2}. \qquad (12.5.58)$$

Each step applies for $t_n < t < t_n + \Delta t$, and time is reset to the initial value, t_n, after completion of the first and second fractional steps. Each step can be carried out using an unconditionally stable implicit method. When Dirichlet or Neumann boundary conditions are prescribed, the discretization results in tridiagonal systems of equations.

Problems

12.5.1 *Du Fort–Frankel method*

Perform a consistency analysis of the Du Fort–Frankel method and show that the corresponding partial differential equation is given by (12.2.23) in the presence of a convection term on the left-hand side.

12.5.2 *Implicit MacCormack method*

Write the finite-difference equations for the implicit MacCormack method.

12.5.3 *Implicit BTBC-CS method*

Write the difference equation for the BTBC-CS method where a backward difference is used for the convection term and a central difference is used for the diffusion term. Assess the stability of the method.

12.5.4 *Generalized upwind differencing*

(*a*) Derive the modified differential equation corresponding to the difference equation (12.5.17). Confirm that the method is consistent and compute the effective diffusivity.

(*b*) Derive the stability criteria shown in (12.5.18).

12.5.5 *Hopscotch method for linear three-dimensional convection–diffusion*

Develop a hopscotch method for the linear convection–diffusion equation in three dimensions.

12.5.6 *Fractional-step method*

Write the finite-difference equations corresponding to the Crank–Nicolson discretization of the three equations in (12.5.58).

 Computer Problems

12.5.7 *Burgers equation*

Compute the solution of the Burgers equation with initial condition $F(x) = \exp(-x^2)$ using (*a*) the FTCS, (*b*) the Du Fort–Frankel, and (*c*) the MacCormack explicit method, with $\kappa = 0.01$, 0.1, and 1.0. Discuss and compare the results of the three simulations.

12.5.8 *Korteweg–de Vries equation*

A standard form of the Korteweg–de Vries equation is

$$\frac{\partial f}{\partial t} + \epsilon f \frac{\partial f}{\partial x} + \mu \frac{\partial^3 f}{\partial x^3} = 0, \qquad (12.5.59)$$

where ϵ and μ are two positive constants.

(*a*) Confirm that an exact solution in an unbounded domain expressing the propagation of a solitary wave is

$$f = \frac{3c}{\cosh^2\left[\alpha\left(x - \epsilon c t - d\right)\right]}, \qquad (12.5.60)$$

where $c \geq 0$ and d are two arbitrary constants and $\alpha = \sqrt{\epsilon c/(4\mu)}$ [98].

(*b*) Develop an explicit finite-difference method and compute the evolution of the solution from the initial state described by (12.5.60). Discuss the reliability and accuracy of your computations.

Finite-difference methods for incompressible Newtonian flow

13

Having discussed finite-difference methods for solving the convection–diffusion equation in its general form, we proceed to develop corresponding methods for solving the equations of steady and unsteady incompressible Newtonian flow. The set of governing equations includes the Navier–Stokes equation and the continuity equation. The primary unknowns are the velocity and the pressure. However, we recall that an arbitrary rotational flow can be described, and therefore computed, in terms of the secondary variables discussed in Chapter 2, including the vorticity, the stream functions, and the vector potential. Numerical methods based on these secondary fields will be outlined.

Considering the evolution of an unsteady flow, we regard the Navier–Stokes equation as an evolution equation for the velocity, providing us with the rate of change of the velocity at a particular point in the flow in terms of the instantaneous velocity and pressure fields. In the absence of the pressure gradient term, the Navier–Stokes equation reduces to a nonlinear convection–diffusion equation that is amenable to the finite-difference methods discussed in Chapter 12. The presence of the pressure gradient in the equation of motion requires special attention due to the absence of an explicit evolution equation for the pressure, as discussed in Section 9.1. In place of this evolution equation, we have the restriction of incompressibility requiring that the pressure evolves so that the rate of expansion is zero and the velocity field remains solenoidal at all times. In Section 9.1, we saw that the restriction of incompressibility can be expressed in terms of a Poisson equation either for the pressure or for the rate of change of the pressure, with a time-dependent forcing function. These equations implicitly determine the evolution of the pressure field.

Computing the evolution of an incompressible Newtonian flow is thus distinguished by the need to solve simultaneously a parabolic equation in time, which is the equation of motion, and an elliptic equation in space, which is the Poisson equation for the pressure or rate of change of the pressure in time. It is interesting to note that the continuity equation for compressible fluids has the form of an evolution equation for the density, which is related to the hydrodynamic pressure through a chosen equation of state. Since all governing equations are parabolic in time, they can be marched forward in time using a standard numerical method. For this reason, computing the evolution of a compressible flow is more straightforward than computing the evolution of an incompressible flow.

An additional concern in computing the structure of a steady incompressible flow or the evolution of an unsteady incompressible flow in primary variables, including the velocity and the pressure, is the derivation and numerical implementation of boundary conditions for the pressure. In the vast majority of fluid dynamics applications, boundary conditions for the pressure are not directly available, but must be derived from the equation of motion subject to specified boundary conditions for the velocity or traction. We will see that the accurate implementation of derived pressure boundary conditions requires special attention.

We will begin this chapter by discussing a class of methods for computing the structure of a steady flow and the evolution of an unsteady flow based on the vorticity transport equation. The numerical procedure involves computing the evolution of the vorticity field, while simultaneously obtaining the simultaneous evolution of the velocity field by inverting the definition of the vorticity, $\omega = \nabla \times \mathbf{u}$, subject to the continuity equation. One advantage of this approach is that the pressure field does not need to considered. One disadvantage is the need to derive boundary conditions for the vorticity.

Methods based on the vorticity transport equation can be regarded as generalizations of the vortex methods for inviscid or weakly viscous flows discussed in Chapter 11. What distinguishes vortex methods from other methods based on vorticity transport is that the velocity field is obtained from the vorticity field efficiently using the Biot–Savart integral or a related contour integral. In the case of viscous flow, because the support of the vorticity is not necessarily compact, and it is more expedient to recover the velocity from the vorticity field by finite-difference or other domain discretization methods.

A variety of finite-difference procedures are available for solving the equations of steady and unsteady incompressible Newtonian flow, and a choice must be made according to the tolerated level of programming complexity and available computational resources. In this chapter, we outline the fundamental principles of several alternative formulations and discuss the basic steps involved their numerical implementation. Extensions and discussion of particular issues and specialized methods can be found in the references cited as well as in general reviews and monographs on finite-difference methods in fluid dynamics dynamics (e.g., [9, 69, 155, 185, 286, 294, 346]). Numerical methods for free-surface and interfacial flow are reviewed in Reference [131].

13.1   Vorticity–stream function formulation for two-dimensional flow

We begin the discussion of finite-difference methods by presenting a classical formulation based on the vorticity transport equation for two-dimensional flow, known as the stream function–vorticity formulation. In Section 13.3, we address the more general case of three-dimensional flow.

13.1.1   Governing equations

In the case of two-dimensional flow, solving for the velocity in terms of the vorticity is done with the least amount of computational effort by introducing the stream function, ψ. The two components of the velocity in the x and y directions are $u_x = \partial\psi/\partial y$ and $u_y = -\partial\psi/\partial x$, and the z component of the vorticity is

$$\omega_z = -\nabla^2\psi, \tag{13.1.1}$$

where ∇^2 is the Laplacian operator in the xy plane. The computation proceeds according to the two fundamental steps of the vortex methods discussed in Chapter 11.

In the first step, the rate of change of the vorticity is computed using the simplified vorticity transport equation for two-dimensional flow written in the vorticity–stream function form

$$\frac{\partial \omega_z}{\partial t} + \frac{\partial \psi}{\partial y}\frac{\partial \omega_z}{\partial x} - \frac{\partial \psi}{\partial x}\frac{\partial \omega_z}{\partial y} = \nu\, \nabla^2 \omega_z, \tag{13.1.2}$$

where ν is the kinematic viscosity of the fluid. The sum of the second and third terms on the left-hand side of (13.1.2) is sometimes designated as the Jacobian, $\mathcal{J}(\omega_z, \psi)$. In the second step, the simultaneous evolution of the stream function is obtained by solving the Poisson equation (13.1.1) for ψ in terms of ω_z. Boundary conditions are requiring in both the integration of (13.1.2) and the inversion of (13.1.1). It is instructive to note that the absence of an explicit evolution equation for the pressure in the original system of governing equations is reflected in the absence of an explicit evolution equation for the stream function.

Pressure field

One important feature of the vorticity–stream function formulation is that computing the pressure is not required. If the instantaneous pressure field is desired, it can be computed *a posteriori* by solving a Poisson equation that emerges by taking the divergence of the Navier–Stokes equation and using the continuity equation to obtain

$$\nabla^2 p = 2\rho \left[\frac{\partial^2 \psi}{\partial x^2}\frac{\partial^2 \psi}{\partial y^2} - \left(\frac{\partial^2 \psi}{\partial x \partial y} \right)^2 \right]. \tag{13.1.3}$$

Boundary conditions for the pressure arise by applying the equation of motion at the boundaries, enforcing the specified boundary conditions, and then projecting the result onto the normal or tangential unit vector, as discussed in Section 13.3 in the context of the velocity–pressure formulation.

13.1.2 Flow in a rectangular cavity

To illustrate the implementation of the finite-difference method, we consider the classical problem of flow in a rectangular cavity driven by a lid that translates parallel to itself with a generally time-dependent velocity, $V(t)$, as illustrated in Figure 13.1.1. The no-penetration condition requires that the normal velocity component is zero at each wall. In terms of the stream function,

$$\psi = c \quad \text{over all walls}, \tag{13.1.4}$$

where c is an arbitrary constant set for simplicity to zero. The no-slip boundary condition requires that the tangential component of the velocity is zero over the bottom, left, and right walls, whereas the tangential velocity at the upper wall is equal to the wall velocity, $V(t)$. In terms of the stream function, we obtain the equivalent statement

$$\frac{\partial \psi}{\partial y} = 0 \quad \text{at the bottom}, \qquad \frac{\partial \psi}{\partial x} = 0 \quad \text{at the sides}, \qquad \frac{\partial \psi}{\partial y} = V \quad \text{at the lid}. \tag{13.1.5}$$

Given these boundary conditions for the velocity, we derive simplified expressions for the boundary values of the vorticity in terms of the stream function. Beginning with (13.1.1) and noting that, for

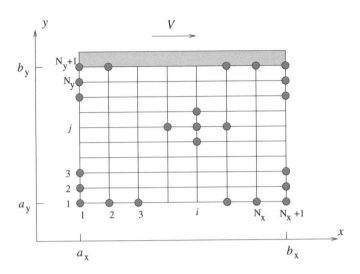

FIGURE 13.1.1   A finite-difference grid for computing two-dimensional flow in a rectangular cavity driven by a moving lid. The pressure and the two components of the velocity are defined at the same grid nodes.

example, $\partial^2\psi/\partial x^2 = -\partial u_y/\partial x = 0$ at the top wall because of the no-penetration condition, we find that

$$\omega_z = -\frac{\partial^2\psi}{\partial y^2} \quad \text{at the top and bottom walls,} \qquad \omega_z = -\frac{\partial^2\psi}{\partial x^2} \quad \text{at the side walls.} \qquad (13.1.6)$$

To implement a finite-difference method, we introduce a two-dimensional grid with $N_x \times N_y$ divisions, as illustrated in Figure 13.1.1. For simplicity, we have assumed that the grid lines are evenly spaced, which means that the grid spacings, Δx and Δy, are uniform throughout the domain of flow. The finite-difference formulation involves assigning discrete values to the stream function and vorticity at all internal and boundary grid points, and replacing the governing differential equations (13.1.1) and (13.1.2) with difference equations, as discussed in Chapter 12. The specific strategy of computation depends on whether we wish to compute a steady or an unsteady flow.

Steady flow

Two distinct but somewhat related approaches are available for computing a steady flow. The first class of methods involves solving the equations of steady flow using an iterative scheme. The second class of methods involves computing the solution of a fictitious transient flow problem governed by a modified set of differential equations from a given initial condition up to the steady state. The solution of the modified problem at steady state satisfies the equations of steady two-dimensional incompressible Newtonian flow.

13.1.3 Direct computation of a steady flow

In one version of the direct approach, the governing equations (13.1.1) and (13.1.2) are regarded as a coupled nonlinear system of Poisson equations for ψ and ω_z,

$$\nabla^2 \psi = -\omega_z, \tag{13.1.7}$$

and

$$\nabla^2 \omega_z = \frac{1}{\nu} \left(u_x \frac{\partial \omega_z}{\partial x} + u_y \frac{\partial \omega_z}{\partial y} \right). \tag{13.1.8}$$

In the special case of Stokes flow, the right-hand side of (13.1.8) vanishes, yielding Laplace's equation for the vorticity, $\nabla^2 \omega_z = 0$. Equation (13.1.7) then shows that the stream function satisfies an inhomogeneous biharmonic equation, $\nabla^4 \psi = -\omega_z$. The computational algorithm in the general case of nonzero Reynolds-number flow involves the following steps:

Step 1: *Guess the vorticity distribution.*

Step 2: *Solve the Poisson equation (13.1.7) for the stream function.*

For boundary conditions, we have the choice between the Dirichlet boundary condition that specifies the boundary distribution of the stream function, and the Neumann boundary condition that specifies the boundary distribution of the normal derivative of the stream function, which is equal to the tangential component of the velocity. A combination of the Dirichlet and Neumann boundary conditions at different boundaries can also be employed. If we use the Neumann condition over all boundaries, the Poisson equation will have a solution only if the following compatibility condition is fulfilled,

$$\oint_{Walls} \mathbf{n} \cdot \nabla \psi \, \mathrm{d}l = \iint_{Flow} \omega_z \, \mathrm{d}x \, \mathrm{d}y, \tag{13.1.9}$$

where $\mathbf{n}$ is the normal unit vector pointing into the flow. Even when (13.1.9) is fulfilled by a fortuitous guess of the vorticity distribution in Step 1, the singular nature of the linear system of equations that arises from the finite-difference discretization of (13.1.7) introduces additional complications. For these reasons, the Dirichlet condition expressed by (13.1.4) is preferred around all boundaries.

Step 3: *Compute the right-hand side of (13.1.8) and the boundary values of the vorticity using the specified boundary conditions for the velocity.*

Step 4: *Solve the Poisson equation (13.1.8) for the vorticity.*

Step 5: *Check whether the computed vorticity field agrees with that guessed in Step 1 at all grid points. If it does not agree within a specified tolerance, replace the guessed with the computed vorticity and return to Step 2.*

Implementation

The details of the numerical implementation will be discussed with reference to flow in a cavity driven by a moving lid, as illustrated in Figure 13.1.1.

Step 1: *Assign initial values for the stream function to all internal and boundary $(N_x + 1) \times (N_y + 1)$ grid points.*

A simple choice consistent with the no-penetration boundary condition is to set the initial stream function to zero, corresponding to a quiescent fluid.

Step 2: *Assign values for the vorticity to all internal $N_x \times N_y$ grid points.*

A simple choice is to set all initial vorticity gird values to zero, corresponding to a quiescent fluid. Note that this choice disregards the motion of the fluid due to the translation of the upper wall.

Step 3: *Solve the Poisson equation (13.1.7) for ψ subject to the Dirichlet boundary condition (13.1.4) at all four walls.*

Since the vorticity on the right-hand side is only an approximation to the exact solution, high accuracy is not necessary. To reduce the computational effort, we solve the Poisson equation iteratively and carry out only a small number of iterations. To perform the iterations, we introduce a fictitious unsteady diffusion–reaction problem governed by the equation

$$\frac{\partial \psi}{\partial t} = \kappa \left(\nabla^2 \psi + \omega_z \right), \tag{13.1.10}$$

where κ is a diffusivity. The solution of this equation at steady state satisfies (13.1.7). For simplicity, we denote $\omega = \omega_z$. Implementing the forward time centered space (FTCS) discretization discussed in Section 12.3.7, we find that

$$\psi_{ij}^{(l+1)} = \psi_{ij}^{(l)} + \alpha_x \, R_{ij}^{(l)}, \tag{13.1.11}$$

where $\alpha_x = \kappa \Delta t / \Delta x^2$ is the diffusion number in the x direction, the superscript (l) denotes the lth iteration level,

$$R_{i,j} \equiv \psi_{i+1,j} - 2 \left(1 + \beta \right) \psi_{i,j} + \psi_{i-1,j} + \beta \psi_{i,j+1} + \beta \psi_{i,j-1} + \Delta x^2 \omega_{i,j} \tag{13.1.12}$$

is the residual, and $\beta = \Delta x^2 / \Delta y^2$. The transient evolution is numerically stable if $\alpha_x (1 + \beta) < \frac{1}{2}$. The iterative method involves computing a time-like sequence of grid values parametrized by the index l, computed using the formula

$$\psi_{i,j}^{(l+1)} = \psi_{i,j}^{(l)} + \frac{\varrho}{2(1 + \beta)} R_{i,j}^{(l)}, \tag{13.1.13}$$

for $l = 1, 2, \ldots$, where ϱ is a relaxation factor used to control the iterations. When the iterations are executed for the first time, the initial guess $\psi^{(0)}$ is set equal to that assigned in Step 1. When the iterations converge, the solution is second-order accurate with respect to Δx and Δy.

Step 4: *Use the simplified expressions (13.1.6) to compute the vorticity at the boundary grid points by one-sided finite differences.*

To compute the vorticity at a grid point along the upper wall, we expand the stream function in a Taylor series with respect to y about a grid point that lies at the lid and evaluate the series at the

N_y level immediately below to obtain

$$\psi_{i,N_y} \simeq \psi_{i,N_y+1} - \Delta y \left(\frac{\partial \psi}{\partial y}\right)_{i,N_y+1} + \frac{1}{2} \Delta y^2 \left(\frac{\partial^2 \psi}{\partial y^2}\right)_{i,N_y+1}. \tag{13.1.14}$$

The no-slip boundary condition (13.1.5) requires that $(\partial \psi/\partial y)_{i,N_y+1} = V$, and the first equation in (13.1.6) requires that $\omega_{i,N_y+1} = -(\partial^2 \psi/\partial y^2)_{i,N_y+1}$. Substituting these expressions into (13.1.14) and solving for ω_{i,N_y+1}, we obtain

$$\omega_{i,N_y+1} = 2 \frac{\psi_{i,N_y+1} - \psi_{i,N_y}}{\Delta y^2} - 2 \frac{V}{\Delta y}, \tag{13.1.15}$$

which is first-order accurate in Δy. Working in a similar fashion for the bottom and side walls, we derive the corresponding expressions

$$\omega_{i,1} = 2 \frac{\psi_{i,1} - \psi_{i,2}}{\Delta y^2}, \qquad \omega_{1,j} = 2 \frac{\psi_{1,j} - \psi_{2,j}}{\Delta y^2}, \qquad \omega_{N_x+1,j} = 2 \frac{\psi_{N_x+1,j} - \psi_{N_x,j}}{\Delta y^2}. \tag{13.1.16}$$

To increase the accuracy of the method to second order, we expand the stream function in a Taylor series about a grid point at the upper wall, evaluate the series at two layers immediately below the upper wall, and retain terms up to third order to find

$$\psi_{i,N_y} \simeq \psi_{i,N_y+1} - \Delta y \left(\frac{\partial \psi}{\partial y}\right)_{i,N_y+1} + \frac{1}{2} \Delta y^2 \left(\frac{\partial^2 \psi}{\partial y^2}\right)_{i,N_y+1} + \frac{1}{6} \Delta y^3 \left(\frac{\partial^3 \psi}{\partial y^3}\right)_{i,N_y+1} \tag{13.1.17}$$

and

$$\psi_{i,N_y-1} \simeq \psi_{i,N_y+1} - 2\Delta y \left(\frac{\partial \psi}{\partial y}\right)_{i,N_y+1} + 2 \Delta y^2 \left(\frac{\partial^2 \psi}{\partial y^2}\right)_{i,N_y+1} - \frac{4}{3} \Delta y^3 \left(\frac{\partial^3 \psi}{\partial y^3}\right)_{i,N_y+1}. \tag{13.1.18}$$

Combining these equations to eliminate the third derivative of the stream function, solving for the second derivative, using the boundary condition (13.1.5), and taking into account (13.1.6), we find that

$$\omega_{i,N_y+1} = \frac{7\psi_{i,N_y+1} - 8\psi_{i,N_y} + \psi_{i,N_y-1}}{2\Delta y^2} - 3 \frac{V}{\Delta y}. \tag{13.1.19}$$

Working in a similar fashion for the bottom and side walls, we find that

$$\omega_{i,1} = \frac{7\psi_{i,1} - 8\psi_{i,2} + \psi_{i,3}}{2\Delta y^2}, \qquad \omega_{1,j} = \frac{7\psi_{1,j} - 8\psi_{2,j} + \psi_{3,j}}{2\Delta x^2},$$

$$\omega_{N_x+1,j} = \frac{7\psi_{N_x+1,j} - 8\psi_{N_x,j} + \psi_{N_x,j}}{2\Delta x^2}. \tag{13.1.20}$$

Step 5: *Differentiate the stream function to compute the velocity at the internal grid points subject to the boundary values (13.1.4).*

Step 6: *Differentiate the vorticity to compute the right-hand side of (13.1.8) at the internal grid points subject to the boundary values computed from (13.1.15) and (13.1.16) or from (13.1.19) and (13.1.20).*

Step 7: *Solve the Poisson equation (13.1.8) for the vorticity subject to the Dirichlet boundary conditions for the vorticity derived in Step 4.*

This is done by iteration using, for example, the forward time–centered space (FTCS) algorithm discussed in Step 3, carrying out only a small number of iterations. Grid values of the forcing function on the right-hand side at the internal nodes are available from Step 6.

Step 8: *If the vorticity computed in Step 7 does not agree with that previously available within a specified tolerance, we return to Step 2 and repeat the computations with the new values of the vorticity.*

This outer iteration is terminated when the absolute value of the difference in the vorticity between two successive iterations at each grid point becomes less than a present threshold, ϵ, or when the sum of the absolute values of the differences in the vorticity over all internal $N_x \times N_y$ grid points becomes less than $N_x \times N_y \times \epsilon$.

One noteworthy feature of the algorithm is that the corner grid points do not enter the computation, which means that the numerical scheme is oblivious to the velocity discontinuity at the two upper corner points. The local jump may cause local oscillations and slow down the convergence of the iterations, but does not have a deleterious effect on the overall numerical method.

Structure of the flow

Vorticity and stream function contour plots for flow in a cavity with aspect ratio $L_x/L_y = 2$, and Reynolds number Re $= VL_x/\nu = 1$ and 100 are shown in Figure 13.1.2. In the notation of Figure 13.1.1, $L_x = b_x - a_x$ and $L_y = b_y - a_y$. Stream function contours are both streamlines and particle paths in a two-dimensional flow.

When Re $= 1$, we obtain a nearly creeping flow whose streamlines are symmetric with respect to the midplane due to reversibility of Stokes flow, as discussed in Section 6.1.7. Small regions of recirculating flow are present at the bottom two corners, requiring increased spatial resolution. The vorticity is singular at the upper two cavity corners due to the discontinuous boundary velocity. A local analysis shows that the shear stress diverges at these corners, and an infinite force is required to slide the lid as a result of the sharp-corner idealization, as discussed in Section 6.2. However, these physical singularities do not deter the performance of the numerical method.

As the Reynolds number increases, the center of the central eddy is shifted toward the upper right corner due to the fluid inertia and the flow becomes unsymmetric. To compute flow at even higher Reynolds numbers, it is helpful to perform parameter continuation where the initial guesses for the stream function and vorticity are identified with the corresponding converged values at a lower Reynolds number. In fact, the Reynolds number can be increased gradually up to a targeted value in the course of the iterations.

Improvements and extensions

The computational procedure discussed in this section can be improved in several ways (e.g., [159]). In one improvement, instead of iterating on the Poisson equation for the vorticity in Step 7, we

(*a*)

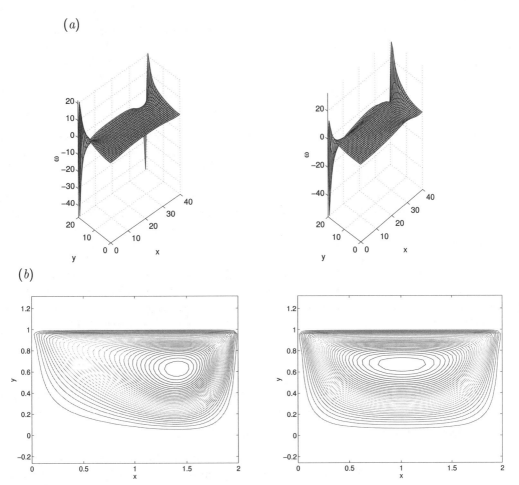

(*b*)

FIGURE 13.1.2   (*a*) Vorticity and (*b*) stream function contour plots for flow in a rectangular cavity with aspect ratio $L_x/L_y = 2$, at Reynolds number $\mathrm{Re} = V L_x/\nu = 1$ (left) and 100 (right).

iterate on the full diffusion–convection equation (13.1.8). This requires that the nonlinear term on the right-hand side is recomputed after each iteration using the updated vorticity. The iterations can be carried out using an explicit or implicit finite-difference method for the convection–diffusion equation in two dimensions discussed in Section 12.7. Because the flow near the center of the cavity is dominated by convection at high Reynolds numbers, using upwind differencing helps improve numerical stability. The rate of convergence of the inner and outer iterations depends on the details of the particular implementation [198].

The method for computing the boundary values of the vorticity described in Step 3 has been the subject of criticism [155]. It has been argued that it is not appropriate to explicitly impose the boundary values of the vorticity. Instead, the boundary distribution of the vorticity should arise implicitly as part of the solution using the natural boundary conditions for the velocity or traction. This issue will be discussed further in Section 13.2 in the more general context of three-dimensional flow.

13.1.4 Modified dynamics for steady flow

A distinguishing feature of the direct approach described in Section 13.1.3 is the solution of a Poisson equation for the stream function, reflecting the elliptic nature of the equations governing the structure of a steady flow. Physically, the velocity at a point in a flow affects the velocity at every other point. We saw that one way to perform the associated inner iterations is to introduce a fictitious unsteady diffusion–reaction problem and then implement the explicit FTCS discretization.

This observation suggests reformulating the problem by retaining the vorticity transport equation (13.1.1) and replacing (13.1.2) with an evolution equation equation for the stream function expressed by (13.1.10), where the diffusivity κ is a free parameter of the numerical method. The idea is to compute the evolution of the flow from a certain initial condition subject to (13.1.1) and (13.1.10) until a steady state has been established. The steady-state solution will satisfy the original equations (13.1.7) and (13.1.8). An advantage of this approach is that all governing equations are parabolic in time. Time-marching schemes similar to those developed in Chapter 12 for convection–diffusion may then be employed.

For the problem of flow in a cavity illustrated in Figure 13.1.1, the method of modified dynamics is implemented according to the following steps:

1. Assign initial values to the stream function and vorticity at all internal and boundary grid points. A simple choice is to set them both to zero.

2. Differentiate the stream function to compute the two components of the velocity at all internal grid points.

3. Compute the vorticity at the boundary grid points.

4. Advance the vorticity at all internal grid points on the basis of (13.1.2) using, for example, the ADI method for the convection–diffusion equation described in Section 12.7, while keeping the vorticity at the boundary grid points constant.

5. Advance the stream function at all internal grid points on the basis of (13.1.10) using, for example, the ADI method for the convection–diffusion equation described in Section 12.7.

6. Return to Step 3 and repeat the computations for another time step.

13.1.5 Unsteady flow

To compute the evolution of an unsteady flow, we combine features of the direct method with features of the method of modified dynamics for steady flow. The algorithm involves computing the evolution of the vorticity field using (13.1.2), while simultaneously describing the evolution of the velocity field in terms of the stream function using (13.1.1). To simplify the notation, we denote $u = u_x$, $v = u_y$, and $\omega = \omega_z$. A simple strategy for computing the evolution of flow in a cavity when the lid starts translating suddenly with constant velocity V involves the following steps:

1. At the initial instant, we set the stream function and velocity equal to zero at all internal and boundary grid points. Then we set the x velocity component at the grid points at the upper wall equal to the lid velocity, V.

2. Now we differentiate the velocity to compute the vorticity. For the internal grid points, we use central differences. For grid points that lie on the lid, we use the definition of the vorticity and apply second-order backward difference to find that

$$\omega_{i,N_y+1} = \frac{-3V + 4\,u_{i,N_y} - u_{i,N_y-1}}{2\Delta y}.$$
(13.1.21)

For grid points that lie at the bottom and side walls, we use the corresponding second-order difference formulas

$$\omega_{i,1} = \frac{-4\,u_{i,2} + u_{i,3}}{2\Delta y}, \quad \omega_{1,j} = \frac{4\,v_{2,j} - v_{3,j}}{2\Delta x}, \quad \omega_{N_x+1,j} = \frac{-4\,v_{N_x,j} + v_{N_x-1,j}}{2\Delta x}.$$
(13.1.22)

3. Next, we integrate (13.1.2) to obtain the vorticity at the next time level at all internal grid points subject to the boundary conditions (13.1.21) and (13.1.22) using an explicit method, such as the FTCS method. At high Reynolds numbers, we use upwind differences.

4. In the fourth step, we solve the Poisson equation (13.1.1) for the stream function at the next time level subject to the Dirichlet boundary condition (13.1.4).

5. Now we differentiate the stream function to compute the velocity at the next time level at all internal grid points.

6. Having advanced the flow by one time step, we return to Step 2 and repeat the computation for another time step.

To improve the temporal accuracy and promote the numerical stability, we may update the vorticity using an implicit or semi-implicit method, such as the ADI method.

ADI method

In a simple implementation of the ADI method, the convection velocity is held constant and equal to initial value at the beginning of the first substep during both substeps. In a more advanced implementation, the Poisson equation is solved for an intermediate stream function after completing the first substep, and the convection velocity is set equal to an intermediate velocity computed by differentiating the intermediate stream function at the second substep. Since the convection velocity is held constant and equal to its value at the beginning of a step or substep, the overall accuracy of both methods is first-order in time.

To achieve second-order accuracy, we use the ADI method with time-dependent velocities described in (12.5.56) and (12.5.57). Collecting the values of the vorticity at all grid points into a vector, $\boldsymbol{\omega}$, we obtain two ADI equations written in the symbolic form

$$\mathbf{A}(\mathbf{u}^{n-1}, \mathbf{u}^n, \mathbf{u}^{n+1}) \cdot \boldsymbol{\omega}^{n+1/2} = \mathbf{B}(\mathbf{u}^{n-1}, \mathbf{u}^n, \mathbf{u}^{n+1}) \cdot \boldsymbol{\omega}^n,$$

$$\mathbf{C}(\mathbf{u}^{n-1}, \mathbf{u}^n, \mathbf{u}^{n+1}) \cdot \boldsymbol{\omega}^n = \mathbf{B}(\mathbf{u}^{n-1}, \mathbf{u}^n, \mathbf{u}^{n+1}) \cdot \boldsymbol{\omega}^{n+1/2},$$
(13.1.23)

where $\mathbf{A}$, $\mathbf{B}$, $\mathbf{C}$, and $\mathbf{D}$ are tridiagonal matrices dependent on their stated arguments. Equations (13.1.23) replace the explicit FTCS equation in Step 3. Steps 3–5 combine into the following inner iterative loop: (a) Guess the velocities $\mathbf{u}^{n+1}$ and solve the two tridiagonal systems (13.1.23) with boundary conditions (13.1.21) and (13.1.22) for $\boldsymbol{\omega}^{n+1/2}$ and $\boldsymbol{\omega}^{n+1}$; ($b$) execute Steps 4 and 5; (c) solve (13.1.23) with the computed values of $\mathbf{u}^{n+1}$ or with a weighted average of old and new values.

If the boundary velocity changes in time, we solve the first tridiagonal system in (13.1.23) subject to the boundary conditions $\boldsymbol{\omega}^{n+1/2} = \frac{1}{2}(\boldsymbol{\omega}^n + \boldsymbol{\omega}^{n+1})$, where $\boldsymbol{\omega}^{n+1}$ has been approximated from the previous inner iteration. To accelerate the convergence, we may replace the boundary conditions for $\boldsymbol{\omega}^{n+1}$ during the inner iterations with a weighted average of old and new values. Further details on the implementation of the method are given in Reference [294] (p. 197).

Problems

13.1.1 *Poisson equation for the pressure*

Take the divergence of the two-dimensional Navier–Stokes equation and introduce the stream function to derive the pressure Poisson equation (13.1.3).

13.1.2 *Axisymmetric flow*

Write the counterparts of equations (13.1.1)–(13.1.3) for axisymmetric flow in terms of the Stokes stream function.

13.1.3 *Boundary condition for the vorticity in unsteady flow*

Derive expressions (13.1.21) and (13.1.22).

Computer Problems

13.1.4 *Steady flow in a cavity*

(a) Write a program that uses the direct approach discussed in the text to compute the steady flow in a square cavity of equal width and depth, L, generated by the steady translation of a lid. The inner iterations should may be carried out using the FTCS method. Carry out computations for a sequence of Reynolds numbers, Re $= VL/\nu = 1, 10, 100, 500, \ldots$, and discuss the changes in the structure of the flow. Study the convergence of the method as a function of the relaxation parameter, ϱ, and number of iterations. Estimate the critical Reynolds number where the spatial resolution appears to be inadequate.

(b) Repeat (a) with the inner iterations carried out using the LSOR method and discuss possible improvements (see Table 12.3.1).

(c) Repeat (a) using the method of modified dynamics.

13.1.5 *Unsteady flow in a cavity*

Write a program that uses the first-order method discussed in the text to solve the unsteady version of Problem 13.1.4 when the lid is set in motion impulsively with constant velocity.

13.2 Vorticity transport equation for three-dimensional flow

Methods based on the vorticity transport equation for three-dimensional flow employ the repetitive application of the two basic steps discussed in Section 13.1, involving the evolution of the vorticity field and the simultaneous evolution of the velocity field.

Evolution of the vorticity field

In the first step, the evolution of the vorticity field is computed based on the vorticity transport equation written in the conservative Eulerian form

$$\frac{\partial \boldsymbol{\omega}}{\partial t} + \nabla \times (\boldsymbol{\omega} \times \mathbf{u}) = \nu \nabla^2 \boldsymbol{\omega}. \tag{13.2.1}$$

Taking the divergence of (13.2.1) and recalling that the divergence of the curl of any twice differentiable vector function is identically zero, we find that the divergence of the vorticity, $\beta \equiv \nabla \cdot \boldsymbol{\omega}$, evolves according to the unsteady heat conduction equation,

$$\frac{\partial \beta}{\partial t} = \nu \nabla^2 \beta. \tag{13.2.2}$$

Accordingly, the computed vorticity field will be solenoidal, as required, provided that it is solenoidal at the initial instant and the boundary distribution of β remains zero at all times (Problem 13.2.1).

Boundary conditions for the vorticity

To integrate the vorticity transport equation (13.2.1) in time, we require the boundary distribution of the vorticity. In the majority of numerical implementations, the boundary vorticity emerges by applying the definition of the vorticity at or near boundaries, $\boldsymbol{\omega} = \nabla \times \mathbf{u}$, and then simplifying the resulting expressions by implementing specified boundary conditions for the velocity. This procedure guarantees that an initially solenoidal vorticity field will remain solenoidal at all times [158, 407]. The numerical procedure is analogous to that involved in the stream function–vorticity formulation discussed in Section 13.1.

It has been argued that it is not entirely appropriate to impose an explicit local boundary condition for the vorticity [155]. Instead, the boundary distribution of the vorticity should emerge as part of the solution in terms of the specified boundary conditions for the velocity or traction. Computational experiments have shown that computing instead of imposing the boundary values of the vorticity promotes the stability of the numerical method at the cost of increased programming complexity and computational effort.

Quartapelle & Valz-Gris [332] replaced the boundary conditions for the vorticity with an integral constraint. In the case of two-dimensional flow with homogeneous boundary conditions for the velocity, the constraint requires that the vorticity is orthogonal to any nonsingular harmonic function defined in the domain of flow, that is, the integral of the vorticity multiplied by any nonsingular harmonic function over the area of the flow is zero. The implementation of this method is discussed by Quartapelle [331] and Anderson [8] in the context of Chorin's vortex sheet method (Section 11.6).

Noninertial frames

When the flow is described in a noninertial frame that translates with velocity $\mathbf{V}(t)$ while rotating with angular velocity $\boldsymbol{\Omega}(t)$ about a chosen point, we work with the modified vorticity $\mathbf{W} = \boldsymbol{\omega} + 2\boldsymbol{\Omega}$, which evolves according to the standard vorticity transport equation (13.2.1) for an inertial frame, as discussed in Section 3.12.5 [383]. Since the acceleration of the frame of reference enters the solution only through the boundary condition, using the modified vorticity simplifies the implementation and reduces the computational cost.

Evolution of the velocity

In the second step of methods based on the vorticity transport equation, the evolution of the velocity field is computed by inverting the definition of the vorticity, $\boldsymbol{\omega} = \nabla \times \mathbf{u}$, subject to the continuity equation, $\nabla \cdot \mathbf{u} = 0$. The inversion can be done in two ways according to the vector potential–vorticity or velocity–vorticity formulation.

13.2.1 Vorticity–vector potential formulation

In this formulation, the velocity field is decomposed into a solenoidal and irrotational component expressed by the gradient of a harmonic potential function, $\nabla\phi$, and a rotational component expressed by the curl of a solenoidal vector potential, $\mathbf{A}$, so that

$$\mathbf{u} = \nabla\phi + \nabla \times \mathbf{A}, \tag{13.2.3}$$

as discussed in Section 2.8 [181]. In the case of flow past a solid boundary, the scalar potential ϕ is computed by solving Laplace's equation subject to the no-penetration boundary condition, $\mathbf{n} \cdot \nabla\phi = \mathbf{n} \cdot \mathbf{u} = 0$. The solution is unique up to an arbitrary but physically irrelevant constant. To compute the vector potential, we write

$$\boldsymbol{\omega} = \nabla \times \nabla \times \mathbf{A} = \nabla(\nabla \cdot \mathbf{A}) - \nabla^2\mathbf{A}, \tag{13.2.4}$$

and stipulate that $\mathbf{A}$ is solenoidal, $\nabla \cdot \mathbf{A} = 0$, to obtain a vectorial Poisson equation,

$$\nabla^2\mathbf{A} = -\boldsymbol{\omega}. \tag{13.2.5}$$

Taking the divergence of (13.2.5) and remembering that the vorticity field is solenoidal, we find that the divergence $\nabla \cdot \mathbf{A}$ satisfies Laplace's equation, which means that the computed $\mathbf{A}$ will be solenoidal provided that $\nabla \cdot \mathbf{A} = 0$ over all boundaries.

Boundary conditions for the vector potential

Consider flow in a simply connected domain. To ensure that $\nabla \cdot \mathbf{A} = 0$ over the boundaries, we require that the tangential components of $\mathbf{A}$ are zero,

$$\mathbf{n} \times (\mathbf{A} \times \mathbf{n}) = \mathbf{0}. \tag{13.2.6}$$

To validate this stipulation in light of the no-slip boundary condition, and also derive a boundary condition for the normal component of $\mathbf{A}$, it is convenient to introduce a local coordinate system

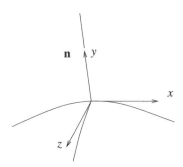

FIGURE 13.2.1   Illustration of a local coordinate system on a solid boundary used to derive a boundary condition for the normal component of the vector potential.

where the x and z axes are tangential to the boundary and the y axis is normal to the boundary at a point, as shown in Figure 13.2.1. The boundary condition (13.2.6) specifies that $A_x = 0$ and $A_z = 0$ at the origin. The normal component of the velocity is

$$(\nabla \times \mathbf{A})_y = \frac{\partial A_x}{\partial z} - \frac{\partial A_z}{\partial x} = \frac{\partial \mathbf{A}}{\partial z} \cdot \mathbf{t}_x - \frac{\partial \mathbf{A}}{\partial x} \cdot \mathbf{t}_z \tag{13.2.7}$$

evaluated at the origin, where $\mathbf{t}_x$ and $\mathbf{t}_z$ are unit tangent vectors along the x and z axes. Rearranging, we obtain

$$(\nabla \times \mathbf{A})_y = \frac{\partial (\mathbf{A} \cdot \mathbf{t}_x)}{\partial z} - \frac{\partial (\mathbf{A} \cdot \mathbf{t}_z)}{\partial x} - \mathbf{A} \cdot \left(\frac{\partial \mathbf{t}_x}{\partial z} - \frac{\partial \mathbf{t}_z}{\partial x} \right). \tag{13.2.8}$$

Now implementing the boundary condition (13.2.6), we obtain

$$(\nabla \times \mathbf{A})_y = -A_y\, \mathbf{n} \cdot \left(\frac{\partial \mathbf{t}_x}{\partial z} - \frac{\partial \mathbf{t}_z}{\partial x} \right) = A_y\, \mathbf{n} \cdot \left(\mathbf{t}_x \frac{\partial \mathbf{n}}{\partial z} - \mathbf{t}_z \frac{\partial \mathbf{n}}{\partial x} \right), \tag{13.2.9}$$

which is zero in light of (1.7.13), consistent with the requirement that $\mathbf{n} \cdot (\nabla \times \mathbf{A}) = 0$, originating from the no-penetration boundary condition.

Now using the condition $\nabla \cdot \mathbf{A} = 0$, we find that, at the origin,

$$\nabla \cdot \mathbf{A} = \mathbf{n} \cdot (\nabla \mathbf{A}) \cdot \mathbf{n} + \frac{\partial \mathbf{A}}{\partial x} \cdot \mathbf{t}_x + \frac{\partial \mathbf{A}}{\partial z} \cdot \mathbf{t}_z, \tag{13.2.10}$$

where the term $\mathbf{n} \cdot (\nabla \mathbf{A}) \cdot \mathbf{n}$ represents the derivative of the normal component of $\mathbf{A}$ in the direction normal to the boundary. Rearranging, we find that

$$\nabla \cdot \mathbf{A} = \mathbf{n} \cdot (\nabla \mathbf{A}) \cdot \mathbf{n} + \frac{\partial (\mathbf{A} \cdot \mathbf{t}_x)}{\partial x} + \frac{\partial (\mathbf{A} \cdot \mathbf{t}_z)}{\partial z} - \mathbf{A} \cdot \left(\frac{\partial \mathbf{t}_x}{\partial x} + \frac{\partial \mathbf{t}_z}{\partial z} \right), \tag{13.2.11}$$

yielding

$$\nabla \cdot \mathbf{A} = \mathbf{n} \cdot (\nabla \mathbf{A}) \cdot \mathbf{n} + 2\kappa_m\, \mathbf{A} \cdot \mathbf{n} = 0, \tag{13.2.12}$$

or

$$\mathbf{n} \cdot (\nabla \mathbf{A}) \cdot \mathbf{n} = -2\kappa_m \, \mathbf{A} \cdot \mathbf{n}, \qquad (13.2.13)$$

where κ_m is the mean curvature of the boundary. In the case of a flat boundary, $\mathbf{n} \cdot (\nabla \mathbf{A}) \cdot \mathbf{n} = 0$.

Boundary conditions for multiply connected domains are available [341]. The numerical implementation of the vector potential–vorticity formulation for three-dimensional flow has been discussed on several occasions (e.g., [11, 20, 252]). In the case of two-dimensional or axisymmetric flow, all components of $\mathbf{A}$ are set to zero, except for the z or azimuthal components that are identified, respectively, with the stream function or with the Stokes stream function divided by the distance from the axis of revolution, σ. The vector potential–vorticity formulation then reduces to the vorticity–stream function formulation discussed in Section 13.1.

13.2.2 Velocity–vorticity formulation

In the most popular version of the velocity–vorticity formulation, the velocity is computed from the vorticity by solving a vectorial Poisson equation,

$$\nabla^2 \mathbf{u} = -\nabla \times \boldsymbol{\omega}, \qquad (13.2.14)$$

which arises by taking the curl of the definition $\boldsymbol{\omega} = \nabla \times \mathbf{u}$ and requiring that the velocity $\mathbf{u}$ is solenoidal, $\nabla \cdot \mathbf{u} = 0$. The solution is subject to specified boundary conditions for the velocity.

To validate the method, we must show that the curl of the velocity computed by solving (13.2.14) is identical to the vorticity, $\boldsymbol{\omega}$, provided that $\boldsymbol{\omega}$ is solenoidal and its boundary values are computed from the definition $\boldsymbol{\omega} = \nabla \times \mathbf{u}$ [100, 407]. Working toward this goal, we use the vector identity $\nabla \times (\nabla \times \mathbf{u}) = \nabla(\nabla \cdot \mathbf{u}) - \nabla^2 \mathbf{u}$ in conjunction with (13.2.14). Defining $\delta \boldsymbol{\omega} \equiv (\nabla \times \mathbf{u}) - \boldsymbol{\omega}$ and rearranging, we obtain

$$\nabla \times \delta \boldsymbol{\omega} = \nabla(\nabla \cdot \mathbf{u}). \qquad (13.2.15)$$

Taking the curl of both sides, we find that

$$\nabla \times \nabla \times \delta \boldsymbol{\omega} = \nabla(\nabla \cdot \delta \boldsymbol{\omega}) - \nabla^2 \delta \boldsymbol{\omega} = \mathbf{0}. \qquad (13.2.16)$$

Since $\boldsymbol{\omega}$ and thus $\delta \boldsymbol{\omega}$ is solenoidal, $\nabla \cdot \delta \boldsymbol{\omega} = 0$, we can be sure that $\nabla^2 \delta \boldsymbol{\omega} = \mathbf{0}$. Because the boundary values of $\delta \boldsymbol{\omega}$ are zero, $\delta \boldsymbol{\omega}$ must vanish identically and the curl of the computed velocity, $\nabla \times \mathbf{u}$, must be equal to the prescribed vorticity $\boldsymbol{\omega}$ throughout the domain of flow.

To show that the velocity computed by solving (13.2.14) is solenoidal, we invoke once again the identity $\nabla \times (\nabla \times \mathbf{u}) = \nabla(\nabla \cdot \mathbf{u}) - \nabla^2 \mathbf{u}$. Using (13.2.14), we find that $\nabla \times (\nabla \times \mathbf{u}) = \nabla(\nabla \cdot \mathbf{u}) + \boldsymbol{\omega}$, which shows that $\nabla(\nabla \cdot \mathbf{u}) = 0$. Integrating in space, we find that $\nabla \cdot \mathbf{u}$ is a constant equal to zero due to mass conservation throughout the domain of flow [100, 407].

Numerical implementations of the velocity–vorticity formulation based on (13.2.14) have been presented by several authors [76, 100, 105, 158, 407]. Guj & Stella [157] developed a method of false transients for steady flow by replacing the elliptic equation (13.2.14) with the parabolic equation

$$\text{Re} \, \frac{\partial \mathbf{u}}{\partial t} = \nabla^2 \mathbf{u} + \nabla \times \boldsymbol{\omega}. \qquad (13.2.17)$$

In another version of the velocity–vorticity formulation, the velocity field is computed directly by solving the Cauchy–Riemann system of equations $\boldsymbol{\omega} = \nabla \times \mathbf{u}$ and $\nabla \cdot \mathbf{u} = 0$ for the velocity, subject to the no-penetration boundary condition [142, 288].

Problem

13.2.1 *Integration of the vorticity transport equation*

Consider the time integration of the vorticity transport equation written in nonconservative form where the vortex stretching term appears explicitly on the right-hand side, subject to the boundary condition $\boldsymbol{\omega} = \nabla \times \mathbf{u}$. Discuss whether the vorticity field will remain solenoidal during the integration [142].

13.3 Velocity–pressure formulation

In the remainder of this chapter, we discuss methods for solving the Navier–Stokes equation for two- or three-dimensional flow in primitive variables, including the velocity and the pressure. The fluid density and viscosity are assumed to be uniform throughout the domain of flow. Relaxing these restrictions requires only straightforward modifications. To simplify the notation, we work with the hydrodynamic pressure, denoted by p, excluding the effect of the body force.

We begin by recasting the Navier–Stokes equation as an evolution equation for the velocity involving the pressure,

$$\frac{\partial \mathbf{u}}{\partial t} = \mathbf{N}(\mathbf{u}) - \frac{1}{\rho} \nabla p + \nu \mathcal{L}(\mathbf{u}), \qquad (13.3.1)$$

where the terms $\mathbf{N}(\mathbf{u})$ and $\mathcal{L}(\mathbf{u})$ are associated, respectively, with the nonlinear inertial force and the linear viscous force,

$$\mathbf{N}(\mathbf{u}) \equiv -\mathbf{u} \cdot \nabla \mathbf{u} = -\nabla \cdot (\mathbf{u} \otimes \mathbf{u}), \qquad \mathcal{L}(\mathbf{u}) \equiv \nabla^2 \mathbf{u} = \nabla(\nabla \cdot \mathbf{u}) - \nabla \times \boldsymbol{\omega}. \qquad (13.3.2)$$

Because the fluid is incompressible, the velocity field is solenoidal and the term $\nabla(\nabla \cdot \mathbf{u})$ in the definition of $\mathcal{L}(\mathbf{u})$ is identically zero. However, because numerical discretization destroys the exact equality, keeping this term helps ensure numerical stability and minimize the numerical error.

The equation of motion is accompanied by the continuity equation for incompressible fluids,

$$\nabla \cdot \mathbf{u} = 0, \qquad (13.3.3)$$

stating that the velocity field is and must remain solenoidal during the evolution.

13.3.1 Pressure Poisson equation

Taking the divergence of (13.3.1) and switching the order of the gradient and temporal derivative on the left-hand side, we obtain an evolution equation for the rate of expansion, $\alpha \equiv \nabla \cdot \mathbf{u}$,

$$\frac{\partial \alpha}{\partial t} = \nabla \cdot \mathbf{N}(\mathbf{u}) - \frac{1}{\rho} \nabla^2 p + \nu \nabla \cdot \mathcal{L}(\mathbf{u}). \qquad (13.3.4)$$

The continuity equation requires that the right-hand side vanishes at all times. Accordingly, the pressure must satisfy the pressure Poisson equation (PPE)

$$\nabla^2 p = \rho \, \nabla \cdot \mathbf{N}(\mathbf{u}) + \mu \nabla \cdot \mathcal{L}(\mathbf{u}). \qquad (13.3.5)$$

One may argue that, since the divergence and the linear operator $\mathcal{L}$ commute, the second term on the right-hand side of (13.3.5) vanishes, yielding the simplified form

$$\nabla^2 p = \rho \, \nabla \cdot \mathbf{N}(\mathbf{u}). \qquad (13.3.6)$$

For reasons that will become evident in our further analysis, equation (13.3.5) is called the consistent PPE and equation (13.3.6) is called the simplified PPE [153].

13.3.2   Alternative systems of governing equations

To this end, we consider replacing the governing equations (13.3.1) and (13.3.3) either with equations (13.3.1) and (13.3.5) or with equations (13.3.1) and (13.3.6). These substitutions will be acceptable, provided that the velocity field remains solenoidal at all times. Substituting (13.3.5) into (13.3.4), we obtain

$$\frac{\partial \alpha}{\partial t} = 0, \qquad (13.3.7)$$

which states that, if the velocity field is solenoidal at the initial time, it will remain solenoidal at all times. Thus, if the initial velocity field is solenoidal, it is permissible to replace the continuity equation with the consistent pressure Poisson equation (13.3.5). However, if the initial rate of expansion is nonzero, the divergence of the velocity will remain nonzero throughout the evolution.

Substituting (13.3.6) into (13.3.4) and interchanging the divergence with the Laplacian, we obtain an unsteady diffusion equation for the rate of expansion,

$$\frac{\partial \alpha}{\partial t} = \nu \, \mathcal{L}(\alpha). \qquad (13.3.8)$$

The general properties of the unsteady diffusion equation in a bounded domain show that the rate of expansion will vanish at all times provided that (a) the initial velocity field is solenoidal and (b) the rate of expansion or its normal derivative are zero over all boundaries of the flow at all times. When these conditions are met, it is permissible to replace the continuity equation with the simplified pressure Poisson equation (13.3.5). The second condition underlines the importance of accurately satisfying mass conservation at the grid points near the boundaries. When the initial rate of expansion is not zero but the boundary distribution is kept zero during the evolution, the magnitude of the divergence of the velocity will keep decreasing and eventually vanish during the evolution.

13.3.3   Boundary conditions for the pressure

The consistent and modified pressure Poisson equations must be solved subject to one scalar boundary condition for the pressure over each boundary of the flow. According to our previous discussion, this condition must ensure that the boundary distribution of the divergence of the velocity is zero at all times. The derivation of proper boundary conditions for the pressure is discussed in two illuminating articles by Orszag, Israeli & Deville [285] and Gresho & Sani [153].

A Neumann boundary condition emerges by applying (13.3.1) at the boundaries of the flow and projecting the resulting equation onto the normal unit vector, $\mathbf{n}$, obtaining

$$\mathbf{n} \cdot \nabla p = \rho \left(-\frac{\partial \mathbf{u}}{\partial t} \cdot \mathbf{n} + \mathbf{N}(\mathbf{u}) \cdot \mathbf{n} \right) + \mu \mathcal{L}(\mathbf{u}) \cdot \mathbf{n}. \tag{13.3.9}$$

Analysis shows that this condition is an acceptable substitute of the condition of incompressibility over the boundaries of the flow. Specifically, using the boundary condition (13.3.9) guarantees that replacing the continuity equation with the simplified PPE is appropriate. The solution for the pressure computed using (13.3.9) will also satisfy the Dirichlet condition that emerges by projecting the equation of motion onto a tangential vector and then integrating it with respect to tangential arc length. In contrast, the pressure field computed by solving the pressure Poisson equation using the Dirichlet condition will satisfy the Neumann condition only when the initial velocity field satisfies certain regularity conditions [153].

Flow over a plane wall

In the case of flow over a plane wall that is either stationary or translates parallel to itself with constant velocity, we use the no-slip and no-penetration boundary conditions to find that the Neumann condition (13.3.9) simplifies into

$$\mathbf{n} \cdot \nabla p = \mu \frac{\partial^2 \mathbf{u}}{\partial \ell^2} \cdot \mathbf{n}, \tag{13.3.10}$$

where $\mathbf{n}$ is the normal unit vector directed into the flow and ℓ is the arc length normal to the wall measured in the direction of the fluid. When the Reynolds number is sufficiently high, the right-hand side of (13.3.10) is small and can be set to zero in order to simplify the implementation.

Implicit implementations

In a certain class of finite-difference methods discussed in Section 13.4, the Neumann condition (13.3.9) is not explicitly enforced. Instead, the numerical method employs an alternative boundary condition that emerges by enforcing the continuity equation at grid nodes adjacent to the boundaries. The consistent implementation of this constraint is equivalent to the Neumann boundary condition (13.3.9). Other implementations have been proposed. For example, Quartapelle & Napolitano [333] replaced the boundary condition for the pressure with an integral constraint involving the projection of the boundary distribution of the pressure onto a nonsingular solution of the vectorial Helmholtz equation defined within the available domain of flow.

13.3.4 Compatibility condition for the PPE

The computational procedure emerging from the previous discussion involves solving a Poisson equation for the pressure,

$$\nabla^2 p = w, \tag{13.3.11}$$

subject to the Neumann boundary condition

$$\mathbf{n} \cdot \nabla p = q, \tag{13.3.12}$$

over all boundaries of the flow, where the function w is identified with the right-hand side of (13.3.5) or (13.3.6), and the function q is identified with the right-hand side of (13.3.9).

Integrating the pressure Poisson equation over the domain of flow and using the divergence theorem, we find that a solution will exist only when a compatibility condition is fulfilled,

$$\iiint_{Flow} \nabla^2 p \, \mathrm{d}V = - \iint_{B} \mathbf{n} \cdot \nabla p \, \mathrm{d}S \tag{13.3.13}$$

or

$$\iiint_{Flow} w \, \mathrm{d}V = - \iint_{B} q \, \mathrm{d}S, \tag{13.3.14}$$

where B stands for the boundaries of the flow and $\mathbf{n}$ is the normal unit vector pointing into the flow. The two-dimensional counterpart of (13.3.14) is

$$\iint_{Flow} w \, \mathrm{d}A = - \oint_{B} q \, \mathrm{d}l, \tag{13.3.15}$$

where l is the arc length around the boundary, B. When the compatibility condition is fulfilled, the pressure can be determined only up to an arbitrary constant. When the compatibility condition is not fulfilled, a solution for the pressure cannot be found.

Discrete compatibility condition

In the discrete formulation of the problem, we obtain a linear system of algebraic equations incorporating the pressure Poisson equation applied at the internal grid points and the associated boundary conditions, written in the symbolic form

$$\mathbf{A} \cdot \boldsymbol{\psi} = \mathbf{b}(w, q), \tag{13.3.16}$$

where $\mathbf{A}$ is a singular matrix. The vector on the right-hand side, $\mathbf{b}(w, q)$, is a function of the source term, w, and boundary pressure flux, q. For the linear system to have a solution, the vector $\mathbf{b}$ must be orthogonal to an eigenvector of the transpose of $\mathbf{A}$ corresponding to the null eigenvalue, denoted by $\mathbf{v}$ and satisfying $\mathbf{v} \cdot \mathbf{A} = 0$. The solvability condition is

$$\mathbf{v} \cdot \mathbf{b} = \mathbf{0}. \tag{13.3.17}$$

In practice, the eigenvector $\mathbf{v}$ can be computed directly or else compiled by inspection in terms of integration quadrature weights.

The solvability condition (13.3.17) is the discrete implementation of the compatibility condition (13.3.14). In this light, the left-hand side of (13.3.17) is recognized as the implementation of a numerical integration quadrature pertinent to the volume or surface integrals on the left- and right-hand sides of (13.3.14). The particular nature of this quadrature depends on the structure of the matrix $\mathbf{A}$, which is determined by the method used to discretize the Laplacian in the pressure Poisson equation. This observation reveals an intimate relation between the numerical differentiation matrix embedded in $\mathbf{A}$ and the singular eigenvector of its transpose of this matrix with reference to numerical integration.

13.3.5   Ensuring compatibility

Although the satisfaction of (13.3.14) is guaranteed in the continuous formulation of the problem, numerical error may destroy the exact equality of the compatibility condition in the discrete formulation. As a result, the linear system (13.3.16) will not have a solution. In solving the Poisson equation by iterative methods, the inconsistency may result in slow convergence or even divergence of the numerical solution. The iterations amount to stepping in time on the verge of numerical instability based on the unsteady diffusion–reaction equation that emerges by adding the time derivative $\partial p/\partial t$ to the right-hand side of (13.3.10). It is clear then that, if the compatibility condition is not fulfilled, a convergent solution corresponding to steady state cannot be found. In most implementations, only a few iterations are carried out and a solution of unknown accuracy is obtained [134]. To produce a numerical solution, we may abandon one linear equation expressing the Poisson equation at a particular grid point at the risk of introducing numerical oscillations. However, more sensible methods are available.

Consistent finite-difference discretization

A custom-made finite-difference method can be developed that coordinates the discretization of the equation of motion, the pressure Poisson equation, and the boundary conditions, so as to automatically satisfy the discrete version of the compatibility condition [1]. This procedure is sometimes called the consistent finite-difference discretization. Extensions to three-dimensions, curvilinear grids, and unstructured finite-volume grids are available [21, 134, 256, 380, 381, 394].

Regularization

In an alternative and more general approach, the right-hand side of the Poisson equation, $\nabla^2 p = w$, is modified by an appropriate amount so as to yield a singular system with an infinite number of solutions [57, 144, 145, 173]. This is done by replacing w with $w - \epsilon f$, where f is a specified function independent of w, and ϵ is a small dimensionless number to be found as part of the solution. The pressure is computed by solving the altered pressure Poisson equation $\nabla^2 p = w - \epsilon f$. The compatibility condition is satisfied if

$$\epsilon = \Big[\iiint_{Flow} w \, dV + \iint_B q \, dS \Big] \Big/ \iiint_{Flow} f \, dV. \qquad (13.3.18)$$

The method can be implemented according to the following steps:

1. Solve the first $N - 1$ equations of the system (13.3.16) for the first $N - 1$ unknowns, set the last unknown to zero, $\psi_N = 0$, and call the solution $\boldsymbol{\psi}^{(1)}$.

2. Compute the residual of the last equation $R^{(1)} \equiv A_{Ni}\psi_i^{(1)} - b_N$.

3. Solve the first $N - 1$ equations of the system (13.3.16) with $q = 0$ for the first $N - 1$ unknowns, set the last unknown to zero, $w_N = 0$, and call the solution $\boldsymbol{\psi}^{(2)}$.

4. Compute the residual of the last equation $R^{(2)} \equiv A_{Ni}\psi_i^{(2)} - b_N$.

5. Set $\epsilon = R^{(1)}/R^{(2)}$ and compute the final solution $\boldsymbol{\psi} = \boldsymbol{\psi}^{(1)} - \epsilon \boldsymbol{\psi}^{(2)}$.

The final solution satisfies all N equations of the perturbed linear system $\mathbf{A} \cdot \boldsymbol{\psi} = \mathbf{b}(w + \epsilon f, q)$.

To formalize and optimize the method, we replace the linear system (13.3.16) with the perturbed system

$$\mathbf{A} \cdot \psi = \mathbf{b} - \epsilon\, \mathbf{c}, \tag{13.3.19}$$

where $\mathbf{c}$ is a suitable vector normalized so that $\mathbf{c} \cdot \mathbf{c} = 1$, and the constant ϵ on the right-hand side is adjusted to ensure the satisfaction of the solvability condition When the adjoint eigenvector $\mathbf{v}$ is available, we may enforce the solvability condition by setting $\epsilon = (\mathbf{v} \cdot \mathbf{b})/(\mathbf{v} \cdot \mathbf{c})$. When the linear system (13.3.16) is solved by iteration, the regularization embodied in (13.3.19) with a uniform vector $\mathbf{c}$ can be implemented by shifting all components of the solution vector $\mathbf{w}$ by the same amount after each iteration, so that one arbitrarily chosen component is anchored at a fixed value.

The best way of regularizing the linear system (13.3.16) is by projecting the right-hand side onto the orthogonal complement of the adjoint eigenvector of the matrix $\mathbf{A}$, denoted by $\mathbf{v}$ and satisfying $\mathbf{v} \cdot \mathbf{A} = \mathbf{0}$, thereby obtaining the regularized system

$$\mathbf{A} \cdot \psi = (\mathbf{I} - \mathbf{v} \otimes \mathbf{v}) \cdot \mathbf{b}, \tag{13.3.20}$$

where $\mathbf{v} \cdot \mathbf{v} = 1$ [312]. By construction then, the solvability condition is fulfilled. Comparing (13.3.19) with (13.3.20), we identify the otherwise arbitrary vector $\mathbf{c}$ with $\mathbf{v}$, and the constant ϵ with the projection $\mathbf{v} \cdot \mathbf{b}$. The regularization expressed by (13.3.20) perturbs the source term in the Poisson equation as well as the boundary conditions in some special way.

Solution of a singular system

When the discrete version of the compatibility condition is fulfilled, the linear system of equations associated with the discrete Poisson equation has a multiplicity of solutions that differ by an arbitrary constant. To render the solution unique, we may either specify the value of the pressure at an arbitrary grid point, or set the average value of the pressure over all grid points at an arbitrary level.

13.3.6   Explicit evolution equation for the pressure

An evolution equation for the pressure can be obtained by differentiating the pressure Poisson equation in time and using the equation of motion to eliminate the time derivatives of the velocity in favor of the velocity and pressure, as discussed in Section 9.1. The result is a Poisson equation for $\partial p/\partial t$ shown in (9.1.5), to be solved subject to a Neumann boundary condition that arises by differentiating (13.3.9) in time and interchanging the order of the normal spatial derivative and temporal derivative on the left-hand side. This method appears to have been untested in practice.

13.3.7   Assessment

Comparing the formulation in primitive variables to the formulations based on the vorticity transport equation discussed in the preceding sections of this chapter, we identify relative strengths and weaknesses. One weakness is the need to derive boundary conditions for the pressure, and one strength is the ease in extending the implementation to multi-fluid and interfacial flows.

Problem

13.3.1 *Compatibility condition for the PPE*

Show that the compatibility condition (13.3.13) for the pressure Poisson equation (13.3.5) with boundary conditions given in (13.3.9) is fulfilled [153].

 Computer Problem

13.3.2 *Solving the Poisson equation in a rectangular domain*

Write a program that solves the Poisson equation in a two-dimensional rectangular domain with an arbitrary source term assigned at the grid points and arbitrary Neumann boundary conditions all around the boundaries. The Laplacian should be approximated using the five-point formula and the normal derivative at the boundaries should be approximated using second-order one-side finite differences (see Section B.5, Appendix B.) A regularization of your choice should be employed to ensure that the discrete problem admits a solution.

13.4   Implementation in primitive variables

Having addressed the fundamental concepts involved in the computation of an incompressible Newtonian flow in terms of the velocity and the pressure, we proceed to discussing specific implementations. The marker-and-cell (MAC) method combines a finite-difference method for solving the equations of incompressible flow with a marker-point tracing method for tracking the evolution of a free surface or interface between two immiscible fluids [168].

The finite-difference method is based on the explicit forward-time discretization of the equation of motion (1.4.32), yielding the velocity at the next time level, $n+1$, from the velocity and pressure at the current time level, n,

$$\mathbf{u}^{n+1} = \mathbf{u}^n + \Delta t \left(\mathbf{N}(\mathbf{u}^n) + \nu \mathcal{L}(\mathbf{u}^n) - \frac{1}{\rho} \nabla p^n \right), \tag{13.4.1}$$

where Δt is the time step. To compute the pressure, p^n, we discretize the evolution equation for the rate of expansion (13.3.4) using a forward difference in time and obtain

$$\frac{(\nabla \cdot \mathbf{u})^{n+1} - (\nabla \cdot \mathbf{u})^n}{\Delta t} = \nabla \cdot \mathbf{N}(\mathbf{u}^n) + \nu \mathcal{L}(\nabla \cdot \mathbf{u}^n) - \frac{1}{\rho} \nabla^2 p^n. \tag{13.4.2}$$

Requiring that the divergence of the velocity is zero at the $n+1$ time level, we obtain a Poisson equation for the pressure,

$$\nabla^2 p^n = \frac{\rho}{\Delta t} (\nabla \cdot \mathbf{u})^n + \rho \, \nabla \cdot \mathbf{N}(\mathbf{u}^n) + \mu \, \mathcal{L}(\nabla \cdot \mathbf{u}^n), \tag{13.4.3}$$

which is a modification of the consistent pressure Poisson equation. Although small, the first and third terms on the right-hand side of (13.4.3) should be retained to prevent the onset of numerical instability. The computational algorithm involves the following steps:

1. Specify an initial solenoidal velocity field that satisfies the prescribed boundary conditions.

2. Compute the pressure field by solving the Poisson equation (13.4.3), subject to the Neumann boundary condition (13.3.9). In practice, this is done using an alternative but equivalent set of boundary conditions to be discussed shortly.

3. Use (13.4.1) to advance the velocity field by one step in time, subject to the prescribed boundary conditions. The spatial derivatives on the right-hand side of (13.4.1) are computed using central differences. The size of time step is kept sufficiently small to suppress the growth of numerical oscillations.

4. Return to Step 2 and repeat the computation for another time step.

The method can be implemented on a staggered grid that decouples velocity from pressure nodes, or on a nonstaggered grid where all variables are defined at the same nodes.

13.4.1   Implementation on a staggered grid

Harlow & Welch [168] employed the staggered grid shown in Figure 13.4.1 for flow in a cavity driven by a translating grid. The staggered grid consists of two interwoven grids parametrized by two pairs of indices, (i, j) and (i', j'), as illustrated in Figure 13.4.1. The grid lines of the primary grid are represented by the solid lines and the grid lines of the secondary grid are represented by the broken lines. The secondary grid conforms with the physical boundaries of the flow. The values of the primed indices are printed in bold.

Discrete values of the pressure are assigned to the primary nodes, (i, j), defined by the intersection of the solid lines, shown as circles. Discrete values of the x component of the velocity are defined at the intersection of horizontal primary grid lines and vertical secondary grid lines, (i', j), shown as horizontal arrows. Discrete values of the y component of the velocity are defined at the intersection of vertical primary grid lines and horizontal secondary grid lines, (i, j'), shown as vertical arrows.

A distinguishing feature of the staggered-grid method is that the unknown functions and governing equations are defined or enforced at different nodes. This decoupling simplifies the numerical implementation and promotes the stability of the numerical method. We will see that distributing the pressure nodes in the interior of the flow excluding the boundaries bypasses the derivation of explicit boundary conditions for the pressure.

Interpolation and extrapolation of the velocity

For simplicity, we denote u_x by u and u_y by v. The velocity components at the vertices of the primary or secondary grids are computed by linear interpolation from the closest nodes,

$$u_{i'+1/2,j} = \frac{1}{2}\left(u_{i',j} + u_{i'+1,j}\right), \qquad u_{i',j+1/2} = \frac{1}{2}\left(u_{i',j} + u_{i',j+1}\right).$$

$$v_{i+1/2,j'} = \frac{1}{2}\left(v_{i,j'} + v_{i+1,j'}\right), \qquad v_{i,j'+1/2} = \frac{1}{2}\left(v_{i,j'} + v_{i,j+1'}\right). \qquad (13.4.4)$$

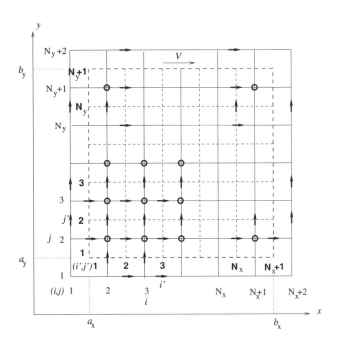

FIGURE 13.4.1   A staggered grid for computing flow in a rectangular domain representing a cavity. The primary grid is drawn with heavy lines and the secondary grid is drawn with broken lines. Note that the secondary grid conforms with the boundaries of the flow. The pressure is defined at the circles, the x and y velocity components are defined at the horizontal or vertical arrows.

Unphysical x velocity components are required at the horizontal lines $j = 1$ and $j = N_y + 2$ for the computation of the second y derivative of u near the top and bottom boundaries. Corresponding unphysical y velocity components are required at the vertical levels $i = 1$ and $N_x + 2$ for the computation of the second x derivative of v near the left and right boundaries. These exterior velocities are computed by extrapolation to satisfy the boundary conditions at the physical levels $i' = 1$, $i' = N_x + 1$, $j' = 1$, and $j' = N_y + 1$.

For example, approximating u with a parabolic function near the lid located at $y = b_y$, and enforcing the no-slip boundary condition $u = V$, we obtain

$$u = V + A\,(y - b_y)^2 + B\,(y - b_y),\tag{13.4.5}$$

where V is the lid velocity, and A, B are unknown coefficients. Applying this expression at three neighboring grid levels, we obtain

$$u_{i',N_y} = V + A\,\frac{9}{4}\,h^2 - B\,\frac{3}{2}\,h, \qquad u_{i',N_y+1} = V + A\,\frac{1}{4}\,h^2 - B\,\frac{1}{2}\,h,$$

$$u_{i',N_y+2} = V + A\,\frac{1}{4}\,h^2 + B\,\frac{1}{2}\,h,\tag{13.4.6}$$

where $h = \Delta y$. Eliminating A and B, we obtain

$$u_{i',N_y+2} = \frac{1}{3}\left(8\,V - 6\,u_{i,N_y+1} + u_{i,N_y}\right). \tag{13.4.7}$$

Similar expressions are derived for the other external velocities.

Advancing the velocity field

To advance the x and y velocity components in time, we apply (13.4.1) at the corresponding interior velocity nodes, express the nonlinear convection term in conservative form, and approximate the spatial derivatives using central differences. Enforcing the x component of the Navier–Stokes equation at the (i',j) x-velocity node, we obtain

$$u_{i'j}^{n+1} = u_{i'j}^{n} + \Delta t\left(-\tilde{u}_{i',j}^{n} + \nu\,\check{u}_{i',j}^{n} - \frac{1}{\rho}\frac{p_{i'+1,j} - p_{i',j}}{\Delta x}\right), \tag{13.4.8}$$

where

$$\tilde{u}_{i',j} = \frac{u_{i'+1/2,j}^2 - u_{i'-1/2,j}^2}{\Delta x} + \frac{(uv)_{i',j+1/2} - (uv)_{i',j-1/2}}{\Delta y} \tag{13.4.9}$$

is the discretized nonlinear term and

$$\check{u}_{i'j} = \frac{u_{i'-1,j} - 2\,u_{i',j} + u_{i'+1,j}}{\Delta x^2} + \frac{u_{i',j-1} - 2\,u_{i',j} + u_{i',j+1}}{\Delta y^2} \tag{13.4.10}$$

is the discretized Laplacian. Working similarly, we enforce the y component of the Navier–Stokes equation at the (i,j') y-velocity node and obtain

$$v_{ij'}^{n+1} = v_{ij'}^{n} + \Delta t\left(-\tilde{v}_{i,j'}^{n} + \check{v}_{i,j'}^{n} - \frac{1}{\rho}\frac{p_{i,j'+1} - p_{i,j'}}{\Delta x}\right), \tag{13.4.11}$$

where

$$\tilde{v}_{i,j'} = \frac{(uv)_{i+1/2,j'} - (uv)_{i-1/2,j'}^2}{\Delta x} + \frac{v_{i,j'+1/2}^2 - v_{i,j'-1/2}}{\Delta y} \tag{13.4.12}$$

and

$$\check{v}_{ij'} = \frac{v_{i-1,j'} - 2\,v_{i,j'} + v_{i+1,j'}}{\Delta x^2} + \frac{v_{i,j'-1} - 2\,v_{i,j'} + v_{i,j'+1}}{\Delta y^2}. \tag{13.4.13}$$

Before we can evaluate the right-hand sides of (13.4.8) and (13.4.11), the nodal pressures must be available.

Pressure Poisson equation

To compute the pressure, we approximate the rate of expansion at the (i,j) pressure node using central differences and obtain

$$D_{i,j} \equiv (\nabla \cdot \mathbf{u})_{i,j} = \frac{u_{i,j} - u_{i-1,j}}{\Delta x} + \frac{v_{i,j} - v_{i,j-1}}{\Delta y} = 0 \tag{13.4.14}$$

for $i = 2, \ldots, N_x + 1$ and $j = 2, \ldots, N_y + 1$. Requiring that the right-hand side of (13.4.14) vanishes at the $n + 1$ time level and expressing the velocity at the $n + 1$ level in terms of the velocity and pressure at the nth level using (13.4.8) and (13.4.11), we obtain

$$\frac{\rho}{\Delta t} D_{ij}^n - \rho \left(\frac{\tilde{u}_{i,j} - \tilde{u}_{i-1,j}}{\Delta x} + \frac{\tilde{v}_{i,j} - \tilde{v}_{i,j-1}}{\Delta y} \right)^n + \mu \left(\frac{\check{u}_{i,j} - \check{u}_{i-1,j}}{\Delta x} + \frac{\check{v}_{i,j} - \check{v}_{i,j-1}}{\Delta y} \right)^n$$
$$- \left(\frac{p_{i+1,j} - 2\,p_{i,j} + p_{i-1,j}}{\Delta x^2} + \frac{p_{i,j+1} - 2\,p_{i,j} + p_{i,j-1}}{\Delta y^2} \right)^n = 0. \quad (13.4.15)$$

Rearranging, we derive the discrete version of the pressure Poisson equation (13.4.3), where the Laplacian is approximated with the five-point formula over intervals equal to Δx and Δy.

Pressure boundary condition

For the pressure nodes adjacent to the boundaries, we follow a slightly different approach that takes into account the prescribed velocity boundary conditions. Considering the second horizontal layer, $j = 2$, hosting the pressure nodes, we require that the discrete form of the rate of expansion given by the right-hand side of (13.4.14) is zero at the $n + 1$ time level. Focusing on the pressure nodes $i = 3, \ldots, N_x$, located away from the corners, we express the velocities at the adjacent nodes at the $n + 1$ level inside the domain of flow in terms of the velocities and pressure at the nth level using (13.4.8) and (13.4.11), and obtain

$$\frac{\rho}{\Delta t} \left(\frac{u_{i,2}^n - u_{i-1,2}^n}{\Delta x} + \frac{v_{i,2}^n - v_{i,1}^{n+1}}{\Delta y} \right) - \rho \left(\frac{\tilde{u}_{i,2} - \tilde{u}_{i-1,2}}{\Delta x} + \frac{\tilde{v}_{i,2}}{\Delta y} \right)^n \quad (13.4.16)$$
$$+ \mu \left(\frac{\check{u}_{i,2} - \check{u}_{i-1,2}}{\Delta x} + \frac{\check{v}_{i,2}}{\Delta y} \right)^n - \left(\frac{p_{i+1,2} - 2\,p_{i,2} + p_{i-1,2}}{\Delta x^2} + \frac{p_{i,3} - p_{i,2}}{\Delta y^2} \right)^n = 0.$$

The v velocity at the $(i, 1)$ node is available from the prescribed boundary conditions. Straightforward modifications are necessary for the two corner nodes, $i = 2$ and $N_x + 1$. Working in a similar fashion for the last horizontal layer corresponding to $j = N_y + 1$ for $i = 3, \ldots, N_x$, we obtain

$$\frac{\rho}{\Delta t} \left(\frac{u_{i,j}^n - u_{i-1,j}^n}{\Delta x} + \frac{v_{i,j}^{n+1} - v_{i,j-1}^n}{\Delta y} \right) - \rho \left(\frac{\tilde{u}_{i,j} - \tilde{u}_{i-1,j}}{\Delta x} - \frac{\tilde{v}_{i,j-1}}{\Delta y} \right)^n \quad (13.4.17)$$
$$+ \mu \left(\frac{\check{u}_{i,j} - \check{u}_{i-1,j}}{\Delta x} - \frac{\check{v}_{i,j-1}}{\Delta y} \right)^n - \left(\frac{p_{i+1,j} - 2\,p_{i,j} + p_{i-1,j}}{\Delta x^2} - \frac{p_{i,j} - p_{i,j-1}}{\Delta y^2} \right)^n = 0,$$

where $j = N_y + 1$. Similar equations can be written for the first and last vertical layers corresponding to $i = 1$ and $N_x + 1$, for $j = 2, \ldots, N_y + 1$. Equations (13.4.16), (13.4.17), and their counterparts for the first and last vertical layers, provide us with boundary conditions for the discrete Poisson equation (13.4.15).

To implement an explicit Neumann boundary condition for the pressure, we introduce the exterior pressure nodes $p_{i,1}$. Using the continuity equation, we find that $\partial v / \partial y$ vanishes at the bottom wall. Applying (13.3.10) at the lower wall located at $y = a_y$, and approximating the derivatives using central differences, we obtain

$$\mathbf{n} \cdot \nabla p = \left(\frac{\partial p}{\partial y} \right)_{y=a_y} \simeq \frac{p_{i,2} - p_{i,1}}{\Delta y} = \mu \left(\frac{\partial^2 u_y}{\partial y^2} \right)_{y=a_y} \simeq 2\mu \frac{v_{i,2}}{\Delta y}. \quad (13.4.18)$$

Rearranging, we find that

$$p_{i,1} = p_{i,2} - 2\mu \frac{v_{i,2}}{\Delta y^2},$$ (13.4.19)

where all variables are evaluated at the nth time level. Working in a similar fashion with the upper, left, and right walls, we find that

$$p_{i,N_y+2} = p_{i,N_y+1} + 2\mu \frac{v_{i,N_y+1}}{\Delta y}, \qquad p_{1,j} = p_{2,j} - 2\mu \frac{u_{2,j}}{\Delta x},$$

$$p_{N_x+2,j} = p_{N_x+1,j} + 2\mu \frac{u_{N_x+1,j}}{\Delta x}.$$ (13.4.20)

Equations (13.4.19) and (13.4.20) provide us with an alternative set of boundary conditions for the Poisson equation (13.4.15), which is now also applied at pressure nodes adjacent to the boundaries, $i = 2, N_x + 1$ and $j = 2, N_y + 1$.

It appears that, by using a staggered grid, we have circumvented the derivation of explicit boundary conditions for the pressure. In fact, equation (13.4.16) and its counterparts for the other three walls amount to the Neumann pressure boundary condition stated in(13.3.9) [153]. To see this, we subtract (13.4.16) from (13.4.15) written for $j = 2$, and obtain

$$\frac{\rho}{\Delta t} \left(u_{i,j-1}^{n+1} - u_{i,j-1}^{n} \right) + \rho \frac{\tilde{v}_{i,j-1}^{n}}{\Delta y} - \mu \frac{\tilde{v}_{i,j-1}^{n}}{\Delta y} + \frac{p_{i,j}^{n} - p_{i,j-1}^{n}}{\Delta y^2} = 0.$$ (13.4.21)

As Δy tends to zero, the first and second terms on the left-hand side vanish and the remaining terms reproduce the Neumann boundary condition (13.4.19).

Assessment

Bypassing the derivation of the boundary conditions for the pressure is a significant advantage of the staggered-grid implementation. Unfortunately, the numerical formulation becomes considerably more involved and prohibitively expensive for grids defined in curvilinear coordinates.

13.4.2 Non-staggered grid

To implement the explicit method on the non-staggered grid shown in Figure 13.1.1, we enforce the discretized equation of motion (13.4.1) and pressure Poisson equation (13.4.3) at all interior nodes and compute all spatial derivatives by central difference approximations. To derive boundary conditions for the pressure, we require that the discretized form of the rate of expansion is zero at grid points located along the four walls. Considering the first horizontal layer, $j = 1$ for $i = 2, \ldots, N_x$, we use central differences in the x direction and second-order one-sided differences in the y direction to write [80]

$$D_{ij} = \frac{u_{i+1,1} - u_{i-1,1}}{2\Delta x} + \frac{-v_{i,3} + 4v_{i,2} - 3v_{i,1}}{2\Delta y}.$$ (13.4.22)

Requiring that the right-hand side of (13.4.22) is zero at the $n + 1$ time level and expressing the velocities at the $(i, 2)$ and $(i, 3)$ nodes in terms of the discrete version of the right-hand side of

(13.4.1), we obtain the counterpart of equation (13.4.16). Working in a similar fashion, we derive corresponding equations for the upper, left, and right walls.

It appears that we have again circumvented the explicit derivation of boundary conditions for the pressure. However, it can be shown that (13.4.22) and its counterparts for the three other walls amount to the Neumann boundary conditions (13.3.9) [153]. This is true even when first-order one-sided differences are used to approximate the second partial derivatives. Unfortunately, solving the Poisson equation by this method is susceptible to numerical instability due to the decoupling of the pressure at adjacent nodes. A small numerical incompressibility must be tolerated in the finite-difference solution.

13.4.3 Second-order methods

The explicit method discussed in this section is first-order accurate in time and conditionally stable. To improve the accuracy and relax the stability constraint, we may resort to a semi-implicit, fully implicit, or predictor–corrector method. Adopting the Crank–Nicolson discretization, we apply the equation of motion at an intermediate $n + \frac{1}{2}$ level and obtain a second-order scheme expressed by the equation

$$\mathbf{u}^{n+1} = \mathbf{u}^n + \frac{\Delta t}{2} \left(\mathbf{N}(\mathbf{u}^{n+1}) + \mathbf{N}(\mathbf{u}^n) + \nu \left[\mathcal{L}(\mathbf{u}^{n+1}) + \mathcal{L}(\mathbf{u}^n) \right] \right) - \frac{\Delta t}{\rho} \nabla p^{n+1/2}. \qquad (13.4.23)$$

Since the method is implicit in the nonlinear terms, carrying out each time step requires solving a system of nonlinear algebraic equations for the velocity at the $n + 1$ time level and for the pressure at the $n + \frac{1}{2}$ time level. In practice, this is done by iteration ([294], p. 167).

Problems

13.4.1 *Pressure boundary conditions*

Derive the counterparts of equations (13.4.16) and (13.4.17) for (*a*) the first and last vertical layers corresponding to $i = 1, N_x$ and $j = 2, \ldots, N_y - 1$, and (*b*) the four corner pressure nodes.

13.4.2 *Neumann boundary conditions for the pressure on a non-staggered grid*

Derive the finite-difference statement of the Neumann boundary condition for the pressure in the case of flow in a cavity using the non-staggered grid shown in Figure 13.1.1.

 *Computer Problem*

13.4.3 *Flow in a rectangular cavity*

(*a*) Write a program that uses the explicit method on a a staggered grid described in the text to compute transient flow in a rectangular cavity due to the impulsive translation of the lid. The pressure Poisson equation should be solved using an iterative method of your choice.

(*b*) Repeat (*a*) for a non-staggered grid.

13.5   Operator splitting, projection, and pressure-correction methods

The Navier–Stokes equation states that the velocity at a particular point in a flow changes because of the simultaneous action of the nonlinear convective term, $\rho \mathbf{u} \cdot \nabla \mathbf{u}$, the linear viscous term, $\mu \nabla \mathbf{u}^2$, and the pressure gradient, ∇p. In the operator-splitting method, each one or groups of these terms are decoupled and considered to operate sequentially during fictitious time intervals of equal length. Since a main issue in the computation of an unsteady incompressible flow is the absence of an evolution equation for the pressure, it is natural to decouple the convective–diffusive term from the pressure gradient term. The Navier–Stokes equation is then split into two component equations,

$$\frac{\partial \mathbf{u}}{\partial t} = \mathbf{N}(\mathbf{u}) + \nu \, \mathcal{L}(\mathbf{u}) \qquad (13.5.1)$$

and

$$\frac{\partial \mathbf{u}}{\partial t} = -\frac{1}{\rho} \nabla p. \qquad (13.5.2)$$

These equations are assumed to apply sequentially, each for the full time interval Δt within each time step. The nonlinear convection–diffusion equation (13.5.1) advances the velocity from the initial state, $\mathbf{u}^n$, to an intermediate state, $\mathbf{u}^*$. The pressure correction equation (13.5.2) advances the intermediate state, $\mathbf{u}^*$, to the final state, $\mathbf{u}^{n+1}$, thus completing the nth time step.

13.5.1   Solenoidal projection and the role of the pressure

To analyze the nature of the fractional-step decomposition expressed by equations (13.5.1) and (13.5.2), it is helpful to introduce the concept of solenoidal projection due to Chorin [80], discussed in Section 9.1. A key idea is to introduce the space of all vector functions, and note that the velocity field of an unsteady incompressible flow must evolve within the subspace of solenoidal functions. Next, we observe that the intermediate velocity, $\mathbf{u}^*$, will not be necessarily solenoidal at the end of the convection–diffusion step (13.5.1). Thus, the convection–diffusion step causes a departure from the subspace of solenoidal functions. The role of the pressure-correction step is to cancel this departure by projecting $\mathbf{u}^*$ into the subspace of solenoidal functions, thus producing the final velocity, $\mathbf{u}^{n+1}$.

Since the pressure p has not been updated in the convection–diffusion step, it loses its physical meaning and must be regarded as an auxiliary function whose main purpose is to project the intermediate velocity, $\mathbf{u}^*$, onto the subspace of solenoidal functions. To indicate this subtle differentiation, we replace the pressure p in (13.5.2) with a projection function, ϕ, writing

$$\frac{\partial \mathbf{u}}{\partial t} = -\frac{1}{\rho} \nabla \phi. \qquad (13.5.3)$$

In the case of a constant density fluid, the right-hand side of this equation is irrotational. The relation between the projection function and the pressure will be discussed later in this section.

13.5.2   Boundary conditions for intermediate variables

An important consideration in the implementation of the fractional-step method is the choice of boundary conditions for the intermediate velocity, $\mathbf{u}^*$, and projection function, ϕ. Ideally, the boundary conditions for $\mathbf{u}^*$ should be chosen to minimize its divergence while ensuring that the stipulated boundary conditions for the physical velocity are observed at the end of a complete time step. The derivation of boundary conditions for $\mathbf{u}^*$ and ϕ has been the subject of extensive investigation [285].

A consistent set of boundary conditions include the Dirichlet boundary condition for the normal and tangential velocity components of the intermediate velocity,

$$\mathbf{u}^* \cdot \mathbf{n} = \mathbf{V}^{n+1} \cdot \mathbf{n}, \qquad \mathbf{u}^* \cdot (\mathbf{I} - \mathbf{n} \otimes \mathbf{n}) = \left(\mathbf{V}^{n+1} + \frac{1}{\rho} \int_{t^n}^{t^n + \Delta t} \nabla \phi \, dt \right) \cdot (\mathbf{I} - \mathbf{n} \otimes \mathbf{n}), \qquad (13.5.4)$$

where $\mathbf{V}$ is the prescribed boundary velocity, $\mathbf{I}$ is the identity matrix, and $\mathbf{I} - \mathbf{n} \otimes \mathbf{n}$ is the tangential projection operator. Correspondingly, the projection function satisfies the homogeneous Neumann condition

$$\mathbf{n} \cdot \int_{t^n}^{t^n + \Delta t} \nabla \phi \, dt = 0. \qquad (13.5.5)$$

Straightforward substitution demonstrates that (13.5.4) and (13.5.5) ensure the required boundary condition at the end of a complete step, $\mathbf{u}^{n+1} = \mathbf{V}^{n+1}$. In expressions (13.5.4) and (13.5.5), the projection function is regarded as a continuous function of time over the interval where equation (13.5.3) applies. In numerical practice, the continuous evolution is replaced by an instantaneous projection.

Optimal and simplified boundary conditions with varying degrees of accuracy and ease of implementation are available [154]. In the simplest scheme, the intermediate velocity, $\mathbf{u}^*$, satisfies the boundary conditions stipulated at the $n + 1$ time level, and the projection function ϕ satisfies the homogeneous Neumann boundary condition $\mathbf{n} \cdot \nabla \phi = 0$. Improved boundary conditions are discussed in Section 13.5.4.

13.5.3 Evolution of the rate of expansion

Evolution equations for the rate of expansion during the convection–diffusion and projection steps arise by taking the divergence of (13.5.1) and (13.5.3), finding

$$\rho \, \frac{\partial \alpha}{\partial t} = \rho \, \nabla \cdot \mathbf{N} + \mu \nabla^2 \alpha, \qquad (13.5.6)$$

and

$$\rho \, \frac{\partial \alpha}{\partial t} = -\nabla^2 \phi, \qquad (13.5.7)$$

where $\alpha \equiv \nabla \cdot \mathbf{u}$ is the rate of expansion. The convection–diffusion equation (13.5.6) receives the presumably vanishing rate of expansion at the t^n time level, and advances it over a time interval Δt to an intermediate nonzero field, α^*. The projection equation (13.5.7) receives α^* and advances it over the same time interval Δt to the final field, α^{n+1}, which is required to vanish throughout the domain of flow and over the boundaries.

13.5.4   First-order projection method

To make matters more specific, we discuss the implementation of a first-order projection method for three-dimensional flow [80]. We begin by splitting the convection–diffusion step (13.5.1) into three sequential convection–diffusion fractional steps according to the equations

$$\frac{\partial \mathbf{u}}{\partial t} + u_x \frac{\partial \mathbf{u}}{\partial x} = \nu \frac{\partial^2 \mathbf{u}}{\partial x^2}, \qquad \frac{\partial \mathbf{u}}{\partial t} + u_y \frac{\partial \mathbf{u}}{\partial y} = \nu \frac{\partial^2 \mathbf{u}}{\partial y^2}, \qquad \frac{\partial \mathbf{u}}{\partial t} + u_z \frac{\partial \mathbf{u}}{\partial z} = \nu \frac{\partial^2 \mathbf{u}}{\partial z^2}. \qquad (13.5.8)$$

Each component equation applies for the full time interval, Δt, and time is reset to the initial value, t_n, at the end of each fractional step. The three intermediate velocity fields, denoted by

$$\mathbf{u}^{n+1/4}, \qquad \mathbf{u}^{n+2/4}, \qquad \mathbf{u}^{n+3/4} \equiv \mathbf{u}^*, \qquad (13.5.9)$$

are not generally solenoidal. In the case of two-dimensional flow, only the first two steps in (13.5.8) are present and the intermediate velocity fields are denoted by

$$\mathbf{u}^{n+1/3}, \qquad \mathbf{u}^{n+2/3} \equiv \mathbf{u}^*. \qquad (13.5.10)$$

The three steps in (13.5.8) can be carried out using the implicit BTCS method or the Crank–Nicolson method with first- or second-order accuracy in time, respectively. Both methods are unconditionally stable and require the easy task of solving tridiagonal systems of equations. To express the constituent convection–diffusion equations in conservative form, we may set the convection velocity in all three steps equal to the presumed solenoidal velocity at the beginning of the step corresponding to the time level, t_n. Boundary conditions for the intermediate velocities will be derived.

Approximating the time derivative in the projection step in (13.5.3) with a difference equation and rearranging, we obtain

$$\mathbf{u}^{n+1} = \mathbf{u}^* - \frac{\Delta t}{\rho} \nabla \phi^{n+1}. \qquad (13.5.11)$$

To compute ϕ^{n+1}, we take the divergence of (13.5.11) and require that $\nabla \cdot \mathbf{u}^{n+1} = 0$ to derive a Poisson equation,

$$\nabla^2 \phi^{n+1} = \frac{\rho}{\Delta t} \nabla \cdot \mathbf{u}^*, \qquad (13.5.12)$$

accompanied by the boundary condition

$$\mathbf{n} \cdot \nabla \phi^{n+1} = 0. \qquad (13.5.13)$$

As in the case of the Poisson equation for the pressure discussed in Section 13.3, the explicit implementation of this boundary condition can be circumvented by using a staggered grid and requiring that $\nabla \cdot \mathbf{u}^{n+1} = 0$ at pressure nodes adjacent to the boundaries, or by using a non-staggered grid and requiring that $\nabla \cdot \mathbf{u}^{n+1} = 0$ at nodes located at the boundaries.

Given an initial velocity field along with a prescribed boundary condition, $\mathbf{u} = \mathbf{V}$, we compute the evolution of the flow according to the following steps:

1. Assign the initial velocity to velocity nodes and provide an estimate for the projection function, ϕ^{n+1}, at the pressure nodes.

2. Advance the velocity field sequentially according to equations (13.5.8) with boundary conditions $\mathbf{u}^* \cdot \mathbf{n} = \mathbf{V} \cdot \mathbf{n}$ for the normal velocity component and

$$\mathbf{u}^* \cdot (\mathbf{I} - \mathbf{n} \otimes \mathbf{n}) = \left(\mathbf{V} + \frac{\Delta t}{\rho} \nabla \phi^{n+1}\right) \cdot (\mathbf{I} - \mathbf{n} \otimes \mathbf{n}) \qquad (13.5.14)$$

for the tangential velocity component, where $\mathbf{I}$ is the identity matrix and $\mathbf{I} - \mathbf{n} \otimes \mathbf{n}$ is the tangential projection operator.

3. Solve the Poisson equation (13.5.12) with the Neumann boundary condition (13.5.13). Since the flow rate of $\mathbf{u}^*$ and thus of $\nabla \phi$ across the boundaries of the flow vanishes because of the imposed boundary conditions on $\mathbf{u}^*$, the compatibility condition is automatically satisfied and the Poisson equation has an infinite number of solutions.

4. Compute the velocity $\mathbf{u}^{n+1}$ at all internal and boundary grid points according to (13.5.11). If the tangential boundary velocity is not equal to the specified velocity, producing a nonzero slip velocity, return to Step 2 and repeat the computations with the current boundary distribution of ϕ. Otherwise, proceed to Step 5.

5. Reset the time to t_{n+1}, return to Step 2, and repeat the computations for another time step.

Consider the familiar problem of two-dimensional flow in a cavity driven by a moving lid, illustrated in Figure 13.1.1. The convection–diffusion equations in the x and y directions can be integrated in time using an implicit method, such as the Crank–Nicolson method. When integrating in the x direction, we use the side wall boundary conditions

$$u_{1,j} = 0, \quad v_{1,j} = \frac{\Delta t}{\rho}\left(\frac{\partial \phi^{n+1}}{\partial y}\right)_{1,j}, \quad u_{N_x+1,j} = 0, \quad v_{N_x+1,j} = \frac{\Delta t}{\rho}\left(\frac{\partial \phi^{n+1}}{\partial y}\right)_{N_x+1,j}, \qquad (13.5.15)$$

where $u = u_x$ and $v = u_y$. Boundary conditions over the upper and lower walls are not required. When integrating in the y direction, we use the lower and upper wall boundary conditions

$$u_{i,1} = \frac{\Delta t}{\rho}\left(\frac{\partial \phi^{n+1}}{\partial y}\right)_{i,1}, \quad v_{i,1} = 0, \quad u_{i,N_y+1} = \frac{\Delta t}{\rho}\left(\frac{\partial \phi^{n+1}}{\partial y}\right)_{i,N_y+1}, \quad v_{i,N_y+1} = 0. \qquad (13.5.16)$$

Boundary conditions over the side walls are not required.

In a more advanced implementation, the convection–diffusion equation (13.5.1) is integrated in time using the alternating direction implicit (ADI) method. This modification reduces the temporal error due to the directional spatial decoupling involved in (13.5.8) and renders the convection–diffusion step, but not necessarily the overall scheme, second-order accurate in time.

13.5.5 Second-order methods

Methods with second-order temporal accuracy can be developed. Discretizing the evolution equations (13.5.1) and (13.5.6) for the velocity and rate of expansion according to the generalized Crank–Nicolson method, we obtain the semi-discrete forms

$$\rho \frac{\mathbf{u}^* - \mathbf{u}^n}{\Delta t} = \rho \mathbf{N}^{n+\gamma} + \mu\left[(1-\gamma)\nabla^2\mathbf{u}^n + \gamma\nabla^2\mathbf{u}^*\right] \qquad (13.5.17)$$

and

$$\rho \frac{\alpha^* - \alpha^n}{\Delta t} = \rho \left[\nabla \cdot \mathbf{N}^{n+\gamma} + \mu (1 - \gamma) \nabla^2 \alpha^n + \gamma \nabla^2 \alpha^* \right], \tag{13.5.18}$$

where the numerical parameter γ varies in the interval $(0, 1]$. The term $\mathbf{N}^{n+\gamma}$ is assumed to be available with $O(\Delta t^m)$ accuracy in terms of the extrapolated or interpolated velocity of the physical velocity field evaluated at time $t^n + \gamma \Delta t$. When $m = 2$, we obtain second-order accuracy.

Now turning to the projection step, we replace the corresponding evolution equations (13.5.3) and (13.5.7) with the difference equations

$$\rho \frac{\mathbf{u}^{n+1} - \mathbf{u}^*}{\Delta t} = -\nabla \phi^{n+1}, \tag{13.5.19}$$

and

$$\rho \frac{\alpha^{n+1} - \alpha^*}{\Delta t} = -\nabla^2 \phi^{n+1}. \tag{13.5.20}$$

Solving (13.5.20) for α^*, we obtain

$$\alpha^* = \alpha^{n+1} + \frac{\Delta t}{\rho} \nabla^2 \phi^{n+1}. \tag{13.5.21}$$

Requiring that $\alpha^{n+1} = 0$, we derive a familiar Poisson equation for ϕ^{n+1},

$$\nabla^2 \phi^{n+1} = \frac{\rho}{\Delta t} \alpha^*, \tag{13.5.22}$$

which is to be solved subject to the homogeneous Neumann condition, $\mathbf{n} \cdot \nabla \phi^{n+1} = 0$, originating from (13.5.5).

Now using (13.5.19) to eliminate the intermediate velocity $\mathbf{u}^*$ from (13.5.17), we obtain the equation

$$\rho \frac{\mathbf{u}^{n+1} - \mathbf{u}^n}{\Delta t} = \rho \mathbf{N}^{n+\gamma} - \nabla (\phi^{n+1} - \gamma \mu \alpha^*) + \mu \left[(1 - \gamma) \nabla^2 \mathbf{u}^n + \gamma \nabla^2 \mathbf{u}^{n+1} \right]. \tag{13.5.23}$$

Comparing this expression with the equation of motion discretized according to the generalized Crank–Nicolson method shows that second-order temporal accuracy for the velocity is achieved when $\gamma = \frac{1}{2}$, the pressure is evaluated according to

$$p^{n+\frac{1}{2}} = \phi^{n+1} - \frac{1}{2} \mu \alpha^* + O(\Delta t^2) = \phi^{n+1} - \frac{1}{2} \nu \Delta t \nabla^2 \phi^{n+1} + O(\Delta t^2), \tag{13.5.24}$$

and the nonlinear term is evaluated with second-order accuracy by extrapolation or interpolation, so that $m = 2$ [59, 206]. Equation (13.5.24) provides us with a relation between the projection function and the pressure, worthy of further investigation.

Relation between the projection function and the pressure

To analyze further the relation between the projection function and the pressure, we rearrange (13.5.18) into an inhomogeneous Helmholtz equation,

$$\alpha^* - \gamma \nu \Delta t \, \nabla^2 \alpha^* = \Delta t \, \nabla \cdot \mathbf{N}^{n+\gamma} + R_+^n, \tag{13.5.25}$$

where

$$R_+^n \equiv \alpha^n + (1 - \gamma) \, \nu \Delta t \, \nabla^2 \alpha^n \tag{13.5.26}$$

is a residual. In the temporal semi-discrete formulation presently considered, R_+ is zero throughout the domain of flow and along the boundaries. In a full time-space discretization, R_+ vanishes to machine accuracy only for a certain class of carefully designed numerical methods [380]. Substituting the right-hand side of (13.5.21) for α^* into (13.5.25) and rearranging, we obtain

$$\nabla^2 \left(\phi^{n+1} - \gamma \nu \Delta t \, \nabla^2 \phi^{n+1} \right) = \rho \, \nabla \cdot \mathbf{N}^{n+\gamma} - \frac{\rho}{\Delta t} \left(R_-^{n+1} - R_+^n \right). \tag{13.5.27}$$

The residual,

$$R_-^{n+1} \equiv \alpha^{n+1} - \gamma \nu \Delta t \, \nabla^2 \alpha^{n+1}, \tag{13.5.28}$$

is precisely zero in the space-continuous formulation presently considered, but only approximately zero in a spatially discrete implementation.

We can infer the relation between the projection function and the pressure by taking the divergence of the equation of motion and enforcing the incompressibility condition to derive the pressure Poisson equation. Applying the Poisson equation at time $t^{n+\gamma}$, we obtain

$$\nabla^2 p^{n+\gamma} = \rho \, \nabla \cdot \mathbf{N}_E^{n+\gamma}, \tag{13.5.29}$$

where the subscript E denotes the exact value. Subtracting (13.5.27) from (13.5.29) and rearranging, we find that

$$\nabla^2 \left(p^{n+\gamma} - \phi^{n+1} + \gamma \nu \Delta t \, \nabla^2 \phi^{n+1} \right) + \rho \left[\nabla \cdot \mathbf{N}^{n+\gamma} - \nabla \cdot \mathbf{N}_E^{n+\gamma} \right] = \frac{\rho}{\Delta t} \left(R_-^{n+1} - R_+^n \right). \tag{13.5.30}$$

By definition, the expression enclosed by the square brackets on the left-hand side is of $O(\Delta t^m)$.

In the absence of spatial discretization or under fortunate circumstances, $\alpha^n = 0$ and $\alpha^{n+1} = 0$, the right-hand side of (13.5.30) vanishes, and the expression enclosed by the first set of parentheses on the left-hand side of (13.5.30) is a harmonic function up to $O(\Delta t^m)$, denoted by χ. Rearranging, we find that

$$p^{n+\gamma} = \phi^{n+1} - \gamma \nu \Delta t \, \nabla^2 \phi^{n+1} + \chi + O(\Delta t^m) = \phi^{n+1} - \gamma \mu \alpha^* + \chi + O(\Delta t^m). \tag{13.5.31}$$

Neumann boundary conditions for χ arise by projecting the gradient of (13.5.31) normal to the boundaries of the flow. Using the equation of motion and the homogeneous Neumann boundary condition for ϕ^{n+1}, we find that

$$\mathbf{n} \cdot \nabla \chi = \mathbf{n} \cdot \left[- \rho \left(\frac{\partial \mathbf{u}_B}{\partial t} \right)^{n+\gamma} + \rho \, \mathbf{N}_E^{n+\gamma} + \mu \nabla^2 \mathbf{u}_E^{n+\gamma} + \gamma \nu \, \Delta t \, \nabla \left(\nabla^2 \phi^{n+1} \right) \right] + O(\Delta t^m). \tag{13.5.32}$$

Next, we substitute into the right-hand side of (13.5.32) the approximation

$$\nabla^2 \mathbf{u}_E^{n+\gamma} = (1 - \gamma)\, \nabla^2 \mathbf{u}^n + \gamma\, \nabla^2 \mathbf{u}^{n+1} + O(\Delta t^l), \tag{13.5.33}$$

where l is the temporal order of the physical velocity. Using (13.5.17) and the Laplacian of (13.5.19) to eliminate the Laplacian of the velocity, we finally obtain

$$\mathbf{n} \cdot \nabla \chi = \rho\, \mathbf{n} \cdot \left[\frac{\mathbf{u}^* - \mathbf{u}^n}{\Delta t} - \left(\frac{\partial \mathbf{u}_B}{\partial t} \right)^{n+\gamma} + O(\Delta t^l) + O(\Delta t^m) \right]. \tag{13.5.34}$$

In light of the first boundary condition in (13.5.4), the right-hand side of (13.5.34) vanishes up to order Δt^m or Δt^l. Consequently, the harmonic function χ is constant within this precision, and equation (13.5.31) yields

$$p^{n+\gamma} = \phi^{n+1} - \gamma \nu \Delta t\, \nabla^2 \phi^{n+1} + O(\Delta t^l) + O(\Delta t^m) \tag{13.5.35}$$

or

$$p^{n+\gamma} = \phi^{n+1} - \gamma \mu\, \alpha^* + O(\Delta t^l) + O(\Delta t^m). \tag{13.5.36}$$

In the case of the Crank–Nicolson discretization corresponding to $\gamma = \frac{1}{2}$ and $l = 2$, and a second-order interpolation or extrapolation method for the nonlinear term corresponding to $m = 2$, we recover (13.5.24).

Boundary layer of the intermediate rate of expansion

Consider an unsteady Stokes flow in the absence of a distributed body force. In the space-continuous formulation, equation (13.5.25) reduces into a homogeneous Helmholtz equation for the intermediate rate of expansion, α^*,

$$\alpha^* - \gamma \nu \Delta t\, \nabla^2 \alpha^* = 0. \tag{13.5.37}$$

Balancing the magnitude of the two terms on the left-hand side, we find that α^* is supported by boundary layers of thickness $\delta \simeq (\gamma \nu \Delta t)^{1/2}$. Since the normal derivative of ϕ at the boundary is zero, the corresponding normal derivative of α^* is comparable to that of the pressure according to (13.5.36). Consequently, α^* is of order δ inside the boundary layer. Correspondingly, the second term on the right-hand side of (13.5.36) makes a leading-order contribution of order δ inside the boundary layer and decays exponentially outside the boundary layer [117, 154, 393]. As γ tends to zero, yielding an explicit discretization, the thickness of the boundary layer vanishes and the numerical method expectedly fails.

Under more general conditions of unsteady Navier–Stokes flow and in the presence of a distributed body force, the right-hand side of (13.5.25) is significant throughout the domain of flow and α^* is of order Δt in the interior of the flow and varies by an amount of order $\Delta t^{1/2}$ across the boundary layers.

In the fully discrete time-space implementation, the right-hand side of (13.5.30) does not generally vanish to machine accuracy, and is of order $\Delta t \times h^m$, where the exponent m is determined by the spatial discretization. Accordingly, the right-hand sides of (13.5.35), (13.5.36), and (13.5.24) include corrections that may lower the nominal order of the numerical method.

A semi-implicit three-level method

Kim & Moin [206] proposed advancing the velocity field according to two fractional steps,

$$\frac{\mathbf{u}^* - \mathbf{u}^n}{\Delta t} = \frac{1}{2}\left[3\,\mathbf{N}(\mathbf{u}^n) - \mathbf{N}(\mathbf{u}^{n-1})\right] + \frac{1}{2}\nu\left[\mathcal{L}(\mathbf{u}^*) + \mathcal{L}(\mathbf{u}^n)\right] \tag{13.5.38}$$

and

$$\frac{\mathbf{u}^{n+1} - \mathbf{u}^*}{\Delta t} = -\frac{1}{\rho}\nabla\phi^{n+1}. \tag{13.5.39}$$

Equation (13.5.38) uses the explicit second-order Adams–Bashforth method for the nonlinear term and the implicit second-order Crank–Nicolson method for the viscous term. The projection function, ϕ, is computed by solving the Poisson equation (13.5.12) to ensure that the velocity field $\mathbf{u}^{n+1}$ is solenoidal. Solving (13.5.39) for $\mathbf{u}^*$ and substituting the result into (13.5.38), we find that

$$\frac{\mathbf{u}^{n+1} - \mathbf{u}^n}{\Delta t} = -\frac{1}{\rho}\nabla\left(\phi^{n+1} + \frac{1}{2}\nu\Delta t\,\nabla^2\phi^{n+1}\right)$$

$$+ \frac{1}{2}\left[3\,\mathbf{N}(\mathbf{u}^n) - \mathbf{N}(\mathbf{u}^{n-1})\right] + \frac{1}{2}\nu\left[\mathcal{L}(\mathbf{u}^{n+1}) + \mathcal{L}(\mathbf{u}^n)\right], \tag{13.5.40}$$

which makes a clear distinction between p and ϕ. Setting

$$p^{n+1/2} \equiv \phi^{n+1} + \frac{1}{2}\nu\Delta t\,\nabla^2\phi^{n+1} \tag{13.5.41}$$

shows that (13.5.39) is second-order accurate in time.

To carry out the convection–diffusion step, we recast (13.5.38) into the form

$$(1 - \frac{1}{2}\nu\Delta t\,\nabla^2)(\mathbf{u}^* - \mathbf{u}^n) = \frac{1}{2}\Delta t\left[3\,\mathbf{N}(\mathbf{u}^n) - \mathbf{N}(\mathbf{u}^{n-1})\right] + \nu\Delta t\,\nabla^2(\mathbf{u}^n), \tag{13.5.42}$$

and then approximately factorize the operator on the left-hand side,

$$(1 - \frac{1}{2}\nu\Delta t\,\nabla^2) \simeq (1 - \frac{1}{2}\nu\Delta t\,\frac{\partial^2}{\partial x^2})(1 - \frac{1}{2}\nu\Delta t\,\frac{\partial^2}{\partial y^2})(1 - \frac{1}{2}\nu\Delta t\,\frac{\partial^2}{\partial z^2}). \tag{13.5.43}$$

To compute the difference $\mathbf{u}^* - \mathbf{u}^n$, we solve three tridiagonal systems of equations using the Thomas algorithm discussed in Section B.1.4, Appendix B (Problem 13.5.2). Boundary conditions for $\mathbf{u}^*$ arise as discussed in Section 13.5.2. The Poisson equation (13.5.12) can be solved on a staggered grid to avoid the explicit implementation of boundary conditions for the projection function, ϕ.

An iterative second-order method

Bell, Colella & Glaz [27] proposed a two-level projection method with second-order accuracy in time. Introducing the velocity and pressure at the integer time levels, n, as well at intermediate time levels, $n + \frac{1}{2}$, we compute the intermediate velocity, $\mathbf{u}^*$, based on a semi-discrete version of a modified equation of motion,

$$\frac{\mathbf{u}^* - \mathbf{u}^n}{\Delta t} = \mathbf{N}(\mathbf{u}^{n+1/2}) - \frac{1}{\rho}\nabla\psi + \frac{1}{2}\nu\left[\mathcal{L}(\mathbf{u}^*) + \mathcal{L}(\mathbf{u}^n)\right], \tag{13.5.44}$$

where ψ is a projection function. The velocity at the next time level derives from a further projection in terms of a new projection function, ϕ,

$$\mathbf{u}^{n+1} = \mathbf{u}^* - \frac{\Delta t}{\rho}\nabla\phi. \tag{13.5.45}$$

Solving (13.5.45) for $\mathbf{u}^*$ and substituting the result into (13.5.44), we obtain

$$\frac{\mathbf{u}^{n+1} - \mathbf{u}^n}{\Delta t} = \mathbf{N}(\mathbf{u}^{n+1/2}) - \frac{1}{\rho}\nabla(\psi + \phi) + \frac{1}{2}\nu\left[\mathcal{L}(\mathbf{u}^{n+1}) + \mathcal{L}(\mathbf{u}^n)\right] + \frac{1}{2}\mu\Delta t\,\nabla\mathcal{L}(\phi), \tag{13.5.46}$$

which implements a second-order discretization of the equation of motion provided that $p^{n+1/2} = \psi + \phi$ and the last term on the right-hand side is zero. The algorithm can be implemented according to the following steps [443]:

1. Given the pressure, $p^{n-1/2}$, and the velocity, $\mathbf{u}^n$, estimate $\mathbf{u}^{n+1/2}$ by extrapolation and set $\psi = p^{n-1/2}$.

2. Calculate the nonlinear term, $\mathbf{N}(\mathbf{u}^{n+1/2})$, compute $\mathbf{u}^*$ from (13.5.44), and call the solution $\mathbf{u}^{*,k}$, where k is an inner iteration number.

3. Introduce the discrete form of the equation of motion

$$\frac{\mathbf{u}^{n+1,k} - \mathbf{u}^n}{\Delta t} = \mathbf{N}(\mathbf{u}^{n+1/2}) - \frac{1}{2}\nabla p^{n+1/2,k} + \frac{1}{2}\nu\left[\mathcal{L}(\mathbf{u}^{n+1,k}) + \mathcal{L}(\mathbf{u}^n)\right], \tag{13.5.47}$$

take its divergence, and require that $\mathbf{u}^{n+1,k}$ is solenoidal to obtain a Poisson equation for $p^{n+1/2,k}$. Solve the Poisson equation subject to appropriate boundary conditions and use (13.5.47) to compute $\mathbf{u}^{n+1,k}$.

4. Set $\psi = p^{n+1/2,k}$, return to Step 2 and repeat the computations with $\mathbf{u}^{n+1} = \mathbf{u}^{n+1,k}$, where $\mathbf{u}^{n+1/2}$ is computed by interpolation or extrapolation, increasing k by one.

Problems

13.5.1 *Solenoidal projection*

(*a*) Consider a nonsolenoidal rotational velocity field defined over the whole three-dimensional space and decaying at infinity. Develop a procedure for removing the nonsolenoidal component while leaving the vorticity unchanged.

(*b*) Repeat (*a*) for a bounded flow and discuss appropriate boundary conditions for the projection function.

13.5.2 *Approximate factorization*

Write the three tridiagonal systems of equations corresponding to the factorized form (13.5.43).

🖥 **Computer Problem**

13.5.3 *Flow in a rectangular cavity*

Write a program that uses the first-order projection method described in the text to compute the transient flow in a rectangular cavity due to the impulsive translation of the lid.

13.6   Methods of modified dynamics or false transients

The methods discussed in the preceding two sections involve advancing the velocity using a time-marching method suitable for parabolic differential equations, and updating the pressure or projection function by solving the elliptic Poisson's equation over the flow domain. The origin of this dual approach can be traced back to the absence of an evolution equation for the pressure in the original system of governing equations. The solution of the elliptic equation consumes most of the computational effort and hinders the development of methods with second-order accuracy in time. If all governing equations were available the form of an evolution equation, a simple time-marching scheme would suffice.

The main idea behind the methods of modified dynamics is to amend either the continuity equation or the equation of motion so as to render all governing equations parabolic in time. In certain cases, the error introduced by modifying the original equations is mild and the transient or steady solution obtained by solving the modified equations describes the physical evolution with acceptable accuracy. In other cases, the transient evolution is purely fictitious and thus physically irrelevant, and the temporal evolution is significant only insofar as to provide us with a venue that leads us to a steady state. The modified equations are designed so that the solution obtained at steady state satisfies the equations of steady incompressible Newtonian flow within a specified tolerance.

13.6.1   Artificial compressibility method for steady flow

To render all governing equations parabolic in time, we may transform the continuity equation into an evolution equation for the pressure. In Chorin's artificial compressibility method, the continuity equation is replaced by the modified evolution equation

$$\frac{\partial p}{\partial t} + \frac{\rho}{\delta} \, \nabla \cdot \mathbf{u} = 0, \qquad (13.6.1)$$

where ρ is the fluid density and δ is a small positive constant called the artificial compressibility [79]. At steady state, the first term on the left-hand side of (13.6.1) vanishes and the steady solution satisfies the equations of steady incompressible flow.

Implementation of a non-staggered grid

On a non-staggered grid, the modified continuity equation (13.6.1) is discretized using standard centered-time and centered-space differences. The divergence of the velocity at boundary nodes is computed using central differences for the velocity component tangential to the boundary or first- or second-order one-sided differences for the velocity component normal to the boundary. The pressure at boundary nodes is computed as discussed previously in this chapter.

To expedite the approach to steady state, it is desirable to use larger times steps. However, a semi-implicit or fully-implicit method must be employed to ensure numerical stability. Chorin [79] implemented a variation of the explicit CTCS method according to the Du Fort–Frankel scheme. The finite-difference equation corresponding to the x component of the equation of motion at the (i, j) grid point is

$$u_{ij}^{n+1} = u_{ij}^{n-1} + 2\Delta t \left(-\frac{1}{\rho} \frac{\partial p^n}{\partial x} + \mathcal{F}_x \right), \tag{13.6.2}$$

where

$$\mathcal{F}_x = N_x(\mathbf{u}^n) + \nu \left(\frac{u_{i+1,j}^n - u_{i,j}^{n+1} - u_{i,j}^{n-1} + u_{i-1,j}^n}{\Delta x^2} + \frac{u_{i,j+1}^n - u_{i,j}^{n+1} - u_{i,j}^{n-1} + u_{i,j-1}^n}{\Delta y^2} \right), \tag{13.6.3}$$

and $u = u_x$. The nonlinear term, $\mathbf{N}$, is discretized using centered differences in its conservative form. In Chapter 12, we saw that the modified differential equation corresponding to the Du Fort–Frankel scheme is not consistent with the original differential equation. However, since the difference is a small term involving second partial derivatives with respect to time, the modified and original equations agree at steady state. The numerical scheme can be extended in a straightforward fashion to three-dimensional flow where the Du Fort–Frankel scheme is applied to one out of the three second spatial derivatives. Chorin [79] finds that, given boundary conditions for the velocity, and provided that the flow remains subsonic with respect to the artificial speed of sound $1/\sqrt{\delta}$, the method is stable as long as the maximum Courant number is less than $2(\delta/n)^{1/2}/(1 + 5^{1/2})$, where $n = 2$ or 3 for two- or three-dimensional flow.

Other explicit or implicit implementations of the artificial compressibility method can be devised. Implementing explicit upwind methods for the convection term at high Reynolds numbers and implicit methods for the viscous term allows us to use of large time steps and thus accelerate the approach to steady state.

Implementation on a staggered grid

Next, we consider the implementation of the method on the staggered grid shown in Figure 13.4.1. Adopting an explicit formulation, we advance the velocity at the x and y velocity nodes according to equations (13.4.8) and (13.4.11), subject to the boundary conditions discussed in Section 13.4. To advance the pressure, we apply (13.6.1) at the pressure nodes and introduce the explicit FTCS discretization to obtain

$$p_{ij}^{n+1} = p_{ij}^n - \frac{\rho \Delta t}{\delta} \left(\frac{u_{i,j} - u_{i-1,j}}{\Delta x} + \frac{v_{i,j} - v_{i,j-1}}{\Delta y} \right), \tag{13.6.4}$$

where $u = u_x$ and $v = u_y$. One notable feature of the artificial compressibility method implemented on a staggered grid is that boundary conditions for the pressure are not required. When steady state has been reached, the pressure satisfies the pressure Poisson equation with Neumann boundary conditions that arise by projecting the equation of motion onto the unit vector normal to the boundaries.

13.6.2   Modified PPE

Sotiropoulos & Abdallah [379] modified the evolution equation (13.3.4) for the rate of expansion, $\alpha \equiv \nabla \cdot \mathbf{u}$, transforming it into an evolution equation for the pressure,

$$\frac{\partial p}{\partial t} = \frac{\rho}{\beta} \left(\frac{1}{\rho} \nabla^2 p - \nabla \cdot \mathbf{N}(\mathbf{u}) + \frac{\partial \alpha}{\partial t} \right), \tag{13.6.5}$$

where β is a positive constant. At steady state, equation (13.6.5) reduces to the familiar Poisson equation for the pressure. The transient solution lacks physical meaning and equation (13.6.5) is significant only insofar as to provide us with a route toward the steady state.

13.6.3   Penalty-function formulation

The penalty-function formulation uses an artificial constitutive equation for the pressure in terms of the rate of expansion,

$$p = -\frac{1}{\epsilon} \nabla \cdot \mathbf{u}, \tag{13.6.6}$$

where ϵ is a small positive constant [401]. Since the pressure is an order-one variable, the rate of expansion is restricted to remain small at all times. Substituting (13.6.6) into the equation of motion yields a modified evolution equation for the velocity alone. In practice, to ensure numerical stability, it is necessary to add to the modified equation of motion an additional small term involving the rate of expansion. The governing equation of motion is

$$\frac{\partial \mathbf{u}}{\partial t} = \mathbf{N}(\mathbf{u}) + \frac{1}{\rho \epsilon} \nabla (\nabla \cdot \mathbf{u}) - \frac{1}{2} \mathbf{u} (\nabla \cdot \mathbf{u}) + \nu \, \mathcal{L}(\mathbf{u}). \tag{13.6.7}$$

The computations proceed by integrating (13.6.7) forward in time from a given initial state subject to the specified velocity boundary conditions. The penalty-function formulation has found extensive applications in numerical procedures based on finite-element methods [196].

Problem

13.6.1 *Artificial compressibility method*

(*a*) Write the counterparts of (13.6.2) and (13.6.3) for the x and y velocity components on a staggered grid.

(*b*) Develop an ADI method for two-dimensional flow.

Mathematical supplement

A

Understanding theoretical and computational fluid dynamics requires familiarity with elementary mathematics summarized in this appendix. The theory of differential, integral, and vector calculus is discussed in two highly mathematical textbooks by Boas [44] and Hildebrand [178]. The equations of fluid mechanics and transport processes in general curvilinear coordinates are discussed in an essential monograph by Aris [14].

A.1 Functions of one variable

A function of a real variable is a device that receives a real number, x, and produces another real number, $f(x)$. We say that a function maps a point, x, to another point, $f(x)$.

Continuity

If the graph of a function does not exhibit sudden jumps inside a certain interval, $a \leq x \leq b$, the function is continuous in that interval.

Intermediate-value theorem

Assume that a function, $f(x)$, is continuous in an interval, $a \leq x \leq b$, and denote the minimum value of the function as $\min \equiv \mathrm{infimum}[f(x)]$, and the maximum value of the function as $\max \equiv \mathrm{supreme}[f(x)]$. The intermediate-value theorem asserts, and geometrical intuition confirms, that, for any number c such that $\min \leq c \leq \max$, there is at least one point, ξ, in the interval $a \leq \xi \leq b$, such that $f(\xi) = c$, as illustrated in Figure A.1.1(a).

Mean-value theorem for the derivative

Assume that a function, $f(x)$, is continuous in a closed interval, $a \leq x \leq b$, and the derivative of the function, $f'(x)$, is finite in the open interval $a < x < b$, as illustrated in Figure A.1.1(b). The mean-value theorem for the derivative ensures that at least one point, ξ, can be found in the closed interval $a \leq \xi \leq b$, where the derivative of the function is equal to the slope of the line subtended between the two end points,

$$f'(\xi) = \frac{f(b) - f(a)}{b - a}. \tag{A.1.1}$$

A consequence of the mean-value theorem is Rolle's theorem stating that, if $f(a) = 0$ and $f(b) = 0$, there is at least one point, ξ, in the interval $a \leq \xi \leq b$, where $f'(\xi) = 0$.

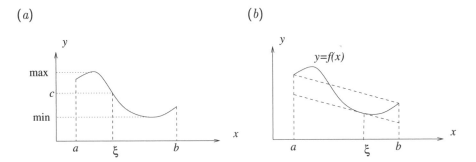

(a) (b)

FIGURE A.1.1   Graphical interpretation of (a) the intermediate-value theorem and (b) the mean-value theorem.

Mean-value theorem for an integral

Let $w(x)$ be a nonnegative function whose definite integral is finite in an interval, $a \leq x \leq b$. If the function $f(x)$ is continuous in that interval, then at least one point, ξ, can be found in the interval $a \leq \xi \leq b$, such that

$$\int_a^b f(x)\, w(x)\, \mathrm{d}x = f(\xi) \int_a^b w(x)\, \mathrm{d}x. \tag{A.1.2}$$

The value $f(\xi)$ is a weighted average of the function $f(x)$. If $w(x)$ is constant, the value $f(\xi)$ is the mean value of the function $f(x)$.

Taylor series expansion

If (a) a point, x, is close to another point, x_0, (b) the function $f(x)$ is continuous in a neighborhood of x_0, and (c) the graph of the function does not become vertical at x_0, it is reasonable to expect that the difference between $f(x)$ and $f(x_0)$ will be proportional to the difference between x and x_0. The Taylor series provides us with an infinite polynomial expression for the difference in the function values, $f(x) - f(x_0)$, in terms of the corresponding difference in the arguments, $x - x_0$.

If the first $N+1$ derivatives of a function $f(x)$ are continuous at every point between and including the points x and x_0, then

$$f(x) = f(x_0) + \sum_{m=1}^N \left(\frac{1}{m!} f^{(m)}(x_0) (x - x_0)^m \right) + R_N(x, x_0), \tag{A.1.3}$$

where the superscript (m) denotes the mth derivative and $m! = 1 \cdots 2 \cdots m$ is the factorial, subject to the convention that $0! = 1$. The remainder is given by

$$R_N(x, x_0) = \frac{1}{(N+1)!} f^{(N+1)}(\xi) (x - x_0)^{N+1}, \tag{A.1.4}$$

where $\xi(x, x_0)$ is a special point; if $x < x_0$, then $x \leq \xi \leq x_0$; whereas, if $x_0 < x$, then $x_0 \leq \xi \leq x$. An alternative expression for the remainder is

$$R_N(x, x_0) = \frac{1}{N!} \int_{x_0}^{x} f^{(N+1)}(\eta) \, (x - \eta)^N \, d\eta. \tag{A.1.5}$$

Expression (A.1.4) arises from (A.1.5) using the mean-value theorem for an integral. As the truncation threshold, N, tends to infinity, we obtain an infinite series representation.

It is important to note that the remainder does not necessarily vanish in the limit $N \to \infty$, but may oscillate or grow unbounded. This means that the infinite Taylor series without the remainder does not necessarily converge to the function, $f(x)$. The radius of convergence is determined by the location of the pole of the complex function $f(z)$ nearest x_0.

Maclaurin series expansion

If the expansion point is set at the origin, $x_0 = 0$, the infinite Taylor series without the remainder reduces to the Maclaurin series. The Maclaurin series expansions of several common functions are presented in Table A.1.1.

Problem

A.1.1 *Maclaurin series*

Derive the Maclaurin series of the function $f(x) = 1/(1+x+x^2)$ and deduce its radius of convergence.

A.2   Functions of two variables

A function of two variables is a device that receives a pair of numbers, x and y, and produces a new number, $f(x, y)$. We may say that the function $f(x, y)$ maps an ordered pair, (x, y), to a single point, $f(x, y)$. Functions of more than two variables are defined in similar ways.

Taylor series expansion

The Taylor series expansion of a function of two variables about a point, (x_0, y_0), takes the form

$$f(x, y) = f(x_0, y_0) + \sum_{m=1}^{N} \left[\frac{1}{m!} \left((x - x_0) \frac{\partial}{\partial x} + (y - y_0) \frac{\partial}{\partial y} \right)^m f(x, y) \right]_{x_0, y_0} + R_N(x, y, x_0, y_0), \tag{A.2.1}$$

where

$$R_N(x, y, x_0, y_0) = \frac{1}{(N+1)!} \left((x - x_0) \frac{\partial}{\partial x} + (y - y_0) \frac{\partial}{\partial y} \right)^{N+1} f(x, y) \Big|_{\xi_x, \xi_y} \tag{A.2.2}$$

is the remainder. If $x < x_0$, $x \leq \xi_x \leq x_0$; whereas, if $x_0 < x$, $x_0 \leq \xi_x \leq x$. Similar relations apply for ξ_y. The operators on the right-hand sides of (A.2.1) and (A.2.2) are defined in terms of the binomial expansion as

$$\left[(x - x_0) \frac{\partial}{\partial x} + (y - y_0) \frac{\partial}{\partial y} \right]^m f(x, y) \Big|_{x_0, y_0} \equiv \sum_{k=0}^{m} \binom{m}{k} \left(\frac{\partial^m f}{\partial x^k \partial y^{m-k}} \right)_{x_0, y_0} (x - x_0)^k (y - y_0)^{m-k}. \tag{A.2.3}$$

Rational	$\frac{1}{1-x} = 1 + x + x^2 + x^3 + \cdots$	for $-1 < x < 1$
	$\frac{1}{1+x} = 1 - x + x^2 - x^3 + \cdots$	for $-1 < x < 1$
Exponential	$\mathrm{e}^x = 1 + x + \frac{1}{2} x^2 + \frac{1}{6} x^3 + \cdots + \frac{1}{i!} x^i + \cdots$	for any real x
	$\sinh x = x + \frac{1}{6} x^3 + \frac{1}{5!} x^5 + \frac{1}{7!} x^7 + \cdots$	for any real x
	$\cosh x = 1 + \frac{1}{2} x^2 + \frac{1}{4!} x^4 + \frac{1}{6!} x^6 + \cdots$	for any real x
	$\tanh x = x - \frac{1}{3} x^3 + \frac{2}{15} x^5 - \frac{17}{315} x^7 + \cdots$	for any real x
	$a^x = 1 + x \ln a + \frac{1}{2} (x \ln a)^2 + \cdots + \frac{1}{i!} (x \ln a)^i + \cdots$	for any real x
Logarithmic	$\ln(1 + x) = x - \frac{1}{2} x^2 + \frac{1}{3} x^3 - \frac{1}{4} x^4 + \ldots$	for $-1 < x < 1$
Trigonometric	$\sin x = x - \frac{1}{6} x^3 - \frac{1}{5!} x^5 + \frac{1}{7!} x^7 + \cdots$	for any real x
	$\cos x = 1 - \frac{1}{2} x^2 + \frac{1}{4!} x^4 - \frac{1}{6!} x^6 + \cdots$	for any real x
	$\tan x = x + \frac{1}{3} x^3 + \frac{2}{15} x^5 + \frac{17}{315} x^7 + \cdots$	for any real x
Trigonometric	$\arcsin x = x + \frac{1}{6} x^3 + \frac{3}{40} x^5 + \frac{5}{112} x^7 + \cdots$	for any real x
	$\arccos x = \frac{\pi}{2} - x - \frac{1}{6} x^3 - \frac{3}{40} x^5 - \frac{5}{112} x^7 + \cdots$	for any real x

TABLE A.1.1   Maclaurin series expansions of common functions. A comprehensive compilation can be found at the CRC Standard Mathematical Tables [444].

The first set of large parentheses on the right-hand side of (A.2.3) denote the combinatorial,

$$\binom{m}{k} \equiv \frac{m!}{k!\,(m-k)!} = \prod_{p=1}^{l} \frac{m-p+1}{p}, \tag{A.2.4}$$

where l is the minimum of k and $m - k$. The combinatorial expresses the number of possible combinations by which k objects can be chosen from a set of m identical objects, leaving behind $m - k$ objects.

As the truncation level, N, tends to infinity, we obtain an infinite series polynomial representation. However, the remainder does not necessarily vanish, and the Taylor series does necessarily converge as $N \to \infty$. The Maclaurin series arises by setting $x_0 = 0$ and $y_0 = 0$.

Problems

A.2.1 *Maclaurin series*

Derive the Maclaurin series of the function $f(x) = e^{x+y}$ and $f(x) = e^{xy}$.

A.2.2 *Taylor series of a function of three variables*

State the Taylor series of a function of three variables, $f(x, y, z)$.

A.2.3 *Binomial distribution*

The binomial distribution is defined as

$$\mathcal{B}_{k|m}(x) \equiv \binom{m}{k} x^k (1-x)^{m-k}, \tag{A.2.5}$$

where $k \leq m$. In statistics, the binomial distribution represents the probability of k successes in m trials, where x is the success probability for each individual trial, called the Bernoulli probability. Show that the binomial distribution reaches a maximum at $x = k/m$.

A.3 Dirac delta function

The Dirac delta function in one dimension, $\delta(x-x_0)$, represents an infinitesimally narrow distribution due to a unit impulse applied at a point, x_0. The induced field is distinguished by the following properties:

1. $\delta(x - x_0)$ vanishes everywhere except at the singular point, $x = x_0$, where it becomes infinite.

2. The integral of $\delta(x - x_0)$ with respect to x over an interval I that contains the point x_0 is equal to unity,

$$\int_I \delta(x - x_0)\, \mathrm{d}x = 1. \tag{A.3.1}$$

   This property reveals that the delta function with argument of length has units of inverse length.

3. The integral of the product of an arbitrary function, $f(x)$, and the delta function over an interval I that contains the point x_0 is equal to value of the function at the singular point,

$$\int_I \delta(x - x_0)\, f(x)\, \mathrm{d}x = f(x_0). \tag{A.3.2}$$

   The integral of the product of an arbitrary function, $f(x)$, and the delta function over an interval I that does *not* contain the singular point x_0 is zero.

4. If a function, $f(x)$, undergoes a discontinuity from a left value f_- to a right value f_+ at the point x_0, the derivative of the function at that point can be expressed in terms of the delta function as

$$f'(x_0) = (f_+ - f_-)\,\delta(x - x_0) + \cdots, \tag{A.3.3}$$

   where the dots denote nonsingular terms.

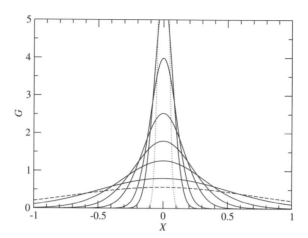

FIGURE A.3.1 A family of dimensionless test functions, $G \equiv gL$, described by (A.3.4), plotted against the dimensionless distance $X \equiv (x - x_0)/L$ for $\lambda = 1$ (dashed line), 2, 5, 10, 20, 50, 100, and 500 (dotted line). As the dimensionless parameter λ tends to infinity, we recover Dirac's delta function in one dimension.

In formal mathematics, $\delta(x - x_0)$ can be obtained from the family of Gaussian test functions,

$$g(x - x_0) = \left(\frac{\lambda}{\pi L^2} \right)^{1/2} \exp \left[-\lambda \left(\frac{x - x_0}{L} \right)^2 \right], \tag{A.3.4}$$

in the limit as the dimensionless parameter λ tends to infinity, where L is an arbitrary length. The mean value of the Gaussian distribution is x_0 and the variance is $\sigma^2 = L^2/(2\lambda)$.

Graphs of the dimensionless function $G \equiv gL$ plotted against the dimensionless distance $X \equiv (x - x_0)/L$ are shown in Figure A.3.1 for $\lambda = 1$ (dashed line), 2, 5, 10, 20, 50, 100, and 500 (dotted line). The maximum height of each graph is inversely proportional to its width, so that the area underneath each graph is equal to unity,

$$\int_{-\infty}^{\infty} g(x) \, \mathrm{d}x = \left(\frac{\lambda}{\pi L^2} \right)^{1/2} \int_{-\infty}^{\infty} \exp(-\lambda \frac{x^2}{L^2}) \, \mathrm{d}x = 1. \tag{A.3.5}$$

To prove this identity, we recall the definite integral associated with the error function,

$$\frac{2}{\sqrt{\pi}} \int_0^{\infty} \exp(-w^2) \, \mathrm{d}w = 1, \tag{A.3.6}$$

and substitute $w = \lambda^{1/2} x/L$.

Dirac delta function in two dimensions

The Dirac delta function in two dimensions, $\delta_2(x - x_0, y - y_0)$, represents an infinitesimally narrow distribution due to a unit impulse applied at a point, (x_0, y_0), in the xy plane. The induced field is distinguished by the following properties:

1. $\delta_2(x - x_0, y - y_0)$ vanishes everywhere except at the point $x = x_0$, $y = y_0$, where it becomes infinite.

2. The integral of the delta function over an area, D, that contains the singular point (x_0, y_0) is equal to unity,

$$\iint_D \delta_2(x - x_0, y - y_0) \, dx \, dy = 1. \tag{A.3.7}$$

This property illustrates that the delta function in two dimensions has units of inverse squared length.

3. The integral of the product of an arbitrary function, $f(x, y)$, and the delta function over an area D that contains the singular point, (x_0, y_0), is equal to the value of the function at the singular point,

$$\iint_D \delta_2(x - x_0, y - y_0) \, f(x, y) \, dx \, dy = f(x_0, y_0). \tag{A.3.8}$$

The integral of the product of an arbitrary function, $f(x, y)$, and the delta function over an area D that does *not* contain the singular point, (x_0, y_0), is equal to zero.

4. We may write

$$\delta_2(x - x_0, y - y_0) = \delta_1(x - x_0)\delta_1(y - y_0), \tag{A.3.9}$$

where δ_1 is the one-dimensional delta function.

In formal mathematics, $\delta_2(x - x_0, y - y_0)$ arises from the family of test functions

$$q(r) = \frac{\lambda}{\pi L^2} \exp\left(-\lambda \frac{r^2}{L^2}\right), \tag{A.3.10}$$

in the limit as the dimensionless parameter λ tends to infinity, where $r \equiv |\mathbf{x} - \mathbf{x}_0|$ and L is an arbitrary length.

Graphs of the dimensionless function $Q \equiv qL^2$ plotted against the dimensionless distance $X \equiv (x - x_0)/L$ are shown in Figure A.3.2 for $y = y_0$, and $\lambda = 1$ (dashed line), 2, 5, 10, 20, 50, 100, and 500 (dotted line). The maximum height of each graph is inversely proportional to the square of its width, so that the area underneath each graph in the xy plane is equal to unity,

$$\int_{-\infty}^{\infty} \int_{-\infty}^{\infty} q(r) \, dx \, dy = \int_0^{2\pi} \int_0^{\infty} q(r) \, r \, dr \, d\theta = 2\pi \int_0^{\infty} q(r) \, r \, dr. \tag{A.3.11}$$

Making substitutions, we find that

$$\int_{-\infty}^{\infty} \int_{-\infty}^{\infty} q(r) \, dx \, dy = \frac{2\lambda}{L^2} \int_0^{\infty} \exp\left(-\lambda \frac{r^2}{L^2}\right) r \, dr = 1, \tag{A.3.12}$$

independent of L.

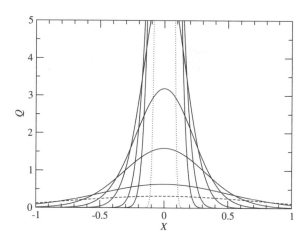

FIGURE A.3.2   A family of dimensionless test functions, $Q \equiv qL^2$, described by (A.3.10), plotted against the dimensionless distance $X \equiv (x - x_0)/L$ for $y = y_0$ and $\lambda = 1$ (dashed line), 2, 5, 10, 20, 50, 100, and 500 (dotted line). As the dimensionless parameter λ tends to infinity, we recover Dirac's delta function in two dimensions.

Dirac delta function in three and higher dimensions

The Dirac delta function can be defined in a space with arbitrary dimensions. The mathematical properties are similar to those in one and two dimensions.

Problems

A.3.1 *Dirac delta function in three dimensions*

(a) State the distinguishing properties of the Dirac delta function in three dimensions.

(b) Devise a family of test functions, similar to those shown in (A.3.4) and (A.3.10), that reduce to the three-dimensional delta function as λ tends to zero.

A.3.2 *Heaviside function*

The dimensionless Heaviside function, $\mathcal{H}(t)$, is defined such that $\mathcal{H}(t) = 0$ for $t < 0$ and $\mathcal{H}(t) = 1$ for $t > 0$. Express the first derivative of the Heaviside function in terms of Dirac's delta function. Discuss the second derivative of the Heaviside function.

A.4   Index notation

In index notation, a Cartesian vector in the Nth dimensional space, $\mathbf{u}$, is denoted as u_i, where $i = 1, \ldots, N$; a two-dimensional matrix $\mathbf{A}$ is denoted as A_{ij}; and an N-dimensional matrix $\mathbf{B}$ is denoted as $B_{ij\cdots m}$, where the number of subscripts, i–m, is equal to N. The same index may not appear more than twice in an index array or across index arrays in a product.

Repeated-index summation convention

Einstein's repeated-index summation convention states that, if a subscript appears twice in an expression involving products, summation over that subscript is implied in its range. Under this convention,

$$u_i \equiv A_{ij} v_j \equiv \sum_j A_{ij} v_j, \qquad C_{ikl} \equiv A_{ij} B_{jkl} \equiv \sum_j A_{ij} B_{jkl},$$

$$u_i v_i \equiv u_1 v_1 + u_2 v_2 + \cdots + u_N v_N, \qquad A_{ii} \equiv A_{11} + A_{22} + \ldots + A_{NN}, \qquad (A.4.1)$$

where N in the last expression is the maximum value of the index i. The third expression defines the inner product of a pair of vectors, **u** and **v**.

Free indices

The vector or matrix nature of an expression is determined by the number of free indices, that is, the indices that appear only once in the expression. For example, $A_{ij} u_j$ is a vector, whereas $u_i u_j$ and $u_i A_{ijl}$ are two-dimensional matrices.

Kronecker's delta

Kronecker's delta, δ_{ij}, represents the identity or unit matrix: $\delta_{ij} = 1$ if $i = j$, and $\delta_{ij} = 0$ if $i \neq j$. Using this definition, we find that

$$u_i \delta_{ij} = u_j, \quad A_{ij} \delta_{jk} = A_{ik}, \quad \delta_{ij} A_{jk} \delta_{kl} = A_{il}, \quad \delta_{ij} \delta_{jk} \delta_{kl} = \delta_{il}, \quad \delta_{ii} = N, \quad (A.4.2)$$

where the index i in the last expression ranges from 1 to N. If x_i is a set of N independent variables, then

$$\frac{\partial x_i}{\partial x_j} = \delta_{ij}, \qquad \frac{\partial x_i}{\partial x_i} = N. \qquad (A.4.3)$$

The Einstein summation convention is implied in the second expression.

Alternating tensor

The alternating tensor ϵ_{ijk}, where all three indices takes the values $1, 2, 3$, is defined such that (a) $\epsilon_{ijk} = 0$ if all three indices or any two indices have the same value, (b) $\epsilon_{ijk} = 1$ if the indices are arranged in cyclic order, 123, 312, or 231, and (c) $\epsilon_{ijk} = -1$ otherwise. Thus, $\epsilon_{132} = -1$, $\epsilon_{122} = 0$, $\epsilon_{ijj} = 0$, and $\epsilon_{ijk} = \epsilon_{jki} = \epsilon_{kij} = -\epsilon_{jik}$. Formally, we define

$$\epsilon_{ijk} = \frac{1}{2} (i - j)(j - k)(k - i). \qquad (A.4.4)$$

Two useful properties of the alternating tensor are

$$\epsilon_{ijk} \epsilon_{ijl} = 2 \delta_{kl}, \qquad \epsilon_{ijk} \epsilon_{ilm} = \delta_{jl} \delta_{km} - \delta_{jm} \delta_{kl}, \qquad (A.4.5)$$

where summation is implied over indices that appear twice.

Antisymmetric matrix

A three-dimensional antisymmetric (skew-symmetric) matrix, A_{ij}, has only three independent components that can be represented in terms of a three-dimensional vector, $\mathbf{v}$, as

$$A_{ij} = \epsilon_{ijk} v_k. \tag{A.4.6}$$

Multiplying this definition by ϵ_{ijl} and using (A.4.5), we obtain the inverse relation

$$v_k = \frac{1}{2} \epsilon_{ijk} A_{ij}. \tag{A.4.7}$$

For example, in fluid mechanics the matrix $\mathbf{A}$ can be the vorticity tensor, $\boldsymbol{\Xi}$, and the vector $\mathbf{v}$ can be the vorticity vector, $\boldsymbol{\omega}$.

Vector and matrix products

It is a standard practice in the theory of matrix calculus and numerical analysis to regard by convention an N-dimensional vector $\mathbf{u}$ as a column vector, which is a contrived two-dimensional $N \times 1$ matrix. To obtain the corresponding row vector, which is a contrived $1 \times N$ matrix, we formulate the transpose of the column vector, $\mathbf{u}^T$, denoted by the superscript T. If $\mathbf{A}$ is an $N \times N$ square matrix, then the products $\mathbf{Au}$ and $\mathbf{u}^T \mathbf{A}$, defined according to the usual rules of matrix multiplication, represent, respectively, an N-dimensional column and an N-dimensional row vector.

In fluid mechanics, we avoid using the superscript T by indicating the one-index-product with a centered dot. For example, $\mathbf{v} = \mathbf{A} \cdot \mathbf{u}$ is equivalent to $v_i = A_{ij} u_j$, where summation is implied over the repeated index, j. Similarly, $\mathbf{w} = \mathbf{u} \cdot \mathbf{A}$ is equivalent to $w_j = u_i A_{ij}$, where summation is implied over the repeated index, i. The scalar

$$s = (\mathbf{A} \cdot \mathbf{u}) \cdot \mathbf{v} = \mathbf{v} \cdot (\mathbf{A} \cdot \mathbf{u}) = (\mathbf{v} \cdot \mathbf{A}) \cdot \mathbf{u} = \mathbf{u} \cdot (\mathbf{v} \cdot \mathbf{A}) \cdot \mathbf{u} \tag{A.4.8}$$

is defined as $s = v_i A_{ij} u_j$, where summation is implied over the repeated indices, i and j.

Inner vector product

The inner product of a pair of vectors with the same dimensions, $\mathbf{u}$ and $\mathbf{v}$, is a scalar defined as the sum of the products of corresponding elements,

$$s \equiv \mathbf{u} \cdot \mathbf{v} = u_i v_i. \tag{A.4.9}$$

If the inner product is zero, the two vectors are orthogonal.

Tensor vector product

If $\mathbf{u}$ and $\mathbf{v}$ are two N-dimensional vectors, then the tensor product

$$\mathbf{A} = \mathbf{u} \otimes \mathbf{v} \tag{A.4.10}$$

is an $N \times N$ matrix with components $A_{ij} = u_i v_j$. By contrast, $\mathbf{u} \cdot \mathbf{v} = u_i v_i$ is a scalar defining the inner product of the vectors $\mathbf{u}$ and $\mathbf{v}$, which is also equal to the sum of the diagonal elements (trace) of the matrix $\mathbf{A}$.

Using the definitions of the inner and tensor products, we obtain

$$\mathbf{w} \cdot (\mathbf{u} \otimes \mathbf{v}) = (\mathbf{w} \cdot \mathbf{u})\, \mathbf{v}, \qquad (\mathbf{u} \otimes \mathbf{v}) \cdot \mathbf{w} = (\mathbf{v} \cdot \mathbf{w})\, \mathbf{u}, \qquad \text{(A.4.11)}$$

where $\mathbf{u}, \mathbf{v}$, and $\mathbf{w}$ are three vectors with matching dimensions. Identifying $\mathbf{w}$ with $\mathbf{u}$ or $\mathbf{v}$, we find that, if the vectors $\mathbf{u}$ and $\mathbf{v}$ are orthogonal, the matrix $\mathbf{u} \otimes \mathbf{v}$ has a zero eigenvalue and is thus singular.

Double-dot scalar matrix product

The double-dot product of a pair of two-dimensional matrices, $\mathbf{A}$ and $\mathbf{B}$, is a scalar defined as

$$s \equiv \mathbf{A} : \mathbf{B} = \mathbf{B} : \mathbf{A} = A_{ij} B_{ij}, \qquad \text{(A.4.12)}$$

where summation is implied over the repeated indices, i and j. If the matrix $\mathbf{A}$ is symmetric, $A_{ij} = A_{ji}$, and the matrix $\mathbf{B}$ is antisymmetric, $B_{ij} = -B_{ji}$, then $\mathbf{A} : \mathbf{B} = 0$.

Problem

A.4.1 *Tensor vector product*

Show that $(\mathbf{u} \otimes \mathbf{v}) \cdot (\mathbf{w} \otimes \mathbf{z}) = \alpha\, (\mathbf{u} \otimes \mathbf{z})$ and evaluate the constant α, where $\mathbf{u}, \mathbf{v}, \mathbf{w}$, and $\mathbf{z}$ are four vectors with matching dimensions.

A.5   Three-dimensional Cartesian vectors

Consider a Cartesian system of axes, (x, y, z), and denote the corresponding unit vectors by $\mathbf{e}_x$, $\mathbf{e}_y$, and $\mathbf{e}_z$, or $\mathbf{e}_1$, $\mathbf{e}_2$, and $\mathbf{e}_3$. The Cartesian components of a three-dimensional vector, $\mathbf{a}$, representing, for example, position, velocity, or acceleration, are hosted by the three entries of $\mathbf{a}$. A similar notation is used in two dimensions.

Inner product

In Section A.4, we defined the inner product of a pair of vectors, $\mathbf{a}$ and $\mathbf{b}$, as the scalar $s \equiv \mathbf{a} \cdot \mathbf{b} = a_i\, b_i$, where summation is implied over the repeated index, i. If the inner product is zero, the two vectors are orthogonal. In two or three dimensions, the inner product of two vectors is equal to the product of the lengths of the vectors, $|\mathbf{a}|$ and $|\mathbf{b}|$, and the cosine of the angle α subtended between the vectors in their plane,

$$\mathbf{a} \cdot \mathbf{b} = |\mathbf{a}|\, |\mathbf{b}|\, \cos \alpha. \qquad \text{(A.5.1)}$$

A similar interpretation applies in two dimensions.

Outer product

The outer or vector product of an ordered pair of three-dimensional vectors, $\mathbf{a}$ and $\mathbf{b}$, is a new vector, $\mathbf{c} \equiv \mathbf{a} \times \mathbf{b}$, given by the determinant of a matrix,

$$\mathbf{c} \equiv \begin{bmatrix} \mathbf{e}_1 & \mathbf{e}_2 & \mathbf{e}_3 \\ a_1 & a_2 & a_3 \\ b_1 & b_2 & b_3 \end{bmatrix} = \mathbf{e}_1\,(a_2 b_3 - a_3 b_2) + \mathbf{e}_2\,(a_3 b_1 - a_1 b_3) + \mathbf{e}_3\,(a_1 b_2 - a_2 b_1). \qquad \text{(A.5.2)}$$

In index notation,

$$c_i = \epsilon_{ijk} a_j b_k. \tag{A.5.3}$$

It is evident from this definition that

$$\mathbf{a} \times \mathbf{b} = -\mathbf{b} \times \mathbf{a}. \tag{A.5.4}$$

The outer product, $\mathbf{c}$, is oriented normal to the plane of $\mathbf{a}$ and $\mathbf{b}$. The direction of the outer product is determined by the right-handed rule applied to the ordered triplet $\mathbf{a}, \mathbf{b}, \mathbf{c}$.

The magnitude of the outer product is equal to the product of the lengths of the two vectors, $|\mathbf{a}|$ and $|\mathbf{b}|$, and the sine of the angle α subtended between these vectors in their plane,

$$|\mathbf{a} \times \mathbf{b}| = |\mathbf{a}|\,|\mathbf{b}|\,\sin\alpha. \tag{A.5.5}$$

This expression demonstrates that, if the outer product is a null vector, the vectors $\mathbf{a}$ and $\mathbf{b}$ are parallel, and *vice versa*.

Triple scalar product

The triple scalar product of an ordered vector triplet, $\mathbf{a}$, $\mathbf{b}$, $\mathbf{c}$, is a scalar representing the volume of a parallelepiped with three sides defined by the vectors $\mathbf{a}$, $\mathbf{b}$, and $\mathbf{c}$,

$$V \equiv [\mathbf{a}, \mathbf{b}, \mathbf{c}] \equiv (\mathbf{a} \times \mathbf{b}) \cdot \mathbf{c} = (\mathbf{c} \times \mathbf{a}) \cdot \mathbf{b} = (\mathbf{b} \times \mathbf{c}) \cdot \mathbf{a}. \tag{A.5.6}$$

Notice two cyclic permutations. The triple scalar product can be computed as the determinant of a matrix,

$$[\mathbf{a}, \mathbf{b}, \mathbf{c}] \equiv \det\left(\begin{bmatrix} a_1 & a_2 & a_3 \\ b_1 & b_2 & b_3 \\ c_1 & c_2 & c_3 \end{bmatrix} \right) = a_1(b_2 c_3 - b_3 c_2) + a_2(b_3 c_1 - b_1 c_3) + a_3(b_1 c_2 - b_2 c_1). \tag{A.5.7}$$

Cyclic permutation preserves the triple scalar product,

$$[\mathbf{a}, \mathbf{b}, \mathbf{c}] = [\mathbf{c}, \mathbf{a}, \mathbf{b}] = [\mathbf{b}, \mathbf{c}, \mathbf{a}]. \tag{A.5.8}$$

Since the determinant of a matrix is equal to the determinant of the matrix transpose,

$$V^2 = \det\left(\begin{bmatrix} a_1 & a_2 & a_3 \\ b_1 & b_2 & b_3 \\ c_1 & c_2 & c_3 \end{bmatrix} \cdot \begin{bmatrix} a_1 & b_1 & c_1 \\ a_2 & b_2 & c_2 \\ a_3 & b_3 & c_3 \end{bmatrix} \right) = \det\left(\begin{bmatrix} \mathbf{a} \cdot \mathbf{a} & \mathbf{a} \cdot \mathbf{b} & \mathbf{a} \cdot \mathbf{c} \\ \mathbf{b} \cdot \mathbf{a} & \mathbf{b} \cdot \mathbf{b} & \mathbf{b} \cdot \mathbf{c} \\ \mathbf{c} \cdot \mathbf{a} & \mathbf{c} \cdot \mathbf{b} & \mathbf{c} \cdot \mathbf{c} \end{bmatrix} \right). \tag{A.5.9}$$

If the vectors $\mathbf{a}, \mathbf{b}, \mathbf{c}$ are mutually orthogonal, the off-diagonal elements of the last matrix are zero.

Triple vector product

The triple vector product of an ordered vector triplet, $\mathbf{a}$, $\mathbf{b}$, $\mathbf{c}$, is a new vector,

$$\mathbf{d} \equiv \mathbf{a} \times (\mathbf{b} \times \mathbf{c}). \tag{A.5.10}$$

Writing

$$d_i = \epsilon_{ijk}\, a_j\, (\epsilon_{klm} b_l c_m) = \epsilon_{ijk}\epsilon_{klm} a_j b_l c_m = \epsilon_{kij}\epsilon_{klm} a_j b_l c_m, \qquad (A.5.11)$$

and then

$$d_i = (\delta_{il}\delta_{jm} - \delta_{im}\delta_{jl})a_j b_l c_m = a_j b_i c_j - a_j b_j c_i, \qquad (A.5.12)$$

we obtain the vector identity

$$\mathbf{a} \times (\mathbf{b} \times \mathbf{c}) = \mathbf{b}\,(\mathbf{a} \cdot \mathbf{c}) - \mathbf{c}\,(\mathbf{a} \cdot \mathbf{b}). \qquad (A.5.13)$$

Working in a similar fashion, we derive the identity

$$\mathbf{a} \times (\mathbf{b} \times \mathbf{c}) + \mathbf{b} \times (\mathbf{c} \times \mathbf{a}) + \mathbf{c} \times (\mathbf{a} \times \mathbf{b}) = \mathbf{0}. \qquad (A.5.14)$$

Notice the cyclic permutation of the vector triplets on the left-hand side.

Directional vector components

The component of a vector, $\mathbf{b}$, in the direction of a unit vector, $\mathbf{e}$, is given by $(\mathbf{b}\cdot\mathbf{e})\,\mathbf{e}$, where $\mathbf{e}\cdot\mathbf{e} = 1$. Applying (A.5.13) with $\mathbf{a} = \mathbf{c} = \mathbf{e}$, we find that the vector

$$\mathbf{e} \times (\mathbf{b} \times \mathbf{e}) = \mathbf{b} - \mathbf{e}\,(\mathbf{e} \cdot \mathbf{b}) \qquad (A.5.15)$$

represents the component of $\mathbf{b}$ normal to $\mathbf{e}$. Using index notation, we find that

$$\mathbf{e} \times (\mathbf{b} \times \mathbf{e}) = \mathbf{b} \cdot (\mathbf{I} - \mathbf{e} \otimes \mathbf{e}), \qquad (A.5.16)$$

where $\mathbf{I}$ is the identity matrix and $\otimes$ denotes the tensor vector product.

Vector–matrix outer product

Given a three-dimensional vector, $\mathbf{u}$, and a square three-dimensional matrix, $\mathbf{A}$, we define their left outer product as a new matrix, $\mathbf{B} = \mathbf{u} \times \mathbf{A}$, with elements

$$B_{ij} = \epsilon_{ikl}\, u_k\, A_{lj}. \qquad (A.5.17)$$

Similarly, we define the right cross product as a new matrix, $\mathbf{C} = \mathbf{A} \times \mathbf{u}$, with elements

$$C_{ij} = A_{ik}\epsilon_{klj}u_l, \qquad (A.5.18)$$

where summation is implied over the repeated indices, k and l.

Problem

A.5.1 *Vector components*
Prove identity (A.5.16) working index notation.

A.6 Vector calculus

The Cartesian components of the *del* or *nabla* vector operator, denoted by ∇, are the partial derivatives with respect to the corresponding coordinates,

$$\nabla \equiv \mathbf{e}_x \frac{\partial}{\partial x} + \mathbf{e}_y \frac{\partial}{\partial y} + \mathbf{e}_z \frac{\partial}{\partial z}, \qquad (A.6.1)$$

where $\mathbf{e}_i$ are unit vectors in the directions of the subscripted axes. In two dimensions, only the x and y derivatives and associated unit vectors appear.

Gradient of a scalar function

The gradient of a scalar function of position, $f(\mathbf{x})$, is a vector defined as

$$\nabla f = \mathbf{e}_x \frac{\partial f}{\partial x} + \mathbf{e}_y \frac{\partial f}{\partial y} + \mathbf{e}_z \frac{\partial f}{\partial z}, \qquad (A.6.2)$$

where $\mathbf{e}_i$ are unit vectors in the directions of the subscripted axes. Physically, the gradient points into the direction where the function f increases most rapidly at a point. To prove this, we consider the directional derivative of the function f, as discussed at the end of this section.

Divergence of a vector field

The divergence of a vector function of position, $\mathbf{f} = (f_x, f_y, f_z)$, is a scalar defined as

$$\nabla \cdot \mathbf{f} = \frac{\partial f_i}{\partial x_i} = \frac{\partial f_x}{\partial x} + \frac{\partial f_y}{\partial y} + \frac{\partial f_z}{\partial z}. \qquad (A.6.3)$$

If $\nabla \cdot \mathbf{f} = 0$ at every point, the vector field $\mathbf{f}$ is called solenoidal. The continuity equation requires that the velocity field of an incompressible fluid is solenoidal throughout the domain of a flow.

Gradient of a vector field

The gradient of a vector field, $\mathbf{f}$, denoted by $\mathbf{L} \equiv \nabla \mathbf{f}$, is a two-dimensional matrix with elements

$$L_{ij} = \frac{\partial f_j}{\partial x_i}. \qquad (A.6.4)$$

The divergence of $\mathbf{f}$ is equal to the trace of the matrix $\mathbf{L}$, which is equal to the sum of the diagonal elements of $\mathbf{L}$,

$$\nabla \cdot \mathbf{f} = \text{trace}(\mathbf{L}). \qquad (A.6.5)$$

In fact, the velocity gradient is the tensor product of the gradient vector operator and the vector field of interest, $\mathbf{f}$. To simplify the notation, we have denoted

$$\nabla \mathbf{f} \equiv \nabla \otimes \mathbf{f}. \qquad (A.6.6)$$

Curl of a vector field

The curl of a vector field, $\mathbf{f}$, denoted by $\nabla \times \mathbf{f}$, is another vector field computed according to the usual rules of the outer vector product, treating the del operator as a regular vector, yielding

$$\nabla \times \mathbf{f} = \det\left(\begin{bmatrix} \mathbf{e}_x & \mathbf{e}_y & \mathbf{e}_z \\ \partial/\partial x & \partial/\partial y & \partial/\partial z \\ f_x & f_y & f_z \end{bmatrix} \right). \tag{A.6.7}$$

Explicitly,

$$\nabla \times \mathbf{f} = \mathbf{e}_x \left(\frac{\partial f_z}{\partial y} - \frac{\partial f_y}{\partial z} \right) + \mathbf{e}_y \left(\frac{\partial f_x}{\partial z} - \frac{\partial f_z}{\partial x} \right) + \mathbf{e}_z \left(\frac{\partial f_y}{\partial x} - \frac{\partial f_x}{\partial y} \right). \tag{A.6.8}$$

In the case of the two-dimensional vector field, $\mathbf{f}$, lying in the xy plane, only the z component of the curl survives.

Laplacian of a scalar function

The Laplacian of a scalar function, f, is a scalar defined as

$$\nabla \cdot (\nabla f) \equiv \nabla^2 f = \frac{\partial^2 f}{\partial x^2} + \frac{\partial^2 f}{\partial y^2} + \frac{\partial^2 f}{\partial z^2}. \tag{A.6.9}$$

The Laplacian is equal to the divergence of the gradient.

Vector identities

Let f be a scalar function and $\mathbf{u}$, $\mathbf{v}$ be two vector functions. The following identities can be shown working in index notation:

$$\nabla \cdot (f\,\mathbf{u}) = f\,\nabla \cdot \mathbf{u} + \mathbf{u} \cdot \nabla f, \tag{A.6.10}$$

$$\nabla (\mathbf{u} \cdot \mathbf{v}) = \mathbf{u} \cdot \nabla \mathbf{v} + \mathbf{v} \cdot \nabla \mathbf{u} + \mathbf{u} \times (\nabla \times \mathbf{v}) + \mathbf{v} \times (\nabla \times \mathbf{u}), \tag{A.6.11}$$

$$\nabla \cdot (\mathbf{u} \times \mathbf{v}) = \mathbf{v} \cdot \nabla \times \mathbf{u} - \mathbf{u} \cdot \nabla \times \mathbf{v}, \tag{A.6.12}$$

$$\nabla \times (\mathbf{u} \times \mathbf{v}) = \mathbf{u}\,\nabla \cdot \mathbf{v} - \mathbf{v}\,\nabla \cdot \mathbf{u} + \mathbf{v} \cdot \nabla \mathbf{u} - \mathbf{u} \cdot \nabla \mathbf{v}, \tag{A.6.13}$$

$$\nabla \times (\nabla f) = \mathbf{0}, \tag{A.6.14}$$

$$\nabla \cdot (\nabla \times \mathbf{u}) = 0, \tag{A.6.15}$$

$$\nabla \times (\nabla \times \mathbf{u}) = \nabla (\nabla \cdot \mathbf{u}) - \nabla^2 \mathbf{u}. \tag{A.6.16}$$

Divergence of a matrix

The divergence of a two-dimensional matrix function of position, $\mathbf{Q}(\mathbf{x})$, is a vector, $\mathbf{v} = \nabla \cdot \mathbf{Q}$, with components

$$v_j = \frac{\partial Q_{ij}}{\partial x_i}, \tag{A.6.17}$$

where summation is implied over the repeated index, i.

Directional derivatives

The rate of change of a scalar function, f, with respect to arc length, l_t, in the direction of a unit vector $\mathbf{t}$, is given by

$$\mathbf{t} \cdot \nabla f = \frac{\partial f}{\partial l_t} = t_x \frac{\partial f}{\partial x} + t_y \frac{\partial f}{\partial y} + t_z \frac{\partial f}{\partial z}. \tag{A.6.18}$$

Using the geometrical interpretation of the inner vector product, we find that $\mathbf{t} \cdot \nabla f$ is maximum when $\mathbf{t}$ is parallel to ∇f at a point.

The corresponding rate of change of a vector function, $\mathbf{f}$, is a vector given by

$$\mathbf{t} \cdot \nabla \mathbf{f} = \frac{\partial \mathbf{f}}{\partial l_t} = t_x \frac{\partial \mathbf{f}}{\partial x} + t_y \frac{\partial \mathbf{f}}{\partial y} + t_z \frac{\partial \mathbf{f}}{\partial z}. \tag{A.6.19}$$

Note that, by convention, the gradient $\nabla \mathbf{f}$ is multiplied *from the left* by $\mathbf{t}$ to produce a directional derivative.

Problem

A.6.1 *Curl*

Compute the curl of the vector function $\mathbf{v}(\mathbf{x}) = f(r)\,\mathbf{x}$, where $r = |\mathbf{x}|$, $\mathbf{x}$ is the position vector, and $f(r)$ is a differentiable function.

A.7   Divergence and Stokes' theorems

The divergence and Stokes' theorems allow us to convert a volume integral into a surface integral or line integral. In practice, these theorems are used to derive differential from integral balances, and *vice versa*, originating from physical laws.

Divergence theorem in space

Let V_c be an arbitrary volume in space bounded by a closed surface, D, as illustrated in Figure A.7.1(a). The Gauss divergence theorem states that the volume integral of the divergence of any differentiable vector function, $\mathbf{f} = (f_x, f_y, f_z)$, over V_c is equal to the flow rate of $\mathbf{f}$ across D,

$$\iiint_{V_c} \nabla \cdot \mathbf{f}\, dV = \iint_D \mathbf{f} \cdot \mathbf{n}\, dS, \tag{A.7.1}$$

where $\mathbf{n}$ be the unit vector normal to D pointing outward from V_c.

Making the three sequential choices $\mathbf{f} = (f, 0, 0)$, $\mathbf{f} = (0, f, 0)$, and $\mathbf{f} = (0, 0, f)$, where f is a differentiable scalar function, we derive the vector form of the divergence theorem,

$$\iiint_{V_c} \nabla f\, dV = \iint_D f\, \mathbf{n}\, dS. \tag{A.7.2}$$

(a)   (b)   (c)

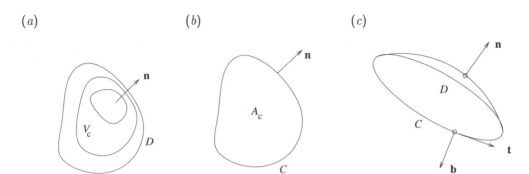

FIGURE A.7.1   Illustration of (a) a volume used to derive the divergence theorem in space and (b) an area used to derive the divergence theorem in the xy plane. (c) Illustration of a closed loop in space, C, bounded by a surface, D, used to derive Stokes' theorem.

The particular choices $f = x$, $f = y$, or $f = z$, provide us with expressions for the volume of V_c in terms of a surface integral of the x, y, or z component of the normal vector.

Setting $\mathbf{f} = \mathbf{a} \times \mathbf{q}$, where $\mathbf{a}$ is a constant vector and $\mathbf{q}$ is a differentiable vector function, and discarding the arbitrary constant $\mathbf{a}$, we obtain the identity

$$\iiint_{V_c} \nabla \times \mathbf{q} \, dV = \iint_D \mathbf{n} \times \mathbf{q} \, dS. \tag{A.7.3}$$

The integral on the right-hand side involves the tangential component of the vector field, $\mathbf{q}$.

Divergence theorem in a plane

Let A_c be an arbitrary control area in the xy plane bounded by a closed contour, C, as illustrated in Figure A.7.1(b). The Gauss divergence theorem states that the areal integral of the divergence of any two-dimensional differentiable vector function, $\mathbf{f} = (f_x, f_y)$, over A_c is equal to the flow rate of $\mathbf{f}$ across C,

$$\iint_{A_c} \nabla \cdot \mathbf{f} \, dA = \oint_C \mathbf{f} \cdot \mathbf{n} \, dl, \tag{A.7.4}$$

where l is the arc length along C and $\mathbf{n}$ is the unit vector normal to C pointing outward from A_c.

Making the sequential choices $\mathbf{f} = (f, 0)$ and $\mathbf{f} = (0, f)$, where f is a differentiable scalar function, we obtain the vector form of the divergence theorem,

$$\iint_{A_c} \nabla f \, dA = \oint_C f \mathbf{n} \, dl. \tag{A.7.5}$$

The particular choices $f = x$ or $f = y$ yield the area of A_c in terms of a line integral of the x or y component of the normal vector.

Divergence theorem on a surface

Consider a surface patch, D, with normal unit vector $\mathbf{n}$, enclosed by a closed loop, C, with tangent unit vector $\mathbf{t}$, as illustrated in Figure A.7.1(c). The normal unit vector, $\mathbf{n}$, is oriented according to the right-handed rule with respect to $\mathbf{t}$ and with reference to a designated side of D. Let $\mathbf{f}$ be a function of position defined over the patch. The Gauss divergence theorem provides us with the identity

$$\iint_P (\mathbf{P} \cdot \nabla) \cdot (\mathbf{P} \cdot \mathbf{f}) \, \mathrm{d}S = \oint_C (\mathbf{t} \times \mathbf{n}) \cdot \mathbf{f} \, \mathrm{d}l, \tag{A.7.6}$$

where $\mathbf{P} = \mathbf{I} - \mathbf{n} \otimes \mathbf{n}$ is the tangential projection operator, l is the arc length along C, and $\mathbf{t} \times \mathbf{n} = \mathbf{b}$ is the tangent unit vector shown in Figure A.7.1(c). The integrand on the left-hand side of (A.7.6) is the surface divergence of the tangential component of $\mathbf{f}$. If $\mathbf{f}$ is normal to D, the surface divergence is identically zero.

Stokes' theorem

Let C be an arbitrary closed loop with tangent unit vector $\mathbf{t}$ and D be an arbitrary surface bounded by C, as illustrated in Figure A.7.1(c). Stokes' theorem states that the circulation of a differentiable vector function $\mathbf{f}$ along C is equal to the flow rate of the curl of $\mathbf{f}$ across D,

$$\oint_C \mathbf{f} \cdot \mathbf{t} \, \mathrm{d}l = \iint_D (\nabla \times \mathbf{f}) \cdot \mathbf{n} \, \mathrm{d}S, \tag{A.7.7}$$

where l is the arc length along C and $\mathbf{n}$ is the unit vector normal to D oriented according to the right-handed rule with respect to $\mathbf{t}$ and with reference to a designated side of D. As we look at a designated side of D, the normal vector points toward us, while the tangent vector describes a counterclockwise path.

Setting $\mathbf{f} = \mathbf{a} \times \mathbf{q}$, where $\mathbf{a}$ is a constant vector and $\mathbf{q}$ is a differentiable vector function of position, expanding the integrand on the right-hand side of (A.7.7), and discarding the arbitrary constant $\mathbf{a}$, we derive the identity

$$\oint_C \mathbf{q} \times \mathbf{t} \, \mathrm{d}l = \iint_D \left[(\nabla \cdot \mathbf{q}) \, \mathbf{n} - (\nabla \mathbf{q}) \cdot \mathbf{n} \right] \mathrm{d}S, \tag{A.7.8}$$

where the matrix $\nabla \mathbf{q}$ is the gradient of $\mathbf{q}$.

Problems

A.7.1 *Volume of a sphere*

Use the divergence theorem to compute the volume of a sphere in terms of a surface integral.

A.7.2 *Curl*

Show that $\oint f(r) \, \mathbf{x} \cdot \mathbf{t} \, \mathrm{d}l = 0$, where $r = |\mathbf{x}|$, $\mathbf{x}$ is the position vector, $f(r)$ is a differentiable function, and $\mathbf{t}$ is the unit vector tangent to a closed contour.

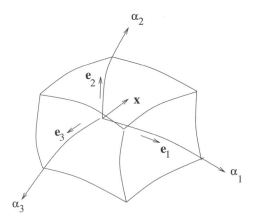

FIGURE A.8.1   Illustration of three-dimensional orthogonal curvilinear coordinates. The variables, α_1, α_2, and α_3 do not necessarily measure physical arc length.

A.8   Orthogonal curvilinear coordinates

Consider three distinct families of surfaces that fill the entire three-dimensional space, where two surfaces of the same family do not intersect at a singular line or point. Next, we introduce three continuous scalar variables, α_1, α_2, and α_3, defined such that two of these variables vary over a chosen surface, while the third variable remains constant over the chosen surface, as shown in Figure A.8.1. The position vector, $\mathbf{x}$, can be regarded as a function of the curvilinear coordinates, α_1, α_2, and α_3, so that $\mathbf{x}(\alpha_1, \alpha_2, \alpha_3)$.

Base vectors and unit vectors

Three surfaces can be identified passing through a chosen point in space, one surface from each family. Two variables are constant along the intersection of a pair of surfaces, and the third variable varies in a continuous fashion along the intersection. The base vectors

$$\mathbf{g}_1 \equiv \frac{\partial \mathbf{x}}{\partial \alpha_1}, \qquad \mathbf{g}_2 \equiv \frac{\partial \mathbf{x}}{\partial \alpha_2}, \qquad \mathbf{g}_3 \equiv \frac{\partial \mathbf{x}}{\partial \alpha_3}, \qquad (A.8.1)$$

are tangential to the intersections, pointing into the directions of increasing α_1, α_2, and α_3. The magnitude and direction of these base vectors generally change with position in space. The corresponding unit vectors are denoted by

$$\mathbf{e}_1 \equiv \frac{\mathbf{g}_1}{|\mathbf{g}_1|}, \qquad \mathbf{e}_2 \equiv \frac{\mathbf{g}_2}{|\mathbf{g}_2|}, \qquad \mathbf{e}_3 \equiv \frac{\mathbf{g}_3}{|\mathbf{g}_3|}, \qquad (A.8.2)$$

as shown in Figure A.8.1. The direction of each base unit vector generally changes with position in space. Since $|\mathbf{e}_i|^2 = \mathbf{e}_i \cdot \mathbf{e}_i = 1$,

$$\frac{\partial (\mathbf{e}_i \cdot \mathbf{e}_i)}{\partial \alpha_j} = 2\,\mathbf{e}_i \cdot \frac{\partial \mathbf{e}_i}{\partial \alpha_j} = 0 \qquad (A.8.3)$$

for any j, where summation is *not* implied over the repeated index, i. This equation demonstrates that the vector $\partial \mathbf{e}_i / \partial \alpha_j$ lacks a component in the direction of $\mathbf{e}_i$.

Orthonormality

If the base unit vectors form a local right-handed system of orthogonal axes, so that

$$\mathbf{e}_p \times \mathbf{e}_q = \epsilon_{pqm}\, \mathbf{e}_m, \tag{A.8.4}$$

then the variables α_1, α_2, and α_3 comprise a system of orthogonal curvilinear coordinates, where summation is implied over the repeated index, m. In that case, the inner product of any two different base vectors $\mathbf{g}_i$ and $\mathbf{g}_j$ is zero, $\mathbf{g}_i \cdot \mathbf{g}_j = 0$ for $i \neq j$. Orthonormality of the base unit vectors requires that

$$\mathbf{e}_i \cdot \mathbf{e}_j = \delta_{ij}, \tag{A.8.5}$$

where δ_{ij} is Kronecker's delta.

Metric tensor

An infinitesimal vector in space can be described as

$$d\mathbf{x} = \mathbf{g}_1\, d\alpha_1 + \mathbf{g}_2\, d\alpha_2 + \mathbf{g}_3\, d\alpha_3, \tag{A.8.6}$$

where $\mathbf{g}_i$ are the local base vectors and $d\alpha_i$ are appropriate differential increments for $i = 1, 2, 3$. Unless the base vectors are constant, equation (A.8.6) cannot be integrated readily to produce an explicit expression for the position vector, $\mathbf{x}(\alpha_1, \alpha_2, \alpha_3)$. Exceptions arise in the case of polar coordinates discussed in Sections A.9–A.11.

The square of the length of the vector is

$$d\mathbf{x} \cdot d\mathbf{x} = g_{ij}\, d\alpha_i\, d\alpha_j, \tag{A.8.7}$$

where summation is implied over the repeated indices, i and j. We have introduced the diagonal metric tensor $\mathbf{g}$ with diagonal components

$$g_{ii} = \mathbf{g}_i \cdot \mathbf{g}_i \equiv h_i^2, \tag{A.8.8}$$

where $h_i = |\mathbf{g}_i|$ are three positive metric coefficients, and summation is *not* implied over the repeated index, i. With these definitions, the base unit vectors are

$$\mathbf{e}_1 \equiv \frac{\mathbf{g}_1}{h_1}, \qquad \mathbf{e}_2 \equiv \frac{\mathbf{g}_2}{h_2}, \qquad \mathbf{e}_3 \equiv \frac{\mathbf{g}_3}{h_3}. \tag{A.8.9}$$

The differential arc length along a line over which α_2 and α_3 are constant but α_1 changes, is $dl_1 = h_1 d\alpha_1$. Similar relations can be written for the other two axes,

$$dl_1 = h_1 d\alpha_1, \qquad dl_2 = h_2 d\alpha_2, \qquad dl_3 = h_3 d\alpha_3. \tag{A.8.10}$$

Properties of the base vectors

Using the definition of the base vectors, we write the identity

$$\frac{\partial(h_i\mathbf{e}_i)}{\partial\alpha_j} = \frac{\partial(h_j\mathbf{e}_j)}{\partial\alpha_i} = \frac{\partial^2\mathbf{x}}{\partial\alpha_i\partial\alpha_j}, \tag{A.8.11}$$

where summation is *not* implied over the repeated indices, i and j. Expanding the first two derivatives, projecting the resulting equation onto $\mathbf{e}_j$, and using (A.8.3), we obtain

$$\frac{\partial h_j}{\partial\alpha_i} = h_i \frac{\partial\mathbf{e}_i}{\partial\alpha_j} \cdot \mathbf{e}_j \tag{A.8.12}$$

for $i \neq j$, where summation is *not* implied over the repeated index, j.

Now using the orthogonality condition $\mathbf{g}_1 \cdot \mathbf{g}_2 = 0$, we write

$$\frac{\partial(\mathbf{g}_1 \cdot \mathbf{g}_2)}{\partial\alpha_3} = \frac{\partial}{\partial\alpha_3}\left(\frac{\partial\mathbf{x}}{\partial\alpha_1} \cdot \frac{\partial\mathbf{x}}{\partial\alpha_2}\right) = \frac{\partial\mathbf{g}_3}{\partial\alpha_1} \cdot \frac{\partial\mathbf{x}}{\partial\alpha_2} + \frac{\partial\mathbf{g}_3}{\partial\alpha_2} \cdot \frac{\partial\mathbf{x}}{\partial\alpha_1} = -2\,\mathbf{g}_3 \cdot \frac{\partial^2\mathbf{x}}{\partial\alpha_1\partial\alpha_2} = 0. \tag{A.8.13}$$

We have demonstrated that

$$\mathbf{g}_3 \cdot \frac{\partial^2\mathbf{x}}{\partial\alpha_1\partial\alpha_2} = \mathbf{g}_3 \cdot \frac{\partial(h_1\mathbf{e}_1)}{\partial\alpha_2} = \mathbf{g}_3 \cdot \frac{\partial(h_2\mathbf{e}_2)}{\partial\alpha_1} = 0, \tag{A.8.14}$$

which shows that

$$\mathbf{e}_3 \cdot \frac{\partial\mathbf{e}_1}{\partial\alpha_2} = 0, \qquad \mathbf{e}_3 \cdot \frac{\partial\mathbf{e}_2}{\partial\alpha_1} = 0. \tag{A.8.15}$$

Similar relations can be written for other combinations, yielding

$$\mathbf{e}_i \cdot \frac{\partial\mathbf{e}_j}{\partial\alpha_k} = 0 \tag{A.8.16}$$

when the indices i, j, and k are all different. This equation demonstrates that the vector $\partial\mathbf{e}_j/\partial\alpha_k$ lacks a component in the direction of $\mathbf{e}_i$, where i, j, and k are all different. Combining this result with (A.8.12), we obtain

$$\frac{\partial\mathbf{e}_i}{\partial\alpha_j} = \frac{1}{h_i}\frac{\partial h_j}{\partial\alpha_i}\mathbf{e}_j \tag{A.8.17}$$

for $i \neq j$, where summation is *not* implied over the repeated index, j.

We may now write

$$\frac{\partial\mathbf{e}_1}{\partial\alpha_1} = \frac{\partial(\mathbf{e}_2 \times \mathbf{e}_3)}{\partial\alpha_1} = \frac{\partial\mathbf{e}_2}{\partial\alpha_1} \times \mathbf{e}_3 - \frac{\partial\mathbf{e}_3}{\partial\alpha_1} \times \mathbf{e}_2 = \frac{1}{h_2}\frac{\partial h_1}{\partial\alpha_2}\mathbf{e}_1 \times \mathbf{e}_3 - \frac{1}{h_3}\frac{\partial h_1}{\partial\alpha_3}\mathbf{e}_1 \times \mathbf{e}_2, \tag{A.8.18}$$

yielding

$$\frac{\partial\mathbf{e}_1}{\partial\alpha_1} = -\frac{1}{h_2}\frac{\partial h_1}{\partial\alpha_2}\mathbf{e}_2 - \frac{1}{h_3}\frac{\partial h_1}{\partial\alpha_3}\mathbf{e}_3. \tag{A.8.19}$$

Similar equations can be written for two other combinations,

$$\frac{\partial \mathbf{e}_2}{\partial \alpha_2} = -\frac{1}{h_3}\frac{\partial h_2}{\partial \alpha_3}\mathbf{e}_3 - \frac{1}{h_1}\frac{\partial h_2}{\partial \alpha_1}\mathbf{e}_1, \qquad \frac{\partial \mathbf{e}_3}{\partial \alpha_3} = -\frac{1}{h_1}\frac{\partial h_3}{\partial \alpha_1}\mathbf{e}_1 - \frac{1}{h_2}\frac{\partial h_3}{\partial \alpha_2}\mathbf{e}_2. \qquad \text{(A.8.20)}$$

Surface and volume metrics

A differential volume in space can be expressed in the form

$$dV(\mathbf{x}) = h_1 h_2 h_3 \, dV(\boldsymbol{\alpha}), \qquad \text{(A.8.21)}$$

where $dV(\mathbf{a}) \equiv d\alpha_1 \, d\alpha_2 \, d\alpha_3$. We see that the product $h_1 h_2 h_3$ serves as a volume metric coefficient for the curvilinear coordinates. The size of a differential surface element that lies in the surface over which α_1, α_2, or α_3 is constant is given, respectively, by

$$dS_1 = h_2 h_3 \, d\alpha_2 \, d\alpha_3, \qquad dS_2 = h_3 h_1 \, d\alpha_3 \, d\alpha_1, \qquad dS_3 = h_1 h_2 \, d\alpha_1 \, d\alpha_2. \qquad \text{(A.8.22)}$$

We see that the products $h_i h_j$ for $i \neq j$ are surface metric coefficients.

Vector components

The component of a vector function of position, $\mathbf{f}(\mathbf{x})$, in the direction of the unit vector, $\mathbf{e}_i$, is given by the scalar product $f_i = \mathbf{f} \cdot \mathbf{e}_i$. By definition,

$$\mathbf{f} = f_1 \mathbf{e}_1 + f_2 \mathbf{e}_2 + f_3 \mathbf{e}_3. \qquad \text{(A.8.23)}$$

To prevent misinterpretation the curvilinear components, f_i, should not be collected into a vector.

Vector and tensor calculus

Vector and tensor calculus in orthogonal curvilinear coordinates are simplifications of those in more general nonorthogonal curvilinear coordinates discussed in Section A.12. However, it is both expedient and instructive to develop concepts and expressions in a specialized framework.

Gradient of a scalar function

The gradient of a scalar function, $f(\mathbf{x})$, is given by

$$\nabla f = \mathbf{e}_1 \frac{1}{h_1}\frac{\partial f}{\partial \alpha_1} + \mathbf{e}_2 \frac{1}{h_2}\frac{\partial f}{\partial \alpha_2} + \mathbf{e}_3 \frac{1}{h_3}\frac{\partial f}{\partial \alpha_3} = \mathbf{e}_1 \frac{\partial f}{\partial l_1} + \mathbf{e}_2 \frac{\partial f}{\partial l_2} + \mathbf{e}_3 \frac{\partial f}{\partial l_3}, \qquad \text{(A.8.24)}$$

where l_i is the arc length measured along the α_i coordinate line.

Gradient operator

The definitions (A.8.24) motivate introducing the gradient operator

$$\nabla \equiv \mathbf{e}_1 \frac{1}{h_1}\frac{\partial}{\partial \alpha_1} + \mathbf{e}_2 \frac{1}{h_2}\frac{\partial}{\partial \alpha_2} + \mathbf{e}_3 \frac{1}{h_3}\frac{\partial}{\partial \alpha_3} = \mathbf{e}_1 \frac{\partial}{\partial l_1} + \mathbf{e}_2 \frac{\partial}{\partial l_2} + \mathbf{e}_3 \frac{\partial}{\partial l_3}. \qquad \text{(A.8.25)}$$

The unit vectors and differentiation operators should not be transposed in these expressions.

Gradient of a vector field

The gradient of a vector function of position, $\mathbf{f}(\mathbf{x})$, is a matrix

$$\mathbf{L} \equiv \nabla \mathbf{f} = \left(\frac{1}{h_1} \mathbf{e}_1 \otimes \frac{\partial}{\partial \alpha_1} + \frac{1}{h_2} \mathbf{e}_2 \otimes \frac{\partial}{\partial \alpha_2} + \frac{1}{h_3} \mathbf{e}_3 \otimes \frac{\partial}{\partial \alpha_3} \right) (f_1 \mathbf{e}_1 + f_2 \mathbf{e}_2 + f_3 \mathbf{e}_3). \quad (A.8.26)$$

The expression enclosed by the first set of parentheses on the right-hand side originates from the gradient operator. Carrying out the differentiation, we find that

$$\mathbf{L} = \frac{1}{h_1} \frac{\partial f_1}{\partial \alpha_1} \mathbf{e}_1 \otimes \mathbf{e}_1 + \frac{f_1}{h_1} \mathbf{e}_1 \otimes \frac{\partial \mathbf{e}_1}{\partial \alpha_1} + \frac{1}{h_2} \frac{\partial f_1}{\partial \alpha_2} \mathbf{e}_2 \otimes \mathbf{e}_1 + \frac{f_1}{h_2} \mathbf{e}_2 \otimes \frac{\partial \mathbf{e}_1}{\partial \alpha_2}$$
$$+ \frac{1}{h_3} \frac{\partial f_1}{\partial \alpha_3} \mathbf{e}_3 \otimes \mathbf{e}_1 + \frac{f_1}{h_3} \mathbf{e}_3 \otimes \frac{\partial \mathbf{e}_1}{\partial \alpha_3} + \cdots, \quad (A.8.27)$$

where the three dots represent similar terms involving f_2 and f_3. Using (A.8.17), (A.8.19), and (A.8.20) to evaluate the derivative of the unit vectors, we obtain

$$\mathbf{L} = \frac{1}{h_1} \frac{\partial f_1}{\partial \alpha_1} \mathbf{e}_1 \otimes \mathbf{e}_1 - \frac{f_1}{h_1 h_2} \frac{\partial h_1}{\partial \alpha_2} \mathbf{e}_1 \otimes \mathbf{e}_2 - \frac{f_1}{h_1 h_3} \frac{\partial h_1}{\partial \alpha_3} \mathbf{e}_1 \otimes \mathbf{e}_3 + \frac{1}{h_2} \frac{\partial f_1}{\partial \alpha_2} \mathbf{e}_2 \otimes \mathbf{e}_1$$
$$+ \frac{f_1}{h_1 h_2} \frac{\partial h_2}{\partial \alpha_1} \mathbf{e}_2 \otimes \mathbf{e}_2 + \frac{1}{h_3} \frac{\partial f_1}{\partial \alpha_3} \mathbf{e}_3 \otimes \mathbf{e}_1 + \frac{f_1}{h_1 h_3} \frac{\partial h_3}{\partial \alpha_1} \mathbf{e}_3 \otimes \mathbf{e}_3 + \cdots, \quad (A.8.28)$$

where the three dots represent similar terms involving f_2 and f_3. Collecting similar terms, we derive the dyadic expansion

$$\mathbf{L} = L_{ij} \, \mathbf{e}_i \otimes \mathbf{e}_j, \quad (A.8.29)$$

where L_{ij} are the components of $\mathbf{L}$ in the chosen curvilinear coordinates, and summation is implied over the repeated indices, i and j. The calculations yield the diagonal components

$$L_{11} = \frac{1}{h_1} \left(\frac{\partial f_1}{\partial \alpha_1} + \frac{f_2}{h_2} \frac{\partial h_1}{\partial \alpha_2} + \frac{f_3}{h_3} \frac{\partial h_2}{\partial \alpha_3} \right), \qquad L_{22} = \frac{1}{h_2} \left(\frac{f_1}{h_1} \frac{\partial h_2}{\partial \alpha_1} + \frac{\partial f_2}{\partial \alpha_2} + \frac{f_3}{h_3} \frac{\partial h_1}{\partial \alpha_3} \right),$$
$$(A.8.30)$$

$$L_{33} = \frac{1}{h_3} \left(\frac{f_1}{h_1} \frac{\partial h_3}{\partial \alpha_1} + \frac{f_2}{h_2} \frac{\partial h_3}{\partial \alpha_2} + \frac{\partial f_3}{\partial \alpha_3} \right), \quad (A.8.31)$$

and the off-diagonal components

$$L_{12} = \frac{1}{h_1} \left(\frac{\partial f_2}{\partial \alpha_1} - \frac{f_1}{h_2} \frac{\partial h_1}{\partial \alpha_2} \right), \qquad L_{21} = \frac{1}{h_2} \left(\frac{\partial f_1}{\partial \alpha_2} - \frac{f_2}{h_1} \frac{\partial h_2}{\partial \alpha_1} \right),$$
$$L_{13} = \frac{1}{h_1} \left(\frac{\partial f_3}{\partial \alpha_1} - \frac{f_1}{h_3} \frac{\partial h_1}{\partial \alpha_3} \right), \qquad L_{31} = \frac{1}{h_3} \left(\frac{\partial f_1}{\partial \alpha_3} - \frac{f_3}{h_1} \frac{\partial h_3}{\partial \alpha_1} \right), \qquad (A.8.32)$$
$$L_{23} = \frac{1}{h_2} \left(\frac{\partial f_3}{\partial \alpha_2} - \frac{f_2}{h_3} \frac{\partial h_2}{\partial \alpha_3} \right), \qquad L_{32} = \frac{1}{h_3} \left(\frac{\partial f_2}{\partial \alpha_3} - \frac{f_3}{h_2} \frac{\partial h_3}{\partial \alpha_2} \right).$$

The trace of $\mathbf{L}$ is the divergence of $\mathbf{f}$, whereas the curl of $\mathbf{f}$ is related to the skew-symmetric part of $\mathbf{L}$, as discussed later in this section.

Divergence of a vector field

The divergence of a vector function of position, $\mathbf{f}(\mathbf{x})$, denoted by $\nabla \cdot \mathbf{f}$, is the trace of the gradient, $\mathbf{L} \equiv \nabla \mathbf{f}$, that is, $\nabla \cdot \mathbf{f} = L_{11} + L_{22} + L_{33}$. Alternatively, we write

$$\nabla \cdot \mathbf{f} \equiv \left(\frac{1}{h_1} \mathbf{e}_1 \cdot \frac{\partial}{\partial \alpha_1} + \frac{1}{h_2} \mathbf{e}_2 \cdot \frac{\partial}{\partial \alpha_2} + \frac{1}{h_3} \mathbf{e}_3 \cdot \frac{\partial}{\partial \alpha_3} \right) (f_1 \mathbf{e}_1 + f_2 \mathbf{e}_2 + f_3 \mathbf{e}_3). \tag{A.8.33}$$

Since the base unit vectors are orthonormal, the trace of the matrix $\mathbf{e}_i \otimes \mathbf{e}_j$ is zero if $i \neq j$, or unity if $i = j$. Using (A.8.27) and recalling that, because $\mathbf{e}_i$ is a unit vector, $\mathbf{e}_i \cdot \partial \mathbf{e}_i / \partial \alpha_j = 0$, where summation is *not* implied over the repeated index, i, we find that

$$\nabla \cdot \mathbf{f} = \frac{1}{h_1} \frac{\partial f_1}{\partial \alpha_1} + f_1 \left(\frac{1}{h_2} \mathbf{e}_2 \cdot \frac{\partial \mathbf{e}_1}{\partial \alpha_2} + \frac{1}{h_3} \mathbf{e}_3 \cdot \frac{\partial \mathbf{e}_1}{\partial \alpha_3} \right) + \cdots, \tag{A.8.34}$$

which also follows from (A.8.33). The three dots on the left-hand side represent similar terms involving f_2 and f_3. A slight rearrangement yields

$$\nabla \cdot \mathbf{f} = \frac{1}{h_1} \frac{\partial f_1}{\partial \alpha_1} + f_1 \left(\frac{1}{h_1 h_2} \mathbf{e}_2 \cdot \frac{\partial (h_1 \mathbf{e}_1)}{\partial \alpha_2} + \frac{1}{h_1 h_3} \mathbf{e}_3 \cdot \frac{\partial (h_1 \mathbf{e}_1)}{\partial \alpha_3} \right) + \cdots. \tag{A.8.35}$$

Now using the relations (A.8.11) and similar relations and simplifying, we obtain

$$\nabla \cdot \mathbf{f} = \frac{1}{h_1 h_2 h_3} \left[\frac{\partial}{\partial \alpha_1} (h_2 h_3 f_1) + \frac{\partial}{\partial \alpha_2} (h_3 h_1 f_2) + \frac{\partial}{\partial \alpha_3} (h_1 h_2 f_3) \right]. \tag{A.8.36}$$

We recall that the products $h_i h_j$ for $i \neq j$ are surface metric coefficients corresponding to coordinate surfaces.

Laplacian of a scalar function

The Laplacian of a scalar function, $f(\mathbf{x})$, is equal to the divergence of its gradient. Using expressions (A.8.24) and (A.8.36), we find that

$$\nabla^2 f = \frac{1}{h_1 h_2 h_3} \left[\frac{\partial}{\partial \alpha_1} \left(\frac{h_2 h_3}{h_1} \frac{\partial f}{\partial \alpha_1} \right) + \frac{\partial}{\partial \alpha_2} \left(\frac{h_3 h_1}{h_2} \frac{\partial f}{\partial \alpha_2} \right) + \frac{\partial}{\partial \alpha_3} \left(\frac{h_1 h_2}{h_3} \frac{\partial f}{\partial \alpha_3} \right) \right]. \tag{A.8.37}$$

We recall that $(1/h_1) \partial f / \partial \alpha_1 = \partial f / \partial l_1$, $(1/h_2) \partial f / \partial \alpha_2 = \partial f / \partial l_2$, and $(1/h_3) \partial f / \partial \alpha_3 = \partial f / \partial l_3$.

Curl of a vector field

The curl of a vector function, $\mathbf{f}$, is a vector defined as

$$\boldsymbol{\omega} \equiv \nabla \times \mathbf{f} = \left(\frac{1}{h_1} \mathbf{e}_1 \times \frac{\partial}{\partial \alpha_1} + \frac{1}{h_2} \mathbf{e}_2 \times \frac{\partial}{\partial \alpha_2} + \frac{1}{h_3} \mathbf{e}_3 \times \frac{\partial}{\partial \alpha_3} \right) (f_1 \mathbf{e}_1 + f_2 \mathbf{e}_2 + f_3 \mathbf{e}_3). \tag{A.8.38}$$

Expanding the derivatives, we find that

$$\begin{aligned}
\boldsymbol{\omega} = &\frac{1}{h_1} \left(f_1 \mathbf{e}_1 \times \frac{\partial \mathbf{e}_1}{\partial \alpha_1} + f_2 \mathbf{e}_1 \times \frac{\partial \mathbf{e}_2}{\partial \alpha_1} + \frac{\partial f_2}{\partial \alpha_1} \mathbf{e}_3 + f_3 \mathbf{e}_1 \times \frac{\partial \mathbf{e}_3}{\partial \alpha_1} - \frac{\partial f_3}{\partial \alpha_1} \mathbf{e}_2 \right) \\
&+ \frac{1}{h_2} \left(f_1 \mathbf{e}_2 \times \frac{\partial \mathbf{e}_1}{\partial \alpha_2} - \frac{\partial f_1}{\partial \alpha_2} \mathbf{e}_3 + f_2 \mathbf{e}_2 \times \frac{\partial \mathbf{e}_2}{\partial \alpha_2} + f_3 \mathbf{e}_2 \times \frac{\partial \mathbf{e}_3}{\partial \alpha_2} + \frac{\partial f_3}{\partial \alpha_2} \mathbf{e}_1 \right) \\
&+ \frac{1}{h_3} \left(f_1 \mathbf{e}_3 \times \frac{\partial \mathbf{e}_1}{\partial \alpha_3} + \frac{\partial f_1}{\partial \alpha_3} \mathbf{e}_2 + f_2 \mathbf{e}_3 \times \frac{\partial \mathbf{e}_2}{\partial \alpha_3} - \frac{\partial f_2}{\partial \alpha_3} \mathbf{e}_1 + f_3 \mathbf{e}_3 \times \frac{\partial \mathbf{e}_3}{\partial \alpha_3} \right).
\end{aligned} \tag{A.8.39}$$

The first component of the curl is

$$\omega_1 \equiv \boldsymbol{\omega} \cdot \mathbf{e}_1 = \frac{1}{h_2} \left(f_1 \, \mathbf{e}_1 \cdot (\mathbf{e}_2 \times \frac{\partial \mathbf{e}_1}{\partial \alpha_2}) + f_2 \, \mathbf{e}_1 \cdot (\mathbf{e}_2 \times \frac{\partial \mathbf{e}_2}{\partial \alpha_2}) + f_3 \, \mathbf{e}_1 \cdot (\mathbf{e}_2 \times \frac{\partial \mathbf{e}_3}{\partial \alpha_2}) + \frac{\partial f_3}{\partial \alpha_2} \right)$$

$$+ \frac{1}{h_3} \left(f_1 \, \mathbf{e}_1 \cdot (\mathbf{e}_3 \times \frac{\partial \mathbf{e}_1}{\partial \alpha_3}) + f_2 \, \mathbf{e}_1 \cdot (\mathbf{e}_3 \times \frac{\partial \mathbf{e}_2}{\partial \alpha_3}) - \frac{\partial f_2}{\partial \alpha_3} + f_3 \, \mathbf{e}_1 \cdot (\mathbf{e}_3 \times \frac{\partial \mathbf{e}_3}{\partial \alpha_3}) \right). \tag{A.8.40}$$

Rearranging the triple scalar products and using the orthogonality of the base vectors, we obtain

$$\omega_1 = \frac{1}{h_2} \left(f_1 \, \mathbf{e}_3 \cdot \frac{\partial \mathbf{e}_1}{\partial \alpha_2} + f_2 \, \mathbf{e}_3 \cdot \frac{\partial \mathbf{e}_2}{\partial \alpha_2} + f_3 \, \mathbf{e}_3 \cdot \frac{\partial \mathbf{e}_3}{\partial \alpha_2} + \frac{\partial f_3}{\partial \alpha_2} \right)$$

$$- \frac{1}{h_3} \left(f_1 \, \mathbf{e}_2 \cdot \frac{\partial \mathbf{e}_1}{\partial \alpha_3} + f_2 \, \mathbf{e}_2 \cdot \frac{\partial \mathbf{e}_2}{\partial \alpha_3} + \frac{\partial f_2}{\partial \alpha_3} + f_3 \, \mathbf{e}_2 \cdot \frac{\partial \mathbf{e}_3}{\partial \alpha_3} \right). \tag{A.8.41}$$

Using relations (A.8.3) and (A.8.15), we obtain

$$\omega_1 = \frac{1}{h_2} \left(f_2 \, \mathbf{e}_3 \cdot \frac{\partial \mathbf{e}_2}{\partial \alpha_2} + \frac{\partial f_3}{\partial \alpha_2} \right) - \frac{1}{h_3} \left(\frac{\partial f_2}{\partial \alpha_3} + f_3 \, \mathbf{e}_2 \cdot \frac{\partial \mathbf{e}_3}{\partial \alpha_3} \right), \tag{A.8.42}$$

which can be rearranged into

$$\omega_1 = \frac{1}{h_2 h_3} \left(h_3 \frac{\partial f_3}{\partial \alpha_2} - f_3 \, h_2 \, \mathbf{e}_2 \cdot \frac{\partial \mathbf{e}_3}{\partial \alpha_3} - h_2 \frac{\partial f_2}{\partial \alpha_3} + f_2 \, h_3 \, \mathbf{e}_3 \cdot \frac{\partial \mathbf{e}_2}{\partial \alpha_2} \right). \tag{A.8.43}$$

Using relations (A.8.20), we obtain

$$\omega_1 = \frac{1}{h_2 h_3} \left(h_3 \frac{\partial f_3}{\partial \alpha_2} + f_3 \frac{\partial h_3}{\partial \alpha_2} - h_2 \frac{\partial f_2}{\partial \alpha_3} - f_2 \frac{\partial h_2}{\partial \alpha_3} \right) = \frac{1}{h_2 h_3} \left(\frac{\partial (h_3 f_3)}{\partial \alpha_2} - \frac{\partial (h_2 f_2)}{\partial \alpha_3} \right). \tag{A.8.44}$$

Similar relations can be written for ω_2 and ω_3. Compiling these results, we find that the curl is given by the determinant of a matrix,

$$\boldsymbol{\omega} \equiv \nabla \times \mathbf{f} = \frac{1}{h_1 h_2 h_3} \det \left(\begin{bmatrix} h_1 \mathbf{e}_1 & h_2 \mathbf{e}_2 & h_3 \mathbf{e}_3 \\ \partial/\partial \alpha_1 & \partial/\partial \alpha_2 & \partial/\partial \alpha_3 \\ h_1 f_1 & h_2 f_2 & h_3 f_3 \end{bmatrix} \right). \tag{A.8.45}$$

The determinant is computed according to the usual rules of the outer vector product, treating the partial derivative operator triplet as a regular vector.

Alternatively, the components of the curl are given by

$$\omega_1 = L_{23} - L_{32}, \qquad \omega_2 = L_{31} - L_{13}, \qquad \omega_3 = L_{12} - L_{21}, \tag{A.8.46}$$

where $\mathbf{L} = \nabla \mathbf{f}$ is the gradient of $\mathbf{f}$.

Problems

A.8.1 *Divergence of a vector function*

Derive expression (A.8.36) working as discussed in the text.

A.8.2 *Curl of a vector field*

Prove the expression for the curl of a vector field given in (A.8.46).

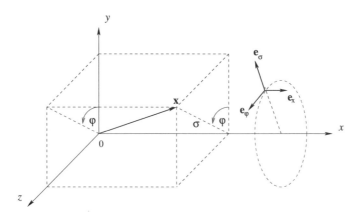

FIGURE A.9.1   Cylindrical polar coordinates, (x, σ, φ), defined with respect to companion Cartesian coordinates, (x, y, z).

A.9   Cylindrical polar coordinates

A point in space can be identified by the values of the ordered triplet (x, σ, φ), as illustrated in Figure A.9.1, where:

- x is the projection of the position vector onto the straight (rectilinear) x axis passing through a designated origin, taking values in the range $(-\infty, +\infty)$.

- σ is the distance of a point of interest from the x axis, taking values in the range $[0, \infty)$.

- φ is the azimuthal angle measured around the x axis, taking values in the range $[0, 2\pi)$.

The value $\varphi = 0$ corresponds to the first and second quadrants, and the value $\varphi = \pi$ corresponds to the third and fourth quadrants of the xy plane. An $x\sigma$ half-plane of constant azimuthal angle φ is called an azimuthal plane. In astronomy, land, and sea navigation, the yz plane lies in the horizon and the y axis points to the north. A star may be located at a point $\mathbf{x}$ in the sky.

Using elementary trigonometry, we derive relations between the Cartesian and associated polar cylindrical coordinates,

$$y = \sigma \cos\varphi, \qquad z = \sigma \sin\varphi, \tag{A.9.1}$$

and the inverse relations between the polar cylindrical and Cartesian coordinates,

$$\sigma = \sqrt{y^2 + z^2}, \qquad \varphi = \arccos\frac{y}{\sigma}. \tag{A.9.2}$$

In computing the inverse cosine function, arccos, care must be taken so that φ is a continuous function of y and σ.

Unit vectors

We consider a point in space and define three vectors of unit length, denoted by $\mathbf{e}_x$, $\mathbf{e}_\sigma$, and $\mathbf{e}_\varphi$, pointing, respectively, in the direction of the x axis, normal to the x axis, and in the direction of the azimuthal angle φ, as depicted in Figure A.9.1. Note that the orientation of the unit vectors $\mathbf{e}_\sigma$ and $\mathbf{e}_\varphi$ changes with position in space, whereas the orientation of $\mathbf{e}_x$ is fixed and independent of position in space. The position vector is

$$\mathbf{x} = x\,\mathbf{e}_x + \sigma\,\mathbf{e}_\sigma. \tag{A.9.3}$$

The dependence of the position vector on the azimuthal angle, φ, is mediated through the unit vector $\mathbf{e}_\sigma$ on the right-hand side. The absence of $\mathbf{e}_\varphi$ from the right-hand side of (A.9.3) is justified by observing that the distance from the origin, expressed by the position vector $\mathbf{x}$, is perpendicular to the third unit vector, $\mathbf{e}_\varphi$.

Vector components

Any vector, $\mathbf{v}$, can be resolved as

$$\mathbf{v} = v_x\,\mathbf{e}_x + v_\sigma\,\mathbf{e}_\sigma + v_\varphi\,\mathbf{e}_\varphi, \tag{A.9.4}$$

where the coefficients v_x, v_σ, and v_φ are the cylindrical polar components of $\mathbf{v}$.

Relation to Cartesian vector components

Using elementary trigonometry, we derive relations between the Cartesian and cylindrical polar unit vectors and the inverse relations,

$$
\begin{aligned}
\mathbf{e}_\sigma &= \cos\varphi\,\mathbf{e}_y + \sin\varphi\,\mathbf{e}_z, & \mathbf{e}_\varphi &= -\sin\varphi\,\mathbf{e}_y + \cos\varphi\,\mathbf{e}_z, \\
\mathbf{e}_y &= \cos\varphi\,\mathbf{e}_\sigma - \sin\varphi\,\mathbf{e}_\varphi, & \mathbf{e}_z &= \sin\varphi\,\mathbf{e}_\sigma + \cos\varphi\,\mathbf{e}_\varphi.
\end{aligned}
\tag{A.9.5}
$$

The corresponding relations for a vector, $\mathbf{v}$, are

$$
\begin{aligned}
v_\sigma &= \cos\varphi\,v_y + \sin\varphi\,v_z, & v_\varphi &= -\sin\varphi\,v_y + \cos\varphi\,v_z, \\
v_y &= \cos\varphi\,v_\sigma - \sin\varphi\,v_\varphi, & v_z &= \sin\varphi\,v_\sigma + \cos\varphi\,v_\varphi.
\end{aligned}
\tag{A.9.6}
$$

All derivatives $\partial\mathbf{e}_\alpha/\partial\beta$ are zero, except for two derivatives,

$$\frac{\partial\mathbf{e}_\sigma}{\partial\varphi} = \mathbf{e}_\varphi, \qquad \frac{\mathrm{d}\mathbf{e}_\varphi}{\mathrm{d}\varphi} = -\mathbf{e}_\sigma, \tag{A.9.7}$$

where Greek variables stand for x, σ, and φ.

Metric coefficients

The cylindrical polar coordinates, (x, σ, φ), comprise a set of orthogonal curvilinear coordinates $(\alpha_1, \alpha_2, \alpha_3)$, as discussed in Section A.8. The associated metric coefficients are

$$h_x = 1, \qquad h_\sigma = 1, \qquad h_\varphi = \sigma. \tag{A.9.8}$$

Expressions (A.9.7) arise by substituting these metric coefficients into the general relations (A.8.17), (A.8.19), and (A.8.20).

Differential operators

The gradient a scalar function, $f(\mathbf{x})$, is

$$\nabla f = \mathbf{e}_x \frac{\partial f}{\partial x} + \mathbf{e}_\sigma \frac{\partial f}{\partial \sigma} + \mathbf{e}_\varphi \frac{1}{\sigma} \frac{\partial f}{\partial \varphi}. \tag{A.9.9}$$

The Laplacian of a scalar function, $f(\mathbf{x})$, is

$$\nabla^2 f = \frac{\partial^2 f}{\partial x^2} + \frac{1}{\sigma} \frac{\partial}{\partial \sigma}\left(\sigma \frac{\partial f}{\partial \sigma}\right) + \frac{1}{\sigma^2} \frac{\partial^2 f}{\partial \varphi^2}. \tag{A.9.10}$$

The divergence of a vector function, $\mathbf{f}(\mathbf{x})$, is

$$\nabla \cdot \mathbf{f} = \frac{\partial f_x}{\partial x} + \frac{1}{\sigma} \frac{\partial(\sigma f_\sigma)}{\partial \sigma} + \frac{1}{\sigma} \frac{\partial f_\varphi}{\partial \varphi}. \tag{A.9.11}$$

The curl of a vector field, $\mathbf{f}(\mathbf{x})$, is

$$\nabla \times \mathbf{f} = \mathbf{e}_x \frac{1}{\sigma}\left(\frac{\partial(\sigma f_\varphi)}{\partial \sigma} - \frac{\partial f_\sigma}{\partial \varphi}\right) + \mathbf{e}_\sigma \left(\frac{1}{\sigma}\frac{\partial f_x}{\partial \varphi} - \frac{\partial f_\varphi}{\partial x}\right) + \mathbf{e}_\varphi \left(\frac{\partial f_\sigma}{\partial x} - \frac{\partial f_x}{\partial \sigma}\right). \tag{A.9.12}$$

The Laplacian of a vector function, $\mathbf{f}(\mathbf{x})$, is

$$\nabla^2 \mathbf{f} = \mathbf{e}_x \nabla^2 f_x + \mathbf{e}_\sigma \left(\nabla^2 f_\sigma - \frac{f_\sigma}{\sigma^2} - \frac{2}{\sigma^2}\frac{\partial f_\varphi}{\partial \varphi}\right) + \mathbf{e}_\varphi \left(\nabla^2 f_\varphi - \frac{f_\varphi}{\sigma^2} + \frac{2}{\sigma^2}\frac{\partial f_\sigma}{\partial \varphi}\right). \tag{A.9.13}$$

The derivative of a vector function, $\mathbf{f}(\mathbf{x})$, with respect to arc length, l, measured in the direction of a unit vector, $\mathbf{t}$, is

$$\mathbf{t} \cdot \nabla \mathbf{f} = \frac{\partial \mathbf{f}}{\partial l} = \mathbf{e}_x \left(\mathbf{t} \cdot \nabla f_x\right) + \mathbf{e}_\sigma \left(\mathbf{t} \cdot \nabla f_\sigma - \frac{t_\varphi f_\varphi}{\sigma}\right) + \mathbf{e}_\varphi \left(\mathbf{t} \cdot \nabla f_\varphi + \frac{t_\varphi f_\sigma}{\sigma}\right), \tag{A.9.14}$$

where the cylindrical polar coordinates of the gradient of the individual components, ∇f_α, are computed from (A.9.9).

Problem

A.9.1 *Laplacian of a vector field*

Derive the expression for the Laplacian of a vector function given in (A.9.13).

A.10 Spherical polar coordinates

A point in space can be identified by the values of the ordered triplet (r, θ, φ), as illustrated in Figure A.10.1, where:

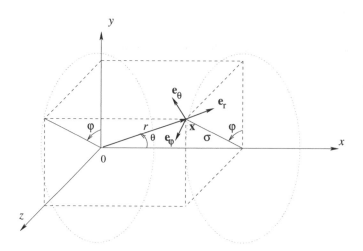

FIGURE A.10.1   Spherical polar coordinates, (r, θ, φ), defined with respect to companion Cartesian coordinates, (x, y, z), and cylindrical polar coordinates, (x, σ, φ).

- r is the distance from the designated origin taking values in the range $[0, \infty)$.

- θ is the meridional angle subtended between the x axis, the origin, and the chosen point, taking values in the range $[0, \pi]$.

- φ is the azimuthal angle measured around the x axis, taking values in the range $[0, 2\pi)$.

The value $\varphi = 0$ corresponds to the first and second quadrants, and the value $\varphi = \pi$ corresponds to the third and fourth quadrants of the xy plane. An $r\theta$ half-plane of constant azimuthal angle φ is called an azimuthal plane. In astronomy, land and sea navigation, the yz plane is the horizon and the y axis points to the north. A star may be located at the point $\mathbf{x}$ in the sky.

Using elementary trigonometry, we derive relations between the Cartesian, cylindrical, and spherical polar coordinates,

$$x = r \cos \theta, \qquad \sigma = r \sin \theta, \tag{A.10.1}$$

and

$$y = \sigma \cos \varphi = r \sin \theta \cos \varphi, \qquad z = \sigma \sin \varphi = r \sin \theta \sin \varphi. \tag{A.10.2}$$

The inverse relations are

$$r = \sqrt{x^2 + y^2 + z^2} = \sqrt{x^2 + \sigma^2}, \qquad \theta = \arccos \frac{x}{r}, \qquad \varphi = \arccos \frac{y}{\sigma}. \tag{A.10.3}$$

In computing the inverse cosine functions, care must be taken so that θ and φ are continuous functions of x, y, r, and σ.

Unit vectors

We consider a point in space and define three vectors of unit length, denoted by $\mathbf{e}_r$, $\mathbf{e}_\theta$, and $\mathbf{e}_\varphi$, pointing, respectively, in the radial, meridional, and azimuthal direction, as illustrated in Figure A.10.1. Note that the orientation of all three unit vectors changes with position in space. In contrast, the orientation of the Cartesian unit vectors, $\mathbf{e}_x$, $\mathbf{e}_y$, and $\mathbf{e}_z$, is fixed. The position vector is proportional to the radial unit vector,

$$\mathbf{x} = r\,\mathbf{e}_r. \tag{A.10.4}$$

The dependence on θ and φ is mediated through the unit vector $\mathbf{e}_r$. The absence of $\mathbf{e}_\theta$ and $\mathbf{e}_\varphi$ from the right-hand side of (A.10.4) is explained by observing that the distance from the origin, expressed by the position vector $\mathbf{x}$, is perpendicular to the unit vectors $\mathbf{e}_\theta$ and $\mathbf{e}_\varphi$.

Vector components

A vector, $\mathbf{v}$, can be resolved in the component form

$$\mathbf{v} = v_r\,\mathbf{e}_r + v_\theta\,\mathbf{e}_\theta + v_\varphi\,\mathbf{e}_\varphi, \tag{A.10.5}$$

where the coefficients v_r, v_θ, and v_φ are the spherical polar components of $\mathbf{v}$.

Relation to Cartesian vector components

Using elementary trigonometry, we derive relations between the spherical polar, cylindrical polar, and Cartesian unit vectors,

$$
\begin{aligned}
\mathbf{e}_r &= \cos\theta\,\mathbf{e}_x + \sin\theta\cos\varphi\,\mathbf{e}_y + \sin\theta\sin\varphi\,\mathbf{e}_z = \cos\theta\,\mathbf{e}_x + \sin\theta\,\mathbf{e}_\sigma, \\
\mathbf{e}_\theta &= -\sin\theta\,\mathbf{e}_x + \cos\theta\cos\varphi\,\mathbf{e}_y + \cos\theta\sin\varphi\,\mathbf{e}_z = -\sin\theta\,\mathbf{e}_x + \cos\theta\,\mathbf{e}_\sigma, \\
\mathbf{e}_\varphi &= -\sin\varphi\,\mathbf{e}_y + \cos\varphi\,\mathbf{e}_z.
\end{aligned}
\tag{A.10.6}
$$

The corresponding relations for a vector, $\mathbf{v}$, are

$$
\begin{aligned}
v_r &= \cos\theta\,v_x + \sin\theta\cos\varphi\,v_y + \sin\theta\sin\varphi\,v_z = \cos\theta\,v_x + \sin\theta\,v_\sigma, \\
v_\theta &= -\sin\theta\,v_x + \cos\theta\cos\varphi\,v_y + \cos\theta\sin\varphi\,v_z = -\sin\theta\,v_x + \cos\theta\,v_\sigma, \\
v_\varphi &= -\sin\varphi\,v_y + \cos\varphi\,v_z.
\end{aligned}
\tag{A.10.7}
$$

All derivatives $\partial\mathbf{e}_\alpha/\partial\beta$ are zero, except for the five derivatives

$$\frac{d\mathbf{e}_r}{d\theta} = \mathbf{e}_\theta, \qquad \frac{d\mathbf{e}_r}{d\varphi} = \sin\theta\,\mathbf{e}_\varphi, \qquad \frac{d\mathbf{e}_\theta}{d\theta} = -\mathbf{e}_r, \qquad \frac{d\mathbf{e}_\theta}{d\varphi} = \cos\theta\,\mathbf{e}_\varphi,$$

$$\frac{d\mathbf{e}_\varphi}{d\varphi} = -\sin\theta\,\mathbf{e}_r - \cos\theta\,\mathbf{e}_\theta, \tag{A.10.8}$$

where Greek variables stand for r, θ, and φ.

Metric coefficients

The spherical polar coordinates, (r, θ, φ), comprise a set of orthogonal curvilinear coordinates identified with the triplet $(\alpha_1, \alpha_2, \alpha_3)$, as discussed in Section A.8. The associated metric coefficients are given by

$$h_r = 1, \qquad h_\theta = r, \qquad h_\varphi = r \sin \theta. \tag{A.10.9}$$

Expressions (A.10.8) arise by substituting these metric coefficients into the general relations (A.8.17), (A.8.19), and (A.8.20).

Differential operators

The gradient a scalar function, f, is given by

$$\nabla f = \mathbf{e}_r \frac{\partial f}{\partial r} + \mathbf{e}_\theta \frac{1}{r} \frac{\partial f}{\partial \theta} + \mathbf{e}_\varphi \frac{1}{r \sin \theta} \frac{\partial f}{\partial \varphi}. \tag{A.10.10}$$

The Laplacian of a scalar function, f, is given by

$$\nabla^2 f = \frac{1}{r^2} \frac{\partial}{\partial r} \left(r^2 \frac{\partial f}{\partial r} \right) + \frac{1}{r^2 \sin \theta} \frac{\partial}{\partial \theta} \left(\sin \theta \frac{\partial f}{\partial \theta} \right) + \frac{1}{r^2 \sin^2 \theta} \frac{\partial^2 f}{\partial \varphi^2}. \tag{A.10.11}$$

The divergence of a vector function, $\mathbf{f}$, is given by

$$\nabla \cdot \mathbf{f} = \frac{1}{r^2} \frac{\partial (r^2 f_r)}{\partial r} + \frac{1}{r \sin \theta} \left(\frac{\partial (\sin \theta f_\theta)}{\partial \theta} + \frac{\partial f_\varphi}{\partial \varphi} \right), \tag{A.10.12}$$

yielding

$$\nabla \cdot \mathbf{f} = \frac{\partial f_r}{\partial r} + 2 \frac{u_r}{r} + \frac{1}{r} \frac{\partial f_\theta}{\partial \theta} + \frac{f_\theta}{r} \cot \theta + \frac{1}{r \sin \theta} \frac{\partial f_\varphi}{\partial \varphi}. \tag{A.10.13}$$

The curl of a vector field, $\mathbf{f}$, is

$$\nabla \times \mathbf{f} = \mathbf{e}_r \frac{1}{r \sin \theta} \left(\frac{\partial (\sin \theta f_\varphi)}{\partial \theta} - \frac{\partial f_\theta}{\partial \varphi} \right) + \mathbf{e}_\theta \frac{1}{r} \left(\frac{1}{\sin \theta} \frac{\partial f_r}{\partial \varphi} - \frac{\partial (r f_\varphi)}{\partial r} \right)$$
$$+ \mathbf{e}_\varphi \frac{1}{r} \left(\frac{\partial (r f_\theta)}{\partial r} - \frac{\partial f_r}{\partial \theta} \right). \tag{A.10.14}$$

The Laplacian of a vector function, $\mathbf{f}$, is given by

$$\nabla^2 \mathbf{f} = q_r \mathbf{e}_r + q_\theta \mathbf{e}_\theta + q_\varphi \mathbf{e}_\varphi, \tag{A.10.15}$$

where

$$q_r = \nabla^2 f_r - 2 \frac{f_r}{r^2} - \frac{2}{r^2} \frac{\partial f_\theta}{\partial \theta} - \frac{2}{r^2} f_\theta \cot \theta - \frac{2}{r^2 \sin \theta} \frac{\partial f_\varphi}{\partial \varphi},$$

$$q_\theta = \nabla^2 f_\theta + \frac{2}{r^2} \frac{\partial f_r}{\partial \theta} - \frac{f_\theta}{r^2 \sin^2 \theta} - \frac{2}{r^2} \frac{\cot \theta}{\sin \theta} \frac{\partial f_\varphi}{\partial \varphi}, \tag{A.10.16}$$

$$q_\varphi = \nabla^2 f_\varphi + \frac{1}{r^2 \sin^2 \theta} \left(- f_\varphi + 2 \frac{\partial f_r}{\partial \varphi} + 2 \cos \theta \frac{\partial f_\theta}{\partial \varphi} \right),$$

and the Laplacian of the individual vector components is computed using (A.10.11).

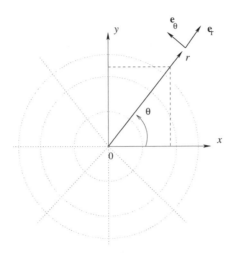

FIGURE A.11.1   Plane polar coordinates, (r, θ), in the xy plane defined with respect to companion Cartesian coordinates, (x, y).

The derivative of a vector function, $\mathbf{f}$, with respect to arc length, l, measured in the direction of a unit vector, $\mathbf{t}$, is given by

$$\mathbf{t} \cdot \nabla \mathbf{f} = \frac{\partial \mathbf{f}}{\partial l} = \mathbf{e}_r \left(\mathbf{t} \cdot \nabla f_r - \frac{t_\theta f_\theta + t_\varphi f_\varphi}{r} \right) + \mathbf{e}_\theta \left(\mathbf{t} \cdot \nabla f_\theta + \frac{t_\theta f_r}{r} - \frac{\cot \theta}{r} t_\varphi f_\varphi \right)$$
$$+ \mathbf{e}_\varphi \left(\mathbf{t} \cdot \nabla f_\varphi + \frac{t_\varphi f_r}{r} + \frac{t_\varphi f_\theta}{r} \cot \theta \right). \tag{A.10.17}$$

The gradients of the individual spherical polar components, ∇f_r, ∇f_θ, and ∇f_φ, are computed according to (A.10.10).

Problem

A.10.1 *Curl of a vector field*

Derive the expression for the curl of a vector field given in (A.10.14).

A.11   Plane polar coordinates

A point in the xy plane can be identified by the doublet (r, θ), where r is the distance from the origin and θ is the polar angle subtended between the x axis, the origin, and the chosen point, measured in the counterclockwise direction, as illustrated in Figure A.11.1. The radial position, r, takes values in the range $[0, \infty)$, and the polar angle, θ, takes values in the range $[0, 2\pi)$. Using elementary trigonometry, we derive the relations between the Cartesian and plane polar coordinates,

$$x = r \cos \theta, \qquad y = r \sin \theta. \tag{A.11.1}$$

The inverse relations are

$$r = \sqrt{x^2 + y^2}, \qquad \theta = \arccos \frac{y}{r}. \qquad (A.11.2)$$

In computing the inverse cosine function, care must be taken so that θ is a continuous function of y and r.

Unit vectors

We consider a point in the xy plane and define two vectors of unit length, denoted by $\mathbf{e}_r$ and $\mathbf{e}_\theta$, pointing in the radial or polar direction, as depicted in Figure A.11.1. Note that the orientation of these unit vectors changes with position in the xy plane, whereas the orientation of the Cartesian unit vectors $\mathbf{e}_x$ and $\mathbf{e}_y$ is fixed. The position vector is proportional to the radial unit vector,

$$\mathbf{x} = r\,\mathbf{e}_r. \qquad (A.11.3)$$

The dependence of the position vector on the polar angle, θ, is mediated through the unit vector $\mathbf{e}_r$.

Vector components

A vector in the xy plane, $\mathbf{v}$, can be expressed in the form

$$\mathbf{v} = v_r\,\mathbf{e}_r + v_\theta\,\mathbf{e}_\theta, \qquad (A.11.4)$$

where the coefficients v_r and v_θ are the plane polar components of $\mathbf{v}$.

Relation to Cartesian vector components

Using elementary trigonometry, we derive the following relations between the Cartesian and plane polar unit vectors and *vice versa*,

$$\begin{aligned}
\mathbf{e}_r &= \cos\theta\,\mathbf{e}_x + \sin\theta\,\mathbf{e}_y, & \mathbf{e}_\theta &= -\sin\theta\,\mathbf{e}_x + \cos\theta\,\mathbf{e}_y, \\
\mathbf{e}_x &= \cos\theta\,\mathbf{e}_r - \sin\theta\,\mathbf{e}_\theta, & \mathbf{e}_y &= \sin\theta\,\mathbf{e}_r + \cos\theta\,\mathbf{e}_\theta.
\end{aligned} \qquad (A.11.5)$$

The corresponding relations for a vector, $\mathbf{v}$, are

$$\begin{aligned}
v_r &= \cos\theta\,v_x + \sin\theta\,v_y, & v_\theta &= -\sin\theta\,v_x + \cos\theta\,v_y, \\
v_x &= \cos\theta\,v_r - \sin\theta\,v_\theta, & v_y &= \sin\theta\,v_r + \cos\theta\,v_\theta.
\end{aligned} \qquad (A.11.6)$$

All derivatives $\partial\mathbf{e}_\alpha/\partial\beta$ are zero, except for two derivatives,

$$\frac{\partial\mathbf{e}_r}{\partial\theta} = \mathbf{e}_\theta, \qquad \frac{\mathrm{d}\mathbf{e}_\theta}{\mathrm{d}\theta} = -\mathbf{e}_r, \qquad (A.11.7)$$

where Greek indices stand for r or θ.

Metric coefficients

The plane polar coordinates, (r, θ), comprise a set of orthogonal curvilinear coordinates (α_1, α_2). The associated metric coefficients are

$$h_r = 1, \qquad h_\theta = r. \qquad (A.11.8)$$

Expressions (A.11.7) arise by substituting these metric coefficients into the general relations (A.8.17), (A.8.19), and (A.8.20).

Differential operators

The differential vector operators in plane polar coordinates, (r, θ), derive from those in cylindrical polar coordinates by discarding the dependence on x and renaming σ as r and φ as θ. The gradient of a scalar function, f, is given by

$$\nabla f = \mathbf{e}_r \frac{\partial f}{\partial r} + \mathbf{e}_\theta \frac{1}{r} \frac{\partial f}{\partial \theta}. \tag{A.11.9}$$

The Laplacian of a scalar function, f, is

$$\nabla^2 f = \frac{1}{r} \frac{\partial}{\partial r} \left(r \frac{\partial f}{\partial r} \right) + \frac{1}{r^2} \frac{\partial^2 f}{\partial \theta^2}. \tag{A.11.10}$$

The divergence of a vector function, $\mathbf{f}$, is given by

$$\nabla \cdot \mathbf{f} = \frac{1}{r} \frac{\partial (r f_r)}{\partial r} + \frac{1}{r} \frac{\partial f_\theta}{\partial \theta}. \tag{A.11.11}$$

The curl of a vector field, $\mathbf{f}$, is given by

$$\nabla \times \mathbf{f} = \frac{1}{r} \left(\frac{\partial (r f_\theta)}{\partial r} - \frac{\partial f_r}{\partial \theta} \right) \mathbf{e}_z, \tag{A.11.12}$$

where $\mathbf{e}_z$ is the unit vector normal to the xy plane. The Laplacian of a vector function, $\mathbf{f}$, is

$$\nabla^2 \mathbf{f} = \mathbf{e}_r \left(\nabla^2 f_r - \frac{f_r}{r^2} - \frac{2}{r^2} \frac{\partial f_\theta}{\partial \theta} \right) + \mathbf{e}_\theta \left(\nabla^2 f_\theta - \frac{f_\theta}{r^2} + \frac{2}{r^2} \frac{\partial f_r}{\partial \theta} \right). \tag{A.11.13}$$

The derivative of a vector function, $\mathbf{f}$, with respect to arc length, l, measured in the direction of a unit vector, $\mathbf{t}$, is given by

$$\mathbf{t} \cdot \nabla \mathbf{f} = \frac{\partial \mathbf{f}}{\partial l} = \mathbf{e}_r \left(\mathbf{t} \cdot \nabla f_r - \frac{t_\theta f_\theta}{r} \right) + \mathbf{e}_\theta \left(\mathbf{t} \cdot \nabla f_\theta + \frac{t_\theta f_r}{r} \right), \tag{A.11.14}$$

where the plane polar coordinates of the gradient of the individual components, f_r and f_θ, are computed from (A.11.9).

Problem

A.11.1 *Laplacian of a vector field*

Derive the expression for the Laplacian of a vector function given in (A.11.13).

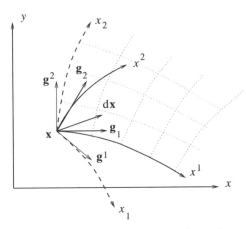

FIGURE A.12.1   Nonorthogonal curvilinear coordinates in a plane; $(\mathbf{g}_1, \mathbf{g}_2)$ are the covariant base vectors, and $(\mathbf{g}^1, \mathbf{g}^2)$ are the contravariant base vectors.

A.12   Nonorthogonal coordinates

Rectilinear and curvilinear nonorthogonal coordinates are employed to accommodate the particular geometry of a solution domain of interest. The main objective is to facilitate the implementation of boundary conditions in solving partial differential equations by analytical or numerical methods.

Base vectors

We begin the discussion of nonorthogonal coordinates by introducing two continuous intersecting families of generally curvilinear axes in the xy plane, (x^1, x^2), as shown in Figure A.12.1. The position vector, $\mathbf{x}$, is regarded as a function of x^1 and x^2, signified by writing $\mathbf{x}(x^1, x^2)$. Next, we introduce a corresponding pair of tangent base vectors,

$$\mathbf{g}_1 = \frac{\partial \mathbf{x}}{\partial x^1}, \qquad \mathbf{g}_2 = \frac{\partial \mathbf{x}}{\partial x^2}, \tag{A.12.1}$$

where the derivative with respect to x^1 is taken keeping x^2 constant, and *vice versa*. The two base vectors, $\mathbf{g}_1$ and $\mathbf{g}_2$, are unit vectors only if the coordinates x^1 and x^2 measure physical arc length. The theory of differential calculus allows us to express an infinitesimal vector in the xy plane at a point, $\mathbf{x}$, as

$$d\mathbf{x} = \mathbf{g}_1 dx^1 + \mathbf{g}_2 dx^2. \tag{A.12.2}$$

In general, equation (A.12.2) cannot be integrated readily to produce an explicit expression for the position vector, $\mathbf{x}(x^1, x^2)$.

The ordered doublet (x^1, x^2) constitutes a pair of nonorthogonal curvilinear coordinates with associated base vectors $(\mathbf{g}_1, \mathbf{g}_2)$. If the lines of constant x^1 and x^2 are straight, we obtain rectilinear coordinates. If the angle subtended between the two vectors $\mathbf{g}_1$ and $\mathbf{g}_2$ is equal to $90°$ at every point in the plane, we obtain orthogonal curvilinear or rectilinear coordinates.

Biorthonormal base vectors

In the general case of nonorthogonal curvilinear coordinates where the variables x^1 and x^2 do not measure physical arc length from a designated origin,

$$\mathbf{g}_1 \cdot \mathbf{g}_2 \neq 0, \qquad \mathbf{g}_1 \cdot \mathbf{g}_1 \neq 1, \qquad \mathbf{g}_2 \cdot \mathbf{g}_2 \neq 1. \tag{A.12.3}$$

The first inequality expressing nonorthogonality prevents us from computing the differential components dx^1 and dx^2 in (A.12.2) by projecting the position vector given in (A.12.2) onto each base vector, that is,

$$dx^1 \neq d\mathbf{x} \cdot \frac{\mathbf{g}_1}{\mathbf{g}_1 \cdot \mathbf{g}_1}, \qquad dx^2 \neq d\mathbf{x} \cdot \frac{\mathbf{g}_2}{\mathbf{g}_2 \cdot \mathbf{g}_2}. \tag{A.12.4}$$

Instead, the differential components must be found by solving a system of two equations for two unknowns originating from (A.12.2),

$$g_{11}\, dx^1 + g_{12}\, dx^2 = d\mathbf{x} \cdot \mathbf{g}_1, \qquad g_{21}\, dx^1 + g_{22}\, dx^2 = d\mathbf{x} \cdot \mathbf{g}_2, \tag{A.12.5}$$

where

$$g_{11} = \mathbf{g}_1 \cdot \mathbf{g}_1, \qquad g_{12} = g_{21} = \mathbf{g}_1 \cdot \mathbf{g}_2, \qquad g_{22} = \mathbf{g}_2 \cdot \mathbf{g}_2 \tag{A.12.6}$$

are the components of the metric tensor, $\mathbf{g}$. Using Cramer's rule, we find the solution

$$dx^1 = d\mathbf{x} \cdot \left(\frac{g_{22}}{g}\, \mathbf{g}_1 - \frac{g_{12}}{g}\, \mathbf{g}_2 \right), \qquad dx^2 = d\mathbf{x} \cdot \left(-\frac{g_{12}}{g}\, \mathbf{g}_1 + \frac{g_{11}}{g}\, \mathbf{g}_2 \right), \tag{A.12.7}$$

where

$$g \equiv g_{11}g_{22} - g_{12}^2 = \det(g_{ij}). \tag{A.12.8}$$

If $\mathbf{g}_1$ and $\mathbf{g}_2$ are unit vectors, $g_{11} = 1$ and $g_{22} = 1$. In the case of orthogonal coordinates, $g_{12} = g_{21} = 0$.

Motivated by this solution, we introduce the new base vectors

$$\mathbf{g}^1 \equiv \frac{g_{22}}{g}\, \mathbf{g}_1 - \frac{g_{12}}{g}\, \mathbf{g}_2, \qquad \mathbf{g}^2 \equiv -\frac{g_{12}}{g}\, \mathbf{g}_1 + \frac{g_{11}}{g}\, \mathbf{g}_2, \tag{A.12.9}$$

and express (A.12.7) in the form

$$dx^1 = d\mathbf{x} \cdot \mathbf{g}^1, \qquad dx^2 = d\mathbf{x} \cdot \mathbf{g}^2. \tag{A.12.10}$$

Direct substitution shows that

$$\mathbf{g}^1 \cdot \mathbf{g}_1 = 1, \qquad \mathbf{g}^2 \cdot \mathbf{g}_1 = 0, \qquad \mathbf{g}^1 \cdot \mathbf{g}_2 = 0, \qquad \mathbf{g}^2 \cdot \mathbf{g}_2 = 1, \tag{A.12.11}$$

which reveals that $\mathbf{g}^2$ is orthogonal to $\mathbf{g}_1$ and $\mathbf{g}^1$ is orthogonal to $\mathbf{g}_2$. We have arrived at the biorthogonality condition

$$\mathbf{g}^i \cdot \mathbf{g}_j = \delta_{ij}, \tag{A.12.12}$$

where δ_{ij} is Kronecker's delta. The vectors $\mathbf{g}^i$ and $\mathbf{g}_j$ constitute a biorthonormal set.

$\mathbf{g}_1, \mathbf{g}_2$	covariant base vectors
x^1, x^2	contravariant coordinates
$\mathbf{g}^1, \mathbf{g}^2$	contravariant base vectors
x_1, x_2	covariant coordinates
$\mathbf{g}_1, \mathbf{g}_2$	covariant base vectors
$g_{ij} = \mathbf{g}_i \cdot \mathbf{g}_j$	covariant metric tensor
$\mathbf{e}_1 = \mathbf{g}_1/\sqrt{g_{11}}, \; \mathbf{e}_2 = \mathbf{g}_2/\sqrt{g_{22}},$	covariant base unit vectors
$\mathbf{g}^1, \mathbf{g}^2$	contravariant base vectors
$g^{ij} = \mathbf{g}^i \cdot \mathbf{g}^j$	contravariant metric tensor
$\mathbf{e}^1 = \mathbf{g}^1/\sqrt{g^{11}}, \; \mathbf{e}^2 = \mathbf{g}^2/\sqrt{g^{22}},$	contravariant base unit vectors
a^1, a^2	contravariant vector components
a_1, a_2	covariant vector components

TABLE A.12.1   Definitions of covariant and contravariant coordinates, base vectors, and vector components in two dimensions. Analogous definitions are made in three dimensions.

Contravariant and covariant coordinates and base vectors

Since the new base vectors $\mathbf{g}^1$ and $\mathbf{g}^2$ form a complete geometrical base, we may also write

$$ \mathrm{d}\mathbf{x} = \mathbf{g}^1 \mathrm{d}x_1 + \mathbf{g}^2 \mathrm{d}x_2, \tag{A.12.13} $$

where (x_1, x_2) are covariant curvilinear coordinates. In contrast, (x^1, x^2) are contravariant curvilinear coordinates. Working as in the case of contravariant coordinates, we find that

$$ \mathrm{d}x_1 = \mathrm{d}\mathbf{x} \cdot \mathbf{g}_1, \qquad \mathrm{d}x_2 = \mathrm{d}\mathbf{x} \cdot \mathbf{g}_2. \tag{A.12.14} $$

The covariant base vectors are defined in (A.12.1). The contravariant base vectors are defined as

$$ \mathbf{g}^1 = \frac{\partial \mathbf{x}}{\partial x_1}, \qquad \mathbf{g}^2 = \frac{\partial \mathbf{x}}{\partial x_2}. \tag{A.12.15} $$

Important definitions are summarized in Table A.12.1.

Metric tensor

Using the definitions of the covariant and contravariant metric tensor, we find that

$$ \mathrm{d}\mathbf{x} \cdot \mathrm{d}\mathbf{x} = g_{ij} \, \mathrm{d}x^i \, \mathrm{d}x^j = g^{ij} \, \mathrm{d}x_i \, \mathrm{d}x_j = \mathrm{d}x^i \, \mathrm{d}x_i, \tag{A.12.16} $$

where summation is implied over the repeated indices, i and j. Combining (A.12.5) with (A.12.14) and juxtaposing the two coordinates, we obtain

$$ g_{ij} \, \mathrm{d}x^j = \mathrm{d}x_i, \qquad g^{ij} \, \mathrm{d}x_j = \mathrm{d}x^i, \tag{A.12.17} $$

where summation is implied over the repeated index, j. These expressions demonstrate that the covariant metric tensor is the inverse of the contravariant metric tensor, and *vice versa*,

$$ [g_{ij}] = [g^{ij}]^{-1}, \qquad [g^{ij}] = [g_{ij}]^{-1}, \tag{A.12.18} $$

where the superscript -1 denotes the matrix inverse.

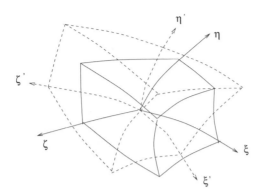

FIGURE A.12.2 Illustration of three-dimensional nonorthogonal curvilinear coordinates, (x^1, x^2, x^3). Contravariant coordinate lines, (ξ, η, ζ), are drawn with solid lines and covariant coordinate lines, (ξ', η', ζ'), are drawn with broken lines.

Because of the biorthonormality of the covariant and contravariant base vectors, the trace of the matrices $\mathbf{g}_i \otimes \mathbf{g}^j$ and $\mathbf{g}^j \otimes \mathbf{g}_j$ satisfy the equations

$$\text{trace}(\mathbf{g}_i \otimes \mathbf{g}^j) = \delta_{ij}, \qquad \text{trace}(\mathbf{g}^i \otimes \mathbf{g}_j) = \delta_{ij}, \tag{A.12.19}$$

where δ_{ij} is Kronecker's delta.

By definition, the traces of the matrices $\mathbf{g}_i \otimes \mathbf{g}_j$ and $\mathbf{g}^i \otimes \mathbf{g}^j$ are

$$\text{trace}(\mathbf{g}_i \otimes \mathbf{g}_j) = g_{ij}, \qquad \text{trace}(\mathbf{g}^i \otimes \mathbf{g}^j) = g^{ij}, \tag{A.12.20}$$

as defined in Table A.12.1.

Three-dimensional coordinates

All notions and equations discussed previously in this section in two dimensions can be extended in a straightforward fashion to three dimensions. A system of three-dimensional nonorthogonal coordinates is shown in Figure A.12.2. Subscripts and superscripts now vary from 1 to 3, and the size of the metric tensor is 3×3. Given the covariant base vectors, the contravariant base vectors can be computed using the formula

$$\mathbf{g}^i = \frac{1}{\mathcal{J}} \epsilon_{ijk} \, \mathbf{g}_j \times \mathbf{g}_k, \qquad \mathbf{g}_j \times \mathbf{g}_k = \mathcal{J} \, \epsilon_{jki} \, \mathbf{g}^i, \tag{A.12.21}$$

where ϵ_{ijk} is the alternating tensor, and summation is implied over the repeated indices j and k. We have introduced the volumetric Jacobian associated with the covariant base,

$$\mathcal{J} \equiv \mathbf{g}_1 \cdot (\mathbf{g}_2 \times \mathbf{g}_3) = \sqrt{g}, \tag{A.12.22}$$

where

$$g \equiv \det(g_{ij}) = \frac{1}{\det(g^{ij})}. \tag{A.12.23}$$

The last equality in (A.12.22) originates from (A.5.9). Explicitly,

$$\mathbf{g}^1 = \frac{1}{\mathcal{J}} \mathbf{g}_2 \times \mathbf{g}_3, \qquad \mathbf{g}^2 = \frac{1}{\mathcal{J}} \mathbf{g}_3 \times \mathbf{g}_1, \qquad \mathbf{g}^3 = \frac{1}{\mathcal{J}} \mathbf{g}_1 \times \mathbf{g}_2. \qquad (A.12.24)$$

The contravariant base vectors can be computed from the covariant base vectors using similar expressions,

$$\mathbf{g}_i = \mathcal{J} \, \epsilon_{ijk} \, \mathbf{g}^j \times \mathbf{g}^k, \qquad \mathbf{g}^j \times \mathbf{g}^k = \frac{1}{\mathcal{J}} \epsilon_{jki} \, \mathbf{g}_i. \qquad (A.12.25)$$

Explicitly,

$$\mathbf{g}_1 = \mathcal{J} \, \mathbf{g}^2 \times \mathbf{g}^3, \qquad \mathbf{g}_2 = \mathcal{J} \, \mathbf{g}^3 \times \mathbf{g}^1, \qquad \mathbf{g}_3 = \mathcal{J} \, \mathbf{g}^1 \times \mathbf{g}^2. \qquad (A.12.26)$$

Problems

A.12.1 *Base vectors in three dimensions*

Derive the counterpart of relations (A.12.7) in three dimensions.

A.12.2 *Orthogonal coordinates*

Derive a relation between the covariant and contravariant base vectors in the case of orthogonal curvilinear coordinates in the plane discussed in Section A.8.

A.13 Vector components in nonorthogonal coordinates

Any vector, $\mathbf{a}$, can be described in terms of its contravariant components, (a^1, a^2, a^3), and associated covariant base vectors, or covariant components, (a_1, a_2, a_3), and associated contravariant base vectors, so that

$$\mathbf{a} = a^1 \, \mathbf{g}_1 + a^2 \, \mathbf{g}_2 + a^3 \, \mathbf{g}_3 = a_1 \, \mathbf{g}^1 + a_2 \, \mathbf{g}^2 + a_3 \, \mathbf{g}^3. \qquad (A.13.1)$$

We may write

$$\mathbf{a} = a^i \, \mathbf{g}_i = a_i \, \mathbf{g}^i, \qquad (A.13.2)$$

where summation is implied over the repeated index, i. Using the biorthonormality of the base vectors, we find that

$$\begin{aligned} a^1 &= \mathbf{a} \cdot \mathbf{g}^1, \quad a^2 = \mathbf{a} \cdot \mathbf{g}^2, \quad a^3 = \mathbf{a} \cdot \mathbf{g}^2, \\ a_1 &= \mathbf{a} \cdot \mathbf{g}_1, \quad a_2 = \mathbf{a} \cdot \mathbf{g}_2, \quad a_3 = \mathbf{a} \cdot \mathbf{g}_3. \end{aligned} \qquad (A.13.3)$$

Taking the inner product of (A.13.1) with each of the covariant vectors, $\mathbf{g}_1$, $\mathbf{g}_2$, and $\mathbf{g}_2$, we find that

$$\begin{aligned} a_1 &= a^1 \, g_{11} + a^2 \, g_{21} + a^3 \, g_{31}, \qquad a_2 = a^1 \, g_{12} + a^2 \, g_{22} + a^3 \, g_{32}, \\ a_3 &= a^1 \, g_{13} + a^2 \, g_{23} + a^3 \, g_{33}. \end{aligned} \qquad (A.13.4)$$

Taking the inner product of (A.13.1) with each one of the contravariant vectors, $\mathbf{g}^1$, $\mathbf{g}^2$, and $\mathbf{g}^2$, we find that

$$a^1 = a_1\,g^{11} + a_2\,g^{21} + a_3\,g^{31}, \qquad a^2 = a_1\,g^{12} + a_2\,g^{22} + a_3\,g^{32}$$
$$a^3 = a_1\,g^{13} + a_2\,g^{23} + a_3\,g^{33}. \tag{A.13.5}$$

We may then write

$$a_i = a^j\,g_{ji}, \qquad a^i = a_j\,g^{ji}, \tag{A.13.6}$$

where summation is implied over the repeated index, j. Expressions (A.13.6) are used to raise or lower the indices.

Inner vector product

Using the biorthogonality of the base vectors, we find that the inner product of two vectors, $\mathbf{a}$ and $\mathbf{b}$, is the scalar

$$\mathbf{a} \cdot \mathbf{b} = a^i b_i = a_i b^i, \tag{A.13.7}$$

where summation is implied over the repeated index, i. The square of the length of a vector, $\mathbf{a}$, is

$$|\mathbf{a}|^2 \equiv \mathbf{a} \cdot \mathbf{a} = a^i a_i, \tag{A.13.8}$$

where summation is implied over the repeated index, i.

Outer vector product

The inner product of two vectors $\mathbf{a}$ and $\mathbf{b}$ is a new vector given by

$$\mathbf{a} \times \mathbf{b} = (a^i \mathbf{g}_i) \times (b^j \mathbf{g}_j) = (a_i \mathbf{g}^i) \times (b_j \mathbf{g}^j), \tag{A.13.9}$$

where summation is implied over the repeated indices, i and j. Using relations (A.12.21) and (A.12.25), we obtain

$$\mathbf{a} \times \mathbf{b} = \mathcal{J}\,\epsilon_{ijk}\,a^i b^j\,\mathbf{g}^k = \frac{1}{\mathcal{J}}\,\epsilon_{ijk}\,a_i b_j\,\mathbf{g}_k, \tag{A.13.10}$$

where ϵ_{ijk} is the alternating tensor.

Gradient of a scalar function

Consider a scalar function of position, $f(\mathbf{x})$. The gradient of the function, ∇f, is a vector defined such that its projection on $\mathbf{g}_i$ or $\mathbf{g}^i$ provides us with the rate of change in the respective direction,

$$\mathbf{g}_i \cdot (\nabla f) = \frac{\partial f}{\partial x^i}, \qquad \mathbf{g}^i \cdot (\nabla f) = \frac{\partial f}{\partial x_i}. \tag{A.13.11}$$

Based on these definitions, we deduce that the derivatives, $\partial f / \partial x^i$, are the covariant components of the gradient, whereas the derivatives, $\partial f / \partial x_i$, are the contravariant components of the gradient,

$$(\nabla f)_i = \frac{\partial f}{\partial x^i}, \qquad (\nabla f)^i = \frac{\partial f}{\partial x_i}. \tag{A.13.12}$$

We thus write

$$\nabla f = \frac{\partial f}{\partial x_1} \mathbf{g}_1 + \frac{\partial f}{\partial x_2} \mathbf{g}_2 + \frac{\partial f}{\partial x_3} \mathbf{g}_3, \qquad \nabla f = \frac{\partial f}{\partial x^1} \mathbf{g}^1 + \frac{\partial f}{\partial x^2} \mathbf{g}^2 + \frac{\partial f}{\partial x^3} \mathbf{g}^3, \qquad (A.13.13)$$

and observe that the derivatives with respect to contravariant coordinates, x^i, produce covariant components. To signify this rule, sometimes the gradient is regrettably called a covariant vector.

Convective derivative

Using the biorthonormality relation between the covariant and contravariant base vectors, we find that the convective derivative in two dimensions can be computed either from the expression

$$\mathbf{u} \cdot \nabla f = (u^1 \mathbf{g}_1 + u^2 \mathbf{g}_2) \cdot (\mathbf{g}^1 \frac{\partial f}{\partial x^1} + \mathbf{g}^2 \frac{\partial f}{\partial x^2}) = u^1 \frac{\partial f}{\partial x^1} + u^2 \frac{\partial f}{\partial x^2}, \qquad (A.13.14)$$

or from the expression

$$\mathbf{u} \cdot \nabla f = (u_1 \mathbf{g}^1 + u_2 \mathbf{g}^2) \cdot (\mathbf{g}_1 \frac{\partial f}{\partial x_1} + \mathbf{g}_2 \frac{\partial f}{\partial x_2}) = u_1 \frac{\partial f}{\partial x_1} + u_2 \frac{\partial f}{\partial x_2}, \qquad (A.13.15)$$

where $\mathbf{u}$ represents the velocity. Similar expressions can be written in three dimensions, yielding

$$\mathbf{u} \cdot \nabla f = u^i \frac{\partial f}{\partial x^i} = u_i \frac{\partial f}{\partial x_i}, \qquad (A.13.16)$$

where summation is implied over the repeated index, i.

Gradient operator

Motivated by (A.13.13), we introduce the gradient operator,

$$\nabla = \mathbf{g}_k \frac{\partial}{\partial x_k} = \mathbf{g}^k \frac{\partial}{\partial x^k}, \qquad (A.13.17)$$

where summation is implied over the repeated index, k. Explicitly in three dimensions,

$$\nabla = \mathbf{g}_1 \frac{\partial}{\partial x_1} + \mathbf{g}_2 \frac{\partial}{\partial x_2} + \mathbf{g}_3 \frac{\partial}{\partial x_3} = \mathbf{g}^1 \frac{\partial}{\partial x^1} + \mathbf{g}^2 \frac{\partial}{\partial x^2} + \mathbf{g}^3 \frac{\partial}{\partial x^3}. \qquad (A.13.18)$$

The base vectors and derivative operators may not be transposed.

Numerical computation

As an application, we consider the computation of the gradient of a scalar function, ∇f, at a point in a plane. We are given the values of the function at nodes distributed along two rectilinear axes passing through that point, as shown in Figure A.13.1. For convenience, we denote $x^1 = \xi$ and $x^2 = \eta$. A suitable algorithm involves the following steps:

1. Compute the covariant base vectors by finite-difference approximations

$$\mathbf{g}_\xi \equiv \frac{\partial \mathbf{x}}{\partial \xi} \simeq \frac{\mathbf{x}_R - \mathbf{x}_L}{\xi_R - \xi_L}, \qquad \mathbf{g}_\eta \equiv \frac{\partial \mathbf{x}}{\partial \eta} \simeq \frac{\mathbf{x}_T - \mathbf{x}_B}{\eta_T - \eta_B}, \qquad (A.13.19)$$

where L, R, T, B stand for left, right, top, and bottom nodes relative to the intersection.

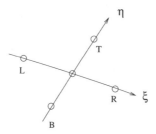

FIGURE A.13.1   Illustration of two curvilinear coordinates supporting data passing through the central point.

2. Compute the corresponding contravariant base vectors, $\mathbf{g}^\xi$, $\mathbf{g}^\eta$, using the formulas discussed in this section.

3. Compute the covariant derivatives by finite-difference approximations,

$$\frac{\partial f}{\partial \xi} \simeq \frac{f_R - f_L}{\xi_R - \xi_L}, \qquad \frac{\partial f}{\partial \eta} \simeq \frac{f_T - f_B}{\eta_T - \eta_B}. \tag{A.13.20}$$

4. Compute the gradient, ∇f, using the second expression in (A.13.13).

Exactly the same results are obtained if we compute the gradient by solving a system of two equations for two unknowns originating from the chain rule,

$$\begin{bmatrix} \partial x/\partial \xi & \partial y/\partial \xi \\ \partial x/\partial \eta & \partial y/\partial \eta \end{bmatrix} \cdot \nabla f = \begin{bmatrix} \partial f/\partial \xi \\ \partial f/\partial \eta \end{bmatrix} \tag{A.13.21}$$

Problem

A.13.1 *Vector components*

Consider a vector in the xy plane. Can this vector be specified by one contravariant and one covariant component?

A.14   Christoffel symbols of the second kind

The derivatives of the covariant base vectors, $\mathbf{g}_i$, with respect to the contravariant coordinates, x_j, are vectors themselves. By definition,

$$\frac{\partial \mathbf{g}_i}{\partial x^j} \equiv \Gamma_{ij}^k \, \mathbf{g}_k, \tag{A.14.1}$$

where Γ_{ij}^k are the Christoffel symbols of the second kind and summation is implied over the repeated index, k. Using the biorthonormality of the covariant and contravariant base vectors, we obtain

$$\Gamma_{ij}^k \equiv \left\{ \begin{matrix} k \\ i \, j \end{matrix} \right\} = \frac{\partial \mathbf{g}_i}{\partial x^j} \cdot \mathbf{g}^k. \tag{A.14.2}$$

Using the biorthonormality of the base vectors once more, we find that $\partial(\mathbf{g}_i \cdot \mathbf{g}^k)/\partial x^j = 0$. Expanding the derivative, we obtain

$$\frac{\partial \mathbf{g}^k}{\partial x^j} \cdot \mathbf{g}_i = -\frac{\partial \mathbf{g}_i}{\partial x^j} \cdot \mathbf{g}^k = -\Gamma^k_{ij}, \tag{A.14.3}$$

which shows that

$$\frac{\partial \mathbf{g}^k}{\partial x^j} = -\Gamma^k_{ij}\,\mathbf{g}^i. \tag{A.14.4}$$

Using the definition (A.14.2), we write

$$\Gamma^k_{ij} = \frac{\partial \mathbf{g}_i}{\partial x^j} \cdot \mathbf{g}^k = \frac{\partial \mathbf{x}}{\partial^i \partial x^j} \cdot \mathbf{g}^k = \frac{\partial \mathbf{g}_j}{\partial x^i} \cdot \mathbf{g}^k, \tag{A.14.5}$$

which reveals the symmetry property

$$\Gamma^k_{ij} = \Gamma^k_{ji}. \tag{A.14.6}$$

Christoffel symbols in terms of the metric tensor

Next, we compute the derivatives

$$\frac{\partial g_{mi}}{\partial x^j} = \frac{\partial(\mathbf{g}_m \cdot \mathbf{g}_i)}{\partial x^j} = \mathbf{g}_m \cdot \frac{\partial \mathbf{g}_i}{\partial x^j} + \mathbf{g}_i \cdot \frac{\partial \mathbf{g}_m}{\partial x^j} = \Gamma^k_{ij}\,\mathbf{g}_m \cdot \mathbf{g}_k + \Gamma^k_{mj}\,\mathbf{g}_i \cdot \mathbf{g}_k, \tag{A.14.7}$$

yielding

$$\frac{\partial g_{mi}}{\partial x^j} = \Gamma^k_{ij}\,g_{mk} + \Gamma^k_{mj}\,g_{ik}, \tag{A.14.8}$$

where summation is implied over the repeated index, k. Multiplying the last equation by g^{mp}, where p is a free index, summing over m, and noting that $g_{mk}g^{mp} = \delta_{kp}$, we obtain

$$g^{mp}\,\frac{\partial g_{mi}}{\partial x^j} = \Gamma^p_{ij} + \Gamma^k_{mj}\,g_{ik}\,g^{mp}. \tag{A.14.9}$$

Switching the indices i and j, we obtain

$$g^{mp}\,\frac{\partial g_{mj}}{\partial x^i} = \Gamma^p_{ij} + \Gamma^k_{mi}\,g_{jk}\,g^{mp}. \tag{A.14.10}$$

Combining the last three equations, we find that the Christoffel symbols of the second kind can be computed in terms of the metric tensor from the expression

$$\Gamma^k_{ij} = \frac{1}{2}\,g^{km}\left(\frac{\partial g_{mi}}{\partial x^j} + \frac{\partial g_{mj}}{\partial x^i} - \frac{\partial g_{ij}}{\partial x^m}\right), \tag{A.14.11}$$

where summation is implied over the repeated index, m.

Gradient of a vector field

Consider a vector function of position, $\mathbf{f}(\mathbf{x})$. The right gradient of this vector field,

$$\mathbf{M} \equiv (\nabla \mathbf{f})^T, \tag{A.14.12}$$

is a matrix defined such that its *projection from the right* on $\mathbf{g}_i$ or $\mathbf{g}^i$ provides us with the rate of change of $\mathbf{f}$ in the respective directions,

$$\mathbf{M} \cdot \mathbf{g}_i = \frac{\partial \mathbf{f}}{\partial x^i}, \qquad \mathbf{M} \cdot \mathbf{g}^i = \frac{\partial \mathbf{f}}{\partial x_i}. \tag{A.14.13}$$

By analogy with (A.13.13) for the gradient of a scalar, we introduce the covariant and contravariant expansions

$$\mathbf{M} = \frac{\partial \mathbf{f}}{\partial x^1} \otimes \mathbf{g}^1 + \frac{\partial \mathbf{f}}{\partial x^2} \otimes \mathbf{g}^2 + \frac{\partial \mathbf{f}}{\partial x^3} \otimes \mathbf{g}^3,$$

$$\mathbf{M} = \frac{\partial \mathbf{f}}{\partial x_1} \otimes \mathbf{g}_1 + \frac{\partial \mathbf{f}}{\partial x_2} \otimes \mathbf{g}_2 + \frac{\partial \mathbf{f}}{\partial x_3} \otimes \mathbf{g}_3. \tag{A.14.14}$$

Expressing the vector field $\mathbf{f}$ in terms of the covariant or contravariant base vectors and carrying out the differentiations, we obtain four combinations. The first combination is

$$\mathbf{M} = \frac{\partial (f^i \, \mathbf{g}_i)}{\partial x^j} \otimes \mathbf{g}^j = \left(\frac{\partial f^i}{\partial x^j} + \Gamma^i_{kj} \, f^k \right) \mathbf{g}_i \otimes \mathbf{g}^j. \tag{A.14.15}$$

The second combination is

$$\mathbf{M} = \frac{\partial (f_i \, \mathbf{g}^i)}{\partial x^j} \otimes \mathbf{g}^j = \left(\frac{\partial f_i}{\partial x^j} - \Gamma^k_{ij} \, f_k \right) \mathbf{g}^i \otimes \mathbf{g}^j. \tag{A.14.16}$$

The third combination is

$$\mathbf{M} = \frac{\partial (f^i \, \mathbf{g}_i)}{\partial x_j} \otimes \mathbf{g}_j = \frac{\partial f^i}{\partial x_j} \, \mathbf{g}_i \otimes \mathbf{g}_j + f^i \, \frac{\partial \mathbf{g}_i}{\partial x_j} \otimes \mathbf{g}_j. \tag{A.14.17}$$

The fourth combination is

$$\mathbf{M} = \frac{\partial (f_i \, \mathbf{g}^i)}{\partial x_j} \otimes \mathbf{g}_j = \frac{\partial f_i}{\partial x_j} \, \mathbf{g}^i \otimes \mathbf{g}_j + f_i \, \frac{\partial \mathbf{g}^i}{\partial x_j} \otimes \mathbf{g}_j. \tag{A.14.18}$$

Summation over the repeated indices i and j is implied in all four combinations. In solid mechanics, continuum mechanics, and applied mathematics, the right gradient $\mathbf{M} = (\nabla \mathbf{f})^T$ is simply called the gradient of the vector field $\mathbf{f}$.

In fluid mechanics, the gradient of the vector field $\mathbf{f}$ is identified with the left gradient,

$$\mathbf{L} = \nabla \mathbf{f} = \mathbf{M}^T. \tag{A.14.19}$$

An example is the velocity gradient $\mathbf{L} = \nabla \mathbf{u}$ discussed in Section 1.1.2. Using (A.14.15), we find that

$$\mathbf{L} = \left(\frac{\partial f^i}{\partial x^j} + \Gamma^i_{kj} \, f^k \right) \mathbf{g}^j \otimes \mathbf{g}_i \equiv f^i_{,j} \, \mathbf{g}^j \otimes \mathbf{g}_i, \tag{A.14.20}$$

where $f^i_{,j}$ is the first covariant derivative. Similar representations can be derived from expressions (A.14.16)–(A.14.18).

As an example, using (A.14.20), we find that the convective derivative of the velocity on the left-hand side on the equation of motion is

$$\mathbf{u} \cdot \nabla \mathbf{u} = (u^m \, \mathbf{g}_m) \cdot \left(\frac{\partial u^i}{\partial x^j} + \Gamma^i_{kj} \, u^k \right) \mathbf{g}^j \otimes \mathbf{g}_i = u^j \left(\frac{\partial u^i}{\partial x^j} + \Gamma^i_{kj} \, u^k \right) \mathbf{g}_i = u^j u^i_{,j} \, \mathbf{g}_i. \tag{A.14.21}$$

The term $u^j u^i_{,j}$ is the ith covariant component of the convective derivative.

Divergence of a vector field

The divergence of a vector field, $\mathbf{f}$, is the trace of the gradient, $\nabla \mathbf{f}$, or its transpose, $(\nabla \mathbf{f})^T$. Using (A.14.15) and (A.14.16), we find that

$$\nabla \cdot \mathbf{f} = \text{trace}(\mathbf{M}) = \frac{\partial f^i}{\partial x^i} + \Gamma^i_{ki} \, f^k = \left(\frac{\partial f_i}{\partial x^j} - \Gamma^k_{ij} \, f_k \right) g^{ij}. \tag{A.14.22}$$

Making substitutions, we obtain

$$\nabla \cdot \mathbf{f} = \frac{\partial f^i}{\partial x^i} + \frac{1}{2} \, g^{mi} \frac{\partial g_{im}}{\partial x^k} \, f^k. \tag{A.14.23}$$

Further manipulation gives

$$\nabla \cdot \mathbf{f} = \frac{1}{\sqrt{g}} \frac{\partial}{\partial x^i} \left(f^i \sqrt{g} \right), \tag{A.14.24}$$

where $g = \det(g_{ij})$.

Invoking the vector representation of the gradient operator in (A.13.17), we write

$$\nabla \cdot \mathbf{f} = \left(\mathbf{g}^i \cdot \frac{\partial}{\partial x^i} \right) (f^j \mathbf{g}_j), \tag{A.14.25}$$

where summation is implied over the repeated indices, i and j. Explicitly,

$$\nabla \cdot \mathbf{f} = \left(\mathbf{g}^1 \cdot \frac{\partial}{\partial x^1} + \mathbf{g}^2 \cdot \frac{\partial}{\partial x^2} + \mathbf{g}^3 \cdot \frac{\partial}{\partial x^3} \right) \left(f^1 \mathbf{g}_1 + f^2 \mathbf{g}_2 + f^3 \mathbf{g}_3 \right). \tag{A.14.26}$$

Carrying out the differentiations and using (A.14.1), we recover (A.14.22).

Laplacian of a scalar field

The Laplacian of a scalar field, f, is the divergence of the gradient, $\nabla^2 f = \nabla \cdot (\nabla f)$. Setting in (A.14.24) $\mathbf{f} = \nabla f$, $f^i = \partial f / \partial x_i$ and $f_i = \partial f / \partial x^i$, we find that

$$\nabla^2 f = \frac{1}{\sqrt{g}} \frac{\partial}{\partial x^i} \left(g^{ki} \frac{\partial f}{\partial x^k} \sqrt{g} \right), \tag{A.14.27}$$

where $g = \det(g_{ij})$.

Curl of a vector field

Invoking the vector representation of the gradient operator in (A.13.17), we compute the curl of a vector field, $\mathbf{f}$, as

$$\boldsymbol{\omega} \equiv \nabla \times \mathbf{f} = \left(\mathbf{g}^i \times \frac{\partial}{\partial x^i} \right)(f_j \, \mathbf{g}^j), \tag{A.14.28}$$

where summation is implied over the repeated indices, i and j. Explicitly,

$$\boldsymbol{\omega} = \left(\mathbf{g}^1 \times \frac{\partial}{\partial x^1} + \mathbf{g}^2 \times \frac{\partial}{\partial x^2} + \mathbf{g}^3 \times \frac{\partial}{\partial x^3} \right) \left(f_1 \mathbf{g}^1 + f_2 \mathbf{g}^2 + f_3 \mathbf{g}^3 \right). \tag{A.14.29}$$

Carrying out the differentiations, we obtain

$$\boldsymbol{\omega} = \frac{\partial f_1}{\partial x^2} (\mathbf{g}^2 \times \mathbf{g}^1) + \frac{\partial f_1}{\partial x^3} (\mathbf{g}^3 \times \mathbf{g}^1) + f^1 \left(\mathbf{g}^1 \times \frac{\partial \mathbf{g}^1}{\partial x^1} + \mathbf{g}^2 \times \frac{\partial \mathbf{g}^1}{\partial x^2} + \mathbf{g}^3 \times \frac{\partial \mathbf{g}^1}{\partial x^3} \right) + \cdots, \tag{A.14.30}$$

where the dots indicate similar terms involving f^2 and f^3. Using (A.12.25) and (A.14.4), we obtain

$$\boldsymbol{\omega} = \frac{1}{\mathcal{J}} \left(-\frac{\partial f_1}{\partial x^2} \mathbf{g}_3 + \frac{\partial f_1}{\partial x^3} \mathbf{g}_2 \right) - f^1 \left[\left(\Gamma^1_{i1} \mathbf{g}^1 + \Gamma^1_{i2} \mathbf{g}^2 + \Gamma^1_{i3} \mathbf{g}^3 \right) \times \mathbf{g}^i \right] + \cdots, \tag{A.14.31}$$

where summation is implied over the repeated index, i. Using (A.12.25), we find that the term enclosed by the square brackets on the right-hand side of (A.14.31) is identically zero, yielding the simple expression

$$\boldsymbol{\omega} = \frac{1}{\mathcal{J}} \epsilon_{ijk} \frac{\partial f_k}{\partial x^j} \mathbf{g}_k. \tag{A.14.32}$$

Laplacian of a vector field

The Laplacian of a vector field, $\mathbf{f}$ is another vector field defined by $\nabla^2 \mathbf{f} = \nabla \cdot (\nabla \mathbf{f})$. Using (A.14.20) and the vector representation of the gradient operator, we find that

$$\nabla^2 \mathbf{f} = \left(\mathbf{g}^m \cdot \frac{\partial}{\partial x^m} \right) \left[\left(\frac{\partial f^i}{\partial x^j} + \Gamma^i_{kj} f^k \right) \mathbf{g}^j \otimes \mathbf{g}_i \right]. \tag{A.14.33}$$

Expanding the derivatives, we obtain

$$\nabla^2 \mathbf{f} = (\mathbf{g}^m \cdot \mathbf{g}^j) \frac{\partial}{\partial x^m} \left(\frac{\partial f^i}{\partial x^j} + \Gamma^i_{kj} f^k \right) \mathbf{g}_i + \left(\frac{\partial f^i}{\partial x^j} + \Gamma^i_{kj} f^k \right) \left(\mathbf{g}^m \cdot \frac{\partial \mathbf{g}^j \otimes \mathbf{g}_i}{\partial x^m} \right). \tag{A.14.34}$$

τ^{ij}	contravariant components
τ_{ij}	covariant components
$\tau^{i}_{\ j}$	mixed right-covariant components
$\tau_{i}^{\ j}$	mixed left-covariant components

TABLE A.15.1   Possible components of a two-index tensor in contravariant, covariant, and mixed forms.

Carrying out the differentiations and simplifying, we obtain

$$\nabla^2 \mathbf{f} = g^{kj} f^{i}_{,jk} \, \mathbf{g}_i, \tag{A.14.35}$$

where

$$f^{i}_{,jk} = \frac{\partial^2 f^i}{\partial x^k \partial x^j} + \Gamma^{i}_{jl} \frac{\partial f^l}{\partial x^k} + \Gamma^{i}_{lk} \frac{\partial f^l}{\partial x^j} - \Gamma^{l}_{jk} \frac{\partial f^i}{\partial x^l} + \Big(\frac{\partial \Gamma^{i}_{jl}}{\partial x^k} + \Gamma^{i}_{mk} \Gamma^{m}_{jl} - \Gamma^{m}_{jk} \Gamma^{i}_{ml} \Big) f^l \tag{A.14.36}$$

is the covariant second derivative. This expression can be used to evaluate the viscous force on the right-hand side of the Navier–Stokes equation (e.g., [417]).

Problem

A.14.1 *Christoffel symbols*

Derive expression (A.14.11).

A.15   Components of a second-order tensor in nonorthogonal coordinates

Expressions (A.14.15)–(A.14.18) for the gradient of a vector field illustrate that an arbitrary second-order tensor, $\boldsymbol{\tau}$, can be represented in four different ways as

$$\begin{aligned} \boldsymbol{\tau} &= \tau^{i}_{\ j} \, \mathbf{g}_i \otimes \mathbf{g}^j, \qquad & \boldsymbol{\tau} &= \tau_{i}^{\ j} \, \mathbf{g}^i \otimes \mathbf{g}_j, \\ \boldsymbol{\tau} &= \tau^{ij} \, \mathbf{g}_i \otimes \mathbf{g}_j, \qquad & \boldsymbol{\tau} &= \tau_{ij} \, \mathbf{g}^i \otimes \mathbf{g}^j. \end{aligned} \tag{A.15.1}$$

The scalar coefficients representing the pure contravariant, the pure covariant, and the mixed components are named as shown in Table A.15.1.

Repeating our previous analysis for vectors, we find that the various tensor components derive from one another by the relations

$$\begin{aligned} \tau^{j}_{\ k} &= g_{ki} \tau^{ij}, \qquad & \tau^{i}_{\ k} &= \tau^{ij} g_{jk}, \qquad & \tau_{lk} &= g_{li} \tau^{ij} g_{jk}, \qquad & \tau^{k}_{\ j} &= g^{ki} \tau_{ij}, \tag{A.15.2} \\ \tau_{i}^{\ k} &= \tau_{ij} g^{jk}, \qquad & \tau^{lk} &= g^{il} \tau_{ij} g^{jk}, \qquad & \tau_{kj} &= g_{ki} \tau^{i}_{\ j}, \qquad & \tau_{il} &= \tau_{i}^{\ j} g_{jl}, \end{aligned}$$

where summation is implied over the repeated indices, k and l. Cursory inspection reveals obvious rules for raising and lowering the indices.

Divergence of a second-order tensor

Departing from the the last expression in (A.15.1) and using (A.13.17), we find that

$$\nabla \cdot \boldsymbol{\tau} = \mathbf{g}^k \cdot \frac{\partial}{\partial x^k} \left(\tau_{ij}\, \mathbf{g}^i \otimes \mathbf{g}^j \right) = \frac{\partial \tau_{ij}}{\partial x^k}\, g^{ik}\, \mathbf{g}^j + \tau_{ij} \left((\mathbf{g}^k \cdot \frac{\partial \mathbf{g}^i}{\partial x^k})\, \mathbf{g}^j + (\mathbf{g}^k \cdot \mathbf{g}^i)\, \frac{\partial \mathbf{g}^j}{\partial x^k} \right). \quad (A.15.3)$$

Computing the derivatives of the base vectors using (A.14.4), we obtain

$$\nabla \cdot \boldsymbol{\tau} = \frac{\partial \tau_{ij}}{\partial x^k}\, g^{ik}\, \mathbf{g}^j - \tau_{ij} \left(g^{mk} \Gamma^i_{mk}\, \mathbf{g}^j + g^{ik}\, \Gamma^j_{mk}\, \mathbf{g}^m \right), \quad (A.15.4)$$

which can be recast into the form

$$\nabla \cdot \boldsymbol{\tau} = \left(\frac{\partial \tau_{ij}}{\partial x^k} - \Gamma^m_{ik}\, \tau_{mj} - \Gamma^m_{jk}\, \tau_{im} \right) g^{ik}\, \mathbf{g}^j. \quad (A.15.5)$$

Working in a similar fashion, we find the alternative forms

$$\nabla \cdot \boldsymbol{\tau} = \left(\frac{\partial \tau^{ij}}{\partial x^i} + \Gamma^i_{im}\, \tau^{mj} + \Gamma^j_{im}\, \tau^{im} \right) \mathbf{g}_j,$$

$$\nabla \cdot \boldsymbol{\tau} = \left(\frac{\partial \tau^i_{\ j}}{\partial x^i} + \Gamma^i_{im}\, \tau^m_{\ j} - \Gamma^m_{ij}\, \tau^i_{\ m} \right) \mathbf{g}^j, \quad (A.15.6)$$

$$\nabla \cdot \boldsymbol{\tau} = \left(\frac{\partial \tau_i^{\ j}}{\partial x^k} - \Gamma^m_{ik}\, \tau_m^{\ j} + \Gamma^j_{km}\, \tau_i^{\ m} \right) g^{ik}\, \mathbf{g}^j.$$

The right-hand sides of the last four equations are vectors expressed in covariant or contravariant form.

Problem

A.15.1 *Divergence of a tensor*
Derive expressions (A.15.6).

A.15.2 *Gradient of a tensor*
Derive expressions for the gradient of a tensor in a suitable base of your choice.

A.16   Rectilinear coordinates with constant base vectors

As an example, we consider rectilinear nonorthogonal coordinates in the plane and assume that the covariant base consists of two dimensionless constant vectors,

$$\mathbf{g}_1 = \begin{bmatrix} 1 \\ 0 \end{bmatrix}, \qquad \mathbf{g}_2 = \begin{bmatrix} 1 \\ 1 \end{bmatrix}, \quad (A.16.1)$$

where the square brackets enclose the Cartesian coordinates, as illustrated in Figure A.16.1. The covariant metric tensor is

$$g_{ij} = \begin{bmatrix} 1 & 1 \\ 1 & 2 \end{bmatrix}. \quad (A.16.2)$$

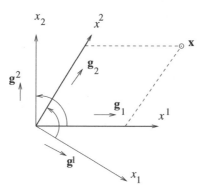

FIGURE A.16.1   Illustration of two-dimensional rectilinear nonorthogonal coordinates in the plane.

The contravariant base vectors computed from (A.12.9) are

$$\mathbf{g}^1 = \begin{bmatrix} 1 \\ -1 \end{bmatrix}, \qquad \mathbf{g}^2 = \begin{bmatrix} 0 \\ 1 \end{bmatrix}. \tag{A.16.3}$$

The contravariant metric tensor is

$$g^{ij} = \begin{bmatrix} 2 & -1 \\ -1 & 1 \end{bmatrix}. \tag{A.16.4}$$

We can easily confirm that g^{ij} is the inverse of g_{ij} and *vice versa*.

Next, we formulate the matrices

$$\mathbf{g}_1 \otimes \mathbf{g}^1 = \begin{bmatrix} 1 & -1 \\ 0 & 0 \end{bmatrix}, \quad \mathbf{g}_1 \otimes \mathbf{g}^2 = \begin{bmatrix} 0 & 1 \\ 0 & 0 \end{bmatrix},$$
$$\mathbf{g}_2 \otimes \mathbf{g}^1 = \begin{bmatrix} 1 & -1 \\ 1 & -1 \end{bmatrix}, \quad \mathbf{g}_2 \otimes \mathbf{g}^2 = \begin{bmatrix} 0 & 1 \\ 0 & 1 \end{bmatrix}, \tag{A.16.5}$$

The traces of $\mathbf{g}_1 \otimes \mathbf{g}^1$ and $\mathbf{g}_2 \otimes \mathbf{g}^2$ are unity, whereas the traces of $\mathbf{g}_1 \otimes \mathbf{g}^2$ and $\mathbf{g}_2 \otimes \mathbf{g}^1$ are zero.

The covariant, contravariant, and Cartesian axes share a common origin. The contravariant and covariant coordinates of the position vector with Cartesian coordinates $\mathbf{x} = [x, y]$ are

$$x^1 = \mathbf{x} \cdot \mathbf{g}^1 = x - y, \quad x^2 = \mathbf{x} \cdot \mathbf{g}^2 = y, \quad x_1 = \mathbf{x} \cdot \mathbf{g}_1 = x, \quad x_2 = \mathbf{x} \cdot \mathbf{g}_2 = x + y. \tag{A.16.6}$$

Conversely,

$$x = x^1 + x^2 = x_1, \qquad y = x^2 = -x_1 + x_2. \tag{A.16.7}$$

If we know that contravariant or covariant coordinates, we can easily compute the Cartesian coordinates.

Gradient of a scalar

Now we consider a scalar field,

$$f(x,y) = 2x + y = 2x^1 + 3x^2 = x_1 + x_2, \tag{A.16.8}$$

where x and y are Cartesian coordinates. In Cartesian coordinates, the gradient of this field is $\nabla f = [2,1]$. Straightforward differentiation yields

$$\frac{\partial f}{\partial x^1} = 2, \qquad \frac{\partial f}{\partial x^2} = 3, \qquad \frac{\partial f}{\partial x_1} = 1, \qquad \frac{\partial f}{\partial x_2} = 1, \tag{A.16.9}$$

which is consistent with (A.13.13).

Vector field

Next, we consider a vector field,

$$\mathbf{u} = \begin{bmatrix} 2x + y \\ x - 2y \end{bmatrix} = (x + 3y)\,\mathbf{g}_1 + (x - 2y)\,\mathbf{g}_2 = (2x + y)\,\mathbf{g}^1 + (3x - y)\,\mathbf{g}^2. \tag{A.16.10}$$

The contravariant and covariant components are

$$u^1 = x + 3y = x^1 + 4x^2, \qquad u^2 = x - 2y = x^1 - x^2,$$
$$u_1 = 2x + y = x_1 + x_2, \qquad u_2 = 3x - y = 4x_1 - x_2. \tag{A.16.11}$$

Using (A.14.22), we find that $\nabla \cdot \mathbf{u} = 0$.

The gradient of the vector field is

$$\nabla \mathbf{u} = \frac{\partial u^1}{\partial x^1}\,\mathbf{g}_1 \otimes \mathbf{g}^1 + \frac{\partial u^1}{\partial x^2}\,\mathbf{g}_1 \otimes \mathbf{g}^2 + \frac{\partial u^2}{\partial x^1}\,\mathbf{g}_2 \otimes \mathbf{g}^1 + \frac{\partial u^2}{\partial x^2}\,\mathbf{g}_2 \otimes \mathbf{g}^2, \tag{A.16.12}$$

yielding

$$\nabla \mathbf{u} = \mathbf{g}_1 \otimes \mathbf{g}^1 + 4\,\mathbf{g}_1 \otimes \mathbf{g}^2 + \mathbf{g}_2 \otimes \mathbf{g}^1 - \mathbf{g}_2 \otimes \mathbf{g}^2 = \begin{bmatrix} 2 & 1 \\ 1 & -2 \end{bmatrix}. \tag{A.16.13}$$

Because $\mathbf{u}$ is solenoidal, the trace of this matrix zero.

Problems

A.16.1 *Rectilinear coordinates*

Repeat the calculations in this section for base vectors

$$\mathbf{g}_1 = \begin{bmatrix} 2 \\ 0 \end{bmatrix}, \qquad \mathbf{g}_2 = \begin{bmatrix} 1 \\ 2 \end{bmatrix}. \tag{A.16.14}$$

A.16.2 *Rectilinear coordinates in three dimensions*

Repeat the calculations in this section for nonorthogonal three-dimensional rectilinear coordinates of your choice.

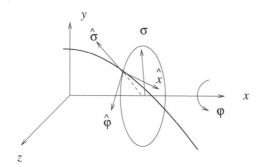

FIGURE A.17.1   Illustration of helical coordinates, $(\hat{x}, \hat{\sigma}, \hat{\varphi})$, in relation to companion Cartesian and cylindrical polar coordinates, (x, σ, φ).

A.17   Helical coordinates

Nonorthogonal helical coordinates are employed when the structure of a scalar or vector field of interest is invariant along a helical path. Examples include fluid flow in a tube with helical corrugations, flow through a tube with a helical centerline, and flow induced by a helical line vortex. In these applications, a point in space is identified by the helical curvilinear coordinates, $(\hat{\sigma}, \hat{\varphi}, \hat{x})$, defined in Figure A.17.1, which are related to the cylindrical polar coordinates, (x, σ, φ), by

$$\sigma = \hat{\sigma}, \qquad \varphi = \hat{\varphi} + \alpha\,\hat{x}, \qquad x = \hat{x}, \qquad (A.17.1)$$

and to the associated Cartesian coordinates by

$$x = \hat{x}, \qquad y = \hat{\sigma}\,\cos(\hat{\varphi} + \alpha\,\hat{x}), \qquad z = \hat{\sigma}\,\sin(\hat{\varphi} + \alpha\,\hat{x}), \qquad (A.17.2)$$

where $\alpha \equiv 2\pi/L$ is the helical wave number and L is the helical pitch. In the limit of infinite pitch, $\alpha \to 0$, the helical coordinates reduce to cylindrical polar coordinates. In problems with helical symmetry, the partial derivative of a variable of interest, f, with respect to $\hat{x}$ is zero, so that $f(\hat{\sigma}, \hat{\varphi})$.

Setting $x^1 = \hat{\sigma}$, $x^2 = \hat{\varphi}$, and $x^3 = \hat{x}$, we compute the base vectors

$$\mathbf{g}_1 \equiv \frac{\partial \mathbf{x}}{\partial x^1} = \frac{\partial \mathbf{x}}{\partial \hat{\sigma}} = \begin{bmatrix} 0 \\ \cos(\hat{\varphi} + \alpha\,\hat{x}) \\ \sin(\hat{\varphi} + \alpha\,\hat{x}) \end{bmatrix}, \qquad \mathbf{g}_2 \equiv \frac{\partial \mathbf{x}}{\partial x^2} = \frac{\partial \mathbf{x}}{\partial \hat{\varphi}} = \hat{\sigma} \begin{bmatrix} 0 \\ -\sin(\hat{\varphi} + \alpha\,\hat{x}) \\ \cos(\hat{\varphi} + \alpha\,\hat{x}) \end{bmatrix},$$

$$\mathbf{g}_3 \equiv \frac{\partial \mathbf{x}}{\partial x^3} = \frac{\partial \mathbf{x}}{\partial \hat{x}} = \begin{bmatrix} 1 \\ -\alpha\hat{\sigma}\sin(\hat{\varphi} + \alpha\,\hat{x}) \\ \alpha\hat{\sigma}\cos(\hat{\varphi} + \alpha\,\hat{x}) \end{bmatrix}. \qquad (A.17.3)$$

The covariant metric tensor is

$$[g_{ij}] = [\mathbf{g}_i \cdot \mathbf{g}_j] = \begin{bmatrix} 1 & 0 & 0 \\ 0 & \hat{\sigma}^2 & \alpha\,\hat{\sigma}^2 \\ 0 & \alpha\,\hat{\sigma}^2 & 1 + \alpha^2\,\hat{\sigma}^2 \end{bmatrix}. \qquad (A.17.4)$$

The contravariant metric tensor is the inverse of the covariant tensor,

$$[g^{ij}] = \begin{bmatrix} 1 & 0 & 0 \\ 0 & \hat{\sigma}^{-2} + \alpha^2 & -\alpha \\ 0 & -\alpha & 1 \end{bmatrix}. \tag{A.17.5}$$

The presence of nondiagonal elements indicates that the helical coordinates are orthogonal only in the limit of infinite pitch, $\alpha \to 0$. The only nonzero Christoffel symbols of the second kind are

$$\Gamma^2_{21} = \frac{1}{\hat{\sigma}}, \qquad \Gamma^2_{31} = \frac{\alpha}{\hat{\sigma}}, \qquad \Gamma^1_{22} = -\hat{\sigma}, \qquad \Gamma^1_{32} = -\alpha\,\hat{\sigma},$$

$$\Gamma^2_{12} = \frac{1}{\hat{\sigma}}, \qquad \Gamma^1_{23} = -\alpha\,\hat{\sigma}, \qquad \Gamma^1_{33} = -\alpha^2\,\hat{\sigma}, \qquad \Gamma^2_{13} = \frac{\alpha}{\hat{\sigma}}. \tag{A.17.6}$$

(e.g., [411]).

A vector field, $\mathbf{u}$, can be resolved into physical components corresponding to polar cylindrical and associated helical coordinates,

$$\mathbf{u} = u_\sigma\,\mathbf{e}_\sigma + u_\varphi\,\mathbf{e}_\varphi + u_x\,\mathbf{e}_x = u_{\hat{\sigma}}\,\mathbf{e}_{\hat{\sigma}} + u_{\hat{\varphi}}\,\mathbf{e}_{\hat{\varphi}} + u_{\hat{x}}\,\mathbf{e}_{\hat{x}}, \tag{A.17.7}$$

where

$$\mathbf{e}_{\hat{\sigma}} = \frac{1}{\sqrt{g_{11}}}\,\mathbf{g}_1 = \mathbf{g}_1, \qquad \mathbf{e}_{\hat{\varphi}} = \frac{1}{\sqrt{g_{22}}}\,\mathbf{g}_2 = \frac{1}{\hat{\sigma}}\,\mathbf{g}_2, \qquad \mathbf{e}_{\hat{x}} = \frac{1}{\sqrt{g_{33}}}\,\mathbf{g}_3 = \frac{1}{\sqrt{1 + \alpha^2\hat{\sigma}^2}}\,\mathbf{g}_3 \tag{A.17.8}$$

are position-dependent unit vectors. Projecting (A.17.7) onto $\mathbf{e}_\sigma$, $\mathbf{e}_\varphi$, or $\mathbf{e}_x$, we find that

$$u_\sigma = u_{\hat{\sigma}}, \qquad u_\varphi = u_{\hat{\varphi}} + \frac{\alpha\,\hat{\sigma}}{\sqrt{1 + \alpha^2\hat{\sigma}^2}}\,u_{\hat{x}} = u_{\hat{\varphi}} + \alpha\,\hat{\sigma}\,u_x, \qquad u_x = \frac{1}{\sqrt{1 + \alpha^2\hat{\sigma}^2}}\,u_{\hat{x}}. \tag{A.17.9}$$

If the vector field $\mathbf{u}$ is helically symmetric, the cylindrical polar as well as the helical components of $\mathbf{u}$ are independent of $\hat{x}$, and only depend on $\hat{\sigma}$ and $\hat{\varphi}$.

The contravariant components of $\mathbf{u}$ corresponding to the helical coordinates, defined such that $\mathbf{u} = u^1\mathbf{g}_1 + u^2\mathbf{g}_2 + u^3\mathbf{g}_3$, are given by

$$u^1 = \frac{u_{\hat{\sigma}}}{\sqrt{g_{11}}} = u_{\hat{\sigma}}, \qquad u^2 = \frac{u_{\hat{\varphi}}}{\sqrt{g_{22}}} = \frac{u_{\hat{\varphi}}}{\hat{\sigma}}, \qquad u^3 = \frac{u_{\hat{x}}}{\sqrt{g_{33}}} = \frac{u_{\hat{x}}}{\sqrt{1 + \alpha^2\hat{\sigma}^2}} = u_x. \tag{A.17.10}$$

In the case of helical symmetry, the contravariant components are independent of $\hat{x}$, and only depend on $\hat{\sigma}$ and $\hat{\varphi}$.

Problem

A.17.1 *Helical coordinates*

Alternative helical coordinates $(\tilde{\sigma}, \tilde{\varphi}, \tilde{x})$ can be defined such that the cylindrical polar coordinates are $\sigma = \tilde{\sigma}$, $\varphi = \tilde{\varphi}$, and $x = \tilde{x} + \frac{1}{\alpha}\tilde{\varphi}$ (e.g., [439]). The associated Cartesian coordinates are

$$x = \tilde{x} + \frac{1}{\alpha}\tilde{\varphi}, \quad y = \tilde{\sigma}\cos\tilde{\varphi}, \quad z = \tilde{\sigma}\sin\tilde{\varphi}. \tag{A.17.11}$$

Sketch the coordinate lines and compute the corresponding covariant and contravariant metric tensors.

B

Primer of numerical methods

A summary of methods employed in numerical and computational fluid dynamics is presented in this appendix. Further information can be found in two highly recommended texts on numerical methods and numerical analysis [18, 317].

B.1 Linear algebraic equations

We seek to compute an N-dimensional vector, $\mathbf{x}$, which, when multiplied by a given $N \times N$ square matrix, $\mathbf{A}$, yields a known vector, $\mathbf{b}$, so that $\mathbf{A} \cdot \mathbf{x} = \mathbf{b}$. In our discussion, summation will not be implied over a repeated index; instead, it will be stated explicitly, as required.

B.1.1 Diagonal and triangular systems

When the matrix $\mathbf{A}$ is diagonal, the unknown vector $\mathbf{x}$ can be computed by the simple algorithm

$$x_i = \frac{b_i}{A_{i,i}} \tag{B.1.1}$$

for $i = 1, \ldots, N$, where summation is *not* implied over the repeated index, i. For clarity, a comma was inserted between the two matrix indices.

When the matrix $\mathbf{A}$ is lower triangular, we use the forward substitution algorithm

$$x_1 = \frac{b_1}{A_{1,1}}, \qquad x_i = \frac{1}{A_{i,i}} \left(b_i - \sum_{j=1}^{i-1} A_{i,j} x_j \right) \tag{B.1.2}$$

for $i = 2, \ldots, N$.

When the matrix $\mathbf{A}$ is upper triangular, we use the backward substitution algorithm

$$x_N = \frac{b_N}{A_{N,N}}, \qquad x_i = \frac{1}{A_{i,i}} \left(b_i - \sum_{j=i+1}^{N} A_{i,j} x_j \right) \tag{B.1.3}$$

for $i = N - 1, \ldots, 1$. Note that the last unknown, x_N, is computed first.

B.1.2   Gauss elimination and LU decomposition

When the matrix $\mathbf{A}$ does not have a recognized structure, the solution can be found by direct or iterative methods. The latter are designated for systems of large size involving sparse matrices with many zeros.

Gauss elimination is the simplest and most popular direct method. The basic idea is to solve the first equation of the given system, $\mathbf{A} \cdot \mathbf{x} = \mathbf{b}$, for the first unknown, x_1, and use the expression thus obtained to eliminate x_1 from all subsequent equations. We then retain the first equation as is, and replace all subsequent equations with their descendants that do not contain x_1. At the second stage, we solve the second equation for the second unknown, x_2, and use the expression thus obtained to eliminate x_2 from all subsequent equations. We then retain the first and second equations, and replace all subsequent equations with their descendants that do not contain x_1 or x_2. Continuing in this fashion, we arrive at the last equation, which contains only the last unknown, x_N.

Having completed the elimination, we compute the unknowns by the method of backward substitution. First, we solve the last equation for x_N, which thus becomes a known. Second, we solve the penultimate equation for x_{N-1}, which also becomes a known. Continuing in the backward direction, we scan the reduced system until we have computed all unknowns.

Pivoting

Immediately before the mth equation has been solved for the mth unknown in the process of elimination, where $m = 1, 2, \ldots, N - 1$, the linear system takes the form

$$
\begin{bmatrix}
A_{1,1}^{(m)} & A_{1,2}^{(m)} & \cdots & & \cdots & & \cdots & A_{1,N}^{(m)} \\
0 & A_{2,2}^{(m)} & \cdots & & \cdots & & \cdots & A_{2,N}^{(m)} \\
0 & 0 & \cdots & & \cdots & & \cdots & \cdots \\
0 & 0 & A_{m-1,m-1}^{(m)} & A_{m-1,m}^{(m)} & \cdots & A_{m-1,N}^{(m)} \\
0 & \cdots & 0 & A_{m,m}^{(m)} & \cdots & A_{m,N}^{(m)} \\
0 & \cdots & 0 & \cdots & \cdots & \cdots \\
0 & \cdots & 0 & A_{N,m}^{(m)} & \cdots & A_{N,N}^{(m)}
\end{bmatrix}
\cdot
\begin{bmatrix}
x_1 \\ x_2 \\ x_3 \\ \vdots \\ x_{N-1} \\ x_N
\end{bmatrix}
=
\begin{bmatrix}
b_1^{(m)} \\ b_2^{(m)} \\ b_3^{(m)} \\ \vdots \\ b_{N-1}^{(m)} \\ b_N^{(m)}
\end{bmatrix},
\tag{B.1.4}
$$

where $A_{i,j}^{(m)}$ are intermediate coefficients and $b_i^{(m)}$ are intermediate right-hand sides. The first equation of (B.1.4) is identical to the first equation of the given linear system for any value of m. Subsequent equations are different, except at the first step corresponding to $m = 1$.

A difficulty arises when the diagonal element, $A_{m,m}^{(m)}$, is nearly or precisely zero, since we are unable to solve the mth equation for x_m, as required. However, the failure of the method does not imply that the system does not have a unique solution. To circumvent this difficulty, we simply rearrange the equations or relabel the unknowns so as to bring the mth unknown to the mth equation using the method of pivoting. If there is no way we can make this happen, the matrix $\mathbf{A}$ is singular and the linear system has either no solution or an infinite number of solutions.

In the method of row pivoting, potential difficulties are bypassed by switching the mth equation in the system (B.1.4) with the subsequent kth equation, where $k > m$; the value of k is chosen so that $|A_{k,m}^{(m)}|$ is the maximum value of the elements in the mth column below the diagonal, $A_{i,m}^{(m)}$, for $i \geq m$. If $A_{i,m}^{(m)} = 0$ for all $i \geq m$, the matrix $\mathbf{A}$ is singular and the system under consideration does not have a unique solution.

Algorithm with row pivoting

To implement the method of Gauss elimination with row pivoting, we work according to the following steps:

Setting up

Formulate the $N \times (N+1)$ partitioned augmented matrix

$$\mathbf{C}^{(1)} \equiv \left[\, \mathbf{A} \,\middle|\, \mathbf{b} \,\right], \tag{B.1.5}$$

and introduce an $N \times N$ matrix, $\mathbf{L}$, whose elements are initialized to zero.

First pass

1. Find the location of the element with the maximum norm in the first column of $\mathbf{C}^{(1)}$. This is done by searching for the maximum norm of the elements $|C_{i,1}^{(1)}|$, for $i = 1, \ldots, N$. Assume that this is equal to $|C_{k,1}^{(1)}|$, corresponding to the kth row.

2. Skip this step if $k = 1$. Otherwise, switch the first row with the kth row of $\mathbf{C}^{(1)}$; repeat for the matrix $\mathbf{L}$.

3. Compute the first column of $\mathbf{L}$ below the diagonal by setting $L_{i,1} = C_{i,1}^{(1)}/C_{1,1}^{(1)}$, for $i = 2, \ldots, N$.

4. Subtract from the ith row of $\mathbf{C}^{(1)}$ the first row multiplied by $L_{i,1}$, for $i = 2, \ldots, N$, to obtain a new augmented matrix,

$$\mathbf{C}^{(2)} \equiv \left[\mathbf{A}^{(2)}|\mathbf{b}^{(2)}\right]. \tag{B.1.6}$$

Second pass

1. Find the location of the element with the maximum norm in the second column of $\mathbf{C}^{(2)}$, below the diagonal. This is done by searching for the maximum norm of the elements $|C_{i,2}^{(2)}|$, for $i = 2, \ldots, N$. Assume that this is equal to $|C_{k,2}^{(2)}|$, corresponding to the kth row.

2. Skip this step if $k = 2$. Otherwise, switch the second row with the kth row of $\mathbf{C}^{(2)}$; repeat for the matrix $\mathbf{L}$.

3. Compute the second column of $\mathbf{L}$ below the diagonal, by setting $L_{i,2} = C_{i,2}^{(2)}/C_{2,2}^{(2)}$, for $i = 3, \ldots, N$.

4. Subtract from the ith row of $\mathbf{C}^{(2)}$ the second row multiplied by $L_{i,2}$, for $i = 3, \ldots, N$, to obtain a new augmented matrix,

$$\mathbf{C}^{(3)} \equiv \left[\mathbf{A}^{(3)}|\mathbf{b}^{(3)}\right]. \tag{B.1.7}$$

. . .

*m*th *pass*

1. Find the location of the element with the maximum norm in the *m*th column of $\mathbf{C}^{(m)}$, below the diagonal. This is done by searching for the maximum norm of the elements $|C_{i,m}^{(m)}|$, for $i = m, \ldots, N$. Assume that this is equal to $|C_{k,m}^{(m)}|$, corresponding to the *k*th row.

2. Skip this step if $k = m$. Otherwise, switch the *m*th row with the *k*th row of $\mathbf{C}^{(m)}$; repeat for the matrix $\mathbf{L}$.

3. Compute the *m*th column of $\mathbf{L}$ below the diagonal, by setting $L_{i,m} = C_{i,m}^{(m)}/C_{m,m}^{(m)}$, for $i = m+1, \ldots, N$.

4. Subtract from the *i*th row of $\mathbf{C}^{(m)}$ the *m*th row multiplied by $L_{i,m}$, for $i = m+1, \ldots, N$, to obtain a new augmented matrix,

$$\mathbf{C}^{(m+1)} \equiv \left[\mathbf{A}^{(m+1)} | \mathbf{b}^{(m+1)}\right]. \tag{B.1.8}$$

...

$(N-1)$ *pass*

At the end of the $N-1$ pass, corresponding to $m = N-1$, the augmented matrix, $\mathbf{C}^{(N)}$, has the form

$$\mathbf{C}^{(N)} = \left[\mathbf{A}^{(N)} \mid \mathbf{b}^{(N)}\right], \tag{B.1.9}$$

where $\mathbf{A}^{(N)} \equiv \mathbf{U}$ is an upper triangular matrix.

Backward substitution

Finally, we solve by backward substitution the upper triangular system

$$\mathbf{U} \cdot \mathbf{x} = \mathbf{b}^{(N)} \tag{B.1.10}$$

to extract the solution of the original system of equations, $\mathbf{A} \cdot \mathbf{x} = \mathbf{b}$. This is done by solving the last equation in (B.1.10) for the last unknown, x_N; once this is available, we solve the penultimate equation for x_{N-1}; continuing backward in this fashion, we finally compute x_1.

Complete the matrix $\mathbf{L}$ *(optional)*

Set the diagonal elements of the matrix $\mathbf{L}$ equal to 1.

LU decomposition

It can be shown by straightforward algebraic manipulations that the matrices $\mathbf{L}$ and $\mathbf{U}$ provide us with the *LU* decomposition of the matrix $\mathbf{A}$. Specifically,

$$\mathbf{L} \cdot \mathbf{U} = \mathbf{A}^{mod}, \tag{B.1.11}$$

where the matrix $\mathbf{A}^{mod}$ is identical to $\mathbf{A}$, except that the rows may have been switched due to pivoting. If pivoting is disabled, $\mathbf{A}^{mod} = \mathbf{A}$.

Determinant

The determinant of the matrix $\mathbf{A}$ follows from the relation

$$\det(\mathbf{A}) = \pm \det(\mathbf{A}^{mod}) = \pm \det(\mathbf{L}) \cdot \det(\mathbf{U}) = \pm U_{1,1} U_{2,2} \cdots U_{N,N}, \tag{B.1.12}$$

where the plus sign applies when an even number of row interchanges have been made due to pivoting, and the minus sign otherwise. In the absence of pivoting, the plus sign is chosen.

Other algorithms for performing the LU decomposition of a matrix directly without reference to a particular system of linear equations are available. Examples are Crout's and Doolittle's algorithms [18, 318]. Once the $\mathbf{L}$ and $\mathbf{U}$ matrices have been obtained, the solution of the linear system $\mathbf{A} \cdot \mathbf{x} = \mathbf{b}$ is found in two stages: first, we solve the lower triangular system $\mathbf{L} \cdot \mathbf{y} = \mathbf{b}$ for $\mathbf{y}$ by forward substitution; second, we solve the upper triangular system $\mathbf{U} \cdot \mathbf{x} = \mathbf{y}$ for $\mathbf{x}$ by backward substitution.

B.1.3   Thomas algorithm for tridiagonal systems

The Thomas algorithm is used for solving an $N \times N$ tridiagonal system,

$$\mathbf{D} \cdot \mathbf{x} = \mathbf{s}, \tag{B.1.13}$$

where $\mathbf{s}$ is a given vector. The matrix $\mathbf{D}$ has the tridiagonal form

$$\mathbf{D} = \begin{bmatrix} a_1 & b_1 & 0 & 0 & 0 & \cdots & 0 & 0 & 0 \\ c_2 & a_2 & b_2 & 0 & 0 & \cdots & 0 & 0 & 0 \\ 0 & c_3 & a_3 & b_3 & 0 & \cdots & 0 & 0 & 0 \\ \cdots & \cdots & \cdots & \cdots & \cdots & \cdots & \cdots & \cdots & \cdots \\ 0 & 0 & 0 & 0 & 0 & \cdots & c_{N-1} & a_{N-1} & b_{N-1} \\ 0 & 0 & 0 & 0 & 0 & \cdots & 0 & c_N & a_N \end{bmatrix}. \tag{B.1.14}$$

The Thomas algorithm proceeds in two stages. At the first stage, the tridiagonal system (B.1.13) is transformed into a bidiagonal system,

$$\mathbf{D}' \cdot \mathbf{x} = \mathbf{y}, \tag{B.1.15}$$

involving the bidiagonal coefficient matrix

$$\mathbf{D}' = \begin{bmatrix} 1 & d_1 & 0 & 0 & 0 & \cdots & 0 & 0 & 0 \\ 0 & 1 & d_2 & 0 & 0 & \cdots & 0 & 0 & 0 \\ 0 & 0 & 1 & d_3 & 0 & \cdots & 0 & 0 & 0 \\ \cdots & \cdots & \cdots & \cdots & \cdots & \cdots & \cdots & \cdots & \cdots \\ 0 & 0 & 0 & 0 & 0 & \cdots & 0 & 1 & d_{N-1} \\ 0 & 0 & 0 & 0 & 0 & \cdots & 0 & 0 & 1 \end{bmatrix}. \tag{B.1.16}$$

At the second stage, the bidiagonal system (B.1.15) is solved by backward substitution. The algorithm involves solving the last equation for the last unknown, x_N, and then moving upward to compute the rest of the unknowns in a sequential fashion. The combined algorithm is listed in Table B.1.1.

Reduction to bidiagonal:

$$\begin{bmatrix} d_1 \\ y_1 \end{bmatrix} = \frac{1}{a_1} \begin{bmatrix} b_1 \\ s_1 \end{bmatrix}$$

Do $i = 1, 2, \ldots, N - 1$

$$\begin{bmatrix} d_{i+1} \\ y_{i+1} \end{bmatrix} = \frac{1}{a_{i+1} - c_{i+1} d_i} \begin{bmatrix} b_{i+1} \\ s_{i+1} - c_{i+1} y_i \end{bmatrix}$$

End Do

Backward substitution:

$x_N = y_N$

Do $i = N - 1, N - 2, \ldots, 1$ (step$=-1$)

$x_i = y_i - d_i\, x_{i+1}$

End Do

TABLE B.1.1   The Thomas algorithm for solving a tridiagonal system of linear equations.

B.1.4   Fixed-point iterations and successive substitutions

Fixed-point iteration or successive substitution algorithms arise by recasting the original linear system $\mathbf{A} \cdot \mathbf{x} = \mathbf{b}$ into the system

$$\mathbf{M} \cdot \mathbf{x} = \mathbf{N} \cdot \mathbf{x} + \mathbf{b}, \tag{B.1.17}$$

where $\mathbf{A} = \mathbf{M} - \mathbf{N}$ is a splitting of the matrix $\mathbf{A}$, We then write

$$\mathbf{x} = \mathbf{P} \cdot \mathbf{x} + \mathbf{c}, \tag{B.1.18}$$

where

$$\mathbf{P} = \mathbf{M}^{-1} \cdot \mathbf{N}, \qquad \mathbf{c} = \mathbf{M}^{-1} \cdot \mathbf{b}. \tag{B.1.19}$$

The algorithm involves computing the sequence of vectors $\mathbf{x}^{(k)}$ based on the equation

$$\mathbf{M} \cdot \mathbf{x}^{(k+1)} = \mathbf{N} \cdot \mathbf{x}^{(k)} + \mathbf{b}, \tag{B.1.20}$$

or on the equation

$$\mathbf{x}^{(k+1)} = \mathbf{P} \cdot \mathbf{x}^{(k)} + \mathbf{c}, \tag{B.1.21}$$

beginning with a certain initial vector, $\mathbf{x}^{(0)}$. If the spectral radius of the projection matrix $\mathbf{P}$ is less than unity, that is, the magnitude of all eigenvalues of $\mathbf{P}$ is less than one, the sequence of vectors will converge to the fixed point of the mapping, denoted by $\mathbf{X}$, that satisfies the equation $\mathbf{X} = \mathbf{P} \cdot \mathbf{X} + \mathbf{c}$ and therefore the given equation $\mathbf{A} \cdot \mathbf{X} = \mathbf{b}$. In that case, the sequence $\mathbf{x}^{(k)}$ will contain increasingly accurate successive approximations to the solution.

Jacobi's method

In Jacobi's method, the iteration matrix, $\mathbf{P}$, and constant vector, $\mathbf{c}$, are constructed by solving the individual scalar equations of the given system, $\mathbf{A} \cdot \mathbf{x} = \mathbf{b}$, for the diagonal unknowns, obtaining the iteration formula

$$x_i^{(k+1)} = \frac{1}{A_{i,i}} \left(b_i - \sum_{j=1}^{N}{}' A_{i,j}\, x_j^{(k)} \right), \tag{B.1.22}$$

where the prime denotes that the term $i = j$ is excluded from the sum. The matrix $\mathbf{M}$ defined in (B.1.17) is the diagonal component of $\mathbf{A}$, and the matrix $\mathbf{N}$ is the negative of $\mathbf{A}$ with the diagonal elements set equal to zero. The components of the projection matrix are $P_{i,j} = -A_{i,j}/A_{i,i}$ if $i \neq j$, $P_{i,j} = 0$ if $i = j$, and $c_i = -b_i/A_{i,i}$, where summation is not implied over the repeated index, i.

A sufficient but not necessary condition for the successive substitutions to converge is that the matrix $\mathbf{A}$ is diagonally dominant,

$$|A_{i,i}| > \sum_{j=1}^{N}{}' |A_{i,j}| \tag{B.1.23}$$

for any i, where summation is *not* implied the repeated index, i, and the prime denotes that the term $i = j$ is excluded from the sum.

Gauss–Seidel method

The Gauss–Seidel method is a variation of Jacobi's method with the new feature that the components of $\mathbf{x}^{(k+1)}$ replace the corresponding components of $\mathbf{x}^{(k)}$ as soon as the former are available. The effective algorithm is

$$x_i^{(k+1)} = \frac{1}{A_{i,i}} \left(b_i - \sum_{j=1}^{i-1} A_{i,j}\, x_j^{(k+1)} - \sum_{j=i+1}^{N} A_{i,j}\, x_j^{(k)} \right). \tag{B.1.24}$$

Let $\mathbf{D}$ be a diagonal matrix, $\mathbf{L}$ be a lower triangular, and $\mathbf{U}$ be an upper triangular matrix, containing, respectively, the diagonal, lower triangular, and upper triangular components of $\mathbf{A}$. The Gauss–Seidel mapping takes the form

$$(\mathbf{D} + \mathbf{L})\, \mathbf{x}^{(k+1)} = -\mathbf{U}\, \mathbf{x}^{(k)} + \mathbf{b}, \tag{B.1.25}$$

where

$$\mathbf{M} = \mathbf{D} + \mathbf{L}, \qquad \mathbf{N} = -\mathbf{U}. \tag{B.1.26}$$

A sufficient but not necessary condition for the successive substitutions to converge is that the matrix $\mathbf{A}$ is symmetric and positive definite.

Successive over-relaxation

The successive over-relaxation method (SOR) is constructed by restating the Gauss–Seidel iteration algorithm in the residual correction form $\mathbf{x}^{(k+1)} = \mathbf{x}^{(k)} + \mathbf{r}^{(k)}$, where

$$\mathbf{r}^{(k)} \equiv (\mathbf{P} - \mathbf{I}) \cdot \mathbf{x}^{(k)} + \mathbf{c} \tag{B.1.27}$$

is the residual correction. The correction is controlled by introducing a relaxation parameter, ω, and setting

$$\mathbf{x}^{(k+1)} = \mathbf{x}^{(k)} + \omega \mathbf{r}^{(k)} = (1 - \omega)\,\mathbf{x}^{(k)} + \omega\big(\mathbf{P} \cdot \mathbf{x}^{(k)} + \mathbf{c}\big). \tag{B.1.28}$$

The method is implemented in terms of the modified Gauss–Seidel algorithm

$$x_i^{(k+1)} = (1 - \omega)\,x_i^{(k)} + \frac{\omega}{A_{i,i}} \left(b_i - \sum_{j=1}^{i-1} A_{i,j}\, x_j^{(k+1)} - \sum_{j=i+1}^{N} A_{i,j}\, x_j^{(k)} \right). \tag{B.1.29}$$

The corresponding iteration formula is

$$(\mathbf{D} + \omega \mathbf{L}) \cdot \mathbf{x}^{(k+1)} = \big[\,(1 - \omega)\,\mathbf{D} - \omega \mathbf{U}\,\big] \cdot \mathbf{x}^{(k)} + \omega \mathbf{b}. \tag{B.1.30}$$

When $\omega = 1$, we obtain the Gauss–Seidel algorithm. A necessary condition for the successive substitutions to converge is that $0 < \omega < 2$. Ideally, ω should take the value that minimizes the spectral radius of the underlying projection matrix, $\mathbf{P}$. The optimal value can be constructed during the iterations.

B.1.5   Minimization and search methods

When the coefficient matrix $\mathbf{A}$ is symmetric and positive definite, computing the solution of the linear system $\mathbf{A} \cdot \mathbf{x} = \mathbf{b}$ is equivalent to finding a vector $\mathbf{x}$ that minimizes the quadratic functional

$$\mathcal{F}(\mathbf{x}) = \frac{1}{2}\,\mathbf{x} \cdot \mathbf{A} \cdot \mathbf{x} - \mathbf{b} \cdot \mathbf{x}. \tag{B.1.31}$$

If the matrix $\mathbf{A}$ is not symmetric, we can multiply the equation $\mathbf{A} \cdot \mathbf{x} = \mathbf{b}$ by the transpose of $\mathbf{A}$ to obtain the preconditioned system $\mathbf{B} \cdot \mathbf{x} = \mathbf{c}$, where $\mathbf{B} = \mathbf{A}^T \cdot \mathbf{A}$ is symmetric and positive definite, and $\mathbf{c} = \mathbf{A}^T \cdot \mathbf{b}$.

Steepest decent search

In the method of steepest decent, an initial guess is made and then improved by stepping in the direction where the quadratic functional appears to change most rapidly. The initial steepest-descent direction is aligned with the residual vector

$$\mathbf{r} \equiv -\nabla \mathcal{F} = -\mathbf{A} \cdot \mathbf{x} + \mathbf{b}, \tag{B.1.32}$$

evaluated at the initial guess, $\mathbf{x}^{(0)}$,

$$\mathbf{r}^{(0)} \equiv -\mathbf{A} \cdot \mathbf{x}^{(0)} + \mathbf{b}. \tag{B.1.33}$$

Thus, the first search direction vector is

$$\mathbf{p}^{(1)} = \mathbf{r}^{(0)}. \tag{B.1.34}$$

To compute the travel distance, we note that, as we travel in the direction of $\mathbf{r}^{(0)}$, the vector $\mathbf{x}$ is described by $\mathbf{x} = \mathbf{x}^{(0)} + \alpha\,\mathbf{r}^{(0)}$, where α is a scalar parameter. The value of the quadratic functional along the travel path is a quadratic function of α,

$$\mathcal{F}(\mathbf{x}) = \frac{1}{2}\,(\mathbf{x}^{(0)} + \alpha\,\mathbf{r}^{(0)}) \cdot \mathbf{A} \cdot (\mathbf{x}^{(0)} + \alpha\,\mathbf{r}^{(0)}) - \mathbf{b} \cdot (\mathbf{x}^{(0)} + \alpha\,\mathbf{r}^{(0)}). \tag{B.1.35}$$

We want to stop traveling when $\mathcal{F}$ has reached a minimum, $\partial\mathcal{F}/\partial\alpha = 0$. Taking the partial derivative of the right-hand side of (B.1.35) with respect to α and setting the resulting expression to zero, we find the optimal value

$$\alpha_1 = \frac{\mathbf{r}^{(0)} \cdot \mathbf{r}^{(0)}}{\mathbf{r}^{(0)} \cdot (\mathbf{A} \cdot \mathbf{r}^{(0)})}, \tag{B.1.36}$$

which yields the improved position

$$\mathbf{x}^{(1)} = \mathbf{x}^{(0)} + \alpha_1\,\mathbf{r}^{(0)}. \tag{B.1.37}$$

The minimization process is subsequently repeated in the search direction vector

$$\mathbf{p}^{(k)} = \mathbf{r}^{(k-1)} \equiv -\mathbf{A} \cdot \mathbf{x}^{(k-1)} + \mathbf{b} \tag{B.1.38}$$

so that $\mathbf{x}^{(k)} = \mathbf{x}^{(k-1)} + \alpha_k\,\mathbf{p}^{(k)}$ with

$$\alpha_k = \frac{\mathbf{p}^{(k)} \cdot \mathbf{p}^{(k)}}{\mathbf{p}^{(k)} \cdot (\mathbf{A} \cdot \mathbf{p}^{(k)})}, \tag{B.1.39}$$

until the minimum has been reached within a specified tolerance. While the method is guaranteed to converge, the rate of convergence can be slow.

Directional search

In this method, the minimization problem is solved by selecting a set of search directions expressed by the N-dimensional vectors $\mathbf{p}^{(1)}, \mathbf{p}^{(2)}, \mathbf{p}^{(3)}, \ldots$, and then stepwise advancing an initial guess, $\mathbf{x}^{(0)}$, by setting

$$\mathbf{x}^{(k)} = \mathbf{x}^{(k-1)} + \alpha_k\,\mathbf{p}^{(k)}, \tag{B.1.40}$$

for $k = 1, \ldots$, where α_k are appropriate coefficients determining the length of each step. The evolved solution at the end of the kth step is

$$\mathbf{x}^{(k)} = \mathbf{x}^{(0)} + \sum_{i=1}^{k} \alpha_i\,\mathbf{p}^{(i)}, \tag{B.1.41}$$

for $k > 0$. Without loss of generality, we can begin the search from the origin, $\mathbf{x}^{(0)} = \mathbf{0}$. Our goal is to compute the search directions so that the exact solution, $\mathbf{X}$, is found exactly after N steps,

$$\mathbf{X} = \alpha_1 \, \mathbf{p}^{(1)} + \alpha_2 \, \mathbf{p}^{(2)} + \cdots + \alpha_N \, \mathbf{p}^{(N)}, \qquad (B.1.42)$$

where N is the system size.

As we travel in the search direction $\mathbf{p}^{(k)}$, the vector $\mathbf{x}$ is described by $\mathbf{x} = \mathbf{x}^{(k-1)} + \alpha_k \, \mathbf{p}^{(k)}$, and the quadratic functional along the path is given by

$$\mathcal{F}(\mathbf{x}) = \frac{1}{2} \left(\mathbf{x}^{(k-1)} + \alpha_k \, \mathbf{p}^{(k)} \right) \cdot \left(\mathbf{A} \cdot \left(\mathbf{x}^{(k-1)} + \alpha_k \, \mathbf{p}^{(k)} \right) \right) - \mathbf{b} \cdot \left(\mathbf{x}^{(k-1)} + \alpha_k \, \mathbf{p}^{(k)} \right), \qquad (B.1.43)$$

which is a quadratic function of α_k. Setting $\partial \mathcal{F} / \partial \alpha_k = 0$ to identify the optimal stopping point, we find that

$$\alpha_k = \frac{\mathbf{p}^{(k)} \cdot \mathbf{r}^{(k-1)}}{\mathbf{p}^{(k)} \cdot \left(\mathbf{A} \cdot \mathbf{p}^{(k)} \right)}, \qquad (B.1.44)$$

which is consistent with (B.1.39).

We have considerable freedom in selecting the search directions. In the steepest descend search discussed earlier in this section, we have stipulated that $\mathbf{p}^{(k)} = \mathbf{r}^{(k-1)}$. However, when the graph of the functional exhibits narrow valleys, the search directions tend to align and successive points bounce off opposite sides. Better methods for selecting the search directions are available.

In the method of conjugate gradients, the search vectors are computed so that they are A-conjugate with each other, that is, $\mathbf{p}^{(i)} \cdot \mathbf{A} \cdot \mathbf{p}^{(j)} = 0$ for $i \neq j$. The direction $\mathbf{p}^{(k)}$ is aligned as much as possible with $\mathbf{p}^{(k-1)}$, subject to the A-conjugation constraint. The basic algorithm is shown in Table B.1.1.

B.1.6 Other methods

Powerful iterative methods based on variational principles and residual minimization techniques are available. In the method of conjugate gradients, the residuals form an orthogonal set in the Krylov space. In the method of generalized minimal residuals (GMRES), an orthonormal base consisting of a set of vectors, $\mathbf{v}^{(i)}$, is constructed explicitly at every step using Gram–Schmidt orthogonalization. Specifically, the GMRES sequence is constructed using the formula

$$\mathbf{x}^{(k)} = \mathbf{x}^{(0)} + \alpha_1 \, \mathbf{v}^{(1)} + \alpha_2 \, \mathbf{v}^{(2)} + \cdots + \alpha_k \, \mathbf{v}^{(k)}, \qquad (B.1.45)$$

where the coefficients, α_i, are chosen to minimize the norm of the residual $|\mathbf{A} \cdot \mathbf{x}^{(k)} - \mathbf{b}|$ at every step.

B.2 Matrix eigenvalues

The set of eigenvalues of a matrix is called the spectrum of the matrix. The eigenvalues of a diagonal, upper triangular, or lower triangular matrix are the diagonal elements. Multiple eigenvalues correspond to repeated elements. The eigenvalues of a tridiagonal matrix whose diagonal, subdiagonal, and superdiagonal elements are constant are known in closed flow, as discussed in Problem 12.2.6.

$\mathbf{x}^{(0)} = \mathbf{0}$
Do $k = 1, \dots, N$
    If $k = 1$
        $\mathbf{p}^{(1)} = \mathbf{r}^{(0)} = \mathbf{b}$
    Else If $k > 1$
        $\beta_k = \dfrac{\mathbf{r}^{(k-1)} \cdot \mathbf{r}^{(k-1)}}{\mathbf{r}^{(k-2)} \cdot \mathbf{r}^{(k-2)}}$
        $\mathbf{p}^{(k)} = \mathbf{r}^{(k-1)} + \beta_k \, \mathbf{p}^{(k-1)}$
    End if
    $\alpha_k = \dfrac{\mathbf{r}^{(k-1)} \cdot \mathbf{r}^{(k-1)}}{\mathbf{p}^{(k)} \cdot (\mathbf{A} \cdot \mathbf{p}^{(k)})}$
    $\mathbf{x}^{(k)} = \mathbf{x}^{(k-1)} + \alpha_k \, \mathbf{p}^{(k)}$
    $\mathbf{r}^{(k)} = \mathbf{r}^{(k-1)} - \alpha_k \, \mathbf{A} \cdot \mathbf{p}^{(k)}$
End Do

TABLE B.1.1 A conjugate gradients algorithm for solving a linear system, $\mathbf{A} \cdot \mathbf{x} = \mathbf{b}$, with a real, symmetric, and positive definite-coefficient matrix, $\mathbf{A}$.

Gerschgorin's theorem locates the eigenvalues of a general $N \times N$ matrix, $\mathbf{A}$, inside the union of N disks in the complex plane. The ith disk is centered at the diagonal element, $A_{i,i}$, and the corresponding radius is equal to the minimum of

$$\sum_{i=1}^{N}{}' |A_{i,j}|, \qquad \sum_{j=1}^{N}{}' |A_{i,j}|, \qquad (\mathrm{B.2.1})$$

where the prime denotes that the term $i = j$ is excluded from the sum. Other methods for locating eigenvalues are available [429]. Three general classes of methods for computing eigenvalues are available.

B.2.1 Roots of the characteristic polynomial

The first class of methods produces the eigenvalues of a matrix, $\mathbf{A}$, by computing the roots of the characteristic polynomial,

$$\mathcal{P}(\lambda) \equiv \det(\mathbf{A} - \lambda \mathbf{I}). \qquad (\mathrm{B.2.2})$$

This can be done using general-purpose numerical methods for solving nonlinear algebraic discussed in Section B.3. To compute $P(\lambda)$, we may perform the LU decomposition of the matrix $\mathbf{A} - \lambda \mathbf{I}$ for a trial value of λ using, for example, the method of Gauss elimination discussed in Section B.2.1.2.

Determinant of a tridiagonal matrix

When the matrix $\mathbf{A}$ is tridiagonal, the shifted matrix, $\mathbf{B} = \mathbf{A} - \lambda\mathbf{I}$, is also tridiagonal. To compute the determinant of $\mathbf{B}$, we denote the nonzero elements as $B_{i,i-1} = \gamma_i$, $B_{i,i} = \alpha_i$, $B_{i,i+1} = \beta_i$, set $P_0 = 1$ and $P_1 = \alpha_1$, and compute the sequence

$$P_i = \alpha_i P_{i-1} - \beta_{i-1}\gamma_i P_{i-2}, \tag{B.2.3}$$

for $i = 2, \ldots, N$. It can be shown by straightforward substitutions that $\mathcal{P} = \det(\mathbf{B})$.

B.2.2   Power method

The power method successively transforms a chosen vector by projecting it onto the matrix $\mathbf{A}$ until it becomes an eigenvector corresponding to the eigenvalue with the maximum norm. In the numerical implementation, we select an initial vector with unit length, $\mathbf{x}^{(0)}$, and compute a Krylov sequence based on the formula

$$\mathbf{x}^{(k+1)} = \mathbf{A} \cdot \mathbf{x}^{(k)} \tag{B.2.4}$$

for $k = 0, 1, \ldots$. The algorithm is:

```
Choose x(0)
for k = 0, 1, ...
    s = |x^(k)|
    x^(k) ← 1/s x^(k)
    x^(k+1) = A · x^(k)
    λ^(k+1) = x^(k) · x^(k+1)
end
```

If the matrix $\mathbf{A}$ is real and the dominant eigenvector is complex, a complex starting vector $\mathbf{x}^{(0)}$ must be provided.

Suppose that the eigenvalue of $\mathbf{A}$ with the maximum norm, λ_1, is available. To obtain a second eigenvalue, we may apply the power method to the singular matrix $\mathbf{B} = \mathbf{A} - \lambda_1\mathbf{I}$. Note that at least one eigenvalue of $\mathbf{B}$ is zero. Since the eigenvalues of $\mathbf{B}$ are shifted with respect to those of $\mathbf{A}$ by λ_1, the dominant eigenvalue has been moved to the origin and does not play a role in the iterations. The power method produces the eigenvalue of $\mathbf{B}$ with the maximum norm, λ_B. The corresponding eigenvalue of $\mathbf{A}$ is $\lambda_A = \lambda_B + \lambda_1$.

To compute the eigenvalue of $\mathbf{A}$ with the minimum norm, we apply the power method to the inverse matrix, $\mathbf{A}^{-1}$, and thus obtain the corresponding eigenvalue with the maximum norm, $\lambda_{A^{-1}}$. The eigenvalue of $\mathbf{A}$ with the minimum norm is $\lambda_A = 1/\lambda_{A^{-1}}$. The success of this method hinges on our ability to compute the inverse $\mathbf{A}^{-1}$ with adequate precision.

To compute the eigenvalue of $\mathbf{A}$ that is closest to a specified complex number, c, we apply the power method to the matrix $\mathbf{B} = (\mathbf{A} - c\mathbf{I})^{-1}$, and obtain the eigenvalue of $\mathbf{B}$ with the maximum norm, λ_B. The theory of eigenvalue shifting ensures that $\lambda_A = c + 1/\lambda_B$ is the eigenvalue of $\mathbf{A}$ closest to c. To avoid the demanding and possibly precarious computation of the inverse matrix,

$(\mathbf{A} - c\,\mathbf{I})^{-1}$, we carry out the iterations by selecting a starting vector, $\mathbf{x}^{(0)}$, and then solve a system of linear equations,

$$(\mathbf{A} - c\,\mathbf{I}) \cdot \mathbf{x}^{(k+1)} = \mathbf{x}^{(k)}. \qquad (\text{B.2.5})$$

In practice, we may perform the LU decomposition of the coefficient matrix $\mathbf{A} - c\,\mathbf{I}$ and carry out iterations by forward and backward substitution, as discussed in Section B.1. The method is reliable even when c is close to an eigenvalue, rendering the matrix $\mathbf{A} - c\,\mathbf{I}$ nearly singular.

Deflation by size reduction

Let us assume that the real eigenvalue of $\mathbf{A}$ with the maximum norm, λ_1, and the corresponding real eigenvector, $\mathbf{u}^{(1)}$, are available. The eigenvector is normalized so that $\mathbf{u}^{(1)} \cdot \mathbf{u}^{(1)} = 1$. Now we introduce the orthogonal Householder matrix $\mathbf{H} = \mathbf{I} - 2\,\mathbf{w} \otimes \mathbf{w}$ with elements

$$H_{i,j} = \delta_{i,j} - 2\,w_i w_j, \qquad (\text{B.2.6})$$

where $\mathbf{w}$ is a real vector with unit length, $\mathbf{w} \cdot \mathbf{w} = 1$. We want to compute the vector $\mathbf{w}$ so that the elements in the first column of the matrix

$$\mathbf{B} = \mathbf{H} \cdot \mathbf{A} \cdot \mathbf{H} \qquad (\text{B.2.7})$$

are zero, except for the first element, $B_{1,1}$, that is equal to λ_1. Analysis shows that this is true when

$$w_1 = \left(\frac{1}{2} \left(1 \pm u_1^{(1)} \right) \right)^{1/2}, \qquad w_i = \pm \frac{1}{2w_1}\,u_i^{(1)} \qquad (\text{B.2.8})$$

for $i = 2, \ldots, N$ (e.g., [317]). The eigenvalues of the bottom $(N-1) \times (N-1)$ diagonal block of $\mathbf{B}$ are also eigenvalues of $\mathbf{A}$. Applying the power method to the reduced matrix $\mathbf{B}$ yields an additional eigenvalue. The deflation can be repeated until the sequentially deflated matrix has been reduced to a scalar.

B.2.3 Similarity transformations

If $\mathbf{P}$ is a nonsingular matrix, the matrix $\mathbf{B}$ computed by the similarity transformation $\mathbf{B} = \mathbf{P}^{-1} \cdot \mathbf{A} \cdot \mathbf{P}$ shares its eigenvalues with $\mathbf{A}$. This observation suggests searching for a transformation matrix, $\mathbf{P}$, that renders the matrix $\mathbf{B}$ as simple as possible, ideally diagonal or triangular.

Jacobi's method

Jacobi's method seeks to reduce the norm of the off-diagonal elements of a symmetric matrix, $\mathbf{A}$, by performing consecutive similarity transformations described as plane rotations by an angle, θ. The algorithm involves the following steps:

1. Compute

$$w \equiv \cot 2\theta = \frac{A_{i,i} - A_{j,j}}{2A_{i,j}}. \qquad (\text{B.2.9})$$

2. Obtain $\tan\theta$ from the equation

$$\tan\theta = \frac{\text{sign}(w)}{|w| + \sqrt{w^2 + 1}}. \tag{B.2.10}$$

3. Set

$$\cos\theta = \frac{1}{\sqrt{\tan^2\theta + 1}}. \tag{B.2.11}$$

4. Compute

$$r = \frac{\sin\theta}{1 + \cos\theta}. \tag{B.2.12}$$

5. Compute the elements of $\mathbf{B}$ corresponding to the amended elements of $\mathbf{A}$ from the equations

$$B_{i,i} = A_{i,i} + A_{i,j}\,\tan\theta, \qquad B_{j,j} = A_{j,j} - A_{i,j}\,\tan\theta, \tag{B.2.13}$$

and

$$B_{i,k} = B_{k,i} = A_{i,k} + (A_{j,k} - rA_{i,k})\sin\theta, \qquad B_{j,k} = B_{k,j} = A_{j,k} - (A_{i,k} + rA_{j,k})\sin\theta, \tag{B.2.14}$$

for $k = 1, \ldots, N$ and $k \neq i, j$.

The algorithm involves sweeping the off-diagonal elements of the upper or lower triangular block of the evolving matrix according to a certain protocol.

As the similarity transformations continue, the transformed matrix tends to become diagonal at a quadratic rate. This means that, after a full sweep has been completed, the sum of the squares of the norms of the off-diagonal elements has been roughly squared. In practice, only a few sweeps are necessary to reduce the magnitude of the off-diagonal elements below a negligible threshold.

QR decomposition

The method of QR decomposition can be applied to an arbitrary matrix that is not necessarily symmetric. The algorithm involves decomposing a matrix $\mathbf{A}$ as $\mathbf{A} = \mathbf{Q}\cdot\mathbf{R}$, where $\mathbf{Q}$ is an orthogonal matrix and $\mathbf{R}$ is an upper triangular matrix whose diagonal elements are equal to unity. Setting $\mathbf{P} = \mathbf{Q}$, we obtain $\mathbf{B} = \mathbf{Q}^T\cdot\mathbf{A}\cdot\mathbf{Q} = \mathbf{R}\cdot\mathbf{Q}$. As the transformations are repeated, the evolving matrix tends to become precisely or nearly upper triangular.

B.3   Nonlinear algebraic equations

Consider a system of N nonlinear algebraic equations for N scalar unknowns,

$$f_i(x_1, x_2, \ldots, x_N) = 0, \tag{B.3.1}$$

where $i = 1, \ldots, N$. It is convenient to introduce the vector of independent variables, $\mathbf{x}$, and the vector function $\mathbf{f}(\mathbf{x})$.

Fixed-point iterations

To compute the solution of the vector equation $\mathbf{f}(\mathbf{x}) = \mathbf{0}$, we recast it into the form $\mathbf{x} = \mathbf{g}(\mathbf{x})$, and perform fixed-point iterations based on the formula

$$\mathbf{x}^{(k+1)} = \mathbf{g}(\mathbf{x}^{(k)}), \tag{B.3.2}$$

beginning with a certain initial guess, $\mathbf{x}^{(0)}$, where the superscript (k) is an iteration index. It is possible that the sequence $\mathbf{x}^{(k)}$ will converge to the fixed point of the mapping function $\mathbf{g}$, denoted by $\mathbf{X}$, which, by definition, satisfies the equation $\mathbf{X} = \mathbf{g}(\mathbf{X})$ or $\mathbf{f}(\mathbf{X}) = \mathbf{0}$. A sufficient condition for the fixed-point iterations to converge is that

$$\sum_{i=1}^{N} |G_{i,j}(\mathbf{X})| \leq 1, \tag{B.3.3}$$

for any j. The Jacobian matrix, $\mathbf{G}$, with elements $G_{i,j} = \partial g_i / \partial x_j$, is the transpose of the gradient of g, $\mathbf{G} \equiv (\nabla g)^T$. A necessary condition for the iterations to converge is that the spectral radius of $\mathbf{G}$ evaluated at the *a priori* unknown fixed point is less than unity.

Newton's and related methods

To implement Newton's method, we introduce the matrix of partial derivatives of the vector function $\mathbf{f}$ with elements $F_{i,j} = \partial f_i / \partial x_j$, and use the iteration function

$$\mathbf{g}(\mathbf{x}) = \mathbf{x} - \mathbf{F}^{-1}(\mathbf{x}) \cdot \mathbf{f}(\mathbf{x}), \tag{B.3.4}$$

where $\mathbf{F}^{-1}$ is the inverse of $\mathbf{F}$. The algorithm involves solving a linear system of equations for a correction vector $\mathbf{e}$,

$$\mathbf{F}(\mathbf{x}^{(k)}) \cdot \mathbf{e} = -\mathbf{f}(\mathbf{x}^{(k)}), \tag{B.3.5}$$

and then setting $\mathbf{x}^{(k+1)} = \mathbf{x}^{(k)} + \mathbf{e}$. The iterations are terminated when $|\mathbf{e}|$ falls below a specified threshold.

In the case of a single equation, $N = 1$, we obtain the Newton–Raphson algorithm

$$x^{(k+1)} = x^{(k)} - \frac{f(x^{(k)})}{f'(x^{(k)})}, \tag{B.3.6}$$

where $f' = \mathrm{d}f/\mathrm{d}x$. In practice, the derivative f' is computed by numerical differentiation setting, for example, $f' = [f(x + \epsilon) - f(x)]/\epsilon$, where ϵ is a small increment, as discussed in Section B.5.

Newton's method is guaranteed to converge so long as the initial guess is sufficiently close to a root. The rate of convergence is quadratic: each time we carry out one iteration, the error is raised to the second power and then multiplied by a constant coefficient. In the case of a multiple root, the rate of convergence is linear: each time we carry out one iteration, the error is multiplied by a constant factor that is less than unity. In the case of a double root, the constant factor is equal to one half. However, if necessary, second-order convergence can be rectified by a judicious modification of the basic algorithm.

Secant method

The secant method is a modification of Newton's method designed to circumvent the computation of derivatives. In the case of one equation, $N = 1$, the algorithm is

$$x^{(k+1)} = x^{(k)} - f^{(k)} \frac{x^{(k)} - x^{(k-1)}}{f^{(k)} - f^{(k-1)}}. \tag{B.3.7}$$

Two guesses are required to initialize the iterations. Each time we carry out one iteration, the error is raised approximately to the 1.6 power.

Quasi-Newton's methods are modifications of Newton's method where the Jacobian matrix is either kept constant or constructed during the iterations.

Bairstow's method for polynomials

This powerful method allows us to compute a pair of roots of an Nth degree polynomial, $P_N(x) = a_1 x_N + a_2 x^{N-1} + a_3 x^{N-2} + \cdots + a_N x + a_{N+1}$. When the coefficients a_i are real, we are able to capture complex conjugate pairs of roots. The idea is to express the polynomial in the form

$$\begin{aligned} P_N(x) &= (x^2 - r\,x - s)\,(b_1\,x^{N-2} + b_2\,x^{N-3} + b_3\,x^{N-4} + \cdots + b^{N-2}\,x + b_{N-1}) \\ &\quad + b_N\,(x - r) + b_{N+1}, \end{aligned} \tag{B.3.8}$$

where b_i are real or complex coefficients. Now we seek to compute the constants r and s so that the coefficients b_N and b_{N+1} are zero and the polynomial has a factorized form. When this has been accomplished, the roots of the binomial $x^2 - r\,x - s$ are also roots of the polynomial, $P_N(x)$, and can be extracted using the quadratic formula. To compute further roots, we apply the method to the $(N - 2)$-degree polynomial enclosed by the second set of parentheses on the right-hand side of (B.3.8).

The problem has been reduced to solving a system of two nonlinear equations,

$$b_N(r, s) = 0, \qquad b_{N+1}(r, s) = 0, \tag{B.3.9}$$

where the coefficients b_N and b_{N+1} are polynomial functions of r and s. The solution is found using Newton's method for two equations. To evaluate the coefficients b_N and b_{N+1} and their partial derivatives with respect to r and s, we construct the sequence

$$\begin{aligned} b_1 &= a_1, \qquad b_2 = a_2 + r\,b_1, \qquad b_3 = a_3 + r\,b_2 + s\,b_1, \qquad \ldots, \\ b_N &= a_N + r\,b_{N-1} + s\,b_{N-2}, \qquad b_{N+1} = a_{N+1} + r\,b_N + s\,b_{N-1}, \end{aligned} \tag{B.3.10}$$

and then the sequence

$$\begin{aligned} c_1 &= 0, \qquad c_2 = b_1, \qquad c_3 = b_2 + r\,c_1, \qquad c_4 = b_3 + r\,c_3 + s\,c_2, \qquad \ldots, \\ c_N &= b_{N-1} + r\,c_{N-1} + s\,c_{N-2}, \qquad c_{N+1} = b_N + r\,c_N + s\,c_{N-1}. \end{aligned} \tag{B.3.11}$$

It can be shown by straightforward substitution that $\partial b_k/\partial r = c_k$ and $\partial b_k/\partial s = c_{k-1}$ for $k = 1, \ldots, N$.

Deflation

Once a root of the equation $\mathbf{f}(\mathbf{x}) = \mathbf{0}$ has been found, denoted by $\mathbf{X}$, it can be screened out by considering the deflated equation $\tilde{\mathbf{f}}(\mathbf{x}) = \mathbf{0}$, where $\tilde{\mathbf{f}}(\mathbf{x}) = \mathbf{f}(\mathbf{x})/|\mathbf{x} - \mathbf{X}|^m$ and m is the multiplicity of $\mathbf{X}$.

Initial guess

A successful estimate of the location of a root can be made by examining whether any terms of the function $\mathbf{f}$ take small or large values for small or large values of the individual scalar components of the vector of independent variables, $\mathbf{x}$. If they do, we either ignore these terms or discard the rest of the terms, compute the solution of the simplified system, and check *a posteriori* whether the approximations that lead us to the approximate solution are justified.

Parameter continuation

A successful initial guess can be found by the method of parameter continuation or embedding. The idea is to perturb the equation $\mathbf{f}(\mathbf{x}) = \mathbf{0}$ to formulate a modified equation, $\mathbf{q}(\mathbf{x}, \epsilon) = \mathbf{0}$, where $\mathbf{q}(\mathbf{x}, \epsilon = 1) = \mathbf{f}(\mathbf{x})$ and the equation $\mathbf{q}(\mathbf{x}, \epsilon = 0) = \mathbf{0}$ is easy to solve. The procedure involves solving a sequence of equations, such as

$$\mathbf{q}(\mathbf{x}, \epsilon = 0) = \mathbf{0}, \qquad \mathbf{q}(\mathbf{x}, \epsilon = 0.1) = \mathbf{0}, \qquad \ldots, \qquad \mathbf{q}(\mathbf{x}, \epsilon = 1.0) = \mathbf{0}, \qquad \text{(B.3.12)}$$

where the initial guess in solving each equation is the converged solution of the previous equation. In fluid mechanics, the parameter ϵ can be identified with the Reynolds number.

B.4　Function interpolation

Given the values of a function, $f(x)$, at $N + 1$ data points, x_i, where $i = 1, \ldots, N + 1$, we wish to compute the value of the function at an intermediate point.

The interpolating polynomial

A popular method of function interpolation replaces the interpolated function, $f(x)$, with an Nth-degree interpolating polynomial passing through the data points,

$$f(x) \simeq P_N(x) = a_1 x^N + a_2 x^{N-1} + a_3 x^{N-2} + \cdots + a_N x + a_{N+1}, \qquad \text{(B.4.1)}$$

so that $f(x_i) = P_N(x_i)$. The error incurred by this approximation is given by

$$e(x) = -\frac{f^{(N+1)}(\xi)}{(N+1)!} \, \Phi_{N+1}(x), \qquad \text{(B.4.2)}$$

where

$$\Phi_{N+1}(x) = (x - x_1)(x - x_2) \cdots (x - x_N)(x - x_{N+1}), \qquad \text{(B.4.3)}$$

is an $(N+1)$-degree polynomial and $f^{(N+1)}(\xi)$ is the $N+1$ derivative of f evaluated at a certain point ξ that is located somewhere between x_1 and x_{N+1}. The actual location of ξ depends on the value of x. The interpolation error does not necessarily vanish as N tends to infinity, especially when the data points are evenly spaced. However, if the data points coincide with the scaled zeros of an Nth degree orthogonal polynomial, the error generally diminishes uniformly as N is increased.

Vandermonde matrix

To compute the interpolating polynomial, we may enforce the interpolation constraint $f(x_i) = P_N(x_i)$ at the $N+1$ data points, and thereby derive a system of $N+1$ equations for the $N+1$ unknown coefficients, a_i. The matrix of the linear system multiplying the unknown vector is the transpose of the Vandermonde matrix. The determinant of this matrix can be shown to be equal to the product of $x_i - x_j$ with $i > j$, and this ensures that the solution of the linear system is unique as long as the data points are distinct. Unfortunately, the Vandermonde matrix is nearly singular for moderate and high values of N, placing limits on the accuracy and practicality of the numerical method.

Lagrange interpolation

In Lagrange's method, the interpolating polynomial is constructed in an expedient manner that circumvents the explicit computation of the polynomial coefficients, by setting

$$P_N(x) = \sum_{i=1}^{N+1} f(x_i)\, l_{N,i}(x),$$ (B.4.4)

where

$$l_{N,i}(x) = \frac{(x-x_1)(x-x_2)\cdots(x-x_{i-1})(x-x_{i+1})\cdots(x-x_{N+1})}{(x_i-x_1)(x_i-x_2)\cdots(x_i-x_{i-1})(x_i-x_{i+1})\cdots(x_i-x_{N+1})}$$ (B.4.5)

are Nth-degree Lagrange polynomials defined in terms of the data points. An alternative compact representation is

$$l_{N,i}(x) = \frac{1}{\Phi'_{N+1}(x_i)}\frac{\Phi_{N+1}(x)}{x-x_i},$$ (B.4.6)

where the polynomial Φ_{N+1} is defined in (B.4.3). The denominators of the fractions defining $l_{N,i}(x)$ are constant, whereas the numerators are Nth-degree polynomials in x.

Local polynomial interpolation

When the number of data points is large or the data carry an appreciable amount of error, global polynomial interpolation may lead to substantial error in certain regions, especially near the ends of the interpolation domain. To avoid this pitfall, we replace the global interpolating polynomial with the union of local interpolating polynomials defined with respect to a small group of data points.

Linear polynomial interpolation employs two consecutive data points and yields the first-degree polynomial

$$P_1^{(i)}(x) = f(x_i) + (x-x_i)\frac{f(x_{i+1})-f(x_i)}{x_{i+1}-x_i},$$ (B.4.7)

or

$$P_1^{(i)}(x) = f(x_i)\frac{x - x_{i+1}}{x_i - x_{i+1}} + f(x_{i+1})\frac{x - x_i}{x_{i+1} - x_i}, \tag{B.4.8}$$

for $x_i < x < x_{i+1}$.

Quadratic interpolation employs three consecutive data points and yields the second-degree polynomial

$$P_2^{(i)}(x) = f(x_i) + [\,(a_i(x - x_i) + b_i\,](x - x_i), \tag{B.4.9}$$

for $x_{i-1} < x < x_{i+1}$. The coefficients a_i and b_i are computed sequentially as

$$a_i = \frac{1}{x_{i+1} - x_{i-1}}\left(\frac{f(x_{i+1}) - f(x_i)}{x_{i+1} - x_i} - \frac{f(x_{i-1}) - f(x_i)}{x_{i-1} - x_i}\right),$$

$$b_i = \frac{f(x_{i+1}) - f(x_i)}{x_{i+1} - x_i} - a_i\,(x_{i+1} - x_i). \tag{B.4.10}$$

Simplifications occur when the points are evenly spaced.

Cubic-spline interpolation

Cubic-spline interpolation fits a third-degree polynomial over each interval between two consecutive data points and matches the first and second derivatives of adjacent polynomials at the data points. Denoting the ith cubic polynomial by

$$P_3^{(i)}(x) = a_i(x - x_i)^3 + b_i(x - x_i)^2 + c_i(x - x_i) + y_i, \tag{B.4.11}$$

we find that

$$a_i = \frac{1}{3}\frac{b_{i+1} - b_i}{h_i}, \qquad c_i = \frac{y_{i+1} - y_i}{h_i} - \frac{1}{3}\,h_i\,(2b_i + b_{i+1}), \tag{B.4.12}$$

where $h_i = h_{i+1} - h_i$ and

$$b_{N+1} \equiv \frac{1}{2}\left(\frac{\mathrm{d}^2 P_3^{(N)}}{\mathrm{d}x^2}\right)_{x = x_{N+1}} = 3\,a_N h_N + b_N. \tag{B.4.13}$$

The coefficients b_i satisfy the $N - 1$ equations

$$h_i\,b_i + 2\,(h_i + h_{i+1})\,b_{i+1} + h_{i+1}\,b_{i+2} = 3\left(\frac{y_{i+2} - y_{i+1}}{h_{i+1}} - \frac{y_{i+1} - y_i}{h_i}\right), \tag{B.4.14}$$

for $i = 1, \ldots, N - 1$. To compute the $N + 1$ unknowns, $b_1, b_2, \ldots, b_{N+1}$, we require two additional conditions. In the case of clamped splines, we specify the slope of the first and last cubics at the first and last data points. If the interpolated function is periodic, we require that the first and second derivatives of the first cubic at the first point are equal to those of the last cubic at the last point.

Polynomial interpolation in two variables

Polynomial interpolation in two variables is similar to that in one variable discussed earlier in this section. The counterpart of local linear interpolation is bilinear interpolation: for each value of x, the interpolating function varies linearly with respect to y; for each value of y, the interpolating function varies linearly with respect to x. Suppose that a point $\mathbf{x}$ lies inside a rectangle confined by the vertical grid lines $x = x_i$, x_{i+1}, and the horizontal grid lines $y = y_j$, y_{j+1}. The interpolated value, $f(x, y)$, is computed as a weighted average of values of f at the four closest grid points,

$$f(x, y) = \mathbf{w}(x, y) : \mathbf{F}. \tag{B.4.15}$$

In terms of the dimensionless coordinates

$$\xi = \frac{x - x_i}{\Delta x}, \qquad \eta = \frac{y - y_j}{\Delta y}, \tag{B.4.16}$$

where $\Delta x = x_{i+1} - x_i$ and $\Delta y = y_{j+1} - y_j$, the weight matrix takes the form

$$\mathbf{w} = \begin{bmatrix} (1 - \xi)\,\eta & \xi\,\eta \\ (1 - \xi)\,(1 - \eta) & \xi\,(1 - \eta) \end{bmatrix}. \tag{B.4.17}$$

Bilinear interpolation generates a continuous function with generally discontinuous first derivatives across the grid lines.

Trigonometric interpolation

Consider a real function, $f(x)$, defined in the interval $[a, b]$. Trigonometric or Fourier approximation and interpolation represents the function in that interval with a truncated complete Fourier series,

$$F_M(x) = \frac{1}{2} a_0 + \sum_{p=1}^{M} a_p \, \cos(pk\widehat{x}) + \sum_{p=1}^{M} b_p \, \sin(pk\widehat{x}), \tag{B.4.18}$$

where $\widehat{x} = x - a$, $k = 2\pi/L$ is the wave number, $L = b - a$ is the length of the interval, M is a chosen truncation level, and a_p, b_p are real Fourier coefficients. Although $\widehat{x}$ can be replaced by x without loss of generality, this complicates the forthcoming algebraic expressions. Outside the interval $[a, b]$, the Fourier series yields the periodic repetition of the portion of f in $[a, b]$.

An alternative complex form of the complete Fourier series is

$$F_M(x) = \sum_{p=-M}^{M} c_p \exp(-\mathrm{i}\,pk\widehat{x}), \tag{B.4.19}$$

where c_p are complex Fourier coefficients and i is the imaginary unit, $\mathrm{i}^2 = -1$. To ensure that the right-hand side of (B.4.19) is real, we require that

$$c_{-p} = c_p^*, \tag{B.4.20}$$

where an asterisk denotes the complex conjugate. Using the Euler decomposition of the complex exponential,

$$\exp(-\mathrm{i}pk\widehat{x}) = \cos(pk\widehat{x}) - \mathrm{i}\sin(pk\widehat{x}), \tag{B.4.21}$$

$$\int_a^b \exp[\mathrm{i}\,(p-q)k\widehat{x}]\,\mathrm{d}x = \int_0^L \exp[\mathrm{i}\,(p-q)k\widehat{x}]\,\mathrm{d}\widehat{x} = L\,\delta_{pq}$$

$$\int_0^L \cos(pk\widehat{x})\,\cos(qk\widehat{x})\,\mathrm{d}\widehat{x} = \begin{cases} L & \text{if} \quad p=q=0 \\ \tfrac{1}{2}L & \text{if} \quad p=q \neq 0 \\ 0 & \text{otherwise} \end{cases}$$

$$\int_0^L \sin(pk\widehat{x})\,\sin(qk\widehat{x})\,\mathrm{d}\widehat{x} = \begin{cases} \tfrac{1}{2}L & \text{if} \quad p=q \neq 0, \\ 0 & \text{otherwise.} \end{cases}$$

$$\int_0^L \cos(pk\widehat{x})\,\sin(qk\widehat{x})\,\mathrm{d}\widehat{x} = 0$$

TABLE B.4.1   Orthogonality of trigonometric functions in the interval $[a,b]$; $L = b - a$, $k = 2\pi/L$, p and q are two integers, and $\widehat{x} = x - a$.

we find that the real and complex Fourier coefficients are related by

$$c_p = \frac{1}{2}\,(a_p + \mathrm{i}\,b_p), \tag{B.4.22}$$

with the understanding that $b_0 = 0$ and therefore $c_0 = \frac{1}{2}a_0$ is real. Because of the preferential treatment of the first term on the right-hand side of (B.4.18), equation (B.4.22) holds true for any value of p, including zero. A graph of $|c_p|^2$ against the index p provides us with the discrete power spectrum of the function $f(x)$.

To compute the Fourier coefficients, we use of the Fourier orthogonality properties shown in Table B.4.1. Multiplying both sides of (B.4.19) by $\exp(\mathrm{i}\,q\widehat{x})$, where q is an integer, using the first orthogonality property in Table B.4.1, and then relabeling $q \to p$, we find that

$$c_p = \frac{1}{L} \int_a^b f(x)\,\exp(\mathrm{i}\,pk\widehat{x})\,\mathrm{d}x, \tag{B.4.23}$$

and thus

$$a_p = \frac{2}{L} \int_a^b f(x)\,\cos(pk\widehat{x})\,\mathrm{d}x, \qquad b_p = \frac{2}{L} \int_a^b f(x)\,\sin(pk\widehat{x})\,\mathrm{d}x. \tag{B.4.24}$$

It can be shown that, when the coefficients are computed in this fashion, in the limit as M tends to infinity, the Fourier representation (B.4.18) becomes exact. The integrals in (B.4.24) can be computed by numerical integration, as discussed in Section B.6.

Let us divide the interval L into N subintervals separated by $N+1$ data points $x_i = a + (i-1)h$, where $i = 1, \ldots, N+1$, $h = L/N$, and $x_{N+1} = b$. Using the trapezoidal rule to evaluate the complex

$$\sum_{j=1}^{N} \exp(\mathrm{i}pk\widehat{x}_j) = \begin{cases} N & \text{if} \quad p = sN \\ 0 & \text{otherwise} \end{cases}$$

$$\sum_{j=1}^{N} \exp[-\mathrm{i}jk(x_p - x_q)] = \begin{cases} N & \text{if} \quad p - q = sN \\ 0 & \text{otherwise} \end{cases}$$

$$\sum_{j=1}^{N} \cos(pk\widehat{x}_j)\cos(qk\widehat{x}_j) = \begin{cases} N & \text{if} \quad p - q = sN \neq 0 \\ \frac{1}{2}N & \text{if} \quad p - q = 0 \\ 0 & \text{otherwise} \end{cases}$$

$$\sum_{j=1}^{N} \sin(pk\widehat{x}_j)\sin(qk\widehat{x}_j) = \begin{cases} \frac{1}{2}N & \text{if} \quad p - q = sN \\ 0 & \text{otherwise} \end{cases}$$

$$\sum_{j=1}^{N} \cos(pk\widehat{x}_j)\sin(qk\widehat{x}_j) = 0$$

TABLE B.4.2   Discrete orthogonality of trigonometric functions, where $x_j = a + (j-1)h$ for $j = 1, \ldots, N+1$, is a sequence of evenly spaced points, $h = L/N$, $L = b - a$, $x_1 = a$, $x_{N+1} = b$, $k = 2\pi/L$, $\widehat{x} = x - a$, i is the imaginary unit, and p, q, s are integers.

Fourier integral in (B.4.23), we find that

$$c_p = \frac{1}{N}\left[\frac{1}{2}f(x_1) + \omega^p f(x_2) + \omega^{2p} f(x_3) + \cdots + \omega^{(N-1)p} f(x_N) + \frac{1}{2}f(x_{N+1})\right], \quad \text{(B.4.25)}$$

where

$$\omega \equiv \exp(\mathrm{i}kh) = \exp(\frac{2\pi\mathrm{i}}{N}). \quad \text{(B.4.26)}$$

Using the identities shown in Table B.4.2, we find that, when the coefficients are computed in this fashion, the truncated Fourier series interpolates through the *interior* data,

$$F_M(x_j) = f(x_j) \quad \text{(B.4.27)}$$

for $j = 2, \ldots, N$. The value of the truncated Fourier series at the first and last data is the end point average,

$$F_M(x_1) = F_M(x_{N+1}) = \frac{1}{2}\left[f(x_1) + f(x_{N+1})\right]. \quad \text{(B.4.28)}$$

Thus, the truncated Fourier series interpolates through the mean of the first and last points.

If the interpolated function $f(x)$ is repeated along the x axis with period L, $f(x_1) = f(x_{N+1})$, formula (B.4.25) takes the simpler form

$$c_p = \frac{1}{N} \sum_{j=1}^{N} \omega^{(j-1)p} f(x_j). \tag{B.4.29}$$

In that case,

$$F_M(x_j) = f(x_j) \tag{B.4.30}$$

for all $j = 1, \ldots, N+1$, that is, the truncated Fourier series interpolates through all $N+1$ data. In practice, when N is large, the sum on the right-hand side of (B.4.29) is computed most efficiently by the method of fast Fourier transform (FFT) requiring $N \log_2 N$ operations.

B.5 Numerical differentiation

Given the values of a function, $f(x)$, at $N+1$ data points, x_i, where $i = 1, \ldots, N+1$, we wish to compute its derivatives at the data points or at an intermediate point. This can be done by approximating the function with a local interpolating polynomial, as discussed in Section B.4, and then differentiating the interpolating polynomial to obtain approximations to the derivatives.

Consider the computation of the derivatives of a function at the ith datum point, x_i. If we use an equal number of data points on either side of x_i to construct the interpolating polynomial, we obtain central differences. Otherwise, we obtain spatially biased, forward or backward differences. Table B.5.1 summarizes finite-difference formulas for computing the derivatives of a function of one variable, x, at the point x_i using the values of the function at a collection of evenly spaced data points separated by distance $\Delta x = h$ (e.g., [185]).

Finite-difference formulas for computing the Laplacian of a function of two variables, x and y at a grid point, (x_i, y_i), can be derived working in a similar fashion. Consider a uniform Cartesian grid with spacings Δx and Δy, and define $\beta = (\Delta x / \Delta y)^2$. The five-point formula yields

$$(\nabla^2 f)_{i,j} = \frac{1}{\Delta x^2} \begin{bmatrix} 0 & \beta & 0 \\ 1 & -2(1+\beta) & 1 \\ 0 & \beta & 0 \end{bmatrix} : \mathbf{F}, \tag{B.5.1}$$

with accuracy of $O(\Delta x^2)$ and $O(\Delta y^2)$, where

$$\mathbf{F} = \begin{bmatrix} f_{i-1,j+1} & f_{i,j+1} & f_{i+1,j+1} \\ f_{i-1,j} & f_{i,j} & f_{i+1,j} \\ f_{i-1,j-1} & f_{i,j-1} & f_{i+1,j-1} \end{bmatrix}. \tag{B.5.2}$$

The nine-point formula with $\Delta x = \Delta y$ yields

$$\left(\nabla^2 f\right)_{i,j} \simeq \frac{1}{6\Delta x^2} \begin{bmatrix} 1 & 4 & 1 \\ 4 & -20 & 4 \\ 1 & 4 & 1 \end{bmatrix} : \mathbf{F}, \tag{B.5.3}$$

with accuracy of $O(\Delta x^2)$.

Backward differences with accuracy $O(h)$

$$
\begin{bmatrix} h\,f_i' \\ h^2\,f_i'' \\ h^3\,f_i''' \\ h^4\,f_i'''' \end{bmatrix}
=
\begin{bmatrix}
0 & 0 & 0 & -1 & 1 \\
0 & 0 & 1 & -2 & 1 \\
0 & -1 & 3 & -3 & 1 \\
1 & -4 & 6 & -4 & 1
\end{bmatrix}
\begin{bmatrix} f_{i-4} \\ f_{i-3} \\ f_{i-2} \\ f_{i-1} \\ f_i \end{bmatrix}
$$

Backward differences with accuracy $O(h^2)$

$$
\begin{bmatrix} 2\,h\,f_i' \\ h^2\,f_i'' \\ 2\,h^3\,f_i''' \\ h^4\,f_i'''' \end{bmatrix}
=
\begin{bmatrix}
0 & 0 & 0 & 1 & -4 & 3 \\
0 & 0 & -1 & 4 & -5 & 2 \\
0 & 3 & -14 & 24 & -18 & 5 \\
-2 & 11 & -24 & 26 & -14 & 3
\end{bmatrix}
\begin{bmatrix} f_{i-5} \\ f_{i-4} \\ f_{i-3} \\ f_{i-2} \\ f_{i-1} \\ f_i \end{bmatrix}
$$

Centered differences with accuracy $O(h^2)$

$$
\begin{bmatrix} 2\,h\,f_i' \\ h^2\,f_i'' \\ h^2\,f_i'' \\ 2\,h^3\,f_i''' \\ h^4\,f_i'''' \end{bmatrix}
=
\begin{bmatrix}
0 & -1 & 0 & 1 & 0 \\
0 & 1 & -2 & 1 & 0 \\
-1 & 2 & 0 & -2 & 1 \\
1 & -4 & 6 & -4 & 1
\end{bmatrix}
\begin{bmatrix} f_{i-2} \\ f_{i-1} \\ f_i \\ f_{i+1} \\ f_{i+2} \end{bmatrix}
$$

Centered differences with accuracy $O(h^4)$

$$
\begin{bmatrix} 12\,h\,f_i' \\ 12\,h^2\,f_i'' \\ 8\,h^3\,f_i''' \\ 6\,h^4\,f_i'''' \end{bmatrix}
=
\begin{bmatrix}
0 & 1 & -8 & 0 & 8 & -1 & 0 \\
0 & -1 & 16 & -30 & 16 & -1 & 0 \\
1 & -8 & 13 & 0 & -13 & 8 & -1 \\
-1 & 12 & -39 & 56 & -39 & 12 & -1
\end{bmatrix}
\begin{bmatrix} f_{i-3} \\ f_{i-2} \\ f_{i-1} \\ f_i \\ f_{i+1} \\ f_{i+2} \\ f_{i+3} \end{bmatrix}
$$

TABLE B.5.1   Finite-difference formulas for the derivatives of a function, $f(x)$, at the grid point, x_i, in terms of values of the function at a set of evenly spaced neighboring points separated by the distance $\Delta x = h$. (*Continuing.*)

Forward differences with accuracy $O(h)$

$$
\begin{bmatrix} h\,f_i' \\ h^2\,f_i'' \\ h^3\,f_i''' \\ h^4\,f_i'''' \end{bmatrix} = \begin{bmatrix} -1 & 1 & 0 & 0 & 0 \\ 1 & -2 & 1 & 0 & 0 \\ -1 & 3 & -3 & 1 & 0 \\ 1 & -4 & 6 & -4 & 1 \end{bmatrix} \begin{bmatrix} f_i \\ f_{i+1} \\ f_{i+2} \\ f_{i+3} \\ f_{i+4} \end{bmatrix}
$$

Forward differences with accuracy $O(h^2)$

$$
\begin{bmatrix} 2\,h\,f_i' \\ h^2\,f_i'' \\ 2\,h^3\,f_i''' \\ h^4\,f_i'''' \end{bmatrix} = \begin{bmatrix} -3 & 4 & -1 & 0 & 0 & 0 \\ 2 & -5 & 4 & -1 & 0 & 0 \\ -5 & 18 & -24 & 14 & -3 & 0 \\ 3 & -14 & 26 & -24 & 11 & -2 \end{bmatrix} \begin{bmatrix} f_i \\ f_{i+1} \\ f_{i+2} \\ f_{i+3} \\ f_{i+4} \\ f_{i+5} \end{bmatrix}
$$

TABLE B.5.1 (*Continued.*)

B.6   Numerical integration

Given the values of a function $f(x)$ at $N+1$ data or base points, x_i, $i = 1, \ldots, N+1$, distributed inside a closed interval, $[a, b]$, we want to compute the integral

$$
I \equiv \int_a^b f(x)\,dx. \tag{B.6.1}
$$

A typical method of numerical integration involves approximating the integrand, $f(x)$, with a global interpolating polynomial or with a set of local interpolating polynomials defined in $[a, b]$, as discussed in Section B.4, and then integrating the interpolating polynomials over their domain of definition.

Trapezoidal rule

Approximating $f(x)$ with a straight line between two consecutive data points, described by a first-degree local interpolating polynomial, we obtain the trapezoidal rule. When the base points are distributed evenly with constant separation $h = (b - a)/N$, we obtain

$$
I_{tr}(h) = h\,(\tfrac{1}{2}\,f_1 + f_2 + \cdots + f_N + \tfrac{1}{2}\,f_{N+1}), \tag{B.6.2}
$$

where $x_1 = a$ and $x_{N+1} = b$. It can be shown that the leading-order numerical error is

$$
E_{tr}(h) = \frac{1}{12}\,(b - a)\,f''(\xi)\,h^2, \tag{B.6.3}
$$

where the point ξ lies somewhere inside the integration domain.

Romberg integration uses the results of two computations with different interval sizes h to improve the accuracy to fourth order in h. For example,

$$I_{Rom}(h = \frac{\epsilon}{2}) = \frac{1}{3}\left(4 I_{tr}(h = \frac{\epsilon}{2}) - I_{tr}(h = \epsilon) \right). \tag{B.6.4}$$

Note that the most accurate result enjoys a higher weight.

Simpson's rule

Approximating $f(x)$ with a parabola subtended across triplets of consecutive data points, described by a second-degree local interpolating polynomial, we obtain Simpson's one-third rule. Assuming that the number of intervals N is even, we obtain

$$I_{smp}(h) = \frac{1}{3}h\left(f_1 + 4f_2 + 2f_3 + 4f_4 + \cdots + 2f_{N-1} + 4f_N + f_{N+1} \right). \tag{B.6.5}$$

The numerical error is

$$E_{Smp}(h) = \frac{1}{180}(b-a)\, f''''(\xi)\, h^4 \simeq \frac{1}{180}(f_{N+1}''' - f_1''')\, h^4, \tag{B.6.6}$$

where the point ξ lies somewhere inside the integration domain. Romberg integration improves the accuracy to sixth order in h by setting

$$I_{Rom}(h = \frac{\epsilon}{2}) = \frac{1}{15}\left(16 I_{Smp}(h = \frac{\epsilon}{2}) - I_{Smp}(h = \epsilon) \right). \tag{B.6.7}$$

Note that the most accurate result is assigned a considerably higher weight.

Gauss quadratures

Gauss quadratures require that the data points are distributed in special ways over the integration domain. Global polynomial interpolation followed by analytical integration leads to the numerical approximation

$$I_{Gauss} = \frac{1}{2}(b-a)\sum_{i=1}^{N_Q} f(x_i)\, w_i, \tag{B.6.8}$$

where N_Q is the chosen number of quadrature base points and w_i are proper weights. The position of the base points, x_i, is determined by the zeros of a properly selected N_Q-degree orthogonal polynomial.

The Gauss–Legendre quadrature is designed for functions that are free of singularities inside the entire integration domain, including the boundaries. The base points where the integrand is evaluated are located at $x_i = \frac{1}{2}[a + b + t_i(b - a)]$, where t_i are the zeros of the N_Q-degree Legendre polynomial defined in the interval $[-1, 1]$. For each chosen N_Q, the weights w_i add up to two, which is necessary for (B.6.8) is to be valid when f is a constant function. Polynomials of degree $2N_Q - 1$ or less are integrated exactly. Values of z_i and w_i for several values of N_Q are given in texts on numerical methods (e.g., [317]).

Other Gauss quadratures applicable to singular integrals and semi-infinite or infinite integration domains are available (e.g., [317]).

Two-dimensional integrals

The choice of a successful numerical method for computing the integral of a function of two variables $f(x, y)$ over a two-dimensional domain depends on the smoothness of the function and geometry of the integration domain.

When the function $f(x, y)$ is smooth and the domain is a rectangular area confined within $a < x < b$ and $c < yd$, the successive application of two Gauss–Legendre quadratures for integration in each direction yields the compound quadrature

$$\iint_{rectangle} f(x, y)\, dx\, dy \simeq \frac{1}{4}(b - a)(c - a) \sum_{i=1}^{N_{Q_x}} \sum_{i=1}^{N_{Q_y}} f(x_i, y_j)\, w_i\, w_j, \tag{B.6.9}$$

where N_{Q_x} and N_{Q_y} are two independent integers determining the order of the quadrature, and x_i, y_j correspond to the Gauss-Legendre base points.

To compute the integral of a smooth function over the surface of a triangle with vertices at the points $\mathbf{v}_1$, $\mathbf{v}_2$, and $\mathbf{v}_3$, we use the quadrature

$$\iint_{triangle} f(x, y)\, dx\, dy \simeq A \sum_{i=1}^{N_Q} f(\mathbf{x}_i)\, w_i, \tag{B.6.10}$$

where $A = |(\mathbf{x}_2 - \mathbf{x}_1) \times (\mathbf{x}_3 - \mathbf{x}_1)|$ is the area of the triangle and

$$\mathbf{x}_i = \alpha_i \mathbf{v}_1 + \beta_i \mathbf{v}_2 + \gamma_i \mathbf{v}_3 \tag{B.6.11}$$

is the position of the base points over the triangle. The coefficients α_i, β_i, and γ_i, and corresponding weights, w_i, are are given in texts on numerical methods [317].

B.7   Function approximation

Given a certain amount of information about a function, $f(x)$, inside an interval, $[a, b]$, we wish to approximate the function with an Nth-degree approximating polynomial, $P_N(x)$. Weierstrass's theorem guarantees that, if $f(x)$ is a continuous function, a polynomial of sufficiently high degree, N, can be found such that $|f(x) - P_N(x)| < \epsilon$ for any x between a and b, where ϵ is an arbitrary small number. Our task is to compute this optimal polynomial.

In the least-squares method, the coefficients of the polynomial are found by minimizing the integral of the squared difference, $|f(x) - P_N(x)|^2$, multiplied by a chosen weighting function, $w(x)$, over $[a, b]$. Unfortunately, when N is higher than about five, this approach results in a nearly singular system of algebraic equations.

Orthogonal polynomials

In an alternative approach, we introduce a family of orthogonal polynomials, $p_i(x)$, for $i = 0, 1, \ldots$, where $p_i(x)$ is an ith degree polynomial with associated weighting function $w(x)$ defined in the interval $[c, d]$. By construction,

$$(p_i, p_j) \equiv \int_c^d p_i(x) \, p_j(x) \, w(x) \, \mathrm{d}x = D_{ij}, \tag{B.7.1}$$

where $\mathbf{D}$ is a diagonal matrix. Two popular families of orthogonal polynomials are the Legendre polynomials and the Chebyshev polynomials, both defined in the interval $[-1, 1]$. An important property of the Chebyshev polynomials is that their magnitude is less than unity for any value of x in the interval $[-1, 1]$.

Next, we introduce the scaled variable $t = c + (d - c)(x - b)/(a - b)$, where $a < b$ and $c < d$. As x increases from a to b, the variable t increases from c to d. Finally, we express the approximating polynomial in terms of a weighted sum of orthogonal polynomials of a chosen class as

$$P_n(t) = \sum_{i=0}^N a_i \, p_i(t), \tag{B.7.2}$$

and compute the coefficients a_i using the orthogonality condition for $p_i(t)$ stated in (B.7.1),

$$a_i = \frac{1}{D_{ii}} \int_c^d f(t) \, p_i(t) \, w(t) \, \mathrm{d}t, \tag{B.7.3}$$

where summation is *not* implied over the repeated index, i. The integral in (B.7.3) can be computed by numerical integration in terms of the values of f at data points, as discussed in Section B.6. To relate t to x, we use the inverse transformation, $x = a + (t - c)(b - a)/(d - c)$.

B.8   Ordinary differential equations

Consider a system of N ordinary differential equations with respect to an independent variable, t,

$$\frac{\mathrm{d}x_i}{\mathrm{d}t} = f_i(x_1, x_2, \ldots, x_N, t), \tag{B.8.1}$$

subject to a given initial condition, $x_i(t = 0)$, for $i = 1, \ldots, N$. In terms of the vector of unknowns functions, $\mathbf{x}$, and velocity vector, $\mathbf{f}$, we obtain

$$\frac{\mathrm{d}\mathbf{x}}{\mathrm{d}t} = \mathbf{f}(\mathbf{x}, t). \tag{B.8.2}$$

The space of scalar variables encapsulated in the vector $\mathbf{x}$ comprise the phase space of the solution. If the function $\mathbf{f}$ does not depend explicitly on t but only implicitly through $\mathbf{x}$, the system is called autonomous; otherwise it is called nonautonomous.

In the following discussion, the superscript (k) denotes the kth time level, t_k, and $\Delta t = t_{k+1} - t_k$ is the time step.

Euler method

In Euler's method, the solution is advanced over a small time interval, Δt, by moving in the phase space by a small distance with the initial phase velocity, so that

$$\mathbf{x}^{(k+1)} = \mathbf{x}^{(k)} + \Delta t \, \mathbf{f}(\mathbf{x}^{(k)}, t_k). \tag{B.8.3}$$

Each time step introduces a numerical error of order Δt^2.

Second-order Runge–Kutta method

The second-order Runge–Kutta method (RK2) is a predictor–corrector scheme requiring two velocity evaluations in each time step. The solution is advanced according to the following five substeps:

1. Compute: $\mathbf{f}^{(k)} \equiv \mathbf{f}(\mathbf{x}^{(k)}, t_k)$

2. Set: $\mathbf{x}^{tmp} = \mathbf{x}^{(k)} + \kappa \, \Delta t \, \mathbf{f}^{(k)}$

3. Compute: $\mathbf{f}^{tmp} = \mathbf{f}(\mathbf{x}^{tmp}, t_k + \kappa \, \Delta t)$

4. Set: $\mathbf{f}^{final} = (1 - \alpha) \, \mathbf{f}^{(k)} + \alpha \, \mathbf{f}^{tmp}$

5. Advance: $\mathbf{x}^{(k+1)} = \mathbf{x}^{(k)} + \Delta t \, \mathbf{f}^{final}$

where the superscript "tmp" denotes a temporary solution, and α and κ are positive numerical parameters satisfying $2\alpha\kappa = 1$.

- Setting $\alpha = \frac{1}{2}$ and $\kappa = 1$ yields the standard version of the RK2 method, also called the modified Euler method.

- Setting $\alpha = 1$ and $\kappa = \frac{1}{2}$ yields the midpoint RK2 method.

- Setting $\alpha = \frac{3}{4}$ and $\kappa = \frac{2}{3}$ yields Heun's method.

Heun's method minimizes the magnitude of the numerical error. In all cases, each complete time step introduces a numerical error of order Δt^3.

Third-order Runge–Kutta method

The third-order Runge–Kutta method (RK3) requires three velocity evaluations in each step. The solution is advanced according to the following seven substeps:

1. Compute: $\mathbf{f}^{(k)} \equiv \mathbf{f}(\mathbf{x}^{(k)}, t_k)$

2. Set: $\mathbf{x}^{tmp1} = \mathbf{x}^{(k)} + \frac{1}{2} \Delta t \, \mathbf{f}^{(k)}$

3. Compute: $\mathbf{f}^{tmp1} = \mathbf{f}(\mathbf{x}^{tmp1}, t_k + \frac{1}{2} \Delta t)$

4. Set: $\mathbf{x}^{tmp2} = \mathbf{x}^{(k)} + \Delta t \left(2 \, \mathbf{f}^{tmp1} - \mathbf{f}^{(k)} \right)$

5. Compute: $\mathbf{f}^{tmp2} = \mathbf{f}(\mathbf{x}^{tmp2}, t_k + \Delta t)$

6. Set: $\qquad\qquad \mathbf{f}^{final} = \frac{1}{6}\left(\mathbf{f}^{(k)} + 4\,\mathbf{f}^{tmp1} + \mathbf{f}^{tmp2} \right)$

7. Advance: $\qquad \mathbf{x}^{(k+1)} = \mathbf{x}^{(k)} + \Delta t\, \mathbf{f}^{final}$

Each complete time step introduces a numerical error of order Δt^4.

Fourth-order Runge–Kutta method

The fourth-order Runge–Kutta method (RK3) requires four velocity evaluations in each step. The solution is advanced according to the following nine substeps:

1. Compute: $\quad \mathbf{f}^{(k)} \equiv \mathbf{f}(\mathbf{x}^{(k)}, t_k)$

2. Set: $\qquad \mathbf{x}^{tmp1} = \mathbf{x}^{(k)} + \frac{1}{2}\,\Delta t\, \mathbf{f}^{(k)}$

3. Compute: $\quad \mathbf{f}^{tmp1} = \mathbf{f}(\mathbf{x}^{tmp1}, t_k + \frac{1}{2}\,\Delta t)$

4. Set: $\qquad \mathbf{x}^{tmp2} = \mathbf{x}^{(k)} + \frac{1}{2}\,\Delta t\, \mathbf{f}^{tmp1}$

5. Compute: $\quad \mathbf{f}^{tmp2} = \mathbf{f}(\mathbf{x}^{tmp2}, t_k + \frac{1}{2}\,\Delta t)$

6. Set: $\qquad \mathbf{x}^{tmp3} = \mathbf{x}^{(k)} + \Delta t\, \mathbf{f}^{tmp2}$

7. Compute: $\quad \mathbf{f}^{tmp3} = \mathbf{f}(\mathbf{x}^{tmp3}, t_k + \Delta t)$

8. Set: $\qquad \mathbf{f}^{final} = \frac{1}{6}\left(\mathbf{f}^{(k)} + 2\,\mathbf{f}^{tmp1} + 2\,\mathbf{f}^{tmp2} + \mathbf{f}^{tmp3} \right)$

9. Advance: $\quad \mathbf{x}^{(k+1)} = \mathbf{x}^{(k)} + \Delta t\, \mathbf{f}^{final}$

Each complete time step introduces a numerical error of order Δt^5.

FDLIB software library

C

The software library FDLIB contains a collection of Fortran 77, C++, Matlab, and other programs that solve a broad range of problems in fluid dynamics and related disciplines by a variety of numerical methods. FDLIB consist of the thirteen main directories listed in Table C.1. Each main directory contains a multitude of nested subdirectories that include main programs, assisting subroutines, and utility subroutines. Linked with drivers, the utility subroutines become stand-alone modules; all drivers are provided. A list of subdirectories and a brief description of their contents are given in this appendix. Further information is available at the FDLIB Internet site: http://dehesa.freeshell.org/FDLIB. Sample results generated by various codes are shown in Figures C.1–C.10.

C.1 Installation

The source code of FDLIB, containing programs and data files, can be obtained from the FDLIB Internet site stated above. The directories have been archived using the **tar** Unix facility into the compressed *FDLIB_*.tgz* file, where the asterisk denotes the version number encoded as year and month (yy_mm). To unravel the directories (folders) in a Unix system, issue the Unix command: **tar xzf FDLIB_*.tgz**. This will generate the *FDLIB_** directory containing nested subdirectories. To unravel the directories in another operating system, double-click on the archived tar file and follow the instructions of the invoked application.

Compilation and execution

Matlab programs are executed as interpreted scripts. The downloaded FDLIB package does not contain Fortran 77 or C++ object or executable files. To compile and link Fortran 77 or C++ programs, follow the instructions of your compiler. An application can be built using the makefile provided in each subdirectory. The makefiles contain scripts interpreted by the *make* utility that instruct the operating system how to compile the main programs and subroutines, and then link the object files into executable binary files.

To compile an application named *nea_krini*, navigate to the subdirectory where the application resides and type: **make nea_krini**. To compile the Fortran 77 programs using a FORTRAN 90 compiler, make appropriate compiler call substitutions in the makefiles. To remove the object files, output files, and executable of an application named *polihni*, navigate to the subdirectory where the application resides and issue the command: **make clean**.

	Subject	Directory
1	Numerical methods	*01_num_meth*
2	Grids	*02_grids*
3	Hydrostatics	*03_hydrostat*
4	Various	*04_various*
5	Lubrication	*05_lub*
6	Stokes flow	*06_stokes*
7	Potential flow	*07_ptf*
8	Hydrodynamic stability	*08_stab*
9	Vortex motion	*09_vortex*
10	Boundary layers	*10_bl*
11	Finite-difference methods	*11_fdm*
12	Boundary-element methods	*12_bem*
13	Turbulence	*13_turbo*

TABLE C.1   The software library FDLIB is arranged in thirteen directories according to physical or numerical classification.

CFDLAB

A subset of FDLIB has been combined with the X11 graphics library *vogle* into the integrated application CFDLAB that visualizes the results of simulations and performs interactive animation. *Vogle* is written in C but offers Fortran 77 and C++ interfaces. The source code of CFDLAB can be downloaded from the Internet site: `http://dehesa.freeshell.org/CFDLAB`.

BEMLIB

A subset of FDLIB containing boundary-element and related codes have been arranged in the library BEMLIB [313]. The source code of BEMLIB can be downloaded from the Internet site: `http://dehesa.freeshell.org/BEMLIB`.

FIGURE C.1   Streamlines of potential flow exiting a two-dimensional channel generated using code *strml* in Directory *04_various* of FDLIB.

C.2   FDLIB directory contents

The public FDLIB directories are listed in the following tables along with a brief description. Further information can be found at the FDLIB Internet site.

01_num_meth

This directory contains a suite of general-purpose programs on general numerical methods and differential equations accompanying Reference [317].

Subdirectory	Topic
01_num_comp	General aspects of numerical computation.
02_lin_calc	Linear algebra and linear calculus.
03_lin_eq	Systems of linear algebraic equations.
04_nl_eq	Nonlinear algebraic equations.
05_eigen	Eigenvalues and eigenvectors of matrices.
06_interp_diff	Function interpolation and differentiation.
07_integration	Function integration.
08_approximation	Function approximation.
09_ode	Ordinary differential equations.
10_ode_ddm	Ordinary differential equations (domain discretization methods).
11_pde_diffusion	Partial differential equations (unsteady diffusion equation).
12_pde_poisson	Partial differential equations (Poisson equation).
13_pde_cd	Partial differential equations (convection--diffusion equation).
14_bem	Boundary-element methods.
15_fem	Finite-element methods.
99_spec_fnc	Special functions.

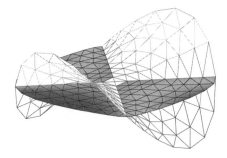

FIGURE C.2   Finite-element solution of Laplace's equation with the Dirichlet boundary condition generated using code *lapl3_d* in Directory of *15_fem* inside Directory *01_num_meth* of FDLIB.

02_grids

This directory contains programs that perform grid generation, adaptive discretization, parametrization, representation, and meshing of planar lines, three-dimensional lines, and three-dimensional surfaces.

Subdirectory	Topic
grid_2d	Discretization of a planar line into a mesh of straight or circular elements.
prd_2d	Adaptive parametrization of a closed line.
prd_3d	Adaptive parametrization of a closed three-dimensional line.
prd_ax	Adaptive parametrization of a planar line representing the trace of an axisymmetric surface in a meridional plane.
rec_2d	Interpolation through a Cartesian grid.
trgl	Triangulation of a closed surface.
trgl_flat	Triangulation of a flat patch.

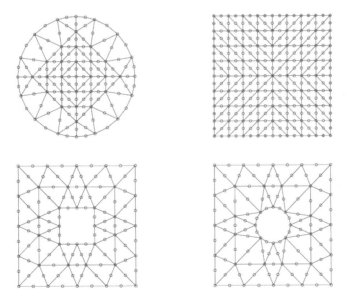

FIGURE C.3 Triangulation of a disk, square, square with a square hole and square with a circular hole into six-node triangular elements generated using codes *trgl6_disk*, *trgl6_sqr*, *trgl6_ss*, and *trgl6_sc* in Directory of *trgl_flat* inside Directory *02_grids* of FDLIB.

03_hydrostat

This directory contains codes that compute interfacial shapes in hydrostatics.

Subdirectory	Topic
drop_2d	Shape of a two-dimensional pendant or sessile drop on a horizontal or inclined plane.
drop_ax	Shape of an axisymmetric pendant or sessile drop on a horizontal plane.
flsphere	Position of a sphere floating at an interface with a curved meniscus.
men_2d	Shape of a two-dimensional meniscus between two parallel plates.
men_2d_plate	Shape of a two-dimensional meniscus attached to an inclined plane.
men_ax	Shape of an axisymmetric meniscus inside a vertical circular tube.
men_axe	Shape of an axisymmetric meniscus outside a vertical circular tube.
men_cc	Shape of a three-dimensional meniscus between two circular cylinders.
men_ell	Shape of a three-dimensional meniscus in the exterior of an ellipse.
men_hexa	Shape of a doubly periodic meniscus with hexagonal cells.

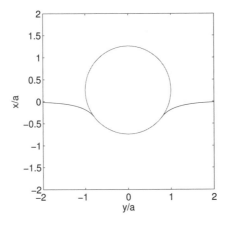

FIGURE C.4 Equilibrium position of a floating sphere generated using code *flsphere* in Directory *03_hydrostat* of FDLIB.

04_various

This directory contains miscellaneous codes that compute the structure and kinematics of various flows.

Subdirectory	Topic
chan_2d	Steady flow in a channel.
chan_2d_2l	Steady two-layer flow in a channel.
chan_2d_imp	Impulsive flow in a channel.
chan_2d_ml	Multi-layer flow in a channel.
chan_2d_osc	Oscillatory flow in a channel.
chan_2d_trans	Transient flow in a channel.
chan_2d_wom	Pulsating flow in a channel.
chan_brush	Steady flow in a brush-like channel.
film	Film flow down an inclined plane.
films	Multi-film flow down an inclined plane.
flow_1d	Steady unidirectional flow in a tube with arbitrary cross-section.
flow_1d_1p	Steady unidirectional flow over a periodic array of cylinders with arbitrary cross-section.
flow_1d_osc	Oscillatory unidirectional flow in a tube with arbitrary cross-section.
path_lines	Computation of path lines.
plate_imp	Flow due to the impulsive motion of a plate.
plate_osc	Flow due to the oscillations of a plate.
spf	Similarity solutions for stagnation-point flows.
strml	Streamline patterns of a broad range of flows offered in a menu.
tube_ann	Steady annular flow.
tube_ann_ml	Steady multi-layer annular flow.
tube_ann_sw	Steady swirling annular flow.
tube_ann_sw_ml	Steady multi-layer swirling annular flow.
tube_crc	Steady flow through a circular tube.
tube_crc_ml	Steady multi-layer flow through a circular tube.
tube_sec	Steady flow through a circular tube due to the translation of a sector.
tube_sw	Transient swirling flow in a circular tube.
tube_sw_trans	Transient flow through a circular tube.
tube_sw_wom	Pulsating flow through a circular tube.
tube_ell	Steady flow through a tube with elliptical cross-section.
tube_rec	Steady flow through a tube with rectangular cross-section.
tube_trgl_eql	Steady flow through a tube with triangular cross-section.

05_lub

This directory contains codes that solve problems involving lubrication flows.

Subdirectory	Topic
bear_2d	Dynamical simulation of the motion of a slider bearing pressing against a flat wall.
chan_21_exp	Dynamical simulation of the evolution of two superposed viscous layers in a horizontal or inclined channel computed by an explicit finite-difference method.
chan_21_imp	Same as chan_21_exp but with an implicit finite-difference method.
films	Evolution of an arbitrary number of superposed films on a horizontal or inclined wall.

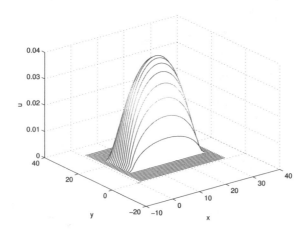

FIGURE C.5   Velocity profiles of unidirectional pressure-driven flow through a rectangular tube computed by a boundary-element method implemented in the code *flow_1d* in Directory *04_various* of FDLIB.

FIGURE C.6   Evolution of two films flowing down an inclined plane computed using code *films* in Directory *05_lub* of FDLIB.

06_stokes

This directory contains codes that compute viscous flows at vanishing Reynolds number.

Subdirectory	Topic
bump_3d	Shear flow over a spherical bump on a plane wall.
drop_ax	Dynamical simulation of the motion of an axisymmetric liquid drop in free space, toward a plane wall, or through a circular tube.
flow_2d	Two-dimensional flow in a domain with arbitrary geometry.
flow_3x	Shear flow over an axisymmetric cavity, orifice, or protrusion.
prtcl_2d	Flow past a fixed bed of two-dimensional particles with arbitrary shapes for a variety of flow configurations computed by a boundary-element method.
prtcl_3d	Flow past or due to the motion of a three-dimensional particle for a variety of flow configurations computed by a boundary-element method.
prtcl_3d_mob	Same as prtcl_3d but for the mobility problem where the force and torque are specified and the particle velocity is computed as part of the solution.
prtcl_ax	Flow past or due to the motion of a collection of axisymmetric particles computed by a boundary-element method.
prtcl_sw	Swirling flow produced by the rotation of an axisymmetric particle computed by a boundary-element method.
rbc_2d	Flow-induced deformation of a two-dimensional red blood cell.
sgf_2d	Green's functions of two-dimensional flow.
sgf_3d	Green's functions of three-dimensional flow.
sgf_ax	Green's functions of axisymmetric flow.

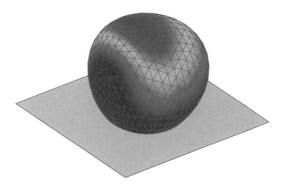

FIGURE C.7   Color-coded (or grayscale) plot of the shear stress over the surface of a particle attached to a wall in simple shear flow computed by a boundary-element method implemented in code *bump_3d* included in Directory *06_stokes* of FDLIB.

07_ptf

This directory contains codes that solve problems involving potential flow.

Subdirectory	Topic
airf_2d	Airfoil shapes.
airf_2d_cdp	Flow past an airfoil computed by the constant-dipole-panel method.
airf_2d_csdp	Flow past an airfoil computed by the constant-source-dipole-panel method.
airf_2d_lvp	Flow past an airfoil computed by the linear-vortex-panel method.
body_2d	Flow past or due to the motion of a two-dimensional body computed by a boundary-element method.
body_ax	Flow past or due to the motion of an axisymmetric body computed by a boundary-element method.
cvt_2d	Flow in a rectangular cavity computed by a finite-difference method.
flow_2d	Two-dimensional flow in an arbitrary domain computed by a boundary-element method.
lgf_2d	Green and Neumann functions of Laplace's equation in two dimensions.
lgf_3d	Green and Neumann functions of Laplace's equation in three dimensions.
lgf_ax	Green and Neumann functions of Laplace's equation in axisymmetric domains.
tank_2d	Dynamical simulation of liquid sloshing in a rectangular tank computed by a boundary-element method.

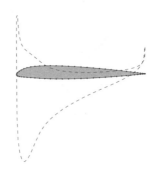

FIGURE C.8 Distribution of the pressure coefficient around the NACA 23012 airfoil computed by a panel method implemented in code *airf_2d_lvp* included in Directory *07_ptf* of FDLIB.

08_stab

This directory contains codes that perform the linear stability analysis of miscellaneous flows. The
fundamental concepts underlying the linear stability analysis are discussed in Chapter 9. The User
Guide of this Directory can be found in Appendix D.

Subdirectory	Topic
ann21	Capillary instability of two annular layers between two concentric cylinders in the presence of an insoluble surfactant.
ann21o	Same as ann21 for Stokes flow.
ann21el	Same as ann21 for an elastic interface.
ann21el0	Same as ann21o for an elastic interface.
ann21vs0	Same as ann21o for a viscous interface.
chan_210	Instability of two-layer Stokes flow in a channel.
chan_210_s	Instability of two-layer Stokes flow in a channel in the presence of an insoluble surfactant.
coat0_s	Stokes flow instability of a liquid film resting on a plane wall in the presence of an insoluble surfactant.
drop_ax	Relaxation of a deformed liquid drop.
film0	Stokes flow instability of a liquid film down an inclined plane.
film0_s	Stokes flow instability of a liquid film down an inclined plane in the presence of an insoluble surfactant.
if0	Stokes flow instability of a horizontal interface between two semi-infinite fluids.
ifsf0_s	Stokes flow instability of a horizontal interface between two semi-infinite fluids in the presence of an insoluble surfactant.
layer0	Stokes flow instability of a horizontal liquid layer resting on a horizontal wall under a semi-infinite fluid.
layersf0_s	Stokes flow instability of a sheared liquid layer coated on a horizontal plane in the presence of an insoluble surfactant.
orr	Solution of the Orr-Sommerfeld equation by a finite-difference method.
prony	Prony fitting of a times series with a sum of complex exponentials.
sf1	Instability of an inviscid shear flow with arbitrary velocity profile.
thread0	Stokes flow instability of an infinite viscous thread suspended in an ambient viscous fluid.
thread1	Instability of an inviscid thread suspended in an inert gas.
vl	Instability of a vortex layer.
vs	Instability of a vortex sheet.

09_vortex

This directory contains codes that compute vortex motion.

Subdirectory	Topic
kirch_stab	Linear stability analysis of Kirchhoff's elliptical vortex by a discrete perturbation method based on the contour dynamics formulation.
lv_lia	Dynamical simulation of the motion of a three-dimensional line vortex computed by the local-induction approximation (LIA).
lvr	Velocity induced by line vortex rings.
lvrm	Dynamical simulation of the motion of a collection of coaxial line vortex rings.
pv	Velocity induced by point vortices.
pvm	Dynamic simulation of the motion of a collection of point vortices.
pvm_pr	Dynamical simulation of the motion of a periodic collection of point vortices.
pvpoly2	Equilibrium of point vortices on two nested polygons.
ring	Self-induced velocity of a vortex ring with core of finite size.
vl_2d	Dynamical simulation of the evolution of compound periodic vortex layers.
vp_2d	Dynamical simulation of the evolution of a collection of two-dimensional vortex patches.
vp_ax	Dynamical simulation of the evolution of a collection of axisymmetric vortex rings and vortex patches with distributed vorticity.

FIGURE C.9   Evolution of a vortex patch computed by the method of contour dynamics implemented in code *vp_2d* in Directory *09_vortex* of FDLIB.

10_bl

This directory contains codes that solve boundary-layers flows.

Subdirectory	Topic
blasius	Computation of the Blasius boundary layer.
falskan	Computation of the Falkner—Skan boundary layer.
kp_cc	Boundary layer around a circular cylinder computed by the Kármán—Pohlhausen method.
pohl_pol	Profiles of the Pohlhausen polynomials.

11_fdm

This directory contains codes that solve problems using finite-difference methods.

Subdirectory	Topic
channel	Unidirectional flow in a channel.
cvt_pm	Transient flow in a rectangular cavity computed by a projection method.
cvt_stag	Steady Stokes flow in a rectangular cavity computed on a staggered grid.
cvt_sv	Steady flow in a rectangular cavity computed by the stream function—vorticity formulation.

12_bem

This directory contains codes that produce solutions to Laplace's equation by boundary-element methods.

Subdirectory	Topic
ldr_3d	Solution of Laplace's equation with the Dirichlet boundary condition in the interior or exterior of a three-dimensional region computed using the boundary-integral formulation.
ldr_3d_2p	Solution of Laplace's equation with the Dirichlet boundary in a semi-infinite domain bounded by a doubly-periodic surface computed using the double-layer formulation.
ldr_3d_ext	Solution of Laplace's equation with the Dirichlet boundary condition in the exterior of a three-dimensional region computed using the double-layer formulation.
ldr_3d_int	Solution of Laplace's equation with the Dirichlet boundary condition in the interior of a three-dimensional region computed using the double-layer formulation.
lnm_3d	Solution of Laplace's equation with the Neumann boundary condition in the interior or exterior of a three-dimensional region computed using the boundary-integral formulation.

13_turbo

This directory contains data and codes pertinent to turbulent flow.

Subdirectory	Topic
stats	Statistical analysis of a turbulent velocity time series.

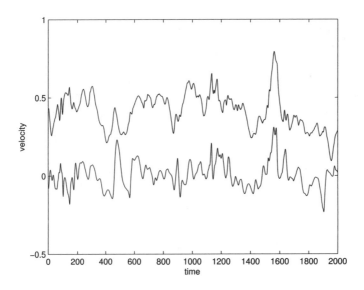

FIGURE C.10   A time series of two velocity components in a turbulent flow contained in the file *keller.dat* included in Directory *stats* inside Directory *13_turbo* of FDLIB.

D

User Guide of directory 08_stab of FDLIB on hydrodynamic stability

This appendix contains the User Guide of the eighth directory of the software library FDLIB on hydrodynamic stability (see Appendix C). The individual subdirectories contain programs that perform the linear stability analysis of several families of inviscid and viscous flows at zero or nonzero Reynolds numbers. The fundamental concepts underlying the linear stability analysis are discussed in Chapter 9. The subdirectories of the main directory *08_stab* are listed in Table D.1.

Subdirectory	Topic
ann21	Rayleigh capillary instability of an annular interface between two concentric cylinders.
ann210	Rayleigh capillary instability of an annular interface between two concentric cylinders under conditions of Stokes flow.
ann21el	Rayleigh capillary instability of an annular elastic interface between two concentric cylinders.
ann21el0	Rayleigh capillary instability of an annular elastic interface between two concentric cylinders under conditions of Stokes flow.
ann21vs0	Rayleigh capillary instability of an annular viscous interface between two concentric cylinders under conditions of Stokes flow.
chan210	Linear stability of two-layer Stokes flow in a channel.
chan210_s	Linear stability of two-layer Stokes flow in a channel in the presence of an insoluble surfactant.

TABLE D.1   Contents of the FDLIB directory *08_stab* on hydrodynamic stability. (*Continuing.*)

Subdirectory	Topic
coat0_s	Instability of a liquid film resting on a horizontal wall in the presence of an insoluble surfactant in Stokes flow.
drop_ax	Relaxation rate of a deformed spherical viscous drop.
film0	Instability of a liquid film flowing down an inclined plane under conditions of Stokes flow.
film0_s	Instability of a liquid film flowing down an inclined plane in the presence of an insoluble surfactant under conditions of Stokes flow.
if0	Instability of a planar interface between two semi-infinite fluids under conditions of Stokes flow.
ifsf0_s	Instability of a planar interface between two semi-infinite fluids in shear flow in the presence of an insoluble surfactant under conditions of Stokes flow.
layer0	Instability of a liquid layer resting on a horizontal wall under a semi-infinite fluid in Stokes flow.
layersf0_s	Instability of a liquid layer resting on a horizontal wall under a semi-infinite fluid undergoing simple shear flow in the presence of an insoluble surfactant under conditions of Stokes flow.
prony	Decomposition of a time series into normal modes expressing exponentially growing or decaying sinusoidal waves. Decomposition of linear spatial waves into exponentially evolving normal modes.
orr	Solution of the Orr–Sommerfeld equation for viscous unidirectional flow.
thread0	Capillary instability of viscous thread immersed in an infinite ambient viscous fluid under conditions of Stokes flow.
sf1	Normal-mode stability analysis of inviscid shear flow with a velocity profile specified in analytical or numerical form.

TABLE D.1 (*Continued.*)

Subdirectory	Topic
thread1	Capillary instability of an inviscid liquid column suspended in an infinite ambient fluid with negligible density.
vl	Kelvin–Helmholtz instability of an inviscid vortex layer with constant vorticity.
vs	Instability of a vortex sheet.

TABLE D.1   (*Continued.*) Contents of the FDLIB directory *08_stab* on hydrodynamic stability.

The User Guide of the individual directories is presented in this appendix, including the definition of the base state, the formulation of the linear stability problem, and the derivation of an algebraic eigenvalue problem determining the complex growth rate. To facilitate the study of the individual programs, some parts of the analysis and a few figures are repeated across the individual directories. This User Guide is intended to be used in concert with Chapter 9 on the fundamentals of hydrodynamic stability.

D.1   Directory ann2l

This directory contains a code that performs the linear stability analysis of the interface between two annular layers confined between two concentric cylinders, as illustrated in Figure D.1.1 [214]. The interface is populated by an insoluble surfactant that is convected and diffuses over the interface but not into the bulk of the fluids.

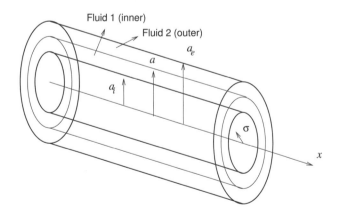

FIGURE D.1.1   Illustration of two annular layers confined between an inner cylindrical tube of radius a_i and an outer cylindrical tube of radius a_e. The interface is occupied by an insoluble surfactant.

Base state

We consider two stationary annular layers separated by a cylindrical interface of radius a, placed between an inner cylindrical tube of radius a_i and an outer cylindrical tube of radius a_e, as illustrated in Figure D.1.1. The inner fluid is labeled 1 and the outer fluid is labeled 2. In the unperturbed configuration, the fluids are quiescent and the surfactant is distributed uniformly over the interface. Several special cases can be recognized:

- When the inner cylinder is absent and the radius of the outer cylinder is infinite, we obtain a thread suspended in an infinite ambient fluid.

- When the inner cylinder is absent, we obtain a core–annular arrangement consisting of an annular layer coated on the interior surface of the inner cylinder.

- When the radius of the outer cylinder is infinite, we obtain an annular layer coated on the exterior surface of the inner cylinder.

Normal-mode analysis

To carry out the normal-mode stability analysis for axisymmetric perturbations, we introduce cylindrical polar coordinates, (x, σ, φ), where the x axis coincides with the axis of revolution of the interface. The radial position of the interface is described by the real or imaginary part of the function

$$\sigma = f(x,t) = a + \epsilon \eta(x,t), \tag{D.1.1}$$

where ϵ is a dimensionless coefficient whose magnitude is much less than unity,

$$\eta(x,t) = A \exp[ik(x - ct)], \tag{D.1.2}$$

is the normal-mode wave form of the perturbation, A is the complex amplitude of the interfacial wave, $k = 2\pi/L$ is the wave number, L is the wavelength, and c is the complex phase velocity.

Stream function, velocity and vorticity

The motion of the fluid on either side of the interface is governed by the continuity equation and the Navier–Stokes equation with appropriate physical constants for each fluid. Taking advantage of the axial symmetry of the flow, we describe the perturbation flow in terms of the Stokes stream functions, $\psi_1^{(1)}$ and $\psi_2^{(1)}$, for the inner and outer fluid, respectively, where the superscript (1) denotes the disturbance. The axial and radial components of the perturbation velocity are given by

$$u_{x_j} = \epsilon \frac{1}{\sigma} \frac{\partial \psi_j^{(1)}}{\partial \sigma}, \qquad u_{\sigma_j} = -\epsilon \frac{1}{\sigma} \frac{\partial \psi_j^{(1)}}{\partial x} \tag{D.1.3}$$

for $j = 1, 2$. The azimuthal component of the vorticity is given by

$$\omega_{\varphi_j} = -\epsilon \frac{1}{\sigma} E^2 \psi_j^{(1)}, \tag{D.1.4}$$

where

$$E^2 \equiv \frac{\partial^2}{\partial x^2} + \frac{\partial^2}{\partial \sigma^2} - \frac{1}{\sigma}\frac{\partial}{\partial \sigma} \qquad (D.1.5)$$

is a second-order differential operator. It is convenient to work with the vorticity transport equation for axisymmetric flow,

$$\frac{D}{Dt}\left(\frac{1}{\sigma}\omega_{\varphi_j}\right) = \frac{\nu_j}{\sigma^2} E^2(\sigma\,\omega_{\varphi_j}), \qquad (D.1.6)$$

where D/Dt is the material derivative and ν_j is the kinematic viscosity of the jth fluid. Substituting expression (D.1.4) into (D.1.6), linearizing with respect to ϵ and rearranging, we obtain

$$\left(E^2 - \frac{1}{\nu_j}\frac{\partial}{\partial t}\right) E^2\,\psi_j^{(1)} = 0 \qquad (D.1.7)$$

for $j = 1, 2$.

To carry out the normal-mode analysis, we express the perturbation stream function in the form

$$\psi_j^{(1)}(x, \sigma, t) = \phi_j(\sigma)\,\exp[ik(x - ct)], \qquad (D.1.8)$$

where $\phi_j(\sigma)$ are eigenfunctions. Substituting this expression into (D.1.7), we obtain

$$\left(\frac{d^2}{d\sigma^2} - \frac{1}{\sigma}\frac{d}{d\sigma} - k^2\right)\left(\frac{d^2}{d\sigma^2} - \frac{1}{\sigma}\frac{d}{d\sigma} - k_j^2\right)\phi_j = 0, \qquad (D.1.9)$$

where

$$k_j^2 \equiv k^2 - i\,c\,\frac{k}{\nu_j}. \qquad (D.1.10)$$

The general solution is

$$\phi_j(\sigma) = \sigma\left[A_{1,j}\,\mathrm{I}_1(\hat{\sigma}) + A_{2,j}\,\mathrm{I}_1(\hat{\sigma}_j) + B_{1,j}\,\mathrm{K}_1(\hat{\sigma}) + B_{2,j}\,\mathrm{K}_1(\hat{\sigma}_j)\right] \qquad (D.1.11)$$

for $j = 1, 2$, where $\hat{\sigma} \equiv k\sigma$, $\hat{\sigma}_j \equiv k_j\sigma$, I_1 and K_1 are Bessel functions, and $A_{i,j}, B_{i,j}$ are complex coefficients to be found as part of the solution. The first and third terms on the right-hand side of (D.1.11) correspond to the first operator, while the second and fourth terms correspond to the second operator on the left-hand side of (D.1.9).

Substituting into (D.1.3) the preceding expressions and using the properties of the Bessel functions

$$\mathrm{I}_0'(z) = \mathrm{I}_1(z), \qquad \mathrm{I}_1'(z) = \mathrm{I}_0(z) - \frac{1}{z}\mathrm{I}_1(z), \qquad \mathrm{K}_0'(z) = -\mathrm{K}_1(z),$$

$$\mathrm{K}_1'(z) = -\mathrm{K}_0(z) - \frac{1}{z}\mathrm{K}_1(z), \qquad (D.1.12)$$

we find that

$$u_{x_j} = \epsilon\, U_{x_j}(\sigma)\, \exp[\mathrm{i}k(x - ct)], \qquad u_{\sigma_j} = \epsilon\, U_{\sigma_j}(\sigma)\, \exp[\mathrm{i}k(x - ct)], \tag{D.1.13}$$

where

$$U_{x_j}(\sigma) = \frac{1}{\sigma}\frac{\mathrm{d}\phi_j}{\mathrm{d}\sigma} = A_{1,j}\, k\, \mathrm{I}_0(\hat\sigma) + A_{2,j}\, k_j\, \mathrm{I}_0(\hat\sigma_j) - B_{1,j}\, k\, \mathrm{K}_0(\hat\sigma) - B_{2,j}\, k_j\, \mathrm{K}_0(\hat\sigma_j), \tag{D.1.14}$$

and

$$U_{\sigma_j}(\sigma) = -\mathrm{i}\frac{k}{\sigma}\phi_j = -\mathrm{i}k\left[\, A_{1,j}\, \mathrm{I}_1(\hat\sigma) + A_{2,j}\, \mathrm{I}_1(\hat\sigma_j) + B_{1,j}\, \mathrm{K}_1(\hat\sigma) + B_{2,j}\, \mathrm{K}_1(\hat\sigma_j)\,\right]. \tag{D.1.15}$$

Pressure field

The normal-mode expression for the pressure is

$$p_j = P_j + \epsilon\, \chi_j(\sigma)\, \exp[\mathrm{i}k(x - ct)], \tag{D.1.16}$$

where P_1 and P_2 are the unperturbed pressures, $P_1 - P_2 = \gamma_0/a$ is the unperturbed capillary pressure, γ_0 is the unperturbed surface tension, and χ_j are eigenfunctions. To compute the disturbance pressure, we consider the x component of the Navier–Stokes equation for axisymmetric flow,

$$\rho_j\frac{\mathrm{D}u_{x_j}}{\mathrm{D}t} = -\frac{\partial p_j}{\partial x} + \mu_j\left(\frac{\partial^2 u_{x_j}}{\partial x^2} + \frac{1}{\sigma}\frac{\partial}{\partial\sigma}(\sigma\frac{\partial u_{x_j}}{\partial\sigma})\right). \tag{D.1.17}$$

Substituting (D.1.16) together with the first equation into (D.1.13) and linearizing, we obtain

$$\chi_j(\sigma) = (c\rho_j + \mathrm{i}\mu_j\, k)\, U_{x_j} - \mathrm{i}\frac{\mu_j}{k}\frac{1}{\sigma}\frac{\mathrm{d}}{\mathrm{d}\sigma}\left(\sigma\frac{\mathrm{d}U_{x_j}}{\mathrm{d}\sigma}\right) = -\mathrm{i}\frac{\mu_j}{k}\left(\frac{1}{\sigma}\frac{\mathrm{d}}{\mathrm{d}\sigma}(\sigma\frac{\mathrm{d}U_{x_j}}{\mathrm{d}\sigma}) - k_j^2\, U_{x_j}\right), \tag{D.1.18}$$

yielding

$$\chi_j(\sigma) = -\mathrm{i}\frac{\mu_j}{k}\left(\frac{\mathrm{d}^2 U_{x_j}}{\mathrm{d}\sigma^2} + \frac{1}{\sigma}\frac{\mathrm{d}U_{x_j}}{\mathrm{d}\sigma} - k_j^2\, U_{x_j}\right). \tag{D.1.19}$$

Making substitutions, we find that

$$\chi_j(\sigma) = -\mathrm{i}\mu_j\, (k^2 - k_j^2)\left[\, A_{1,j}\, \mathrm{I}_0(\hat\sigma) - B_{1,j}\, \mathrm{K}_0(\hat\sigma)\,\right] = kc\rho_j\left[\, A_{1,j}\, \mathrm{I}_0(\hat\sigma) - B_{1,j}\, \mathrm{K}_0(\hat\sigma)\,\right]. \tag{D.1.20}$$

Note that only two terms contribute to the pressure field.

Continuity of velocity at the interface

Requiring that the x and σ velocity components are continuous at the interface and using expressions (D.1.14) and (D.1.15), we derive the linearized kinematic interfacial conditions

$$A_{1,1}\, k\, \mathrm{I}_0(\hat{k}) + A_{2,1}\, k_1\, \mathrm{I}_0(\hat{k}_1) - B_{1,1}\, k\, \mathrm{K}_0(\hat{k}) - B_{2,1}\, k_1\, \mathrm{K}_0(\hat{k}_1)$$
$$= A_{1,2}\, k\, \mathrm{I}_0(\hat{k}) + A_{2,2}\, k_2\, \mathrm{I}_0(\hat{k}_2) - B_{1,2}\, k\, \mathrm{K}_0(\hat{k}) - B_{2,2}\, k_2\, \mathrm{K}_0(\hat{k}_2) \tag{D.1.21}$$

and

$$A_{1,1}\, \mathrm{I}_1(\hat{k}) + A_{2,1}\, \mathrm{I}_1(\hat{k}_1) + B_{1,1}\, \mathrm{K}_1(\hat{k}) + B_{2,1}\, \mathrm{K}_1(\hat{k}_1)$$
$$= A_{1,2}\, \mathrm{I}_1(\hat{k}) + A_{2,2}\, \mathrm{I}_1(\hat{k}_2) + B_{1,2}\, \mathrm{K}_1(\hat{k}) + B_{2,2}\, \mathrm{K}_1(\hat{k}_2), \tag{D.1.22}$$

where $\hat{k} = ka$, $\hat{k}_1 = k_1 a$, and $\hat{k}_2 = k_2 a$.

Kinematic compatibility

Kinematic compatibility requires that $\mathrm{D}[f(x,t) - \sigma]/\mathrm{D}t = 0$ at the interface, where $\mathrm{D}/\mathrm{D}t$ is the material derivative and the function f describes the position of the interface according to (D.1.1). Carrying out the differentiation, we obtain

$$\frac{\partial f}{\partial t} + u_{x_j}\frac{\partial f}{\partial x} - u_{\sigma_j} = 0, \tag{D.1.23}$$

where the velocity is evaluated at the interface. Substituting the preceding expressions and linearizing, we find that

$$A = \frac{\mathrm{i}}{ck}\,U_{\sigma_j}(\sigma = a) = \frac{1}{c}\,[\,A_{1,j}\,\mathrm{I}_1(\hat{k}) + A_{2,j}\,\mathrm{I}_1(\hat{k}_j) + B_{1,j}\,\mathrm{K}_1(\hat{k}) + B_{2,j}\,\mathrm{K}_1(\hat{k}_j)\,]. \tag{D.1.24}$$

Boundary conditions at the cylinders

The no-slip and no-penetration conditions at the inner and outer cylinder require that

$$U_{x_j}(a_i) = U_{\sigma_j}(a_i) = 0, \qquad U_{x_j}(a_e) = U_{\sigma_j}(a_e) = 0. \tag{D.1.25}$$

Making substitutions, we find that

$$\begin{aligned} A_{1,1}\,k\,\mathrm{I}_0(\hat{k}_i) + A_{2,1}\,k_1\,\mathrm{I}_0(\hat{k}_{i_1}) - B_{1,1}\,k\,\mathrm{K}_0(\hat{k}_i) - B_{2,1}\,k_1\,\mathrm{K}_0(\hat{k}_{i_1}) &= 0, \\ A_{1,1}\,\mathrm{I}_1(\hat{k}_i) + A_{2,1}\,\mathrm{I}_1(\hat{k}_{i_1}) + B_{1,1}\,\mathrm{K}_1(\hat{k}_i) + B_{2,1}\,\mathrm{K}_1(\hat{k}_{i_1}) &= 0, \end{aligned} \tag{D.1.26}$$

and

$$\begin{aligned} A_{1,2}\,k\,\mathrm{I}_0(\hat{k}_e) + A_{2,2}\,k_2\,\mathrm{I}_0(\hat{k}_{e_2}) - B_{1,2}\,k\,\mathrm{K}_0(\hat{k}_e) - B_{2,2}\,k_2\,\mathrm{K}_0(\hat{k}_{e_2}) &= 0, \\ A_{1,2}\,\mathrm{I}_1(\hat{k}_e) + A_{2,2}\,\mathrm{I}_1(\hat{k}_{e_2}) + B_{1,2}\,\mathrm{K}_1(\hat{k}_e) + B_{2,2}\,\mathrm{K}_1(\hat{k}_{e_2}) &= 0, \end{aligned} \tag{D.1.27}$$

where $\hat{k}_i = ka_i$, $\hat{k}_{i_1} = k_1 a_i$, $\hat{k}_e = ka_e$, and $\hat{k}_{e_2} = k_2 a_e$.

Surfactant transport

The interface is occupied by an insoluble surfactant with surface concentration Γ. For a normal-mode disturbance, $\Gamma = \Gamma_0 + \epsilon\,\Gamma^{(1)}$, where Γ_0 is the unperturbed surfactant concentration,

$$\Gamma^{(1)} \equiv \Gamma_1\,\exp[\mathrm{i}k(x - ct)] \tag{D.1.28}$$

is the normal-mode perturbation, and Γ_1 is the complex amplitude of the perturbation. Linearizing the surfactant transport equation in the absence of base flow, we derive the evolution equation

$$\epsilon\,\frac{\partial \Gamma^{(1)}}{\partial t} + \Gamma_0\,\frac{\partial u_x}{\partial x} = -\frac{\Gamma_0}{a}\,u_\sigma + \epsilon D_s\,\frac{\partial^2 \Gamma^{(1)}}{\partial x^2}, \tag{D.1.29}$$

where the velocity is evaluated at the unperturbed position, $\sigma = a$. Substituting the normal-mode forms and rearranging, we find that

$$\begin{aligned} \frac{\Gamma_1}{\Gamma_0} &= -\frac{1}{s + D_s k^2}\left(\mathrm{i}k\,U_{x_j} + \frac{U_{\sigma_j}}{\sigma}\right)_{\sigma=a} = -\mathrm{i}\,\frac{k}{a\,(s + D_s\,k^2)}\Big(A_{1,j}\,[\hat{k}\,\mathrm{I}_0(\hat{k}) - \mathrm{I}_1(\hat{k})] \\ &\quad + A_{2,j}\,[\hat{k}_j\,\mathrm{I}_0(\hat{k}_j) - \mathrm{I}_1(\hat{k}_j)] - B_{1,j}\,[\hat{k}\,\mathrm{K}_0(\hat{k}) + \mathrm{K}_1(\hat{k})] - B_{2,j}\,[\hat{k}_j\,\mathrm{K}_0(\hat{k}_j) + \mathrm{K}_1(\hat{k}_j)]\Big), \end{aligned} \tag{D.1.30}$$

where $s \equiv -ikc$. If c is imaginary, as expected in the absence of a base flow, $s = kc_I$ is the real growth rate.

Surface equation of state

Surfactant concentration inhomogeneities over the interface cause corresponding variations in the surface tension. For a normal-mode perturbation,

$$\gamma = \gamma_0 + \epsilon\,\gamma_1 \exp[ik(x - ct)], \tag{D.1.31}$$

where γ_0 is the surface tension in the unperturbed state corresponding to the surfactant concentration Γ_0, and γ_1 is the complex amplitude of the perturbation. In the linearized approximation, the surface tension depends linearly on the surfactant concentration,

$$\frac{\gamma_1}{\Gamma_1} = -\mathrm{Ma}\,\frac{\gamma_0}{\Gamma_0}, \tag{D.1.32}$$

where Ma is the Marangoni number. Substituting (D.1.30) and rearranging, we obtain

$$\frac{\gamma_1}{\gamma_0} = i\,\mathrm{Ma}\,\frac{k}{a\,(s + D_s k^2)} \Big(A_{1,j}\,[\hat{k}\,\mathrm{I}_0(\hat{k}) - \mathrm{I}_1(\hat{k})] + A_{2,j}\,[\hat{k}_j\,\mathrm{I}_0(\hat{k}_j) - \mathrm{I}_1(\hat{k}_j)]$$
$$- B_{1,j}\,[\hat{k}\,\mathrm{K}_0(\hat{k}) + \mathrm{K}_1(\hat{k})] - B_{2,j}\,[\hat{k}_j\,\mathrm{K}_0(\hat{k}_j) + \mathrm{K}_1(\hat{k}_j)] \Big). \tag{D.1.33}$$

Next, we select $j = 1$ and recast this equation into the form

$$\gamma_1 = i\,(\delta_1 A_{1,1} + \delta_2 A_{2,1} - \delta_3 B_{1,1} - \delta_4 B_{2,1}), \tag{D.1.34}$$

where

$$\delta_1 = \Pi\,[\hat{k}\,\mathrm{I}_0(\hat{k}) - \mathrm{I}_1(\hat{k})], \qquad \delta_2 = \Pi\,[\hat{k}_1 \mathrm{I}_0(\hat{k}_1) - \mathrm{I}_1(\hat{k}_1)], \qquad \delta_3 = \Pi\,[\hat{k}\mathrm{K}_0(\hat{k}) + \mathrm{K}_1(\hat{k})],$$
$$\delta_4 = \Pi\,[\hat{k}_1\mathrm{K}_0(\hat{k}_1) + \mathrm{K}_1(\hat{k}_1)], \tag{D.1.35}$$

and

$$\Pi \equiv \mathrm{Ma}\,\gamma_0\,\frac{k}{a\,(s + D_s k^2)} \tag{D.1.36}$$

is a dimensionless number.

Shear stress interfacial condition

Balancing the tangential forces exerted on an infinitesimal portion of the interface and the interfacial tension γ, we derive the linearized shear stress balance

$$\Big(\sigma_{x\sigma_1} - \sigma_{x\sigma_2}\Big)_{\sigma=a} = \mu_1 \Big(\frac{\partial u_{x_1}}{\partial \sigma} + \frac{\partial u_{\sigma_1}}{\partial x}\Big)_{\sigma=a} - \mu_2 \Big(\frac{\partial u_{x_2}}{\partial \sigma} + \frac{\partial u_{\sigma_2}}{\partial x}\Big)_{\sigma=a} = \frac{\partial \gamma}{\partial x}. \tag{D.1.37}$$

Substituting the normal-mode forms, we obtain

$$ik\,\gamma_1 = \mu_1 \Big(\frac{dU_{x_1}}{d\sigma} + ik\,U_{\sigma_1}\Big)_{\sigma=a} - \mu_2 \Big(\frac{dU_{x_2}}{d\sigma} + ik\,U_{\sigma_2}\Big)_{\sigma=a}. \tag{D.1.38}$$

Now using the expression

$$\frac{dU_{x_j}(\sigma)}{d\sigma} = A_{1,j}\, k^2\, \mathrm{I}_1(\hat\sigma) + A_{2,j}\, k_j^2\, \mathrm{I}_1(\hat\sigma_j) + B_{1,j}\, k^2\, \mathrm{K}_1(\hat\sigma) + B_{2,j}\, k_j^2\, \mathrm{K}_1(\hat\sigma_j) \tag{D.1.39}$$

along with (D.1.15), we obtain

$$\mu_1\left[A_{1,1}2\,k^2\,\mathrm{I}_1(\hat k) + A_{2,1}(k^2+k_1^2)\,\mathrm{I}_1(\hat k_1) + B_{1,1}2\,k^2\,\mathrm{K}_1(\hat k) + B_{2,1}(k^2+k_1^2)\,\mathrm{K}_1(\hat k_1)\right]$$
$$-\mu_2\left[A_{1,2}2\,k^2\,\mathrm{I}_1(\hat k) + A_{2,2}(k^2+k_2^2)\,\mathrm{I}_1(\hat k_2) + B_{1,2}2\,k^2\,\mathrm{K}_1(\hat k) + B_{2,2}(k^2+k_2^2)\,\mathrm{K}_1(\hat k_2)\right]$$
$$= \mathrm{i}k\,\gamma_1 = -k\,(\delta_1 A_{1,1} + \delta_2 A_{2,1} - \delta_3 B_{1,1} - \delta_4 B_{2,1}). \tag{D.1.40}$$

Consolidating the various terms, we derive a linear algebraic equation,

$$F_1\,A_{1,1} + F_2\,A_{2,1} + F_3\,B_{1,1} + F_4\,B_{2,1} + F_5\,A_{1,2} + F_6\,A_{2,2} + F_7\,B_{1,2} + F_8\,B_{2,2} = 0, \tag{D.1.41}$$

where

$$\begin{aligned}
F_1 &= -2\mu_1\,k^2\,\mathrm{I}_1(\hat k) - k\,\delta_1, & F_2 &= -\mu_1\,(k^2+k_1^2)\,\mathrm{I}_1(\hat k_1) - k\,\delta_2,\\
F_3 &= -2\mu_1\,k^2\,\mathrm{K}_1(\hat k) + k\,\delta_3, & F_4 &= -\mu_1\,(k^2+k_1^2)\,\mathrm{K}_1(\hat k_1) + k\,\delta_4,\\
F_5 &= 2\mu_2\,k^2\,\mathrm{I}_1(\hat k), & F_6 &= \mu_2\,(k^2+k_2^2)\,\mathrm{I}_1(\hat k_2),\\
F_7 &= 2\mu_2\,k^2\,\mathrm{K}_1(\hat k), & F_8 &= \mu_2\,(k^2+k_2^2)\,\mathrm{K}_1(\hat k_2).
\end{aligned} \tag{D.1.42}$$

Normal stress interfacial condition

Balancing the normal forces exerted on an infinitesimal patch of the interface with the interfacial tension, γ, and linearizing, we derive the normal stress balance

$$\left(\sigma_{\sigma\sigma 1} - \sigma_{\sigma\sigma 2}\right)_{\sigma=a} = \left(-p_1 + 2\mu_1\frac{\partial u_{\sigma 1}}{\partial\sigma}\right)_{\sigma=a} - \left(-p_2 + 2\mu_2\frac{\partial u_{\sigma 2}}{\partial\sigma}\right)_{\sigma=a} = \gamma\,2\kappa_m, \tag{D.1.43}$$

where κ_m is the mean curvature. In the linearized approximation,

$$2\kappa_m = -\frac1a + \epsilon\left(\frac{\eta}{a^2} + \eta_{xx}\right) = -\frac1a + \epsilon\frac{A}{a^2}\,(1-\hat k^2)\,\exp[\mathrm{i}k(x-ct)]. \tag{D.1.44}$$

Using (D.1.24) with $j=1$ to eliminate the interfacial amplitude, A, we obtain

$$2\kappa_m = -\frac1a + \epsilon\,2\,\kappa_m^{(1)}\,\exp[\mathrm{i}k(x-ct)], \tag{D.1.45}$$

where

$$2\kappa_m^{(1)} = \frac{1-\hat k^2}{ca^2}\left(A_{1,1}\,\mathrm{I}_1(\hat k) + A_{2,1}\,\mathrm{I}_1(\hat k_1) + B_{1,1}\,\mathrm{K}_1(\hat k) + B_{2,1}\,\mathrm{K}_1(\hat k_1)\right). \tag{D.1.46}$$

Next, we substitute into (D.1.43) the normal-mode expansions and obtain

$$\left(-\chi_1 + 2\mu_1\frac{\partial U_{\sigma 1}}{\partial\sigma}\right)_{\sigma=a} - \left(-\chi_2 + 2\mu_2\frac{\partial U_{\sigma 2}}{\partial\sigma}\right)_{\sigma=a}$$
$$= -\frac{\gamma_1}{a} + \gamma_0\frac{1-\hat k^2}{ca^2}\left(A_{1,1}\,\mathrm{I}_1(\hat k) + A_{2,1}\,\mathrm{I}_1(\hat k_1) + B_{1,1}\,\mathrm{K}_1(\hat k) + B_{2,1}\,\mathrm{K}_1(\hat k_1)\right). \tag{D.1.47}$$

Substituting into (D.1.47) the expression

$$\frac{\mathrm{d}U_{\sigma_j}}{\mathrm{d}\sigma} = -\mathrm{i}\,k\left[A_{1,j}\,k\left(\mathrm{I}_0(\hat{\sigma}) - \frac{1}{\hat{\sigma}}\,\mathrm{I}_1(\hat{\sigma})\right) + A_{2,j}\,k_j\left(\mathrm{I}_0(\hat{\sigma}_j) - \frac{1}{\hat{\sigma}_j}\,\mathrm{I}_1(\hat{\sigma}_j)\right)\right.$$
$$\left. - B_{1,j}\,k\left(\mathrm{K}_0(\hat{\sigma}) + \frac{1}{\hat{\sigma}}\,\mathrm{K}_1(\hat{\sigma})\right) - B_{2,j}\,k_j\left(\mathrm{K}_0(\hat{\sigma}_j) + \frac{1}{\hat{\sigma}_j}\,\mathrm{K}_1(\hat{\sigma}_j)\right)\right], \qquad (D.1.48)$$

together with the last expression in (D.1.20) for the pressure, we obtain

$$\alpha_{1,1}A_{1,1} + \alpha_{2,1}A_{2,1} + \beta_{1,1}B_{1,1} + \beta_{2,1}B_{2,1} + \alpha_{1,2}A_{1,2} + \alpha_{2,2}A_{2,2} + \beta_{1,2}B_{1,2} + \beta_{2,2}B_{2,2}$$
$$= -\mathrm{i}\,\frac{\gamma_1}{a} = \frac{1}{a}\left(\delta_1\,A_{1,1} + \delta_2\,A_{2,1} - \delta_3\,B_{1,1} - \delta_4\,B_{2,1}\right), \qquad (D.1.49)$$

where

$$\alpha_{1,1} = s\,\rho_1\,\mathrm{I}_0(\hat{k}) + 2\,\mu_1\,k^2\left(\mathrm{I}_0(\hat{k}) - \frac{1}{\hat{k}}\,\mathrm{I}_1(\hat{k})\right) - \frac{\gamma_0 k^3}{s}\,\frac{1-\hat{k}^2}{\hat{k}^2}\,\mathrm{I}_1(\hat{k}),$$

$$\alpha_{2,1} = 2\,\mu_1\,k\,k_1\left(\mathrm{I}_0(\hat{k}_1) - \frac{1}{\hat{k}_1}\,\mathrm{I}_1(\hat{k}_1)\right) - \frac{\gamma_0 k^3}{s}\,\frac{1-\hat{k}^2}{\hat{k}^2}\,\mathrm{I}_1(\hat{k}_1),$$

$$\beta_{1,1} = -s\,\rho_1\,\mathrm{K}_0(\hat{k}) - 2\,\mu_1\,k^2\left(\mathrm{K}_0(\hat{k}) + \frac{1}{\hat{k}}\,\mathrm{K}_1(\hat{k})\right) - \frac{\gamma_0 k^3}{s}\,\frac{1-\hat{k}^2}{\hat{k}^2}\,\mathrm{K}_1(\hat{k}), \qquad (D.1.50)$$

$$\beta_{2,1} = -2\,\mu_1\,k\,k_1\left(\mathrm{K}_0(\hat{k}_1) + \frac{1}{\hat{k}_1}\,\mathrm{K}_1(\hat{k}_1)\right) - \frac{\gamma_0 k^3}{s}\,\frac{1-\hat{k}^2}{\hat{k}^2}\,\mathrm{K}_1(\hat{k}_1),$$

and

$$\alpha_{1,2} = -s\,\rho_2\,\mathrm{I}_0(\hat{k}) - 2\,\mu_2\,k^2\left(\mathrm{I}_0(\hat{k}) - \frac{1}{\hat{k}}\,\mathrm{I}_1(\hat{k})\right), \qquad \alpha_{2,2} = -2\,\mu_2\,k\,k_2\left(\mathrm{I}_0(\hat{k}_2) - \frac{1}{\hat{k}_2}\,\mathrm{I}_1(\hat{k}_2)\right),$$

$$\beta_{1,2} = s\,\rho_2\,\mathrm{K}_0(\hat{k}) + 2\,\mu_2\,k^2\left(\mathrm{K}_0(\hat{k}) + \frac{1}{\hat{k}}\,\mathrm{K}_1(\hat{k})\right), \qquad \beta_{2,2} = 2\,\mu_2\,k\,k_2\left(\mathrm{K}_0(\hat{k}_2) + \frac{1}{\hat{k}_2}\,\mathrm{K}_1(\hat{k}_2)\right).$$

$$(D.1.51)$$

Consolidating the various terms, we obtain a linear algebraic equation,

$$G_1 A_{1,1} + G_2 A_{2,1} + G_3 B_{1,1} + G_4 B_{2,1} + G_5 A_{1,2} + G_6 A_{2,2} + G_7 B_{1,2} + G_8 B_{2,2} = 0, \qquad (D.1.52)$$

where

$$G_1 = \alpha_{1,1} - \frac{\delta_1}{a}, \qquad G_2 = \alpha_{2,1} - \frac{\delta_2}{a}, \qquad G_3 = \beta_{1,1} + \frac{\delta_3}{a}, \qquad G_4 = \beta_{2,1} + \frac{\delta_4}{a},$$
$$G_5 = \alpha_{1,2}, \qquad G_6 = \alpha_{2,2}, \qquad G_7 = \beta_{1,2}, \qquad G_8 = \beta_{2,2}. \qquad (D.1.53)$$

Formulation of an eigenvalue problem

Collecting equations (D.1.21), (D.1.22), (D.1.26), (D.1.27), (D.1.41), and (D.1.52), we formulate a homogeneous linear system, $\mathbf{M} \cdot \mathbf{q} = \mathbf{0}$, where

$$\mathbf{q} = \left(A_{1,1}, A_{2,1}, B_{1,1}, B_{2,1}, A_{1,2}, A_{2,2}, B_{1,2}, B_{2,2}\right) \qquad (D.1.54)$$

$$\mathbf{M} = \begin{bmatrix} \mathrm{I}_1(\hat{k}) & \mathrm{I}_1(\hat{k}_1) & \mathrm{K}_1(\hat{k}) & \mathrm{K}_1(\hat{k}_1) & -\mathrm{I}_1(\hat{k}) \\ k\,\mathrm{I}_0(\hat{k}) & k_1\,\mathrm{I}_0(\hat{k}_1) & -k\,\mathrm{K}_0(\hat{k}) & -k_1\,\mathrm{K}_0(\hat{k}_1) & -k\,\mathrm{I}_0(\hat{k}) \\ \mathrm{I}_1(\hat{k}_i) & \mathrm{I}_1(\hat{k}_{i_1}) & \mathrm{K}_1(\hat{k}_i) & \mathrm{K}_1(\hat{k}_{i_1}) & 0 \\ k\,\mathrm{I}_0(\hat{k}_i) & k_1\,\mathrm{I}_0(\hat{k}_{i_1}) & -k\,\mathrm{K}_0(\hat{k}_i) & -k_1\,\mathrm{K}_0(\hat{k}_{i_1}) & 0 \\ 0 & 0 & 0 & 0 & \mathrm{I}_1(\hat{k}_e) \\ 0 & 0 & 0 & 0 & k\,\mathrm{I}_0(\hat{k}_e) \\ F_1 & F_2 & F_3 & F_4 & F_5 \\ G_1 & G_2 & G_3 & G_4 & G_5 \end{bmatrix}$$

$$\begin{bmatrix} -\mathrm{I}_1(\hat{k}_2) & -\mathrm{K}_1(\hat{k}) & -\mathrm{K}_1(\hat{k}_2) \\ -k_2\mathrm{I}_0(\hat{k}_2) & k\,\mathrm{K}_0(\hat{k}) & k_2\mathrm{K}_0(\hat{k}_2) \\ 0 & 0 & 0 \\ 0 & 0 & 0 \\ \mathrm{I}_1(\hat{k}_{e_2}) & \mathrm{K}_1(\hat{k}_e) & \mathrm{K}_1(\hat{k}_{e_2}) \\ k_2\,\mathrm{I}_0(\hat{k}_{e_2}) & -k\,\mathrm{K}_0(\hat{k}_e) & -k_2\,\mathrm{K}_0(\hat{k}_{e_2}) \\ F_6 & F_7 & F_8 \\ G_6 & G_7 & G_8 \end{bmatrix}$$

TABLE D.1.1 Setting the determinant of the matrix displayed to zero provides us with an algebraic equation for the complex growth rate.

and the coefficient matrix, $\mathbf{M}$, is shown in Table D.1.1. For a nontrivial solution to exist, the determinant of the coefficient matrix must be zero. In the numerical method implemented in the code, the roots of the secular equation $\det(\mathbf{M}) = 0$ are found by Newton's method with a suitable initial guess. The determinant of the matrix is calculated by *LU* decomposition using Gauss elimination.

Thread surrounded by an annular layer

When the inner cylinder is absent, we obtain an annular layer surrounding a liquid thread. For small values of their arguments, the modified Bessel functions become singular and the general expression for the stream function (D.1.11) exhibits singular terms. For the velocity to be regular at the x axis, the constants $B_{1,1}$ and $B_{2,1}$ must be set to zero, yielding a simplified expression for the inner fluid stream function,

$$\phi_j(\sigma) = \sigma\left[A_{1,j}\,\mathrm{I}_1(\hat{\sigma}) + A_{2,j}\,\mathrm{I}_1(\hat{\sigma}_j)\right], \tag{D.1.55}$$

where $j = 1$. The linear system $\mathbf{M}\cdot\mathbf{q} = \mathbf{0}$ undergoes analogous simplifications.

Annular layer coated on a tube

When the outer cylinder is absent, we obtain an annular layer coated on the exterior surface of a cylindrical tube, surrounded by an infinite outer fluid. For large values of their arguments, the modified Bessel functions become unbounded. To ensure a regular behavior, we set the coefficient

$A_{1,2}$ and $A_{2,2}$ to zero and obtain a simplified expression for the exterior stream function,

$$\phi_j(\sigma) = \sigma\left[B_{1,j}\,\mathrm{K}_1(\hat\sigma) + B_{2,j}\,\mathrm{K}_1(\hat\sigma_j)\right], \tag{D.1.56}$$

where $j = 2$. The linear system $\mathbf{M}\cdot\mathbf{q} = \mathbf{0}$ undergoes analogous simplifications.

Suspended thread

In the simplest configuration, both the internal and external cylinders are absent, yielding an infinite thread suspended in an infinite ambient fluid. The stream functions for the internal and external flows are described, respectively, by equations (D.1.55) and (D.1.56). The linear system $\mathbf{M}\cdot\mathbf{q} = \mathbf{0}$ undergoes analogous simplifications. When the interface is devoid of surfactants, the coefficient matrix reduces to that derived by Tomotika [410], as discussed in Section 9.13.2.

Program files:

1. `ann21`
   Computes the growth rate.

2. `matrix`
   Formulates the matrix $\mathbf{M}$.

3. `bess_I01K0`
   Computes the Bessel functions.

4. `gel`
   Computes the determinant of a matrix by LU decomposition using Gauss elimination.

Input files:

1. `ann21.dat`
   Problem selection and specification of input parameters.

Output files:

1. `ann21.out`
   Growth rates.

D.2   Directory ann2l0

This directory contains a code that performs the linear stability analysis of two annular layers confined between two concentric cylinders, as illustrated in Figure D.2.1. The instability occurs under conditions of Stokes flow. The interface is populated by an insoluble surfactant that diffuses and is convected over the interface but does not enter the bulk of the fluids.

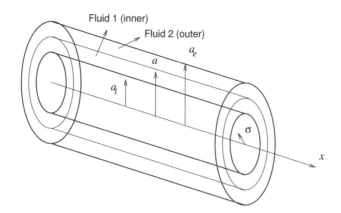

FIGURE D.2.1   Illustration of two concentric annular layers confined between an inner cylindrical tube of radius a_i, and an outer cylindrical tube of radius a_e.

Base state

We consider the instability of the interface between two annular layers separated by a cylindrical interface of radius a, confined between an inner cylindrical tube of radius a_i and an outer cylindrical tube of radius a_e, as illustrated in Figure D.2.1. The inner fluid is labeled 1 and the outer fluid is labeled 2. The viscosity of the inner fluid is μ_1 and the viscosity of outer fluid is $\mu_2 = \lambda\mu_1$, where λ is the viscosity ratio. The interface is occupied by an insoluble surfactant that diffuses and is convected over the interface, but does not enter the bulk of the fluids. In the unperturbed configuration, the fluids are quiescent and the surfactant is distributed uniformly over the cylindrical interface. Several special cases can be recognized:

- When the inner cylinder is absent and the radius of the outer cylinder is infinite, we obtain a thread suspended in an infinite ambient fluid.

- When the inner cylinder is absent, we obtain an annular layer coated on the interior surface of the outer cylinder.

- When the radius of the outer cylinder is infinite, we obtain an annular layer coated on the exterior surface of the inner cylinder.

Normal-mode analysis

To carry out the normal-mode stability analysis for axisymmetric perturbations, we introduce cylindrical polar coordinates, (x, σ, φ), where the x axis coincides with the axis of revolution of the unperturbed interface. Next, we describe the radial position of the interface in terms of the real or imaginary part of the function

$$\sigma = f(x, t) = a + \epsilon\eta(x, t), \tag{D.2.1}$$

where ϵ is a dimensionless coefficient whose magnitude is much less than unity,

$$\eta(x,t) = A \exp[ik(x - ct)] \tag{D.2.2}$$

is the normal-mode wave form of the perturbation, A is the complex amplitude of the interfacial wave, $k = 2\pi/L$ is the wave number, L is the wavelength, and c is the complex phase velocity.

Stream function, velocity and vorticity

The motion of the fluid on either side of the interface is governed by the continuity equation and the Stokes equation with appropriate physical constants for each fluid. Taking advantage of the axial symmetry of the flow, we describe the perturbation flow in terms of the Stokes stream functions, $\psi_1^{(1)}$ and $\psi_2^{(1)}$, for the inner and outer fluid, respectively, where the superscript (1) denotes the disturbance. The axial and radial components of the perturbation velocity are given by

$$u_{x_j} = \epsilon \frac{1}{\sigma} \frac{\partial \psi_j^{(1)}}{\partial \sigma}, \qquad u_{\sigma_j} = -\epsilon \frac{1}{\sigma} \frac{\partial \psi_j^{(1)}}{\partial x} \tag{D.2.3}$$

for $j = 1, 2$. The azimuthal component of the vorticity is given by

$$\omega_{\varphi_j} = -\epsilon \frac{1}{\sigma} E^2 \psi_j^{(1)}, \tag{D.2.4}$$

where

$$E^2 \equiv \frac{\partial^2}{\partial x^2} + \frac{\partial^2}{\partial \sigma^2} - \frac{1}{\sigma} \frac{\partial}{\partial \sigma} \tag{D.2.5}$$

is a second-order differential operator. Far from the axis of revolution, E^2 reduces to the Laplacian in a meridional plane of constant azimuthal angle, φ.

The vorticity transport equation for axisymmetric Stokes flow requires that

$$E^4 \psi_j \equiv E^2 \left(E^2 \psi_j^{(1)} \right) = 0, \tag{D.2.6}$$

which can be decomposed into two second-order constituent equations,

$$E^2 \psi_j^{(1)} = \tilde{\psi}_j^{(1)}, \qquad E^2 \tilde{\psi}_j^{(1)} = 0, \tag{D.2.7}$$

where a tilde denotes an intermediate solution.

Next, we express the stream function in the normal-mode form

$$\psi_j^{(1)}(x, \sigma, t) = \phi_j(\sigma) \exp[ik(x - ct)], \tag{D.2.8}$$

where $\phi_j(\sigma)$ are eigenfunctions. Substituting this expression into (D.2.7), we find that

$$\left(\frac{d^2}{d\sigma^2} - \frac{1}{\sigma} \frac{d}{d\sigma} - k^2 \right) \phi_j = \tilde{\phi}_j, \tag{D.2.9}$$

where

$$\left(\frac{d^2}{d\sigma^2} - \frac{1}{\sigma} \frac{d}{d\sigma} - k^2 \right) \tilde{\phi}_j = 0. \tag{D.2.10}$$

The general solution is

$$\phi_j(\sigma) = \sigma \left[A_{1,j} \, I_1(\hat{\sigma}) + A_{2,j} \sigma \, I_0(\hat{\sigma}) + B_{1,j} \, K_1(\hat{\sigma}) + B_{2,j} \, \sigma \, K_0(\hat{\sigma}) \right], \tag{D.2.11}$$

where I_0, I_1, K_0, and K_1 are Bessel functions, $A_{i,j}, B_{i,j}$ are complex constants, and $\hat{\sigma} \equiv k\sigma$. The first and third terms on the right-hand side of (D.2.11) satisfy (D.2.9) with $\tilde{\phi} = 0$. The second and fourth terms satisfy (D.2.9) with $\tilde{\phi}$ being a nontrivial solution of (D.2.10).

Substituting the preceding expressions into (D.2.3) and using the properties of the Bessel functions

$$I_0'(z) = I_1(z), \qquad I_1'(z) = I_0(z) - \frac{1}{z} I_1(z), \qquad K_0'(z) = -K_1(z),$$

$$K_1'(z) = -K_0(z) - \frac{1}{z} K_1(z), \tag{D.2.12}$$

we find that

$$u_{x_j} = \epsilon \, U_{x_j}(\sigma) \, \exp[ik(x - ct)], \qquad u_{\sigma_j} = \epsilon \, U_{\sigma_j}(\sigma) \, \exp[ik(x - ct)], \tag{D.2.13}$$

where

$$U_{x_j}(\sigma) = \frac{1}{\sigma} \frac{d\phi_j}{d\sigma} = A_{1,j} \, k \, I_0(\hat{\sigma}) + A_{2,j} \left[2 \, I_0(\hat{\sigma}) + \hat{\sigma} \, I_1(\hat{\sigma}) \right]$$
$$- B_{1,j} \, k \, K_0(\hat{\sigma}) + B_{2,j} \left[2 \, K_0(\hat{\sigma}) - \hat{\sigma} \, K_1(\hat{\sigma}) \right], \tag{D.2.14}$$

and

$$U_{\sigma_j}(\sigma) = -i \frac{k}{\sigma} \phi_j = -ik \left[A_{1,j} \, I_1(\hat{\sigma}) + A_{2,j} \, \sigma \, I_0(\hat{\sigma}) + B_{1,j} \, K_1(\hat{\sigma}) + B_{2,j} \, \sigma \, K_0(\hat{\sigma}) \right]. \tag{D.2.15}$$

Pressure field

The normal-mode expression for the pressure is

$$p_j = P_j + \epsilon \chi_j(\sigma) \, \exp[ik(x - ct)], \tag{D.2.16}$$

where P_1 and P_2 are the uniform unperturbed pressures, the pressure jump $P_1 - P_2 = \gamma_0/a$ is the unperturbed capillary pressure, γ_0 is the unperturbed surface tension, and $\chi_j(\sigma)$ are pressure eigenfunctions. To compute the disturbance pressure, we consider the x component of the Stokes equation for axisymmetric flow,

$$\frac{\partial p_j}{\partial x} = \mu_j \left[\frac{\partial^2 u_{x_j}}{\partial x^2} + \frac{1}{\sigma} \frac{\partial}{\partial \sigma} \left(\sigma \frac{\partial u_{x_j}}{\partial \sigma} \right) \right]. \tag{D.2.17}$$

Substituting (D.2.16) together with the first equation in (D.2.13) and linearizing, we obtain

$$\chi_j(\sigma) = \mathrm{i}\mu_j\, k\, U_{x_j} - \mathrm{i}\,\frac{\mu_j}{k}\,\frac{1}{\sigma}\,\frac{\mathrm{d}}{\mathrm{d}\sigma}\Big(\sigma\frac{\mathrm{d}U_{x_j}}{\mathrm{d}\sigma}\Big) = -\mathrm{i}\,\frac{\mu_j}{k}\,\Big(\frac{1}{\sigma}\,\frac{\mathrm{d}}{\mathrm{d}\sigma}(\sigma\frac{\mathrm{d}U_{x_j}}{\mathrm{d}\sigma}) - k^2\,U_{x_j}\Big), \qquad (\mathrm{D.2.18})$$

yielding

$$\chi_j(\sigma) = -\mathrm{i}\,\frac{\mu_j}{k}\,\Big(\frac{\mathrm{d}^2U_{x_j}}{\mathrm{d}\sigma^2} + \frac{1}{\sigma}\,\frac{\mathrm{d}U_{x_j}}{\mathrm{d}\sigma} - k^2\,U_{x_j}\Big) = -\mathrm{i}\,2\,\mu_j\,k\,\big[A_{2,j}\,\mathrm{I}_0(\hat\sigma) + B_{2,j}\,\mathrm{K}_0(\hat\sigma)\big]. \qquad (\mathrm{D.2.19})$$

Note that only two terms contribute to the pressure field.

Continuity of velocity at the interface

Requiring that the x and σ velocity components are continuous at the interface and using expressions (D.2.14) and (D.2.15), we derive the linearized kinematic interfacial conditions

$$A_{1,1}k\,\mathrm{I}_0(\hat k) + A_{2,1}[2\,\mathrm{I}_0(\hat k) + \hat k\,\mathrm{I}_1(\hat k)] - B_{1,1}k\,\mathrm{K}_0(\hat k) + B_{2,1}[2\,\mathrm{K}_0(\hat k) - \hat k\,\mathrm{K}_1(\hat k)]$$
$$= A_{1,2}\,k\,\mathrm{I}_0(\hat k) + A_{2,2}\,[2\,\mathrm{I}_0(\hat k) + \hat k\,\mathrm{I}_1(\hat k)] - B_{1,2}\,k\,\mathrm{K}_0(\hat k) + B_{2,2}\,[2\,\mathrm{K}_0(\hat k) - \hat k\,\mathrm{K}_1(\hat k)] \qquad (\mathrm{D.2.20})$$

and

$$A_{1,1}\,\mathrm{I}_1(\hat k) + A_{2,1}\,a\,\mathrm{I}_0(\hat k) + B_{1,1}\,\mathrm{K}_1(\hat k) + B_{2,1}\,a\,\mathrm{K}_0(\hat k)$$
$$= A_{1,2}\,\mathrm{I}_1(\hat k) + A_{2,2}\,a\,\mathrm{I}_0(\hat k) + B_{1,2}\,\mathrm{K}_1(\hat k) + B_{2,2}\,a\,\mathrm{K}_0(\hat k), \qquad (\mathrm{D.2.21})$$

where $\hat k = ka$ is a scaled wave number.

Kinematic compatibility

Kinematic compatibility requires that $\mathrm{D}[f(x,t) - \sigma]/\mathrm{D}t = 0$ evaluated at the interface, where $\mathrm{D}/\mathrm{D}t$ is the material derivative and the function f describes the position of the interface according to (D.2.1). Carrying out the differentiation, we obtain

$$\frac{\partial f}{\partial t} + u_x\,\frac{\partial f}{\partial x} - u_\sigma = 0, \qquad (\mathrm{D.2.22})$$

where the velocities u_x and u_σ are evaluated at the interface. Substituting the preceding expressions and linearizing, we find that

$$A = \frac{\mathrm{i}}{ck}\,U_{\sigma_j}(\sigma=a) = \frac{1}{c}\,\big[A_{1,j}\,\mathrm{I}_1(\hat k) + A_{2,j}\,a\,\mathrm{I}_0(\hat k) + B_{1,j}\,\mathrm{K}_1(\hat k) + B_{2,j}\,a\,\mathrm{K}_0(\hat k)\big]. \qquad (\mathrm{D.2.23})$$

Boundary conditions at the cylinders

The no-slip and no-penetration conditions at the inner and outer cylinders require that

$$U_{x_j}(a_i) = 0, \qquad U_{\sigma_j}(a_i) = 0, \qquad U_{x_j}(a_e) = 0, \qquad U_{\sigma_j}(a_e) = 0. \qquad (\mathrm{D.2.24})$$

Making substitutions, we find that

$$A_{1,1}\,k\,\mathrm{I}_0(\hat k_i) + A_{2,1}\,[2\,\mathrm{I}_0(\hat k_i) + \hat k_i\,\mathrm{I}_1(\hat k_i)] - B_{1,1}\,k\,\mathrm{K}_0(\hat k_i) + B_{2,1}\,[2\,\mathrm{K}_0(\hat k_i) - \hat k_i\,\mathrm{K}_1(\hat k_i)] = 0,$$
$$A_{1,1}\,\mathrm{I}_1(\hat k_i) + A_{2,1}\,a_i\,\mathrm{I}_0(\hat k_i) + B_{1,1}\,\mathrm{K}_1(\hat k_i) + B_{2,1}\,a_i\,\mathrm{K}_0(\hat k_i) = 0, \qquad (\mathrm{D.2.25})$$

where $\hat{k}_i = ka_i$, and

$$A_{1,2}\, k\, \mathrm{I}_0(\hat{k}_e) + A_{2,2}\, [2\,\mathrm{I}_0(\hat{k}_e) + \hat{k}_e\, \mathrm{I}_1(\hat{k}_e)] - B_{1,2}\, k\, \mathrm{K}_0(\hat{k}_e) + B_{2,2}\, [2\,\mathrm{K}_0(\hat{k}_e) - \hat{k}_e\, \mathrm{K}_1(\hat{k}_e)] = 0,$$

$$A_{1,2}\, \mathrm{I}_1(\hat{k}_e) + A_{2,2}\, a_e\, \mathrm{I}_0(\hat{k}_e) + B_{1,2}\, \mathrm{K}_1(\hat{k}_e) + B_{2,2}\, a_e\, \mathrm{K}_0(\hat{k}_e) = 0, \tag{D.2.26}$$

where $\hat{k}_e = ka_e$.

Surfactant transport

The interface is occupied by an insoluble surfactant with surface concentration Γ. For a normal-mode perturbation, $\Gamma = \Gamma_0 + \epsilon\, \Gamma^{(1)}$, where Γ_0 is the unperturbed surfactant concentration,

$$\Gamma^{(1)} \equiv \Gamma_1 \exp[ik(x - ct)] \tag{D.2.27}$$

is the normal-mode surfactant perturbation, and Γ_1 is the complex amplitude of the perturbation. Linearizing the surfactant transport equation in the absence of a base flow, we derive the evolution equation

$$\epsilon\, \frac{\partial \Gamma^{(1)}}{\partial t} + \Gamma_0\, \frac{\partial u_x}{\partial x} = -\frac{\Gamma_0}{a}\, u_\sigma + \epsilon\, D_s\, \frac{\partial^2 \Gamma^{(1)}}{\partial x^2}, \tag{D.2.28}$$

where the velocity is evaluated at the unperturbed position $\sigma = a$. Substituting the normal-mode forms and rearranging, we find that

$$\frac{\Gamma_1}{\Gamma_0} = -\frac{1}{s + D_s k^2}\left(ik\, U_{x_j} + \frac{1}{a}\, U_{\sigma_j}\right)_{\sigma = a} = -\mathrm{i}\, \frac{k}{a\,(s + D_s k^2)}\left(A_{1,j}\, [\hat{k}\, \mathrm{I}_0(\hat{k}) - \mathrm{I}_1(\hat{k})]\right.$$

$$\left. + A_{2,j}\, a\, [\mathrm{I}_0(\hat{k}) + \hat{k}\, \mathrm{I}_1(\hat{k})] - B_{1,j}\, [\hat{k}\, \mathrm{K}_0(\hat{k}) + \mathrm{K}_1(\hat{k})] + B_{2,j}\, a\, [\mathrm{K}_0(\hat{k}) - \hat{k}\, \mathrm{K}_1(\hat{k})]\right), \tag{D.2.29}$$

where $s \equiv -\mathrm{i}kc$. If c is imaginary, as expected in the absence of a base flow, $s = kc_I$ is the real growth rate.

Surface equation of state

Surfactant concentration inhomogeneities over the interface cause corresponding variations in the surface tension. For a normal-mode perturbation,

$$\gamma = \gamma_0 + \epsilon\, \gamma_1 \exp[ik(x - ct)], \tag{D.2.30}$$

where γ_0 is the surface tension in the unperturbed state corresponding to the surfactant concentration Γ_0, and γ_1 is the complex amplitude of the perturbation. Consistent with the linearized approximation, we assume that the surface tension is a linear function of the surfactant concentration,

$$\frac{\gamma_1}{\Gamma_1} = -\mathrm{Ma}\, \frac{\gamma_0}{\Gamma_0}, \tag{D.2.31}$$

where Ma is the Marangoni number. Substituting (D.2.29) and rearranging, we obtain

$$\frac{\gamma_1}{\gamma_0} = \mathrm{i}\, \mathrm{Ma}\, \frac{k}{a\,(s + D_s k^2)}\left(A_{1,j}\, [\hat{k}\, \mathrm{I}_0(\hat{k}) - \mathrm{I}_1(\hat{k})] + A_{2,j}\, a\, [\mathrm{I}_0(\hat{k}) + \hat{k}\, \mathrm{I}_1(\hat{k})]\right.$$

$$\left. - B_{1,j}\, [\hat{k}\, \mathrm{K}_0(\hat{k}) + \mathrm{K}_1(\hat{k})] + B_{2,j}\, a\, [\mathrm{K}_0(\hat{k}) - \hat{k}\, \mathrm{K}_1(\hat{k})]\right). \tag{D.2.32}$$

Selecting $j = 1$, corresponding to the inner fluid, we recast this equation into the form

$$\gamma_1 = \mathrm{i}\,(\,\delta_1 A_{1,1} + \delta_2 A_{2,1} + \delta_3 B_{1,1} + \delta_4 B_{2,1}),\qquad (\mathrm{D.2.33})$$

where

$$\delta_1 = \frac{\Pi}{a}\,[\hat{k}\,\mathrm{I}_0(\hat{k}) - \mathrm{I}_1(\hat{k})], \qquad \delta_2 = \Pi\,[\mathrm{I}_0(\hat{k}) + \hat{k}\,\mathrm{I}_1(\hat{k})], \qquad \delta_3 = -\frac{\Pi}{a}\,[\hat{k}\,\mathrm{K}_0(\hat{k}) + \mathrm{K}_1(\hat{k})],$$

$$\delta_4 = \Pi\,[\mathrm{K}_0(\hat{k}) - \hat{k}\,\mathrm{K}_1(\hat{k})],\qquad (\mathrm{D.2.34})$$

and

$$\Pi \equiv \mathrm{Ma}\,\frac{\gamma_0}{s + D_s k^2}\qquad (\mathrm{D.2.35})$$

is a dimensionless number pertinent to the surfactant.

Tangential interfacial force balance

Balancing the tangential hydrodynamic forces exerted on an infinitesimal portion of the interface and the interfacial tension, γ, we derive the linearized shear stress condition

$$\left(\sigma_{x\sigma_1} - \sigma_{x\sigma_2}\right)_{\sigma=a} = \mu_1\left(\frac{\partial u_{x_1}}{\partial \sigma} + \frac{\partial u_{\sigma_1}}{\partial x}\right)_{\sigma=a} - \mu_2\left(\frac{\partial u_{x_2}}{\partial \sigma} + \frac{\partial u_{\sigma_2}}{\partial x}\right)_{\sigma=a} = \frac{\partial \gamma}{\partial x}.\qquad (\mathrm{D.2.36})$$

Substituting the normal-mode forms, we find that

$$\mu_1\left(\frac{\mathrm{d}U_{x_1}}{\mathrm{d}\sigma} + \mathrm{i}k\,U_{\sigma_1}\right)_{\sigma=a} - \mu_2\left(\frac{\mathrm{d}U_{x_2}}{\mathrm{d}\sigma} + \mathrm{i}k\,U_{\sigma_2}\right)_{\sigma=a} = \mathrm{i}k\gamma_1.\qquad (\mathrm{D.2.37})$$

Using the expression

$$\frac{\mathrm{d}U_{x_j}(\sigma)}{\mathrm{d}\sigma} = A_{1,j}\,k^2\,\mathrm{I}_1(\hat{\sigma}) + A_{2,j}k[2\,\mathrm{I}_1(\hat{\sigma}) + \hat{\sigma}\,\mathrm{I}_0(\hat{\sigma})]$$
$$+ B_{1,j}k^2\,\mathrm{K}_1(\hat{\sigma}) - B_{2,j}k\,[2\,\mathrm{K}_1(\hat{\sigma}) - \hat{\sigma}\,\mathrm{K}_0(\hat{\sigma})]\qquad (\mathrm{D.2.38})$$

along with (D.2.15) and (D.2.33), we obtain

$$2\mu_1 k\left(A_{1,1}k\,\mathrm{I}_1(\hat{k}) + A_{2,1}[\mathrm{I}_1(\hat{k}) + \hat{k}\,\mathrm{I}_0(\hat{k})] + B_{1,1}k\,\mathrm{K}_1(\hat{k}) - B_{2,1}[\mathrm{K}_1(\hat{k}) - \hat{k}\,\mathrm{K}_0(\hat{k})]\right)$$
$$-2\mu_2 k\left(A_{1,2}k\,\mathrm{I}_1(\hat{k}) + A_{2,2}[\mathrm{I}_1(\hat{k}) + \hat{k}\,\mathrm{I}_0(\hat{k})] + B_{1,2}k\,\mathrm{K}_1(\hat{k}) - B_{2,2}[\mathrm{K}_1(\hat{k}) - \hat{k}\,\mathrm{K}_0(\hat{k})]\right)$$
$$= \mathrm{i}k\,\gamma_1 = -k\left(\delta_1 A_{1,1} + \delta_2 A_{2,1} + \delta_3 B_{1,1} + \delta_4 B_{2,1}\right).\qquad (\mathrm{D.2.39})$$

Consolidating the various terms, we obtain a linear algebraic equation,

$$F_1 A_{1,1} + F_2 A_{2,1} + F_3 B_{1,1} + F_4 B_{2,1} + F_5 A_{1,2} + F_6 A_{2,2} + F_7 B_{1,2} + F_8 B_{2,2} = 0,\qquad (\mathrm{D.2.40})$$

where

$$\begin{aligned}
F_1 &= 2\mu_1 k\,\mathrm{I}_1(\hat{k}) + \delta_1, & F_2 &= 2\mu_1[\mathrm{I}_1(\hat{k}) + \hat{k}\,\mathrm{I}_0(\hat{k})] + \delta_2,\\
F_3 &= 2\mu_1\,k\,\mathrm{K}_1(\hat{k}) + \delta_3, & F_4 &= 2\mu_1\,[-\mathrm{K}_1(\hat{k}) + \hat{k}\,\mathrm{K}_0(\hat{k})] + \delta_4,\\
F_5 &= -2\mu_2\,k\,\mathrm{I}_1(\hat{k}), & F_6 &= -2\mu_2\,[\mathrm{I}_1(\hat{k}) + \hat{k}\,\mathrm{I}_0(\hat{k})],\\
F_7 &= -2\mu_2\,k\,\mathrm{K}_1(\hat{k}), & F_8 &= -2\mu_2\,[-\mathrm{K}_1(\hat{k}) + \hat{k}\,\mathrm{K}_0(\hat{k})].
\end{aligned}\qquad (\mathrm{D.2.41})$$

Normal stress interfacial condition

Balancing the normal forces exerted on an infinitesimal patch of the interface and the interfacial tension γ, and linearizing, we derive the normal stress balance

$$\left(\sigma_{\sigma\sigma_1} - \sigma_{\sigma\sigma_2}\right)_{\sigma=a} = \left(-p_1 + 2\mu_1 \frac{\partial u_{\sigma_1}}{\partial \sigma}\right)_{\sigma=a} - \left(-p_2 + 2\mu_2 \frac{\partial u_{\sigma_2}}{\partial \sigma}\right)_{\sigma=a} = \gamma\, 2\,\kappa_m, \quad \text{(D.2.42)}$$

where κ_m is the mean curvature. In the linearized approximation,

$$2\kappa_m = -\frac{1}{a} + \epsilon\left(\frac{\eta}{a^2} + \eta_{xx}\right) = -\frac{1}{a} + \epsilon\,\frac{A}{a^2}\,(1 - \hat{k}^2)\,\exp[ik(x - ct)]. \quad \text{(D.2.43)}$$

Using (D.2.23) with $j = 1$ to eliminate the interfacial amplitude, A, we obtain

$$2\kappa_m = -\frac{1}{a} + \epsilon\,2\kappa_m^{(1)}\,\exp[ik(x - ct)], \quad \text{(D.2.44)}$$

where

$$2\kappa_m^{(1)} = \frac{1 - \hat{k}^2}{ca^2}\left[A_{1,1}\,\mathrm{I}_1(\hat{k}) + A_{2,1}\,a\,\mathrm{I}_0(\hat{k}) + B_{1,1}\,\mathrm{K}_1(\hat{k}) + B_{2,1}\,a\,\mathrm{K}_0(\hat{k})\right]. \quad \text{(D.2.45)}$$

Substituting into (D.2.42) the normal-mode expansions, we obtain

$$\left(-\chi_1 + 2\mu_1 \frac{\partial U_{\sigma_1}}{\partial \sigma}\right)_{\sigma=a} - \left(-\chi_2 + 2\mu_2 \frac{\partial U_{\sigma_2}}{\partial \sigma}\right)_{\sigma=a} = -\frac{\gamma_1}{a} \quad \text{(D.2.46)}$$

$$+\gamma_0\,\frac{1 - \hat{k}^2}{ca^2}\left(A_{1,1}\,\mathrm{I}_1(\hat{k}) + A_{2,1}\,a\,\mathrm{I}_0(\hat{k}) + B_{1,1}\,\mathrm{K}_1(\hat{k}) + B_{2,1}\,a\,\mathrm{K}_0(\hat{k})\right).$$

Substituting into the left-hand side the expression

$$\frac{\mathrm{d}U_{\sigma_j}}{\mathrm{d}\sigma} = -\mathrm{i}\,k\left[A_{1,j}\,k\left[\mathrm{I}_0(\hat{\sigma}) - \frac{1}{\hat{\sigma}}\mathrm{I}_1(\hat{\sigma})\right] + A_{2,j}\left[\mathrm{I}_0(\hat{\sigma}) + \hat{\sigma}\,\mathrm{I}_1(\hat{\sigma})\right]\right.$$

$$\left.-B_{1,j}\,k\left[\mathrm{K}_0(\hat{\sigma}) + \frac{1}{\hat{\sigma}}\mathrm{K}_1(\hat{\sigma})\right] + B_{2,j}\left[\mathrm{K}_0(\hat{\sigma}) - \hat{\sigma}\,\mathrm{K}_1(\hat{\sigma})\right]\right] \quad \text{(D.2.47)}$$

together with the last expression in (D.2.19) for the pressure, we find that

$$\alpha_{1,1}A_{1,1} + \alpha_{2,1}A_{2,1} + \beta_{1,1}B_{1,1} + \beta_{2,1}B_{2,1} + \alpha_{1,2}A_{1,2} + \alpha_{2,2}A_{2,2}$$

$$+\beta_{1,2}B_{1,2} + \beta_{2,2}B_{2,2} = -\mathrm{i}\,\frac{\gamma_1}{a} = \frac{1}{a}\left(\delta_1\,A_{1,1} + \delta_2\,A_{2,1} + \delta_3\,B_{1,1} + \delta_4\,B_{2,1}\right), \quad \text{(D.2.48)}$$

where

$$\alpha_{1,1} = 2\mu_1\,k^2\left[\mathrm{I}_0(\hat{k}) - \frac{1}{\hat{k}}\mathrm{I}_1(\hat{k})\right] - \frac{\gamma_0 k^3}{s}\,\frac{1 - \hat{k}^2}{\hat{k}^2}\,\mathrm{I}_1(\hat{k}),$$

$$\alpha_{2,1} = 2\mu_1\,k\,\hat{k}\,\mathrm{I}_1(\hat{k}) - \frac{\gamma_0 k^3}{s}\,\frac{1 - \hat{k}^2}{\hat{k}^2}\,a\,\mathrm{I}_0(\hat{k}),$$

$$\beta_{1,1} = -2\mu_1\,k^2\left[\mathrm{K}_0(\hat{k}) + \frac{1}{\hat{k}}\mathrm{K}_1(\hat{k})\right] - \frac{\gamma_0 k^3}{s}\,\frac{1 - \hat{k}^2}{\hat{k}^2}\,\mathrm{K}_1(\hat{k}), \quad \text{(D.2.49)}$$

$$\beta_{2,1} = -2\mu_1\,k\,\hat{k}\,\mathrm{K}_1(\hat{k}) - \frac{\gamma_0 k^3}{s}\,\frac{1 - \hat{k}^2}{\hat{k}^2}\,a\,\mathrm{K}_0(\hat{k}),$$

and

$$\alpha_{1,2} = -2\,\mu_2\,k^2 \left[\, \mathrm{I}_0(\hat{k}) - \frac{1}{\hat{k}}\,\mathrm{I}_1(\hat{k})\,\right], \qquad \alpha_{2,2} = -2\,\mu_2\,k\,\hat{k}\,\mathrm{I}_1(\hat{k}),$$

$$\beta_{1,2} = 2\,\mu_2\,k^2 \left[\, \mathrm{K}_0(\hat{k}) + \frac{1}{\hat{k}}\,\mathrm{K}_1(\hat{k})\,\right], \qquad \beta_{2,2} = 2\,\mu_2\,k\,\hat{k}\,\mathrm{K}_1(\hat{k}). \tag{D.2.50}$$

Consolidating the various terms, we derive a linear algebraic equation,

$$G_1\,A_{1,1} + G_2\,A_{2,1} + G_3\,B_{1,1} + G_4\,B_{2,1} + G_5\,A_{1,2} + G_6\,A_{2,2} + G_7\,B_{1,2} + G_8\,B_{2,2} = 0, \tag{D.2.51}$$

where

$$G_1 = \alpha_{1,1} - \frac{\delta_1}{a}, \qquad G_2 = \alpha_{2,1} - \frac{\delta_2}{a}, \qquad G_3 = \beta_{1,1} - \frac{\delta_3}{a}, \qquad G_4 = \beta_{2,1} - \frac{\delta_4}{a},$$

$$G_5 = \alpha_{1,2}, \qquad G_6 = \alpha_{2,2}, \qquad G_7 = \beta_{1,2}, \qquad G_8 = \beta_{2,2}. \tag{D.2.52}$$

Formulation of an eigenvalue problem

To formulate the algebraic eigenvalue problem, we collect equations (D.2.20), (D.2.21), (D.2.25), (D.2.26), (D.2.40), and (D.2.51) into a linear homogeneous system, $\mathbf{M} \cdot \mathbf{q} = \mathbf{0}$, where

$$\mathbf{q} = \left(\, A_{1,1}, A_{2,1}, B_{1,1}, B_{2,1}, A_{1,2}, A_{2,2}, B_{1,2}, B_{2,2}\,\right), \tag{D.2.53}$$

and the coefficient matrix, $\mathbf{M}$, is given in Table D.2.1 [214]. When both the inner and outer cylinders are absent and the interface is devoid of surfactants, the coefficient matrix is in agreement with that derived by Tomotika [410], as discussed in Section 9.13.2.

For a nontrivial solution to exist, the determinant of the matrix $\mathbf{M}$ must be zero, $\det(\mathbf{M}) = 0$. To compute the growth rate, $s = -ikc$, we eliminate the denominators from all entries of the matrix $\mathbf{M}$ by multiplying corresponding rows by them, thereby obtaining a new matrix, $\mathbf{Q}$. Setting the determinant of $\mathbf{Q}$ to zero, we obtain a quadratic equation for the shifted growth rate,

$$P_2(s) = \det(\mathbf{Q}) = \tilde{a}\,(s - s_r)^2 + \tilde{b}\,(s - s_r) + \tilde{c} = 0, \tag{D.2.54}$$

where s_r is an arbitrary reference value. In the numerical method, the binomial coefficients, $\tilde{a}, \tilde{b}, \tilde{c}$, are computed from the values $P_2(s_r - \delta)$, $P_2(s_r)$, $P_2(s_r + \delta)$, using finite-difference formula with zero error, where δ is an arbitrary offset. The growth rate is then computed using the quadratic formula, yielding two normal modes. In the absence of surfactant, $\tilde{a} = 0$, yielding one normal mode.

Program files:

1. ann210
   Solves the secular equation. The determinant of the matrix is calculated by LU decomposition using Gauss elimination.

2. ann210_dr
   Driver for ann210.

$$
\mathbf{M} =
\begin{bmatrix}
I_1(\hat{k}) & a\,I_0(\hat{k}) & K_1(\hat{k}) & a\,K_1(\hat{k}) \\
k\,I_0(\hat{k}) & 2\,I_0(\hat{k}) + \hat{k}\,I_1(\hat{k}) & -k\,K_0(\hat{k}) & 2\,K_0(\hat{k}) - \hat{k}\,K_1(\hat{k}) \\
I_1(\hat{k}_i) & a_i\,I_0(\hat{k}_i) & K_1(ka_i) & a_i\,K_0(\hat{k}_i) \\
k\,I_0(\hat{k}_i) & 2\,I_0(\hat{k}_i) + \hat{k}_i\,I_1(\hat{k}_i) & -k\,K_0(\hat{k}_i) & 2\,K_0(\hat{k}_i) - \hat{k}_i\,K_1(\hat{k}_i) \\
0 & 0 & 0 & 0 \\
0 & 0 & 0 & 0 \\
F_1 & F_2 & F_3 & F_4 \\
G_1 & G_2 & G_3 & G_4
\end{bmatrix}
$$

$$
\begin{bmatrix}
-I_1(\hat{k}) & -a\,I_0(\hat{k}) & -K_1(\hat{k}) & -a\,K_0(\hat{k}) \\
-k\,I_0(\hat{k}) & -2\,I_0(\hat{k}) - \hat{k}\,I_1(\hat{k}) & k\,K_0(\hat{k}) & -2\,K_0(\hat{k}) + \hat{k}\,K_1(\hat{k}) \\
0 & 0 & 0 & 0 \\
0 & 0 & 0 & 0 \\
I_1(\hat{k}_e) & a_e\,I_0(\hat{k}_e) & K_1(\hat{k}_e) & a_e\,K_0(\hat{k}_e) \\
k\,I_0(\hat{k}_e) & 2\,I_0(\hat{k}_e) + \hat{k}_e\,I_1(\hat{k}_e) & -k\,K_0(\hat{k}_e) & 2\,K_0(\hat{k}_e) - \hat{k}_e\,K_1(\hat{k}_e) \\
F_5 & F_6 & F_7 & F_8 \\
G_5 & G_6 & G_7 & G_8
\end{bmatrix}
$$

TABLE D.2.1   Setting the determinant of the matrix displayed to zero provides us with a quadratic equation for the complex growth rate.

3. `matrix`
   Formulates the secular matrix, $\mathbf{M}$.

4. `bess_I01K0`
   Computes the Bessel functions.

5. `gel`
   Computes the determinant of a matrix by *LU* decomposition using Gauss elimination.

Input files:

1. `ann210.dat`
   Problem selection and specification of input parameters.

Output files:

1. `ann210.out`
   Growth rates.

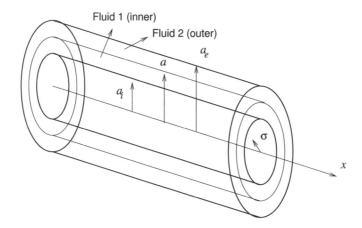

FIGURE D.3.1 Illustration of two annular layers separated by an elastic interface between an inner cylindrical tube of radius a_i and an outer cylindrical tube of radius a_e.

D.3 Directory ann2lel

This directory contains a code that performs the linear stability analysis of the elastic interface between two annular layers confined between two concentric cylinders, as illustrated in Figure D.3.1, subject to axisymmetric perturbations [310].

Base state

We consider two stationary annular layers separated by a cylindrical interface of radius a, confined between an inner cylindrical tube of radius a_i and an outer cylindrical tube of radius a_e, as illustrated in Figure D.3.1. The inner fluid is labeled 1 and the outer fluid is labeled 2. The viscosity of the inner fluid is μ_1 and the viscosity of outer fluid is $\mu_2 = \lambda \mu_1$, where λ is the viscosity ratio. Several special cases can be recognized:

- When the inner cylinder is absent and the radius of the outer cylinder is infinite, we obtain a thread suspended in an infinite ambient fluid.

- When the inner cylinder is absent, we obtain a core-annular arrangement consisting of an annular layer coated on the interior surface of the inner cylinder.

- When the radius of the outer cylinder is infinite, we obtain an annular layer coated on the exterior surface of the inner cylinder.

In the unperturbed configuration, the fluids are quiescent.

Normal-mode analysis

To carry out the normal-mode stability analysis for axisymmetric perturbations, we introduce cylindrical polar coordinates, (x, σ, φ), where the x axis coincides with the axis of revolution of the interface. The radial position of the interface is described in terms of the real or imaginary part of the right-hand side of the equation

$$\sigma = f(x,t) = a + \epsilon\eta(x,t), \tag{D.3.1}$$

where ϵ is a dimensionless coefficient whose magnitude is much less than unity,

$$\eta(x,t) = A\exp[ik(x-ct)] \tag{D.3.2}$$

is the normal-mode wave form of the perturbation, A is the complex amplitude of the interfacial wave, $k = 2\pi/L$ is the wave number, L is the wavelength, and c is the complex phase velocity.

Stream function, velocity and vorticity

The motion of the fluid on either side of the interface is governed by the continuity equation and the Navier–Stokes equation with appropriate physical constants for fluid. Taking advantage of the axial symmetry of the flow, we describe the perturbation flow in terms of the Stokes stream functions, $\psi_1^{(1)}$ and $\psi_2^{(1)}$, for the inner and outer fluid, respectively, where the superscript (1) denotes the disturbance. The axial and radial components of the perturbation velocity are

$$u_{x_j} = \epsilon\,\frac{1}{\sigma}\,\frac{\partial\psi_j^{(1)}}{\partial\sigma}, \qquad u_{\sigma_j} = -\epsilon\,\frac{1}{\sigma}\,\frac{\partial\psi_j^{(1)}}{\partial x} \tag{D.3.3}$$

for $j = 1,2$, and the azimuthal component of the vorticity is given by

$$\omega_{\varphi_j} = -\epsilon\,\frac{1}{\sigma}\,E^2\psi_j^{(1)}, \tag{D.3.4}$$

where

$$E^2 \equiv \frac{\partial^2}{\partial x^2} + \frac{\partial^2}{\partial\sigma^2} - \frac{1}{\sigma}\,\frac{\partial}{\partial\sigma} \tag{D.3.5}$$

is a second-order differential operator.

Next, we recall the vorticity transport equation for axisymmetric flow,

$$\frac{\mathrm{D}}{\mathrm{D}t}\left(\frac{1}{\sigma}\,\omega_{\varphi_j}\right) = \frac{\nu_j}{\sigma^2}\,E^2(\sigma\,\omega_{\varphi_j}), \tag{D.3.6}$$

where $\mathrm{D}/\mathrm{D}t$ is the material derivative and ν_j is the kinematic viscosity of the jth fluid. Substituting (D.3.4) into (D.3.6), linearizing with respect to ϵ and rearranging, we obtain

$$\left(E^2 - \frac{1}{\nu_j}\,\frac{\partial}{\partial t}\right)E^2\,\psi_j^{(1)} = 0 \tag{D.3.7}$$

for $j = 1,2$.

To carry out the normal-mode analysis, we express the stream function in the form

$$\psi_j^{(1)}(x,\sigma,t) = \phi_j(\sigma)\,\exp[ik(x-ct)], \tag{D.3.8}$$

where $\phi_j(\sigma)$ are eigenfunctions. Substituting this expression into (D.3.7), we obtain

$$\left(\frac{\mathrm{d}^2}{\mathrm{d}\sigma^2} - \frac{1}{\sigma}\frac{\mathrm{d}}{\mathrm{d}\sigma} - k^2\right)\left(\frac{\mathrm{d}^2}{\mathrm{d}\sigma^2} - \frac{1}{\sigma}\frac{\mathrm{d}}{\mathrm{d}\sigma} - k_j^2\right)\phi_j = 0, \tag{D.3.9}$$

where

$$k_j^2 \equiv k^2 - \mathrm{i}\,c\,\frac{k}{\nu_j}. \tag{D.3.10}$$

The general solution is

$$\phi_j(\sigma) = \sigma\left[A_{1,j}\,\mathrm{I}_1(\hat{\sigma}) + A_{2,j}\,\mathrm{I}_1(\hat{\sigma}_j) + B_{1,j}\,\mathrm{K}_1(\hat{\sigma}) + B_{2,j}\,\mathrm{K}_1(\hat{\sigma}_j)\right] \tag{D.3.11}$$

for $j = 1, 2$, where $\hat{\sigma} \equiv k\sigma$, $\hat{\sigma}_j \equiv k_j\sigma$, I_1 and K_1 are Bessel functions, and $A_{i,j}, B_{i,j}$ are complex constants. The first and third terms on the right-hand side of (D.3.11) correspond to the first operator, while the second and fourth terms correspond to the second operator on the left-hand side of (D.3.9).

Substituting the preceding expressions into (D.3.3) and using the properties of the Bessel functions

$$\mathrm{I}_0'(z) = \mathrm{I}_1(z), \qquad \mathrm{I}_1'(z) = \mathrm{I}_0(z) - \frac{1}{z}\mathrm{I}_1(z), \qquad \mathrm{K}_0'(z) = -\mathrm{K}_1(z),$$

$$\mathrm{K}_1'(z) = -\mathrm{K}_0(z) - \frac{1}{z}\mathrm{K}_1(z), \tag{D.3.12}$$

we find that

$$u_{x_j} = \epsilon\,U_{x_j}(\sigma)\,\exp[\mathrm{i}k(x - ct)], \qquad u_{\sigma_j} = \epsilon\,U_{\sigma_j}(\sigma)\,\exp[\mathrm{i}k(x - ct)], \tag{D.3.13}$$

where

$$U_{x_j}(\sigma) = \frac{1}{\sigma}\frac{\mathrm{d}\phi_j}{\mathrm{d}\sigma} = A_{1,j}\,k\,\mathrm{I}_0(\hat{\sigma}) + A_{2,j}\,k_j\,\mathrm{I}_0(\hat{\sigma}_j) - B_{1,j}\,k\,\mathrm{K}_0(\hat{\sigma}) - B_{2,j}\,k_j\,\mathrm{K}_0(\hat{\sigma}_j) \tag{D.3.14}$$

and

$$U_{\sigma_j}(\sigma) = -\mathrm{i}\frac{k}{\sigma}\phi_j = -\mathrm{i}k\left[A_{1,j}\,\mathrm{I}_1(\hat{\sigma}) + A_{2,j}\,\mathrm{I}_1(\hat{\sigma}_j) + B_{1,j}\,\mathrm{K}_1(\hat{\sigma}) + B_{2,j}\,\mathrm{K}_1(\hat{\sigma}_j)\right]. \tag{D.3.15}$$

Pressure field

The normal-mode expression for the pressure is

$$p_j = P_j + \epsilon\,\chi_j(\sigma)\,\exp[\mathrm{i}k(x - ct)], \tag{D.3.16}$$

where P_1 and P_2 are the uniform unperturbed pressures, $P_1 - P_2 = \gamma_0/a$ is the unperturbed capillary pressure, and $\chi_j(\sigma)$ are pressure eigenfunctions. To compute the disturbance pressure, we resort to the x component of the Navier–Stokes equation for axisymmetric flow,

$$\rho_j\frac{\mathrm{D}u_{x_j}}{\mathrm{D}t} = -\frac{\partial p_j}{\partial x} + \mu_j\left(\frac{\partial^2 u_{x_j}}{\partial x^2} + \frac{1}{\sigma}\frac{\partial}{\partial\sigma}\left(\sigma\frac{\partial u_{x_j}}{\partial\sigma}\right)\right). \tag{D.3.17}$$

Substituting (D.3.16) together with the first equation in (D.3.13) and linearizing, we obtain

$$\chi_j(\sigma) = (c\rho_j + \mathrm{i}\mu_j\,k)\,U_{x_j} - \mathrm{i}\frac{\mu_j}{k}\frac{1}{\sigma}\frac{\mathrm{d}}{\mathrm{d}\sigma}(\sigma\frac{\mathrm{d}U_{x_j}}{\mathrm{d}\sigma}) = -\mathrm{i}\frac{\mu_j}{k}\left(\frac{1}{\sigma}\frac{\mathrm{d}}{\mathrm{d}\sigma}(\sigma\frac{\mathrm{d}U_{x_j}}{\mathrm{d}\sigma}) - k_j^2\,U_{x_j}\right) \quad (\text{D.3.18})$$

or

$$\chi_j(\sigma) = -\mathrm{i}\frac{\mu_j}{k}\left(\frac{\mathrm{d}^2 U_{x_j}}{\mathrm{d}\sigma^2} + \frac{1}{\sigma}\frac{\mathrm{d}U_{x_j}}{\mathrm{d}\sigma} - k_j^2\,U_{x_j}\right). \quad (\text{D.3.19})$$

Making substitutions, we find that

$$\chi_j(\sigma) = -\mathrm{i}\mu_j\,(k^2 - k_j^2)\big[A_{1,j}\,\mathrm{I}_0(\hat{\sigma}) - B_{1,j}\,\mathrm{K}_0(\hat{\sigma})\big] = ck\rho_j\big[A_{1,j}\,\mathrm{I}_0(\hat{\sigma}) - B_{1,j}\,\mathrm{K}_0(\hat{\sigma})\big]. \quad (\text{D.3.20})$$

Note that only two terms contribute to the pressure field.

Continuity of velocity at the interface

Requiring that the x and σ velocity components are continuous at the interface and using expressions (D.3.14) and (D.3.15), we derive the linearized kinematic interfacial conditions

$$A_{1,1}\,k\,\mathrm{I}_0(\hat{k}) + A_{2,1}\,k_1\,\mathrm{I}_0(\hat{k}_1) - B_{1,1}\,k\,\mathrm{K}_0(\hat{k}) - B_{2,1}\,k_1\,\mathrm{K}_0(\hat{k}_1)$$
$$= A_{1,2}\,k\,\mathrm{I}_0(\hat{k}) + A_{2,2}\,k_2\,\mathrm{I}_0(\hat{k}_2) - B_{1,2}\,k\,\mathrm{K}_0(\hat{k}) - B_{2,2}\,k_2\,\mathrm{K}_0(\hat{k}_2) \quad (\text{D.3.21})$$

and

$$A_{1,1}\,\mathrm{I}_1(\hat{k}) + A_{2,1}\,\mathrm{I}_1(\hat{k}_1) + B_{1,1}\,\mathrm{K}_1(\hat{k}) + B_{2,1}\,\mathrm{K}_1(\hat{k}_1)$$
$$= A_{1,2}\,\mathrm{I}_1(\hat{k}) + A_{2,2}\,\mathrm{I}_1(\hat{k}_2) + B_{1,2}\,\mathrm{K}_1(\hat{k}) + B_{2,2}\,\mathrm{K}_1(\hat{k}_2), \quad (\text{D.3.22})$$

where $\hat{k} = ka$, $\hat{k}_1 = k_1 a$, and $\hat{k}_2 = k_2 a$.

Kinematic compatibility

Kinematic compatibility requires $\mathrm{D}[f(x,t) - \sigma]/\mathrm{D}t = 0$ at the interface, where $\mathrm{D}/\mathrm{D}t$ is the material derivative, and the function f describes the position of the interface according to (D.3.1). Carrying out the differentiation, we find that

$$\frac{\partial f}{\partial t} + u_x\frac{\partial f}{\partial x} - u_\sigma = 0, \quad (\text{D.3.23})$$

where the velocity is evaluated at the interface. Substituting the preceding expressions and linearizing, we obtain

$$A = \mathrm{i}\frac{1}{kc}\,U_{\sigma_j}(\sigma = a) = \frac{1}{c}\big[A_{1,j}\,\mathrm{I}_1(\hat{k}) + A_{2,j}\,\mathrm{I}_1(\hat{k}_j) + B_{1,j}\,\mathrm{K}_1(\hat{k}) + B_{2,j}\,\mathrm{K}_1(\hat{k}_j)\big]. \quad (\text{D.3.24})$$

Boundary conditions at the cylinders

The no-slip and no-penetration conditions at the inner and outer cylinder surface require that

$$U_{x_j}(a_i) = 0, \qquad U_{\sigma_j}(a_i) = 0, \qquad U_{x_j}(a_e) = 0, \qquad U_{\sigma_j}(a_e) = 0. \quad (\text{D.3.25})$$

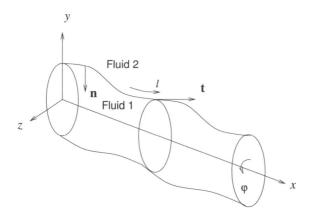

FIGURE D.3.2 Illustration of an interfacial deformation subject to an axisymmetric normal-mode perturbation.

Making substitutions, we obtain

$$A_{1,1}\, k\, \mathrm{I}_0(\hat{k}_i) + A_{2,1}\, k_1\, \mathrm{I}_0(\hat{k}_{i_1}) - B_{1,1}\, k\, \mathrm{K}_0(\hat{k}_i) - B_{2,1}\, k_1\, \mathrm{K}_0(a_{i_1}) = 0,$$

$$A_{1,1}\, \mathrm{I}_1(\hat{k}_i) + A_{2,1}\, \mathrm{I}_1(\hat{k}_{i_1}) + B_{1,1}\, \mathrm{K}_1(\hat{k}_i) + B_{2,1}\, \mathrm{K}_1(\hat{k}_{i_1}) = 0, \qquad (\mathrm{D.3.26})$$

and

$$A_{1,2}\, k\, \mathrm{I}_0(\hat{k}_e) + A_{2,2}\, k_2\, \mathrm{I}_0(\hat{k}_{e_2}) - B_{1,2}\, k\, \mathrm{K}_0(\hat{k}_e) - B_{2,2}\, k_2\, \mathrm{K}_0(\hat{k}_{e_2}) = 0,$$

$$A_{1,2}\, \mathrm{I}_1(\hat{k}_e) + A_{2,2}\, \mathrm{I}_1(\hat{k}_{e_2}) + B_{1,2}\, \mathrm{K}_1(\hat{k}_e) + B_{2,2}\, \mathrm{K}_1(\hat{k}_{e_2}) = 0, \qquad (\mathrm{D.3.27})$$

where $\hat{k}_i = ka_i$, $\hat{k}_{i_1} = k_1 a_i$, $\hat{a}_e = ka_e$, and $\hat{k}_{e_2} = k_2 a_e$.

Interfacial force balance

The dynamic interfacial condition requires that the jump in the hydrodynamic traction is balanced by the stress resultants due to the surface tension, γ, as well as by the elastic interfacial tensions,

$$\Delta \mathbf{f} \equiv (\boldsymbol{\sigma}^{(1)} - \boldsymbol{\sigma}^{(2)}) \cdot \mathbf{n} = \gamma\, 2\kappa_m\, \mathbf{n} + \Delta \mathbf{f}^{els}, \qquad (\mathrm{D.3.28})$$

where $\boldsymbol{\sigma}^{(1)}$ and $\boldsymbol{\sigma}^{(2)}$ are the hydrodynamic stress tensors for the inner and outer fluid, $\mathbf{n}$ is the unit vector normal to the interface pointing into the inner fluid labeled 1, as shown in Figure D.3.2, κ_m is the mean curvature of the interface, and $\Delta \mathbf{f}^{els}$ is an elastic component.

It will be necessary to introduce the unit vector tangent to the interface in an azimuthal plane, $\mathbf{t}$, and the arc length l in the direction of $\mathbf{t}$, as shown in Figure D.3.2. A force balance over a small section of the interface shows that the jump in the hydrodynamic traction due to the elastic tensions is given by

$$\Delta \mathbf{f}^{els} = (\kappa_l \tau_l + \kappa_\varphi \tau_\varphi)\, \mathbf{n} - \left(\frac{\partial \tau_l}{\partial l} + \frac{1}{\sigma} \frac{\partial \sigma}{\partial l} (\tau_l - \tau_\varphi) \right) \mathbf{t}, \qquad (\mathrm{D.3.29})$$

where κ_l and κ_φ are the principal curvatures of the interface in an azimuthal and its conjugate plane, τ_l is the principal elastic tension in the direction of the tangential vector $\mathbf{t}$, and τ_φ is the principal elastic tension in the azimuthal direction (e.g., [306], pp. 152–153). All functions in equations (D.3.28) and (D.3.29) are evaluated at the position of the perturbed interface. Consistent with the normal-mode analysis, we express the principal elastic tensions as

$$\tau_l = \epsilon\gamma_l \exp[ik(x-ct)], \qquad \tau_\varphi = \epsilon\gamma_\varphi \exp[ik(x-ct)], \tag{D.3.30}$$

where γ_l and γ_φ are complex coefficients to be determined as part of the solution.

Using (D.3.1) and standard expressions for the normal unit vector and curvatures, we derive linearized forms for the normal unit vector,

$$\mathbf{n} = -\mathbf{e}_\sigma + \epsilon ikA \exp[ik(x-ct)]\,\mathbf{e}_x, \tag{D.3.31}$$

principal curvatures,

$$\kappa_l = -\epsilon Ak^2 \exp[ik(x-ct)], \qquad \kappa_\varphi = -\frac{1}{a} + \epsilon\,\frac{A}{a^2}\exp[ik(x-ct)], \tag{D.3.32}$$

and mean curvature,

$$2\kappa_m = \kappa_l + \kappa_\varphi = -\frac{1}{a} + \epsilon\Big(\frac{\eta}{a^2}+\eta_{xx}\Big) = -\frac{1}{a} + \epsilon\,\frac{A}{a^2}(1-\hat{k}^2)\exp[ik(x-ct)], \tag{D.3.33}$$

where $\mathbf{e}_x$ and $\mathbf{e}_\sigma$ are unit vectors in the axial and radial directions.

Linearizing the tangential and normal components of the interfacial force balance expressed by (D.3.28), we obtain

$$\big(\sigma_{x\sigma_1}-\sigma_{x\sigma_2}\big)_{\sigma=a} = \mu_1\Big(\frac{\partial u_{x_1}}{\partial\sigma}+\frac{\partial u_{\sigma_1}}{\partial x}\Big)_{\sigma=a} - \mu_2\Big(\frac{\partial u_{x_1}}{\partial\sigma}+\frac{\partial u_{\sigma_1}}{\partial x}\Big)_{\sigma=a} = \epsilon i\kappa\gamma_l \exp[ik(x-ct)], \tag{D.3.34}$$

and

$$\big(\sigma_{\sigma\sigma_1}-\sigma_{\sigma\sigma_2}\big)_{\sigma=a} = \Big(-p_1+2\,\mu_1\frac{\partial u_{\sigma_1}}{\partial\sigma}\Big)_{\sigma=a} - \Big(-p_2+2\,\mu_2\frac{\partial u_{\sigma_2}}{\partial\sigma}\Big)_{\sigma=a}, \tag{D.3.35}$$

yielding

$$\big(\sigma_{\sigma\sigma_1}-\sigma_{\sigma\sigma_2}\big)_{\sigma=a} = \epsilon\Big[\gamma\,\frac{A}{a^2}(1-\hat{k}^2)-\frac{\gamma_\varphi}{a}\Big]\exp[ik(x-ct)]. \tag{D.3.36}$$

Interface constitutive equations

A constitutive equation relating the elastic tensions to the interfacial deformation is required. Following standard practice, we introduce the principal extension ratios

$$\lambda_l \equiv \frac{\partial l}{\partial l_R}, \qquad \lambda_\varphi \equiv \frac{\sigma}{\sigma_R}, \tag{D.3.37}$$

where the subscript R denotes the resting state. Assuming that the interface behaves like a thin elastic sheet of an incompressible material that obeys the Mooney constitutive law (e.g., [152, 236, 262]), we write

$$\tau_l = \frac{2}{3} E \left(2\hat{\lambda}_l + \hat{\lambda}_\varphi\right), \qquad \tau_\varphi = \frac{2}{3} E \left(\hat{\lambda}_l + 2\hat{\lambda}_\varphi\right), \tag{D.3.38}$$

where $\hat{\lambda}_l = \lambda_l - 1$ and $\hat{\lambda}_\varphi = \lambda_\varphi - 1$. Introducing the normal-mode forms

$$\lambda_l = 1 + \epsilon \chi_l \, \exp[ik(x - ct)], \qquad \lambda_\varphi = 1 + \epsilon \chi_\varphi \, \exp[ik(x - ct)], \tag{D.3.39}$$

we find that

$$\gamma_l = \frac{2}{3} E \left(2\chi_l + \chi_\varphi\right), \qquad \gamma_\varphi = \frac{2}{3} E \left(\chi_l + 2\chi_\varphi\right). \tag{D.3.40}$$

Evolution of the extension ratios

It remains to derive evolution equations for the principal extension ratios. An evolution equation for λ_φ arises by substituting (D.3.1) into the second equation of (D.3.37) and setting σ_R equal to the unperturbed radius of the interface, a, to obtain

$$\chi_\varphi = \frac{A}{a}. \tag{D.3.41}$$

An evolution equation for χ_l arises from the equation of motion of a material vector lying along the trace of the interface in an azimuthal plane,

$$\frac{1}{\chi_l} \frac{D\chi_l}{Dt} = \mathbf{t} \cdot \mathbf{L} \cdot \mathbf{t}, \tag{D.3.42}$$

where D/Dt is the material derivative and $\mathbf{L}$ is the velocity gradient tensor. Linearizing, we obtain

$$\chi_l = -\frac{1}{ca} \left(\frac{\partial \phi_j}{\partial \sigma}\right)_{\sigma=a}. \tag{D.3.43}$$

Interfacial displacement

Setting the right-hand side of the first equation in (D.3.37) equal to the right-hand side of the first equation in (D.3.39) and substituting (D.3.43), we obtain

$$\frac{\partial l}{\partial l_R} = 1 + \epsilon \chi_l \, \exp[ik(x - ct)] = 1 - \frac{\epsilon}{ca} \left(\frac{\partial \phi_j}{\partial \sigma}\right)_{\sigma=a} \exp[ik(x - ct)]. \tag{D.3.44}$$

Integrating with respect to l_R, we find that the axial displacement of a material membrane point which at the unstressed state was located at x, is given by

$$\Delta X = i\,\frac{\epsilon}{kca} \left(\frac{\partial \phi_j}{\partial \sigma}\right)_{\sigma=a} \exp[ik(x - ct)], \tag{D.3.45}$$

where $j = 1$ or 2. The corresponding axial displacement given by (D.3.41) is

$$\Delta\Sigma = \epsilon A \exp[ik(x - ct)]. \tag{D.3.46}$$

We anticipate that, because of the absence of a mean flow, the phase velocity c will be purely imaginary. Selecting the real parts of the last two equations, we obtain the normal-mode parametrization

$$X = x + \epsilon X_1 \sin(kx) \exp[ik(x - ct)], \qquad \Sigma = a + \epsilon \Sigma_1 \cos(kx) \exp[ik(x - ct)], \tag{D.3.47}$$

where

$$X_1 = -i \frac{1}{kca} \left(\frac{\partial \phi_j}{\partial \sigma} \right)_{\sigma = a}, \qquad \Sigma_1 = A. \tag{D.3.48}$$

The ratio $\delta \equiv X_1/\Sigma_1$ determines the direction of displacement of the individual material points corresponding to a normal mode.

Formulation of an eigenvalue problem

To formulate the algebraic eigenvalue problem, we collect equations (D.3.21), (D.3.22), (D.3.26), (D.3.27), (D.3.34), and (D.3.36) into a linear homogeneous system, $\mathbf{M} \cdot \mathbf{q} = \mathbf{0}$, where

$$\mathbf{q} = \left(A_{1,1}, A_{2,1}, B_{1,1}, B_{2,1}, A_{1,2}, A_{2,2}, B_{1,2}, B_{2,2} \right), \tag{D.3.49}$$

and the coefficient matrix, $\mathbf{M}$, is given in Table D.3.1. The functions F_i and G_i are coded in the function `matrix`. For a nontrivial solution to exist, the determinant of the coefficient matrix must be zero, $\det(\mathbf{M}) = 0$. In practice, the roots are found numerically using Newton's method with a suitable initial guess.

Program files:

1. `ann2lel`
   Computes the roots of the secular equation $\det(\mathbf{M}) = 0$ using Newton's method. The determinant of the matrix is calculated by LU decomposition using Gauss elimination.

2. `matrix`
   Formulates the matrix $\mathbf{M}$.

3. `bess_I01K0`
   Computes the Bessel functions.

4. `gel`
   Computes the determinant of a matrix by LU decomposition through Gauss elimination.

Input files:

1. `ann2lel.dat`
   Problem selection and specification of input parameters.

Output files:

1. `ann2lel.out`
   Growth rates.

$$
\mathbf{M} = \begin{bmatrix}
I_1(\hat{k}) & I_1(\hat{k}_1) & K_1(\hat{k}) & K_1(\hat{k}_1) & -I_1(\hat{k}) \\
k\,I_0(\hat{k}) & k_1 I_0(\hat{k}_1) & -k\,K_0(\hat{k}) & -k_1 K_0(\hat{k}_1) & -k\,I_0(\hat{k}) \\
I_1(\hat{k}_i) & I_1(\hat{k}_{i_1}) & K_1(\hat{k}_i) & K_1(\hat{k}_{i_1}) & 0 \\
k\,I_0(\hat{k}_i) & k_1 I_0(\hat{k}_{i_1}) & -k\,K_0(\hat{k}_i) & -k_1 K_0(\hat{k}_{i_2}) & 0 \\
0 & 0 & 0 & 0 & I_1(\hat{k}_e) \\
0 & 0 & 0 & 0 & k\,I_0(\hat{k}_e) \\
F_1 & F_2 & F_3 & F_4 & F_5 \\
G_1 & G_2 & G_3 & G_4 & G_5
\end{bmatrix}
$$

$$
\begin{bmatrix}
-I_1(\hat{k}_2) & -K_1(\hat{k}) & -K_1(\hat{k}_2) \\
-k_2 I_0(\hat{k}_2) & k\,K_0(\hat{k}) & k_2 K_0(\hat{k}_2) \\
0 & 0 & 0 \\
0 & 0 & 0 \\
I_1(\hat{k}_{e_2}) & K_1(\hat{k}_e) & K_1(\hat{k}_{e_2}) \\
k_2 I_0(\hat{k}_{e_2}) & -k\,K_0(\hat{k}_e) & -k_2 K_0(\hat{k}_{e_2}) \\
F_6 & F_7 & F_8 \\
G_6 & G_7 & G_8
\end{bmatrix}
$$

TABLE D.3.1 Setting the determinant of the matrix displayed to zero provides us with an algebraic equation for the complex growth rate.

D.4   Directory ann2lel0

This directory contains a code that performs the linear stability analysis of an elastic interface separating two annular layers confined between two concentric cylinders, as shown in Figure D.4.1, subject to axisymmetric perturbations. The instability occurs under conditions of Stokes flow [310].

Base state

We consider the instability of an elastic interface separating two annular layers confined between two concentric cylinders, as shown in Figure D.4.1. The inner fluid is labeled 1 and the outer fluid is labeled 2. The viscosity of the inner fluid is μ_1 and the viscosity of outer fluid is $\mu_2 = \lambda\mu_1$, where λ is the viscosity ratio. In the unperturbed configuration, the fluids are quiescent. Several special cases can be recognized:

- When the inner cylinder is absent and the radius of the outer cylinder is infinite, we obtain a thread suspended in an infinite ambient fluid.

- When the inner cylinder is absent, we obtain an annular layer coated on the interior surface of the outer cylinder.

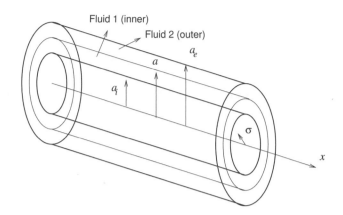

Fluid 1 (inner)

Fluid 2 (outer)

FIGURE D.4.1 Illustration of two annular layers separated by an elastic interface between an inner cylindrical tube with radius a_i and an outer cylindrical tube with radius a_e.

- When the radius of the outer cylinder is infinite, we obtain an annular layer coated on the exterior surface of the inner cylinder.

Normal-mode analysis

To carry out the normal-mode stability analysis for axisymmetric perturbations, we introduce cylindrical polar coordinates, (x, σ, φ), where the x axis coincides with the axis of revolution of the unperturbed interface. Next, we describe the radial position of the interface in terms of the real or imaginary part of the function

$$\sigma = f(x, t) = a + \epsilon\, \eta(x, t), \tag{D.4.1}$$

where ϵ is a dimensionless coefficient whose magnitude is much less than unity,

$$\eta(x, t) = A \exp[ik(x - ct)] \tag{D.4.2}$$

is the normal-mode wave form of the perturbation, A is the complex amplitude of the interfacial wave, $k = 2\pi/L$ is the wave number, L is the wavelength, and c is the complex phase velocity.

Stream function, velocity, and vorticity

The motion of the fluid on either side of the interface is governed by the continuity equation and the Stokes equation with appropriate physical constants for each fluid. Taking advantage of the axial symmetry of the flow, we describe the perturbation flow in terms of the Stokes stream functions, $\psi_1^{(1)}$ and $\psi_2^{(2)}$, for the inner and outer fluid, respectively, where the superscript (1) denotes the disturbance. The axial and radial components of the perturbation velocity are given by

$$u_{x_j} = \epsilon\, \frac{1}{\sigma}\, \frac{\partial \psi_j^{(1)}}{\partial \sigma}, \qquad u_{\sigma_j} = -\epsilon\, \frac{1}{\sigma}\, \frac{\partial \psi_j^{(1)}}{\partial x} \tag{D.4.3}$$

for $j = 1, 2$. The azimuthal component of the vorticity is given by

$$\omega_{\varphi_j} = -\epsilon \frac{1}{\sigma} E^2 \psi_j^{(1)}, \tag{D.4.4}$$

where

$$E^2 \equiv \frac{\partial^2}{\partial x^2} + \frac{\partial^2}{\partial \sigma^2} - \frac{1}{\sigma} \frac{\partial}{\partial \sigma} \tag{D.4.5}$$

is a second-order differential operator.

The vorticity transport equation for axisymmetric Stokes flow requires that

$$E^4 \psi_j \equiv E^2 \left(E^2 \psi_j^{(1)} \right) = 0, \tag{D.4.6}$$

which can be decomposed into two second-order constituent equations,

$$E^2 \psi_j = \tilde{\psi}_j^{(1)}, \qquad E^2 \tilde{\psi}_j^{(1)} = 0, \tag{D.4.7}$$

where a tilde denotes an intermediate solution.

Next, we express the stream function in the normal-mode form

$$\psi_j^{(1)}(x, \sigma, t) = \phi_j(\sigma) \, \exp[ik(x - ct)], \tag{D.4.8}$$

where $\phi_j(\sigma)$ are eigenfunctions. Substituting this expression into (D.4.7), we find that

$$\left(\frac{d^2}{d\sigma^2} - \frac{1}{\sigma} \frac{d}{d\sigma} - k^2 \right) \phi_j = \tilde{\phi}_j, \tag{D.4.9}$$

where

$$\left(\frac{d^2}{d\sigma^2} - \frac{1}{\sigma} \frac{d}{d\sigma} - k^2 \right) \tilde{\phi}_j = 0. \tag{D.4.10}$$

The general solution is

$$\phi_j(\sigma) = \sigma \left[A_{1,j} I_1(\hat{\sigma}) + A_{2,j} \sigma I_0(\hat{\sigma}) + B_{1,j} K_1(\hat{\sigma}) + B_{2,j} \sigma K_0(\hat{\sigma}) \right], \tag{D.4.11}$$

where I_0, I_1, K_0, and K_1 are Bessel functions, $A_{i,j}, B_{i,j}$ are complex constants, and $\hat{\sigma} \equiv k\sigma$. The first and third terms on the right-hand side of (D.4.11) satisfy (D.4.9) with $\tilde{\phi} = 0$. The second and fourth terms satisfy (D.4.9) with $\tilde{\phi}$ being a nontrivial solution of (D.4.10).

Substituting the preceding expressions into (D.4.3) and using the properties of the Bessel functions

$$I_0'(z) = I_1(z), \qquad I_1'(z) = I_0(z) - \frac{1}{z} I_1(z), \qquad K_0'(z) = -K_1(z),$$

$$K_1'(z) = -K_0(z) - \frac{1}{z} K_1(z), \tag{D.4.12}$$

we find that

$$u_{x_j} = \epsilon\, U_{x_j}(\sigma)\, \exp[\mathrm{i}k(x-ct)], \qquad u_{\sigma_j} = \epsilon\, U_{\sigma_j}(\sigma)\, \exp[\mathrm{i}k(x-ct)], \tag{D.4.13}$$

where

$$U_{x_j}(\sigma) = \frac{1}{\sigma}\frac{\mathrm{d}\phi_j}{\mathrm{d}\sigma} = A_{1,j}\, k\, \mathrm{I}_0(\hat\sigma) + A_{2,j}\,[\,2\,\mathrm{I}_0(\hat\sigma) + \hat\sigma\,\mathrm{I}_1(\hat\sigma)\,]$$
$$- B_{1,j}\, k\, \mathrm{K}_0(\hat\sigma) + B_{2,j}\,[\,2\,\mathrm{K}_0(\hat\sigma) - \hat\sigma\,\mathrm{K}_1(\hat\sigma)\,] \tag{D.4.14}$$

and

$$U_{\sigma_j}(\sigma) = -\mathrm{i}\,\frac{k}{\sigma}\,\phi_j = -\mathrm{i}k\,[\,A_{1,j}\,\mathrm{I}_1(\hat\sigma) + A_{2,j}\,\sigma\,\mathrm{I}_0(\hat\sigma) + B_{1,j}\,\mathrm{K}_1(\hat\sigma) + B_{2,j}\,\sigma\,\mathrm{K}_0(\hat\sigma)\,]. \tag{D.4.15}$$

Pressure field

The normal-mode expression for the pressure is

$$p_j = P_j + \epsilon\,\chi_j(\sigma)\,\exp[\mathrm{i}k(x-ct)], \tag{D.4.16}$$

where P_1 and P_2 are the uniform unperturbed pressures, $P_1 - P_2 = \gamma/a$ is the unperturbed capillary pressure, γ is the surface tension, and $\chi_j(\sigma)$ are pressure eigenfunctions. To compute the disturbance pressure field, we consider the x component of the Stokes equation for axisymmetric flow,

$$\frac{\partial p_j}{\partial x} = \mu_j\left(\frac{\partial^2 u_{x_j}}{\partial x^2} + \frac{1}{\sigma}\frac{\partial}{\partial\sigma}\left(\sigma\frac{\partial u_{x_j}}{\partial\sigma}\right)\right). \tag{D.4.17}$$

Substituting (D.4.16) together with the first equation in (D.4.13) and linearizing, we obtain

$$\chi_j(\sigma) = \mathrm{i}\mu_j k U_{x_j} - \mathrm{i}\frac{\mu_j}{k}\frac{1}{\sigma}\frac{\mathrm{d}}{\mathrm{d}\sigma}\left(\sigma\frac{\mathrm{d}U_{x_j}}{\mathrm{d}\sigma}\right) = -\mathrm{i}\frac{\mu_j}{k}\left[\frac{1}{\sigma}\frac{\mathrm{d}}{\mathrm{d}\sigma}\left(\sigma\frac{\mathrm{d}U_{x_j}}{\mathrm{d}\sigma}\right) - k^2\,U_{x_j}\right], \tag{D.4.18}$$

yielding

$$\chi_j(\sigma) = -\mathrm{i}\frac{\mu_j}{k}\left(\frac{\mathrm{d}^2 U_{x_j}}{\mathrm{d}\sigma^2} + \frac{1}{\sigma}\frac{\mathrm{d}U_{x_j}}{\mathrm{d}\sigma} - k^2\,U_{x_j}\right) = -\mathrm{i}\,2\,\mu_j k\,[\,A_{2,j}\,\mathrm{I}_0(\hat\sigma) + B_{2,j}\,\mathrm{K}_0(\hat\sigma)\,]. \tag{D.4.19}$$

Note that only two terms contribute to the pressure field.

Continuity of velocity at the interface

Requiring that the x and σ velocity components are continuous at the interface and using expressions (D.4.14) and (D.4.15), we derive the linearized kinematic interfacial conditions

$$A_{1,1}\, k\, \mathrm{I}_0(\hat k) + A_{2,1}\,[2\,\mathrm{I}_0(\hat k) + \hat k\,\mathrm{I}_1(\hat k)] - B_{1,1}\, k\, \mathrm{K}_0(\hat k) + B_{2,1}\,[2\,\mathrm{K}_0(\hat k) - \hat k\,\mathrm{K}_1(\hat k)]$$
$$= A_{1,2}\, k\, \mathrm{I}_0(\hat k) + A_{2,2}\,[2\,\mathrm{I}_0(\hat k) + \hat k\,\mathrm{I}_1(\hat k)] - B_{1,2}\, k\, \mathrm{K}_0(\hat k) + B_{2,2}\,[2\,\mathrm{K}_0(\hat k) - \hat k\,\mathrm{K}_1(\hat k)] \tag{D.4.20}$$

and

$$A_{1,1}\,\mathrm{I}_1(\hat k) + A_{2,1}\, a\,\mathrm{I}_0(\hat k) + B_{1,1}\,\mathrm{K}_1(\hat k) + B_{2,1}\, a\,\mathrm{K}_0(\hat k)$$
$$= A_{1,2}\,\mathrm{I}_1(\hat k) + A_{2,2}\, a\,\mathrm{I}_0(\hat k) + B_{1,2}\,\mathrm{K}_1(\hat k) + B_{2,2}\, a\,\mathrm{K}_0(\hat k), \tag{D.4.21}$$

where $\hat k = ka$ is a scaled wave number.

Kinematic compatibility

Kinematic compatibility requires $D[f(x,t) - \sigma]/Dt = 0$ at the interface, where D/Dt is the material derivative and the function f describes the position of the interface according to (D.4.1). Carrying out the differentiation, we find that

$$\frac{\partial f}{\partial t} + u_x \frac{\partial f}{\partial x} - u_\sigma = 0, \tag{D.4.22}$$

where the velocity is evaluated at the interface. Substituting the preceding expressions and linearizing, we find

$$A = i\,\frac{1}{ck}\,U_{\sigma_j}(\sigma = a) = \frac{1}{c}\,[\,A_{1,j}\,\mathrm{I}_1(\hat{k}) + A_{2,j}\,a\,\mathrm{I}_0(\hat{k}) + B_{1,j}\,\mathrm{K}_1(\hat{k}) + B_{2,j}\,a\,\mathrm{K}_0(\hat{k})\,]. \tag{D.4.23}$$

Boundary conditions at the cylinders

The no-slip and no-penetration conditions at the inner and outer cylinder surface require that

$$U_{x_j}(a_i) = 0, \qquad U_{\sigma_j}(a_i) = 0, \qquad U_{x_j}(a_e) = 0, \qquad U_{\sigma_j}(a_e) = 0. \tag{D.4.24}$$

Making substitutions, we obtain

$$A_{1,1}\,k\,\mathrm{I}_0(\hat{k}_i) + A_{2,1}\,[2\,\mathrm{I}_0(\hat{k}_i) + \hat{k}_i\,\mathrm{I}_1(\hat{k}_i)] - B_{1,1}\,k\,\mathrm{K}_0(\hat{k}_i) + B_{2,1}\,[2\,\mathrm{K}_0(\hat{k}_i) - \hat{k}_i\,\mathrm{K}_1(\hat{k}_i)] = 0,$$
$$A_{1,1}\,\mathrm{I}_1(\hat{k}_i) + A_{2,1}\,a_i\,\mathrm{I}_0(\hat{k}_i) + B_{1,1}\,\mathrm{K}_1(\hat{k}_i) + B_{2,1}\,a_i\,\mathrm{K}_0(\hat{k}_i) = 0, \tag{D.4.25}$$

and

$$A_{1,2}\,k\,\mathrm{I}_0(\hat{k}_e) + A_{2,2}\,[2\,\mathrm{I}_0(\hat{k}_e) + \hat{k}_e\,\mathrm{I}_1(\hat{k}_e)] - B_{1,2}\,k\,\mathrm{K}_0(\hat{k}_e) + B_{2,2}\,[2\,\mathrm{K}_0(\hat{k}_e) - \hat{k}_e\,\mathrm{K}_1(\hat{k}_e)] = 0,$$
$$A_{1,2}\,\mathrm{I}_1(\hat{k}_e) + A_{2,2}\,a_e\,\mathrm{I}_0(\hat{k}_e) + B_{1,2}\,\mathrm{K}_1(\hat{k}_e) + B_{2,2}\,a_e\,\mathrm{K}_0(\hat{k}_e) = 0, \tag{D.4.26}$$

where $\hat{k}_i = ka_i$ and $\hat{k}_e = ka_e$.

Interfacial force balance

The dynamic interfacial condition requires that the jump in the hydrodynamic traction is balanced by the capillary pressure due to the surface tension, γ, as well as by the normal and tangential elastic interfacial tensions,

$$\Delta\mathbf{f} \equiv (\boldsymbol{\sigma}^{(1)} - \boldsymbol{\sigma}^{(2)}) \cdot \mathbf{n} = \gamma\,2\kappa_m\,\mathbf{n} + \Delta\mathbf{f}^{els}, \tag{D.4.27}$$

where $\boldsymbol{\sigma}^{(1)}$ and $\boldsymbol{\sigma}^{(2)}$ are the hydrodynamic stress tensors in the inner and outer fluid, $\mathbf{n}$ is the unit vector normal to the interface pointing into the inner fluid labeled 1, as shown in Figure D.4.2, γ is the surface tension, κ_m is the mean curvature of the interface, and $\Delta\mathbf{f}^{els}$ is an elastic component.

It will be necessary to introduce the unit vector tangent to the interface in an azimuthal plane, $\mathbf{t}$, and the arc length l in the direction of $\mathbf{t}$, as shown in Figure D.4.2. A force balance over a small

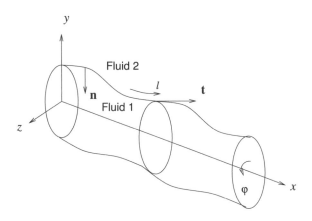

FIGURE D.4.2   Illustration of an interfacial deformation subject to an axisymmetric normal-mode perturbation.

section of the interface shows that the jump in the hydrodynamic traction due to the elastic tensions is given by

$$\Delta \mathbf{f}^{els} = (\kappa_l \tau_l + \kappa_\varphi \tau_\varphi) \, \mathbf{n} - \Big(\frac{\partial \tau_l}{\partial l} + \frac{1}{\sigma} \frac{\partial \sigma}{\partial l} (\tau_l - \tau_\varphi) \Big) \mathbf{t}, \tag{D.4.28}$$

where κ_l and κ_φ are the principal curvatures of the interface in a meridional and its conjugate plane, τ_l is the principal elastic tension in the direction of $\mathbf{t}$, and τ_φ is the principal elastic tension in the azimuthal direction of increasing angle φ (e.g., [306], pp. 152–153). All functions in equations (D.4.27) and (D.4.28) are evaluated at the position of the perturbed interface.

Consistent with the normal-mode analysis, we express the principal elastic tensions as

$$\tau_l = \epsilon \gamma_l \, \exp[ik(x - ct)], \qquad \tau_\varphi = \epsilon \gamma_\varphi \, \exp[ik(x - ct)], \tag{D.4.29}$$

where γ_l and γ_φ are complex coefficients to be determined as part of the solution.

Using (D.4.1) and standard expressions for the normal unit vector and curvatures, we derive linearized forms for the normal unit vector,

$$\mathbf{n} = -\mathbf{e}_\sigma + \epsilon A i k \, \exp[ik(x - ct)] \, \mathbf{e}_x, \tag{D.4.30}$$

principal curvatures,

$$\kappa_l = -\epsilon A k^2 \exp[ik(x - ct)], \qquad \kappa_\varphi = -\frac{1}{a} + \epsilon \frac{A}{a^2} \exp[ik(x - ct)], \tag{D.4.31}$$

and mean curvature,

$$2\kappa_m = \kappa_l + \kappa_\varphi = -\frac{1}{a} + \epsilon \Big(\frac{\eta}{a^2} + \eta_{xx} \Big) = -\frac{1}{a} + \epsilon \frac{A}{a^2} (1 - \hat{k}^2) \exp[ik(x - ct)], \tag{D.4.32}$$

where $\mathbf{e}_x$ and $\mathbf{e}_\sigma$ are unit vectors in the axial and radial directions.

Linearizing the tangential and normal components of the interfacial force balance expressed by (D.4.27), we find that

$$\left(\sigma_{x\sigma_1} - \sigma_{x\sigma_2}\right)_{\sigma=a} = \mu_1 \left(\frac{\partial u_{x_1}}{\partial \sigma} + \frac{\partial u_{\sigma_1}}{\partial x}\right)_{\sigma=a} - \mu_2 \left(\frac{\partial u_{x_2}}{\partial \sigma} + \frac{\partial u_{\sigma_2}}{\partial x}\right)_{\sigma=a} = \epsilon\, \mathrm{i}\, \kappa \gamma_l \, \exp[\mathrm{i}k(x - ct)]$$

(D.4.33)

and

$$\left(\sigma_{\sigma\sigma_1} - \sigma_{\sigma\sigma_2}\right)_{\sigma=a} = \left(-p_1 + 2\mu_1 \frac{\partial u_{\sigma_1}}{\partial \sigma}\right)_{\sigma=a} - \left(-p_2 + 2\mu_2 \frac{\partial u_{\sigma_2}}{\partial \sigma}\right)_{\sigma=a},$$

(D.4.34)

yielding

$$\left(\sigma_{\sigma\sigma_1} - \sigma_{\sigma\sigma_2}\right)_{\sigma=a} = \epsilon\left[\gamma A \frac{1}{a^2}(1 - \hat{k}^2) - \frac{\gamma_\varphi}{a}\right]\exp[\mathrm{i}k(x - ct)].$$

(D.4.35)

Both sides are evaluated at the location of the unperturbed interface, $\sigma = a$.

Interface constitutive equations

A constitutive equation relating the elastic tensions to the interfacial deformation is required. Following standard practice, we introduce the principal extension ratios

$$\lambda_l \equiv \frac{\partial l}{\partial l_R}, \qquad \lambda_\varphi \equiv \frac{\sigma}{\sigma_R},$$

(D.4.36)

where the subscript R signifies the resting state. Assuming that the membrane behaves like a thin elastic sheet consisting of an incompressible material that obeys the Mooney constitutive law (e.g., [152, 236, 262]), we write

$$\tau_l = \frac{2}{3}E\,(2\hat{\lambda}_l + \hat{\lambda}_\varphi), \qquad \tau_\varphi = \frac{2}{3}E\,(\hat{\lambda}_l + 2\hat{\lambda}_\varphi),$$

(D.4.37)

where $\hat{\lambda}_l = \lambda_l - 1$ and $\hat{\lambda}_\varphi = \lambda_\varphi - 1$. Introducing the normal-mode forms

$$\lambda_l = 1 + \epsilon\chi_l\,\exp[\mathrm{i}k(x - ct)], \qquad \lambda_\varphi = 1 + \epsilon\chi_\varphi\,\exp[\mathrm{i}k(x - ct)],$$

(D.4.38)

we find that

$$\gamma_l = \frac{2}{3}E\,(2\chi_l + \chi_\varphi), \qquad \gamma_\varphi = \frac{2}{3}E\,(\chi_l + 2\chi_\varphi).$$

(D.4.39)

Evolution of the extension ratios

It remains to derive evolution equations for the principal extension ratios. An evolution equation for λ_φ arises by substituting (D.4.1) into the second equation of (D.4.36) and setting σ_R equal to the unperturbed radius of the interface, a, to obtain

$$\chi_\varphi = \frac{A}{a}.$$

(D.4.40)

An evolution equation for χ_l arises from the equation of motion of a material vector lying along the trace of the interface in an azimuthal plane,

$$\frac{1}{\chi_l} \frac{\mathrm{D}\chi_l}{\mathrm{D}t} = \mathbf{t} \cdot \mathbf{L} \cdot \mathbf{t}, \tag{D.4.41}$$

where $\mathrm{D}/\mathrm{D}t$ is the material derivative and $\mathbf{L}$ is the velocity gradient tensor. Linearizing, we find

$$\chi_l = -\frac{1}{ca} \left(\frac{\partial \phi_j}{\partial \sigma} \right)_{\sigma=a}, \tag{D.4.42}$$

where $j = 1$ or 2.

Interfacial displacement

Setting the right-hand side of the first equation in (D.4.36) equal to the right-hand side of the first equation in (D.4.38) and substituting (D.4.42), we find

$$\frac{\partial l}{\partial l_R} = 1 + \epsilon \chi_l \exp[\mathrm{i}k(x - ct)] = 1 - \frac{\epsilon}{ca} \left(\frac{\partial \phi_j}{\partial \sigma} \right)_{\sigma=a} \exp[\mathrm{i}k(x - ct)]. \tag{D.4.43}$$

Integrating with respect to l_R, we find that the axial displacement of a material point which at the unstressed state was located at x, is given by

$$\Delta X = \mathrm{i} \frac{\epsilon}{kca} \left(\frac{\partial \phi_j}{\partial \sigma} \right)_{\sigma=a} \exp[\mathrm{i}k(x - ct)], \tag{D.4.44}$$

where $j = 1$ or 2. The corresponding axial displacement given by (D.4.40) is

$$\Delta \Sigma = \epsilon A \exp[\mathrm{i}k(x - ct)]. \tag{D.4.45}$$

We anticipate that, because of the absence of a base flow, the phase velocity c will be purely imaginary and select the real parts of the last two equations to obtain the normal-mode parametrization

$$X = x + \epsilon X_1 \sin(kx) \exp[\mathrm{i}k(x - ct)], \qquad \Sigma = a + \epsilon \Sigma_1 \cos(kx) \exp[\mathrm{i}k(x - ct)], \tag{D.4.46}$$

where

$$X_1 = -\mathrm{i} \frac{1}{kca} \left(\frac{\partial \phi_j}{\partial \sigma} \right)_{\sigma=a}, \qquad \Sigma_1 = A. \tag{D.4.47}$$

The ratio $\delta \equiv X_1/\Sigma_1$ determines the direction of displacement of the individual material points corresponding to the normal modes.

Formulation of an eigenvalue problem

To formulate an algebraic eigenvalue problem, we collect equations (D.4.20), (D.4.21), (D.4.25), (D.4.26), (D.4.33), and (D.4.35) into a linear homogeneous system, $\mathbf{M} \cdot \mathbf{q} = \mathbf{0}$, where

$$\mathbf{q} = \left(A_{1,1}, A_{2,1}, B_{1,1}, B_{2,1}, A_{1,2}, A_{2,2}, B_{1,2}, B_{2,2} \right) \tag{D.4.48}$$

$$\mathbf{M} = \begin{bmatrix}
I_1(\hat{k}) & a\,I_0(\hat{k}) & K_1(\hat{k}) & a\,K_1(\hat{k}) \\
k\,I_0(\hat{k}) & 2\,I_0(\hat{k}) + \hat{k}\,I_1(\hat{k}) & -k\,K_0(\hat{k}) & 2\,K_0(\hat{k}) - \hat{k}\,K_1(\hat{k}) \\
I_1(\hat{k}_i) & a_i\,I_0(\hat{k}_i) & K_1(\hat{k}_i) & a_i\,K_0(\hat{k}_i) \\
k\,I_0(\hat{k}_i) & 2\,I_0(\hat{k}_i) + \hat{k}_i\,I_1(\hat{k}_i) & -k\,K_0(\hat{k}_i) & 2\,K_0(\hat{k}_i) - \hat{k}_i\,K_1(\hat{k}_i) \\
0 & 0 & 0 & 0 \\
0 & 0 & 0 & 0 \\
F_1 & F_2 & F_3 & F_4 \\
G_1 & G_2 & G_3 & G_4
\end{bmatrix}$$

$$\begin{bmatrix}
-I_1(\hat{k}) & -a\,I_0(\hat{k}) & -K_1(\hat{k}) & -a\,K_0(\hat{k}) \\
-k\,I_0(\hat{k}) & -2\,I_0(\hat{k}) - \hat{k}\,I_1(\hat{k}) & k\,K_0(\hat{k}) & -2\,K_0(\hat{k}) + \hat{k}\,K_1(\hat{k}) \\
0 & 0 & 0 & 0 \\
0 & 0 & 0 & 0 \\
I_1(\hat{k}_e) & a_e\,I_0(\hat{k}_e) & K_1(\hat{k}_e) & a_e\,K_0(\hat{k}_e) \\
k\,I_0(\hat{k}_e) & 2\,I_0(\hat{k}_e) + \hat{k}_e\,I_1(\hat{k}_e) & -k\,K_0(\hat{k}_e) & 2\,K_0(\hat{k}_e) - \hat{k}_e\,K_1(\hat{k}_e) \\
F_5 & F_6 & F_7 & F_8 \\
G_5 & G_6 & G_7 & G_8
\end{bmatrix}$$

TABLE D.4.1   Setting the determinant of the matrix displayed to zero provides us with an algebraic equation for the complex growth rate.

and the coefficient matrix, $\mathbf{M}$, is given in Table D.4.1 [310]. The functions F_1–F_8 in the seventh row and the functions G_1–G_8 in the eighth row are defined in the computer code. For a nontrivial solution to exist, the determinant of the coefficient matrix must be zero, $\det(\mathbf{M}) = 0$. Expanding the determinant, we obtain a quadratic equation for the growth rate.

Program files:

1. `ann21e10`
   Computes the roots of the secular equation $\det(\mathbf{M}) = 0$. The determinant of the matrix is calculated by LU decomposition using Gauss elimination.

2. `matrix`
   Formulates the matrix $\mathbf{M}$.

3. `bess_I01K0`
   Computes the Bessel functions.

4. `gel`
   Computes the determinant of a matrix by LU decomposition using Gauss elimination.

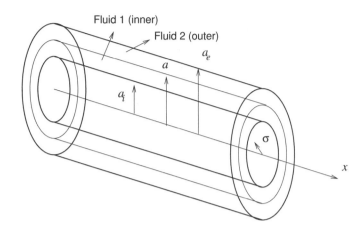

FIGURE D.5.1 Illustration of two annular layers separated by a viscous interface confined between an inner cylindrical tube with radius a_i and an outer cylindrical tube with radius a_e.

Input files:

1. `ann2lel.dat`
   Problem selection and specification of input parameters.

Output files:

1. `ann2lel.out`
   Growth rates.

D.5 Directory ann2lvs0

This directory contains a code that performs the linear stability analysis of a viscous interface separating two annular layers confined between two concentric cylinders, as shown in Figure D.5.1 [326]. The instability occurs under conditions of Stokes flow.

Base state

We consider the capillary instability of an infinite, cylindrical, viscous interface separating two concentric annular layers confined between two concentric cylinders, as shown in Figure D.5.1. The inner fluid is labeled 1 and the outer fluid is labeled 2. The viscosity of the inner fluid is μ_1 and the viscosity of outer fluid is $\mu_2 = \lambda\mu_1$, where λ is the viscosity ratio. In the unperturbed configuration, the fluids are quiescent. Several special cases can be recognized:

- When the inner cylinder is absent and the radius of the outer cylinder is infinite, we obtain a thread suspended in an infinite ambient fluid.

- When the inner cylinder is absent, we obtain an annular layer coated on the interior surface of the outer cylinder.

- When the radius of the outer cylinder is infinite, we obtain an annular layer coated on the exterior surface of the inner cylinder.

Normal-mode analysis

To carry out the normal-mode stability analysis for axisymmetric perturbations, we introduce cylindrical polar coordinates, (x, σ, φ), where the x axis coincides with the axis of revolution of the unperturbed interface. Next, we describe the radial position of the interface in terms of the real or imaginary part of the function

$$\sigma = f(x, t) = a + \epsilon \eta(x, t), \tag{D.5.1}$$

where ϵ is a dimensionless coefficient whose magnitude is much less than unity,

$$\eta(x, t) = A \exp[\mathrm{i} k(x - ct)] \tag{D.5.2}$$

is the normal-mode wave form of the perturbation, A is the complex amplitude of the interfacial wave, i is the imaginary unit satisfying $\mathrm{i}^2 = -1$, $k = 2\pi/L$ is the wave number, L is the wavelength, and c is the complex phase velocity.

Stream function, velocity and vorticity

The motion of the fluid on either side of the interface is governed by the continuity equation and the Stokes equation with appropriate physical constants for each fluid. Taking advantage of the axial symmetry of the flow, we describe the perturbation in terms of the Stokes stream functions, $\psi_1^{(1)}$ and $\psi_2^{(1)}$, for the inner and outer fluid, respectively, where the superscript (1) denotes the disturbance. The axial and radial components of the perturbation velocity are given by

$$u_{x_j} = \epsilon \, \frac{1}{\sigma} \, \frac{\partial \psi_j^{(1)}}{\partial \sigma}, \qquad u_{\sigma_j} = -\epsilon \, \frac{1}{\sigma} \, \frac{\partial \psi_j^{(1)}}{\partial x} \tag{D.5.3}$$

for $j = 1, 2$. The azimuthal component of the vorticity is given by

$$\omega_{\varphi_j} = -\epsilon \, \frac{1}{\sigma} \, E^2 \psi_j^{(1)}, \tag{D.5.4}$$

where

$$E^2 \equiv \frac{\partial^2}{\partial x^2} + \frac{\partial^2}{\partial \sigma^2} - \frac{1}{\sigma} \frac{\partial}{\partial \sigma} \tag{D.5.5}$$

is a second-order differential operator.

The vorticity transport equation for axisymmetric Stokes flow requires that

$$E^4 \psi_j \equiv E^2 \left(E^2 \psi_j^{(1)} \right) = 0, \tag{D.5.6}$$

which can be decomposed into two second-order constituent equations,

$$E^2 \psi_j^{(1)} = \tilde{\psi}_j, \qquad E^2 \tilde{\psi}_j^{(1)} = 0, \tag{D.5.7}$$

where a tilde denotes an intermediate solution.

The stream function is now expressed in the normal-mode form

$$\psi_j^{(1)}(x, \sigma, t) = \phi_j(\sigma) \, \exp[ik(x - ct)], \tag{D.5.8}$$

where $\phi_j(\sigma)$ are eigenfunctions. Substituting this expression into (D.5.7), we obtain

$$\left(\frac{d^2}{d\sigma^2} - \frac{1}{\sigma} \frac{d}{d\sigma} - k^2 \right) \phi_j = \tilde{\phi}_j, \tag{D.5.9}$$

where

$$\left(\frac{d^2}{d\sigma^2} - \frac{1}{\sigma} \frac{d}{d\sigma} - k^2 \right) \tilde{\phi}_j = 0. \tag{D.5.10}$$

The general solution is

$$\phi_j(\sigma) = \sigma \left[A_{1,j} \, I_1(\hat{\sigma}) + A_{2,j} \, \sigma \, I_0(\hat{\sigma}) + B_{1,j} \, K_1(\hat{\sigma}) + B_{2,j} \, \sigma \, K_0(\hat{\sigma}) \right], \tag{D.5.11}$$

where $I_0, I_1, K_0,$ and K_1 are Bessel functions, $A_{i,j}, B_{i,j}$ are complex constants, and $\hat{\sigma} \equiv k\sigma$. The first and third terms on the right-hand side of (D.5.11) satisfy (D.5.9) with $\tilde{\phi}_j = 0$. The second and fourth terms satisfy (D.5.9) with $\tilde{\phi}_j$ being a nontrivial solution of (D.5.10).

Substituting the preceding expressions in (D.5.3) and using the following properties of the Bessel functions,

$$I_0'(z) = I_1(z), \qquad I_1'(z) = I_0(z) - \frac{1}{z} I_1(z), \qquad K_0'(z) = -K_1(z),$$

$$K_1'(z) = -K_0(z) - \frac{1}{z} K_1(z), \tag{D.5.12}$$

we find that

$$u_{x_j} = \epsilon \, U_{x_j}(\sigma) \, \exp[ik(x - ct)], \qquad u_{\sigma_j} = \epsilon \, U_{\sigma_j}(\sigma) \, \exp[ik(x - ct)], \tag{D.5.13}$$

where

$$U_{x_j}(\sigma) = \frac{1}{\sigma} \frac{d\phi_j}{d\sigma} = A_{1,j} \, k \, I_0(\hat{\sigma}) + A_{2,j} \left[2 \, I_0(\hat{\sigma}) + \hat{\sigma} \, I_1(\hat{\sigma}) \right]$$

$$-B_{1,j} \, k \, K_0(\hat{\sigma}) + B_{2,j} \left[2 \, K_0(\hat{\sigma}) - \hat{\sigma} \, K_1(\hat{\sigma}) \right], \tag{D.5.14}$$

and

$$U_{\sigma_j}(\sigma) = -i \frac{k}{\sigma} \phi_j = -ik \left[A_{1,j} \, I_1(\hat{\sigma}) + A_{2,j} \, \sigma \, I_0(\hat{\sigma}) + B_{1,j} \, K_1(\hat{\sigma}) + B_{2,j} \, \sigma \, K_0(\hat{\sigma}) \right]. \tag{D.5.15}$$

Pressure field

The normal-mode pressure field is

$$p_j = P_j + \epsilon \chi_j(\sigma) \, \exp[ik(x - ct)], \tag{D.5.16}$$

where P_1 and P_2 are the unperturbed uniform pressures, $P_1 - P_2 = \gamma/a$ is the unperturbed capillary pressure, γ is the surface tension, and $\chi_j(\sigma)$ are pressure eigenfunctions. To compute the disturbance pressure field, we consider the x component of the Stokes equation for axisymmetric flow,

$$\frac{\partial p_j}{\partial x} = \mu_j \left[\frac{\partial^2 u_{x_j}}{\partial x^2} + \frac{1}{\sigma} \frac{\partial}{\partial \sigma} \left(\sigma \frac{\partial u_{x_j}}{\partial \sigma} \right) \right]. \tag{D.5.17}$$

Substituting (D.5.16) along with the first equation into (D.5.13) and linearizing, we obtain

$$\chi_j(\sigma) = \mathrm{i}\mu_j k \, U_{x_j} - \mathrm{i} \frac{\mu_j}{k} \frac{1}{\sigma} \frac{\mathrm{d}}{\mathrm{d}\sigma} \left(\sigma \frac{\mathrm{d}U_{x_j}}{\mathrm{d}\sigma} \right) = -\mathrm{i} \frac{\mu_j}{k} \left[\frac{1}{\sigma} \frac{\mathrm{d}}{\mathrm{d}\sigma} \left(\sigma \frac{\mathrm{d}U_{x_j}}{\mathrm{d}\sigma} \right) - k^2 U_{x_j} \right] \tag{D.5.18}$$

or

$$\chi_j(\sigma) = -\mathrm{i} \frac{\mu_j}{k} \left(\frac{\mathrm{d}^2 U_{x_j}}{\mathrm{d}\sigma^2} + \frac{1}{\sigma} \frac{\mathrm{d}U_{x_j}}{\mathrm{d}\sigma} - k^2 U_{x_j} \right). \tag{D.5.19}$$

Making substitutions, we obtain

$$\chi_j(\sigma) = -\mathrm{i}\, 2\, \mu_j k \left[A_{2,j} \, \mathrm{I}_0(\hat{\sigma}) + B_{2,j} \, \mathrm{K}_0(\hat{\sigma}) \right]. \tag{D.5.20}$$

Note that only two terms contribute to the pressure field.

Continuity of velocity at the interface

Requiring that the x and σ velocity components are continuous at the interface, and using expressions (D.5.14) and (D.5.15), we derive the linearized kinematic interfacial conditions

$$A_{1,1}k \, \mathrm{I}_0(\hat{k}) + A_{2,1}[2\, \mathrm{I}_0(\hat{k}) + \hat{k}\, \mathrm{I}_1(\hat{k})] - B_{1,1}k \, \mathrm{K}_0(\hat{k}) + B_{2,1}[2\, \mathrm{K}_0(\hat{k}) - \hat{k}\, \mathrm{K}_1(\hat{k})]$$
$$= A_{1,2}\, k \, \mathrm{I}_0(\hat{k}) + A_{2,2}[2\, \mathrm{I}_0(\hat{k}) + \hat{k}\, \mathrm{I}_1(\hat{k})] - B_{1,2}\, k \, \mathrm{K}_0(\hat{k}) + B_{2,2}\, [2\, \mathrm{K}_0(\hat{k}) - \hat{k}\, \mathrm{K}_1(\hat{k})], \tag{D.5.21}$$

and

$$A_{1,1}\, \mathrm{I}_1(\hat{k}) + A_{2,1}\, a\, \mathrm{I}_0(\hat{k}) + B_{1,1}\, \mathrm{K}_1(\hat{k}) + B_{2,1}\, a\, \mathrm{K}_0(\hat{k})$$
$$= A_{1,2}\, \mathrm{I}_1(\hat{k}) + A_{2,2}\, a\, \mathrm{I}_0(\hat{k}) + B_{1,2}\, \mathrm{K}_1(\hat{k}) + B_{2,2}\, a\, \mathrm{K}_0(\hat{k}), \tag{D.5.22}$$

where $\hat{k} = ka$ is a scaled wave number.

Kinematic compatibility

Kinematic compatibility requires that $\mathrm{D}[f(x,t) - \sigma]/\mathrm{D}t = 0$ at the interface, where $\mathrm{D}/\mathrm{D}t$ is the material derivative and the function f describes the position of the interface according to (D.5.1). Carrying out the differentiation, we obtain

$$\frac{\partial f}{\partial t} + u_x \frac{\partial f}{\partial x} - u_\sigma = 0, \tag{D.5.23}$$

where the velocity is evaluated at the interface. Substituting the preceding expressions and linearizing, we find that

$$A = \mathrm{i}\,\frac{1}{kc}\,U_{\sigma_j}(a) = \frac{1}{c}\,\big[\,A_{1,j}\,\mathrm{I}_1(\hat{k}) + A_{2,j}\,a\,\mathrm{I}_0(\hat{k}) + B_{1,j}\,\mathrm{K}_1(\hat{k}) + B_{2,j}\,a\,\mathrm{K}_0(\hat{k})\,\big]. \tag{D.5.24}$$

Boundary conditions at the cylinders

The no-slip and no-penetration condition at the inner and outer cylinder require that

$$U_{x_j}(a_i) = U_{\sigma_j}(a_i) = 0, \qquad U_{x_j}(a_e) = U_{\sigma_j}(a_e) = 0. \tag{D.5.25}$$

Making substitutions, we obtain

$$A_{1,1}\,k\,\mathrm{I}_0(\hat{k}_i) + A_{2,1}\,[2\,\mathrm{I}_0(\hat{k}_i) + \hat{k}_i\,\mathrm{I}_1(\hat{k}_i)] - B_{1,1}\,k\,\mathrm{K}_0(\hat{k}_i) + B_{2,1}\,[2\,\mathrm{K}_0(\hat{k}_i) - \hat{k}_i\,\mathrm{K}_1(\hat{k}_i)] = 0,$$
$$A_{1,1}\,\mathrm{I}_1(\hat{k}_i) + A_{2,1}\,a_i\,\mathrm{I}_0(\hat{k}_i) + B_{1,1}\,\mathrm{K}_1(\hat{k}_i) + B_{2,1}\,a_i\,\mathrm{K}_0(\hat{k}_i) = 0, \tag{D.5.26}$$

and

$$A_{1,2}\,k\,\mathrm{I}_0(\hat{k}_e) + A_{2,2}\,[2\,\mathrm{I}_0(\hat{k}_e) + \hat{k}_e\,\mathrm{I}_1(\hat{k}_e)] - B_{1,2}\,k\,\mathrm{K}_0(\hat{k}_e) + B_{2,2}\,[2\,\mathrm{K}_0(\hat{k}_e) - \hat{k}_e\,\mathrm{K}_1(\hat{k}_e)] = 0,$$
$$A_{1,2}\,\mathrm{I}_1(\hat{k}_e) + A_{2,2}\,a_e\,\mathrm{I}_0(\hat{k}_e) + B_{1,2}\,\mathrm{K}_1(\hat{k}_e) + B_{2,2}\,a_e\,\mathrm{K}_0(\hat{k}_e) = 0, \tag{D.5.27}$$

where $\hat{k}_i = ka_i$ and $\hat{k}_e = ka_e$.

Interfacial force balance

The dynamic interfacial condition requires that the jump in the hydrodynamic traction is balanced by the normal stress due to the surface tension, γ, as well as by viscous interfacial tensions generated by the fluid motion,

$$\Delta\mathbf{f} \equiv (\boldsymbol{\sigma}^{(1)} - \boldsymbol{\sigma}^{(2)})\cdot\mathbf{n} = \gamma\,2\kappa_m\,\mathbf{n} + \Delta\mathbf{f}^v, \tag{D.5.28}$$

where $\boldsymbol{\sigma}^{(1)}$ and $\boldsymbol{\sigma}^{(2)}$ are the hydrodynamic stress tensors in the inner and outer fluid, $\mathbf{n}$ is the unit vector normal to the interface pointing into the inner fluid labeled 1, as shown in Figure D.5.2, κ_m is the mean curvature of the interface, and $\Delta\mathbf{f}^v$ is a viscous component.

A force balance over a small section of the interface shows that the jump in the hydrodynamic traction due to the interfacial tensions is given by

$$\Delta\mathbf{f}^v = (\kappa_l\tau_l + \kappa_\varphi\tau_\varphi)\,\mathbf{n} - \Big[\frac{\partial\tau_l}{\partial l} + \frac{1}{\sigma}\frac{\partial\sigma}{\partial l}\,(\tau_l - \tau_\varphi)\Big]\,\mathbf{t}, \tag{D.5.29}$$

where κ_l and κ_φ are the principal curvatures of the interface in a meridional and its conjugate plane, τ_l and τ_φ are the principal viscous tensions referring to orthogonal curvilinear axes corresponding to the unit tangential vector $\mathbf{t}$ and to the azimuthal angle, φ, and l is the arc length measured in the direction of $\mathbf{t}$, as shown in Figure D.5.2 (e.g., [306], pp. 152–153). All functions in equations (D.5.28) and (D.5.29) are evaluated at the position of the perturbed interface.

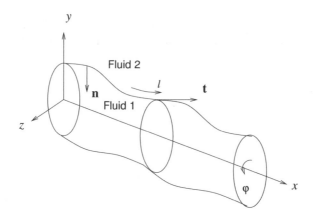

FIGURE D.5.2   Illustration of an interfacial deformation subject to an axisymmetric normal-mode perturbation.

Consistent with the normal-mode analysis, we express the principal viscous tensions as

$$\tau_l = \epsilon \gamma_l \, \exp[ik(x - ct)], \qquad \tau_\varphi = \epsilon \gamma_\varphi \, \exp[ik(x - ct)], \tag{D.5.30}$$

where γ_l and γ_φ are complex coefficients to be determined as part of the solution.

Using (D.5.1) and standard expressions for the normal and tangent unit vectors and curvatures, we derive linearized forms for the normal and tangent unit vectors,

$$\mathbf{n} = -\mathbf{e}_\sigma + \epsilon Aik \, \exp[ik(x - ct)]\mathbf{e}_x, \qquad \mathbf{t} = \mathbf{e}_x + \epsilon Aik \, \exp[ik(x - ct)]\mathbf{e}_\sigma \tag{D.5.31}$$

principal curvatures,

$$\kappa_l = -\epsilon Ak^2 \exp[ik(x - ct)], \qquad \kappa_\varphi = -\frac{1}{a} + \epsilon \frac{A}{a^2} \, \exp[ik(x - ct)], \tag{D.5.32}$$

and mean curvature,

$$2\kappa_m = \kappa_l + \kappa_\varphi = -\frac{1}{a} + \epsilon \left(\frac{\eta}{a^2} + \eta_{xx} \right) = -\frac{1}{a} + \epsilon \frac{A}{a^2} \left(1 - \hat{k}^2 \right) \exp[ik(x - ct)], \tag{D.5.33}$$

where $\mathbf{e}_x$ and $\mathbf{e}_\sigma$ are the unit vectors in the axial and radial directions.

To evaluate the viscous interfacial tensions, we introduce the principal extension ratios

$$\lambda_l \equiv \frac{\partial l}{\partial l_R}, \qquad \lambda_\varphi \equiv \frac{\sigma}{\sigma_R}, \tag{D.5.34}$$

where the subscript R signifies the resting state.

The viscous tensions developing in a three-dimensional interface can be expressed in global Cartesian form as [366]

$$\tau^v = \mu_e \, \theta \, \mathbf{P} + \mu_s \left(2 \, \mathbf{E}^s - \theta \, \mathbf{P} \right) = (\mu_e - \mu_s) \, \theta \, \mathbf{P} + \mu_s \, 2 \, \mathbf{E}^s, \tag{D.5.35}$$

where μ_e is the interface expansion viscosity, μ_s is the interface shear viscosity, $\mathbf{P} \equiv \mathbf{I} - \mathbf{n} \otimes \mathbf{n}$ is the tangential projection operator, $\mathbf{n}$ is the unit vector normal to the interface, $\mathbf{E}^s$ is the surface rate-of-deformation tensor given by

$$\mathbf{E}^s = \frac{1}{2} \mathbf{P} \cdot \left[\nabla \mathbf{u} + (\nabla \mathbf{u})^T \right] \cdot \mathbf{P} = \mathbf{P} \cdot \nabla \mathbf{u} \cdot \mathbf{P}, \tag{D.5.36}$$

the superscript T denotes the transpose, and θ is the surface rate of dilatation given by

$$\theta = \mathbf{P} : \nabla \mathbf{u} = \mathrm{trace}(\mathbf{E}^s). \tag{D.5.37}$$

In the case of axisymmetric deformation, $\mathbf{P} = \mathbf{t} \otimes \mathbf{t} + \mathbf{e}_\varphi \otimes \mathbf{e}_\varphi$,

$$\mathbf{E}^s = (\mathbf{t} \cdot \nabla \mathbf{u} \cdot \mathbf{t}) \mathbf{t} \otimes \mathbf{t} + (\mathbf{e}_\varphi \cdot \nabla \mathbf{u} \cdot \mathbf{e}_\varphi) \mathbf{e}_\varphi \otimes \mathbf{e}_\varphi, \qquad \boldsymbol{\tau}^v = \tau_l \, \mathbf{t} \otimes \mathbf{t} + \tau_\varphi \otimes \mathbf{e}_\varphi \mathbf{e}_\varphi, \tag{D.5.38}$$

where

$$\tau_l = (\mu_e - \mu_s) \theta + 2\mu_s \mathbf{t} \cdot \nabla \mathbf{u} \cdot \mathbf{t}, \qquad \tau_\varphi = (\mu_e - \mu_s) \theta + 2\mu_s \, \mathbf{e}_\varphi \cdot \nabla \mathbf{u} \cdot \mathbf{e}_\varphi, \tag{D.5.39}$$

and $\mathbf{e}_\varphi$ is the unit vector in the azimuthal direction. To interpret these expressions, we introduce the principal extension ratios

$$\mathbf{t} \cdot \nabla \mathbf{u} \cdot \mathbf{t} = \frac{\partial \mathbf{u}}{\partial l} \cdot \mathbf{t} = \frac{\partial l_R}{\partial l} \frac{\partial \mathbf{u}}{\partial l_R} \cdot \mathbf{t} = \frac{1}{\lambda_l} \frac{\mathrm{D}\lambda_l}{\mathrm{D}t} = \frac{\partial u_t}{\partial l} + \kappa_l \, u_n,$$

$$\mathbf{e}_\varphi \cdot \nabla \mathbf{u} \cdot \mathbf{e}_\varphi = \frac{1}{\sigma} \frac{\partial \mathbf{u}}{\partial \varphi} \cdot \mathbf{e}_\varphi = \frac{1}{\lambda_\varphi} \frac{\mathrm{D}\lambda_\varphi}{\mathrm{D}t} = \frac{u_\sigma}{\sigma}, \tag{D.5.40}$$

where $\mathrm{D}/\mathrm{D}t$ is the material derivative, $u_t \equiv \mathbf{u} \cdot \mathbf{t}$ is the tangential velocity, and $u_n \equiv \mathbf{u} \cdot \mathbf{n}$ is the normal velocity (e.g., [107]). The radial velocity can be expressed in terms of the tangential and normal velocities as

$$u_\sigma = u_t \frac{\partial \sigma}{\partial l} - u_n \frac{\partial x}{\partial l}, \tag{D.5.41}$$

yielding

$$\mathbf{e}_\varphi \cdot \nabla \mathbf{u} \cdot \mathbf{e}_\varphi = \frac{u_\sigma}{\sigma} = \frac{u_t}{\sigma} \frac{\partial \sigma}{\partial l} - \frac{u_n}{\sigma} \frac{\partial x}{\partial l} = \frac{u_t}{\sigma} \frac{\partial \sigma}{\partial x} + \kappa_\varphi \, u_n. \tag{D.5.42}$$

The rate of surface dilatation is then

$$\theta = \mathbf{t} \cdot \nabla \mathbf{u} \cdot \mathbf{t} + \mathbf{e}_\varphi \cdot \nabla \mathbf{u} \cdot \mathbf{e}_\varphi = \frac{1}{\sigma} \frac{\partial(\sigma u_t)}{\partial l} + 2\kappa_m \, u_n. \tag{D.5.43}$$

The first term on the right-hand side expresses dilatation due to in-plane axisymmetric motion, and the second term expresses dilatation due to normal motion.

In the linear approximation,

$$\mathbf{t} \cdot \nabla \mathbf{u} \cdot \mathbf{t} = \left(\frac{\partial u_x}{\partial x} \right)_{\sigma=a} = \epsilon \, ik U_x(a) \exp[ik(x - ct)], \qquad \mathbf{e}_\varphi \cdot \nabla \mathbf{u} \cdot \mathbf{e}_\varphi = \epsilon \frac{U_\sigma(a)}{a} \exp[ik(x - ct)],$$

$$\theta = \epsilon i \left(kU_x - i \frac{U_\sigma}{a} \right)_{\sigma=a} \exp[ik(x - ct)]. \tag{D.5.44}$$

Substituting these expressions into the surface constitutive equation, we find that

$$\gamma_l = i\,(\delta_1 A_{1,1} + \delta_2 A_{2,1} + \delta_3 B_{1,1} + \delta_4 B_{2,1}), \tag{D.5.45}$$

where

$$\delta_1 = k^2\,[(\mu_e + \mu_s)\,\mathrm{I}_0(\hat{k}) - (\mu_e - \mu_s)\,\frac{1}{\hat{k}}\,\mathrm{I}_1(\hat{k})],$$

$$\delta_2 = k\,[(\mu_e + \mu_s)[2\,\mathrm{I}_0(\hat{k}) + \hat{k}\,\mathrm{I}_1(\hat{k})] - (\mu_e - \mu_s)\,\mathrm{I}_0(\hat{k})],$$

$$\delta_3 = k^2\,[-(\mu_e + \mu_s)\,\mathrm{K}_0(\hat{k}) - (\mu_e - \mu_s)\,\frac{1}{\hat{k}}\,\mathrm{K}_1(\hat{k})], \tag{D.5.46}$$

$$\delta_4 = k\,[\,(\mu_e + \mu_s)[2\,\mathrm{K}_0(\hat{k}) - \hat{k}\,\mathrm{K}_1(\hat{k})] - (\mu_e - \mu_s)\,\mathrm{K}_0(\hat{k})\,],$$

and

$$\gamma_\varphi = i\,(\delta_1' A_{1,1} + \delta_2' A_{2,1} + \delta_3' B_{1,1} + \delta_4' B_{2,1}), \tag{D.5.47}$$

where

$$\delta_1' = k^2\,[(\mu_e - \mu_s)\,\mathrm{I}_0(\hat{k}) - (\mu_e + \mu_s)\,\frac{1}{\hat{k}}\,\mathrm{I}_1(\hat{k})],$$

$$\delta_2' = k\,[(\mu_e - \mu_s)[2\,\mathrm{I}_0(\hat{k}) + \hat{k}\,\mathrm{I}_1(\hat{k})] - (\mu_e + \mu_s)\,\mathrm{I}_0(\hat{k})],$$

$$\delta_3' = k^2\,[-(\mu_e - \mu_s)\,\mathrm{K}_0(\hat{k}) - (\mu_e + \mu_s)\,\frac{1}{\hat{k}}\,\mathrm{K}_1(\hat{k})], \tag{D.5.48}$$

$$\delta_4' = k\,[\,(\mu_e - \mu_s)[2\,\mathrm{K}_0(\hat{k}) - \hat{k}\,\mathrm{K}_1(\hat{k})] - (\mu_e + \mu_s)\,\mathrm{K}_0(\hat{k})\,].$$

Tangential interfacial balance

Linearizing the tangential component of the interfacial force balance (D.5.28), we obtain

$$\left(\sigma_{x\sigma_1} - \sigma_{x\sigma_2}\right)_{\sigma=a} = \mu_1\left(\frac{\partial u_{x_1}}{\partial\sigma} + \frac{\partial u_{\sigma_1}}{\partial x}\right)_{\sigma=a} - \mu_2\left(\frac{\partial u_{x_2}}{\partial\sigma} + \frac{\partial u_{\sigma_2}}{\partial x}\right)_{\sigma=a}$$

$$= \frac{\partial\tau_l}{\partial l} = \epsilon i k \gamma_l\,\exp[ik(x - ct)]. \tag{D.5.49}$$

Substituting the normal-mode expression for the velocity, we find that

$$\mu_1\left(\frac{dU_{x_1}}{d\sigma} + i k\,U_{\sigma_1}\right)_{\sigma=a} - \mu_2\left(\frac{dU_{x_2}}{d\sigma} + i k\,U_{\sigma_2}\right)_{\sigma=a} = i k \gamma_l. \tag{D.5.50}$$

Using the expression

$$\frac{dU_{x_j}(\sigma)}{d\sigma} = A_{1,j}\,k^2\,\mathrm{I}_1(\hat{\sigma}) + A_{2,j}\,k\,[2\,\mathrm{I}_1(\hat{\sigma}) + \hat{\sigma}\,\mathrm{I}_0(\hat{\sigma})]$$

$$+ B_{1,j}\,k^2\,\mathrm{K}_1(\hat{\sigma}) - B_{2,j}\,k\,[2\,\mathrm{K}_1(\hat{\sigma}) - \hat{\sigma}\,\mathrm{K}_0(\hat{\sigma})] \tag{D.5.51}$$

along with (D.5.15) and (D.5.45), we obtain

$$
2\mu_1\,k\left(A_{1,1}\,k\,\mathrm{I}_1(\hat{k})+A_{2,1}\left[\mathrm{I}_1(\hat{k})+\hat{k}\,\mathrm{I}_0(\hat{k})\right]+B_{1,1}\,k\,\mathrm{K}_1(\hat{k})-B_{2,1}\left[\mathrm{K}_1(\hat{k})-\hat{k}\,\mathrm{K}_0(\hat{k})\right]\right)
$$
$$
-2\mu_2\,k\left(A_{1,2}\,k\,\mathrm{I}_1(\hat{k})+A_{2,2}\left[\mathrm{I}_1(\hat{k})+\hat{k}\,\mathrm{I}_0(\hat{k})\right]+B_{1,2}\,k\,\mathrm{K}_1(\hat{k})-B_{2,2}\left[\mathrm{K}_1(\hat{k})-\hat{k}\,\mathrm{K}_0(\hat{k})\right]\right)
$$
$$
=\mathrm{i}k\,\gamma_l=-k\left(\delta_1 A_{1,1}+\delta_2 A_{2,1}+\delta_3 B_{1,1}+\delta_4 B_{2,1}\right). \tag{D.5.52}
$$

Consolidating the various terms, we obtain a linear algebraic equation,

$$
F_1 A_{1,1}+F_2 A_{2,1}+F_3 B_{1,1}+F_4 B_{2,1}+F_5 A_{1,2}+F_6 A_{2,2}+F_7 B_{1,2}+F_8 B_{2,2}=0, \tag{D.5.53}
$$

where

$$
\begin{aligned}
&F_1=2\,\mu_1 k\,\mathrm{I}_1(\hat{k})+\delta_1, && F_2=2\,\mu_1\left[\mathrm{I}_1(\hat{k})+\hat{k}\,\mathrm{I}_0(\hat{k})\right]+\delta_2,\\
&F_3=2\,\mu_1 k\,\mathrm{K}_1(\hat{k})+\delta_3, && F_4=2\,\mu_1\left[-\mathrm{K}_1(\hat{k})+\hat{k}\mathrm{K}_0(\hat{k})\right]+\delta_4,\\
&F_5=-2\,\mu_2 k\,\mathrm{I}_1(\hat{k}), && F_6=-2\,\mu_2\left[\mathrm{I}_1(\hat{k})+\hat{k}\,\mathrm{I}_0(\hat{k})\right],\\
&F_7=-2\,\mu_2 k\,\mathrm{K}_1(\hat{k}), && F_8=-2\,\mu_2\left[-\mathrm{K}_1(\hat{k})+\hat{k}\,\mathrm{K}_0(\hat{k})\right].
\end{aligned} \tag{D.5.54}
$$

Normal interfacial balance

Next, we linearize the normal component of the interfacial force balance expressed by (D.5.28), and find that

$$
\left(\sigma_{\sigma\sigma_1}-\sigma_{\sigma\sigma_2}\right)_{\sigma=a}=\left(-p_1+2\,\mu_1\frac{\partial u_{\sigma_1}}{\partial\sigma}\right)_{\sigma=a}-\left(-p_2+2\,\mu_2\frac{\partial u_{\sigma_2}}{\partial\sigma}\right)_{\sigma=a}, \tag{D.5.55}
$$

yielding

$$
\sigma_{\sigma\sigma_1}\Big|_{\sigma=a}-\sigma_{\sigma\sigma_2}\Big|_{\sigma=a}=-\frac{\gamma}{a}-\epsilon\left[-\gamma\frac{A}{a^2}\left(1-\hat{k}^2\right)+\frac{\gamma_\varphi}{a}\right]\exp[\mathrm{i}k(x-ct)]. \tag{D.5.56}
$$

Both sides are evaluated at the location of the unperturbed interface, $\sigma=a$. Substituting the normal-mode expansions, we obtain

$$
\left(-\chi_1+2\,\mu_1\frac{\partial U_{\sigma_1}}{\partial\sigma}\right)_{\sigma=a}-\left(-\chi_2+2\,\mu_2\frac{\partial U_{\sigma_2}}{\partial\sigma}\right)_{\sigma=a}=\gamma\frac{A}{a^2}\left(1-\hat{k}^2\right)-\frac{\gamma_\varphi}{a}. \tag{D.5.57}
$$

Substituting into the left-hand side of (D.5.57) the expression

$$
\begin{aligned}
\frac{\mathrm{d}U_{\sigma_j}}{\mathrm{d}\sigma}=-\mathrm{i}\,k\Big(&A_{1,j}\,k\left[\mathrm{I}_0(\hat{\sigma})-\frac{1}{\hat{\sigma}}\,\mathrm{I}_1(\hat{\sigma})\right]+A_{2,j}\left[\mathrm{I}_0(\hat{\sigma})+\hat{\sigma}\,\mathrm{I}_1(\hat{\sigma})\right]\\
&-B_{1,j}\,k\left[\mathrm{K}_0(\hat{\sigma})+\frac{1}{\hat{\sigma}}\,\mathrm{K}_1(\hat{\sigma})\right]+B_{2,j}\left[\mathrm{K}_0(\hat{\sigma})-\hat{\sigma}\,\mathrm{K}_1(\hat{\sigma})\right]\Big)
\end{aligned} \tag{D.5.58}
$$

along with the last expression in (D.5.20) for the pressure and expression (D.5.24) for the interfacial amplitude, A, we obtain

$$
\alpha_{1,1}A_{1,1}+\alpha_{2,1}A_{2,1}+\beta_{1,1}B_{1,1}+\beta_{2,1}B_{2,1}+\alpha_{1,2}A_{1,2}+\alpha_{2,2}A_{2,2}+\beta_{1,2}B_{1,2}+\beta_{2,2}B_{2,2}
$$
$$
=-\mathrm{i}\frac{\gamma_\varphi}{a}=\frac{1}{a}\left(\delta_1' A_{1,1}+\delta_2' A_{2,1}+\delta_3' B_{1,1}+\delta_4' B_{2,1}\right), \tag{D.5.59}
$$

where

$$
\alpha_{1,1} = 2\,\mu_1\,k^2\left[\,\mathrm{I}_0(\hat{k}) - \frac{1}{\hat{k}}\,\mathrm{I}_1(\hat{k})\,\right] - \frac{\gamma k^3}{s}\,\frac{1 - \hat{k}^2}{\hat{k}^2}\,\mathrm{I}_1(\hat{k}),
$$

$$
\alpha_{2,1} = 2\,\mu_1\,k\,\hat{k}\,\mathrm{I}_1(\hat{k}) - \frac{\gamma\,k^3}{s}\,\frac{1 - \hat{k}^2}{\hat{k}^2}\,a\,\mathrm{I}_0(\hat{k}),
$$

$$
\beta_{1,1} = -2\,\mu_1\,k^2\left[\,\mathrm{K}_0(\hat{k}) + \frac{1}{\hat{k}}\,\mathrm{K}_1(\hat{k})\,\right] - \frac{\gamma\,k^3}{s}\,\frac{1 - \hat{k}^2}{\hat{k}^2}\,\mathrm{K}_1(\hat{k}), \tag{D.5.60}
$$

$$
\beta_{2,1} = -2\,\mu_1\,k\,\hat{k}\,\mathrm{K}_1(\hat{k}) - \frac{\gamma k^3}{s}\,\frac{1 - \hat{k}^2}{\hat{k}^2}\,a\,\mathrm{K}_0(\hat{k}),
$$

and

$$
\alpha_{1,2} = -2\,\mu_2\,k^2\left[\,\mathrm{I}_0(\hat{k}) - \frac{1}{\hat{k}}\,\mathrm{I}_1(\hat{k})\,\right], \qquad \alpha_{2,2} = -2\,\mu_2\,k\,\hat{k}\,\mathrm{I}_1(\hat{k}),
$$

$$
\beta_{1,2} = 2\,\mu_2\,k^2\left[\,\mathrm{K}_0(\hat{k}) + \frac{1}{\hat{k}}\,\mathrm{K}_1(\hat{k})\,\right], \qquad \beta_{2,2} = 2\,\mu_2\,k\,\hat{k}\,K_1(\hat{k}). \tag{D.5.61}
$$

Consolidating the various terms, we obtain a linear algebraic equation,

$$
G_1 A_{1,1} + G_2 A_{2,1} + G_3 B_{1,1} + G_4 B_{2,1} + G_5 A_{1,2} + G_6 A_{2,2} + G_7 B_{1,2} + G_8 B_{2,2} = 0, \tag{D.5.62}
$$

where

$$
G_1 = \alpha_{1,1} - \frac{\delta_1'}{a}, \quad G_2 = \alpha_{2,1} - \frac{\delta_2'}{a}, \quad G_3 = \beta_{1,1} - \frac{\delta_3'}{a}, \quad G_4 = \beta_{2,1} - \frac{\delta_4}{a},
$$

$$
G_5 = \alpha_{1,2}, \qquad G_6 = \alpha_{2,2}, \qquad G_7 = \beta_{1,2}, \qquad G_8 = \beta_{2,2}. \tag{D.5.63}
$$

Formulation of an eigenvalue problem

To formulate the algebraic eigenvalue problem, we collect equations (D.5.21), (D.5.22), (D.5.26), (D.5.27), (D.5.53), and (D.5.62) into a liner homogeneous system $\mathbf{M} \cdot \mathbf{q} = \mathbf{0}$, where

$$
\mathbf{q} = \left(\,A_{1,1}, A_{2,1}, B_{1,1}, B_{2,1}, A_{1,2}, A_{2,2}, B_{1,2}, B_{2,2}\,\right), \tag{D.5.64}
$$

and the coefficient matrix $\mathbf{M}$ is given by in Table D.5.1. For a nontrivial solution to exist the determinant of $\mathbf{M}$ must be zero, $\det(\mathbf{M}) = 0$. Expanding the determinant, we obtain a linear algebraic equation for the growth rate.

Thread surrounded by an annular layer

When the inner cylinder is absent, we obtain an annular layer surrounding a liquid thread. For small values of their arguments, the modified Bessel functions become singular and the general expression for the stream function (D.5.11) exhibits divergent terms. For the velocity to be regular at the x axis, the constants $B_{1,1}$ and $B_{2,1}$ must be set to zero, yielding a simplified expression for the inner fluid stream function,

$$
\phi_j(\sigma) = \sigma\left[\,A_{1,j}\,\mathrm{I}_1(\hat{\sigma}) + A_{2,j}\,\mathrm{I}_1(\hat{\sigma}_j)\,\right], \tag{D.5.65}
$$

where $j = 1$. The linear system $\mathbf{M} \cdot \mathbf{q} = \mathbf{0}$ undergoes analogous simplifications.

$$
\mathbf{M} =
\begin{bmatrix}
\mathrm{I}_1(\hat{k}) & a\,\mathrm{I}_0(\hat{k}) & \mathrm{K}_1(\hat{k}) & a\,\mathrm{K}_1(\hat{k}) \\
k\,\mathrm{I}_0(\hat{k}) & 2\,\mathrm{I}_0(\hat{k}) + \hat{k}\,\mathrm{I}_1(\hat{k}) & -k\,\mathrm{K}_0(\hat{k}) & 2\,\mathrm{K}_0(\hat{k}) - \hat{k}\,\mathrm{K}_1(\hat{k}) \\
\mathrm{I}_1(\hat{k}_i) & a_i\,\mathrm{I}_0(\hat{k}_i) & \mathrm{K}_1(\hat{k}_i) & a\,\mathrm{K}_1(\hat{k}) \\
k\,\mathrm{I}_0(\hat{k}) & 2\,\mathrm{I}_0(\hat{k}) + \hat{k}\,\mathrm{I}_1(\hat{k}) & -k\,\mathrm{K}_0(\hat{k}) & 2\,\mathrm{K}_0(\hat{k}) - \hat{k}\,\mathrm{K}_1(\hat{k}) \\
\mathrm{I}_1(\hat{k}_i) & a_i\,\mathrm{I}_0(\hat{k}_i) & \mathrm{K}_1(\hat{k}_i) & a_i\,\mathrm{K}_0(\hat{k}_i) \\
k\,\mathrm{I}_0(\hat{k}_i) & 2\,\mathrm{I}_0(\hat{k}_i) + \hat{k}_i\,\mathrm{I}_1(\hat{k}_i) & -k\,\mathrm{K}_0(\hat{k}_i) & 2\,\mathrm{K}_0(\hat{k}_i) - \hat{k}_i\,\mathrm{K}_1(\hat{k}_i) \\
0 & 0 & 0 & 0 \\
0 & 0 & 0 & 0 \\
F_1 & F_2 & F_3 & F_4 \\
G_1 & G_2 & G_3 & G_4 \\
\end{bmatrix}
$$

$$
\begin{bmatrix}
-\mathrm{I}_1(\hat{k}) & -a\,\mathrm{I}_0(\hat{k}) & -\mathrm{K}_1(\hat{k}) & -a\,\mathrm{K}_0(\hat{k}) \\
-k\,\mathrm{I}_0(\hat{k}) & -2\,\mathrm{I}_0(\hat{k}) - \hat{k}\,\mathrm{I}_1(\hat{k}) & k\,\mathrm{K}_0(\hat{k}) & -2\,\mathrm{K}_0(\hat{k}) + \hat{k}\,\mathrm{K}_1(\hat{k}) \\
0 & 0 & 0 & 0 \\
0 & 0 & 0 & 0 \\
\mathrm{I}_1(\hat{k}_e) & a_e\,\mathrm{I}_0(\hat{k}_e) & \mathrm{K}_1(\hat{k}_e) & a_e\,\mathrm{K}_0(\hat{k}_e) \\
k\,\mathrm{I}_0(\hat{k}_e) & 2\,\mathrm{I}_0(\hat{k}_e) + \hat{k}_e\,\mathrm{I}_1(\hat{k}_e) & -k\,\mathrm{K}_0(\hat{k}_e) & 2\,\mathrm{K}_0(\hat{k}_e) - \hat{k}_e\,\mathrm{K}_1(\hat{k}_e) \\
F_5 & F_6 & F_7 & F_8 \\
G_5 & G_6 & G_7 & G_8 \\
\end{bmatrix}
$$

TABLE D.5.1   Setting the determinant of the matrix displayed to zero provides us with an algebraic equation for the complex growth rate.

Annular layer coated on a tube

When the outer cylinder is absent, we obtain an annular layer coated on the exterior surface of a cylindrical tube and surrounded by an infinite outer fluid. The modified Bessel functions become singular as their argument tends to infinity. To ensure a regular behavior, we set the coefficient $A_{1,2}$ and $A_{2,2}$ to zero and obtain a simplified expression for the exterior stream function,

$$
\phi_j(\sigma) = \sigma \left[B_{1,j}\,\mathrm{K}_1(\hat{\sigma}) + B_{2,j}\,\mathrm{K}_1(\hat{\sigma}_j) \right], \tag{D.5.66}
$$

where $j = 2$. The linear system $\mathbf{M} \cdot \mathbf{q} = \mathbf{0}$ undergoes analogous simplifications.

Suspended thread

In the simplest configuration, both the internal and external cylinder are absent, yielding an infinite thread suspended in an infinite ambient fluid. The stream functions for the internal and external flow are described, respectively, by equations (D.5.65) and (D.5.66). The linear system $\mathbf{M} \cdot \mathbf{q} = \mathbf{0}$ undergoes analogous simplifications. In the absence of interfacial viscosity, the coefficient matrix reduces to that derived by Tomotika [410].

Program files:

1. `ann2lvs0`
   Computes the roots of the secular equation. The determinant of the matrix is calculated by LU decomposition using Gauss elimination.

2. `matrix`
   Formulates the matrix **M**.

3. `bess_I01K0`
   Computes the Bessel functions.

4. `gel`
   Computes the determinant of a matrix by LU decomposition using Gauss elimination.

Input files:

1. `ann2lvs.dat`
   Problem selection and specification of input parameters.

Output files:

1. `ann2lvs.out`
   Growth rates.

D.6   Directory chan2l0

This directory contains a code that performs the linear stability analysis of two-layer flow in a horizontal or inclined channel, as illustrated in Figure D.6.1. The instability occurs under conditions of Stokes flow.

Base state

In the unperturbed configuration, the interface is flat and the fluids execute unidirectional motion parallel to the channel walls. In the inclined system of coordinates depicted in Figure D.6.1, the unperturbed interface is located at $y = 0$, the lower wall is located at $y = -h_1$, and the upper wall is located at $y = h_2 = 2h - h_1$, where $2h$ is the channel width. For convenience, the x and y velocity components are denoted by $u = u_x$ and $v = u_y$.

The velocity profile of the base flow is

$$u_j^{(0)} = -\frac{\chi + \rho_j g_x}{2\mu_j}\, y^2 + \xi_j y + u_I, \qquad v_j^{(0)} = 0 \tag{D.6.1}$$

for $j = 1, 2$, corresponding to the lower or upper fluid, where the superscript (0) denotes the unperturbed base state, subject to the following definitions: χ is the negative of the axial pressure gradient; $g_x = g \sin \beta$ is the x component of the acceleration of gravity; g is the magnitude of the

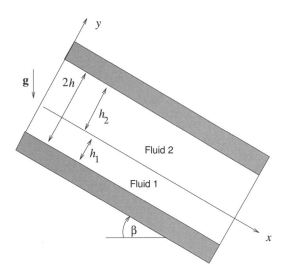

FIGURE D.6.1   Two-layer flow in an inclined channel confined between two parallel plane walls inclined
at angle β.

acceleration of gravity; ρ_j are the fluid densities; μ_j are the fluid viscosities; $u_I \equiv u_j^{(0)}(y = 0)$ is the
common interfacial velocity. The coefficients

$$\xi_1 \equiv \left(\frac{\partial u_{x_1}^{(0)}}{\partial y}\right)_{y=0}, \qquad \xi_2 \equiv \left(\frac{\partial u_{x_2}^{(0)}}{\partial y}\right)_{y=0}, \tag{D.6.2}$$

are the interfacial shear rates of the unperturbed flow. Continuity of shear stress across the interface
requires that $\mu_1 \xi_1 = \mu_2 \xi_2$ or

$$\xi_1 = \lambda \xi_2, \tag{D.6.3}$$

where $\lambda = \mu_2/\mu_1$ is the viscosity ratio.

The corresponding pressure distributions are given by

$$p_j^{(0)}(y) = -\chi\, x + \rho_j\, g_y\, y + p_0 \tag{D.6.4}$$

for $j = 1, 2$, where $g_y = -g \cos \beta$ is the y component of the acceleration of gravity and p_0 is an
unspecified reference pressure.

Enforcing the no-slip boundary condition at the walls, $u_1 = U_1$ at $y = -h_1$ and $u_2 = U_2$ at
$y = h_2$, we find that

$$\xi_1 = \frac{\Delta U}{h_1}\frac{\lambda}{\lambda + r} - \frac{h}{\mu_1}\left(\chi + \rho_1 g_x \frac{\lambda - \delta\, r^2}{\lambda - r^2}\right)\frac{\lambda - r^2}{(1 + r)(\lambda + r)}, \qquad \xi_2 = \frac{\xi_1}{\lambda} \tag{D.6.5}$$

and

$$u_I = \frac{r\,U_1 + \lambda\,U_2}{\lambda + r} + \frac{h^2}{\mu_1}\left(\chi + \rho_1 g_x \frac{1 + \delta\,r}{1 + r}\right)\frac{2r}{(1 + r)(\lambda + r)},\qquad\text{(D.6.6)}$$

where $\Delta U = U_2 - U_1$ is the difference in the wall velocities, $r = h_2/h_1$ is the layer thickness ratio, and $\delta = \rho_2/\rho_1$ is the density ratio. When the densities of the two liquids are matched, $\delta = 1$ and $\rho_1 = \rho_2 = \rho$, the two terms enclosed by the large parentheses on the right-hand side of (D.6.5) and (D.6.6) combine into an effective negative pressure gradient, $\chi + \rho\,g_x$.

Normal-mode analysis

A normal-mode perturbation displaces the interface to a position given by the real or imaginary part of the function

$$y = f(x,t) = \epsilon\eta(x,t),\qquad\text{(D.6.7)}$$

where ϵ is a dimensionless coefficient whose magnitude is much less than unity,

$$\eta(x,t) = A\exp[ik(x - ct)]\qquad\text{(D.6.8)}$$

is the normal-mode wave form of the perturbation, A is the complex amplitude of the interfacial wave, $k = 2\pi/L$ is the wave number, L is the wavelength, and c is the complex phase velocity.

Stream function

The stream function, ψ, is defined by the equations $u = \partial\psi/\partial y$ and $v = -\partial\psi/\partial x$. In the linear analysis, we set $\psi_j = \psi_j^{(0)} + \epsilon\psi_j^{(1)}$ and introduce the normal-mode form

$$\psi_j^{(1)}(x,y,t) = \phi_j(\hat{y})\,\exp[ik(x - ct)]\qquad\text{(D.6.9)}$$

for $j = 1, 2$, where $\hat{y} = ky$ and the superscript (1) denotes the disturbance. The perturbation velocity components are $u^{(1)} = \partial\psi^{(1)}/\partial y$ and $v^{(1)} = -\partial\psi^{(1)}/\partial x$.

The Reynolds number of the flow is so small that the motion of the fluid is governed by the equations of Stokes flow. Requiring that the stream function satisfies the biharmonic equation, $\nabla^4\psi_j^{(1)} = 0$, we find that

$$\phi_j(\hat{y}) = a_{1j}\cosh\hat{y} + a_{2j}\,\hat{y}\cosh\hat{y} + a_{3j}\sinh\hat{y} + a_{4j}\,\hat{y}\sinh\hat{y},\qquad\text{(D.6.10)}$$

where a_{ij} are eight complex coefficients for $i = 1\text{--}4$ and $j = 1, 2$.

Pressure field

The linearized pressure field in the jth fluid is $p_j = p_j^{(0)} + \epsilon p_j^{(1)}$. The disturbance pressure field is expressed by the normal-mode form

$$p_j^{(1)}(x,y,t) = \mu_j\,q_j(\hat{y})\,\exp[ik(x - ct)]\qquad\text{(D.6.11)}$$

for $j = 1, 2$, where $q_j(\hat{y})$ are pressure eigenfunctions. Substituting this expression into the x component of the Stokes equation, we find that

$$p_j^{(1)}(y) = -\frac{\mathrm{i}}{k}\left(\frac{\partial^3\psi_j}{\partial x^2\partial y} + \frac{\partial^3\psi_j}{\partial y^3}\right), \tag{D.6.12}$$

yielding

$$q_j(\hat{y}) = \mathrm{i}\,k^2\left(\frac{\mathrm{d}\phi_j}{\mathrm{d}\hat{y}} - \frac{\mathrm{d}^3\phi_j}{\mathrm{d}\hat{y}^3}\right). \tag{D.6.13}$$

Continuity of y velocity at the interface

In the linear approximation, continuity of y velocity at the interface requires that

$$\psi_1^{(1)}(x, y = 0, t) = \psi_2^{(1)}(x, y = 0, t), \tag{D.6.14}$$

yielding $a_{11} = a_{12}$.

Kinematic compatibility

Kinematic compatibility requires that $\mathrm{D}[f(x,t) - y]/\mathrm{D}t = 0$, where $\mathrm{D}/\mathrm{D}t$ is the material derivative and the function $f(x,t)$ describe the shape of the interface according to (D.6.7). In the linear approximation,

$$\frac{\partial\eta}{\partial t} + u_I\frac{\partial\eta}{\partial x} - v^{(1)}(y = 0) = 0, \tag{D.6.15}$$

where $v^{(1)} = -\partial\psi^{(1)}/\partial x$. Substituting the preceding expressions, we find that $-\mathrm{i}kcA + u_I\mathrm{i}kA + \mathrm{i}ka_{11} = 0$, yielding $a_{11} = (c - u_I)\,A$ or

$$A = \zeta a_{11}, \tag{D.6.16}$$

where $\zeta \equiv 1/(c - u_I)$.

Continuity of x velocity at the interface

In the linear approximation, continuity of x velocity at the interface requires that

$$\xi_1\eta(x,y) + \left(\frac{\partial\psi_1^{(1)}}{\partial y}\right)_{y=0} = \xi_2\eta(x,y) + \left(\frac{\partial\psi_2^{(1)}}{\partial y}\right)_{y=0}, \tag{D.6.17}$$

yielding

$$\xi_1 A + k\left(\frac{\mathrm{d}\phi_1}{\mathrm{d}\hat{y}}\right)_{\hat{y}=0} = \xi_2 A + k\left(\frac{\mathrm{d}\phi_2}{\mathrm{d}\hat{y}}\right)_{\hat{y}=0}. \tag{D.6.18}$$

Rearranging and using (D.6.16), we find that

$$\zeta\,(\xi_1 - \xi_2)\,a_{11} + k\left(\frac{\mathrm{d}\phi_1}{\mathrm{d}\hat{y}}\right)_{\hat{y}=0} - k\left(\frac{\mathrm{d}\phi_2}{\mathrm{d}\hat{y}}\right)_{\hat{y}=0} = 0, \tag{D.6.19}$$

yielding

$$\zeta \left(\Xi_1 - \Xi_2 \right) a_{11} + a_{21} + a_{31} - a_{22} - a_{32} = 0, \qquad (D.6.20)$$

where $\Xi_i \equiv \xi_i/k$, for $i = 1, 2$, are coefficients with dimensions of velocity.

Wall boundary conditions

The no-slip and no-penetration conditions at the lower and upper wall require that

$$\phi_1(\hat{y} = -\hat{k}_1) = 0, \qquad \phi_1'(\hat{y} = -\hat{k}_1) = 0, \qquad \phi_2(\hat{y} = \hat{k}_2) = 0, \qquad \phi_2'(\hat{y} = \hat{k}_2) = 0, \qquad (D.6.21)$$

where $\hat{k}_1 \equiv k\hat{h}_1$, $\hat{k}_2 \equiv k\hat{h}_2$, and a prime denotes a derivative with respect to $\hat{y}$. Making substitutions, we obtain

$$\begin{bmatrix} \cosh \hat{k}_1 & -\hat{k}_1 \cosh \hat{k}_1 & -\sinh \hat{k}_1 & \hat{k}_1 \sinh \hat{k}_1 \\ -\sinh \hat{k}_1 & \cosh \hat{k}_1 + \hat{k}_1 \sinh \hat{k}_1 & \cosh \hat{k}_1 & -\sinh \hat{k}_1 - \hat{k}_1 \cosh \hat{k}_1 \end{bmatrix} \cdot \mathbf{w}_1 = \mathbf{0}, \qquad (D.6.22)$$

and

$$\begin{bmatrix} \cosh \hat{k}_2 & \hat{k}_2 \cosh \hat{k}_2 & \sinh \hat{k}_2 & \hat{k}_2 \sinh \hat{k}_2 \\ \sinh \hat{k}_2 & \cosh \hat{k}_2 + \hat{k}_2 \sinh \hat{k}_2 & \cosh \hat{k}_2 & \sinh \hat{k}_2 + \hat{k}_2 \cosh \hat{k}_2 \end{bmatrix} \cdot \mathbf{w}_2 = \mathbf{0}, \qquad (D.6.23)$$

where $\mathbf{w}_1 = [a_{11}, a_{21}, a_{31}, a_{41}]$ and $\mathbf{w}_2 = [a_{12}, a_{22}, a_{32}, a_{42}]$.

Tangential component of the interfacial force balance

The linearized tangential component of the interfacial force balance requires that

$$\mu_1 \left(\frac{\partial u_1^{(1)}}{\partial y} + \frac{\partial v_1^{(1)}}{\partial x} \right)_{y=0} = \mu_2 \left(\frac{\partial u_2^{(1)}}{\partial y} + \frac{\partial v_2^{(1)}}{\partial x} \right)_{y=0} + \Delta\rho\, g_x A, \qquad (D.6.24)$$

where $\Delta\rho = \rho_1 - \rho_2$. Expressing the velocity in terms of the stream function and using (D.6.16) to eliminate the interfacial amplitude, A, we find that

$$\left(\frac{\partial^2 \psi_1^{(1)}}{\partial y^2} - \frac{\partial^2 \psi_1^{(1)}}{\partial x^2} \right)_{y=0} = \lambda \left(\frac{\partial^2 \psi_2^{(1)}}{\partial y^2} - \frac{\partial^2 \psi_2^{(1)}}{\partial x^2} \right)_{y=0} + \frac{\Delta\rho\, g_x}{\mu_1} \zeta\, a_{11}, \qquad (D.6.25)$$

yielding

$$\left(\frac{\mathrm{d}^2 \phi_1}{\mathrm{d}\hat{y}^2} + \phi_1 \right)_{\hat{y}=0} = \lambda \left(\frac{\mathrm{d}^2 \phi_2}{\mathrm{d}\hat{y}^2} + \phi_2 \right)_{\hat{y}=0} + \frac{\Delta\rho\, g_x}{\mu_1 k^2} \zeta\, a_{11}. \qquad (D.6.26)$$

Substituting the preceding expressions for ϕ_j and simplifying, we find that

$$\left(1 - \zeta\, \mathcal{B}_x \right) a_{11} + a_{41} - \lambda \left(a_{12} + a_{42} \right) = 0, \qquad (D.6.27)$$

where

$$\mathcal{B}_x \equiv \frac{\Delta\rho\, g_x}{2\mu_1 k^2}. \qquad (D.6.28)$$

We recall that $a_{11} = a_{12}$, and obtain

$$a_{41} = -a_{12}\left(1 - \lambda - \zeta\,\mathcal{B}_x\right) + \lambda\,a_{42}. \tag{D.6.29}$$

Normal component of the interface force balance

In the linear approximation, the normal component of the interface force balance reads

$$\left(-p_1^{(1)} + 2\mu_1\frac{\partial v_1^{(1)}}{\partial y}\right)_{y=0} - \left(-p_2^{(1)} + 2\mu_2\frac{\partial v_2^{(1)}}{\partial y}\right)_{y=0} - \Delta\rho\,g_y\eta(x) = \gamma\frac{\partial^2\eta}{\partial x^2}, \tag{D.6.30}$$

where γ is the surface tension. Substituting the preceding expressions, we obtain

$$\left(-\mu_1 q_1 + \mu_2 q_2 - 2\mathrm{i}k^2\mu_1\frac{\mathrm{d}\phi_1}{\mathrm{d}\hat{y}} + 2\mathrm{i}k^2\mu_2\frac{\mathrm{d}\phi_2}{\mathrm{d}\hat{y}}\right)_{\hat{y}=0} - \Delta\rho\,g_y\,A = -\gamma\,A\,k^2. \tag{D.6.31}$$

Substituting $q_j(\hat{y}=0) = -2\mathrm{i}\,k^2 a_{2j}$, we obtain

$$2\,\mathrm{i}\,k^2(\mu_1 a_{21} - \mu_2 a_{22}) - 2\,\mathrm{i}\,k^2\mu_1\,(a_{21} + a_{31}) + 2\,\mathrm{i}\,k^2\mu_2\,(a_{22} + a_{32}) - \Delta\rho\,g_y\,A = -\gamma\,A\,k^2, \tag{D.6.32}$$

which can be simplified into

$$-2\mathrm{i}k^2\mu_1\,a_{31} + 2\mathrm{i}k^2\mu_2\,a_{32} - \Delta\rho\,g_y A = -\gamma\,Ak^2. \tag{D.6.33}$$

Rearranging and using (D.6.16), we find that

$$a_{31} = \lambda\,a_{32} - \mathrm{i}\,\Pi\,\zeta\,a_{12}, \tag{D.6.34}$$

where

$$\Pi \equiv \frac{1}{2\mu_1}\left(-\frac{\Delta\rho\,g_y}{k^2} + \gamma\right) = -\mathcal{B}_y + \frac{\gamma}{2\mu_1} \tag{D.6.35}$$

is a property group with dimensions of velocity, and

$$\mathcal{B}_y \equiv \frac{\Delta\rho\,g_y}{2\mu_1 k^2}. \tag{D.6.36}$$

Formulation of an eigenvalue problem

Substituting (D.6.34) into (D.6.20) and rearranging, we find that

$$a_{21} = \zeta\left(\mathrm{i}\,\Pi + \Xi_2 - \Xi_1\right)a_{12} + a_{22} + (1 - \lambda)\,a_{32}. \tag{D.6.37}$$

Finally, we substitute (D.6.29), (D.6.34), and (D.6.37) into the linear system (D.6.22) and obtain the equivalent system

$$\left[\begin{array}{cc|c} C_{11} - \zeta\,C_{12} & -\hat{k}_1\cosh\hat{k}_1 & -\lambda\sinh\hat{k}_1 - (1-\lambda)\,\hat{k}_1\cosh\hat{k}_1 \\ C_{21} + \zeta\,C_{22} & \cosh\hat{k}_1 + \hat{k}_1\sinh\hat{k}_1 & \cosh\hat{k}_1 + (1-\lambda)\,\hat{k}_1\sinh\hat{k}_1 \end{array}\right.$$

$$\left.\begin{array}{c} \lambda\,\hat{k}_1\sinh\hat{k}_1 \\ -\lambda\,(\sinh\hat{k}_1 + \hat{k}_1\cosh\hat{k}_1) \end{array}\right] \cdot \mathbf{w}_2 = \mathbf{0}, \tag{D.6.38}$$

where

$$C_{11} = \cosh \hat{k}_1 - (1 - \lambda)\,\hat{k}_1\,\sinh \hat{k}_1,$$
$$C_{12} = \hat{k}_1 \left[(\Xi_2 - \Xi_1)\cosh \hat{k}_1 - \mathcal{B}_x \sinh \hat{k}_1 \right] + i\,\Pi\,(\hat{k}_1 \cosh \hat{k}_1 - \sinh \hat{k}_1),$$
$$C_{21} = -\sinh \hat{k}_1 + (1 - \lambda)(\sinh \hat{k}_1 + \hat{k}_1 \cosh \hat{k}_1), \tag{D.6.39}$$
$$C_{22} = \left[(\Xi_2 - \Xi_1)(\cosh \hat{k}_1 + \hat{k}_1 \sinh \hat{k}_1) - \mathcal{B}_x (\sinh \hat{k}_1 + \hat{k}_1 \cosh \hat{k}_1) \right] + i\,\Pi\,\hat{k}_1 \sinh \hat{k}_1.$$

Appending equations (D.6.38) to equations (D.6.23), we obtain a linear homogeneous system, $\mathbf{Q} \cdot \mathbf{w}_2 = \mathbf{0}$, where

$$\mathbf{Q} = \begin{bmatrix} \cosh \hat{k}_2 & \hat{k}_2 \cosh \hat{k}_2 \\ \sinh \hat{k}_2 & \cosh \hat{k}_2 + \hat{k}_2 \sinh \hat{k}_2 \\ C_{11} - \zeta\,C_{12} & -\hat{k}_1 \cosh \hat{k}_1 \\ C_{21} + \zeta\,C_{22} & \cosh \hat{k}_1 + \hat{k}_1 \sinh \hat{k}_1 \end{bmatrix}$$

$$\begin{matrix} \sinh \hat{k}_2 & \hat{k}_2 \sinh \hat{k}_2 \\ \cosh \hat{k}_2 & \sinh \hat{k}_2 + \hat{k}_2 \cosh \hat{k}_2 \\ -\lambda \sinh \hat{k}_1 - (1-\lambda)\hat{k}_1 \cosh \hat{k}_1 & \lambda\,\hat{k}_1 \sinh \hat{k}_1 \\ \cosh \hat{k}_1 + (1-\lambda)\hat{k}_1 \sinh \hat{k}_1 & -\lambda(\sinh \hat{k}_1 + \hat{k}_1 \cosh \hat{k}_1) \end{matrix} \Bigg]. \tag{D.6.40}$$

Setting the determinant of $\mathbf{Q}$ to zero provides us with a secular equation for the computation of the complex phase velocity, c.

Performing the Laplace expansion with respect to the first column and rearranging, we obtain

$$c = u_I + \frac{C_{12}\,M_{31} + C_{22}\,M_{41}}{\cosh \hat{k}_2\,M_{11} - \sinh \hat{k}_2\,M_{21} + C_{11}\,M_{31} - C_{21}\,M_{41}}, \tag{D.6.41}$$

where M_{11}, M_{21}, M_{31}, and M_{41} are the determinants of 3×3 minor matrices defined in Table D.6.1.

Separating the complex phase velocity into its real and imaginary parts, we obtain the real phase velocity

$$c_R = u_I + \frac{N}{D} \tag{D.6.42}$$

where

$$N = \hat{k}_1 \left[(\Xi_2 - \Xi_1)\cosh \hat{k}_1 - \mathcal{B}_x \sinh \hat{k}_1 \right] M_{31} \tag{D.6.43}$$
$$+ \left[(\Xi_2 - \Xi_1)(\cosh \hat{k}_1 + \hat{k}_1 \sinh \hat{k}_1) - \mathcal{B}_x (\sinh \hat{k}_1 + \hat{k}_1 \cosh \hat{k}_1) \right] M_{41},$$

and

$$D = \cosh \hat{k}_2\,M_{11} - \sinh \hat{k}_2\,M_{21} + C_{11}\,M_{31} - C_{21}\,M_{41}. \tag{D.6.44}$$

$$M_{11} = \det(\begin{bmatrix} \cosh \hat{k}_2 + \hat{k}_2 \sinh \hat{k}_2 & \cosh \hat{k}_2 \\ -\hat{k}_1 \cosh \hat{k}_1 & -\lambda \sinh \hat{k}_1 + (\lambda - 1) \hat{k}_1 \cosh \hat{k}_1 \\ \cosh \hat{k}_1 + \hat{k}_1 \sinh \hat{k}_1 & \cosh \hat{k}_1 - (\lambda - 1) \hat{k}_1 \sinh \hat{k}_1 \end{bmatrix}$$

$$\begin{bmatrix} \sinh \hat{k}_2 + \hat{k}_2 \cosh \hat{k}_2 \\ \lambda \hat{k}_1 \sinh \hat{k}_1 \\ -\lambda (\sinh \hat{k}_1 + \hat{k}_1 \cosh \hat{k}_1) \end{bmatrix})$$

$$M_{21} = \det(\begin{bmatrix} \hat{k}_2 \cosh \hat{k}_2 & \sinh \hat{k}_2 \\ -\hat{k}_1 \cosh \hat{k}_1 & -\lambda \sinh \hat{k}_1 + (\lambda - 1) \hat{k}_1 \cosh \hat{k}_1 \\ \cosh \hat{k}_1 + \hat{k}_1 \sinh \hat{k}_1 & \cosh \hat{k}_1 - (\lambda - 1) \hat{k}_1 \sinh \hat{k}_1 \end{bmatrix}$$

$$\begin{bmatrix} \hat{k}_2 \sinh \hat{k}_2 \\ \lambda \hat{k}_1 \sinh \hat{k}_1 \\ -\lambda (\sinh \hat{k}_1 + \hat{k}_1 \cosh \hat{k}_1) \end{bmatrix})$$

$$M_{31} = \det(\begin{bmatrix} \hat{k}_2 \cosh \hat{k}_2 & \sinh \hat{k}_2 \\ \cosh \hat{k}_2 + \hat{k}_2 \sinh \hat{k}_2 & \cosh \hat{k}_2 \\ \cosh \hat{k}_1 + \hat{k}_1 \sinh \hat{k}_1 & \cosh \hat{k}_1 - (\lambda - 1) \hat{k}_1 \sinh \hat{k}_1 \end{bmatrix}$$

$$\begin{bmatrix} \hat{k}_2 \sinh \hat{k}_2 \\ \sinh \hat{k}_2 + \hat{k}_2 \cosh \hat{k}_2 \\ -\lambda (\sinh \hat{k}_1 + \hat{k}_1 \cosh \hat{k}_1) \end{bmatrix})$$

$$M_{41} = \det(\begin{bmatrix} \hat{k}_2 \cosh \hat{k}_2 & \sinh \hat{k}_2 \\ \cosh \hat{k}_2 + \hat{k}_2 \sinh \hat{k}_2 & \cosh \hat{k}_2 \\ -\hat{k}_1 \cosh \hat{k}_1 & -\lambda \sinh \hat{k}_1 + (\lambda - 1) \hat{k}_1 \cosh \hat{k}_1 \end{bmatrix}$$

$$\begin{bmatrix} \hat{k}_2 \sinh \hat{k}_2 \\ \sinh \hat{k}_2 + \hat{k}_2 \cosh \hat{k}_2 \\ \lambda \hat{k}_1 \sinh \hat{k}_1 \end{bmatrix})$$

TABLE D.6.1 Matrices involved in the secular equation determining hydrodynamic stability. The vertical lines separate the three matrix columns.

The growth rate is

$$s_I = kc_I = \frac{\Pi}{D} \left[(\hat{k}_1 \cosh \hat{k}_1 - \sinh \hat{k}_1) M_{31} + \hat{k}_1 \sinh \hat{k}_1 M_{41} \right], \tag{D.6.45}$$

where Π is defined in (D.6.35). Note that the growth rate is independent of the shear rate at the interface, that is, it is independent of the overall structure of the velocity profile of the base flow and is the same as that prevailing in quiescent fluids.

Program files:

1. chan210
   Evaluates the complex phase velocity.

2. chan210_dr
   Driver for the subroutine chan210

3. det_33c
   Determinant of a 3×3 complex matrix.

4. det_44c
   Determinant of a 4×4 complex matrix.

Input file:

1. chan210.dat
   Specification of input parameters.

Output file:

1. chan210.out
   Recording of the computed phase velocity and growth rate.

D.7   Directory chan2l0_s

This directory contains a code that performs the linear stability analysis of two-layer flow in a horizontal or inclined channel, as illustrated in Figure D.7.1. The interface is occupied by an insoluble surfactant and the instability occurs under conditions of Stokes flow.

Base state

In the inclined system of coordinates depicted in Figure D.7.1, the unperturbed interface is located at $y = 0$, the lower wall is located at $y = -h_1$ and the upper wall is located at $y = h_2 = 2h - h_1$, where $2h$ is the channel width. The surfactant is distributed uniformly over the interface. For convenience, the x and y velocity components are denoted by $u = u_x$ and $v = u_y$.

The velocity profile of the base flow is

$$u_j^{(0)} = -\frac{\chi + \rho_j g_x}{2\mu_j} y^2 + \xi_j y + u_I, \qquad v_j^{(0)} = 0 \tag{D.7.1}$$

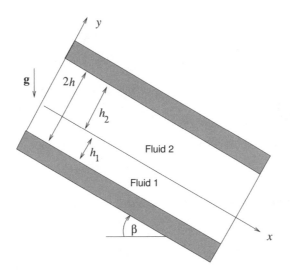

FIGURE D.7.1   Two-layer flow in an inclined channel confined between two parallel walls. The interface is occupied by an insoluble surfactant.

for $j = 1, 2$, corresponding to the lower and upper fluid, where the superscript (0) denotes the unperturbed base state, subject to the following definitions: χ is the negative of the axial pressure gradient; $g_x = g \sin \beta$ is the x component of the acceleration of gravity; g is the magnitude of the acceleration of gravity; ρ_j are the fluid densities; μ_j are the fluid viscosities; $u_I \equiv u_j^{(0)}(y = 0)$ is the common interfacial velocity. The coefficients

$$\xi_1 \equiv \left(\frac{\partial u_{x_1}^{(0)}}{\partial y} \right)_{y=0}, \qquad \xi_2 \equiv \left(\frac{\partial u_{x_2}^{(0)}}{\partial y} \right)_{y=0}, \tag{D.7.2}$$

are the interfacial shear rates of the unperturbed flow. Continuity of shear stress across the interface requires that $\mu_1 \xi_1 = \mu_2 \xi_2$ or

$$\xi_1 = \lambda \xi_2, \tag{D.7.3}$$

where $\lambda = \mu_2 / \mu_1$ is the viscosity ratio. The corresponding pressure distributions are

$$p_j^{(0)}(y) = -\chi \, x + \rho_j g_y \, y + p_0 \tag{D.7.4}$$

for $j = 1, 2$, where $g_y = -g \cos \beta$ is the y component of the acceleration of gravity and p_0 is a reference pressure.

Enforcing the no-slip wall boundary conditions, $u_1 = U_1$ at $y = -h_1$ and $u_2 = U_2$ at $y = h_2$, we find that

$$\xi_1 = \frac{\Delta U}{h_1} \frac{\lambda}{\lambda + r} - \frac{h}{\mu_1} \left(\chi + \rho_1 g_x \frac{\lambda - \delta \, r^2}{\lambda - r^2} \right) \frac{\lambda - r^2}{(1 + r)(\lambda + r)}, \qquad \xi_2 = \frac{\xi_1}{\lambda}, \tag{D.7.5}$$

and

$$u_I = \frac{r\,U_1 + \lambda\,U_2}{\lambda + r} + \frac{h^2}{\mu_1}\left(\chi + \rho_1 g_x \frac{1 + \delta\,r}{1 + r}\right)\frac{2r}{(1+r)(\lambda+r)},\tag{D.7.6}$$

where $\Delta U = U_2 - U_1$ is the difference in the wall velocities, $r = h_2/h_1$ is the layer thickness ratio, and $\delta = \rho_2/\rho_1$ is the density ratio. When the densities of the two liquids are matched, $\delta = 1$ and $\rho_1 = \rho_2 = \rho$, the two terms enclosed by the large parentheses on the right-hand side of (D.7.5) and (D.7.6) combine into an effective negative pressure gradient, $\chi + \rho\,g_x$.

Normal-mode analysis

A normal-mode perturbation displaces the interface to a position given by the real or imaginary part of the function

$$y = f(x,t) = \epsilon\eta(x,t),\tag{D.7.7}$$

where ϵ is a dimensionless coefficient whose magnitude is much less than unity,

$$\eta(x,t) = A\exp[ik(x - ct)]\tag{D.7.8}$$

is the normal-mode wave form of the perturbation, A is the complex amplitude of the interfacial wave, $k = 2\pi/L$ is the wave number, L is the wavelength, and c is the complex phase velocity.

Stream function

The stream function, ψ, is defined by the equations $u = \partial\psi/\partial y$ and $v = -\partial\psi/\partial x$. In the linear analysis, we set $\psi_j = \psi_j^{(0)} + \epsilon\psi_j^{(1)}$ and introduce the normal-mode form

$$\psi_j^{(1)}(x,y,t) = \phi_j(\hat{y})\,\exp[ik(x - ct)],\tag{D.7.9}$$

where $\hat{y} = ky$ and the superscript (1) denotes the perturbation. The perturbation velocity components are $u^{(1)} = \partial\psi^{(1)}/\partial y$ and $v^{(1)} = -\partial\psi^{(1)}/\partial x$.

The Reynolds number of the flow is so small that the motion of the fluid is governed by the equations of Stokes flow. Requiring that the stream function satisfies the biharmonic equation, $\nabla^4\psi_j^{(1)} = 0$, we find that

$$\phi_j(\hat{y}) = a_{1j}\cosh\hat{y} + a_{2j}\,\hat{y}\cosh\hat{y} + a_{3j}\sinh\hat{y} + a_{4j}\,\hat{y}\sinh\hat{y},\tag{D.7.10}$$

where a_{ij} are eight complex coefficients for $i = 1\text{–}4$ and $j = 1,2$.

Pressure field

The linearized pressure field in the jth fluid is $p_j = p_j^{(0)} + \epsilon p_j^{(1)}$. The disturbance pressure field is expressed by the normal-mode form

$$p_j^{(1)}(x,y,t) = \mu_j q_j(\hat{y})\,\exp[ik(x - ct)]\tag{D.7.11}$$

for $j = 1, 2$, where $q_j(\hat{y})$ are pressure eigenfunctions. Substituting this expression into the x component of the Stokes equation, we find that

$$p_j^{(1)}(y) = -\frac{i}{k}\left(\frac{\partial^3 \psi_j}{\partial x^2 \partial y} + \frac{\partial^3 \psi_j}{\partial y^3}\right), \tag{D.7.12}$$

yielding

$$q_j(\hat{y}) = i\,k^2\left(\frac{d\phi_j}{d\hat{y}} - \frac{d^3\phi_j}{d\hat{y}^3}\right). \tag{D.7.13}$$

Continuity of y velocity at the interface

In the linear approximation, continuity of y velocity at the interface requires that

$$\psi_1^{(1)}(x, y = 0, t) = \psi_2^{(1)}(x, y = 0, t), \tag{D.7.14}$$

yielding $a_{11} = a_{12}$.

Kinematic compatibility

Kinematic compatibility requires that $D[f(x, t) - y]/Dt = 0$, where D/Dt is the material derivative. In the linear approximation,

$$\frac{\partial \eta}{\partial t} + u_I\frac{\partial \eta}{\partial x} - v^{(1)} = 0, \tag{D.7.15}$$

where the velocity is evaluated at $y = 0$. Substituting the preceding expressions, we find that $-ikcA + u_I ikA + ika_{11} = 0$, yielding $a_{11} = (c - u_I)\,A$ or

$$A = \zeta a_{11}, \tag{D.7.16}$$

where $\zeta \equiv 1/(c - u_I)$ is the inverse of the complex phase velocity shifted by the real interfacial velocity.

Continuity of x velocity at the interface

In the linear approximation, continuity of the x velocity at the interface requires that

$$\xi_1 \eta(x, y) + \left(\frac{\partial \psi_1^{(1)}}{\partial y}\right)_{y=0} = \xi_2 \eta(x, y) + \left(\frac{\partial \psi_2^{(1)}}{\partial y}\right)_{y=0}, \tag{D.7.17}$$

yielding

$$\xi_1 A + k\left(\frac{d\phi_1}{d\hat{y}}\right)_{\hat{y}=0} = \xi_2 A + k\left(\frac{d\phi_2}{d\hat{y}}\right)_{\hat{y}=0}. \tag{D.7.18}$$

Rearranging and using (D.7.16), we obtain

$$(\xi_1 - \xi_2)\,\zeta\,a_{11} + k\left(\frac{d\phi_1}{d\hat{y}}\right)_{\hat{y}=0} - k\left(\frac{d\phi_2}{d\hat{y}}\right)_{\hat{y}=0} = 0, \tag{D.7.19}$$

yielding

$$(\Xi_1 - \Xi_2)\,\zeta\,a_{11} + a_{21} + a_{31} - a_{22} - a_{32} = 0, \tag{D.7.20}$$

where $\Xi_1 \equiv \xi_1/k$ and $\Xi_2 \equiv \xi_2/k$ are coefficients with dimensions of velocity.

Wall conditions

The no-slip and no-penetration conditions at the lower and upper wall require that

$$\phi_1(\hat{y} = -\hat{k}_1) = 0, \qquad \phi_1'(\hat{y} = -\hat{k}_1) = 0, \qquad \phi_2(\hat{y} = \hat{k}_2) = 0, \qquad \phi_2'(\hat{y} = \hat{k}_2) = 0, \tag{D.7.21}$$

where $\hat{k}_1 \equiv k\hat{h}_1$, $\hat{k}_2 \equiv k\hat{h}_2$, and a prime denotes a derivative with respect to $\hat{y}$. Making substitutions, we obtain

$$\left[\begin{array}{cccc} \cosh\hat{k}_1 & -\hat{k}_1\cosh\hat{k}_1 & -\sinh\hat{k}_1 & \hat{k}_1\sinh\hat{k}_1 \\ -\sinh\hat{k}_1 & \cosh\hat{k}_1 + \hat{k}_1\sinh\hat{k}_1 & \cosh\hat{k}_1 & -\sinh\hat{k}_1 - \hat{k}_1\cosh\hat{k}_1 \end{array} \right] \cdot \mathbf{w}_1 = \mathbf{0} \tag{D.7.22}$$

and

$$\left[\begin{array}{cccc} \cosh\hat{k}_2 & \hat{k}_2\cosh\hat{k}_2 & \sinh\hat{k}_2 & \hat{k}_2\sinh\hat{k}_2 \\ \sinh\hat{k}_2 & \cosh\hat{k}_2 + \hat{k}_2\sinh\hat{k}_2 & \cosh\hat{k}_2 & \sinh\hat{k}_2 + \hat{k}_2\cosh\hat{k}_2 \end{array} \right] \cdot \mathbf{w}_2 = \mathbf{0}, \tag{D.7.23}$$

where $\mathbf{w}_1 = \left[a_{11}, a_{21}, a_{31}, a_{41}\right]$ and $\mathbf{w}_2 = \left[a_{12}, a_{22}, a_{32}, a_{42}\right]$.

Surfactant concentration and surface tension

The distribution of the interfacial surfactant concentration and surface tension are described by the companion functions

$$\Gamma(x,t) = \Gamma_0 + \epsilon\,\Gamma_1 \exp[ik(x - ct)], \qquad \gamma(x,t) = \gamma_0 + \epsilon\,\gamma_1 \exp[ik(x - ct)], \tag{D.7.24}$$

where Γ_0, γ_0 are uniform values corresponding to the flat interface, and Γ_1, γ_1 are complex amplitudes. Since the perturbations are small, we can adopt the linear constitutive equation

$$\frac{\gamma_1}{\gamma_0} = -\mathrm{Ma}\,\frac{\Gamma_1}{\Gamma_0}, \tag{D.7.25}$$

where Ma is the Marangoni number.

Surfactant transport

Surfactant transport over the interface is governed by the equation

$$\frac{\mathrm{D}\Gamma}{\mathrm{D}t} + \Gamma\,\mathbf{t}\cdot\frac{\partial\mathbf{u}}{\partial l} = \frac{\partial}{\partial l}\left(D_s\frac{\partial\Gamma}{\partial l}\right), \tag{D.7.26}$$

where $\mathrm{D}/\mathrm{D}t$ is the material derivative. Regarding Γ as a function of x and t, we obtain

$$\frac{\partial\Gamma}{\partial t} + u\frac{\partial\Gamma}{\partial x} + \Gamma\,\mathbf{t}\cdot\left(\frac{\partial\mathbf{u}}{\partial x} + \frac{\partial\mathbf{u}}{\partial y}\frac{\partial\Gamma}{\partial x}\right)\frac{\partial x}{\partial l} = \frac{\partial}{\partial l}\left(D_s\frac{\partial\Gamma}{\partial l}\right). \tag{D.7.27}$$

The unit tangential vector can be linearized as

$$\mathbf{t} \simeq \mathbf{e}_x + \epsilon \frac{\partial \eta}{\partial x} \mathbf{e}_y, \tag{D.7.28}$$

where $\mathbf{e}_x$ and $\mathbf{e}_y$ are unit vectors along the x and y axes. To leading order, $\partial x / \partial l = 1$, and the linearized form of the surfactant transport equation becomes

$$\frac{\partial \Gamma^{(1)}}{\partial t} + u^{(0)} \frac{\partial \Gamma^{(1)}}{\partial x} + \Gamma_0 \left(\frac{\partial u^{(1)}}{\partial x} + \frac{\partial u^{(0)}}{\partial y} \frac{\partial \eta}{\partial x} \right) = D_s \frac{\partial^2 \Gamma^{(1)}}{\partial x^2}, \tag{D.7.29}$$

where all terms are evaluated at the unperturbed position, $y = 0$, the velocity on the left-hand side is evaluated on either side of the interface, and $\Gamma^{(1)}(x,t) = \Gamma_1 \exp[ik(x-ct)]$. Substituting the preceding expressions for the upper fluid velocity, we find that

$$\frac{\Gamma_1}{\Gamma_0} = \frac{a_{22} + a_{32} + \Xi_2 \, \zeta \, a_{11}}{1 + iD_s k \zeta} \, k\zeta, \tag{D.7.30}$$

where we recall that $\Xi_2 = \xi_2 / k$.

Tangential component of the interfacial force balance

The linearized tangential component of the interfacial force balance requires that

$$\mu_1 \left(\frac{\partial u_1^{(1)}}{\partial y} + \frac{\partial v_1^{(1)}}{\partial x} \right)_{y=0} = \mu_2 \left(\frac{\partial u_2^{(1)}}{\partial y} + \frac{\partial v_2^{(1)}}{\partial x} \right)_{y=0} + \Delta \rho \, g_x A + \frac{\partial \gamma^{(1)}}{\partial x}, \tag{D.7.31}$$

where $\gamma^{(1)}(x,t) = \gamma_1 \exp[ik(x-ct)]$ and $\Delta\rho = \rho_1 - \rho_2$. Expressing the velocity in terms of the stream function and using (D.7.16) to eliminate A in favor of a_{11}, we obtain

$$\left(\frac{\partial^2 \psi_1^{(1)}}{\partial y^2} - \frac{\partial^2 \psi_1^{(1)}}{\partial x^2} \right)_{y=0} = \lambda \left(\frac{\partial^2 \psi_2^{(1)}}{\partial y^2} - \frac{\partial^2 \psi_2^{(1)}}{\partial x^2} \right)_{y=0} + \frac{\Delta \rho \, g_x}{\mu_1} \zeta a_{11} + \frac{1}{\mu_1} \frac{\partial \gamma^{(1)}}{\partial x}, \tag{D.7.32}$$

yielding

$$\left(\frac{d^2 \phi_1}{d\hat{y}^2} + \phi_1 \right)_{\hat{y}=0} = \lambda \left(\frac{d^2 \phi_2}{d\hat{y}^2} + \phi_2 \right)_{\hat{y}=0} + \frac{\Delta \rho \, g_x}{\mu_1 k^2} \zeta \, a_{11} + i \frac{\gamma_1}{k\mu_1}. \tag{D.7.33}$$

Substituting the preceding expressions for ϕ_j and simplifying, we obtain

$$(1 - \zeta \mathcal{B}_x) \, a_{11} + a_{41} - \lambda \, (a_{12} + a_{42}) = i \frac{\gamma_1}{2\mu_1 k}, \tag{D.7.34}$$

where

$$\mathcal{B}_x \equiv \frac{\Delta \rho \, g_x}{2\mu_1 k^2}. \tag{D.7.35}$$

We recall that $a_{11} = a_{12}$ and $\gamma_1 / \gamma_0 = -\text{Ma} \, \Gamma_1 / \Gamma_0$, and use (D.7.30) to obtain

$$a_{41} = -a_{12} \, (1 - \lambda - \zeta \mathcal{B}_x) + \lambda \, a_{42} + \Lambda \, (a_{22} + a_{32} + \Xi_2 \, a_{12} \, \zeta), \tag{D.7.36}$$

where

$$\Lambda \equiv -\text{Ma}\,\frac{\gamma_0}{2\mu_1}\,\frac{\mathrm{i}\zeta}{1+D_s k\mathrm{i}\zeta} \tag{D.7.37}$$

is a dimensionless group. In the absence of surfactant, $\Lambda = 0$.

Normal component of the interface force balance

In the linear approximation, the normal component of the interfacial force balance requires that

$$\left(-p_1^{(1)} + 2\mu_1\frac{\partial v_1^{(1)}}{\partial y}\right)_{y=0} - \left(-p_2^{(1)} + 2\mu_2\frac{\partial v_2^{(1)}}{\partial y}\right)_{y=0} - \Delta\rho\, g_y \eta(x) = \gamma_0\frac{\partial^2\eta}{\partial x^2}. \tag{D.7.38}$$

Substituting the preceding expressions, we find that

$$-\mu_1 q_1 + \mu_2 q_2 - 2\mathrm{i}k^2\mu_1\frac{\mathrm{d}\phi_1}{\mathrm{d}\hat{y}} + 2\mathrm{i}k^2\mu_2\frac{\mathrm{d}\phi_2}{\mathrm{d}\hat{y}} - \Delta\rho\, g_y\, A = -\gamma_0\, A\, k^2, \tag{D.7.39}$$

where the left-hand side is evaluated at $\hat{y} = 0$. Substituting $q_j(y=0) = -2\mathrm{i}\,k^2 a_{2j}$, we obtain

$$2\mathrm{i}\,k^2(\mu_1 a_{21} - \mu_2 a_{22}) - 2\,\mathrm{i}\,k^2\mu_1\,(a_{21}+a_{31}) + 2\,\mathrm{i}\,k^2\mu_2\,(a_{22}+a_{32}) + (-\Delta\rho\, g_y + \gamma_0\, k^2\,)\, A = 0, \tag{D.7.40}$$

which simplifies into

$$-2\mathrm{i}\,k^2\mu_1\, a_{31} + 2\mathrm{i}k^2\mu_2\, a_{32} - (\rho_1 - \rho_2)\, g_y A\,(-\Delta\rho\, g_y + \gamma_0\, k^2\,)\, A = 0. \tag{D.7.41}$$

Rearranging and using (D.7.16), we obtain

$$a_{31} = \lambda\, a_{32} - \mathrm{i}\,\Pi\,\zeta a_{12}, \tag{D.7.42}$$

where

$$\Pi \equiv \frac{1}{2\mu_1}\left(-\frac{\Delta\rho\, g_y}{k^2} + \gamma_0\right) = -\mathcal{B}_y + \frac{\gamma_0}{2\mu_1} \tag{D.7.43}$$

is a property group with dimensions of velocity, and

$$\mathcal{B}_y \equiv \frac{\Delta\rho\, g_y}{2\mu_1 k^2}. \tag{D.7.44}$$

Formulation of an eigenvalue problem

Substituting (D.7.42) into (D.7.20) and rearranging, we find that

$$a_{21} = \zeta\,(\mathrm{i}\,\Pi + \Xi_2 - \Xi_1)\, a_{12} + a_{22} + (1-\lambda)\, a_{32}. \tag{D.7.45}$$

Finally, we substitute (D.7.36), (D.7.42), and (D.7.45) into the linear system (D.7.22) and derive the equivalent system $\mathbf{A} \cdot \mathbf{w}_2 = \mathbf{0}$, where

$$
\mathbf{A} = \left[\begin{array}{c|c}
\begin{array}{l} C_{11} - \zeta\,(C_{12} - \Lambda\,C_{13}) \\ C_{21} + \zeta\,(C_{22} + \Lambda\,C_{23}) \end{array} & \begin{array}{l} \hat{k}_1\,(\Lambda\sinh\hat{k}_1 - \cosh\hat{k}_1) \\ (1 - \Lambda\,\hat{k}_1)\cosh\hat{k}_1 + (\hat{k}_1 - \Lambda)\sinh\hat{k}_1 \end{array} \\[2ex]
\begin{array}{l} -(1 - \lambda)\,\hat{k}_1\,\cosh\hat{k}_1 + (\Lambda\,\hat{k}_1 - \lambda)\sinh\hat{k}_1 \\ (1 - \Lambda\,\hat{k}_1)\cosh\hat{k}_1 + [(1 - \lambda)\,\hat{k}_1 - \Lambda]\sinh\hat{k}_1 \end{array} & \begin{array}{l} \lambda\,\hat{k}_1\sinh\hat{k}_1 \\ -\lambda\,(\sinh\hat{k}_1 + \hat{k}_1\cosh\hat{k}_1) \end{array}
\end{array} \right],
$$ (D.7.46)

and

$$
\begin{aligned}
C_{11} &= \cosh\hat{k}_1 - (1 - \lambda)\,\hat{k}_1\,\sinh\hat{k}_1, \\
C_{12} &= \hat{k}_1\left[(\Xi_2 - \Xi_1)\cosh\hat{k}_1 - \mathcal{B}_x\,\sinh\hat{k}_1 \right] + i\,\Pi\,(\hat{k}_1\,\cosh\hat{k}_1 - \sinh\hat{k}_1), \\
C_{13} &= \Xi_2\,\hat{k}_1\,\sinh\hat{k}_1, \\
C_{21} &= -\sinh\hat{k}_1 + (1 - \lambda)(\sinh\hat{k}_1 + \hat{k}_1\cosh\hat{k}_1), \\
C_{22} &= \left[(\Xi_2 - \Xi_1)(\cosh\hat{k}_1 + \hat{k}_1\sinh\hat{k}_1) - \mathcal{B}_x\,(\sinh\hat{k}_1 + \hat{k}_1\cosh\hat{k}_1) \right] + i\,\Pi\,\hat{k}_1\sinh\hat{k}_1, \\
C_{23} &= -\Xi_2\,(\sinh\hat{k}_1 + \hat{k}_1\cosh\hat{k}_1).
\end{aligned}
$$ (D.7.47)

Note that the coefficients C_{13} and C_{23} are real.

Appending equations (D.7.46) to equations (D.7.23), we obtain a linear homogeneous system $\mathbf{Q} \cdot \mathbf{w}_2 = \mathbf{0}$, where

$$
\mathbf{Q} = \left[\begin{array}{c|c}
\begin{array}{l} \cosh\hat{k}_2 \\ \sinh\hat{k}_2 \\ C_{11} - \zeta\,(C_{12} - \Lambda C_{13}) \\ C_{21} + \zeta\,(C_{22} + \Lambda C_{23}) \end{array} & \begin{array}{l} \hat{k}_2\cosh\hat{k}_2 \\ \cosh\hat{k}_2 + \hat{k}_2\sinh\hat{k}_2 \\ \hat{k}_1\,(\Lambda\sinh\hat{k}_1 - \cosh\hat{k}_1) \\ (1 - \Lambda\,\hat{k}_1)\cosh\hat{k}_1 + (\hat{k}_1 - \Lambda)\sinh\hat{k}_1 \end{array} \\[4ex]
\begin{array}{l} \sinh\hat{k}_2 \\ \cosh\hat{k}_2 \\ -(1 - \lambda)\,\hat{k}_1\,\cosh\hat{k}_1 + (\Lambda\,\hat{k}_1 - \lambda)\sinh\hat{k}_1 \\ (1 - \Lambda\,\hat{k}_1)\cosh\hat{k}_1 + [(1 - \lambda)\,\hat{k}_1 - \Lambda]\sinh\hat{k}_1 \end{array} & \begin{array}{l} \hat{k}_2\sinh\hat{k}_2 \\ \sinh\hat{k}_2 + \hat{k}_2\cosh\hat{k}_2 \\ \lambda\,\hat{k}_1\sinh\hat{k}_1 \\ -\lambda\,(\sinh\hat{k}_1 + \hat{k}_1\cosh\hat{k}_1) \end{array}
\end{array} \right].
$$ (D.7.48)

In the absence of surfactant, $\Lambda = 0$. Setting the determinant of $\mathbf{Q}$ to zero provides us with a secular equation for the complex phase velocity, c.

Numerical method

The growth rates of the two normal modes are identified with the roots of a quadratic equation with complex coefficients,

$$
P_2(\zeta) = (1 + D_s ki\zeta)\det(\mathbf{Q}) = a\,\zeta^2 + b\,\zeta + c = 0.
$$ (D.7.49)

The factor $1 + D_s ki\zeta$ on the left-hand side involving the surface surfactant diffusivity arises from the definition of the parameter Λ. In the numerical method, the binomial coefficients are obtained by an exact finite-difference method based on the equations

$$P_2(1) = a + b + c, \qquad P_2(-1) = a - b + c, \qquad P_2(0) = c. \qquad (D.7.50)$$

The roots are found analytically using the quadratic formula.

Program files:

1. `chan210_s`
   Evaluates the complex phase velocity.

2. `chan210_s_dr`
   Driver for the subroutine `chan210_s`

3. `det_33c`
   Determinant of a 3×3 complex matrix.

4. `det_44c`
   Determinant of a 4×4 complex matrix.

5. `quadc`
   Computes the roots of a quadratic equation with complex coefficients.

Input file:

1. `chan210_s.dat`
   Specification of input parameters.

Output file:

1. `chan210_s.out`
   Recording of the computed phase velocity and growth rate.

D.8 Directory coat0_s

This directory contains a code that performs the linear stability analysis of a liquid film resting on a horizontal plane wall underneath a constant-pressure medium. The formulation of the linear stability problem is discussed in Section 9.11.

Program files:

1. `coat0_s`
   Evaluates the growth rate.

2. `cramer_33c`
   Solves a system of three complex equations.

3. det_33c
   Computes the determinant of a 3×3 complex matrix.

4. det_44c
   Computes the determinant of a 4×4 complex matrix.

5. matrix
   Evaluates the secular matrix.

Input file:

1. coat0_s.dat
   Specification of input parameters.

Output file:

1. coat0_s.out
   Recording of the computed phase velocity and growth rate.

D.9 Directory drop_ax

This directory contains a code that computes the rate of relaxation of a viscous drop immersed in an ambient fluid with the same viscosity under conditions of Stokes flow. The formulation of the linear stability problem is discussed in Section 9.3.5. Results are presented in Figure 9.3.2.

Program files:

1. drop_ax
   Evaluates the rate of relaxation by discrete perturbations.

2. drop_slp_splines
   Evaluates the single-layer potential on cubic splines.

3. ell_int
   Evaluates complete elliptic integrals.

4. gauss_leg
   Gauss–Legendre base points and weights.

5. sgf_ax_fs
   Free-space Green's function of axisymmetric Stokes flow.

6. splc_clm
   Cubic-spline interpolation with clamped boundary conditions.

7. splc_geo
   Geometry of interpolated interface.

8. thomas
   Thomas algorithm for tridiagonal systems.

D.10   Directory film0

This directory contains a code that performs the linear stability analysis of a liquid film flowing down an inclined plane under conditions of Stokes flow. Expressions for the phase velocity and growth rate are derived in directory `film0_s` in the more general context of flow in the presence of an insoluble surfactant (Section D.11).

Program files:

1. `det_33c`
   Determinant of a complex 3×3 matrix.

2. `film0`
   Evaluates the complex phase velocity.

3. `film0_dr`
   Driver for `film0`.

4. `matrix`
   Evaluates the secular matrix.

5. `cramer_33c`
   Solution of a complex 3×3 linear system.

Input file:

1. `film0.dat`
   Specification of input parameters.

Output file:

1. `film0.out`
   Recording of the computed growth rate.

D.11   Directory film0_s

This directory contains a code that performs the linear stability analysis of a liquid film flowing down an inclined plane, as illustrated in Figure D.11.1. The film surface is occupied by an insoluble surfactant that is convected and diffuses over the film surface but not into the bulk of the film fluid [314]. The instability occurs under conditions of Stokes flow.

Base state

The flow is described in the inclined coordinates defined in Figure D.11.1. The unperturbed configuration is described by the flat-film Nusselt solution designated by the superscript (0). The velocity, stream function, and pressure are given by

$$u_x^{(0)} = \frac{\rho g}{2\mu} \sin \beta \, y(2h - y), \qquad u_y^{(0)} = 0, \qquad \psi^{(0)} = \frac{\rho g}{2\mu} \sin \beta \, y^2 \left(h - \frac{1}{3}y\right),$$

$$p^{(0)} = \rho g \cos \beta (h - y) + p_a, \tag{D.11.1}$$

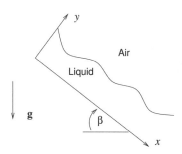

FIGURE D.11.1   Instability of a liquid film flowing down an inclined plane. The surface of the film is occupied by an insoluble surfactant.

where $g = |\mathbf{g}|$ is the magnitude of the acceleration of gravity, β is the plane inclination angle, and p_a is the ambient atmospheric pressure. The free-surface velocity is

$$u_s = u_x^{(0)}(y = h) = \frac{\rho g h^2}{2\mu} \sin\beta. \tag{D.11.2}$$

Normal-mode analysis

To carry out the normal-mode stability analysis for two-dimensional perturbations, we describe the position of the film surface by the function

$$y = f(x, t) = h + \epsilon\eta(x, t), \tag{D.11.3}$$

where ϵ is a dimensionless coefficient, whose magnitude is small compared to unity,

$$\eta(x, t) = A \exp[ik(x - ct)] \tag{D.11.4}$$

is the normal-mode wave form of the perturbation, A is the complex amplitude of the interfacial wave, $k = 2\pi/L$ is the wave number, L is the wavelength, and c is the complex phase velocity.

Stream function

The stream function, ψ, is defined by the equations $u = \partial\psi/\partial y$ and $v = -\partial\psi/\partial x$. In the linear analysis, we set $\psi = \psi^{(0)} + \epsilon\psi^{(1)}$ and introduce the normal-mode form

$$\psi^{(1)}(x, y, t) = \phi(\hat{y}) \exp[ik(x - ct)], \tag{D.11.5}$$

where the superscript (1) denotes the perturbation and $\hat{y} \equiv ky$.

The Reynolds number of the flow is so small that the motion of the fluid is governed by the equations of Stokes flow. Requiring that the stream function satisfies the biharmonic equation, $\nabla^4\psi_j^{(1)} = 0$, we obtain

$$\phi(\hat{y}) = a_1\, e^{\hat{y}} + a_2\, \hat{y}\, e^{\hat{y}} + a_3\, e^{-\hat{y}} + a_4\, \hat{y}\, e^{-\hat{y}}, \tag{D.11.6}$$

where a_1–a_4 are four complex coefficients.

Wall conditions

The no-penetration and no-slip boundary conditions over the wall require that $\phi(0) = 0$ and $\phi'(0) = 0$. Making substitutions, we obtain

$$a_3 = -a_1, \qquad 2a_1 + a_2 + a_4 = 0. \tag{D.11.7}$$

Kinematic compatibility

Kinematic compatibility requires that $\mathrm{D}[f(x,t) - y]/\mathrm{D}t = 0$, where $\mathrm{D}/\mathrm{D}t$ is the material derivative, yielding

$$\frac{\partial f}{\partial t} + u_x \frac{\partial f}{\partial x} - u_y = 0, \tag{D.11.8}$$

evaluated at the film surface. Linearizing, we obtain

$$\frac{\partial \eta}{\partial t} + u_s \frac{\partial \eta}{\partial x} + \frac{\partial \psi^{(1)}}{\partial x} = 0. \tag{D.11.9}$$

Substituting the preceding expressions, we find that $-\mathrm{i}Akc + u_s\mathrm{i}kA + \mathrm{i}k\phi(\hat{k}) = 0$, where $\hat{k} = kh$. Rearranging, we obtain

$$A = \zeta\phi(\hat{k}), \tag{D.11.10}$$

where $\zeta = 1/(c - u_s)$ is the inverse of the complex phase velocity shifted by the real interfacial velocity.

Surfactant concentration and surface tension

Surfactant concentration inhomogeneities over the film surface cause corresponding variations in surface tension. The distribution of the interfacial surfactant concentration, Γ, and surface tension, γ, are described by the companion functions

$$\Gamma(x,t) = \Gamma_0 + \epsilon\Gamma_1 \exp[\mathrm{i}k(x - ct)], \qquad \gamma(x,t) = \gamma_0 + \epsilon\gamma_1 \exp[\mathrm{i}k(x - ct)], \tag{D.11.11}$$

where Γ_0, γ_0 are uniform values corresponding to the flat film, and Γ_1, γ_1 are complex amplitudes. Since the perturbations are small, we can write

$$\frac{\gamma_1}{\gamma_0} = -\mathrm{Ma}\frac{\Gamma_1}{\Gamma_0}, \tag{D.11.12}$$

where Ma is the Marangoni number.

Surfactant transport

The linearized form of the surfactant transport equation is

$$\frac{\partial \Gamma^{(1)}}{\partial t} + u_x^{(0)} \frac{\partial \Gamma^{(1)}}{\partial x} + \Gamma^{(0)} \left(\frac{\partial u_x^{(1)}}{\partial x} + \frac{\partial u_x^{(0)}}{\partial y} \frac{\mathrm{d}\eta}{\mathrm{d}x} \right) = D_s \frac{\partial^2 \Gamma^{(1)}}{\partial x^2}, \tag{D.11.13}$$

evaluated at the unperturbed interface, $y = h$. The second term inside the parentheses on the left-hand side arises from the derivative $\partial u_x^{(0)}/\partial l \simeq (\partial u_x^{(0)}/\partial y)(\mathrm{d}f/\mathrm{d}x) = \epsilon\,(\partial u_x^{(0)}/\partial y)(\mathrm{d}\eta/\mathrm{d}x)$. However, because the shear stress and thus slope of the Nusselt velocity profile vanishes at the film surface, this term does not make a contribution. In terms of the stream function,

$$\frac{\partial \Gamma^{(1)}}{\partial t} + u_s\,\frac{\partial \Gamma^{(1)}}{\partial x} + \Gamma_0\,\frac{\partial^2 \psi^{(1)}}{\partial x \partial y} = D_s\,\frac{\partial^2 \Gamma^{(1)}}{\partial x^2}. \tag{D.11.14}$$

Substituting the normal-mode forms and rearranging, we find that the complex amplitude of the surfactant concentration is

$$\frac{\Gamma_1}{\Gamma_0} = k\,\frac{\phi'(\hat{k})}{c - u_s + \mathrm{i}k D_s}, \tag{D.11.15}$$

where a prime denotes a derivative with respect to $\hat{k}$. Correspondingly, the complex amplitude of the surface tension is

$$\frac{\gamma_1}{\gamma_0} = k\,\frac{\mathrm{Ma}}{c - u_s + \mathrm{i}k D_s}\,\phi'(\hat{k}). \tag{D.11.16}$$

Evaluating the derivative and remembering that $a_3 = -a_1$, we obtain

$$\frac{\gamma_1}{\gamma_0} = -k\,\mathrm{e}^{\hat{k}}\,\frac{\mathrm{Ma}}{c - u_s + \mathrm{i}k D_s}\,\big[\,a_1(1 + q) + a_2(1 + \hat{k}) + a_4(1 - \hat{k})\,q\,\big], \tag{D.11.17}$$

where $q = \exp(-2\hat{k})$.

Tangential component of the interfacial force balance

The linearized tangential component of the interfacial force balance requires that

$$\mu\left(\frac{\partial u_x^{(1)}}{\partial y} + \frac{\partial u_y^{(1)}}{\partial x}\right)_{y=h} = 2\,\mu\,\frac{u_s}{h^2}\,\eta + \frac{\partial \gamma^{(1)}}{\partial x}. \tag{D.11.18}$$

The last term on the right-hand side represents the Marangoni traction. Expressing the velocity in terms of the stream function and rearranging, we obtain

$$\left(\frac{\partial^2 \psi^{(1)}}{\partial y^2} - \frac{\partial \psi^{(1)}}{\partial x^2}\right)_{y=h} = 2\,\frac{u_s}{h^2}\,\eta + \frac{1}{\mu}\,\frac{\partial \gamma^{(1)}}{\partial x}, \tag{D.11.19}$$

Substituting the normal-mode forms and using (D.11.10), we obtain

$$\hat{k}^2\left(\frac{\mathrm{d}^2\phi}{\mathrm{d}\hat{y}^2} + \phi\right)_{\hat{y}=\hat{k}} = 2\,u_s\,\zeta\,\phi(\hat{k}) + \mathrm{i}\,\hat{k}\,\frac{h}{\mu}\,\gamma_1. \tag{D.11.20}$$

Computing the derivatives, recalling that $a_3 = -a_1$, and rearranging, we find that

$$\eta\hat{k}\,\big[\,(1 - q)\,a_1 + (\hat{k} + 1)\,a_2 + (\hat{k} - 1)\,q\,a_4\,\big] = -\Big(\frac{1 - q}{\hat{k}}\,a_1 + a_2 + q\,a_4\Big) + \mathrm{i}\,\chi h\,\frac{\gamma_1}{2\mu}\,\mathrm{e}^{-\hat{k}}, \tag{D.11.21}$$

where $q = \exp(-2\hat{k})$ and $\chi = -1/(u_s\zeta) = 1 - c/u_s$.

Normal component of the interfacial force balance

The linearized pressure field is $p_j = p^{(0)} + \epsilon p^{(1)}$. The linearized normal component of the interfacial stress balance provides us with an expression for the perturbation pressure,

$$p^{(1)} = 2\mu \frac{\partial u_y^{(1)}}{\partial y} + \rho g \cos\beta\, \eta - \gamma^{(0)} \frac{\partial^2 \eta}{\partial x^2} + \gamma^{(1)} \kappa_0, \tag{D.11.22}$$

where κ_0 is the interfacial curvature and all terms are evaluated at the unperturbed film surface, $y = h$. Because in the unperturbed configuration the film surface is flat, $\kappa_0 = 0$, the last term on the right-hand side is identically zero. Consequently, surface tension variations do not affect the normal force balance.

Differentiating (D.11.22) with respect to x and using the tangential projection of the equation of motion to evaluate the pressure derivative on the left-hand side, we obtain the preferred pressure-free form

$$\frac{\partial p^{(1)}}{\partial x} = \mu\, \nabla^2 u_x^{(1)} = 2\mu \frac{\partial^2 u_y^{(1)}}{\partial x \partial y} + \rho g \cos\beta \frac{\partial \eta}{\partial x} - \gamma_0 \frac{\partial^3 \eta}{\partial x^3}, \tag{D.11.23}$$

where all terms are evaluated at the unperturbed film surface, $y = h$. Introducing the stream function, we obtain

$$\mu \left(3 \frac{\partial^3 \psi^{(1)}}{\partial x^2 \partial y} + \frac{\partial^3 \psi^{(1)}}{\partial y^3} \right) = \rho g \cos\beta \frac{\partial \eta}{\partial x} - \gamma_0 \frac{\partial^3 \eta}{\partial x^3}. \tag{D.11.24}$$

Substituting the normal-mode forms and using (D.11.10), we obtain

$$\mu \left[-3\, \phi'(\hat{k}) + \phi'''(\hat{k}) \right] = \mathrm{i}\zeta \left(\frac{\rho g}{k^2} \cos\beta + \gamma_0 \right) \phi(\hat{k}), \tag{D.11.25}$$

where a prime denotes a derivative with respect to $\hat{y}$. Evaluating the derivatives, remembering that $a_3 = -a_1$, and setting $\zeta = -1/(u_s\chi)$, we find that

$$\chi\left[(1+q)\, a_1 + a_2 \hat{k} - a_4 \hat{k}\, q \right] = \mathrm{i}\, \frac{1}{2\mu u_s} \left(\frac{\rho g}{k^2} \cos\beta + \gamma_0 \right) \left[a_1(1-q) + a_2\, \hat{k} + a_4\, \hat{k}\, q \right], \tag{D.11.26}$$

where $q = \exp(-2\hat{k})$.

Formulation of an eigenvalue problem

Collecting the second equation in (D.11.7), the shear stress balance (D.11.21), the normal stress balance (D.11.26), and equation (D.11.17), we derive a linear system of homogeneous, $\mathbf{M} \cdot \mathbf{w} = \mathbf{0}$, where $\mathbf{w} = [a_1, a_2, a_4, \gamma_1/(\mu k)]$ is the unknown vector. The coefficient matrix is given by

$$\mathbf{M} = \begin{bmatrix} 2 & 1 & 1 & 0 \\ (1-q)(\chi\hat{k} + 1/\hat{k}) & \chi\hat{k}(\hat{k}+1) + 1 & \chi\hat{k}(\hat{k}-1)q + q & -\mathrm{i}\frac{1}{2}\chi\hat{k}e^{-\hat{k}} \\ \chi\hat{k}(1+q) - \mathrm{i}\tau(1-q)/\hat{k} & \chi\hat{k}^2 - \mathrm{i}\tau & -(\chi\hat{k}^2 + \mathrm{i}\tau)q & 0 \\ \mathrm{Ma}\,(1+q) & \mathrm{Ma}\,(1+\hat{k}) & \mathrm{Ma}\,(1-\hat{k})q & \mu\,(-u_s\chi + \mathrm{i}kD_s)e^{-\hat{k}} \end{bmatrix}, \tag{D.11.27}$$

where $q = \exp(-2\hat{k})$,

$$\tau \equiv \frac{1}{2\mu u_s}(\rho g h^2 \cos\beta + \gamma_0 \hat{k}^2) = \cot\beta + \frac{2\pi^2}{\text{Ca}}, \qquad \text{Ca} \equiv \frac{\mu u_s}{\gamma_0}\left(\frac{L}{h}\right)^2 = \frac{1}{2}\frac{\rho g L^2}{\gamma_0}\sin\beta, \qquad (D.11.28)$$

are a dimensionless group and a capillary number defined with respect to the wavelength of the perturbation. Setting the determinant of $\mathbf{M}$ to zero provides us with a third-order algebraic equation for the reduced and shifted complex phase velocity, χ. One trivial root is $\chi = 0$ corresponding to $c = u_s$. The other two roots can be computed in terms of the coefficients of the remainder binomial using the quadratic formula. In practice, the coefficients are extracted by solving a system of complex linear equations for three trial values of χ. Thus, in the presence of surfactant, the flow admits two normal modes.

Absence of surfactant

In the absence of surfactant, $\text{Ma} = 0$, we set $\gamma_1 = 0$ and retain the first three equations in (D.11.27). Yih [435] derived analytical expressions for the phase velocity c_R and rate of decay of surface waves $s_I = kc_I$, where the subscripts R and I denote the real and imaginary part. The dimensionless phase velocity and dimensionless growth rate are given by

$$\frac{c_R}{u_s} = 1 + \frac{1}{\cosh^2 \hat{k} + \hat{k}^2}, \qquad \hat{s} \equiv \frac{s_I h}{u_s} = \frac{c_I \hat{k}}{u_s} = -\frac{\tau}{2\hat{k}}\frac{\sinh(2\hat{k}) - 2\hat{k}}{\cosh^2 \hat{k} + \hat{k}^2}. \qquad (D.11.29)$$

Since the fraction on the right-hand side of the expression for the growth rate is positive for any $\hat{k}$, the growth rate is negative and the flow is stable.

Program files:

1. `cramer_33c`
   Solves a system of three complex equations.

2. `det_33c`
   Computes the determinant of a 3×3 complex matrix.

3. `det_44c`
   Computes the determinant of a 4×4 complex matrix.

4. `film0_s`
   Evaluates the growth rate.

5. `matrix`
   Evaluates the secular matrix.

Input file:

1. `film0_s.dat`
   Specification of input parameters.

Output file:

1. `film0_s.out`
   Recording of the computed phase velocity and growth rate.

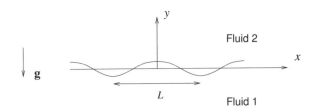

FIGURE D.12.1   Instability of the interface between two semi-infinite fluids in Stokes flow.

D.12   Directory if0

This directory contains a code that performs the linear stability analysis of an infinite horizontal interface separating two semi-infinite quiescent fluids, as illustrated in Figure D.12.1. When the upper fluid is heavier than the lower fluid, we obtain the Rayleigh–Taylor instability. The flow due to the instability is assumed to occur under conditions of Stokes flow.

Base state

The lower fluid is labeled 1 and the upper fluid is labeled 2. The viscosity of the lower fluid is μ_1 and the viscosity of the upper fluid is $\mu_2 = \lambda\mu_1$, where λ is the viscosity ratio. The origin of the y axis is set at the location of the unperturbed interface. For convenience, the x and y velocity components are denoted by $u = u_x$ and $v = u_y$.

In the unperturbed configuration, the interface is flat, the fluids are quiescent, and the pressure assumes the hydrostatic distribution

$$p_j^{(0)}(y) = -\rho_j g y + p_0, \tag{D.12.1}$$

where $1, 2$ for the lower or upper fluid, ρ_j are the fluid densities, g is the acceleration of gravity, p_0 is an unspecified reference pressure, and the superscript (0) denotes the unperturbed base state.

Normal-mode analysis

A normal-mode perturbation displaces the interface to a position given by the real or imaginary part of the function

$$y = f(x,t) = \epsilon\eta(x,t), \tag{D.12.2}$$

where ϵ is a dimensionless coefficient whose magnitude is much less than unity,

$$\eta(x,t) = A\exp[ik(x - ct)] \tag{D.12.3}$$

is the normal-mode wave form of the perturbation, A is the complex amplitude of the interfacial wave, $k = 2\pi/L$ is the wave number, L is the wavelength, and c is the complex phase velocity of the perturbation.

Stream function

The stream function, ψ, is defined by the equations $u = \partial\psi/\partial y$ and $v = -\partial\psi/\partial x$. In the absence of base flow, we set $\psi_j = \epsilon\psi_j^{(1)}$ and introduce the normal-mode form

$$\psi_j^{(1)}(x,y,t) = \phi_j(\hat{y})\,\exp[ik(x-ct)], \tag{D.12.4}$$

where $j = 1,2$ for the lower or upper fluid, and $\hat{y} = ky$. The perturbation velocity components are $u^{(1)} = \partial\psi^{(1)}/\partial y$ and $v^{(1)} = -\partial\psi^{(1)}/\partial x$.

The Reynolds number of the flow is so small that the motion of the fluid is governed by the equations of Stokes flow. Requiring that the stream function satisfies the biharmonic equation, $\nabla^4\psi_j^{(1)} = 0$, we obtain

$$\phi_1(\hat{y}) = a_1\,e^{\hat{y}} + b_1\,\hat{y}\,e^{\hat{y}}, \qquad \phi_2(\hat{y}) = c_2\,e^{-\hat{y}} + d_2\,\hat{y}\,e^{-\hat{y}}, \tag{D.12.5}$$

where a_1, b_1, c_2, and d_2 are four complex coefficients.

Pressure field

The linearized pressure field in the jth fluid is $p_j = p_j^{(0)} + \epsilon p_j^{(1)}$. The perturbation pressure is described by the normal-mode form

$$p_j^{(1)}(x,t) = \mu_j\,q_j(\hat{y})\,\exp[ik(x-ct)] \tag{D.12.6}$$

for $j = 1,2$, where $q_j(\hat{y})$ are pressure eigenfunctions. Substituting this expression into the x component of the Stokes equation, we obtain

$$p_j^{(1)} = -\frac{i}{k}\left(\frac{\partial^3\psi_j^{(1)}}{\partial x^2\partial y} + \frac{\partial^3\psi_j^{(1)}}{\partial y^3}\right), \tag{D.12.7}$$

yielding

$$q_j(\hat{y}) = i\,k^2\left(\frac{d\phi_j}{d\hat{y}} - \frac{d^3\phi_j}{d\hat{y}^3}\right). \tag{D.12.8}$$

Continuity of y velocity at the interface

In the linear approximation, continuity of y velocity at the interface requires that

$$\psi_1(x,y=0,t) = \psi_2(x,y=0,t), \tag{D.12.9}$$

yielding $a_1 = c_2$.

Continuity of x velocity at the interface

In the linear approximation, continuity of x velocity at the interface requires that

$$\left(\frac{\partial\psi_1}{\partial y}\right)_{y=0} = \left(\frac{\partial\psi_2}{\partial y}\right)_{y=0}, \tag{D.12.10}$$

yielding

$$a_1 + b_1 = -c_2 + d_2. \tag{D.12.11}$$

Kinematic compatibility

Kinematic compatibility requires that $D[f(x,t) - y]/Dt = 0$ at the interface, where D/Dt is the material derivative. In the linearized approximation,

$$\frac{\partial \eta}{\partial t} - v^{(1)}(y = 0) = 0. \tag{D.12.12}$$

Substituting the preceding expressions, we find that $-ikcA + ika_1 = 0$ or

$$a_1 = cA. \tag{D.12.13}$$

Tangential component of the interfacial force balance

The linearized tangential component of the interfacial force balance requires that

$$\mu_1 \left(\frac{\partial u_1}{\partial y} + \frac{\partial v_1}{\partial x} \right)_{y=0} = \mu_2 \left(\frac{\partial u_2}{\partial y} + \frac{\partial v_2}{\partial x} \right)_{y=0} \tag{D.12.14}$$

or

$$\left(\frac{\partial^2 \psi_1}{\partial y^2} - \frac{\partial^2 \psi_1}{\partial x^2} \right)_{y=0} = \lambda \left(\frac{\partial^2 \psi_2}{\partial y^2} - \frac{\partial^2 \psi_2}{\partial x^2} \right)_{y=0}, \tag{D.12.15}$$

where $\lambda = \mu_2/\mu_1$ is the viscosity ratio. Substituting the preceding expressions for the stream function and simplifying, we find that

$$a_1 + b_1 = \lambda (c_2 - d_2). \tag{D.12.16}$$

Combining (D.12.16) with (D.12.11), we obtain

$$d_2 = c_2 = a_1, \qquad b_1 = -a_1. \tag{D.12.17}$$

Normal component of the interfacial force balance

The linearized normal component of the interfacial force balance requires that

$$\left(-p_1^{(1)} + 2\mu_1 \frac{\partial v_1^{(1)}}{\partial y} \right)_{y=0} - \left(-p_2^{(1)} + 2\mu_2 \frac{\partial v_2^{(1)}}{\partial y} \right)_{y=0} + \Delta\rho\, g\, \eta = \gamma \frac{\partial^2 \eta}{\partial x^2}, \tag{D.12.18}$$

where γ is the surface tension and $\Delta\rho = \rho_1 - \rho_2$. Substituting the preceding expressions, we obtain

$$-\mu_1 q_1 + \mu_2 q_2 - 2ik^2 (\mu_1 - \mu_2) \frac{d\phi_1}{d\hat{y}} + \Delta\rho\, g A = -\gamma A\, k^2, \tag{D.12.19}$$

where all terms are evaluated at $y = 0$. Substituting also

$$q_1(y = 0) = -2\mathrm{i}\, k^2\, b_1, \qquad q_2(y = 0) = -2\mathrm{i}\, k^2\, d_2, \tag{D.12.20}$$

we obtain

$$2\mathrm{i}\, k^2(\mu_1 b_1 - \mu_2 d_2) - 2\mathrm{i}k^2\,(\mu_1 - \mu_2)(a_1 + b_1) + \Delta\rho\, g\, A = -\gamma A k^2. \tag{D.12.21}$$

Using (D.12.17) to simplify and rearranging, we find that

$$a_1 = -\frac{\mathrm{i}}{2\mu_1(1 + \lambda)}\left(\frac{\Delta\rho\, g}{k^2} + \gamma\right) A. \tag{D.12.22}$$

Growth rate

Substituting expression (D.12.13) into (D.12.22), we obtain an imaginary phase velocity,

$$c = -\frac{\mathrm{i}}{2\mu_1(1 + \lambda)}\left(\frac{\Delta\rho\, g}{k^2} + \gamma\right). \tag{D.12.23}$$

The growth rate is

$$\sigma_I \equiv k c_I = -\frac{1}{2\mu_1(1 + \lambda)}\left(\frac{\Delta\rho\, g}{k} + k\gamma\right). \tag{D.12.24}$$

When the fluids are unstably stratified, $\rho_2 > \rho_1$ or $\Delta\rho < 0$, the first term inside the large parentheses on the right-hand side is responsible for the Rayleigh–Taylor instability. The second term expresses the stabilizing influence of surface tension.

Program files:

1. if0
   Evaluates expression (D.12.24).

2. if0_dr
   Driver for if0.

Input file:

1. if0.dat
   Specification of input parameters.

Output file:

1. if0.out
   Recording of the growth rate.

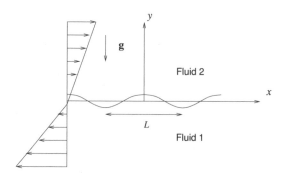

FIGURE D.13.1   Instability of the interface between two semi-infinite fluids in simple shear flow. The interface is occupied by an insoluble surfactant.

D.13   Directory ifsf0_s

This directory contains a code that performs the linear stability analysis of the horizontal interface between two semi-infinite fluids undergoing simple shear flow, as illustrated in Figure D.13.1. The interface is occupied by an insoluble surfactant and the fluid motion occurs under conditions of Stokes flow.

Base state

The lower fluid is labeled 1 and the upper fluid is labeled 2. The viscosity of the lower fluid is μ_1 and the viscosity of the upper fluid is $\mu_2 = \lambda\mu_1$, where λ is the viscosity ratio. The origin of the y axis is set at the unperturbed interface. For convenience, the x and y velocity components are denoted as $u = u_x$ and $v = u_y$.

In the unperturbed configuration, the interface is flat and the fluids undergo simple shear flow with a linear velocity profile parallel to the interface. The velocity profiles in the lower and upper fluids are given by

$$u_1^{(0)} = \lambda\xi y, \qquad u_2^{(0)} = \xi y, \tag{D.13.1}$$

where ξ is the shear rate in the upper fluid and the superscript (0) denotes the base state. The corresponding pressure distributions are

$$p_j^{(0)}(y) = -\rho_j g\, y + p_0 \tag{D.13.2}$$

for $j = 1, 2$, where ρ_j are the fluid densities, g is the gravitational acceleration, and p_0 is an inconsequential interfacial pressure.

Normal-mode analysis

A normal-mode perturbation displaces the interface to a position given by the real or imaginary part of the function

$$y = f(x,t) = \epsilon \eta(x,t),\tag{D.13.3}$$

where ϵ is a dimensionless number whose magnitude is much less than unity,

$$\eta(x,t) = A\exp[\mathrm{i}k(x-ct)]\tag{D.13.4}$$

is the normal-mode wave form of the perturbation, A is the complex amplitude of the interfacial wave, $k = 2\pi/L$ is the wave number, L is the wavelength, and c is the complex phase velocity of the perturbation.

Stream function

The stream function, ψ, is defined by the equations $u = \partial\psi/\partial y$ and $v = -\partial\psi/\partial x$. In the linear analysis, we set $\psi_j = \psi_j^{(0)} + \epsilon\psi_j^{(1)}$ and introduce the normal-mode form

$$\psi_j^{(1)}(x,y,t) = \phi_j(\hat{y})\,\exp[\mathrm{i}k(x-ct)]\tag{D.13.5}$$

for $j = 1,2$, where $\hat{y} = ky$ and the superscript (1) denotes the perturbation. The perturbation velocity components are $u^{(1)} = \partial\psi^{(1)}/\partial y$ and $v^{(1)} = -\partial\psi^{(1)}/\partial x$.

The Reynolds number of the flow is so small that the motion of the fluid is governed by the equations of Stokes flow. Requiring that the stream function satisfies the biharmonic equation, $\nabla^4\psi_j^{(1)} = 0$, and decays far from the interface, we obtain

$$\phi_1(\hat{y}) = a_1\,\mathrm{e}^{\hat{y}} + b_1\hat{y}\,\mathrm{e}^{\hat{y}}, \qquad \phi_2(\hat{y}) = c_2\,\mathrm{e}^{-\hat{y}} + d_2\hat{y}\,\mathrm{e}^{-\hat{y}},\tag{D.13.6}$$

where a_1, b_1, c_2, and d_2 are four complex coefficients.

Pressure field

The linearized pressure field in the jth fluid is $p_j = p_j^{(0)} + \epsilon p_j^{(1)}$. The perturbation pressure is described by the normal-mode form

$$p_j^{(1)}(x,y,t) = \mu_j\,q_j(\hat{y})\,\exp[\mathrm{i}k(x-ct)]\tag{D.13.7}$$

for $j = 1,2$, where $q_j(\hat{y})$ are pressure eigenfunctions. Substituting this expression into the x component of the Stokes equation, we obtain

$$p_j^{(1)} = -\frac{\mathrm{i}}{k}\Big(\frac{\partial^3\psi_j^{(1)}}{\partial x^2\partial y} + \frac{\partial^3\psi_j^{(1)}}{\partial y^3}\Big),\tag{D.13.8}$$

yielding

$$q_j(y) = \mathrm{i}\,k^2\Big(\frac{\mathrm{d}\phi_j}{\mathrm{d}\hat{y}} - \frac{\mathrm{d}^3\phi_j}{\mathrm{d}\hat{y}^3}\Big).\tag{D.13.9}$$

Continuity of the y velocity at the interface

In the linear approximation, continuity of the y velocity component at the interface requires that

$$\psi_1^{(1)}(x, y = 0, t) = \psi_2^{(1)}(x, y = 0, t), \tag{D.13.10}$$

yielding $a_1 = c_2$.

Kinematic compatibility

Kinematic compatibility requires that $D[f(x,t) - y]/Dt = 0$ evaluated at the interface, where D/Dt is the material derivative. In the linearized approximation,

$$\frac{\partial \eta}{\partial t} - v_j^{(1)} = 0, \tag{D.13.11}$$

evaluated at $y = 0$ for either fluid, $j = 1, 2$. Substituting the preceding expressions for the lower fluid velocity, we find that $-ikcA + ika_1 = 0$ or

$$A = \zeta a_1, \tag{D.13.12}$$

where $\zeta = 1/c$ is the inverse of the complex phase velocity.

Continuity of the x velocity component at the interface

In the linear approximation, continuity of the x velocity component at the interface requires that

$$\lambda \xi \eta + \left(\frac{\partial \psi_1^{(1)}}{\partial y}\right)_{y=0} = \xi \eta + \left(\frac{\partial \psi_2^{(1)}}{\partial y}\right)_{y=0}, \tag{D.13.13}$$

yielding

$$\lambda \xi A + k \left(\frac{d\phi_1}{d\hat{y}}\right)_{\hat{y}=0} = \xi A + k \left(\frac{d\phi_2}{d\hat{y}}\right)_{\hat{y}=0}. \tag{D.13.14}$$

Rearranging and using (D.13.12), we obtain

$$-\zeta \xi (1 - \lambda) a_1 + k \left(\frac{d\phi_1}{d\hat{y}}\right)_{\hat{y}=0} = k \left(\frac{d\phi_2}{d\hat{y}}\right)_{\hat{y}=0}, \tag{D.13.15}$$

yielding

$$\left[1 - \zeta \Xi (1 - \lambda)\right] a_1 + b_1 + c_2 - d_2 = 0, \tag{D.13.16}$$

where $\Xi \equiv \xi/k$ is a coefficient with dimensions of velocity. Substituting $c_2 = a_1$ and rearranging, we find that

$$b_1 = d_2 - \left[2 - \Xi \zeta (1 - \lambda)\right] c_2. \tag{D.13.17}$$

Surfactant concentration and surface tension

The distribution of the surfactant surface concentration and surface tension are described by the companion functions

$$\Gamma(x,t) = \Gamma_0 + \epsilon\,\Gamma_1 \exp[ik(x-ct)], \qquad \gamma(x,t) = \gamma_0 + \epsilon\,\gamma_1 \exp[ik(x-ct)], \qquad (D.13.18)$$

where Γ_0, γ_0 are the uniform unperturbed values corresponding to the flat interface, and Γ_1, γ_1 are the complex amplitudes of the perturbation. Since the perturbations are small, we may use the linearized form

$$\frac{\gamma_1}{\gamma_0} = -\mathrm{Ma}\,\frac{\Gamma_1}{\Gamma_0}, \qquad (D.13.19)$$

where Ma is the Marangoni number.

Surfactant transport

Interfacial surfactant transport is governed by the convection–diffusion equation

$$\frac{D\Gamma}{Dt} + \Gamma\mathbf{t}\cdot\frac{\partial\mathbf{u}}{\partial l} = \frac{\partial}{\partial l}\left(D_s\frac{\partial\Gamma}{\partial l}\right), \qquad (D.13.20)$$

where D/Dt is the material derivative, $\mathbf{t}$ is the unit tangent vector pointing in the direction of increasing arc length, l, and D_s is the surface surfactant diffusivity. Regarding Γ as a function of x and t, we obtain

$$\frac{\partial\Gamma}{\partial t} + u_x\frac{\partial\Gamma}{\partial x} + \Gamma\mathbf{t}\cdot\left(\frac{\partial\mathbf{u}}{\partial x} + \frac{\partial\mathbf{u}}{\partial y}\frac{\partial\Gamma}{\partial x}\right)\frac{\partial x}{\partial l} = \frac{\partial}{\partial l}\left(D_s\frac{\partial\Gamma}{\partial l}\right). \qquad (D.13.21)$$

The unit tangential vector can be linearized as

$$\mathbf{t} \simeq \mathbf{e}_x + \epsilon\frac{\partial\eta}{\partial x}\mathbf{e}_y, \qquad (D.13.22)$$

where $\mathbf{e}_x$ and $\mathbf{e}_y$ are unit vectors along the x and y axes. To first order in ϵ, $\partial x/\partial l = 1$. The linearized surfactant transport equation becomes

$$\frac{\partial\Gamma^{(1)}}{\partial t} + u^{(0)}\frac{\partial\Gamma^{(1)}}{\partial x} + \Gamma_0\left(\frac{\partial u^{(1)}}{\partial x} + \frac{\partial u^{(0)}}{\partial y}\frac{\partial\eta}{\partial x}\right) = D_s\frac{\partial^2\Gamma^{(1)}}{\partial x^2}, \qquad (D.13.23)$$

where $\Gamma^{(1)}(x,t) = \Gamma_1\exp[ik(x-ct)]$, all terms are evaluated at the unperturbed position, $y=0$, and the velocity on the left-hand side is evaluated on either side of the interface. Choosing the upper fluid velocity, we obtain

$$-\Gamma_1 ikc + \Gamma_0\left[ik^2(a_1+b_1) + \xi Aik\right] = -D_s k^2\Gamma_1. \qquad (D.13.24)$$

Substituting $A = \zeta a_1$ and rearranging, we obtain

$$\frac{\Gamma_1}{\Gamma_0} = \frac{a_1 + b_1 + \Xi\zeta a_1}{1 + iD_s k\zeta}\,k\zeta, \qquad (D.13.25)$$

where we recall that $\Xi = \xi/k$.

Tangential component of the interfacial force balance

The linearized tangential component of the interfacial force balance requires that

$$\mu_1 \left(\frac{\partial u_1^{(1)}}{\partial y} + \frac{\partial v_1^{(1)}}{\partial x} \right)_{y=0} = \mu_2 \left(\frac{\partial u_2^{(1)}}{\partial y} + \frac{\partial v_2^{(1)}}{\partial x} \right)_{y=0} + \frac{\partial \gamma^{(1)}}{\partial x}, \tag{D.13.26}$$

where $\gamma^{(1)}(x, t) = \gamma_1 \exp[ik(x - ct)]$. Expressing the velocity in terms of the stream function, we find that

$$\left(\frac{\partial^2 \psi_1^{(1)}}{\partial y^2} - \frac{\partial^2 \psi_1^{(1)}}{\partial x^2} \right)_{\hat{y}=0} = \lambda \left(\frac{\partial^2 \psi_2^{(1)}}{\partial y^2} - \frac{\partial^2 \psi_2^{(1)}}{\partial x^2} \right)_{\hat{y}=0} + \frac{1}{\mu_1} \frac{\partial \gamma^{(1)}}{\partial x}, \tag{D.13.27}$$

yielding

$$\left(\frac{d^2 \phi_1}{d\hat{y}^2} + \phi_1 \right)_{y=0} = \lambda \left(\frac{d^2 \phi_2}{d\hat{y}^2} + \phi_2 \right)_{y=0} + i \frac{\gamma_1}{k\mu_1}. \tag{D.13.28}$$

Substituting the preceding expressions for ϕ_j, simplifying, and using (D.13.19), we obtain an algebraic equation,

$$a_1 + b_1 - \lambda (c_2 - d_2) = i \frac{\gamma_1}{2\mu_1 k} = -i \, \text{Ma} \, \frac{\Gamma_1}{2\mu_1 k} \frac{\gamma_0}{\Gamma_0}. \tag{D.13.29}$$

Substituting (D.13.25), we find that

$$a_1 + b_1 - \lambda (c_2 - d_2) = -i \, \text{Ma} \, \frac{\gamma_0}{2\mu_1} \frac{a_1 + b_1 + \xi\zeta a_1/k}{1 + iD_s k \zeta} \zeta, \tag{D.13.30}$$

which can be rearranged into

$$a_1 + b_1 - \lambda (c_2 - d_2) = [\, a_1(1 + \Xi \zeta) + b_1 \,] \Lambda, \tag{D.13.31}$$

where

$$\Lambda \equiv -i \, \text{Ma} \, \frac{\gamma_0}{2\mu_1} \frac{\zeta}{1 + D_s k \, i \zeta} \tag{D.13.32}$$

is a dimensionless group. In the absence of surfactant, $\Lambda = 0$. Setting $a_1 = c_2$ and rearranging, we find that

$$(1 - \Lambda) b_1 = [\lambda - 1 + \Lambda(1 + \Xi \zeta)] c_2 - \lambda d_2. \tag{D.13.33}$$

Substituting the expression for b_1 from (D.13.17) and rearranging, we obtain

$$(1 + \lambda - \Lambda) d_2 = \Big[(1 - \Lambda) [2 - \Xi\zeta(1 - \lambda)] + \lambda - 1 + \Lambda(1 + \Xi \zeta) \Big] c_2. \tag{D.13.34}$$

Normal component of the interface force balance

The linearized normal component of the interfacial force balance requires that

$$\left(-p_1^{(1)} + 2\mu_1 \frac{\partial v_1^{(1)}}{\partial y}\right)_{y=0} - \left(-p_2^{(1)} + 2\mu_2 \frac{\partial v_2^{(1)}}{\partial y}\right)_{y=0} + \Delta\rho\, g\,\eta = \gamma_0 \frac{\partial^2 \eta}{\partial x^2}, \qquad \text{(D.13.35)}$$

where $\Delta\rho = \rho_1 - \rho_2$. Substituting the preceding expressions, we find that

$$\left(-\mu_1 q_1 + \mu_2 q_2 - 2\mathrm{i}k^2\,\mu_1 \frac{\mathrm{d}\phi_1}{\mathrm{d}\hat{y}} + 2\mathrm{i}k^2\,\mu_2 \frac{\mathrm{d}\phi_2}{\mathrm{d}\hat{y}}\right)_{\hat{y}=0} + \Delta\rho\, g A = -\gamma_0 A k^2. \qquad \text{(D.13.36)}$$

Substituting $q_1(y = 0) = -2\mathrm{i}\,k^2 b_1$, $q_2(y = 0) = -2\mathrm{i}\,k^2 d_2$, and using (D.13.17), we obtain

$$2\mathrm{i}k^2 \big[\mu_1 b_1 - \mu_2 d_2 - (\mu_1 - \mu_2)(a_1 + b_1) - \mu_2\, \Xi\, \zeta(1 - \lambda)\, c_2\big] + \big(\Delta\rho\, g + \gamma_0\, k^2\big) A = 0. \qquad \text{(D.13.37)}$$

Setting $a_1 = c_2$ and rearranging, we obtain

$$\mu_2\,(b_1 - d_2) - (\mu_1 - \mu_2)\, c_2 - \mu_2\, \Xi\, \zeta(1 - \lambda)\, c_2 - \mathrm{i}\frac{1}{2}\left(\frac{\Delta\rho\, g}{k^2} + \gamma_0\right)\zeta\, c_2 = 0. \qquad \text{(D.13.38)}$$

Finally, we substitute the expression for b_1 from (D.13.17) and simplify to obtain

$$\left(\mu_1 + \mu_2 + \mathrm{i}\frac{1}{2}\left(\frac{\Delta\rho\, g}{k^2} + \gamma_0\right)\zeta\right)c_2 = 0. \qquad \text{(D.13.39)}$$

First normal mode

Assuming that $c_2 \neq 0$, we set the expression inside the large parentheses in (D.13.39) to zero and obtain the imaginary phase velocity of the first normal mode,

$$c = -\mathrm{i}\frac{1}{2\mu_1(1 + \lambda)}\left(\frac{\Delta\rho\, g}{k^2} + \gamma_0\right). \qquad \text{(D.13.40)}$$

The constant d_2 arises from (D.13.34) in terms of an arbitrary c_2. We note that the complex phase velocity is independent of the Marangoni number and is thus unaffected by the surfactant. The growth rate is always negative and this mode is stable.

Second normal mode

Assuming that $c_2 = 0$ and $d_2 \neq 0$, we set the coefficient multiplying d_2 in (D.13.34) to zero and obtain

$$\Lambda \equiv -\mathrm{i}\,\mathrm{Ma}\,\frac{\gamma_0}{2\mu_1}\,\frac{\zeta}{1 + D_s k\mathrm{i}\zeta} = 1 + \lambda. \qquad \text{(D.13.41)}$$

Rearranging, we derive the imaginary phase velocity of the second normal mode,

$$c = -\mathrm{i}\left(\mathrm{Ma}\,\frac{\gamma_0}{2\mu_1(1 + \lambda)} + k D_s\right). \qquad \text{(D.13.42)}$$

Since the growth rate $\varrho_I = k c_I$ is always negative, this mode is stable. Since $c_2 = 0$, the interface is flat and the perturbation flow is driven by in-plane interfacial motion.

Program files:

1. `ifsf0_s`
   Evaluates the analytical expressions for the phase velocity and growth rates.

2. `ifsf0_s_dr`
   Driver for `ifsf0_s`.

Input file:

1. `ifsf0_s.dat`
   Specification of input parameters.

Output file:

1. `ifsf0_s.out`
   Recording of the growth rate.

D.14   Directory layer0

This directory contains a code that performs the linear stability analysis of a liquid layer resting on a horizontal wall underneath a semi-infinite fluid, as illustrated in Figure D.14.1. The flow is assumed to occur under conditions of Stokes flow [280].

Base state

The layer is designated as fluid 1 and the upper fluid is designated as fluid 2. The viscosity of the layer fluid is μ_1 and the viscosity of the upper fluid is $\mu_2 = \lambda \mu_1$, where λ is the viscosity ratio. The origin of the y axis is set at the location of the unperturbed interface. For convenience, the x and y velocity components are denoted by $u = u_x$ and $v = u_y$.

In the unperturbed configuration, the interface is flat, the fluids are quiescent, and the pressure distribution assumes the hydrostatic profile

$$p_j^{(0)}(y) = -\rho_j g y + p_0, \tag{D.14.1}$$

where $j = 1$ or 2 for the film or upper fluid, ρ_j are the fluid densities, g is the acceleration of gravity, p_0 is an unspecified reference pressure, and the superscript (0) denotes the unperturbed base state.

Normal-mode analysis

A normal-mode perturbation displaces the interface to a position described by the real or imaginary part of the function

$$y = f(x,t) = \epsilon \eta(x,t), \tag{D.14.2}$$

where ϵ is a dimensionless coefficient whose magnitude is much less than unity,

$$\eta(x,t) = A \exp[ik(x - ct)] \tag{D.14.3}$$

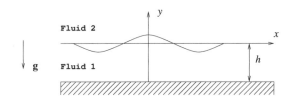

FIGURE D.14.1   Gravitational (Rayleigh–Taylor) instability and leveling of a liquid layer resting on a
horizontal wall underneath a semi-infinite fluid.

is the normal-mode wave form of the perturbation, A is the complex amplitude of the interfacial
wave, $k = 2\pi/L$ is the wave number, L is the wavelength, and c is the complex phase velocity.

Stream function

The stream function, ψ, is defined by the equations $u = \partial\psi/\partial y$ and $v = -\partial\psi/\partial x$. In the absence of
base flow, we set $\psi_j = \epsilon\psi_j^{(1)}$ and introduce the normal-mode form

$$\psi_j^{(1)}(x,y,t) = \phi_j(\hat{y})\,\exp[ik(x-ct)],\tag{D.14.4}$$

where $j = 1$ or 2 for the film or upper fluid, $\hat{y} = ky$, and the superscript (1) denotes the perturbation.
The perturbation velocity components are $u^{(1)} = \partial\psi^{(1)}/\partial y$ and $v^{(1)} = -\partial\psi^{(1)}/\partial x$.

The Reynolds number of the flow is so small that the motion of the fluid is governed by
the equations of Stokes flow. Requiring that the stream function satisfies the biharmonic equation,
$\nabla^4\psi_j^{(1)} = 0$, we obtain

$$\phi_1(\hat{y}) = a_1\,e^{\hat{y}} + b_1\,\hat{y}\,e^{\hat{y}} + c_1\,e^{-\hat{y}} + d_1\,\hat{y}\,e^{-\hat{y}}, \qquad \phi_2(\hat{y}) = c_2\,e^{-\hat{y}} + d_2\,\hat{y}\,e^{-\hat{y}},\tag{D.14.5}$$

where a_1, b_1, c_1, d_1, a_2, and b_2 are six complex coefficients. Note that the disturbance flow decays
exponentially with distance from the interface into the overlying fluid.

Pressure field

The linearized pressure field in the jth fluid is $p_j = p_j^{(0)} + \epsilon p_j^{(1)}$. The perturbation pressure field is
described by the normal-mode form

$$p_j^{(1)}(x,y,t) = \mu_j q_j(\hat{y})\,\exp[ik(x-ct)]\tag{D.14.6}$$

for $j = 1, 2$, where $q_j(\hat{y})$ are pressure eigenfunctions. Substituting this expression into the x component of the Stokes equation, we obtain

$$p_j^{(1)} = -\frac{i}{k}\left(\frac{\partial^3\psi_j^{(1)}}{\partial x^2\partial y} + \frac{\partial^3\psi_j^{(1)}}{\partial y^3}\right),\tag{D.14.7}$$

yielding

$$q_j(\hat{y}) = i\,k^2\Big(\frac{\mathrm{d}\phi_j}{\mathrm{d}\hat{y}} - \frac{\mathrm{d}^3\phi_j}{\mathrm{d}\hat{y}^3}\Big). \tag{D.14.8}$$

Wall conditions

The no-slip and no-penetration conditions at the wall, located at $y = -h$, require that

$$\phi_1(\hat{y} = -\hat{k}) = 0, \qquad \frac{\mathrm{d}\phi_1}{\mathrm{d}\hat{y}}(\hat{y} = -\hat{k}) = 0, \tag{D.14.9}$$

where $\hat{k} = kh$ is a dimensionless wave number. Making substitutions, we find that

$$a_1\,e^{-\hat{k}} - b_1\,\hat{k}\,e^{-\hat{k}} + c_1\,e^{\hat{y}} - d_1\,\hat{k}\,e^{\hat{k}} = 0,$$

$$a_1\,e^{-\hat{k}} + b_1\,(1 - \hat{y})\,e^{-\hat{k}} - c_1\,e^{\hat{y}} + d_1\,(1 + \hat{y})\,e^{\hat{k}} = 0. \tag{D.14.10}$$

Continuity of y velocity at the interface

In the linear approximation, continuity of y velocity at the interface requires that

$$\psi_1^{(1)}(x, y = 0, t) = \psi_2^{(1)}(x, y = 0, t), \tag{D.14.11}$$

yielding

$$a_1 + c_1 = c_2. \tag{D.14.12}$$

Kinematic compatibility

Kinematic compatibility requires that $\mathrm{D}[f(x,t) - y]/\mathrm{D}t = 0$ evaluated at the interface, where $\mathrm{D}/\mathrm{D}t$ is the material derivative. In the linear approximation,

$$\frac{\partial\eta}{\partial t} - v^{(1)}(y = 0) = 0. \tag{D.14.13}$$

Substituting the preceding expressions and choosing the upper fluid interfacial velocity, we find that $-ikc\,A + ikc_2 = 0$. Simplifying, we obtain

$$c_2 = c\,A. \tag{D.14.14}$$

Continuity of x velocity at the interface

In the linear approximation, continuity of x velocity at the interface requires that

$$\Big(\frac{\partial\psi_1^{(1)}}{\partial y}\Big)_{y=0} = \Big(\frac{\partial\psi_2^{(1)}}{\partial y}\Big)_{y=0}, \tag{D.14.15}$$

yielding

$$a_1 + b_1 - c_1 + d_1 = -c_2 + d_2. \tag{D.14.16}$$

Tangential component of the interface force balance

The linearized tangential component of the interfacial force balance requires that

$$\mu_1 \left(\frac{\partial u_1^{(1)}}{\partial y} + \frac{\partial v_1^{(1)}}{\partial x} \right)_{y=0} = \mu_2 \left(\frac{\partial u_2^{(1)}}{\partial y} + \frac{\partial v_2^{(1)}}{\partial x} \right)_{y=0} \tag{D.14.17}$$

or

$$\left(\frac{\partial^2 \psi_1^{(1)}}{\partial y^2} - \frac{\partial^2 \psi_1^{(1)}}{\partial x^2} \right)_{y=0} = \lambda \left(\frac{\partial^2 \psi_2^{(1)}}{\partial y^2} - \frac{\partial^2 \psi_2^{(1)}}{\partial x^2} \right)_{y=0}, \tag{D.14.18}$$

where $\lambda = \mu_2/\mu_1$ is the viscosity ratio. Substituting the preceding expressions and simplifying, we find that

$$a_1 + b_1 + c_1 - d_1 = \lambda (c_2 - d_2). \tag{D.14.19}$$

Normal component of the interface force balance

The linearized normal component of the interfacial force balance requires that

$$\left(-p_1^{(1)} + 2\mu_1 \frac{\partial v_1^{(1)}}{\partial y} \right)_{y=0} - \left(-p_2^{(1)} + 2\mu_2 \frac{\partial v_2^{(1)}}{\partial y} \right)_{y=0} + \Delta\rho\, g\, \eta = \gamma \frac{\partial^2 \eta}{\partial x^2}, \tag{D.14.20}$$

where γ is the surface tension and $\Delta\rho = \rho_1 - \rho_2$. Substituting the preceding expressions, we find that

$$-\mu_1 q_1 + \mu_2 q_2 - 2\mathrm{i} k^2 (\mu_1 - \mu_2) \frac{\mathrm{d}\phi_1}{\mathrm{d}\hat{y}} + \Delta\rho\, g\, A = -\gamma A\, k^2, \tag{D.14.21}$$

evaluated at $y = 0$. Substituting

$$q_1(y = 0) = -2\mathrm{i}\, k^2 (b_1 + d_1), \qquad q_2(y = 0) = -2\mathrm{i}\, k^2 d_2, \tag{D.14.22}$$

we obtain

$$2\mathrm{i}\, k^2 c\, \mu_1 \left[b_1 + d_1 - \lambda\, d_2 - (1 - \lambda)(-c_2 + d_2) \right] + \Delta\rho\, g\, c_2 = -\gamma\, k^2 c_2. \tag{D.14.23}$$

Simplifying and rearranging, we obtain

$$b_1 + d_1 + \Phi\, c_2 - d_2 = 0, \tag{D.14.24}$$

where

$$\Phi \equiv 1 - \lambda - \frac{\mathrm{i}}{2\mu_1 c} \left(\frac{\Delta\rho\, g}{k^2} + \gamma \right) \tag{D.14.25}$$

is a dimensionless group. Rearranging the definition of Φ, we derive an expression for the growth rate,

$$\varrho_I \equiv kc_I = \frac{1}{2\mu_1(\Phi - 1 + \lambda)k} \left(\frac{\Delta\rho\, g}{k^2} + \gamma \right). \tag{D.14.26}$$

Formulation of an eigenvalue problem

Compiling equations (D.14.10), (D.14.12) (D.14.16), (D.14.19), and (D.14.24), we obtain a linear homogeneous system $\mathbf{M} \cdot \mathbf{w} = \mathbf{0}$, where $\mathbf{w} = (a_1, b_1, c_1, d_1, c_2, d_2)$,

$$\mathbf{M} = \begin{bmatrix} q & (1-\hat{k})q & -1 & 1+\hat{k} & 0 & 0 \\ q & -q\hat{k} & 1 & -\hat{k} & 0 & 0 \\ 1 & 0 & 1 & 0 & -1 & 0 \\ 1 & 1 & -1 & 1 & 1 & -1 \\ 1 & 1 & 1 & -1 & -\lambda & \lambda \\ 0 & 1 & 0 & 1 & \Phi & -1 \end{bmatrix}, \tag{D.14.27}$$

and $q \equiv \exp(-2\hat{k})$. Setting the determinant of the matrix $\mathbf{M}$ to zero provides us with a secular equation for the computation of Φ. Once Φ is available, the growth rate follows from (D.14.26). After some algebra, we derive the dimensionless growth rate

$$\varrho_I = -\frac{1}{2\mu_1(1+\lambda)k} \left(\frac{\Delta\rho\, g}{k^2} + \gamma \right) \mathcal{F}(\hat{k}, \lambda), \tag{D.14.28}$$

where

$$\mathcal{F}(\hat{k}, \lambda) = (1+\lambda) \frac{\frac{1}{2}\sinh(2\hat{k}) - \hat{k} + \lambda\,(\sinh^2\hat{k} - \hat{k}^2)}{(1-\lambda^2)\,\hat{k}^2 + (\cosh\hat{k} + \lambda\sinh\hat{k})^2}. \tag{D.14.29}$$

As $\hat{k} \to \infty$, the function $\mathcal{F}(\hat{k}, \lambda)$ tends to unity, yielding the growth rate of perturbations at the interface between two semi-infinite fluids. In the limit $\lambda \to 0$, the function $\mathcal{F}(\hat{k}, \lambda)$ reduces the second fraction on the right-hand side of (9.11.35) for a liquid film underneath a constant-pressure ambient gas.

When the fluids are unstably stratified, $\rho_2 > \rho_1$ and $\Delta\rho < 0$, the first term inside the parentheses on the right-hand side of (D.14.28) is responsible for the Rayleigh–Taylor instability. The second term inside the parentheses expresses the stabilizing influence of the surface tension.

Program files:

1. `layer0`
   Computes the growth rate by two methods.

2. `layer0_dr`
   Driver for `layer0`

3. `crout`
   Crout decomposition for computing the determinant.

Input file:

1. `layer0.dat`
   Specification of input parameters.

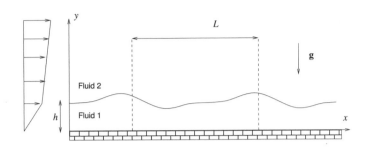

FIGURE D.15.1 Instability of a liquid layer coated on a horizontal wall underneath a semi-infinite fluid undergoing simple shear flow. The interface is occupied by an insoluble surfactant.

Output file:

1. `layer0.out`
   Recording of the growth rate.

D.15 Directory layersf0_s

This directory contains a code that performs the linear stability analysis of a liquid layer resting on a horizontal wall underneath a semi-infinite fluid undergoing simple shear flow, as shown in Figure D.15.1. The interface is occupied by an insoluble surfactant and the motion of the fluid occurs under conditions of Stokes flow [325].

Base state

The liquid layer is designated as fluid 1 and the overlying semi-infinite fluid is designated as fluid 2. The viscosity of the layer is μ_1 and the viscosity of the upper fluid is $\mu_2 = \lambda\mu_1$, where λ is the viscosity ratio. The origin of the y axis is set at the wall. For convenience, the x and y velocity components are denoted by $u = u_x$ and $v = u_y$.

In the unperturbed configuration, the interface is flat and the fluids undergo unidirectional flow parallel to the wall. The velocity profiles inside the layer and upper fluid are given by

$$u_1^{(0)} = \lambda\xi y, \qquad u_2^{(0)} = \xi\left[y + (\lambda - 1)h\right], \tag{D.15.1}$$

where ξ is the shear rate in the upper fluid, h is the layer thickness, and the superscript (0) denotes the unperturbed unidirectional flow. These expressions ensure continuity of velocity and shear stress at the interface for any viscosity ratio. The corresponding pressure distributions are given by

$$p_j^{(0)}(y) = -\rho_j g\left(y - h\right) + p_0 \tag{D.15.2}$$

for $j = 1, 2$, where p_0 is an inconsequential interfacial pressure.

Normal-mode analysis

A normal-mode perturbation displaces the interface to a position given by the real or imaginary part of the function

$$y = f(x,t) = h + \epsilon \eta(x,t), \tag{D.15.3}$$

where ϵ is a dimensionless coefficient whose magnitude is much less than unity,

$$\eta(x,t) = A \exp[ik(x - ct)] \tag{D.15.4}$$

is the normal-mode wave form of the perturbation, A is the complex amplitude of the interfacial wave, $k = 2\pi/L$ is the wave number, L is the wavelength, and c is the complex phase velocity.

Stream function

The stream function, ψ, is defined by the equations $u = \partial \psi / \partial y$ and $v = -\partial \psi / \partial x$. In the linear analysis, we set $\psi_j = \psi^{(0)} + \epsilon \psi_j^{(1)}$ and introduce the normal-mode form

$$\psi_j^{(1)}(x,y,t) = \phi_j(\hat{y}) \exp[ik(x - ct)], \tag{D.15.5}$$

where the superscript (1) denotes the perturbation, $\hat{y} \equiv k(y - h)$, and $j = 1, 2$ for the liquid film or semi-infinite fluid. The perturbation velocity components are $u^{(1)} = \partial \psi^{(1)}/\partial y$ and $v^{(1)} = -\partial \psi^{(1)}/\partial x$.

The Reynolds number of the flow is so small that the motion of the fluid is governed by the equations of Stokes flow. Requiring that the stream function satisfies the biharmonic equation, $\nabla^4 \psi_j^{(1)} = 0$, we obtain

$$\phi_j(\hat{y}) = a_{1j} \cosh \hat{y} + a_{2j} \, \hat{y} \cosh \hat{y} + a_{3j} \, \sinh \hat{y} + a_{4j} \, \hat{y} \sinh \hat{y}, \tag{D.15.6}$$

where a_{ij} are eight complex coefficients for $i = 1$–4 and $j = 1, 2$. To ensure that the perturbation velocity decays in the upper fluid far from the wall, we set

$$a_{32} = -a_{12}, \qquad a_{42} = -a_{22}. \tag{D.15.7}$$

Pressure field

The linearized pressure field in the jth fluid is $p_j = p_j^{(0)} + \epsilon p_j^{(1)}$. The disturbance pressure field takes the normal-mode form

$$p_j^{(1)}(x,y,t) = \mu_j q_j(\hat{y}) \exp[ik(x - ct)] \tag{D.15.8}$$

for $j = 1, 2$, where $q_j(\hat{y})$ are pressure eigenfunctions. Substituting this expression into the x component of the Stokes equation, we obtain

$$p_j^{(1)} = -\frac{i}{k} \left(\frac{\partial^3 \psi_j^{(1)}}{\partial x^2 \partial y} + \frac{\partial^3 \psi_j^{(1)}}{\partial y^3} \right), \tag{D.15.9}$$

yielding

$$q_j(y) = \mathrm{i}\, k^2 \left(\frac{\mathrm{d}\phi_j}{\mathrm{d}\hat{y}} - \frac{\mathrm{d}^3\phi_j}{\mathrm{d}\hat{y}^3} \right). \tag{D.15.10}$$

Continuity of the y velocity component at the interface

In the linear approximation, continuity of the y velocity component at the interface requires that

$$\psi_1^{(1)}(x, y = h, t) = \psi_2^{(1)}(x, y = h, t), \tag{D.15.11}$$

yielding $a_{11} = a_{12}$. Combining this equation with (D.15.7), we obtain $a_{32} = -a_{12} = -a_{11}$.

Kinematic compatibility

Kinematic compatibility requires that $\mathrm{D}[f(x,t) - y]/\mathrm{D}t = 0$, where $\mathrm{D}/\mathrm{D}t$ is the material derivative. In the linear approximation,

$$\frac{\partial\eta}{\partial t} + u_I \frac{\partial\eta}{\partial x} - v_j^{(1)}(y = h) = 0 \tag{D.15.12}$$

for $j = 1, 2$, where $u_I = \lambda\xi h$ is the unperturbed interfacial velocity. Substituting the preceding expressions, we obtain $-\mathrm{i}kcA + u_I\mathrm{i}kA + \mathrm{i}ka_{11} = 0$, yielding $a_{11} = (c - u_I)A$ and

$$A = \zeta\, a_{11}, \tag{D.15.13}$$

where $\zeta \equiv 1/(c - u_I)$ is the inverse of the complex phase velocity shifted by the real interfacial velocity.

Continuity of the x velocity component at the interface

In the linear approximation, continuity of the x velocity component at the interface requires that

$$\lambda\xi\eta(x,t) + \left(\frac{\partial\psi_1^{(1)}}{\partial y} \right)_{y=h} = \xi\eta(x,t) + \left(\frac{\partial\psi_2^{(1)}}{\partial y} \right)_{y=h}, \tag{D.15.14}$$

yielding

$$\lambda\xi A + k \left(\frac{\mathrm{d}\phi_1}{\mathrm{d}\hat{y}} \right)_{\hat{y}=\hat{k}} = \xi A + k \left(\frac{\mathrm{d}\phi_2}{\mathrm{d}\hat{y}} \right)_{\hat{y}=\hat{k}}, \tag{D.15.15}$$

where $\hat{k} = kh$. Rearranging and using (D.15.13), we obtain

$$-\xi\zeta\,(1 - \lambda)\, a_{11} + k \left(\frac{\mathrm{d}\phi_1}{\mathrm{d}\hat{y}} \right)_{\hat{y}=h} - k \left(\frac{\mathrm{d}\phi_2}{\mathrm{d}\hat{y}} \right)_{\hat{y}=h} = 0, \tag{D.15.16}$$

yielding

$$-\Xi\,\zeta\,(1 - \lambda)\, a_{11} + a_{21} + a_{31} - a_{22} - a_{32} = 0, \tag{D.15.17}$$

where $\Xi = \xi/k$. Substituting $a_{32} = -a_{11}$, we obtain

$$[1 - \Xi\zeta\,(1 - \lambda)]\, a_{11} + a_{21} + a_{31} - a_{22} = 0. \tag{D.15.18}$$

Wall conditions

The no-slip and no-penetration conditions at the horizontal wall require that

$$\phi_1(\hat{y} = -\hat{k}) = 0, \qquad \phi_1'(\hat{y} = -\hat{k}) = 0, \tag{D.15.19}$$

where a prime denotes a derivative with respect to $\hat{y}$. Making substitutions, we obtain

$$\begin{bmatrix} \cosh \hat{k} & -\hat{k}\cosh\hat{k} & -\sinh\hat{k} & \hat{k}\sinh\hat{k} \\ -\sinh\hat{k} & \cosh\hat{k}+\hat{k}\sinh\hat{k} & \cosh\hat{k} & -\sinh\hat{k}-\hat{k}\cosh\hat{k} \end{bmatrix} \cdot \mathbf{w}_1 = \mathbf{0}, \tag{D.15.20}$$

where $\mathbf{w}_1 = \big(a_{11}, a_{21}, a_{31}, a_{41}\big)$ is a collection of unknown coefficients.

Surfactant concentration and surface tension

The distribution of the interfacial surfactant concentration, Γ, and surface tension, γ, are described by the companion functions

$$\Gamma(x,t) = \Gamma_0 + \epsilon \Gamma_1 \exp[ik(x - ct)], \qquad \gamma(x,t) = \gamma_0 + \epsilon \gamma_1 \exp[ik(x - ct)], \tag{D.15.21}$$

where Γ_0, γ_0 are uniform values corresponding to the flat interface, and Γ_1, γ_1 are complex amplitudes. Since the perturbations are small, we can write

$$\frac{\gamma_1}{\gamma_0} = -\mathrm{Ma}\,\frac{\Gamma_1}{\Gamma_0}, \tag{D.15.22}$$

where Ma is the Marangoni number.

Surfactant transport

Surfactant transport over the interface is governed by the equation

$$\frac{D\Gamma}{Dt} + \Gamma \mathbf{t} \cdot \frac{\partial \mathbf{u}}{\partial l} = \frac{\partial}{\partial l}\Big(D_s \frac{\partial \Gamma}{\partial l}\Big), \tag{D.15.23}$$

where D/Dt is the material derivative, l is the arc length along the interface measured in the direction of the unit tangent vector, $\mathbf{t}$, and D_s is the surface surfactant diffusivity. Regarding Γ as a function of x and t, we obtain

$$\frac{\partial \Gamma}{\partial t} + u_x \frac{\partial \Gamma}{\partial x} + \Gamma \mathbf{t} \cdot \Big(\frac{\partial \mathbf{u}}{\partial x} + \frac{\partial \mathbf{u}}{\partial y}\frac{\partial \Gamma}{\partial x}\Big)\frac{\partial x}{\partial l} = \frac{\partial}{\partial l}\Big(D_s \frac{\partial \Gamma}{\partial l}\Big). \tag{D.15.24}$$

The unit tangential vector can be linearized as

$$\mathbf{t} \simeq \mathbf{e}_x + \epsilon \frac{\partial \eta}{\partial x}\mathbf{e}_y, \tag{D.15.25}$$

where $\mathbf{e}_x$ and $\mathbf{e}_y$ are unit vectors along the x and y axes. To leading order, $\partial x / \partial l = 1$. The emerging linearized form of the surfactant transport equation is

$$\frac{\partial \Gamma^{(1)}}{\partial t} + u^{(0)}\frac{\partial \Gamma^{(1)}}{\partial x} + \Gamma_0\Big(\frac{\partial u^{(1)}}{\partial x} + \frac{\partial u^{(0)}}{\partial y}\frac{\partial \eta}{\partial x}\Big) = D_s \frac{\partial^2 \Gamma^{(1)}}{\partial x^2}, \tag{D.15.26}$$

where all terms are evaluated at the unperturbed position, $y = h$, and the velocity on the left-hand side is evaluated on either side of the interface, and $\Gamma^{(1)}(x, t) = \Gamma_1 \exp[\mathrm{i}k(x - ct)]$. Substituting the preceding expressions for the upper fluid velocity, we obtain

$$\frac{\Gamma_1}{\Gamma_0} = \frac{a_{22} + a_{32} + \Xi\,\zeta a_{11}}{1 + D_s k\,\mathrm{i}\zeta}\,k\zeta = \frac{(\Xi\,\zeta - 1)\,a_{11} + a_{22}}{1 + D_s k\,\mathrm{i}\zeta}\,k\zeta, \tag{D.15.27}$$

where we recall that $\Xi = \xi/k$.

Tangential component of the interfacial force balance

The linearized tangential component of the interfacial force balance requires that

$$\mu_1 \Big(\frac{\partial u_1^{(1)}}{\partial y} + \frac{\partial v_1^{(1)}}{\partial x}\Big)_{y=h} = \mu_2 \Big(\frac{\partial u_2^{(1)}}{\partial y} + \frac{\partial v_2^{(1)}}{\partial x}\Big)_{y=h} + \frac{\partial \gamma^{(1)}}{\partial x}, \tag{D.15.28}$$

where $\gamma^{(1)}(x, t) = \gamma_1 \exp[\mathrm{i}k(x - ct)]$. Expressing the velocity in terms of the stream function and rearranging, we obtain

$$\Big(\frac{\partial^2 \psi_1^{(1)}}{\partial y^2} - \frac{\partial^2 \psi_1^{(1)}}{\partial x^2}\Big)_{y=h} = \lambda\,\Big(\frac{\partial^2 \psi_2^{(1)}}{\partial y^2} - \frac{\partial^2 \psi_2^{(1)}}{\partial x^2}\Big)_{y=h} + \frac{1}{\mu_1}\,\frac{\partial \gamma^{(1)}}{\partial x}. \tag{D.15.29}$$

Substituting the normal-mode forms, we find that

$$\Big(\frac{\mathrm{d}^2 \phi_1}{\mathrm{d}\hat{y}^2} + \phi_1\Big)_{\hat{y}=0} = \lambda\,\Big(\frac{\mathrm{d}^2 \phi_2}{\mathrm{d}\hat{y}^2} + \phi_2\Big)_{\hat{y}=0} + \mathrm{i}\,\frac{\gamma_1}{\mu_1 k}. \tag{D.15.30}$$

Substituting the preceding expressions for ϕ_j and simplifying, we obtain

$$a_{11} + a_{41} - \lambda\,(a_{12} + a_{42}) = \mathrm{i}\,\frac{\gamma_1}{2\mu_1 k}, \tag{D.15.31}$$

yielding

$$(1 - \lambda)\,a_{11} + a_{41} + \lambda\,a_{22} - \Lambda\,[\,(\Xi\,\zeta - 1)a_{11} + a_{22}\,] = 0, \tag{D.15.32}$$

where

$$\Lambda \equiv -\mathrm{i}\,\mathrm{Ma}\,\frac{\gamma_0}{2\mu_1}\,\frac{\zeta}{1 + D_s k\,\mathrm{i}\zeta} \tag{D.15.33}$$

is a dimensionless group. In the absence of surfactant, $\Lambda = 0$.

Normal component of the interfacial force balance

The linearized normal component of the interfacial force balance requires that

$$\Big(-p_1^{(1)} + 2\mu_1\,\frac{\partial v_1^{(1)}}{\partial y}\Big)_{y=h} - \Big(-p_2^{(1)} + 2\mu_2\,\frac{\partial v_2^{(1)}}{\partial y}\Big)_{y=h} + \Delta\rho\,g\,\eta = \gamma_0\,\frac{\partial^2 \eta}{\partial x^2}, \tag{D.15.34}$$

where $\Delta\rho = \rho_1 - \rho_2$. Substituting the preceding expressions, we obtain

$$-\mu_1 q_1 + \mu_2 q_2 - 2ik^2\mu_1\frac{d\phi_1}{d\hat{y}} + 2ik^2\mu_2\frac{d\phi_2}{d\hat{y}} + \Delta\rho\, g\, A = -\gamma_0\, Ak^2, \qquad (D.15.35)$$

where all terms are evaluated at $\hat{y} = 0$. Substituting further $q_j(0) = -2i\,k^2 a_{2j}$, we obtain

$$2ik^2(\mu_1 a_{21} - \mu_2 a_{22}) - 2ik^2\mu_1(a_{21} + a_{31}) + 2ik^2\mu_2(a_{22} + a_{32}) + \Delta\rho\, g\, A = -\gamma_0 Ak^2, \quad (D.15.36)$$

which simplifies into

$$-2ik^2\mu_1 a_{31} + 2ik^2\mu_2 a_{32} + (\Delta\rho\, g + \gamma_0 k^2)\, A = 0. \qquad (D.15.37)$$

Rearranging, setting $a_{32} = -a_{11}$, and using (D.7.16), we find that

$$(\lambda + i\Pi\zeta)\, a_{11} + a_{31} = 0, \qquad (D.15.38)$$

where

$$\Pi \equiv \frac{1}{2\mu_1}\left(\frac{\Delta\rho\, g}{k^2} + \gamma_0\right) = \frac{\gamma_0}{2\mu_1}(\mathrm{Bo} + 1) \qquad (D.15.39)$$

is a property group with dimensions of velocity, and $\mathrm{Bo} \equiv \Delta\rho\, g/(\gamma_0 k^2)$ is a Bond number,

Formulation of an eigenvalue problem

Collecting equations (D.15.20), (D.15.18), (D.15.32), and (D.15.38), we obtain a linear homogeneous system, $\mathbf{M}\cdot\mathbf{w} = \mathbf{0}$, where $\mathbf{w} = (a_{11}, a_{21}, a_{31}, a_{41}, a_{22})$ and

$$\mathbf{M} = \begin{bmatrix} \cosh\hat{k} & -\hat{k}\cosh\hat{k} & -\sinh\hat{k} & \hat{k}\sinh\hat{k} & 0 \\ -\sinh\hat{k} & \cosh\hat{k} + \hat{k}\sinh\hat{k} & \cosh\hat{k} & -\sinh\hat{k} - \hat{k}\cosh\hat{k} & 0 \\ 1 - \Xi\zeta(1-\lambda) & 1 & 1 & 0 & -1 \\ 1 - \lambda - \Lambda(\Xi\zeta - 1) & 0 & 0 & 1 & \lambda - \Lambda \\ \lambda + i\Pi\zeta & 0 & 0 & 1 & 0 \end{bmatrix}. (D.15.40)$$

Using the last equation to eliminate a_{31} in favor of a_{11}, we derive a smaller system, $\mathbf{M}'\cdot\mathbf{w}' = \mathbf{0}$, where $\mathbf{w}' = (a_{11}, a_{21}, a_{41}, a_{22})$ and

$$\mathbf{M}' = \begin{bmatrix} \cosh\hat{k} + (\lambda + i\Pi\zeta)\sinh\hat{k} & -\hat{k}\cosh\hat{k} & \hat{k}\sinh\hat{k} & 0 \\ -\sinh\hat{k} - (\lambda + i\Pi\zeta)\cosh\hat{k} & \cosh\hat{k} + \hat{k}\sinh\hat{k} & -\sinh\hat{k} - \hat{k}\cosh\hat{k} & 0 \\ 1 - \lambda - \zeta[\Xi(1-\lambda) + i\Pi] & 1 & 0 & -1 \\ 1 - \lambda - \Lambda(\Xi\zeta - 1) & 0 & 1 & \lambda - \Lambda \end{bmatrix}. (D.15.41)$$

Setting the determinant of the matrix $\mathbf{M}'$ to zero provides us with a quadratic equation for two possible growth rates,

$$P_2(\zeta) = (1 + D_s k\, i\zeta)\det(\mathbf{M}') = a\zeta^2 + b\zeta + c = 0. \qquad (D.15.42)$$

The factor $1 + D_s k \, \mathrm{i} \zeta$ on the left-hand side involving the surface surfactant diffusivity arises from the definition of the dimensionless parameter Λ. In the numerical method, the binomial coefficients are computed by an exact finite-difference based on the equations

$$P_2(1) = a + b + c, \qquad P_2(-1) = a - b + c, \qquad P_2(0) = c. \qquad (D.15.43)$$

The roots are found analytically using the quadratic formula.

As the dimensionless wave number kh increases, the effect of the wall becomes decreasingly important, yielding shear flow past an interface separating two semi-infinite fluids. Our analysis in directory `ifsf0_s` has revealed two normal modes with complex phase velocities given in (D.13.42) and (D.13.42),

$$c = u_I - \mathrm{i} \, \frac{1}{2\mu_1(1+\lambda)} \Big(\frac{\Delta \rho \, g}{k^2} + \gamma \Big), \qquad c = u_I - \mathrm{i} \Big(\mathrm{Ma} \, \frac{\gamma_0}{2\mu_1(1+\lambda)} + kD_s \Big). \qquad (D.15.44)$$

Both modes are stable with negative growth rates and phase velocity equal to the undisturbed interfacial velocity.

Program files:

1. `layersf0_s`
   Evaluates the complex phase velocity.

2. `layersf0_s_dr`
   Driver for the subroutine `layersf0_s`

3. `det_33c`
   Determinant of a 3×3 complex matrix.

4. `det_44c`
   Determinant of a 4×4 complex matrix.

5. `quadc`
   Computes the roots of a quadratic equation with complex coefficients.

Input file:

1. `layersf0_s.dat`
   Specification of input parameters.

Output file:

1. `layersf0_s.out`
   Recording of the growth rate.

D.16 Directory orr

This directory contains a code that computes the complex phase velocity of normal-mode perturbations in a viscous unidirectional shear flow using a finite-difference method. The mathematical formulation and numerical method are discussed in Section 9.7.1

D.17   Directory prony

This directory contains programs that decompose a time series into a linear superposition of normal modes. The mathematical formulation and numerical method are discussed in Section 8.9 of Reference [317].

D.18   Directory sf1

This directory contains a code that computes the complex phase velocity of normal-mode perturbations in an inviscid unidirectional shear flow using a finite-difference method. The mathematical formulation and numerical method are discussed in Section 9.7.1

D.19   Directory thread0

This directory contains a code that performs the linear stability analysis of a viscous thread suspended in an infinite ambient viscous fluid. The instability occurs under conditions of Stokes flow. The formulation of the linear stability problem is discussed in Section 9.13.2.

Program files:

1. thread0
   Evaluates the growth rate from analytical expression derived in Section 9.13.2.

2. bess_I01K01
   Computes the Bessel functions.

3. thread0_dr
   Driver calling the main function.

Input files:

1. thread0.dat
   Specification of input parameters.

Output files:

1. thread0.out
   Growth rates.

D.20   Directory thread1

This directory contains a code that performs the linear stability analysis of an inviscid liquid column suspended in an infinite constant-pressure ambient gas. The formulation of the linear stability problem is discussed in Section 9.13.1.

Program files:

1. thread1
   Evaluates the phase velocity and growth rate from analytical expressions.

2. bess_I01K01
   Computes Bessel functions.

3. thread1_dr
   Driver calling the main function.

D.21   Directory vl

This directory contains a code that computes the phase velocity and growth rate of periodic disturbances on a vortex layer from analytical expressions derived in Section 9.6.

D.22   Directory vs

This directory contains a code that produces the phase velocity and growth rate of periodic disturbances on a vortex sheet separating two streams merging with different velocities from analytical expressions derived in Section 9.10.

References

[1] ABDALLAH, S. (1987) Numerical solutions for the incompressible Navier–Stokes equations in primitive variables using a non-staggered grid II. *J. Comp. Phys.* **70**, 193–202.

[2] ABRAMOWITZ, M. & STEGUN, I. A. (1972) *Handbook of Mathematical Functions.* Dover.

[3] ADAMSON, A. W. (1990) *Physical Chemistry of Surfaces.* Fifth Edition. Wiley.

[4] AHLFORS, L. A. (1979) *Complex Analysis.* McGraw–Hill.

[5] AMES, W. F. (1977) *Numerical Methods for Partial Differential Equations.* Academic.

[6] ANDERSON, C. R. (1986) A method of local corrections for computing the velocity field due to a distribution of vortex blobs. *J. Comp. Phys.* **62**, 111–123.

[7] ANDERSSON, C. R. (2009) Modified Schwarz–Christoffel mappings using approximate curve factors. *J. Comp. Appl. Math.* **233**, 1117–1127.

[8] ANDERSON, C. R. (1989) Vorticity boundary conditions and boundary vorticity generation for two-dimensional viscous incompressible flows. *J. Comp. Phys.* **80**, 72–97.

[9] ANDERSON, P. A., TANNEHILL, J. C. & PLETCHER, R. H. (1984) *Computational Fluid Dynamics and Heat Transfer.* Taylor & Francis.

[10] ANSHUS, B. E. & GOREN, S. L. (1966) A method of getting approximate solutions to the Orr-Sommerfeld equation for flow on a vertical wall. *AIChE J.* **12**, 1004–1008.

[11] ARAGBESOLA, Y. A. S. & BURLEY, D. M. (1977) The vector and scalar potential method for the numerical solution of two- and three-dimensional Navier–Stokes equations. *J. Comp. Phys.* **24**, 398–415.

[12] AREF, H. (1983) Integrable, chaotic, and turbulent vortex motion in two-dimensional flows. *Annu. Rev. Fluid Mech.* **15**, 345–389.

[13] AREF, H., NEWTON, P. K., STREMLER, M. A., TOKIEDA, T. & VAINCHTEIN, D. L. (1983) Vortex Crystals. *Arch. Appl. Mech.* **39**, 2–81.

[14] ARIS, R. (1962) *Vectors, Tensors, and the Basic Equations of Fluid Mechanics.* Prentice-Hall.

[15] ATKINSON, K. E. (1973) Iterative variants of the Nÿstrom method for the numerical solution of integral equations. *Numer. Math.* **22**, 17–31.

[16] ATKINSON, K. E. (1976) *A Survey of Numerical Methods for the Solution of Fredholm Integral Equations of the Second Kind.* SIAM.

[17] ATKINSON, K. E. (1980) The numerical solution of Laplace's equation in three dimensions - II. In *Numerical Treatment of Integral Equations*, J. Albrecht & L. Collatz (Eds.), 1–23, Birkhäuser, Basel, Switzerland.

[18] ATKINSON, K. E. (1989) *An Introduction to Numerical Analysis.* Wiley.

[19] ATKINSON, K. E. (1990) A Survey of Boundary Integral Methods for the Numerical Solution of Laplace's Equation in Three Dimensions. In *Numerical Solution of Integral Methods*, Goldberg, M. A. (Ed.), Plenum Press.

[20] AZIZ, K. & HELLUMS, J. D. (1967) Numerical solution of the three-dimensional equations of motion for laminar natural convection. *Phys. Fluids* **10**, 314–324.

[21] BABU, V. & KORPELA, S. (1994) Numerical solution of the incompressible, three-dimensional Navier–Stokes equations. *Comput. & Fluids* **23**, 675–691.

[22] BAKER, G. R. (1983) Generalized vortex methods for free-surface flows. In: *Waves in Fluid Interfaces.* Academic Press.

[23] BATCHELOR, G. K. (1954) The skin friction on infinite cylinders moving parallel to their length. *Quart. J. Mech. Appl. Math.* **7**, 179–192.

[24] BATCHELOR, G. K. (1967) *An Introduction to Fluid Dynamics.* Cambridge University Press.

[25] BATCHELOR, G. K. (1970) Slender-body theory for particles of arbitrary cross-section in Stokes flow. *J. Fluid Mech.* **44**, 419–440 .

[26] BAZANT, M. Z. (2004) Conformal mapping of some non-harmonic functions in transport theory. *Proc. R. Soc. Lond.* A **460**, 1433–1452.

[27] BELL, J., COLELLA, P. & GLAZ, H. (1989) A second-order projection method for the incompressible Navier–Stokes equations. *J. Comp. Phys.* **85**, 257–283.

[28] BENJAMIN, T. B. (1957) Wave formation in laminar flow down an inclined plane. *J. Fluid Mech.* **2**, 554–574.

[29] BENNEY, D. J. (1966) Long waves in liquid films. *J. Math. Phys.* **45**, 150–155.

[30] BENTON, E. R. & PLATZMAN, G. W. (1972) A table of solutions of the one-dimensional Burgers equation. *Quart. Appl. Math.* **30**, 195–212.

[31] BERKER, R. (1963) Intégration des équations du mouvement d'un fluide visqueux incompressible. In *Handbuch der Physik*, Flugge, S. (Ed.), **8**(2), 1–384, Springer, Berlin.

[32] BERTSEKAS, D. P. & TSITSIKLIS, J. N. (1989) *Parallel and Distributed Computation.* Prentice-Hall.

[33] BETCHOV, R. (1965) On the curvature and torsion of an isolated vortex filament. *J. Fluid Mech.* **22**, 471–479.

[34] BETCHOV, R. & CRIMINALE, W. O. (1967) *Stability of Parallel Flows*. Academic Press.

[35] BETCHOV, R. & SZEWCZYK, A. (1963) Stability of a shear layer between parallel streams. *Phys. Fluids* **6**, 1391–1396.

[36] BIRD, R. B., STEWART, W. E. & LIGHTFOOT, E. N. (2006) *Transport Phenomena*. Second Edition. Wiley.

[37] BLAKE, J. R. (1971) A note for the image system for a Stokeslet in a no-slip boundary. *Proc. Camb. Phil. Soc.* **70**, 303–310.

[38] BLAKE, J. R. & GIBSON, D. C. (1987) Cavitation bubbles near boundaries. *Annu. Rev. Fluid Mech.* **19**, 99–123.

[39] BLASIUS, H. (1908) Grenzschichten in Flüssigkeiten mit kleiner reibung. Mathematik und Physik 56, 1-37. *Z. Math. Phys.* **56**, 1–37.

[40] BLASIUS, H. (1910) Funktionentheoretische methoden in der hydrodynakik. *Z. Math. Phys.* **58**, 90–110.

[41] BLOTTNER, F. G. (1970) Finite difference methods of solution of the boundary layer equations. *AIAA J.* **8**, 193–205.

[42] BLYTH, M. G. & POZRIKIDIS, C. (2005) Stagnation-point flow against a liquid film on a plane wall. *Acta Mech.* **180**, 203–219.

[43] BLYTH, M. G. & POZRIKIDIS, C. (2008) Particle encapsulation due to thread breakup in Stokes flow. *J. Fluid Mech.* **617**, 141–166.

[44] BOAS, M. L. (2005) *Mathematical Methods in the Physical Sciences*. Third Edition. Wiley.

[45] BODEWIG, E. (1959) *Matrix Calculus*. North-Holland.

[46] BOGY, D. B. (1979) Drop formation in a circular liquid jet. *Annu. Rev. Fluid Mech.* **11**, 207–228.

[47] BURSTEIN, S. Z. (1967) Finite-difference calculations for hydrodynamic flows containing discontinuities. *J. Comp. Phys.* **2**, 198–222.

[48] GUPTA N. R. & BORHAN, A. (2003) Capsule motion and deformation in tube and channel flow. In *Modeling and Simulation of Capsules and Biological Cells*, Pozrikidis C. (Ed.), Chapman & Hall /CRC.

[49] BORIS, J. P. & BOOK, D. L. (1973) Flux-corrected transport. I. SHASTA, A fluid transport algorithm that works. *J. Comp. Phys.* **11**, 38–69.

[50] BORIS, J. P., BOOK, D. L. & HAIN, K. (1975) Flux-corrected transport II: Generalizations of the method. *J. Comp. Phys.* **18**, 248–283.

[51] BORIS, J. P. & BOOK, D. L. (1976) Flux-corrected transport. III. Minimal-error FCT algorithms. *J. Comp. Phys.* **20**, 397–431.

[52] BOUCHER, E. A. (1980) Capillary phenomena: properties of systems with fluid/fluid interfaces. *Rep. Prog. Phys.* **43**, 497–545.

[53] BRADY, M. & POZRIKIDIS, C. (1993) Diffusive transport across irregular and fractal walls. *Proc. R. Soc. Lond. A* **442**, 571–583.

[54] BRENNER, H. (1963) The Stokes resistance of an arbitrary particle. *Chem. Eng. Sci.* **18**, 1–25.

[55] BRENNER, H. (1964) The Stokes resistance of an arbitrary particle–II: An extension. *Chem. Eng. Sci.* **19**, 599–629.

[56] BRENNER, H. (1964) The Stokes resistance of an arbitrary particle–IV: Arbitrary fields of flow. *Chem. Eng. Sci.* **19**, 703–727.

[57] BRILEY, R. W. (1974) Numerical method for predicting three-dimensional steady viscous flow in ducts. *J. Comp. Phys.* **14**, 8–28.

[58] BROWN, R. A. (1979) Finite-element methods for the calculation of capillary surfaces. *J. Comp. Phys.* **33**, 217–235.

[59] BROWN, D. L., CORTEZ, R. & MINION, M. L. (2001) Accurate projection methods for the incompressible Navier–Stokes equations. *J. Comp. Phys.* **168**, 464–499.

[60] BUCKMASTER, J. (1973) Viscous-gravity spreading of an oil slick. *J. Fluid Mech.* **59**, 481–491.

[61] BURGERS, J. M. (1938) On the motion of small particles of elongated form, suspended in a viscous liquid. *Second Report on Viscosity and Plasticity.* Kon. Ned. Akad. Wet. Verhand. (Eerste Sectie) **16**: 113.

[62] BURGERS, J. M. (1948) A mathematical model illustrating the theory of turbulence. *Adv. Appl. Mech.* **1**, 171–199.

[63] BUTLER, S. F. J (1953) A note on Stokes's stream function for motion with a spherical boundary, *Proc. Camb. Phil. Soc.* **49**, 169–174.

[64] CAFLISCH, R. E., LI, X. & SHELLEY, M. J. (1993) The collapse of an axi-symmetric, swirling vortex sheet. *Nonlinearity* **6**, 843–867.

[65] CARRIER, G. F., KROOK, M. & PEARSON, C. E. (1983) *Functions of a Complex Variable: Theory and Technique.* Hod Books.

[66] CARSLAW, H. S. & JAEGER, J. C. (1959) *Conduction of Heat in Solids.* Second Edition. Oxford University Press.

[67] CASE, K. M. (1960) Stability of inviscid plane Couette flow. *Phys. Fluids* **3**, 143–148.

[68] CASE, K. M. (1960) Stability of an idealized atmosphere I. Discussion of results. *Phys. Fluids* **3**, 149–154.

[69] CEBECI, T. (1982) *Numerical and Physical Aspects of Aerodynamic Flows.* Springer-Verlag.

[70] CEBECI, T. & BRADSHAW, P. (1977) *Momentum Transfer in Boundary Layers.* Hemisphere.

[71] CEBECI, T. & BRADSHAW, P. (1984) *Physical and Computational Aspects of Convective Heat Transfer* Springer–Verlag.

[72] CERCIGNANI, C. (2000) *Rarefied Gas Dynamics.* Cambridge University Press.

[73] CHANDRASEKHAR, S. (1961) *Hydrodynamic and Hydromagnetic Stability.* Dover.

[74] CHANG, H.-C. (1994) Wave evolution on a falling film. *Annu. Rev. Fluid Mech.* **26**, 103–136.

[75] CHEBBI, R. (2001) Viscous–gravity spreading of oil on water. *AIChE J.* **47**, 288–294.

[76] CHIEN, J. C. (1976) A general finite-difference formulation with applications to Navier–Stokes equations. *J. Comp. Phys.* **20**, 268–278.

[77] CHIN, R. W., ABERNATHY, F. H. & BERTSCHY, J. R. (1986) Gravity and shear wave stability of free surface flows. Part 1. Numerical calculations. *J. Fluid Mech.* **168**, 501–513.

[78] CHIU, C. & NICOLAIDES, R. A. (1988) Convergence of a higher-order vortex method for two-dimensional Euler's equations. *Math. Comp.* **51**, 507–534.

[79] CHORIN, A. J. (1967) A numerical method for solving incompressible viscous flow problems. *J. Comp. Phys.* **2**, 12–26.

[80] CHORIN, A. J. (1968) Numerical solution of the Navier–Stokes equations. *Math. Comp.* **22**, 745–762.

[81] CHORIN, A. J. (1973) Numerical study of slightly viscous flow. *J. Fluid Mech.* **57**, 785–796.

[82] CHORIN, A. J. & BERNARD, P. S. (1973) Discretization of a vortex sheet, with an example of roll-up. *J. Comp. Phys.* **13**, 423–429.

[83] CHORIN, A. J. (1978) Vortex sheet approximation of boundary layers. *J. Comp. Phys.* **27**, 428–442.

[84] CHORIN, A. J. (1980) Vortex models and boundary layer instability. *SIAM J. Sci. Stat. Comput.* **1**, 1–21.

[85] CHRISTIANSEN, J. P. (1973) Numerical simulation of hydrodynamics by the method of point vortices. *J. Comp. Phys.* **13**, 363–379.

[86] CHWANG, A. T. & WU, T. Y.-T. (1974) Hydromechanics of low-Reynolds-number flow. Part 1, Rotation of axisymmetric prolate bodies. *J. Fluid Mech.* **63**, 607–622.

[87] CHWANG, A. T. & WU, T. Y.-T. (1975) Hydromechanics of low-Reynolds-number flow. Part 2, Singularity methods for Stokes flows. *J. Fluid Mech.* **67**, 787–815.

[88] CLEMENTS, R. R. & MAULL, D. J. (1975) The representation of sheets of vorticity by discrete vortices. *Prog. Aerospace Sci.* **16**, 129–146.

[89] CLIFT, R., GRACE, J. R. & WEBER, M. E. (1978) *Bubbles, Drops, and Particles*. Academic Press.

[90] COCHRAN, W. G. (1934) The flow due to a rotating disk. *Proc. Camb. Phil. Soc.* **30**, 365–375.

[91] COLTON, D. & KRESS, R. (1983) *Integral Equation Methods in Scattering Theory*. Wiley.

[92] CONCUS, P. (1968) Static menisci in a vertical right circular cylinder. *J. Fluid Mech.* **34**, 481–495.

[93] CONCUS, P. & FINN, R. (1974) On capillary free surfaces in a gravitational field. *Acta Math.* **132**, 207–223.

[94] COUETTE, M. (1890) Etudes sur le frottement des liquides. *Ann. Chim. Phys.* **21**, 433–510.

[95] CRAIK, A. D. D. (1985) *Wave Interactions and Fluid Flows*. Cambridge University Press.

[96] CRANE, L. J. (1970) Flow past a stretching plate. *ZAMP* **21**, 645–647.

[97] CRANK, J. & NICOLSON, P. (1947) A practical method for numerical evaluation of solutions of partial differential equations of the heat-conduction type. *Proc. Camb. Phil. Soc.* **43**, 50–67.

[98] GREIG, I. S. & MORRIS, J. L. (1976) A hopscotch method for the Korteweg-de-Vries equation. *J. Comp. Phys.* **20**, 64–80.

[99] CRIGHTON, D. G. (1985) The Kutta condition in unsteady flow. *Annu. Rev. Fluid Mech.* **17**, 411–445.

[100] DAUBE, O. (1992) Resolution of the two-dimensiona Navier–Stokes equations in velocity-vorticity form by means of an influence matrix technique. *J. Comp. Phys.* **103**, 402–414.

[101] DAVEY, A. (1977) On the numerical solution of difficult eigenvalue problems. *J. Comput. Phys.* **24**, 331–338.

[102] DAVID, T. & BLYTH, G. (1992) Parallel algorithms for panel methods. *Int. J. Num. Meth. Fluids* **14**, 95–108.

[103] DEAN, W. R. & MONTAGNON, P. E. (1949) On the steady motion of viscous liquid in a corner. *Proc. Camb. Phil. Soc.* **45**, 389–394.

[104] DELVES, L. M. & MOHAMED, J. L. (1985) *Computational Methods for Integral Equations*. Cambridge University Press.

[105] DENNIS, S. C. R., INGHAM, D. B., & COOK, R. N. (1979) Finite-difference methods for calculating steady incompressible flows in three dimensions. *J. Comp. Phys.* **33**, 325–339.

[106] DETTMAN, J. W. (1965) *Applied Complex Variables*. Dover.

[107] DIAZ, A., BARTHÈS-BIESEL, D. & PELEKASIS, N. (2001) Effect of membrane viscosity on the dynamic response of an axisymmetric capsule. *Phys Fluids* **13**, 3835–3838.

[108] DORREPAAL, J. M. (1986) An exact solution of the Navier–Stokes equation which describes non-orthogonal stagnation-point flow in two dimensions. *J. Fluid Mech.* **163**, 141–147.

[109] DOUGLAS, J. JR. (1955) On the numerical integration of $\partial^2 u/\partial x^2 + \partial^2 u/\partial y^2 = \partial u/\partial t$ by implicit methods. *J. Soc. Indust. Appl. Math.* **3**, 42–65.

[110] DOUGLAS, J. JR. & GUNN, J. E. (1964) A general formulation of alternating direction methods. I. Parabolic and hyperbolic problems. *Numerische Mathematik* **6**, 428–453.

[111] DOUGLAS, J. JR. (1962) Alternating direction methods for three space variables. *Numerische Mathematik* **4**, 41–63.

[112] DRAZIN, P. G. & REID, W. H. (2004) *Hydrodynamic Stability.* Second Edition. Cambridge University Press.

[113] DRITSCHEL, D. G. (1989) Contour dynamics and contour surgery: numerical algorithms for extended, high-resolution modelling of vortex dynamics in two-dimensional, inviscid, incompressible flows. *Comp. Phys. Rep.* **10**, 77–146.

[114] DUSSAN V., E. B. (1979) On the spreading of liquids on solid surfaces: static and dynamic contact lines *Ann. Rev. Fluid Mech.* **11**, 371–400.

[115] DUSSAN V., E. B. (1985) On the ability of drops or bubbles to stick to non-horizontal surfaces of solids. Part 2. Small drops or bubbles having contact angles of arbitrary size. *J. Fluid Mech.* **151**, 1–20.

[116] DU FORT, E. C. & FRANKEL, S. P. (1953) Stability conditions in the numerical treatment of parabolic differential equations. *Mathematical Tables and Other Aids to Computation* **7**, 135–152.

[117] E, W. & LIU, J.-G. (1995) Projection method I: Convergence and numerical boundary layers, *SIAM J. Num. Anal.* **32** 1017–1057.

[118] EDWARDS, D. A., BRENNER, H. & WASAN D. T. (1991) *Interfacial Transport Processes and Rheology.* Butterworth-Heinemann.

[119] ELIASSEN, A., HØILLAND, E. & RIIS, E. (1953) Two-dimensional perturbation of a flow with constant shear of a stratified fluid. *Publ. 1, Inst. Weather & Climate Res.*, Norwegian Acad. Sci. Letters.

[120] EVANS, H. (1968) *Laminar Boundary Layers.* Addison-Wesley.

[121] FALKNER, V. M. & SKAN, S. W. (1931) Solutions of the boundary layer equations. *Phil. Mag.* **12**, 865–896.

[122] FAXÉN, H. (1924) Der widerstand gegen die bewegung einer starren kugel in einer zähen flüssigkeit, die zwischen zwei parallelen, ebener Wänden eingeschlossen ist. *Ark. Mat. Astr. Fys.* **18**(29), 3.

[123] FELDMAN, S. (1957) On the hydrodynamic stability of two viscous incompressible fluids in parallel uniform shearing motion. *J. Fluid Mech.* **2**, 343–370.

[124] FERZIGER, J. H. (1981) *Numerical Methods for Engineering Application.* Wiley.

[125] FETTER, A. L. (1974) Translational velocity of a classical vortex ring. *Phys. Rev. A* **10**, 1724–1727.

[126] FINN, R. (1985) *Equilibrium Capillary Surfaces.* Springer–Verlag.

[127] FINK, P. T. & SOH, W. K. (1974) Calculation of vortex sheets in unsteady flow and applications in ship hydrodynamics. *Proc. 10th Symp. Naval Hydrodynamics*, 463–491.

[128] FJØRTOFT, R. (1950) Application of integral theorems in deriving criteria of stability for laminar flows and for the baroclinic circular vortex. *Geophys. Publ.* **17**(5).

[129] FLORYAN, J. M. (1985) Conformal-mapping-based coordinate generation method for channel flows. *J. Comp. Phys.* **62**, 221–247.

[130] FLORYAN, J. M. (1986) Conformal-mapping-based coordinate generation methods for flows in periodic configurations. *J. Comp. Phys.* **62**, 221–247.

[131] FLORYAN, J. M. & RASMUSSEN, H. (1989) Numerical methods for viscous flows with moving boundaries. *Appl. Mech. Rev.* **42**, 323–341.

[132] FLORYAN, J. M. & ZEMACH, C. (1988) Quadrature rules for singular integrals with application to Schwartz–Christoffel mappings. *J. Comp. Phys.* **75**, 15–30.

[133] FLORYAN, J. M. & ZEMACH, C. (1993) Schwarz–Christoffel methods for conformal mapping of regions with a periodic boundary. *J. Comp. Appl. Math.* **46**, 77–102.

[134] DE FOY, B. & DAWES, W. (2000) Unstructured pressure-correction solver based on a consistent discretization of the Poisson equation. *Int. J. Num. Meth. Fluids* **34**, 463–478.

[135] FODA, M. & COX, R. G. (1980) The spreading of thin liquid films on a water-air interface. *J. Fluid Mech.* **101**, 33–51.

[136] FRAENKEL, L. E. (1962) Laminar flow in symmetrical channels with slightly curved walls I. On the Jeffery-Hamel solutions for flow between plane walls. *Proc. R. Soc. Lond. A* **267**, 119–138.

[137] FRAENKEL, L. E. (1962) Laminar flow in symmetrical channels with slightly curved walls II. An asymptotic series for the stream function. *Proc. R. Soc. Lond. A* **272**, 406–428.

[138] FUENTES, Y. O. & KIM, S. (1992) Parallel computational microhydrodynamics: communication scheduling strategies. *AIChE J.* **38**, 1059–1078.

[139] FULFORD, G. D. (1964) The flow of liquids in thin films. *Adv. Chem. Eng.* **5**, 151–236.

[140] FUNG, Y. C. (1984) *Biodynamics: Circulation.* Springer.

[141] GASTER, M. (1962) A note on the relation between temporally-increasing and spatially-increasing disturbances in hydrodynamic stability. *J. Fluid Mech.* **14**, 222–224.

[142] GATSKI, T. B., GROSCH, C. E. & ROSE, M. E. (1989) The numerical solution of the Navier–Stokes equations for three-dimensional, unsteady, incompressible flows by compact schemes. *J. Comp. Phys.* **82**, 298–329.

[143] GERSTING, J. M. & JANKOWSKI, D. F. (1972) Numerical methods for Orr-Sommerfeld problems. *Int. J. Num. Meth. Eng.* **4**, 195–206.

[144] GHIA, U., GHIA, K. N. & STUDERUS, C. J. (1976) A study of three-dimensional laminar incompressible flow in ducts. *AIAA Paper 76-424.*

[145] GHIA, K. N., HANKEY, W. L. & HODGE, J. K. (1977) Study of incompressible Navier–Stokes equations in primitive variables using implicit numerical technique. *AIAA Paper 77-648*, 156–165.

[146] GOLDSHTIK, M. A. (1990) Viscous-flow paradoxes. *Annu. Rev. Fluid Mech.* **22**, 441–472.

[147] GÖLTER, H. (1940) On the three-dimensional instability of laminar boundary layers on concave walls. *Tech. Mem. Nat. Adv. Comm. Aero. Wash. No 1375.*

[148] GÖLTER, H. (1957) A new series for the calculation of steady laminar boundary layer flows. *J. Math. Mech.* **6**, 1–66.

[149] GOURLAY, A. R. (1970) Hopscotch: A fast second-order partial differential equation solver. *J. Inst. Maths. Applics.* **6**, 375–390.

[150] GRADSHTEYN, I. S. & RYZHIK, I. M. (1980) *Table of Integrals, Series, and Products.* Academic Press.

[151] GRAY, J. & HANCOCK, G. J. (1955) The propulsion of sea-urchin spermatozoa. *J. Exp. Biol.* **32**, 802–814.

[152] GREEN, A. E. & ADKINS, J. E. (1960) *Large Elastic Deformations and Non-linear Continuum Mechanics.* Oxford University Press.

[153] GRESHO, P. M. & SANI, R. L. (1987) On pressure boundary conditions for the incompressible Navier–Stokes equations. *Int. J. Num. Meth. Fluids* **7**, 1111–1145.

[154] GRESHO, P. M. (1990) On the theory of semi-implicit projection methods for viscous incompressible flow and its implementation via a finite-element method that also introduces a nearly consistent mass matrix. Part 1: Theory. *Int. J. Num. Meth. Fluids* **11**, 587–620.

[155] GRESHO, P. M. (1991) Incompressible fluid dynamics: some fundamental formulation issues. *Annu. Rev. Fluid Mech.* **23**, 413–453.

[156] GUERON, S. & LIRON, N. (1992) Ciliary motion modeling, and dynamic multicilia interactions. *Biophys. J.* **63**, 1045–1058.

[157] GUJ, G. & STELLA, F. (1988) Numerical solutions of high-Re recirculating flows in vorticity-velocity form. *Int. J. Num. Meth. Fluids.* **8**, 405–416.

[158] GUJ, G. & STELLA, F. (1993) A vorticity–velocity method for the numerical solution of 3D incompressible flows. *J. Comp. Phys.* **106**, 286–298.

[159] GUPTA, M. M. (1991) High-accuracy solutions of incompressible Navier–Stokes equations. *J. Comp. Phys.* **93**, 343–359.

[160] HAMADICHE, M., SCOTT, J. & JEANDEL, D. (1994) Temporal stability of Jeffery-Hamel flow. *J. Fluid Mech.* **268**, 71–88.

[161] HAMEL, G. (1916) Spiralförmige bewegungen zäher Flüssigkeiten. *Jahresb. Deutsch. Math.-Verein.* **25**, 34–60.

[162] HAMA, F. R. (1962) Progressive deformation of a curved vortex filament by its own induction. *Phys. Fluids* **5**, 1156–1162.

[163] HAMA, F. R. (1963) Progressive deformation of a perturbed line vortex filament by its own induction. *Phys. Fluids* **6**, 526–534.

[164] HAMROCK, B. J. (1994) *Fundamentals of Fluid Film Lubrication.* McGraw–Hill.

[165] HANCOCK, G. J. (1953) The self-propulsion of microscopic organisms through liquids. *Proc. R. Soc. Lond. A* **217**, 96–121.

[166] HAPPEL, J. & BRENNER, H. (1973) *Low Reynolds Number Hydrodynamics.* Martinus Nijhoff.

[167] HARDIN, J. C. (1982) The velocity field induced by a helical vortex filament. *Phys. Fluids* **25**, 1949–1952.

[168] HARLOW, H. H. & WELCH, J. E. (1965) Numerical calculation of time-dependent viscous incompressible flow of fluid with free surface. *Phys. Fluids* **8**, 2182–2189.

[169] HARPER, J. F. & MOORE, D. W. (1968) The motion of a spherical liquid drop at high Reynolds number *J. Fluid Mech.* **32**, 367–391.

[170] HARTREE, D. R. (1937) On an equation occurring in Falkner and Skan's approximate treatment of the equations of the boundary layer. *Proc. Camb. Phil. Soc.* **33**, 223–239.

[171] HASIMOTO, H. (1972) A soliton on a vortex filament. *J. Fluid Mech.* **51**, 477–485.

[172] HAVELOCK, T. H. (1931) The stability of motion of rectilinear vortices in ring formation. *Phil. Mag. Ser. 7.* **11**(70), Suppl. 617–633.

[173] HENSHAW, W. D. (1994) A fourth-order accurate method for the incompressible Navier–Stokes equations on overlapping grids. *J. Comp. Phys.* **113**, 13–25.

[174] HESS, J. L. (1990) Panel methods in computational fluid dynamics. *Annu. Rev. Fluid Mech.* **22**, 255–274.

[175] HICKS, W. M. (1923) On the mutual threading of vortex rings. *Proc. R. Soc. Lond. A* **102**, 111–131.

[176] HIEMENZ, K. (1911) Die Grenzschicht an einem in den gleichförmigen flüssigkeitsstrom einge-tauchten geraden kreiszylinder. *Dinglers Polyt. J.* **326**(21), 321–410.

[177] HIGDON, J. J. L. (1979) A hydrodynamic analysis of flagellar propulsion. *J. Fluid Mech.* **90**, 685–711.

[178] HILDEBRAND, F. B. (1976) *Advanced Calculus for Applications.* Prentice-Hall.

[179] HILL, A. I. & POZRIKIDIS, C. (2011) On the shape of a hydrostatic meniscus attached to a corrugated plate or wavy cylinder *J. Colloid Interf. Sci.* **356**, 763–774.

[180] HINDMARSH, A. C., GRESHO, P. M. & GRIFFITHS, D. F. (1984) The stability of explicit Euler time-integration for certain finite difference approximations of the multi-dimensional advection-diffusion equation. *Int. J. Num. Meth. Fluids* **4**, 853–897.

[181] HIRASAKI, G. J. & HELLUMS, J. D. (1970) Boundary conditions on the vector and scalar potentials in viscous three-dimensional hydrodynamics. *Quart. Appl. Math.* **28**, 293–296.

[182] HIRSCH, C. (1991) *Numerical Computation of Internal and External Flows.* Vols. I and II. Wiley.

[183] HOCQUART, R. & HINCH, E. J. (1983) The long-time tail of the angular-velocity autocorre-lation function for a rigid Brownian particle of arbitrary centrally symmetric shape. *J. Fluid Mech.* **137**, 217–220.

[184] HOFFMAN, J. D. (1992) *Numerical Methods for Engineers and Scientists.* McGraw–Hill.

[185] HOFFMANN, K. A. & CHIANG, S. T. (1993) *Computational Fluid Dynamics for Engineers.* Vol I. Engineering Education System.

[186] HOFFMANN, K. A. & CHIANG, S. T. (1993) *Computational Fluid Dynamics for Engineers.* Vol II. Engineering Education System.

[187] HOMANN, F. (1936) Einfluß großer zähigkeit bei strömung um zylinder. *Forschg. Ing.-Wes.* **7**, 1–10.

[188] HOMANN, F. (1936) Der einfluß großer zähigkeit bei der strömung um den zylinder und um die kugel. *Z. Angew. Math. Mech.* **16**(3), 153–164.

[189] HOOPER, A. (1989) The stability of two superposed viscous fluids in a channel. *Phys. Fluids A* **1**, 1133–1142.

[190] HOOPER, A., DUFFY, B. R. & MOFFATT, H. K. (1982) Flow of fluid of non-uniform viscosity in converging and diverging channels. *J. Fluid Mech.* **117**, 283–304.

[191] HORNUNG, U. & MITTELMANN, H. D. (1990) A finite element method for capillary surfaces with volume constraints. *J. Comp. Phys.* **87**, 126–136.

[192] HOWARD, L. N. (1961) Note on a paper by John W. Miles. *J. Fluid Mech.* **10**, 509–512.

[193] HOWARTH, L. (1951) The boundary layer in three-dimensional flow. Part II: The flow near a stagnation point. *Phil. Mag. Ser. 7* **42**, 1433–1440.

[194] HOWELL, L. H. (1993) Numerical conformal mapping of circular arc polygons. *J. Comp. Appl. Math.* **46**, 7–28.

[195] HUERRE, P. & MONKEWITZ, P. A. (1990) Local and global instabilities in spatially developing flows. *Annu. Rev. Fluid Mech.* **22**, 473–537.

[196] HUGHES, T. J. R., LIU, W. K. & BROOKS, A. (1979) Finite element analysis of incompressible viscous flows by the penalty function formulation. *J. Comp. Phys.* **30**, 1–60.

[197] ISENBERG, C. (1992) *The Science of Soap Films and Soap Bubbles.* Dover.

[198] ISRAELI, M. (1972) On the evaluation of iteration parameters for the boundary vorticity. *Stud. Appl. Math.* **51**, 67–71.

[199] JEFFERY, G. B. (1915) The two-dimensional steady motion of a viscous fluid. *Phil. Mag.* **29**, 455–465.

[200] JOHNSON, R. E. (1980) An improved slender-body theory for Stokes flow. *J. Fluid Mech.* **99**, 411–431.

[201] JOSEPH, D. D., LIAO, T. Y. & HU, H. H. (1993) Drag and moment in viscous potential flow. *Eur. J. Mech. B/Fluids* **12**, 97–106.

[202] KANEDA, Y. (1990) On the three-dimensional motion of an infinitely thin vortex sheet in an ideal fluid. *Phys. Fluids A* **2**, 1817–1826.

[203] KANEKO, A. & HONJI, H. (1979) Double structures of steady streaming in the oscillatory viscous flow over a wavy wall. *J. Fluid Mech.* **93**, 727–736.

[204] KELLOGG, O. D. (1953) *Foundations of Potential Theory.* Dover.

[205] KERNER, W. (1989) Large-scale complex eigenvalue problems. *J. Comput. Phys.* **85**, 1–85.

[206] KIM, J. & MOIN, P. (1985) Application of a fractional-step method to incompressible Navier–Stokes equations. *J. Comp. Phys.* **59**, 308–323.

[207] KIM, S. (1985) A note on Faxén laws for nonspherical particles. *Int. J. Multiph. Flow* **11**, 713–719.

[208] KIM, S. & KARRILA, S. J. (1991) *Microhydrodynamics: Principles and Selected Applications.* Butterworth.

[209] KIM, S. & LU, S.-Y. (1987) The functional similarity between Faxén relations and singularity solutions for fluid-fluid, fluid-solid and solid-solid dispersions. *Int. J. Multiph. Flow* **13**, 837–844.

[210] KIM, S. K. & TROESCH, A. W. (1989) Streaming flows generated by high-frequency small-amplitude oscillations of arbitrarily shaped cylinders. *Phys. Fluids A* **1**, 975–985.

[211] KINO, O. M. & GHONIEM, A. F. (1990) Numerical study of a three-dimensional vortex method. *J. Comp. Phys.* **86**, 75–106.

[212] KOVASZNAY, L. I. G. (1948) Laminar flow behind a two-dimensional grid. *Proc. Camb. Phil. Soc.* **44**, 58–62.

[213] KOZLOV, S. V., LIFANOV, I. K. & MIKHAILOV, A. A. (1991) A new approach to mathematical modelling of flow of ideal fluid around bodies. *Sov. J. Num. Anal. Mod.* **6**, 209–222.

[214] KWAK, S. & POZRIKIDIS, C. (2001) Effect of surfactants on the instability of a liquid thread of annular layer, Part I: Quiescent fluids. *Int. J. Multiph. Flow* **27**, 1–37.

[215] KRASNY, R. (1986) A study of singularity formation in a vortex sheet by the point-vortex approximation. *J. Fluid Mech.* **167**, 65–93.

[216] KRASNY, R. (1986) Desingularization of periodic vortex sheet roll-up. *J. Comp. Phys.* **65**, 65–93.

[217] LADYZHENSKAYA, O. A. (1975) Mathematical analysis of Navier–Stokes equations for incompressible liquids. *Annu. Rev. Fluid Mech.* **7**, 249–272.

[218] LAGESTROM, P. A. (1964) *Laminar Flow Theory.* In *Theory of Laminar Flows*, Moore, F. K. (Ed.), Princeton University Press.

[219] LAMB, H. (1911) On the uniform motion of a sphere through a viscous fluid. *Phil. Mag.* **21**, 112–121.

[220] LAMB, H. (1932) *Hydrodynamics.* Dover.

[221] LANDAU, L. (1944) An exact solution of Navier–Stokes equations. *Doklady Acad. Nauk. SSSR* **43**(7), 299–301.

[222] LANDAU, L. & LEVICH, V. G. (1942) Dragging of a liquid by a moving plate. *Acta Physicochim.* (USSR) **17**, 42–54.

[223] LANGLOIS, W. E. (1964) *Slow Viscous Flow.* MacMillan.

[224] LASSEIGNE, D. G. & JACKSON, T. L. (1992) Stability of a nonorthogonal stagnation flow to three-dimensional disturbances. *Theor. Comput. Fluid Dyn.* **3**, 207–218.

[225] LAURA, P. A, ROMANELLI, E. & MAURIZI, M. J. (1972) On the analysis of waveguides of doubly-connected cross-section by the method of conformal mapping. *J. Sound Vibr.* **20**, 27–38.

[226] LAX, P. D. (1954) Weak solutions of nonlinear hyperbolic equations and their computation. *Comm. Pure Appl. Math.* **7**, 159–193.

[227] LAX, P. D. & RICHTMYER, R. D. (1956) Survey of the stability of linear finite difference equations. Part I. An equivalence theorem. *Comm. Pure Appl. Math.* **9**, 267–293.

[228] LAX, P. D. & WENDROFF, B. (1960) Systems of conservation laws. *Comm. Pure Appl. Math.* **13**, 217–237.

[229] LEES, M. (1962) Alternating direction methods for hyperbolic differential equations. *J. Soc. Indust. Appl. Math.* **10**, 610–616.

[230] LEE, S. H. & LEAL, L. G. (1986) Low-Reynolds-number flow past cylindrical bodies of arbitrary cross-sectional shape. *J. Fluid Mech.* **164**, 401–427.

[231] LEIDER, P. J. & BIRD, R. B. (1974) Squeezing flow between parallel disks. I. Theoretical analysis. *Ind. Eng. Che. Fundam.* **13**, 336–341.

[232] LEONARD, B. P., LESCHZINER, M. A. & McGUIRK, J. (1978) Third-order finite-difference method for steady two-dimensional convection. *Numerical Methods in Laminar and Turbulent Flow*, Proc. First Int. Conf. Swansea, 807–819, Pentech.

[233] LEONARD, A. (1980) Vortex methods for flow simulation. *J. Comp. Phys.* **37**, 289–335.

[234] LEONARD, A. (1985) Computing three-dimensional incompressible flows with vortex elements. *Annu. Rev. Fluid Mech.* **17**, 523–559.

[235] LEVICH, V. G. (1962) *Physicochemical Hydrodynamics*. Prentice-Hall.

[236] LI, X. Z., BARTHÈS-BIESEL, D. & HELMY, A. (1988) Large deformation and burst of a capsule freely suspended in an elongational flow. *J. Fluid Mech.* **187**, 179–196.

[237] LI, X. & POZRIKIDIS, C. (1996) Shear flow over a liquid drop adhering to a solid surface. *J. Fluid Mech.* **307**, 167–190.

[238] LIGHTHILL, M. J. (1963) Attachment and separation in three-dimensional flow. In *Laminar Boundary Layer Theory*, Rosenhead, L. (Ed.), Chap. II, Sec. 2.6, pp. 72–82, Oxford University Press.

[239] LIGHTHILL, M. J. (1975) *Mathematical Biofluiddynamics*. SIAM.

[240] LIGHTHILL, M. J. (1976) Flagellar hydrodynamics, *SIAM Rev.* **18**, 161–230.

[241] LIN, C. C. (1955) *The Theory of Hydrodynamic Stability*. Cambridge University Press.

[242] LO, L. L. (1983) The meniscus on a needlea lesson in matching. *J. Fluid Mech.* **132**, 65–78.

[243] LOEWENBERG, M. (1994) Axisymmetric unsteady Stokes flow past an oscillating finite-length cylinder. *J. Fluid Mech.* **265**, 265–288.

[244] LONGUET-HIGGINS, M. S. & COKELET, E. D. (1976) The deformation of steep surface waves on water. I. A numerical method of computation. *Proc. R. Soc. Lond. A* **350**, 1–26.

[245] LORENTZ, H. A. (1907) Ein allgemeiner Satz, die bewegung einer reibenden flüssigkeit betreffend, nebst einigen anwendungen desselben. *Abhand. Theor. Phys. (Leipzig)* **1**, 23–42. Translated into English: *Lorentz, H. A. Collected Papers Vol. IV, 7–14*, Martinus Nijhoff.

[246] LOVE, A. E. H. (1893) On the stability of certain vortex motions. *Proc. Lond. Math. Soc.* **25**, 18–42.

[247] LOVE, A. E. H. (1944) *A Treatise on the Mathematical Theory of Elasticity*. Dover.

[248] Lugt, H. J. (1987) Local flow properties at a viscous free surface. *Phys. Fluids* **30**, 3647–3652.

[249] Luo, H. & Pozrikidis, C. (2008) Effect of surface slip on Stokes flow past a spherical particle in infinite fluid and near a plane wall. *J. Eng. Math.* **62**, 1–21.

[250] MacCormack, R. W. (1969) The effect of viscosity in hypervelocity impact cratering. *AIAA Paper 69-354.*

[251] Malik, M. R. (1986) The neutral curve for stationary disturbances in rotating-disk flow. *J. Fluid Mech.* **164**, 275–287.

[252] Mallinson, G. D. & de Vahl Davis, G. (1977) Three-dimensional natural convection in a box: a numerical study. *J. Fluid Mech.* **83**, 1–31.

[253] Mager, A. (1964) Three-dimensional laminar boundary layers. In: *Theory of Laminar Flows*, Moore F. K. (Ed.), Princeton University Press.

[254] Mangler, W. (1948) Zusammenhang zwischen ebenen und rotationssymmetrischen grenzschichten in kompressiblen Flüssigkeiten. *Z. Angew. Math. Mech.* **28**, 97–103.

[255] Mansour, N. N. & Lundgren, T. S. (1990) Satellite formation in capillary jet breakup. *Phys. Fluids A* **2**, 1141–1144.

[256] Mansour, M. L. & Hamed, A. (1990) Implicit solution of the incompressible Navier–Stokes equations on a non-staggered grid. *J. Comp. Phys.* **86**, 147–167.

[257] Maslowe, S. A. (1981) Shear flow instabilities and transition. In *Hydrodynamic Instabilities and the Transition to Turbulence*, Swinney & Gollub (Eds.), Springer–Verlag.

[258] Maxey, M. R. & Riley, J. J. (1983) Equation of motion for a small rigid sphere in a nonuniform flow. *Phys. Fluids* **26**, 883–889.

[259] Moore, D. W. (1963) The boundary layer on a spherical gas bubble. *J. Fluid Mech.* **16**, 161-176.

[260] Moore, F. K. (1964) *Theory of Laminar Flows.* Princeton University Press.

[261] McCraken, M. F. & Peskin, C. S. (1980) A vortex method for blood flow through heart valves. *J. Comp. Phys.* **35**, 183–205.

[262] McDonald, P. (1996) *Continuum Mechanics.* PWS Press.

[263] Mei, C. C. (1966) Nonlinear gravity waves in a thin sheet of viscous fluid. *J. Math. Phys.* **45**, 266–288.

[264] Michalke, A. (1964) On the inviscid instability of the hyperbolic-tangent velocity profile. *J. Fluid Mech.* **19**, 543–556.

[265] Milinazzo, F. & Saffman, P. G. (1977) The calculation of large Reynolds number two-dimensional flow using discrete vortices with random walks. *J. Comp. Phys.* **23**, 380–392.

[266] MILINAZZO, F. & SHINBROT, M. (1988) A numerical study of a drop on a vertical wall. *J. Colloid Interf. Sci.* **121**, 254–264.

[267] MILNE-THOMSON, L. M. (1940) Hydrodynamic images. *Proc. Camb. Phil. Soc.* **36**, 246–247.

[268] MILNE-THOMSON, L. M. (1968) *Theoretical Hydrodynamics.* MacMillan.

[269] MITCHELL, A. R. & GRIFFITHS, D. F. (1980) *The Finite-Difference Method in Partial Differential Equations.* Wiley.

[270] MITCHELL, T. B. & ROSSI L. F. (2008) The evolution of Kirchhoff elliptic vortices. *Phys. Fluids* **449**, 054103.

[271] MOFFATT, H. K. (1964) Viscous and resistive eddies near a sharp corner. *J. Fluid Mech.* **18**, 1–18.

[272] MOFFATT, H. K. & DUFFY, B. R. (1980) Local similarity solutions and their limitations. *J. Fluid Mech.* **96**, 299–313.

[273] MOFFATT, H. K. & TSINOBER, A. (1992) Helicity in laminar and turbulent flow. *Annu. Rev. Fluid Mech.* **24**, 281–312.

[274] MURDOCK, J. N. & STEWARTSON, K. (1977) Spectra of the Orr-Sommerfeld equation. *Phys. Fluids* **20**, 1404–1411.

[275] NAKAYAMA, T. (1990) A computational method for simulating transient motions of an incompressible inviscid fluid with a free surface. *Int. J. Num. Meth. Fluids* **10**, 683–695.

[276] NEHARI, Z. (1952) *Conformal Mapping.* Dover.

[277] NAYFEH, A. H. (1985) *Problems in Perturbation.* Wiley .

[278] NEWTON, P. K. (2001) *The N-Vortex Problem: Analytical Techniques.* Springer.

[279] NG, B. S. & REID, W. H. (1979) An initial value method for eigenvalue problems using compound matrices. *J. Comp. Phys.* **30**, 125–136.

[280] NEWHOUSE, L. A. & POZRIKIDIS, C. (1990) The Rayleigh-Taylor instability of a viscous liquid layer resting on a plane wall. *J. Fluid Mech.* **217**, 615–638.

[281] NEWHOUSE, L. A. & POZRIKIDIS, C. (1992) The capillary instability of annular layers and liquid threads. *J. Fluid Mech.* **242**, 193–209.

[282] NORBURY, J. (1973) A family of steady vortex rings. *J. Fluid Mech.* **57**, 417–431.

[283] ORR, W. M. F. (1907) The stability or instability of the steady motions of a perfect liquid and of a viscous liquid; Part I: A perfect liquid; Part II: A viscous liquid. *Proc. R. Irish Acad. A* **27**, 9–138.

[284] ORSZAG, S. A. (1971) Accurate solution of the Orr-Sommerfeld stability equation. *J. Fluid Mech.* **50**, 689–703.

[285] ORSZAG, S. A., ISRAELI, M. & DEVILLE, M. O. (1986) Boundary conditions for incompressible flows. *J. Sci. Comp.* **1**, 75–111.

[286] ORSZAG, S. A. & ISRAELI, M. (1974) Numerical simulation of viscous incompressible flows. *Annu. Rev. Fluid Mech.* **5**, 281–318.

[287] OSEEN, C. W. (1927) *Hydrodynamik.* Leipzig.

[288] OSSWALD, G. A., GHIA, K. N. & GHIA, U. (1987) A direct algorithm for solution of incompressible three-dimensional unsteady Navier–Stokes equations. *AIAA Paper 87–1139*, 408–421.

[289] OSWATITSCH, K. (1958) Die Ablösungsbedingung von Grenzschichten. In: *IUTAM Symposium on Boundary Layer Research, Freiburg, 1957*, Görtler, H. (Ed.), 357–367, Springer–Berlin.

[290] PADMANABHAN, N. & PEDLEY, T. J. (1987) Three-dimensional steady streaming in a uniform tube with an oscillating elliptical cross-section. *J. Fluid Mech.* **178**, 325–343.

[291] PEACEMAN, D. W. & RACHFORD, H. H. JR. (1955) The numerical solution of parabolic and elliptic differential equations. *J. Soc. Indust. Appl. Math.* **3**, 28–41.

[292] PEREGRINE, D. H. (1981) The fascination of fluid mechanics. *J. Fluid Mech.* **106**, 59–80.

[293] PETRILA, T. & TRIF, D. (2005) *Introduction to Computational Fluid Dynamics.* Springer.

[294] PEYRET, R. & TAYLOR, T. D. (1983) *Computational Methods for Fluid Flow.* Springer–Verlag.

[295] PHAN-THIEN, N. & KIM, S. (1994) *Microstrutures in Elastic Media.* Oxford University Press.

[296] PHILLIPS, W. R. C. (1996) On a class of unsteady boundary layers of finite extent. *J. Fluid Mech.* **319**, 151–170.

[297] PHILLIPS, W. R. C. (1997) On the spreading radius of surface tension driven oil on deep water. *Appl. Sci. Res.* **57**, 67–80.

[298] PIERSON, F. W. & WHITAKER, S. (1977) Some theoretical and experimental observations of the wave structure of falling liquid films. *Ind. Eng. Chem. Fund.* **16**, 401–408.

[299] PLESSET, M. S. & PROSPERETTI, A. (1977) Bubble dynamics and cavitation. *Annu. Rev. Fluid Mech.* **9**, 145–185.

[300] POGORZELSKI, W. (1966) *Integral Equations and their Applications.* Pergamon Press.

[301] POHLHAUSEN, E. (1921) Zur näherungsweisen integration der differentialgleichung der laminaren grenzschicht. *Z. Angew. Math. Mech.* **1**, 252–268.

[302] POZRIKIDIS, C. (1986) The nonlinear instability of Hill's spherical vortex. *J. Fluid Mech.* **168**, 337–367.

[303] POZRIKIDIS, C. (1989) A singularity method for unsteady linearized flow. *Phys. Fluids A* **1**, 1508–1520.

[304] POZRIKIDIS, C. (1989) A study of linearized oscillatory flow past particles by the boundary integral method. *J. Fluid Mech.* **202**, 17–41.

[305] POZRIKIDIS, C. (1990) The instability of moving viscous drops. *J. Fluid Mech.* **210**, 1–21.

[306] POZRIKIDIS, C. (1992) *Boundary Integral and Singularity Methods for Linearized Viscous Flow.* Cambridge University Press, Cambridge.

[307] POZRIKIDIS, C. (1993) Unsteady viscous flow over irregular boundaries. *J. Fluid Mech.* **255**, 11–34.

[308] POZRIKIDIS, C. (1995) A bibliographical note on the unsteady force on a spherical drop. *Phys. Fluids* **6**, 3209.

[309] POZRIKIDIS, C. (1997) Shear flow over a protuberance on a plane wall. *J. Eng. Math.* **31**, 29–42.

[310] POZRIKIDIS, C. (2000) Instability of two annular layers or a liquid thread bounded by an elastic membrane. *J. Fluid Mech.* **405**, 211–241.

[311] POZRIKIDIS, C. (2001) Theoretical and computational aspects of the motion of three-dimensional vortex sheets. *J. Fluid Mech.* **425**, 335–366.

[312] POZRIKIDIS, C. (2001) A note on the regularization of the Poisson–Neumann problem. *J. Comp. Phys.* **172**, 1917–1923.

[313] POZRIKIDIS, C. (2002) *A Practical Guide to Boundary-Element Methods with the Software Library BEMLIB.* Francis & Taylor/CRC, Boca Raton.

[314] POZRIKIDIS, C. (2003) Effect of surfactants on film flow down a periodic wall. *J. Fluid Mech.* **496**, 105–127.

[315] POZRIKIDIS, C. (2003) Numerical simulation of the flow-induced deformation of red blood cells. *Annals Biomed. Eng.* **31**, 1194–1205.

[316] POZRIKIDIS, C. (2003) On the relation between the pressure and the projection function for the numerical computation of incompressible flow. *Eur. J. Mech. B/Fluids* **22**, 105–121.

[317] POZRIKIDIS, C. (2008) *Numerical Computation in Science and Engineering.* Second Edition. Oxford University Press.

[318] POZRIKIDIS, C. (2009) *Fluid Dynamics: Theory, Computation, and Numerical Simulation.* Second Edition. Springer.

[319] POZRIKIDIS, C. (2009) Shape of hexagonal hydrostatic menisci. *Int. J. Num. Meth. Fluids* DOI 10.1002/fld.2200.

[320] POZRIKIDIS, C. (2010) Computation of three-dimensional hydrostatic menisci. *IMA J. Appl. Math.* **75**, 418–438.

[321] POZRIKIDIS, C. (2010) Hydrostatic meniscus between two eccentric circular cylinders. *J. Colloid Interf. Sci.* **349**, 366–373.

[322] POZRIKIDIS, C. (2011) Stability of sessile and pendant liquid drops. *J. Eng. Math.* DOI 10.1007/s10665-011-9459-3.

[323] POZRIKIDIS, C. & HIGDON, J. J. L. (1985) Nonlinear Kelvin-Helmholtz instability of a finite vortex layer. *J. Fluid Mech.* **157**, 225–263.

[324] POZRIKIDIS, C. & HIGDON, J. J. L. (1987) Instability of compound vortex layers and wakes. *Phys. Fluids* **30**, 2965–2975.

[325] POZRIKIDIS, C. & HILL, A. I. (2011) Surfactant-induced instability of a sheared liquid layer *IMA J. Appl. Math.* doi: 10.1093/imamat/hxq067.

[326] POZRIKIDIS, C. & JUNEJA, V. (2010) Effect of surface viscosity on the capillary instability of an annular layer or viscous thread. *IMA J. Appl. Math.* **75**, 932–950.

[327] PRANDTL, L. (1904) Über Flüssigkeitsbewegung bei sehr kleiner Reibung, Verh. III. Int. Math. Kongr., Heidelburg, 484–491. (Translated into English in NACA Tech. Mem. 452.)

[328] PUCKETT, E. G. (1993) Vortex methods: An introduction and survey of selected research topics. In *Incompressible Computational Fluids Dynamics: Trends and Advances.* Cambridge University Press.

[329] PULLIN, D. I. (1981) The nonlinear behaviour of a constant vorticity layer at a wall. *J. Fluid Mech.* **108**, 401–421.

[330] PULLIN, D. I. (1992) Contour dynamics methods. *Annu. Rev. Fluid Mech.* **24**, 89–115.

[331] QUARTAPELLE, L. (1981) Vorticity conditioning in the computation of two-dimensional viscous flows. *J. Comp. Phys.* **40**, 453–477.

[332] QUARTAPELLE, L. & VALZ-GRIS, F. (1981) Projection conditions on the vorticity in viscous incompressible flows. *Int. J. Num. Meth. Fluids.* **1**, 129–144.

[333] QUARTAPELLE, L. & NAPOLITANO, M. (1986) Integral conditions for the pressure in the computation of incompressible viscous flows. *J. Comp. Phys.* **62**, 340–348.

[334] RAYLEIGH, L. (1879) On the instability of jets. *Proc. Lond. Math. Soc.* **10**, 4–13.

[335] RAYLEIGH, L. (1880) On the stability or instability of certain fluid motions. *Proc. Lond. Math. Soc.* **11**, 57–70.

[336] RAYLEIGH, L. (1883) Investigation of the character of the equilibrium of an incompressible heavy fluid of variable density. *Proc. Lond. Math. Soc.* **14**, 170–177.

[337] RAYLEIGH, L. (1894) *The Theory of Sound.* Macmillan (Reprinted by Dover 1945).

[338] RAYLEIGH, L. (1916) On the dynamics of revolving fluids. *Proc. R. Soc. Lond. A* **93**, 148–154.

[339] RICCA, R. L. (1991) Rediscovery of Da Rios equations. *Nature* **352**, 561–562.

[340] RICHARDSON, S. (1973) On the no-slip boundary condition. *J. Fluid Mech.* **59**, 707–719.

[341] RICHARDSON, S. M. & CORNISH, A. R. H. (1977) Solution of three-dimensional incompressible flow problems. *J. Fluid Mech.* **82**, 109–319.

[342] RICHTMYER, R. D. (1963) A survey of difference methods for non-steady fluid dynamics. *NCAR Technical Notes* 63-2.

[343] RICHTMYER, R. D. & MORTON, K. W. (1967) *Difference Methods for Initial-Value Problems.* Interscience.

[344] RIERA, G., CARRASCO, H. & PREISS R. (2008) The Schwarz–Christoffel conformal mapping for "Polygons" with infinitely many sides. *Int. J. Maths. Math. Sci.*, No 350326.

[345] RILEY, N. (1992) Acoustic streaming about a cylinder in orthogonal beams. *J. Fluid Mech.* **242**, 387–394.

[346] ROACH, P. J. (1982) *Computational Fluid Dynamics.* Hermosa Publishers.

[347] ROBERTS, S. G. (1985) Accuracy of the random vortex methods for a problem with non-smooth initial conditions. *J. Comp. Phys.* **58**, 29–43.

[348] ROSENHEAD, L. (1930) The spread of vorticity in the wake behind a cylinder. *Proc. R. Soc. Lond. A* **127**, 590–612.

[349] ROSENHEAD, L. (1931) The formation of vortices from a surface of discontinuity. *Proc. R. Soc. Lond. A* **134**, 170–192.

[350] ROSENHEAD, L. (1940) The steady two-dimensional radial flow of viscous fluid between two inclined plane walls. *Proc. R. Soc. Lond. A* **175**, 436–467.

[351] ROSENHEAD, L. (1954) A discussion on the first and second viscosities of fluids. *Proc. R. Soc. Lond. A* **226**, 1–65.

[352] ROSENHEAD, L. (1963) *Laminar Boundary Layers.* Oxford University Press.

[353] ROTT, N. (1956) Unsteady viscous flow in the vicinity of a stagnation point. *Quart. Apl. Math.* **13**, 444–451.

[354] ROTT, N. (1958) On the viscous core of a line vortex. *ZAMP* **9**, 543–553.

[355] ROTT, N. (1964) Time-dependent solutions of the Navier–Stokes equations. In *Theory of Laminar Flows*, Moore, F. K. (Ed.), Princeton University Press.

[356] RYNHART, P. R., MCLACHLAN, R., JONES, J. R. & MCKIBBIN, R. (2003) Solution of the Young-Laplace equation for three particles. *Res. Lett. Math. Sci.* **5**, 119–127.

[357] SAFFMAN, P. G. (1995) *Vortex Dynamics.* Oxford University Press.

[358] SAFFMAN, P. G. & BAKER, G. R. (1979) Vortex interactions. *Annu. Rev. Fluid Mech.* **11**, 95–122.

[359] SAFFMAN, P. G. & TAYLOR, G. I. (1958) The penetration of a fluid into a porous medium or Hele–Shaw cell containing a more viscous liquid. *Proc. R. Soc. Lond. A* **245**, 312–329.

[360] SAKIADIS, B. C. (1961) Boundary-layer behavior on continuous solid surfaces: I Boundary-layer equations for two-dimensional and axisymmetric flow. *AIChE J.* **7**, 26–28.

[361] SAKIADIS, B. C. (1961) Boundary-layer behavior on continuous solid surfaces: II The boundary layer on a continuous flat surface. *AIChE J.* **7**, 221–225.

[362] SCHAAF, S. A. & CHAMBRE. P. L. (1961) *Flow of Rarefied Gases.* Princeton University Press.

[363] SCHLICHTING, H. (1968) *Boundary Layer Theory.* McGraw–Hill.

[364] SCHMID, P. J. & HENNINGSON, S. (2001) *Stability and Transition in Shear Flows.* Springer–Verlag.

[365] SCHOWALTER, W. R. (1978) *Mechanics of non-Newtonian Fluids.* Pergamon Press.

[366] SECOMB, T. W. & SKALAK, R. (1982) Surface flow of viscoelastic membranes in viscous fluids. *Q. J. Mech. Appl. Math.* **35**, 233–247.

[367] SHARIFF, K. & LEONARD, A. (1992) Vortex rings. *Annu. Rev. Fluid Mech.* **24**, 235–279.

[368] SHARIFF, K., LEONARD, A. & FERZIGER, J. H. (1989) Dynamics of a class of vortex rings. *NASA TM-102257.*

[369] SHEN, S. F. (1954) Calculated amplified oscillations in plane Poisseuille and Blasius flows. *J. Aero. Sc.* **21**, 62–64.

[370] SHERCLIFF, J. A. (1975) A note on vorticity. *J. Inst. Maths. Applics.* **16**, 259–262.

[371] SHERMAN, F. S. (1990) *Viscous Flow.* McGraw–Hill.

[372] SHTERN, V. & HUSSAIN, F. (1993) Azimuthal instability of divergent flows. *J. Fluid Mech.* **256**, 535–560.

[373] SINGH, P. & HESLA, T. I. (2004) The interfacial torque on a partially submerged sphere. *J. Colloid Interf. Sci.* **280**, 542–543.

[374] SMITH, A. M. O. (1956) Rapid laminar boundary-layer calculations by piecewise application of similar solutions. *J. Aeronaut. Sci.* **23**, 901–912.

[375] SMITH, A. M. O. & CLUTTER, D. W. (1963) Solution of the incompressible laminar boundary-layer equations. *AIAA J.* **1**, 2062–2071.

[376] SMITH, F. T. (1982) On the high Reynolds number theory of laminar flows. *IMA J. Appl. Math.* **28**, 207–281.

[377] SOD, G. A. (1985) *Numerical Methods in Fluid Fynamics. Initial and Initial Boundary-Value Problems.* Cambridge University Press.

[378] SOMMERFELD, A. (1908) Ein Beitrag zur hydrodynamischen Erklärung der turbulenten Flüssigkeitsbewegung. *Proc. Fourth Int. Cong. Math. Rome*, 116–124.

[379] SOTIROPOULOS, F. & ABDALLAH, S. (1990) Coupled fully implicit solution procedure for the steady incompressible Navier–Stokes equation. *J. Comp. Phys.* **87**, 328–348.

[380] SOTIROPOULOS, F. & ABDALLAH, S. (1991) The discrete continuity equation in primitive variable solutions of incompressible flow. *J. Comp. Phys.* **95**, 212–227.

[381] SOTIROPOULOS, F. & ABDALLAH, S. (1992) A primitive variable method for the solution of three-dimensional incompressible viscous flow. *J. Comp. Phys.* **103**, 336–349.

[382] SPARROW, E. M. (1962) Laminar flow in isosceles triangular ducts. *AIChE J.* **8**, 599–604.

[383] SPEZIALE, C. G. (1987) On the advantages of the vorticity–velocity formulation of the equations of fluid dynamics. *J. Comp. Phys.* **73**, 476–480.

[384] SQUIRES, H. B. (1933) On the stability for three-dimensional disturbances of viscous flow between parallel walls. *Proc. R. Soc. Lond. A* **142**, 621–628.

[385] SQUIRES, H. B. (1951) The round laminar jet. *Quart. J. Mech. Appl. Math.* **4**, 321–329.

[386] STAKGOLD, I. (1967) *Boundary Value Problems of Mathematical Physics*. Volume I. Macmillan.

[387] STAKGOLD, I. (1968) *Boundary Value Problems of Mathematical Physics*. Volume II. Macmillan.

[388] STEWARTSON, K. (1954) Further solutions of the Falkner-Skan equation. *Proc. Camb. Phil. Soc.* **50**, 454–465.

[389] STRUIK, D. J. (1961) *Lectures on Classical Differential Geometry*. Dover.

[390] STUART, J. T. (1959) The viscous flow near a stagnation point when the external flow has uniform vorticity. *J. Aero/Space Sc.* **26**, 310–311.

[391] STUART, J. T. (1963) Unsteady boundary layers. In *Laminar Boundary Layers*, Rosenhead, L. (Ed.), Oxford University Press.

[392] SUTERA, S. P. & SKALAK, R. (1993) The history of Poiseuille's law. *Annu. Rev. Fluid. Mech.* **25**, 1–19.

[393] STRIKWERDA, J. C. & LEE, Y. S. (1999) The accuracy of the fractional step method, *SIAM J. Numer. Anal.* **37**, 37–47.

[394] TAFTI, D. (1995) Alternate formulations for the pressure equation Lplacian in a collocated grid for solving the unsteady incompressible Navier–Stokes equations. *J. Comp. Phys.* **116**, 143–153.

[395] TANI, I. (1977) History of boundary-layer theory. *Annu. Rev. Fluid Mech.* **9**, 87–111.

[396] TANNER, R. I. (1988) *Engineering Rheology*. Oxford University Press.

[397] TAYLOR, G. I. (1923) On the decay of vortices in a viscous fluid. *Phil. Mag.* **46**, 671–674.

[398] TAYLOR, G. I. (1923) Stability of a viscous liquid contained contained between two rotating cylinders. *Phil. Trans. R. Soc. Lond. A* **223**, 289–343.

[399] TAYLOR, G. I. (1950) The instability of liquid surfaces when accelerated in a direction perpendicular to their planes. *Proc. R. Soc. Lond. A* **201**, 192–196.

[400] TAYLOR, G. I. (1960) Deposition of a viscous fluid on a plane surface. *J. Fluid Mech.* **9**, 218–224.

[401] TEMAM, R. (1968) Une méthode d'approximation de la solution des équations de Navier–Stokes. *Bull. Soc. Math. France* **96**, 115–152.

[402] THWAITES, B. (1949) Approximate calculation of the laminar boundary layer. *Aeronaut. Quart.* **1**, 245–280.

[403] THOMAS, L. H. (1953) The stability of plane Poseuille flow. *Phys. Rev.* **91**, 780–783.

[404] THOMSON, W. (1880) On the vibrations of a columnar vortex. *Phil. Mag.* **10**, 155–168.

[405] TOBAK, M. & PEAKE, D. J. (1982) Topology of three-dimensional separated flows. *Ann. Rev. Fluid Mech.* **14**, 61–85.

[406] TREFETHEN, L. N. (1980) Numerical computation of the Schwarz-Christoffel transformation. *SIAM J. Sci. Stat. Comput.* **1**, 82–102.

[407] TRUJILLO, J. R. (1994) *Spectral Element Vorticity-Velocity Algorithm for the Incompressible Navier–Stokes Equations*. Doctoral dissertation, Princeton University.

[408] TRUESDELL, C. (1974) The meaning of viscometry in fluid dynamics. *Annu. Rev. Fluid. Mech.* **6**, 111–146.

[409] TRYGGVASON, G. (1988) Numerical simulations of the Rayleigh-Taylor instability. *J. Comp. Phys.* **75**, 253–283.

[410] TOMOTIKA, S. (1935) On the instability of a cylindrical thread of a viscous liquid surrounded by another viscous liquid, *Proc. R. Soc. Lond. A* **150**, 322–337.

[411] TUNG, T. T. & LAURENCE, R. L. (1975) A coordinate frame for helical flows. *Polym. Eng. Sci.* **15**(6), 401–405.

[412] VAN DYKE, M. D. (1975) *Perturbation Methods in Fluid Mechanics*. Academic Press.

[413] VOJIR, D. R. & MICHAELIDES, E. E. (1994) Effect of the history term on the motion of rigid spheres in a viscous fluid. *Int. J. Multiph. Flow* **20**, 547–556.

[414] VON KÁRMÁN, T. (1934) Some aspects of the turbulence problem. *Proc. Fourth Int. Congr. Appl. Mech.*, Cambridge, England.

[415] VON KÁRMÁN, T. (1921) Über laminare und turbulente Reibung. *Z. Angew. Math. Mech.* **1**, 233–252.

[416] WALLACE, D. & REDEKOPP, L. G. (1992) Linear instability characteristics of wake-shear layers. *Phys. Fluids A* **4**, 189–191.

[417] WANG, C. Y. (1981) On the low-Reynolds-number flow in a helical pipe. *J. Fluid Mech.* **108**, 185–194.

[418] WANG, C. Y. (1984) The three-dimensional flow due to a stretching flat surface. *Phys. Fluids* **27**, 1915–1917.

[419] WANG, C. Y. (1989) Exact solutions of the unsteady Navier–Stokes equations. *Appl. Mech. Rev.* **42**, S269–S282.

[420] WANG, C. Y. (1991) Exact solutions of the steady-state Navier–Stokes equations. *Annu. Rev. Fluid Mech.* **23**, 159–177.

[421] WANG, C. Y. (2008) Similarity stagnation point solutions of the Navier–Stokes equations-review and extension. *Eur. J. Mech. B/Fluids* **27**, 678–683.

[422] WANG, C. Y. (2010) The rounded triangular cross section–Exact solutions for torsion, flow and heat transfer. *Z. Angew. Math. Mech.* **90**, 522–527.

[423] WANG, C. Y. (2011) Review of similarity stretching exact solutions of the Navier–Stokes equations. *Eur. J. Mech. B/Fluids* **30**, 475–479.

[424] WARMING, R. F. & HYETT, B. J. (1974) The modified equation approach to the stability and accuracy analysis of finite-difference methods. *J. Comp. Phys.* **14**, 159–179.

[425] WATSON, J. (1959) The two-dimensional laminar flow near the stagnation point of a cylinder which has an arbitrary transverse motion. *Quart. J. Mech.* **12**, 175–190.

[426] WEISS, P. (1945) On hydrodynamic images, arbitrary irrotational flow disturbed by a sphere. *Proc. Camb. Phil. Soc.* **3**, 259–261.

[427] WHITE, F. M. (1974) *Viscous Fluid Flow.* McGraw–Hill.

[428] WHITHAM, G. B. (1963) The Navier–Stokes equations of motion. In *Laminar Boundary Layers*, Rosenhead, L. (Ed.), Oxford University Press.

[429] WILKINSON, J. H. (1965) *The Algebraic Eigenvalue Problem.* Oxford University Press.

[430] WILSON, S. D. R. (1982) The drag-out problem in film coating theory. *J. Eng. Math.* **16**, 209–221.

[431] WINCKELMANS, G. S. & LEONARD, A. (1993) Contributions to vortex particle methods for the computation of three-dimensional incompressible unsteady flow. *J. Comp. Phys.* **109**, 247–273.

[432] WU, J. Z. & WU, J. M. (1996) Vorticity dynamics on boundaries. *Adv. Appl. Mech.* **32**, 119–275.

[433] Wu, J.-Z., Yang, Y.-T, Luo, Y.-B. & Pozrikidis, C. (2005) Fluid kinematics on a deformable surface. *J. Fluid Mech.* **541**, 371–381.

[434] Yanenko, N. N. (1970) *The Method of Fractional Steps.* Springer–Verlag.

[435] Yih, C.-S. (1963) Stability of liquid flow down an inclined plane. *Phys. Fluids A* **6**, 321–334.

[436] Yih, C.-S. (1967) Instability due to viscosity stratification. *J. Fluid Mech.* **27**, 337–352.

[437] Yih, C.-S. (1979) *Fluid Mechanics.* West River Press.

[438] Yon, S. & Pozrikidis, C. (1998) A finite-volume / boundary-element method for interfacial flow in the presence of surfactants, with applications to shear flow past a viscous drop. *Computers & Fluids* **27**, 879–902.

[439] Yu, Q. & Hu, G.-H. (1997) Development of a helical coordinate system and its application to analysis of polymer flow in screw extruders, Part I. The balance equations in a helical coordinate system. *J. Non-Newt. Fluid Mech.* **69**, 155–167.

[440] Zabusky, N. J., Hughes, M. H. & Roberts, K. V. (1979) Contour dynamics for the Euler equations in two dimensions. *J. Comp. Phys.* **30**, 96–106.

[441] Zheng, Q.-S., Yu, Y. & Zhao, Z.-H. (2005) Effects of hydraulic pressure on the stability and transition of wetting modes of superhydrophobic surfaces. *Langmuir* **21**, 12207–12212.

[442] Zhou, H. & Pozrikidis, C. (1995) Adaptive singularity method for Stokes flow past particles. *J. Comp. Phys.* **117**, 78–89.

[443] Zhu, J. & Sethian, J. (1992) Projection methods coupled to level set interface techniques. *J. Comp. Phys.* **102**, 128–138.

[444] Zwillinger, D. (2002) *CRC Standard Mathematical Tables and Formulae.* 31st Edition. Chapman & Hall/CRC.

Index